BIOLOGY

Of related interest from the Benjamin/Cummings Series in the Life Sciences

GENERAL BIOLOGY

N. A. Campbell, L. G. Mitchell, and J. B. Reece
Biology: Concepts and Connections (1994)

J. Dickey
Laboratory Investigations for Biology (1995)

R. J. Kosinski
Fish Farm: Simulation Software (1993)

J. G. Morgan and M. E. B. Carter
Investigating Biology: A Laboratory Manual, Second Edition (1996)

PLANT BIOLOGY

M. G. Barbour, J. H. Burk, and W. D. Pitts
Terrestrial Plant Ecology, Second Edition (1987)

J. Mauseth
Plant Anatomy (1988)

BIOCHEMISTRY AND CELL BIOLOGY

W. M. Becker, J. B. Reece, and M. F. Poenie
The World of the Cell, Third Edition (1996)

C. Mathews and K. H. van Holde
Biochemistry, Second Edition (1996)

G. L. Sackheim
Chemistry for Biology Students, Fifth Edition (1996)

W. B. Wood, J. H. Wilson, R. M. Benbow, and L. E. Hood
Biochemistry: A Problems Approach, Second Edition (1981)

MOLECULAR BIOLOGY AND GENETICS

F. J. Ayala and J. A. Kiger, Jr.
Modern Genetics, Second Edition (1984)

M. V. Bloom, G. A. Freyer, and D. A. Micklos
Laboratory DNA Science (1996)

L. E. Hood, I. L. Weissman, W. B. Wood, and J. H. Wilson
Immunology, Second Edition (1984)

R. Schleif
Genetics and Molecular Biology (1986)

J. D. Watson, N. H. Hopkins, J. W. Roberts, J. A. Steitz, and A. M. Weiner
Molecular Biology of the Gene, Fourth Edition (1987)

G. Zubay
Genetics (1987)

MICROBIOLOGY

I. E. Alcamo
Fundamentals of Microbiology, Fourth Edition (1994)

R. M. Atlas and R. Bartha
Microbial Ecology: Fundamentals and Applications, Third Edition (1992)

J. Cappuccino and N. Sherman
Microbiology: A Laboratory Manual, Fourth Edition (1996)

T. R. Johnson and C. L. Case
Laboratory Experiments in Microbiology, Brief Edition, Fourth Edition (1995)

G. J. Tortora, B. R. Funke, and C. L. Case
Microbiology: An Introduction, Fifth Edition (1995)

EVOLUTION, ECOLOGY, AND BEHAVIOR

M. Lerman
Marine Biology: Environment, Diversity, and Ecology (986)

R. Trivers
Social Evolution (1985)

ANIMAL BIOLOGY

H. E. Evans
Insect Biology: A Textbook of Entomology (1984)

E. N. Marieb
Essentials of Human Anatomy and Physiology, Fourth Edition (1994)

E. N. Marieb
Human Anatomy and Physiology, Third Edition (195)

E. N. Marieb and J. Mallatt
Human Anatomy (1992)

A. P. Spence
Basic Human Anatomy, Third Edition (1991)

BIOLOGY

FOURTH EDITION

NEIL A. CAMPBELL

UNIVERSITY OF CALIFORNIA, RIVERSIDE

THE BENJAMIN/CUMMINGS PUBLISHING COMPANY, INC.

Menlo Park, California ■ Reading, Massachusetts ■ New York ■ Don Mills, Ontario
Wokingham, U.K. ■ Amsterdam ■ Bonn ■ Paris ■ Milan ■ Madrid ■ Sydney
Singapore ■ Tokyo ■ Seoul ■ Taipei ■ Mexico City ■ San Juan

Sponsoring Editor: Don O'Neal
Senior Developmental Editor: Susan Weisberg
Executive Editor: Johanna Schmid
Consulting Developmental Editors: Betsy Dilernia, Joanne Fraser
Editorial–Production Coordinator: Laura Kenney
Assistant Editor: Hilair Chism
Editorial Assistant: Tabinda Khan
Production Editors: Rani Cochran, Laura Kenney
Art Supervisor: Donna Kalal
Developmental Artist: Carla Simmons
Principal Artists for the Fourth Edition: Precision Graphics, Carla Simmons
Photo Editor: Cecilia Mills
Photo Researcher: Laurel Anderson/Photosynthesis
Text Designer: Bruce Kortebein, Design Office
Cover Designer: Yvo Riezebos
Layout Artists: Bruce Kortebein, Peter Martin, Design Office; Curtis Boyer; Chad Colburn
Art and Design Manager: Don Kesner
Copyeditor: Betsy Dilernia
Proofreaders: Roy Zitting, John Hammett
Indexer: Shane-Armstrong Information Systems
Compositor: PC&F, Inc.
Film House: Colotone Imaging, Inc.
Composition and Film Manager: Lillian Hom
Manufacturing Supervisor: Merry Free Osborn
Marketing Manager: Larry Swanson
Marketing Specialist: Erika Nelson
Executive Marketing Manager: Anne Emerson

ABOUT THE COVER:

Figure acknowledgments begin on page C-1.

Library of Congress Cataloging-in-Publication Data
Campbell, Neil A., 1946–
Biology/Neil A. Campbell—4th ed.
p. cm.
Includes bibliographical references and index.
ISBN 0-8053-1940-9
1. Biology. I. Title.
QH308.2.C34 1996
574—dc20

2 3 4 5 6 7 8 9 10—DO—00 99 98 97 96

High School Binding distributed by
Addison-Wesley School Division,
Menlo Park, California
ISBN 0-8053-1957-3

The Benjamin/Cummings Publishing Company, Inc.
2725 Sand Hill Road
Menlo Park, California 94025

To Rochelle and Allison, with love

ABOUT THE AUTHOR

BIOLOGY is the product of 27 years of teaching experience and many years of intensive writing and revision by Dr. Neil A. Campbell. This textbook is a natural outgrowth of Dr. Campbell's broad interest in his science. He earned his M.A. in Zoology from UCLA, where he studied the control of protein synthesis during animal development, and went on to the University of California, Riverside, where he earned a Ph.D. in plant biology. Dr. Campbell's research efforts on salt transport in plants and the cellular basis of leaf movements have resulted in publications in *Science, The Proceedings of the National Academy of Sciences,* and *Plant Physiology,* among other journals.

In addition to his accomplishments as a research scientist, Dr. Campbell has earned a reputation as an outstanding classroom teacher with a strong commitment to improving undergraduate education. After 10 years of teaching general biology and cell biology at San Bernardino Valley College, he took an academic leave and accepted a faculty position at Cornell University, where he reorganized a two-semester general biology course. After three successful years at Cornell, Dr. Campbell returned to California to reassume his teaching position at San Bernardino Valley College, where in 1986 he received the college's first Outstanding Professor Award for excellence in classroom instruction. He frequently returned to Cornell to teach the summer general biology course to advanced-placement high school students and Cornell undergraduates on a six-week schedule. In 1988, Dr. Campbell accepted an invitation to teach a one-semester general biology course at Pomona College.

During his many years of teaching general biology—most frequently as the sole lecturer—Dr. Campbell's teaching sensibilities have been honed in both large lecture and small classroom environments and with a diverse group of students. His textbooks have helped introduce biology to another million students. Dr. Campbell is currently a Visiting Scholar in the Department of Plant Sciences at the University of California, Riverside.

PREFACE

This fourth edition of *BIOLOGY* is true to the earlier versions' dual objectives: to explain the key concepts of biology clearly and accurately within a context of unifying themes, and to help students develop positive and realistic impressions of science as a human activity. These two teaching values evolved in the classroom, and I am pleased that the book's conceptual approach and its emphasis on science as a process have appealed to the educators and students who have made *BIOLOGY* the most widely used textbook in its field. With that acceptance, however, comes the responsibility to serve the biology community even better. Thus, in 1993, as I began to plan this new edition, I visited dozens of campuses to listen to what students and their professors had to say about their biology courses and textbooks. The most common criticism of biology texts, including this one, was that the most important concepts in each chapter did not pop out enough from the background of supporting details. Those conversations with faculty and students inspired the fourth edition's most visible and pervasive improvement.

■ The major improvement in this new edition of *BIOLOGY* is a much sharper focus on key concepts

Classroom experience convinces me that when students seem overwhelmed by the amount of biological information, it is usually because they are having difficulty structuring what they learn into a hierarchical scheme in which the main ideas rise above the supporting details, examples, and terminology. For instance, one key concept in the study of photosynthesis can be phrased like this: "The light reactions of photosynthesis transform light energy to chemical energy." But a beginning student trying to understand photosynthesis may have difficulty distilling this concept from a sea of details such as the names of electron carriers. And a student who is frustrated with what seems like a formless flood of information too often resorts to rote memorization as a substitute for conceptual comprehension. In trying to help students construct a framework of concepts, many colleagues share my experience that *over*simplifying photosynthesis or any other topic is counterproductive; without sufficient depth, key concepts themselves collapse into meaningless factual statements to be memorized instead of organizing principles to be investigated.

My mentors are biology teachers who keep the spotlight on a topic's starring concepts, even as they introduce a supporting cast of carefully chosen details, useful terms, reinforcing examples, and validating evidence. That is the teaching philosophy that has always been at the heart of *BIOLOGY*, and my vision for this edition was to do an even better job of keeping each chapter's key concepts in the limelight. Just as every classroom presentation has its main take-home lessons, so does every chapter in this new edition of *BIOLOGY*.

There are many excellent general biology textbooks, and I am familiar with those that accent key concepts by listing them at the beginning of a chapter or highlighting them as sidebars within a chapter. But that is not what I had in mind for this edition of *BIOLOGY*. The mission was to completely rebuild each chapter around a manageable number of key concepts, usually about ten, which serve as the titles and focal points of the chapter's major sections. That meant: first, deciding which concepts are most important to teach and then phrasing those concepts as clear, concise, accurate sentences; second, arranging those key concepts so that they build in a logical sequence and interlock into a cohesive chapter; and third, rethinking the text and figures for each section so that their function is to explain and reinforce the key concept stated at the beginning of that section.

With this sharper focus on concepts, it then made sense to use the key concept statements to help students both preview and review the chapter. Thus, on the first page of each chapter, you will find a list of key concepts as a preview of the chapter's main ideas. The second time a student sees one of those conceptual statements, it will be in the body of the chapter as the banner for a section that blends text and illustrations to teach that concept. These main ideas appear for a third time at the end of the chapter, their encore in the Review of Key Concepts section. A blue square (■) marks the key concepts in all three locations, making it easier for students to relate the sections of a chapter to the overall conceptual framework.

After students study a chapter, they should be able to take one more pass through the list of key concepts and explain them in their own words. Understanding biology through its key concepts will stick with students even as the knowledge explosion refines our view of life. *BIOLOGY* should continue to serve students as a general reference after they succeed in their introductory course and continue their education.

■ The spotlight on key concepts complements *BIOLOGY*'s other hallmarks

By reconstructing each chapter of *BIOLOGY* to enable students to keep the key concepts in clear view, I have built on the pedagogical foundation of the first three editions.

Unifying Themes. A thematic approach continues to distinguish *BIOLOGY* from an "encyclopedia of life science." Chapter 1 introduces twelve themes that resurface throughout the text to help students synthesize connections in their study of life. The themes complement the key concepts in giving form to the vastness of biology: the key concepts apply at the chapter level as that subject's most important ideas; the themes cut across all fifty chapters as unifying features of life. For example, "The light reactions of photosynthesis transform light energy to chemical energy" is a key concept in the subject area of photosynthesis. But "Organisms are open systems that interact continuously with their environments" is a theme that unifies diverse biological concepts, including the idea that plants and other photosynthetic organisms are open systems that depend on transfusions of solar energy to make sugar. In the fourth edition of *BIOLOGY*, the focus on key concepts gives form to each chapter; the application of unifying themes gives form to the whole book.

Evolution as the Core Theme of *BIOLOGY*. If the function of *BIOLOGY*'s themes is to help students integrate their study of life, then evolution is the theme of all themes—the thread that ties together the other unifying features of biology. For example, evolution accounts for the unity and diversity of life (one of *BIOLOGY*'s themes) and lends meaning to the correlation of biological structure and function (another theme). As the overarching theme of *BIOLOGY*, evolution is built into every chapter.

Science as a Process. Chapter 1 includes a thorough introduction to the power and limitations of science as one of the book's themes, but *BIOLOGY*'s commitment to showcasing science as a human activity does not end there. Case studies, each announced by the subtitle *"science as a process"* following the concept statement it supports, enrich many chapters throughout the book by balancing "what we know" with "how we know" and "what we don't yet know."

BIOLOGY also features many Methods Boxes, which demystify science by explaining laboratory and field methods in the context of experiments. For example, a new Methods Box in Chapter 49 ("Ecosystems") describes how paleoecologists are studying fossil pollen to reconstruct how past climate changes affected biological communities and to make predictions about the future consequences of global warming. And eight new interviews with influential biologists, which introduce the eight units of the book, personalize science and portray it as a social activity of creative men and women, rather than an impersonal collection of facts.

Science, Technology, and Society. Biology and its applications have a profound impact on culture—on our perspective of nature, on our environmental awareness, and on our health and quality of life. It is important for students to appreciate that ethics has a place in science, even in basic research, and that technology brings with it the need to examine values and make choices. This interrelatedness of science, technology, and society is one of the themes of *BIOLOGY*. For example, environmental problems, such as the destruction of tropical rain forests, are presented as complex issues with cultural, political, and economic considerations as well as biological ones. At the end of each chapter, Science, Technology, and Society questions encourage students to incorporate the biological concepts they have learned into their broader view of the world.

A Marriage of Text and Illustrations. Biology is a visual science, and many students are visual learners. As a teacher trying to help students learn biological concepts, I have always authored the illustration program of *BIOLOGY* side-by-side with the text. Beginning with the first draft of each edition, the artists, photo researchers, editors, and I begin working together to embed the illustrations and their self-contained legends into the story line of each chapter.

With this commitment to a marriage of words and pictures, *BIOLOGY* has pioneered many breakthroughs that enhance the pedagogy of textbooks. For example, many chapters of *BIOLOGY* use a sequence of orientation diagrams as road signs to help students keep track of where they are going as they navigate through a biological process such as photosynthesis. In figures that illustrate stepwise processes, circled numbers in the text or figure legends match the numbered steps in the diagrams in order to walk students through the process. Another important navigation aid is the consistent use of color coding and icons to help students connect concepts as they move from chapter to chapter. For example, proteins are always color-coded purple, and ATP always appears in illustrations as a yellow sunburst. In this new edition, text and illustrations have continued to co-evolve, and most of the figures have been refined to improve their teaching effectiveness.

An "Overview–Closer Look" Teaching Style. *BIOLOGY* begins the development of many complex topics such as cellular respiration (Chapter 9) and protein synthesis (Chapter 16) with a panoramic view—*an overview*—of what the overall process accomplishes. Text and figures then invite the student inside the process for *a closer look* at how it works. The orientation diagrams, miniature versions of the overview illustration with appropriate parts highlighted, help students keep the larger process in sight even as they dissect it for a closer look. The "overview–closer look" teaching strategy is another example of how the hallmarks of *BIOLOGY* complement this new edition's sharper focus on key concepts to help students find their way through the multidisciplinary landscape of biology.

I have thoroughly updated the content of each chapter while preserving *BIOLOGY*'s versatile organization

BIOLOGY makes no pretense that there is one "correct" sequence for the major topics in a general biology course: the individuality of biology professors is one of the strengths of science education. Therefore, I built *BIOLOGY* to be versatile enough to support instructors' diverse courses, whether they choose to start with molecules or ecosystems, or somewhere in between. The eight units of the book are self-contained, and most of the chapters can be assigned in a different sequence without substantial loss of continuity. For example, professors who integrate plant and animal physiology can merge chapters from Units Six and Seven to fit their courses. And instructors who begin their courses with ecology can assign Unit Eight ("Ecology") right after students have read Chapter 1, which introduces the themes that give each unit of chapters a general context.

Although specific updates and pedagogical improvements in this new edition of *BIOLOGY* are too numerous to list here, a brief survey of the eight units is a useful roadmap to the book's content.

Unit One: The Chemistry of Life. Many students struggle in general biology courses because they are uncomfortable with basic chemistry. Chapters 2–4 help those students by developing, in carefully paced steps, the concepts of chemistry that are essential for success in biology. I designed the chapters so that students of diverse backgrounds can use them for self-study, reducing the amount of valuable class time instructors need to spend on chemical review before they can get on to biology. However, Chapter 5 ("The Structure and Function of Macromolecules") and Chapter 6 ("An Introduction to Metabolism") provide important orientation even for those students with solid chemistry backgrounds. The role of chaperones in the building of proteins (Chapter 5) is one specific example of what is new in this edition.

Unit Two: The Cell. Chapters 7–11 build the study of cells around the theme of the correlation between structure and function. Throughout the unit, for example, I have accented the role of membranes in ordering cell physiology. Among the changes in this edition are greater emphasis on the structure and functions of the extracellular matrix of animal cells in Chapter 7 and a new section in Chapter 8 introducing signal-transduction pathways, a topic that is reinforced in later chapters on plant and animal physiology.

Unit Three: The Gene. Chapters 12–19 trace the history of genetics, from Mendel to DNA technology, with "science as a process" as a theme. *BIOLOGY*'s extensive coverage of human genetics is not artificially collected into a single chapter; it appears throughout the unit, integrated with general concepts that are applied to human genetics. New sections on emerging viruses and prions in Chapter 17 are examples of how *BIOLOGY* is keeping pace with current research.

Unit Four: Mechanisms of Evolution. As the core theme of *BIOLOGY*, evolution figures prominently in every unit, but Chapters 20–23 are where students will learn *how* life evolves and how biologists study evolution and test evolutionary hypotheses. Chapter 20 ("Descent with Modification: A Darwinian View of Life") sets the stage for the unit by grounding evolutionary biology in the process of science. Students will then find many examples throughout the unit of research and debate about mechanisms of evolution. New examples of natural selection in action and a comparison of different definitions of species are among the improvements in this edition. Chapter 23 bridges this unit to the survey of biological diversity in Unit Five by highlighting how modern methods of systematics, including applications of molecular biology, are helping biologists trace the history of life.

Unit Five: The Evolutionary History of Biological Diversity. Chapters 24–30 consider the diversity of life in the context of key evolutionary junctures, such as the origin of prokaryotes, the evolution of the eukaryotic cell, the genesis of multicellular life, and the adaptive radiation of plants, fungi, and animals. The evolutionary theme of this unit contrasts with a "parade of phyla" approach. Recent discoveries of important fossils, improvements in molecular systematics, and the growing consensus for cladistic classification are transforming our view of biological history and the diversity of life. Thus, this is the most extensively revised unit in the fourth edition of *BIOLOGY*. A few examples of what is new are phylogenetic classification of bacteria, a reevaluation of protistan taxonomy, new hypotheses on the origin of plants, evolutionary relationships of fungi to other kingdoms, and the ongoing debates about human origins. The rationale for alternatives to the classical five-kingdom system of classification is thoroughly evaluated, but the unit is organized so that it also supports courses that favor the five-kingdom scheme.

Unit Six: Plants: Form and Function. Chapters 31–35 introduce students to plants in the evolutionary context of adaptation to terrestrial environments. The correlation between structure and function is also a prominent theme throughout the unit. The chapters emphasize how plant cell biologists and molecular biologists are reshaping our understanding of the morphology, physiology, and development of plants. An example of how the unit has been updated is a new section in Chapter 35 ("Control Systems in Plants") on the responses of plants to environmental stress.

Unit Seven: Animals: Form and Function. The interaction between organisms and their environment is the focus of Chapters 36–45, which take a comparative approach in exploring the diverse adaptations that have evolved in different animal groups. Humans fit into this comparative format as an important mammalian example. The connection of bioenergetics to other animal functions is much more prevalent in this new edition, beginning with an introduction that relates bioenergetics to animal form and physiology in Chapter 36 ("An Introduction to Animal Structure and Function"). Chapter 39 ("The Body's Defenses") has been updated to reflect progress in immunology. And Chapter 43 ("Animal Development") now features invertebrate models for the study of pattern formation, especially the developmental genetics of *Drosophila.*

Unit Eight: Ecology. Chapters 46–50 feature stronger connections to evolution, including an updated section on the evolution of life histories. The unit also reflects the urgent need for basic ecological research in an era when the exploding human population and its technology are treading blindly and carelessly throughout the biosphere. After careful consideration, I decided not to collect environmental issues into a separate chapter on human ecology. Given the relevance of general ecological concepts to our current environmental crises, I opted for thorough examination of environmental issues throughout the ecology unit so that students could evaluate those complex problems in the context of the basic concepts that apply. For example, an improved section in Chapter 48 ("Community Ecology") links what students have learned about community structure to strategies for setting up biodiversity preserves. The last chapter ("Behavior") is based on the evolutionary perspective of behavioral ecology, an orientation that fits the chapter into the ecology unit. Chapter 50 also serves as a capstone for the entire book, relating behavior and ecology to the other fields of biology, to the other natural sciences, and to the student's general education.

Learning tools at the end of each chapter help students review and apply *BIOLOGY*'s key concepts

In the spirit of this edition's major improvement, the earlier editions' Study Outline has been replaced with a **Review of Key Concepts.** Each entry in the review restates a concept and provides an abridged explanation of it. Along with the concept are page numbers to direct students if they need to return to where that concept is developed in more depth. In most cases, each entry in the review also refers students to a particular illustration (or illustrations) in the chapter that provides the most useful visual summary of the concept.

A **Self-Quiz** at the end of each chapter helps students test their comprehension of key concepts, but many of these questions also require students to apply concepts or solve problems. Students will find the answers to the Self-Quiz questions in Appendix One. **Challenge Questions** encourage students to verbalize their interpretations of concepts, to extrapolate from what they have learned to new situations, to think critically about complex debates in biology, to apply quantitative skills in the context of biological problems, and to generate testable hypotheses of their own. The **Science, Technology, and Society** questions ask students to think about biology's place in culture and about the consequences of applied biology. A short **Further Reading** list completes the tools at the end of each chapter.

Students will also find learning tools at the back of the book, including an extensive **Glossary** of biological terms and an improved **Index.**

Carefully developed supplements add value to the *BIOLOGY* package

Student Study Guide by Martha Taylor, Cornell University.

Investigating Biology, Second Edition by Judith Morgan, Emory University, and Eloise Carter, Oxford College of Emory University. A laboratory manual, with accompanying **Annotated Instructor's Edition and Preparation Guide.**

Instructor's Guide by Nina Caris and Harold Underwood, both of Texas A&M University.

Test Bank edited by Daniel Wivagg, Baylor University, with contributors Richard Duhrkopf, Baylor University; Richard Storey, Colorado College; Gary Fabris, Red Deer College; Rebecca Pyles, East Tennessee State University; and Kurt Redborg, Coe College. This test bank is available on Microtest, a microcomputer test-generation program. The test bank is available to qualified college and university adopters.

Overhead Transparencies A set of 300 color acetates of illustrations and micrographs from *BIOLOGY,* Fourth Edition, is available to qualified college and university adopters.

35-mm Slides The same 300 illustrations available as acetates are available in 35-mm slides to qualified college and university adopters.

Transparency Masters A set of 300 black-and-white transparency masters from *BIOLOGY,* Fourth Edition, is available to qualified college and university adopters.

BioShow II: The Videodisc BioShow II is a videodisc of illustrations from the text, original animations, and motion sequences. Side one accompanies *BIOLOGY.* Art conversion from *BIOLOGY* to still figures, stepped figures, and animations was developed and executed by Tom Dallman, Ph.D. BioShow II is available to qualified adopters.

The Art CD-ROM for *BIOLOGY,* Fourth Edition A new CD-ROM containing text illustrations for projection during lecture is available to qualified adopters.

Fish Farm: Simulation Software by Robert J. Kosinski, Clemson University, with accompanying **Student Workbook** and **Instructor's Guide.**

* * *

The real test of any textbook is how well it helps instructors teach and students learn. I welcome comments from students and professors who use *BIOLOGY.* Please address your suggestions for improving the next edition directly to me:

Neil A. Campbell
Department of Plant Sciences
University of California
Riverside, California 92521

ACKNOWLEDGMENTS

Like science itself, building a science textbook is a social process. Though *BIOLOGY* is in my voice, its text and illustrations are a synthesis of what I continue to learn from students, teachers, research scientists, artists, and editors. My name alone goes on *BIOLOGY*'s cover, but what you find between the covers is the result of many committed people working together toward the shared goal of improving biology education.

The reviewers listed on p. xi suggested many ways to improve *BIOLOGY*'s scientific accuracy and pedagogy. Many other professors and their students offered suggestions by writing directly to me. Those correspondents include: Peter Atsatt (University of California, Irvine), Karl Aufderheide (Texas A&M University), Howard Berg (Memphis State University), Robert Cleland (University of Washington), Morris Cline (The Ohio State University), Marshall Darley (University of Georgia), Barbara Demming-Adams (University of Colorado), Robert Eaton (University of Colorado), Steven Eiger (Montana State University), Richard Ellis (Bucknell University), John W. Evans (Memorial University, Newfoundland), Joseph Frankel (University of Iowa), Roger Gordon (Memorial University, Newfoundland), Lane Graham (University of Manitoba), Gary Grimes (Hofstra University), Jack Hailman (University of Wisconsin), Jill Hendrickson (Carleton College), Harvey Hinsz (International School of Kenya), Douglas Hunter (Oakland University), Grahame Kelly (Queensland University of Technology), Arlene Larsen (University of Colorado, Denver), Rodger Lloyd (Florida Community College, Jacksonville), Karen Mainer (El Rancho High School), Len Millis (Douglas College, British Columbia), William Moore (Wayne State University), Joseph Pelliccia (Bates College), Carl Pike (Franklin and Marshall College), Bob Ross (Linn-Benton Community College),Walter Sakai (Santa Monica City College), Ethyl Stanley (Millikin University), Lloyd Stark (University of Nevada, Las Vegas), Cyril Thong (Simon Fraser University), Gordon Ultsch (The University of Alabama), F. Vella (University of Saskatchewan), Kerry Walsh (University of Central Queensland), and Reid Wiseman (College of Charleston). Although I am responsible for any errors that remain, they are all the fewer because of the dedication of the reviewers and correspondents.

Several biologists participated in this edition by helping to revise text or by submitting early drafts of new material. These contributors are: Antonie Blackler (Cornell University), who helped plan the revision of the animal development chapter (Chapter 43); David Bourgaize (Colby College), who helped revise the chapters on the eukaryotic genome and biotechnology (Chapters 18 and 19); Lawrence Mitchell (University of Montana), the major contributor for the animal physiology unit (Unit Seven); Mary Jane Niles (University of San Francisco), who collaborated on the immunology chapter (Chapter 39); Karen Oberhauser (University of Minnesota), who made major improvements in the ecology chapters (Chapters 46–49); and Eric Strauss (University of Massachusetts, Boston), who helped with the behavior chapter (Chapter 50). Barbara Beitch (Hamden Hall Country Day School), and Dan Wivagg (Baylor University) contributed several new questions for the ends of chapters. I thank these contributors for helping me make this new edition more correct, current, and clear.

Numerous U.C.-Riverside colleagues continue to help shape *BIOLOGY* by discussing their research fields and exchanging ideas about biology education. In particular, I would like to thank Katharine Atkinson, Richard Cardullo, Mark Chappell, Darleen DeMason, Norman Ellstrand, Leah Haimo, Robert Heath, Anthony Huang, Bradley Hyman, Tracy Kahn, Elizabeth Lord, Carol Lovatt, Leonard Nunney, John Oross, Kathryn Platt, Mary Price, David Reznick, Rodolfo Ruibal, Clay Sassaman, Vaughan Shoemaker, William Thomson, Linda Walling, Nickolas Waser, and John Moore (whose "Science as a Way of Knowing" essays have had such an important influence on the evolution of *BIOLOGY*). I am also grateful to Pius Horner, who was my mentor during the many years we taught general biology together at San Bernardino Valley College.

One of my incentives for revising *BIOLOGY* is the opportunity to conduct new interviews to open the eight units of the text. For the fourth edition, it was my pleasure to interview Eloy Rodriguez, Shinya Inoué, David Satcher, John Maynard Smith, Edward O. Wilson, Adrienne Clarke, Patricia Churchland, and Margaret Davis. I thank them for helping *BIOLOGY* to communicate the human face of science.

Many publishing professionals welcomed the challenge to help reinvent *BIOLOGY* so that it has a much sharper focus on key concepts. A new design, ambitious revision of the art and photo program, and an editorial eye toward the greater emphasis on key concepts are three examples of what our goals for this edition meant for the publishing team. Bruce Kortebein of Design Office experimented with us until we had the right design to make the conceptual framework of each chapter visible to students. Bruce was also one of the layout artists, along with Peter Martin, Curtis Boyer, and Chad Colburn. Yvo Riezebos designed a beautiful cover that is at once fresh and true to the elegant simplicity that distinguished the covers of earlier editions. Donna Kalal supervised production of the very complex art revision, and Don Kesner, art and design manager, played a crucial role in coordinating with Precision Graphics to develop computer art of the highest quality. Photo editor Cecilia Mills, with the help of photo researcher Laurel Anderson of Photosynthesis, were patient partners in searching for just the right photos to reinforce key concepts. Joanne Fraser and Betsy Dilernia, as consulting developmental editors, made many helpful suggestions for improving chapters. Betsy was also the copyeditor, and I thank her for her perfectionism and consistency. Editorial assistant Tabinda Khan and assistant editor Hilair Chism were essential members of the editorial team. Hilair also edited most of the interviews. Kim Johnson worked with the authors of the supplements to provide students and professors with the best package of support materials ever to accompany *BIOLOGY.* Arlene Cowan and Curt Cowan worked hundreds of hours at the keyboard to produce clean "galleys" from my first- and second-draft manuscripts. Roy Zitting and John Hammett were vigilant proofreaders for the fourth edition, and Charlotte Shane created an index that works much better than earlier versions. Composition and film manager Lillian Hom and manufacturing supervisor Merry Free Osborn worked wonders to transform piles of folders to a bound book. And I am grateful for the dedication and flexibility of Laura Kenney, who coordinated the editorial and production efforts and served as our final production editor. I also thank Rani Cochran and Donna Linden, production editors. The entire publishing group worked together to craft a book that teaches biological concepts even better than earlier editions.

I am also fortunate to have the support of the Benjamin/Cummings marketing department, which keeps *BIOLOGY* in touch with the students and professors it serves. Erika Nelson, Nathalie Mainland, and biology marketing manager Larry Swanson got inside the fourth edition to understand its improvements and then launched the book with an informative promotion.

The field staff that represents *BIOLOGY* on campus is my living link to the students and professors who use the text. The field representatives tell me what you like and don't like about the book, and they provide prompt service to biology departments. The field reps are good allies in science education, and I thank them for their professionalism in communicating the merits of our book without denigrating other publishers and their competing texts.

BIOLOGY originated from a 1979 meeting with Jim Behnke. Jim was my editor for the first edition, and it took us eight years to craft the new kind of biology text we envisioned. Robin Heyden took over as sponsoring editor of the second edition and inspired many improvements in the book. Edith Brady, editor for the third edition, joined in an ambitious revision that made the book more successful than ever. I am also grateful to former editorial director Barbara Piercecchi and executive editor Johanna Schmid. And I am indebted to Benjamin/Cummings president Sally Elliott for her sustaining confidence in *BIOLOGY* and its author.

Don O'Neal was sponsoring editor for this fourth edition. I thank Don for supporting my drive to rebuild each chapter of *BIOLOGY* around a framework of key concepts. When it would have been safe for us all to be complacent with the book's success and do a cosmetic revision, Don shared the commitment to take the book to a new level of teaching effectiveness.

Although the *BIOLOGY* team changes somewhat from edition to edition, five key veterans have been especially important in the book's long-term success. They are Carla Simmons, Anne Emerson, Susan Weisberg, Jane Reece, and Larry Mitchell. Carla Simmons, the finest illustrator in biology publishing, has helped craft the art for all four editions. Carla played an especially large role in this new edition, serving as developmental artist and bringing her sense of clarity and aesthetics to the entire art program. Anne Emerson, executive marketing manager, has been my *BIOLOGY* colleague through three editions. Anne's creativity, common sense, fairness, and commitment to science education set standards for everyone in college publishing. Susan Weisberg, senior developmental editor, was my main partner for this very demanding revision of *BIOLOGY*. Susan also worked with me on the development of the first edition of the book. She understands what I am trying to accomplish, and she helps me do it better. Susan's solid publishing values show on every page of our book, and I thank her so much for the years of teamwork and friendship.

Jane Reece is an editor/author with a Ph.D. in bacterial genetics. In various roles, Jane has worked with me on *BIOLOGY* for 15 years. She has had an especially important impact on the genetics unit and on the book's entire illustration program. Jane is also a coauthor, along with Larry Mitchell and me, of our nonmajors biology textbook. The vision the three of us shared for that book had a major influence on the improvements in this fourth edition of *BIOLOGY*. Larry is a gifted writer, careful scientist, and dedicated educator. I have learned so much from my work with Jane and Larry, and I want them to know how much I value our partnership.

Most of all, I thank my family and friends for their encouragement and for enduring my obsession with making *BIOLOGY* a better learning tool for students.

FOURTH EDITION REVIEWERS

Kenneth Able, *State University of New York, Albany*
Richard Almon, *State University of New York, Buffalo*
Tania Beliz, *College of San Mateo*
Werner Bergen, *Michigan State University*
Annalisa Berta, *San Diego State University*
Theodore A. Bremner, *Howard University*
Danny Brower, *University of Arizona*
Carole Browne, *Wake Forest University*
Linda Butler, *University of Texas, Austin*
Iain Campbell, *University of Pittsburgh*
Richard Cardullo, *University of California, Riverside*
Lynwood Clemens, *Michigan State University*
Bruce Criley, *Illinois Wesleyan University*
Norma Criley, *Illinois Wesleyan University*
John Drees, *Temple University School of Medicine*
Charles Drewes, *Iowa State University*
William Fixsen, *Harvard University*
Kerry Foresman, *University of Montana*
Chris George, *California Polytechnic State University, San Luis Obispo*
Frank Gilliam, *Marshall University*
Anne Good, *University of California, Berkeley*
Judith Goodenough, *University of Massachusetts, Amherst*
Robert Grammer, *Belmont University*
Joseph Graves, *Arizona State University*
Serine Gropper, *Auburn University*
Leah Haimo, *University of California, Riverside*
Rebecca Halyard, *Clayton State College*
Helmut Hirsch, *State University of New York, Albany*
James Hoffman, *University of Vermont*
Ron Hoy, *Cornell University*
Donald Humphrey, *Emory University School of Medicine*
Bradley Hyman, *University of California, Riverside*
Dan Johnson, *East Tennessee State University*
Wayne Johnson, *Ohio State University*
Greg Kopf, *University of Pennsylvania School of Medicine*
Diane Lavett, *Emory University*
John Lewis, *Loma Linda University*
Kenneth Mitchell, *Tulane University School of Medicine*
William Moore, *Wayne State University*
Michael Mote, *Temple University*
Elliot Myerowitz, *California Institute of Technology*
John Olsen, *Rhodes College*
Sharman O'Neill, *University of California, Davis*
Georgiandra Ostarello, *Diablo Valley College*
Bob Pittman, *Michigan State University*
Rebecca Pyles, *East Tennessee State University*
Brian Reeder, *Morehead State University*
Steve Rothstein, *University of California, Santa Barbara*
Don Sakaguchi, *Iowa State University*
Mark Sanders, *University of California, Davis*
Lisa Shimeld, *Crafton Hills College*
Eric Strauss, *University of Massachusetts, Boston*
John Sullivan, *Southern Oregon State University*
William Thomas, *Colby-Sawyer College*
William Wade, *Dartmouth Medical College*
Charles Webber, *Loyola University of Chicago*
Peter Webster, *University of Massachusetts, Amherst*
Patrick Woolley, *East Central College*

PREVIOUS EDITION REVIEWERS

John Alcock, *Arizona State University,* Katherine Anderson, *University of California, Berkeley,* Richard J. Andren, *Montgomery County Community College,* J. David Archibald, *Yale University,* Leigh Auleb, *San Francisco State University,* P. Stephen Baenziger, *University of Nebraska,* Katherine Baker, *Millersville University,* William Barklow, *Framingham State College,* Steven Barnhart, *Santa Rosa Junior College,* Ron Basmajian, *Merced College,* Tom Beatty, *University of British Columbia,* Wayne Becker, *University of Wisconsin, Madison,* Jane Beiswenger, *University of Wyoming,* Anne Bekoff, *University of Colorado, Boulder,* Marc Bekoff, *University of Colorado, Boulder,* Adrianne Bendich, *Hoffman-La Roche, Inc.,* Barbara Bentley, *State University of New York, Stony Brook,* Darwin Berg, *University of California, San Diego,* Gerald Bergstrom, *University of Wisconsin, Milwaukee,* Anna W. Berkovitz, *Purdue University,* Dorothy Berner, *Temple University,* Paulette Bierzychudek, *Pomona College,* Charles Biggers, *Memphis State University,* Robert Blystone, *Trinity University,* Robert Boley, *University of Texas, Arlington,* Eric Bonde, *University of Colorado, Boulder,* Richard Boohar, *University of Nebraska, Omaha,* Carey L. Booth, *Reed College,* James L. Botsford, *New Mexico State University,* J. Michael Bowes, *Humboldt State University,* Richard Bowker, *Alma College,* Barry Bowman, *University of California, Santa Cruz,* Jerry Brand, *University of Texas, Austin,* James Brenneman, *University of Evansville,* Donald P. Briskin, *University of Illinois, Urbana,* Herbert Bruneau, *Oklahoma State University,* Gary Brusca, *Humboldt State University,* Alan H. Brush, *University of Connecticut, Storrs,* Meg Burke, *University of North Dakota,* Edwin Burling, *De Anza College,* William Busa, *Johns Hopkins University,* John Bushnell, *University of Colorado,* Deborah Canington, *University of California, Davis,* Gregory Capelli, *College of William and Mary,* Nina Caris, *Texas A&M University,* Doug Cheeseman, *De Anza College,* Shepley Chen, *University of Illinois, Chicago,* Henry Claman, *University of Colorado Health Science Center,* William P. Coffman, *University of Pittsburgh,* J. John Cohen, *University of Colorado Health Science Center,* John Corliss, *University of Maryland,* Stuart J. Coward, *University of Georgia,* Charles Creutz, *University of Toledo,* Richard Cyr, *Pennsylvania State University,* Marianne Dauwalder, *University of Texas, Austin,* Bonnie J. Davis, *San Francisco State University,* Jerry Davis, *University of Wisconsin, La Crosse,* Thomas Davis, *University of New Hampshire,* John Dearn, *University of Canberra,* James Dekloe, *University of California, Santa Cruz,* T. Delevoryas, *University of Texas, Austin,* Diane C. DeNagel, *Northwestern University,* Jean DeSaix, *University of North Carolina,* Marvin Druger, *Syracuse University,* Betsey Dyer, *Wheaton College,* Robert Eaton, *University of Colorado,* Robert S. Edgar, *University of California, Santa Cruz,* Betty J. Eidemiller, *Lamar University,* David Evans, *University of Florida,* Robert C. Evans, *Rutgers University, Camden,* Sharon Eversman, *Montana State University,* Lincoln Fairchild, *Ohio State University,* Bruce Fall, *University of Minnesota,* Lynn Fancher, *College of DuPage,* Larry Farrell, *Idaho State University,* Jerry F. Feldman, *University of California, Santa Cruz,* Russell Fernald, *University of Oregon,* Milton Fingerman, *Tulane University,* Barbara Finney, *Regis College,* David Fisher, *University of Hawaii at Manoa,* Abraham Flexer, *Manuscript Consultant, Boulder, Colorado,* Norma Fowler, *University of Texas, Austin,* David Fox, *University of Tennessee, Knoxville,* Otto Friesen, *University of Virginia,* Virginia Fry, *Monterey Peninsula College,* Alice Fulton, *University of Iowa,* Sara Fultz, *Stanford University,* Berdell Funke, *North Dakota State University,* Anne Funkhouser, *University of the Pacific,* Arthur W. Galston, *Yale University,* Carl Gans, *University of Michigan,* John Gapter, *University of Northern Colorado,* Reginald Garrett, *University of Virginia,* Patricia Gensel, *University of North Carolina,* Robert George, *University of Wyoming,* Todd Gleeson, *University of Colorado,* William Glider, *University of Nebraska,* Elizabeth A. Godrick, *Boston University,* Lynda Goff, *University of California, Santa Cruz,* Paul Goldstein, *University of Texas, El Paso,* Judith Goodenough, *University of Massachusetts, Amherst,* Ester Goudsmit, *Oakland University,* A. J. F. Griffiths, *University of British Columbia,* William Grimes, *University of Arizona,* Mark Gromko, *Bowling Green State University,* Katherine L. Gross, *Ohio State University,* Gary Gussin, *University of Iowa,* R. Wayne Habermehl, *Montgomery County Community College,* Mac Hadley, *University of Arizona,* Jack P. Hailman, *University of Wisconsin,* Leah Haimo, *University of California, Riverside,* Penny Hanchey-Bauer, *Colorado State University,* Laszlo Hanzely, *Northern Illinois University,* Richard Harrison, *Cornell University,* H. D. Heath, *California State University, Hayward,* George Hechtel, *State University of New York at Stony Brook,* Jean Heitz-Johnson, *University of Wisconsin, Madison,* Caroll Henry, *Chicago State University,* Frank Heppner, *University of Rhode Island,* Paul E. Hertz, *Barnard College,* Ralph Hinegardner, *University of California, Santa Cruz,* William Hines, *Foothill College,* Tuan-hua David Ho, *Washington University,* Carl Hoagstrom, *Ohio Northern University,* James Holland, *Indiana State University, Bloomington,* Laura Hoopes, *Occidental College,* Nancy Hopkins, *Massachusetts Institute of Technology,* Kathy Hornberger, *Widener University,* Pius F. Horner, *San Bernardino Valley College,* Margaret Houk, *Ripon College,* Ronald R. Hoy, *Cornell University,* Robert J. Huskey, *University of Virginia,* Alice Jacklet, *State University of New York, Albany,* John Jackson, *North Hennepin Community College,* Kenneth C. Jones, *California State University, Northridge,* Russell Jones, *University of California, Berkeley,* Alan Journet, *Southeast Missouri State University,* Thomas Kane, *University of Cincinnati,* E. L. Karlstrom, *University of Puget Sound,* George Khoury, *National Cancer Institute,* Robert Kitchen, *University of Wyoming,* Attila O. Klein, *Brandeis University,* Thomas Koppenheffer, *Trinity University,* Janis Kuby, *San Francisco State University,* J. A. Lackey, *State University of New York at Oswego,* Lynn Lamoreux, *Texas A&M University,* Carmine A. Lanciani, *University of Florida,* Kenneth Lang, *Humboldt State University,* Allan Larson, *Washington University,* Diane K. Lavett, *State University of New York, Cortland,* and *Emory University,* Charles Leavell, *Fullerton College,* C. S. Lee, *University of Texas,* Robert Leonard, *University of California, Riverside,* Joseph Levine, *Boston College,* Bill Lewis, *Shoreline Community College,* Lorraine Lica, *California State University, Hayward,* Harvey Lillywhite, *University of Florida, Gainesville,* Sam Loker, *University of New Mexico,* Jane Lubchenco, *Oregon State University,* James MacMahon, *Utah State University,* Charles Mallery, *University of Miami,* Lynn Margulis, *Boston University,* Edith Marsh, *Angelo State University,* Karl Mattox, *Miami University of Ohio,* Joyce Maxwell, *California State University, Northridge,* Richard McCracken, *Purdue University,* John Merrill, *University of Washington,* Ralph Meyer, *University of Cincinnati,* Roger Milkman, *University of Iowa,* Helen Miller, *Oklahoma State University,* John Miller, *University of California, Berkeley,* Kenneth R. Miller, *Brown University,* John E. Minnich, *University of Wisconsin, Milwaukee,* Russell Monson, *University of Colorado, Boulder,* Frank Moore, *Oregon State University,* Randy Moore, *Wright State University,* Carl Moos, *Veterans Administration Hospital, Albany, New York,* John Mutchmor, *Iowa State University,* John Neess, *University of Wisconsin, Madison,* Todd Newbury, *University of California, Santa Cruz,* Harvey Nichols, *University of Colorado, Boulder,* Deborah Nickerson, *University of South Florida,* Bette Nicotri, *University of Washington,* Charles R. Noback, *College of Physicians and Surgeons,*

Columbia University, Mary C. Nolan, *Irvine Valley College,* David O. Norris, *University of Colorado, Boulder,* Cynthia Norton, *University of Maine, Augusta,* Bette H. Nybakken, *Hartnell College,* Brian O'Conner, *University of Massachusetts, Amherst,* Gerard O'Donovan, *University of North Texas,* Eugene Odum, *University of Georgia,* Wan Ooi, *Houston Community College,* Gay Ostarello, *Diablo Valley College,* Barry Palevitz, *University of Georgia,* Peter Pappas, *County College of Morris,* Bulah Parker, *North Carolina State University,* Stanton Parmeter, *Chemeketa Community College,* Robert Patterson, *San Francisco State University,* Crellin Pauling, *San Francisco State University,* Kay Pauling, *Foothill Community College,* Patricia Pearson, *Western Kentucky University,* James Platt, *University of Denver,* Scott Poethig, *University of Pennsylvania,* Jeffrey Pommerville, *Texas A&M University,* Warren Porter, *University of Wisconsin,* Donald Potts, *University of California, Santa Cruz,* David Pratt, *University of California, Davis,* Halina Presley, *University of Illinois, Chicago,* Scott Quackenbush, *Florida International University,* Ralph Quatrano, *Oregon State University,* Charles Ralph, *Colorado State University,* C. Gary Reiness, *Pomona College,* Charles Remington, *Yale University,* David Reznick, *University of California, Riverside,* Fred Rhoades, *Western Washington State University,* Christopher Riegle, *Irvine Valley College,* Donna Ritch, *Pennsylvania State University,* Thomas Rodella, *Merced College,* Rodney Rogers, *Drake University,* Wayne Rosing, *Middle Tennessee State University,* Thomas Rost, *University of California, Davis,* Stephen I. Rothstein, *University of California, Santa Barbara,* John Ruben, *Oregon State University,* Albert Ruesink, *Indiana University,* Mark F. Sanders, *University of California, Davis,* Ted Sargent, *University of Massachusetts, Amherst,* Carl Schaefer, *University of Connecticut,* David Schimpf, *University of Minnesota, Duluth,* William H. Schlesinger, *Duke University,* Erik P. Scully, *Towson State University,* Stephen Sheckler, *Virginia Polytechnic Institute and State University,* James Shinkle, *Trinity University,* Barbara Shipes, *Hampton University,* Peter Shugarman, *University of Southern California,* Alice Shuttey, *DeKalb Community College,* James Sidie, *Ursinus College,* Daniel Simberloff, *Florida State University,* John Smarrelli, *Loyola University,* Andrew T. Smith, *Arizona State University,* Andrew J. Snope, *Essex Community College,* Susan Sovonick-Dunford, *University of Cincinnati,* Karen Steudel, *University of Wisconsin,* Barbara Stewart, *Swarthmore College,* Cecil Still, *Rutgers University, New Brunswick,* John Stolz, *California Institute of Technology,* Richard D. Storey, *Colorado College,* Stephen Strand, *University of California, Los Angeles,* Russell Stullken, *Augusta College,* Gerald Summers, *University of Missouri,* Marshall Sundberg, *Louisiana State University,* Daryl Sweeney, *University of Illinois, Urbana-Champaign,* Samuel S. Sweet, *University of California, Santa Barbara,* Lincoln Taiz, *University of California, Santa Cruz,* Samuel Tarsitano, *Southwest Texas State University,* David Tauck, *Santa Clara University,* James Taylor, *University of New Hampshire,* Roger Thibault, *Bowling Green State University,* John Thornton, *Oklahoma State University,* Robert Thornton, *University of California, Davis,* James Traniello, *Boston University,* Robert Tuveson, *University of Illinois, Urbana,* Maura G. Tyrrell, *Stonehill College,* Gordon Uno, *University of Oklahoma,* James W. Valentine, *University of California, Santa Barbara,* Joseph Vanable, *Purdue University,* Theodore Van Bruggen, *University of South Dakota,* Frank Visco, *Orange Coast College,* Laurie Vitt, *University of California, Los Angeles,* Susan D. Waaland, *University of Washington,* John Waggoner, *Loyola Marymount University,* Dan Walker, *San Jose State University,* Robert L. Wallace, *Ripon College,* Jeffrey Walters, *North Carolina State University,* Margaret Waterman, *University of Pittsburgh,* Terry Webster, *University of Connecticut, Storrs,* Peter Wejksnora, *University of Wisconsin, Milwaukee,* Kentwood Wells, *University of Connecticut,* Stephen Williams, *Glendale Community College,* Christopher Wills, *University of California, San Diego,* Fred Wilt, *University of California, Berkeley,* Robert T. Woodland, *University of Massachusetts Medical School,* Joseph Woodring, *Louisiana State University,* Philip Yant, *University of Michigan,* Hideo Yonenaka, *San Francisco State University,* John Zimmerman, *Kansas State University,* Uko Zylstra, *Calvin College.*

THE CAMPBELL INTERVIEWS

BRIEF CONTENTS

DETAILED CONTENTS

CHAPTER 1

Introduction: Themes in the Study of Life 2

UNIT ONE

THE CHEMISTRY OF LIFE 22

CHAPTER 2

The Chemical Context of Life 25

CHAPTER 3

Water and the Fitness of the Environment 41

CHAPTER 8

Membrane Structure and Function 140

CHAPTER 9

Cellular Respiration: Harvesting Chemical Energy 159

CHAPTER 10

Photosynthesis 182

CHAPTER 11

The Reproduction of Cells 204

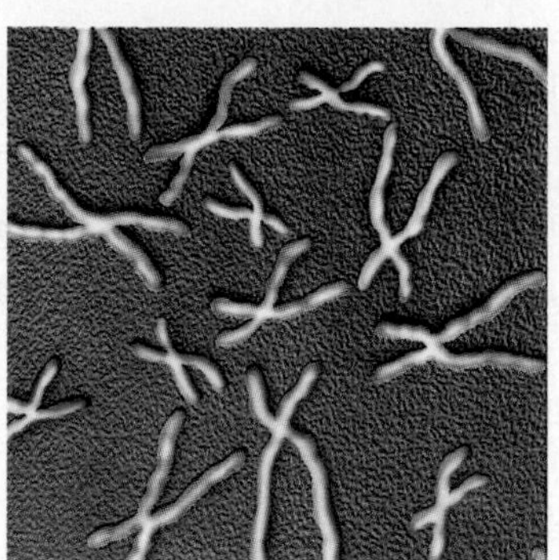

UNIT THREE

THE GENE 222

CHAPTER 12

Meiosis and Sexual Life Cycles 225

CHAPTER 13

Mendel and the Gene Idea 238

CHAPTER 14

The Chromosomal Basis of Inheritance 262

CHAPTER 15

The Molecular Basis of Inheritance 281

CHAPTER 16

From Gene to Protein 297

CHAPTER 17

Microbial Models: The Genetics of Viruses and Bacteria 324

CHAPTER 18

Genome Organization and Expression in Eukaryotes 351

CHAPTER 19

DNA Technology 369

UNIT FOUR

MECHANISMS OF EVOLUTION 396

CHAPTER 20

Descent with Modification: A Darwinian View of Life 399

CHAPTER 25

Prokaryotes and the Origins of Metabolic Diversity 498

CHAPTER 26

The Origins of Eukaryotic Diversity 518

CHAPTER 27

Plants and the Colonization of Land 547

CHAPTER 28

Fungi 573

CHAPTER 29

Invertebrates and the Origin of Animal Diversity 589

CHAPTER 30

The Vertebrate Genealogy 628

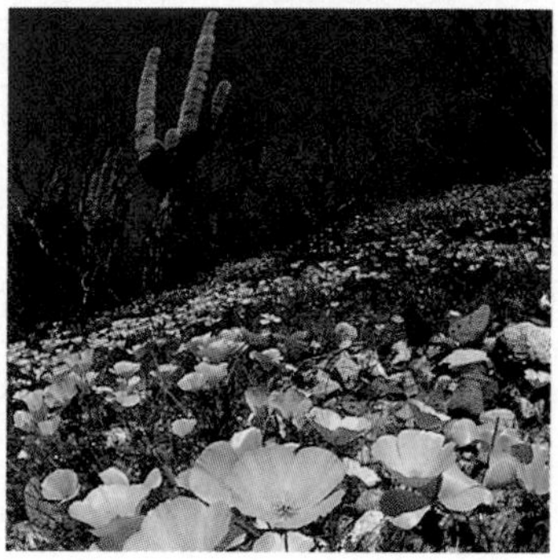

CHAPTER 35

Control Systems in Plants 750

UNIT SEVEN

ANIMALS: FORM AND FUNCTION 776

CHAPTER 36

An Introduction to Animal Structure and Function 779

CHAPTER 37

Animal Nutrition 796

CHAPTER 38

Circulation and Gas Exchange 819

CHAPTER 43

Animal Development 963

CHAPTER 44

Nervous Systems 993

CHAPTER 45

Sensory and Motor Mechanisms 1026

UNIT EIGHT

ECOLOGY 1058

CHAPTER 46

An Introduction to Ecology: Distribution and Adaptations of Organisms 1061

CHAPTER 47

Population Ecology 1093

CHAPTER 48

Community Ecology 1118

CHAPTER 49

Ecosystems 1145

CHAPTER 50

Behavior 1172

APPENDIX ONE

Self-Quiz Answers A-1

APPENDIX TWO

The Metric System A-3

APPENDIX THREE

Classification of Life A-4

APPENDIX FOUR

A Comparison of the Light Microscope and the Electron Microscope A-6

CHAPTER 1

INTRODUCTION: THEMES IN THE STUDY OF LIFE

- Life is organized on many structural levels
- Each level of biological organization has emergent properties
- Cells are an organism's basic units of structure and function
- The continuity of life is based on heritable information in the form of DNA
- A feeling for organisms enriches the study of life
- Structure and function are correlated at all levels of biological organization
- Organisms are open systems that interact continuously with their environments
- Diversity and unity are the dual faces of life on Earth
- Evolution is the core theme of biology
- Science as a process of inquiry often involves hypothetico-deductive thinking
- Science and technology are functions of society
- Biology is a multidisciplinary adventure

Biology, the study of life, is rooted in the human spirit. People keep pets, nurture houseplants, invite avian visitors with backyard birdhouses, and visit zoos and nature parks. This behavior expresses what Harvard biologist E. O. Wilson calls biophilia, *an innate attraction to life in its diverse forms. Biology is the scientific extension of this human tendency to feel connected to and curious about all forms of life. It is a science for adventurous minds. It takes us, personally or vicariously, into jungles, deserts, seas, and other environments, where a variety of living forms and their physical surroundings are interwoven into complex webs called ecosystems. Studying life leads us into laboratories to examine more closely how living things, which biologists call organisms, work. Biology draws us into the microscopic world of the fundamental units of organisms known as cells, and into the submicroscopic realm of the molecules that make up those cells. Our intellectual journey also takes us back in time, for biology encompasses not only contemporary life, but also a history of ancestral forms stretching nearly four billion years into the past. The scope of biology is immense. The purpose of this book is to introduce you to this multifaceted science* (Figure 1.1). *It is unlikely that any biology course could or should cover all 50 chapters, but this textbook is designed to help you succeed in your general biology course* and *to serve as a durable reference in your continuing education.*

You are becoming involved with biology during its most exciting era. Using fresh approaches and new research methods, biologists are beginning to unravel some of life's most engaging mysteries. Though stimulating, the information explosion in biology is also intimidating. Most of the biologists who have ever lived are alive today, and they add about a half-million new research articles to the scientific literature annually. Each of biology's many subfields changes continuously, and it is very difficult for a professional biologist to remain current in more than one narrowly defined specialty. How, then, can beginning biology students hope to keep their heads above water in this deluge of data and discovery? The key is to recognize unifying themes that pervade all of biology—themes that will still apply decades from now, when much of the specific information presented in any textbook will be obsolete. This chapter introduces some broad, enduring themes in the study of life. The list on this page previews these themes.

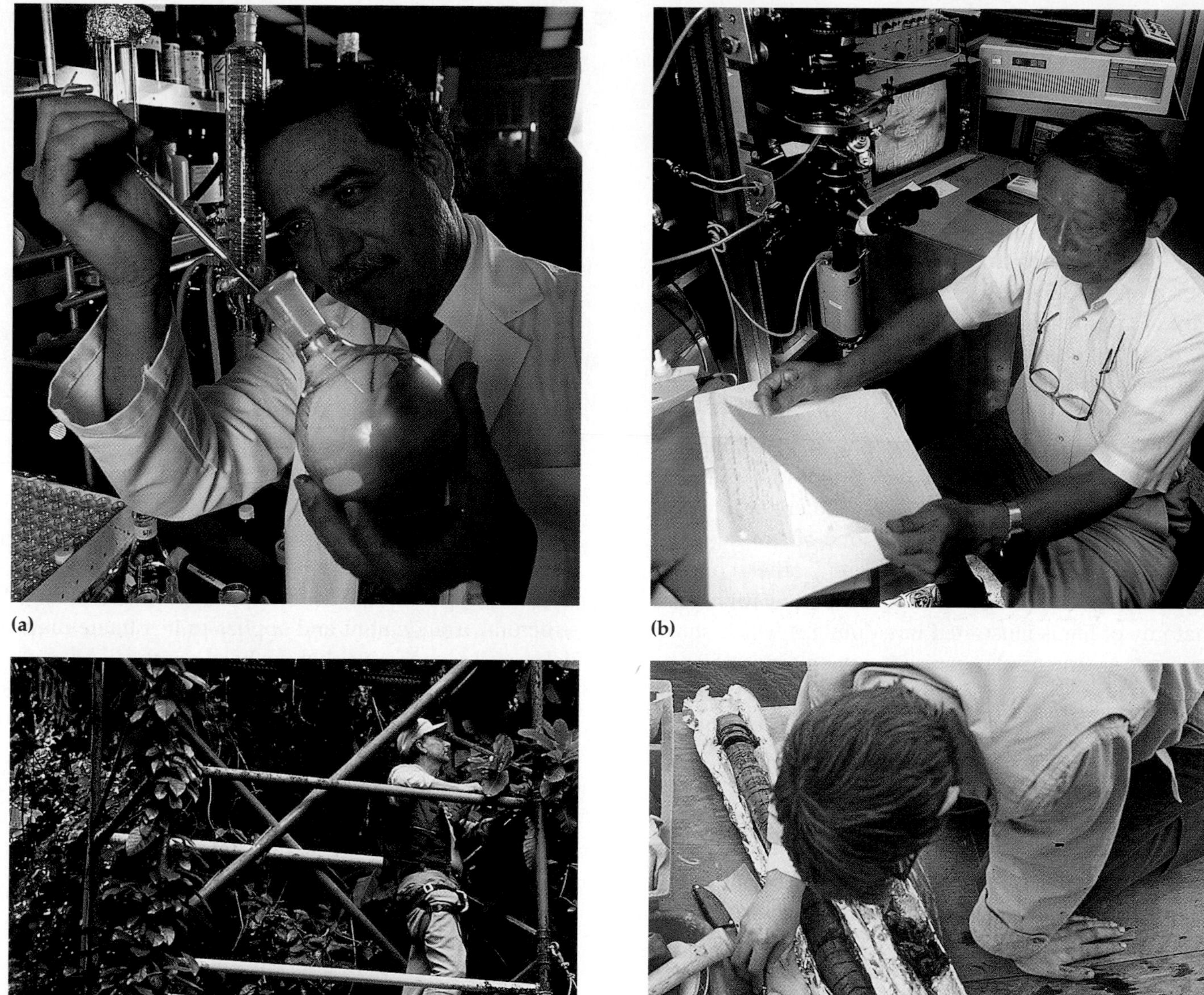

(a) (b) (c) (d)

FIGURE 1.1

Biologists study life on many different scales of size and time. (**a**) Investigating life at the molecular level, Eloy Rodriquez has identified many of the unique chemical substances produced by plants he collects in the tropics. (**b**) Shinya Inoué builds state-of-the-art microscopes, which he uses to explore the world of living cells. (**c**) Edward O. Wilson, at work here on a scaffold built to study the canopy in a Panamanian rain forest, focuses his research on whole organisms. (**d**) Biologists also trace the history of life over the scale of geologic time. Here, Margaret Davis examines a column of sediment bored from the bottom of a lake. The sediments contain a record of ancient pollen, providing clues about how the ice ages changed forests. You will learn more about the diverse research of these and other scientists in the interviews that precede each unit of chapters in your textbook.

Life is organized on many structural levels

A basic characteristic of life is a high degree of order. You can see it in the intricate pattern of veins throughout a leaf or in the colorful pattern of a bird's plumage. If you were to examine the vein of a leaf or the feather of a bird under a microscope, you would discover that biological order also exists at levels below what the unaided eye can see.

Biological organization is based on a hierarchy of structural levels, each level building on the levels below it. Atoms, the chemical building blocks of all matter, are ordered into complex biological molecules such as proteins. The molecules of life are arranged into minute structures called organelles, which are in turn the components of cells. Some organisms consist of single cells, but others, including plants and animals, are aggregates of many specialized types of cells. In such multicellular organisms, similar cells are grouped into tissues, and specific arrangements of different tissues form organs. For example, the nervous impulses that coordinate your movements are transmitted along specialized cells called neurons. The nervous tissue within your brain has billions of neurons organized into a communications network of spectacular complexity. The brain, however, is not pure nervous tissue; it is an organ built of many different tissues, including a type called connective tissue that forms the protective covering of the brain. The brain is itself part of the nervous system, which also includes the spinal cord and the many nerves that transmit messages between the spinal cord and other parts of the body. The nervous system is only one of several organ systems characteristic of humans and other complex animals. Another example of this parts-within-parts anatomy of life is illustrated in FIGURE 1.2, which shows the levels of structural organization in an aspen grove.

In the hierarchy of biological organization, there are tiers beyond the individual organism. A population is a localized group of organisms belonging to the same species; populations of species living in the same area make up a biological community; and community interactions that include nonliving features of the environment, such as soil and water, form an ecosystem.

Unfolding biological organization at its many levels is fundamental to the study of life. This text essentially follows such an organization, beginning by looking at the chemistry of life and ending with the study of ecosystems and the biosphere, the sum of all Earth's ecosystems. However, we will also see that biological processes transcend this hierarchy, with causes and effects at several organizational levels. For example, when a rattlesnake explodes from its coiled posture and strikes a desert mouse, the predator's coordinated movements result from complex interactions at the molecular, cellular, tissue, and organ levels within the snake. But there are also causes and effects of this behavior that operate on the level of the biological community where the snake and its prey live; the feeding response is triggered when the snake senses and locates the nearby mouse. And such episodes of predation have an important cumulative impact on the sizes of both the mouse and the rattlesnake populations. Most biologists specialize in the study of life at a particular level, but they gain broader perspective when they integrate their discoveries with processes occurring at lower or higher levels. A narrow focus on a single level of biological organization depreciates the fun and the power of biology.

Each level of biological organization has emergent properties

With each step upward in the hierarchy of biological order, novel properties emerge that were not present at the simpler levels of organization. These emergent properties result from interactions between components. A molecule such as a protein has attributes not exhibited by any of its component atoms, and a cell is certainly much more than a bag of molecules. If the intricate organization of the human brain is disrupted by a head injury, that organ will cease to function properly, even though all its parts may still be present. And an organism is a living whole greater than the sum of its parts.

The theme of emergent properties may seem, at first, to support a doctrine known as vitalism, which views life as a supernatural phenomenon beyond the bounds of physical and chemical laws. However, the concept of emergent properties merely accents the importance of structural arrangement and applies to inanimate material as well as to life. Neither the head nor the handle of a hammer alone is very useful for driving nails, but put these parts together in a certain way, and the functional properties of a hammer emerge. Diamonds and graphite have different properties because their carbon atoms are arranged differently. Unique properties of organized matter arise from how parts are arranged and interact, not from supernatural powers. And life is driven not by "vital forces" that defy explanation, but by principles of physics and chemistry extended into a new territory. The emergent properties of life simply reflect a hierarchy of structural organization without counterpart among inanimate objects.

Life resists a simple, one-sentence definition because it is associated with numerous emergent properties. Yet almost any child perceives that a dog or a bug or a tree is alive and a rock is not. We can recognize life without defining it, and we recognize life by what living things do. FIGURE 1.3 on p. 6 illustrates and describes some of the properties and processes we associate with the state of being alive.

Because the properties of life emerge from complex organization, scientists seeking to understand biological processes confront a dilemma. One horn of the dilemma is that we cannot fully explain a higher level of order by breaking it down into its parts. A dissected animal no longer functions; a cell dismantled to its chemical ingredients is no longer a cell. According to a principle known as holism, disrupting a living system interferes with the meaningful explanation of its processes. The other horn of the dilemma is the futility of trying to analyze something as complex as an organism or a cell without taking

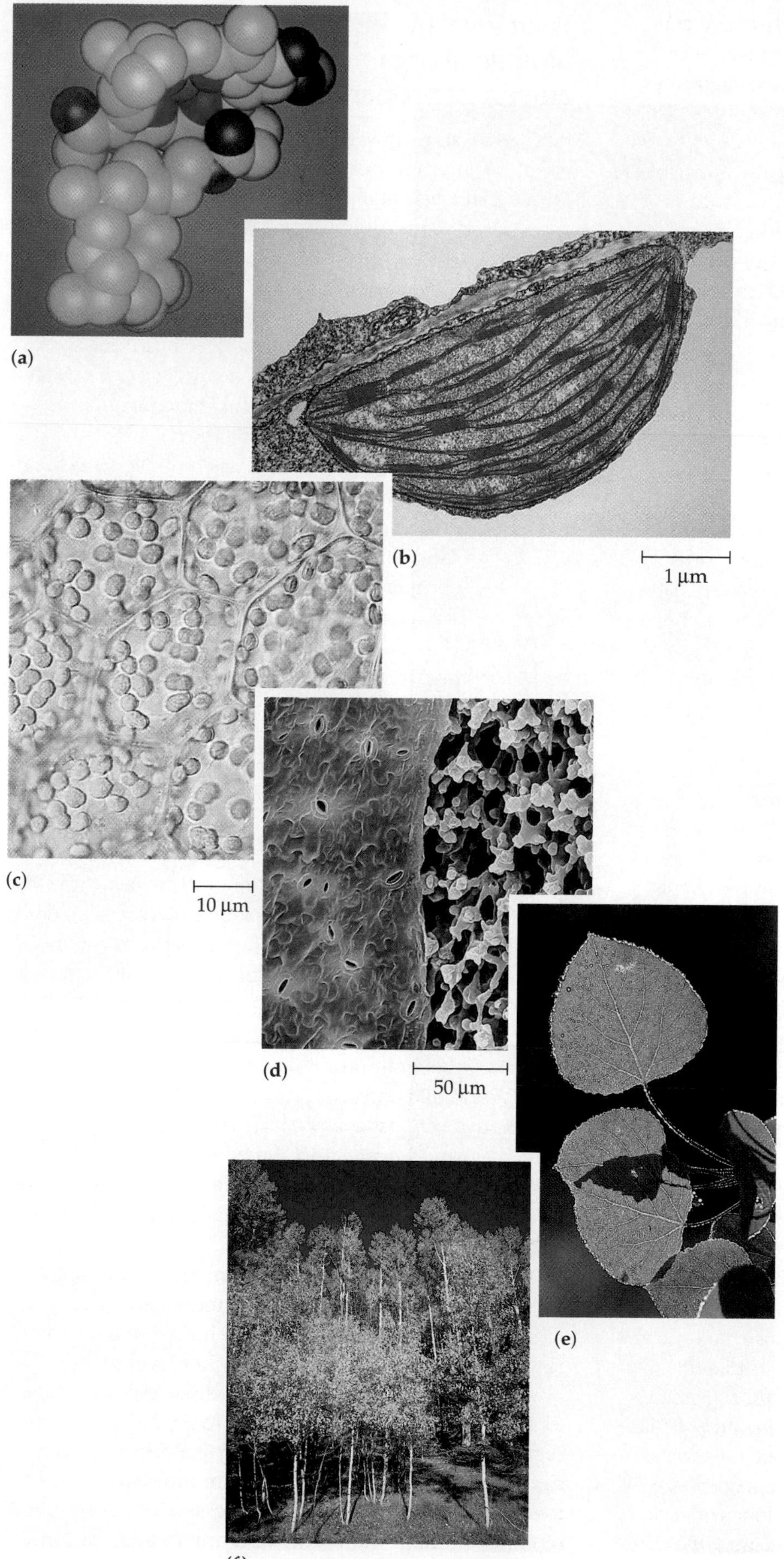

FIGURE 1.2

The hierarchy of biological organization. This sequence of images takes us all the way from atoms to a biological community of many interacting species. (**a**) Chlorophyll, represented here by a computer graphic model, is a molecule built from many atoms. This molecule in the leaves of plants absorbs sunlight as a source of energy for driving photosynthesis, the manufacture of food in the leaf. (**b**) The process of photosynthesis requires the participation of many other molecules organized within the cellular organelle called the chloroplast. (**c**) Many organelles cooperate in the functioning of the living unit we call a cell. Chloroplasts are evident in these leaf cells. (**d**) In multicellular organisms, cells are usually organized into tissues, groups of similar cells forming a functional unit. The leaf in this micrograph (a photograph taken with a microscope) has been cut obliquely, revealing two different specialized tissues. The honeycomblike tissue on the right consists of photosynthetic cells within the leaf. The tissue on the left with the small pores is the epidermis, the "skin" of the plant. The pores in the epidermis allow carbon dioxide, a raw material that is converted to sugar by photosynthesis, to enter the leaf. (**e**) The aspen leaf, a plant organ, has a specific organization of many different tissues, including photosynthetic tissue, epidermis, and the vascular tissue that transports water from the roots to the leaves. (**f**) These aspens are members of a biological community that includes many other species of organisms.

FIGURE 1.3

Some properties of life. (a) **Order:** All other characteristics of life emerge from an organism's complex organization, which is apparent in this closeup of a sunflower. (**b**) **Reproduction:** Organisms reproduce their own kind. Life comes only from life, an axiom known as biogenesis. Here, a Japanese macaque protects its offspring. (**c**) **Growth and development:** Heritable programs in the form of DNA direct the pattern of growth and development, producing an organism that is characteristic of its species. Shown here are embryos of a Costa Rican species of frog. (**d**) **Energy utilization:** Organisms take in energy and transform it to do many kinds of work. This bat obtains fuel in the form of nectar from the saguaro cactus. The bat will use energy stored in the molecules of its food to power flight and other work. (**e**) **Response to the environment:** This soon-to-be-digested cricket "tripped" the Venus flytrap when it stimulated hair cells on the surface of the modified leaves that make up the trap. The plant responded to this environmental stimulus with a rapid closure of the trap. (**f**) **Homeostasis:** Regulatory mechanisms maintain an organism's internal environment within tolerable limits, even though the external environment may fluctuate. This regulation is called homeostasis. For example, regulation of the amount of blood flowing through the blood vessels of this jackrabbit's large ears constantly adjusts heat loss to the surroundings. This contributes to homeostasis of the animal's body temperature. (**g**) **Evolutionary adaptation:** Life evolves as a result of the interaction between organisms and their environments. One consequence of evolution is the adaptation of organisms to their environment. The white fur of the arctic fox makes it nearly invisible against the animal's snowy surroundings.

(a)

(b)

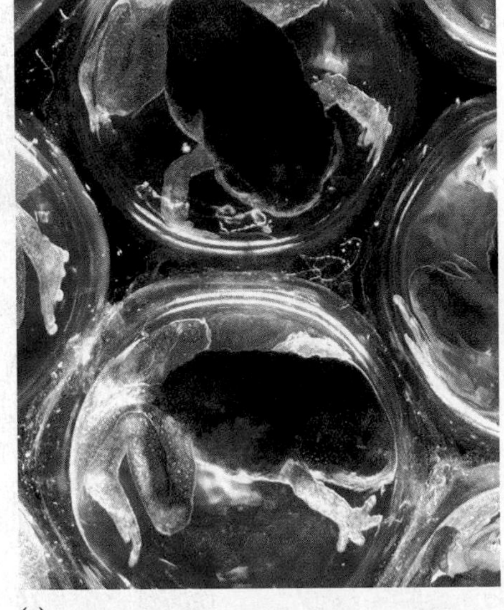

(c)

(d)

(e)

(f)

(g)

it apart. Reductionism—reducing complex systems to simpler components that are more manageable to study—is a powerful strategy in biology. For example, by studying the molecular structure of a substance called DNA that had been extracted from cells, James Watson and Francis Crick deduced, in 1953, how this molecule could serve as the chemical basis of inheritance. The central role of DNA was better understood, however, when it was possible to study its interactions with other substances in the cell. Biology balances the pragmatic reductionist strategy with the longer-range objective of understanding how the parts of cells and organisms are functionally integrated.

Cells are an organism's basic units of structure and function

As the lowest level of structure capable of performing *all* the activities of life, the cell has a special place in the hierarchy of biological organization. All organisms are composed of cells. They occur singly as a great variety of unicellular organisms, and they occur as the subunits of organs and tissues in plants, animals, and other multicellular organisms. In either case, the cell is an organism's basic unit of structure and function.

Robert Hooke, an English scientist, first described and named cells in 1665, when he observed a slice of cork (bark from an oak tree) with a microscope that magnified 30 times (30×). Apparently believing that the tiny boxes, or "cells," that he saw were unique to cork, Hooke never realized the significance of his discovery. His contemporary, a Dutchman named Anton van Leeuwenhoek, discovered organisms we now know to be single-celled. Using grains of sand that he had polished into magnifying glasses as powerful as 300×, Leeuwenhoek discovered a microbial world in droplets of pond water and also observed the blood cells and sperm cells of animals. In 1839, nearly two centuries after the discoveries of Hooke and Leeuwenhoek, cells were finally acknowledged as the ubiquitous units of life by Matthias Schleiden and Theodor Schwann, two German biologists. In a classic case of inductive reasoning—reaching a generalization based on many concurring observations—Schleiden and Schwann summarized their own microscopic studies and those of others by concluding that all living things consist of cells. This generalization forms the basis of what is known as the cell theory. This theory was later expanded to include the idea that all cells come from other cells. The ability of cells to divide to form new cells is the basis for all reproduction and for the growth and repair of multicellular organisms, such as humans.

Over the past 40 years, a powerful instrument called the electron microscope has revealed the complex structure of cells. All cells are enclosed by a membrane that regulates the passage of materials between the cell and its surroundings. Every cell, at some stage in its life, contains DNA, the heritable material that directs the cell's many activities.

Two major kinds of cells—prokaryotic cells and eukaryotic cells—can be distinguished based on structural organization (FIGURE 1.4). The cells of the microorganisms known as bacteria are prokaryotic. All other forms of life are composed of eukaryotic cells.

The eukaryotic cell, by far the more complex, is subdivided by internal membranes into many different

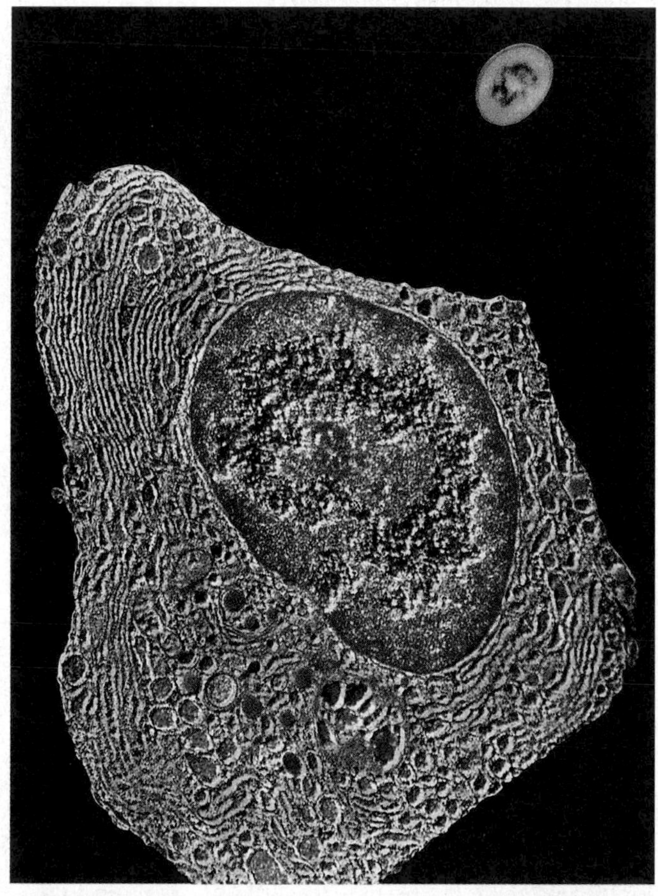

FIGURE 1.4

Two types of cells, as viewed with the electron microscope at the same scale. The eukaryotic cell, found in plants, animals, and all other organisms except bacteria, is characterized by an extensive subdivision into many different compartments, or organelles. In this micrograph of a human cell, the relatively large organelle at the right is known as the nucleus. Various organelles can also be seen in the cytoplasm outside the nucleus. The prokaryotic cell (above right), unique to bacteria, is much simpler, lacking most of the organelles found in eukaryotic cells. Compared to eukaryotic cells, prokaryotic cells are also generally much smaller. (Micrographs are artificially colored.)

functional compartments, or organelles. In eukaryotic cells, the DNA is organized along with proteins into structures called chromosomes contained within a nucleus, the largest organelle of most cells. Surrounding the nucleus is the cytoplasm, a thick liquid in which are suspended the various organelles that perform most of the cell's functions. Some eukaryotic cells, including those of plants, have tough walls external to their membranes. Animal cells lack walls.

In the much simpler prokaryotic cell, the DNA is not separated from the rest of the cell into a nucleus. Prokaryotic cells also lack the cytoplasmic organelles typical of eukaryotic cells. Almost all prokaryotic cells (bacteria) have tough external cell walls.

Although eukaryotic and prokaryotic cells contrast sharply in complexity, we will see that they have some key similarities. Cells vary widely in size, shape, and specific structural features, but all are highly ordered structures that carry out complicated processes necessary for maintaining life.

The continuity of life is based on heritable information in the form of DNA

Order implies information; instructions are required to arrange parts or processes in a nonrandom way. Biological instructions are encoded in the molecule known as DNA (deoxyribonucleic acid). DNA is the substance of genes, the units of inheritance that transmit information from parents to offspring (FIGURE 1.5).

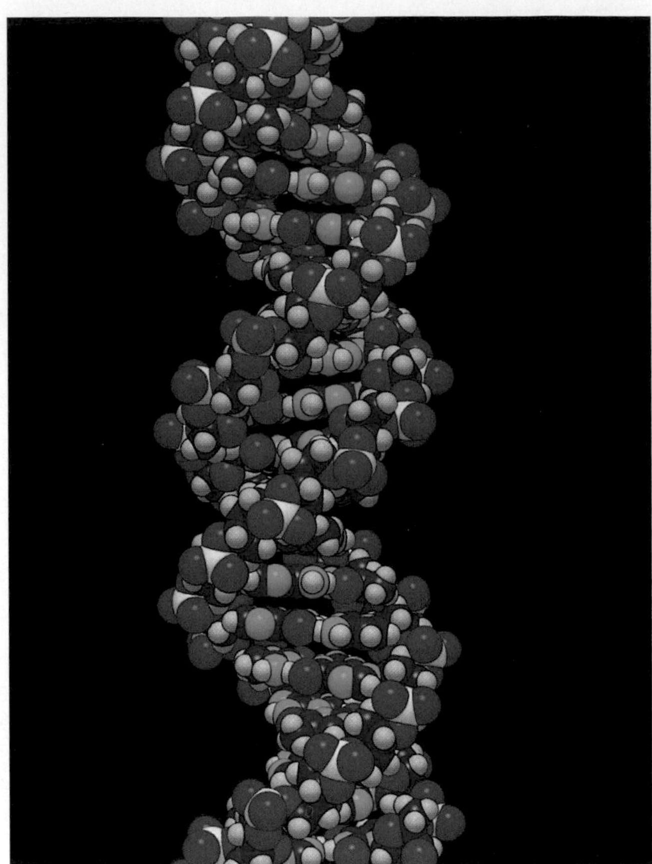

FIGURE 1.5
The genetic material: DNA. The molecule that carries biological information from one generation to the next takes the three-dimensional form of a double helix. Along the length of the molecule, information is encoded in specific sequences of nucleotides, the chemical building blocks of DNA.

Each DNA molecule is a long chain made up of four chemical building blocks called nucleotides. The way DNA conveys information is analogous to the way we arrange the letters of the alphabet into precise sequences with specific meanings. The word *rat*, for example, conjures up an image of a rodent; *tar* and *art*, which contain the same letters, mean quite different things. Libraries are filled with books containing information encoded in varying sequences of only 26 letters. We can think of nucleotides as the alphabet of inheritance. Specific sequential arrangements of these four chemical letters encode the precise information in a gene. If the entire library of genes stored within the microscopic nucleus of a single human cell were written in letters the size of those you are now reading, the information would fill more than a hundred books as large as this one. The complex structural organization of an organism is thus specified by an inherited script conveying an enormous amount of coded information.

All forms of life employ essentially the same genetic code. A particular sequence of nucleotides says the same thing to one organism as it does to another; differences between organisms reflect genetic programs of different nucleotide sequences. The diverse forms of life are different expressions of a common language for programming biological order.

Inheritance itself depends on a mechanism for copying DNA and passing its sequence of chemical letters on to offspring. As a cell prepares to divide to form two cells, it copies its DNA. A mechanical system that moves chromosomes then distributes the genetic material equally to the two "daughter" cells. In species that reproduce sexually, offspring inherit copies of DNA present in the parents' sperm and egg cells. The continuity of life over the generations and over the eons has its molecular basis in the reproduction of DNA.

A feeling for organisms enriches the study of life

While cells are the units of organisms, it is organisms that are the units of life. It's an important distinction. Except for unicellular life, "cell" does not equal "organism." A single-celled organism such as an amoeba is analogous not to one of your cells, but to your whole body. What the amoeba accomplishes with a single cell—the uptake and processing of nutrients, excretion of wastes, response to environmental stimuli, reproduction, and other functions—a human or other multicellular organism accomplishes with a division of labor among specialized tissues, organs, and organ systems. Unlike the amoeba, none of your cells could live for long on its own. The organism we recognize as an animal or plant is not a collection of unicells, but a multicellular cooperative with the emergent properties of "whole organism."

We may go on to study life at different levels of organization, but it is organisms we must keep in mind if our exploration of cells or ecosystems is to be meaningful. The story of Barbara McClintock is a classic example. In 1983, McClintock (1902–1992) was awarded a Nobel Prize for discoveries she had made three decades earlier (FIGURE 1.6). Investigating the inheritance of kernel color in Indian corn, McClintock uncovered a type of mutation, a change in DNA, that did not conform to genetic theory of the 1950s. In fact, few geneticists understood the importance of McClintock's experimental results until molecular biologists discovered the same phenomenon in bacteria in the 1970s. This mechanism of mutation turns out to be one of the types of genetic rearrangement that can cause cancer. You will learn about these important discoveries in Chapters 17 and 18. What is important to appreciate now is that Barbara McClintock was only able to see the signs of this unexpected behavior of DNA because she was so familiar with corn, the organism with which she worked for over 60 years. Author Evelyn Keller highlighted this asset in *A Feeling for the Organism*, her biography of McClintock (New York: W. H. Freeman, 1983). Keller tells how McClintock knew the individual corn plants in her experiments and how she enjoyed her close involvement with them. It is a common misconception that scientific objectivity demands that biologists be detached from the organisms they study. In fact, as Keller concluded from her conversations with Barbara McClintock, "Good science cannot proceed without a deep emotional investment on the part of a scientist."

FIGURE 1.6
Barbara McClintock, legendary corn geneticist. In the 1930s, McClintock, far right, and her fellow graduate students at Cornell University were breeding corn in order to trace variations among the plants to specific genes. McClintock later continued her research at the Cold Spring Harbor Laboratory on Long Island, New York. McClintock's most famous discovery, a type of genetic mutation, had little impact until new methods made it possible for other scientists to observe these changes in the DNA of microorganisms.

A "feeling for organisms" is an emotion you and the best research specialists in biology have in common. The exploration of cells and their molecular components has revealed much about life. However, most of us do not arrive at biology from an interest in these minute structures, but from our personal experiences with organisms—from observing spiders weaving webs, plant seedlings poking through sidewalk cracks, birds courting mates, and the leaves of trees marking the seasons. Biophilia is manifest in every child.

There are a few questions we should ask about any organism we choose to study. What kind of organism is it? Where does it live? How does it acquire its nutrients and other resources from the environment? How is the organism equipped for its way of life? A feeling for organisms is a theme that enriches our study of life at all levels.

Structure and function are correlated at all levels of biological organization

Given a choice of tools, you would not loosen a screw with a hammer or pound a nail with a screwdriver. How a device works is correlated with its structure: Form fits function. Applied to biology, this theme is a guide to the anatomy of life at its many structural levels, from molecules to organisms. Analyzing a biological structure gives us clues about what it does and how it works. Conversely, knowing the function of a structure provides insight about its construction.

An example of this structure-function theme is the aerodynamically efficient shape of a bird's wing. Beneath the external contours, the skeleton of the bird also has structural qualities that contribute to flight, with bones that have a strong but light honeycombed structure. The flight muscles of a bird are controlled by neurons. Long extensions of the neurons transmit nervous impulses, making these cells especially well structured for communication. As an example of functional anatomy at the

(a)

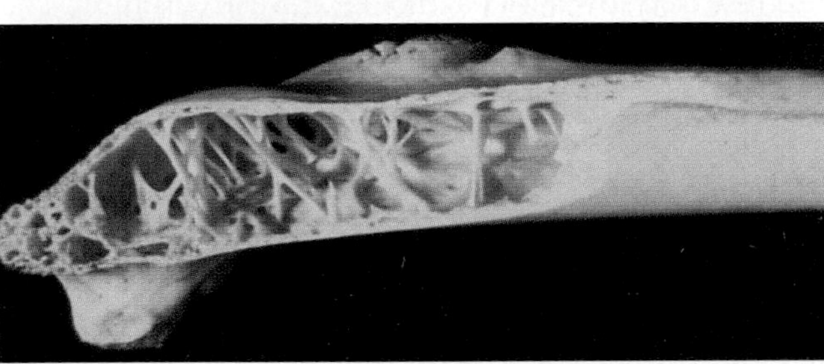

(b)

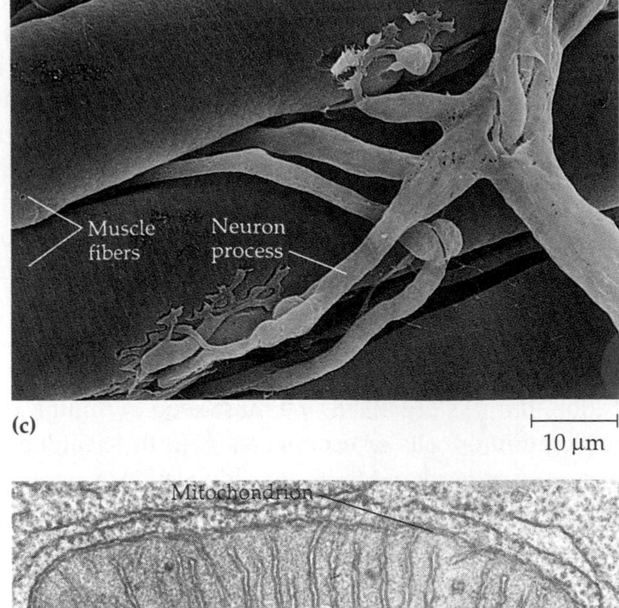

(c)

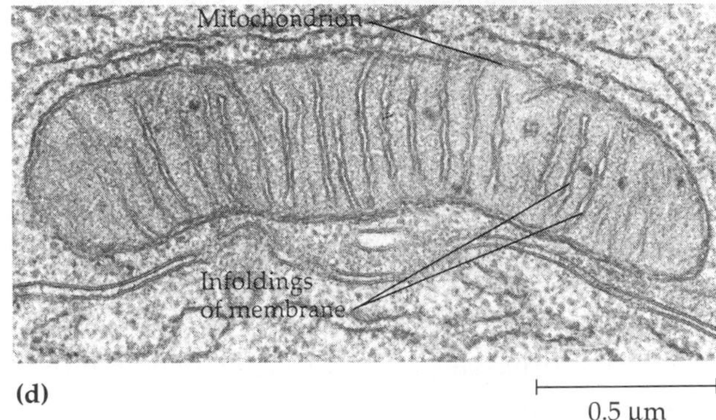

(d)

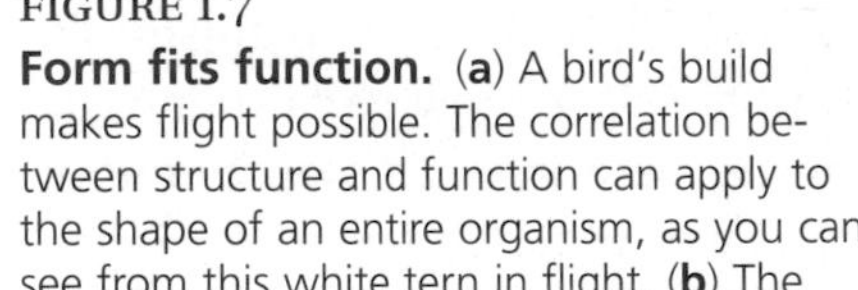

FIGURE 1.7
Form fits function. (**a**) A bird's build makes flight possible. The correlation between structure and function can apply to the shape of an entire organism, as you can see from this white tern in flight. (**b**) The structure-function theme also applies to organs and tissues. For example, the honeycombed construction of a bird's bones provides a lightweight skeleton of great strength. (**c**) The form of a cell fits its specialized function. Nerve cells, or neurons, have long extensions (processes) that transmit nervous impulses—here, to muscle cells. (**d**) Functional beauty is also apparent at the subcellular level. This organelle, called a mitochondrion, has an inner membrane that is extensively folded, a structural solution to the problem of packing a relatively large amount of this membrane into a very small container.

subcellular level, consider the organelles called mitochondria. They are the sites of cellular respiration, the chemical process that powers the cell by using oxygen to help tap the energy stored in sugar and other food molecules. A mitochondrion is surrounded by an outer membrane, but it also has an inner membrane with many infoldings. Molecules embedded in the inner membrane carry out many of the steps in cellular respiration, and the infoldings pack a large amount of this membrane into a minute container. In exploring life on its different structural levels, we will discover functional beauty at every turn (FIGURE 1.7).

Organisms are open systems that interact continuously with their environments

Life does not exist in a vacuum. An organism is an example of what scientists call an open system, with "open" referring to the exchange of materials and energy between the system and its surroundings. As an open system, each organism interacts continuously with its environment, which includes other organisms as well as nonliving factors. The roots of a tree, for example, absorb water and minerals from the soil, and the leaves take in carbon dioxide from the air. Sunlight absorbed by chlorophyll, the green pigment of leaves, drives photosynthesis, which converts water and carbon dioxide to sugar and oxygen. The tree releases oxygen to the air, and its roots change the soil by breaking up rocks into smaller particles, secreting acid, and absorbing minerals. Both organism and environment are affected by the interaction between them. The tree also interacts with other life, including soil microorganisms associated with its roots and animals that eat its leaves and fruit.

The many interactions between organisms and their environment are interwoven to form the fabric of an ecosystem. The dynamics of any ecosystem include two major processes. One is the cycling of nutrients. For example, minerals acquired by plants will eventually be returned to the soil by microorganisms that decompose leaf litter, dead roots, and other organic debris. The second major process in an ecosystem is the flow of energy from sunlight to photosynthetic life (producers) to organisms that feed on plants (consumers) (FIGURE 1.8). The theme of organisms as open systems that exchange materials and energy with their surroundings is essential to understanding life on all levels of organization.

Diversity and unity are the dual faces of life on Earth

Diversity is a hallmark of life. Biologists have identified and named about 1.5 million species, including over 260,000 plants, almost 50,000 vertebrates (animals with backbones), and more than 750,000 insects. Thousands of newly identified species are added to the list each year. Estimates of the total diversity of life range from about 5 million to over 30 million species.

Biological diversity is something to relish and preserve, but it can also be a bit overwhelming. To make the diversity somewhat more comprehensible, people have devised ways of grouping species that are similar. Thus, we may speak of squirrels and pine trees without distinguishing the many different species belonging to each group. Taxonomy, the branch of biology concerned with naming and classifying species, groups organisms according to a more formal scheme. The scheme consists of different levels of classification, each more comprehensive than those below it (FIGURE 1.9). The broadest units of classification are the kingdoms. For the past two decades, most biologists divided the diversity of life into five kingdoms, though many now prefer classification schemes with six or more kingdoms. The rationale for these classifications will be introduced in Unit Five of the text. For now, it is mainly the immense range of biological diversity that is important to appreciate, not so much the ways we try to order the diverse forms of life.

The microorganisms known as bacteria are either all placed in the kingdom Monera (five-kingdom scheme) or divided into two major groups, the kingdoms

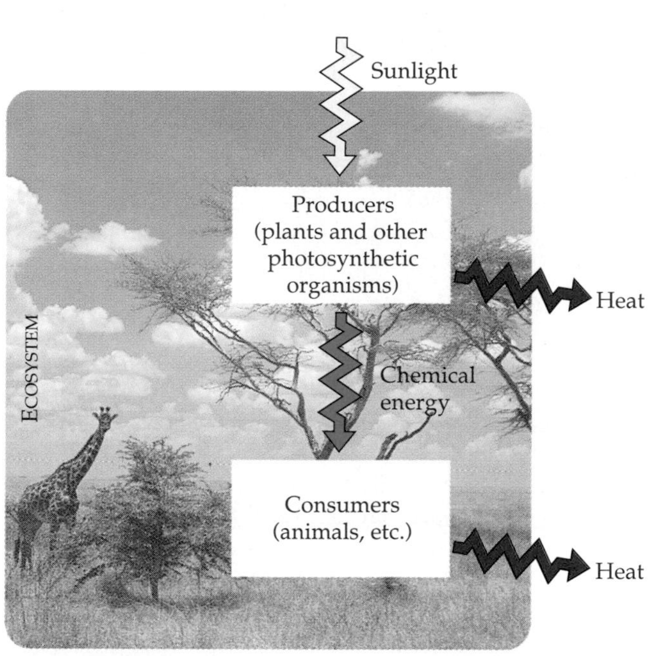

FIGURE 1.8

An introduction to energy flow in an ecosystem. Living is work, and work requires that organisms obtain and use energy. Most ecosystems are solar-powered. Plants and other photosynthetic organisms convert light energy to the chemical energy stored in sugar and other complex molecules. By breaking these fuel molecules down to simpler molecules, organisms can harvest the stored energy and put it to work. In this schematic diagram of energy flow in an ecosystem, photosynthetic organisms are called producers because the entire ecosystem depends on their photosynthetic products. Animals and other consumers acquire their energy in chemical form by eating plants, by eating animals that ate plants, or by decomposing organic refuse, such as leaf litter and dead animals. The energy that enters an ecosystem in the form of light exits in the form of heat, which all organisms dissipate to their surroundings whenever they perform work.

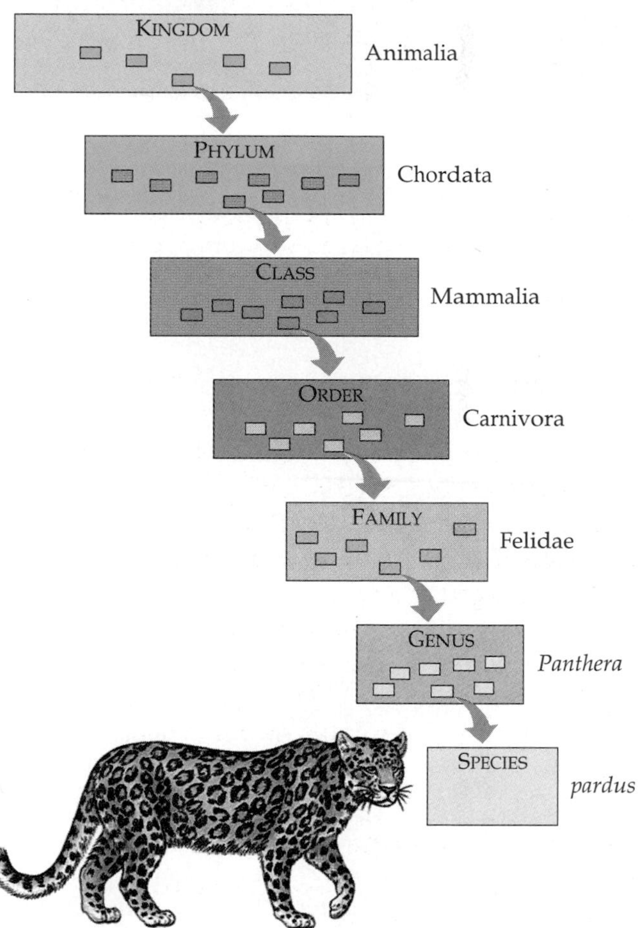

FIGURE 1.9

Classifying life. The taxonomic scheme classifies species into groups subordinate to more comprehensive groups. Species that are very similar are placed in the same genus, genera are grouped into families, and so on, each level of classification being more comprehensive than those it includes. This example classifies a leopard.

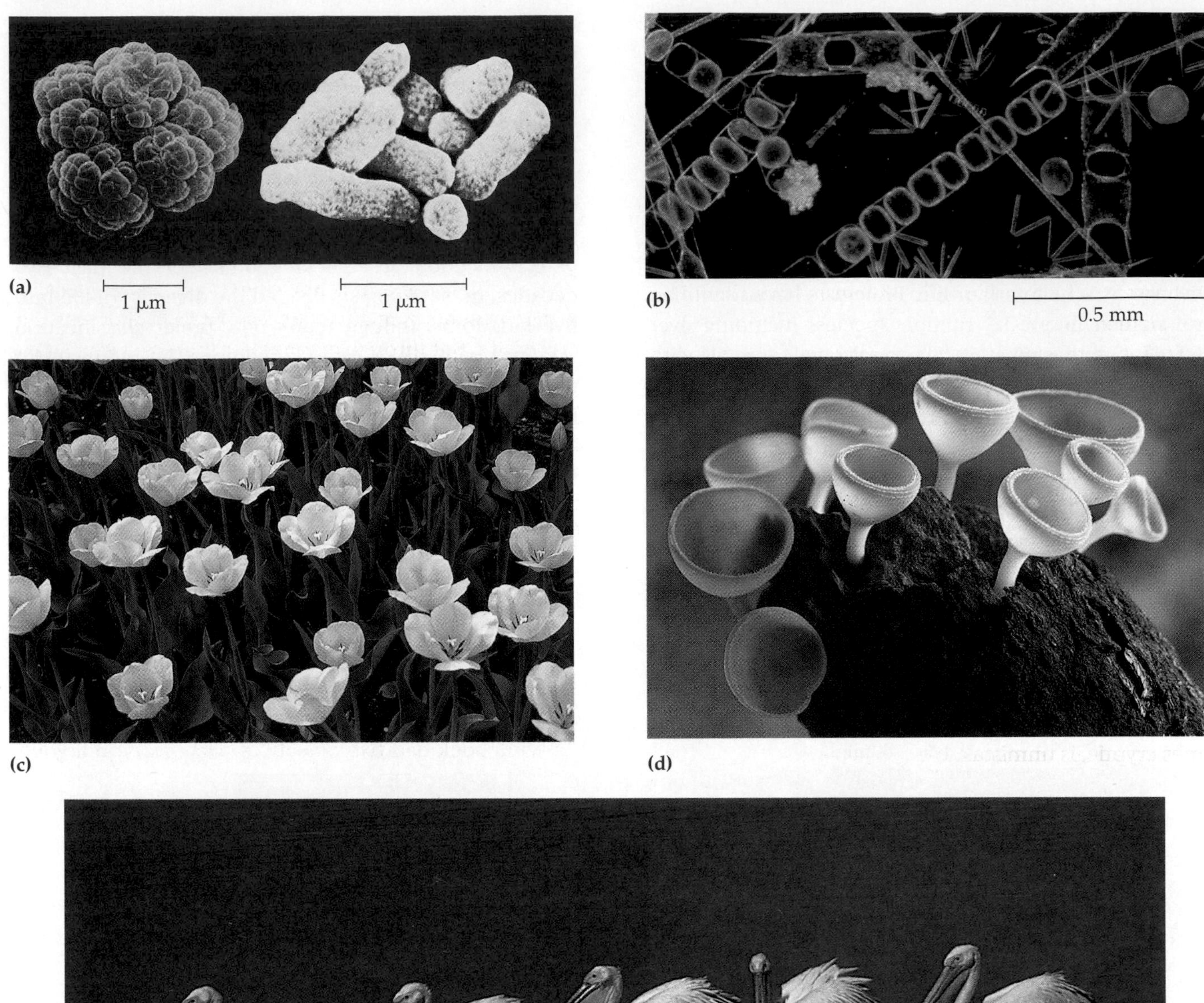

FIGURE 1.10

Kingdoms of life. (**a**) The simpler structure of the prokaryotic cell distinguishes bacteria from the eukaryotic kingdoms. In the five-kingdom classification scheme, all prokaryotes are placed in the kingdom Monera. A six-kingdom scheme recognizes two distinct groups of prokaryotes, the kingdoms Archaebacteria (left) and Eubacteria (right). In Chapter 25, you'll learn how these two prokaryotic groups differ. (**b**) Kingdom Protista consists of unicellular eukaryotes and their relatively simple multicellular relatives. This is an assortment of protists inhabiting pond water. (**c**) Kingdom Plantae, represented by these tulips, consists of multicellular eukaryotes that carry out photosynthesis. (**d**) Kingdom Fungi is defined, in part, by the nutritional mode of its members, organisms that absorb nutrients after decomposing organic refuse. These are Costa Rican cup fungi. (**e**) Kingdom Animalia consists of multicellular eukaryotes that ingest other organisms. These are rose pelicans in Kenya.

Archaebacteria and Eubacteria (six-kingdom scheme). Although there are important differences between these two groups, they have a key feature in common: All bacterial cells are prokaryotic, the simpler of the two types of cells (see FIGURE 1.4). This distinguishes the bacteria from the four kingdoms of eukaryotic organisms: protists, fungi, plants, and animals (FIGURE 1.10).

Kingdom Protista consists mostly of eukaryotic organisms that are unicellular—for example, amoebas and the other microscopic organisms known as protozoa. Many biologists extend the boundaries of the kingdom Protista to include certain multicellular forms such as seaweeds that seem to be more closely related to unicellular species than to plants, fungi, or animals.

The remaining three kingdoms—Plantae, Fungi, and Animalia—consist of multicellular eukaryotes. Plants are characterized by photosynthesis. Fungi are mostly decomposers that absorb nutrients by breaking down the complex molecules of dead organisms and waste, such as leaf litter and feces. Animals obtain food by ingestion, which is the eating and digesting of other organisms. Thus, plants, fungi, and animals are distinguished partly by their contrasting modes of nutrition. Each kingdom, however, has many other unique features, which will be discussed in Unit Five.

If life is so diverse, how can biology have any unifying themes at all? What, for instance, can a mold, a tree, and a human possibly have in common? As it turns out, a great deal! Underlying the diversity of life is a striking unity, especially at the lower levels of organization. We can see it, for example, in the universal genetic code shared by all organisms. Unity is also evident in certain similarities of cell structure (FIGURE 1.11). Above the cellular level, however, organisms are so variously adapted to their ways of life that describing biological diversity remains an essential goal of biology. But few biologists view taxonomy as a mere cataloging of seemingly unrelated living objects. In fact, the kinship of all life, though sometimes cryptic, is unmistakable.

Evolution is the core theme of biology

The history of life is a chronicle of a restless Earth billions of years old, inhabited by a changing cast of living forms (FIGURE 1.12). Life evolves. Just as an individual has a family history, each species is one tip on a branching tree of life extending back in time through ancestral species more and more remote. Species that are very similar, such as the horse and zebra, share a common ancestor that represents a relatively recent branch point on the tree of life. But through an ancestor that lived much farther back in time, horses and zebras are also related to rabbits, humans, and all other mammals. And mammals, reptiles, birds, and all other vertebrates share a common ancestor even more ancient. Trace evolution back far enough, and there are only the primeval prokaryotes that inhabited Earth more than three billion years ago. All of life is connected. Evolution, the processes that have transformed life on Earth from its earliest beginnings to its apparently unending diversity today, is the one biological theme that ties together all others.

Charles Darwin brought biology into focus in 1859 when he published *The Origin of Species* (FIGURE 1.13). Darwin's book had two objectives. First, Darwin argued

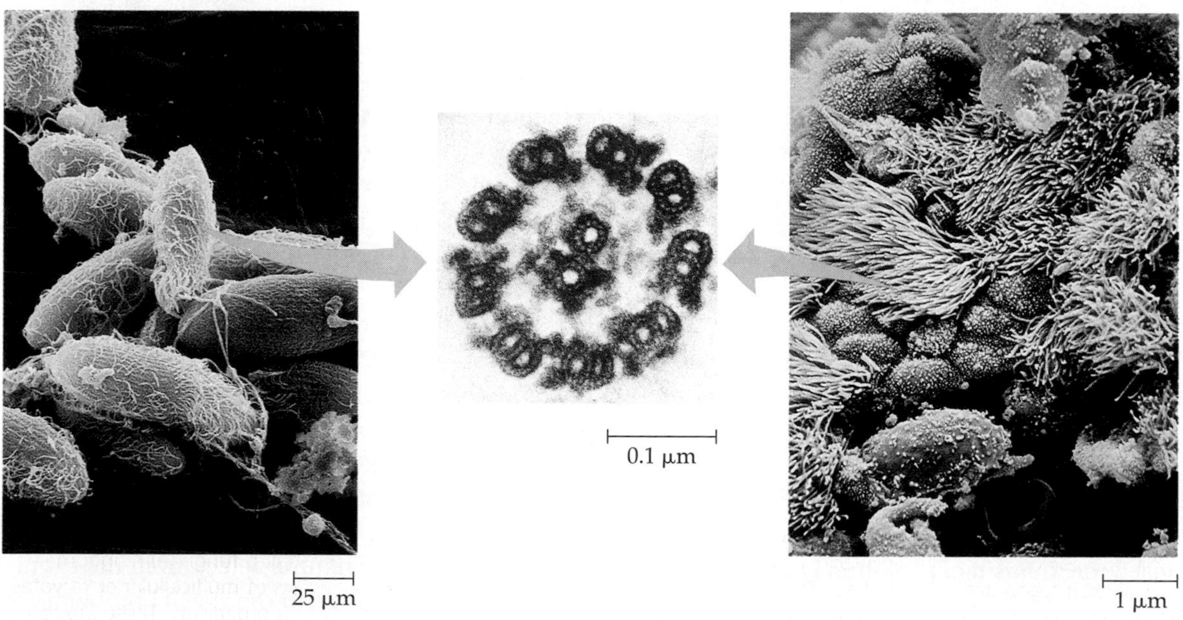

FIGURE 1.11

An example of unity underlying the diversity of life: the architecture of eukaryotic cilia. Eukaryotic organisms as diverse as protozoa (kingdom Protista) and animals possess cilia, locomotory "hairs" that extend from cells. The cilia of *Paramecium* (left micrograph), a protozoan, propel the cell through pond water. The cells that line the human windpipe (right micrograph) are also equipped with cilia, which help keep the lungs clean by moving a film of debris-trapping mucus upward. Comparing cross sections of cilia from diverse eukaryotes reveals a common structural organization (center micrograph). Such striking similarity in complex components contributes to the evidence that organisms as different as protozoa and humans are, to some degree, related.

FIGURE 1.12

The fossil record is one type of historical documentation that chronicles evolution. *Archaeopteryx,* the animal documented in this fossil, lived about 150 million years ago. It had feathers, the exclusive trademark of birds. But *Archaeopteryx* was much more reptilian than any modern bird. It had teeth, claws on its wings, and a bony tail—all traits of reptiles. The fossil record is compatible with other types of evidence supporting the hypothesis that birds evolved from reptiles. Evolutionary biology is solid science because its hypotheses can be tested.

FIGURE 1.13

Charles Darwin (1809–1882). Darwin and his son William posed for this photograph in 1842. The author of numerous books and monographs on topics as diverse as barnacles, plant movements, and island geology, Darwin would be remembered as one of the greatest naturalists of the nineteenth century, even if he had never published on the topic of evolution. But it was *The Origin of Species* that established Darwin's place as the most influential scientist in the development of modern biology. He is buried next to Isaac Newton in London's Westminster Abbey.

convincingly from several lines of evidence that contemporary species arose from a succession of ancestors through a process of "descent with modification," his phrase for evolution. (The evidence for evolution is discussed in detail in Chapter 20.) The second objective of Darwin's book was to present his theory for *how* life evolves. This proposed mechanism of evolution is called natural selection.

Darwin synthesized the concept of natural selection from observations that by themselves were neither new nor profound. Others had the pieces of the puzzle, but Darwin saw how they fit together. He inferred natural selection by connecting two observations:

OBSERVATION #1: *Individual variation.* Individuals in a population of any species vary in many heritable traits.

OBSERVATION #2: *Struggle for existence.* Any population of a species has the potential to produce far more offspring than the environment can possibly support with food, space, and other resources. This overproduction makes a struggle for existence among the variant members of a population inevitable.

INFERENCE: *Differential reproductive success.* Those individuals with traits best suited to the local environment generally leave a disproportionately large number of surviving, fertile offspring. This selective reproduction favors the representation of certain heritable variations in the next generation. It is this differential reproductive success that Darwin called natural selection, and he envisioned it as the cause of evolution.

We see the products of natural selection in the exquisite adaptations of organisms to the special problems of their environments (FIGURE 1.14). Notice, however, that natural selection does not *create* adaptations; rather, it screens the heritable variations in each generation, increasing the frequencies of some variations and decreasing the frequencies of others over the generations. Adaptation is an editing process, with heritable variations exposed to environmental factors that favor the reproductive success of some individuals over others. The

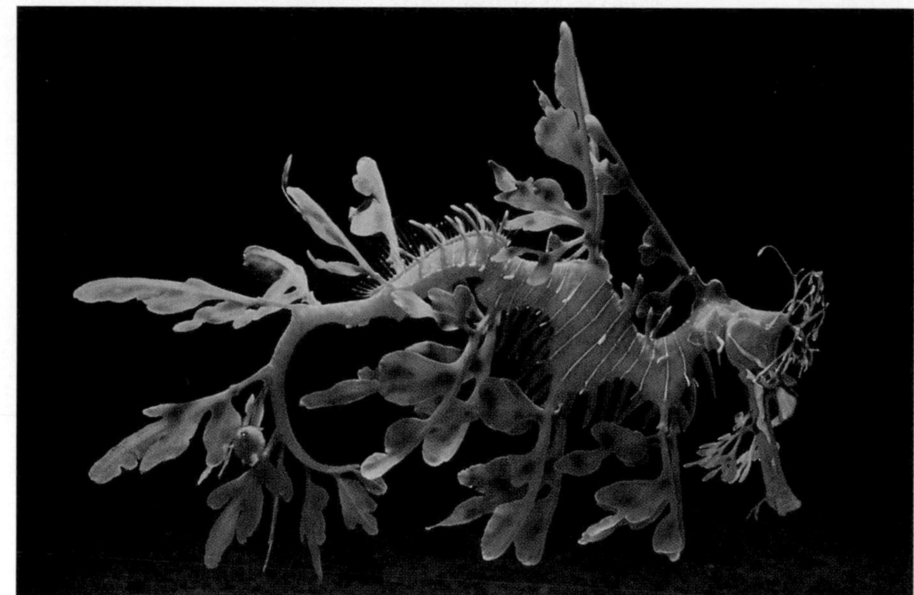

FIGURE 1.14
Evolutionary adaptation is a product of natural selection. This sea horse lives among kelp (seaweed). The fish looks so much like a seaweed that it lures prey into the seeming safety of the kelp forest and then eats them. In the Darwinian view of life, the best-camouflaged members of the sea horse population have the greatest probability of obtaining food and escaping predators—and thus the greatest probability of surviving and leaving offspring. By this mechanism of differential reproductive success, or natural selection, the interaction between environment and the heritable variation among members of the sea horse population gradually refined and maintained the camouflage over the generations.

camouflage of the sea horse in FIGURE 1.14 did not result from individuals changing during their lifetimes to look more like their backgrounds and then passing that improvement on to offspring. The adaptation evolved gradually over many generations by the greater reproductive success in each generation of individuals who were innately better camouflaged than the average member of the sea horse population.

Darwin proposed that natural selection, by its cumulative effects over vast spans of time, could produce new species from ancestral species. This would occur, for example, when a population fragments into several populations isolated in different environments. In these various arenas of natural selection, what begins as one species may gradually diversify into many, as the geographically isolated populations adapt over many generations to different sets of environmental problems. Descent with modification accounts for both the unity and the diversity we observe in life. Features shared by two species are due to their descent from common ancestors, and differences between the species are due to natural selection modifying the ancestral equipment in different environmental contexts. Evolution is the core theme of biology—a unifying thread that will tie together every chapter of this text.

Science as a process of inquiry often involves hypothetico-deductive thinking

Biology is classified as a natural science. Having identified unifying themes that apply specifically to the study of life, we now examine some general features of science as a process.

Like life, science is better understood by observing it than by trying to create a precise definition. The word *science* is derived from a Latin verb meaning "to know." Science is a way of knowing. It emerges from our curiosity about ourselves, the world, and the universe. Striving to understand seems to be one of our basic drives. At the heart of science are people asking questions about nature and believing that those questions are answerable. Scientists tend to be quite passionate in their quest for discovery. Max Perutz, a Nobel Prize–winning biochemist, puts it this way: "A discovery is like falling in love and reaching the top of a mountain after a hard climb all in one, an ecstasy induced not by drugs but by the revelation of a face of nature that no one has seen before."

A process known as the scientific method outlines a series of steps for answering questions, but few scientists adhere rigidly to this prescription. Science is a less structured process than most people realize. Like other intellectual activities, the best science is a process of minds that are creative, intuitive, imaginative, and social. Perhaps science is distinguished by its conviction that natural phenomena, including the processes of life, have natural causes—and by its obsession with evidence. Scientists are generally skeptics. As you begin each unit of study in this text, you will meet such scientists through personal interviews and come to understand a bit more about how they think and why they enjoy their work.

Although it is counterproductive to reduce science to a stereotyped method, we *can* identify a common theme of the scientific process: *hypothetico-deductive thinking*. The first part of this term refers to *hypothesis*, which is a tentative answer to some question—an explanation on trial. Consider, for example, this imaginary scenario: Scott and Ian, identical twins, are sleepy every day in their 1:00 history class; they often doze and want to know why, so they can prevent the drowsiness and improve their history grades. Maybe eating a big lunch just before going to history every day makes them sleepy. Or maybe the classroom is too warm. Maybe Scott and Ian are listless because they sit in the back of their history class and are less involved than they are in their other classes, where they always sit in front. Perhaps their history professor is really boring. Or maybe it's just the time of day. These are all hypotheses, possible explanations for this daily behavior of sleeping through history class. It is possible to test these hypotheses.

The *deductive* in hypothetico-deductive thinking refers to how deductive reasoning is used to test hypotheses. Deduction contrasts with induction. Recall that induction is reasoning from a set of specific observations to reach a general conclusion (as in "All organisms are composed of cells"). In deduction, the reasoning flows in the reverse direction, from the general to the specific. From general premises, we extrapolate to specific results we should expect if the premises are true: If all organisms are made of cells (premise #1) and humans are organisms (premise #2), then humans are composed of cells (prediction about a specific case). In the scientific process, deduction usually takes the form of predictions about the results of experiments or observations we should expect *if* a particular hypothesis (premise) is correct. We then test the hypothesis by performing the experiments or making observations to see whether or not the predicted results occur. This deductive testing takes the form of *"If . . . then"* logic:

HYPOTHESIS #1: *If* the twins get sleepy because they eat lunch before their 1:00 class,

EXPERIMENT: and Scott postpones lunch until class ends at 2:00 (but Ian still eats at noon),

PREDICTED RESULT #1: *then* Scott should be less sleepy than Ian in history class.

PREDICTED RESULT #2: *then* Scott should get sleepy in his 3:00 class instead.

OR

HYPOTHESIS #2: *If* the twins get sleepy in their 1:00 class because they sit in the back of the room,

EXPERIMENT: and Scott (but not Ian) moves to the front of the class,

PREDICTED RESULT: *then* Scott should be more alert than Ian during the 1:00 class.

Now, suppose the result of the first experiment is that Scott is still as sleepy as Ian (but hungrier) in their 1:00 class. The twins try the second experiment, with the result that now Scott dozes in the front row of class instead of the back row. The twins continue to test and reject various hypotheses. And then they are the subjects of an experiment they did not even plan—daylight savings time begins. Scott and Ian (and most of their classmates) notice that for a day or two, they have more trouble than usual waking up in the morning. However, the twins find that they are more alert than usual in their 1:00 class and instead hit their low point during the 2:00–3:00 break in their schedule. Within a few days, patterns are back to normal, and Scott and Ian are drowsy during their 1:00 class again. The twins conclude that they are just naturally sleepy at this time of day and decide not to schedule 1:00 classes in future semesters.

Although this example is fanciful and there are flaws in the experiments, five important points about hypotheses are evident:

1. *Hypotheses are possible causes.* A generalization based on inductive reasoning is not a hypothesis. Based on the twins' daily experience in history class, they may conclude, "We are sleepy every day at about 1:00," but that conclusion merely summarizes a set of observations. A hypothesis, remember, is a tentative *explanation* for what we have observed. "The warm temperature of the room causes Scott and Ian to sleep through class" is a hypothesis.

2. *Hypotheses reflect past experience with similar questions.* Sometimes hypotheses are described as *educated* propositions about cause. For example, the hypothesis that a warm classroom causes drowsiness may be based on a general experience of sleepiness under such conditions. Although we should consider any hypothesis that is a possible explanation for what we have observed, the hypotheses we test first should be those that seem the most reasonable, based on what we already know.

3. *Multiple hypotheses should be proposed whenever possible.* Proposing alternative explanations that can answer a question is good science. If we operate with a single hypothesis, especially one we favor, we may direct our investigation toward a hunt for evidence in support of this hypothesis.

4. *Hypotheses should be testable via the hypothetico-deductive approach.* Hypotheses should be phrased in a way that enables us to make predictions that can be tested by experiments or further observation. "Scott falls asleep in history class because the devil makes him do it" is not a testable hypothesis and therefore does not lend itself to the scientific process. The requirement that hypotheses be testable limits the scope of questions that science can answer.

5. *Hypotheses can be eliminated but not confirmed with absolute certainty.* If moving to the front of his 1:00 class

does not reduce Ian's drowsiness, this casts doubt on one hypothesis. A hypothesis can be falsified by experimental tests, especially if the experiments are repeated with the same results. The onset of daylight savings time supported the "sleepy time of day" hypothesis, and our confidence in this explanation grows if the hypothesis continues to stand up to various kinds of experimental tests. But we can never *prove* that this hypothesis is the true explanation. It is impossible to repeat an experiment enough times to be absolutely certain that the results will *always* be the same. And some false hypotheses make accurate predictions. Consider this hypothesis: "Night and day are caused by the sun orbiting around Earth in an east-west direction." This hypothesis predicts that the sun will rise each morning in the east, move across the sky, and set in the west, which is exactly what we observe. However, the "geocentric universe" hypothesis makes many other predictions that enable us to falsify the hypothesis. Of the many hypotheses proposed to answer a particular question, the correct explanation may not even be included. Even the most thoroughly tested hypotheses are accepted only conditionally, pending further investigation.

The "sleepy twins" scenario introduces another important feature of the scientific process: the controlled experiment. In a *controlled experiment,* the subjects (Scott and Ian, in this case) are divided into two groups, an *experimental group* and a *control group.* Ideally, the two groups are treated exactly alike except for the one variable the experiment is designed to test. This provides a basis for comparison, enabling us to draw conclusions about the effects of our experimental manipulation. Scott and Ian attempted to control their experiments. When Scott moved to the front of the 1:00 class as the *experimental,* Ian remained in the back of the room as the *control.*

There are, of course, serious deficiencies in this experiment. If the twins were trying to test the effect of seating locale on sleepiness in class generally, then a single individual is an inadequate sample size for a control group or experimental group. Note also that there were uncontrolled variables in this experiment. Temperature and other unknown factors may have varied between the front and back of the classroom. Some experiments are easier to control than others, but setting up the best possible controls is characteristic of good experimental design.

Now that we have analyzed the hypothetico-deductive approach by using an imaginary example, you should be able to recognize the process in an elegant study actually reported in the scientific literature. For many years, David Reznick of the University of California, Riverside, and John Endler of the University of California, Santa Barbara, have been investigating differences between populations of guppies in Trinidad, a Caribbean island (FIGURE 1.15). Guppies *(Poecilia reticulata)* are small freshwater fishes you probably recognize as common aquarium pets. In the Aripo River system of Trinidad, guppies live in small pools as populations that are relatively isolated from one another. In some cases, two populations inhabiting the same stream live less than 100 meters (m) apart, but they are separated by a waterfall that impedes the migration of guppies between the two pools.

FIGURE 1.15
David Reznick conducting field experiments on guppy evolution in Trinidad.

When Reznick and Endler compared guppy populations, they observed differences in what are called *life history characteristics.* These characteristics included the average age and size of guppies when they reach sexual maturity and begin to reproduce, as well as the average number of offspring per brood (a litter of baby guppies). The researchers were able to correlate variations in these life history characteristics with the types of predators present in different locations. In some pools, the main predator is a small fish called a killifish, which preys predominately on small, juvenile guppies. In other locations, a larger predator called a pike-cichlid preys more intensely on guppies and mainly eats relatively large, sexually mature individuals (FIGURE 1.16). Guppies in populations exposed to these pike-cichlids have larger broods, reproduce at a younger age, and are smaller at maturity, on average, than guppies that coexist with killifish.

What causes these life history differences between the guppy populations? Correlation with the type of predator present is suggestive, but a correlation does not necessarily imply a cause-and-effect relationship. The type of predator present and the life history characteristics of the guppy populations in a particular location may be independent consequences of some third factor. In fact, Reznick and Endler tested the hypothesis that the life

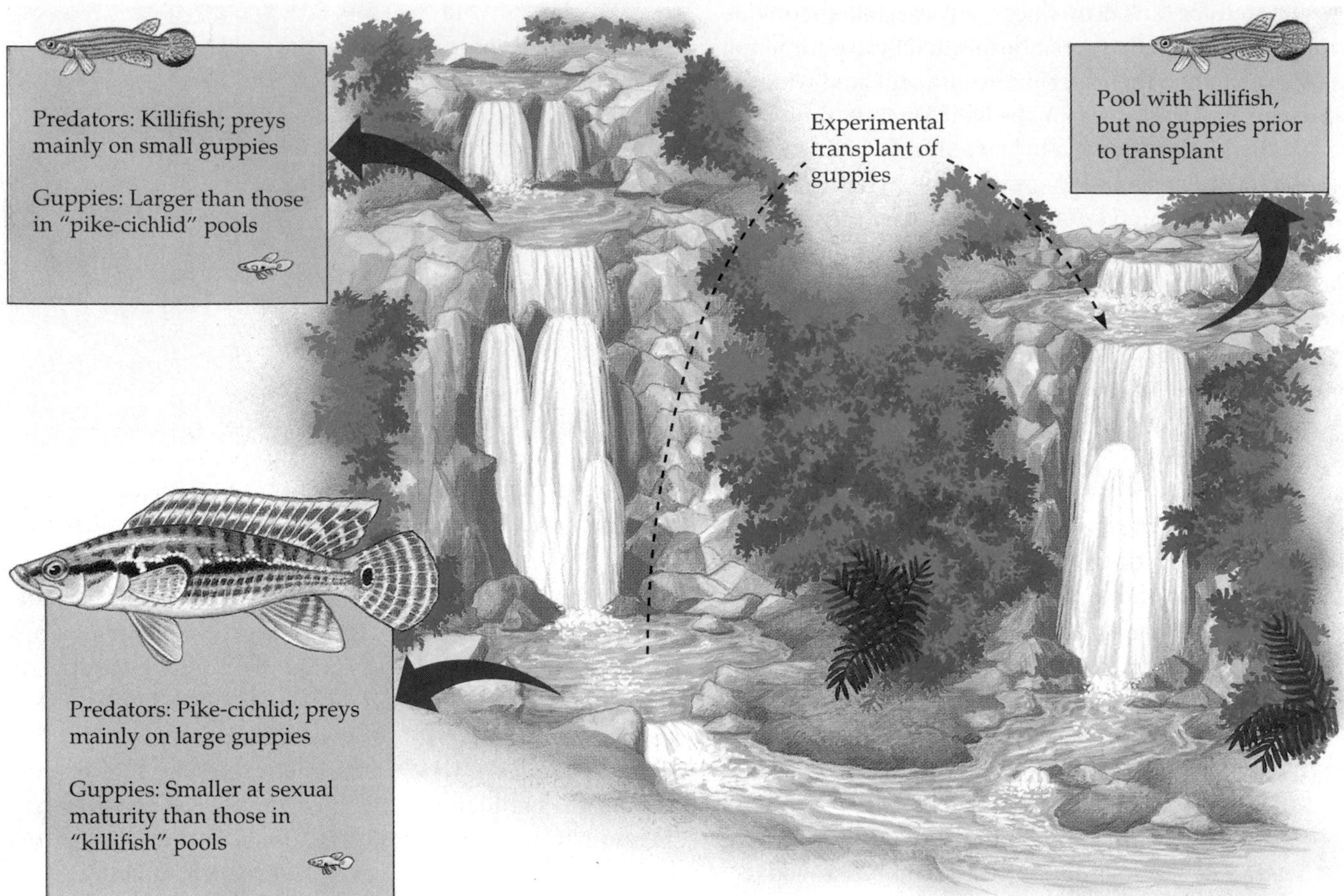

FIGURE 1.16

Testing the hypothesis that selective predation affects the evolution of guppy populations. This drawing symbolizes three pools along streams of Trinidad's Aripo River system. In the pool at the lower left, pike-cichlids prey intensively on guppies, mainly eating relatively large individuals. The predators in the upper-left pool are killifish, which prey less intensively than pike-cichlids and feed mainly on relatively small guppies. In this and other killifish pools, guppies are larger and older at sexual maturity than guppies in pike-cichlid pools. This observation led to Reznick and Endler's hypothesis that selective predation was affecting the evolution of life history characteristics of the guppy populations. One way the researchers tested this hypothesis was to transplant guppies from pike-cichlid pools to pools that contained killifish but had no natural guppy populations (upper right). Reznick and Endler then tracked the evolution of life history in the experimental guppy populations for 11 years. They compared average age and size of mature guppies in the experimental pools to these life history characteristics of guppies in control pools inhabited by pike-cichlids. FIGURE 1.17 summarizes the results of these experiments.

history variations were due to differences in water temperature or other features of the physical environment. Notice the *"If . . . then"* logic characteristic of the hypothetico-deductive approach:

HYPOTHESIS #1: *If* differences in physical environments cause variations in the life histories of guppy populations,

EXPERIMENT: and samples from different wild guppy populations are collected and cultured for several generations in identical environments in predator-free aquaria,

PREDICTED RESULT: *then* the laboratory populations should become more similar in their life history characteristics.

When the researchers performed this experiment, the differences persisted for many generations. This result eliminates hypothesis #1 and also indicates that the life history differences in guppy populations are inherited. Based on the assumption that natural selection can lead to genetic differences in populations, Reznick and Endler tested the following explanation:

HYPOTHESIS #2: *If* the feeding preferences of different predators caused contrasting life histories in different guppy populations to evolve by natural selection,

EXPERIMENT: and guppies are transplanted from locations with pike-cichlids (predators of mature guppies) to guppy-free sites inhabited by killifish (predators of juvenile guppies),

PREDICTED RESULT: *then* the transplanted guppy populations should show a generation-to-generation trend toward later maturation, larger size, and smaller broods—life history characteristics typical of natural guppy populations that coexist with killifish.

In 1976, Reznick and Endler introduced guppies from locations with pike-cichlids to sites that had killifish but no guppies (see FIGURE 1.16). These transplanted populations were the researchers' experimental groups, and they studied them for 11 years, measuring age and size at maturity, brood size, and other life history characteristics. The scientists compared these measurements to data collected over the same period on control groups, guppies that remained in the original locations inhabited by pike-cichlids. To be certain that only heritable differences were counted, the measurements were made after samples from the experimental and control populations had been reared for two generations in identical aquarium environments (FIGURE 1.17). Over 11 years, or 30 to 60 generations, the average weight at maturity for guppies in the introduced (experimental) populations increased by about 14% compared to the control populations. Other life history characteristics also changed in the direction predicted by hypothesis #2.

Without a control group for comparison, there would be no way to tell whether it was the killifish or some *other* factor that caused the transplanted guppy population to change. But since control sites and experimental sites were often nearby pools of the same stream, the main variable was probably the presence of different predators. And these careful researchers observed similar results when guppy populations were reared in artificial streams that were identical except for the type of predator.

Of the several hypotheses Reznick and Endler have tested (we examined only two), they are left with natural selection due to differential predation on larger versus smaller guppies as the most likely explanation for the observed differences between guppy populations. Apparently, when predators such as pike-cichlids prey mainly on reproductively mature adults, the chance that a guppy will survive to reproduce several times is relatively low. The guppies with greatest reproductive success should then be the individuals that mature at a young age and small size and produce at least one brood before growing to a size preferred by the local predator.

The popular press gave Reznick and Endler's research a lot of attention because it documents evolution in a natural setting over a relatively short time (just 11 years). We have examined the guppy experiments in some detail because they also provide a fine example of the key role of hypothetico-deductive thinking in science.

Another key feature of science is its progressive, self-correcting quality. A succession of scientists working on the same problem build on what has been learned earlier. It is also common for scientists to check on the conclusions of others by attempting to repeat observations and experiments. Among contemporary scientists working on the same question, there are both cooperation and competition. Scientists share information through publications, seminars, meetings, and personal communication. They also subject one another's work to careful scrutiny. Nobel laureate James Watson, of DNA fame, recently highlighted this social nature of science: "Constantly exposing your ideas to informed criticism is very important. . . . It's very hard to succeed in science if you don't want to be with other scientists."

Many people associate the word *discovery* with science. Often, what they have in mind is the discovery of new facts. But accumulating facts is not really what science is about; a telephone book is a catalog of facts, but

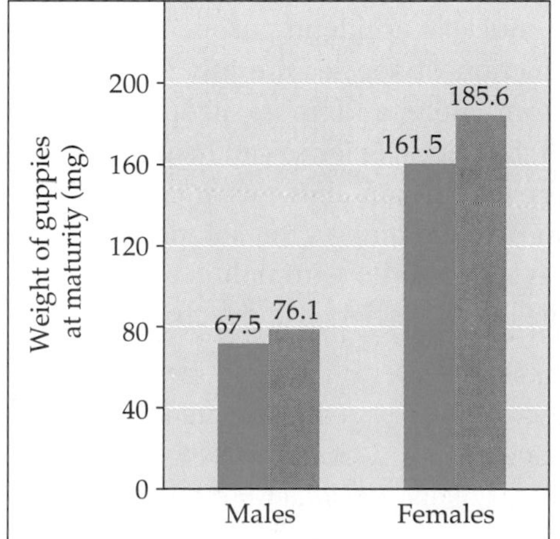

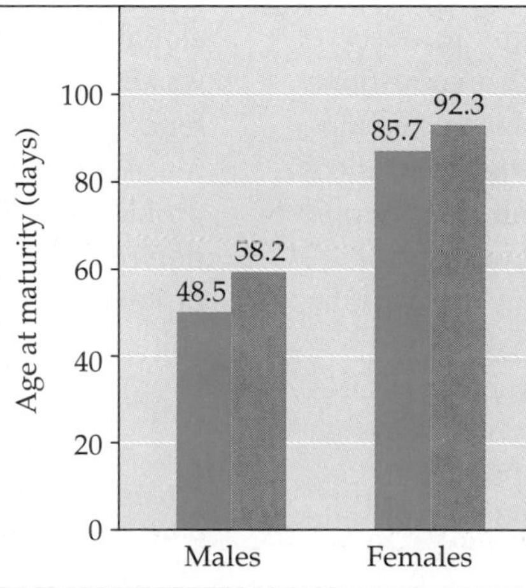

FIGURE 1.17
Experimental evidence for natural selection in action: results of the guppy transplant experiments. These data represent average measurements for two life history characteristics in guppy populations: weight at sexual maturity (left) and age at sexual maturity (right). The vertical bars of the histograms compare control guppy populations (■) to experimental populations (■). Controls were populations native to pools where the main predator is the pike-cichlid, which feeds predominately on large, sexually mature guppies. The experimental populations consisted of guppies that were removed from pike-cichlid pools and transplanted to guppy-free pools inhabited by killifish (see FIGURE 1.16). Killifish prey mainly on small, immature guppies. After just 11 years, the transplanted guppy populations had evolved measurably.

it has little to do with science. It is true that facts, in the form of observations and experimental results, are the prerequisites of science. What really advances science, however, is a new idea that collectively explains a number of observations that previously seemed to be unrelated. The most exciting ideas in science are those that explain the greatest variety of phenomena. People like Newton, Darwin, and Einstein stand out in the history of science not because they discovered a great many facts, but because they synthesized ideas with great explanatory power. Such ideas, much broader in scope than the hypotheses that pose possible causes for one set of observations, are known as theories. Because theories are comprehensive, they only become widely accepted if they are supported by a large body of evidence. "Natural selection" qualifies as a theory because of its broad application to so many situations. And the theory of natural selection is widely accepted because researchers working in the many fields of biology continue to validate the theory with new observations and experiments, including those of Reznick and Endler.

This book is only partly about the current state of biological knowledge. It is also important for you to learn, by example and by practice, how the process of science works. The power of hypothetico-deductive thinking is one of the themes of this text, and you will read many examples of how biologists have applied it in their research. However, it is mainly your own experience in the laboratory and in the field that will teach you how to do science. And your practice with the hypothetico-deductive approach will help you think more critically in general.

Science and technology are functions of society

Science and technology are associated. You have just learned that science is a process, usually involving the hypothetico-deductive approach, that curious people use to answer their questions about nature. Technology, especially in the form of new instruments (electron microscopes, for example), extends our ability to observe and measure and enables scientists to work on questions that were previously unapproachable. In turn, technological inventions often apply the discoveries of science. For example, the inventors of the electron microscope borrowed electromagnetic theory from physics.

We can think of at least some technology as scientific discoveries applied to the development of goods and services. Watson and Crick discovered the structure of DNA through the process of science. This breakthrough sparked an explosion of scientific activity that led to better understanding of DNA chemistry and the genetic code. These discoveries eventually made it possible to manipulate DNA, enabling genetic technologists to transplant foreign genes into microorganisms and produce such valuable products as human insulin. The new biotechnology is revolutionizing the pharmaceutical industry, and DNA technology has also had a enormous impact in other areas, including the legal profession (FIGURE 1.18). Perhaps Watson and Crick envisioned that their discovery would someday have technological applications, but that was probably not what motivated their research, nor could they have predicted exactly what the applications would be. Scientist/writer Lewis Thomas put it this way in his essay "Making Science Work": "We cannot say to ourselves, we need this or that sort of technology, therefore we should be doing this or that sort of science. . . . Science is useful, indispensable sometimes, but whenever it moves forward it does so by producing a surprise; you cannot specify the surprise you'd like. Technology should be watched closely, monitored, criticized, even voted in or out by the electorate, but science itself must be given its head if we want it to work."*

Not all technology can be described as applied science. In fact, technology predates science, driven by inventive humans who built tools, crafted pots, mixed paints, designed musical instruments, and made clothing—all without necessarily understanding why their inventions worked. As Cecily Selby of the Department of Teaching and Learning at New York University writes, "know-how most often comes before know-why." Science catalyzes certain technologies by complementing trial and error with more informed design. But the direction technology takes depends less on science than it does on the needs of humans and the values of society; science could presently enable many technologies that do not exist because there is no demand.

Technology has improved our standard of living in many ways, but it is a double-edged sword. Technology, especially technology that keeps people healthier, has enabled the human population to grow more than tenfold in the past three centuries. The environmental consequences are monstrous. Acid rain, deforestation, global warming, nuclear accidents, ozone holes, toxic wastes, and extinction of species are just a few of the repercussions from more and more people wielding more and more technology. Science can help us identify problems and provide insight about what course of action may prevent further damage. But solutions to these problems have as much to do with politics, economics, culture, and values as with science and technology.

*Thomas, L. "Making Science Work." *Late Night Thoughts on Listening to Mahler's Ninth Symphony.* New York: Viking Press, 1983, p. 28.

(a)

(b)

FIGURE 1.18

Two examples of DNA technology. (**a**) The employees in this biotechnology plant are packaging human growth hormone, which physicians can prescribe to treat children with a hormonal disorder called pituitary dwarfism. Genetic engineering makes mass production of the hormone possible by inserting the gene for human growth hormone into bacteria. Because the genetic code is essentially universal, the bacteria then translate the transplanted gene and produce the human hormone. (**b**) Forensic technicians can use traces of DNA extracted from a blood sample or other body fluid to produce a molecular "fingerprint." The stained bands visible in this photograph represent fragments of DNA, and the pattern of bands varies from one person to another. The legal applications of DNA technology became very public in 1995 during the O. J. Simpson double-murder trial. You will learn more about DNA technology in Chapter 19.

Science and technology are partners, but they originated in very different ways. And now that both science and technology have become such powerful functions of society, it is more important than ever to distinguish "what we would like to understand" from "what we would like to build." Scientists should not distance themselves from technology but instead try to influence how technology applies the discoveries of science. And scientists have a responsibility to help educate politicians, bureaucrats, corporate leaders, and voters about how science works and about the potential benefits and hazards of specific technologies. The crucial relationship between science, technology, and society is a theme that increases the significance of our study of life.

Biology is a multidisciplinary adventure

In some ways, biology is the most demanding of all sciences, partly because living systems are so complex and partly because biology is a multidisciplinary science that requires a knowledge of chemistry, physics, and mathematics. Modern biology is the decathlon of natural science. If you are a biology major or a preprofessional student, you have an opportunity to become a versatile scientist. If you are a physical science major or an engineering student, you will discover in the study of life many applications for what you have learned in your other science courses. If you are a nonscience student enrolled in biology as part of a liberal arts education, you have selected a course in which you can sample many scientific disciplines. And of all the sciences, biology is the most connected to the humanities and social sciences.

No matter what brings you to biology, you will find the study of life to be challenging and uplifting. Do not let the details of biology spoil a good time. The complexity of life is inspiring, but it can be overwhelming. To help you keep from "getting lost in the forest because of all the trees," each chapter of this book is constructed from a manageable number of key concepts. The concepts are first listed at the beginning of the chapter, they head the main sections within the chapter, and then they reappear in the review at the end of the chapter. The details of a chapter enrich your understanding of the concepts and how they fit together. This introductory chapter is the exception; instead of presenting the key concepts of a particular area of biology, this chapter introduced themes that cut across all biological fields—ways of thinking about biology. These themes, along with the key concepts in each chapter, will provide you with a framework for fitting together the many things you will learn in your multidisciplinary exploration of life—and will encourage you to begin asking important questions of your own.

UNIT ONE

THE CHEMISTRY OF LIFE

AN INTERVIEW WITH

ELOY RODRIGUEZ

Eloy Rodriguez works at the interface of biology and chemistry. As one of the world's leading plant chemists, Dr. Rodriguez searches for molecules in plants and insects that function in defense. He and his colleagues have also invented the field of zoopharmacognosy, the study of how wild animals use plants for possible medicinal purposes. We visited Dr. Rodriguez in his research greenhouses at the University of California, Irvine. He has since moved to Cornell University, where he is James A. Perkins Professor of plant biochemistry and natural products biology. His comments in this interview highlight the multidisciplinary character of biology.

When we came in, we noticed that your facility is called Phytochemistry and Toxicology. Let's start with the phytochemistry part. Tell us about this field.

The simplest translation of phytochemistry is plant chemistry (*phyto* meaning plant). Phytochemistry is the study of the chemicals produced by plants, especially the so-called secondary chemicals, or secondary compounds. Unlike primary compounds, molecules such as sugars and chlorophyll that all plants produce and require for normal growth, secondary compounds are molecules without known functions in the basic processes of life. Each secondary compound is unique to certain plant species. Secondary compounds were originally thought of as waste products, but as evolutionary biologists got involved in studying the function of these chemicals, they began to see them as very important to the survival of plants. Some secondary plant compounds are very familiar to us. Aspirin, for instance, was originally derived from willow trees before it was synthetically made. Caffeine is not only a stimulant for humans; it also has physiological effects on fungi. These substances are small molecular compounds that function mainly in defense for the plants.

How do secondary chemicals help defend plants?

Natural selection favors those plants with the right kinds of chemical compounds to ward off fungi, viruses, insects, and large herbivores. There is a large array of organisms that eat plants, and a plant can't just get up and run. A plant's ability to survive is really due to chemistry—the ability to have an effect on the physiology of an organism. An animal will eat a plant and then either become paralyzed or otherwise impaired by the secondary compounds, or it will escape and learn to avoid that plant in the future. Dead plants don't reproduce, so obviously the presence of defensive compounds allows for fitness and reproduction within a given population.

Secondary compounds aren't just randomly produced and packaged in the plant. Flower chemistry is very different from leaf chemistry, which is very different from root chemistry. In flowers you wouldn't expect to find a lot of poisonous chemicals; you would expect to find chemicals that attract pollinators. And that's what flowers have—nice perfumes and pigments. They generally aren't very toxic. But when you get into the leaf, that's where you begin to see a large diversity of alkaloids and other toxic compounds. There are different kinds of chemistry in different parts of a plant, which makes sense from an evolutionary perspective.

The creosote bush is a really interesting example. This is a desert plant that's found all the way from South America to Utah and north. If you ever see a creosote bush, it just glistens in the sun because it's coated with a very thick resin that protects against extreme radiation, something you have to worry about if you're a desert plant. It also helps prevent critical water loss through transpiration. It's an incredible plant as far as the ability to make compounds; we've calculated that this plant has over 500 secondary chemicals! Chiles are another intriguing case. Chiles are South American in origin and became an important food for the Aztecs and indigenous groups of Mexico, and now Asia is the biggest consumer of chiles. Studies have shown that chiles are very good antibiotics. In fact, I had some school kids do an experiment where they put chiles on bread and found that it took longer for fungus to grow on the chile-coated bread than on plain bread. I'm still studying the chemistry of chiles; the alkaloid chemistry is quite fascinating.

The chemistry part of what you do sounds like lab work. Does your research get you out into the field as well?

I was trained as a plant biologist and as an organic chemist. A lot of people can get pretty focused and pretty reductionist in their approach, but I have always had a very broad biological and chemical background. When I do chemistry, I don't just do it for the sake of chemistry. It's exciting to look at new chemical structures, but I have always wanted to understand the reason for their existence. To do that you have to get out in the field. You have to go out, make observations, see the system, and ask the question, "What is the function of this chemical?" A lot of my research starts from going out in the field and observing.

Can you give an example?

One project that I got involved with has now developed into a discipline. It's called zoopharmacognosy, the study of how animals possibly medicate themselves with plants. This came about from real observations. It started with looking at general feeding behavior in great apes and the family of chimps. My colleagues and I observed some very peculiar things going on that had to do with apes selecting certain plants and swallowing the leaves. I don't think we ever would have discovered this if we hadn't been out there looking at these animals. The fieldwork was essential. I still go down to the rain forest every summer. It is an integral part of the way I do science.

Tell us more about finding medicinal plants by following animals.

I got involved with Richard Wrangham, a primatologist from Harvard who was studying the feeding behavior of primates in the Kibale Forest of Uganda. We were interested in the notion that monkeys are able to eat more poisonous plants than chimps. That was the original general hypothesis—that monkeys have a greater physiological ability to deal with toxins. If you go out in the rain forest, you'll see monkeys eating all kinds of stuff that chimps will avoid. If you're ever starving to death in the rain forest, don't follow the monkeys, follow the chimps!

We followed chimpanzees that were sick. Apes get every illness that we do, and you can tell when they are sick. Animals that are very sick will focus on one particular plant species and stop eating everything else. In one observation, researchers followed a particular animal for several days, and for two whole days this animal concentrated on one plant species. It wouldn't eat the whole plant. It would take off the leaves, crack the stalk, and then suck out the juice. The scientists then did some beautiful chemistry that showed that the leaf chemistry was totally different from the stem chemistry. It was a clear example that animals were selecting a certain part of the plant that had the proper chemicals. I think it's pretty clear that these plants we are calling the pharmacopia of the apes are very good ones.

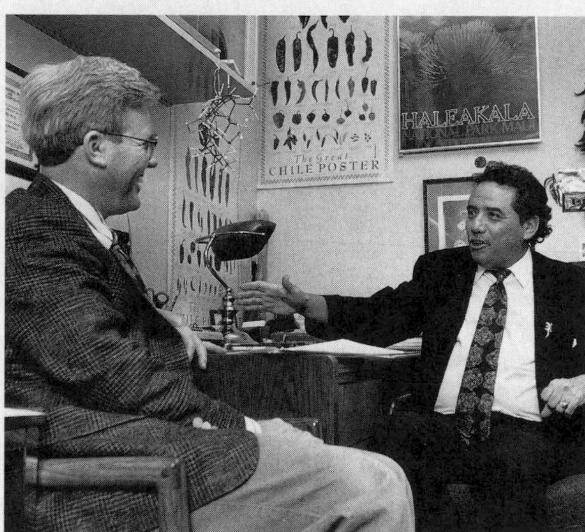

Richard pointed out to me that sometimes chimps seem to select young leaves from certain plant species. From his earlier studies in Africa, he showed that these animals get up in the morning, make a beeline toward these plants, and take a certain amount of the young leaves. We calculated that they were more or less getting a set dosage of the drug or drugs in the leaf. Once we put that together, we published a paper. The response we got was incredible, from both scientists and nonscientists. It's like everyone who owns a dog or cat was telling you that they should be part of this discovery. I think science should always do that—involve people.

This kind of science really catches the imagination of everybody. It's kind of interesting how this field has really exploded in the sense that a lot of scientists from other disciplines are asking questions. Behavioral scientists are no longer just going out, looking at animals, and saying, "Well, the plants are there for food." They now have to look at another component indicating that plants might also have an effect on reducing the parasite load. In the area of phytochemistry, a result of this new field is that there is now a more direct way to go after plant parts to look at the chemistry. A lot of collection is done randomly—you go out and collect a bunch of bark, for example. It is a very tedious and not exactly a very scientific way to go about getting drugs. It's nice when you have animals basically telling you, "Here, try these leaves."

Of the 15 plant species that we have looked at while studying these chimps, we have figured out the chemistry of five of them. When you figure out a chemical structure, you are breaking a code of nature. That code gives you all the information. It says, "Look at me, look at my triple bonds. Do you see what I can do with these triple bonds?"

Have apes and other animals led you to any compounds that have potential medicinal value in humans?

Oh, clearly. I certainly see some potential in modern medicine. For example, some of the compounds we've identified by zoopharmacognosy kill parasitic worms, and some of these chemicals may be useful against tumors. There is no question that the templates for most drugs are in the natural world.

How do you think people of ancient cultures, including Native Americans, first identified medicinal plants?

That is really still the sixty-four-thousand-dollar question! Besides trial and error, there must have been some experimentation taking place and some very keen observation. Now, when I go out to the tropical rain forest I always ask the people, "Do you look at animals?" and they will tell me, "Oh yeah, sure we look at animals." Some of them even keep animals in captivity and feed them plants. I am not implying that they use animal experimentation the way we do, but pets are an interesting idea. I mean if you keep pets, you may as well learn something from them.

I think the use of medicinal plants started when the first humans got sick. When you are sick, you try to figure out how to take care of yourself. We are talking about a long history of drug development and drug use. When the Europeans came to the New World, they not only brought a lot of destruction, they also brought a lot of disease. Indigenous people tried to use some of the medicinal plants to deal with these new diseases. For example, malaria was probably brought to the Americas from Africa, through the African slave trade. Quinine was discovered in Peru, and native people there were probably already using it to treat fever related to other diseases when malaria was introduced.

Anthropologists have documented that native people extract plant materials, grind them up, filter them, and treat them with burned leaves (which act as a base)—almost the same process that I use. They get out materials that are relatively pure for use as medicine. This is not a crude process. These people don't run around grabbing leaves and sticking them in their mouth or just boiling them and then drinking the tea. It is a lot more sophisticated than that. Indigenous people have been extremely successful! They have really filtered out some of the best medicines from plants. A lot of the drug companies started out thanks to those folks.

Is the medicinal potential of tropical plants among the incentives for conserving the biological diversity of rain forests?

There is no question that diversity, any kind of diversity, is worth keeping. In the tropical rain forest we are talking about the ultimate diversity in plants, animals, fungi—life! The arguments that have been made for preserving the forests are excellent, ranging from production of oxygen to the fact that we now have good evidence that the rain forest plays a very important role in the uptake of excess CO_2. The more we search the rain forest, the more in awe of it we are. We see how important it is.

Then there is the argument that, "Well, you know, we can't get rid of this genetic material because how are we going to cure cancer?" But the importance of the rain forest goes beyond human concerns. We are talking about the health of the planet.

One way I am trying to increase awareness in college students is by taking a few undergraduates each summer into the rain forest. The students come back with a passion like you have never seen before. I wish there was a major undertaking among all universities to do this. I think it would give a big boost to ecology. Students are very good communicators, so they will spread the word!

What experiences did you have as a student that led you to a career in science?

I grew up in South Texas, in a very poor but typically large family. I had 67 cousins. Sixty-four of us got undergraduate degrees,

and ten of us got master's degrees or Ph.D.s, which was unique. I grew up in a predominately Mexican-American community so I am bilingual. I never lived in Mexico, but my parents maintained my Spanish. It is the greatest thing that happened to me as a researcher because it allows me to communicate with all of South America, which is over 400 million people.

I grew up in a strong community where the value of education was obviously very important. We knew we couldn't get good jobs unless we got an education. The community was so poor that you couldn't be a criminal. There was no profit! What are you going to do, steal from the poor? That didn't make sense. If you stole a car, you probably got into greater debt trying to fix the darn thing! We saw that the people who seemed to be the best off were teachers. They were the ones with Ramblers and brick homes.

My interest in science started with visits to my grandfather in Mexico. He had a ranch—a small farm, actually. He would always take me out and show me animals and plants. I was intrigued with them, and I think it just stayed with me. I always wanted to be a scientist. I never thought about medical school. Maybe it has to do with the fact that we had a pretty lousy medical doctor. That may have been the reason—he never seemed to cure us!

Certainly my study of medicinal plants has to do with my roots. The only kind of medication that we used to get, I remember very clearly, was people bringing in herbs (primarily women did this). My aunt always maintained a little medicinal garden in the backyard. I was never negative about this because it was the only kind of medication that we used. You can't be negative about what cures you, and it obviously cured me because I am here! The more I think about it now, the more I realize the value of those herbal treatments. I know now that the reason these plants work is because they have certain chemicals in them that work.

Is that what led you to phytochemistry?

I went to the University of Texas, and as an undergraduate I felt very lonely. I was one of the very small number of minority students. There were no programs for minority students. We were kind of abandoned. It was like shopping at a 7-11 store—you know, just get your stuff and get out. That was my experience until I got hooked.

I was in a work-study program and they needed somebody to mop the labs. That's how I got started in the laboratory. There was a post-doc from Switzerland who was a nice guy, but kind of a lazy guy. He wanted someone to help him in his lab performing chemical separations. He said, "Can you do it?" and I said, "Sure." I was really good at it. He always used to be out with his coffee reading his newspaper at the desk. One day the professor in charge of the lab came in and asked, "What's going on?" That is how I got my first publication. I published two more papers as an undergraduate and fifteen more as a graduate student. I worked hard and loved what I was doing. The research bug got to me, the passion, the excitement of science. Once it gets to you, you can't let go of it.

Based on your own experiences in science, what advice can you share with first-year biology students?

I always tell undergraduates that it's okay to become a specialist, but when you're an undergraduate, give yourselves some breadth of experience. As an undergraduate, I started as a zoology major, then I went into botany, and then organic chemistry. But I never really could separate them. Never in a lifetime would I have stumbled across the idea of zoopharmacognosy if I had just been stuck within an isolated field of organic chemistry. But if you talk to organic chemists now, they talk constantly about biology. This area is wide open today; there's a rebirth of excitement in looking at natural products. You can't separate the chemistry from the biology. I say this so young students realize that the interdisciplinary approach is the way things are going.

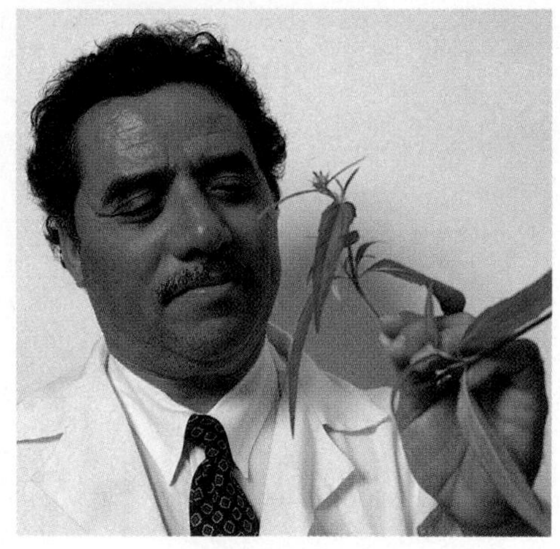

CHAPTER 2

THE CHEMICAL CONTEXT OF LIFE

KEY CONCEPTS

- Matter consists of chemical elements in pure form and in combinations called compounds
- Life requires about 25 chemical elements
- Atomic structure determines the behavior of an element
- Atoms combine by chemical bonding to form molecules
- Weak chemical bonds play important roles in the chemistry of life
- A molecule's biological function is related to its shape
- Chemical reactions change the composition of matter
- Chemical conditions on the early Earth set the stage for the origin and evolution of life

Eloy Rodriguez, the scientist you met in the preceding interview, holds a branch of Aspilia *in the photograph that opens this chapter. He uses his chemistry training to extract and purify substances from the leaves so that the medicinal potential of these chemicals can be tested. Does this research fit into the category of* biology *or* chemistry? *It's a trick question! In contrast to a college catalog of courses, nature is not packaged into biology, chemistry, physics, and the other natural sciences. Biologists are scientists who specialize in the study of life, but organisms are natural systems to which basic concepts of chemistry and physics apply. Biology is a multidisciplinary science.*

This unit of chapters introduces key concepts of chemistry that will apply throughout our study of life. We will make many connections to the themes introduced in Chapter 1. One of those themes is the organization of life on a hierarchy of structural levels, with additional properties emerging at each successive level (Figure 2.1, p. 26)*. In this unit, we will see how the theme of emergent properties applies to the lowest levels of biological organization—to the ordering of atoms into molecules and to the interactions of those molecules within cells. Somewhere in the transition from molecules to cells, we will cross the blurry boundary between nonlife and life.*

Matter consists of chemical elements in pure form and in combinations called compounds

Organisms are composed of **matter,** which is anything that takes up space and has mass.* Matter exists in many diverse forms, each with its own characteristics. Rocks, metals, wood, glass, and you and I are just a few examples of what seems an endless assortment of matter.

Some of the ancient Greek philosophers believed that the great variety of matter arises from four basic ingredients, or elements. They imagined these elements to be air, water, fire, and earth—supposedly pure substances that could not be decomposed to other forms of matter. All other substances were thought to be formed by blending various proportions of two or more of the elements. Even though the classical philosophers proposed the wrong elements, their basic idea was correct.

*Sometimes we substitute the term *weight* for *mass,* although the two are not equivalent. We can think of mass as the amount of matter an object represents. The weight of an object measures how strongly that mass is pulled by gravity. An astronaut in a space shuttle is weightless, but the astronaut's mass is the same as it would be on Earth. However, as long as we are earthbound, the weight of an object is a measure of its quantity of matter; so for our purposes, we can use the terms interchangeably.

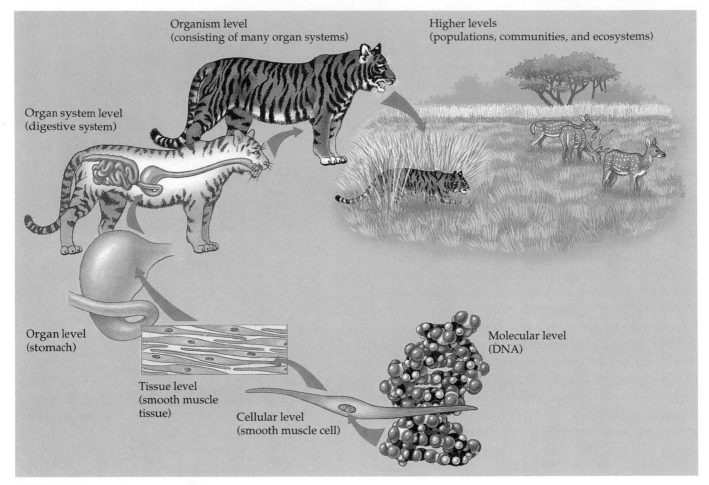

FIGURE 2.1
The hierarchy of biological order.

An **element** is a substance that cannot be broken down to other substances by chemical reactions. Today, chemists recognize 92 elements occurring in nature—for example, gold, copper, carbon, and oxygen. Each element has a symbol, usually the first letter or two of its name. Some of the symbols are derived from Latin or German names; for instance, the symbol for sodium is Na, from the Latin, *natrium.*

Two or more elements may combine in a fixed ratio to produce a **compound.** Table salt, for example, is sodium chloride (NaCl), a compound composed of the elements sodium (Na) and chlorine (Cl). Pure sodium is a metal and pure chlorine is a poisonous gas. Chemically combined, however, sodium and chlorine form an edible compound. This is a simple example of organized matter having emergent properties: A compound has characteristics beyond those of its combined elements (FIGURE 2.2).

FIGURE 2.2
The emergent properties of a compound. The metal sodium combines with the poisonous gas chlorine to form the edible compound sodium chloride, or table salt.

Life requires about 25 chemical elements

About 25 of the 92 natural elements are known to be essential to life. Just four of these—carbon (C), oxygen (O), hydrogen (H), and nitrogen (N)—make up 96% of living matter. Phosphorus (P), sulfur (S), calcium (Ca), potassium (K), and a few other elements account for most of the remaining 4% of an organism's weight. TABLE 2.1 lists by percentage the elements found in the human body. FIGURE 2.3 illustrates a deficiency of an essential element in plants.

Trace elements are those required by an organism in minute quantities. However, because they are mandatory for good health, trace elements are not nutrients of marginal importance to the organism. Some trace elements, such as iron (Fe), are needed by all forms of life; others are required only by certain species. For example, in vertebrates (animals with backbones), the element iodine (I) is a necessary ingredient of a hormone produced by the thyroid gland. A daily intake of only 0.15 milligram (mg) of iodine is adequate for normal activity of the human thyroid. An iodine deficiency in the diet causes the thyroid gland to grow to abnormal size, producing a deformity called goiter (FIGURE 2.4). Where it is available, iodized salt has reduced the incidence of goiter.

FIGURE 2.3
Nitrogen deficiency in corn. In this controlled experiment, the plants on the left are growing in soil that was fertilized with compounds containing nitrogen, an essential element. The soil on the right is deficient in nitrogen. Even if the poorly nourished crop growing in this soil can be harvested, it will yield less food and its chemical deficiencies will be passed on to livestock or human consumers.

FIGURE 2.4
Goiter. The enlarged thyroid gland of this Malaysian woman is due to an iodine deficiency.

Atomic structure determines the behavior of an element

The units of matter are called **atoms.** They are so small that it would take about a million of them to stretch across the period printed at the end of this sentence. Each element consists of a certain kind of atom, which is different from the atoms of any other element. We symbolize atoms with the same abbreviation used for the element made up of those atoms; thus, C stands for both

TABLE 2.1

Naturally Occurring Elements in the Human Body

SYMBOL	ELEMENT	ATOMIC NUMBER (SEE P. 28)	PERCENTAGE OF HUMAN BODY WEIGHT
O	Oxygen	8	65.0
C	Carbon	6	18.5
H	Hydrogen	1	9.5
N	Nitrogen	7	3.5
Ca	Calcium	20	1.5
P	Phosphorus	15	1.0
K	Potassium	19	0.4
S	Sulfur	16	0.3
Na	Sodium	11	0.2
Cl	Chlorine	17	0.2
Mg	Magnesium	12	0.1

Trace elements (less than 0.01%): boron (B), chromium (Cr), cobalt (Co), copper (Cu), fluorine (F), iodine (I), iron (Fe), manganese (Mn), molybdenum (Mo), selenium (Se), silicon (Si), tin (Sn), vanadium (V), and zinc (Zn).

the element carbon and a single carbon atom. An atom is the smallest possible amount of an element that retains that element's properties.

Subatomic Particles

Although the atom is the smallest unit having the physical and chemical properties of its element, these tiny bits of matter are composed of even smaller parts, called subatomic particles. Physicists have split the atom into more than a hundred types of particles, but only three kinds of particles are stable enough to be of relevance here: **neutrons, protons,** and **electrons.** Neutrons and protons are packed together tightly to form a dense core, or nucleus, at the center of the atom. The electrons move about this nucleus at nearly the speed of light (FIGURE 2.5).

Electrons and protons are electrically charged. Each electron has one unit of negative charge, and each proton has one unit of positive charge. A neutron, as its name implies, is electrically neutral. Protons give the nucleus a positive charge, and it is the attraction between opposite charges that keeps the rapidly moving electrons in the vicinity of the nucleus.

The neutron and proton are almost identical in mass, each about 1.7×10^{-24} grams (g). Grams and other conventional units are not very useful for describing the mass of objects so minuscule. Thus, for atoms and subatomic particles, we use a unit of measurement called the **dalton,** in honor of John Dalton, the British scientist who helped develop atomic theory around 1800. Neutrons and protons have a mass of almost exactly 1 dalton apiece (actually 1.009 and 1.007, respectively). Because the mass of an electron is only about $\frac{1}{2000}$ that of a neutron or proton, we can ignore electrons when computing the total mass of an atom.

Atomic Number and Atomic Weight

Atoms of the various elements differ in their number of subatomic particles. All atoms of a particular element have the same number of protons in their nuclei. This number, which is unique to that element, is referred to as the **atomic number** and is written as a subscript to the left of the symbol for the element. The abbreviation ${}_{2}He$, for example, tells us that an atom of the element helium has two protons in its nucleus. Unless otherwise indicated, an atom is neutral in electrical charge, which means that its protons must be balanced by an equal number of electrons. Therefore, the atomic number tells us the number of protons *and* the number of electrons in an electrically neutral atom.

We can deduce the number of neutrons from a second quantity, the **mass number,** which is the sum of protons plus neutrons in the nucleus of an atom. The mass number is written as a superscript to the left of an element's symbol. For example, we can use this shorthand to write an atom of helium as ${}^{4}_{2}He$. Since the atomic number indicates how many protons there are, we can determine the quantity of neutrons by subtracting the atomic number from the mass number: A ${}^{4}_{2}He$ atom has 2 neutrons. An atom of sodium, ${}^{23}_{11}Na$, has 11 protons, 11 electrons, and 12 neutrons. The simplest atom is hydrogen, ${}^{1}_{1}H$, which has no neutrons—a lone proton with a single electron moving about it constitutes a hydrogen atom.

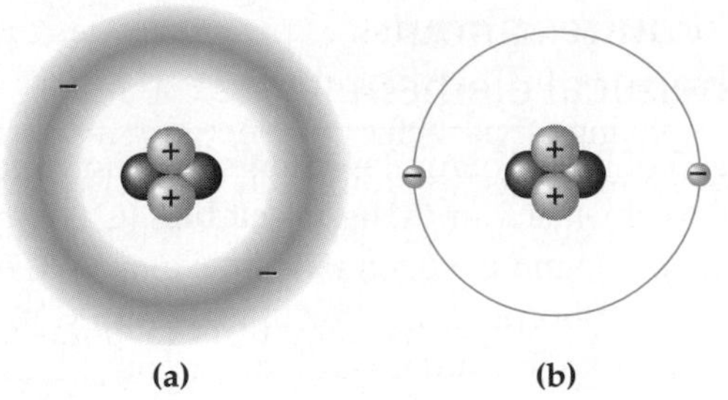

FIGURE 2.5
Two simplified models of a helium (He) atom. The helium nucleus consists of two neutrons (dark gray) and two protons (light gray). (**a**) The nucleus is surrounded by a cloud of negative charge (blue), owing to the rapid movement of two electrons. (**b**) The circle indicates the *average* distance of the electrons from the nucleus, although this distance is not drawn to scale compared to the size of the nucleus. (Our model of an atom will be refined as the chapter progresses.)

Essentially all of an atom's mass is concentrated in its nucleus, because the contribution of electrons to mass is negligible. Since neutrons and protons each have a mass that is very close to 1 dalton, the mass number tells us the approximate mass of the whole atom. The term **atomic weight** is often used to refer to what is technically the total atomic mass, or mass number. Thus, the atomic weight of helium (${}^{4}_{2}He$) is 4 daltons (4.003, to be exact).

Isotopes

All atoms of a given element have the same number of protons, but some atoms have more neutrons than other atoms of the same element and therefore weigh more. These different atomic forms are referred to as **isotopes** of the element. In nature, an element occurs as a mixture of its isotopes. For example, consider the three isotopes of the element carbon, which has the atomic number 6. The most common isotope is carbon-12, ${}^{12}_{6}C$, which accounts for about 99% of the carbon in nature. It has 6 neutrons. Most of the remaining 1% of carbon consists of atoms of the isotope ${}^{13}_{6}C$, with 7 neutrons. A third isotope, ${}^{14}_{6}C$, has 8 neutrons; it is present in the environment in minute quantities. Notice that all three isotopes of carbon have 6 protons—otherwise, they would not be carbon.

Both ${}^{12}C$ and ${}^{13}C$ are stable isotopes, meaning that their nuclei do not have a tendency to lose particles. The isotope ^{14}C, however, is unstable, or radioactive. A **radioactive isotope** is one in which the nucleus decays spontaneously, giving off particles and energy. A loss of nuclear particles transforms the atom to an atom of a different element. For example, radioactive carbon decays to form nitrogen.

Radioactive isotopes have many useful applications in biology. In Chapter 23, you will learn how researchers use the amount of radioactivity in fossils to date those relics of past life. Radioactive isotopes are also useful as tracers to follow atoms through metabolism, the chemical processes of an organism (see the Methods Box, p. 30). Cells use the radioactive atoms as they would nonradioactive isotopes of the same element, but the radioactive tracers can be detected. Radioactive tracers have thus become important diagnostic tools in medicine. For example, certain kidney disorders can be diagnosed by injecting small doses of substances containing radioactive isotopes into the blood and then measuring the amount of tracer excreted in the urine.

Although radioactive isotopes are very useful in biological research and medicine, radiation from these decaying isotopes also poses a hazard to life by damaging cellular molecules. The severity of this damage depends on the type and amount of radiation an organism absorbs. One of the most serious environmental threats is radioactive fallout from nuclear accidents (FIGURE 2.6).

FIGURE 2.6
The Chernobyl accident. During the 1986 disaster at this nuclear power plant at Chernobyl, Ukraine, giant clouds of radioactive materials were belched into the air, producing widespread contamination downwind. Photographic censors removed smoke from this photo before it was shown on Moscow television a few days after the explosion.

Energy Levels

The simplified models of the atom in FIGURE 2.5 distort the size of the nucleus relative to the volume of the whole atom. If the nucleus were the size of a golf ball, the electrons would be moving about the nucleus at an average distance of approximately 1 kilometer (km). Atoms are mostly empty space.

When two atoms approach each other during a chemical reaction, their nuclei do not come close enough to interact. Of the three kinds of subatomic particles we have discussed, only electrons are directly involved in the chemical reactions between atoms.

An atom's electrons vary in the amount of energy they possess. **Energy** is defined as the ability to do work. **Potential energy** is the energy that matter stores because of its position or location. For example, because of its altitude, water in a reservoir on a hill has potential energy. When the gates of the dam are opened and the water runs downhill, the energy is taken out of storage to do work, such as turning generators. Since potential energy has been expended, the water stores less energy at the bottom of the hill than it did in the reservoir. Matter has a natural tendency to move to the lowest possible state of potential energy; in this example, water runs downhill. To restore the potential energy of a reservoir, work must be done to elevate the water against gravity.

The electrons of an atom also have potential energy because of their position in relation to the nucleus. The negatively charged electrons are attracted to the positively charged nucleus; the more distant the electrons are from the nucleus, the greater their potential energy. Unlike the continuous flow of water downhill, changes in the potential energy of electrons can occur only in steps of fixed amounts. An electron having a certain discrete amount of energy is analogous to a ball on a staircase. The ball can have different amounts of potential energy, depending on which step it is on, but it cannot spend much time between the steps.

The different states of potential energy for electrons in an atom are called **energy levels,** or **electron shells.** The first shell is closest to the nucleus, and electrons in this shell have the lowest energy. Electrons in the second shell have more energy, electrons in the third shell more energy still, and so on. An electron can change its shell, but only by absorbing or losing an amount of energy equal to the difference in potential energy between the old shell and the new shell. To move to a shell farther out from the nucleus, the electron must absorb energy. For example, light can excite an electron to a higher energy level. (Indeed, this is the first step when plants harness light energy for photosynthesis, the process that produces

METHODS: THE USE OF RADIOACTIVE TRACERS IN BIOLOGY

Radioactive isotopes are among the most versatile tools in biological research. These isotopes serve as "spies" within an organism: They are used to label certain chemical substances in order to follow the steps of a metabolic process or to determine the location of the substance within an organism. Organisms do not generally discriminate between radioactive and stable isotopes of the same element; thus, they assimilate and process the labeled substance normally.

The experiments being conducted here were designed to determine how temperature affects the rate at which the DNA replicates in a population of dividing animal cells, and to locate the newly synthesized DNA within the cells. The experiment begins by culturing cells in an artificial medium that contains, among other things, the specific chemical ingredients used by cells to make new DNA. One of those ingredients, a molecule named thymidine, is labeled with ^{3}H, a radioactive isotope of hydrogen. We can use the radioactive label to trace the incorporation of thymidine into new DNA.

After a certain amount of time has elapsed, samples of cells grown at various temperatures in the presence of the radioactive tracer are killed, and their DNA is precipitated onto pieces of filter paper. The papers are then placed in vials containing scintillation fluid, which emits flashes of light whenever certain chemicals in the fluid are excited by radiation from the decay of the tracer in the DNA. The frequency of flashes, proportional to the amount of radioactive material present, is measured in counts per minute by placing the vials in a scintillation counter (FIGURE a). The effect of temperature on the rate of DNA synthesis can be determined by plotting the counts per minute for the various DNA samples against the temperatures at which the cells were grown (FIGURE b).

A technique known as autoradiography can be used to determine the location of the radioactively labeled DNA within the cells. The cells are washed free of any radioactive material that was not incorporated into the DNA. Then they are fixed (preserved), and thin sections (slices) of them are placed on glass slides and covered by a layer of photographic emulsion. The slides are then kept for some time in the dark. Wherever DNA is located in the cells, radiation from the radioactive tracer will expose the photographic emulsion. The emulsion is developed, and the slide is examined with a microscope (FIGURE c). Black grains of the exposed emulsion are superimposed on the nuclei of the cells, the sites of DNA. The nucleus of the cell on the left has been radioactively labeled.

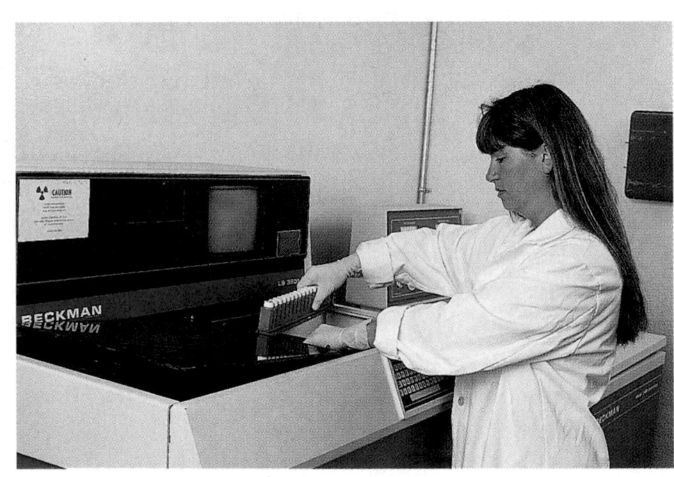

(a)

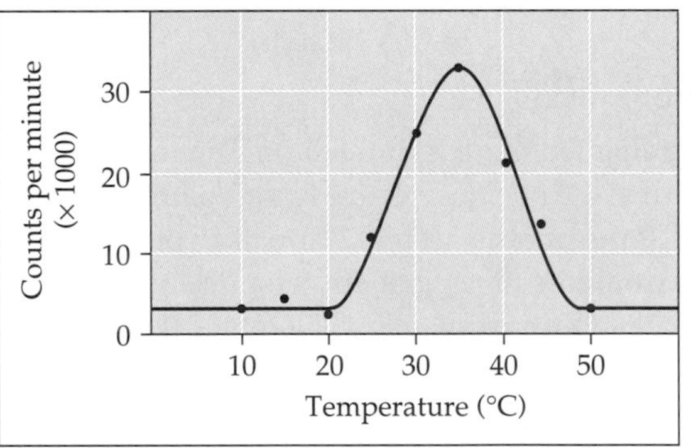

(b)

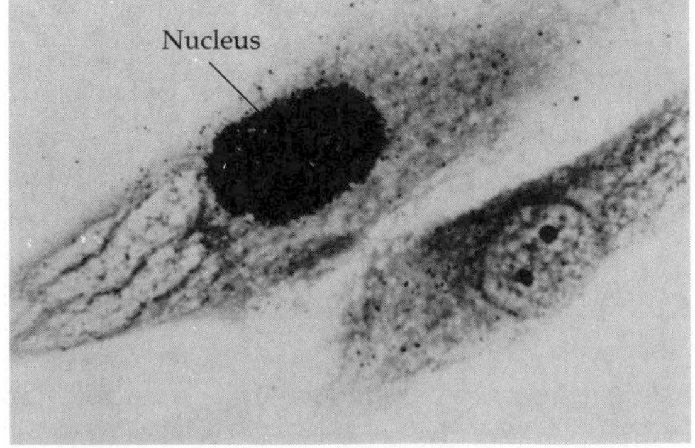

(c)

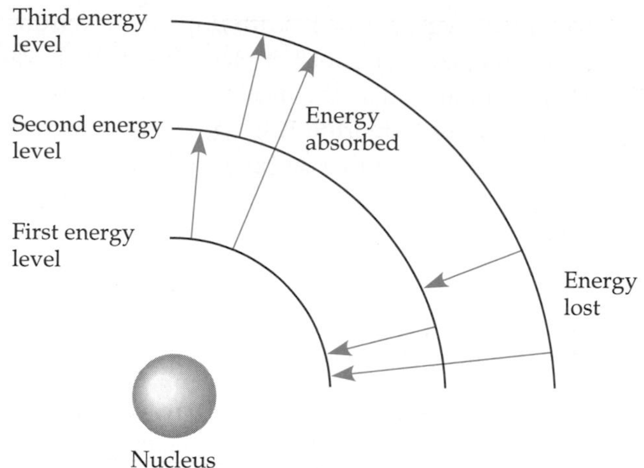

FIGURE 2.7

Energy levels of electrons. Electrons exist only at fixed levels of potential energy. An electron can move from one level to another only if the energy it gains or loses is exactly equal to the difference in energy between the two levels. Arrows indicate some of the stepwise changes in potential energy that are possible for electrons. Energy levels are also called electron shells.

food from carbon dioxide and water.) To move to a shell closer in, an electron must lose energy, which is usually released to the environment in the form of heat (FIGURE 2.7). Thus, when the energy of sunlight excites electrons in the paint of an automobile roof to higher energy levels, the roof heats up as the electrons "fall" back to their original levels.

Electron Orbitals

Earlier in this century, the electron shells of an atom were visualized as concentric paths of electrons orbiting the nucleus, something like planets orbiting the sun (see FIGURE 2.5). The atom is not this simple, however. In fact, we can never know the exact trajectory of an electron. What we can do instead is describe the volume of space in which an electron spends most of its time. The three-dimensional space where an electron is found 90% of the time is called an **orbital** (FIGURE 2.8).

No more than two electrons can occupy the same orbital. The first energy shell has a single orbital and can thereby accommodate a maximum of two electrons. This single orbital, which is spherical in shape, is designated the 1*s* orbital. The lone electron of a hydrogen atom occupies the 1*s* orbital, as do the two electrons of a helium atom. Electrons, like all matter, tend to exist in the lowest available state of potential energy, which they have in the first shell. An atom with more than two electrons must use higher shells because the first shell is full.

The second electron shell can hold eight electrons, two in each of four orbitals. Electrons in the four different orbitals all have virtually the same energy, but they move in different volumes of space. There is a 2*s* orbital, spherical in shape like the 1*s* orbital, but with a slightly greater diameter. The other three orbitals, called 2*p* orbitals, are dumbbell-shaped, each oriented at right angles to the other two. (At higher energy levels, the orbitals are referred to as 3*s*, 3*p*, and so on.)

Electron Configuration and Chemical Properties

The chemical behavior of an atom is determined by its electron configuration—that is, the distribution of electrons in the atom's electron shells. Beginning with hydrogen, the simplest atom, we can imagine building the

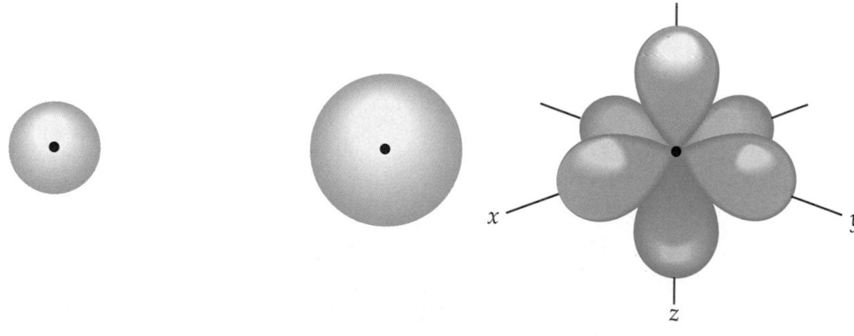

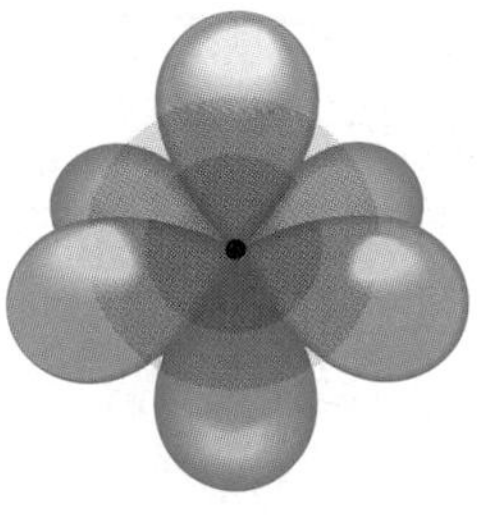

(a) 1*s* orbital

(b) 2*s* and 2*p* orbitals

(c) Electron orbitals of neon ($_{10}$Ne)

FIGURE 2.8

Electron orbitals. These three-dimensional shapes represent the volumes of space where the constantly moving electrons are most likely to be found. Each orbital holds a maximum of two electrons. (**a**) The first electron shell has one spherical (*s*) orbital, designated 1*s*. Only this orbital is occupied in hydrogen, which has one electron, and helium, which has two electrons. (**b**) The second and all higher shells each have one larger *s* orbital (designated 2*s*, in the case of the second shell) plus three dumbbell-shaped orbitals called *p* orbitals (2*p* for the second shell). The three 2*p* orbitals are arranged at right angles to one another along imaginary *x*, *y*, and *z* axes of the atom. The third and higher electron shells can hold additional electrons in orbitals of more complex shapes. (**c**) To symbolize the electron orbitals of the element neon, which has a total of ten electrons, we superimpose the 1*s* orbital of the first shell and the 2*s* and three 2*p* orbitals of the second shell. (Each orbital, remember, can hold two electrons.

atoms of other elements by adding one proton and one electron at a time (along with an appropriate number of neutrons). FIGURE 2.9, an abbreviated version of what is called a periodic table, shows this for the first 18 elements, from hydrogen ($_1H$) to argon ($_{18}Ar$). The elements are arranged in three tiers, or periods, corresponding to the sequential addition of electrons to orbitals in the first three electron shells. The main point is that the chemical properties of an atom depend mostly on the number of electrons in its *outermost* shell. We refer to those outer electrons as **valence electrons,** and to the outermost energy shell as the **valence shell.**

An atom with a complete valence shell is unreactive; that is, it will not interact readily with other atoms it encounters. At the far right of the periodic table are helium, neon, and argon, the only three elements shown that have full valence shells. They are termed inert elements because they are chemically unreactive. All other atoms shown in FIGURE 2.9 are chemically reactive because they have incomplete valence shells with unpaired electrons. Atoms with the same number of electrons in their valence shells exhibit similar chemical behavior. For example, fluorine (F) and chlorine (Cl) both have seven valence electrons, and both combine with the element sodium to form compounds.

Atoms combine by chemical bonding to form molecules

Now that we have looked at the structure of atoms, we will move up in the hierarchy of organization and see how atoms combine to form molecules. Atoms with incomplete valence shells will interact with certain other atoms in such a way that each partner completes its valence shell. Atoms do this by either sharing or completely transferring valence electrons. These interactions usually result in atoms staying close together, held by attractions

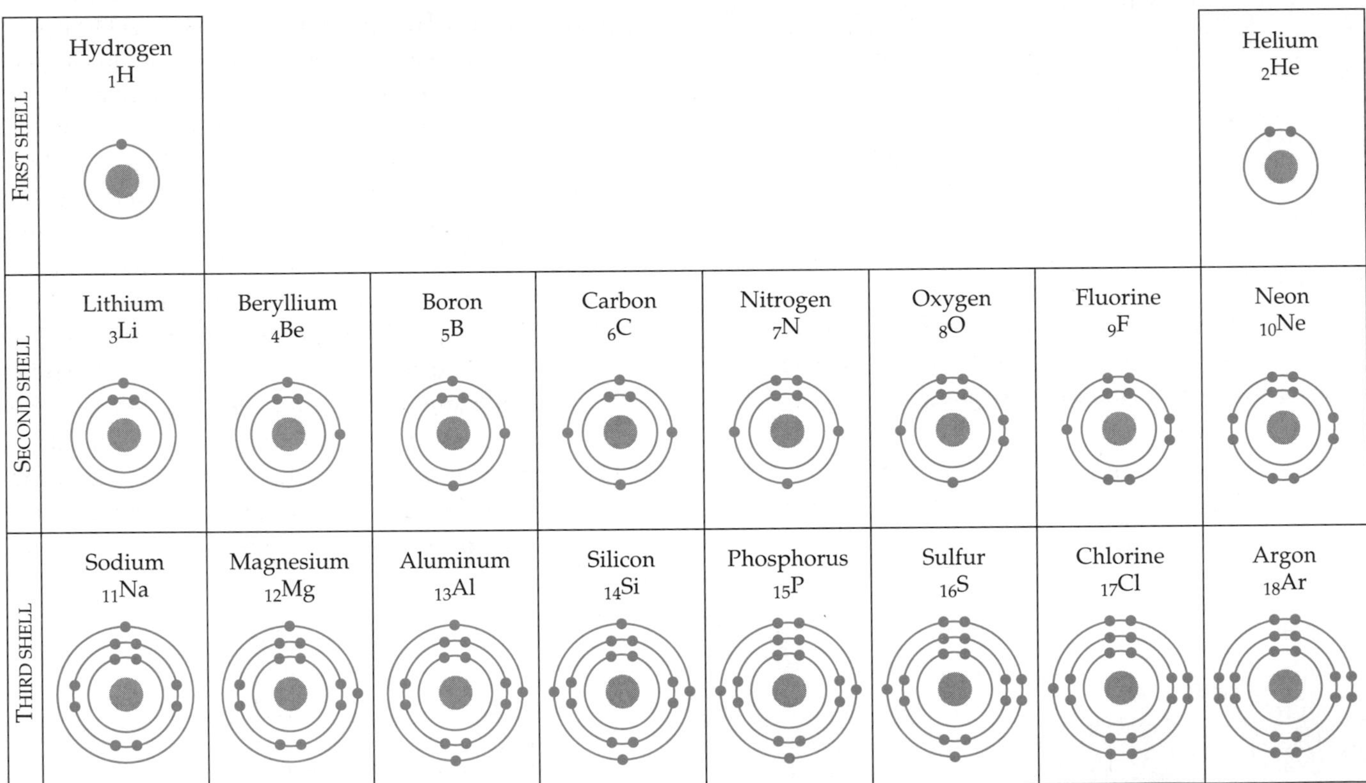

FIGURE 2.9

Electron configurations of the first 18 elements. The number of electrons in each energy level (shell) is diagrammed as dots on concentric rings. (These symbolic rings do not portray the three-dimensional freedom of electrons in their orbitals.) The elements are arranged in rows, each representing the filling of an electron shell. As electrons are added, they occupy the lowest available energy level. Hydrogen's one electron and helium's two electrons are located in the first level. The next element, lithium, has three electrons. Two electrons fill the first energy level, while the third electron occupies the second energy level, not some level more distant from the nucleus. This behavior of electrons illustrates the general tendency for matter to exist in its lowest state of potential energy. The outermost energy level occupied by electrons is called the valence shell. Of these first 18 elements, only helium, neon, and argon are made of atoms with full valence shells; such elements are called inert because they are unreactive. All other elements in this figure consist of atoms with incomplete valence shells, and the presence of unpaired electrons makes these atoms chemically reactive. Those elements with the same number of valence electrons—fluorine and chlorine, for instance—have similar chemical properties.

called **chemical bonds.** The strongest kinds of chemical bonds are covalent bonds and ionic bonds.

Covalent Bonds

A **covalent bond** is the sharing of a pair of valence electrons by two atoms. For example, let's see what happens when two hydrogen atoms approach each other. Recall that hydrogen has one valence electron in the first shell, but the shell's capacity is for two electrons. When the two hydrogen atoms come close enough for their 1*s* orbitals to overlap, they share their electrons. Each hydrogen atom now has two electrons associated with it in what amounts to a completed valence shell. In this case, we have formed a hydrogen molecule consisting of two hydrogen atoms held together by a covalent bond (FIGURE 2.10a). We abbreviate this molecule by writing H—H, where the line represents a covalent bond, that is, a pair of shared electrons. This type of notation, which represents both atoms and bonding, is called a **structural formula.** We can abbreviate even further by writing H_2, a **molecular formula** indicating simply that the molecule consists of two atoms of hydrogen.

With six electrons in its second electron shell, oxygen needs *two* more electrons to complete this valence shell. Two oxygen atoms form a molecule by sharing *two* pairs of valence electrons (FIGURE 2.10b). The atoms are thus joined by what is called a **double covalent bond.**

Notice that each atom sharing electrons has a bonding capacity: a certain number of covalent bonds that must be formed for the atom to have a full complement of valence electrons. This bonding capacity is called the atom's **valence.** The valence of hydrogen is 1; the valence of oxygen is 2. You should be able to deduce the valences of other elements from the electron configurations in FIGURE 2.9. For example, can you explain why nitrogen atoms usually form three covalent bonds?

The molecules we have looked at so far—H_2 and O_2—are pure elements, not compounds. (Recall that a compound is a combination of two or more different elements.) An example of a compound is water, with the molecular formula H_2O; it takes two atoms of hydrogen to satisfy the valence of one oxygen atom. FIGURE 2.10c shows the structure of a water molecule. This molecule is so important to life that the next chapter is devoted entirely to its structure and behavior.

Another molecule that is also a compound is methane, the main component of natural gas, with the molecular formula CH_4 (FIGURE 2.10d). Carbon ($_6C$) has four valence electrons, so its bonding capacity, or valence, is 4. It takes four hydrogen atoms, each with a valence of 1, to complement one atom of carbon. You now know the valences of the four most abundant elements in life: Hydrogen forms one bond, oxygen forms two bonds, nitrogen forms three bonds (usually), and carbon forms four bonds.

Nonpolar and Polar Covalent Bonds. The attraction of an atom for the electrons of a covalent bond is called **electronegativity.** The more electronegative an atom, the more strongly it pulls shared electrons toward itself. In a covalent bond between two atoms of the same element, the outcome of the tug-of-war for common electrons is a standoff; the two atoms are equally electronegative. In a **nonpolar covalent bond,** electrons are shared equally. The covalent bond of H_2 is nonpolar, as is the double bond of O_2. The bonds of methane (CH_4) are also nonpolar; although the partners are different elements, carbon and hydrogen do not differ substantially in electronegativity. This is not always the case in a compound where covalent bonds join atoms of different elements. If one atom is more electronegative than the other, electrons of the bond will not be shared equally. In such cases, the bond is called a **polar covalent bond.** In a water molecule, the bonds between oxygen and hydrogen are

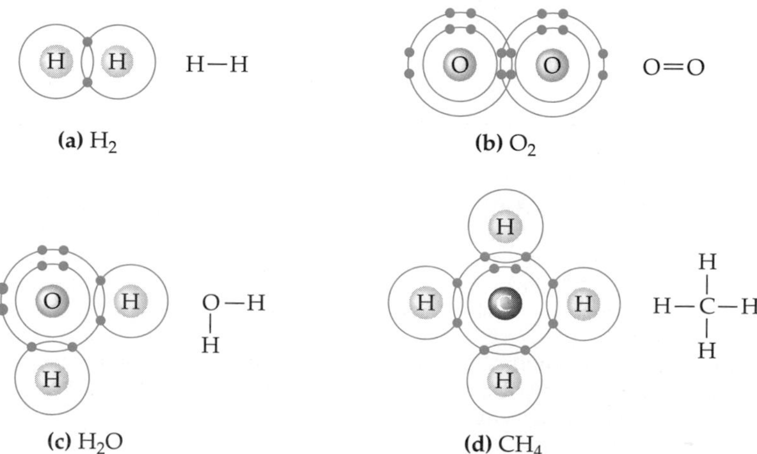

FIGURE 2.10

Covalent bonds. Each covalent bond consists of a pair of shared electrons. The number of electrons required to complete an atom's valence shell determines how many bonds that atom will form. (**a**) If two unattached hydrogen atoms meet, they will form a single covalent bond. (**b**) Two oxygen atoms form a molecule by sharing two pairs of valence electrons; the atoms are joined by a double covalent bond. (**c**) Two hydrogen atoms can be joined to one oxygen atom by covalent bonds to produce a molecule of water. (**d**) Four hydrogen atoms satisfy the valence of one carbon atom, forming methane.

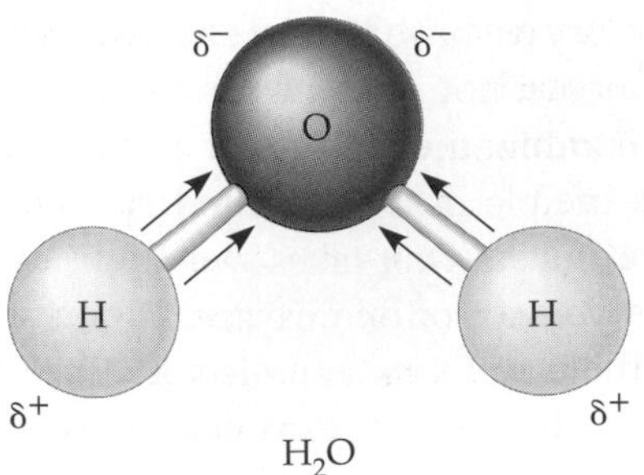

FIGURE 2.11
Polar covalent bonds in a water molecule. Oxygen, being much more electronegative than hydrogen, pulls the shared electrons of the bond toward itself, as indicated by the arrows in this diagram. This unequal sharing of electrons gives the oxygen a slight negative charge and the hydrogens a small amount of positive charge. The Greek symbol delta (δ) indicates that the charges are less than full units.

polar. Oxygen is one of the most electronegative of the 92 elements, attracting shared electrons much more strongly than hydrogen does. In a covalent bond between oxygen and hydrogen, the electrons spend more time around the oxygen atom than they do around the hydrogen atom. Since electrons have a negative charge, the unequal sharing of electrons in water causes the oxygen atom to have a slight negative charge and each hydrogen atom a slight positive charge (FIGURE 2.11).

Ionic Bonds

In some cases, two atoms are so unequal in their attraction for valence electrons that the more electronegative atom strips an electron completely away from its partner. This is what happens when an atom of sodium ($_{11}Na$) encounters an atom of chlorine ($_{17}Cl$) (FIGURE 2.12). A sodium atom has a total of 11 electrons, with its single valence electron in the third electron shell. A chlorine atom has a total of 17 electrons, with 7 electrons in its valence shell. When these two atoms meet, the lone valence electron of sodium is transferred to the chlorine atom, and both atoms end up with their valence shells complete. (Since sodium no longer has an electron in the third shell, the second shell is now outermost.)

The electron transfer between the two atoms moves one unit of negative charge from sodium to chlorine. Sodium, now with 11 protons but only 10 electrons, has a net electrical charge of +1. A charged atom (or molecule) is called an **ion.** When the charge is positive, the ion is specifically called a **cation.** Conversely, the chlorine atom, having gained an extra electron, now has 17 protons and 18 electrons, giving it a net electrical charge of −1. It has become a chloride ion—specifically, an **anion,** or negatively charged ion. Because of their opposite charges, cations and anions attract each other in what is called an **ionic bond.**

Ionic compounds are called salts. We know the compound sodium chloride (NaCl) as table salt. Salts are often found in nature as crystals of various sizes and shapes, each an aggregate of vast numbers of cations and anions bonded by their electrical attraction and arranged in a three-dimensional lattice (FIGURE 2.13). A salt crystal does not really consist of molecules in the same sense that a covalent compound does because a covalently bonded molecule has a definite size and number of

FIGURE 2.13
A sodium chloride crystal. The sodium ions (Na^+) and chloride ions (Cl^-) are held together by ionic bonds.

FIGURE 2.12
Electron transfer and ionic bonding. A valence electron is transferred from sodium (Na) to chlorine (Cl), giving both atoms completed valence shells. The electron transfer leaves the sodium atom with a net charge of +1 (cation) and the chlorine atom with a net charge of −1 (anion). The attraction between the oppositely charged atoms, or ions, is an ionic bond. Ions can bond not only to the atom they reacted with, but to any other ion of opposite charge.

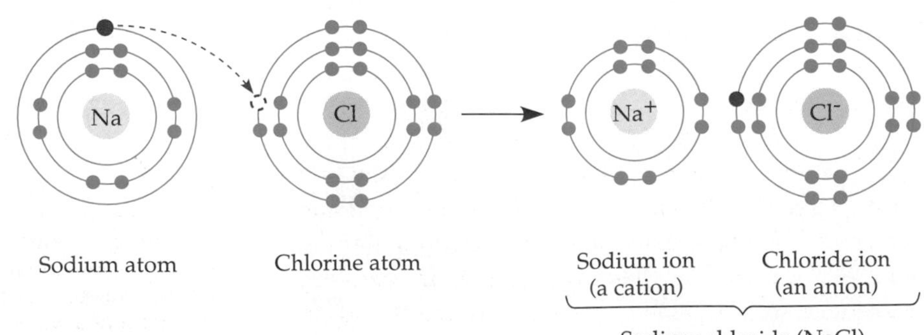

atoms. The formula for an ionic compound, such as NaCl, indicates only the ratio of elements in a crystal of the salt.

Not all salts have equal numbers of cations and anions. For example, the ionic compound magnesium chloride ($MgCl_2$) has two chloride ions for each magnesium ion. Magnesium ($_{12}Mg$) must lose two outer electrons if the atom is to have a complete valence shell. One magnesium atom can supply valence electrons to two chlorine atoms. After losing two electrons, the magnesium atom is a cation with a net charge of +2 (Mg^{2+}).

The term *ion* also applies to entire covalent molecules that are electrically charged. In the salt ammonium chloride (NH_4Cl), for instance, the anion is a single chloride ion (Cl^-), but the cation is ammonium (NH_4^+), a nitrogen atom with four covalently bonded hydrogen atoms. The whole ammonium ion has an electrical charge of +1 because it is one electron short.

Environment affects the strength of ionic bonds. In a dry salt crystal, the bonds are so strong that it takes a hammer and chisel to break enough of them to crack the crystal in two. Place the same salt crystal in water, however, and the salt dissolves as the attractions between its ions decrease. In the next chapter, you will learn how water dissolves salts.

There is no distinct line between covalent bonding and ionic bonding. A nonpolar covalent bond and an ionic bond are opposite extremes in a range of situations where atoms share electrons. In the middle zone is the polar covalent bond, in which electrons are shared, but unequally. We might think of an ionic bond as a covalent bond that is so polar that one atom has pulled an electron completely away from its less electronegative partner. Indeed, some compounds spend part of their time in a polar covalent state and the rest of their time as ions.

Weak chemical bonds play important roles in the chemistry of life

Covalent bonds are important in living matter because they link the atoms of a cell's molecules. But bonding *between* molecules is also important, especially in the chemistry of the cell, where the properties of life emerge from molecular interactions. When two molecules in the cell associate, they may adhere temporarily by types of chemical bonds that are much weaker than covalent bonds. The advantage of weak bonds is that the contact between the molecules can be brief; the molecules come together, respond to one another in some way, and then separate. An example is chemical signaling in the brain (see FIGURE 2.16). The signal molecule uses weak bonds to dock on the receptor molecule just long enough to trigger a momentary response by the receiving cell. If the signal molecule attached by stronger covalent bonds, the receiving cell would continue to respond long after the transmitting cell ceased dispatching the message. (Imagine, for instance, if your brain continued to perceive the ringing sound of a bell for hours after nerve cells transmitted the information from the ears to the brain.)

Several types of weak chemical bonds enable the molecules of the cell to associate. One is the ionic bond, which is relatively weak in the presence of water. Another type of weak bond that functions in living matter is known as a hydrogen bond.

Hydrogen Bonds

Among the various kinds of weak chemical bonds, hydrogen bonds are so important in the chemistry of life that they deserve special attention. A **hydrogen bond** occurs when a hydrogen atom covalently bonded to one electronegative atom is also attracted to another electronegative atom (FIGURE 2.14). In living cells, the electronegative partners involved are usually oxygen or nitrogen atoms.

Let's examine the relatively simple case of hydrogen bonding between water (H_2O) and ammonia (NH_3). You have seen how the polar covalent bonds of water result in the oxygen atom having a slight negative charge and the hydrogen atoms having a slight positive charge. A similar situation arises in the ammonia molecule where an electronegative nitrogen atom has a small amount of negative charge because of its pull on the electrons it shares covalently with hydrogen. If a water molecule and an ammonia molecule are close to each other, there will be a weak attraction between the negatively charged nitrogen atom and a positively charged hydrogen atom of the

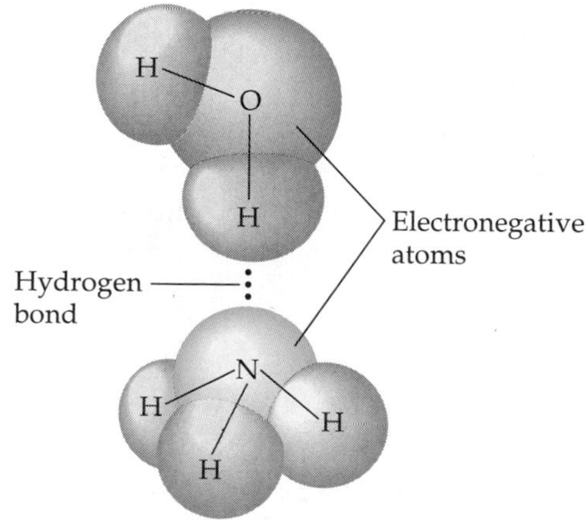

FIGURE 2.14
A hydrogen bond. Through a weak electrical attraction, one electronegative atom shares a hydrogen atom with another electronegative atom. In this diagram, a hydrogen bond joins a hydrogen atom of a water molecule (H_2O) with the nitrogen atom of an ammonia molecule (NH_3).

adjacent water molecule. This attraction is a hydrogen bond.

Hydrogen bonds, ionic bonds, and other weak bonds form not only between molecules; they also may form between different regions of a single large molecule, such as a protein. Although these bonds are individually weak, their cumulative effect is to reinforce the three-dimensional shape of a large molecule. You will learn more about the biological roles of weak bonds in Chapter 5.

A molecule's biological function is related to its shape

A molecule has a characteristic size and shape (FIGURE 2.15). The water molecule, for example, is shaped roughly like a right angle, its two covalent bonds spread apart by 104.5°. The methane molecule is shaped like a tetrahedron, a pyramid with a three-sided base. The nucleus of the carbon atom is at the center, with its four covalent bonds radiating to the hydrogen nuclei at the corners of the tetrahedron. Larger molecules, including many of the molecules that make up living matter, have more complex shapes. Molecular shape is important to biologists because it is the basis for how most molecules of life recognize and respond to one another. For example, one nerve cell in the brain signals another by dispatching specific molecules that have unique shapes. On the surface of the receiving cell are receptor molecules that fit a certain type of signal molecule something like a lock and key (FIGURE 2.16). Some of these signals even affect our emotions. For instance, the human brain has molecular signals called endorphins that contribute to feelings of elation. Morphine, heroin, and the other opiate drugs mimic endorphins and artificially produce euphoria by binding to endorphin receptors in the brain (FIGURE 2.17).

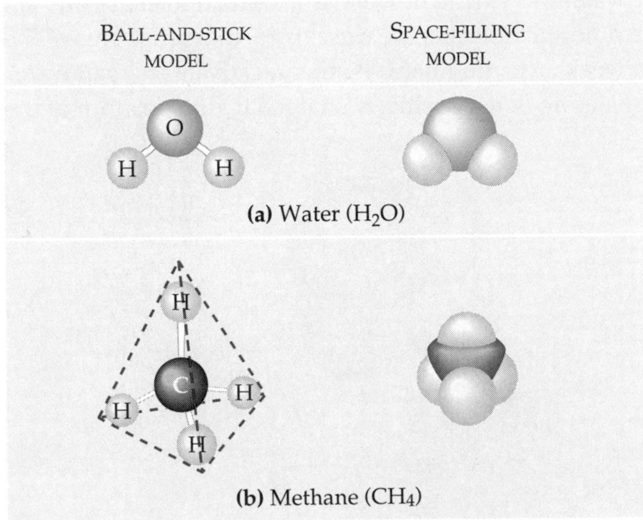

FIGURE 2.15
Molecular shapes of water and methane. This figure introduces two types of models that represent three-dimensional shapes of molecules. A ball-and-stick model emphasizes the bond angles of the molecule. A space-filling model portrays a molecule's shape more accurately. (**a**) The two covalent bonds of water are angled at 104.5°. (**b**) In methane, carbon's four covalent bonds angle toward the corners of an imaginary tetrahedron, outlined here with a dotted magenta line.

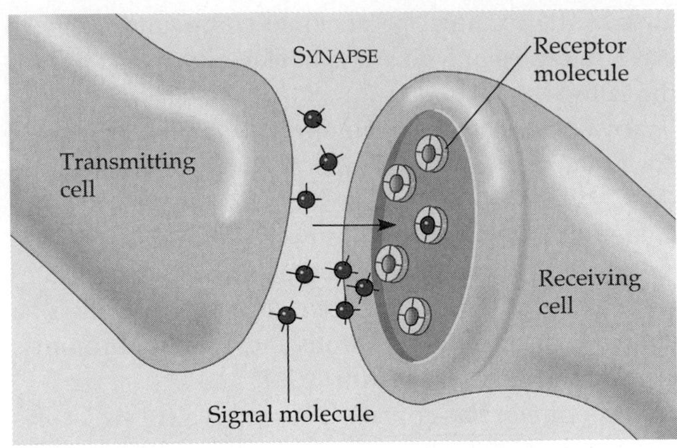

FIGURE 2.16
Molecular shape and brain chemistry. One nerve cell in the brain signals another by releasing a chemical messenger that fits a receptor molecule on the membrane of the receiving cell. The transmitting cell contains numerous vesicles, not visible in this exterior view. A signal is transmitted when these vesicles release specific molecules into the synapse, a tiny gap between the nerve cell and its neighbor. The signal molecules pass across the synapse and stimulate the receiving cell. In the drawing, we see a highly simplified version of the chemical communication that occurs between two cells. The signal molecule and the receptor molecule are shown here as simple objects of specific shapes that will fit together; the actual molecules have much more complex geometries. The role of molecular shape in brain chemistry is an example of the correlation between structure and function, one of biology's unifying themes.

Chemical reactions change the composition of matter

The making and breaking of chemical bonds, leading to changes in the composition of matter, are called **chemical reactions.** An example is the reaction between hydrogen and oxygen to form water:

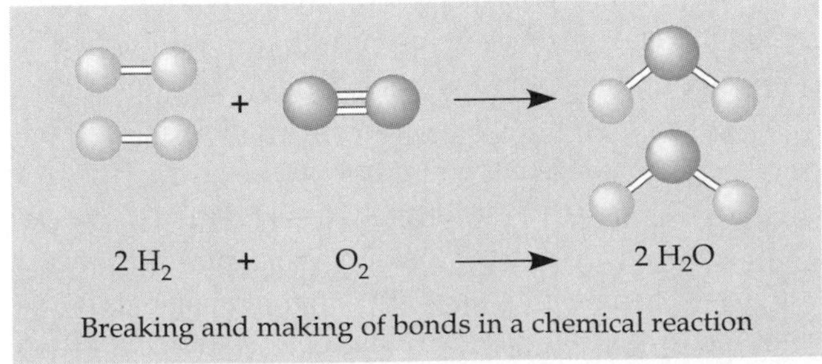

$2 H_2$ + O_2 → $2 H_2O$
Breaking and making of bonds in a chemical reaction

This reaction breaks the covalent bonds of H_2 and O_2, forming the new bonds of a water molecule. When we

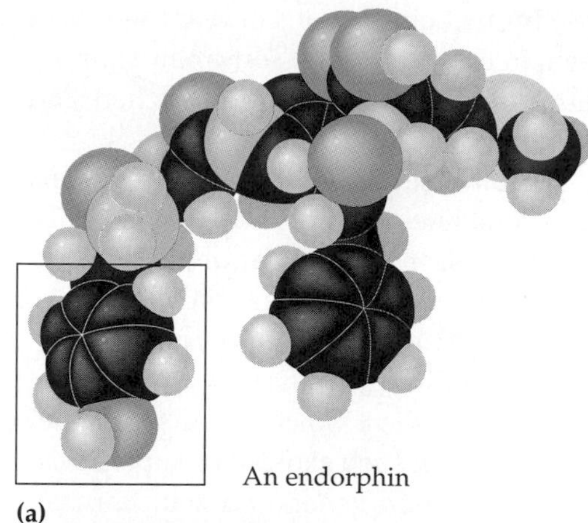

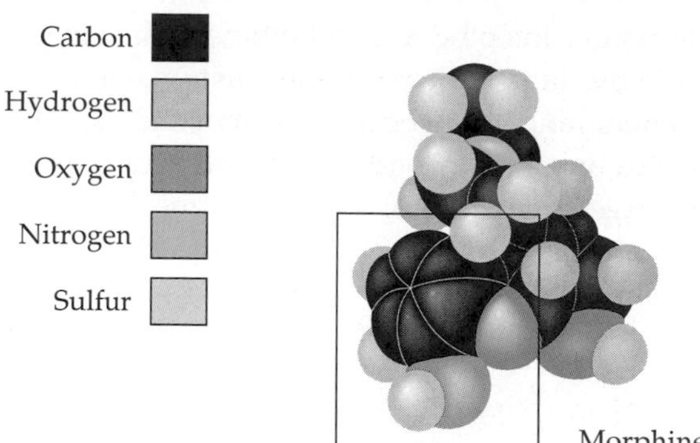

FIGURE 2.17
A molecular copycat. (**a**) An endorphin is a natural brain signal that helps us feel good. The boxed portion of the endorphin molecule is the shape that is recognized by receptor molecules on appropriate target cells in the brain. The boxed portion of the morphine molecule, an opiate drug, is a close match. Morphine, a chemical imposter, affects emotional state by mimicking the brain's natural endorphins. (**b**) It was Candace Pert, working in the lab of Solomon Snyder at Johns Hopkins University, who helped revolutionize neurochemistry in 1972 when she discovered that opiates bind to receptors in the brain that normally recognize the brain's own endorphins.

write a chemical reaction, we use an arrow to indicate the conversion of the starting materials, called **reactants,** to the **products.** The coefficients indicate the number of molecules involved; for example, the 2 in front of the H_2 means that the reaction starts with two molecules of hydrogen. Notice that all atoms of the reactants must be accounted for in the products. Matter is conserved in a chemical reaction: Reactions cannot create or destroy matter but can only rearrange it.

Below is the chemical shorthand that summarizes the process of photosynthesis, a particularly important example of chemical reactions rearranging matter:

$$6\ CO_2 + 6\ H_2O \longrightarrow\longrightarrow\longrightarrow C_6H_{12}O_6 + 6\ O_2$$

Photosynthesis takes place within chloroplasts, green organelles in leaf cells. The raw materials are carbon dioxide (CO_2), which is taken from the air, and water (H_2O), which is absorbed from the soil. Within the chloroplasts, sunlight powers the conversion of these ingredients to a sugar called glucose ($C_6H_{12}O_6$) and oxygen (O_2), which the leaves release into the air (FIGURE 2.18). Although photosynthesis is actually a sequence of many chemical reactions, we still end up with the same number and kinds of atoms we had when we started. Matter has been rearranged but conserved.

FIGURE 2.18
Photosynthesis: a solar-powered rearrangement of matter. This *Elodea,* a freshwater plant, produces sugars and other food by rearranging the atoms of carbon dioxide and water in the chemical process known as photosynthesis. Sunlight powers this chemical transformation. The oxygen bubbles escaping from the leaves are by-products of photosynthesis. Humans and other animals ultimately depend on photosynthesis for food and oxygen. Photosynthesis, the process at the foundation of almost all ecosystems, is an important example of how chemistry applies to the study of life.

Some chemical reactions go to completion; that is, all the reactants are converted to products. But most reactions are reversible, the products of the forward reaction becoming the reactants for the reverse reaction. For example, hydrogen and nitrogen molecules combine to form ammonia, but ammonia can also decompose to regenerate hydrogen and nitrogen:

$$3\ H_2 + N_2 \rightleftharpoons 2\ NH_3$$

The opposite-headed arrows indicate that the reaction is reversible.

One of the factors affecting the rate of reaction is the concentration of reactants. The greater the concentration of reactant molecules, the more frequently they collide with one another and have an opportunity to react to form products. The same holds true for the products. As products accumulate, collisions resulting in the reverse reaction become increasingly frequent. Eventually, the forward and reverse reactions occur at the same rate, and the relative concentrations of products and reactants remain fixed. The point at which the reactions offset one another exactly is called **chemical equilibrium.** This is a dynamic equilibrium; reactions are still going on, but with no net effect on the concentrations of reactants and products. Equilibrium does *not* mean that the reactants and products are equal in concentration, but only that their concentrations have stabilized. For the above reaction involving ammonia, equilibrium is reached when ammonia decomposes as rapidly as it forms. In this case, the forward reaction is much more favorable than the reverse reaction; thus, at equilibrium, there will be far more ammonia than hydrogen and nitrogen.

We will end this chapter by placing the chemical basis of life in an evolutionary context.

Chemical conditions on the early Earth set the stage for the origin and evolution of life

Chemical reactions and physical processes on the early Earth created an environment that made life possible. And life, once it began, transformed the planet's chemistry. Biological and geological history are inseparable.

The formation of planet Earth and its life is a fragment of a much bigger story. Earth is one of nine planets orbiting the sun, one of billions of stars in the Milky Way, which is one of millions of galaxies in the universe. The star closest to our sun, Proxima Centauri, is four light years—40 trillion km—away; we see it by the light it emitted four years ago. Some stars are so distant that even if they burned out millions of years ago, we would still see them in the sky tonight; some new stars are invisible because their light has not yet reached Earth. Gazing at stars, we look back in time.

The universe has not always been so spread out. Based on several lines of evidence, most astronomers now believe that all matter was at one time concentrated in a single mass that blew apart with a "big bang" sometime between 10 and 20 billion years ago. The universe has been expanding ever since.

Some stars die a violent death when they explode; astronomers call such explosions supernovae. Our sun is a second- or third-generation star, born about 5 billion years ago from the fallout of defunct stars. Compared with the overall universe, our solar system is relatively rich in the heavier elements that formed by fusion from smaller atoms in the crucibles of ancestral stars. Most of the swirling matter in the disk-shaped cloud of dust that formed our solar system condensed in the center as the sun. Peripheral material was left spinning around the infant sun in several rings. The planets, including Earth, formed about 4.6 billion years ago from kernels that used gravity to draw together the dust and ice in their zones.

Most geologists believe that Earth began as a cold world that later melted from the heat produced by compaction, radioactive decay, and the impact of meteorites. Molten material sorted into layers of varying density. Most of the nickel and iron sank to the center and formed a core. Less dense material became concentrated in a mantle, and the least dense material solidified into a thin crust. The present continents are attached to plates of crust that float on the flexible mantle.

The first atmosphere, which was probably composed mostly of hot hydrogen gas (H_2), escaped because the gravity of Earth was not strong enough to hold such small molecules. Volcanoes belched gases that formed a new atmosphere. Based on analysis of gases vented by modern volcanoes, scientists have speculated that the second early atmosphere consisted mostly of water vapor (H_2O), carbon monoxide (CO), carbon dioxide (CO_2), nitrogen (N_2), methane (CH_4), and ammonia (NH_3). The first seas formed from torrential rains that began when Earth had cooled enough for water in the atmosphere to condense. In addition to an atmosphere very different from the one we know, lightning, volcanic activity, and ultraviolet radiation were much more intense when Earth was young. In such a world, life began.

In upcoming chapters, you will learn more about how chemical evolution on the early Earth made the origin of life possible. To begin, in the next chapter we will examine the life-promoting properties of water.

REVIEW OF KEY CONCEPTS (with page numbers and key figures)

- Matter consists of chemical elements in pure form and in combinations called compounds (pp. 25–26, FIGURE 2.2)
  - The basic ingredients of matter are the elements, which cannot be broken down to other substances.
  - A compound contains two or more elements in a fixed ratio and has emergent properties very different from those of its constituent elements.
- Life requires about 25 chemical elements (p. 27, TABLE 2.1)
  - Carbon, oxygen, hydrogen, and nitrogen make up 96% of living matter.
  - The remaining 4% of living matter includes trace elements, which are required in minute amounts.
- Atomic structure determines the behavior of an element (pp. 27–32, FIGURE 2.9)
  - An atom is the smallest unit of an element.
  - An atom consists of three types of subatomic particles. Uncharged neutrons and positively charged protons are tightly bound in a nucleus; negatively charged electrons move rapidly about the nucleus.
  - The number of protons in an atom is called the atomic number. The number of electrons in an electrically neutral atom is equal to the number of protons.
  - The mass number of an element indicates the sum of the protons and neutrons and approximates the atomic weight of an atom in daltons.
  - Most elements consist of two or more isotopes, different in neutron number and mass. Some isotopes are unstable and give off particles and energy as radioactivity. Radioactivity has important uses in science and medicine but can also harm organisms.
  - Electron configuration determines the chemical behavior of an atom—that is, the way it reacts with other atoms.
  - Electrons move within orbitals, three-dimensional spaces located within successive electron shells (energy levels) surrounding the nucleus.
  - Chemical properties depend on the number of valence electrons, those in the outermost shell. An atom with an incomplete valence shell is reactive.
- Atoms combine by chemical bonding to form molecules (pp. 32–35, FIGURES 2.10 and 2.12)
  - Chemical bonds form when atoms interact and complete their valence shells.
  - A covalent bond forms when two atoms share a pair of valence electrons.
  - Molecules consist of two or more bonded atoms.
  - A structural formula shows the atoms and bonds in a molecule. A molecular formula indicates only the number and types of atoms.
  - A nonpolar covalent bond forms when both atoms are equally electronegative. Electrons of a polar covalent bond are pulled closer to the more electronegative atom.
  - An ionic bond is created when two atoms differ so much in electronegativity that one or more electrons are actually transferred from one atom to the other. The recipient atom becomes a negatively charged anion. The donor atom becomes a positively charged cation. Cations and anions attract each other in an ionic bond.
- Weak chemical bonds play important roles in the chemistry of life (pp. 35–36, FIGURE 2.14)
  - Hydrogen bonds are relatively weak bonds between a partially positive hydrogen atom of one polar molecule and a partially negative atom of another polar molecule.
  - Hydrogen bonds and other weak bonds function in adhesion between molecules and help reinforce the shapes of large biological molecules.
- A molecule's biological function is related to its shape (p. 36, FIGURE 2.15)
  - Molecules have characteristic sizes and shapes symbolized with ball-and-stick and space-filling models.
  - In many cases, shape is the basis for one biological molecule recognizing another when they interact.
- Chemical reactions change the composition of matter (pp. 36–38)
  - Chemical reactions break or form chemical bonds to change reactants into products. During a reaction, matter is conserved.
  - Most chemical reactions are reversible. Chemical equilibrium is reached when the forward and backward reaction rates are equal.
- Chemical conditions on the early Earth set the stage for the origin and evolution of life (p. 38)
  - The chemical composition of Earth reflects the history of the solar system, galaxy, and universe.
  - Conditions on the early Earth, very different from those of today, made the origin of life possible.

SELF-QUIZ

1. An element is to a (an) ____________ as a tissue is to a (an) ____________.
   a. atom; organism
   b. compound; organ
   c. molecule; cell
   d. atom; organ
   e. compound; organelle

2. In the term *trace element,* the modifier *trace* means
   a. the element is required in very small amounts
   b. the element can be employed as a label to trace atoms through an organism's metabolism
   c. the element is very rare on Earth
   d. the element enhances health but is not essential for the organism's long-term survival
   e. the element passes rapidly through the organism

3. Compared to ^{31}P, the radioactive isotope ^{32}P has
   a. a different atomic number
   b. one more neutron
   c. one more proton
   d. one more electron
   e. a different charge

4. What do the four elements most abundant in life—carbon, oxygen, hydrogen, and nitrogen—have in common?
   a. They all have the same number of valence electrons.
   b. Each element exists in only one isotopic form.

c. They are all relatively light elements, near the top of the periodic table.
d. They are all about equal in electronegativity.
e. They are elements produced only by living cells.

5. The atomic number of sulfur is 16. Sulfur combines with hydrogen by covalent bonding to form a compound, hydrogen sulfide. Based on the electron configuration of sulfur, we can predict that the molecular formula of the compound will be (explain your answer)

a. HS b. HS_2 c. H_2S d. H_3S_2 e. H_4S

6. Review the valences of carbon, oxygen, hydrogen, and nitrogen, and then determine which of the following molecules is most likely to exist.

a. O═C—H

c. H—O—C(H)(H)—C(H)═O (structure: H—O—C—C═O, with H above and H below the first C, and H above the second C)

b.

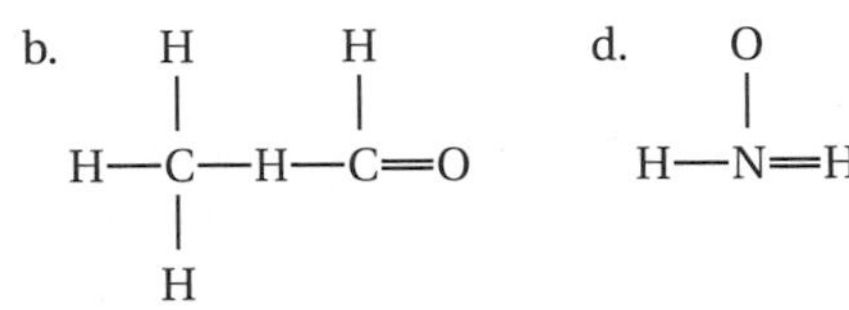

d. H—N═H, with O bonded above N

7. Which orientation is most likely for two adjacent water molecules? Explain your answer.

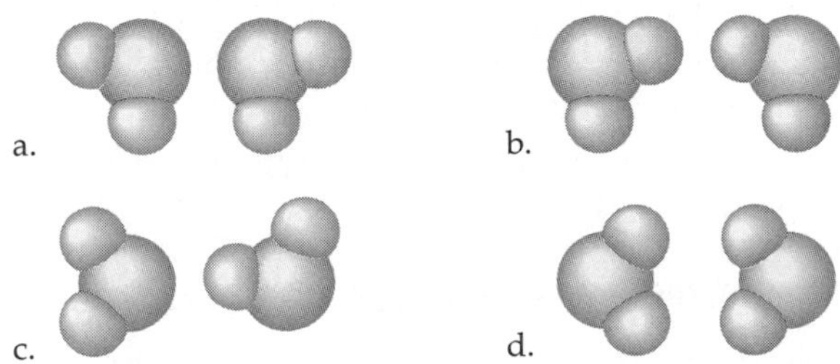

8. Which of these statements is true of all anionic atoms?
   a. The atom has more electrons than protons.
   b. The atom has more protons than electrons.
   c. The atom has fewer protons than does a neutral atom of the same element.
   d. The atom has more neutrons than protons.
   e. The net charge is –1.

9. What coefficients must be placed in the blanks to balance this chemical reaction?

$$C_6H_{12}O_6 \longrightarrow \underline{?}\, C_2H_6O + \underline{?}\, CO_2$$

a. 1; 2 b. 2; 2 c. 1; 3 d. 1; 1 e. 3; 1

10. Which of the following statements correctly describes any chemical reaction that has reached equilibrium?
    a. The concentration of products equals the concentration of reactants.
    b. The rate of the forward reaction equals the rate of the reverse reaction.
    c. Both forward and reverse reactions have halted.
    d. The reaction is now irreversible.
    e. No reactants remain.

CHALLENGE QUESTIONS

1. Recall from Chapter 1 that vitalism is the belief that life possesses supernatural forces that cannot be explained by physical and chemical principles. Explain why the concept of emergent properties does not lend credibility to vitalism.

2. Female silkworm moths (Bombyx mori) attract males by emitting chemical signals that spread through the air. A male hundreds of meters away can detect these molecules and fly toward their source. The sensory organs responsible for this behavior are the comblike antennae visible in the photograph below. Each filament of an antenna is equipped with thousands of receptor cells that detect the sex attractant. Based on what you learned in this chapter, propose a hypothesis to account for the ability of the male moth to detect a specific molecule in the presence of many other molecules in the air. What predictions does your hypothesis make? Design an experiment to test one of these predictions.

SCIENCE, TECHNOLOGY, AND SOCIETY

1. The use of radioactive isotopes as tracers in biochemical research is based on cells using the tracers in place of nonradioactive isotopes. Explain why this ability of radioactive isotopes to infiltrate the chemical processes of the cell also compounds the threat posed by radioactive contaminants in air, soil, and water.

2. While waiting at an airport, your author once overheard this claim: "It's paranoid and ignorant to worry about industry or agriculture contaminating the environment with their chemical wastes. After all, this stuff is just made of the same atoms that were already present in our environment." How would you counter this argument?

FURTHER READING

Allègre, C. J., and S. H. Schneider. "The Evolution of the Earth." *Scientific American*, October 1994.

Atkins, P. W. *Molecules*. New York: Scientific American Library, 1987. Beautifully illustrated tour of the world of molecules.

Kahn, P. "A Grisly Archive of Key Cancer Data." *Science*, January 22, 1993. Alarming health problems among uranium workers in Germany.

Pennisi, E. "Natureworks." *Science News*, May 16, 1992. How organisms make minerals.

Sackheim, G. *Introduction to Chemistry for Biology Students*, 4th ed. Menlo Park, CA: Benjamin/Cummings, 1991. A programmed review of the fundamentals of chemistry.

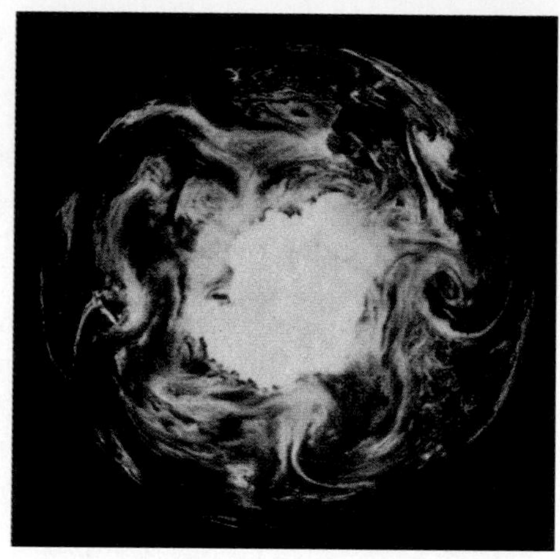

CHAPTER 3

WATER AND THE FITNESS OF THE ENVIRONMENT

KEY CONCEPTS

- The polarity of water molecules results in hydrogen bonding
- Organisms depend on the cohesion of water molecules
- Water contributes to Earth's habitability by moderating temperatures
- Oceans and lakes don't freeze solid because ice floats
- Water is the solvent of life
- Organisms are sensitive to changes in pH
- Acid precipitation threatens the fitness of the environment

If we could cruise the universe in a quest for life, we would do well to search for worlds with water. We might not recognize life on dry planets even if it existed. All organisms familiar to us are made mostly of water and live in a world where water dominates climate and many other features of the environment. Here on Earth, water is the biological medium—the substance that makes possible life as we know it.

Life began in water and evolved there for three billion years before spreading onto land. Modern life, even terrestrial (land-dwelling) life, remains tied to water. Most cells are surrounded by water, and cells contain from about 70% to 95% water. Earth's surface is also wet, with water covering three-quarters of our planet. Although most of this water—enough to cover the United States to a depth of 130 km—is in liquid form, water is also present on Earth as ice and vapor. Water is the only common substance to exist in the natural environment in all three physical states of matter: solid, liquid, and gas. These three states of water are visible in the photograph on this page, a view of Earth's South Pole.

The abundance of water is a major reason Earth is habitable. In a classic book called The Fitness of the Environment, *Lawrence Henderson highlights the importance of water to life. While acknowledging that life adapts to its environment through natural selection, Henderson emphasizes that for life to exist at all in a particular location, the environment must first be a suitable abode. Your objective in this chapter is to develop a conceptual understanding of how water contributes to the fitness of Earth for life.*

Water is so common that it is easy to overlook the fact that it is an exceptional substance with many extraordinary qualities. Following the theme of emergent properties, we can trace water's unique behavior to the structure and interactions of its molecules.

■ The polarity of water molecules results in hydrogen bonding

Studied in isolation, the water molecule is deceptively simple. Its two hydrogen atoms are joined to the oxygen atom by single covalent bonds (see FIGURE 2.11). Because oxygen is more electronegative than hydrogen, the electrons of the polar bonds spend fractionally more time closer to the oxygen atom. This results in the oxygen region of the molecule having a slight negative charge and the hydrogens having a slight positive charge. The water molecule, shaped something like a right angle, is a **polar**

molecule, meaning that it has opposite charges on opposite ends.

The anomalous properties of water arise from attractions among these polar molecules. The attraction is electrical; a slightly positive hydrogen of one molecule is attracted to the slightly negative oxygen of a nearby molecule. The molecules are thus held together by a hydrogen bond (FIGURE 3.1). Each water molecule can form hydrogen bonds to a maximum of four neighbors. The extraordinary qualities of water are emergent properties resulting from the hydrogen bonding that orders molecules into a higher level of structural organization.

We will examine four of water's properties that contribute to the fitness of Earth as an environment for life: water's cohesive behavior, its ability to stabilize temperature, its expansion upon freezing, and its versatility as a solvent for life.

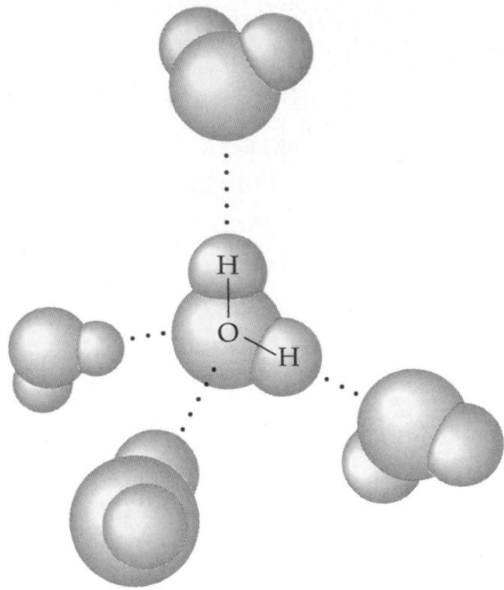

FIGURE 3.1
Hydrogen bonds between water molecules. The charged regions of a polar water molecule are attracted to oppositely charged parts of neighboring molecules. (The oxygen has a slight negative charge; the hydrogens have a slight positive charge.) Each molecule can hydrogen-bond to a maximum of four partners. At any instant in liquid water at 37°C (human body temperature), about 15% of the molecules are bonded to four partners in short-lived clusters.

Organisms depend on the cohesion of water molecules

Water molecules stick together as a result of hydrogen bonding. When water is in its liquid form, its hydrogen bonds are very fragile, about one-twentieth as strong as covalent bonds. They form, break, and re-form with great frequency. Each hydrogen bond lasts only a few trillionths of a second, but the molecules are constantly forming new bonds with a succession of partners. Thus, at any instant, a substantial percentage of all the water molecules are bonded to their neighbors, giving water more structure than most other liquids. Collectively, the hydrogen bonds hold the substance together, a phenomenon called **cohesion.**

Cohesion due to hydrogen bonding contributes to the transport of water against gravity in plants (FIGURE 3.2). Water reaches the leaves through microscopic vessels that extend upward from the roots. Water that evaporates from a leaf is replaced by water from the vessels in the veins of the leaf. Hydrogen bonds cause water molecules leaving the veins to tug on molecules farther down in the vessel, and the upward pull is transmitted along the vessel all the way down to the root. **Adhesion,** the clinging of one substance to another, also plays a role. Adhesion of water to the walls of the vessels helps counter the downward pull of gravity.

Related to cohesion is **surface tension,** a measure of how difficult it is to stretch or break the surface of a liquid.

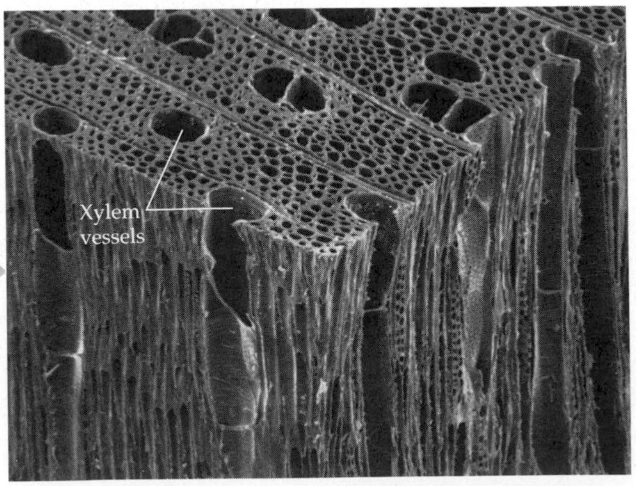

FIGURE 3.2
Water transport in plants. Evaporation from leaves pulls water upward from the roots through microscopic conduits called xylem vessels, in this case located in the trunk of a tree. Cohesion due to hydrogen bonding helps hold together the column of water within a vessel. Adhesion of the water to the vessel wall also helps in resisting the downward pull of gravity. (SEM from *Scanning Electron Microscopy in Biology,* by R. G. Kessel and C. Y. Shih, Springer-Verlag, New York, 1974, p. 147.)

Water has a greater surface tension than most other liquids. At the interface between water and air is an ordered arrangement of water molecules, hydrogen-bonded to one another and to the water below. This makes the water behave as though it were coated with an invisible film. We can observe the surface tension of water by slightly overfilling a drinking glass; the water will stand above the rim. Water's surface tension also allows us to skip rocks on a pond. In a more biological example, some animals can stand, walk, or run on water without breaking the surface (FIGURE 3.3).

Water contributes to Earth's habitability by moderating temperatures

Water stabilizes air temperatures by absorbing heat from air that is warmer and releasing the stored heat to air that is cooler. Water is effective as a heat bank because a slight change in its own temperature is accompanied by the absorption or release of a relatively large amount of heat. To understand this quality of water, we must first look briefly at heat and temperature.

Heat and Temperature

Anything that moves has **kinetic energy,** the energy of motion. Atoms and molecules have kinetic energy because they are always moving, although in no particular direction. The faster a molecule moves, the greater its kinetic energy. **Heat** is a measure of the *total* quantity of kinetic energy due to molecular motion in a body of matter. **Temperature** measures the intensity of heat due to the *average* kinetic energy of the molecules. When the average speed of the molecules increases, a thermometer records this as a rise in temperature. Heat and temperature are related, but they are not the same. A swimmer crossing the English Channel has a higher temperature than the water, but the ocean contains far more heat because of its volume.

Whenever two objects of different temperature are brought together, heat passes from the warmer to the cooler body until the two are the same temperature. Molecules in the cooler object speed up at the expense of the kinetic energy of the warmer object. An ice cube cools a drink not by adding coldness to the liquid, but by absorbing heat as the ice melts.

Throughout this book, we will use the **Celsius scale** to indicate temperature (Celsius degrees are abbreviated as °C). At sea level, water freezes at 0°C and boils at 100°C. The temperature of the human body averages 37°C, and comfortable room temperature is about 20°C to 25°C.

One convenient unit of heat used in this book is the **calorie (cal).** A calorie is the amount of heat energy it takes to raise the temperature of 1 g of water by 1°C. Conversely, a calorie is also the amount of heat that 1 g of

FIGURE 3.3
Running on water. This basilisk lizard, a native of tropical rain forests in Central and South America, can escape from predators by running across streams and small ponds. Although the lizard is denser than water, its membranous feet spread the animal's weight over enough surface to prevent breaking the surface tension of the water, as long as the trip is speedy.

water releases when it cools down by 1°C. A **kilocalorie (kcal),** 1000 cal, is the quantity of heat required to raise the temperature of 1 kilogram (kg) of water by 1°C. (The "calories" on food packages are actually kilocalories.) Another energy unit used in this book is the **joule (J).** One joule equals 0.239 cal; a calorie equals 4.184 J.

Water's High Specific Heat

The ability of water to stabilize temperature depends on its relatively high specific heat. The **specific heat** of a substance is defined as the amount of heat that must be absorbed or lost for 1 g of that substance to change its temperature by 1°C. We already know water's specific heat because we have defined a calorie as the amount of heat that causes water to change its temperature by 1°C. Therefore, the specific heat of water is 1 calorie per gram per degree Celsius, abbreviated as 1 cal/g/°C. Compared with most other substances, water has an unusually high specific heat. For example, ethyl alcohol, the type of alcohol in alcoholic beverages, has a specific heat of 0.6 cal/g/°C.

Because of the high specific heat of water relative to other materials, water will change its temperature less when it absorbs or loses a given amount of heat. The reason you can burn your fingers by touching the metal handle of a pot on the stove when the water in the pot is still lukewarm is that the specific heat of water is ten times greater than that of iron. In other words, it will take only 0.1 cal to raise the temperature of 1 g of iron 1°C. Specific heat can be thought of as a measure of how well a substance resists changing its temperature when it absorbs or releases heat. Water resists changing its temperature; when it does change its temperature, it absorbs or loses a relatively large quantity of heat for each degree of change.

We can trace water's high specific heat, like many of its other properties, to hydrogen bonding. Heat must be absorbed in order to break hydrogen bonds, and heat is released when hydrogen bonds form. A calorie of heat causes a relatively small change in the temperature of water because much of the heat energy is used to disrupt hydrogen bonds before the water molecules can begin moving faster. And when the temperature of water drops slightly, many additional hydrogen bonds form, releasing a considerable amount of energy in the form of heat.

What is the relevance of water's high specific heat to life on Earth? By warming up only a few degrees, a large body of water can absorb and store a huge amount of heat from the sun in the daytime and during summer. At night and during winter, the gradually cooling water can warm the air. This is the reason coastal areas generally have milder climates than inland regions. The high specific heat of water also makes ocean temperatures quite stable, creating a favorable environment for marine life. Thus, because of its high specific heat, the water that covers most of planet Earth keeps temperature fluctuations within limits that permit life. Also, because organisms are made primarily of water, they are more able to resist changes in their own temperatures than if they were made of a liquid with a lower specific heat.

Evaporative Cooling

Molecules of any liquid stay close together because they are attracted to one another. Molecules moving fast enough to overcome these attractions can depart from the liquid and enter the air as gas. This transformation from a liquid to a gas is called vaporization, or evaporation. The speed of molecular movement varies. Recall that temperature is the *average* kinetic energy of molecules. Even at low temperatures, the speediest molecules can escape into the air. Some evaporation occurs at any temperature; a glass of water, for example, will eventually evaporate at room temperature. If a liquid is heated, the average kinetic energy of molecules increases and the liquid evaporates more rapidly.

Heat of vaporization is the quantity of heat a liquid must absorb for 1 g of it to be converted from the liquid to the gaseous state. Compared with most other liquids, water has a high heat of vaporization. To evaporate each gram of water at room temperature, about 580 cal of heat are needed—nearly double the amount needed to vaporize a gram of alcohol or ammonia. Water's high heat of vaporization is another emergent property caused by hydrogen bonds, which must be broken before the molecules can make their exodus from the liquid state.

Water's high heat of vaporization helps moderate Earth's climate. A considerable amount of solar heat absorbed by tropical seas is consumed during the evaporation of surface water. Then, as moist tropical air circulates poleward, it releases heat as it condenses to form rain.

As a substance evaporates, the surface of the liquid that remains behind cools down. This **evaporative cooling** occurs because the "hottest" molecules, those with the greatest kinetic energy, are the most likely to leave as gas. It is as if the 100 fastest runners at a college transferred to another school; the average speed of the remaining students would decline.

Evaporative cooling of water contributes to the stability of temperature in lakes and ponds and also provides a mechanism that prevents terrestrial organisms from overheating. For example, evaporation of water from the leaves of a plant helps keep the tissues in the leaves from becoming too warm in the sunlight. Evaporation of sweat from human skin dissipates body heat and helps prevent overheating on a hot day or when excess heat is generated by strenuous activity (FIGURE 3.4). High humidity on a hot day increases discomfort because the concentration of water vapor in the air inhibits the evaporation of sweat from the body.

Oceans and lakes don't freeze solid because ice floats

Water is one of the few substances that are less dense as a solid than as a liquid. In other words, ice floats. While other materials contract when they solidify, water expands. The cause of this exotic behavior is, once again, hydrogen bonding. At temperatures above 4°C, water behaves like other liquids, expanding as it warms and contracting as it cools. Water begins to freeze when its molecules are no longer moving vigorously enough to break their hydrogen bonds. As the temperature reaches 0°C, the water becomes locked into a crystalline lattice, each water molecule bonded to the maximum of four partners (FIGURE 3.5). The hydrogen bonds keep the molecules at "arm's length," far enough apart to make ice about 10% less dense (10% fewer molecules for the same volume) than liquid water at 4°C. When ice absorbs enough heat for its temperature to increase to above 0°C, hydrogen bonds between molecules are disrupted. As the crystal collapses, the ice melts, and molecules are free to slip closer together. Water reaches its greatest density at 4°C

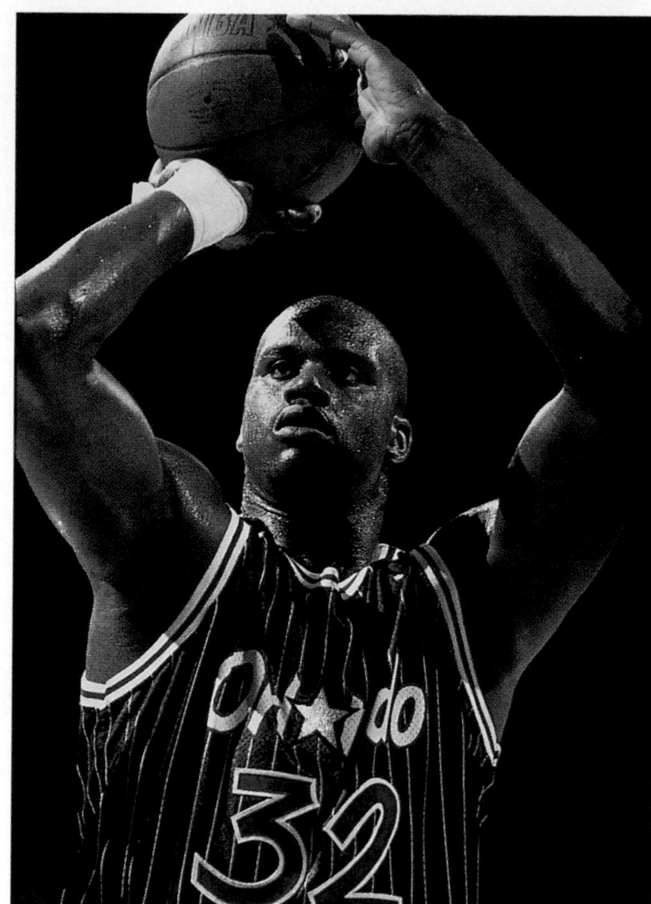

FIGURE 3.4
Evaporative cooling. Because of water's high heat of vaporization, evaporation of sweat cools the surface of the body.

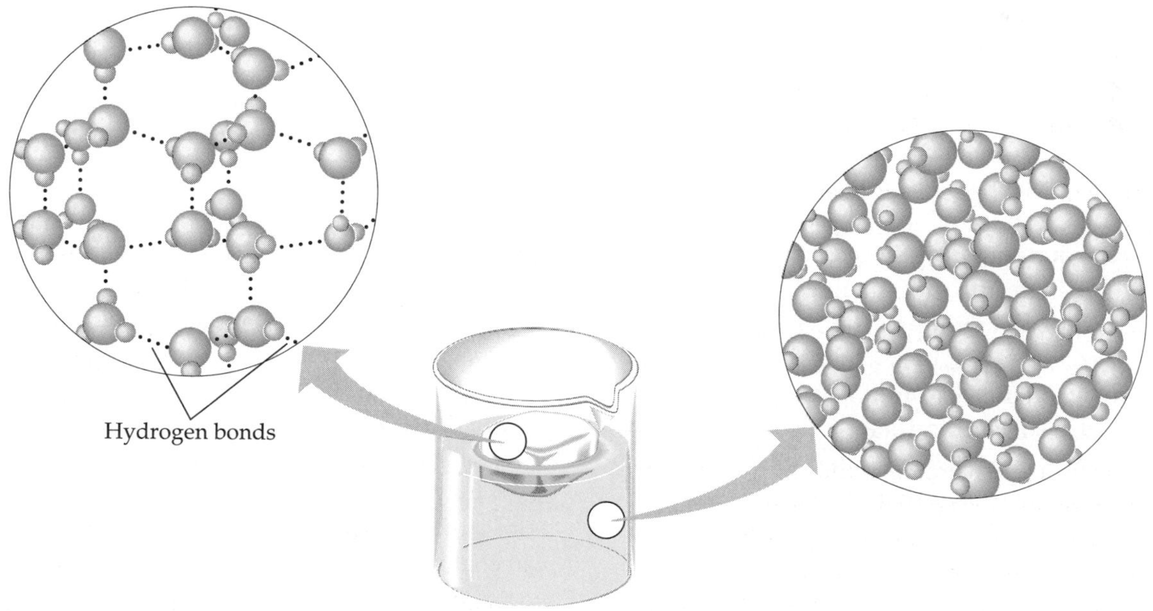

FIGURE 3.5
The structure of ice. Each molecule is hydrogen-bonded to four neighbors in a three-dimensional crystal with open channels. Because the hydrogen bonds make the crystal spacious, ice has fewer molecules than an equal volume of liquid water. In other words, ice is less dense than liquid water.

FIGURE 3.6
Floating ice and the fitness of the environment. Floating ice becomes a barrier that protects the liquid water below from the colder air. These are invertebrates called krill, photographed beneath Antarctic ice.

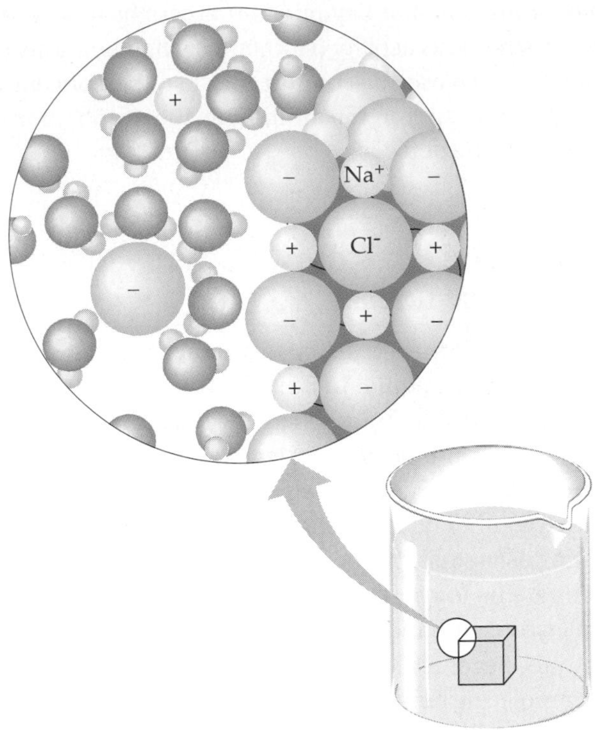

FIGURE 3.7
A crystal of table salt dissolving in water. The positive hydrogen regions of the polar water molecules are attracted to the chloride anions (green), whereas the negative oxygen regions cling to the sodium cations (beige).

and then begins to expand as the molecules move faster. Keep in mind, however, that even liquid water is semistructured because of transient hydrogen bonds.

The ability of ice to float because of the expansion of water as it solidifies is an important factor in the fitness of the environment. If ice sank, then eventually all ponds, lakes, and even the oceans would freeze solid, making life as we know it impossible on Earth. During summer, only the upper few inches of the ocean would thaw. Instead, when a deep body of water cools, the floating ice insulates the liquid water below, preventing it from freezing and allowing life to exist under the frozen surface (FIGURE 3.6).

Water is the solvent of life

A sugar cube placed in a glass of water will dissolve. The glass will then contain a uniform mixture of sugar and water; the concentration of dissolved sugar will be the same everywhere in the mixture. A liquid that is a homogeneous mixture of two or more substances is called a **solution.** The dissolving agent of a solution is the **solvent,** and the substance that is dissolved is the **solute.** In this case, water is the solvent and sugar is the solute. An **aqueous solution** is one in which water is the solvent.

The medieval alchemists tried to find a universal solvent, one that would dissolve anything. They learned that nothing works better than water. However, water is not a universal solvent; if it were, it could not be stored in any container, including our cells. But water is a very versatile solvent, a quality we can trace to the polarity of the water molecule.

Suppose, for example, that a crystal of the ionic compound sodium chloride is placed in water (FIGURE 3.7). At the surface of the crystal, the sodium and chloride ions are exposed to the solvent. The ions and water molecules have a mutual affinity through electrical attraction. The oxygen regions of the water molecules are negatively charged and cling to sodium cations. The hydrogen regions of the water molecules are positively charged and are attracted to chloride anions. Water surrounds the individual ions, separating the sodium from the chloride and shielding the ions from one another. Working inward from the surface of the salt crystal, water eventually dissolves all the ions. This produces a solution of two solutes, sodium and chloride, homogeneously mixed with water, the solvent. Other ionic compounds also dissolve in water. Seawater, for instance, contains a great variety of dissolved ions, as do living cells.

A compound does not need to be ionic in order to dissolve in water; polar compounds are also water-soluble. For example, sugars are soluble because water molecules can coat the polar sugar molecules. Even such large molecules as certain proteins dissolve in water because of ionic and polar regions on the surface of the large solute (FIGURE 3.8). Many different kinds of polar compounds are dissolved (along with ions) in the water of such biological fluids as blood, the sap of plants, and the liquid within all cells. Water is the solvent of life.

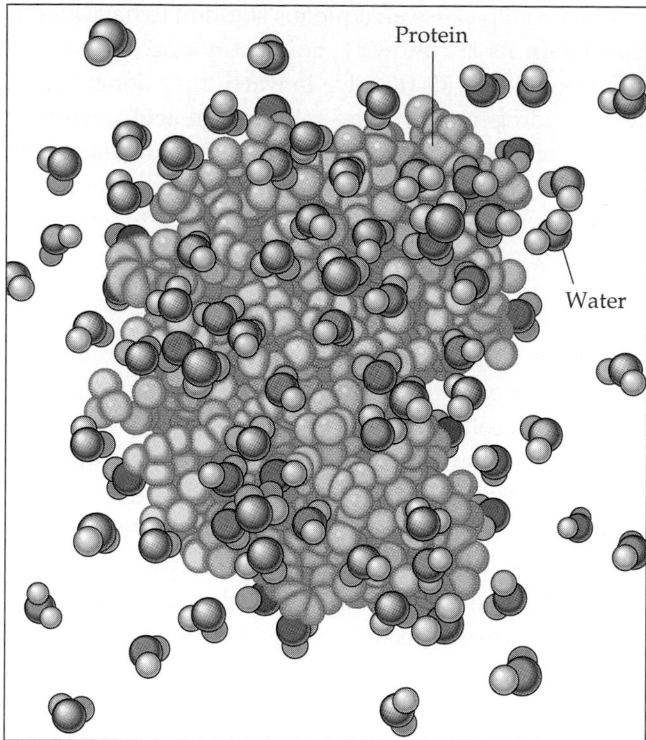

FIGURE 3.8
Hydration of a soluble protein. The polar water molecules have an affinity for ionic and polar regions of the large protein molecule.

Hydrophilic and Hydrophobic Substances

Whether ionic or polar, any substance that has an affinity for water is said to be **hydrophilic** (from the Greek *hydro,* "water," and *philios,* "loving"). This term is used even if the substance does not dissolve—because the molecules are too large, for instance. Cotton, a plant product, is an example of a hydrophilic substance that absorbs water without dissolving. Cotton consists of giant molecules of a compound called cellulose. With numerous regions of partial positive and partial negative charges associated with polar bonds, the cellulose fibers adhere to water. Thus, a cotton towel does a great job of drying the body, yet does not dissolve in the washing machine. Cellulose is also present in the walls of water-conducting vessels in a plant; you learned earlier how the adhesion of water to these hydrophilic walls functions in water transport.

There are, of course, substances that neither dissolve in water nor have an affinity for water. Such substances, which actually seem to repel water, are termed **hydrophobic** (Gr. *phobos,* "fearing"). They are nonionic, nonpolar substances. For example, vegetable oil and water do not mix. The hydrophobic behavior of the oil molecules results from a prevalence of nonpolar bonds, particularly bonds between carbon and hydrogen, which share electrons almost equally. Hydrophobic molecules related to oils are major ingredients of cell membranes. (Imagine what would happen to a cell if its membrane dissolved.)

Solute Concentration in Aqueous Solutions

Biological chemistry is "wet" chemistry. Most of the chemical reactions that occur in life involve solutes dissolved in water. To understand the chemistry of life, it is important to learn how to calculate the concentrations of the solutes dissolved in aqueous solutions.

Suppose we wanted to prepare an aqueous solution of table sugar having a specified concentration of sugar molecules (a certain number of solute molecules in a certain volume of solution). Because counting or weighing individual molecules is not practical, we instead usually measure substances in units called moles. A **mole (mol)** is equal in number to the molecular weight of a substance, but upscaled to units of grams rather than daltons. Imagine weighing out 1 mol of table sugar (sucrose), which has the molecular formula $C_{12}H_{22}O_{11}$. A carbon atom weighs 12 daltons, a hydrogen atom weighs 1 dalton, and an oxygen atom weighs 16 daltons. **Molecular weight** is the sum of the weights of all the atoms in a molecule; thus, the molecular weight of sucrose is 342 daltons. To obtain 1 mol of sucrose, we weigh out 342 g, the molecular weight of sucrose expressed in grams.

The practical advantage of measuring a quantity of chemicals in moles is that a mole of one substance has exactly the same number of molecules as a mole of any other substance. If substance A has a molecular weight of 10 daltons and substance B has a molecular weight of 100 daltons, then 10 g of A will have the same number of molecules as 100 g of B. The number of molecules in a mole, called Avogadro's number, is 6.02×10^{23}. A mole of table sugar contains 6.02×10^{23} sucrose molecules and weighs 342 g. A mole of ethyl alcohol (C_2H_6O) also contains 6.02×10^{23} molecules, but it weighs only 46 g because the molecules are smaller than those of sucrose. The mole concept rescales the weighing of molecules from daltons, used for single molecules, to grams, which are more practical units of weight for the laboratory. Measuring in moles also enables scientists to combine substances in fixed ratios of molecules.

How would we make a liter (L) of solution consisting of 1 mol of sugar dissolved in water? To obtain this concentration, we would weigh out 342 g of sucrose and then gradually add water, while stirring, until the sugar was completely dissolved. We would then add enough water to bring the total volume of the solution up to 1 L. At that point, we would have a one-molar (1 *M*) solution of sucrose. **Molarity**—the number of moles of solute per liter of solution—is the unit of concentration most often used by biologists for aqueous solutions.

Organisms are sensitive to changes in pH

Biologists use something called the pH scale to measure how acidic or basic (the opposite of acidic) a solution is. In this section you will learn about acids, bases, pH, and why changes in pH can adversely affect organisms.

Dissociation of Water Molecules

Occasionally, a hydrogen atom shared between two water molecules in a hydrogen bond shifts from one molecule to the other. When this happens, the hydrogen atom leaves its electron behind, and what is actually transferred is a **hydrogen ion,** a single proton with a charge of +1. The water molecule that lost a proton is now a **hydroxide ion** (OH^-), which has a charge of −1. The proton binds to the second water molecule, making that molecule a hydronium ion (H_3O^+). We can picture the chemical reaction this way:

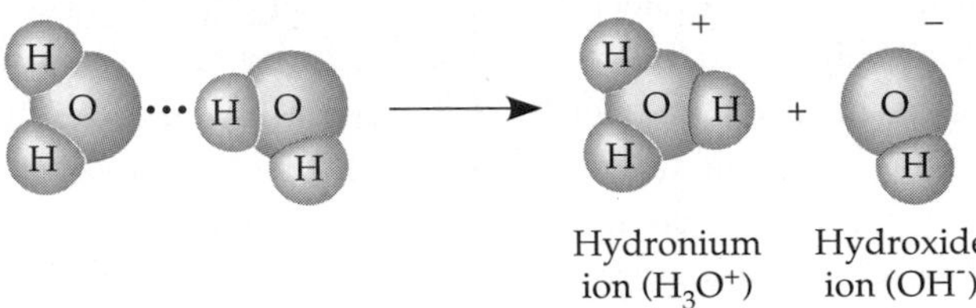

Although this is technically what happens, it is more useful to think of the process as the dissociation (separation) of a water molecule into a hydrogen ion and a hydroxide ion:

$$H_2O \rightleftharpoons \underset{\text{Hydrogen ion}}{H^+} + \underset{\text{Hydroxide ion}}{OH^-}$$

As the double arrows indicate, this is a reversible reaction that will reach a state of dynamic equilibrium when water dissociates at the same rate that it is being reformed from H^+ and OH^-. At this equilibrium point, the concentration of water molecules greatly exceeds the concentrations of H^+ and OH^-. In fact, in pure water, only one water molecule in every 554 million is dissociated. Although the dissociation of water is reversible and statistically rare, it is exceedingly important in the chemistry of life. Hydrogen and hydroxide ions are very reactive. Even slight changes in their concentrations can affect a cell's proteins and other complex molecules.

Acids and Bases

Since the dissociation of water produces one H^+ for every OH^-, the concentrations of these ions will be equal in pure water. The concentration of each ion is 10^{-7} *M* (at a temperature of 25°C). This means that there is only one ten-millionth of a mole of hydrogen ions per liter of pure water, and an equal number of hydroxide ions.

What would cause an aqueous solution to have an imbalance in its H^+ and OH^- concentrations? When substances called acids dissolve in water, they donate additional hydrogen ions to the solution. An **acid,** according to the definition most biologists use, is a substance that increases the H^+ concentration of a solution. For example, when hydrochloric acid (HCl) is added to water, hydrogen ions dissociate from chloride ions:

$$HCl \longrightarrow H^+ + Cl^-$$

Now there are two sources of H^+ in the solution (dissociation of water is the other), resulting in more H^+ than OH^-. Such a solution is known as an acidic solution.

A substance that reduces the hydrogen ion concentration in a solution is called a **base.** Some bases reduce the H^+ concentration indirectly by dissociating to form hydroxide ions, which then combine with hydrogen ions to form water. An example of a base that acts this way is sodium hydroxide (NaOH), which in water dissociates into its ions:

$$NaOH \longrightarrow Na^+ + OH^-$$

Other bases reduce H^+ concentration directly by accepting hydrogen ions. Ammonia (NH_3), for instance, acts as a base by binding a hydrogen ion from the solution, resulting in an ammonium ion (NH_4^+):

$$NH_3 + H^+ \rightleftharpoons NH_4^+$$

In either case, the base reduces the H^+ concentration. Solutions with a higher concentration of OH^- than H^+ are known as basic solutions. A solution in which the H^+ and OH^- concentrations are equal is said to be neutral.

Notice that single arrows were used in the reactions for HCl and NaOH. These compounds dissociate completely when mixed with water. Hydrochloric acid is called a strong acid and sodium hydroxide a strong base because they dissociate completely. In contrast, ammonia is a relatively weak base. The double arrows in the reaction for ammonia indicate that the binding and release of the hydrogen ion are reversible. When the reaction reaches equilibrium, there will be a fixed ratio of NH_4^+ to NH_3.

There are also weak acids, which dissociate reversibly to release and reaccept hydrogen ions. An example is carbonic acid:

$$\underset{\text{Carbonic acid}}{H_2CO_3} \rightleftharpoons \underset{\text{Bicarbonate ion}}{HCO_3^-} + \underset{\text{Hydrogen ion}}{H^+}$$

The equilibrium so favors the reaction in the left direction that when carbonic acid is added to water, only 1% of the molecules are dissociated at any particular time. Still, that is enough to shift the balance of H^+ and OH^- from neutrality.

The pH Scale

In any solution, the *product* of the H^+ and the OH^- concentrations is constant at 10^{-14} *M*. This can be written

$$[H^+][OH^-] = 10^{-14}\ M^2$$

In such an equation, brackets indicate molar concentration for the substance enclosed within them. In a neutral solution at room temperature (25°C), $[H^+] = 10^{-7}$ and $[OH^-] = 10^{-7}$, so the product is $10^{-14}\ M^2$ ($10^{-7} \times 10^{-7}$). If enough acid is added to a solution to increase $[H^+]$ to 10^{-5} *M*, then $[OH^-]$ will decline by an equivalent amount to 10^{-9} *M* ($10^{-5} \times 10^{-9} = 10^{-14}$). An acid not only adds hydrogen ions to a solution, but also removes hydroxide ions because of the tendency for H^+ to combine with OH^- to form water. A base has the opposite effect, increasing OH^- concentration but also reducing H^+ concentration by the formation of water. If enough of a base is added to raise the OH^- concentration to 10^{-4} *M*, the H^+ concentration will drop to 10^{-10} *M*. Whenever we know the concentration of either H^+ or OH^- in a solution, we can deduce the concentration of the other ion.

Because the H^+ and OH^- concentrations of solutions can vary by a factor of 100 trillion or more, scientists have developed a way to express this variation more conveniently, by use of the **pH scale,** which ranges from 0 to 14. The pH scale compresses the range of H^+ and OH^- concentrations by employing a common mathematical device: logarithms. The pH of a solution is defined as the negative logarithm (base 10) of the hydrogen ion concentration:

$$pH = -\log [H^+]$$

For a neutral solution, $[H^+]$ is 10^{-7} *M*, giving us

$$-\log 10^{-7} = -(-7) = 7$$

Notice that pH *declines* as H^+ concentration *increases.* Notice, too, that although the pH scale is based on H^+ concentration, it also implies OH^- concentration. A solution of pH 10 has a hydrogen ion concentration of 10^{-10} *M* and a hydroxide ion concentration of 10^{-4} *M*.

The pH of a neutral solution is 7, the midpoint of the scale (FIGURE 3.9). A pH value less than 7 denotes an acidic solution; the lower the number, the more acidic the solution. The pH for basic solutions is above 7. Most biological fluids are within the range pH 6 to pH 8. There are a few exceptions, however, including the strongly acidic digestive juice of the human stomach, which has a pH of about 1.5.

It is important to remember that each pH unit represents a tenfold difference in H^+ and OH^- concentrations. It is this mathematical feature that makes the pH scale so compact. A solution of pH 3 is not twice as acidic as a solution of pH 6, but a thousand times more acidic.

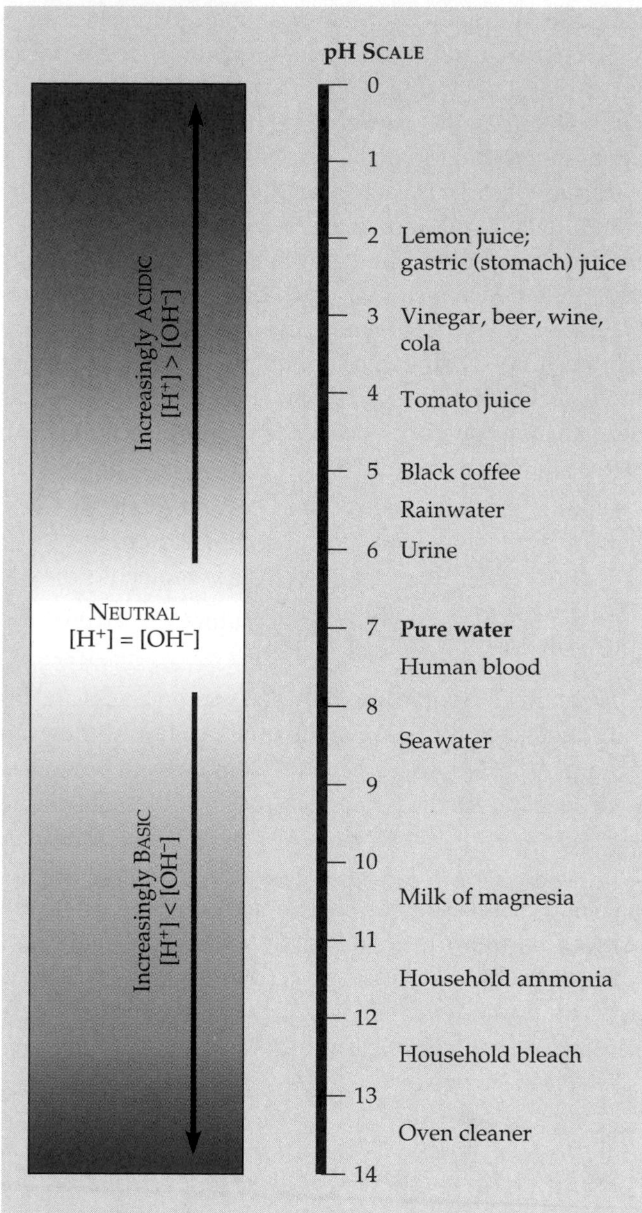

FIGURE 3.9
The pH of some aqueous solutions.

When the pH of a solution changes slightly, the actual concentrations of H^+ and OH^- in the solution change substantially.

Buffers

The internal pH of most living cells is close to 7. Even a slight change in pH can be harmful, because the chemical processes of the cell are very sensitive to the concentrations of hydrogen and hydroxide ions.

Biological fluids resist changes to their own pH when acids or bases are introduced because of the presence of **buffers,** substances that minimize changes in the concentrations of H^+ and OH^-. Buffers in human blood, for example, normally maintain the blood pH very close to 7.4. A person cannot survive for more than a few minutes if the blood pH drops to 7 or rises to 7.8. Under normal circumstances, the buffering capacity of the blood prevents such swings in pH.

A buffer works by accepting hydrogen ions from the solution when they are in excess and donating hydrogen ions to the solution when they have been depleted. Most buffers are weak acids or weak bases that combine reversibly with hydrogen ions. One of the buffers that contributes to pH stability in human blood and many other biological solutions is carbonic acid (H_2CO_3), which, as already mentioned, dissociates to yield a bicarbonate ion (HCO_3^-) and a hydrogen ion (H^+):

Response to a rise in pH

$$H_2CO_3 \rightleftharpoons HCO_3^- + H^+$$

H^+ donor (acid) — Response to a drop in pH — H^+ acceptor (base) — Hydrogen ion

The chemical equilibrium between carbonic acid and bicarbonate acts as a pH regulator, the reaction shifting left or right as other processes in the solution add or remove hydrogen ions. If the H^+ concentration in blood begins to fall (that is, if pH rises), more carbonic acid dissociates, replenishing hydrogen ions. But when H^+ concentration in blood begins to rise (pH drops), the bicarbonate ion acts as a base and removes the excess hydrogen ions from the solution. Thus, the carbonic acid-bicarbonate buffering system actually consists of an acid *and* a base in equilibrium with each other. Most other buffers are also acid-base pairs.

Acid precipitation threatens the fitness of the environment

Considering the dependence of all life on water, contamination of rivers, lakes, and seas is a dire environmental problem. One of the most serious assaults on water quality is acid precipitation.

Uncontaminated rain has a pH of about 5.6, slightly acidic, owing to the formation of carbonic acid from carbon dioxide and water. **Acid precipitation** refers to rain, snow, or fog more acidic than pH 5.6. The questions to consider are: What causes acid precipitation, what are its effects on the fitness of the environment, and what can be done to reduce the problem?

Acid precipitation is caused primarily by the presence in the atmosphere of sulfur oxides and nitrogen oxides, gaseous compounds that react with water in the air to form acids, which fall to Earth with rain or snow. A major source of these oxides is the combustion of fossil fuels by factories and automobiles. Acid precipitation from air pollution is as old as the Industrial Revolution, but the problem has escalated and become more widespread in the past three decades. One practice that has increased the occurrence of acid precipitation is the construction of taller smokestacks designed to reduce local pollution by dispersing factory exhaust. Ironically, prevailing winds simply move the problem, and acid rain and snow fall hundreds or thousands of miles away from industrial centers, often in formerly pristine regions. In the Adirondack Mountains of upstate New York, the pH of rainfall averages 4.2, about 25 times more acidic than normal rain. Acid precipitation falls on many other regions, including the Cascade Mountains of the Pacific Northwest and certain parts of Europe and Asia. One West Virginia storm dropped rain having a pH of 1.5, as acidic as the digestive juices in our stomachs!

Experiments and observations have confirmed that acid precipitation is harming both terrestrial and freshwater ecosystems. Acid rain and snow that fall on land lower the pH of the soil solution, which affects the solubility of minerals. Some mineral nutrients required by plants are washed out of the topsoil, while other minerals, such as aluminum, reach toxic concentrations when acidification increases their solubility. The effects of acid rain on soil chemistry have contributed to the decline of European forests and are taking a toll on some North American forests (FIGURE 3.10). Acid precipitation has also lowered the pH of lakes and ponds in some regions, and the accumulation of certain minerals leached from the soil by acid rain further contaminates freshwater habitats. The acidic assault has adversely affected many species of fish, amphibians, and aquatic invertebrates. More than half the lakes at the higher elevations of the western Adirondacks are now more acidic than pH 5, and fish have completely disappeared from nearly all those lakes. Like the coal mine canary, the fish—or their

FIGURE 3.10
The effects of acid rain on a forest. Rain and snow bearing the acidic products of coal and oil combustion have been blamed for the demise of this forest in the Erzgebirge mountain range in the Czech Republic.

absence—are a warning that something has gone awry in the environment.

If there is reason for optimism about the future quality of water resources, it is our progress in reducing certain kinds of pollution in some locations. For example, acid precipitation can be reduced through industrial controls and antipollution devices. In fact, there is evidence that emissions of sulfur oxides have declined by about 30% since 1985. Continued progress can come only from business leaders, voters, consumers, and politicians who are concerned about environmental quality. An essential part of their education is to understand the crucial role that water plays in the environmental fitness for continued life on Earth.

REVIEW OF KEY CONCEPTS (with page numbers and key figures)

- The polarity of water molecules results in hydrogen bonding (pp. 41–42, FIGURE 3.1)
  - Water is a polar molecule. A hydrogen bond is formed when the oxygen of one water molecule is electrically attracted to the hydrogen of an adjacent molecule.
  - Hydrogen bonding between water molecules is the basis for water's emergent properties.
- Organisms depend on the cohesion of water molecules (pp. 42–43)
  - The formation and re-formation of hydrogen bonds makes liquid water cohesive. For example, water is pulled upward in the microscopic vessels of plants.
  - Hydrogen bonding of water molecules on the surface of liquid water is responsible for water's surface tension.
- Water contributes to Earth's habitability by moderating temperatures (pp. 43–45)
  - Heat is the total kinetic energy of molecules in a body of matter. Temperature measures the average kinetic energy of those molecules. A calorie is the heat energy required to raise the temperature of 1 g of water by 1°C.
  - Hydrogen bonding gives water a high specific heat. Heat is absorbed when hydrogen bonds break and is released when hydrogen bonds form, minimizing temperature fluctuations to within limits that permit life.
  - Evaporative cooling is based on water's high heat of vaporization. Water molecules must have a relatively high kinetic energy to overcome hydrogen bonds. The evaporative loss of these energetic water molecules cools a surface.
- Oceans and lakes don't freeze solid because ice floats (pp. 45–46, FIGURE 3.5)
  - Ice is less dense than liquid water, owing to more organized hydrogen bonding, which forces water to expand into a characteristic crystal. Floating ice allows life to exist under the frozen surfaces of lakes and polar seas.
- Water is the solvent of life (pp. 46–47, FIGURE 3.7)
  - Water is an unusually versatile solvent because its polarity attracts it to charged and polar substances. When ions or polar substances are surrounded by water molecules, they dissolve and are called solutes.
  - Hydrophilic substances have an affinity for water. Hydrophobic substances seem to repel water.
  - Biologists usually use molarity as a measure of solute concentration in solutions. Molarity is the number of moles per liter of solution. A mole is the number of grams of a substance equal to its molecular weight in daltons.
- Organisms are sensitive to changes in pH (pp. 48–50, FIGURE 3.9)
  - Water can dissociate into H^+ and OH^-.
  - The concentration of H^+ is measured in pH units as follows: $pH = -\log [H^+]$.
  - Acids donate additional H^+ in aqueous solutions; bases donate OH^- or accept H^+ in solutions.
  - In a neutral solution, $[H^+] = [OH^-] = 10^{-7}$, and the $pH = 7$. In an acidic solution, $[H^+]$ is greater than $[OH^-]$, and the pH is less than 7. In a basic solution, $[H^+]$ is less than $[OH^-]$, and the pH is greater than 7.
  - Buffers in biological fluids resist changes in pH. A buffer consists of an acid-base pair that combines reversibly with hydrogen ions.
- Acid precipitation threatens the fitness of the environment (pp. 50–51)
  - Acid precipitation occurs when water in the air reacts with sulfur oxides and nitrogen oxides that result from the combustion of fossil fuels. The acids that form give rain and snow a pH less than 5.6, sometimes causing serious environmental consequences.

SELF QUIZ

1. The main thesis of Lawrence Henderson's *The Fitness of the Environment* is
   a. Earth's environment is constant
   b. it is the physical environment, not life, that has evolved
   c. the environment of Earth has adapted to life
   d. life as we know it depends on certain environmental qualities on Earth
   e. water and other aspects of Earth's environment exist because they make the planet more suitable for life

2. Air temperature often increases slightly as clouds begin to drop rain or snow. Which behavior of water is *most directly* responsible for this phenomenon?
   a. Water's change in density when it condenses.
   b. Water's reactions with other atmospheric compounds.
   c. Release of heat by the formation of hydrogen bonds.
   d. Release of heat by the breaking of hydrogen bonds.
   e. Water's high surface tension.

3. For two bodies of matter in contact, heat always flows from
   a. the body with greater heat to the one with less heat
   b. the body of higher temperature to the one of lower temperature

c. the more dense to the less dense body
d. the body with more water to the one with less water
e. the larger to the smaller body

4. A slice of pizza has 500 kcal. If we could burn the pizza and use all the heat to warm a 50-L container of cold water, what would be the approximate increase in the temperature of the water? (*Note:* A liter of cold water weighs about 1 kg.)
a. 50°C
b. 5°C
c. 10°C
d. 100°C
e. 1°C

5. The bonds that are broken when water vaporizes are
a. ionic bonds
b. bonds between water molecules
c. bonds between atoms of individual water molecules
d. polar covalent bonds
e. nonpolar covalent bonds

6. Which of the following is an example of a hydrophobic material?
a. paper
b. table salt
c. wax
d. sugar
e. pasta

7. We can be sure that a mole of table sugar and a mole of vitamin C are equal in their
a. weight in daltons
b. weight in grams
c. number of molecules
d. number of atoms
e. volume

8. How many grams of acetic acid ($C_2H_4O_2$) would you use to make 10 L of a 0.1 *M* aqueous solution of the acetic acid? (*Note*: The atomic weights, in daltons, are approximately 12 for carbon, 1 for hydrogen, and 16 for oxygen.)
a. 10 g
b. 0.1 g
c. 6 g
d. 60 g
e. 0.6 g

9. Acid rain has lowered the pH of a particular lake to 4.0. What is the hydrogen ion concentration of the lake?
a. 4.0 *M*
b. 10^{-10} *M*
c. 10^{-4} *M*
d. 10^{4} *M*
e. 4%

10. What is the *hydroxide* ion concentration of the lake described in question 9?
a. 10^{-7} *M*
b. 10^{-4} *M*
c. 10^{-10} *M*
d. 10^{-14} *M*
e. 10 *M*

CHALLENGE QUESTIONS

1. Explain how panting helps regulate a dog's body temperature.
2. Design a controlled experiment to test the hypothesis that acid precipitation inhibits the growth of *Elodea*, a common freshwater plant.
3. Write a paragraph explaining how the insect in this photograph, a water strider, can walk on water.

SCIENCE, TECHNOLOGY, AND SOCIETY

1. Discuss the special political obstacles to reducing acid precipitation (as compared with environmental issues confined to a more localized region).
2. Agriculture, industry, and the growing populations of cities all compete, through political influence, for water. If you were in charge of water resources in an arid region, what would your priorities be for allocating the limited water supply for various uses? How would you defend your position and try to build consensus among the different special interest groups?

FURTHER READING

Henderson, L. J. *The Fitness of the Environment.* New York: Macmillan, 1913. A classic book highlighting the importance of water and carbon to life.
Lehninger, A. L., D. L. Nelson, and M. M. Cox. *Principles of Biochemistry.* New York: Worth, 1992. A readable biochemistry text with an excellent discussion of water in Chapter 4.
Mohner, V. A. "The Challenge of Acid Rain." *Scientific American,* August 1988. Analysis of a complex environmental problem.
Pennisi, E. "Water, Water Everywhere." *Science News,* February 20, 1993. How water stabilizes the structure of proteins and other large biological molecules.
Pennisi, E. "Water: The Power, Promise, and Turmoil of North America's Fresh Water." *National Geographic.* A special 1993 issue examining how we use and abuse our water resources.
Raloff, J. "When Nitrate Reigns." *Science News,* February 11, 1995. A link between air pollution and water pollution.

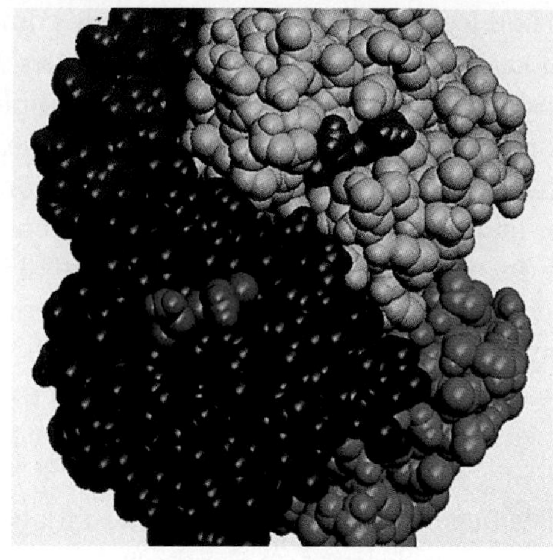

CHAPTER 4

CARBON AND THE MOLECULAR DIVERSITY OF LIFE

KEY CONCEPTS

- Organic chemistry is the study of carbon compounds
- Carbon atoms are the most versatile building blocks of molecules
- Variation in carbon skeletons contributes to the diversity of organic molecules
- Functional groups also contribute to the molecular diversity of life
- The chemical elements of life: *a review*

From the origin of the first cells to the current variety of organisms, carbon has played a prominent role in the evolution of life on Earth. Biological diversity reflects molecular diversity; and carbon, of all chemical elements, is unparalleled in its ability to form molecules that are large, complex, and diverse. An example of this molecular complexity is the blood protein hemoglobin, illustrated in the computer graphic model that heads this chapter. Your objective in this chapter is to learn a few concepts of molecular architecture that highlight carbon's importance to life. These concepts will further illustrate the theme that emergent properties arise from the organization of life's matter.

Organic chemistry is the study of carbon compounds

Although a cell is composed of 70% to 95% water, most of the rest consists of carbon-based compounds. Proteins, DNA, carbohydrates, and other molecules that distinguish living matter from inanimate material are all composed of carbon atoms bonded to one another and to atoms of other elements. Hydrogen (H), oxygen (O), nitrogen (N), sulfur (S), and phosphorus (P) are other common ingredients of these compounds, but it is carbon (C) that accounts for the endless diversity of biological molecules.

Compounds containing carbon are said to be organic, and the branch of chemistry that specializes in the study of carbon compounds is called **organic chemistry.** Organic compounds range from simple molecules such as methane (CH_4) to colossal ones such as hemoglobin and other proteins with thousands of atoms and molecular weights in excess of 100,000 daltons. The percentages of the major elements of life—C, O, H, N, S, and P—are quite uniform from individual to individual, and even from species to species. The atoms of organic molecules, however, can be arranged so many different ways that the uniqueness of each organism is ensured. Carbon's versatility allows a limited assortment of atomic building blocks, taken in roughly the same proportions, to be used to build an inexhaustible variety of organic molecules.

Eloy Rodriguez, the phytochemist you met in the interview preceding this unit, continues a centuries-old tradition of scientists investigating other organisms as sources of valued substances—everything from wine and food to medicines and fabrics. Organic chemistry originated in attempts to purify and improve the yield of these products. By the early nineteenth century, chemists had learned to make many simple compounds in the laboratory by combining elements under the right conditions, but artificial synthesis of the complex molecules extracted from living matter seemed impossible. It was at that time that the Swedish Chemist Jöns Jakob Berzelius first made the distinction between organic compounds, those that seemingly could arise only within living organisms, and inorganic compounds, those that were found in the nonliving world. The new discipline of organic chemistry was first built on a foundation of vitalism, the belief in a life force outside the jurisdiction of physical and chemical laws.

Chemists began to chip away at the foundation of vitalism when they learned to synthesize organic compounds in their laboratories. In 1828, Friedrich Wöhler, a German chemist who had studied with Berzelius, attempted to make an inorganic salt, ammonium cyanate, by mixing solutions of ammonium (NH_4^+) and cyanate (CNO^-) ions. Wöhler was astonished to find that instead of the expected product, he had made urea, an organic compound present in the urine of animals. Wöhler challenged the vitalists when he wrote, "I must tell you that I can prepare urea without requiring a kidney or an animal, either man or dog." However, one of the ingredients used in the synthesis, the cyanate, had been extracted from animal blood, and the vitalists were not swayed by Wöhler's discovery. A few years later, Hermann Kolbe, a student of Wöhler's, made the organic compound acetic acid from inorganic substances that could themselves be prepared directly from pure elements.

The foundation of vitalism was shaking. It finally crumbled after several more decades of laboratory synthesis of increasingly complex organic compounds. In 1953, Stanley Miller, a graduate student at the University of Chicago, helped place this abiotic (nonliving) synthesis of organic compounds into the context of evolution. Miller used a laboratory simulation of chemical conditions on the primitive Earth to demonstrate that the spontaneous synthesis of organic compounds may have been an early stage in the origin of life (FIGURE 4.1).

The pioneers of organic chemistry helped shift the mainstream of biological thought from vitalism to **mechanism,** the belief that all natural phenomena, including the processes of life, are governed by physical and chemical laws. Organic chemistry was redefined as the study of carbon compounds, regardless of their origin. Most naturally occurring organic compounds are produced by organisms, and these molecules present a diversity and range of complexity unrivaled by inorganic compounds. However, the same rules of chemistry apply to inorganic and organic molecules alike. The foundation of organic chemistry is not some intangible life force, but the unique chemical versatility of the element carbon.

FIGURE 4.1
Abiotic synthesis of organic compounds under "early Earth" conditions. Stanley Miller recreates his 1953 experiment, a laboratory simulation demonstrating that environmental conditions on the lifeless, primordial Earth favored the synthesis of some organic molecules. Miller used electrical discharges (simulated lightning) to trigger reactions in a primitive "atmosphere" of H_2O, H_2, NH_3 (ammonia), and CH_4 (methane)—some of the gases that are belched into the air by volcanoes. From these ingredients, Miller's apparatus made a variety of organic compounds that play key roles in living cells. Similar chemistry may have set the stage for the origin of life on Earth, a hypothesis we will explore in more detail in Chapter 24.

Carbon atoms are the most versatile building blocks of molecules

The key to the chemical characteristics of an atom, as you learned in Chapter 2, is in its distribution of electrons; electron distribution determines the kinds and number of bonds an atom will form with other atoms. Carbon has a total of six electrons, with two in the first electron shell and four in the second shell. Having four valence electrons in a shell that holds eight, carbon has little tendency to gain or lose electrons and form ionic bonds; it would have to donate or accept four electrons to do so. Instead, a carbon atom completes its valence shell by sharing electrons with other atoms in four covalent bonds. Each carbon atom thus acts as an intersec-

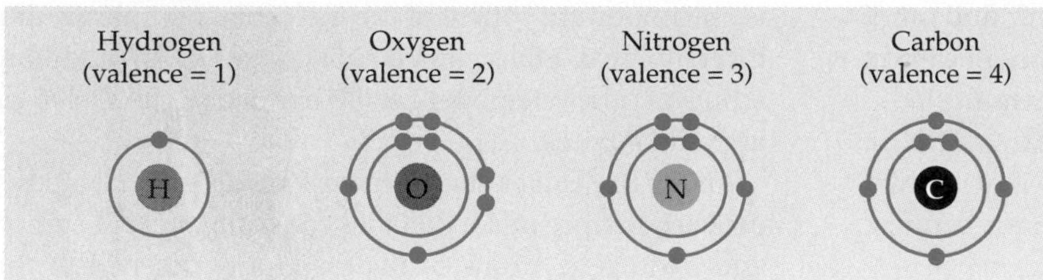

FIGURE 4.2
Valences for the major elements of organic molecules. Valence is the number of bonds an atom will usually form, equal to the number of electrons required to complete the outermost (valence) electron shell.

tion point from which a molecule can branch off in up to four directions. This *tetravalence* is one facet of carbon's versatility that makes large, complex molecules possible.

The electron distribution of carbon also gives it covalent compatibility with many different elements. FIGURE 4.2 reviews the valences of the four major atomic components of organic molecules: carbon and its most frequent partners, oxygen, hydrogen, and nitrogen. We can think of these valences as the rules of covalent bonding in organic chemistry—the building codes that govern the architecture of organic molecules.

In Chapter 2, you learned that when a carbon atom forms single covalent bonds, the bonds angle toward the corners of an imaginary tetrahedron (see FIGURE 2.15b). The bond angles in methane (CH_4) are 109°, and they would be approximately the same in any molecule where carbon has four single bonds. For example, ethane (C_2H_6) is shaped like two tetrahedrons joined at their apexes (FIGURE 4.3). It is convenient to write structural formulas as though molecules were flat, but molecules are three-dimensional, and the shape of an organic molecule can determine its function in a living cell.

A couple of examples will demonstrate the rules of covalent bonding in organic molecules. In the carbon dioxide molecule (CO_2), a single carbon atom is joined to two atoms of oxygen by double covalent bonds. The structural formula for CO_2 is O═C═O. Each line (bond) in a structural formula represents a pair of shared electrons. Notice that the carbon atom in CO_2 is involved in a total of four covalent bonds, two with each oxygen atom. The arrangement completes the valence shells of all atoms in the molecule. Carbon dioxide is such a simple molecule that it is generally considered inorganic, even though it contains carbon. Whether we call CO_2 organic or inorganic is an arbitrary distinction, but there is no ambiguity about its importance to the living world. Taken from

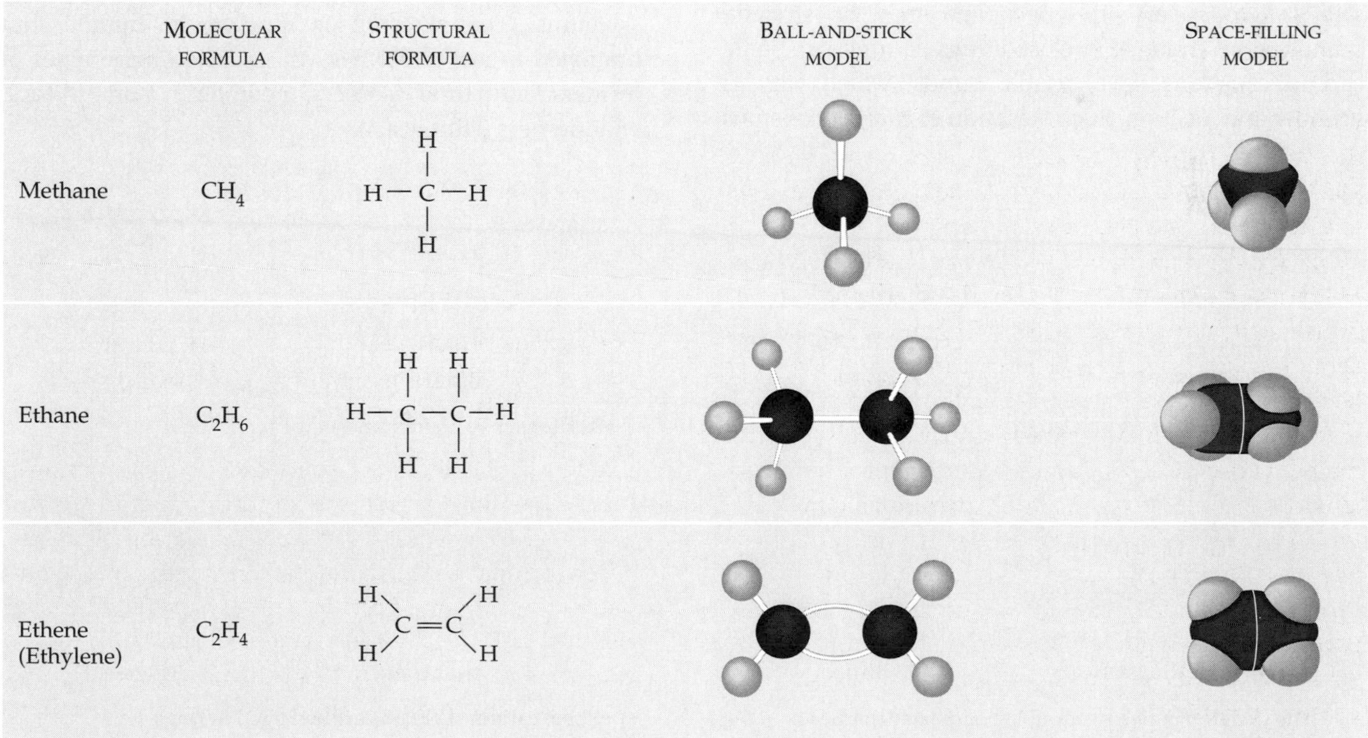

	Molecular formula	Structural formula	Ball-and-stick model	Space-filling model
Methane	CH_4	$\begin{array}{c}H\\ \vert\\ H-C-H\\ \vert\\ H\end{array}$		
Ethane	C_2H_6	$\begin{array}{c}H\ \ H\\ \vert\ \ \ \vert\\ H-C-C-H\\ \vert\ \ \ \vert\\ H\ \ H\end{array}$		
Ethene (Ethylene)	C_2H_4	$\begin{array}{c}H\diagdown\ \ \diagup H\\ C=C\\ H\diagup\ \ \diagdown H\end{array}$		

FIGURE 4.3
The shapes of three simple organic molecules. Whenever a carbon atom has four single bonds, the bonds angle toward the corners of an imaginary tetrahedron. When two carbons are joined by a double bond, all bonds around those atoms are in the same plane. (Notice, for example, that ethylene is a flat molecule, its atoms all in the same plane.)

the air by plants and incorporated into sugar and other foods during photosynthesis, CO_2 is the source of carbon for all the organic molecules found in organisms.

Another relatively simple molecule is urea, $CO(NH_2)_2$. This is the organic compound from urine that Wöhler learned to synthesize in the early nineteenth century. The structural formula for urea is:

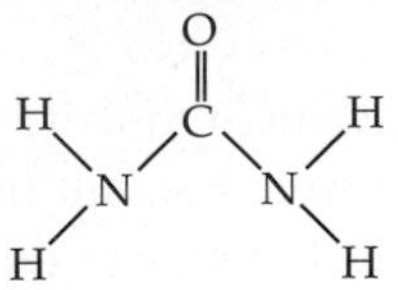

Again, each atom has the required number of covalent bonds. In this case, one carbon atom is involved in both single and double bonds.

Both urea and carbon dioxide are simple molecules with only one carbon atom. But a carbon atom can also use one or more of its valence electrons to form covalent bonds to other carbon atoms, making it possible to link the atoms into chains of seemingly infinite variety.

Variation in carbon skeletons contributes to the diversity of organic molecules

Carbon chains form the skeletons of organic molecules. The skeletons vary in length and may be straight, branched, or arranged in closed rings (FIGURE 4.4). Some carbon skeletons have double bonds, which vary in number and location. Such variation in carbon skeletons is one important source of the molecular complexity and diversity that characterize living matter. In addition, atoms of other elements can be bonded to the skeletons at available sites.

All the molecules shown in FIGURES 4.3 and 4.4 are **hydrocarbons,** organic molecules consisting only of carbon and hydrogen. Atoms of hydrogen are attached to the carbon skeleton wherever electrons are available for covalent bonding. Hydrocarbons are the major components of petroleum, which is called a fossil fuel because it consists of the partially decomposed remains of organisms that lived millions of years ago. Although hydrocarbons are not prevalent in living organisms, many of a cell's organic molecules have regions consisting of only carbon and hydrogen. For example, the molecules known as fats have long hydrocarbon tails attached to a nonhydrocarbon component (FIGURE 4.5). Neither petroleum nor fat mixes with water; both are hydrophobic compounds because the bonds between the carbon and hydrogen atoms are nonpolar.

Isomers

Variation in the architecture of organic molecules can be seen in **isomers,** compounds that have the same molecular formula but different structures and hence different properties. Compare, for example, the two butanes in FIGURE 4.4. Both have the molecular formula C_4H_{10}, but they differ in the covalent arrangement of their carbon skeletons. The skeleton is straight in butane, but branched in isobutane. We will examine three types of isomers: structural isomers, geometric isomers, and enantiomers (FIGURE 4.6).

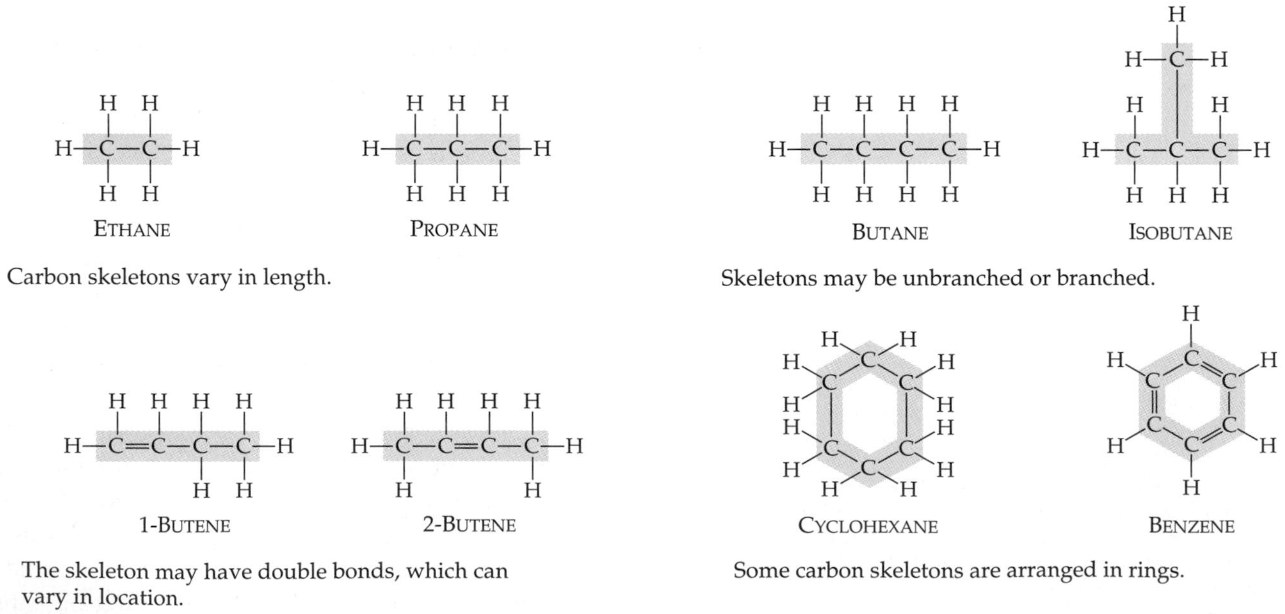

ETHANE

PROPANE

Carbon skeletons vary in length.

BUTANE

ISOBUTANE

Skeletons may be unbranched or branched.

1-BUTENE

2-BUTENE

The skeleton may have double bonds, which can vary in location.

CYCLOHEXANE

BENZENE

Some carbon skeletons are arranged in rings.

FIGURE 4.4
Variations in carbon skeletons. Hydrocarbons, organic molecules consisting only of carbon and hydrogen, illustrate the diversity of the carbon skeletons of organic molecules.

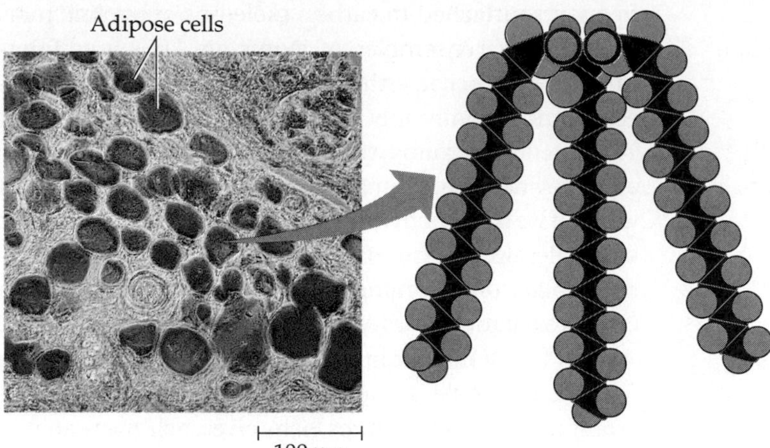

FIGURE 4.5

The role of hydrocarbons in the characteristics of fats. Humans and other mammals store fats in specialized cells called adipose cells. Each cell is almost completely filled with a large fat droplet (stained red in this light micrograph), which stockpiles an enormous number of fat molecules. The drawing is a space-filling model of one fat molecule (black = carbon; gray = hydrogen; pink = oxygen). Three hydrocarbon tails are attached to a headpiece containing oxygen as well as carbon and hydrogen. Hydrocarbon bonds are nonpolar, which accounts for the hydrophobic behavior of fats. Another characteristic of hydrocarbons is that they store a relatively large amount of energy. The gasoline that fuels a car consists of hydrocarbons, and the hydrocarbon tails of fat molecules act as stored fuel for your body.

(a) Structural isomers: variation in covalent arrangement, as shown in the example of butane and isobutane.

(b) Geometric isomers: variation in arrangement about a double bond. (In these diagrams, X represents an atom or group of atoms attached to a double-bonded carbon.)

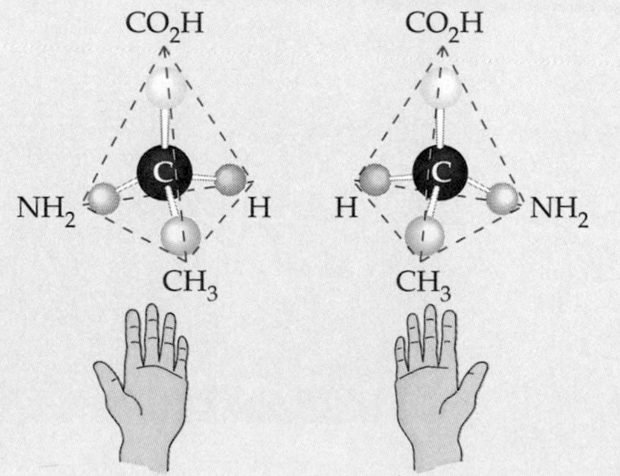

(c) Enantiomers: variation in spatial arrangement around an asymmetric carbon, resulting in molecules that are mirror images, like left and right hands. Enantiomers cannot be superimposed on each other.

FIGURE 4.6

Three types of isomers. Compounds with the same molecular formula but different structures, isomers are a source of diversity in organic molecules.

Structural isomers differ in the covalent arrangements of their atoms. The number of possible isomers increases tremendously as carbon skeletons increase in size. There are only two butanes, but there are 18 variations of C_8H_{18} and 366,319 possible structural isomers of $C_{20}H_{42}$. Structural isomers may also differ in the location of double bonds.

Geometric isomers of a molecule have all the same covalent partnerships, but they differ in their spatial arrangements. Geometric isomers arise from the inflexibility of double bonds, which, unlike single bonds, will not allow the atoms they join to rotate freely about the axis of the bonds. The subtle difference in shape between geometric isomers can dramatically affect the biological activities of organic molecules. For example, the biochemistry of vision involves a light-induced change of rhodopsin, a chemical compound in the eye, from one geometric isomer to another.

Enantiomers are molecules that are mirror images of each other. In the ball-and-stick models shown in FIGURE 4.6c, the middle carbon is called an *asymmetric carbon* because it is attached to four different atoms or groups of atoms. The four groups can be arranged in space about the asymmetric carbon in two different ways that are mirror images. They are, in a way, left-handed and right-handed versions of the molecule. A cell can distinguish these isomers based on their different shapes. Usually, one isomer is biologically active and the other is inactive.

The concept of enantiomers is important in the pharmaceutical industry, because the two enantiomers of a drug may not be equally effective. In fact, one of the isomers could even produce harmful effects. This was the case with thalidomide, a drug prescribed for thousands of pregnant women in the early 1960s. The drug was a mixture of two enantiomers. One enantiomer produced the desired effect by acting as a sedative, but the other caused birth defects. Organisms are sensitive to even the most

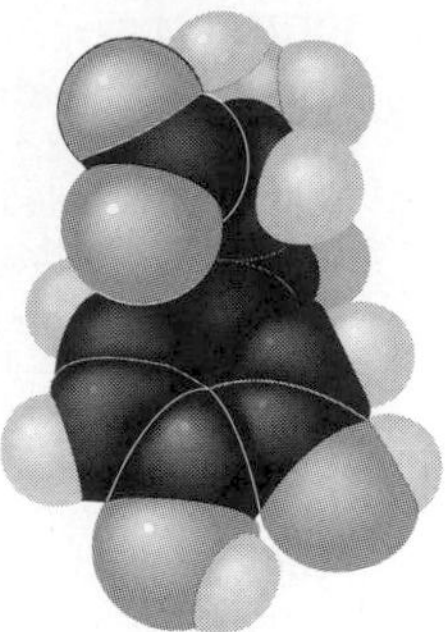

L-DOPA (effective against Parkinson's disease)

D-DOPA (biologically inactive)

FIGURE 4.7

The pharmacological importance of enantiomers. L-dopa is a drug used to treat Parkinson's disease, a disorder of the central nervous system. The drug's enantiomer, the mirror-image molecule designated D-dopa, has no effect on patients. Pharmaceutical companies are developing better techniques for synthesizing such drugs in pure isometric form rather than as mixtures of enantiomers.

subtle variations in molecular architecture (FIGURE 4.7). Once again, we see that molecules have emergent properties that depend on the specific arrangement of their atoms.

Functional groups also contribute to the molecular diversity of life

The distinctive properties of an organic molecule depend not only on the arrangement of its carbon skeleton, but also on the molecular components attached to that skeleton. We will now examine certain groups of atoms that are frequently attached to carbon skeletons of organic molecules. These ensembles of atoms are known as **functional groups** because they are the regions of organic molecules most commonly involved in chemical reactions. The functional groups we will consider are all hydrophilic and thus increase the solubility of organic compounds in water. If we think of hydrocarbons as the simplest organic molecules, we can view functional groups as attachments that replace one or more of the hydrogens bonded to the carbon skeleton of the hydrocarbon.

Each functional group behaves consistently from one organic molecule to another, and the number and arrangement of the groups help give each molecule its unique properties. Consider the differences between estradiol and testosterone, female and male sex hormones, respectively, in humans and other vertebrates (FIGURE 4.8). Both are steroids, organic molecules with a carbon skeleton in the form of four interlocking rings. These sex hormones differ only in the attachment of certain functional groups to the common skeleton. The different actions of these two molecules on many targets throughout the body help produce the contrasting features of females and males. Thus, even our sexuality has its biological basis in the variations of molecular architecture.

The six functional groups most important in the chemistry of life are the hydroxyl, carbonyl, carboxyl, amino, sulfhydryl, and phosphate groups (TABLE 4.1).

The Hydroxyl Group

In a **hydroxyl group,** a hydrogen atom is bonded to an oxygen atom, which in turn is bonded to the carbon

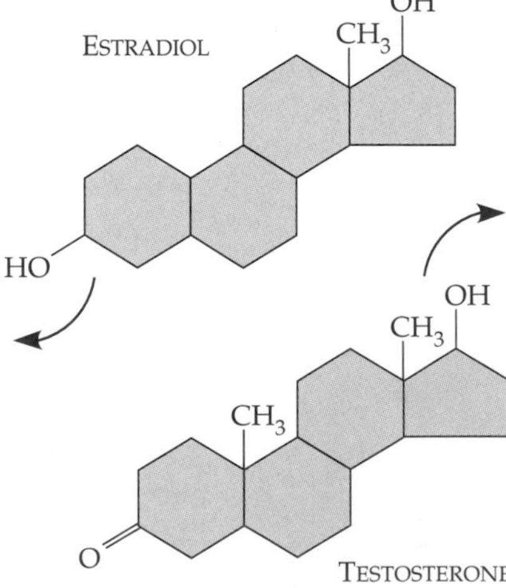

FIGURE 4.8

A comparison of functional groups of female (estradiol) and male (testosterone) sex hormones. The two molecules differ in the attachment of functional groups to a common carbon skeleton (which has been simplified here by omitting the double bonds and hydrogens of the four fused rings). This subtle variation in molecular architecture influences the development of the anatomical and physiological differences between female and male vertebrates. Compare, for example, the plumage of the female (left) and male (right) wood ducks.

TABLE 4.1

Functional Groups of Organic Compounds

FUNCTIONAL GROUP	FORMULA*	NAME OF COMPOUNDS	EXAMPLE
Hydroxyl	R—OH	Alcohols	CH_3—CH_2—OH Ethanol (the drug of alcoholic beverages)
Carbonyl	R—C(=O)—H	Aldehydes	CH_3—CH_2—C(=O)—H Propanal
	R—C(=O)—R	Ketones	CH_3—C(=O)—CH_3 Acetone
Carboxyl	R—C(=O)—OH (non-ionized) R—C(=O)—O^- (ionized)	Carboxylic acids	CH_3—C(=O)—OH Acetic acid** (the acid of vinegar)
Amino	R—N(H)—H (non-ionized) R—^+N(H)(H)—H (ionized)	Amines	HO—C(=O)—CH_2—N(H)—H Glycine** (an amino acid)
Sulfhydryl	R—SH	Thiols	CH_3—CH_2—SH Mercaptoethanol
Phosphate	R—O—P(=O)(O^-)—O^-	Organic phosphates	H—C(OH)(H)—C(OH)(H)—C(H)(H)—O—P(=O)(O^-)—O^- Glycerol phosphate

*The letter R symbolizes the carbon skeleton to which the functional group is attached.
**The ionized forms of the carboxyl and amino groups prevail in cells. However, acetic acid and gylcine are represented here in their non-ionized forms.

skeleton of the organic molecule. Organic compounds containing hydroxyl groups are called **alcohols,** and their specific names usually end in *-ol,* as in ethanol, the drug present in alcoholic beverages. In a structural formula, the hydroxyl group is abbreviated by omission of the covalent bond between the oxygen and hydrogen, and is written as —OH or HO—. (Do not confuse this functional group with the hydroxide ion, OH^-, formed by the dissociation of bases such as sodium hydroxide.) The hydroxyl group is polar as a result of the electronegative oxygen atom drawing electrons toward itself. Consequently, water molecules are attracted to the hydroxyl group, and this helps dissolve organic compounds containing such groups. Sugars, for example, owe their solubility in water to the presence of hydroxyl groups.

The Carbonyl Group

The **carbonyl group** (>CO) consists of a carbon atom joined to an oxygen atom by a double bond. If the carbonyl group is on the end of a carbon skeleton, the organic compound is called an **aldehyde;** otherwise the compound is called a **ketone.** For the latter to be the case, the chain must be at least three carbons long, as it is in acetone, the simplest ketone (see TABLE 4.1). Acetone has different properties from propanal, a three-carbon aldehyde (acetone and propanal are structural isomers). Thus, variation in locations of functional groups along carbon skeletons is another source of molecular diversity in life.

The Carboxyl Group

When an oxygen atom is double-bonded to a carbon atom that is also bonded to a hydroxyl group, the entire assembly of atoms is called a **carboxyl group** (—COOH). Compounds containing carboxyl groups are known as **carboxylic acids,** or organic acids. The simplest is the one-carbon compound called formic acid (HCOOH), the substance some ants inject when they sting. Acetic acid, which has two carbons, gives vinegar its sour taste. (In general, acids, including carboxylic acids, taste sour.)

Why does a carboxyl group have acidic properties? A carboxyl group is a source of hydrogen ions. The covalent bond between the oxygen and the hydrogen is so polar that the hydrogen tends to dissociate reversibly from the molecule as an ion (H^+). In the case of acetic acid, we have:

$$CH_3COOH \rightleftharpoons CH_3COO^- + H^+$$

Acetic acid — Acetate ion — Hydrogen ion

Dissociation occurs as a result of the two electronegative oxygen atoms of the carboxyl group pulling shared electrons away from hydrogen. If the double-bonded oxygen and the hydroxyl group are attached to *separate* carbon atoms, there is less tendency for the —OH group to dissociate because the second oxygen is farther away. Here is another example of how emergent properties result from a specific arrangement of building components.

The Amino Group

The **amino group** (—NH_2) consists of a nitrogen atom bonded to two hydrogen atoms and to the carbon skeleton. Organic compounds with this functional group are called **amines.** An example is glycine, illustrated in TABLE 4.1. Because glycine *also* has a carboxyl group, it is both an amine and a carboxylic acid. Most of the cell's organic compounds have two or more different functional groups attached to their carbon skeletons. Glycine belongs to a group of organic compounds named amino acids, which are the molecular building blocks of proteins.

The amino group acts as a base. You learned in Chapter 3 that ammonia (NH_3) can pick up a proton from the surrounding solution. Amino groups of organic compounds can do the same:

$$-NH_2 + H^+ \rightleftharpoons -{}^{+}NH_3$$

This process gives the amino group a charge of +1, its most common state within the cell.

The Sulfhydryl Group

Sulfur is grouped with oxygen in the periodic table; both have six valence electrons and form two covalent bonds. The organic functional group known as the **sulfhydryl group** (—SH), which consists of a sulfur atom bonded to an atom of hydrogen, resembles a hydroxyl group in shape (see TABLE 4.1). Organic compounds containing sulfhydryls are called **thiols.** In the next chapter, you will learn how sulfhydryl groups can interact to help stabilize the intricate structure of many proteins.

The Phosphate Group

Phosphate is an anion formed by dissociation of an inorganic acid called phosphoric acid (H_3PO_4). The loss of hydrogen ions by dissociation leaves the phosphate with a negative charge. Organic compounds containing **phosphate groups** have a phosphate ion covalently attached by one of its oxygen atoms to the carbon skeleton (see TABLE 4.1). One function of phosphate groups is the transfer of energy between organic molecules. In Chapter 6, you will learn how cells harness the transfer of phosphate groups to perform work, such as the contraction of muscle cells.

The chemical elements of life: *a review*

Living matter, as you have learned, consists mainly of carbon, oxygen, hydrogen, and nitrogen, with smaller amounts of sulfur and phosphorus. These elements share the characteristic of forming strong covalent bonds, a quality that is essential in the architecture of complex organic molecules. Of all these elements, carbon is the virtuoso of the covalent bond. The chemical behavior of carbon makes it exceptionally versatile as a building block in molecular architecture: It can form four covalent bonds, link together into intricate molecular skeletons, and join with several other elements. The versatility of carbon makes possible the great diversity of organic molecules, each with special properties that emerge from the unique arrangement of its carbon skeleton and the functional groups appended to that skeleton. At the foundation of all biological diversity lies this variation at the molecular level.

Now that we have examined the basic principles of organic chemistry, we can move on to the next chapter, where we will explore the specific structures and functions of the most elegant molecules made by living cells: carbohydrates, lipids, proteins, and nucleic acids.

REVIEW OF KEY CONCEPTS (with page numbers and key figures)

- Organic chemistry is the study of carbon compounds (pp. 53–54)
  - Carbon is unparalleled in its ability to form the large, complex, and diverse molecules that characterize living matter.
  - The belief in vitalism was challenged when chemists were able to synthesize organic compounds from inorganic ones. Organic chemistry is now defined as the chemistry of carbon compounds.
- Carbon atoms are the most versatile building blocks of molecules (pp. 54–56, FIGURE 4.3)
  - A bonding capacity of four contributes to carbon's ability to form complex and diverse molecules.
  - Carbon can bond covalently to a variety of atoms, including O, H, N, and S.
  - Carbon atoms can also bond to other carbon atoms, forming the carbon skeletons of organic compounds.
- Variation in carbon skeletons contributes to the diversity of organic molecules (pp. 56–58, FIGURE 4.4)
  - The carbon skeletons of organic molecules vary in length and shape and possess bonding sites for atoms of other elements.
  - Hydrocarbons consist only of carbon and hydrogen.
  - Carbon's versatile bonding is the basis for isomers, molecules with the same molecular formula but different structures and thus different properties. Three types of isomers are structural isomers, geometric isomers, and enantiomers.
- Functional groups also contribute to the molecular diversity of life (pp. 58–61, TABLE 4.1)
  - Functional groups consist of specific groups of atoms that covalently bond to carbon skeletons and give the overall molecule distinctive chemical properties.
  - The hydroxyl group (—OH), found in alcohols, has a polar covalent bond, which helps alcohols dissolve in water.
  - The carbonyl group ($>$CO) can be either at the end of a carbon skeleton (aldehyde) or within the skeleton (ketone).
  - The carboxyl group (—COOH) is found in carboxylic acids. The hydrogen of this group can dissociate to some extent, making the molecule a weak acid.
  - The amino group (—NH_2) can accept an H^+, thereby acting as a base.
  - The sulfhydryl group (—SH) helps stabilize the structure of some proteins.
  - The phosphate group can bond to the carbon skeleton by one of its oxygen atoms and has an important role in the transfer of cellular energy.
- The chemical elements of life: *a review* (p. 61)
  - Living matter is made mostly of carbon, oxygen, hydrogen, and nitrogen, with some sulfur and phosphorus.
  - Biological diversity has its molecular basis in carbon's ability to form an incredible array of molecules with characteristic shapes and chemical properties.

SELF QUIZ

1. Organic chemistry is currently defined as
   a. the study of compounds that can be made only by living cells
   b. the study of carbon compounds
   c. the study of vital forces
   d. the study of natural (as opposed to synthetic) compounds
   e. the study of hydrocarbons

2. Choose the pair of terms that completes this sentence: Hydroxyl is to __________ as __________ is to aldehyde.
   a. carbonyl; ketone
   b. oxygen; carbon
   c. alcohol; carbonyl
   d. amine; carboxyl
   e. alcohol; ketone

3. Which of these hydrocarbons has a double bond in its carbon skeleton?
   a. C_3H_8
   b. C_2H_6
   c. CH_4
   d. C_2H_4
   e. C_2H_2

4. The gasoline consumed by an automobile is a fossil fuel consisting mostly of
   a. aldehydes
   b. amino acids
   c. alcohols
   d. hydrocarbons
   e. thiols

5. Choose the term that correctly describes the relationship between these two sugar molecules:

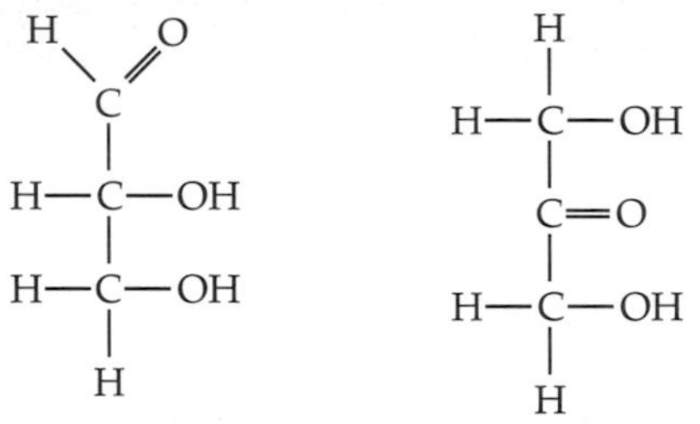

```
 H     O              H
  \   //              |
    C              H—C—OH
    |                 |
 H—C—OH              C=O
    |                 |
 H—C—OH            H—C—OH
    |                 |
    H                 H
```

   a. structural isomers
   b. geometric isomers
   c. enantiomers
   d. isotopes

6. Identify the asymmetric carbon in this molecule:

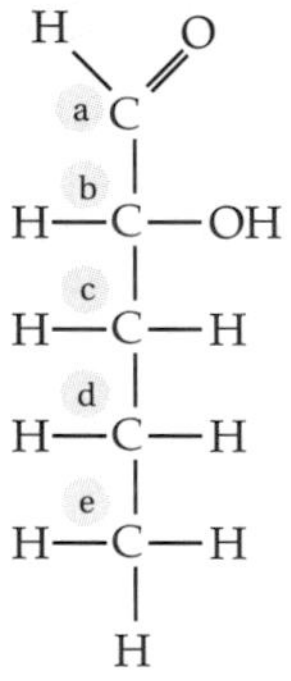

```
     H    O
      \  //
      a C
        |
  b  H—C—OH
        |
  c  H—C—H
        |
  d  H—C—H
        |
  e  H—C—H
        |
        H
```

7. Which functional group is *not* present on this molecule?

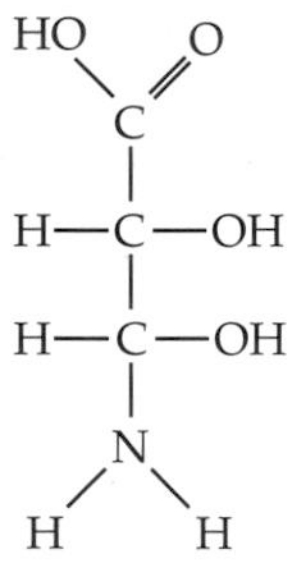

```
   HO    O
     \  //
       C
       |
    H—C—OH
       |
    H—C—OH
       |
       N
      / \
     H   H
```

   a. carboxyl
   b. sulfhydryl
   c. hydroxyl
   d. amino

8. An organic chemist would classify the molecule in question 7 as a (an)
   a. ketone
   b. aldehyde
   c. hydrocarbon
   d. amino acid
   e. thiol

9. Which functional group is most responsible for some organic molecules behaving as bases?
   a. hydroxyl
   b. carbonyl
   c. carboxyl
   d. amino
   e. phosphate

10. Which of the following molecules would be the strongest acid? Explain your answer.

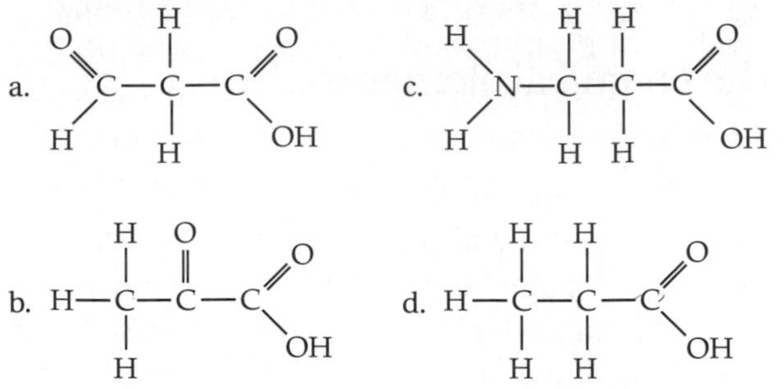

```
        H                            H  H
  O     |      O              H      |  |     O
   \\   |    //                \     |  |   //
a.    C—C—C               c.     N—C—C—C
   /    |    \                 /     |  |   \
  H     H     OH              H      H  H    OH

     H  O                         H  H
     |  ||    O                   |  |    O
     |  ||  //                    |  |  //
b. H—C—C—C                   d. H—C—C—C
     |      \                     |  |  \
     H       OH                   H  H   OH
```

CHALLENGE QUESTIONS

1. Draw an organic molecule having all six functional groups described in this chapter.

2. Draw three structural isomers of the hydrocarbon pentane (C_5H_{12}).

3. Alice, in Lewis Carroll's classic *Alice in Wonderland,* poses this question: "Is looking-glass milk good to drink?" Respond, and justify your answer based on what you have learned about the structure of organic molecules, the importance of isomers, and the biological relevance of molecular shape.

SCIENCE, TECHNOLOGY, AND SOCIETY

1. How would you respond to a modern-day vitalist who argues as follows: "Biologists have been unable to discover a physical or chemical basis for human thoughts or ideas because such abstract aspects of being human *have* no basis in natural processes."

2. The role of the Food and Drug Administration (FDA) in monitoring the testing of new drugs continues to be controversial. The thalidomide tragedy (p. 57) is often cited by those who argue that the FDA should not approve any drug until no doubt remains that the drug can be safely prescribed. Others would like to see FDA standards less rigid for drugs that may help people suffering from terminal diseases, such as AIDS. Where do you stand on this issue? Defend your position.

3. Each year, industrial chemists synthesize and test thousands of new organic compounds for use as insecticides and weed killers. In what ways are these chemicals useful and important to us? In what ways can they be harmful? What influences have shaped your opinions about these chemicals?

FURTHER READING

Asimov, I. *The World of Carbon,* 2nd ed. New York: Macmillan, 1962. A primer on the basics of organic chemistry by one of America's most popular science writers.

Bradley, D. "Frog Venom Cocktail Yields a One-Handed Painkiller." *Science,* August 27, 1993. Organic chemists synthesize a new drug based on the toxic compound secreted by the "poison arrow frog."

Bradley, D. "A New Twist in the Tale of Nature's Asymmetry." *Science,* May 13, 1994. The pharmaceutical importance of isomers.

Kessler, D. A., and K. L. Feiden. "Faster Evaluation of Vital Drugs." *Scientific American,* March 1995. When is it ethical to trade safety for speed in order to make new drugs available?

Ourisson, G., P. Albrecht, and M. Rohmer. "The Microbial Origin of Fossil Fuels." *Scientific American,* August 1984. The biology behind our important energy resources.

CHAPTER 5

THE STRUCTURE AND FUNCTION OF MACROMOLECULES

KEY CONCEPTS

- Most macromolecules are polymers
- A limitless variety of polymers can be built from a small set of monomers
- Organisms use carbohydrates for fuel and building material
- Lipids are mostly hydrophobic molecules with diverse functions
- Proteins are the molecular tools for most cellular functions
- A polypeptide is a polymer of amino acids connected in a specific sequence
- A protein's function depends on its specific conformation
- Nucleic acids store and transmit hereditary information
- A DNA strand is a polymer with an information-rich sequence of nucleotides
- Inheritance is based on precise replication of DNA
- We can use DNA and proteins as tape measures of evolution

We have applied the concept of emergent properties to our study of water and relatively simple organic molecules. These substances are central to life, each one having unique behavior arising from the orderly arrangement of its atoms. Another level in the hierarchy of biological organization is attained when cells join together small organic molecules to form larger molecules that belong to four classes: carbohydrates, lipids, proteins, and nucleic acids. Many of these cellular molecules are, on the molecular scale, huge. For example, a protein may consist of thousands of covalently connected atoms that form a molecular colossus weighing more than 100,000 daltons. Biologists use the term ***macromolecule*** *for such giant molecules of living matter.*

*Considering the size and complexity of macromolecules, it is remarkable that biochemists have determined the detailed structures of so many of them (*Figure 5.1, p. 64*). Understanding the architecture of a particular macromolecule helps explain how that molecule works. For example, the structure of the silk protein from which the orb spider in the photograph on this page weaves its web gives the fibers their strength and springiness. In molecular biology, as in the study of life at all levels, form and function are inseparable. To help us understand the relationship between the structure and function of life's macromolecules, we begin with a key generalization about how cells build large molecules from smaller ones.*

Most macromolecules are polymers

Cells make macromolecules by linking relatively small molecules together, forming chains called polymers (from the Greek *polys,* "many," and *meris,* "part"). A **polymer** is a large molecule consisting of many identical or similar building blocks linked by bonds, much as a train consists of a chain of cars. The subunits that serve as the building blocks of a polymer are called **monomers.** Some of the small molecules that serve as monomers also have other functions of their own.

The classes of macromolecules differ in the nature of their monomers, but the chemical mechanisms that cells use to make and break polymers are basically the same for all macromolecules. Monomers are connected by **condensation reactions,** also called **dehydration reactions,** in which two molecules become covalently bonded to each other through loss of a small molecule,

(a)

(b)

FIGURE 5.1

Building models to study the structure and function of macromolecules. (**a**) Linus Pauling (1901–1994) posed with a model of part of a protein. In the 1950s, Pauling discovered several of the basic structural features of proteins. (**b**) Today, scientists use computers to help build molecular models. Though methods have improved, the goal remains to correlate the structure of macromolecules with their functions.

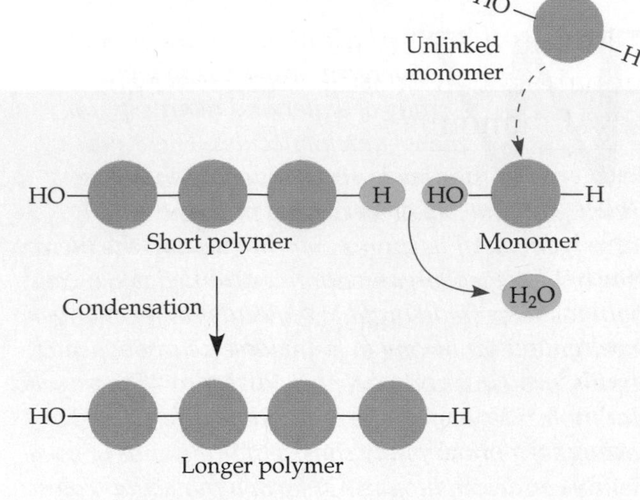

(**a**) Condensation synthesis (dehydration) of a polymer

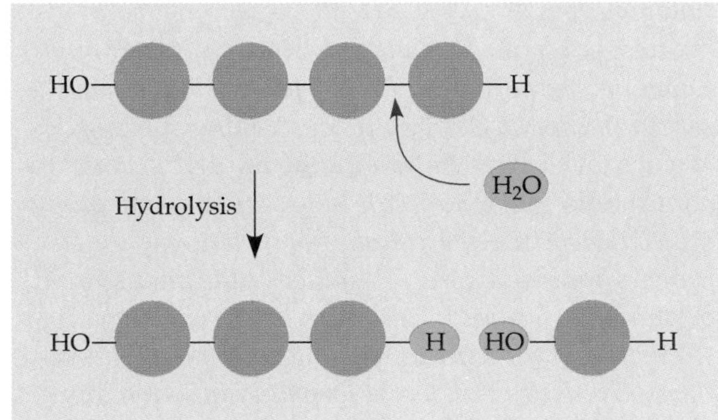

(**b**) Hydrolysis of a polymer

FIGURE 5.2

The synthesis and breakdown of polymers. (**a**) Monomers are joined by condensation reactions (dehydration reactions). The net effect is the removal of a water molecule. (**b**) The reverse of this process, hydrolysis, breaks bonds between monomers by adding water molecules.

usually water (FIGURE 5.2a). When a bond forms between two monomers, each monomer contributes part of the water molecule that is lost: One molecule provides a hydroxyl (—OH), while the other provides a hydrogen (—H). To make a polymer, this condensation reaction is repeated over and over as each monomer is added to the chain. The cell must expend energy to carry out these condensation reactions, and the process occurs only with the help of enzymes, specialized proteins that speed up chemical reactions in cells.

Polymers are disassembled to monomers by **hydrolysis,** a process that is essentially the reverse of condensation (FIGURE 5.2b). Hydrolysis means to break with water (from the Greek *hydro,* "water," and *lysis,* "break"). Bonds between monomers are broken by the addition of water molecules, a hydrogen from the water attaching to one monomer, and a hydroxyl attaching to the adjacent monomer. An example of hydrolysis working in our bodies is the process of digestion. The bulk of the organic material in our food is in the form of polymers that are much too large to enter our cells. Within the digestive tract, various enzymes attack the polymers, speeding up hydrolysis. The released monomers are then absorbed into the bloodstream for distribution to all body cells. Those cells can then use condensation reactions to assemble the monomers into new polymers that differ from the ones that were digested.

A limitless variety of polymers can be built from a small set of monomers

Each cell has thousands of different kinds of macromolecules, many of which vary from one type of cell to another in the same organism. The inherent differences between human siblings reflect variations in polymers, particularly DNA and proteins. Molecular differences between unrelated individuals are more extensive, and between species greater still. The diversity of macromolecules in the living world is vast—and the potential variety is essentially infinite.

What is the basis for such diversity in life's polymers? All macromolecules are constructed from only 40 to 50 common monomers and some others that occur rarely. Building a limitless variety of polymers from such a limited list of monomers is analogous to constructing hundreds of thousands of words from only 26 letters of the alphabet. The key is arrangement—variation in the linear sequence in which the subunits are strung together. However, this analogy falls far short of describing the great diversity of macromolecules, because most biological polymers are much longer than the longest word. Proteins, for example, are built from 20 kinds of amino acids arranged in chains that are typically more than 100 amino acids long. The molecular logic of life is simple but elegant: Small molecules common to all organisms are ordered into unique macromolecules.

Having developed an overview of macromolecules as polymers assembled from monomers, we are now prepared to investigate the specific structures and functions of the four major classes of organic compounds found in cells. For each class, we will see that the macromolecules have emergent properties not found in their individual monomers.

Organisms use carbohydrates for fuel and building material

Carbohydrates include sugars and their polymers. The simplest carbohydrates are the monosaccharides, single sugars also known as simple sugars (FIGURE 5.3). Disaccharides are double sugars, consisting of two

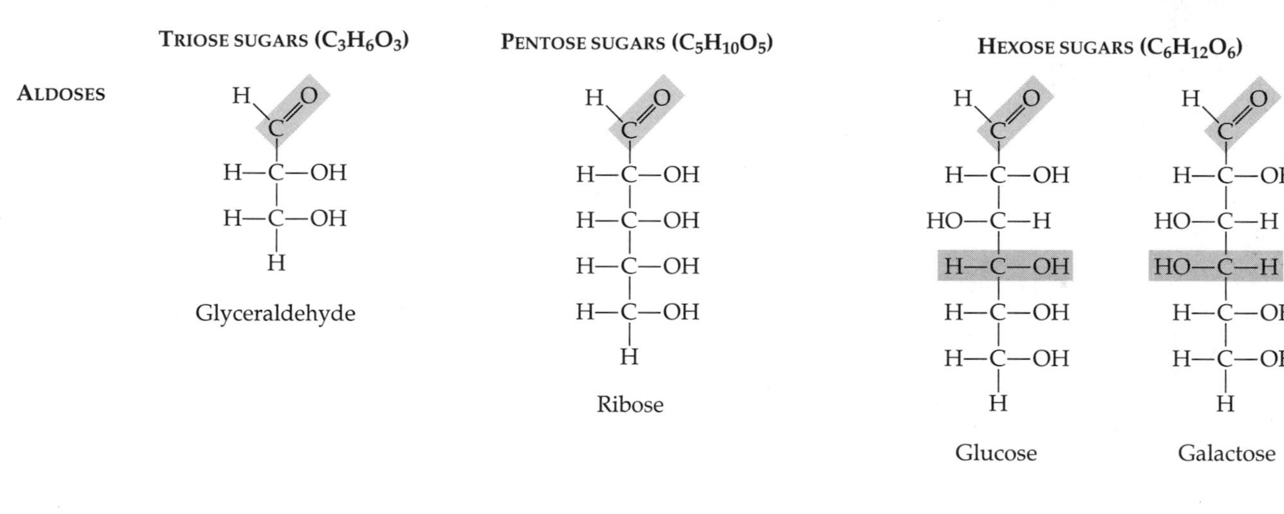

FIGURE 5.3

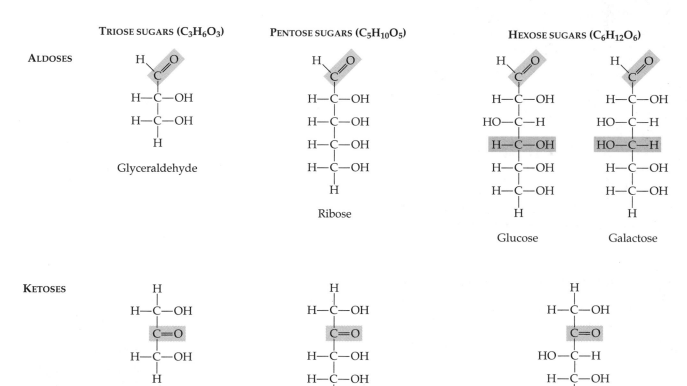

The structure and classification of some monosaccharides. Sugars may be aldoses (aldehydes) or ketoses (ketones), depending on the location of the carbonyl group (pink). Sugars are also classified according to the length of their carbon skeletons. A third point of variation is in the spatial arrangement around asymmetric carbons (compare, for example, the gray portions of glucose and galactose).

monosaccharides joined by condensation. Carbohydrates also include macromolecules in the form of polysaccharides, polymers of many sugars.

Monosaccharides

Monosaccharides (from the Greek *monos*, "single," and *sacchar*, "sugar") generally have molecular formulas that are some multiple of CH_2O (see FIGURE 5.3). Glucose ($C_6H_{12}O_6$), the most common monosaccharide, is of central importance in the chemistry of life. In the structure of glucose, we can see all the trademarks of a sugar. A hydroxyl group is attached to each carbon except one, which is double-bonded to an oxygen to form a carbonyl group. Depending on the location of the carbonyl group, a sugar is either an aldose (aldehyde sugar) or a ketose (ketone sugar). Glucose, for example, is an aldose; fructose, a structural isomer of glucose, is a ketose. (Most names for carbohydrates end in *-ose*.) Another criterion for classifying sugars is the size of the carbon skeleton, which ranges from three to seven carbons long. Glucose, fructose, and other sugars that have six carbons are called hexoses. Trioses and pentoses are also common.

Still another source of diversity for simple sugars is in the spatial arrangement of their parts around asymmetric carbons. (Recall from Chapter 4 that an asymmetric carbon is one attached to four different kinds of covalent partners.) Glucose and galactose, for example, differ only in the placement of parts around one asymmetric carbon (see the gray boxes in FIGURE 5.3). What may seem at first a small difference is significant enough to give the two sugars distinctive shapes, and shape is a major means by which molecules within cells recognize and interact with one another.

Although it is convenient to draw glucose as though its carbon skeleton were linear, this representation is not accurate. In aqueous solutions, glucose molecules, as well as most other sugars, form rings (FIGURE 5.4).

Monosaccharides, particularly glucose, are major nutrients for cells. In the process known as cellular respiration, cells extract the energy stored in glucose molecules. Not only are simple sugar molecules a major fuel for cellular work, but their carbon skeletons serve as raw material for the synthesis of other types of small organic molecules, including amino acids and fatty acids. Sugar molecules that are not immediately used by cells as fuel or as a source of carbon skeletons are generally incorporated as monomers into disaccharides or polysaccharides.

Disaccharides

A **disaccharide,** or double sugar, consists of two monosaccharides joined by a **glycosidic linkage,** a covalent bond formed between two monosaccharides (FIGURE 5.5). For example, maltose is a disaccharide formed by linking two molecules of glucose. Also known as malt sugar, maltose is an ingredient for brewing beer. Lactose, the sugar present in milk, is another disaccharide, consisting of a glucose molecule joined to a galactose molecule. The most prevalent disaccharide is sucrose, better known as table sugar. Its two monomers are glucose and fructose. Plants generally transport carbohydrates from leaves to roots and other nonphotosynthetic organs in the form of sucrose.

Polysaccharides

Polysaccharides are macromolecules. They are polymers in which a few hundred to a few thousand monosaccharides are linked together. Some polysaccharides are storage material, hydrolyzed as needed to provide sugar for cells. Other polysaccharides serve as building material for structures protecting the cell or the whole organism.

Storage Polysaccharides. Starch (FIGURE 5.6a), a storage polysaccharide of plants, is a polymer consisting

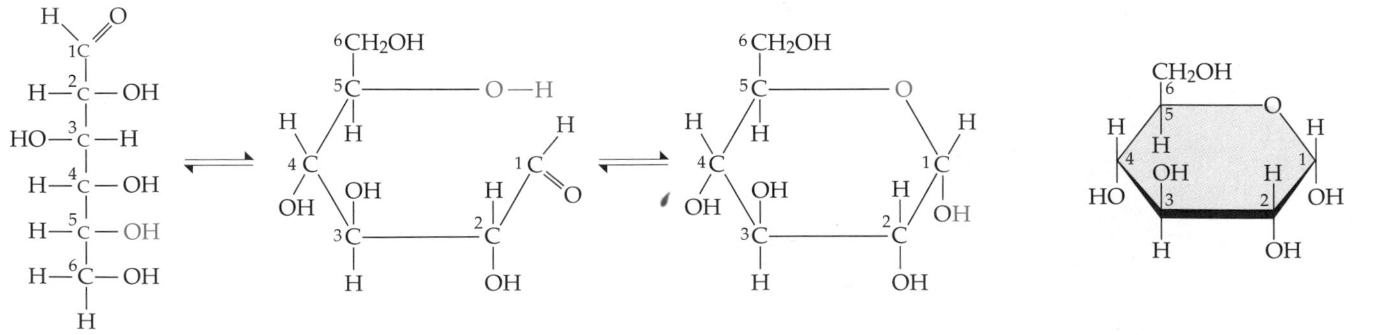

(a) Linear and ring forms

(b) Abbreviated ring structure

FIGURE 5.4
Linear and ring forms of glucose.
(**a**) Chemical equilibrium between the linear structures and rings greatly favors the formation of rings. To form the glucose ring, carbon number 1 bonds to the oxygen attached to carbon number 5. (**b**) In this abbreviated ring formula, the thicker edge indicates that you are looking at the ring edge-on; that is, the components attached to the ring by the vertical lines lie above or below the plane of the ring.

GLUCOSE GLUCOSE MALTOSE

1–4 glycosidic linkage

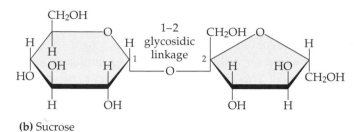

(a) Condensation synthesis of maltose

1–2 glycosidic linkage

(b) Sucrose

FIGURE 5.5

Examples of disaccharides. (**a**) The bonding of two glucose units forms maltose. Notice that the glycosidic link joins the number 1 carbon of one glucose to the number 4 carbon of the second glucose. Joining the glucose monomers at different places would result in different disaccharides. (**b**) Sucrose is a disaccharide formed from glucose and fructose. (Notice that fructose forms a five-sided ring rather than a six-membered ring.)

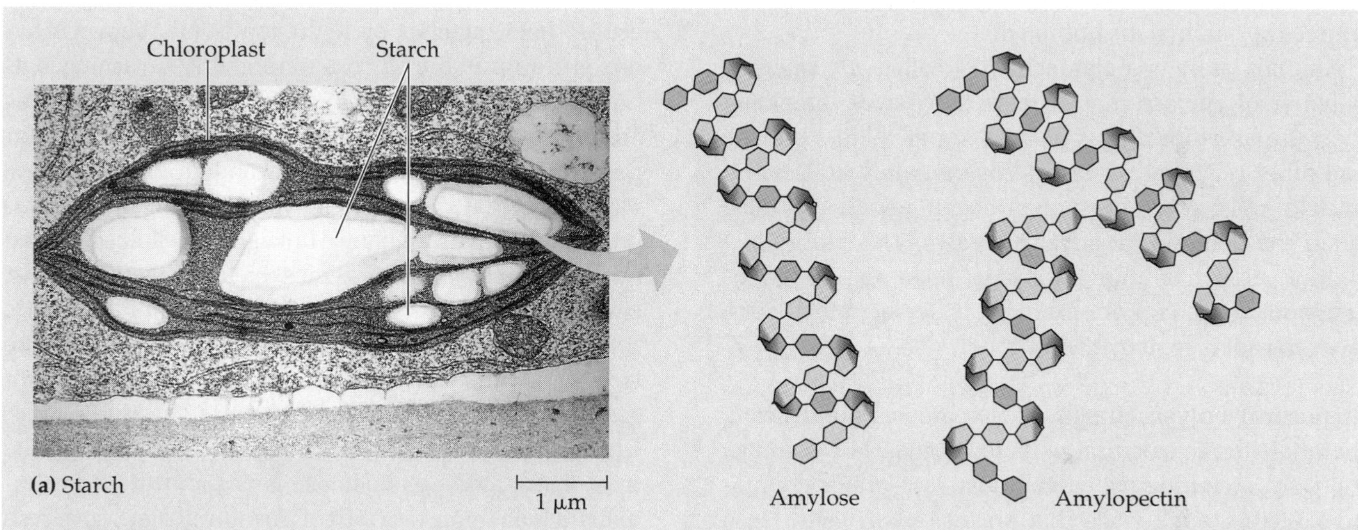

(a) Starch

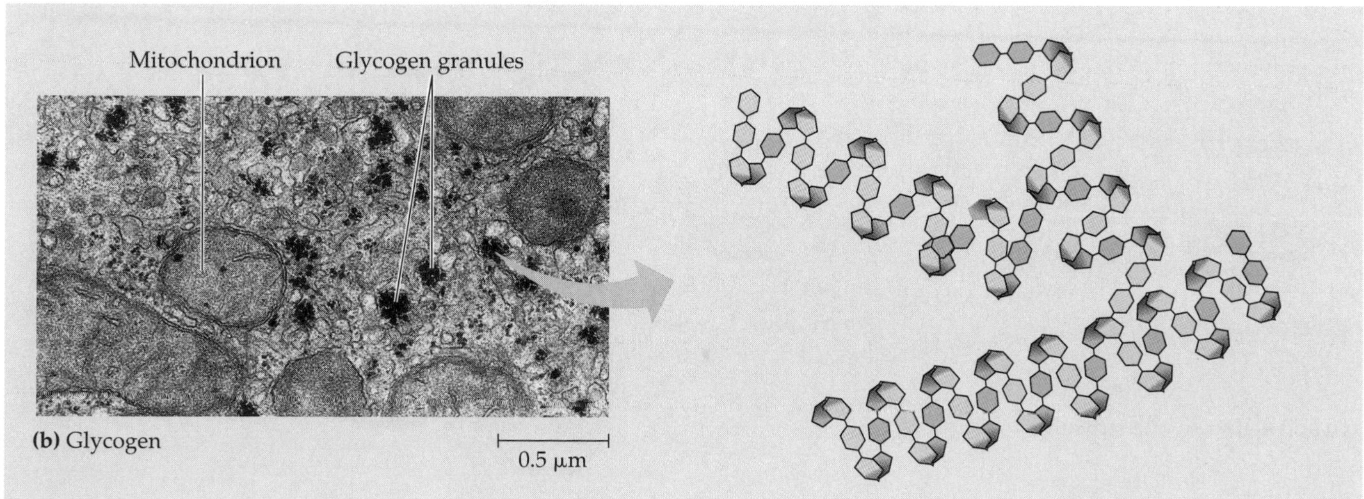

(b) Glycogen

FIGURE 5.6

Storage polysaccharides. These examples are composed entirely of glucose monomers, which are abbreviated here as hexagons. Starch and glycogen have helical shapes. (**a**) Two forms of starch are amylose (unbranched) and amylopectin (branched). The light ovals in the micrograph are granules of starch within the chloroplasts of a plant cell (TEM). (**b**) Glycogen is more extensively branched than amylopectin. Animal cells stockpile the polysaccharide glycogen as dense clusters of granules within liver and muscle cells. Hydrolysis withdraws the glucose from storage (TEM, portion of a liver cell).

entirely of glucose monomers. These monomers are joined by 1–4 linkages (number 1 carbon to number 4 carbon), like the glucose monomers in maltose (see FIGURE 5.5a). The angle of these bonds results in the polymer's helical shape. The simplest form of starch, amylose, is unbranched. Amylopectin, a more complex form of starch, is a branched polymer.

Plants store starch as granules within cellular structures called plastids, including chloroplasts (see FIGURE 5.6a). By synthesizing starch, the plant can stockpile surplus sugar. Because glucose is a major cellular fuel, starch represents stored energy. The sugar can later be withdrawn from this carbohydrate bank by hydrolysis, which breaks the bonds between the glucose monomers. Most animals, including humans, also have enzymes that can hydrolyze plant starch, making glucose available as a nutrient for cells. Potato tubers and grains—the fruits of wheat, corn, rice, and other grasses—are the major sources of starch in the human diet.

Animals store a polysaccharide called **glycogen,** a polymer of glucose that is more extensively branched than the amylopectin of plants (FIGURE 5.6b). Humans and other vertebrates store glycogen mainly in liver and muscle cells. Hydrolysis of glycogen releases glucose when the demand for sugar increases. This stored fuel cannot sustain an animal for long, however. The glycogen bank of humans, for example, is depleted in about a day unless it is replenished by food.

Structural Polysaccharides. Organisms build strong materials from structural polysaccharides. For example, the polysaccharide called **cellulose** is a major component of the tough walls that enclose plant cells. On a global scale, plants produce almost 10^{11} (100 billion) tons of cellulose per year; it is the most abundant organic compound on Earth. Like starch, cellulose is a polymer of glucose, but the glycosidic linkages in these two polymers differ. The difference is based on two possible ring structures for glucose. When the carbon chain of glucose forms a ring, the hydroxyl group attached to the number 1 carbon at the site where the ring closes is locked into one of two alternative positions: lying either below or above the plane of the ring. These two ring forms for glucose are called alpha (α) and beta (β), respectively (FIGURE 5.7a). In starch, all the glucose monomers are in the α configuration (FIGURE 5.7b). In contrast, the glucose monomers of cellulose are all in the β configuration (FIGURE 5.7c). This variation in the geometry of the glycosidic links results in starch and cellulose having different three-dimensional shapes, and therefore very different properties.

In the cell wall of a plant, many parallel cellulose molecules, held together by hydrogen bonds between hydroxyl groups of the glucose monomers, are arranged as units called microfibrils (FIGURE 5.8). These strong cables are an excellent building material for plants as well as for humans, who use cellulose-rich wood for lumber.

Enzymes that digest starch by hydrolyzing the α bonds are unable to hydrolyze the β linkages of cellulose. In fact, few organisms possess enzymes that can digest cellulose. Humans do not; the cellulose fibrils in our food pass through the digestive tract and are eliminated with the feces. Along the way, the fibrils abrade the wall of the digestive tract and stimulate the lining to secrete mucus, which aids in the smooth passage of food through the tract. Thus, although cellulose is not a nutrient for humans, it is an important part of a healthful diet. Most fresh fruits, vegetables, and grains are rich in cellulose, or fiber.

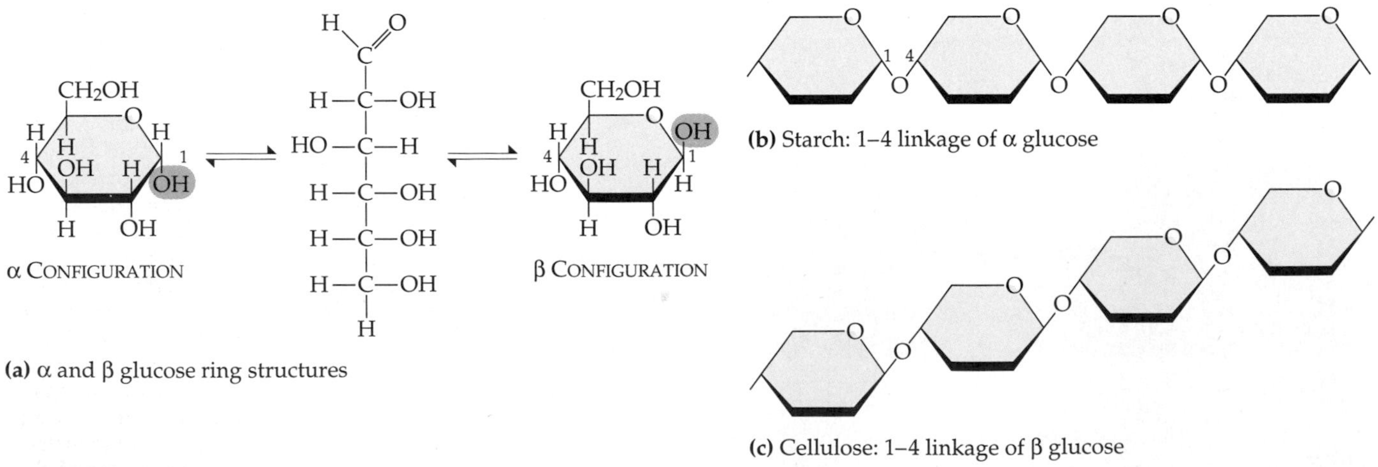

α CONFIGURATION

β CONFIGURATION

(a) α and β glucose ring structures

(b) Starch: 1–4 linkage of α glucose

(c) Cellulose: 1–4 linkage of β glucose

FIGURE 5.7

Starch and cellulose structures compared. (**a**) Glucose forms two interconvertible ring structures, designated α and β. These two forms differ in the placement of the hydroxyl group attached to the number 1 carbon. (**b**) The α ring form is the monomer for starch. (**c**) Cellulose consists of glucose monomers in the β configuration.

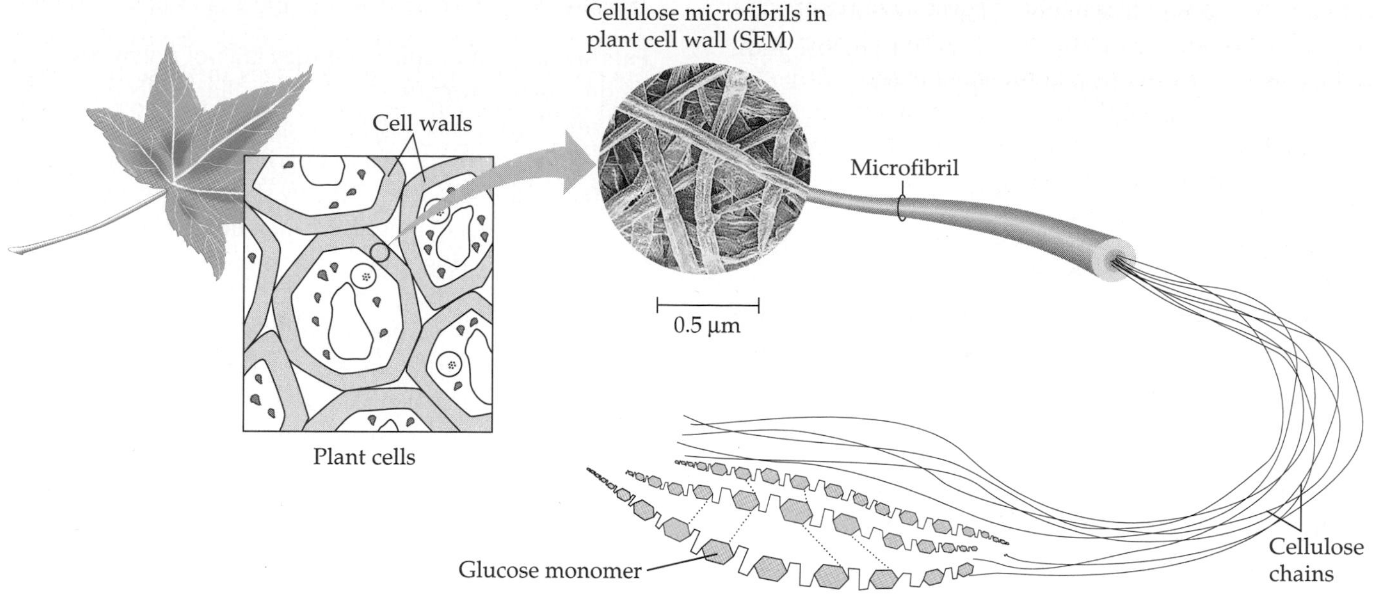

FIGURE 5.8
Cellulose and plant cell walls. Cellulose is an unbranched polysaccharide. Parallel cellulose molecules are held together by hydrogen bonds (dotted lines) between hydroxyl groups projecting from both sides. About 80 cellulose molecules associate to form a microfibril, the main architectural unit of the plant cell wall.

Some bacteria and other microorganisms can digest cellulose, breaking it down to glucose monomers. A cow harbors cellulose-digesting bacteria in the rumen, a pouch attached to its stomach. The bacteria hydrolyze the cellulose of hay and grass, converting the glucose molecules to other nutrients that nourish the cow. Similarly, a termite, which is unable to digest cellulose for itself, has unicellular organisms living in its gut that can make a meal of wood. Some molds (fungi) can also digest cellulose, thereby serving as decomposers that are crucial in recycling chemical elements within Earth's ecosystems.

Another important structural polysaccharide is **chitin,** the carbohydrate used by arthropods (insects, spiders, crustaceans, and related animals) to build their exoskeletons (FIGURE 5.9). An exoskeleton is a hard case that surrounds the soft parts of the animal. Pure chitin is leathery, but it becomes hardened when encrusted with calcium carbonate, a salt. Chitin is also found in many fungi, which use this polysaccharide rather than

(a)

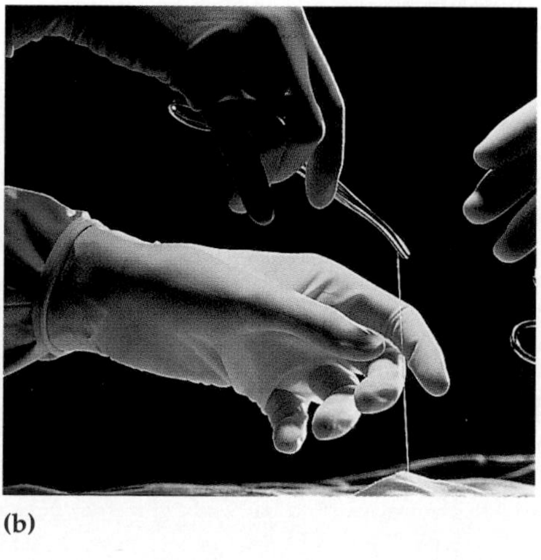

(b)

FIGURE 5.9
Chitin. (**a**) A structural polysaccharide, chitin forms the exoskeleton of arthropods. This cicada is molting, shedding its old exoskeleton and emerging in adult form. Chitin is similar in structure to cellulose, except the monomer in chitin is an amino sugar (a sugar with nitrogen). (**b**) The medical industry uses chitin to make one type of surgical thread, taking advantage of the material's combination of strength and flexibility. There is no need to remove chitin threads after a wound heals because they slowly decompose.

cellulose as the building material for their cell walls. The monomer of chitin, an amino sugar, is a glucose molecule with a nitrogen-containing appendage:

Lipids are mostly hydrophobic molecules with diverse functions

The compounds called **lipids** are grouped together because they share one important trait: They have little or no affinity for water. The hydrophobic behavior of lipids is based on their molecular structure. Although they may have some polar bonds associated with oxygen, lipids consist mostly of hydrocarbon. Three important families of lipids are fats, phospholipids, and steroids.

Fats

Fats are large molecules, but they are not polymers. A fat is constructed from two kinds of smaller molecules: glycerol and fatty acids (FIGURE 5.10). Glycerol is an alcohol with three carbons, each bearing a hydroxyl group. A **fatty acid** has a long carbon skeleton, usually 16 or 18 carbon atoms in length. At one end of the fatty acid is a "head" consisting of a carboxyl group, the functional group that gives these molecules the name fatty *acids.* Attached to the carboxyl group is a long hydrocarbon "tail." The nonpolar C—H bonds in the tails of fatty acids are the reason fats are hydrophobic. Fats separate from water because the water molecules hydrogen-bond to one another and exclude the fats. A common example of this phenomenon is the separation of vegetable oil from the aqueous vinegar solution in a bottle of salad dressing.

Three fatty acids can each be joined to glycerol by an ester linkage, a bond between a hydroxyl group and a carboxyl group. The product is a fat, or **triacylglycerol,** which consists of three fatty acids linked to one glycerol molecule. (Still another name for a fat is triglyceride, a word often found in the list of ingredients on packaged

FATTY ACID (PALMITIC ACID)

GLYCEROL

(a)

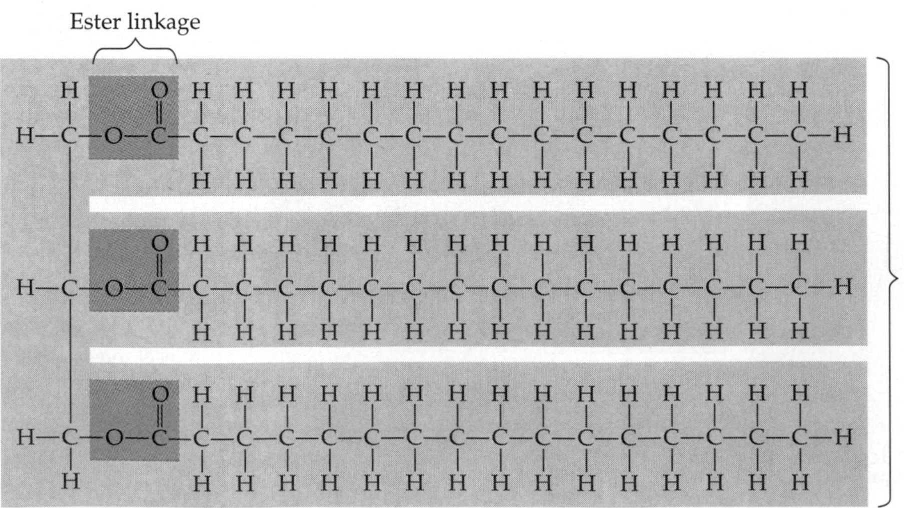

(b)

FIGURE 5.10
The structure of a fat, or triacylglycerol. The molecular building blocks of a fat are one molecule of glycerol and three molecules of fatty acids. (**a**) One water molecule is removed for each fatty acid joined to the glycerol. (**b**) The result is a fat, such as the one shown here.

foods.) All the fatty acids in a fat can be the same (FIGURE 5.10b), or they can be of two or three different kinds.

Fatty acids vary in length and in the number and locations of double bonds. The terms *saturated fats* and *unsaturated fats* are commonly used in the context of nutrition. These terms derive from the structure of the hydrocarbon tails of the fatty acids. If there are no double bonds between the carbon atoms composing the tail, then as many hydrogen atoms as possible are bonded to the carbon skeleton, creating a **saturated fatty acid** (FIGURE 5.11a). An **unsaturated fatty acid** has one or more double bonds, formed by the removal of hydrogen atoms from the carbon skeleton. The fatty acid will have a kink in its shape wherever a double bond occurs (FIGURE 5.11b).

Most animal fats are saturated; they have fatty acids that lack double bonds. Saturated animal fats—such as bacon grease, lard, and butter—solidify at room temperature. In contrast, the fats of plants and fishes are generally unsaturated. Usually liquid at room temperature, plant and fish fats are referred to as oils—for instance, corn oil, olive oil, and cod liver oil. The kinks where the double bonds are located prevent the molecules from packing together closely enough to solidify at room temperature. The phrase "hydrogenated vegetable oils," often found on food labels, means that unsaturated fats have been synthetically converted to saturated fats by adding hydrogen. Peanut butter, margarine, and many other products are hydrogenated to prevent lipids from separating out in liquid (oil) form.

A diet rich in saturated fats is one of several factors that may contribute to the human cardiovascular disease known as atherosclerosis. In this condition, deposits called plaques develop on the internal lining of blood vessels, impeding blood flow and reducing the resilience of the vessels.

Fat has come to have such a negative connotation in our culture that you might wonder whether fats serve any useful purpose. The major function of fats is energy storage. The hydrocarbons of fats are similar to gasoline molecules and are just as rich in energy. A gram of fat stores more than twice as much energy as a gram of a polysaccharide, such as starch. Because plants are relatively immobile, they can function with bulky energy storage in the form of starch. (Vegetable oils are generally obtained from seeds, where more compact storage is an asset to the plant.) Animals, on the other hand, must carry their energy baggage with them, so there is an advantage to having a more compact reservoir of fuel—fat. Humans and other mammals stock their long-term food reserves in adipose cells, which swell and shrink as fat is deposited and withdrawn from storage (see FIGURE 4.5). In addition to storing energy, adipose tissue also cushions such vital organs as the kidneys, and a layer of fat beneath the skin insulates the body. This subcutaneous

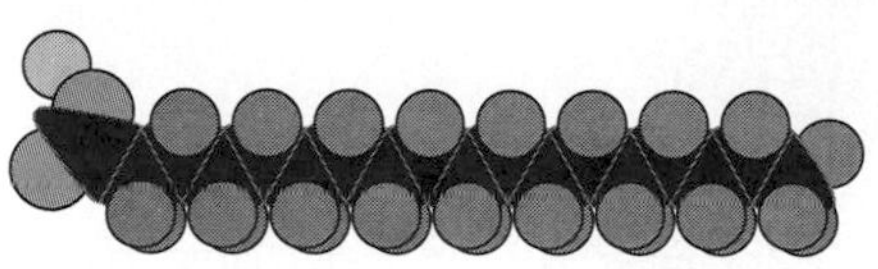

(a) Saturated fatty acid: stearic acid

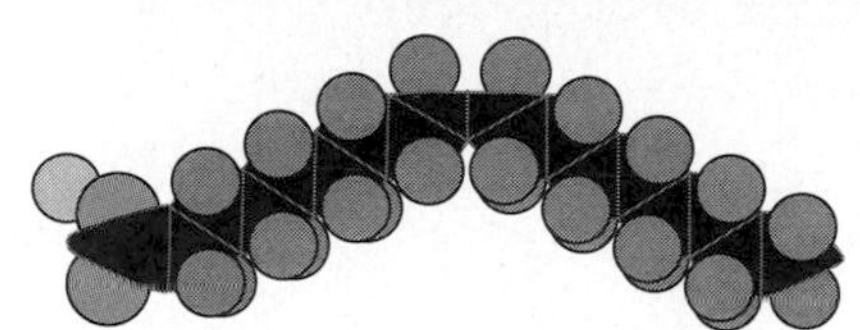

(b) Unsaturated fatty acid: oleic acid

FIGURE 5.11

Saturated and unsaturated fats compared. (**a**) If its fatty acids are saturated (with hydrogen), a fat is also said to be saturated. Most animal fats, such as those in butter, are saturated. They are solids at room temperature. (**b**) Unsaturated fatty acids, such as oleic acid, have one or more double bonds between carbons. The fatty acid bends where double bonds are located. Unsaturated fats have unsaturated fatty acids. Most vegetable fats are unsaturated, and they are called oils because they form liquids at room temperature. The kinks in the fatty acids prevent the fats from packing together closely enough to solidify.

layer is especially thick in whales, seals, and most other marine mammals.

Phospholipids

Phospholipids are structurally related to fats, but they have only two fatty acids rather than three (FIGURE 5.12). The third hydroxyl group of glycerol is joined to a phosphate group, which is negative in electrical charge. Additional small molecules, usually charged or polar, can be linked to the phosphate group to form a variety of phospholipids.

Phospholipids show ambivalent behavior toward water. Their tails, which consist of hydrocarbons, are hydrophobic and are excluded from water. However, the phosphate group and its attachments form a hydrophilic head that has an affinity for water.

When phospholipids are added to water, they self-assemble into aggregates that shield their hydrophobic portions from water. One such cluster is a micelle, a phospholipid droplet with the phosphate heads on the outside, in contact with water. The hydrocarbon tails are restricted to the water-free interior of the micelle (FIGURE 5.13a).

At the surface of a cell, phospholipids are arranged in a bilayer, or double layer (FIGURE 5.13b). The hydrophilic heads of the molecules are on the outside of the bilayer, in contact with the aqueous solutions inside and outside the cell. The hydrophobic tails point toward the interior of the membrane, away from the water. The phospholipid bilayer forms a boundary between the cell and its external environment; in fact, phospholipids are major components of cell membranes. This behavior provides another example of how form fits function at the molecular level.

Steroids

Steroids are lipids characterized by a carbon skeleton consisting of four interconnected rings (FIGURE 5.14). Different steroids vary in the functional groups attached to this ensemble of rings. An important steroid is **cholesterol,** a common component of the membranes of animal cells. Cholesterol is also the precursor from which

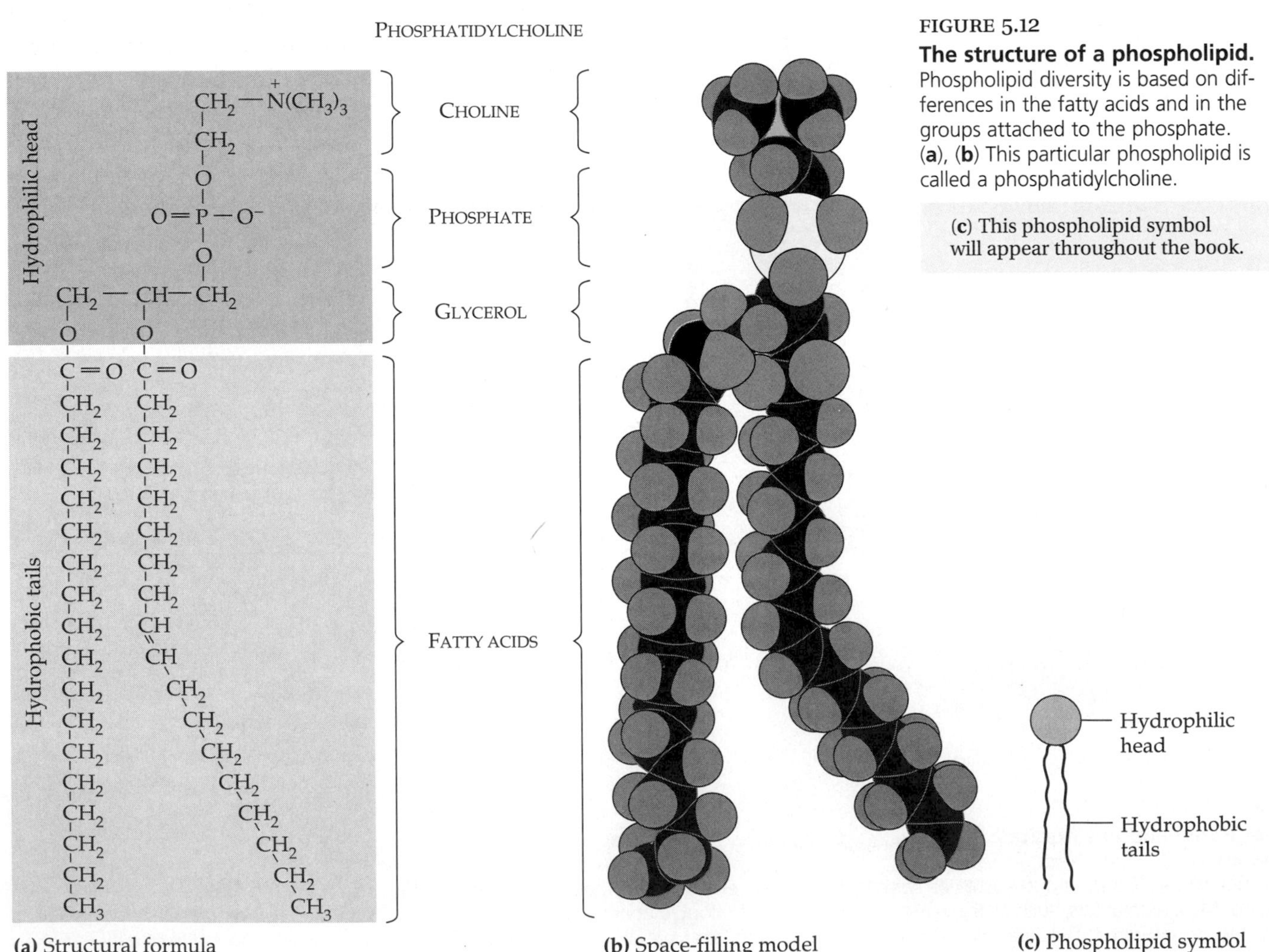

(a) Structural formula

(b) Space-filling model

(c) Phospholipid symbol

FIGURE 5.12
The structure of a phospholipid. Phospholipid diversity is based on differences in the fatty acids and in the groups attached to the phosphate. (**a**), (**b**) This particular phospholipid is called a phosphatidylcholine.

(**c**) This phospholipid symbol will appear throughout the book.

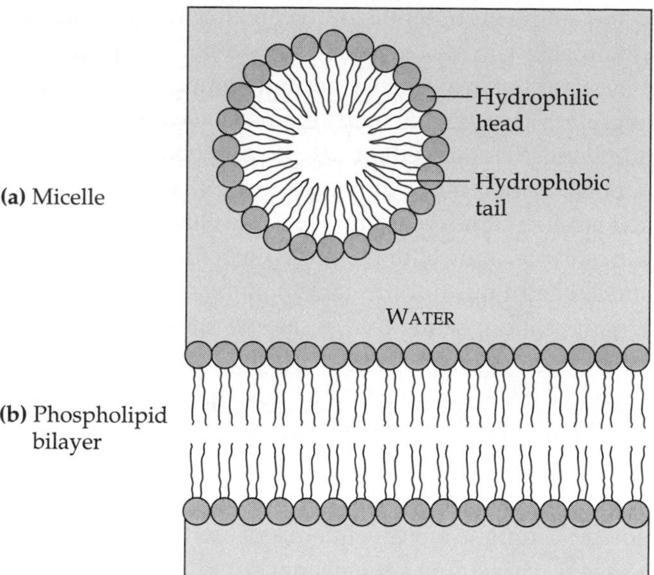

FIGURE 5.13
Two structures formed by self-assembly of phospholipids in aqueous environments. (**a**) A micelle, in cross section. (**b**) A cross section of a phospholipid bilayer between two aqueous compartments. Such bilayers are the main fabric of biological membranes. The hydrophilic heads (spheres) of the phospholipids are in contact with water, while the hydrophobic tails are in contact with each other and remote from water.

most other steroids are synthesized. For example, many hormones, including the sex hormones of vertebrates, are steroids produced from cholesterol (see FIGURE 4.8). Thus, cholesterol has important functions in animals, although a high concentration of cholesterol in the blood may contribute to atherosclerosis.

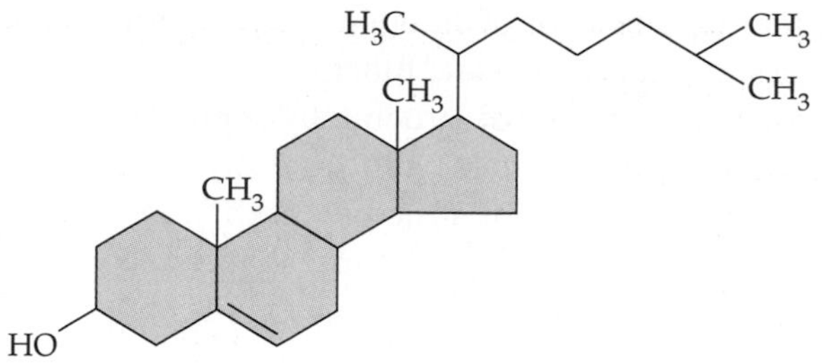

FIGURE 5.14
Cholesterol: a steroid. Cholesterol is the molecule from which other steroids, including the sex hormones, are synthesized. Steroids vary in the functional groups attached to their four interconnected rings (shown in gold).

In addition to fats, phospholipids, and steroids, other families of lipids include waxes and certain pigments in plants and animals.

Proteins are the molecular tools for most cellular functions

The importance of proteins is implied by their name, which comes from the Greek word *proteios,* meaning "first place." **Proteins** account for more than 50% of the dry weight of most cells, and they are instrumental in almost everything organisms do. Proteins are used for structural support, storage, transport of other substances, signaling from one part of the organism to another, movement, and defense against foreign substances (TABLE 5.1). In addition, as enzymes, proteins regulate metabolism by selectively accelerating chemical reactions in the cell. A human

TABLE 5.1

An Overview of Protein Functions

TYPE OF PROTEIN	FUNCTION	EXAMPLES
Structural proteins	Support	Insects and spiders use silk fibers to make their cocoons and webs, respectively. Collagen and elastin provide a fibrous framework in animal connective tissues, such as tendons and ligaments. Keratin is the protein of hair, horns, feathers, and other skin appendages.
Storage proteins	Storage of amino acids	Ovalbumin is the protein of egg white, used as an amino acid source for the developing embryo. Casein, the protein of milk, is the major source of amino acids for baby mammals. Plants store proteins in seeds.
Transport proteins	Transport of other substances	Hemoglobin, the iron-containing protein of vertebrate blood, transports oxygen from the lungs to other parts of the body. Other proteins transport molecules across cell membranes.
Hormonal proteins	Coordination of an organism's activities	Insulin, a hormone secreted by the pancreas, helps regulate the concentration of sugar in the blood of vertebrates.
Receptor proteins	Response of cell to chemical stimuli	Receptors built into the membrane of a nerve cell detect chemical signals released by other nerve cells.
Contractile proteins	Movement	Actin and myosin are responsible for the movement of muscles. Contractile proteins are responsible for the undulations of cilia and flagella, which propel many cells.
Defensive proteins	Protection against disease	Antibodies combat bacteria and viruses.
Enzymatic proteins	Selective acceleration of chemical reactions	Digestive enzymes hydrolyze the polymers in food.

has tens of thousands of different kinds of proteins, each with a specific structure and function.

Proteins are the most structurally sophisticated molecules known. Consistent with their diverse functions, they vary extensively in structure, each type of protein having a unique three-dimensional shape, or **conformation.** Diverse though proteins may be, they are all polymers constructed from the same set of 20 amino acids, the universal monomers of proteins. Polymers of amino acids are called polypeptide chains. A protein consists of one or more polypeptide chains folded and coiled into specific conformations.

A polypeptide is a polymer of amino acids connected in a specific sequence

Amino acids are organic molecules possessing both carboxyl and amino groups (see Chapter 4). Cells build their thousands of different proteins from just 20 kinds of amino acids. There are some other amino acids with important functions in organisms, but they are not incorporated into proteins.

Most amino acids consist of an asymmetric carbon bonded to four different covalent partners. Three of these are a hydrogen atom, a carboxyl group, and an amino group. The amino acids that make up proteins differ only in what is attached by the fourth bond to the asymmetric carbon (FIGURE 5.15). This variable part of the amino acid is symbolized by the letter R. The R group, also called the side chain, may be as simple as a hydrogen atom, as in the amino acid glycine, or it may be a carbon skeleton with various functional groups attached, as in glutamine.

The physical and chemical properties of the side chain determine the unique characteristics of a particular amino acid. In FIGURE 5.15, the amino acids are grouped according to the properties of their side chains. One group consists of amino acids with nonpolar side chains, which are hydrophobic. Another group consists of amino acids with polar side chains, which are hydrophilic. Acidic amino acids are those with side chains that are generally negative in charge, owing to the presence of a carboxyl group, which is usually dissociated at cellular pH. Basic amino acids have amino groups in their side chains that are generally positive in charge. (Notice that *all* amino acids have carboxyl groups and amino groups; the terms *acidic* and *basic* in this context refer only to the nature of the side chains.) Because they are ionic, acidic and basic side chains are hydrophilic.

Now that we have examined amino acids, let's see how they are linked to form polymers. When two amino acids are arranged so that the carboxyl group of one is adjacent to the amino group of the other, an enzyme can join the amino acids by means of condensation. This results in a covalent linkage called a **peptide bond.** Thus, a **polypeptide** is a polymer of many amino acids linked by peptide bonds (FIGURE 5.16, p. 76). At one end of the polypeptide chain is a free amino group, and at the opposite end is a free carboxyl group. Thus, the chain has polarity, with an N-terminus (for the nitrogen of the amino group) and a C-terminus (for the carbon of the carboxyl group). The repeating sequence of atoms along the chain (—N—C—C—N—C—C—) is referred to as the polypeptide backbone. Attached to this repetitive backbone are different kinds of appendages, the side chains of the amino acids. Polypeptides range in length from a few monomers to a thousand or more. Each specific polypeptide has a unique linear sequence of amino acids. This is a specific case of the general concept introduced earlier, that cells can make a limitless variety of polymers by linking monomers into diverse sequences.

A protein's function depends on its specific conformation

"Polypeptide chain" is not quite synonymous with "protein." The relationship is somewhat analogous to that between a long strand of yarn and a sweater of particular size and shape that one can knit from the yarn strand. A functional protein is not *just* a polypeptide chain, but one or more polypeptides precisely twisted, folded, and coiled into a molecule of unique shape (FIGURE 5.17, p. 76). A polypeptide has information in the form of its amino acid sequence, and it is that information that determines what three-dimensional conformation the protein will take. Many proteins are globular (roughly spherical), while others are fibrous in shape. However, within these broad categories, countless variations are possible.

A protein's specific conformation determines how it works. In almost every case, the function of a protein depends on its ability to recognize and bind to some other molecule. For instance, an antibody binds to a particular foreign substance that has invaded the body, and an enzyme recognizes and binds to its substrate, the substance the enzyme works on. In Chapter 2, you learned that one nerve cell in the brain signals another by dispatching specific molecules that have a unique shape. The receptor molecules on the surface of the receiving cell are proteins that fit the signal molecules something like a lock and key (see FIGURE 2.16).

Four Levels of Protein Structure

When a cell synthesizes a polypeptide, the chain folds spontaneously to assume the functional conformation for that protein. As this occurs, the protein's structure is

NONPOLAR

GLYCINE (Gly) ALANINE (Ala) VALINE (Val) LEUCINE (Leu) ISOLEUCINE (Ile)

METHIONINE (Met) PHENYLALANINE (Phe) TRYPTOPHAN (Trp) PROLINE (Pro)

POLAR

SERINE (Ser) THREONINE (Thr) CYSTEINE (Cys) TYROSINE (Tyr) ASPARAGINE (Asn) GLUTAMINE (Gln)

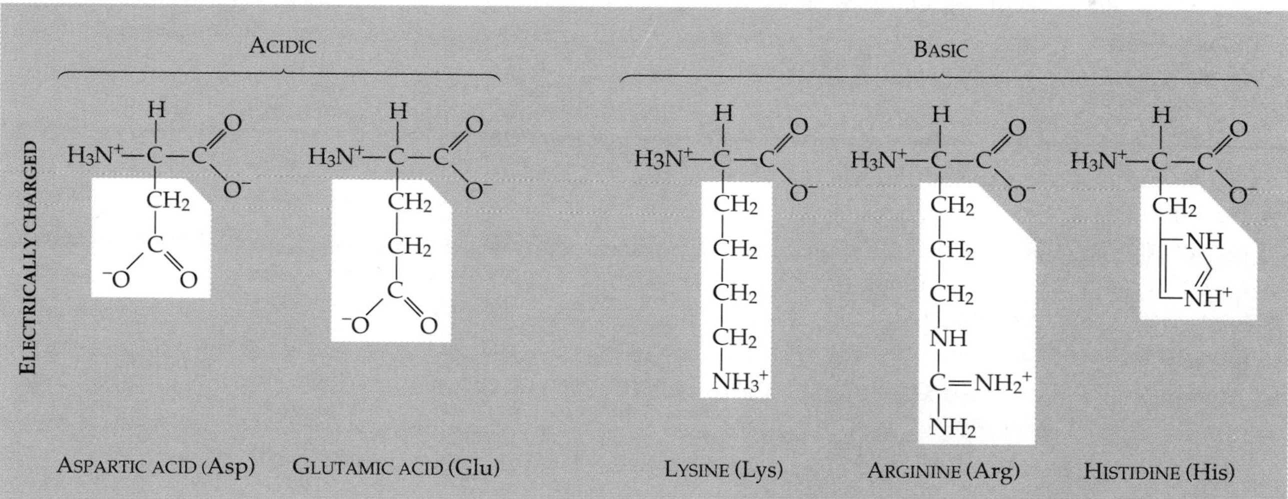

FIGURE 5.15

The 20 amino acids of proteins. The amino acids are grouped here according to the properties of their side chains (R groups), highlighted in white. The amino acids are shown in their prevailing ionic forms at pH 7, the approximate pH within a cell. In parentheses are the three-letter abbreviations for the amino acids.

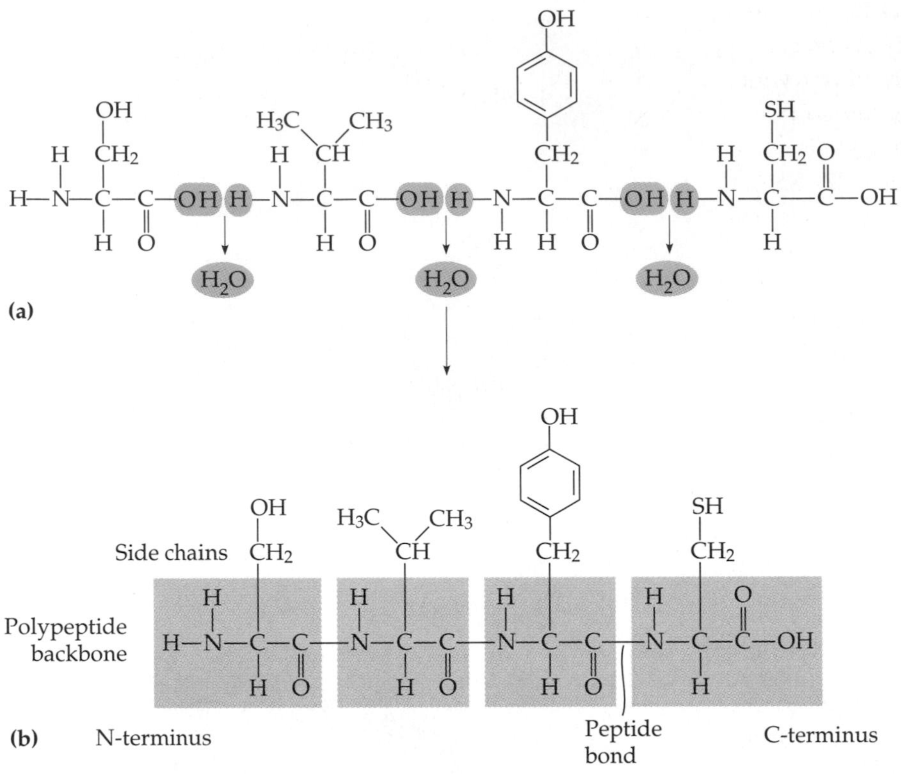

(a)

H—N—C—C—OH H—N—C—C—OH H—N—C—C—OH H—N—C—C—OH

H_2O H_2O H_2O

(b)

Side chains

Polypeptide backbone

N-terminus

Peptide bond

C-terminus

FIGURE 5.16

Polypeptide chains. (**a**) Peptide bonds formed by condensation reactions link the carboxyl group of one amino acid to the amino group of the next. (**b**) The polypeptide has a repetitive backbone (purple) to which the amino acid side chains are attached.

reinforced by a variety of chemical bonds between parts of the chain. Thus, the function of a protein—the ability of a receptor protein to identify and associate with a particular chemical messenger in the brain, for instance—is an emergent property resulting from exquisite molecular order. In this complex architecture of a protein, we can recognize three superimposed levels of structure, known as primary, secondary, and tertiary structure. A fourth level, quaternary structure, occurs when a protein consists of two or more polypeptide chains.

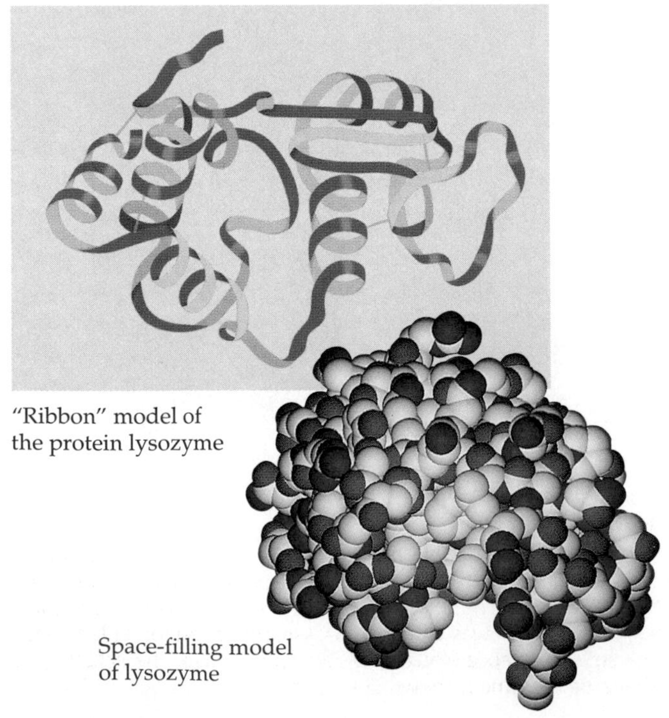

FIGURE 5.17

Functional conformation of a protein, the enzyme lysozyme. Present in our sweat, tears, and saliva, lysozyme is an enzyme that helps prevent infection by binding to and destroying specific molecules on the surface of many kinds of bacteria. You can see from the space-filling model that lysozyme's shape is roughly spherical (globular), as are many other proteins. This *exact* conformation, however, is unique to lysozyme, and the protein's specialized function emerges from this shape. A groove on the surface of lysozyme is the part of the protein that recognizes and binds to the target molecules on bacterial walls. A "ribbon" model of lysozyme makes it easier to see how the single polypeptide chain, represented by the ribbon, forms the basis of the functional protein. In this simplification, the yellow lines symbolize one type of chemical bond that stabilizes the protein's shape.

Primary Structure. The **primary structure** of a protein is its unique sequence of amino acids. As an example, we will examine the primary structure of lysozyme, the antibacterial enzyme illustrated in its three-dimensional form in FIGURE 5.17. Lysozyme is a relatively small protein, its single polypeptide chain only 129 amino acids long. In FIGURE 5.18, the polypeptide chain is unraveled for a closer look at its primary structure. A specific one of the 20 amino acids occupies each of the 129 positions along the chain. The primary structure is like the order of letters in a very long word. If left to chance, there would be 20^{129} different ways of arranging amino acids into a polypeptide chain of this length. However, the precise primary structure of a protein is determined not by the random linking of amino acids, but by inherited genetic information.

Even a slight change in primary structure can affect a protein's conformation and ability to function. For instance, sickle-cell disease is an inherited blood disorder in which one amino acid is substituted for another in a single position in the primary structure of hemoglobin, the protein that carries oxygen in red blood cells (FIGURE 5.19).

Biochemists have determined the primary structures of hundreds of proteins. The pioneer in this work was Frederick Sanger, who, with his colleagues at Cambridge University in England, worked out the amino acid sequence of the hormone insulin in the late 1940s and early 1950s. His approach was to use protein-digesting enzymes and other catalysts that break polypeptides at specific places rather than completely hydrolyzing the chain. Treatment with one of these agents would cleave the polypeptide into fragments that could be separated by a technique called chromatography. Hydrolysis with another agent would break the polypeptide at different sites, yielding a second group of fragments. Sanger used chemical methods to determine the sequence of amino acids in these small fragments. Then he searched for overlapping regions among the pieces obtained by hydrolyzing with the different agents. Consider, for instance, two fragments with the following sequences:

Cys-Ser-Leu-Tyr-Gln-Leu
Tyr-Gln-Leu-Glu-Asn

We can deduce from the overlapping regions that the intact polypeptide contains in its primary structure the following segment:

Cys-Ser-Leu-Tyr-Gln-Leu-Glu-Asn

Just as we could reconstruct this sentence from a collection of fragments with overlapping sequences of letters, Sanger and his co-workers were able, after years of effort, to reconstruct the complete primary structure of insulin. Since then, most of the steps involved in sequencing a polypeptide have been automated. However,

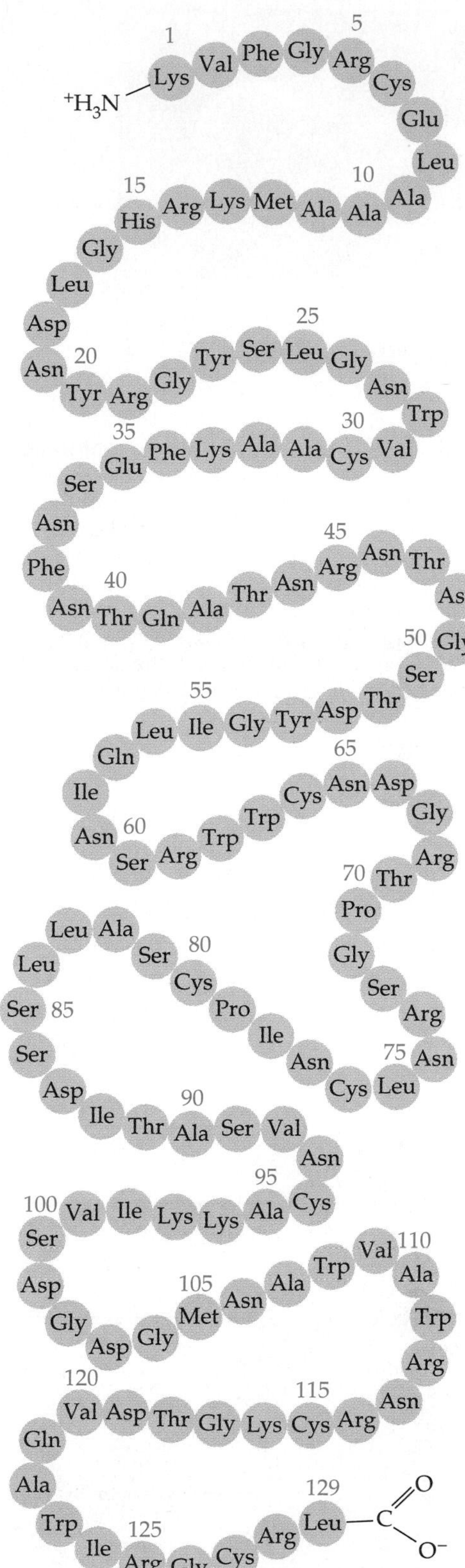

FIGURE 5.18

The primary structure of a protein. This is the unique amino acid sequence, or primary structure, of the enzyme lysozyme. The names of the amino acids are given as their three-letter abbreviations. (The chain was drawn in this serpentine fashion only so that the entire sequence would be visible on the page. The actual shape of lysozyme is shown in FIGURE 5.17.)

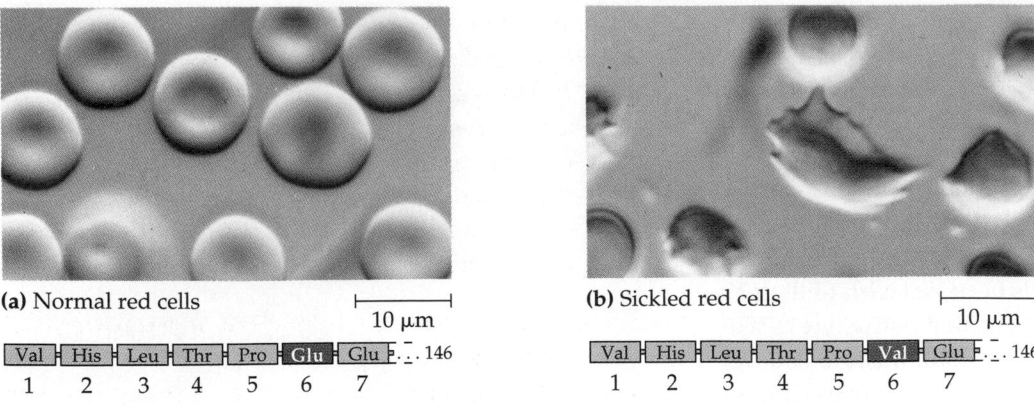

FIGURE 5.19

Sickle-cell disease is caused by a single amino acid substitution in a single protein. (**a**) The function of normal human red blood cells, which are disk-shaped, is to transport oxygen from the lungs to the other organs of the body. Each cell contains millions of molecules of hemoglobin, the protein that carries the oxygen (LM). (**b**) A slight change in the primary structure of hemoglobin—an inherited substitution of one amino acid—causes sickle-cell disease. The substitution occurs in the number 6 position of a polypeptide chain that is 146 amino acids long. The abnormal hemoglobin molecules tend to link together and crystallize, deforming the cells. The disease is named for the "sickle" shape of some of these deformed cells. The life of someone with the disease is punctuated by "sickle-cell crises," which occur when the angular cells clog tiny blood vessels, impeding blood flow (LM).

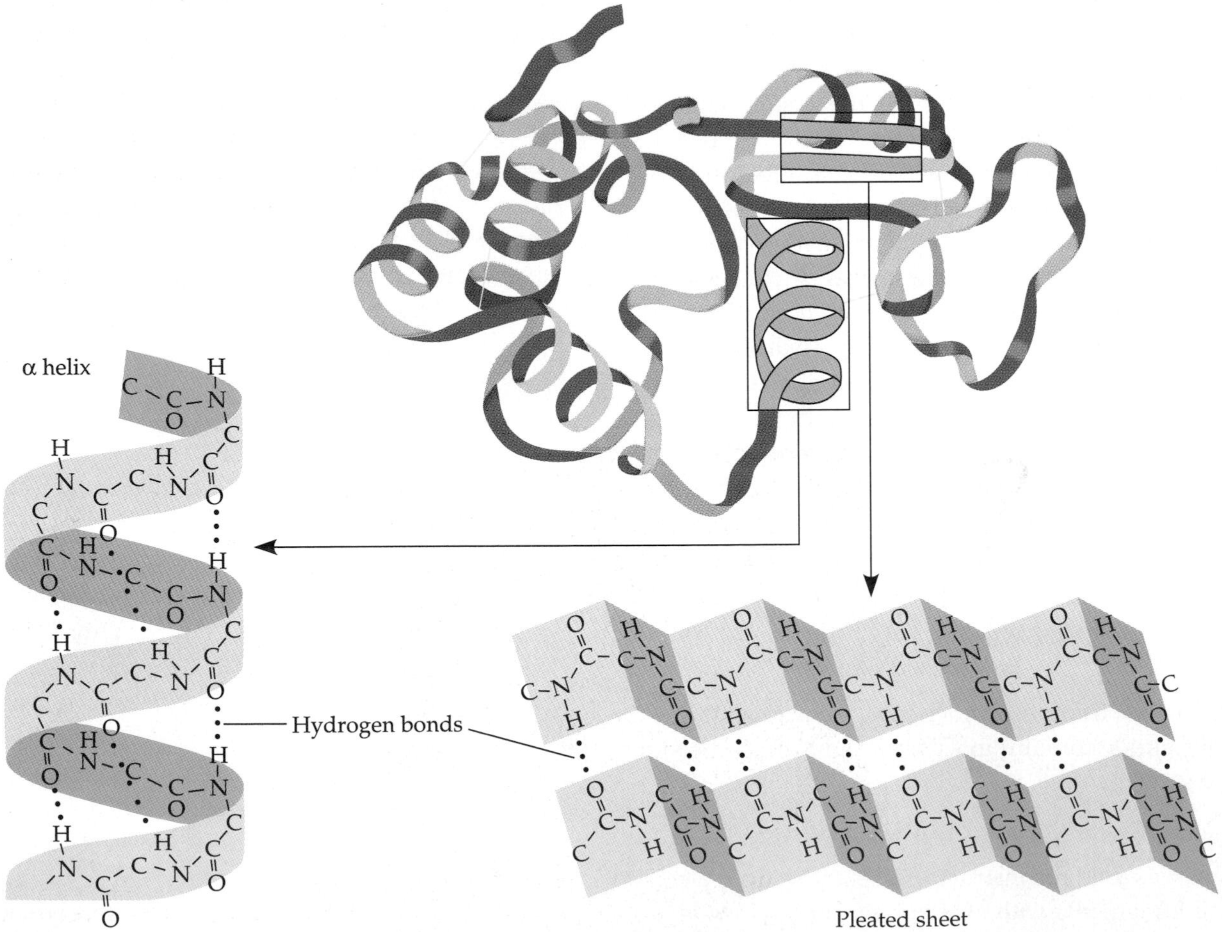

α helix

Hydrogen bonds

Pleated sheet

FIGURE 5.20

The secondary structure of a protein. Two types of secondary structure, the α helix and the pleated sheet, can both be found in the protein lysozyme. Both patterns depend on hydrogen bonding along the polypeptide chain. The R groups of the amino acids are omitted in these diagrams.

it was Sanger's analysis of insulin that first demonstrated what is now a fundamental axiom of molecular biology: Each type of protein has a unique primary structure, a precise sequence of amino acids.

Secondary Structure. Most proteins have segments of their polypeptide chain repeatedly coiled or folded in patterns that contribute to the protein's overall conformation. These coils and folds, collectively referred to as **secondary structure,** are the result of hydrogen bonds at regular intervals along the polypeptide backbone (FIGURE 5.20, facing page, bottom). Because they are electronegative, both the oxygen and the nitrogen atoms of the backbone have weak negative charges (see Chapter 2). The weakly positive hydrogen atom attached to the nitrogen atom has an affinity for the oxygen atom of a nearby peptide bond. Individually, these hydrogen bonds are weak, but because they are repeated many times over a relatively long region of the polypeptide chain, they can support a particular shape for that part of the protein. One such secondary structure is the **alpha (α) helix,** a delicate coil held together by hydrogen bonding between every fourth peptide bond. The regions of α helix in the enzyme lysozyme are evident in FIGURE 5.20, where one α helix is enlarged to show the hydrogen bonds. Lysozyme is fairly typical of a globular protein in that it has a few stretches of α helix separated by nonhelical regions. In contrast, some fibrous proteins, such as α-keratin, the structural protein of hair, have the α-helix formation over most of their entire length.

Another type of secondary structure is the **pleated sheet,** in which the polypeptide chain folds back and forth, or where two regions of the chain lie parallel to each other. Hydrogen bonds between the parallel regions hold the structure together. Pleated sheets make up the dense core of many globular proteins, and we can recognize one such region in lysozyme (see FIGURE 5.20). Also, pleated sheets dominate some fibrous proteins, including the structural protein of silk produced by many insects and spiders (FIGURE 5.21).

Tertiary Structure. Superimposed on the patterns of secondary structure is a protein's **tertiary structure,** consisting of irregular contortions from bonding between side chains (R groups) of the various amino acids. (In contrast, recall that secondary structure results from hydrogen bonds formed at regular intervals along the polypeptide's backbone.) One of the factors that contributes to tertiary structure is called a **hydrophobic interaction.** As a polypeptide folds into its functional conformation, amino acids with hydrophobic (nonpolar) side chains usually congregate at the core of the protein, out of contact with water. Their mutual exclusion from water keeps the hydrophobic side chains together in localized clusters. Thus, what we call a hydrophobic interaction is actually caused by the behavior of water molecules, which exclude nonpolar substances as the water molecules hydrogen-bond to one another and to hydrophilic molecules. In addition, hydrogen bonds between side chains of certain amino acids and ionic bonds between positively and negatively charged side chains also help stabilize tertiary structure. These are all weak interactions, but their cumulative effect helps give the protein a specific shape.

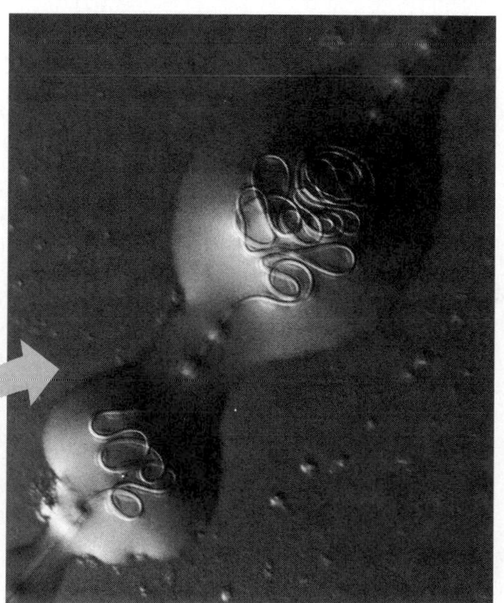

100 μm

FIGURE 5.21
Spider silk: a structural protein. Abdominal glands of this spider secrete silk fibers. Secreted in liquid form, the protein solidifies upon contact with air. The silk protein owes its strength to its secondary structure. The polypeptides are folded back and forth as pleated sheets. The teamwork of so many hydrogen bonds makes each silk fiber stronger than steel. The strands that radiate out from the web's center are made of dry silk protein; their strength maintains the basic shape of the web. The elastic strands forming the spiral, called capture strands, stretch and coil in response to wind, rain, and insects that are unfortunate enough to fly into the web. The light micrograph, right, helps us understand this resilience. A capture strand consists of a coiled silk fiber coated by a sticky fluid. Force on the strand unwinds the silk fiber, absorbing the shock. Then the fluid's surface tension—the tendency for the aqueous solution to "bead"—rewinds the fiber.

The conformation of a protein may be reinforced further by strong covalent bonds called **disulfide bridges.** Disulfide bridges form where two cysteine monomers, amino acids with sulfhydryl groups (—SH) on their side chains, are brought close together by the folding of the protein. The sulfur of one cysteine bonds to the sulfur of a second, and the disulfide bridge (—S—S—) rivets parts of the protein together. (The yellow lines in FIGURES 5.17 and 5.20 represent disulfide bridges.) All of these different kinds of bonds can occur in one protein, as shown diagramatically in FIGURE 5.22.

Quaternary Structure. As mentioned previously, some proteins consist of two or more polypeptide chains aggregated into one functional macromolecule. **Quaternary structure** is the overall protein structure that results from the aggregation of these polypeptide subunits. For example, collagen is a fibrous protein that has helical subunits supercoiled into a larger triple helix (FIGURE 5.23a). This supercoiled organization of collagen, which is similar to the construction of a rope, gives the long fibers great strength. This suits collagen fibers to their function as the girders of connective tissue, such as tendons and ligaments. Hemoglobin, the oxygen-binding protein of red blood cells, is an example of a globular protein with quaternary structure (FIGURE 5.23b). It consists of two kinds of polypeptide chains, with two of each kind per hemoglobin molecule.

We have taken the reductionist approach in dissecting proteins to their four levels of structural organization. However, it is the overall product, a macromolecule with a unique shape, that works in a cell. The specific function of a protein is an emergent property that arises from the architecture of the molecule (FIGURE 5.24).

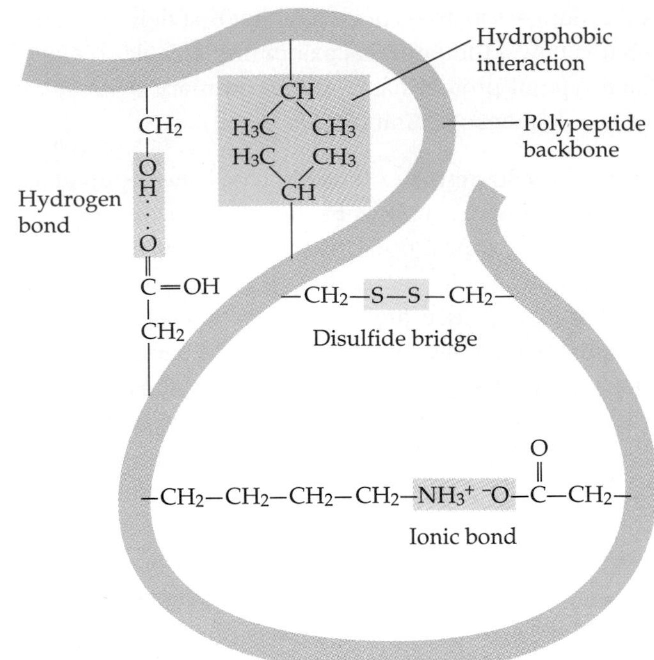

FIGURE 5.22
Examples of bonds contributing to the tertiary structure of a protein. Hydrogen bonds, ionic bonds, and hydrophobic interactions are weak bonds between side chains that collectively hold the protein in a specific conformation. Much stronger are the disulfide bridges, covalent bonds between the side chains of cysteine pairs.

What Determines Protein Conformation?

You've learned that unique conformation endows each protein with a specific function, but what are the key factors determining conformation? You already know part of the answer: A polypeptide chain of a given amino acid sequence will spontaneously arrange itself into a three-dimensional shape maintained by the interactions re-

(a) Collagen

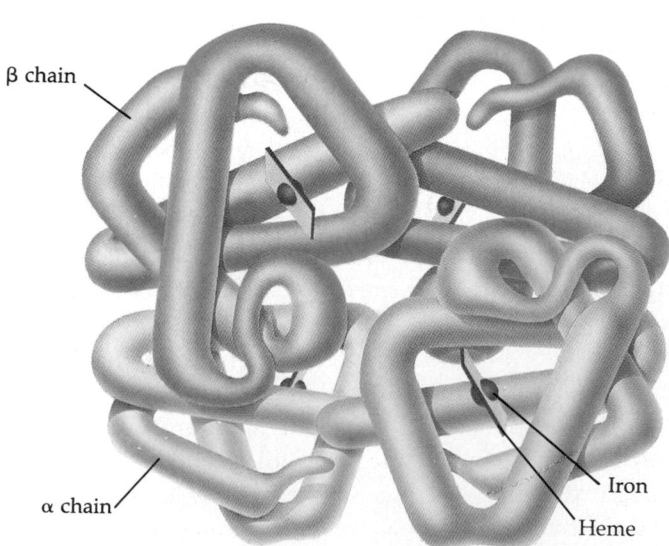

(b) Hemoglobin

FIGURE 5.23
The quaternary structure of a protein. At this level of structure, two or more polypeptide subunits associate to form a functional protein. (**a**) Collagen is a fibrous protein consisting of three helical polypeptides that are supercoiled to form a ropelike structure of great strength. Accounting for 40% of the protein in the human body, collagen strengthens connective tissue in our skin, bones, ligaments, tendons, and other body parts. (**b**) Hemoglobin is a globular protein with four subunits, two of one kind (α chains) and two of another kind (β chains). (Each subunit has a nonpolypeptide component, called heme, with an iron atom that binds oxygen.)

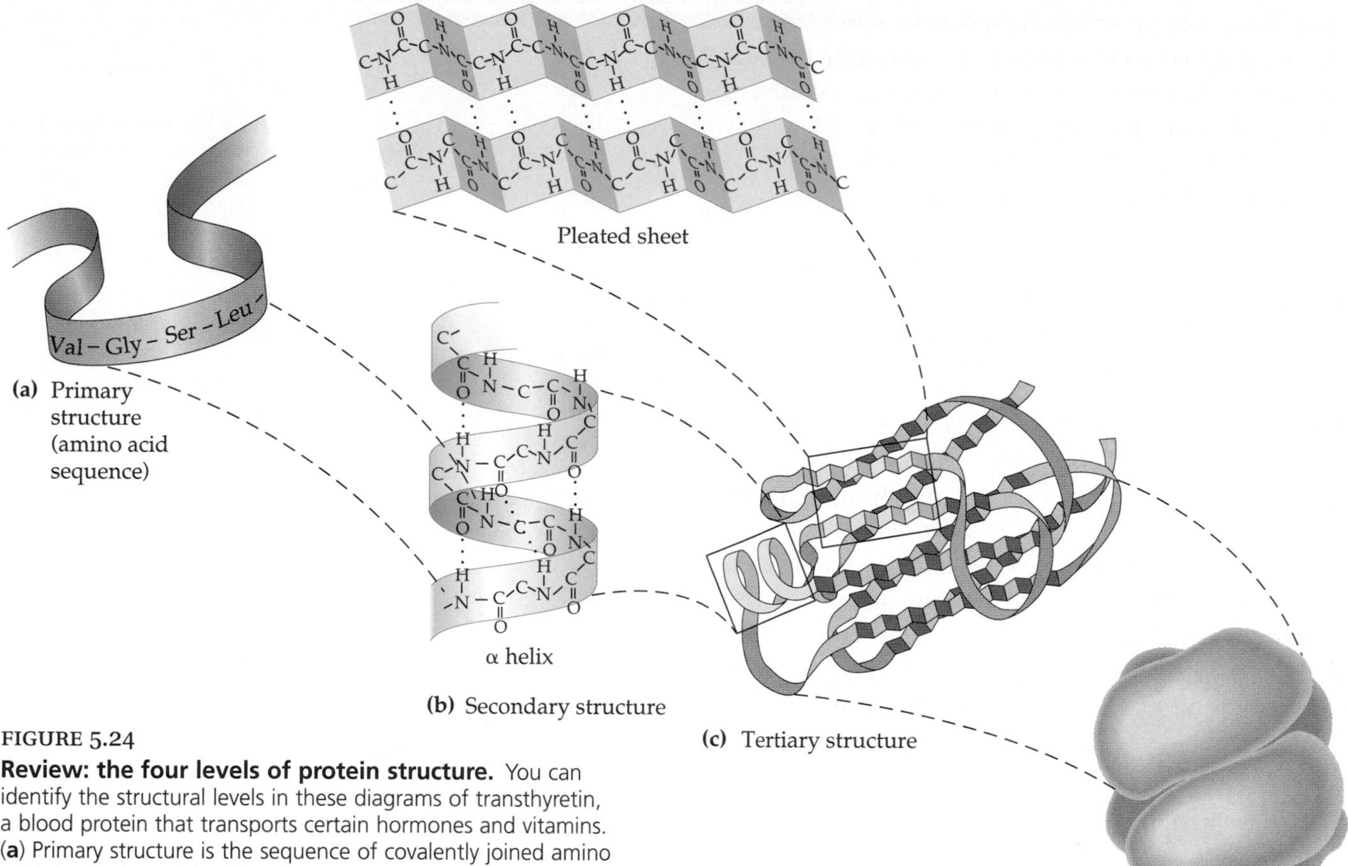

FIGURE 5.24
Review: the four levels of protein structure. You can identify the structural levels in these diagrams of transthyretin, a blood protein that transports certain hormones and vitamins. (**a**) Primary structure is the sequence of covalently joined amino acids in a polypeptide. (**b**) Secondary structure is the bending and hydrogen bonding of a polypeptide backbone to form α helices and pleated sheets. (**c**) Tertiary structure is the overall conformation (shape) of a polypeptide, as reinforced by interactions between the side chains (R groups) of amino acids. (**d**) Quaternary structure is the association between two or more polypeptides that make up a protein. In the case of transthyretin, the whole protein consists of four identical polypeptide subunits.

sponsible for secondary and tertiary structure. This normally occurs when the protein is being synthesized within the cell. However, protein conformation also depends on the physical and chemical conditions of the protein's environment. If the pH, salt concentration, temperature, or other aspects of its environment are altered, the protein may unravel and lose its native conformation, a change called **denaturation** (FIGURE 5.25, p. 83). Misshapen, the denatured protein becomes biologically inactive. Most proteins become denatured if they are transferred from an aqueous environment to an organic solvent, such as ether or chloroform; the protein turns inside out, its hydrophobic regions changing places with its hydrophilic portions. Other agents of denaturation include chemicals that disrupt the hydrogen bonds, ionic bonds, and disulfide bridges that maintain a protein's shape. Denaturation can also result from excessive heat, which agitates the polypeptide chain enough to overpower the weak interactions that stabilize conformation. The white of an egg becomes opaque during cooking because the denatured proteins are insoluble and solidify.

When a protein is denatured, it may re-form to its functional shape when returned to its normal environment. We can conclude that the information for building specific shape is intrinsic in the protein's primary structure. The sequence of amino acids determines conformation—where an α helix can form, where pleated sheets can occur, where disulfide bridges are located, and so on.

The Protein-Folding Problem

Biochemists have determined the amino acid sequences of hundreds of proteins, and the three-dimensional shapes of many of those proteins are also known (see the Methods Box, p. 82). One would think that by correlating the primary structures of many proteins with their conformations, it would be possible to discover the rules of protein folding, especially with the help of computers. Unfortunately, the protein-folding problem is not that simple. Most proteins probably go through several intermediate states on their way to a stable conformation, and looking at the "mature" conformation does not reveal the stages of folding required to achieve that form. However, biochemists have recently developed methods for tracking a protein through its intermediate stages of folding. Researchers have also discovered **chaperone proteins,** molecules that function as temporary braces in assisting

METHODS: MOLECULAR MODELS AND COMPUTER GRAPHICS

Three-dimensional structures of biological macromolecules provide important insights into molecular function. Determining the structures of macromolecules as complex as proteins, each made up of thousands of atoms, is a formidable task. Pauling, Watson and Crick, and the other pioneers of molecular biology built models from wood, wire, and plastic model sets. Computers have made it possible to build models much more quickly.

In the illustrations in this box (from the Department of Biochemistry at the University of California, Riverside), we follow the development of a computer model for the structure of an enzymatic protein called ribonuclease, whose function involves binding to a nucleic acid molecule. The first step is to crystallize the protein. (In this case the protein is combined with a short strand of nucleic acid.) Then, using a method called X-ray crystallography, an instrument aims an X-ray beam through the crystal. The regularly spaced atoms of the crystal diffract (deflect) the X-rays into an orderly array (FIGURE a). The diffracted X-rays expose photographic film, producing a pattern of spots (FIGURE b). From such diffraction patterns, computer programs generate electron density maps of successive, cross-sectional slices through the protein (FIGURE c).

By combining the information from electron density maps with the primary structure of the protein, as determined by chemical methods, it is possible to plot the three-dimensional (*x, y, z*) coordinates of each atom. Finally, graphics software generates a picture showing the position of each atom in the molecule (FIGURE d). The scientist can move the image around on the screen to simulate the molecule's appearance from various angles—even from *inside* the molecule. Thus, computers have expanded the ways we can view a molecule while retaining the advantages of being able to manipulate the model by hand.

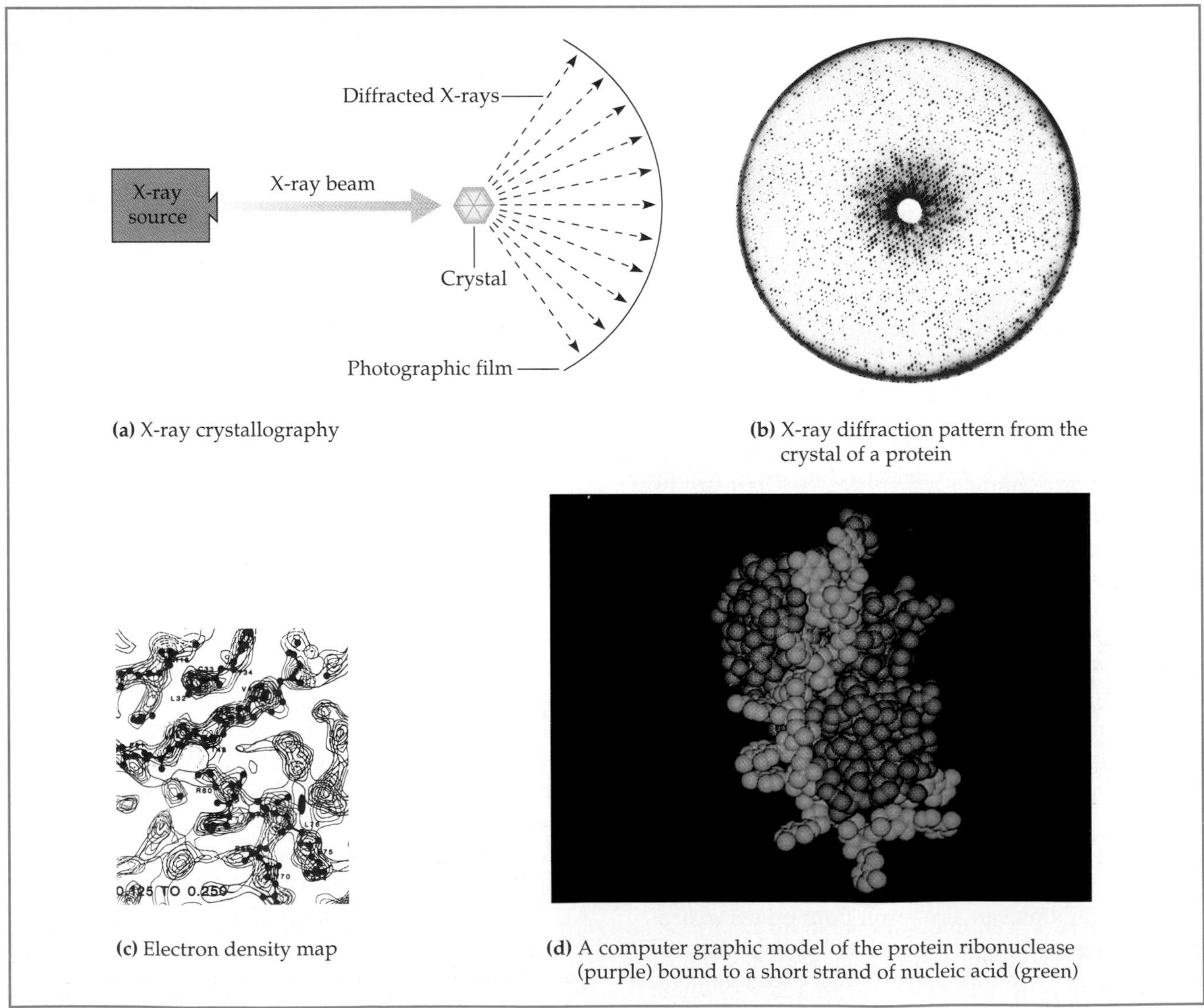

(a) X-ray crystallography

(b) X-ray diffraction pattern from the crystal of a protein

(c) Electron density map

(d) A computer graphic model of the protein ribonuclease (purple) bound to a short strand of nucleic acid (green)

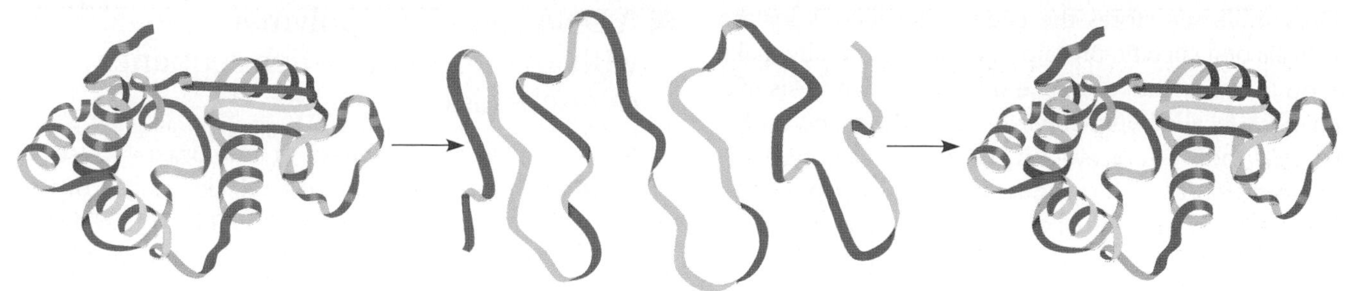

FIGURE 5.25

Denaturation and renaturation of a protein. High temperatures or various chemical treatments will denature a protein, causing it to lose its conformation and ability to function. If the denatured protein remains dissolved, it may renature when the environment is restored to normal.

the folding of other proteins. These breakthroughs will accelerate our understanding of protein folding.

The protein-folding problem is as important as it is challenging. Once the rules of protein folding are known, it should be possible to design proteins that will carry out specific tasks by making polypeptide chains with appropriate amino acid sequences. Based on what has already been learned, a University of Colorado research team recently made an enzyme that digests proteins.

Nucleic acids store and transmit hereditary information

If primary structure determines the conformation of a protein, what determines primary structure? The amino acid sequence of a polypeptide is programmed by a unit of inheritance known as a **gene.** Genes consist of DNA, which is a polymer belonging to the class of compounds known as **nucleic acids.**

There are two types of nucleic acids: **deoxyribonucleic acid (DNA)** and **ribonucleic acid (RNA).** These are the molecules that enable living organisms to reproduce their complex equipment from one generation to the next. Unique among molecules, DNA provides directions for its own replication; this molecular reproduction is the basis for the continuity of all life.

DNA is the genetic material that organisms inherit from their parents. A DNA molecule is very long and consists of hundreds or thousands of genes, each occupying a specific position along the single molecule. When a cell reproduces itself by dividing, its DNA is copied and passed along from one generation of cells to the next. Encoded in the structure of DNA is the information that programs all the cell's activities. The DNA, however, is not directly involved in running the operations of the cell, any more than computer software by itself can print a bank statement or read the bar code on a box of cereal. Just as a printer is needed to print out a statement and a scanner is needed to read a bar code, proteins are required to implement genetic programs. Proteins are the molecular hardware of the cell—the tools for most biological functions. For example, it is the protein hemoglobin that carries oxygen in the blood, not the DNA that specifies the structure of hemoglobin.

FIGURE 5.26 illustrates how RNA, the other type of nucleic acid, fits into the flow of genetic information from DNA to proteins. Although each gene along the length of

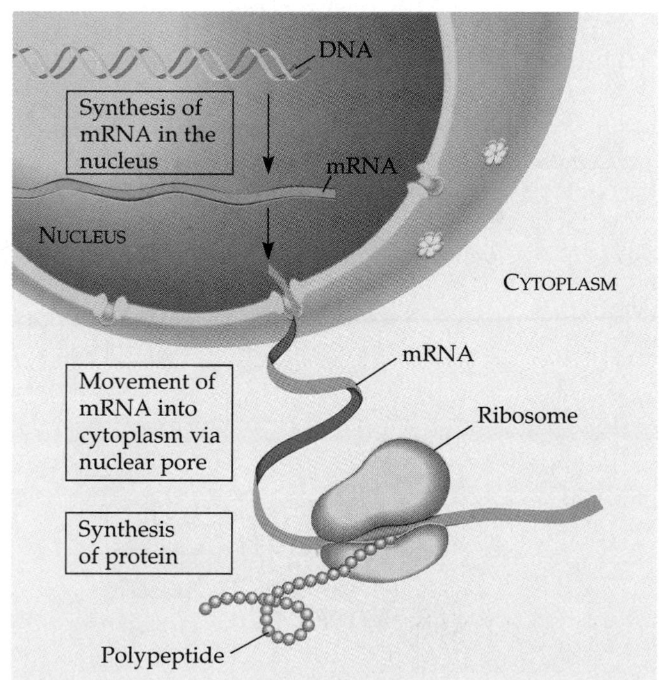

FIGURE 5.26

DNA⟶RNA⟶protein: a diagrammatic overview of information flow in a cell. In a eukaryotic cell, DNA in the nucleus programs protein production in the cytoplasm by producing messenger RNA (mRNA), which travels to the cytoplasm and binds to ribosomes. As a ribosome (greatly enlarged in this drawing) moves along the mRNA, the genetic message conveyed from the nucleus is translated into a polypeptide of specific amino acid sequence.

a DNA molecule stores the coded instructions for the synthesis of a specific protein, it does not actually make the protein. Instead, the gene directs the synthesis of a type of RNA called messenger RNA (mRNA). The mRNA molecule then interacts with the protein-synthesizing machinery to direct the production of a polypeptide. We can summarize the flow of genetic information this way: DNA⟶RNA⟶protein. The actual sites of protein synthesis are cellular structures called ribosomes. In a eukaryotic cell, ribosomes are located in the cytoplasm, but DNA resides in the nucleus. Messenger RNA conveys the genetic instructions for building proteins from the nucleus to the cytoplasm. Prokaryotic cells (bacteria) lack nuclei, but they still use RNA to send a message from the DNA to the ribosomes and other equipment of the cell that translate the coded information into amino acid sequences. We will defer further study of RNA until Chapter 16. For now, we will take a closer look at the structure and function of DNA.

A DNA strand is a polymer with an information-rich sequence of nucleotides

Nucleic acids are polymers of monomers called **nucleotides.** Each nucleotide is itself composed of three parts: a phosphate group, which is joined to a pentose (five-carbon sugar), which in turn is bonded to an organic molecule called a nitrogenous base (FIGURE 5.27). In the case of DNA, the pentose is a sugar named deoxyribose, hence the designation *deoxyribo*nucleic acid. Phosphate and deoxyribose are common to all DNA nucleotides. It is the third component of a nucleotide, the nitrogenous base, that varies. There are two families of bases: pyrimidines and purines. A **pyrimidine** is characterized by a six-membered ring made up of carbon and nitrogen atoms. **Purines** are larger, with the six-membered ring fused to a five-membered ring. Within each of these families are specific nitrogenous bases that differ

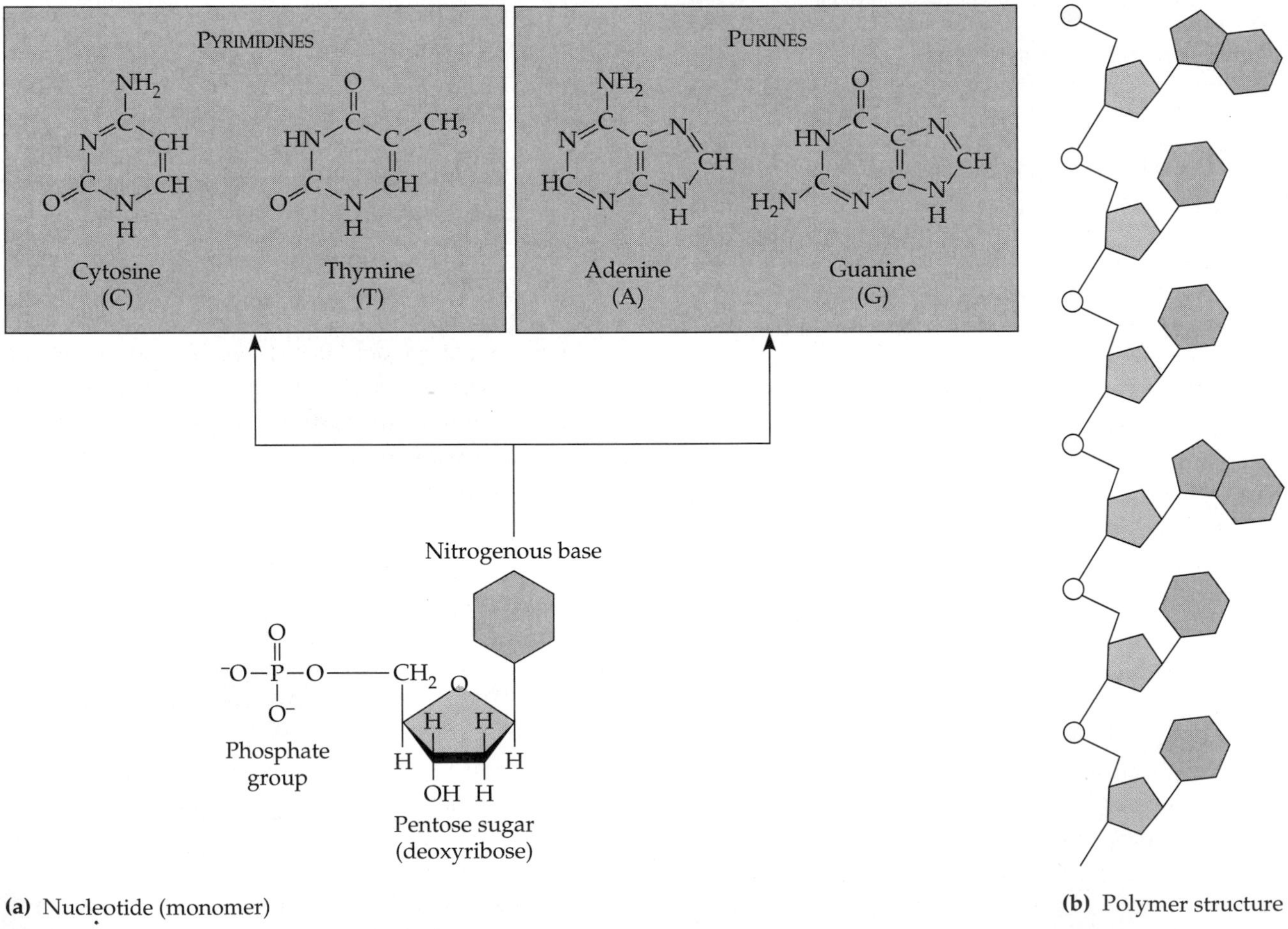

FIGURE 5.27
The structures of nucleotides and DNA polymers. (**a**) Nucleotides, the monomers of nucleic acids, are themselves composed of three smaller molecular building blocks: a phosphate group, a pentose sugar, and a nitrogenous base, either a purine or a pyrimidine. In DNA, the pentose is deoxyribose and the possible bases are cytosine, thymine, adenine, and guanine. (**b**) In a DNA strand, each nucleotide monomer has its phosphate group bonded to the sugar of the next nucleotide. The polymer has a regular sugar-phosphate backbone with variable appendages, the four kinds of nitrogenous bases.

only in the functional groups attached to the rings. Altogether in DNA, there are four different bases: two pyrimidines named cytosine (abbreviated C) and thymine (T), and two purines named adenine (A) and guanine (G). Thus, each nucleotide consists of a phosphate group, deoxyribose, and A, G, C, or T as its nitrogenous base.

In a DNA polymer, nucleotides are joined by covalent bonds called phosphodiester linkages between the phosphate of one nucleotide and the sugar of the next monomer. This bonding results in a backbone with a repeating pattern of sugar-phosphate-sugar-phosphate (FIGURE 5.27b). All along this sugar-phosphate backbone are appendages consisting of the nitrogenous bases. Unlike the regular backbone, the sequence of bases along a DNA polymer is unique for each gene. Because genes are typically hundreds of nucleotides long, the number of possible base sequences is essentially limitless. A gene's meaning to the cell is encoded in its specific sequence of the four bases. For example, the sequence AGGTAACTT means one thing, while the sequence CGCTTTAAC has a different translation. (Real genes, of course, are much longer.) The linear order of bases encoded in a gene specifies the amino acid sequence—the primary structure—of a protein, which then specifies that protein's three-dimensional conformation and function in the cell.

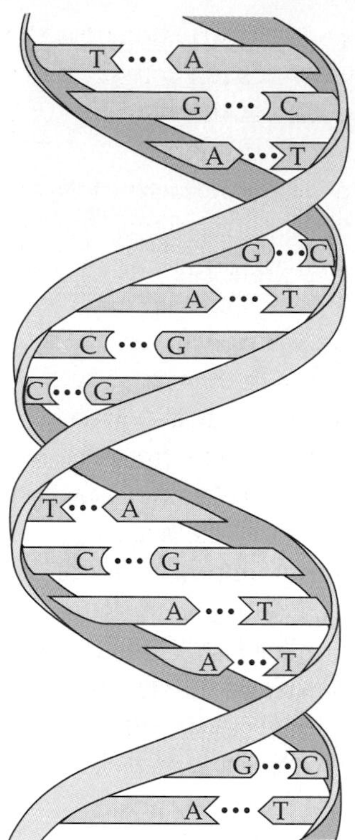

FIGURE 5.28
The double helix. The DNA molecule is usually double-stranded, with the sugar-phosphate backbone of the polynucleotides (abbreviated here by blue ribbons) on the outside of the helix. In the interior are pairs of nitrogenous bases, holding the two strands together by hydrogen bonds. Hydrogen bonding between the bases is specific. As illustrated here with symbolic shapes for the bases, adenine (A) can pair only with thymine (T), and guanine (G) can pair only with cytosine (C). As a ell prepares to divide, the two strands of the double helix separate, and each serves as a template for the precise ordering of nucleotides into new complementary strands.

Inheritance is based on precise replication of DNA

The DNA molecules of cells actually consist of two chains of nucleotides that spiral around an imaginary axis to form a **double helix.** James Watson and Francis Crick, working at Cambridge University, first proposed the double helix as the three-dimensional structure of DNA in 1953 (FIGURE 5.28). The two sugar-phosphate backbones are on the outside of the helix, and the nitrogenous bases are paired in the interior of the helix. The two chains of nucleotides, or strands, as they are called, are held together by hydrogen bonds between the paired bases. Most DNA molecules are very long—some several millimeters—with thousands or even millions of base pairs holding the two chains together. One long DNA molecule represents a large number of genes, each one a particular segment of the double helix.

Only certain bases in the double helix are compatible with each other. Adenine (A) always pairs with thymine (T), and guanine (G) always pairs with cytosine (C). If we were to read the sequence of bases along one strand as we traveled the length of the double helix, we would know the sequence of bases along the other strand. If a stretch of one strand has the base sequence AGGTCCG, then the base-pairing rules tell us that the same stretch of the other strand must have the sequence TCCAGGC. The two strands of the double helix are complementary, each the predictable counterpart of the other. It is this feature of DNA that makes possible the precise copying of genes that is responsible for inheritance. As a cell prepares to divide, the two strands of each gene separate. Each existing strand then serves as a template to order nucleotides into a new complementary strand. The identical copies of each gene are then distributed to the two cells formed by the division of one cell. Thus, the structure of DNA accounts for its function in transmitting genetic information whenever a cell reproduces.

We can use DNA and proteins as tape measures of evolution

Genes (DNA) and their products (proteins) document the hereditary background of an organism. The linear sequences of nucleotides in DNA molecules are passed from parents to offspring, and these DNA sequences determine the amino acid sequences of proteins. Siblings have greater similarity in their DNA and proteins than do unrelated individuals of the same species. If the evolutionary view of life is valid, we should be able to extend this concept of "molecular genealogy" to relationships *between* species: We should expect two species that appear to be closely related based on fossil and anatomical evidence to also share a greater proportion of their DNA and proteins than do more distantly related species. That

TABLE 5.2

Evolutionary Relationships and Similarities in a Polypeptide Chain of Hemoglobin

SPECIES	NUMBER OF AMINO ACID DIFFERENCES IN THE β CHAIN OF HEMOGLOBIN, COMPARED TO HUMAN HEMOGLOBIN (TOTAL CHAIN LENGTH = 146 AMINO ACIDS)
Human	0
Gorilla	1
Gibbon	2
Rhesus monkey	8
Mouse	27
Frog	67

is the case. For example, TABLE 5.2 compares a polypeptide chain of the protein hemoglobin in humans with the hemoglobin in eight other vertebrates. In this chain of 146 amino acids, humans and gorillas differ in just one amino acid. More distantly related species have chains that are less similar. Molecular biology has added a new tape measure to the toolkit biologists use to assess evolutionary kinship.

* * *

We have concluded our survey of macromolecules, but not our study of the chemistry of life. Applying the reductionist strategy, we have examined the architecture of molecules, but we have yet to explore the dynamic interactions between molecules that result in the chemical changes in cells collectively referred to as metabolism. Chapter 6, the concluding chapter of this unit, will take us another step up the hierarchy of biological order by introducing the fundamental principles of metabolism.

REVIEW OF KEY CONCEPTS (with page numbers and key figures)

- Most macromolecules are polymers (pp. 63–64, FIGURE 5.2)
  - Carbohydrates, lipids, proteins, and nucleic acids are the four major classes of organic compounds in cells. Some of these compounds are *very* large and are called macromolecules.
  - Cells can combine small organic molecules into large macromolecules to form a higher level in the hierarchy of biological order.
  - Most macromolecules are polymers, chains of identical or similar building blocks called monomers.
  - Monomers form larger molecules by condensation reactions (dehydration), in which one monomer donates a hydroxyl group and the other a hydrogen atom, forming a water molecule.
  - Polymers can be disassembled to monomers by the reverse process, called hydrolysis.
- A limitless variety of polymers can be built from a small set of monomers (p. 65)
  - Although there is a limited number of monomers common to organisms, each organism is unique because of the specific arrangement of these monomers into polymers.
- Organisms use carbohydrates for fuel and building material (pp. 65–70, FIGURES 5.3 and 5.7)
  - Carbohydrates are sugars and their derivatives.
  - Monosaccharides are the simplest carbohydrates. They are used directly for fuel, converted to other types of organic molecules, or used as monomers for polymers.
  - Disaccharides consist of two monosaccharide monomers connected by a glycosidic bond.
  - Polysaccharides are macromolecules that may consist of thousands of monosaccharide monomers connected by glycosidic bonds. Starch in plants and glycogen in animals are both storage polymers of glucose. Cellulose is an important structural polysaccharide in the cell walls of plants.
- Lipids are mostly hydrophobic molecules with diverse functions (pp. 70–73, FIGURE 5.10)
  - Lipids all share the property of being wholly or partly hydrophobic.
  - Fats are high-energy, compact storage molecules also known as triacylglycerols. They are constructed by joining a glycerol molecule to three fatty acids.
  - Saturated fatty acids have the maximum number of hydrogen atoms because of single bonding between all the carbons. Unsaturated fatty acids (present in oils) have one or more double bonds between the carbons.
  - Phospholipids substitute the third fatty acid of fats with a negatively charged phosphate group, which may be joined, in turn, to another small molecule. Such bonding introduces hydrophilic behavior to this part of the molecule. Phospholipids are ideally suited for construction of cell membranes.
  - Steroids, such as cholesterol and the sex hormones, are also classified as lipids.
- Proteins are the molecular tools for most cellular functions (pp. 73–74, TABLE 5.1)
  - A protein consists of one or more polypeptide chains folded into a specific three-dimensional conformation.

- A polypeptide is a polymer of amino acids connected in a specific sequence (p. 74, FIGURE 5.16)
  - Proteins are constructed from 20 different amino acids.
  - Each amino acid is defined by its side chain, the one variable part of an amino acid.
  - The carboxyl and amino groups of adjacent amino acids link together in a peptide bond, forming long polymers.
- A protein's function depends on its specific conformation (pp. 74–83, FIGURE 5.24)
  - Protein conformation can be described by three or four hierarchical levels. Primary structure is the first level and describes the unique sequence of amino acids.
  - Secondary structure describes how the primary structure is folded into localized configurations, the α helix and the pleated sheet, which result from hydrogen bonding between peptide linkages.
  - Tertiary structure describes the less regular contortions of the molecule caused by the involvement of side groups in hydrophobic interactions, hydrogen bonds, ionic bonds, and disulfide bridges.
  - Proteins made of more than one polypeptide chain also show a specific arrangement of their constituent subunits in a quaternary level of structure.
  - The structure and function of a protein are sensitive to conditions such as pH, salt concentration, and temperature. Changing these conditions can cause denaturation, or alteration of the protein's shape.
  - Protein shape is ultimately determined by its primary structure. Biochemists are looking for rules that will predict protein folding.
- Nucleic acids store and transmit hereditary information (pp. 83–84, FIGURE 5.26)
  - DNA stores information for the synthesis of specific proteins. RNA links this genetic information to the protein-synthesizing machinery.
- A DNA strand is a polymer with an information-rich sequence of nucleotides (pp. 84–85, FIGURE 5.27)
  - DNA is a polymer of nucleotides, monomers consisting of a pentose (five-carbon sugar) covalently bonded to a phosphate group and to one of four different kinds of nitrogenous bases (A, G, C, or T).
  - In the formation of a DNA strand, nucleotides join by phosphodiester linkages to form a sugar-phosphate backbone from which the nitrogenous bases project.
  - The sequence of bases along a DNA strand specifies the amino acid sequence of a particular protein.
- Inheritance is based on precise replication of DNA (p. 85, FIGURE 5.28)
  - DNA is a helical, double-stranded macromolecule with bases projecting into the interior of the molecule. Because A always hydrogen-bonds to T, and C to G, the nucleotide sequences of the two strands are complementary. One strand can serve as a template for the formation of the other. This unique feature of DNA provides a mechanism for the continuity of life.
- We can use DNA and proteins as tape measures of evolution (pp. 85–86, TABLE 5.2)
  - Molecular comparisons are helping biologists sort out the evolutionary connections among species.

SELF QUIZ

1. Which of the following terms includes all others in the list?
   a. monosaccharide
   b. disaccharide
   c. starch
   d. carbohydrate
   e. polysaccharide

2. The molecular formula for glucose is $C_6H_{12}O_6$. What would be the molecular formula for a polymer made by linking ten glucose molecules together by condensation reactions? Explain your answer.
   a. $C_{60}H_{120}O_{60}$
   b. $C_6H_{12}O_6$
   c. $C_{60}H_{102}O_{51}$
   d. $C_{60}H_{100}O_{50}$
   e. $C_{60}H_{111}O_{51}$

3. There are two ring forms of glucose (alpha and beta) because
   a. the two forms are made from two structural isomers of glucose
   b. they arise from different linear (nonring) formulas for glucose
   c. different carbons of the linear structure join to form the rings
   d. when the ring closes, the hydroxyl group at the point of closure can be trapped in either one of two possible positions
   e. one is an aldose and the other is a ketose

4. Choose the pair of terms that completes this sentence: Nucleotides are to __________ as __________ are to proteins.
   a. nucleic acids; amino acids
   b. amino acids; polypeptides
   c. glycosidic linkages; polypeptide linkages
   d. genes; enzymes
   e. polymers; polypeptides

5. Which of the following statements concerning *unsaturated* fats is correct?
   a. They are more common in animals than in plants.
   b. They have double bonds in the carbon chains of their fatty acids.
   c. They generally solidify at room temperature.
   d. They contain more hydrogen than saturated fats having the same number of carbon atoms.
   e. They have fewer fatty acid molecules per fat molecule.

6. Human sex hormones are classified as
   a. proteins
   b. steroids
   c. amino acids
   d. triacylglycerols
   e. carbohydrates

7. For a protein to have a quaternary structure, it *must*
   a. have four amino acids
   b. consist of two or more polypeptide subunits
   c. consist of four polypeptide subunits
   d. have at least four disulfide bridges
   e. exist in several alternative conformational states

8. What does a protein lose when it denatures?
   a. its primary structure
   b. its three-dimensional shape
   c. its peptide bonds
   d. its sequence of amino acids
   e. its amino acid side chains

9. Which of the following is a complication contributing to the difficulty of the protein-folding problem?
   a. A specific protein has several alternative amino acid sequences.
   b. There are no methods for revealing the three-dimensional shape of a protein.
   c. Intermediate stages in folding affect the final shape.
   d. It is impossible to determine the precise primary structure of a protein.
   e. One must identify the gene that codes for a particular protein before the protein's folding pattern can be determined.

10. Which of these terms includes all others in the list?
   a. nucleic acid
   b. nucleotide
   c. nitrogenous base
   d. purine
   e. pyrimidine

CHALLENGE QUESTIONS

1. Most amino acids can exist in two forms:

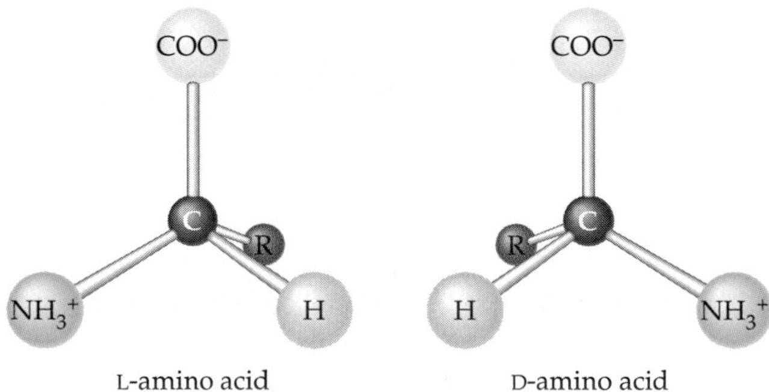

   a. Apply what you learned in Chapter 4 to explain the structural difference between these two molecules.
   b. When an organic chemist synthesizes amino acids, a mixture of these two forms is made; the result was probably the same when the first amino acids formed by abiotic synthesis in the "primordial soup" of the early Earth. However, with very few exceptions, only the L form of amino acids occurs in proteins. This is one example of molecular "handedness" in life. What are the possible advantages of life's becoming locked into the exclusive use of one of these two amino acid versions? Speculate on the possible evolutionary significance of this "handedness."

2. A particular small polypeptide is nine amino acids long. Using three different enzymes to hydrolyze the polypeptide at various sites, we obtain the following five fragments (N denotes the amino terminus of the chain): Ala-Leu-Asp-Tyr-Val-Leu; Tyr-Val-Leu; N-Gly-Pro-Leu; Asp-Tyr-Val-Leu; N-Gly-Pro-Leu-Ala-Leu. Determine the primary structure of this polypeptide.

SCIENCE, TECHNOLOGY, AND SOCIETY

1. Some amateur and professional athletes take anabolic steroids to help them "bulk up" or build strength. The health risks of this practice are extensively documented. Apart from health considerations, how do you feel about the use of chemicals to enhance athletic performance? Is an athlete who takes anabolic steroids cheating, or is his or her use of such chemicals just part of the preparation required to succeed in a competitive sport? Defend your answer.

2. Molecular biologists have learned how to splice the genes of one species into the DNA of a different species. For example, the gene that codes for the protein insulin can be transferred from a human cell to a bacterial cell, resulting in a culture of bacteria that produce human insulin, which is then used to treat diabetes. Some critics think we should not manipulate Mother Nature and refer to this new biotechnology as "playing God." Given that our ability to modify nature is nothing new, do you feel that there is a fundamental ethical difference between genetic engineering and other technologies? In your view, what determines whether a particular technology is acceptable?

FURTHER READING

Arthur, C. "Hunt Is on for Money-Spinning Spider Gene." *New Scientist,* May 7, 1994. Molecular biology applied to silk technology.

Birge, R. R. "Protein-Based Computers." *Scientific American,* March 1995. Imitating nature to build better machines.

Craig, E. A. "Chaperones: Helpers Along the Pathways to Protein Folding." *Science,* June 25, 1993. A step closer to understanding how a functional protein is built.

Dushesne, L. C., and D. W. Larson. "Cellulose and the Evolution of Plant Life." *Bioscience,* April 1989. The chemistry and natural history of the most abundant organic molecule in the biosphere.

Gibbons, A. "Geneticists Trace the DNA Trail of the First Americans." *Science,* January 15, 1993. Applying the "DNA tape measure" to a problem in anthropology.

Hoberman, J. M., and C. E. Yesalis. "The History of Synthetic Testosterone." *Scientific American,* February 1995. Anabolic steroids and sports.

Horgan, J. "Stubbornly Ahead of His Time." *Scientific American,* March 1993. A profile of Linus Pauling, a controversial giant of twentieth-century science.

Johnson, J. "First Fold Your Protein." *New Scientist,* December 10, 1994. Potential applications of solving the protein-folding problem.

Mathews, C. K., and K. E. van Holde. *Biochemistry,* 2nd ed. Menlo Park, CA: Benjamin/Cummings, 1996. Excellent explanations and beautiful art.

The living cell is a chemical industry in miniature, where thousands of reactions occur within a microscopic space. Sugars are converted to amino acids, and vice versa. Small molecules are assembled into polymers, which may later be hydrolyzed as the needs of the cell change. Many cells export chemical products that are used in other parts of the organism. The chemical process known as cellular respiration drives the cellular economy by extracting the energy stored in sugars and other fuels. Cells apply this energy to perform various types of work. For example, bacteria in the "headlight" of the fish (Photoblepharon palpebratus) *in the photograph at the right convert the energy stored in food to light, a process called bioluminescence. In this metabolic function and all others, the myriad reactions going on in the cell are precisely coordinated. In its complexity, its efficiency, its integration, and its responsiveness to subtle changes, the cell is peerless as a chemical institution. The concepts of metabolism you learn in this chapter will help you understand how organisms work.*

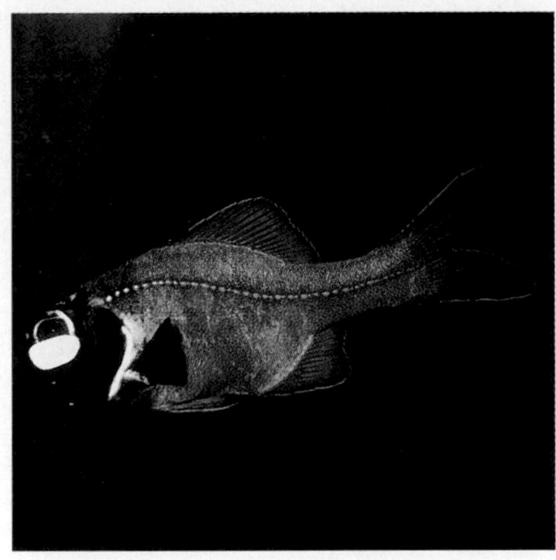

CHAPTER 6

AN INTRODUCTION TO METABOLISM

KEY CONCEPTS

- The chemistry of life is organized into metabolic pathways
- Organisms transform energy
- The energy transformations of life are subject to two laws of thermodynamics
- Organisms live at the expense of free energy
- ATP powers cellular work by coupling exergonic to endergonic reactions
- Enzymes speed up metabolic reactions by lowering energy barriers: *an overview*
- Enzymes are substrate-specific: *a closer look*
- The active site is an enzyme's catalytic center: *a closer look*
- A cell's chemical and physical environment affects enzyme activity: *a closer look*
- Metabolic order emerges from the cell's regulatory systems and structural organization
- The theme of emergent properties is manifest in the chemistry of life: *a review*

The chemistry of life is organized into metabolic pathways

Metabolism (from the Greek *metabole*, "change") is the totality of an organism's chemical processes. It is an emergent property of life that arises from specific interactions between molecules within the orderly environment of the cell. We can think of a cell's metabolism as an elaborate road map of the thousands of reactions that occur in that cell (FIGURE 6.1). These reactions are arranged in intricately branched metabolic pathways, which alter molecules by a series of steps. Enzymes route matter through the metabolic pathways by selectively accelerating each step. Analogous to the red, green, and yellow lights that control the flow of traffic and prevent snarls, mechanisms that regulate enzymes balance metabolic supply and demand, averting deficits and surpluses of chemicals.

As a whole, metabolism is concerned with managing the material and energy resources of the cell. Some metabolic pathways release energy by breaking down complex molecules to simpler compounds. These degradative processes are called **catabolic pathways.** A major thoroughfare of catabolism is cellular respiration, in which the sugar glucose and other organic fuels are broken down to carbon dioxide and water. Energy that was stored in the organic molecules becomes available to do the work of the cell, such as the bioluminescence

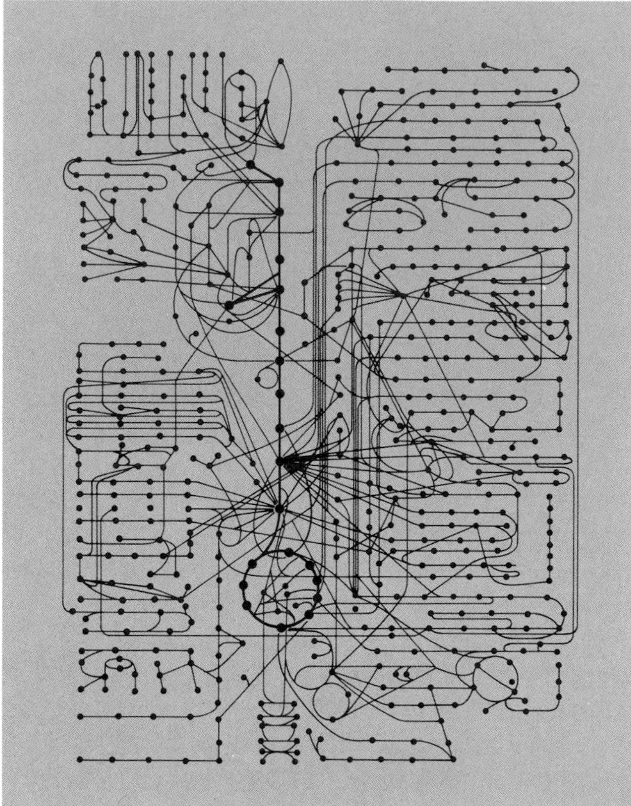

FIGURE 6.1
The metabolic map: an overview of the complexity of metabolism. This schematic diagram traces only a few hundred of the thousands of metabolic reactions that occur in a cell. The dots represent molecules, and the lines represent the chemical reactions that transform them. The reactions proceed in stepwise sequences called metabolic pathways, each step catalyzed by a specific enzyme.

illustrated in the photograph on p. 89. There are also **anabolic pathways,** which consume energy to build complicated molecules from simpler ones. An example of anabolism is the synthesis of a protein from amino acids. Catabolic and anabolic pathways are the downhill and uphill avenues of the metabolic map. The metabolic pathways intersect in such a way that energy released from the "downhill" reactions of catabolism can be used to drive the "uphill" reactions of the anabolic pathways. This transfer of energy from catabolism to anabolism is called energy coupling.

In this chapter, we will focus on the mechanisms common to metabolic pathways. Since energy is fundamental to all metabolic processes, a basic knowledge of energy is necessary to understand how the living cell works. We will use some nonliving examples to study energy. But keep in mind that the same concepts demonstrated by these examples also apply to **bioenergetics,** the study of how organisms manage their energy resources. An understanding of energy is as important for students of biology as it is for students of physics, chemistry, and engineering.

Organisms transform energy

Energy is the capacity to do work—that is, to move matter against opposing forces, such as gravity and friction. Put another way, energy is the ability to rearrange a collection of matter. For example, you expend energy to turn the pages of this book. Energy exists in various forms, and the work of life depends on the ability of cells to transform energy from one type into another.

Anything that moves possesses a form of energy called **kinetic energy**—the energy of motion. Moving objects perform work by imparting motion to other matter: A pool player uses the motion of the cue stick to push the cue ball, which in turn moves the other balls; water gushing through a dam turns turbines; electrons flowing along a wire run household appliances; the contraction of leg muscles pushes bicycle pedals. Heat, or thermal energy, is kinetic energy that results from the random movement of molecules. Light also represents kinetic energy, which can be harnessed to perform work, such as powering photosynthesis in green plants.

A resting object not presently at work may still possess energy, which, remember, is the *capacity* to do work. Stored energy, or **potential energy,** is energy that matter possesses because of its location or structure. Water behind a dam, for instance, stores energy because of its altitude. Chemical energy, a form of potential energy especially important to biologists, is stored in molecules because of the structural arrangement of the atoms in those molecules.

Energy can be converted from one form to another. Consider, for example, the playground scene in FIGURE 6.2. The girl at the bottom of the slide transformed kinetic energy to potential energy when she climbed the ladder up to the slide. This stored energy was converted back to kinetic energy as she slid down. It was another source of potential energy, the chemical energy in the food she ate for breakfast, that enabled the girl to climb the ladder in the first place.

Chemical energy can be tapped when chemical reactions rearrange the atoms of molecules in such a way that potential energy stored in the molecules is converted to kinetic energy. This transformation occurs, for example, in the engine of an automobile when the hydrocarbons of gasoline react explosively with oxygen, releasing the energy that pushes the pistons. Similarly, chemical energy fuels organisms. Cellular respiration and other catabolic pathways unleash energy stored in sugar and other complex molecules and make that energy available for cellular work. Each child who climbed the ladder in FIGURE 6.2 transformed some of the chemical energy that was stored in the organic molecules of food to the kinetic energy of movements. The chemical energy stored in

FIGURE 6.2
Transformations between kinetic and potential energy. The children in this playground scene have more potential energy at the top of the slide (because of the effect of gravity) than they do at the bottom. They convert kinetic energy to potential energy when they climb the ladder up to the slide, and convert stored (potential) energy back to kinetic energy during their descent.

these fuel molecules had itself been converted from light energy by plants during photosynthesis. Organisms are energy transformers.

The energy transformations of life are subject to two laws of thermodynamics

The study of the energy transformations that occur in a collection of matter is called **thermodynamics.** Scientists use the word *system* to denote the collection of matter under study and refer to the rest of the universe—everything outside the system—as the *surroundings.* A *closed system,* such as that approximated by liquid in a thermos bottle, is isolated from its surroundings. In an *open system,* energy can be transferred between the system and its surroundings. Organisms are open systems. They absorb light energy or chemical energy in the form of organic molecules and release heat and metabolic waste products, such as carbon dioxide, to the surroundings. Two laws of thermodynamics govern energy transformations in organisms and all other collections of matter.

According to the **first law of thermodynamics,** the energy of the universe is constant. *Energy can be transferred and transformed, but it can be neither created nor destroyed.* The first law is also known as *conservation of energy.* The electric company does not manufacture energy, but merely converts it to a form that is convenient to use. By converting light to chemical energy, a green plant acts as an energy transformer, not an energy producer. What happens to energy after it has performed work in a machine or an organism? If energy cannot be destroyed, what prevents organisms from behaving like closed systems and recycling their energy? The second law answers these questions.

The **second law of thermodynamics** can be stated many ways. Let's begin with the following interpretation: Every energy transfer or transformation makes the universe more disordered. Scientists use a quantity called **entropy** as a measure of disorder, or randomness. The more random a collection of matter is, the greater its entropy. We can now restate the second law as follows: *Every energy transfer or transformation increases the entropy of the universe.* There is an unstoppable trend toward randomization. In many cases, increased entropy is evident in the physical disintegration of a system's organized structure. For example, you can observe this increasing entropy in the gradual decay of an unmaintained building. Much of the increasing entropy of the universe is less apparent, however, because it takes the form of an increasing amount of heat, which is the energy of random molecular motion.

In most energy transformations, ordered forms of energy are at least partly converted to heat. Only about 25% of the chemical energy stored in the fuel tank of an automobile is transformed into the motion of the car; the remaining 75% is lost from the engine as heat, which dissipates rapidly through the surroundings. Similarly, the children in FIGURE 6.2 convert only a fraction of the energy stored in their food to the kinetic energy of ladder climbing and other play. In performing various kinds of work, living cells unavoidably convert organized forms of energy to heat. (This can make a room crowded with people uncomfortably warm.)

In machines and organisms, even energy that performs useful work is eventually converted to heat. The organized energy of an automobile's forward movement becomes heat when the friction of the brakes and tires stops the car. Conversion to heat is the fate of *all* the chemical energy a child uses to climb a slide: Metabolic breakdown of food generates heat during the climb, and the fraction of energy temporarily stored as gravitational potential energy is converted to heat on the way down, as friction between child and slide warms the surrounding air.

Conversion of other forms of energy to heat does not violate the first law of thermodynamics. Energy has been conserved, because heat is a form of energy, though energy in its most random state. By combining the first and second laws of thermodynamics, we can conclude that the *quantity* of energy in the universe is constant, but its *quality* is not. In a sense, heat is the lowest grade of energy. It is the uncoordinated movement of molecules, which many systems cannot harness in order to perform work. A system can only put heat to work when there is a temperature difference that results in the heat flowing from a warmer location to a cooler one. If temperature is uniform throughout a system, as it is in a living cell, then the only use for heat energy is to warm a body of matter, such as an organism.

Thus, an organism takes in organized forms of matter and energy from the surroundings and replaces them with less ordered forms. For example, an animal obtains starch, proteins, and other complex molecules from the food it eats. As catabolic pathways break these molecules down, the animal releases carbon dioxide and water, relatively small, simple molecules that store less chemical energy than the food. The depletion of chemical energy is accounted for by heat generated during metabolism. On a larger scale, energy flows into an ecosystem in the form of light and leaves in the form of heat. Living systems increase the entropy of their surroundings, as predicted by thermodynamic law.

How can we reconcile the second law of thermodynamics—the unstoppable increase in the entropy of the universe—with the orderliness of life, which is one of this book's themes? The key is to remember another theme: Organisms are open systems that exchange energy and materials with their surroundings. Cells create ordered structures from less organized starting materials. For example, amino acids are ordered into the specific sequences of polypeptide chains. During the early history of life, complex organisms evolved from simpler ancestors (FIGURE 6.3). However, this high degree of organization in no way violates the second law. The entropy of a particular system, such as an organism, may actually decrease, as long as the total entropy of the *universe*—the system plus its surroundings—increases. Thus, organisms are islands of low entropy in an increasingly random universe. The evolution of biological order is perfectly consistent with the laws of thermodynamics.

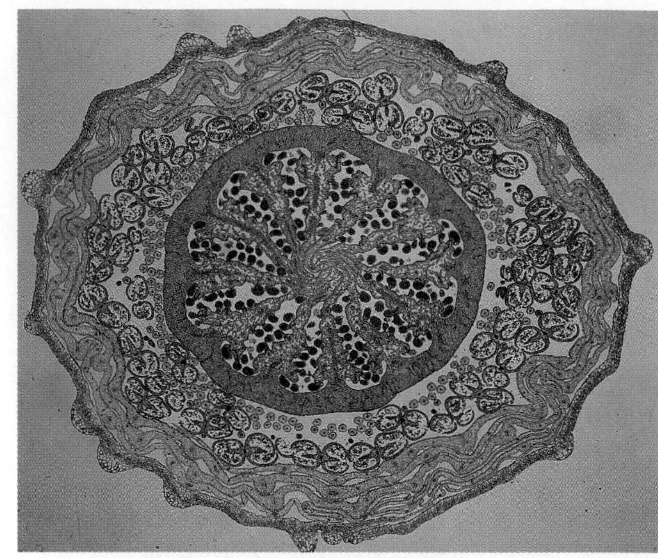

(a)

(b)

FIGURE 6.3

The evolution of biological order is consistent with thermodynamic law. (**a**) Order is a characteristic of life. It is evident, for example, in this cross section of a poppy blossom (LM). Organisms decrease their entropy when they order raw materials, such as organic monomers, into macromolecules and then organize these macromolecules into cellular structures. (**b**) Biological order has, at times, also increased over the grander scale of geological time.This horsetail fossil represents one group of complex plants that evolved from simpler ancestors. An evolutionary interpretation of the fossil record does not violate thermodynamic law. The second law requires only that processes increase the entropy of the universe. Open systems can increase their order at the expense of the order of their surroundings.

Organisms live at the expense of free energy

How can we predict what can and cannot occur in nature? How can we distinguish the possible from the impossible? We know from experience that certain events occur spontaneously and others do not. For instance, we know that water flows downhill, that objects of opposite charge move toward each other, that an ice cube melts at room temperature, and that a sugar cube dissolves in water. Explaining *why* these processes occur spontaneously is tricky.

Let's begin by defining a spontaneous process as a change that can occur without outside help. A spontaneous change can be harnessed in order to perform work. The downhill flow of water can be used to turn a turbine in a power plant, for example. A process that cannot occur on its own is said to be nonspontaneous; it will happen only if an external energy source is added. Water moves uphill only when a windmill or some other machine pumps the water against gravity, and a cell must expend energy to synthesize a protein from amino acids.

When a spontaneous process occurs in a system, the stability of that system increases. Unstable systems tend to change in such a way that they become more stable. A body of elevated water, such as a reservoir, is less stable than the same water at sea level. A system of charged particles is less stable when opposite charges are apart than when they are together. In situations less familiar to us, how can we predict which changes lead to greater stability in a system? That is, which changes are spontaneous? You have already learned that a process can only occur spontaneously if it increases the disorder (entropy) of the universe. This principle is helpful in theory, but it does not give us a practical criterion to apply to biological systems because it requires that we measure changes in the surroundings. We need some standard for spontaneity that is based on the system alone. That criterion is called free energy.

Free Energy: A Criterion for Spontaneous Change

The concept of free energy is not easy to grasp, but the effort is worthwhile because we can apply the idea to many biological problems. ***Free energy** is the portion of a system's energy that can perform work when temperature is uniform throughout the system, as in a living cell.* It is called *free* energy because it is available for work, not because it can be spent without cost to the universe. In fact, you will soon understand that organisms can only live at the expense of free energy acquired from the surroundings.

A system's quantity of free energy is symbolized by the letter G. There are two components to G: the system's total energy (symbolized by H) and its entropy (symbolized by S). Free energy is related to these factors in the following way:

$$G = H - TS$$

T stands for absolute temperature (in Kelvin units, K, equal to °C + 273; see Appendix Three). Notice that temperature amplifies the entropy term of the equation. This makes sense if you remember that temperature measures the intensity of random molecular motion (heat), which tends to disrupt order. What does this equation tell us about free energy? Not all the energy stored in a system (H) is available for work. The system's disorder, the entropy factor, is subtracted from total energy in computing the maximum capacity of the system to perform useful work. We are then left with free energy, which is somewhat less than the system's total energy.

How does the concept of free energy help us determine whether a particular process can occur spontaneously? Think of free energy, G in the above equation, as a measure of a system's instability—its tendency to change to a more stable state. Systems that are rich in energy, such as stretched springs or separated charges, are unstable; so are highly ordered systems, such as complex molecules. Thus, those systems that tend to change spontaneously to a more stable state are those that have high energy, low entropy, or both. The free-energy equation weighs these two factors, which are consolidated in the system's G content. Now we can state a versatile criterion for spontaneous change: *In any spontaneous process, the free energy of a system decreases.*

The change in free energy as a system goes from a starting state to a different state is represented by ΔG:

$$\Delta G = G_{final\ state} - G_{starting\ state}$$

Or, put another way:

$$\Delta G = \Delta H - T\Delta S$$

For a process to occur spontaneously, the system must either give up energy (a decrease in H), give up order (an increase in S), or both. When these changes in H and S are tallied, ΔG must have a negative value. The greater this decrease in free energy, the greater the maximum amount of work the spontaneous process can perform. This is a formal, mathematical way of stating the obvious: Nature runs "downhill" (downhill in the metaphorical sense of a loss in useful energy—the capacity to perform work).

Free Energy and Equilibrium

There is an important relationship between free energy and equilibrium, including chemical equilibrium. Recall from Chapter 2 that most chemical reactions are reversible and proceed until the forward and backward reactions occur at the same rate. The reaction is then said to be at chemical equilibrium, and there is no further change in the concentration of products or reactants. As a reaction proceeds toward equilibrium, the free energy of the mixture of reactants and products decreases. Free energy increases when a reaction is somehow pushed away from equilibrium. For a reaction at equilibrium, $\Delta G = 0$, because there is no net change in the system. We can think of equilibrium as an energy valley. A chemical reaction or physical process at equilibrium performs no work. A process is spontaneous and can perform work when sliding toward equilibrium. Movement away from equilibrium is nonspontaneous; it can occur only with the help of an outside energy source. The relationships between free energy, equilibrium, and work are summarized in FIGURE 6.4. We can now apply the free-energy concept more specifically to the chemistry of life.

Free Energy and Metabolism

Exergonic and Endergonic Reactions in Metabolism. Based on their free-energy changes, chemical reactions can be classified as either exergonic (meaning "energy outward") or endergonic (meaning "energy inward"). An **exergonic reaction** proceeds with a net release of free energy. Since the chemical mixture loses free energy, ΔG is negative for an exergonic reaction. In other words, exergonic reactions are those that occur spontaneously. The magnitude of ΔG for an exergonic reaction is the maximum amount of work the reaction can perform. We can use cellular respiration as an example:

$$C_6H_{12}O_6 + 6\,O_2 \longrightarrow 6\,CO_2 + 6\,H_2O$$
$$\Delta G = -686 \text{ kcal/mol } (-2870 \text{ kJ/mol})$$

For each mole (180 g) of glucose broken down by respiration, 686 kilocalories (or 2870 kilojoules) of energy are made available for work. Because energy must be conserved, the chemical products of respiration store 686 kcal less free energy than the reactants. The products are, in a sense, the spent exhaust of a process that tapped most of the free energy stored in the sugar molecules (FIGURE 6.4c).

An **endergonic reaction** is one that absorbs free energy from its surroundings. Because this kind of reaction stores more free energy in the molecules, ΔG is positive. Such reactions are nonspontaneous, and the magnitude of ΔG is the minimum quantity of energy required to drive the reaction. If a chemical process is exergonic (downhill) in one direction, then the reverse process must be endergonic (uphill). A reversible process cannot travel downhill in both directions. If $\Delta G = -686$ kcal/mol for respiration, for photosynthesis to produce sugar from carbon dioxide and water, $\Delta G = +686$ kcal/mol. Sugar production in the leaf cells of a plant is steeply endergonic, an uphill process powered by the absorption of light energy from the sun.

Metabolic Disequilibrium. The chemical reactions of metabolism are reversible and would reach equilibrium if they occurred in the isolation of a test tube. Because chemical systems at equilibrium have a ΔG of zero and can do no work, a cell that has reached equilibrium is dead! In fact, metabolic disequilibrium is one of the defining features of life. We'll use the catabolic pathway of respiration as an example. Some of the reversible reactions of respiration are pulled in one direction and kept

More free energy
Less stable
Greater work capacity

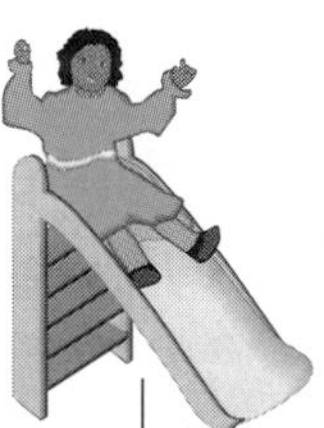

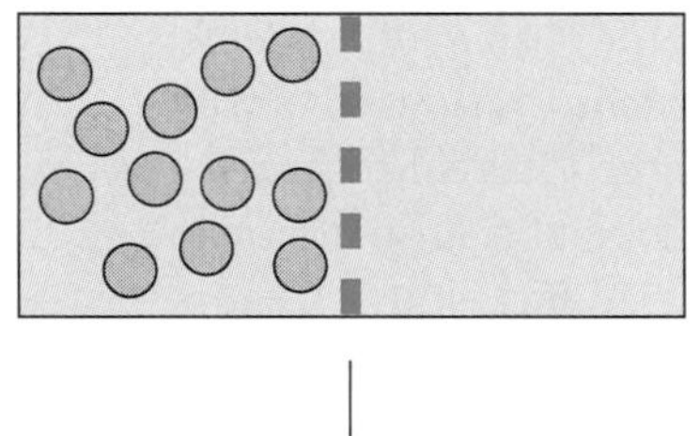

Free energy decreases; direction of spontaneous change; $\Delta G < 0$; change toward equilibrium; change can be harnessed to perform work

Less free energy
More stable
Less work capacity

(a)

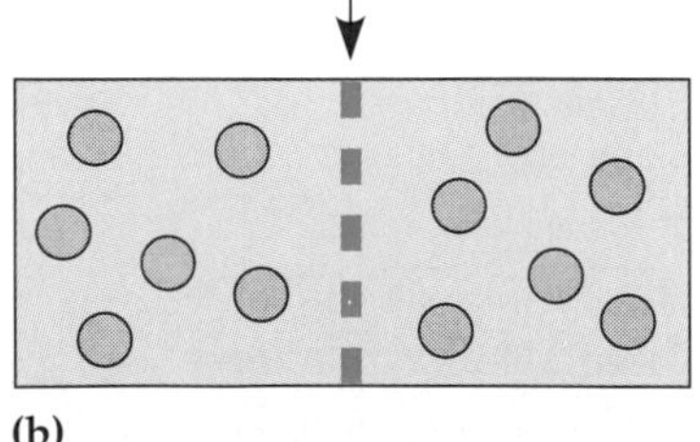

(b)

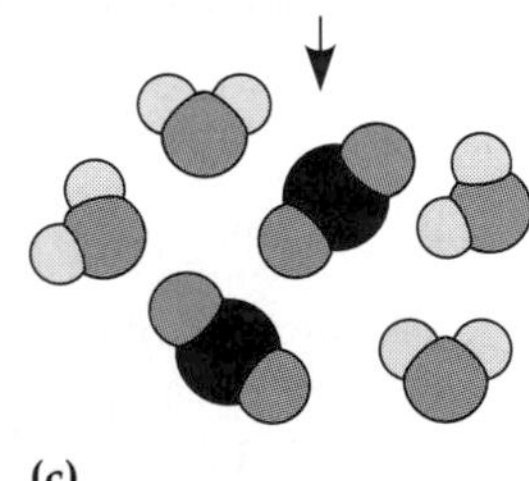

(c)

FIGURE 6.4
The relationship of free energy to stability, spontaneous change, equilibrium, and work. An unstable system is rich in free energy. It has a tendency to change spontaneously to a more stable state, and it is possible to harness this "downhill" change in order to perform work. (**a**) In this case, free energy is proportional to the girl's altitude. (**b**) The free-energy concept also applies on the molecular scale, in this case to the physical movement of molecules known as diffusion. Here, a membrane separates two aqueous compartments. Molecules of a particular solute are distributed unequally across the membrane. This ordered state is unstable; it is rich in free energy. If the solute molecules can cross the membrane, there will be a net movement (diffusion) of the molecules until they are equally concentrated in both compartments. (**c**) Chemical reactions also involve free energy. The sugar molecule on top is less stable than the simpler molecules below. When catabolic pathways break down complex organic molecules, a cell can harness the free energy stored in the molecules to perform work.

out of equilibrium. The key to sustaining this disequilibrium is that the product of one reaction does not accumulate, but instead becomes a reactant in the next step along the metabolic pathway (FIGURE 6.5). The overall sequence of reactions is pulled by the huge free-energy difference between glucose at the uphill end of respiration and carbon dioxide and water at the downhill end of the pathway. As long as the cell has a steady supply of glucose or other fuels and is able to expel the CO_2 waste to the surroundings, the cell does not reach equilibrium and continues to do the work of life.

We see once again how important it is to think of organisms as open systems. Sunlight provides a daily source of free energy for an ecosystem's plants and other photosynthetic organisms. Animals and other nonphotosynthetic organisms in an ecosystem depend on free-energy transfusions in the form of the organic products of photosynthesis.

Now that we have applied the free-energy concept to metabolism, we are ready to see how a cell actually performs the work of life. A key strategy in bioenergetics is **energy coupling,** the use of an exergonic process to drive an endergonic one. A molecule called ATP is responsible for driving most energy coupling in cells.

ATP powers cellular work by coupling exergonic to endergonic reactions

A cell does three main kinds of work:

1. *Mechanical work,* such as the beating of cilia, the contraction of muscle cells, the flow of cytoplasm within cells, and the movement of chromosomes during cellular reproduction.
2. *Transport work,* the pumping of substances across membranes against the direction of spontaneous movement.
3. *Chemical work,* the pushing of endergonic reactions that would not occur spontaneously, such as the synthesis of polymers from monomers.

In most cases, the immediate source of energy that powers cellular work is ATP.

The Structure and Hydrolysis of ATP

ATP (adenosine triphosphate) is closely related to one type of nucleotide found in nucleic acids. ATP consists of the nitrogenous base adenine bonded to ribose, a five-carbon sugar. In a nucleotide, one phosphate group is attached to the sugar (see FIGURE 5.27a). Adenosine

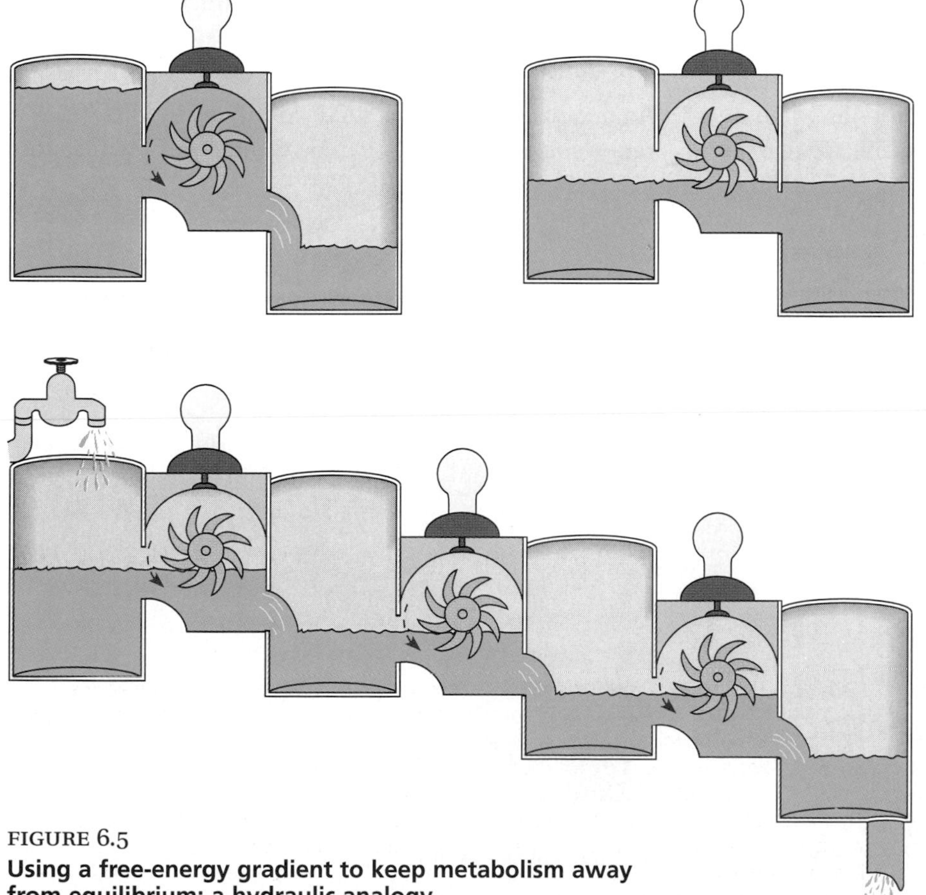

(a) This diagram portrays a system in which water generates electrical energy only while it is falling. Once the levels in the two containers are equal, the turbine ceases to turn and the light goes out. The individual steps of respiration, in isolation, would also come to equilibrium, and cellular work would cease.

(b) If there is a series of drops in water level, electrical energy can be generated at each drop. In respiration, there is a series of drops in free energy between glucose, the starting material, and the metabolic wastes at the end (carbon dioxide and water). The overall process never reaches equilibrium as long as the organism lives. A cell continues to acquire free energy in the form of glucose. The product of each reaction becomes the reactant for the next, and the metabolic wastes are expelled from the cell.

FIGURE 6.5
Using a free-energy gradient to keep metabolism away from equilibrium: a hydraulic analogy.

*tri*phosphate differs by having a chain of *three* phosphate groups attached to the ribose (FIGURE 6.6a).

The triphosphate tail of ATP is unstable, and the bonds between the phosphate groups can be broken by hydrolysis. When water hydrolyzes the terminal phosphate bond, a molecule of inorganic phosphate (abbreviated Ⓟ$_i$ throughout this book) is removed from ATP, which then becomes adenosine diphosphate, or ADP (FIGURE 6.6b). The reaction is exergonic and, under laboratory conditions, releases 7.3 kcal of energy per mole of ATP hydrolyzed:

$$\text{ATP} + H_2O \longrightarrow \text{ADP} + Ⓟ_i$$

$$\Delta G = -7.3 \text{ kcal/mol } (-31 \text{ kJ/mol})$$

This is the free-energy change measured at what are called standard conditions. However, the chemical and physical conditions in the cell do not conform to standard conditions. When the reaction occurs in the cellular environment rather than in a test tube, the actual ΔG is about −13 kcal/mole, 77% greater than the energy released by ATP hydrolysis under standard conditions.

Because their hydrolysis releases energy, the phosphate bonds of ATP are sometimes referred to as high-energy phosphate bonds, but the term is misleading. The phosphate bonds of ATP are not strong bonds, as the words "high-energy" imply. In fact, these bonds are relatively weak, and it is *because* they are unstable that their hydrolysis yields energy. The products of hydrolysis (ADP and Ⓟ$_i$) are more stable than ATP. When a system changes in the direction of greater stability—as when a compressed spring relaxes, for instance—the change is exergonic. Thus, the release of energy during the hydrolysis of ATP comes from the chemical change to a more stable condition, and not from the phosphate bonds themselves. Why are the phosphate bonds so fragile? If we reexamine the ATP molecule in FIGURE 6.6a, we can see that all three phosphate groups are negatively charged. These like charges are crowded together, and their repulsion contributes to the instability of this region of the ATP molecule. The triphosphate tail of ATP is the chemical equivalent of a loaded spring.

How ATP Performs Work

When ATP is hydrolyzed in a test tube, the release of free energy merely heats the surrounding water. In the cell, that would be an inefficient and dangerous use of a valuable energy resource. With the help of specific enzymes, the cell is able to couple the energy of ATP hydrolysis directly to endergonic processes by transferring a phosphate group from ATP to some other molecule. The recipient of the phosphate group is then said to be phosphorylated. The key to the coupling is the formation of this **phosphorylated intermediate,** which is more reactive (less stable) than the original molecule (FIGURE 6.7). Nearly all cellular work depends on ATP's energizing of other molecules by transferring phosphate groups. For instance, ATP powers the movement of muscles by transferring phosphate to contractile proteins.

The Regeneration of ATP

An organism at work uses ATP continuously, but ATP is a renewable resource that can be regenerated by the addi-

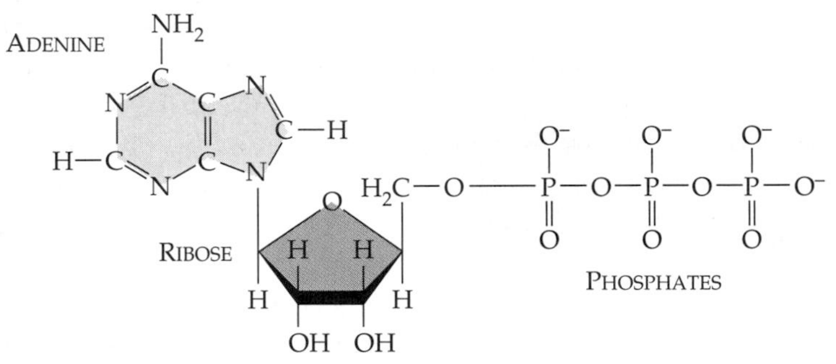

Adenine

NH_2

C—H

H—C

H_2C—O—P—O—P—O—P—O^-

O^- O^- O^-

O O O

Ribose

H H

H H

OH OH

Phosphates

(a) Adenosine triphosphate (ATP)

H_2O

Ⓟ—Ⓟ—Ⓟ

Ⓟ—Ⓟ + Ⓟ$_i$

(b)

ATP

Adenosine diphosphate (ADP) + Inorganic phosphate

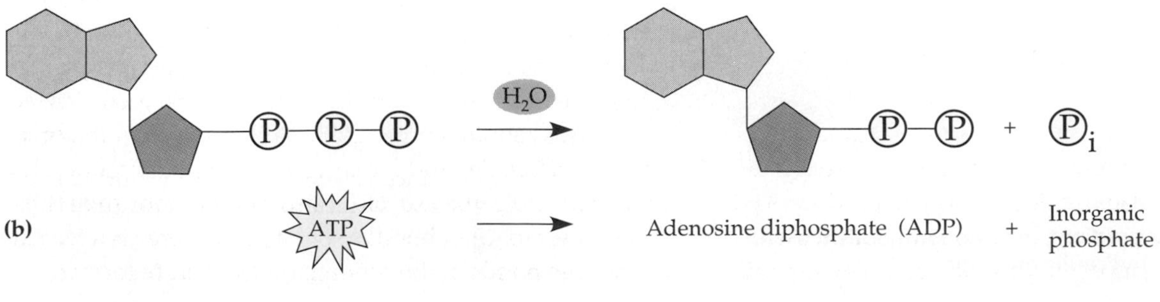

FIGURE 6.6

ATP. This figure illustrates (**a**) the structure of ATP and (**b**) the hydrolysis of ATP to yield ADP and inorganic phosphate. In the cell, most hydroxyl groups of phosphates are ionized (—O^-).

The "sunburst" symbol for ATP introduced in this figure will reappear throughout the book.

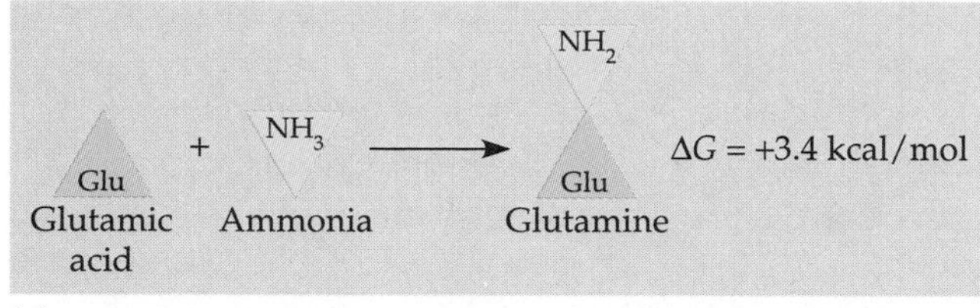

(a)

① Glu + ATP ⟶ Glu—Ⓟ + ADP

② Glu—Ⓟ + NH_3 ⟶ Glu—NH_2 + Ⓟi

(b)

Glu + NH_3	⟶	Glu—NH_2	$\Delta G = +3.4$ kcal/mol
ATP	⟶	ADP + Ⓟi	$\Delta G = -7.3$ kcal/mol
			Net $\Delta G = -3.9$ kcal/mol

(c)

FIGURE 6.7

Energy coupling by phosphate transfer. In this example, ATP hydrolysis is used to drive an endergonic reaction, the conversion of the amino acid glutamic acid (Glu) to another amino acid, glutamine (Glu—NH_2). (**a**) Without the help of ATP, the conversion is nonspontaneous. (**b**) As it actually occurs in the cell, the synthesis of glutamine is a two-step reaction driven by ATP. The formation of a phosphorylated intermediate couples the two steps. ① In the first step, ATP phosphorylates glutamic acid, transferring chemical instability to the amino acid. ② In the second step, ammonia displaces the phosphate group from the phosphorylated intermediate, forming glutamine. (**c**) We can calculate the free-energy change for the overall reaction by adding together ΔG for each step of the two-step reaction. Because the overall process is exergonic (has a negative ΔG), it occurs spontaneously.

tion of phosphate to ADP (FIGURE 6.8). The ATP cycle moves at an astonishing pace. For example, a working muscle cell recycles its entire pool of ATP about once each minute. That turnover represents 10 million molecules of ATP consumed and regenerated per second per cell. If ATP could not be regenerated by the phosphorylation of ADP, humans would consume nearly their body weight in ATP each day.

Since a reversible process cannot go downhill both ways, the regeneration of ATP from ADP is necessarily endergonic:

$$\text{ADP} + \text{Ⓟ}_i \longrightarrow \text{ATP} + H_2O$$
$$\Delta G = +7.3 \text{ kcal/mol (standard conditions)}$$

Catabolic (exergonic) pathways, especially cellular respiration, provide the energy to make ATP, an endergonic process. Plants can also use light energy to produce ATP.

Cellular respiration is a stepwise pathway by which enzymes decompose glucose and other complex organic

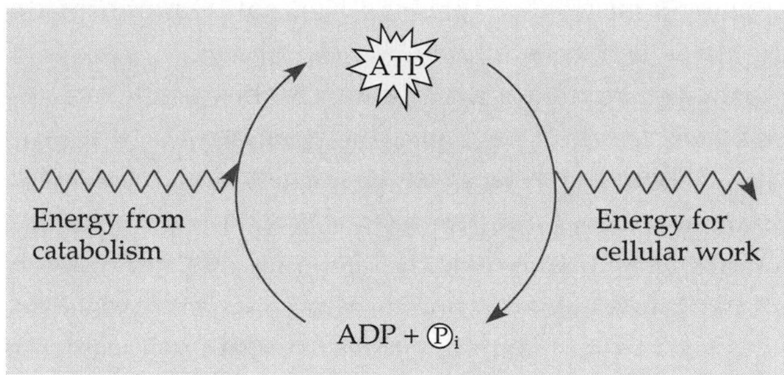

FIGURE 6.8

The ATP cycle. Energy released by breakdown reactions (catabolism) in the cell is used to phosphorylate ADP, regenerating ATP. Energy stored in ATP drives most cellular work. Thus, ATP couples the cell's energy-yielding processes to the energy-consuming ones.

molecules. The process is overwhelmingly exergonic, and the energy it releases drives phosphorylation of ADP to regenerate ATP. The ATP cycle is a turnstile through which energy passes during its transfer from catabolic to anabolic pathways.

Enzymes speed up metabolic reactions by lowering energy barriers: *an overview*

The laws of thermodynamics tell us what can and cannot happen but say nothing about the speed of these processes. A spontaneous chemical reaction may occur so slowly that it is imperceptible. For example, the hydrolysis of sucrose (table sugar) to glucose and fructose is exergonic, occurring spontaneously with a release of free energy ($\Delta G = -7$ kcal/mol). Yet a solution of sucrose dissolved in sterile water will sit for years at room temperature with no appreciable hydrolysis. However, if we add a small amount of the enzyme sucrase to the solution, then all the sucrose may be hydrolyzed within seconds. **Enzymes** are catalytic proteins. (Another class of biological catalysts, ribozymes, made of RNA, are discussed in Chapters 16 and 24.) A **catalyst** is a chemical agent that changes the rate of a reaction without being consumed by the reaction. In the absence of enzymes, chemical traffic through the pathways of metabolism would become hopelessly congested. What impedes a spontaneous reaction, and how does an enzyme lower the barrier?

A chemical reaction involves bond breaking and bond forming. When a reaction rearranges the atoms of molecules, existing bonds in the reactants must be broken and the new bonds of the products formed. These processes require exchanges of energy between the mixture of molecules and the surrounding environment. The reactant molecules must absorb energy from their surroundings for their bonds to break, and energy is released when the bonds of the product molecules are formed.

The initial investment of energy for starting a reaction—the energy required to break bonds in the reactant molecules—is known as the **free energy of activation,** or **activation energy,** abbreviated E_A in this book. It is usually provided in the form of heat absorbed by the reactant molecules from the surroundings. If the reaction is exergonic, E_A will be repaid with dividends, as the formation of new bonds releases more energy than was invested in the breaking of old bonds. FIGURE 6.9 graphs these energy changes for a hypothetical reaction that swaps portions of two reactant molecules:

$$AB + CD \longrightarrow AC + BD$$

The bonds of the reactants break only when the molecules have absorbed enough energy to become unstable. (Recall that systems rich in free energy are intrinsically unstable, and unstable systems are reactive.) The activation energy is represented by the uphill portion of the graph, with the free-energy content of the reactants increasing. The absorption of thermal energy increases the speed of the reactants, so they are colliding more often and more forcefully. Also, thermal agitation of the atoms in the molecules makes the bonds more fragile and more likely to break. At the summit, the reactants are in an unstable condition known as the *transition state;* they are primed, and the reaction can occur. As the molecules settle into their new bonding arrangements, energy is released to the surroundings. This phase of the reaction corresponds to the downhill portion of the curve, which indicates a loss of free energy by the molecules. The difference in the free energy of the products and reactants is ΔG for the overall reaction, which is negative for an exergonic reaction.

As FIGURE 6.9 shows, even for an exergonic reaction, which is energetically downhill overall, the barrier of activation energy must be scaled before the reaction can occur. For some reactions, E_A is modest enough that even at room temperature, there is sufficient thermal energy for many of the reactants to reach the transition state. In most cases, however, the E_A barrier is loftier, and the reaction will occur at a noticeable rate only if the reactants are heated. The spark plugs in an automobile engine heat the gasoline-oxygen mixture so that the molecules reach the transition state and react; only then can there be the explosive release of energy that pushes the pistons. Without a spark, the hydrocarbons of gasoline are too stable to react with oxygen.

The barrier of activation energy is essential to life. Proteins, DNA, and other complex molecules of the cell are rich in free energy and have the potential to decompose spontaneously; that is, the laws of thermodynamics favor their breakdown. These molecules exist only because at temperatures typical for cells, few molecules

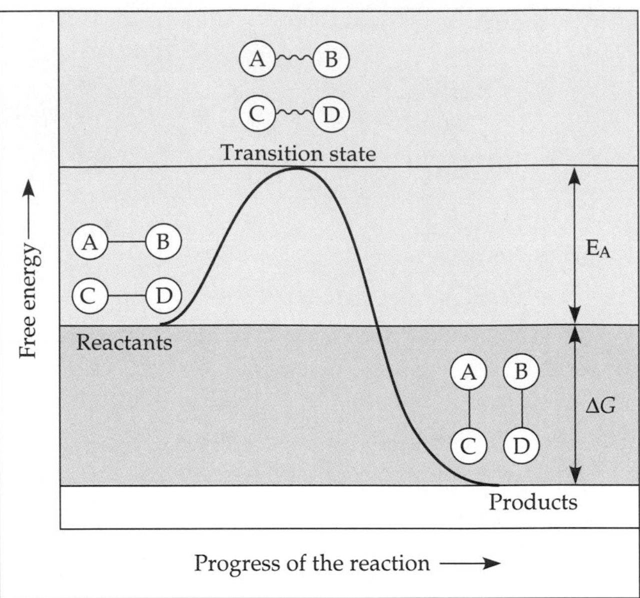

FIGURE 6.9
An energy profile of a reaction. In this hypothetical reaction, the reactants AB and CD must absorb enough energy from the surroundings to surmount the hill of activation energy (E_A) and reach the transition state. The bonds can then break, and as the reaction proceeds, energy is released to the surroundings during the formation of new bonds. This particular graph profiles the energy inputs and outputs of an exergonic reaction, which has a negative ΔG; the products have less free energy than the reactants.

can make it over the hump of activation energy. Occasionally, however, the barrier for selected reactions must be surmounted, or else the cell would be metabolically stagnant. Heat speeds a reaction, but high temperature kills cells. Organisms must therefore use an alternative: a catalyst.

An enzyme speeds a reaction by lowering the E_A barrier, so that the precipice of the transition state is within reach even at moderate temperatures (FIGURE 6.10). An enzyme cannot change the ΔG for a reaction. It cannot make an endergonic reaction exergonic. Enzymes can only hasten reactions that would occur eventually anyway, but this function makes it possible for the cell to have a dynamic metabolism. Further, because enzymes are very selective in the reactions they catalyze, they determine which chemical processes will be going on in the cell at any particular time.

Enzymes are substrate-specific: *a closer look*

The reactant an enzyme acts on is referred to as the enzyme's **substrate.** The enzyme binds to its substrate (or substrates, when there are two or more reactants). While enzyme and substrate are joined, the catalytic action of the enzyme converts the substrate to the product (or

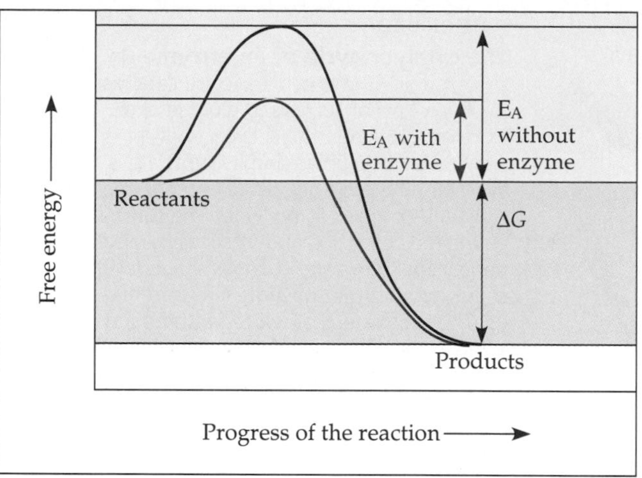

FIGURE 6.10
Enzymes: lowering the barrier of activation energy. Without affecting the free-energy change (ΔG) for the reaction, an enzyme speeds the reaction by reducing the uphill climb to the transition state. The black curve shows the course of the reaction without an enzyme; the magenta curve shows the course of the reaction with an enzyme.

products) of the reaction. This process can be generalized in this way:

$$\text{Substrate} \xrightarrow{\text{Enzyme}} \text{Product}$$

For example, the enzyme sucrase (many enzyme names end in *-ase*) breaks the disaccharide sucrose into its two monosaccharides, glucose and fructose:

$$\text{Sucrose} + H_2O \xrightarrow{\text{Sucrase}} \text{Glucose} + \text{Fructose}$$

An enzyme can distinguish its substrate from even closely related compounds, such as isomers, so that each type of enzyme catalyzes a particular reaction. For instance, sucrase will act only on sucrose and will reject other disaccharides, such as maltose. What accounts for this molecular recognition? Recall that enzymes are proteins, and proteins are macromolecules with unique three-dimensional conformations. The specificity of an enzyme is based on its shape.

Only a restricted region of the enzyme molecule actually binds to the substrate. This region, called the **active site,** is typically a pocket or groove on the surface of the protein (FIGURE 6.11). Usually, the active site is formed by only a few of the enzyme's amino acids, with the rest of the protein molecule providing a framework that reinforces the configuration of the active site.

The specificity of an enzyme is attributed to a compatible fit between the shape of its active site and the shape of the substrate. The active site, however, is not a rigid receptacle for the substrate. As the substrate enters the active site, it induces the enzyme to change its shape slightly so that the active site fits even more snugly around the substrate. This **induced fit** is like a clasping handshake. Induced fit brings chemical groups of the active site into positions that enhance their ability to work on the substrate and catalyze the chemical reaction.

The active site is an enzyme's catalytic center: *a closer look*

In an enzymatic reaction, the substrate binds to the active site to form an enzyme-substrate complex (FIGURE 6.12). In most cases, the substrate is held in the active site by weak interactions, such as hydrogen bonds and ionic bonds. Side chains (R groups) of a few of the amino acids that make up the active site catalyze the conversion of substrate to product, and the product departs from the active site. The enzyme is then free to take another substrate molecule into its active site. The entire cycle happens so fast that a single enzyme molecule typically converts about a thousand substrate molecules per second. Some enzymes are much faster. Enzymes, like other catalysts, emerge from the reaction in their original form. Therefore, very small amounts of enzyme can have a huge metabolic impact by functioning over and over again in catalytic cycles.

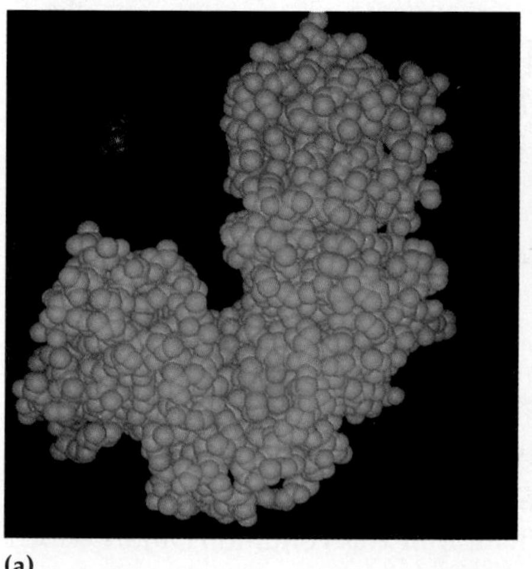

(a)

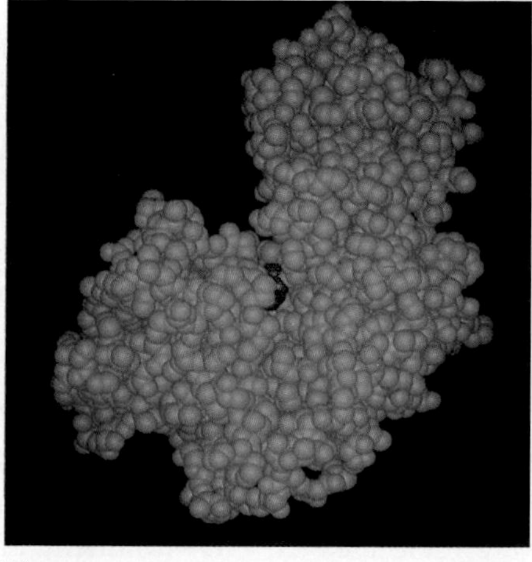

(b)

FIGURE 6.11
The induced fit between an enzyme and its substrate. (**a**) The active site of this enzyme, called hexokinase, can be seen in this computer graphic model as a groove on the surface of the blue protein. (**b**) On entering the active site, the substrate, which is glucose (red), induces a slight change in the shape of the protein, causing the active site to embrace the substrate.

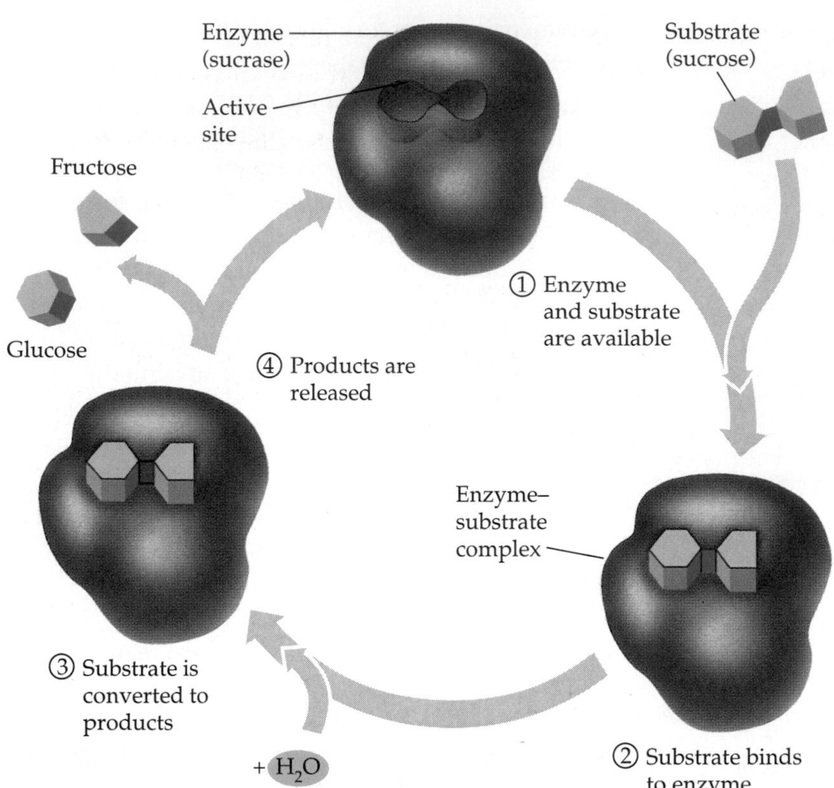

FIGURE 6.12
The catalytic cycle of an enzyme. In this example, the enzyme sucrase catalyzes the hydrolysis of sucrose to glucose and fructose. ① When the active site of an enzyme is unoccupied and its substrate is available, the cycle begins. ② An enzyme-substrate complex forms when the substrate enters the active site and attaches by weak bonds. The active site changes shape to fit snugly around the substrate (induced fit). ③ The substrate is converted to products while in the active site. ④ The enzyme releases the products, and ① its active site is then available for another molecule of substrate. Most metabolic reactions are reversible, and an enzyme can catalyze both the forward and the reverse reactions. Which reaction prevails depends mainly on the relative concentrations of reactants and products; that is, the enzyme catalyzes the reaction in the direction of equilibrium.

Enzymes use a variety of mechanisms that lower activation energy and speed up a reaction. In reactions involving two or more reactants, the active site provides a template for the substrates to come together in the proper orientation for a reaction to occur between them. As induced fit causes the active site to clinch the substrates, the enzyme may stress the substrate molecules, stretching and bending critical chemical bonds that must be broken during the reaction. Since E_A is proportional to the difficulty of breaking bonds, distorting the substrate reduces the amount of thermal energy that must be absorbed in order to achieve a transition state.

The active site may also provide a microenvironment that is conducive to a particular type of reaction. For example, if the active site has amino acids with acidic side chains (R groups), the active site may be a pocket of low pH in an otherwise neutral cell. In such cases, an acidic amino acid facilitates H^+ transfer to the substrate as a key step in catalyzing the reaction. Still another mechanism of catalysis is the direct participation of the active site in the chemical reaction. Sometimes this process even involves brief covalent bonding between the substrate and a side chain of an amino acid of the enzyme. Subsequent steps of the reaction restore the side chains to their original states, so the active site is the same after the reaction as it was before.

The rate at which a given amount of enzyme converts substrate to product is partly a function of the initial concentration of substrate: The more substrate molecules available, the more frequently they access the active sites of the enzyme molecules. However, there is a limit to how fast the reaction can be pushed by adding more substrate to a fixed concentration of enzyme. At some point, the concentration of substrate will be high enough that all enzyme molecules have their active sites engaged. As soon as the product exits an active site, another substrate molecule enters. At this substrate concentration, the enzyme is said to be saturated, and the rate of the reaction is determined by the speed at which the active site can convert substrate to product. When an enzyme population is saturated, the only way to increase productivity is to add more enzyme. Cells sometimes do this by making more enzyme molecules.

A cell's chemical and physical environment affects enzyme activity: *a closer look*

The activity of an enzyme is affected by general environmental factors, such as temperature and pH, and also by particular chemicals that specifically influence that enzyme.

Effects of Temperature and pH

Recall from Chapter 5 that the three-dimensional structures of enzymes and other proteins are sensitive to their environment. Each enzyme has conditions under which it works optimally, because that environment favors the most active conformation for the enzyme molecule.

Temperature is one environmental factor important in the activity of an enzyme (FIGURE 6.13a). Up to a point, the velocity of an enzymatic reaction increases with increasing temperature, partly because substrates collide with active sites more frequently when the molecules move rapidly. At some point on the temperature scale, however, the speed of the enzymatic reaction drops sharply with additional temperature increase. At that point, the thermal agitation of the enzyme molecule disrupts the hydrogen bonds, ionic bonds, and other weak interactions that stabilize the active conformation, and the protein molecule denatures. Each type of enzyme has an optimal temperature at which its reaction rate is fastest. This temperature allows the greatest number of molecular collisions without denaturing the enzyme. Most human enzymes have optimal temperatures of about 35°C to 40°C (close to human body temperature). Bacteria that live in hot springs contain enzymes with optimal temperatures of 70°C or higher.

Just as each enzyme has an optimal temperature, it also has a pH at which it is most active (FIGURE 6.13b). The optimal pH values for most enzymes fall in the range of 6 to 8, but there are exceptions. For example, pepsin, a digestive enzyme in the stomach, works best at pH 2. Such an acidic environment denatures most enzymes, but the active conformation of pepsin is adapted to the acidic environment of the stomach. In contrast, trypsin, a digestive enzyme residing in the alkaline environment of the intestine, has an optimal pH of 8.

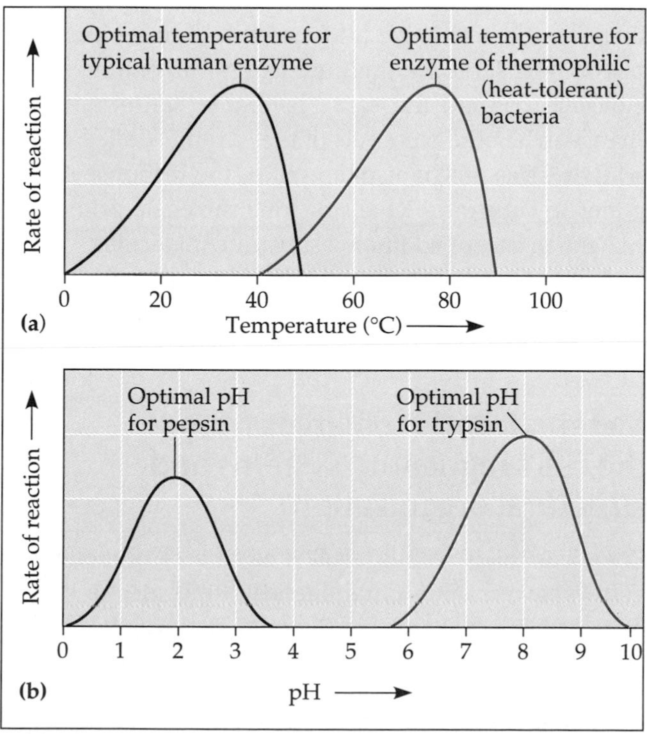

FIGURE 6.13
Environmental factors affecting enzymes. Each enzyme has an optimal (**a**) temperature and (**b**) pH that favor the active conformation of the protein molecule.

Cofactors

Many enzymes require nonprotein helpers for catalytic activity. These adjuncts, called **cofactors,** may be bound tightly to the active site as permanent residents, or they may bind loosely and reversibly along with the substrate. The cofactors of some enzymes are inorganic, such as the metal atoms zinc, iron, and copper. If the cofactor is an organic molecule, it is more specifically called a **coenzyme.** Most vitamins are coenzymes or raw materials from which coenzymes are made. Cofactors function in various ways, but in all cases they are necessary for catalysis to take place.

Enzyme Inhibitors

Certain chemicals selectively inhibit the action of specific enzymes (FIGURE 6.14). If the inhibitor attaches to the enzyme by covalent bonds, inhibition is usually irreversible. The inactivation is reversible, however, if the inhibitor binds to the enzyme by weak bonds.

Some inhibitors resemble the normal substrate molecule and compete for admission into the active site. These mimics, called **competitive inhibitors,** reduce the productivity of enzymes by blocking the substrate from entering active sites. If the inhibition is reversible, it can be overcome by increasing the concentration of substrate so that as active sites become available, more substrate molecules than inhibitor molecules are around to gain entry to the sites.

Noncompetitive inhibitors impede enzymatic reactions by binding to a part of the enzyme away from the active site. This interaction causes the enzyme molecule to change its shape, rendering the active site unreceptive to substrate, or leaving the enzyme less effective at catalyzing the conversion of substrate to product.

Some enzyme inhibitors absorbed from an organism's environment act as metabolic poisons. For example, the pesticides DDT and parathion are inhibitors of key enzymes in the nervous system. Many antibiotics are inhibitors of specific enzymes in bacteria. For instance, penicillin blocks the active site of an enzyme that many bacteria use to make their cell walls.

Allosteric Regulation

Examples of enzyme inhibitors as metabolic poisons may give the impression that enzyme inhibition is generally abnormal and harmful. In fact, selective

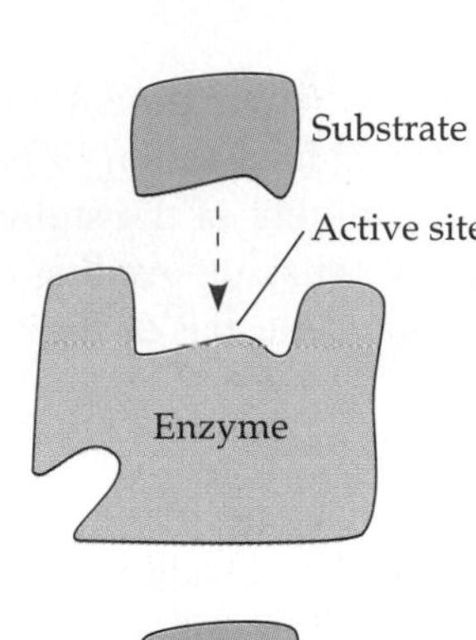

(a) A substrate can normally bind to the active site of an enzyme.

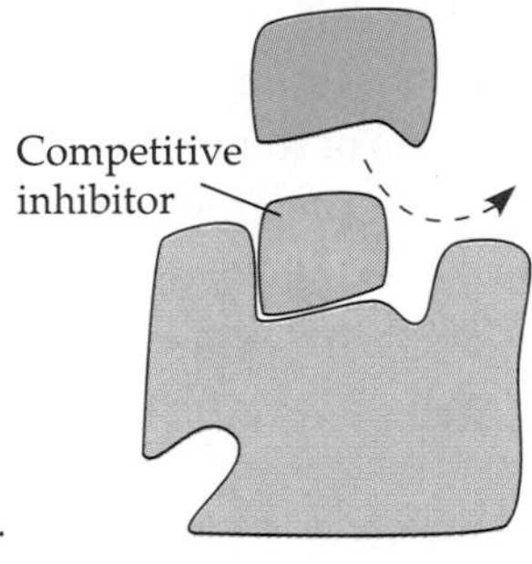

(b) A competitive inhibitor mimics the substrate and competes for the active site.

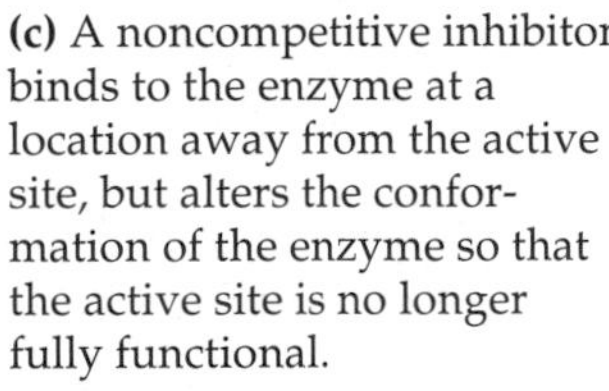

(c) A noncompetitive inhibitor binds to the enzyme at a location away from the active site, but alters the conformation of the enzyme so that the active site is no longer fully functional.

FIGURE 6.14
Enzyme inhibition.

inhibition and activation of enzymes by molecules naturally present in the cell are essential mechanisms in metabolic control.

In most cases, the molecules that naturally regulate enzyme activity bind to an **allosteric site,** a specific receptor site on some part of the enzyme molecule remote from the active site. Most enzymes having allosteric sites are proteins constructed from two or more polypeptide chains, or subunits. Each subunit has its own active site, and allosteric sites are usually located where subunits are joined (FIGURE 6.15a). The entire complex oscillates between two conformational states, one catalytically active and the other inactive. The binding of an activator to an allosteric site stabilizes the conformation that has a functional active site, while the binding of an allosteric inhibitor stabilizes the inactive form of the enzyme (FIGURE 6.15b). (Because they bind away from the active site, allosteric inhibitors fit our definition of noncompetitive inhibitors.) The areas of contact between the subunits of an allosteric enzyme articulate in such a way that a conformational change in one subunit is transmitted to all others. Through this interaction of subunits, a single activator or inhibitor molecule that binds to one allosteric site will affect the active sites of all subunits.

Because allosteric regulators attach to an enzyme by weak bonds, the activity of the enzyme changes in response to fluctuating concentrations of the regulators. In some cases, an inhibitor and an activator are similar enough in shape to compete for the same allosteric site. For example, an enzyme that catalyzes a step in a catabolic pathway, such as respiration, may have an allosteric site that fits both ATP and ADP. The enzyme is inhibited by ATP and activated by ADP. This control seems logical because a major function of catabolism is to regenerate ATP from ADP. If ATP production lags behind its use, ADP accumulates and activates key enzymes that speed up catabolism. If the supply of ATP exceeds demand, then catabolism slows down as ATP molecules outnumber ADP molecules in competition for allosteric sites. In this way, allosteric enzymes act as valves that control the rates of key reactions in metabolic pathways.

Cooperativity

By a mechanism that resembles allosteric activation, substrate molecules may stimulate the catalytic powers of an enzyme (FIGURE 6.15c). Recall that the binding of a substrate to an enzyme induces a favorable change in the shape of the active site (induced fit). If an enzyme has two or more subunits, this interaction with one substrate molecule triggers the same favorable conformational change in all other subunits of the enzyme. Called **cooperativity,** this mechanism amplifies the response of enzymes to substrates: One substrate molecule primes an enzyme to accept additional substrate molecules.

Next we will see how the regulation of enzymes fits into the overall metabolic economy of the cell.

Metabolic order emerges from the cell's regulatory systems and structural organization

Chemical chaos would result if all of a cell's metabolic pathways were open simultaneously. Imagine, for example, a substance synthesized by one pathway being immediately broken down by another. If the two pathways were to run at the same time, the cell would be spinning its metabolic wheels. Actually, the operation of each metabolic pathway is tightly regulated. Pathways are switched on and off by controlling enzyme activity.

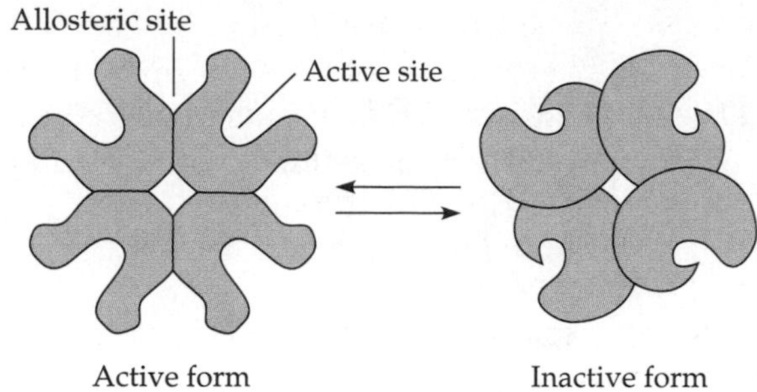

(a) Conformational changes in allosteric enzymes

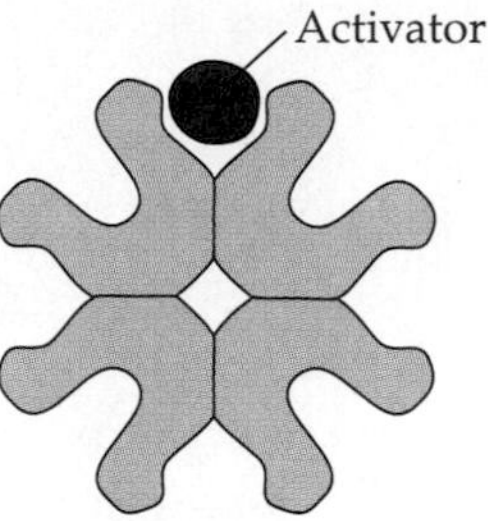

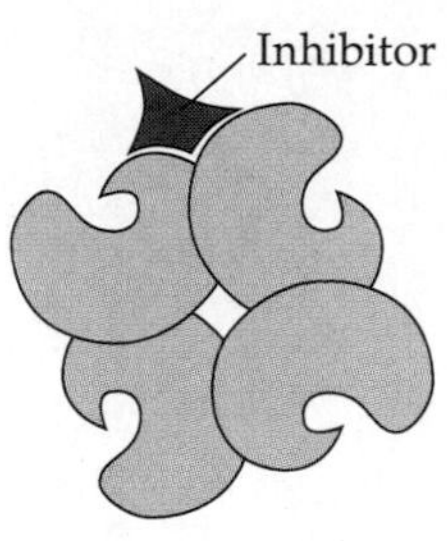

(b) Allosteric regulation

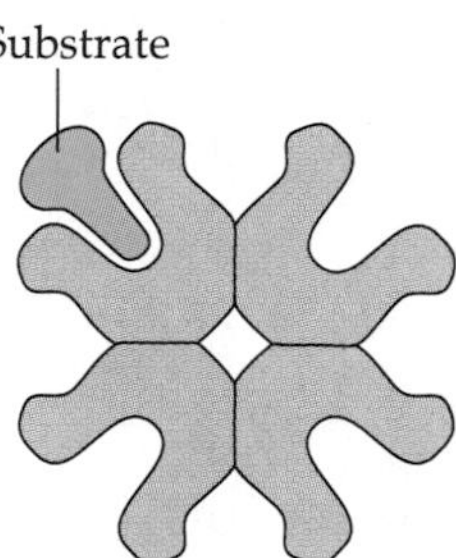

(c) Cooperativity

FIGURE 6.15
Allosteric regulation and cooperativity. (**a**) Most allosteric enzymes are constructed from two or more subunits, each having its own active site. The enzyme oscillates between two conformational states, one active and the other inactive. Remote from the active sites are allosteric sites, specific receptors for regulators of the enzyme, which may be activators or inhibitors. (**b**) Here we see the opposing effects of an allosteric inhibitor and activator on the conformation of all four subunits of an enzyme. (**c**) Similarly, through a phenomenon called cooperativity, one substrate molecule can activate all subunits of the enzyme by the mechanism of induced fit.

Feedback Inhibition

One of the most common methods of metabolic control is **feedback inhibition.** This occurs when a metabolic pathway is switched off by its end-product, which acts as an inhibitor of an enzyme within the pathway. A specific example of feedback inhibition will reveal the logic of this control mechanism. Some cells use a pathway of five steps to synthesize the amino acid isoleucine from threonine, another amino acid (FIGURE 6.16). As isoleucine, the end-product of the pathway, accumulates, it slows down its own synthesis. This happens because isoleucine is an allosteric inhibitor of the enzyme that catalyzes the very first step of the pathway, the enzyme for which threonine is the substrate. Feedback inhibition thereby prevents the cell from wasting chemical resources to synthesize more isoleucine than is necessary.

Structural Order and Metabolism

The cell is not just a bag of chemicals with thousands of different kinds of enzymes and substrates wandering about randomly. The complex structure of the cell orders metabolic pathways in space and time. In some cases, a team of enzymes for several steps of a metabolic pathway is assembled together as a multienzyme complex. The arrangement orders the sequence of reactions, as the product from the first enzyme becomes substrate for the adjacent enzyme in the complex, and so on, until the end-product is released. Some enzymes have fixed locations within the cell because they are incorporated into the structure of a specific membrane. Membranes also partition the cell into many kinds of compartments, with each organelle containing its own internal chemical environment and special blend of enzymes. For example, the enzymes for cellular respiration reside within mitochondria (FIGURE 6.17). If the cell had the same number of enzymes for respiration but they were diluted throughout the entire volume of the cell, respiration would be very inefficient.

By examining the structural basis of metabolic order, we have returned to the theme with which this unit of chapters began.

The theme of emergent properties is manifest in the chemistry of life: *a review*

Recall that life is organized along a hierarchy of structural levels. With each increase in the level of order, new properties emerge in addition to those of the component parts. In these chapters, we have dissected the chemistry of life using the strategy of the reductionist. But we have

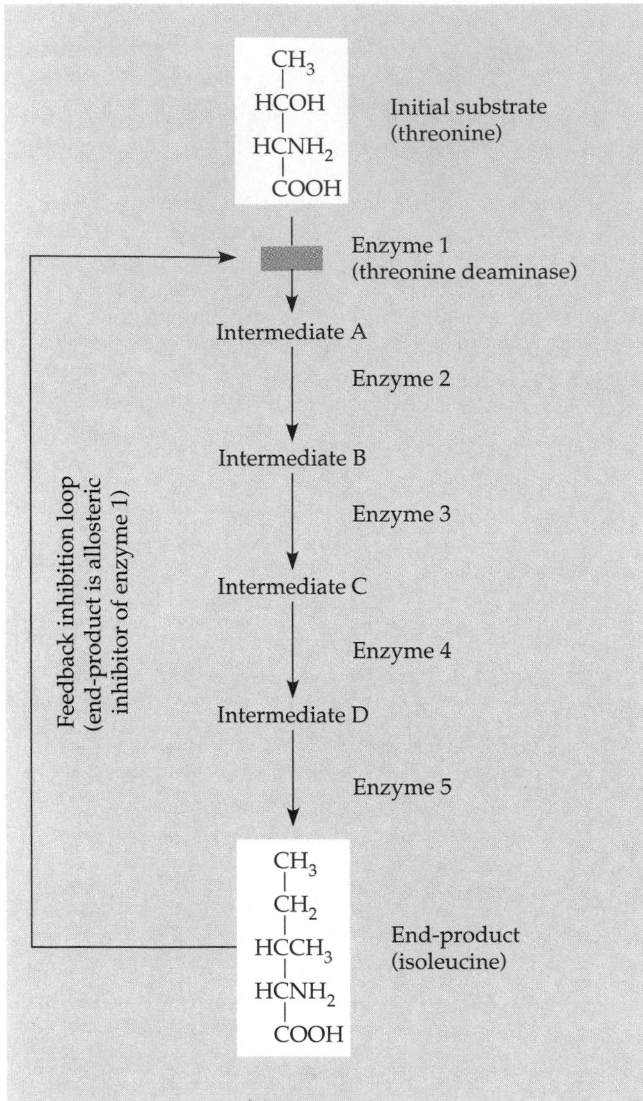

FIGURE 6.16
Feedback inhibition. The initial reactions of many metabolic pathways are switched off by the end-products of the metabolic sequence, which act as allosteric inhibitors of the first enzyme of the pathway.

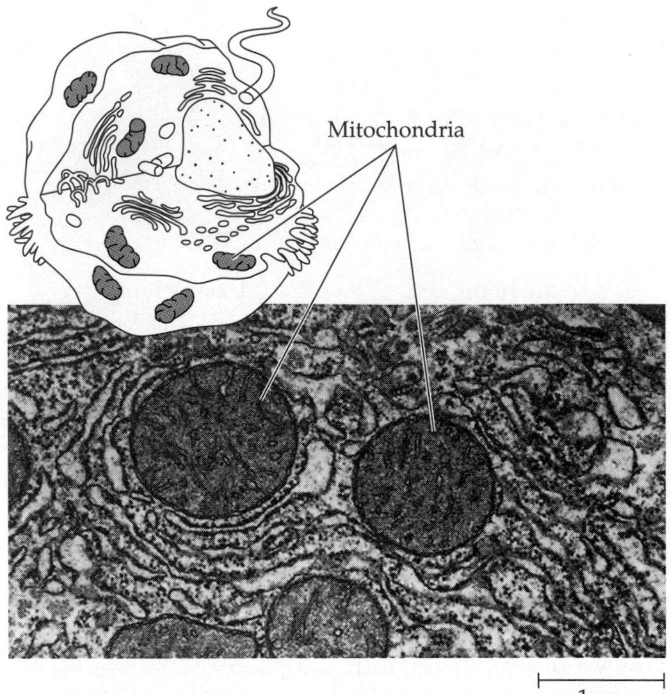

FIGURE 6.17
Structural order and metabolism. Membranes partition a eukaryotic cell into various metabolic compartments, or organelles, each with a corps of enzymes that carry out specific functions. The organelles called mitochondria, for instance, are the sites of cellular respiration (TEM).

also begun to develop a more integrated view of life as we have seen how properties emerge with increasing order.

We have seen that the peculiar behavior of water, so essential to life on earth, results from interactions of the water molecules, themselves an ordered arrangement of hydrogen and oxygen atoms. We reduced the great complexity and diversity of organic compounds to the chemical characteristics of carbon, but we also saw that the unique properties of organic compounds are related to the specific structural arrangements of carbon skeletons and their appended functional groups. We learned that small organic molecules are often assembled into giant molecules, but we also discovered that a macromolecule does not behave like a simple composite of its monomers. For example, the unique form and function of a protein are consequences of a hierarchy of primary, secondary, and tertiary structures. And in this chaper we have seen that metabolism, that orderly chemistry that characterizes life, is a concerted interplay of thousands of different kinds of molecules in an organized cell.

By completing our overview of metabolism with an introduction to its structural basis in the compartmentalized cell, we have built a bridge to Unit Two, where we will study the cell's structure and function in more depth. We will maintain our balance between the need to reduce life to a conglomerate of simpler processes and the ultimate satisfaction of viewing those processes in their integrated context.

REVIEW OF KEY CONCEPTS (with page numbers and key figures)

- The chemistry of life is organized into metabolic pathways (pp. 89–90, FIGURE 6.1)
  - Metabolism is the sum of all the chemical reactions occurring in the cells of an organism.
  - Aided by enzymes, metabolism proceeds by steps along intersecting pathways.
  - Catabolic pathways, such as those of cellular respiration, break complex molecules into simpler compounds, releasing energy in the process. Anabolic pathways build up complex molecules from simpler compounds, requiring the energy input usually provided by catabolism.
- Organisms transform energy (pp. 90–91)
  - Energy is the capacity to do work by moving matter against an opposing force.
  - A moving object has kinetic energy.
  - Potential energy is stored energy based on the specific location or structure of matter. Chemical energy is potential energy stored in molecular structure.
  - Energy can be changed from one form to another, governed by the laws of thermodynamics.
- The energy transformations of life are subject to two laws of thermodynamics (pp. 91–92, FIGURE 6.4)
  - The first law of thermodynamics, conservation of energy, states that energy cannot be created or destroyed.
  - The second law of thermodynamics states that every time energy changes form, there is an increase in the entropy (S), or disorder, of the universe. Whenever matter becomes more ordered, it does so only as a result of a process that increases the disorder of the surroundings.
- Organisms live at the expense of free energy (pp. 92–95, FIGURE 6.5)
  - A system's free energy is the amount of energy that can actually be put to work under cellular conditions—that is, in the absence of temperature gradients. Free energy (G) is directly related to total energy (H) and inversely related to entropy (S): $\Delta G = \Delta H - T\Delta S$.
  - Every spontaneous change in a system proceeds with a decrease in free energy ($-\Delta G$).
  - A spontaneous chemical reaction, one in which the products have less free energy than the reactants, is termed an exergonic reaction ($-\Delta G$). Endergonic (nonspontaneous) reactions are those that occur only with a supply of energy from the surroundings ($+\Delta G$).
  - In metabolism, exergonic reactions are used to power endergonic reactions. This is called energy coupling.
  - The removal of metabolic end-products prevents metabolism from reaching equilibrium.
- ATP powers cellular work by coupling exergonic to endergonic reactions (pp. 95–97, FIGURE 6.6)
  - ATP (adenosine triphosphate) serves as the main energy shuttle in cells. Hydrolysis of one of its phosphate bonds releases ADP (adenosine diphosphate) and inorganic phosphate. This is an exergonic reaction that releases free energy.
  - ATP drives endergonic reactions in the cell by the enzymatic transfer of the phosphate group to specific reactants. The phosphorylated intermediates formed are more reactive than the original molecules. In this way, cells can carry out work, such as movement, pumping solutes across membranes, and anabolism.
  - The regeneration of ATP from ADP and phosphate is driven by catabolic pathways such as cellular respiration.
- Enzymes speed up metabolic reactions by lowering energy barriers: *an overview* (pp. 97–98, FIGURE 6.10)
  - Enzymes, which are proteins, are biological catalysts.
  - Enzymes enable molecules to react during metabolism by lowering activation energy (E_A), which allows bonds to break at the moderate body temperatures characteristic of most organisms.
- Enzymes are substrate-specific: *a closer look* (pp. 98–99, FIGURE 6.11)
  - Each type of enzyme has a uniquely shaped active site, giving it specificity in combining with its particular substrate, the reactant molecule on which an enzyme acts.
  - The binding of substrate to an active site causes the enzyme to change shape slightly (induced fit).
- The active site is an enzyme's catalytic center: *a closer look* (pp. 99–100, FIGURE 6.12)
  - The active site of an enzyme can lower activation energy in a number of ways: by providing a template for substrates to come together in proper orientation, by binding to the substrate in such a way that critical bonds of the substrate are strained, and by providing suitable microenvironments.
- A cell's chemical and physical environment affects enzyme activity: *a closer look* (pp. 100–102, FIGURES 6.13–6.15)
  - As proteins, enzymes are very sensitive to environmental conditions that influence their three-dimensional structure. Each enzyme has optimal conditions of temperature and pH.
  - Cofactors are nonprotein ions or molecules required for the function of some enzymes. If the cofactor is organic, it is known as a coenzyme.
  - Enzyme inhibitors selectively reduce enzyme function. A competitive inhibitor is structurally similar to the substrate and can bind to the active site in its place. A noncompetitive inhibitor binds to a place on the enzyme other than the active site, disrupting the enzyme's shape and function.
  - Some enzymes change shape when regulatory molecules, either activators or inhibitors, bind to specific allosteric sites. Induced fit by the binding of substrate activates other attached subunits by the mechanism known as cooperativity.

- Metabolic order emerges from the cell's regulatory systems and structural organization (pp. 102–103, FIGURE 6.16)
  - A common method of regulating metabolism is feedback inhibition, in which the end-product of a metabolic pathway inhibits the first enzyme in that pathway.
  - Membranes partition the cell into metabolic compartments (organelles).
- The theme of emergent properties is manifest in the chemistry of life: *a review* (pp. 103–104)
  - In this unit, we have seen how increasing levels of organization result in the emergence of properties that are different from those of lower levels. Organization is the key to the chemistry of life.

SELF QUIZ

1. Choose the pair of terms that completes this sentence: Catabolism is to anabolism as __________ is to __________.
   a. exergonic; spontaneous
   b. exergonic; endergonic
   c. free energy; entropy
   d. work; energy
   e. entropy; order
2. Most cells cannot harness heat in order to perform work because
   a. heat is not a form of energy
   b. cells do not have much heat; they are relatively cool
   c. temperature is usually uniform throughout a cell
   d. there are no mechanisms in nature that can use heat to do work
   e. heat denatures enzymes
3. According to the first law of thermodynamics,
   a. matter can be neither created nor destroyed
   b. energy is conserved in all processes
   c. all processes increase the order of the universe
   d. systems rich in energy are intrinsically stable
   e. the universe constantly loses energy because of friction
4. Which of the following metabolic processes can occur without a net influx of energy from some other process?
   a. $ADP + ⓟ_i \longrightarrow ATP + H_2O$
   b. $C_6H_{12}O_6 + 6\,O_2 \longrightarrow 6\,CO_2 + 6\,H_2O$
   c. $6\,CO_2 + 6\,H_2O \longrightarrow C_6H_{12}O_6 + 6\,O_2$
   d. amino acids $\longrightarrow$ protein
   e. glucose + fructose $\longrightarrow$ sucrose
5. Which molecule binds to the active site of an enzyme?
   a. allosteric activator
   b. allosteric inhibitor
   c. noncompetitive inhibitor
   d. competitive inhibitor
6. If an enzyme solution is saturated with substrate, the most effective way to obtain an even faster yield of products would be to
   a. add more of the enzyme
   b. heat the solution to 90°C
   c. add more substrate
   d. add an allosteric inhibitor
   e. add a noncompetitive inhibitor
7. An enzyme accelerates a metabolic reaction by
   a. altering the overall free-energy change for the reaction
   b. making an endergonic reaction occur spontaneously
   c. lowering the activation energy
   d. pushing the reaction away from equilibrium
   e. making the substrate molecule more stable
8. Some bacteria are metabolically active in hot springs because
   a. they are able to maintain an internal temperature much cooler than that of the surrounding water
   b. the high temperatures facilitate active metabolism without the need of catalysis
   c. their enzymes have high optimal temperatures
   d. their enzymes are insensitive to temperature
   e. they use molecules other than proteins as their main catalysts
9. Which metabolic process in bacteria is directly inhibited by the antibiotic penicillin?
   a. cellular respiration
   b. ATP hydrolysis
   c. synthesis of fats
   d. synthesis of chemical components of the cell wall
   e. replication of DNA, the genetic material
10. In the following branched metabolic pathway, a dotted arrow with a minus sign symbolizes inhibition of a metabolic step by an end-product:

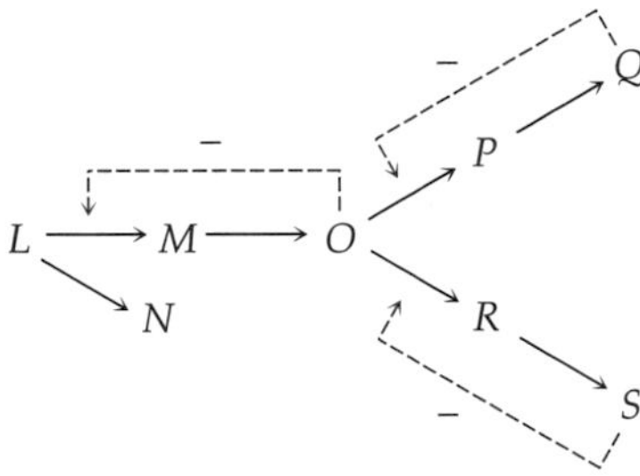

Which reaction would prevail if both Q and S are present in the cell in high concentrations?
   a. $L \longrightarrow M$
   b. $M \longrightarrow O$
   c. $L \longrightarrow N$
   d. $O \longrightarrow P$
   e. $R \longrightarrow S$

CHALLENGE QUESTIONS

1. A particular enzyme has an optimal temperature of 37°C and begins to denature at 45°C. During denaturation, entropy increases (the protein loses much of its organization). The protein also increases its energy content (energy must be absorbed from the surroundings to break the numerous weak bonds that reinforce the native conformation). Thus, for denaturation, ΔS and ΔH are both positive. Using the free-energy equation ($\Delta G = \Delta H - T\Delta S$), explain why denaturation becomes spontaneous at a certain temperature.

2. In an experiment, the enzyme sucrase is mixed with various concentrations of its substrate, sucrose. Each test tube begins with a certain sucrose concentration, and all tubes contain the same concentration of the enzyme. The rate of the reaction—conversion of substrate to product—is measured for each of the samples, and the results are plotted on the graph below. Explain the shape of the curve in the graph.

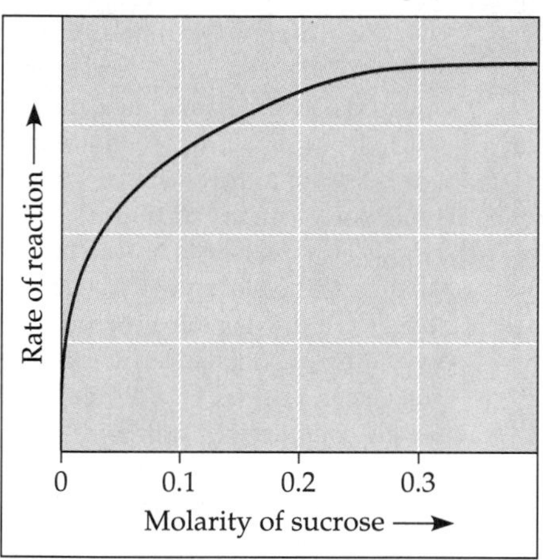

SCIENCE, TECHNOLOGY, AND SOCIETY

1. Biologists often hear the following type of argument: "Evolutionists claim that the complexity of organisms has increased during the history of life. Such evolution of greater biological order contradicts the second law of thermodynamics, which is known to be unbreakable. Therefore, biological evolution is a scientifically invalid concept." How would you respond to this argument?

2. In the early 1970s, the United States was forced to take its energy consumption more seriously when a cartel of oil-exporting countries began to regulate their production in order to control prices. "Energy crisis" became part of our everyday vocabulary, and our dependence on imported oil still affects our economy and foreign policies. Your author remembers a faculty colleague who questioned the very premise of an energy crisis by invoking the first law of thermodynamics: "If energy can't be destroyed," he said, "then how can there be an energy crisis? We just need to develop some clever ways of recycling our energy." What are the scientific flaws in this wishful thinking? How would you explain thermodynamic reality to this professor?

FURTHER READING

Adams, S. "No Way Back." *New Scientist,* October 22, 1994. An entertaining article about the second law of thermodynamics.

Becker, W. M., J. B. Reece, and M. F. Poenie. *The World of the Cell,* 3rd ed. Menlo Park, CA: Benjamin/Cummings, 1996. A lucid explanation of cellular energetics and enzymes in Chapters 5 and 6.

Harold, F. M. *The Vital Force: A Study of Bioenergetics.* New York: W. H. Freeman, 1986. A challenging but clear introduction to energy and life in Chapters 1–3.

Kauffman, S. A. *The Origins of Order.* New York: Oxford University Press, 1993. An influential book applying thermodynamics and chaos theory to life.

Lewin, R. "A Simple Matter of Complexity." *New Scientist,* February 5, 1994. Our interpretation of life's history depends in part on how we define complexity.

Mathews, C. K., and K. E. van Holde. *Biochemistry,* 2nd ed. Menlo Park, CA: Benjamin/Cummings, 1996. An introduction to enzymes and metabolism in Chapters 10 and 12.

UNIT TWO

THE CELL

AN INTERVIEW WITH

SHINYA INOUÉ

Part biologist, part engineer, Shinya Inoué has helped reinvent the use of light microscopes in the study of cells. Building his own microscopes, Professor Inoué invented optical technology that reveals the structure and movement of minute organelles without having to kill and stain cells. For example, such technical virtuosity enabled Dr. Inoué to confirm the existence of structures called spindle fibers, which separate chromosomes during cell division. In 1992, the American Society of Cell Biology recognized Dr. Inoué's contributions with its highest honor, the E. B. Wilson Award. Professor Inoué was elected to the prestigious National Academy of Sciences in 1993. He presently holds the title of Distinguished Scientist at the Marine Biological Laboratory (MBL) in Woods Hole, Massachusetts. Each summer, scientists from around the world converge at MBL in order to share ideas and collaborate on their research. Science is social, and the Woods Hole culture has helped shape modern biology. At least 34 Nobel laureates have been part of the MBL community. In this interview, Shinya Inoué shares some of his thoughts about MBL and about the microscopic world of cells.

What is the Marine Biological Laboratory, and how did it start?

It started in 1888 as an offshoot of the Women's Educational Association of Boston. U.S. fisheries researchers first recognized that the coast off Woods Hole is a place where the warm current and the cold current mix, and therefore it is very rich in life. MBL came shortly after. It was started essentially as a summer research station with a collection of scientists from different universities. Scientists came here not just to study marine life per se, but to use marine organisms as appropriate models for studying various basic aspects of life.

Why do you think the lab has been such a magnet for top scientists who seem to migrate here each summer?

The ambiance certainly contributes. But a tradition has built up over the years of people coming together and exchanging ideas. All the lectures are open to everybody in the community. The summer courses are taught by top-notch scientists. And we have a world-famous library.

Team research seems to be common at the MBL.

Well, none of us are experts in enough fields. Collaborations are extremely important. I think in order to have a fruitful collaboration, you need to have people who know not only their own field but enough of another field to have a feeling for it. Biologists, physicists, and chemists collaborate naturally at the MBL, where they are free from university chores.

Can you tell us a little bit more about your own career? What attracted you to science?

My father was in the foreign service, so I grew up in different countries. I was born in England, and then we went to Japan briefly, then China, the United States, Australia. From my high school years on, I lived in Japan. Science was one of the things that spanned the different countries—it provided cultural continuity.

Technology and science interested me a lot when I was very young. Biology as such didn't interest me very much in high school, except for one class where the teacher showed us how birds' feathers stayed as neat as they are with simple barbs. The rest was so rote, and memorizing things is not my forte. There was so much memorization in the Japanese school system. In college, which is between high school and university in Japan, I had very good science teachers who were researchers, and one, Professor Katsuma Dan, really sparked my interest. He challenged us with interesting projects in the biology lab, so it was no longer memorizing things or repeating what others had done, but finding out how you discover things. I think that is when I really got interested in science.

As a cell biologist, you have engineered many improvements in light microscopes. What are the advantages of light microscopes compared to electron microscopes?

About the time I started studying cells, a lot of people went into electron microscopy because of the very high resolution. Certainly we learned a lot from electron microscopes, but in electron microscopy you have to kill, fix (preserve), and slice the cell. You get a still image of a part of a cell, which may or may not be modified during the preparation. With light microscopes, on the other hand, you have a limited resolution but you can see the dynamics of living cells.

With so many types of light microscopes available from optical companies, why do you build your own?

None of them did the job that we needed to get done. The polarizing microscope does an excellent job of identifying the optical characteristics of crystals, but these microscopes as they came wouldn't work in the small areas of living cells, so I started building. For example, this special objective lens increases sensitivity for detecting where molecules are lined up in the cell.

When did you build your first microscope?

The first one was while I was still in Japan after the second world war. Emperor Hirohito came visiting to Dan's laboratory at the Misaki Marine Station. Before the war, the Emperor had published volumes of biology, but he had to publish under somebody else's name because the Emperor was considered to be a god. After the war, he got released from being a god and became an ordinary person and a biologist. Anyhow, when he came I was told, "Why don't you show the Emperor some of the specimens swimming around?" I had different bits and pieces of microscopes that were stacked on books to align their axes, but it was too unstable. So I tied them together on a sturdy base and got started on the first version of my microscopes.

And most recently you have been a pioneer in video microscopy.

With video microscopy you can use the light microscope the way it was supposed to work but never did because the contrast became too low at high resolution and shallow depth of field. When you put that image through video and a computer, out pops all kinds of images that weren't possible, such as molecular filaments that are much thinner than the microscope's resolution limit. Another advantage of video microscopy is the time dimension. With video you can have immediate playback of what you have been recording, so you can plan the experiments based on what the recording shows.

Your inventiveness with light microscopy dates back to your studies in the 1950s of spindle fibers and their role in cell division. Tell us a little bit about that work.

In those studies I was able to see the arrangement of molecules inside living cells. By being able to see regions where the molecules were lined up, I could show that chromosomes were moved by fibers attached to them. In textbooks, you see these spindle fibers attached to chromosomes and it looks convincing. But people who were studying mitosis seriously at that time said, "You see those fibers when you fix a cell and stain it, but that doesn't mean they are there in the living cell." In fact if you fix the cell very carefully, you don't see any fibers. There was a 50-year battle raging back and forth as to whether chromosomes were moved by those fibers or had some other means of moving. I was able to show, by improving the polarizing microscope, that these were real structures.
I was also able to show why there was such

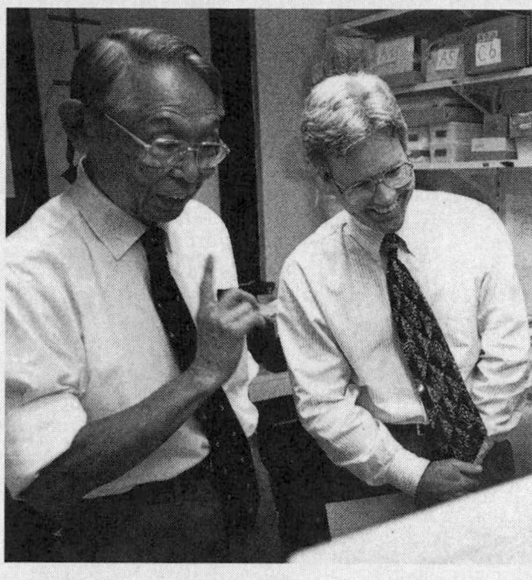

a controversy. Essentially, those fibers are very unstable; if you say "boo" to a cell, they disappear.

How do you think these spindle fibers work—how do the chromosomes move apart?

We don't have a definitive answer yet. The fibers that I saw were bundles of structures called microtubules. I showed that the molecules making up the microtubles were assembling and disassembling. Furthermore, I said that the chromosomes move as the microtubules fall apart. Scientists said, "Well, that's crazy! You can't have a string that is both falling apart and pulling something." I showed these fibers and their behavior in 1950 or 1951, but it took about 15 years before the microtubules could be isolated from cells. When they were isolated, the microtubules showed none of the characteristics I predicted they would. It took another 10 years or so until somebody figured out why the isolated fibers would not behave the way the live fibers behave. It turns out that these fibers are terribly sensitive to calcium levels. When you grind up the cell, a huge amount of calcium leaks from stored compartments and destroys the microtubules. A young researcher figured out that you have to sequester calcium that leaks out so it can't attack the spindle. Then he was able to isolate the spindles and show that the molecules do assemble and disassemble. Now, disassembling and shortening microtubules can be made to pull chromosomes even outside the cell.

But there was also very good work showing that there are molecular motors that move on top of microtubules in cilia and flagella. When that came into vogue, everybody wanted to say that chromosomes move because there are molecular motors on the chromosomes that make them crawl along microtubules. How to combine these two different concepts or observations is still up in the air.

What are some of the other important questions about how cells divide?

By using extracts from frog eggs and clam eggs, scientists have been able to show where mitosis can be interrupted. They also have identified chemical compounds produced by the cells that control stages of cell division, for example, when the cell goes into mitosis, when the nuclear envelope breaks down and cell division takes place, and so on. That obviously has an immediate significance in cancer research because in cancer cells, cell division goes wild. Why do some cells go wild and not others? What controls the division?

How does your research on cell division connect to other questions in basic cell biology?

Well, if we look at this example of the microtubules, it turns out that the concept of the microtubules assembling and disassembling applies not just to cell division but also to nerve growth, white blood cells moving around and capturing bacteria, and transport of organelles in cells. All of these things are related. Once we start understanding the basics of how the cell and its components behave, then we have a better chance of understanding the kind of things that many scientists are asking questions about but are having difficulty finding answers to.

What advice do you have for students as they begin to think about the structure and function of cells?

I continue to worry about science being learned as a collection of facts and theories. One needs to have a certain body of knowledge—but in addition, one needs to understand how the knowledge is acquired—that really is at the heart of science.

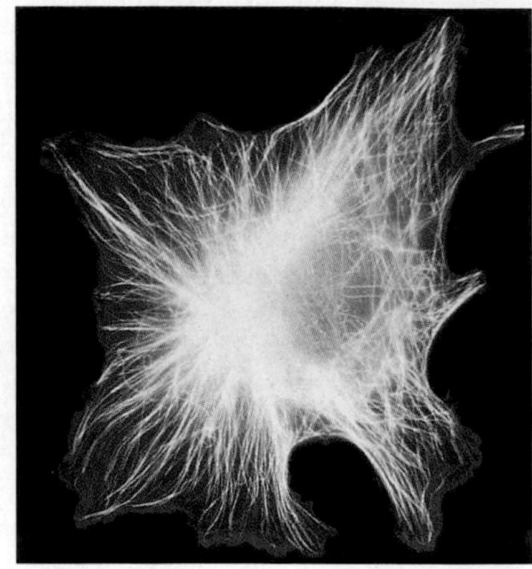

CHAPTER 7

A TOUR OF THE CELL

KEY CONCEPTS

- Microscopes provide windows to the world of the cell
- Cell biologists can isolate organelles to study their functions
- A panoramic view of the cell
- The nucleus contains a cell's genetic library
- Ribosomes build a cell's proteins
- Many organelles are related through the endomembrane system
- The endoplasmic reticulum manufactures membranes and performs many other biosynthetic functions
- The Golgi apparatus finishes, sorts, and ships many products of the cell
- Lysosomes are digestive compartments
- Vacuoles have diverse functions in cell maintenance
- Peroxisomes consume oxygen in various metabolic functions
- Mitochondria and chloroplasts are the main energy transformers of cells
- The cytoskeleton provides structural support and functions in cell motility
- Plant cells are encased by cell walls
- The extracellular matrix (ECM) of animal cells functions in support, adhesion, movement, and development
- Intercellular junctions integrate cells into higher levels of structure and function
- The cell is a living unit greater than the sum of its parts

The cell is as fundamental to biology as the atom is to chemistry: All organisms are made of cells. In the hierarchy of biological organization, the cell is the simplest collection of matter that can live. Indeed, there are diverse forms of life existing as single-celled organisms. More complex organisms, including plants and animals, are multicellular; their bodies are cooperatives of many kinds of specialized cells that could not survive for long on their own. However, even when they are arranged into higher levels of organization, such as tissues and organs, cells can be singled out as the organism's basic units of structure and function. The contraction of muscle cells moves your eyes as you read this sentence; when you decide to turn this page, nerve cells will transmit that decision from your brain to the muscle cells of your hand. Everything an organism does is fundamentally occurring at the cellular level. This chapter introduces the microscopic world of the cell.

This text takes a thematic approach to the study of life, and the cell is a microcosmic model of many of the themes introduced in Chapter 1. We will see that life at the cellular level arises from structural order, reinforcing the themes of emergent properties and the correlation between structure and function. For example, the movement of an animal cell depends on an intricate interplay of structures that make up a cellular skeleton (visible in the above light micrograph of a type of cell called a fibroblast). Another recurring theme in biology is the interaction of organisms with their environment. Cells sense and respond to environmental fluctuations. As open systems, they continuously exchange both materials and energy with their surroundings. And keep in mind the one biological theme that unifies all others: evolution. All cells are related by their descent from earlier cells, but they have been modified in various ways during the long evolutionary history of life on Earth. For example, if one unicellular organism lives in fresh water and another inhabits the sea, we can expect these cells to be somewhat differently equipped as a result of their divergent adaptations to disparate environments.

Perhaps the greatest obstacle to becoming acquainted with the cell is imagining how something too small to be seen by the unaided eye can be so complex. How can cell biologists possibly dissect so small a package to investigate its inner workings? Before we actually tour the cell, it will be helpful to learn how cells are studied.

Microscopes provide windows to the world of the cell

The evolution of a science often parallels the invention of instruments that extend human senses to new limits. The discovery and early study of cells progressed with the invention and improvement of microscopes in the seventeenth century. Microscopes of various types are still indispensable tools for the study of cells.

The microscopes first used by Renaissance scientists, as well as the microscopes you are likely to use in the laboratory, are all **light microscopes (LMs)**. Visible light is passed through the specimen and then through glass lenses. The lenses refract (bend) the light in such a way that the image of the specimen is magnified as it is projected into the eye. (See the Appendix at the back of the book that diagrams microscope structure.)

Two important values in microscopy are magnification and resolving power, or resolution. Magnification is how much larger the object appears compared to its real size. **Resolving power** is a measure of the clarity of the image; it is the minimum distance two points can be separated and still be distinguished as two separate points. For example, what appears to the unaided eye as one star in the sky may be resolved as twin stars with a telescope.

Just as the resolving power of the human eye is limited, the resolving power of telescopes and microscopes is limited. Microscopes can be designed to magnify objects as much as desired, but the light microscope can never resolve detail finer than about 0.2 µm, the size of a small bacterium (FIGURE 7.1). This resolution is limited by the wavelength of the visible light used to illuminate the specimen. Light microscopes can magnify effectively to about 1000 times the size of the actual specimen; greater magnifications increase blurriness. Most of the improvements in light microscopy since the beginning of this century have involved new methods for enhancing contrast, which makes the details that can be resolved stand out better to the eye (TABLE 7.1). Shinya Inoué, the scientist you met in the interview that precedes this chapter, has pioneered videomicroscopy and other improvements in light microscopy.

Although cells were discovered by Robert Hooke in 1665, the geography of the cell was largely uncharted until the past few decades. Most subcellular structures, or **organelles,** are too small to be resolved by the light microscope. Cell biology advanced rapidly in the 1950s with the introduction of the electron microscope. Instead of using visible light, the **electron microscope (EM)** focuses a beam of electrons through the specimen. Resolving power is inversely related to the wavelength of radiation a microscope uses, and electron beams have wavelengths much shorter than the wavelengths of

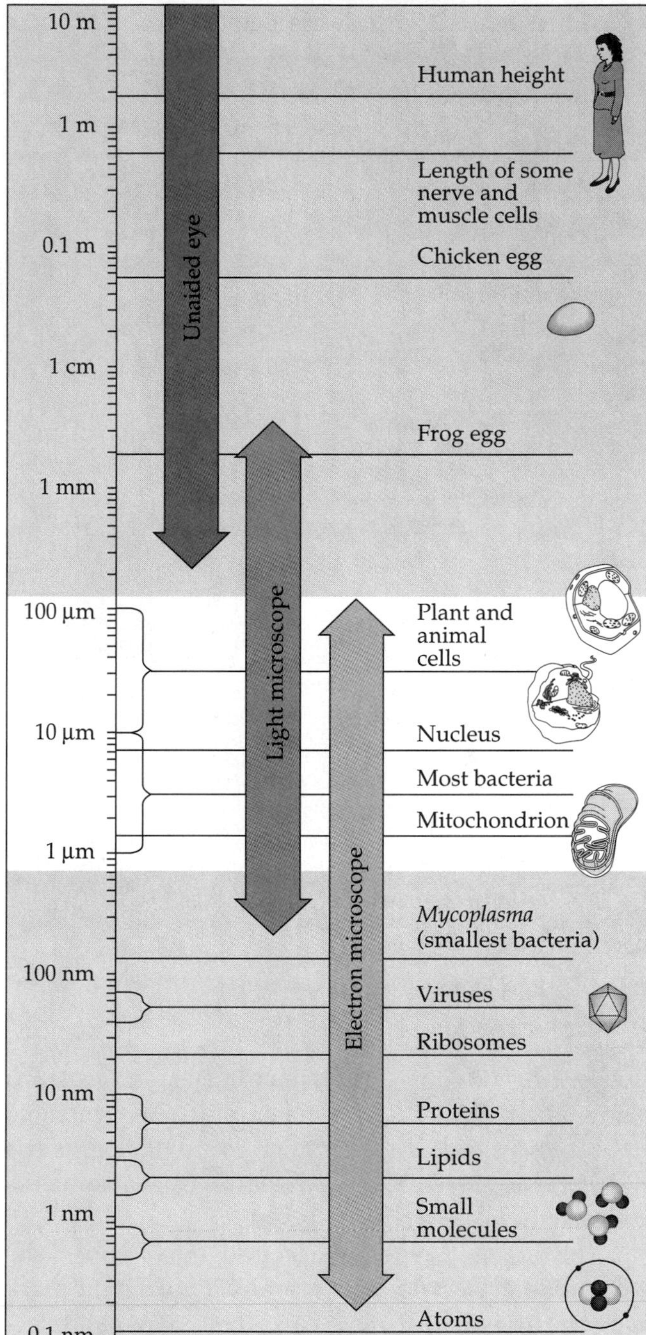

MEASUREMENTS
1 centimeter (cm) = 10^{-2} meter = 0.4 inch
1 millimeter (mm) = 10^{-3} meter
1 micrometer (µm) = 10^{-3} mm
1 nanometer (nm) = 10^{-3} µm

FIGURE 7.1
The size range of cells. Most cells are between 1 and 100 µm in diameter and are therefore visible only under a microscope. Notice that the scale is logarithmic to accommodate the range of sizes shown. Starting at the top of the scale with 10 meters and going down, each reference measurement along the left side marks a *tenfold* decrease in size.

TABLE 7.1

Different Types of Light Microscopy: A Comparison

TYPE OF MICROSCOPY	LIGHT MICROGRAPHS OF HUMAN CHEEK EPITHELIAL CELLS		TYPE OF MICROSCOPY
Brightfield (unstained specimen): Passes light directly through specimen; unless cell is naturally pigmented or artificially stained, image has littte contrast.			**Phase-contrast:** Enhances contrast in unstained cells by amplifying variations in density within specimen; especially useful for examining living, unpigmented cells.
Brightfield (stained specimen): Staining with various dyes enhances contrast, but most staining procedures require that cells be fixed (preserved).			**Differential-interference-contrast (Nomarski):** Also uses optical modifications to exaggerate differences in density.
Darkfield: Passes light through specimen obliquely, and only light scattered by particles can be seen.			**Confocal:** Uses lasers and special optics for "optical sectioning." Only those regions within a narrow depth of focus are imaged. Regions above and below the selected plane of view appear black rather than blurry.

50 μm

visible light. Modern electron microscopes can achieve a resolution of about 0.2 nanometer (nm), a thousandfold improvement over the light microscope. Biologists use the term cell ultrastructure to refer to a cell's anatomy as resolved by an electron microscope.

There are two types of electron microscopes: the **transmission electron microscope (TEM)** and the **scanning electron microscope (SEM).** The TEM aims an electron beam through a thin section of the specimen, similar to the way the light microscope transmits light through a slide. However, instead of using glass lenses, which are opaque to electrons, the TEM uses electromagnets as lenses to focus and magnify the image by bending the trajectories of the charged electrons. The image is ultimately focused onto a screen for viewing or onto photographic film. To enhance contrast in the image, very thin sections of preserved cells are stained with atoms of heavy metals, which attach to certain places in the cells. Cell biologists use the TEM mainly to study the internal ultrastructure of cells (FIGURE 7.2a).

The SEM is especially useful for detailed study of the surface of the specimen (FIGURE 7.2b). The electron beam scans the surface of the sample, which is usually coated with a thin film of gold. The beam excites electrons on the sample surface itself, and these secondary electrons are collected and focused onto a screen. This forms an image of the topography of the specimen. An important attribute of the SEM is its great depth of field, which results in an image that appears three-dimensional.

Electron microscopes reveal many organelles that are impossible to resolve with the light microscope. But the light microscope offers many advantages, especially for the study of live cells. A disadvantage of electron microscopy is that the chemical and physical methods used to prepare the specimen not only kill cells but also may introduce artifacts, structural features seen in micrographs that do not exist in the living cell.

Microscopes of various kinds are the most important tools of cytology, the study of cell structure. But simply describing the diverse organelles within the cell reveals little about their function. Modern cell biology developed from an integration of cytology with biochemistry, the study of life's chemical processes, or metabolism. A biochemical approach called cell fractionation has been particularly important in this multidisciplinary synthesis of cell biology.

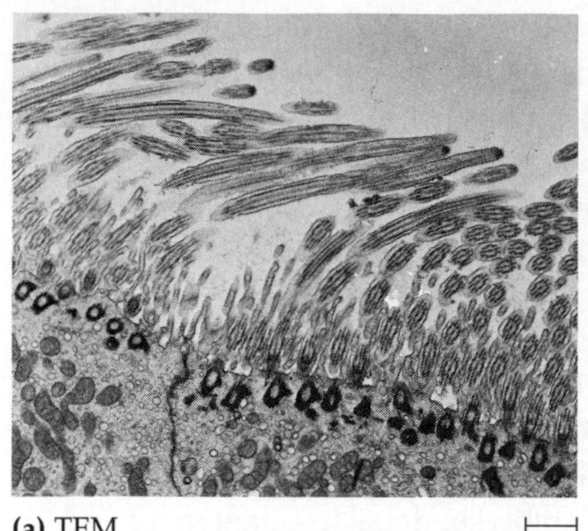

(a) TEM

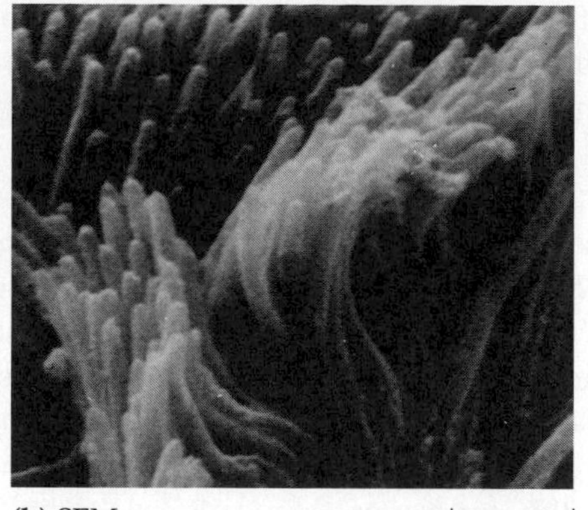

(b) SEM

FIGURE 7.2
Electron micrographs: photographs taken with electron microscopes. (**a**) This micrograph, taken with a transmission electron microscope (TEM), profiles a thin section of a cell from a rabbit trachea (windpipe), revealing its ultrastructure. (**b**) The scanning electron microscope (SEM) produces a three-dimensional image of the surface of the same type of cell. Both micrographs show motile organelles called cilia. Beating of the cilia that line the windpipe helps move inhaled debris upward back toward the pharynx (throat).

Throughout this book, micrographs are identified by the type of microscopy: LM for a light micrograph, TEM for a transmission electron micrograph, and SEM for a scanning electron micrograph.

Cell biologists can isolate organelles to study their functions

The objective of **cell fractionation** is to take cells apart, separating the major organelles so that their individual functions can be studied (FIGURE 7.3). The instrument used to fractionate cells is the centrifuge, a merry-go-round for test tubes capable of spinning at various speeds. The most powerful machines, called **ultracentrifuges,** can spin as fast as 80,000 revolutions per minute (rpm) and apply forces on particles of up to 500,000 times the force of gravity (500,000 *g*).

Fractionation begins with homogenization, the disruption of cells. The objective is to break the cells without severely damaging their organelles. Spinning the soupy homogenate in a centrifuge separates the parts of the cell into two fractions: the pellet, consisting of the larger structures that become packed at the bottom of the test tube; and the supernatant, consisting of smaller parts of the cell suspended in the liquid above the pellet. The supernatant is decanted into another tube and centrifuged again. The process is repeated, increasing the speed with each step, collecting smaller and smaller components of the homogenized cells in each successive pellet.

Cell fractionation enables the researcher to prepare specific components of cells in bulk quantity in order to

FIGURE 7.3
Cell fractionation. Disrupted cells are centrifuged at various speeds and durations to isolate components of different sizes. The process begins with homogenization, the disruption of a tissue and its cells with the help of such instruments as kitchen blenders or ultrasound devices. The homogenate, a soupy mixture of organelles, bits of membrane, and molecules from the broken cells, is then fractionated by a series of spins in a centrifuge. After each centrifugation, the unpelleted portion, or supernatant, is decanted and centrifuged again, at a higher speed. By determining which cell fractions are associated with particular metabolic processes, those functions can be tied to certain organelles.

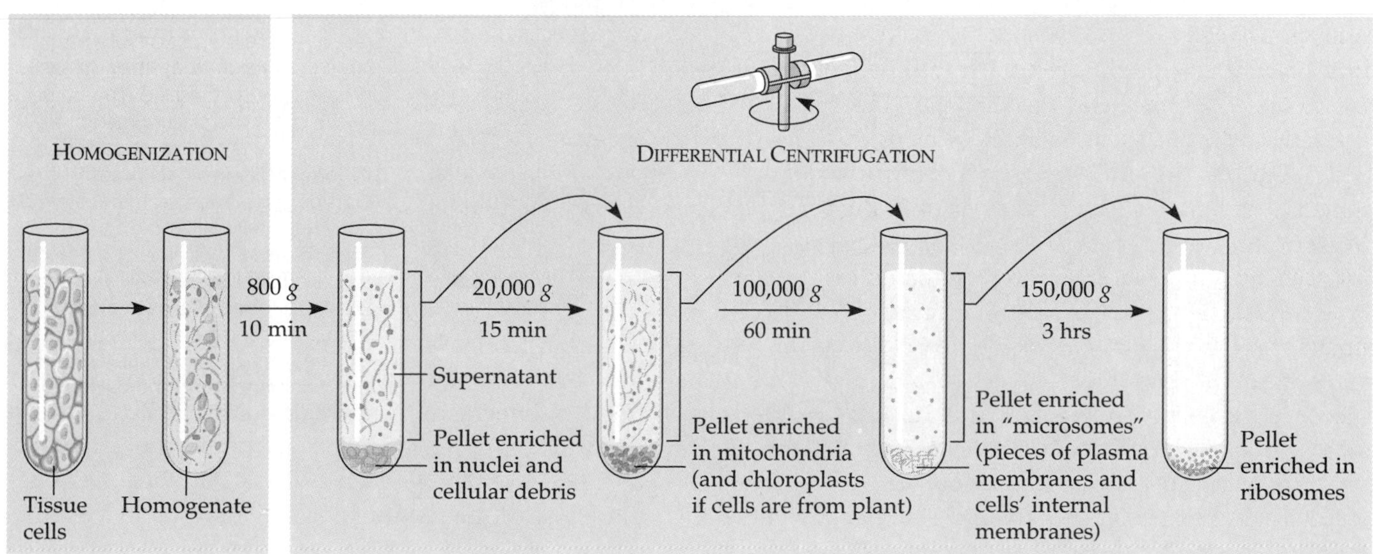

study their composition and functions. By following this approach, biologists have been able to assign various functions of the cell to the different organelles, a task that would be far more difficult if they had to study intact cells. For example, one cellular fraction collected by centrifugation has enzymes that function in the metabolic process known as cellular respiration. The prevalent organelles in that fraction match the structures called mitochondria, as visualized in the electron microscope. This helped cell biologists determine that mitochondria are the sites of cellular respiration. Cytology and biochemistry complement each other in the correlation between cellular structure and function.

A panoramic view of the cell

Prokaryotic and Eukaryotic Cells

Every organism is composed of one of two structurally different types of cells: prokaryotic cells or eukaryotic cells. Prokaryotic cells are found only in the kingdom Monera, the bacteria. Protists, plants, fungi, and animals—four of the five kingdoms of life—are all eukaryotes.

The two types of cells differ markedly in their internal organization. One difference is denoted by their names. The **prokaryotic cell** has no nucleus (FIGURE 7.4). (The word *prokaryote* is from the Greek *pro,* "before," and *karyon,* "kernel," referring here to the nucleus.) Its genetic material (DNA) is concentrated instead in a region called the **nucleoid,** but no membrane separates this region from the rest of the cell. In contrast, the **eukaryotic cell** (Gr. *eu,* "true," and *karyon*) has a true nucleus enclosed by a membranous nuclear envelope. The entire region between the nucleus and the membrane bounding the cell is called the **cytoplasm.** It consists of a semifluid medium called the **cytosol,** in which are suspended organelles of specialized form and function, most of them absent in prokaryotic cells. Thus, the presence or absence of a true nucleus is just one example of the disparity in structural complexity between the two types of cells. The prokaryotic cell will be described in detail in Chapters 17 and 25, and the possible evolutionary relationships between the two types of cells will be discussed in Chapter 26. Most of the discussion of cell structure that follows in this chapter applies to eukaryotes.

Cell Size

Size is a general feature of cell structure that relates to function. The logistics of carrying out metabolism set limits on the size range of cells. The smallest cells known are bacteria called mycoplasmas, which have diameters of between 0.1 and 1.0 micrometer (μm). These are perhaps the smallest packages with enough DNA to program metabolism and enough enzymes and other cellular equipment to carry out the activities necessary for a cell to sustain itself and reproduce. Most bacteria are 1 to 10 μm in diameter, about ten times larger than myco-

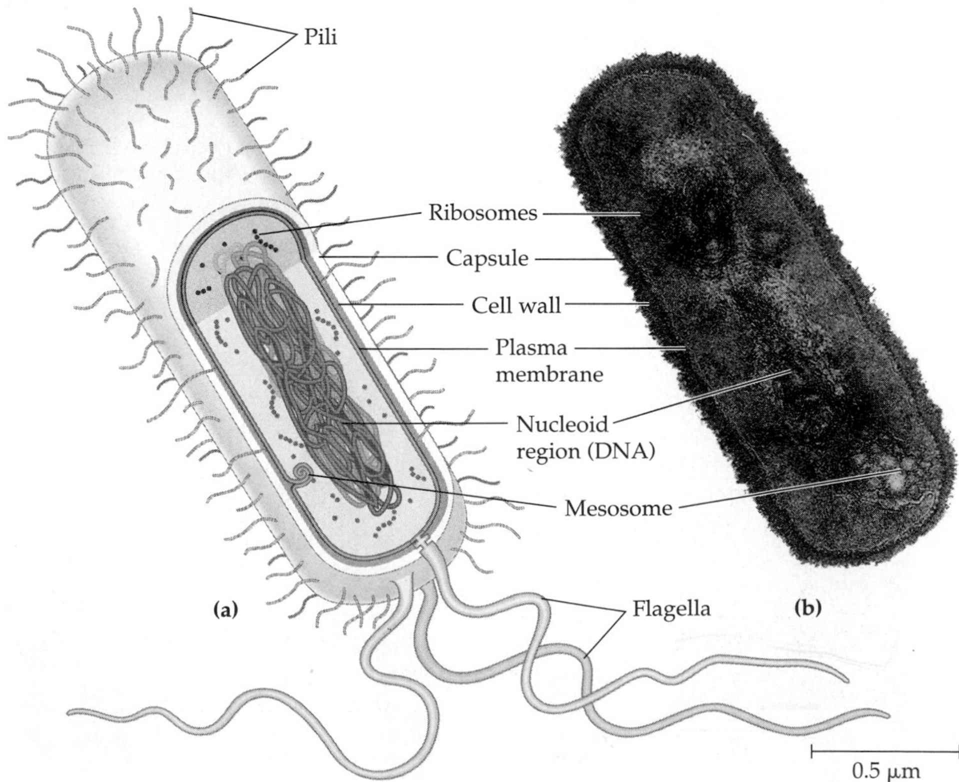

FIGURE 7.4
A prokaryotic cell. Prokaryotes are bacteria, including cyanobacteria. (**a**) The drawing depicts a typical rod-shaped bacterium. Lacking the membrane-enclosed organelles of a eukaryote, the prokaryote is much simpler in structure. The DNA is in the nucleoid region, and no membrane separates the DNA from the rest of the cell. A prokaryote has a large number of ribosomes, where proteins are synthesized. The border of the cell is the plasma membrane, which in some prokaryotes folds in to form structures called mesosomes. Outside the plasma membrane are a fairly rigid cell wall and often an outer capsule, usually jellylike. Some bacteria have flagella (locomotion organelles), pili (attachment structures), or both projecting from their surface. (**b**) The electron micrograph shows a thin section through the bacterium *Bacillus coagulans* (TEM).

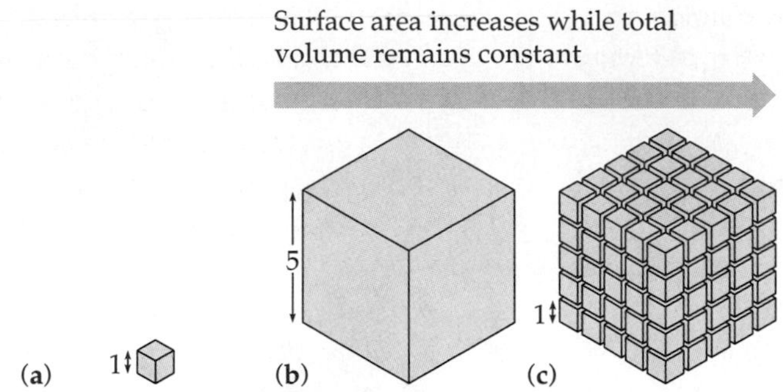

Surface area increases while total volume remains constant

	(a) 1	(b) 5	(c) 1
Total surface area (height × width × number of sides × number of boxes)	6	150	750
Total volume (height × width × length × number of boxes)	1	125	125
Surface-area-to-volume ratio (area ÷ volume)	6	1.2	6

FIGURE 7.5

Why are most cells microscopic? In this diagram, cells are represented as boxes. (**a**) In arbitrary units of length, this small cell measures 1 on each side. We can calculate the cell's surface area (in square units), volume (in cubic units), and surface-area-to-volume ratio. (**b**) As the size of the cell increases to 5 units of length per side, the ratio of surface area to volume decreases compared to the smaller cell. Rates of chemical exchange with the extracellular environment might be inadequate to maintain the cell because most of its cytoplasm is relatively far from the outer membrane. (**c**) By dividing the large cell into many smaller cells, we can restore a surface-area-to-volume ratio that can serve each cell's need for acquiring nutrients and expelling waste products. These geometric relationships explain why most cells are microscopic, and why larger organisms do not generally have *larger* cells than smaller organisms, but *more* cells.

plasmas (see FIGURE 7.1). Eukaryotic cells are typically 10 to 100 μm in diameter, ten times larger than bacteria.

Metabolic requirements also impose upper limits on the size that is practical for a single cell. As an object of a particular shape increases in size, its volume grows proportionately more than its surface area. (Area is proportional to a linear dimension squared, whereas volume is proportional to the linear dimension cubed.) For objects of the same shape, the smaller the object, the greater its ratio of surface area to volume (FIGURE 7.5).

At the boundary of every cell, the **plasma membrane** functions as a selective barrier that allows sufficient passage of oxygen, nutrients, and wastes to service the entire volume of the cell (FIGURE 7.6). For each square micrometer of membrane, only so much of a particular substance can cross per second. The need for a surface sufficiently large to accommodate its volume helps explain the microscopic size of most cells.

The Importance of Compartmental Organization

The complexity of a eukaryotic cell is another structural feature correlated with function. Internal membranes partition the cell into compartments and also participate directly in much of the cell's metabolism; many enzymes are built right into the membranes. Because the compartments of the eukaryotic cell provide different local environments that facilitate specific metabolic functions, processes that are incompatible can go on simultaneously in separate subcellular compartments.

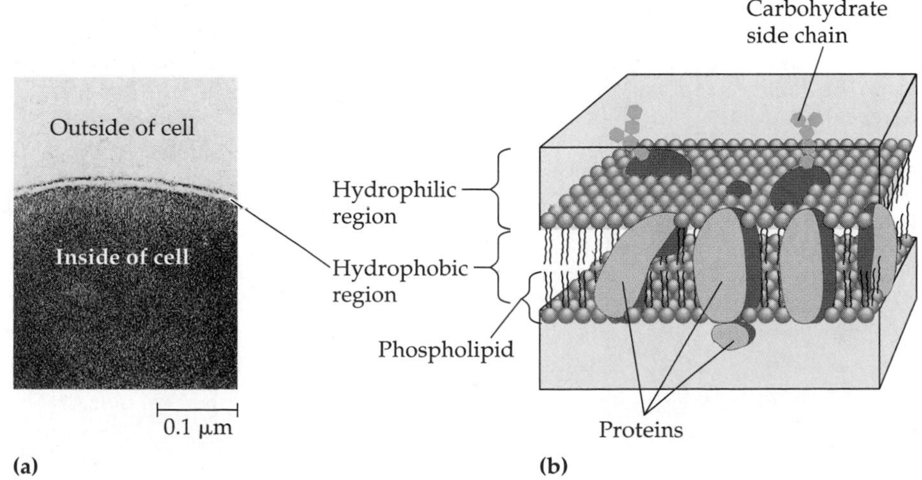

FIGURE 7.6

The plasma membrane. (**a**) In electron micrographs of sufficient magnification, the plasma membrane appears as a pair of dark bands separated by a light band. This is the membrane of a red blood cell (TEM). (**b**) The plasma membrane and the various internal membranes of cells consist of proteins of diverse functions attached to or embedded in a double layer (bilayer) of phospholipids. (See Chapter 5 to review the dual hydrophilic/hydrophobic behavior of phospholipids.) The specific functions of the plasma membrane and the various types of membranes within the cell depend on the kinds of phospholipids and proteins present. The plasma membrane also has carbohydrates attached to its outer surface.

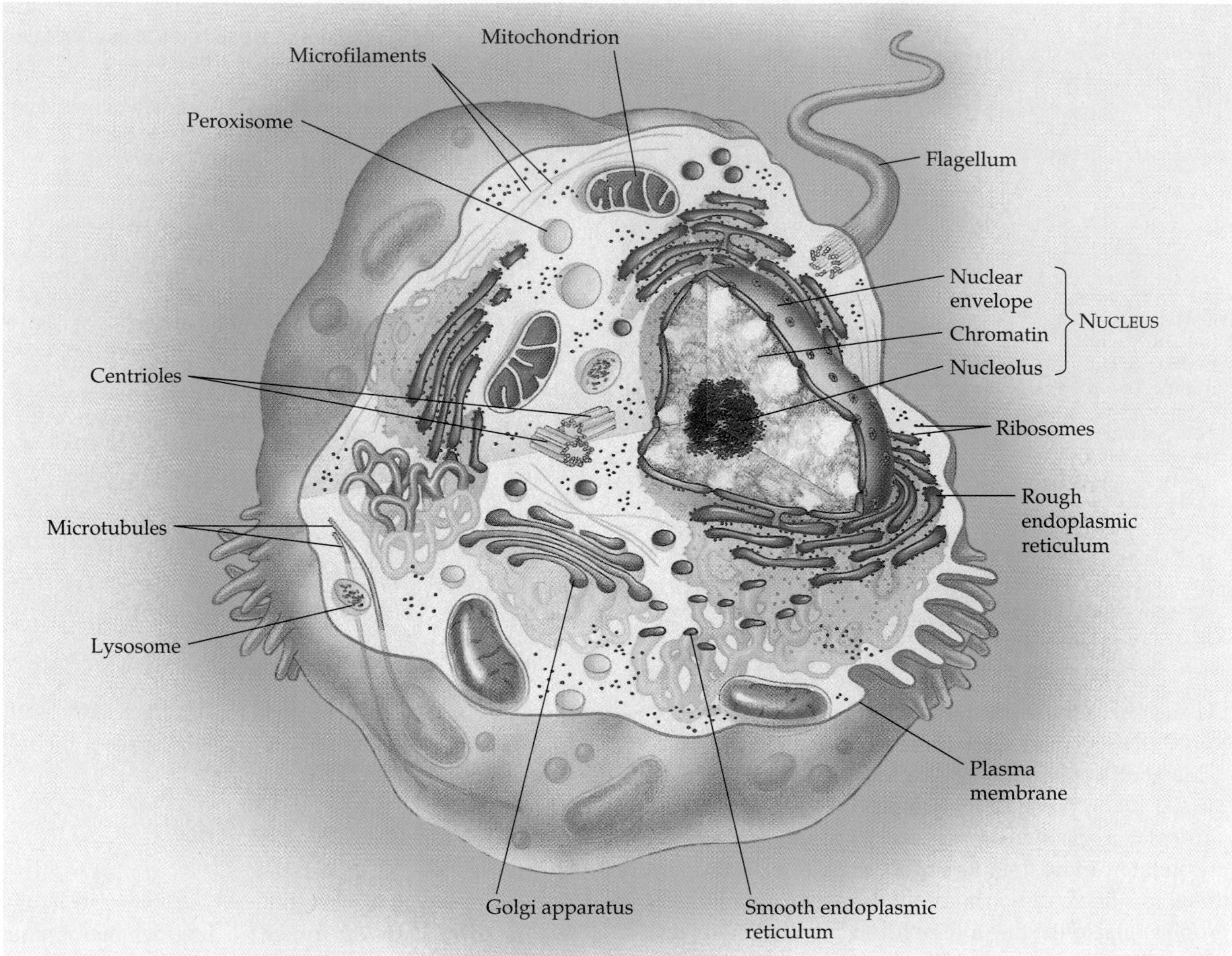

FIGURE 7.7

Overview of an animal cell. This drawing of a generalized animal cell combines the most common structures found in animal cells. No single cell looks like this. Within the cell are a variety of components collectively called organelles ("little organs"). Some organelles are bounded by membranes, while others lack membranes. The most prominent organelle in an animal cell is usually the nucleus, in which inherited genes reside in the form of DNA. The chromatin consists of this DNA along with proteins. The chromatin is actually a collection of separate structures called chromosomes, which are visible as separate units only in a dividing cell. Also present in the nucleus are one or more nucleoli (singular, nucleolus). Nucleoli are involved in the production of particles called ribosomes, which function in protein synthesis. The nucleus is bordered by a porous envelope consisting of two membranes.

Most of the cell's metabolic activities occur in the cytoplasm, the entire region between the nucleus and the plasma membrane surrounding the cell. The cytoplasm is full of specialized organelles suspended in a semifluid medium called the cytosol. Pervading much of the cytoplasm is the endoplasmic reticulum (ER), a labyrinth of membranes forming flattened sacs and tubes that segregate the contents of the ER from the cytosol. The ER takes two forms: rough (studded with ribosomes) and smooth. Many types of proteins are made by ribosomes attached to ER membranes, and the ER also plays a major role in assembling the other membranes of the cell. The Golgi apparatus, another type of membranous organelle in the cytoplasm, consists of stacks of flattened sacs that play an active role in the synthesis, refinement, storage, sorting, and secretion of chemical products by the cell.

Other classes of organelles enclosed by membranes are: lysosomes, which contain mixtures of digestive enzymes that hydrolyze macromolecules; peroxisomes, a diverse group of organelles containing specialized enzymes that perform specific metabolic processes; and vacuoles, which have a variety of storage and metabolic functions. The mitochondria (singular, mitochondrion) are organelles that generate ATP from organic fuels such as sugar in the process of cellular respiration.

Nonmembranous organelles within the cells include microtubules and microfilaments. They form a framework called the cytoskeleton, which reinforces the cell's shape and functions in cell movement. The cell in the drawing has a flagellum, an organelle of locomotion, which is an assembly of microtubules. Also made of microtubules are centrioles, located near the nucleus. These play a role in cell division.

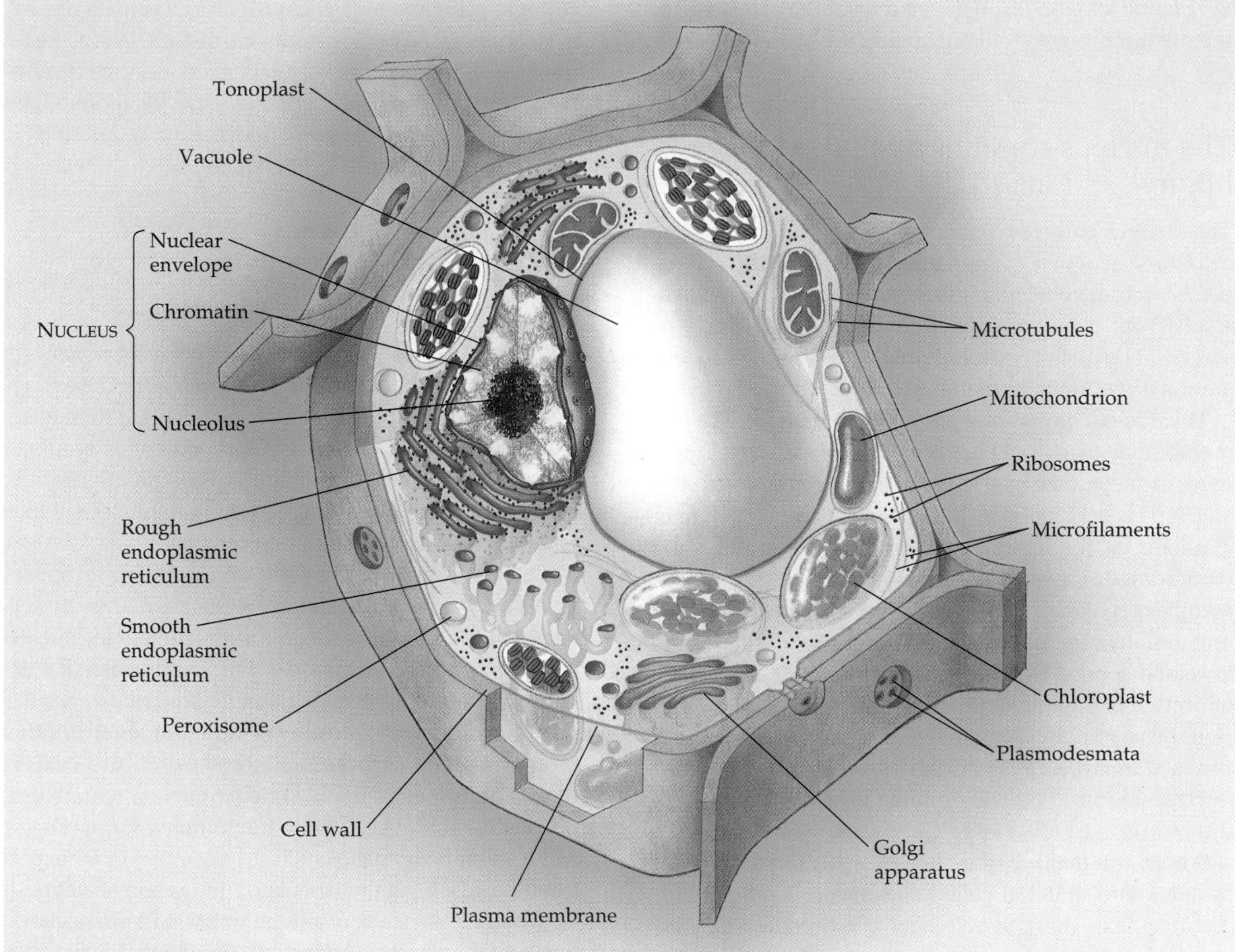

FIGURE 7.8

Overview of a plant cell. This drawing of a generalized plant cell reveals the similarities and differences between an animal cell and a plant cell. Like the animal cell, the plant cell is surrounded by a plasma membrane and contains a nucleus, ribosomes, ER, Golgi apparatus, mitochondria, peroxisomes, and microfilaments and microtubules. However, a plant cell also contains membrane-enclosed organelles called plastids. The most important type of plastid is the chloroplast, which carries out photosynthesis, converting sunlight to chemical energy stored in sugar and other organic molecules. Another prominent organelle in many plant cells, especially older ones, is a large central vacuole. The vacuole stores chemicals, breaks down macromolecules, and, by enlarging, plays a major role in plant growth. The membrane of the vacuole is called the tonoplast. Outside a plant cell's plasma membrane (as well as in fungi and some protists) is a thick cell wall, which helps maintain the cell's shape and protects the cell from mechanical damage. The cytosol of adjacent cells connects through trans-wall channels called plasmodesmata.

If you preview the rest of the chapter now, you'll see Figures 7.7 and 7.8 repeated in miniature as orientation diagrams. In each case, a particular organelle is highlighted, color-coded to its appearance in Figures 7.7 and 7.8. As we take a closer look at individual organelles, the orientation diagrams will help you place those structures in the context of the whole cell.

Membranes of various kinds are fundamental to the organization of the cell. In general, biological membranes consist of a double layer of phospholipids and other lipids. Embedded in this lipid bilayer or attached to its surfaces are diverse proteins (see FIGURE 7.6). However, each membrane has a unique composition of lipids and proteins suited to that membrane's specific functions. For example, enzymes that function in cellular respiration are embedded in the membranes of organelles called mitochondria.

Before continuing with this chapter, examine the overviews of eukaryotic cells in FIGURES 7.7 and 7.8. These figures and their legends introduce the various organelles and provide a map of the cell for the detailed tour upon which we will soon embark. FIGURES 7.7 and 7.8 also contrast animal and plant cells. As eukaryotic cells, they have much more in common with each other than either has with any prokaryote. As you will see, however, there are important differences between plant and animal cells.

The first stop on our detailed tour of the cell is one of the membrane-enclosed organelles, the nucleus.

The nucleus contains a cell's genetic library

The **nucleus** contains most of the genes that control the cell (some genes are located in mitochondria and chloroplasts). It is generally the most conspicuous organelle in a eukaryotic cell, averaging about 5 μm in diameter (FIGURE 7.9). The nuclear envelope encloses the nucleus, separating its contents from the cytoplasm.

The nuclear envelope is a double membrane. The two membranes, each a lipid bilayer with associated proteins, are separated by a space of about 20 to 40 nm. The envelope is perforated by pores that are about 100 nm in diameter. At the lip of each pore, the inner and outer membranes of the nuclear envelope are fused. The pore complex regulates the entrance and exit of certain large macromolecules and particles. The nuclear side of the envelope is lined by the **nuclear lamina,** a netlike array of protein filaments that maintains the shape of the nucleus. There is also growing evidence for a nuclear matrix, a framework of fibers distributed throughout the nuclear interior. (We will examine possible functions of this matrix in Chapter 18.)

Within the nucleus, the DNA is organized along with proteins into material called **chromatin.** Stained chromatin appears through both light microscopes and electron microscopes as a diffuse mass. As a cell prepares to divide (reproduce), the stringy, entangled chromatin condenses, becoming thick enough to be discerned as separate structures called **chromosomes.** Each eukaryotic species has a characteristic number of chromosomes. A human cell, for example, has 46 chromosomes in its nucleus; the exceptions are the sex cells—eggs and sperm—which have only 23 chromosomes in humans.

The most visible structure within the nondividing nucleus is the nucleolus, which synthesizes molecular ingredients of ribosomes. These ribosomal components pass through the nuclear pores to the cytoplasm, where their assembly is completed. Sometimes there are two or more nucleoli; the number depends on the species and the stage in the cell's reproductive cycle. The nucleolus is roughly spherical, and through the electron microscope it appears as a mass of densely stained granules and fibers.

The nucleus controls protein synthesis in the cytoplasm by sending molecular messengers in the form of ribonucleic acid (RNA). This messenger RNA (mRNA), as it is called, is synthesized in the nucleus according to instructions provided by the DNA. The mRNA then conveys the genetic messages to the cytoplasm via the nuclear pores. Once in the cytoplasm, the mRNA attaches to ribosomes, the sites where the genetic message is translated into the primary structure of a specific protein. This process of translating genetic information is described in detail in Chapter 16.

Ribosomes build a cell's proteins

Ribosomes are the sites where the cell assembles proteins. Cells that have high rates of protein synthesis have a particularly great number of ribosomes, another example of cell structure fitting function. For example, a human liver cell has a few million ribosomes. Cells active in protein synthesis also have prominent nucleoli, which function in ribosome production.

Ribosomes build proteins in two cytoplasmic locales (FIGURE 7.10). *Free* ribosomes are suspended in the cytosol, while *bound* ribosomes are attached to the outside of a membranous network called the endoplasmic reticulum. Most of the proteins made by free ribosomes will function within the cytosol; examples are enzymes that catalyze metabolic processes localized in the cytosol. Bound ribosomes generally make proteins that are destined either for inclusion into membranes, for packaging within certain organelles such as lysosomes, or for export from the cell. Cells that specialize in protein secretion—for instance, the cells of the pancreas and other glands that secrete digestive enzymes—frequently have a high proportion of bound ribosomes. Bound and free ribosomes are structurally identical and interchangeable, and the cell can adjust the relative numbers of each as its metabolism changes. You will learn more about ribosome structure and function in Chapter 16.

Many organelles are related through the endomembrane system

Many of the different membranes of the eukaryotic cell are part of an **endomembrane system.** These membranes are related either through direct physical continuity or by the transfer of membrane segments through the movement of tiny vesicles (membrane-enclosed sacs). These relationships, however, do not mean that the various membranes are alike in structure and function. The thickness, molecular composition, and metabolic behavior of a membrane are not fixed, but may be modified several times during the membrane's life. The endomembrane system includes the nuclear envelope,

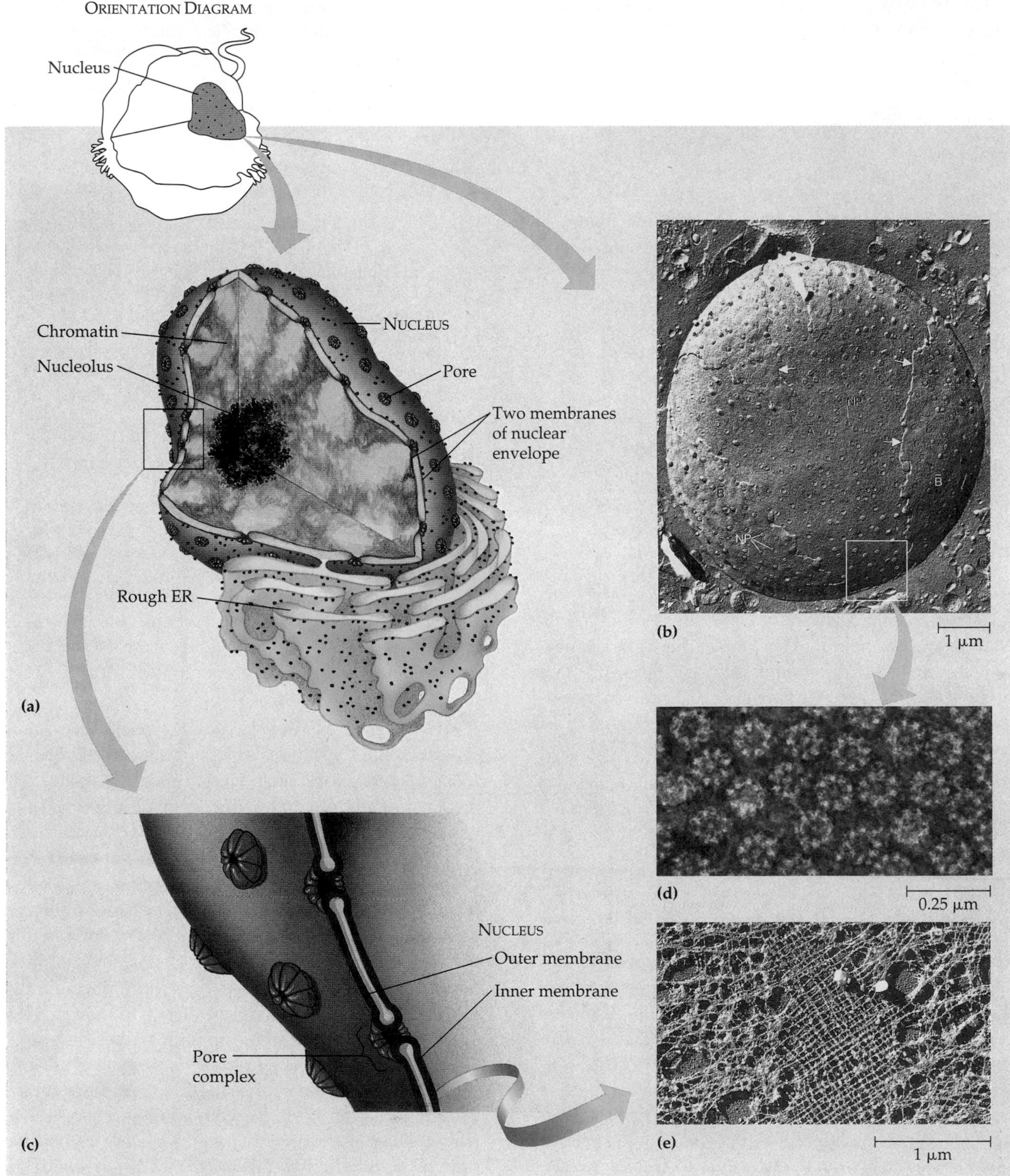

FIGURE 7.9

The nucleus and its envelope.
(**a**) Within the nucleus is chromatin, consisting of DNA and proteins. When a cell prepares to divide, individual chromosomes become visible as the chromatin condenses. The nucleolus functions in ribosome synthesis. The nuclear envelope, which consists of two membranes separated by a narrow space, is perforated with pores. (**b**) Numerous nuclear pores (NP) through the envelope are evident in this electron micrograph, prepared by a method called freeze-fracture (TEM). (The A and B labels distinguish the two membranes of the envelope.) (**c**) The nuclear envelope. (**d**) An electron micrograph of the outer surface of the envelope reveals that each pore is bordered by a ring of protein particles (TEM). (**e**) The netlike nuclear lamina lines the inner surface of the envelope and reinforces the shape of the nucleus (TEM).

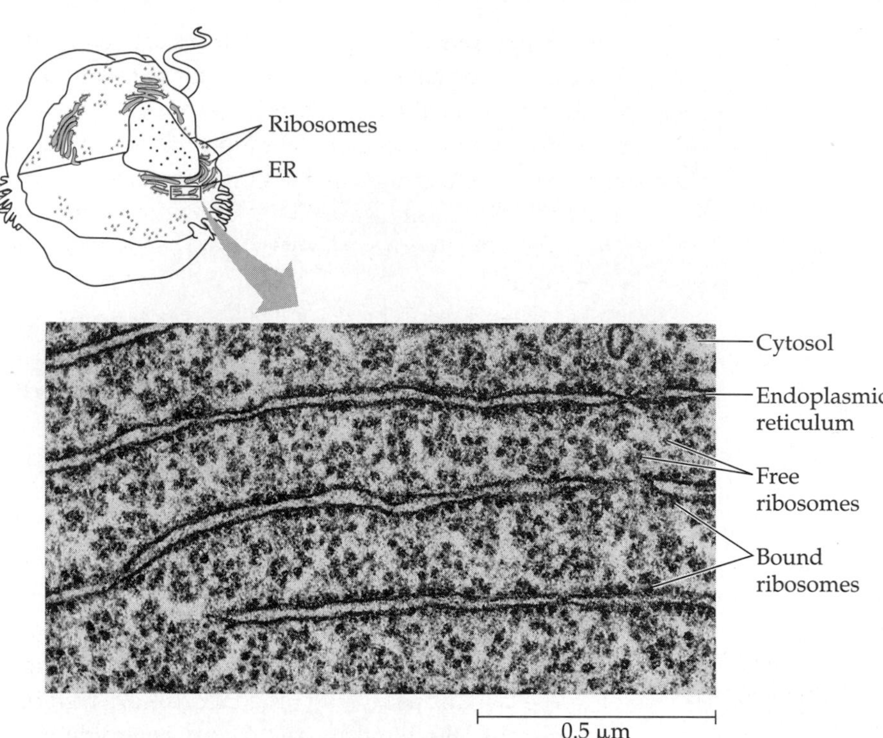

FIGURE 7.10
Ribosomes. Both free and bound ribosomes are abundant in this electron micrograph of a cell from the pancreas (TEM). The pancreas is a gland specialized for the secretion of proteins. It secretes hormones, including the protein insulin, into the bloodstream, and secretes digestive enzymes, which are proteins, into the intestine. Bound ribosomes, those presently attached to the endoplasmic reticulum (ER), produce secretory proteins. Free ribosomes mainly make proteins that will remain dissolved in the cytosol. Bound and free ribosomes are identical and can alternate between these two roles.

endoplasmic reticulum, Golgi apparatus, lysosomes, various kinds of vacuoles, and the plasma membrane (not actually an *endo*membrane in physical location, but nevertheless related to the endoplasmic reticulum and other internal membranes). We have already discussed the nuclear envelope and will now focus on the endoplasmic reticulum and the other endomembranes to which it gives rise.

The endoplasmic reticulum manufactures membranes and performs many other biosynthetic functions

The **endoplasmic reticulum (ER)** is a membranous labyrinth so extensive that it accounts for more than half the total membrane in many eukaryotic cells. (The word *endoplasmic* means "within" the *cytoplasm*, and *reticulum* is derived from the Latin for "network.") The ER consists of a network of membranous tubules and sacs called cisternae (L. *cisterna*, "box" or "chest"). The ER membrane separates its internal compartment, the cisternal space, from the cytosol. And because the ER membrane is continuous with the nuclear envelope, the space between the two membranes of the envelope is continuous with the cisternal space of the ER (FIGURE 7.11).

There are two distinct, though connected, regions of ER that differ in structure and function: smooth ER and rough ER. **Smooth ER** is so named because its cytoplasmic surface lacks ribosomes. **Rough ER** appears rough through the electron microscope because ribosomes stud the cytoplasmic surface of the membrane. Ribosomes are also attached to the cytoplasmic side of the nuclear envelope's outer membrane, which is confluent with rough ER.

Functions of Smooth ER

The smooth ER of various cell types functions in diverse metabolic processes, including synthesis of lipids, metabolism of carbohydrates, and detoxification of drugs and other poisons.

Enzymes of the smooth ER are important to the synthesis of fatty acids, phospholipids, steroids, and other lipids. Among the steroids produced by smooth ER are the sex hormones of vertebrates and the various steroid hormones secreted by the adrenal glands. The cells that actually synthesize and secrete these hormones—in the testes and ovaries, for example—are rich in smooth ER, a structural feature that fits the function of these cells.

Liver cells provide one example of the role of smooth ER in carbohydrate metabolism. Liver cells store carbohydrate in the form of glycogen, a polysaccharide. The hydrolysis of glycogen leads to the release of glucose from the liver cells, which is important in the regulation of sugar concentration in the blood. However, the first product of glycogen hydrolysis is glucose phosphate, an ionic form of the sugar that cannot exit the cell and enter the blood. An enzyme embedded in the membrane of the liver cell's smooth ER removes the phosphate from the glucose, which can then leave the cell and elevate blood sugar concentration.

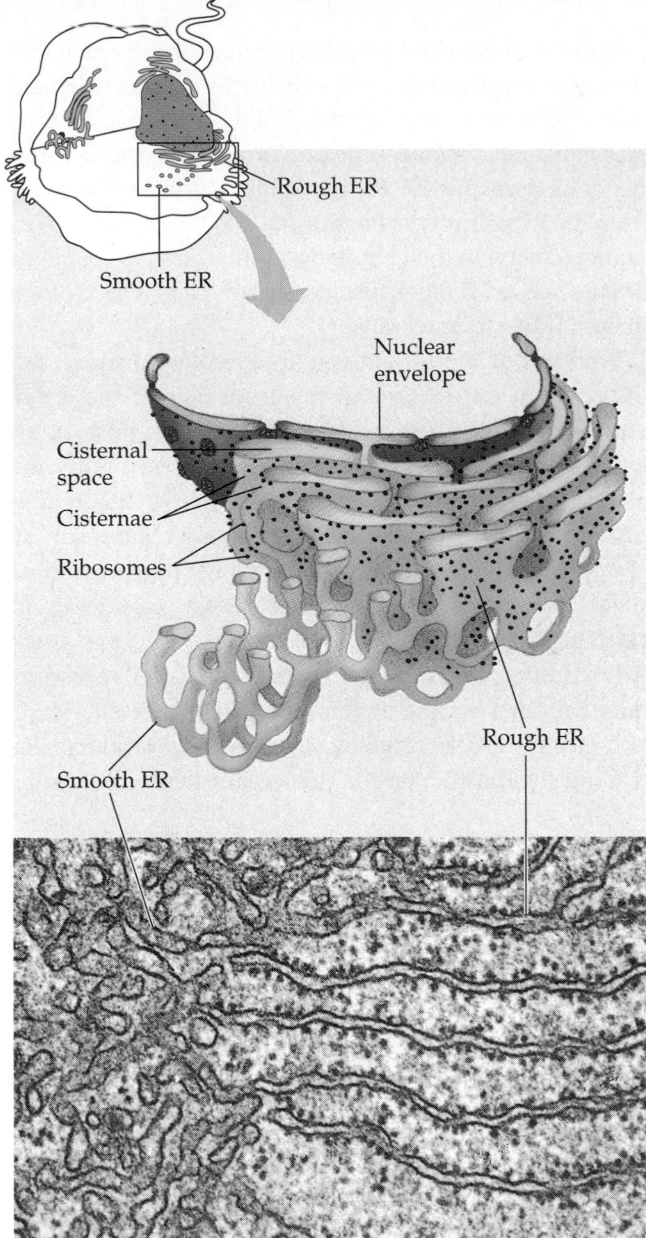

FIGURE 7.11
Endoplasmic reticulum (ER). A membranous system of interconnected tubules and flattened sacs called cisternae, the ER is also continuous with the nuclear envelope. (The drawing is a cutaway view.) The membrane of the ER encloses a compartment called the cisternal space. Rough ER, which is studded on its cytoplasmic surface with ribosomes, can be distinguished from smooth ER in the electron micrograph (TEM).

Enzymes of the smooth ER help detoxify drugs and other poisons, especially in liver cells. Detoxification usually involves adding hydroxyl groups to drugs, increasing their solubility and making it easier to flush the compounds from the body. The sedative phenobarbital and other barbiturates are examples of drugs metabolized in this manner by smooth ER in liver cells. In fact, barbiturates, alcohol, and many other drugs induce the proliferation of smooth ER and its associated detoxification enzymes. This in turn increases tolerance to the drugs, meaning that higher doses are required to achieve a particular effect, such as sedation. Also, because some of the detoxification enzymes have relatively broad action, the proliferation of smooth ER in response to one drug can increase tolerance to other drugs as well. Barbiturate abuse, for example, may decrease the effectiveness of certain antibiotics and other useful drugs.

Muscle cells exhibit still another specialized function of smooth ER. The ER membrane pumps calcium ions from the cytosol into the cisternal space. When a muscle cell is stimulated by a nerve impulse, calcium rushes back across the ER membrane into the cytosol and triggers contraction of the muscle cell.

Rough ER and Protein Synthesis

Many types of specialized cells secrete proteins produced by ribosomes attached to rough ER. For example, certain cells in the pancreas secrete the protein insulin, a hormone, into the bloodstream. As a polypeptide chain grows from a bound ribosome, it is threaded through the ER membrane into the cisternal space, possibly through a pore. As it enters the cisternal space, the protein folds into its native conformation. Most secretory proteins are **glycoproteins,** proteins that are covalently bonded to carbohydrates. In the cisternal space, the carbohydrate is attached to the protein by specialized molecules built into the ER membrane. The carbohydrate appendage of a glycoprotein is an oligosaccharide, the term for a relatively small polymer of sugar units.

Once the secretory proteins are formed, the ER membrane keeps them separate from the proteins produced by free ribosomes that will remain in the cytosol. Secretory proteins depart from the ER wrapped in the membranes of vesicles budded like bubbles from a specialized region called transitional ER. Such vesicles in transit from one part of the cell to another are called **transport vesicles,** and we will soon learn their fate.

Rough ER and Membrane Production

In addition to making secretory proteins, rough ER is a membrane factory that grows in place by adding proteins and phospholipids. As membrane proteins elongate from the ribosomes, they are inserted into the ER membrane itself and are anchored there by hydrophobic portions of the proteins. The rough ER also makes its own membrane phospholipids; enzymes built into the ER membrane assemble phospholipids from precursors in the cytosol. The ER membrane expands and can be transferred in the form of transport vesicles targeted for other components of the endomembrane system.

The Golgi apparatus finishes, sorts, and ships many products of the cell

After leaving the ER, many transport vesicles travel to the **Golgi apparatus.** We can think of the Golgi as a center of manufacturing, warehousing, sorting, and shipping. Here, products of the ER are modified and stored, and then sent to other destinations. Not surprisingly, the Golgi apparatus is especially extensive in cells specialized for secretion.

The Golgi apparatus consists of flattened membranous sacs, looking like a stack of pita bread (FIGURE 7.12). A cell may have several of these stacks, all interconnected. Each cisterna in a stack consists of a membrane that separates its internal space from the cytosol. Vesicles concentrated in the vicinity of the Golgi apparatus are engaged in the transfer of material between the Golgi and other structures.

The Golgi apparatus has a distinct polarity, with the membranes of cisternae at opposite ends of a stack differing in thickness and molecular composition. The two poles of a Golgi stack are referred to as the *cis* face and the *trans* face; these act, respectively, as the receiving and shipping departments of the Golgi apparatus. The *cis* face is usually located near ER. Transport vesicles move material from the ER to the Golgi. A vesicle that buds from the ER will add its membrane and the contents of its lumen (cavity) to the *cis* face by fusing with a Golgi membrane. The *trans* face gives rise to vesicles, which pinch off and travel to other sites.

Products of the ER are usually modified during their transit from the *cis* pole to the *trans* pole of the Golgi. Proteins and phospholipids of membranes may be altered. For example, various Golgi enzymes modify the oligosaccharide portions of glycoproteins. When first added to proteins in the ER, the oligosaccharides of all glycoproteins are identical. The Golgi removes some sugar monomers and substitutes others, producing diverse oligosaccharides.

In addition to its finishing work, the Golgi apparatus manufactures certain macromolecules by itself. Many polysaccharides secreted by cells are Golgi products, including hyaluronic acid, a sticky substance that helps

FIGURE 7.12
The Golgi apparatus. A Golgi apparatus consists of stacks of flattened membranous sacs. (The drawing is a cutaway view.) The apparatus receives and dispatches transport vesicles and the products they contain. Materials received from the ER are modified and stored in the Golgi and eventually shipped to the cell surface or other destinations. Note the vesicles forming at the edges of the stack and also the free vesicles that have just arisen. The stacks have a structural and functional polarity, with a *cis* face that receives vesicles and a *trans* face that dispatches vesicles (right, TEM).

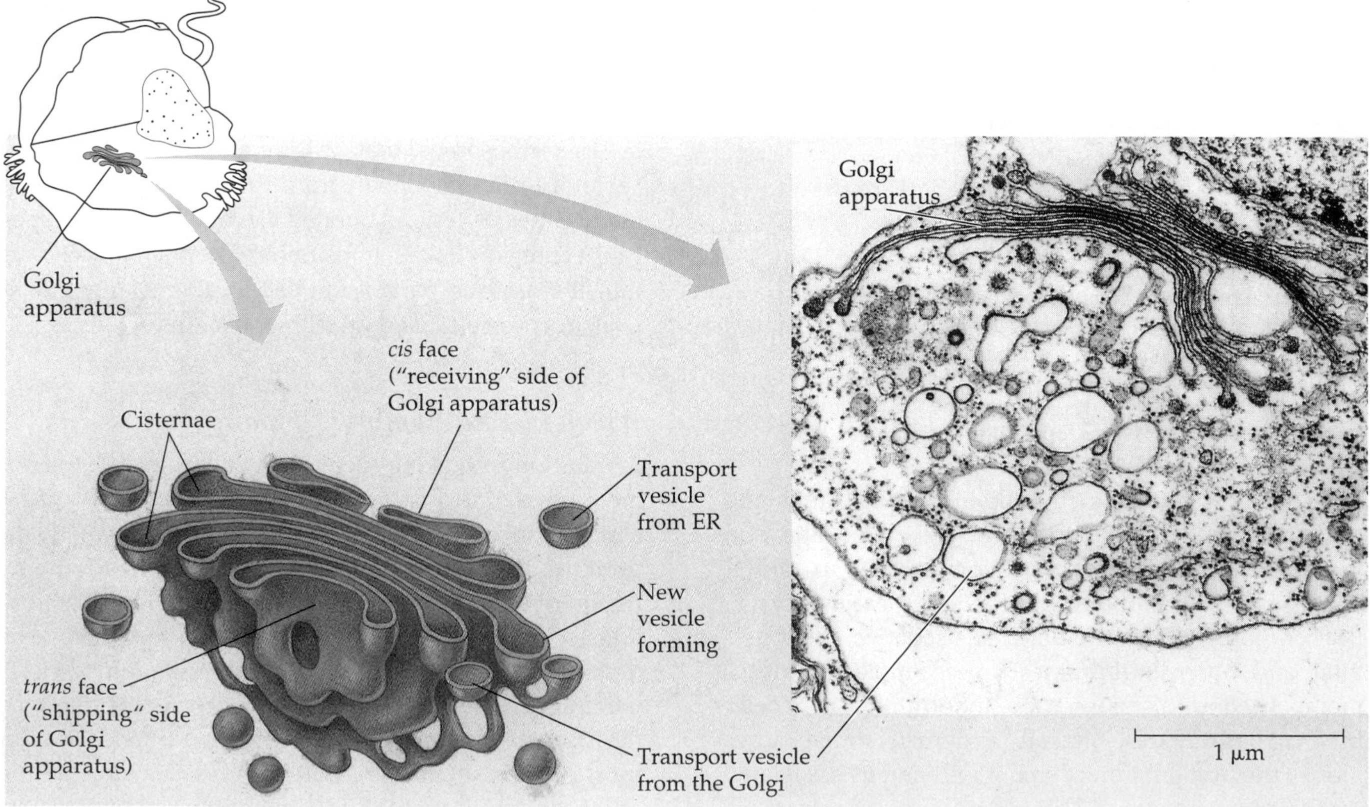

glue animal cells together. Golgi products that will be secreted depart from the *trans* faces of Golgi in the lumen of transport vesicles that eventually fuse with the plasma membrane.

The Golgi manufactures and refines its products in stages, with different cisternae between the *cis* and *trans* ends containing unique teams of enzymes. Products in various stages of processing are transferred from one cisterna to the next by vesicles.

Before the Golgi apparatus dispatches its products by budding vesicles from the *trans* face, it sorts these products and targets them for various parts of the cell. Molecular identification tags, such as phosphate groups that have been added to the Golgi products, aid in sorting. And vesicles budded from the Golgi may have external molecules on their membranes that recognize "docking sites" on the surface of specific organelles.

Lyosomes are digestive compartments

A lysosome is a membrane-enclosed sac of hydrolytic enzymes that the cell uses to digest macromolecules (FIGURE 7.13). There are lysosomal enzymes that can hydrolyze proteins, polysaccharides, fats, and nucleic acids—all the major classes of macromolecules. These enzymes work best in an acidic environment, at about pH 5. The lysosomal membrane maintains this low internal pH by pumping hydrogen ions from the cytosol into the lumen of the lysosome. If the lysosome should break open or leak its contents, the enzymes would not be very active in the neutral environment of the cytosol. However, excessive leakage from a large number of lysosomes can destroy a cell by autodigestion. From this example, we can see once again how important compartmental organization is to the functions of the cell: The lysosome provides a space where the cell can digest macromolecules safely, without the general destruction that would occur if hydrolytic enzymes roamed at large.

Hydrolytic enzymes and lysosomal membrane are made by rough ER and then transferred to a Golgi apparatus for further processing. Lysosomes probably arise by budding from the *trans* face of the Golgi apparatus. Proteins of the inner surface of the lysosomal membrane and the digestive enzymes themselves are probably spared from self-destruction by having three-dimensional conformations that protect vulnerable bonds from enzymatic attack.

Lysosomes function in intracellular digestion in a variety of circumstances. *Amoeba* and many other protists eat by engulfing smaller organisms or other food particles, a process called **phagocytosis** (Gr. *phagein*, "to eat," and *kytos*, "vessel," referring here to the cell). The food vacuole formed in this way then fuses with a lysosome, whose enzymes digest the food (FIGURE 7.14). Some human cells also carry out phagocytosis. Among them are macrophages, cells that help defend the body by destroying bacteria and other invaders.

Lysosomes also use their hydrolytic enzymes to recycle the cell's own organic material, a process called autophagy. This occurs when a lysosome engulfs another organelle or a small parcel of cytosol (see FIGURE 7.13b). The lysosomal enzymes dismantle the ingested material,

FIGURE 7.13

Lysosomes. (**a**) In this white blood cell from a rat, the lysosomes are very dark because of a specific stain that reacts with one of the products of digestion within the lysosome. This type of white blood cell ingests bacteria and viruses and destroys them in the lysosomes (TEM). (**b**) In the cytoplasm of this rat liver cell, an autophagic lysosome has engulfed two disabled organelles, a mitochondrion and a peroxisome (TEM).

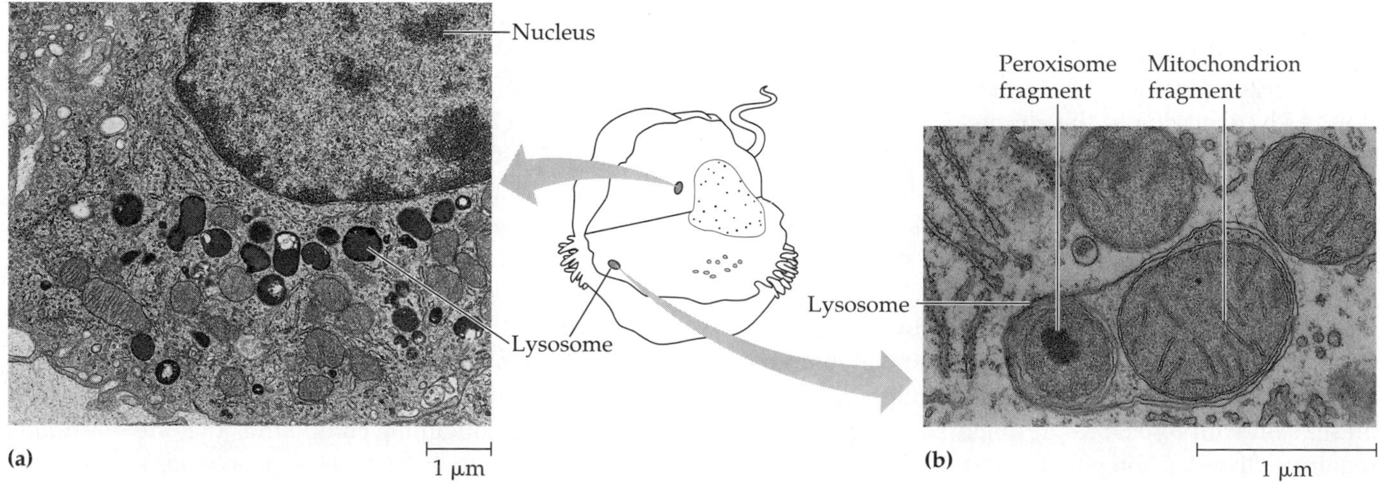

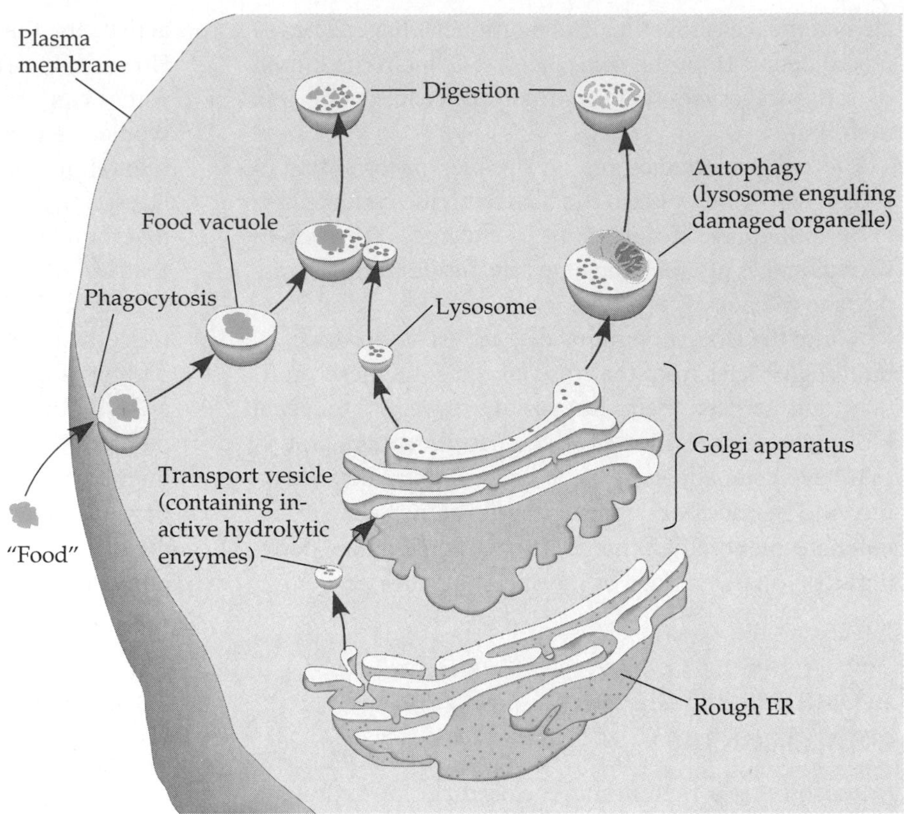

FIGURE 7.14
The formation and functions of lysosomes. Lysosomes digest materials taken into the cell and recycle materials from intracellular refuse. During phagocytosis, the cell encloses food in a vacuole with a membrane that pinches off internally from the plasma membrane. This food vacuole fuses with a lysosome, and hydrolytic enzymes digest the food. After hydrolysis, simple sugars, amino acids, and other monomers pass across the lysosomal membrane into the cytosol as nutrients for the cell. By the process of autophagy, lysosomes recycle the molecular ingredients of organelles. The ER and Golgi may cooperate in the production of lysosomes containing active enzymes, although some lysosomes may bud directly from specialized regions of the ER.

and the organic monomers are returned to the cytosol for reuse. With the help of lysosomes, the cell continually renews itself. A human liver cell, for example, recycles half of its macromolecules each week.

Programmed destruction of cells by their own lysosomal enzymes is important in the development of many organisms. During the transforming of a tadpole into a frog, for instance, lysosomes destroy the cells of the tail. And the hands of human embryos are webbed until lysosomes digest the tissue between the fingers.

A variety of inherited disorders called storage diseases affect lysosomal metabolism. A person afflicted with a storage disease lacks one of the active hydrolytic enzymes normally present in lysosomes. The lysosomes become engorged with indigestible substrates, which begin to interfere with other cellular functions. In Pompe's disease, for example, the liver is damaged by an accumulation of glycogen due to the absence of a lysosomal enzyme needed to break down the polysaccharide. In Tay-Sachs disease, a lipid-digesting enzyme is missing or inactive, and the brain becomes impaired by an accumulation of lipids in the cells. Fortunately, storage diseases are rare in the general population. In the future, it might be possible to treat storage diseases by injecting the missing enzymes into the blood along with adaptor molecules that target the enzymes for engulfment by cells and fusion with lysosomes. It might also be possible to repair a disorder directly by inserting genes (DNA) for the missing enzyme into the appropriate cells (see Chapter 19).

Vacuoles have diverse functions in cell maintenance

Vacuoles and vesicles are both membrane-enclosed sacs within the cell, but vacuoles are larger than vesicles. Vacuoles have various functions. **Food vacuoles,** formed by phagocytosis, have already been mentioned (see FIGURE 7.14). Many freshwater protists have **contractile vacuoles** that pump excess water out of the cell. Mature plant cells generally contain a large **central vacuole** enclosed by a membrane called the **tonoplast,** which is part of their endomembrane system (FIGURE 7.15).

The plant cell vacuole is a versatile compartment. It is a place to store organic compounds, such as the proteins that are stockpiled in the vacuoles of storage cells in seeds. The vacuole is also the plant cell's main repository of inorganic ions, such as potassium and chloride. Many plant cells use their vacuoles as disposal sites for metabolic by-products that would endanger the cell if they accumulated in the cytoplasm. Some vacuoles are enriched in pigments that color the cells, such as the red and blue pigments of petals that help attract pollinating insects to flowers. Vacuoles may also help protect the plant against predators by containing compounds that are poisonous or unpalatable to animals. The vacuole has a major role

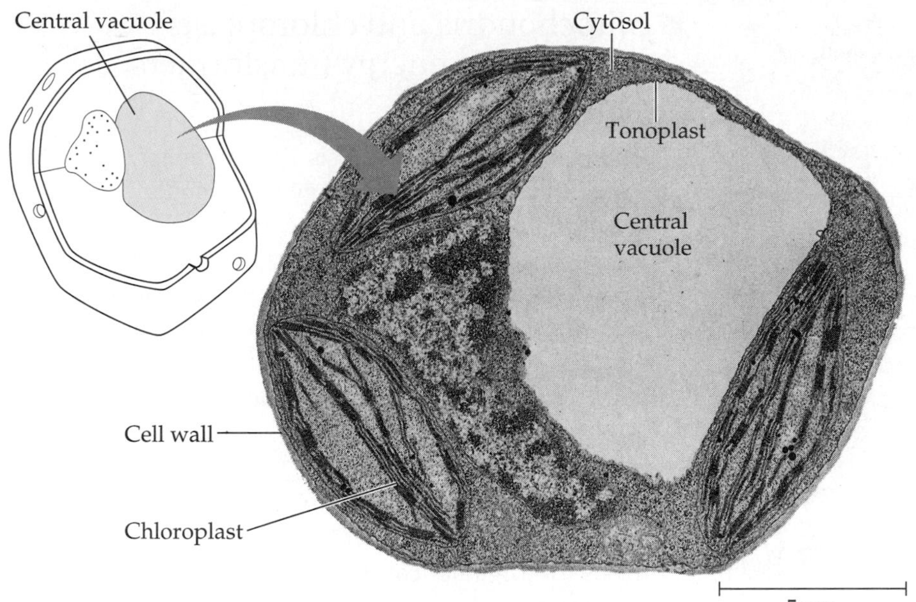

FIGURE 7.15

The plant cell vacuole. The central vacuole is usually the largest compartment in a plant cell, comprising 80% or more of a mature cell. The cytoplasm is generally confined to a narrow zone between the vacuole and the plasma membrane. The membrane enclosing the vacuole, the tonoplast, separates the cytosol from the solution inside the vacuole, which is called cell sap. Like all cellular membranes, the tonoplast is selective in transporting solutes; therefore, cell sap differs in composition from the cytosol. Functions of the vacuole include storage, waste disposal, protection, and growth (TEM).

in the growth of plant cells, which elongate as their vacuoles absorb water, enabling the cell to become larger with a minimal investment in new cytoplasm. And because the cytoplasm occupies a thin shell between the plasma membrane and the tonoplast, the ratio of membrane surface to cytoplasmic volume is great, even for a large plant cell.

The large vacuole of a plant cell develops by the coalescence of smaller vacuoles, themselves derived from the endoplasmic reticulum and Golgi apparatus. Through these relationships, the vacuole is an integral part of the endomembrane system. FIGURE 7.16 reviews the endomembrane system. We'll continue our tour of the cell with three organelles that are *not* closely related to the endomembrane system: peroxisomes, mitochondria, and chloroplasts.

Peroxisomes consume oxygen in various metabolic functions

The **peroxisome** is a specialized metabolic compartment bounded by a single membrane (FIGURE 7.17).

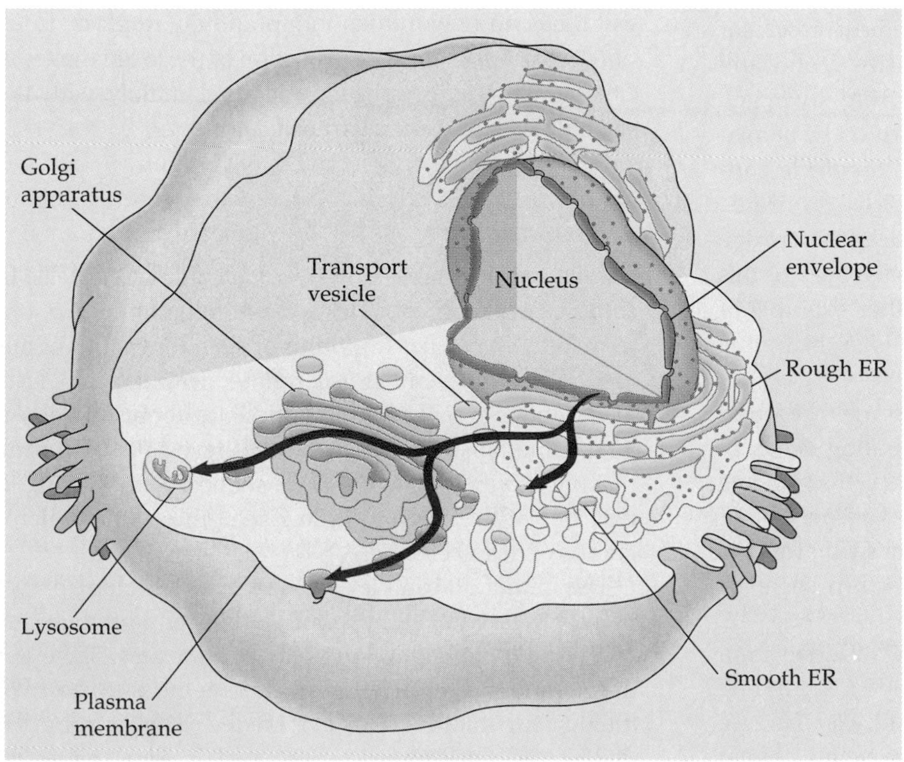

FIGURE 7.16

Review: relationships among endomembranes. The nuclear envelope is an extension of the rough ER, which is also confluent with smooth ER. Membrane produced by the ER flows in the form of transport vesicles to the Golgi. The Golgi, in turn, pinches off vesicles that give rise to lysosomes and vacuoles. Even the plasma membrane expands by the fusion of vesicles born in the ER and Golgi. (Coalescence of vesicles with the plasma membrane also releases secretory proteins and other products to the outside of the cell.) As membranes of the system flow from ER to the Golgi and then elsewhere, their molecular compositions and metabolic functions are modified. The endomembrane system is a complex and dynamic player in the cell's compartmental organization.

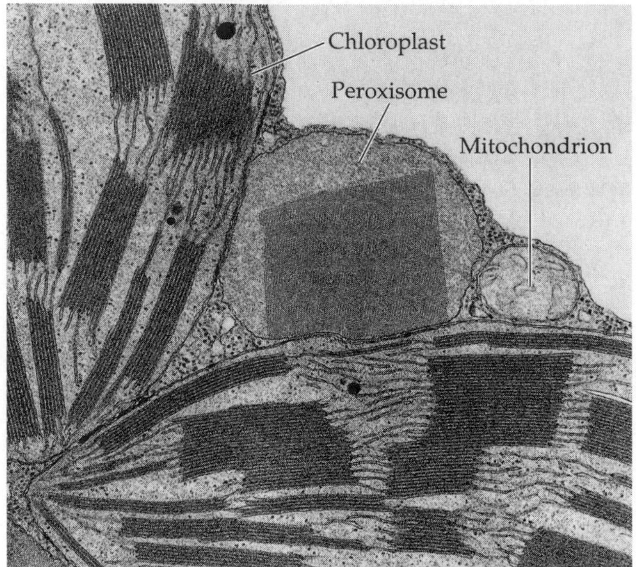

FIGURE 7.17
Peroxisomes. Peroxisomes are roughly spherical and often have a granular or crystalline core that is probably a dense collection of enzymes. This peroxisome is in a leaf cell. Notice its close relationship with mitochondria and choloroplasts, which cooperate with peroxisomes in certain metabolic functions (TEM).

Peroxisomes contain enzymes that transfer hydrogen from various substrates to oxygen, producing hydrogen peroxide (H_2O_2) as a by-product, from which the organelle derives its name. These reactions may have many different functions. Some peroxisomes use oxygen to break fatty acids down into smaller molecules that can then be transported to mitochondria as fuel for cellular respiration. Peroxisomes in the liver detoxify alcohol and other harmful compounds by transferring hydrogen from the poisons to oxygen. The H_2O_2 formed by peroxisome metabolism is itself toxic, but the organelle contains an enzyme that converts the H_2O_2 to water. Packaging the enzymes that produce hydrogen peroxide in an enclosure where an enzyme that disposes of this by-product is also concentrated is another example of how the cell's compartmental structure is crucial to its functions.

Specialized peroxisomes called glyoxysomes are found in the fat-storing tissues of the germinating seeds of plants. These organelles contain enzymes that initiate the conversion of fatty acids to sugar, a process that makes the energy stored in the oils of the seed available until the seedling is able to produce its own sugar by photosynthesis.

Unlike lysosomes, peroxisomes are not budded from the endomembrane system. They grow by incorporating proteins and lipids produced in the cytosol, and increase in number by splitting in two when they reach a certain size.

Mitochondria and chloroplasts are the main energy transformers of cells

One of this book's themes is that organisms are open systems that transform energy they acquire from their surroundings. In eukaryotic cells, mitochondria and chloroplasts are the organelles that convert energy to forms that cells can use for work. **Mitochondria** (singular, **mitochondrion**) are the sites of cellular respiration, the catabolic process that generates ATP by extracting energy from sugars, fats, and other fuels with the help of oxygen. **Chloroplasts,** found only in plants and eukaryotic algae (kingdom Protista), are the sites of photosynthesis. They convert solar energy to chemical energy by absorbing sunlight and using it to drive the synthesis of organic compounds from carbon dioxide and water.

Although mitochondria and chloroplasts are enclosed by membranes, they are not considered part of the endomembrane system. Their membrane proteins are made not by the ER, but by free ribosomes in the cytosol and by ribosomes contained within the mitochondria and chloroplasts themselves. Not only do these organelles have ribosomes, but they also contain a small amount of DNA that programs the synthesis of some of their own proteins (although most proteins in these organelles are made in the cytosol, programmed by messenger RNA sent by nuclear genes). Mitochondria and chloroplasts are semiautonomous organelles that grow and reproduce within the cell. In Chapters 9 and 10, we will focus on how mitochondria and chloroplasts function. We will consider the evolution of these organelles in Chapter 26. Here, we are concerned mainly with the structure of these energy transformers.

Mitochondria

Mitochondria are found in nearly all eukaryotic cells. In some cases, there is a single large mitochondrion, but more often, a cell has hundreds or even thousands of mitochondria; the number is correlated with the cell's level of metabolic activity. Mitochondria are about 1 to 10 µm long. Time-lapse films of living cells reveal mitochondria moving around, changing their shapes, and dividing in two, unlike the static cylinders seen in electron micrographs of dead cells.

The mitochondrion is enclosed in an envelope of two membranes, each a phospholipid bilayer with a unique collection of embedded proteins (FIGURE 7.18). The outer membrane is smooth, but the inner membrane is convoluted, with infoldings called **cristae.** The membranes divide the mitochondrion into two internal compartments. The first is the intermembrane space, the narrow

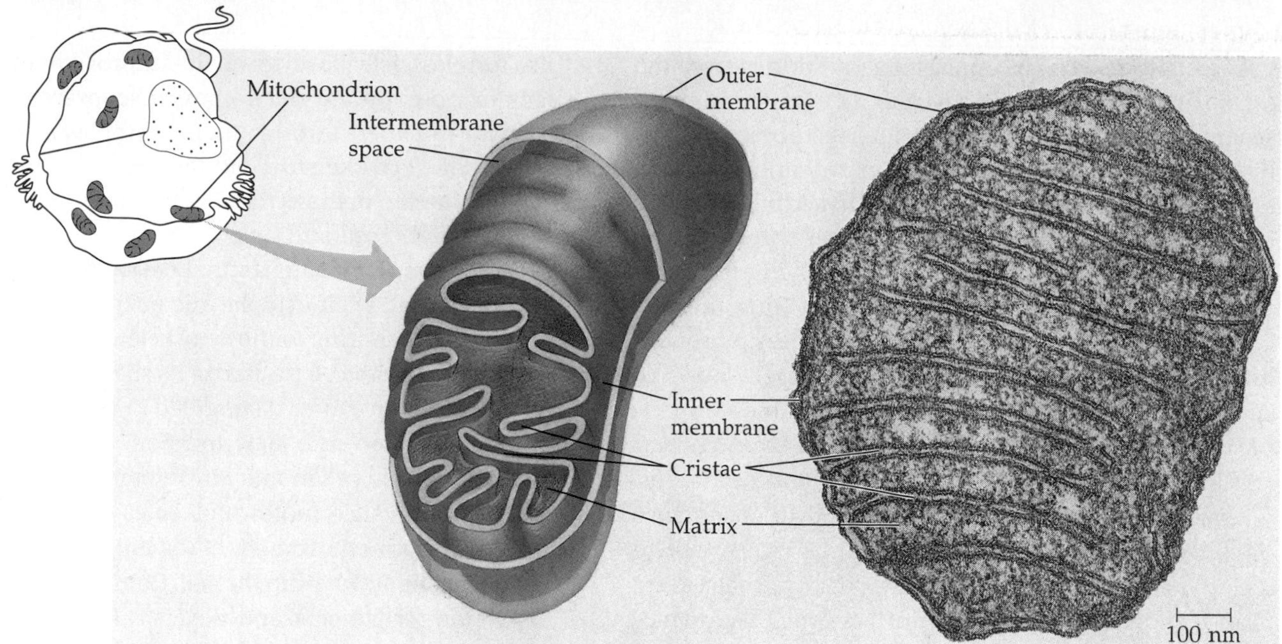

FIGURE 7.18

The mitochondrion, site of cellular respiration. The double membrane of the mitochondrion is evident in the drawing and the micrograph (TEM). The cristae are infoldings of the inner membrane. The cutaway drawing emphasizes the relationships between the two membranes and the compartments they bound: the intermembrane space and the mitochondrial matrix.

region between the inner and outer membranes. The second compartment, the **mitochondrial matrix,** is enclosed by the inner membrane. Some of the metabolic steps of cellular respiration occur in the matrix, where many different enzymes are concentrated. Other proteins that function in respiration, including the enzyme that makes ATP, are built into the inner membrane. The cristae give the inner mitochondrial membrane a large surface area that enhances the productivity of cellular respiration, another example of the correlation between structure and function.

Chloroplasts

The chloroplast is a specialized member of a family of closely related plant organelles called **plastids.** Amyloplasts are colorless plastids that store starch (amylose), particularly in roots and tubers. Chromoplasts are enriched in pigments that give fruits and flowers their orange and yellow hues. Chloroplasts contain the green pigment chlorophyll along with enzymes and other molecules that function in the photosynthetic production of food. These lens-shaped organelles, measuring about 2 μm by 5 μm, are found in leaves and other green organs of plants and in eukaryotic algae (FIGURE 7.19).

FIGURE 7.19

The chloroplast, site of photosynthesis. Chloroplasts, like mitochondria, are enclosed by two membranes separated by a narrow intermembrane space. The inner membrane encloses fluid called stroma. The stroma surrounds a third compartment, delineated by its own membrane, the thylakoid membrane. Throughout the chloroplast, thylakoid sacs are stacked to form structures called grana. Individual thylakoids of one granum are continuous with other grana through extensions that traverse the stroma (TEM).

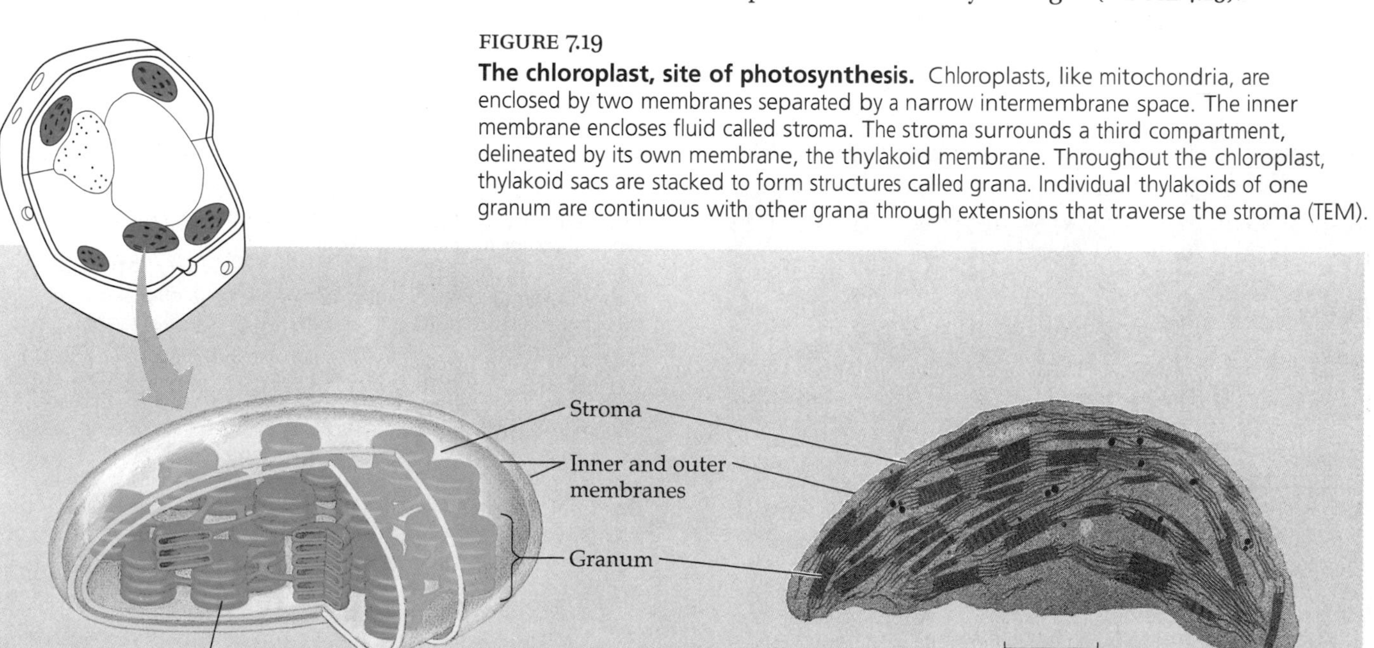

The contents of a chloroplast are partitioned from the cytosol by an envelope consisting of two membranes separated by a very narrow intermembrane space. Inside the chloroplast is another membranous system, arranged into flattened sacs called **thylakoids.** In some regions, thylakoids are stacked like poker chips, forming structures called **grana** (singular, **granum**). The fluid outside the thylakoids is called the **stroma.** Thus, the thylakoid membrane divides the interior of the chloroplast into two compartments: the thylakoid space and the stroma. In Chapter 10, you will learn how this compartmental organization enables the chloroplast to convert light energy to chemical energy during photosynthesis.

As with mitochondria, the static and rigid appearance of chloroplasts in electron micrographs is not true to their dynamic behavior in the living cell. Their shapes are plastic, and they occasionally pinch in two. They are mobile and move around the cell with mitochondria and other organelles along tracks of the cytoskeleton, the next stop on our tour of the cell.

The cytoskeleton provides structural support and functions in cell motility

In the early days of electron microscopy, the cell seemed to consist of a variety of organelles suspended or floating in a formless, jellylike cytosol. But improvements in both light microscopy and electron microscopy have revealed a network of fibers throughout the cytoplasm. This mesh is called the **cytoskeleton** (FIGURE 7.20).

One function of the cytoskeleton is to give mechanical support to the cell and help maintain its shape. This is especially important for animal cells, which lack walls. Organelles and even cytoplasmic enzymes may be held in place by anchoring to the cytoskeleton. The cytoskeleton also enables a cell to change its shape; like a scaffold, the cytoskeleton can be dismantled in one part of the cell and reassembled in a new location. The cytoskeleton also functions in cell motility by interacting with specialized proteins called motor molecules (FIGURE 7.21). This motility includes movement of the entire cell and movement of organelles within the cell. Components of the cytoskeleton wiggle cilia and flagella and enable muscle

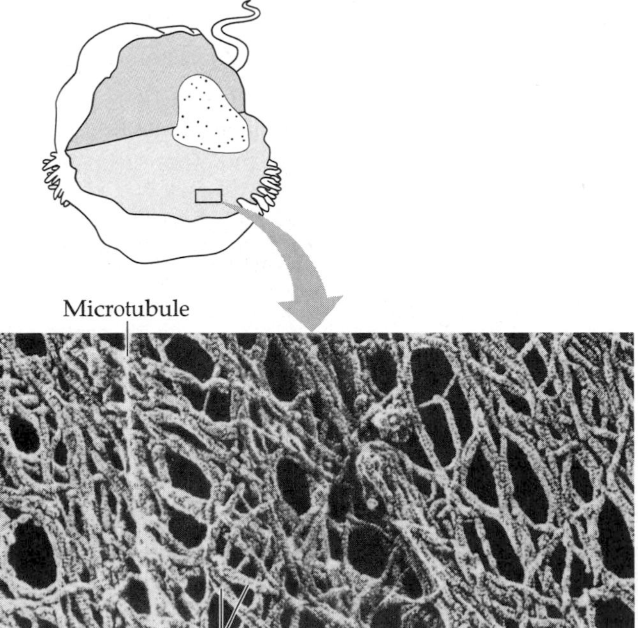

FIGURE 7.20
The cytoskeleton. The cytoskeleton gives the cell shape, anchors some organelles and directs the movement of others, and may enable the entire cell to change shape or move. In this electron micrograph prepared by a method known as deep-etching, microtubules and microfilaments are visible. A third component of the cytoskeleton, intermediate filaments, is not evident (TEM).

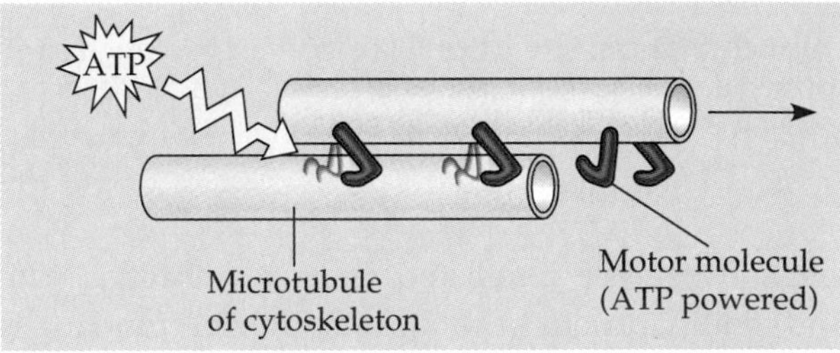

(a)

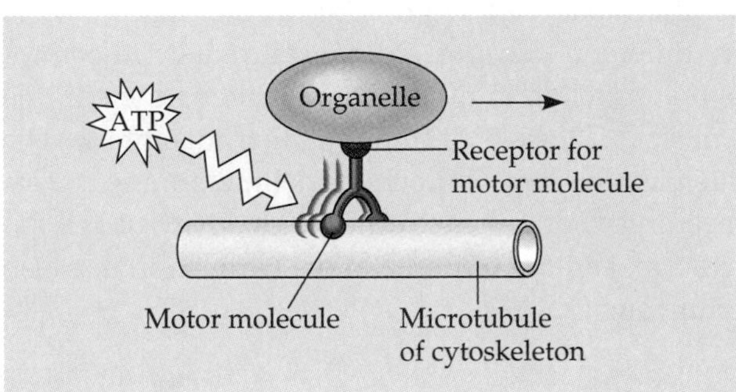

(b)

FIGURE 7.21
Motor molecules and the cytoskeleton. The microtubules and microfilaments of the cytoskeleton function in motility by interacting with protein complexes called motor molecules. Various types of motor molecules all work by changing their shapes, moving back and forth something like microscopic legs. ATP powers these conformational changes. With each cycle of shape change, the motor releases at its free end and grips at a site farther along a microtubule or microfilament. (**a**) In some types of cell motility, motor molecules attached to one element of the cytoskeleton cause it to slide over another cytoskeletal element. For example, a sliding of neighboring microtubules moves cilia and flagella. A similar mechanism causes muscle cells to contract, but in this case motor molecules slide microfilaments rather than microtubules. (**b**) Motor molecules can also attach to receptors on organelles such as vesicles and enable the organelles to "walk" along microtubules of the cytoskeleton. For example, this is how vesicles containing neurotransmitters migrate to the tips of axons, the long extensions of nerve cells that release transmitter molecules as chemical signals to adjacent nerve cells.

TABLE 7.2

The Structure and Function of the Cytoskeleton

PROPERTY	MICROTUBULES	MICROFILAMENTS (ACTIN FILAMENTS)	INTERMEDIATE FILAMENTS
Structure	Hollow tubes; wall consists of 13 columns of tubulin proteins	Two intertwined strands of actin	Fibrous proteins supercoiled into thicker cables
Diameter	25 nm with 15-nm lumen	7 nm	8–12 nm
Monomers	α-tubulin, β-tubulin } form dimer*	Actin	One of several different proteins of the keratin family, depending on cell type
Functions	Cell motility (as in cilia or flagella) Chromosome movements Movement of organelles Maintenance of cell shape	Muscle contraction Cytoplasmic streaming Cell motility (as in pseudopodia) Cell division (cleavage furrow formation) Maintenance of cell shape Changes in cell shape	Structural support Maintenance of cell shape

*A dimer consists of two similar molecular units bonded together.

Source: Adapted from W. M. Becker, J .B. Reece, and M. F. Poenie, The World of the Cell, 3rd ed. Menlo Park, CA: Benjamin/Cummings, 1996, p. 555.

cells to contract. The cytoskeleton extends the pseudopodia of *Amoeba* and also functions in the streaming of cytoplasm that circulates materials within many large plant cells. Vesicles may travel to their destinations in the cell along "monorails" provided by the cytoskeleton, and contractile components of the cytoskeleton manipulate the plasma membrane to form food vacuoles during phagocytosis.

The cytoskeleton is constructed from at least three types of fibers (TABLE 7.2). **Microtubules** are the thickest of the three types; **microfilaments** (also called actin filaments) are the thinnest. **Intermediate filaments** are collections of fibers whose diameters fall in a middle range.

Microtubules

Microtubules are found in the cytoplasm of all eukaryotic cells. They are straight, hollow rods measuring about 25 nm in diameter and from 200 nm to 25 μm in length. The wall of the hollow tube is constructed from globular proteins called tubulins, of which there are two closely related kinds, α-tubulin and β-tubulin. A microtubule elongates by adding tubulin molecules to its ends.

Microtubules can be disassembled and their tubulin used to build microtubules elsewhere in the cell.

Microtubules shape and support the cell and also serve as tracks along which organelles equipped with motor molecules can move (see FIGURE 7.21). For example, microtubules probably help guide secretory vesicles from the Golgi apparatus to the plasma membrane. Microtubules are also involved in the separation of chromosomes during cell division, discussed in Chapter 11.

In many cells, microtubules radiate from a **centrosome,** a region located near the nucleus (see TABLE 7.2). These microtubules function as girders that support the cell. Within the centrosome of an animal cell are a pair of **centrioles.** Each centriole is composed of nine sets of triplet microtubules arranged in a ring (FIGURE 7.22). When a cell divides, the centrioles replicate. Although centrioles may help organize microtubule assembly, they are not mandatory for this function in all eukaryotes; centrosomes of most plants lack centrioles altogether.

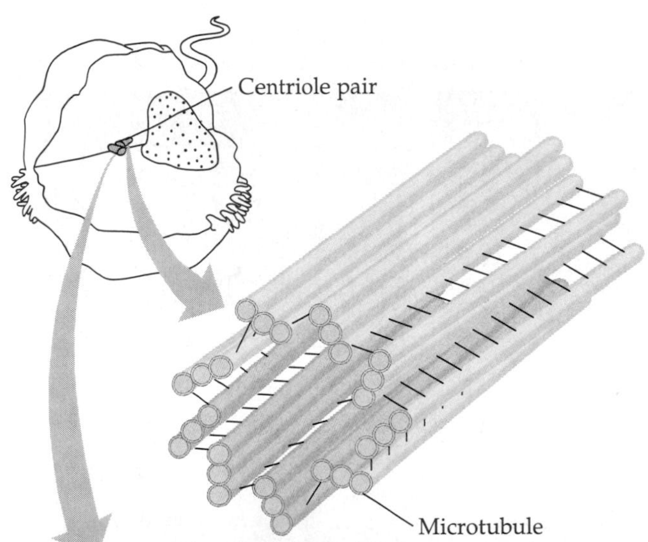

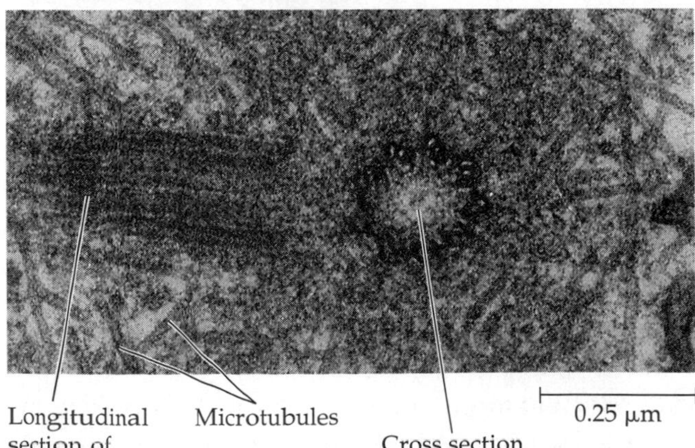

FIGURE 7.22

Centrioles. An animal cell has a pair of centrioles within its centrosome, the region near the nucleus where the cell's microtubules are initiated. The centrioles, each about 250 nm (0.25 µm) in diameter, are arranged at right angles to each other, and each is made up of nine sets of three microtubules (TEM).

Cilia and Flagella. In eukaryotes, a specialized arrangement of microtubules is responsible for the beating of **flagella** and **cilia,** locomotive appendages that protrude from some cells. Many unicellular organisms (kingdom Protista) are propelled through water by cilia or flagella, and the sperm of animals, algae, and some plants are flagellated. If cilia or flagella extend from cells that are held in place as part of a tissue layer, then they function to draw fluid over the surface of the tissue. For example, the ciliated lining of the windpipe sweeps mucus with trapped debris out of the lungs (see FIGURE 7.2).

Cilia usually occur in large numbers on the cell surface. They are about 0.25 µm in diameter and about 2 to 20 µm in length. Flagella are the same diameter but longer than cilia, measuring 10 to 200 µm in length. Also, flagella are usually limited to just one or a few per cell.

Flagella and cilia also differ in their patterns of beating. A flagellum has an undulating motion that generates force in the same direction as the flagellum's axis. In contrast, cilia work more like oars, with alternating power and recovery strokes generating force in a direction perpendicular to the cilium's axis (FIGURE 7.23).

Though different in length, number per cell, and beating pattern, cilia and flagella actually share a common ultrastructure. A cilium or flagellum has a core of microtubules ensheathed in an extension of the plasma membrane (FIGURE 7.24). Nine doublets of microtubules, the members of each pair sharing part of their walls, are arranged in a ring. In the center of the ring are two single microtubules. This arrangement, referred to as the "9 + 2" pattern, is found in nearly all eukaryotic flagella and cilia. (The flagella of motile prokaryotes, which will be discussed in Chapter 25, are entirely different.) The doublets of the outer ring are connected to the center of the cilium or flagellum by radial spokes that terminate near the central pair of microtubules. Each doublet of the outer ring also has pairs of arms evenly spaced along its length and reaching toward the neighboring doublet of microtubules. The microtubule assembly of a cilium or flagellum is anchored in the cell by a **basal body,** which is structurally identical to a centriole.

The arms extending from each microtubule doublet to the next are the motors responsible for the bending movements of cilia and flagella. The motor molecule that makes up these arms is a very large protein called **dynein.** A dynein arm performs a complex cycle of movements caused by changes in the conformation (shape) of the protein, with ATP providing the energy for these changes (see FIGURE 7.21). The mechanics of dynein "walking" are reminiscent of a cat climbing a tree by

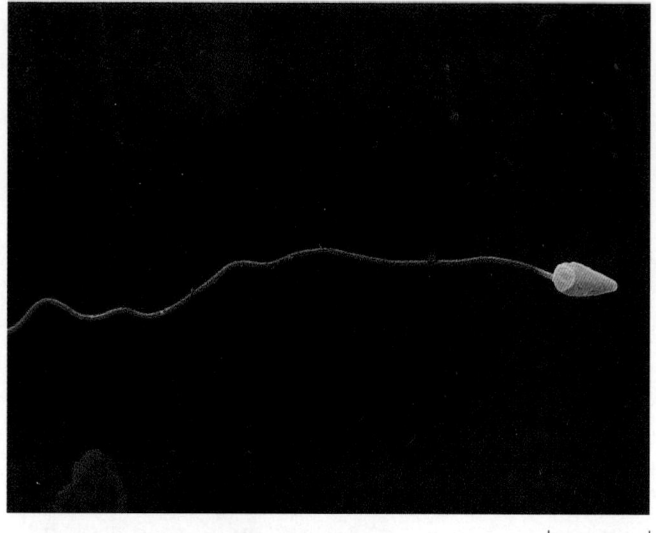

(a) Motion of flagella

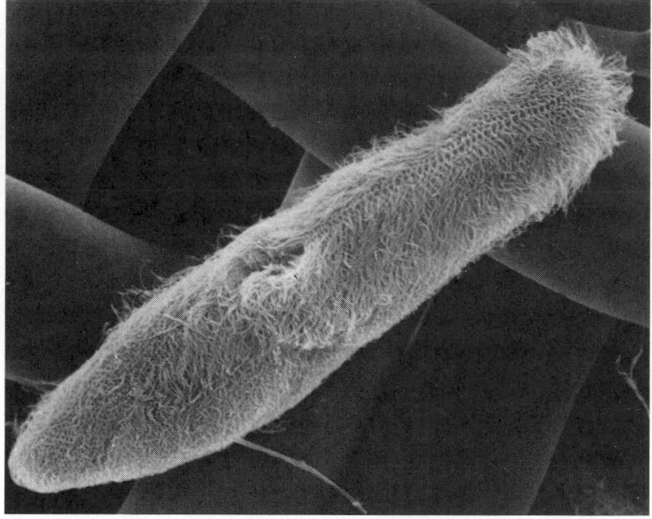

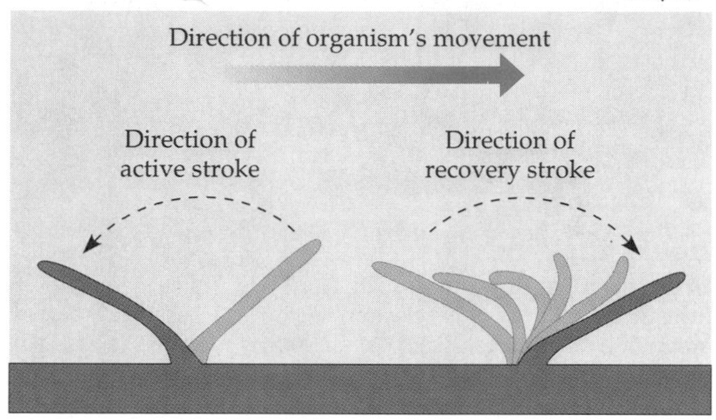

(b) Motion of cilia

FIGURE 7.23

A comparison of the beating of flagella and cilia. (**a**) A flagellum usually undulates, its snakelike motion driving a cell in the same direction as the axis of the flagellum. Propulsion of a sperm cell is an example of flagellate locomotion (SEM). (**b**) A dense nap of beating cilia covers this *Paramecium,* a motile protist (SEM). The cilia beat at a rate of about 40 to 60 strokes per second. Cilia have a back-and-forth motion, alternating active strokes with recovery strokes. This moves the cell, or moves a fluid over the surface of a stationary cell, in a direction perpendicular to the axis of the cilium.

attaching its claws, moving its legs, releasing its claws, and grabbing again farther up the tree. Similarly, the dynein arms of one doublet attach to an adjacent doublet and pull so that the doublets slide past each other in opposite directions. The arms then release from the other doublet and reattach a little farther along its length. Without any restraints on this movement, one doublet would continue to "walk" along the surface of the other, elongating the cilium rather than bending it. For lateral movement of a cilium, the dynein "walking" must be restrained; that is, it must have something to pull against, as when the muscles in your leg pull against your bones to move your knee. In cilia and flagella, the microtubule doublets are held in place, perhaps by the radial spokes or other structural elements. Thus neighboring doublets cannot slide past each other very far. Instead, the forces exerted by the dynein arms cause the doublets to curve, bending the cilium or flagellum (FIGURE 7.25). In the beating mechanism of cilia and flagella, we see once again that structure fits function.

Microfilaments (Actin Filaments)

Microfilaments are solid rods about 7 nm in diameter. They are also called actin filaments because they are built from molecules of **actin,** a globular protein. The actin molecules are linked into chains; two of these chains twisted about each other in a helix form the microfilament (see TABLE 7.2).

Microfilaments are part of the contractile apparatus in muscle cells. Thousands of actin filaments are arranged parallel to one another along the length of a muscle cell, interdigitated with thicker filaments made of a protein called **myosin** (FIGURE 7.26a). Contraction of the cell results from the actin and myosin filaments sliding past one another, which shortens the cell. Extending from the myosin filaments to the actin filaments are movable

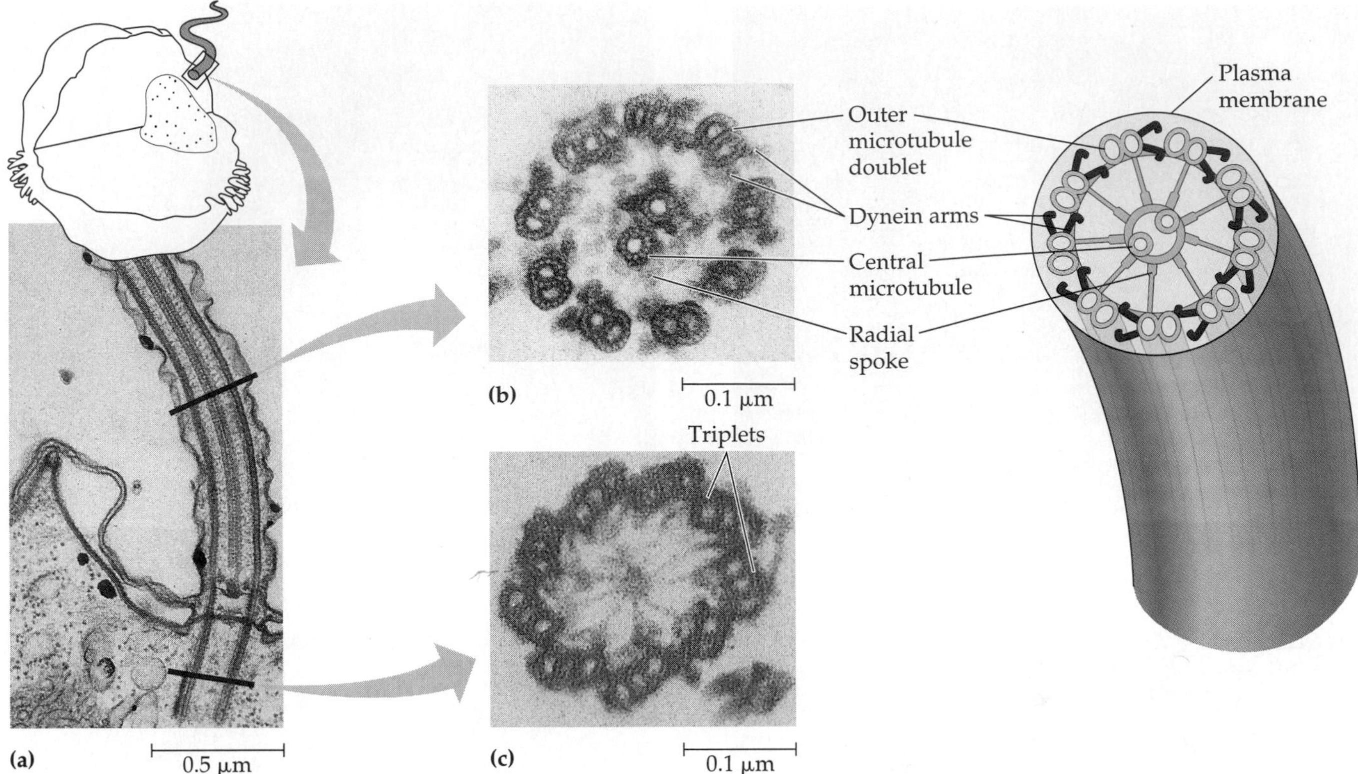

FIGURE 7.24
Ultrastructure of a eukaryotic flagellum or cilium. (**a**) In this electron micrograph of a longitudinal section of a cilium, microtubules can be seen running the length of the structure (TEM). (**b**) A cross section through the cilium shows the "9 + 2" arrangement of microtubules (TEM). (**c**) The basal body anchoring the cilium or flagellum to the cell has a ring of nine microtubule triplets (the basal body is structurally identical to a centriole). The nine doublets of the cilium extend into the basal body, where each doublet joins another microtubule to form the ring of nine triplets. The two central microtubules terminate above the basal body (TEM).

arms, regions of the myosin proteins that function as motor molecules. Powered by ATP, the conformational changes of these arms slide the adjacent actin filaments.

Although microfilaments are especially concentrated and well ordered in muscle cells, they seem to be present to some extent in all eukaryotic cells. Along with the rest of the cytoskeleton, microfilaments function in support. For example, bundles of microfilaments make up the core of microvilli, delicate projections that increase the surface area of cells specialized for the transport of material across the plasma membrane (FIGURE 7.26b).

In some parts of the cell, actin filaments are associated with myosin in miniature versions of the arrangement found in muscle cells. These actin–myosin aggregates are responsible for localized contractions of cells.

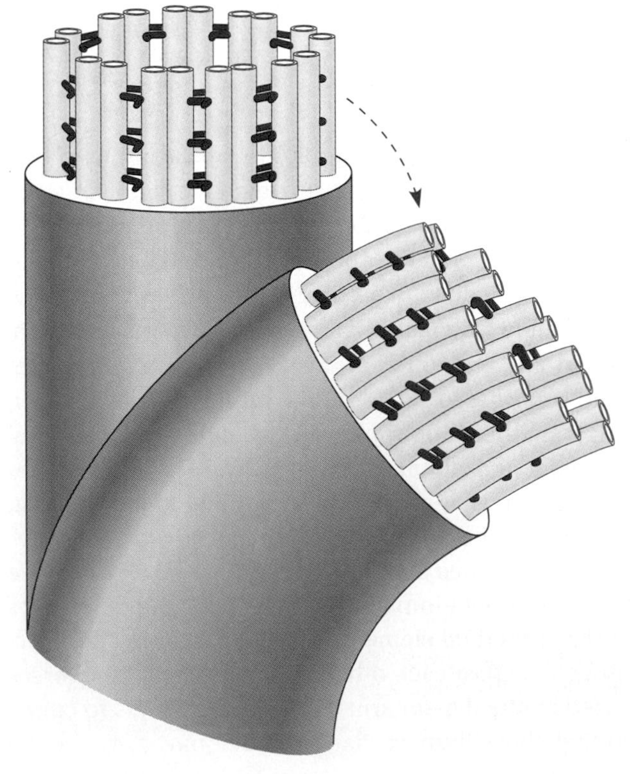

FIGURE 7.25
How dynein "walking" moves cilia and flagella. The dynein arms of one microtubule doublet grip the adjacent doublet, pull, release, and then grip again. This cycle of the dynein motors is powered by ATP. The doublets cannot slide far because they are physically restrained within the cilium. Thus, the action of the dynein arms causes the doublets to bend.

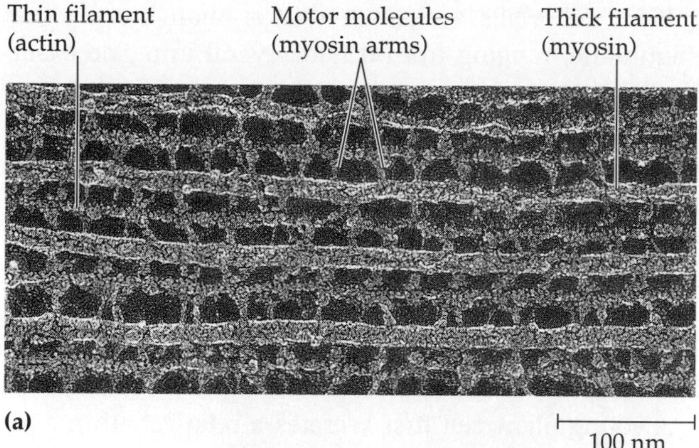

(a)

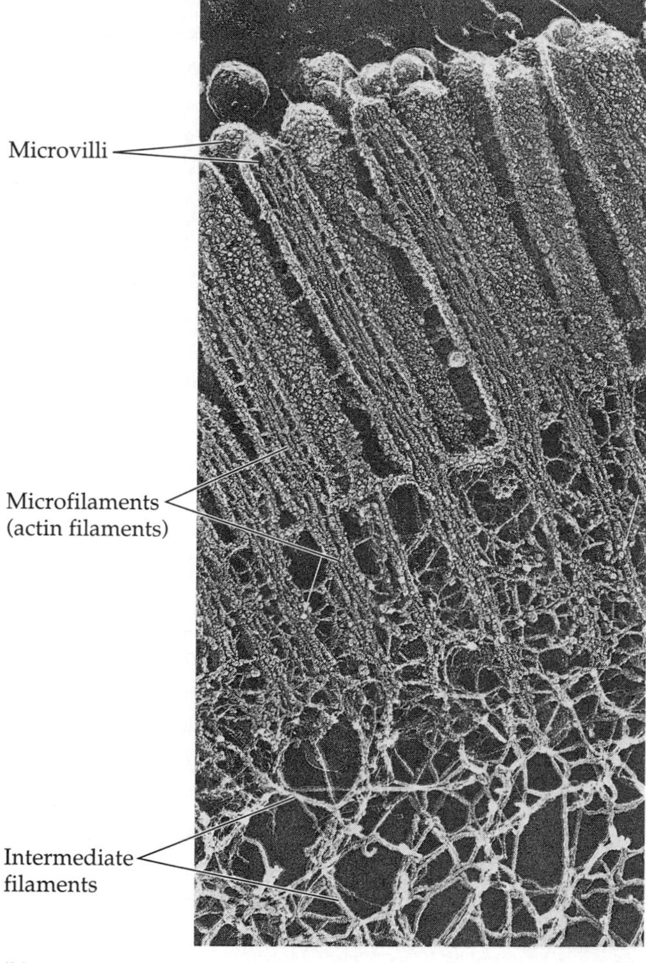

(b)

FIGURE 7.26

Functions of microfilaments. (**a**) In muscle cells, actin filaments interdigitate with thicker filaments made of the protein myosin. Myosin acts as a motor molecule by means of arms that "walk" the two types of filaments past each other. The teamwork of many such sliding filaments enables the entire muscle cell to shorten (TEM). (**b**) The surface area of this nutrient-absorbing intestinal cell is increased by its many microvilli, cellular extensions reinforced by bundles of microfilaments. These actin filaments are anchored to a network of intermediate filaments (TEM, Hirokawa et al. 1982. J. Cell Biol. 94, pp. 425–443, Fig. 1).

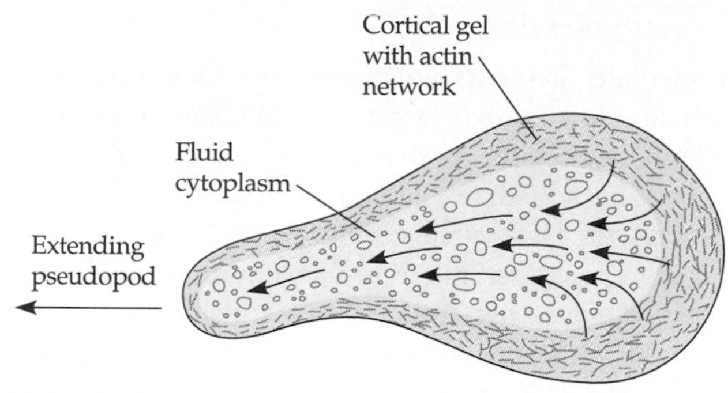

(**a**) Ameboid movement

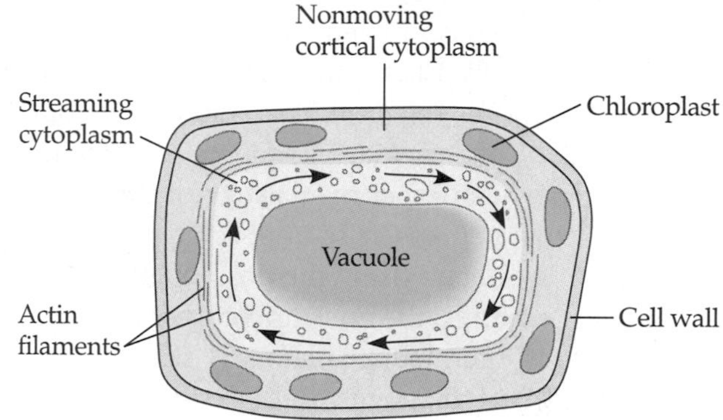

(**b**) Cytoplasmic streaming

FIGURE 7.27

Microfilaments and motility in nonmuscle cells. (**a**) In ameboid movement, the inner portion of cytoplasm is fluid and can move into an extending pseudopod. The cortex, the outer layer of cytoplasm, is in a gel state due to a network of actin filaments held together by binding proteins. This drawing presents one hypothesis of the way the actin behaves during ameboid movement. Actin filaments at the trailing end of the cell interact with myosin, causing the actin network in the cortex to contract. Like squeezing on a toothpaste tube, contraction of the thick cortex at the trailing end forces the fluid cytoplasm into the pseudopod, where the cortex is much thinner. The actin is also redistributed. Microfilaments at the trailing end disassemble, and the actin subunits are carried by the cytoplasmic current. The pseudopod stops extending when these actin subunits reassemble into microfilaments that thicken the cortical gel in that region of the cell. (**b**) In cytoplasmic streaming, a fluid layer of cytoplasm cycles around the cell, moving over a carpet of actin filaments. According to one hypothesis, myosin motors attached to organelles in the fluid cytoplasm drive the streaming by interacting with the actin.

For example, when an animal cell divides, it is pinched in two by a contracting belt of microfilaments. In addition, microfilaments function in ameboid movement (FIGURE 7.27a), in which a cell moves along a surface by extending and flowing into cellular extensions called **pseudopodia** (Gr. *pseudes*, "false," and *pod*, "foot"). Microfilaments in plant cells function in **cytoplasmic streaming,** a circular flow of cytoplasm within cells (FIGURE 7.27b). This movement, which is especially common in large plant cells, speeds the distribution of materials within the cell.

Intermediate Filaments

Intermediate filaments are named for their diameter, which, at 8 to 12 nm, is larger than the diameter of microfilaments but smaller than that of microtubules (see TABLE 7.2). Intermediate filaments actually include a diverse class of cytoskeletal elements. Each type is constructed from a different molecular subunit belonging to a diverse family of proteins called keratins. Microtubules and microfilaments, in contrast, are consistent in diameter and composition in all eukaryotic cells.

Intermediate filaments are also more permanent fixtures of cells than are microfilaments and microtubules, which are often disassembled and reassembled in various parts of a cell. Chemical treatments that remove microfilaments and microtubules from the cytoplasm leave a web of intermediate filaments that retains its original shape. Such experiments suggest that intermediate filaments are especially important in reinforcing the shape of a cell and fixing the position of certain organelles. For example, the nucleus commonly sits within a cage made of intermediate filaments, fixed in location by branches of the filaments that extend into the cytoplasm. Other intermediate filaments make up the nuclear lamina that lines the interior of the nuclear envelope (see FIGURE 7.9). In cases where the shape of the entire cell is correlated with function, intermediate filaments support that shape. For instance, the long extensions (axons) of nerve cells that transmit impulses are strengthened by one class of intermediate filament. Specialized for bearing tension, the various kinds of intermediate filaments may function as the framework of the entire cytoskeleton.

Having criss-crossed the interior of the cell to explore various organelles, we complete our tour of the cell by returning to the surface of this microscopic world, where there are additional structures with important functions. Although the plasma membrane is usually regarded as the boundary of the living cell, most cells synthesize and secrete coats of one kind or another that are external to the plasma membrane.

Plant cells are encased by cell walls

The **cell wall** is one of the features of plant cells that distinguishes them from animal cells. The wall protects the plant cell, maintains its shape, and prevents excessive uptake of water. On the level of the whole plant, the strong walls of specialized cells hold the plant up, against the force of gravity. Prokaryotes, fungi, and some protists also have cell walls, but we will postpone discussion of them until Unit Five.

Plant cell walls are much thicker than the plasma membrane, ranging from 0.1 to several μm. The exact chemical composition of the wall varies from species to species and from one cell type to another in the same plant, but the basic design of the wall is consistent (see FIGURE 5.8). Microfibrils made of the polysaccharide cellulose are embedded in a matrix of other polysaccharides and protein. This combination of materials, strong fibers in a "ground substance" (matrix), is the same basic architectural design found in steel-reinforced concrete and in fiberglass.

A young plant cell first secretes a relatively thin and flexible wall called the **primary cell wall** (FIGURE 7.28). Between primary walls of adjacent cells is the **middle lamella,** a thin layer rich in sticky polysaccharides called pectins. The middle lamella glues the cells together (pectin is used as a thickening agent in jams and jellies). When the cell matures and stops growing, it strengthens its wall. Some cells do this simply by secreting hardening substances into the primary wall. Other plant cells add a **secondary cell wall** between the plasma membrane and the primary wall. The secondary wall, often deposited in several laminated layers, has a strong and durable matrix that affords the cell protection and support. Wood, for example, consists mainly of secondary walls.

The extracellular matrix (ECM) of animal cells functions in support, adhesion, movement, and development

Although animal cells lack walls akin to those of plant cells, they do have an elaborate **extracellular matrix (ECM).** The main ingredients of the ECM are glycoproteins secreted by the cells. (Recall that glycoproteins are proteins covalently bonded to carbohydrate.) The most abundant glycoprotein in the ECM of most animal cells is **collagen,** which forms strong fibers outside the cells. In fact, collagen accounts for about half of the total protein in the human body. The collagen fibers are embedded in a network woven from another class of glycoproteins called **proteoglycans** (FIGURE 7.29). These molecules are especially rich in carbohydrate—up to 95%. Some cells are attached to the ECM by still another class of glycoproteins, most commonly **fibronectins**. Fibronectins bind to receptor proteins called **integrins** that are built into the plasma membrane. Integrins span the membrane and bind on their cytoplasmic side to microfilaments of the cytoskeleton. Thus, integrins integrate responses of the cytoskeleton to changes in the ECM, and vice versa.

In addition to providing support and anchorage for cells, the ECM functions in a cell's dynamic behavior.

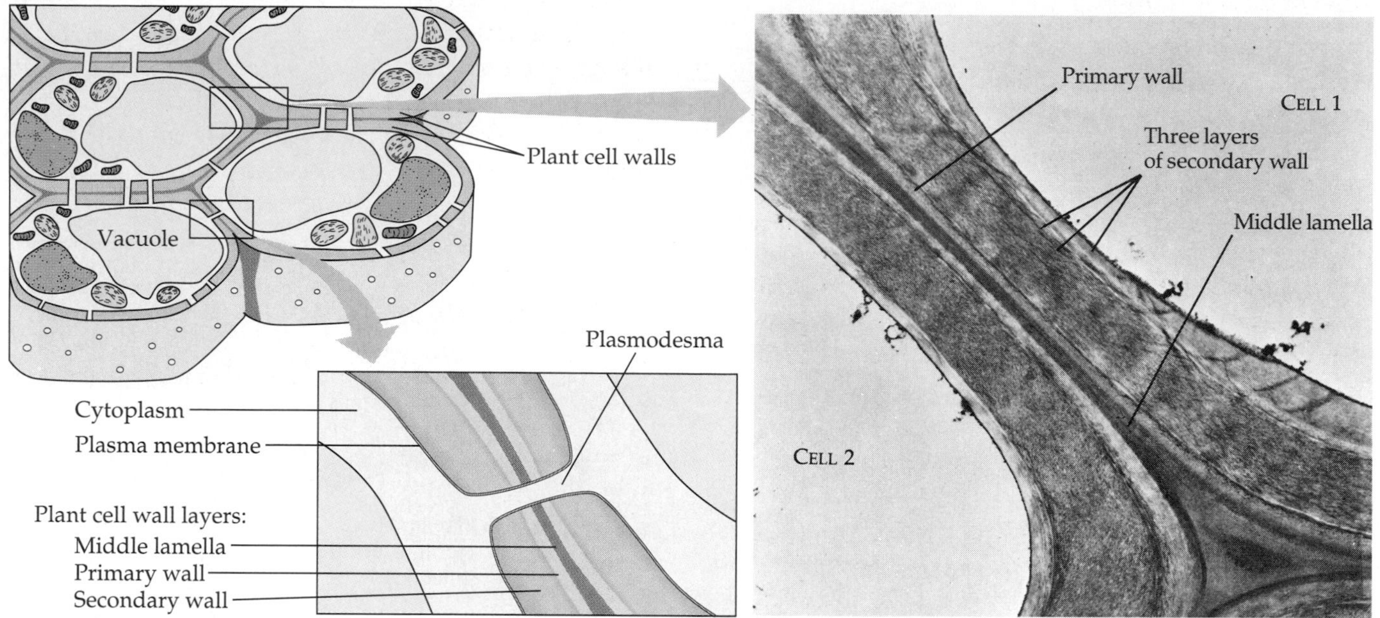

FIGURE 7.28

Plant cell walls. Young cells first construct thin primary walls, often adding stronger secondary walls to the inside of the primary wall when growth ceases. A sticky middle lamella cements adjacent cells together. Thus, the multilayered partition between these cells consists of adjoining walls individually secreted by the cells. The walls do not isolate the cells: The cytoplasm of one cell is continuous with the cytoplasm of its neighbors via plasmodesmata, channels through the walls (TEM).

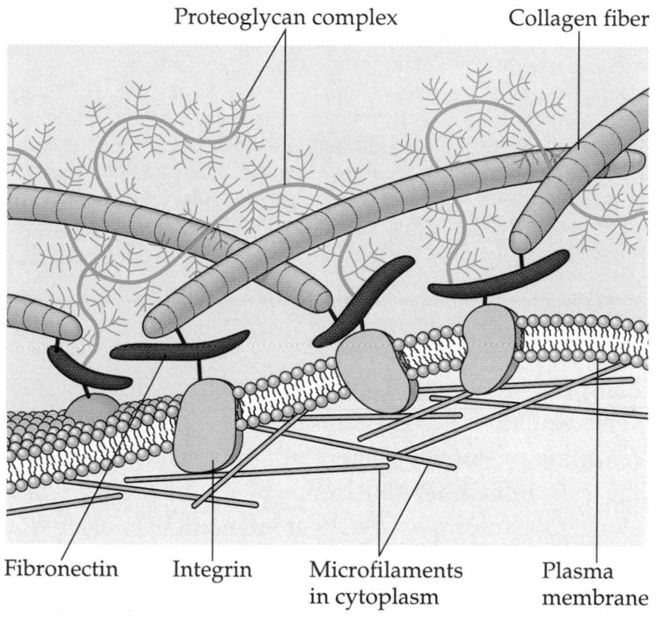

FIGURE 7.29

Extracellular matrix (ECM) of an animal cell. The molecular composition and structure of the ECM varies from one cell type to another. In this example, three different types of glycoproteins are present. Fibers made of the glycoprotein collagen are embedded in a web of proteogylcans, which can be as much as 95% carbohydrate. The third glycoprotein is fibronectin, the adhesive that attaches the ECM to the plasma membrane of the cell. Membrane proteins called integrins are bound to the ECM on one side and the cytoskeleton on the other. This linkage can transmit mechanical stimuli between the cell's extracellular environment and its interior.

For example, some cells in a developing embryo migrate along specific pathways by matching the orientation of their microfilaments to the "grain" of fibers in the extracellular matrix. Researchers are also learning that a cell's contacts with its ECM help control the activity of genes in the nucleus—that is, the production of messenger RNA by specific genes. Perhaps integrins transmit mechanical stimuli from the ECM to the cytoskeleton, which in turn triggers production of chemical signals within the cell that relay the information into the nucleus. Biologists are just beginning to understand the important functions of the extracellular matrix in the lives of cells.

Intercellular junctions integrate cells into higher levels of structure and function

The many cells of an animal or plant are integrated into one functional organism. Neighboring cells often adhere, interact, and communicate through special patches of direct physical contact.

It might seem that the nonliving cell walls of plants would isolate cells from one another. In fact, the walls are perforated with channels called **plasmodesmata** (singular, **plasmodesma;** Gr. *desmos,* "to bind"). Strands of cytoplasm pass through the plasmodesmata and connect the living contents of adjacent cells (see FIGURE 7.28).

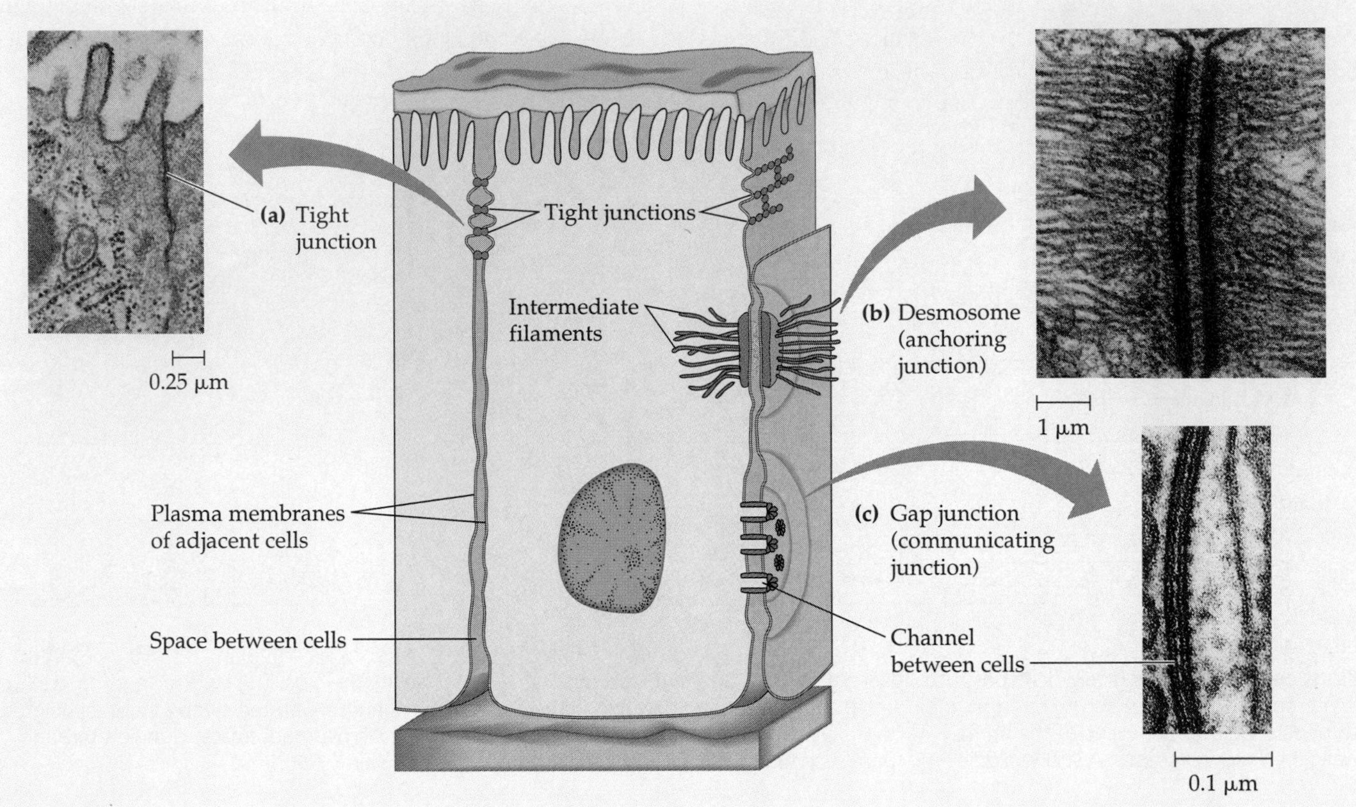

FIGURE 7.30
Intercellular junctions in animals. These specialized connections are especially common in epithelial tissue, which lines the internal surfaces of the body. Here, we use epithelial cells lining the intestine to describe three kinds of intercellular junctions, each with a structure well adapted for its function. (**a**) Tight junctions. These connections form continuous belts around the cell. The membranes of neighboring cells are actually fused at a tight junction, forming a seal that prevents leakage of extracellular fluid across a layer of epithelial cells. For example, the tight junctions of the intestinal epithelium keep the contents of the intestine separate from the body fluid on the opposite side of the epithelium (TEM). (**b**) Desmosomes (anchoring junctions). These junctions function like rivets, fastening cells together into strong epithelial sheets. Intermediate filaments made of the sturdy protein keratin reinforce desmosomes (TEM). (**c**) Gap junctions (communicating junctions). These connections provide cytoplasmic channels between adjacent cells. Special membrane proteins surround each pore, which is wide enough for salts, sugars, amino acids, and other small molecules to pass (TEM). In the muscle tissue of the heart, the flow of ions through gap junctions coordinates the contractions of the cells. Gap junctions are especially common in animal embryos, where chemical communication between cells is essential for development.

This unifies most of the plant into one living continuum. The plasma membranes of adjacent cells are continuous through a plasmodesma; the membrane lines the channel. Water and small solutes can pass freely from cell to cell, a transport that is enhanced by cytoplasmic streaming.

In animals, there are three main types of intercellular junctions: **tight junctions, desmosomes,** and **gap junctions.** These are illustrated and described in detail in FIGURE 7.30.

The cell is a living unit greater than the sum of its parts

From our panoramic view of the cell's overall compartmental organization to our closeup inspection of each organelle's architecture, this tour of the cell has provided many opportunities to correlate structure with function. (This would be a good time to review cell structure by returning to FIGURES 7.7 and 7.8.) But even as we dissect the cell, remember that none of its organelles works alone. As an example of cellular integration, consider the microscopic scene in FIGURE 7.31. The large cell is a macrophage. It helps defend the body against infections by ingesting bacteria (the smaller green cells). The macrophage creeps along a surface and reaches out with cellular extensions that help it catch the bacteria. Actin filaments interact with other elements of the cytoskeleton in these movements. After the macrophage engulfs the bacteria, they are destroyed by lysosomes. The elaborate endomembrane system, which includes the ER and the Golgi apparatus, produces the lysosomes. The digestive enzymes of the lysosomes and the proteins of the cytoskeleton are all made on ribosomes. And the synthesis of these proteins is programmed by genetic messages dispatched from the DNA in the nucleus. All these processes require energy, which mitochondria supply in the form of ATP. Cellular functions arise from cellular order: The cell is a living unit greater than the sum of its parts.

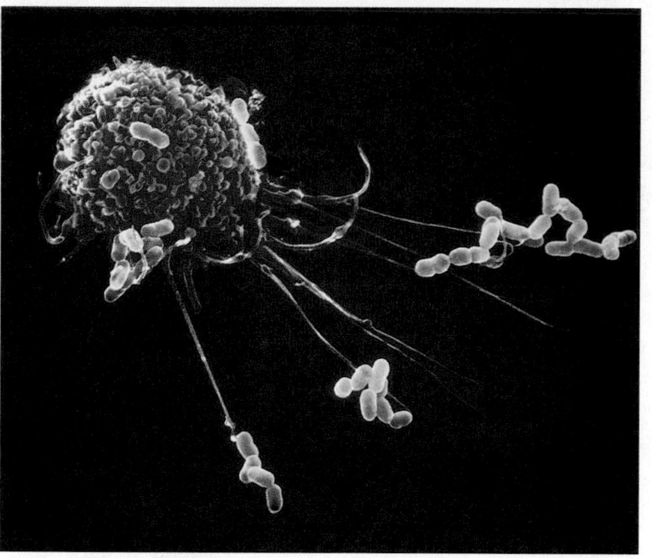

FIGURE 7.31
The emergence of cellular functions from the cooperation of many organelles. The ability of this macrophage (orange) to recognize, apprehend, and destroy bacteria (green) is a coordinated activity of the whole cell. The cytoskeleton, lysosomes, and plasma membrane are among the components that function in phagocytosis. Other cellular functions are also emergent properties that depend on interactions of the cell's parts (colorized SEM).

REVIEW OF KEY CONCEPTS

- Microscopes provide windows to the world of the cell (pp. 111–112, FIGURE 7.1)
  - Electron microscopes and improvements in light microscopes have catalyzed progress in the study of cell structure.
- Cell biologists can isolate organelles to study their function (pp. 113–114, FIGURE 7.3)
  - After homogenizing cells, researchers use the ultracentrifuge to fractionate the cells into pellets enriched in specific organelles.
- A panoramic view of the cell (pp.114–118, FIGURES 7.7, 7.8)
  - Prokaryotic cells lack nuclei and membrane-enclosed organelles; bacteria are prokaryotes. All other organisms are made up of eukaryotic cells with membrane-enclosed nuclei surrounded by cytoplasm, in which are suspended specialized organelles not found in prokaryotic cells.
  - The requirement for a favorable ratio of membrane surface to cell volume sets upper limits on cell size.
  - Eukaryotic cells are surrounded by a plasma membrane and are partitioned into various compartments by internal membranes. These internal membranes provide local environments for specific metabolic processes.
  - All membranes consist of phospholipids and proteins. The diversity of membrane function reflects the variation in a membrane's specific molecular composition.
- The nucleus contains a cell's genetic library (p. 118, FIGURE 7.9)
  - The trademark of a eukaryotic cell is its distinctive nucleus, enclosed in the nuclear envelope. Pores in the envelope allow for exchange of macromolecules between the nucleus and the cytoplasm.
  - The nucleus contains the genetic material, DNA, organized with proteins in a characteristic number of chromosomes in each eukaryotic species.
  - The nucleolus is the nuclear site where the parts of ribosomes are produced.
- Ribosomes build a cell's proteins (p. 118, FIGURE 7.10)
  - Ribosomes carry out protein synthesis in the cytosol (free ribosomes) or attached to the outside of the membranous endoplasmic reticulum (bound ribosomes).
- Many organelles are related through the endomembrane system (pp. 118–120, FIGURE 7.16)
  - Many of the membranes of a eukaryotic cell are interrelated directly through physical continuity or indirectly through transport vesicles, pinched-off portions of membrane in transit from one membrane site to another.

- The endoplasmic reticulum manufactures membranes and performs many other biosynthetic functions (pp. 120–121, FIGURE 7.11)
  - The endoplasmic reticulum (ER) is a network of membrane-enclosed compartments called cisternae. Smooth ER, so named because it lacks ribosomes, synthesizes steroids, metabolizes carbohydrates, stores calcium in muscle cells, and detoxifies poisons in liver cells.
  - Rough ER, that portion of the ER with bound ribosomes, is continuous with the nuclear envelope and functions in producing cell membrane and manufacturing proteins for secretion. Membrane and secretory proteins can be transferred to other locations in the cell by the budding of transport vesicles from ER.
- The Golgi apparatus finishes, sorts, and ships many products of the cell (pp. 122–123, FIGURE 7.12)
  - The Golgi apparatus consists of stacks of membranous sacs that synthesize various macromolecules and also modify, store, sort, and export products of the ER.
  - One side of a Golgi stack, the cis face, receives secretory proteins from the ER through transport vesicles. Once inside, these proteins can be chemically modified and sorted before release from the trans face of the Golgi in vesicles.
- Lyosomes are digestive compartments (pp. 123–124, FIGURE 7.14)
  - A lysosome is a membrane-enclosed sac of hydrolytic enzymes. Its acidic microenvironment is optimal for the functioning of its enzymes in recycling monomers from cell macromolecules and in digesting substances ingested by phagocytosis.
- Vacuoles have diverse functions in cell maintenance (pp. 124–125, FIGURE 7.15)
  - The central vacuole of plant cells functions in storage, waste disposal, cell elongation, and protection. The vacuole's membrane is called the tonoplast.
  - Food vacuoles of animal cells and contractile vacuoles of freshwater protists are other examples of vacuoles.
- Peroxisomes consume oxygen in various metabolic functions (pp. 125–126, FIGURE 7.17)
  - Peroxisomes function in a variety of metabolic processes that produce hydrogen peroxide as a waste product. An enzyme in the peroxisome then converts the peroxide to water.
- Mitochondria and chloroplasts are the main energy transformers of cells (pp. 126–128, FIGURE 7.18, 7.19)
  - Mitochondria are sites of cellular respiration in eukaryotic cells. Energy is released, with the help of oxygen, from chemical fuels such as sugars and fats and is used to restock the cellular supply of ATP.
  - Mitochondria are compartmentalized by an outer membrane and an inner membrane folded into convolutions called cristae. Some of the metabolic reactions of respiration take place in the space enclosed by the inner membrane, the mitochondrial matrix. Enzymes built into the inner membrane also function in respiration.
  - Chloroplasts, specialized members of a family of plant organelles called plastids, contain chlorophyll and other pigments, which function in photosynthesis. Chloroplasts are enclosed by two membranes surrounding the fluid called stroma, in which are embedded the membranous thylakoids. These flattened sacs are stacked in some regions, forming grana.
- The cytoskeleton provides structural support and functions in cell motility (pp. 128–134, TABLE 7.2)
  - The cytoskeleton is constructed from microtubules, microfilaments, and intermediate filaments.
  - Microtubules are hollow cylinders. In many cells, microtubules radiate out from the centrosome, an area near the nucleus that surrounds the centrioles in animal cells. Microtubules shape and support the cell, guide the movement of organelles, and participate in chromosome separation during cell division.
  - Cilia and flagella are motile cellular appendages consisting of a "9 + 2" arrangement of microtubules. Movement of cilia and flagella occurs when arms consisting of the protein dynein move the microtubule doublets past each other.
  - Microfilaments, thinner than microtubules, are solid rods built from the protein actin. Microfilaments in muscle cells interact with the protein myosin to cause contraction. They also function in ameboid movement, cytoplasmic streaming, and support for cellular projections, such as microvilli.
  - In addition to microtubules and microfilaments, most cells have a variety of intermediate filaments that are important in supporting cell shape and fixing various organelles in place.
- Plant cells are encased by cell walls (p. 134, FIGURE 7.28)
  - The cells of plants, prokaryotes, fungi, and some protists are reinforced by cell walls external to the plasma membrane. Plant cell walls are composed of cellulose fibers embedded in other polysaccharides and protein.
- The extracellular matrix (ECM) of animal cells functions in support, adhesion, movement, and development (pp. 134–135, FIGURE 7.29)
  - Animal cells secrete glycoproteins that form the ECM.
  - The ECM of many cells consists of collagen fibers and a proteoglycan complex all attached to the plasma membrane by fibronectins.
- Intercellular junctions integrate cells into higher levels of structure and function (pp. 135–136, FIGURE 7.30)
  - Plants have plasmodesmata, cytoplasmic channels that pass through adjoining cell walls.
  - Cell-to-cell contact in animals is provided by tight junctions, desmosomes, and gap junctions.
- The cell is a living unit greater than the sum of its parts (pp. 136–137, FIGURE 7.31)
  - Organelles do not function in isolation; they cooperate with other organelles. At the cellular level, life emerges from these complex interactions of a cell's parts.

SELF QUIZ

1. As a spherical cell grows in diameter, the μm^2 of plasma membrane surface area per μm^3 of cellular volume
   a. increases
   b. decreases
   c. stays the same
   d. adjusts by adding infoldings
   e. adjusts by becoming thinner

2. From the following, choose the statement that correctly characterizes bound ribosomes.
   a. Bound ribosomes are enclosed in their own membrane.
   b. Bound ribosomes are structurally different from free ribosomes.
   c. Bound ribosomes generally synthesize membrane proteins and secretory proteins.
   d. The most common location for bound ribosomes is the cytoplasmic surface of the plasma membrane.
   e. Bound ribosomes are concentrated in the cisternal space of rough ER.

3. Which of the following organelles is least closely associated with the endomembrane system?
   a. nuclear envelope c. Golgi apparatus e. ER
   b. chloroplast d. plasma membrane

4. Cells of the pancreas will incorporate radioactively labeled amino acids into proteins. This "tagging" of newly synthesized proteins enables a researcher to track the location of these proteins in a cell. In this case, we are tracking an enzyme that is eventually secreted by pancreatic cells. Which of the following is the most likely pathway for movement of this protein in the cell?
   a. ER ⟶ Golgi ⟶ nucleus
   b. Golgi ⟶ ER ⟶ lysosome
   c. nucleus ⟶ ER ⟶ Golgi
   d. ER ⟶ Golgi ⟶ vesicles that fuse with plasma membrane
   e. ER ⟶ lysosomes ⟶ vesicles that fuse with plasma membrane

5. A certain poison disrupts the cytoskeleton of cells. Which of the following functions would be affected most directly by this drug?
   a. cell division d. protein synthesis
   b. cellular respiration e. digestion within lysosomes
   c. photosynthesis

6. Which of the following organelles is common to plant *and* animal cells?
   a. chloroplasts c. tonoplast e. centrioles
   b. wall made of cellulose d. mitochondria

7. Which component is present in a prokaryotic cell?
   a. mitochondria c. nuclear envelope e. ER
   b. ribosomes d. chloroplasts

8. The cytoskeleton senses mechanical disturbance of the extracellular matrix via
   a. collagen c. integrin e. cellulose
   b. myosin d. proteoglycan

9. Which type of cell would probably provide the best opportunity to study lysosomes? Explain your answer.
   a. muscle cell d. leaf cell of a plant
   b. nerve cell e. bacterial cell
   c. phagocytic white blood cell

10. Which of the following structure-function pairs is *mismatched?*
   a. nucleolus—ribosome production
   b. lysosome—intracellular digestion
   c. ribosome—protein synthesis
   d. Golgi—secretion of cell products
   e. microtubules—muscle contraction

CHALLENGE QUESTIONS

1. An inherited disorder in humans results in the absence of dynein in flagella and cilia. The disease causes respiratory problems and, in males, sterility. What is the ultrastructural connection between these two symptoms?

2. When very small viruses infect a plant cell by crossing its membrane, the viruses often spread rapidly throughout the entire plant without crossing additional membranes. Explain how this occurs.

3. Write a short essay describing similarities and differences between plant cells and animal cells.

SCIENCE, TECHNOLOGY, AND SOCIETY

Doctors at a California university removed a man's spleen, standard treatment for a type of leukemia. The disease did not recur. Researchers kept some of the spleen cells alive in a nutrient medium. They found that some of the cells produced a blood protein called GM-CSF, which they are now testing to fight cancer and AIDS. The researchers patented the cells. The patient sued, claiming a share in profits from any products derived from his cells. In 1988, the California Supreme Court ruled against the plaintiff (patient), stating that his suit "threatens to destroy the economic incentive to conduct important medical research." The U.S. Supreme Court agreed. The plaintiff's attorney argued that the ruling left patients "vulnerable to exploitation at the hands of the state." Do you think the plaintiff was treated fairly? Is there anything else you would like to know about this case that might help you make up your mind?

FURTHER READING

Alberts, B., D. Bray, J. Lewis, M. Raff, K. Roberts, and J. D. Watson. *Molecular Biology of the Cell,* 3rd ed. New York: Garland, 1994. A popular text; lucidly written, well illustrated, and comprehensive.

Becker, W. M., J. B. Reece, and M. F. Poenie. *The World of the Cell,* 3rd ed. Menlo Park, CA: Benjamin/Cummings, 1996. A very readable and student-oriented text.

Christensen, D. "Of Craters and Crevices." *Science News,* April 23, 1994. What determines the shape of a cell?

DeDuve, C. A Guided Tour of the Living Cell. New York: *Scientific American* Books, 1986. A beautifully illustrated introduction to the cell by the discoverer of lysosomes.

Lichtman, J.W. "Confocal Microscopy." *Scientific American,* August 1994. A powerful new window to the cell.

Murray, M. "Life on the Move." *Discover,* March 1991. How cellular motors work.

Nowak, R. "Matrix Work Wins Acclaim." *Science,* January 7, 1994. How signals from the extracellular matrix are transmitted to the nucleus.

Pennisi, E. "Piecing Together the Ribosome." *Science News,* November 5, 1994. How cell biologists are dissecting the protein machine.

Stossel, T.P. "The Machinery of Cell Crawling." *Scientific American,* September 1994. Cells on the move.

CHAPTER 8

MEMBRANE STRUCTURE AND FUNCTION

KEY CONCEPTS

- Membrane models have evolved to fit new data: *science as a process*
- A membrane is a fluid mosaic of lipids, proteins, and carbohydrates
- A membrane's molecular organization results in selective permeability
- Passive transport is diffusion across a membrane
- Osmosis is the passive transport of water
- Cell survival depends on balancing water uptake and loss
- Specific proteins facilitate the passive transport of selected solutes
- Active transport is the pumping of solutes against their gradients
- Some ion pumps generate voltage across membranes
- In cotransport, a membrane protein couples the transport of one solute to another
- Exocytosis and endocytosis transport large molecules
- Specialized membrane proteins transmit extracellular signals to the inside of the cell.

The plasma membrane is the edge of life, the boundary that separates the living cell from its nonliving surroundings. A remarkable film only about 8 nm thick—you would have to stack over 8000 such membranes to equal the thickness of this page—the plasma membrane controls traffic into and out of the cell it surrounds. Like all biological membranes, the plasma membrane has ***selective permeability;*** *that is, it allows some substances to cross it more easily than others. One of the earliest episodes in the evolution of life may have been the formation of a membrane that could enclose a solution of different composition from the surrounding solution, while still permitting the selective uptake of nutrients and elimination of waste products. This ability of the cell to discriminate in its chemical exchanges with the environment is fundamental to life, and it is the plasma membrane that makes this selectivity possible.*

In this chapter, you will learn how biological membranes control the passage of substances. We will concentrate on the plasma membrane, the outermost membrane of the cell, represented by the drawing on this page. However, the general principles of membrane traffic also apply to the many varieties of internal membranes that partition the eukaryotic cell. To understand how membranes work, we begin by examining their architecture.

Membrane models have evolved to fit new data: *science as a process*

Lipids and proteins are the staple ingredients of membranes, although carbohydrates are also present. Currently, the most widely accepted model for the arrangement of these molecules in membranes is the fluid mosaic model. We will trace the evolution of this model in some detail as an example of how scientists build on earlier observations and ideas and how they construct models as working hypotheses.

Scientists began building molecular models of the membrane decades before membranes were first resolved by the electron microscope in the 1950s. In 1895, Charles Overton postulated that membranes are made of lipids, based on his observations that substances that dissolve in lipids enter cells much more rapidly than substances that are insoluble in lipids. Twenty years later, membranes isolated from red blood cells were chemically analyzed and found to be composed of lipids and proteins.

Phospholipids are the most abundant lipids in most membranes. The ability of phospholipids to form membranes is built into their molecular structure. A phospholipid is an **amphipathic** molecule, meaning it has both a hydrophilic region and a hydrophobic region (see FIGURE 5.13). Other types of membrane lipids are also amphipathic.

In 1917, I. Langmuir made artificial membranes by adding phospholipids dissolved in benzene (an organic solvent) to water. After the benzene evaporated, the phospholipids remained as a film covering the surface of the water, with only the hydrophilic heads of the phospholipids immersed in the water (FIGURE 8.1a). In 1925, two Dutch scientists, E. Gorter and F. Grendel, reasoned that cell membranes are actually phospholipid bilayers, two molecules thick. Such a bilayer could exist as a stable boundary between two aqueous compartments because the molecular arrangement shelters the hydrophobic tails of the phospholipids from water, while exposing the hydrophilic heads to water (FIGURE 8.1b). Gorter and Grendel measured the phospholipid content of membranes isolated from red blood cells and found just enough of the lipid to cover the cells with two layers. (Ironically, Gorter and Grendel underestimated both the phospholipid content and the surface area of the cells, but the two errors canceled each other. Thus, what turned out to be a correct conclusion was based on flawed measurements.)

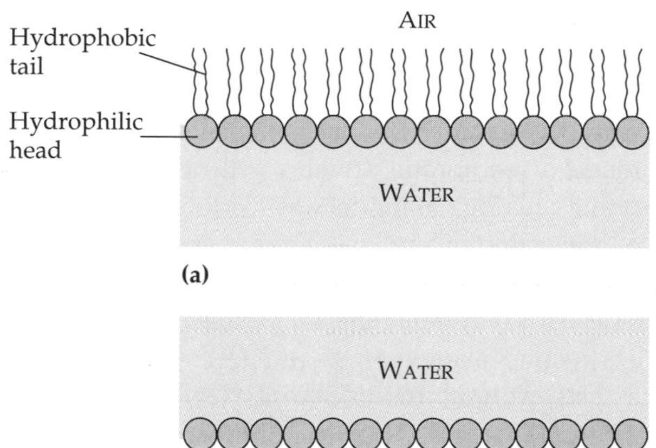

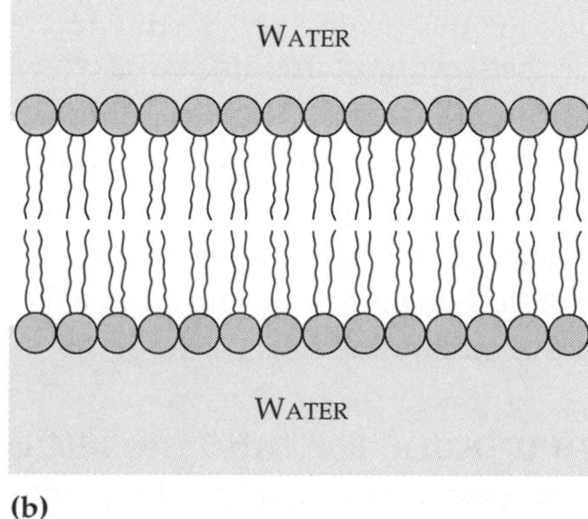

FIGURE 8.1

Artificial membranes (cross sections). (**a**) Water can be coated with a single layer of phospholipids. The hydrophilic heads of the phospholipids are immersed in water, and the hydrophobic tails are excluded from water. (**b**) A bilayer of phospholipids forms a stable boundary between two aqueous compartments. The arrangement exposes the hydrophilic parts of the molecules to water and shields the hydrophobic parts from water.

If we assume that a phospholipid bilayer is the main fabric of the membrane, where do we place the proteins? Although the heads of phospholipids are hydrophilic, the surface of an artificial membrane consisting of a phospholipid bilayer adheres less strongly to water than does the surface of an actual biological membrane. This difference could be accounted for if the membrane were coated on both sides with hydrophilic proteins. In 1935, H. Davson and J. Danielli incorporated this hypothesis into a molecular model of the membrane. The Davson-Danielli model was a sandwich: a phospholipid bilayer between two layers of globular protein (FIGURE 8.2a).

Not until the 1950s did biologists finally see membranes with the help of electron microscopes. With a thickness of only 7 to 8 nm, the plasma membrane was a little too thin to be the molecular sandwich that Davson and Danielli had predicted. Modifying the model by replacing the globular proteins with continuous layers of protein in the pleated-sheet configuration fit the model to the observed thickness of the actual membrane.

In electron micrographs of cells stained with atoms of heavy metals, the plasma membrane is triple-layered, having two dark ("stained") bands separated by an unstained layer (see FIGURE 7.6, p. 115). This, too, was interpreted as evidence for the Davson-Danielli model. Most early electron microscopists assumed that the stain adhered to the proteins and hydrophilic heads of the phospholipids, leaving the hydrophobic core of the membrane unstained. In a somewhat circular manner, the electron micrographs and the membrane model became more and more associated as explanations for each other. By the 1960s, the Davson-Danielli sandwich had become widely accepted as the structure not only of the plasma membrane, but of all the internal membranes of the cell. However, by the end of that decade, many cell biologists recognized two problems with the model.

First, the generalization that all membranes of the cell are identical was challenged. Not all membranes look alike in the electron microscope. For example, whereas the plasma membrane is 7 to 8 nm thick and has the three-layered structure, the inner membrane of the mitochondrion is only 6 nm thick and in electron micrographs looks like a row of beads. Mitochondrial membranes also have a substantially greater percentage of proteins than plasma membranes, and there are differences in the specific kinds of phospholipids and other lipids. Membranes with different functions differ in chemical composition and structure.

The second problem with the sandwich model is in the placement of the proteins. Unlike proteins dissolved in the cytosol, membrane proteins are not very soluble in

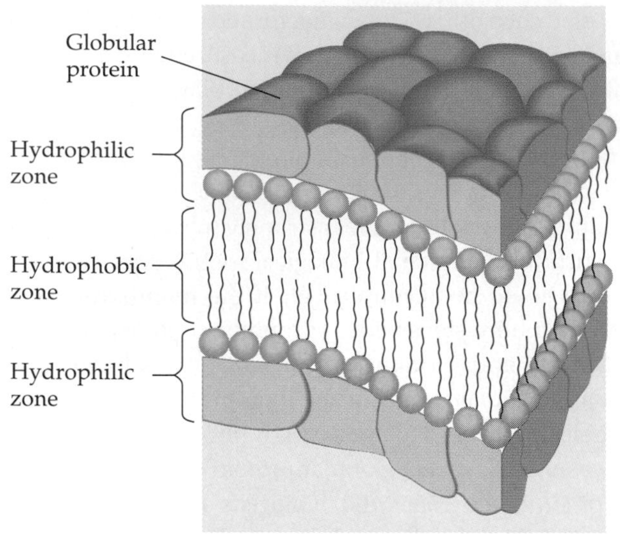

(a) Original Davson-Danielli model

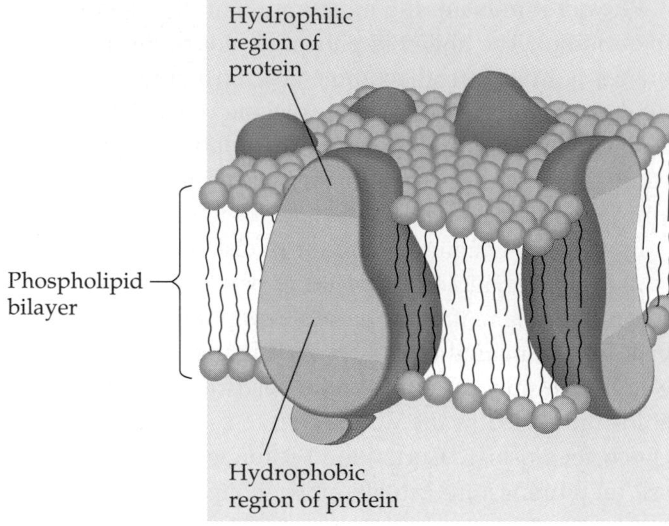

(b) Current fluid mosaic model

FIGURE 8.2

Two generations of membrane models. (**a**) The Davson-Danielli model, proposed in 1935, sandwiched the phospholipid bilayer between two protein layers. With later modifications, this model was widely accepted until about 1970. (**b**) The fluid mosaic model disperses the proteins and immerses them in the phospholipid bilayer, which is in a fluid state. This is our present working model of the membrane.

water. Membrane proteins have hydrophobic and hydrophilic regions; they are amphipathic, as are their phospholipid partners in membranes. If proteins were layered on the surface of the membrane, their hydrophobic parts would be in an aqueous environment, and the proteins would separate the hydrophilic heads of the phospholipids from water.

In 1972, S. J. Singer and G. Nicolson advocated a revised membrane model that placed the proteins in a location compatible with their amphipathic character. Instead of seeing the phospholipid bilayer as coated with solid sheets of protein, Singer and Nicolson proposed that membrane proteins are dispersed and individually inserted into the phospholipid bilayer, with only their hydrophilic regions protruding far enough from the bilayer to be exposed to water. This molecular arrangement would maximize contact of hydrophilic regions of proteins and phospholipids with water, while providing their hydrophobic parts with a nonaqueous environment. According to this model, the membrane is a mosaic of protein molecules bobbing in a fluid bilayer of phospholipids; hence the term **fluid mosaic model** (FIGURE 8.2b).

A method of preparing cells for electron microscopy called freeze-fracture has provided convincing evidence that proteins are embedded in the phospholipid bilayer of the membrane, rather than being spread upon the surface. Freeze-fracture can delaminate a membrane along the middle of the bilayer, splitting the membrane into outer and inner faces (see the Methods Box). When the halves of the fractured membrane are viewed in the electron microscope, the interior of the bilayer appears cobblestoned, with protein particles interspersed in a smooth matrix. Proteins penetrate into the hydrophobic interior of the membrane, which would not be the case if the Davson-Danielli sandwich were correct.

We have examined the evolution of our understanding of membrane structure as a case history of how science works. Models are proposed by scientists as ways of organizing and explaining existing information. Replacing one model of membrane structure with another does not imply that the original model was worthless. The acceptance or rejection of a model depends on how well it fits observations and explains experimental results. A good model also makes predictions that shape future research. Models inspire experiments, and few models survive these tests without modification. New findings may make a model obsolete; but even then, it may not be totally scrapped, just revised to incorporate the new observations. Like its predecessor, which endured for 35 years, the fluid mosaic model may eventually be retailored to fit new data. For now, it is our most acceptable working model of membrane structure.

A membrane is a fluid mosaic of lipids, proteins, and carbohydrates

What exactly does it mean to describe a membrane as a fluid mosaic? Let's begin with the word *fluid.*

The Fluid Quality of Membranes

Membranes are not static sheets of molecules locked rigidly in place. A membrane is held together primarily by

METHODS: FREEZE-FRACTURE AND FREEZE-ETCH

(**a**) A researcher freezes the specimen at the temperature of liquid nitrogen, then fractures the cells with a cold knife.

(**b**) The knife does not cut cleanly through the frozen cells; instead, it cracks the specimen with the fracture plane following the path of least resistance. The fracture plane often follows the hydrophobic interior of a membrane, splitting the lipid bilayer down the middle into a P (protoplasmic) face and an E (exterior) face. The membrane proteins are not split but go with one or the other of the phospholipid layers. The topography of the fractured surface may be enhanced by etching, the removal of water by sublimation (direct evaporation of frozen water to water vapor).

(**c**) A fine mist of platinum is sprayed from an angle onto the fractured surface of the cell. There will be "shadows" where elevated regions of the fractured cell block the platinum. Adding a film of carbon strengthens the platinum coat.

The original specimen is digested away with bleach, acids, and enzymes, leaving the platinum-carbon film as a replica of the fractured surface. It is this replica, not the membrane itself, that is examined in the electron microscope.

(**d**) The electron micrographs have been superimposed on a drawing of a delaminated membrane. Notice the protein particles (the "bumps").

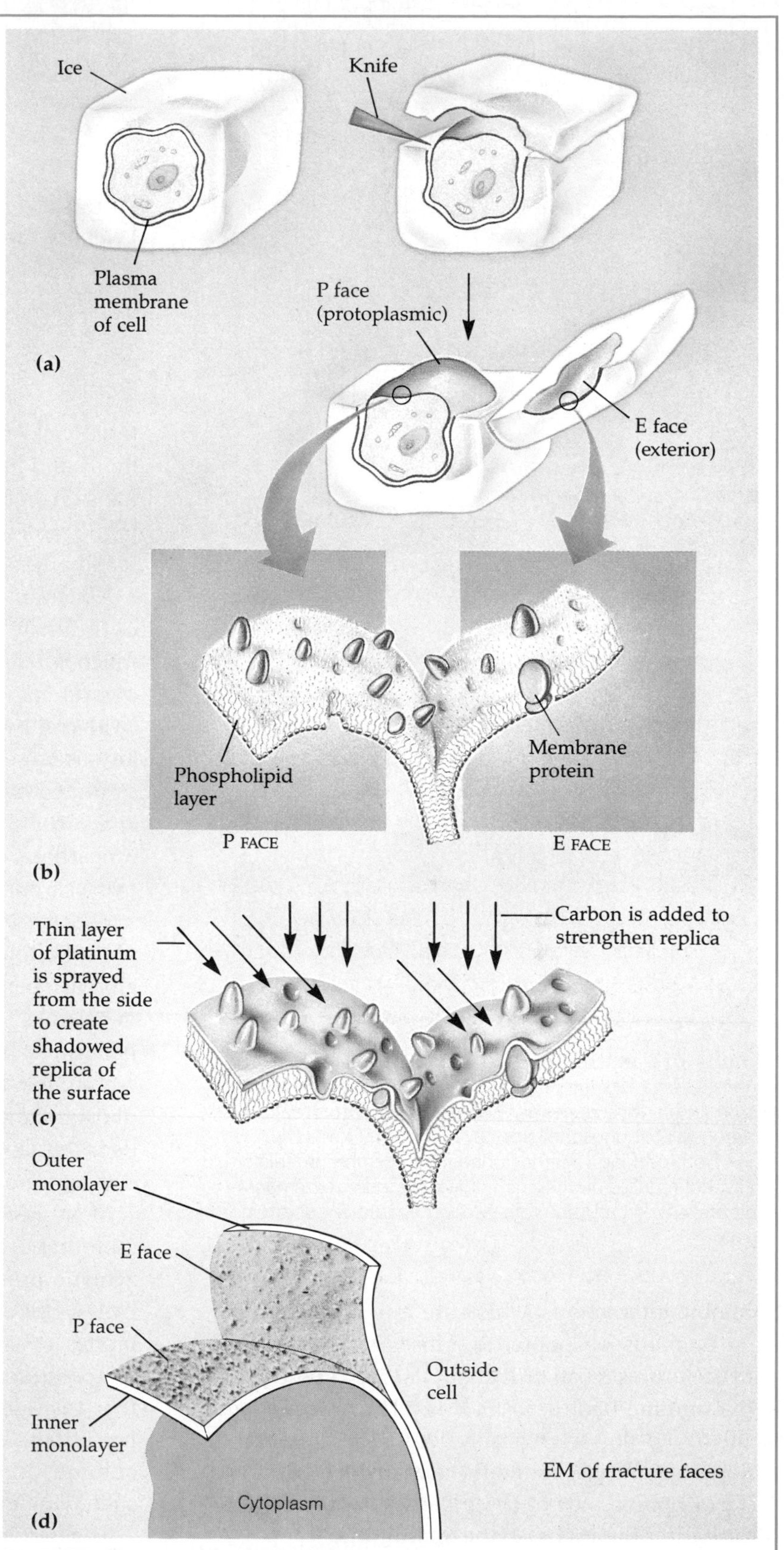

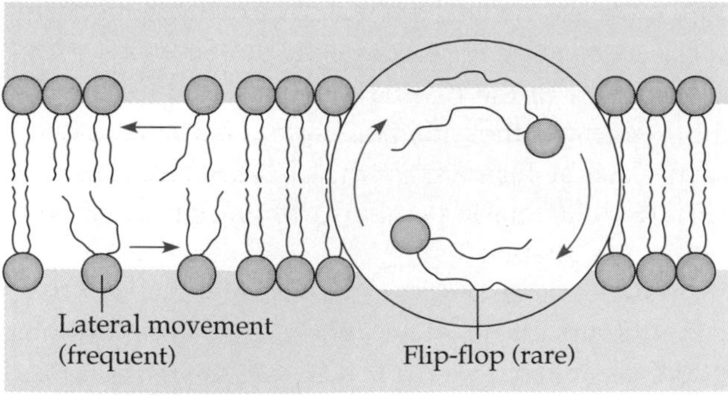

(a)

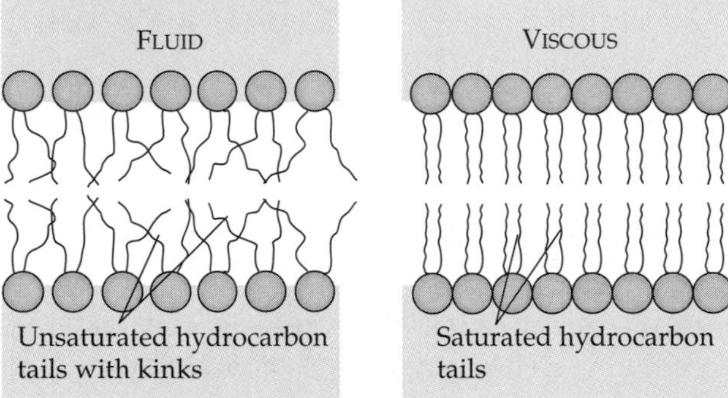

(b)

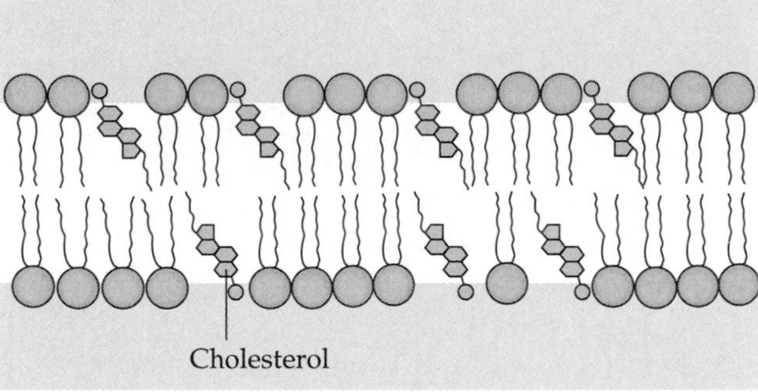

(c)

FIGURE 8.3

The fluidity of membranes. (**a**) Lipids move laterally (that is, in two dimensions) in a membrane, but flip-flopping across the membrane (in the third dimension) is rare. (**b**) Unsaturated hydrocarbon tails of phospholipids have kinks that keep the molecules from packing together, enhancing membrane fluidity. (**c**) Cholesterol reduces membrane fluidity at moderate temperatures but prevents membrane solidification at cold temperatures.

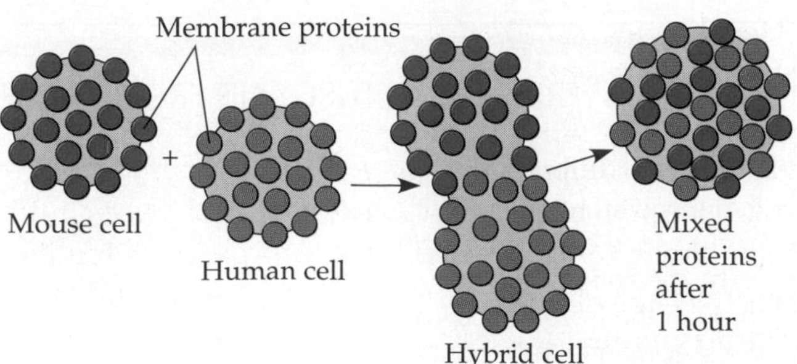

FIGURE 8.4

Evidence for the drifting of membrane proteins. When researchers fuse a human cell with a mouse cell, it takes less than an hour for the membrane proteins of the two species to completely intermingle in the membrane of the hybrid cell.

hydrophobic interactions, which are much weaker than covalent bonds (see Chapter 5). Most of the lipids and some of the proteins can drift about laterally in the plane of the membrane (FIGURE 8.3a). It is rare, however, for a molecule to flip-flop transversely across the membrane, switching from one phospholipid layer to the other; to do so, the hydrophilic part of the molecule would have to cross the hydrophobic core of the membrane.

Phospholipids move along the plane of the membrane rapidly, averaging about 2 μm—the length of a large bacterial cell—per second. Proteins are much larger than lipids and move more slowly, but some membrane proteins do, in fact, drift (FIGURE 8.4). However, many membrane proteins are unable to move far because of their attachment to the cytoskeleton.

A membrane remains fluid as temperature decreases, until finally, at some critical temperature, it solidifies, much as bacon grease forms lard when it cools. The temperature at which a membrane solidifies depends on its lipid composition. The membrane remains fluid to a lower temperature if it is rich in phospholipids with unsaturated hydrocarbon tails (see Chapter 5). Because of kinks where double bonds are located, unsaturated hydrocarbons do not pack together as closely as saturated hydrocarbons (FIGURE 8.3b).

The steroid cholesterol, which is wedged between phospholipid molecules in the plasma membranes of animals, helps stabilize membrane fluidity (FIGURE 8.3c). At relatively warm temperatures—37°C, the body temperature of humans, for example—cholesterol makes the membrane less fluid by restraining the movement of phospholipids. However, because cholesterol also hinders the close packing of phospholipids, it lowers the temperature required for the membrane to solidify.

Membranes must be fluid to work properly. When a membrane solidifies, its permeability changes, and enzymatic proteins in the membrane may become inactive. A cell can alter the lipid composition of its membranes to some extent as an adjustment to changing temperature. For instance, in many varieties of plants that tolerate extreme cold, such as winter wheat, the percentage of unsaturated phospholipids increases in autumn, an adaptation that keeps the membranes from solidifying during winter.

The functioning membrane, then, is a liquid film consisting of proteins dissolved in a lipid bilayer. It is normally about as fluid as salad oil.

Membranes as Mosaics of Structure and Function

Now we come to the word *mosaic.* A membrane is a collage of many different proteins embedded in the fluid matrix of the lipid bilayer (FIGURE 8.5). The lipid bilayer is the main fabric of the membrane, but proteins determine most of the membrane's specific functions (FIGURE 8.6). The plasma membrane and the membranes of the various organelles have their unique collections of proteins. More than 50 kinds of proteins have been found in the plasma membrane of red blood cells, for example, and there are probably many more that have not yet been detected.

Notice in FIGURE 8.5 that there are two major populations of membrane proteins. **Integral proteins** penetrate far enough into the membrane for their hydrophobic regions to be surrounded by the hydrocarbon tails of lipids. Some integral proteins reach only partway across the membrane. Others completely span the membrane. These transmembrane proteins have hydrophobic midsections between hydrophilic ends exposed to the aqueous solutions on both sides of the membrane. **Peripheral proteins** are not embedded in the lipid bilayer at all; they are appendages attached to the surface of the membrane, often to the exposed parts of integral proteins. On the cytoplasmic side of the plasma membrane, some peripheral proteins and their integral protein partners may be held in place by filaments of the cytoskeleton. Fibers of the extracellular matrix also adhere to specific membrane proteins (see Chapter 7).

Membranes have distinct inside and outside faces. The two lipid layers may differ in specific lipid composition, and each protein has directional orientation in the membrane. The plasma membrane also has carbohydrates, which are restricted to the exterior surface. This asymmetrical distribution of proteins, lipids, and carbohydrates is determined as the membrane is being built by the endoplasmic reticulum (FIGURE 8.7).

Membrane Carbohydrates and Cell-Cell Recognition

Cell-cell recognition, a cell's ability to distinguish one type of neighboring cell from another, is crucial to the functioning of an organism. It is important, for example, in the

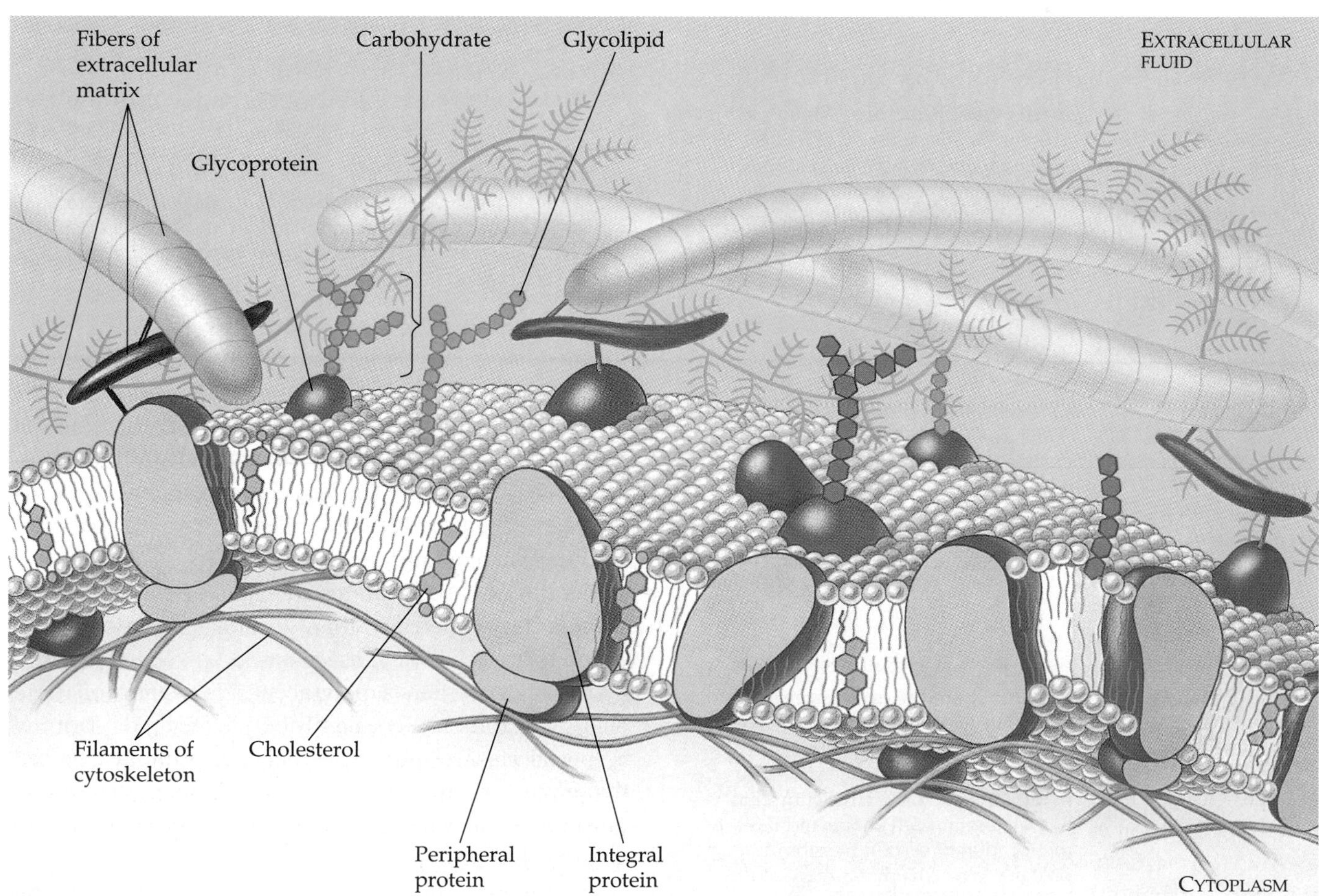

FIGURE 8.5
The detailed structure of an animal cell's plasma membrane (cross section).

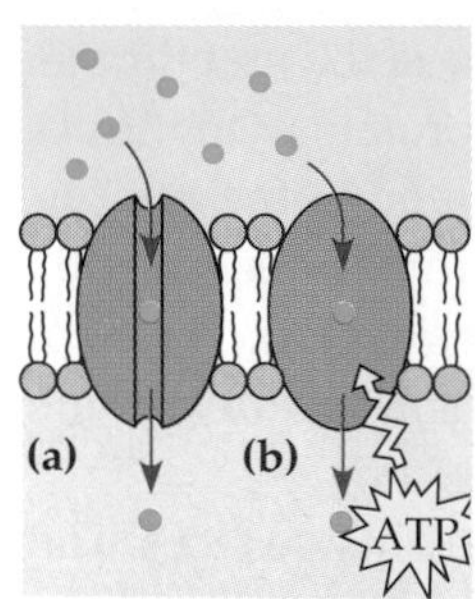

Transport proteins **(a)** A protein that spans the membrane may provide a hydrophilic channel across the membrane that is selective for a particular solute. **(b)** Some transport proteins hydrolyze ATP as an energy source to actively pump substances across the membrane.

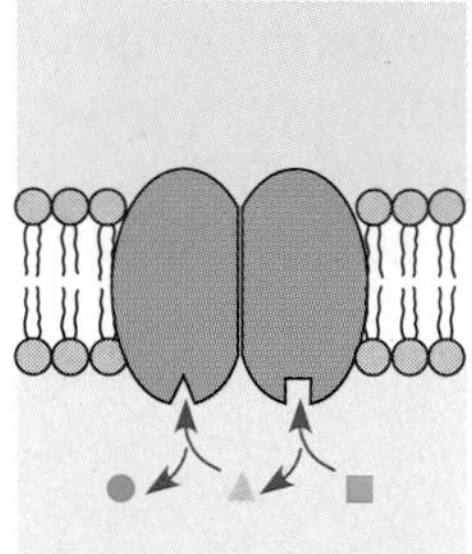

Enzymes A protein built into the membrane may be an enzyme with its active site exposed to substances in the adjacent solution. In some cases, several enzymes are ordered in a membrane as a team that carries out sequential steps of a metabolic pathway.

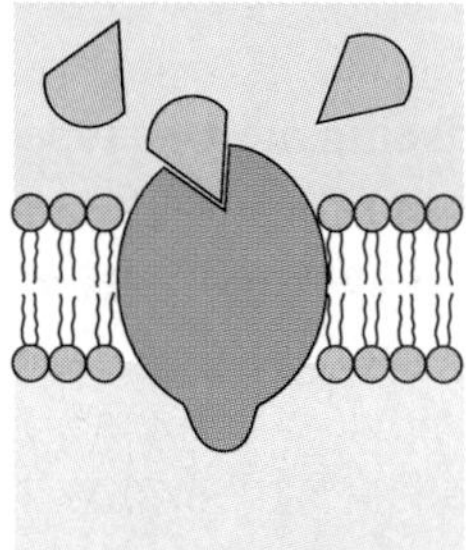

Proteins as receptor sites
A membrane may have a binding site with a specific shape that fits the shape of a chemical messenger, such as a hormone. The external signal may cause a conformational change in the protein that relays the message to the inside of the cell.

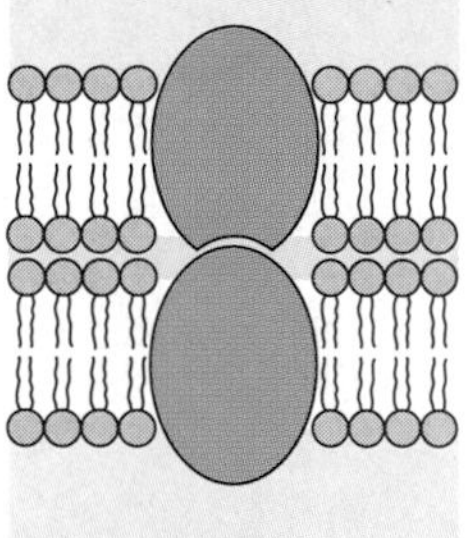

Intercellular junctions Membrane proteins of adjacent cells may be hooked together in various kinds of junctions (see Figure 7.30).

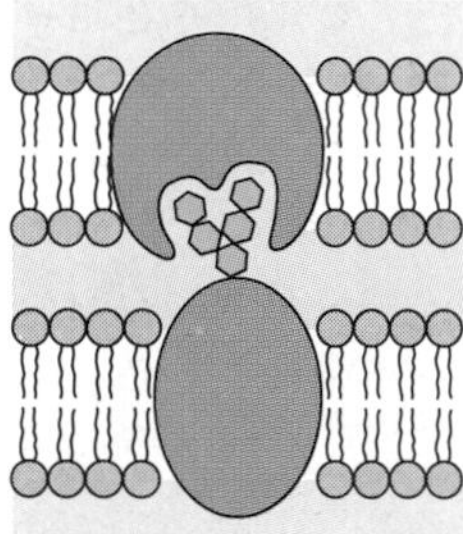

Cell-cell recognition
Some glycoproteins (proteins bonded to short chains of sugars) serve as identification tags that are specifically recognized by other cells.

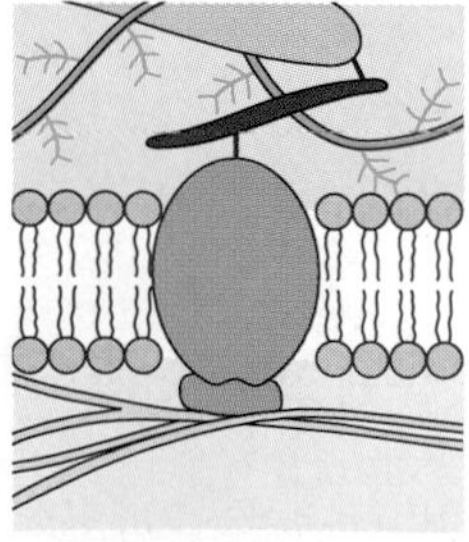

Attachment to the cytoskeleton and extracellular matrix (ECM)
Actin filaments or other elements of the cytoskeleton may be bonded to membrane proteins, a function that helps maintain cell shape and fixes the location of certain membrane proteins. Proteins that adhere to the ECM can coordinate extracellular and intracellular changes.

FIGURE 8.6
Some functions of membrane proteins. A single protein may perform some combination of these tasks.

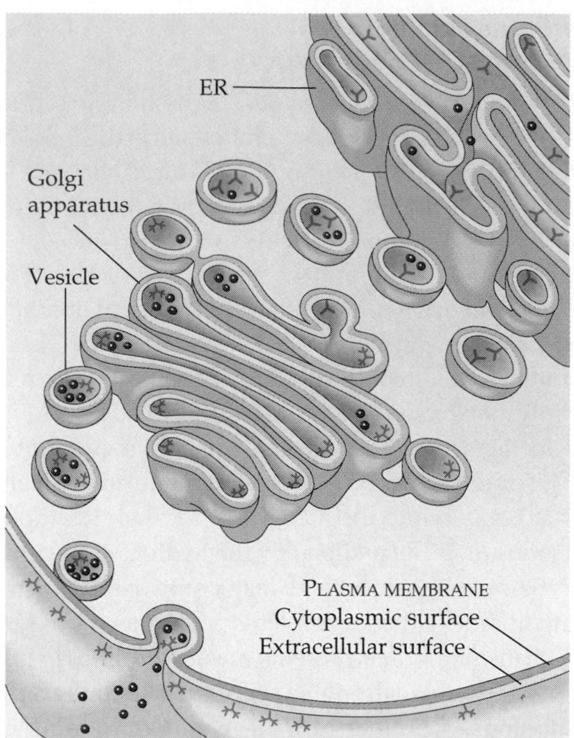

FIGURE 8.7
Sidedness of the plasma membrane. The membrane has distinct cytoplasmic and extracellular sides. This bifacial quality is determined when the membrane is first synthesized and modified by the ER and Golgi. The diagram color-codes the two sides of the membrane, to illustrate that the one facing the lumen (cavity) of the ER, Golgi, and vesicles is topologically equivalent to the extracellular surface of the plasma membrane. The other side faces the cytoplasm, from the time the membrane is produced by the ER to the time it is added to the plasma membrane by fusion of a vesicle. The small green "trees" represent membrane carbohydrates that are synthesized in the ER and modified in the Golgi. Vesicle fusion with the plasma membrane is also responsible for secretion of some cell products (purple).

sorting of cells into tissues and organs in an animal embryo. It is also the basis for the rejection of foreign cells (including those of transplanted organs) by the immune system, an important line of defense in vertebrate animals (see Chapter 39). The way cells recognize other cells is by keying on surface molecules, often carbohydrates, on the plasma membrane.

Membrane carbohydrates are usually branched oligosaccharides with fewer than 15 sugar units. (*Oligo* is Greek for "few"; an oligosaccharide is a sugar polymer that is shorter than a polysaccharide.) Some of these oligosaccharides are covalently bonded to lipids, forming a special class of molecules called glycolipids. (Recall that *glyco* refers to the presence of carbohydrates.) Most, however, are covalently bonded to proteins, which are thereby glycoproteins.

The oligosaccharides on the external side of the plasma membrane vary from species to species, among individuals of the same species, and even from one cell type to another in a single individual. The diversity of the molecules and their location on the cell's surface enable oligosaccharides to function as markers that distinguish one cell from another. For example, the four human blood groups—designated A, B, AB, and O—reflect variation in the oligosaccharides on the surfaces of red blood cells.

The biological membrane is an exquisite example of a supramolecular structure—many molecules ordered into a higher level of organization—with emergent properties beyond those of the individual molecules. The remainder of this chapter addresses one of the most important of those properties: the ability to regulate transport across cellular boundaries, a function essential to the cell's existence as an open system. We will see once again that form fits function: The fluid mosaic model helps explain how membranes regulate the cell's molecular traffic.

A membrane's molecular organization results in selective permeability

There is steady traffic of small molecules across the plasma membrane. Consider the chemical exchanges between a human muscle cell and the extracellular fluid that bathes it. Sugars, amino acids, and other nutrients enter the cell, and metabolic waste products leave. The cell takes in oxygen for cellular respiration and expels carbon dioxide. It also regulates its concentrations of inorganic ions, such as Na^+, K^+, Ca^{2+}, and Cl^-, by shuttling them one way or the other across the plasma membrane. Although traffic through the membrane is extensive, biological membranes are selectively permeable, and substances do not cross the barrier indiscriminately. The cell is able to take up many varieties of small molecules and exclude others. Moreover, substances that move through the membrane do so at different rates.

Permeability of the Lipid Bilayer

The hydrophobic core of the membrane impedes the transport of ions and polar molecules, which are hydrophilic. Hydrophobic molecules, such as hydrocarbons and oxygen, can dissolve in the membrane and cross it with ease. Very small molecules that are polar but uncharged can also pass through the membrane rapidly. Examples are water and carbon dioxide, which are tiny enough to pass between the lipids of the membrane. The lipid bilayer is not very permeable to larger, uncharged polar molecules, such as glucose and other sugars. It is also relatively impermeable to all ions, even such small ones as H^+ and Na^+. A charged atom or molecule and its shell of water find the hydrophobic layer of the membrane difficult to penetrate. However, the lipid bilayer is only part of the story of a membrane's selective permeability. Proteins built into the membrane play a key role in regulating transport.

Transport Proteins

Biological membranes are permeable to specific ions and certain polar molecules. These hydrophilic substances avoid contact with the lipid bilayer by passing through **transport proteins** that span the membrane (see FIGURE 8.6, top). Some transport proteins function by having a channel that certain molecules use as a hydrophilic tunnel through the membrane. Other transport proteins bind to their passengers and physically move them across the membrane. In any case, each transport protein is very specific for the substances it translocates, allowing only a certain molecule or class of closely related molecules to cross the membrane. For example, glucose carried in blood to the human liver enters liver cells rapidly through specific transport proteins inserted in the plasma membrane. The protein is so selective that it even rejects fructose, a structural isomer of glucose.

The selective permeability of a membrane depends on both the discriminating barrier of the lipid bilayer and the specific transport proteins built into the membrane. But what determines the *direction* of traffic across a membrane? At a given time, will a particular substance enter the cell through the membrane, or leave? And what mechanisms actually *drive* molecules across membranes? The next few sections explore these questions by contrasting two modes of membrane traffic: passive transport and active transport.

Passive transport is diffusion across a membrane

Molecules have intrinsic kinetic energy called thermal motion, or heat. One result of thermal motion is **diffusion,** the tendency for molecules of any substance to spread out into the available space. Each molecule moves randomly, yet diffusion of a *population* of molecules may be directional. For example, imagine a membrane separating pure water from a solution of a dye dissolved in water. Assume that this membrane is permeable to the dye molecules (FIGURE 8.8a). Each dye molecule wanders randomly, but there will be a *net* movement of the dye molecules across the membrane to the side that began as pure water. The spreading of the dye across the membrane will continue until both solutions have equal concentrations of the dye. Once that point is reached, there will be a dynamic equilibrium, with as many dye molecules moving per second across the membrane in one direction as in the other.

We can now state a simple rule of diffusion: In the absence of other forces, a substance will diffuse from where it is more concentrated to where it is less concentrated. Put another way, any substance will diffuse down its **concentration gradient.** No work must be done to make this happen; diffusion is a spontaneous process because it decreases free energy (see FIGURE 6.4). Recall that in any system there is a tendency for entropy, or disorder, to increase. Diffusion of a solute in water increases entropy by producing a more random mixture than exists when there are localized concentrations of the solute. It is important to note that each substance diffuses down its *own* concentration gradient, unaffected by the concentration differences of other substances (FIGURE 8.8b).

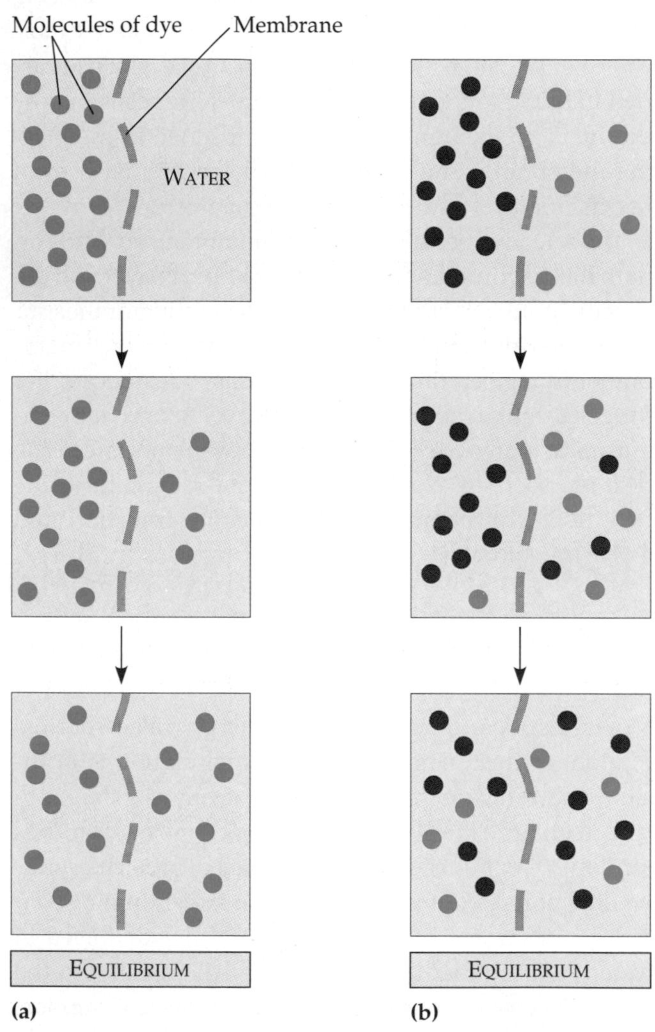

FIGURE 8.8
The diffusion of solutes across membranes. (**a**) A substance will diffuse from where it is more concentrated to where it is less concentrated. The membrane, viewed here in cross section, has pores large enough for molecules of dye to pass. Diffusion down the concentration gradient leads to a dynamic equilibrium; the solute molecules continue to cross the membrane, but at equal rates in both directions. (**b**) In this case, solutions of two different dyes are separated by a membrane that is permeable to both dyes. Each dye diffuses down its own concentration gradient. There will be a net diffusion of the green dye toward the left, even though the total solute concentration was initially greater on the left side.

Much of the traffic across cell membranes occurs by diffusion. When a substance is more concentrated on one side of a membrane than on the other, there is a tendency for the substance to diffuse across the membrane down its concentration gradient (assuming that the membrane is permeable to that substance). One important example is the uptake of oxygen by a cell performing cellular respiration. Dissolved oxygen diffuses into the cell across the plasma membrane. As long as cellular respiration consumes the O_2 as it enters, diffusion into the cell will continue, because the concentration gradient favors movement in that direction.

The diffusion of a substance across a biological membrane is called **passive transport,** because the cell does not have to expend energy to make it happen. The concentration gradient itself represents potential energy and drives diffusion. Remember, however, that membranes are selectively permeable and therefore affect the rates of diffusion of various molecules. One molecule that diffuses freely across most membranes is water, a fact that has important consequences for cells.

Osmosis is the passive transport of water

In comparing two solutions of unequal solute concentration, the solution with a higher concentration of solutes is said to be **hypertonic.** The solution with a lower solute concentration is **hypotonic.** (*Hyper* and *hypo* mean "more" and "less," respectively, referring here to solute concentration.) These are relative terms that are meaningful only in a comparative sense. For example, tap water is hypertonic to distilled water but hypotonic to seawater. In other words, tap water has a higher solute concentration than distilled water, but a lower concentration than seawater. Solutions of equal solute concentration are said to be **isotonic** (*iso* means "the same").

Picture a U-shaped vessel with a selectively permeable membrane separating two sugar solutions of different concentrations (FIGURE 8.9). Pores in this synthetic membrane are too small to allow the sugar molecules to pass but large enough for water to cross the membrane. In effect, the solution with higher solute concentration (hypertonic) has a lower water concentration.* Thus, the water will diffuse across the membrane from the hypotonic solution to the hypertonic solution. This diffusion

*The effect of solutes on water concentration is actually not so simple. At any instant, a fraction of the water molecules in a solution lose their freedom of independent movement by being bound to solute molecules in hydration shells. It is not really a difference in *total* water concentration that causes osmosis, but a difference in the concentration of *unbound* water that is free to cross the membrane.

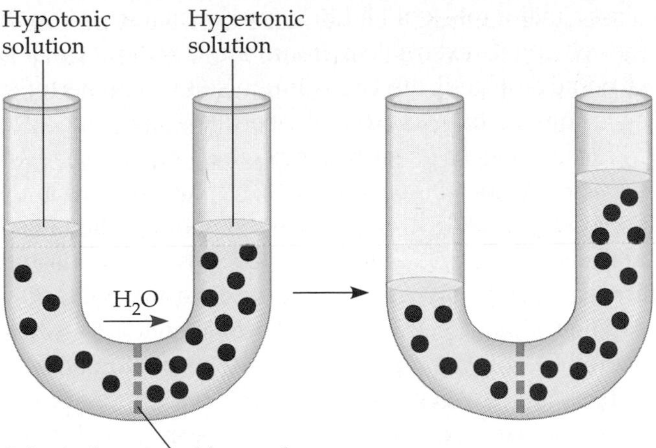

FIGURE 8.9
Osmosis. Two sugar solutions of different concentration are separated by a porous membrane that is permeable to the solvent (water) but impermeable to the solute (sugar). Water will diffuse from the hypotonic solution to the hypertonic solution. This passive transport of water, or osmosis, reduces the difference in sugar concentrations.

of water across a selectively permeable membrane is a special case of passive transport called **osmosis.**

The direction of osmosis is determined only by a difference in *total* solute concentration, not by the nature of the solutes. Water will move from a hypotonic to a hypertonic solution even if the hypotonic solution has more *kinds* of solutes. Seawater, which has a great variety of solutes, will lose water to a very concentrated sugar solution, because the total solute concentration of the seawater is lower. Water moves across a membrane separating isotonic solutions at an equal rate in both directions; that is, there is no net osmosis between isotonic solutions. (In Units Six and Seven of the text, we will examine the direction and rate of osmosis with more quantitative measures called *water potential* and *osmotic pressure.*)

Cell survival depends on balancing water uptake and loss

The movement of water across cell membranes and the balance of water between the cell and its environment are crucial to organisms. Let's now apply to living cells what we have learned about osmosis in artificial systems.

Water Balance of Cells Without Walls

If an animal cell is immersed in an environment that is isotonic to the cell, there will be no net movement of water across the plasma membrane. Water is flowing across the membrane, but at the same rate in both directions. In an isotonic environment, the volume of an animal cell is stable (FIGURE 8.10). Now let's transfer the cell to a solution that is hypertonic to the cell. The cell will lose water to its environment, shrivel, and probably die. This is one reason why an increase in the salinity (saltiness) of a lake can kill the animals there. However, taking up too much water can be just as hazardous to an animal cell as losing water. If we place the cell in a solution that is hypotonic to the cell, water will enter faster than it leaves, and the cell will swell and lyse (burst) like an overfilled water balloon.

A cell without rigid walls can tolerate neither excessive uptake nor excessive loss of water. This problem of water balance is automatically solved if such a cell lives in isotonic surroundings. Seawater is isotonic to many marine invertebrates. The cells of most terrestrial (land-dwelling) animals are bathed in an extracellular fluid that is isotonic to the cells. Animals and other organisms without rigid cell walls living in hypertonic or hypotonic

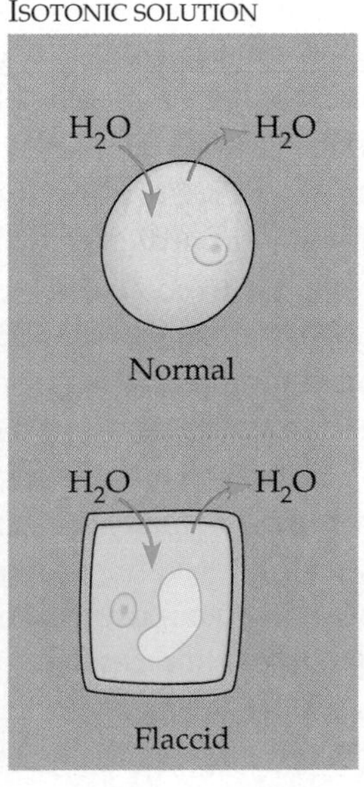

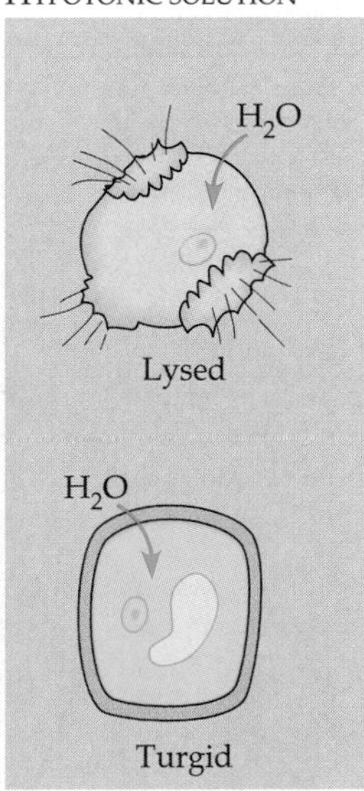

FIGURE 8.10
The water balance of living cells. How living cells react to changes in the solute concentrations of their environments depends on whether or not they have cell walls. Animal cells do not have cell walls; plant cells do. Unless it has special adaptations to offset the osmotic uptake or loss of water, an animal cell fares best in an isotonic environment. Plant cells are turgid (firm) and generally healthiest in a hypotonic environment, where the tendency for continued uptake of water is balanced by the elastic wall pushing back on the cell. (Arrows indicate net water movement when the cells are *first* placed in these solutions.)

environments must have special adaptations for **osmoregulation,** the control of water balance. For example, the protist *Paramecium* lives in pond water, which is hypotonic to the cell. Water continually tends to enter the cell, but *Paramecium* has a plasma membrane that is much less permeable to water than the membranes of most other cells. Also, *Paramecium* is equipped with a contractile vacuole, an organelle that functions as a bilge pump to force water out of the cell as fast as it enters by osmosis (FIGURE 8.11). We will examine other adaptations for osmoregulation in Chapter 40.

Water Balance of Cells with Walls

The cells of plants, prokaryotes, fungi, and some protists have walls. Under certain conditions, the wall plays a major role in maintaining water balance between the cell and its external environment. However, a wall is of no advantage if the cell is immersed in a hypertonic environment. In this case, a plant cell, like an animal cell, will lose water to its surroundings and shrink (see FIGURE 8.10). As the plant cell shrivels, its plasma membrane pulls away from the wall. This phenomenon, called **plasmolysis,** is usually lethal. The walled cells of bacteria and fungi also plasmolyze in hypertonic environments.

When a plant cell is in a hypotonic solution—when bathed by rainwater, for example—the wall functions in water balance. Again like an animal cell, the plant cell swells as water enters by osmosis. However, the elastic wall will expand only so much before it exerts a back pressure on the cell that offsets the tendency for further water uptake from the hypotonic surroundings. At this point, the cell is **turgid** (very firm). This is the healthy state for most plant cells. Plants that are not woody, such as most house plants, depend on turgid cells for mechanical support. This requires that the cells be hypertonic to the solution on the outside of their plasma membranes. If a plant cell and its surroundings are isotonic, there is no net tendency for water to enter, and the cell is **flaccid** (limp). A plant wilts when its cells are flaccid.

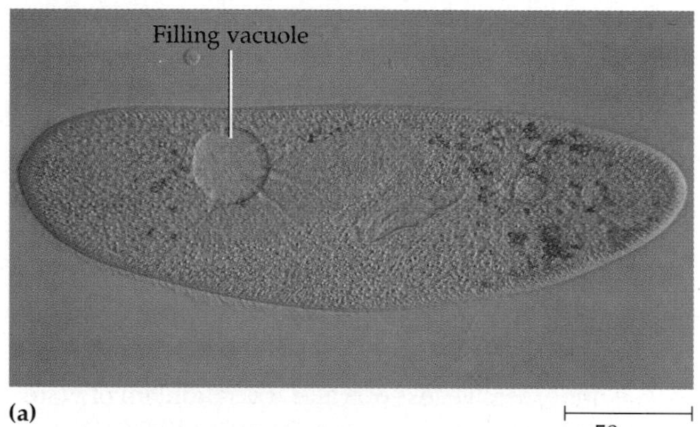

(a)

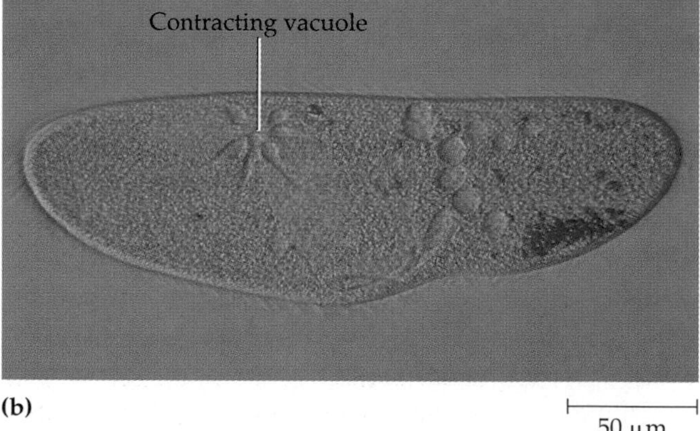

(b)

FIGURE 8.11
Evolutionary adaptations for osmoregulation in *Paramecium.* *Paramecium* is a genus of freshwater protists. Pond water, which is hypotonic to the cell, surrounds a *Paramecium*, and water tends to enter the cell by osmosis. What keeps the cell, which lacks a cell wall, from bursting? Compared to the plasma membranes of most other organisms, the *Paramecium* membrane is less permeable to water. However, this adaptation only slows the uptake of water. The contractile vacuole offsets osmosis by bailing water out of the cell. (**a**) A contractile vacuole fills with fluid that enters from a system of canals radiating throughout the cytoplasm (LM). (**b**) When full, the vacuole and canals contract, expelling fluid from the cell (LM).

Specific proteins facilitate the passive transport of selected solutes

Let's now turn our attention from the transport of water across a membrane to the traffic of specific solutes dissolved in the water. As mentioned earlier, many polar molecules and ions impeded by the lipid bilayer of the membrane diffuse with the help of transport proteins that span the membrane. This phenomenon is called **facilitated diffusion.**

A transport protein has many of the properties of an enzyme. Just as an enzyme is specific for its substrate, a membrane protein is specialized for the solute it transports and has a specific binding site akin to the active site of an enzyme. Like enzymes, transport proteins can be saturated. There are only so many molecules of each type of transport protein built into the plasma membrane, and when these molecules are binding and translocating passengers as fast as they can, transport is occurring at a maximum rate. Also like enzymes, transport proteins can be inhibited by molecules that resemble the normal "substrate." This occurs when the imposter competes with the normally transported solute by binding to the transport protein. Unlike enzymes, however, transport proteins do not usually catalyze chemical reactions. Their function is to catalyze a *physical* process—the faster transport of a molecule across a membrane that would otherwise be relatively impermeable to the substance.

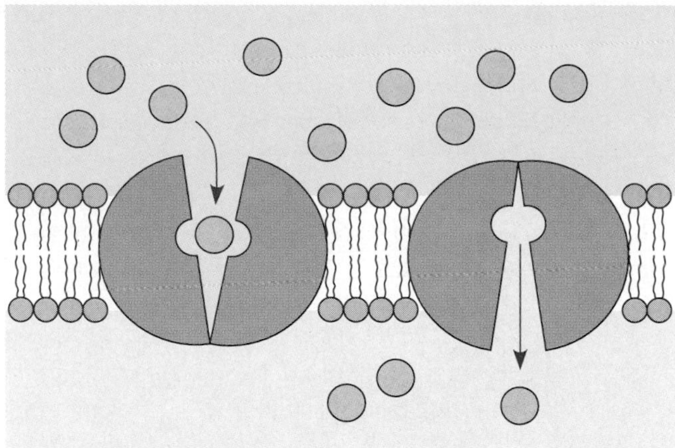

FIGURE 8.12

One model for facilitated diffusion. The transport protein (purple) alternates between two conformations, moving a solute across the membrane as the shape of the protein changes. The protein can transport the solute in either direction, with the net movement being down the concentration gradient of the solute.

Cell biologists are still trying to learn how various transport proteins facilitate diffusion. In many cases, the protein probably undergoes a subtle change in shape that translocates the solute-binding site from one side of the membrane to the other (FIGURE 8.12). The changes in shape could be triggered by the binding and release of the transported molecule. Other transport proteins simply provide selective corridors allowing a specific solute across the membrane (see FIGURE 8.6, top left). Some of these proteins function as **gated channels;** an electrical or chemical stimulus causes them to open. For example, stimulation of a nerve cell opens gated channels that facilitate the diffusion of sodium ions into the cell.

In certain inherited diseases, specific transport systems are either defective or missing altogether. An example is cystinuria, a human disease characterized by the absence of a protein that transports cystine and other amino acids across the membranes of kidney cells. Kidney cells normally reabsorb these amino acids from the urine and return them to the blood, but an individual afflicted with cystinuria develops painful stones from amino acids that accumulate and crystallize in the kidneys.

Active transport is the pumping of solutes against their gradients

Despite the help of a transport protein, facilitated diffusion is still considered passive transport because the solute is moving down its concentration gradient. Facilitated diffusion speeds the transport of a solute by providing a specific corridor through the membrane, but it does not alter the direction of transport. Some transport proteins, however, *can* move solutes against their concentration gradients, across the plasma membrane from the side where they are less concentrated to the side where they are more concentrated. This transport is "uphill"; it goes against the tendency for substances to diffuse down their concentration gradients, and therefore requires work. To pump a molecule across a membrane against its gradient, the cell must expend its own metabolic energy; therefore, this type of membrane traffic is called **active transport.**

Active transport is a major factor in the ability of a cell to maintain internal concentrations of small molecules that differ from concentrations in the surrounding environment. For example, compared to its surroundings, an animal cell has a much higher concentration of potassium ions and a much lower concentration of sodium ions. The plasma membrane helps maintain these steep gradients by pumping sodium out of the cell and potassium into the cell.

The work of active transport is performed by specific proteins embedded in membranes. As in other types of cellular work, ATP supplies the energy for most active transport. One way ATP can power active transport is by transferring its terminal phosphate group directly to the transport protein. This may induce the protein to change its conformation in a manner that translocates a solute bound to the protein across the membrane. One transport system that works this way is the **sodium-potassium pump,** which exchanges sodium (Na^+) for potassium (K^+) across the plasma membrane of animal cells (FIGURE 8.13). FIGURE 8.14 (p. 153) reviews the distinction between passive transport and active transport.

Some ion pumps generate voltage across membranes

All cells have voltages across their plasma membranes. Voltage is electrical potential energy—a separation of opposite charges. The cytoplasm of a cell is negative in charge compared to the extracellular fluid because of an unequal distribution of anions and cations on opposite sides of the membrane. The voltage across a membrane, called a **membrane potential,** ranges from about −50 to −200 millivolts. (The minus sign indicates that the inside of the cell is negative compared to the outside.)

The membrane potential acts like a battery, an energy source that affects the traffic of all charged substances across the membrane. Because the inside of the cell is negative compared to the outside, the membrane potential favors the passive transport of cations into the cell and anions out of the cell. Thus, *two* forces drive the diffusion of ions across a membrane: a chemical force (the ion's concentration gradient) and an electrical force (the effect of the membrane potential on the ion's movement). This combination of forces acting on an ion

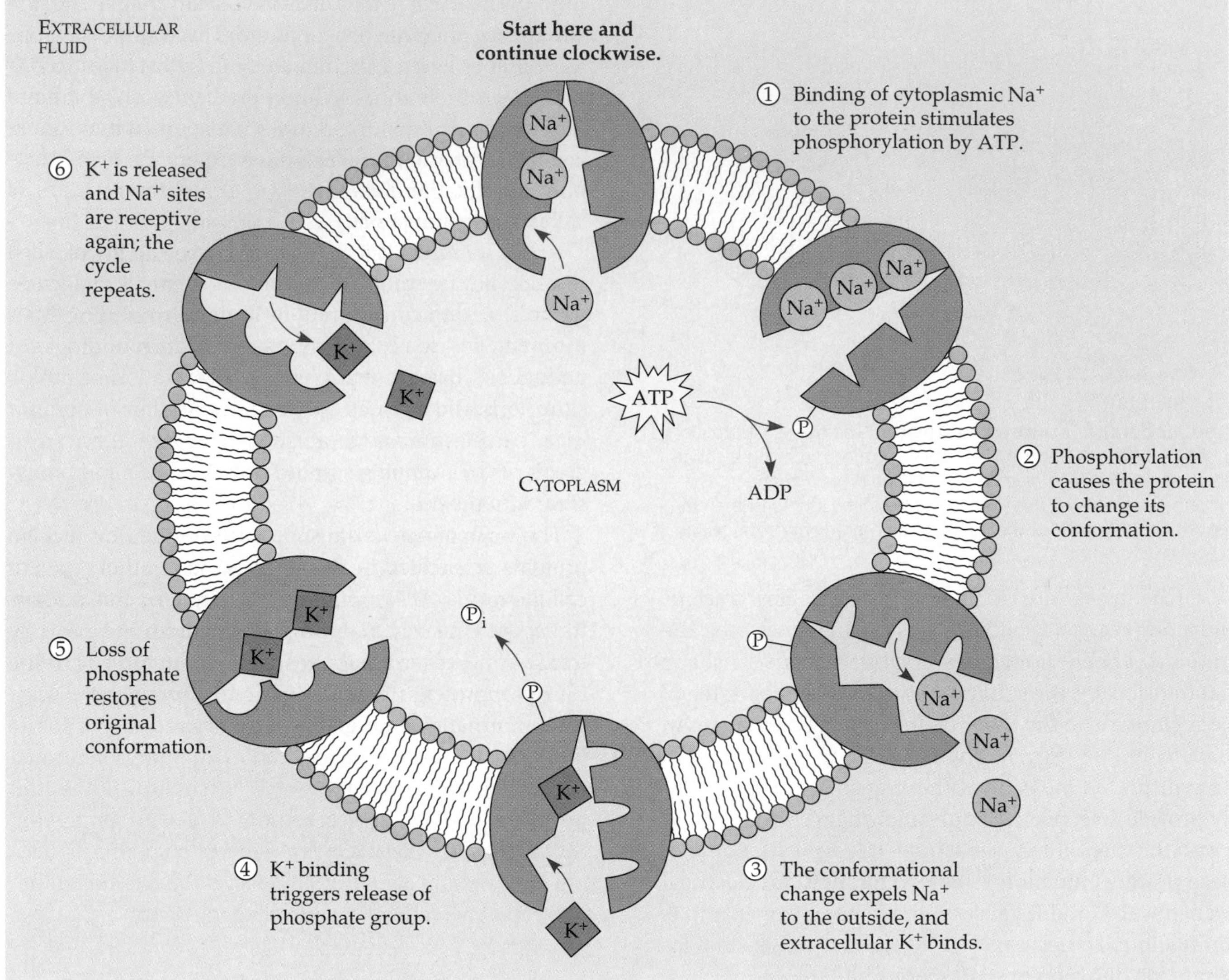
EXTRACELLULAR FLUID
Start here and continue clockwise.
① Binding of cytoplasmic Na+ to the protein stimulates phosphorylation by ATP.
② Phosphorylation causes the protein to change its conformation.
③ The conformational change expels Na+ to the outside, and extracellular K+ binds.
④ K+ binding triggers release of phosphate group.
⑤ Loss of phosphate restores original conformation.
⑥ K+ is released and Na+ sites are receptive again; the cycle repeats.
CYTOPLASM
ATP
ADP
P
Pi
Na+
K+

FIGURE 8.13

The sodium-potassium pump: a specific case of active transport. This transport system pumps ions against steep concentration gradients. The pump oscillates between two conformational states in a pumping cycle that translocates three Na^+ ions out of the cell for every two K^+ ions pumped into the cell. ATP powers the changes in conformation by phosphorylating the transport protein (that is, by transferring a phosphate group to the protein).

is called the **electrochemical gradient.** In the case of ions, we must refine our concept of passive transport: An ion does not simply diffuse down its *concentration* gradient, but diffuses down its *electrochemical* gradient. For example, a resting nerve cell has a much higher concentration of sodium ions (Na^+) outside the cell than inside. When the cell is stimulated, gated channels that facilitate Na^+ diffusion open. Sodium ions then "fall" down their electrochemical gradient, driven by the concentration gradient of Na^+ and by the tendency of cations to cross to the negative side of the membrane.

Some membrane proteins that actively transport ions contribute to the membrane potential. An example is the Na^+-K^+ pump. Notice in FIGURE 8.13 that the pump does not translocate Na^+ and K^+ ions one for one, but actually pumps three sodium ions out of the cell for every two potassium ions it pumps into the cell. With each crank of the pump, there is a net transfer of one positive charge from the cytoplasm to the extracellular fluid, a process that stores energy in the form of voltage. A transport protein that generates voltage across a membrane is called an **electrogenic pump.** The sodium-potassium pump seems to be the major electrogenic pump of animal cells. The main electrogenic pump of plants, bacteria, and fungi is a **proton pump,** which actively transports hydrogen ions (protons) out of the cell. The pumping of H^+ transfers positive charge from the cytoplasm to the extracellular solution (FIGURE 8.15).

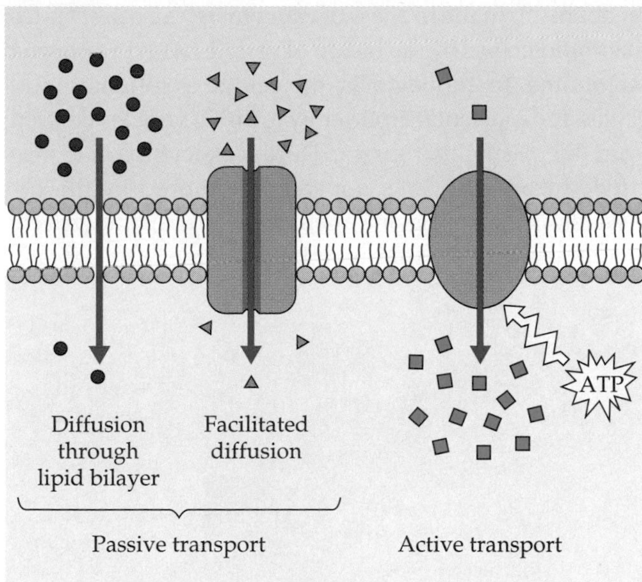

FIGURE 8.14
Review: passive and active transport compared.

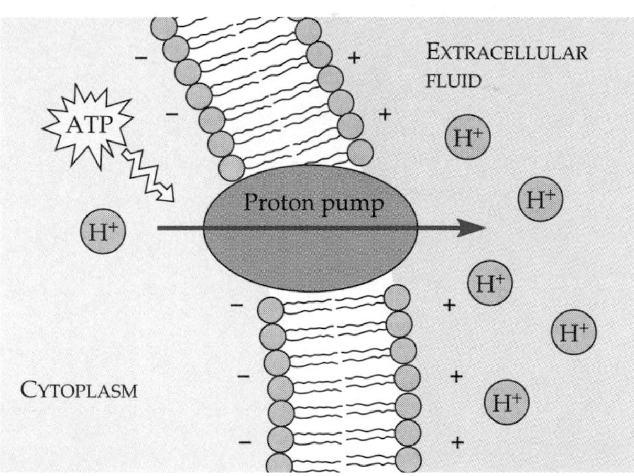

FIGURE 8.15
An electrogenic pump. Proton pumps are examples of membrane proteins that store energy by generating voltage (charge separation) across membranes. Using ATP for power, a proton pump translocates positive charge in the form of hydrogen ions. The voltage and H^+ gradient represent a dual energy source that can be tapped by the cell to drive other processes, such as the uptake of sugar and other nutrients. Proton pumps are the main electrogenic pumps of plants, fungi, and bacteria.

By generating voltage across membranes, electrogenic pumps store energy that can be tapped for cellular work, including a type of membrane traffic called cotransport.

In cotransport, a membrane protein couples the transport of one solute to another

A single ATP-powered pump that transports a specific solute can indirectly drive the active transport of several other solutes in a mechanism called **cotransport.** A substance that has been pumped across a membrane can do work as it leaks back by diffusion, analogous to water that has been pumped uphill and performs work as it flows back down. Another specialized transport protein, separate from the pump, can couple the "downhill" diffusion of this substance to the "uphill" transport of a second substance against its own concentration gradient. For example, a plant cell uses the gradient of hydrogen ions generated by its proton pumps to drive the active transport of amino acids, sugars, and several other nutrients into the cell. One specific transport protein couples the return of hydrogen ions to the transport of sucrose into the cell (FIGURE 8.16). The protein can translocate sucrose into the cell against a concentration gradient, but only if the sucrose molecule travels in the company of a hydrogen ion. The sucrose rides on the coattails of the hydrogen ion, which uses the common transport protein as an avenue to diffuse down the concentration gradient maintained by the proton pump. Plants use this mechanism to load sucrose produced by photosynthesis into specialized cells in the veins of leaves. The sugar can then be distributed by the vascular tissue of the plant to nonphotosynthetic organs, such as roots.

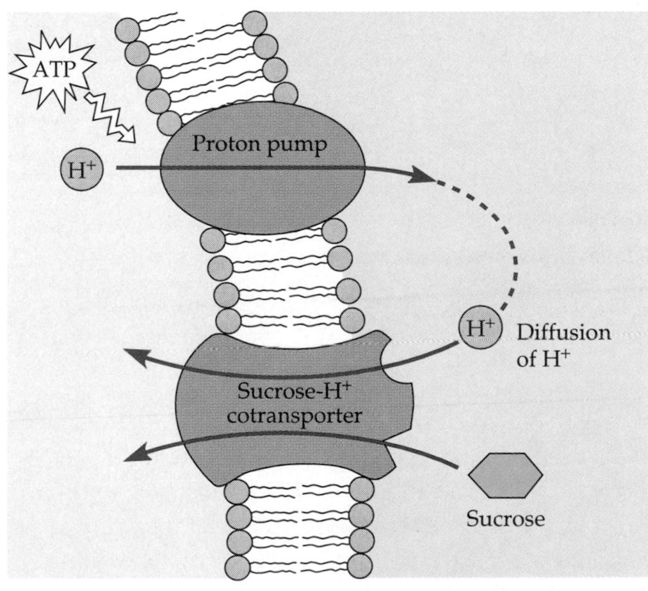

FIGURE 8.16
Cotransport. An ATP-driven pump stores energy by concentrating a substance (H^+, in this case) on one side of the membrane. As the substance leaks back across the membrane through specific transport proteins, it escorts other substances. In this case, the proton pump of the membrane is indirectly driving sucrose accumulation by a plant cell, with the help of a protein that cotransports the two solutes.

Exocytosis and endocytosis transport large molecules

Water and small solutes enter and leave the cell by passing through the lipid bilayer of the membrane, or they are pumped or carried across the membrane by transport proteins. Large molecules, such as proteins and polysaccharides, generally cross the membrane by a different mechanism. In the process of **exocytosis,** the cell secretes macromolecules by the fusion of vesicles with the plasma membrane. In **endocytosis,** the cell takes in macromolecules and particulate matter by forming vesicles derived from the plasma membrane. During exocytosis, a vesicle budded from the Golgi apparatus is moved by the cytoskeleton to the plasma membrane. When the vesicle membrane and plasma membrane come into contact,

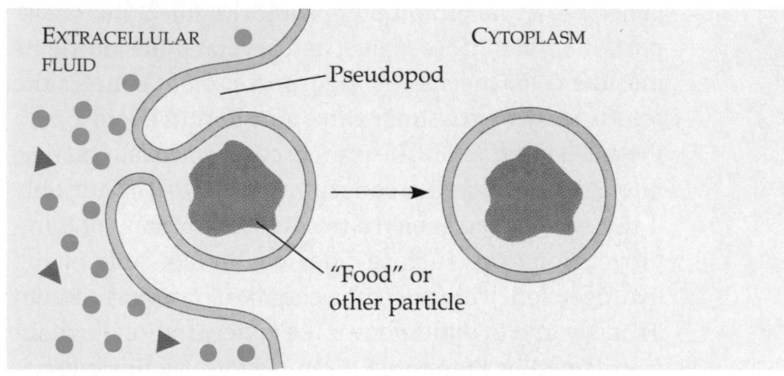

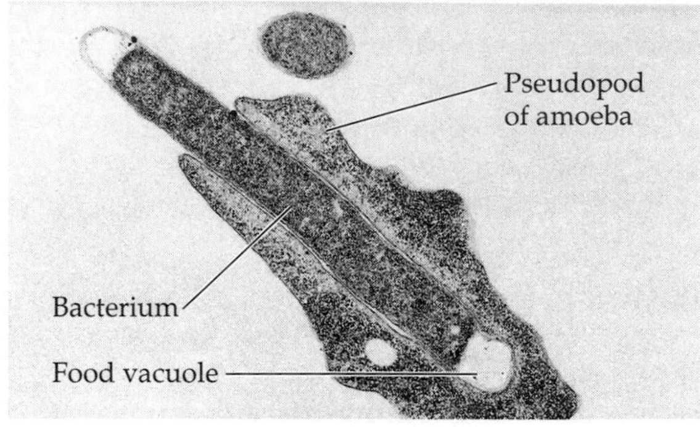

(a) Phagocytosis

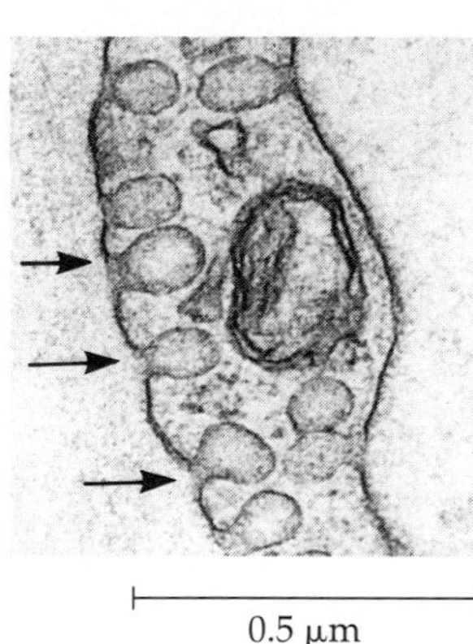

(b) Pinocytosis

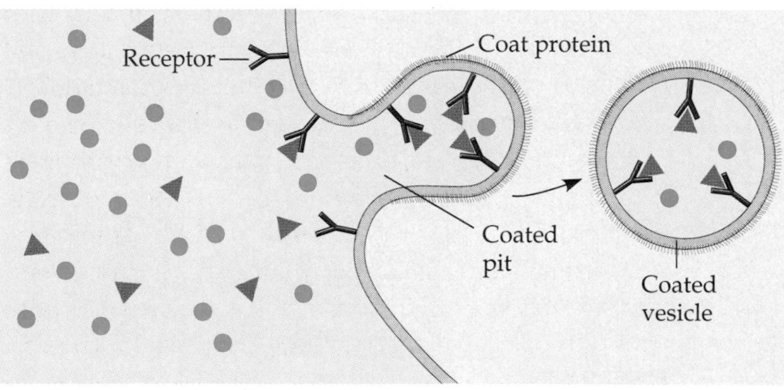

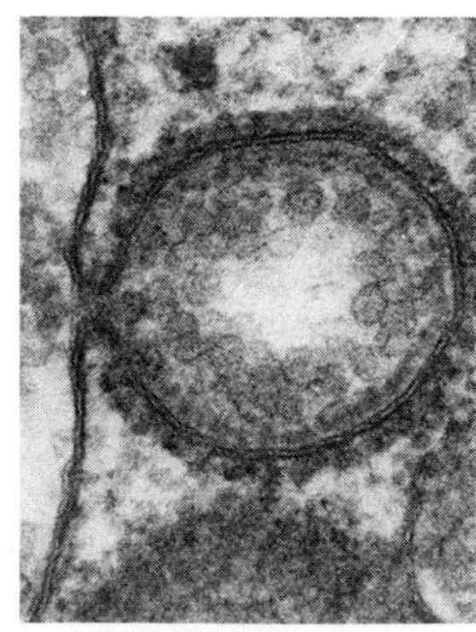

(c) Receptor-mediated endocytosis

FIGURE 8.17

The three types of endocytosis in animal cells. (**a**) In phagocytosis, pseudopodia engulf a particle and package it in a vacuole. The micrograph shows an amoeba engulfing a bacterium (TEM). (**b**) In pinocytosis, droplets of extracellular fluid are incorporated into the cell in small vesicles. The micrograph shows pinocytotic vesicles forming (arrows) in a cell lining a small blood vessel (TEM). (**c**) In receptor-mediated endocytosis, coated pits form vesicles when specific molecules (ligands) bind to receptors on the cell surface. Notice that there is a greater relative proportion of bound molecules (purple) inside the vesicles, though other molecules (green) are also present. The micrographs show two progressive stages of receptor-mediated endocytosis (TEMs). After the ingested material is liberated from the vesicle for metabolism, the receptors are recycled to the plasma membrane.

the lipid molecules of the two bilayers rearrange themselves. The two membranes then fuse to become continuous, and the contents of the vesicle spill to the outside of the cell (see FIGURE 8.7). The steps are basically reversed during endocytosis. A localized region of the plasma membrane sinks inward to form a pocket. As the pocket deepens, it pinches into the cytoplasm from the plasma membrane, forming a vesicle containing material that had been outside the cell.

Many secretory cells use exocytosis to export their products. For example, certain cells in the pancreas manufacture the hormone insulin and secrete it into the blood by exocytosis. Another example is the neuron, or nerve cell, which uses exocytosis to release chemical signals that stimulate other neurons or muscle cells (see FIGURE 2.15). When plant cells are making walls, exocytosis delivers carbohydrates from Golgi vesicles to the outside of the cell.

There are three types of endocytosis: phagocytosis ("cellular eating"), pinocytosis ("cellular drinking"), and receptor-mediated endocytosis. In **phagocytosis,** a cell engulfs a particle by wrapping pseudopodia around it and packaging it within a membrane-enclosed sac large enough to be classified as a vacuole (FIGURE 8.17a). The particle is digested after the vacuole fuses with a lysosome containing hydrolytic enzymes. In **pinocytosis,** the cell "gulps" droplets of extracellular fluid in tiny vesicles (FIGURE 8.17b). Because any and all solutes dissolved in the droplet are taken into the cell, pinocytosis is unspecific in the substances it transports. In contrast, **receptor-mediated endocytosis** is very specific (FIGURE 8.17c). Embedded in the membrane are proteins with specific receptor sites exposed to the extracellular fluid. The extracellular substances that bind to the receptors are called **ligands,** a general term for any molecule that binds specifically to a receptor site of another molecule (from the Latin *ligare,* "to bind"). The receptor proteins are usually clustered in regions of the membrane called coated pits, which are lined on their cytoplasmic side by a fuzzy layer of protein. These coat proteins probably help deepen the pit and form the vesicle.

Receptor-mediated endocytosis enables the cell to acquire bulk quantities of specific substances, even though those substances may not be very concentrated in the extracellular fluid. For example, human cells use the process to take in cholesterol for use in the synthesis of membranes and as a precursor for the synthesis of other steroids. Cholesterol travels in the blood in particles called low-density lipoproteins (LDLs), complexes of lipids and proteins. These particles bind to LDL receptors on membranes, and then enter the cell by endocytosis. In humans with familial hypercholesterolemia, an inherited disease characterized by a very high level of cholesterol in the blood, the LDL receptor proteins are defective, and the LDL particles cannot enter cells. Cholesterol accumulates in the blood, where it contributes to early atherosclerosis (the buildup of fat deposits on blood vessel linings).

Vesicles not only transport substances between the cell and its surroundings, they also provide a mechanism for rejuvenating or remodeling the plasma membrane. Endocytosis and exocytosis occur continually to some extent in most eukaryotic cells, yet the amount of plasma membrane in a nongrowing cell remains fairly constant over the long run. Apparently, the addition of membrane by one process offsets the loss of membrane by the other. The plasma membrane is a component of the cell's dynamic endomembrane system.

Specialized membrane proteins transmit extracellular signals to the inside of the cell

Although this chapter has emphasized the role of membrane proteins in transport, certain membrane proteins have other important functions. You learned in Chapter 7, for example, that specific membrane proteins bound to both the extracellular matrix and the cytoskeleton help regulate cell shape and cell movement. Other specialized proteins built into the membrane transmit chemical signals from the extracellular environment to the inside of the cell. Binding of a specific extracellular molecule is the first step in a chain of molecular interactions, known as a **signal-transduction pathway,** that leads to various responses within the cell. FIGURE 8.18 illustrates a simplified version of a signal-transduction pathway involving three membrane proteins. The extracellular molecule that binds to a receptor is the pathway's **first messenger.** The binding activates the receptor protein, which in turn activates a relay protein. The relay protein stimulates still another membrane protein, which functions as the effector (because it effects changes within the cell). The effector protein is an enzyme that produces a **second messenger,** a cytoplasmic molecule that triggers metabolic and structural responses within the cell. An example of such a signal-transduction pathway is the hormone epinephrine (adrenaline) stimulating liver cells to hydrolize stored glycogen and release glucose to the blood. The hormone, acting as a first messenger, binds to a specific receptor in the plasma membrane of a liver cell. Like falling dominoes, the hormone-activated receptor activates another protein, which activates another protein, and so on, until the enzyme that hydrolyzes glycogen is activated.

We will examine specific examples of how such signal-transduction pathways coordinate cellular responses in Chapters 35 and 41. For now, the important concept is the role of the plasma membrane in a cell's ability to sense and respond to environmental change.

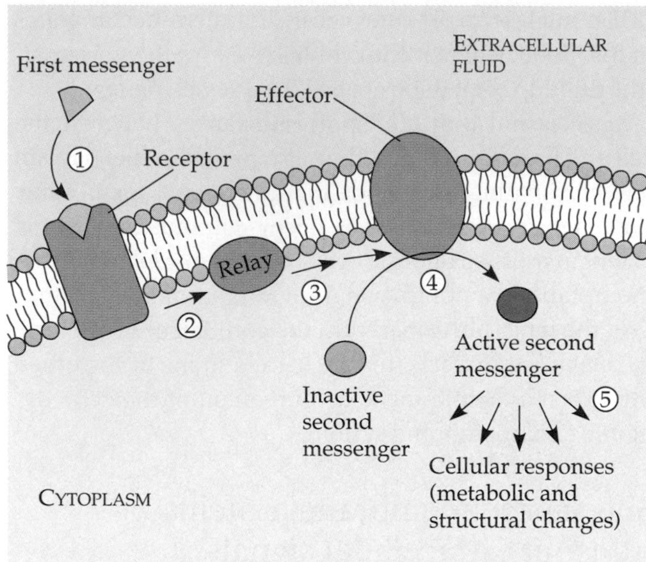

FIGURE 8.18
The role of membranes in signal-transduction pathways. In this simplified diagram, ① an extracellular molecule is a first messenger that binds to a specific receptor protein. ② The receptor activates a relay protein that ③ stimulates an effector protein. ④ The effector is an enzyme that produces a second messenger on the cytoplasmic side of the membrane. ⑤ The second messenger triggers various metabolic and structural responses of the cell.

Energy and cellular work have figured prominently in our study of membranes. We have seen, for example, that active transport is powered by ATP. In the next two chapters, you will learn more about how cells acquire and harvest chemical energy to do the work of life. We'll call membranes back for an encore in these chapters, as they play a major role in how mitochondria and chloroplasts make energy available to cells.

REVIEW OF KEY CONCEPTS (with page numbers and key figures)

- Membrane models have evolved to fit new data: *science as a process* (pp. 140–142, FIGURE 8.2)
  - The Davson-Danielli model of the membrane placed layers of proteins on either side of the lipid bilayer.
  - In the current membrane model, the fluid mosaic model, the membrane is a mosaic of dispersed individual proteins floating laterally in a fluid bilayer of phospholipids.
- A membrane is a fluid mosaic of lipids, proteins, and carbohydrates (pp. 142–147, FIGURE 8.5)
  - Proteins with diverse functions are either embedded in the lipid bilayer (integral proteins) or attached to the surface (peripheral proteins).
  - Membranes have specific inside and outside faces arising from differences in the lipid composition of the two bilayers and the directional orientation of proteins and any attached carbohydrates.
  - Carbohydrates linked to the proteins and lipids are important for cell-cell recognition.
- A membrane's molecular organization results in selective permeability (p. 147)
  - The cell requires an extensive interchange of small nutrient and waste molecules, respiratory gases, and inorganic ions. The plasma membrane regulates the passage of these substances.
  - Hydrophobic substances pass through membranes rapidly because of their solubility in the lipid bilayer. Small polar molecules, such as H_2O and CO_2, can also pass through the membrane. Larger polar molecules and ions require specific transport proteins, which provide channels.
- Passive transport is diffusion across a membrane (pp. 147–148, FIGURE 8.8)
  - Diffusion is the spontaneous movement of a substance down its concentration gradient.
- Osmosis is the passive transport of water (pp. 148–149, FIGURE 8.9).
  - Water flows across a membrane from the side with a lesser concentration of solute (hypotonic) to the side with the greater solute concentration (hypertonic). No net osmosis occurs across membranes separating solutions of equal concentration (isotonic).
- Cell survival depends on balancing water uptake and loss (pp. 149–150, FIGURE 8.10)
  - Cells lacking walls (as in animals and some protists) are either isotonic with their environments or else have adaptations for osmoregulation.
  - Plants, prokaryotes, fungi, and some protists have an elastic wall around their cells, which keeps the cells from bursting in a hypotonic environment. Under such conditions, these cells are turgid.
- Specific proteins facilitate the passive transport of selected solutes (pp. 150–151, FIGURE 8.12)
  - In facilitated diffusion, transport proteins hasten the movement of certain substances across a membrane down their concentration gradients.
  - Diffusion, osmosis, and facilitated diffusion are all passive transport processes that do not require the input of energy from the cell.

- Active transport is the pumping of solutes against their gradients (p. 151, FIGURE 8.13)
  - Energy, usually in the form of ATP, is harnessed by specific membrane proteins that perform the active transport.
- Some ion pumps generate voltage across membranes (pp. 151–153, FIGURE 8.15)
  - The diffusion of uncharged solutes depends only on their concentration gradients. However, ions have both a concentration (chemical) gradient and an electric gradient (voltage). These two forces are combined in an overall force called the electrochemical gradient, which determines the net direction of ionic diffusion.
  - Electrogenic pumps, such as the sodium-potassium pump and proton pump, are transport proteins that generate voltage across a membrane. This voltage, or membrane potential, contributes to the electrochemical gradients driving ion transport.
- In cotransport, a membrane protein couples the transport of one solute to another (p. 153, FIGURE 8.16)
  - Special membrane proteins can cotransport two solutes, coupling the "downhill" diffusion of one to the "uphill" transport of the other.
- Exocytosis and endocytosis transport large molecules (pp. 154–155, FIGURE 8.17)
  - In exocytosis, intracellular vesicles migrate to the plasma membrane, fuse with it, and release their contents.
  - In endocytosis, large molecules enter cells within vesicles pinched inward from the plasma membrane. The three types of endocytosis are: phagocytosis, the ingestion of large particles or whole cells; pinocytosis, the intake of tiny droplets of extracellular fluid with all its contained solutes; and receptor-mediated endocytosis, the ingestion of specific substances that bind to receptor proteins located in coated pits on the membrane.
- Specialized membrane proteins transmit extracellular signals to the inside of the cell (pp. 155–156, FIGURE 8.18)
  - In a signal-transduction pathway, a team of membrane proteins converts a first messenger, an external signal such as a hormone, to a second messenger, a cytoplasmic chemical that triggers the cell's responses.

SELF-QUIZ

1. In what way do the various membranes of a eukaryotic cell differ?
   a. Phospholipids are only found in certain membranes.
   b. Certain proteins are unique to each membrane.
   c. Only certain membranes of the cell are selectively permeable.
   d. Only certain membranes are constructed from amphipathic molecules.
   e. Some membranes have hydrophobic surfaces exposed to the cytoplasm, while others have hydrophilic surfaces facing the cytoplasm.

2. According to the fluid mosaic model of membrane structure, proteins of the membrane are
   a. spread in a continuous layer over the inner and outer surfaces of the membrane
   b. confined to the hydrophobic core of the membrane
   c. embedded in a lipid bilayer
   d. randomly oriented in the membrane, with no fixed inside-outside polarity
   e. free to depart from the fluid membrane and dissolve in the surrounding solution

3. Which of the following factors would tend to increase membrane fluidity?
   a. a greater proportion of unsaturated phospholipids
   b. a lower temperature
   c. a relatively high protein content in the membrane
   d. a greater proportion of relatively large glycolipids compared to lipids having smaller molecular weights
   e. a high membrane potential

4. Which of the following processes includes all others in the list?
   a. osmosis
   b. diffusion of a solute across a membrane
   c. facilitated diffusion
   d. passive transport
   e. transport of an ion down its electrochemical gradient

5. Based on the model of sucrose uptake in FIGURE 8.16, which of the following experimental treatments would increase the rate of sucrose transport into the cell?
   a. decreasing extracellular sucrose concentration
   b. decreasing extracellular pH
   c. decreasing cytoplasmic pH
   d. adding an inhibitor that blocks the regeneration of ATP
   e. adding a substance that makes the membrane more permeable to hydrogen ions

Questions 6–10

An artificial cell with an aqueous solution enclosed in a selectively permeable membrane has just been immersed in a beaker containing a different solution.

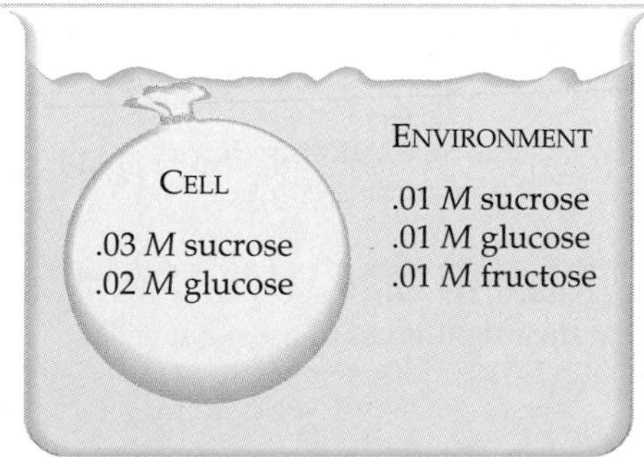

The membrane is permeable to water and to the simple sugars glucose and fructose but completely impermeable to the disaccharide sucrose.

6. Which solute(s) will exhibit a net diffusion into the cell?
7. Which solute(s) will exhibit a net diffusion out of the cell?

8. Which solution—the *cell contents* or the *environment*—is hypertonic?

9. In which direction will there be a net osmotic movement of water?

10. After the cell is placed into the beaker, which of the following changes would occur?
    a. The artificial cell would become more flaccid.
    b. The artificial cell would become more turgid.
    c. The entropy of the system (cell plus surrounding solution) would decrease.
    d. The overall free energy stored in the system would increase.
    e. The membrane potential would decrease.

CHALLENGE QUESTIONS

1. An experiment is designed to study the mechanism of sucrose uptake by plant cells. Cells are immersed in a sucrose solution, and the pH of the surrounding solution is monitored with a pH meter. The measurements show that sucrose uptake by the plant cells raises the pH of the surrounding solution. The magnitude of the pH change is proportional to the starting concentration of sucrose in the extracellular solution. A metabolic poison that blocks the ability of the cells to regenerate ATP also inhibits the pH change in the surrounding solution. Propose a hypothesis accounting for these results. Suggest an additional experiment to test your hypothesis.

2. In an adaptation of the preceding experiment, the rates of sucrose uptake from solutions of different sucrose concentrations are compared.

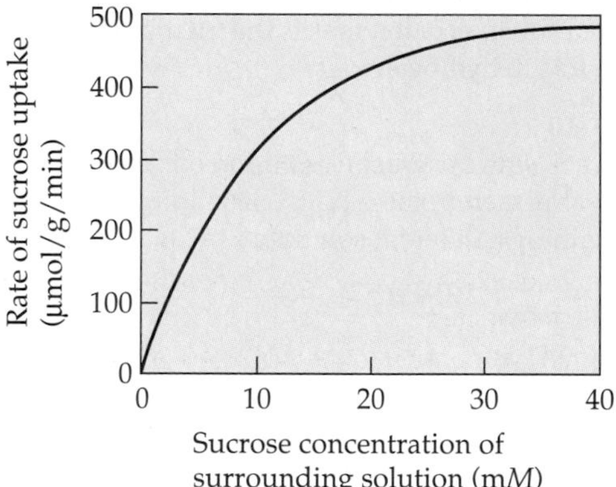

Explain the shape of the curve above in terms of what is happening at the membranes of the plant cells.

3. The cells of plant seeds store oils in the form of droplets enclosed by membranes. Unlike the membranes you studied in this chapter, the oil droplet membrane probably consists of a single layer of phospholipids rather than a bilayer. Draw a model for a membrane around an oil droplet, and explain why this arrangement is more stable than a bilayer.

SCIENCE, TECHNOLOGY, AND SOCIETY

1. A U.S. government panel has recommended that all people over age 20 should have their blood cholesterol level measured, and that those with high cholesterol levels should be put on a special diet or drug therapy. The annual cost of screening and treating everyone in the United States with elevated cholesterol could be \$10–\$50 billion. Research suggests that relatively few people benefit from drug treatment, mainly those suffering from abnormal conditions like familial hypercholesterolemia. Do you think it is worth the effort and expense of testing and medicating everyone with high cholesterol to help only 1% or 2% of the population? Why or why not?

2. Extensive irrigation in arid regions causes salts to accumulate in the soil. (The water contains low concentrations of salts, but when the water evaporates from the fields, the salts are left behind to concentrate in the soil.) Based on what you have learned about water balance in plant cells, explain why increasing soil salinity (saltiness) has an adverse effect on agriculture. Suggest some ways to minimize this damage. What costs are attached to your solutions?

FURTHER READING

Alberts, B., D. Bray, J. Lewis, M. Raff, K. Roberts, and J. D. Watson. *Molecular Biology of the Cell*, 3rd ed. New York: Garland, 1994. A description of the structure and functions of the plasma membrane in Chapter 6.
Barinaga, M. "Genetic Disease: Novel Functions Discovered for the Cystic Fibrosis Gene." *Science*, April 24, 1992. How a defective Cl^- pump impairs health.
Bretscher, M. S., and S. Munro. "Cholesterol and the Golgi Apparatus." *Science*, September 3, 1993. Tracing the pathway of a membrane molecule.
Changeux, J. P. "Chemical Signaling in the Brain." *Scientific American*, November 1993. How the membranes of synapses function.
Sharon, N., and H. Lis. "Carbohydrates in Cell Recognition." *Scientific American*, January 1993. How do membrane markers enable an organism to distinguish "self" cells from "nonself" cells?
Wickner, W. T. "How ATP Drives Proteins Across Membranes." *Science*, November 18, 1994. Transporting huge molecules requires special mechanisms.

CHAPTER 9

CELLULAR RESPIRATION: HARVESTING CHEMICAL ENERGY

KEY CONCEPTS

- Cellular respiration and fermentation are catabolic (energy-yielding) pathways
- Cells must recycle the ATP they use for work
- Redox reactions release energy when electrons move closer to electronegative atoms
- Electrons "fall" from organic molecules to oxygen during cellular respiration
- The "fall" of electrons during respiration is stepwise, via NAD^+ and an electron transport chain
- Respiration is a cumulative function of glycolysis, the Krebs cycle, and electron transport: *an overview*
- Glycolysis harvests chemical energy by oxidizing glucose to pyruvate: *a closer look*
- The Krebs cycle completes the energy-yielding oxidation of organic molecules: *a closer look*
- The inner mitochondrial membrane couples electron transport to ATP synthesis: *a closer look*
- Cellular respiration generates many ATP molecules for each sugar molecule it oxidizes: *a review*
- Fermentation enables some cells to produce ATP without the help of oxygen
- Glycolysis and the Krebs cycle connect to many other metabolic pathways
- Feedback mechanisms control cellular respiration

*Living is work. A cell organizes small organic molecules into polymers, such as proteins and DNA. It pumps substances across membranes. Many cells move or change their shapes. They grow and reproduce. A cell must work just to maintain its complex structure, because order is intrinsically unstable. To perform their many tasks, cells require transfusions of energy from outside sources. Energy enters most ecosystems in the form of sunlight, the energy source for plants and other photosynthetic organisms (*Figure 9.1, *p. 160). Animals, such as the desert locust in the photograph to the right, obtain fuel by eating plants, or by eating other organisms that eat plants. In this chapter, you will learn how cells harvest the chemical energy stored in organic molecules and use it to regenerate ATP, the molecule that drives most cellular work.*

Cellular respiration and fermentation are catabolic (energy-yielding) pathways

Organic compounds store energy in their arrangements of atoms. With the help of enzymes, a cell systematically degrades complex organic molecules that are rich in potential energy to simpler waste products that have less energy. Some of the energy taken out of chemical storage can be used to do work; the rest is dissipated as heat. As you learned in Chapter 6, metabolic pathways that release stored energy by breaking down complex molecules are called catabolic pathways. One catabolic process, called **fermentation,** is a partial degradation of sugars that occurs without the help of oxygen. However, the most prevalent and efficient catabolic pathway is **cellular respiration,** in which oxygen is consumed as a reactant along with the organic fuel. In eukaryotic cells, mitochondria house most of the metabolic equipment for cellular respiration.

Although very different in mechanism, respiration is in principle similar to the combustion of gasoline in an automobile engine after oxygen is mixed with the fuel (hydrocarbons). Food is the fuel for respiration, and the exhaust is carbon dioxide and water. The overall process can be summarized as follows:

$$\text{Organic compounds} + \text{Oxygen} \longrightarrow \text{Carbon dioxide} + \text{Water} + \text{Energy}$$

Although carbohydrates, fats, and proteins can all be processed and consumed as fuel, it is traditional to learn

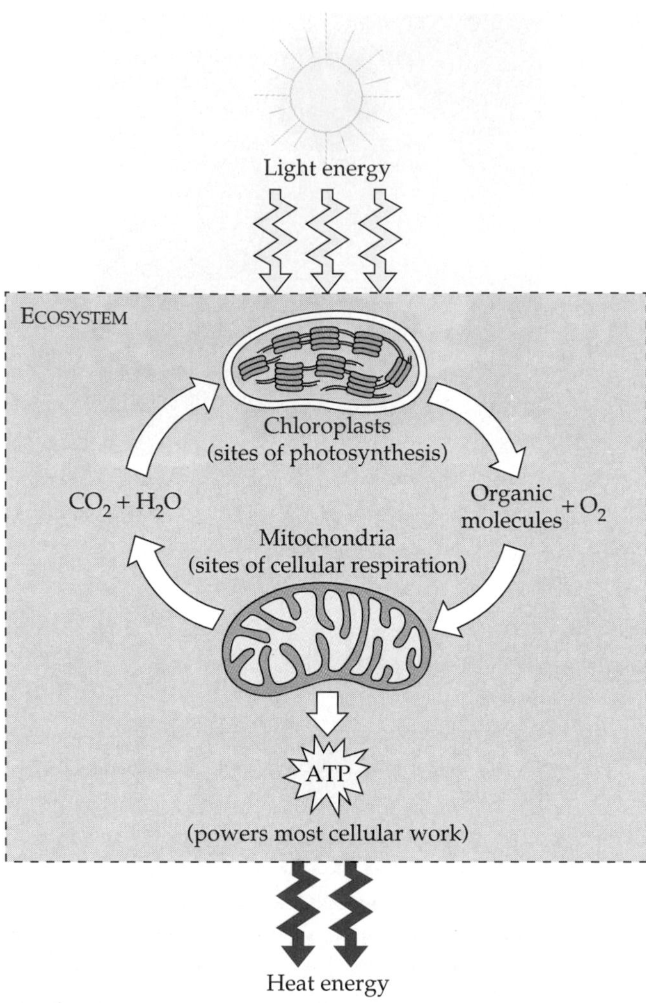

FIGURE 9.1
Energy flow and chemical recycling in ecosystems. The mitochondria of eukaryotes (including plants) use the organic products of photosynthesis as fuel for cellular respiration, which also consumes the oxygen produced by photosynthesis. Respiration harvests the energy stored in organic molecules to generate ATP, which powers most cellular work. The waste products of respiration, carbon dioxide and water, are the very substances that chloroplasts use as raw materials for photosynthesis. Thus, the chemical elements essential to life are recycled. But energy is not: It flows into an ecosystem as sunlight and back out as heat.

the steps of cellular respiration by tracking the degradation of the sugar glucose ($C_6H_{12}O_6$):

$$C_6H_{12}O_6 + 6\,O_2 \longrightarrow 6\,CO_2 + 6\,H_2O + \text{Energy (ATP + Heat)}$$

This breakdown of glucose is exergonic, having a free-energy change of −686 kcal (−2870 kJ) per mole of glucose decomposed ($\Delta G = -686$ kcal/mol; recall that a negative ΔG indicates that the products of the chemical process store less energy than the reactants).

Catabolic pathways do not directly move flagella, pump solutes, polymerize monomers, or perform other cellular work. Catabolism is linked to work by a chemical drive shaft: ATP. The processes of cellular respiration and fermentation are complex and challenging to learn. Throughout the chapter, it is important to keep in mind the main objective: discovering how cells use the energy stored in food molecules to make ATP.

Cells must recycle the ATP they use for work

The molecule known as ATP, short for adenosine triphosphate, is the central character in bioenergetics. Recall from Chapter 6 that the triphosphate tail of ATP is the chemical equivalent of a loaded spring; the close packing of the three negatively charged phosphate groups is an unstable, energy-storing arrangement (like charges repel). The chemical "spring" tends to "relax" from the loss of the terminal phosphate (see FIGURE 6.6). The cell taps this energy source by using enzymes to transfer phosphate groups from ATP to other compounds, which are then said to be phosphorylated. Phosphorylation primes a molecule to undergo some kind of change that performs work, and the molecule loses its phosphate group in the process (FIGURE 9.2). The price of most cellular work, then, is the conversion of ATP to ADP and inorganic phosphate (abbreviated Ⓟ$_i$ in this book), products that store less energy than ATP. To keep working, the cell must regenerate its supply of ATP from ADP and inorganic phosphate. A working muscle cell, for example, recycles its ATP at a rate of about 10 million molecules per second. To understand how cellular respiration regenerates ATP, we need to examine the fundamental chemical processes known as oxidation and reduction.

Redox reactions release energy when electrons move closer to electronegative atoms

Just what happens when cellular respiration decomposes glucose and other organic fuels? And why does this metabolic pathway yield energy? The answers are based on the transfer of electrons during the chemical reactions. The relocation of electrons releases the energy stored in food molecules, and this energy is used to synthesize ATP.

In many chemical reactions, there is a transfer of one or more electrons (e^-) from one reactant to another. These electron transfers are called oxidation-reduction reactions, or **redox reactions** for short. During a redox reaction, the loss of electrons from one substance is called **oxidation,** and the addition of electrons to another substance is known as **reduction.*** Consider, for example, the

*This term defies intuition; *adding* electrons is called *reduction.* The term was derived from the electrical effects of adding electrons. Negatively charged electrons added to a cation reduce the amount of positive charge of the cation.

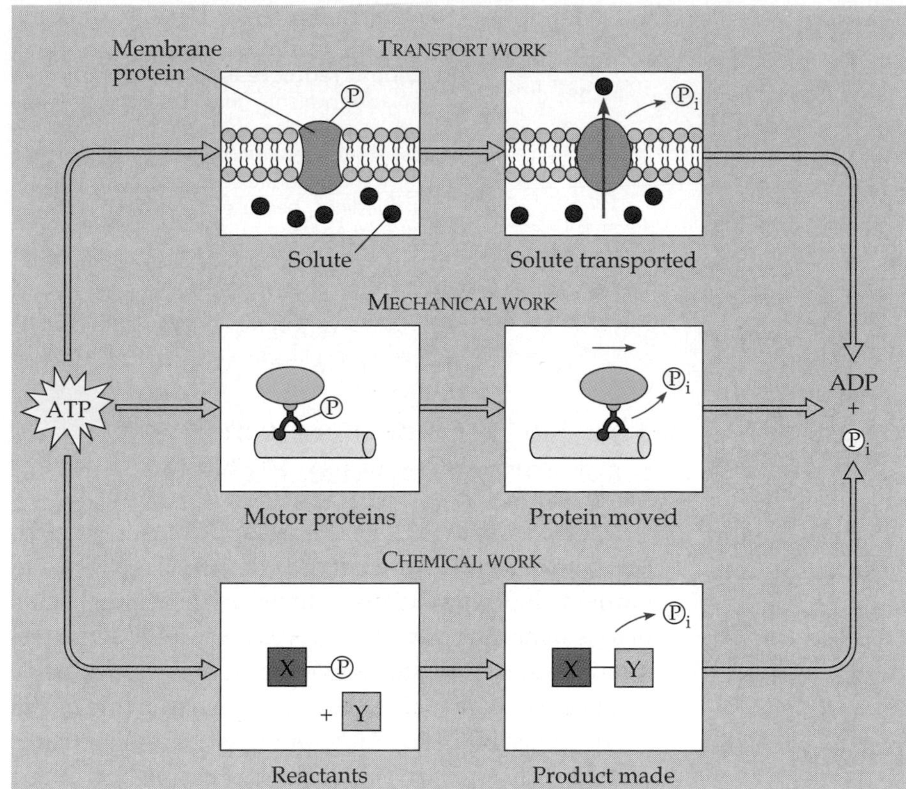

FIGURE 9.2
A review of how ATP drives cellular work. Phosphate group transfer is the mechanism responsible for most types of cellular work. Enzymes shift a phosphate group (Ⓟ) from ATP to some other molecule, and this phosphorylated molecule undergoes a change that performs work. For example, ATP drives active transport by phosphorylating specialized proteins built into membranes; drives mechanical work by phosphorylating motor proteins, such as the dynein responsible for the sliding of microtubules in cilia and flagella; and drives chemical work by phosphorylating key reactants. The phosphorylated molecules lose the phosphate groups as work is performed, leaving ADP and inorganic phosphate ($Ⓟ_i$) as products. Cellular respiration replenishes the ATP supply by powering the phosphorylation of ADP.

reaction between sodium and chlorine to form table salt:

$$\text{Na} + \text{Cl} \longrightarrow \text{Na}^+ + \text{Cl}^-$$

(Oxidation: Na → Na^+; Reduction: Cl → Cl^-)

Or we could generalize a redox reaction this way:

$$Xe^- + Y \longrightarrow X + Ye^-$$

(Oxidation: Xe^- → X; Reduction: Y → Ye^-)

In the generalized reaction, substance X, the electron donor, is called the **reducing agent;** it reduces Y. Substance Y, the electron acceptor, is the **oxidizing agent;** it oxidizes X. Since an electron transfer requires both a donor and an acceptor, oxidation and reduction always go together.

Not all redox reactions involve the complete transfer of electrons from one substance to another; some change the *degree* of electron sharing in covalent bonds. The reaction between methane and oxygen to form carbon dioxide and water, shown in FIGURE 9.3, is an example. As explained in Chapter 2, the covalent electrons in methane are shared equally between the bonded atoms because carbon and hydrogen have about the same affinity for valence electrons; they are about equally electronegative. But when methane reacts with oxygen to form carbon dioxide, electrons are shifted away from the carbon atoms to their new covalent partner, oxygen, which is very electronegative. Methane has thus been oxidized. The two atoms of the oxygen molecule also share their electrons equally. But when the oxygen reacts with the hydrogen from methane to form water, the electrons of the covalent bonds are drawn closer to the oxygen; the oxygen molecule has been reduced. Because oxygen is so electronegative, it is one of the most potent of all oxidizing agents.

Energy must be added to pull an electron away from an atom, just as energy must be added to push a large ball uphill. The more electronegative the atom (the stronger its pull on electrons), the more energy required to keep the electron away from it, just as more energy is required to push a ball up a steeper hill. An electron *loses* potential energy when it shifts from a less electronegative atom *toward* a more electronegative one, just as a ball loses potential energy when it rolls downhill. A redox reaction that relocates electrons closer to oxygen, such as the burning of methane, releases chemical energy that can be put to work.

Electrons "fall" from organic molecules to oxygen during cellular respiration

The oxidation of methane by oxygen is the main combustion reaction that occurs at the burner of a gas stove.

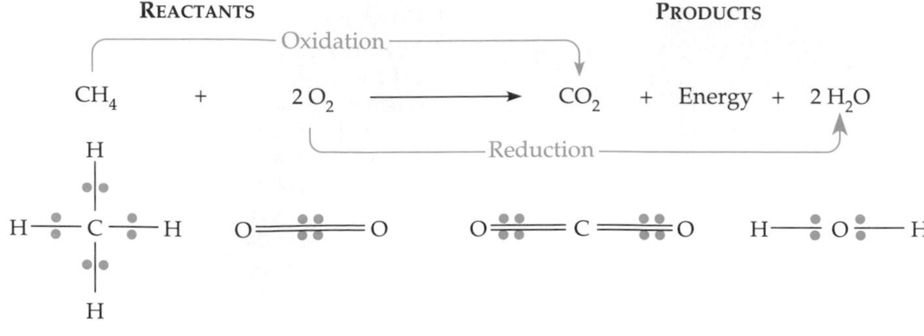

FIGURE 9.3
Methane combustion as an energy-yielding redox reaction. During the reaction, covalently shared electrons move away from carbon and hydrogen atoms and closer to oxygen, which is very electronegative. The reaction releases energy to the surroundings, because the electrons lose potential energy as they move closer to electronegative atoms.

Combustion of gasoline in an automobile engine is also a redox reaction, and the energy released pushes the pistons. But the energy-yielding redox process of greatest interest here is respiration: the oxidation of glucose and other fuel molecules in food. Examine again the summary equation for cellular respiration, but this time think of it as a redox process:

Oxidation ($C_6H_{12}O_6$ → $6 CO_2$)

$$C_6H_{12}O_6 + 6\,O_2 \longrightarrow 6\,CO_2 + 6\,H_2O$$

Reduction ($6 O_2$ → $6 H_2O$)

As in the combustion of methane or gasoline, the fuel (sugar) is oxidized and oxygen is reduced, and the electrons lose potential energy along the way.

In general, organic molecules that have an abundance of hydrogen are excellent fuels because their bonds are a source of "hilltop" electrons with the potential to "fall" closer to oxygen. The summary equation for respiration indicates that hydrogen is transferred from glucose to oxygen. But the important point, not visible in the summary equation, is that the change in the covalent status of electrons as hydrogen is transferred to oxygen is what liberates energy. By oxidizing glucose, respiration takes energy out of storage and makes it available for ATP synthesis.

The main energy foods, carbohydrates and fats, are reservoirs of electrons associated with hydrogen. Only the barrier of activation energy holds back the flood of electrons to a lower energy state (see Chapter 6). Without this barrier, a food substance like glucose would combine spontaneously with oxygen. When we supply the activation energy by igniting glucose, it burns in air, releasing 686 kcal (2870 kJ) of heat per mole (about 180 g). Body temperature is not high enough to initiate burning, which is the rapid oxidation of fuel accompanied by an enormous release of energy as heat. But swallow some glucose in the form of a sugar cube, and when the molecules reach your cells, enzymes will lower the barrier of activation energy, allowing the sugar to be oxidized slowly.

The "fall" of electrons during respiration is stepwise, via NAD^+ and an electron transport chain

The wholesale release of energy from a fuel is difficult to harness efficiently for constructive work: The explosion of a gasoline tank cannot drive a car very far. Cellular respiration does not oxidize glucose in a single explosive step that would transfer all the hydrogen from the fuel to the oxygen at one time. Rather, glucose and other organic fuels are broken down gradually in a series of steps, each one catalyzed by an enzyme. At key steps, hydrogen atoms are stripped from the glucose, but they are not transferred directly to oxygen. They are usually passed first to a coenzyme called **NAD^+** (nicotinamide adenine dinucleotide). Thus, NAD^+ functions as an oxidizing agent during respiration.

How does NAD^+ trap electrons from glucose and other fuel molecules? Enzymes called dehydrogenases remove a pair of hydrogen atoms from the substrate, a sugar or some other fuel. We can think of this as the removal of two electrons and two protons (the nuclei of hydrogen atoms). The enzyme delivers the *two* electrons along with *one* proton to its coenzyme, NAD^+ (FIGURE 9.4). The other proton is released as a hydrogen ion (H^+) into the surrounding solution:

$$\text{H—}\overset{|}{\underset{|}{\text{C}}}\text{—OH} + NAD^+ \xrightarrow{\text{Dehydrogenase}} \overset{|}{\underset{|}{\text{C}}}\text{═O} + NADH + H^+$$

While the oxidized form, NAD^+, has a positive charge, the reduced form, NADH, is electrically neutral. By receiving two negatively charged electrons but only one positively charged proton, NAD^+ has its charge neutralized. The name NADH for the reduced form shows the hydrogen that has been received in the reaction. Since NAD^+ gains electrons, it is an electron acceptor (a synonym for oxidizing agent). The most versatile electron acceptor in cellular respiration, NAD^+ functions in many of the redox steps during the breakdown of sugar.

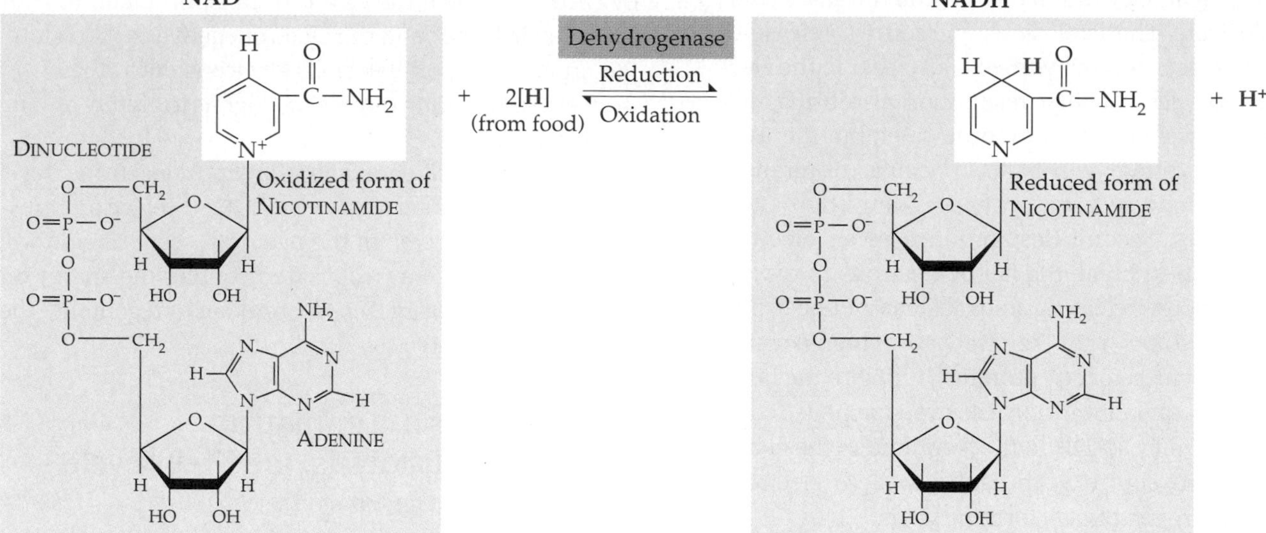

FIGURE 9.4
NAD$^+$ as an electron shuttle. The full name for NAD$^+$, nicotinamide adenine dinucleotide, describes its structure; the molecule consists of two nucleotides joined together. The enzymatic transfer of two electrons and one proton from some organic substrate to NAD$^+$ reduces the NAD$^+$ to NADH. Most of the electrons removed from food are transferred initially to NAD$^+$.

Electrons lose very little of their potential energy when they are transferred from food to NAD$^+$. Each NADH molecule formed during respiration represents stored energy that can be tapped to make ATP when the electrons complete their "fall" from NADH to oxygen.

How do electrons extracted from food and stored by NADH finally reach oxygen? It will help to compare this complex redox chemistry of cellular respiration to a much simpler reaction: the reaction between hydrogen and oxygen to form water (FIGURE 9.5a). Mix H_2 and O_2,

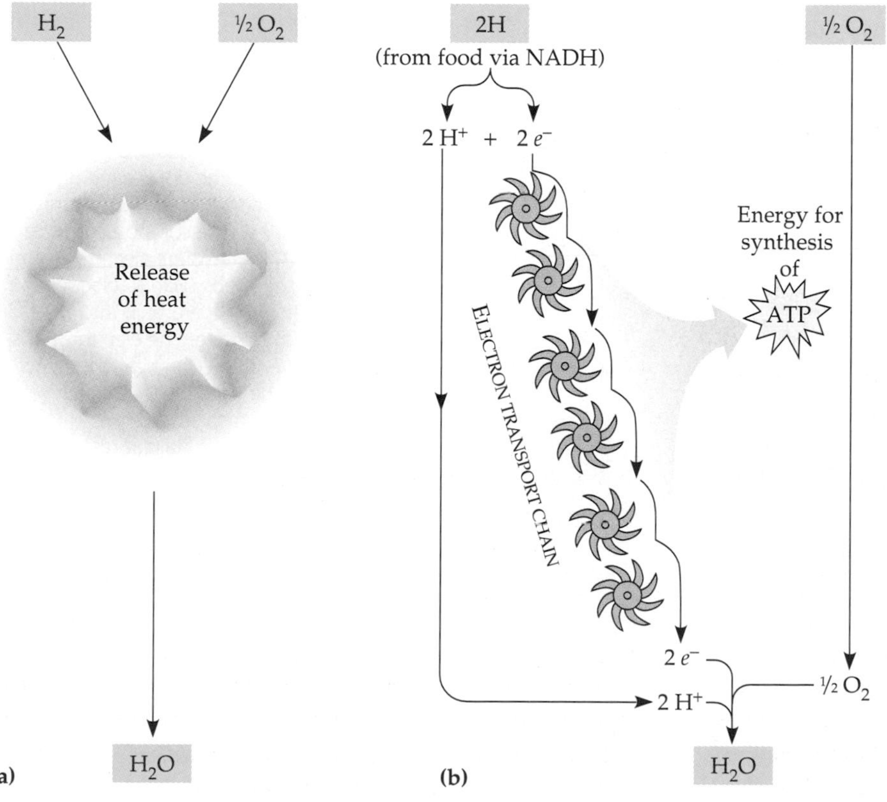

FIGURE 9.5
An introduction to electron transport chains. (**a**) The exergonic reaction of hydrogen with oxygen to form water releases a large amount of energy in the form of heat and light: an explosion. (**b**) In cellular respiration, an electron transport chain breaks the "fall" of electrons in this reaction into a series of smaller steps and stores some of the released energy in a form that can be used to make ATP (the rest of the energy is released as heat).

provide a spark for activation energy, and the gases combine explosively. The explosion represents a release of energy as the electrons of hydrogen fall closer to the electronegative oxygen. Cellular respiration also brings hydrogen and oxygen together to form water, but there are two important differences. First, in cellular respiration, the hydrogen that reacts with oxygen is derived from organic molecules. Second, respiration uses an **electron transport chain** to break the fall of electrons to oxygen into several energy-releasing steps instead of one explosive reaction (FIGURE 9.5b). The transport chain consists of several molecules, mostly proteins, built into the inner membrane of a mitochondrion. Electrons removed from food are shuttled by NADH to the "top" end of the chain. At the "bottom" end, oxygen captures these electrons along with hydrogen nuclei, forming water.

Electron transfer from NADH to oxygen is an exergonic reaction with a free energy change of −53 kcal/mol (−222 kJ/mol). Instead of this energy being released and wasted in a single explosive step, electrons cascade down the chain from one carrier molecule to the next, losing a small amount of energy with each step until they finally reach oxygen, the terminal electron acceptor. What keeps the electrons moving is that each carrier is more electronegative than its "uphill" neighbor in the chain. At the bottom of the chain is oxygen, which has a very great affinity for electrons. Thus, electrons removed from food by NAD^+ fall down the electron transport chain to a far more stable location in the electronegative oxygen atom. Put another way, oxygen pulls electrons down the chain in an energy-yielding tumble analogous to gravity pulling objects downhill.

Thus, during cellular respiration, most electrons travel this "downhill" route: food ⟶ NADH ⟶ electron transport chain ⟶ oxygen. In the next four sections, you will learn more about how respiration couples the energy released from this exergonic electron fall to regenerate the cell's supply of ATP.

Respiration is a cumulative function of glycolysis, the Krebs cycle, and electron transport: *an overview*

Now that we have covered the basic redox mechanisms of respiration, let's look at the entire process. Respiration is a cumulative function of three metabolic stages, which are diagrammed in FIGURE 9.6:

1. Glycolysis (color-coded blue-green throughout the chapter).
2. The Krebs cycle (color-coded salmon).
3. The electron transport chain and oxidative phosphorylation (color-coded violet).

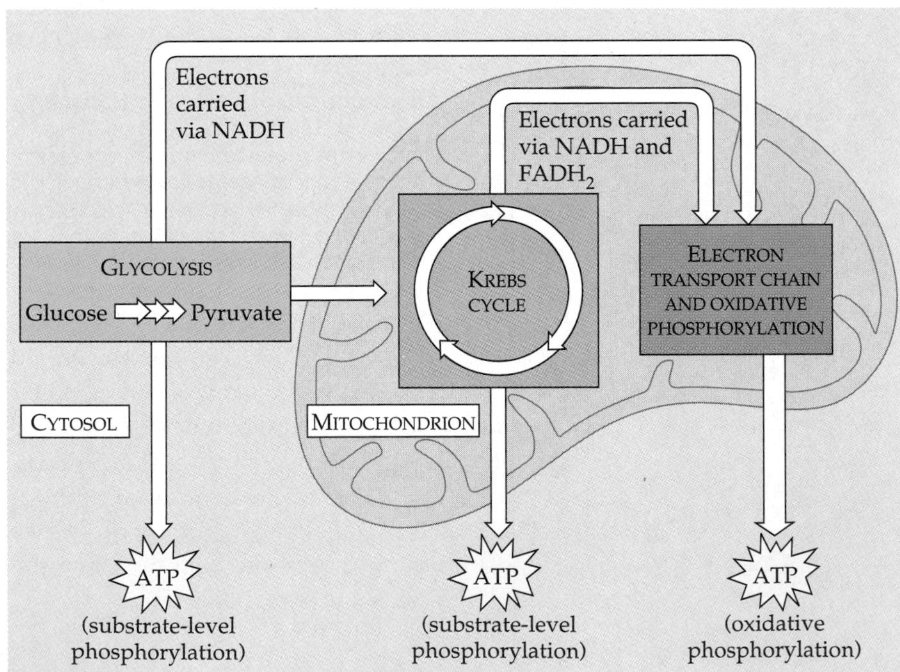

FIGURE 9.6
An overview of cellular respiration. In a eukaryotic cell, glycolysis occurs outside the mitochondria in the cytosol. The Krebs cycle and the electron transport chains are located inside the mitochondria. During glycolysis, each glucose molecule is broken down into two molecules of the compound pyruvate. The pyruvate crosses the double membrane of the mitochondrion to enter the matrix, where the Krebs cycle decomposes it to carbon dioxide. NADH transfers electrons from glycolysis and the Krebs cycle to electron transport chains, which are built into the membrane of the cristae. The electron transport chain converts the chemical energy to a form that can be used to drive oxidative phosphorylation, which accounts for most of the ATP generated by cellular respiration. A smaller amount of ATP is formed directly during glycolysis and the Krebs cycle by substrate-level phosphorylation.

A miniature version of this diagram accompanies many upcoming figures. In each case, the appropriate metabolic component—glycolysis, Krebs cycle, or electron transport chain—is highlighted. These orientation diagrams and the color coding will help you learn how the details of cellular respiration fit into the overall process.

The first two stages, glycolysis and the Krebs cycle, are the catabolic pathways that decompose glucose and other organic fuels. **Glycolysis,** which occurs in the cytosol, begins the degradation by breaking glucose into two molecules of a compound called pyruvate. The **Krebs cycle,** which takes place within the mitochondrial matrix, completes the job by decomposing a derivative of pyruvate to carbon dioxide.

Thus, the carbon dioxide produced by respiration represents fragments of oxidized organic molecules. Some of the steps of glycolysis and the Krebs cycle are redox reactions in which dehydrogenase enzymes transfer electrons from substrates to NAD^+, forming NADH. The third stage of respiration, the electron transport chain, accepts electrons from the breakdown products of the first two stages (usually via NADH) and passes these electrons from one molecule to another. At the end of the chain, the electrons are combined with hydrogen ions and molecular oxygen to form water (see FIGURE 9.5). The energy released at each step of the chain is stored in a form the mitochondrion can use to make ATP. This mode of ATP synthesis is called **oxidative phosphorylation** because it is powered by the redox reactions that transfer electrons from food to oxygen.

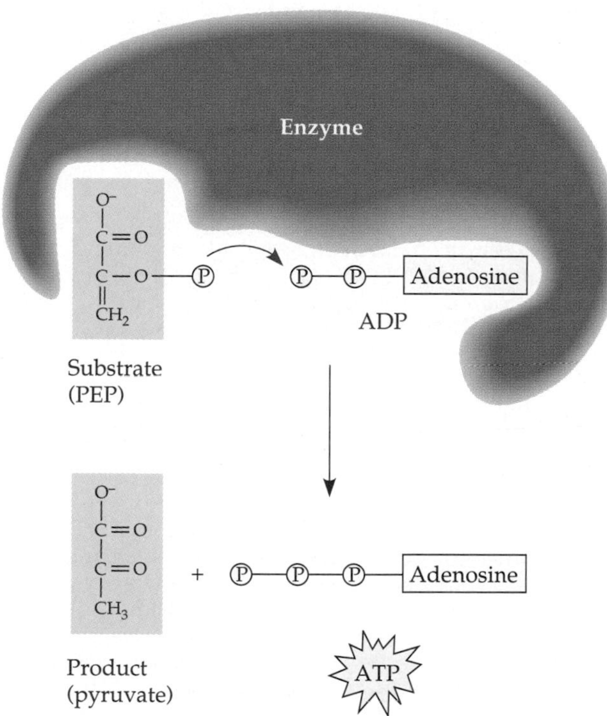

FIGURE 9.7
Substrate-level phosphorylation. Some ATP is made by direct enzymatic transfer of a phosphate group from a substrate to ADP. The phosphate donor in this case is phosphoenolpyruvate (PEP), which is formed from the breakdown of sugar during glycolysis.

The site of electron transport and oxidative phosphorylation is the inner membrane of the mitochondrion (see FIGURE 7.18). Oxidative phosphorylation accounts for almost 90% of the ATP generated by respiration. A smaller amount of ATP is formed directly in a few reactions of glycolysis and the Krebs cycle by a mechanism called **substrate-level phosphorylation** (FIGURE 9.7). This mode of ATP synthesis occurs when an enzyme transfers a phosphate group from a substrate to ADP. ("Substrate" here refers to an organic molecule generated by the sequential catabolism of glucose.)

Respiration cashes in the large denomination of energy banked in glucose for the small change of ATP, which is more practical for the cell to spend on its work. For each molecule of glucose degraded to carbon dioxide and water by respiration, the cell makes up to 36 molecules of ATP.

This overview introduced how glycolysis, the Krebs cycle, and electron transport fit into the overall process of cellular respiration. We are now ready to take a closer look at each of these three stages of respiration.

Glycolysis harvests chemical energy by oxidizing glucose to pyruvate: *a closer look*

The word *glycolysis* means "splitting of sugar," and that is exactly what happens during this pathway. Glucose, a six-carbon sugar, is split into two three-carbon sugars. These smaller sugars are then oxidized, and their remaining atoms are rearranged to form two molecules of pyruvate.

The catabolic pathway of glycolysis consists of ten steps, each catalyzed by a specific enzyme. We can divide these ten steps into two phases. The energy-investment phase includes the first five steps and the energy-payoff phase includes the next five steps. During the energy-investment phase, the cell actually spends ATP to phosphorylate the fuel molecules. This investment is repaid with dividends during the energy-payoff phase, when ATP is produced by substrate-level phosphorylation and NAD^+ is reduced to NADH by oxidation of the food. On the next two pages, you will find a detailed diagram of glycolysis (FIGURE 9.8). Do not try to memorize the chemical structures, and do not let the details overcome the key point: glycolysis as a source of ATP and NADH.

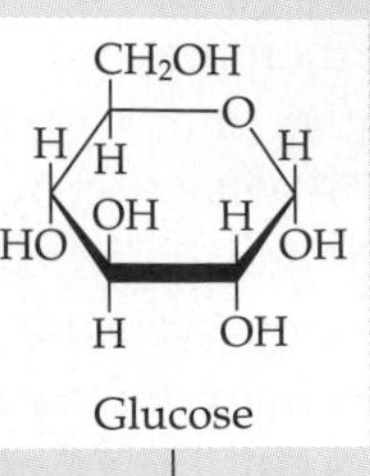

FIGURE 9.8

A closer look at glycolysis. The orientation diagram at the upper right relates glycolysis to the whole process of respiration. Do not let the chemical detail in this diagram block your view of glycolysis as a source of ATP and NADH.

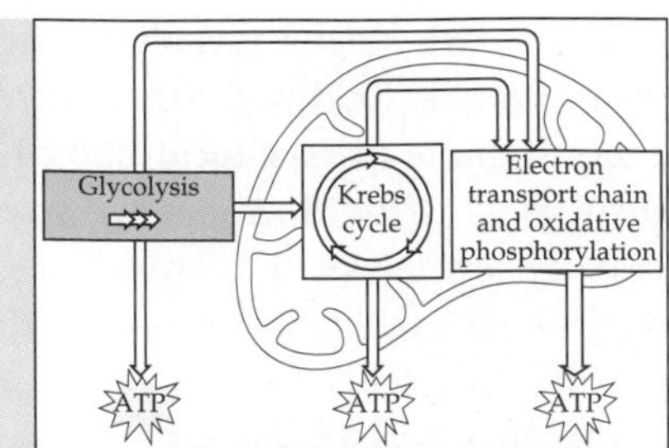

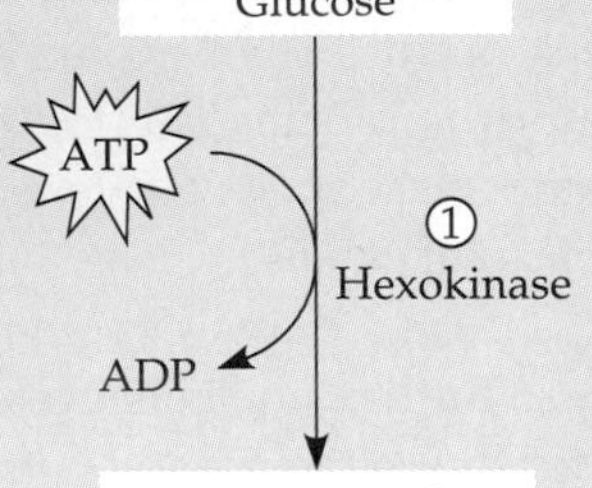

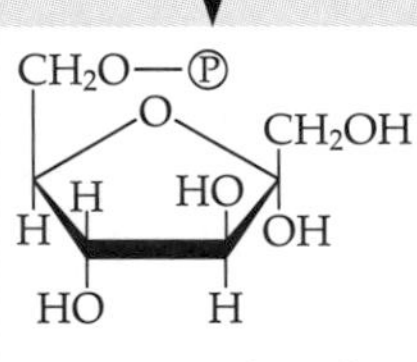

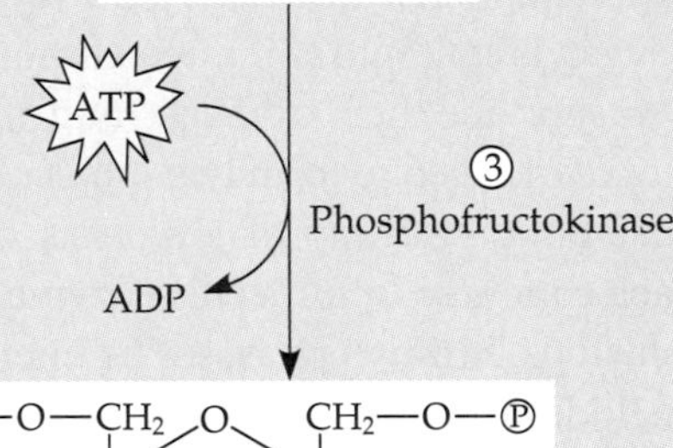

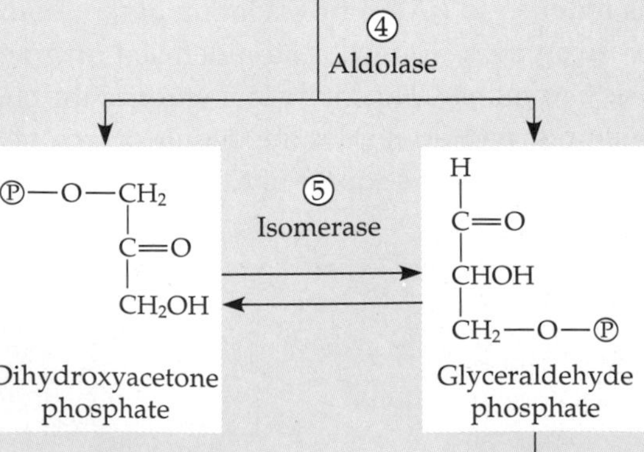

ENERGY-INVESTMENT PHASE

Step ① Glucose enters the cell and is phosphorylated by the enzyme hexokinase, which transfers a phosphate group from ATP to the sugar. The electrical charge of the phosphate group traps the sugar in the cell because of the impermeability of the plasma membrane to ions. Phosphorylation of glucose also makes the molecule more chemically reactive. In this diagram, coupled arrows (↩) indicate the transfer of a phosphate group or pair of electrons from one reactant to another.

Step ② Glucose 6-phosphate is rearranged to convert it to its isomer, fructose 6-phosphate.

Step ③ In this step, still another molecule of ATP is invested in glycolysis. An enzyme transfers a phosphate group from ATP to the sugar. So far, the ATP ledger shows a debit of 2. With phosphate groups on its opposite ends, the sugar is now ready to be split in half.

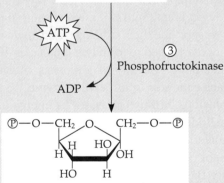

Step ④ This is the reaction from which glycolysis gets its name. An enzyme cleaves the sugar molecule into two different three-carbon sugars: glyceraldehyde phosphate and dihydroxyacetone phosphate. These two sugars are isomers of each other.

Step ⑤ Another enzyme catalyzes the reversible conversion between the two three-carbon sugars, and if left alone in a test tube, the reaction reaches equilibrium. This does not happen in the cell, however, because the next enzyme in glycolysis uses only glyceraldehyde phosphate as its substrate and is unreceptive to dihydroxyacetone phosphate. This pulls the equilibrium between the two three-carbon sugars in the direction of glyceraldehyde phosphate, which is removed as fast as it forms. Thus, the net result of steps 4 and 5 is cleavage of a six-carbon sugar into two molecules of glyceraldehyde phosphate; each will progress through the remaining steps of glycolysis.

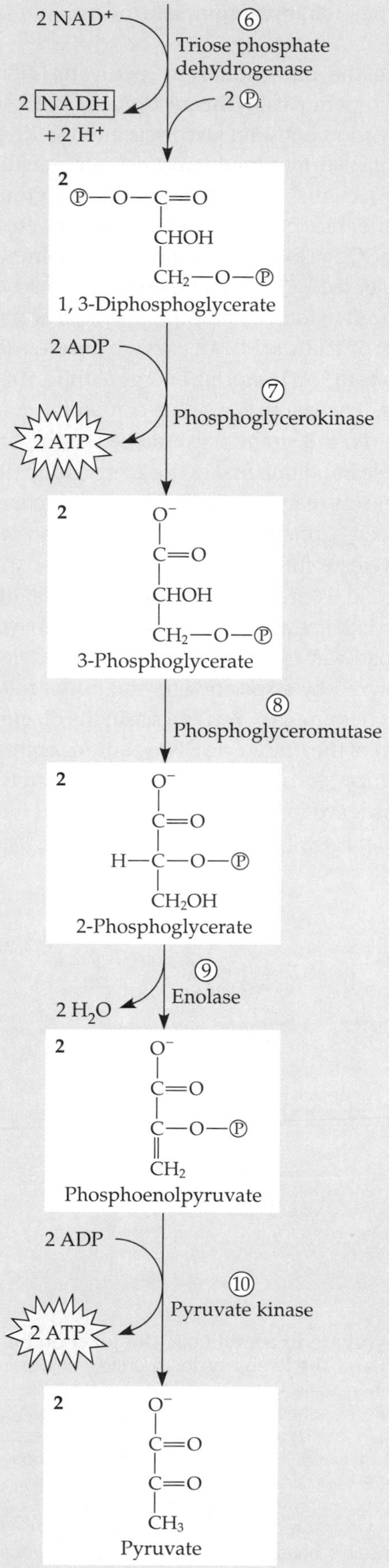

2 NAD^+

⑥ Triose phosphate dehydrogenase

2 NADH + 2 H^+ ← 2 Ⓟi

2 Ⓟ—O—C=O | CHOH | CH_2—O—Ⓟ

1, 3-Diphosphoglycerate

ENERGY-YIELDING PHASE

Step ⑥ An enzyme now catalyzes two sequential reactions while it holds glyceraldehyde phosphate in its active site. First, the sugar is oxidized by the transfer of electrons and H^+ to NAD^+, forming NADH. Here we see in metabolic context the type of redox reaction described earlier. This reaction is very exergonic (ΔG = –10.3 kcal/mol), and the enzyme capitalizes by attaching a phosphate group to the oxidized substrate. The source of the phosphate is inorganic phosphate, which is always present in the cytosol. Notice that the coefficient 2 precedes all molecules in the energy-yielding phase; these steps occur after glucose is split into two three-carbon sugars.

2 ADP

⑦ Phosphoglycerokinase

2 ATP

2 O^- | C=O | CHOH | CH_2—O—Ⓟ

3-Phosphoglycerate

Step ⑦ Finally, glycolysis produces some ATP. The phosphate group added in the previous step is transferred to ADP. For each glucose molecule that began glycolysis, step 7 produces two molecules of ATP, since every product after the sugar-splitting step (step 4) is doubled. Of course, two ATPs were invested to get sugar ready for splitting. The ATP ledger now stands at zero. By the end of step 7, glucose has been converted to two molecules of 3-phosphoglycerate. This compound is not a sugar. The carbonyl group that characterizes a sugar has been oxidized to a carboxyl group, the hallmark of an organic acid. The sugar was oxidized in step 6, and now the energy made available by that oxidation has been used to make ATP.

⑧ Phosphoglyceromutase

2 O^- | C=O | H—C—O—Ⓟ | CH_2OH

2-Phosphoglycerate

Step ⑧ Next, an enzyme relocates the remaining phosphate group. This prepares the substrate for the next reaction.

⑨ Enolase

2 H_2O

2 O^- | C=O | C—O—Ⓟ ‖ CH_2

Phosphoenolpyruvate

Step ⑨ An enzyme forms a double bond in the substrate by extracting a water molecule to form phosphoenolpyruvate, or PEP. This results in the electrons of the substrate being rearranged in such a way that the remaining phosphate bond becomes very unstable, preparing the substrate for the next reaction.

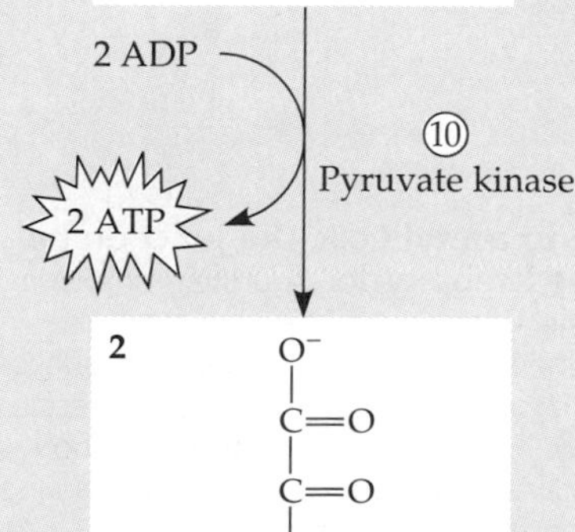

2 O^- | C=O | C=O | CH_3

Pyruvate

Step ⑩ The last reaction of glycolysis produces more ATP by transferring the phosphate group from PEP to ADP. Since this step occurs twice for each glucose molecule, the ATP ledger now shows a net gain of two ATPs. Steps 7 and 10 each produce two ATPs for a total credit of four, but a debt of two ATPs was incurred from steps 1 and 3. Glycolysis has repaid the ATP investment with 100% interest. Additional energy was stored by step 6 in NADH, which can be used to make ATP by oxidative phosphorylation (if oxygen is present). In the meantime, glucose has been broken down and oxidized to two molecules of pyruvate, the end-product of the glycolytic pathway.

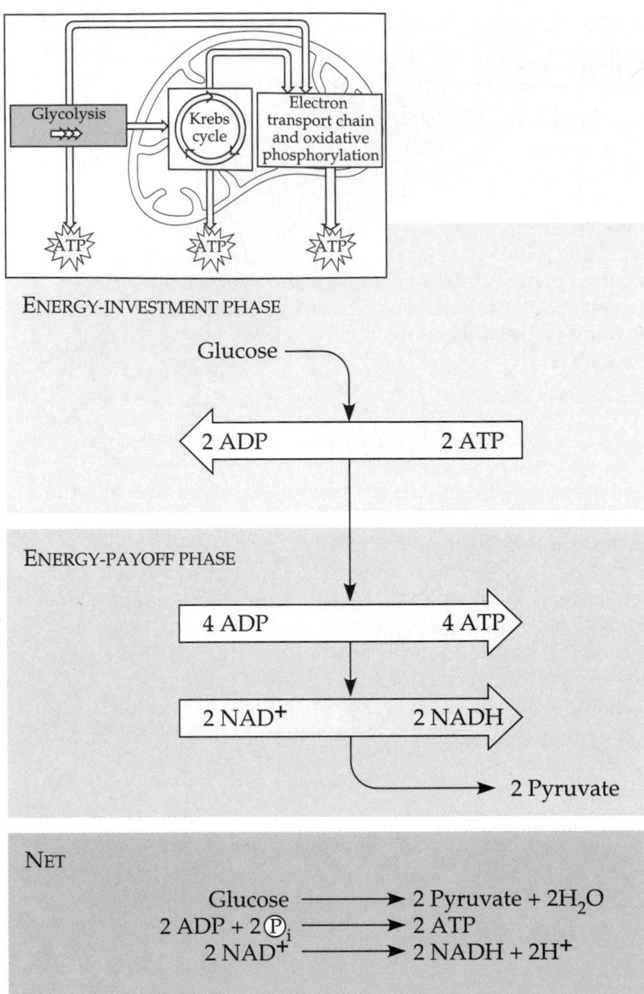

FIGURE 9.9
A summary of glycolysis.

FIGURE 9.9 summarizes the inputs and outputs of glycolysis. The net energy yield from glycolysis, per glucose molecule, is 2 ATP plus 2 NADH. Notice that all of the carbon originally present in glucose is accounted for in the two molecules of pyruvate; no CO_2 is released during glycolysis. Also notice that glycolysis occurs whether or not oxygen is present. However, if oxygen *is* present, the energy stored in NADH can be converted to ATP energy when the electron transport chain drives oxidative phosphorylation. And in the presence of oxygen, the chemical energy left in pyruvate can be extracted by the Krebs cycle.

The Krebs cycle completes the energy-yielding oxidation of organic molecules: *a closer look*

Glycolysis releases less than a quarter of the chemical energy stored in glucose; most of the energy remains stocked in the two molecules of pyruvate. If oxygen is present, the pyruvate enters the mitochondrion, where the enzymes of the Krebs cycle complete the oxidation of the organic fuel.

Upon entering the mitochondrion, pyruvate is first converted to a compound called **acetyl CoA** (FIGURE 9.10). This step, the junction between glycolysis and the Krebs cycle, is accomplished by a multienzyme complex that catalyzes three reactions: ① Pyruvate's carboxyl group, which has little chemical energy, is removed and given off as a molecule of CO_2. (This is the first step in respiration where CO_2 is released.) ② The remaining two-carbon fragment is oxidized to form a compound named acetate (the ionized form of acetic acid). An enzyme transfers the extracted electrons to NAD^+, storing energy in the form of NADH. ③ Finally, coenzyme A, a sulfur-containing compound derived from a B vitamin, is attached to the acetate by an unstable bond that makes the acetyl group (the attached acetate) very reactive. The product of this chemical grooming, acetyl CoA, is now ready to feed its acetate into the Krebs cycle for further oxidation.

The Krebs cycle is named after Hans Krebs, the German-British scientist who was largely responsible for elucidating the pathway in the 1930s. The cycle has eight steps, each catalyzed by a specific enzyme in the mitochondrial matrix (FIGURE 9.11). You can see in the diagram that for each turn of the Krebs cycle, two carbons enter in the relatively reduced form of acetate (step ①), and two different carbons leave in the completely oxidized form of CO_2 (steps ③ and ④). The acetate joins the cycle by its

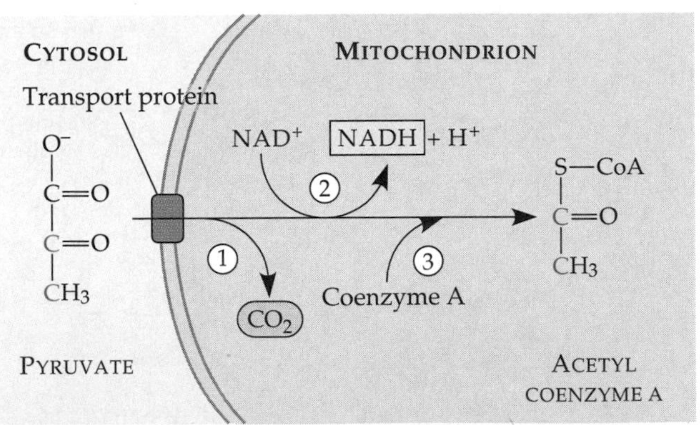

FIGURE 9.10
Conversion of pyruvate to acetyl CoA, the junction between glycolysis and the Krebs cycle. A protein built into the mitochondrial membrane translocates pyruvate from the cytosol into the mitochondrion. Then ① the carboxyl group of pyruvate, already fully oxidized, is removed as a CO_2 molecule, which diffuses out of the cell. ② The remaining two-carbon fragment is oxidized while NAD^+ is reduced to NADH. ③ Finally, the two-carbon acetyl group is attached to coenzyme A (CoA). The coenzyme has a sulfur atom, which attaches to the acetyl fragment by an unstable bond. This activates the acetyl group for the first reaction of the Krebs cycle.

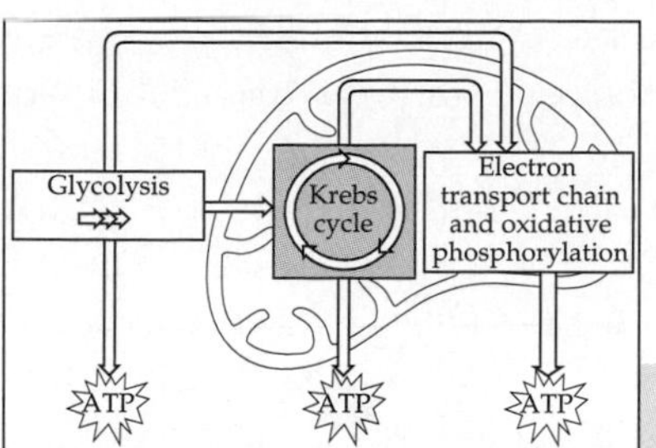

FIGURE 9.11

A closer look at the Krebs cycle. Red type traces the fate of the two carbon atoms that enter the cycle via acetyl CoA (step ①); blue type indicates the two carbons that exit the cycle as CO_2 in steps ③ and ④. (Notice that carboxylic acids are represented in their ionized forms, as —COO⁻. For example, citrate is the ionized form of citric acid.)

Step ① Acetyl CoA adds its two-carbon fragment to oxaloacetate, a four-carbon compound. The unstable bond of acetyl CoA is broken as oxaloacetate displaces the coenzyme and attaches to the acetyl group. The product is the six-carbon citrate. CoA is then free to prime another two-carbon fragment derived from pyruvate. Notice that oxaloacetate is regenerated by the last step of the cycle.

Step ② A molecule of water is removed and another is added back. The net result is the conversion of citrate to its isomer, isocitrate.

Step ③ The substrate loses a CO_2 molecule, and the remaining five-carbon compound is oxidized, reducing NAD^+ to NADH.

Step ④ This step is catalyzed by a multienzyme complex very similar to the one that converts pyruvate to acetyl CoA. CO_2 is lost; the remaining four-carbon compound is oxidized by the transfer of electrons to NAD^+ to form NADH, and is then attached to coenzyme A by an unstable bond.

Step ⑤ Substrate-level phosphorylation occurs in this step. CoA is displaced by a phosphate group, which is then transferred to GDP to form guanosine triphosphate (GTP). GTP is similar to ATP, which is formed when GTP donates a phosphate group to ADP.

Step ⑥ In another oxidative step, two hydrogens are transferred to FAD to form $FADH_2$.

Step ⑦ Bonds in the substrate are rearranged in this step by the addition of a water molecule.

Step ⑧ The last oxidative step produces another molecule of NADH and regenerates oxaloacetate, which accepts a two-carbon fragment from acetyl CoA for another turn of the cycle.

KREBS CYCLE

Acetyl CoA; Oxaloacetate; Citrate; Isocitrate; α-Ketoglutarate; Succinyl CoA; Succinate; Fumarate; Malate

enzymatic addition to the compound oxaloacetate, forming citrate. Subsequent steps decompose the citrate back to oxaloacetate, giving off CO_2 as "exhaust." It is this regeneration of oxaloacetate that accounts for the "cycle" in the Krebs cycle.

Most of the energy harvested by the oxidative steps of the cycle is conserved in NADH. For each acetate that enters the cycle, three molecules of NAD^+ are reduced to NADH—steps ③, ④, and ⑧. In one oxidative step, ⑥, electrons are transferred not to NAD^+, but to a different electron acceptor, FAD (flavin adenine dinucleotide, derived from riboflavin, a B vitamin). The reduced form, $FADH_2$, donates its electrons to the electron transport chain, as does NADH. (However, as we will see in the next section, $FADH_2$ "feeds" its electrons to the transport chain at a lower energy level than does NADH.) There is also a step in the Krebs cycle, step ⑤, that forms an ATP molecule directly by substrate-level phosphorylation, similar to the ATP-generating steps of glycolysis. But most of the ATP output of respiration results from oxidative phosphorylation, when the NADH and $FADH_2$ produced by the Krebs cycle relay the electrons extracted from food to the electron transport chain. Use FIGURE 9.12 to review the inputs and outputs of the Krebs cycle before proceeding to the electron transport chain.

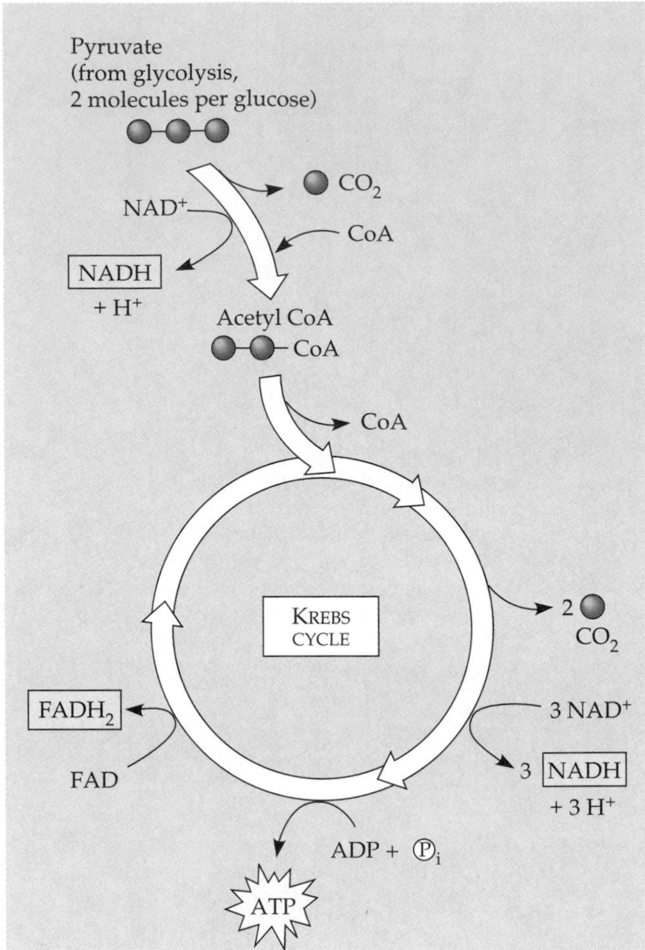

FIGURE 9.12
A summary of the Krebs cycle. The cycle functions as a metabolic "furnace" that oxidizes organic fuel derived from pyruvate, the product of glycolysis. This diagram summarizes the inputs and outputs as pyruvate is broken down to 3 molecules of CO_2. (The diagram includes the pre-Krebs cycle conversion of pyruvate to acetyl CoA.) The cycle generates 1 ATP per turn by substrate phosphorylation, but most of the chemical energy is transferred during the redox reactions to NAD^+ and FAD. The reduced coenzymes, NADH and $FADH_2$, shuttle their cargo of high-energy electrons to the electron transport chain, which uses the energy to synthesize ATP by oxidative phosphorylation. (To calculate the inputs and outputs on a "per-glucose" basis, multiply by 2 because each glucose molecule is split during glycolysis into 2 pyruvate molecules.)

The inner mitochondrial membrane couples electron transport to ATP synthesis: *a closer look*

Our main objective in this chapter is to learn how cells harvest the energy of food to make ATP. But the metabolic components of respiration we have dissected so far, glycolysis and the Krebs cycle, produce only four molecules of ATP per glucose molecule by substrate-level phosphorylation: two net ATPs from glycolysis and two ATPs from the Krebs cycle. At this point, molecules of NADH (and $FADH_2$) account for most of the energy extracted from food. These electron escorts link glycolysis and the Krebs cycle to the machinery for oxidative phosphorylation, which uses energy released by the electron transport chain to power ATP synthesis. In this section, you will first learn how the electron transport chain works, then how the inner membrane of the mitochondrion couples ATP synthesis to electron flow down the chain.

The Pathway of Electron Transport

You learned earlier that the electron transport chain is a collection of molecules embedded in the inner membrane of the mitochondrion. The folding of the inner membrane to form cristae provides space for thousands of copies of the chain in each mitochondrion—once again, we see that structure fits function. Most components of the chain are proteins. Tightly bound to these proteins are prosthetic groups, nonprotein components essential for the catalytic functions of certain enzymes. During electron transport along the chain, these prosthetic groups alternate between reduced and oxidized states as they accept and donate electrons.

FIGURE 9.13 traces the sequence of electron transfers along the electron transport chain. Electrons removed from food during glycolysis and the Krebs cycle are transferred by NADH to the first molecule of the electron

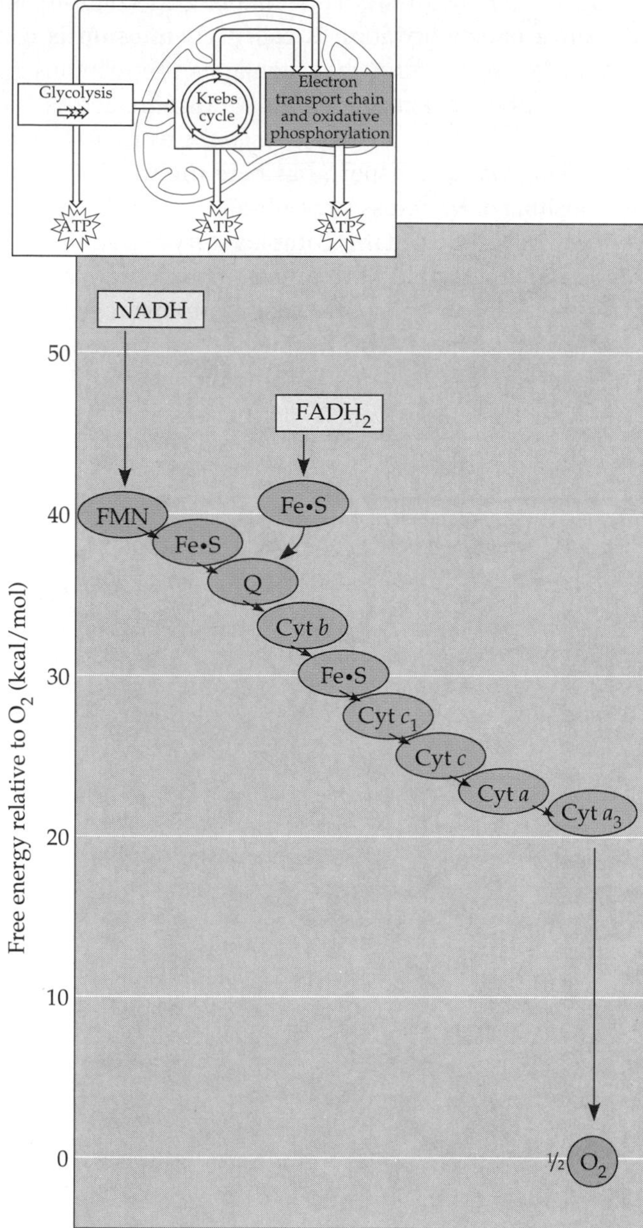

FIGURE 9.13

A closer look at the electron transport chain. Each member of the chain oscillates between a reduced state and an oxidized state. A component of the chain becomes reduced when it accepts electrons from its "uphill" neighbor (which has a lower affinity for the electrons). Each member of the chain returns to its oxidized form as it passes electrons to its "downhill" neighbor (which has a greater affinity for the electrons). At the bottom of the chain is oxygen, which is *very* electronegative. The overall energy drop for electrons traveling from NADH to oxygen is 53 kcal/mol, but this fall is broken up into a series of smaller steps by the electron transport chain.

transport chain. This molecule is a flavoprotein, so named because it has a prosthetic group called flavin mononucleotide (FMN in FIGURE 9.13). In the next redox reaction, the flavoprotein returns to its oxidized form as it passes electrons to an iron-sulfur protein (Fe•S in FIGURE 9.13), one of a family of proteins with both iron and sulfur tightly bound. The iron-sulfur protein then passes the electrons to a compound called ubiquinone (Q in FIGURE 9.13). This electron carrier is a lipid, the only member of the electron transport chain that is not a protein.

Most of the remaining electron carriers between Q and oxygen are proteins called **cytochromes (cyt).** Their prosthetic group, called a heme group, has four organic rings surrounding a single iron atom. It is similar to the iron-containing prosthetic group found in hemoglobin, the red protein of blood that transports oxygen. But the iron of cytochromes transfers electrons, not oxygen. The electron transport chain has several types of cytochromes, each a different protein with a heme group. The last cytochrome of the chain, cyt a_3, passes its electrons to oxygen, which also picks up a pair of hydrogen ions from the aqueous solution to form water. (An oxygen atom is represented in FIGURE 9.13 as $\frac{1}{2}O_2$ to emphasize that the electron transport chain reduces molecular oxygen, O_2, not individual oxygen atoms. For every two NADH molecules, one O_2 molecule is reduced to two molecules of water.)

Another source of electrons for the transport chain is $FADH_2$, the other reduced product of the Krebs cycle. Notice in FIGURE 9.13 that $FADH_2$ adds its electrons to the electron transport chain at a lower energy level than NADH does. Consequently, the electron transport chain provides about one-third less energy for ATP synthesis when the electron donor is $FADH_2$ rather than NADH.

The electron transport chain makes no ATP directly. Its function is to ease the fall of electrons from food to oxygen, breaking a large free-energy drop into a series of smaller steps that release energy in manageable amounts. How does the mitochondrion couple this electron transport to ATP synthesis? The answer is a mechanism called chemiosmosis.

Chemiosmosis: The Energy-Coupling Mechanism

Populating the inner membrane of the mitochondrion are many copies of a protein complex called an **ATP synthase,** the enzyme that actually makes ATP. It works like an ion pump running in reverse. Recall from Chapter 8 that ion pumps use ATP as an energy source to transport ions against their gradients. In the reverse of that process, an ATP synthase uses the energy of an existing ion gradient to power ATP synthesis. The ion gradient that drives oxidative phosphorylation is a proton (hydrogen ion) gradient; that is, the power source for the ATP synthase is a difference in the concentration of H^+ on opposite sides of the inner mitochondrial membrane. We can also think of this gradient as a difference in pH, since pH is a measure of H^+ concentration.

How does the mitochondrial membrane generate and maintain an H^+ gradient? That is the function of the electron transport chain. The chain is an energy converter that uses the exergonic flow of electrons to pump H^+ across the membrane, from the matrix into the intermembrane space (FIGURE 9.14). The H^+ leaks back across the membrane, diffusing down its gradient. But the ATP synthases are the only patches of the membrane that are freely permeable to H^+. The ions pass through a channel in an ATP synthase, and the complex of proteins functions as a mill that harnesses the exergonic flow of H^+ to drive the phosphorylation of ADP. Thus, an H^+ gradient couples the redox reactions of the electron transport chain to ATP synthesis. This coupling mechanism for oxidative phosphorylation is called **chemiosmosis** (Gr. *osmos*, "push"), a term that highlights the relationship between chemical reactions and transport across the membrane. We have previously used the word *osmosis* in discussing water transport, but here the word refers to the pushing of H^+ across a membrane.

If you have followed this complex story of chemiosmosis so far, you should have at least two questions. How does the electron transport chain pump hydrogen ions? And how does the ATP synthase use H^+ backflow to make ATP? Researchers have made some progress on the first problem. Certain members of the electron transport

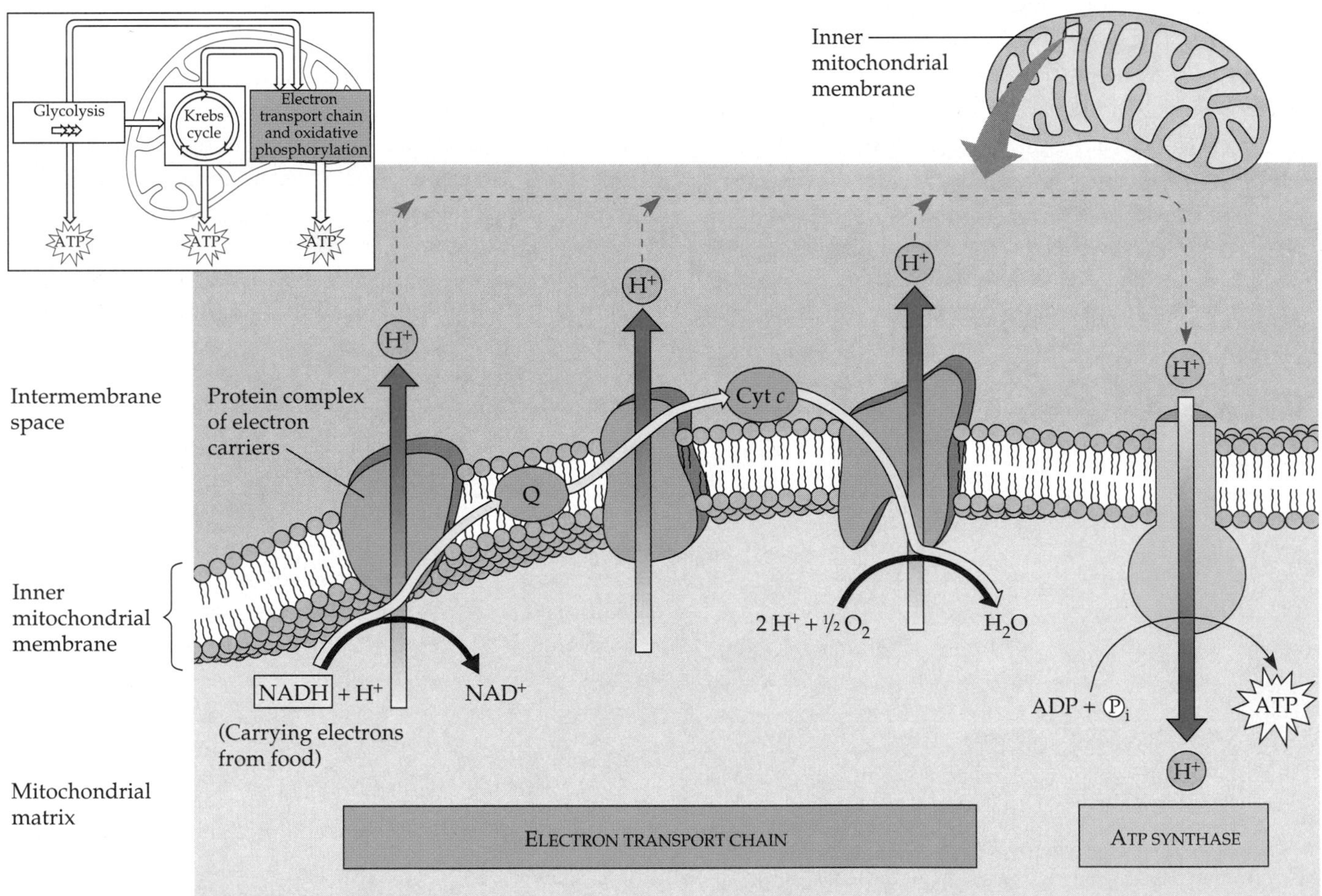

FIGURE 9.14

Chemiosmosis: How the mitochondrial membrane couples electron transport to oxidative phosphorylation. NADH shuttles high-energy electrons extracted from food during the Krebs cycle to an electron transport chain, which is built into the inner mitochondrial membrane. The yellow arrow in this diagram traces the transport of electrons, which pass to oxygen at the "downhill" end of the chain to form water. Most of the cytochromes and other electron carriers of the chain (see FIGURE 9.13) are collected into three complexes, each represented here by a purple "blob" embedded in the membrane. Two mobile carriers, ubiquinone (Q) and cytochrome *c*, move rapidly along the membrane, ferrying electrons between the three large complexes. As each complex of the chain accepts and then donates electrons, it pumps hydrogen ions (protons) from the mitochondrial matrix into the space between the inner and outer membranes (magenta arrows trace H^+ transport). Thus, chemical energy harvested from food is transformed to a proton-motive force, a gradient of H^+ across the membrane. The hydrogen ions complete their circuit by flowing down their gradient through an H^+ channel in an ATP synthase, another protein complex built into the membrane. The ATP synthase harnesses the proton-motive force to phosphorylate ADP, forming ATP. (This is called oxidative phosphorylation because it is driven by the exergonic transfer of electrons from food to oxygen.) This mechanism for energy coupling—the use of an H^+ gradient (proton-motive force) to transfer energy from redox reactions to cellular work (ATP synthesis, in this case)—is called chemiosmosis.

chain must accept and release protons (H^+) along with electrons, while other carriers transport only electrons. Therefore, at certain steps along the chain, electron transfers cause H^+ to be taken up and released back into the surrounding solution. The electron carriers are spatially arranged in the membrane in such a way that H^+ is accepted from the mitochondrial matrix and deposited in the intermembrane space (see FIGURE 9.14). The H^+ gradient that results is referred to as a **proton-motive force,** emphasizing the capacity of the gradient to perform work. The force drives H^+ back across the membrane through the specific H^+ channels provided by ATP synthase complexes. How the ATP synthase uses the downhill H^+ current to attach inorganic phosphate to ADP is not yet known. The hydrogen ions may participate directly in the reaction, or they may induce a conformation (shape) change of the ATP synthase that facilitates phosphorylation. Research has revealed the general mechanism of energy coupling by chemiosmosis, but many details of the process are still uncertain.

Let's review the key feature of chemiosmosis: It is an energy-coupling mechanism that uses exergonic redox reactions to store energy in the form of an H^+ gradient, which then drives other kinds of work, including ATP synthesis. Chemiosmosis is not unique to mitochondria. Chloroplasts also use the mechanism to generate ATP during photosynthesis; the main difference is that light drives electrons along an electron transport chain. Bacteria, which lack both mitochondria and chloroplasts, generate H^+ gradients across their plasma membranes. They then tap the proton-motive force to make ATP, to pump nutrients and waste products across the membrane, and even to move by rotating their flagella. In 1961, British biochemist Peter Mitchell first proposed chemiosmosis as an energy-coupling mechanism based on his experiments with bacteria. Nearly two decades later, after many scientists had confirmed the centrality of chemiosmosis in energy conversions within bacteria, mitochondria, and chloroplasts, Mitchell was awarded the Nobel Prize. Chemiosmosis has helped unify the study of bioenergetics.

Cellular respiration generates many ATP molecules for each sugar molecule it oxidizes: *a review*

Now that we have looked more closely at the key processes of cellular respiration, let's return to its overall function: harvesting the energy of food for ATP synthesis (FIGURE 9.15).

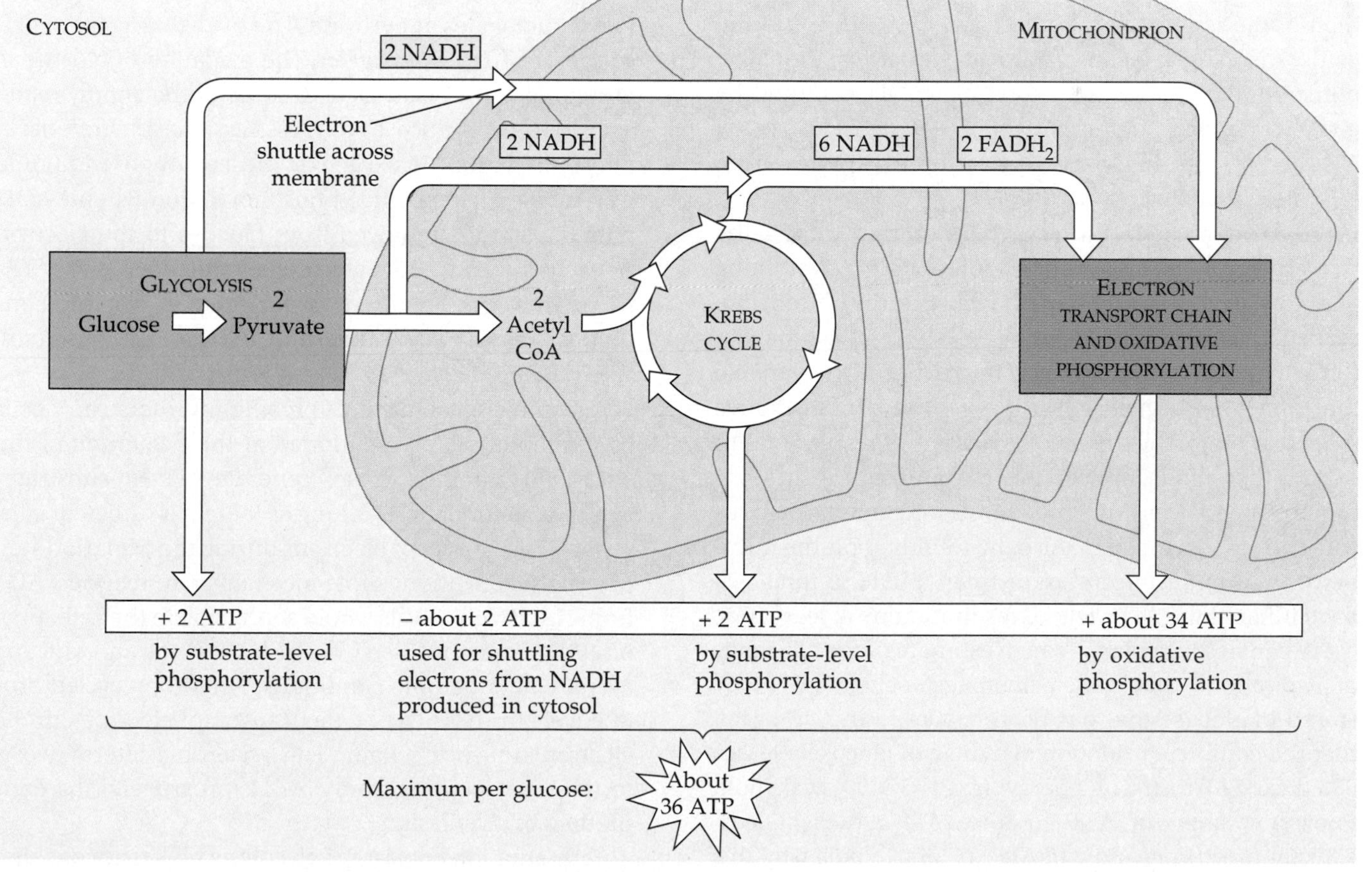

FIGURE 9.15

Review: each molecule of glucose yields many ATP molecules during cellular respiration. The text explains why the yield of 36 ATP per glucose is only an estimate of the maximum output.

During respiration, most energy flows in this sequence: Glucose ⟶ NADH ⟶ Electron transport chain ⟶ proton-motive force ⟶ ATP.

We can do some bookkeeping to calculate the net ATP profit when cellular respiration oxidizes a molecule of glucose to six molecules of carbon dioxide. The three main departments of this metabolic enterprise are glycolysis, the Krebs cycle, and the electron transport chain, which drives oxidative phosphorylation. FIGURE 9.15, a follow-up to the overview of cellular respiration presented in FIGURE 9.6, gives a detailed accounting of the ATP yield per glucose molecule oxidized. The tally adds the few molecules of ATP produced directly by substrate-level phosphorylation during glycolysis and the Krebs cycle to the many more molecules of ATP generated by oxidative phosphorylation. Each NADH that transfers a pair of electrons from food to the electron transport chain contributes enough to the proton-motive force to generate a maximum of about three ATPs. (The average ATP yield per NADH is probably between two and three; we are rounding off to three here to simplify the bookkeeping.) The Krebs cycle also supplies electrons to the electron transport chain via $FADH_2$, but each molecule of this electron carrier is worth a maximum of only about two molecules of ATP. In most eukaryotic cells, this lower ATP yield per electron pair also applies to the NADH produced by glycolysis in the cytosol. The mitochondrial membrane is impermeable to NADH, so NADH in the cytosol is segregated from the machinery of oxidative phosphorylation. The electrons of NADH captured by glycolysis must be relayed, at the expense of some ATP, across the membrane to electron acceptors within the mitochondrion.

Subtracting the debit of 2 ATP incurred during the preparatory steps of glycolysis, and doubling everything after the sugar-splitting step of glycolysis, the bottom line reads 36 ATP. Our bookkeeping gives only an estimate of the maximum ATP yield from respiration. One variable that affects ATP yield is the use of the proton-motive force (generated by the redox reactions of respiration) to drive other kinds of work. For example, the proton-motive force powers the mitochondrion's uptake of pyruvate from the cytoplasm. Also, by rounding off the number of ATP molecules produced per NADH to three, we have inflated the ATP yield of respiration by at least 10%.

We can now calculate a rough estimate of the efficiency of respiration—that is, the percentage of chemical energy stored in glucose that has been restocked in ATP. Recall that the complete oxidation of a mole of glucose releases 686 kcal (2870 kJ) of energy ($\Delta G = -686$ kcal/mol). Phosphorylation of ADP to form ATP stores at least 7.3 kcal (31 kJ) per mole of ATP (p. 96 explains why this number is probably higher under cellular conditions). Therefore, the efficiency of respiration is 7.3 times 36 (maximum ATP yield per glucose) divided by 686, or about 38%. The rest of the stored energy is lost as heat. We use some of this heat to maintain our relatively high body temperature (37°C), and we dissipate the rest through sweating and other cooling mechanisms. Cellular respiration is remarkably efficient in its energy conversion. (By comparison, the most efficient automobile converts about 25% of the energy stored in gasoline to movement of the car.)

Because most of the ATP generated by cellular respiration is the work of oxidative phosphorylation, our estimate of ATP yield from respiration is contingent upon an adequate supply of oxygen to the cell. Without the electronegative oxygen to pull electrons down the transport chain, oxidative phosphorylation ceases. However, fermentation provides a mechanism by which some cells can oxidize organic fuel and generate ATP *without* the help of oxygen.

Fermentation enables some cells to produce ATP without the help of oxygen

How can food be oxidized without oxygen? Remember, oxidation refers to the loss of electrons to *any* electron acceptor, not just to oxygen. Glycolysis oxidizes glucose to two molecules of pyruvate. The oxidizing agent of glycolysis is NAD^+, *not* oxygen. The oxidation of glucose is exergonic, and glycolysis uses some of the energy made available to produce two ATPs (net) by substrate-level phosphorylation. If oxygen *is* present, then additional ATP is made by oxidative phosphorylation when NADH passes electrons removed from glucose to the electron transport chain. But glycolysis generates two ATPs whether oxygen is present or not—that is, whether conditions are **aerobic** or **anaerobic** (Gr. *aer,* "air," and *bios,* "life"; the prefix *an-* means "without").

Anaerobic catabolism of organic nutrients can occur by fermentation, as mentioned at the beginning of the chapter. Fermentation can generate ATP by substrate-level phosphorylation, as long as there is a sufficient supply of NAD^+ to accept electrons during the oxidation step of glycolysis. Without some mechanism to recycle NAD^+ from NADH, glycolysis would soon deplete the cell's pool of NAD^+ and shut itself down for lack of an oxidizing agent. Under aerobic conditions, NAD^+ is recycled productively from NADH by the transfer of electrons to the electron transport chain. The anaerobic alternative is to transfer electrons from NADH to pyruvate, the end-product of glycolysis.

Fermentation consists of glycolysis plus reactions that regenerate NAD^+ by transferring electrons from NADH to pyruvate or derivatives of pyruvate. There are many types of fermentation, differing in the waste products formed from pyruvate. Two common types are alcohol fermentation and lactic acid fermentation.

In **alcohol fermentation** (FIGURE 9.16a), pyruvate is converted to ethanol, or ethyl alcohol, in two steps. The first step releases carbon dioxide from the pyruvate, which is converted to the two-carbon compound acetaldehyde. In the second step, acetaldehyde is reduced by NADH to ethyl alcohol. This regenerates the supply of NAD^+ needed for glycolysis. Alcohol fermentation by yeast, a fungus, is used in brewing and wine making (FIGURE 9.17). Many bacteria also carry out alcohol fermentation under anaerobic conditions.

During **lactic acid fermentation** (FIGURE 9.16b) pyruvate is reduced directly by NADH to form lactate as a waste product, with no release of CO_2. (Lactate is the ionized form of lactic acid.) Lactic acid fermentation by certain fungi and bacteria is used in the dairy industry to make cheese and yogurt. Acetone and methyl alcohol are among the by-products of other types of microbial fermentation that are commercially important.

Human muscle cells make ATP by lactic acid fermentation when oxygen is scarce. This occurs during the early stages of strenuous exercise, when sugar catabolism for ATP production outpaces the muscle's supply of oxygen from the blood. Under these conditions, the cells switch from aerobic respiration to fermentation. The lactate that accumulates as a waste product may cause muscle fatigue and pain, but it is gradually carried away

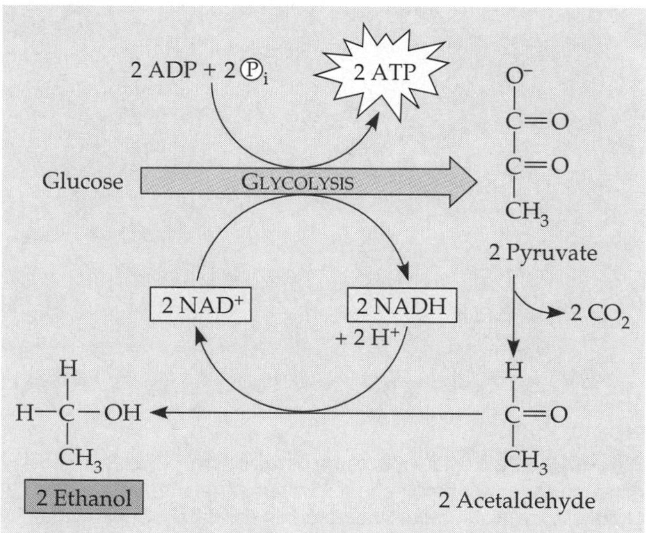

(a) Alcohol fermentation

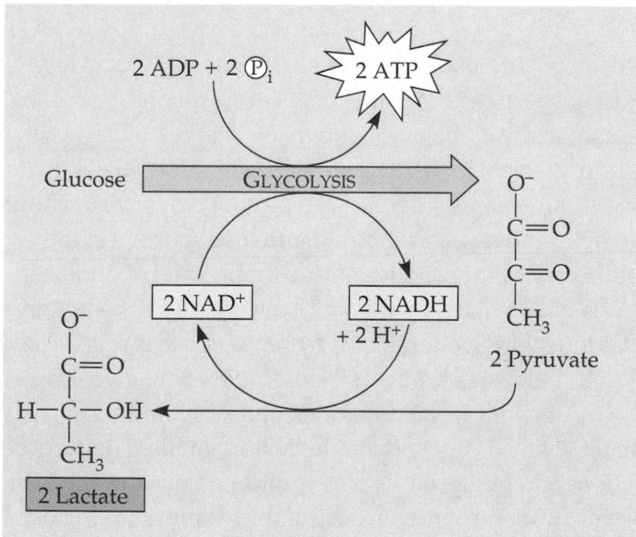

(b) Lactic acid fermentation

FIGURE 9.16

Fermentation. Pyruvate, the end-product of glycolysis, serves as an electron acceptor for oxidizing NADH back to NAD^+. The NAD^+ can then be reused to oxidize sugar during glycolysis, which yields two net molecules of ATP by substrate-level phosphorylation. Two of the common waste products formed from fermentation are (**a**) ethanol and (**b**) lactate.

(a)

(b)

FIGURE 9.17

Commercial applications of alcohol fermentation. The basic steps used to make wine from grapes were the same in (**a**) ancient times as they are (**b**) today. The process employs yeast to convert some of the sugar in fruit juice to alcohol. One-way gas valves allow carbon dioxide to escape from the ferment without letting air in. To make a sparkling wine, such as champagne, the CO_2 is left dissolved in the wine.

by the blood to the liver. Lactate is converted back to pyruvate by liver cells.

Fermentation and Respiration Compared

Fermentation and cellular respiration are anaerobic and aerobic alternatives, respectively, for producing ATP by harvesting the chemical energy of food. Both pathways use glycolysis to oxidize glucose and other organic fuels to pyruvate, with a net production of 2 ATP by substrate phosphorylation. And in both fermentation and respiration, NAD^+ is the oxidizing agent that accepts electrons from food during glycolysis. A key difference is the contrasting mechanisms for oxidizing NADH back to NAD^+, which is required to sustain glycolysis. In fermentation, the final electron acceptor is an organic molecule such as pyruvate (lactic acid fermentation) or acetaldehyde (alcohol fermentation). In respiration, by contrast, the final acceptor for electrons from NADH is oxygen. This not only regenerates the NAD^+ required for glycolysis but pays an ATP bonus when the stepwise electron transport from NADH to oxygen drives oxidative phosphorylation. An even bigger ATP payoff comes from the oxidation of pyruvate in the Krebs cycle, which is unique to respiration. Without oxygen, the energy still stored in pyruvate is unavailable to the cell. Thus, cellular respiration harvests much more energy from each sugar molecule than a cell can tap by fermentation. In fact, respiration yields as much as 18 times more ATP per glucose molecule than does fermentation—36 ATP for respiration, compared to 2 ATP produced by substrate-level phosphorylation in fermentation.

Some organisms, including yeasts and many bacteria, can make their ATP by either fermentation or respiration. Such species are called **facultative anaerobes.** On the cellular level, our muscle cells behave as facultative anaerobes. In a facultative anaerobe, pyruvate is a fork in the metabolic road that leads to two alternative catabolic routes (FIGURE 9.18). Under aerobic conditions, pyruvate may be converted to acetyl CoA, and oxidation continues in the Krebs cycle. Under anaerobic conditions, pyruvate is diverted from the Krebs cycle, serving instead as an electron acceptor to recycle NAD^+. To make the same amount of ATP, a facultative anaerobe would have to consume sugar at a much faster rate when fermenting than it would when respiring.

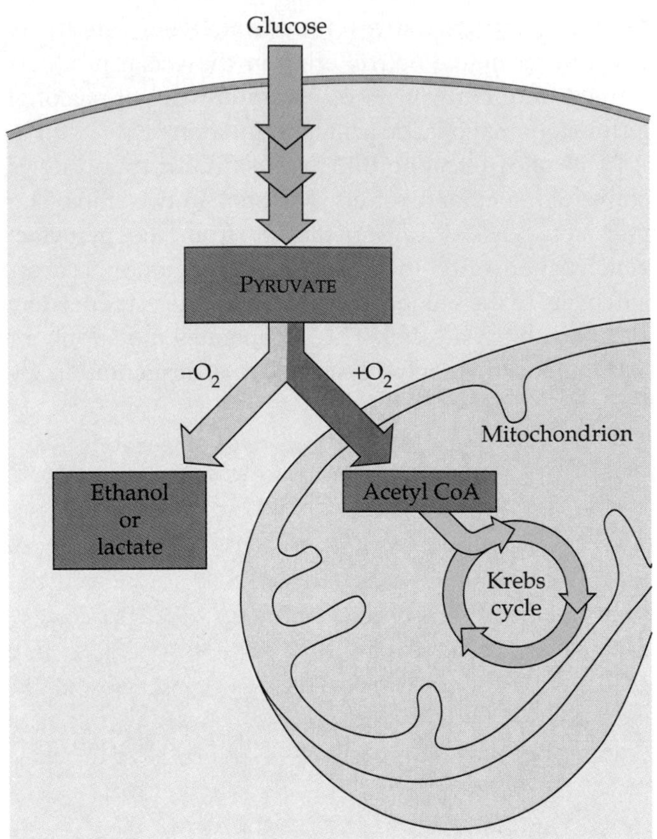

FIGURE 9.18
Pyruvate as a key juncture in catabolism. Glycolysis is common to fermentation and respiration. The end-product of glycolysis, pyruvate, represents a fork in the catabolic pathways of glucose oxidation. In a cell capable of both respiration and fermentation, pyruvate is committed to one of those two pathways, usually depending on whether or not oxygen is present.

The Evolutionary Significance of Glycolysis

Notice again in FIGURE 9.18 that glycolysis is the one catabolic pathway common to fermentation and respiration, a similarity with an evolutionary basis. Ancient prokaryotes probably used glycolysis to make ATP long before oxygen was present in the Earth's atmosphere. The oldest known fossils of bacteria date back over 3.5 billion years, but appreciable quantities of oxygen probably did not begin to accumulate in the atmosphere until about 2.5 billion years ago. (According to fossil evidence, the cyanobacteria that produce O_2 as a by-product of photosynthesis had evolved by then.) Therefore, the first prokaryotes must have generated ATP exclusively from glycolysis, which does not require oxygen. In addition, glycolysis is the most widespread metabolic pathway, which suggests that it evolved very early in the history of life. The cytosolic location of glycolysis also implies great antiquity; the pathway does not require any of the membrane-enclosed organelles of the eukaryotic cell, which evolved nearly 2 billion years after the prokaryotic cell. Glycolysis is a metabolic heirloom from the earliest cells that continues to function in fermentation and as the first stage in the breakdown of organic molecules by respiration.

Glycolysis and the Krebs cycle connect to many other metabolic pathways

So far, we have treated the oxidative breakdown of glucose in isolation from the cell's overall metabolic economy. In this section, you will learn that glycolysis and the Krebs cycle are major intersections of various catabolic and anabolic (biosynthetic) pathways.

The Versatility of Catabolism

Throughout this chapter, we have used glucose as the fuel for cellular respiration. But free glucose molecules are not common in the diets of humans and other animals. We obtain most of our calories in the form of fats, proteins, sucrose and other disaccharides, and starch, a polysaccharide. All these food molecules can be used by cellular respiration to make ATP (FIGURE 9.19).

Glycolysis can accept a wide range of carbohydrates for catabolism. In the digestive tract, starch is hydrolyzed to glucose, which can then be broken down in the cells by glycolysis and the Krebs cycle. Similarly, glycogen, the polysaccharide that humans and many other animals store in their liver and muscle cells, can be hydrolyzed to glucose between meals as fuel for respiration. The digestion of disaccharides, including sucrose, provides glucose and other monosaccharides as fuel for respiration.

Proteins can also be used for fuel, but first they must be digested to their constituent amino acids. Many of the amino acids, of course, are used by the organism to build new proteins. Amino acids present in excess are converted by enzymes to intermediates of glycolysis and the Krebs cycle. Before amino acids can feed into glycolysis or the Krebs cycle, their amino groups must be removed, a process called deamination. The nitrogenous refuse is excreted from the animal in the form of ammonia, urea, or other waste products.

Catabolism can also harvest energy stored in fats obtained either from food or from storage cells in the body. After fats are digested, the glycerol is converted to glyceraldehyde phosphate, an intermediate of glycolysis. Most of the energy of a fat is stored in the fatty acids. A metabolic sequence called **beta oxidation** breaks the fatty acids down to two-carbon fragments, which enter the Krebs cycle as acetyl CoA. Fats make excellent fuel. A gram of fat oxidized by respiration produces more than twice as much ATP as a gram of carbohydrate. Unfortunately, this also means that a dieter must be patient while using fat stored in the body, because so many calories are stockpiled in each gram of fat.

Biosynthesis (Anabolic Pathways)

Cells need substance as well as energy. Not all the organic molecules of food are destined to be oxidized as fuel to make ATP. In addition to calories, food must also provide the carbon skeletons that cells require to make their own molecules. Some organic monomers obtained from digestion can be used directly. For example, amino acids from the hydrolysis of proteins in food can be incorporated into the organism's own proteins. Often, however, the body needs specific molecules that are not present as such in food. Compounds formed as intermediates of glycolysis and the Krebs cycle can be diverted into anabolic pathways as precursors from which the cell can synthesize the molecules it requires. For example, humans can make about half of the 20 amino acids by modifying compounds siphoned away from the Krebs cycle. Also, glucose can be made from pyruvate, and fatty acids

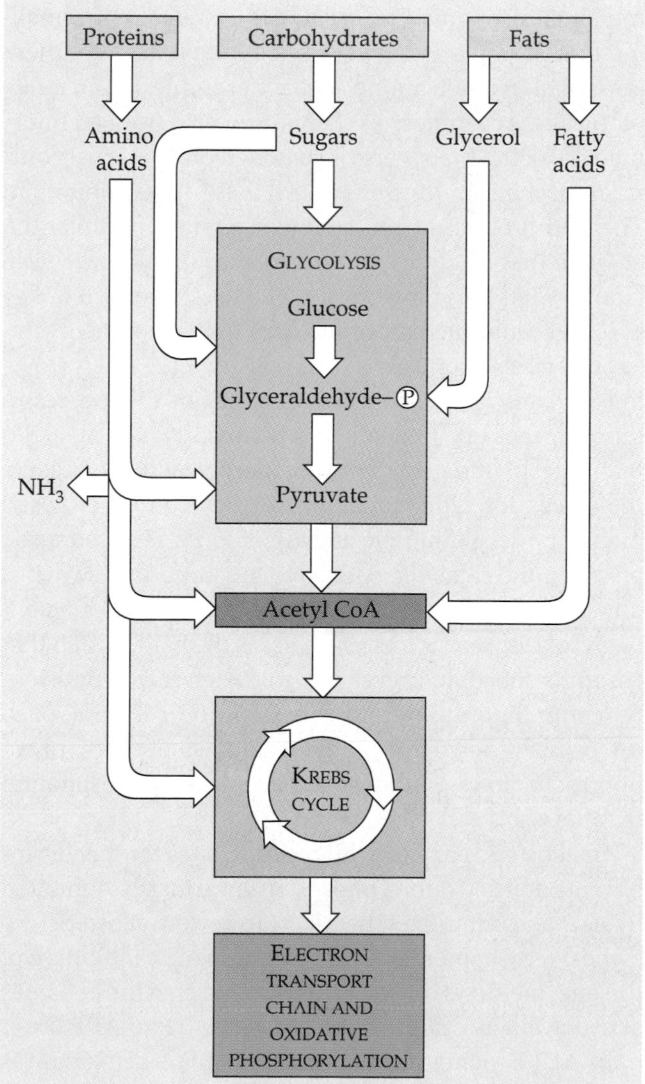

FIGURE 9.19
The catabolism of various food molecules. Carbohydrates, fats, and proteins can all be used as fuel for cellular respiration. Monomers of these food molecules enter glycolysis or the Krebs cycle at various points. Glycolysis and the Krebs cycle are catabolic funnels through which electrons from all kinds of food molecules flow on their exergonic fall to oxygen.

can be synthesized from acetyl CoA. Of course, these anabolic or biosynthetic pathways do not generate ATP, but consume it instead.

In addition, glycolysis and the Krebs cycle function as metabolic interchanges, enabling our cells to convert some molecules to others as we need them. For instance, carbohydrates and proteins can be converted to fats through intermediates of glycolysis and the Krebs cycle. If we eat more food than we need, we will store fat even if our diet is fat-free. Metabolism is remarkably versatile and adaptable.

Feedback mechanisms control cellular respiration

Basic principles of supply and demand regulate the metabolic economy. The cell does not waste energy making more of a particular substance than it needs. If there is a glut of a certain amino acid, for example, the anabolic pathway that synthesizes that amino acid from an intermediate of the Krebs cycle is switched off. The most common mechanism for this control is feedback inhibition: The end-product of the anabolic pathway inhibits the enzyme that catalyzes the first step of the pathway (see FIGURE 6.16). This prevents the needless diversion of key metabolic intermediates from uses that are more urgent.

The cell also controls its catabolism. If the cell is working hard and its ATP concentration begins to drop, respiration speeds up. When there is plenty of ATP to meet demand, respiration slows down, sparing valuable organic molecules for other functions. Again, control is based mainly on regulating the activity of enzymes at strategic points in the catabolic pathway. One important switch is phosphofructokinase, the enzyme that catalyzes step 3 of glycolysis (see FIGURE 9.8). That is the earliest step that commits substrate irreversibly to the glycolytic pathway. By controlling the rate of this step, the cell can speed up or slow down the entire catabolic process; phosphofructokinase is thus the pacemaker of respiration (FIGURE 9.20).

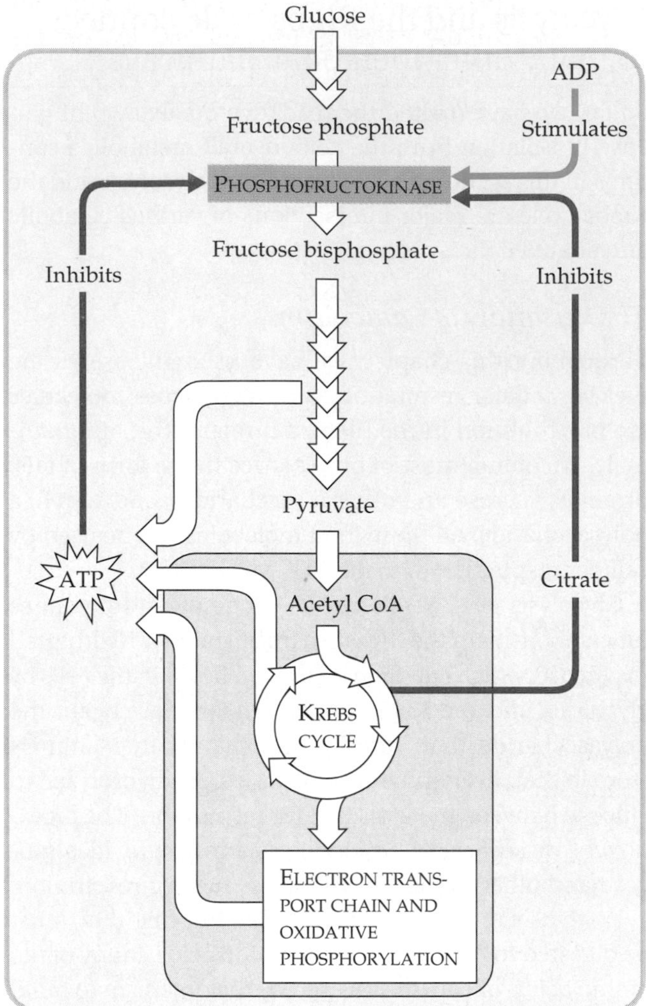

FIGURE 9.20
The control of cellular respiration. Allosteric enzymes at certain points in the respiratory pathway respond to inhibitors and activators to set the pace of glycolysis and the Krebs cycle. Phosphofructokinase, the enzyme that catalyzes step 3 of glycolysis, is one such enzyme. It is stimulated by ADP but inhibited by ATP and by citrate. This feedback regulation adjusts the rate of respiration as the cell's catabolic and anabolic demands change.

An allosteric enzyme with receptor sites for specific inhibitors and activators, phosphofructokinase is inhibited by ATP and stimulated by ADP. (To review allosteric enzymes, see Chapter 6.) As ATP accumulates, inhibition of the enzyme slows down glycolysis. The enzyme becomes active again as cellular work converts ATP to ADP faster than ATP is being regenerated. Phosphofructokinase is also sensitive to citrate, the first product of the Krebs cycle. If citrate accumulates in mitochondria, some of it passes into the cytosol and inhibits phosphofructokinase. This mechanism helps synchronize the rates of glycolysis and the Krebs cycle. As citrate accumulates, glycolysis slows down, and the supply of acetate to the Krebs cycle decreases. If citrate consumption increases, either because of a demand for more ATP or because anabolic pathways are draining off intermediates of the Krebs cycle, glycolysis accelerates and meets the demand. Metabolic balance is augmented by the control of other enzymes at other key locations in glycolysis and the Krebs cycle. Cells are thrifty, expedient, and responsive in their metabolism.

* * *

Examine FIGURE 9.1 again to put cellular respiration into the broader context of energy flow and chemical cycling in ecosystems. The energy that keeps us alive is *released,* but not *produced,* by cellular respiration. We are tapping energy that was stored in food by photosynthesis. In the next chapter, you will learn how photosynthesis captures light and converts it to chemical energy.

REVIEW OF KEY CONCEPTS (with page numbers and key figures)

- Cellular respiration and fermentation are catabolic (energy-yielding) pathways (pp. 159–160, FIGURE 9.1)
  - The metabolic breakdown of glucose and other organic fuels to simpler products is exergonic, yielding energy for ATP synthesis.
- Cells must recycle the ATP they use for work (p. 160, FIGURE 9.2)
  - ATP drives cellular work by transferring phosphate groups to various substrates, priming them to undergo change.
  - To keep on working, a cell must regenerate ATP.
  - Starting with glucose or another organic fuel and using oxygen, cellular respiration yields water, carbon dioxide, and energy in the form of ATP and heat.
- Redox reactions release energy when electrons move closer to electronegative atoms (pp. 160–161, FIGURE 9.3)
  - Food molecules store energy in their arrangement of electrons. The cell taps this energy through oxidation-reduction, or redox, reactions, in which one substance (reducing agent) partially or totally shifts electrons to another (oxidizing agent). The substance receiving electrons is reduced; the substance losing electrons is oxidized.
- Electrons "fall" from organic molecules to oxygen during cellular respiration (pp. 161–162)
  - In cellular respiration, glucose ($C_6H_{12}O_6$) is oxidized to CO_2, and O_2 is reduced to H_2O. Electrons lose potential energy during their transfer from organic compounds to oxygen, and this energy drives ATP synthesis.
- The "fall" of electrons during respiration is stepwise, via NAD^+ and an electron transport chain (pp. 162–164, FIGURE 9.5)
  - Electrons extracted from food are usually transferred first to NAD^+, reducing the compound to NADH.
  - NADH passes these electrons to an electron transport chain. The chain then conducts the electrons to oxygen in a series of steps that release energy. Oxidative phosphorylation uses this energy to make ATP.
- Respiration is a cumulative function of glycolysis, the Krebs cycle, and electron transport: *an overview* (pp. 164–165, FIGURE 9.6)
  - Glycolysis and the Krebs cycle supply electrons to the transport chain (via NADH), and the transport chain drives oxidative phosphorylation.
  - Glycolysis occurs in the cytosol. The Krebs cycle occurs in the mitochondrial matrix. The electron transport chain is built into the inner mitochondrial membrane.
- Glycolysis harvests chemical energy by oxidizing glucose to pyruvate: *a closer look* (pp. 165–168, FIGURE 9.9)
  - Glycolysis nets two molecules of ATP, produced by substrate-level phosphorylation, and two molecules of NADH.
- The Krebs cycle completes the energy-yielding oxidation of organic molecules: *a closer look* (pp. 168–170, FIGURE 9.12)
  - The link between glycolysis and the Krebs cycle is the conversion of pyruvate to acetyl CoA.
  - The acetate of acetyl CoA joins a four-carbon molecule, oxaloacetate, to form the six-carbon citrate molecule, which is subsequently degraded back to oxaloacetate in a series of steps constituting one turn of the cycle. In the process, carbon dioxide is given off, one molecule of ATP is formed by substrate-level phosphorylation, and electrons are passed to three molecules of NAD^+ and one molecule of FAD, another electron acceptor.
- The inner mitochondrial membrane couples electron transport to ATP synthesis: *a closer look* (pp. 170–173, FIGURE 9.14)
  - Most of the ATP created from the energy stored in glucose is produced by oxidative phosphorylation when NADH and $FADH_2$ donate their electrons to a system of electron carriers embedded in the inner mitochrondrial membrane.
  - The electron transport chain receives electrons from NADH and $FADH_2$. At the end of the chain, electrons are passed to oxygen, reducing it to water.
  - Electron transport is coupled to ATP synthesis by a mechanism called chemiosmosis. The structural arrangement of the electron carriers causes electron transfers at certain steps along the chain to translocate H^+ from the matrix to the intermembrane space, storing energy as a proton-motive force (H^+ gradient). As hydrogen ions diffuse back into the matrix through ATP synthase, the exergonic passage of H^+ drives the endergonic phosphorylation of ADP.
- Cellular respiration generates many ATP molecules for each sugar molecule it oxidizes: *a review* (pp. 173–174, FIGURE 9.15)
  - The complete oxidation of glucose to carbon dioxide during respiration in eukaryotes produces a maximum net yield of about 36 molecules of ATP.

- Fermentation enables some cells to produce ATP without the help of oxygen (pp. 174–176, FIGURE 9.16)
  - Fermentation is an anaerobic catabolism of organic nutrients. It yields ATP from glycolysis.
  - The electrons from NADH are transferred to pyruvate or some derivative of that glycolytic end-product. This restores the NAD^+ required to sustain glycolysis.
  - Yeast and certain bacteria are facultative anaerobes, capable of making ATP by either aerobic respiration or fermentation. Of the two pathways, respiration is the more efficient in terms of ATP yield per glucose.
  - Glycolysis occurs in nearly all organisms and probably evolved in ancient prokaryotes before oxygen was available in the atmosphere.
- Glycolysis and the Krebs cycle connect to many other metabolic pathways (pp. 177–178, FIGURE 9.19)
  - Fats, proteins, and carbohydrates can all be consumed by cellular respiration to form ATP. Thus, glycolysis and the Krebs cycle are catabolic pathways that funnel electrons from all kinds of food molecules into the electron transport chain, which powers ATP synthesis.
  - Carbon skeletons for anabolism come either directly from digestion or from glycolysis and the Krebs cycle, which donate intermediates for use in biosynthesis.
- Feedback mechanisms control cellular respiration (pp. 178–179, FIGURE 9.20)
  - Cellular respiration is controlled by allosteric enzymes at key points in glycolysis and the Krebs cycle. This control strikes a moment-to-moment balance between catabolism and anabolism.

SELF-QUIZ

1. The *direct* energy source that drives ATP synthesis during oxidative phosphorylation is
   a. the oxidation of glucose and other organic compounds
   b. the exergonic flow of electrons down the electron transport chain
   c. the affinity of oxygen for electrons
   d. a difference of H^+ concentration on opposite sides of the inner mitochondrial membrane
   e. the transfer of phosphate from Krebs cycle intermediates to ADP

2. What is the oxidizing agent in the following reaction?

$$\text{phosphoenolpyruvate} + NAD^+ \longrightarrow \text{pyruvate} + NADH + H^+$$

   a. oxygen
   b. NAD^+
   c. NADH
   d. phosphoenolpyruvate
   e. pyruvate

3. Which metabolic pathway is common to both fermentation and cellular respiration?
   a. Krebs cycle
   b. electron transport chain
   c. glycolysis
   d. synthesis of acetyl CoA from pyruvate
   e. reduction of pyruvate to lactate

4. In a eukaryotic cell, most of the enzymes of the Krebs cycle are located in the
   a. plasma membrane
   b. cytosol
   c. inner mitochondrial membrane
   d. mitochondrial matrix
   e. intermembrane space

5. The *final* electron acceptor of the electron transport chain that functions in oxidative phosphorylation is
   a. oxygen
   b. water
   c. NAD^+
   d. pyruvate
   e. ADP

6. When electrons flow along the electron transport chains of mitochondria, which of the following changes occurs? (Explain your answer)
   a. The pH of the matrix increases.
   b. The ATP synthase pumps protons by active transport.
   c. The electrons gain free energy.
   d. The cytochromes of the chain phosphorylate ADP to form ATP.
   e. NAD^+ is oxidized.

7. In the presence of a metabolic poison that specifically inhibits the mitochondrial ATP synthase, you would expect (explain your answer)
   a. a decrease in the pH difference across the mitochondrial membrane
   b. an increase in the pH difference across the mitochondrial membrane
   c. increased synthesis of ATP
   d. oxygen consumption to cease
   e. proton pumping by the electron transport chain to cease

8. Most of the ATP made during cellular respiration is generated by
   a. glycolysis
   b. oxidative phosphorylation
   c. substrate-level phosphorylation
   d. direct synthesis of ATP by the Krebs cycle
   e. transfer of phosphate from glucose-phosphate to ADP

9. Which of the following is a true distinction between fermentation and cellular respiration?
   a. Only respiration oxidizes glucose.
   b. NADH is oxidized by the electron transport chain only in respiration.
   c. Fermentation, but not respiration, is an example of a catabolic pathway.
   d. Substrate-level phosphorylation is unique to fermentation.
   e. NAD^+ functions as an oxidizing agent only in respiration.

10. Most CO_2 from catabolism is released during
    a. glycolysis
    b. the Krebs cycle
    c. lactate fermentation
    d. electron transport
    e. oxidative phosphorylation

CHALLENGE QUESTIONS

1. A century ago, Louis Pasteur, the great French biochemist, investigated the metabolism of yeast, a facultative anaerobe. He observed that the yeast consumed sugar at a much faster rate under anaerobic conditions than it did under aerobic conditions. Explain this Pasteur effect, as the observation is called.

2. In the 1940s, some physicians prescribed low doses of a drug called dinitrophenol (DNP) to help patients lose weight. This unsafe method was abandoned after a few patients died. DNP uncouples the chemiosmotic machinery by making the lipid bilayer of the inner mitochondrial membrane leaky to H^+. Explain how this causes weight loss.

SCIENCE, TECHNOLOGY, AND SOCIETY

1. Nearly all human societies use fermentation to produce alcoholic drinks such as beer and wine. The practice dates back to the earliest days of agriculture. How do you suppose this use of fermentation was first discovered? Why did wine prove to be a more useful beverage, especially to a preindustrial culture, than the grape juice from which it was made?

2. Trematol is a metabolic poison produced by white snakeroot, the plant in this photograph:

Cows that stray from pastures into wooded areas sometimes ingest white snakeroot, a common inhabitant of North American forests. Trematol becomes concentrated in the milk of these cows and may poison humans who drink the milk. In the early 1800s, "milk sickness" killed thousands of settlers in the U.S. Midwest, including Abraham Lincoln's mother, Nancy Hanks Lincoln, who died when the future president was only 7 years old. Milk sickness is very rare today because white snakeroot is removed from grazing areas, and commercial dairies mix the milk of many cows, a practice that dilutes any toxins present in one or a few cows. The biochemical basis of the disease is now understood: Trematol inhibits an enzyme that helps convert lactate to other compounds in the liver. Why does physical exertion intensify the symptoms of milk sickness? Why does the pH of blood decrease in a person afflicted with milk disease?

FURTHER READING

Becker, W. M., J. B. Reece, and M. F. Poenie. *The World of the Cell,* 3rd ed. Menlo Park, CA: Benjamin/Cummings, 1996. A particularly clear and comprehensive coverage of how cells harvest energy in Chapters 11 and 12.

Duffy, D. C. "Land of Milk and Poison" *Natural History,* July 1990. How nineteenth-century medical sleuths solved the mystery of milk sickness.

Harold, F. M. *The Vital Force: A Study of Bioenergetics.* New York: W. H. Freeman, 1986. An examination of the proton-motive force.

Mathews, C., and K. van Holde. *Biochemistry,* 2nd ed. Redwood City, CA: Benjamin/Cummings, 1996. Effective diagrams on catabolism.

Weiss, R. "Blazing Blossoms." *Science News,* June 24, 1989. How skunk cabbage and other plants use their catabolism to generate heat.

CHAPTER 10

PHOTOSYNTHESIS

KEY CONCEPTS

- Plants and other autotrophs are the producers of the biosphere
- Chloroplasts are the sites of photosynthesis in plants
- Evidence that chloroplasts split water molecules enabled researchers to track atoms through photosynthesis: *science as a process*
- The light reactions and the Calvin cycle cooperate in converting light energy to the chemical energy of food: *an overview*
- The light reactions convert solar energy to the chemical energy of ATP and NADPH: *a closer look*
- The Calvin cycle uses ATP and NADPH to convert CO_2 to sugar: *a closer look*
- Alternative mechanisms of carbon fixation have evolved in hot, arid climates
- Photosynthesis is the biosphere's metabolic foundation: *a review*

Life on Earth is solar-powered. The chloroplasts of plants capture light energy that has traveled 160 million kilometers from the sun and convert it to chemical energy stored in sugar and other organic molecules. The process is called ***photosynthesis.*** *In this chapter, you will learn how photosynthesis works. We begin by placing photosynthesis in its ecological context.*

Plants and other autotrophs are the producers of the biosphere

Photosynthesis nourishes almost all of the living world directly or indirectly. An organism acquires the organic compounds it uses for energy and carbon skeletons by one of two major modes: autotrophic or heterotrophic nutrition. At first, the term autotrophic (Gr. *autos,* "self," and *trophos,* "feed") may seem to contradict the principle that organisms are open systems, taking in resources from their environment. **Autotrophs** are not totally self-sufficient, however; they are self-feeders only in the sense that they sustain themselves without eating or decomposing other organisms. They make their organic molecules from inorganic raw materials obtained from the environment. It is for this reason that biologists refer to autotrophs as the *producers* of the biosphere.

Plants are autotrophs; the only nutrients they require are carbon dioxide from the air, and water and minerals from the soil. Specifically, plants are *photo*autotrophs, organisms that use light as a source of energy to synthesize carbohydrates, lipids, proteins, and other organic substances. Photosynthesis also occurs in algae, including certain protists, and in some prokaryotes (FIGURE 10.1). In this chapter, our emphasis will be on plants. Variations in photosynthesis that occur in algae and bacteria will be discussed in Unit Five. A much rarer form of self-feeding is unique to those bacteria that are *chemo*autotrophs. These organisms produce their organic compounds without the help of light, obtaining their energy by oxidizing inorganic substances, such as sulfur or ammonia. (We will postpone further discussion of this type of autotrophic nutrition until Chapter 25.)

Heterotrophs obtain their organic material by the second major mode of nutrition. Unable to make their own food, they live on compounds produced by other organisms; heterotrophs are the biosphere's *consumers.* The most obvious form of this "other-feeding" (*hetero* means "other, different") is when an animal eats plants or other animals. But heterotrophic nutrition may be more subtle. Some heterotrophs do not kill prey, but instead de-

FIGURE 10.1
Photoautotrophs: producers for most ecosystems. These organisms use light energy to drive the synthesis of organic molecules from carbon dioxide and (usually) water. They feed not only themselves, but the entire living world. (**a**) On land, plants are the predominant producers of food. Three major groups of land plants—mosses, ferns, and flowering plants—are represented in this scene. In oceans, ponds, lakes, and other aquatic environments, photosynthetic organisms include (**b**) multicellular algae, such as kelp; (**c**) some unicellular protists, such as *Euglena;* (**d**) the prokaryotes called cyanobacteria; and (**e**) other photosynthetic prokaryotes, such as these purple sulfur bacteria (c, d, e: LMs).

compose and feed on organic litter—such as carcasses, feces, and fallen leaves—and thus are known as decomposers. Most fungi and many types of bacteria get their nourishment this way. Almost all heterotrophs, including humans, are completely dependent on photoautotrophs for food, and also for oxygen, a by-product of photosynthesis. Thus, we can trace the food we eat and the oxygen we breathe to the chloroplast.

Chloroplasts are the sites of photosynthesis in plants

All green parts of a plant, including green stems and unripened fruit, have chloroplasts, but the leaves are the major sites of photosynthesis in most plants (FIGURE 10.2). There are about half a million chloroplasts per square millimeter of leaf surface. The color of the leaf is from **chlorophyll,** the green pigment located within the chloroplasts. It is the light energy absorbed by chlorophyll that drives the synthesis of food molecules in the chloroplast. Chloroplasts are found mainly in the cells of the **mesophyll,** the tissue in the interior of the leaf. Carbon dioxide enters the leaf, and oxygen exits, by way of microscopic pores called **stomata** (singular, **stoma;** Gr. "mouth"). Water absorbed by the roots is delivered to the leaves in veins. Leaves also use veins to export sugar to roots and other nonphotosynthetic parts of the plant.

A typical mesophyll cell has about 30 to 40 chloroplasts, each a lens-shaped organelle measuring about 2–4 μm by 4–7 μm. An envelope of two membranes encloses the stroma, the dense fluid within the chloroplast. An elaborate system of interconnected thylakoid membranes segregates the stroma from another compartment, the thylakoid space. In some places, thylakoid sacs are layered in dense stacks called grana. Chlorophyll resides in the thylakoid membranes. (Photosynthetic prokaryotes lack chloroplasts, but, as you will see in Chapter 25, they do have membranes that function in a manner similar to the thylakoid membranes of chloroplasts.) Now that we have identified chloroplasts as the sites of photosynthesis in plants, we are ready to see how these organelles convert the light energy absorbed by chlorophyll to chemical energy.

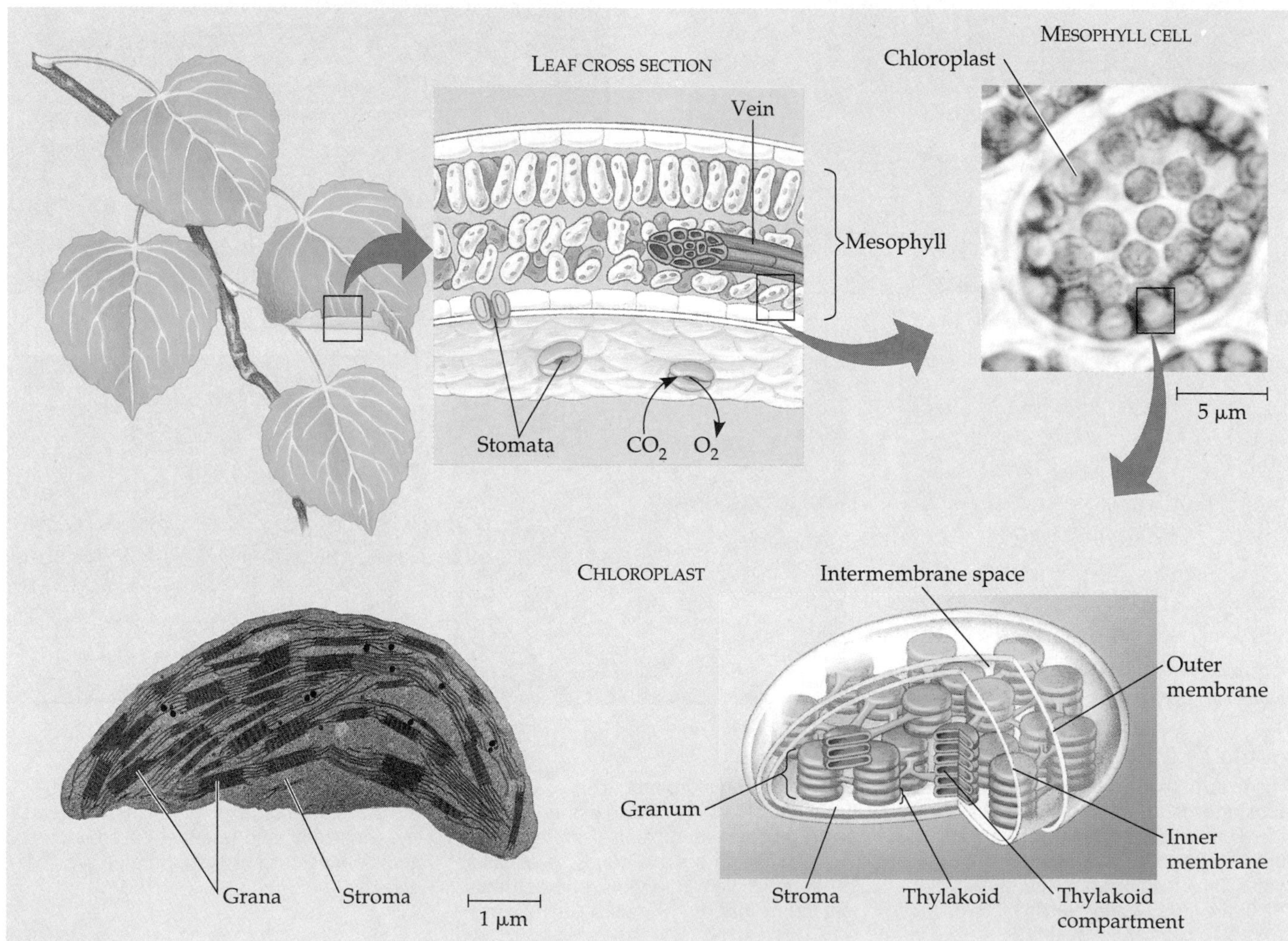

FIGURE 10.2

The site of photosynthesis in a plant. Leaves are the major organs of photosynthesis in plants. Gas exchange between the mesophyll and the atmosphere occurs through microscopic pores called stomata. Chloroplasts, found mainly in the mesophyll, are bounded by a double membrane that encloses the stroma, the dense fluid content of the chloroplast. Membranes of the thylakoid system separate the stroma from the thylakoid space. Thylakoids are concentrated in stacks called grana. (Top right, LM; bottom left, TEM.)

Evidence that chloroplasts split water molecules enabled researchers to track atoms through photosynthesis: *science as a process*

Scientists have tried for centuries to piece together the process by which plants make food. Although some of the steps are still not understood, the overall photosynthetic equation has been known since the 1800s: In the presence of light, the green parts of plants produce organic material and oxygen from carbon dioxide and water. Using molecular formulas, we can summarize photosynthesis with this chemical equation:

$$6\,CO_2 + 12\,H_2O + \text{Light energy} \longrightarrow C_6H_{12}O_6 + 6\,O_2 + 6\,H_2O$$

The carbohydrate $C_6H_{12}O_6$ is glucose.* Water appears on both sides of the equation, because 12 molecules are consumed and 6 molecules are newly formed during photosynthesis. We can simplify the equation by indicating the net consumption of water:

$$6\,CO_2 + 6\,H_2O + \text{Light energy} \longrightarrow C_6H_{12}O_6 + 6\,O_2$$

Writing the equation in this form, we can see that the chemical change during photosynthesis is the reverse of the one that occurs during cellular respiration. Both of these metabolic processes occur in plant cells. However, it will soon become apparent that plants do not make food by simply reversing the steps of respiration.

Now let's divide the photosynthetic equation by 6 to put it in its simplest possible form:

$$CO_2 + H_2O \longrightarrow CH_2O + O_2$$

Here, CH_2O is not a true sugar but symbolizes the general formula for a carbohydrate. In other words, we are imagining the synthesis of a sugar molecule one carbon at a time. Six repetitions would produce a glucose molecule. Let's now use this simplified formula to see how researchers tracked the chemical elements (C, H, and O) from the reactants of photosynthesis to the products.

The Splitting of Water

One of the first clues to the mechanism of photosynthesis came from the discovery that the oxygen given off by plants is derived from water and not from carbon dioxide. The chloroplast splits water into hydrogen and oxygen. Before this discovery, the prevailing hypothesis was that photosynthesis split carbon dioxide and then added water to the carbon:

$$\text{Step 1: } CO_2 \longrightarrow C + O_2$$
$$\text{Step 2: } C + H_2O \longrightarrow CH_2O$$

*The direct product of photosynthesis is actually a three-carbon sugar. Glucose is used here only to simplify the relationship between photosynthesis and respiration.

This hypothesis predicted that the O_2 released during photosynthesis came from CO_2. The idea was challenged in the 1930s by C. B. van Niel of Stanford University. Van Niel was investigating photosynthesis in bacteria that make their carbohydrate from CO_2 but do not release O_2. Van Niel concluded that, at least in these bacteria, CO_2 is not split into carbon and oxygen. One group of bacteria required hydrogen sulfide (H_2S) rather than water for photosynthesis, forming yellow globules of sulfur as a waste product (these globules are visible in FIGURE 10.1e). Here is the chemical equation:

$$CO_2 + 2\,H_2S \longrightarrow CH_2O + H_2O + 2\,S$$

Van Niel reasoned that the bacteria split H_2S and used the hydrogen to make sugar. He generalized that all photosynthetic organisms require a hydrogen source, but that the source varies:

$$\text{General: } CO_2 + 2\,H_2X \longrightarrow CH_2O + H_2O + 2\,X$$
$$\text{Sulfur bacteria: } CO_2 + 2\,H_2S \longrightarrow CH_2O + H_2O + 2\,S$$
$$\text{Plants: } CO_2 + 2\,H_2O \longrightarrow CH_2O + H_2O + O_2$$

Thus, van Niel hypothesized that plants split water as a source of hydrogen, releasing oxygen as a by-product.

Nearly 20 years later, scientists confirmed van Niel's hypothesis by using oxygen-18 (^{18}O), a heavy isotope, as a tracer to follow the fate of oxygen atoms during photosynthesis. The O_2 that came from plants was labeled with ^{18}O *only* if water was the source of the tracer. If the ^{18}O was introduced to the plant in the form of CO_2, the label did not turn up in the released O_2. In the following summary of these experiments, red denotes labeled atoms of oxygen:

$$\text{Experiment 1: } CO_2 + 2\,H_2O \longrightarrow CH_2O + H_2O + O_2$$
$$\text{Experiment 2: } CO_2 + 2\,H_2O \longrightarrow CH_2O + H_2O + O_2$$

The most important result of the shuffling of atoms during photosynthesis is the extraction of hydrogen from water and its incorporation into sugar. The waste product of photosynthesis, O_2, restores the atmospheric oxygen consumed during cellular respiration. FIGURE 10.3 traces the fate of all atoms during photosynthesis.

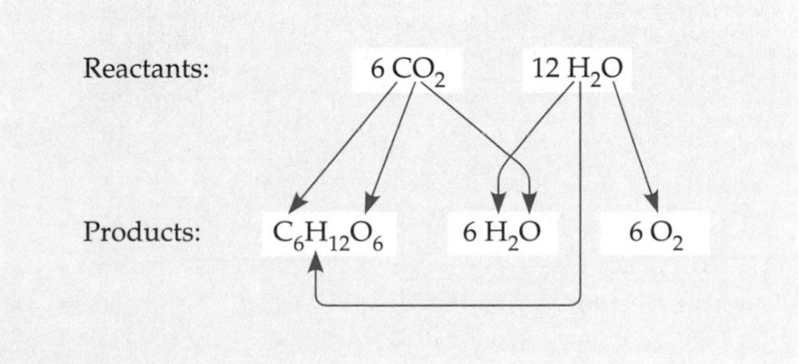

FIGURE 10.3
Tracking atoms through photosynthesis.

Photosynthesis as a Redox Process

Let's briefly contrast photosynthesis with cellular respiration. During respiration, energy is released from sugar when electrons associated with hydrogen are transported by carriers to oxygen, forming water as a by-product. The electrons lose potential energy as electronegative oxygen pulls them down the electron transport chain, and the mitochondrion uses the energy to synthesize ATP. Photosynthesis, also a redox process, reverses the direction of electron flow. Water is split, and electrons are transferred along with hydrogen ions from the water to carbon dioxide, reducing it to sugar. Since electrons cannot travel downhill in both directions, they must increase their potential energy when moved from water to sugar. The energy boost is provided by light.

The light reactions and the Calvin cycle cooperate in converting light energy to the chemical energy of food: *an overview*

The equation for photosynthesis is a deceptively simple summary of a very complex process. Actually, photosynthesis is not a single process, but two, each with multiple steps. These two stages of photosynthesis are known as the **light reactions** (the *photo* part of photosynthesis) and the **Calvin cycle** (the *synthesis* part) (FIGURE 10.4).

The light reactions are the steps of photosynthesis that convert solar energy to chemical energy. Light absorbed by chlorophyll drives a transfer of electrons and hydrogen from water to an acceptor called **NADP$^+$** (nicotinamide adenine dinucleotide phosphate), which temporarily stores the energized electrons. Water is split in the process, and thus it is the light reactions of photosynthesis that give off O_2 as a by-product. The electron acceptor of the light reactions, NADP$^+$, is first cousin to NAD$^+$, which functions as an electron carrier in cellular respiration; the two molecules differ only by the presence of an extra phosphate group in the NADP$^+$ molecule. The light reactions use solar power to reduce NADP$^+$ to NADPH by adding a pair of electrons along with a hydrogen nucleus, or H$^+$. The light reactions also generate ATP by powering the addition of a phosphate group to ADP, a process called **photophosphorylation.** Thus, light energy is initially converted to chemical energy in the form of two compounds: NADPH, a source of energized electrons ("reducing power"); and ATP, the versatile energy currency of cells. Notice that the light reactions produce no sugar; that happens in the second stage of photosynthesis, the Calvin cycle.

The Calvin cycle is named for Melvin Calvin, who began to elucidate its steps along with his colleagues in the late 1940s. The cycle begins by incorporating CO_2 from the air into organic molecules already present in the chloroplast. This initial incorporation of carbon into organic compounds is known as **carbon fixation.** The Calvin cycle then reduces the fixed carbon to carbohydrate by the addition of electrons. The reducing power is provided by NADPH, which acquired energized electrons in the light reactions. To convert CO_2 to carbohydrate, the

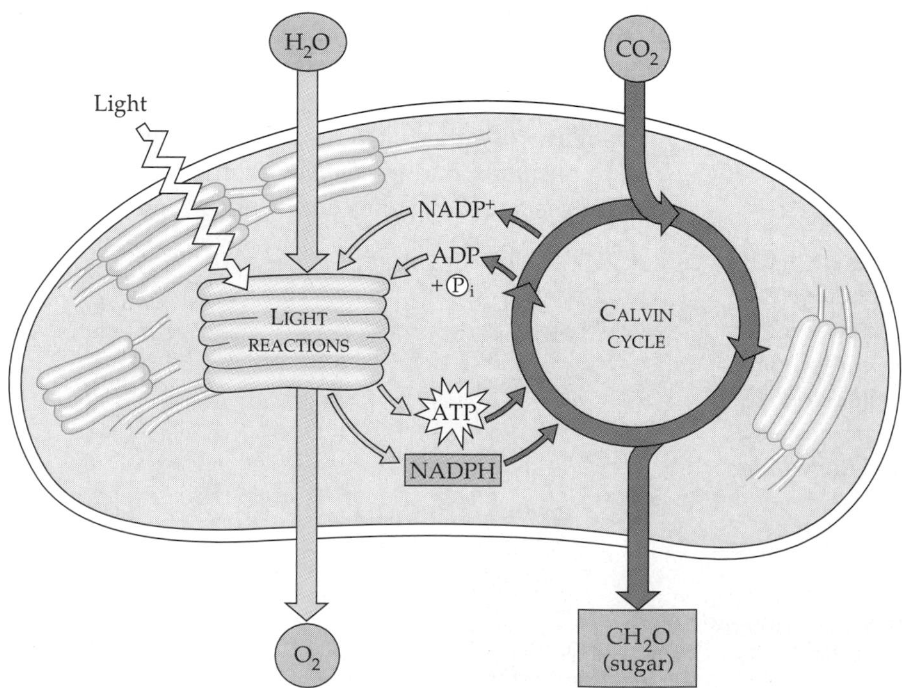

FIGURE 10.4
An overview of photosynthesis: cooperation of the light reactions and the Calvin cycle. The light reactions use solar energy to make ATP and NADPH, which function as chemical energy and reducing power, respectively, in the Calvin cycle. (Notice that in contrast to ATP generated by cellular respiration, ATP produced in the light reactions of photosynthesis is usually dedicated to a single kind of cellular work, driving the Calvin cycle.) The Calvin cycle incorporates CO_2 into organic molecules. Thylakoid membranes, especially those of the grana, are the sites of the light reactions, whereas the Calvin cycle occurs in the stroma.

A smaller version of this diagram will reappear in several subsequent figures as a reminder of whether the events being described occur in the light reactions or in the Calvin cycle.

Calvin cycle also requires chemical energy in the form of ATP, which is also generated by the light reactions. Thus, it is the Calvin cycle that makes sugar, but it can do so only with the help of the NADPH and ATP produced by the light reactions. The metabolic steps of the Calvin cycle are sometimes referred to as the dark reactions, or light-independent reactions, because none of the steps requires light *directly.* Nevertheless, the Calvin cycle in most plants occurs during daylight, for only then can the light reactions regenerate the NADPH and ATP spent in the reduction of CO_2 to sugar. In essence, the chloroplast uses light energy to make sugar by coordinating the two stages of photosynthesis.

As FIGURE 10.4 shows, the thylakoids of the chloroplast are the sites of the light reactions, while the Calvin cycle occurs in the stroma. As molecules of $NADP^+$ and ADP bump into the thylakoid membrane, they pick up electrons and phosphate, respectively, and then transfer their high-energy cargo to the Calvin cycle. The two stages of photosynthesis are treated in this figure as metabolic modules that take in ingredients and crank out products. Our next step toward understanding photosynthesis is to look more closely at how the two stages work, beginning with the light reactions.

The light reactions convert solar energy to the chemical energy of ATP and NADPH: *a closer look*

Chloroplasts are chemical factories powered by the sun. Their thylakoids transform light energy into the chemical energy of ATP and NADPH. To understand this conversion better, it is first necessary to learn some important properties of light.

The Nature of Sunlight

Light is a form of energy known as electromagnetic energy, also called radiation. Electromagnetic energy travels in rhythmic waves analogous to those created by dropping a pebble into a puddle of water. Electromagnetic waves, however, are disturbances of electrical and magnetic fields rather than disturbances of a material medium such as water.

The distance between the crests of electromagnetic waves is called the **wavelength.** Wavelengths range from less than a nanometer (for gamma rays) to more than a kilometer (for radio waves). This entire range of radiation is known as the **electromagnetic spectrum** (FIGURE 10.5). The segment most important to life is the narrow band that ranges from about 380 to 750 nm in wavelength. This radiation is known as **visible light,** because it is detected as various colors by the human eye.

The model of light as waves explains many of light's properties, but in certain respects, light behaves as though it consists of discrete particles called **photons.** Photons are not tangible objects, but they act like objects in that each of them has a fixed quantity of energy. The amount of energy is inversely related to the wavelength of the light; the shorter the wavelength, the greater the energy of each photon of that light. Thus, a photon of violet light packs nearly twice as much energy as a photon of red light.

Although the sun radiates the full spectrum of electromagnetic energy, the atmosphere acts like a selective window, allowing visible light to pass through while screening out a substantial fraction of other radiation. The same part of the spectrum we can see—visible light—is also the radiation that drives photosynthesis. Blue and red, the two wavelengths most effectively absorbed by chlorophyll, are the colors most useful as energy for the light reactions.

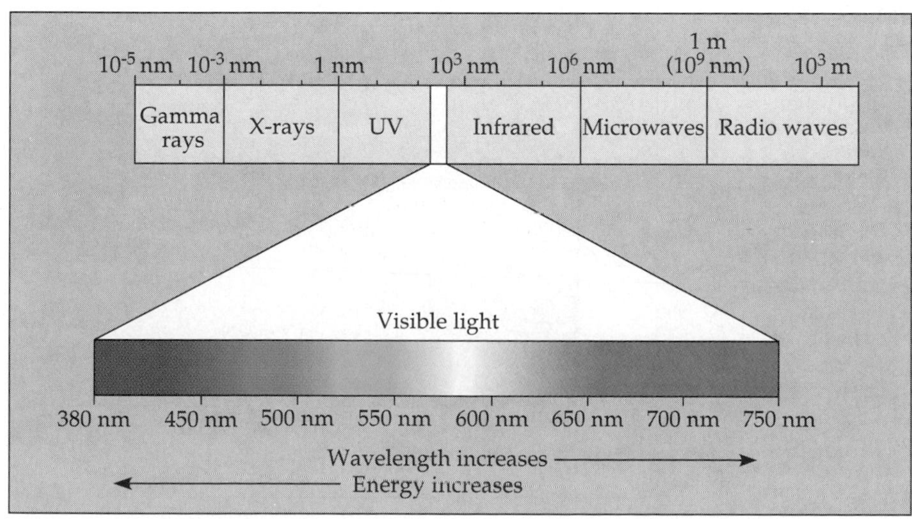

FIGURE 10.5

The electromagnetic spectrum. Visible light and other forms of electromagnetic energy radiate through space as waves of various lengths. We perceive different wavelengths of visible light as different colors. White light is a mixture of wavelengths. A prism can sort white light into its component colors by bending light of different wavelengths varying degrees. Visible light drives photosynthesis.

Photosynthetic Pigments: The Light Receptors

As light meets matter, it may be reflected, transmitted, or absorbed (FIGURE 10.6). Substances that absorb visible light are called pigments. Different pigments absorb light of different wavelengths, and the wavelengths that are absorbed disappear. If a pigment is illuminated with white light, the color we see is the color most reflected or transmitted by the pigment. (If a pigment absorbs all wavelengths, it appears black.) We see green when we look at a leaf, because chlorophyll absorbs red and blue light while transmitting and reflecting green light. The ability of a pigment to absorb various wavelengths of light can be measured by placing a solution of the pigment in a **spectrophotometer** (see the Methods Box). A graph plotting a pigment's light absorption versus wavelength is called an **absorption spectrum.**

FIGURE 10.7a compares the absorption spectrum of a type of chlorophyll called **chlorophyll *a*** to other pig-

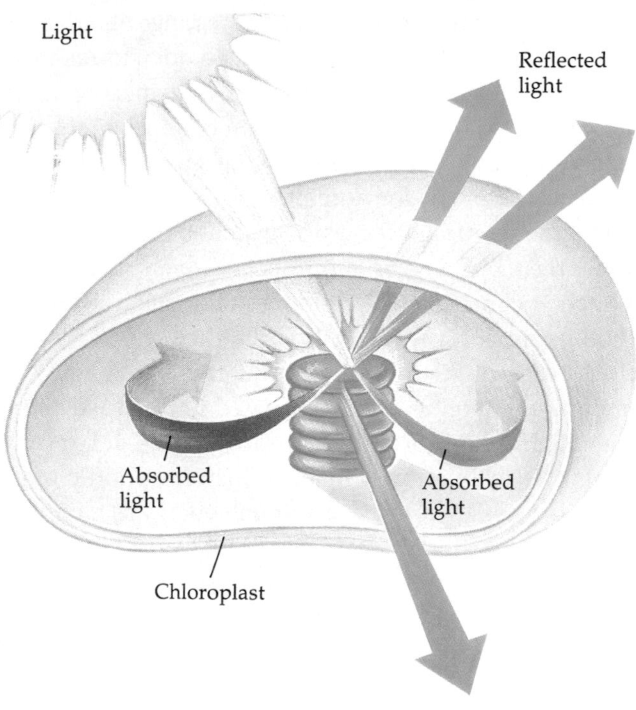

FIGURE 10.6
Interactions of light with matter. The pigments of chloroplasts absorb blue and red light, the colors most effective in photosynthesis. The pigments reflect or transmit green light, which is why leaves appear green.

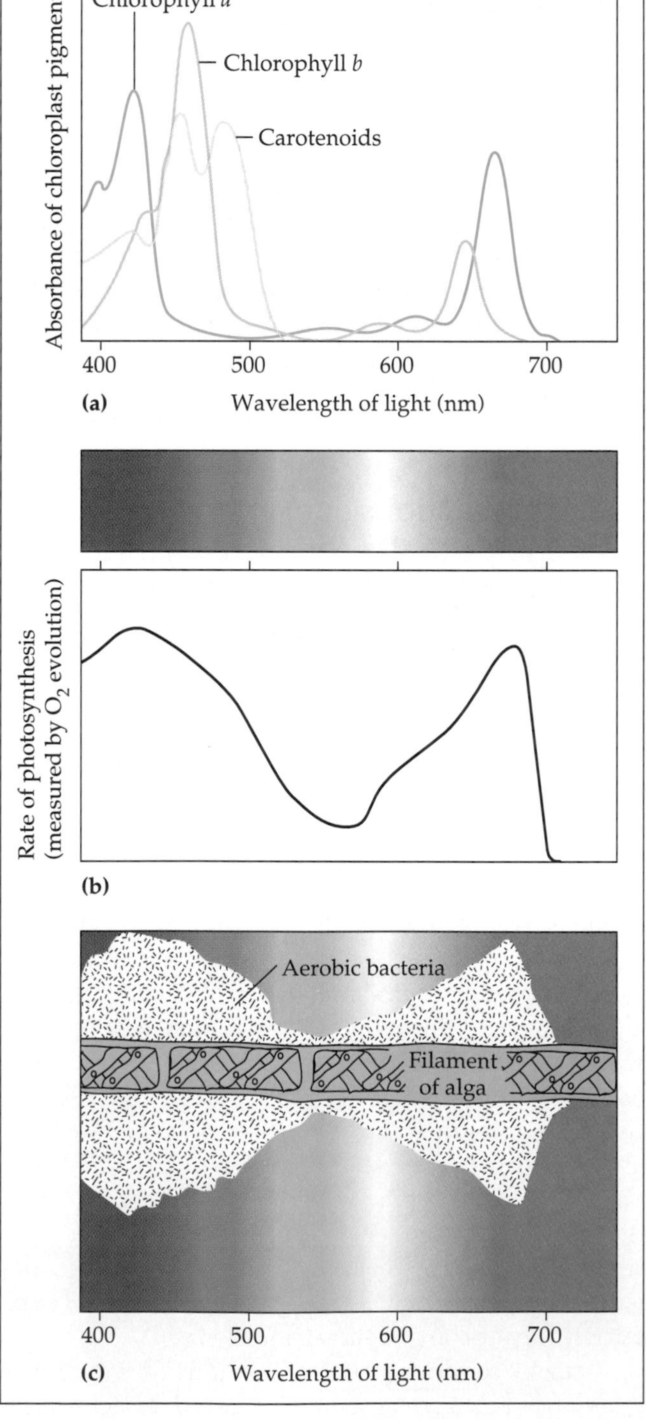

FIGURE 10.7
Absorption and action spectra for photosynthesis. (**a**) A comparison of the absorption spectra for chlorophyll *a* and accessory pigments extracted from chloroplasts. (**b**) An action spectrum, profiling the effectiveness of different wavelengths of light in driving photosynthesis. Compared to the peaks in the absorption spectrum for chlorophyll *a* (blue line), the peaks in the action spectrum are broader, and the valley is narrower and not as deep. This is partly due to the absorption of light by accessory pigments, which broaden the spectrum of colors that can be used for photosynthesis. (**c**) An elegant experiment demonstrating the action spectrum for photosynthesis was first performed in 1883 by Thomas Engelmann, a German botanist. He illuminated a filamentous alga with light that had been passed through a prism, thus exposing different segments of the alga to different wavelengths of light. Engelmann used aerobic bacteria, which concentrate near an oxygen source, to determine which segments of the alga were releasing the most O_2. Bacteria congregated in greatest density around the parts of the alga illuminated with red and blue light. Notice the close match of the bacterial distribution to the action spectrum in part b.

METHODS: DETERMINING AN ABSORPTION SPECTRUM

Spectrophotometers are among the most widely used research instruments in biology. A spectrophotometer measures the proportions of light of different wavelengths absorbed and transmitted by a pigment solution. Inside the spectrophotometer, white light is separated into colors (wavelengths) by a prism. Then, one by one, the different colors of light are passed through the sample. The transmitted light strikes a photoelectric tube, which converts the light energy to electricity, and the electrical current is measured by a meter. Each time the wavelength of light is changed, the meter indicates the proportion of light transmitted through the sample or, conversely, the proportion of light absorbed. A graph that profiles absorbance at different wavelengths is called an absorption spectrum. For example, the absorption spectrum for chlorophyll *a*, the form of chlorophyll most important in photosynthesis, has two peaks, corresponding to blue and red light. These are the colors chlorophyll *a* absorbs best (see FIGURE 10.7a). The absorption spectrum has a valley in the green region because the pigment transmits that color.

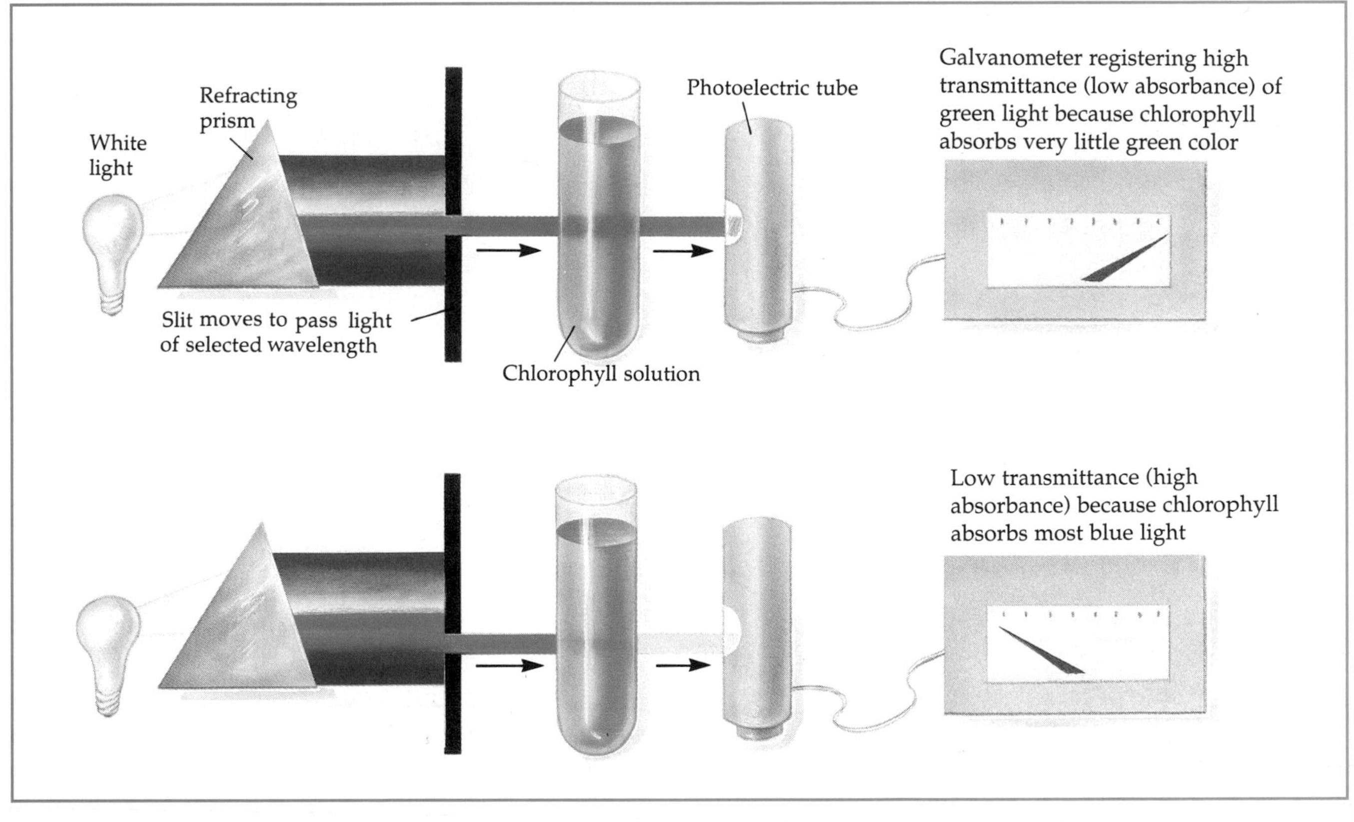

ments in the chloroplast. The absorption spectrum for chlorophyll *a* provides clues to the relative effectiveness of different wavelengths for driving photosynthesis, since light can perform work in chloroplasts only if it is absorbed. As previously mentioned, blue and red light work best for photosynthesis, while green is the least effective color. An **action spectrum** (FIGURE 10.7b) profiles the relative performance of the different wavelengths more accurately than an absorption spectrum. An action spectrum is prepared by illuminating chloroplasts with different colors of light and then plotting wavelength against some measure of photosynthetic rate, such as oxygen release or carbon dioxide consumption (FIGURE 10.7c).

Notice by comparing FIGURE 10.7a to FIGURE 10.7b that the action spectrum for photosynthesis does not exactly match the absorption spectrum of chlorophyll *a*. The absorption spectrum underestimates the effectiveness of certain wavelengths in driving photosynthesis. This is partly because chlorophyll *a* is not the only pigment in chloroplasts important to photosynthesis. Only chlorophyll *a* can participate directly in the light reactions, which convert solar energy to chemical energy. But other pigments can absorb light and transfer the energy to chlorophyll *a*, which then initiates the light reactions. One of these accessory pigments is another form of chlorophyll, **chlorophyll *b*.** Chlorophyll *b* is almost identical to chlorophyll *a* (FIGURE 10.8), but the slight structural difference between them is enough to give the two pigments slightly different absorption spectra and hence different colors. Chlorophyll *a* is blue-green, whereas chlorophyll *b* is

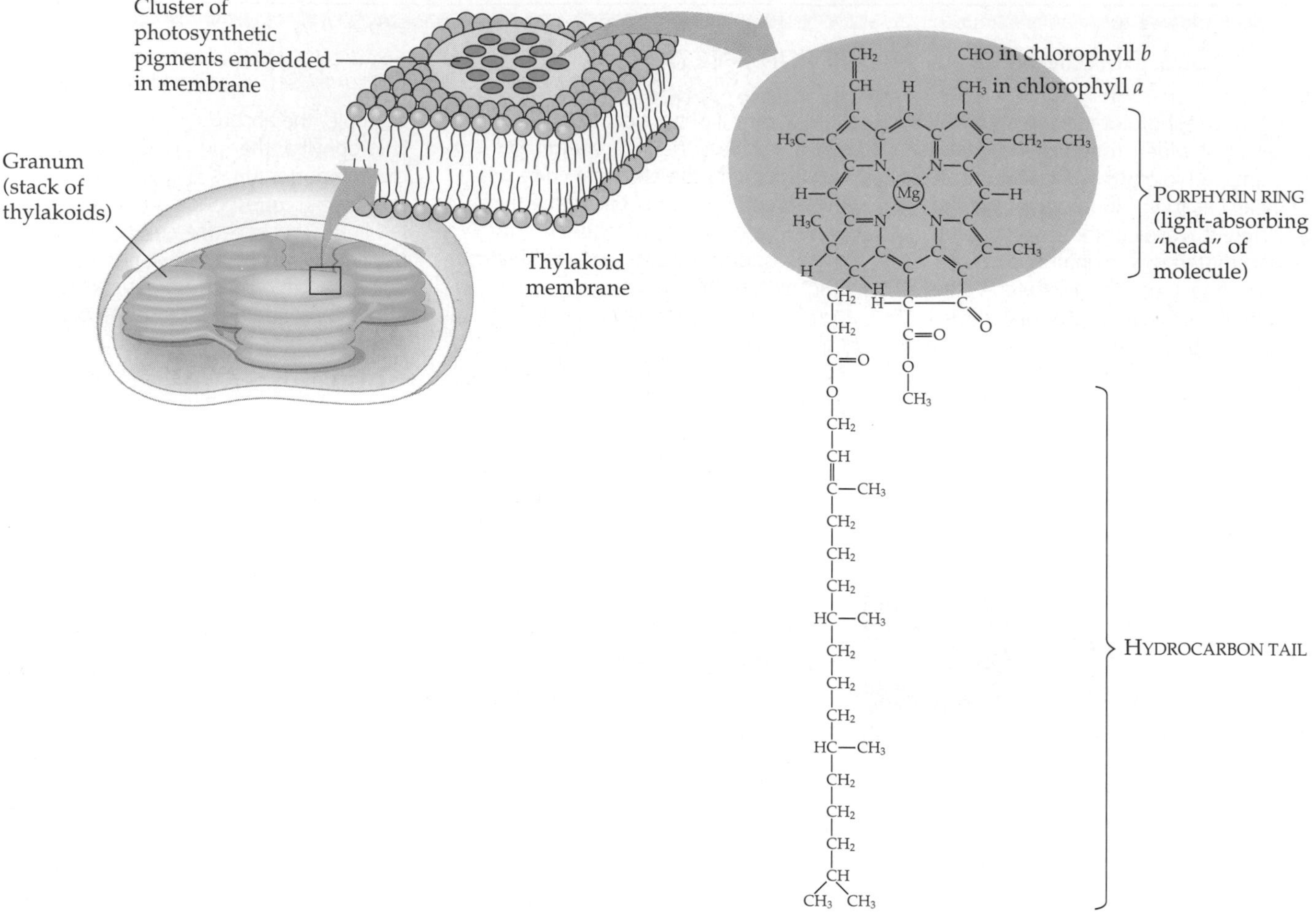

FIGURE 10.8

Chlorophyll. Chlorophyll *a,* the pigment that participates directly in the light reactions of photosynthesis, has a "head" called a porphyrin ring with a magnesium atom at its center. Attached to the porphyrin is a hydrocarbon tail, which interacts with hydrophobic regions of proteins in the thylakoid membrane. Chlorophyll *b* differs from chlorophyll *a* only in one of the functional groups bonded to the porphyrin. This diagram simplifies by placing chlorophyll at the surface of the membrane; most of the molecule is actually immersed in the hydrophobic core of the membrane.

yellow-green. If a photon of sunlight is absorbed by chlorophyll *b,* energy is conveyed to chlorophyll *a,* which then behaves just as though it had absorbed the photon. The chloroplast also has a family of accessory pigments called **carotenoids,** which are various shades of yellow and orange (see FIGURE 10.7a). These hydrocarbons are built into the thylakoid membrane along with the two kinds of chlorophyll. Carotenoids can absorb wavelengths of light that chlorophyll cannot, and this may broaden the spectrum of colors that can drive photosynthesis. However, excessive light intensity can damage chlorophyll. Instead of transmitting energy *to* chlorophyll, some carotenoids can accept energy *from* chlorophyll, thus providing a function known as photoprotection.

Photoexcitation of Chlorophyll

What exactly happens when chlorophyll and other pigments absorb photons? The colors corresponding to the absorbed wavelengths disappear from the spectrum of the transmitted and reflected light, but energy cannot disappear. When a molecule absorbs a photon, one of the molecule's electrons is elevated to an orbital where it has more potential energy. When the electron is in its normal orbital, the pigment molecule is said to be in its ground state. After absorption of a photon boosts an electron to an orbital of higher energy, the pigment molecule is said to be in an excited state. The only photons absorbed are those whose energy is exactly equal to the energy difference between the ground state and an excited state, and this energy difference varies from one kind of atom or molecule to another. Thus, a particular compound absorbs only photons corresponding to specific wavelengths, which is why each pigment has a unique absorption spectrum.

The energy of an absorbed photon is converted to the potential energy of an electron raised from the ground state to an excited state. But the electron cannot remain

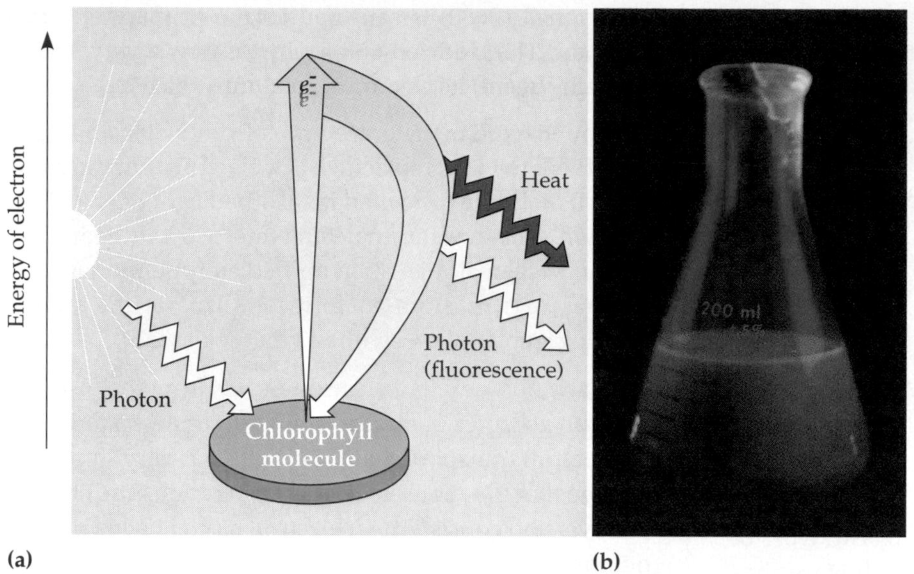

FIGURE 10.9
Photoexcitation of isolated chlorophyll. (**a**) Absorption of a photon causes a transition of the chlorophyll molecule from its ground state to its excited state. The photon boosts an electron to an orbital where it has more potential energy. If isolated chlorophyll is illuminated, its excited electron immediately drops back down to the ground-state orbital, giving off its excess energy as heat and fluorescence (light). (**b**) A chlorophyll solution excited with ultraviolet light will fluoresce, giving off a red-orange glow.

there long; the excited state, like all high-energy states, is unstable. Generally, when pigments absorb light, their excited electrons drop back down to the ground-state orbital in a billionth of a second, releasing their excess energy as heat. This conversion of light energy to heat is what makes the top of an automobile so hot on a sunny day. (White cars are coolest because their paint reflects all wavelengths of visible light, although it may absorb ultraviolet and other invisible radiation.) Some pigments, including chlorophyll, emit light as well as heat after absorbing photons. The electron jumps to a state of greater energy, and as it falls back to ground state, a photon is given off. This afterglow is called fluorescence. If a solution of chlorophyll isolated from chloroplasts is illuminated, it will fluoresce in the red part of the spectrum and also give off heat (FIGURE 10.9).

Photosystems: Light-Harvesting Complexes of the Thylakoid Membrane

In contrast to the behavior of isolated chlorophyll when it absorbs light energy, the excitation of chlorophyll in intact chloroplasts produces very different results. In its native environment of the thylakoid membrane, chlorophyll is organized along with other molecules into **photosystems.**

A photosystem has a light-gathering "antenna complex" consisting of a cluster of a few hundred chlorophyl *a*, chlorophyll *b*, and carotenoid molecules (FIGURE 10.10). The number and variety of pigment molecules enable a photosystem to harvest light over a larger surface and larger portion of the spectrum than any single pigment molecule could harvest. When any antenna molecule absorbs a photon, the energy is transmitted from pigment molecule to pigment molecule until it reaches a particular chlorophyll *a*. What is special about *this* chlorophyll *a* molecule is not its molecular structure, but its position. Only this chlorophyll molecule is located in the region of the photosystem called the **reaction center,** where the first light-driven chemical reaction of photosynthesis occurs.

Sharing the reaction center with the chlorophyll *a* molecule is a specialized molecule called the **primary electron acceptor.** In an oxidation-reduction reaction, the chlorophyll *a* molecule at the reaction center loses one of its electrons to the primary electron acceptor. This

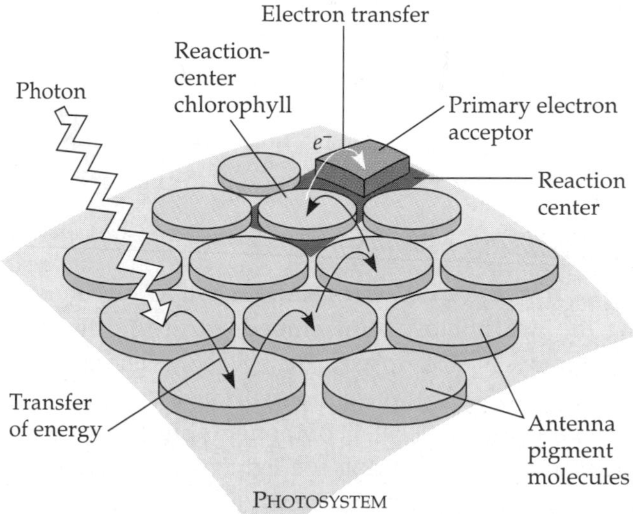

FIGURE 10.10
How a photosystem harvests light. Photosystems are the light-harvesting units of the thylakoid membrane. Each photosystem has an antenna of a few hundred pigment molecules. When a photon strikes a pigment molecule, the energy is passed from molecule to molecule until it reaches the reaction center. At the reaction center, the energy drives an oxidation-reduction reaction. An excited electron from the reaction-center chlorophyll is captured by a specialized molecule called the primary electron acceptor.

redox reaction occurs when light excites the electron to a higher energy level in chlorophyll and the electron acceptor traps the high-energy electron before it can return to the ground state in the cholorophyll molecule. Isolated chlorophyll fluoresces because there is no electron acceptor to prevent electrons of photoexcited chlorophyll from dropping right back to the ground state. In a chloroplast, the acceptor molecule functions like a dam that prevents this immediate plunge of high-energy electrons back to the ground state. Thus, each photosystem—reaction-center chlorophyll and electron acceptor surrounded by an antenna complex—functions in the chloroplast as a light-harvesting unit. The solar-powered transfer of electrons from chlorophyll to the primary electron acceptor is the first step of the light reactions.

The thylakoid membrane is populated by two types of photosystems that cooperate in the light reactions of photosynthesis. They are called **photosystem I** and **photosystem II,** in order of their discovery. Each has a characteristic reaction center—a particular kind of primary electron acceptor next to a chlorophyll *a* molecule associated with specific proteins. The reaction-center chlorophyll of photosystem I is known as P700 because this pigment is best at absobing light having a wavelength of 700 nm (the far-red part of the spectrum). The chlorophyll at the reaction center of photosystem II is called P680 because its absorption spectrum has a peak of 680 nm (also in the red part of the spectrum). These two pigments, P700 and P680, are actually identical chlorophyll *a* molecules. However, their association with different proteins in the thylakoid membrane affects the electron distribution in the chlorophyll molecules and accounts for the slight differences in light-absorbing properties. Let's now see how the two photosystems work together in using light energy to generate ATP and NADPH, the two main products of the light reactions.

Noncyclic Electron Flow

Light drives the synthesis of NADPH and ATP by energizing the two photosystems embedded in the thylakoid membranes of chloroplasts. The key to this energy transformation is a flow of electrons through the photosystems and other molecular components built into the thylakoid membrane. During the light reactions there are two possible routes for electron flow: cyclic and noncyclic. In **noncyclic electron flow,** which predominates during the light reactions of photosynthesis, electrons ejected from chlorophyll molecules do not cycle back to the ground state in chlorophyll. FIGURE 10.11 tracks the path of noncyclic electron flow, with the numbers below corresponding to the numbered steps in the figure.

① When photosystem II absorbs light, an electron excited to a higher energy level in the reaction-center chlorophyll (P680) is captured by the primary electron acceptor. The oxidized chlorophyll is now a very strong oxidizing agent; its electron "hole" must be filled.

② An enzyme extracts electrons from water and supplies them to P680, replacing each electron that the chlorophyll molecule lost when it absorbed light energy. This reaction splits a water molecule into two hydrogen ions and an oxygen atom, which immediately combines with another oxygen atom to form O_2. This is the water-splitting step of photosynthesis that releases O_2.

③ Each photoexcited electron passes from the primary electron acceptor of photosystem II to photosystem I via an electron transport chain. This chain is very similar to the one that functions in cellular respiration. The chloroplast version consists of an electron carrier called plastoquinone (Pq); a complex of two cytochromes (closely related to the cytochromes of mitochondria); and a copper-containing protein called plastocyanin (Pc).

④ As electrons cascade down the chain, their exergonic "fall" to a lower energy level is harnessed by the thylakoid membrane to produce ATP. This ATP synthesis is called photophosphorylation because it is driven by light energy. Specifically, the production of ATP during noncyclic electron flow is called **noncyclic photophosphorylation.** (The mechanism for photophosphorylation is chemiosmosis, the same basic process that generates ATP during respiration. We'll soon examine chemiosmosis in chloroplasts.) This ATP generated by the light reactions will provide chemical energy for the synthesis of sugar during the Calvin cycle, the second major stage of photosynthesis.

⑤ When an electron reaches the "bottom" of the electron transport chain, it fills an electron "hole" in P700, the chlorophyll *a* molecule in the reaction center of photosystem I. This replaces the electron that light energy drives from the chlorophyll to the primary electron acceptor of photosystem I.

⑥ The primary electron acceptor of photosystem I passes the photoexcited electrons to ferredoxin (Fd), an iron-containing protein. An enzyme called $NADP^+$ reductase then transfers the electrons from Fd to $NADP^+$. This is the redox reaction that stores the high-energy electrons in NADPH, the molecule that will provide reducing power for the synthesis of sugar in the Calvin cycle.

The energy changes of electrons as they flow through the light reactions is analogous to the cartoon in FIGURE 10.12. As complicated as the scheme is, do not lose track of its functions: The light reactions use solar power to generate ATP and NADPH, which provide chemical energy and reducing power, respectively, to the sugar-making reactions of the Calvin cycle.

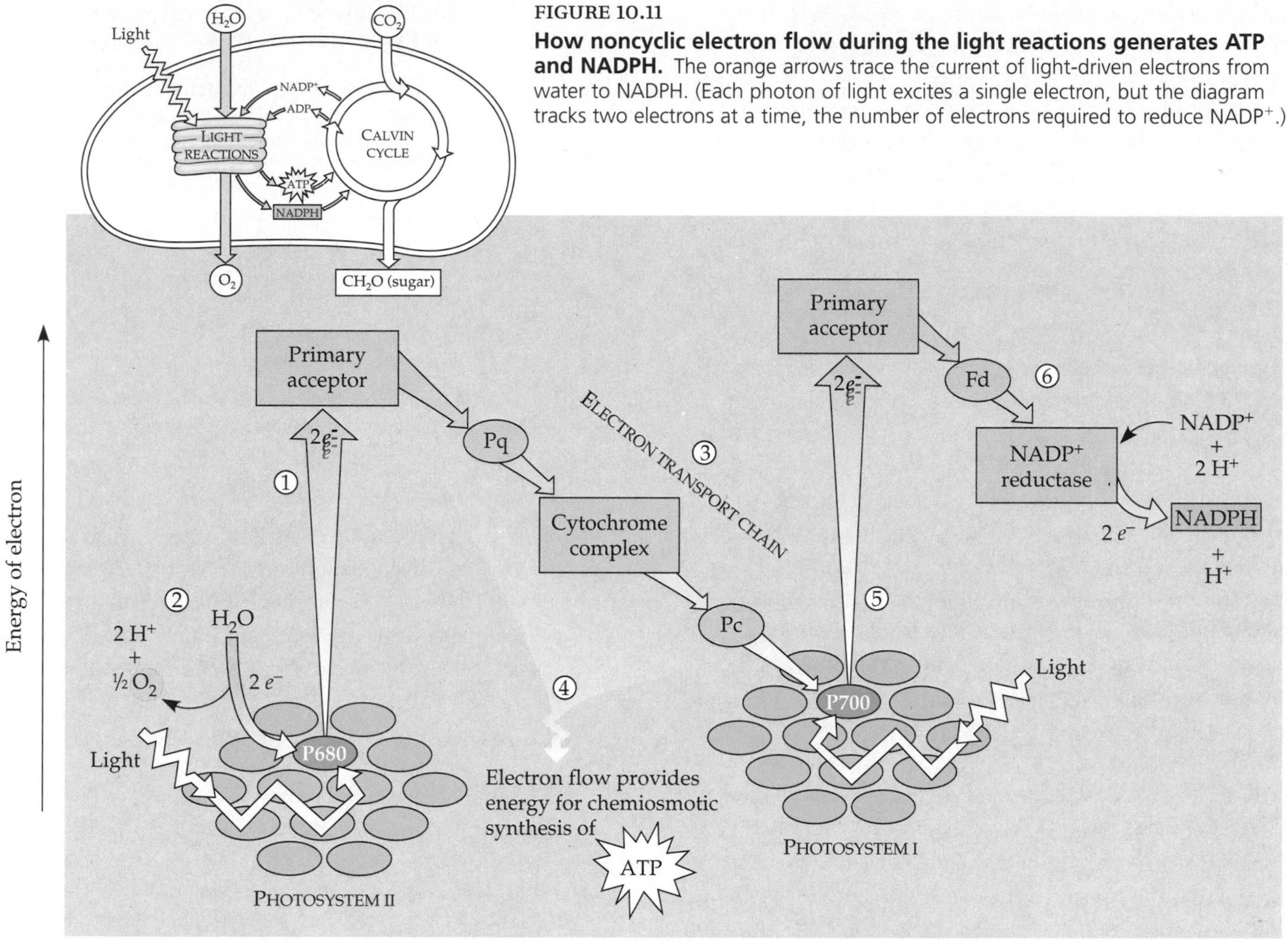

FIGURE 10.11
How noncyclic electron flow during the light reactions generates ATP and NADPH. The orange arrows trace the current of light-driven electrons from water to NADPH. (Each photon of light excites a single electron, but the diagram tracks two electrons at a time, the number of electrons required to reduce $NADP^+$.)

FIGURE 10.12
A mechanical analogy of the light reactions.

Cyclic Electron Flow

Under certain conditions, photoexcited electrons take an alternative path called **cyclic electron flow,** which uses photosystem I but not photosystem II. You can see in FIGURE 10.13 that cyclic flow is a short circuit: the electrons cycle back to the P700 chlorophyll via the same electron transport chain that functions in noncyclic flow. There is no production of NADPH and no release of oxygen. Cyclic flow does, however, generate ATP. This is called **cyclic photophosphorylation,** to distinguish it from noncyclic photophosphorylation.

What is the function of cyclic electron flow? Noncyclic electron flow produces ATP and NADPH in roughly equal quantities, but the Calvin cycle consumes more ATP than NADPH. Cyclic electron flow makes up the difference. The concentration of NADPH in the chloroplast may help regulate which pathway, cyclic versus noncyclic, electrons take through the light reactions. If the chloroplast runs low on ATP for the Calvin cycle, NADPH will begin to accumulate as the Calvin cycle slows down. The rise in NADPH may stimulate a temporary shift from

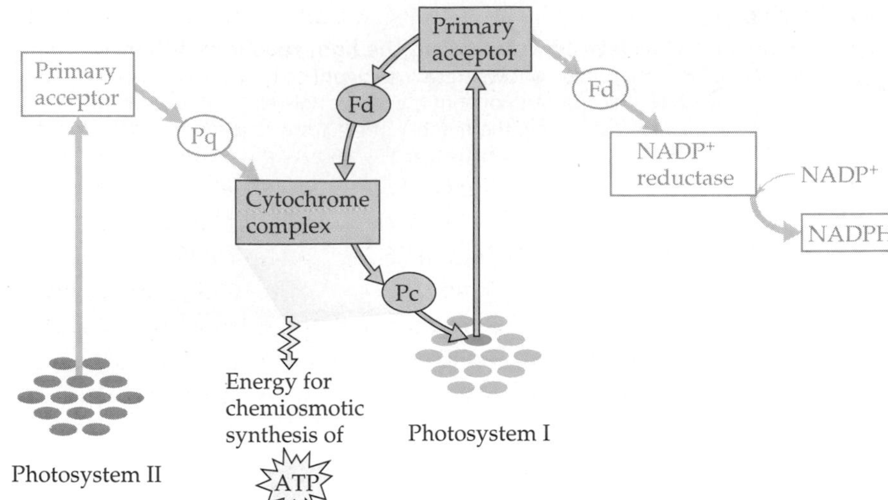

FIGURE 10.13
Cyclic electron flow. Photoexcited electrons from photosystem I are occasionally shunted back to chlorophyll via the electron transport chain. This cyclic electron flow supplements the supply of ATP but produces no NADPH. (The "shadow" of noncyclic electron flow is included in this diagram for comparison.)

noncyclic to cyclic electron flow until ATP supply catches up with demand.

Whether photophosphorylation is driven by noncyclic or cyclic electron flow, the actual mechanism for ATP synthesis is the same. This is a good time to review chemiosmosis, the basic process that uses membranes to couple redox reactions to ATP production.

A Comparison of Chemiosmosis in Chloroplasts and Mitochondria

Chloroplasts and mitochondria generate ATP by the same basic mechanism: chemiosmosis. An electron transport chain assembled in a membrane pumps protons across the membrane as electrons are passed through a series of carriers that are progressively more electronegative. Thus, electron transport chains transform redox energy to a proton-motive force, potential energy stored in the form of an H^+ gradient across a membrane. Built into the same membrane is an ATP synthase complex that couples the diffusion of hydrogen ions down their gradient to the phosphorylation of ADP. Some of the electron carriers, including the iron-containing proteins called cytochromes, are very similar in chloroplasts and mitochondria. The ATP synthase complexes of the two organelles are also very much alike. But there are noteworthy differences between oxidative phosphorylation in mitochondria and photophosphorylation in chloroplasts. In mitochondria, the high-energy electrons dropped down the transport chain are extracted by the oxidation of food molecules. Chloroplasts do not need food to make ATP; their photosystems capture light energy and use it to drive electrons to the top of the transport chain. In other words, mitochondria transfer chemical energy from food molecules to ATP, while chloroplasts transform light energy into chemical energy.

The spatial organization of chemiosmosis also differs in chloroplasts and mitochondria (FIGURE 10.14). The inner membrane of the mitochondrion pumps protons from the matrix out to the intermembrane space, which then serves as a reservoir of hydrogen ions that powers

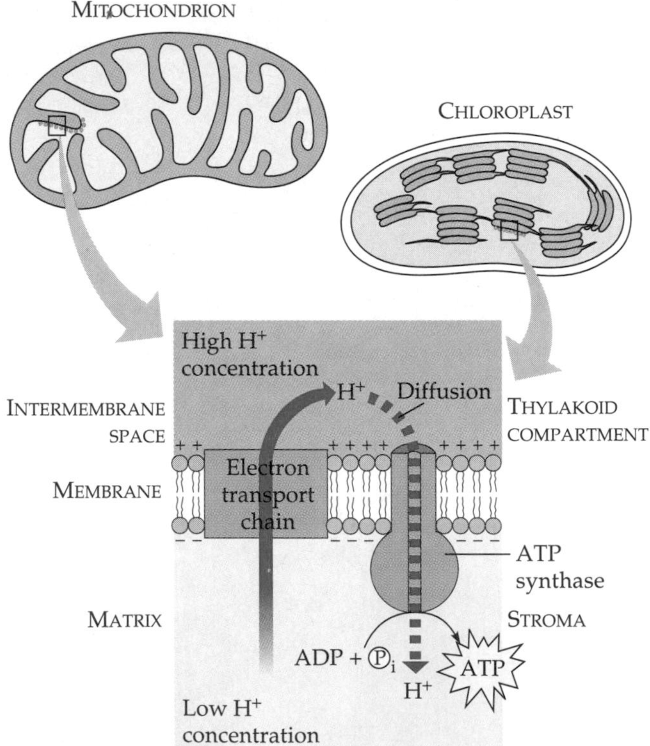

FIGURE 10.14
The logistics of chemiosmosis in mitochondria and chloroplasts. The inner membrane of the mitochondrion pumps protons (H^+) from the matrix into the intermembrane space (darker brown). ATP is made on the matrix side of the membrane as hydrogen ions diffuse through ATP synthase complexes. In chloroplasts, the thylakoid membrane pumps protons from the stroma into the thylakoid compartment. As the hydrogen ions leak back across the membrane through the ATP synthase, phosphorylation of ADP occurs on the stroma side of the membrane.

the ATP synthase. The thylakoid membrane of the chloroplast pumps protons from the stroma into the thylakoid compartment, which functions as the H^+ reservoir. The membrane makes ATP as the hydrogen ions diffuse from the thylakoid compartment back to the stroma through ATP synthase complexes, whose catalytic heads are on the stroma side of the membrane. Thus, ATP forms in the stroma, where it is used to help drive sugar synthesis during the Calvin cycle.

The proton gradient, or pH gradient, across the thylakoid membrane is substantial. When chloroplasts are illuminated, the pH in the thylakoid compartment drops to about 5, and the pH in the stroma increases to about 8. This gradient of three pH units corresponds to a thousandfold difference in H^+ concentration. When the lights are turned off, the pH gradient is abolished, but it can quickly be restored by turning the lights back on. Such experiments add to the evidence described in Chapter 9 in support of the chemiosmotic model.

Considerably more research is required before the precise organization of the thylakoid membrane can be discerned. FIGURE 10.15 shows a tentative model based on studies in several laboratories. Notice that NADPH, like ATP, is produced on the side of the membrane facing the stroma, where sugar is synthesized by the Calvin cycle.

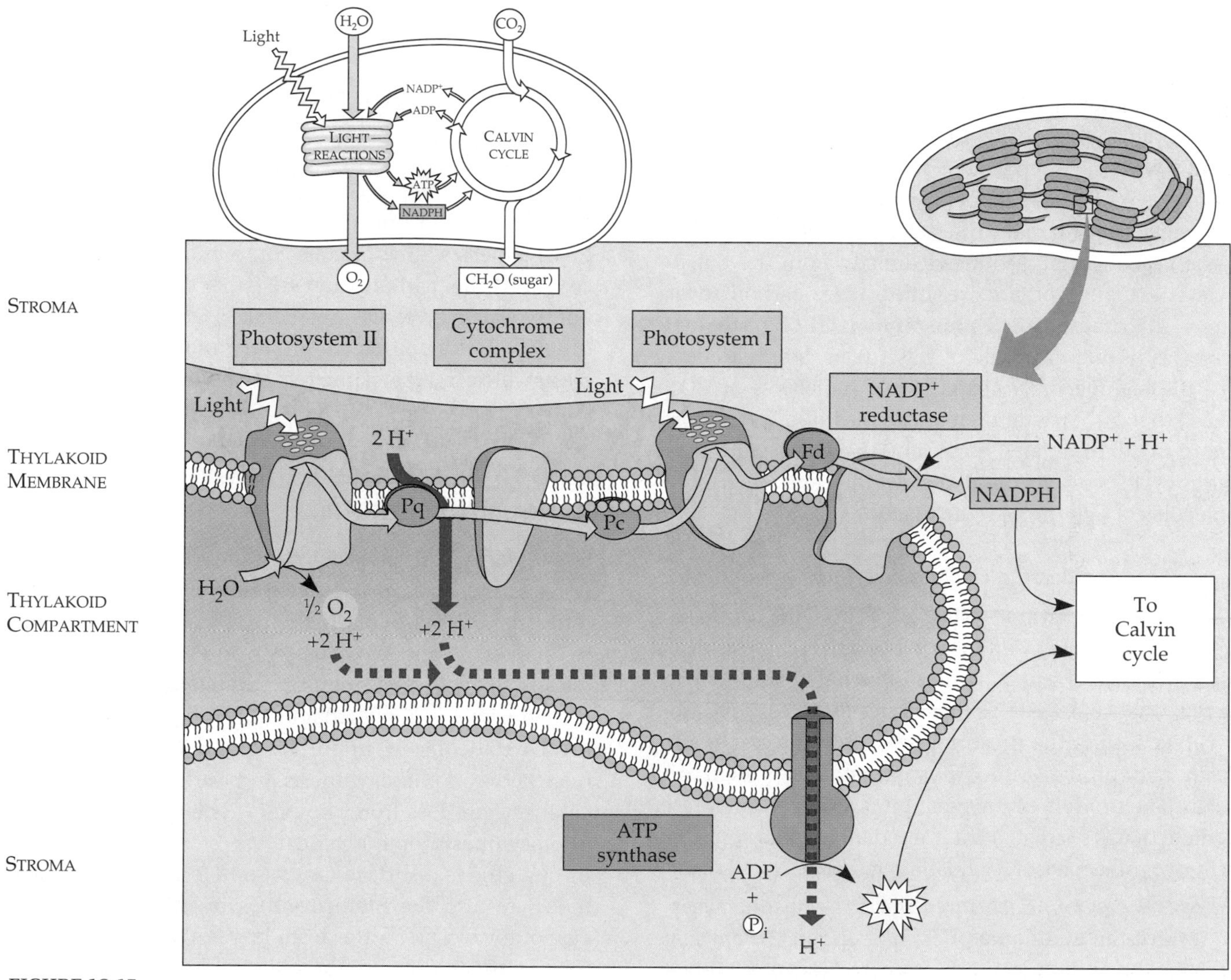

FIGURE 10.15

A tentative model for the organization of the thylakoid membrane. The orange arrows track electron flow. As electrons pass from carrier to carrier during redox reactions, hydrogen ions removed from the stroma are deposited in the thylakoid compartment, storing energy as a proton-motive force (H^+ gradient). There are at least three steps in the light reactions that contribute to the proton gradient: Water is split by photosystem II on the side of the membrane facing the thylakoid compartment; as plastoquinone (Pq), a mobile carrier, transfers electrons to the cytochrome complex, protons are translocated across the membrane; and a hydrogen ion in the stroma is taken up by $NADP^+$ when it is reduced to NADPH. The diffusion of H^+ from the thylakoid compartment to the stroma (along the H^+ concentration gradient) powers the ATP synthase. These light-driven reactions store chemical energy in NADPH and ATP, which shuttle the energy to the sugar-producing Calvin cycle.

Let's summarize the light reactions. Noncyclic electron flow pushes electrons from water, where they are at a low state of potential energy, to NADPH, where they are stored at a high state of potential energy. The light-driven electron current also generates ATP. Thus, the equipment of the thylakoid membrane converts light energy to the chemical energy stored in NADPH and ATP. (Oxygen is a by-product.) Let's now see how the Calvin cycle uses the products of the light reactions to synthesize sugar from CO_2.

The Calvin cycle uses ATP and NADPH to convert CO_2 to sugar: *a closer look*

The Calvin cycle is a metabolic pathway similar to the Krebs cycle in that a starting material is regenerated after molecules enter and leave the cycle. Carbon enters the Calvin cycle in the form of CO_2 and leaves in the form of sugar. The cycle spends ATP as an energy source and consumes NADPH as reducing power for adding high-energy electrons to make the sugar (FIGURE 10.16).

The carbohydrate produced directly from the Calvin cycle is actually not glucose, but a three-carbon sugar named **glyceraldehyde 3-phosphate (G3P).** For the net synthesis of one molecule of this sugar, the cycle must take place three times, fixing three molecules of CO_2. (Recall that carbon fixation refers to the initial incorporation of CO_2 into organic material.) As we trace the steps of the cycle, keep in mind that we are following three molecules of CO_2 through the reactions.

FIGURE 10.16 divides the Calvin cycle into three phases:

Phase 1: Carbon fixation. The Calvin cycle incorporates each CO_2 molecule by attaching it to a five-carbon sugar named ribulose bisphosphate (abbreviated RuBP). The enzyme that catalyzes this first step is RuBP carboxylase, or **rubisco.** (It is the most abundant protein in chloroplasts and probably the most abundant protein on Earth.) The product of the reaction is a six-carbon intermediate that is so unstable it immediately splits in half to form two molecules of 3-phosphoglycerate.

Phase 2: Reduction. Each molecule of 3-phosphoglycerate receives an additional phosphate group. An enzyme transfers the phosphate group from ATP, forming 1,3-bisphosphoglycerate as a product. Next, a pair of electrons donated from NADPH reduces 1,3-bisphosphoglycerate to G3P. Specifically, the electrons from NADPH reduce the carboyxl group of 3-phosphoglycerate to the carbonyl group of G3P, which stores more potential energy. G3P is a sugar—the same three-carbon sugar formed in glycolysis by the splitting of glucose. Notice in FIGURE 10.16 that for every *three* molecules of CO_2, there are *six* molecules of G3P. But only one molecule of this three-carbon sugar can be counted as a net gain of carbohydrate. The cycle began with 15 carbons' worth of carbohydrate in the form of three molecules of the five-carbon sugar RuBP. Now there are 18 carbons' worth of carbohydrate in the form of six molecules of G3P. One molecule exits the cycle to be used by the plant cell, but the other five molecules must be recycled to regenerate the three molecules of RuBP.

Phase 3: Regeneration of CO_2 acceptor (RuBP). In a complex series of reactions, the carbon skeletons of five molecules of G3P are rearranged by the last steps of the Calvin cycle into three molecules of RuBP. To accomplish this, the cycle spends three more molecules of ATP. The RuBP is now prepared to receive CO_2 again, and the cycle continues.

For the net synthesis of one G3P molecule, the Calvin cycle consumes a total of nine molecules of ATP and six molecules of NADPH. The light reactions regenerate the ATP and NADPH. The G3P spun off from the Calvin cycle becomes the starting material for metabolic pathways that synthesize other organic compounds, including glucose and other carbohydrates. Neither the light reactions nor the Calvin cycle alone can make sugar from CO_2. Photosynthesis is an emergent property of the intact chloroplast, which integrates the two stages of photosynthesis.

Alternative mechanisms of carbon fixation have evolved in hot, arid climates

Since plants first moved onto land about 425 million years ago, they have been adapting to the problems of terrestrial life, particularly the problem of dehydration. In Chapters 27 and 32, we will consider anatomical adaptations that help plants conserve water. Here, we are concerned with metabolic adaptations. The solutions often involve tradeoffs. An important example is the compromise between photosynthesis and the prevention of excessive water loss from the plant. The CO_2 required for photosynthesis enters a leaf via stomata, the pores through the leaf surface (see FIGURE 10.2). However, stomata are also the main avenues of transpiration, the evaporative loss of water from leaves. On a hot, dry day, most plants close their stomata, a response that conserves water. This response also reduces photosynthetic yield by limiting access to CO_2. With stomata even partially closed, CO_2 concentrations begin to decrease in the air spaces within the leaf, and the concentration of O_2 released from photosynthesis begins to increase. These conditions within the leaf favor a seemingly wasteful process called photorespiration.

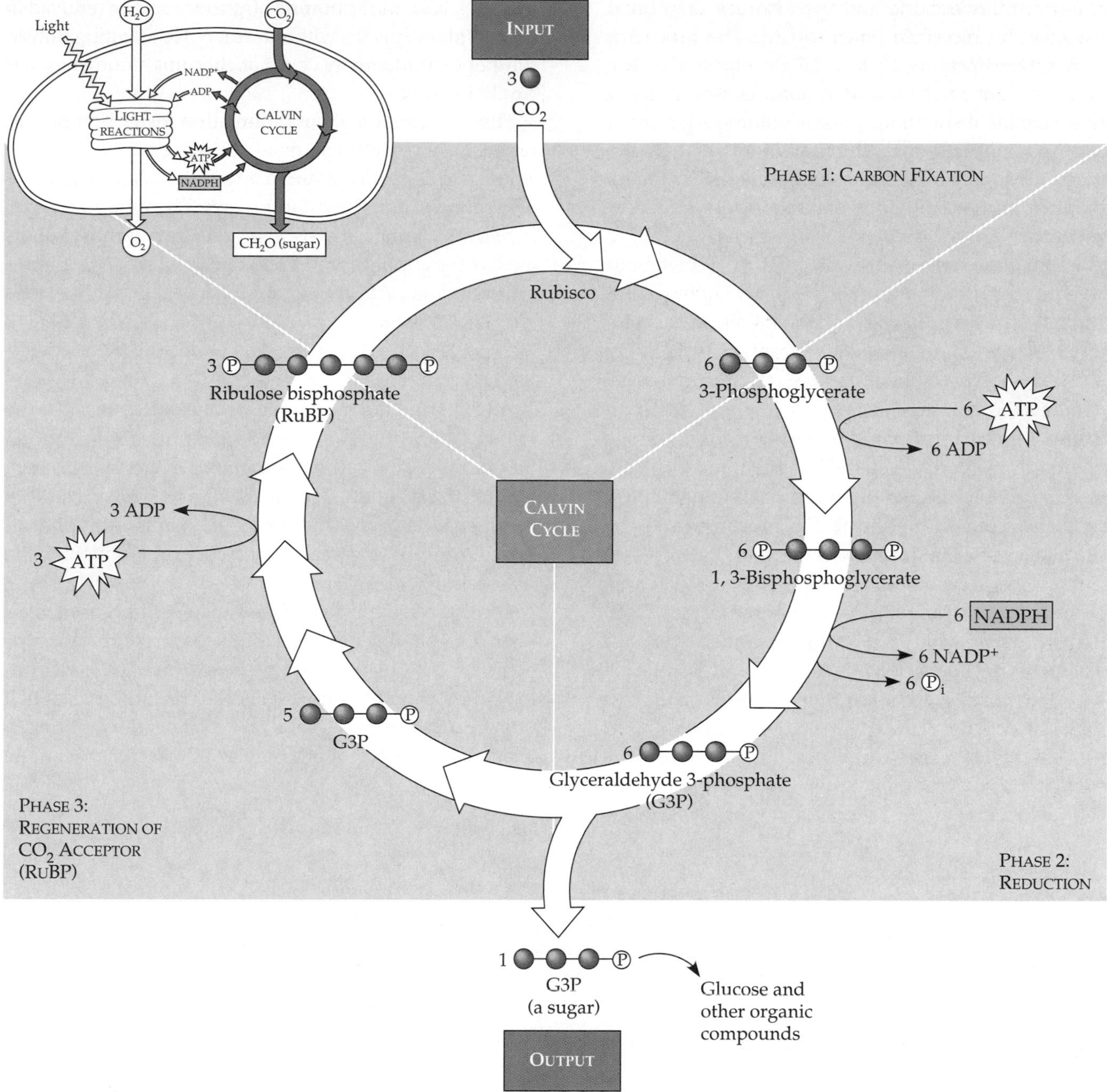

FIGURE 10.16

The Calvin cycle. This diagram tracks carbon atoms (gray balls) through the cycle. The three phases of the cycle correspond to the phases discussed in the text. For every three molecules of CO_2 that enter the cycle, the net output is one molecule of glyceraldehyde 3-phosphate (G3P), a three-carbon sugar. For each G3P synthesized, the cycle spends nine molecules of ATP and six molecules of NADPH. The light reactions sustain the Calvin cycle by regenerating the ATP and NADPH.

Photorespiration: An Evolutionary Relic?

In most plants, initial fixation of carbon occurs via rubisco, the Calvin cycle enzyme that adds CO_2 to ribulose bisphosphate. Such plants are called **C_3 plants** because the first organic product of carbon fixation is a three-carbon compound, 3-phosphoglycerate (see FIGURE 10.16). Rice, wheat, and soybeans are among the C_3 plants that are important in agriculture. These plants produce less food when their stomata close on hot, dry days. The declining level of CO_2 in the leaf starves the Calvin cycle. Making matters worse, rubisco can accept O_2 in place of CO_2. As O_2 concentrations overtake CO_2 concentrations within the air spaces of the leaf, rubisco adds O_2 to the Calvin cycle instead of CO_2. The product splits, and one piece, a two-carbon compound, is exported from the chloroplast. Mitochondria and peroxisomes then break the two-carbon molecule down to CO_2. The process is called **photorespiration** because it occurs in the light (*photo*) and consumes O_2 (*respiration*). However, unlike normal cellular respiration, photorespiration generates no ATP. And unlike photosynthesis, photorespiration produces no food. In fact, photorespiration *decreases* photosynthetic output by siphoning organic material from the Calvin cycle.

How can we explain the existence of a metabolic process that seems to be counterproductive to the plant? According to one hypothesis, photorespiration is evolutionary baggage—a metabolic relic from a much earlier time, when the atmosphere had less O_2 and more CO_2 than it does today. In that ancient atmosphere, when rubisco first evolved, the inability of the enzyme's active site to exclude O_2 would have made little difference. The hypothesis speculates that modern rubisco retains some of its ancestral affinity for O_2, which is so concentrated in the present atmosphere that a certain amount of photorespiration is inevitable.

It is not known whether photorespiration is beneficial to plants in any way. It *is* known that in many types of plants, including some of agricultural importance, such as soybeans, photorespiration drains away as much as 50% of the carbon fixed by the Calvin cycle. As heterotrophs that depend on carbon fixation in chloroplasts for our food, we naturally view photorespiration as wasteful. Indeed, if photorespiration could be reduced in certain plant species without otherwise affecting photosynthetic productivity, crop yields and food supplies would increase.

The environmental conditions that foster photorespiration are hot, dry, bright days—the conditions that cause stomata to close. In certain plant species, alternate modes of carbon fixation that minimize photorespiration—even in hot, arid climates—have evolved. The two most important of these photosynthetic adaptations are C_4 photosynthesis and CAM.

C_4 Plants

The C_4 **plants** are so named because they preface the Calvin cycle with an alternate mode of carbon fixation that forms a four-carbon compound as its first product. Several thousand species in at least 19 plant families use the C_4 pathway. Among the C_4 plants important to agriculture are sugarcane and corn, members of the grass family.

A unique leaf anatomy is correlated with the mechanism of C_4 photosynthesis (FIGURE 10.17a; compare to FIGURE 10.2). In C_4 plants, there are two distinct types of photosynthetic cells: bundle-sheath cells and mesophyll cells. **Bundle-sheath cells** are arranged into tightly packed sheaths around the veins of the leaf. Between the bundle sheath and the leaf surface are the more loosely

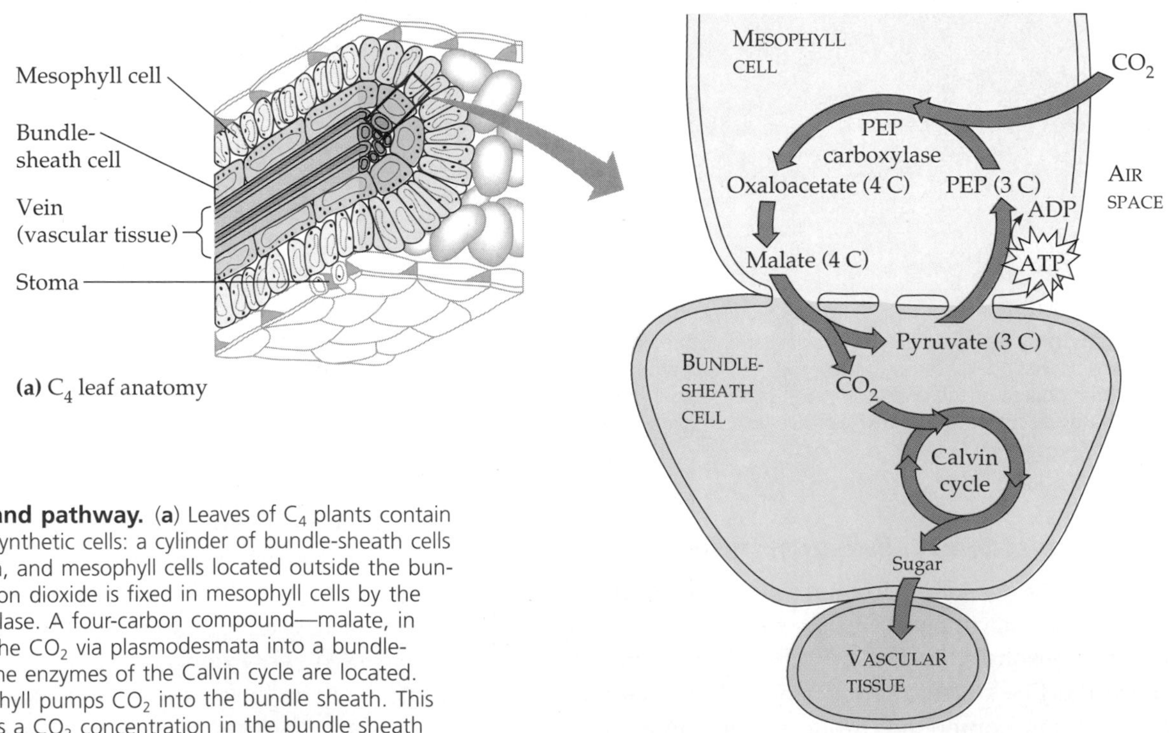

FIGURE 10.17
The C_4 anatomy and pathway. (**a**) Leaves of C_4 plants contain two types of photosynthetic cells: a cylinder of bundle-sheath cells surrounding the vein, and mesophyll cells located outside the bundle sheath. (**b**) Carbon dioxide is fixed in mesophyll cells by the enzyme PEP carboxylase. A four-carbon compound—malate, in this case—conveys the CO_2 via plasmodesmata into a bundle-sheath cell, where the enzymes of the Calvin cycle are located. In effect, the mesophyll pumps CO_2 into the bundle sheath. This adaptation maintains a CO_2 concentration in the bundle sheath that favors photosynthesis over photorespiration.

arranged **mesophyll cells.** The Calvin cycle is confined to the chloroplasts of the bundle sheath. However, the cycle is preceded by incorporation of CO_2 into organic compounds in the mesophyll (FIGURE 10.17b). The first step is the addition of CO_2 to phosphoenolpyruvate (PEP) to form the four-carbon product oxaloacetate. The enzyme **PEP carboxylase** adds CO_2 to PEP. Compared to rubisco, PEP carboxylase has a much higher affinity for CO_2. Therefore, PEP carboxylase can fix CO_2 efficiently when rubisco cannot—that is, when it is hot and dry and stomata are partially closed, causing CO_2 concentration in the leaf to fall and O_2 concentration to rise. After the C_4 plant fixes CO_2, the mesophyll cells export their four-carbon products to bundle-sheath cells through plasmodesmata (see FIGURE 7.28). Within the bundle-sheath cells, the four-carbon compounds release CO_2, which is reassimilated into organic material by rubisco and the Calvin cycle.

In effect, the mesophyll cells of a C_4 plant pump CO_2 into the bundle sheath, keeping the CO_2 concentration in the bundle-sheath cells high enough for rubisco to accept carbon dioxide rather than oxygen. In this way, C_4 photosynthesis minimizes photorespiration and enhances sugar production. This adaptation is especially advantageous in hot regions with intense sunlight, and it is in such environments that C_4 plants evolved and thrive today.

CAM Plants

A second photosynthetic adaptation to arid conditions has evolved in succulent (water-storing) plants (including ice plants), many cacti, pineapples, and representatives of several other plant families. These plants open their stomata during the night and close them during the day, just the reverse of how other plants behave. Closing stomata during the day helps desert plants conserve water, but it also prevents CO_2 from entering the leaves. During the night, when their stomata are open, these plants take up CO_2 and incorporate it into a variety of organic acids. This mode of carbon fixation is called **crassulacean acid metabolism,** or **CAM,** after the plant family Crassulaceae, the succulents in which the process was first discovered. The mesophyll cells of **CAM plants** store the organic acids they make during the night in their vacuoles until morning, when the stomata close. During the day, when the light reactions can supply ATP and NADPH for the Calvin cycle, CO_2 is released from the organic acids made the night before to become incorporated into sugar in the chloroplasts (FIGURE 10.18).

FIGURE 10.18

C_4 and CAM photosynthesis compared. Both adaptations are characterized by preliminary incorporation of CO_2 into organic acids, followed by transfer of the CO_2 to the Calvin cycle. In C_4 plants, such as sugarcane, these two steps are separated spatially; they are segregated into two cell types. In CAM plants, such as pineapple, the two steps are separated temporally; carbon fixation into organic acids occurs at night, and the Calvin cycle operates during the day. C_4 and CAM are two evolutionary solutions to the problem of maintaining photosynthesis with stomata partially or completely closed on hot, dry days.

Notice in FIGURE 10.18 that the CAM pathway is similar to the C_4 pathway in that carbon dioxide is first incorporated into organic intermediates before it enters the Calvin cycle. The difference is that in C_4 plants, the initial steps of carbon fixation are separated structurally from the Calvin cycle, whereas in CAM plants the two steps occur at separate times. Keep in mind that CAM, C_4, and C_3 plants all eventually use the Calvin cycle to make sugar from carbon dioxide.

Photosynthesis is the biosphere's metabolic foundation: *a review*

In this chapter, we have followed photosynthesis from photons to food (FIGURE 10.19). The light reactions capture solar energy and use it to make ATP and transfer electrons from water to $NADP^+$. The Calvin cycle uses the ATP and NADPH to produce sugar from carbon dioxide. The energy that entered the chloroplasts as sunlight becomes stored as chemical energy in organic compounds.

What are the fates of photosynthetic products? The sugar made in the chloroplasts supplies the entire plant with chemical energy and carbon skeletons to synthesize all the major organic molecules of cells. About 50% of the organic material made by photosynthesis is consumed as fuel for cellular respiration in the mitochondria of the plant cells. Sometimes there is a loss of photosynthetic products to photorespiration.

Technically, green cells are the only autotrophic parts of the plant. The rest of the plant depends on organic molecules exported from leaves in veins. In most plants, carbohydrate is transported out of the leaves in the form of sucrose, a disaccharide. After arriving at nonphotosynthetic cells, the sucrose provides raw material for cellular respiration and a multitude of anabolic pathways that synthesize proteins, lipids, and other products. A considerable amount of sugar in the form of glucose is linked together to make the polysaccharide cellulose, especially in plant cells that are still growing and maturing. Cellulose, the main ingredient of cell walls, is the most abundant organic molecule in the plant—and probably on the surface of the planet.

Most plants manage to make more organic material each day than they need to use as respiratory fuel and precursors for biosynthesis. They stockpile the extra sugar by synthesizing starch, storing some in the chloroplasts themselves and some in storage cells of roots, tubers, seeds, and fruits. In accounting for the consumption of the food molecules produced by photosynthesis, let's not forget that most plants lose leaves, roots, stems, fruits, and sometimes their entire bodies to heterotrophs, including humans.

On a global scale, the collective productivity of the minute chloroplasts is prodigious; it is estimated that photosynthesis makes about 160 billion metric tons of carbohydrate per year (a metric ton is 1000 kg, about 1.1 tons). That's organic matter equivalent to a stack of about 60 trillion copies of this textbook—17 stacks of books reaching from Earth to the Sun! No other chemical process on the planet can match the output of photosynthesis. And no process is more important than photosynthesis to the welfare of life on Earth.

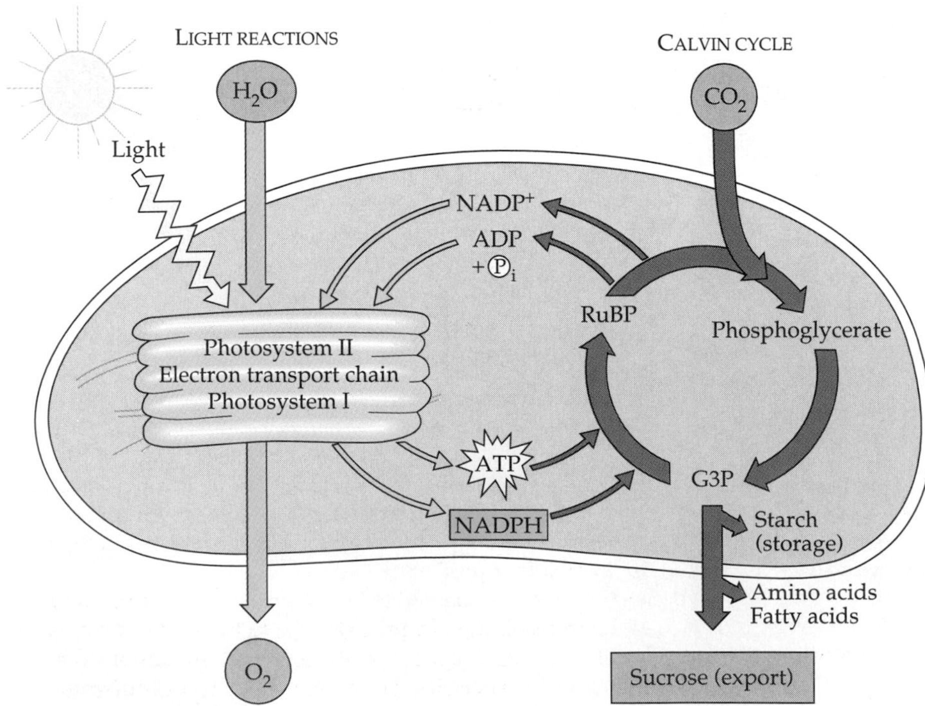

FIGURE 10.19
A review of photosynthesis. This diagram outlines the main reactants and products of photosynthesis as it occurs in the chloroplasts of plant cells. The light reactions convert light energy to the chemical energy of ATP and NADPH. The pigment and protein molecules that carry out the light reactions are found in the thylakoid membranes and include the molecules of two photosystems and an electron transport chain. The light reactions split H_2O and release O_2 to Earth's atmosphere. The Calvin cycle, which takes place in the stroma of the chloroplast, uses ATP and NADPH to convert CO_2 to carbohydrate (three key compounds of the cycle are shown). The direct product of the Calvin cycle is the three-carbon sugar glyceraldehyde 3-phosphate (G3P). Enzymes in the chloroplast and cytoplasm convert this small sugar to a diversity of other organic compounds. The Calvin cycle returns ADP, inorganic phosphate, and $NADP^+$ to the light reactions. The entire ordered operation depends on the structural integrity of the chloroplast and its membranes.

REVIEW OF KEY CONCEPTS (with page numbers and key figures)

- Plants and other autotrophs are the producers of the biosphere (pp. 182–183, FIGURE 10.1)
  - Autotrophs sustain themselves without ingesting organic molecules. Photoautotrophs use the energy of sunlight to synthesize organic molecules from CO_2 and H_2O.
  - Heterotrophs must ingest other organisms or their by-products to obtain energy and carbon skeletons.
- Chloroplasts are the sites of photosynthesis in plants (p. 184, FIGURE 10.2)
  - In autotrophic eukaryotes, photosynthesis occurs inside chloroplasts, organelles containing an elaborate system of thylakoid membranes that are stacked in places as grana. These membranes separate the thylakoid compartment from the chloroplast's stroma.
- Evidence that chloroplasts split water molecules enabled researchers to track atoms through photosynthesis: *science as a process* (pp. 185–186, FIGURE 10.3)
  - Photosynthesis can be summarized by the following equation:

$$6\ CO_2 + 12\ H_2O + \frac{\text{Light}}{\text{energy}} \longrightarrow C_6H_{12}O_6 + 6\ O_2 + 6\ H_2O$$

  - Experiments reveal that the chloroplast splits water into hydrogen and oxygen, incorporating the electrons of hydrogen into the bonds of sugar molecules. Photosynthesis is thus a redox process in which water is oxidized and carbon dioxide is reduced.
- The light reactions and the Calvin cycle cooperate in converting light energy to the chemical energy of food: *an overview* (pp. 186–187, FIGURE 10.4)
  - The light reactions in the grana produce ATP and split water, releasing oxygen and forming NADPH by transferring electrons from water to $NADP^+$.
  - The Calvin cycle occurs in the stroma and uses ATP for energy and NADPH for reducing power to form sugar from CO_2.
- The light reactions convert solar energy to the chemical energy of ATP and NADPH: *a closer look* (pp. 187–196, FIGURE 10.11)
  - Light is a form of electromagnetic energy, which travels in waves. The range of wavelengths of this radiation constitutes the electromagnetic spectrum, part of which we detect as the colors of visible light.
  - Light behaves as discrete energy packets called photons.
  - A pigment is a substance that absorbs specific wavelengths of light. A spectrophotometer can be used to produce an absorption spectrum for a specific pigment. The action spectrum of photosynthesis and the absorption spectrum of chlorophyll *a* do not directly correspond. Accessory pigments—chlorophyll *b* and various carotenoids—absorb different wavelengths of light and pass their energy on to chlorophyll *a*.
  - A pigment goes from a ground state to an excited state when a photon boosts one of its electrons to a higher-energy orbital. In isolated pigments, the electron immediately returns to the ground state, releasing the energy as light (fluorescence) and/or heat.
  - The pigments of chloroplasts are built into the thylakoid membrane near molecules known as primary electron acceptors, which trap the high-energy electrons before they return to the ground state.
  - Accessory pigments in chloroplasts are clustered in an antenna complex of a few hundred molecules surrounding a chlorophyll *a* molecule at the reaction center. Photons absorbed anywhere in the antenna can pass their energy along to energize this chlorophyll *a*, which then passes an electron to a nearby primary electron acceptor. The antenna complex, the reaction-center chlorophyll, and the primary electron acceptor make up a photosystem, a light-harvesting unit built into the thylakoid membrane.
  - There are two kinds of photosystems. Photosystem I contains P700, and photosystem II contains P680, chlorophyll *a* molecules at the reaction centers.
  - Noncyclic electron flow involves both photosystems and produces NADPH, ATP, and oxygen. Electrons from P700 in photosystem I are trapped by $NADP^+$, which stores them in the form of NADPH. Electrons from P680 in photosystem II restore the electrons lost from P700. These electrons travel between the two photosystems via an electron transport chain that powers ATP synthesis. The electron "holes" in P680 are filled by electrons from water, which is split into hydrogen ions and oxygen. The overall process powers the transfer of electrons from water to $NADP^+$ and the phosphorylation of ADP to form ATP.
  - Cyclic electron flow employs only photosystem I, producing ATP but no NADPH or O_2.
  - ATP production during the light reactions is called photophosphorylation. The mechanism is chemiosmosis. The redox reactions of the electron transport chain generate an H^+ gradient across the thylakoid membrane. An ATP synthase harnesses this proton-motive force to make ATP.
- The Calvin cycle uses ATP and NADPH to convert CO_2 to sugar: *a closer look* (p. 196, FIGURE 10.16)
  - The Calvin cycle is a cyclic metabolic pathway in the chloroplast stroma. An enzyme (rubisco) combines carbon dioxide with ribulose bisphosphate (RuBP), a five-carbon sugar. Then, using electrons from NADPH and energy from ATP, the cycle synthesizes the three-carbon sugar glyceraldehyde 3-phosphate in a series of reactions. Most of the G3P is reused in the cycle as an intermediate for reconversion to RuBP, but some exits the cycle and is converted to other essential organic molecules.
- Alternative mechanisms of carbon fixation have evolved in hot, arid climates (pp. 196–200, FIGURE 10.18)
  - On dry, hot days, plants close their stomata, which conserves water. Oxygen from the light reactions builds up. When O_2 substitutes for CO_2 in the active site of rubisco, an intermediate is formed that leaves the cycle and is oxidized to CO_2 and H_2O in the peroxisomes and mitochondria. This process, called photorespiration, consumes organic fuel without producing ATP.
  - C_4 plants are adapted to hot, dry conditions. They avert photorespiration by prefacing the Calvin cycle with a series of reactions that incorporate CO_2 into four-carbon compounds in specialized mesophyll cells. These four-carbon

compounds are exported to photosynthetic bundle-sheath cells, where they release carbon dioxide for use in the Calvin cycle.

- CAM plants open their stomata during the night and incorporate the CO_2 that enters into organic acids, which are stored in the vacuoles of mesophyll cells. During the day, the stomata close, conserving water, and the CO_2 is released from the organic acids for use in the Calvin cycle.

- Photosynthesis is the biosphere's metabolic foundation: *a review* (p. 200, FIGURE 10.19)
  - Veins export sucrose made in green cells to nonphotosynthetic parts of the plant. Cellular respiration in a plant's mitochondria degrade about 50% of the carbohydrate made by photosynthesis. Much of the remaining carbohydrate is converted to a variety of molecules, including cellulose.
  - Excess organic material is stockpiled as starch, oils, and proteins and stored in leaves, roots, tubers, seeds, and fruit. Heterotrophs consume much of this organic material.

SELF-QUIZ

1. The light reactions of photosynthesis supply the Calvin cycle with
   a. light energy
   b. CO_2 and ATP
   c. H_2O and NADPH
   d. ATP and NADPH
   e. sugar and O_2

2. Which sequence correctly portrays the flow of electrons during photosynthesis?
   a. NADPH $\longrightarrow$ O_2 $\longrightarrow$ CO_2
   b. H_2O $\longrightarrow$ NADPH $\longrightarrow$ Calvin cycle
   c. NADPH $\longrightarrow$ chlorophyll $\longrightarrow$ Calvin cycle
   d. H_2O $\longrightarrow$ photosystem I $\longrightarrow$ photosystem II
   e. NADPH $\longrightarrow$ electron transport chain $\longrightarrow$ O_2

3. Which of the following conclusions does *not* follow from studying the absorption spectrum for chlorophyll *a* and the action spectrum for photosynthesis?
   a. Not all wavelengths are equally effective for photosynthesis.
   b. There must be accessory pigments that broaden the spectrum of light that contributes energy for photosynthesis.
   c. The red and blue areas of the spectrum are most effective in driving photosynthesis.
   d. Chlorophyll owes its color to the absorption of green light.
   e. Chlorophyll *a* has two absorption peaks.

4. Cooperation of the *two* photosystems of the chloroplast is required for
   a. ATP synthesis
   b. reduction of $NADP^+$
   c. cyclic photophosphorylation
   d. oxidation of the reaction center of photosystem I
   e. generation of a proton-motive force

5. In *mechanism*, photophosphorylation is most similar to
   a. substrate-level phosphorylation in glycolysis
   b. oxidative phosphorylation in cellular respiration
   c. the Calvin cycle
   d. carbon fixation
   e. reduction of $NADP^+$

6. In what respect are the photosynthetic adaptations of C_4 plants and CAM plants similar?
   a. In both cases, the stomata normally close during the day.
   b. Both types of plants make their sugar without the Calvin cycle.
   c. In both cases, an enzyme other than rubisco carries out the first step in carbon fixation.
   d. Both types of plants make most of their sugar in the dark.
   e. Neither C_4 plants nor CAM plants have grana in their chloroplasts.

7. The stage of photosynthesis that actually produces sugar is
   a. the Calvin cycle
   b. photosystem I
   c. photosystem II
   d. the light reactions
   e. the splitting of water

8. For every CO_2 molecule fixed by photosynthesis, how many molecules of O_2 are released?
   a. 1 d. 6
   b. 2 e. 12
   c. 3

9. Which of the following statements is a correct distinction between autotrophs and heterotrophs?
   a. Only heterotrophs require chemical compounds from the environment.
   b. Cellular respiration is unique to heterotrophs.
   c. Only heterotrophs have mitochondria.
   d. Autotrophs, but not heterotrophs, can nourish themselves beginning with nutrients that are entirely inorganic.
   e. Only heterotrophs require oxygen.

10. Which of the following processes could still occur in a chloroplast in the presence of an inhibitor that prevents H^+ from passing through ATP synthase complexes? (Explain your answer.)
    a. sugar synthesis
    b. generation of a proton-motive force
    c. photophosphorylation
    d. the Calvin cycle
    e. oxidation of NADPH

CHALLENGE QUESTIONS

1. The photosynthetic rate of aquatic plants in a test tube can be determined by collecting and measuring the amount of oxygen that gases out of the water. If bicarbonate, the source of CO_2 for aquatic plants, is added to the water, the rate of oxygen evolution increases. If CO_2 is fixed by the Calvin cycle but oxygen is evolved by the light reactions, how can an increase in CO_2 supply increase the rate of oxygen evolution?

2. Compare and contrast the mechanisms of photosynthesis and cellular respiration. How has the chemiosmotic model helped unify current ideas on the mechanisms of these two processes?

3. The graph below tracks a bean plant's stomata over 24 hours. The dashed line indicates partial closing of stomata on a hot, dry afternoon. How would this affect CO_2 and O_2 concentrations within the leaf of a C_3 plant? What would be the impact on photosynthesis and photorespiration? How would these effects differ for a C_4 plant? How would the behavior of a CAM plant's stomata differ from the behavior recorded by this graph?

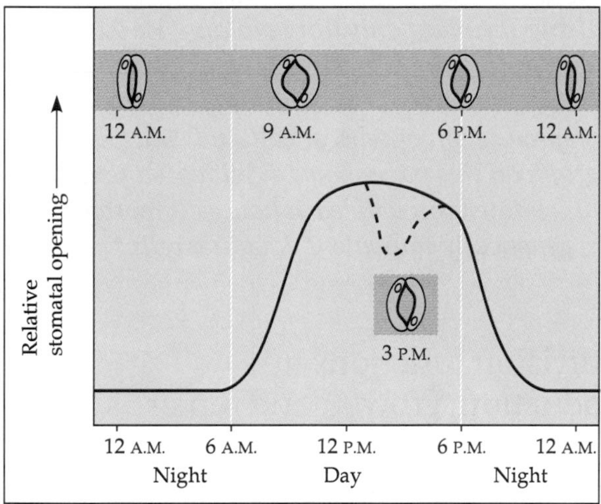

4. The diagram below represents an experiment with isolated chloroplasts. The chloroplasts were first made acidic by soaking them in a solution at pH 4. After the thylakoid compartment reached pH 4, the chloroplasts were transferred to a basic solution with pH 8. This caused the chloroplasts to make ATP, even if placed in the dark. Explain this result.

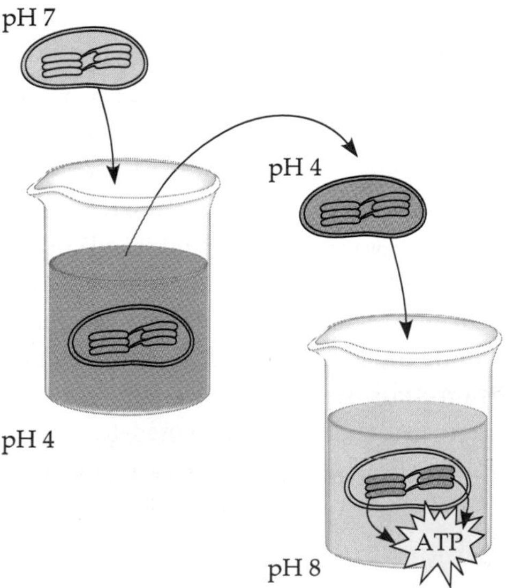

SCIENCE, TECHNOLOGY, AND SOCIETY

1. Tropical rain forests cover only about 3% of Earth's surface, but they are estimated to be responsible for more than 20% of global photosynthesis. It seems reasonable to expect that the lush growth of jungle foliage would produce large amounts of oxygen and reduce global warming by consuming carbon dioxide. But in fact, many experts now believe that rain forests make little or no *net* contribution to global oxygen production or reduction of global warming. Using your knowledge of photosynthesis and cellular respiration, explain what the basis of this hypothesis might be. What happens to the food produced by a rainforest tree when it is eaten by animals or the tree dies?

2. Atmospheric levels of CO_2 are increasing. We know most of the sources of atmospheric CO_2, such as fossil fuel combustion, deforestation, and the oxidation of organic matter in farm soils. Thus, we can estimate world CO_2 production with some accuracy. However, we do not know where all this CO_2 goes. Some is retained in the atmosphere and some dissolves in the ocean, but there is a significant amount of CO_2 that is not accounted for. Where do you think this CO_2 is going? How would you test your hypothesis?

FURTHER READING

Bazzazz, F. A., and E. D. Fajer. "Plant Life in a CO_2-Rich World." *Scientific American,* January 1992. How will increasing atmospheric CO_2 and global warming affect the relative success of C_3 and C_4 plants?

Becker, W. M., J. B. Reece, and M. F. Poenie. *The World of the Cell,* 3rd ed. Menlo Park, CA: Benjamin/Cummings, 1996. Chapter 13.

Galston, A. W. "Photosynthesis as a Basis for Life Support on Earth and in Space." *Bioscience,* July/August 1992. Plants in space.

Govindjee and W. J. Coleman. "How Plants Make Oxygen." *Scientific American,* February 1990.

Hendry, G. "Oxygen, the Great Destroyer." *Natural History,* August 1992. Photorespiration and other problems associated with an oxygen-rich atmosphere.

Walker, D. *Energy, Plants, and Man.* Mill Valley, CA: University Science Books, 1992. Cartoons and stories that make bioenergetics fun.

CHAPTER 11

THE REPRODUCTION OF CELLS

KEY CONCEPTS

- Cell division functions in reproduction, growth, and repair
- Bacteria reproduce by binary fission
- The genome of a eukaryotic cell is organized into multiple chromosomes
- Mitosis alternates with interphase in the cell cycle: *an overview*
- The mitotic spindle distributes chromosomes to daughter cells: *a closer look*
- Cytokinesis divides the cytoplasm: *a closer look*
- External and internal cues control cell division
- Cyclical changes in regulatory proteins function as a mitotic clock
- Cancer cells escape from the controls on cell division

Photograph courtesy of J. M. Murray, University of Pennsylvania Medical School.

Life arises only from life. The ability of organisms to reproduce their kind is the one phenomenon that best distinguishes life from nonliving matter. (Although the analogy of making photocopies is frequently used, the reproduction of an organism is actually more analogous to a photocopier making more photocopiers!) This unique capacity to procreate, like all biological functions, has a cellular basis. Rudolf Virchow, a German physician, put it this way in 1855: "Where a cell exists, there must have been a preexisting cell, just as the animal arises only from an animal and the plant only from a plant." He summarized with the axiom, "Omnis cellula e cellula," *meaning "All cells from cells." The continuity of life is based on the reproduction of cells, or* ***cell division.*** *(The micrograph on this page shows a cell reproducing.)*

*In this chapter, you will learn how cells reproduce to form genetically equivalent daughter cells.**

Cell division functions in reproduction, growth, and repair

In some cases, the division of one cell to form two reproduces an entire organism, as when a unicellular organism, such as *Amoeba,* divides to form duplicate offspring (FIGURE 11.1a). But cell division also enables a multicellular organism, such as a human, to grow and develop from a single cell—the fertilized egg (FIGURE 11.1b). Even after the organism is fully grown, cell division continues to function in renewal and repair, replacing cells that die from normal wear and tear or accidents. For example, dividing cells in your bone marrow continuously supply new blood cells (FIGURE 11.1c).

The reproduction of an ensemble as complex as a cell cannot occur by a mere pinching in half; the cell is not like a soap bubble that simply grows and splits in two. Cell division involves the distribution of identical genetic material—DNA—to the two daughter cells. A cell's total hereditary endowment of DNA is called its **genome.** The genome consists of one or more very long DNA molecules. Arranged along the length of each DNA molecule are hundreds or thousands of genes, the hereditary units that specify an organism's traits. In eukaryotes, the compact coiling and folding of DNA molecules form chromo-

*Although the terms *daughter cells* and *sister chromatids* (a term you will encounter later in the chapter) are traditional and will be used throughout this book, the structures they refer to are actually genderless.

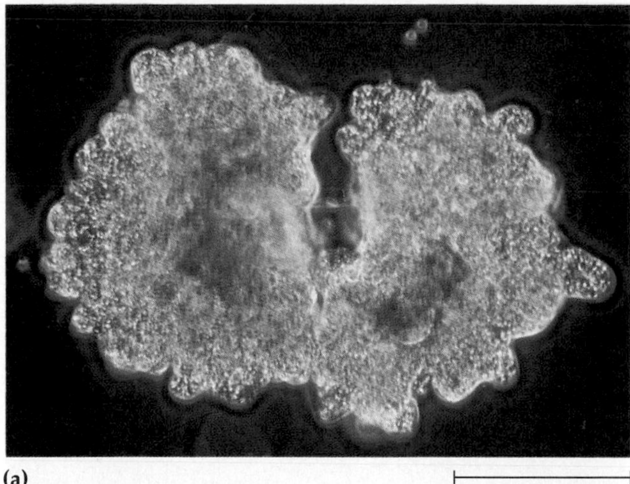

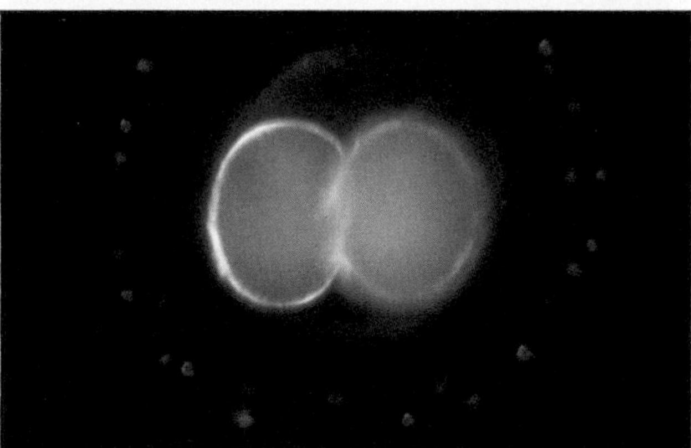

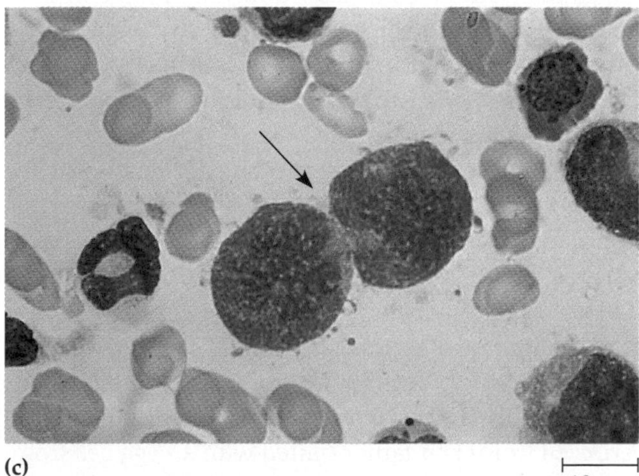

FIGURE 11.1

The functions of cell division. (**a**) *Amoeba,* a single-celled eukaryote, divides to form two cells, each an individual organism. In this case, cell division functions in reproduction. (**b**) For multicellular organisms, the division of embryonic cells is important in growth and development. This darkfield micrograph shows a sand dollar embryo shortly after the fertilized egg divided to form two cells. (**c**) Even in a mature organism, cell division continues to function in the renewal and repair of tissues. For example, these dividing cells in bone marrow (arrow) give rise to new blood cells. (All LMs.)

somes, visible with a light microscope just before and during cell division. What is most remarkable about cell division is the fidelity with which the genome is passed along without dilution from one generation of cells to the next. A cell preparing to divide first copies all its genes, allocates them equally to opposite ends of the cell, and then separates into two daughter cells.

Although the emphasis in this chapter will be on the division of eukaryotic cells, we will begin with a brief look at prokaryotic cell division.

Bacteria reproduce by binary fission

Prokaryotes (bacteria) reproduce by a type of cell division called **binary fission,** meaning literally "division in half" (FIGURE 11.2). Most bacterial genes are carried on a single chromosome that consists of a circular DNA molecule

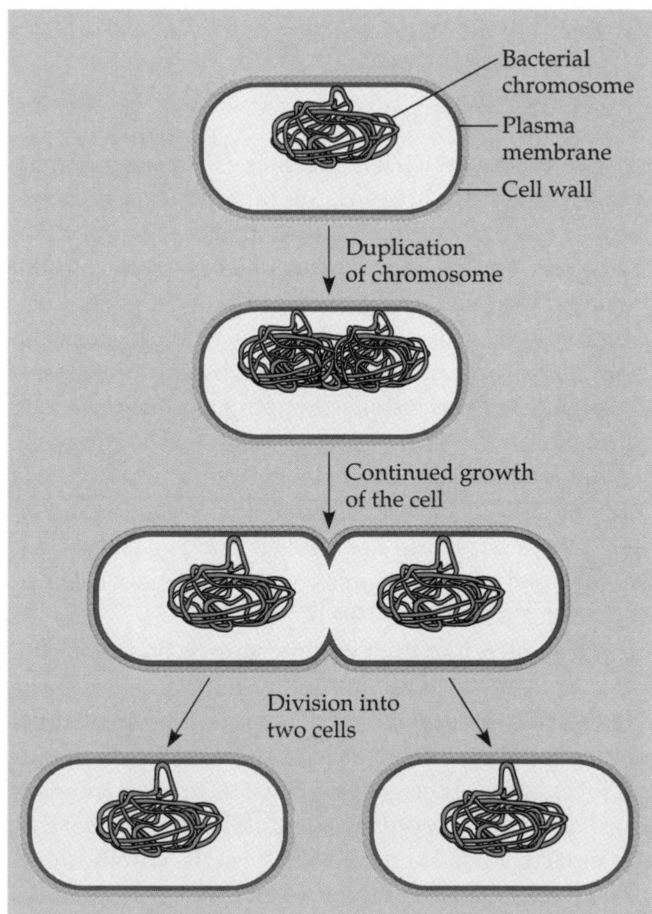

FIGURE 11.2

Bacterial cell division (binary fission). Allocation of equal genomes to daughter cells depends on the attachment of duplicated chromosomes to the plasma membrane of the parent cell. Continued growth of the cell gradually separates the chromosomes. Eventually, the plasma membrane pinches inward to divide the cell in two, as a new cell wall is deposited between the daughter cells.

and associated proteins. Although bacteria are smaller and simpler than eukaryotic cells, the problem of replicating their genomes in an orderly fashion and distributing the copies equally to two daughter cells is still formidable. Consider, for example, the chromosome of the bacterium *Escherichia coli;* when fully stretched out, this chromosome is about 500 times longer than the length of the cell. Clearly, such a chromosome must be highly folded within the cell.

The bacterial chromosome is attached to the plasma membrane. After a bacterial cell replicates its chromosome in preparation for fission, the two copies remain attached to the membrane at adjacent sites. Growth of the membrane between the two attachment sites separates the two copies of the chromosome. When the bacterium has reached about twice its initial size, its plasma membrane pinches inward, and a cell wall forms between the two chromosomes, dividing the parent cell into two daughter cells. Each cell inherits a complete genome.

The genome of a eukaryotic cell is organized into multiple chromosomes

Eukaryotes have much larger genomes than prokaryotes. Yet each of the tens of thousands of genes in a typical eukaryotic cell must be copied, and then the mechanics of cell division must apportion complete genomes to the two daughter cells. The problem of replicating and distributing so many genes is only manageable because the genes are grouped into multiple chromosomes (FIGURE 11.3). Every species of eukaryote has a characteristic number of chromosomes in each cell nucleus. For example, there are 46 chromosomes in human **somatic cells** (all body cells except the reproductive cells). Reproductive cells, or **gametes**—sperm cells and egg cells—have half as many chromosomes as do somatic cells, or 23 chromosomes in the case of humans.

Incorporated into each chromosome is one very long, linear DNA molecule representing thousands of genes. The DNA is associated with various proteins, which maintain the structure of the chromosome and help control the activity of the genes. The DNA-protein complex, called **chromatin,** is organized into a long, thin fiber that is compactly folded and coiled to form the chromosome.

As part of its preparation for division, a cell copies its entire genome by duplicating each chromosome. After replication, each chromosome consists of two **sister chromatids.** The two chromatids have matching copies of the DNA molecule that was present in the single copy

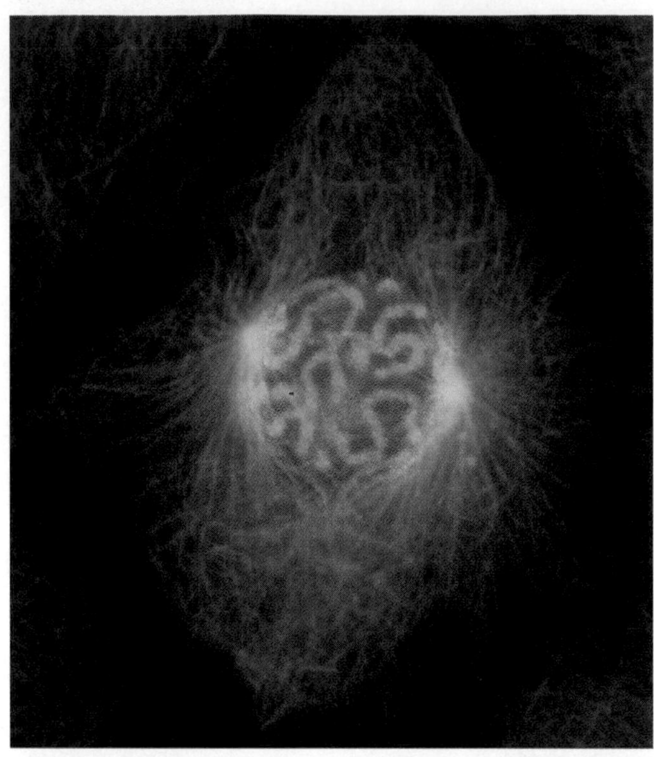

FIGURE 11.3
Eukaryotic chromosomes. A tangle of threadlike chromosomes (orange) is visible within the nucleus of this kangaroo rat epithelial cell, which is preparing to divide (LM). Chromosomes are so named because they take up certain dyes used in microscopy (*chromo,* "colored," and *somes,* "bodies"). (Photo courtesy of J. M. Murray, University of Pennsylvania Medical School.)

prior to replication. A specialized region of the chromosome called the **centromere** holds the two sister chromatids together. Then, in the process of **mitosis,** the sister chromatids are pulled apart and repackaged as complete chromosome sets in two nuclei, one at each end of the cell (FIGURE 11.4). Mitosis, the division of the nucleus, is usually followed immediately by **cytokinesis,** the division of the cytoplasm. Where there was one cell, there are now two, each the genetic equivalent of the parent cell.

Let's see what happens to chromosome number as we follow the human life cycle through the generations. You inherited 46 chromosomes, 23 from each parent. They were combined in the nucleus of a single cell when a sperm cell from your father united with an egg cell from your mother to form a fertilized egg, or zygote. Mitosis produced the trillions of somatic cells that now make up your body, and the same process continues to generate new cells to replace dead and damaged ones. In contrast, your gametes—egg or sperm cells—are produced by a variation of cell division called **meiosis,** which yields daughter cells that have half as many chromosomes as

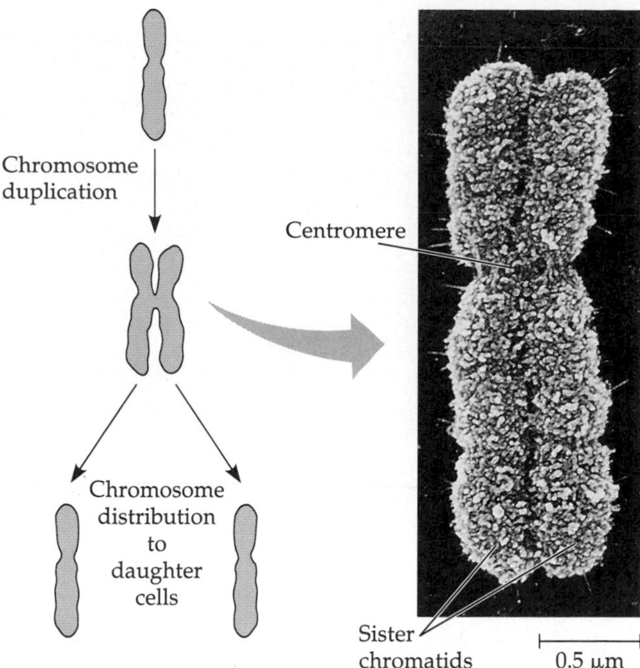

FIGURE 11.4
Chromosome duplication and distribution during mitosis. As a cell prepares to divide, it duplicates each of its multiple chromosomes, one of which is represented in this diagram. A duplicated chromosome consists of two sister chromatids joined at their centromeres. The micrograph shows a human chromosome in this duplicated state (SEM). The chromosome has a "hairy" appearance because it consists of a very long chromatin fiber folded and coiled in a compact arrangement. One long DNA molecule associated with a variety of protein molecules makes up the chromatin fiber. The DNA molecules of the sister chromatids are identical, products of the precise process that replicates the genetic material. As mitosis continues, mechanical processes separate the sister chromatids and distribute them to the two daughter cells.

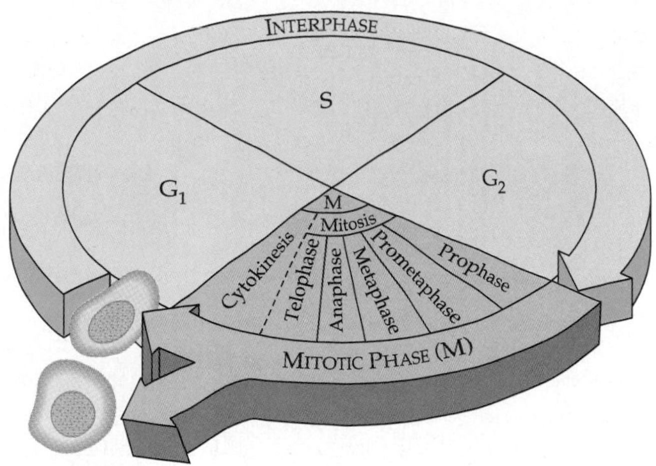

FIGURE 11.5
The cell cycle. In a dividing cell, the mitotic (M) phase alternates with an interphase, or growth period. The first part of interphase is called G_1, followed by the S phase, during which the chromosomes replicate; the last part of interphase is called G_2. Next, mitosis divides the nucleus and its chromosomes. Finally, cytokinesis divides the cytoplasm, producing two daughter cells. The stages of mitosis are described in FIGURE 11.6.

the parent cell. Meiosis occurs only in your gonads (your ovaries or testes). In each generation of humans, meiosis reduces the chromosome number from 46 to 23. Fertilization joins gametes and doubles the chromosome number to 46 again, and mitosis conserves that number in every somatic cell of the new individual. In Chapter 12, we will examine the role of meiosis in reproduction and inheritance in much more detail. In the rest of this chapter, we will focus on mitosis.

Mitosis alternates with interphase in the cell cycle: *an overview*

Mitosis is just one part of a cell division cycle, or **cell cycle** (FIGURE 11.5). In fact, the **mitotic (M) phase,** when mitosis and cytokinesis actually divide the nucleus and cytoplasm, is the shortest part of the cell cycle. Successive mitotic divisions alternate with a much longer **interphase,** which usually accounts for about 90% of the time that elapses during each cell cycle. It is during interphase that the cell grows and copies its chromosomes in preparation for cell division. Interphase consists of three periods of growth. These subphases of interphase are called, in order, the **G_1 phase** (for first "gap"), the **S phase,** and the **G_2 phase** (for second "gap"). During all three subphases, the cell grows by synthesizing proteins and producing cytoplasmic organelles. However, chromosomes are duplicated only during the S phase (S stands for synthesis, as in DNA synthesis). Thus, a cell grows (G_1), continues to grow as it copies its chromosomes (S), grows more as it completes preparations for cell division (G_2), and divides (M). The daughter cells may then repeat the cycle.

Time-lapse films of living, dividing cells reveal the dynamics of mitosis as a continuum of changes. For purposes of description, however, mitosis is conventionally described as occurring in five subphases: **prophase, prometaphase, metaphase, anaphase,** and **telophase.** FIGURE 11.6 on the following pages describes these stages as they occur in an animal cell. Be sure to study this figure thoroughly before progressing to the next section, which examines mitosis more closely.

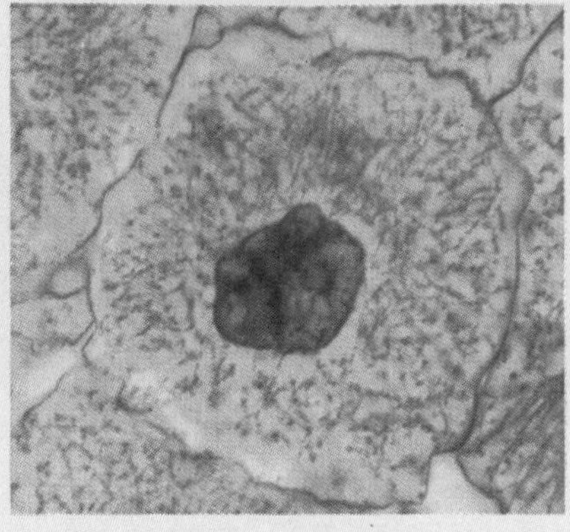

G_2 OF INTERPHASE

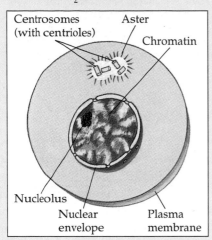

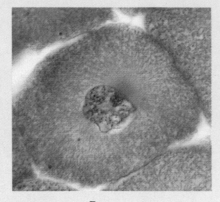

Centrosomes (with centrioles)
Aster
Chromatin
Nucleolus
Nuclear envelope
Plasma membrane

During late interphase, the nucleus is well defined and bounded by the nuclear envelope. It contains one or more nucleoli. Just outside the nucleus are two centrosomes, formed during early interphase by replication of a single centrosome. In animal cells, each centrosome features a pair of centrioles. Microtubules extend from the centrosomes in radial arrays called **asters** ("stars"). The chromosomes have already duplicated (during the S phase), but at this stage they cannot be distinguished individually because they are still in the form of loosely packed chromatin fibers.

PROPHASE

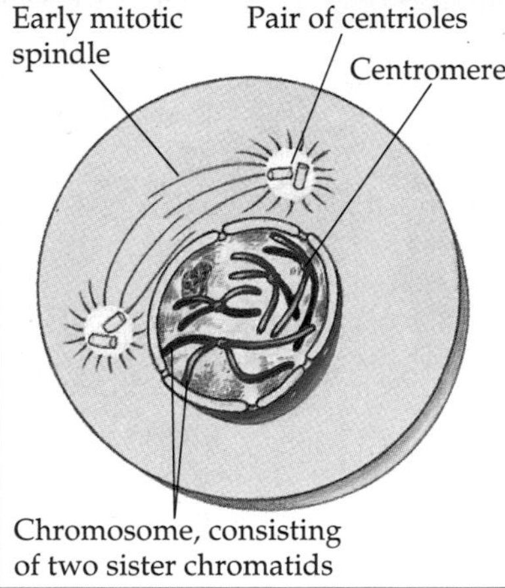

During prophase, changes occur in both the nucleus and the cytoplasm. In the nucleus, the nucleoli disappear. The chromatin fibers become more tightly coiled as they condense into discrete chromosomes observable with a light microscope. Each duplicated chromosome appears as two identical sister chromatids joined at the centromere. In the cytoplasm, the mitotic spindle begins to form; it is made of microtubules arranged between the two centrosomes. During prophase, the centrosomes move away from each other, apparently propelled along the surface of the nucleus by the lengthening bundles of microtubules between them.

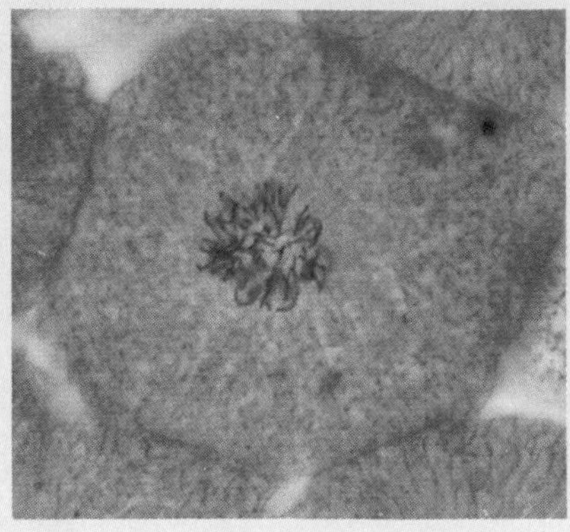

PROMETAPHASE

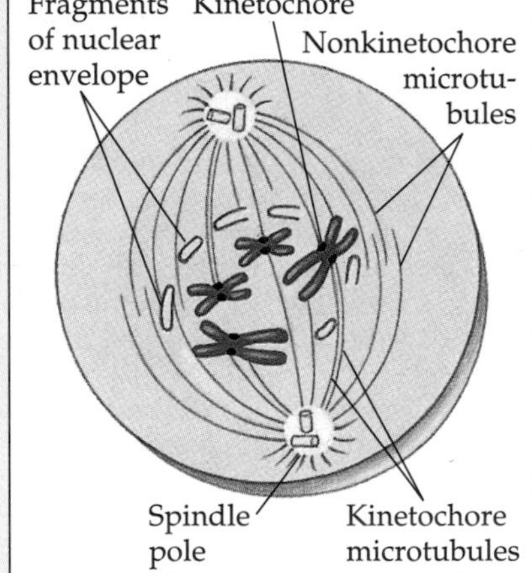

During prometaphase, the nuclear envelope fragments. The microtubules of the spindle can now invade the nucleus and interact with the chromosomes, which have become even more condensed. Bundles of microtubules extend from each pole toward the equator of the cell. Each of the two chromatids of a chromosome now has a specialized structure called the **kinetochore,** located at the centromere region. Some of the microtubules, called kinetochore microtubules, attach to the kinetochores. This interaction causes the chromosomes to begin jerky movements. Nonkinetochore microtubules overlap with those from the opposite pole of the cell.

FIGURE 11.6

The stages of mitotic cell division in an animal cell. The light micrographs, similar to slides you are likely to see in lab, show dividing cells from a fish embryo. (In plant cells, centrioles are lacking and cytokinesis occurs by a different mechanism; see FIGURES 11.9 and 11.10.) The schematic drawings show details not visible in the micrographs. For the sake of simplicity, only four chromosomes are drawn. A dividing cell does not "click" from one mitotic stage to the next, as it appears to in these static micrographs. Cell division is a dynamic continuum of changes.

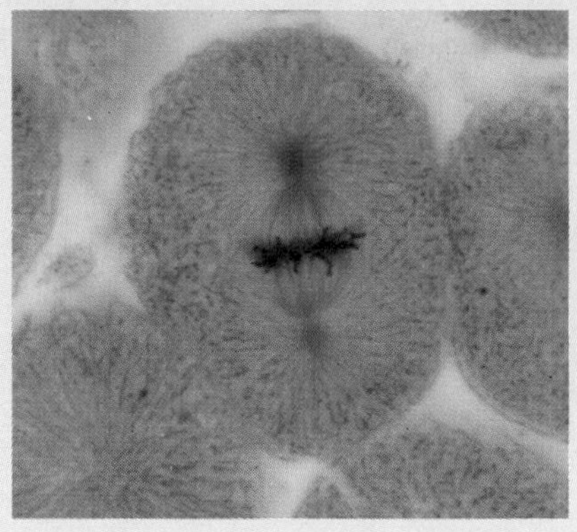

METAPHASE

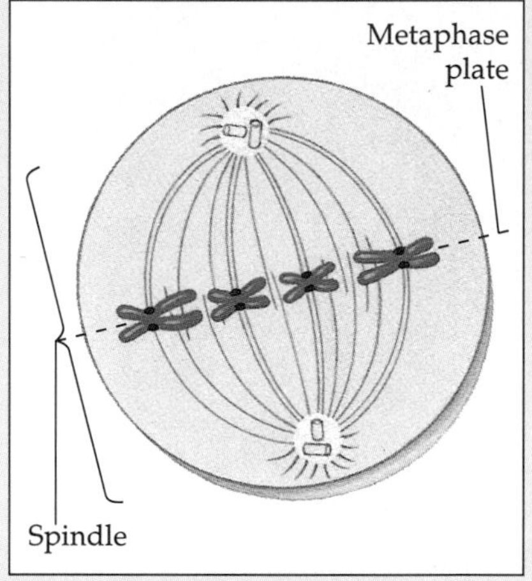

The centrosomes are now at opposite poles of the cell. The chromosomes convene on the **metaphase plate,** an imaginary plane that is equidistant between the spindle's two poles. The centromeres of all the chromosomes are aligned with one another, and sister chromatids of each chromosome straddle the metaphase plate. For each chromosome, the kinetochores of the sister chromatids face opposite poles of the cell. Thus, the identical chromatids of each chromosome are attached to kinetochore microtubules radiating from opposite ends of the parent cell. The entire apparatus of nonkinetochore microtubules plus kinetochore microtubules is called the spindle because of its shape.

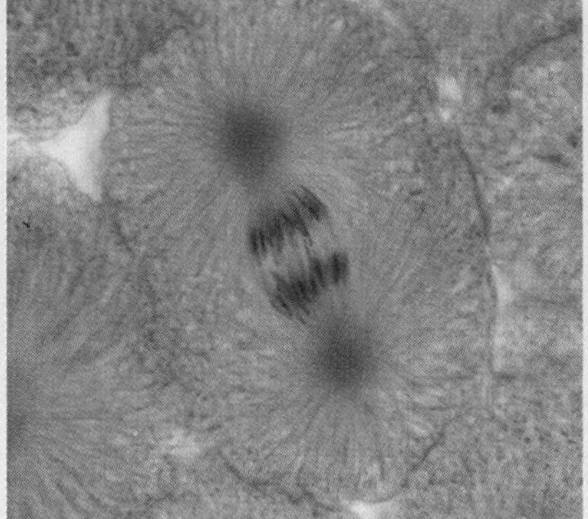

ANAPHASE

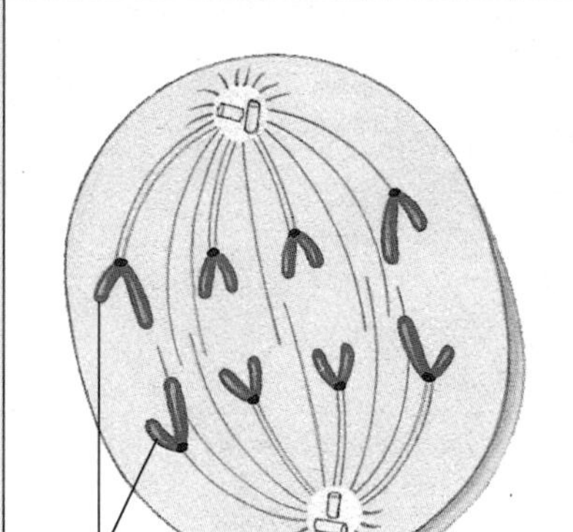

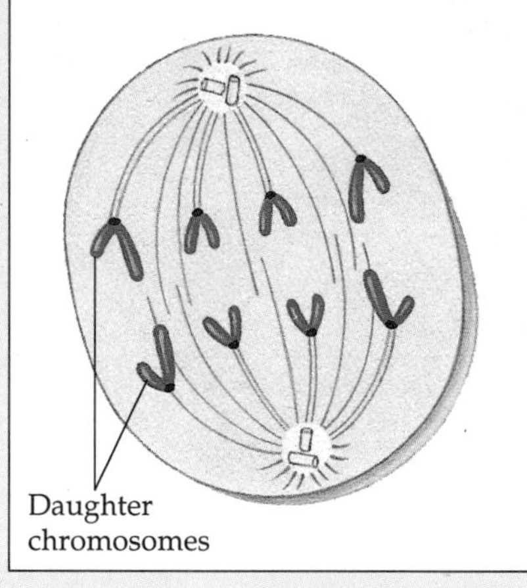

Anaphase begins when the paired centromeres of each chromosome separate, liberating the sister chromatids from each other. Each chromatid is now considered a full-fledged chromosome. The once-joined sisters then begin moving along microtubules toward opposite poles of the cell. Because the kinetochore microtubules are attached to the centromere, the chromosomes move centromere first (their pace is about 1 μm/min). At the same time, the poles of the cell also move farther apart. By the end of anaphase, the two poles of the cell have equivalent—and complete—collections of chromosomes.

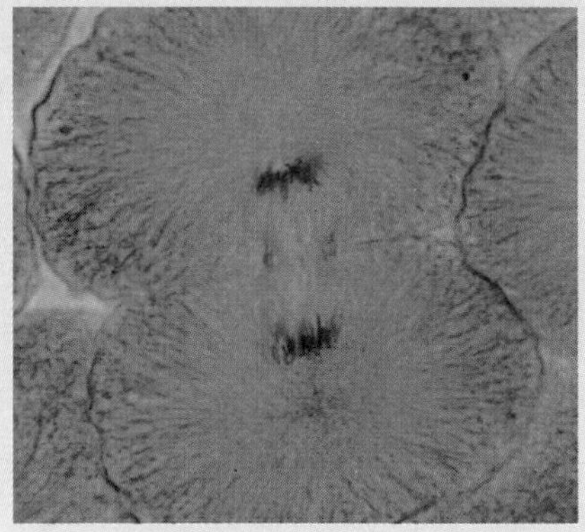

TELOPHASE AND CYTOKINESIS

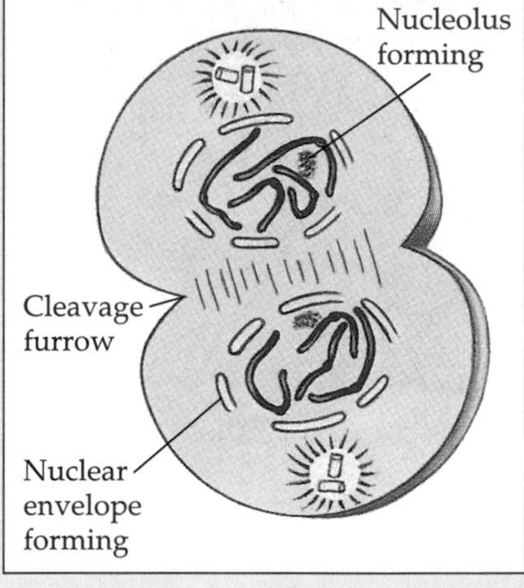

At telophase, the nonkinetochore microtubules elongate the cell still more, and the daughter nuclei begin to form at the two poles of the cell where the chromosomes have gathered. Nuclear envelopes are formed from the fragments of the parent cell's nuclear envelope and other portions of the endomembrane system. In a further reversal of prophase and prometaphase events, the nucleoli reappear, and the chromatin fiber of each chromosome becomes less tightly coiled. Mitosis, the equal division of one nucleus into two genetically identical nuclei, is now complete. Cytokinesis, the division of the cytoplasm, is usually well under way by this time, so the appearance of two separate daughter cells follows shortly after the end of mitosis. In animal cells, cytokinesis involves the formation of a cleavage furrow, which pinches the cell in two.

The mitotic spindle distributes chromosomes to daughter cells: *a closer look*

Many of the events of mitosis depend on a structure called the **mitotic spindle,** which begins to form in the cytoplasm during prophase. This structure consists of fibers made of microtubules and associated proteins (FIGURE 11.7). While the mitotic spindle is being assembled, the microtubules of the cytoskeleton are partially disassembled, probably providing the material used to construct the spindle. The spindle microtubules elongate by incorporating more subunits of the protein tubulin (see TABLE 7.2). Several parallel microtubules form bun-

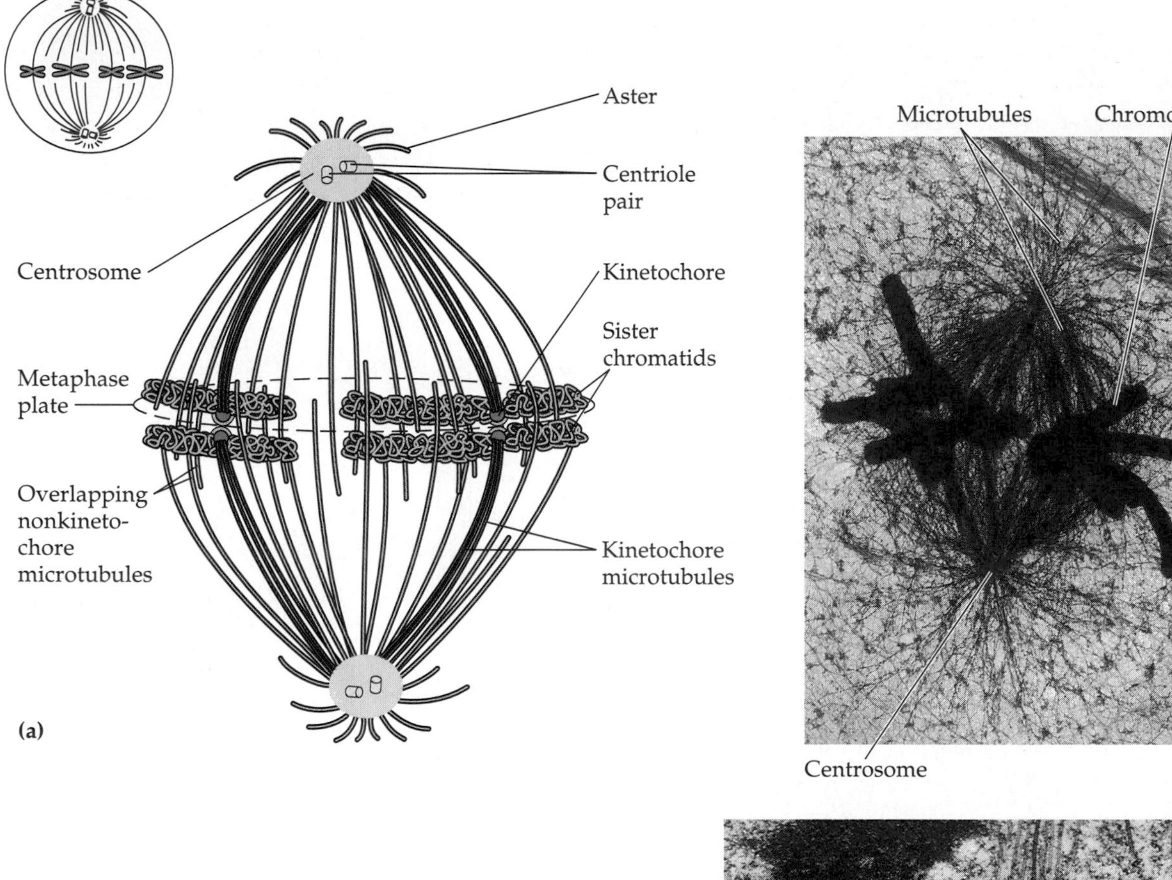

FIGURE 11.7

The mitotic spindle at metaphase. (**a**) The chromosomes are arranged on the metaphase plate, attached to kinetochore microtubules radiating from the centrosomes at the poles of the cell. Nonkinetochore microtubules, not attached to chromosomes, overlap at the metaphase plate. The orientation diagram in the upper left corner will help you relate this closer look at the mitotic spindle to the overview of mitosis in FIGURE 11.6. (**b**) You can see in these transmission electron micrographs that the kinetochores of a chromosome's two sister chromatids face opposite poles of the cell.

dles called spindle fibers that are large enough to see with a light microscope.

The assembly of spindle microtubules is initiated in the centrosome (see Chapter 7), also called the microtubule organizing center. In animal cells, a pair of centrioles is located at the center of the centrosome, but these structures are not essential for cell division. In fact, the centrosomes of most plants lack centrioles. And if the centrioles of animal cells are destroyed with a laser microbeam, spindles nevertheless form and function during mitosis.

During interphase of the cell cycle, the single centrosome replicates to form two centrosomes located just outside the nucleus (see FIGURE 11.6). As the two centrosomes move farther apart during prophase and prometaphase, spindle microtubules radiate from them, their tips away from the centrosomes. By the end of prometaphase, the two centrosomes are arranged at opposite poles of the cell.

Each of the two joined chromatids of a chromosome has a protein structure called a **kinetochore** located at the centromere region. Some microtubules called **kinetochore microtubules** attach to the kinetochores. Other **nonkinetochore microtubules** overlap with nonkinetochore microtubules from the opposite pole of the cell.

During prometaphase, kinetochore microtubules from one pole of the cell may attach to a kinetochore first, and the chromosome begins to move toward that pole. However, this movement is checked as soon as microtubules from the opposite pole attach to the chromosome's other kinetochore. What happens next is like a tug-of-war that ends in a draw. The chromosome moves first in one direction, then the other, back and forth, finally settling at the cell's midpoint. Apparently, microtubules can only remain attached to a kinetochore when there is a force exerted on the chromosome from the opposite pole of the cell. If microtubules from just one pole attach to a chromosome, they lose their grip. Attachment to one kinetochore is stabilized only when microtubules from the opposite pole hook to the other kinetochore. This check-and-balance system equalizes the number of microtubules attached to the two kinetochores of a duplicated chromosome and moves the chromosome to the midline of the cell. At metaphase, all the duplicated chromosomes are aligned on the cell's midline, or metaphase plate (see FIGURE 11.7).

Let's now see how the structure of the entire spindle apparatus is correlated with its function during mitosis. Anaphase commences when each chromosome's centromeres separate and the sister chromatids, now separate chromosomes, move toward opposite poles of the cell. How do the kinetochore microtubules function in this poleward movement of chromosomes? There is experimental evidence that kinetochore microtubules shorten during anaphase by depolymerizing at their kinetochore ends (FIGURE 11.8). A chromosome creeps poleward, as its kinetochore somehow hangs onto the remaining tips of microtubules just ahead of the zone of depolymerization. The exact mechanism of this interaction between kinetochores and microtubules is still unresolved. However, there is increasing evidence that the kinetochore is equipped with motor proteins that "walk" a chromosome along the shortening microtubules. (To review the relationship between motor proteins and the cytoskeleton, see Chapter 7.)

What is the function of the nonkinetochore microtubules? In a dividing animal cell, these tubules overlap at the middle of the cell and are responsible for elongating the whole cell along the polar axis during anaphase (see FIGURE 11.7). This occurs when the interdigitating tubules slide past each other away from the cell's equator. The

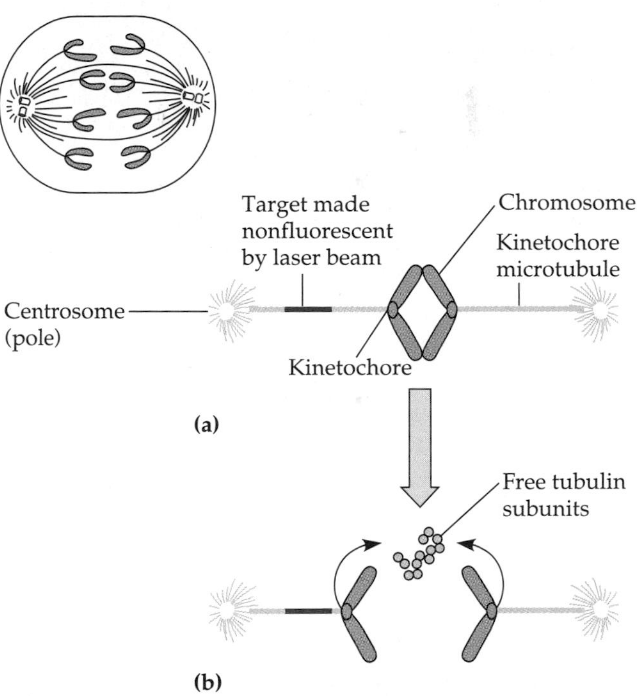

FIGURE 11.8

Testing a hypothesis for chromosome separation during anaphase. (**a**) In this experiment, the kinetochore microtubules were labeled with a fluorescent dye that glows in the microscope. During early anaphase, the researchers aimed a laser microbeam at microtubules about midway between the pole and the kinetochore. The laser eliminated fluorescence at the target region, but the microtubules still functioned. This treatment marked a fixed point on a microtubule, making it possible to monitor changes in the length of the microtubule on either side of the mark. (**b**) As anaphase proceeded and the chromosomes moved toward the poles, the segment of microtubule on the kinetochore side of the laser mark shortened. The portion of the tubule on the centrosome side of the mark retained its length. This is one of the experiments supporting the hypothesis that a chromosome tracks along shortening microtubules as tubulin (disproportionately large in this drawing) depolymerizes.

movement may be quite similar to the sliding of neighboring microtubules in a flagellum. According to this hypothesis, motor proteins drive nonkinetochore fibers past one another. ATP provides the energy.

At the end of anaphase, then, duplicate sets of chromosomes are clustered at opposite poles of the parent cell, which has elongated in its polar axis. Nuclei reform during telophase. Concurrent with this last stage of mitosis, the process of cytokinesis generally divides the cell's cytoplasm.

Cytokinesis divides the cytoplasm: *a closer look*

In animal cells, cytokinesis occurs by a process known as cleavage. The first sign of cleavage is the appearance of a **cleavage furrow,** which begins as a shallow groove in the cell surface near the old metaphase plate (FIGURE 11.9a). On the cytoplasmic side of the furrow is a contractile ring of microfilaments made of the protein actin, the same protein that plays a key role in muscle contraction, as well

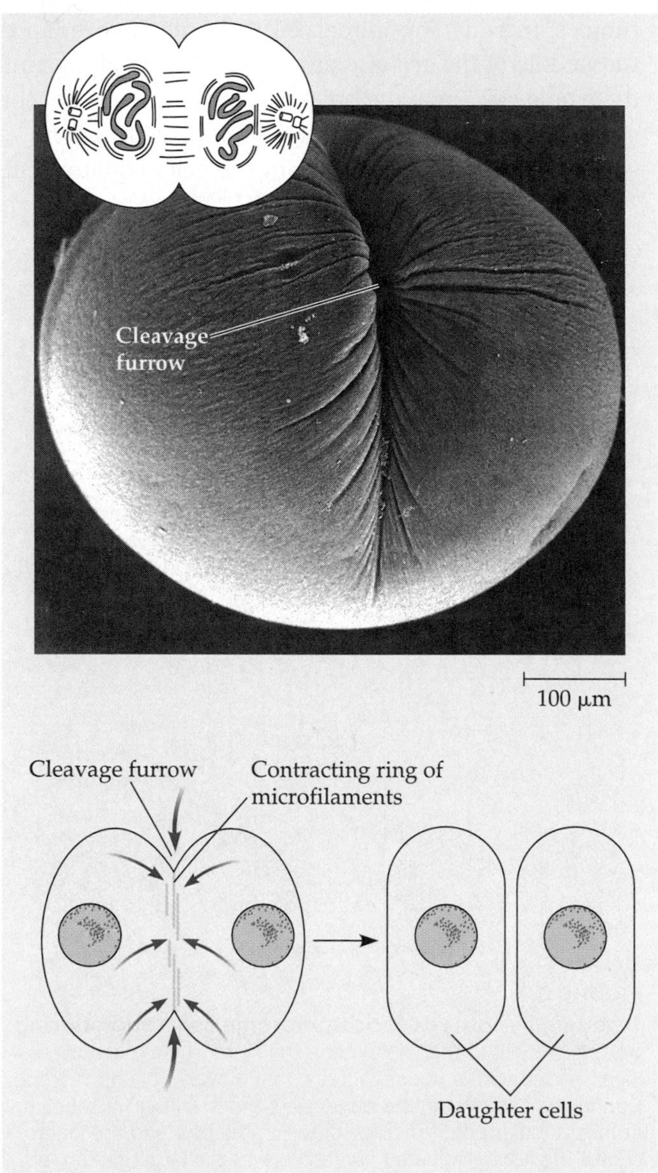

(a) Cleavage of an animal cell

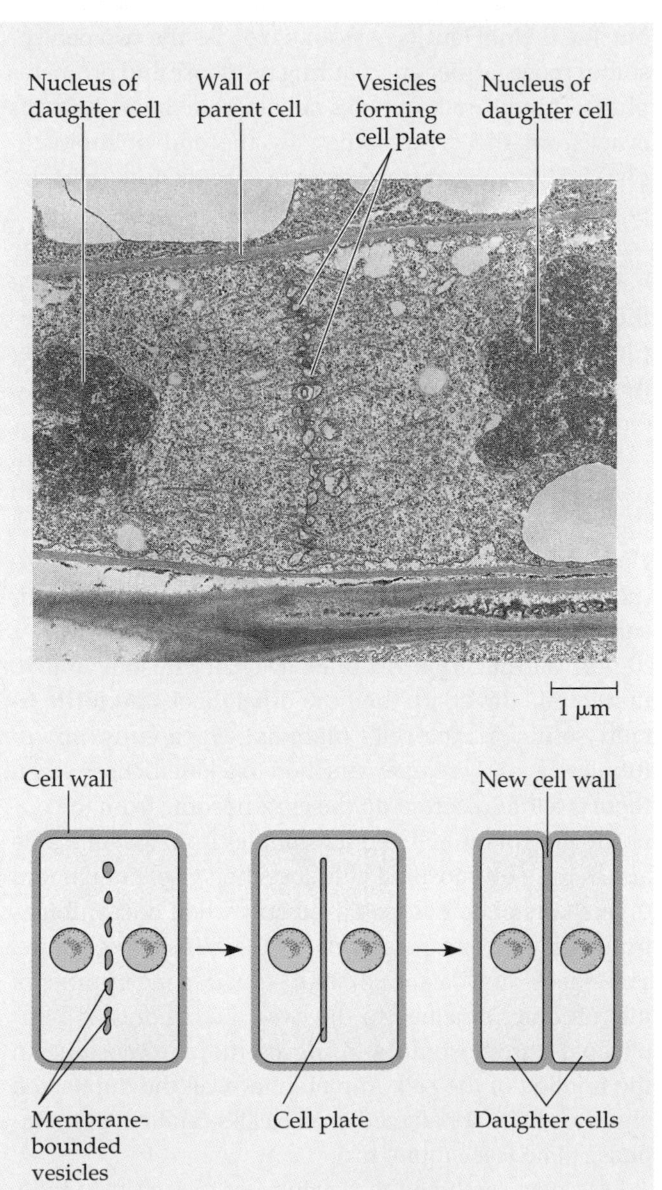

(b) Cell plate formation in a plant cell

FIGURE 11.9

A comparison of cytokinesis in animal and plant cells. (**a**) In the micrograph, we see the cleavage furrow of a dividing animal cell as it appears on the cell surface (SEM). Microfilaments form a ring just inside the plasma membrane at the location of furrowing. These microfilaments consist of actin and myosin, which cause the cleavage furrow to deepen until the cell is pinched in two. (**b**) In this micrograph of a soybean root cell at telophase, you can see the nuclei that form at the two poles (TEM). Meanwhile, membrane-bounded vesicles fuse to form a double membrane, which encloses the cell plate at the equator of the cell. Materials are secreted into the space between the membranes to form the new cell wall.

as many other kinds of cell movement (see Chapter 7). As the dividing cell's ring of microfilaments contracts and its diameter shrinks, the effect is like the pulling of drawstrings. The cleavage furrow deepens until the parent cell is pinched in two. The last bridge between the two daughter cells, containing the remains of the mitotic spindle, finally breaks, producing two completely separated cells.

Cytokinesis in plant cells, which have walls, is markedly different. There are no cleavage furrows. Instead, a structure called the **cell plate** forms during telophase across the midline of the parent cell where the old metaphase plate was located (FIGURE 11.9b). Vesicles derived from the Golgi apparatus move along microtubules to the middle of the cell, where they coalesce to form the cell plate. The fusion of vesicles forms two membranes, which eventually unite laterally with the existing plasma membrane. This results in the formation of two daughter cells, each with its own plasma membrane. A new cell wall forms between the two membranes of the cell plate. FIGURE 11.10 shows one complete turn of the cell division cycle for a

FIGURE 11.10

Mitosis in a plant cell. These micrographs show cells of *Tradescantia* as viewed by differential-interference-contrast microscopy.

Nucleus
Nucleoli

INTERPHASE

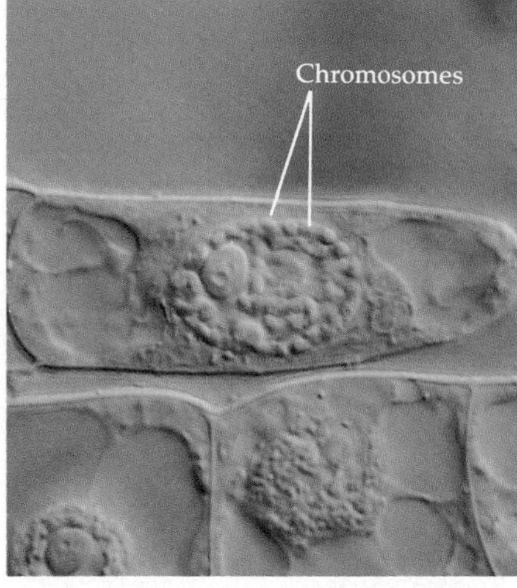

PROPHASE

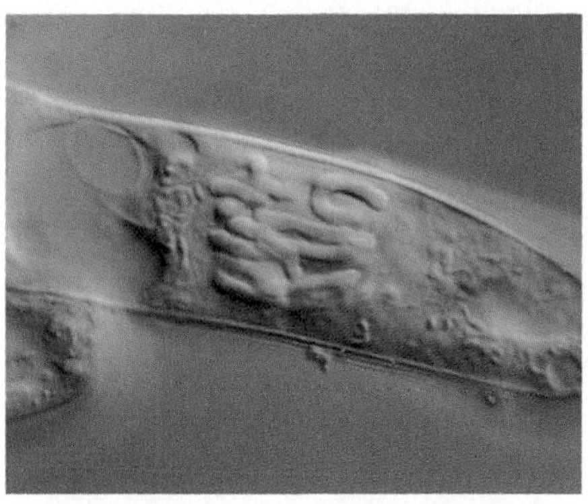

METAPHASE

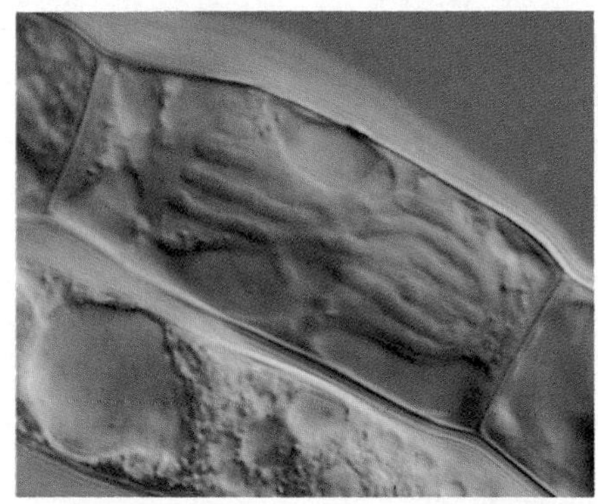

ANAPHASE

LATE ANAPHASE

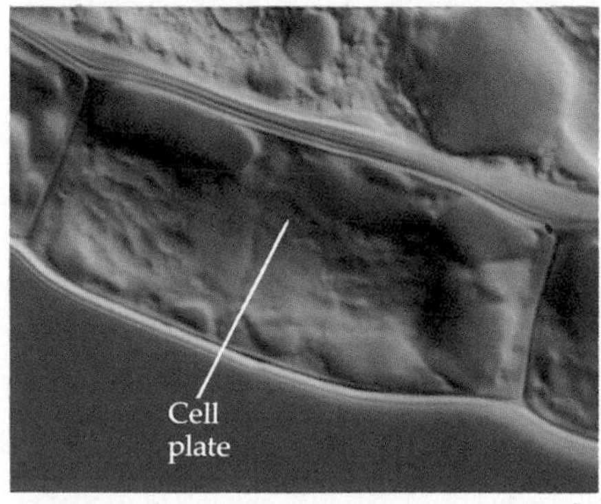

TELOPHASE

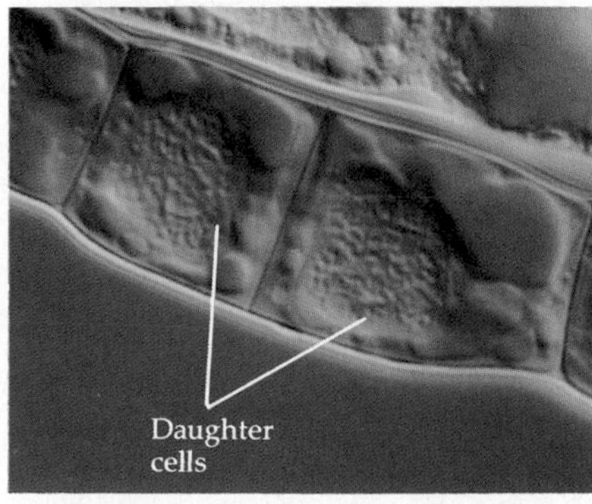

CYTOKINESIS

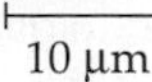

plant cell. Examining this figure will help you review the process of mitosis.

Researchers still disagree about many of the details of mitotic mechanics, and only additional experiments will resolve these debates. Almost all cell biologists *would* agree, however, that the complex cellular choreography of mitosis represents an evolutionary breakthrough that solved the problem of perpetuating the large genomes of eukaryotes. FIGURE 11.11 traces a hypothesis for a stepwise origin of mitosis.

External and internal cues control cell division

The timing and rate of cell division in different parts of a plant or animal are critical to normal growth, development, and maintenance. We can observe several patterns of cell division in different types of cells. For example, human skin cells divide frequently throughout life, whereas liver cells maintain the ability to divide but keep it in reserve until an appropriate situation arises—say, to

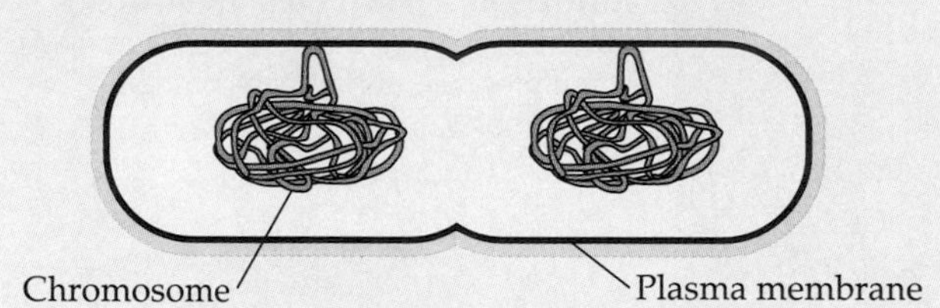

During binary fission of bacteria, the daughter chromosomes are attached to the plasma membrane. Elongation of the cell separates the chromosomes.

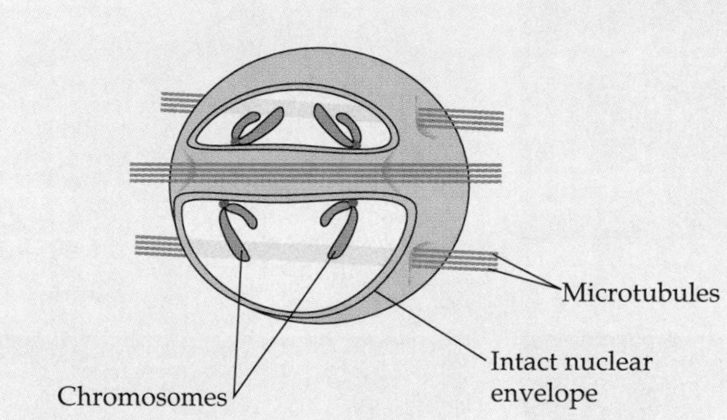

A nucleus contains the multiple chromosomes of a eukaryotic cell. In unicellular algae called dinoflagellates, the nuclear envelope remains intact during cell division. Chromosomes attach to the nuclear envelope, much as the bacterial chromosome attaches to the plasma membrane. Microtubules pass through cytoplasmic tunnels through the nucleus. The microtubules do not attach to chromosomes. They function instead to reinforce the spatial orientation of the nucleus, which then divides in a fission process reminiscent of bacterial reproduction.

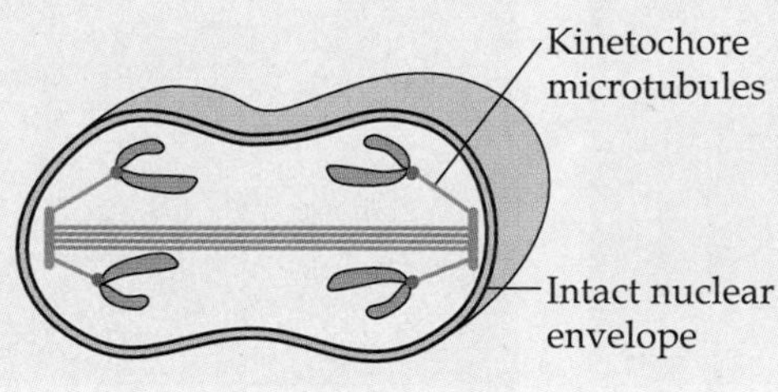

In another group of unicellular algae called diatoms, the nuclear envelope also remains intact during cell division. But in these organisms, the microtubules form a spindle *within* the nucleus. The microtubules separate the chromosomes, and the nucleus splits into two daughter nuclei.

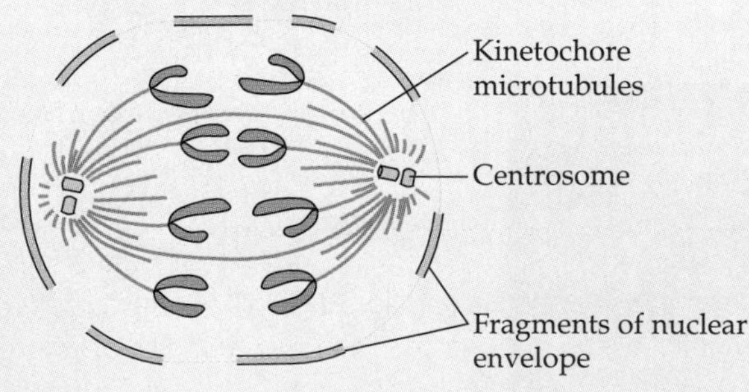

In most other eukaryotes, including plants and animals, the spindle forms outside the nucleus, and the nuclear envelope breaks down during mitosis. Microtubules separate the chromosomes, and the nuclear envelope re-forms.

FIGURE 11.11

The evolution of mitosis. Mitotic cell division is unique to eukaryotes. Prokaryotes (bacteria) have much smaller genomes and reproduce by the simpler process of binary fission. As eukaryotes, with their larger genomes, evolved, their ancestral process of binary fission somehow gave rise to mitosis. Researchers interested in this evolutionary problem have observed what they believe are mechanisms of cell division intermediate between binary fission and mitosis. Two of these variations in cell division are illustrated here, along with bacterial fission and mitosis as it occurs in plants and animals.

METHODS: CELL CULTURE

Many types of animal and plant cells can be removed from an organism and cultured in an artificial environment. This makes it possible to study the cell cycle and many other activities of cells in controlled conditions.

In the experiment illustrated here, we test the effect of a specific growth factor on the ability of human fibroblast cells to divide. Fibroblasts are the cells of connective tissue that secrete collagen, the protein making up the extracellular fibers that strengthen our tissues and organs. Our source of fibroblasts is a small sample of tissue removed during a biopsy. Two steps are required to isolate these cells. First, small pieces of connective tissue are dissected from the tissue sample. Second, specific enzymes are added to digest the collagen fibers and other components of the extracellular matrix, resulting in a suspension of free fibroblast cells. The isolated fibroblasts are placed in a culture vessel, a flat flask that sits on its side. The cells adhere to the glass on the inside of the vessel, and they are bathed in a liquid culture medium of known composition. It is essential to sterilize the vessels and the culture media before use, in order to kill microorganisms that would contaminate the culture. The culture medium is a complex recipe of glucose, amino acids, salts, and antibiotics (as a further precaution to inhibit bacterial growth). The cultures are incubated at an optimal temperature, 37°C for human fibroblast cultures.

In our experiment, some culture vessels contain only this basic medium, while others contain the basic medium plus a protein called platelet-derived growth factor, or PDGF. In the intact animal, blood cells called platelets release this growth factor at the site of an injury, stimulating nearby fibroblasts to proliferate. This is an important step in the healing of wounds. Our experiment confirms that PDGF stimulates the cell division of cultured fibroblasts.

Source: Art adapted from R. I. Freshney, *Culture of Animal Cells: A Manual of Basic Technique*, 2nd ed., fig. 1.2, p. 5. © 1989 John Wiley and Sons. Reprinted by permission of Wiley-Liss, a division of John Wiley and Sons, Inc.

DISSECTION

Obtain suspension of free cells by using enzymes to digest extracellular matrix.

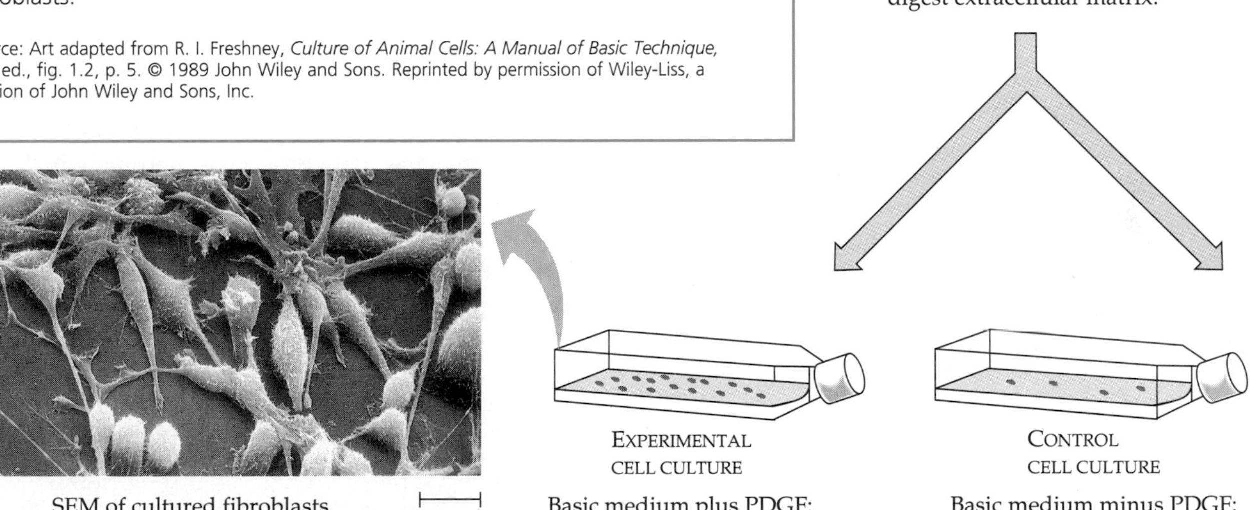

SEM of cultured fibroblasts

Basic medium plus PDGF: Cells proliferate.

Basic medium minus PDGF: Cells fail to divide.

repair a wound. Some of the most specialized cells, such as nerve cells and muscle cells, do not divide at all in a mature human.

By growing cells in artificial cultures, researchers have been able to identify many chemical and physical factors that can stimulate or inhibit cell division (see the Methods Box). For example, cells fail to divide if an essential nutrient is left out of the culture medium. Even if all other conditions are favorable, some types of mammalian cells will only divide in cultures if the growth medium includes specific regulatory substances called **growth factors.** For example, fibroblasts, the main cells of connective tissue, require platelet-derived growth factor, or PDGF. Fibroblast cells have receptors for PDGF on their plasma membranes, and binding of the growth factor to the cell stimulates cell division. This regulation occurs not only in the artificial conditions of cell culture, but in the body of the animal as well. Blood cells called platelets fragment and release PDGF in the vicinity of an injury. This triggers cell division of fibroblasts in that location, a response that helps heal the wound. Researchers have discovered several other growth factors. Each cell type probably responds specifically to a certain growth factor or combination of factors.

The density of cells is another important condition regulating cell division (FIGURE 11.12). Cultured cells normally divide until they form a single layer of cells on the inner surface of the culture container, at which point the cells stop dividing. If some cells are removed to create a space, those bordering the open space begin dividing again until the vacancy is filled. Crowding inhibits cell division. This phenomenon is called **density-dependent inhibition** of cell division. A population of cells competes for nutrients and for minute quantities of growth regulators. Apparently, when cells reach a certain density, the amount of these required substances per cell is insufficient to allow continued growth of the cell population. To divide, most cells also require adhesion to a substratum: the inside of a culture jar or the extracellular matrix of a tissue. Cells normally stop dividing if they lose their anchorage. Density-dependent inhibition probably functions in the body's tissues as well as in cell culture, checking the proliferation of cells at some optimal population density. Cancer cells, which we will discuss later in the chapter, do not exhibit density-dependent inhibition.

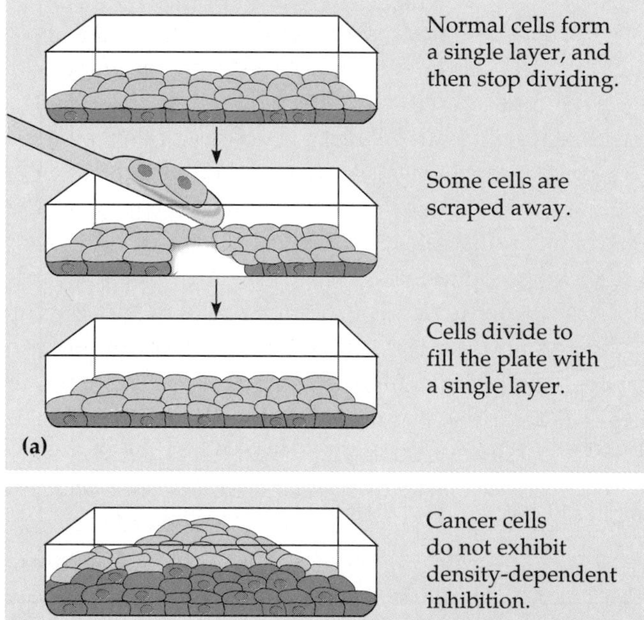

FIGURE 11.12

Density-dependent inhibition of cell division. (**a**) When grown in cell culture, normal cells will multiply only until they form a single layer. The availability of nutrients, growth factors, and substratum for attachments limits the density of the cell population. If some cells are removed, those at the border will divide until the gap is filled with a single layer. (**b**) In contrast, cancer cells usually continue to divide after they are crowded, forming a clump of overlapping cells. (Individual cells are shown disproportionately large in this figure.)

The G_1 phase of the cell cycle is a key period in the control of cell division. A crucial checkpoint occurs late in G_1, just before DNA synthesis (the S phase) would begin. Called the **restriction point,** this is when the "go/no-go decision" to divide is made. If all systems are "go" at this point—if all the internal and external cues are favorable—the cell proceeds to copy its DNA. Then the cell divides. Alternatively, the cell may "exit" the cell cycle at the restriction point and switch to a nondividing state called the **G_0 phase.** Most cells of the human body are actually in the G_0 phase. The most specialized cells, such as nerve and muscle cells, will never divide again. Other cells, such as liver cells, can be "called back" to the cell cycle by certain environmental cues, such as growth factors released during injury.

For cells that *are* actively proliferating in favorable conditions, cell size seems to be one important criterion for passing the restriction point. The cell must grow to a certain size during G_1 before DNA synthesis begins. How does the cell sense its size? The ratio of cytoplasmic volume to genome size seems to be an important indicator. As a cell grows by adding cytoplasm, the amount of DNA in the nucleus remains constant. The cell may pass the restriction point and copy its DNA once the cell's volume-to-genome ratio reaches a certain threshold value. Without some regulatory mechanism based on cell size, we can imagine daughter cells getting smaller and smaller with each cell cycle.

Cyclical changes in regulatory proteins function as a mitotic clock

The restriction point is a point of no return. The onset of the S phase commits ("restricts") the cell to continue through the G_2 and M phases and divide, regardless of external conditions, such as nutrient supply. This does not mean, however, that the cell is out of control. The precise sequence of events required for successful division seems to depend on the completion of each task before the cell can progress to the next stage. For example, chromosomes do not condense until they have replicated, the nuclear envelope does not break down until the chromosomes are condensed, the spindle does not separate sister chromatids until all the chromosomes have arrived at the metaphase plate, and cytokinesis does not occur until anaphase moves chromosomes toward opposite poles of the cell. Cell biologists are beginning to work out the "switches" that control this precise sequence.

Rhythmic fluctuations in regulatory proteins function as a molecular clock that paces the sequential events of the cell cycle. Some of these molecules belong to a family of proteins called **protein kinases,** enzymes that control the activities of other proteins. A protein kinase activates or inactivates another protein by catalyzing the transfer of a phosphate group from ATP to the target protein. This phosphorylation of the target protein typically switches the protein "on" by changing its shape to a more active conformation. (In some cases, phosphorylation switches a protein "off" by changing its conformation to a less active form.)

If the action of a particular protein kinase marks the transition from one phase of the cell cycle to another, then what causes the cyclical change in activity of that kinase? The answer is another class of regulatory proteins called **cyclins,** so named because their concentration fluctuates cyclically. A protein kinase that helps control the cell cycle is active only when it is attached to a particular cyclin. Because of this requirement, these kinases are called **cyclin-dependent kinases,** or **Cdks.** A particular Cdk is present in uniform concentration throughout the cell cycle, but its activity in phosphorylating target proteins rises and falls with changes in the concentration of its cyclin "adaptor."

Let's examine the specific case of the cyclin-Cdk complex called **MPF.** The initials stand for "maturation promoting factor," but we can think of it as "M-phase promoting factor," because it is the master switch for a cell's passage from interphase to mitosis (FIGURE 11.13). When cyclins that accumulate during G_2 of interphase associate with the protein kinase molecules, the resulting MPF complex initiates prophase and activates diverse proteins that function in mitosis. For example, phosphorylation of certain chromatin proteins causes chromosomes to condense during prophase. The nuclear envelope disperses during prometaphase when kinases phosphorylate proteins of the nuclear lamina, the internal lining of the nuclear envelope (see FIGURE 7.9). Acting directly on target proteins, or indirectly by phosphorylating other protein kinases, MPF is probably required for correct performance of the entire mitotic sequence.

Near the end of the mitotic phase, MPF *switches itself off* by activating an enzyme that destroys cyclin. The other part of MPF, the Cdk, persists in the cell in inactive form until it associates with new cyclin synthesized during interphase of the next cell cycle. Fluctuating activities of different cyclin-Cdk complexes time other stages of the cell cycle. For example, a particular cyclin-Cdk complex is required for a cell to pass the restriction point, the transition from G_1 to S of interphase. The rhythmic changes in activity of regulatory proteins keeps the sequential changes in a dividing cell on schedule.

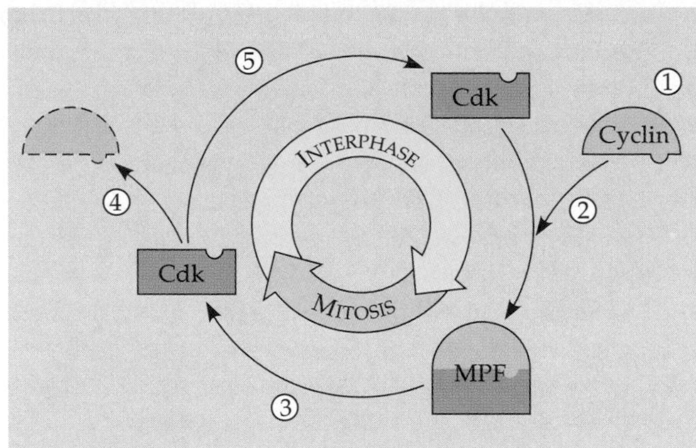

FIGURE 11.13
Protein kinases and the control of cell division. The stepwise processes of the cell cycle are timed by rhythmic fluctuations in the activity of protein kinases, enzymes that switch target proteins on or off by phosphorylating them. One such protein kinase is MPF, a complex of a cyclin (a regulatory protein whose concentration changes cyclically) and a Cdk (a cyclin-dependent kinase). ① The cyclin is synthesized throughout the cycle and accumulates during interphase. ② Cyclin attaches to the Cdk, and the protein complex is activated at the end of interphase. ③ The active complex, MPF, coordinates mitosis by phosphorylating various proteins, including other protein kinases. ④ One of the proteins activated by MPF is a cyclin-degrading enzyme that destroys MPF activity. ⑤ The Cdk component of MPF is recycled, its kinase activity restored by association with new cyclin that accumulates during interphase.

Let's review control of the cell cycle. G_1 is the most variable phase, both in duration and in the variety of external and internal controls over cell division. Nutritional status, growth factors, the density of the cell population, and the developmental state of the cell all affect the length of G_1 and whether the cell will pass the restriction point and divide. A cell that does not pass the restriction point will diverge from the cell cycle, temporarily or permanently, as a nondividing G_0 cell. If all other conditions favor cell division, the cell will proceed through the restriction point when it grows enough to achieve a certain volume-to-genome ratio. This cell is now irreversibly committed to dividing. Protein kinases, activated by cyclins at key points in the cell cycle, sequence the steps of cell division. Thus, the cell cycle has a "divide/don't-divide" decision point (restriction point) and a control system that choreographs division once the cell is committed to reproducing.

Cancer cells escape from the controls on cell division

Cancer cells do not respond normally to the body's control mechanisms. They divide excessively and invade other tissues. If unchecked, they can kill the whole organism.

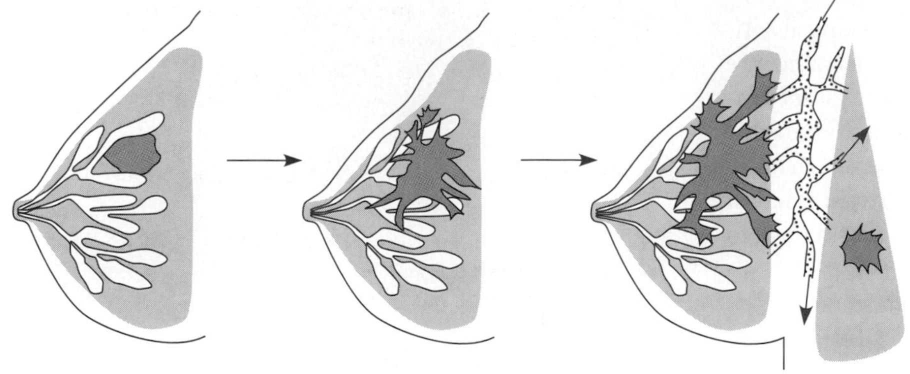

Tumor grows from a single cancer cell.

Cancer cells invade neighboring tissue.

Cancer cells spread through lymph and blood vessels to other parts of the body.

FIGURE 11.14
The growth and metastasis of a malignant breast tumor. The cells of malignant (cancerous) tumors grow in an uncontrolled way and can spread to neighboring tissues and, via the circulatory system, to other parts of the body. The spread of cancer cells beyond their original sites is called metastasis.

By studying cancer cells in culture, researchers have learned that they do not heed the normal signals that stop growth. In particular, they ignore density-dependent inhibition when growing in culture (see FIGURE 11.12b). The cultured cells continue to multiply even after contacting one another, accumulating until the nutrients in the growth medium are exhausted.

There are other important differences between normal cells and cancer cells that reflect derangements of the cell cycle. If and when they stop dividing, cancer cells seem to do so at random points in the cycle, rather than just at the restriction point of G_1. Moreover, in culture, cancer cells can go on dividing indefinitely if they have a continual supply of nutrients; thus they are said to be "immortal." A striking example is a cell line that has been reproducing in culture since 1951. (Cells of this line are called HeLa cells because their original source was a tumor removed from a woman named Henrietta Lacks.) By contrast, nearly all normal mammalian cells growing in culture divide only about 20 to 50 times before they stop dividing, age, and die.

The abnormal behavior of cancer cells can be catastrophic when it occurs in the body. The potential problem begins when a single cell in a tissue is transformed, transformation being the term for the conversion of a normal cell to a cancer cell. The body's defense system—the immune system—normally destroys these insurgent cells. However, if the cancer cell somehow evades destruction, it may proliferate to form a **tumor,** a mass of cancer cells within an otherwise normal tissue (FIGURE 11.14). If the cells remain at this original site, the lump is called a **benign tumor.** Benign tumors do not usually cause serious problems, and they can be completely removed by surgery. In contrast, a **malignant tumor** becomes invasive enough to impair the functions of one or more organs. An individual with a malignant tumor is said to have cancer.

The cells of malignant tumors are abnormal in many ways besides their lack of self-control over cell division. They may have unusual numbers of chromosomes. Their metabolism may be deranged, and they cease to function in any constructive way. Also, due to abnormal changes of the cells' surfaces, they lose their attachments to neighboring cells and the extracellular matrix. This enables the cancer cells to spread into other tissues surrounding the malignant tumor. Cancer cells may also separate from the tumor and enter the blood and lymph vessels of the circulatory system. These cells can invade other parts of the body and proliferate to form more tumors. This spread of cancer cells beyond their original site is called **metastasis.** If a tumor metastasizes, it is usually treated with high-energy radiation and poisonous chemicals that are especially harmful to actively dividing cells.

Researchers are beginning to understand how a normal cell is transformed into a cancer cell. Though the causes of cancer are diverse, most involve the transformation of cells by altering genes that control cell division. (We will consider the role of DNA rearrangements in cancer in Chapter 18.) Our knowledge of how changes in the genome lead to the various abnormalities of cancer cells is still rudimentary. Perhaps the reason we have so many unanswered questions about cancer cells is that there is still so much to learn about how normal cells function. The cell, life's basic unit of structure and function, holds enough secrets to engage researchers well into the future.

*   *   *

With this chapter on the reproduction of cells, we have built a bridge to the next unit, which features genes and their role in inheritance.

REVIEW OF KEY CONCEPTS (with page numbers and key figures)

- Cell division functions in reproduction, growth, and repair (pp. 204–205, FIGURE 11.1)
  - The perpetuation of life has its basis in cell division. Unicellular organisms reproduce by cell division. Multicellular organisms depend on cell division for development, growth, and repair.
- Bacteria reproduce by binary fission (pp. 205–206, FIGURE 11.2)
  - Prokaryotes undergo binary fission, in which the cell splits in two after replication of the single chromosome.
- The genome of a eukaryotic cell is organized into multiple chromosomes (pp. 206-207, FIGURE 11.4)
  - Eukaryotic cell division consists of mitosis (division of the nucleus) and cytokinesis (division of the cytoplasm).
  - The grouping of genes in chromosomes makes it possible for a eukaryotic cell to reproduce and distribute an enormous number of genes.
  - Chromosomes are composed of chromatin, a threadlike complex of DNA and protein that becomes more condensed during mitosis.
  - When chromosomes replicate, they form identical sister chromatids joined by a centromere. These chromatids separate during mitosis, thereby becoming the chromosomes of the new daughter cells.
- Mitosis alternates with interphase in the cell cycle: *an overview* (pp. 207–209, FIGURES 11.5, 11.6)
  - Mitosis and cytokinesis make up the M (mitotic) phase of the cell cycle, a sequence of events in dividing cells.
  - Between divisions, cells are in interphase, consisting of the G_1, S, and G_2 phases. The cell grows throughout interphase, but DNA is replicated only during the S (synthesis) phase.
  - Mitosis is a dynamic continuum of sequential changes, often described by the five stages of prophase, prometaphase, metaphase, anaphase, and telophase.
- The mitotic spindle distributes chromosomes to daughter cells: *a closer look* (pp. 210–212, FIGURE 11.7)
  - The mitotic spindle is a complex of microtubules that orchestrates chromosome movement during mitosis. During prophase, the spindle begins to form from the centrosome, a region near the nucleus associated with centrioles in animal cells.
  - The spindle includes kinetochore microtubules, which attach to the kinetochores of chromatids.
  - Kinetochore microtubules move all the chromosomes to the metaphase plate.
  - During anaphase, sister chromatids separate and move toward opposite poles of the cell. The kinetochore somehow moves along the shortening microtubules. At the same time, a sliding of nonkinetochore microtubules elongates the whole cell in the polar axis.
  - During telophase, daughter nuclei form at opposite ends of the dividing cell.
- Cytokinesis divides the cytoplasm: *a closer look* (pp. 212–214, FIGURE 11.9)
  - In most cases, mitosis is followed by cytokinesis, the formation of cleavage furrows in animals or cell plates in plants.
- External and internal cues control cell division (pp. 214–216, FIGURE 11.12)
  - Cell culture is a powerful method for studying the cell cycle. It has enabled researchers to determine the nutrients, growth factors, and other requirements for cell division.
  - The restriction point, near the end of the G_1 phase, is a critical time in the cell cycle. At this checkpoint, the cell either becomes a nondividing (G_0) cell or commits to DNA synthesis and cell division. The cell's volume-to-genome ratio must reach a certain threshold value for the cell to pass the restriction point, and other environmental and developmental conditions must also favor cell division.
- Cyclical changes in regulatory proteins function as a mitotic clock (pp. 216–217, FIGURE 11.13)
  - Protein kinases, enzymes that regulate other proteins by phosphorylating them, control the sequential processes of the cell cycle.
  - These protein kinases are themselves regulated by their requirement for cyclins, proteins whose concentrations fluctuate rhythmically during the cell cycle.
- Cancer cells escape from the controls on cell division (pp. 217–218, FIGURE 11.14)
  - Cancer cells elude normal regulation and divide out of control, forming tumors.
  - Malignant tumors are those that spread to surrounding tissues or export cancer cells via the circulatory system to other parts of the body, in a process called metastasis.

SELF-QUIZ

1. Which process is not associated with the reproduction of bacteria?
   a. replication of DNA
   b. binary fission
   c. mitosis
   d. synthesis of new cell wall material
   e. elongation of the parent cell

2. Through a microscope, you can see a cell plate beginning to develop across the middle of the cell and nuclei re-forming at opposite poles of the cell. This cell is most likely a (an)
   a. animal cell in the process of cytokinesis
   b. plant cell in the process of cytokinesis
   c. animal cell in the S phase of the cell cycle
   d. bacterial cell dividing
   e. plant cell in metaphase

3. In a typical cell cycle, cytokinesis generally overlaps in time with which stage?
   a. S phase
   b. prophase
   c. telophase
   d. anaphase
   e. metaphase

4. A particular cell has half as much DNA as some of the other cells in a mitotically active tissue. The cell in question is most likely in
   a. G_1
   b. G_2
   c. prophase
   d. metaphase
   e. anaphase

5. One difference between a cancer cell and a normal cell is that
   a. the cancer cell is unable to synthesize DNA
   b. the cell cycle of the cancer cell is arrested at the S phase
   c. cancer cells continue to divide even when they are tightly packed
   d. cancer cells cannot function properly because they suffer from density-dependent inhibition
   e. cancer cells are always in the M phase of the cell cycle

6. The decline of MPF at the end of mitosis is caused by
   a. the destruction of the protein kinase (Cdk)
   b. decreased synthesis of cyclin
   c. the enzymatic destruction of the cyclin
   d. synthesis of DNA
   e. an increase in the cell's volume-to-genome ratio

7. In the light micrograph below of dividing cells near the tip of an onion root, identify a cell in interphase, prophase, metaphase, and anaphase. Describe the major events occurring at each stage.

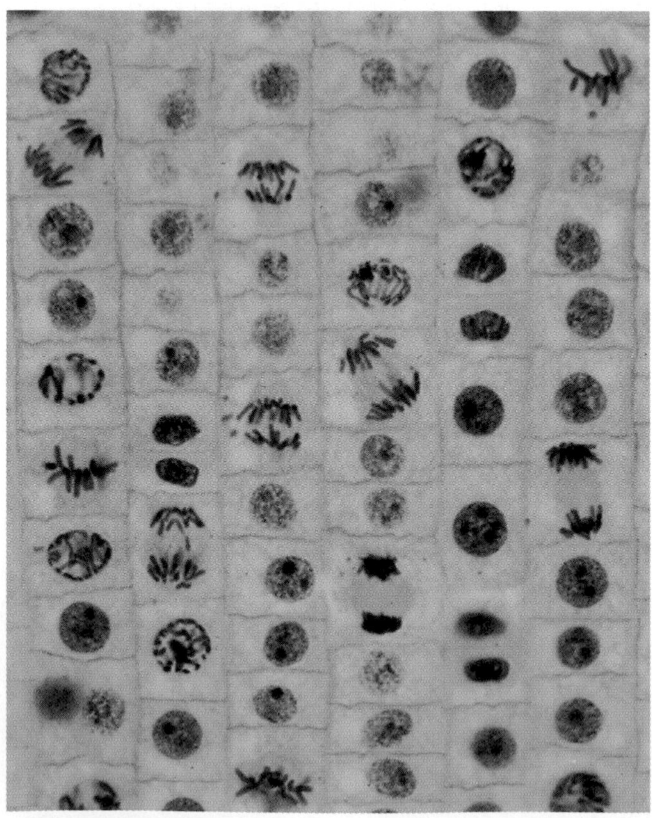

8. Which of the following is *not* a function of mitosis in humans?
   a. growth
   b. wound repair
   c. embryonic development
   d. production of gametes (sex cells)
   e. replacement of blood cells

9. In *function,* the plant cell structure that is analogous to an animal cell's cleavage furrow is the
   a. chromosome
   b. cell plate
   c. nucleus
   d. centrosome
   e. spindle apparatus

10. In some organisms, mitosis occurs without cytokinesis occurring. This will result in
   a. cells with more than one nucleus
   b. cells that are unusually small
   c. cells lacking nuclei
   d. destruction of chromosomes
   e. cell cycles lacking an S phase

CHALLENGE QUESTIONS

1. When a population of cells is examined with a microscope, the percentage of the cells in the M phase is called the mitotic index. The greater the proportion of cells that are dividing, the higher the mitotic index. In a particular study, cells from a cell culture are spread on a slide, preserved and stained, and then inspected in the microscope. A hundred cells are examined: 9 cells are in prophase; 5 cells are in metaphase; 2 cells are in anaphase; 4 cells are in telophase; the remainder, 80 cells, are in interphase. Answer the following questions:
   a. What is the mitotic index for this cell culture?
   b. The average duration for the cell cycle in this culture is known to be 20 hours. What is the duration of interphase? Of metaphase?
   c. Going back to the living culture of these cells, the average quantity of DNA per cell is measured. Of the cells in interphase, 50% contain 10 ng (1 nanogram = 10^{-9} g) of DNA per cell, 20% contain 20 ng DNA per cell, and the remaining 30% of the interphase cells have amounts of DNA between 10 and 20 ng. Based on these data, determine the duration of the G_1, S, and G_2 portions of the cell cycle.

2. About a day after a human egg is fertilized by a sperm cell, the zygote (fertilized egg) divides for the first time. The two daughter cells usually stick together, and their repeated cell divisions give rise to a multicellular embryo. On rare occasions, however, the two daughter cells formed by the first division of the zygote separate. Each of these cells can go on to form a normal embryo—not a half-embryo or an otherwise defective embryo. Based on what you have learned in this chapter, explain why these "monozygotic twins" are essentially genetically identical.

3. Vinblastine is a drug that interferes with the assembly of microtubules. It is widely used for chemotheraphy in treating cancer patients. Suggest a hypothesis to explain how vinblastine slows tumor growth.

4. Red blood cells, which carry oxygen to body tissues, live for only about 120 days. Replacement cells are produced by cell division in bone marrow. How many cell divisions must occur each second in your bone marrow just to replace red blood cells? Here is some information to use in calculating your answer. There are about 5 million red blood cells per mm^3 of blood. An average adult has about 5 L (5000 cm^3) of blood.

SCIENCE, TECHNOLOGY, AND SOCIETY

1. Imagine that the federal government has awarded you a $100,000 research grant from funds budgeted for cancer research. However, you do not work on cancer at all. Your research is very basic: You culture cells from mouse embryos in order to track the concentrations of key proteins through the normal cell cycle. A friend who cannot make the connection between cancer research and your work kids you about squandering tax dollars on dishes of mouse cells. How would you convince your friend that this is money well spent? Does public interest in how research funds are used threaten basic science? How does science benefit from skillful educators who can communicate the excitement of current scientific research to general audiences?

2. A biotechnology company has succeeded in extracting natural vanilla from cells grown in cell culture. Vanilla extracted from vanilla beans costs about $1200 per pound. Vanillin, a synthetic flavor used in many foods, costs about $7 per pound but lacks the many trace compounds that give natural vanilla its complex flavor. Cell culture vanilla costs less than $200 per pound, which might make it economical for use in more foods. Can you think of other products that might be produced by means of cell culture? What would be the advantages and disadvantages of producing them this way?

3. Every year about a million Americans are diagnosed as having cancer. This means that about 75 million Americans now living will eventually have cancer, and one in five will die of the disease. There are many kinds of cancers and many causes of the disease. For example, smoking causes most lung cancers. Overexposure to ultraviolet rays in sunlight causes most skin cancers. There is evidence that a high-fat, low-fiber diet is a factor in breast, colon, and prostate cancers. And agents in the workplace such as asbestos and vinyl chloride are also implicated as causes of cancer. Hundreds of millions of dollars are spent each year in the search for effective treatments for cancer; far less money is spent on preventing cancer. Why might this be the case? What kinds of lifestyle changes could we make to help prevent cancer? What kinds of prevention programs could be initiated or strengthened to encourage these changes? What factors might impede such changes and programs?

FURTHER READING

Alberts, B., D. Bray, T. Lewis, M. Raff, K. Roberts, and J. D. Watson. *Molecular Biology of the Cell,* 3rd ed. New York: Garland, 1994. Chapter 13.

Baringa, M. "A New Twist to the Cell Cycle." *Science,* August 4, 1995. How the timed destruction of cyclins regulates cell division.

Beardsley, T. "A War Not Won." *Scientific American,* January 1994. So far, the overall rate of death due to cancer has not decreased in the U.S.

Becker, W. M., J. Reece, and M. F. Poenie. *The World of the Cell,* 3rd ed. Menlo Park, CA: Benjamin/Cummings, 1996. Chapter 15.

Douglas, K. "Making Friends with Death-Wish Genes." *New Scientist,* July 30, 1994. Using natural regulatory mechanism to kill cancer cells.

Glover, D. M., C. Gonzalez, and J. W. Raff. "The Centrosome." *Scientific American,* June 1993. Structure and function of the mitotic spindle.

Marx, J. "How Cells Cycle Toward Cancer." *Science,* January 21, 1994. The normal functions and pathology of cyclins.

McIntosh, J. R., and K. L. McDonald. "The Mitotic Spindle." *Scientific American,* October 1989.

Murray, A.W., and T. Hunt. *The Cell Cycle: An Introduction.* New York: W. H. Freeman, 1993. A thorough overview of current research.

Murray, A. W., and M. W. Kirschner. "What Controls the Cell Cycle." *Scientific American,* March 1991.

UNIT THREE

THE GENE

AN INTERVIEW WITH DAVID SATCHER

David Satcher began his career as a medical geneticist. In 1972, he helped develop the King-Drew Sickle-Cell Research Center in Los Angeles and served as its director for 6 years. In addition to his abilities as a researcher and physician, Dr. Satcher began to distinguish himself as a gifted leader and administrator in a succession of positions, leading in 1982 to the Presidency of Meharry Medical College in Nashville. In 1993, when Dr. Satcher was appointed director of the Centers for Disease Control and Prevention (CDC), he became one of the most visible and important health scientists in the world. In this interview, Dr. Satcher talks about the responsibilities of the CDC and explains how a childhood experience inspired his interest in medicine and his continuing commitment to community service.

Dr. Satcher, you started your career in the field of genetics, specifically cytogenetics. Could you explain what cytogenetics is and what its medical significance is?

I was actually in the M.D.-Ph.D. Program at Case Western Reserve School of Medicine and my Ph.D. was in cytogenetics. Cytogenetics is the study of chromosomes located in cells. I studied the effects that X-radiation and radioactive iodine, I-131, had on chromosomes and was able to show a linear relationship between radiation dosage—and, to a certain extent, I-131 dosage—and chromosome damage. That was some of the earlier work looking directly at the impact of radiation on chromosomes.

Much of your early work focused on sickle-cell disease. What did you do, and what accomplishments were especially satisfying?

I directed one of the ten national sickle-cell research centers starting back in 1972. The center I directed, the King-Drew Sickle-Cell Research Center, at the King-Drew Medical Center in Los Angeles, emphasized community education, early diagnosis, screening, and counseling programs as well as treatment. I think we made major contributions in all those areas.

What is the objective of screening? How is it done? And what role do counselors play?

That's a very important issue. In the absence of community education, screening for sickle-cell disease can be a problem. The sickle-cell gene is recessive, which means that a person can be a carrier—have "sickle-cell trait"—but not be affected with the disease. Because communities didn't understand that in the early days of screening programs, there were laws saying that people with sickle-cell trait couldn't go to public schools because it was thought to be contagious. This is one of the best examples of what's necessary if you're going to develop a mass screening program. We don't recommend mass screening, but if you are going to do it, you have to make sure you have the education first so that you are prepared to deal with the results. Although the cause is not genetic, we have a similar problem right now with HIV and AIDS. You might ask why only 15% of the people in this country have been tested for HIV. Part of the answer is that people are afraid to find out the results. Similarly, if you don't have the appropriate support mechanisms in place, genetic screening programs can do more harm than good. You also have to have trained people other than geneticists, for example physician's assistants and nurse practitioners, to do genetic counseling. That is happening now, all over the country.

You mentioned that at Case Western Reserve you earned dual medical and Ph.D. degrees. Why did you select such a tough course of study, and how does one actually go about an M.D.-Ph.D.?

Let me first answer the question, "Why?" When I came to Morehouse College here in Atlanta in 1959, my goal was to do premedicine and then go to medical school to be a physician. When I was at Morehouse, I became interested in research. Some of my professors supported me, and in addition to working and being a preceptor in the laboratory, I actually got involved in doing research. By the time I graduated from Morehouse, I thought maybe I could make a contribution in the area of research since I really enjoyed it. So I went to Case Western as part of the program where you get both the M.D. and the Ph.D. It took 7 years, but it was enjoyable. I attribute a lot of my success to the fact that I have been able to relate to the basic scientist as well as to the physician.

It sounds like a good case for getting undergraduates into the laboratory and involved in research.

There's something about research that provides a certain amount of discipline and a certain amount of vision; it allows you to create.

So when you started as an undergrad at Morehouse, you already wanted to become a physician. How did you decide that?

That's an interesting story. I grew up in

Anniston, Alabama. Neither of my parents had finished elementary school. When I was 2 years old, I came down with a severe case of pertussis, or whooping cough, and it led to pneumonia. I was severely ill. The situation was such that we couldn't go to the hospital, but my parents were able to get Dr. Jackson, the only black physician in Anniston at the time, to come out to the farm where we lived and check on me. He came on one of his off days and spent almost the whole day there working with me and my parents. When he got ready to leave, he did not want to get my parents' hopes up yet, so he told them I probably wouldn't live out the week. But he did take the time to show my mother how to keep my chest clear and my temperature down. My mother worked very hard, and I pulled through. And my mother told me that story from the time I was old enough to understand words. By the time I was 8 years old, I was already telling people that I was going to be a doctor, like Dr. Jackson. All through elementary school and middle school and high school, I had this drive: I'm going to be a physician. At that time, I was going to come back to Anniston and be a family doctor. As time went on, my vision grew and I realized there were a lot of Anniston, Alabamas, in the world, and that I could probably contribute to a lot of them. That's how it started. And that motivation still drives me.

As your vision grew, you made the transition to health science administration at some point. What motivated that?

It really started back when I went out to Los Angeles to King-Drew in 1972. That was a fairly new institution, one that King Hospital had just opened three months earlier, and they were developing the school. I pulled people together from throughout the institution—basic scientists as well as clinicians—to get our first major grant for the Sickle-Cell Research Center. Within a few months, I was asked to become director of the center, which meant that I had a major administrative responsibility, not just a scientific or medical one. I had to make the team go. I guess I did a good job because people kept asking me to do more things administratively. When I was asked to be the president of Meharry Medical College, I didn't want to be a college president. But after I visited Meharry and learned that this institution had sent more of its graduates to serve in inner-city communities than any other medical school in the country, I thought I could make a difference. So I finally said, "Yes." I ended up spending 12 years there.

Then you resigned to become director of the CDC. Was that a hard decision?

It was. I was really not ready to leave Meharry. I struggled with it and decided to do it.

Tell us a little bit about the CDC.

It's interesting how it came to be. During World War II malaria was a major cause of death for American troops, so the public

health service decided to put together a program to teach our troops to combat malaria, how to rid the swamps of the mosquitoes that were causing the malaria. They put this program in the south, of course, where there are more swamps. When the war ended, the decision was made to continue a Communicable Disease Control Center in Atlanta. So in 1946, we became the CDC and have been here in Atlanta ever since. We are the nation's prevention agency. Highlights of the CDC's 40-year history include things like eradicating smallpox in the world, working with Legionnaire's disease, working to eradicate polio. We haven't had a case of wild-virus polio in this country since 1979, and we haven't had a case in the whole Western Hemisphere, including Latin America, since 1991. We believe that we can eradicate polio in the world by the year 2000. Those are just some examples.

In recent years, we've gotten much more involved in the prevention of chronic diseases, such as heart disease and cancer. Our newest center deals with injury control, things like automobile accidents, and we've made a lot of progress there. HIV/AIDS and violence are probably our toughest two problems right now.

We have about 6800 employees. About 4000 of them are here in Atlanta, the rest scattered throughout the country and throughout the world. Our budget is about $2.5 billion a year. And we have seven different centers. The CDC is known all over the world. People often look to us before they look to the World Health Organization to help with epidemics, such as the recent plague in India.

Is it becoming increasingly important for the CDC to have a global perspective?

It is, yes. We view health as a global issue. A lot of things have helped us realize it but nothing more than emerging infections. During the last 15 years or so, we've had several new infections, like HIV/AIDS, Lyme disease, Legionnaire's disease, and Ebola fever. In addition, we've had reemerging or resurgent infections like tuberculosis, which we thought we had essentially gotten rid of. About 30% of the new cases of TB in this country since 1985 have come from people traveling, and immigrating. HIV/AIDS probably started on another continent. Ebola fever was actually transmitted to this country through an experimental monkey in Africa. During the recent plague in India, we were only one plane flight away from that plague. Public health is definitely a global issue.

You mentioned that HIV is one of the biggest problems at the CDC now. What role did the CDC play in recognizing AIDS and the viral cause of the disease?

When the five original cases in this country were discovered in San Francisco and Los Angeles, the CDC was called right away. We sent public health advisors to the scene. We got blood samples. We were able to determine after a period of time that the cause was a virus that was being transmitted through the blood, through sexual relations, or other body fluid transmission. The CDC has been involved for the last 13 years, and it accounts for over 25% of our budget now. We have over 1000 employees who work just with HIV/AIDS. Just to show you how serious this problem is, there was a 60% increase in the number of new AIDS cases in the world last year. In this country, we believe we're beginning to make some progress; in the last 2 years, it's been increasing only about 3% annually. We are very cautious about saying this, however, because we want to keep the intensity up until we really have this disease under control. It doesn't look as if we are going to have a cure or even a vaccine soon, which means that we really must do everything we can to prevent the spread of this virus by helping people modify their behavior. And that's our major role right now. We're working with people at every level to try to do that.

The public probably thinks of the CDC mainly in connection with epidemiology and communicable diseases, but you've been advocating greater emphasis on some of its broader charges in terms of the health of the country. I'd like to get your reactions to a few of these broader views of public health—the health impact of smoking, for example. What role does the CDC have there?

Our mission is to promote health and quality of life and to prevent disease. So when we find that something is impacting the health of the American people and the quality of life, we try to find out whether there are ways to prevent it. We have about 2 million deaths in this country every year. About 20% of them are related to smoking, 420,000 deaths a year. That's a lot of deaths. And they represent the most preventable causes of death that we deal with, whether they're from cancer or respiratory disease or heart disease. So the CDC has taken very seriously its responsibility to try to control tobacco usage. Many people don't realize this, but 85% to 90% of the new smokers in this country are teenagers. We have interviewed some of them recently, and a lot of them, at the age of 16, are talking about how they wished they had not started. They wished they could quit because regardless of what some people say, smoking is addictive. So we feel that smoking is clearly a public health problem.

The same thing applies to violence. It is a little more difficult to make that argument to some people. There are people who argue that violence is a criminal justice problem: You just put the criminals in jail, and you won't have a problem. There are people who argue on the other side that violence is a social problem, that it's due to unemployment, poverty, and racism. And we don't disagree with any of that. We believe that violence is a criminal justice problem, and we believe that criminals have to be dealt with. We believe that violence is a social problem, that we must continue to improve the social environment where people live and work and grow. When we say it is a public health problem, we're saying that we believe that we can identify the risk factors for violence, that we can develop intervention strategies and decrease the rate of violence. In the African-American population, homicide is the leading cause of death in men and women between the ages of 15 and 34. So right now, the major problem of violence is with young people, the same group of people who are starting to smoke, the same group of people who are at risk from their sexual behavior. So it's very clear that we need to target teenagers more in this society.

What role is the CDC playing in women's health issues?

Throughout the Department of Health & Human Services—which includes the CDC, the NIH, and other agencies—we've acknowledged that women's health has been neglected in many ways. A lot of clinical trials have not included women. A lot of the research has targeted men. There are a lot of areas for prevention that have just not targeted women. For example, every year we have about 46,000 women dying from breast cancer and about 6000 dying from cervical cancer. All 6000 of those cervical cancer deaths and about 30% to 40% of the breast cancer deaths are preventable. So we are really intensifying our efforts to do just that. Another good example is the fact that every year about 150,000 women in this country become infertile because they have undiagnosed and untreated sexually transmitted diseases like chlamydia. In states where we've had demonstration projects, the CDC has reduced that problem by 50% in the last 2 years. We would like to expand that program nationwide. We are developing our office of women's health to really begin focusing on those kinds of problems.

Your whole career is marked by public service, and much of that has been community-based. How did that commitment to community develop, and how do we encourage young physicians and others to dedicate time to better the community?

Well, as I mentioned, my motivation for going into medicine in the first place was to make a difference at the community level. I had some good experiences early, even when I was in medical school. We were required to visit patients in their homes during our first year of medical school, to really get to know the situation in the communities. Then when I was an intern and resident at Strong Memorial Hospital in Rochester, New York, I had an opportunity to work at migrant health clinics one night a week and in neighborhood health centers. When I left there I went to Watts and worked for Al Haynes, one of the best epidemiologists in this country, who had a scientific approach to community. His attitude was that you bring the best science to bear on problems in communities. So I was able to bring together strong science and strong commitment to community. And that's what drives me. I think the way we get young people interested is that we find more mentors, more faculty members, who are willing to go with young people and to show them that they can make a difference in communities. We grew up thinking that we could change things, we could make things better. A lot of our young people today don't have that attitude. As adults, we've got to give young people meaningful, successful experiences very early so that they can feel the same way.

What other advice can you offer to undergraduates who are considering careers in the health sciences?

In the first place, I think it's important to really examine your motivation for going into health sciences. If you care about people and you're comfortable with science, I don't think there is any better way to go than health sciences. I do think it requires a real commitment to service to people. When you say medicine is a profession, you're not just saying it's good science, you're saying it's good humanity. We have to stress that it is a way to really make a difference in the lives of people, a very positive difference. We want to get young people excited about doing that.

Life's most exclusive distinction is the ability of organisms to reproduce their kind. Like begets like. Only oak trees produce oaks, and only elephants can make more elephants. Furthermore, offspring resemble their parents more than they do less closely related individuals of the same species. This continuity of traits from one generation to the next is called ***heredity*** *(L.* heres, *"heir"). Along with inherited similarity, there is also* ***variation:*** *Offspring exhibit individuality, differing somewhat in appearance from parents and siblings. These observations have been exploited for the thousands of years that people have bred plants and animals. Curiosity about human similarities and differences is just as ancient. The photo on this page illustrates family resemblance and individuality in the Tilly family. But the mechanisms of heredity and variation eluded biologists until the development of genetics in this century.* ***Genetics,*** *the scientific study of heredity and variation, is the subject of this unit. You will learn how biologists are answering questions about life that have endured for centuries; how discoveries in genetics are catalyzing progress in every other biological field, including physiology, evolutionary biology, ecology, and the behavioral sciences; how modern genetics is revolutionizing the pharmaceutical industry and other technologies; and how all these achievements are affecting society and raising new philosophical and ethical questions. In this chapter, we begin our study of genetics by learning how sexual reproduction passes chromosomes from parents to offspring.*

CHAPTER 12

MEIOSIS AND SEXUAL LIFE CYCLES

KEY CONCEPTS

- Offspring acquire genes from parents by inheriting chromosomes
- Like begets like, more or less: a comparison of asexual and sexual reproduction
- Fertilization and meiosis alternate in sexual life cycles: *an overview*
- Meiosis reduces chromosome number from diploid to haploid: *a closer look*
- Sexual life cycles produce genetic variation among offspring
- Evolutionary adaptation depends on a population's genetic variation

Offspring acquire genes from parents by inheriting chromosomes

Friends sometimes tell my daughter that she has her father's freckles, but *I* still *have* mine. Parents do not, in any literal sense, give their children freckles, eyes, hair, or any other traits. What, then, actually *is* inherited? Parents endow their offspring with coded information in the form of hereditary units called **genes.** The tens of thousands of genes we inherit from our mothers and fathers constitute our genome. Our genetic link to our parents accounts for family resemblance. My daughter's genome includes the gene for freckles, which she inherited from her father. Our genes program the emergence of specific traits as we develop from fertilized eggs into adults.

Genes are segments of DNA. You learned in Chapters 1 and 5 that DNA is a polymer of four different kinds of monomers called nucleotides. Inherited information is passed on in the form of each gene's specific sequence of nucleotides, much as printed information is communicated in the form of meaningful sequences of letters. Language is abstract. The brain translates words and sentences into mental images and ideas; for example, the object you imagine when you read "apple" looks nothing like the word itself. Analogously, cells translate genetic "sentences" into freckles and other features that bear no resemblance to genes. Most genes program cells to synthesize specific enzymes and other proteins, and it is the cumulative action of these proteins that produces an organism's inherited traits.

Inheritance has its chemical basis in the precise replication of DNA, which produces copies of genes that can be passed along from parents to offspring. The cellular vehicles for these genes are sperm and ova (unfertilized eggs). After a sperm cell unites with an ovum (a single egg), genes from both parents are present in the nucleus of the fertilized egg. Biological order based on heritable programs in the form of DNA is one of the unifying themes of biology.

The DNA of a eukaryotic cell is subdivided into chromosomes located within the nucleus. Every living species has a characteristic number of chromosomes. For example, humans have 46 chromosomes (except in their reproductive cells). Each chromosome consists of a single DNA molecule that is much longer than the chromosome itself. Along with various kinds of proteins, the DNA is elaborately folded and coiled, making up the structure of the chromosome. One chromosome includes hundreds or thousands of genes, each of which is a specific region of the DNA molecule. A gene's specific location along the length of a chromosome is called the gene's **locus** (plural, **loci**).

The physical mechanism of inheritance—the actual transmission of genes from parents to offspring—depends on the behavior of chromosomes. Our genetic endowment consists of whatever genes happened to be part of the chromosomes we inherited from our parents.

Like begets like, more or less: a comparison of asexual and sexual reproduction

Strictly speaking, "Like begets like" applies only to organisms that reproduce asexually. In **asexual reproduction,** a single individual is the sole parent and passes copies of all its genes on to its offspring. For example, single-celled organisms can reproduce asexually by mitotic cell division, in which DNA is copied and allocated equally to two daughter cells. The genomes of the offspring are exact copies of the parent's genome. Some multicellular organisms are also capable of reproducing asexually. *Hydra,* a relative of the jellyfish, can reproduce by budding (FIGURE 12.1). Since the cells of the bud were derived by mitosis in the parent, the "chip off the old block" is usually genetically identical to its parent. Occasional genetic differences are due to relatively rare changes in the DNA called mutations, which will be discussed in Chapter 16. An individual that reproduces asexually gives rise to a **clone,** a group of genetically identical individuals.

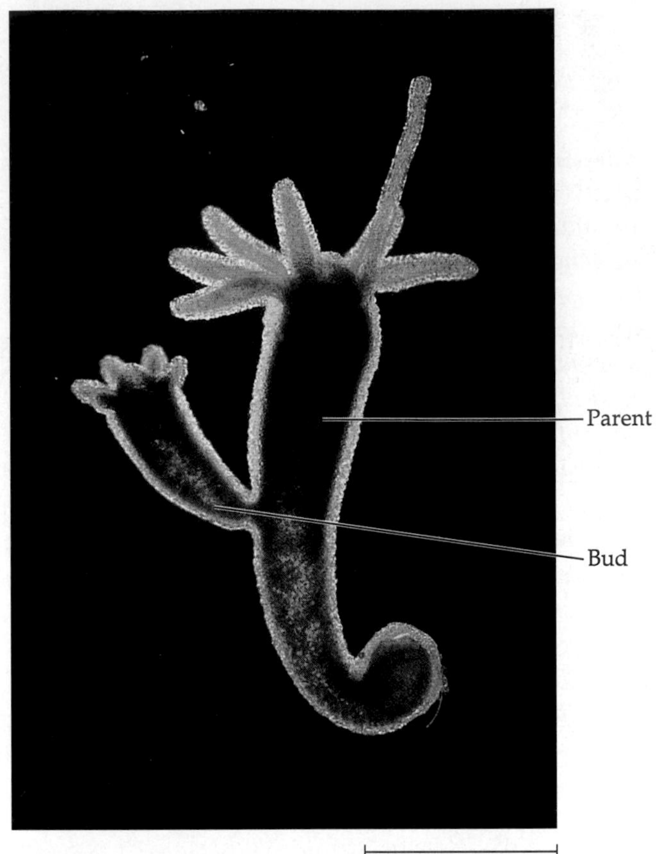

FIGURE 12.1
The asexual reproduction of *Hydra.* This relatively simple multicellular animal reproduces by budding. The bud, a localized mass of mitotically dividing cells, develops into a small *Hydra,* which detaches from the parent (LM).

Compared to asexual reproduction, **sexual reproduction** usually results in greater variation; two parents give rise to offspring that have unique combinations of genes inherited from both parents. In contrast to a clone, offspring of sexual reproduction vary genetically from their siblings and both parents (FIGURE 12.2). What mechanisms generate this genetic variation? The key is the activity of chromosomes during the sexual life cycle.

Fertilization and meiosis alternate in sexual life cycles: *an overview*

A **life cycle** is the generation-to-generation sequence of stages in the reproductive history of an organism, from conception to production of its own offspring. In this

FIGURE 12.2
Two families. Two sets of parents make up the top row, but the photos are randomly arranged. Each couple has two children represented among the four photos in the bottom row, also randomly arranged. (All individuals were photographed at about the same age; these are senior pictures from high school annuals.) Can you match the offspring with their parents? (See the bottom of the page for the answers.*) "Like begets like" in the general sense of family resemblance, but notice that each offspring is unique in her or his appearance, differing from parents and siblings. This genetic variation is an important consequence of sexual reproduction.

section, we track the behavior of chromosomes through sexual life cycles.

The Human Life Cycle

In humans, each **somatic cell**—any cell other than a sperm or ovum—has 46 chromosomes. With a light microscope, chromosomes can be distinguished from one another by their appearance. The sizes of chromosomes and the positions of their centromeres differ. Each chromosome also has a distinctive pattern of bands, which is visible after staining with certain dyes.

Careful examination of a micrograph of the 46 human chromosomes reveals that there are two of each type. This becomes clear when the chromosomes are arranged in pairs, starting with the longest chromosomes. The resulting display is called a **karyotype** (see the Methods Box on p. 234). The chromosomes that make up a pair—that have the same length, centromere position, and staining pattern—are called **homologous chromosomes,** or homologues. The two chromosomes of each pair carry genes controlling the same inherited characters. For example, if a gene for eye color is situated at a particular locus on a certain chromosome, then the homologue of that chromosome will also have a gene specifying eye color at the equivalent locus.

There is an important exception to the rule of homologous chromosomes for human somatic cells: the two distinct chromosomes referred to as *X* and *Y*. Human females have a homologous pair of *X* chromosomes (*XX*), but males have one *X* and one *Y* chromosome (*XY*). Because they determine an individual's sex, the *X* and *Y* chromosomes are called **sex chromosomes.** The other chromosomes are called **autosomes.**

The occurrence of pairs of chromosomes in our karyotype is a consequence of our sexual origins. We inherit one member of each chromosome pair from each parent. So the 46 chromosomes in our somatic cells are actually two sets of 23 chromosomes—a maternal set (from our mother) and a paternal set (from our father).

Sperm cells and ova are distinct from somatic cells in their chromosome count. Each of these reproductive cells, or **gametes,** has a single set of the 22 autosomes plus a single sex chromosome, either *X* or *Y*. A cell with a single chromosome set is called a **haploid cell.** For humans, the haploid number, abbreviated n, is 23 ($n = 23$).

By means of sexual intercourse, a haploid sperm cell from the father reaches and fuses with a haploid ovum of the mother. This union of gametes is called **fertilization,** or **syngamy.** The resulting fertilized egg, or **zygote,** contains the two haploid sets of chromosomes bearing genes representing the maternal and paternal family lines. The zygote and all other cells having two sets of chromosomes are called **diploid cells.** For humans, the diploid number, abbreviated $2n$, is 46 ($2n = 46$).

As a human develops from a zygote to a sexually mature adult, the zygote's genes are passed on with precision to all somatic cells of the body by the process of mitosis. Thus, somatic cells, like the zygote from which they are derived, are diploid.

The only cells of the human body *not* produced by mitosis are the gametes, which develop in the gonads (ovaries in females and testes in males). Imagine what would happen if gametes *were* made by mitosis: They

*Answer: The boy and girl on either end of the bottom row are children of the parents at top right. The boy and girl in the center of the bottom row are children of the parents at top left.

would be diploid like the somatic cells. At the next round of fertilization, when two gametes fused, the normal chromosome number of 46 would double to 92, and each subsequent generation would double the number of chromosomes yet again. But sexually reproducing organisms carry out a process that halves the chromosome number in the gametes, compensating for the doubling that occurs at fertilization. This process is a form of cell division called **meiosis,** and it occurs only in the ovaries or testes. While mitosis conserves chromosome number, meiosis reduces the chromosome number by half. As a result, human sperm and ova have haploid sets of the 23 different chromosomes. Fertilization restores the diploid condition, and the human life cycle goes on, generation after generation (FIGURE 12.3).

The processes of fertilization and meiosis are the unique trademarks of sexual reproduction. Fertilization and meiosis alternate in sexual life cycles, offsetting each other's effects on the chromosome number and thus perpetuating a species' chromosome count from one generation to the next. The life cycles of *all* sexually reproducing organisms follow a basic pattern of alternation between the diploid and haploid conditions, even though the details of the life cycles differ.

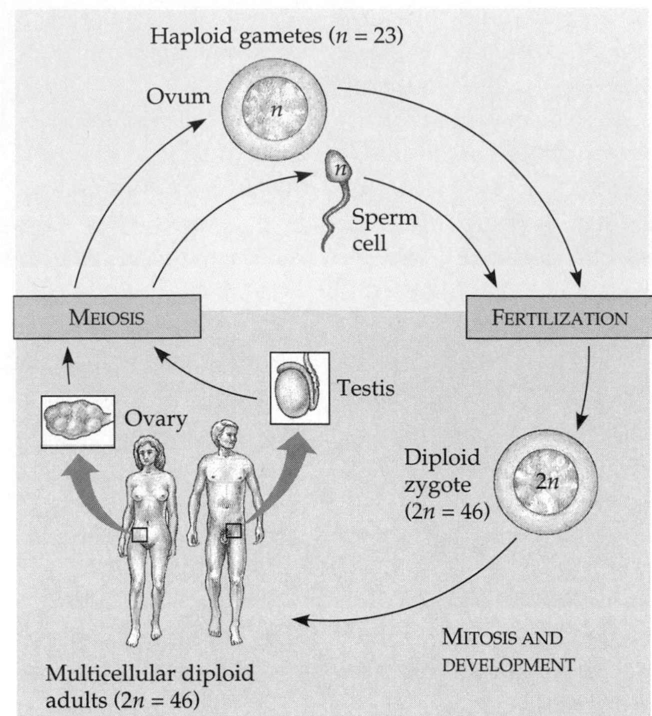

FIGURE 12.3

The human life cycle. In each generation, the doubling of chromosome number that results from fertilization is offset by the halving of chromosome number that results from meiosis. For humans, the number of chromosomes in a haploid cell is 23 ($n = 23$); the number of chromosomes in the diploid zygote and all somatic cells arising from it is 46 ($2n = 46$).

This figure introduces a color code that will be used for all life cycles throughout the book. A blue-green background represents haploid stages of a life cycle, and a tan background represents diploid stages.

The Variety of Sexual Life Cycles

Although the alternation of meiosis and fertilization is common to all organisms that reproduce sexually, the timing of these two events in the life cycle varies, depending on the species. These variations can be grouped into three main types of life cycles (FIGURE 12.4). The human life cycle is an example of one type, characteristic of most animals. Gametes are the only haploid cells. Meiosis occurs during the production of gametes, which

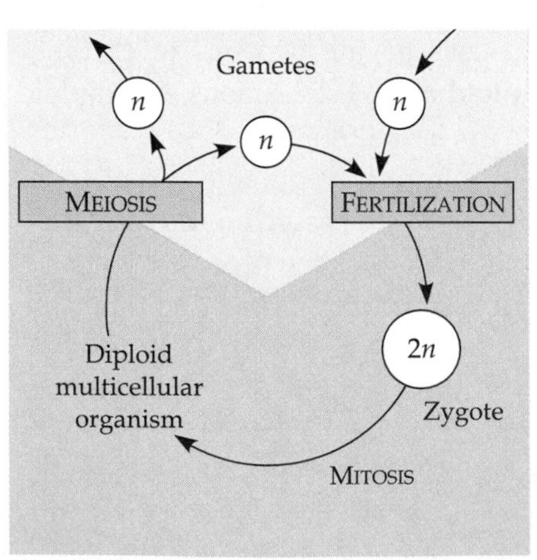

(a) Animals

Haploid multicellular organism
MITOSIS
MITOSIS
Gametes
MEIOSIS
FERTILIZATION
$2n$
Zygote

(b) Some fungi and some algae

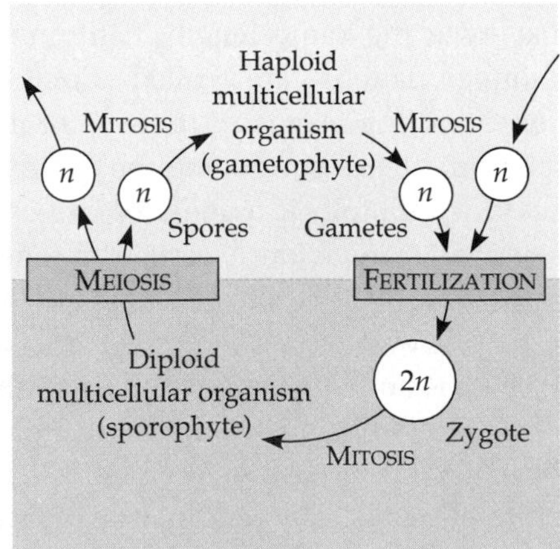

(c) Plants and some algae

FIGURE 12.4

Three sexual life cycles differing in the timing of meiosis and fertilization (syngamy). The common feature of all three cycles is the alternation of these two key events, which contribute to genetic variation among offspring.

HAPLOID
DIPLOID

undergo no further cell division prior to fertilization. The diploid zygote divides by mitosis, producing a multicellular organism that is diploid (FIGURE 12.4a).

A second type of life cycle occurs in many fungi and some protists (including some algae). After gametes fuse to form a diploid zygote, meiosis occurs before offspring develop. This produces haploid cells, which then divide by mitosis to give rise to a multicellular adult organism that is haploid. Subsequently, gametes are produced from the haploid organism by mitosis rather than by meiosis. The only diploid stage is the zygote (FIGURE 12.4b). (Note that *either* haploid or diploid cells can divide by mitosis depending on the type of life cycle. Only diploid cells, however, can undergo meiosis.)

Plants and some species of algae exhibit a third type of life cycle called **alternation of generations.** In this type of life cycle, there are both diploid and haploid multicellular stages. The multicellular diploid stage is called the **sporophyte.** Meiosis in the sporophyte produces haploid cells called **spores.** Unlike a gamete, a spore gives rise to a multicellular individual without fusing with another cell. A spore divides mitotically to generate a multicellular haploid stage called the **gametophyte.** The haploid gametophyte makes gametes by mitosis. Fertilization results in a diploid zygote, which develops into the next sporophyte generation. In this type of life cycle, therefore, the sporophyte and gametophyte generations take turns reproducing each other (FIGURE 12.4c).

Though the three types of sexual life cycles differ in the timing of meiosis and fertilization, they share a fundamental result: Each cycle of chromosome halving and doubling contributes to genetic variation among offspring. A closer look at meiosis will reveal the source of this variation.

Meiosis reduces chromosome number from diploid to haploid: *a closer look*

Many of the steps of meiosis closely resemble corresponding steps in mitosis. Meiosis, like mitosis, is preceded by the replication of chromosomes. However, this single replication is followed by *two* consecutive cell divisions, called **meiosis I** and **meiosis II** (FIGURE 12.5). These divisions result in four daughter cells (rather than the two daughter cells of mitosis), each with only half as many chromosomes as the parent. The drawings and text in FIGURE 12.6 describe in some detail the two divisions of meiosis for an animal cell whose diploid number is 4. Study FIGURE 12.6 thoroughly before going on to the next section.

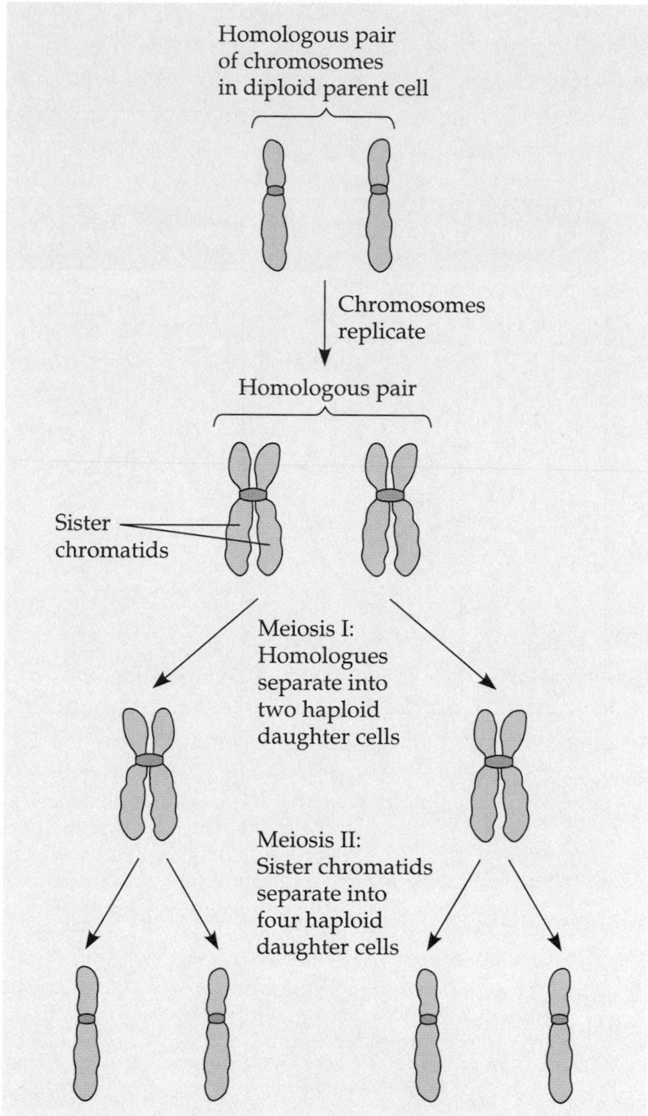

FIGURE 12.5
How meiosis reduces chromosome number. Chromosomes replicate *once,* but the diploid parent cell divides *twice* to form four haploid daughter cells. The first division (meiosis I) segregates the two chromosomes of each homologous pair, packaging them in separate daughter cells. The second division (meiosis II) separates the sister chromatids of each chromosome. Before proceeding to the more detailed drawings of meiosis in FIGURE 12.6, be sure you understand the difference between homologous chromosomes and sister chromatids. The two chromosomes of a homologous pair are individual chromosomes inherited from different parents. Homologues *appear* alike in the microscope, but they have different versions of genes at some of their corresponding loci (for example, a gene for freckles on one chromosome and a gene for the absence of freckles at the same locus of the homologue). Before meiosis, *each* of these chromosomes replicates to form sister chromatids that remain attached until meiosis II. (This simplified diagram tracks just one homologous chromosome in the diploid parent cell.)

Meiosis I:
Separates homologues

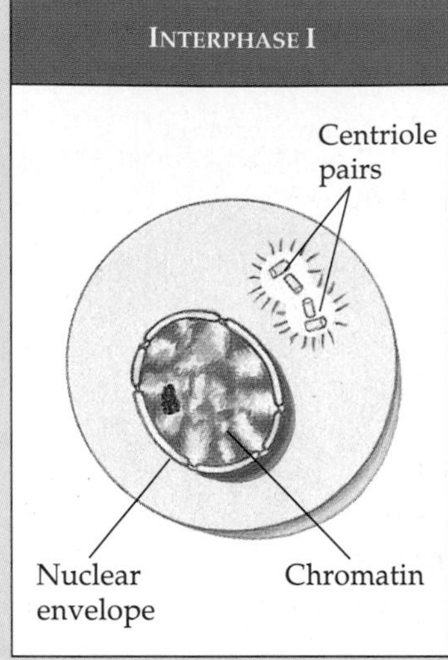

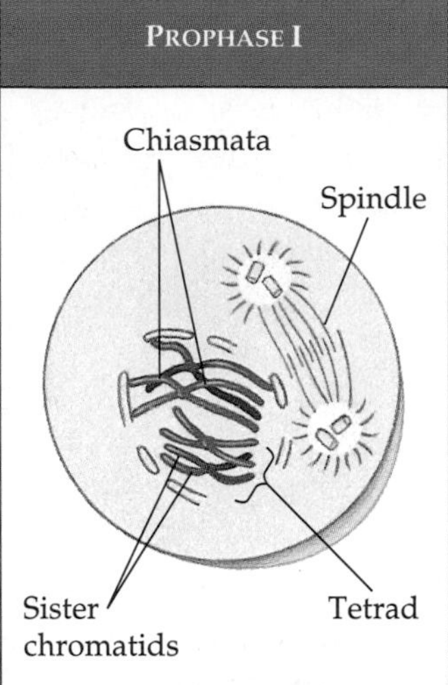

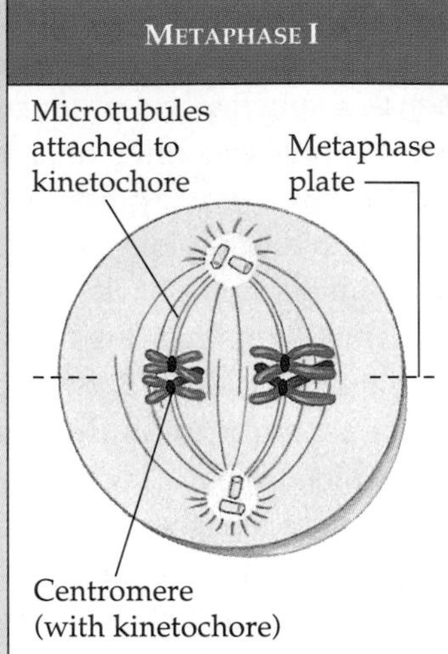

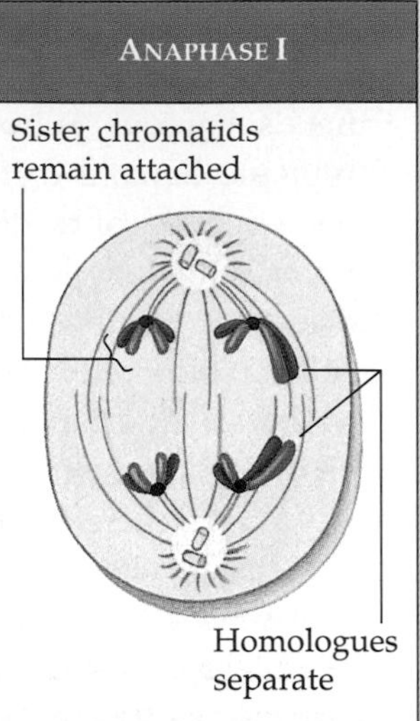

Interphase I

Meiosis is preceded by an interphase, during which each of the chromosomes replicates. This process is similar to the chromosome replication preceding mitosis. For each chromosome, the result is two genetically identical sister chromatids attached at their centromeres. The centriole pairs (in an animal cell) also replicate to form the two pairs represented in this drawing.

Prophase I

Meiotic prophase I lasts longer and is more complex than prophase in mitosis. The chromosomes begin to condense. In a process called synapsis, homologous chromosomes, each made up of two chromatids, come together as pairs. Each chromosome pair is now visible in the microscope as a tetrad, a complex of four chromatids. At numerous places along their length, chromatids of homologous chromosomes are criss-crossed. These crossings, which help hold homologous chromosomes togther, are called chiasmata (singular, chiasma). (We will examine the genetic significance of these crossings later in the chapter.)

As prophase I continues, the cell prepares for the division of the nucleus in a manner similar to that observed during mitosis. The centriole pairs move away from each other, and spindle microtubules form between them. The nuclear envelope and nucleoli disperse. Finally, the chromosomes begin their migration to the metaphase plate, midway between the two poles of the spindle apparatus. Prophase I, which can last for days or even longer, typically occupies more than 90% of the time required for meiosis.

Metaphase I

Chromosomes are now arranged on the metaphase plate, still in homologous pairs. Spindle fibers from one pole of the cell attach to one chromosome of each pair, while spindle fibers from the opposite pole attach to the homologue.

Anaphase I

As in mitosis, the spindle apparatus moves the chromosomes toward the poles. However, sister chromatids remain attached at their centromeres and move as a single unit toward the same pole. The homologous chromosome moves toward the opposite pole. This contrasts with the behavior of chromosomes during mitosis. In mitosis, chromosomes align individually on the metaphase plate rather than in pairs, and the spindle separates sister chromatids of each chromosome.

FIGURE 12.6

The stages of meiotic cell division. These drawings show meiotic cell division for an animal cell with a diploid number of 4 ($2n = 4$). The behavior of the chromosomes is emphasized. For a discussion about spindle formation and other features common to mitosis and meiosis, see FIGURE 11.7.

MEIOSIS II:
Separates sister chromatids

TELOPHASE I AND CYTOKINESIS

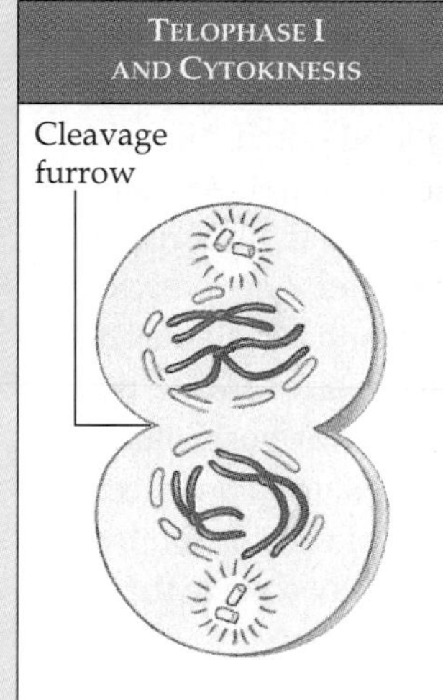

PROPHASE II

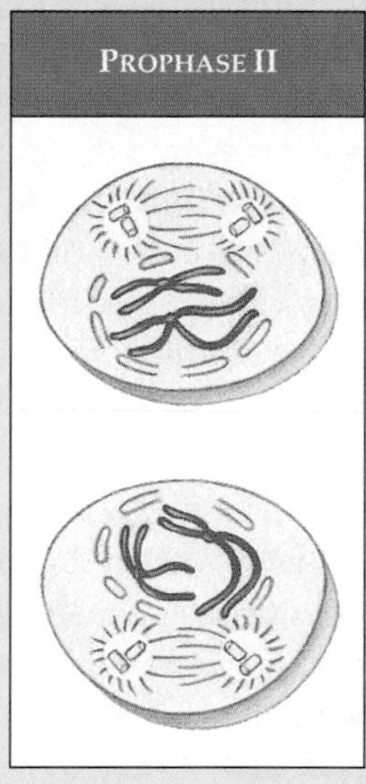

METAPHASE II

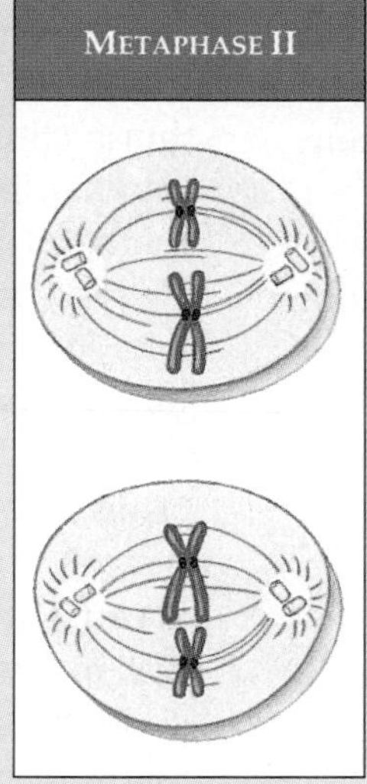

ANAPHASE II

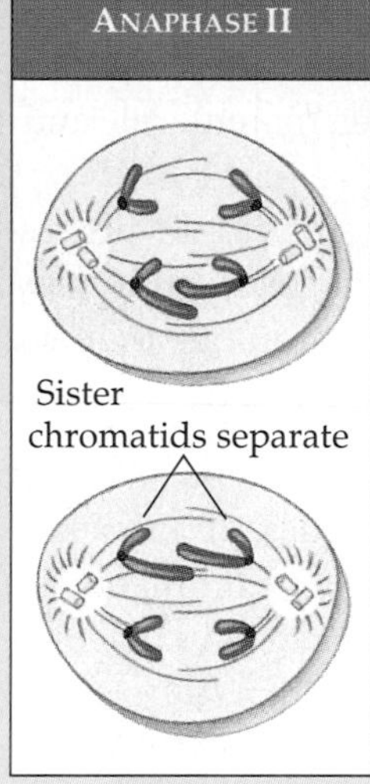

TELOPHASE II AND CYTOKINESIS

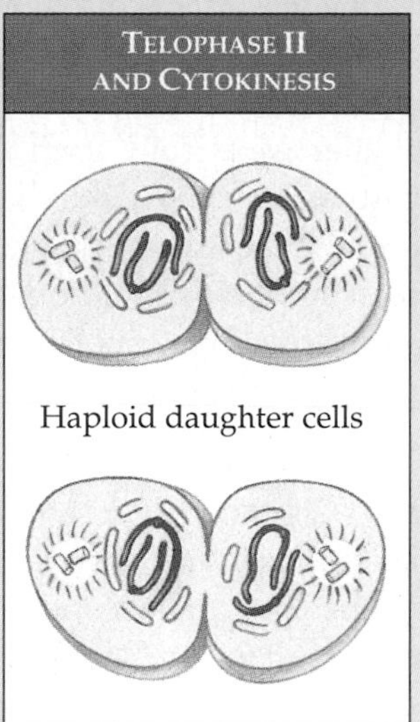

TELOPHASE I AND CYTOKINESIS

The spindle apparatus continues to separate the homologous chromosome pairs until the chromosomes reach the poles of the cell. Each pole now has a haploid chromosome set, but each chromosome still has two chromatids. Usually cytokinesis (division of the cytoplasm) occurs simultaneously with telophase I, forming two daughter cells. Cleavage furrows form in animal cells, and cell plates appear in plant cells. In some species, nuclear membranes and nucleoli re-form, and there is a period of time before meiosis II. In other species, daughter cells of telophase I immediately begin preparation for the second meiotic division. There is no further replication of the genetic material prior to the second division of meiosis.

PROPHASE II

A spindle apparatus forms, and the chromosomes progress toward the metaphase II plate.

METAPHASE II

The chromosomes align on the metaphase plate in mitosis-like fashion, with the kinetochores of sister chromatids of each chromosome pointing toward opposite poles.

ANAPHASE II

The centromeres of sister chromatids finally separate, and the sister chromatids of each pair, now individual chromosomes, move toward opposite poles of the cell.

TELOPHASE II AND CYTOKINESIS

Nuclei begin to form at opposite poles of the cell, and cytokinesis occurs. There are now four daughter cells, each with the haploid number of chromosomes.

Mitosis and Meiosis Compared

Now that we have followed chromosomes through meiosis in FIGURE 12.6, let's summarize the key differences between meiosis and mitosis. The chromosome number is reduced by half in meiosis but not in mitosis. The genetic consequences of this difference are important. Whereas mitosis produces daughter cells genetically identical to their parent cell and to each other, meiosis produces cells that differ genetically from their parent cell and from each other.

FIGURE 12.7 compares the key steps in the processes of mitosis and meiosis. Although meiosis involves two cell divisions, the three events that are unique to meiosis all occur during the first division, meiosis I:

1. During prophase I of meiosis, the duplicated chromosomes pair with their homologues, a process called **synapsis.** The four closely associated chromatids are visible in the light microscope as tetrads. Also visible in the light microscope are X-shaped regions called **chiasmata** (singular, **chiasma**). They represent a crossing of *nonsister* chromatids, which are two chromatids belonging to separate but homologous chromosomes. Chiasmata are the physical manifestations of a genetic rearrangement called crossing over, discussed in the next section. Neither synapsis nor chiasmata occur during mitosis.
2. At metaphase I of meiosis, homologous pairs of chromosomes, rather than individual chromosomes, align on the metaphase plate.
3. At anaphase I of meiosis, centromeres do not divide and sister chromatids do not separate, as they do in mitosis. Rather, the two sister chromatids of each chromosome remain attached and go to the same pole of the cell. *Meiosis I separates homologous pairs of chromosomes, not sister chromatids of individual chromosomes.*

The second meiotic division, meiosis II, separates sister chromatids and is virtually identical in mechanism to mitosis. However, since the chromosomes do not replicate between meiosis I and meiosis II, the final outcome of meiosis is a halving of the number of chromosomes per cell.

Sexual life cycles produce genetic variation among offspring

How do we account for the genetic variation apparent in FIGURE 12.2? In species that reproduce sexually, the behavior of chromosomes during meiosis and fertilization is responsible for most of the variation that arises each generation. Let's examine three sexual mechanisms that contribute to genetic variation: independent assortment of chromosomes, crossing over, and random fertilization.

Independent Assortment of Chromosomes

One way sexual reproduction generates genetic variation is shown in FIGURE 12.8 (p. 235), which color-codes the chromosomes so that we can track them as they are packaged in gametes. The two colors are used to distinguish chromosomes in a diploid cell inherited from the mother from those inherited from the father. At metaphase of meiosis I, each homologous pair of chromosomes, consisting of one maternal and one paternal chromosome, is situated on the metaphase plate. The orientation of the homologous pair relative to the two poles of the cell is random; there are two possibilities. Thus, there is a fifty-fifty chance that a particular daughter cell of meiosis I will get the maternal chromosome of a certain homologous pair, and a fifty-fifty chance that it will receive the paternal chromosome.

Because each homologous pair of chromosomes orients independently of the other pairs at metaphase I—it is as random as the flip of a coin—the first meiotic division results in independent assortment of maternal and paternal chromosomes into daughter cells. Each gamete represents one outcome of all possible combinations of maternal and paternal chromosomes. The number of combinations possible for gametes formed by meiosis starting with two homologous pairs of chromosomes ($2n = 4$, $n = 2$) is four, as shown in FIGURE 12.8. In the case of $n = 3$, there are eight combinations of chromosomes possible for gametes. More generally, the number of combinations possible when meiosis packages chromosomes into gametes by independent assortment is 2^n, where n is the haploid number.

In the case of humans, the haploid number (n) in the formula is 23. Thus, the number of possible combinations of maternal and paternal chromosomes in the resulting gametes is 2^{23}, or about 8 million. The variations are analogous to the 8 million combinations of heads and tails possible for the simultaneous tossing of 23 coins. Thus, each gamete that a human produces contains one of 8 million possible assortments of chromosomes inherited from that individual's mother and father.

Crossing Over

Because of the independent assortment of chromosomes during meiosis, each of us produces gametes containing diverse combinations of the chromosomes we inherited from our two parents. But from what you have learned so far, it would seem that each *individual* chromosome in a gamete would be exclusively maternal or paternal in origin; that is, it would consist of DNA derived from the mother or the father, but not from both. In fact,

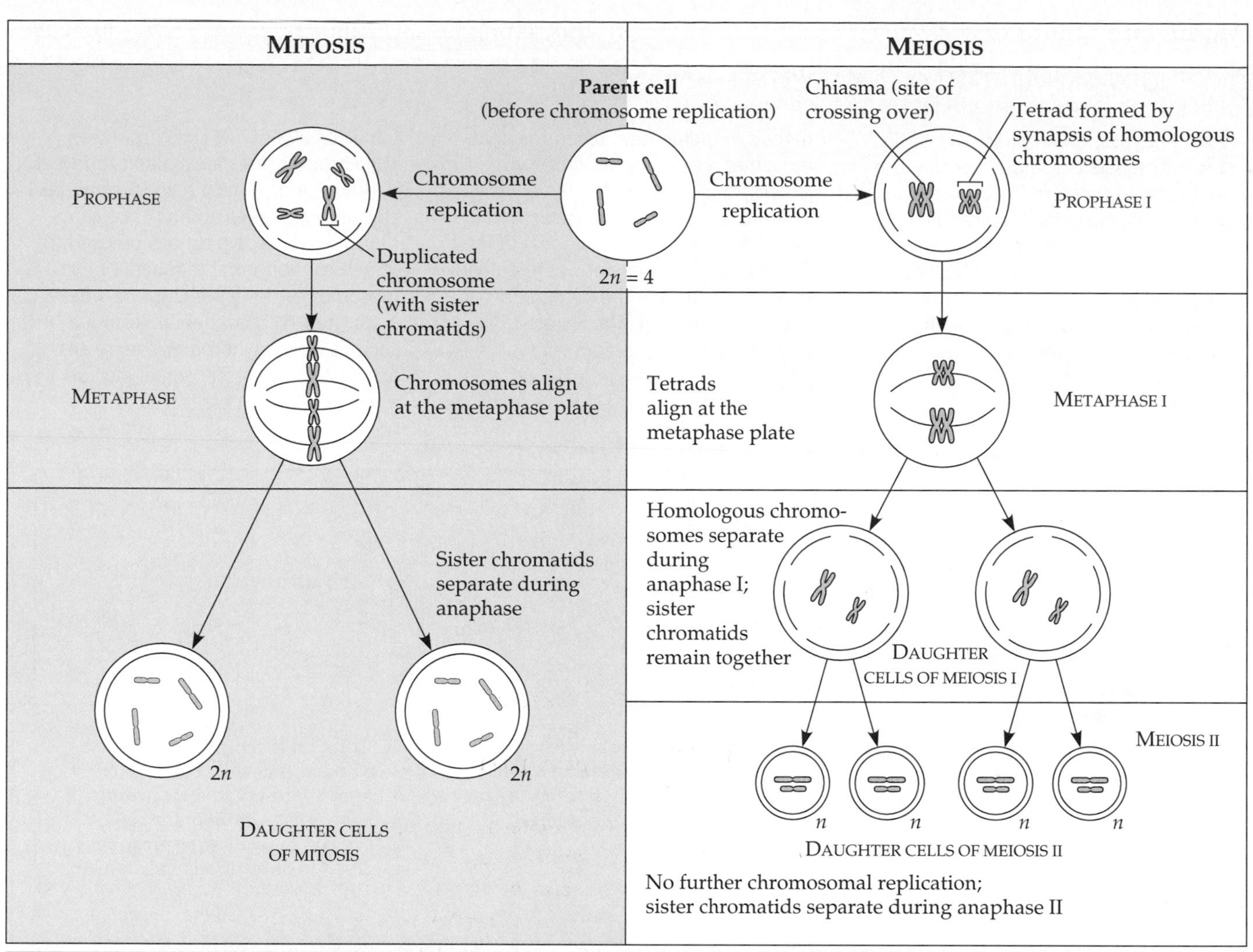

SUMMARY		
Event	Mitosis	Meiosis
DNA replication	Occurs during interphase before nuclear division begins.	Occurs during interphase before nuclear division begins.
Number of divisions	One, consisting of prophase, metaphase, anaphase, and telophase.	Two, each consisting of prophase, metaphase, anaphase, and telophase; DNA replication does not occur between the two nuclear divisions (meiosis I and meiosis II); an event unique to meiosis is that during meiosis I, the homologous chromosomes synapse (join along their length), forming tetrads (groups of four chromatids).
Number of daughter cells and genetic composition	Two, each diploid ($2n$) and genetically identical to the mother cell.	Four, each containing half as many chromosomes as the mother cell (haploid, or n); genetically nonidentical to the mother cell and to each other.
Importance in the animal body	Development of multicellular adult from zygote; production of cells for growth and tissue repair.	Production of gametes; reduces chromosome number by half and introduces genetic variability in the gametes.

FIGURE 12.7
A comparison of mitosis and meiosis in animals.

METHODS: PREPARATION OF A KARYOTYPE

Karyotypes, ordered displays of an individual's chromosomes, are useful in identifying certain abnormalities in the chromosomes. Medical technicians usually prepare karyotypes by using lymphocytes, a type of white blood cell.

After the cells are treated with a drug to stimulate mitosis, they are grown in culture for several days. Another drug is then applied to arrest mitosis at metaphase, when the chromosomes, each consisting of two joined sister chromatids, are very condensed. This is the stage when chromosomes are easiest to identify in the microscope. The drawings here outline the further steps in the preparation of a karyotype from lymphocytes. The light micrograph in step 5 shows the karyotype of a human male, with one *X* and one *Y* chromosome. (Human females have two *X* chromosomes.) The chromosomes are stained to reveal band patterns, which help identify specific chromosomes and parts of chromosomes. Karyotyping can be used to screen for abnormal numbers of chromosomes or defective chromosomes associated with congenital disorders, such as Down syndrome. The causes and effects of chromosomal disorders are discussed in Chapter 14.

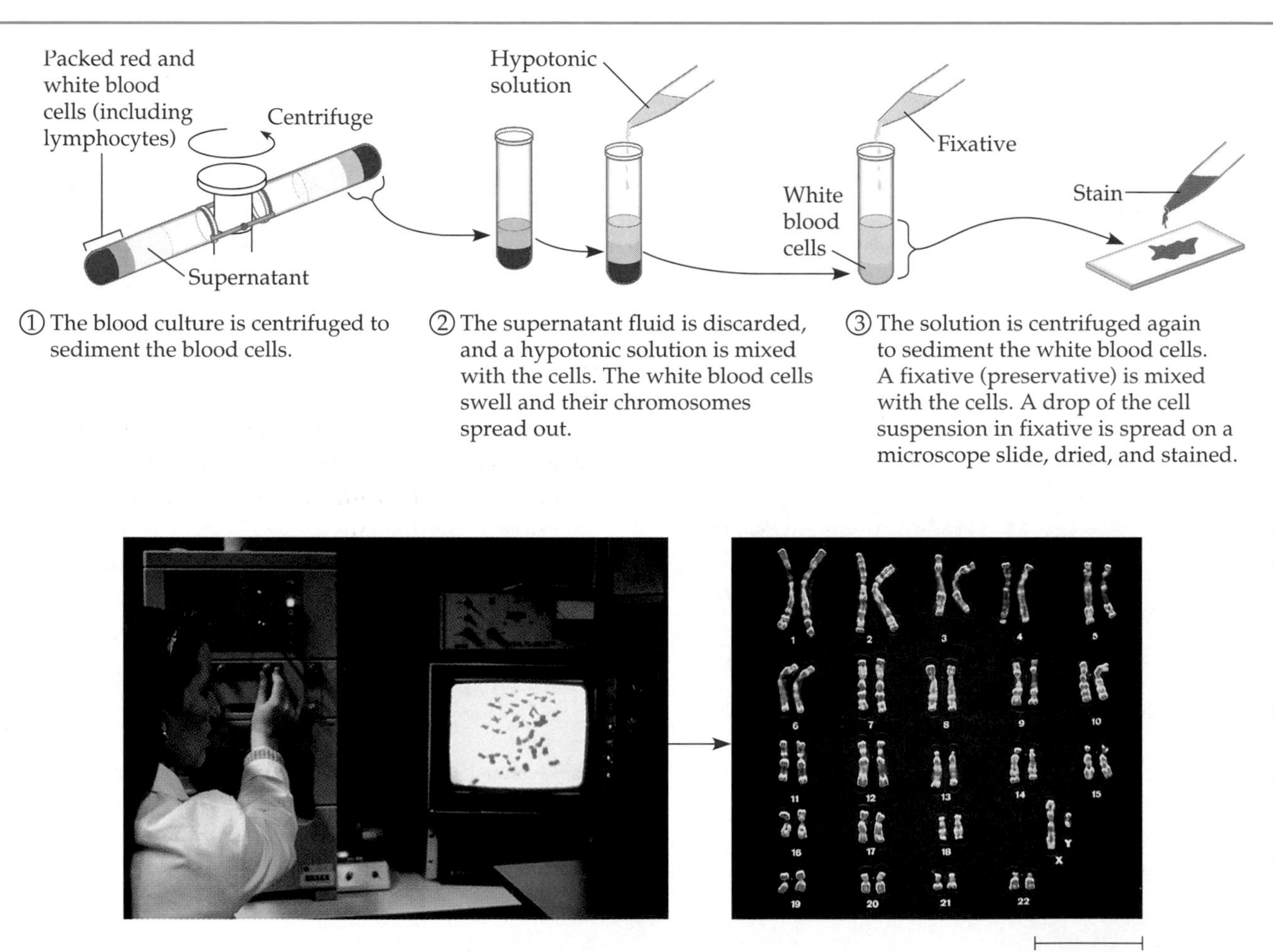

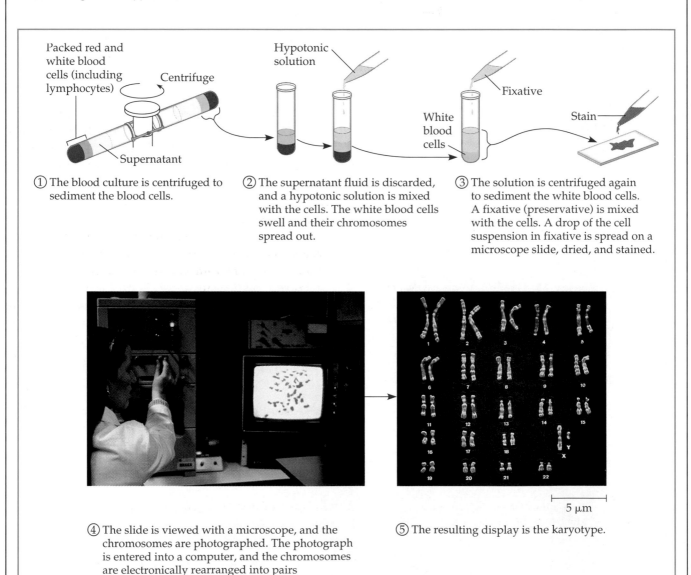

① The blood culture is centrifuged to sediment the blood cells.

② The supernatant fluid is discarded, and a hypotonic solution is mixed with the cells. The white blood cells swell and their chromosomes spread out.

③ The solution is centrifuged again to sediment the white blood cells. A fixative (preservative) is mixed with the cells. A drop of the cell suspension in fixative is spread on a microscope slide, dried, and stained.

④ The slide is viewed with a microscope, and the chromosomes are photographed. The photograph is entered into a computer, and the chromosomes are electronically rearranged into pairs according to size and shape.

⑤ The resulting display is the karyotype.

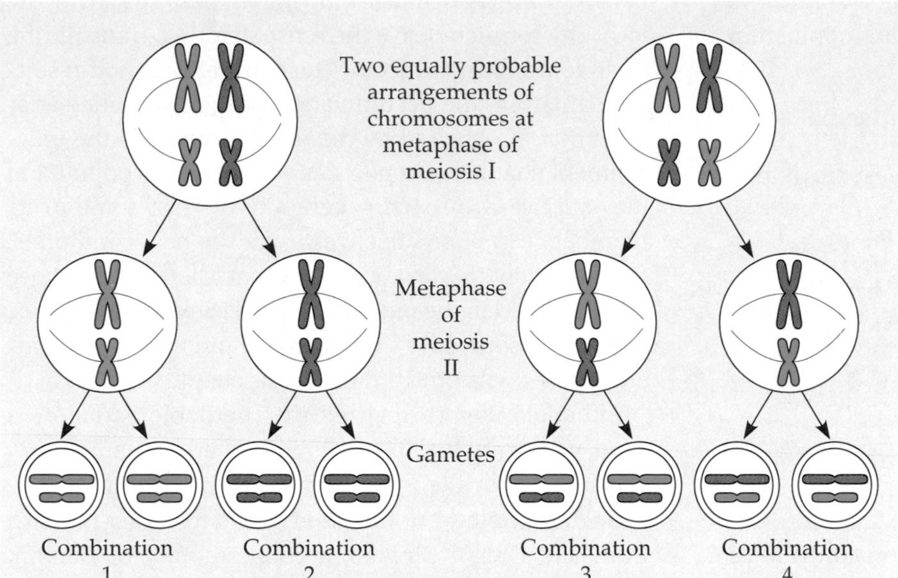

FIGURE 12.8
Alternative arrangements of two homologous chromosome pairs on the metaphase plate in meiosis I. In this figure, we consider the consequences of meiosis in a hypothetical organism with a diploid chromosome number of 4 ($2n = 4$). The parental origins of the chromosomes are symbolized with color coding, blue representing chromosomes inherited from one parent, red for chromosomes from the other parent. The positioning of each homologous pair of chromosomes at metaphase of meiosis I is a matter of chance, like the flip of a coin. The arrangement of chromosomes at metaphase I determines which chromosomes will be packaged together in the haploid daughter cells.

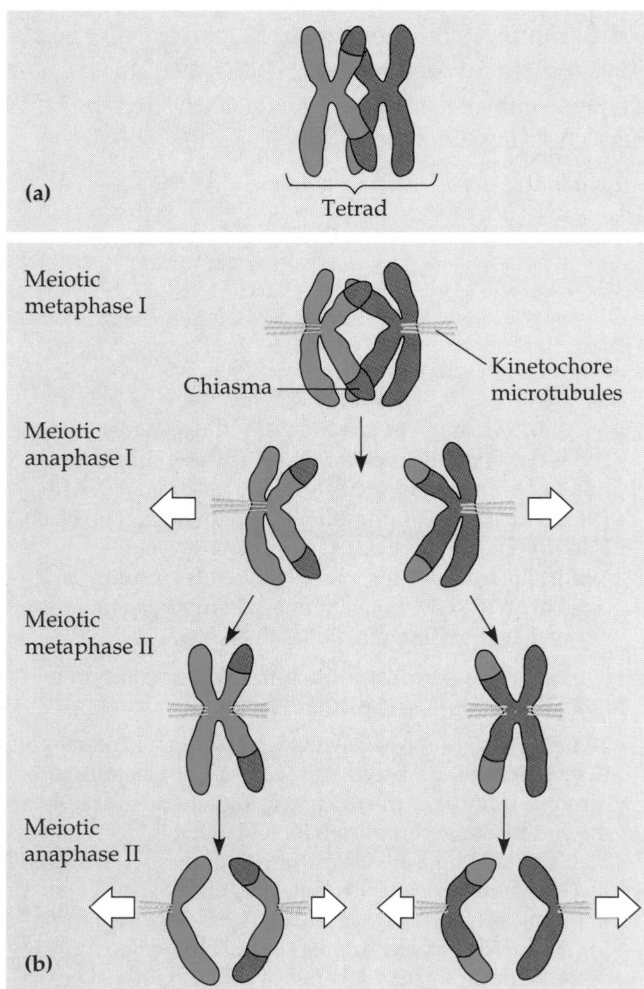

FIGURE 12.9
Crossing over. (**a**) During prophase of meiosis I, nonsister chromatids of homologous chromosomes exchange corresponding segments. (**b**) Following these chromosomes through meiosis, we can see that crossing over gives rise to individual chromosomes that have some combination of DNA originally derived from two different parents.

this is *not* the case. A process called **crossing over** produces individual chromosomes that combine genes inherited from the two parents. It occurs during prophase of meiosis I. Recall that during prophase I, homologous chromosomes come together as pairs (see FIGURE 12.6). A protein apparatus called the synaptonemal complex functions something like a zipper to bring the chromosomes into close association, but the exact mechanism of synapsis is not yet known. The pairing is precise, the homologues aligning with each other gene by gene.

Crossing over occurs when homologous portions of two nonsister chromatids trade places (FIGURE 12.9). In the case of humans, an average of two or three such crossover events occur per chromosome pair. The locations of these genetic exchanges are visible in light micrographs as chiasmata. The mechanism of crossing over will be discussed in more detail in Chapter 14. The important point for now is that crossing over, by combining DNA inherited from two parents into a single chromosome, is an important source of genetic variation in sexual life cycles.

Random Fertilization

The random nature of fertilization adds to the genetic variation arising from meiosis. Consider a zygote resulting from a mating between a woman and a man. A human ovum, representing one of approximately 8 million possible chromosome combinations, is fertilized by a single sperm cell, which represents one of 8 million *different* possibilities. Thus, even without considering crossing over, any two parents will produce a zygote with any of about 64 trillion (8 million × 8 million) diploid combinations. No wonder brothers and sisters can be so different. You really *are* unique.

So far, we have seen that there are three sources of genetic variability in a sexually reproducing population of organisms:

- Independent assortment of homologous chromosome pairs during meiosis I.
- Crossing over between homologous chromosomes during prophase of meiosis I.
- Random fertilization of an ovum by a sperm.

All three mechanisms reshuffle the various genes carried by the individual members of a population. However, as you will learn in later chapters, mutations are what ultimately create a population's diversity of genes.

Evolutionary adaptation depends on a population's genetic variation

Having considered how sexual reproduction contributes to genetic variation in a population, we should bridge these concepts to evolution, biology's core theme. Darwin recognized the importance of genetic variation in the evolutionary mechanism he called natural selection. Recall from Chapter 1 that a population evolves through the differential reproductive success of its variant members. On average, those individuals best suited to the local environment leave the most offspring, transmitting their genes in the process. This natural selection results in adaptation, the accumulation of those genetic variations that are favored by the environment. As the environment changes or a population moves, the population may survive if in each generation, at least some of its members can cope effectively with the new conditions. Different genetic variations may work better than those that prevailed in the old time or place. Sex and mutations are the two sources of this variation, and we have considered the sexual contribution in this chapter.

Although Darwin realized that heritable variation is what makes evolution possible, he had no satisfactory explanation for the precise mechanism that makes offspring resemble—but not be identical to—their parents. Ironically, Gregor Mendel, a contemporary of Darwin, published a theory of inheritance that helps explain genetic variation, but his discoveries had no impact on biologists until 1900, more than 15 years after Darwin (1809–1882) and Mendel (1822–1884) died. In the next chapter, you will learn how Mendel discovered the basic rules governing the inheritance of specific traits.

REVIEW OF KEY CONCEPTS (with page numbers and key figures)

- Offspring acquire genes from parents by inheriting chromosomes (pp. 225–226)
  - Genetics is the study of heredity and variation.
  - Genetic material consists of DNA organized into genes, each with a specific locus on a certain chromosome.
- Like begets like, more or less: a comparison of asexual and sexual reproduction (p. 226, FIGURE 12.2)
  - In asexual reproduction, a single parent gives rise to genetically identical offspring by mechanisms involving mitosis.
  - Sexual reproduction combines genes from two different parents to form offspring that are genetically diverse.
- Fertilization and meiosis alternate in sexual life cycles: *an overview* (pp. 226–229, FIGURE 12.4)
  - All sexually reproducing organisms alternate diploid ($2n$) and haploid (n) states through fertilization and meiosis.
  - Normal human somatic cells contain 46 chromosomes, half inherited from the father and half from the mother.
  - Each of 22 autosomes in the maternal set has a corresponding homologous chromosome in the paternal set. The twenty-third pair, the sex chromosomes, determines whether the person is a female (*XX*) or a male (*XY*).
  - Single, haploid (n) sets of chromosomes in ovum and sperm unite during fertilization to produce a diploid ($2n$), single-celled zygote. The zygote develops into a multicellular individual by mitosis.
  - At sexual maturity, ovaries and testes (the gonads) produce haploid gametes by meiosis, which reduces chromosome number from diploid to haploid.
  - Differences in the timing of meiosis with respect to fertilization characterize a variety of sexual life cycles. Multicellular organisms may be diploid (as in animals), haploid (as in some fungi), or may alternate generations between haploid and diploid (as in plants).
- Meiosis reduces chromosome number from diploid to haploid: *a closer look* (pp. 229–232, FIGURE 12.7)
  - Meiosis consists of two cell divisions, meiosis I and meiosis II, resulting in four daughter cells, each with half the chromosome number of the original cell. Thus, meiosis reduces the chromosome count from diploid to haploid.
  - Meiosis is distinguished from mitosis by a series of distinctive events that occur during meiosis I.
  - In prophase I of meiosis, replicated homologous chromosomes, each with two chromatids, undergo synapsis. This association allows the exchange of genetic material by the crossing over of homologous segments of nonsister chromatids. The crossing-over sites are visible as chiasmata.
  - The paired chromosomes (tetrads) align on the metaphase plate, and at anaphase I, the two chromosomes of each homologous pair (rather than sister chromatids) are pulled toward separate poles. This halves the number of chromosomes in the daughter cells.

- Meiosis II separates the sister chromatids to form four haploid daughter cells.

- Sexual life cycles produce genetic variation among offspring (pp. 232–236, FIGURE 12.9)
  - The sexual processes that contribute to genetic variation in a population are independent assortment of chromosomes during meiosis I, crossing over between homologous chromosones during meiosis I, and random fertilization of an ovum by a sperm.
- Evolutionary adaptation depends on a population's genetic variation (p. 236)
  - Genetic variation among a population's members is the raw material for evolution by natural selection. Sex and mutations are the two processes that generate this variation.

SELF-QUIZ

1. A human cell containing 22 autosomes and a *Y* chromosome is
   a. a somatic cell of a male
   b. a zygote
   c. a somatic cell of a female
   d. a sperm cell
   e. an ovum

2. Homologous chromosomes segregate toward opposite poles of a dividing cell during
   a. mitosis
   b. meiosis I
   c. meiosis II
   d. fertilization
   e. binary fission

3. Meiosis II is similar to mitosis in that
   a. homologous chromosomes synapse
   b. DNA replicates before the division
   c. the daughter cells are diploid
   d. sister chromatids separate during anaphase
   e. the chromosome number is reduced

4. The DNA content of a diploid cell in the G_1 phase of the cell cycle is measured (see Chapter 11). If this DNA content is X, then the DNA content of the same cell at metaphase of meiosis I would be
   a. $0.25X$
   b. $0.5X$
   c. X
   d. $2X$
   e. $4X$

5. If we continued to follow the cell lineage from question 4, then the DNA content at metaphase of meiosis II would be
   a. $0.25X$
   b. $0.5X$
   c. X
   d. $2X$
   e. $4X$

6. How many different combinations of maternal and paternal chromosomes can be packaged in gametes made by an organism with a diploid number of 8 ($2n = 8$)?
   a. 2
   b. 4
   c. 8
   d. 16
   e. 32

7. The most direct product of meiosis in a plant is a
   a. spore
   b. gamete
   c. zygote
   d. sporophyte
   e. gametophyte

8. The following choices indicate chromosome number before and after a process. Which choice corresponds to the process of fertilization?
   a. $2n \longrightarrow n$
   b. $n \longrightarrow 2n$
   c. $2n \longrightarrow 2n$
   d. $n \longrightarrow n$
   e. $2n \longrightarrow 4n$

9. Crossing over contributes to genetic variation when it exchanges chromosomal regions between
   a. sister chromatids of a chromosome
   b. chromatids of nonhomologues
   c. nonsister chromatids of homologues
   d. nonhomologous loci of the genome
   e. autosomes and sex chromosomes

10. In comparing the typical life cycles of plants and animals, a stage found in plants but not in animals is a
    a. gamete
    b. zygote
    c. multicellular diploid
    d. multicellular haploid

CHALLENGE QUESTIONS

1. In domestic turkeys, viable offspring are sometimes produced by the development of an unfertilized egg cell, a process called parthenogenesis. Such offspring, like the mother, are diploid. What variation in meiosis could produce a diploid organism without fertilization?

2. Many species can reproduce either asexually or sexually. It is often when the environment changes in some way that is unfavorable to an existing population that the organisms begin to reproduce sexually. Speculate about the evolutionary significance of this switch from asexual to sexual reproduction.

SCIENCE, TECHNOLOGY, AND SOCIETY

1. Studies of human characteristics indicate that most variation in the human population is due to the differences *within* racial groups. Only a small amount of variation is due to differences *between* what have traditionally been considered races. Based on these observations, write a paragraph or two evaluating whether the concept of "race" is valid.

2. It is possible to grow seedlings of pine trees from short pieces of their needles. A few of the straightest, fastest-growing trees are selected for this treatment. By this method, thousands of genetically identical trees can be grown and transplanted to create a forest that is a superior producer of lumber. What are the short-term and long-term advantages and disadvantages of this technology?

FURTHER READING

Anderson, A. "The Evolution of Sexes." *Science,* July 17, 1992. Why do humans and many other organisms exist in only two sexes?

Becker, W. M., J. B. Reece, and M. F. Poenie. *The World of the Cell,* 3rd ed. Menlo Park, CA: Benjamin/Cummings, 1996. Chapter 16.

Cunningham, P. "The Genetics of Thoroughbred Horses." *Scientific American,* May 1991. The history of how humans have exploited heritable variation in one species.

Gould, S. J. "Ghosts of Bell Curves Past." *Natural History,* February 1995. A famous evolutionary biologist argues against a best-selling book about the genetic contribution to intelligence.

Griffiths, A. J. F., J. H. Miller, D. T. Suzuki, R. C. Lewontin, and W. M. Gelbart. *An Introduction to Genetic Analysis,* 5th ed. New York: W. H. Freeman, 1993. A good undergraduate genetics text.

Ridley, M. "Is Sex Good For Anything?" *New Scientist,* December 4, 1993. Did disease play an important role in the evolution of sex?

CHAPTER 13

MENDEL AND THE GENE IDEA

KEY CONCEPTS

- Mendel brought an experimental and quantitative approach to genetics: *science as a process*
- According to the law of segregation, the two alleles for a character are packaged into separate gametes
- According to the law of independent assortment, each pair of alleles segregates into gametes independently
- Mendelian inheritance reflects rules of probability
- Mendel discovered the particulate behavior of genes: *a review*
- The relationship between genotype and phenotype is rarely simple
- Pedigree analysis reveals Mendelian patterns in human inheritance
- Many human disorders follow Mendelian patterns of inheritance
- Technology is providing new tools for genetic testing and counseling

A person's eyes can be blue, brown, green, gray, or hazel; a person's hair can be different shades of brown, red, blond, or black; a parakeet's feathers can be green, blue, or yellow, with black or gray markings. What causes these biological spectra of colors? We can frame the question in more general terms: What is the genetic basis of variation among a population's individuals? Or, what principles account for the transmission of these variations from parents to offspring?

One possible explanation of heredity is a "blending hypothesis," the idea that genetic material contributed by the two parents mixes in a manner analogous to the way blue and yellow paints blend to make green. This hypothesis predicts that mating a blue parakeet with a yellow one would result in green offspring, and once blended, the hereditary material of the two parents would be as inseparable as the colors of mixed paint. If the blending hypothesis were correct, over many generations a freely mating population of blue and yellow parakeets would give rise to a uniform population of green birds. The actual results of parakeet breeding, however, contradict such a prediction. The blending hypothesis also fails to explain other phenomena of inheritance, such as traits skipping a generation.

An alternative to the blending model is a "particulate model" of inheritance: the gene idea. According to this model, parents pass on discrete heritable units—genes—that retain their separate identities in offspring. An organism's collection of genes is more like a bucket of marbles than a pail of paint. Like marbles, genes can be sorted and passed along, generation after generation, in undiluted form.

Modern genetics had its genesis in an abbey garden when a monk named Gregor Mendel documented a particulate mechanism of inheritance. In the painting on this page, Mendel works with his experimental organism, garden peas. In this chapter, you will learn how Mendel developed his theory and how the Mendelian model applies to the inheritance of human variations.

Mendel brought an experimental and quantitative approach to genetics: *science as a process*

Gregor Mendel discovered the basic principles of heredity by breeding garden peas in carefully planned experiments. As we retrace Mendel's work in this and the following sections, we will be able to recognize the key elements of the scientific process that were introduced in Chapter 1.

Johann Mendel (he took the name Gregor when he entered the Augustinian brotherhood) grew up on his parents' small farm in a region of Austria that is now part of the Czech Republic. In this agricultural area, crops and orchards were of great local interest, and at school Mendel and the other children received agricultural training along with basic education. Later, Mendel overcame financial hardships and a series of illnesses to excel in high school and at the Olmutz Philosophical Institute.

Mendel entered the Augustinian monastery in 1843. After three years of theological studies, he was assigned to a school as a temporary teacher but failed the teacher's examination. An administrator sent Mendel to the University of Vienna, where he studied from 1851 to 1853. These were very important years for Mendel's development as a scientist. Two professors were especially influential. One was the physicist Doppler, who encouraged his students to learn science through experimentation and trained Mendel to apply mathematics to help explain natural phenomena. The second was a botanist named Unger, who aroused Mendel's interest in the causes of variation in plants. These influences came together in Mendel's subsequent experiments with garden peas.

After attending the university, Mendel was assigned to teach at the Brünn Modern School, where several teachers shared his enthusiasm for scientific research. At the monastery where Mendel lived, there were also stimulating colleagues, many of them university professors and active researchers. There had also been a long tradition of interest in the breeding of plants, including peas, at the monastery. Thus, it was probably not extraordinary when, around 1857, Mendel began breeding garden peas in the abbey garden to study inheritance. What *was* extraordinary was Mendel's fresh approach to very old questions about heredity.

Mendel probably chose to work with peas because they are available in many varieties. For example, one variety has purple flowers, while a contrasting variety has white flowers. Geneticists use the term **character** for a heritable feature, such as flower color, that varies among individuals. Each variant for a character, such as purple or white flowers, is called a **trait.**

The use of peas also gave Mendel strict control over which plants mated with which. The petals of the pea flower almost completely enclose the female and male parts (carpel and stamens, respectively); normally, the plants self-fertilize after pollen grains released from the stamens land on the carpel. When Mendel wanted cross-pollination (fertilization between different plants), he would remove the immature stamens of a plant before they produced pollen and then dust pollen from another plant onto the emasculated flowers (FIGURE 13.1). Whether ensuring self-pollination or executing artificial cross-pollination, Mendel could always be sure of the parentage of new seeds.

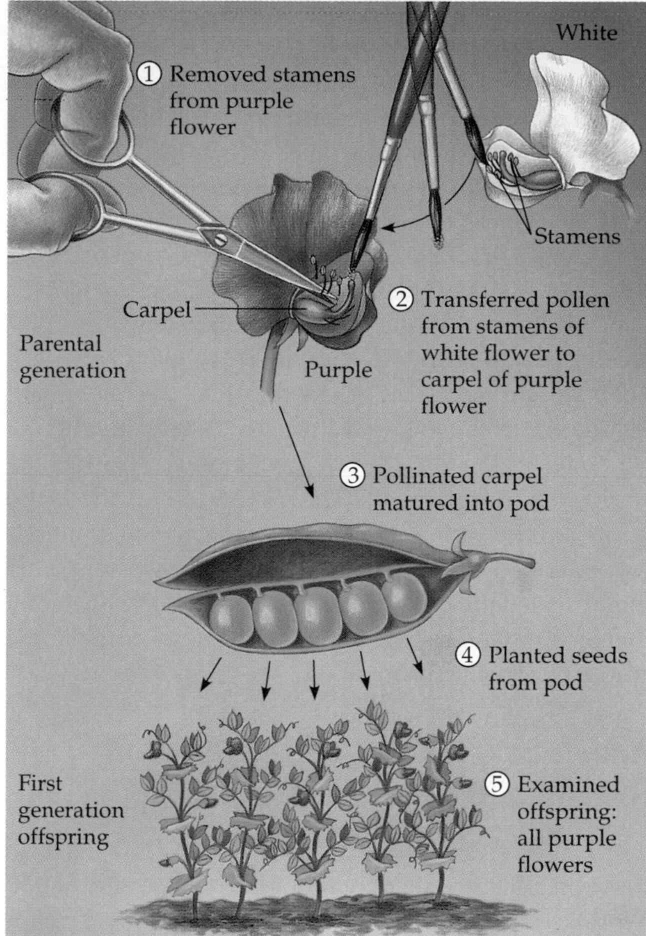

FIGURE 13.1

A genetic cross. To hybridize between pea varieties, Mendel used an artist's brush to transfer pollen. In this case, the character of interest is flower color, and the two varieties are purple-flowered and white-flowered. Seeds develop within the female organ, or carpel, which develops into the fruit (pod). Germination of the seeds produces the first generation hybrids, which all have purple flowers. The result is the same for the reciprocal cross, the transfer of pollen from purple flowers to white flowers.

Mendel was careful to track the inheritance of only categorical variations, that is, inherited characters that varied in an "either-or" rather than a "more-or-less" manner. For example, his plants had either purple or white flowers; there was nothing intermediate between these two varieties. Had Mendel focused instead on characters that vary in a continuum among individuals—seed weight, for example—he would not have discovered the particulate nature of inheritance.

Mendel also made sure he started his experiments with varieties that were **true-breeding,** which means that when the plants self-pollinate, all their offspring are of the same variety. For example, a plant with purple flowers is true-breeding if its seeds produced by self-pollination all give rise to plants that also have purple flowers.

In a typical breeding experiment, Mendel would cross-pollinate between two contrasting, true-breeding pea varieties—between purple-flowered plants and white-flowered plants, for example (see FIGURE 13.1). This mating, or crossing, of two varieties is called **hybridization.** Our example is specifically a **monohybrid cross,** the term for a cross that tracks the inheritance of a single character—flower color, in this case. The true-breeding parents are referred to as the **P generation** (for parental), and their hybrid offspring are the **F_1 generation** (for first filial, referring to offspring). Allowing these F_1 hybrids to self-pollinate produces an **F_2 generation** (second filial). Mendel generally followed traits for at least these three generations: the P, F_1, and F_2 generations. Had Mendel stopped his experiments with the F_1 generation, the basic patterns of inheritance would have eluded him. It was mainly Mendel's quantitative analysis of F_2 plants that revealed the two fundamental principles of heredity that are now known as the law of segregation and the law of independent assortment.

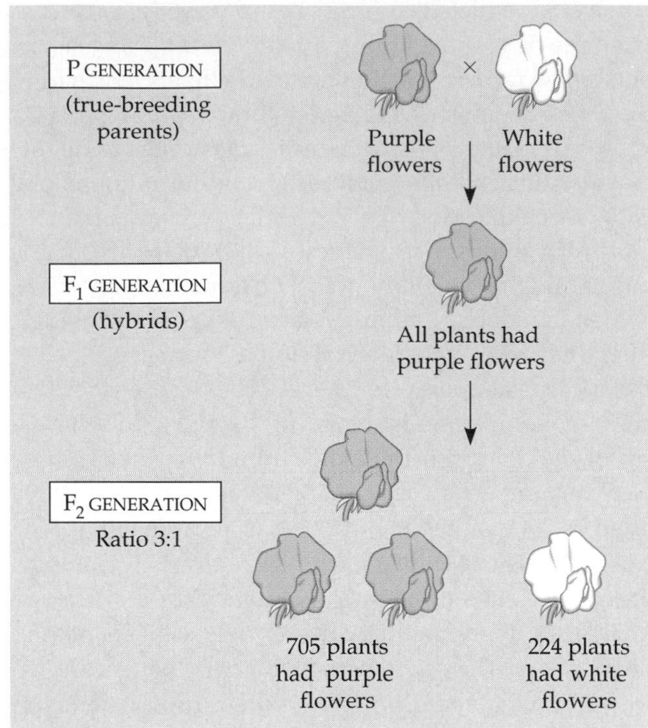

FIGURE 13.2
Mendel tracked heritable characters for three generations. When F_1 hybrids were allowed to self-pollinate, or when they were cross-pollinated with other F_1 hybrids, a 3:1 ratio of the two varieties occurred in the F_2 generation. An "×" sign symbolizes a genetic cross, or mating.

According to the law of segregation, the two alleles for a character are packaged into separate gametes

If the blending model of inheritance were correct, the F_1 hybrids from a cross between purple-flowered and white-flowered pea plants would have pale purple flowers, intermediate between the two varieties of the P generation. Notice in FIGURE 13.1 that the experiment produced a very different result: The F_1 offspring all had flowers just as purple as the purple-flowered parents. What happened to the white-flowered plants' genetic contribution to the hybrids? If it were lost, then the F_1 plants could produce only purple-flowered offspring in the F_2 generation. But when Mendel allowed the F_1 plants to self-pollinate and planted their seeds, the white-flower trait reappeared in the F_2 generation. Mendel used very large sample sizes and kept accurate records of his results: 705 of the F_2 plants had purple flowers, and 224 had white flowers. These data fit a ratio of about 3 purple to 1 white (FIGURE 13.2). Mendel reasoned that the heritable factor for white flowers did not disappear in the F_1 plants, but only the purple-flower factor was affecting flower color in these hybrids. In Mendel's terminology, purple flower is a dominant trait and white flower is a recessive trait. The occurrence of white-flowered plants in the F_2 generation was evidence that the heritable factor causing that recessive trait had not been diluted in any way by coexisting with the purple-flower factor in the F_1 hybrids.

Mendel observed the same pattern of inheritance in six other characters, each represented by two contrasting varieties (TABLE 13.1). For example, the parental pea seeds either had a smooth, round shape, or they were wrinkled. In a monohybrid cross for this character, all the F_1 hybrids produced round seeds; this is the dominant trait. In the F_2 generation, 75% of the seeds were round and 25% were wrinkled—the typical 3:1 ratio. How did Mendel explain this pattern, which he consistently observed in his monohybrid crosses? He developed a hypothesis that we can break down into four related ideas. (We will replace some of Mendel's original terms with modern words; for example, "gene" will be used in place of Mendel's "heritable factors.")

1. *Alternative versions of genes (different alleles) account for variations in inherited characters.* The gene for flower color, for example, exists in two versions, one for purple flowers and the other for white. These alternative versions of a gene are now called **alleles.** Today, we can relate this concept to chromosomes and DNA. As we saw in Chapter 12, each gene resides at a specific locus on a specific chromosome. The DNA at that locus, however, can vary somewhat in its sequence of nucleotides, and hence in its information content. The purple-flower allele and the white-flower allele are two DNA variations possible at the flower-color locus on one of a pea plant's chromosomes.

2. *For each character, an organism inherits two alleles, one from each parent.* Mendel made this deduction without

TABLE 13.1

The Results of Mendel's F_1 Crosses for Seven Characters in Pea Plants

CHARACTER	DOMINANT TRAIT	×	RECESSIVE TRAIT	F_2 GENERATION DOMINANT:RECESSIVE	RATIO
Flower color	Purple	×	White	705:224	3.15:1
Flower position	Axial	×	Terminal	651:207	3.14:1
Seed color	Yellow	×	Green	6022:2001	3.01:1
Seed shape	Round	×	Wrinkled	5474:1850	2.96:1
Pod shape	Inflated	×	Constricted	882:299	2.95:1
Pod color	Green	×	Yellow	428:152	2.82:1
Stem length	Tall	×	Dwarf	787:277	2.84:1

knowing about the role of chromosomes, but what you learned about chromosomes in Chapter 12 will help you understand Mendel's idea. Recall that a diploid organism has homologous pairs of chromosomes, one chromosome of each pair inherited from each parent. Thus, a genetic locus is actually represented twice in a diploid cell. These homologous loci may have matching alleles, as in the true-breeding plants of Mendel's P generation. Or, the two alleles may differ, as in the F_1 hybrids. In the flower-color example, the hybrids inherited a purple-flower allele from one parent and a white-flower allele from the other parent (FIGURE 13.3). This brings us to the third part of Mendel's hypothesis.

3. *If the two alleles differ, then one, the* ***dominant allele,*** *is fully expressed in the organism's appearance; the other, the* ***recessive allele,*** *has no noticeable effect on the organism's appearance.* According to this part of the hypothesis, Mendel's F_1 plants had purple flowers because the allele for that variation is dominant and the allele for white flowers is recessive.

4. *The two alleles for each character segregate during gamete production.* Thus, an ovum and a sperm each receive only one of the alleles that are present in two copies in the somatic cells of the organism. (In the case of peas, "sperm" refers to a cell in a pollen grain.) In terms of

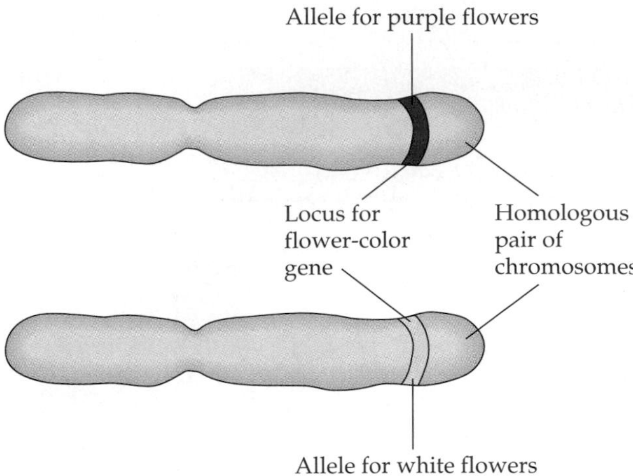

FIGURE 13.3

Alleles, contrasting versions of a gene. The gene for a particular inherited character, such as flower color in garden peas, resides at a specific locus (position) on a certain chromosome. Alleles are variants of that gene. In this case, the flower-color gene exists in two versions: the allele for purple flowers and the allele for white flowers. The homologous pair of chromosomes illustrated here represents an F_1 hybrid, which inherited the allele for purple flowers from one parent and the allele for white flowers from the other parent.

chromosomes, this segregation corresponds to meiotic reduction of chromosome count from the diploid to the haploid number. Note that if an organism has matching alleles for a particular character—that is, the organism is true-breeding for that character—then that allele exists in a single copy in all gametes. But if contrasting alleles are present, as in the F_1 hybrids, then 50% of the gametes receive the dominant allele, while 50% receive the recessive allele. It is this last part of the hypothesis, the separation of allelles into separate gametes, for which Mendel's **law of segregation** is named.

One test of Mendel's segregation hypothesis is whether or not it can account for the 3:1 ratio he observed in the F_2 generation of his numerous monohybrid crosses. The hypothesis predicts that the F_1 hybrids will produce two classes of gametes. When alleles segregate, half the gametes receive a purple-flower allele, while the other half get a white-flower allele. During self-pollination, these two classes of gametes unite randomly. An ovum with a purple-flower allele has an equal chance of being fertilized by a sperm with a purple-flower allele or one with a white-flower allele. Since the same is true for an ovum with a white-flower allele, there are a total of four equally likely combinations of sperm and ovum. FIGURE 13.4 illustrates these combinations using a type of diagram called a Punnett square, a handy device for predicting the results of a genetic cross. Notice that a capital letter symbolizes a dominant allele and a lowercase letter represents a recessive allele. In our example, *P* is the purple-flower allele, and *p* is the white-flower allele.

What will be the physical appearance of these F_2 offspring? One-fourth of the plants have two alleles specifying purple flowers; clearly, these plants will have purple flowers. But one-half of the F_2 offspring have inherited

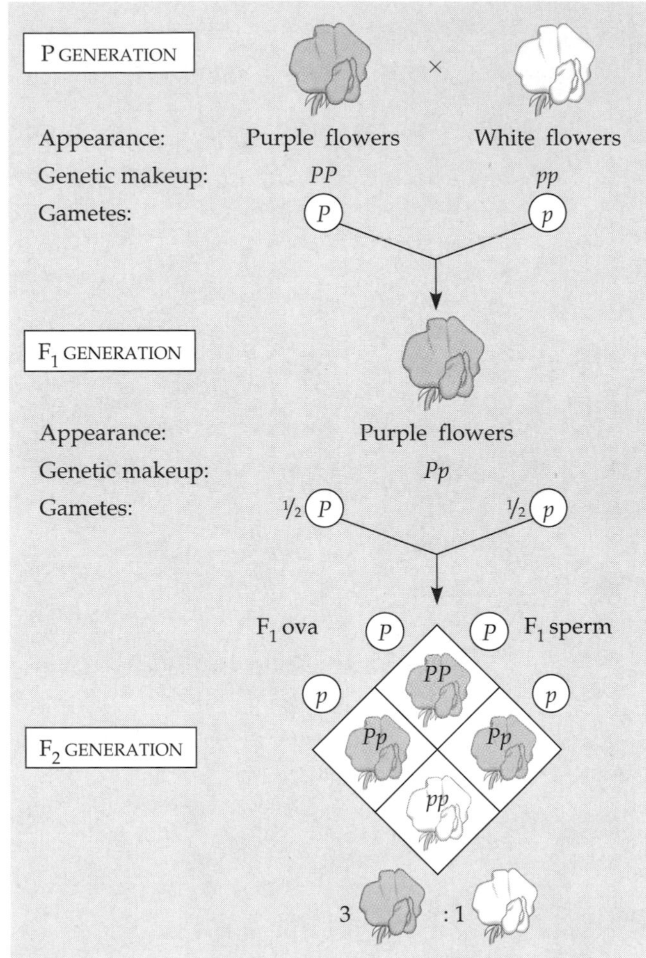

FIGURE 13.4

Mendel's law of segregation. A more detailed version of FIGURE 13.2, this diagram illustrates Mendel's model for monohybrid inheritance. The purple-flower allele (*P*) is dominant, and the white-flower allele (*p*) is recessive. Each plant has two alleles for the gene controlling flower color. One allele for the gene is inherited from each parent. Each true-breeding plant of the parental generation has matching alleles, either *PP* (purple-flower parentals) or *pp* (white-flower parentals). Gametes, symbolized with circles, contain only one allele for the flower color gene. Union of the parental gametes produces F_1 hybrids having nonmatching alleles, a *Pp* combination. Because the purple-flower allele is dominant, all these hybrids have purple flowers. When the hybrids produce gametes, the two alleles segregate, half the gametes receiving the *P* allele and the other half receiving the *p* allele. Random combination of these gametes results in the 3:1 ratio that Mendel observed in the F_2 generation. The box at the bottom of the figure is a Punnett square, a useful tool for showing all possible combinations of alleles in offspring. Each square represents an equally probable product of fertilization. For example, the box at the right of the Punnett square shows the genetic combination resulting from a (*p*) sperm fertilizing a (*P*) ovum.

one allele for purple flowers and one allele for white flowers; like the F_1 plants, these plants will also have purple flowers, the dominant trait. Finally, one-fourth of the F_2 plants have inherited two alleles specifying white flowers and will, in fact, express the recessive trait. Thus, Mendel's model accurately explains the 3:1 ratio that he observed in the F_2 generation.

Some Useful Genetic Vocabulary

An organism having a pair of identical alleles for a character is said to be **homozygous** for that gene. A pea plant that is true-breeding for purple flowers (*PP*) is an example. Pea plants with white flowers are also homozygous, but for the recessive allele (*pp*). If we cross dominant homozygotes with recessive homozygotes, as in the parental cross (P generation) of FIGURE 13.4, all the offspring will have a nonmatching combination of alleles—*Pp* in the case of the F_1 hybrids of our flower-color experiment. Organisms having two different alleles for a character are said to be **heterozygous** for that character. Unlike homozygotes, heterozygotes are not true-breeding because they produce gametes having one *or* the other of the different alleles. We have seen that a *Pp* plant of the F_1 generation will produce both purple-flowered and white-flowered offspring when it self-pollinates.

Because of dominance and recessiveness, an organism's appearance does not always reveal its genetic composition. Therefore, we have to distinguish between an organism's appearance, called its **phenotype,** and its genetic makeup, its **genotype.** In the case of flower color in peas, *PP* and *Pp* plants have the same phenotype (purple), but different genotypes. FIGURE 13.5 reviews these terms. (Phenotype refers to physiological traits as well as physical traits. For example, one pea variety is incapable of the normal trait of self-pollination.)

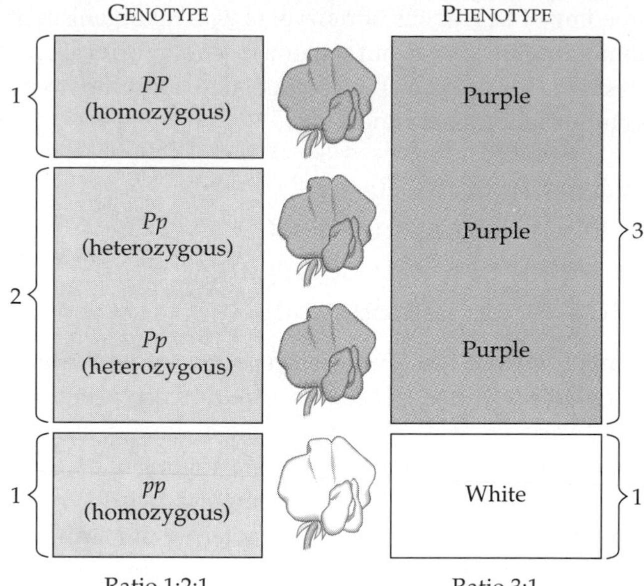

FIGURE 13.5

Genotype versus phenotype. Grouping F_2 offspring from a monohybrid cross for flower color according to phenotype results in the typical 3:1 ratio. In terms of genotype, however, there are actually two categories of purple-flowered plants: *PP* (homozygous) and *Pp* (heterozygous).

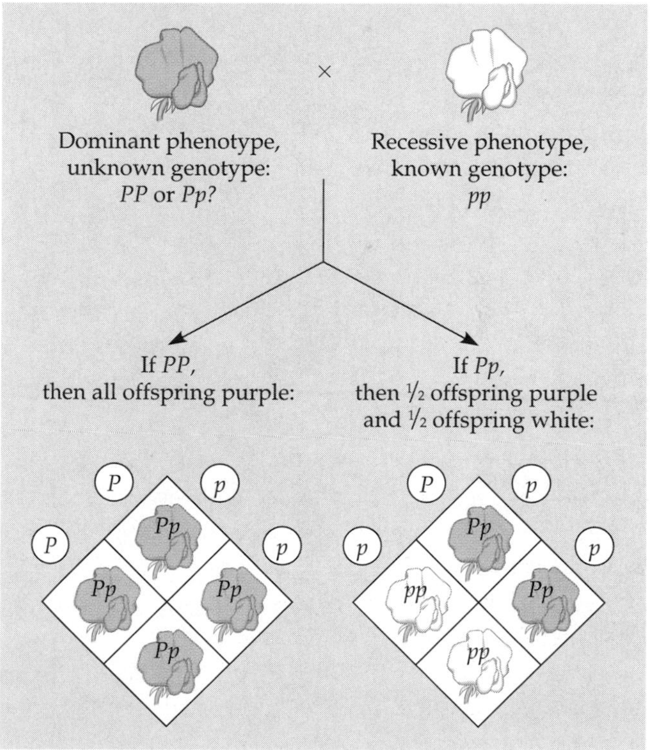

FIGURE 13.6

A testcross. A testcross is designed to reveal the genotype of an organism that exhibits a dominant trait, such as purple flowers in pea plants. Such an organism could be either homozygous for the dominant allele or heterozygous. The most efficient way to resolve the genotype is to cross the organism with an individual expressing the recessive trait. Since the genotype of the white-flowered parent must be homozygous, we can deduce the genotype of the purple-flowered parent by observing the phenotypes of the offspring.

The Testcross

Suppose we have a pea plant that has purple flowers. We cannot tell from its flower color if this plant is homozygous or heterozygous because the genotypes *PP* and *Pp* result in the same phenotype. If we cross this pea plant with one having white flowers, the appearance of the offspring will reveal the genotype of the purple-flowered parent. The genotype of the plant with white flowers is known: Because this is the recessive trait, the plant must be homozygous. If all the offspring of the cross have purple flowers, then the other parent is homozygous for the dominant allele; a $PP \times pp$ cross produces nothing but *Pp* offspring. But if both the purple and the white phenotypes appear among the offspring, the purple-flowered parent must be heterozygous. The offspring of a $Pp \times pp$ cross will have a 1:1 phenotypic ratio (FIGURE 13.6). This breeding of a recessive homozygote with an organism of dominant phenotype, but unknown genotype, is called a **testcross.** It was devised by Mendel and continues to be an important tool of geneticists.

According to the law of independent assortment, each pair of alleles segregates into gametes independently

Mendel derived the law of segregation by performing monohybrid crosses, breeding experiments using parental varieties that differ in a single character, such as flower color. What would happen in a mating of parental varieties differing in *two* characters—a **dihybrid cross?** For instance, two of the seven characters Mendel studied were seed color and seed shape. Seeds may be either yellow or green. They also may be either round (smooth) or wrinkled. From monohybrid crosses, Mendel knew that the allele for yellow seeds is dominant (*Y*), and the allele for green seeds is recessive (*y*). For the seed-shape character, round is dominant (*R*), and wrinkled is recessive (*r*). Imagine breeding two pea varieties differing in *both* of these characters—a parental cross between a plant with yellow-round seeds (*YYRR*) and a plant with green-wrinkled seeds (*yyrr*). Are these two characters, seed color and seed shape, transmitted from parents to offspring as a package? Put another way, will the *Y* and *R* alleles always stay together, generation after generation? Or, are seed color and seed shape inherited independently of each other? FIGURE 13.7 illustrates how a dihybrid cross can determine which of these two hypotheses is correct.

In the F_1 generation of this dihybrid cross, the genotype is *YyRr,* and the plants exhibit both dominant phenotypes, yellow seeds with round shapes. The key step in

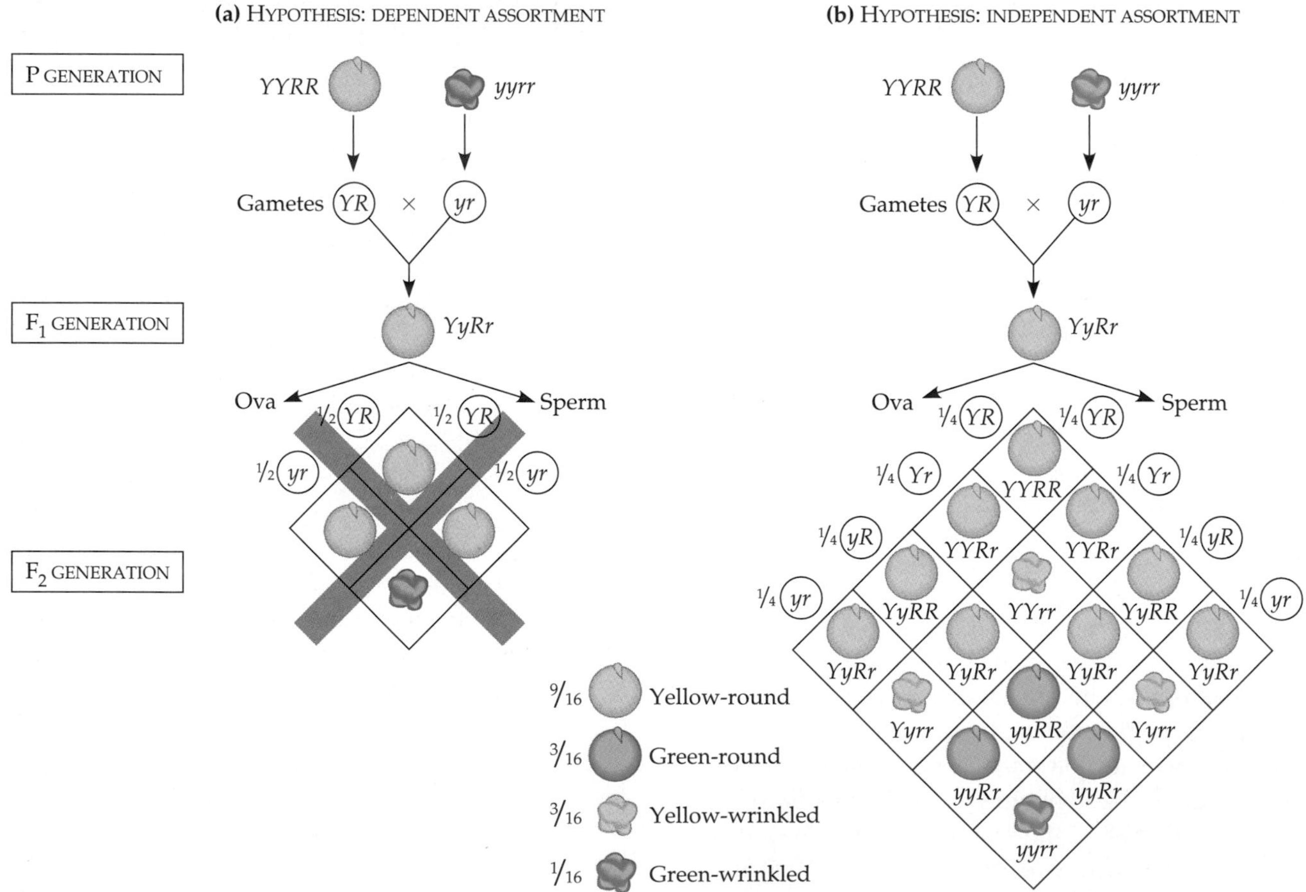

FIGURE 13.7
Testing two hypotheses for segregation in a dihybrid cross. A cross between true-breeding parent plants that differ in two characters produces F_1 hybrids that are heterozygous for both characters. In this example, the two characters are seed color and seed shape. Yellow color (*Y*) and round shape (*R*) are dominant. (**a**) If the two characters segregate dependently (together), the F_1 hybrids can only produce the same two classes of gametes that they received from the parents, and the F_2 offspring will show a 3:1 phenotypic ratio. (**b**) If the two characters segregate independently, four classes of gametes will be produced by the F_1 generation, and there will be a 9:3:3:1 phenotypic ratio in the F_2 generation. Mendel's results supported this latter hypothesis, called independent assortment.

the experiment is to see what happens when F_1 plants self-pollinate to produce F_2 offspring. If the hybrids must transmit their alleles in the same combinations in which they were inherited from the P generation, then there will only be two classes of gametes: *YR* and *yr*. This hypothesis predicts that the phenotypic ratio of the F_2 generation will be 3:1, just as in a monohybrid cross (FIGURE 13.7a).

The alternative hypothesis is that the two pairs of alleles segregate independently of each other. In other words, genes are packaged into gametes in all possible allelic combinations, as long as each gamete has one allele for each gene. In our example, four classes of gametes would be produced in equal quantities: *YR, Yr, yR,* and *yr*. If four classes of sperm are mixed with four classes of ova, there will be 16 (4 × 4) equally probable ways in which the alleles can combine in the F_2 generation, as shown in FIGURE 13.7b. These combinations make up four phenotypic categories with a ratio of 9:3:3:1 (nine yellow-round to three green-round to three yellow-wrinkled to one green-wrinkled). When Mendel did the experiment and "scored" (classified) the F_2 offspring, he obtained a ratio of 315:108:101:32, which is approximately 9:3:3:1.

The experimental results supported the hypothesis that each character is independently inherited; that in the dihybrids (*YyRr*), the two alleles for seed color segregate independently of the two alleles for seed shape. Mendel tried his seven pea characters in various dihybrid combinations and always observed a 9:3:3:1 phenotypic ratio in the F_2 generation. Notice in FIGURE 13.7b, however, that there remains a 3:1 phenotypic ratio for each character: three yellow to one green; three round to one wrinkled. As far as an individual character is concerned, the segregation behavior is the same as if this were a monohybrid cross. The independent segregation of each pair of alleles during gamete formation is now called Mendel's **law of independent assortment.** FIGURE 13.8 reinforces the concept of independent assortment by applying it to a mating of budgies (parakeets) that are heterozygous for two characters.

Mendelian inheritance reflects rules of probability

Mendel's laws of segregation and independent assortment are specific applications of the same general rules of probability that apply to tossing coins, rolling dice, or drawing cards. A basic understanding of these rules of chance is essential for genetic analysis.

The probability scale ranges from 0 to 1. An event that is certain to occur has a probability of 1, while an event that is certain *not* to occur has a probability of 0. With a two-headed coin, the probability of tossing heads is 1, and the probability of tossing tails is 0. With a normal coin, the chance of tossing heads is ½ and the chance of tossing tails is ½. The probability of rolling the number 3 with a die, which is six-sided, is ⅙, and the chance of drawing the queen of spades from a full deck of cards is 1/52. The probabilities of all possible outcomes for an event must add up to 1. With a die, the chance of rolling a number other than 3 is ⅚. In a deck of cards, the chance of drawing a card other than the queen of spades is 51/52.

We can learn an important lesson about probability from tossing a coin. For every toss, the probability of heads is ½. The outcome of any particular toss is unaffected by what has happened on previous trials. We refer to phenomena such as successive coin tosses as independent events. (The term also applies to simultaneous tosses of several coins.) It is entirely possible that five successive tosses of a normal coin will produce five successive heads. Before the sixth toss, an observer might predict, "A tail is due to come up, because there have already been so many heads." But on the sixth toss, the chance that the outcome will again be heads is still ½.

Two basic laws of probability that can help us in games of chance and in solving genetic problems are the rule of multiplication and the rule of addition.

The Rule of Multiplication

If two coins are tossed simultaneously, the outcome for each coin is independent of what happens with the other coin. What is the chance that both coins will land heads up? How do we determine the chance that two or more independent events will occur together in some specific combination? The solution is in computing the probability for each independent event, then multiplying these individual probabilities to obtain the overall probability of these events occurring in combination. According to this rule of multiplication, the probability that both coins will land heads up is ½ × ½ = ¼. An F_1 monohybrid cross is analogous to this game of chance. With flower color as the heritable character, the genotype of a given F_1 plant is *Pp*. What is the probability that a particular F_2 plant will have white flowers? For this to happen, both the ovum and the sperm must carry the *p* allele, so we invoke the rule of multiplication. Segregation in the heterozygous plant is like flipping a coin. The probability that an ovum will have the *p* allele is ½. The chance that a sperm will have the *p* allele is ½. Thus, the probability that two *p* alleles will come together at fertilization is ½ × ½ = ¼, equivalent to the probability that two independently tossed coins will land heads up (FIGURE 13.9, p. 247).

We can also apply the rule of multiplication to dihybrid crosses. In FIGURE 13.7b, the independent assortment of the two pairs of alleles during the F_1 dihybrid cross is analogous to flipping a dime and a quarter simultaneously.

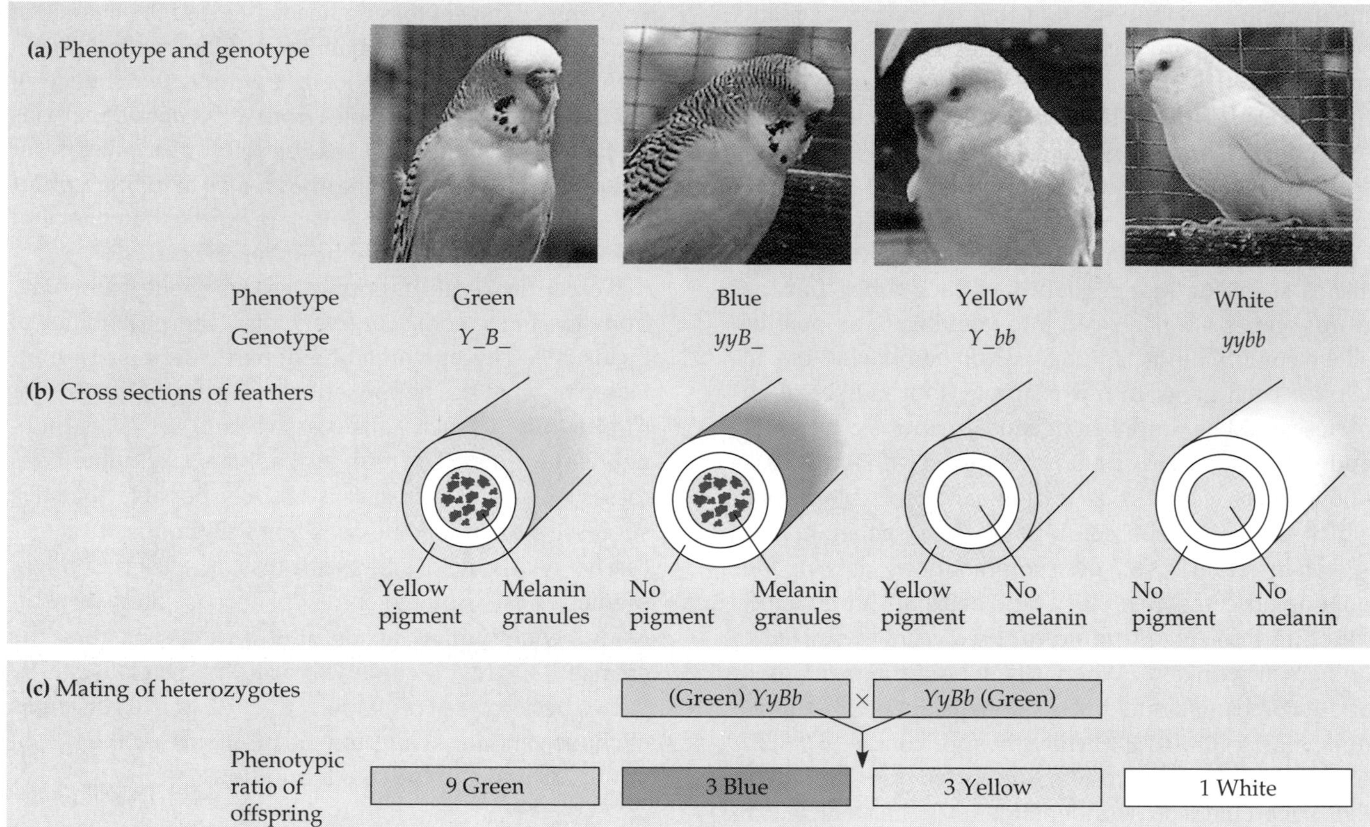

FIGURE 13.8

Independent assortment and variations in the feather color of budgies. (**a**) Two different and independently inherited genes affect the color of plumage of the budgies. One gene determines pigmentation in the outer region of a feather, and the second gene determines pigmentation of the feather's core. Two alleles exist for each gene. For the outer zone, the allele for yellow color is dominant (*Y*) and the allele for a colorless condition is recessive (*y*). For the core of the feather, an allele for the presence of a pigment called melanin is dominant (*B*) and the allele for absence of melanin is recessive (*b*). Various combinations of these alleles result in the four phenotypes illustrated by the photos. (The blanks in the genotypes indicate that homozygous dominant and heterozygous conditions result in the same phenotype.) (**b**) Cross sections of feathers reveal the anatomical and biochemical bases for the four phenotypes. A dihybrid cross between true-breeding green budgies (*YYBB*) and white birds (*yybb*) results in green offspring that are heterozygous for both genes (*YyBb*). (**c**) When these dihybrids mate, a Mendelian F_1 cross, the four possible phenotypes occur in the F_2 generation in a 9:3:3:1 ratio. If you construct a Punnett square for the results of the F_1 cross, you can derive this ratio by grouping genotypes that produce the same phenotype.

For a parent with the genotype *YyRr*, the probability that a gamete will carry the *Y* and *R* alleles is ¼, equivalent to the chance of a quarter and dime both coming up heads. We can extend the rule of multiplication to determine the probability of specific genotypes in the F_2 generation, avoiding the time-consuming need to construct a 16-part Punnett square. For example, the probability of an F_2 plant having the genotype *YYRR* is 1/16 (¼ chance for a *YR* ovum × ¼ chance for a *YR* sperm). This corresponds to the top box in the Punnett square of FIGURE 13.7b.

The Rule of Addition

What is the probability that an F_2 plant from a monohybrid cross will be heterozygous? Notice in FIGURE 13.9 that there are two ways F_1 gametes can combine to produce a heterozygous result. The dominant allele can come from the ovum and the recessive allele from the sperm, or vice versa. According to the rule of addition, the probability of an event that can occur in two or more different ways is the sum of the separate probabilities of those ways. Using this rule, we can calculate the probability of an F_2 heterozygote as ¼ + ¼ = ½.

Using Rules of Probability to Solve Genetics Problems

We can combine the rules of multiplication and addition to solve complex problems in Mendelian genetics. For instance, Mendel crossed pea varieties that differed in three characters—*trihybrid* crosses. Imagine a cross of garden peas in which a trihybrid with purple flowers and yellow, round seeds (heterozygous for all three genes) is crossed with a plant with purple flowers and green, wrin-

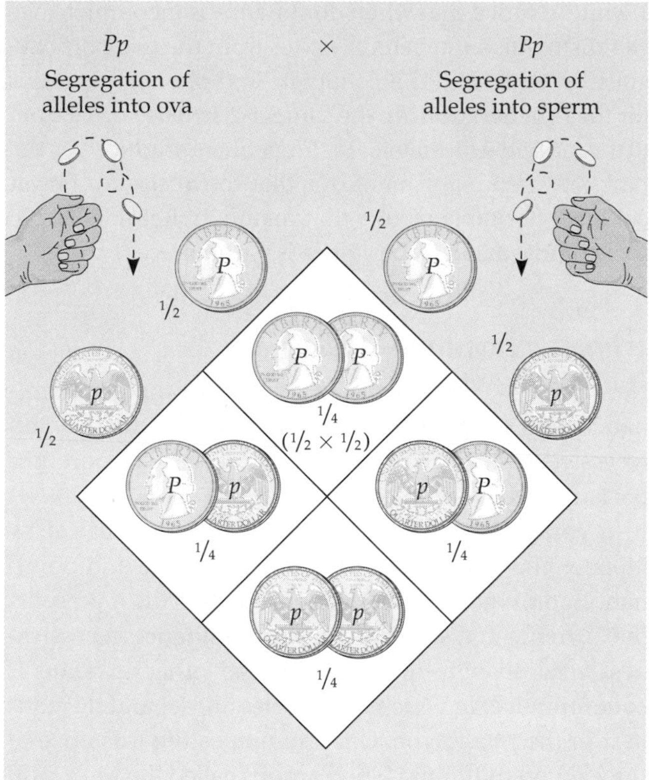

FIGURE 13.9

Segregation of alleles and fertilization as chance events. When a heterozygote (*Pp*) forms gametes, segregation of alleles is like the toss of a coin. An ovum has a 50% chance of receiving the dominant allele and a 50% chance of receiving the recessive allele. The same odds apply to a sperm cell. Like two separately tossed coins, segregation during sperm and ovum formation occurs as two independent events. To determine the probability that an individual offspring will inherit the dominant allele from both parents, we multiply the probabilities of each required event: $\frac{1}{2} \times \frac{1}{2} = \frac{1}{4}$. Similarly, we can use the rule of multiplication to predict the probability for any genotype among offspring, as long as we know the genotypes of the parents.

kled seeds (heterozygous for flower color but homozygous recessive for the other two characters). Using Mendelian symbols, our cross looks like this:

PpYyRr × *Ppyyrr*

Let's use the rules of probability to calculate the fraction of offspring that would exhibit the recessive phenotypes for *at least two* of the three traits. We can start by listing all genotypes that fulfill this condition: *ppyyRr, ppYyrr, Ppyyrr, PPyyrr,* and *ppyyrr.* (Because the condition is *at least* two recessive traits, this last genotype, which produces all three recessive phenotypes, counts.) Next, use the rule of multiplication to calculate the individual probabilities for each of these genotypes from our *PpYyRr* × *Ppyyrr* cross. Finally, use the rule of addition to pool the probabilities for fulfilling the condition of at least two recessive traits:

ppyyRr	$\frac{1}{4}$ (probability of *pp*) × $\frac{1}{2}$ (*yy*) × $\frac{1}{2}$ (*Rr*)	= $\frac{1}{16}$
ppYyrr	$\frac{1}{4} \times \frac{1}{2} \times \frac{1}{2}$	= $\frac{1}{16}$
Ppyyrr	$\frac{1}{2} \times \frac{1}{2} \times \frac{1}{2}$	= $\frac{2}{16}$
PPyyrr	$\frac{1}{4} \times \frac{1}{2} \times \frac{1}{2}$	= $\frac{1}{16}$
ppyyrr	$\frac{1}{4} \times \frac{1}{2} \times \frac{1}{2}$	= $\frac{1}{16}$
Chance of *at least two* recessive characters		= $\frac{6}{16}$ or $\frac{3}{8}$

With practice, you'll be able to apply the rules of probability to solve such genetics problems faster than you could by filling in Punnett squares.

Mendel discovered the particulate behavior of genes: *a review*

If we plant a seed from the F_2 generation of FIGURE 13.4, we cannot predict with absolute certainty that the plant will yield white flowers, any more than we can predict with certainty that two tossed coins will both come up heads. What we *can* say is that there is exactly a $\frac{1}{4}$ chance that the plant will have white flowers. Stated another way, among a large sample of F_2 plants, one-fourth (25%) will have white flowers. Usually, the larger the sample size, the closer the results will conform to our predictions. The fact that Mendel counted so many offspring from his crosses suggests that he understood this statistical feature of inheritance and had a keen sense of the rules of chance.

Thus, Mendel's two laws, segregation and independent assortment, explain heritable variations in terms of alternate forms of genes (hereditary "particles") that are passed along, generation after generation, according to simple rules of probability. This particulate theory of inheritance, first discovered in garden peas, is equally valid for figs, flies, fish, birds, and human beings. Mendel's impact endures, not only on genetics, but on all of science, as a case study of the power of hypothetico-deductive thinking.

The relationship between genotype and phenotype is rarely simple

In this century, geneticists have extended Mendelian principles not only to diverse organisms, but also to patterns of inheritance more complex than Mendel actually described. It was brilliant (or lucky) that Mendel chose pea plant characters that turned out to have a relatively simple genetic basis: Each character he studied is determined by one gene, for which there are only two alleles, one completely dominant to the other. But these conditions are not met by all heritable characters, not even in garden peas. The relationship between genotype and phenotype is rarely so simple. This does not diminish the utility of Mendelian genetics, for the basic principles of segregation and independent assortment apply even to more complex patterns of inheritance. In this section, we will extend Mendelian genetics to hereditary patterns that were not reported by Mendel.

Incomplete Dominance

The F_1 offspring of Mendel's classic pea crosses always looked like one of the two parental varieties because of the complete dominance of one allele over another. But for some traits, there is **incomplete dominance,** where the F_1 hybrids have an appearance somewhere in between the phenotypes of the two parental varieties. For instance, when red snapdragons are crossed with white snapdragons, all the F_1 hybrids have pink flowers (FIGURE 13.10). This third phenotype results from flowers of the heterozygotes having less red pigment than the red homozygotes. We should not, however, regard incomplete dominance as evidence of the blending theory, which would predict that the red or white traits could never be retrieved from the pink hybrids. In fact, breeding the F_1 hybrids produces F_2 offspring with a phenotypic ratio of 1 red to 2 pink to 1 white. (Notice that when dominance is incomplete, we can distinguish the heterozygotes from the two homozygous varieties, and the genotypic and phenotypic ratios for the F_2 generation are the same, 1:2:1.) The segregation of the red and white alleles in the gametes produced by the pink-flowered plants confirms that the alleles for flower color are heritable factors that maintain their identity in the hybrids; that is, inheritance is particulate.

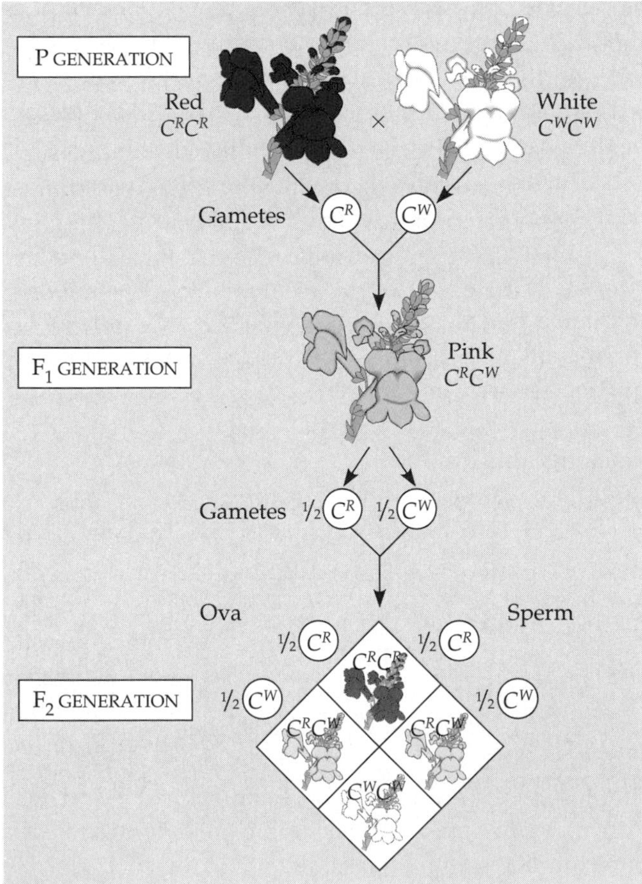

FIGURE 13.10
Incomplete dominance in snapdragon color. When red snapdragons are crossed with white ones, the F_1 hybrids have pink flowers. Segregation of alleles into the gametes of the F_1 plants results in an F_2 generation with a 1:2:1 ratio for both genotype and phenotype. C^R = allele for red flower color; C^W = allele for white flower color.

What Is *a Dominant Allele?*

Now that you have learned about incomplete dominance, let's reexamine the meaning of dominance and recessiveness. What *is* a dominant allele? Or, more importantly, what is it *not*?

In **complete dominance,** the situation described by Mendel, the phenotypes of the heterozygote and dominant homozygote are indistinguishable. This represents one extreme of a spectrum in the dominance/recessiveness relationships of alleles. At the other extreme is **codominance,** in which *both* alleles are separately manifest in the phenotype. One example is the existence of three different human blood groups called the M, N, and MN blood groups. These groupings are based on two specific molecules located on the surfaces of blood cells. People of group M have one of these two types of molecules, and people of group N have the other type. Group MN is characterized by the presence of *both* molecules on blood cells. What is the genetic basis of these phenotypes? A single gene locus, at which two allelic variations are possible, determines these blood groups. M individuals are homozygous for one allele; N individuals are homozygous for the other allele. A heterozygous condition results in the blood of the MN group. Notice that the MN phenotype is *not* intermediate between M and N phenotypes, but that both of these phenotypes are individually expressed by the presence of the two types of molecules on blood cells. In contrast, incomplete dominance is characterized by an intermediate phenotype, as in the pink flowers of snapdragon hybrids. Thus, the range of relationships between alleles includes complete dominance, codominance, and different degrees of incomplete dominance. These variations are reflected in the phenotypes of heterozygotes.

For any character, the dominance/recessiveness relationship we observe depends on the level at which we examine phenotype. For example, consider Tay-Sachs disease, an inherited disorder in humans. The brain cells of a baby with Tay-Sachs disease are unable to metabolize gangliosides, a type of lipid, because a crucial enzyme does not work properly. As the lipids accumulate in the brain, the brain cells gradually cease to function normally, leading to death. Only children who inherit two copies of the Tay-Sachs allele (homozygotes) have the disease. Thus, on the *organismal* level of normal versus Tay-Sachs phenotype, the Tay-Sachs allele qualifies as a recessive. At the *biochemical* level, however, we observe an intermediate phenotype characteristic of incomplete dominance: The enzyme deficiency that causes Tay-Sachs disease can be detected in heterozygotes, who have an activity level of the lipid-metabolizing enzyme that is intermediate between individuals homozygous for the normal allele and individuals with Tay-Sachs disease. Heterozygotes lack symptoms of the disease, apparently because half the normal amount of functional enzyme is sufficient to prevent lipid accumulation in the brain. In fact, heterozygous individuals produce equal numbers of normal and dysfunctional enzymes. Thus, at the *molecular* level, the normal allele and the Tay-Sachs allele are codominant. As you can see, dominance/recessiveness relationships are rarely as straightforward as Mendel reported.

It is also important to understand that an allele is not termed *dominant* because it somehow subdues and mutes a recessive allele. Recall that alleles are simply variations in a gene's nucleotide sequence. When a dominant allele coexists with a recessive allele in a heterozygous genotype, they do not actually interact at all. It is in the pathway from genotype to phenotype that dominance and recessiveness come into play. We can use one of Mendel's characters—round versus wrinkled pea seed shape—as an example. The dominant allele codes for the synthesis of an enzyme that helps convert sugar to starch in the seed. The recessive allele codes for a defective form of this enzyme. Thus, in a recessive homozygote, sugar accumulates in the seed because it is not converted to starch. As the seed develops, the high sugar concentration causes the osmotic uptake of water, and the seed swells. When the mature seed dries, it has wrinkles, analogous to the stretch marks of a person who has lost a lot of weight. In contrast, if a dominant allele is present, sugar is converted to starch, and the seeds do not wrinkle when they dry. One dominant allele results in enough of the enzyme to convert sugar to starch, and thus dominant homozygotes and heterozygotes have the same phenotype: round seeds. By exploring the mechanisms responsible for phenotype, we can demystify the concepts of dominance and recessiveness.

There is another important lesson about the meaning of the term *dominance*. Because an allele for a particular character is dominant does not necessarily mean that it is more common in a population than the recessive allele for that character. For example, about one baby out of 400 in the United States is born with extra fingers or toes, a condition known as polydactyly. The allele for polydactyly is *dominant* to the allele for five digits per appendage. In other words, 399 out of every 400 people are recessive homozygotes for this character; the recessive allele is far more prevalent than the dominant allele in the population. In Chapter 21, you will learn how the relative frequencies of alleles in a population are affected by natural selection.

Let's summarize three important points about dominance/recessiveness relationships:

1. They range from complete dominance, through various degrees of incomplete dominance, to codominance.
2. They reflect the mechanisms by which specific alleles are expressed in phenotype and do not involve the ability of one allele to subdue another at the level of the DNA.
3. They do not determine the relative abundance of alleles in a population.

Multiple Alleles

Most genes actually exist in more than two allelic forms. The ABO blood groups in humans are one example of multiple alleles of a single gene. There are four phenotypes for this character. A person's blood group may be either A, B, AB, or O. These letters refer to two carbohydrates, the A substance and the B substance, which may be found on the surface of red blood cells. (These groups are based on different molecular markers on blood cells than those used for the MN classification discussed earlier.) A person's blood cells may be coated with one substance or the other (type A or B), with both (type AB), or with neither (type O). Matching compatible blood groups is critical for blood transfusions. If the donor's blood has a factor (A or B) that is foreign to the recipient, specific proteins called antibodies produced by the recipient bind to the foreign molecules and cause the donated blood cells to agglutinate (clump together). This agglutination can kill the recipient.

The four blood groups result from various combinations of three different alleles of one gene, symbolized as I^A (for the A carbohydrate), I^B (for B), and i (giving rise to neither A nor B). Six genotypes are possible (FIGURE 13.11, p. 250). Both the I^A and the I^B alleles are dominant to the i allele. Thus, I^AI^A and I^Ai individuals have type A blood, and I^BI^B and I^Bi individuals have type B. Recessive homozygotes, ii, have type O blood, because neither the A nor the B substance is produced. The I^A and I^B alleles are codominant; both are expressed in the phenotype of the I^AI^B heterozygote, who has type AB blood.

Pleiotropy

So far, we have treated Mendelian inheritance as though each gene affects one phenotypic character. Most genes, however, have multiple phenotypic effects. The ability of a gene to affect an organism in many ways is called **pleiotropy** (Gr. *pleion*, "more"). For example, alleles that

Phenotype (Blood Group)	Genotypes	Antibodies Present in Blood Serum	Reacts (Clumps) When Red Blood Cells from Groups Below Are Added to Serum from Groups at Left?			
			O	A	B	AB
O	ii	Anti-A Anti-B	No	Yes	Yes	Yes
A	$I^A I^A$ or $I^A i$	Anti-B	No	No	Yes	Yes
B	$I^B I^B$ or $I^B i$	Anti-A	No	Yes	No	Yes
AB	$I^A I^B$	—	No	No	No	No

FIGURE 13.11
Multiple alleles for the ABO blood groups. There are three alleles. Because each person carries two alleles, six genotypes are possible. Whenever the I^A or the I^B allele is present, the corresponding factor (A or B, respectively) is present on the surface of red blood cells. Both of these alleles, which are codominant, are dominant to the i allele, which does not code for any surface factor. A person produces antibodies against foreign blood factors, causing clumping of red blood cells if a transfusion is performed with incompatible blood.

are responsible for certain hereditary diseases in humans, including sickle-cell disease, usually cause multiple symptoms (see FIGURE 13.16, p. 255). Considering the intricate molecular and cellular interactions responsible for an organism's development, it is not surprising that a gene can affect many of the organism's characteristics.

Epistasis

In some cases, a gene at one locus alters the phenotypic expression of a gene at a second locus, a condition known as **epistasis** (Gr. for "standing upon"). An example will help clarify this concept. In mice and many other mammals, black coat color is dominant to brown. The two alleles for this character are designated *B* and *b*. For a mouse to have brown fur, its genotype must be *bb*. But there is more to the story. A second gene locus determines whether or not pigment will be deposited in the hair. For this second gene, the dominant allele, symbolized by *C* (for color), results in the deposition of pigment. This allows either black or brown color, depending on the genotype at the first locus. But if the mouse is homozygous recessive for the second locus (*cc*), then the coat is white (albino), regardless of the genotype at the black/brown locus.

What happens if we mate black mice that are heterozygous for both genes (*BbCc*)? Although the two genes affect the same phenotypic character (coat color), they follow the law of independent assortment (the two genes are inherited separately). Thus, our breeding experiment represents an F_1 dihybrid cross, which produced a 9:3:3:1 ratio in Mendel's experiments. In the case of coat color, however, the ratio of phenotypes among F_2 offspring is 9 black to 3 brown to 4 white. FIGURE 13.12 uses a Punnett square to account for this ratio in terms of epistasis. Other types of epistatic interactions produce different ratios.

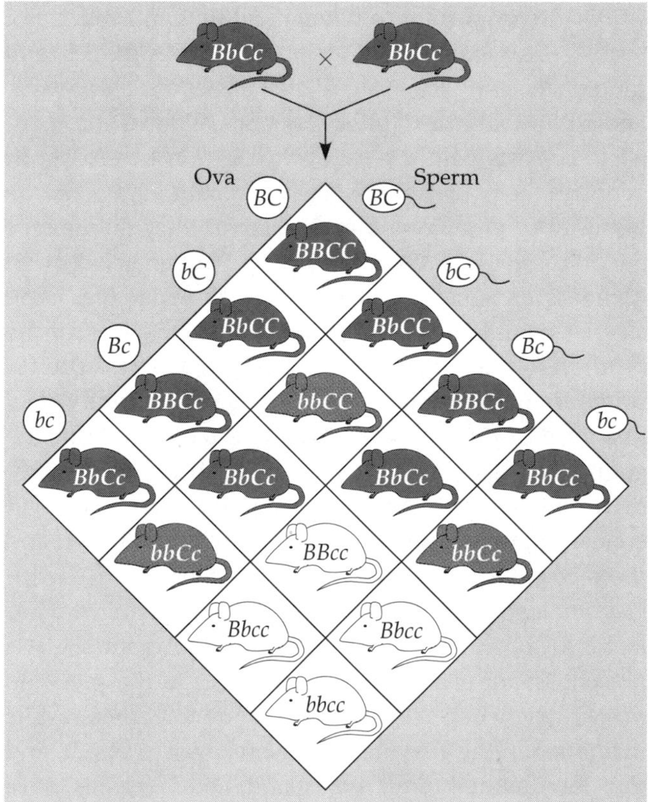

FIGURE 13.12
An example of epistasis. This Punnett square illustrates the genotypes and phenotypes for offspring of matings between two black mice. The parents are heterozygous for two genes, which assort independently of each other. Thus, our breeding experiment corresponds to a Mendelian F_1 cross between dihybrids. One gene determines whether the coat will be black (dominant, *B*) or brown (recessive, *b*). The second gene controls whether or not pigment of any color will be deposited in the hair, with the allele for the presence of color (*C*) dominant to the allele for the absence of color (*c*). All offspring of the *cc* genotype are white (albino), regardless of the genotype for the black/brown genetic locus. This epistatic relationship of the color gene to the black/brown gene results in an F_2 phenotypic ratio of 9 black to 3 brown to 4 white.

Polygenic Inheritance

Mendel studied characters that could be classified on an either-or basis, such as purple versus white flower color. There are many characters, however, including human skin color and height, for which an either-or classification is impossible, because the characters vary in the population along a continuum (in gradations). These are called **quantitative characters.** Quantitative variation usually indicates **polygenic inheritance,** an additive effect of two or more genes on a single phenotypic character (the converse of pleiotropy, where a single gene affects many phenotypic characters).

There is evidence, for instance, that skin pigmentation in humans is controlled by at least three separately inherited genes (probably more, but we will simplify). Let's consider three genes, with a dark-skin allele for each gene (*A, B, C*) contributing one "unit" of darkness to the phenotype and being incompletely dominant to the other alleles (*a, b, c*). An *AABBCC* person would be very dark, while an *aabbcc* individual would be very light. An *AaBbCc* person would have skin of an intermediate shade. Because the alleles have a cumulative effect, the genotypes *AaBbCc* and *AABbcc* would make the same genetic contribution (three units) to skin darkness. FIGURE 13.13 shows how this polygenic inheritance could result in a bell-shaped curve, called a normal distribution, for skin darkness among the members of a hypothetical population. (You are probably familiar with the concept of a normal distribution for class curves of test scores.) Environmental factors, such as exposure to the sun, also affect the skin-color phenotype and help make the graph a smooth curve rather than a stairlike histogram.

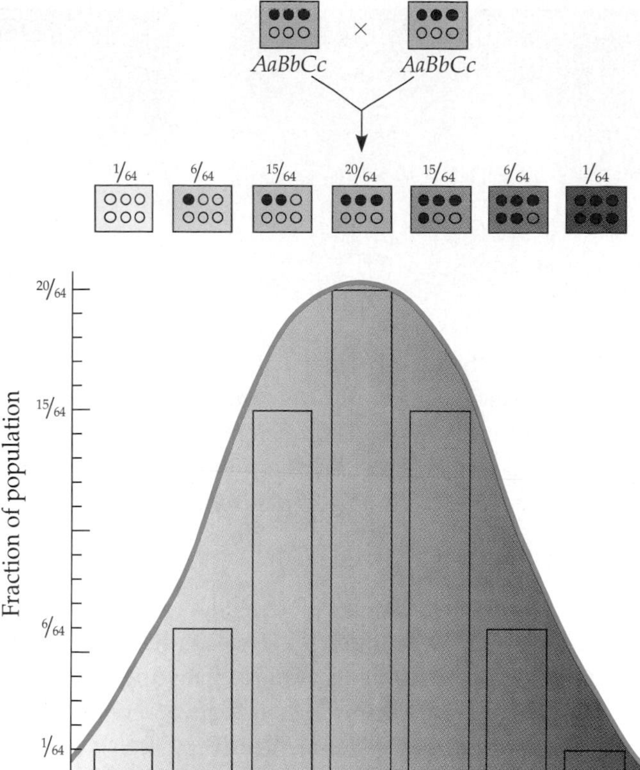

FIGURE 13.13
A simplified model for polygenic inheritance of skin color. According to this model, three separately inherited genes affect the darkness of skin. For each gene, an allele for dark skin is incompletely dominant to an allele for light skin. Thus, an individual who is heterozygous for all three genes (*AaBbCc*) has inherited three "units" of darkness, indicated in this figure by the three dots in the small square. An individual with the genotype *AABbcc* would also have a total of three units for dark skin. Imagine a large number of matings between individuals who are heterozygous for all three genes. Along the top of the graph are the variations that can occur among offspring. The *y*-axis represents the fractions of these variations among offspring of such trihybrid matings. The resulting histogram is smoothed into a bell-shaped curve by environmental factors, such as exposure to the sun, that affect skin color. This model is an oversimplification, but it should help clarify how polygenic inheritance can contribute to phenotypic characters that vary in gradations among the members of a population.

Nature Versus Nurture: The Environmental Impact on Phenotype

Phenotype depends on environment as well as on genes. A single tree, locked into its inherited genotype, has leaves that vary in size, shape, and greenness, depending on exposure to wind and sun. For humans, nutrition influences height, exercise alters build, sun-tanning darkens the skin, and experience increases performance on intelligence tests. Even identical twins, who are genetic equals, accumulate phenotypic differences as a result of their unique experiences.

Whether it is genes or the environment—nature or nurture—that most influences human characteristics is a very old and hotly contested debate that we will not attempt to settle here. We can say, however, that the product of a genotype is generally not a rigidly defined phenotype, but a range of phenotypic possibilities over which there may be variation due to environmental influence. This phenotypic range is called the **norm of reaction** for a genotype (FIGURE 13.14). There are cases where the norm of reaction has no breadth whatsoever; that is, a given genotype mandates a very specific phenotype. An example is the gene locus that determines a person's ABO blood group. In contrast, a person's blood count of red and white cells varies, depending on such factors as the altitude of one's home, the person's customary level of physical activity, and the presence of infectious agents.

Generally, norms of reaction are broadest for polygenic characters, including behavioral traits. Environment

FIGURE 13.14
The effect of environment on phenotype. The effect of a genotype is defined by its norm of reaction, a phenotypic range that depends on the environment in which the genotype is expressed. For example, hydrangea flowers of the same genetic variety range in color from blue-violet to pink, depending on the acidity of the soil.

contributes to the quantitative nature of these characters, as we have seen in the continuous variation of skin color. Geneticists refer to such characters as **multifactorial,** meaning that many factors, both genetic and environmental, collectively influence phenotype.

Integrating a Mendelian View of Heredity and Variation

Over the past several pages, we have broadened our view of Mendelian inheritance by exploring incomplete dominance and other variations in dominance/recessiveness relationships, multiple alleles, pleiotropy, polygenic inheritance, and the phenotypic impact of the environment. How can we integrate these refinements into a comprehensive theory of Mendelian genetics? The key is to make the transition from the reductionist emphasis on single genes and phenotypic characters to the idea of the organism as a whole, one of the themes of this book. In fact, the term *phenotype* does double duty. We have been using the word in the context of specific characters, such as flower color and blood group. But phenotype is also used to describe the organism in its entirety—*all* aspects of its physical appearance, internal anatomy, physiology, and behavior. Similarly, the term *genotype* can also refer to an organism's entire genetic makeup, not just its alleles for a single genetic locus. In most cases, a gene's impact on phenotype is affected by other genes and by the environment. In this integrated view of heredity and variation, an organism's phenotype reflects its overall genotype and unique environmental history.

Considering all that can occur in the pathway from genotype to phenotype, it is indeed impressive that Mendel could simplify the complexities to reveal the fundamental principles governing the transmission of individual genes from parents to offspring. By extending the principles of segregation and independent assortment to help explain such hereditary patterns as epistasis and quantitative characters, we begin to see how broadly Mendelism applies. From Mendel's abbey garden came a theory of particulate inheritance that anchors modern genetics. In the last two sections of this chapter, we will apply Mendelian genetics to human inheritance, especially the transmission of hereditary diseases.

Pedigree analysis reveals Mendelian patterns in human inheritance

Whereas peas are convenient subjects for genetic research, humans are not. The human generation span is about 20 years, and human parents produce relatively few offspring (compared to peas and most other species). Furthermore, well-planned breeding experiments like the ones Mendel performed are impossible (or at least socially unacceptable) with humans. In spite of these difficulties, the study of human genetics continues to advance, powered by the incentive to understand our own inheritance. New techniques in molecular biology have led to many breakthrough discoveries, as we will see in Chapter 19, but basic Mendelism endures as the foundation of human genetics.

Unable to manipulate the mating patterns of people, geneticists must analyze the results of matings that have already occurred. As much information as possible is collected about a family's history for a particular trait, and this information is assembled into a family tree describing the interrelationships of parents and children across the generations—the family **pedigree.** The pedigree in FIGURE 13.15a traces the occurrence of widow's peak (a pointed contour of the hairline on the forehead). The trait is due to a dominant allele, which we will symbolize as *W.*

We know that all individuals in this family who lack a widow's peak are homozygous recessive, and thus we can fill their genotypes on the pedigree (*ww*). We also know that both grandparents with widow's peaks are heterozygous (*Ww*); if they were homozygous dominant (*WW*), then all of their offspring would have widow's peaks. Those offspring in this second generation who *do* have widow's peaks must also be heterozygous, since they are the products of *Ww* × *ww* matings. The third generation in this pedigree consists of two sisters. The one who has a widow's peak could be either homozygous (*WW*) or heterozygous (*Ww*), given what we know about the genotypes of her parents (both *Ww*).

FIGURE 13.15b is a pedigree of the same family, but this time we are tracing a recessive trait, attached earlobes. We'll use *f* for the recessive allele and *F* for the dominant

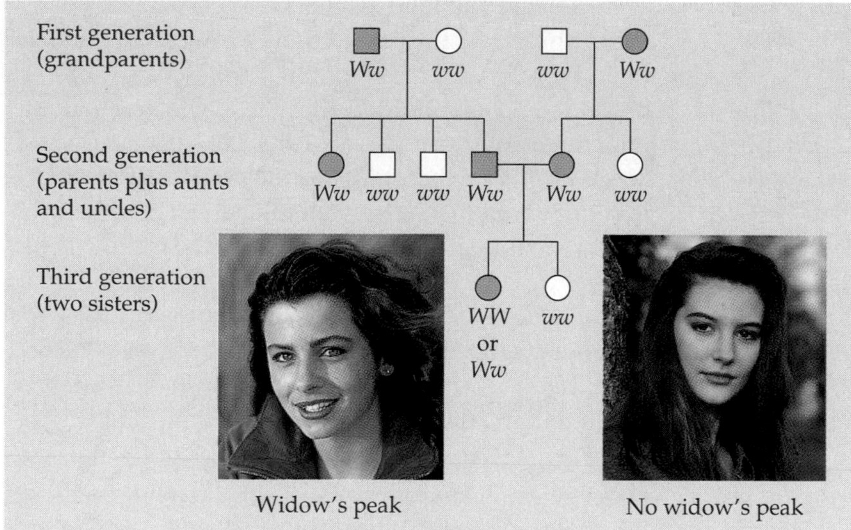

(a) Pedigree tracing a dominant trait (widow's peak)

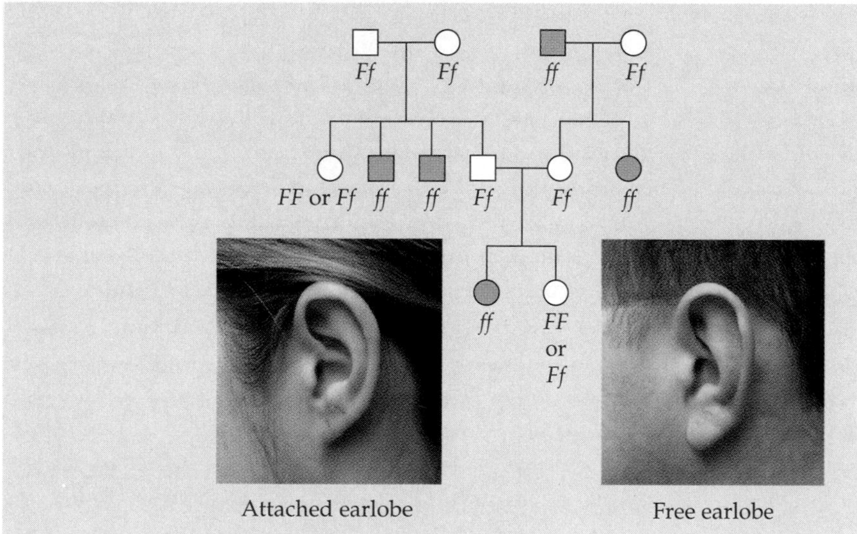

(b) Pedigree tracing a recessive trait (attached earlobes)

FIGURE 13.15
Pedigree analysis. In these family trees, squares symbolize males and circles represent females. A horizontal line connecting a male and female (□—○) indicates a mating, with offspring listed below in their order of birth, from left to right. Shaded symbols stand for individuals with the trait being traced. (**a**) **Dominant trait.** This pedigree traces a trait called widow's peak through three generations of a family. Notice in the third generation that the second-born daughter lacks a widow's peak, although both of her parents had the trait. Such a pattern of inheritance supports the hypothesis that the trait is due to a dominant allele. (If the trait were due to a *recessive* allele, and both parents have the recessive phenotype, then *all* of their offspring should also have the recessive phenotype.) (**b**) **Recessive trait.** This is the same family, but in this case we are tracing the inheritance of a recessive trait, attached earlobes. Notice that the first-born daughter in the third generation has attached lobes, although both of her parents lack that trait (they have free earlobes). Such a pattern is easily explained if the attached-lobe phenotype is due to a recessive allele. (If it were due to a *dominant* allele, then at least one parent would also have the trait.)

allele, which results in free earlobes. As you work your way through the pedigree, notice once again that you can apply what you have learned about Mendelian inheritance to fill in the genotypes for most individuals.

A pedigree not only helps us understand the past; it also helps us predict the future. Suppose that the couple represented in the second generation of FIGURE 13.15 decide to have one more child. What is the probability that the child will have a widow's peak? This is like a Mendelian F_1 cross (*Ww* × *Ww*), and thus the probability that a child will exhibit the dominant phenotype (widow's peak) is ¾. What is the probability that the child will have attached earlobes? Again, we can treat this as an F_1 monohybrid cross (*Ff* × *Ff*), but this time we want to know the chance that the offspring would be homozygous recessive. The probability is ¼. What is the chance that the child will have a widow's peak *and* attached earlobes? If the two pairs of alleles assort independently in this dihybrid problem (*WwFf* × *WwFf*), then we can use the rule of multiplication: ¾ (chance of widow's peak) × ¼ (chance of attached earlobes) = 3⁄16 (chance of widow's peak and attached earlobes).

Many human disorders follow Mendelian patterns of inheritance

Pedigree analysis is much more significant when the alleles in question cause disabling or deadly hereditary diseases instead of innocuous human variations such as hairline or earlobe configuration. However, for disorders that are inherited as simple Mendelian traits, these same techniques of pedigree analysis apply.

Recessively Inherited Disorders

Several thousand genetic disorders are known to be inherited as simple recessive traits. These disorders range in severity from traits that are usually nonlethal, such as albinism (lack of skin pigmentation), to life-threatening conditions, such as cystic fibrosis.

How can we account for the recessive behavior of the alleles causing these disorders? Recall that genes code for proteins of specific function. An allele that causes a genetic disorder codes either for a malfunctional protein or for no protein at all. In the case of disorders classified as recessive, heterozygotes are normal in phenotype because one copy of the "normal" allele produces a sufficient amount of the specific protein. Thus, a recessively inherited disorder shows up only in the homozygous individuals who inherit one recessive allele from each parent. We can symbolize the genotype of such people as *aa*, with individuals lacking the disorder being either *AA* or *Aa*. The heterozygotes (*Aa*), who are phenotypically normal, are called *carriers* of the disorder because they may transmit the recessive allele to their offspring.

Most of the people who have recessive disorders are born to parents of normal phenotype who are both carriers. A mating between two carriers corresponds to a Mendelian F_1 cross ($Aa \times Aa$), with the zygote having a ¼ chance of inheriting a double dose of the recessive allele. A child of normal phenotype from such a cross has a ⅔ chance of being a carrier. (Genotypic ratio for the offspring is 1*AA*:2*Aa*:1*aa*. Thus, two out of three individuals of *normal* phenotype—*AA* or *Aa*—are heterozygous carriers.) Recessive homozygotes could also result from $Aa \times aa$ and $aa \times aa$ matings, but if the disorder is lethal before reproductive age or results in sterility, no *aa* individuals will reproduce. Even if recessive homozygotes are able to reproduce, such individuals will still account for a much smaller percentage of the population than heterozygous carriers, for reasons we will examine in Chapter 21.

In general, a genetic disorder is not evenly distributed among all groups of humans. These disparities result from the different genetic histories of the world's peoples during less technological times, when populations were more geographically, hence genetically, isolated. We will now examine three examples of such recessively inherited disorders.

The most common lethal genetic disease in the United States is **cystic fibrosis,** which strikes one out of every 2500 Caucasians but is much rarer in other groups. One out of 25 Caucasians (4%) is a carrier (heterozygote). The normal allele for this gene codes for a membrane protein that functions in chloride transport between certain cells and the extracellular fluid. These chloride channels are defective or absent in the plasma membranes of children who have inherited two of the recessive alleles that cause cystic fibrosis. This results in an abnormal concentration of extracellular chloride, which in turns causes the mucus that coats certain cells to become thicker and stickier than normal. The mucus builds up in the pancreas, lungs, digestive tract, and other organs, a condition that favors pneumonia and other infections. Untreated, most children with cystic fibrosis die before their fifth birthday. A special diet, daily doses of antibiotics to prevent infection, and other preventive treatments can prolong life. In the United States, the average longevity of individuals with cystic fibrosis is now about 27 years.

Another lethal disorder inherited as a recessive allele is **Tay-Sachs disease,** described earlier in the chapter. Recall that the disease is caused by a dysfunctional enzyme that fails to break down a class of brain lipids. The symptoms of Tay-Sachs disease usually become manifest a few months after birth. The infant begins to suffer seizures, blindness, and degeneration of motor and mental performance. Inevitably, the child dies within a few years. There is a disproportionately high incidence of Tay-Sachs disease among Ashkenazic Jews, Jewish people whose ancestors lived in central Europe. In that population, the frequency of the disease is one case out of 3600 births, about 100 times greater than the incidence among non-Jews or Mediterranean (Sephardic) Jews.

The most common inherited disease among blacks is **sickle-cell disease,** which affects one out of 400 African-Americans. This disease is caused by the substitution of a single amino acid in the hemoglobin protein of red blood cells (see FIGURE 5.19). When the oxygen content of an affected individual's blood is low (at high altitudes or under physical stress, for instance), the sickle-cell hemoglobin can deform the red cells to a sickle shape. Sickling of the cells, in turn, can lead to other symptoms. The multiple effects of the sickle-cell allele exemplify pleiotropy (FIGURE 13.16).

Individuals who are heterozygous for the sickle-cell allele are said to have sickle-cell trait. These carriers are usually healthy, although a fraction of the heterozygotes suffer some symptoms of sickle-cell disease when there is an extended reduction of blood oxygen. (The two alleles are codominant at the molecular level; both normal and abnormal hemoglobins are made.) About one out of ten African-Americans has sickle-cell trait, an unusually high frequency of heterozygotes for an allele with severe detrimental effects in homozygotes. The reason for the prevalence of this allele appears to be that while only individuals who are homozygous for the sickle-cell allele suffer from the disease, in certain environments, heterozygotes have an advantage over people who carry no copies of the sickle-cell allele. A single copy of the sickle-cell allele increases resistance to malaria. Thus, in tropical Africa, where malaria is common, the sickle-cell allele is both boon and bane. The relatively high frequency of

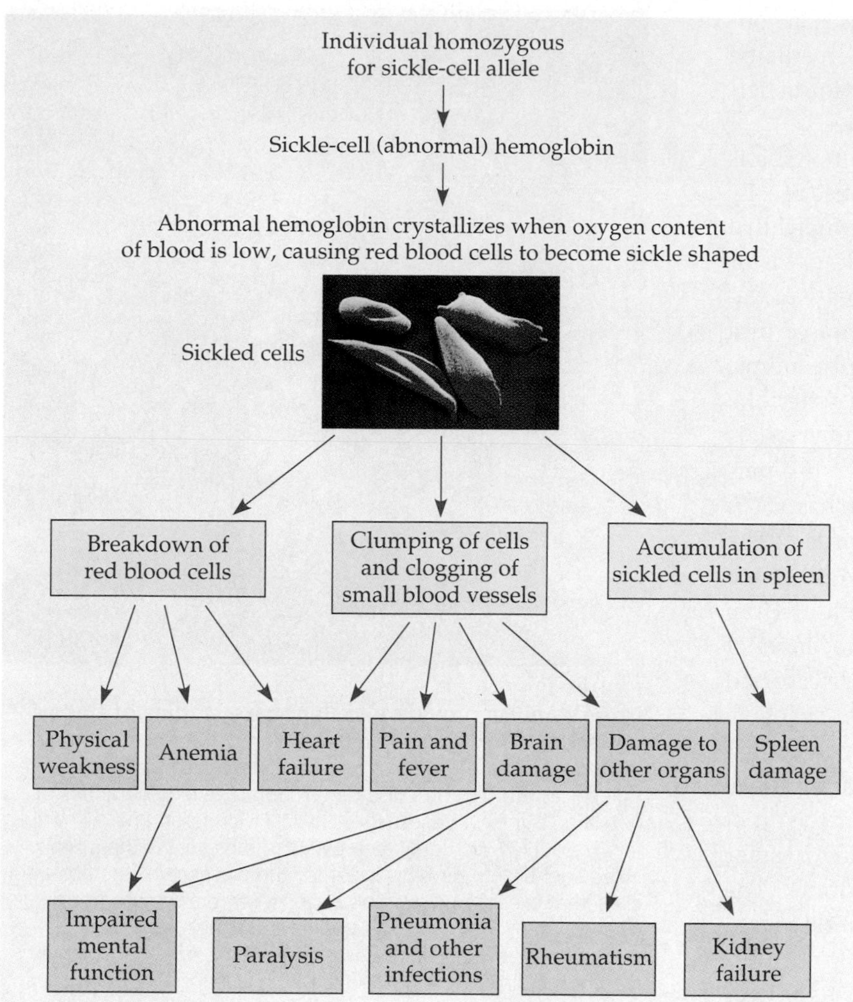

FIGURE 13.16
Pleiotropic effects of the sickle-cell allele. At the molecular level, the recessive allele responsible for sickle-cell disease has a single direct effect: It causes red blood cells to produce an abnormal version of the protein hemoglobin. If sickle-cell alleles are inherited from both parents, then all the hemoglobin is of the abnormal variety. The abnormal hemoglobin deforms the red blood cells, starting a cascade of symptoms throughout the body. Thus, at the level of the whole organism, the sickle-cell allele has multiple phenotypic effects.

African-Americans with sickle-cell trait is a vestige of their African roots.

Although it is relatively unlikely that two carriers of the same rare harmful allele will meet and mate, the probability increases greatly if the man and woman are close relatives (for example, siblings or first cousins). These are called consanguineous ("same blood") matings, and they are indicated in pedigrees with double lines. Because people with recent common ancestors are more likely to carry the same recessive alleles than are unrelated people, it is more likely that a mating of close relatives will produce offspring homozygous for a harmful recessive trait. Such effects can be observed in many types of domesticated and zoo animals that have become inbred.

There is debate among geneticists about the extent to which human consanguinity increases the risk of inherited diseases. Many deleterious alleles have such severe effects that a homozygous embryo spontaneously aborts long before birth. Most societies and cultures have laws or taboos forbidding marriages between close relatives. These rules may have evolved out of empirical observation that in most populations, stillbirths and birth defects are more common when parents are closely related. But social and economic factors have also influenced the development of customs and laws against consanguineous marriages, which could concentrate wealth in a few families.

Dominantly Inherited Disorders

Although most harmful alleles are recessive, several human disorders are due to dominant alleles. One example is achondroplasia, a form of dwarfism with an incidence of one case among every 10,000 people. Heterozygous individuals have the dwarf phenotype. Therefore, all people who are not achondroplastic dwarfs—99.99% of the population—are homozygous for the recessive allele.

Lethal dominant alleles are much less common than lethal recessives. One reason for this difference is that the effects of lethal dominant alleles are not masked in heterozygotes. Many lethal dominant alleles are the result of new changes (mutations) in a gene of the sperm or egg that subsequently kill the developing offspring. (In achondroplasia, for example, homozygosity for this dominant allele is usually lethal, causing spontaneous abortion.) An individual who does not survive to reproductive maturity will not pass on the new form of the gene. This is in contrast to lethal recessive mutations, which are perpetuated from generation to generation by the reproduction of heterozygous carriers who have normal phenotypes.

A lethal dominant allele can escape elimination if it is late-acting, causing death at a relatively advanced age. By the time the symptoms become evident, the individual may have already transmitted the lethal allele to his or her children. For example, **Huntington's disease,** a degenerative disease of the nervous system, is caused by a lethal dominant allele that has no obvious phenotypic effect until the individual is about 35 to 45 years old. Once the deterioration of the nervous system begins, it is irreversible and inevitably fatal. Any child born to a parent who has the allele for Huntington's disease has a 50% chance of inheriting the allele and the disorder. (The mating can be symbolized as $Aa \times aa$, with A being the dominant allele that causes Huntington's disease.) Until recently, it was impossible to tell before the onset of symptoms if a person at risk for Huntington's disease had actually inherited the allele, but that has changed. Analyzing DNA samples from a large family with a high incidence of the disorder, molecular geneticists have tracked the Huntington's allele to a locus near the tip of chromosome 4 (FIGURE 13.17). It is now possible to test for the presence of the allele in an individual's genome. (The methods that make such tests possible will be discussed in Chapter 19.) For those with a family history of Huntington's disease, the availability of this test poses an agonizing dilemma: Under what circumstances is it beneficial for a presently healthy person to find out whether he or she has inherited a fatal and not yet curable disease?

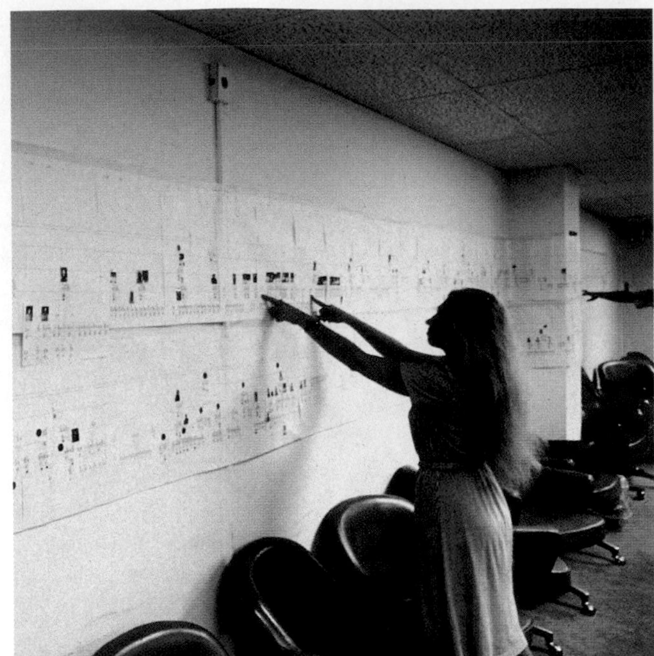

FIGURE 13.17
Large families provide excellent case studies of human genetics. Here, Nancy Wexler, of Columbia University and the Hereditary Disease Foundation, stands in front of a huge pedigree that traces Huntington's disease through several generations of one large family in Venezuela. Classical Mendelian analysis of this family, coupled with new techniques of molecular biology, enabled scientists to develop a test for the presence of the dominant allele that causes Huntington's disease—before symptoms are evident. Dr. Wexler, who has studied the Venezuelan family, is herself at risk for developing Huntington's disease. Her mother died of the disorder, and there is a 50% chance that Dr. Wexler inherited the dominant allele that causes the disease.

Multifactorial Disorders

The hereditary diseases we have studied so far are sometimes described as simple Mendelian disorders because they result from certain alleles at a single genetic locus. Many more people are susceptible to diseases that have a multifactorial basis—a genetic component plus a significant environmental influence. The long list of multifactorial diseases includes heart disease, diabetes, cancer, alcoholism, and certain mental illnesses, such as schizophrenia and manic-depressive psychosis. In many cases, the hereditary component is polygenic. For example, many genes affect our cardiovascular health, making some of us more prone than others to heart attacks and strokes. But our lifestyle intervenes tremendously between genotype and phenotype for cardiovascular health and other multifactorial characters. Exercise, a healthful diet, abstinence from smoking, and an ability to put stressful situations in perspective all reduce our risk of heart disease and some types of cancer.

At present, so little is understood about the genetic contributions to most multifactorial diseases that the best public-health strategy is to educate people about the importance of environmental factors and to promote healthful behavior.

Technology is providing new tools for genetic testing and counseling

A preventive approach to simple Mendelian disorders is sometimes possible, because in some cases the risk that a particular genetic disorder will occur can be assessed before a child is conceived or in the early stages of the pregnancy. Many hospitals have genetic counselors who can provide information to prospective parents concerned about a family history for a specific disease.

Let's consider the example of an imaginary couple, John and Carol, who are planning to have their first child and are seeking genetic counseling because of family histories of a lethal disease known to be recessively inherited. John and Carol each had a brother who died of the disorder, so they want to determine the risk of their having a child with the disease. From the information about their brothers, we know that both parents of John and both parents of Carol must have been carriers of the recessive allele. Thus, John and Carol are both products of *Aa* and *Aa* crosses, where *a* symbolizes the allele that causes this particular disease. We also know that John and Carol are not homozygous recessive (*aa*) because they do not have the disease. Therefore, their genotypes are either *AA* or *Aa*. Given a genotypic ratio of 1*AA*:2*Aa*:1*aa* for offspring of an *Aa* × *Aa* cross, John and Carol each have a ⅔ chance of being carriers (*Aa*). Using the rule of multiplication, we can determine that the overall probability of their firstborn having the disorder is ⅔ (the chance that John is a carrier) multiplied by ⅔ (the chance that Carol is a carrier) multiplied by ¼ (the chance of two carriers having a child with the disease), which equals ⅑. Suppose that Carol and John decide to take the risk and have a child; after all, there is an ⁸⁄₉ chance that their baby will not have the disorder. But their child is born with the disease. We no longer have to guess about John's and Carol's genotypes. We now know that both John and Carol are, in fact, carriers. If the couple decides to have another child, they now know there is a ¼ chance that the second child will have the disease.

When we use Mendel's laws to predict possible outcomes of matings, it is important to recall that chance has no memory: Each child represents an independent event in the sense that its genotype is unaffected by the genotypes of older siblings. Suppose that John and Carol have three more children, and *all three* have the hypothetical hereditary disease. This is an unfortunate family, for there is only one chance in 64 (¼ × ¼ × ¼) that such an outcome will occur. But this run of misfortune will in no way affect the result if John and Carol decide to have still another child. There is still a ¼ chance that the additional child will have the disease and a ¾ chance that it will not. Mendel's laws, remember, are simply rules of probability applied to heredity.

Carrier Recognition

Because most children with recessive disorders are born to parents with normal phenotypes, the key to assessing the genetic risk for a particular disease is determining whether the prospective parents are heterozygous carriers of the recessive trait. For some heritable disorders, there are tests that can distinguish individuals of normal phenotype who are dominant homozygotes from those who are heterozygotes, and the number of such tests increases each year. Examples are tests that can identify carriers of the alleles for Tay-Sachs disease, sickle-cell disease, and the most common form of cystic fibrosis. On one hand, these tests enable people with family histories of genetic disorders to make informed decisions about having children. On the other hand, these new methods for genetic screening could be abused. If confidentiality is breached, will carriers be stigmatized? Will they be denied health or life insurance, even though they are themselves healthy? Will misinformed employers equate "carrier" with disease? And will sufficient genetic counseling be available to help a large number of individuals understand their test results? David Satcher addresses some of these issues in the interview that precedes this unit.

New biotechnology offers possibilities of reducing human suffering, but not before key ethical issues are resolved. The dilemmas posed by human genetics reinforce one of this book's themes: the immense social implications of biology.

Fetal Testing

Suppose a couple learns that they are both Tay-Sachs carriers, but they decide to have a child anyway. Tests performed in conjunction with a technique known as **amniocentesis** can determine, beginning at the fourteenth to sixteenth week of pregnancy, whether the developing fetus has Tay-Sachs disease (FIGURE 13.18). To perform this procedure, a physician inserts a needle into the uterus and extracts about 10 milliliters of amniotic fluid, the liquid that bathes the fetus. Some genetic disorders can be detected from the presence of certain chemicals in the amniotic fluid itself. Tests for other disorders, including Tay-Sachs disease, are performed on cells grown in the laboratory from the fetal cells that had been sloughed off into the amniotic fluid. These cultured cells can also be used for karyotyping to identify certain chromosomal defects (see the Methods Box in Chapter 12, p. 234).

In an alternative technique called **chorionic villus sampling (CVS),** the physician suctions off a small amount of fetal tissue from the placenta. Because the cells of the chorionic villi are proliferating rapidly, enough cells are undergoing mitosis to allow karyotyping to be carried out immediately, giving results within 24 hours. This is an advantage over amniocentesis, in which the cells must be cultured for several weeks before karyotyping. Another advantage of CVS is that it can be performed as early as the eighth to tenth week of pregnancy.

Other techniques allow a physician to examine a fetus directly for major abnormalities. One such technique is ultrasound, which uses sound waves to produce an image of the fetus by a simple noninvasive procedure. This procedure has no known risk to either fetus or mother. With another technique, fetoscopy, a needle-thin tube containing a viewing scope and fiber optics (to

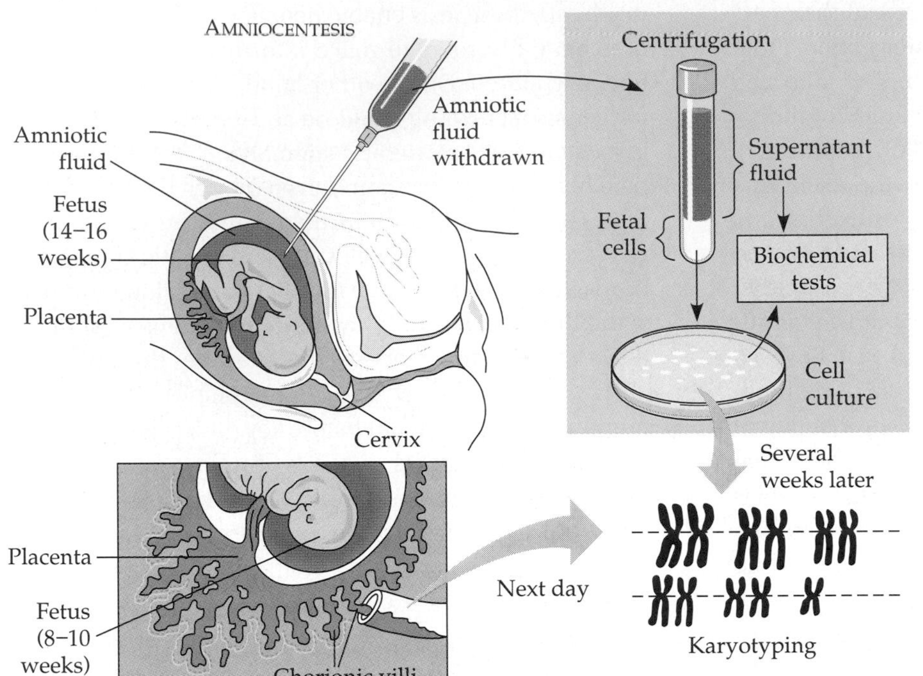

FIGURE 13.18
Fetal diagnosis. In amniocentesis, ultrasound is used to locate the fetus, and a small amount of amniotic fluid is extracted for testing. Physicians can diagnose some disorders from chemicals in the fluid itself, while other disorders may show up in tests performed on cells cultured from fetal cells present in the fluid. This analysis includes biochemical tests, to detect the presence of certain enzymes, and karyotyping, to determine whether the chromosomes of the fetal cells are normal in number and microscopic appearance. In chorionic villus sampling, a physician inserts a narrow tube through the cervix and uses suction to extract a tiny sample of fetal tissue from the placenta, the organ that transmits nutrients and wastes between the fetus and the mother. The tissue sample is used for immediate karyotyping.

transmit light) is inserted into the uterus. Fetoscopy enables the physician to examine the fetus for certain anatomical deformities.

In about 1% of the cases, amniocentesis or fetoscopy causes complications, such as maternal bleeding or fetal death. Thus, these techniques are usually reserved for cases in which the risk of a genetic disorder or other type of birth defect is relatively great. If the fetal tests reveal a serious disorder, the parents face the difficult choice between terminating the pregnancy or preparing to care for a child with a genetic disorder.

Newborn Screening

Some genetic disorders can be detected at birth by simple tests that are now routinely performed in most hospitals in the United States. One screening program is for the recessively inherited disorder called phenylketonuria (PKU), which occurs in about one out of every 10,000 births in the United States. Children with this disease cannot properly break down the amino acid phenylalanine. This compound and its by-product, phenylpyruvate, can accumulate to toxic levels in the blood, causing mental retardation. However, if the deficiency is detected in the newborn, retardation can usually be prevented with a special diet, low in phenylalanine, that allows normal development. Thus, screening newborns for PKU and other treatable disorders is vitally important. Unfortunately, very few genetic disorders are treatable at the present time.

*   *   *

In this chapter, you have learned about the Mendelian model of inheritance and its application to human genetics. We owe the "gene idea," the concept of heritable factors that are transmitted according to simple rules of chance, to the elegant experiments of Gregor Mendel. Mendel's quantitative approach was foreign to the biology of his era, and even the few biologists who read his papers apparently missed the importance of his discoveries. It wasn't until the beginning of the twentieth century that Mendelian genetics was rediscovered by biologists who studied the role of chromosomes in inheritance. In the next chapter, you will learn how Mendel's laws have their physical basis in the behavior of chromosomes during sexual life cycles, and how the synthesis of Mendelism and a chromosome theory of inheritance catalyzed progress in genetics.

REVIEW OF KEY CONCEPTS (with page numbers and key figures)

- Mendel brought an experimental and quantitative approach to genetics: *science as a process* (pp. 238–240, FIGURE 13.2)
  - In the 1860s, Gregor Mendel developed a particulate theory of inheritance based on his experiments with garden peas.
  - Mendel demonstrated that parents pass on to their offspring discrete genes that retain their identity generation after generation.
- According to the law of segregation, the two alleles for a character are packaged into separate gametes (pp. 240–244, FIGURE 13.4)
  - By producing hybrid offspring and allowing them to self-pollinate, Mendel arrived at the law of segregation. The hybrids (F_1) exhibited the dominant trait. In the next generation (F_2), 75% of offspring had the dominant trait and 25% had the recessive trait, for a 3:1 ratio.
  - To account for these results, Mendel postulated that genes have alternative forms (now called alleles) and that each organism inherits one allele for each gene from each parent. These separate (segregate) from each other during gamete formation, so that a sperm or an egg carries only one allele. After fertilization, if the two alleles of the pair are different, one (the dominant allele) is fully expressed in the offspring and the other (the recessive allele) is completely masked.
  - Homozygous individuals have two identical alleles for a given character and are true-breeding. Heterozygous individuals have two different alleles for a given character.
- According to the law of independent assortment, each pair of alleles segregates into gametes independently (pp. 244–245, FIGURE 13.7)
  - Mendel proposed the law of independent assortment based on dihybrid crosses between plants contrasting in two or more characters, such as flower color and seed shape. Alleles for each character segregate into gametes independently of alleles for other characters.
  - The four possible phenotypes are represented in a 9:3:3:1 ratio in the F_2 generation of a dihybrid cross.
- Mendelian inheritance reflects rules of probability (pp. 245–247, FIGURE 13.9)
  - Mendel's laws operate according to the rules of probability. According to the rule of multiplication, the probability of a compound event is equal to the product of the separate probabilities of the independent single events. The rule of addition states that the probability of an event that can occur in two or more independent ways is the sum of the separate probabilities.
- Mendel discovered the particulate behavior of genes: *a review* (p. 247)
  - Mendel's quantitative analysis of carefully planned experiments provides a case study in the process of science.
- The relationship between genotype and phenotype is rarely simple (pp. 247–252, FIGURES 13.10–13.14)
  - Some heterozygous genotypes result in incomplete dominance. Such individuals have an appearance that is intermediate between the phenotypes of the two parents.
  - In codominance, a heterozygous organism expresses *both* phenotypes of its two alleles.
  - Many genes have *multiple* (more than two) alleles.
  - Pleiotropy is the ability of a single gene to affect multiple phenotypic traits.
  - In epistasis, one gene interferes with the expression of another gene.
  - Certain characters are quantitative characters that vary in a continuous fashion, indicating polygenic inheritance, an additive effect of two or more genes on a single phenotypic character.
  - Environment also influences quantitative characters. Such characters are said to be multifactorial.
- Pedigree analysis reveals Mendelian patterns in human inheritance (pp. 252–253, FIGURE 13.15)
  - Family pedigrees can be used to deduce the possible genotypes of individuals and make predictions about future offspring. Any predictions are usually statistical probabilities rather than absolute statements.
- Many human disorders follow Mendelian patterns of inheritance (pp. 253–256)
  - Certain genetic disorders are inherited as simple recessive traits from phenotypically normal, heterozygous carriers.
  - Although they are far less common than the recessive type, some human disorders are due to dominant alleles.
  - Medical researchers are just beginning to sort out the genetic and environmental components of multifactorial disorders, such as heart disease and cancer.
- Technology is providing new tools for genetic testing and counseling (pp. 256–258, FIGURE 13.18)
  - Using family histories, genetic counselors help couples determine the odds that their children will have genetic disorders. For certain diseases, tests that can identify carriers can more accurately define those odds.
  - Once a child is conceived, the techniques of amniocentesis and chorionic villus sampling can help determine whether a suspected genetic disorder is present.

GENETICS PROBLEMS

1. A rooster with gray feathers is mated with a hen of the same phenotype. Among their offspring, 15 chicks are gray, 6 are black, and 8 are white. What is the simplest explanation for the inheritance of these colors in chickens? What offspring would you predict from the mating of a gray rooster and a black hen?

2. In some plants, a true-breeding, red-flowered strain gives all pink flowers when crossed with a white-flowered strain: *RR* (red) × *rr* (white) ⟶ *Rr* (pink). If flower position (axial or terminal) is inherited as it is in peas (see TABLE 13.1), what will be the ratios of genotypes and phenotypes of the generation resulting from the following cross: axial-red (true-breeding) × terminal-white? What will be the ratios in the F_2 generation?

3. Flower position, stem length, and seed shape were three characters that Mendel studied. Each is controlled by an independently assorting gene and has dominant and recessive expression as follows:

Character	*Dominant*	*Recessive*
Flower position	Axial (*A*)	Terminal (*a*)
Stem length	Tall (*L*)	Dwarf (*l*)
Seed shape	Round (*R*)	Wrinkled (*r*)

If a plant that is heterozygous for all three characters were allowed to self-fertilize, what proportion of the offspring would be expected to be as follows? (*Note:* Use the rules of probability instead of a huge Punnett square.)

a. homozygous for the three dominant traits
b. homozygous for the three recessive traits
c. heterozygous
d. homozygous for axial and tall, heterozygous for seed shape

4. A black guinea pig crossed with an albino guinea pig produced 12 black offspring. When the albino was crossed with a second black one, 7 blacks and 5 albinos were obtained. What is the best explanation for this genetic situation? Write genotypes for the parents, gametes, and offspring.

5. In sesame plants, the one-pod condition (*P*) is dominant to the three-pod condition (*p*), and normal leaf (*L*) is dominant to wrinkled leaf (*l*). Pod type and leaf type are inherited independently. Determine the genotypes for the two parents for all possible matings producing the following offspring:

a. 318 one-pod normal, 98 one-pod wrinkled
b. 323 three-pod normal, 106 three-pod wrinkled
c. 401 one-pod normal
d. 150 one-pod normal, 147 one-pod wrinkled, 51 three-pod normal, 48 three-pod wrinkled
e. 223 one-pod normal, 72 one-pod wrinkled, 76 three-pod normal, 27 three-pod wrinkled

6. A man with group A blood marries a woman with group B blood. Their child has group O blood. What are the genotypes of these individuals? What other genotypes, and in what frequencies, would you expect in offspring from this marriage?

7. Color pattern in a species of duck is determined by one gene with three alleles. Alleles *H* and *I* are codominant, and allele *i* is recessive to both. How many phenotypes are possible in a flock of ducks that contains all the possible combinations of these three alleles?

8. Phenylketonuria (PKU) is an inherited disease caused by a recessive allele. If a woman and her husband are both carriers, what is the probability of each of the following?

a. all three of their children will be of normal phenotype
b. one *or* more of the three children will have the disease
c. all three children will have the disease
d. *at least* one child will be phenotypically normal

(*Note:* Remember that the probabilities of all possible outcomes always add up to 1.)

9. The genotype of F_1 individuals in a tetrahybrid cross is *AaBbCcDd*. Assuming independent assortment of these four genes, what are the probabilities that F_2 offspring would have the following genotypes?

a. *aabbccdd*
b. *AaBbCcDd*
c. *AABBCCDD*
d. *AaBBccDd*
e. *AaBBCCdd*

10. In 1981, a stray black cat with unusual rounded, curled-back ears was adopted by a family in California. Hundreds of descendants of the cat have since been born, and cat fanciers hope to develop the "curl" cat into a show breed. Suppose you owned the first curl cat and wanted to develop a true-breeding variety. How would you determine whether the curl allele is dominant or recessive? How would you select for true-breeding cats? How would you know they are true-breeding?

11. What is the probability that each of the following pairs of parents will produce the indicated offspring (assume independent assortment of all gene pairs)?

a. *AABBCC* × *aabbcc* ⟶ *AaBbCc*
b. *AABbCc* × *AaBbCc* ⟶ *AAbbCC*
c. *AaBbCc* × *AaBbCc* ⟶ *AaBbCc*
d. *aaBbCC* × *AABbcc* ⟶ *AaBbCc*

12. Karen and Steve each have a sibling with sickle-cell disease. Neither Karen, Steve, nor any of their parents has the disease, and none of them has been tested to reveal sickle-cell trait. Based on this incomplete information, calculate the probability that if this couple has a child, the child will have sickle-cell disease.

13. Imagine that a newly discovered, recessively inherited disease is expressed only in individuals with type O blood, although the disease and blood group are independently inherited. A normal man with type A blood and a normal woman with type B blood have already had one child with the disease. The woman is now pregnant for a second time. What is the probability that the second child will also have the disease? Assume both parents are heterozygous for the "disease" gene.

14. In tigers, a recessive allele causes an absence of fur pigmentation (a "white tiger") and a cross-eyed condition. If two phenotypically normal tigers that are heterozygous at this locus are mated, what percentage of their offspring will be cross-eyed? What percentage will be white?

15. In corn plants, a dominant allele *I* inhibits kernal color, while the recessive allele *i* permits color when homozy-

gous. At a different locus, the dominant gene *P* causes purple kernel color, while the homozygous recessive genotype *pp* causes red kernels. If plants heterozygous at both loci are crossed, what will be the phenotypic ratio of the F_1 generation?

16. The pedigree below traces the inheritance of alkaptonuria, a biochemical disorder. Affected individuals, indicated here by the filled-in circles and squares, are unable to break down a substance called alkapton, which colors the urine and stains body tissues. Does alkaptonuria appear to be caused by a dominant or recessive allele? Fill in the genotypes of the individuals whose genotypes you know. What genotypes are possible for each of the other individuals?

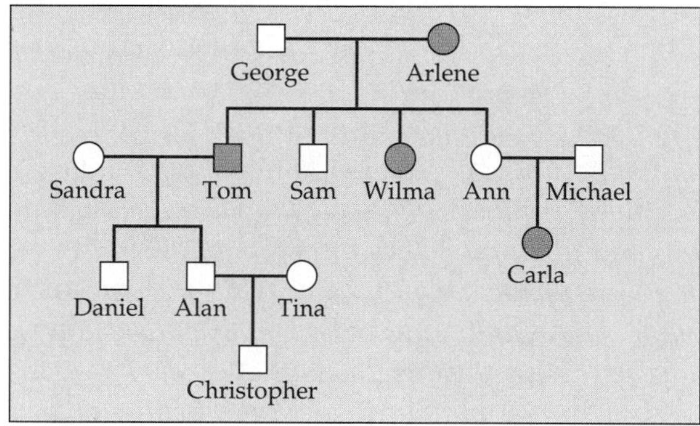

17. A man has six fingers on each hand and six toes on each foot. His wife and their daughter have the normal number of digits. Extra digits is a dominant trait. What fraction of this couple's children would be expected to have extra digits?

18. Imagine you are a genetic counselor, and a couple planning to start a family came to you for information. Charles was married once before, and he and his first wife had a child who has cystic fibrosis. The brother of his current wife Elaine died of cystic fibrosis. What is the probability that Charles and Elaine will have a baby with cystic fibrosis? (Neither Charles nor Elaine has cystic fibrosis.)

19. In mice, black color (*B*) is dominant to white (*b*). At a different locus, a dominant allele (*A*) produces a band of yellow just below the tip of each hair in mice with black fur. This gives a frosted appearance known as agouti. Expression of the recessive allele (*a*) results in a solid coat color. If mice that are heterozygous at both loci are crossed, what will be the expected phenotypic ratio of their offspring?

20. The pedigree below traces the inheritance of a vary rare biochemical disorder in humans. Affected individuals are indicated by filled-in circles and squares. Is the allele for this disorder dominant or recessive? What genotypes are possible for the individuals marked 1, 2, and 3?

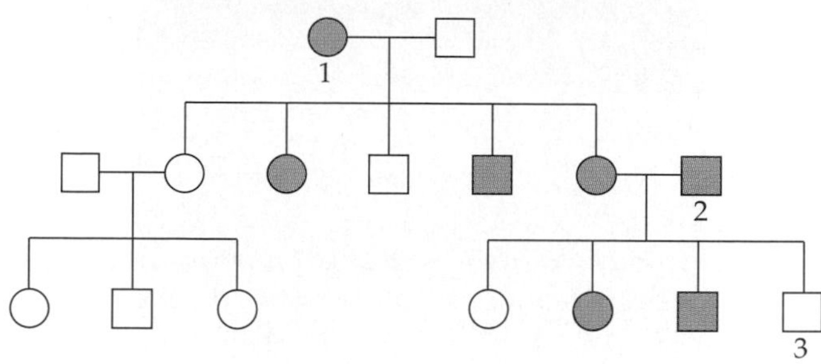

SCIENCE, TECHNOLOGY, AND SOCIETY

1. Imagine that one of your parents had Huntington's disease. What would be the probability that you, too, would someday manifest the disease? There is no cure for Huntington's. Would you want to be tested for the Huntington's allele? Why or why not?

2. Proponents of the discredited eugenics movement believe that the human population would be improved if individuals suffering from genetic disorders, such as cystic fibrosis or Tay-Sachs disease, could be sterilized or persuaded not to have children. Would implementation of this strategy eliminate these defects? Why or why not? How might new tests that identify carriers of harmful alleles bolster eugenics? Are such tests an opportunity or a danger for society? Why?

3. In the near future, gene therapy may be an option for the treatment and cure of many inherited disorders. What do you think are the most serious ethical issues that must be dealt with before human gene therapy is used on a large scale? Why do you think these issues are important?

FURTHER READING

Baringa, M. "New Alzheimer's Gene Found." *Science,* June 30, 1995. Located on chromosome 14, this allele may cause up to 80% of early-onset Alzheimer's cases of the inherited type.

Diamond, J. "Curse and Blessing of the Ghetto." *Discover,* March 1991. The biology and sociology of Tay-Sachs disease.

Fackelmann, K. "Drug Wards Off Sickle-Cell Attacks." *Science News,* February 4, 1995. Progress in treating a relatively common genetic disorder.

Griffiths, A. J. F., J. H. Miller, D. T. Suzuki, R. C. Lewontin, and W. M. Gelbart. *An Introduction to Genetic Analysis,* 5th ed. New York: W. H. Freeman, 1993. A good basic undergraduate genetics text.

Horgan, J. "Eugenics Revisited." *Scientific American,* June 1993. Controversies centered on the genetics of human behavior.

Mann, C. C. "Behavioral Genetics in Transition." *Science,* June 17, 1994. To what extent do genes influence how we behave?

Revkin, A. "Hunting Down Huntington's." *Discover,* December 1993. A team of genetic sleuths locate the gene for Huntington's disease.

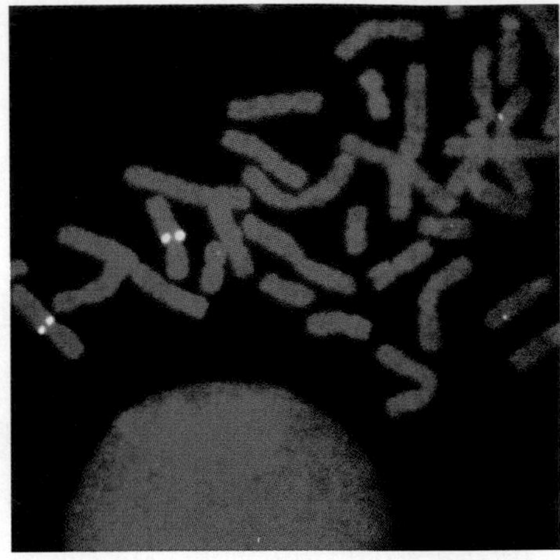

CHAPTER 14

THE CHROMOSOMAL BASIS OF INHERITANCE

KEY CONCEPTS

- Mendelian inheritance has its physical basis in the behavior of chromosomes during sexual life cycles
- Morgan traced a gene to a specific chromosome: *science as a process*
- Linked genes tend to be inherited together because they are located on the same chromosome
- Independent assortment of chromosomes and crossing over cause genetic recombination
- Geneticists can use recombination data to map a chromosome's genetic loci
- The chromosomal basis of sex produces unique patterns of inheritance
- Alterations of chromosome number or structure cause some genetic disorders
- The phenotypic effects of some genes depend on whether they were inherited from the mother or the father
- Extranuclear genes exhibit a non-Mendelian pattern of inheritance

It was not until the year 1900 that biology finally caught up with Gregor Mendel. At that time, three botanists, working independently on plant breeding experiments, reproduced Mendel's results. By searching the literature, the German Karl Correns, the Austrian Erich von Tschermak, and the Dutchman Hugo de Vries all found that Mendel had explained the same results 35 years before. During the intervening years, biology had grown more experimental and quantitative and thus more receptive to Mendelism. Nevertheless, many biologists remained incredulous about Mendel's laws of segregation and independent assortment until evidence had mounted that these principles of heredity had a physical basis in the behavior of chromosomes. Mendel's hereditary factors—genes—are located on chromosomes. For example, the yellow dots (a fluorescent dye) in the light micrograph on this page mark the locus of a specific gene on a homologous pair of human chromosomes (which have already replicated, hence the two dots per chromosome). This chapter integrates and extends what you have learned in the previous two chapters by describing the chromosomal basis for the transmission of genes from parents to offspring.

Mendelian inheritance has its physical basis in the behavior of chromosomes during sexual life cycles

Cytologists worked out the process of mitosis in 1875 and the process of meiosis in the 1890s. Then, around the turn of the century, cytology and genetics converged as biologists began to see parallels between the behavior of chromosomes and the behavior of Mendel's factors. For example, chromosomes and genes are both present in pairs in diploid cells, homologous chromosomes separate and alleles segregate during meiosis, and fertilization restores the paired condition for both chromosomes and genes. Around 1902, Walter S. Sutton, Theodor Boveri, and others independently noted these parallels, and a **chromosome theory of inheritance** began to take form. According to this theory, Mendelian genes have specific loci on chromosomes, and it is the chromosomes that undergo segregation and independent assortment (FIGURE 14.1).

Morgan traced a gene to a specific chromosome: *science as a process*

Thomas Hunt Morgan, an embryologist at Columbia University, first associated a specific gene with a specific

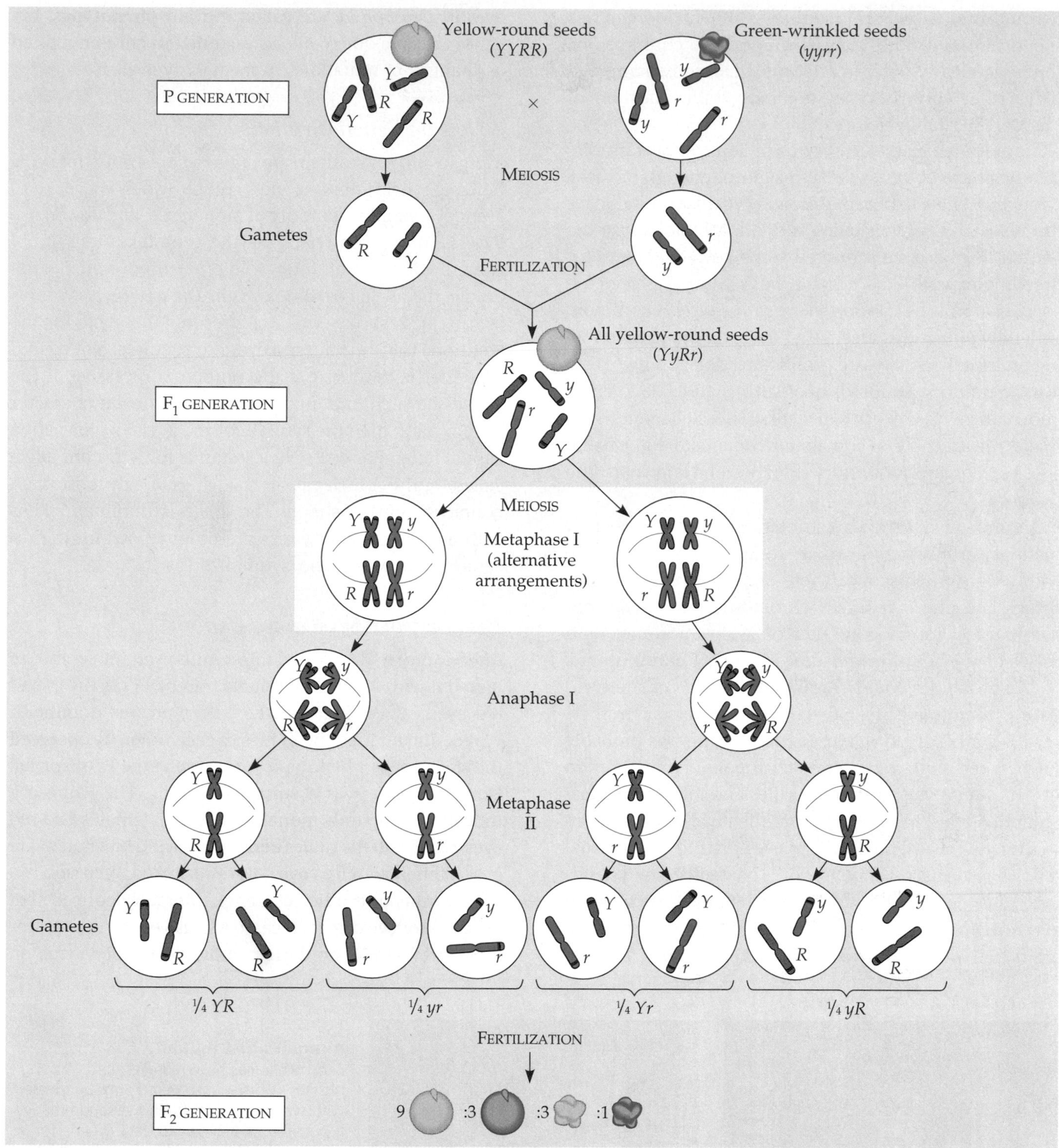

FIGURE 14.1

The chromosomal basis of Mendel's laws. Here, we correlate the results of one of Mendel's dihybrid crosses with the behavior of chromosomes. The two characters we are tracking are the color and shape of pea seeds. The two genes are on different chromosomes, and black bands symbolize the loci in these drawings. (Peas actually have a total of seven chromosome pairs, but only two are illustrated here.) The movements of chromosomes during metaphase and anaphase of meiosis I account for the segregation and independent assortment of the alleles for seed color and shape. The two alleles for each character segregate when homologous chromosomes move toward opposite poles of the cell and become packaged in different cells during the first meiotic division. Mendel's law of independent assortment has its physical basis in the random arrangement of chromosomes on the metaphase plate during meiosis I. For each homologous pair of chromosomes, the polar orientation of the maternal and paternal versions (color-coded red and blue, respectively) of the chromosome is unrelated to the orientation of the other chromosome pair. This results in the alternative metaphase I arrangements illustrated in the yellow box. If we continue to trace the results of these alternative alignments of chromosomes to the F_2 generation, we see the physical explanation for Mendel's 9:3:3:1 ratio of phenotypes.

chromosome early in this century. Although Morgan was skeptical about both Mendelism and the chromosome theory, his early experiments provided convincing evidence that chromosomes are indeed the location of Mendel's heritable factors.

Many times in the history of biology, important discoveries have come to those insightful enough or lucky enough to choose an experimental organism suitable for the research problem being tackled. Mendel chose the garden pea because it offered some key advantages for breeding experiments. For his work, Morgan selected a species of fruit fly, *Drosophila melanogaster,* a common, generally innocuous insect that feeds on the fungi growing on fruit. Fruit flies are prolific breeders; a single mating will produce hundreds of offspring, and a new generation can be bred every two weeks. These characteristics make the fruit fly a convenient organism for genetic studies. Morgan's laboratory soon became known as "the fly room."

Another advantage of the fruit fly is that it has only four pairs of chromosomes, which are easily distinguishable with a light microscope. There are three pairs of autosomes and one pair of sex chromosomes. Female fruit flies have a homologous pair of *X* chromosomes, and males have one *X* chromosome and one *Y* chromosome.

While Mendel could readily obtain different pea varieties, there were no convenient suppliers of fruit fly varieties for Morgan to employ. Indeed, he was probably the first person to want different varieties of this common insect. After a year of breeding flies and looking for variant individuals, Morgan was rewarded with the discovery of a single male fly with white eyes instead of the usual red. The normal phenotype for a character (the phenotype most common in natural populations), such as red eyes in *Drosophila,* is called the **wild type** (FIGURE 14.2). Traits that are alternatives to the wild type, such as white eyes in *Drosophila,* are called **mutant phenotypes,** because they are due to alleles assumed to have originated as changes, or mutations, in the wild-type allele.

A Note on Genetic Symbols

Morgan and his students invented a convention for symbolizing alleles that is now more widely used than Mendel's simple notation of uppercase and lowercase letters. For a given character, the gene takes its symbol from the first mutant (non–wild type) discovered. For example, the allele for white eyes in *Drosophila* is symbolized by *w.* A superscript + represents the allele for the wild-type trait—w^+ for red eyes, for example. Notice that the letter is lowercase if the mutant is recessive. If the first-discovered mutant version of a particular character is dominant, then the symbol for the allele begins with a capital letter. For example, *Cy* stands for a mutant allele called "curly," which causes the wings of a fly to form an abnormal, curled shape. The allele is dominant. Flies with normal, straight wings are homozygous for the recessive, wild-type allele, symbolized as Cy^+.

Discovery of a Sex-Linked Gene

After Morgan discovered his white-eyed male fly, he mated it with a red-eyed female. All the F_1 offspring had red eyes, suggesting that the wild type was dominant. When Morgan bred the F_1 flies to each other, he observed the classical 3:1 phenotypic ratio among the F_2 offspring. However, there was a surprising result: The white-eye trait showed up only in males. All the F_2 females had red eyes, while half the males had red eyes and half had white eyes. Somehow, a fly's eye color was linked to its sex.

From this and other evidence, Morgan deduced that the gene for eye color is located exclusively on the *X* chromosome; there is no corresponding eye-color locus on

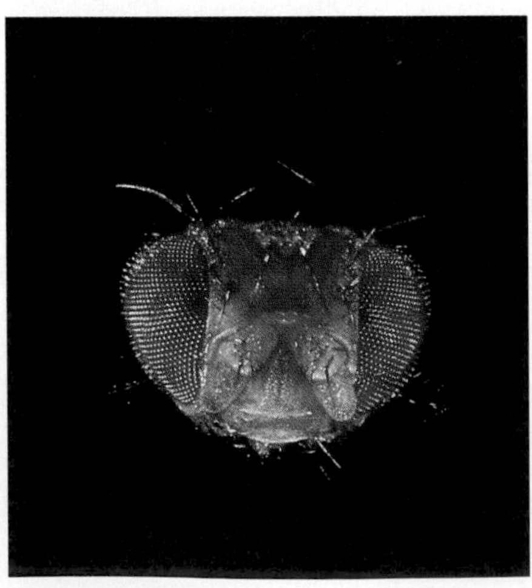

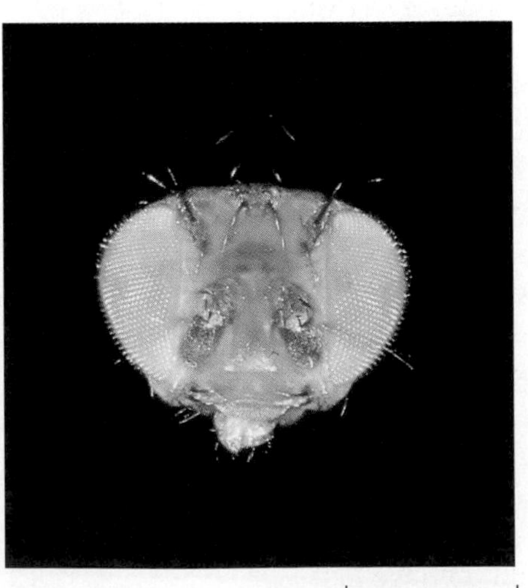

FIGURE 14.2
Morgan's first mutant. Wild-type *Drosophila* flies have red eyes (left). Among his flies, Morgan discovered a mutant male with white eyes (right). This variation made it possible for Morgan to trace a gene for eye color to a specific chromosome (LMs).

the *Y* chromosome (FIGURE 14.3). Thus, females (*XX*) carry two copies of the gene for this character, while males (*XY*) carry only one. Because the mutant allele is recessive, a female will have white eyes only if she receives that allele on both *X* chromosomes—an impossibility for the F_2 females in Morgan's experiment. For a male, on the other hand, a single copy of the mutant allele confers white eyes. Since a male has only one *X* chromosome, there can be no wild-type allele present to offset the recessive allele.

Genes located on a sex chromosome are called **sex-linked genes.** Morgan's evidence that a specific gene is carried on the *X* chromosome added credibility to the chromosome theory of inheritance. Recognizing the importance of this work, many bright students were attracted to Morgan's fly room, and his laboratory dominated genetics research for the next three decades. We will see the influence of Morgan and his colleagues as we consider some other important aspects of the chromosomal basis of inheritance.

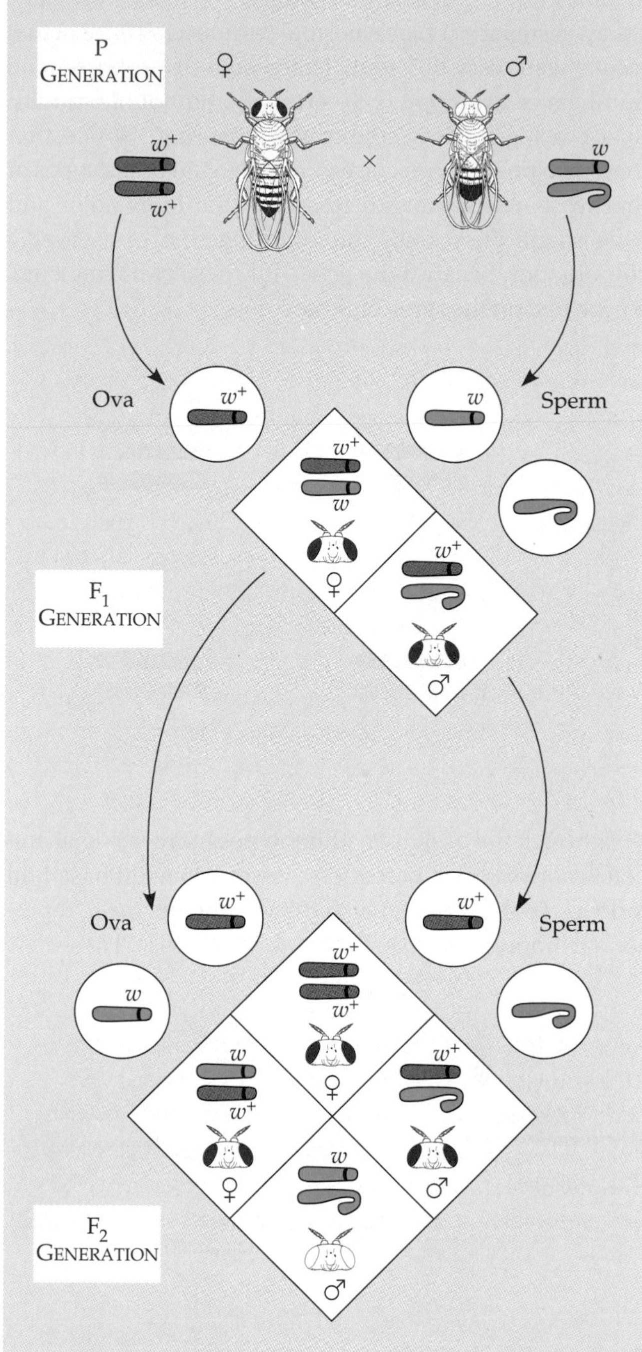

FIGURE 14.3
Sex-linked inheritance. When Morgan bred his white-eyed male to a wild-type female, all F_1 offspring had red eyes. The F_2 generation showed a typical Mendelian 3:1 ratio of traits, but the recessive trait—white eyes—was linked to sex. All females had red eyes, but half the males had white eyes. Morgan hypothesized that the gene responsible was located on the *X* chromosome (symbolized as a straight chromosome here) and that there was no corresponding locus on the *Y* chromosome (hooked, in this diagram). In this figure, the dominant allele (for red eyes) is symbolized by w^+, and the recessive allele (for white eyes) is symbolized by *w*. The symbols ♀ and ♂ stand for female and male, respectively.

Linked genes tend to be inherited together because they are located on the same chromosome

The number of genes in a cell is far greater than the number of chromosomes; in fact, each chromosome has hundreds or thousands of genes. Genes located on the same chromosome tend to be inherited together in genetic crosses because they are part of a single chromosome that is passed along as a unit. Such genes are said to be **linked genes.** (Note that the use of the word *linked* in this way is different from its use in the term *sex-linked.*) When geneticists follow linked genes in breeding experiments, the results deviate from those expected according to the Mendelian principle of independent assortment.

Let's examine another of Morgan's *Drosophila* experiments to see how linkage affects the inheritance of two different characters. In this case, the two characters are body color and wing size. Wild-type flies have gray bodies and normal wings. Mutant phenotypes for these characters are black body and vestigial wings, which are much smaller than normal wings. The alleles for these traits are represented by the following symbols: b^+ = gray, b = black; vg^+ = normal wings, vg = vestigial wings. (Neither gene is sex-linked; the loci are on an autosome.) Morgan crossed female dihybrids ($b^+ b\, vg^+ vg$) with males that had both mutant phenotypes, black bodies and vestigial wings ($bb\, vgvg$). Notice that this corresponds to a Mendelian testcross rather than to an F_1 cross between two dihybrids. According to Mendel's law of independent assortment, Morgan's *Drosophila* testcross would produce four phenotypic classes of offspring, approximately equal in number: 1 gray-normal (wings) : 1 black-vestigial : 1 gray-vestigial : 1 black-normal (FIGURE 14.4). The actual results were very different. There were disproportionate numbers of wild-type (gray-normal) and double mutant (black-vestigial) flies among the offspring. Notice that these two phenotypes corresponded to the phenotypes of the two parents. Morgan reasoned that body color and wing shape are usually inherited together in a specific combination because the genes for these two characters are located on the same chromosome:

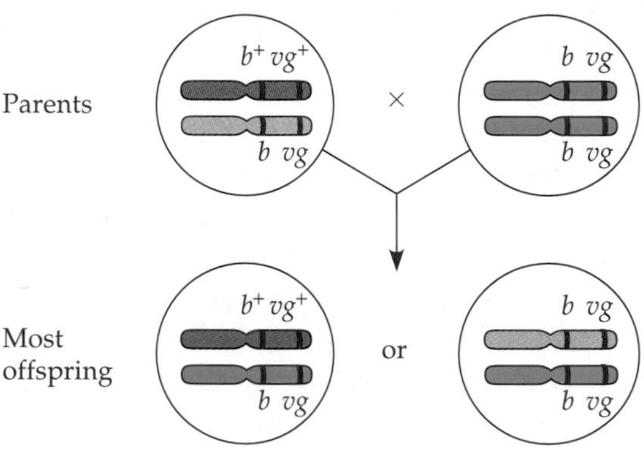

Although the other two phenotypes (gray-vestigial and black-normal) numbered fewer than expected based on independent assortment, these phenotypes *were* represented among the offspring of Morgan's cross. These new phenotypic variations resulted from crossing over, a source of genetic variation discussed in the next section.

Independent assortment of chromosomes and crossing over cause genetic recombination

In Chapter 12, we saw that meiosis and random fertilization generate genetic variation among offspring of sexually reproducing organisms. The general term for the production of offspring with new combinations of traits inherited from two parents is **genetic recombination.** Here, we will examine the chromosomal basis of recombination in more detail.

The Recombination of Unlinked Genes: Independent Assortment of Chromosomes

Mendel learned from his dihybrid crosses that some offspring have combinations of traits that do not match either parent. For example, draw a Punnett square for a testcross between a pea plant with yellow-round seeds that is heterozygous for both traits (*YyRr*) and a plant with green-wrinkled seeds (homozygous for both recessive traits, *yyrr*). The gene loci for these two characters are on separate chromosomes: Seed shape and seed color are unlinked, their alleles assorting indepedently. Notice that in your Punnett square, one-half of the offspring of the testcross inherit a phenotype that matches one of the

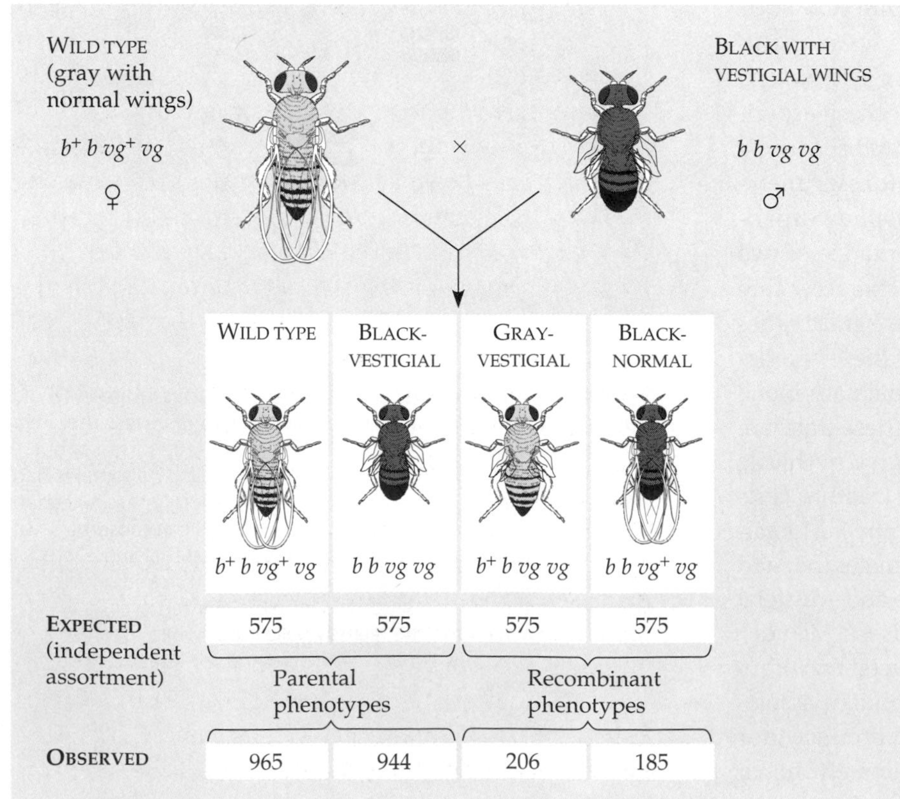

FIGURE 14.4
Evidence for linked genes in *Drosophila.* This is a testcross between flies differing in two characters, body color and wing size. The females are heterozygous for both genes and their phenotypes are wild type (gray bodies and normal wings). The males are homozygous recessive and express the mutant phenotypes for both characters (black bodies and vestigial wings). Morgan "scored" (classified according to phenotype) 2300 offspring from such matings. The top row of numbers (Expected) represents the 1:1:1:1 ratio of phenotypes we would expect if the two genes assort independently of each other. The bottom row (Observed) shows the actual results. The parental phenotypes are disproportionately represented among offspring. Morgan concluded that genes for body color and wing size are usually transmitted together from parents to offspring because they are located on the same chromosome—that is, they are linked genes. Crossing over accounts for the recombinant phenotypes, offspring that have combinations of traits different from either parent.

parental phenotypes—either yellow-round seeds or wrinkled-green seeds. These offspring are called **parental types.** But other phenotypes are also represented: one-fourth of the offspring have green-round seeds, and one-fourth have yellow-wrinkled seeds. Because these offspring have different combinations of seed shape and color than either parent, they are called **recombinants.** When half of all offspring are recombinants, geneticists say that there is a 50% frequency of recombination.

A 50% frequency of recombination is observed for any two genes that are located on different chromosomes. The physical basis of recombination between unlinked genes is the random alignment of homologous chromosomes during metaphase I of meiosis, which leads to the independent assortment of alleles (see FIGURE 14.1).

The Recombination of Linked Genes: Crossing Over

Linked genes do not assort independently because they are located on the same chromosomes and tend to move together through meiosis and fertilization. We would not expect linked genes to recombine into assortments of alleles not found in the parents. But, in fact, recombination between linked genes *does* occur. To see how, let's return to Morgan's fly room.

How can we explain the results of the *Drosophila* cross illustrated in FIGURE 14.4? The offspring of the testcross for body color and wing shape did not conform to the 1:1:1:1 phenotypic ratio we would expect if the genes for these two characters were on different chromosomes and assorted independently. But if the two genes were *completely* linked because their loci are on the same chromosome, then we should observe a 1:1 ratio, with only the parental phenotypes represented among offspring. The actual results satisfy neither of these expectations. Most of the offspring had parental phenotypes, suggesting linkage between the two genes, but about 17% of the flies were recombinants (FIGURE 14.5). Although there was linkage, it appeared incomplete. Morgan proposed that some mechanism that exchanges segments between homologous chromosomes must occasionally break the linkage between the two genes. Subsequent experiments have demonstrated that such an exchange—crossing over—accounts for the recombination of linked genes (see Chapter 12). While homologous chromosomes are paired in synapsis during prophase of meiosis I, nonsister chromatids may break at corresponding points and switch fragments. A crossover between homologous chromosomes breaks linkages in the parental chromosomes to form recombinant chromosomes that may bring together alleles in new combinations. The subsequent events of meiosis distribute the recombinant chromosomes to gametes.

Geneticists can use recombination data to map a chromosome's genetic loci

The discovery of linked genes and recombination due to crossing over led Morgan's research team to a method for constructing genetic maps. A genetic map lists the sequence of genetic loci along a particular chromosome. In this section, you will learn how to use crossover data to construct a type of genetic map called a linkage map. Newer techniques for mapping chromosomes, based on methods in molecular biology, will be introduced in Chapter 19.

As discussed earlier, a gene for *Drosophila* body color—gray (b^+) versus black (b)—is linked to a gene for wing size—normal (vg^+) versus vestigial (vg). The recombination frequency between these two genetic loci is about 17%; that is, crossing over results in 17% of the offspring having a combination of body color and wing size unlike either parent (see FIGURE 14.5). Carried on the same chromosome as these two genes is a third gene called cinnabar (*cn*), one of the many *Drosophila* genes affecting eye color. Cinnabar eyes, a mutant phenotype, are brighter red than the wild-type color. The recombination frequency between the cinnabar gene (*cn*) and the *b* locus is 9%. Thus, crossovers between the *b* and *vg* loci are about twice as frequent as crossovers between *b* and *cn* (17% versus 9%). In 1917, Alfred H. Sturtevant, one of Morgan's students, reasoned that different recombination frequencies reflect different distances between genes on a chromosome; that is, if two genes are far apart on a chromosome, there is a higher probability that a crossover event will separate them than if the two genes are close together. If we assume that the probability of crossing over between two genes is directly proportional to the distance between them, then the distance along the *Drosophila* chromosome between *b* and *vg* must be about twice as great as the distance between *b* and *cn.*

Sturtevant began using recombination data to assign genes positions on a map. He defined one "map unit" as the equivalent of a 1% recombination frequency. (Today, the word *centimorgan* is often used in place of map unit, in honor of Morgan.) Thus, the *b* and *cn* loci are separated by 9 map units, while the *b* and *vg* genes are 17 map units apart. But what is the sequence of the three genes? We can eliminate the sequence *b-vg-cn,* since we know that *cn* is closer to *b* than is *vg.* This leaves two possible sequences: *cn-b-vg* and *b-cn-vg.* The frequency of recombination between *cn* and *vg* should reveal the correct sequence of the three genes. The first possible sequence predicts that *cn* and *vg* are about 26 map units (17 + 9) apart, while the second sequence predicts a separation of about 8 map units (17 − 9). Sturtevant found that the

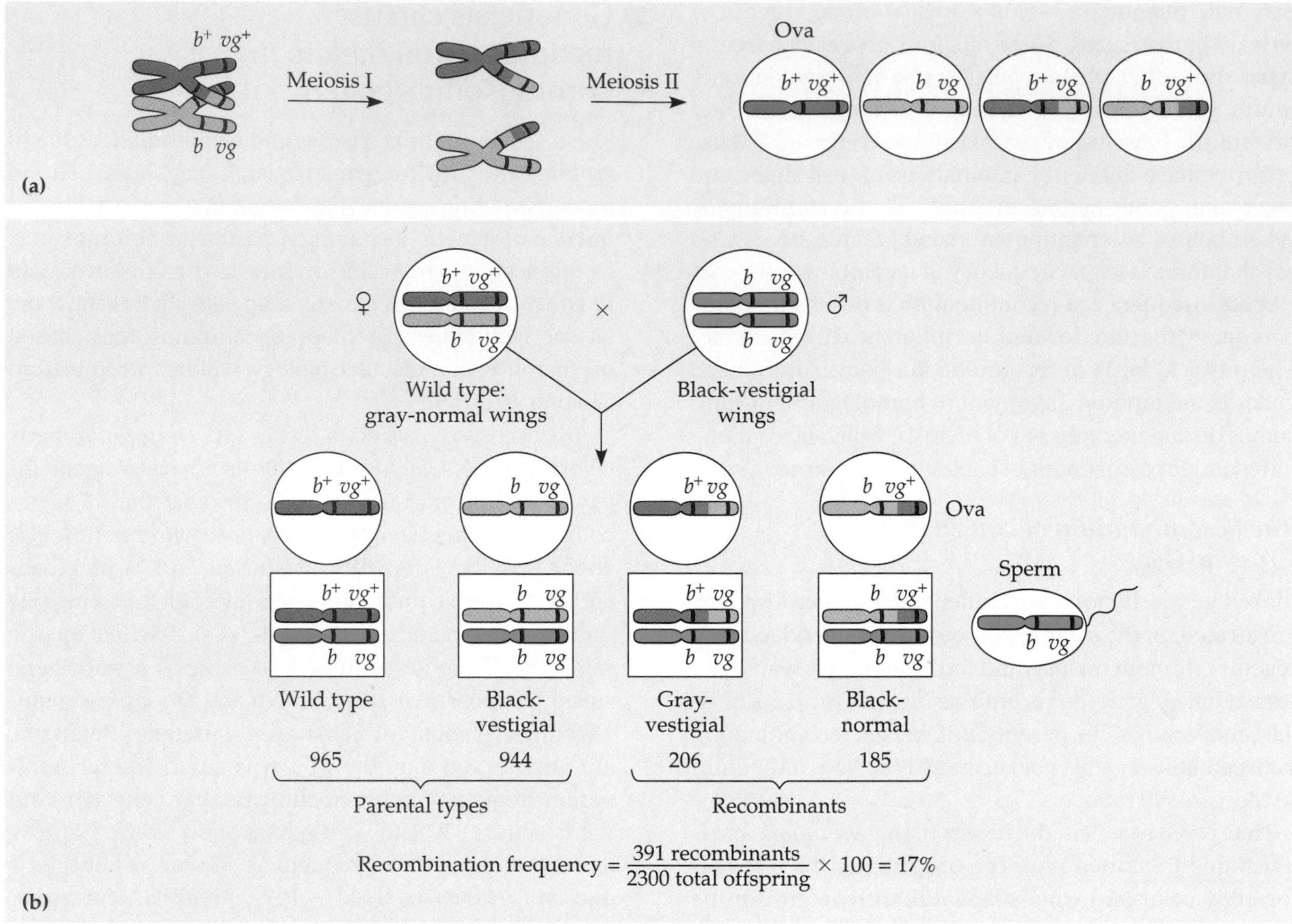

FIGURE 14.5

Recombination due to crossing over. These diagrams recreate the testcross in FIGURE 14.4, but this time we track chromosomes as well as genes. The *b* and *vg* loci are linked; they are on the same chromosome. The maternal chromosomes are color-coded with two shades of red so that we can distinguish one homologue from the other. (**a**) Crossing over occurs when homologous chromosomes pair during prophase of meiosis I. Nonsister chromatids break, and the fragments join the homologous chromosome. In this case, the crossover occurs in the region between the *b* and *vg* loci. If we follow the chromosomes through meiosis, we see that crossing over results in ova that, when fertilized, give rise to the recombinant offspring. (**b**) Linkage of the genes for body color and wing size explains the prevalence of parental phenotypes among offspring. Recombinant offspring, those with genotypes and phenotypes different from either parent, result from crossing over. We can calculate the recombination frequency from the proportion of recombinant flies out of the total pool of offspring.

frequency of recombination between *vg* and *cn* was 9.5%; he therefore proposed that the genes were arranged along a chromosome in the sequence *b-cn-vg* (FIGURE 14.6). This method was soon extended to map the other identified *Drosophila* genes in linear arrays.

Some genes on a chromosome are so far apart from each other that crossovers between them occur very often. The frequency of recombination measured between such genes can have a maximum value of 50%, a result indistinguishable from that for genes on different chromosomes. In fact, the seven characters that Mendel studied in his peas are not all on separate chromosomes, although the pea coincidentally has seven chromosome pairs. Seed color and flower color, for instance, are now known to be on chromosome 1. But they are so far apart on that chromosome that linkage is not observed in genetic crosses. Genes located far apart on a chromosome are mapped by adding the recombination frequencies from crosses involving each of the distant genes and an intermediate gene. Only for one pair of the genes Mendel studied, the genes for plant height and pod shape, do modern biologists observe linkage. Although Mendel observed segregation of alleles for each of these characters in monohybrid crosses, he did not report the results of dihybrid crosses for this particular combination of characters.

Using crossover data, Sturtevant and his colleagues were able to cluster the known mutations (and hence the wild-type alleles) of *Drosophila* into four groups of linked genes. Because microscopists had found four

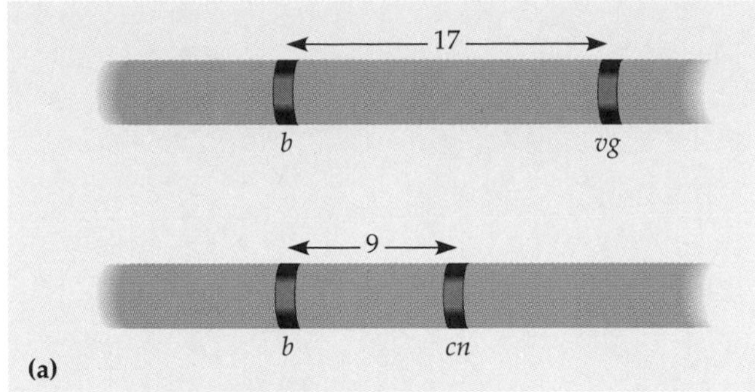

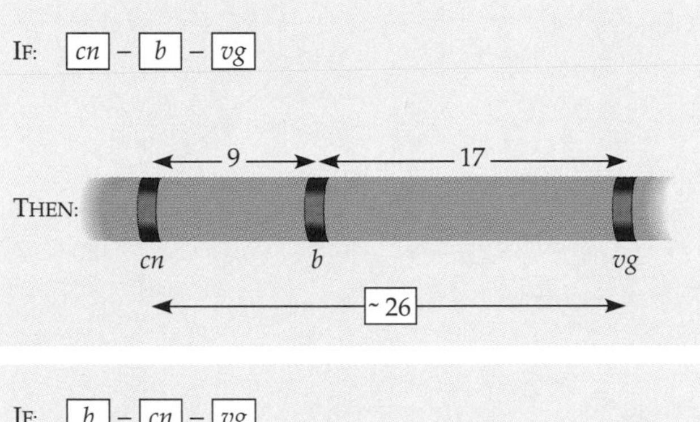

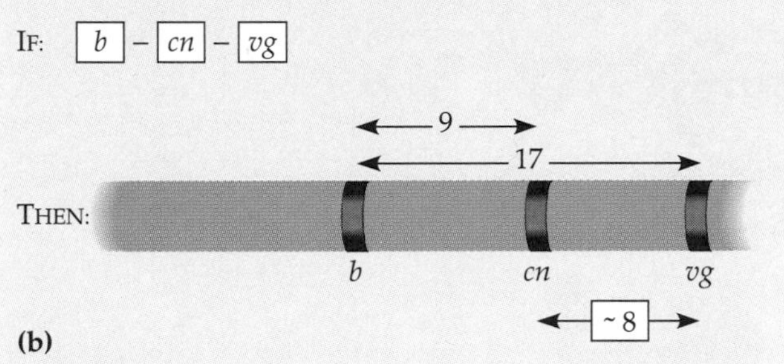

FIGURE 14.6
Using crossover data to construct a genetic map. A genetic map represents the linear sequence of genes along a chromosome. One method for constructing a genetic map is to determine how frequently crossovers occur in the region between two genes. This method is based on the assumption that the probability of a crossover between two genetic loci is proportional to the distance separating the loci. In this example, we use crossover data (recombination frequencies) to sequence three *Drosophila* genes: *b, vg,* and *cn.* (**a**) The *b* and *vg* loci are separated by 17 map units, each map unit being equal to a 1% recombination frequency. The *b* and *cn* loci are separated by 9 map units. (**b**) In order to decide which of two possible sequences of genes is correct, we must determine the frequency of crossing over between the *vg* and *cn* genes. Experiments reveal this recombination frequency to be 9.5%, which best fits the sequence *b-cn-vg.*

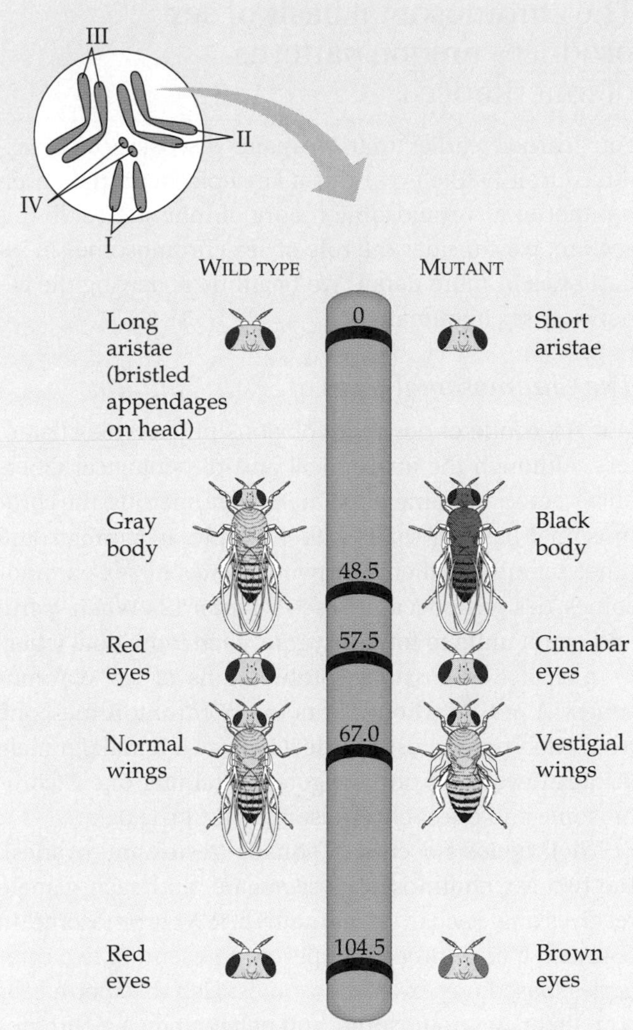

FIGURE 14.7
A partial genetic map of a *Drosophila* chromosome. *Drosophila* has four pairs of chromosomes, and genetic cartographers have been able to map many loci along each chromosome. The simplified map shown here represents just a few of the genes that have been mapped on chromosome II. Included are the *b* (black body), *cn* (cinnabar eyes), and *vg* (vestigial wings) loci we mapped in FIGURE 14.6. (Notice that more than one gene can affect a given character, such as eye color.)

pairs of chromosomes in *Drosophila* cells, this clustering of genes was additional evidence that genes are located on chromosomes. Each chromosome has a linear array of specific gene loci (FIGURE 14.7).

A **linkage map**—a genetic map based on recombination frequencies—is *not* a picture of an actual chromosome. The frequency of crossing over is not uniform over the length of the chromosome, and thus map units do not have absolute size (in nanometers, for instance). Thus, a linkage map portrays the sequence of genes along a chromosome, but it does not give the precise locations of genes. Additional methods enable geneticists to construct **cytological maps** of chromosomes, which actually pinpoint genes. One of these methods locates a gene by associating a mutant phenotype with the position of a chromosomal defect or other feature that can be seen in the microscope. In a comparison of a linkage map with a cytological map of the same chromosome, the sequence of genetic loci matches, but the spacing between loci does not. One reason a 1% recombination frequency does not correspond exactly to a fixed length of chromosome is that crossovers are more common for some chromosomal regions than for others.

The chromosomal basis of sex produces unique patterns of inheritance

You learned earlier that Morgan's discovery of a sex-linked trait (white eyes) was a key episode in the development of a chromosome theory of inheritance. In this section, we consider the role of sex chromosomes in inheritance in more detail. We begin by reviewing the genetics of sex in humans.

The Chromosomal Basis of Sex in Humans

Our sex is one of our more obvious phenotypic characters. Although the anatomical and physiological differences between women and men are numerous, the chromosomal basis of sex is rather simple. In humans and other mammals, there are two varieties of sex chromosomes, designated *X* and *Y* (see Chapter 12). When sperm and ovum unite to form a zygote, each individual inherits one of two possible combinations of sex chromosomes. A person who inherits two *X* chromosomes, one from each parent, usually develops as a female. A male usually develops from a zygote containing one *X* chromosome and one *Y* chromosome (FIGURE 14.8a).

When meiosis occurs in gonads (testes and ovaries), the two sex chromosomes segregate, and each gamete receives one. Each ovum contains one *X* chromosome. In contrast, a male produces sperm representing two categories, based on sex chromosomes: Half the sperm cells contain an *X* chromosome, and half contain a *Y* chromosome. We can trace the sex of each offspring to the moment of conception: If a sperm cell bearing an *X* chromosome happens to fertilize an ovum, the zygote is *XX;* if a sperm cell containing a *Y* chromosome fertilizes an ovum, the zygote is *XY.* Sex is a matter of chance—a fifty-fifty chance. FIGURE 14.8 compares this *X-Y* system of mammals to chromosomal systems of sex determination in other animal groups.

In humans, the anatomical evidence of sex begins to emerge when the embryo is about two months old. Before then, the rudiments of the gonads are generic—that is, they can develop into either ovaries or testes, depending on hormonal conditions within the embryo. Which of these two possibilities occurs depends on whether or not a *Y* chromosome is present, and it is probably just a small region of the *Y* chromosome that confers masculinity. In 1990, a British research team identified a single gene that seems to be required for the development of testes. They named the gene *Sry,* for sex-determining region of *Y.* In the absence of *Sry,* the gonads develop into ovaries. The researchers emphasized that the presence (or absence) of *Sry* is just a trigger. The biochemical, physiological, and anatomical features of sex

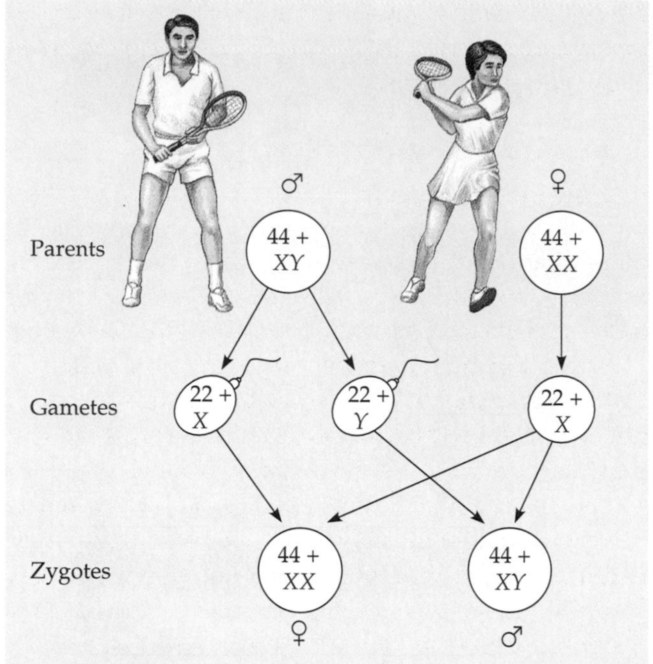

(a) The *X-Y* system

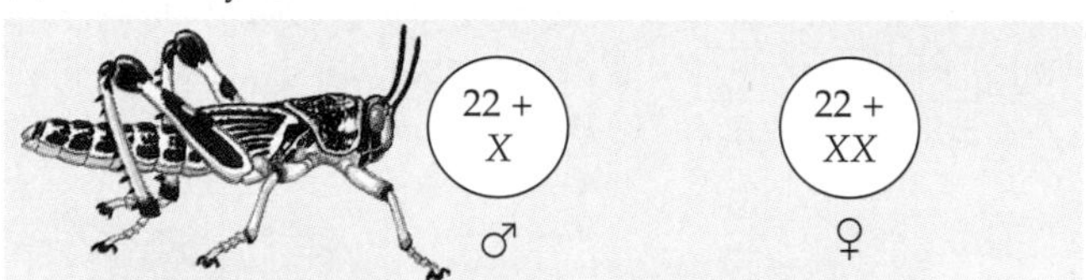

(b) The *X-0* system

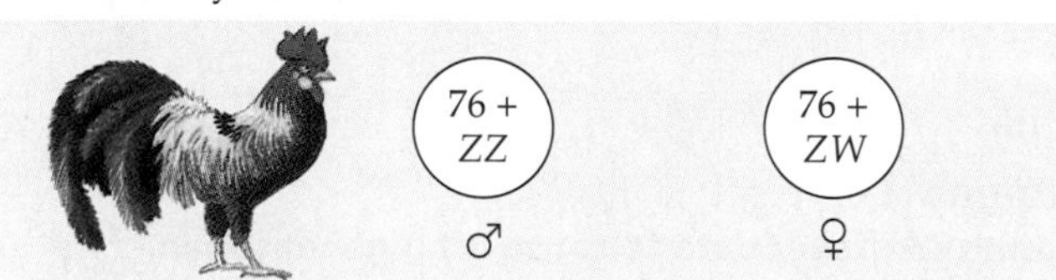

(c) The *Z-W* system

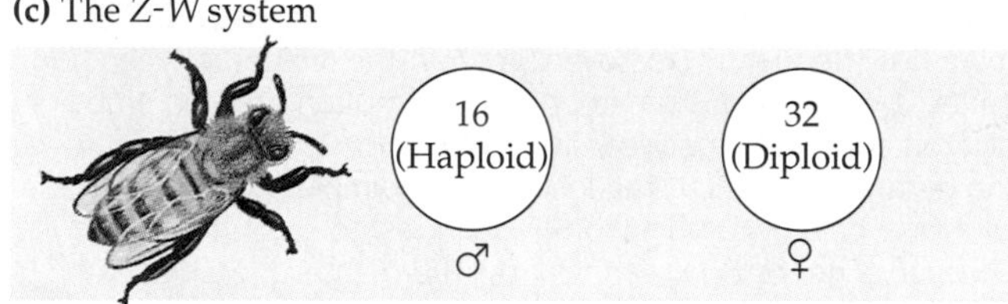

(d) The haplo-diploid system

FIGURE 14.8

Some chromosomal systems of sex determination. (**a**) In humans and other mammals, the sex of an offspring depends on whether the sperm cell carries an *X* chromosome or a *Y* chromosome. (Numbers in circles indicate the number of autosomes, the nonsex chromosomes.) (**b**) In grasshoppers, crickets, roaches, and some other insects, there is only one type of sex chromosome, the *X.* Females are *XX;* males are *X*0 (the 0 is a zero; males have only one sex chromosome). Sex of the offspring is determined by whether the sperm cell contains an *X* chromosome or no sex chromosome. (**c**) In birds, some fishes, and some insects, including butterflies and moths, the variable that determines sex is the sex chromosome present in the ovum (not the sperm, as is the case in the *X-Y* and *X-0* systems). The sex chromosomes are designated *Z* and *W* to avoid confusions with the *X-Y* system. Males are *ZZ* and females are *ZW.* (**d**) There are no sex chromosomes in most species of bees and ants. Females develop from fertilized ova and are thus diploid. Males develop from unfertilized eggs and are haploid; they are fatherless.

are complex, and many genes are involved in their development. It is likely that *Sry* codes for a protein that regulates many other genes.

Sex-Linked Disorders in Humans

In addition to their role in determining sex, the sex chromosomes, especially *X* chromosomes, have genes for many characters unrelated to sex. In humans, the term *sex-linked* usually refers to *X*-linked characters. These traits all follow the same pattern of inheritance that Morgan observed for the white-eye locus in *Drosophila*. Fathers pass *X*-linked alleles to all their daughters but to none of their sons (FIGURE 14.9). In contrast, mothers can pass sex-linked alleles to both sons and daughters.

If a sex-linked trait is due to a recessive allele, a female will express the phenotype only if she is a homozygote. Because males have only one locus, the terms *homozygous* and *heterozygous* lack meaning for describing their sex-linked genes (the term *hemizygous* is used in such cases). Any male receiving the recessive allele from his mother will express the trait. For this reason, far more males than females have disorders that are inherited as sex-linked recessives. However, even though the chance of a female inheriting a double dose of the mutant allele is much less than the probability of a male inheriting a single dose, there *are* females with sex-linked disorders. For instance, color blindness is a mild disorder inherited as a sex-linked trait. A color-blind daughter may be born to a color-blind father whose mate is a carrier (see FIGURE 14.9c). However, because the sex-linked allele for color blindness is rare, the probability that such a man and woman will come together is very low.

An example of a sex-linked disorder much more serious than color blindness is **Duchenne muscular dystrophy,** which affects about one out of every 3500 males born in the United States. People with Duchenne muscular dystrophy rarely live past their early twenties. The disease is characterized by a progressive weakening of the muscles and loss of coordination. Researchers have traced the disorder to the absence of a key muscle protein called dystrophin and have tracked the gene for this protein to a specific locus on the *X* chromosome. Perhaps in the future this new information will lead to treatments that will prevent the disease from progressing.

Hemophilia is a sex-linked recessive trait defined by absence of a certain protein required for blood clotting.

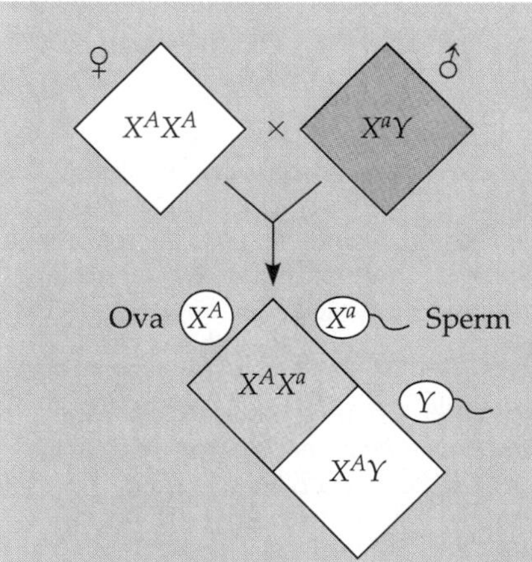

(a) A father with the trait will transmit the mutant allele to all daughters but to no sons. When the mother is a dominant homozygote, the daughters will have the normal phenotype but will be carriers of the mutation.

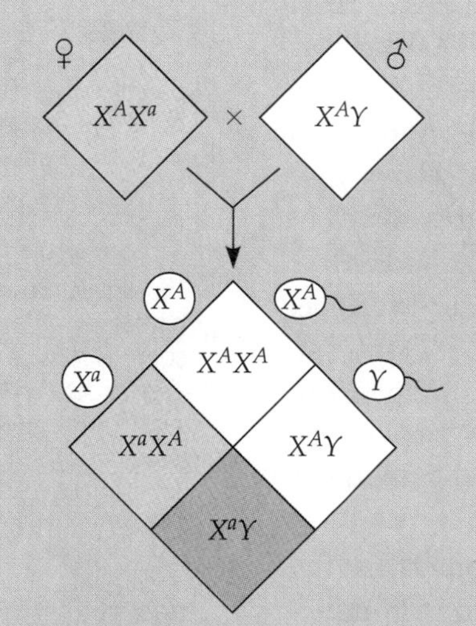

(b) A carrier who mates with a male of normal phenotype will pass the mutation to half her sons and half her daughters. The sons with the mutation will have the disorder. The daughters who have inherited the mutation in single dose will have the normal phenotype but will be carriers like their mother.

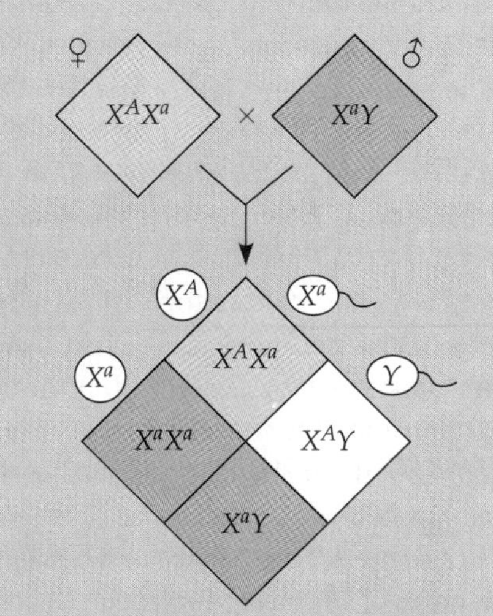

(c) If a carrier mates with a male who has the trait, there is a 50% chance that each child born to them will have the trait, regardless of sex. Daughters who do not have the trait will be carriers, whereas males without the trait will be completely free of the harmful recessive allele.

FIGURE 14.9

The transmission of sex-linked recessive traits. In this diagram, *X* and *Y* symbolize the sex chromosomes. The superscript *A* represents a dominant allele carried on the *X* chromosome, and the superscript *a* represents a recessive allele. Imagine that this recessive allele is a mutation that causes a sex-linked disease. White boxes indicate unaffected individuals, light-colored boxes indicate carriers, and dark-colored boxes indicate individuals with the sex-linked disorder.

Hemophiliacs bleed excessively when injured: the most seriously affected individuals may bleed to death after relatively minor skin abrasions, bruises, or cuts.

Hemophilia has an interesting history. The ancient Hebrews must have had some understanding of its hereditary pattern because sons born to women having a family history of hemophilia were exempted from circumcision. A high frequency of sex-linked hemophilia has plagued the royal families of Europe. The first hemophiliac in the royal line seems to have been Leopold, son of Queen Victoria (1819–1901) of England. The recessive allele for hemophilia was probably introduced to the royal family through a mutation in one of the sex cells of Victoria's mother or father, making Victoria a heterozygote, or carrier, of the deadly allele. Leopold survived to father a daughter who was also a carrier, transmitting hemophilia to one of her sons. Hemophilia was eventually introduced to the royal families of Prussia, Russia, and Spain through the marriages of two of Victoria's daughters, Alice and Beatrice, both carriers. The age-old practice of strengthening international alliances by having royalty marry royalty effectively spread hemophilia through the royal families of several European kingdoms.

X-*Inactivation in Females*

Although female mammals, including humans, inherit two *X* chromosomes, one *X* chromosome in each cell becomes almost completely inactivated during embryonic development. As a result, the cells of females and males have the same effective dose (one copy) of genes with loci on the *X* chromosome. The inactive *X* in each cell of a female condenses into a compact object, called a **Barr body,** which lies along the inside of the nuclear envelope. Most of the genes of the *X* chromosome that forms the Barr body are not expressed, although small regions of that chromosome remain active. (Barr bodies are "reactivated" in the cells of gonads that undergo meiosis to form gametes.)

British geneticist Mary Lyon has demonstrated that the selection of which of the two *X*s will form the Barr body occurs randomly and independently in each of the embryonic cells present at the time of *X*-inactivation. As a consequence, females consist of a mosaic of two types of cells: those with the active *X* derived from the father and those with the active *X* derived from the mother. After an *X* chromosome is inactivated in a particular cell, all mitotic descendants of that cell have the same inactive *X*. Therefore, if the female is heterozygous for a sex-linked trait, approximately half her cells will express one allele, while the others will express the alternate allele. This mosaicism can be seen graphically in the coloration of a calico cat (FIGURE 14.10). In humans, there is a recessive *X*-linked mutation that prevents the development of sweat glands. A woman who is heterozygous for this trait will have patches of normal skin and patches of skin lacking sweat glands.

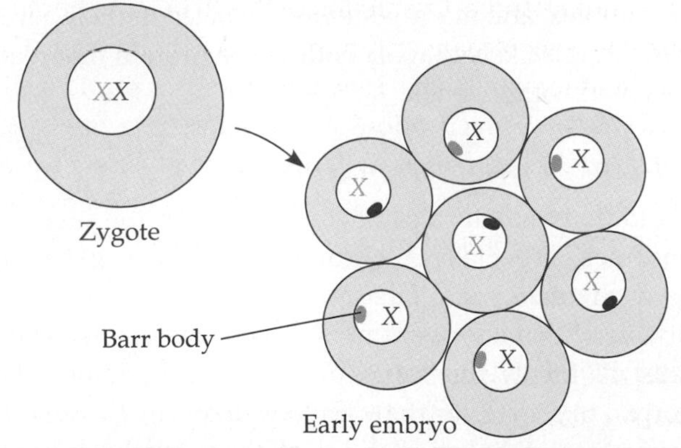

FIGURE 14.10
X-inactivation and the calico cat. On the *X* chromosome is a gene controlling fur color, with one allele causing black fur and another causing orange fur. (A separate gene is responsible for a patchlike pattern of colored and white fur.) A male (*XY*) can inherit one of these alleles, but not both. Thus, calicos, which have both black and orange patches, are almost always females. A female calico is heterozygous for the patch-color locus, inheriting an allele for black on one *X* chromosome and an allele for orange on the other *X*. This is symbolized in the diagram by color-coding the *X* chromosomes according to the allele they carry. During the cat's early embryonic development, one or the other *X* chromosome is randomly inactivated in each cell. The inactivated *X* condenses as a Barr body located just inside the nuclear envelope. As the embryonic cells divide by mitosis, each gives rise to a clone of cells having a specific *X* chromosome active and the other inactive. The patchwork coat of the calico results from these dual cell populations.

Inactivation of an *X* chromosome involves attachment of methyl groups (—CH_3) to cytosine, one of the nitrogenous bases of DNA nucleotides. (The regulatory role of DNA methylation is discussed in more detail in Chapter 18.) But what determines which of the two *X* chromosomes is targeted for methylation? Researchers have recently discovered a gene that is active *only* on the Barr body. The gene is called *XIST,* for *X*-inactive specific transcript. The gene's product, or "specific transcript," is an RNA molecule that remains in close physical contact with the *X* chromosome. Interaction of this RNA with the chromosome may initiate and maintain *X*-inactivation. This leads to more questions. How does the RNA produced by *XIST* initiate *X*-inactivation? And what determines which of the two *X* chromosomes in each of a female's cells will have an active *XIST* gene and become the Barr body? Our understanding of *X*-inactivation is still rudimentary.

Alterations of chromosome number or structure cause some genetic disorders

Physical and chemical disturbances, as well as errors during meiosis, can damage chromosomes or alter their number in a cell. In this section, we survey these chromosomal alterations and apply this information to some important disorders in humans.

Alterations of Chromosome Number: Aneuploidy and Polyploidy

Ideally, the meiotic spindle distributes chromosomes to daughter cells without error. But there is an occasional mishap, called a **nondisjunction,** in which the members of a pair of homologous chromosomes do not move apart properly during meiosis I, or in which sister chromatids fail to separate during meiosis II. In these cases, one gamete receives two of the same type of chromosome and another gamete receives no copy (FIGURE 14.11). The other chromosomes are usually distributed normally. If either of these aberrant gametes unites with a normal one, the offspring will have an abnormal chromosome number, known as **aneuploidy.** If the chromosome is present in triplicate in the fertilized egg (so that the cell has a total of $2n + 1$ chromosomes), the aneuploid cell is said to be **trisomic** for that chromosome. If a chromosome is missing (so that the cell has $2n - 1$ chromosomes), the aneuploidy is **monosomic** for that chromosome. Mitosis will subsequently transmit the anomaly to all embryonic cells. If the organism survives, it usually has a set of symptoms caused by the abnormal dose of genes located on the extra or missing chromosome. Nondisjunction can also occur during mitosis. If such an error takes place early in embryonic development, then the aneuploid condition is passed along by mitosis to a large number of cells and is likely to have a substantial effect on the organism.

Some organisms have more than two complete chromosome sets. The general term for this chromosomal alteration is **polyploidy,** with the specific terms triploidy ($3n$) and tetraploidy ($4n$) indicating the number of chromosomal sets. One way a triploid cell may be produced is by the fertilization of an abnormal diploid egg produced

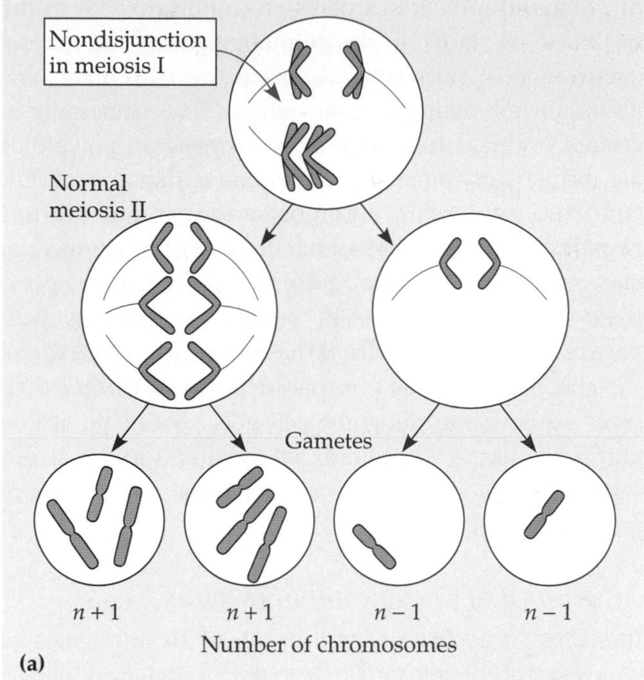

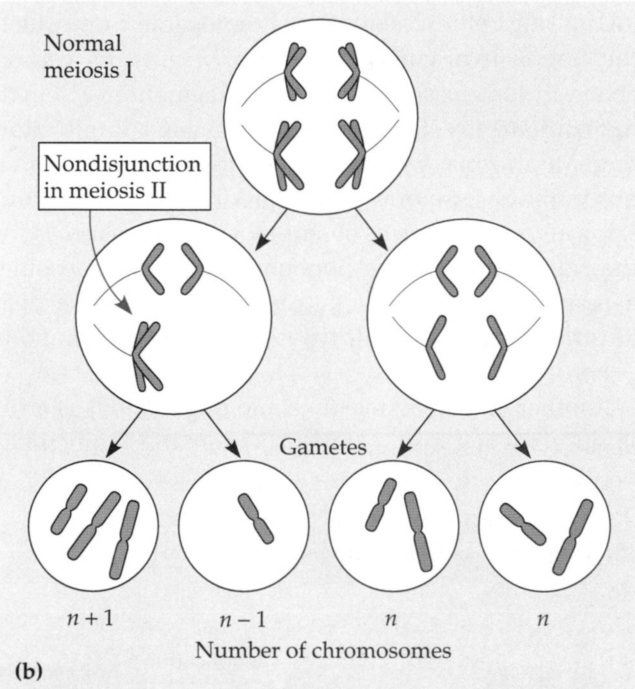

FIGURE 14.11
Meiotic nondisjunction. (**a**) Homologues may fail to separate during anaphase of meiosis I, or (**b**) chromatids may fail to separate during anaphase of meiosis II. Either type of meiotic error will produce gametes with an anomalous chromosome number.

by nondisjunction of all its chromosomes. An example of an accident that would result in tetraploidy is the failure of a $2n$ zygote to divide after replicating its chromosomes. Subsequent mitosis would then produce a $4n$ embryo.

Polyploidy is relatively common in the plant kingdom, and we will see in Chapter 22 that the spontaneous origin of polyploid individuals plays an important role in the evolution of plants. In the animal kingdom, the natural occurrence of polyploids seems to be extremely rare, although polyploidy can be induced experimentally in certain animals, including frogs. In general, polyploids are more nearly normal in appearance than aneuploids. One extra (or missing) chromosome apparently disrupts genetic balance more than having an entire extra set of chromosomes does. More common than complete polyploid animals are mosaic polyploids, animals with patches of polyploid cells. If the sister chromatids for all the chromosomes fail to separate during a *mitotic* division, so that one daughter cell gets all the replicated chromosomes, a tetraploid cell results. Additional mitotic divisions of this cell produce a localized clone of tetraploid cells.

Alterations of Chromosome Structure

Breakage of a chromosome can lead to four types of changes in chromosome structure. A **deletion** occurs when a chromosomal fragment lacking a centromere is lost during cell division. The chromosome from which the fragment originated will then be missing certain genes. In some cases, however, the fragment may join to the homologous chromosome, producing a **duplication** there. It also may reattach to the original chromosome but in the reverse orientation, producing an **inversion.** A fourth possible result of chromosomal breakage is for the fragment to join a nonhomologous chromosome, a rearrangement called a **translocation.** FIGURE 14.12 illustrates these different types of structural alterations of chromosomes.

Another source of deletions and duplications is error during crossing over. Nonsister chromatids sometimes break at different places, and one partner consequently gives up more genes than it receives. The products of such a nonreciprocal crossover are one chromosome with a deletion and one chromosome with a duplication.

An organism that inherits a homozygous deletion (or a single *X* chromosome with a deletion, in a male) is missing certain genes, resulting in a genetic imbalance that is usually lethal. Duplications and translocations also tend to have harmful effects. In reciprocal translocations, in which segments are exchanged between nonhomologous chromosomes, and in inversions, the balance of genes is not abnormal—all genes are present in their normal doses. Nevertheless, inversions and translocations can alter phenotype because a gene's expression can be influenced by its location among neighboring genes.

Human Disorders Due to Chromosomal Alterations

Alterations of chromosome number and structure are associated with a number of serious human disorders. When nondisjunction occurs in meiosis, the result is aneuploidy, an abnormal number of chromosomes in the gamete produced and, later, in the zygote. Although the frequency of aneuploid zygotes may be quite high in humans, most of these chromosomal alterations are so disastrous to development that the embryos are spontaneously (naturally) aborted long before birth. However, some types of aneuploidy appear to upset the genetic balance less than others, with the result that individuals with certain aneuploid conditions can survive to birth and beyond. These individuals have a set of symptoms—a syndrome—characteristic of the type of aneuploidy. Genetic disorders caused by aneuploidy can be diagnosed before birth by by fetal testing (see Chapter 13).

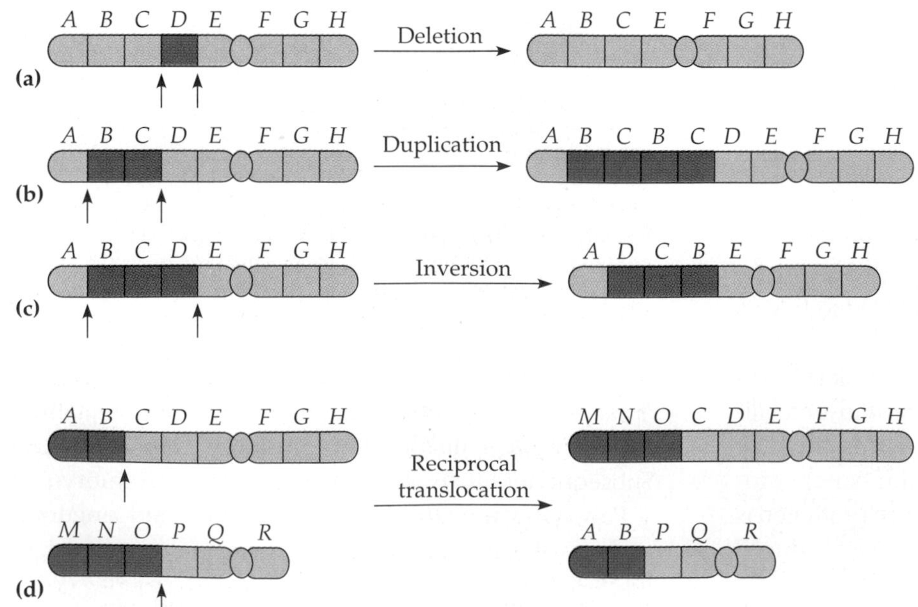

FIGURE 14.12
Alterations of chromosome structure. Arrows indicate where chromosomes break. Dark purple symbolizes the genes affected by the chromosomal rearrangements. (**a**) A deletion removes a chromosomal segment. (**b**) A duplication repeats a segment. (**c**) An inversion reverses a segment within a chromosome. (**d**) A translocation moves a segment from one chromosome to another, nonhomologous one. The most common type of translocation is reciprocal, in which nonhomologues exchange fragments. Nonreciprocal translocations, in which a chromosome transfers a fragment without receiving a fragment in return, also occur.

One aneuploid condition, **Down syndrome,** affects approximately one out of every 700 children born in the United States. Down syndrome is usually the result of an extra chromosome 21, so that each body cell has a total of 47 chromosomes (FIGURE 14.13). In chromosomal terms, the cells are trisomic for chromosome 21. Although chromosome 21 is the smallest human chromosome, its trisomy severely alters the individual's phenotype. Down syndrome includes characteristic facial features, short stature, heart defects, susceptibility to respiratory infection, and mental retardation. Furthermore, individuals with Down syndrome are prone to developing leukemia and Alzheimer's disease. (It is probably not a coincidence that alleles associated with the latter two diseases are located on chromosome 21.) Although people with Down syndrome, on average, have a lifespan much shorter than normal, some live to middle age or beyond. Most are sexually underdeveloped and sterile. Thus, most cases of Down syndrome result from nondisjunction during gamete production in one of the parents.

The frequency of Down syndrome correlates with the age of the mother. Down syndrome occurs in 0.04% of children born to women under age 30. The risk climbs to 1.25% for mothers in their early thirties and is even higher for older mothers. Because of this relatively high risk, pregnant women who are over 35 are candidates for fetal testing in order to check for trisomy 21 in the embryo. The correlation of Down syndrome with maternal age has not yet been explained. One hypothesis is that older women are more likely than younger women to carry Down babies to term rather than spontaneously aborting the trisomic embryos. No other chromosomal disorder is known to follow this pattern of increased incidence with maternal age.

Far rarer than Down syndrome are several other human disorders caused by autosomal aneuploidy. Patau syndrome, or trisomy 13, is characterized by serious eye, brain, and circulatory defects, as well as cleft palate. Patau syndrome occurs once in every 5000 live births. Edwards syndrome, trisomy 18, occurs once in every 10,000 live births and affects almost every body organ. In both of these syndromes, the babies rarely survive more than a year.

Nondisjunction of sex chromosomes produces a variety of aneuploid conditions in humans. Most of these conditions appear to upset genetic balance less than aneuploid conditions involving autosomes. This may be because the *Y* chromosome carries relatively few genes and because extra copies of the *X* chromosome become inactivated as Barr bodies in the somatic cells.

An extra *X* chromosome in a male, producing *XXY*, occurs approximately once in every 2000 live births. People with this disorder, called Klinefelter syndrome, have male sex organs, but the testes are abnormally small and

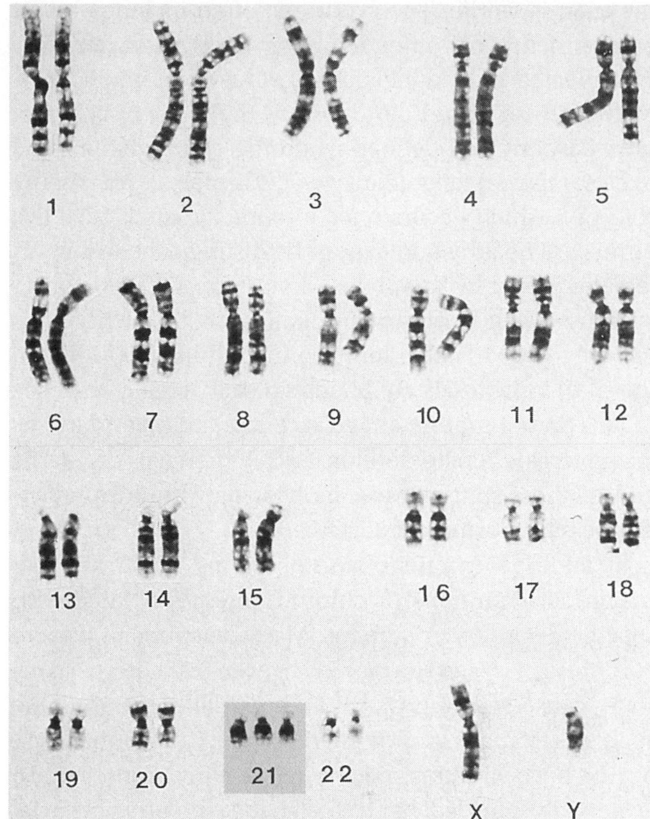

(a)

(b)

FIGURE 14.13

Down syndrome. (**a**) The karyotype shows trisomy 21. (**b**) Thousands of individuals with Down syndrome are among those who participate nationwide in the Special Olympics.

the man is sterile. The syndrome often includes breast enlargement and other feminine body characteristics. The affected individual is usually of normal intelligence. Males with an extra *Y* chromosome (*XYY*) are not characterized by any well-defined syndrome, although they tend to be somewhat taller than average. Females with trisomy *X* (*XXX*), which occurs once in approximately 1000 live births, are healthy and cannot be distinguished from *XX* females except by karyotype. Monosomy *X*, called Turner syndrome, occurs about once in every 5000 births and is the only known viable monosomy in humans. Although these *X*0 individuals are phenotypically female, their sex organs do not mature at adolescence, and secondary sex characteristics fail to develop. Such individuals are sterile and of short stature. Most individuals with Turner syndrome have normal intelligence.

In addition to aneuploid conditions, there are also structural alterations of chromosomes associated with specific disorders in humans. Many deletions in human chromosomes (see FIGURE 14.12a), even in a heterozygous state, cause severe physical and mental problems. One such syndrome is known as *cri du chat* ("cry of the cat"). A child born with this specific deletion in chromosome 5 is mentally retarded, has a small head with unusual facial features, and has a cry that sounds like the mewing of a distressed cat. Such individuals usually die in infancy or early childhood.

Another type of chromosomal structural alteration associated with human disorders is chromosomal translocation, the attachment of a fragment from one chromosome to another, nonhomologous chromosome (see FIGURE 14.12d). Chromosomal translocations have been implicated in certain cancers. One example is chronic myelogenous leukemia (CML). Leukemia is a cancer affecting the cells that give rise to white blood cells, and in the cancerous cells of CML patients, a reciprocal translocation has occurred. A portion of chromosome 22 has switched places with a small fragment from a tip of chromosome 9. (How such a switch might cause cancer will be discussed in Chapter 18.)

A small fraction of individuals with Down syndrome have a chromosomal translocation of a different sort. All the cells of such people have the normal number of chromosomes, 46. Close inspection of the karyotype, however, shows the presence of part or all of a third chromosome 21 attached to another chromosome by translocation.

The phenotypic effects of some genes depend on whether they were inherited from the mother or the father

Throughout our discussions of Mendelian genetics and the chromosomal basis of inheritance, we have been assuming that a specific allele will have the same effect regardless of whether it was inherited from the mother or the father. This is probably a safe assumption most of the time. For example, when Mendel crossed purple-flowered peas with white-flowered peas, he observed the same results regardless of whether the purple-flowered parent supplied the ova or pollen. Recently, however, geneticists have identified some traits, including some inherited disorders in humans, that seem to depend on which parent passed along the alleles for those traits.

Genomic Imprinting

Consider two different disorders called Prader-Willi syndrome and Angelman syndrome. The symptoms are different. Prader-Willi syndrome is characterized by mental retardation, obesity, short stature, and unusually small hands and feet. People with Angelman syndrome exhibit spontaneous (uncontrollable) laughter, jerky movements, and other motor and mental symptoms. For both disorders, the genetic cause seems to be the same: deletion of a particular segment of chromosome 15. If a child inherits the abnormal chromosome from the father, the result is Prader-Willi syndrome. If the abnormal chromosome is inherited from the mother, the result is Angelman syndrome. Put another way, an individual with Prader-Willi lacks those genes that have been deleted from the paternal version of chromosome 15; an individual with Angleman lacks those genes that have been deleted from the maternal version of the chromosome. The implication is that the genes of the deleted region normally behave differently in offspring, depending on whether they belong to the maternal or the paternal chromosome.

A process called **genomic imprinting** can explain the Prader-Willi/Angelman enigma and some similar phenomena. Certain genes are imprinted in some way each generation, and the imprint is different depending on whether the genes reside in females or in males (FIGURE 14.14). In other words, the same alleles may have different effects on offspring depending on whether they arrive in the zygote via the ovum or via the sperm. In the new generation, both maternal and paternal imprints are apparently "erased" in gamete-producing cells, and all the chromosomes are recoded according to the sex of the individual in which they now reside. There is evidence that a parental imprint consists of methyl ($—CH_3$) groups that are added to nucleotides at specific loci on chromosomes (see Chapter 18).

Fragile X and Triplet Repeats

Genomic imprinting may help explain another disorder called **fragile *X* syndrome.** The disorder is named for the physical appearance of an abnormal *X* chromosome, the tip of which hangs on to the rest of the chromosome by a thin thread of DNA. Children with fragile *X* syndrome—

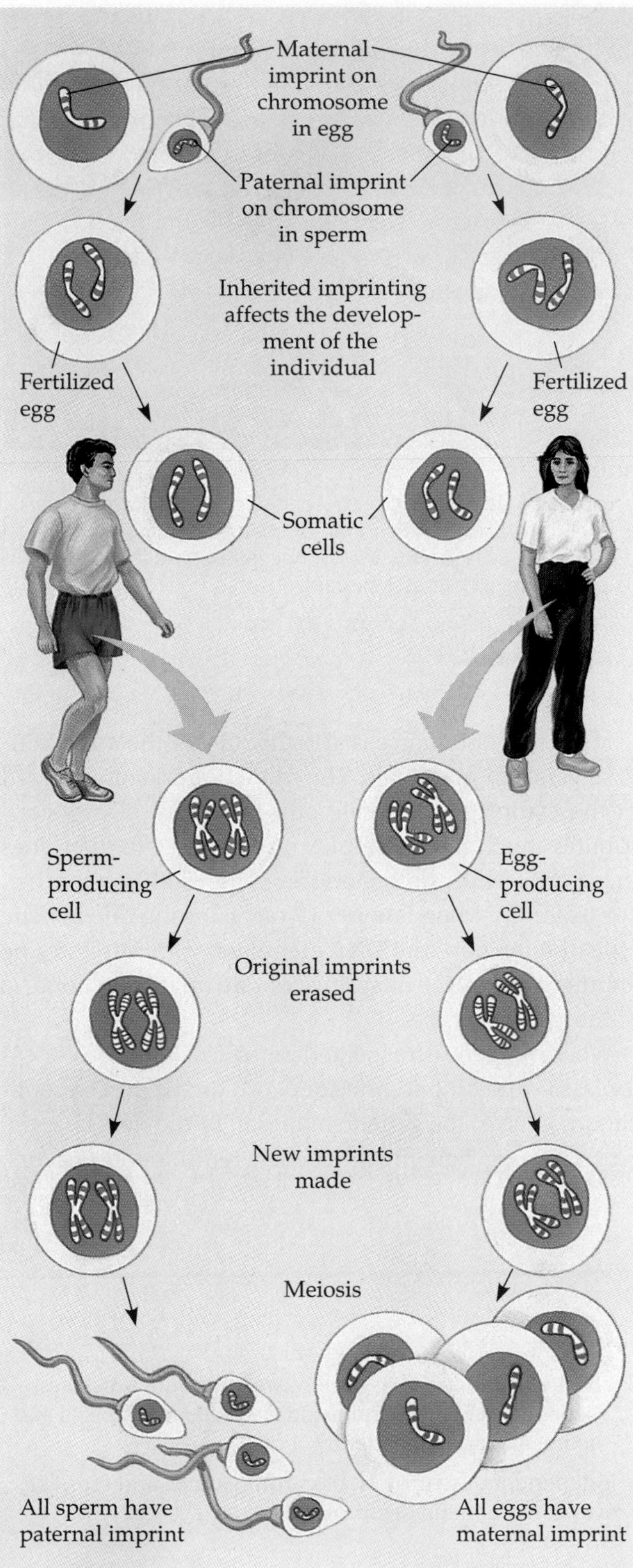

FIGURE 14.14
Genomic imprinting. Sperm and ova may convey chromosomes that are differently imprinted. The same allele may have a different phenotypic effect in offspring, depending on whether it has the maternal or the paternal imprint. Each generation, the old imprints are "erased" when sperm or ova are produced, and all the chromosomes are newly imprinted according to the sex of the individual.

about one in every 1500 males and about one in every 2500 females—are mentally retarded. Of all forms of mental retardation with a genetic basis, fragile *X* is the most common.

Inheritance of fragile *X* has a complex pattern, but researchers are beginning to put the pieces of the puzzle together. The region near the tip of a normal *X* chromosome that is altered in fragile *X* syndrome is an example of what geneticists call **triplet repeats.** These are sections of DNA where a specific triplet of nucleotides (the monomers of DNA; see Chapter 5) are repeated many times. At one tip of a normal *X* chromosome, the triplet of nucleotide bases CGG (cytosine, guanine, guanine) is repeated up to 50 times. This region of the *X* chromosome is imprinted in the female parent by the addition of methyl groups. The same triplet of bases is repeated more than 200 times in the fragile *X* chromosome. The alteration seems to occur in steps, with "pre–fragile *X*" forms of the chromosome having about 50 to 200 CGG repeats.

An individual who inherits the pre–fragile *X* chromosome is phenotypically normal, but a generation-to-generation addition of CGG triplets can eventually alter the chromosome enough to cause fragile *X* syndrome in offspring. This is where genomic imprinting seems to play a role. Extension of the CGG region is more likely to occur if the pre–fragile *X* chromosome is present in cells that give rise to ova rather than in cells that form sperm. And the excessive methylation of this part of the chromosome during genomic imprinting may prevent one or more genes in that region of the *X* chromosome from functioning normally in offspring. This maternal imprinting accounts for the observation that fragile *X* syndrome is more commonly expressed if the abnormal chromosome is inherited from the mother rather than the father. The hypothesis can also explain why the fragile *X* disorder is more common in males than in females: If a male (*XY*) inherits the fragile *X* chromosome, it *has* to be from his mother—a maternally imprinted version of the abnormal chromosome. In contrast, a female with the fragile *X* chromosome could have inherited it either from her mother (and therefore have the syndrome) *or* from her father (and be free of the syndrome because the copy of the fragile *X* chromosome is not maternally imprinted).

Triplet repeats occur normally in many places in human genomes. It is the progressive addition to those repeats that can lead to a disorder, such as fragile *X* syndrome. Another example is the allele that causes Huntington's disease, which has an extended version of a CAG (cytosine, adenine, guanine) triplet repeat at a locus near the tip of chromosome 4. Genomic imprinting also seems to be important. In contrast to fragile *X*, an addition to the triplet repeat of the Huntington's locus is more likely to occur if the allele is inherited from the father rather than the mother. Genomic imprinting and mutations that extend triplet repeats are important new concepts in genetics.

Extranuclear genes exhibit a non-Mendelian pattern of inheritance

Although our focus in this chapter has been on the chromosomal basis of inheritance, we end with an important amendment: Not all of a eukaryotic cell's genes are located on nuclear chromosomes, or even in the nucleus. Extranuclear genes are found in some cytoplasmic organelles: mitochondria and, in plants, plastids, including chloroplasts. Both mitochondria and plastids replicate and transmit their genes to daughter organelles. These cytoplasmic genes do not display Mendelian inheritance because they are not distributed to offspring according to the same rules that direct the distribution of nuclear chromosomes during meiosis.

Cytoplasmic genes were first observed in plants. In 1909, Karl Correns studied the inheritance of yellow or white patches on the leaves of an otherwise green ornamental plant (FIGURE 14.15). He found that the coloration of the offspring was determined only by the maternal parent (the source of seeds that germinate to give rise to the offspring) and not by the paternal parent (the pollen source). Subsequent research has shown that such variegated (striped or spotted) colorations of leaves are due to genes in the plastids that control pigmentation. In most plants, a zygote receives all its plastids from the cytoplasm of the ovum and none from pollen. Thus, as the zygote of these plants develops, its pattern of leaf coloration depends only on maternal cytoplasmic genes.

FIGURE 14.15
Cytoplasmic inheritance. Variegated leaves result from genes located in the plastids rather than on the nuclear chromosomes of plant cells. Because only the ovum contributes plastids to a plant zygote, all plastid genes are inherited from the maternal parent.

Maternal inheritance is also the rule for the mitochondrial genes in mammals. The mitochondria are situated in the cytoplasm of a cell, and the ovum always contributes much more cytoplasm to the zygote than does the sperm. Thus, mitochondria come from the mother. For example, when laboratory rats carrying one genetic type of mitochondrial DNA are mated with rats carrying another type, all the offspring contain only mitochondria of the maternal type.

Wherever genes are located in the cell—nucleus or cytoplasm—their inheritance depends on the precise replication of DNA, the genetic material. In the next chapter, you will learn how this molecular reproduction occurs.

REVIEW OF KEY CONCEPTS (with page numbers and key figures)

- Mendelian inheritance has its physical basis in the behavior of chromosomes during sexual life cycles (p. 262, FIGURE 14.1)
  - In the early 1900s, geneticists demonstrated that chromosomal movements during meiosis account for Mendel's laws.
- Morgan traced a gene to a specific chromosome: *science as a process* (pp. 262–265, FIGURE 14.3)
  - Morgan discovered that a gene for eye color in *Drosophila* is carried on the *X* chromosome, giving support to the chromosome theory of inheritance.
- Linked genes tend to be inherited together because they are located on the same chromosome (pp. 265–266, FIGURE 14.4)
  - Each chromosome has hundreds or thousands of genes. Genes on the same chromosome are said to be linked and do not assort independently.
- Independent assortment of chromosomes and crossing over cause genetic recombination (pp. 266–267, FIGURE 14.5)
  - The events of meiosis and random fertilization are responsible for genetic recombination, the production of offspring with new combinations of traits inherited from two parents.
  - Offspring that have the same phenotype for specific characters as one or the other parent are called parental types. The recombinant offspring, however, have combinations of traits that do not match either of the parents,

owing partly to independent assortment of alleles during the first meiotic division.
  - A recombination frequency of less than 50% indicates that the genes are linked but that crossing over has occurred. In this process, homologous chromosomes in synapsis during prophase of meiosis I break at corresponding points and switch fragments, thereby creating new combinations of alleles that are subsequently passed on to the gametes.

- Geneticists can use recombination data to map a chromosome's genetic loci (pp. 267–269, FIGURE 14.6)
  - One way to map genes is to deduce relative distances between them based on crossover data. Genes that are far apart on a chromosome are more likely to be separated during crossover than are genes that are close together. One map unit is defined as the equivalent of a 1% recombination frequency.
  - Cytological mapping is a technique that pinpoints the physical locus of a gene by associating a mutant phenotype with a chromosomal defect seen in the microscope.

- The chromosomal basis of sex produces unique patterns of inheritance (pp. 270–273, FIGURES 14.8, 14.9)
  - Sex is an inherited phenotypic character usually determined by the presence or absence of special chromosomes, but the exact mechanism for sex determination varies among different species.
  - In humans and other mammals, an *X-Y* system is operative. The *XY* males apportion either an *X* or a *Y* chromosome to a given sperm, which combine with an ovum containing an *X* chromosome from an *XX* female. Thus, the sex of the offspring is determined at conception by whether the sperm cell bears an *X* or a *Y* chromosome.
  - Certain genes for traits that are unrelated to maleness or femaleness are located on the sex chromosomes. Hemophilia is an example of a sex-linked recessive disorder whose gene is carried on the *X* chromosome.
  - In mammalian females, one of the two *X* chromosomes in each cell is randomly inactivated during early embryonic development.

- Alterations of chromosome number or structure cause some genetic disorders (pp. 273–276, FIGURES 14.11, 14.12)
  - Chromosomal mutations, including errors in meiosis, can change the number of chromosomes per cell or the structure of individual chromosomes. Such alterations can significantly affect phenotype.
  - Aneuploidy is an abnormal chromosome number. It can arise when a normal parental gamete unites with its counterpart that contains, for example, either two copies or no copies of a particular chromosome as a result of nondisjunction during meiosis. The resulting zygote will be trisomic or monosomic for the given chromosome.
  - Polyploidy, a condition in which there are more than two entire sets of chromosomes, can occur as a result of complete nondisjunction during gamete formation.
  - A variety of rearrangements can result from breakage of a chromosome. A lost fragment leaves the original chromosome with a deletion and may produce a duplication, translocation, or inversion by reattaching to a chromosome.
  - Chromosomal alterations cause a variety of human disorders. For example, trisomy of chromosome 21 is responsible for Down syndrome.

- The phenotypic effects of some genes depend on whether they were inherited from the mother or the father (pp. 276–278, FIGURE 14.14)
  - Individuals apparently imprint chromosomes in their gamete-producing cells with either a male or a female "stamp." This affects the way some genes are expressed in offspring.
  - Genomic imprinting is helping explain some baffling hereditary disorders, including fragile *X* syndrome. Another factor in fragile *X* is the addition of nucleotides to triple repeats.

- Extranuclear genes exhibit a non-Mendelian pattern of inheritance (p. 278, FIGURE 14.15)
  - Mitochondria and chloroplasts contain some of their own genes.
  - The zygote receives almost all its cytoplasm from the ovum. Such inheritance causes certain features of the offspring's phenotype to depend solely on maternal cytoplasmic genes.

GENETICS PROBLEMS

1. A man with hemophilia (a recessive, sex-linked condition) has a daughter of normal phenotype. She marries a man who is normal for the trait. What is the probability that a daughter of this mating will be a hemophiliac? A son? If the couple has four sons, what is the probability that all four will be born with hemophilia?

2. Pseudohypertrophic muscular dystrophy is a disorder that causes gradual deterioration of the muscles. It is seen only in boys born to apparently normal parents and usually results in death in the early teens. Is pseudohypertrophic muscular dystrophy caused by a dominant or recessive allele? Is its inheritance sex-linked or autosomal? How do you know? Explain why this disorder is seen only in boys and never in girls.

3. Red-green color blindness is caused by a sex-linked recessive allele. A color-blind man marries a woman with normal vision whose father was color-blind. What is the probability that they will have a color-blind daughter? What is the probability that their first son will be color-blind? (*Note:* The two questions are worded a bit differently.)

4. A wild-type fruit fly (heterozygous for gray body color and normal wings) was mated with a black fly with vestigial wings. The offspring had the following phenotypic distribution: wild type, 778; black-vestigial, 785; black-normal, 158; gray-vestigial, 162. What is the recombination frequency between these genes for body color and wing type?

5. In another cross, a wild-type fruit fly (heterozygous for gray body color and red eyes) was mated with a black fruit fly with purple eyes. The offspring were as follows: wild type, 721; black-purple, 751; gray-purple, 49; black-red, 45. What is the recombination frequency between these genes for body color and eye color? Following up on this problem and problem 4, what fruit flies (genotypes and phenotypes) would you mate to determine the sequence of the body color, wing shape, and eye color genes on the chromosome?

6. A space probe discovers a planet inhabited by creatures who reproduce with the same hereditary patterns as those in humans. Three phenotypic characters are height (T = tall, t = dwarf), head appendages (A = antennae, a = no antennae), and nose morphology (S = upturned snout, s = downturned snout). Since the creatures were not "intelligent," Earth scientists were able to do some controlled breeding experiments, using various heterozygotes in testcrosses. For a tall heterozygote with antennae, the offspring were tall-antennae, 46; dwarf-antennae, 7; dwarf-no antennae, 42; tall-no antennae, 5. For a heterozygote with antennae and an upturned snout, the offspring were antennae-upturned snout, 47; antennae-downturned snout, 2; no antennae-downturned snout, 48; no antennae-upturned snout, 3. Calculate the recombination frequencies for both experiments.

7. Using the information from problem 6, a further testcross was done using a heterozygote for height and nose morphology. The offspring were tall-upturned nose, 40; dwarf-upturned nose, 9; dwarf-downturned nose, 42; tall-downturned nose, 9. Calculate the recombination frequency from these data; then use your answer from problem 6 to determine the correct sequence of the three linked genes.

8. Imagine that a geneticist has identified two disorders that appear to be caused by the same chromosomal defect and are affected by genomic imprinting: blindness and numbness of the hands and feet. A blind woman (whose mother suffered from numbness) has four children, two of whom, a son and daughter, have inherited the chromosomal defect. If this defect works like Prader-Willi and Angelman syndromes, what disorders do this son and daughter display? What disorders would be seen in *their* sons and daughters?

9. What pattern of inheritance would lead a geneticist to suspect that an inherited disorder of cell metabolism is due to a defective mitochondrial gene?

10. An aneuploid person is obviously female, but her cells have two Barr bodies. What is the probable complement of sex chromosomes in this individual?

11. Determine the sequence of genes along a chromosome based on the following recombination frequencies: $A-B$, 8 map units; $A-C$, 28 map units; $A-D$, 25 map units; $B-C$, 20 map units; $B-D$, 33 map units.

12. About 5% of individuals with Down syndrome are the result of chromosomal translocation. In most of these cases, one copy of chromosome 21 becomes attached to chromosome 14. How does this translocation lead to children with Down syndrome?

13. Assume genes A and B are linked and are 50 map units apart. An individual heterozygous at both loci is crossed with an individual who is homozygous recessive at both loci. What percentage of the offspring will show phenotypes resulting from crossovers? If you did not know genes A and B were linked, how would you interpret the results of this cross?

14. In *Drosophila,* the gene for white eyes and the gene that produces "hairy" wings have both been mapped to the same chromosome and have a crossover frequency of 1.5%. A geneticist doing some crosses involving these two mutant characteristics noticed that in a particular stock of flies, these two genes assorted independently; that is, they behaved as though they were on different chromosomes. What explanation can you offer for this observation?

SCIENCE, TECHNOLOGY, AND SOCIETY

1. A chromosome test is used to determine whether women Olympic competitors are actually female. In the past, the test checked for the presence of a Barr body. The method used in the 1992 Olympics checked for genes found on the *Y* chromosome. Athletes who fail the test—about one in 500—are barred from competition. Some *XY* individuals are anatomically female, although they lack ovaries and a uterus; their *Y* chromosome fails to cause the development of testes, or their cells are insensitive to the effects of the male hormone testosterone. Athletes and physicians argue that the chromosome test unfairly bars *XY* females from competition. What is the purpose of the test? Is it fair? Can you suggest an alternative?

2. Gregor Mendel never saw a gene, yet he concluded that these "heritable factors" were responsible for inheritance in peas. Similarly, Morgan and Sturtevant never actually saw linked genes on chromosomes; their maps were deduced from patterns of inheritance. Is it legitimate science for biologists to claim the existence of things and processes they cannot actually see? Why or why not?

3. Opinions differ about whether children with learning disorders should be tested by karyotyping for the presence of a fragile *X* chromosome. Some argue that it's always better to know the cause of the problem so that education specialized for that disorder can be prescribed. Others counter that attaching a specific biological cause to a learning disability stigmatizes a child and limits his or her opportunities. What is your evaluation of these arguments?

FURTHER READING

Balter, M. "Filtering a River of Cancer Data." *Science,* February 24, 1995. Russian researchers are investigating chromosomal damage caused by radiation from nuclear accidents.

Gillis, A. M. "Turning Off the *X* Chromosome." *BioScience,* March 1994. How is one *X* chromosome in the cells of females inactivated?

Miller, K. R. "Whither the *Y*?" *Discover,* February 1995. How did sex chromosomes evolve?

Patterson, D. "The Causes of Down Syndrome." *Scientific American,* August 1987. How genes considered to be responsible for the symptoms of Down syndrome are being identified and mapped on chromosome 21.

Sapienza, C. "Parental Imprinting of Genes." *Scientific American,* October 1990.

Webb, J. "A Fragile Case For Screening?" *New Scientist,* January 1, 1994. The debate about screening children with learning disabilities for the fragile *X* chromosome.

CHAPTER 15

THE MOLECULAR BASIS OF INHERITANCE

KEY CONCEPTS

- The search for the genetic material led to DNA: *science as a process*
- Watson and Crick discovered the double helix by building models to conform to X-ray data: *science as a process*
- During DNA replication, base pairing enables existing DNA strands to serve as templates for new complementary strands
- A team of enzymes and other proteins functions in DNA replication
- Enzymes proofread DNA during its replication and repair damage to existing DNA

In 1953, James Watson and Francis Crick won a very competitive race to discover the molecular structure of DNA. The photograph that introduces this chapter captures Watson and Crick admiring their model of deoxyribonucleic acid, or DNA. DNA is the most celebrated molecule of our time, for it is the substance of inheritance. Mendel's heritable factors and Morgan's genes on chromosomes are, in fact, composed of DNA. Chemically speaking, your genetic endowment consists of the DNA inherited from your mother and father.

Of all nature's molecules, nucleic acids are unique in their ability to direct their own replication. Indeed, the resemblance of offspring to their parents has its molecular basis in the precise replication and transmission of DNA from one generation to the next. In other words, DNA is the substance behind the adage "Like begets like." Hereditary information is encoded in the chemical language of DNA and reproduced in all the cells of your body. It is this DNA program that directs the development of your biochemical, anatomical, physiological, and, to some extent, behavioral traits. In this chapter, you will learn how biologists deduced that DNA is the genetic material, how Watson and Crick discovered its structure, and how cells replicate their DNA—the molecular basis of inheritance.

The search for the genetic material led to DNA: *science as a process*

Today, molecular biologists can make or alter DNA in the laboratory and insert it into a cell, changing the cell's heritable characteristics. Earlier in this century, however, no one realized the relationship between DNA and heredity, and identification of the molecules of inheritance then loomed as a major challenge to biologists. As with the work of Mendel and Morgan, a key factor in meeting this challenge was the choice of appropriate experimental organisms. Because microscopic organisms—bacteria and the viruses that infect them—are far simpler than pea plants, fruit flies, or humans, the role of DNA in heredity was first worked out by studying such microbes.

By the 1940s, scientists realized that chromosomes, which were known to carry hereditary information, consisted of two substances, DNA and protein. Most researchers thought it was the protein that was the material of genes. The case for proteins seemed strong, especially since biochemists had identified them as a class of macromolecules with great heterogeneity and specificity of function, essential requirements for the hereditary material. Moreover, little was known about nucleic acids, whose physical and chemical properties seemed far too uniform to account for the multitude of specific inherited traits expressed by every organism. This view gradually changed, as experiments with microorganisms yielded unexpected results. In this section, we will trace the search for the genetic material in some detail, as a case study of the scientific process.

Evidence That DNA Can Transform Bacteria

The first evidence that genes are specific molecules was found in 1928. Frederick Griffith, a British medical officer, was studying *Streptococcus pneumoniae,* a bacterium that causes pneumonia in mammals. When Griffith grew colonies of the bacteria in petri dishes, he could distinguish between two genetic varieties, or strains. One strain produced colonies that appeared smooth, while colonies of the other strain appeared rough. Cells of the smooth strain (abbreviated S) synthesized a polysaccharide that surrounded the cells with a mucous coat, or capsule, which was not formed by the rough cells (R). These alternative phenotypes were inherited: Each strain reproduced its own kind.

When Griffith injected the bacteria into mice, he found that only the S strain was pathogenic (disease-causing). Mice injected with S cells died of pneumonia, while those injected with R cells survived (FIGURE 15.1). But it was not the polysaccharide of the coat that caused pneumonia, for Griffith found that S cells that had been killed by heat were harmless to the mice.

Then Griffith tried something remarkable. He mixed heat-killed S cells with live R cells and injected the mixture into mice. Although neither the dead S cells nor the live R cells alone were pathogenic, mice injected with the mixture developed pneumonia and died. More startling, Griffith found live S cells in blood samples taken from the dead mice, although only dead S cells had been injected. Somehow, some of the R cells had acquired from the dead S cells the ability to make polysaccharide coats and cause disease. Furthermore, this new-found ability was heritable: When Griffith cultured S cells taken from the dead mice, the dividing bacteria produced daughter cells with coats. The phenomenon that Griffith discovered is now called **transformation,** a change in phenotype due to the assimilation of external genetic material by a cell.* (Many bacteria live in a "broth" enriched with molecules derived from decomposing organic material. In Chapter 17, you will learn how bacteria take up macromolecules, including genetic material, from such surroundings.) Although Griffith did not know the chemical nature of the transforming agent, his observations spurred other scientists to search intensively for the elusive genetic material. Griffith's use of heat to inactivate the S cells hinted that protein might not be the genetic material. Heat denatures most proteins, yet the S-cell genetic material retained its ability to transform R cells into S cells.

For a decade, American bacteriologist Oswald Avery tried to identify Griffith's transforming agent. He purified various chemicals from the heat-killed S cells, then tested each substance with live R cells to see if it could transform

*This usage of *transformation* should not be confused with the conversion of a normal animal cell to a cancerous one; see Chapter 11.

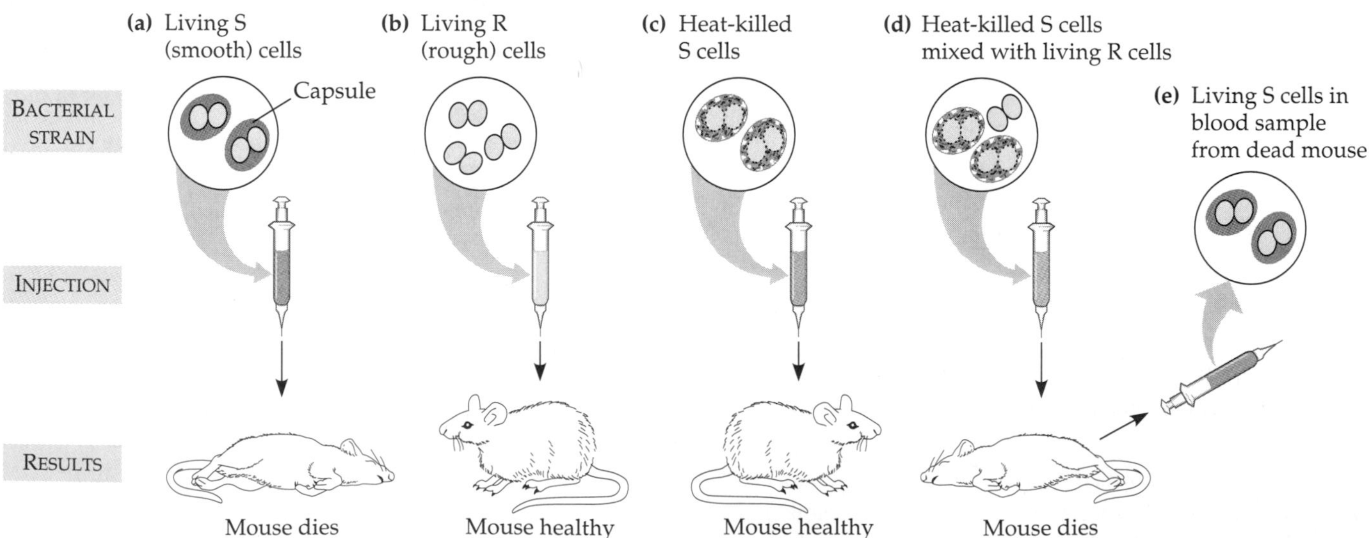

FIGURE 15.1
Transformation of bacteria. Griffith discovered that (**a**) the S strain of the bacterium *Streptococcus pneumoniae,* which was protected from a mouse's defensive system by a capsule, was pathogenic; (**b**) the R strain, a mutant lacking the coat, was nonpathogenic; (**c**) heat-killed S cells were harmless; but (**d**) a mixture of heat-killed S cells and live R cells caused pneumonia and death. (**e**) Live S bacteria could be retrieved from the dead mice injected with the mixture. Griffith concluded that a molecule from the dead S cells had genetically transformed some of the living R bacteria into S bacteria.

the bacteria. Only DNA worked. Thus, in 1944, Avery and his colleagues Maclyn McCarty and Colin MacLeod announced that the transforming agent was DNA. Their discovery was greeted with considerable skepticism, in part because of the lingering belief that proteins were better candidates for the genetic material. Moreover, many biologists were not convinced that the genes of bacteria would be similar in composition and function to those of more complex organisms. But the major reason for the continued doubt was that so little was known about DNA. No one could imagine how DNA could carry genetic information.

Evidence That Viral DNA Can Program Cells

Additional evidence for DNA as the genetic material came from studies of a virus that infects bacteria. Viruses are much simpler than cells. A virus is little more than DNA (or sometimes RNA) enclosed by a protective coat of protein. To reproduce, a virus must infect a cell and take over the cell's metabolic machinery.

Viruses that infect bacteria are widely used in laboratory research. These viruses are called bacteriophages (meaning "bacteria-eaters"), or just **phages,** for short. In 1952, Alfred Hershey and Martha Chase discovered that DNA is the genetic material of a phage known as T2. This is one of many phages that infect the bacterium *Escherichia coli* (*E. coli*), which normally lives in the intestines of mammals. At that time, biologists already knew that T2, like other viruses, was composed almost entirely of DNA and protein. They also knew that the phage could quickly turn an *E. coli* cell into a T2-producing factory that released phages when the cell ruptured. Somehow, T2 could reprogram its host cell to produce viruses, but which viral component—protein or DNA—was responsible?

Hershey and Chase answered this question by devising an experiment to determine which substance was transferred from the phage to the *E. coli* during infection (FIGURE 15.2). They used different radioactive isotopes to tag the molecules of DNA and protein. First, they grew T2

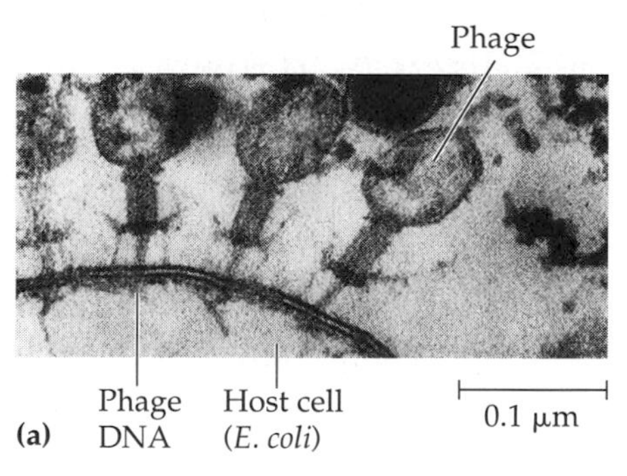
Phage
Phage DNA
Host cell (E. coli)
0.1 μm
(a)

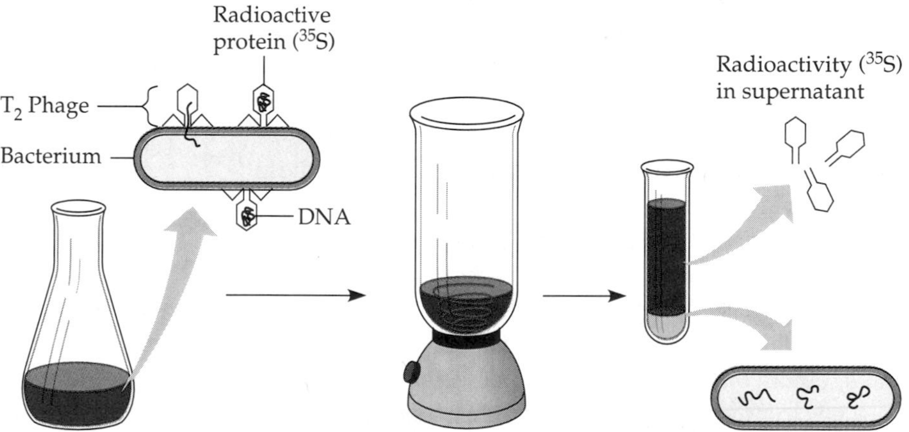
Radioactive protein (^{35}S)
T_2 Phage
Bacterium
DNA
Radioactivity (^{35}S) in supernatant
Mix radioactively labeled phage with bacteria. The phage infects the bacterial cells.
Agitate in a blender to separate phage outside the bacteria from the cells and their contents.
Centrifuge and measure the radioactivity in the pellet and supernatant.

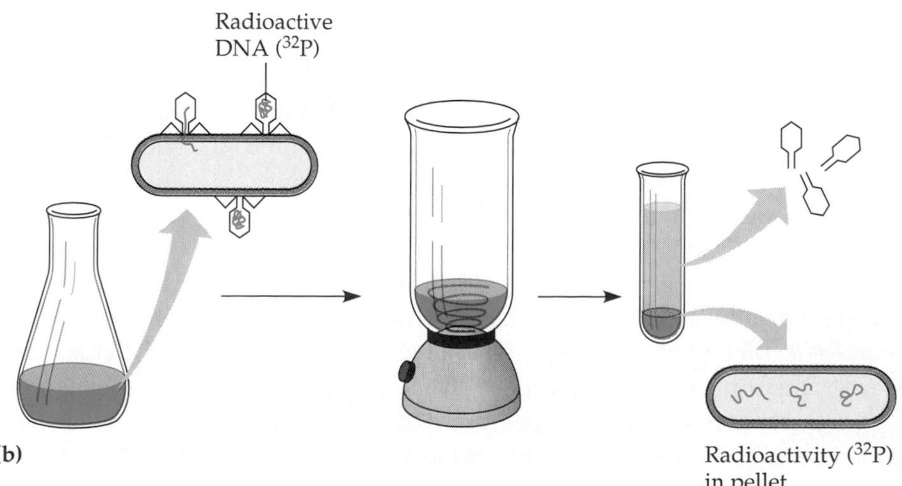
Radioactive DNA (^{32}P)
(b)
Radioactivity (^{32}P) in pellet

FIGURE 15.2
The Hershey-Chase experiment.
(**a**) Phages are viruses that infect bacteria. They use their tail pieces to attach to the host cell and inject their genetic material (TEM). (**b**) In their famous 1952 experiment, Hershey and Chase demonstrated that it was DNA, not protein, that functioned as the phages' genetic material. Viral proteins, labeled with radioactive sulfur, remained outside the host cell during infection. In contrast, viral DNA, labeled with radioactive phosphorus, entered the bacterial cell.

with *E. coli* in the presence of radioactive sulfur. Because protein, but not DNA, contains sulfur, the radioactive atoms were incorporated only into the protein of the phage. Next, in a similar way, the DNA of a separate batch of phage was labeled with atoms of radioactive phosphorus; because nearly all the phage's phosphorus is in its DNA, this procedure left the phage protein unlabeled. Then the protein-labeled and DNA-labeled batches of T2 were each allowed to infect separate samples of nonradioactive *E. coli* cells. Shortly after the onset of infection, the cultures were agitated in a kitchen blender to shake loose any parts of the phages that remained outside the bacterial cells. The mixtures were then spun in a centrifuge, forcing the heavier bacterial cells to form a pellet at the bottom of the centrifuge tubes, but allowing the lighter viral parts to remain suspended in the liquid, or supernatant. The scientists then measured and compared radioactivity in the pellet and supernatant.

Hershey and Chase found that when the bacteria had been infected with the T2 containing labeled proteins, most of the radioactivity was found in the supernatant, which contained viral particles (but not bacteria). This suggested that the phage protein did not enter the host cells. But when the bacteria had been infected with T2 phage whose DNA was tagged with radioactive phosphorus, then the pellet of mainly bacterial material contained most of the radioactivity. Moreover, when these bacteria were returned to culture medium, the infection ran its course; the *E. coli* released phages containing radioactive phosphorus.

Hershey and Chase concluded that the DNA of the virus is injected into the host cell, while most of the proteins remain outside. The injected DNA molecules cause the cells to produce additional viral DNA and proteins—indeed, additional intact viruses—providing powerful evidence that nucleic acids, rather than proteins, are the hereditary material, at least in viruses.

Additional Evidence That DNA Is the Genetic Material of Cells

Additional circumstantial evidence pointed to DNA as the genetic material in eukaryotes. Prior to mitosis, a eukaryotic cell doubles its DNA content, and during mitosis, this DNA is distributed equally to the two daughter cells. Also, diploid sets of chromosomes have twice as much DNA as the haploid sets found in the gametes of the same organism.

Still more compelling evidence came from the laboratory of biochemist Erwin Chargaff. Recall from Chapter 5 that DNA is a polymer of monomers called nucleotides, each consisting of three components: a nitrogenous (nitrogen-containing) base, a pentose sugar called deoxyribose, and a phosphate group (FIGURE 15.3). The base of each nucleotide can be any one of four different bases: adenine (A), thymine (T), guanine (G), or cytosine (C). Chargaff analyzed the base composition of DNA from a number of different organisms. In 1947, he reported that DNA composition is species-specific: In the DNA of any one organism, the amounts of the four nitrogenous bases are not all equal, and the ratios of nitrogenous bases vary from one species to another. Such evidence of molecular diversity, which had been presumed absent from DNA, made DNA a more credible candidate for the genetic material.

Chargaff also found a peculiar regularity in the ratios of nucleotide bases. In the DNA of each species he studied, the number of adenine components approximately equaled the number of thymines, and the number of guanines approximately equaled the number of cytosines. In human DNA, for example, the four bases are present in these percentages: A = 30.9% and T = 29.4%; G = 19.9% and C = 19.8%. The A = T and G = C equalities, later known as Chargaff's rules, remained unexplained until the discovery of the double helix.

Watson and Crick discovered the double helix by building models to conform to X-ray data: *science as a process*

Once most biologists were convinced that DNA was the genetic material, the race was under way to determine how the structure of DNA could account for its role in inheritance. By the beginning of the 1950s, the arrangement of covalent bonds in a nucleic acid polymer was well established (see FIGURE 15.3), and the competition focused on discovering the three-dimensional structure of DNA. Among the scientists working on the problem were Linus Pauling, in California, and Maurice Wilkins and Rosalind Franklin, in London. First to the finish line, however, were two scientists who were relatively unknown at the time—the American James Watson and the Englishman Francis Crick.

The brief but celebrated partnership that solved the DNA puzzle began soon after the young Watson journeyed to Cambridge University, where Crick was studying protein structure with a technique called X-ray crystallography (see the Methods Box in Chapter 5, p. 82). While visiting the laboratory of Maurice Wilkins at King's College in London, Watson saw an X-ray photograph of DNA, produced by Wilkins' colleague, Rosalind Franklin (FIGURE 15.4a, p. 286). Photographs produced by the X-ray crystallography method are not actually "pictures" of molecules. The spots and smudges in FIGURE 15.4b were produced by X-rays that were diffracted (deflected) as

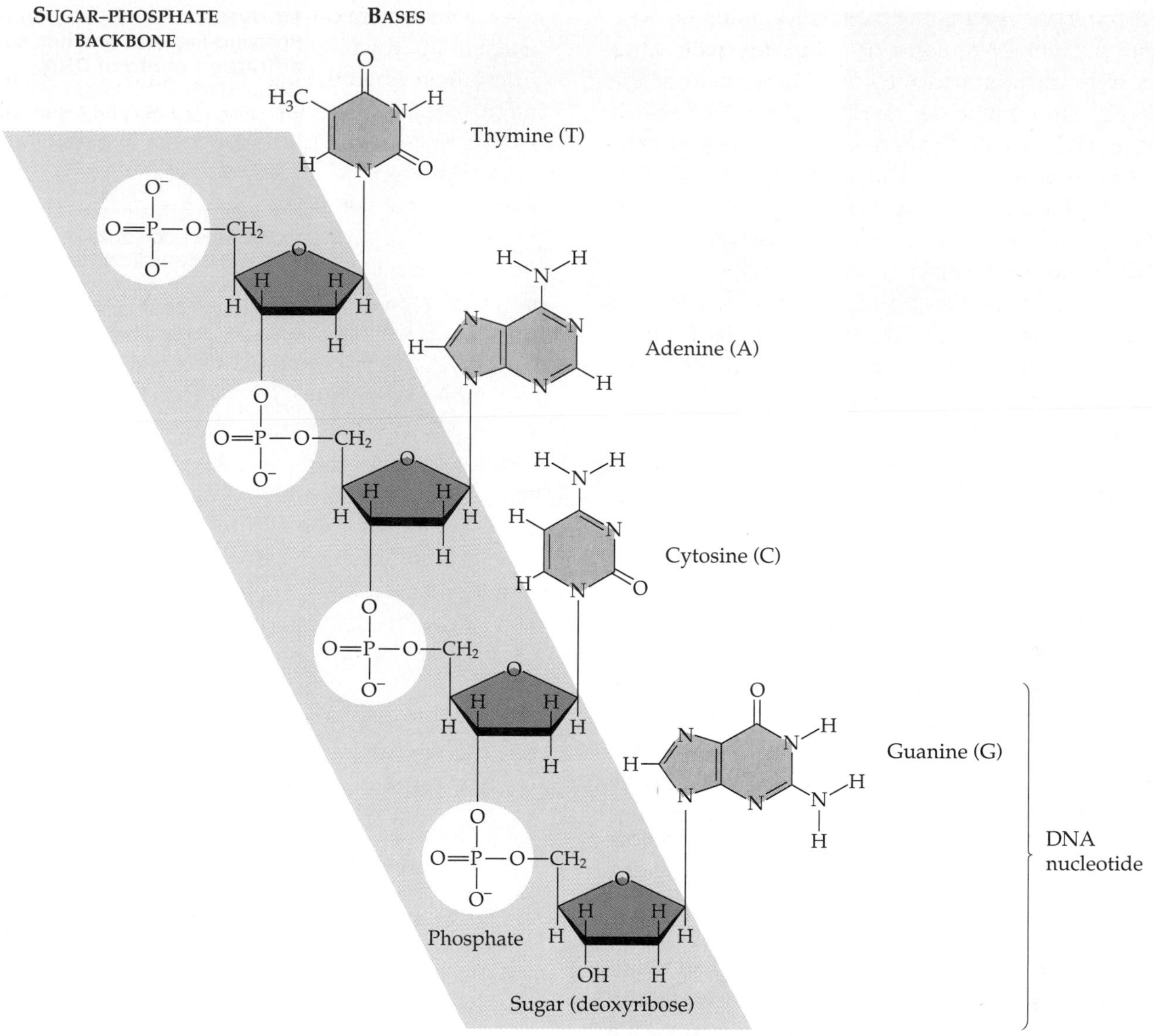

FIGURE 15.3

The structure of a DNA strand. Each nucleotide unit of the polynucleotide chain consists of a nitrogenous base (T, A, C, or G), the sugar deoxyribose, and a phosphate group. The phosphate of one nucleotide is attached to the sugar of the next nucleotide in line. The result is a "backbone" of alternating phosphates and sugars, from which the bases project.

they passed through crystallized DNA. Crystallographers use mathematical equations to translate such patterns of spots into information about the three-dimensional shapes of molecules. Watson and Crick based their model of DNA on data they were able to extract from Franklin's X-ray diffraction photo. They interpreted the pattern of spots on the X-ray photograph to mean that DNA was helical in shape. Based on Watson's recollection of the photograph, he and Crick deduced that the helix had a uniform width of 2 nanometers (nm), with its nitrogenous bases stacked 0.34 nm apart. The width of the helix suggested that it was made up of two strands, contrary to a three-stranded model that Linus Pauling had recently proposed. The presence of two strands accounts for the now-familiar term **double helix.**

Using molecular models made of wire, Watson and Crick began building scale models of a double helix that would conform to the X-ray measurements and what was then known about the chemistry of DNA. After failing to make a satisfactory model that placed the sugar-phosphate chains on the inside of the molecule, Watson tried putting them on the outside and forcing the nitrogenous bases to swivel to the interior of the double helix. Imagine this double helix as a rope ladder having rigid rungs, with the ladder twisted into a spiral. The side ropes are the equivalent of the sugar-phosphate backbones, and the rungs represent pairs of nitrogenous bases. Franklin's X-ray data indicated that the helix makes one full turn every 3.4 nm along its length. Because the bases are stacked just 0.34 nm apart, there are ten layers of base

(a)

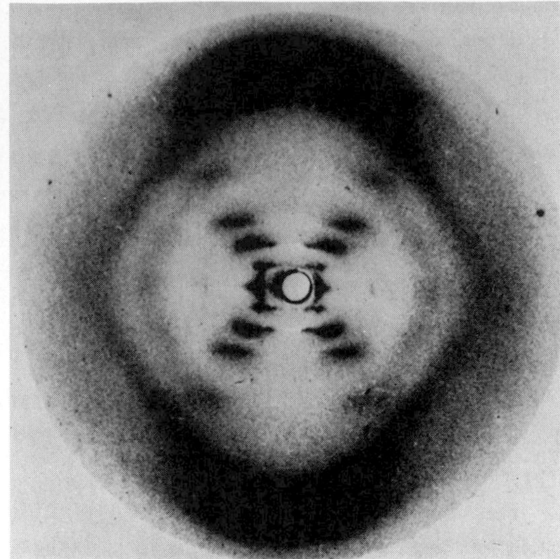

(b)

FIGURE 15.4
Rosalind Franklin and her X-ray diffraction photo of DNA.
(**a**) Franklin was the X-ray crystallographer who took the photo that enabled Watson and Crick to deduce the double-helical structure of DNA. Franklin died of cancer when she was only 38. Her colleague, Maurice Wilkins, received the Nobel Prize in 1962 along with Watson and Crick. Because the Nobel Prize is not awarded posthumously, science historians can only speculate about whether the committee would have recognized Franklin's contribution to the discovery of the double helix. (**b**) Franklin's X-ray diffraction photograph of DNA.

pairs, or rungs on the ladder, in each turn of the helix (FIGURE 15.5). This arrangement was appealing because it put the relatively hydrophobic nitrogenous bases in the molecule's interior and thus away from the surrounding aqueous medium.

Notice in FIGURE 15.5 that the nitrogenous bases of the double helix are paired in specific combinations: adenine (A) with thymine (T), and guanine (G) with cytosine (C). It was mainly by trial and error that Watson and Crick arrived at this key feature of DNA. At first, Watson imagined that the bases paired like-with-like—for example, A with A and C with C. But this model did not fit with the X-ray data, which suggested that the double helix had a uniform diameter. Why is this requirement inconsistent with like-with-like pairing of bases? Adenine and guanine are purines, nitrogenous bases with two organic rings. In contrast, cytosine and thymine belong to the family of nitrogenous bases known as pyrimidines, which have a single ring. Thus, purines (A and G) are about twice as wide as pyrimidines (C and T). A purine-purine pair is too wide and a pyrimidine-pyrimidine pair too narrow to account for the 2-nm diameter of the double helix. The solution is to always pair a purine with a pyrimidine:

Purine + purine: too wide

Pyrimidine + pyrimidine: too narrow

Purine + pyrimidine: width consistent with X-ray data

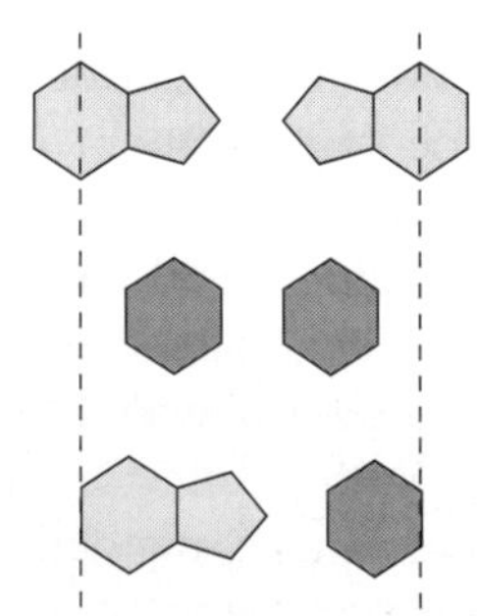

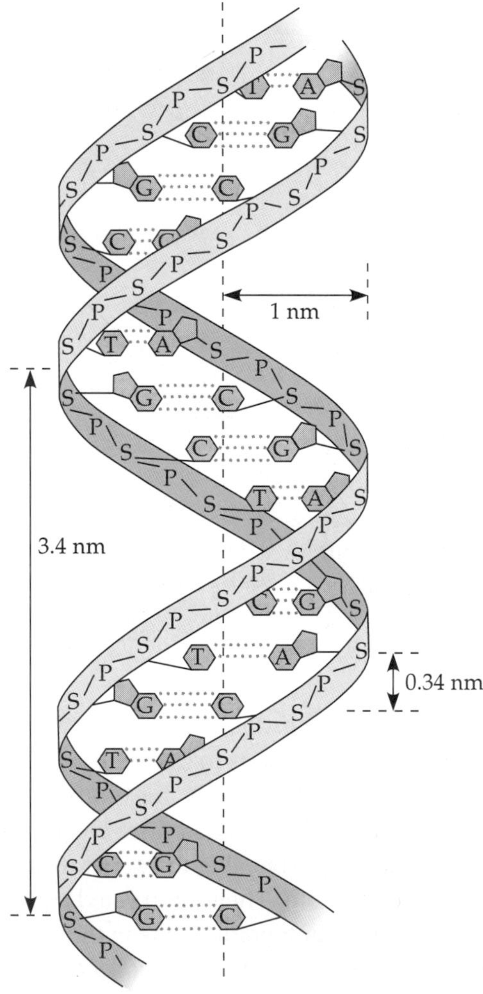

FIGURE 15.5
The double helix. The "ribbons" in this diagram represent the sugar-phosphate backbones of the two DNA strands. The two strands are held together by hydrogen bonds (dotted lines) between the nitrogenous bases, which are paired in the interior of the double helix. The base pairs are 0.34 nm apart; there are ten pairs per turn of the helix.

Watson and Crick reasoned that there must be additional specificity of pairing dictated by the structure of the bases. Each base has chemical side groups that can form hydrogen bonds with its appropriate partner: Adenine can form two hydrogen bonds with thymine and only thymine; guanine forms three hydrogen bonds with cytosine and only cytosine. In shorthand, A pairs with T, and G pairs with C:

ADENINE (A) THYMINE (T)

GUANINE (G) CYTOSINE (C)

The Watson-Crick model explained Chargaff's rules. Wherever one strand of a DNA molecule has an A, the partner strand has a T. And a G in one strand is always paired with a C in the complementary strand. Therefore, in the DNA of any organism, the amount of adenine equals the amount of thymine, and the amount of guanine equals the amount of cytosine.

In April 1953, Watson and Crick surprised the scientific world with a succinct, one-page paper in the British journal *Nature*. The paper reported a new molecular model for DNA: the double helix, which has since become the symbol of molecular biology. The beauty of the model was that its structure suggested the basic mechanism of DNA replication.

During DNA replication, base pairing enables existing DNA strands to serve as templates for new complementary strands

The relationship between structure and function, one of the themes that drives biology, is manifest in the double helix. The molecular architecture of DNA provides the mechanism for the replication of genes (FIGURE 15.6). The idea that there is specific pairing of nitrogenous bases in

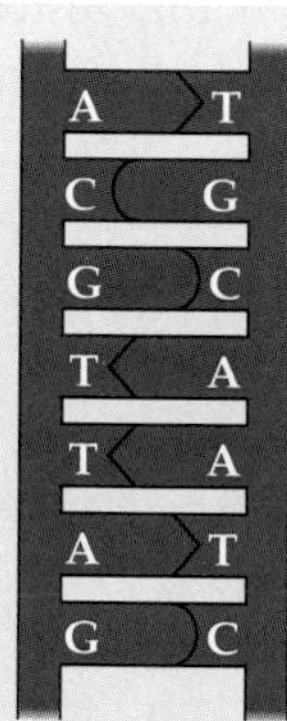

(a) Before replication, the parent molecule has two complementary strands of DNA. Each base is paired with its specific partner, A with T and G with C.

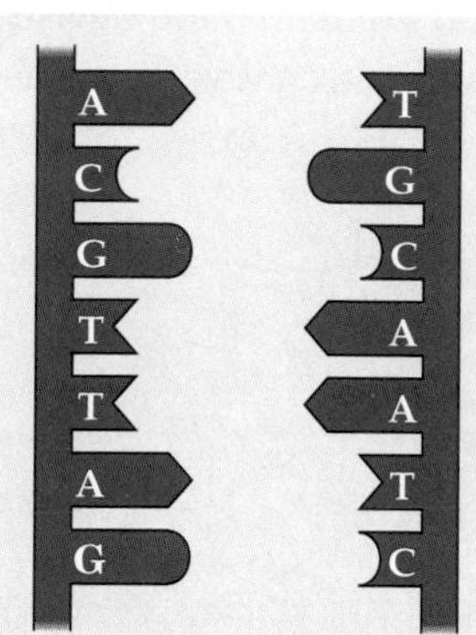

(b) The first step in replication is separation of the two DNA strands.

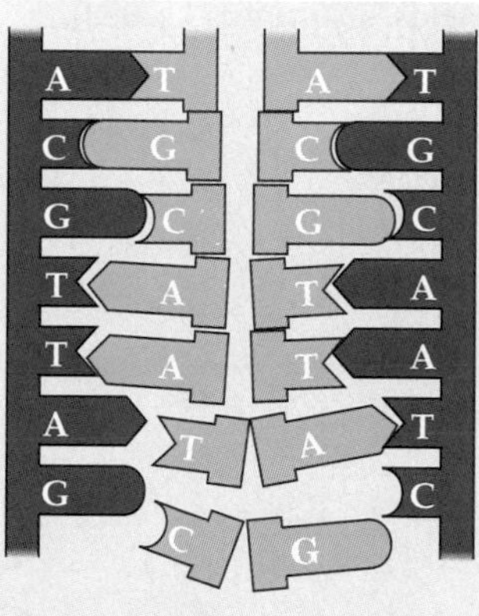

(c) Each "old" strand now serves as a template that directs synthesis of "new" complementary strands. Nucleotides plug into specific sites along the template surface according to the base-pairing rules.

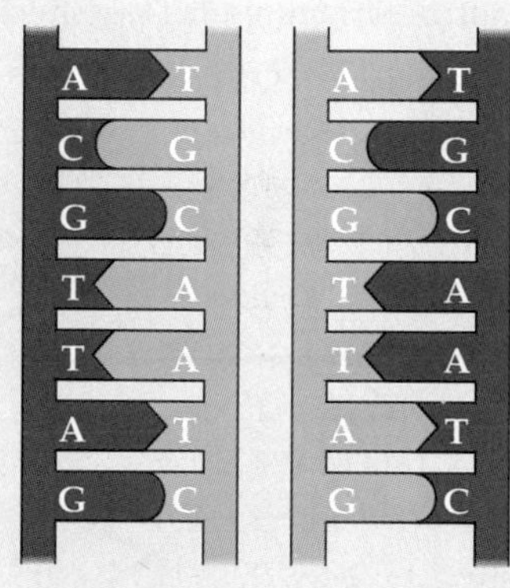

(d) The nucleotides are connected to form the sugar–phosphate backbones of the new strands. Each DNA molecule now consists of one "old" strand and one "new" strand, resulting in two copies identical to the one DNA molecule with which we started.

FIGURE 15.6
DNA replication: the basic concept. In this simplification, a short segment of DNA has been untwisted to convert the double helix to a two-dimensional version of the molecule that resembles a ladder. The rails of the ladder are the sugar-phosphate backbones of the two DNA strands. The rungs are the pairs of nitrogenous bases. Simple shapes are used to symbolize the four kinds of bases. Dark blue represents DNA strands originally present in the parent cell. Newly synthesized DNA is represented by light blue.

DNA was the flash of inspiration that led Watson and Crick to the double helix. At the same time, they understood the functional significance of the base-pairing rules. They ended their classic paper with this wry statement: "It has not escaped our notice that the specific pairing we have postulated immediately suggests a possible copying mechanism for the genetic material."* In this section, you will learn about this basic mechanism of DNA replication. Some important details of the process will be presented in the next section.

Although the base-pairing rules dictate the combinations of nitrogenous bases that form the "rungs" of the double helix, they do not restrict the sequence of nucleotides *along* each DNA strand. Thus, the linear sequence of the four bases can be varied in countless ways, and each gene has a unique order, or base sequence. FIGURE 15.6 untwists a short section of double helix so that it is easier to follow replication of the DNA. Our objective here is to visualize how a cell copies this genetic information. Cover one of the two DNA strands of FIGURE 15.6a with a piece of paper, and you can still determine its linear sequence of bases by referring to the unmasked strand and applying the base-pairing rules. The two strands are complementary; each stores the information necessary to reconstruct the other. When a cell copies a DNA molecule, the two strands separate, and each strand then serves as a template (mold) for ordering nucleotides into a new complementary strand. One at a time, nucleotides line up along the template strand according to the base-pairing rules. Enzymes link the nucleotides to form the new strands. Where there was one double-stranded DNA molecule at the beginning of the process, there are now two, each an exact replica of the "parent" molecule. The copying mechanism is analogous to using a photographic negative to make a positive image, which can in turn be used to make another negative, and so on.

This model of gene replication remained untested for several years following publication of the DNA structure. The requisite experiments were simple in concept but difficult to perform. Watson and Crick's model predicts that when a double helix reproduces, each of the two daughter molecules will have one old strand derived from the parent molecule and one newly made strand. This **semiconservative model** can be distinguished from a conservative model of replication, in which the parent molecule remains intact (is conserved) and the new molecule is formed entirely from scratch. In a third model, called the dispersive model, all four strands of DNA, after the double helix is replicated, have a mixture of "old" and "new" DNA (FIGURE 15.7). In the late 1950s, Matthew Meselson and Franklin Stahl devised experiments that tested these three alternative hypotheses for DNA replication. Their

*Watson, J. D., and F. H. C. Crick. "Molecular Structure of Nucleic Acids: A Structure for Deoxynucleic Acids." *Nature* 171 (1953), p. 738.

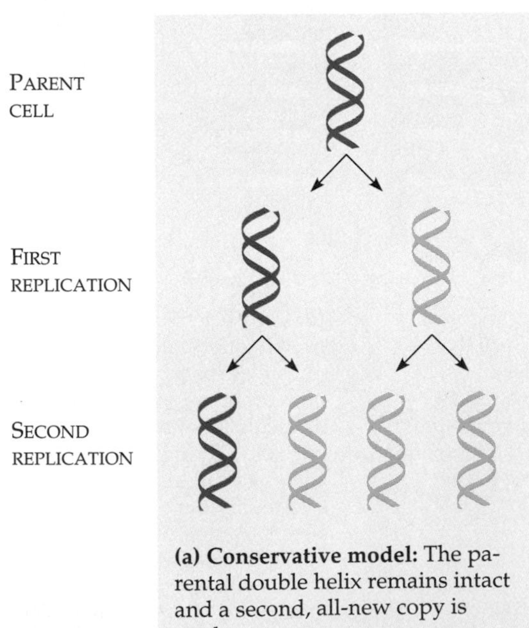

(a) Conservative model: The parental double helix remains intact and a second, all-new copy is made.

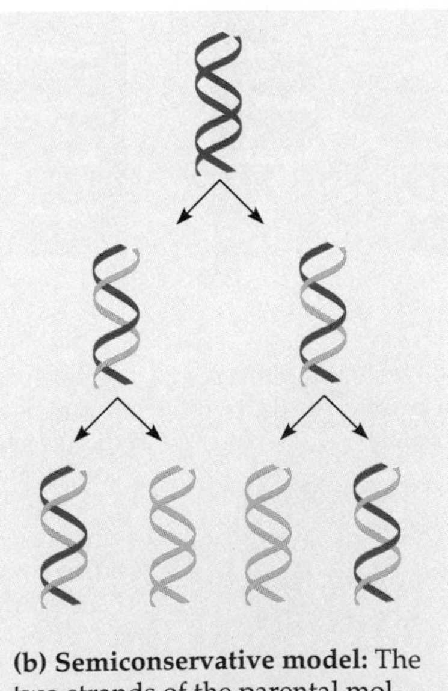

(b) Semiconservative model: The two strands of the parental molecule separate, and each functions as a template for synthesis of a new complementary strand.

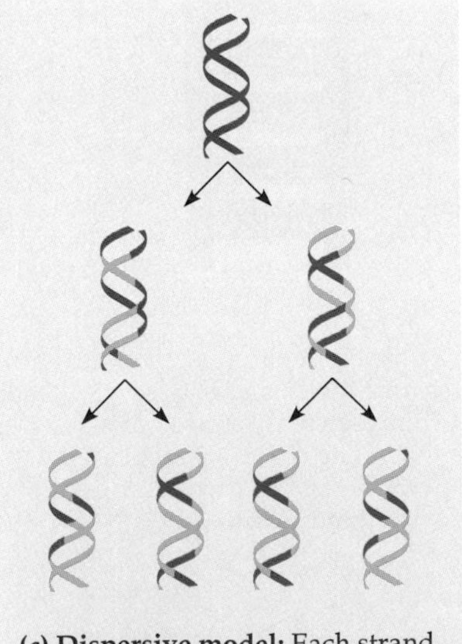

(c) Dispersive model: Each strand of *both* daughter molecules would contain a mixture of old and newly synthesized parts.

FIGURE 15.7
Three models of DNA replication. The short segments of double helix illustrated here symbolize the genetic material within a cell. Beginning with a parent cell, we follow the DNA for two generations of cells—two replications of the genetic material.

experiments supported the semiconservative model, as predicted by the Watson-Crick model (FIGURE 15.8).

In a second paper that followed their announcement of the double helix, Watson and Crick summarized their model of DNA replication:

> Now our model for deoxyribonucleic acid is, in effect, a pair of templates, each of which is complementary to the other. We imagine that prior to duplication the hydrogen bonds are broken, and the two chains unwind and separate. Each chain then acts as a template for the formation onto itself of a new companion chain, so that eventually we shall have two pairs of chains, where we only had one before. Moreover, the sequence of the pairs of bases will have been duplicated exactly.*

The copying of the genetic material is, in basic concept, elegantly simple. However, the actual process involves complex biochemical gymnastics, as we will now see.

*Crick, F. H. C., and J. D. Watson. "The Complementary Structure of Deoxyribonucleic Acid." *Proc. Roy. Soc.* (A) 223 (1954), p. 80.

A team of enzymes and other proteins functions in DNA replication

The bacterium *E. coli* has a single chromosome of about 5 million base pairs. In a favorable environment, an *E. coli* cell can copy all this DNA and divide to form two genetically identical daughter cells in less than an hour. Each one of *your* cells has 46 DNA molecules, one giant molecule per chromosome (see Chapter 11). In all, that represents about 6 billion base pairs, or over a thousand times more DNA than is found in a bacterial cell. If we were to print the one-letter symbols for these bases (A, G, C, and T) the size of the letters you are reading, the 6 billion bases of a single human cell would fill about 900 books as thick as this text. Yet it takes a cell just a few hours to copy all this DNA. This replication of an enormous amount of genetic information is achieved with very few errors—only about one per billion nucleotides. The copying of DNA is remarkable in its speed and accuracy.

More than a dozen enzymes and other proteins participate in DNA replication. Much more is known about how

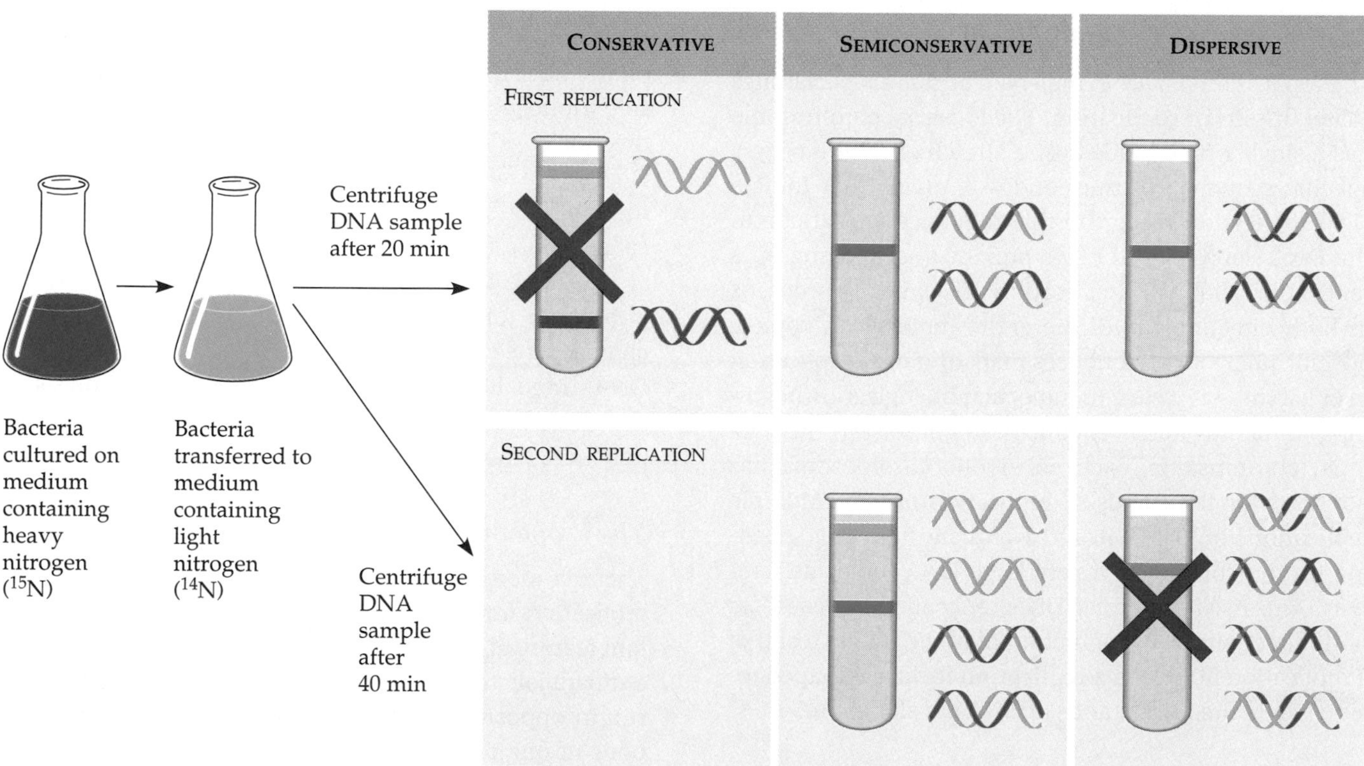

FIGURE 15.8
The Meselson-Stahl experiment tested three hypotheses of DNA replication. Meselson and Stahl cultured *E. coli* for several generations on a medium containing a heavy isotope of nitrogen, ^{15}N. The bacteria incorporated the heavy nitrogen into their nucleotides and then into their DNA. The scientists then transferred the bacteria to a medium containing ^{14}N, the lighter, more common isotope of nitrogen. Thus, any new DNA that the bacteria synthesized would be lighter than the "old" DNA made in the ^{15}N medium. Meselson and Stahl could distinguish DNA of different densities by centrifuging DNA extracted from the bacteria. The centrifuge tubes in this drawing represent the results predicted by the three hypotheses in FIGURE 15.7. The first replication in the ^{14}N medium produced a band of hybrid (^{15}N-^{14}N) DNA. This result eliminated the conservative hypothesis. A second replication produced both light and hybrid DNA, a result that eliminated the dispersive hypotheses and supported the semiconservative hypothesis.

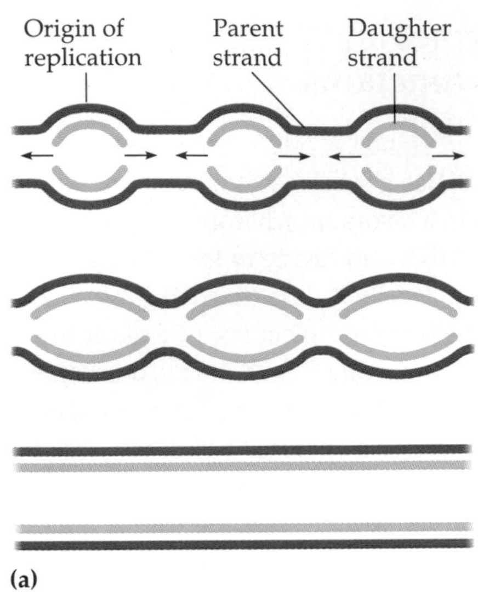

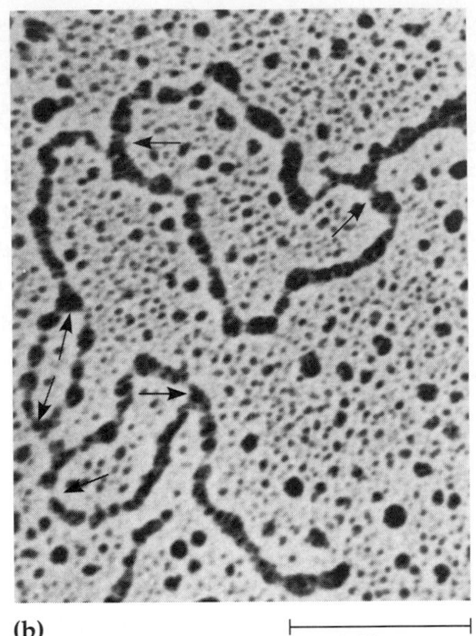

FIGURE 15.9
Origins of replication. (**a**) DNA replication begins at specific sites where the two parental strands of DNA separate to form replication "bubbles" (top). In eukaryotes, there are hundreds or thousands of origin sites along the giant DNA molecule of each chromosome. A replication bubble expands laterally, as DNA replication proceeds in both directions. Eventually, the replication bubbles fuse (center), and synthesis of the "daughter" strands of DNA is complete (bottom). (**b**) In this micrograph, three replication bubbles are visible along the DNA of cultured Chinese hamster cells. The arrows indicate the directions of DNA replication at the two ends of each bubble (TEM).

this "replication machine" works in bacteria than in eukaryotes. However, most of the basic steps of replication seem to be similar for prokaryotes and eukaryotes. In this section, we take a closer look at some of these steps.

Getting Started: Origins of Replication

The replication of a DNA molecule begins at special sites called **origins of replication.** The bacterial chromosome has a single origin marked by a stretch of DNA having a specific sequence of nucleotides. Proteins that initiate DNA replication recognize this sequence and attach to the DNA, separating the two strands and opening up a replication "bubble." Replication of DNA then proceeds in both directions, until the entire molecule is copied. Specific nucleotide sequences mark origins of replication in eukaryotes as well, and molecular biologists are beginning to identify such sequences. In contrast to the bacterial chromosome, each eukaryotic chromosome has hundreds or thousands of replication origins. Multiple replication bubbles form and eventually fuse, thus speeding up the copying of a very large DNA molecule (FIGURE 15.9). As in bacteria, DNA replication proceeds in both directions from each point of origin. At each end of a replication bubble is a **replication fork,** a Y-shaped region where the new strands of DNA are elongating.

Elongating a New DNA Strand

Elongation of new DNA at a replication fork is catalyzed by enzymes called **DNA polymerases.** As nucleotides align with complementary bases along the "old" template strand of DNA, they are added by polymerase, one by one, to the growing end of the new DNA strand. The rate of elongation is about 500 nucleotides per second in bacteria and 50 per second in human cells.

What is the source of energy that drives the polymerization of nucleotides to form new DNA strands? The substrates for DNA polymerase are not actually nucleotides but related compounds called nucleoside triphosphates (FIGURE 15.10). Instead of the one phosphate group characteristic of nucleotides, nucleoside triphosphates have three, just like ATP. (In fact, ATP *is* a nucleoside triphosphate. The only difference between ATP and the nucleoside triphosphate that supplies adenine to DNA is the sugar component, which is deoxyribose for the building block of DNA, but a closely related sugar called ribose for ATP.) Like ATP, the monomers for DNA synthesis are chemically reactive, partly because their triphosphate tails have an unstable cluster of negative charge. As each monomer joins the growing end of a DNA strand, it loses two phosphate groups. Hydrolysis of the phosphate is the exergonic reaction that drives the polymerization of nucleotides to form DNA.

The Problem of Antiparallel DNA Strands

There is more to the scenario of DNA synthesis at the replication fork. Until now, we have ignored an important feature of the double helix: The two DNA strands are antiparallel; that is, their sugar-phosphate backbones run in opposite directions. In FIGURE 15.11, the five carbons of one deoxyribose sugar of each DNA strand are numbered, from 1′ to 5′. (The prime sign is used to distinguish the numbered carbons of the sugar from the numbers representing the atoms of the nitrogenous bases.) Notice in FIGURE 15.11 that a nucleotide's phosphate group is attached to the 5′ carbon of deoxyribose. Notice also that the phosphate group of one nucleotide is joined to the 3′ carbon of the adjacent nucleotide. The result is a DNA strand of distinct polarity. At one end, de-

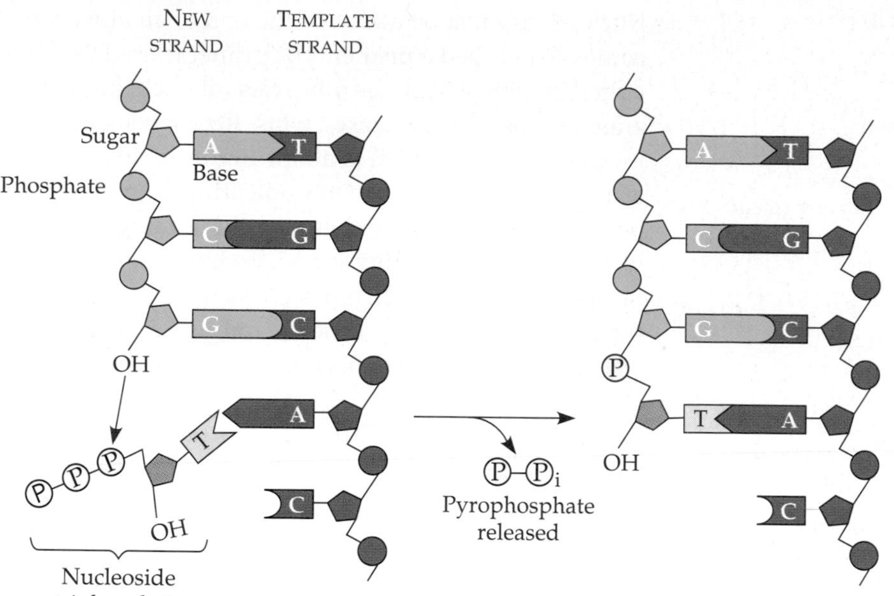

FIGURE 15.10
Incorporation of a nucleotide into a DNA strand. When a nucleoside triphosphate links to the sugar-phosphate backbone of a growing DNA strand, it loses two of its phosphates as a pyrophosphate molecule. The enzyme DNA polymerase catalyzes the reaction, and hydrolysis of the bonds between the phosphate groups provides the energy.

noted the 3′ end, a hydroxyl group is attached to the 3′ carbon of the terminal deoxyribose. At the opposite end, the 5′ end, the sugar-phosphate backbone terminates with the phosphate group attached to the 5′ carbon of the last nucleotide. In the double helix, the two sugar-phosphate backbones are essentially upside-down (antiparallel) relative to each other.

How does the antiparallel structure of the double helix affect replication? The enzyme DNA polymerase can add nucleotides only to the free 3′ end of a growing DNA strand, never to the 5′ end. Thus, a new DNA strand can elongate only in the 5′ ⟶ 3′ direction. With this in mind, let's reexamine a replication fork (FIGURE 15.12). Along one template strand, DNA polymerase can synthesize a continuous complementary strand by elongating the new DNA in the mandatory 5′ ⟶ 3′ direction. The polymerase simply nestles in the replication fork and moves along the template strand as the fork progresses. The DNA strand made by this mechanism is called the **leading strand.**

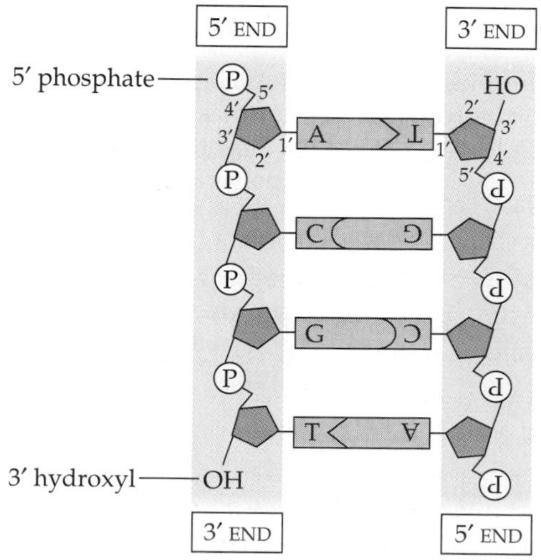

FIGURE 15.11
The two strands of DNA are antiparallel. The 5′ ⟶ 3′ direction of one strand runs counter to the other strand. The carbon atoms of the deoxyribose sugars at the top are numbered for orientation.

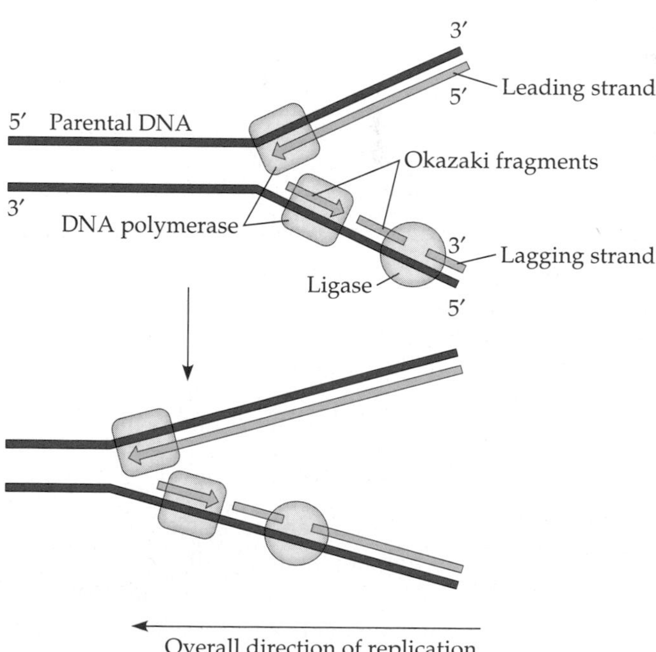

FIGURE 15.12
Synthesis of leading and lagging strands during DNA replication. DNA polymerase elongates strands only in the 5′ ⟶ 3′ direction. One new strand, called the leading strand, can therefore elongate continually in the 5′ ⟶ 3′ direction as the replication fork progresses. But the other new strand, the lagging strand, must grow in an overall 3′ ⟶ 5′ direction by the addition of short segments, Okazaki fragments, that individually grow 5′ ⟶ 3′. An enzyme called ligase connects the fragments.

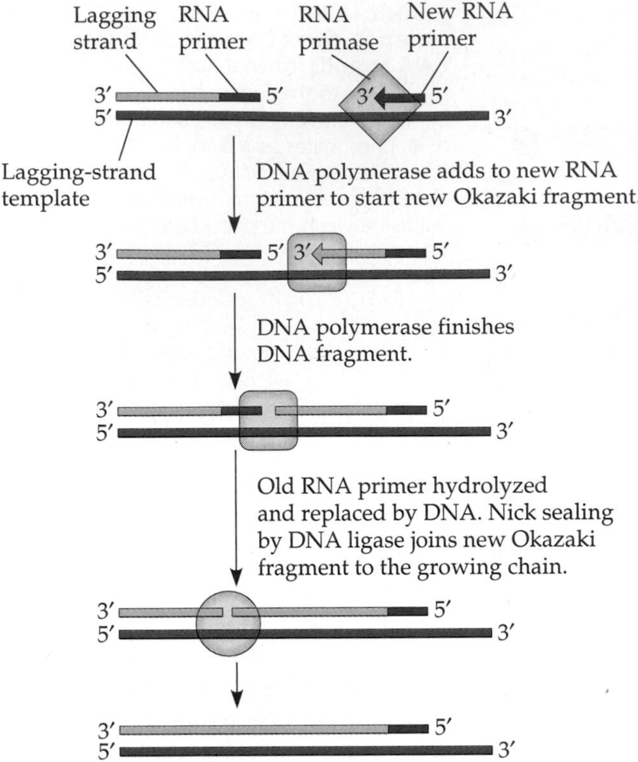
Lagging strand
RNA primer
RNA primase
New RNA primer
3′
5′
5′
3′
5′
3′
Lagging-strand template
DNA polymerase adds to new RNA primer to start new Okazaki fragment.
DNA polymerase finishes DNA fragment.
Old RNA primer hydrolyzed and replaced by DNA. Nick sealing by DNA ligase joins new Okazaki fragment to the growing chain.

FIGURE 15.13
Priming DNA synthesis. DNA polymerase cannot initiate a polynucleotide strand; it can only add to the 3′ end of an already-started strand. The primer is a short segment of RNA synthesized by the enzyme primase.

To elongate the other new strand of DNA, polymerase must work along the template *away from* the replication fork. The DNA synthesized in this direction is called the **lagging strand.** The process is analogous to a sewing method called back-stitching. As a replication bubble opens, polymerase can work its way away from a replication fork and synthesize a short segment of DNA. As the bubble widens, another short segment of the lagging strand can be made by a polymerase working away from the fork. In contrast to the leading strand, which can be elongated continuously, the lagging strand is first synthesized as a series of segments. These pieces are called Okazaki fragments, after the Japanese scientist who discovered them. The fragments are about 100 to 200 nucleotides long in eukaryotes. Another enzyme, **DNA ligase,** joins the Okazaki fragments into a single DNA strand.

Priming DNA Synthesis

There is another important restriction for DNA polymerase. It can add a nucleotide only to a polynucleotide that is already correctly paired with the complementary strand. (This requirement is evident in FIGURE 15.10.) This means that DNA polymerase cannot actually *initiate* synthesis of a DNA strand by joining the first nucleotides. Nucleotides must be added to the end of an already existing chain called a **primer.** The primer is not DNA, but a short stretch of RNA, the other class of nucleic acid. Still another enzyme, **primase,** joins RNA nucleotides to make the primer, which is about 10 nucleotides long in eukaryotes (FIGURE 15.13). Only one primer is required for polymerase to begin synthesizing the leading strand of new DNA. For the lagging strand, each fragment must be primed. An enzyme then replaces the RNA nucleotides of the primers with DNA versions, and ligase joins all the DNA fragments into a strand.

Other Proteins Assisting DNA Replication

You have learned about three of the proteins that function in DNA synthesis: DNA polymerase, ligase, and primase. Many other proteins also participate; two of them are helicase and single-strand binding proteins. **Helicase** is an enzyme that works at the crotch of the replication fork, untwisting the double helix and separating the two "old" strands. **Single-strand binding proteins** then attach in chains along the unpaired DNA strands, holding these templates straight until new complementary strands can be synthesized.

The diagram below summarizes the functions of the proteins that cooperate in DNA replication. FIGURE 15.14, on the facing page, is a visual summary of DNA replication.

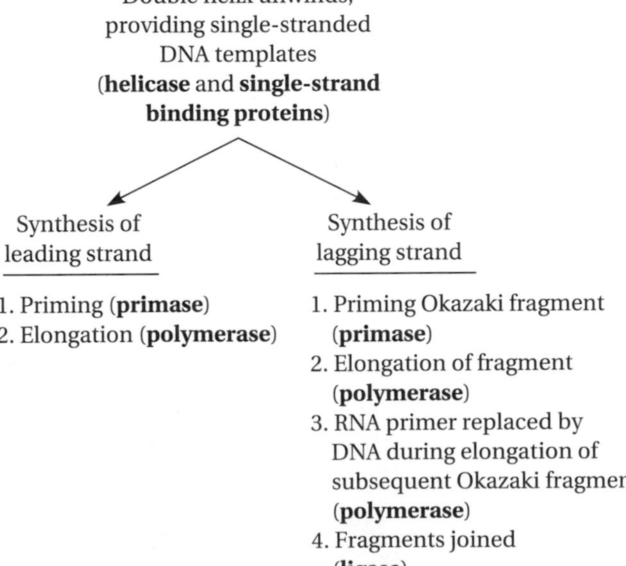
Double helix unwinds, providing single-stranded DNA templates (**helicase** and **single-strand binding proteins**)
Synthesis of leading strand
1. Priming (**primase**)
2. Elongation (**polymerase**)
Synthesis of lagging strand
1. Priming Okazaki fragment (**primase**)
2. Elongation of fragment (**polymerase**)
3. RNA primer replaced by DNA during elongation of subsequent Okazaki fragment (**polymerase**)
4. Fragments joined (**ligase**)

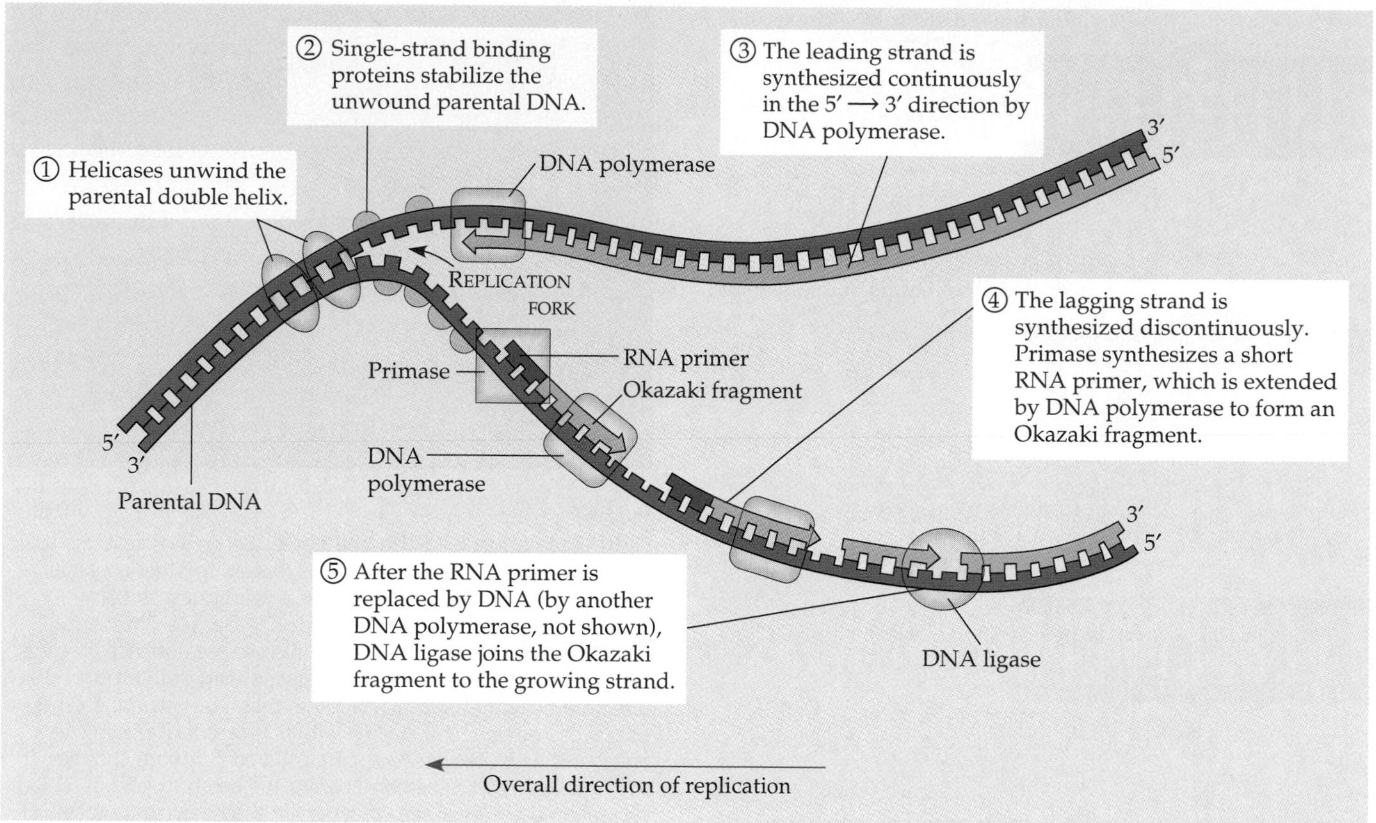
① Helicases unwind the parental double helix.
② Single-strand binding proteins stabilize the unwound parental DNA.
③ The leading strand is synthesized continuously in the 5′ ⟶ 3′ direction by DNA polymerase.
DNA polymerase
3′
5′
REPLICATION FORK
④ The lagging strand is synthesized discontinuously. Primase synthesizes a short RNA primer, which is extended by DNA polymerase to form an Okazaki fragment.
Primase
RNA primer
Okazaki fragment
5′
3′
Parental DNA
DNA polymerase
3′
5′
⑤ After the RNA primer is replaced by DNA (by another DNA polymerase, not shown), DNA ligase joins the Okazaki fragment to the growing strand.
DNA ligase
Overall direction of replication

FIGURE 15.14
A summary of DNA replication.

Enzymes proofread DNA during its replication and repair damage in existing DNA

We cannot attribute the accuracy of DNA replication solely to the specificity of base pairing. Although errors in the completed DNA molecule amount to only one in 1 billion nucleotides, initial pairing errors between incoming nucleotides and those in the template strand are 100,000 times more common—an error rate of one in 10,000 base pairs. One DNA repair mechanism, called **mismatch repair,** fixes mistakes when DNA is copied. In bacteria, DNA polymerase itself functions in mismatch repair. The polymerase proofreads each nucleotide against its template as soon as it is added to the strand. Upon finding an incorrectly paired nucleotide, the polymerase backs up, removes the incorrect nucleotide, and replaces it before continuing with synthesis. (This action resembles a command in a word-processing program, in which you backspace to an error, then delete and replace the error before continuing.) In eukaryotes, several proteins in addition to polymerase are required for mismatch repair. In 1994, researchers discovered that a hereditary defect in one of these proteins is associated with one form of colon cancer. Apparently, errors in the DNA accumulate without adequate proofreading.

In addition to proofreading, precise maintenance of the genetic information encoded in DNA also requires a way of repairing accidental changes that occur in existing DNA. DNA molecules are constantly subjected to many potentially damaging physical and chemical agents. Reactive chemicals, radioactive emissions, X-rays, and ultraviolet light can change nucleotides in ways that can affect encoded genetic information, usually adversely. Fortunately, these changes, or mutations, are usually corrected. Each cell continuously monitors and repairs its genetic material. Biochemists have identified more than 50 different types of DNA repair enzymes. Most mechanisms of DNA repair take advantage of the base-paired structure of DNA. For example, in **excision repair,** a segment of the strand containing the damage is cut out by one repair enzyme, and the resulting gap is filled in with nucleotides properly paired with the nucleotides in the undamaged strand. The enzymes involved in filling the gap are a DNA polymerase and DNA ligase (FIGURE 15.15).

One function of the DNA repair enzymes in our skin cells is to repair genetic damage caused by the ultraviolet rays of sunlight. The importance of this function to healthy people is underscored by the skin disorder xeroderma pigmentosum, caused by an inherited defect in an excision repair enzyme. Individuals with this disorder are hypersensitive to sunlight; mutations in skin cells caused by ultraviolet light are left uncorrected and cause skin cancer.

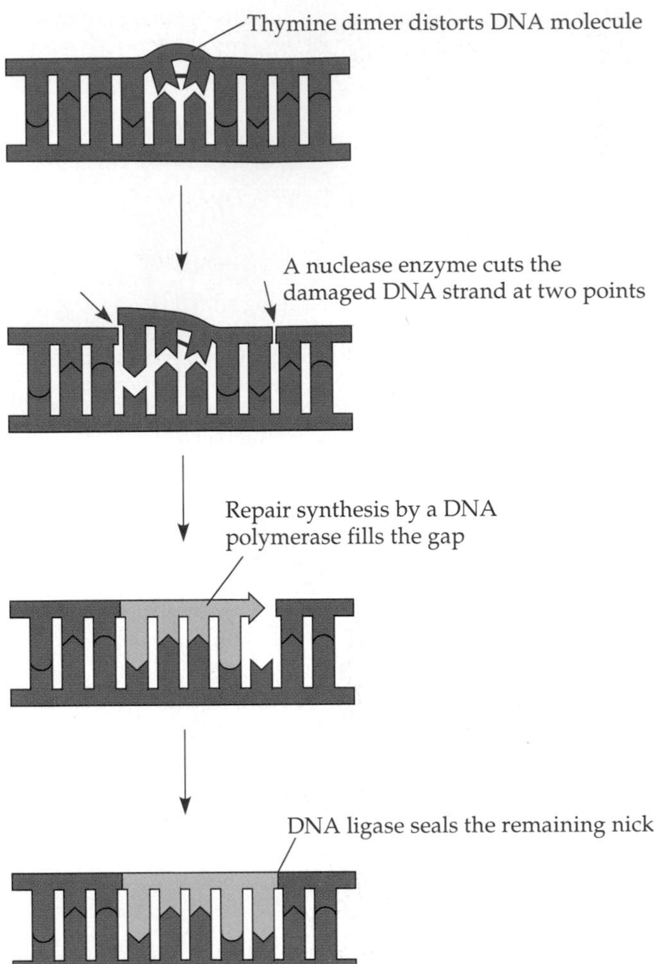

Thymine dimer distorts DNA molecule
A nuclease enzyme cuts the damaged DNA strand at two points
Repair synthesis by a DNA polymerase fills the gap
DNA ligase seals the remaining nick

FIGURE 15.15
Excision repair of DNA. A team of enzymes detects and repairs damaged DNA. One type of damage, shown here, is the covalent linking of thymine bases that are adjacent on a DNA strand. Such thymine dimers, induced by ultraviolet radiation, cause the DNA to buckle and will lead to errors during DNA replication. Repair enzymes can excise the damaged region from the DNA and replace it with a normal DNA segment.

10 µm

FIGURE 15.16
DNA replication and the cell cycle. Before we leave the topic of DNA replication, let's connect this process to other concepts you have learned about cells. Before a cell divides, its DNA is replicated during the S phase of the cell cycle (see Chapter 11). Duplication of chromosomes is the microscopic evidence that the genetic material has been copied. Mitosis then distributes duplicated chromosomes to daughter cells. The cell in the center of this micrograph (LM, stained with fluorescent dye) is in late anaphase, with the spindle apparatus (green) moving chromosomes (blue) toward the two poles of the parent cell. Thus, DNA replication provides copies of genes, and the mechanics of cell division transmit these genes from one cellular generation to the next—and from one organismal generation to the next when the genes' cellular vehicles are gametes.

In this chapter, we have concentrated on the structure of DNA and how this genetic material is copied. Mitosis distributes these copies to daughter cells every time one of our cells divides (FIGURE 15.16). And DNA replication provides the copies of genes that parents pass to offspring via gametes. However, it is not enough that genes be copied and transmitted; they must also be expressed. How can genes manifest themselves in such phenotypic characters as eye color, for instance? In the next chapter, we will examine the molecular basis of gene expression—how the cell translates genetic information encoded in DNA.

REVIEW OF KEY CONCEPTS (with page numbers and key figures)

- The search for the genetic material led to DNA: *science as a process* (pp. 281–284, FIGURE 15.3)
  - The ability of DNA from a pathogenic strain of bacteria to transform harmless bacteria into pathogens provided the first evidence that the genetic material was DNA.
  - The ability of phages to take over bacterial cells by injecting DNA was further evidence that DNA is the genetic material.
  - The correlation between a cell's DNA content and replication of its chromosomes before cell division added more evidence for DNA as the genetic material.
- Watson and Crick discovered the double helix by building models to conform to X-ray data: *science as a process* (pp. 284–287, FIGURE 15.5)
  - Basing their model on data from Franklin's X-ray diffraction photo of DNA, Watson and Crick discovered that DNA is a double helix. Two antiparallel sugar-phosphate chains wind around the outside of the molecule; the nitrogenous bases project into the interior, where they hydrogen-bond in specific pairs, A with T and G with C.
- During DNA replication, base pairing enables existing DNA strands to serve as templates for new complemen-

tary strands (pp. 287–289, FIGURE 15.6)

- Meselson and Stahl demonstrated that DNA replication is semiconservative, confirming Watson and Crick's hypothesis that the parent molecule unwinds and each strand then serves as a template for the synthesis of a new half-molecule according to base-pairing rules.

- A team of enzymes and other proteins functions in DNA replication (pp. 289–292, FIGURE 15.14)
  - Replication begins at special sites called origins of replication. Y-shaped replication forks form at opposite ends of a replication bubble, where the two DNA strands separate.
  - DNA polymerases catalyze the synthesis of the new DNA strand, working in the $5' \longrightarrow 3'$ direction.
  - Simultaneous $5' \longrightarrow 3'$ synthesis of antiparallel strands at a replication fork yields a continuous leading strand and short, discontinuous segments of lagging strand. The fragments are later joined together with the help of DNA ligase.
  - DNA synthesis must start on the end of a primer, a short segment of RNA synthesized by the enzyme primase.

- Enzymes proofread DNA during its replication and repair damage to existing DNA (pp. 293–294, FIGURE 15.15).
  - In mismatch repair, proteins proofread replicating DNA and correct errors in base pairing.
  - In excision repair, repair enzymes fix DNA damaged by physical and chemical agents, including ultraviolet light.

SELF-QUIZ

1. In his work with pneumonia-causing bacteria and mice, Griffith found that
   a. the protein coat from smooth (S) cells was able to transform rough (R) cells
   b. heat-killed S cells were able to cause pneumonia only when they were transformed by the DNA of R cells
   c. some chemical from S cells was transferred to R cells, transforming them into S cells
   d. the polysaccharide coat of R cells caused pneumonia
   e. bacteriophages injected DNA from S cells into R cells

2. *E. coli* cells grown on ^{15}N medium are transferred to ^{14}N medium and allowed to grow for two generations (two cell replications). DNA extracted from these cells is centrifuged. What density distribution of DNA would you expect in this experiment? Explain your answer.
   a. one high-density and one low-density band
   b. one intermediate-density band
   c. one high-density and one intermediate-density band
   d. one low-density and one intermediate-density band
   e. one low-density band

3. A biochemist isolated and purified molecules needed for DNA replication. When she added some DNA, replication occurred, but the DNA molecules formed were defective. Each consisted of a normal DNA strand paired with numerous segments of DNA a few hundred nucleotides long. What had she probably left out of the mixture? Explain your answer.
   a. DNA polymerase
   b. ligase
   c. nucleotides
   d. Okazaki fragments
   e. primers

4. What is the basis for the difference in the synthesis of the leading and lagging strands of DNA molecules?
   a. The origins of replication occur only at the 5′ end of the molecule.
   b. Helicases and single-strand binding proteins work at the 5′ end.
   c. DNA polymerase can join new nucleotides only to the 3′ end of the growing strand.
   d. DNA ligase works only in the $3' \longrightarrow 5'$ direction.
   e. Polymerase can only work on one strand at a time.

5. In analyzing the number of different bases in a DNA sample, which result would be consistent with the base-pairing rules? Explain your answer.
   a. $A = G$
   b. $A + G = C + T$
   c. $A + T = G + T$
   d. $A = C$
   e. $G = T$

6. The primer required to initiate synthesis of a new DNA strand consists of
   a. RNA
   b. DNA
   c. an Okazaki fragment
   d. a structural protein
   e. a thymine dimer

7. A particular gene measures about 1 μm in length along a double-stranded DNA molecule. What is the approximate number of base pairs in this gene (see FIGURE 15.5)?
   a. 3
   b. 10
   c. 1000
   d. 3000
   e. 30,000

8. The elongation of the *leading* strand during DNA synthesis
   a. progresses away from the replication fork
   b. occurs in the $3' \longrightarrow 5'$ direction
   c. produces Okazaki fragments
   d. depends on the action of DNA polymerase
   e. does not require a template strand

9. Why does cytosine pair with guanine and not adenine?
   a. A C-A pair would be too wide to fit in the double helix.
   b. C and A are both polar.
   c. A C-A pair would not reach across the double helix.
   d. The functional groups that form hydrogen bonds are not complementary between C and A.
   e. C and A are both purines.

10. Of the following, the most reasonable inference from the observation that defects in DNA repair enzymes contribute to some forms of cancer is that
    a. cancer is generally inherited
    b. uncorrected changes in DNA can cause cancer
    c. cancer cannot occur when DNA repair enzymes work properly
    d. mutations generally lead to cancer
    e. cancer is caused by environmental factors that damage DNA repair enzymes

CHALLENGE QUESTIONS

1. Mature nerve cells no longer divide, so they do not replicate their DNA. A cell biologist found that there was X amount of DNA in a human nerve cell. The biologist then measured the amount of DNA in four other types of human cells; the

results are recorded in the chart below. Complete the chart by filling in the type of cell from these choices: (a) sperm cell; (b) bone marrow cell just beginning interphase of the cell cycle; (c) skin cell in the S phase of the cell cycle; (d) intestinal cell beginning mitosis.

Cell	Amount of DNA	Type of Cell
A	$2X$	
B	$1.6X$	
C	$0.5X$	
D	X	

2. Cells proofread and repair errors in the genetic information encoded in DNA. But some lasting mistakes, or mutations, still happen. The most error-prone stage in DNA replication occurs after parent DNA strands have separated but before new complementary strands are in place. Why might permanent errors be more likely then?
3. Write a short essay explaining the evidence that DNA is the genetic material.

SCIENCE, TECHNOLOGY, AND SOCIETY

1. DNA molecules with any desired base sequence can be synthesized and replicated in large quantities in the laboratory. It is also possible to determine the base sequence of a DNA molecule. For these reasons, it has been suggested that oil shipments could be tagged with DNA of known sequences. If an oil slick were found, the DNA could be sequenced and the offending ship tracked down. Suggest some other potential applications for "DNA labeling."
2. Cooperation and competition are both common in science. What roles did these two social behaviors play in Watson and Crick's discovery of the double helix? How might competition between scientists accelerate progress in a scientific field? How might it slow progress?

FURTHER READING

Becker, W. M., J. R. Reece, and M. F. Poenie. *The World of the Cell,* 3rd ed. Menlo Park, CA: Benjamin/Cummings, 1996. Chapter 15 integrates DNA replication, the cell cycle, and mitosis.

Culotta, E., and D. E. Koshland, Jr. "DNA Repair Works Its Way to the Top." *Science,* December 23, 1994. How enzymes detect and fix damaged DNA.

Hall, S. S. "Old School Ties: Watson, Crick, and 40 years of DNA." *Science,* March 11, 1993. Reunion of key players in the origin of molecular biology.

Judson, H. F. *The Eighth Day of Creation: Makers of the Revolution in Biology.* New York: Simon & Schuster, 1979. An engaging history of molecular biology.

Rennie, J. "DNA's New Twists." *Scientific American,* March 1993. How new discoveries are changing our view of molecular biology.

Watson, J. D. *The Double Helix.* New York: Atheneum, 1968. The brash, controversial best-seller by the codiscoverer of the double helix.

Watson, J. D., and F. H. C. Crick. "Molecular Structure of Nucleic Acids: A Structure for Deoxynucleic Acids." *Nature* 171 (1953): 737–738. Watson and Crick's classic paper.

CHAPTER 16

FROM GENE TO PROTEIN

KEY CONCEPTS

- The study of metabolic defects provided evidence that genes specify proteins: *science as a process*
- Transcription and translation are the two main steps from gene to protein: *an overview*
- In the genetic code, a particular triplet of nucleotides specifies a certain amino acid: *a closer look*
- Transcription is the DNA-directed synthesis of RNA: *a closer look*
- Translation is the RNA-directed synthesis of a polypeptide: *a closer look*
- Some polypeptides have signal sequences that target them to specific destinations in the cell
- Comparing protein synthesis in prokaryotes and eukaryotes: *a review*
- Eukaryotic cells modify RNA after transcription
- A point mutation can affect the function of a protein
- What is a gene?

Written in the DNA of a human zygote's nucleus are the basic instructions for how to build a person—not a generic human, but one with a unique combination of traits. Each individual's phenotype is the cumulative product of a one-of-a-kind genetic makeup combined with environmental influences. The information content of DNA, the genetic material, is in the form of specific sequences of nucleotides along the DNA strands. But how is this information related to an organism's inherited traits? Put another way, what does a gene actually say? And how is its message translated by cells into a specific trait, such as brown hair or type A blood?

Consider, once again, Mendel's peas. One of the characters Mendel studied was stem length. Variation in a single gene accounts for the difference between the tall and dwarf varieties of pea plants in the drawing on this page. Mendel did not know the physiological basis of this phenotypic difference, but plant scientists have since worked out the explanation: Dwarf peas lack growth hormones called gibberellins, which stimulate the normal elongation of stems. A dwarf plant treated with gibberellins grows to normal height. Why do dwarf peas fail to make their own gibberellins? They are missing a key protein, an enzyme required for gibberellin synthesis. This example illustrates the main point of this chapter: The DNA inherited by an organism leads to specific traits by dictating the synthesis of certain proteins. Proteins are the links between genotype and phenotype. This chapter explores the steps in the flow of information from genes to proteins.

The study of metabolic defects provided evidence that genes specify proteins: *science as a process*

In 1909, British physician Archibald Garrod first suggested that genes dictate phenotypes through enzymes that catalyze specific chemical processes in the cell. Garrod postulated that the symptoms of an inherited disease reflect a person's inability to make a particular enzyme. He referred to such diseases as "inborn errors of metabolism." Garrod gave as one example the hereditary condition called alkaptonuria, in which the urine appears black because it contains the chemical alkapton, which darkens upon exposure to air. Garrod reasoned that normal individuals have an enzyme that breaks down alkapton, whereas alkaptonuric individuals have inherited an inability to make the enzyme that metabolizes alkapton.

How Genes Control Metabolism

Garrod's idea was ahead of its time, but research conducted several decades later supported his hypothesis that the function of a gene is to dictate the production of a specific enzyme. Biochemists accumulated much evidence that cells synthesize and degrade most organic molecules via metabolic pathways, in which each chemical reaction in a sequence is catalyzed by a specific enzyme. Such metabolic pathways lead, for instance, to the synthesis of the pigments that give *Drosophila* their eye color (see FIGURE 14.2). In the 1930s, George Beadle and Boris Ephrussi speculated that each of the various mutations affecting eye color in *Drosophila* blocks pigment synthesis at a specific step by preventing production of the enzyme that catalyzes that step. However, neither the chemical reactions nor the enzymes that catalyze them were known at the time.

A breakthrough in demonstrating the relationship between genes and enzymes came a few years later, after Beadle and Edward Tatum began to search for mutants of a bread mold, *Neurospora crassa.* They discovered mutants that differed from the wild-type mold in their nutritional needs (FIGURE 16.1). Wild-type *Neurospora* has modest food requirements. It can survive in the laboratory on agar (a moist support medium) mixed only with inorganic salts, sucrose, and the vitamin biotin. From this *minimal medium,* the mold uses its metabolic pathways to produce all the other molecules it needs. Beadle and Tatum identified mutants that could not survive on minimal medium, apparently because they were unable to synthesize certain essential molecules from the minimal ingredients. Such nutritional mutants are called **auxotrophs** (Gr. *auxely,* "to increase"; *trophikos,* "nourishment"). Most auxotrophs *can* survive on a *complete growth medium,* minimal medium supplemented with all 20 amino acids and some other nutrients.

To pinpoint an auxotroph's metabolic defect, Beadle and Tatum took samples from the mutant growing on complete medium and distributed them to several different vials. Each vial contained minimal medium plus a single additional nutrient. The particular supplement that allowed growth indicated the metabolic defect. For example, if the only supplemented vial that supported growth of the mutant was the one fortified with the amino acid arginine, the researchers could conclude that the mutant was defective in the pathway that normally synthesizes arginine.

With further experimentation, a defect could be described even more specifically. For instance, consider three of the steps in the synthesis of arginine: A precursor nutrient is converted to ornithine, which is converted to citrulline, which is converted to arginine (see FIGURE 16.1). Beadle and Tatum could distinguish among the various arginine auxotrophs they found. Some required arginine, others required either arginine or citrulline, and still others could grow when any of the three compounds—arginine, citrulline, or ornithine—was provided. These three classes of mutants, Beadle and Tatum reasoned, must be blocked at different steps in the pathway that synthesizes arginine. They concluded that each mutant lacked a different enzyme. Assuming that each mutant was defective in a single gene, they formulated the one gene–one enzyme hypothesis, which states that the function of a gene is to dictate the production of a specific enzyme.

One Gene–One Polypeptide

As researchers learned more about proteins, they made minor revisions in the one gene–one enzyme hypothesis. Not all proteins are enzymes. Keratin, the structural protein of animal hair, and the hormone insulin are two examples of nonenzyme proteins. Because proteins that are not enzymes are nevertheless gene products, molecular biologists began to think in terms of one gene–one protein. However, many proteins are constructed from two or more different polypeptide chains (see Chapter 5), and each subunit is specified by its own gene. For example, hemoglobin, the oxygen-transporting protein of vertebrate blood, is built from two kinds of polypeptides, and thus *two* genes code for this protein. We can now restate Beadle and Tatum's idea as the **one gene–one polypeptide hypothesis.** Because most proteins consist of single polypeptides, in this chapter we will sometimes simplify by referring to proteins as the gene products.

Transcription and translation are the two main steps from gene to protein: *an overview*

Genes are the instructions for making specific proteins. But a gene does not build a protein directly. The bridge between genetic information and protein synthesis is ribonucleic acid, or RNA. You learned in Chapter 5 that RNA and DNA are both nucleic acids, polymers of nucleotides. There are two important structural differences between the nucleotides of DNA and RNA (FIGURE 16.2). One is a subtle difference in the five-carbon (pentose) sugar component of the nucleotides. **Deoxyribose,** the sugar component of DNA, has one less hydroxyl group than **ribose,** the sugar component of RNA. The second difference is in the selection of nitrogenous bases. Adenine (A), guanine (G), and cytosine (C) are common to both classes of nucleic acids. However, thymine (T) is unique to DNA. A different base called **uracil (U),** a pyrimidine very similar to thymine, is unique to RNA. Thus,

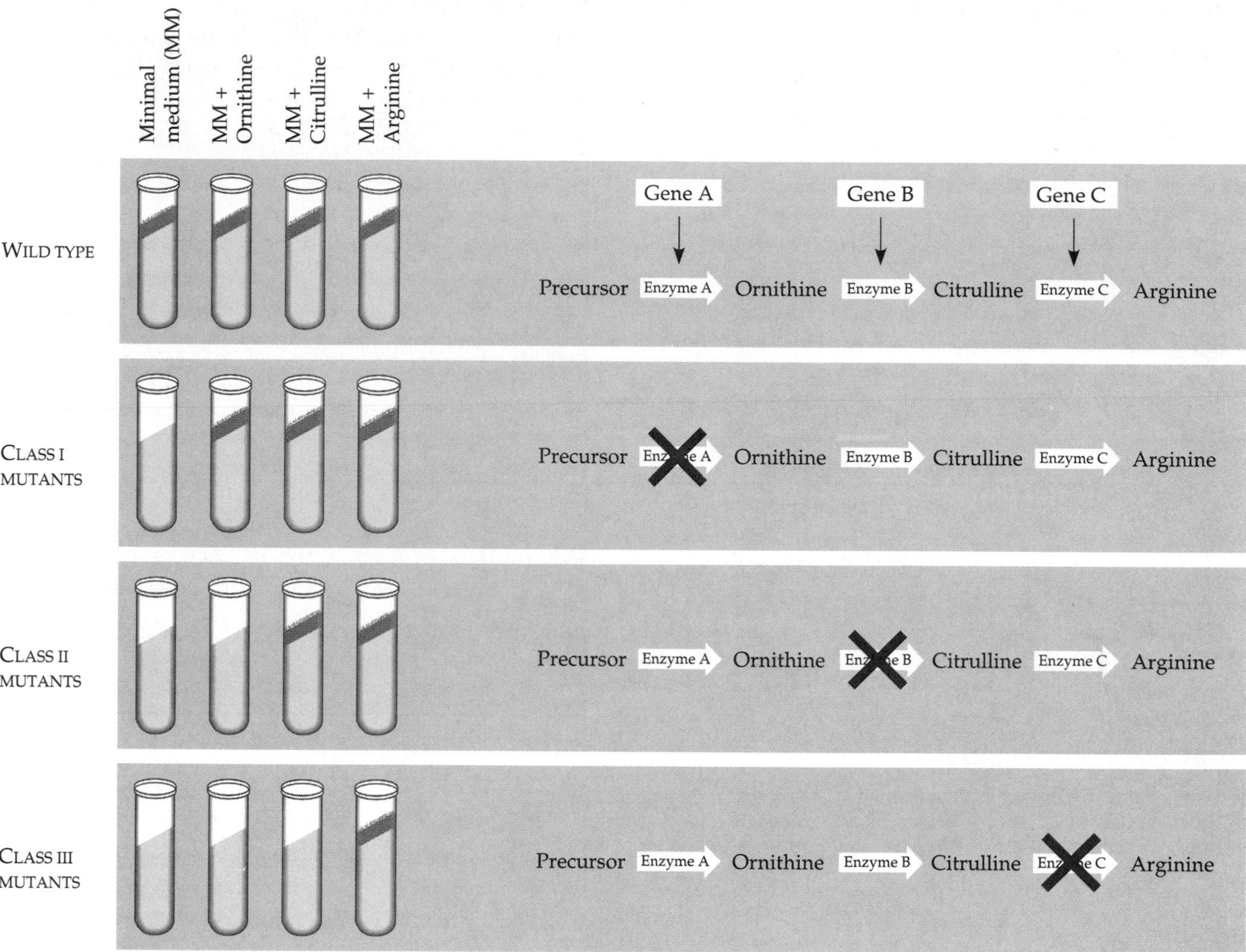

FIGURE 16.1
The one gene–one enzyme hypothesis. Beadle and Tatum based this idea on their study of nutritional mutants of the red bread mold *Neurospora crassa.* The wild-type strain requires only a minimal nutritional medium containing sucrose, essential minerals (inorganic salts), and one vitamin. The mold uses a multistep pathway to synthesize the amino acid arginine from a precursor. Beadle and Tatum identified three classes of mutants unable to synthesize arginine for three different reasons. Each mutant had a metabolic block (X in this diagram) at a different step in the pathway. For example, class II mutants failed to grow on minimal medium or minimal medium supplemented with ornithine. Adding either citrulline or arginine to the nutritional medium enabled these mutants to grow. Beadle and Tatum deduced that class II mutants lacked the enzyme that converts ornithine to citrulline. Adding citrulline to the medium bypasses the metabolic block and allows the mold to survive. The other classes of mutants lacked different enzymes. Beadle and Tatum concluded that various mutations were abnormal variations of different genes, each gene dictating the production of one enzyme; hence, the one gene–one enzyme hypothesis.

each nucleotide along a DNA strand has deoxyribose as its sugar and A, G, C, or T as its base; each nucleotide along an RNA strand has ribose as its sugar and A, G, C, or U as its base.

It is customary to describe the flow of information from gene to protein in linguistic terms because both nucleic acids and proteins have specific sequences of monomers that confer information, much as specific sequences of letters communicate information in a written language. In DNA or RNA, the monomers are the four types of nucleotides, which differ in their nitrogenous bases. Genes are typically hundreds or thousands of nucleotides long, each gene having a specific sequence of bases. A protein also has monomers arranged in a particular linear order (the protein's primary structure; see Chapter 5), but its monomers are the 20 amino acids. Thus, nucleic acids and proteins contain information written in two different chemical languages. To get from one to the other requires two major steps, called transcription and translation.

Transcription is the synthesis of RNA under the direction of DNA. Both nucleic acids use the same language, and the information is simply transcribed, or copied, from one molecule to the other. A gene's unique

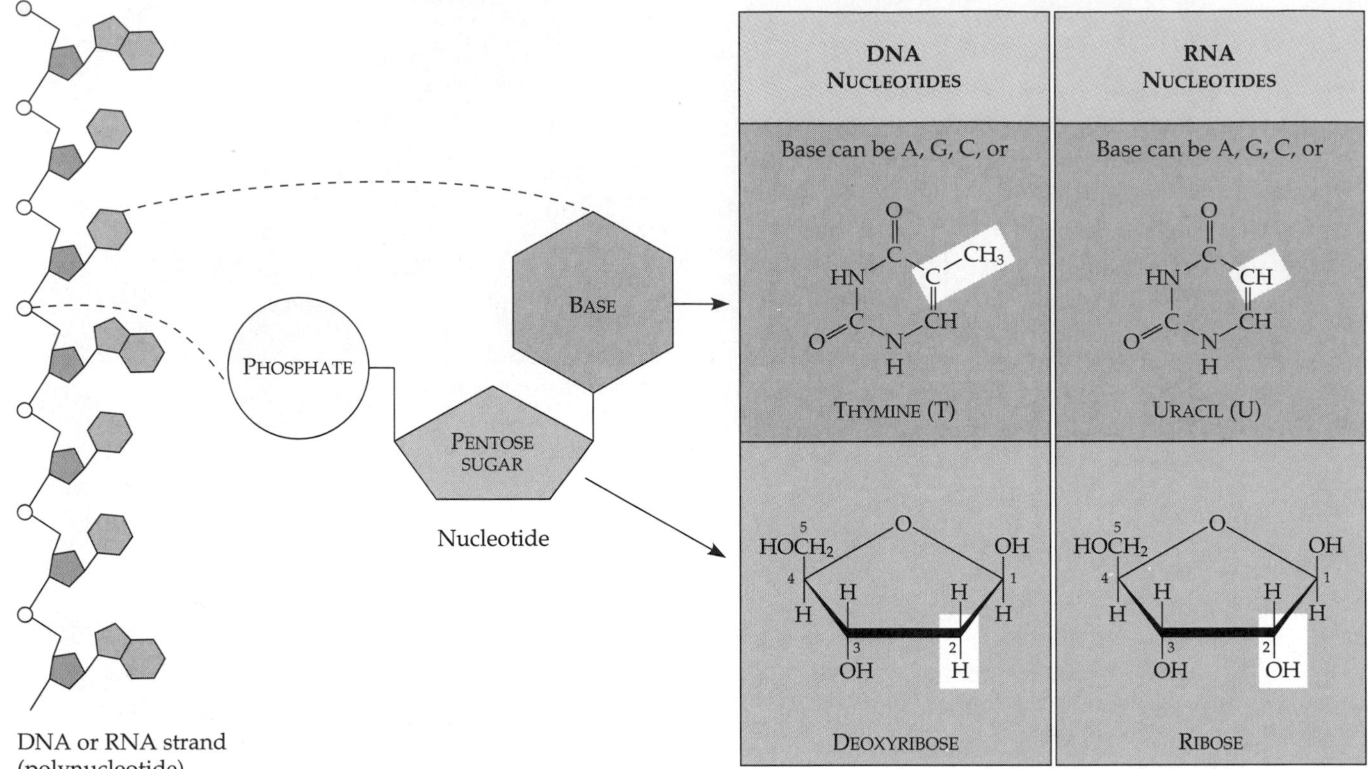

FIGURE 16.2
Two structural differences between the nucleotides of DNA and RNA. The two types of nucleic acid are named for their pentose sugars: deoxyribose in the case of DNA (deoxyribonucleic acid); ribose in the case of RNA (ribonucleic acid). A second difference is that RNA has the nitrogenous base uracil in place of thymine.

sequence of DNA nucleotides provides a template for assembling a unique sequence of RNA nucleotides, much as one DNA strand provides a template for the synthesis of a new complementary strand during DNA replication. The RNA molecule made according to the DNA template is a transcript of the gene's protein-building instructions. This type of RNA molecule is called **messenger RNA (mRNA)** because it functions as a genetic message from DNA to the protein-synthesizing machinery of the cell (FIGURE 16.3a). (There are also RNA molecules with different functions, which we will examine later in the chapter.) **Translation** is the actual synthesis of a polypeptide, which occurs under the direction of mRNA. In this step there is a change in language: The cell must translate the base sequence of an mRNA molecule into the amino acid sequence of a polypeptide. The sites of translation are ribosomes, complex particles with many enzymes and other agents that facilitate the orderly linking of amino acids into polypeptide chains.

Although the basic mechanics of transcription and translation are similar for prokaryotes and eukaryotes, there is an important difference in how the overall process of protein synthesis is organized within the cells. Because bacteria lack nuclei, their DNA is not segregated from ribosomes and the other protein-synthesizing equipment. As a result, transcription and translation are coupled, with ribosomes attaching to the "leading" end of an mRNA molecule while transcription is still in progress (see FIGURE 16.3a). In a eukaryotic cell, by contrast, the nuclear envelope separates transcription from translation in space and time. Transcription occurs in the nucleus, and mRNA is dispatched to the cytoplasm, where translation occurs (see FIGURE 16.3b). The compartmentalization of transcription and translation in eukaryotes provides an opportunity to modify mRNA in various ways before it leaves the nucleus. This modification, called **RNA processing,** occurs only in eukaryotes. You will learn more about RNA processing later in the chapter.

Let's summarize the main point of our overview of protein synthesis: Genes program protein synthesis via genetic messages in the form of messenger RNA. Put another way, cells are governed by a molecular chain of command: DNA ⟶ RNA ⟶ protein. The next sections discuss in more detail how the instructions for assembling amino acids into a specific order are encoded in the language of DNA.

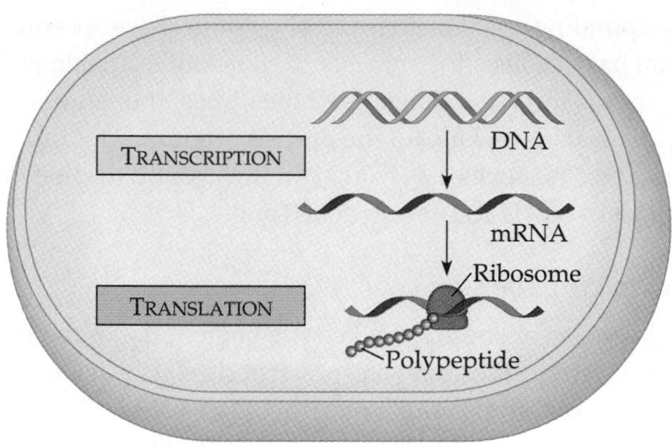

(a) Prokaryotic cell

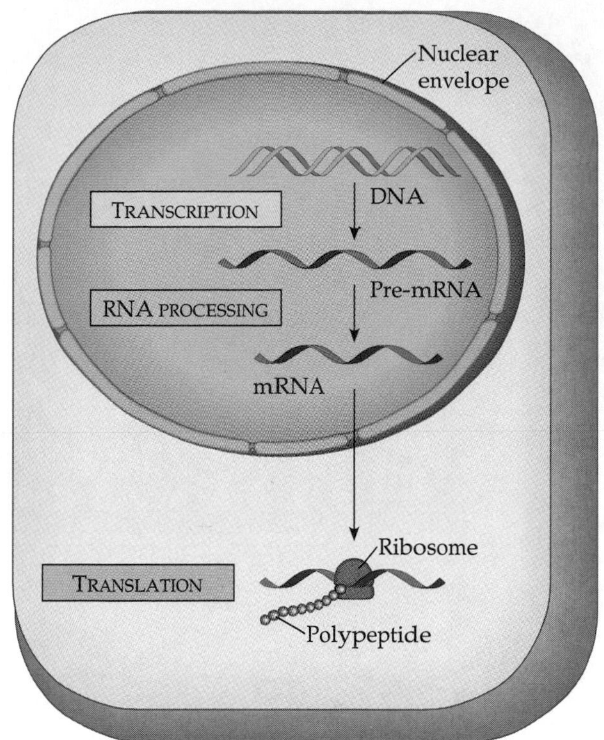

(b) Eukaryotic cell

FIGURE 16.3
Overview: the roles of transcription and translation in the flow of genetic information. In a cell's chain of command, inherited information flows from DNA to RNA to protein. The two major steps in this passing of commands are called transcription and translation. During transcription, a gene provides the instructions for synthesizing a messenger RNA (mRNA) molecule. During translation, the genetic message encoded in mRNA orders amino acids into a protein of specific amino acid sequence. Ribosomes are the sites of translation. (**a**) In a prokaryotic cell, which lacks a nucleus, mRNA produced by transcription is immediately translated without additional processing. (**b**) In a eukaryotic cell, the two main steps of protein synthesis occur in separate compartments: transcription in the nucleus and translation in the cytoplasm. Thus, mRNA must be translocated from nucleus to cytoplasm via pores in the nuclear envelope. The RNA is first synthesized as pre-mRNA, which is processed by enzymes before leaving the nucleus as mRNA. A miniature version of this illustration accompanies several figures later in the chapter as an orientation diagram to help you see how the details of protein synthesis fit into the overview.

In the genetic code, a particular triplet of nucleotides specifies a certain amino acid: *a closer look*

When biologists began to suspect that the instructions for protein synthesis are encoded in DNA, they recognized a problem: There are only 4 nucleotides to specify 20 amino acids. Thus, the genetic code cannot be a language like Chinese, where each written symbol corresponds to a single word. If each nucleotide base were translated into an amino acid, only 4 of the 20 amino acids could be specified. Would a language of two-letter code words suffice? The base sequence AG, for example, could specify one amino acid, and GT could specify another. Since there are four bases, this would give us 16 (4^2) possible arrangements—still not enough to code for all 20 amino acids.

Triplets of nucleotide bases are the smallest units of uniform length that can code for all the amino acids. If each arrangement of 3 consecutive bases specifies an amino acid, there can be 64 (4^3) possible code words—more than enough to specify all the amino acids. Experiments have verified that the flow of information from gene to protein is based on a **triplet code:** The genetic instructions for a polypeptide chain are written in the DNA as a series of three-nucleotide words. For example, the base triplet AGT at a particular position along a DNA strand says to place the amino acid serine at the corresponding position of the polypeptide to be produced.

A cell cannot directly translate a gene's base triplets into amino acids. The intermediate step is transcription, during which the gene determines the sequence of base triplets along the length of an mRNA molecule. For each gene, only one of the two DNA strands is transcribed. This strand is called the *template strand,* because it provides the template for ordering the sequence of bases in an RNA transcript. The other strand provides the instructions for making a new template strand when the DNA replicates. A given DNA strand can be the template strand in some regions of a DNA molecule, while in other

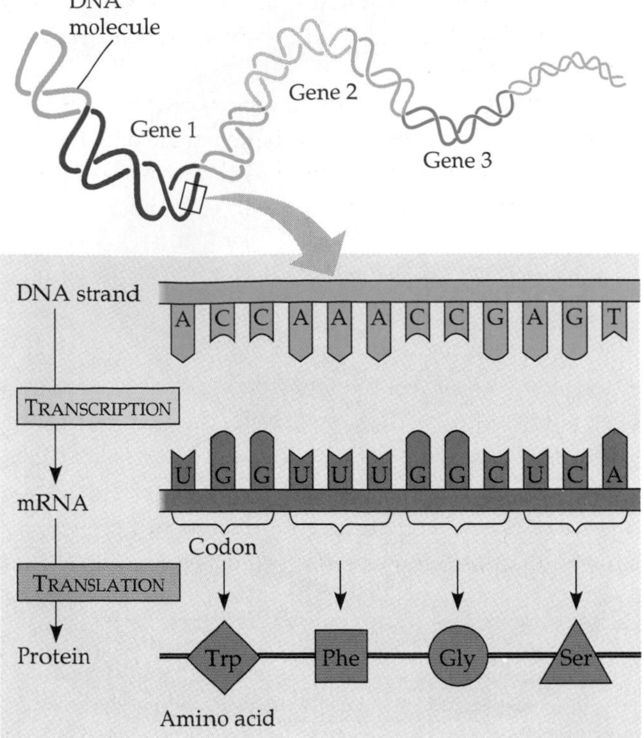

FIGURE 16.4
The triplet code. For each gene, one of the two strands of DNA functions as a template for transcription—the synthesis of an mRNA molecule of complementary sequence. The same base-pairing rules that apply to DNA synthesis also guide transcription, but the base uracil (U) takes the place of thymine (T) in RNA. During translation, the genetic message (mRNA) is read as a sequence of base triplets, analogous to three-letter code words. Each of these triplets, called a codon (bracketed in the figure), specifies the amino acid to be added at the corresponding position along a growing protein chain. All of these polymers—the gene, its mRNA transcript, and the protein product—are typically much longer than the segments shown here.

regions along the double helix it is the complementary strand that functions as the template for RNA synthesis.

An mRNA molecule is complementary rather than identical to its DNA template because RNA bases are assembled on the template according to base-pairing rules (FIGURE 16.4). The principle is the same as the specific base pairing that occurs during DNA replication. For example, when a DNA strand is transcribed, the base triplet CCG in DNA provides a template for ordering the complementary triplet GGC in the mRNA molecule. The mRNA base triplets are called **codons.** Notice in FIGURE 16.4 that U, the RNA substitute for T, pairs with A. Thus, the DNA triplet AGT is transcribed as an mRNA codon UCA, the code word for the amino acid serine.

During translation, the sequence of codons along a genetic message (mRNA) is decoded, or translated, into a sequence of amino acids making up a polypeptide chain. Each codon along the mRNA molecule specifies which one of the 20 amino acids will be incorporated at the corresponding position along a polypeptide. Because codons are base triplets, the number of nucleotides making up a genetic message must be three times the number of amino acids making up the protein product. For example, it takes 300 nucleotides along an RNA strand to code for a protein that is 100 amino acids long.

Cracking the Genetic Code

Molecular biologists cracked the code of life in the early 1960s, when a series of elegant experiments disclosed the amino acid translations of each of the RNA codons. The first codon was deciphered in 1961 by Marshall Nirenberg, of the National Institutes of Health. Nirenberg had synthesized an artificial mRNA by linking identical RNA nucleotides containing uracil as their base. No matter where this message started or stopped, it could contain only one codon in repetition: UUU. Nirenberg added this "poly U" to a test-tube mixture containing amino acids, ribosomes, and the other components required for protein synthesis. His artificial system translated the poly U into a polypeptide containing a single amino acid, phenylalanine (Phe), strung together as a long polyphenylalanine chain. Thus, Nirenberg determined that the mRNA codon UUU specifies the amino acid phenylalanine. Soon, the amino acids specified by the codons AAA, GGG, and CCC were also determined.

Although more elaborate techniques were required to decode mixed triplets such as AUA and CGA, all 64 codons were deciphered by the mid-1960s. As FIGURE 16.5 shows, 61 of the 64 triplets code for amino acids. Notice that the codon AUG has a dual function: It not only codes for the amino acid methionine (Met), but also functions as a "start" signal, or an initiation codon. Genetic messages begin with the mRNA codon AUG, which signals the protein-synthesizing machinery to begin translating the mRNA at that location. (Since AUG also stands for methionine, polypeptide chains begin with methionine when they are synthesized. However, an enzyme may subsequently remove this "starter" amino acid from a chain.) The remaining three codons do not designate amino acids. Instead, they are "stop" signals, or termination codons, marking the end of a genetic message.

Notice in FIGURE 16.5 that there is redundancy in the genetic code, but no ambiguity. For example, although codons GAA and GAG both specify glutamic acid (redundancy), neither of them ever specifies any other amino acid (no ambiguity). The redundancy in the code is not altogether random. In many cases, codons that are "synonyms" for a particular amino acid differ only in the third base of the triplet. We will consider a possible explanation for this redundancy later in the chapter.

Our ability to extract the intended message from a written language depends on reading the symbols in the

First base	Second base: U	C	A	G	Third base
U	UUU Phe	UCU Ser	UAU Tyr	UGU Cys	U
	UUC	UCC	UAC	UGC	C
	UUA Leu	UCA	UAA Stop	UGA Stop	A
	UUG	UCG	UAG Stop	UGG Trp	G
C	CUU Leu	CCU Pro	CAU His	CGU Arg	U
	CUC	CCC	CAC	CGC	C
	CUA	CCA	CAA Gln	CGA	A
	CUG	CCG	CAG	CGG	G
A	AUU Ile	ACU Thr	AAU Asn	AGU Ser	U
	AUC	ACC	AAC	AGC	C
	AUA	ACA	AAA Lys	AGA Arg	A
	AUG Met or start	ACG	AAG	AGG	G
G	GUU Val	GCU Ala	GAU Asp	GGU Gly	U
	GUC	GCC	GAC	GGC	C
	GUA	GCA	GAA Glu	GGA	A
	GUG	GCG	GAG	GGG	G

FIGURE 16.5
The dictionary of the genetic code. The three bases of an mRNA codon are designated here as the first, second, and third bases. Practice using this dictionary by finding the codon UGG, the first codon in FIGURE 16.4. This is the only codon for the amino acid tryptophan, but most amino acids are specified by two or more codons. For example, both UUU and UUC stand for the amino acid phenylalanine (Phe). When either of these codons is read along an mRNA molecule, phenylalanine will be incorporated into the growing polypeptide chain. We can think of UUU and UUC as synonyms in the genetic code. Notice that the codon AUG not only stands for the amino acid methonine (Met) but also functions as a "start" signal for ribosomes, the cell organelles that actually assemble proteins, to begin translating the mRNA at that location. Three of the 64 codons function as "stop" signals. Any one of these termination codons marks the end of a genetic message, and the completed polypeptide chain is released from the ribosome.

correct sequence and groupings. This ordering is called the **reading frame.** Consider this statement: "The red dog ate the cat." Read the words out of order, and you may get the unintended message, "The red cat ate the dog." Group the letters incorrectly, as in "Thereddogatethecat," and you may think that the first word of the message is "There." The reading frame is also important in the molecular language of cells. The amino acids shown in FIGURE 16.4, for instance, can only be assembled in the correct order if the mRNA codons UGGUUUGGCUCA are read from start to finish in the correct sequence and groups of three. Although a genetic message is written with no spaces between the codons, the cell's protein-synthesizing machinery reads the message in the correct frame as a series of nonoverlapping three-letter words. The message is *not* read as a series of overlapping words—UGGUUU, and so on—which would convey a very different message.

Let's summarize what we have just covered. Genetic information is encoded as a sequence of nonoverlapping base triplets, or codons, each of which is translated into a specific amino acid during protein synthesis.

The Evolutionary Significance of a Common Genetic Language

The genetic code is nearly universal, shared by organisms as diverse as bacteria and humans. The RNA codon CCG, for instance, is translated as the amino acid proline in all organisms whose genetic code has been examined. In laboratory experiments, genes can be transcribed and translated after they are transplanted from one species to another (FIGURE 16.6). One important application is that

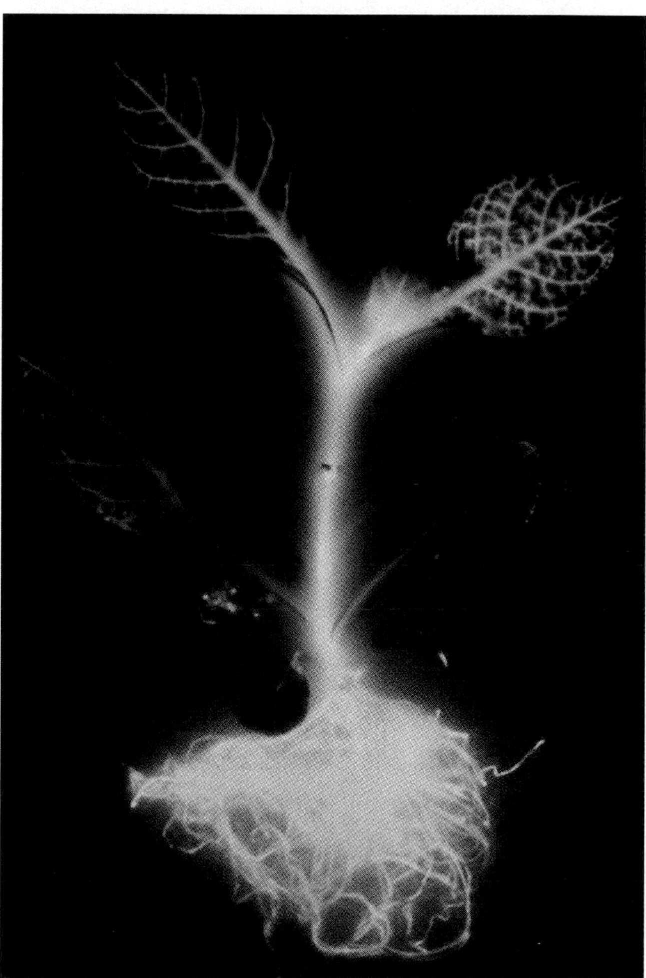

FIGURE 16.6
A tobacco plant expressing a firefly gene. Because diverse forms of life share a common genetic code, it is possible to program one species to produce proteins characteristic of another species by transplanting DNA. In this experiment, researchers were able to incorporate a gene from a firefly into the DNA of a tobacco plant. The gene codes for the firefly enzyme that catalyzes the chemical reaction that releases energy in the form of light.

bacteria can be programmed by the insertion of a human gene to synthesize the protein insulin, a product that can be used to treat diabetes. Such applications have produced many exciting developments in biotechnology, which you will learn about in Chapter 19.

There are some interesting exceptions to the universality of the genetic code. In several single-celled eukaryotes called ciliates (including *Paramecium,* an organism you may know from the lab), biologists have found a variation from the standard code. In these organisms, the RNA codons UAA and UAG are not stop signals, as they are in other organisms; instead, they code for the amino acid glutamine. Researchers have discovered other exceptions to the standard genetic code in mitochondria and chloroplasts, organelles that contain DNA coding for some of their own proteins.

Although biologists do not yet understand how these variations of the genetic code evolved, they generally agree on the evolutionary significance of the code's *near* universality. A language shared by all living things must have been operating very early in the history of life—early enough to be present in the organisms that were the common ancestors of all modern organisms, from the simplest bacteria to the most complex plants and animals. A shared genetic vocabulary is a reminder of the kinship that bonds all life on Earth.

Now that we have considered the linguistic logic and evolutionary significance of the genetic code, we can reexamine, in more detail, the key processes of transcription and translation.

Transcription is the DNA-directed synthesis of RNA: *a closer look*

Messenger RNA, the carrier of information from DNA to protein-synthesizing machinery, is transcribed from the template strand of a gene. Enzymes called **RNA polymerases** pry the two strands of DNA apart and hook together the RNA nucleotides as they base-pair along the DNA template. Like the DNA polymerases that function in DNA replication, RNA polymerases can add nucleotides only to the 3′ end of the growing polymer. Thus, an RNA molecule elongates in its 5′ ⟶ 3′ direction. Specific sequences of nucleotides along the DNA mark the initiation and termination sites, where transcription of a gene begins and ends. Including these "bookends" and the hundreds or thousands of nucleotides in between, the entire stretch of DNA that is transcribed into a single RNA molecule is called a **transcription unit** (FIGURE 16.7).

In eukaryotes, a transcription unit represents a single gene; mRNA codes for the synthesis of one polypeptide. In prokaryotes, one transcription unit may include a few genes that code for proteins of related function—enzymes catalyzing sequential steps of a metabolic pathway, for example. In these cases, the mRNA is punctuated with start and stop codons that signal the ends of each segment of mRNA coding for one protein.

Bacteria have a single type of RNA polymerase that synthesizes not only mRNA but also other types of RNA that function in protein synthesis (discussed in the next section). In contrast, eukaryotes are equipped with three types of RNA polymerase. The one specialized for mRNA synthesis is called RNA polymerase II. About 40,000 RNA polymerase II molecules are found in the nucleus of a human cell.

The three key steps in transcription are polymerase binding and initiation, elongation, and termination. Let's examine each of these processes in greater detail.

RNA Polymerase Binding and Initiation of Transcription

RNA polymerases bind to regions of DNA known as **promoters** (FIGURE 16.8, p. 306). A promoter includes the initiation site, where transcription (RNA synthesis) actually begins, and dozens of nucleotides "upstream" from the initiation site. Certain regions within the promoter are especially important for recognition by RNA polymerases. For example, the RNA polymerase II of eukaryotes keys on a region called the *TATA box,* so named because it is enriched with the nitrogenous bases thymine (T) and adenine (A). TATA boxes are centered at locations about 25 nucleotides upstream from initiation sites.

RNA polymerase II cannot recognize and bind to a promoter on its own. Other proteins, called **transcription factors,** aid the polymerases in their search for promoter regions along DNA molecules (see FIGURE 16.8). One type of transcription factor is a protein that must attach to a promoter before an RNA polymerase II molecule can bind. Apparently, RNA polymerase II recognizes the complex between this protein and DNA as its binding site. Once active RNA polymerase is bound to a promoter region, the enzyme begins to separate the two DNA strands at the initiation site, and transcription is under way.

Elongation of the RNA Strand

As RNA polymerase II moves along the DNA, it untwists one turn of the double helix at a time, separating the strands and exposing about ten DNA bases for pairing with RNA nucleotides (see FIGURE 16.7). The enzyme adds nucleotides to the 3′ end of the growing RNA molecule as it continues along the double helix. In the wake of this advancing wave of RNA synthesis, the mRNA molecule peels away from its DNA template. Transcription progresses at a rate of about 60 nucleotides per second.

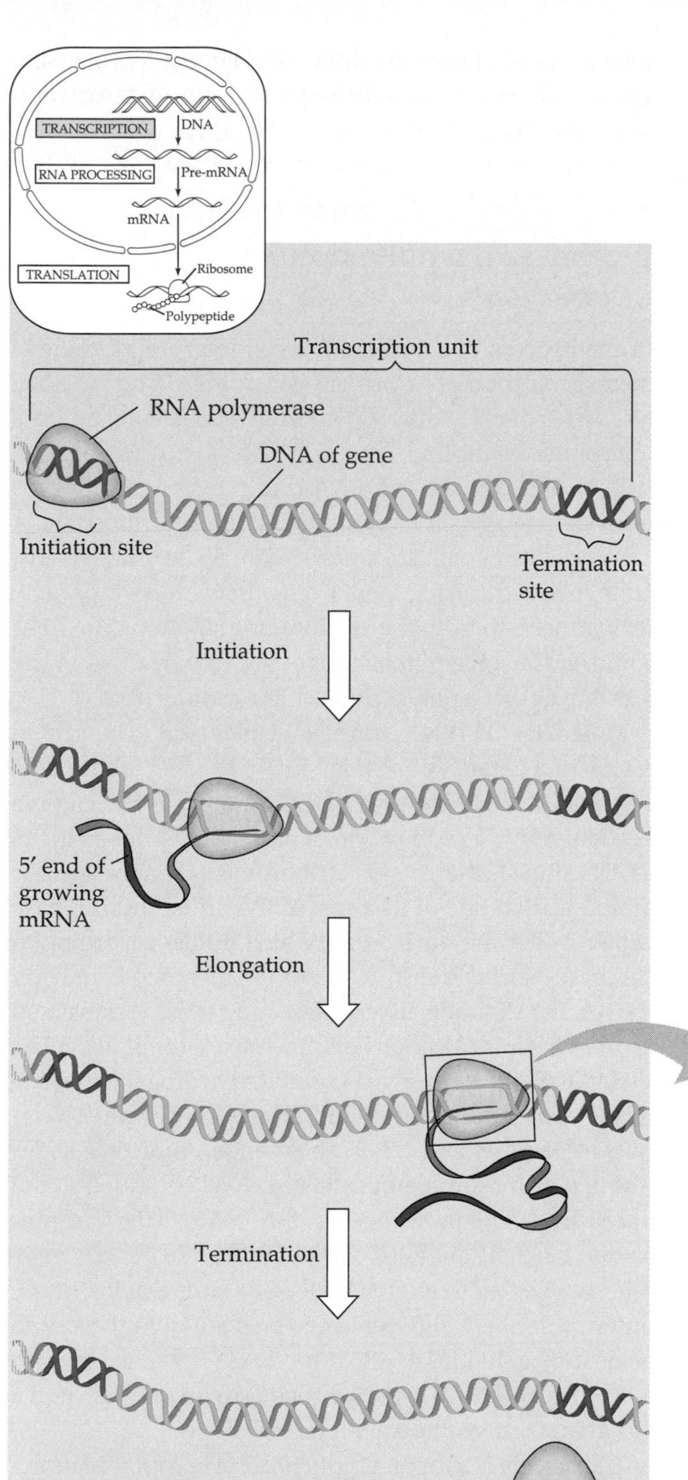

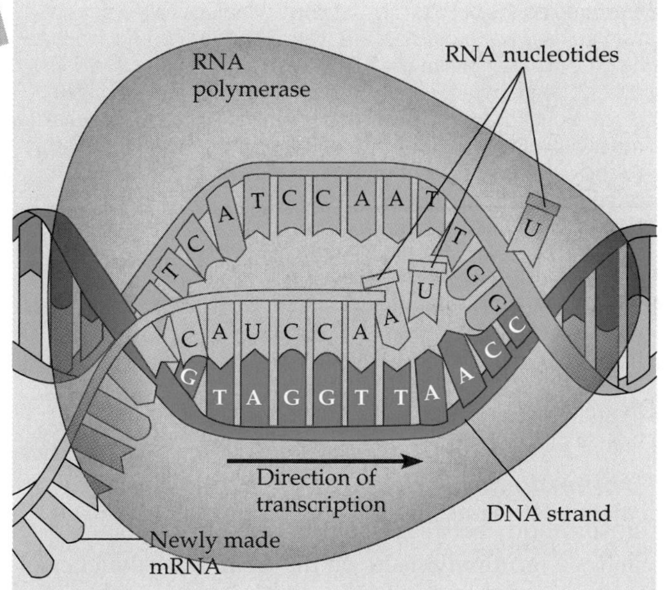

FIGURE 16.7
A closer look at transcription. As an RNA polymerase molecule moves along a gene from the initiation site to the termination site, it synthesizes an RNA molecule that consists of the nucleotide sequence determined by the template strand of the gene. The entire stretch of DNA that is transcribed is called a transcription unit. Transcription begins at the initiation site when the polymerase separates the two DNA strands and exposes the template strand for base pairing with RNA nucleotides. The RNA polymerase works its way "downstream" from the initiation site, prying apart the two strands of DNA and elongating the mRNA in the 5′ ⟶ 3′ direction. In the wake of transcription, the two DNA strands re-form the double helix. The RNA polymerase continues to elongate the RNA molecule until it reaches the termination site, a specific sequence of nucleotides along the DNA that signals the end of the transcription unit. The mRNA, a transcript of the gene, is released, and the polymerase subsequently dissociates from the DNA.

A single gene can be transcribed simultaneously by several molecules of RNA polymerase II, following each other like trucks in a convoy. The growing strands of RNA trail off from each polymerase, with the length of each new strand reflecting how far along the template the enzyme has traveled from the initiation site. The congregation of many polymerase molecules simultaneously transcribing a single gene increases the number of mRNA molecules and allows a cell to produce a particular protein in large amounts.

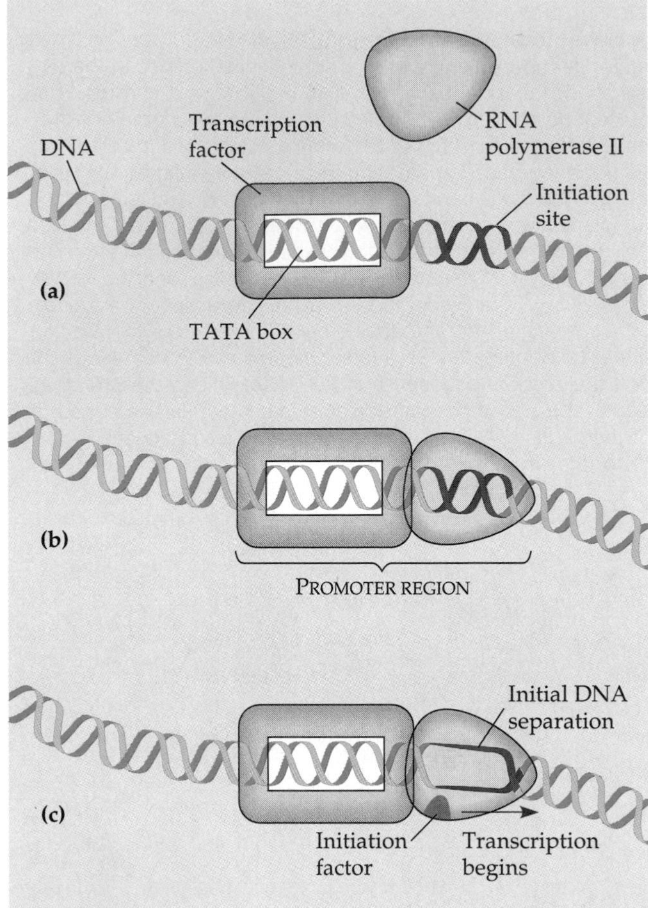

FIGURE 16.8
Promoters and the initiation of transcription. RNA polymerase molecules bind to DNA at special regions called promoters. In eukaryotes, promoters are typically about 100 nucleotides long. They consist of the actual initiation site and a few regions of DNA that identify the promoter to proteins that help initiate transcription. For example, RNA polymerase II, the enzyme that synthesizes mRNA in eukaryotes, binds to a promoter that has a TATA box, a short segment of thymines (T) and adenines (A) located about 25 nucleotides upstream from the initiation site. (**a**) An RNA polymerase cannot recognize the TATA box and other landmarks of the promoter region on its own. Another protein, a transcription factor that recognizes the TATA box, binds to the DNA before the RNA polymerase can do so. (**b**) Once RNA polymerase attaches to the promoter region, (**c**) it associates with other transcription factors before RNA synthesis begins (only one of these additional transcription factors, an initiation factor, is illustrated here).

Termination of Transcription

Transcription proceeds until the RNA polymerase reaches a termination site on the DNA. The sequence of nitrogenous bases that marks this site signals RNA polymerase to stop adding nucleotides to the RNA strand and release the RNA molecule. In eukaryotes, the most common termination sequence is AATAAA.

In bacteria, mRNAs are ready for translation as soon as they peel away from their DNA templates. In contrast, the RNA products of transcription in eukaryotes are processed before they leave the nucleus as mRNA molecules. We will postpone discussing this RNA processing until after we have studied the process of translation more closely.

Translation is the RNA-directed synthesis of a polypeptide: *a closer look*

In the process of translation, a cell interprets a genetic message and builds a protein accordingly. The message is a series of codons along an mRNA molecule, and the interpreter is another type of RNA called **transfer RNA (tRNA).** The function of tRNA is to transfer amino acids from the cytoplasm's amino acid pool to a ribosome. A cell keeps its cytoplasm stocked with all 20 amino acids, either by synthesizing them from other compounds or by taking them up from the surrounding solution. The ribosome adds each amino acid brought to it by tRNA to the growing end of a polypeptide chain (FIGURE 16.9).

Molecules of tRNA are not all identical. The key to translating a genetic message into a specific amino acid sequence is that each type of tRNA molecule associates a particular mRNA codon with a particular amino acid. As a tRNA molecule arrives at a ribosome, it bears a specific amino acid at one of its ends. At the other end is a base triplet called the **anticodon,** which binds, according to the base-pairing rules, to a complementary codon on mRNA. For example, the mRNA codon UUU is translated as the amino acid phenylalanine (see FIGURE 16.5). The tRNA that plugs into this codon by hydrogen bonding has AAA as its anticodon and always carries phenylalanine at its other end. Thus, as an mRNA molecule slides through a ribosome, phenylalanine will be added to the polypeptide chain whenever the codon UUU is presented for translation. Codon by codon, the genetic message is translated, as tRNAs deposit amino acids in the order prescribed, and ribosomal enzymes join the amino acids into a chain. The tRNA molecule is like a flashcard with a "nucleic acid word" (anticodon) on one side and a "protein word" (amino acid) on the other.

Translation is simple in principle but complex in its actual biochemistry and mechanics. To dissect the process, let's take a closer look at some of the major players in this cellular drama of protein synthesis, then see how they act together to make a polypeptide.

The Structure and Function of Transfer RNA

Transfer RNA molecules, like mRNA and other types of RNA, are transcribed from DNA templates within the nucleus of a eukaryotic cell. Like mRNA, tRNA must travel from the nucleus to the cytoplasm, where translation occurs. Each tRNA molecule can then be used repeatedly,

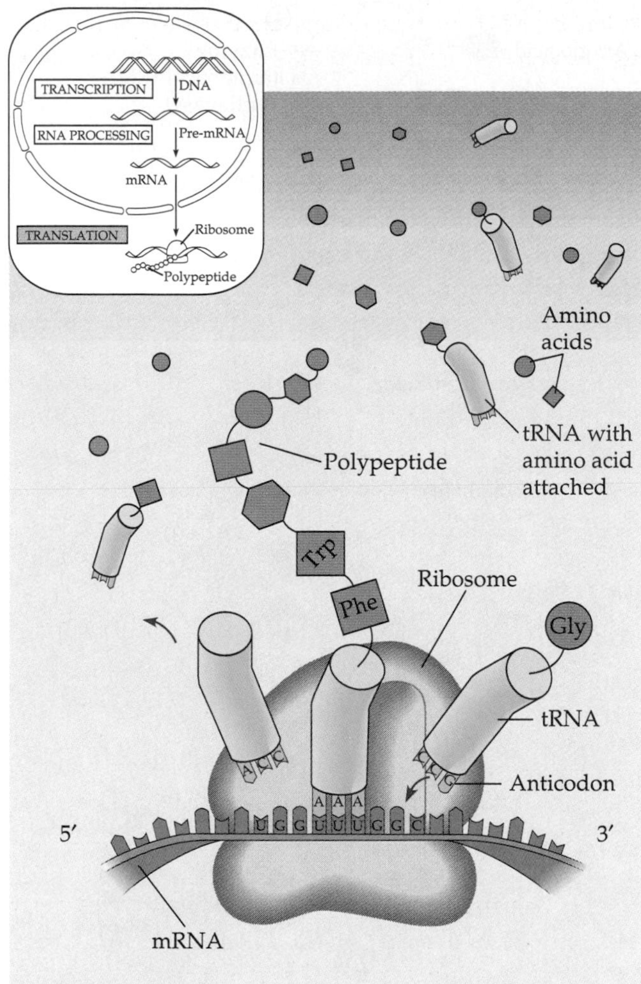

FIGURE 16.9
Translation: the basic concept. As a strand of mRNA slides through a ribosome, codons are translated into amino acids, one by one. The interpreters are tRNA molecules, each type with a specific anticodon at one end and a certain amino acid at the other end. A tRNA adds its amino acid cargo to a growing polypeptide chain when the anticodon binds to a complementary codon on the mRNA.

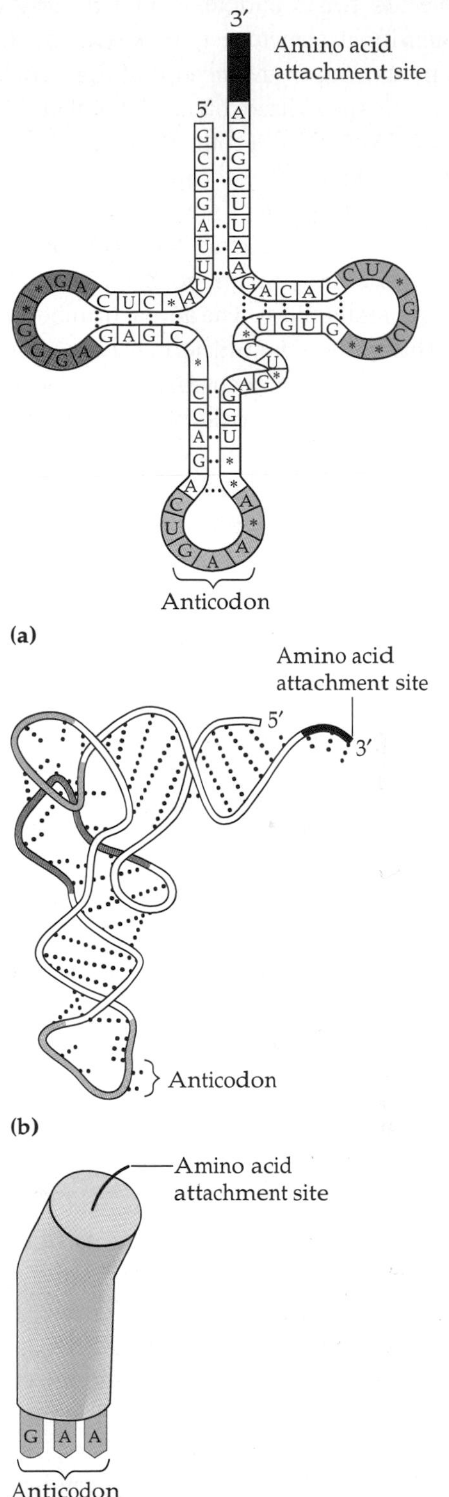

FIGURE 16.10
The structure of transfer RNA. (**a**) The two-dimensional structure of a tRNA molecule specific for the amino acid leucine. Notice the four base-paired regions and three loops characteristic of all tRNAs. At one end of the molecule is the amino acid attachment site, which has the same base sequence for all tRNAs; within the middle loop is the anticodon triplet, which is unique to each tRNA type. (The asterisks mark unusual bases that are unique to tRNAs.) (**b**) A diagram of the three-dimensional structure of an L-shaped tRNA molecule. (**c**) In the figures that follow, tRNA is represented by this simplified shape.

picking up its designated amino acid in the cytosol, depositing this cargo at the ribosome, and leaving the ribosome to pick up another load. The structure of a tRNA molecule fits its function as a shuttle for a specific amino acid (FIGURE 16.10).

A tRNA molecule consists of a single RNA strand that is only about 80 nucleotides long (compared to hundreds of nucleotides for most mRNA molecules). This single RNA strand folds back upon itself to form a molecule with a secondary structure, a three-dimensional structure reinforced by interactions between different parts of the nucleotide chain. Certain regions of the tRNA strand form hydrogen bonds with complementary bases of other regions. Flattened into one plane to reveal these regions of hydrogen bonding, a tRNA molecule has a cloverleaf shape. The two-dimensional cloverleaf of the typical tRNA molecule twists and folds into a fairly compact three-dimensional structure that is roughly L-shaped. The loop protruding from one end of the L includes the anticodon, the specialized base triplet that binds to a specific mRNA codon. From the other end of the L-shaped tRNA molecule protrudes its 3′ end, which is the attachment site for an amino acid.

If one tRNA variety existed for each of the mRNA codons that specifies an amino acid, there would be 61 tRNAs (see FIGURE 16.5). The actual number is smaller: about 45. This number is sufficient, because some tRNAs have anticodons that can recognize two or more different codons. Such versatility is possible because the rules for base pairing between the third base of a codon and the corresponding base of a tRNA anticodon are not as strict as those for DNA and mRNA codons. For example, the base U of a tRNA anticodon can pair with either A or G in the third position of an mRNA codon. This relaxation of the base-pairing rules is called *wobble.* The most versatile tRNAs are those with inosine (I), a modified base, in the wobble position of the anticodon. Inosine is formed by enzymatic alteration of adenine after tRNA is synthesized. When anticodons associate with codons, the base I can hydrogen-bond with any one of three bases: U, C, or A. Thus, the tRNA molecule that has CCI as its anticodon can bind to the codons GGU, GGC, and GGA, all of which code for the amino acid glycine. Wobble explains why the synonymous codons for a given amino acid can differ in their third base, but usually not in their other bases.

Aminoacyl-tRNA Synthetases

Codon-anticodon bonding is actually the second of two recognition steps required for the accurate translation of a genetic message. It must be preceded by a correct match between tRNA and an amino acid. A tRNA that binds to an mRNA codon specifying a particular amino acid must carry *only* that amino acid to the ribosome. Each amino acid is matched with the correct tRNA by a specific enzyme called an **aminoacyl-tRNA synthetase.** There is a whole family of these enzymes, one enzyme for each amino acid. The active site of each type of aminoacyl-tRNA synthetase fits only a specific combination of amino acid and tRNA. The synthetase catalyzes the attachment of the amino acid to its tRNA in a two-step process driven by the hydrolysis of ATP (FIGURE 16.11). The resulting amino acid–tRNA complex is released from the enzyme and delivers its amino acid to a growing polypeptide chain on a ribosome.

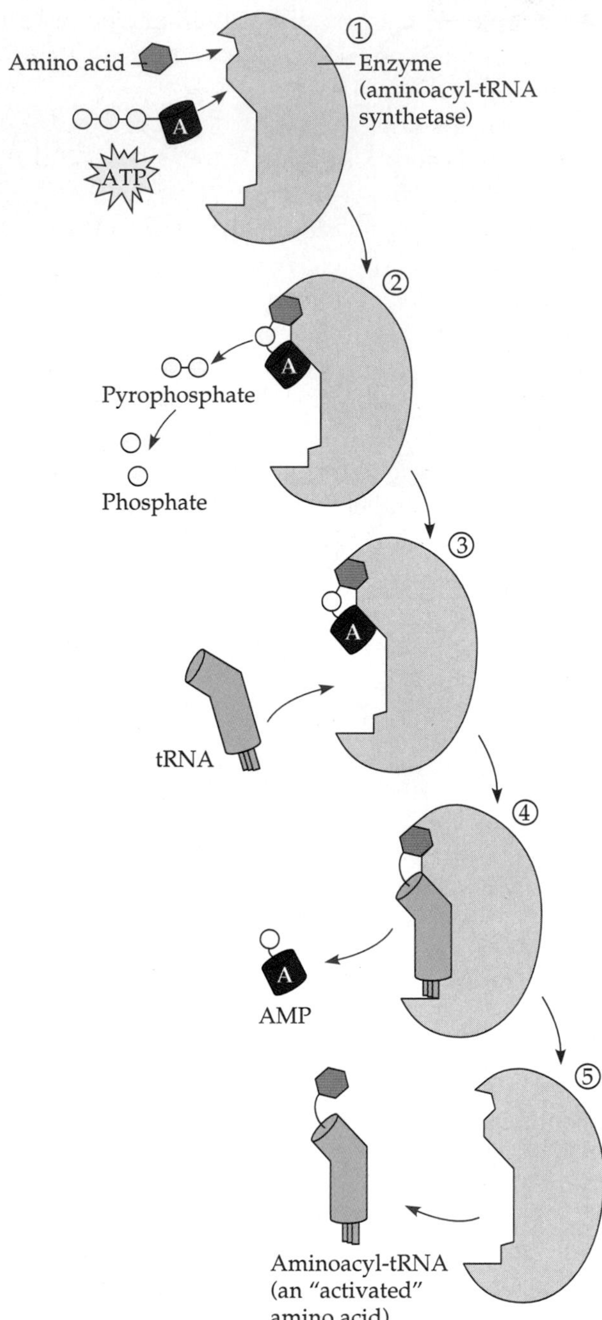

FIGURE 16.11
An aminoacyl-tRNA synthetase joins a tRNA to an amino acid in specific combination. Linkage of the tRNA to the amino acid is an endergonic reaction that occurs at the expense of ATP. ① The active site of the enzyme binds the amino acid and an ATP molecule. ② The ATP loses two phosphate groups and joins to the amino acid as AMP (adenosine monophosphate). ③ The appropriate tRNA covalently bonds to the amino acid, ④ displacing the AMP from the enzyme's active site. ⑤ The enzyme releases the aminoacyl-tRNA, the "activated" amino acid.

Ribosomes

Ribosomes facilitate the specific coupling of tRNA anticodons with mRNA codons during protein synthesis. A ribosome, which can be seen with the electron microscope, is made up of two subunits, termed the large and small subunits (FIGURE 16.12a). In eukaryotes, the ribosomal subunits are constructed in the nucleolus (see Chapter 7). The subunits are exported via nuclear pores to the cytoplasm. A large and small subunit join to form a functional ribosome only when they attach to a mRNA molecule. Each ribosomal subunit is an aggregate of numerous proteins and yet another form of specialized RNA called **ribosomal RNA (rRNA).** About 60% of the weight of each ribosome is rRNA. Because most cells contain thousands of ribosomes, rRNA is the most abundant type of RNA.

Although the ribosomes of prokaryotes and eukaryotes are very similar in structure and function, those of prokaryotes are slightly smaller and differ somewhat from eukaryotic ribosomes in their molecular composition. The differences are medically significant. Certain drugs can paralyze prokaryotic ribosomes without inhibiting the ability of eukaryotic ribosomes to make proteins. These drugs, including tetracycline and streptomycin, are used as antibiotics to combat bacterial infection.

The structure of a ribosome reflects its function of bringing mRNA together with amino acid–bearing tRNAs. In addition to a binding site for mRNA, each ribosome has two binding sites for tRNA (FIGURE 16.12b). The **P site** (peptidyl-tRNA site) holds the tRNA carrying the growing polypeptide chain, while the **A site** (aminoacyl-tRNA site) holds the tRNA carrying the next amino acid to be added to the chain. Acting like a vise, the ribosome holds the tRNA and mRNA molecules close together and catalyzes the addition of an amino acid to the carboxyl end of the growing polypeptide chain (FIGURE 16.12c).

Building a Polypeptide

We can divide translation, the synthesis of a polypeptide chain, into three stages: chain initiation, chain elongation, and chain termination. All three stages require protein factors (mostly enzymes) that aid mRNA, tRNA, and ribosomes in the translation process. For chain initiation and elongation, energy is also required. It is provided by GTP (guanosine triphosphate), a molecule closely related to ATP.

Initiation. The initiation stage of translation brings together mRNA, a tRNA bearing the first amino acid of the polypeptide, and the two subunits of a ribosome. First, a small ribosomal subunit binds to both mRNA and a special initiator tRNA (FIGURE 16.13). The small ribosomal subunit attaches to a specific sequence of nucleotides at the 5′ (upstream) end of the mRNA. Just downstream from this loading site is the initiation codon, AUG, where translation actually begins. The initiator tRNA, which carries the amino acid methionine, attaches to the initiation codon.

The union of mRNA, initiator tRNA, and small ribosomal subunit is followed by the attachment of a large ribosomal subunit to form a functional ribosome. Proteins called *initiation factors* are required to bring all these components together. The cell also spends energy in the form of one GTP to form the initiation complex. At the completion of the initiation process, the initiator tRNA sits in the P site of the ribosome, and the vacant A site is ready for the next tRNA.

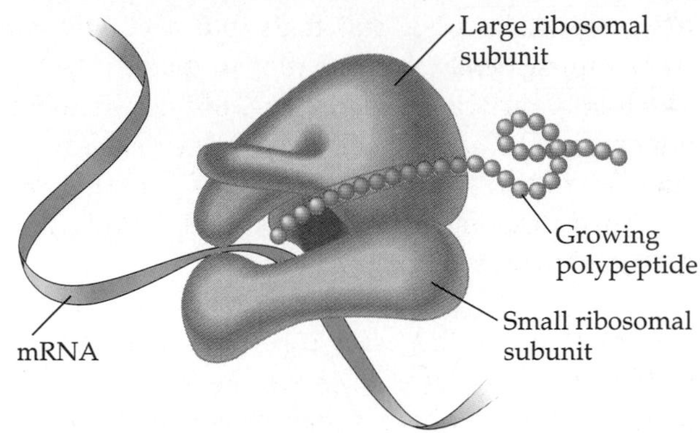

(a)

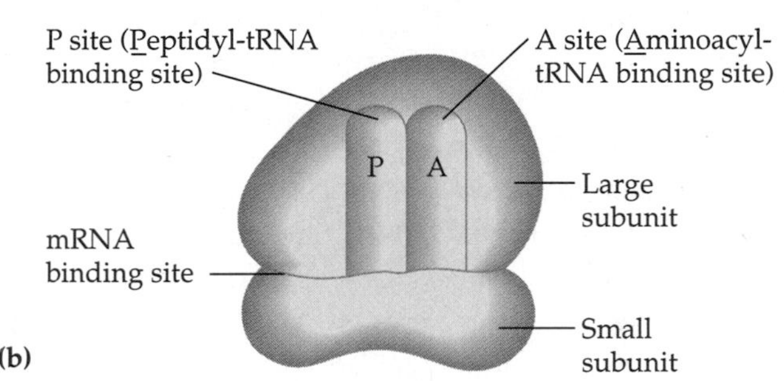

(b)

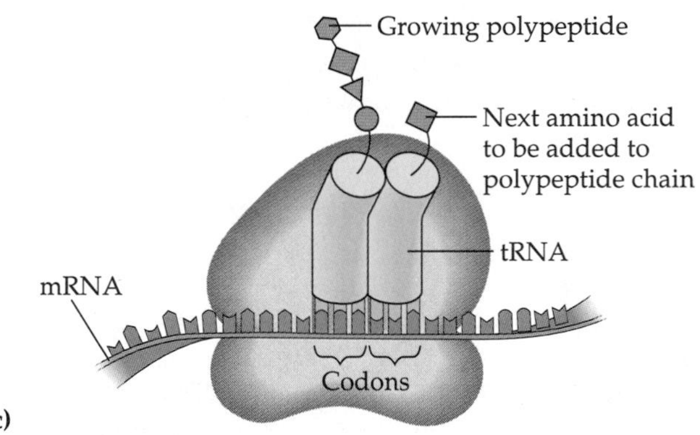

(c)

FIGURE 16.12
The anatomy of a ribosome. (**a**) A functional ribosome consists of two subunits, each an aggregate of ribosomal RNA and many proteins. This is a model of a bacterial ribosome. The eukaryotic ribosome is similar in shape, but larger, with more proteins and rRNA molecules. (**b**) A ribosome has an mRNA-binding site and two tRNA-binding sites, known as the P and A sites. This is a simplified version of the ribosomal shape that will appear in the next several figures. (**c**) A tRNA fits into a binding site when its anticodon base-pairs with an mRNA codon. The P site holds the tRNA attached to the growing polypeptide. The A site holds the tRNA carrying the next amino acid to be added to the polypeptide chain.

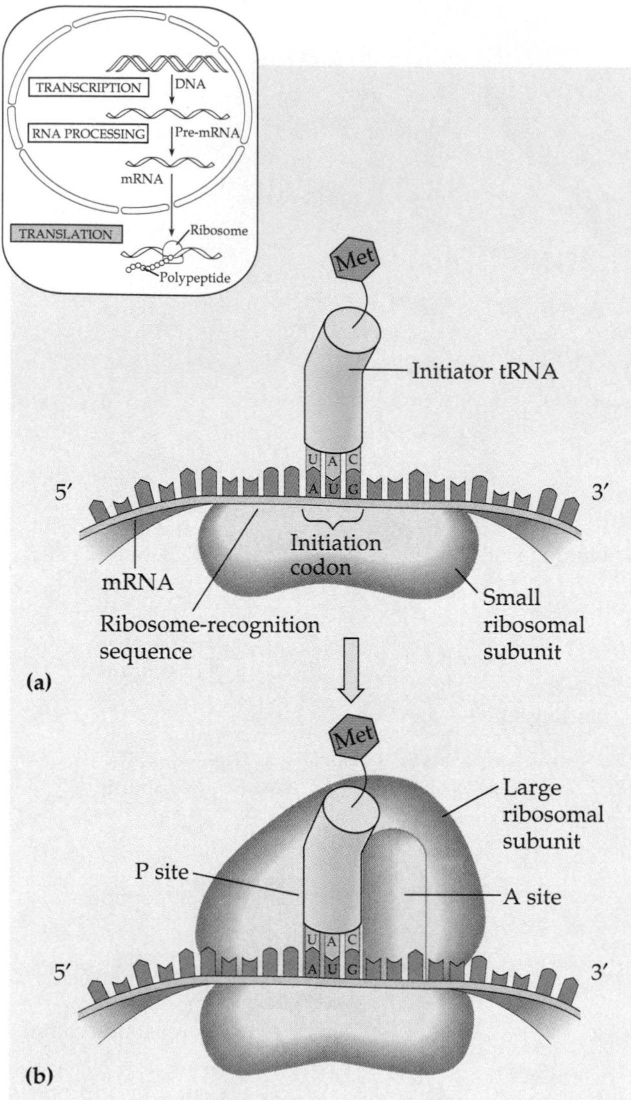

FIGURE 16.13
Initiation of translation. (**a**) A small ribosomal subunit binds to a molecule of mRNA. The positioning of the mRNA is signaled by a ribosome recognition sequence on the mRNA. At the same time, the initiator tRNA, with the anticodon UAC, base-pairs with the initiation codon AUG. This tRNA carries the amino acid methionine (Met). (**b**) The large ribosomal subunit joins the initiation complex. The initiator tRNA is in the P site. The A site is available to the tRNA bearing the next amino acid. Proteins called initiation factors are also required to bring these translation components together. GTP, closely related to ATP, provides the energy for the initiation process.

Elongation. In the elongation stage of translation, amino acids are added one by one to the initial amino acid. Each addition, which involves the participation of several proteins called *elongation factors,* occurs in a three-step cycle (FIGURE 16.14, facing page):

① *Codon recognition.* The mRNA codon in the A site of the ribosome forms hydrogen bonds with the anticodon of an incoming molecule of tRNA carrying its appropriate amino acid. An elongation factor ushers the tRNA into the A site. This step also requires the hydrolysis of a phosphate bond from GTP.

② *Peptide bond formation.* A component of the large ribosomal subunit catalyzes the formation of a peptide bond between the polypeptide extending from the P site and the newly arrived amino acid in the A site. In this step, the polypeptide separates from the tRNA to which it was bound and is transferred to the amino acid carried by the tRNA in the A site.

③ *Translocation.* The tRNA in the P site dissociates from the ribosome. The tRNA in the A site, now attached to the growing polypeptide, is translocated to the P site. As the tRNA changes sites, its anticodon remains hydrogen-bonded to the mRNA codon, allowing the mRNA and tRNA molecules to move as a unit. This movement, in turn, brings the next codon to be translated into the A site. The translocation step requires energy, which is provided by hydrolysis of a GTP molecule. The mRNA is moved through the ribosome in the 5′ ⟶ 3′ direction only, much as a ratchet allows a mechanical device to be turned in only one direction; or perhaps it is the *ribosome* that moves. The important point is that the ribosome and the mRNA move relative to each other, unidirectionally, codon by codon.

The elongation cycle takes only about 60 milliseconds and is repeated as each amino acid is added to the chain until the polypeptide is completed.

Termination. The final stage of translation is termination (FIGURE 16.15, p. 312). Elongation continues until a termination codon reaches the A site of the ribosome. These special base triplets—UAA, UAG, and UGA (see FIGURE 16.5)—do not code for amino acids but instead act as signals to stop translation. A protein called a *release factor* binds directly to the termination codon in the A site. The release factor causes the ribosome to add a water molecule instead of an amino acid to the polypeptide chain. This reaction hydrolyzes the completed polypeptide from the tRNA that is in the P site, thereby freeing the polypeptide from the ribosome. The ribosome then separates into its small and large subunits.

Polyribosomes

A single ribosome can make an average-sized polypeptide in less than a minute. Typically, however, a single mRNA is used to make many copies of a polypeptide simultaneously, because several ribosomes work on translating the message at the same time. Once a ribosome moves past the initiation codon, a second ribosome can attach to the mRNA, and thus several ribosomes may trail along the

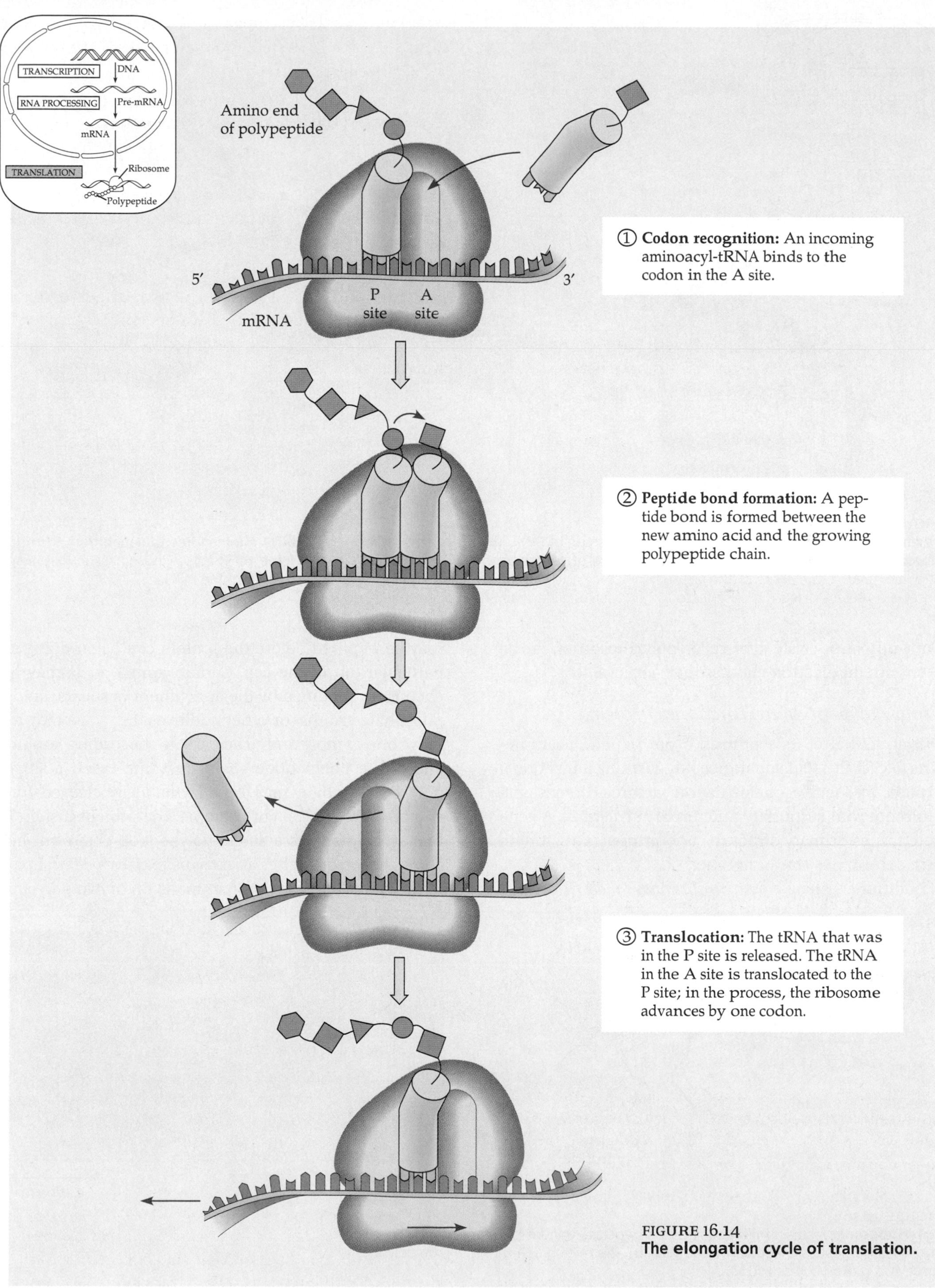

FIGURE 16.14
The elongation cycle of translation.

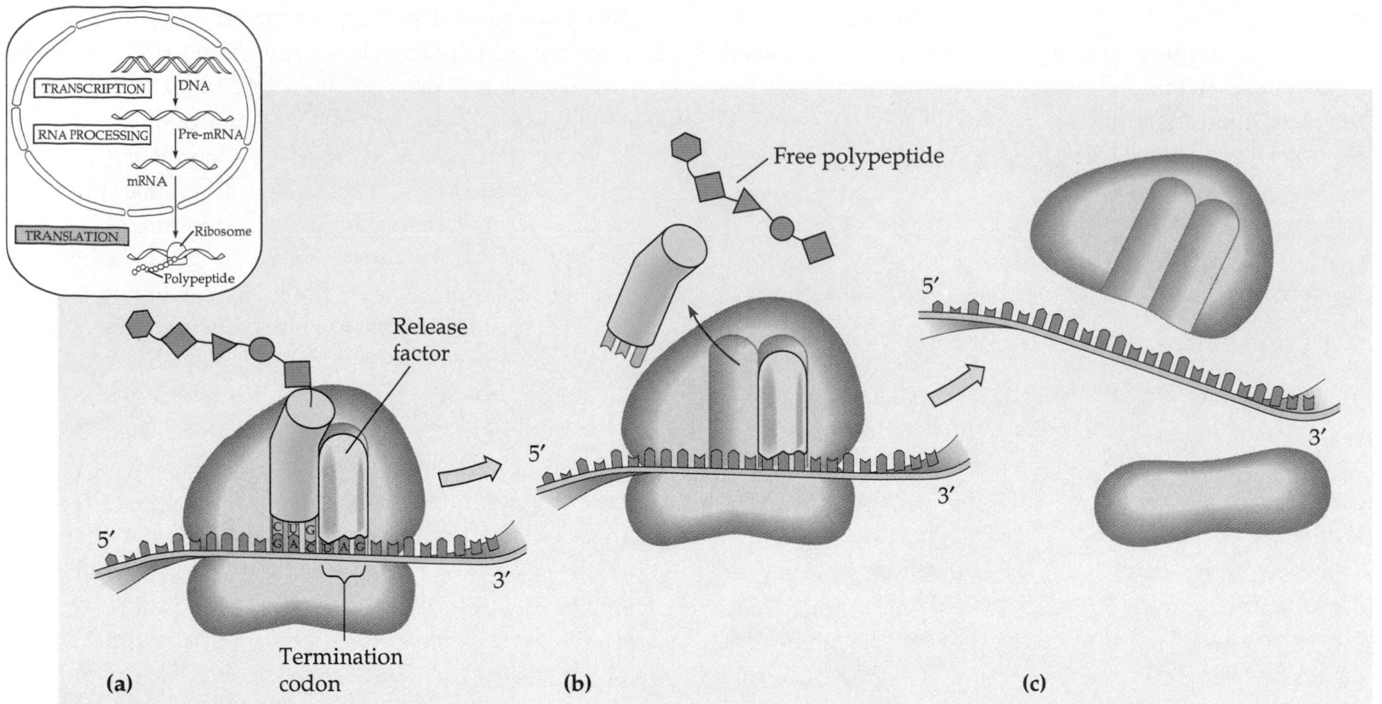

FIGURE 16.15
Termination of translation. (**a**) When a ribosome reaches a termination codon on a strand of mRNA, the A site of the ribosome accepts a protein called a release factor instead of tRNA. (**b**) The release factor hydrolyzes the bond between the tRNA in the P site and the last amino acid of the polypeptide chain. Both polypeptide and tRNA are then free to depart from the ribosome. (**c**) The two ribosomal subunits dissociate from the mRNA.

same mRNA. Such clusters, called **polyribosomes,** can be seen with the electron microscope (FIGURE 16.16).

From Polypeptide to Functional Protein

During and after its synthesis, a polypeptide chain begins to coil and fold spontaneously, forming a functional protein of specific conformation: a three-dimensional molecule with secondary and tertiary structures. A gene determines primary structure, and primary structure in turn determines conformation.

Additional steps—*posttranslational modifications*—may be required before the protein can begin doing its particular job in the cell. Certain amino acids may be chemically modified by the attachment of sugars, lipids, phosphate groups, or other additives. Enzymes may remove one or more amino acids from the leading (amino) end of the polypeptide chain. In some cases, a single polypeptide chain may be enzymatically cleaved into two or more pieces. For example, the protein insulin is first synthesized as a single polypeptide chain but becomes active only after an enzyme excises a central part of the chain, leaving a protein made up of two polypep-

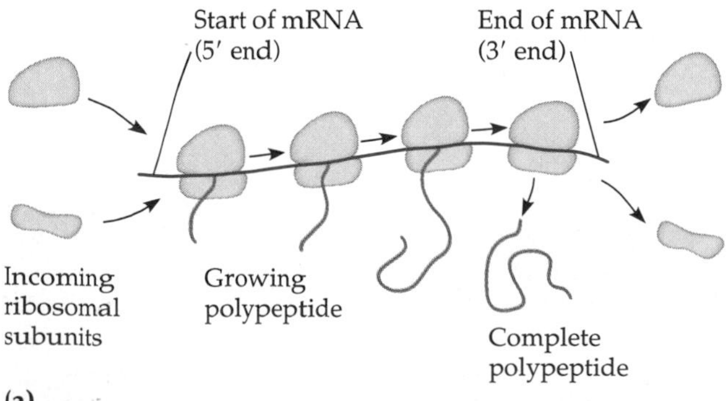

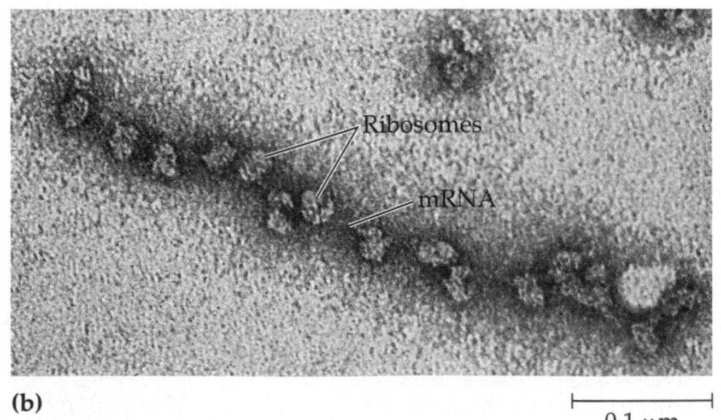

FIGURE 16.16
Polyribosomes. (**a**) An mRNA molecule is generally translated simultaneously by several ribosomes in clusters called polyribosomes. (**b**) This micrograph shows a large polyribosome in a prokaryotic cell (TEM).

tide chains connected by disulfide bridges. In other cases, two or more polypeptides that are synthesized separately may join to become the subunits of a protein that has quaternary structure. (To review the levels of protein structure, see Chapter 5.)

Some polypeptides have signal sequences that target them to specific destinations in the cell

In electron micrographs of eukaryotic cells active in protein synthesis, two populations of ribosomes (and polyribosomes) are evident: free and bound (see FIGURE 7.11). Free ribosomes are suspended in the cytosol and mostly synthesize proteins that dissolve in the cytosol and function there. In contrast, bound ribosomes are attached to the cytosolic side of the endoplasmic reticulum (ER). They make proteins of the endomembrane system (the nuclear envelope, ER, Golgi apparatus, lysosomes, vacuoles, and plasma membrane) and proteins that are secreted from the cell. Insulin is an example of a secretory protein. The ribosomes themselves are identical and can switch their status from free to bound.

What determines whether a ribosome will be free in the cytosol or bound to rough ER at any particular time? The synthesis of all proteins begins in the cytosol when a ribosome starts to translate a messenger RNA molecule. The growing polypeptide chain itself cues the ribosome to either remain in the cytosol or attach to the ER. Secretory proteins are marked by a **signal sequence** of about 20 amino acids (FIGURE 16.17). The signal sequence, usually the first part of the polypeptide made, is recognized by a protein complex called a *signal-recognition particle.* This particle functions as an adaptor that attaches the ribosome to a receptor protein built into the ER membrane. Synthesis of the protein continues there, and as the growing polypeptide snakes across the membrane into the cisternal space, the signal sequence is removed by an enzyme. In contrast, if the mRNA molecule lacks a segment that programs synthesis of the ER signal sequence, the ribosome translating that RNA remains free in the cytosol, where the finished protein will be released. Thus, the function of the signal sequence is to dispatch proteins to their target; in this case, the ER.

The use of signal sequences to target secretory proteins to the ER is only one example of a general mechanism for dispatching proteins to specific sites. Other signal sequences target proteins for mitochondria or chloroplasts after the proteins are released from free ribosomes. Signal sequences function like ZIP codes, addressing proteins to certain locations in the cell.

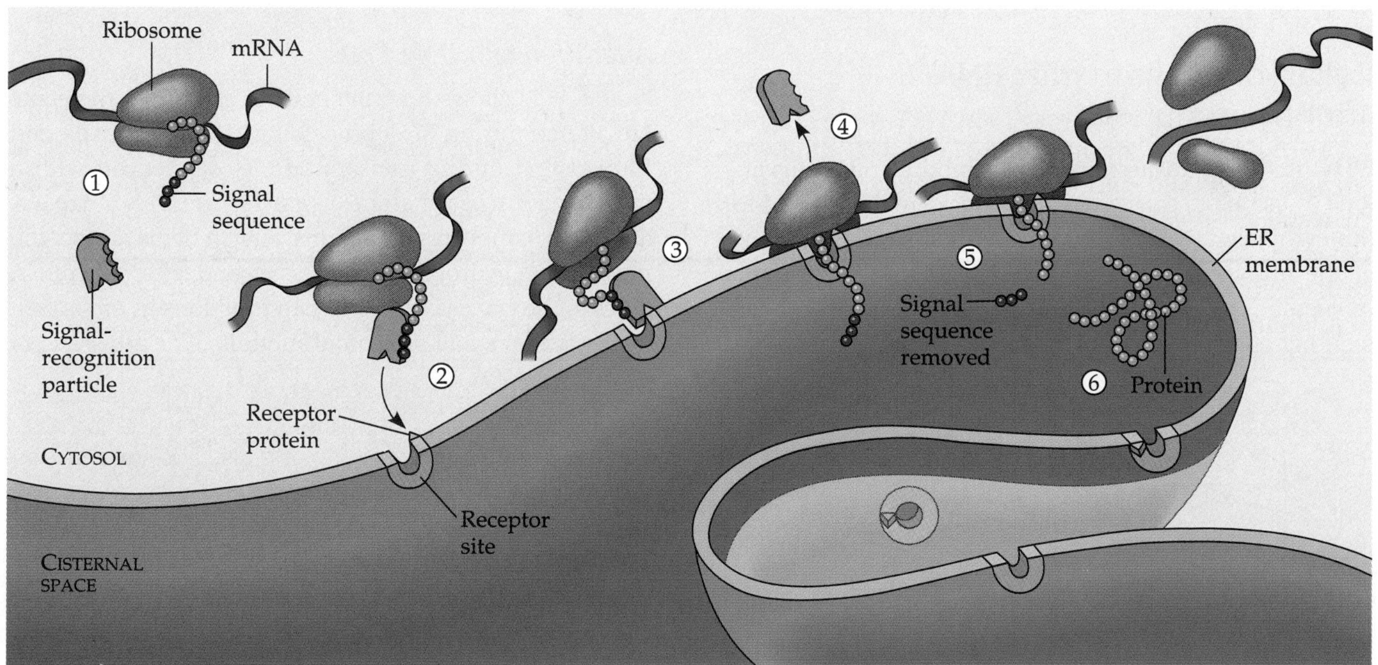

FIGURE 16.17
The signal mechanism for targeting proteins. Many polypeptide chains begin with a signal sequence, a stretch of amino acids that targets the protein for some organelle in the cell. ① In this example, a secretory protein, one that will eventually be secreted from the cell, is targeted for the endoplasmic reticulum (ER). ② A signal-recognition particle, a complex of proteins, identifies and binds to the signal sequence. ③ The signal-recognition particle then binds to a receptor protein that is built into the ER membrane. ④ The signal-recognition particle is released, while the elongating polypeptide snakes across the ER membrane. ⑤ An enzyme removes the signal sequence. ⑥ Released from the ribosome, the completed polypeptide folds into the conformation of a specific protein.

Comparing protein synthesis in prokaryotes and eukaryotes: *a review*

Bacteria and eukaryotes carry out transcription and translation in very similar ways, although some details of each step in the overall process of protein synthesis differ. There are also differences in some of the basic equipment, such as RNA polymerases and ribosomes. But these distinctions are not essential for a basic understanding of the pathway from gene to protein. What is important to learn is the significance of a eukaryotic cell's compartmental organization of protein synthesis. This point was made earlier in the chapter (see FIGURE 16.3), but it is worth reinforcing now that you know more about transcription and translation. These two processes are coupled in prokaryotes. In fact, bacterial ribosomes may attach to a growing mRNA molecule, and translation may begin before transcription has even been completed (FIGURE 16.18). In contrast, the nuclear envelope of a eukaryotic cell separates transcription from translation. This provides the time for RNA processing, an extra step between transcription and translation that does not occur in prokaryotes. Let's examine the mechanisms and functions of RNA processing in eukaryotic cells.

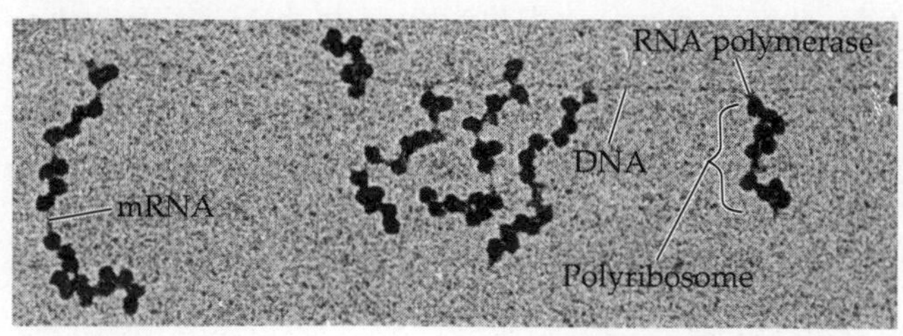

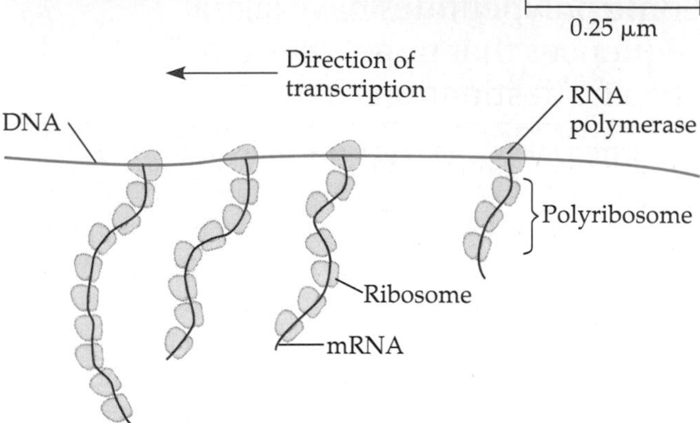

FIGURE 16.18
Coupled transcription and translation in bacteria. In prokaryotic cells, transcription and translation are not segregated by a nuclear envelope. The translation of mRNA can begin as soon as the leading end (5′) of the mRNA molecule peels away from the DNA template. The micrograph shows a strand of *E. coli* DNA being transcribed by RNA polymerase molecules. Attached to each RNA polymerase molecule is a growing strand of mRNA, which is already being translated by ribosomes. The newly synthesized polypeptides are not visible here (TEM).

Eukaryotic cells modify RNA after transcription

Enzymes in the eukaryotic nucleus modify mRNA in various ways before the genetic messages are dispatched to the cytoplasm. During this RNA processing, both ends of the mRNA molecule are altered. In some cases, the molecule is then cut apart, and parts of it are spliced together again.

Alteration of mRNA Ends

FIGURE 16.19 shows how the ends of an mRNA molecule are altered during RNA processing. The 5′ end, the end formed first during transcription, is capped off with a modified form of a guanine (G) nucleotide. This **5′ cap** has at least two important functions. First, it helps protect the mRNA from hydrolytic enzymes. Second, after the mRNA reaches the cytoplasm, the 5′ cap functions as an "attach here" sign for small ribosomal subunits. The other end of

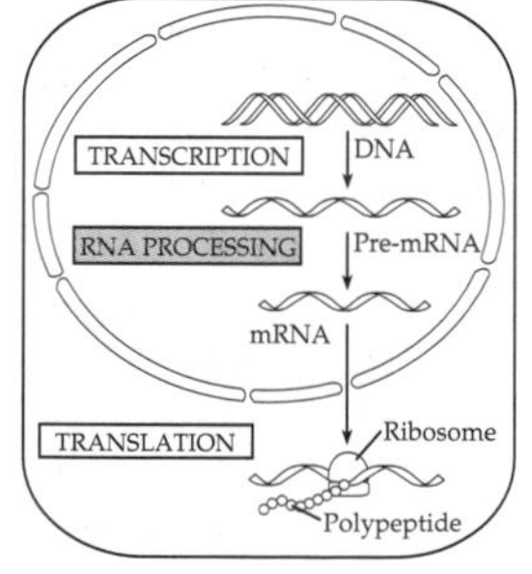

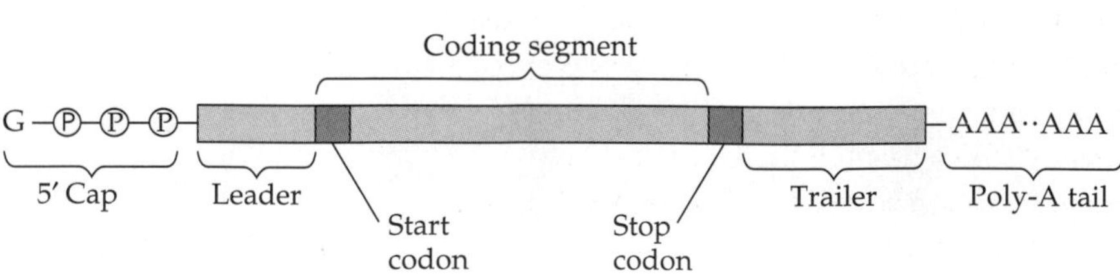

FIGURE 16.19
RNA processing: addition of the 5′ cap and poly-A tail. Enzymes alter the two ends of an mRNA molecule after transcription. A cap consisting of a modified guanosine triphosphate is added to the 5′ end. A poly-A tail consisting of up to 200 adenine nucleotides is attached to the 3′ end of the mRNA. These modified ends probably help protect the molecule from degradation. Also, the 5′ cap, along with a leader segment of RNA that is not translated into amino acids, functions as an "attach here" sign for ribosomes. The poly-A tail may be required for the mRNA to leave the nucleus. The tail is not attached directly to the stop codon, but to a trailer segment of RNA that is not translated.

an mRNA molecule, the 3′ end, is also modified before the message exits the nucleus. To the 3′ end, which is synthesized last during transcription, an enzyme adds a **poly-A tail** consisting of 30 to 200 adenine nucleotides. Like the 5′ cap, the poly-A tail helps inhibit degradation of the RNA. Addition of the tail may also play a regulatory role in protein synthesis by somehow facilitating the export of mRNA from the nucleus to the cytoplasm.

Split Genes and RNA Splicing

The most remarkable stage of RNA processing in the eukaryotic nucleus is the removal of a large portion of the molecule that is initially synthesized during transcription: a cut-and-paste job called **RNA splicing** (FIGURE 16.20). The average length of a transcription unit along a DNA molecule is about 8000 nucleotides, so the RNA product of transcription is also about that long. But it takes only about 1200 nucleotides to code for an average-sized protein of 400 amino acids. (Remember, each amino acid is coded for by a *triplet* of nucleotide bases.) This means that most eukaryotic genes and their RNA transcripts have long noncoding stretches of nucleotides, regions that are not translated. Even more surprising is that these noncoding sequences are interspersed between coding segments of the gene, and thus between coding segments of the mRNA transcript. In other words, the sequence of nucleotides that codes for a protein does not occur as an unbroken continuum. The noncoding segments of DNA are called intervening sequences, or **introns** for short. The coding regions are called **exons,** because they are eventually expressed (translated into protein). Richard Roberts and Phillip Sharp, who independently found evidence of "split genes" in 1977, shared the 1993 Nobel Prize in physiology or medicine for their surprising discovery.

Both introns and exons are transcribed to form an oversized RNA molecule. This pre-mRNA is more technically called *heterogeneous nuclear RNA (hnRNA),* a reference to the wide size range of these molecules. The hnRNA never leaves the nucleus; the mRNA molecule that enters the cytoplasm is an edited version of the original transcript. Enzymes excise the introns from the molecule and join the exons to form an mRNA molecule with a continuous coding sequence. This RNA splicing also occurs during the posttranscriptional processing of transfer RNA and ribosomal RNA.

The details of RNA splicing are still being worked out. Researchers have learned that the signals for RNA splicing are sets of a few nucleotides located at either end of each intron. Particles called *small nuclear ribonucleoproteins,* or *snRNPs* (pronounced "snurps"), play a key role in RNA splicing. As the name implies, these small particles are located in the cell nucleus and are composed of RNA and seven or more protein molecules. The RNA in a snRNP particle is called *small nuclear RNA (snRNA),* and it is typically a single molecule about 150 nucleotides long. Several snRNPs join to form an even larger assembly called a **spliceosome.** The spliceosome interacts with the ends of an RNA intron. It cuts at specific points to release the intron, then immediately joins the two exons that were adjacent to the intron (FIGURE 16.21).

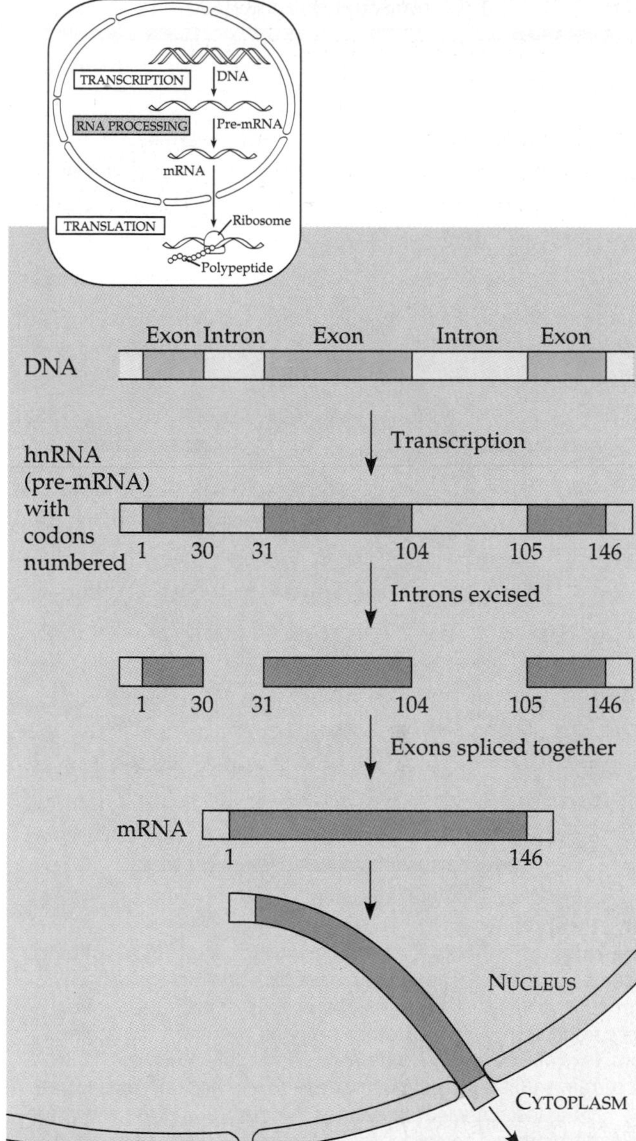

FIGURE 16.20
RNA processing: mRNA splicing. The gene depicted here codes for β-globin, one of the polypeptides of hemoglobin. β-globin is 146 amino acids long. Its gene has three segments containing coding regions, called exons, that are separated by noncoding regions, called introns. The entire gene is transcribed to form an hnRNA (heterogeneous nuclear RNA) molecule. However, before the molecule leaves the nucleus as mRNA, the introns are excised and the exons are spliced together. The mRNA also contains noncoding sequences at either end (the leader and trailer; see FIGURE 16.19).

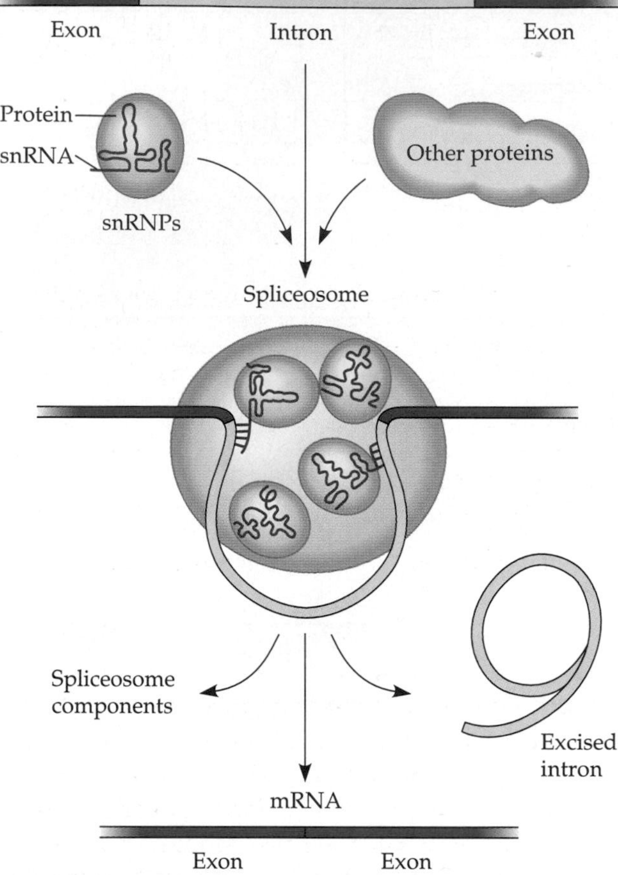

FIGURE 16.21
The roles of snRNPs and spliceosomes in mRNA splicing. After a eukaryotic gene containing exons and introns is transcribed, the RNA transcript combines with small nuclear ribonucleoproteins (snRNPs) and other proteins to form a molecular complex called a spliceosome. Within the spliceosome, the RNA of certain snRNPs forms base pairs with the ends of each intron, the RNA transcript is cut to release the intron, and the exons are spliced together. The spliceosome then comes apart, releasing mRNA, which now contains only exons.

Ribozymes

The splicing of other kinds of RNA transcripts, such as those for tRNA and rRNA, occurs in several different ways. However, as with mRNA splicing, RNA is often involved in catalyzing the reactions. In some cases—for example, in the splicing of rRNA in the ciliated protozoan *Tetrahymena*—the splicing occurs completely without proteins or even extra RNA molecules: The intron RNA itself catalyzes the process! RNA molecules that function as catalysts are called **ribozymes.**

The discovery of ribozymes rendered the statement "All biological catalysts are enzymatic proteins" obsolete. We have just seen that ribozymes catalyze reactions during RNA splicing. In addition, molecular biologists have recently learned that ribosomal RNA has catalytic functions during the translation process. For example, the ribosomal component that catalyzes the bonding of an amino acid to a growing polypeptide may be a ribozyme. TABLE 16.1 summarizes the diverse roles of RNA molecules during protein synthesis. The ability of RNA to perform so many different functions is based on variations in the three-dimensional shapes of different types of RNA. DNA may be the genetic material of cells, but RNA is much more versatile. As you will learn in Chapter 17, many viruses even use RNA as their genetic material. Thus, we can also add "All genes consist of DNA" to the list of overstated generalizations. Molecular biology has produced many reminders that the phrase "scientific dogma" is an oxymoron.

TABLE 16.1

The Major Types of RNA in a Eukaryotic Cell

TYPE OF RNA	FUNCTION
Messenger RNA (mRNA)	Carries information specifying amino acid sequences of proteins from DNA to ribosomes.
Transfer RNA (tRNA)	Serves as adaptor molecule in protein synthesis; translates mRNA codons into amino acids.
Ribosomal RNA (rRNA)	Plays structural and catalytic roles (as ribozymes) in ribosomes.
Heterogeneous nuclear RNA (hnRNA)	Pre-mRNA containing noncoding regions (introns) that separate coding region (exons). RNA splicing converts the pre-mRNA to mRNA.
Small nuclear RNA (snRNA)	Plays structural and catalytic roles (as ribozymes) in spliceosomes, the molecular complexes that splice mRNA.

The Functional and Evolutionary Importance of Introns

What are the biological functions of introns and gene splicing? One hypothesis is that introns play a regulatory role in the cell. Perhaps intron DNA includes sequences that control gene activity in some way, or perhaps the splicing process itself is part of a mechanism that regulates the passage of mRNA from the nucleus to the cytoplasm. Introns may also enable different kinds of cells in the same organism to make different proteins from a common gene. This can occur if all introns are removed from a particular transcript in one cell type, but one or more of the introns are left in place in the same transcript in another cell type. The maintained introns are then translated along with the mRNA's exons to produce a protein different from the one formed by a cell type where all introns were removed.

Introns also play an important role in the evolution of protein diversity. Many proteins have a modular architecture, consisting of structural and functional components called **domains.** One domain of an enzymatic protein, for instance, might include the active site, while another might attach the protein to some cellular membrane. In many cases, the exons of a "split gene" code for the different domains of a protein. It is possible for genetic recombination to modify the function of a protein by changing just one of its domains without altering the other domains. Because coding regions for a particular protein can be separated by considerable distances along the DNA, the frequency of recombination *within* a split gene can be higher than for a continuous coding region lacking introns. Thus, introns facilitate the recombination of exons between different alleles of a gene by increasing the probability that a crossover will switch one variation of an exon for another variation found on the homologous chromosome. This exon shuffling can lead to a novel protein with a single one of its multiple domains altered.

Mutations in one or more of a gene's exons can also contribute to protein evolution. Mutation, the ultimate source of genetic diversity, is the topic of the next section.

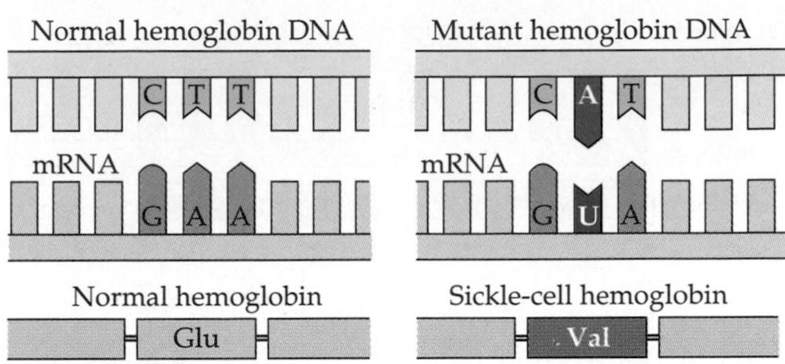

FIGURE 16.22
The molecular basis of sickle-cell disease. The allele that causes sickle-cell disease differs from the normal allele by a change in a single nucleotide—a point mutation. Where the normal gene for one of the polypeptides making up the protein hemoglobin has the base thymine, the sickle-cell allele has adenine. This alters one of the codons (base triplets) in the messenger RNA transcribed from the gene, placing the amino acid valine in the polypeptide in place of the glutamic acid found in the normal polypeptide. In individuals who are homozygous for the mutant allele, the sickling of red blood cells caused by the altered hemoglobin produces the multiple symptoms associated with sickle-cell disease (see FIGURE 13.16).

A point mutation can affect the function of a protein

Mutations are changes in the genetic makeup of a cell. In Chapter 14, we considered mutations that alter the structure of chromosomes. Now that you have learned about the genetic code and its translation, we can examine **point mutations,** which are chemical changes in just one nucleotide in a single gene.

If a point mutation occurs in a gamete, or in a cell that gives rise to gametes, it may be transmitted to offspring and to a succession of future generations. If the mutation has a noticeably adverse effect on the phenotype, the mutant condition is referred to as a genetic disorder, or hereditary disease. For example, we can trace the genetic basis of sickle-cell disease to a mutation affecting a single nucleotide in the gene that codes for one of the polypeptides of hemoglobin (FIGURE 16.22). Let's see how different types of point mutations translate into altered proteins.

Types of Point Mutations

Point mutations within a gene can be divided into two general categories: base-pair substitutions and base-pair insertions or deletions. While reading about how these mutations affect proteins, refer to the appropriate parts of FIGURE 16.23.

Substitutions. A **base-pair substitution** is the replacement of one nucleotide and its partner from the complementary DNA strand with another pair of nucleotides. Some substitutions are called *silent mutations* because, due to the redundancy of the genetic code, they have no effect on the protein coded for. In other words, a change in a base pair may transform one codon into another that is translated into the same amino acid. For example, if CCG mutated to CCA, the mRNA codon that used to be GGC would become GGU, and a glycine would still be inserted at the proper location in the protein (see FIGURE 16.5). Other changes of a single nucleotide pair may switch an amino acid but have little effect on the encoded protein. The new amino acid may have properties similar to those of the amino acid it replaces, or it may be in a region of the protein where the exact sequence of amino acids is not essential to the protein's function.

However, the base-pair substitutions of greatest interest are those that cause a readily detectable change in a protein. The alteration of a single amino acid in a crucial area of a protein—in the active site of an enzyme, for example—will significantly alter protein activity. Occasionally, such a mutation leads to an improved protein or one with novel capabilities that enhance the success of the mutant organism and its descendants. But much more often, such mutations are detrimental, creating a useless or less active protein that impairs cellular function.

Substitution mutations are usually **missense mutations;** that is, the altered codons still code for amino acids and thus make sense, although not necessarily the *right* sense. But if a point mutation changes a codon for an amino acid into a stop codon, translation will be terminated prematurely, and the resulting polypeptide will be shorter than the polypeptide encoded by the normal gene. Alterations that change an amino acid codon to a

WILD TYPE

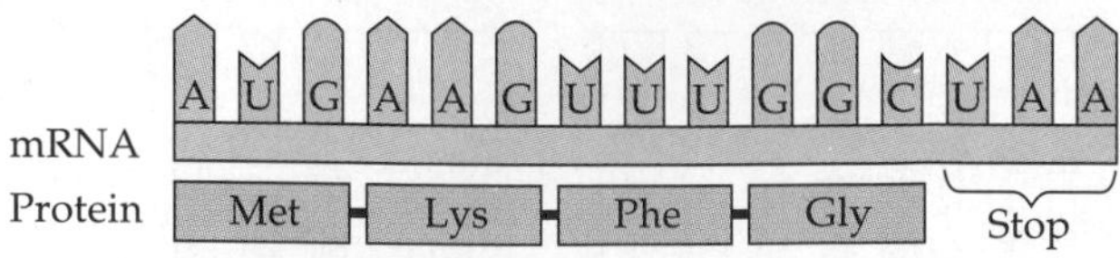

BASE-PAIR SUBSTITUTION

No effect on amino acid sequence

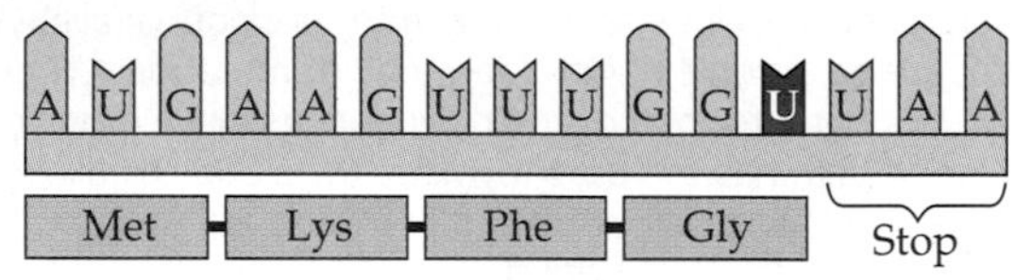

Missense

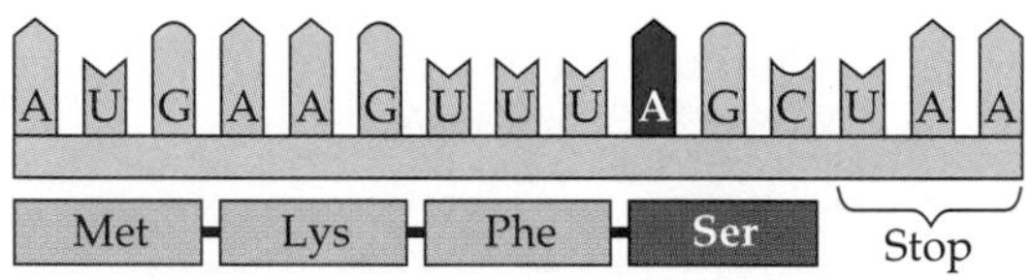

Nonsense

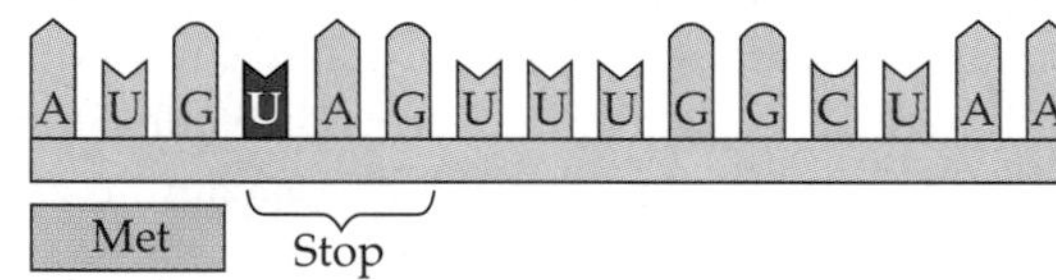

BASE-PAIR INSERTION OR DELETION

Frameshift causing extensive missense

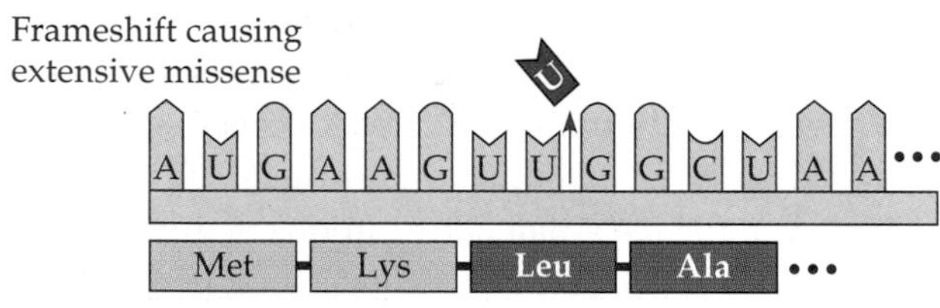

Frameshift causing immediate nonsense

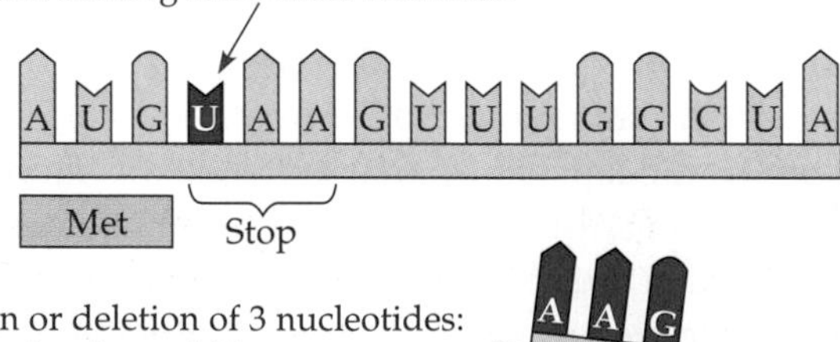

Insertion or deletion of 3 nucleotides: no extensive frameshift

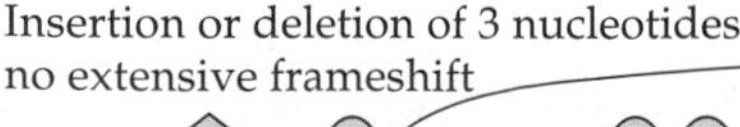
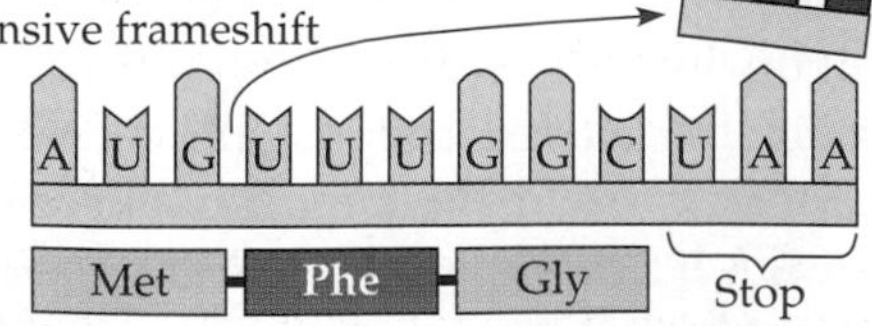

FIGURE 16.23
Categories and consequences of point mutations.
Mutations are changes in DNA, but they are represented here as they are reflected in mRNA and its protein product.

stop signal are called **nonsense mutations,** and nearly all nonsense mutations lead to nonfunctional proteins.

Insertions and Deletions. **Insertions** and **deletions** are additions or losses of one or more nucleotide pairs in a gene. These mutations usually have a more disastrous effect than substitutions on the resulting protein. Because mRNA is read as a series of nucleotide triplets during translation, the insertion or deletion of nucleotides may alter the reading frame (triplet grouping) of the genetic message. Such a mutation, called a **frameshift mutation,** will occur whenever the number of nucleotides inserted or deleted is not a multiple of 3. All the nucleotides that are downstream of the deletion or insertion will be improperly grouped into codons, and the result will be extensive missense ending sooner or later in nonsense—premature termination. Unless the frameshift is very near the end of the gene, it will produce a protein that is almost certain to be nonfunctional.

Mutagens

In the 1920s, Hermann Muller discovered that if he subjected fruit flies to X-rays, genetic changes increased in frequency. Using this method, Muller was able to obtain mutant *Drosophila* that he could use in his genetic studies. But he also recognized an alarming implication of his discovery: X-rays and other forms of radiation pose hereditary hazards to people as well as to laboratory animals. Since then, many other causes of mutations have been discovered.

The production of mutations can occur in a number of ways. Errors during DNA replication, repair, or recombination can lead to base-pair substitutions, insertions, or deletions. Mutations resulting from such errors are called *spontaneous mutations.*

A number of physical and chemical agents, called **mutagens,** interact with DNA to cause mutations. X-rays and ultraviolet (UV) light are examples of physical mutagens. For example, the UV of sunlight can produce thymine dimers in DNA (see FIGURE 15.15). Chemical mutagens fall into several categories, including base analogues, chemicals that are similar to normal DNA bases but that pair incorrectly. Researchers have developed various methods to test the mutagenic activity of different chemicals. One of the simplest and most popular methods is the Ames test (see the Methods Box).

■ What is a gene?

Our definition of a gene has evolved over the past few chapters. We began with the Mendelian concept of a gene as a discrete unit of inheritance that affects a phenotypic character (Chapter 13). We saw that Morgan and

METHODS: THE AMES TEST

The Ames test, named for its developer, microbiologist Bruce Ames, measures the mutagenic strength of various chemicals. The suspected mutagen is mixed with a culture of bacteria. It is also necessary to add a rat liver extract, which contains enzymes that convert certain chemicals from nonmutagenic to mutagenic forms. (These liver enzymes normally function to metabolize toxic substances, but unfortunately the chemical modification sometimes makes a bad situation worse by making the substances more mutagenic.) The bacteria, *Salmonella,* represent a mutant strain unable to produce the amino acid histidine. These bacteria fail to survive when plated on a growth medium lacking histidine. Some of the bacteria, however, undergo a back-mutation that restores the ability to make histidine. Such revertant bacteria give rise to colonies on the histidine-free medium. A mutagen will increase the frequency of these back-mutations, thereby resulting in more colonies on the histidine-free medium. The mutagenic activity of a chemical can be tested by comparing the number of colonies that grow after chemical treatment to control samples not treated with the suspected mutagen. One application of the Ames test is the screening of chemicals to identify those that can cause cancer. This works because most carcinogens (cancer-causing chemicals) are mutagenic and, conversely, most mutagens are carcinogenic.

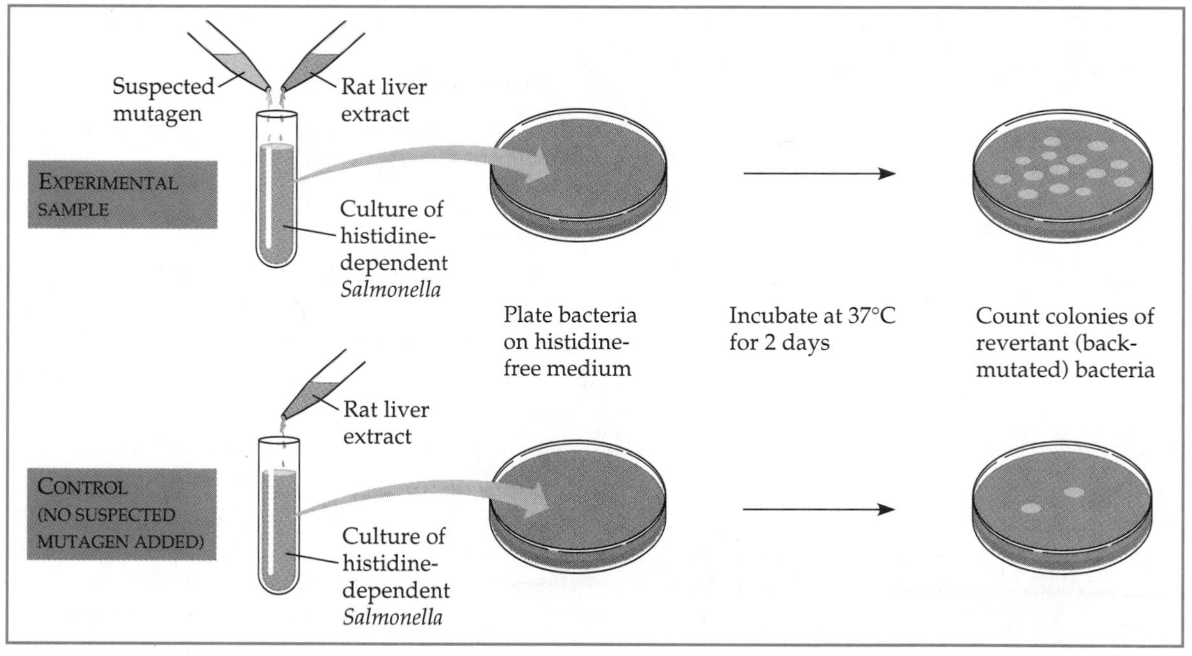

his colleagues assigned such genes to specific loci on chromosomes and that geneticists sometimes use the term *locus* as a synonym for gene (Chapter 14). We went on to view a gene as a region of specific nucleotide sequence along the length of a DNA molecule that includes thousands of genes (Chapter 15). Finally, in this chapter, we have moved toward a functional definition of a gene as a DNA sequence coding for a specific polypeptide chain. All these definitions are useful, depending on the context in which genes are being studied.

Even the one gene–one polypeptide definition must be refined and applied selectively. Most eukaryotic genes contain noncoding segments (introns), so large portions of these genes have no corresponding segments in polypeptides. Most molecular biologists also include promoters and other regulatory regions of DNA within the boundaries of a gene. These DNA sequences are not transcribed, but they can be considered part of the functional gene because they must be present for transcription to occur. Our molecular definition of a gene must also be broad enough to include the DNA that codes for rRNA, tRNA, and snRNA. These genes have no polypeptide products. The following definition of a gene applies more generally at the molecular level than does the one

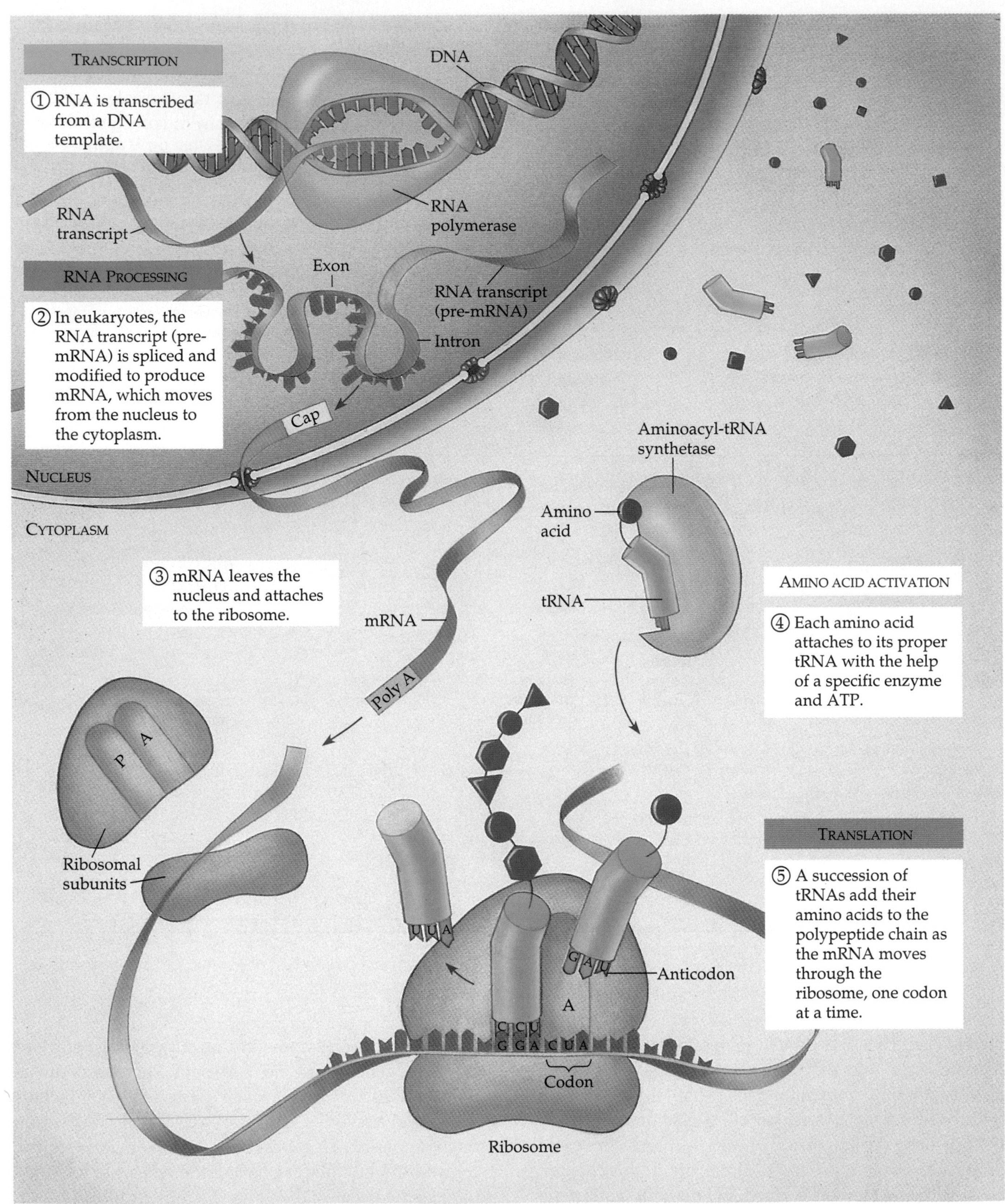

FIGURE 16.24
A summary of transcription and translation in eukaryotes. In general, the processes in prokaryotic and eukaryotic cells are similar. The major difference is the occurrence of RNA processing in the eukaryotic nucleus.

gene—one polypeptide idea: A gene is a region of DNA that is required for the production of an RNA molecule.

* * *

You have learned in this chapter how genetic information is expressed by its translation into proteins of specific structure and function, which in turn bring about an organism's phenotype. FIGURE 16.24 summarizes this path from gene to protein.

Genes are subject to regulation. The control of gene expression enables a bacterium, for example, to vary the amounts of particular enzymes as the metabolic needs of the cell change. In eukaryotes, the control of gene expression makes it possible for cells with the same DNA to diverge during their development into different cell types, such as muscle and nerve cells. The regulation of gene expression in eukaryotes is the main topic of Chapter 18. In the next chapter, Chapter 17, we begin our discussion of gene regulation by focusing on the simpler molecular biology of bacteria and viruses.

REVIEW OF KEY CONCEPTS (with page numbers and key figures)

- The study of metabolic defects provided evidence that genes specify proteins: *science as a process* (pp. 297–299, FIGURE 16.1)
  - DNA controls metabolism by commanding cells to make specific enzymes and other proteins.
  - The early idea that inherited diseases were a result of "inborn errors of metabolism" was supported by data obtained by Beadle and Tatum on mutant strains of *Neurospora* bread mold. These classic experiments gave rise to the one gene–one enzyme hypothesis.
  - Beadle and Tatum's hypothesis was later modified to one gene–one polypeptide. The amino acid sequence of each polypeptide chain is determined by a gene.
- Transcription and translation are the two main steps from gene to protein: *an overview* (pp. 299–300; FIGURES 16.3, 16.24)
  - Both nucleic acids and proteins are informational polymers assembled from linear sequences of nucleotides and amino acids, respectively.
  - Messenger RNA (mRNA) is the intermediate in the flow of information from DNA to proteins.
  - The nucleotide-to-nucleotide transfer of information from DNA to RNA is called transcription. The informational transfer from nucleotide sequence in RNA to amino acid sequence in a polypeptide is called translation.
- In the genetic code, a particular triplet of nucleotides specifies a certain amino acid: *a closer look* (pp. 301–304, FIGURE 16.4)
  - Genetic instructions from DNA are written in three-nucleotide units. In an mRNA molecule transcribed from a gene, these base triplets are called codons.
  - Of the 64 codons, 61 code for amino acids. In most cases, 2 or more codons are synonyms for the same amino acid. A few codons function as start and stop signals that mark the ends of a genetic message.
  - The near universality of the genetic code suggests that the code had already evolved in ancestors common to all kingdoms of life.
- Transcription is the DNA-directed synthesis of RNA: *a closer look* (pp. 304–306, FIGURE 16.7)
  - RNA synthesis on a DNA template is catalyzed by RNA polymerase. It follows the base-pairing rules governing DNA replication, except that in RNA, uracil substitutes for thymine.
  - Promoters, specific nucleotide sequences at the start of a gene, signal the initiation of mRNA synthesis. Transcription factors (proteins) help RNA polymerase recognize promoter sequences and bind to the DNA. Transcription continues until the RNA polymerase reaches the termination sequence of nucleotides on the DNA template.
- Translation is the RNA-directed synthesis of a polypeptide: *a closer look* (pp. 306–313, FIGURE 16.14)
  - Transfer RNA (tRNA) molecules pick up specific amino acids and line up by means of their anticodon triplets at complementary codons on the mRNA molecule.
  - The binding of a specific amino acid to its particular tRNA is a precise, ATP-driven process catalyzed by a family of aminoacyl-tRNA synthetase enzymes.
  - Ribosomes coordinate the coupling of tRNAs to mRNA codons. They provide a site for the binding of mRNA, as well as P and A sites for holding adjacent tRNAs, as amino acids are linked in the growing polypeptide chain. Each ribosome is composed of two subunits made of aggregates of protein and ribosomal RNA (rRNA).
  - The translation process comprises three stages: initiation, elongation, and termination.
  - Several ribosomes often read a single mRNA, forming polyribosome clusters.
  - A protein often undergoes one or more alterations during and after translation that affect its three-dimensional structure and hence its final activity in the cell.
- Some polypeptides have signal sequences that target them to specific destinations in the cell (p. 313, FIGURE 16.17)
  - In eukaryotic cells, proteins destined for membranes or for export from the cell are synthesized on ribosomes bound to the endoplasmic reticulum. A signal-recognition particle binds to a signal sequence on the leading end of the growing polypeptide, enabling the ribosome to bind to the ER.

- Proteins that will remain in the cytosol lack this signal and are manufactured on free ribosomes.
- Other signal sequences target proteins for mitochondria or chloroplasts.

- Comparing protein synthesis in prokaryotes and eukaryotes: *a review* (p. 314)
  - In comparing protein synthesis in prokaryotes and eukaryotes, the most important difference is in the spatial and temporal relationships of transcription and translation. In a bacterial cell, which lacks a nuclear envelope, translation of an mRNA can begin while transcription is still in progress. In a eukaryotic cell, the nuclear envelope separates transcription from translation, making RNA processing possible.
- Eukaryotic cells modify RNA after transcription (pp. 314–317, FIGURES 16.19, 16.20)
  - Eukaryotic mRNA molecules are processed before leaving the nucleus by modification of their ends and by RNA splicing.
  - The mRNA molecule receives a modified nucleotide for a cap at the 5′ end and a poly-A tail of nucleotides at the 3′ end. The cap and tail probably protect the molecule from degradation and enhance translation.
  - Most eukaryotic genes are interrupted by long noncoding regions, called introns, interspersed among coding regions, known as exons. RNA splicing involves removing the introns and joining the exons.
  - RNA splicing is catalyzed by small nuclear ribonucleoproteins (snRNPs), consisting of small nuclear RNA (snRNA) and proteins; they operate within larger assemblies called spliceosomes.
  - In some cases, only RNA is needed to catalyze RNA splicing. Catalytic RNA molecules are called ribozymes.
  - The shuffling of exons due to recombination contributes to the evolution of protein diversity.
- A point mutation can affect the function of a protein (pp. 317–318, FIGURE 16.23)
  - Point mutations are changes in a single nucleotide pair.
  - Base-pair substitutions within a gene have a variable effect, depending on whether or not an amino acid is actually altered, and if so, whether the alteration has any effect on the function of the protein. Many substitutions are detrimental, causing missense or nonsense mutations.
  - Base-pair insertions or deletions are almost always disastrous, often resulting in frameshift mutations that disrupt the codon messages downstream of the mutation.
  - Spontaneous mutations can occur during DNA replication or repair. In addition, various chemical and physical mutagens can alter the gene.
- What is a gene? (pp. 318–321)
  - Different definitions of a gene suit different situations. At the molecular level, a gene can be defined as a region of DNA required for the production of an RNA molecule.

SELF-QUIZ

1. The synthesis of an RNA molecule from a DNA template is known as
   a. transcription
   b. translation
   c. RNA splicing
   d. replication
   e. recombination

2. Which of the following statements about RNA polymerase is correct?
   a. It functions in translation.
   b. It transcribes both introns and exons.
   c. It is a ribozyme.
   d. It starts transcribing at an AUG triplet on one DNA strand.
   e. It can produce several polypeptide chains at one time through the creation of polyribosomes.

3. Which of the following is *not* true of a codon?
   a. It consists of three nucleotides.
   b. It may code for the same amino acid as another codon does.
   c. It never codes for more than one amino acid.
   d. It extends from one end of a tRNA molecule.
   e. It is the basic unit of the genetic code.

4. Beadle and Tatum discovered several classes of *Neurospora* mutants that were able to grow on minimal medium with arginine added. Class I mutants were also able to grow on medium supplemented with either ornithine or citrulline, whereas class II mutants could grow on citrulline medium but not on ornithine medium. The metabolic pathway of arginine synthesis is as follows:

$$\text{Precursor} \xrightarrow{} \underset{A}{} \text{Ornithine} \xrightarrow{} \underset{B}{} \text{Citrulline} \xrightarrow{} \underset{C}{} \text{Arginine}$$

   From these growth results, they could conclude that
   a. one gene codes for the entire metabolic pathway
   b. the genetic code of DNA is a triplet code
   c. class I mutants have their mutations later in the nucleotide chain than do class II mutants, and thus have more functional enzymes
   d. class I mutants have a nonfunctional enzyme at step A, and class II mutants have a nonfunctional enzyme at step B
   e. class I mutants have a nonfunctional enzyme at step B, and class II mutants have a nonfunctional enzyme at step C

5. The anticodon of a particular tRNA molecule is
   a. complementary to the corresponding mRNA codon
   b. complementary to the corresponding triplet in DNA
   c. the part of tRNA that bonds to a specific amino acid
   d. changeable, depending on the amino acid that attaches to the tRNA
   e. catalytic, making the tRNA a ribozmye

6. Which of the following is *not* true of RNA processing?
   a. Exons are excised and hydrolyzed before mRNA moves out of the nucleus.
   b. The existence of exons and introns may facilitate crossing over between regions of a gene that code for polypeptide domains.
   c. Ribozymes function in RNA splicing.
   d. RNA splicing may be catalyzed by spliceosomes.
   e. An initial RNA transcript is often much longer than the final RNA molecule that may leave the nucleus.

7. Which of the following terms includes all others in the list?
   a. ribozyme
   b. enzyme
   c. catalyst
   d. snRNP
   e. aminoacyl-tRNA synthetase
8. Using the genetic code in FIGURE 16.5, identify a possible sequence of nucleotides in the *DNA template* for an mRNA coding for the polypeptide sequence Phe-Pro-Lys.
   a. AAA-GGG-UUU
   b. TTC-CCC-AAG
   c. TTT-CCA-AAA
   d. AAG-GGC-TTC
   e. UUU-CCC-AAA
9. Which of the following mutations would be *most* likely to have a harmful effect on an organism? Explain your answer.
   a. a base-pair substitution
   b. a deletion of three bases near the middle of the gene
   c. a single base deletion near the middle of an intron
   d. a single base deletion close to the end of the coding sequence
   e. a single base insertion near the start of the coding sequence
10. Which component is *not directly* involved in the process known as translation?
   a. mRNA
   b. DNA
   c. tRNA
   d. ribosomes
   e. GTP

CHALLENGE QUESTIONS

1. A biologist inserted a gene from a human liver cell into the chromosome of a bacterium. The bacterium then transcribed this gene into mRNA and translated the mRNA into protein. The protein produced was useless; it contained many more amino acids than the protein made by the eukaryotic cell, and the amino acids were in a different sequence. Explain why.
2. The base sequence of the gene coding for a short polypeptide is CTACGCTAGGCGATTATC. What would be the base sequence of the mRNA transcribed from this gene? Using the genetic code chart (FIGURE 16.5), give the amino acid sequence of the polypeptide translated from this mRNA.
3. Choose one catalyst, either an enzyme or a ribozyme, and write a paragraph explaining how it functions in the pathway from gene to protein.

SCIENCE, TECHNOLOGY, AND SOCIETY

1. As part of the Human Genome Project (discussed in Chapter 19), researchers are determining the nucleotide sequences of human genes and identifying the proteins coded by the genes. Labs of the U.S. National Institutes of Health (NIH), for example, have worked out thousands of sequences, and similar analysis is being carried out by many private companies. Knowing the nucleotide sequence of a gene and identifying its product can be useful; this information might be used to treat genetic defects or produce life-saving medicines. U.S. law allows the first person or research group to isolate a pure protein or a gene to patent it, whether or not a practical use for the discovery has been demonstrated. The NIH and biotechnology companies have applied for patents on their discoveries. What are the purposes of a patent? How might the discoverer of a gene benefit from a patent? How might the public benefit? What kinds of negative impacts might result from patenting genes? Do you think individuals and companies should be able to patent genes and gene products? Why or why not? Under what conditions should such patenting be permitted?
2. Our civilization generates many potentially mutagenic chemicals (pesticides, for example) and modifies the environment in ways that increase exposure to other mutagens, notably UV radiation. What role should government play in identifying mutagens and regulating their industrial causes?

FURTHER READING

Barinaga, M. "Ribozymes: Killing the Messenger." *Science,* December 3, 1993. Is it possible to design RNA catalysts (ribozymes) to destroy specific RNA targets, including the RNA of the virus that causes AIDS?

Goldberg, A. L. "Functions of the Proteasome: The Lysis at the End of the Tunnel." *Science,* April 28, 1995. Controlled degradation of proteins as a regulatory mechanism in cells.

Radetsky, P. "Genetic Heretic." *Discover,* November 1990. The story of how the discovery of ribozymes led to a Nobel Prize.

Tijan, R. "Molecular Machines That Control Genes." *Scientific American,* February 1995. Transcription is a team effort.

Waldrop, M. "The Structure of the Second Genetic Code." *Science,* December 1, 1989. How enzymes match tRNAs and amino acids.

Wise, J. "Guides to the Heart of the Spliceosome." *Science,* December 24, 1993. How does the RNA splicing machinery recognize where to cut and paste?

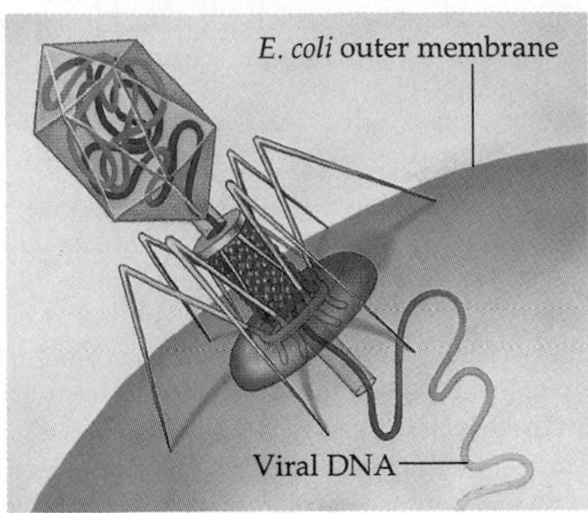

CHAPTER 17

MICROBIAL MODELS: THE GENETICS OF VIRUSES AND BACTERIA

KEY CONCEPTS

- Researchers discovered viruses by studying a plant disease: *science as a process*
- Most viruses consist of a genome enclosed in a protein shell
- Viruses can only reproduce within a host cell
- Phages exhibit two reproductive cycles: the lytic and lysogenic cycles
- Animal viruses are diverse in their modes of infection and mechanisms of replication
- Plant viruses are serious agricultural pests
- Viroids and prions are infectious agents even simpler than viruses
- Viruses may have evolved from other mobile genetic elements
- The short generation span of bacteria facilitates their evolutionary adaptation to changing environments
- Genetic recombination and transposition produce new bacterial strains
- The control of gene expression enables individual bacteria to adjust their metabolism to environmental change

The drawing that opens this chapter dramatizes the first step of one of the most remarkable events in biology: the genetic takeover of a cell by a virus. Infection begins when the virus injects its DNA into the host cell. In this case, the cell is the bacterium E. coli *and the virus is T4, which looks something like a miniature lunar landing craft. Molecular biology was born in the laboratories of microbiologists studying such viruses and bacteria. Microbiologists were the ones who provided most of the evidence that DNA is the genetic material (see Chapter 15). And microbiologists were the ones who outlined the major steps of replication, transcription, and translation, the three key processes in the flow of genetic information. Viruses and bacteria are the simplest biological systems—microbial models where scientists find life's fundamental molecular mechanisms in their most basic, accessible forms.*

The value of viruses and bacteria as model systems in biological research is just one reason to learn about these microbes. While microbial models have helped biologists understand the molecular genetics of more complex organisms, viruses and bacteria also have unique features that make microbial genetics interesting in its own right. These specialized mechanisms have important applications for understanding how viruses and bacteria cause disease. In addition, new techniques enabling scientists to manipulate genes and transfer them from one organism to another have emerged from the study of the genetic mechanisms of microorganisms. These techniques are having an important impact on both basic research and biotechnology (see Chapter 19).

In this chapter, we will explore the genetics of viruses and bacteria. Recall that bacteria are prokaryotic organisms, characterized by cells much smaller and more simply organized than those of eukaryotes, such as plants and animals. Viruses are smaller and simpler still, lacking the structures and most of the metabolic machinery found in cells (Figure 17.1). *In fact, most viruses are little more than aggregates of nucleic acids and proteins—genes packaged in protein coats. It is with these simplest of all genetic systems that we begin.*

Researchers discovered viruses by studying a plant disease: *science as a process*

Microbiologists were able to observe viruses indirectly long before they were actually able to see them. The story of how viruses were discovered begins in 1883 with

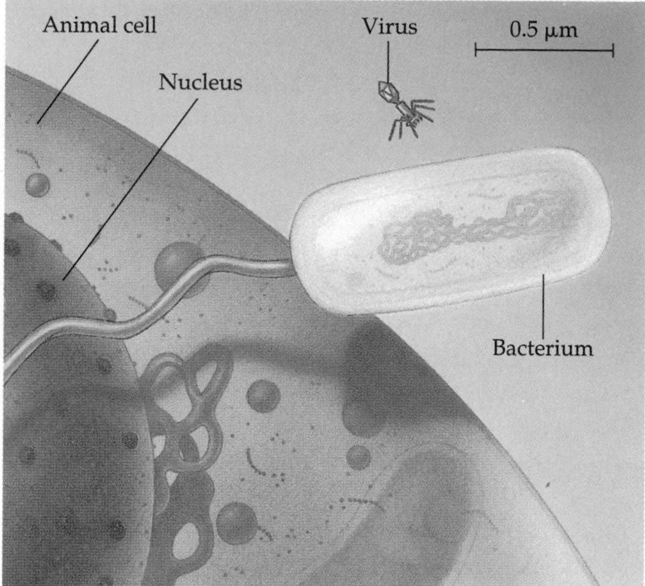

FIGURE 17.1
Comparing the sizes of a virus, a bacterium, and a eukaryotic cell. By studying viruses and bacteria, the simplest biological systems, scientists caught their first glimpses of the elegant molecular mechanisms of heredity.

A. Mayer, a German scientist who was seeking the cause of tobacco mosaic disease. This disease stunts the growth of tobacco plants and gives their leaves a mottled, or mosaic, coloration (see FIGURE 17.8a). Mayer discovered that the disease was contagious when he found he could transmit it from plant to plant by spraying sap extracted from diseased leaves onto healthy plants. He searched for a microbe in the infectious sap but found none. Mayer concluded that the disease was caused by unusually small bacteria that could not be seen with the microscope. This hypothesis was tested a decade later by D. Ivanowsky, a Russian, who passed sap from infected tobacco leaves through a filter designed to remove bacteria. After filtering, the sap still produced mosaic disease.

Ivanowsky clung to the hypothesis that bacteria caused tobacco mosaic disease. Perhaps, he reasoned, the pathogenic bacteria were so small they could pass through the filter. Or perhaps the bacteria made a filterable toxin that caused the disease. This latter possibility was ruled out in 1897 when Dutch microbiologist M. Beijerinck discovered that the infectious agent in the filtered sap could reproduce. Beijerinck sprayed plants with the filtered sap, and after these plants developed mosaic disease, he used their sap to infect more plants, continuing this process through a series of infections. The pathogen must have been reproducing, for its ability to cause disease was undiluted after several transfers from plant to plant.

In fact, the pathogen could reproduce only within the host it infected. Unlike bacteria, the mysterious agent of mosaic disease could not be cultivated on nutrient media in test tubes or petri dishes. Also, the pathogen was not inactivated by alcohol, which is generally lethal to bacteria. Beijerinck imagined a reproducing particle much smaller and simpler than bacteria. His suspicions were confirmed in 1935, when American scientist Wendell Stanley crystallized the infectious particle, now known as tobacco mosaic virus (TMV). Subsequently, TMV and many other viruses were actually seen with the help of the electron microscope.

Most viruses consist of a genome enclosed in a protein shell

The tiniest viruses are only 20 nm in diameter—smaller than a ribosome. Millions could easily fit on a pinhead. Even the largest viruses can barely be resolved with the light microscope. Stanley's discovery that viruses could be crystallized was exciting and puzzling news. Not even the simplest of cells can aggregate into regular crystals. But if viruses are not cells, then what are they? They are infectious particles consisting usually of only the viral genes enclosed in a shell made of proteins. Let's examine these two components of viruses more closely, then take a preliminary look at how they function in viral replication.

Viral Genomes

We usually think of genes as being made of double-stranded DNA—the conventional double helix—but viruses often defy this convention. Their genomes (sets of genes) may consist of double-stranded DNA, single-stranded DNA, double-stranded RNA, or single-stranded RNA, depending on the specific type of virus. A virus is called a DNA virus or an RNA virus, according to the type of nucleic acid that makes up its genome. In either case, the genome is usually organized as a single linear or circular molecule of nucleic acid. The smallest viruses have only four genes, while the largest have several hundred.

Capsids and Envelopes

The protein shell that encloses the viral genome is called a **capsid.** Depending on the type of virus, the capsid may be rod-shaped (more precisely, helical), polyhedral, or more complex in shape. Capsids are built from a large number of protein subunits called capsomeres, but the number of different *kinds* of proteins is usually small. Tobacco mosaic virus, for example, has a rigid, rod-shaped capsid made from over a thousand molecules of a single type of protein (FIGURE 17.2a). Adenoviruses, which infect the respiratory tracts of animals, have 252 identical protein molecules arranged into a polyhedral capsid with 20 triangular facets—an icosahedron (FIGURE 17.2b).

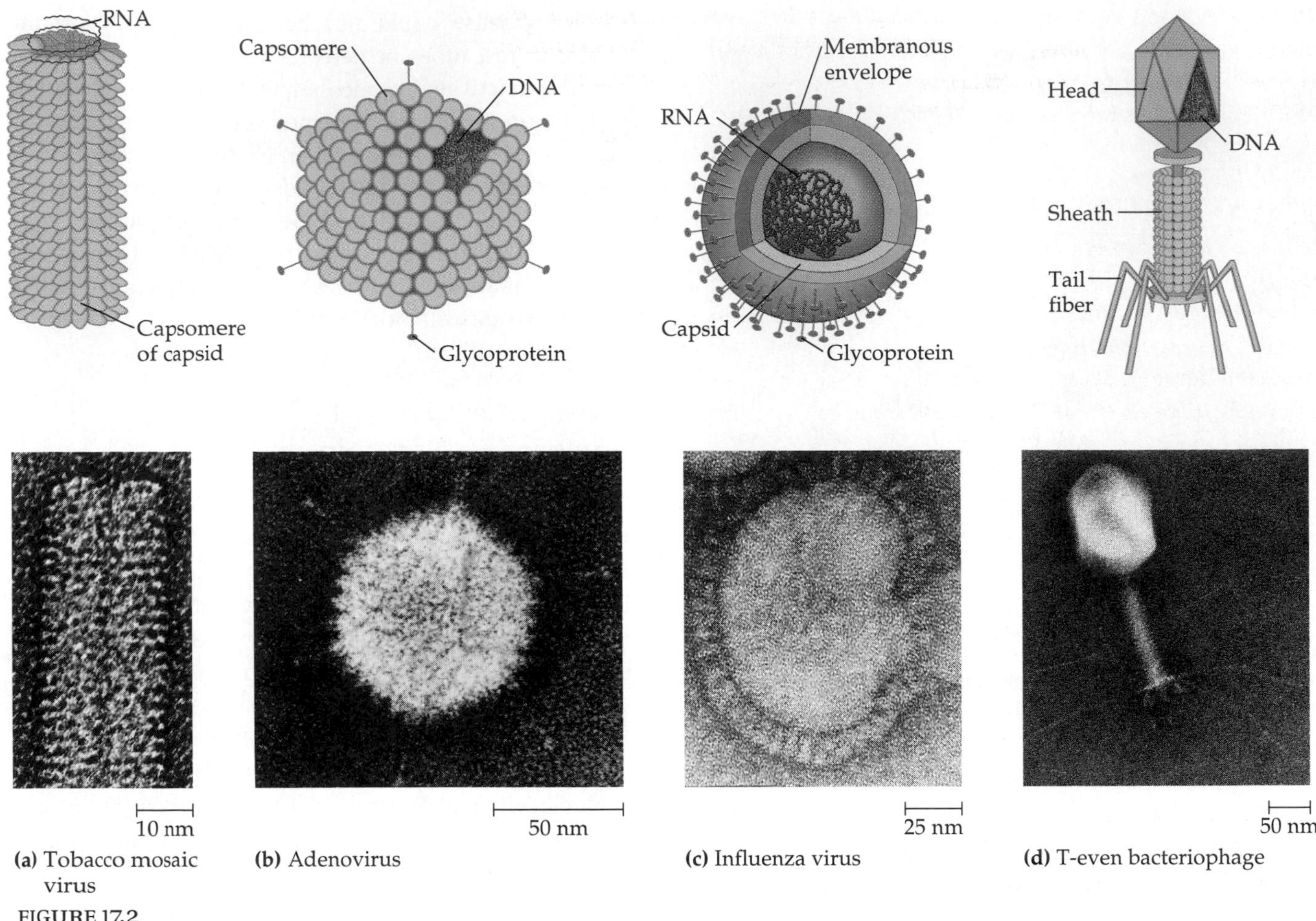

(a) Tobacco mosaic virus **(b)** Adenovirus **(c)** Influenza virus **(d)** T-even bacteriophage

FIGURE 17.2

Viral structure. Viruses are made up of nucleic acid (DNA or RNA) enclosed in a protein coat (the capsid) and sometimes further wrapped in a membranous envelope. The individual protein subunits making up the capsid are called capsomeres. Although viruses are diverse in size and shape, there are common structural motifs, most of which appear in the four examples shown here. (**a**) Tobacco mosaic virus has a helical capsid with the overall shape of a rigid rod. (**b**) Adenovirus has a polyhedral capsid with a protein spike at each vertex. Some adenoviruses cause upper respiratory infections in humans. (**c**) Influenza virus has an outer viral envelope studded with glycoprotein spikes. (**d**) Phages are viruses that infect bacteria. A T-even phage, such as T4, has a complex capsid consisting of a polyhedral head and a tail apparatus. DNA is stored in the head, and the tail piece functions in the injection of this DNA into a bacterium (see the drawing on p. 324). (All TEMs.)

Some viruses have accessory structures that help them infect their hosts. Influenza viruses, and many other viruses found in animals, have **viral envelopes,** membranes cloaking their capsids (FIGURE 17.2c). These envelopes are derived from membrane of the host cell, but in addition to host cell phospholipids and proteins, they also contain proteins and glycoproteins (proteins with carbohydrate covalently attached) of viral origin.

The most complex capsids are found among viruses that infect bacteria. Bacterial viruses are called bacteriophages, or simply **phages** (see Chapter 15). The first phages studied included seven that infect the bacterium *Escherichia coli.* These seven phages were named type 1 (T1), type 2 (T2), and so forth, in the order of their discovery. By coincidence, the three T-even phages—T2, T4, and T6—turned out to be very similar in structure. Their capsids have icosahedral heads that enclose the genetic material. Attached to the head is a protein tail piece with tail fibers that the phage uses to attach to a bacterium (FIGURE 17.2d).

Viruses can only reproduce within a host cell

Viruses are obligate intracellular parasites; they can only reproduce within a host cell. An isolated virus is unable to replicate itself—or do anything else, for that matter, except infect an appropriate host cell. Viruses lack the enzymes for metabolism and have no ribosomes or other equipment for making their own proteins. Thus, isolated viruses are merely protein-coated sets of genes in transit from one host cell to another.

Each type of virus can infect and parasitize only a limited range of host cells, called its **host range.** This host specificity depends on the evolution of recognition systems by the virus. Viruses identify their host cells by a "lock-and-key" fit between proteins on the outside of the virus and specific receptor molecules on the surface of the cell. Some viruses have host ranges broad enough to include several species. Swine flu virus, for example, can infect both hogs and humans, and the rabies virus can infect a number of mammalian species, including rodents, dogs, and humans. In other cases, viruses have host ranges so narrow that they infect a single species or a single type of tissue within a species. For instance, there are several phages that can parasitize only the bacterium *E. coli.* Human cold viruses usually infect only the cells lining the human upper respiratory tract, ignoring other tissues. And the virus that causes AIDS binds to a specific receptor on certain types of white blood cells.

A viral infection begins when the genome of a virus makes its way into a cell. The mechanism by which this nucleic acid enters the host varies, depending on the type of virus. For example, the T-even phages use their elaborate tail apparatus to inject DNA into a bacterium (see the drawing on p. 324). Once inside, the viral genome can commandeer its host, reprogramming the cell to copy the viral genes and manufacture capsid proteins (FIGURE 17.3). Most DNA viruses use the DNA polymerases of the host cell to synthesize new genomes along the templates provided by the viral DNA. In contrast, RNA viruses usually contain enzymes of their own to initiate replication of their genomes within the host. A cell has no native enzymes for copying RNA; it never produces its own RNA by transcribing one RNA molecule from another. We will describe the replication of DNA and RNA viruses in more detail later in the chapter, when we discuss specific viral infections.

Regardless of the type of viral genome, the parasite diverts its host's resources for viral production. The host provides the nucleotides for nucleic acid synthesis. It also uses its enzymes, ribosomes, tRNAs, amino acids, ATP, and other components to produce the viral proteins dictated by mRNA transcribed from the parasite's genes.

After the viral nucleic acid molecules and capsomeres are produced, their assembly into new viruses is often a spontaneous process, a process of self-assembly. In fact, the RNA and capsomeres of TMV can be separated in the laboratory and then reassembled to form complete viruses simply by mixing the components together again.

The simplest type of viral life cycle is completed when hundreds or thousands of viruses emerge from the infected host cell. The cell is often destroyed in the process. In fact, some of the symptoms of human viral infections, such as colds and influenza, result from cellular damage and death and from the body's responses to this destruction. The viral offspring that exit the cell that produced them have the potential to infect additional cells, spreading the viral infection.

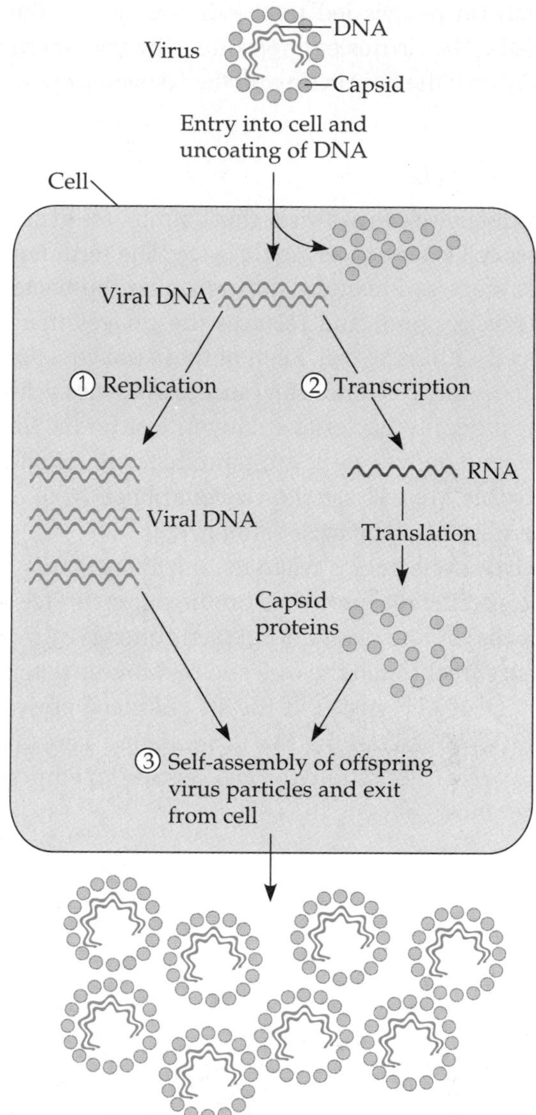

FIGURE 17.3
A simplified viral life cycle. A virus is an obligate intracellular parasite that uses the equipment of its host cell to reproduce. In this simplest of all possible viral life cycles, the parasite is a DNA virus with a capsid consisting of a single type of protein. ① After entering the cell, the viral DNA uses host nucleotides and enzymes to replicate itself. ② It uses other host materials and machinery to produce its own capsid proteins. ③ Viral DNA and capsid proteins then assemble into new virus particles.

There are many variations on the simplified life cycle we have traced in this overview, a few of which we will see as we take a closer look at some bacterial viruses (phages), animal viruses, and plant viruses.

Phages exhibit two reproductive cycles: the lytic and lysogenic cycles

The phages are the best understood of all viruses, although some of them are also among the most complex.

Research on phages led to the discovery that double-stranded DNA viruses can reproduce by two alternative mechanisms: the lytic cycle and the lysogenic cycle.

The Lytic Cycle

A reproductive cycle of a virus that culminates in death of the host cell is known as a **lytic cycle.** The term refers to the last stage of infection, during which the bacterium lyses (breaks open) and releases the phages that were produced within the cell. Each of these phages can then infect a healthy cell, and a few successive lytic cycles can destroy an entire bacterial colony in just hours. Viruses that depend on lytic cycles to reproduce are called **virulent viruses.** We will use the virulent phage T4 to examine the steps of a lytic cycle (FIGURE 17.4).

The lytic cycle begins when the tail fibers of a T4 virus stick to specific receptor sites on the outer surface of an *E. coli* cell. The sheath of the tail then contracts, thrusting a hollow core through the wall and membrane of the cell. Molecules of ATP stored in the T4 tail piece power this penetration. Functioning like a miniature syringe, the phage injects its DNA into the cell, leaving an empty capsid as a "ghost" outside the cell.

Once infected by the phage DNA, the *E. coli* cell quickly begins to transcribe and translate the viral genes. Phage T4 has about 100 genes, and most of their functions are known. One of the first phage genes translated by the *E. coli* cell codes for an enzyme that chops up the host cell's own DNA. The phage DNA itself is protected because it contains a modified form of cytosine that is not recognized by the enzyme. The cell now submits completely to the genetic commands of its invader.

Once the phage genome gains control of the cell, it commands the host's metabolic machinery to produce phage components. Nucleotides salvaged from the cell's degraded DNA are recycled to make many copies of the phage genome. Three separate sets of capsid proteins are made and assembled into phage tails, tail fibers, and polyhedral heads. The phage completes its subversion of the cell when one of its genes directs production of an enzyme (lysozyme) that digests the bacterial cell wall. With the cell wall damaged, osmotic uptake of water causes the cell to swell and finally to burst. The lysed bacterium releases 100 to 200 phage particles, which can then infect other cells nearby. The entire lytic cycle, from the phage's contact with the cell surface to lysis, takes only 20 to 30 minutes at 37°C.

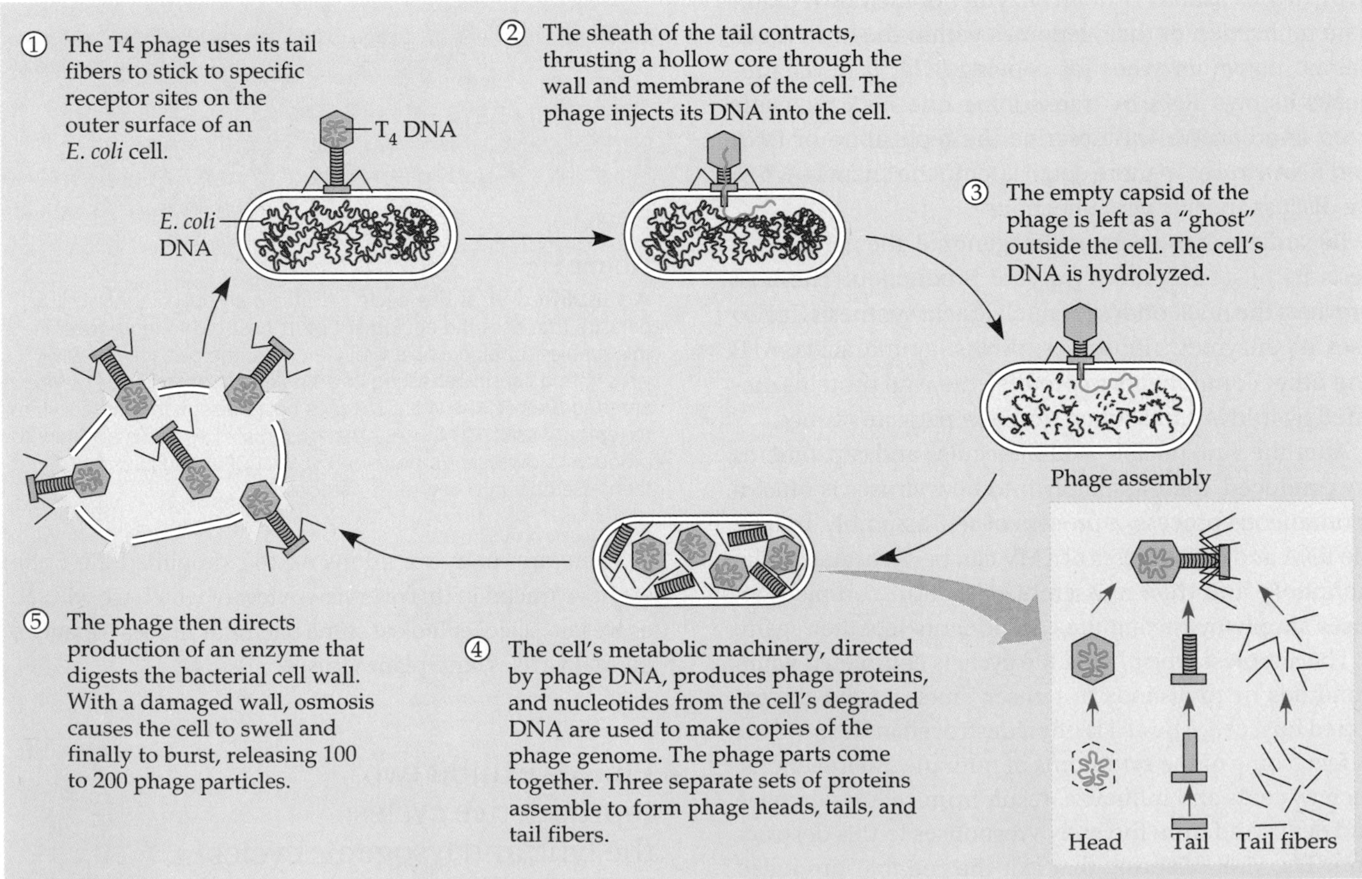

① The T4 phage uses its tail fibers to stick to specific receptor sites on the outer surface of an E. coli cell.
T_4 DNA
E. coli DNA
② The sheath of the tail contracts, thrusting a hollow core through the wall and membrane of the cell. The phage injects its DNA into the cell.
③ The empty capsid of the phage is left as a "ghost" outside the cell. The cell's DNA is hydrolyzed.
Phage assembly
④ The cell's metabolic machinery, directed by phage DNA, produces phage proteins, and nucleotides from the cell's degraded DNA are used to make copies of the phage genome. The phage parts come together. Three separate sets of proteins assemble to form phage heads, tails, and tail fibers.
⑤ The phage then directs production of an enzyme that digests the bacterial cell wall. With a damaged wall, osmosis causes the cell to swell and finally to burst, releasing 100 to 200 phage particles.
Head
Tail
Tail fibers

FIGURE 17.4
The lytic cycle of phage T4.

After reading about the lytic cycle, you may wonder why phages haven't exterminated all bacteria. Actually, bacteria are not defenseless. Natural selection favors bacterial mutants with receptor sites that are no longer recognized by a particular type of phage. And when phage DNA successfully enters a bacterium, various types of degradative enzymes may break it down. Enzymes called restriction enzymes, for example, recognize and cut up DNA that is foreign to the cell, including certain phage DNA. The bacterial cell's own DNA is chemically modified in a way that prevents attack by restriction enzymes. But just as natural selection favors bacteria with effective restriction enzymes, natural selection favors phage mutants that are resistant to these enzymes. Thus, the parasite-host relationship is in a constant evolutionary flux.

There is still another important reason bacteria have been spared from extinction as a result of phage activity. Many phages can check their own destructive tendencies and, instead of lysing their host cells, coexist with them in what is called the lysogenic cycle.

The Lysogenic Cycle

In contrast to the lytic cycle, which kills the host cell, the **lysogenic cycle** reproduces the viral genome without destroying the host. Viruses that are capable of the two different modes of reproducing within a bacterium are called **temperate viruses.** To compare the lytic and lysogenic cycles, we will examine a temperate virus called lambda, abbreviated with the Greek letter λ. Phage λ resembles T4 but lacks tail fibers.

Infection of an *E. coli* cell by λ begins when the phage binds to the surface of the cell and injects its DNA (FIGURE 17.5). Within the host, the λ DNA molecule forms a circle. What happens next depends on the type of reproductive mode: lytic cycle or lysogenic cycle. During a lytic cycle, the viral genes immediately turn the host cell into a λ-producing factory, and the cell soon lyses and releases its viral products. The viral genome behaves differently during a lysogenic cycle. The λ DNA molecule is incorporated into a specific site of the host cell's chromosome, and it is then known as a **prophage.** One prophage gene codes for a protein that represses most of the other prophage genes. Thus, the phage genome is mostly silent within the bacterium. How, then, does the phage reproduce? Every time the *E. coli* cell prepares to divide, it replicates the phage DNA along with its own and passes the copies on to daughter cells. A single infected cell can soon give rise to a large population of bacteria carrying the

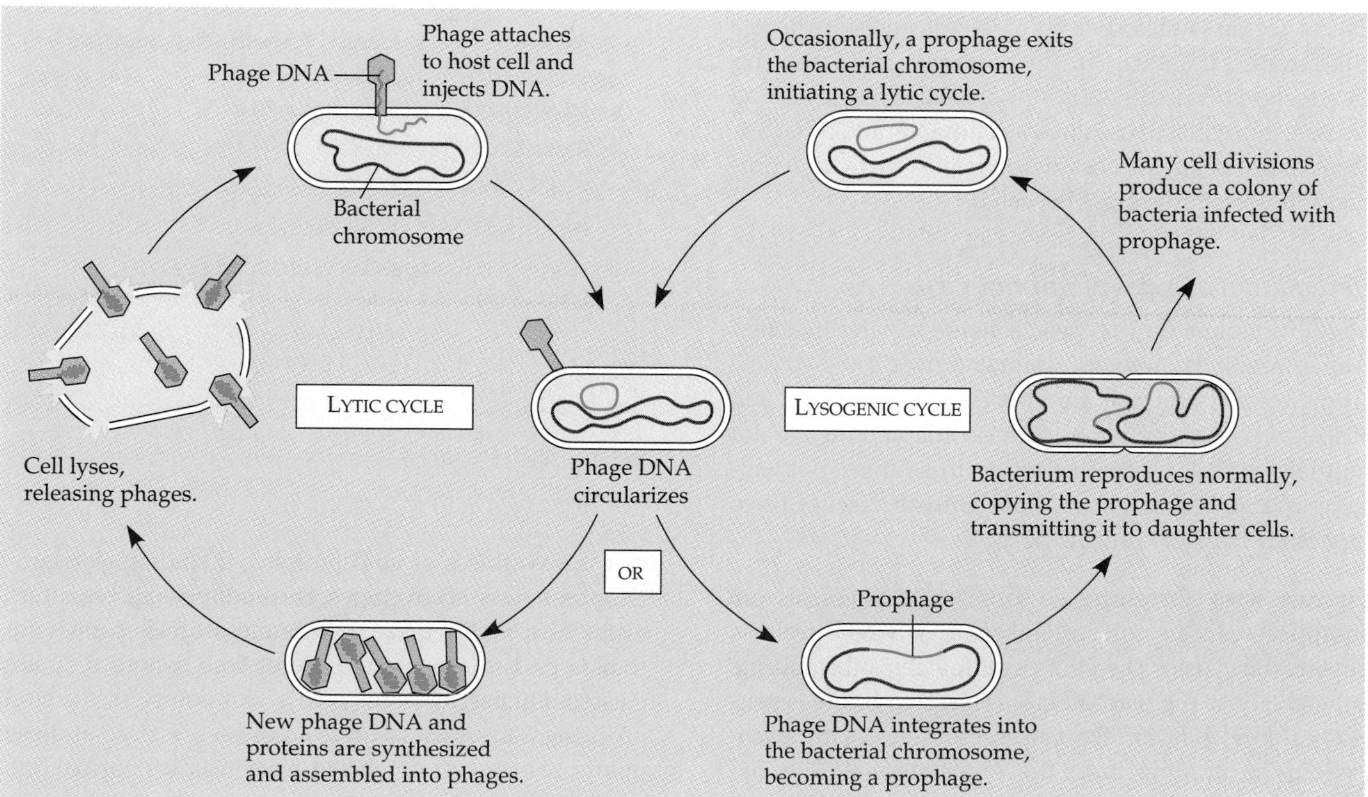
Phage DNA
Phage attaches to host cell and injects DNA.
Bacterial chromosome
Occasionally, a prophage exits the bacterial chromosome, initiating a lytic cycle.
Many cell divisions produce a colony of bacteria infected with prophage.
LYTIC CYCLE
LYSOGENIC CYCLE
Cell lyses, releasing phages.
Phage DNA circularizes
Bacterium reproduces normally, copying the prophage and transmitting it to daughter cells.
OR
Prophage
New phage DNA and proteins are synthesized and assembled into phages.
Phage DNA integrates into the bacterial chromosome, becoming a prophage.

FIGURE 17.5
The lysogenic and lytic reproductive cycles of phage λ. After entering the bacterial cell, the λ DNA can either integrate into the bacterial chromosome (lysogenic cycle) or immediately initiate the production of a large number of offspring (lytic cycle). In most cases, the lytic pathway is followed, but once a lysogenic cycle is started, the prophage may be carried in the host cell's chromosome for many generations.

virus in prophage form. This mechanism enables viruses to propagate without eliminating the host cells upon which they depend.

The term *lysogenic* implies that prophages can, at some point, give rise to active phages that lyse their host cells. This occurs when the λ genome exits the bacterial chromosome. At this time, the λ genome commands the host cell to manufacture phages and then self-destruct, releasing the infectious phages. It is usually an environmental trigger, such as radiation or the presence of certain chemicals, that switches the virus from the lysogenic to the lytic mode.

In addition to the gene for the repressor protein, a few other prophage genes may also be expressed during lysogenic cycles, and these genes may alter the phenotype of the host bacteria. This can have important medical significance. For example, the bacteria that cause the human diseases diphtheria, botulism, and scarlet fever would be harmless if it were not for certain prophage genes that induce bacterial production of toxins.

Animal viruses are diverse in their modes of infection and mechanisms of replication

Everyone has suffered from viral infections, whether chicken pox, influenza, or the common cold. TABLE 17.1 lists some important classes of animal viruses. Like all viruses, those that cause illness in humans and other animals are obligate intracellular parasites that can reproduce only after infecting host cells.

TABLE 17.1

Classes of Animal Viruses, Grouped by Type of Nucleic Acid

CLASS*		EXAMPLES/DISEASES
I. dsDNA**		
	Papovavirus	Papilloma (human warts, cervical cancer); polyoma (tumors in certain animals)
	Adenovirus	Respiratory disease; some cause tumors in certain animals
	Herpesvirus	Herpes simplex I (cold sores); herpes simplex II (genital sores); varicella zoster (chicken pox, shingles); Epstein-Barr virus (mononucleosis, Burkitt's lymphoma)
	Poxvirus	Smallpox; vaccinia; cowpox
II. ssDNA (parvovirus)		
		Roseola; most parvoviruses depend on coinfection with adenoviruses for growth
III. dsRNA (reovirus)		
		Diarrhea viruses
IV. ssRNA that can serve as mRNA		
	Picornavirus	Poliovirus; rhinovirus (common cold); enteric (intestinal) viruses
	Togavirus	Rubella virus; yellow fever virus; encephalitis viruses
V. ssRNA that is a template for mRNA		
	Rhabdovirus	Rabies
	Paramyxovirus	Measles, mumps
	Orthomyxovirus	Influenza viruses
VI. ssRNA that is a template for DNA synthesis (retrovirus)		
		RNA tumor viruses (e.g., leukemia); HIV (AIDS virus)

*The subclasses within each class differ mainly in capsid structure and in the presence or absence of a membrane envelope.
**ds = double-stranded; ss = single-stranded.

Reproductive Cycles of Animal Viruses

Many variations on the basic scheme of viral infection are represented among the animal viruses. The two variations we will examine are viruses with envelopes and viruses with RNA genomes. (These adaptations are not mutually exclusive—that is, some viruses have both envelopes and RNA genomes—but we will discuss them separately here for convenience.)

Viruses with Envelopes. Some animal viruses are equipped with an outer membrane, or viral envelope, outside the capsid. The viral envelope helps the parasite enter the host cell (FIGURE 17.6). This membrane is generally a lipid bilayer, like cell membranes, with glycoproteins protruding from the outer surface. The glycoprotein spikes bind to specific receptor molecules on the surface of the host cell. The viral envelope then fuses with the host's plasma membrane, transporting the capsid and viral genome into the cell. After cellular enzymes remove the capsid, the viral genome can replicate and direct the synthesis of viral proteins, including glycoproteins for new viral envelopes. The endoplasmic reticulum of the host makes these membrane proteins, which are transported to the plasma membrane, where they are clustered in patches that serve as exit points for the viral offspring. The viruses bud from the cell surface at these points, in a process much like exocytosis in a normal cell, wrapping themselves in membrane as they go. In other words, the viral envelope is derived from the host cell's membrane, although some of the molecules of this membrane are specified by viral genes. The enveloped viruses are now free to spread the infection to other cells.

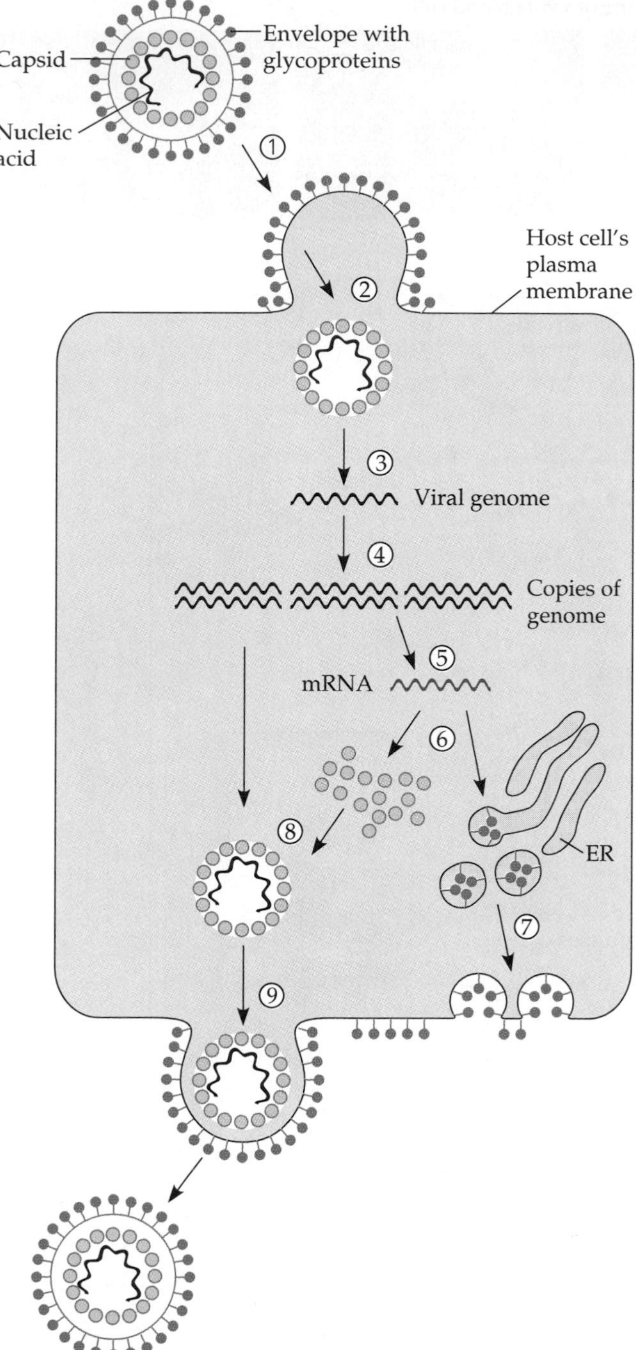

FIGURE 17.6

The reproductive cycle of an enveloped virus. ① Glycoproteins projecting from the viral envelope recognize and bind to specific receptor molecules on the surface of the host cell. The viral envelope fuses with the cell's plasma membrane. ② The capsid and viral genome enter the cell. ③ Cellular enzymes remove the capsid. ④ The viral genome is replicated, and ⑤ the copies are transcribed into messenger RNA. ⑥ The mRNA is translated into both capsid proteins and glycoproteins characteristic of the viral envelope. The host cell's endoplasmic reticulum (ER) synthesizes the glycoproteins. ⑦ Vesicles transport the glycoproteins to the cell's plasma membrane. ⑧ Capsids assemble around the viral nucleic acid molecules. ⑨ The virus buds from the cell. Its envelope, studded with glycoproteins, is derived from the cell's plasma membrane.

Notice that this reproductive cycle does not necessarily kill the host cell, in contrast to the lytic cycles of phages.

Other viruses have envelopes that are not derived from plasma membrane. Herpesviruses, for example, have envelopes derived from the nuclear membrane of the host. The genomes of herpesviruses are double-stranded DNA, and these viruses reproduce within the cell nucleus, using a combination of viral and cellular enzymes to replicate and transcribe their DNA. While within the nucleus, herpesvirus DNA may become integrated into the cell's genome as a **provirus,** similar to a bacterial prophage. Once acquired, herpes infections (including cold sores and genital sores) tend to recur throughout a person's life. Between these episodes, the virus apparently remains latent within the host cells' nuclei. From time to time, physical or emotional stress may cause the herpes proviruses to be excised from the host's genome and reproduce, resulting in the blisters of active infections.

RNA Viruses. The full range of viral genomes is represented among animal viruses. Particularly interesting are the RNA viruses, and we will discuss their molecular biology here, even though some phages and most plant viruses are also RNA viruses.

The RNA viruses with the most complicated reproductive cycles are the **retroviruses.** *Retro,* meaning "backward," refers to the reverse direction in which genetic information flows in these viruses. Retroviruses are equipped with a unique enzyme called **reverse transcriptase,** which can transcribe DNA from an RNA template, providing an RNA ⟶ DNA information flow. The newly formed DNA then integrates as a provirus into a chromosome within the nucleus of the animal cell. The host's RNA polymerase transcribes the viral DNA into RNA molecules, which can function both as mRNA for the synthesis of viral proteins and as new genomes for viral offspring released from the cell. A retrovirus of particular importance is **HIV (human immunodeficiency virus),** the virus that causes **AIDS (acquired immunodeficiency syndrome).** FIGURE 17.7 traces the reproductive cycle of HIV as an example of a retrovirus. We will postpone a detailed discussion of AIDS until Chapter 39.

Important Viral Diseases in Animals

The link between a viral infection and the symptoms it produces is often obscure. Some viruses damage or kill cells by causing the release of hydrolytic enzymes from lysosomes. Some viruses cause the infected cells to produce toxins that lead to disease symptoms, and some have toxic components themselves, such as envelope proteins. How much damage a virus causes depends partly on the ability of the infected tissue to regenerate by cell division. We usually recover completely from colds

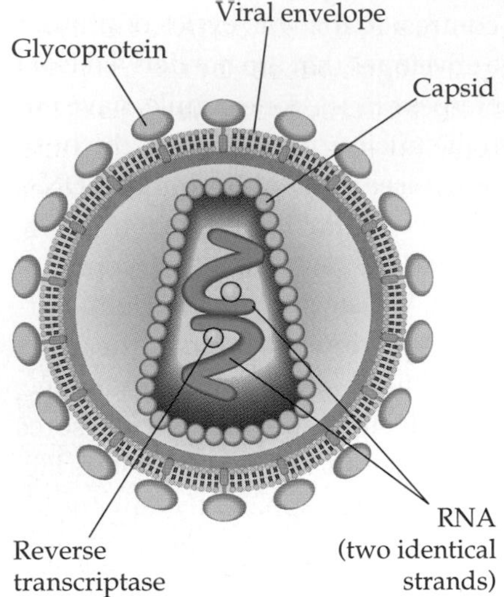

(a)

FIGURE 17.7

The life cycle of HIV, a retrovirus.

(**a**) The structure of HIV, the infectious agent that causes AIDS. The glycoproteins of the envelope enable the virus to bind to specific receptors on the surface of certain white blood cells. Although there are two RNA molecules, they are identical, not complementary, strands. (**b**) The reproductive cycle of HIV. ① The genome enters a host cell when the virus fuses with the plasma membrane and the proteins of the capsid are enzymatically removed. ② Reverse transcriptase then catalyzes the synthesis of DNA complementary to the RNA template provided by the viral genome. ③ The new DNA strand then serves as a template for the synthesis of a complementary DNA strand, and ④ the double-stranded DNA is incorporated as a provirus into the host cell's genome. ⑤ Proviral genes are transcribed into mRNA molecules that are ⑥ translated in the cytoplasm into HIV proteins. The RNA transcribed from the provirus also provides the genomes for the next generation of viruses. ⑦ The assembly of capsids around genomes is followed by ⑧ budding of the new viruses from the host cell. Transmission electron micrographs (artificially colored) depict HIV entering (top) and leaving (bottom) a human white blood cell.

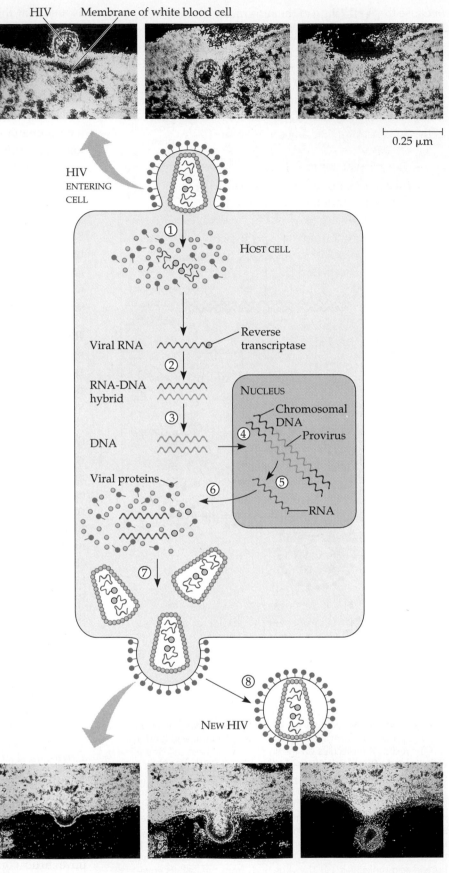

HIV
Membrane of white blood cell
0.25 μm
HIV ENTERING CELL
① HOST CELL
Viral RNA
Reverse transcriptase
②
RNA-DNA hybrid
③
DNA
NUCLEUS
Chromosomal DNA
④
Provirus
⑤
RNA
Viral proteins
⑥
⑦
⑧
NEW HIV

(b)

because the epithelium of the respiratory tract, which the viruses infect, can efficiently repair itself. In contrast, the poliovirus attacks nerve cells, which do not divide and cannot be replaced. Polio's damage to such cells, unfortunately, is permanent. Many of the temporary symptoms associated with viral infections, such as fever, aches, and inflammation, may actually result from the body's own efforts at defending itself against the infection.

As we will see in Chapter 39, the immune system is a complex and critical part of the body's natural defense mechanisms. The immune system is also the basis for the major medical weapon for preventing viral infections—vaccines. **Vaccines** are harmless variants or derivatives of pathogenic microbes that stimulate the immune system to mount defenses against the actual pathogen.

The term *vaccine* is derived from *vacca,* the Latin word for cow; the first vaccine, against smallpox, consisted of cowpox virus. In the late 1700s, Edward Jenner, an English physician, learned from his patients in farm country that milkmaids who had contracted cowpox (a milder disease that usually infects cows) were resistant to subsequent smallpox infections. In his famous experiment of 1796, Jenner scratched a farmboy with a needle bearing fluid from a sore of a milkmaid who had cowpox. When the boy was later exposed to smallpox, he resisted the disease.

The cowpox and smallpox viruses are so similar that the immune system cannot distinguish them. Vaccination with the mimic, the cowpox virus, sensitizes the immune system to react vigorously in defending the body, if it is ever exposed to the actual smallpox virus. This strategy has eradicated smallpox, which was once a devastating scourge in many parts of the world. There are also effective vaccines against many other viral diseases, including polio, rubella, measles, and mumps.

Although vaccines can prevent illnesses caused by certain viruses, medical technology can do little, at present, to cure most viral infections once they occur. The antibiotics that help us recover from bacterial infections are powerless against viruses. Antibiotics kill bacteria by inhibiting enzymes or biosynthetic processes specific to the pathogens, but viruses have few or no enzymes of their own; they use those of their hosts. However, a few antiviral drugs interfere with viral nucleic acid synthesis. One such drug is adenine arabinoside (also called Ara-A or vidarabine), which acts on several human viruses at concentrations well below those that inhibit the synthesis of host nucleic acid. Another is acyclovir, which seems to inhibit herpesvirus DNA synthesis.

Emerging Viruses

HIV, the AIDS virus, seemed to make a sudden appearance in the early 1980s. In 1993, dozens of people in the southwestern states of the U.S. died from hantavirus infection, which the news media first described as a "new" disease. Epidemiologists are still puzzled by the 1995 outbreak of deadly Ebola virus infections in Zaire. Each year, new strains of influenza virus cause millions to miss work or classes. From where or what do these and other "emerging viruses" arise?

Three processes contribute to the emergence of viral diseases:

1. *An existing virus can evolve and cause disease in individuals who had developed immunity to the ancestral virus.* For example, annual or biennial influenza episodes are caused by viruses that have evolved into new genetic strains. The "new model" is different enough from last year's virus that the human immune system must develop defenses as though it were responding to the virus for the first time. Between flu outbreaks, the evolving viruses are maintained within other animal hosts, especially ducks and other water fowl.

2. *An existing virus can spread from one host species to another.* For example, monkeypox virus, which causes pocking of the host's skin, spread from African monkeys to Asian monkeys in the 1950s, when monkeys from those two continents were transported together under crowded conditions to laboratories that were using the animals to develop and test polio vaccines. The World Health Organization documented the first cases of monkeypox in humans in forest villages of Zaire (in central Africa) in the 1970s. It is likely that humans did not acquire the virus directly from monkeys, but rather from squirrels and other game animals that had harbored the virus without becoming ill.

3. *An existing virus can disseminate from a small population to become more widespread.* The hantavirus that made news in 1993 was not really a new virus, even to humans. Small rodents called deer mice are the reservoirs for the virus, and it is probable that occasional infections of humans in the southwestern U.S. have occurred for decades. Indeed, to the Navajo, mice are taboo, bearers of strange illnesses. In 1993, the population of deer mice exploded after a wet year increased their food supply. Humans became infected when they inhaled dust containing hantavirus deposited in the urine and feces of deer mice. AIDS is another example of a disease that may have been unnoticed and unnamed for decades before it became widespread. In the case of HIV, technological and social factors—including affordable travel, blood transfusion technology, sexual promiscuity, and intravenous drug abuse—allowed a rare disease to become a global epidemic.

Thus, emerging viruses are generally not new, but are existing viruses that expand their "host territory" by evolving, by spreading to new host species, or by disseminating to a larger proportion of the host species.

Environmental change, natural or caused by humans, can increase the viral traffic responsible for emerging diseases. For example, new roads through remote areas can allow viruses to spread between previously isolated human populations. Another problem is the destruction of forests to expand cropland, an environmental disturbance that brings humans into contact with other animals that may host viruses capable of infecting humans.

Viruses and Cancer

For many years, scientists have recognized that some viruses can cause cancer in animals. Research on these tumor viruses, which include members of the retrovirus, papovavirus, adenovirus, and herpesvirus groups (see TABLE 17.1), has been facilitated by techniques for growing them in cell cultures. When certain tumor viruses infect animal cells growing in laboratory culture, the cells undergo transformation into a cancerous state. They assume the rounded shapes characteristic of cancer cells and abandon their orderly growth.

In a few cases, there is strong evidence that viruses cause certain types of human cancer. The virus responsible for hepatitis B also seems to cause liver cancer in individuals with chronic hepatitis. And the Epstein-Barr virus, the herpesvirus that causes infectious mononucleosis, has been linked to several types of cancer prevalent in parts of Africa, notably Burkitt's lymphoma. Papilloma viruses (of the papovavirus group) have been associated with cancer of the cervix. Among the retroviruses, one called HTLV-1 causes a type of adult leukemia. All tumor viruses transform cells through the integration of viral nucleic acid into host cell DNA.

Scientists have identified a number of viral genes directly involved in triggering cancerous characteristics in cells. Many of these genes, called oncogenes, are not unique to tumor viruses or tumor cells; they are also found in normal cells of many species. The oncogenes identified thus far all seem to code for cellular growth factors or for proteins involved in growth factor action (for example, growth factor receptors). In some cases, the tumor virus lacks oncogenes and transforms the cell simply by turning on or increasing the expression of one or more of the cell's own oncogenes. Whatever the mechanism by which a particular virus causes cancer, there is evidence that more than one oncogene must usually be activated to transform a cell into a fully cancerous state. It is likely that infection with most cancer-causing viruses is effective only in combination with other events, such as exposure to chemical mutagens. This leads us to suspect that **carcinogens** (*non*viral cancer-causing agents) also act by affecting the activity of cellular oncogenes. (Oncogenes and cancer will be covered more thoroughly in Chapter 18.)

Plant viruses are serious agricultural pests

Plant viruses can stunt plant growth and diminish crop yields (FIGURE 17.8a). Most plant viruses discovered thus far are RNA viruses. Many of them, including the tobacco mosaic virus, have rod-shaped capsids with protein arranged in a spiral.

There are two major routes by which a plant viral disease can spread. By the first route, called *horizontal transmission,* a plant is infected from an external source of the virus. Since the invading virus must get past the plant's outer protective layer of cells (the epidermis), the plant becomes more susceptible to viral infections if it has been damaged by wind, chilling, injury, or insects. Insects are a double threat, because they often also act as carriers of viruses, transmitting disease from plant to plant. Farmers and gardeners themselves may transmit plant viruses inadvertently on pruning shears and other tools. The other route of viral infection is *vertical trans-*

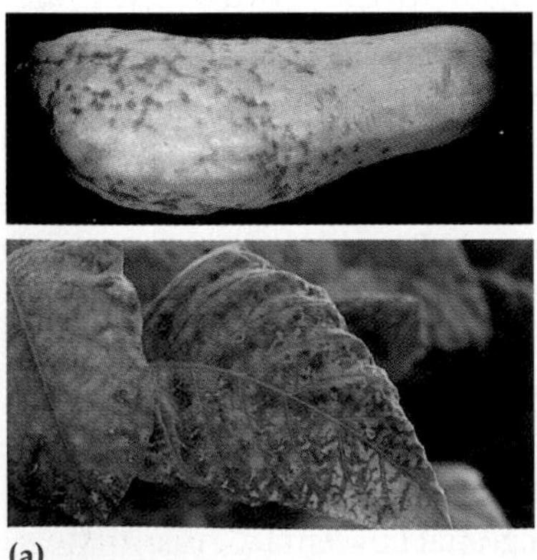

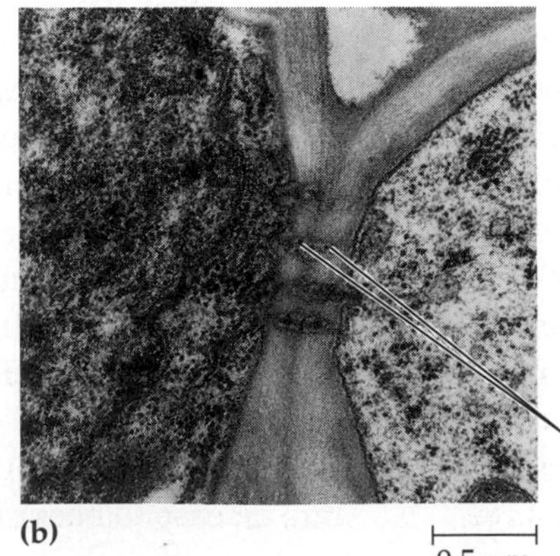

FIGURE 17.8
Viral infection of plants. (**a**) Mosaic viruses caused the mottling of this summer squash (top) and tobacco leaf (bottom). (**b**) Viruses that infect a plant, such as these rice yellow mottle viruses, can spread throughout the plant body via the plasmodesmata that connect cells.

mission, in which a plant inherits a viral infection from a parent. Vertical transmission can occur in asexual propagation (for example, by taking cuttings) or in sexual reproduction via infected seeds.

Once a virus enters a plant cell and begins reproducing, virus particles can spread throughout the plant by passing through plasmodesmata, the cytoplasmic connections that penetrate the walls between adjacent plant cells (FIGURE 17.8b). Agricultural scientists have not yet devised cures for most viral diseases of plants. Therefore, their efforts have focused largely on reducing the incidence and transmission of such diseases and on breeding genetic varieties of crop plants that are relatively resistant to certain viruses.

Viroids and prions are infectious agents even simpler than viruses

As small and simple as viruses are, they dwarf another class of plant pathogens called **viroids.** These are tiny molecules of naked RNA, only several hundred nucleotides long. Somehow, these RNA molecules can disrupt the metabolism of a plant cell and stunt the growth of the whole plant. One viroid disease has killed over ten million coconut palms in the Philippines. Clues to what viroids do within cells may emerge from an intriguing discovery that relates viroids to certain normal eukaryotic genes, including rRNA genes. The nucleotide sequences of viroid RNA turn out to be similar to the sequences of introns found within those genes, introns that can catalyze their own excision from a larger RNA molecule (see Chapter 16). Perhaps viroids originated as "escaped introns."

It seems likely that viroids cause errors in the regulatory systems that control the genes of the cell. Indeed, the symptoms that are typically associated with viroid diseases are abnormal development and stunted growth.

An important lesson from viroids is that a *molecule* can be an infectious agent that spreads a disease. Viroids are nucleic acids, and thus their self-directed replication prevents them from being diluted when they are transmitted from host to host. More difficult to explain are infectious agents called **prions,** which are proteins. Prions cause a disease called scrapie in sheep and may be responsible for some degenerative diseases of the nervous system in humans. How can a protein, which cannot replicate itself, be an infectious pathogen? According to one hypothesis, a prion is a defective form of a protein normally present in specific cells. When the prion infects a cell with the normal protein, the prion catalyzes conversion of the normal protein to the prion version. If this hypothesis is correct, a prion molecule could trigger a chain reaction that increases the number of prions. This would enable prions to spread through a host population without dilution. Several labs are testing this and other hypotheses that attempt to explain how proteins can function as infectious pathogens.

Viruses may have evolved from other mobile genetic elements

Viruses are in the semantic fog between life and nonlife. Do we think of them as nature's most complex molecules or as the simplest forms of life? Either way, we must bend our usual definitions. An isolated virus is biologically inert, unable to replicate its genes or regenerate its own supply of ATP. Yet it has a genetic program written in the universal language of life. Although viruses are obligate intracellular parasites that cannot reproduce independently, it is hard to deny their evolutionary connection to the living world.

How did viruses originate? Because they depend on cells for their own propagation, it is reasonable to assume that viruses are not the descendants of precellular prototypes of life, but that they evolved *after* the first cells. Most molecular biologists favor the hypothesis that viruses originated from fragments of cellular nucleic acids that could move from one cell to another. Consistent with this idea is the observation that a viral genome usually has more in common with the host cell's genome than with the genomes of viruses infecting other hosts. Indeed, some viral genes are essentially identical to genes of the host, as in the case of oncogenes, for example. Perhaps the earliest viruses were naked bits of nucleic acid, similar to plant viroids, that made it from one cell to another via injured cell surfaces. The evolution of genes coding for capsid proteins may have facilitated the infection of undamaged cells.

The most likely candidates for potential sources of viral genomes are two genetic elements of cells called plasmids and transposons. Plasmids are small, circular DNA molecules that are separate from chromosomes. They are found in bacteria and yeasts, which are unicellular eukaryotes. Plasmids, like most viruses, can replicate independently of the rest of the cell's genome and are occasionally transferred between cells. Transposons are DNA segments that can move from one location to another in a cell's genome. Thus, plasmids, transposons, and viruses all share an important feature: They are mobile genetic elements. (We will discuss plasmids and transposons in more detail later in the chapter.)

It is the evolutionary relationship between viruses and the genomes of their host cells that makes viruses such useful model systems in molecular biology. By studying how the replication of viruses is controlled, researchers are learning more about the mechanisms that regulate DNA replication and gene expression (transcription and translation) in cells. Bacteria are equally valuable as microbial models in genetics research, but for different reasons. Unlike viruses, bacteria are true cells. But as prokaryotic cells, bacteria provide researchers with the opportunity to investigate molecular genetics in the simplest organisms. In fact, *E. coli,* the bacterium sometimes called "the laboratory rat of molecular biology," is the most completely understood of all organisms at the molecular level. Let's move from the topic of viruses and learn more about the genetics of bacteria.

The short generation span of bacteria facilitates their evolutionary adaptation to changing environments

Bacteria are very adaptable, in both the evolutionary sense of adaptation via natural selection and the physiological sense of adjustment to changes in the environment by individual bacteria. These sections on bacterial genetics will help clarify how these microbes can be so malleable.

The major component of the bacterial genome is one double-stranded DNA molecule arranged in a circle. Although we will refer to this structure as the bacterial chromosome, it is very different from eukaryotic chromosomes, which have linear DNA molecules associated with a considerable amount of protein. In the case of the common intestinal bacterium *E. coli,* the chromosome consists of approximately 4 million base pairs representing about 3000 genes. This is 100 times more DNA than is found in a typical virus, but only about one-thousandth as much DNA as in an average eukaryotic cell. Still, this is a lot of DNA to be packaged in such a small container. Stretched out, the DNA of an *E. coli* cell would measure about a millimeter in length, 500 times longer than the cell. Within a bacterium, however, the chromosome is so tightly packed that it does not even fill the whole cell but forms a structure something like a long loop of yarn tangled into a ball. This dense region of DNA, called the **nucleoid,** is not bounded by a membrane like the true nucleus of a eukaryotic cell. In addition to the chromosome, many bacteria also have plasmids, much smaller circles of DNA. Each plasmid has only a small number of genes, from just a few to about two dozen. You will learn about the structure and function of plasmids in the next section.

Bacterial cells divide by binary fission, which is preceded by replication of the bacterial chromosome (see Chapter 11). From a single origin of replication, the copying of DNA progresses in both directions around the circular chromosome (FIGURE 17.9).

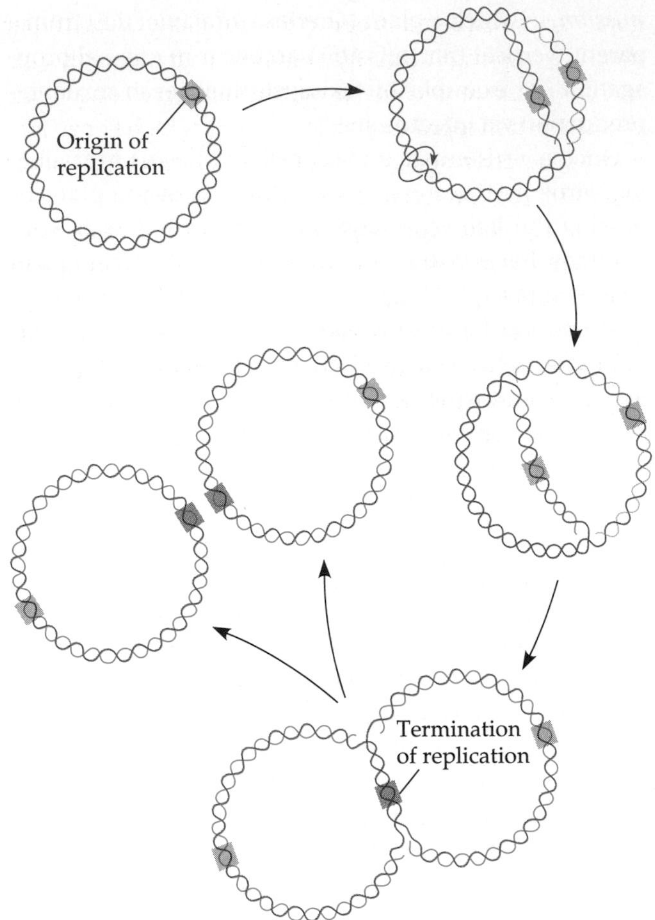

FIGURE 17.9
Replication of the bacterial chromosome. From one origin, DNA replication progresses in both directions around the circular chromosome until the entire chromosome has been reproduced. Enzymes that cut, twirl, and reseal the double helix prevent the DNA from knotting (magenta arrow).

Bacteria can proliferate very rapidly in a favorable environment, whether in a natural habitat or in a laboratory culture. For example, *E. coli,* growing under optimal conditions, can divide every 20 minutes. A laboratory culture started with a single cell can produce a colony of 10^7 to 10^8 bacteria overnight (12 hours). Reproductive rates in the organism's natural habitat, the large intestines (colons) of mammals, are just as impressive. For example, in the human colon, *E. coli* reproduces rapidly enough to replace the 2×10^{10} bacteria lost each day in feces.

Since fission is an asexual process—the production of offspring from a single parent—most of the bacteria in a colony are genetically identical to the parent cell. As a result of mutation, however, some of the offspring *do* differ slightly in genetic makeup. For a given *E. coli* gene, the probability of a mutation averages only about 1×10^{-7} per cell division. But since about 2×10^{10} new *E. coli* cells are produced daily in a human colon, mutations in each gene will give rise to approximately 2000 ($2 \times 10^{10}/1 \times 10^{-7}$) *E. coli* mutants for that gene each day in just one human host. The total number of mutations when all 3000 *E. coli* genes are considered is about 6×10^6 (3000×2000) per day. The important point is that new mutations, though individually rare, can have a significant impact on genetic diversity when reproductive rates are very high because of short generation spans. This diversity, in turn, affects the evolution of bacterial populations: Individual bacteria that are genetically well equipped for the local environment clone themselves more prolifically than do less fit individuals.

In contrast, new mutations make a relatively small contribution to genetic variation among individuals of a population of slowly reproducing organisms, such as humans. Most of the heritable variation we observe in a human population is due not to the creation of novel alleles by *new* mutations, but to the sexual recombination of existing alleles (see Chapter 14). Even in bacteria, where new mutations *are* a major source of individual variation, genetic recombination adds more diversity to a population, as we will now see.

Genetic recombination and transposition produce new bacterial strains

Natural selection depends on heritable variation among the individuals of a population (see Chapter 1). In addition to mutations, genetic recombination generates diversity within bacterial populations. We will define recombination here as the combining of genetic material from two individuals into the genome of a single individual.

How can we detect genetic recombination in bacteria? Consider two *E. coli* strains (genetic varieties), each unable to synthesize one of its required amino acids. Wild-type *E. coli* can grow on a minimal medium containing only glucose as a source of organic carbon. Our two mutant strains cannot grow on this culture medium of minimal nutrients because one of them cannot synthesize tryptophan and the other cannot synthesize arginine (see the Methods Box).

METHODS: DETECTING GENETIC RECOMBINATION IN BACTERIA

Each of the two mutant strains of *E. coli* in this experiment is unable to synthesize a particular amino acid—either arginine or tryptophan—from glucose. This is indicated by the genetic symbols for the two strains: arg^+ trp^- (capable of synthesizing arginine but not tryptophan) and arg^- trp^+ (capable of synthesizing tryptophan but not arginine). These mutant bacteria cannot grow on minimal medium, but only on a medium supplemented with the amino acid they are unable to make for themselves. In the middle tube of the diagram, a mixture of the two mutant strains is cultured together on a medium supplemented with both arginine and tryptophan. Samples containing equal numbers of bacteria from the three cultures are then plated onto agar (a gelatinous substance) containing minimal medium (MM) in petri dishes. The original mutant cells fail to produce colonies. However, some of the cells in the sample from the mixed culture are able to grow on minimal medium. Those cells are recombinant bacteria, an arg^+ trp^+ strain produced by gene exchange between the two types of mutant cells.

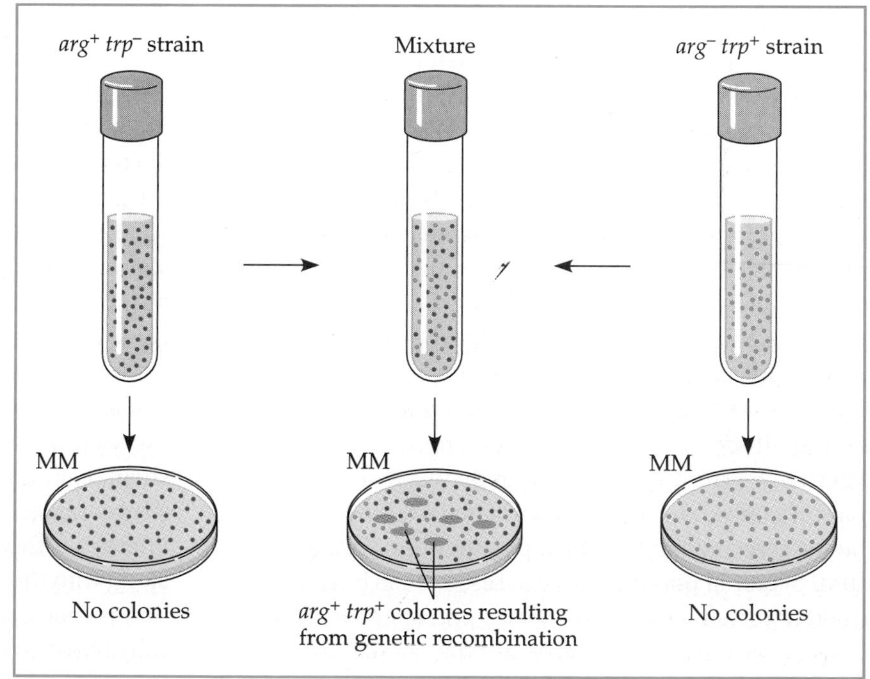

Let's suppose we have grown the two strains together on a medium containing both tryptophan and arginine. After a few hours, we transfer a small sample of this culture to a petri dish containing minimal medium and incubate this dish overnight. The next morning we observe numerous colonies of bacteria on the minimal medium. Each of these colonies must have started with a cell capable of making *both* tryptophan *and* arginine, but their number far exceeds what can be accounted for by mutation. Most of the cells with genes for making both amino acids must have acquired these genes from two different cells, one from each strain. Genetic recombination has occurred.

In mechanism, genetic recombination in bacteria is different from recombination in eukaryotes. The sexual processes of meiosis and fertilization combine genetic material from two individuals in a single zygote in eukaryotes (see Chapter 12). Sex, in this eukaryotic manner, is absent in prokaryotes: Meiosis and fertilization do not occur. The bacterial mechanisms of genetic recombination are three processes called transformation, transduction, and conjugation.

Transformation

Transformation is the alteration of a bacterial cell's genotype by the uptake of naked, foreign DNA from the surrounding environment. For example, we saw in Chapter 15 that harmless *Streptococcus pneumoniae* bacteria lacking coats could be transformed to pneumonia-causing individuals by the uptake of naked DNA from a medium containing dead cells of the coated strain. This transformation occurs when a live, coatless cell absorbs a piece of DNA that happens to include the gene for producing the protective coat. The foreign allele is then incorporated into the bacterial chromosome by replacing the native allele (for the "coatless" condition, in this case) in a DNA exchange similar to crossing over in eukaryotic meiosis. The transformed cell now has a chromosome containing DNA derived from two different cells, which fits our definition of genetic recombination.

For many years after transformation was discovered in laboratory cultures, most biologists believed the process to be too rare and haphazard to play an important role in natural bacterial populations. But researchers have since learned that many bacterial species possess proteins on their surfaces that are specialized for the uptake of naked DNA from the surrounding solution. These proteins specifically recognize and transport only DNA from closely related species of bacteria. Not all bacteria have such membrane proteins. For instance, *E. coli* does not seem to have any specialized mechanism for the uptake of foreign DNA. However, placing *E. coli* in a culture medium containing a relatively high concentration of calcium ions will artificially stimulate the cells to take up small pieces of DNA. In biotechnology, this technique is applied to introduce foreign genes into bacteria—genes coding for valuable proteins, such as human insulin and growth hormone.

Transduction

In the recombination mechanism known as **transduction,** phages (the viruses that infect bacteria) transfer bacterial genes from one host cell to another. There are two forms of transduction: generalized transduction and specialized transduction (FIGURE 17.10). Both result from aberrations in the infection cycles of phages.

First let's consider *generalized* transduction, shown in FIGURE 17.10a. Recall that at the end of a phage's lytic cycle, the viral nucleic acid molecules are packaged within capsids, and the completed phages are released when the host cell lyses. Occasionally, a small piece of the host cell's degraded DNA is packaged within a phage capsid in place of the phage genome. Such a virus is defective because it lacks its own genetic material. However, after its release from the lysed host, the defective phage can attach to another bacterium and inject the piece of bacterial DNA acquired from the first cell. This piece of DNA can subsequently replace the homologous region of the second cell's chromosome by a molecular rearrangement similar to crossing over. The cell's chromosome now has a combination of genetic material derived from two cells; recombination has occurred. This mode of recombination is called generalized transduction because it does not selectively transfer certain genes. Whatever piece of bacterial DNA happens to get packaged within a phage is the genetic material that is moved between cells.

Let's contrast this result with *specialized* transduction, shown in FIGURE 17.10b. This form of transduction requires infection by a temperate phage. Recall that in the lysogenic cycle, the genome of a temperate phage integrates as a prophage into the host bacterium's chromosome, usually at a specific site. Later, when the phage genome is excised from the chromosome, it sometimes takes with it small regions of the bacterial DNA that were adjacent to the prophage. When such a virus carrying bacterial DNA infects another host cell, the bacterial genes are injected along with the phage's genome. This mode of recombination is called specialized transduction because it specifically transfers genes that are near the chromosomal location where the prophage is incorporated.

Conjugation and Plasmids

Conjugation is the direct transfer of genetic material between two bacterial cells that are temporarily joined. This mechanism of genetic recombination, the bacterial version of sex, has been studied most extensively in

E. coli. The DNA transfer is one-way: one cell donating DNA, and its "mate" receiving the genes. The DNA donor, referred to as the "male," uses appendages called sex pili to attach to the DNA recipient, the "female" (FIGURE 17.11). Then a temporary cytoplasmic bridge forms between the two cells, providing an avenue for DNA transfer. "Maleness," the ability to form sex pili and donate DNA during conjugation, requires the presence of a special plasmid called the F plasmid. This is one of several types of plasmids that have been discovered in bacteria. Before learning about the specialized role of the F plasmid in conjugation, we should examine plasmids more generally.

General Characteristics of Plasmids. A plasmid is a small, circular DNA molecule separate from the bacterial chromosome. Plasmids replicate independently, but some do so in synchrony with the chromosome. Other plasmids replicate on their own schedule, and this can alter the number of these plasmids present in the cell at a particular time.

Certain plasmids have another interesting behavior: They can undergo a reversible incorporation into the cell's chromosome. Such a genetic element, which can replicate either as an extrachromosomal molecule or as part of the main bacterial chromosome, is called an **episome.** In addition to some plasmids, temperate viruses, such as phage λ, also qualify as episomes. Recall that the genomes of these phages replicate separately in the cytoplasm during a lytic cycle, and as an integral part of the host's chromosome during a lysogenic cycle. The hypothesis that some viruses evolved from plasmids was introduced earlier. Of course, there are important differences between

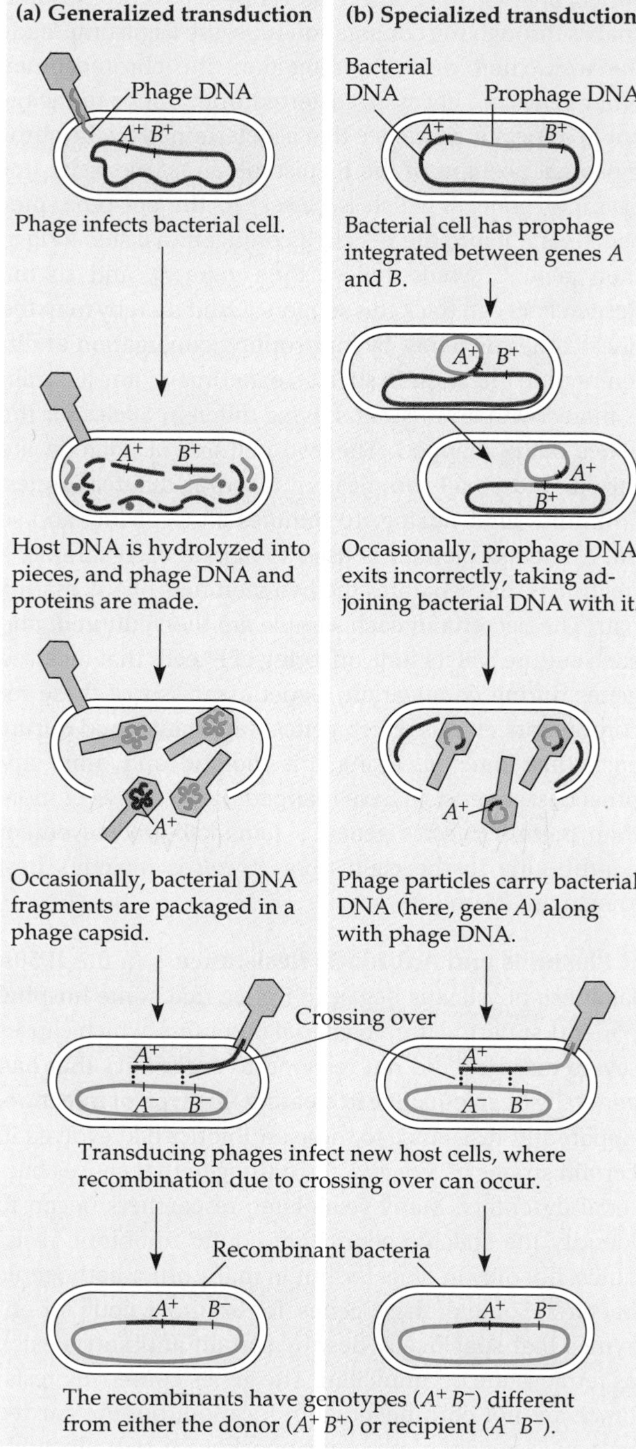
(a) Generalized transduction
Phage DNA
A^+ B^+
Phage infects bacterial cell.
A^+ B^+
Host DNA is hydrolyzed into pieces, and phage DNA and proteins are made.
A^+
Occasionally, bacterial DNA fragments are packaged in a phage capsid.
(b) Specialized transduction
Bacterial DNA
Prophage DNA
A^+ B^+
Bacterial cell has prophage integrated between genes *A* and *B*.
A^+ B^+
A^+
B^+
Occasionally, prophage DNA exits incorrectly, taking adjoining bacterial DNA with it.
A^+
Phage particles carry bacterial DNA (here, gene *A*) along with phage DNA.
Crossing over
A^+
A^- B^-
A^+
A^- B^-
Transducing phages infect new host cells, where recombination due to crossing over can occur.
Recombinant bacteria
A^+ B^-
A^+ B^-
The recombinants have genotypes ($A^+ B^-$) different from either the donor ($A^+ B^+$) or recipient ($A^- B^-$).

FIGURE 17.10

Transduction. Phages occasionally carry bacterial genes from one cell to another. (**a**) In generalized transduction, random pieces of the host chromosome are packaged within a phage capsid. (**b**) In specialized transduction, a prophage exits the chromosome in such a way that it carries adjacent bacterial genes along with it. In both types of transduction, some of the transferred DNA may recombine with the genome of the new host cell.

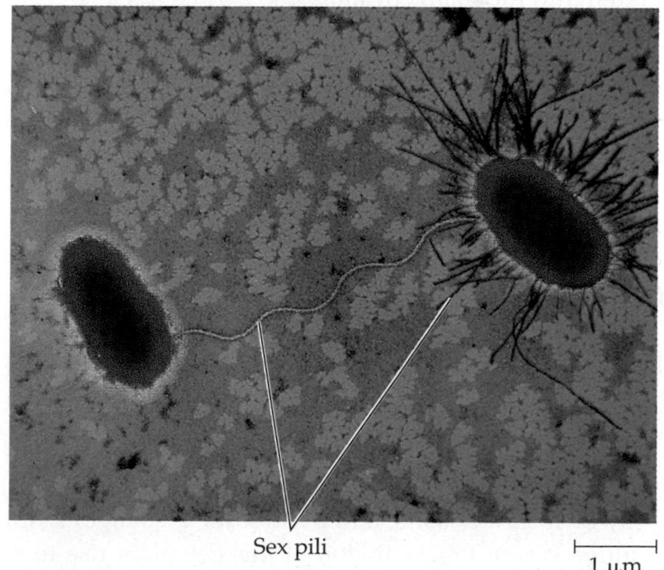
Sex pili
1 μm

FIGURE 17.11

Bacterial mating. The *E. coli* "male" (right) extends sex pili, one of which is attached to a "female" cell. Later, the mates will briefly join by a cytoplasmic bridge through which the "male" transfers DNA to the "female." This mechanism of DNA transfer is called conjugation (colorized TEM).

plasmids and viruses. Plasmids, unlike viruses, lack an extracellular stage. And plasmids are generally beneficial to the bacterial cell, while viruses are parasites that usually harm their hosts.

Each plasmid has only a few genes, and these genes are not required for the survival and reproduction of the bacterium under normal conditions. However, the genes of plasmids can confer advantages in bacteria in stressful environments. For example, the F plasmid facilitates genetic recombination, which may be advantageous in a changing environment that no longer favors existing strains in a bacterial population.

The F Plasmid and Conjugation. The **F plasmid** (F for fertility) consists of about 25 genes, most required for the production of sex pili. Geneticists use the symbol F^+ to denote a cell that contains the F plasmid (a "male" cell). The F^+ condition is heritable: The F plasmid replicates in synchrony with the chromosomal DNA, and division of an F^+ cell usually gives rise to two offspring that are both F^+. Cells lacking the F plasmid are designated F^-, and they function as DNA recipients ("females") during conjugation. The F^+ condition is "contagious" in the sense that an F^+ cell can convert an F^- cell to F^+ when the two cells conjugate. The F plasmid replicates within the "male" cell, and a copy is transferred to the "female" through the conjugation tube joining the cells (FIGURE 17.12a). In such an $F^+ \times F^-$ mating, only an F plasmid is transferred.

Under what circumstances is DNA of the main bacterial chromosome transferred during conjugation? The F plasmid is an episome that occasionally becomes integrated into the main chromosome (FIGURE 17.12b). A cell with the F genes built into its chromosome is called an Hfr cell (for high frequency of recombination). An Hfr cell continues to function as a male during conjugation, transferring the F genes to its F^- partner. But now, the F episome takes chromosomal DNA along with it (FIGURE 17.12c). The Hfr cell's chromosome is replicating as DNA is transferred, so this donor cell retains all of its own genes. Random movements of the bacteria usually disrupt conjugation before an entire copy of the Hfr chromosome can be passed to the F^- cell. Temporarily, the recipient cell is a partial diploid, containing its own chromosome plus DNA copied from part of the donor's chromosome. Recombination occurs when the newly acquired DNA aligns with the homologous region of the cell's own chromosome and a crossover exchanges DNA (FIGURE 17.12d). Binary fission of this cell gives rise to a colony of recombinant bacteria with genes derived from two different cells.

Interrupting Conjugation to Map Bacterial Chromosomes. Microbial geneticists have used Hfr $\times$ F^- matings to map the sequence of genes around the *E. coli* chromosome. When the F plasmid joins a chromosome to form an Hfr cell, it always inserts into the chromosome at a specific site (though this site differs from one genetic variety of *E. coli* to another). As a copy of the chromosome snakes through the conjugation tube, the F episome leads the way. Thus, during conjugation, the chromosome's other genes are always transferred from donor to recipient in a specific sequence that is determined by the chromosomal position of the F episome. In FIGURE 17.12c, for example, gene *A*, which is closest to the episome, precedes gene *B* into the F^- cell. If conjugation lasted longer, then gene C would follow, then gene *D*, and so on. Researchers can track this sequence, and thereby map the linear order of genes, by interrupting conjugation at different time intervals. In such an experiment, an Hfr strain is mated with an F^- strain having different alleles for the genes being mapped. The two cultures of bacteria are mixed, and small samples are taken at different times: 5 minutes after mixing, 10 minutes after mixing, and so on. A kitchen blender is used to agitate each sample, a treatment that separates the two members of any mating pair. The bacteria in each sample are then cultured, and each culture will include offspring of F^- cells that received genes during conjugation. Genetic analysis of these recombinants reveals which genes were transferred during each time interval. Using this method and other approaches, geneticists have mapped the locations of more than half of *E. coli*'s genes, a remarkable achievement contributing to the claim that *E. coli* is biology's best known organism.

R Plasmids and Antibiotic Resistance. In the 1950s, Japanese physicians began to notice that some hospital patients suffering from bacterial dysentery, which causes severe diarrhea, did not respond to antibiotics that had generally been effective in treating this type of infection. Apparently, resistance to these antibiotics had evolved in certain strains of *Shigella*, the pathogen that causes bacterial dysentery. Many years later, researchers began to identify the specific genes that confer antibiotic resistance, not only in *Shigella*, but in many other pathogenic bacteria. Some of these genes, for example, code for enzymes that specifically destroy certain antibiotics, such as tetracycline or ampicillin. The genes conferring resistance are not chromosomal in location, but are carried by a special class of plasmids now known as **R plasmids** (R for resistance).

Exposure of a bacterial population to a specific antibiotic—whether in a laboratory culture or within a host organism, such as a human—will kill antibiotic-sensitive bacteria but not those that happen to have R plasmids that counter that antibiotic. The theory of natural selec-

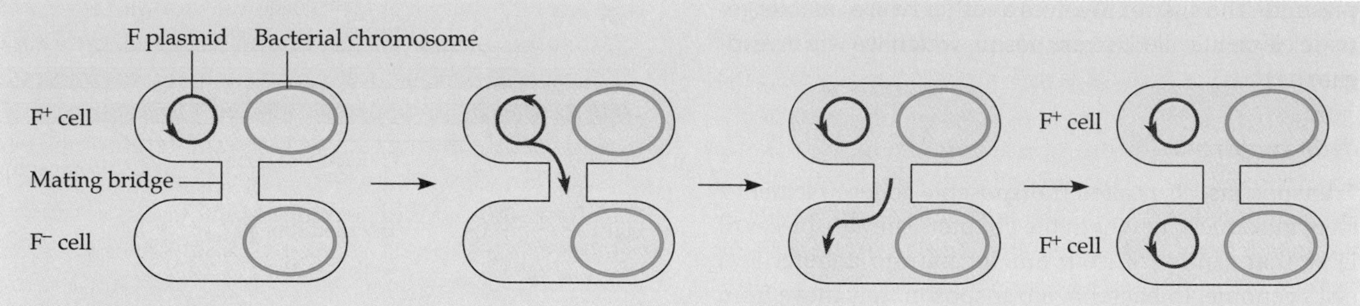

(a) Conjugation between an F⁺ (male) and an F⁻ (female) bacterium. Cells that carry an extrachromosomal fertility factor, the F plasmid, are called F⁺ cells. They are "male" in that they can transfer an F plasmid to a "female" F⁻ cell during conjugation. In this way, an F⁻ cell can become F⁺. The F factor replicates as it is transferred, so that the donor cell remains F⁺. The arrowhead marks the point where replication begins.

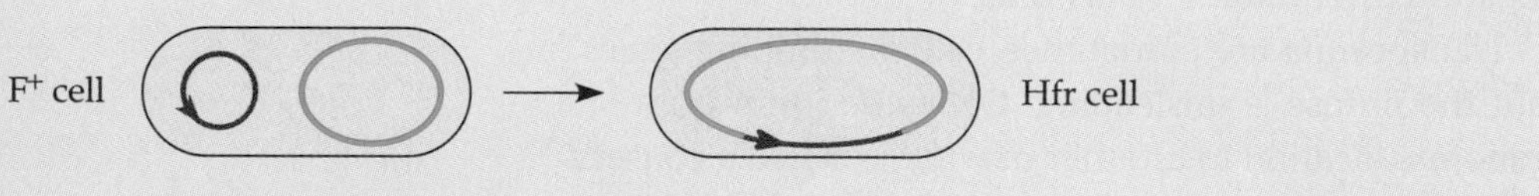

(b) Conversion of an F⁺ male into an Hfr male by integration of the F plasmid (an episome) into the chromosome. This process is similar to viral DNA joining the host chromosome as a prophage.

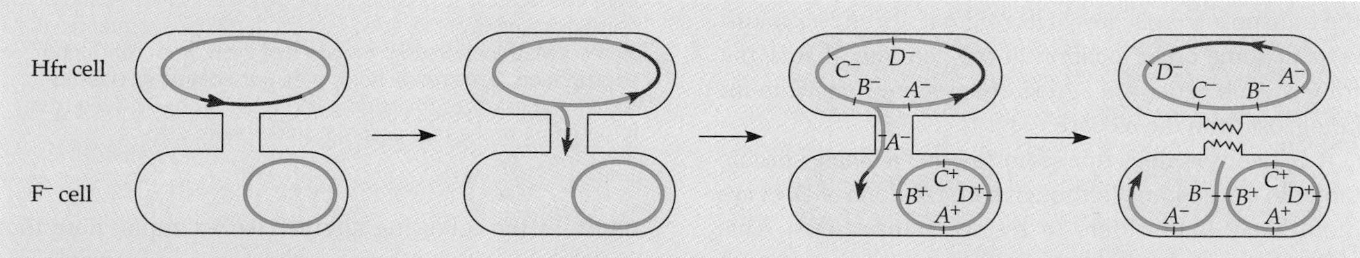

(c) Conjugation between an Hfr and an F⁻ bacterium. Replication of the "male's" chromosome begins at a fixed point (arrowhead) within the F episome (the location where the episome inserts into the chromosome varies from one genetic strain of *E. coli* to another). The location of this point determines the sequence in which genes are transferred to the "female" during conjugation. For example, in this *E. coli* strain, the transfer sequence for four genes is symbolized as *A-B-C-D*. The conjugation bridge usually breaks before the entire chromosome and the tail end of the F episome are transferred.

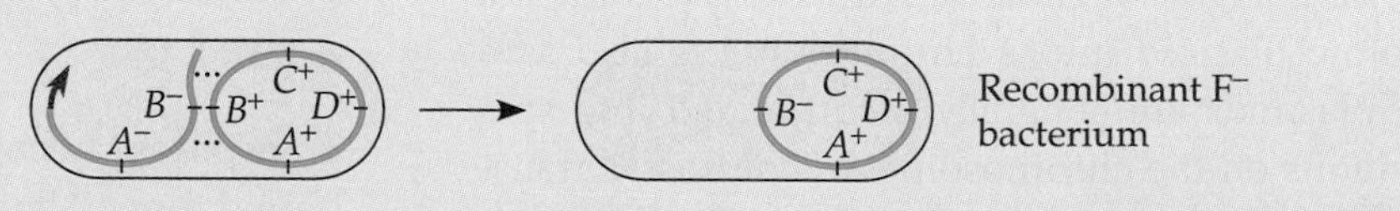

(d) Recombination between the Hfr chromosome fragment and the F⁻ chromosome. Crossing over can occur between genes on the fragment of bacterial chromosome transferred from the Hfr cell and the same (homologous) genes on the recipient (F⁻) cell's chromosome. A recombinant F⁻ cell will result. Pieces of DNA ending up outside the bacterial chromosome will eventually be degraded by the cell's enzymes or lost in cell division.

FIGURE 17.12
Conjugation and recombination in *E. coli.*

tion predicts that, under these circumstances, an increasing number of bacteria will inherit genes for antibiotic resistance, and that is exactly what happens. The medical consequences are also predictable: Resistant strains of pathogens are becoming more common, making the treatment of certain bacterial infections more difficult. The problem is compounded by the fact that R plasmids, like F plasmids, can be transferred from one bacterial cell to another during conjugation. Making the problem still worse, some plasmids carry as many as ten genes for resistance to that many antibiotics. How do so many antibiotic-resistant genes become part of a single plasmid? The answer involves another type of mobile genetic element called a transposon, which we will investigate next.

Transposons

Transposons, also called transposable genetic elements, were introduced earlier in the chapter. They are pieces of DNA that can move from one location to another in a cell's genome. In bacteria, a transposon may move from one locus to another within the chromosome; from a plasmid to the chromosome, or vice versa; or from one plasmid to another. For example, transposons are responsible for combining several genes for antibiotic resistance into a single plasmid by moving the genes to that location from different plasmids.

Transposons are sometimes called "jumping genes," but the phrase is misleading. Genes *do* jump from one genomic location to another in what is called a conservative transposition. In this case, the transposon's genes are not replicated before the move, so the number of copies of these genes is conserved. However, in another type of transposition, called a replicative transposition, the transposon replicates at its original site and a *copy* inserts at some other location in the genome; that is, the transposon's genes are added at some new site without being lost from the old site.

A transposon does not seem to have a single specific target in the genome (although some regions of DNA are more likely than others to receive transposons). This ability to scatter certain genes throughout the genome makes transposition fundamentally different from all other mechanisms of genetic shuffling. Crossing over in eukaryotic meiosis and the three mechanisms of recombination in prokaryotes—transformation, transduction, and conjugation—all depend on the exchange of alleles between homologous regions of DNA. The integration of an episomic plasmid into a chromosome is also site-specific; plasmids are generally incorporated into specific positions on the chromosome. In contrast, a transposon may move genes to a site where no such genes have ever before existed.

Insertion Sequences. The simplest transposons are called **insertion sequences.** They consist of only the DNA necessary for the act of transposition itself; no other genes are present. The one gene found in an insertion sequence codes for transposase, an enzyme that catalyzes transposition. The transposase gene is bracketed by a pair of DNA sequences called inverted repeats, noncoding sequences about 20 to 40 nucleotides long (FIGURE 17.13a). They are called inverted repeats because these two regions of DNA on the ends of an insertion sequence are upside-down, backward versions of each

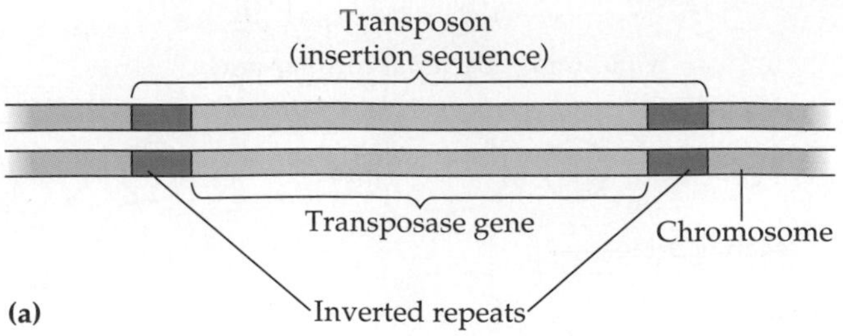

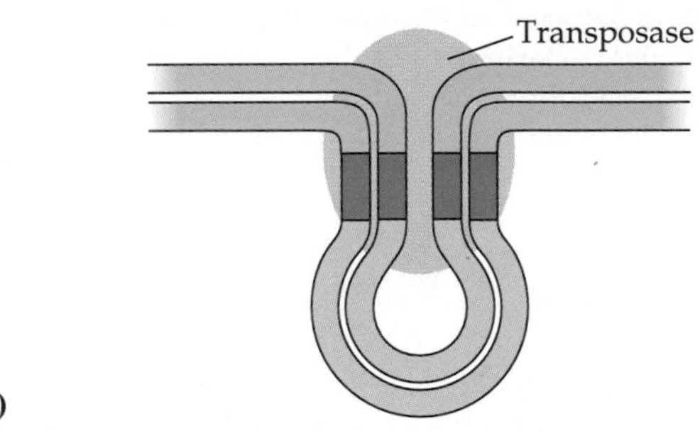

FIGURE 17.13
Insertion sequences, the simplest transposons. (**a**) The one and only gene of an insertion sequence codes for transposase, the enzyme that catalyzes movement of the transposon from one location to another in the genome. On either end of the transposase gene are inverted repeats, nucleotide sequences that are backward, upside-down versions of each other. (**b**) During transposition, transposase holds the inverted repeats close together and catalyzes the cutting and resealing of DNA required for insertion of the transposon at some target site.

other. In the following abbreviated example, note that each base sequence is repeated, in reverse, along the opposite DNA strand of the inverted repeat at the other end of the transposon:

	5′	3′
DNA strand #1	. . . ATCCGGT. . .	. . .ACCGGAT. . .
DNA strand #2	. . . TAGGCCA. . .	. . . .TGGCCTA. . .
	3′	5′

Transposase recognizes these inverted repeats as the boundaries of the transposon. The enzyme binds to these two regions, bringing them close together and catalyzing the DNA cutting and resealing required for transposition (FIGURE 17.13b). Other enzymes are also required for transposition. For example, DNA polymerase helps form identical regions of DNA, called direct repeats, which flank a transposon in its new target site (FIGURE 17.14).

When an insertion sequence invades a gene, it usually impairs the functioning of that gene by interrupting the coding sequence that specifies a protein. In other words, insertion sequences cause mutations when they move. But notice that this mechanism of mutation is intrinsic to the cell, in contrast to mutagenesis caused by extrinsic factors, such as radiation and foreign chemicals. In ad-

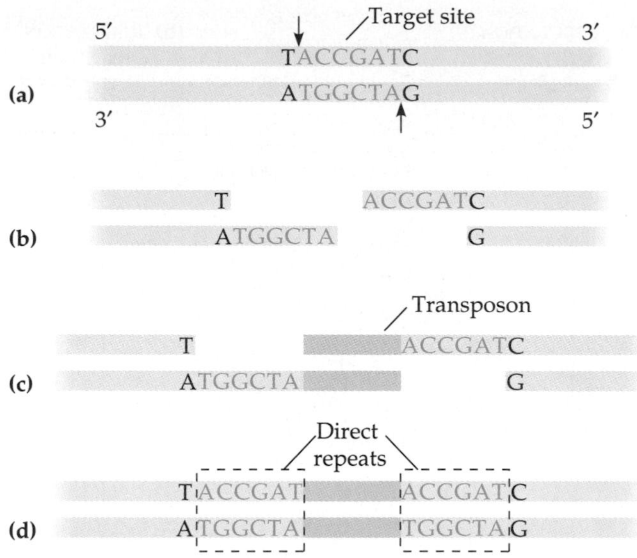

FIGURE 17.14

Insertion of a transposon. (**a**) An enzyme, probably transposase, cuts the two DNA strands at the target site in a staggered fashion (arrows). (**b**) This leaves short segments of unpaired DNA at each end of the cut chromosome or plasmid. (**c**) Transposase inserts the transposon into the opening of the target DNA. (**d**) The gaps in the two DNA strands are filled in when nucleotides pair with the single-stranded regions and DNA polymerase joins the nucleotides. This results in direct repeats, identical segments of DNA on either side of the transposon. Notice that the direct repeats did not exist at the target site until they were created by the act of transposition.

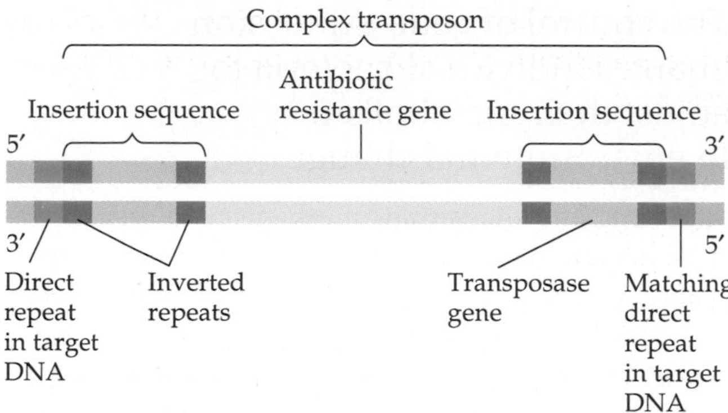

FIGURE 17.15

Anatomy of a complex transposon. A complex transposon consists of one or more genes located between twin insertion sequences. In this case, a gene for resistance to a specific antibiotic will be carried along as part of the complex transposon when this DNA inserts at some new site in the genome. For example, such transposition can add a gene for antibiotic resistance to a plasmid that already carries genes for resistance to other antibiotics. When such composite plasmids are transmitted to bacteria by either cell division or conjugation, the microbes can resist a variety of antibiotics.

dition to inserting within the coding regions of genes, insertion sequences may also increase or decrease the production of a protein by inserting within regulatory regions of DNA that control transcription rates. Wherever they land in the genome, insertion sequences are likely to alter the phenotype of the cell in some way. Insertion sequences account for about 1.5% of the *E. coli* genome, but they transpose and cause mutations rarely—only about once in every ten million generations. This is about the same as the mutation rate due to extrinsic causes. Given the rapid proliferation of bacteria, the transposition of insertion sequences probably plays a significant role in bacterial evolution as a source of genetic variation.

Complex Transposons and R Plasmids. Transposons larger and more complex than insertion sequences also move about in the bacterial genome. In addition to the DNA required for transposition, complex transposons include other genes that go along for the ride, such as genes for antibiotic resistance. These genes are sandwiched between two insertion sequences (FIGURE 17.15). It is as though two insertion sequences happened to land relatively close together in the genome and now travel together, along with all the DNA between them, as a single transposon. In contrast to simple insertion sequences, which are not known to benefit bacteria in any specific way, complex transposons may help bacteria adapt to new environments. We have already mentioned the example of transposition packaging several genes for resistance to different antibiotics into a single plasmid. In an antibiotic-rich environment, natural selection favors bacterial clones that have built up these composite R plasmids through a series of transpositions.

Transposable genetic elements are not unique to bacteria but are important components of eukaryotic genomes as well. In fact, the first evidence for such wandering genes came from American geneticist Barbara McClintock's breeding experiments with Indian corn in the 1940s and 1950s (see Chapter 1). McClintock identified changes in the color of corn kernels that only made sense if she postulated the existence of mobile genetic elements capable of moving from other locations in the genome to the genes for kernel color. She called these mobile elements "controlling elements" because they seemed to insert next to the genes responsible for kernel color, either activating or inactivating those genes. McClintock's discovery received little attention until transposons were discovered in bacteria many years later, and microbial geneticists learned more about the molecular basis of transposition. In 1983, more than 30 years after she discovered transposable genetic elements, Barbara McClintock was awarded a Nobel Prize, at age 81. McClintock continued her experiments at Cold Spring Harbor Laboratory in New York until her death in 1992.

You will learn more about transposable elements in eukaryotes in Chapter 18. We end this chapter by examining how bacterial genes are switched on and off in different environments.

The control of gene expression enables individual bacteria to adjust their metabolism to environmental change

Mutations and the various mechanisms of genetic recombination and transposition we have been studying generate the genetic variation that makes natural selection possible. And natural selection, acting over many generations, can increase the proportion of individuals in a bacterial population that are adapted to some new environmental condition, such as the introduction of a specific antibiotic. But how can an individual bacterium, locked into the genome it has inherited, cope with environmental fluctuation?

Think, for instance, of an *E. coli* cell living in the erratic environment of a human colon, dependent for its nutrients on the whimsical eating habits of its host. If the bacterium is deprived of the amino acid tryptophan, which it needs in order to survive, it responds by activating a metabolic pathway to make its own tryptophan from another compound. Later, if the human host eats a tryptophan-rich meal, the bacterial cell stops producing tryptophan for itself, thus saving the cell from squandering its resources to produce a substance that is available from the surrounding solution in prefabricated form. This is just one example of how bacteria tune their metabolism to changing environments.

Metabolic control occurs on two levels (FIGURE 17.16). First, cells can vary the numbers of specific enzyme molecules; that is, they can regulate the expression of a gene. Second, cells can vary the activities of enzymes already present. The latter mode of control, which is more immediate, depends on the sensitivity of many enzymes to chemical cues that increase or decrease their catalytic activity (see Chapter 6). For example, activity of the first enzyme of the tryptophan-synthesis pathway is inhibited by the pathway's end-product. Thus, if tryptophan accumulates in a cell, it shuts down its own synthesis. Such feedback inhibition, typical of anabolic (biosynthetic) pathways, allows a cell to adapt to short-term fluctuations in levels of a substance it needs.

If, in our example, the environment continues to provide all the tryptophan the cell needs, the regulation of gene expression also comes into play: The cell stops making enzymes of the tryptophan pathway. This control of enzyme number occurs at the level of transcription, the synthesis of messenger RNA coding for these enzymes. More generally, many genes of the bacterial genome are switched on or off by changes in the metabolic status of the cell. The basic mechanism for this control of gene expression, described as the operon model, was discovered in 1961 by François Jacob and Jacques Monod at the Pasteur Institute in Paris. Let's see what an operon is and how it works, using the control of tryptophan synthesis as an example.

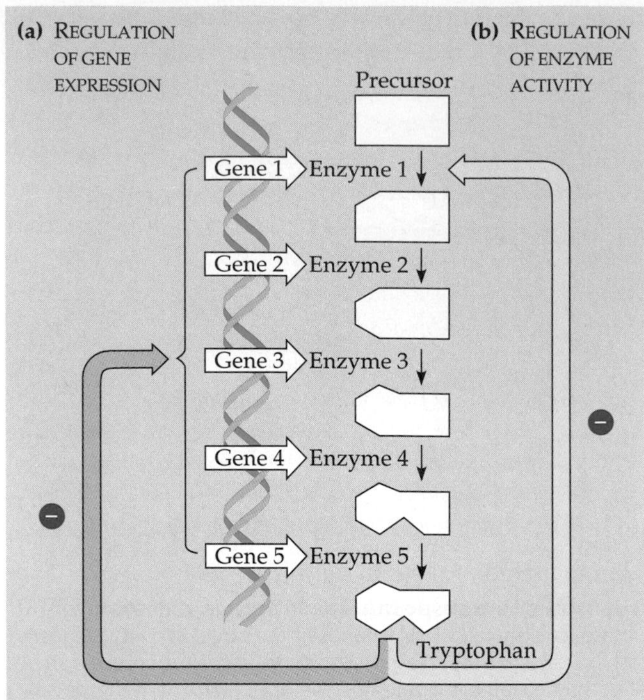

FIGURE 17.16
Regulation of a metabolic pathway. Cells can adjust the rates of specific metabolic pathways by regulating gene expression (the synthesis of new enzyme molecules) or by regulating the catalytic activity of existing enzymes. In the pathway for tryptophan synthesis, an abundance of tryptophan can (**a**) repress expression of the genes for all the enzymes needed for the pathway, and (**b**) inhibit the activity of the first enzyme in the pathway (feedback inhibition).

Operons: The Basic Concept

E. coli synthesizes tryptophan from a precursor molecule in a series of steps, each reaction catalyzed by a specific enzyme (see FIGURE 17.16). The five genes coding for the polypeptide chains that make up these enzymes are clustered together on the chromosome. A single promoter serves all five genes, which constitute a transcription unit. (Recall from Chapter 16 that a promoter is a site where RNA polymerase can bind to DNA and begin transcribing genes.) Thus, transcription gives rise to one long mRNA molecule representing all five genes for the tryptophan pathway. The cell can translate this transcript into separate polypeptides because the mRNA is punctuated with start and stop codons signaling where the coding sequence for each polypeptide begins and ends.

Genes that code for polypeptides are called **structural genes.** A key advantage of grouping structural genes of related function into one transcription unit is that a single "on-off switch" can control the whole cluster of functionally related genes. When an *E. coli* cell must make tryptophan for itself because the nutrient medium lacks this amino acid, all the enzymes for the metabolic pathway are synthesized at one time. The switch is a segment of DNA called an **operator.** Its location and name both suit its function: Positioned within the promoter or between the promoter and the structural genes, the operator controls the access of RNA polymerase to the structural genes. All together, the structural genes, the operator, and the promoter—the entire stretch of DNA required for enzyme production for the tryptophan pathway—is called an **operon** (FIGURE 17.17). Here we are dissecting one of many operons that have been discovered in *E. coli:* the *trp* operon (*trp* for tryptophan).

If the operator is the control point for transcription, what determines whether the operator is in the on or off mode? By itself, the operator is on; RNA polymerase can bind to the promoter and transcribe the structural genes. The operon must be switched off by a protein called the **repressor.** The repressor binds to the operator and blocks attachment of RNA polymerase to the promoter, stopping transcription of the structural genes. Repressor proteins are specific; that is, they recognize and bind only to the operator of a certain operon. The repressor that switches off the *trp* operon has no effect on other operons in the *E. coli* genome.

The repressor is the product of a gene called a **regulatory gene.** The regulatory gene for the *trp* repressor is located some distance away from the operon it controls (see FIGURE 17.17). Transcription of the regulatory gene produces an mRNA molecule that is translated into the repressor protein, which can then reach the operator of the *trp* operon by diffusion. Regulatory genes are transcribed continuously, although at a slow rate, and a few repressor molecules are always present in the cell. Why, then, is the *trp* operon not switched off permanently? First of all, the binding of repressors to operators is

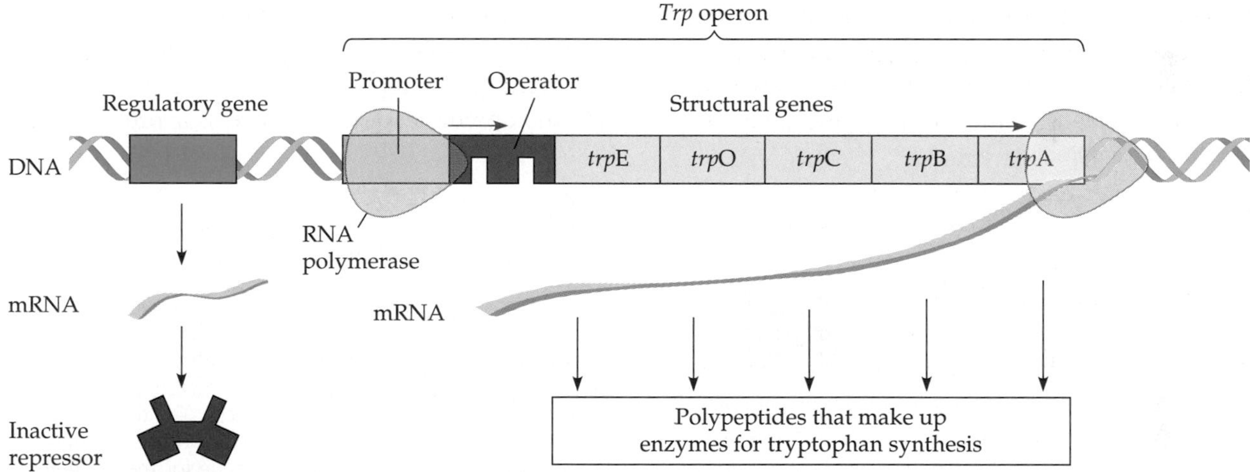

(a) Tryptophan absent, repressor inactive, operon on

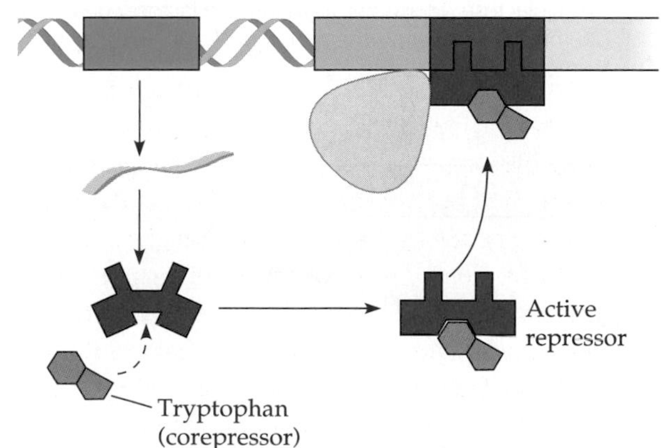

(b) Tryptophan present, repressor active, operon off

FIGURE 17.17

The *trp* operon: regulated synthesis of repressible enzymes. (**a**) Tryptophan is an amino acid produced by an anabolic pathway catalyzed by repressible enzymes. Accumulation of tryptophan, the end-product of the pathway, represses synthesis of the enzymes. The mechanism for this regulation in an *E. coli* cell is shown here. Five structural genes, which code for the polypeptides that make up the enzymes of the pathway, are grouped into an operon. The operon also includes a promoter and an operator. (The operator region is actually located within the promoter, but it is illustrated as a separate region in this simplified diagram of the *trp* operon.) When the operon is in the "on" mode, RNA polymerase molecules attach to the DNA at the promoter region and transcribe the structural genes. A regulatory gene, located outside the operon, codes for a repressor protein. The repressor can switch the *trp* operon off by binding to the operator and blocking access of RNA polymerase to the promoter. The repressor protein is synthesized in an inactive form, and remains inactive in the absence of tryptophan. With no repressor bound to the operator, the operon is on, producing mRNA for the enzymes that synthesize tryptophan. (**b**) As tryptophan accumulates in the cell, it inhibits its own production by activating the repressor protein. The repressor can now bind to the operator and switch the operon off.

reversible. An operator vacillates between the on and off modes, with the relative duration of each state depending on the number of active repressor molecules around. Secondly, the *trp* repressor is first synthesized in an inactive form, which has little affinity for the *trp* operator. It assumes its active conformation and attaches to the operator only if it first binds to a molecule of tryptophan. Tryptophan functions in this regulatory system as a **corepressor,** a small molecule that cooperates with a repressor protein to switch an operon off. As tryptophan levels rise, more tryptophan molecules associate with *trp* repressors, which can then bind to operators and shut down tryptophan production. If the cell's tryptophan concentration drops, transcription of the operon's structural genes resumes. This is one example of how gene expression responds rapidly to changes in the cell's internal and external environment.

Repressible Versus Inducible Enzymes: Two Types of Negative Gene Regulation

The enzymes of the tryptophan pathway are said to be *repressible enzymes* because their synthesis is inhibited by a metabolic end-product (tryptophan, in this case). In contrast, the synthesis of *inducible enzymes* is stimulated, rather than inhibited, by specific small molecules. Let's investigate an example.

The disaccharide lactose (milk sugar) is available to *E. coli* if the host human drinks milk. The bacteria can absorb the lactose and break it down for energy or use it as a source of organic carbon for synthesizing other compounds. Lactose metabolism begins with hydrolysis of the disaccharide into its two component monosaccharides, glucose and galactose. The enzyme that catalyzes this reaction is called β-galactosidase. Only a few molecules of this enzyme are present in an *E. coli* cell that has been growing in the absence of lactose—in the intestines of a person who does not drink milk, for example. But if lactose is added to the bacterium's nutrient medium, it takes only about fifteen minutes for the number of β-galactosidase molecules in the cell to increase a thousandfold.

The gene for β-galactosidase is part of an operon, the *lac* operon (*lac* for lactose metabolism), that includes two other structural genes coding for proteins that function in lactose metabolism (FIGURE 17.18). This entire transcription unit is under the command of a single operator and promoter. The regulatory gene, located outside the operon, codes for a repressor protein that can

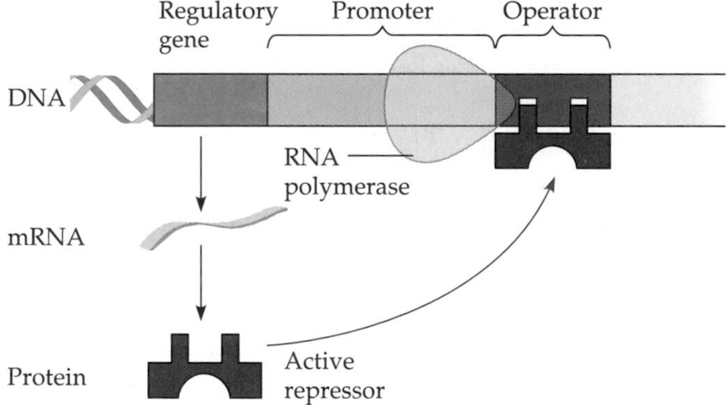

(a) Lactose absent, repressor active, operon off

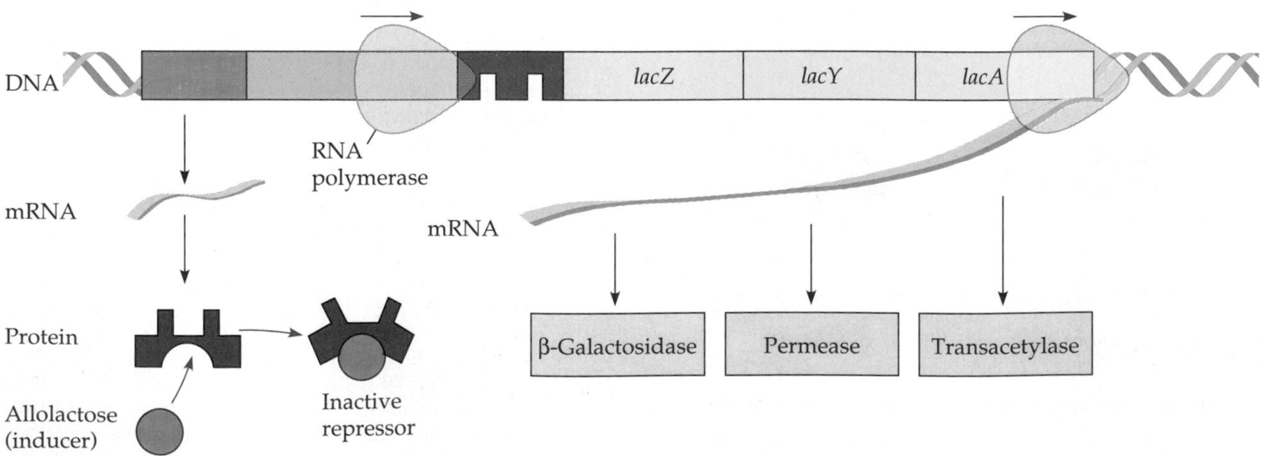

(b) Lactose present, repressor inactive, operon on

FIGURE 17.18

The *lac* operon: regulated synthesis of inducible enzymes. *E. coli* uses three enzymes to take up and metabolize lactose. The structural genes for these three enzymes are clustered in a single operon, the *lac* operon. One gene, *lacZ,* codes for β-galactosidase, which hydrolyzes lactose to glucose and galactose. Another gene, *lacY*, codes for a permease, the membrane protein that transports lactose into the cell. The third gene, *lacA,* codes for an enzyme called transacetylase, whose function in lactose metabolism is still uncertain. The gene for the *lac* repressor happens to be adjacent to the *lac* operon, an unusual situation. (**a**) The *lac* repressor is innately active, and in the absence of lactose, switches off the operon by binding to the operator region. (**b**) Allolactose, an isomer formed from lactose, derepresses the operon by inactivating the repressor. Thus, the enzymes for lactose metabolism are induced.

switch off the *lac* operon by binding to the operator. So far, this sounds just like regulation of the *trp* operon, but there is one important difference. Recall that the *trp* repressor was innately inactive and required tryptophan as a corepressor in order to bind to the operator. The *lac* repressor, in contrast, is active all by itself, binding to the operator and switching the *lac* operon off. In this case, a specific small molecule, called an **inducer,** *inactivates* the repressor. For the *lac* operon, the inducer is allolactose, an isomer of lactose formed in small amounts from lactose that enters the cell. In the absence of lactose (and hence allolactose), the *lac* repressor is in its active configuration, and the structural genes of the *lac* operon are silent. If lactose is added to the cell's nutrient medium, allolactose binds to the *lac* repressor and alters its conformation, nullifying the repressor's ability to attach to the operator. Now, on demand, the *lac* operon produces mRNA for the enzymes of the lactose pathway. In the context of gene regulation, these enzymes are referred to as inducible enzymes because their synthesis is induced by a chemical signal (allolactose, in this case).

Let's contrast repressible enzymes and inducible enzymes in terms of the metabolic economy of the *E. coli* cell. Repressible enzymes generally function in anabolic pathways, which synthesize essential end-products from raw materials (precursors). By suspending production of this end-product when it is already present in sufficient quantity, the cell can allocate its organic precursors and energy for other uses. In contrast, inducible enzymes usually function in catabolic pathways, which break a nutrient down to simpler molecules. By producing the appropriate enzymes only when the nutrient is available, the cell avoids making proteins that have nothing to do. Why, for example, bother to make the enzymes that break down milk sugar when no milk is present?

In comparing repressible and inducible enzymes, there is one more important point: Both systems are examples of the *negative* control of genes, because the operons are switched off by the active form of the repressor protein. It may be easier to see this in the case of the *trp* operon, but it is true for the *lac* operon as well. Allolactose induces enzyme synthesis not by acting directly on the genome, but by freeing the *lac* operon from the negative effect of the repressor. Technically, allolactose is more of a *derepressor* than an inducer of genes. Gene regulation is termed positive only when an activator molecule interacts directly with the genome to switch transcription on. Let's look at an example, again involving the *lac* operon.

An Example of Positive Gene Regulation

For the enzymes that break down lactose to be synthesized in appreciable quantity, it is not enough that lactose be present in the bacterial cell. The other requirement is that the simple sugar glucose be absent. Given a choice of substrates for glycolysis and other catabolic pathways, *E. coli* preferentially uses glucose, the sugar most reliably present in the nutrient medium.

How does the *E. coli* cell sense the glucose concentration, and how is this information relayed to the genome? The key is a protein called **catabolite activator protein (CAP).** (Catabolites are molecules such as lactose that can be consumed by catabolic pathways.) CAP accelerates transcription of an operon, such as the *lac* operon, by adhering to the promoter and facilitating the binding of RNA polymerase (FIGURE 17.19). Since CAP associates

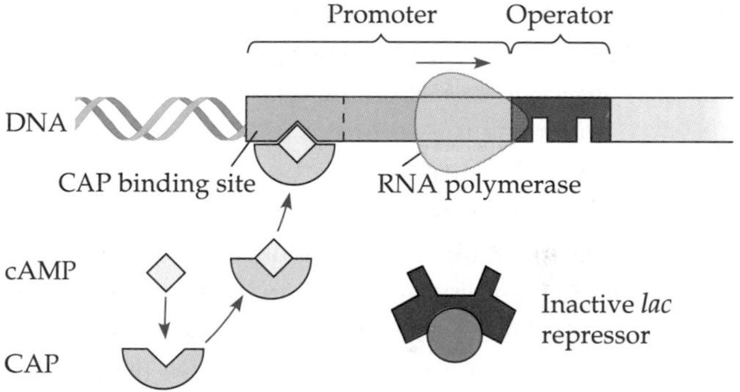

(a) Lactose present, glucose absent (cAMP level high): abundant *lac* mRNA synthesized

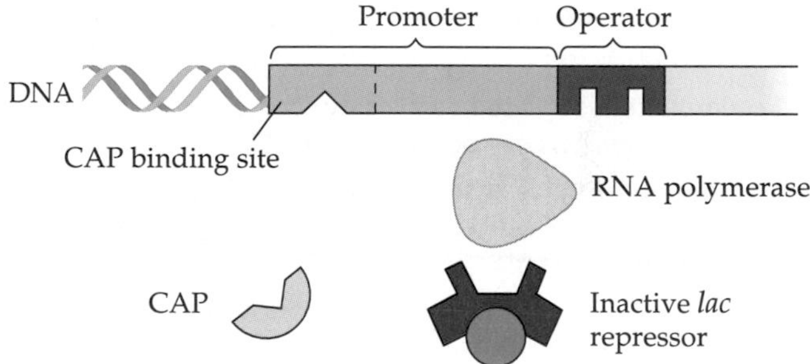

(b) Lactose present, glucose present (cAMP level low): little *lac* mRNA synthesized

FIGURE 17.19

Positive control: catabolite activator protein. RNA polymerase has a low affinity for the promoter of the *lac* operon unless helped by catabolite activator protein (CAP), which binds to the DNA. The CAP molecule can attach to the DNA only when associated with cyclic AMP (cAMP), whose concentration in the cell is inversely proportional to the glucose concentration. (**a**) If glucose is scarce, cAMP activates CAP, and the *lac* operon produces abundant mRNA for the lactose pathway. (**b**) But when glucose is present, cAMP is scarce, and CAP is unable to stimulate transcription. Thus, even if lactose is available, the cell will preferentially catabolize glucose, using enzymes that are always present. This regulatory system assures that *E. coli* will gear up for consumption of lactose and other secondary catabolites only when glucose is unavailable.

directly with the DNA to stimulate gene expression, this mechanism qualifies as positive regulation.

The activity of CAP is sensitive to the cell's glucose concentration. The absence of glucose results in the accumulation of a molecule called **cyclic AMP (cAMP),** which is derived from ATP. The CAP molecule has a binding site for cAMP, and it is actually the CAP-cAMP complex that attaches to the *lac* promoter and stimulates transcription of the genes for lactose catabolism. If glucose is added, the cAMP concentration falls, and CAP molecules disengage from the *lac* promoters. Thus, the *lac* operon is under dual control: negative control by the *lac* repressor, described earlier, and positive control by CAP. The state of the *lac* repressor (active or inactive) determines whether or not transcription of the *lac* operon's structural genes can occur; the state of CAP (plus or minus cAMP) controls the rate of transcription if the operon is repressor-free. It is as though the operon has both an on-off switch and a volume control.

Although we have used the *lac* operon as an example, CAP, unlike specific repressor proteins, works on several different operons. When glucose is present and CAP is inactive, there is a general slowdown in the synthesis of enzymes required for the utilization of all catabolites except glucose. The cell's ability to use alternative catabolites, such as lactose, provides backup systems that enable the cell deprived of glucose to survive. The specific catabolites present at the moment determine which operons are switched on. These elaborate contingency mechanisms suit an organism that cannot control what its host eats. Bacteria are remarkable in their ability to adapt—over the longer term by evolutionary changes in their genetic makeup, and over the shorter term by the control of gene expression in individual cells. Of course, the individual control mechanisms are also evolutionary products that exist because they have been favored by natural selection.

* * *

Molecular genetics was founded on the study of viruses and bacteria, the microbial models that have been the subjects of this chapter. Eukaryotic organisms are much more complex, and researchers are only beginning to learn how the control of gene expression can bring about this complexity. How, for example, can the genome present in a human zygote program the development of so many different kinds of cells in the adult organism? This is one of the problems we will investigate next, in Chapter 18.

REVIEW OF KEY CONCEPTS (with page numbers and key figures)

- Researchers discovered viruses by studying a plant disease: *science as a process* (pp. 324–325)
  - In the late 1800s, researchers discovered that tobacco mosaic disease is caused by an infectious agent much smaller than bacteria.
- Most viruses consist of a genome enclosed in a protein shell (pp. 325–326, FIGURE 17.2)
  - Viruses are not cells, but generally consist only of nucleic acid enclosed in a protein shell called a capsid.
  - The viral genome may be single- or double-stranded DNA, or single- or double-stranded RNA.
- Viruses can only reproduce within a host cell (pp. 326–327, FIGURE 17.3)
  - Viruses are obligate intracellular parasites that use the enzymes, ribosomes, and small molecules of host cells to synthesize multiple copies of themselves.
  - Each type of virus has a characteristic host range, determined by specific receptor sites on host cells.
- Phages exhibit two reproductive cycles: the lytic and lysogenic cycles (pp. 327–330, FIGURE 17.5)
  - In the lytic cycle of phage replication, injection of a phage genome into a bacterium programs the destruction of host DNA, the production of new viruses, and digestion of the bacterial cell wall, which bursts and releases the new virus.
  - In a lysogenic cycle, temperate viruses insert their genome into the bacterial chromosome as a prophage. In this innocuous form, the virus can be passed on to host daughter cells until it is stimulated to leave the bacterial chromosome and initiate a lytic cycle.
- Animal viruses are diverse in their modes of infection and mechanisms of replication (pp. 330–334, FIGURE 17.7)
  - Animal viruses are often equipped with an envelope acquired from host cell membrane. The envelope allows entry and exit through the plasma membrane of host cells.
  - All types of viral genomes are represented among animal viruses. RNA viruses called retroviruses (such as HIV) exhibit the most complex reproductive cycles. They use an enzyme called reverse transcriptase to synthesize DNA from their RNA template. The DNA can then integrate into the host genome as a provirus.
  - Vaccines against specific viruses stimulate the immune system to defend the host against an infection.
  - Emerging viruses that cause new outbreaks of disease are usually not new viruses but existing viruses that manage to expand their "host territory."
  - Tumor viruses insert viral DNA into host cell DNA, triggering subsequent cancerous changes through their own or host cell oncogenes.

- Plant viruses are serious agricultural pests (pp. 334–335, FIGURE 17.8)
  - Most plant viruses are RNA viruses that seriously compromise plant growth and development.
- Viroids and prions are infectious agents even simpler than viruses (p. 335)
  - Plant diseases can also be caused by viroids, molecules of naked RNA that interfere with plant growth and development.
  - Prions are infectious proteins that may increase their number by converting related proteins to more prions.
- Viruses may have evolved from other mobile genetic elements (pp. 335–336)
  - Evidence points to the origin of viruses as fragments of cellular nucleic acid that acquired specialized packaging.
- The short generation span of bacteria facilitates their evolutionary adaptation to changing environments (pp. 336–337)
  - The bacterial chromosome is a circular DNA molecule with few associated proteins. Accessory genes are carried on smaller rings of DNA called plasmids.
  - Because bacteria proliferate rapidly and have a short generation span, new mutations can affect a population's genetic variation in a relatively short period of time.
- Genetic recombination and transposition produce new bacterial strains (pp. 337–343, FIGURES 17.10, 17.12, 17.15)
  - Bacteria have three mechanisms of genetic recombination: transformation, transduction, and conjugation.
  - In transformation, naked DNA enters the cell from the surroundings.
  - In transduction, bacterial DNA is carried from one cell to another by phages.
  - In conjugation, a primitive kind of mating, an F^+ or Hfr cell transfers DNA to an F^- cell. The transfer is brought about by a plasmid called the F (fertility) plasmid, which carries genes for the sex pili and other functions needed for mating. In an Hfr cell, the F episome is integrated into the bacterial chromosome, and the Hfr cell will transfer chromosomal DNA along with the F episome DNA in conjugation.
  - R plasmids confer resistance to various antibiotics. Their transfer between bacterial cells poses serious medical problems.
  - Transposons, DNA segments that can insert at multiple sites in the genome or move from one site to another, also contribute to genetic shuffling in bacteria.
  - Insertion sequences, the simplest transposons, may affect gene function as they move about. They consist of inverted repeats of DNA flanking a gene for transposase.
  - Complex transposons include additional genes, such as genes for antibiotic resistance.
- The control of gene expression enables individual bacteria to adjust their metabolism to environmental change (pp. 344–348, FIGURES 17.17, 17.18, 17.19)
  - Cells control metabolism by regulating enzyme activity or by regulating enzyme synthesis through the activation or inactivation of selected genes.
  - In bacteria, regulated genes are often clustered into units called operons, consisting of a single promoter serving adjacent structural genes. A region called the operator serves as the on-off switch controlling the operon. Binding of a specific repressor protein to the operator shuts off transcription by blocking the attachment of RNA polymerase.
  - A repressible operon is switched off in the presence of a key molecule, usually an end-product of an anabolic pathway; the small molecule acts as a corepressor by activating the repressor protein.
  - In an inducible operon, binding of a small molecule (inducer) to the innately active repressor prevents its attachment to the operator, thereby turning on structural genes only when necessary. Inducible enzymes usually function in catabolic pathways.
  - Operons can also involve positive control via a stimulatory activator protein. For example, catabolite activator protein (CAP) stimulates transcription by binding to the promoter and enhancing its ability to associate with RNA polymerase.

SELF-QUIZ

1. Scientists have discovered how to put together a bacteriophage with the protein coat of phage T2 and the DNA of phage T4. If this composite phage were allowed to infect a bacterium, the phages produced in the host cell would have
   a. the protein of T2 and the DNA of T4
   b. the protein of T4 and the DNA of T2
   c. a mixture of the DNA and proteins of both phages
   d. the protein and DNA of T2
   e. the protein and DNA of T4

2. Horizontal transmission of a plant viral disease could be caused by
   a. the movement of viral particles through plasmodesmata
   b. the inheritance of an infection from a parent plant
   c. the spread of an infection by vegetative (asexual) propagation
   d. insects as vectors carrying viral particles between plants
   e. the transmission of proviruses via cell division

3. RNA viruses require their own supply of certain enzymes because
   a. the viruses are rapidly destroyed by host cell defenses
   b. host cells do not have RNA $\longrightarrow$ RNA or RNA $\longrightarrow$ DNA enzymes
   c. the enzymes translate viral mRNA into proteins
   d. the viruses use these enzymes to penetrate host cell membranes
   e. these enzymes cannot be made in host cells

4. A microbiologist found that some bacteria infected by phages had developed the ability to make a particular amino acid that they could not make before. This new ability was probably a result of
   a. transformation
   b. induction
   c. conjugation
   d. transduction
   e. transposition

5. Transposition differs from other mechanisms of genetic shuffling because it
   a. occurs only in bacteria
   b. moves genes between homologous regions of the DNA
   c. plays little or no role in evolution
   d. occurs only in eukaryotes
   e. scatters genes to new loci in the genome

6. A particular operon produces enzymes that manufacture an important amino acid. If this process works as it does in other operons,
   a. the amino acid inactivates the repressor
   b. the enzymes produced are called inducible enzymes
   c. the repressor binds to the operator in the absence of the amino acid
   d. the amino acid acts as a corepressor
   e. the amino acid "turns on" enzyme synthesis

7. A mutation that renders the regulatory gene of a repressible operon nonfunctional would result in
   a. continuous transcription of the structural genes
   b. continuous synthesis of an inducer
   c. accumulation of large quantities of a substrate for the catabolic pathway controlled by the operon
   d. irreversible binding of the repressor to the promoter
   e. excessive synthesis of a catabolic activator protein

8. Which of the following information transfers is catalyzed by reverse transcriptase?
   a. RNA ⟶ RNA
   b. DNA ⟶ RNA
   c. RNA ⟶ DNA
   d. protein ⟶ DNA
   e. RNA ⟶ protein

9. Which of the following characteristics or processes is common to *both* bacteria *and* viruses?
   a. nucleic acid as the genetic material
   b. binary fission
   c. mitosis
   d. ribosomes in the cytoplasm
   e. conjugation

10. Which of the following processes would not contribute to genetic variation within a bacterial population?
    a. transduction
    b. transformation
    c. conjugation
    d. mutation
    e. meiosis

CHALLENGE QUESTIONS

1. When bacteria infect an animal, the number of bacteria in the body increases gradually. A graph of the growth of the bacterial population is a smoothly increasing curve, as shown in graph a at the top of the next column. A viral infection shows a different pattern. For a while, there is no evidence of infection, and then there is a sudden rise in the number of viruses. The number stays constant for a time, then there is another sudden increase. The curve of viral population growth looks like a series of steps, as shown in graph b below. Explain the difference in these growth curves.

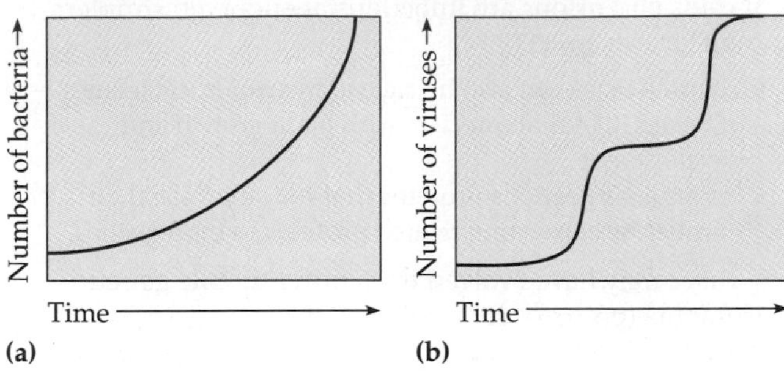

2. Mutations can alter the function of an operon; in fact, it was the effects of various mutations that enabled Jacob and Monod to figure out how the *lac* operon works. Predict how the following mutations would affect *lac* operon function in the presence and absence of allolactose.
   a. Mutation of regulatory gene; repressor will not bind to lactose.
   b. Mutation of operator; repressor will not bind to operator.
   c. Mutation of regulatory gene, repressor will not bind to operator.
   d. Mutation of promoter; RNA polymerase will not attach to promoter.

SCIENCE, TECHNOLOGY, AND SOCIETY

1. AIDS is not a new disease, just new to the Western world. HIV probably originated in central Africa and may have been infecting humans for decades. In what ways might modern technology and social changes contribute to the emergence of viruses that make new epidemics like AIDS possible?

2. Explain how the excessive or inappropriate use of antibiotics poses a health hazard for a population.

FURTHER READING

Beardsley, T. "Better Than a Cure." *Scientific American,* January 1995. The importance of immunizing children.

Diamond, J. "The Arrow of Disease." *Discover,* October 1992. Columbus and other explorers imported European pathogenic viruses and bacteria to the New World.

Eigen, M. "Viral Quasispecies." *Scientific American,* July 1993. How should we classify viruses?

Gibbons, A. "Where Are New Viruses Born?" *Science,* August 6, 1993. The problem of emerging viruses.

Maynard Smith, J. "Bacteria Break the Antibiotic Bank." *Natural History,* June 1994. The threat of drug-resistant bacteria.

Pennisi, E. "HIV Alters DNA, Causing Rare Cancer." *Science News,* April 16, 1994.

Prusiner, S. B. "The Prion Disease." *Scientific American,* January 1995. Infectious proteins.

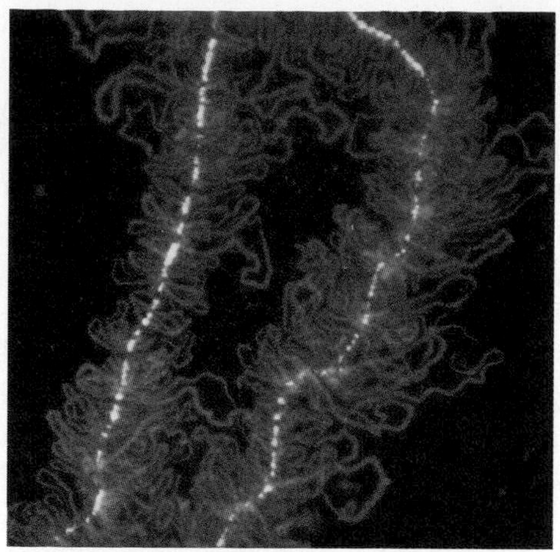

CHAPTER 18

GENOME ORGANIZATION AND EXPRESSION IN EUKARYOTES

Eukaryotic cells face the same challenges as prokaryotic cells in expressing their genes, with two main differences: the vastly greater size of the typical eukaryotic genome and the importance of cell specialization in eukaryotes. These two complications present a formidable information-processing task for the eukaryotic cell.

Consider the genome of a human cell, which has an estimated 50,000 to 100,000 genes—about fifty times more than a typical bacterium. The genomes of humans and other eukaryotes also include a large amount of DNA that does not program for the synthesis of RNA or protein. The entire mass of DNA must be precisely replicated with each turn of the cell cycle, as we saw in Chapter 11. Managing such a large amount of DNA requires that the eukaryotic genome be more complexly organized than prokaryotic DNA. In both prokaryotes and eukaryotes, DNA is associated with proteins, but in eukaryotes, the DNA-protein complex, called chromatin, is ordered into higher structural levels. The light micrograph that introduces this chapter gives a sense of the complex organization of chromatin in a developing salamander ovum. You can see chromatin both actively synthesizing RNA during transcription (red stained loops) and packed into the main axis of each of the chromosome's two chromatids (stained white). In this chapter, you will learn more about how eukaryotic genomes are organized and how expression of their genes is controlled.

KEY CONCEPTS

- Each cell of a multicellular eukaryote expresses only a small fraction of its genome
- The structural organization of chromatin sets coarse controls on gene expression
- Noncoding sequences and gene duplications account for much of a eukaryotic genome
- The control of gene expression can occur at any step in the pathway from gene to functional protein
- Chemical signals that help control gene expression include hormones
- Chemical modification or relocation of DNA within a genome can alter gene expression
- Cancer can result from the abnormal expression of genes that regulate cell growth and division

Each cell of a multicellular eukaryote expresses only a small fraction of its genome

One result of genomic complexity is that cell division is more elaborate in eukaryotes than in prokaryotes (see Chapter 11). At the same time, the challenges the eukaryotic cell faces during interphase may be even greater than those of cell division, especially in multicellular organisms. Interphase is when most genes are expressed—transcribed and translated into the products that enable each cell to carry out its role in the organism. Like unicellular organisms, the cells of multicellular organisms must continually turn certain genes on and off in response to signals from their external and internal environments. In addition, gene expression must be controlled on a long-term basis for **cellular differentiation**—the divergence in structure and function of different types of cells as they become specialized during an organism's development and remain that way. Highly specialized cells, such as those of muscle or nervous tissue, express only a tiny fraction of their genes. In fact, a typical human cell expresses only 3% to 5% of its genes at any given time. The enzymes that transcribe DNA must locate the right genes at the right time, which is like finding—and threading—a needle in a haystack. When the expression or activity of genes goes awry, serious imbalances and diseases, including cancer, can arise. Thus, the question of how eukaryotic genes are regulated is paramount for medical research as well as for basic biology.

Only twenty years ago, an understanding of the mechanisms that control gene expression in eukaryotes seemed almost hopelessly out of reach. Since then, new research methods have empowered molecular biologists to begin solving some of these once impenetrable mysteries. Equipped with DNA technology for cloning genes and increasingly rapid methods of sequencing DNA (discussed in Chapter 19), biologists are uncovering many of the details about how genes are organized within the eukaryotic genome and how their expression is regulated.

Despite the extra "nuts and bolts" it requires, the control of gene activity in eukaryotes involves the same basic principles as prokaryotic gene regulation. In all organisms, gene activity is regulated by DNA-binding proteins that also interact with other proteins and with environmental factors. In most cases, it is the transcription of the DNA that is specifically controlled. However, the greater complexity of genomic organization in eukaryotes and the greater complexity of eukaryotic cell structure do offer additional opportunities for controlling gene expression. So before examining eukaryotic gene expression and its control, we will look at how eukaryotic DNA is physically packaged into chromatin and chromosomes, and how genes and other DNA sequences are organized within the genome. We will also consider how the dynamic nature of the eukaryotic genome affects the availability of its genes for expression.

The structural organization of chromatin sets coarse controls on gene expression

Although both prokaryotic and eukaryotic cells contain hereditary material in the form of double-stranded DNA, their genomes are organized differently. Prokaryotic DNA is usually circular, and the nucleoid it forms is so small that it can be seen only with an electron microscope. The "chromosome" of *E. coli*, for example, contains only 4.3×10^6 nucleotide pairs of DNA; and while it is associated with various proteins to form a sort of chromatin, the protein molecules are relatively few. In contrast, eukaryotic chromatin consists of DNA precisely complexed with a large amount of protein. During interphase, the chromatin fibers are usually highly extended and tangled. When interphase cells are stained, the chromatin appears as a diffuse, colored mass. You learned in Chapter 11, however, that as a cell prepares for mitosis, its chromatin coils and folds up ("condenses") to form a number of short, thick, discrete chromosomes that, when stained, are clearly visible with a light microscope.

Eukaryotic chromosomes contain an enormous amount of DNA relative to their length. Each chromosome contains a single uninterrupted DNA double helix that, in humans, typically has about 2×10^8 nucleotide pairs. If extended, such a DNA molecule would be about 6 cm long, thousands of times longer than the diameter of a cell nucleus. All this DNA—and the DNA of the other 45 human chromosomes, as well—fits into the nucleus through an elaborate, multilevel system of packing, which is diagrammed in FIGURE 18.1.

Nucleosomes, or "Beads on a String"

Proteins called **histones** are responsible for the first level of DNA packing in chromatin. In fact, the amount of histone in chromatin is approximately equal to the amount of DNA. Histones have a high proportion of positively charged amino acids (lysine and arginine), and they bind tightly to the negatively charged DNA, forming chromatin. There are five types of histone found in most eukaryotic cells. Histones are very similar from one eukaryote to another, suggesting that evolution has been relatively conservative with the histone genes.

In electron micrographs, unfolded chromatin has the appearance of beads on a string, as you can see in FIGURE 18.1a. Each "bead" is a **nucleosome,** the basic unit of DNA packing. The nucleosome consists of DNA wound around a protein core composed of two molecules each of four different types of histone: H2A, H2B, H3, and H4. A molecule of the fifth histone, called H1, may be attached to the outside of the "bead." Later, we will consider how the specific organization of DNA in nucleosomes may influence gene expression by limiting the access of transcription proteins to DNA, or by directing them to specific DNA regions.

Higher Levels of DNA Packing

The beaded string undergoes higher-order packing. This is most strikingly evident when the extended interphase chromatin coils and folds further to produce the thickened, compact chromosomes we see during mitosis. In the laboratory, mitotic chromosomes can be isolated and unraveled to reveal several orders of chromatin coiling. FIGURE 18.1 illustrates the various structures in order of

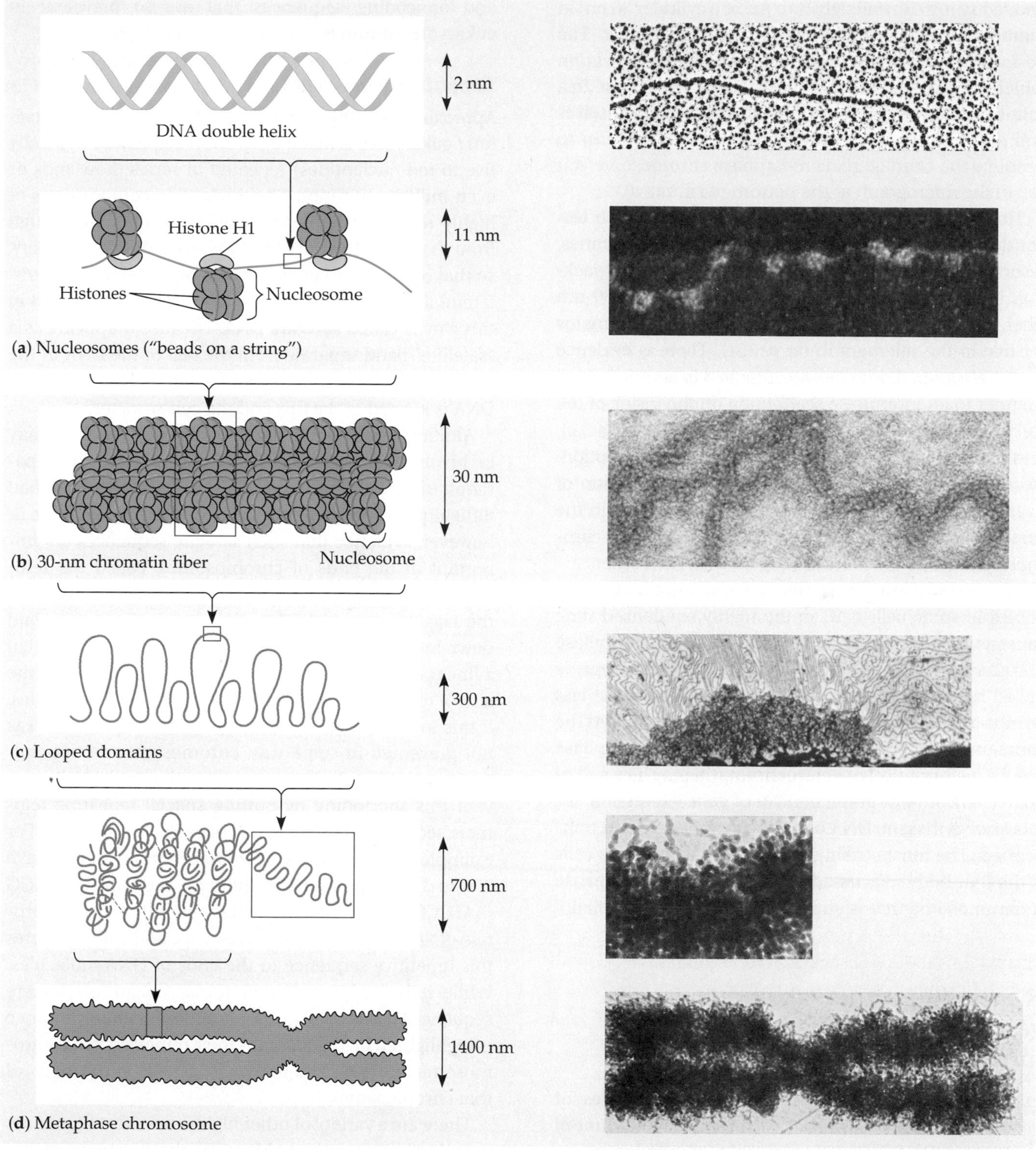

FIGURE 18.1

Levels of chromatin packing. This series of diagrams and transmission electron micrographs shows a current model for the progressive stages of DNA coiling and folding, which culminate in the highly condensed metaphase chromosome. (**a**) DNA, in association with histone, forms "beads on a string," consisting of nucleosomes in an extended configuration. Each nucleosome has two molecules each of four types of histone. The fifth histone (called H1) may be present on DNA adjacent to the "bead." (**b**) The 30-nm chromatin fiber is a tightly wound coil with six nucleosomes per turn. (**c**) Looped domains of 30-nm fibers are visible here because a compact chromosome has been experimentally unraveled. (**d**) These multiple levels of chromatin packing form the compact chromosome, visible at metaphase.

increasing compactness. With the aid of histone H1, the beaded string can coil tightly to make a cylinder 30 nm in diameter, known as the *30-nm chromatin fiber.* The 30-nm fiber, in turn, forms loops called *looped domains,* which are attached to a nonhistone protein scaffold. In a mitotic chromosome, the looped domains themselves coil and fold, further compacting all the chromatin to produce the characteristic metaphase chromosome you see in the micrograph at the bottom of FIGURE 18.1.

Though interphase chromatin is generally much less condensed than the chromatin of mitotic chromosomes, it shows several of these same levels of higher-order packing. Its "beaded string" is usually coiled into a 30-nm fiber, and the 30-nm fiber is folded into looped domains (visible in the micrograph on p. 351). There is evidence that these looped domains are attached in some directed manner to an interphase scaffolding on the inside of the nuclear envelope (the nuclear lamina; see FIGURE 7.9), and that this helps organize areas of active transcription. Recent experiments also suggest that the chromatin of each chromosome occupies a restricted area within the nucleus, and that the chromatin fibers of different chromosomes do not become entangled with each other.

Even during interphase, portions of certain chromosomes in some cells exist in the highly condensed state represented at the bottom of FIGURE 18.1. Such interphase chromatin, which is visible with a light microscope, is called **heterochromatin,** to distinguish it from the less compacted **euchromatin** ("true chromatin"). What is the function of this selective condensation in interphase cells? The formation of heterochromatin may be a sort of coarse adjustment in the control of gene expression, for it is known that the DNA of heterochromatin is not transcribed. The most striking example in mammalian cells is the Barr body (discussed in Chapter 14), an interphase *X* chromosome that is almost entirely heterochromatin.

Noncoding sequences and gene duplications account for much of a eukaryotic genome

In prokaryotes, most of the DNA in a genome codes for protein (or tRNA and rRNA), with the small amount of noncoding DNA consisting mainly of regulatory sequences, such as promoters. The coding sequence of nucleotides along a prokaryotic gene proceeds from start to finish without interruption. In eukaryotic genomes, in contrast, most of the DNA does *not* encode protein or RNA. Moreover, certain DNA sequences may be present in multiple copies in a eukaryotic genome, and coding sequences may be interrupted by long stretches of noncoding DNA (introns; see Chapter 16). In this section, we will examine some examples of the DNA duplications and noncoding sequences that are so prevalent in eukaryotic genomes.

Repetitive Sequences

Approximately 10% to 25% of the total DNA of multicellular eukaryotes is made up of short sequences (typically five to ten nucleotides) repeated in series thousands or even millions of times. The nucleotide compositions of highly repetitive sequences are often different enough from the rest of the cell's DNA to have a different density, so that researchers can isolate the repetitive DNA by differential ultracentrifugation. DNA that can be isolated in this way is called **satellite DNA,** because it appears as a "satellite" band separate from the rest of the DNA in the centrifuge tube. In chromosomes, much of the satellite DNA is located at the tips and the centromeres.

Although its location suggests that satellite DNA may be involved in the replication of chromosomes or separation of chromatids in cell division, scientists have had difficulty finding direct evidence for such a role. There is, however, evidence that such satellite sequences are important at the ends of chromosomes, called the **telomeres.** Recall from Chapter 15 that during replication of the lagging strand of DNA, RNA primers must be laid down in front of the region to be replicated. At the end of a linear chromosome, this is impossible, resulting in the lack of replication of the very ends of the DNA molecules. If this loss of a DNA tip through lack of replication was not prevented in some way, chromosomes would continue to shorten during each replication cycle. Cells prevent this shortening by putting special repeating telomere sequences onto the ends of each chromosome. For example, the telomeres of human chromosomes have 250 to 1500 repetitions of the base sequence TTAGGG (AATCCC on the complementary stand). A special protein-RNA complex, telomerase, periodically restores this repetitive sequence to the ends of DNA molecules. While it is unclear whether these repeating telomere sequences perform any other function, their importance is highlighted by the fact that their removal from a chromosome in yeast cells leads to disintegration of that chromosome.

There are a variety of other highly repetitive sequences in eukaryotic genomes. Many of these sequences are transposons, described in Chapter 17 and later in this chapter. While generally regarded as nonfunctional, transposons have been associated with a number of diseases, including neurofibromatosis-1 (elephant man's disease) and a number of different forms of cancer. Some genes normally have repetitive sequences within their boundaries, but mutations can extend the number of repeats and cause the genes to malfunction. Examples are the elongations of base triplet repeats that lead to fragile *X* syndrome and Huntington's disease (see Chapter 14).

Multigene Families

As in prokaryotes, the eukaryotic DNA sequences that code for proteins or RNA—the genes—are usually present as unique sequences, single copies in the genome. However, some genes are represented by more than one copy, and others closely resemble each other in nucleotide sequence. A collection of identical or very similar genes is called a **multigene family.** It is likely that the members of each family evolved from a single ancestor gene.

The members of a multigene family may be clustered or dispersed in the genome. Some multigene families exist as multiple identical genes, usually clustered together. Although the genes for histone proteins are a notable exception, multigene families usually consist of genes for RNA products. An example is the family of identical genes for the major ribosomal RNA (rRNA) molecules (FIGURE 18.2). These genes are repeated in series (tandemly) hundreds to thousands of times in the genomes of multicellular eukaryotes, forming huge tandem arrays of genes that enable the cells to make the millions of ribosomes needed for active protein synthesis.

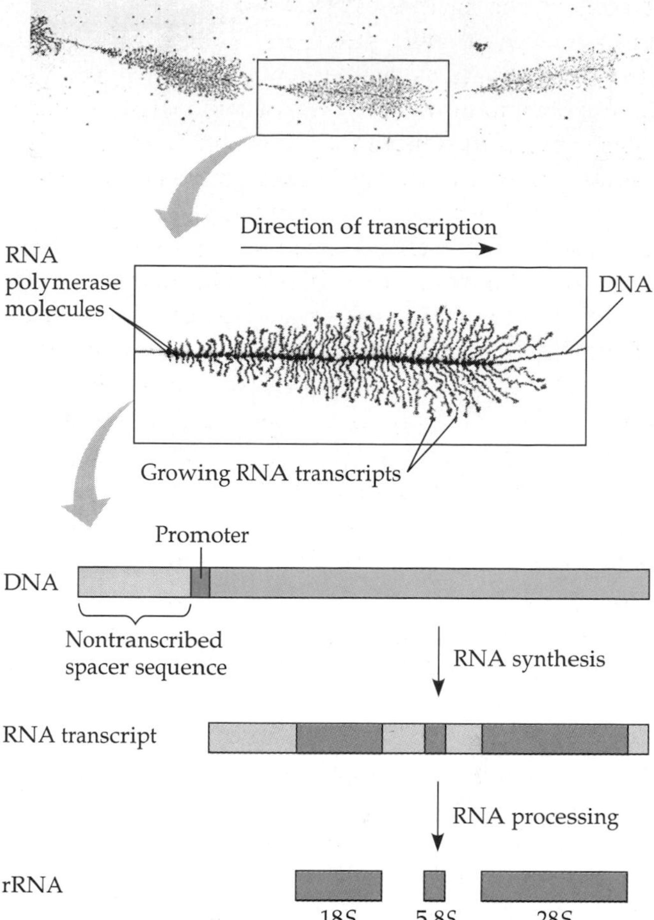

FIGURE 18.2
Part of a family of identical genes for ribosomal RNA. The transmission electron micrograph at the top shows three of the hundreds of copies of rRNA genes in a salamander genome. Each "feather" corresponds to transcript (RNA) produced as about 100 molecules of RNA polymerase (the dark dots along the DNA) move left to right. The growing RNA transcripts extend out from the DNA. The genes are arranged in tandem, each one separated from the next by a spacer sequence of nontranscribed DNA. The RNA transcripts are processed to yield three kinds of rRNA molecules: 18*S*, 5.8*S*, and 28*S*. (The *S* designations refer to their sedimentation rates in the ultracentrifuge.)

The classic examples of multigene families of *nonidentical* genes are the two related families of genes that encode globins, the α and β polypeptide subunits of hemoglobin (see FIGURE 5.23b). One family, located on chromosome 16 in humans, encodes various versions of α-globin; the other, on chromosome 11, encodes versions of β-globin (FIGURE 18.3). Similarities in the sequences of the various globin genes indicate that the α-like globins and β-like globins all evolved from a common ancestral globin. The different versions of each globin subunit are expressed at different times in development, allowing the hemoglobin variations to function effectively in the changing environment of the developing animal. In humans, for example, the embryonic and fetal forms of hemoglobin have a higher affinity for oxygen than the adult forms, ensuring the efficient transfer of oxygen from mother to developing fetus (see FIGURE 18.3b).

How do families of genes arise from a single gene? The most likely explanation for families of identical genes is that they arise by repeated gene duplication. This occurrence, called tandem gene duplication, results from mistakes made in DNA replication and recombination. Families of nonidentical genes probably arise from mutations that accumulate in duplicated genes over a long period of time. The existence of DNA segments called pseudogenes is evidence for this process of gene duplication and mutation. **Pseudogenes** have sequences very similar to real (functional) genes but lack the sites (for example, promoters) necessary for gene expression. The globin gene families include several pseudogenes within the stretches of noncoding DNA between the functional genes.

Pseudogenes and repetitive DNA do not account for all of the enormous amount of noncoding DNA in the genomes of eukaryotes. Significant amounts of additional noncoding DNA are found *within* genes, as introns. While these sequences are also generally regarded as nonfunctional, there are a growing number of examples of RNA molecules produced from introns that regulate the expression of certain genes.

Now that we have considered the organization of chromatin and some unique features of the eukaryotic genome, let's consider how gene expression is regulated in eukaryotic organisms.

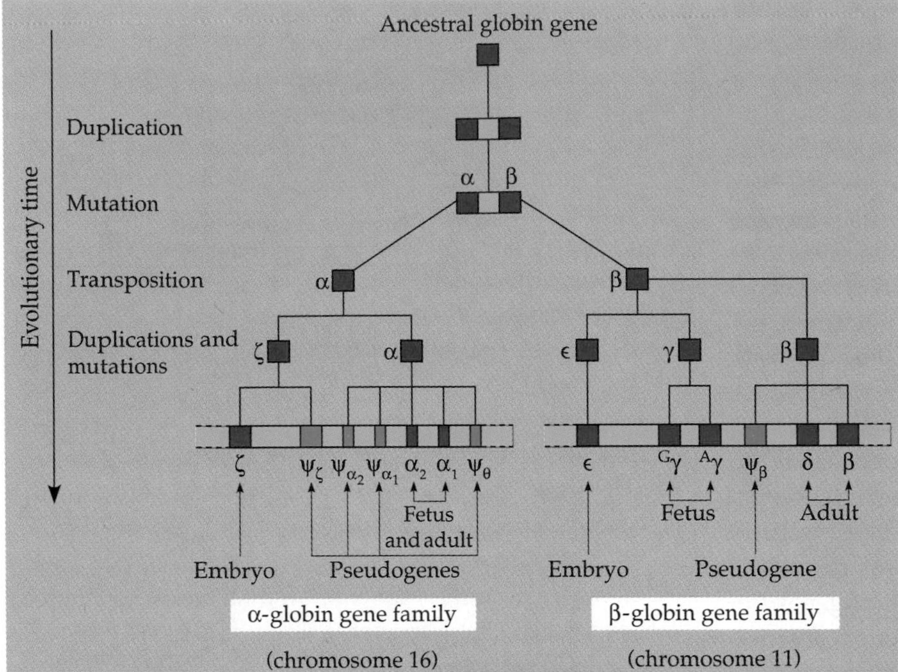

(a)

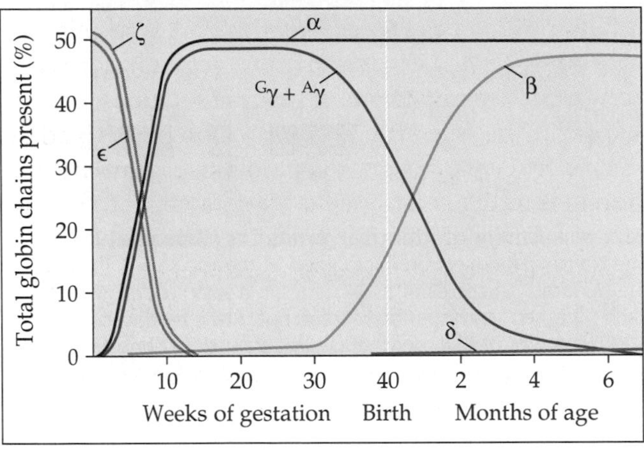

(b)

FIGURE 18.3
The evolution of α-globin and β-globin. (**a**) Each of the two gene families consists of a group of similar, but not identical, genes clustered together on a chromosome. The various genes (red) are represented by different Greek letters. In each gene family, the genes are arranged in order of their expression during development; the expression of individual genes is turned on or off in response to the organism's changing environment as it develops from an embryo into a fetus, and then into an adult. At all times during development, functional hemoglobin consists of a total of four polypeptide subunits, two from the α family and two from the β family. Separating the functional genes within each family cluster are long stretches of noncoding DNA, which include pseudogenes (green), nonfunctional nucleotide sequences very similar to the functional genes. Presumably, the different genes and pseudogenes in each family arose from gene duplication of an original α or β gene followed by mutation. In fact, the original α and β genes themselves undoubtedly arose in much the same way from a common ancestral globin gene. Transposition (see Chapter 17) put the α-globin and β-globin families on different chromosomes, probably early in their evolution. (**b**) This graph traces the utilization of different globin genes during human development. During embryogenesis, the ζ and ϵ forms predominate. About 10 weeks following fertilization, the products of the α-globin genes replace that of the ζ gene, and the G_γ and A_γ forms of the β-globin genes predominate. Just prior to birth, the β-gene product begins to replace the G_γ and A_γ forms, such that within 6 months of birth, the adult forms (β and δ) are being used.

The control of gene expression can occur at any step in the pathway from gene to functional protein

The organization of chromatin discussed earlier in the chapter serves a dual purpose. Certainly one function is to package the DNA into a compact form suitable for containment within the nucleus of a cell. At the same time, it is becoming clear that such packaging is not random; the physical state of DNA within and around a gene is important in helping control which regions of the DNA are available for transcription. Thus, condensed heterochromatin is not expressed, and location relative to scaffold attachment sites and nucleosomes can determine whether a specific gene is expressed. But simply being available does not ensure that a gene will be expressed. Cells must be able to alter gene expression in response to changing environments, external signals, or new demands following differentiation. Thus, even genes available for transcription need to be tightly regulated. FIGURE 18.4 gives an overview of the steps in gene expression in eukaryotes, highlighting the stages at which the expression of a given gene can be regulated. In this section, we will first review the organization of a typical protein-coding gene, and then consider each level of regulation, starting with the control of transcription.

Organization of a Typical Eukaryotic Gene

The DNA that makes up a gene and its control regions is typically organized as shown in FIGURE 18.5 (p. 358), which reviews and extends what you learned about eukaryotic

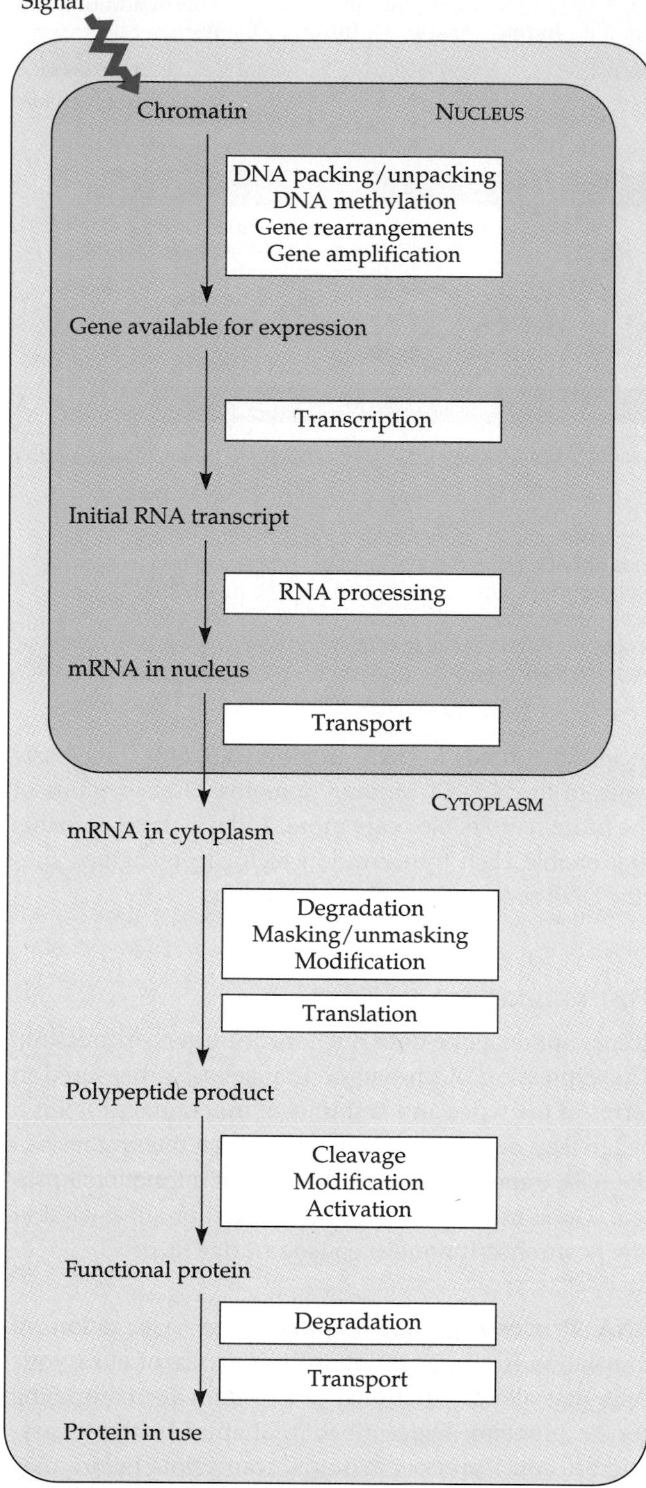

FIGURE 18.4

Opportunities for the control of gene expression in eukaryotic cells: an overview. Unlike a prokaryotic cell, a eukaryotic cell has a nuclear envelope that separates transcription and translation in both space and time. This feature offers greater opportunity for posttranscriptional control in the form of RNA processing. In addition, eukaryotes have a greater variety of control mechanisms at the gene level (before transcription) and at the protein level (after translation). In this diagram, the processes that offer opportunities for regulation are highlighted by white boxes.

genes in Chapter 16. The most striking difference between this gene and a prokaryotic gene is the presence of introns, noncoding sequences that are interspersed within the coding sequence. Recall from Chapter 16 that RNA polymerase attaches to a promoter sequence at the "upstream" end of the gene and proceeds to transcribe the introns along with the coding sequences, called exons. The introns are removed in later RNA processing, so that they do not appear in the mature mRNA. The processing of the initial RNA transcript also includes the addition of a modified guanosine triphosphate cap at the 5′ end and a poly-A tail at the 3′ end.

Another distinctive feature of the eukaryotic gene illustrated in FIGURE 18.5 is the presence of other non-coding control sequences. Some may be located in close proximity to the promoter, while others may be located thousands of bases away. These sequences, called **enhancers,** can have a powerful influence on transcription of the associated gene, an important control mechanism we will examine later in the chapter.

Transcriptional Control

In both prokaryotes and eukaryotes, RNA synthesis begins with an RNA polymerase enzyme acting in concert at a promoter with numerous proteins called transcription factors. The polymerase and transcription factors bind to specific sequences within the promoter region just upstream from the coding sequence of a gene. Once the appropriate initiation complex forms, the polymerase moves along the DNA template, producing the complementary strand of RNA. In eukaryotes, additional transcription factors bind selectively to other regulatory elements, including enhancer regions of DNA that may be thousands of nucleotides away from the promoter and coding sequence.

The specific associations between transcription factors and enhancer sites in the genome play an important role in the control of gene expression in eukaryotes. How do sequences of DNA so far away affect transcription at specific promoters? Transcription is probably enhanced when a loop in the DNA brings the transcription factor attached to the enhancer into contact with the transcription factors and polymerase at the promoter (FIGURE 18.6). Diverse transcription factors keyed to different enhancer sequences in the genome may selectively activate the expression of specific genes at appropriate stages in the development of a cell.

Thus, the control of transcription depends on regulatory proteins that bind selectively to DNA and to other proteins. Hundreds of transcription factors, including those that bind to enhancers, have been discovered so far in eukaryotes. As diverse as these proteins are, they share a few basic structures. Each has a domain (structural

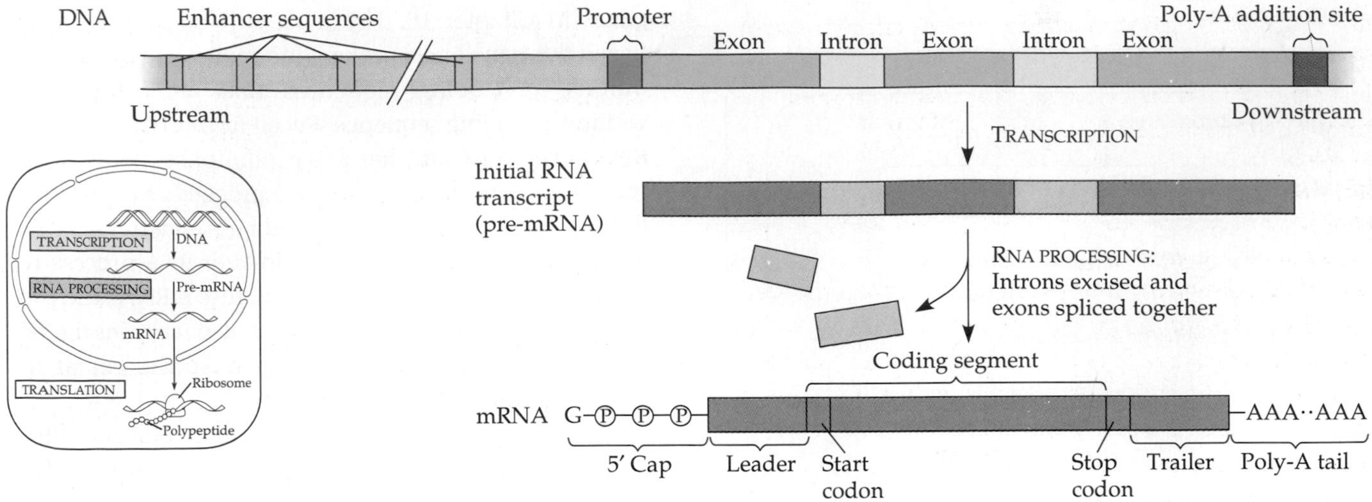

FIGURE 18.5

A review of the molecular anatomy of a eukaryotic gene and its transcript. The promoter and associated neighboring sequences function in the initiation of transcription, which proceeds "downstream," elongating the RNA. Located as many as thousands of nucleotides upstream or downstream from the promoter are enhancer sequences, sites where proteins that regulate transcription of the gene can bind. After the initial RNA transcript is made, processing enzymes excise introns and add the methylated guanosine 5′ cap and poly-A tail. The mRNA is now ready for export to the cytoplasm.

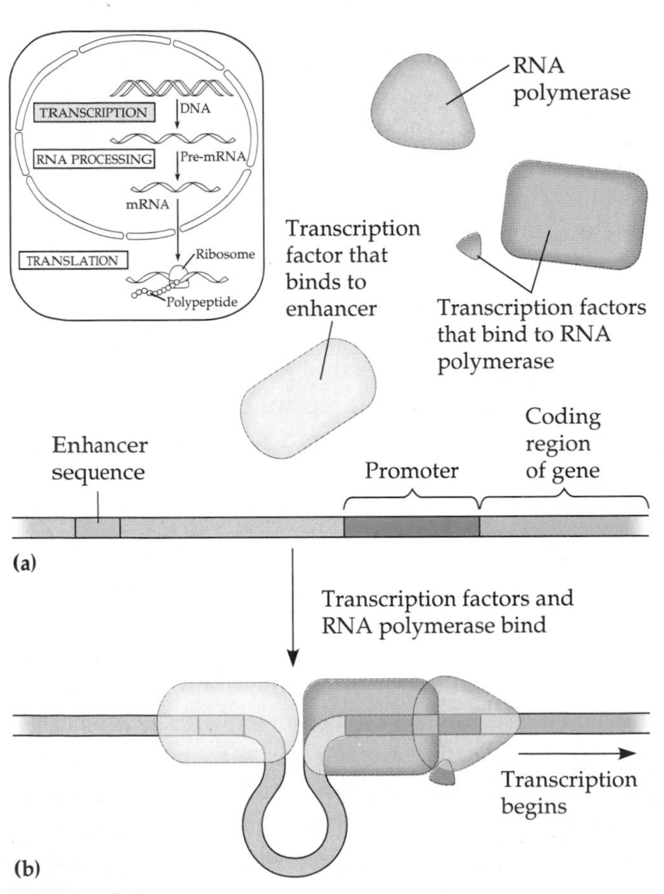

FIGURE 18.6

The function of enhancers in the control of eukaryotic gene expression. (**a**) RNA polymerase requires additional proteins called transcription factors in order to recognize and bind to the promoter region at the upstream end of a gene. Enhancer regions are often found within DNA quite distant from the promoter. (**b**) Binding of a specific transcription factor to the enhancer stimulates the formation of the necessary complex of polymerase and transcription factors.

region) that binds to DNA, and there are only three basic types of those DNA-binding domains. Other regions of the protein molecules vary more, and it is these domains that enable each transcription factor to recognize specific DNA sequences and other proteins.

Posttranscriptional Control

Transcription alone does not constitute gene expression. The expression of protein-coding genes is measured in terms of the types and amounts of functional proteins a cell makes, and much happens between the synthesis of the RNA transcript and the creation of a functional protein. Gene expression may be blocked or stimulated at any posttranscriptional step (see FIGURE 18.4).

RNA Processing and Export. The segregation of translation from transcription is a feature of eukaryotic cells that allows additional possibilities for controlling gene expression. As described in Chapter 16, the eukaryotic cell must process its initial transcripts before they can act as mRNA, tRNA, or rRNA. An mRNA transcript must receive a 5′ cap and a poly-A tail, and the RNA segments representing the introns of the genes must be removed and the exons (coding segments) spliced together (see FIGURE 18.5). Such processed mRNAs, most likely with proteins attached, are transported out of the nucleus through nuclear pores. In the cytoplasm, the mRNA interacts with a number of specific proteins and may associate with ribosomes to undergo translation. Each step in RNA processing and in the mechanisms responsible for the export of mRNA from the nucleus and translation in the cytoplasm represents an opportunity for controlling gene expression. However, little is known about how these posttranscriptional steps are regulated.

Regulation of mRNA Degradation. The lifespan of an mRNA molecule in the cytoplasm is also an important factor in controlling the pattern of protein synthesis in a cell. Prokaryotic mRNA molecules typically have very short lives; they are degraded by enzymes after only a few minutes. This is one reason bacteria can vary their patterns of protein synthesis so quickly in response to environmental changes. In contrast, the mRNA molecules of eukaryotes can have lifetimes of hours, days, or even weeks. A striking example of long-lived mRNA is found in mammalian red blood cells, which are "factories" for the production of the protein hemoglobin. The mRNAs for hemoglobin are unusually stable and are translated repeatedly in the developing red blood cells. If two species of mRNA molecules differ in how rapidly they are broken down by enzymes in the cytoplasm, they may differ in how much protein synthesis each directs.

In cases where mRNAs are synthesized in large quantities and stored, not only translation must be blocked, but degradation as well. In at least some cases, proteins that bind to mRNA and block translation also block degradation. In other cases, modification of the length of the poly-A tail affects the rate of degradation.

Translational and Posttranslational Control

Translation in eukaryotic cells involves many more protein factors, especially initiation factors, than in prokaryotic cells (see Chapter 16). Thus, there are ample opportunities for the control of gene expression at the level of translation.

Most translational control mechanisms block the initiation phase of protein synthesis. Translational repression of specific mRNAs can be accomplished by regulatory proteins that bind to specific sequences or structures at the 5′ ends of particular messages, preventing the attachment of ribosomes. Alternatively, the translation of all mRNAs can be blocked by mechanisms that inactivate certain initiation factors. Such global controls of translation are found in a number of developmental situations. For example, the mRNA used for the active protein synthesis that occurs during the first stage of embryonic development (cleavage) of many organisms has all been synthesized by the egg cell nucleus prior to fertilization. It is stored in the cytoplasm of the unfertilized egg as inactive mRNA and is not translated until fertilization, with the sudden activation of an initiation factor that must be present for ribosomes and mRNA to begin interacting. By synthesizing large quantities of specific mRNAs, stocking them in the cytoplasm, and delaying their translation until a signal is given, a developing cell can respond to a stimulus with an explosive burst of synthesis of particular proteins. Some plants and algae also store mRNAs in this way during periods of darkness; light then triggers the reactivation of the translational apparatus.

The last opportunities for controlling gene expression occur after translation. Often, eukaryotic polypeptides must be processed to yield the active final products. The posttranslational processing of the hormone insulin is an example (see Chapter 16). In addition, many polypeptides must be modified by the addition of chemical groups, such as chains of sugars, in order to be active. For many proteins to reach their final destinations in the cell, the signal mechanism must target these proteins (see FIGURE 16.17). Regulation might occur at any of these steps involved in modifying or transporting a protein. Finally, the extent of gene expression could be controlled by the cell's selective degradation of particular proteins. Abnormal targeting or degradation of a protein can have serious consequences, as, for example, in cystic fibrosis (described in Chapter 13). This disease results from mutations in the gene for a protein that functions as a chloride channel. The defective protein does not reach its final destination in the cell, the plasma membrane, and is rapidly degraded.

The Arrangement of Coordinately Controlled Genes

In some cases, different genes of related function need to be turned on or off at the same time. In Chapter 17, you learned that in prokaryotes, such coordinately controlled genes are often clustered into an operon; they are adjacent to each other in the DNA molecule and share the regulatory sites located at one end of the cluster. All the genes of the operon are transcribed sequentially into a single mRNA molecule and are translated together. Such operons have not been found in eukaryotic cells. Genes coding for the enzymes of a metabolic pathway, for example, are often scattered over different chromosomes in the eukaryotic genome. Even when genes for related functions are located near one another on the same chromosome, each gene has its own promoter and is individually transcribed. Nevertheless, scattered collections of eukaryotic genes are often coordinately expressed.

Coordinate gene expression in eukaryotes probably involves the presence of a specific regulatory element or enhancer associated with every gene of a scattered group. This sequence would be recognized by a single type of transcription factor (just as a prokaryotic operator sequence is recognized by a repressor or activator). Three examples of such coordinate control in eukaryotes are the induction of a set of proteins following exposure to high temperatures (the heat shock response), the changes in transcription that occur as a result of the action of steroid hormones, and the coordinate expression of specific genes during cellular differentiation.

Gene Expression and Differentiation

In multicellular organisms, there is usually a high degree of cellular differentiation, resulting in the formation of different tissues. Such differentiation, often yielding major changes in the structure and function of cells, can only occur as a result of the presence or absence of tissue-specific proteins. While a more complete discussion of cellular differentiation and development is presented in Chapters 34 and 43, it is worthwhile at this point to consider briefly how tissue-specific gene expression might be controlled.

The cytoplasmic contents and position of a cell within an embryo determine the developmental fate of each cell. Both of these influences can provide chemical signals that will affect only certain cells or groups of cells. The function of these signals is to activate transcription factors, which result in the expression of genes for other regulatory proteins (typically more transcription factors). Cascades of regulatory proteins ultimately result in whole sets of tissue-specific genes being expressed as a result of developmental signals.

Chemical signals that help control gene expression include hormones

How does the cell's external or internal environment influence gene expression? In bacteria (see Chapter 17), small organic metabolites such as tryptophan affect transcription by combining with regulatory proteins, such as the repressor protein of the *trp* operon. The regulatory protein's shape is changed in such a way that its ability to bind to DNA is altered. In eukaryotes, too, many small organic molecules influence transcription by combining with regulatory proteins. Among the extracellular molecules that alter gene expression by this mechanism are the steroid hormones of animals. Carried in the blood, these molecules serve as signals from cells in one part of the animal to cells elsewhere in the body. The effect of steroid hormones on gene expression in target cells was first clearly recognized in insects.

Chromosome Puffs: Evidence for the Regulatory Role of Steroid Hormones in Insects

Studies of giant (polytene) chromosomes in the salivary gland and other tissues of certain insect larvae have provided evidence for both the control of gene expression at the transcriptional level and the role of steroids in the process. Polytene chromosomes consist of hundreds of parallel chromatids resulting from multiple rounds of DNA replication in nondividing cells. At characteristic stages in the larva's development, chromosome puffs appear at specific sites on the polytene chromosomes (FIGURE 18.7). A puff forms when DNA loops out from the chromosome axis, perhaps making the DNA in that region more accessible to the transcription machinery. Analysis using autoradiography (see Chapter 2) shows that the puffs indeed correspond to regions of intense transcription.

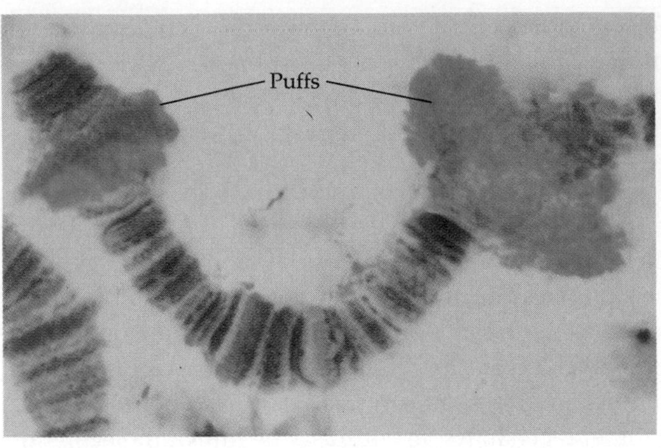

FIGURE 18.7
Chromosome puffs. This light micrograph shows two puffs in a polytene chromosome of an insect, *Trichosia pubescens*. The locations of puffs along the chromosomes change as a cell develops. Puffs are a visible indication of gene activity—sites of active transcription. Experimental addition of ecdysone, the steroid hormone that stimulates molting, causes a pattern of puffs characteristic of natural molting.

The locations of chromosome puffs along a chromosome change as the larva develops. When the larva prepares to molt, certain puffs disappear and others form at new sites. The shifting puffs are visual indicators of the selective switching on and off of specific genes during development. These changes in puffing patterns can be induced by ecdysone, the steroid hormone that initiates molting. This effect demonstrates that gene regulation in the insect is responsive to specific chemical signals, and in particular to a steroid.

The Action of Steroid Hormones in Vertebrates

Researchers are identifying the key steps by which sex hormones and other steroids alter gene expression in target cells of vertebrates. Steroids are soluble in lipids. When a cell is exposed to a steroid, the hormone diffuses across the plasma membrane into the cytoplasm. It then enters the nucleus, where it encounters a soluble recep-

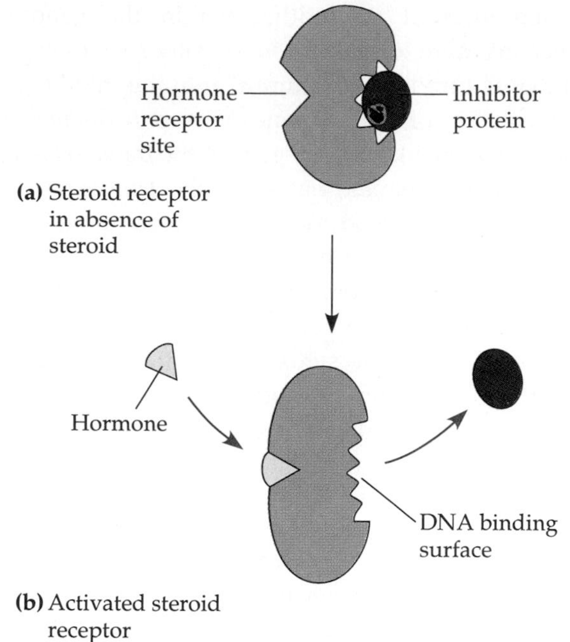

FIGURE 18.8
Activation of a steroid receptor. (**a**) The steroid receptor is a DNA-binding protein that stimulates transcription of specific genes. In the absence of the steroid, an inhibitor protein prevents the receptor from binding to DNA. (**b**) Arrival of the steroid causes the receptor protein to release the inhibitor or alter its structure, and the receptor can now bind to specific regulatory sites on DNA. By this mechanism, steroid hormones activate the expression of certain genes in target cells.

tor protein. (In some cases the hormone binds to its receptor in the cytoplasm, and the hormone-receptor complex then enters the nucleus.) In the absence of the steroid, the receptor protein is usually associated with an inhibitory protein that prevents the receptor from binding to DNA (FIGURE 18.8a). Binding of the steroid hormone to the receptor causes the release of the inhibitory protein, and the activated receptor protein can now attach to specific sites on the DNA (FIGURE 18.8b). These sites are within enhancer or other regulatory regions that control steroid-responsive genes. Binding of the receptor protein to the enhancer activates transcription, and in this way, a steroid acts as a chemical signal to switch on specific genes in certain cells.

Other hormones besides steroid hormones can affect transcription. Since these nonsteroid hormones are not lipid soluble, they cannot cross cell membranes, and thus must operate through a receptor protein on the outside of the cell. The binding of such a hormone to its receptor initiates a signal transduction pathway (see Chapter 8) that ultimately leads to the activation of specific transcription factors and the resulting transcription of specific genes. You will learn more about the pathways by which hormones affect cells in Chapter 41.

This is a good place to summarize five key points about the control of gene expression in eukaryotes:

1. The various cell types of a multicellular organism express different genes.
2. The physical organization of chromatin makes certain genes available for expression and other genes unavailable.
3. For genes that are available for expression, regulatory opportunities exist at each step in the pathway from gene to functional protein.
4. Control of transcription is especially important in determining which genes are expressed; in eukaryotes, the selective binding of transcription factors to enhancer sequences in DNA stimulates transcription of specific genes.
5. The regulatory activity of some of these DNA-binding proteins is sensitive to certain hormones and other chemical signals.

So far, we have examined mechanisms that control gene expression without actually altering the DNA itself. We now turn to some chemical processes that *can* modify a cell's genome.

Chemical modification or relocation of DNA within a genome can alter gene expression

We are accustomed to the idea that, except for rare mutations, the chemical composition and nucleotide sequence of an organism's DNA is constant during its lifetime. So it may come as a surprise that there are important exceptions, enough of them to allow us to say that a cell's genome is somewhat plastic. Chemical modification of DNA and movement of DNA within the genome can have a major influence on gene expression.

Gene Amplification and Selective Gene Loss

Sometimes the number of copies of a gene or gene family may temporarily increase in some tissues during a particular stage of development. For instance, consider the genes for ribosomal RNA (rRNA) in amphibians. As in most eukaryotes, multiple copies of these genes are built into the genome of every cell. A developing ovum, however, synthesizes a million or more additional copies of the rRNA genes, which exist in nucleoli, separate from the chromosomes. This selective replication of certain genes, or **gene amplification,** is a potent way of increasing expression of the rRNA genes, enabling the developing egg cell to make enormous numbers of ribosomes. These ribosomes make possible a burst of protein synthesis once the egg is fertilized. The extra copies of the rRNA genes are hydrolyzed during early embryonic development.

In other instances, especially in certain insects, genes are selectively lost in certain tissues (although not in the cells that give rise to gametes, of course). In fact, whole chromosomes or parts of chromosomes may be eliminated from certain cells early in the embryonic development of insects.

Gene amplification has also been observed in cancer cells exposed to high concentrations of chemotherapeutic drugs. While such drugs may kill a great many cells in a tumor, invariably some cells are resistant. These cells contain amplified regions of DNA with genes conferring drug resistance. In the laboratory, increasing concentrations of such drugs lead to increasing resistance in the cell population by selecting for cells that are amplifying these genes. There is evidence that the resistance of some parasites to specific drugs also occurs through selective gene amplification, as in the rise of drug-resistant *Plasmodia*, the parasites responsible for malaria.

Rearrangements in the Genome

A surprisingly common change in the genome is the shuffling of substantial stretches of DNA. Here we are talking not about the genetic recombination that goes on in meiosis but about rearrangements that change the loci of genes in somatic cells of an organism. Such rearrangements may have powerful effects on gene expression.

Transposons. All organisms seem to have transposons, stretches of DNA that are particularly prone to moving from one location to another within the genome. Transposons were discussed in detail in Chapter 17. Recall that if a transposon "jumps" into the middle of a coding sequence of another gene, it prevents the normal functioning of the interrupted gene. If the transposon inserts within a sequence that is involved in regulating transcription, the transposition may increase or decrease the production of one or more proteins. In some cases, the transposon itself carries a gene that is activated when it is inserted just downstream from an active promoter. Barbara McClintock (see FIGURE 1.6) was the first to discover transposons when she found evidence for mobile genetic elements that affect the color of developing corn kernels. Transposons can also alter flower color in various plants (FIGURE 18.9).

Immunoglobulin Genes. In vertebrates, at least one set of genes undergoes permanent rearrangements of DNA segments during cellular differentiation, and the function of these rearrangements is well known. These changes occur in the genes that encode antibodies, or **immunoglobulins,** proteins that specifically recognize and help combat viruses, bacteria, and other invaders of the body.

Immunoglobulins are made by cells of the immune system called B lymphocytes, a type of white blood cell.

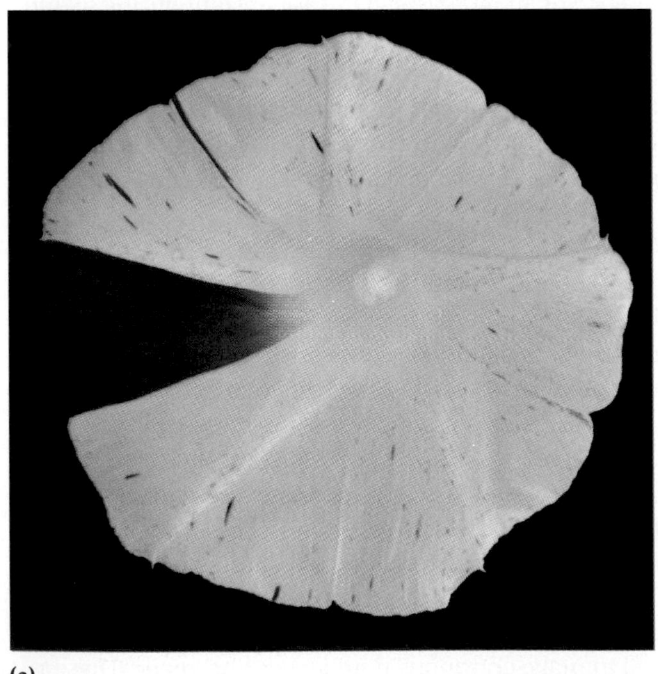

(a)

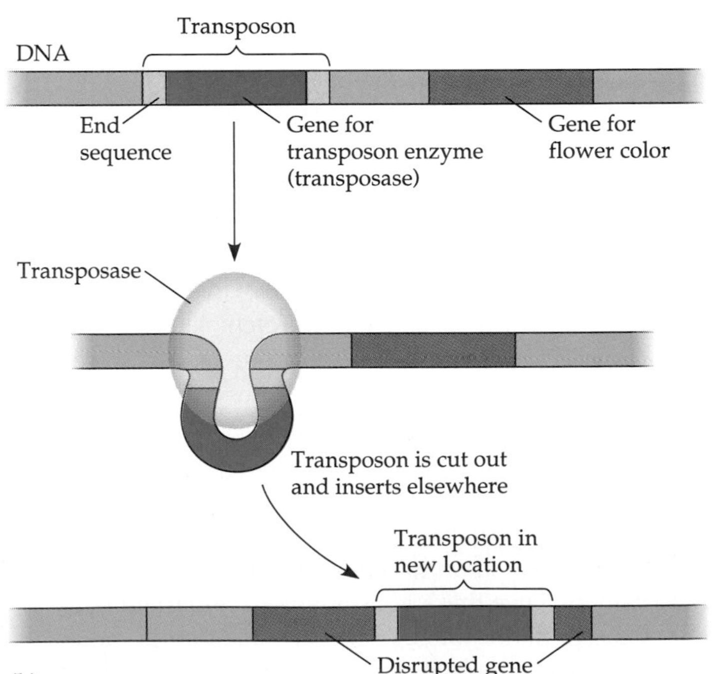

(b)

FIGURE 18.9

A transposon that affects flower color. (**a**) The two-tone coloration of this morning glory flower resulted from a transposon that moved from one location in the genome to a locus that determines flower color and made the gene for flower color nonfunctional. Cells in the colored region of the flower express the allele for purple. The white portion of the flower, which appears blue in the photo here, consists of the clone of cells that have had their gene expression altered by the transposon. (**b**) This simplified diagram shows how a transposon can interrupt the coding sequence of a gene. To review the structure of a simple transposon (insertion sequence), see FIGURE 17.13.

B lymphocytes are highly specialized, with each differentiated cell and its mitotic descendants producing one specific type of antibody that attacks a specific invader. As an unspecialized cell of the immune system differentiates into a B lymphocyte, segments of antibody genes are pieced together randomly from several DNA regions that are physically separated in the genome of an embryonic cell (FIGURE 18.10). In this way, the human immune system, with its millions of subpopulations of B lymphocytes, can make millions of different kinds of antibody molecules.

The basic immunoglobulin (antibody) molecule is shown at the bottom right of FIGURE 18.10. It consists of four polypeptide chains held together by disulfide bridges. Each chain has two major parts: a constant region (C), which is the same for all antibodies of a particular class, and a variable region (V), which gives a particular antibody its unique function—the ability to recognize and bind to a specific foreign molecule. In the genome of an embryonic cell, the DNA region coding for the constant part of each type of antibody polypeptide is separated by a long stretch of DNA from a location containing hundreds of variable region–coding segments. As a B lymphocyte differentiates, a specific variable segment of the DNA is connected to a constant segment by the deletion of intervening DNA. The joined segments form the continuous sequence of nucleotides that functions as the gene for one of the immunoglobulin polypeptides. Thus, much antibody variation arises from different combinations of variable and constant regions in immunoglobulin polypeptides, as well as from different combinations of polypeptides that form the complete antibody molecules. The formation of functional antibody genes shows how a chemical rearrangement of DNA can produce individual somatic cells with distinctive genomes.

DNA Methylation

In another type of chemical change of the genome, the nitrogenous bases of DNA are sometimes modified by a process called methylation. **DNA methylation** is the addition of methyl groups ($—CH_3$) to bases of DNA after DNA synthesis. The DNA of most plants and animals has methylated bases, usually cytosine. About 5% of the cytosine bases in eukaryotic DNA are methylated. Inactive DNA, such as that of Barr bodies, is generally highly methylated compared to DNA that is actively transcribed, although there are important exceptions. Comparison of the same genes in different types of cells (from different tissues, for example) shows that the genes are usually more heavily methylated in the cells where they are not

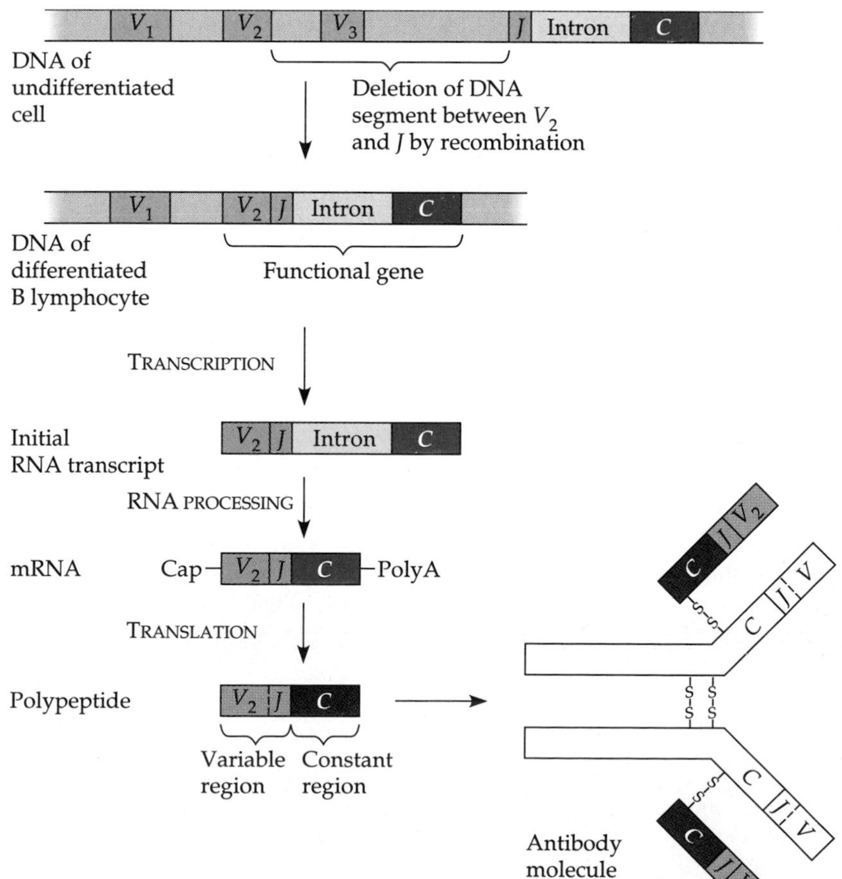

FIGURE 18.10

DNA rearrangement in the maturation of an antibody gene. The DNA of antibody genes in undifferentiated cells carries coding segments for hundreds of different antibody variable (V) regions (only three are shown here), for several different junction (J) regions, and for one or more different constant (C) regions. During differentiation of white blood cells called B lymphocytes, a long segment of DNA, from the end of one of the V segments to the beginning of one of the J segments, is deleted. This deletion brings a V segment (in this case V_2) adjacent to a J segment and produces a gene that can be transcribed. The RNA transcript is processed in the usual way to remove introns (and any extra J segments), and the resulting mRNA is translated into one of the polypeptide chains for an antibody molecule. The amino acids coded by the J segment are considered part of the variable region of the polypeptide. By bringing together V, J, and C regions of DNA in random combinations (many more than would be possible from the simplified version in this diagram), this genome plasticity helps arm the immune system with diverse antibody-producing lymphocytes, each keyed to a particular foreign invader (see Chapter 39). Different combinations of the polypeptide chains that make up each antibody are another source of diversity in these proteins.

expressed. In addition, drugs that inhibit methylation can induce gene reactivation, even in Barr bodies. Is DNA methylation a cellular mechanism for the long-term control of gene expression? There is evidence that DNA methylation in many cells reinforces, or makes permanent, the differential expression of genes that characterizes cellular differentiation. In some cases, methylation patterns are inherited through cell division, assuring that clones of cells forming specialized tissues have a "chemical record" of regulatory events that occurred during embryonic development.

Cancer can result from the abnormal expression of genes that regulate cell growth and division

In Chapter 11, we considered cancer as a variety of diseases in which cells escape from the control mechanisms that normally limit their growth and division. Certain genes normally regulate cell growth and division, and mutations that alter the expression of those genes in somatic cells can lead to cancer. The agent of such change can be random spontaneous mutation. However, it is likely that most of these mutations occur as a result of an environmental influence, such as exposure to carcinogens, physical mutagens such as X-rays, or certain viruses. In fact, a breakthrough in understanding cancer resulted from the study of tumors induced by specific viruses. This led to the discovery of cancer-causing genes, or **oncogenes** (*onco-* comes from the Greek for "tumor"). Harold Varmus and Michael Bishop, who shared a Nobel Prize for their discovery, found oncogenes in certain RNA viruses (retroviruses) that cause uncontrolled growth of infected cells in culture. Other researchers then found close counterparts of these viral genes in the genomes of humans and other animals. These normal cellular genes, called **proto-oncogenes,** code for protein products that normally regulate cell growth, cell division, and cell adhesion.

How might a proto-oncogene—a gene that has an essential function in normal cells—become an oncogene, a cancer-causing gene? In general, the answer seems to be that the product of the gene becomes more active, either as a result of changes leading to increased gene expression or through the production of a more active version of the protein product. Four types of mutations can convert a proto-oncogene to an oncogene: gene amplification; chromosome translocation; gene transposition; and point mutation, a change in a single nucleotide (FIGURE 18.11). In the case of amplification, oncogenes are present in more copies per cell than is normal. Alternatively, malignant cells are frequently found to contain translocated fragments of chromosomes, chromosomes that have broken and rejoined, juxtaposing pieces of different chromosomes (see FIGURE 14.12). An oncogene found in the new joint region may now be adjacent to active promoters or other control regions that increase transcription. In transposition, a transposon may move an oncogene to a new locus where it is controlled by an especially active promoter. Conversely, the oncogene may stay put, but a transposition may move a promoter or other regulatory DNA sequence to the upstream end of the oncogene. A point mutation may change the proto-oncogene to an oncogene that encodes a growth-stimulating protein that is more active or more resistant to degradation than the normal protein. All these mechanisms can increase the activity of regulatory proteins and put the cell on the path to the aberrant growth and division characteristic of tumor cells.

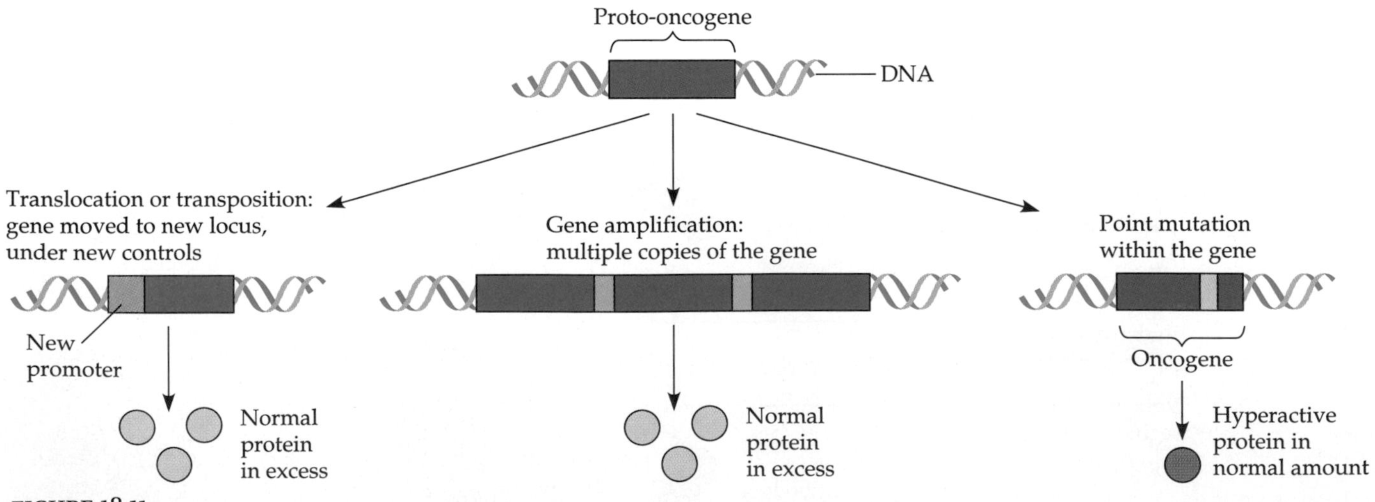

FIGURE 18.11
Mutations that can turn proto-oncogenes into oncogenes.

In addition to mutations affecting growth-stimulating proteins, changes in genes whose products *inhibit* cell division can also be involved in cancer. Such genes are called **tumor-suppressor genes** because the proteins they encode normally help prevent uncontrolled cell growth. Any mutation that decreases the normal activity of tumor-supressor proteins may contribute to the onset of cancer, in effect stimulating growth through the absence of suppression. The first mutation in a tumor-suppressor gene to be recognized was the retinoblastoma allele, which in homozygous dosage allows the formation of a malignant tumor in the eyes of young children. When a normal allele for this locus is inserted into a cell with mutated versions of the gene, the aberrant growth of the cancerous cells in culture is abated. Mutations at this same tumor-suppressor locus are associated with a wide variety of tumors.

More than one somatic mutation is generally needed to produce all the changes characteristic of a full-fledged cancer cell. This may help explain why the incidence of cancer increases greatly with age. If cancer results from an accumulation of mutations, and if mutations occur throughout life, then the longer we live, the more mutations we will accumulate. Thus, the older we are, the more likely it is that any given cell will acquire those multiple mutations necessary to transform it into a cancer cell.

The notion that more than a single mutation is required to produce a cancer cell also helps explain the genetic predispositions to certain types of cancer that run in some families. Consider the hereditary patterns of oncogenes and tumor-suppressor genes. Since oncogenes stimulate growth and division through their increased activity, the mutation of a single proto-oncogene allele can cause such a change in the cell; in other words, the mutation is dominant. Alternatively, since the activity of tumor-suppressor genes must be decreased, or lost altogether, to result in the loss of growth control, it is likely that mutant alleles in tumor-suppressor genes will behave as recessive alleles; that is, both tumor-suppressor alleles must be defective in order to cause cellular growth defects. Thus, a genetic predisposition to cancer might be inherited as a defect in a single suppressor gene. An individual inheriting such a recessive mutant allele will be one step closer to accumulating the necessary mutations that will result in abnormal cell growth and division. Much effort is now being devoted to finding such inherited mutant alleles, so that predisposition to certain cancers can be detected early in life. Researchers have recently identified such mutant tumor-suppressor alleles associated with inherited forms of colorectal cancer (FIGURE 18.12) and breast cancer.

Recent progress in understanding the genetic basis of breast cancer is particularly important because it is the

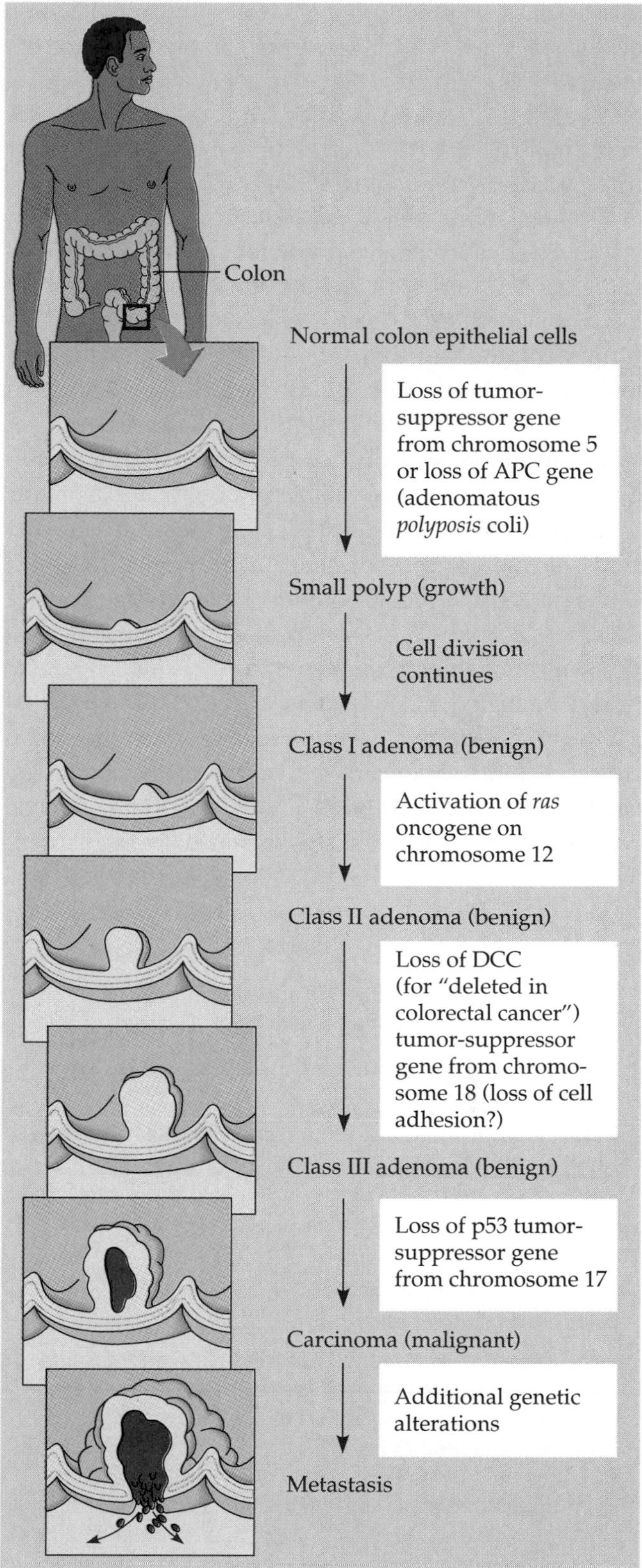

FIGURE 18.12

A multi-step model for the development of colorectal cancer. Changes in the tumor parallel a series of genetic alterations, including mutations in tumor-suppressor genes on chromosomes 5, 18, and 17, as well as a mutation of an oncogene called *ras*. Other mutation sequences can also cause cancer. In each case, the mutations change genes from forms that normally regulate cell growth and division into forms that cause a breakdown in regulation.

second most common type of cancer in women, killing over 45,000 each year in the United States. In 5% to 10% of breast cancer cases, there is evidence of a strong inherited predisposition. Women from these families are much more likely to develop breast cancer than are other women, and at a much earlier age; many often develop ovarian cancer as well. More than half of these hereditary cases are associated with mutant alleles on chromosome 17. In 1994, researchers identified this genetic locus, called BRCA1 (for BReast CAncer), as a tumor-suppressor gene. In that same year, researchers identified another locus, BRCA2, on chromosome 13, which accounts for most of the remaining half of breast cancer cases linked to family history. Even in breast cancer patients with no family history for the disease, new somatic mutations in BRCA1 and BRCA2 are usually present. The study of these and other genes associated with breast cancer may lead to new methods for early diagnosis and treatment of many breast cancers, not just the potential cases in which the mutant genes are inherited.

The multistep path to cancer is well supported by studies of one of the best understood types of human cancer, colorectal cancer (see FIGURE 18.12). Like many cancers, the development of a metastasizing colorectal cancer is gradual. The first sign is unusually frequent division of apparently normal cells in the colon lining. Later, a benign tumor (polyp) appears, and eventually a malignant tumor may develop. These cellular changes are paralleled by a gradual accumulation of oncogenes and tumor-suppressor gene mutations, and only after a number of genes have changed is a full-blown malignant tumor present. In most cases, at least four changes occur at the DNA level: the activation of an oncogene and the inactivation of three tumor-suppressor genes. Approximately 15% of colorectal cancers involve defective DNA repair mechanisms. Such a defect, which is hereditary, probably allows more rapid accumulation of the mutations required for cancer.

How are oncogenes involved in virus-associated cancers, which may contribute to 15% of human cancer cases worldwide? The virus might add an oncogene to the cell (gene amplification) or disrupt the cell's DNA in a way that affects a proto-oncogene or a tumor-suppressor gene.

Unfortunately, increased knowledge about the genetic basis of cancer has not yet led to significant new methods in cancer treatment; presumably these will come as continuing basic research improves our understanding of eukaryotic genomes and the control of gene expression. The DNA technologies you will learn about in the next chapter figure prominently in this research.

REVIEW OF KEY CONCEPTS (with page numbers and key figures)

- Each cell of a multicellular eukaryote expresses only a small fraction of its genome (pp. 351–352)
  - Eukaryotic genomes are more complexly organized than prokaryotic genomes. The control of gene expression is also more elaborate in eukaryotes than in prokaryotes.
  - In particular, selective control of genes is required for cellular differentiation.
- The structural organization of chromatin sets coarse controls on gene expression (pp. 352–354, FIGURE 18.1)
  - The enormous amount of DNA in a eukaryotic chromosome is compacted by a multilevel system of folding.
  - Chromatin is composed of DNA and five types of histone proteins that bind to the DNA and dictate its first level of folding into nucleosomes, basic units of DNA packing.
  - Chromatin fibers fold to form looped domains. Further folding coils the DNA into its most highly compacted form, the metaphase chromosome. In certain cases interphase chromatin also exists in a highly condensed form, called heterochromatin. Active transcription occurs on euchromatin, the more open, unfolded form of the chromosome.
- Noncoding sequences and gene duplications account for much of a eukaryotic genome (pp. 354–355; FIGURES 18.2, 18.3)
  - Much of the DNA in a genome does not encode protein.
  - Some DNA sequences may be present in hundreds or thousands of copies. Highly repetitive sequences of five to ten nucleotides tandemly repeated thousands of times contribute to the enormous amount of noncoding DNA in the eukaryotic genome.
  - Repetitive sequences at telomeres help conserve chromosome tips.
  - Introns also contribute to noncoding DNA.
  - A multigene family is a collection of genes that are similar or identical in nucleotide sequence.
- The control of gene expression can occur at any step in the pathway from gene to functional protein (pp. 356–360, FIGURE 18.4)
  - Binding of transcription factors to enhancer regions and other regulatory regions in the DNA selectively stimulates transcription of genes.
  - Before mRNA can leave the nucleus, it must undergo processing, which presents opportunities for controlling gene expression.
  - The degradation of mRNA is regulated, influencing how much of a specific protein may be made.
  - Translation itself is regulated. Most translational control mechanisms block the initiation of protein synthesis.
  - After translation, proteins may be extensively modified and targeted for specific sites in the cell.
  - Unlike prokaryotic operons, eukaryotic genes of related function are often scattered throughout the genome. Coordinated control may be mediated by specific nucleotide sequences common to all the genes of the group.

- Chemical signals that help control gene expression include hormones (pp. 360–361, FIGURE 18.8)
  - Some small molecules in the cell affect gene expression by associating with DNA-binding proteins and altering the conformations of these regulatory proteins.
  - Ecdysone, a steroid hormone in insects, selectively activates genes during development, a regulation evident in the changing pattern of chromosome puffs.
  - Steroid hormones in vertebrates interact with intracellular receptor proteins, which are thereby activated to bind to enhancers and influence transcription.
- Chemical modification or relocation of DNA within a genome can alter gene expression (pp. 361–364; FIGURES 18.9, 18.10)
  - Gene amplification may occur in some tissues during development.
  - The cells of some species show selective gene loss, in which entire chromosomes or parts of chromosomes of certain cells are eliminated.
  - Rearrangements of DNA can activate or inactivate specific genes. By moving from one location to another within the genome, transposons can affect gene expression.
  - In vertebrates, a rearrangement and selective deletion of DNA segments in differentiating B lymphocytes accounts for antibody diversity.
  - The attachment of methyl groups to DNA may diminish transcription of that DNA.
- Cancer can result from the abnormal expression of genes that regulate cell growth and division (pp. 364–366, FIGURE 18.11)
  - Proto-oncogenes control cell growth and differentiation. When such genes mutate or escape normal control mechanisms and become more active, they are called oncogenes. Oncogenes allow the formation of tumors. Physical agents, chemical carcinogens, or viruses may activate oncogenes in various ways.
  - Tumor-suppressor genes encode proteins that keep cell division in a normal state. The loss or mutation of these genes has effects similar to the activation of oncogenes.
  - Tumors show progressive accumulation of defects in proto-oncogenes and tumor-suppressor genes. Some of these mutations can be inherited, resulting in a predisposition to developing certain types of cancer.

SELF-QUIZ

1. In a nucleosome, the DNA is wrapped around
   a. polymerase molecules
   b. ribosomes
   c. histones
   d. the nucleolus
   e. satellite DNA

2. Apparently, our muscle cells are different from our nerve cells mainly because
   a. they express different genes
   b. they contain different genes
   c. they use different genetic codes
   d. they have unique ribosomes
   e. they have different chromosomes

3. Chromosome puffs on the giant chromosomes of *Drosophila* probably represent regions where
   a. genes are inactivated by repressor proteins
   b. hormones are produced
   c. genes have been damaged
   d. genes are especially active in transcription
   e. ribosomes are being synthesized

4. The function of enhancers is an example of
   a. a transcriptional control that affects gene expression
   b. a posttranscriptional mechanism for editing mRNA
   c. initiation factors that stimulate translation
   d. posttranslational control activating proteins
   e. a eukaryotic equivalent of the prokaryotic promoter

5. Multigene families are
   a. groups of enhancers that control transcription
   b. usually clustered at the telomeres
   c. equivalent to the operons of prokaryotes
   d. collections of genes whose expression is controlled by the same regulatory proteins
   e. identical or similar genes produced by duplication

6. Which of the following statements about the DNA in one of your brain cells is true?
   a. Some DNA sequences may be present in multiple copies.
   b. Most of the DNA codes for protein.
   c. The majority of genes are likely to be transcribed.
   d. Each gene lies next to an enhancer that helps control transcription.
   e. Many genes are grouped into operonlike clusters.

7. Permanent rearrangement of DNA segments is known to be involved in genes coding for
   a. ribosomal RNA
   b. most proteins in eukaryotes
   c. hemoglobin
   d. histone proteins
   e. antibodies

8. Which of the following is an example of a possible step for the posttranscriptional control of gene expression?
   a. the addition of methyl groups to cytosine bases of DNA
   b. the binding of transcription factors to a promoter region
   c. the removal of introns and splicing of exons
   d. gene amplification during a particular stage in development
   e. the folding of DNA to form heterochromatin

9. The amount of protein made from a given mRNA molecule depends partly on
   a. the degree of DNA methylation
   b. the rate at which the mRNA is degraded
   c. the presence of certain transcription factors and enhancers
   d. the number of introns present in the mRNA
   e. the types of ribosomes present in the cytoplasm

10. All our cells contain proto-oncogenes, which can change into oncogenes that cause cancer. Which of the following is the best explanation for the presence of these potential time bombs in our cells?
    a. Proto-oncogenes first arose from viral infections.
    b. Proto-oncogenes normally help regulate cell growth and division.
    c. Proto-oncogenes are genetic "junk" with no known function.
    d. Proto-oncogenes are mutant versions of normal genes.
    e. Cells produce proto-oncogenes as a by-product of the aging process.

CHALLENGE QUESTIONS

1. The amino acid sequences encoded by the genes for certain DNA-binding proteins have been found to be remarkably similar between vertebrates and invertebrates, despite the evolutionary divergence of these animals over 500 million years ago. Speculate on why evolution has been so conservative with genes that control development.

2. The presence of a nucleus in eukaryotic cells allows many things to happen that cannot happen in prokaryotic cells, but it also presents transport problems. Consider the traffic that must flow into and out of the nucleus. Molecules of mRNA must move out, but only after they are properly processed. Ribosomal subunits are assembled in nucleoli inside the nucleus, meaning ribosomal proteins must be moved into the nucleus. All the histones, transcription factors, and other DNA-binding proteins also need to function within the nucleus. Suggest hypotheses for how the cell regulates this nuclear traffic.

3. Write a short essay integrating the roles of inheritance, viruses, mutations, and environmental factors in causing cancer.

SCIENCE, TECHNOLOGY, AND SOCIETY

A chemical called dioxin, or TCDD, is produced as a contaminant of some chemical manufacturing processes. Trace amounts of this substance were present in Agent Orange, a defoliant sprayed on vegetation during the Vietnam War. There has been a continuing controversy over its effects on soldiers exposed to it during the war. Animal tests have shown that dioxin can be lethal and can cause birth defects, cancer, liver and thymus damage, and immune system suppression. But its effects on humans are unclear, and even animal tests are equivocal; a hamster is not affected by a dose that can kill a guinea pig. Researchers have discovered that dioxin exerts its effects like some hormones. It enters a cell and binds to a receptor protein, which in turn attaches to the cell's DNA. How might this mechanism help explain the variety of dioxin's effects on different body systems and in different animals? How might you determine whether a type of illness is related to exposure to dioxin or whether a particular individual became ill as a result of exposure to dioxin? Which would be more difficult to demonstrate? Why?

FURTHER READING

Cavenee, W. K., and R. L. White. "The Genetic Basis of Cancer." *Scientific American,* March 1995. A well-illustrated update.

Cech, T. R. "Chromosome End Games." *Science,* October 21, 1994. How telomeres are conserved.

Gilbert, S. F. *Developmental Biology,* 4th ed. Sunderland, MA: Sinauer Associates, 1994. An excellent textbook used in upper-division developmental biology courses.

Grunstein, M. "Histones as Regulators of Genes." *Scientific American,* October 1992. How one type of chromosomal protein helps control gene expression.

Kaiser, J. "Breast Cancer: Hope for a Genetic Test." *Science News,* February 25, 1995.

Marx, J. "How Cells Cycle Toward Cancer." *Science,* January 21, 1994. The roles of various oncogenes and tumor-suppressor genes within the cell cycle.

Nowak, R. "Mining Treasures from 'Junk DNA.'" *Science,* February 4, 1994. Recent developments in understanding noncoding DNA.

Rennie, J. "DNA's New Twists." *Scientific American,* March 1993. Reviews advances in our understanding of transposons, methylation, and other complexities of the eukaryotic genome.

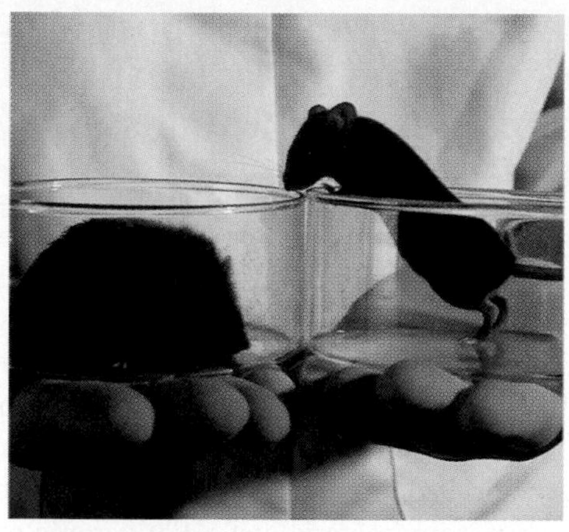

CHAPTER 19

DNA TECHNOLOGY

KEY CONCEPTS

- DNA technology makes it possible to clone genes for basic research and commercial applications: *an overview*
- The toolkit for DNA technology includes restriction enzymes, DNA vectors, and host organisms
- Recombinant DNA technology provides a means to transplant genes from one species into the genome of another
- Additional methods for analyzing and cloning nucleotide sequences increase the power of DNA technology
- DNA technology is catalyzing progress in many fields of biology
- The Human Genome Project is an enormous collaborative effort to map and sequence DNA
- DNA technology is reshaping the medical and pharmaceutical industries
- DNA technology offers forensic, environmental, and agricultural applications
- DNA technology raises important safety and ethical questions

In the photograph that opens this chapter, a researcher at a biotechnology company compares a normal mouse (right) to an obese one. In 1995, the company demonstrated that a protein produced by genetically engineered bacteria can be used to treat such obese mice and trim their weight to more normal size. The bacteria were engineered by splicing the mouse gene that codes for the weight-regulating protein of normal mice into the bacterial DNA.

Today, hundreds of useful products are produced by **genetic engineering,** *the manipulation of genetic material for practical purposes. In the last decade, it has become routine to combine genes from different sources—often different species—in test tubes, and then transfer this recombinant DNA into living cells, where it can be replicated and expressed.* E. coli *is often used as a host organism because it is easy to grow and its biochemistry is well understood.*

DNA technology has launched an industrial revolution in biotechnology. In broad terms, **biotechnology** *is the manipulation of living organisms or their components to perform practical tasks or provide useful products. Practices that go back centuries, such as the use of microorganisms to make wine and cheese and the selective breeding of livestock and field crops, are examples of biotechnology, as are the production of antibiotics from microorganisms and the synthesis of monoclonal antibodies using modern techniques of immunology (see Chapter 39). Biotechnology based on the manipulation of DNA in vitro (outside of living cells) is different from earlier biotechnology in that it is more precise, more rapid, and much more powerful, allowing genes to be moved between organisms as distinct as bacteria, plants, and animals.*

Although the use of genetic engineering to create valuable new products is important, the most impressive achievements resulting from recombinant DNA technology so far have been advances in our understanding of eukaryotic molecular biology and in the development of basic research techniques. Today, biologists have a DNA technology toolkit more powerful than they could have imagined even a decade ago. The new methods have influenced almost every field of biology.

In this chapter, we will examine the main techniques and applications of genetic engineering. We will see that DNA technology is revolutionizing biological research, medicine, criminal law, and agriculture. Finally, we will consider some of the social and ethical issues that we face as our ability to manipulate genes becomes increasingly powerful.

DNA technology makes it possible to clone genes for basic research and commercial applications: *an overview*

There are now many ways to manipulate the genetic material of virtually any organism, but almost all methods share certain general features. As an overview of the process, we will consider a general approach to genetic engineering utilizing bacterial cells and their plasmids (FIGURE 19.1). You learned in Chapter 17 that plasmids are small, circular DNA molecules that can replicate autonomously within bacterial cells. For genetic engineering, plasmids are first isolated from bacterial cells. The tools and techniques of genetic engineering then make it possible to insert foreign genes into the isolated plasmids. These plasmids are now examples of **recombinant DNA** molecules, meaning they have a combination of DNA from two sources. The plasmids are then returned to bacterial cells, which then reproduce, cloning the recombinant DNA as the cells replicate their plasmids. Under suitable conditions, the bacterial culture will produce the protein encoded by the foreign gene.

The potential uses of such cloned DNA or cloned genes vary but fall into three general categories. The goal may be to produce a protein product, such as human growth hormone (used to treat stunted growth), or tissue plasminogen activator (which aids in dissolving blood clots following heart attacks). Alternatively, the goal may be to endow a particular organism with a metabolic capability it did not previously possess, such as transferring pest resistance into crops or engineering bacteria to degrade oil spills. The third goal may be to create more copies of the gene itself, so that it can be studied further. Most genes exist in only one copy per genome—something on the order of one part in a million of DNA. The ability to clone such rare DNA fragments has become a valuable tool in biological research. In the next few sections, we will take a closer look at the steps in FIGURE 19.1 and at related methods for manipulating, cloning, and analyzing DNA.

The toolkit for DNA technology includes restriction enzymes, DNA vectors, and host organisms

The discovery of enzymes that cut DNA molecules in precise locations made genetic engineering possible. In its simplest form, genetic engineering requires three biological "tools": the enzymes to cut DNA, a vector (transfer agent) such as a plasmid to allow recombinant DNA to enter a cell and replicate, and an appropriate host organism for the recombinant DNA.

Restriction Enzymes

One of the major tools of recombinant DNA technology is a group of bacterial enzymes called **restriction enzymes,** which were first discovered in the late 1960s. In nature, these enzymes protect bacteria against intruding DNA from other organisms, such as viruses or other bacterial cells. They work by cutting up the foreign DNA, a process called *restriction.* Most restriction enzymes are very specific, recognizing short, specific nucleotide sequences in DNA molecules and cutting at specific points within these sequences. The bacterial cell protects its own DNA from restriction by adding methyl groups ($—CH_3$) to adenines or cytosines within the sequences recognized by the restriction enzyme. Hundreds of different restriction enzymes have been identified and isolated, and many are available commercially.

The top of FIGURE 19.2 (p. 372) is a diagram of a molecule of DNA containing two recognition sequences for a particular restriction enzyme. As shown in this example, a restriction-enzyme recognition sequence (darker blue) is symmetrical: The same sequence of four to eight nucleotides (here, six) is found on both strands but running in opposite directions. Restriction enzymes cut covalent phosphodiester bonds of both strands, often in a staggered way, as indicated in the diagram. The resulting **restriction fragments** are double-stranded DNA fragments with single-stranded ends, called **sticky ends.** These short extensions will form hydrogen-bonded base pairs with complementary single-stranded stretches on other DNA molecules cut with the same enzyme.

The sticky ends of restriction fragments can be used in the laboratory to join DNA pieces originating from different sources. Such unions are only temporary, because only a few hydrogen bonds hold the fragments together. The DNA fusions can be made permanent, however, by the enzyme DNA ligase, which seals the strands by catalyzing the formation of phosphodiester bonds. (Recall from Chapter 15 that DNA ligase is a key enzyme in DNA replication and repair.) We now have recombinant DNA, DNA that has been cut and spliced from two different sources.

Vectors

DNA engineered in the test tube must be returned to a cell in order to function. Most genetic engineering procedures utilize carriers—**cloning vectors**—for moving recombinant DNA from test tubes back into cells. The two most popular types of vectors are bacterial plasmids and viruses. Recombinant plasmids produced by splicing restriction fragments from foreign DNA into plasmids isolated from bacteria can then be returned relatively easily to bacteria. The bacteria replicate (clone) the recombinant plasmids as the original cell produces a colony.

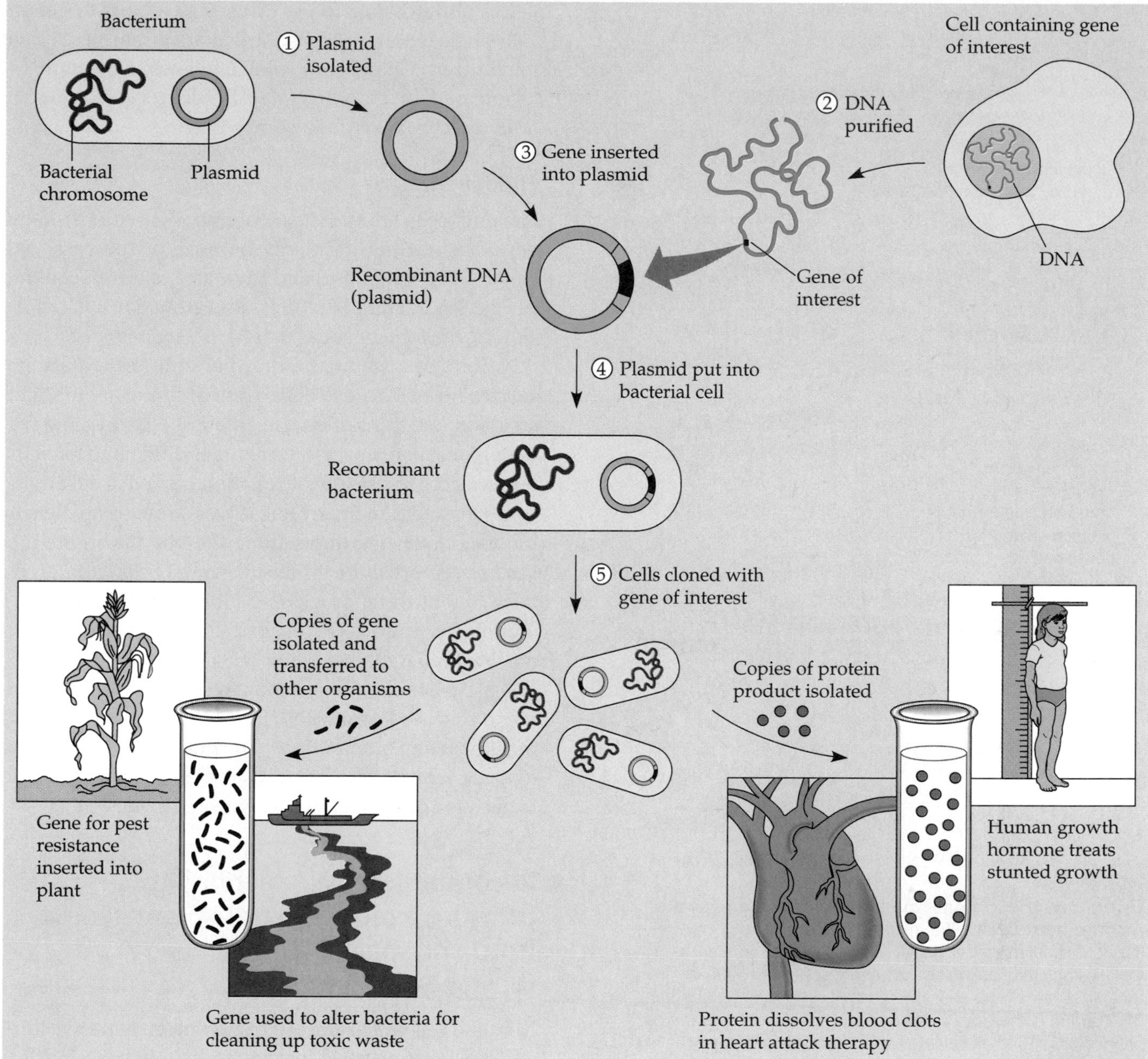

FIGURE 19.1

An overview of how plasmids are used in genetic engineering. ① The genetic engineer isolates plasmid DNA from bacteria and ② purifies DNA containing a gene of interest from another cell. This cell might be a bacterium or a plant or animal cell, and the gene of interest might be, for instance, a plant gene that confers resistance to pest insects or a human gene that encodes a hormone. ③ A piece of DNA containing the gene is inserted into the plasmid, producing recombinant DNA, and ④ the plasmid is put back into a bacterial cell. ⑤ This genetically engineered bacterium is then cloned (grown in culture). Since the foreign DNA spliced into the plasmid does not impair the plasmid's ability to replicate within a bacterial cell, the result is the cloning of the gene of interest. The bacterial culture now contains many copies of the gene (one per cell in this example). The drawings at the bottom represent some current applications of genetically engineered bacteria. In the examples on the left, either the gene itself or its effects on metabolism are the desired product. In the cases on the right, useful protein products are harvested in large quantities from the bacterial cultures.

Certain types of bacteriophage can also serve as cloning vectors in bacterial cells. Fragments of foreign DNA can be spliced into a phage genome by using restriction enzymes and ligase. The recombinant phage DNA is then introduced into an *E. coli* cell through the normal process of infection. Once inside the cell, the phage DNA replicates and produces new phage particles, each carrying the foreign DNA. These phages, in turn, can carry the foreign DNA into other bacterial cells by infection.

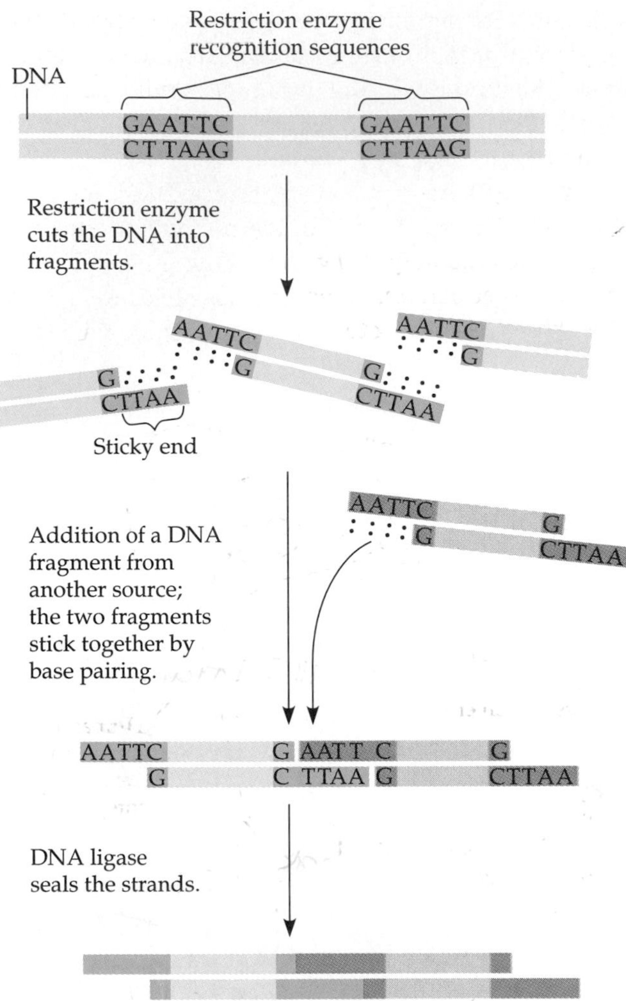

FIGURE 19.2
Using a restriction enzyme and DNA ligase to make recombinant DNA. The restriction enzyme in this example (called *Eco*RI) recognizes a specific six-base-pair sequence and makes staggered cuts in the sugar-phosphate backbone within this sequence. Notice that the recognition sequence along one DNA strand is the exact reverse of the sequence along the complementary strand. Because of this symmetry of sequences, the result of enzyme action is fragments of DNA with single-stranded "sticky" ends. Complementary ends will stick to each other by hydrogen bonding, transiently rejoining fragments in their original combinations or in new, recombinant arrangements. When restriction fragments come together by base pairing, the enzyme DNA ligase can catalyze the formation of covalent bonds joining their ends. If the fragments are from two different sources, the result is recombinant DNA.

As we will see later, it is sometimes desirable to clone DNA in eukaryotic cells rather than in bacteria. Yeast cells, single-celled fungi, offer the advantage of having plasmids, a rarity among eukaryotes. Genetic engineers have even constructed recombinant plasmids that combine yeast and bacterial DNA and can replicate in either type of cell. Scientists have also constructed artificial chromosomes that combine yeast DNA and foreign DNA, and cell division clones these chromosomes.

It is also possible to use viruses as vectors to genetically engineer animal cells. Particularly useful are certain retroviruses (Chapter 17), which, following infection, integrate into the chromosome. This delivers recombinant DNA directly to a chromosome.

Host Organisms

Bacterial cells are the most commonly used host in genetic engineering, primarily because of the ease with which DNA can be isolated from and reintroduced into such cells. Bacterial cultures also grow quickly, rapidly cloning the foreign genes that have been inserted.

Nevertheless, there are drawbacks to using bacterial cells as hosts. Because eukaryotic and prokaryotic transcription and translation use different enzymes and regulatory mechanisms, it is sometimes difficult to force the bacterial cell to use the information carried in eukaryotic genes. In addition, many eukaryotic proteins are heavily modified following translation, often by the addition of lipid or carbohydrate groups. Bacterial cells cannot perform any of these processing functions. Therefore, if a modified protein is the desired end-product, a eukaryotic host cell is often a better choice.

Yeast cells offer many advantages, but it is also possible to transfer foreign DNA into cultured plant and animal (including human) cells. The most serious limitation to using many eukaryotic cells is that it is often difficult to get such cells to take up engineered DNA.

Recombinant DNA technology provides a means to transplant genes from one species into the genome of another

Now that we have examined restriction enzymes, vectors, and host cells as basic tools of genetic engineering, let's take a closer look at how these tools are combined to clone specific genes that have been incorporated into recombinant DNA.

Steps for Using Bacteria and Plasmids to Clone Genes

FIGURE 19.3 is a more detailed version of the gene-cloning method outlined in FIGURE 19.1. Step ① is the isolation of two kinds of DNA: the bacterial plasmid and the DNA containing the gene of interest. In this example, the DNA comes from human tissue cells that have been growing in laboratory culture. The plasmid comes from the bacterium *E. coli* and carries two genes: *amp*^{R} confers resistance to the antibiotic ampicillin on its *E. coli* host cell; *lacZ* encodes the enzyme β-galactosidase, which hydrolyzes the sugar lactose. The plasmid has a single recognition sequence for the restriction enzyme used, and the sequence lies within the *lacZ* gene.

In step ②, both the plasmid and the human DNA are digested with the same restriction enzyme. The enzyme cuts the plasmid DNA at its single recognition sequence, or **restriction site,** disrupting the *lacZ* gene. It also cuts the human DNA, generating many thousands of fragments; one of these fragments carries the gene of interest. In making the cuts, the restriction enzyme creates sticky ends on both the human DNA fragments and the plasmid. For simplicity, the figure here shows the step-by-step processing of one human DNA fragment and one plasmid, but actually millions of plasmids and a heterogeneous mixture of millions of human DNA fragments are treated simultaneously.

In step ③, the human DNA is mixed with the clipped plasmid. The sticky ends of the plasmid base-pair with the complementary sticky ends of the human DNA fragment. Other combinations are possible as well, such as two plasmids pairing together, or a plasmid pairing with

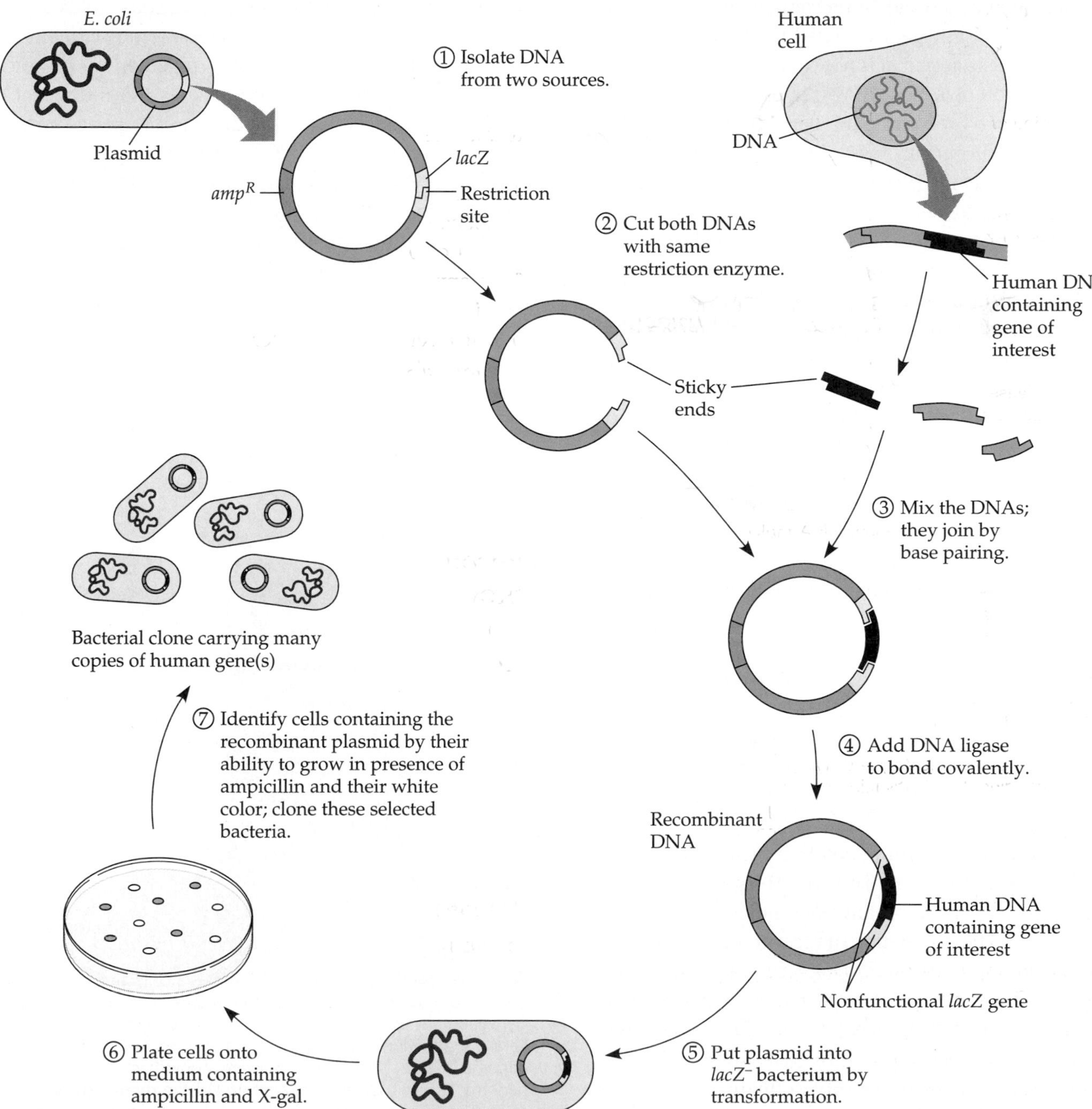

FIGURE 19.3

Cloning a gene in a bacterial plasmid. Unlike in FIGURE 19.2, the two strands of DNA are not depicted separately in this figure. The numbered steps in the text correspond to the circled numbers in this diagram.

several DNA fragments. In step ④, the enzyme DNA ligase joins the DNA molecules by covalent bonds. The result is a mixture of recombinant DNA molecules, some of which are like the one shown. In step ⑤, the recombinant plasmid is introduced into a bacterial cell by the process of transformation (the uptake of naked DNA from the environment; see Chapter 17). Some of the bacteria will take up the desired plasmid DNA from solution; many other cells will take up other recombinant molecules.

Step ⑥ begins the process of selecting the desired transformed cell. The bacterial culture, with its recombinant plasmids, is allowed to reproduce. As the bacteria form cell clones, any genes carried by the recombinant plasmids are also cloned. It is now that we take advantage of the plasmid's own genes. We can identify colonies of bacteria that carry recombinant plasmids by the fact that their cells are ampicillin-resistant (since they carry the amp^R gene). Cells are plated onto a medium containing ampicillin. Only those cells that have taken up the plasmid with its amp^R gene will grow and form colonies. The culture medium also contains a modified sugar called X-gal. The X-gal can be cleaved by β-galactosidase, yielding a blue product. In Step ⑦, colonies of bacteria that have a foreign DNA fragment inserted into the plasmid will appear white because the vector cannot produce β-galactosidase. Bacterial colonies containing plasmids without an insert will show up as blue, since they contain an intact β-galactosidase gene and will produce the enzyme. Thus, the desired cells will form white colonies on medium containing ampicillin and X-gal.

The net result of steps 1–7 is to identify a bacterial colony carrying a recombinant plasmid that contains a foreign DNA fragment.

Sources of Genes for Cloning

Where do biologists get genes for cloning? There are two main sources: DNA isolated directly from an organism and complementary DNA made in the laboratory from mRNA templates.

Scientists isolate DNA directly by starting with all the DNA from cells of an organism with the gene they want and constructing recombinant DNA molecules as shown in FIGURE 19.3. The population of recombinant molecules formed is then introduced into bacterial cells. The resulting set of thousands of plasmid clones, each carrying copies of a particular segment from the foreign genome, is referred to as a **genomic library.** Because this gene-cloning procedure uses a mixture of fragments from the entire genome of an organism, it is often called a "shotgun" approach—no single gene is targeted for cloning. If phages instead of plasmids are used as vectors, then the genomic library is stored as phage particles containing recombinant DNA with different fragments from the foreign genome. FIGURE 19.4 illustrates both a plasmid library and a phage library.

One problem with cloning DNA directly from a eukaryotic genome is that because eukaryotic genes often contain long noncoding regions (introns), bacterial cells will not be able to express these genes. To avoid this problem, scientists can sometimes make an artificial gene that lacks introns. The starting material for this procedure, mRNA, is present in the eukaryotic cell, as shown in the first two steps of FIGURE 19.5. ① Transcription of an intron-containing gene in the cell nucleus yields a pre-RNA molecule. ② Spliceosomes then remove the intron RNA and splice the exon RNA together to produce mRNA. ③ The biologist then isolates the mRNA molecules from the cell and uses them as templates to synthesize a complementary DNA strand. This synthesis of DNA on an RNA template is the reverse of transcription. It is catalyzed by the enzyme reverse transcriptase, which is obtained from retroviruses (see Chapter 17). ④ After a single strand of DNA is synthesized, the RNA is degraded. ⑤ The second DNA strand is made using DNA poly-

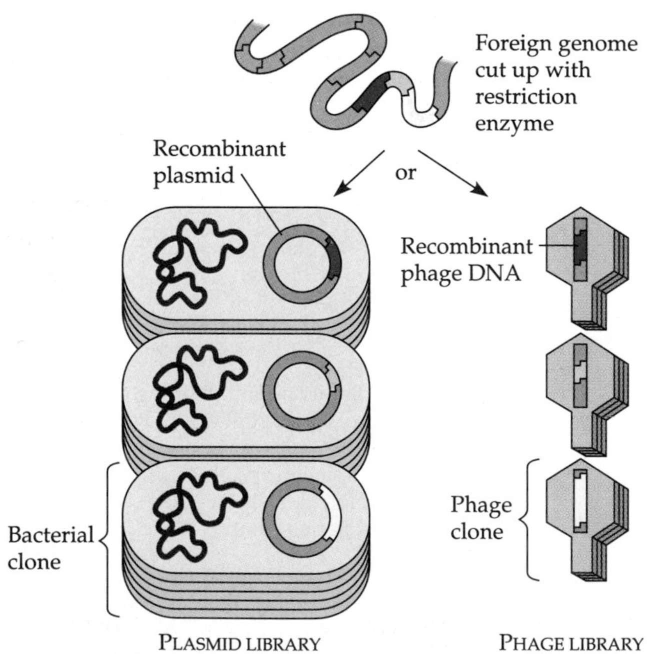

FIGURE 19.4

Genomic libraries. A genomic library is a collection of a large number of bacterial or phage clones, each containing copies of a DNA segment from a foreign genome. In a complete genomic library, the foreign DNA segments represented cover the entire genome of an organism. This diagram shows parts of two genomic libraries. On the left are three of the thousands of "books" in a plasmid library. Each "book" is a bacterial clone containing one particular variety of foreign genome fragment—red, orange, or yellow here—in its recombinant plasmid. On the right, the same three foreign genome fragments are shown in three "books" of a phage library.

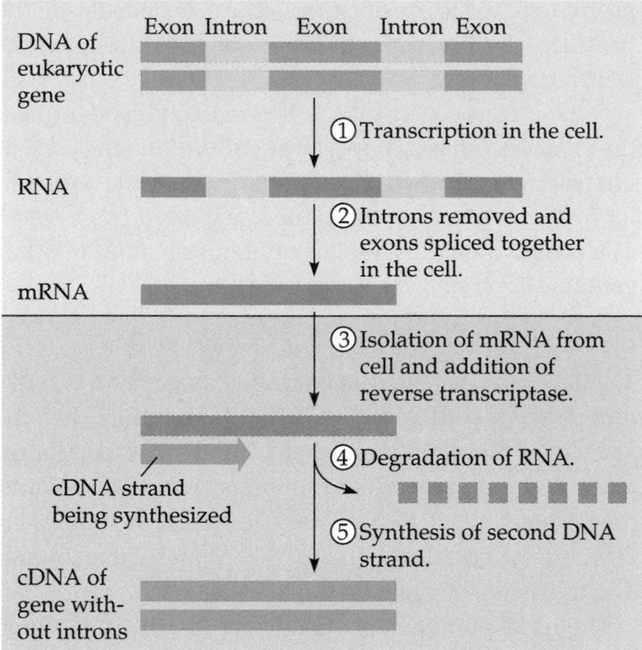

Exon Intron Exon Intron Exon
DNA of eukaryotic gene
① Transcription in the cell.
RNA
② Introns removed and exons spliced together in the cell.
mRNA
③ Isolation of mRNA from cell and addition of reverse transcriptase.
④ Degradation of RNA.
cDNA strand being synthesized
⑤ Synthesis of second DNA strand.
cDNA of gene without introns

FIGURE 19.5

Making complementary DNA (cDNA) for a eukaryotic gene. Complementary DNA is DNA made in the laboratory using mRNA as a template and the enzyme reverse transcriptase. Complementary DNA lacks introns and is therefore smaller than the original gene and easier to clone. It is also much more likely to be functional in bacterial cells, which lack the machinery for removing introns from RNA transcripts. However, to be transcribed, the cDNA will have to be joined to an appropriate bacterial promoter because no promoter will be present in the cDNA copy of the gene.

merase and the first strand as a template. The result is a double-stranded molecule of DNA carrying the coding sequence of the gene but no introns. This is called **complementary DNA** or **cDNA.**

The cDNA gene created by this method has the potential to be transcribed and translated by bacterial cells, which do not have RNA-splicing machinery. However, it contains none of the control sequences for transcription and translation normally associated with genes. These signals must be provided within the vector DNA.

Because the mRNA molecules for a particular gene cannot usually be isolated from the other mRNA molecules of a cell, the cDNA method, like the shotgun method of cloning DNA obtained directly from cells, produces gene libraries, usually representing all mRNAs of the cell. The cDNA libraries represent only part of a cell's genome—only the genes expressed (transcribed) in the cell used. This is of particular advantage if a researcher is trying to identify genes responsible for specialized functions within certain cell types, or if the researcher is trying to identify changes in the expression of genetic information, such as during the different developmental stages of an organism.

Inserting DNA into Cells

A variety of methods is available for introducing DNA into cells. The choice of which method to use depends on both the vector and the type of host cell. In the most common genetic engineering procedures, DNA is transferred into bacterial cells through transformation, the absorption of DNA from the surrounding solution. If phage vectors are used, the recombinant phage DNA is packaged inside its protein coat and then infects the host cell.

Yeast cells can be transformed by the uptake of recombinant plasmids, and they can also be coaxed into taking up linear pieces of DNA, which subsequently become incorporated into a chromosome through recombination. Other eukaryotic cells can be transformed by DNA fragments as well.

A variety of more aggressive techniques for getting DNA into eukaryotic cells have also been developed. In **electroporation,** a brief electrical pulse applied to a solution containing cells causes temporary holes to open in the plasma membrane, through which DNA can enter. Alternatively, DNA can be injected directly into single eukaryotic cells using microscopically thin needles. And, in a technique used primarily for plant cells, DNA can be attached to microscopic particles of metal and fired into cells with a gun (see FIGURE 34.13). In all cases, DNA becomes incorporated into the host cell's DNA through recombination.

Selection: Finding a Gene of Interest

How is it possible to select the clone that carries a particular gene of interest from the many thousands of genes represented in a cDNA library, and perhaps millions of DNA fragments in a shotgun library?

If the clones containing a specific gene actually translate the gene into protein, then it is sometimes possible to identify them by screening all clones for the presence of the protein. Detection of the protein can be based on either its activity (as with an enzyme) or its structure, using antibodies that combine specifically with it. Such an approach relies on translation occurring—which may not be happening in many cases. Most often, screening techniques rely on detecting the gene itself, not the gene product.

Methods for detecting a gene directly all depend on base pairing between the gene and a complementary sequence on another nucleic acid molecule, a process called *hybridization.* The complementary molecule, which can be either RNA or DNA, is called a **probe.** When at least part of the nucleotide sequence of the gene is known, or can be deduced from knowledge of the amino acid sequence of the gene's protein product, short probes complementary to it can be chemically synthesized. The probe, which will hydrogen-bond specifically to the desired gene, can be traced by labeling it with a radioactive isotope or a fluorescent tag. FIGURE 19.6 shows how several bacterial clones, growing as colonies on a plate of solid medium, can be simultaneously screened for the presence of DNA complementary to a specific DNA probe. The radioactive probe tags the correct clone—finds the needle in the haystack—by hydrogen bonding, or hybridizing, to its DNA complement.

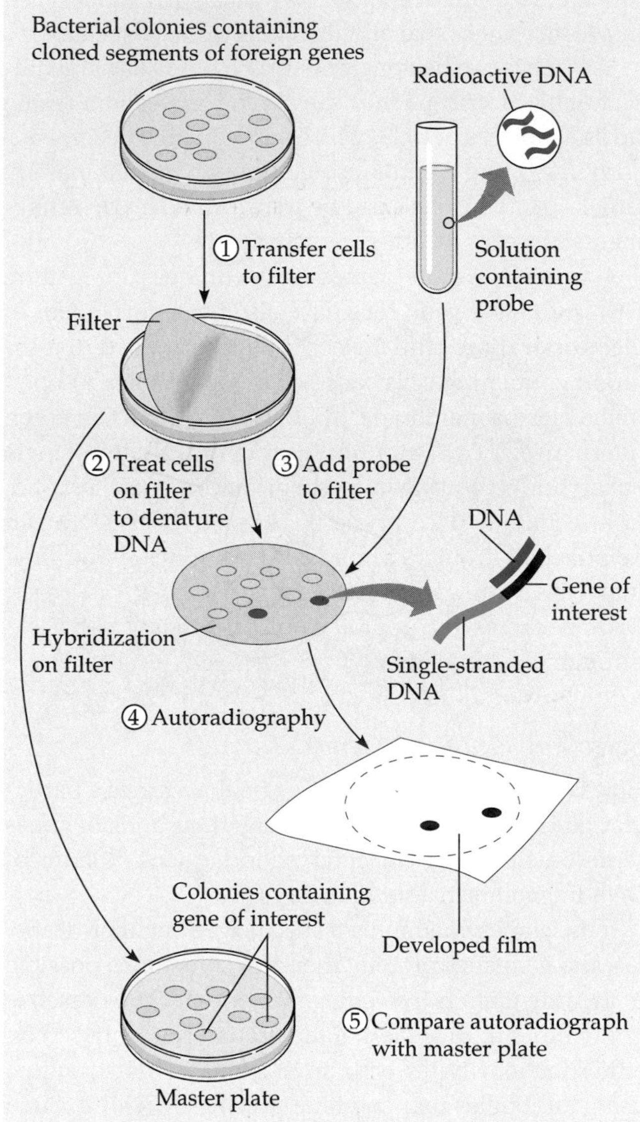

FIGURE 19.6

Using a nucleic acid probe to identify a cloned gene. This technique depends on the fact that complementary nucleotide sequences will base-pair, bind together by hydrogen bonds. Here the cloned gene of interest is carried on a bacterial plasmid, and the probe is a short length of synthetic DNA complementary to part of the gene. ① Bacterial colonies on agar are pressed against special filter paper, transferring cells to the filter. ② The filter is treated to break open the cells and denature their DNA; the resulting single-stranded DNA molecules stick to the filter. ③ A solution of probe molecules is incubated with the filter. The DNA hybridizes (base-pairs) with any complementary DNA on the filter; excess DNA is rinsed off. ④ The filter is laid on photographic film, allowing any radioactive areas to expose the film (autoradiography). ⑤ The developed film, an autoradiograph, is compared with the master culture plate to determine which colonies carry the desired gene.

Once a clone carrying the desired gene is identified, the clone can be grown in a large culture, and the gene of interest can easily be isolated in large amounts and used for further study. Also, the cloned gene itself can be used as a probe to identify similar or identical genes in other genomic libraries.

Achieving the Expression of Cloned Genes

You have already read that getting a cloned gene to function in a new setting can be difficult. Except for the universality of the genetic code, most details of transcription and translation are different in prokaryotes and eukaryotes. Therefore, it is necessary to devise recombinant DNA molecules that contain the proper transcription and translation signals for the host cell.

The usual approach is to engineer the requisite signal sequences directly into the cloning vector. By choosing restriction enzymes to cut and splice at strategic locations in the plasmid or other vector, the foreign gene can be inserted adjacent to the host cell's versions of promoters and other signal sequences. The host cell will then recognize the recombinant molecule as a legitimate gene that can be expressed. Such advanced cloning vectors allow the expression of many eukaryotic proteins in bacterial cells.

Other methods can maximize gene expression. For example, the cloned eukaryotic gene can be attached to a bacterial gene encoding a protein that is produced in large quantities. Enzymes are later used to clip off the unwanted, bacterial portion of the resulting protein. Bacterial cells can also be engineered to secrete a protein as it is made, thereby simplifying the task of purifying it.

Additional methods for analyzing and cloning nucleotide sequences increase the power of DNA technology

Gel Electrophoresis

One of the most important analytical procedures in DNA technology involves separating and visualizing DNA fragments, often following digestion with restriction enzymes. The best way to accomplish this is through **gel electrophoresis** (described in the Methods Box on the facing page). Electrophoresis separates macromolecules—either nucleic acids or proteins—on the basis of size, electrical charge, and other physical properties.

One use is the identification of particular DNA molecules by the band patterns they yield in gel electrophoresis after being cut with various restriction enzymes. Viral

METHODS: GEL ELECTROPHORESIS OF MACROMOLECULES

Gel electrophoresis separates macromolecules on the basis of their rate of movement through a gel under the influence of an electric field. As FIGURE a indicates, mixtures of nucleic acids or proteins are placed in wells near one end of a thin slab of a polymeric gel. The gel is supported by glass plates and bathed in an aqueous solution. Electrodes are attached to both ends, and voltage is applied. Each macromolecule then migrates toward the electrode of opposite charge at a rate determined mostly by the molecule's charge and size. Usually several different samples, each a mixture of molecules, are run simultaneously in multiple "lanes" on a slab gel, as shown here.

For nucleic acids, the rate of migration—how far a molecule travels while the current is on—is inversely proportional to molecular size. Nucleic acids carry negative charges (phosphate groups) proportionate to their lengths, but the gel impedes longer fragments more than it does shorter ones. The gel in FIGURE b has been treated after electrophoresis with a DNA-binding dye that fluoresces pink in ultraviolet light. In each lane, the top of which is closest to the scientist, you can see a number of pink bands, which correspond to DNA molecules of different sizes. The larger molecules move more slowly through the gel and are located toward the bottom in the photograph. The bands contain DNA restriction fragments. Twenty samples were run, each a mixture of fragments from a DNA sample digested with a different restriction enzyme.

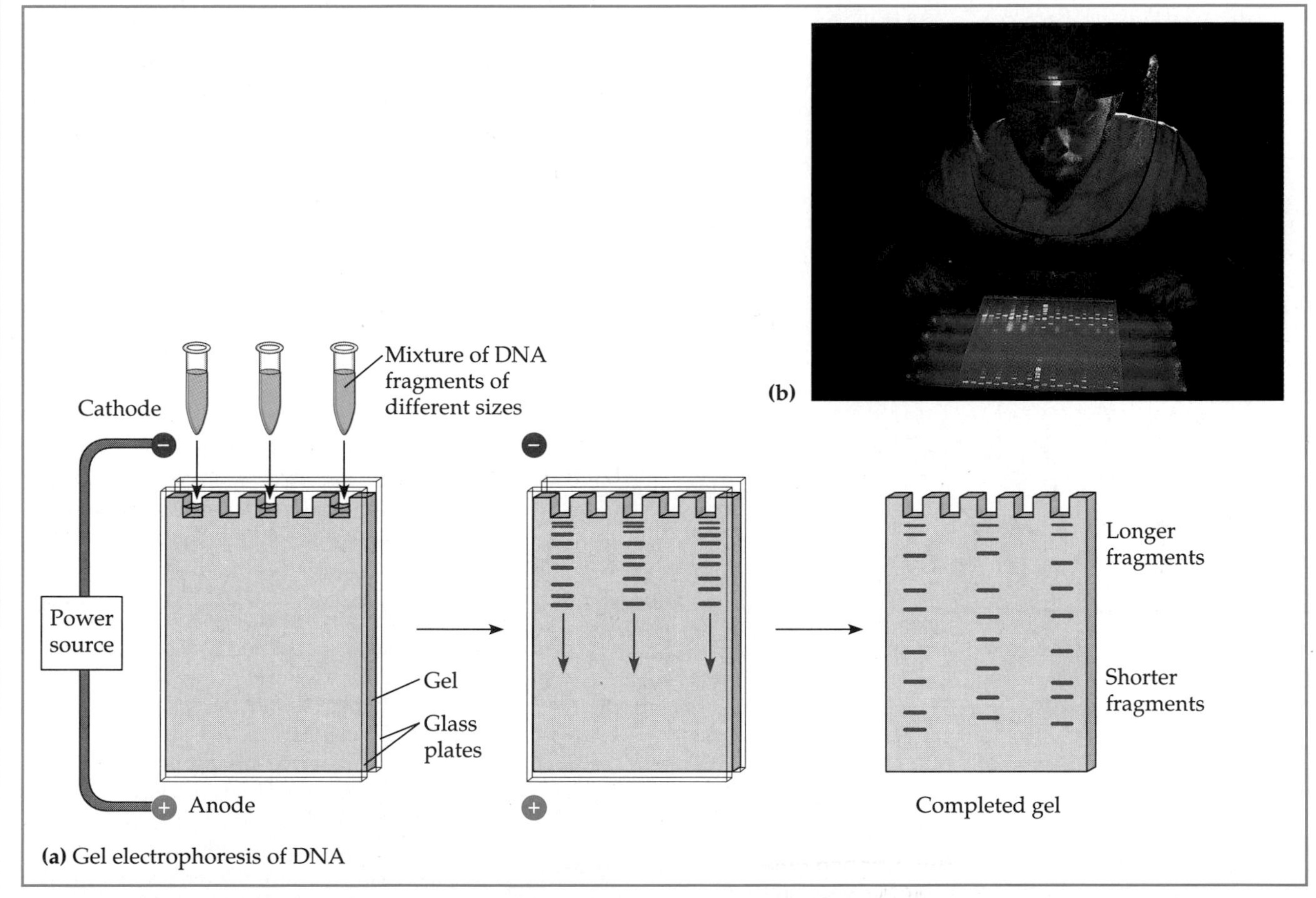

(a) Gel electrophoresis of DNA

(b)

DNA, plasmid DNA, and particular segments of chromosomal DNA can all be identified in this way. Another use is the isolation and purification of individual fragments containing interesting genes, which can be recovered from the gel with full biological activity. We will see still other uses for gel electrophoresis later in the chapter.

DNA Synthesis and Sequencing

It is now possible to determine the nucleotide sequence of a gene in a few days or less. A common method, described in the Methods Box on p. 378, combines base-specific chemical or enzymatic reactions with high-resolution gel electrophoresis.

The sequence of a gene can provide a great deal of information. Analyzing the nucleotide sequence of a gene is often the fastest method for determining the amino acid sequence of its polypeptide. In addition, the gene sequence provides the locations of restriction sites within a piece of DNA, making further manipulation of the gene easier. Most sequences that are determined are being

METHODS: SEQUENCING OF DNA BY THE SANGER METHOD

Developed by British scientist Frederick Sanger, this method for determining the nucleotide sequence of DNA molecules involves synthesizing in vitro DNA strands complementary to one of the strands of the DNA being sequenced. The method is based on the random incorporation of a modified nucleotide (a *di*deoxyribonucleotide, which lacks *two* oxygen atoms) that blocks further DNA synthesis. A series of DNA fragments is synthesized that reflects all the positions of the modified nucleotides, and thus ultimately the sequence of the DNA. Before beginning the synthesis procedure, the DNA to be sequenced is cut up into restriction fragments. Then the procedure illustrated here is carried out with each fragment.

Most of the work of DNA sequencing is now automated. As an alternative to labeling the fragments with radioactive primers, some of the newest sequencing machines use fluorescent dyes to tag the dideoxyribonucleotides, one color for each of the four types of nucleotide. This approach allows the four reactions to be performed in a single tube; the dideoxy ends of the reaction-product DNA strands can be distinguished by the color of their fluorescence.

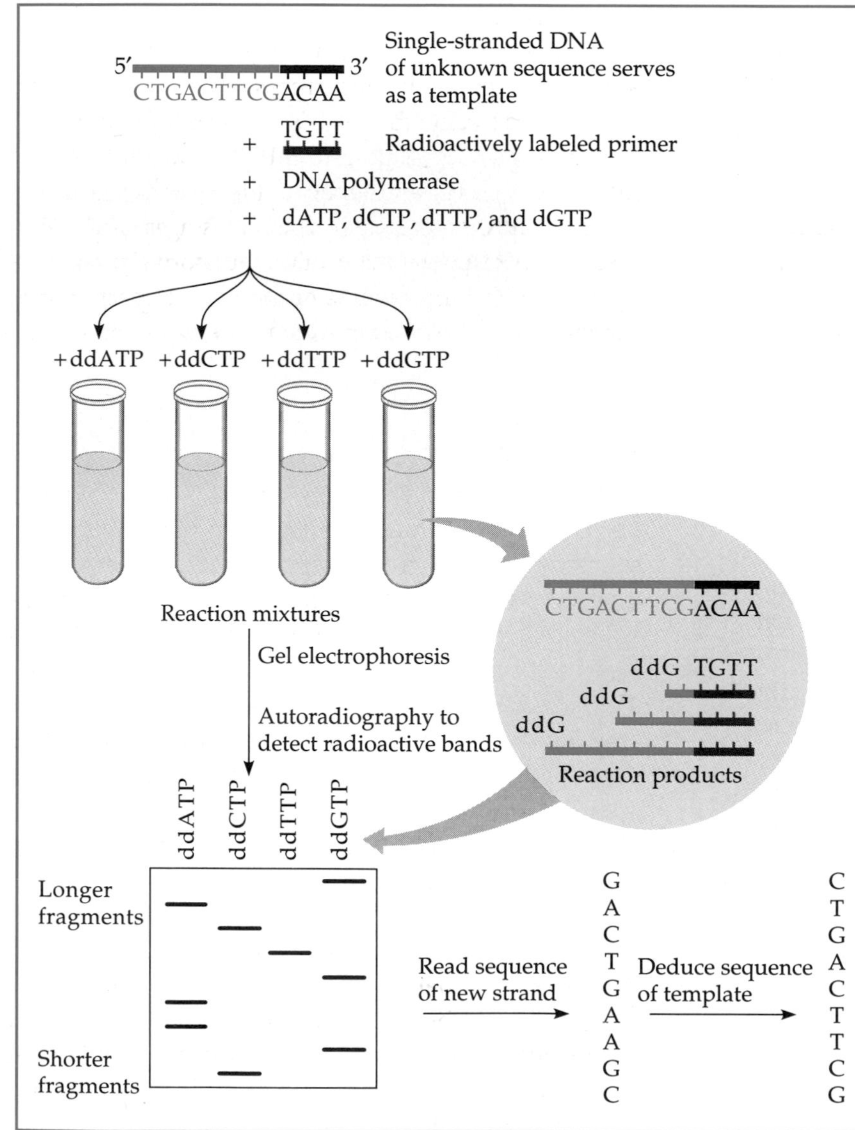

A preparation of one of the strands of the DNA fragment is divided into four portions, and each portion is incubated with all the ingredients needed for the synthesis of complementary strands: a primer (radioactively labeled), DNA polymerase, and the four deoxyribonucleoside triphosphates. In addition, each reaction mixture contains a different *one* of the four nucleotides in the modified, dideoxy (dd) form.

Synthesis of the new strands starts with the primer and continues until a dideoxyribonucleotide is incorporated, which prevents further synthesis. Since the reaction mixture contains both deoxy and dideoxy forms of one nucleotide, the two forms "compete" for incorporation into the strand. Eventually, a set of radioactive strands of various lengths will be generated. This is shown here only for the reaction mixture with ddGTP.

The new DNA strands in each reaction mixture are separated by electrophoresis on a polyacrylamide gel, which can separate strands differing by as little as one nucleotide in length. The sequence of the newly synthesized strands can be read directly from the bands produced in the gel, and from that, the sequence of the original template strand is deduced. In this example, since the longest fragment terminates with ddG, this means that G is the last base in the new DNA strand. Notice that the second longest fragment terminates with ddA, meaning that A is the second to last base, and so on.

collected in computer data banks that are valuable for understanding genes and genetic control elements, as well as for biotechnology. With the help of computers, long sequences can be scanned for shorter segments known to be control sequences, such as promoters or enhancers. It is also possible to scan a DNA sample for similarities to known sequences in other genes or other organisms. Such sequence comparisons provide a new tool for classifying organisms and determining evolutionary relationships.

The Polymerase Chain Reaction (PCR)

The **polymerase chain reaction (PCR)** is a technique by which any piece of DNA can be quickly amplified (copied many times) in vitro (in test tubes). The DNA is incubated under appropriate conditions with the enzyme DNA polymerase and special short pieces of nucleic acid called primers. Billions of copies of a segment of DNA can be made in a few hours, whereas it usually takes weeks to clone a piece of DNA by using recombinant DNA methods and replication within host cells. The Methods Box on p. 380 describes the PCR procedure. Devised in 1985, PCR has revolutionized research in molecular biology and also has many commercial applications, as we will see later in the chapter.

Just as impressive as the speed of PCR is its specificity. Because the primers determine the DNA sequence that is amplified, there is no need to isolate the desired segment of DNA from the starting material before carrying out the reaction. This provides some distinct advantages in certain strategies that combine PCR with cloning in host cells. Recall that one of the more difficult steps in cloning is being able to identify a rare desired clone among thousands of others. By using PCR to amplify a gene prior to additional cloning in host cells, the target DNA becomes by far the most abundant DNA fragment, making the selection of a desired clone a much simpler task.

Another advantage of PCR is that there is no need to prepare a purified sample of DNA. Only minute amounts of DNA need be present in the starting material, and this DNA can be in a partially degraded state. Furthermore, because DNA is an unusually stable biological molecule, it can often be amplified by PCR from sources thousands and even millions of years old.

PCR has been used to amplify DNA from a wide variety of sources: fragments of ancient DNA from a 40,000-year-old frozen woolly mammoth; DNA from tiny amounts of blood, tissue, or semen found at the scenes of violent crimes; DNA from single embryonic cells for rapid prenatal diagnosis of genetic disorders; and DNA of viral genes from cells infected with such difficult-to-detect viruses as HIV (the AIDS virus).

Hybridization

We briefly examined nucleic acid hybridization in the context of probes used to identify specific clones. The technique can be applied to a wider variety of problems. Consider an example: A researcher has just identified and cloned a gene that plays an important regulatory role during mitosis in yeast cells. Do other organisms—for example, humans, or mice, or frogs—also contain this gene? Hybridization can be used to address this question. The basis of the technique is to allow a labeled probe, a specific piece of DNA complementary to the gene of interest, to bind to the DNA from various cells being tested. If such complementary sequences are present, the probe will base-pair to them. Cells carrying the gene can then be identified because the probe has been labeled.

With hybridization, researchers can learn more about the complementary sequence than simply whether it is present or not. The Methods Box on p. 381 describes in more detail a particular kind of hybridization method called a Southern hybridization, or **Southern blotting.** This technique enables a researcher to determine not only whether a particular sequence is present within a sample of DNA, but how many such sequences there are, and the size of the restriction fragments that contain these sequences.

Messenger RNA can also be subjected to hybridization analysis, in an analogous process known as **Northern blotting.** Typically, the goal is to determine whether a particular gene is made into mRNA, how much of that mRNA is present, and whether the abundance of that specific mRNA changes at different stages of development or in response to certain regulatory signals. Such Northern hybridization has become the mainstay of research focusing on the control of gene expression.

RFLP Analysis

Earlier in the chapter, we saw that the DNA fragments that result from cutting a particular piece of DNA with a specific restriction enzyme give a characteristic pattern of bands upon gel electrophoresis. Each band corresponds to a DNA restriction fragment of a certain length. Using this technique to examine homologous segments of DNA known to carry different alleles of a gene, researchers found that DNA of different alleles would show different band patterns after treatment with certain restriction enzymes. This result was not surprising because differences in nucleotide sequence would be expected to result in differences in the number and locations of restriction sites. FIGURE 19.7 (p. 382) shows why this is so. The diagram illustrates the DNA of two different alleles of the same gene that differ in sequence by a single base pair (outlined in black in FIGURE 19.7a). As a consequence of this difference, allele 1 has one more restriction site than is found in allele 2. Therefore, when the DNA segments representing these two alleles are treated with a restriction enzyme, they are cut into fragments that differ in both number (three versus two) and length. Upon gel electrophoresis (FIGURE 19.7b, p. 382), the DNA from the two alleles shows different patterns of bands.

Biologists were excited to discover similar results when homologous but noncoding segments of DNA were used as starting material. In fact, differences in band patterns

METHODS: THE POLYMERASE CHAIN REACTION (PCR)

The polymerase chain reaction (PCR) is a method for making many copies of a specific segment of DNA. It is much faster than gene cloning with plasmid or phage DNA and is performed completely in vitro. The starting material for PCR (FIGURE a) is a solution of double-stranded DNA containing the nucleotide sequence that is "targeted" for copying (it may be either a gene or a noncoding DNA sequence). The scientist then adds DNA polymerase (which catalyzes the reaction), a supply of all four nucleotides (for assembly into new DNA), and primers. The primers are required for the DNA polymerase enzyme to initiate DNA synthesis (see Chapter 16). The particular primers used in PCR are chemically synthesized with sequences that are complementary to the ends of the targeted segment of DNA.

FIGURE b outlines the PCR procedure. ① The DNA is briefly heated to separate its strands and then ② cooled to allow the primers to bind by hydrogen bonding to the ends of the target sequence, one primer on each strand. Then ③ the DNA polymerase extends the primers by adding nucleotides, using the longer DNA strands as templates. Within a short time, the amount of the target DNA sequence has been doubled. The solution is then heated again, starting another cycle of strand separation, primer binding, and DNA synthesis. Again and again the cycle is repeated, at about 5 minutes per cycle, until the targeted sequence has been duplicated enough times. There are practical limits to the number of copies that can be made, often imposed by the accumulation of relatively rare errors in the base sequence of the DNA copies (in vitro versions of mutations).

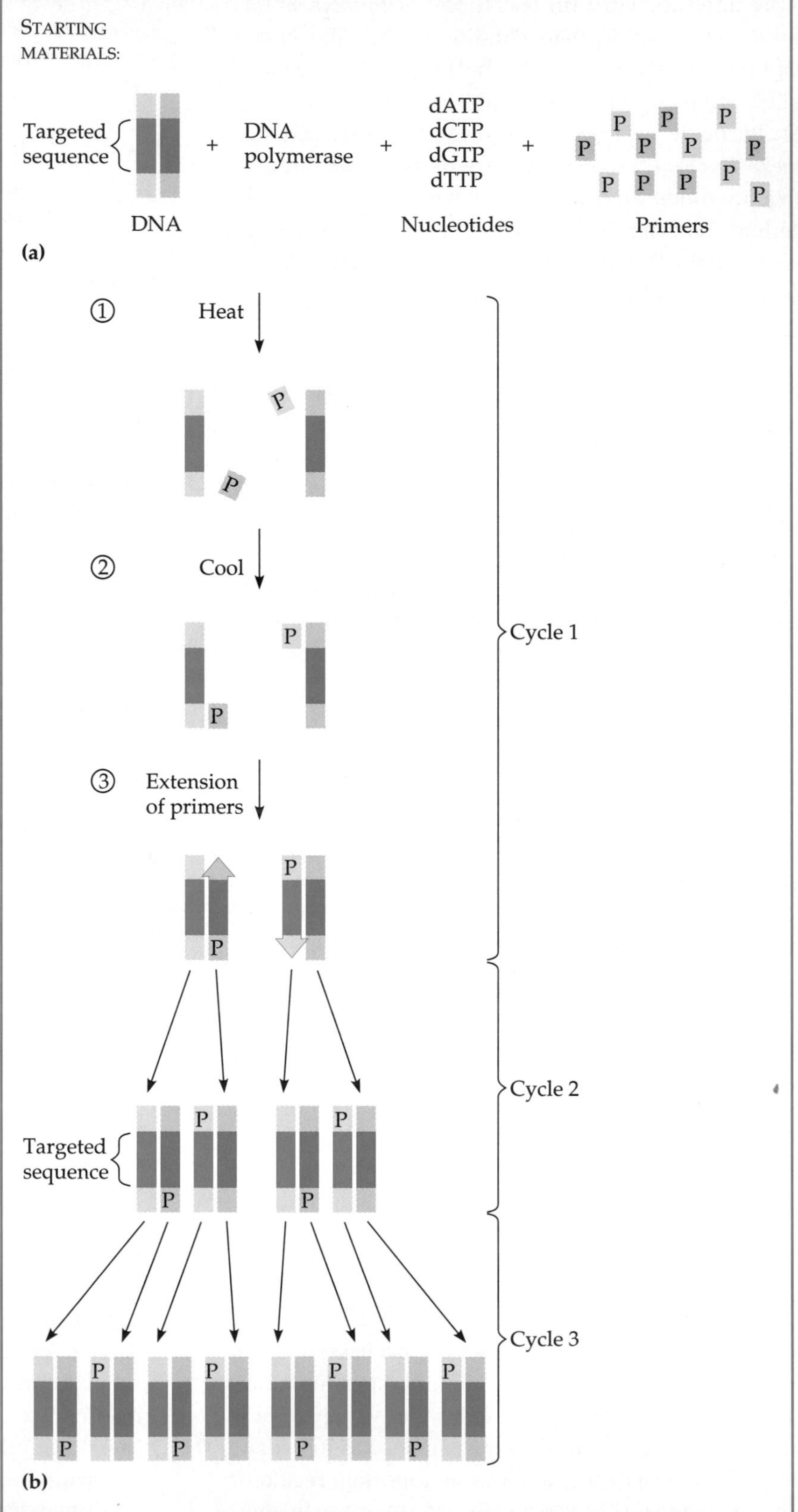

METHODS: NUCLEIC ACID HYBRIDIZATION AND SOUTHERN BLOTTING

Nucleic acid hybridization is a general method for detecting and analyzing sequences of homologous DNA. It can be used simply to determine whether such homologous sequences exist in different samples (perhaps from different organisms), or it can be used to determine the number of such homologous sequences that exist within a genome and the size of the restriction fragments that contain such sequences. In this way, it is possible to study genetic differences between organisms or individuals.

As indicated in these diagrams, nucleic acid hybridization analysis uses five laboratory techniques, some of which have already been discussed: restriction enzyme digestion of DNA, gel electrophoresis, transfer of DNA to a solid support through blotting, the use of DNA probes, and autoradiography (see the Methods Box in Chapter 2, p. 30). The starting material for nucleic acid hybridization is often DNA from the entire genome of an organism. This huge length of DNA will produce so many restriction fragments that, if all are made visible (with a dye, for example), they would appear as a smear in gel electrophoresis, rather than as discrete bands. However, the DNA bands of interest can be selectively visualized by using a labeled probe. The probe consists of multiple copies of a radioactively or fluorescently labeled piece of DNA that will base-pair with the DNA of interest. In order for such hybridization to occur, the DNA being tested must be transferred from the gel to a solid support, typically nitrocellulose or nylon membranes. Such a transfer, which is often accomplished simply by capillary action, is known as Southern blotting (after E. M. Southern, who developed this method of hybridization in 1975).

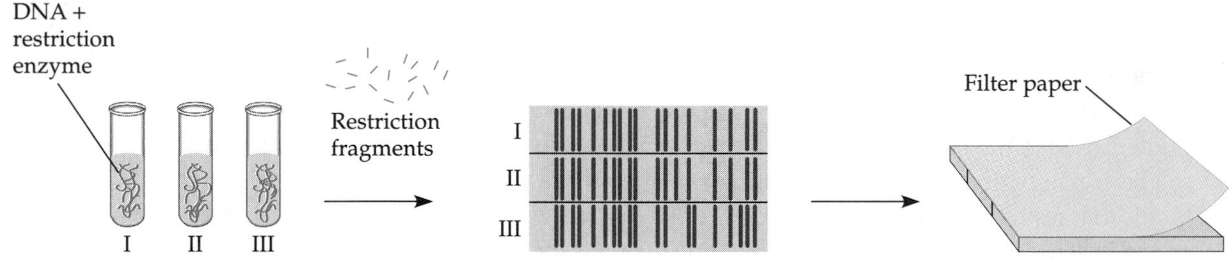

① **Restriction fragment preparation.** DNA samples to be tested (in this case identified as samples I, II, and III) are prepared from the appropriate sources. A restriction enzyme is added to the three samples of DNA to produce restriction fragments.

② **Electrophoresis.** The mixtures of restriction fragments from each sample are separated by electrophoresis. Each sample forms a characteristic pattern of bands. (There would be many more bands than are shown here.)

③ **Blotting.** After the DNA on the gel is denatured by chemical treatment, the single strands are transferred onto special paper or nylon membranes through capillary action. (This is called Southern blotting.)

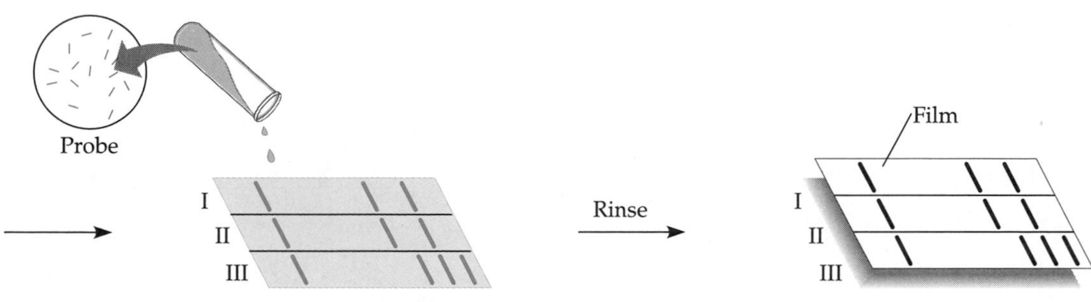

④ **Radioactive probe.** A radioactive probe is added to the DNA bands. The probe is a single-stranded DNA molecule that is complementary to the gene of interest. The probe attaches by base pairing to those restriction fragments that are complementary to it.

⑤ **Autoradiography.** After the probe hybridizes and excess probe is rinsed off, a sheet of photographic film is laid over the gel. The radioactivity in the bound probe exposes the film to form an image corresponding to specific DNA bands—the bands containing DNA that base pairs with the probe.

were observed much more often than they had expected. Differences in DNA sequence on homologous chromosomes that result in different restriction fragment patterns turn out to be scattered abundantly throughout genomes, including the human genome. Such differences have been named **restriction fragment length polymorphisms (RFLPs,** pronounced "riflips"). This type of difference in sequence is conceptually the same as a difference in coding sequence; it too can serve as a genetic marker for a particular location (locus) in the genome. A given RFLP marker frequently occurs in numerous variants in a population (the word *polymorphisms* comes from the Greek for "many forms").

RFLPs are detected and analyzed through Southern hybridization, using nucleic acid probes that are radioactively labeled. The example shown in the Methods Box on p. 381 could just as easily represent the detection of RFLPs in three different individuals. For such a comparison, the entire genome can be used as the DNA starting material. In the case of humans, such samples are typically obtained from white blood cells. Samples I, II, and III might be DNA isolated from different individuals. The DNA is then subjected to Southern hybridization, and a particular RFLP marker is used as the probe, a labeled DNA fragment that will base-pair with its complement among the numerous bands that result from electrophoresis. The example illustrated in the Methods Box indicates that individuals I and II carry the same version(s) of the RFLP marker, but that individual III carries a different version.

Since RFLP markers are inherited in a Mendelian fashion, they can serve as genetic markers for making linkage maps. The geneticist uses the same reasoning you learned about in Chapter 14; the more often a particular RFLP marker and a certain allele for a gene are inherited together, the closer the two loci on a chromosome. The discovery of RFLPs has greatly increased the number of markers available for mapping the human genome. No longer are human geneticists limited to genetic variations that lead to obvious phenotypic differences (such as genetic diseases) or even to differences in protein products. RFLP analysis is also important in the diagnosis of genetic disorders and in forensic (legal) applications. These are among the many applications of DNA technology that we will sample in the next three sections.

DNA technology is catalyzing progress in many fields of biology

DNA technology has triggered research advances in almost all fields of biology by enabling biologists to tackle more specific questions with finer tools. As mentioned earlier, the new techniques have opened up the study of the molecular details of gene structure and function. The beauty of these methods is that they generally do not depend on the expression of genes. Geneticists can now study genes directly, without having to infer genotype from phenotype as in classical genetics, and can begin to answer several important basic research questions: What is the organization of a gene? What regulatory sequences are located near the gene? How is the gene regulated?

At least as important is the opposite problem: determining the function of a gene's product—that is, inferring phenotype from genotype. Very often, researchers clone and identify genes without knowing what the pro-

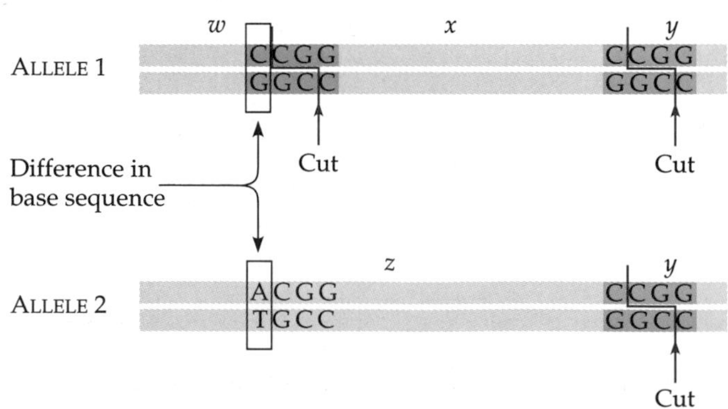

(a) DNA from two alleles

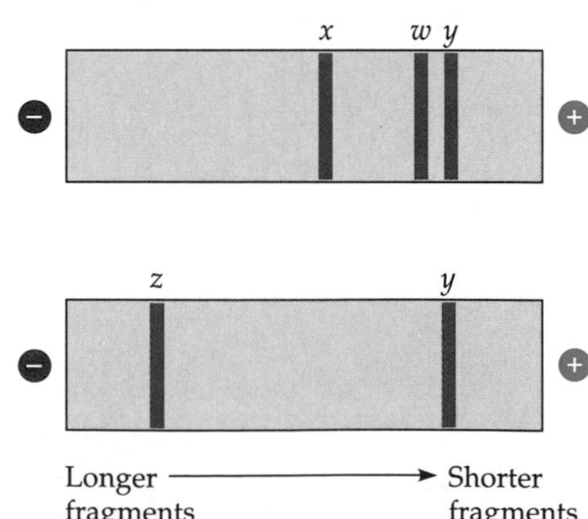

(b) Electrophoresis of restriction fragments

FIGURE 19.7

Using restriction fragment patterns to distinguish DNA from different alleles. (**a**) Restriction sites differ in the DNA of these two alleles (only the relevant bases are shown). The particular restriction enzyme for which the recognition sequences are shown cuts the DNA from allele 1 into three pieces (*w*, *x*, and *y*) but cuts the DNA from allele 2 into only two pieces (*z* and *y*). (**b**) Electrophoresis separates the restriction fragments formed from each allele. A clear difference between the two alleles is revealed by their band patterns on the gel. Allele 1 has three bands, corresponding to fragments *w*, *x*, and *y*; allele 2 has two bands, corresponding to *z* and *y*.

METHODS: IN VITRO MUTAGENESIS

Researchers can use in vitro mutagenesis to introduce mutations into specific genes. It is first necessary to add the wild-type allele (lighter blue) to a plasmid or other appropriate vector, and clone the DNA. ① After cloning, the double-stranded plasmid is denatured to obtain a single-stranded DNA template that includes the wild-type version of the DNA of interest. ② A mutagenic primer is allowed to base-pair with the single-stranded template. The primer consists of two regions that are complementary to template regions on either side of a mismatched region—a region where the base sequence of the mutagenic primer is not complementary and will not pair with the base sequence in the template DNA. ③ Addition of DNA polymerase elongates the primer strand to produce a double-stranded plasmid, one strand representing the original genetic information and the other strand containing the new mutant DNA. ④ The vectors (plasmids) are then used to transform bacterial cells, which replicate both strands of the plasmid. When such a transformed cell first divides, one daughter cell gets a mutant version of the plasmid, while the other cell gets the wild-type version. ⑤ Thus, half the resulting colonies will provide a source of the cloned mutant allele and its protein product. The resulting defective protein provides a comparison to help identify the function of the normal protein. (The colonies are artificially color-coded in this diagram.)

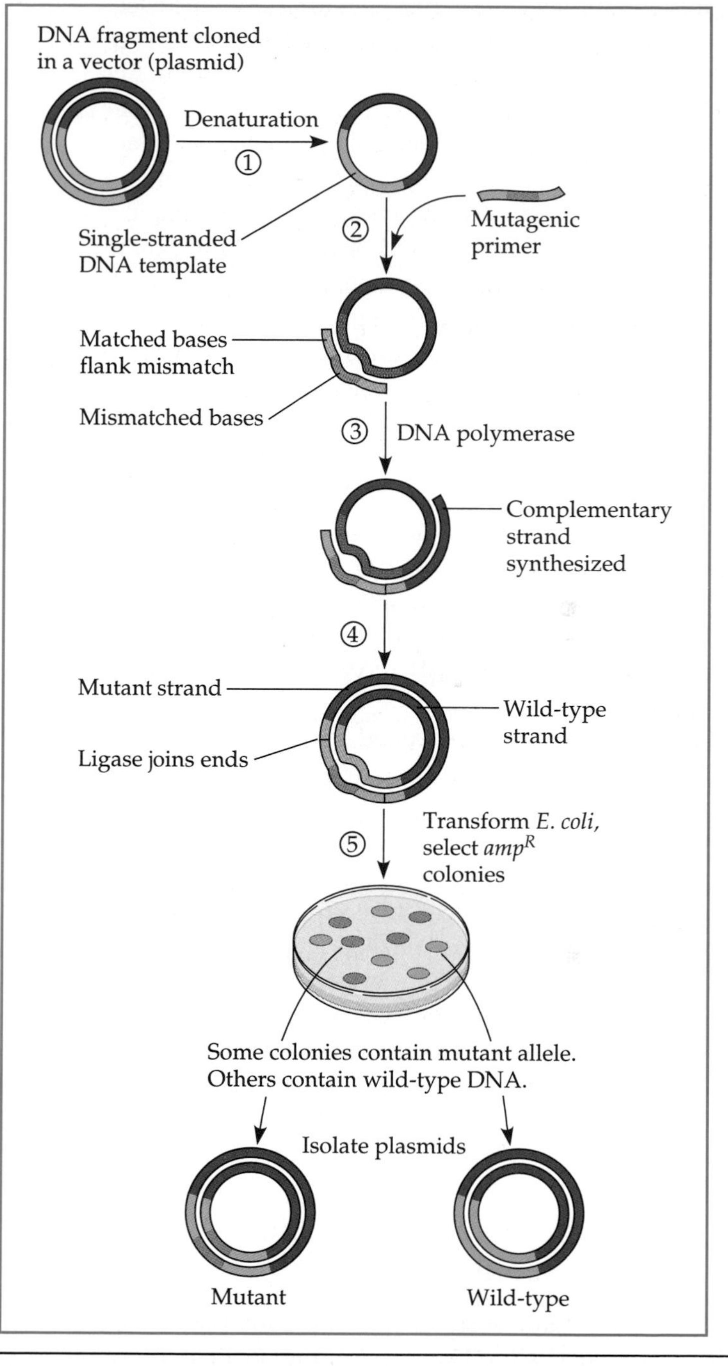

tein product does. For example, it is possible to identify mRNAs that accumulate at particular developmental stages. Cloning of cDNA can be used to obtain the genes corresponding to these mRNAs, but how can the developmental functions of the proteins be determined?

One powerful approach is **in vitro mutagenesis,** a technique that can be used to introduce specific changes into the sequence of a cloned gene (see the Methods Box above). Such mutations often alter the function of the protein product. Thus, when the mutated gene is returned to the host cell, it is sometimes possible to determine the function of the missing normal protein by examining what changes occur in the cell physiology or developmental pattern of the mutant.

Hybridization allows the detection of DNA sequences similar to that of any cloned gene. These sequences can provide information about the evolutionary relationships between the gene of interest and other genes of the same organism or of other organisms. They can also allow the biologist to learn about the natural form of the gene (the probe may have been cDNA), including regulatory sequences and other noncoding sequences adjacent to the gene or within it. Such research is greatly increasing our understanding of how genes are organized and controlled.

A DNA probe can even be used to map a gene on a eukaryotic chromosome. In the technique called **in situ hybridization,** a radioactive DNA probe is allowed to base-pair (hybridize) with complementary sequences on intact chromosomes on a microscopic slide (*in situ* means "in place"). Autoradiography and chromosome staining are then used to reveal which band on which chromosome the probe has attached to (see the photograph on p. 262). In a similar way, the binding of a probe to mRNA can be used to identify cells that are expressing a specific gene, a technique that again finds great usefulness in the study of developmental biology.

In addition to allowing the production of large amounts of particular genes, DNA technology enables the production of large quantities of proteins that are present naturally in only minute amounts. This is important because many molecules crucial for controlling cell metabolism and development are scarce and therefore cannot be purified and characterized by traditional biochemical methods.

Some researchers, emboldened by the power of these new techniques, are attempting to catalog all the proteins made by particular organs, such as the brain. The task has become feasible now that biologists are no longer limited to studying proteins produced in large enough quantities to detect with assays for enzymatic activity or antibody binding. Instead, cDNA from all the mRNA molecules present in a given organ can be prepared, cloned in bacteria, and used to prepare enough of each protein for study. Early estimates are that the brain catalog includes tens of thousands of different proteins.

The Human Genome Project is an enormous collaborative effort to map and sequence DNA

The most ambitious research project made possible by DNA technology is the **Human Genome Project,** an effort to determine the nucleotide sequence of the entire human genome. Four complementary approaches are being used:

1. *Genetic (linkage) mapping of the human genome.* The initial goal is to locate at least 3000 genetic markers (genes or other identifiable loci on the DNA) spaced evenly throughout the chromosomes. This approach is made feasible by the abundance of RFLPs in the human genome; many of the markers used are RFLP markers. The resulting map will make it easier for researchers to find the loci of other markers, including genes, by testing for genetic linkage to known markers.
2. *Physical mapping of the human genome.* This is done by cutting each chromosome into a number of identifiable fragments, and then determining their actual order in the chromosome. A method called **chromosome walking,** which combines several of the techniques we have already examined, has accelerated this chromosome mapping (FIGURE 19.8).
3. *Sequencing the human genome.* This is the process of determining the exact order of the nucleotide pairs of each chromosome. Since a haploid set of human chromosomes contains approximately 3 billion nucleotide pairs, this will be the most time-consuming part of the project.
4. *Analyzing the genomes of other species.* The project also includes similar analysis of the genomes of other species important in genetic research, such as *E. coli,* yeast, mouse, and the plant *Arabadopsis* (see Chapter 31). These "sample" genomes, which are of great interest in their own right, will allow the development of strategies, methods, and new technologies necessary for the daunting task of sequencing the entire human genome.

These four approaches are yielding data that together will give a complete map of the human genome and an understanding of how the human genome compares to those of other organisms. Work is well under way to map the genome of the yeast *Saccharomyces cerevisiae.* In 1992, an international team sequenced one entire chromosome (chromosome 3, consisting of 315,357 nucleotide pairs), and by the end of the decade, the sequencing of most of the other 15 yeast chromosomes will probably be completed. Interestingly, more than half the genes on chromosome 3 encode proteins whose functions are unknown. Researchers predict that many of these genes will also turn up in other eukaryotes.

The potential benefits of genomic information are huge. In the area of health care, the identification and mapping of the genes responsible for genetic diseases will surely aid in the diagnosis, treatment, and prevention of those conditions. In the area of basic science, detailed knowledge of the genomes of humans and other species will give insight into fundamental questions about genome organization, the control of gene expression, cellular growth and differentiation, and evolutionary biology. Analysis of the human genome data and comparison with data from other species will proceed throughout all phases of the project, which involves researchers in many countries.

Much of the progress in completing the Human Genome Project will come through combining all the molecular techniques described earlier in this chapter with classical genetic techniques such as pedigree analy-

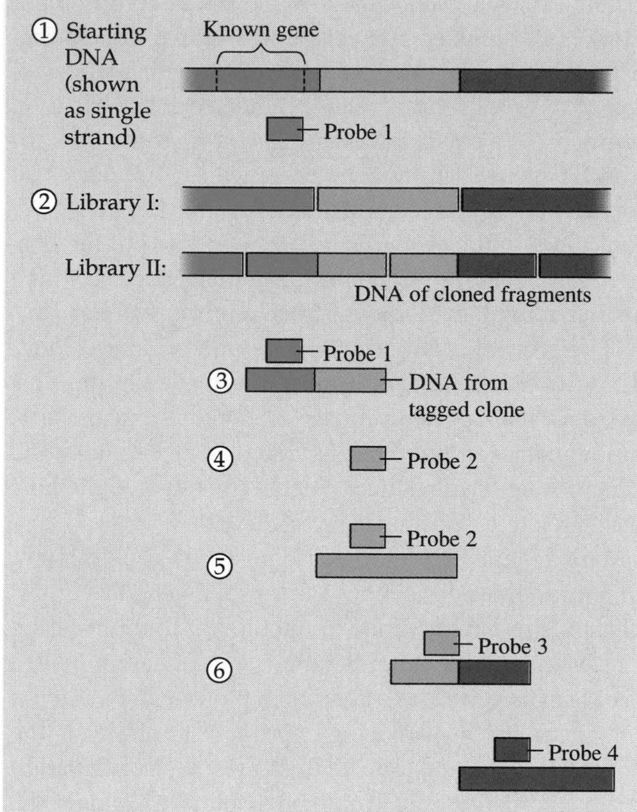

FIGURE 19.8

Chromosome walking. In this method, the researcher starts with a known gene and "walks" along the chromosomal DNA from that locus, producing a map of multiple, overlapping DNA segments. The method follows the steps shown: ① Prepare a probe from the end of the known gene (probe 1). ② Cut the starting DNA with two different restriction enzymes. ③ Use probe 1 to screen library II for DNA fragments that overlap the known gene. ④ Isolate DNA from the clone tagged by probe 1 and prepare a new probe from the far end of that fragment (probe 2). ⑤ Use probe 2 to screen library I for an overlapping fragment farther along. ⑥ Repeat steps 4 and 5, with new probes and alternating libraries, to "walk" down the original DNA.

sis. To mention one example, PCR can amplify specific portions of DNA from individual sperm cells. In this way, researchers can obtain the immediate products of meiotic recombination in amounts sufficient for study, and they can analyze as large a sample as they need, even thousands of sperm. Thus, based on the frequencies of crossovers between genes, researchers can perform human linkage mapping without having to find large families for pedigree analysis.

A significant part of the technological power to achieve the goals of the Human Genome Project will come from advances in automation and utilization of the latest electronic technology, including computer software. Many of these new advances will be developed as the smaller genomes of yeast, *Arabidopsis,* and mouse are analyzed.

DNA technology is reshaping the medical and pharmaceutical industries

Modern biotechnology has made enormous contributions to medicine. Major advances continue in the diagnosis of human genetic disorders and other diseases, in the first applications of gene therapy, and in the development of vaccines and other pharmaceutical products.

Diagnosis of Diseases

A new chapter in the diagnosis of infectious diseases is being opened by DNA technology, in particular the use of PCR and labeled DNA probes to track down elusive pathogens. For example, because the base sequence of HIV DNA is known, PCR can be used to amplify, and thus detect, HIV DNA in blood or tissue samples. This is often the best way to detect an otherwise elusive infection.

The use of DNA technology to diagnose genetic diseases is proceeding even faster. Medical scientists can now diagnose more than 200 human genetic disorders using DNA technology. Increasingly, it is possible to identify individuals with genetic diseases before the onset of symptoms, even before birth. It is also possible to identify symptomless carriers of potentially harmful recessive alleles. Genes have been cloned for a number of human diseases, including hemophilia, phenylketonuria (PKU), cystic fibrosis, and Duchenne muscular dystrophy (see Chapter 14).

Hybridization analysis makes it possible to detect abnormal allelic forms of genes present in DNA samples. Even in cases where the gene has not yet been cloned, the presence of an abnormal allele can be diagnosed with reasonable accuracy if a closely linked RFLP marker has been found. If a mapped RFLP marker is co-inherited at high frequency with a disease, it is probable that the defective gene is located close to the RFLP marker on the chromosomal DNA. Blood samples from relatives of the person at risk must be studied in order to determine which variant of the RFLP marker is linked to the abnormal allele in that family, and this variant must be different from the RFLP variant(s) linked to that family's normal allele (FIGURE 19.9). If these conditions are met, the RFLP marker variants found in the genome of the person at risk reveal whether the abnormal allele is also likely to be present. Alleles for Huntington's disease and a number of other genetic diseases were previously detected in this indirect way. Once chromosome walking or other approaches are able to map a gene more precisely, it can be cloned for study and for use as a probe for identical or similar DNA, as is now the case for Huntington's disease, cystic fibrosis, and many other diseases.

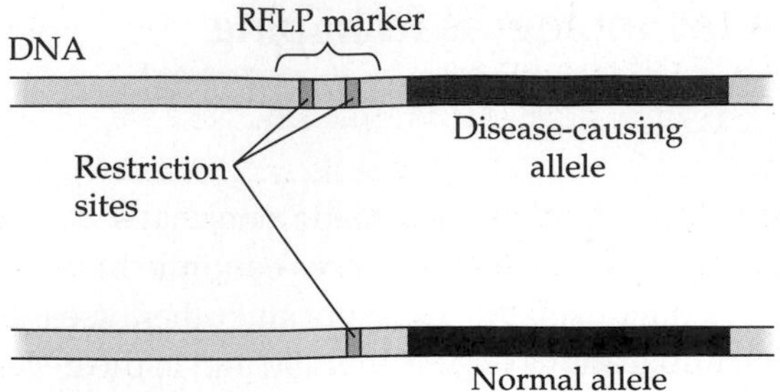

FIGURE 19.9
RFLP markers close to a gene. Even if a disease-causing allele has not been cloned and its precise locus is unknown, its presence can sometimes be detected with high (though not perfect) accuracy by testing for the presence of RFLP markers that are very close to the gene in question. This diagram shows homologous segments of DNA from a family in which some members have a genetic disease. In this family, different versions of a RFLP marker are associated with the different alleles, allowing the test to be applied. If a family member has inherited the version of the RFLP marker with two restriction sites (rather than one), there is a high probability that the individual has also inherited the disease-causing allele.

Human Gene Therapy

Genetic engineering has the potential to actually correct some genetic disorders in individuals. For any genetic disorder traceable to a single defective allele, it should theoretically be possible to replace or supplement the defective allele with a functional, normal allele using recombinant DNA techniques. The new allele could be inserted into the somatic cells of a child or adult, or into germ (gamete-producing) cells or embryonic cells. For example, in 1994 medical researchers reported some success in treating cystic fibrosis patients by using a nasal spray containing a vector to introduce the normal allele of the cystic fibrosis gene into lung cells.

Gene therapy is especially suited to the treatment of single-enzyme deficiency diseases, where an introduced normal allele would provide the missing enzyme. In this case, normal alleles must be introduced into a patient's own somatic cells that actively reproduce in the body (so that the normal allele will be replicated in the individual) and whose normal protein products can correct the disorder. It may not matter precisely how many engineered cells are introduced into the body as long as a minimum amount of the product is produced.

The most successful of the fledgling gene therapy trials involve the treatment of an immunodeficiency disease (see Chapter 39) that results from a lack of the enzyme adenosine deaminase (ADA). In the first attempts at this therapy, patients received intravenous injections of their own T lymphocytes (a type of white blood cell) that were engineered to carry a normal ADA allele. In order to accomplish this, T lymphocytes were earlier removed from the patients and cultured in vitro with harmless retroviral vectors carrying the normal ADA allele. The recombinant retroviruses infected the cells and inserted the normal ADA allele into the cells' genomes along with a DNA version of the viral genome. The engineered cells were then returned to the patient's body. This represents one approach to gene therapy, although since T lymphocytes do not reproduce, patients must undergo this treatment periodically. A more permanent approach is to put the normal ADA allele into patients' bone marrow cells, which are then reimplanted into the bone marrow (FIGURE 19.10). This approach could result in a broader, more long-lasting treatment because bone marrow cells multiply and give rise to all the cells of the immune system. Although it is too early to conclude that such therapy will achieve long-term results, patients thus treated are able to lead relatively normal lives.

Many technical questions are posed by gene therapy. An important one is how the proper genetic control mechanisms can be made to operate on the transferred gene. Consider the potential treatment of hemophilia, a disease caused by a deficiency of a blood-clotting factor. It might be just as damaging to produce too much of the clotting factor as too little. In this case, precise regulation of the therapeutic gene would be essential to achieve the desired result, something that is technically impossible at this time.

Other questions include: At what stage of biological development is intervention most effective, and when is it too late? Is it best to introduce the gene into cells that have not finished differentiating, such as those in the bone marrow, or into specific tissue cells, such as T lymphocytes in the case of ADA therapy? How can we get the correct gene or gene product to the tissues where it is required? How do we treat developmental diseases, where a gene product may be needed for only a brief period during development? How can we be sure that the insertion of the therapeutic gene does not alter some other necessary cell function?

Gene therapy also raises some difficult ethical and social questions. At present, the procedures are extremely expensive and require expertise and equipment found only in major medical centers. Once treatment is available for a particular genetic disorder, how many patients will actually have access to it? It is possible that researchers will eventually develop vectors—perhaps engineered viruses—that can be injected directly into patients, thus reducing costs by eliminating the need for engineering cells in culture.

The most difficult ethical question is whether we should try to treat human germ cells in the hope of correcting the defect in future generations. In mice at least, transferring foreign genes into the germ line (egg cells) is now a routine procedure in experiments. For example,

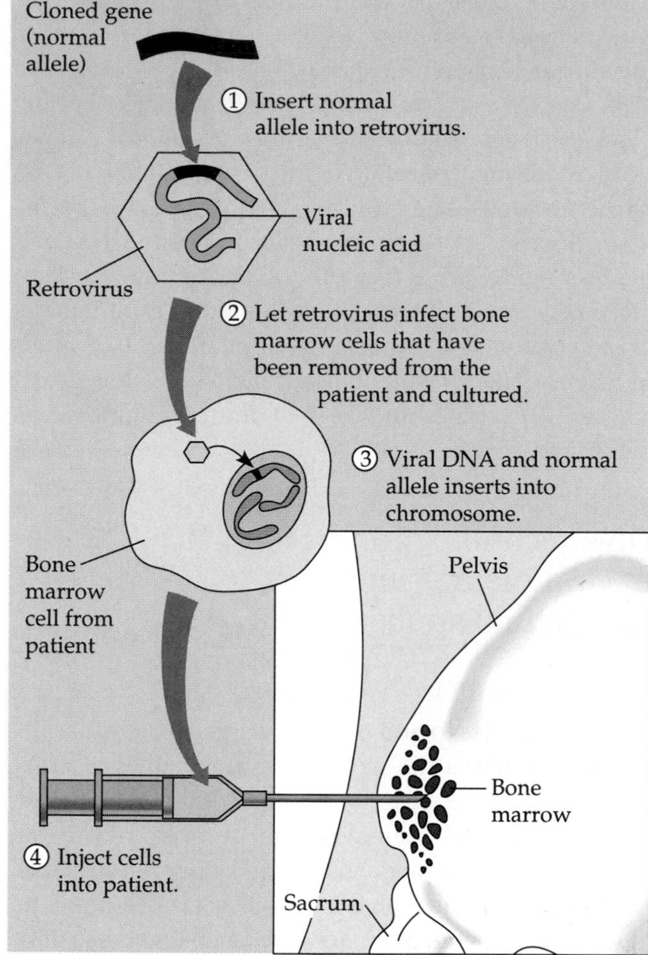
Cloned gene
(normal
allele)
① Insert normal
allele into retrovirus.
Viral
nucleic acid
Retrovirus
② Let retrovirus infect bone
marrow cells that have
been removed from the
patient and cultured.
③ Viral DNA and normal
allele inserts into
chromosome.
Bone
marrow
cell from
patient
Pelvis
Bone
marrow
④ Inject cells
into patient.
Sacrum

FIGURE 19.10

One type of gene therapy. In this procedure, a disarmed retrovirus is used as a vector to introduce a normal allele of a gene into the cells of a patient who lacks it. The method takes advantage of the fact that a retrovirus inserts a copy of its nucleic acid (actually a DNA transcript of its RNA genome; see Chapter 17) into the chromosomal DNA of its host cell. If its nucleic acid includes a foreign gene and that gene is expressed, the cell—and its mitotic descendants—will possess the product of the gene, and will potentially be cured. Cells that continue to reproduce throughout life, such as bone marrow cells, are ideal for gene therapy. (The insertion of such cells is shown here in a side view of a pelvis.)

scientists have successfully transplanted a human gene for one of the polypeptides of hemoglobin into the germ line of mice defective in the corresponding mouse gene. In many of the recipient mice and their descendants, the human gene is active in producing the polypeptide, not only in the correct location (red blood cells) but also at the correct time in development (at the appropriate fetal stage). So far, getting good expression of foreign genes in live animals has been much less successful than in other experimental systems, but it is clear that the technical problems will eventually be solved. Thus, we may soon have to face the question of whether it is advisable, under any circumstances, to intervene in the genetics of human germ lines and embryos.

Some critics suggest that tampering with human genes in any way, even in somatic cells and even to treat individuals who have life-threatening diseases, is wrong. They argue that it will inevitably lead to the practice of eugenics, a deliberate effort to control the genetic makeup of human populations. Other observers see no fundamental difference between genetic engineering of somatic cells and other conventional medical interventions to save lives. They compare transplanting genes to the transplantation of organs.

Vaccines and Other Pharmaceutical Products

For many viral diseases for which there are no effective drug treatments, prevention by vaccination is virtually the only way to fight the disease. Traditional vaccines for viral diseases are of two types: particles of a virulent virus that have been inactivated by chemical or physical means, and active virus particles of an attenuated (nonpathogenic) viral strain. In both cases, the virus particles are similar enough to the active pathogen to trigger an immune response that will produce antibodies against the invading pathogens (see Chapter 39).

There are several ways in which the new biotechnology is being used to modify current vaccines or provide new vaccines against previously recalcitrant diseases. First, recombinant DNA techniques can generate large amounts of a specific protein molecule from the protein coat of a particular disease-causing virus, bacterium, or other microbe. If the protein, referred to as a subunit, is one that triggers an immune response against the intact pathogen, it can be used as a vaccine. Second, genetic engineering methods can be used to modify the genome of the pathogen in order to attenuate it. Vaccination with a live but attenuated organism is often more effective than a subunit vaccine, because a small amount of material triggers a greater response by the immune system. Pathogens attenuated by gene-splicing techniques may be safer than the natural mutants traditionally used.

Another scheme that uses a live vaccine employs vaccinia, the virus that is the basis of the smallpox vaccine. With recombinant DNA techniques, the viral genes that induce immunity to smallpox can be replaced with genes that induce immunity to other diseases. In fact, the vaccinia virus can be made to carry the genes required to vaccinate against several diseases simultaneously. In the future, a single live vaccinia inoculation could conceivably protect people from as many as a dozen diseases.

DNA technology has also been used to create other pharmaceutical products. One of the first practical applications of gene splicing was producing mammalian hormones and other mammalian proteins in bacteria.

Insulin, growth hormone, and several proteins of the immune system, such as potential anticancer molecules called interferons, were among the early examples. Insulin was the first polypeptide hormone made by recombinant DNA procedures to be approved for use in treating human patients in the United States; human growth hormone was the second.

About two million diabetics in the United States depend on insulin treatment to help control their disease. Before 1982, the principal sources of therapeutic insulin were pig and cattle pancreatic tissues obtained from slaughterhouses. Although the insulin extracted from pigs and cattle closely resembles the human version, it is not identical and causes adverse reactions in some people. Now there are several ways of producing insulin in genetically engineered bacteria, and the insulin produced is chemically identical to that made in the human pancreas.

Human growth hormone (HGH) has almost 200 amino acids, making it far larger than insulin. Moreover, growth hormones are more species-specific than insulin, meaning that growth hormone from other animals is generally not an effective growth stimulator if administered to humans. Before genetically engineered HGH became available in 1985, children born with hypopituitarism, a form of dwarfism caused by inadequate amounts of HGH, had to rely on scarce supplies of the hormone obtained from human cadavers. Genetic engineering made much more HGH available for treating children. It also gave researchers a chance to learn about other potential uses for the hormone. In the future, HGH may be used to help heal wounds and broken bones, to treat severe burns, and to retard, if not reverse, the loss of muscle mass that often accompanies the aging process.

In 1989, a biotechnology firm used recombinant DNA methods to produce still another medically important human hormone, erythropoietin (EPO). This protein is normally produced by the kidney as a hormone that stimulates the production of red blood cells in the bone marrow. Biotechnology now provides a source of EPO to treat anemia due to a variety of causes, including kidney damage.

The commercial success of tissue plasminogen activator (TPA) provides another example of a successful venture. TPA helps dissolve blood clots and reduces the risk of secondary heart attacks if administered very shortly after an initial attack. However, TPA illustrates a problem with many genetically engineered products: Because the development costs are high and the application is limited, it is extremely expensive.

The most recent developments in pharmaceutical products involve truly novel ways to fight some diseases that are much less responsive to traditional drug treatments. Current research may soon lead to drugs or gene therapy methods that prevent specific messenger RNAs from being translated into proteins in virus-infected or cancerous cells. For example, one approach is synthesizing **antisense nucleic acid,** single-stranded molecules of DNA or RNA that base-pair with mRNA molecules and block their translation. Interfering with crucial mRNAs involved in viral replication or the transformation of cells into a cancerous state could prevent the spread of the diseases. Another approach is using recombinant DNA techniques to make drugs that either block or mimic surface receptors on cell membranes. One such experimental drug substitutes for a receptor protein that HIV binds to when it attacks white blood cells. The idea is to decoy the AIDS virus and keep it from infecting white blood cells.

DNA technology offers forensic, environmental, and agricultural applications

Forensic Uses of DNA Technology

In violent crimes, blood or small fragments of other tissue may be left at the scene or on the clothes or other possessions of the victim or assailant. If rape is involved, small amounts of semen may be recovered from the victim's body. If enough tissue or semen is available, forensic laboratories can perform tests to determine the blood type or tissue type. However, such tests have limitations. First, they require fairly fresh tissue in a sufficient amount for testing. Second, because there are many people in the population with the same blood type or tissue type, this approach can only exclude a suspect; it is not evidence of guilt.

DNA testing, on the other hand, can identify the guilty individual with a much higher degree of certainty, since the DNA base sequence of every person is unique (except for identical twins). For this forensic application, one common DNA technology is RFLP analysis. The results of such an analysis display selected DNA restriction fragments as bands separated by electrophoresis. The method is used to compare DNA samples from the suspect (a murder suspect, for example), the victim, and a small amount of semen, blood, or other tissue found at the scene of the crime. Radioactive probes mark the bands that contain certain RFLP markers. Even a small set of RFLP markers from an individual can provide a **DNA fingerprint,** or specific pattern of bands, that is of forensic use, since the probability that two people (who are not identical twins) would have the exact same set of RFLP markers is very small. The autoradiograph in FIGURE 19.11 is the type of evidence presented (with explanation) to juries. It shows a match between a victim's DNA fingerprint and that of blood found on the defendant's clothes. In support of such

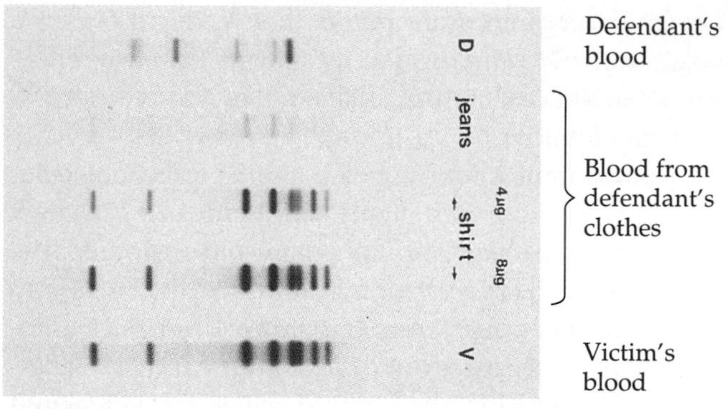

FIGURE 19.11

DNA fingerprints from a murder case. As revealed by RFLP analysis, DNA from bloodstains on the defendant's clothes matches the DNA fingerprint of the victim but differs from the DNA fingerprint of the defendant. This is evidence that the blood on the defendant's clothes came from the victim, not the defendant. (Cellmark Diagnostics.)

evidence, the prosecution presents statistical calculations of the chance that more than one person's RFLP markers would match those in the blood sample.

In most forensic laboratories, standard RFLP analysis is now augmented by DNA fingerprinting based on variations in satellite DNA. Recall from Chapter 18 that satellite DNA consists of tandemly repeated base sequences within the genome. The most useful satellite DNA in forensic applications, known as microsatellite sequences, are about 20 to 200 base pairs long, with the repeating unit 1 to 4 base pairs. For example, a particular microsatellite sequence may have the base doublet AC repeated 50 times at one locus in a person's genome, 65 times at a second locus, and so on. The exact number of repeats at each microsatellite locus varies within the human population. These variations at particular sites in the genome are called **variable number tandem repeats (VNTRs).** Restriction fragments containing VNTRs vary in size among individuals because of differences in the length of the repeating sequence between two restriction sites cut by the enzyme. For example, one person may have AC repeated 65 times in a particular restriction fragment, while another person may have 62 AC repeats. This contrasts with standard RFLP analysis, which depends on differences in the number of restriction sites within a region of the genome. The greater the number of VNTRs or RFLPs examined in a sample, the more likely it is that the DNA fingerprint is unique to a particular individual. VNTR analysis is often coupled with PCR, which amplifies satellite DNA from a minute quantity of blood or other tissue.

Just how reliable is DNA fingerprinting? When we say that the DNA fingerprint from each individual is absolutely unique, we are speaking of the theoretical situation where restriction fragment analysis is done on the entire genome. In practice, forensic DNA tests focus on only five or ten tiny regions of the genome. However, the DNA regions chosen are ones known to be highly variable from one person to another. So the probability is very small that two people (other than identical twins) will have exactly the same sequences, and hence matching DNA fingerprints, for these tested regions. In most legal cases, the probability is between one chance in 100,000 and one in a billion, depending on the number of genomic sites compared. Is this probability small enough for a jury to convict a suspect? Some legal experts contend that juries are so heavily influenced by this kind of scientific data, without appreciating its statistical basis, that the evidence should be virtually flawless before it is admissible. Other experts counter that DNA fingerprints are more reliable than eyewitnesses in placing a suspect at the scene of a crime. The O. J. Simpson murder trial of 1995 made "DNA fingerprint" a household term, and this type of evidence will have an increasing forensic impact.

Environmental Uses of DNA Technology

Increasingly, genetic engineering is being applied to environmental work. The ability of microorganisms to transform chemicals is remarkable, and scientists are now engineering these metabolic capabilities into organisms that will yield economic benefit or help cope with some environmental problems. For example, many microorganisms are able to extract heavy metals, such as copper, lead, and nickel, from their environments and incorporate the metals into compounds such as copper sulfate or lead sulfate. Chemical processing can then recover the metal compounds. Because these metals have been relatively abundant, there has been little economic incentive to engineer more efficient microorganisms to mine them. But as metal reserves are depleted, genetically engineered microbes may become important in mining and, in the process, cleaning up mining waste.

The metabolic diversity of microbes is also employed in the recycling of wastes and detoxification of toxic chemicals. Sewage treatment plants rely on the ability of microbes to degrade many organic compounds into nontoxic form. However, an increasing number of compounds being released into the environment cannot be easily degraded and are not inert—most notably, chlorinated hydrocarbons. Biotechnologists are attempting to engineer microorganisms that can degrade these compounds and can be used in waste water treatment plants. If such microbes can be developed, they might be incorporated directly into manufacturing processes, preventing toxic chemicals from ever being released as waste.

A related research area is the identification and engineering of microbes capable of detoxifying specific toxic wastes found in spills and waste dumps. For example, bacterial strains have been developed that can degrade some of the compounds released during oil spills. Other bacteria have been isolated that transform highly reactive, toxic metals (such as chromium) to much less reactive forms. The ability to move the genes responsible for these transformations into different organisms allows the development of strains that can survive the harsh conditions of these environmental disasters and still help detoxify them.

Agricultural Uses of DNA Technology

Scientists are working to learn more about the genetics of the plants and animals important to agriculture, and they have begun using genetic engineering to improve agricultural productivity.

Animal Husbandry. Farm animals are already being treated with products that have been made by recombinant DNA methods. These products include new or redesigned vaccines, antibodies, and growth hormones. For example, some milk cows are being injected with bovine growth hormone (BGH), made by *E. coli,* in order to raise milk production (it usually increases by about 10%). BGH also improves weight gain in beef cattle. BGH has passed all safety tests so far, and it is now being used extensively in dairy herds. However, some states require that milk from BGH-treated cows be so labeled, and some countries presently refuse to import such milk. Another protein made by engineered *E. coli* that is useful for agriculture is the enzyme cellulase, which hydrolyzes cellulose and makes it possible for virtually all parts of a plant to be used for animal feed.

A number of **transgenic organisms,** organisms that contain genes from another species, have been developed for potential agricultural use. Transgenic animals, including beef and dairy cattle, hogs, sheep, and several species of commercially raised fishes, are being produced by injecting foreign DNA into the nuclei of egg cells or early embryos. For instance, rainbow trout and salmon, widely raised for human food, that are genetically engineered with a cloned growth hormone gene can reach in one year a size that usually requires two or three years of growth. The use of gene-splicing techniques to engineer fishes and other animals for human food is still in the experimental stages, but many biotechnologists and farmers are optimistic about future applications.

Manipulating Plant Genes. In one striking way, plant cells have thus far proved easier to engineer than the cells of many animals. For many plant species, an adult plant can be regenerated from a single cell grown in tissue culture (see FIGURE 34.12). This is an important advantage because many genetic manipulations, such as the introduction of genes from other species, are easier to perform and assess on single cells than on whole organisms. Commercially important plants that readily grow from single somatic cells include asparagus, cabbage, citrus fruits, sunflowers, carrots, alfalfa, millet, tomatoes, potatoes, and tobacco.

Like researchers working with animal cells, molecular biologists working with plants commonly use DNA vectors to move genes from one organism to another. The best-developed DNA vector is a plasmid of the bacterium *Agrobacterium tumefaciens.* In nature, *A. tumefaciens* infects plants and causes tumors called crown galls. The tumors are induced by the plasmid, called the **Ti plasmid** (Ti for tumor-inducing). The Ti plasmid integrates a segment of its DNA, known as T DNA, into the chromosomal DNA of its host plant cells. Researchers have developed ways to eliminate the plasmid's disease-causing properties while maintaining its ability to move genetic material into plant cells.

Foreign genes can be inserted into the Ti plasmid using recombinant DNA techniques. The recombinant plasmid is either put back into *Agrobacterium,* which can then be used to infect plant cells growing in culture, or introduced directly into plant cells, where it incorporates into the plant's chromosomes. Then, taking advantage of the capacity of those cells to regenerate whole plants, it is possible to produce plants that contain and express the foreign gene and pass it on to their offspring (FIGURE 19.12).

A major drawback to using the Ti plasmid as a vector is that only dicots (plants with two seed leaves) are susceptible to infection by *Agrobacterium.* Monocots, including agriculturally important grasses such as corn and wheat, cannot be infected by *Agrobacterium.* Fortunately, newer techniques, such as electroporation and DNA guns, are helping plant scientists overcome this limitation. The DNA gun is being used for a wide variety of plants, especially monocots. In this technique, tiny metal pellets coated with DNA are shot at recipient cells, typically by a burst of gas (see FIGURE 34.13). The pellets are driven through the cell walls into the cytoplasm. The cells quickly repair the tiny punctures in the cell wall and plasma membrane, the metal pellets remain in the cytoplasm but cause no harm, and the DNA becomes integrated into the DNA of the host cell.

Even with these new techniques, engineering plant genomes is a formidable challenge. The cloning of plant DNA is straightforward, but identifying genes of interest may be very difficult. Furthermore, many agriculturally desirable plant traits, such as high crop yield, are polygenic, involving many genes.

Some Early Successes of Genetic Engineering in Plants. Despite its difficulty, genetic engineering of plants is already yielding some positive results, mostly in cases where useful traits are determined by one or only a

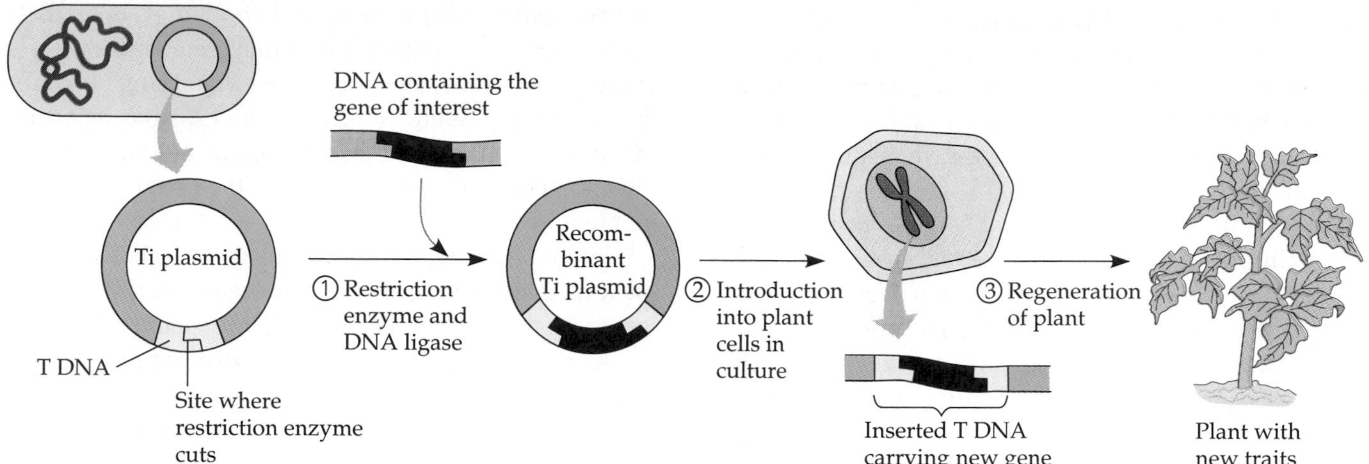

FIGURE 19.12
Using the Ti plasmid as a vector for genetic engineering in plants. ① The Ti plasmid is isolated from the bacterium *Agrobacterium tumefaciens,* and a fragment of foreign DNA is inserted into its T region by standard recombinant DNA techniques. ② When the recombinant plasmid is introduced into cultured plant cells, the T DNA integrates into the plant's chromosomal DNA. ③ As the plant cell divides, each of its descendants receives a copy of the T DNA and any foreign genes it carries. If an entire plant is regenerated, all its cells will carry—and may express—the new genes.

few genes. For example, several chemical companies have developed strains of wheat, cotton, and soybeans carrying a bacterial gene that makes the plants resistant to the herbicides used by many farmers to control weeds. This gene would make it easier to grow crops while still ensuring that weeds are destroyed. The first gene-spliced fruits approved by the FDA for human consumption were tomatoes engineered with antisense genes that retard spoilage (see FIGURE 35.8). After cloning the tomato gene that codes for the enzyme mainly responsible for ripening, researchers cloned a complementary (antisense) gene—one with the complementary sequence of bases. When spliced into the DNA of a tomato plant, the antisense gene transcribes messenger RNA that is complementary to the ripening gene's mRNA. When the ripening gene transcribes the normal mRNA, the complementary (antisense) mRNA binds to the normal RNA, blocking the synthesis of the ripening enzyme. The engineered tomatoes produce only about 1% of the normal amount of ripening enzyme.

A number of crop plants are being engineered to resist infectious pathogens and pest insects. Tomato and tobacco plants, for example, have been engineered to carry and express certain genes of viruses that can normally infect and damage plants. With these versions of the viral genes, however, the plants are resistant to—in effect, vaccinated against—attack by the viruses. Other crop plants have been engineered to resist insect attack. Growing insect-resistant plants will reduce the need to apply chemical insecticides to crops.

These early successes in plant engineering point to an agricultural revolution just around the corner. It is likely that many crop plants will soon be made more productive by enlarging their agriculturally valuable parts—whether they be roots, leaves, flowers, or stems. Researchers are also making strides toward improving the food value of plants, as in engineering corn or wheat to produce storage proteins containing a mix of amino acids more suitable for the human diet. At present, more than 30 different crop plants developed with recombinant DNA techniques are being tested in field trials.

The Nitrogen-Fixation Challenge. Perhaps the most exciting potential use of DNA technology in agriculture involves nitrogen fixation, a bacterial process that benefits plants and their consumers. Nitrogen fixation is the conversion of atmospheric, gaseous nitrogen (N_2), which cannot be utilized by plants, into nitrogen-containing compounds that plants can take up from the soil and use for making essential organic molecules, such as amino acids and nucleotides (see Chapter 33). The bacteria that fix nitrogen live either in the soil or within the roots of certain plants, such as beans and alfalfa. Since the availability of appropriate nitrogen compounds is often the limiting factor in plant growth and crop yield, modern agriculture makes extensive use of chemically synthesized nitrogen fertilizers to supplement bacterial nitrogen fixation. Gene-splicing techniques offer ways to reduce the use of expensive fertilizers by increasing the nitrogen-fixing capability of bacteria. In the future, it may also be possible to design nitrogen-fixing bacteria that can live in the tissues of nitrogen-demanding plants, such as corn and wheat. The ultimate challenge is to engineer crop plants that fix nitrogen themselves.

DNA technology raises important safety and ethical questions

As soon as its potential power became apparent, DNA technology raised questions about possibly dangerous consequences. The earliest concerns were that genetic manipulations of microorganisms could create hazardous new pathogens, which might escape from the laboratory. In response to these concerns, scientists developed a self-monitoring approach—an honor system whereby researchers would adhere to a set of voluntary and self-imposed guidelines to ensure safety. This early approach soon led to formal regulatory programs administered by federal agencies. Today, governments and regulatory agencies throughout the world are grappling with how to promote the potential industrial, medical, and agricultural revolution engendered by biotechnology while ensuring that new products are safe. In the United States, the FDA, the National Institutes of Health Recombinant DNA Advisory Committee, the Department of Agriculture (USDA), and the Environmental Protection Agency (EPA) share responsibility for setting policies and regulating new developments in genetic engineering.

In the past decade, hundreds of genetically designed products and new strains of organisms have been developed. Many have been exhaustively tested, and these, at least, seem to pose little or no threat to humans or the environment. The tests also confirm that genetic engineering holds enormous potential for improving human health and increasing agricultural productivity. Nevertheless, safety remains an important question. The benefits of gene therapy, for example, must be weighed against the need to assure that gene vectors are safe. In the case of environmental problems—oil spills, for example, and chemical wastes that threaten our soil, water, and air—genetically engineered organisms may be part of the solution, but their own impact on the environment must be considered before they are widely used.

With new medical products, the main cause for concern is the potential for harmful side effects, both short-term and long-term. Hundreds of new genetically engineered diagnostic products, vaccines, and drugs, including some designed to treat AIDS and certain forms of cancer, currently await federal approval. Before the FDA considers a new medical product for general marketing, the substance must pass exhaustive tests in laboratory animals and humans.

There is heated debate about genetically engineered agricultural products because of the potential dangers of introducing new organisms into the environment. Some scientists argue that producing transgenic organisms by gene splicing is only an extension of traditional cross-breeding, or hybridization—the procedure that has given us the tangelo (a tangerine-grapefruit hybrid) and beefalo (a cow-buffalo hybrid). Since no hybrid crops or animals were ever tested for safety before they were marketed, it is argued that genetically engineered organisms need not be treated differently. In general, the FDA has held that if the result of genetic engineering is not significantly different from a product already on the market, testing is not required.

On the other side of the argument are scientists who believe that creating transgenic organisms by splicing genes from bacteria or animals into plants, and vice versa, is radically different from hybridizing closely related species of plants or animals. There is a concern that foods produced by gene splicing will contain new proteins that are toxic or that cause severe allergies in some people. There is also concern that genetically engineered crop plants could become "superweeds." For example, plants that have genetically engineered resistance to herbicides or microbial diseases and pest insects could escape into the wild and overgrow native species. They could also colonize croplands where other plant species are grown and become difficult to control there.

Another consideration is that engineered crop plants may pass their new genes to close relatives in neighboring wild areas. Lawn and crop grasses, for example, commonly exchange genes with wild relatives via pollen transfer, and there is little doubt that if some species received new genes, they would pass them to wild plants. If domestic plants pollinate wild ones, the offspring could become superweeds, exceedingly difficult to control because they have genes that confer resistance to herbicides, natural diseases, and insect pests. Researchers are looking for ways to prevent the escape of engineered plant genes. Among the possibilities are crop isolation and the engineering of plants so that they cannot hybridize.

As with all new technologies, developments in DNA technology have ethical overtones. Obtaining a complete map of the human genome, for example, will open the door for significant advances in gene therapy. It will also raise significant ethical questions. Who should have the right to examine someone else's genes? How should that information be used? Should a person's genome be a factor in suitability for a job or eligibility for insurance? Ethical considerations, as well as concerns about potential environmental and health hazards, will likely slow the application of the products of the new biotechnology. There is always a danger that too much regulation will stifle basic research and its potential benefits. However, the power of genetic engineering—our ability to profoundly and rapidly alter species that have been evolving for millennia—demands that we proceed with humility and caution.

REVIEW OF KEY CONCEPTS (with page numbers and key figures)

- DNA technology makes it possible to clone genes for basic research and commercial applications: *an overview* (p. 370, FIGURE 19.1)
  - DNA technology is a powerful set of techniques that enables biologists to manipulate and analyze genetic material.
  - Genetic engineering is the creation of useful new products and organisms using techniques of gene manipulation.
- The toolkit for DNA technology includes restriction enzymes, DNA vectors, and host organisms (pp. 370–372, FIGURE 19.2)
  - A variety of bacterial restriction enzymes recognize short, specific nucleotide sequences in DNA and cut the sequences at specific points on both strands to yield a set of double-stranded DNA fragments with single-stranded sticky ends. The sticky ends readily form base pairs with complementary single-stranded segments on other DNA molecules.
  - Either plasmids or bacteriophages can serve as vectors (carriers) to introduce recombinant DNA molecules into host cells. Recombinant DNA is made by inserting a piece of DNA containing a gene of interest into the plasmid or phage DNA that has been clipped by restriction enzymes. In either case, gene cloning results when the foreign genes replicate inside the host bacterium or other host cell.
  - Although bacteria are the most common host organisms for cloning, DNA can be introduced directly into certain eukaryotic cells as well.
- Recombinant DNA technology provides a means to transplant genes from one species into the genome of another (pp. 372–376, FIGURE 19.3)
  - The two major sources of desired genes for insertion into vectors are genomic libraries that contain all the plasmid-carried or phage-carried DNA segments isolated directly from an organism, and complementary DNA (cDNA) synthesized from mRNA templates. The latter source is especially useful for manipulating and expressing intron-rich eukaryotic DNA.
  - The selection of a desired gene in recombinant DNA can be accomplished using radioactively labeled nucleic acid segments of complementary sequence called probes.
  - In order to achieve expression of a foreign gene in the host organism, the appropriate transcription and translation signals must be engineered into the recombinant molecule.
- Additional methods for analyzing and cloning nucleotide sequences increase the power of DNA technology (pp. 376–382, the chapter's five Methods Boxes)
  - Gel electrophoresis makes it possible to separate and isolate DNA restriction fragments.
  - Determining the nucleotide sequence of a DNA molecule can now be automated, using restriction enzymes and gel electrophoresis as key tools.
  - The polymerase chain reaction (PCR) is a technique for quickly making many copies of DNA in vitro.
  - Hybridization is a general technique for detecting a specific gene or mRNA, by allowing a labeled probe with the complementary sequence to bind to the target molecule.
  - Restriction fragment length polymorphisms (RFLPs) are differences in DNA sequence on homologous chromosomes that result in different patterns of restriction fragment lengths. These patterns are visualized as bands on gel electrophoresis or Southern blots. RFLP analysis has many applications, including genetic mapping for basic research and diagnosing genetic disorders.
- DNA technology is catalyzing progress in many fields of biology (pp. 382–384)
  - In vitro mutagenesis can alter a specific gene and make it possible for researchers to deduce the normal gene's function.
  - Recombinant DNA technology has enabled investigators to answer questions about molecular evolution, probe details of gene organization and control, and produce and catalog proteins of interest.
- The Human Genome Project is an enormous collaborative effort to map and sequence DNA (pp. 384–385)
  - An international research effort, the Human Genome Project involves linkage mapping, physical mapping, and sequencing of the entire human genome.
- DNA technology is reshaping the medical and pharmaceutical industries (pp. 385–388; FIGURES 19.9, 19.10)
  - Medical applications of recombinant DNA technology include the development of diagnostic tests for detecting mutations that cause genetic disease; the design of safer, more effective vaccines; the large-scale production of many new, and some previously scarce, pharmaceutical products; and the ultimate prospect of curing and preventing genetic disorders caused by defective genes.
- DNA technology offers forensic, environmental, and agricultural applications (pp. 388–391; FIGURES 19.11, 19.12)
  - DNA "fingerprints" obtained from RFLP or VNTR (satellite DNA) analysis of blood, hair, or other materials found at the scenes of violent crimes provide evidence in trials.
  - Genetic engineering can be used to modify the metabolism of microorganisms such that they can extract minerals from the environment or degrade waste materials.
  - In agriculture, transgenic plants and animals (those containing genes from other species) are being designed to improve productivity and food quality. Transgenic plants are engineered by regenerating plants following the insertion of recombinant DNA molecules.
- DNA technology raises important safety and ethical questions (p. 392)
  - Several U.S. government agencies are responsible for setting policies and regulating recombinant DNA technology.
  - The potential benefits of genetic engineering must be carefully weighed against the potential hazards of creating products or developing procedures that are harmful to humans or the environment.

SELF-QUIZ

1. Which of the following tools of recombinant DNA technology is *incorrectly* paired with its use?
   a. restriction enzyme—production of RFLPs
   b. DNA ligase—enzyme that cuts DNA, creating the sticky ends of restriction fragments
   c. DNA polymerase—used in a polymerase chain reaction to amplify sections of DNA
   d. reverse transcriptase—production of cDNA from mRNA
   e. electrophoresis—DNA sequencing

2. Which of the following is *not* true of complementary DNA?
   a. It can be amplified by a polymerase chain reaction.
   b. It can be used to create a complete genomic library.
   c. It is produced from mRNA using reverse transcriptase.
   d. It can be used as a probe to locate a gene of interest.
   e. It eliminates the introns of eukaryotic genes and thus is more easily introduced into and cloned by bacterial cells.

3. Plants are more readily manipulated by genetic engineering than are animals because
   a. plant genes do not contain introns
   b. more vectors are available for transferring recombinant DNA into plant cells
   c. a somatic plant cell can give rise to a complete plant
   d. recombinant genes can be inserted into plant cells by microinjection
   e. plant cells have larger nuclei

4. A paleontologist has recovered a bit of tissue from the 400-year-old preserved skin of an extinct dodo (bird). The researcher would like to compare DNA from the sample with DNA from living birds. Which of the following would be most useful for increasing the amount of DNA available for testing?
   a. RFLP analysis
   b. polymerase chain reaction (PCR)
   c. electroporation
   d. gel electrophoresis
   e. Southern hybridization

5. The Human Genome Project includes all of the following goals *except*
   a. the location of RFLP markers
   b. the nucleotide sequencing of the entire human genome
   c. the physical mapping of the chromosomes
   d. the analysis of the genomes of other species
   e. altering the human genome

6. Recombinant DNA technology has many medical applications. Which of the following has *not yet* been attempted or achieved?
   a. production of hormones for treating diabetes and dwarfism
   b. production of subunits of viruses that may serve as vaccines
   c. introduction of genetically engineered genes into human gametes
   d. prenatal identification of genetic disease genes
   e. genetic testing for carriers of harmful alleles

7. RFLP analysis is being used as evidence to link suspects with blood and tissues found at crime scenes. DNA fingerprints look something like bar codes. The pattern of bars in a DNA fingerprint shows
   a. the order of bases in a particular gene
   b. the individual's genotype
   c. the order of genes along particular chromosomes
   d. the presence of dominant or recessive alleles for particular traits
   e. the presence of certain DNA restriction fragments

8. Which of the following sequences along a double-stranded DNA molecule may be recognized as a cutting site for a particular restriction enzyme?
   a. A A G G
   T T C C
   b. A G T C
   T C A G
   c. G G C C
   C C G G
   d. A C C A
   T G G T
   e. A A A A
   T T T T

9. In recombinant DNA methods, the term *vector* refers to
   a. the enzyme that cuts DNA into restriction fragments
   b. the sticky end of a DNA fragment
   c. a RFLP marker
   d. a plasmid used to transfer DNA into a living cell
   e. a DNA probe used to identify a particular gene

10. The template used to make cDNA is
    a. DNA
    b. mRNA
    c. a plasmid
    d. a DNA probe
    e. a restriction fragment

CHALLENGE QUESTIONS

1. A neurophysiologist hopes to study a gene that codes for a neurotransmitter protein in human brain cells. She knows the amino acid sequence of the protein. Briefly explain how she might (a) identify only the genes that are expressed in a specific type of brain cell, (b) identify the gene that codes for the neurotransmitter, (c) produce multiple copies of the gene for study, and (d) produce a quantity of the neurotransmitter for evaluation as a potential medication.

2. Hemophilia A is a hereditary disease characterized by a shortage of clotting factor VIII, a blood protein normally synthesized by liver cells. Without the clotting factor, blood fails to clot normally, and a hemophiliac can bleed to death from a minor injury. Tay-Sachs disease results from a lack of the enzyme hexosaminidase A, which normally breaks down a fatty substance called ganglioside GM2 inside nerve cells. The accumulation of ganglioside GM2 in brain cells causes blindness, seizures, the breakdown of sensory and motor function, and eventually death. Which of these diseases do you think would be easiest to treat with future gene therapy? Why? Describe how this might be done.

SCIENCE, TECHNOLOGY, AND SOCIETY

1. Which uses of DNA technology do you think will prove to be of most value in the coming decades? Why do you think so? What safety and ethical issues do you think are most important? Why? How do you think we should deal with these concerns?

2. Do you think there is a potential in our society for genetic discrimination based on testing for "harmful" genes? What policies can you suggest that would prevent abuses of genetic testing?

3. Under pressure from the biotechnology industry, the U.S. government recently loosened up some regulations affecting DNA technology in order to make the possible introduction of new products and procedures more rapid and less costly. What trade-offs do you see between government regulation of biotechnology and the desire of U.S. industry to remain competitive with other nations in this field?

FURTHER READING

Aldhous, P. "Safer Gene Therapy in Sight for Cystic Fibrosis." *New Scientist,* January 7, 1995. Liposomes, artificial vesicles with lipid membranes, may be safer than viruses as vectors for transferring genes.

Barinaga, M. "Gene Therapy for Clogged Arteries Passes Test in Pig." *Science,* August 5, 1994. A case study in the development of a gene therapy procedure.

Beardsley, T. "Big-Time Biology." *Scientific American,* November 1994. Explores some of the business aspects of molecular biology.

Beardsley, T. "Genes in the Not So Public Domain." *Scientific American,* April 1995. Who owns the base sequences in DNA data bases?

Glick, B. R., and Pasternak, J. J. *Molecular Biotechnology: Principles and Applications of Recombinant DNA.* Washington, DC: ASM Press, 1994. An excellent detailed description of all of the procedures and applications covered in this chapter.

Marshall, E. "A Strategy for Sequencing the Genome 5 Years Early." *Science,* February 10, 1995. Advancing the agenda for the Human Genome Project.

Rennie, J. "Grading the Gene Tests." *Scientific American,* June 1994. Discusses genetic testing and its ethical dilemmas.

Schmidt, K. "Whatever Happened to the Gene Revolution?" *New Scientist,* January 7, 1995. Is DNA technology living up to the hype?

Watzman, H. "DNA to Unravel Secrets of the Dead Sea Scrolls." *New Scientist,* January 14, 1995. DNA technology as a tool in historical research.

UNIT FOUR

MECHANISMS OF EVOLUTION

AN INTERVIEW WITH

JOHN MAYNARD SMITH

My recent trip to southern England was a Darwinian adventure. The itinerary included a pilgrimage to the town of Downe to tour Darwin's home (now a museum) and a visit to the University of Sussex to meet John Maynard Smith, one of the most important evolutionary biologists since Darwin. In fact, the Royal Society of London awarded the Darwin Medal to Dr. Maynard Smith for his research on the evolution of sex. Professor Maynard Smith has also pioneered the application of game theory to evolutionary biology. One of this book's themes is the interdisciplinary character of biology, and John Maynard Smith's career is a grand example. He arrived at biology as an engineer and mathematician, and his research extends biology to the social sciences and humanities.

How and why did you make the transition from aircraft engineer to evolutionary biologist?

I didn't realize when I was at school that being a biologist was something you could do. I didn't learn any science at school at all, formally. I went to read engineering at university and then the Second World War came along. So I finished my engineering degree, and was told to design airplanes. Mind you, I quite enjoyed that. It really wasn't until after the war that I decided I didn't want to spend the rest of my life designing airplanes, and by that time, of course, I was grown up. I had always been interested in biology, so I went back and took a second degree in biology.

You say you were always interested in biology. Tell us about that.

I lived in London until I was eight years old. There is not a lot of natural history you can do in London, but I remember going to the Natural History Museum and the zoo. When I was eight my father died, and I went to live in the country. I started bird watching, beetle collecting—you know, those things that kids do. I have been obsessed by natural history ever since. The real trick was to be able to fit together the two things that I like doing—natural history and mathematics. Fitting them together has been very, very fortunate.

And how did you fit them together? How does your background in math and engineering affect your approach to biology?

I planned my education quite deliberately, with my eyes open. I was twenty-eight when I went back to school after the war. I chose University College, London because I knew that J. B. S. Haldane was there. He was one of the great figures in the application of mathematics to biology, and he was one of the people who really transformed the way we see evolution by showing that you can make a marriage of evolution and genetics. He did so effectively by using very simple-minded mathematics. I sort of learned by copying him, I suppose.

What do you think are the most interesting questions in evolutionary biology today?

At the moment I don't think we are looking for a completely new way of understanding evolution. I think the basic theory is fine, but there are lots of particular problems. The two that I am really interested in are at opposite ends of evolution, oddly enough. One is the origin and evolution of human language, and the other is the evolution of bacteria. Bacterial evolution is practically important right now, and it is also nice to study because bacteria have a short generation time—they get on with it. You don't have to wait for millions of years for something to happen.

One of the problems you work on in bacterial evolution is drug resistance. What are some of the medical implications of this work?

The medical implications are very serious and in some ways tragic. More and more bacteria are becoming resistant to the antibiotics we use to control them. I believe that in the United States, you now have a problem with tuberculosis being drug-resistant. A number of other diseases which until recently were thought very readily controlled by antibiotics are ceasing to be so. This is because the bacteria have evolved a resistance to the drugs that physicians are using, and we can't go on indefinitely finding new drugs that bacteria aren't resistant to. It is a very serious practical problem.

Can you tell us a little bit more about your interest in the origin of language?

It arose because a young Hungarian colleague and I are trying to write a book about what we call "the major transitions." These include all the very big changes, like the origin of life itself, the origin of cells, the origin of chromosomes, the origin of multicellular organisms. The last big transition is the origin of humans. The one thing that really separates us from other animals is our ability to talk. There is no other fundamental difference between us and chimpanzees.

It is a very odd fact that until relatively recently, although humans had become erect and physically like us, and their brains had gotten bigger, and so on, they were incredibly conservative in the tools they used. The same kind of hand ax was made for a quarter of a million years. It is an unbelievable fact, but it's true. Then suddenly, about 50,000 years ago, people became very inventive. They refined fish hooks, needles, painting on the walls, burial of the dead, figurines, and spears. All sorts of new things suddenly started happening. What on Earth happened 50,000 years ago to make this possible? I think most of the people who have thought about this seriously have come to the conclusion that the only answer that makes sense is language—it helps you to think. Just try thinking without words.

Another issue you have thought a lot about is the origin and maintenance of sex. Why is sex such an interesting problem in evolutionary biology?

I think my contribution to the study of the evolution of sex has largely been to ask questions about the points I feel are puzzling rather than to find answers. One can see what's puzzling if one asks, "What is essential about sex?" One talks about sex for reproduction, but in fact sex and reproduction are the precise opposites of one another. Reproduction is a process by which one cell turns into two, and sex is the process whereby two cells turn into one. Now if you are a Darwinist and think that those organisms that multiply most successfully will in some sense outcompete the ones that don't multiply successfully, what on Earth is the point in stopping multiplying and actually reducing the number of individuals by two individuals fusing? It is a very puzzling phenomenon when one looks at it that way. Except for a few parthenogenetic organisms, sex is universal in the sense that all existing eukaryotes are either themselves sexual or descended from something sexual and became asexual later in their evolution. In other words, sexual reproduction seems to be the primitive condition for all eukaryotes.

If not reproduction, then what *does* sex accomplish?

I take the simple-minded view, which is that the great advantage of sex is an advantage to a population rather than to an individual. The advantage is that the population can evolve more. If two favorable mutations occur in two different members of the population and they have sex, those two favorable mutations can come together in a single descendant. If they don't have sex, the mutations can't come together. And when you do calculations, it turns out that sex can have a very considerable effect in accelerating the rate at which populations evolve in response to changing circumstances. I think the data support this view in that very few species of eukaryotes have been asexual for a very long time, in evolutionary terms. Some species may have been asexual for perhaps thousands or maybe even tens of thousands of years, but there probably aren't very many organisms that have been asexual for longer than that. Asexuality seems to work fine for a time and then the world changes, they can't change, and the population goes extinct. I think that is what maintains sex—that's why asexual populations are rare and relatively young.

I don't think the maintenance of sex is as difficult a problem as its origin. I suspect that sex originated just once.

Can you speculate on how sex began?

The most plausible guess that I can make about the origin of sex goes something like this.

The first stage was a population of organisms somewhere that had a haploid/diploid life cycle. Perhaps part of the year they were diploid (with two sets of chromosomes), and then they became haploid (having a single set of chromosomes). There are circumstances in which it is beneficial for an organism to be diploid—notably, if there is a lot of ionizing radiation about. If there are two copies of the chromosome set, an organism can repair DNA damage by comparing one set with the other. Haploid organisms are at a disadvantage because they have only one copy, and if it is damaged, they haven't got another copy to use to tell them how to repair it. However, haploids do have advantages; for example, they can grow faster.

I imagine an organism that had a haploid/diploid life cycle but no sex. When it had to go from being diploid to haploid, it would just undergo what we now call meiosis, which halves the number of chromosomes. When it had to become diploid again, it would just double its chromosomes without the cell dividing. I can then see it evolving into a sexual organism by replacing the process whereby it doubled its chromosomes and became diploid with the fusion of two haploid cells. The advantage is that if two different haploid cells have different mutation damage, when they fuse, they cover up each other's weaknesses. It's what I call the engine and gearbox theory of sex—you combine two clapped-out motor cars and make one functional motor car out of them by taking the engine out of one and the gearbox out of the other. But it's all terribly speculative.

Why are there only two sexes in most species? Why not three or four or ten different sexes?

I think to answer that, one has to ask, "How would you have three sexes?" There are two possible things that might happen. One would be that the new individual is formed from the genetic material from only two of the sexes, and the third sex is some kind of nurse. That's what happens in an ant's nest. The other alternative would be that every new individual has to start from the fusion of genetic material from three kinds of parents: a male, a female, and a zombie if you like. Sex is hard enough. In order to get sex to work, an individual has to produce a gamete with exactly half the chromosome number. Imagine having three sets of chromosomes and producing a gamete with only one of them! Imagine evolving that! There is nothing to be gained by having three sexes that I can see, and it would be incredibly difficult to evolve the cell division that gave you one-third of the genes, and exactly the right third.

Why, in most species, is the male-to-female ratio close to one-to-one?

This is an interesting question. Clearly if you wanted to arrange for the population to have as many offspring as possible, you would have ten, twenty times as many females as males. Each male could then easily fertilize maybe 20 females. But let's say, using humans as an example, that women were allowed to choose the sex of each child with the requirement that they would choose so as to maximize the number of their own grandchildren. What would happen? Clearly, if the other women were mostly choosing to have daughters you should have sons because individuals of whichever is the rarer sex will have more children. If everybody else is choosing to have sons, you would choose daughters. Supposing there are 100 females and only 50 males. On average, each male gets to mate with two females, and he could have twice as many children as each female. The only stable state is when there is an equal number of both boys and girls.

You've also worked on the problem of sexual selection. Tell us about that research.

The idea of sexual selection goes back to Darwin, who said that it would really take two forms. One form would be the members of a single sex, usually males, competing with one another for access to the members of the other sex. Clearly this happens. The form that has puzzled people a lot more is the idea of female choice, which is that females choose to mate with one male instead of another. That has puzzled people because they couldn't understand how a female could choose and what would be in it for her if she did.

Oddly enough, one of the experiments I did in the 1950s answered both these questions for *Drosophila melanogaster,* the common fruit fly in Britain. The females I found mate with healthy males and do not mate with inbred males or old males, sick males or males that can't see very well. It really pays them to do this because they have better offspring if they're choosy. How do they do it? They do it by dancing. What happens is the male sees the female and he comes and faces her. When he is facing her, the female darts quickly to one side and the male has to dart sideways and keep on facing her. There is a very quick dance with the female going from side to side and the male following her. If he succeeds in facing her for a few passages, she stands still, and mating takes place. If he is inbred or aged he falls behind, he can't keep up, and the female just flies away. So the female is selecting which male she mates with, she is doing it by dancing, and it pays off in healthier offspring.

The theory of punctuated equilibrium has captured much attention among evolutionary biologists. What is punctuated equilibrium, and does it pose a problem for the basic Darwinian view of life?

This is a difficult question to answer because I don't think the proponents of punctuated equilibrium are particularly lucid in telling us just what they think. The factual argument, which most of us probably would now accept. is that if you look at fossil records, you do not see continuous gradual change. Instead you see what they call stasis, which is populations changing very little indeed for long periods of time and then changing in really rather sudden transitions. But understand when we talk

about a sudden transition, we are not talking about 10 years or 1000 years. Something that takes 100,000 years is sudden in geological terms. On this scale of time, there is nothing wildly new about the idea of relatively rapid origin of species—you find a lot of this idea in George Gaylord Simpson's work in the 1940s and 1950s. It certainly does not, as I see it, challenge Darwinism. I am not saying it is uninteresting, but it just doesn't make a big difference in the way we see evolution. What *would* make a big difference would be if when those transitions took place, they took place by processes other than by natural selection operating on random variation within populations. It has never been clear to me whether the proponents of punctuated equilibrium actually believe there is some profoundly different process that happens during those punctuational events. Until they answer that question, I have no doubt that, basically, Darwin got it right.

We speak of natural selection and many other wide-ranging sets of ideas in science as "theories." What does *theory* mean in science?

Well, the first thing to say is that all science consists of theory; it's not peculiar to evolutionary biology. The difference between a scientific theory and an idea like a political philosophy or a religious belief is that you can test scientific theories; they have consequences. You can say, "If my theory is correct, then so-and-so," and you can go out and see whether the so-and-so is the case. If it isn't, you've got to forget about your theory. There are plenty of false theories, as well as correct ones. The real point is that theories should be testable.

You wrote that Darwin made it possible for us to see nature simultaneously with the eyes of a child and a philosopher. Can you explain?

Darwin was a supreme naturalist, you understand. Darwin came to this business not originally out of any deep philosophical concerns but simply because he was fascinated by natural history. All through his life he did beautiful work in natural history—on earthworms, on orchids, on coral reefs—you name it, he thought about it. He had, I take it, a child's wonder at nature. To look at nature is astonishing, it really is. If you can look at birds or wildflowers and don't feel a sense of wonder, then something's a bit wrong. And yet you can also think about nature like a scientist, a philosopher, or, in my case, a mathematician: You can reason about nature, and you can deduce things about it. It is that combination I find in my life, and I think Darwin made that possible.

CHAPTER 20

DESCENT WITH MODIFICATION: A DARWINIAN VIEW OF LIFE

KEY CONCEPTS

- Western culture resisted evolutionary views of life
- Theories of geological gradualism helped clear the path for evolutionary biologists
- Lamarck placed fossils in an evolutionary context
- Field research helped Darwin frame his view of life: *science as a process*
- *The Origin of Species* developed two main points: the occurrence of evolution and natural selection as its mechanism
- Evidence from many fields of biology validates the evolutionary view of life
- What is theoretical about the Darwinian view of life?

Biology came of age on November 24, 1859, the day Charles Darwin published On the Origin of Species by Means of Natural Selection. *(The portrait to the right was painted 20 years earlier, when Darwin was 31.) Darwin's book, which presented a convincing case for evolution, connected what had previously seemed a bewildering array of unrelated facts into a cohesive view of life. In biology,* ***evolution*** *refers to the processes that have transformed life on Earth from its earliest forms to the vast diversity that characterizes it today. Darwin addressed the sweeping issues of biology: the great diversity of organisms, their origins and relationships, their similarities and differences, their geographical distribution, and their adaptations to the surrounding environment. Thus, evolution is the most pervasive principle in biology, and a thematic thread woven throughout this book. This unit focuses specifically on the mechanisms by which life evolves.*

Darwin made two points in The Origin of Species. *First, he argued from evidence that species were not specially created in their present forms but had evolved from ancestral species. Second, he proposed a mechanism for evolution, which he termed* ***natural selection.*** *According to Darwin's concept of natural selection, a population of organisms can change over time as a result of individuals with certain heritable traits leaving more offspring than other individuals. Darwin's first point—that evolution occurs—can stand on its own, whether or not natural selection is the cause.*

There is a recurrent theme throughout the chapters of this unit: Evolutionary change is based mainly on the interactions between populations of organisms and their environments. This first chapter defines the Darwinian view of life and traces its historical development.

Western culture resisted evolutionary views of life

To put the Darwinian view in perspective, we must compare it to earlier ideas about Earth and its life. The impact of an intellectual revolution such as Darwinism depends as much on timing as on logic. The timeline in FIGURE 20.1 places Darwin's ideas in a historical context.

The Origin of Species was truly radical; not only did it challenge prevailing scientific views, it also shook the deepest roots of Western culture. Darwin's view of life contrasted sharply with the conventional paradigm of an

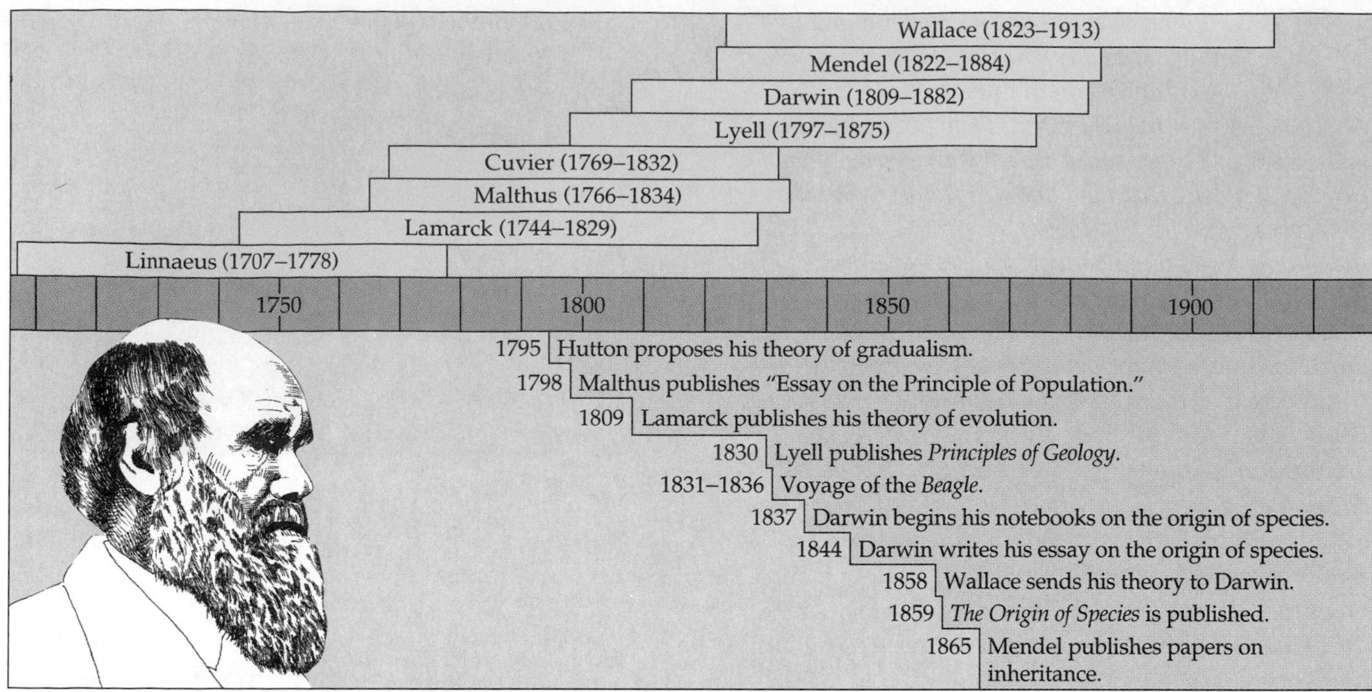

FIGURE 20.1
Darwinism in historical context.

Earth only a few thousand years old, populated by unchanging forms of life that had been individually made during the single week in which the Creator formed the entire universe. Darwin's book subverted a world view that had been taught for centuries.

The Scale of Life and Natural Theology

A number of classical Greek philosophers believed in the gradual evolution of life. But the philosophers who influenced Western culture most, Plato (427–347 B.C.) and his student Aristotle (384–322 B.C.), held opinions that opposed any concept of evolution. Plato believed in two worlds: a real world, ideal and eternal, and an illusory world of imperfection that we perceive through our senses. The variations we see in plant and animal populations were to Plato merely imperfect representatives of ideal forms, or essences, and only the perfect forms were real. Plato's philosophy, known as idealism, or **essentialism,** ruled out evolution, which would be counterproductive in a world where ideal organisms were already perfectly adapted to their environments.

Although Aristotle questioned the Platonic philosophy of dual worlds, his own beliefs also precluded evolution. A careful student of nature, Aristotle recognized that organisms ranged from relatively simple to very complex. He believed that all living forms could be arranged on a scale of increasing complexity, later called the *scala naturae* ("scale of nature"). Each form along this ladder of life had its allotted rung, and every rung was taken. In this view of life, which prevailed for over 2000 years, species are fixed, or permanent, and do not evolve.

Prejudice against evolution was fortified in Judeo-Christian culture by the Old Testament account of creation. The creationist-essentialist dogma that species were individually designed and permanent became firmly embedded in Western thought. There were many evolutionists before Darwin, but none was able to topple the doctrine of fixed species. Even as Darwinism emerged, biology in Europe and America was dominated by **natural theology,** a philosophy dedicated to discovering the Creator's plan by studying nature. Natural theologians saw the adaptations of organisms as evidence that the Creator had designed each and every species for a particular purpose. A major objective of natural theology was to classify species in order to reveal the steps of the scale of life that God had created.

In the eighteenth century, Carolus Linnaeus (1707–1778), a Swedish physician and botanist, sought order in the diversity of life *ad majorem Dei gloriam*—"for the greater glory of God." Linnaeus was the founder of **taxonomy,** the branch of biology concerned with naming and classifying the diverse forms of life. He developed the two-part, or binomial, system of naming organisms according to genus and species that is still used today. In addition, Linnaeus adopted a filing system for grouping species into a hierarchy of increasingly general categories (see

FIGURE 1.9). For example, similar species are grouped in the same genus, similar genera (plural of genus) are grouped in the same family, and so on.

Linnaeus sought and found order in the diversity of life with his hierarchy of taxonomic categories. But clustering certain species under taxonomic banners implied no evolutionary kinship to Linnaeus. As a natural theologian, he believed that species were permanent creations, and he developed his classification scheme only to reveal God's plan. Or, as Linnaeus himself put it, *Deus creavit, Linnaeus disposuit*—"God creates, Linnaeus arranges." Ironically, a century later, the taxonomic system of Linnaeus would become a focal point in Darwin's arguments for evolution.

Cuvier, Fossils, and Catastrophism

Fossils are relics or impressions of organisms from the past, hermetically sealed in rock. Most fossils are found in **sedimentary rocks,** rocks formed from the sand and mud that settle to the bottom of seas, lakes, and marshes. New layers of sediment cover older ones and compress them into rock, such as sandstone and shale. In places where shorelines repeatedly advance and retreat, sedimentary rock will be deposited in many superimposed layers called strata. Later, erosion may scrape or carve through upper (younger) strata and reveal more ancient strata that had been buried (FIGURE 20.2). The fossil record thus displays graphic evidence that Earth has had a succession of floras (plant life) and faunas (animal life).

Paleontology, the study of fossils, was largely developed by Georges Cuvier (1769–1832), the great French anatomist. Realizing that the history of life is recorded in strata containing fossils, he documented the succession of fossil species in the Paris Basin. He noted that each stratum is characterized by a unique suite of fossil species, and the deeper (older) the stratum, the more dissimilar the flora and fauna are from modern life. Cuvier even understood that extinction had been a common occurrence in the history of life. From stratum to stratum, new species appear and others disappear. Yet Cuvier was a staunch and effective opponent to the evolutionists of his day. How did he reconcile the dynamic story told by the fossil record with the concept that species are immutable? Cuvier speculated that the boundaries between the fossil strata corresponded in time to catastrophic events, such as floods or drought, that had destroyed many of the species that had lived at that location at that time. Where there were multiple strata, there had been many catastrophes. This view of Earth's history is known as **catastrophism.**

If Cuvier believed that species were fixed, how did he account for the appearance of species in younger strata that were not present in older rocks? He proposed that the periodic catastrophes that caused mass extinctions were usually confined to local geographical regions. After the extinction of much of the native flora and fauna, the ravaged region would be repopulated by foreign species immigrating from other areas.

Some of Cuvier's followers had more extreme theories of catastrophism. One theory held that the catastrophes were global, and after each holocaust, God created life anew. Although Cuvier himself left religion out of his writing, his aversion to evolution came through loud and clear. But even as Cuvier was winning debates against

FIGURE 20.2
Stratification of sedimentary rock at the Grand Canyon. The Colorado River cut through 2000 m of sedimentary rock and unearthed many strata of varying colors and thicknesses. Each stratum, or layer, represents a particular period in Earth's history and is characterized by a collection of fossils of organisms that lived at that time.

advocates of evolution, a theory of Earth's history that would help pave the way for Darwin was gaining popularity among geologists.

Theories of geological gradualism helped clear the path for evolutionary biologists

Competing with Cuvier's theory of catastrophism was a very different idea of how geological processes had shaped the Earth's crust. In 1795, Scottish geologist James Hutton proposed that it was possible to explain the various land forms by looking at mechanisms currently operating in the world. For example, canyons were cut by rivers running down their lengths, and sedimentary rocks with marine fossils were built of particles that had eroded from the land and been carried by rivers to the sea. Hutton explained the state of the Earth by applying the principle of **gradualism,** which holds that profound change is the cumulative product of slow but continuous processes.

The leading geologist of Darwin's era, Charles Lyell (1797–1875), incorporated Hutton's gradualism into a theory known as **uniformitarianism.** The term refers to Lyell's idea that geological processes are so uniform that their rates and effects must balance out through time. For example, processes that build mountains are eventually balanced by the erosion of mountains. Darwin rejected this extreme version of uniformity in geological processes, but he was strongly influenced by two conclusions that followed directly from the observations of Hutton and Lyell. First, if geological change results from slow, continuous actions rather than sudden events, then Earth must be very old, certainly much older than the 6000 years assigned by many theologians on the basis of biblical inference. Second, very slow and subtle processes persisting over a long period of time can cause substantial change. Darwin was not the first to apply the principle of gradualism to biological evolution, however.

Lamarck placed fossils in an evolutionary context

Toward the end of the eighteenth century, several naturalists suggested that life had evolved along with the evolution of Earth. But only one of Darwin's predecessors developed a comprehensive model that attempted to explain how life evolves: Jean Baptiste Lamarck (1744–1829).

Lamarck published his theory of evolution in 1809, the year Darwin was born. Lamarck was in charge of the invertebrate collection at the Natural History Museum in Paris. By comparing current species to fossil forms, Lamarck could see what appeared to be several lines of descent, each a chronological series of older to younger fossils leading to a modern species.

Where Aristotle saw one ladder of life, Lamarck saw many, and they were more analogous to escalators. On the ground floor were the microscopic organisms, which Lamarck believed were continually generated spontaneously from inanimate material. At the top of the evolutionary escalators were the most complex plants and animals. Evolution was driven by an innate tendency toward greater and greater complexity, which Lamarck seemed to equate with perfection. As organisms attained perfection, they became better and better adapted to their environments. Thus, Lamarck believed that evolution responded to organisms' *sentiments interieurs,* or "felt needs."

Lamarck is remembered most for the mechanism he adopted to explain how specific adaptations evolve. It incorporates two ideas that were popular during Lamarck's era. The first was use and disuse, the idea that those parts of the body used extensively to cope with the environment become larger and stronger, while those that are not used deteriorate. Among the examples Lamarck cited were the blacksmith developing a bigger bicep in the arm that works the hammer and a giraffe stretching its neck to new lengths in pursuit of leaves to eat. The second idea Lamarck adopted was called the inheritance of acquired characteristics. In this concept of heredity, the modifications an organism acquires during its lifetime can be passed along to its offspring. The long neck of the giraffe, Lamarck reasoned, evolved gradually as the cumulative product of a great many generations of ancestors stretching higher and higher. There is, however, no evidence that acquired characteristics can be inherited. Blacksmiths may increase strength and stamina by a lifetime of pounding with a heavy hammer, but these acquired traits do not change genes transmitted by gametes to offspring.

The Lamarckian theory of evolution is ridiculed by some today because of its erroneous assumption that acquired characteristics are inherited; but in Lamarck's times, that concept of inheritance was generally accepted (and, indeed, Darwin could offer no acceptable alternative). To most of Lamarck's contemporaries, however, the mechanism of evolution was an irrelevant issue. In the creationist-essentialist view that still prevailed, species were fixed, and no theory of evolution could be taken seriously. Lamarck was vilified, especially by Cuvier, who would have no part of evolution. In retrospect, Lamarck deserves much credit for his theory, which was visionary in many respects: in its claim that evolution is the best explanation for both the fossil record and the current diversity of life, in its emphasis on the great age of Earth, and in its stress on adaptation to the environment as a primary product of evolution.

Field research helped Darwin frame his view of life: *science as a process*

We have set the scene for the Darwinian revolution. Natural theology, with its view of an ordered world where each living form fit its environment perfectly because it had been specially created, still dominated the intellectual climate as the nineteenth century dawned. A few clouds of doubt about the permanence of species were beginning to gather, but no one could have forecast the thundering storm just over the horizon.

Charles Darwin was born in Shrewsbury, in western England, in 1809. Even as a boy, Darwin's consuming interest in nature was evident. When he was not reading nature books, he was in the fields and forests fishing, hunting, and collecting insects. His father, an eminent physician, could see no future for a naturalist and sent Charles to the University of Edinburgh to study medicine. Only 16 years old at the time, Charles found medical school boring and distasteful, although he managed decent grades. He left Edinburgh without a degree and shortly thereafter enrolled at Christ College at Cambridge University, with the intent of becoming a clergyman. At that time in Great Britain, most naturalists and other scientists belonged to the clergy, and nearly all saw the world in the context of natural theology. Darwin became the protégé of the Reverend John Henslow, professor of botany at Cambridge. Soon after Darwin received his B.A. degree in 1831, Professor Henslow recommended the young graduate to Captain Robert FitzRoy, who was preparing the survey ship *Beagle* for a voyage around the world.

The Voyage of the Beagle

Darwin was 22 years old when he sailed from Great Britain with HMS *Beagle* in December 1831. The primary mission of the voyage was to chart poorly known stretches of the South American coastline (FIGURE 20.3a). While the crew of the ship surveyed the coast, Darwin

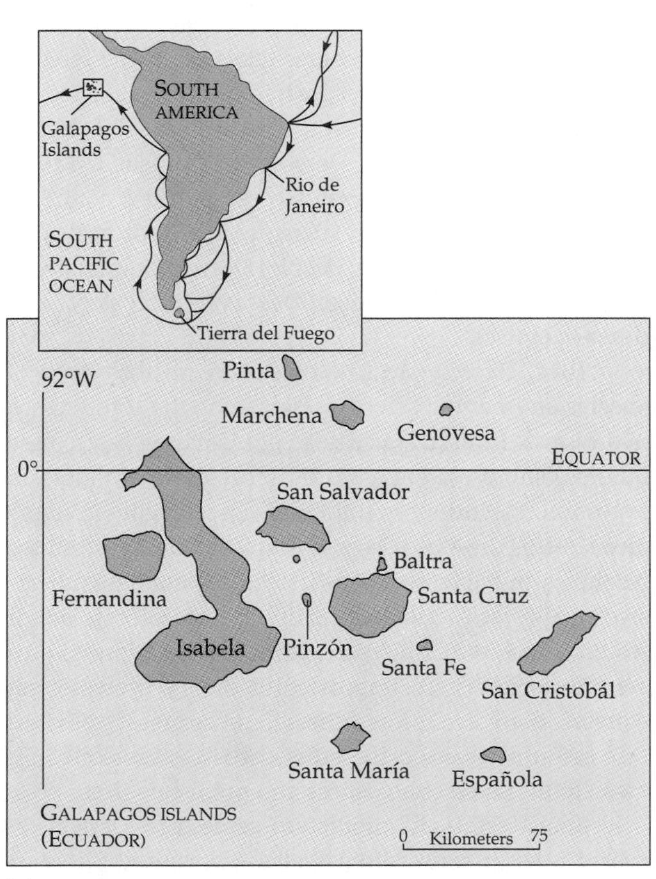

(a)

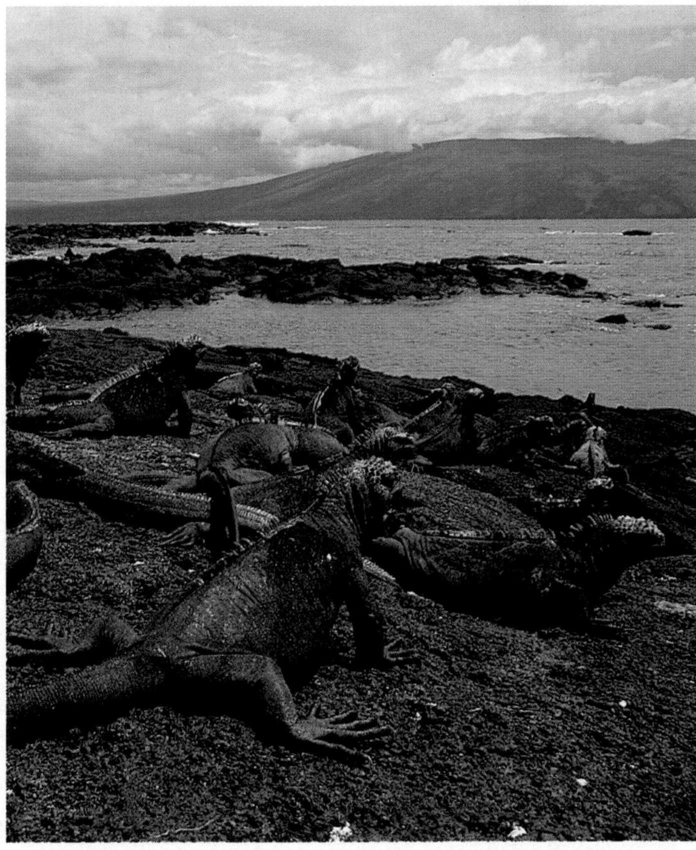

(b)

FIGURE 20.3

The Voyage of the *Beagle*. (**a**) On its round-the-world journey, the *Beagle* spent time at the Galapagos Islands. These islands, located about 900 km off the west coast of Ecuador, are inhabited by many plant and animal species found nowhere else in the world. (**b**) These marine iguanas, on Fernandina Island, are among the relatively young species that evolved from mainland ancestors that colonized the volcanic islands.

spent most of his time on shore, observing and collecting thousands of specimens of the exotic and exceedingly diverse faunas and floras of South America. As the ship worked its way around the continent, Darwin was able to observe the various adaptations of plants and animals that inhabited such diverse environments as the Brazilian jungles, the expansive grasslands of the Argentine pampas, the desolate lands of Tierra del Fuego near Antarctica, and the towering heights of the Andes Mountains.

Their unique adaptations notwithstanding, the fauna and flora of the different regions of the continent all had a definite South American stamp, very distinct from the life forms of Europe. That in itself may not have been surprising. But the plants and animals living in temperate regions of South America were taxonomically closer to species living in tropical regions of that continent than to species in temperate regions of Europe. Furthermore, the South American fossils that Darwin found, though clearly different from modern species, were distinctly South American in their resemblance to the living plants and animals of that continent. Darwin was perplexed by the peculiarities of the geographical distribution of species.

A particularly puzzling case of geographical distribution was the fauna of the Galapagos, islands of relatively recent volcanic origin that lie on the equator about 900 km west of the South American coast (see FIGURE 20.3). Most of the animal species on the Galapagos live nowhere else in the world, although they resemble species living on the South American mainland. It was as though the islands were colonized by plants and animals that strayed from the South American mainland and then diversified on the different islands. Among the birds Darwin collected on the Galapagos were 13 types of finches that, although quite similar, seemed to be different species. Some were unique to individual islands, while other species were distributed on two or more islands that were close together. However, Darwin did not keep careful records of which birds came from which islands, apparently because he did not yet appreciate the full significance of the Galapagos fauna and flora.

By the time the *Beagle* sailed from the Galapagos, Darwin had read Lyell's *Principles of Geology.* Lyell's ideas, together with his experiences on the Galapagos, had Darwin doubting the church's position that the Earth was static and had been created only a few thousand years ago. By acknowledging that Earth was very old and constantly changing, Darwin had taken an important step toward recognizing that life on Earth had also evolved.

Darwin Focuses on Adaptation

At the time Darwin collected the Galapagos finches, he was not sure whether they were actually different species or merely varieties of a single species. Soon after returning to Great Britain in 1836, he learned from ornithologists (bird specialists) that the finches were indeed separate species. He began to reassess all that he had observed during the voyage of the *Beagle,* and in 1837 began the first of several notebooks on the origin of species.

Darwin began to perceive the origin of new species and adaptation as closely related processes. A new species would arise from an ancestral form by the gradual accumulation of adaptations to a different environment. For example, if one species became fragmented into several localized populations isolated in different environments by geographical barriers, the populations would diverge more and more in appearance as each adapted to local conditions. This hypothesis for the origin of species predicted that over many generations, the two populations could become dissimilar enough to be designated separate species. This is apparently what happened to the Galapagos finches. Among the differences between the birds are their beaks, which are adapted to the specific foods available on their home islands (FIGURE 20.4). Darwin anticipated that explaining how such adaptations arise was essential to understanding evolution.

By the early 1840s, Darwin had worked out the major features of his theory of natural selection as the mechanism of evolution. However, he had not yet published his ideas. He was in poor health, and he rarely left home. Despite his reclusiveness, Darwin was not isolated from the scientific community. Already famous as a naturalist because of the letters and specimens he sent to Britain during the voyage of the *Beagle,* Darwin had frequent correspondence and visits from Lyell, Henslow, and other scientists.

In 1844, Darwin wrote a long essay on the origin of species and natural selection. Realizing the significance of this work, he asked his wife to publish the essay in case he died before writing a more thorough dissertation on evolution. Evolutionary thinking was emerging in many areas by this time, but Darwin was reluctant to introduce his theory publicly. Apparently, he understood its subversive quality and quite correctly anticipated the stir it would cause. While he procrastinated, he continued to compile evidence in support of his theory. Lyell, not yet convinced of evolution himself, nevertheless advised Darwin to publish on the subject before someone else came to the same conclusions and published first.

In June 1858, Lyell's prediction came true. Darwin received a letter from Alfred Wallace, a young naturalist working in the East Indies. The letter was accompanied by a manuscript in which Wallace developed a theory of natural selection essentially identical to Darwin's. Wallace asked Darwin to evaluate the paper and forward it to Lyell if it merited publication. Darwin complied, writing to Lyell: "Your words have come true with a vengeance. . . . I

(a)

(b)

(c)

FIGURE 20.4
Galapagos finches. The Galapagos Islands have a total of 13 species of closely related finches, some found on only a single island. The most striking difference among species is in their beaks, which are adapted for specific diets. (**a**) The large ground finch (*Geospiza magnirostris*) has a large beak adapted for cracking seeds. (**b**) The small tree finch (*Camarhynchus parvulus*) uses its beak to grasp insects. (**c**) The woodpecker finch (*Camarhynchus pallidus*) uses a cactus spine or small twig as a tool to probe for termites and other wood-boring insects.

never saw a more striking coincidence . . . so all my originality, whatever it may amount to, will be smashed." That was not to be Darwin's fate, however. Lyell and a colleague presented Wallace's paper, along with extracts from Darwin's unpublished 1844 essay, to the Linnaean Society of London on July 1, 1858. Darwin quickly finished *The Origin of Species* and published it the next year. Although Wallace wrote up his ideas for publication first, Darwin developed and supported the theory of natural selection so much more extensively than Wallace that he is known as its main author. Darwin's notebooks also prove that he formulated his theory of natural selection 15 years before reading Wallace's manuscript.

Within a decade, Darwin's book and its proponents had convinced the majority of biologists that biological diversity was the product of evolution. Darwin succeeded where previous evolutionists had failed, partly because science was beginning to shift away from natural theology, but mainly because he convinced his readers with immaculate logic and an avalanche of evidence in support of evolution.

The Origin of Species developed two main points: the occurrence of evolution and natural selection as its mechanism

Darwinism has a dual meaning. One facet is recognition of evolution as the explanation for life's unity and diversity. The second facet is the Darwinian concept of natural selection as the cause of adaptive evolution. This section highlights these two main points of Darwin's book.

Descent with Modification

In the first edition of *The Origin of Species,* Darwin did not use the word evolution until the last paragraph, referring instead to **descent with modification,** a term that condensed his view of life. Darwin perceived unity in life, with all organisms related through descent from some unknown prototype that lived in the remote past. As the descendants of that inaugural organism spilled into various habitats over millions of years, they accumulated diverse modifications, or adaptations, that fit them to specific ways of life. In the Darwinian view, the history of life is like a tree, with multiple branching and rebranching from a common trunk all the way to the tips of the living twigs, symbolic of the current diversity of organisms. At each fork of the evolutionary tree is an ancestor common to all lines of evolution branching from that fork. Species that are closely related, such as lions and tigers, share many characteristics because their lineage of common descent extends to the smallest branches of the tree of life. Most branches of evolution, even some major ones, are dead ends; about 99% of all species that have ever lived are extinct.

Ironically, Linnaeus, who apparently believed that species are fixed, provided Darwin with a connection to evolution by recognizing that the great diversity of organisms could be ordered into "groups subordinate to groups" (Darwin's phrase). The major taxonomic categories were introduced in Chapter 1, but let's review them here: kingdom > phylum > class > order > family > genus > species.

To Darwin, the natural hierarchy of the Linnaean scheme reflected the branching genealogy of the tree of life, with organisms at the different taxonomic levels related through descent from common ancestors. If we acknowledge that lions and tigers are more closely related than are lions and horses, then we have recognized that evolution has left signs in the form of different degrees of kinship among modern species. Because taxonomy is a human invention, it cannot, by itself, confirm common descent. But taken along with the many other types of evidence, the evolutionary implication of taxonomy is unmistakable. For example, genetic analysis reveals that such species as lions and tigers, thought to be closely related on the basis of anatomical features and other criteria, are indeed "blood" relatives with a common hereditary background.

Natural Selection and Adaptation

Despite the title of his book, Darwin actually devoted little space to the origin of species, concentrating instead on how populations of individual species become better adapted to their local environments through natural selection (FIGURE 20.5).

Evolutionary biologist Ernst Mayr has dissected the logic of Darwin's theory of natural selection into three inferences based on five observations*:

OBSERVATION #1: All species have such great potential fertility that their population size would increase exponentially if all individuals that are born reproduced successfully.

OBSERVATION #2: Most populations are normally stable in size, except for seasonal fluctuations.

OBSERVATION #3: Natural resources are limited.

INFERENCE #1: Production of more individuals than the environment can support leads to a struggle for existence among individuals of a population, with only a fraction of offspring surviving each generation.

OBSERVATION #4: Individuals of a population vary extensively in their characteristics; no two individuals are exactly alike.

OBSERVATION #5: Much of this variation is heritable.

INFERENCE #2: Survival in the struggle for existence is not random, but depends in part on the hereditary constitution of the surviving individuals. Those individuals whose inherited characteristics fit them best to their environment are likely to leave more offspring than less fit individuals.

(a)

(b)

(c)

FIGURE 20.5
Camouflage as an example of evolutionary adaptation shaped by natural selection. Related species of insects called mantids have diverse shapes and colors that evolved in different environments. (**a**) A flower mantid in Malaya. (**b**) A Trinidad tree mantid that mimics dead leaves. (**c**) A Central American mantid that resembles a green leaf.

---

*Adapted from Mayr, E. *The Growth of Biological Thought: Diversity, Evolution and Inheritance.* Cambridge, MA: Harvard University Press, 1982.

INFERENCE #3: This unequal ability of individuals to survive and reproduce will lead to a gradual change in a population, with favorable characteristics accumulating over the generations.

Natural selection is this differential success in reproduction, and its product is adaptation of organisms to their environment. Even if the advantages of some variations over others are slight, the favorable variations will accumulate in the population after many generations of being disproportionately perpetuated by natural selection.

Thus, natural selection occurs through an interaction between the environment and the variability inherent in any population. Although Darwin did not understand the genetic basis of individual variation, we now know that variations arise by the chance mechanisms of mutation and genetic recombination (see Chapter 14). However, natural selection is *not* a chance phenomenon. Environmental factors affect which hereditary traits enhance reproductive success.

A struggle for life is ensured by the excessive production of new individuals. Darwin was already aware of the struggle for existence when he read an influential essay on human population written by the Reverend Thomas Malthus in 1798. Malthus contended that much of human suffering—disease, famine, homelessness, and war—was the inescapable consequence of the potential for the human population to increase faster than food supplies and other resources. Although humans are not the best example, the capacity to overproduce seems to be characteristic of all species. Of the many eggs laid, young born, and seeds spread, only a tiny fraction complete their development and leave offspring of their own. The rest are eaten, frozen, starved, diseased, unmated, or unable to reproduce for some other reason.

Each generation, environmental factors screen heritable variations, favoring some over others. Differential reproduction results in the favored traits being disproportionately represented in the next generation. But can selection actually cause substantial change in a population? Darwin found evidence in **artificial selection,** the breeding of domesticated plants and animals. Humans have modified other species over many generations by selecting individuals with the desired traits as breeding stock. The plants and animals we grow for food bear little resemblance to their wild ancestors (FIGURE 20.6). The

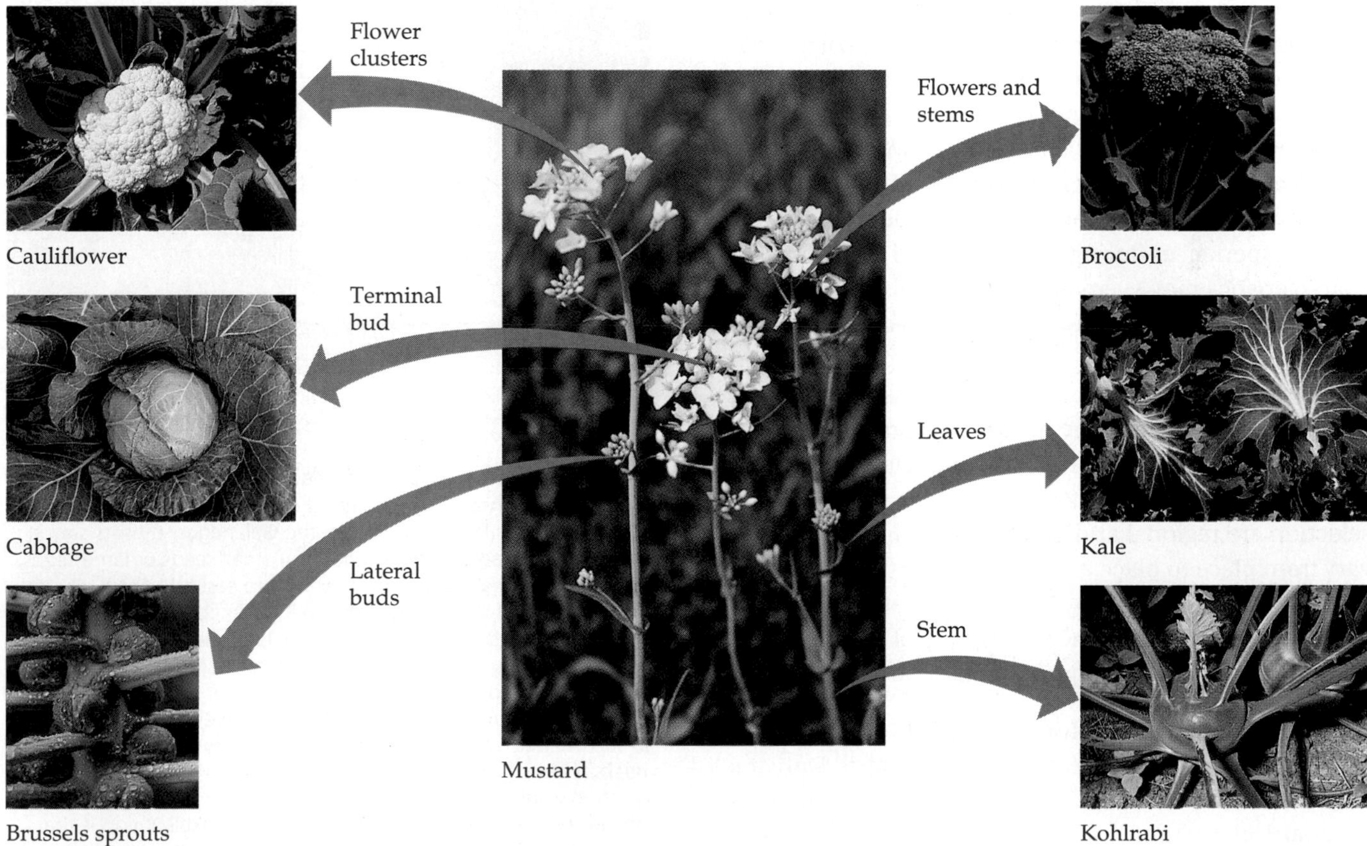

FIGURE 20.6
Artificial selection. These vegetables all have a common ancestor in one species of wild mustard. By selecting different parts of the plant to accentuate, breeders have obtained these divergent results.

power of selective breeding is especially apparent in our pets, which have been bred more for fancy than for utility.

If so much change can be achieved by artificial selection in a relatively short period of time, Darwin reasoned, then natural selection should be capable of considerable modification of species over hundreds or thousands of generations. He postulated that natural selection operating in varying contexts over vast spans of time could account for the entire diversity of life. Darwin envisioned life evolving by a gradual accumulation of minute changes. Gradualism, a concept so important in Lyell's geology, was also incorporated into Darwin's view of evolution.

We can now summarize the two features of the Darwinian view of life: The diverse forms of life have arisen by descent with modification from ancestral species, and the mechanism of modification has been natural selection working continuously over enormous tracts of time.

Some Subtleties of Natural Selection. There are some subtleties of natural selection that require clarification. One is the importance of populations in evolution. For now, we will define a population as a group of interbreeding individuals belonging to a particular species and sharing a common geographic area. A population is the smallest unit that can evolve. Natural selection involves interactions between individual organisms and their environment, but individuals do not evolve. Evolution can be measured only as changes in relative proportions of variations in a population over a succession of generations. Furthermore, natural selection can amplify or diminish only those variations that are heritable. As we have seen, an organism may become modified through its own experiences during its lifetime, and such acquired characteristics may even adapt the organism to its environment, but there is no evidence that acquired characteristics can be inherited. We must distinguish between adaptations an organism acquires by its own actions and inherited adaptations that evolve in a population over many generations as a result of natural selection.

It must also be emphasized that the specifics of natural selection are regional and timely; environmental factors vary from place to place and from time to time. An adaptation in one situation may be useless or even detrimental in different circumstances. A couple of examples will reinforce this situational quality of natural selection.

Natural Selection in Action: Two Examples. One investigation of natural selection in action tests Darwin's hypothesis that the beaks of Galapagos finches are evolutionary adaptations to different food sources. For over twenty years, Peter and Rosemary Grant of Princeton University have been studying the population of medium ground finches (*Geospiza fortis*) on Daphne Major, a tiny islet of the Galapagos. These birds use their strong beaks to crush seeds. They feed preferentially on small seeds, which are produced in abundance by certain plant species during wet years. The finches resort to larger seeds, which are harder to crush, only when small seeds are in short supply. Such shortages occur during dry years, when the plants produce fewer seeds, both small and large. The Grants discovered that the average depth of beaks (the dimension from the top to bottom of the beak) in this finch population changes over the years (FIGURE 20.7). During droughts, beak depth increases, only to decrease again during wet periods. This is a heritable trait. The Grants attribute the change to the relative availability of small seeds from year to year. Birds with stronger beaks may have an advantage during dry periods, when survival and reproduction depend on cracking large seeds. In contrast, a smaller beak is apparently more efficient equipment for feeding on the small seeds that are abundant during wet periods.

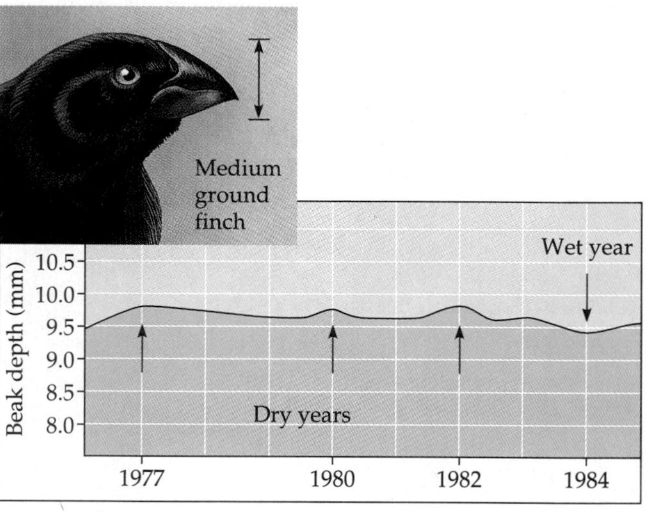

FIGURE 20.7

Natural selection in action: beak evolution in one of Darwin's finches. The medium ground finch (*Geospiza fortis*), one of the birds Darwin found on the Galapagos, uses its strong beak to crush seeds. Given a choice of small seeds or large seeds, the birds eat mostly small ones, which are easier to crush. During wet years, small seeds are produced in such abundance that ground finches consume relatively few large seeds. This changes during dry years, when both small and large seeds are in short supply and the birds resort to eating a greater than usual proportion of large seeds. The change in diet is correlated with a change in the average depth (top-to-bottom dimension) of the birds' beaks. Field studies comparing offspring to parents confirm that this trait is inherited rather than acquired (by exercising the beak on large seeds, for example). The most likely explanation is that those birds that happen to have stronger beaks have a feeding advantage during droughts and pass the genes for this trait on to their offspring.

The Grants' studies of beak evolution reinforce the point that natural selection is situational: What works best in one environmental context may be less suitable in some other situation. It is also important to understand that beak evolution on Daphne Major does not result from the inheritance of acquired characteristics. The environment did not *create* beaks specialized for larger or smaller seeds, depending on annual rainfall. The environment only acted upon the inherited variations manifest in any population, favoring the survival and reproductive success of some individuals over others. Natural selection edits populations. The proportion of thicker-beaked finches increased during dry periods because, on average, those individuals with thicker beaks transmitted their genes to more offspring than did thinner-beaked birds.

In another investigation of natural selection in action, Michael Singer and Camille Parmesan of the University of Texas have documented rapid evolutionary adaptation in a population of a butterfly species called Edith's checkerspot (*Edith editha*) living in a meadow near Carson City, Nevada. Female butterflies lay their eggs on plants, which provide food for the larvae after they hatch. Until recently, the mothers deposited their eggs mainly on a native plant named *Collinsia torreyi*, using tasting cells on their feet to identify the plant (FIGURE 20.8a). For example, in 1983, checkerspots laid about 80% of their eggs on *Collinsia*. During the next decade, a weed named *Plantago lanceolata* invaded the meadow by spreading from surrounding cattle ranches. By 1993—after just 10 years—the butterflies were laying about 70% of their eggs on the invading plant instead of *Collinsia* (FIGURE 20.8b). The switch in plant preference is genetic: The researchers demonstrated in the lab that daughters of butterflies that deposited their eggs on *Plantago* inherited the taste for that plant, choosing it over *Collinsia* when they laid *their* eggs. In only a decade, reproductive behavior in the checkerspot population was apparently adapting to changing vegetation.

Among the other examples of natural selection in action presented in earlier chapters are the evolution of antibiotic resistance in bacteria (Chapter 17) and change in the body size of guppies exposed to different predators (Chapter 1). Researchers have published more than 100 other accounts of natural selection in the wild. And hundreds of other studies have documented natural selection in laboratory populations of such organisms as *Drosophila*.

Thus, scientists do not accept natural selection solely on the face of its logic but because it has been confirmed repeatedly by the hypothetico-deductive approach, in which predictions based on hypotheses are tested by observation and experimentation (see Chapter 1). Ironically,

(a)

(b)

FIGURE 20.8

Natural selection in action: a change in taste by a Western butterfly. (**a**) A decade ago, females of this population of checkerspots (*Edith editha*) deposited about 80% of their eggs on *Collinsia torreyi*, a common native plant in this Nevada meadow. The larva in the inset is feeding on the plant. (**b**) After the weed *Plantago lanceolata* invaded the meadow, it took only seven generations for a "preferred taste" for that plant to evolve in the checkerspot population. The butterflies now lay about 70% of their eggs on *Plantago*.

Darwin himself thought that natural selection always operated too slowly to be observed. He was also unable to satisfactorily account for the genetics of variation (a problem we will address in Chapter 21). For these and other reasons, the theory of natural selection as the mechanism of evolution won relatively few advocates during Darwin's time. In contrast, within just a few years after Darwin published *The Origin of Species,* most biologists were convinced that evolution does, in fact, occur—whatever the mechanism.

Evidence from many fields of biology validates the evolutionary view of life

Evolution leaves observable signs. Such data are the clues to the past, essential to any historical science. For example, historians of human civilization can study written records from earlier times, but they can also piece together the evolution of societies by recognizing vestiges of the past in modern cultures. Even if we did not know from written documents and archaeological evidence that Spaniards colonized the Americas, we would deduce this from the Hispanic stamp on much of Latin America. Similarly, biological evolution has left marks—in the fossil record and in the historical vestiges evident in modern life. In this section, we briefly survey some of these signs of evolution (which will be discussed more thoroughly in Chapter 23). Darwin documented evolution mainly with evidence from the geographical distribution of species and from the fossil record, but we will not limit ourselves to these two categories of evidence. As biology progresses, new discoveries, including the revelations of molecular biology, continue to validate the evolutionary view of life.

Biogeography

It was the geographical distribution of species—**biogeography**—that first suggested evolution to Darwin. Islands have many species of plants and animals that are endemic (native, found nowhere else) but closely related to species of the nearest mainland or neighboring island. Some revealing questions arise. Why are two islands with similar environments in different parts of the world *not* populated by closely related species but by species taxonomically affiliated with the plants and animals of the nearest mainland, where the environment is often quite different? Why are the tropical animals of South America more closely related to species of South American deserts than to species of the African tropics? Why is Australia home to a great diversity of pouched mammals (marsupials) but relatively few placental mammals (those in which embryonic development is completed in the uterus)? It is not because Australia is inhospitable to placental mammals; in recent years, humans have introduced rabbits to Australia, and the rabbit population has exploded. The unique Australian fauna evolved on that island continent in isolation from places where the ancestors of placental mammals lived.

Although such biogeographical patterns are incongruous if one imagines that species were individually placed in suitable environments, they make sense in the historical context of evolution. The Darwinian interpretation is that we find modern species where they are because they evolved from ancestors that inhabited those regions. Consider armadillos, the armored mammals that live only in the Americas. The evolutionary view of biogeography predicts that contemporary armadillos are modified descendants of earlier species that occupied these continents, and the fossil record confirms that such ancestors existed. This example brings us to the more general importance of the fossil record as a chronicle of evolution.

The Fossil Record

The succession of fossil forms is compatible with what is known from other types of evidence about the major branches of descent in the tree of life. For instance, evidence from biochemistry, molecular biology, and cell biology places prokaryotes as the ancestors of all life, and predicts that bacteria should precede all eukaryotic life in the fossil record. Indeed, the oldest known fossils are prokaryotes. Another example is the chronological appearance of the different classes of vertebrate animals in the fossil record. Fossil fishes predate all other vertebrates, with amphibians next, followed by reptiles, then mammals and birds. This sequence is consistent with the history of vertebrate descent as revealed by many other types of evidence. In contrast, the idea that all species were individually created at about the same time predicts that all vertebrate classes would make their first appearance in the fossil record in rocks of the same age, a prediction at odds with what one actually observes.

The Darwinian view of life also predicts that evolutionary transitions should leave signs in the fossil record. Paleontologists have discovered many transitional forms that link even older fossils to modern species. For example, a series of fossils documents the changes in skull shape and size that occurred as mammals evolved from reptiles. Every year, paleontologists turn up other important links between contemporary forms and their ancestors. In the past few years, for instance, researchers have found fossilized whales that link these aquatic mammals to their terrestrial predecessors (FIGURE 20.9).

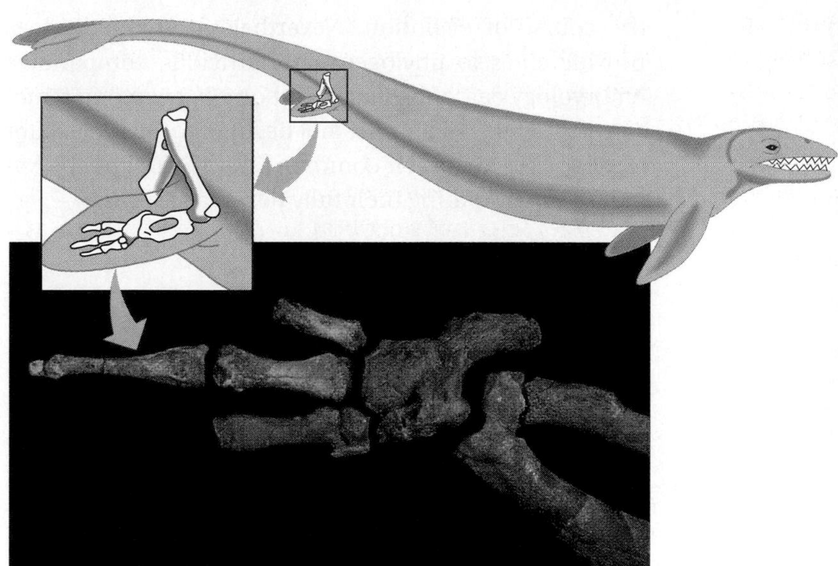

FIGURE 20.9
Transitional fossils linking past and present. Whales evolved from terrestrial ancestors, an evolutionary transition that left many signs, including fossil evidence. Paleontologists digging in Egypt and Pakistan have recently identified extinct whales that had hind limbs. Shown here are the fossilized leg bones of *Basilosaurus,* one of those ancient whales. These whales were already aquatic animals that no longer used their legs to support their weight and walk. The leg bones of an even older fossilized whale named *Ambulocetus* are heftier. *Ambulocetus* may have been amphibian, dividing its time between land and water.

Comparative Anatomy

Descent with modification is evident in anatomical similarities between species grouped in the same taxonomic category. For example, the same skeletal elements make up the forelimbs of humans, cats, whales, bats, and all other mammals, although these appendages have very different functions (FIGURE 20.10). Surely, the best way to construct the infrastructure of a bat's wing is not also the best way to build a whale's flipper. Such anatomical peculiarities make no sense if the structures are uniquely engineered and unrelated. A more likely explanation, especially in view of corroborative evidence, is that the basic similarity of these forelimbs is the consequence of the descent of all mammals from a common ancestor. The forelegs, wings, flippers, and arms of different mammals are variations on a common anatomical theme that has been modified for divergent functions.

Similarity in characteristics resulting from common ancestry is known as **homology,** and such anatomical signs of evolution are called **homologous structures.** Comparative anatomy is consistent with all other evidence in testifying that evolution is a remodeling process in which ancestral structures that functioned in one capacity become modified as they take on new functions.

The oddest homologous structures are **vestigial organs,** rudimentary structures of marginal, if any, use to the organism. Vestigial organs are historical remnants of structures that had important functions in ancestors but are no longer essential. For instance, the skeletons of some snakes retain vestiges of the pelvis and leg bones of walking ancestors.*

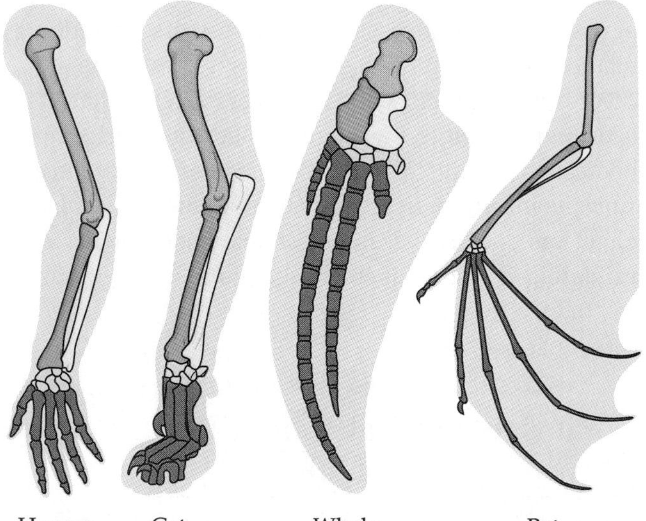

FIGURE 20.10
Homologous structures: anatomical signs of evolution. The forelimbs of all mammals are constructed from the same skeletal elements, an architectural relationship we would expect if a common ancestral forelimb became modified for many different functions.

Comparative Embryology

Closely related organisms go through similar stages in their embryonic development. For example, all vertebrate embryos go through a stage in which they have gill

*Vestigial organs may seem to support the Lamarckian concept of use and disuse, but they can be explained by natural selection. It would be wasteful to continue providing blood, nutrients, and space to an organ that no longer has a major function. Individuals with reduced versions of those organs would be favored, and natural selection operating over thousands of generations would tend to phase out obsolete structures.

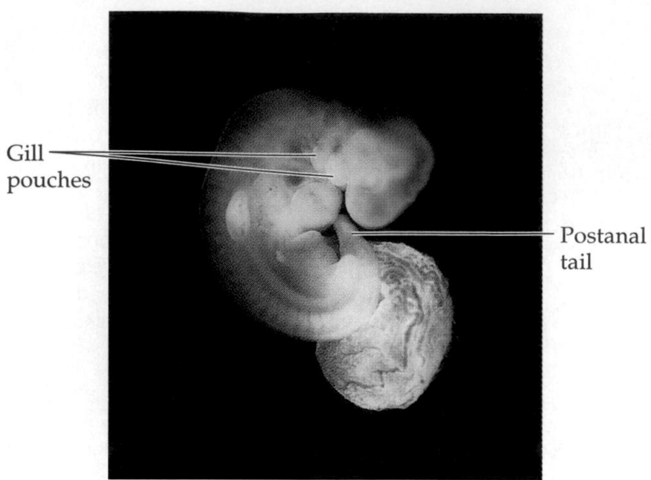

FIGURE 20.11

Evolutionary signs from comparative embryology. At this early stage of development, the kinship of vertebrates is unmistakable. This 4-week-old human embryo has gill pouches and a postanal tail, two of the trademarks of *all* vertebrate embryos. Comparative embryology helps biologists identify anatomical homology that is less apparent in adults because the structures are extensively modified in different ways during later development of the organisms.

pouches on the sides of their throats (FIGURE 20.11). Indeed, at this stage of development, similarities between fishes, frogs, snakes, birds, humans, and all other vertebrates are much more apparent than differences. As development progresses, the various vertebrates diverge more and more, taking on the distinctive characteristics of their classes. In fish, for example, the gill pouches develop into gills; in terrestrial vertebrates, these embryonic structures become modified for other functions, such as the Eustachian tubes that connect the middle ear with the throat in humans.

Inspired by the Darwinian principle of descent with modification, many embryologists in the late nineteenth century proposed the extreme view that "ontogeny recapitulates phylogeny." This notion holds that the development of an individual organism, **ontogeny,** is a replay of the evolutionary history of the species, **phylogeny.** The theory of recapitulation is an overstatement. What recapitulation does occur is a replay of embryonic stages, not a sequence of adultlike stages of ever more advanced vertebrates. Although vertebrates share many features of embryonic development, it is not as though a mammal first goes through a "fish stage," then an "amphibian stage," and so on. Also, because embryonic processes ultimately affect the functioning of the adult organism, they are subject to natural selection. Thus, even relatively early stages of development may become modified over the course of evolution. Nevertheless, ontogeny does provide clues to phylogeny. In particular, comparative embryology can often establish homology among structures, such as gill pouches, that become so altered in later development that their common origin would not be apparent by comparing their fully developed forms.

Molecular Biology

Molecular signs of evolution were discussed in Chapter 5. The main point was that evolutionary relationships among species are reflected in their DNA and proteins—in their genes and gene products. If two species have libraries of genes and proteins with sequences of monomers that match closely, the sequences must have been copied from a common ancestor. If two long paragraphs were identical except for the substitution of a letter here and there, we would surely attribute them both to a single source.

Darwin's boldest speculation—that *all* forms of life are related to some extent through branching descent from the earliest organisms—has also been substantiated by molecular biology. Even taxonomically remote organisms, such as humans and bacteria, have some proteins in common. An example is cytochrome *c*, a respiratory protein found in all aerobic species (see Chapter 9). Mutations have substituted amino acids at some places in the protein during the long course of evolution, but the cytochrome *c* molecules of all species are clearly akin in structure and function. Similarly, a comparison of the number of amino acid differences in hemoglobin in several vertebrates corroborates evidence from paleontology and comparative anatomy about the evolutionary relationships between these species (FIGURE 20.12).

A common genetic code is further evidence that all life is related. Evidently, the language of the genetic code has been passed along through all branches of evolution ever since its inception in an early form of life. Molecular biology has thus added the latest chapter to the evidence validating evolution as the basis for the unity and diversity of life.

What is theoretical about the Darwinian view of life?

Some people dismiss Darwinism as "just a theory." This tactic for nullifying the evolutionary view of life has two flaws. First, it fails to separate Darwin's two claims: that modern species evolved from ancestral forms, and that natural selection is the main mechanism for this evolution. The conclusion that life has evolved is based on his-

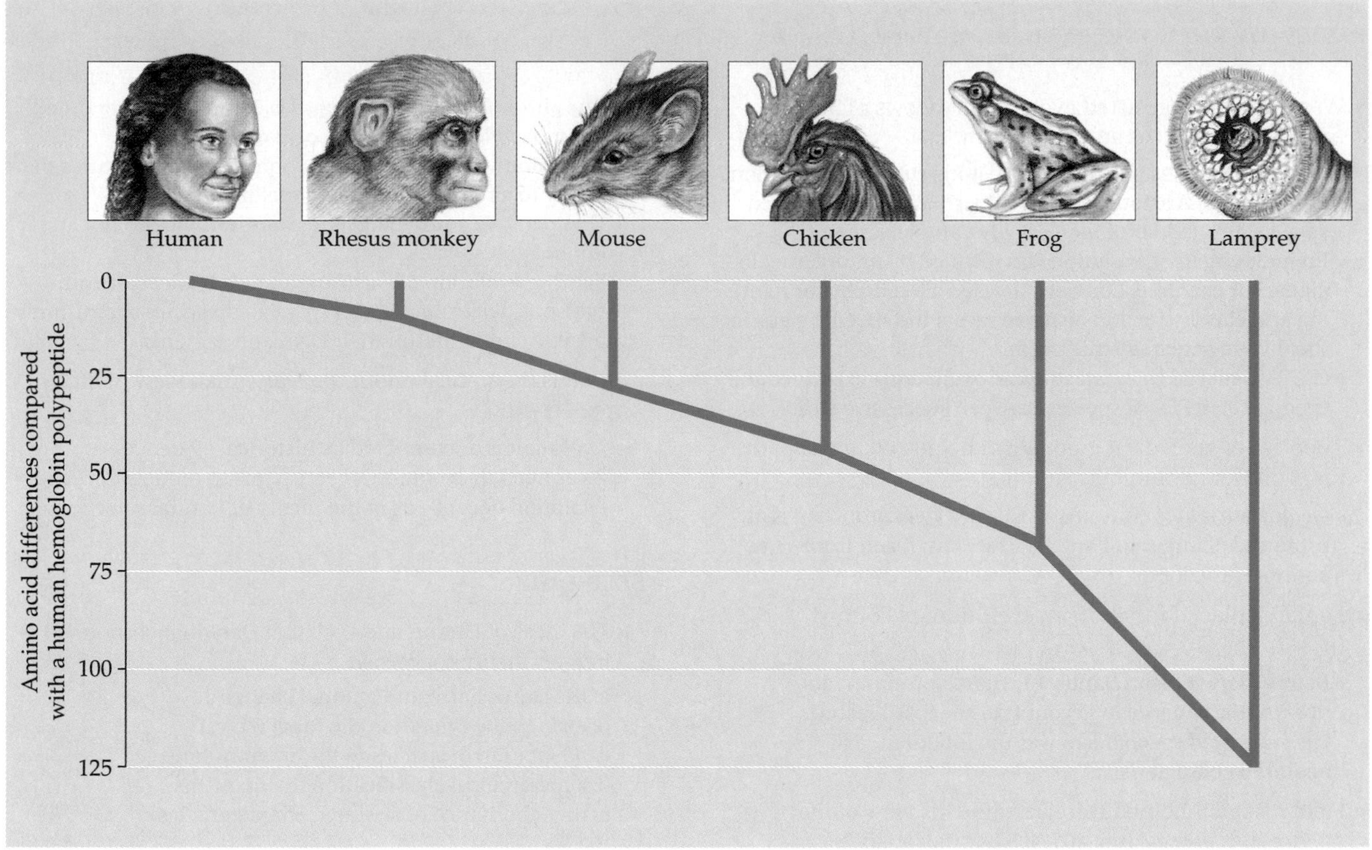

FIGURE 20.12

Molecular data and the evolutionary relationships of vertebrates. Branch points in this vertebrate tree are defined by the number of amino acid differences in a polypeptide of hemoglobin, a protein that functions in oxygen transport. The molecular data are compatible with vertebrate relationships deduced from the fossil record and comparative anatomy.

torical evidence—the signs of evolution discussed in the previous section.

What, then, is theoretical about evolution? Theories are our attempts to explain facts and integrate them with overarching concepts. To biologists, "Darwin's theory of evolution" is natural selection—the mechanism Darwin proposed to explain the historical facts of evolution documented by fossils, biogeography, and other types of evidence.

So the "just a theory" argument concerns Darwin's second claim, his theory of natural selection. This brings us to the second flaw in the "just a theory" case. The term *theory* has a very different meaning in science compared to our general use of the word. The colloquial use of "theory" comes close to what scientists mean by a "hypothesis." In science, a theory is more comprehensive than a hypothesis. A theory, such as Newton's theory of gravitation or Darwin's theory of natural selection, accounts for many facts and attempts to explain a great variety of phenomena. Such a unifying theory does not become widely accepted in science unless its predictions stand up to thorough and continuous testing by experiments and observations. Even then, good scientists do not allow theories to become dogma. For example, many evolutionary biologists now question whether natural selection alone accounts for the evolutionary history observed in the fossil record. The study of evolution is more robust and lively than ever, and we will evaluate some of the current debates in the next three chapters. But these questions about *how* life evolves in no way imply that most biologists consider evolution itself to be "just a theory." Debates about evolutionary theory are like arguments over competing theories about gravity; we know that objects keep right on falling while we debate the cause.

Darwin gave biology a sound scientific basis by attributing the diversity of life to natural causes rather than supernatural creation. Nevertheless, the diverse products of evolution are elegant and inspiring. As Darwin said in the closing paragraph of *The Origin of Species,* "There is grandeur in this view of life."

REVIEW OF KEY CONCEPTS (with page numbers and key figures)

- Western culture resisted evolutionary views of life (pp. 399–402, FIGURE 20.1)
  - The philosophies of Plato and Aristotle ruled out evolution. In particular, Aristotle envisioned a *scala naturae* of fixed species on a ladder of increasingly complex organisms.
  - Prejudice against evolution was fortified by natural theologians. For example, Linnaeus devised a hierarchy for naming and classifying organisms to reveal the specific steps in the divinely created scale of life.
  - Cuvier believed that catastrophic extinctions explained the unique sets of fossil species between successive strata.
- Theories of geological gradualism helped clear the path for evolutionary biologists (p. 402)
  - Geologists James Hutton and Charles Lyell proposed that profound changes in Earth's surface can result from slow, continuous actions.
- Lamarck placed fossils in an evolutionary context (p. 402)
  - Before Darwin, Jean Baptiste Lamarck proposed an evolutionary theory to account for the fossil record. His suggested mechanism was the inheritance of acquired characteristics.
- Field research helped Darwin frame his view of life: *science as a process* (pp. 403–405, FIGURE 20.3)
  - Darwin's view of life apparently began to change when he served as naturalist on HMS *Beagle*. Darwin was impressed by the peculiar geographical distribution and distinctive interrelationships among species, including those of the Galapagos Islands. This experience eventually led him to the idea that new species originate from ancestral forms by the gradual accumulation of adaptations.
  - The long-delayed publication of Darwin's *On the Origin of Species by Means of Natural Selection* in 1859 was catalyzed by Alfred Wallace, who independently arrived at the theory of natural selection.
- *The Origin of Species* developed two main points: the occurrence of evolution and natural selection as its mechanism (pp. 405–410)
  - The Darwinian view of life can be dissected into two separate claims: the evolution of new species by descent with modification from ancestral species, and natural selection as the mechanism of this evolution.
  - Natural selection is based on differential success in reproduction, made possible because of variation among the individuals of any population and the tendency for a population to produce more offspring than the environment can support. On average, the individuals best adapted to the local environment leave the most offspring and thereby pass on their adaptive characteristics.
- Evidence from many fields of biology validates the evolutionary view of life (pp. 410–412, FIGURES 20.9–20.12)
  - The biogeography of species first suggested common descent to Darwin. For instance, he noticed that island species were more closely related to those on the mainland than to those on distant islands with similar environments.
  - The chronological fossil record is compatible with other lines of evidence in support of evolution.
  - Comparative anatomy reveals homologous structures that testify to an evolutionary remodeling process.
  - Comparative embryology uncovers homologies not apparent in adult species.
  - Signs of evolution are also apparent in molecular comparisons. For example, closely related species show unmistakable similarities in their DNA and proteins.
- What is theoretical about the Darwinian view of life? (pp. 412–413)
  - Evolution is documented by historical evidence.
  - As applied to evolution, *theory* refers to models for *how* evolution occurs—as in the theory of natural selection.

SELF-QUIZ

1. The ideas of Hutton and Lyell that Darwin incorporated into his theory concerned
   a. the age of Earth and gradual change
   b. extinctions evident in the fossil record
   c. adaptation of species to the environment
   d. a hierarchical classification of organisms
   e. the inheritance of acquired characteristics

2. Which of the following is *not* an observation or inference upon which natural selection is based?
   a. There is heritable variation among individuals.
   b. Poorly adapted individuals never leave offspring.
   c. Since only a fraction of offspring survive, there is a struggle for limited resources.
   d. Individuals whose inherited characteristics best fit them to the environment will leave more offspring.
   e. Unequal reproductive success leads to adaptations.

3. The gill pouches of mammal and bird embryos are
   a. vestigial structures
   b. support for "ontogeny recapitulates phylogeny"
   c. homologous structures
   d. used by the embryos to breathe
   e. evidence for the degeneration of unused body parts

4. The best evidence for a common origin of *all* life is
   a. comparative anatomy
   b. comparative embryology
   c. biogeography
   d. molecular biology
   e. the fossil record

5. Darwin's theory, as presented in *The Origin of Species,* mainly concerned
   a. how new species arise
   b. the origin of life
   c. how adaptations evolve
   d. how extinctions occur
   e. the genetics of evolution

6. In science, the term *theory* generally applies to an idea that
   a. is a speculation lacking supportive observations or experiments
   b. attempts to explain many related phenomena
   c. is synonymous with what biologists mean by a hypothesis
   d. is so widely accepted that it is considered a fact
   e. cannot be tested

7. Which person is *incorrectly* matched with a term or idea?
   a. Plato—essentialism
   b. Linnaeus—acquired characteristics
   c. Malthus—overpopulation
   d. Lyell—uniformitarianism
   e. Aristotle—*scala naturae*

8. The smallest biological unit that can evolve over time is
   a. a particular cell
   b. an individual organism
   c. a population
   d. a species
   e. an ecosystem

9. Which of the following ideas is common to both Darwin's and Lamarck's theories of evolution?
   a. Adaptation results from differential reproductive success.
   b. Evolution drives organisms to greater and greater complexity.
   c. Evolutionary adaptation results from interactions between organisms and their environments.
   d. Adaptation results from the use and disuse of anatomical structures.
   e. The fossil record supports the view that species are fixed.

10. Which pair of structures is least likely to represent homology?
    a. the wings of a bat and the forelimbs of a human
    b. the cytochrome *c* protein of a bacterium and cytochrome *c* of a cat
    c. the mitochondria of a plant and those of an animal
    d. the bark of a tree and the protective covering of a lobster
    e. the brain of a frog and the brain of a dog

CHALLENGE QUESTIONS

1. Some opponents of evolution have exclaimed, "I just can't believe we came from a chimpanzee!" What is their misconception about evolution?

2. An orange grower discovered that most of his trees were infested with destructive mites. He sprayed the trees with insecticide, which killed 99% of the mites. Five weeks later, most of the trees were infested again, so he sprayed again, using the same quantity of the same insecticide. This time, only about half the mites were killed. Explain why the spray did not work as well the second time.

3. There are two main groups of bats: Smaller "microbats" navigate by using sonar, and larger "megabats" rely on vision. Mammalogists once thought that both kinds of bats evolved from insectivorous mammals. But similarities between the visual systems of megabats and primates have led some researchers to think that megabats may have evolved from primates, perhaps lemurs. Explain how molecular biology could help resolve the question of bat ancestry. What results would support the hypothesis that the two groups of bats have a common origin? Separate origins?

SCIENCE, TECHNOLOGY, AND SOCIETY

1. To what extent are humans in a technological society exempt from natural selection? Explain your answer.

2. Is the concept of natural selection relevant in a political or economic context? In other words, if a particular nation or corporation achieves success or dominance, does this mean that it is more fit than its competitors and that unregulated dominance is justified? Why or why not?

FURTHER READING

Darwin, C. *On the Origin of Species by Means of Natural Selection, or The Preservation of Favored Races in the Struggle for Life.* New York: New American Library, 1963. A modern printing of the historical book that revolutionized biology.

Desmond, A., and J. Moore. *Darwin.* New York: Warner, 1992. A recent biography.

Futuyma, D. J. "The Uses of Evolutionary Biology." *Science,* January 6, 1995. Harnessing the special adaptations of plants and other organisms to help solve environmental problems.

Gould, S. J. "The First Unmasking of Nature." *Natural History,* April 1993. How Linnaeus and his taxonomy help make Darwinism possible.

Gould, S. J. "Hooking Leviathan by Its Past." *Natural History,* May 1994. Newly discovered whale fossils with legs are making a big splash in paleontology.

Landman, O. E. "Inheritance of Acquired Characteristics Revisited." *BioScience,* November 1993. *Are* there examples of Lamarckian evolution?

Weiner, J. "Evolution Made Visible." *Science,* January 6, 1995. Observing evolution in action.

CHAPTER 21

THE EVOLUTION OF POPULATIONS

KEY CONCEPTS

- The modern evolutionary synthesis integrated Darwinism and Mendelism: *science as a process*
- A population has a genetic structure defined by its gene pool's allele and genotype frequencies
- The Hardy-Weinberg theorem describes a nonevolving population
- Microevolution is a generation-to-generation change in a population's allele or genotype frequencies: *an overview*
- Genetic drift can cause evolution via chance fluctuation in a small population's gene pool: *a closer look*
- Gene flow can cause evolution by transferring alleles between populations: *a closer look*
- Mutations can cause evolution by substituting one allele for another in a gene pool: *a closer look*
- Nonrandom mating can cause evolution by shifting the frequencies of genotypes in a gene pool: *a closer look*
- Natural selection can cause evolution via differential reproductive success among varying members of a population: *a closer look*
- Genetic variation is the substrate for natural selection
- Natural selection is the mechanism of adaptive evolution
- Does evolution fashion perfect organisms?

One obstacle to understanding evolution is the common misconception that individual organisms evolve, in the Darwinian sense, during their lifetimes. In fact, natural selection does *act on individuals; their characteristics affect their chances of survival and reproductive success. But the evolutionary impact of this natural selection is only apparent in tracking how a* population *of organisms changes over time. Consider, for example, the African swallowtail butterflies (*Papilio dardanus*) in the photograph on this page. These female specimens were all collected from one population; their different patterns of coloration represent genetic variations within that population. If birds prey preferentially on* Papilio *butterflies having a particular coloration, then the proportion of individuals with that coloration will decline from one generation to the next because such butterflies will produce fewer offspring. Thus, it is the population, not its individuals, that evolves, as some heritable variations become more common at the expense of others.* Figure 21.1 *illustrates another example. Evolution on the smallest scale, or microevolution, can be defined as a change in the genetic makeup of a population. We will refine this definition as we examine in more detail how natural selection and other mechanisms cause populations to evolve. We begin by tracing how biologists finally began to understand Darwin's theory of natural selection during the first half of this century.*

The modern evolutionary synthesis integrated Darwinism and Mendelism: *science as a process*

The Origin of Species convinced most biologists that species are products of evolution, but Darwin was not nearly as successful in gaining acceptance for natural selection as the mechanism of evolution. Natural selection requires hereditary processes that Darwin could not explain. His theory was based on what seems like a paradox of inheritance: Like begets like—but not exactly. What was missing in Darwin's book was an understanding of inheritance that could explain how chance variations arise in a population, while also accounting for the precise transmission of these variations from parents to offspring. Although Gregor Mendel and Charles Darwin were contemporaries, Mendel's discoveries were unappreciated at the time, and apparently no one noticed that he had elucidated the very principles of inheritance that could have resolved Darwin's paradox and given credibility to natural selection.

FIGURE 21.1

Individuals are selected, but populations evolve. The bent grass (*Agrostis tenuis*) in the foreground is growing on the tailings of an abandoned mine in Wales. These plants tolerate a concentration of heavy metals that is toxic to other plants of the same species growing just meters away, in the pasture on the other side of the fence. Each year, many seeds land on the mine tailings, but most are unable to grow successfully there. The only plants that germinate, grow, and reproduce are those that inherited genes making it possible to tolerate metallic soil. Thus, this adaptation does not evolve by individual plants becoming more metal-tolerant during their lifetimes. We can only observe the adaptive evolution as an increasing proportion of tolerant individuals from one generation to the next. Natural selection works by favoring the survival and reproductive success of certain individuals over others among the varying members of a population. But the impact of this individual selection is a generation-to-generation change in the prevalence of certain traits on the level of the whole population.

Ironically, when Mendel's research article was rediscovered and reassessed at the beginning of this century, many geneticists believed that the laws of inheritance were at odds with Darwin's theory of natural selection. As the raw material for natural selection, Darwin emphasized characters that vary along a continuum in a population, such as the fur length of mammals or the speed with which an animal can flee from a predator. We know today that such quantitative characters are influenced by multiple genetic loci. (To review polygenic inheritance and quantitative characters, see Chapter 13.) But Mendel, and later the geneticists of the early twentieth century, recognized only discrete "either-or" traits, such as purple or white flowers in pea plants, as heritable. Thus, there seemed to be no genetic basis for natural selection to work on the more subtle variations within a population that were central to Darwin's theory.

During the 1920s, genetic research focused on mutations. As an alternative to Darwin's theory of natural selection, a widely accepted hypothesis held that evolution occurred in rapid leaps as a result of radical changes in phenotype caused by mutations. This idea contrasted sharply with Darwin's view of gradual evolution due to environmental selection acting on continuous (quantitative) variations among individuals of a population. There was also considerable sentiment for *orthogenesis,* the idea that evolution has been a predictable progression to more and more elite forms of life. This notion of goal-oriented evolution, a throwback to Lamarck's "felt needs," opposed Darwin's mechanistic view that evolution simply reflected differential reproductive success extrapolated over many generations.

An important turning point for evolutionary theory was the birth of **population genetics,** which emphasizes the extensive genetic variation within populations and recognizes the importance of quantitative characters. With progress in population genetics in the 1930s, Mendelism and Darwinism were reconciled, and the genetic basis of variation and natural selection was worked out.

A comprehensive theory of evolution that became known as the **modern synthesis** was forged in the early 1940s. It is called a synthesis because it integrated discoveries and ideas from many different fields, including paleontology, taxonomy, biogeography, and, of course, population genetics. Among the architects of the modern synthesis were geneticist Theodosius Dobzhansky, biogeographer and taxonomist Ernst Mayr, paleontologist George Gaylord Simpson, and botanist G. Ledyard Stebbins. The modern synthesis emphasizes the importance of populations as the units of evolution, the central role of natural selection as the most important mechanism of evolution, and the idea of gradualism to explain how large changes can evolve as an accumulation of small changes occurring over long periods of time. No scientific paradigm is likely to endure without modification for half a century. In the next two chapters, you will learn that many evolutionary biologists are now challenging some of the claims and assumptions of the modern synthesis, and we will evaluate the need for a new evolutionary synthesis at the end of this unit. Still, twentieth-century biology has been profoundly affected by the modern synthesis, which shaped most of the ideas about how populations evolve that are introduced in this chapter.

A population has a genetic structure defined by its gene pool's allele and genotype frequencies

A **population** is a localized group of individuals belonging to the same species. For now, we will define a **species** as a group of populations that have the potential to interbreed in nature (this definition will be examined more

FIGURE 21.2

Population distribution. Populations are localized groups of individuals belonging to the same species. (**a**) Here, two dense populations of Douglas fir (*Pseudotsuga menziesii*) are separated by a river bottom where firs are uncommon. The two populations are not totally isolated; interbreeding occurs when wind blows pollen between the populations. Nevertheless, trees are more likely to interbreed with members of the same population than with trees on the other side of the river. (**b**) Humans also tend to concentrate in localized populations. In this nighttime satellite view of the United States, the lights of major population centers, or cities, are visible. These populations are, of course, not isolated; people move around, and there are low-density suburban and rural communities between cities. But city dwellers are most likely to choose mates who live in the same city, often in the same neighborhood.

(a)

(b)

critically in Chapter 22). Each species has a geographical range within which individuals are not spread out evenly, but are usually concentrated in several localized populations. A population may be isolated from others of the same species, exchanging genetic material only rarely. Such isolation is particularly common for populations confined to widely separated islands, unconnected lakes, or mountain ranges separated by lowlands. However, populations are not always isolated, nor do they necessarily have sharp boundaries. One dense population center may blur into another in an intermediate region where members of the species occur but are less numerous. Although the populations are not isolated, individuals are still concentrated in centers and are more likely to interbreed with members of the same population than with members of other populations. Therefore, individuals near a population center are, on average, more closely related to one another than to members of other populations (FIGURE 21.2).

The total aggregate of genes in a population at any one time is called the population's **gene pool.** It consists of all alleles at all gene loci in all individuals of the population. For a diploid species, each locus is represented twice in the genome of an individual, who may be either homozygous or heterozygous for those homologous loci. If all members of a population are homozygous for the same allele, that allele is said to be *fixed* in the gene pool. More often, there are two or more alleles for a gene, each having a relative frequency (proportion) in the gene pool.

An example will make the concept of allele frequency in a gene pool less abstract. Imagine a wildflower population with two varieties contrasting in flower color. An allele for pink flowers, which we will symbolize by *A*, is completely dominant to an allele for white flowers, symbolized by *a*. For our simplified situation, these are the only two alleles for this locus in the population. Our imaginary population has 500 plants. Twenty have white flowers because they are homozygous for the recessive allele; their genotype is *aa*. Of the 480 plants with pink flowers, 320 are homozygous (*AA*) and 160 are heterozygous (*Aa*). Since these are diploid organisms, there are a total of 1000 copies of genes for flower color in the population. The dominant allele accounts for 800 of these genes ($320 \times 2 = 640$ for *AA* plants, plus $160 \times 1 = 160$ for *Aa* individuals). Thus, the frequency of the *A* allele in the gene pool of this population is 80%, or 0.8. And since there are only two allelic forms of the gene, the *a* allele must have a frequency of 20%, or 0.2. Related to these allele frequencies are the frequencies of genotypes. In our model wildflower population, these frequencies are: $AA = 0.64$ (320 out of 500 plants), $Aa = 0.32$ (160/500), and $aa = 0.04$ (20/500). Population geneticists use the term **genetic structure** to refer to a population's frequencies of alleles and genotypes.

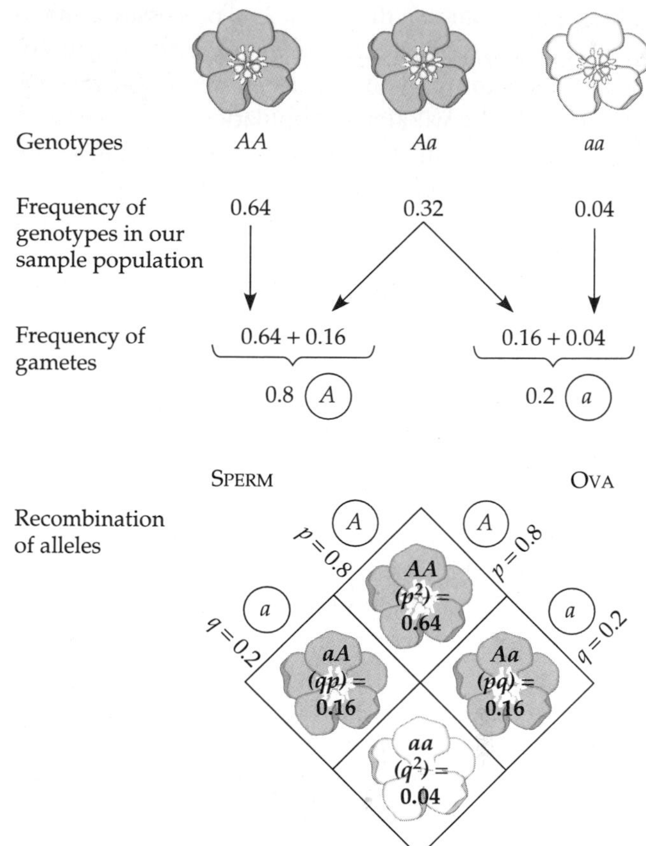

FIGURE 21.3
The Hardy-Weinberg theorem. The genetic structure of a nonevolving population remains constant over the generations. Sexual recombination alone will not alter the relative frequencies of alleles or genotypes. (p = frequency of *A*; q = frequency of *a*)

The Hardy-Weinberg theorem describes a nonevolving population

Before we consider the mechanisms that cause a population to evolve, it will be helpful to examine, for comparison, the genetics of a nonevolving population. Such a gene pool in evolutionary stasis is described by the **Hardy-Weinberg theorem,** named for the two scientists who derived the principle independently in 1908. It states that the frequencies of alleles and genotypes in a population's gene pool remain constant over the generations unless acted upon by agents other than sexual recombination. Put another way, the sexual shuffling of alleles due to meiosis and random fertilization has no effect on the overall genetic structure of a population.

To apply the Hardy-Weinberg theorem, let's return to our imaginary wildflower population of 500 plants (FIGURE 21.3). Recall that 80% (0.8) of the flower-color loci in the gene pool have the *A* allele and 20% (0.2) have the *a* allele. How will genetic recombination during sexual reproduction affect the frequencies of the two alleles in the next generation of our wildflower population? We will assume that the union of sperm and ova in the population is completely random; that is, all male-female mating combinations are equally likely. The situation is analogous to mixing all gametes in a sack and then drawing them randomly, two at a time, to determine the genotype for each zygote. Each gamete has one allele for flower color, and the *A* and *a* alleles will occur in the same frequencies in which they had occurred in the population that made the gametes. Every time a gamete is drawn from the pool at random, the chance that the gamete will bear an *A* allele is 0.8, and the chance that an *a* allele will be present is 0.2.

Using the rule of multiplication (see Chapter 13), we can calculate the frequencies of the three possible genotypes in the next generation of the population. The probability of picking two *A* alleles from the pool of gametes is 0.64 (0.8×0.8). Thus, about 64% of the plants in the next generation will have the genotype *AA*. The frequency of *aa* individuals will be about 4%, or 0.04 (0.2×0.2). And 32%, or 0.32, of the plants will be heterozygous—that is, *Aa* or *aA*, depending on whether it is the sperm or ovum that supplies the dominant allele ($0.8 \times 0.2 = 0.16$, $\times$ 2 ways $= 0.32$).

Notice in FIGURE 21.3 that the sexual processes of meiosis and random fertilization have maintained the same allele and genotype frequencies that existed in the previous generation of the wildflower population. For the flower-color locus, the population's genetic structure is in a state of equilibrium—referred to as Hardy-Weinberg equilibrium. (In this example, the wildflower population was at equilibrium initially. If we had started with the population not yet at equilibrium, only a single generation would be required for equilibrium to be attained. You will have a chance to prove this in Challenge Question #2 at the end of the chapter.)

From the specific case of the wildflower population, we can derive a general formula, called the Hardy-Weinberg equation, for calculating the frequencies of alleles and genotypes in populations at equilibrium. We will restrict our analysis to the simplest case of only two alleles, one dominant over the other. However, the Hardy-Weinberg equation can be adapted to situations in which there are three or more alleles for a particular locus, and there is no clear-cut dominance.

For a gene locus where only two alleles occur in a population, population geneticists use the letter p to represent the frequency of one allele and the letter q to represent the frequency of the other allele (see FIGURE 21.3). In the imaginary wildflower population, $p = 0.8$ and $q = 0.2$. Note that $p + q = 1$; the combined frequencies of all possible alleles must account for 100% of the genes for that locus in the population. If there are only two alleles and we know the frequency of one, the frequency of the other can be calculated:

$$\text{If } p + q = 1, \text{ then } 1 - p = q, \text{ or } 1 - q = p$$

When gametes combine their alleles to form zygotes, the probability of generating an *AA* genotype is p^2 (an application of the rule of multiplication). In the wildflower population, $p = 0.8$, and $p^2 = 0.64$, the probability of an *A* sperm fertilizing an *A* ovum to produce an *AA* zygote. The frequency of individuals homozygous for the other allele (*aa*) is q^2, or $0.2 \times 0.2 = 0.04$ for the wildflower population. There are two ways in which an *Aa* genotype can arise, depending on which parent contributes the dominant allele. Therefore, the frequency of heterozygous individuals in the population is $2pq$ ($2 \times 0.8 \times 0.2 = 0.32$, in our example). If we have calculated the frequencies of all possible genotypes correctly, they should add up to 1:

$$\underset{\text{Frequency of } AA}{p^2} + \underset{\text{Frequency of } Aa \text{ plus } aA}{2pq} + \underset{\text{Frequency of } aa}{q^2} = 1$$

For our wildflowers, this is $0.64 + 0.32 + 0.04 = 1$.*

*Recalling your algebra courses, you may recognize the Hardy-Weinberg equation as a binomial expansion: in this case, $(p + q)^2 = p^2 + 2pq + q^2 = 1$.

The Hardy-Weinberg equation enables us to calculate frequencies of alleles in a gene pool if we know frequencies of genotypes, and vice versa. One application is to calculate the percentage of the human population that carries the allele for a particular inherited disease. For instance, one out of approximately 10,000 babies in the United States is born with phenylketonuria (PKU), a metabolic disorder that, untreated, results in mental retardation and other problems. The disease is caused by a recessive allele, and thus the frequency of individuals in the U.S. population born with PKU corresponds to q^2 in the Hardy-Weinberg equation. Given one PKU occurrence per 10,000 births, $q^2 = 0.0001$. Therefore, the frequency of the recessive allele for PKU in the population is $q = \sqrt{0.0001}$, or 0.01. And the frequency of the dominant allele is $p = 1 - q$, or 0.99. The frequency of carriers, heterozygous people who do not have PKU but may pass the PKU allele on to offspring, is

$$2pq = 2 \times 0.99 \times 0.01 = 0.0198$$

About 2% of the U.S. population carries the PKU allele.

Microevolution is a generation-to-generation change in a population's allele or genotype frequencies: *an overview*

If the Hardy-Weinberg theorem describes a gene pool in equilibrium—that is, a *nonevolving* population—then how is it relevant to our study of evolution? The concept of Hardy-Weinberg equilibrium tells us what to expect if a population is *not* evolving. The equilibrium values for allele and genotype frequencies we calculate from the Hardy-Weinberg equation provide a baseline for tracking the genetic structure of a population over a succession of generations. If the frequencies of alleles or genotypes deviate from values expected from Hardy-Weinberg equilibrium, then the population is evolving. We can now refine our definition of evolution at the population level: *Evolution is a generation-to-generation change in a population's frequencies of alleles or genotypes—a change in a population's genetic structure.* Because such change in a gene pool is evolution on the smallest scale, it is referred to more specifically as **microevolution.**

To meet our criterion for microevolution, it is enough that the frequencies of alleles are changing for a single genetic locus. If we track allele and genotype frequencies in a population over a succession of generations, some loci may be at equilibrium, while frequencies of alleles at other loci may be changing. Such a population is evolving. For example, our imaginary wildflower population is evolving if the frequencies of the pink-flower and white-flower alleles are changing, even if Hardy-Weinberg equilibrium is maintained for all other genetic loci.

If we think of microevolution as a departure from Hardy-Weinberg equilibrium, then it is important to understand the five conditions that are required for Hardy-Weinberg equilibrium to be maintained in a population:

1. *Very large population size.* In a small population, genetic drift, which is chance fluctuation in the gene pool, can alter the frequencies of alleles.
2. *Isolation from other populations.* Gene flow, the transfer of alleles between populations due to the movement of individuals or gametes, can change gene pools.
3. *No net mutations.* By changing one allele into another, mutations alter the gene pool.
4. *Random mating.* If individuals select mates having certain heritable traits, then the random mixing of gametes required for Hardy-Weinberg equilibrium does not occur.
5. *No natural selection.* Differential survival and differential reproductive success alter a gene pool by favoring the transmission of some alleles at the expense of others.

These five mandatory conditions for maintaining Hardy-Weinberg equilibrium provide a framework for understanding the processes that cause microevolution. There are five potential agents of microevolution: genetic drift, gene flow, mutation, nonrandom mating, and natural selection (TABLE 21.1). Each is a departure from one of the five conditions for Hardy-Weinberg equilibrium. Of all the causes of microevolution, only natural selection generally adapts a population to its environment. The other agents of microevolution are sometimes called non-Darwinian because of their usually nonadaptive nature.* Let's take a closer look at the five causes of microevolution.

TABLE 21.1

Causes of Microevolution: An Overview

MECHANISM	ACTION ON GENE POOL	USUALLY ADAPTIVE?*
Genetic drift	Random change in small gene pool due to sampling errors in propagation of alleles.	No
Gene flow	Change in gene pools due to migration of individuals between populations.	No
Mutation	Change in allele frequencies due to net mutation.	No
Nonrandom mating	Inbreeding or selection of mates for specific phenotypes (assortative mating) reduces frequency of heterozygous individuals.	Unknown
Natural selection	Differential reproductive success increases frequencies of some alleles and diminishes others.	Yes

Genetic drift can cause evolution via chance fluctuation in a small population's gene pool: *a closer look*

Flip a coin a thousand times, and a result of 700 heads and 300 tails would make you very suspicious about that coin. Flip a coin ten times, and an outcome of seven heads and three tails is within reason. The smaller a sample, the greater the chance deviations from an idealized result—an equal number of heads and tails, in the case of a sample of coin tosses. This disproportion of results in a small sample is known as sampling error, and it is an important factor in the genetics of small populations of organisms. If a new generation draws its alleles at random, then the larger the sample size, the better it will represent the gene pool of the previous generation. If a population of organisms is small, its existing gene pool may not be accurately represented in the next generation because of sampling error. In the small wildflower population of FIGURE 21.4, for example, the frequencies of the alleles for pink (*A*) and white (*a*) flowers fluctuate over several generations. Only a fraction of the plants in the population manage to leave offspring, and over successive generations, genetic variation has been reduced. This is a case of microevolution caused by **genetic drift,** changes in the gene pool of a small population due to chance. Only luck could result in random drift improving the population's adaptiveness to its environment.

Ideally, a population must be infinitely large for genetic drift to be ruled out completely as an agent of evolution. Although that is impossible, many populations are so large that drift may be negligible. However, some populations are small enough for significant genetic drift to occur; chance certainly plays a major role in the microevolution of populations having fewer than 100 or so individuals. Two situations that can lead to populations small enough for genetic drift to occur are known as the bottleneck effect and the founder effect.

*"Nonadaptive" does not mean "maladaptive." Microevolution due to causes other than natural selection may affect populations in positive, negative, or neutral ways. In contrast, the effects of natural selection are almost always positive because selection favors the disproportionate propagation of favorable traits.

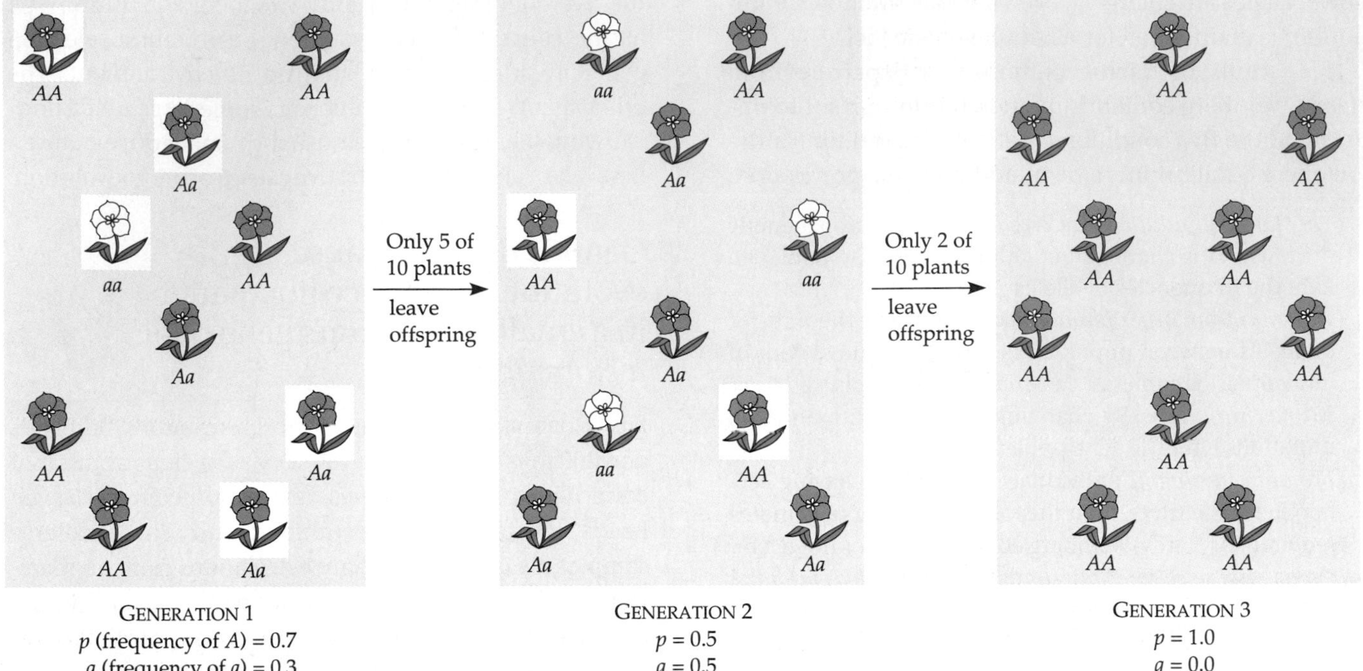

FIGURE 21.4
Genetic drift. This small wildflower population has a stable size of about ten plants. For generation 1, only the five boxed plants produce seeds that germinate and give rise to offspring that survive to flowering. Only two of the plants of generation 2 manage to leave fertile offspring. The random change in the frequencies of alleles and genotypes due to sampling error in a small population is called genetic drift. Eventually, genetic drift usually reduces genetic variability by fixing alleles (as is the case for the *A* allele in generation 3 of this imaginary wildflower population).

The Bottleneck Effect

Disasters such as earthquakes, floods, and fires may reduce the size of a population drastically, killing victims rather unselectively. The result is that the small surviving population is unlikely to be representative of the original population in its genetic makeup—a situation known as the **bottleneck effect.** By chance, certain alleles will be overrepresented among survivors, other alleles will be underrepresented, and some alleles may be eliminated completely (FIGURE 21.5). Genetic drift may continue to affect the population for many generations, until the population is again large enough for sampling errors to be insignificant.

Bottlenecking and the genetic drift that follows usually reduce the overall genetic variability in a population because alleles for at least some loci are likely to be lost from the gene pool. One example concerns the population of northern elephant seals, which passed through a bottleneck in the 1890s when hunters reduced it to about 20 individuals. Since then, the animal has become a protected species, and the population has grown to over 30,000. Researchers have examined 24 gene loci in many individuals of the northern elephant seal population, and no genetic variation has been found; a single allele has been fixed at each of the 24 loci, probably due in large part to genetic drift. By comparison, genetic variation abounds

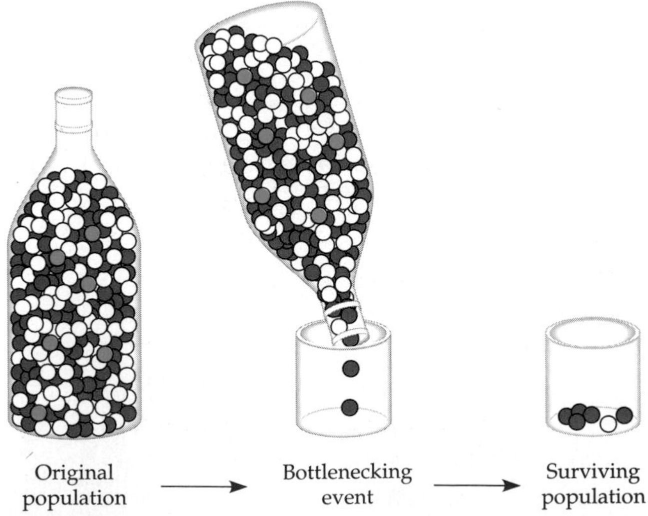

FIGURE 21.5
The bottleneck effect. A gene pool can begin to drift by chance after the population is drastically reduced by a disaster that kills unselectively. In this analogy, a population of magenta, green, and white marbles in a bottle is reduced in the bottleneck. Notice that the composition of the surviving population emerging from the bottleneck is not representative of the makeup of the larger, original population. By chance, the magenta marbles are overrepresented in the new population, and green marbles are absent.

in populations of the southern elephant seal, which have not been bottlenecked. Bottlenecking may also explain why the South African population of cheetahs displays less genetic variation than do inbred strains of laboratory mice. The cheetah population was probably severely reduced during the last ice age about 10,000 years ago, and a second time when the animals were hunted to near extinction at the beginning of this century.

The Founder Effect

Genetic drift is also likely whenever a few individuals colonize an isolated island, lake, or some other new habitat. The smaller the sample size, the less the genetic makeup of the colonists will represent the gene pool of the larger population they left. The most extreme case would be the founding of a new population by one pregnant animal or a single plant seed. If the colony is successful, random drift will continue to affect the frequency of alleles in the gene pool until the population is large enough for sampling errors from generation to generation to be minimal. Genetic drift in a new colony is known as the **founder effect.** The effect undoubtedly contributed to the evolutionary divergence of Darwin's finches after strays from the South American mainland reached the remote Galapagos Islands.

The founder effect is probably responsible for the relatively high frequency of certain inherited disorders among human populations established by a small number of colonists. In 1814, fifteen people founded a British colony on Tristan da Cunha, a group of small islands in the Atlantic Ocean midway between Africa and South America. Apparently, one of the colonists carried a recessive allele for retinitis pigmentosa, a progressive form of blindness that afflicts homozygous individuals. Of the 240 descendants who still lived on the island in the late 1960s, 4 had retinitis pigmentosa, and at least 9 others were known to be carriers, based on pedigree analysis. The frequency of this allele is much higher on Tristan da Cunha than in the populations from which the founders came. Although inherited diseases provide striking examples of the founder effect, this cause of genetic drift also alters the frequencies of many alleles in the gene pool that affect more subtle characteristics.

Gene flow can cause evolution by transferring alleles between populations: *a closer look*

Hardy-Weinberg equilibrium requires the gene pool to be a closed system, but most populations are not completely isolated. A population may gain or lose alleles by **gene flow,** genetic exchange due to the migration of fertile individuals or gametes between populations. Perhaps, for example, a population neighboring our hypothetical wildflower population consists entirely of white-flowered individuals. A wind storm may blow pollen to our wildflowers from this population with a different genetic structure, and allele frequencies may change.

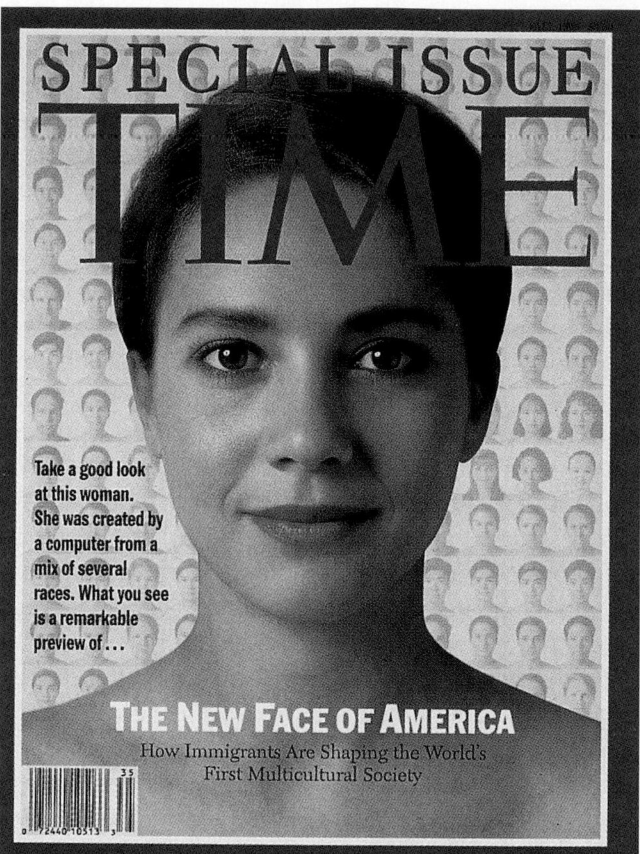

FIGURE 21.6
Gene flow and human evolution. The migration of people from one part of the world to another is transferring alleles between populations that were once more isolated.

Gene flow tends to reduce differences between populations that have accumulated because of natural selection or genetic drift. If it is extensive enough, gene flow can eventually amalgamate neighboring populations into a single population with a common genetic structure. As humans began to move about the world more freely, gene flow undoubtedly became an important agent of microevolutionary change in populations that were previously quite isolated (FIGURE 21.6).

Mutations can cause evolution by substituting one allele for another in a gene pool: *a closer look*

A new mutation that is transmitted in gametes immediately changes the gene pool of a population by substituting one allele for another. For example, a mutation that causes a white-flowered plant in our hypothetical wildflower population to produce gametes bearing the dominant allele for pink flowers would decrease the frequency of the *a* allele in the population and increase the frequency of the *A* allele. By itself, mutation does not have much quantitative effect on a large population in a single generation. This is because a mutation at any given gene locus is a very rare event; although mutation rates vary, depending on the species and the gene locus, rates of one mutation per locus per 10^5 to 10^6 gametes are typical. If an allele has a frequency of 0.50 in the gene pool and mutates to another allele at a rate of 10^{-5} mutations per generation, it would take 2000 generations to reduce the frequency of the original allele from 0.50 to 0.49. The gene pool would be affected even less if the mutation were reversible, as most are. If some new allele produced by mutation increases its frequency by a significant amount in a population, it is not because mutation is generating the allele in abundance, but because individuals carrying the mutant allele are producing a disproportionate number of offspring as a result of natural selection or genetic drift. Over the long run, however, mutation is, in itself, very important to evolution because it is the original source of the genetic variation that serves as raw material for natural selection.

Nonrandom mating can cause evolution by shifting the frequencies of genotypes in a gene pool: *a closer look*

For Hardy-Weinberg equilibrium to hold, an individual of any genotype must choose its mates at random from the population. In actuality, individuals usually mate more often with close neighbors than with more distant members of the population, especially in species that do not disperse far. Other individuals in the same "neighborhood" within a larger population tend to be closely related. This promotes **inbreeding,** mating between closely related partners. The most extreme case of inbreeding is self-fertilization ("selfing"), particularly common in plants.

Inbreeding causes the relative frequencies of genotypes to deviate from that which is expected from Hardy-Weinberg equilibrium. For example, in our imaginary wildflower population, self-pollination would tend to increase the frequencies of homozygous genotypes at the expense of heterozygotes. If *AA* individuals and *aa* individuals "self," then their offspring must also be homozygous. If *Aa* plants "self," however, only half their offspring will be heterozygous. With each generation, the proportion of heterozygotes decreases and the proportions of dominant and recessive homozygotes increase. Even in less extreme cases of inbreeding without selfing, the decline of heterozygosity occurs, though more slowly. One visible effect of this change in genotype frequencies is a greater proportion of individuals expressing recessive phenotypes; the frequency of white-flowered individuals in our wildflower population would be greater than the Hardy-Weinberg equation predicts. Regardless of the impact of inbreeding on the ratio of genotypes and phenotypes in the population, the values of p and q, the frequencies of the two alleles, remain the same. It is just that a smaller proportion of recessive alleles are "masked" in heterozygous individuals.

Another type of nonrandom mating is **assortative mating,** in which individuals select partners that are like themselves in certain phenotypic characters. For example, blister beetles (*Lytta magister*) that live in the Sonoran Desert of Arizona most commonly mate with individuals of the same size (FIGURE 21.7). To some extent,

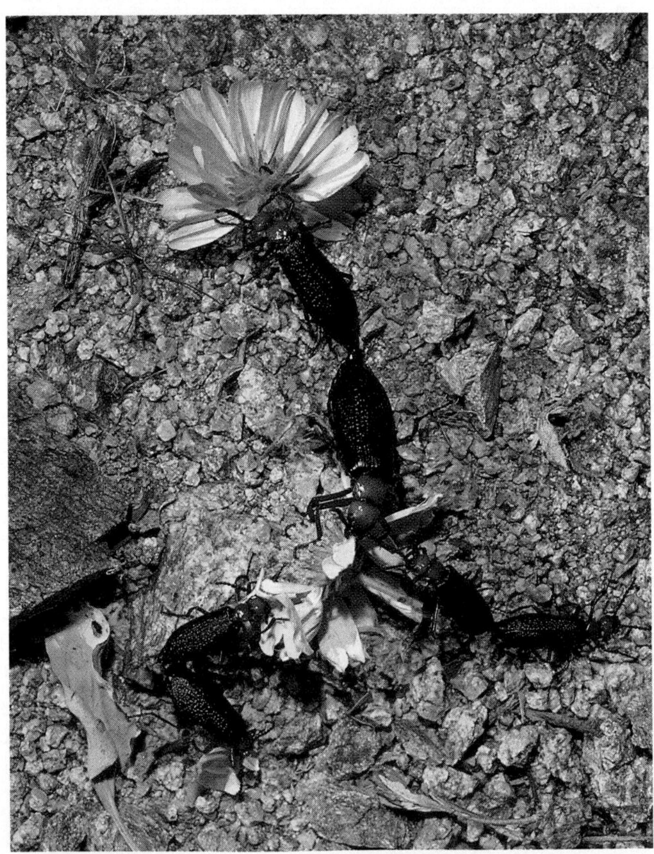

FIGURE 21.7
Assortative mating in blister beetles. These insects (*Lytta magister*) of the Sonoran Desert in Arizona pair off, posterior to posterior, according to size. The animals may continue to feed on brittlebush flowers during copulation, which usually lasts several hours.

humans also use size as a criterion for assortative mating; for example, tall women commonly (but not always) pair with tall men.

Notice again that nonrandom mating—inbreeding or assortative mating—increases the number of gene loci in the population that are homozygous, but nonrandom mating does not in itself alter the overall frequencies of alleles in a population's gene pool. Remember, however, that a population's genetic structure is defined by its frequencies of alleles *and genotypes.* Any change in a population's inbreeding or assortative mating behaviors will shift the frequencies of different genotypes. Thus, nonrandom mating can cause a population to evolve.

Natural selection can cause evolution via differential reproductive success among varying members of a population: *a closer look*

Hardy-Weinberg equilibrium requires that all individuals in a population be equal in their ability to survive and produce viable, fertile offspring. This condition is probably never completely met. Populations consist of varied individuals, and on average, some variants leave more offspring than others. This differential success in reproduction is natural selection. Selection results in alleles being passed along to the next generation in numbers disproportionate to their relative frequencies in the present generation. For example, in our imaginary wildflower population, plants with pink flowers (*AA* or *Aa* genotypes) may for some reason produce more offspring on average than plants having white flowers (*aa*); perhaps white flowers are more visible to herbivorous insects that eat the flowers. This would disturb Hardy-Weinberg equilibrium; the frequency of the *A* allele would increase, and the frequency of the *a* allele would decline in the gene pool.

Of all agents of microevolution that change a gene pool, only selection is likely to be adaptive. Natural selection accumulates and maintains favorable genotypes in a population. If the environment should change, selection responds by favoring genotypes adapted to the new conditions. But the degree of adaptation can be extended only within the realm of the genetic variability present in the population. Before we examine the process of adaptation by natural selection more closely, let's look at the genetic basis of the variation that makes it possible for populations to evolve.

Genetic variation is the substrate for natural selection

Heritable variation is at the heart of Darwin's theory of evolution, for variation provides the raw material—the substrate—on which natural selection works. In what ways do members of a population vary genetically? How extensive is genetic variation? What mechanisms generate and maintain variations in a population? Do all variations function as raw material for selection? These are the questions we will try to answer as we look at the genetic variations so crucial to the process of natural selection.

How Extensive Is Genetic Variation Within and Between Populations?

You have no trouble recognizing your friends in a crowd. Each person has a unique genome, reflected in individual variations of appearance and temperament. Individual variation occurs in populations of all species of sexually reproducing organisms. We are very conscious of human diversity; we are less sensitive to individuality in populations of other animals and of plants, and the diversity may escape our notice because the variations are subtle. But these slight differences between individuals in a population are the variations Darwin wrote most about as the raw material for natural selection.

Not all the variation we observe in a population is heritable. Phenotype is the cumulative product of an inherited genotype and a multitude of environmental influences. For example, bodybuilders alter their phenotypes dramatically. It is important to remember that only the genetic component of variation can have evolutionary consequences as a result of natural selection, because it is the only component that transcends generations.

Both discrete and quantitative characters contribute to variation within a population. Most heritable variation consists of polygenic characters that vary quantitatively within a population. For example, plant height may vary continuously in our hypothetical wildflower population, from very short individuals to very tall individuals and everything in between. Discrete characters, such as pink versus white flowers, vary categorically, usually because they are determined by a single gene locus with different alleles that produce distinct phenotypes.

Polymorphism. In cases of discrete variation when two or more forms of a Mendelian character are represented in a population, the contrasting forms are called *morphs*—as in the pink-flowered and white-flowered morphs of our wildflower population. A population is said to be *polymorphic* for a character if two or more distinct morphs are each represented in high enough frequencies

FIGURE 21.8
Polymorphism. Some populations consist of two or more discrete varieties of individuals. These four garter snakes (*Thamnophis ordinoides*), which differ markedly in their patterns of coloration, were captured in the same Oregon field. Edmund Brodie of the University of Chicago has discovered that the behavior of each morph is keyed to its coloration. Spotted snakes generally blend into their background better than striped snakes, but stripes make it harder to judge the speed of the snake in motion. When approached, spotted garter snakes usually freeze, while snakes of the striped morph move rapidly away.

to be readily noticeable. (Obviously, this definition is arbitrary, but a population is not termed polymorphic if it consists almost exclusively of a single morph, with other morphs extremely rare.) FIGURE 21.8 illustrates a striking example of **polymorphism** (the existence of polymorphic characters) in Oregon garter snakes. Polymorphism is extensive in human populations, both in physical characters, such as the presence or absence of freckles, and in biochemical characters, such as ABO blood groups (for which there are four morphs: type A, type B, type AB, and type O). Polymorphism only applies to such discrete characters, not to characters such as human height that vary continuously in a population.

Measuring Genetic Variation. Population biologists use several quantitative definitions of genetic variation. Two common measures are the percentage of gene loci represented by two or more alleles in a population (loci at which a single allele is not fixed), and the average percentage of loci that are heterozygous in the individuals of a population.

By these two measures of genetic diversity, the reservoir of heritable variation in a population is much more extensive than Darwin realized. Although much of the variation is invisible, it is manifest in molecular differences that can be detected by biochemical methods. For example, some evolutionary biologists use electrophoresis to study variations in the protein products of specific gene loci among individuals in a population. Such variations represent different alleles at a locus. Scores of loci have been studied in many different species. For example, in populations of the fruit fly *Drosophila*, the gene pool typically has two or more alleles for about 30% of the loci examined, and each fly is heterozygous at about 12% of its loci; that amounts to 700 to 1200 heterozygous loci per fly. Expressed another way, any two flies in a *Drosophila* population differ in genotype at about 25% of their loci. The extent of genetic variation in human populations, as revealed by electrophoresis, is comparable.

Geographical Variation. Most species exhibit **geographical variation,** differences in genetic structure between populations. Because at least some environmental factors are likely to be different from one place to another, natural selection can contribute to geographical variation. For example, one population of our now-familiar wildflower species may have a higher frequency of recessive alleles at the flower-color locus than other populations, perhaps because of a local prevalence of pollinators that key on white flowers (recessive homozygotes). Genetic drift can also cause chance variations among different populations. On a more local scale, geographical variation can also occur within a population, either because the environment has patchlike diversity or because the population is differentiated into subpopulations resulting from the limited dispersal of individuals.

One particular type of geographical variation, called a **cline,** is a graded change in some trait along a geographic axis. In some cases, a cline may represent a graded region of overlap where individuals of neighboring populations are interbreeding. In other cases, a gradation in some environmental variable may produce a cline. For example, the average body size of many North American species of birds and mammals increases gradually with increasing latitude. Presumably, the reduced ratio of surface area to volume that accompanies larger size is an adaptation that helps animals living in cold environments conserve body heat. Experimental studies of some clines, such as geographical variation in the height of yarrow plants that grow on the slopes of mountains, confirm the role of genetic variation in the spatial differences of phenotype (FIGURE 21.9).

How Is Genetic Variation Generated?

Mutation and sexual recombination (see Chapter 14) are the two processes that generate genetic variation.

Mutation. New alleles originate only by mutation. A mutation affecting any gene locus is an accident that

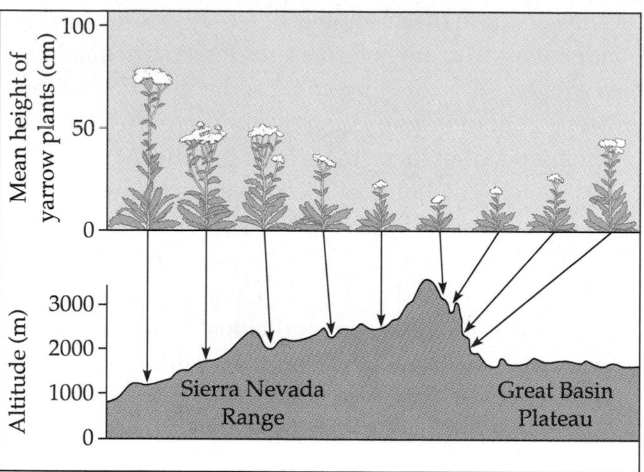

FIGURE 21.9

A cline. Yarrow plants on the slopes of the Sierra Nevada mountains of California and Nevada gradually decrease in average size at higher and higher elevations. Although the environment affects growth rates directly to some extent, some of the variation has a genetic basis. Researchers collected seeds at different elevations and grew plants in a common garden; the average sizes of the plants were correlated with the altitude at which the seeds were collected.

is rare and random. Most mutations occur in somatic cells and die with the individual. Only mutations that occur in cell lines that produce gametes can be passed along to offspring. In humans, there is about one mutation for every 100,000 to 200,000 gene replications. Humans have about 100,000 gene loci, or a diploid number of 200,000 genes. Thus on average, each person bears one or two mutant alleles not present in either parent.

A mutation is a shot in the dark. Chance determines where it will strike and how it will alter a gene. Most point mutations, those affecting a single base in DNA, are probably relatively harmless. Much of the DNA in the eukaryotic genome does not code for protein products, and it is uncertain how a change of a single nucleotide base in this silent DNA will affect the well-being of the organism. Even mutations of structural genes, which do code for proteins, may occur with little or no effect on the organism, partly because of redundancy in the genetic code. Of course, a single point mutation can have a significant impact on phenotype, as in sickle-cell disease, for example.

A mutation that alters a protein enough to affect its function is more often harmful than beneficial. Organisms are the refined products of thousands of generations of past selection, and a random change is not likely to improve the genome any more than firing a gunshot blindly through the hood of a car is likely to improve engine performance. On rare occasions, however, a mutant allele may actually fit its bearer to the environment better and enhance the reproductive success of the individual. This is not especially likely in a stable environment, but becomes more probable when the environment is changing and mutations that were once selected against are now favorable under the new conditions. For example, some mutations that happen to endow house flies with resistance to DDT also reduce growth rate and were deleterious before the pesticide was introduced. A new environmental factor, DDT, tilted the balance in favor of the mutant alleles, and they spread through fly populations by natural selection.

Because chromosomal mutations usually affect many gene loci, they are almost certain to disrupt the development of the organism. But even rearrangements of chromosomes may in rare instances bring benefits. For example, the translocation of a chromosomal piece from one chromosome to another could link alleles that affect the organism in some positive way when they are inherited together as a package.

Duplications of chromosome segments, like other chromosomal mutations, are nearly always harmful. But if the repeated segment does not disrupt genetic balance severely, it can persist over the generations and provide an expanded genome with superfluous loci that may eventually take on new functions by mutation, while the original genes continue to function at their old locations in the genome. New genes may also arise from existing DNA sequences by the shuffling of exons within the genome, either within a single locus or between loci (see Chapter 18).

For bacteria and other microorganisms that have very short generation spans, mutation can have a noticeable effect on a population's variation in a relatively short time. Bacteria reproduce asexually by dividing as often as once every 20 minutes; a single cell can potentially give rise to a billion descendants in just 10 hours. A new mutation that happens to be beneficial can increase its frequency in a bacterial population very rapidly. Imagine, for example, exposing a bacterial population to an antibiotic. If a single individual in the population happens to harbor a mutation that renders it resistant to the poison, in just a few hours there may be millions of resistant bacteria, while bacteria sensitive to the antibiotic may have been almost completely eliminated. Bacterial populations can evolve, one mutation at a time, by the explosive asexual expansion of clones favored by the local environment. On a generation-to-generation time scale, animals and plants depend mainly on sexual recombination for the genetic variation that makes adaptation possible. But even most bacteria increase genetic variation by occasionally exchanging and recombining genes through processes that resemble sex (see Chapter 17).

Recombination. Although mutations are the source of new alleles, they are so infrequent at any one locus that generation to generation, their contribution to genetic variation in a large population is negligible. Members of a population owe nearly all their genetic differences to the unique recombinations of existing alleles each individual draws from the gene pool. (Of course, this allele variation has its ultimate basis in past mutations.)

Sex shuffles alleles and deals them at random to determine individual genotypes. During meiosis, homologous chromosomes, one inherited from each parent, trade some of their genes by crossing over, and then the homologous chromosomes and the alleles they carry segregate randomly into separate gametes. Gametes from one individual vary extensively in their genetic makeup, and each zygote made by a mating pair has a unique assortment of alleles resulting from the random union of a sperm and an ovum. A population, of course, contains a vast number of possible mating combinations, each bringing together the gametes of individuals that are likely to have different genetic backgrounds. Sexual reproduction recombines old alleles into fresh assortments every generation.

How Is Genetic Variation Preserved?

What prevents natural selection from extinguishing a population's variation by culling unfavorable genotypes? The tendency for natural selection to reduce variation is countered by several mechanisms that preserve or restore variation.

Diploidy. The diploid nature of most eukaryotes hides a considerable amount of genetic variation from selection in the form of recessive alleles in heterozygotes. Recessive alleles that are less favorable than their dominant counterparts, or even harmful in the present environment, can persist in a population through their propagation by heterozygous individuals. This latent variation is exposed to selection only when both parents carry the same recessive allele and combine two copies in one zygote. This happens only rarely if the frequency of the recessive allele is very low. For example, if the frequency of the recessive allele is 0.01 and the frequency of the dominant allele is 0.99, then 99% of the copies of that recessive allele are protected from selection in heterozygotes, and only 1% of the recessive alleles are present in homozygotes. The rarer the recessive allele, the greater the degree of protection afforded by heterozygosity. Heterozygote protection maintains a huge pool of alleles that may not be suitable for present conditions but that could bring new benefits when the environment changes.

Balanced Polymorphism. Selection itself may preserve variation at some gene loci. This ability of natural selection to maintain diversity in a population is called **balanced polymorphism.** One of the mechanisms for this preservation of variation is **heterozygote advantage.** If individuals who are heterozygous for a particular locus have greater survivorship and reproductive success than any type of homozygote, then two or more alleles will be maintained at that locus by natural selection.

An example of heterozygote advantage involves the locus in humans for one chain of hemoglobin, the protein of red blood cells that transports oxygen. As we have seen, a specific recessive allele at that locus causes sickle-cell disease in homozygous individuals. Heterozygotes, however, are resistant to malaria, an important advantage in tropical regions where that disease is a major cause of death. The environment in these regions favors the heterozygotes over both homozygous dominant individuals, who are susceptible to malaria, and homozygous recessive individuals, who are harmed by sickle-cell disease. The frequency of the sickle-cell allele in Africa is generally highest in areas where the malaria parasite is most common. In some tribes, the recessive allele accounts for 20% of the hemoglobin loci in the gene pool, a very high frequency for an allele that is disastrous in homozygotes. But at this frequency ($q = 0.2$), 32% of the population consists of heterozygotes resistant to malaria ($2pq$), and only 4% of the population suffers from sickle-cell disease (q^2).

Another example of heterozygote advantage is found in the crossbreeding of crop plants. When corn, for instance, is highly inbred, the number of homozygous gene loci increases, and the corn may gradually become stunted in growth and increasingly sensitive to a variety of diseases. Crossbreeding between two different inbred varieties often produces hybrids that are much more vigorous than either parent stock. This **hybrid vigor** is probably due to two factors: the segregation of deleterious recessives that were homozygous in the inbred varieties, and the heterozygote advantage at many loci in the hybrids.

A patchy environment, where natural selection favors different phenotypes in different subregions within a population's geographical boundaries, can also result in balanced polymorphism. For example, protective coloration suited to different backgrounds may also help explain the morphs of garter snakes in FIGURE 21.8. FIGURE 21.10 illustrates another example of a polymorphism that may be due to environmental patchiness—in this case, specialization for slightly different foods.

Still another cause of balanced polymorphism is **frequency-dependent selection,** in which the reproductive success of any one morph declines if that phenotypic form becomes too common in the population. A particularly intricate example is a balanced polymorphism that has been observed in populations of *Papilio dardanus*, the African swallowtail butterfly in the photograph on p. 416. The males all have similar coloration, but the

FIGURE 21.10

A balanced polymorphism in a finch population. Two different bill sizes occur in a single population of black-bellied seedcrackers, a species of finch that lives in Cameroon, West Africa. There are no individuals with bills of intermediate size. The small-billed individuals (left) feed mainly on soft seeds, while the large-billed birds specialize in cracking hard seeds. Each morph shows a higher feeding efficiency on each respective seed. Thomas Smith of San Francisco State University has been studying this population in the field. His hypothesis is that natural selection maintains the polymorphism by selecting against intermediate bills, which crack both classes of seeds relatively inefficiently.

females occur in several different morphs, each resembling another butterfly species that is noxious to predators. *Papilio* females are not noxious, but birds avoid them because they look so much like the distasteful butterflies. This mimicry would be less effective if all *Papilio* females copied the same noxious species, because birds would be slow to associate a particular pattern of coloration with bad taste if they encountered good-tasting mimics as often as the noxious models. FIGURE 21.11 illustrates another example of frequency-dependent selection.

Does All Genetic Variation Affect Survival and Reproductive Success?

Some of the genetic variations observed in populations are probably trivial in their impact on reproductive success. The diversity of human fingerprints is an example of what is called **neutral variation,** which seems to confer no selective advantage for some individuals over others. Much of the protein variation detectable by electrophoresis may represent chemical "fingerprints" that are neutral in their adaptive qualities. For instance, 99 known mutations affect 71 of the 146 amino acids in the β chain of human hemoglobin, one of two kinds of polypeptide chains that make up that protein. Some

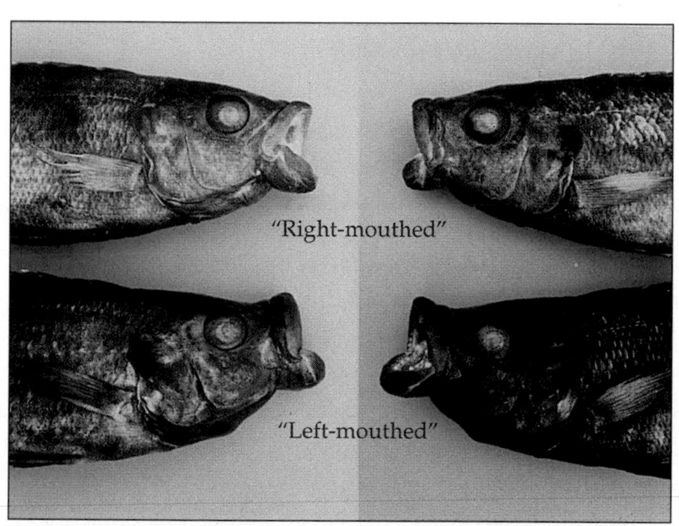

(a)

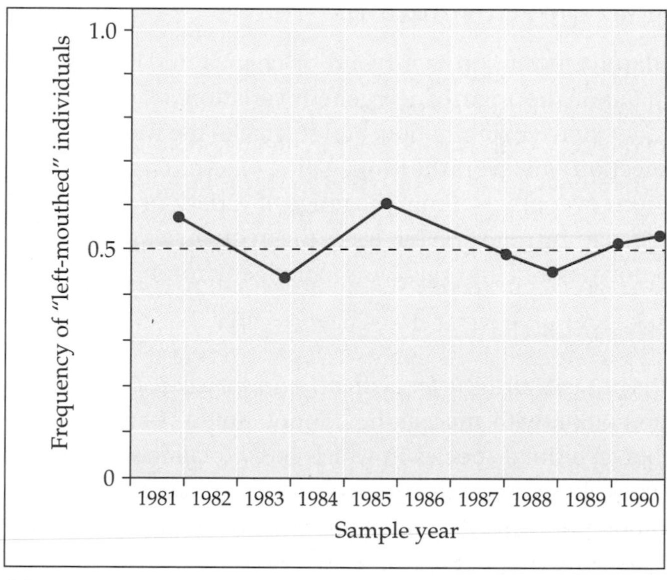

(b)

FIGURE 21.11

Frequency-dependent selection. Michio Hori of Wakayama Medical College in Japan has correlated a balanced polymorphism in mouth anatomy with the feeding habits of *Perissodus microlepis,* a small cichlid fish that lives in Africa's Lake Tanganyika. These fish feed on the scales of other fishes by approaching their prey from behind and using their mouths to snatch scales from the flanks of the prey. (**a**) *P. microlepis* has an asymmetrical mouth oriented either toward the right (top) or left (bottom). The two morphs attack opposite sides of their prey. Simple Mendelian inheritance determines this character. (**b**) The two morphs, "right-mouthed" and "left-mouthed," occur in about equal numbers among individuals in a population of *P. microlepis,* fluctuating only slightly in relative frequency from year to year, as we can conclude from this graph of left-mouthed frequency over a 10-year period. Hori offers this hypothesis: Prey fish guard more effectively against attack from either the left or the right side, depending on which *P. microlepis* morph is more common at a particular time. Thus, the polymorphism is balanced by the less common morph having a feeding advantage that enhances survival and reproductive success.

of those mutations, including the allele for sickle-cell disease, certainly affect the reproductive potential of the individual. However, according to the *theory of neutral evolution,* many of the variant alleles at this locus and others may confer no selective advantage or disadvantage. The relative frequencies of neutral variations will not be affected by natural selection; some neutral alleles will increase in the gene pool, and others will decrease by the chance effects of genetic drift.

There is no consensus among evolutionary biologists on how much genetic variation is neutral, or even if any variation can be considered truly neutral. Variations appearing to be neutral may, in fact, influence survival and reproductive success in ways that are difficult to measure. It is possible to show that a particular allele is detrimental, but it is impossible to demonstrate that an allele brings no benefits at all to an organism. Furthermore, a variation may be neutral in one environment but not in another. We can never know the degree to which genetic variation is neutral. But we can be certain that even if only a fraction of the extensive variation in a gene pool significantly affects the organisms, that is still an enormous resource of raw material for natural selection and the adaptive evolution it causes.

Natural selection is the mechanism of adaptive evolution

Adaptive evolution is a blend of chance and sorting—chance in the origin of new genetic variations by mutation and sexual recombination, and sorting in the workings of selection as it favors the propagation of some chance variations over others. From the range of variations available to it, natural selection increases the frequencies of certain genotypes and fits organisms to their environments.

Fitness

The phrases *struggle for existence* and *survival of the fittest* are loaded with misleading connotations. There are, of course, animal species in which individuals, usually the males, lock horns or otherwise do combat that determines mating privilege. But direct and violent confrontations between members of a population are rare; success is generally subtle and passive. A barnacle may produce more eggs than its neighbors because it is more efficient at collecting food from the water. In a population of moths, certain variants may average more offspring than others because their coloration hides them from predators better. Plants in a wildflower population may differ in reproductive success because some are better able to attract pollinators, owing to slight variations in flower color, shape, or fragrance. ***Darwinian fitness,*** *the measure that is critical to selection, is the relative contribution an individual makes to the gene pool of the next generation.*

In a more quantitative approach to natural selection, population geneticists define **relative fitness** as the contribution of a genotype to the next generation compared to the contributions of alternative genotypes for the same locus. For example, consider our wildflower population, in which *AA* and *Aa* plants have pink flowers and *aa* plants have white flowers. Let's assume that, on average, individuals with pink flowers produce more offspring than those with white flowers. The relative fitness of the most reproductively successful variants is set at 1 as a basis for comparison; so in this case, the relative fitness of an *AA* or *Aa* plant is 1. If plants with white flowers average only 80% as many offspring, their relative fitness is 0.8.

Survival alone does not guarantee reproductive success. Relative fitness is zero for a sterile plant or animal, even if it is robust and outlives other members of the population. But, of course, survival is a prerequisite for reproducing, and longevity increases fitness if it results in certain individuals leaving disproportionately high numbers of descendants. Then again, an individual that matures quickly and becomes fertile at an early age may have a greater reproductive potential than individuals that live longer but mature late. Thus, the components of selection are the many factors that affect both survival and fertility.

What Does Selection Act On?

An organism exposes its phenotype—its physical traits, metabolism, physiology, and behavior—not its genotype, to the environment. Acting on phenotypes, selection indirectly adapts a population to its environment by increasing or maintaining favorable genotypes in the gene pool.

Throughout this chapter, we have been looking at a wildflower population with alleles for flower color that affect phenotype unambiguously, but the connection between genotype and phenotype is rarely that simple or definite. A genotype at a particular locus may have multiple effects, especially if it influences the development or growth of the organism. This ability of genes to influence many phenotypic characters is called pleiotropy (see Chapter 13). The overall fitness of a genotype depends on whether its positive effects outweigh any harmful effects it may have on the survival and reproductive success of the organism. Another complication in the translation of genotype into phenotype arises when many gene loci influence the same characteristic. These are polygenic, or quantitative, characters; human height is an example. Individuals in a population do not usually fit into exclusive categories for a polygenic character, but instead vary continuously over a phenotypic range. This chapter has been simplified by using discrete characters for most of the examples of microevolution, but quantitative characters actually account for most of the variation exposed to natural selection.

The finished organism subjected to natural selection is an integrated composite of its many phenotypic features, not a collage of individual parts. The relative fitness of a genotype at any one locus depends on the entire genetic context in which it works. For example, alleles that enhance the growth rate of the trunk and limbs of a tree may be useless or even detrimental in the absence of alleles at other loci that enhance the growth rate of roots required to support the tree. On the other hand, alleles that contribute nothing to an organism's success, or may even be slightly maladaptive, may be perpetuated because they are present in individuals whose overall fitness is high. A whole baseball team wins the league pennant, even the player with the worst batting average and most errors.

Modes of Natural Selection

Natural selection can affect the frequency of a heritable trait in a population in three different ways, depending on which phenotypes in a varying population are favored. These three modes of selection are called stabilizing selection, directional selection, and diversifying selection. They can be depicted with graphs that show how the frequencies of different phenotypes change with time (FIGURE 21.12). This is most meaningful for quantitative traits that depend on many gene loci.

Stabilizing selection acts against extreme phenotypes and favors the more common intermediate variants. This mode of selection reduces variation and maintains the

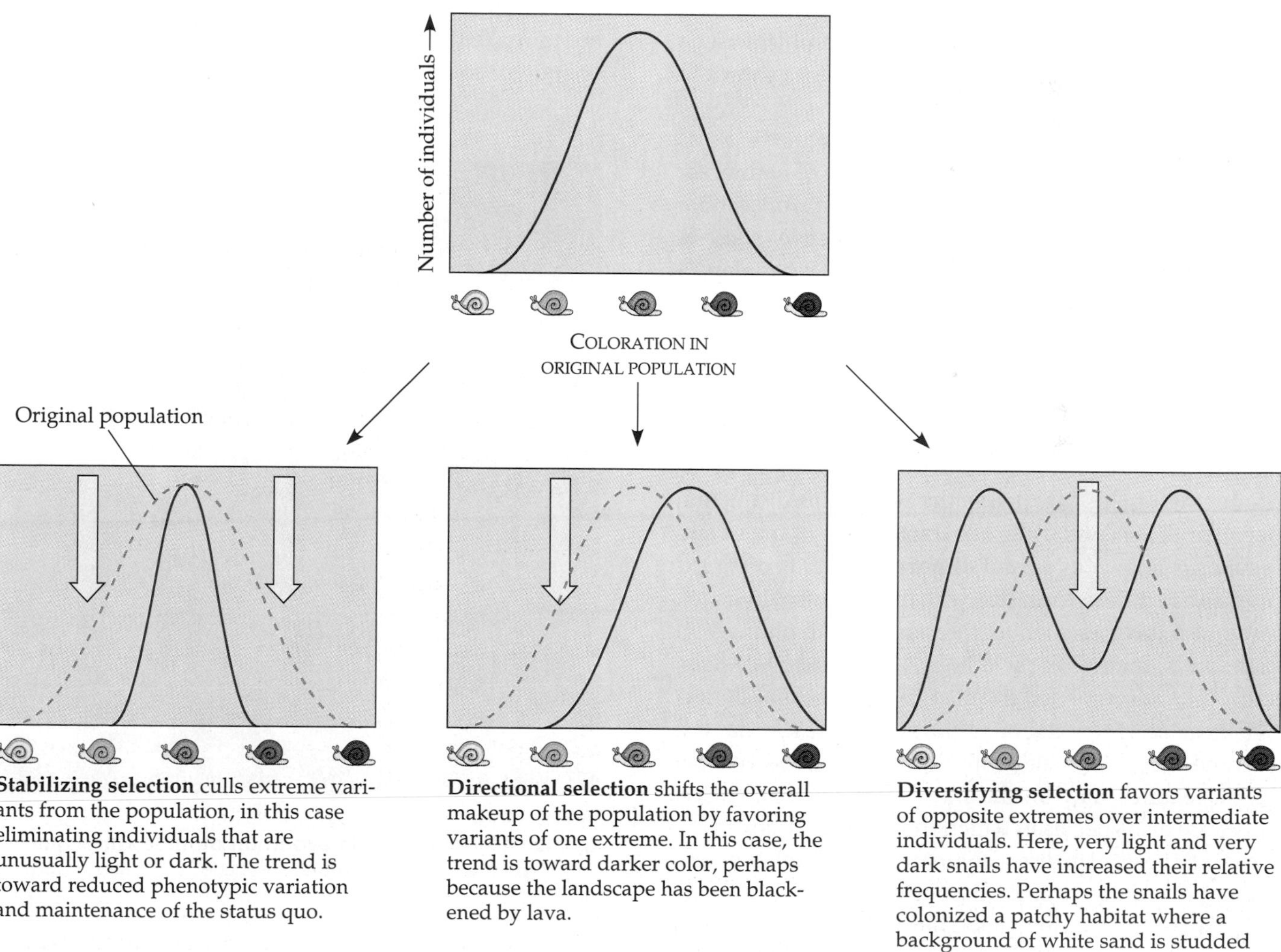

Stabilizing selection culls extreme variants from the population, in this case eliminating individuals that are unusually light or dark. The trend is toward reduced phenotypic variation and maintenance of the status quo.

Directional selection shifts the overall makeup of the population by favoring variants of one extreme. In this case, the trend is toward darker color, perhaps because the landscape has been blackened by lava.

Diversifying selection favors variants of opposite extremes over intermediate individuals. Here, very light and very dark snails have increased their relative frequencies. Perhaps the snails have colonized a patchy habitat where a background of white sand is studded with lava rocks.

FIGURE 21.12
Modes of selection. These cases describe the possible microevolution of a snail population in which there is quantitative variation in coloration. The graphs show how the frequencies of individuals of varying darkness change with time. The large arrows symbolize natural selection working against certain phenotypes.

status quo for a particular phenotypic character. For example, stabilizing selection keeps the majority of human birth weights in the 3–4 kg range. For babies much smaller or larger than this, infant mortality is greater.

Directional selection is most common during periods of environmental change or when members of a population migrate to some new habitat with different environmental conditions. Directional selection shifts the frequency curve for variations in some phenotypic character in one direction or the other by favoring what are initially relatively rare individuals that deviate from the average for that character. For instance, there is fossil evidence that the average size of black bears in Europe increased with each glacial period of the ice ages, only to decrease again during the warmer interglacial periods.

Diversifying selection, probably a rare mode of selection, occurs when environmental conditions are varied in a way that favors individuals on both extremes of a phenotypic range over intermediate phenotypes. Diversifying selection can result in balanced polymorphism, such as the finch population with the two bill sizes shown in FIGURE 21.10.

Although we refer to these three selection trends as "modes of selection," the basic mechanism of natural selection is the same for each case. Selection favors certain heritable traits via differential reproductive success. What determines the mode is which phenotypes along a range of individual variation are favored.

Sexual Selection

Males and females of many animal species exhibit marked differences in addition to the differences in the reproductive organs that define the sexes. This distinction between the secondary sex characteristics of males and females is known as **sexual dimorphism.** It is often expressed as a difference in size, with the male usually larger, but it also involves such features as colorful plumage in male birds, manes on male lions, antlers on male deer, and other adornments. In most cases of sexual dimorphism, at least for vertebrates, the male is the showier sex. In some cases, the males with the most impressive masculine features may be the most attractive to females. There are also species in which the secondary sex structures may be used in direct competition with other males; this is particularly common in species where a single male garners a harem of females. These males may succeed because they defeat smaller, weaker, or less fierce males in combat; more often, however, they are effective in ritualized displays that discourage would-be competitors (see Chapter 50).

Darwin was intrigued by **sexual selection,** which he saw as a separate selection process leading to sexual dimorphism. Many secondary sex features do not seem to be adaptive in the general sense; showy plumage probably does not help male birds cope with their environment and may even attract predators. If such accoutrements give the individual an edge in gaining a mate, however, they will be favored for the most Darwinian of reasons: because they enhance reproductive success. Thus, in many cases, the ultimate evolutionary outcome is a compromise between the two selection forces. In some species, the line between sexual selection and ordinary natural selection blurs because the sexual feature does double duty as an adaptation to the environment. For example, a stag may use his antlers to defend himself against a predator.

Many researchers who study sexual selection and dimorphism have recently shifted their attention to the roles of females in the evolution of these characters. Every time a female chooses a mate based on particular phenotypic traits, she perpetuates the alleles that caused her to make that choice and allows a male of particular phenotype to perpetuate his alleles (FIGURE 21.13).

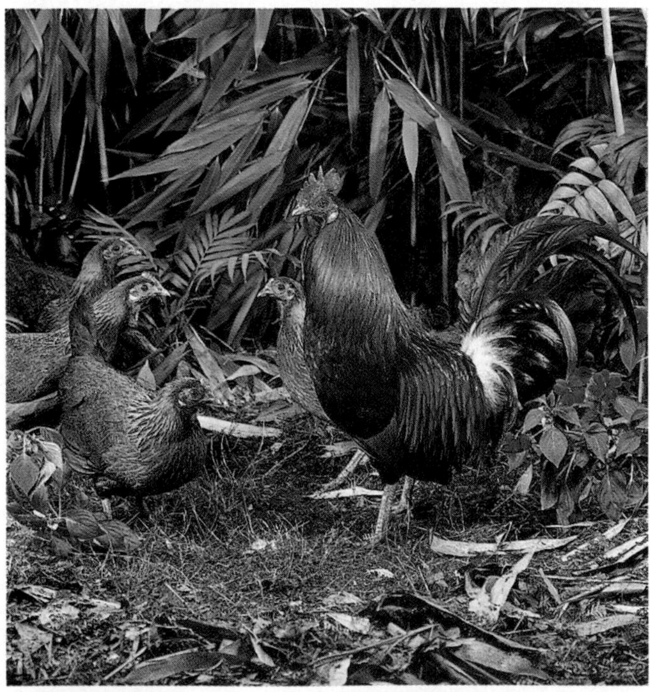

FIGURE 21.13
Sexual selection and the evolution of male appearance in jungle fowl (*Gallus gallus*). Marlene Zuk of the University of California, Riverside, has been studying mate selection in Asian jungle fowl, the ancestors of domesticated chickens. Hens generally choose roosters, such as the one in this photograph, with bright eyes and large red combs and wattles (the flap of skin draped from the throat). These traits are correlated with good health and resistance to pathogens. Thus, according to Dr. Zuk's hypothesis, a hen keys on rooster features that advertise vitality. Over the generations, such choices by females shape the appearance of males and may play an active role in favoring less visible genotypes that contribute to healthy offspring.

Does evolution fashion perfect organisms?

The answer to this question, in a word, is *no.* There are at least four reasons why natural selection cannot breed perfection.

1. *Organisms are locked into historical constraints.* As we saw in Chapter 20, each species has a legacy of descent with modification from a long line of ancestral forms. Evolution does not scrap ancestral anatomy and build each new complex structure from scratch, but co-opts existing structures and adapts them to new situations. For example, the excruciating back problems some humans endure result in part because the skeleton and musculature modified from the anatomy of four-legged ancestors are not fully compatible with upright posture.

2. *Adaptations are often compromises.* Each organism must do many different things. A seal spends part of its time on rocks; it could probably walk better if it had legs instead of flippers, but it would not swim nearly as well. We humans owe much of our versatility and athleticism to our prehensile hands and flexible limbs, which also make us prone to sprains, torn ligaments, and dislocations; structural reinforcement has been compromised for agility.

3. *Not all evolution is adaptive.* Chance probably affects the genetic structure of populations to a greater extent than was once believed. For instance, when a storm blows insects hundreds of miles over an ocean to an island, the wind does not necessarily pick up the specimens that are best suited to the new environment. And not all alleles fixed by genetic drift in the gene pool of the small founding population are better suited to the environment than alleles that are lost. Similarly, the bottleneck effect can cause nonadaptive or even maladaptive evolution.

4. *Selection can only edit variations that exist.* Natural selection favors only the most fit variations from what is available, which may not be the ideal traits. New alleles do not arise on demand.

With all these constraints, we cannot expect evolution to craft perfect organisms. Natural selection operates on a "better than" basis. We can see evidence for evolution in the subtle imperfections of the organisms it produces.

* * *

Natural selection is usually thought of as an agent of change, but it can also act to maintain the status quo. Stabilizing selection probably prevails most of the time, resisting change that may be maladaptive. Evolutionary spurts occur when a population is stressed by a change in the environment, migration to a new place, or a change in the genome. When challenged with a new set of problems, a population either adjusts through natural selection or becomes extinct. The fossil record indicates that extinction is the more common outcome. Those populations that do survive crises often change enough to become new species, as we will see in the next chapter.

REVIEW OF KEY CONCEPTS (with page numbers and key figures)

- The modern evolutionary synthesis integrated Darwinism and Mendelism: *science as a process* (pp. 416–417)
  - The development of population genetics, with its emphases on quantitative inheritance and variation, brought Darwinism and Mendelism together.
  - In the 1940s, the modern synthesis provided a comprehensive theory of evolution that focused on populations as units of evolution.
- A population has a genetic structure defined by its gene pool's allele and genotype frequencies (pp. 417–419)
  - A population, a localized group of organisms belonging to the same species, is united by its gene pool, the aggregate of all alleles in the population.
  - At any given time, different alleles and genotypes are represented in a gene pool in certain frequencies—the population's genetic structure.
- The Hardy-Weinberg theorem describes a nonevolving population (pp. 419–420, FIGURE 21.3)
  - According to the Hardy-Weinberg theorem, the frequencies of alleles in a population will remain constant if sexual reproduction is the only process that affects the gene pool.
  - The mathematical expression for such a nonevolving population is the Hardy-Weinberg equation, which states that for a two-allele locus, $p^2 + 2pq + q^2 = 1$, where p and q represent the relative frequencies of the dominant and recessive alleles, respectively, p^2 and q^2 the frequencies of the homozygous genotypes, and $2pq$ the frequency of the heterozygous genotype.
- Microevolution is a generation-to-generation change in a population's allele or genotype frequencies: *an overview* (pp. 420–421, TABLE 21.1)
  - For Hardy-Weinberg equilibrium to apply, the population must be very large, be totally isolated, have no net mutations, show random mating, and have equal reproductive success for all individuals.
  - Microevolution, a change in allele or genotype frequencies in a population, can occur when the conditions required for Hardy-Weinberg equilibrium are not met.

- Genetic drift can cause evolution via chance fluctuation in a small population's gene pool: *a closer look* (pp. 421–423, FIGURE 21.4)
  - Genetic drift is the change in allele frequencies observed in small populations due to sampling error or chance events. When large segments of a population are destroyed by disasters (the bottleneck effect) or when a small sample of a population colonizes a new habitat (the founder effect), the new, small population is unlikely to be representative of the parent population. Genetic drift will continue until the population grows larger.
- Gene flow can cause evolution by transferring alleles between populations: *a closer look* (p. 423)
  - Gene flow is the exchange of alleles between two populations due to migration.
- Mutations can cause evolution by substituting one allele for another in a gene pool: *a closer look* (pp. 423–424)
  - Mutation can theoretically affect allele frequencies in a gene pool, but it is usually insignificant over the short term for large populations. Mutation is mainly important in evolution because it generates new variations.
- Nonrandom mating can cause evolution by shifting the frequencies of genotypes in a gene pool: *a closer look* (pp. 424–425)
  - Nonrandom mating, such as inbreeding or assortative mating, does not ordinarily affect allele frequencies, but does affect the ratio of genotypes in populations.
- Natural selection can cause evolution via differential reproductive success among varying members of a population: *a closer look* (p. 425)
  - Natural selection, differential success in reproduction, is the only agent of microevolution that tends to be adaptive.
- Genetic variation is the substrate for natural selection (pp. 425–430, FIGURES 21.8–21.11)
  - Genetic variation in a population includes individual variation in discrete and quantitative characters as well as geographical variation.
  - Much genetic variation can be detected only at the molecular level.
  - Mutation and sexual recombination generate genetic variation.
  - Populations remain variable despite natural selection. Diploidy maintains a reservoir of latent variation in heterozygotes. Balanced polymorphism may maintain variation at some gene loci as a result of heterozygote advantage or frequency-dependent selection.
  - Biologists debate what fraction of genetic variation is actually acted on by natural selection and what fraction is neutral in its impact on reproductive success.
- Natural selection is the mechanism of adaptive evolution (pp. 430–432, FIGURE 21.12)
  - Darwinian fitness is measured only by reproductive success. One genotype has a greater relative fitness than another if it contributes to a greater number of viable, fertile offspring.
  - Selection favors certain genotypes in a population by acting on the phenotype of individual organisms. The whole organism is the object of selection.
  - Natural selection can affect the frequency of a phenotype in three different ways: stabilizing selection, which discriminates against extreme phenotypes; directional selection, which favors relatively rare individuals on one end of the phenotypic range; and diversifying selection, which favors individuals at both extremes of a range over intermediate phenotypes.
  - Sexual selection leads to the evolution of secondary sex characteristics, which give the individual an advantage in mating.
- Does evolution fashion perfect organisms? (p. 433)
  - Evolution by natural selection does not fashion perfect organisms for several reasons: Structures result from modified ancestral anatomy, adaptations are often compromises, the gene pool can be affected by genetic drift, and natural selection can act only on available variation.

SELF-QUIZ

1. A gene pool consists of
   a. all the alleles exposed to natural selection
   b. the total of all alleles present in a population
   c. the entire genome of a reproducing individual
   d. the frequencies of the alleles for a gene locus within a population
   e. all the gametes in a population

2. In a population with two alleles for a particular locus, *B* and *b*, the allele frequency of *B* is 0.7. What would be the frequency of heterozygotes if the population is in Hardy-Weinberg equilibrium?
   a. 0.7 d. 0.42
   b. 0.49 e. 0.09
   c. 0.21

3. In a population that is in Hardy-Weinberg equilibrium, 16% of the individuals show the recessive trait. What is the frequency of the dominant allele in the population?
   a. 0.84 d. 0.4
   b. 0.36 e. 0.48
   c. 0.6

4. The average length of jackrabbit ears decreases the farther north the rabbits live. This variation is an example of
   a. a cline d. genetic drift
   b. discrete variation e. diversifying selection
   c. polymorphism

5. Which of the following is an example of a polymorphism in humans?
   a. variation in height
   b. variation in intelligence
   c. the presence or absence of a widow's peak (see FIGURE 13.15)
   d. variation in the number of fingers
   e. variation in fingerprints

6. Selection acts *directly* on

a. phenotype
b. genotype
c. the entire genome
d. each allele
e. the entire gene pool

7. As a mechanism of microevolution, natural selection can be most closely equated with

a. assortative mating
b. genetic drift
c. differential reproductive success
d. mutation
e. gene flow

8. Most of the variation we see in coat coloration and pattern in a population of wild mustangs in any generation is probably due to

a. new mutations that occurred in the preceding generation
b. sexual recombination of alleles
c. genetic drift due to the small size of the population
d. geographical variation within the population
e. environmental effects

9. In terms of the algebraic symbols used in the Hardy-Weinberg equation (p and q), the most likely effect of assortative mating on the frequencies of alleles and genotypes for a gene locus would be

a. a decrease in p^2 compared to q^2
b. a trend toward zero for q^2
c. convergence of p^2 and q^2 toward equal values
d. a change in p and q, the relative frequencies of the two alleles in the gene pool
e. a decrease in $2pq$ below the value expected by the Hardy-Weinberg theorem

10. A founder event favors microevolution in the founding population mainly because

a. mutations are more common in a new environment
b. a small founding population is subject to extensive sampling error in the composition of its gene pool
c. the new environment is likely to be patchy, favoring diversifying selection
d. gene flow increases
e. members of a small population tend to migrate

CHALLENGE QUESTIONS

1. Some species have been rescued from near extinction by conservationists. In terms of evolutionary theory, what problems do such species face as their populations rebound from a small size?

2. Let's return to the wildflowers with which we derived the Hardy-Weinberg theorem. The frequency of *A*, the dominant allele for pink flowers, is 0.8, and the frequency of *a*, the recessive allele for white flowers, is 0.2. In one starting population, the frequencies of genotypes do not conform to Hardy-Weinberg equilibrium: 60% of the plants are *AA* and 40% of the plants are *Aa*. (At this point, the population has no plants with white flowers.) Assuming that all conditions for the Hardy-Weinberg theorem are met, prove that genotypes will reach equilibrium in the next generation.

3. Write a short essay explaining how natural selection adapts a population to its local environment.

SCIENCE, TECHNOLOGY, AND SOCIETY

1. Some scientists have suggested that the increased appreciation of the importance of females in sexual selection is one consequence of increasing numbers of women making contributions to animal behavior and evolutionary biology. Do you think gender makes a difference in the kinds of questions scientists ask and information they uncover? Why or why not?

2. How might modern technologies and lifestyles be altering human gene pools and reshaping human populations and the human species as a whole? Consider the five agents of microevolution discussed in this chapter.

3. Give a few historical examples of how gene flow has affected human populations.

FURTHER READING

Dawkins, R. *The Blind Watchmaker.* New York: Norton, 1986. A master of metaphor explains how complexity can arise in the absence of design.

Diamond, J. "Founding Fathers and Mothers." *Natural History,* June 1988. The importance of genetic drift in human evolution.

Gibbons, A. "The Mystery of Humanity's Missing Mutations." *Science,* January 6, 1995. Evidence for a bottleneck in our history.

Gillis, A. M. "Getting a Picture of Human Diversity." *BioScience,* January 1994. The application of population genetics to anthropology.

Hoffman, P. "The Science of Race." *Discover,* November 1994. The concept of human races may lack a biological basis.

Jones, S. "A Brave, New, Healthy World?" *Natural History,* June 1994. Will amalgamation of human populations decrease the occurrence of hereditary diseases?

Pennisi, E. "Cheetah Countdown." *Science News,* September 25, 1993. Inbreeding is a serious threat to the survival of this cat species.

CHAPTER 22

THE ORIGIN OF SPECIES

KEY CONCEPTS

- The biological species concept emphasizes reproductive isolation
- Reproductive barriers separate species
- Geographical isolation can lead to the origin of species: allopatric speciation
- A new species can originate in the geographical midst of the parent species: sympatric speciation
- Population genetics can account for speciation
- The theory of punctuated equilibrium has stimulated research on the tempo of speciation

When Darwin saw that the geologically young Galapagos Islands had already become populated with many plants and animals known nowhere else in the world, he realized that he was visiting a place of genesis. The islands are named for the giant tortoises, such as the Geochelone elephantopus *shown here, that are among the unique inhabitants.* (Galápago *is the Spanish word for tortoise.) After visiting the Galapagos, Darwin wrote in his diary: "Both in space and time, we seem to be brought somewhat near to that great fact—that mystery of mysteries—the first appearance of new beings on this Earth." The beginning of new forms of life—the origin of species—is at the focal point of evolutionary theory, for it is in new species that biological diversity arises. It is not enough to explain how adaptations evolve in populations, a topic covered in Chapter 21. Evolutionary theory must also explain the multiplication of species, the radiation of an existing species that gives rise to two or more new species.*

The fossil record chronicles two patterns of ***speciation*** *(the origin of new species): anagenesis and cladogenesis* (Figure 22.1). ***Anagenesis*** *(Gr.* ana, *"up," and* genesis, *"origin"), also known as* ***phyletic evolution,*** *is the transformation of one entire species into another.* ***Cladogenesis*** *(Gr.* clados, *"branch"), also called* ***branching evolution,*** *is the budding of one or more new species from a parent species that continues to exist. Cladogenesis is more important than anagenesis in the history of life, not only because it seems to be the more common pattern, but also because only cladogenesis can promote biological diversity by increasing the number of species.*

Our objective in this chapter is to examine and evaluate possible mechanisms of speciation. The first step is to appraise the assumption that species actually exist in nature as discrete biological units distinct from all others.

The biological species concept emphasizes reproductive isolation

In 1927, a young biologist named Ernst Mayr led an expedition into the remote Arafak Mountains of New Guinea to study the wildlife and collect specimens. He found a great diversity of birds, identifying 138 separate species on the basis of differences in their appearance. Mayr was surprised to learn that the local tribe of Papuan natives, who hunted the animals for food and feathers, had given names of their own to 137 birds (two birds assigned to separate species by Mayr are extremely similar, and the Papuans did not distinguish between them). Although

FIGURE 22.1
Two patterns of speciation. (**a**) In anagenesis (phyletic evolution), a single population is transformed enough to be designated a new species. (**b**) Cladogenesis is branching evolution, in which a new species arises from a small population that buds from a parent species. Most new species probably evolve by cladogenesis, the branching evolution that is the basis for biological diversity.

their motives and training could not have been more different, Mayr and the indigenous people agreed almost exactly in their inventory of the local birdlife. There are many such cases of university-trained taxonomists giving specific names to organisms in a particular locale, only to find that the discrete forms they identified correspond to the folk taxonomy of the region. From this we might conclude that species exist in nature as discrete units, demarcated from other species. Yet devising a formal definition of a species poses a formidable challenge.

Species is a Latin word meaning "kind" or "appearance." Indeed, we learn to distinguish between the kinds of plants or animals—between dogs and cats, for instance—from differences in their appearance. Linnaeus, the founder of modern taxonomy, described individual species in terms of their physical form, or morphology, and this is still the method most often used to characterize species. Most of the species recognized by taxonomists have been designated as separate species based on measurable morphological features, an approach that is practical to apply in the field, even on fossils. Modern taxonomists also consider differences in physiology, biochemistry, behavior, and genetic makeup. From the standpoint of evolutionary theory, however, dividing organisms into different species based on comparative

(a) Similarity between different species

(b) Diversity within a species

FIGURE 22.2
The biological species concept is based on interfertility rather than physical similarity. (**a**) The eastern meadowlark (*Sturnella magna,* left) and the western meadowlark (*Sturnella neglecta*) have very similar body shapes and colorations, but they represent different species. Their songs are distinct, a behavioral difference that helps prevent interbreeding between the two species. (**b**) In contrast, all humans, as seemingly diverse as we are, belong to a single species (*Homo sapiens*), defined by our capacity to interbreed.

data begs the question: How does each species remain distinct from other species? Put another way, why is biological diversity not a continuum of variation, but instead broken up into separate forms that we can identify as species? This is the question addressed by a species definition known as the biological species concept, first enunciated by Ernst Mayr in 1942.

The **biological species concept** defines a species as a population or group of populations whose members have the potential to interbreed with one another in nature to produce viable, fertile offspring, but who cannot successfully interbreed with members of other species (FIGURE 22.2). In other words, a biological species is the largest unit of population in which genetic exchange is possible, and that is genetically isolated from other such populations. Put still another way, each species is circumscribed by reproductive barriers that preserve its integrity as a species by blocking genetic mixing with other species. Members of a species, said to be **conspecific,** are united by being reproductively compatible, at least potentially. A businesswoman in Manhattan has little probability of producing offspring with a dairyman in Outer Mongolia, but if the two should get together, they could have viable babies that develop into fertile adults. All humans belong to the same biological species. In contrast, humans and chimpanzees remain distinct species even where they share territory, because the two species do not successfully interbreed and produce hybrid offspring.

Remember that biological species are defined by their reproductive isolation from other species in *natural* environments. In the laboratory or in zoos, hybrids can often be produced between two species that do not interbreed in nature.

The biological species concept does not work in all situations. The criterion of interbreeding is useless for organisms that are completely asexual in their reproduction, as are all prokaryotes, some protists, some fungi, and even some plants (such as the commercial banana) and some animals (including certain lizards and other vertebrates). Many bacteria do transfer genes on a limited scale by conjugation and other processes, but there is nothing akin to the equal contribution of genetic material from two parents that occurs in sexual reproduction. Different lineages of descent give rise to clones, which, genetically speaking, represent single individuals. Asexual organisms can be assigned to species only by the grouping of clones that have the same morphology and biochemical characteristics. Most organisms, however, are sexual.

The biological species concept is also inadequate as a criterion for grouping extinct forms of life, the fossils of which must be classified according to morphology.

There is still another situation where the biological species concept does not work. If two populations are geographically segregated from each other, they do not interbreed, although they may be so much alike that they are placed in the same species on morphological grounds. In classifying these remote populations in the same species, taxonomists are making an educated guess that the two populations could interbreed if they came into contact. This is why *potential to interbreed* is part of the biological species concept.

Even for populations that are sexual, contemporaneous, and geographically contiguous, there are cases where the biological species concept cannot be applied in a straightforward way. Consider, for example, four subspecies of the deer mouse (*Peromyscus maniculatus*) found in the Rocky Mountains (FIGURE 22.3). A subspecies is a population or a group of populations that live in one area and have minor differences from conspecific populations found elsewhere in the species' geographical range. The deer mouse subspecies overlap at certain locations, and some interbreeding occurs in these zones of cohabitation; we would therefore consider the populations to belong to the same species according to the biological species concept. The exception is the subspecies *Peromyscus maniculatus artemisiae* and the subspecies *P. m. nebrascensis.* Though their ranges overlap, these two subspecies do not interbreed. Nevertheless, their gene pools are not completely isolated from each other, because each subspecies interbreeds with its other neighboring subspecies, and genes could conceivably flow between the *artemisiae* and the *nebrascensis* populations. If the corridors for gene flow were eliminated by extinction

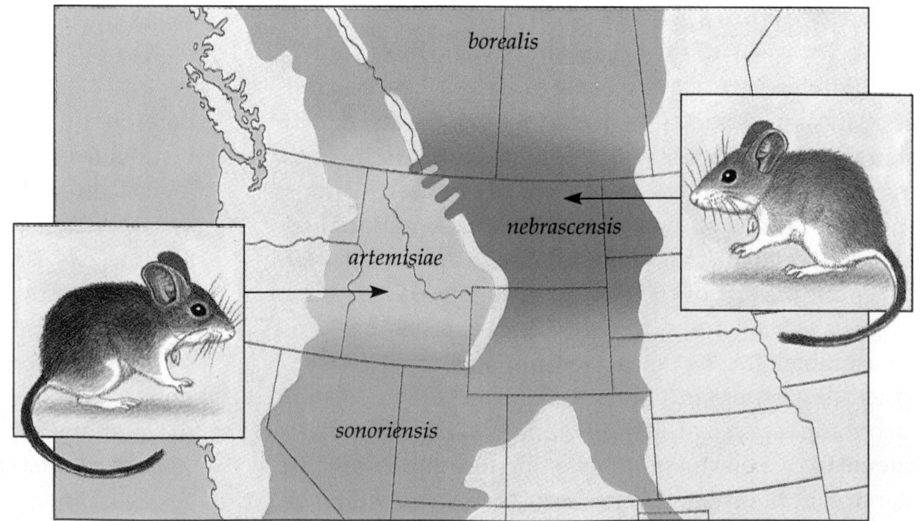

FIGURE 22.3
Speciation in progress? These four subspecies of the deer mouse (*Peromyscus maniculatus*) display geographical variation. Interbreeding between subspecies occurs where their ranges overlap, except for the subspecies *artemisiae* and *nebrascensis* (which come into contact where the light and dark green merge). These two subspecies will not interbreed, yet gene flow between them is possible via the other neighboring subspecies. Such cases may represent speciation in progress; for example, *artemisiae* and *nebrascensis* could legitimately be called different species if, in the future, the populations connecting them become extinct.

of the other two subspecies, then *artemisiae* and *nebrascensis* could be named separate species.

Population biologists are discovering more and more cases where the distinction between subspecies with limited genetic exchange and full biological species with segregated gene pools blurs. It is as though we are catching populations at different stages in their evolutionary descent from common ancestors. This is to be expected if new species usually arise by the gradual divergence of populations.

Many evolutionary biologists now question the usefulness of the biological species concept, and additional species concepts have been proposed in recent years (we will examine one of them later in the chapter). But the species problem may have no completely satisfactory resolution; it is unlikely that any single definition of a species can be stretched to cover all cases. These theoretical problems aside, the use of morphological criteria to distinguish species and the biological species concept usually result in recognizing the same units as species. Ernst Mayr and the Papuan tribe arrived at the same set of species of Arafak birds based on differences in appearance, and it is reproductive isolation that preserves species boundaries.

Reproductive barriers separate species

Any factor that impedes two species from producing viable, fertile hybrids contributes to reproductive isolation. No single barrier may be completely impenetrable to genetic exchange, but most species are genetically sequestered by more than one type of barrier. Here, we are considering only biological barriers to reproduction, which are intrinsic to the organisms. Of course, if two species are geographically segregated, they cannot possibly interbreed, but a geographical barrier is not considered equivalent to reproductive isolation because it is not intrinsic to the organisms themselves. Reproductive isolation prevents populations belonging to different species from interbreeding, even if their ranges overlap.

Clearly, a fly will not mate with a frog or a fern, but what prevents species that are very similar—that is, closely related—from interbreeding? The various reproductive barriers that isolate the gene pools of species can be categorized as prezygotic or postzygotic, depending on whether they function before or after the formation of zygotes, or fertilized eggs.

Prezygotic Barriers

Prezygotic barriers impede mating between species or hinder the fertilization of ova if members of different species attempt to mate.

Habitat Isolation. Two species that live in different habitats within the same area may encounter each other rarely, if at all, even though they are not technically geographically isolated. For example, two species of garter snakes belonging to the genus *Thamnophis* occur in the same areas, but one lives mainly in water and the other is primarily terrestrial. Habitat isolation also affects parasites, which are generally confined to certain plant or animal host species. Two species of parasites living on different hosts will not have a chance to mate.

Behavioral Isolation. Special signals that attract mates, as well as elaborate behavior unique to a species, are probably the most important reproductive barriers among closely related animals. Male fireflies of different species signal to females of their kind by blinking their lights in particular patterns. The females respond only to signals characteristic of their own species, flashing back and attracting the males.

The eastern and western meadowlarks are almost identical in morphology and habitat, and their ranges overlap in the central United States (see FIGURE 22.2a). Yet they remain two separate species, partly because of the difference in their songs, which enables them to recognize individuals of their own kind. Still another form of behavioral isolation is courtship ritual specific to a species (FIGURE 22.4).

Temporal Isolation. Two species that breed during different times of the day, different seasons, or different years cannot mix their gametes. The geographical ranges

FIGURE 22.4
Courtship ritual as a behavioral barrier between species. These blue-footed boobies, inhabitants of the Galapagos, will mate only after a specific ritual of courtship displays. Part of the "script" calls for the male to high-step, a behavior that advertises the bright blue feet characteristic of the species.

of the western spotted skunk (*Spilogale gracilis*) and the eastern spotted skunk (*Spilogale putorius*) overlap, but these two very similar species do not interbreed because *S. gracilis* mates in late summer and *S. putorius* mates in late winter. Three species of the orchid genus *Dendrobium* living in the same rain forest do not hybridize because they flower on different days. Pollination of each species is limited to a single day, because the flowers open in the morning and wither that evening.

Mechanical Isolation. Closely related species may attempt to mate, but fail to consummate the act because they are anatomically incompatible. For example, mechanical barriers contribute to reproductive isolation of flowering plants that are pollinated by insects or other animals. Floral anatomy is often adapted to certain pollinators that transfer pollen only among plants of the same species.

Gametic Isolation. Even if the gametes of different species meet, they rarely fuse to form a zygote. For animals whose eggs are fertilized within the female reproductive tract (internal fertilization), the sperm of one species may not be able to survive in the environment of the female reproductive tract of another species. Many aquatic animals release their gametes into the surrounding water, where the eggs are fertilized (external fertilization). Even when two closely related species release their gametes at the same time in the same place, cross-specific fertilization is uncommon. Gamete recognition may be based on the presence of specific molecules on the coats around the egg, which adhere only to complementary molecules on sperm cells of the same species. A similar mechanism of molecular recognition enables a flower to discriminate between pollen of the same species and pollen of different species.

Postzygotic Barriers

If a sperm cell from one species does fertilize an ovum of another species, then **postzygotic barriers** prevent the hybrid zygote from developing into a viable, fertile adult.

Reduced Hybrid Viability. When prezygotic barriers are crossed and hybrid zygotes are formed, genetic incompatibility between the two species may abort development of the hybrid at some embryonic stage. Of the numerous species of frogs belonging to the genus *Rana*, some live in the same regions and habitats, where they may occasionally hybridize. But the hybrids generally do not complete development, and those that do are frail.

Reduced Hybrid Fertility. Even if two species mate and produce hybrid offspring that are vigorous, reproductive isolation is intact if the hybrids are completely or largely sterile. Since the infertile hybrid cannot backbreed with either parental species, genes cannot flow freely between the species. One cause of this barrier is a failure of meiosis to produce normal gametes in the hybrid if chromosomes of the two parent species differ in number or structure. The most familiar case of a sterile hybrid is the mule, a robust cross between a horse and a donkey; horses and donkeys remain distinct species because, except very rarely, mules cannot backbreed with either species (FIGURE 22.5).

FIGURE 22.5
Hybrid sterility, a postzygotic barrier. A horse (left) can mate with a donkey (center) to produce a viable hybrid offspring in the form of a mule (right). This does not blur the distinction between horses and donkeys as separate species because the mule is sterile and thus genes cannot flow between the two parental species.

Hybrid Breakdown. In some cases when species cross-mate, the first-generation hybrids are viable and fertile, but when these hybrids mate with one another or with either parent species, offspring of the next generation are feeble or sterile. For example, different cotton species can produce fertile hybrids, but breakdown occurs in the next generation when offspring of the hybrids die in their seeds or grow into weak and defective plants.

FIGURE 22.6 summarizes the reproductive barriers between closely related species.

Introgression

Alleles may occasionally seep through all reproductive barriers and pass between the gene pools of closely related species when fertile hybrids mate successfully with one of the parent species. This transplantation of alleles between species is called **introgression.** For example, corn (*Zea mays*) has some alleles that can be traced to a closely related wild grass called teosinte (*Zea mexicana*). Introgression occurs when the two species hybridize and a fraction of the hybrids manage to cross with corn plants. The transplantation of alleles increases the reservoir of genetic variation that can be exploited by breeders

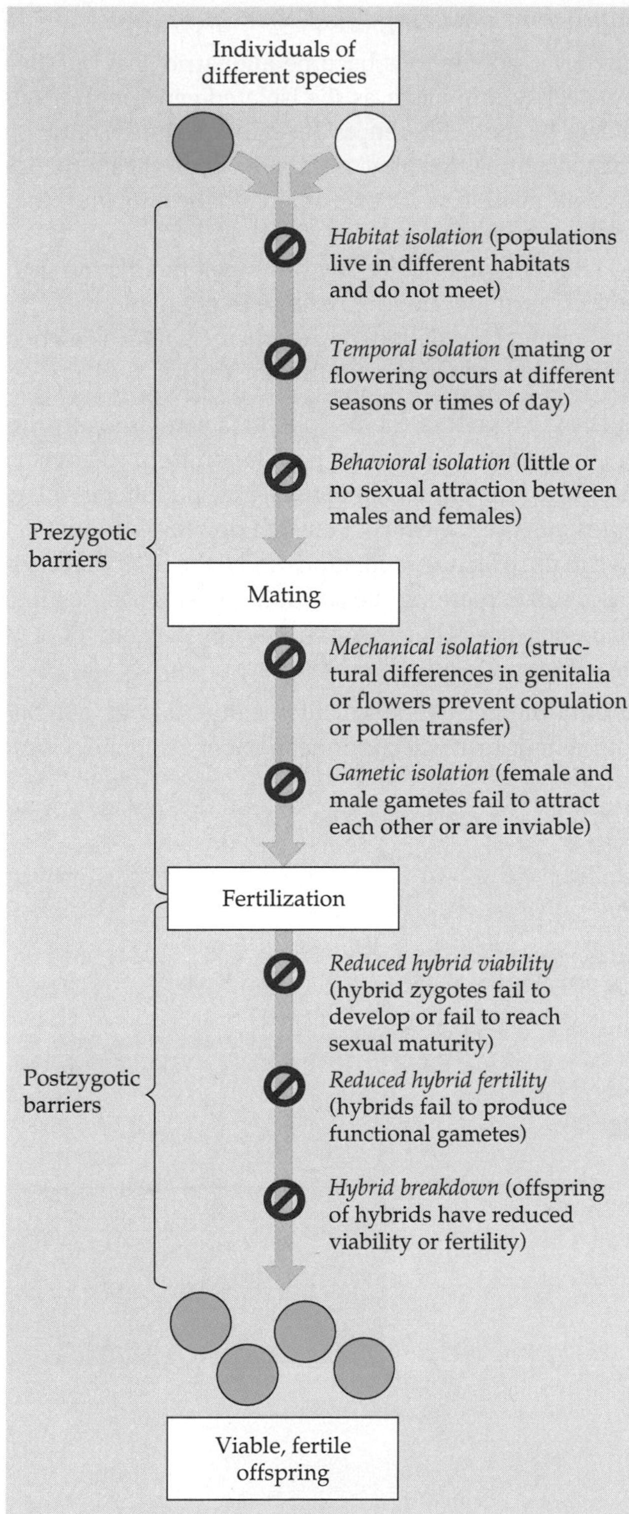

FIGURE 22.6
A summary of reproductive barriers between closely related species.

trying to produce new corn varieties by artificial selection. But occasional hybridization does not erase the boundary between corn and teosinte. As long as reproductive barriers hold introgression to a trickle, the isolation of the two gene pools is not seriously breached, and the two species remain distinct.

If the reproductive barriers we have surveyed form the boundaries around species, then the evolution of these barriers is the key biological event in the origin of new species. Let's now examine situations that bring about reproductive isolation, and hence speciation.

Geographical isolation can lead to the origin of species: allopatric speciation

The crucial episode in the origin of a species occurs when the gene pool of a population is severed from other populations of the parent species and gene flow no longer occurs. With its gene pool isolated, the splinter population can follow its own evolutionary course as changes in allele frequencies caused by selection, genetic drift, and mutations occur undiluted by gene flow from other populations. Speciation episodes can be classified into two modes based on the geographical relationship of a new species to its ancestral species. The initial block to gene flow may be a geographical barrier that physically isolates the population. This speciation mode is termed **allopatric speciation** (Gr. *allos,* "other," and L. *patria,* "homeland"), and populations segregated by a geographical barrier are known as allopatric populations. In the second speciation mode, a subpopulation becomes reproductively isolated in the midst of its parent population; this is **sympatric speciation** (Gr. *sym,* "together"). Populations are said to be sympatric if their ranges overlap. This section focuses on allopatric speciation. We'll examine mechanisms of sympatric speciation in the next section.

Geographical Barriers

Geological processes can fragment a population into two or more isolated populations. A mountain range may emerge and gradually split a population of organisms that can inhabit only lowlands; a creeping glacier may gradually divide a population; a land bridge, such as the Isthmus of Panama, may form and separate the marine life on either side; or a large lake may subside until there are several smaller lakes with their populations now isolated. Alternatively, a small population may become geographically isolated when individuals from the parent population colonize to a new location.

Just how formidable a geographical barrier must be to keep allopatric populations apart depends on the ability of the organisms to disperse based on the mobility of animals or the dispersibility of spores, pollen, and seeds of plants. The Grand Canyon is easily crossed by hawks and many other birds, but it is an impassable barrier to populations of small rodents confined to either the north or south rim of the canyon. Indeed, the same bird species populate both rims of the canyon, but each rim has several unique species of rodents (FIGURE 22.7).

Let's consider an example of how geographical isolation can lead to allopatric speciation. About 50,000 years ago, during an ice age, what is now the Death Valley region of California and Nevada had a very rainy climate and a system of interconnected lakes and rivers. A drying trend began about 10,000 years ago, and by 4000 years ago the region had become a desert. Today, all that is left of the network of lakes and rivers is isolated springs scattered in the desert, mostly in deep clefts between rocky walls. The springs vary extensively in water temperature and salinity. Living in many of the springs are tiny fishes called pupfishes, belonging to the genus *Cyprinodon*. Each inhabited spring, often no more than a few meters in diameter, is home to its own species of pupfish adapted to that pool and found nowhere else in the world. The various pupfishes probably descended from a single ancestral species whose range was broken up when the region became arid, cloistering several small populations that diverged in their evolution as they adapted to their home springs.

Conditions Favoring Allopatric Speciation

Whenever populations become allopatric, it is possible for speciation to occur as the isolated gene pools accumulate genetic differences by microevolution. But an isolated population that is small is more likely than a large population to change substantially enough to become a new species.

The geographical isolation of a small population usually occurs at the fringe of the parent population's range. The splinter population, or *peripheral isolate,* is a good candidate for speciation for three reasons:

1. The gene pool of the peripheral isolate probably differs from that of the parent population from the outset. Living near the border of the range, the peripheral isolate represents the extremes of any genotypic clines that existed in the original population. And if the peripheral isolate is small, there will be a founder effect resulting in a gene pool that is not representative of the gene pool of the parent population.

2. Until the peripheral isolate becomes a large population, genetic drift will continue to change its gene pool at

FIGURE 22.7
Allopatric speciation of squirrels in the Grand Canyon. One example of allopatric speciation is the two species of antelope squirrels that inhabit opposite rims of the Grand Canyon. On the south rim is Harris's antelope squirrel (*Ammospermophilus harrisi*). A few miles away on the north rim is the closely related white-tailed antelope squirrel (*Ammospermophilus leucurus*), which is slightly smaller, with a shorter tail that is white underneath. Birds and other organisms that can disperse easily across the canyon have not diverged into different species on opposite rims.

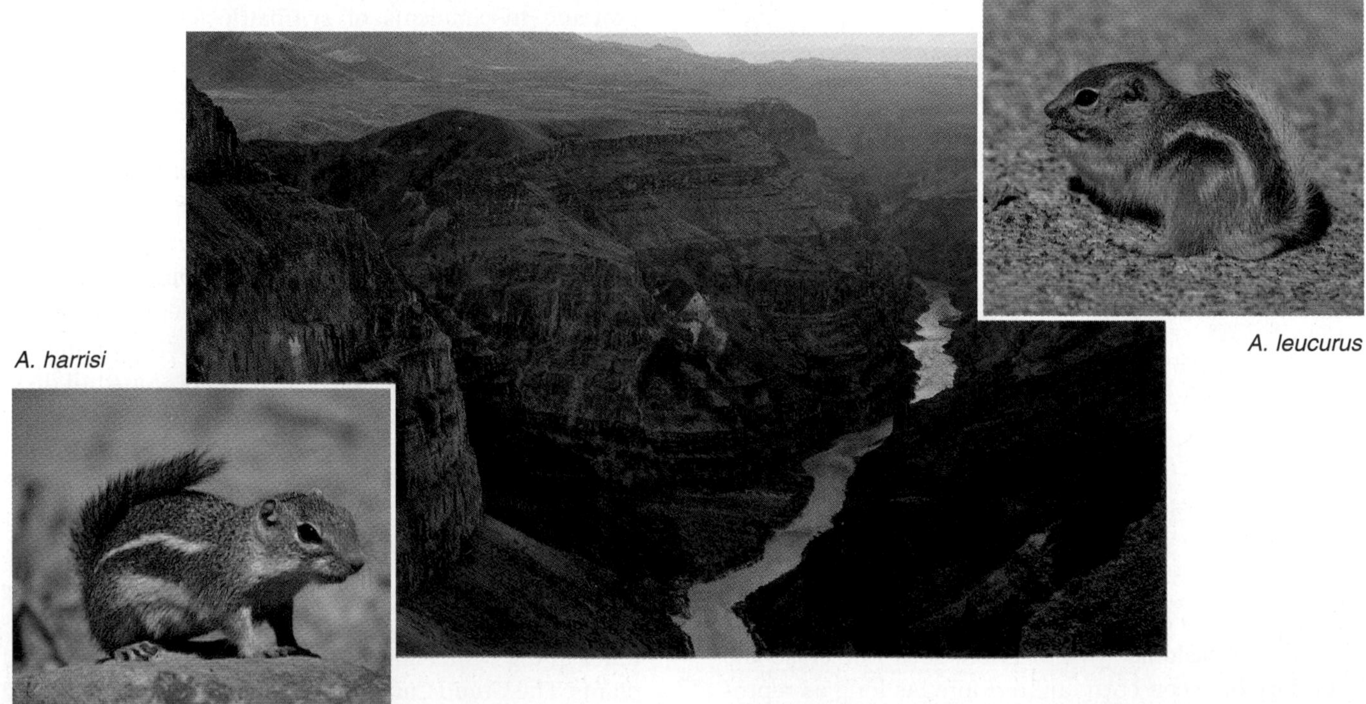

random. New mutations or combinations of existing alleles that are neutral in adaptive value may become fixed in the population by chance alone, causing genotypic and phenotypic divergence from the parent population.

3. Evolution caused by natural selection may take a different direction in the peripheral isolate than in the parent population. Because the peripheral isolate inhabits a frontier, where the environment is somewhat different, the peripheral isolate will probably encounter selection factors that are different from, and generally more severe than, those operating on the parent population.

These factors may cause peripheral isolates to follow an evolutionary course that diverges from that of the parent population as long as the gene pools remain isolated. This does not mean that all peripheral isolates persist long enough or change enough to become new species. Life on the frontier is usually harsh, and most pioneer populations probably become extinct. As evolutionary biologist Stephen Jay Gould puts it: "Status as a peripheral isolate merely gives a lottery ticket to a small population. A population can't win (speciate) without a ticket, but there are very few winning tickets."

There is ample evidence that allopatric speciation is much faster in small populations than in very large ones. The North American sycamore tree and the European sycamore represent large populations that have been allopatric for at least 30 million years, but specimens that are brought together still produce fertile hybrids. In comparison, the relatively small number of organisms that managed to reach the Galapagos Islands has given rise to hundreds of new species in less than two million years.

Adaptive Radiation on Island Chains

Islands are living laboratories for the study of speciation. Flurries of allopatric speciation have occurred on island chains where organisms that have strayed or become passively dispersed from their parent populations have founded new populations that evolved in isolation. The many endemic species of the Galapagos descended from stragglers that floated, flew, or were blown over the sea from the South American mainland. For example, consider Darwin's finches. A single dispersal event may have seeded one island with a small population of the ancestral finch, and the peripheral isolate formed a new species. Later, a few individuals of this island species may have reached neighboring islands, where geographical isolation permitted more speciation episodes (FIGURE 22.8). After diverging on one of these other islands, a new species could recolonize the island from which its founding population emigrated and coexist there with its parent species, or form still another species. Multiple invasions of islands by peripheral isolates of species from neighboring islands would eventually lead to the coexistence of several species on each island. The islands are far enough apart to permit populations to evolve in isolation, but close enough together for occasional dispersion events to occur.

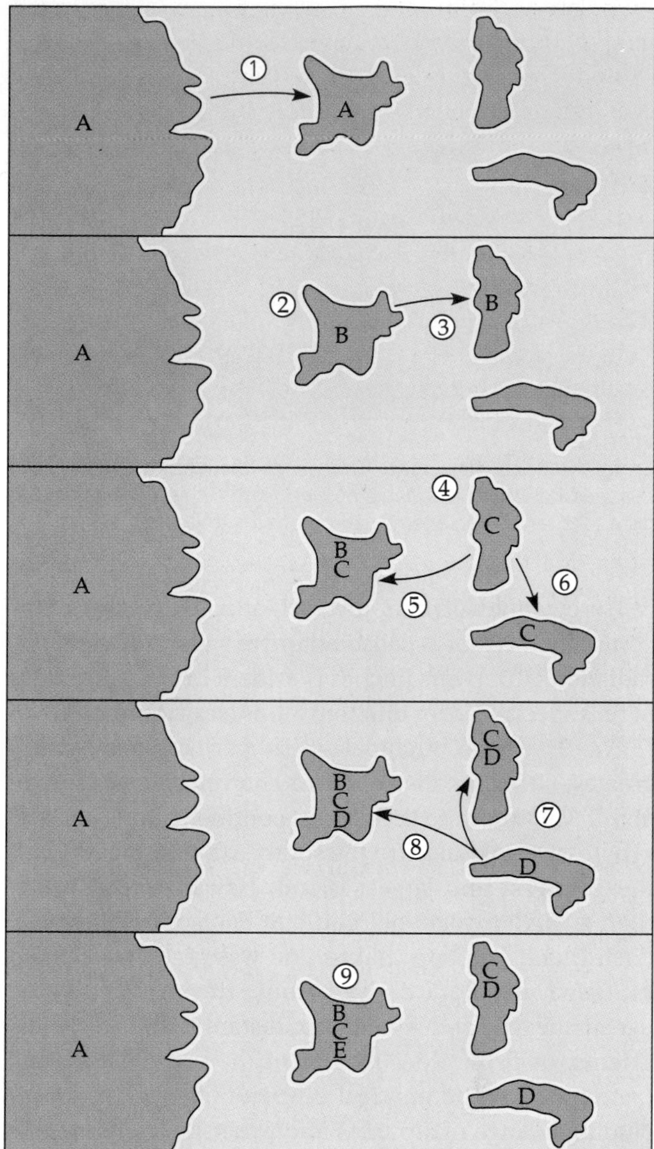

FIGURE 22.8
A model for adaptive radiation on island chains. ① One island in this cluster of three is seeded by a small colony founded by individuals of species A, blown over from a mainland population. ② Its gene pool isolated from the parent species, the island population evolves into species B as it adapts to its new environment. ③ Storms or other agents of dispersion spread species B to a second island, ④ where the isolated colony evolves into species C. ⑤ Later, individuals from species C recolonize the first island and cohabit with species B, but reproductive barriers keep the species distinct. ⑥ A colony of species C may also populate a third island, ⑦ where it adapts and forms species D. ⑧ Species D is dispersed to the two islands of its ancestors, ⑨ forming a new species, E, on one of those islands. The story could go on and on, with a series of allopatric speciation episodes made possible by the combination of isolation and occasional dispersal.

FIGURE 22.9
Long-distance dispersal. Plant seeds (the black dots) cling to this sea bird, which is capable of long flights. This is one mechanism that can disperse terrestrial organisms to isolated islands.

The evolution of many diversely adapted species from a common ancestor is called **adaptive radiation.** Adaptive radiation of Darwin's finches is evident in the many types of bills specialized for different foods (see FIGURE 20.4).

The Hawaiian Archipelago is perhaps the world's greatest showcase of evolution. The volcanic islands are about 3500 km from the nearest continent. They become progressively younger to the southeast, terminating with the youngest and largest island, Hawaii, which is less than a million years old and still has active volcanoes. Each island was born naked and was gradually clothed by fauna and flora derived from strays that rode the ocean currents and winds from distant islands and continents, or from older islands of the archipelago itself (FIGURE 22.9). The physical diversity of each island, including a range of altitudes and extensive differences in rainfall, provides many environmental opportunities for evolutionary divergence by natural selection. Multiple invasions and allopatric speciations have ignited an explosion of adaptive radiation; most of the thousands of species of plants and animals that now inhabit the islands are found nowhere else in the world. In contrast, there are no endemic species on the Florida Keys. Those islands are so close to the mainland that founding populations are not sequestered long enough for the origin of intrinsic reproductive barriers that block their gene pools from the steady stream of immigrants from the parent populations on the mainland.

A new species can originate in the geographical midst of the parent species: sympatric speciation

In sympatric speciation, new species arise within the range of parent populations; reproductive isolation evolves without geographical isolation. This can occur in a single generation if some genetic change results in a reproductive barrier between the mutants and the parent population.

Many plant species have their origins in accidents during cell division that result in extra sets of chromosomes, a mutant condition called **polyploidy.** Examine the causes of these chromosomal changes in FIGURE 22.10. An **autopolyploid** is an individual that has more than two chromosome sets, all derived from a single species. For example, an accident during cell division can double chromosome number from the diploid count ($2n$) to a tetraploid number ($4n$). The tetraploid can then fertilize itself (self-pollinate) or mate with other tetraploids. However, the mutants cannot interbreed successfully with diploid plants of the original population. The offspring, which would be triploid ($3n$), would be sterile because unpaired chromosomes result in abnormal meiosis. In just one generation, a postzygotic barrier has caused reproductive isolation and interrupted gene flow between a tiny population of tetraploids (maybe only a single plant initially) and the parent diploid population that surrounds it.

Sympatric speciation by autopolyploidy was first discovered early in this century by geneticist Hugo de Vries while he was studying the genetics of the evening primrose, *Oenothera lamarckiana,* a diploid species with 14 chromosomes. One day, de Vries noticed an unusual variant that had appeared among his plants, and microscopic inspection revealed that it was a tetraploid with 28 chromosomes. He found that the plant was unable to breed with the diploid primrose, and he named the new species *Oenothera gigas.*

Another type of polyploid species, much more common than autopolyploids, is called an **allopolyploid,** referring to the contribution of two different species to a polyploid hybrid. The potential origin of an allopolyploid begins when two different species interbreed and combine their chromosomes. Interspecific hybrids are usually sterile because the haploid set of chromosomes from one species cannot pair during meiosis with the haploid set from the other species. Though infertile, a hybrid may actually be more vigorous than its parents and propagate itself asexually (which many plants can do). At least two mechanisms, illustrated in FIGURE 22.10b, can transform the sterile hybrids into fertile polyploids.

Some allopolyploids are especially vigorous, apparently because they combine the best qualities of their two parent species. Speciation of polyploids, especially allopolyploids, accounts for 25% to 50% of plant species. Some of these new polyploid species have originated and spread within historic times. An example is a new species of salt-marsh grass that originated along the coast of southern England in the 1870s. It is an allopolyploid

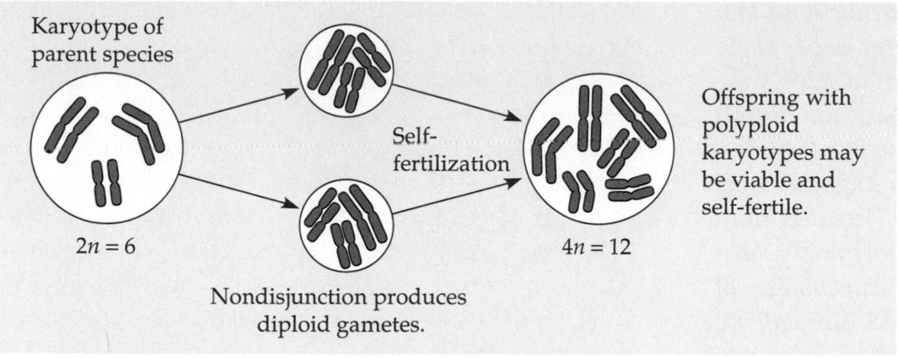

(**a**) Autopolyploidy

Gametes
$n = 2$
Species *A*
$2n = 4$
Species *B*
$2n = 6$
$n = 3$
Hybrid
Mitosis
Nondisjunction
produces polyploid
condition.
$2n = 10$
Meiosis
Gametes
$n = 5$
$n = 5$
Self-
fertilization
Viable
fertile hybrid
$2n = 10$

Species *A*
$2n = 4$
Unreduced gamete
$2n = 4$
Species *B*
$2n = 6$
Normal gamete
$n = 3$
Triploid hybrid
$3n = 7$
Unreduced gamete
$3n = 7$
Normal gamete
$n = 3$
Viable
fertile hybrid
$2n = 10$

(**b**) Allopolyploidy

FIGURE 22.10

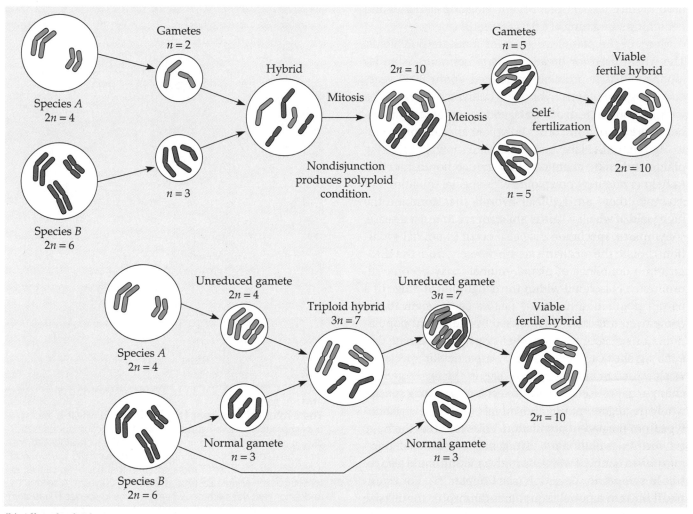

Sympatric speciation by polyploidy in plants. (**a**) **Autopolyploidy.** Nondisjunction in the germ (reproductive) cell line during either mitosis or meiosis results in diploid gametes. Self-fertilization produces a tetraploid zygote. (**b**) **Allopolyploidy.** A hybrid between two species is normally sterile because chromosomes are not homologous and cannot pair during meiosis. However, the hybrids may be able to reproduce asexually. Two different mechanisms can produce allopolyploid species from such hybrids. (Top) At some instant in the history of the hybrid clone, mitotic nondisjunction affecting the reproductive tissue of an individual may double chromosome number. The hybrid will then be able to make gametes because each chromosome has a homologue with which to synapse during meiosis. Union of gametes from this hybrid may give rise to a new species of interbreeding plants, reproductively isolated from both parent species. (Bottom) This diagram illustrates a more complex, but probably more common, origin of allopolyploid species. Both mechanisms produce new species having a chromosome number equal to the sum of the chromosomes in the two parent species. These various kinds of accidents during cell division make it possible for a new species to arise without geographical isolation from parent species.

derived from a European species (*Spartina maritima*) and an American species (*Spartina alternaflora*). Seeds of the American species, stowaways in the ballast of ships, were accidently introduced to England in the early nineteenth century. The invaders hybridized with the local species, and eventually a third species (*Spartina anglica*), morphologically distinct and reproductively isolated from its two parent species, evolved as an allopolyploid. Chromosome numbers are consistent with this mechanism of speciation: For *S. maritima*, $2n = 60$; for *S. alternaflora*, $2n = 62$; and for the new species, *S. anglica*, $2n = 122$. Since its inception, the new marsh grass has spread around the coast of Great Britain, clogging estuaries and becoming something of a pest (a weed).

Many of the plants we grow for food are polyploids. The wheat used for bread, *Triticum aestivum*, is an allopolyploid that probably originated about 8000 years ago as a spontaneous hybrid of a cultivated wheat and a wild grass (FIGURE 22.11). Oats, cotton, potatoes, and tobacco are among the other polyploid species important to agriculture. Plant geneticists are now hybridizing plants and using chemicals that induce nondisjunction to help create new polyploids with special qualities. For example, there are artificial hybrids that combine the high yield of wheat with the ability of rye to resist disease.

Sympatric speciation may also occur in animal evolution, though the mechanisms are different from the chromosome doublings of plants. Animals may become reproductively isolated within the geographical range of a parent population if genetic factors cause them to become fixed on resources not used by the parent population. Consider, for example, the wasps that pollinate figs. Each fig species is pollinated by a particular species of wasp, which mates and lays its eggs in the figs. A genetic change that caused wasps to select a different fig species would segregate mating individuals of this new phenotype from the parent population. This would set the stage for further evolutionary divergence. A balanced polymorphism coupled with assortative mating could also result in sympatric speciation (see Chapter 21). For example, if birds in a population that is dimorphic for bill size began to mate selectively with birds of the same morph, then eventual speciation is possible. Lake Victoria in Africa, which is less than a million years old, is home to almost 200 species of closely related fishes belonging to the cichlid family. Subdivision of populations into groups specialized for exploiting different food sources and other resources in the lake is one process that probably contributed to the explosive radiation of these fishes. However, most evolutionary biologists agree that allopatric speciation is far more common than sympatric speciation in the animal kingdom, while both speciation modes have important roles in plant evolution.

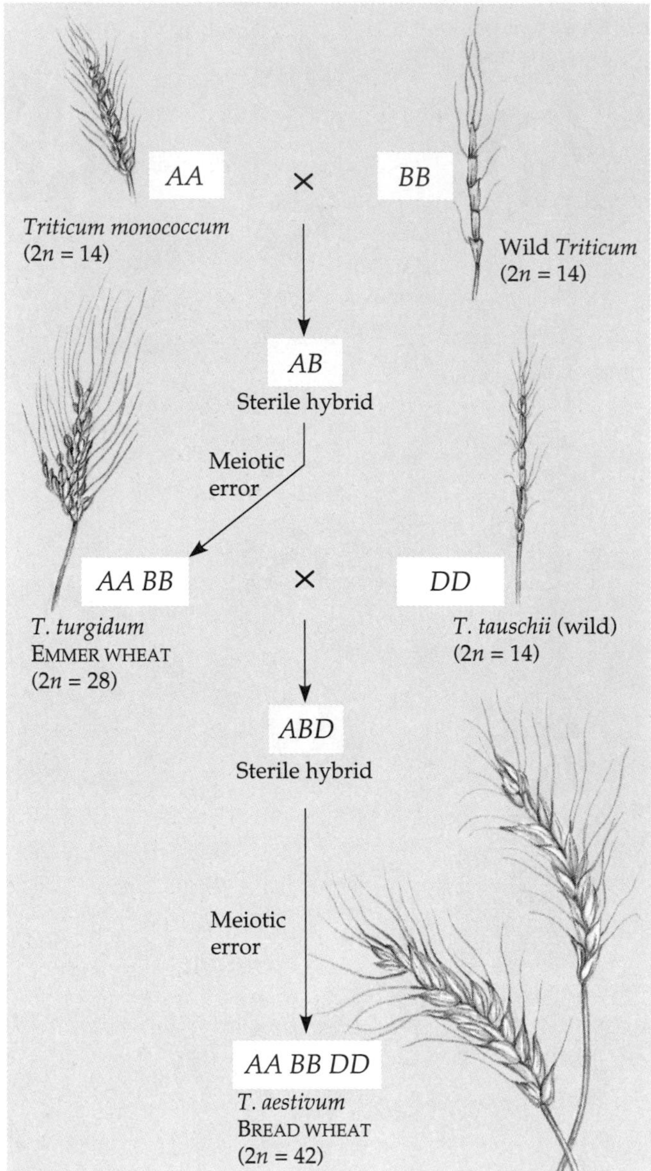

FIGURE 22.11
The evolution of wheat (*Triticum*). Bread wheat is the evolutionary product of two hybridization/nondisjunction episodes producing allopolyploids. The first sympatric speciation produced emmer wheat, derived from hybridization between wild *Triticum* and a domesticated species (*T. monococcum*) that has been cultivated for food in the Middle East for at least 11,000 years. A second hybridization, between emmer wheat and a wild species (*T. tauschii*), gave rise to bread wheat about 8000 years ago. The uppercase letters in this diagram represent sets of chromosomes that can be traced to a particular species. Thus, bread wheat has chromosome sets derived from three different ancestral species.

Population genetics can account for speciation

Classifying speciation modes as allopatric or sympatric accentuates the biogeographical factors of speciation but does not emphasize the actual genetic mechanisms that differentiate populations into separate species. In this section, we examine the changes required for the origin of species in the context of the population genetics presented in Chapter 21.

Speciation by Adaptive Divergence

When two populations adapt to disparate environments, they accumulate differences in genetic structure—differences in the frequencies of alleles and genotypes. In the course of this gradual adaptive divergence of two gene pools, reproductive barriers between the two populations may evolve coincidentally, differentiating the populations into two species.

A key idea in evolution by divergence is that reproductive barriers can arise without being favored directly by natural selection—there is no drive toward speciation for its own sake. Reproductive isolation is usually a secondary consequence of divergence of the two populations as they adapt to their separate environments. Postzygotic barriers may be pleiotropic effects of interspecific differences in genes controlling development. (To review pleiotropy, see Chapter 13). For example, in laboratory hybrids between two very similar species of *Drosophila, D. melanogaster* and *D. simulans,* only one of the two sets of genes for the synthesis of ribosomal RNA is active. This results in a very low viability for the hybrids. Prezygotic barriers can also evolve as by-products of the gradual genetic divergence of two populations. For instance, if a change in genetic structure enables one population of an insect species to adapt to a different host plant than that of other populations, then a habitat barrier to interbreeding with the other populations is a side effect.

There *are* cases where reproductive isolation evolves more directly by sexual selection in isolated populations. Such selection may have helped differentiate *Drosophila* on the Hawaiian Islands into hundreds of species. For example, the wide head of the *D. heteroneura* male enhances reproductive success with females of the same species, but makes successful courtship with females of other species very unlikely (FIGURE 22.12). However, even when sexual selection results in reproductive barriers, they evolve as adaptations that enhance reproductive success within a single population, not as safeguards against interbreeding with other populations. After all, reproductive barriers usually evolve when populations are allopatric, so they cannot possibly be functioning directly to isolate the gene pools of populations. For this reason, one criticism of the biological species concept is its emphasis on reproductive isolating mechanisms. One alternative is the **recognition concept of species,** which defines a species by its set of characteristics that maximize successful mating with members of the same population—the molecular, morphological, and behavioral characteristics that enable an individual to recognize a "good" mate of the same species. Reproductive isolation from other populations would be a spin-off. Whether a species definition emphasizes "reproductive isolating mechanisms" or "mate-recognition mechanisms" may seem like stressing opposite sides of the same coin. But the recognition concept may help focus attention on those characteristics that are actually subject to natural selection in an isolated population in the process of speciation.

FIGURE 22.12
The role of sexual selection in the radiation of Hawaiian *Drosophila* species. The Hawaiian Islands are home to about 800 *Drosophila* species, one-third of all known species of these fruit flies. Colonization and island hopping made this adaptive radiation possible (see FIGURE 22.8). Sexual selection probably played a major role in differentiating populations into separate species. Unique courtship rituals, usually dependent on specific morphological features of the males, contribute to prezygotic barriers between closely related species. In *D. heteroneura,* shown here as an example, males have "hammerheads," with eyes set far apart. This trademark probably functions in species recognition as the male faces the female during a courtship dance. The "pictures" on the wings, species-unique patterns of dots, may also contribute to mate recognition.

Speciation by Shifts in Adaptive Peaks

In the 1930s, Sewell Wright crafted an evolutionary metaphor known as the adaptive landscape. This symbolic landscape has many adaptive peaks separated by valleys (FIGURE 22.13). An **adaptive peak** represents an equilibrium state where the gene pool has allele frequencies that optimize the population's success in that environment. Even in a stable environment, several adaptive peaks for a given population are possible, but natural selection will tend to maintain the population at a single peak. To reach an alternative adaptive peak by some change in the overall gene pool, a population must go through a period corresponding to a valley on the adaptive landscape, where

FIGURE 22.13
The adaptive landscape and peak shifts. For any population in a stable environment, there are many alternative states of genetic adaptation symbolized as peaks on an adaptive landscape. The peaks are separated by valleys representing genetic combinations of relatively low average fitness. In a new environment, the adaptive landscape is redefined, and survival of the population may depend on natural selection moving the gene pool to a new adaptive peak. However, population geneticists reserve the term *peak shift* for speciation episodes triggered by nonadaptive changes in the genetic system, such as genetic drift or the origin of a polyploid population. Once a small population is genetically destabilized and dislodged from its original adaptive peak, natural selection may cause a generation-by-generation climb to some new adaptive peak.

the average fitness of individuals is low. Thus, if some slight change in the frequency of alleles at one or more loci drives a population off an adaptive peak, natural selection will usually push the population back to its original peak. However, if the environment should change, then the adaptive landscape is redefined. A population that survives in this new environment must climb another adaptive peak through microevolution of its gene pool. But this is just another way of looking at speciation by adaptive divergence. What population geneticists call a *peak shift* is triggered *not* by a new physical environment, but by nonadaptive changes in the genetic structure of a population.

Peak shifts can be caused by a founder effect or a bottleneck. By randomly changing allele frequencies in the gene pool, genetic drift can knock a small population off its original adaptive peak. If the gene pool is sufficiently destabilized, new adaptive peaks may be within reach. If the population survives, it will be natural selection that pushes it to a new adaptive peak as the generations pass. Thus, adaptive evolution plays a major role in a peak shift, but it is genetic drift that makes the shift possible. A peak shift can occur in a bottlenecked population even if its environment is stable, because many adaptive peaks are possible under the same environmental conditions. Once a small population is genetically destabilized and dislodged from its original adaptive peak, natural selection may cause a generation-to-generation climb to some new adaptive peak. In the case of a founder effect, the small population is not only subject to genetic drift moving the gene pool randomly over the adaptive landscape, but a new set of adaptive peaks now exists. This combination of genetic drift followed by natural selection in a new environment may be responsible for the relatively rapid radiation of island species.

Hybrid Zones and the Cohesion Concept of Species

What happens when two closely related populations that have been allopatric for some time come back into contact? Three outcomes are possible. The two populations may interbreed freely, in which case their gene pools will be amalgamated; speciation has not occurred during the period of geographical isolation. A second possible outcome is that evolutionary divergence during the period of allopatry resulted in reproductive barriers that keep the gene pools of the two populations separate even if sympatry is restored; speciation has occurred. Between these two extremes is a third outcome, the establishment of a hybrid zone.

A **hybrid zone** is a region where two related populations that diverged after becoming geographically isolated make secondary contact and interbreed where their geographical ranges overlap. For example, two phenotypically distinct populations of woodpeckers, the red-shafted flicker in western North America and the yellow-shafted flicker in central North America, interbreed along a hybrid zone stretching through the Great Plains from the Texas panhandle to southern Alaska. The two populations probably became separated during the ice ages, renewing contact at least a few centuries ago. Although the two flicker populations have been interbreeding in that hybrid zone for at least two hundred years, the introgression of alleles between the populations does not penetrate far beyond the hybrid zone. The hybrid zone is relatively stable, rather than expanding. Put another way, the frequencies of genotypes and phenotypes that distinguish the red-shafted flicker from the yellow-shafted flicker grade sharply in the hybrid zone as steep clines (see Chapter 21). But the two flicker populations remain distinct deep in their ranges, away from the hybrid zone. Evolutionary biologists debate whether such populations should be considered subspecies or full, separate species.

Some researchers who favor species status for populations that remain distinct in spite of hybrid zones also argue that such stable hybrid zones pose a problem for the biological species concept. If two species can maintain their taxonomic identity even though they hybridize, there must be cohesive forces other than reproductive isolation that hold a species together and prevent its complete merger with a closely related species.

Thus, the **cohesion concept of species** is another alternative to the biological species concept. In part, the cohesion may involve a distinctive, integrated set of adaptations that has been refined during the evolutionary history of a population. Stabilizing selection would restrict phenotypic variation to a range narrow enough to define the species as separate from other species. To use the metaphor of the adaptive landscape, the populations of red-shafted and yellow-shafted flickers are clustered around two different adaptive peaks. The genetic basis for this cohesion of phenotype may involve specific combinations of alleles and specific linkages between gene loci on chromosomes. The clinal change of genetic structure and phenotype in the hybrid zone may be correlated with a transition in the environmental factors that help shape the two distinct populations (subspecies or species) on either side of the zone. As a counterpoint to the cohesion concept of species, Ernst Mayr argues that the biological species concept, which emphasizes reproductive barriers as the main boundaries between species, can be stretched to account for limited hybridization: "A population does not lose its species status when an individual belonging to it makes a mistake."

TABLE 22.1 reviews the four species concepts that have come up in this chapter. Keep in mind that each concept may have utility in different situations, depending on the kinds of questions we are asking about species and speciation.

TABLE 22.1

Four Concepts of Species Compared

Morphological concept of species	Emphasizes measurable anatomical differences between species. Most species recognized by taxonomists have been designated as separate species based on morphological criteria.
Biological concept of species	Emphasizes reproductive isolation, the potential of members of a species to interbreed with each other but not with members of other species.
Recognition concept of species	Emphasizes mating adaptations that become fixed in a population, as individuals "recognize" certain characteristics of suitable mates.
Cohesion concept of species	Emphasizes cohesion of phenotype as the basis of species integrity, with each species defined by its integrated complex of genes and set of adaptations.

How Much Genetic Change Is Required for Speciation?

It is not possible to generalize about the "genetic distance" between closely related species. In some cases, reproductive isolation may result from the cumulative divergence of populations at many gene loci. In other cases, changes at only a few loci produce reproductive barriers. For example, two species of Hawaiian *Drosophila, D. silvestris* and *D. heteroneura,* differ in alleles at a gene locus determining head shape, a character important in mate recognition by these flies (see FIGURE 22.12). However, the phenotypic effect of different alleles at this locus is amplified by at least ten other gene loci that interact in an epistatic system. (For a review of the genetic interactions known as epistasis, see Chapter 13.) Thus, it took more than one mutation to differentiate these two species of *Drosophila.* However, it is also clear from such examples that massive genetic change involving hundreds of loci is not mandatory for speciation. This conclusion is relevant to a debate among evolutionary biologists about the tempo of speciation, an issue we will now examine.

The theory of punctuated equilibrium has stimulated research on the tempo of speciation

The traditional evolutionary tree that diagrams the descent of species from ancestral forms sprouts branches that diverge gradually, each new species evolving continuously over long spans of time (FIGURE 22.14a). The theory behind the tree is the extrapolation of the processes of microevolution, changes in the frequencies of alleles in gene pools, to the divergence of species—that is, big changes occur by the accumulation of many small ones. However, paleontologists rarely find gradual transitions of fossil forms. Instead, they often observe species appearing as new forms rather suddenly (in geological terms) in a layer of rocks, persisting essentially unchanged for their tenure on Earth, and then disappearing from the record of the rocks as suddenly as they appeared. To Darwin, the origin of species was an extension of adaptation by natural selection, with isolated populations from common ancestral stock evolving differences gradually as they adapted to their local environments. But Darwin himself was bewildered by the dearth of connecting fossils and wrote: "Although each species must have passed through numerous transitional stages, it is probable that the periods during which each underwent modification, though many and long as measured by years, have been short in comparison with the periods during which each remained in an unchanged condition."

Advocates of a theory known as **punctuated equilibrium** have redrawn the evolutionary tree to represent the fossil evidence for evolution occurring in spurts of relatively rapid change, instead of a gradual divergence of species (FIGURE 22.14b). This theory, first proposed in 1972 by Niles Eldredge and Stephen Jay Gould, depicts species undergoing most of their morphological modification as they first bud from parent species, and then changing little, even as they produce additional species. The name of the theory is derived from the idea of long periods of stasis punctuated by episodes of speciation. A change in the genome, such as occurs in the origin of new polyploid plants, would be one mechanism of sudden speciation. Punctuationalists, as the proponents of this model of evolution are called, point out that allopatric speciation of a splinter population isolated from its parent population by a geographical barrier can also be quite rapid. Genetic drift and natural selection can cause significant change in just a few hundred to a few thousand generations in the gene pool of a population cloistered in a challenging new environment.

How can speciation in a few thousand generations, which may require several thousand years, be called an abrupt episode? The fossil record indicates that successful species last for a few million years, on average. Suppose that a particular species survives for 5 million years, but most of its morphological changes occurred during the first 50,000 years of its existence. In this case, the evolution of the species-defining characteristics was compressed into just 1% of the lifetime of the species. On the time scale that can generally be determined in fossil strata, the species will appear suddenly in rocks of a certain age and then linger with little or no change before becoming extinct. During its formative millennia, the species may have accumulated its modifications gradually, but relative to the overall history of the species, its inception was abrupt. This scenario of an evolutionary spurt preceding a much longer period of morphological stasis would explain why paleontologists find relatively few smooth transitions in the fossil record of species.

Once it is acknowledged that "sudden" may be many thousands of years on the vast scale of geological time, the debate between punctuationalists and gradualists over the rate of speciation is muted somewhat. The degree to which a species changes after its origin is another issue. If the species is adapted to an environment that stays the same, then natural selection would counter changes in the gene pool. In this view, the tendency for stabilizing selection to hold a population at one adaptive peak results in long periods of stasis.

Some gradualists retort that stasis is an illusion. Many species may continue to change after they come into existence, but in ways that cannot be detected from fossils. By necessity, paleontologists base their theories of descent almost entirely on external anatomy and skeletons. Changes in internal anatomy may go unnoticed, as would modifications in physiology or behavior. Population geneticists point out that many of the effects of microevolution occur at the molecular level without overtly affecting morphology.

Even the claim of long periods of morphological stasis in the history of species is contested. Peter Sheldon of Trinity College in Dublin has analyzed about 15,000 fossils of trilobites from shale deposits in Wales (trilobites

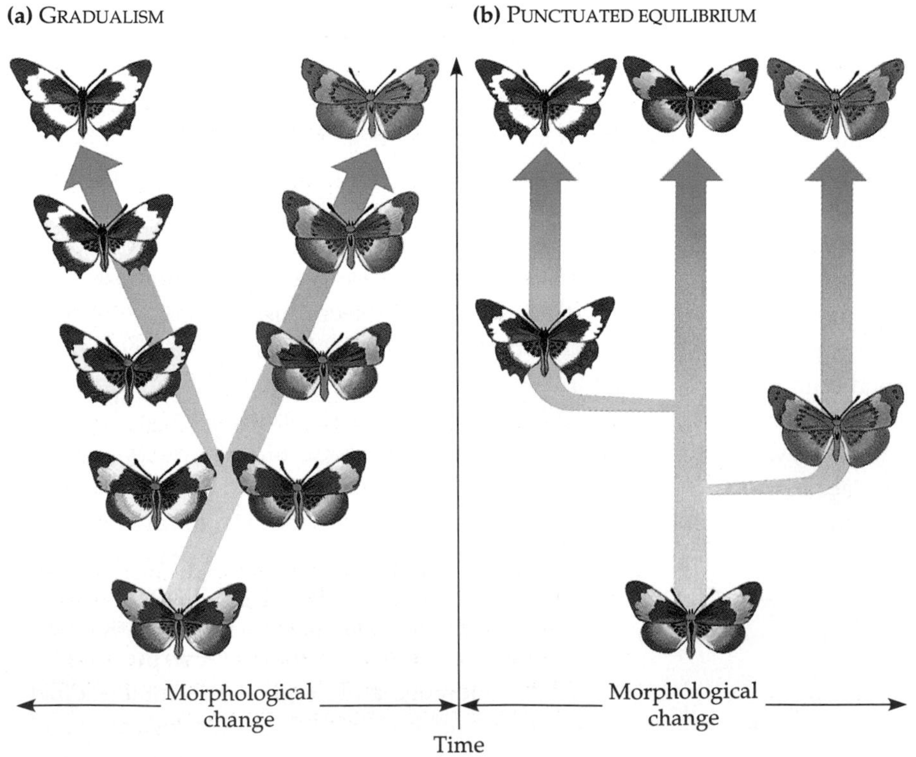

FIGURE 22.14
Models of two theories for the tempo of speciation. A new species buds from its parent species as a small isolated population, indicated in this figure by the narrow base of the branching arrow. (**a**) In gradualism, species descended from a common ancestor diverge more and more in morphology as they acquire unique adaptations. (**b**) According to the theory of punctuated equilibrium, a new species changes most as it buds from a parent species, and then changes little for the rest of its existence.

are extinct arthropods; see Chapter 29). At this study site, the fossil record of trilobites is unusually complete. Paleontologists have arranged these fossils into several evolutionary lineages. The youngest and oldest fossils of each lineage have been classified as different species because of morphological differences, including a different number of ridges on the tail sections of their shells. However, after Sheldon's extensive study of the fossils, he finds it impossible to draw such species boundaries. In each evolutionary lineage of the trilobites, the average number of tail ridges in the fossil populations changes gradually in the succession of rock layers. Sheldon argues that the requirement for paleontologists to give their fossils species names can artificially lead to a punctuated interpretation of a fossil series that actually changes gradually. In contrast to Sheldon's trilobite research, Alan Cheetham of the Smithsonian Institution reported in 1995 that his exhaustive study of fossilized bryozoans (invertebrates) supports punctuated equilibrium. Long periods of stasis in the morphology of these animals are punctuated by the relatively rapid appearance of new species. What is needed, both of these researchers agree, are many more exhaustive studies of fossil morphology where specific lineages are well preserved. Only then will we be able to assess the relative importance of gradual and punctuated tempos in the origin of new species.

Whatever the outcome of research on the pace of speciation, there is no question that the theory of punctuated equilibrium has stimulated research and catalyzed a new interest in paleontology. In the next chapter, we will see how this investigation of the fossil record extends beyond the issue of speciation to major patterns in the history of life.

REVIEW OF KEY CONCEPTS (with page numbers and key figures)

- The biological species concept emphasizes reproductive isolation (pp. 436–439, TABLE 22.1)
  - Biological species are groups of populations that have the potential to interbreed with each other but not with other species.
- Reproductive barriers separate species (pp. 439–441, FIGURE 22.6)
  - Any intrinsic factor that prevents two species from producing viable, fertile hybrids is a reproductive barrier that maintains the integrity of a species.
  - Prezygotic barriers prevent cross-specific mating or fertilization. In particular, species that occupy the same geographic area often live in separate habitats (habitat isolation); breed at different times (temporal isolation); possess unique, exclusive mating signals and courtship behaviors (behavioral isolation); and/or have anatomically incompatible reproductive organs (mechanical isolation) or incompatible sex cells (gametic isolation).
  - Even if two different species manage to mate, postzygotic barriers usually prevent the interspecific hybrids from developing into adults, breeding with either parent species, or producing viable, fertile offspring.
- Geographical isolation can lead to the origin of species: allopatric speciation (pp. 441–444, FIGURE 22.8)
  - Allopatric speciation occurs when a splinter population evolutionarily diverges from its parent population after becoming geographically isolated.
  - Small splinter populations are better candidates for allopatric speciation than large ones because genetic drift and natural selection can change a small gene pool faster.
  - Adaptive radiation is the evolution of numerous species with diverse adaptations from a common ancestor. Island chains are showcases of radiation.
- A new species can originate in the geographical midst of the parent species: sympatric speciation (pp. 444–446, FIGURE 22.10)
  - Sympatric speciation occurs without geographic separation when a segment of the population experiences a genetic change that results in reproductive isolation.
  - Sympatric speciation is most common in plants, in which the mutation is usually the doubling of the chromosome number. Autopolyploids are species derived this way from one ancestral species. Allopolyploids are species with multiple sets of chromosomes derived from two different species.
- Population genetics can account for speciation (pp. 446–449)
  - Reproductive barriers may arise coincidentally as two populations genetically diverge while adapting to different environments.
  - Sexual selection may lead directly to reproductive barriers. According to the recognition concept of species, natural selection would amplify adaptations that enhance reproductive success with members of the same species.
  - According to Wright's evolutionary metaphor, a population's gene pool is perched on an adaptive peak, one of several possible peaks on an adaptive landscape. Natural selection will tend to maintain a population at an adaptive peak that maximizes fitness. Peak shifts, associated with nonadaptive changes in the gene pool, may be initiated by a founder effect or bottleneck and driven by genetic drift. Adaptive evolution can push a destabilized gene pool to a new adaptive peak.
  - According to the cohesion concept of species, reproductive criteria alone cannot account for species integrity. For example, two species may interbreed freely at a hybrid zone without losing their distinctiveness away from the zone.
  - Speciation may result from a change at just a few gene loci or by the cumulative divergence at many loci.

- The theory of punctuated equilibrium has stimulated research on the tempo of speciation (pp. 449–451, FIGURE 22.14)
  - One view of speciation is that it usually occurs gradually by an accumulation of microevolutionary changes in gene pools.
  - According to the theory of punctuated equilibrium, a species changes most when it buds from an ancestral species and then remains fairly static in morphology for the rest of its duration.

SELF-QUIZ

1. Most of biological diversity has probably arisen by
   a. anagenesis
   b. cladogenesis
   c. phyletic evolution
   d. hybridization
   e. sympatric speciation

2. The largest unit in which gene flow is possible is a
   a. population
   b. species
   c. genus
   d. subspecies
   e. phylum

3. Bird guides once listed the myrtle warbler and Audubon's warbler (above) as distinct species, but recent books show them as eastern (left) and western forms of a single species, the yellow-rumped warbler. Experts must have found that the two kinds of warblers
   a. live in the same areas
   b. successfully interbreed in nature
   c. look enough alike to be considered one species
   d. are reproductively isolated from each other
   e. are allopatric

4. Among allopatric species of *Anopheles* mosquito, some live in brackish water, some in running fresh water, and others in stagnant water. What type of reproductive barrier is most obviously separating these different species?
   a. habitat isolation
   b. temporal isolation
   c. behavioral isolation
   d. gametic isolation
   e. postzygotic barriers

5. An eight-lane superhighway is constructed across a grassland. For the next ten years, a researcher compares the genetic structure of several populations of organisms on the two sides of the highway. The most likely organisms to show evolutionary divergence are
   a. crows
   b. mice
   c. grasses with windblown pollen
   d. snails
   e. butterflies

6. According to advocates of the punctuated equilibrium theory,
   a. natural selection is unimportant as a mechanism of evolution
   b. given enough time, most existing species will branch gradually into new species
   c. a new species accumulates most of its unique features as it comes into existence and changes little for the rest of its duration as a species
   d. most evolution is anagenic
   e. speciation is usually due to a single mutation

7. The biological species concept cannot be applied to two putative species that are
   a. sympatric
   b. nearly indistinguishable in morphology
   c. reproductively isolated
   d. capable of forming viable hybrids
   e. exclusively asexual

8. Future cladogenesis of human populations to form new hominid (human) species is probably unlikely because
   a. the environment has stabilized
   b. humans are already perfectly adapted organisms
   c. only one adaptive peak is possible for hominids
   d. most human populations are very large and are incompletely isolated from surrounding populations
   e. human variation is not very extensive

9. Plant species A has a diploid number of 12. Plant species B has a diploid number of 16. A new species, C, arises as an allopolyploid from hybridization of A and B. The diploid number of C would probably be
   a. 12
   b. 14
   c. 16
   d. 28
   e. 56

10. The speciation episode described in question 9 is most likely a case of
    a. allopatric speciation
    b. sympatric speciation
    c. speciation based on sexual selection
    d. adaptive radiation
    e. anagenic speciation

CHALLENGE QUESTIONS

1. According to one hypothesis, two closely related species should be most distinct from each other where their ranges overlap—that is, in regions of sympatry. This phenomenon is called character displacement. The hypothesis goes on to propose that character displacement reduces interspecific competition for food, habitat, and other resources, and also minimizes hybridization between the species in their region of sympatry. Character displacement has been difficult to demonstrate in nature. An experiment has nevertheless addressed the problem by comparing bill sizes of two species of Darwin's ground finches (*Geospiza fortis* and *G. fuliginosa*) in sympatry and allopatry. The experiment controlled for any possible effect on morphology by variation in food supply among locations. Why do you think this control was necessary? After taking this control into account, how would you expect the relative bill sizes of *G. fortis* and *G. fuliginosa* on Santa Cruz Island, the only island they cohabit, to compare with the relative bill sizes of *G. fortis* on the islands of Daphne Major and *G. fuliginosa* on Los Hermanos if character displacement were operative?

2. In 1990, J. Jackson and A. Cheetham, of the Smithsonian Institution, published a study on the reliability of morphology as a taxonomic tool for classifying invertebrates called bryozoans into species. Skeleton morphology alone was sufficient to distinguish modern species. Relate this research to the criticism that the theory of punctuated equilibrium is an inaccurate interpretation of the fossil record because it is based only on measuring morphological change.

3. Write a paragraph explaining why remote islands have a proportionately greater number of endemic species than do islands close to a mainland.

4. Explain why horses and donkeys remain separate species even though they produce viable hybrids.

SCIENCE, TECHNOLOGY, AND SOCIETY

1. Thousands of Snake River sockeye salmon used to spawn in Redfish Lake, Idaho, but in 1990 not a single sockeye returned to the spawning grounds. The salmon are impeded on their journeys up and down the Columbia River and Snake River by eight hydroelectric dams. In 1991, the U.S. government added the Snake River sockeye salmon to the list of endangered species. This is one of several sockeye populations in the river system. The Endangered Species Act grants protection not just to a whole biological species, but also to "distinct population segments." How would you define "distinct" in this context, and how would you try to find out whether a salmon population meets your definition? What measures would you be willing to implement to protect the Snake River sockeye salmon population? What would be the advantages and disadvantages of these protective measures?

2. Through genetic engineering, it is possible to modify some individuals in a population so that they can no longer exchange genes with other individuals from the same population. Would you consider these "engineered" organisms a new species? Why or why not? It is acceptable for *Homo sapiens* to create new species? If so, under what circumstances?

3. What is the biological basis for assigning all human populations to a single species? Explain why it is unlikely that a second human species could arise by cladogenesis in the future.

FURTHER READING

Culotta, E. "How Many Genes Had to Change to Produce Corn?" *Science,* June 28, 1991. The evolution of an important food source.

Dorozynski, A. "A Family Tree of European Bears." *Science,* January 14, 1994. The legal and political issues surrounding the distinction between species and populations.

Kerr, R. A. "Did Darwin Get It All Right?" *Science,* March 10, 1995. Studies of invertebrate fossils are adding support for punctuated equilibrium.

Kluger, T. "Go Fish." *Discover,* March 1992. Rapid speciation in Lake Victoria.

Knowlton, N. "A Tale of Two Seas." *Natural History,* June 1994. Allopatric speciation resulting from the Panamanian land bridge separating the Caribbean from the Pacific Ocean.

Rennie, J. "Darwin's Current Bulldog." *Scientific American,* August 1994. A profile of Ernst Mayr, one of the architects of the modern evolutionary synthesis.

Ryan, M. J. "Signals, Species, and Sexual Selection." *American Scientist,* January/February 1990. Frog songs in the context of the mate-recognition concept of species.

CHAPTER 23

TRACING PHYLOGENY: MACROEVOLUTION, THE FOSSIL RECORD, AND SYSTEMATICS

KEY CONCEPTS

- The fossil record documents macroevolution
- Paleontologists use a variety of methods to date fossils
- What are the major questions about macroevolution? *an overview*
- Some evolutionary novelties are modified versions of older structures
- Genes that control development play a major role in evolutionary novelty
- Recognizing trends in the fossil record does not mean that macroevolution is goal-oriented
- Macroevolution has a biogeographical basis in continental drift
- The history of life is punctuated by mass extinctions followed by adaptive radiations of the survivors
- Systematics connects biological diversity to phylogeny
- Molecular biology provides powerful new tools for systematics
- Cladistics highlights the phylogenetic significance of systematics
- Is a new evolutionary synthesis necessary?

The term macroevolution implies substantial change in organisms. Here is a more specific definition: ***Macroevolution*** *is the origin of taxonomic groups higher than the species level. In other words, macroevolutionary change is substantial enough that we view its products as new genera, new families, or even new phyla. The origin of mammals from their reptilian ancestors is an example: Reptiles and mammals are related, but different enough that zoologists place them in separate vertebrate classes. In broader terms, macroevolution is the story of the major events in the history of life that are revealed by the fossil record. (This chapter opens with a photograph of fossilized fishes.) Evolution on this grand scale encompasses the origin of novel designs, such as the feathers and wings of birds and the upright posture of humans; evolutionary trends, such as increasing brain size in mammals; the explosive diversification of certain groups of organisms following some evolutionary breakthrough, such as the adaptive radiation of flowering plants; and mass extinctions, which cleared the way for new episodes of adaptive radiation, such as the radiation of mammals following the disappearance of the dinosaurs.*

This chapter describes how biologists study macroevolution, discusses theories on the origin of new biological designs, and examines a few important episodes in the history of life. These topics are also related to systematics, the study of biological diversity, both past and present. The last section, the capstone of our unit on mechanisms of evolution, considers one of the fundamental questions of evolutionary biology: Is macroevolution mainly the cumulative product of microevolution working over vast spans of time? Put another way, is the gradual modification of populations due to natural selection the main cause of the larger changes we observe in the fossil record as macroevolution? Throughout our study of macroevolution, the theme that evolution is a consequence of interactions between organisms and their environments will be extended to environmental and biological change of global proportions.

The fossil record documents macroevolution

Fossils are historical documents of biology. They are collected and interpreted by specialists called paleontologists (Gr. *palaios*, "old," a reference to the fossils, not the collectors). In this section, you will learn more about the contributions and limitations of the fossil record in the study of macroevolution.

How Fossils Form

A fossil is any preserved remnant or impression left by an organism that lived in the past (L. *fossilis*, "dug up"). Sedimentary rocks are the richest sources of fossils. Sand and silt weathered and eroded from the land are carried by rivers to seas and swamps, where the particles settle to the bottom. Deposits pile up and compress the older sediments below into rock—sand into sandstone and mud into shale. Aquatic organisms, and terrestrial ones swept into the seas and swamps, settle when they die, along with the sediments. A tiny fraction of them is then preserved as fossils.

The organic substances of a dead organism buried in sediments usually decay rapidly. However, hard parts that are rich in minerals, such as the bones and teeth of vertebrates and the shells of many invertebrates and protists, may remain as fossils (FIGURE 23.1a and b, p. 456). Paleontologists have unearthed nearly complete skeletons of dinosaurs and other forms, but more often, the finds consist of parts of skulls, bone fragments, or teeth. Many of these relics are hardened even more and preserved by a process called petrification. Under the right conditions, minerals dissolved in groundwater seep into the tissues of a dead organism and replace organic material. The plant or animal turns to stone. For example, bizarre forests of petrified trees can be explored in parts of the southwestern United States that are now desert (FIGURE 23.1c).

Rarer than mineralized fossils are those that retain organic material. They are sometimes discovered as thin films pressed between layers of sandstone or shale. For example, paleontologists have discovered plant leaves millions of years old that are still green with chlorophyll and well enough preserved for their organic composition to be analyzed and the ultrastructure of their cells to be explored with the electron microscope (FIGURE 23.1d). One research team even managed to clone a minuscule sample of DNA extracted from an ancient magnolia leaf. The most common fossilized plant material is pollen, which has a hard organic case that resists degradation.

The fossils that paleontologists find in many of their digs are not the actual remnants of organisms at all, but rocks that form as replicas of the organisms. These fossils result when a dead organism captured in sediments decays and leaves an empty mold that becomes filled with minerals dissolved in water. The minerals may subsequently crystallize, forming a cast in the shape of the organism (FIGURE 23.1e).

Casts called trace fossils form in footprints, animal burrows, or other impressions left in sediments by the activities of animals. These rocks are fossilized behavior; they tell paleontologists something about how the animals that left trace fossils lived. For example, dinosaur tracks provide clues about the animal's locomotion—its gait (pattern of leg movements), stride length, and speed (FIGURE 23.1f).

Although relics of past life locked in sedimentary rocks are the most common fossils, they are not the only kind. If an organism happens to die in a place where bacteria and fungi cannot decompose the corpse, the entire body, including soft parts, may be preserved as a fossil. For example, the arthropod in FIGURE 23.1g got stuck in a drop of resin from a tree about 30 million years ago. The resin eventually hardened into amber, entombing the animal much like a biological specimen encased in a plastic block. There are other mechanisms that preserve whole organisms. Explorers have discovered mammoths, bison, and other extinct mammals frozen in arctic ice. You have probably read about Bronze Age humans preserved in acid bogs where conditions retard decomposition. Such rare discoveries make the news, but biologists rely mainly on more common sedimentary fossils to reconstruct the history of life.

Limitations of the Fossil Record

The discovery of a fossil is the culmination of a sequence of improbable coincidences. First, the organism had to die in the right place at the right time for burial conditions to favor fossilization. Then the rock layer containing the fossil had to escape geological processes that destroy or severely distort rocks, such as erosion, pressure from superimposed strata, or the melting of rocks that occurs at some locations. If the fossil *was* preserved, there is only a slight chance that a river carving a canyon or some other process will expose the rock containing the fossil. There is an even more remote chance that someone will find the fossil, although discovery is more probable for people who are purposefully looking for fossils. No wonder the fossil record is incomplete. A substantial fraction of species that have lived probably left no fossils, most fossils that formed have been destroyed, and only a fraction of the existing fossils has been discovered. The fossil record, far from being a complete sampling of organisms of the past, is slanted in favor of species that existed for a long time, were abundant and widespread, and had shells or hard skeletons. Paleontologists, like all historians, must reconstruct the past from incomplete records. Even with

(a)

(b)

(c)

(d)

(e)

(f)

(g)

FIGURE 23.1

A gallery of fossils. (**a**) Sedimentary rocks are the richest hunting grounds for paleontologists. Numerous invertebrate shells and other fossils about 15 million years old are visible in the sediments exposed on the face of this cliff. (**b**) The hard parts of organisms, such as this skull of *Australopithecus africanus,* an ancestor of humans that lived about 2.5 million years ago, are the most common fossils. The fossils may become even harder by the process of petrification, which replaces organic material with minerals. (**c**) These petrified trees in Arizona are about 190 million years old. (**d**) Some sedimentary fossils, such as this 40-million-year-old leaf, retain organic material. (**e**) Buried organisms, such as these invertebrates called brachiopods, which lived about 375 million years ago, decay and leave molds that may be filled by minerals dissolved in water. The casts that form when the minerals harden are replicas of the organisms. (**f**) Trace fossils are footprints, burrows, and other remnants of an ancient organism's behavior. This boy is standing in a 150-million-year-old dinosaur track in Colorado. (**g**) This 30-million-year-old arthropod called a pseudoscorpion, embedded in amber (hardened resin from a tree), is almost perfectly preserved.

its limitations, however, the fossil record is a remarkably detailed document of macroevolution over the vast scale of geological time.

Paleontologists use a variety of methods to date fossils

Fossils are reliable historical data only if we can determine their ages. Here, we examine the methods for determining where fossils fit on the geological time scale.

Relative Dating

The trapping of dead organisms in sediments freezes fossils in time. At any particular location, sedimentation is not continuous but occurs in intervals when the sea level changes or lakes and swamps dry up and refill. Even when a region is submerged, the rate of sedimentation and the types of sedimentary particles vary with time. As a result of these different periods of sedimentation, the rock forms in layers, or strata (see FIGURE 20.2). The fossils in each layer are a local sampling of the organisms that existed at the time that sediment was deposited. Younger sediments are superimposed upon older ones. Thus, this book of sedimentary pages tells the relative ages of fossils.

The strata at one location can often be correlated with strata at another location by the presence of similar fossils, known as *index fossils.* The best index fossils for correlating strata that are far apart are the shells of sea animals that were widespread. At any one location where a roadcut or canyon wall reveals layered rocks, there are likely to be gaps in the sequence. That area may have been above sea level during different periods, and thus no sedimentation occurred. Some of the sedimentary layers that were deposited when the area *was* submerged may have been scraped away by subsequent periods of erosion.

By studying many different sites, geologists have established a **geological time scale** with a consistent sequence of historical periods (TABLE 23.1). These periods are grouped into four eras: the Precambrian, Paleozoic, Mesozoic, and Cenozoic eras. Each era represents a distinct age in the history of Earth and its life; the boundaries are marked in the fossil record by explosive radiations of many new forms of life following mass extinctions. For example, the beginning of the Paleozoic era is recorded by a great diversity of fossilized invertebrates that are absent in rocks of the late Precambrian era. And most of the animals that lived during the late Precambrian became extinct at the end of that era. The boundaries between periods within each era also mark major transitions in the forms of life fossilized in rocks, though these changes are less radical than the geological and biological revolutions that divide the eras. The periods within each era are further subdivided into finer intervals called epochs (only the epochs of the current era, the Cenozoic, appear in TABLE 23.1). The timeline to the right of TABLE 23.1 will help you understand that the geological eras were not equal in duration, nor were the periods. For example, the Jurassic period lasted almost twice as long as the Triassic period. Remember that scientists have not divided geological time in an arbitrary manner, but have located boundaries in the record of the rocks that correspond to times of great change.

The record of the rocks is a serial that chronicles the *relative* ages of fossils; it tells us the order in which groups of species present in a sequence of strata evolved. However, the series of sedimentary rocks does not tell the *absolute* ages of the embedded fossils. The difference is analogous to peeling the layers of wallpaper from the walls of a very old house that has been inhabited by many owners. You could determine the sequence in which the papers had been applied, but not the year that each layer was added.

Absolute Dating

"Absolute" dating does not mean errorless dating, but only that age is given in years instead of relative terms such as *before* and *after, early* and *late.* **Radiometric dating** is the method most often used to determine the ages of rocks and fossils on a scale of absolute time. Fossils contain isotopes of elements that accumulated in the organisms when they were alive. Because each radioactive isotope has a fixed rate of decay, it can be used to date a specimen (see Chapter 2). An isotope's **half-life,** the number of years it takes for 50% of the original sample to decay, is unaffected by temperature, pressure, and other environmental variables. For example, carbon-14 has a half-life of 5600 years, a reliable rate of decay that can be used to date relatively young fossils (see the Methods Box page 459). Paleontologists use radioactive isotopes with longer half-lives to date older fossils. For example, in 1993, a research team used uranium-238, which has a half-life of 4.5 billion years, as a radiometric clock to date the oldest known fossil-containing rocks of the Cambrian period. Their measurements place the base of the Cambrian at 544 million years ago (see TABLE 23.1).

Methods other than radiometric dating can be used on some fossils. Amino acids exist in two isomers with either left-handed or right-handed symmetry, designated the L and D forms, respectively. Organisms synthesize only L-amino acids, which are incorporated into proteins. After an organism dies, however, its population of L-amino acids is slowly converted, resulting in a mixture of L- and D-amino acids. In a fossil, the ratio of L- and D-amino acids can be measured. Knowing the rate at which this chemical conversion, called racemization, takes place, we can determine how long the organism has

TABLE 23.1 **The Geological Time Scale**

ERA	PERIOD	EPOCH	AGE (MILLIONS OF YEARS)	SOME IMPORTANT EVENTS IN THE HISTORY OF LIFE
Cenozoic	Quaternary	Recent	0.01	Historic time
		Pleistocene	1.8	Ice ages; humans appear
	Tertiary	Pliocene	5	Apelike ancestors of humans appear
		Miocene	23	Continued radiation of mammals and angiosperms
		Oligocene	34	Origins of most modern mammalian orders, including apes
		Eocene	57	Angiosperm dominance increases; further increase in mammalian diversity
		Paleocene	65	Major radiation of mammals, birds, and pollinating insects
Mesozoic	Cretaceous		144	Flowering plants (angiosperms) appear; dinosaurs and many groups of organisms become extinct at end of period
	Jurassic		208	Gymnosperms continue as dominant plants; dinosaurs dominant; first birds
	Triassic		245	Gymnosperms dominate landscape; first dinosaurs and mammals
Paleozoic	Permian		286	Radiation of reptiles; origins of mammal-like reptiles and most modern orders of insects; extinction of many marine invertebrates
	Carboniferous		360	Extensive forests of vascular plants; first seed plants; origin of reptiles; amphibians dominant
	Devonian		408	Diversification of bony fishes; first amphibians and insects
	Silurian		438	Diversity of jawless vertebrates; colonization of land by plants and arthropods; origin of vascular plants
	Ordovician		505	First vertebrates (jawless fishes); marine algae abundant
	Cambrian		544	Origin of most invertebrate phyla; diverse algae
Precambrian			700	Origin of first animals
			1500	Oldest eukaryotic fossils
			2500	Oxygen begins accumulating in atmosphere
			3500	Oldest definite fossils known (prokaryotes)
			4600	Approximate origin of Earth

RELATIVE TIME SPAN OF ERAS

- Cenozoic
- Mesozoic
- Paleozoic
- Pre-Cambrian

been dead. Archeologists have recently used this method to date ostrich eggshells found along with fossils of early humans. The humans probably ate the eggs and used the shells for water bowls. Racemization, unlike radioactive decay, is temperature-sensitive, meaning that past changes in climate made the racemization clock run faster or slower. But for fossils found in locations where climate apparently has not changed significantly since the fossils formed, the two dating methods agree closely on the age of the fossils.

The order of sedimentary strata records the sequence of biological change, and dating methods tell us how long ago these episodes of macroevolution occurred. We now shift our attention from the record of macroevolution to the mechanisms that cause these transformations of life.

METHODS: RADIOMETRIC DATING

Radioactive isotopes can be detected and their amounts measured by the radiation they emit as they decompose to more stable atoms (see Chapter 2). Paleontologists use the clocklike decay of radioactive isotopes to date fossils and rocks. Carbon-14, for example, has a half-life of 5600 years, meaning that half the carbon-14 in a specimen will be gone in 5600 years, half the remainder will be gone in another 5600 years, and so on. The graph below traces this exponential decline in radioactivity. In the example illustrated below, we use carbon-14 dating to determine the vintage of a fossilized clam shell. Because the half-life of carbon-14 is relatively short, this isotope is only reliable for dating fossils less than about 50,000 years old. To date older fossils, paleontologists use radioactive isotopes with longer half-lives. For instance, uranium-238, a radioactive isotope with a half-life of 4.5 billion years, can be used to date rocks hundreds or thousands of millions of years old and infer the age of fossils embedded in those rocks. Radiometric dating has an error factor of less than 10%.

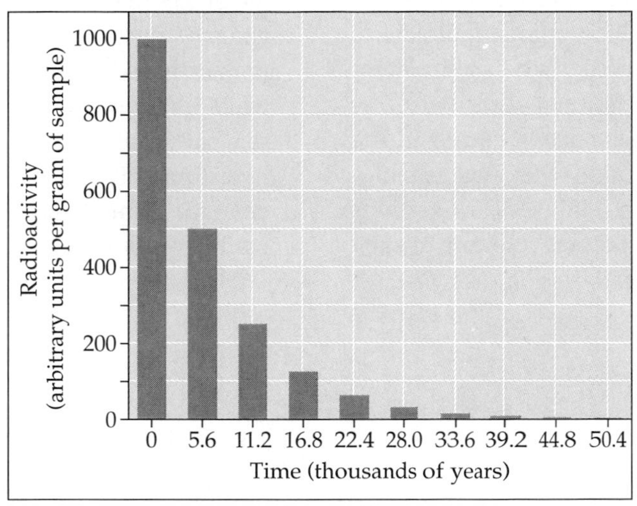

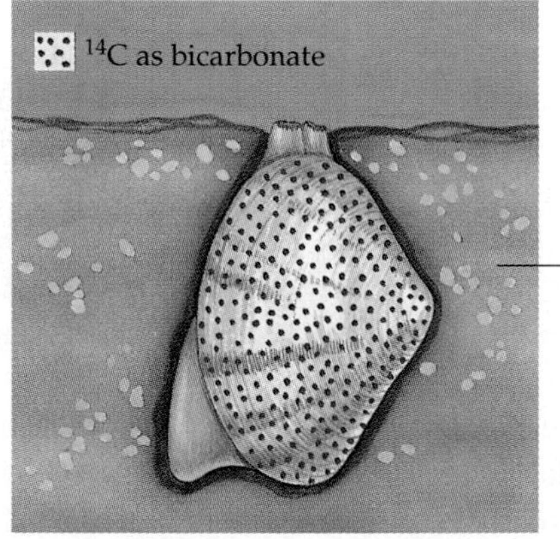

While an organism, in this case a clam, is alive, it assimilates the different isotopes of each element in proportions determined by their relative abundances in the environment. Carbon-14 is taken up in trace quantities, along with much larger quantities of the more common carbon-12.

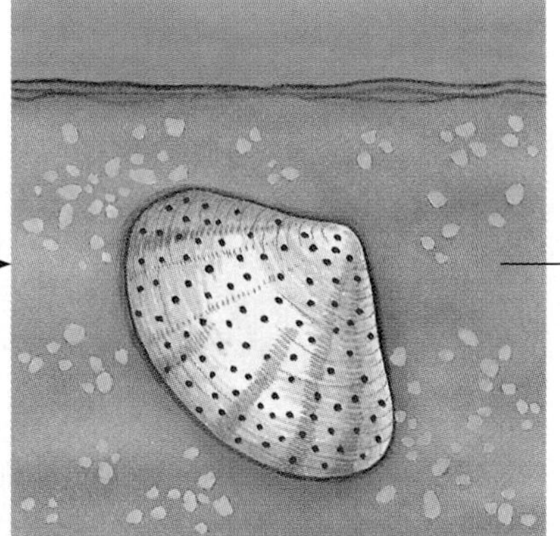

After the clam dies, it is covered with sediment, and its shell eventually becomes consolidated into a layer of rock as the sediment is compressed. From the time the clam dies and ceases to assimilate carbon, the amount of carbon-14 relative to carbon-12 in its remains declines due to radioactive decay.

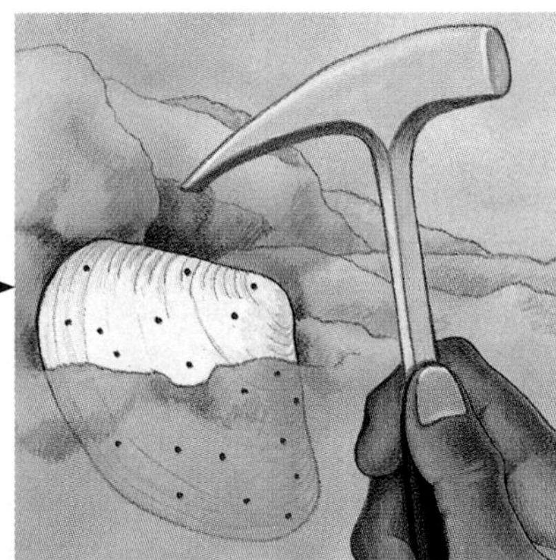

After the clam fossil is found, its age can be determined by measuring the ratio of the two isotopes to learn how many half-life reductions have occurred since it died. For example, if the ratio of carbon-14 to carbon-12 in this fossil clam was found to be one-fourth that of a living organism, this fossil would be about 11,200 years old.

What are the major questions about macroevolution? *an overview*

What processes actually cause the large-scale evolutionary changes that we can trace through the fossil record? How, for instance, do the novel features that define taxonomic groups above the species level, such as the flight adaptations of birds, arise? What accounts for evolutionary trends that appear from the fossil record to be progressive, such as the general size increase in reptiles of several families during the age of dinosaurs, or the increase in brain size during human evolution? How have global geological changes affected macroevolution? And how can we explain the major fluctuations in biological diversity evident in the fossil record, such as a proliferation of animal diversity at the beginning of the Paleozoic era or the mass extinctions that have occurred in the past? These are some of the questions that engage biologists who study macroevolution. The next five sections focus on macroevolutionary processes relevant to such questions about the history of life.

Some evolutionary novelties are modified versions of older structures

Birds evolved from dinosaurs; their wings are homologous to the forelimbs of their modern reptilian cousins. How could flying vertebrates evolve from earthbound ancestors? What creates the evolutionary novelties that define the higher taxonomic groups, such as families and classes? Put another way, how do new designs for living evolve? One mechanism is the gradual refinement of existing structures for new functions.

Most biological structures have an evolutionary plasticity that makes alternative functions possible. From a retrospective vantage point, evolutionists use the term **preadaptation** for a structure that evolved in one context and became co-opted for another function. This concept does not imply that a structure somehow evolves in anticipation of future use. Natural selection cannot predict the future and can only improve a structure in the context of its current utility. The light, honeycombed bones of birds (see FIGURE 1.7b) could not have evolved in earthbound reptilian ancestors as an adaptation for upcoming flights. If these honeycombed bones predated flight, as clearly indicated by the fossil record, then they must have had some function on the ground. The probable ancestors of birds were agile, bipedal dinosaurs that also would have benefited from a light frame. It is possible that winglike forelimbs, as well as feathers, which increased the surface area of these forelimbs, were also co-opted for flight after functioning in some other capacity, such as "netting" insects and other small prey chased by small, fleet-footed dinosaurs. Or perhaps feathers functioned mainly in social displays—in courtship, for example. The first flight may have been only a glide down from a tree or an extended hop in pursuit of prey or escape from a predator. Once flight itself became an advantage, natural selection would have remodeled feathers and wings to better fit their additional function.

Preadaptation offers one explanation for how novel designs can arise gradually through a series of intermediate stages, each of which has some function in the organism's current context. Harvard zoologist Karel Liem puts it this way: "Evolution is like modifying a machine while it's running." The concept that evolutionary novelties can evolve by the remodeling of old structures for new functions is in the Darwinian tradition of large changes being an accumulation of many small changes crafted by natural selection.

Genes that control development play a major role in evolutionary novelty

The evolution of complex structures, such as wings and feathers, from their antecedents requires so much remodeling that changes are probably involved at a large number of gene loci. In other cases, relatively few changes in the genome can apparently cause major modifications of morphology, such as some of the differences between humans and chimpanzees. How can slight genetic divergence become magnified into major differences between organisms? Scientists working at the interface of developmental biology and evolutionary biology are trying to answer this question.

Genes that program development control the rate, timing, and spatial pattern of changes in an organism's form as it is transfigured from a zygote into an adult. For example, **allometric growth,** a difference in the relative rates of growth of various parts of the body, helps shape an organism (Gr. *allos,* "other," and *metron,* "measure"). FIGURE 23.2a tracks how allometric growth alters human body proportions during development. Change these relative rates of growth even slightly, and you change the adult form substantially. For example, different allometric patterns contribute to the contrasting shapes of human and chimpanzee skulls (FIGURE 23.2b). Allometry is one mechanism by which a subtle alteration of development becomes compounded in its effects on the adult.

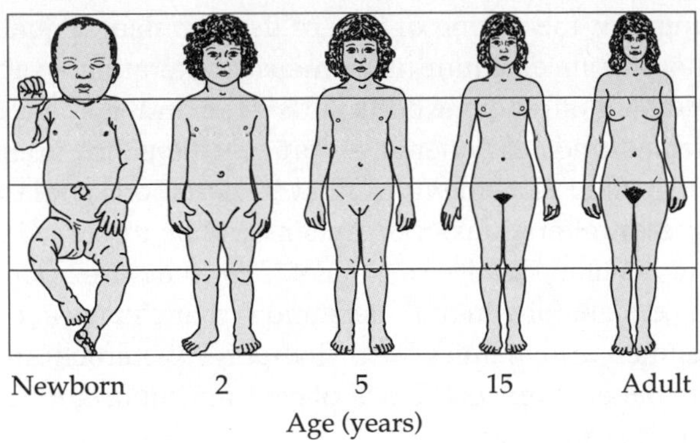

(a)

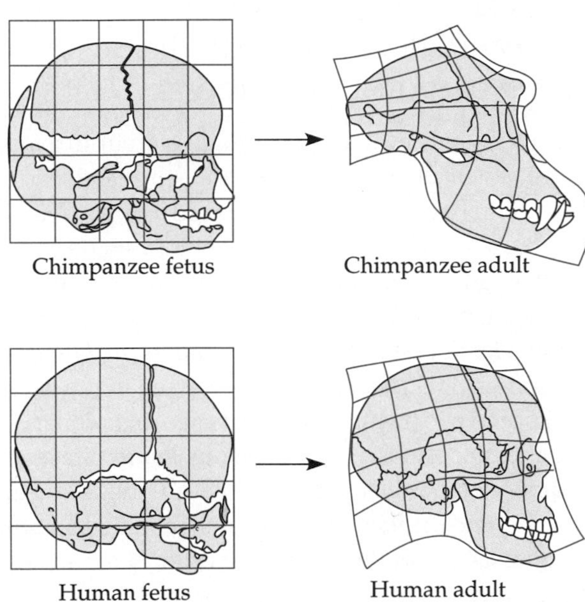

(b)

FIGURE 23.2
Allometric growth. Difference in the growth rates for some parts of the body compared to others determines body proportions. (**a**) During growth of a human, the arms and legs grow faster than the head and trunk, as can be seen in this conceptualization of different-aged individuals all rescaled to the same height. (**b**) The fetal skulls of humans and chimpanzees are similar in shape. Allometric growth of the bones transforms the rounded skull of a newborn chimpanzee to the sloping skull characteristic of adult apes. The same allometric pattern occurs in humans, but it is attenuated, and the adult human skull has departed less than the chimpanzee skull from the fetal shape common to primates.

Plains zebra

Grevy's zebra

FIGURE 23.3
The effect of developmental timing on zebra stripes. The stripes of the plains zebra, *Equus burchelli* (top), are wider and fewer in number than those of Grevy's zebra, *Equus grevyi.* This variation probably reflects the different times at which stripes start to develop in the zebra embryos, three weeks after fertilization in the plains zebra and five weeks after fertilization in Grevy's zebra. To picture this process, imagine painting stripes of equivalent width and spacing around two balloons, one less inflated than the other. If we continue inflating the two balloons to much larger but equal sizes, the one that was smaller when we applied the stripes will have stripes that are fewer and wider than those of the other balloon.

In addition to affecting developmental rates, genetic changes can also alter the timing of developmental events—the sequence in which different body parts start and stop developing. For example, the variation in stripes between two zebra species probably results from a difference in the time at which the stripes begin to take form in the embryos (FIGURE 23.3). Of more general importance are changes in developmental timing that result in **paedomorphosis** (Gr. *paid,* "child," and *morphos,* "form"), in which a sexually mature adult retains features

FIGURE 23.4
Paedomorphosis. Some species retain features as adults that were juvenile in ancestors. This salamander is an axolotl, which grows to full size, becomes sexually mature, and reproduces while retaining certain larval (tadpole) characteristics, including gills. This figure and FIGURES 23.2 and 23.3 illustrate examples of heterochrony, evolutionary change in the timing or rate of development.

that were juvenile structures in its evolutionary ancestors. For example, most salamander species have a larval stage that undergoes metamorphosis to become an adult. But some species grow to adult size and become sexually mature while retaining gills and certain other larval features (FIGURE 23.4). Such an evolutionary alteration of developmental timing can produce animals that appear very different from their ancestors, even though the overall genetic change may be small.

Changes in developmental chronology have also been important in human evolution. Humans and chimpanzees are very closely related through descent from a common ancestor. Of the numerous anatomical differences between these two modern primates that can be attributed to different allometric properties and variations in developmental timing, the most significant contrast is brain size. The human brain is proportionately larger than the chimpanzee brain because growth of the organ is switched off much later in human development. Compared to the brain of chimpanzees, our brain continues to grow for several more years, which can be interpreted as the prolonging of a juvenile process. We owe our culture to this evolutionary novelty, coupled with an extended childhood during which parents and teachers can influence what the young, growing brain stores.

All these examples of temporal changes in development that create evolutionary novelties fit into the category of **heterochrony,** a general term for evolutionary changes in the timing or rate of development. Equally important in evolution is **homeosis,** alteration in the placement of different body parts—the arrangement of different kinds of appendages on an animal, for example, or the placement of flower parts on a plant. Homeotic alterations of organisms will be discussed in more detail in Chapters 34 and 43. For now, the important concept for you to understand is that change in developmental dynamics, both temporal (heterochrony) and spatial (homeosis), has played an important role in macroevolution.

Recognizing trends in the fossil record does not mean that macroevolution is goal-oriented

Extracting a single evolutionary progression from a fossil record that is likely to be incomplete is misleading; it is like describing a bush as growing toward a single point by tracing the system of branches that leads from the base of the bush to one particular twig. A case in point is the evolution of the modern horse, which is a descendant of a much smaller ancestor named *Hyracotherium.* About the size of a large dog, *Hyracotherium* browsed in the woods of the Eocene epoch 40 million years ago. In comparison to this ancestor, modern horses (genus *Equus*) are larger, the number of toes has been reduced from four on each foot to one, and the teeth have become modified for grazing on grasses rather than browsing on shrubs and trees. By selecting certain species from the available fossils, it is possible to arrange a succession of animals intermediate between *Hyracotherium* and modern horses that shows trends toward increased size, reduced number of toes, and modification of teeth for grazing (yellow line in FIGURE 23.5). We might interpret this series of fossils as an unbranched lineage leading directly from *Hyracotherium* to modern horses through a continuum of intermediate stages. If we include all fossil horses known today, however, the illusion of coherent, progressive evolution leading directly to modern horses vanishes. The genus *Equus* just happens to be the only surviving twig of a phylogenetic tree that is so branched it is more like a bush. *Equus* descended through a series of speciation episodes that included several adaptive radiations, not all of which led to large, one-toed, grazing horses. Had *Equus* become extinct and some other horse persisted, we might perceive different evolutionary trends.

Evolution *has* produced many trends that seem from the fossil record to be genuine. For example, a family of

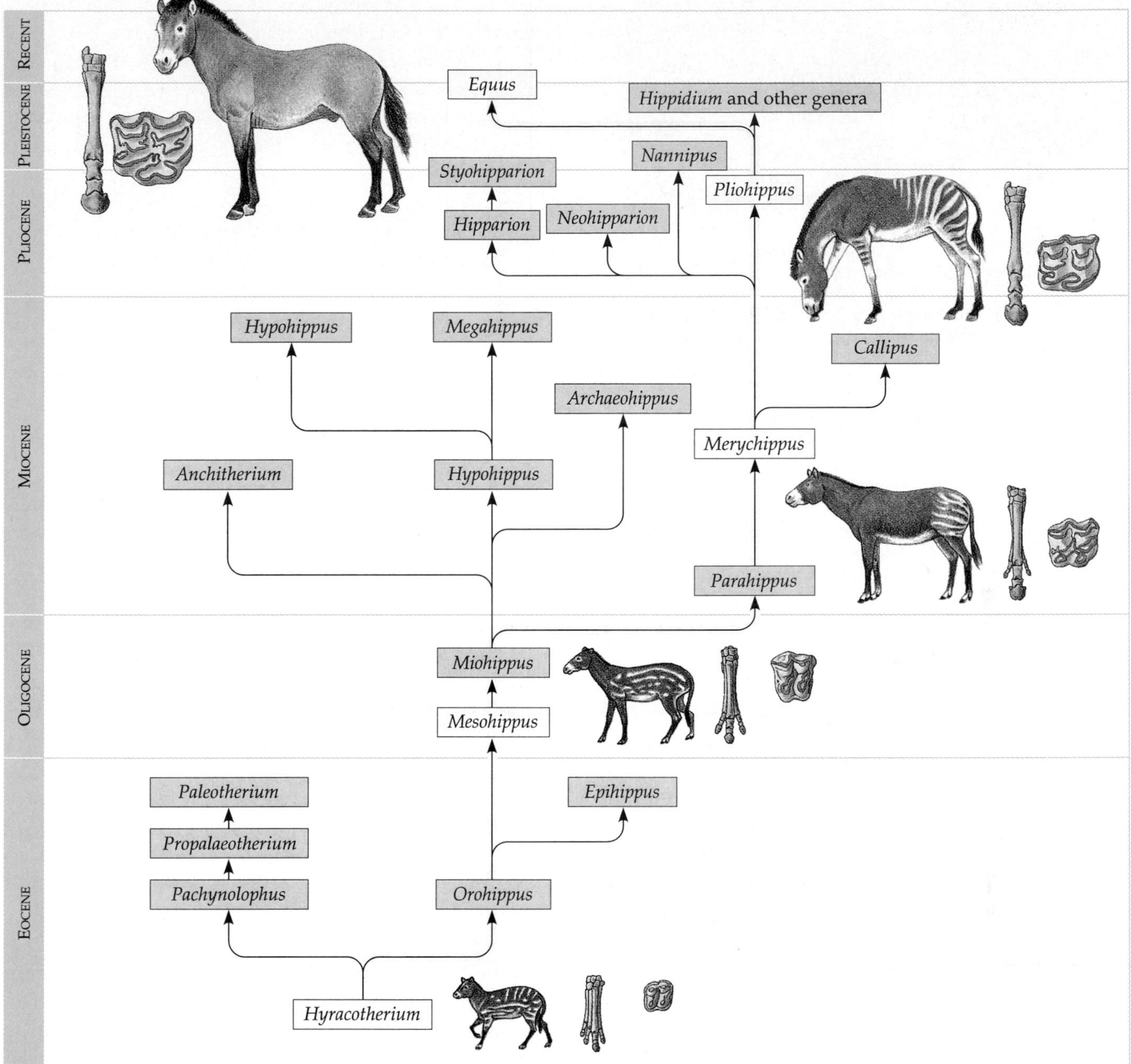

FIGURE 23.5

The branched evolution of horses. If we use a yellow highlighter to trace one sequence of fossil horses that are intermediate in form between the modern horse and its Eocene ancestor *Hyracotherium,* we create the illusion of a progressive trend toward larger size, reduced number of toes, and teeth modified for grazing. In fact, the modern horse is the only surviving twig of an evolutionary bush with many divergent trends.

now-extinct giant mammals called titanotheres, as large as elephants, evolved from a mouse-sized ancestor that lived during the early Cenozoic era (FIGURE 23.6). There is no evidence of a smooth continuum of increasing size in the various lineages of titanotheres; there is a succession of progressively larger species, but each species remains the same size for its duration in the fossil record. Punctuated equilibrium seems to be the mode of evolution in this case (see Chapter 22). According to this interpretation of the fossil record, evolutionary trends in most

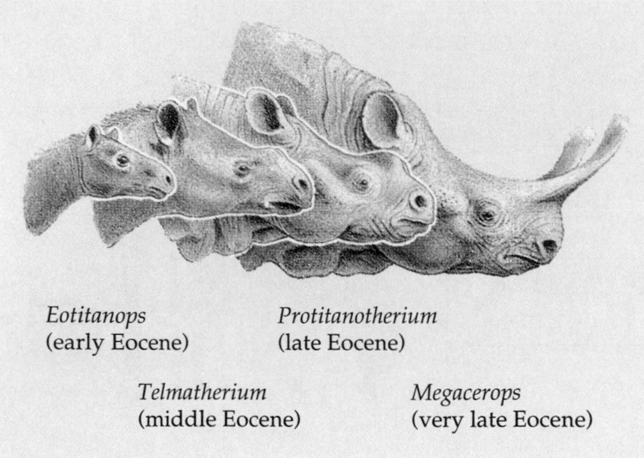

FIGURE 23.6
Trends in the evolution of titanotheres. In each of the genera of this extinct family of mammals, species became larger and their horns bigger during the Eocene epoch. However, no individual species changes significantly in size for its duration in the fossil record. Some paleontologists interpret this to mean that the evolutionary trends were produced not by gradual phyletic evolution (anagenesis) of individual populations, but by a series of speciation episodes, with the average size of the species increasing with time.

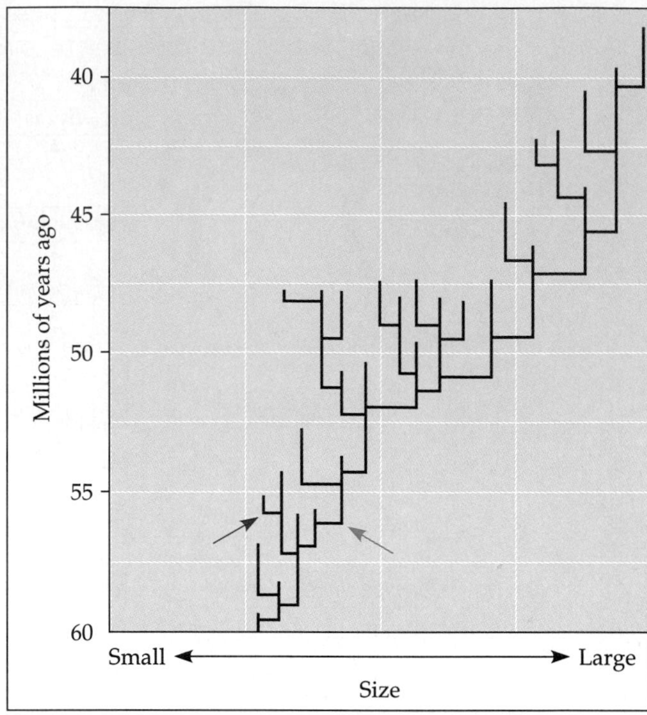

FIGURE 23.7
The species selection hypothesis of macroevolution. This model explains evolutionary trends as the products of the differential longevity and budding of daughter species. The example shows a trend toward species of larger body size. Species smaller than their parent (red arrow) bud from the evolutionary bush as frequently as species that are larger than their parent (blue arrow). Nevertheless, according to this model, there is an overall trend because large species endure longer and leave the most descendant species, because of either a faster speciation rate, a slower extinction rate, or both.

groups of organisms are produced not by a gradual slurring of forms (phyletic evolution, or anagenesis), but by staccato changes occurring in increments during the time that new species branch from ancestral ones.

Branching evolution (cladogenesis) can produce a trend even if some new species counter the trend. During the Mesozoic era, there was an overall trend in reptilian evolution toward largeness. This evolutionary trend produced the dinosaurs, which became the dominant animals of that era. The trend was sustained even though some new species were smaller than their parent species. In fact, even if the descendant species were smaller as often as they were larger than the species from which they budded, an overall trend would still develop if larger reptiles speciated more often than smaller ones or lasted longer than smaller species before becoming extinct (FIGURE 23.7). In this view of macroevolution, enunciated by Steven Stanley of Johns Hopkins University, species are analogous to individuals. Speciation is their birth and extinction is their death. New species are their offspring. According to this model, an evolutionary trend is produced by **species selection,** which is analogous to the production of a trend within a population by natural selection. The species that endure the longest and generate the greatest number of new species determine the direction of major evolutionary trends. This concept suggests that differential speciation plays a role in macroevolution similar to the role of differential reproduction in microevolution.

To the extent that speciation rates and species longevity reflect success, the analogy to natural selection is even stronger. But qualities unrelated to the overall success of organisms in specific environments may be equally important in species selection. For example, the ability of a species to disperse to new locations may contribute to its giving rise to a large number of "daughter species." The species selection model has many critics who argue that evolutionary trends more commonly result from the gradual modification of populations in response to environmental change. The value of such debates is that they stimulate research, and many paleontologists and other evolutionary biologists are now focusing on the question of what produces evolutionary trends.

Whatever its cause, the appearance of an evolutionary trend does not imply that there is some intrinsic drive toward a preordained state of being. Evolution is a response to interactions between organisms and their current environments. If conditions change, an evolutionary trend may cease or even reverse itself. For example, the Mesozoic world favored giant reptiles, but by the end of that era, smaller species prevailed.

Macroevolution has a biogeographical basis in continental drift

Macroevolution has dimension in space as well as in time. Indeed, it was biogeography even more than fossils that first nudged Darwin and Wallace toward an evolutionary view of life. The history of Earth helps explain the current geographical distribution of species. For example, the emergence of volcanic islands such as the Galapagos opens new environments for founders that reach the outposts, and adaptive radiation fills many of the available niches with new species. On a global scale, the drifting of continents is the major geographical factor correlated with the spatial distribution of life and with such macroevolutionary episodes as mass extinctions and explosive increases of biological diversity.

The continents are not fixed, but drift about Earth's surface like passengers on great plates of crust floating on the molten mantle (FIGURE 23.8a). Unless two land masses are embedded in the same plate, their positions relative to each other change. For example, North America and Europe are presently drifting apart at a rate of about

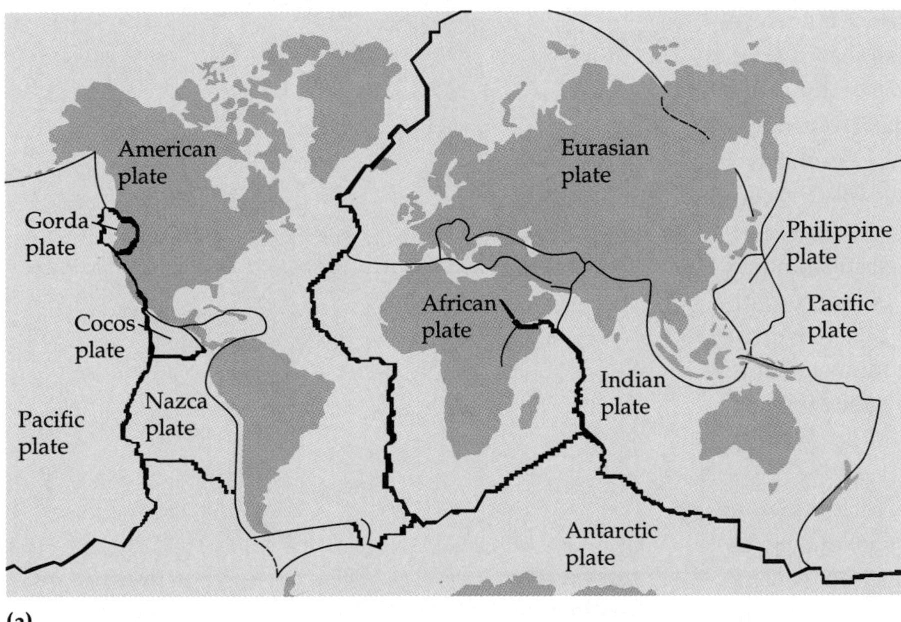

FIGURE 23.8
Earth's crustal plates and plate tectonics (geological processes resulting from plate movements). (**a**) The modern continents are passengers on crustal plates that are swept across Earth's surface by convection currents of the molten mantle below. This map identifies only the major plates. (**b**) At some plate boundaries, such as oceanic ridges, the plates separate, and molten rock wells up in the gap. The rock solidifies and adds crust symmetrically to both plates, a phenomenon called seafloor spreading. There are also areas known as subduction zones, where plates move toward each other, with the denser plate diving below the less dense one and creating a trench. The Marianas Trench of the South Pacific is a subduction zone over 11,000 m deep. The abrasion at subduction zones causes earthquakes and volcanic eruptions. When continents riding on different plates collide, they pile up and build mountains.

2 cm per year. Many important geological phenomena, including mountain building, volcanism, and earthquakes, happen at plate boundaries (FIGURE 23.8b). California's infamous San Andreas Fault is part of a border where two plates slide past each other. The Philippines, where Mount Pinatubo erupted in 1991, sits right on another plate boundary.

Plate movements rearrange geography incessantly, but two chapters in the continuing saga of continental drift must have been especially significant in their influence on life. About 250 million years ago, near the end of the Paleozoic era, plate movements brought all the land masses together into a supercontinent that has been named **Pangaea,** meaning "all land" (FIGURE 23.9). Imagine some of the possible effects on life. Species that had been evolving in isolation came together and competed. When the land masses coalesced, the total amount of shoreline was reduced, and there is evidence that the ocean basins increased in depth, which lowered sea level and drained much of the shallow coastal seas that remained. Then, as now, most marine species inhabited shallow waters, and the formation of Pangaea destroyed a considerable amount of that habitat. It was probably a long, traumatic period for terrestrial life as well. The continental interior, which has a drier and more erratic climate than coastal regions, increased in area substantially when the land came together. Changing ocean currents also would have affected land life as well as sea life. The formation of Pangaea surely had a tremendous environmental impact that reshaped biological diversity by causing extinctions and providing new opportunities for taxonomic groups that survived the crisis.

Another dramatic chapter in the history of continental drift was written about 180 million years ago, during the early Mesozoic era. Pangaea began to break up, causing geographical isolation of colossal proportions. As the continents drifted apart, each became a separate evolutionary arena, and the faunas and floras of the different biogeographical realms diverged.

The pattern of continental separations is the solution to many biogeographical puzzles. For example, paleontologists have discovered matching fossils of Triassic reptiles in Ghana (West Africa) and Brazil. These two parts of the world, now separated by 3000 km of ocean, were contiguous during the early Mesozoic era. Continental drift also explains much about the current distribution of organisms, such as why the Australian fauna and flora contrast so sharply with the rest of the world. The great diversity of marsupials (pouched mammals), which fill ecological roles in Australia analogous to those filled by placental mammals on other continents, is just one example of Australia's unique collection of species. Marsupials probably evolved first in what is now North America and reached Australia via South America and Antarctica while the continents were still joined. The subsequent breakup of the southern continents set Australia "afloat" like a great ark of marsupials, while placental mammals evolved and diversified on other continents. Australia has been

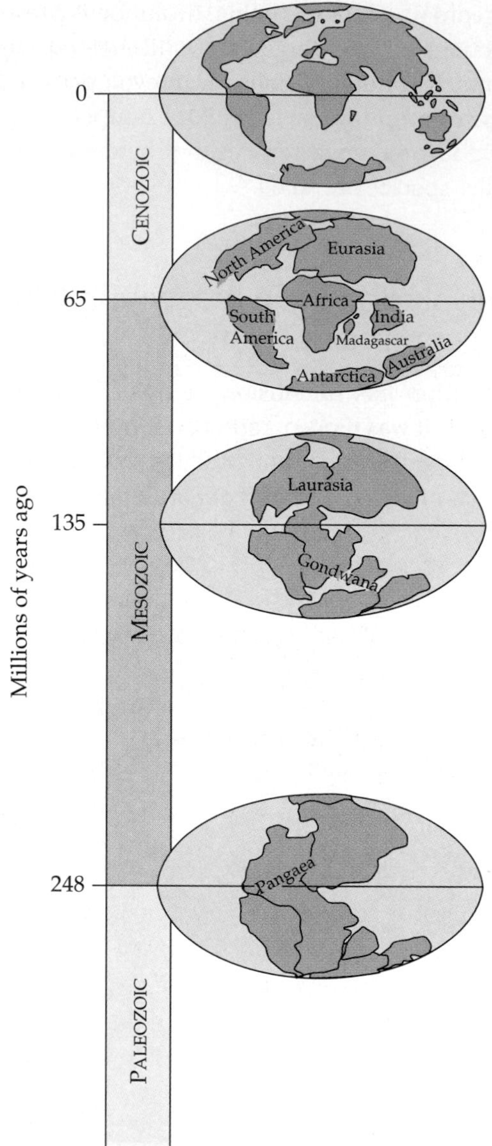

FIGURE 23.9
Continental drift. About 200 to 250 million years ago, all of Earth's land masses were locked together in a supercontinent named Pangaea. About 180 million years ago, Pangaea began to split into northern (Laurasia) and southern (Gondwana) land masses, which later separated into the modern continents. The continents continue to drift. India collided with Eurasia just 10 million years ago, forming the Himalayas, the tallest and youngest of Earth's mountain ranges.

completely isolated for 50 million years. Bats, rats, mice, and humans (and their domesticated animals) are the only placental mammals that have managed to populate the island continent.

The history of life is punctuated by mass extinctions followed by adaptive radiations of the survivors

The evolutionary byways from ancient to modern life have not been smooth. The fossil record reveals an episodic history, with long, relatively quiescent periods punctuated by briefer intervals when the turnover in species composition was much more extensive. The episodes include explosive adaptive radiations of major taxonomic groups as well as mass extinctions.

Examples of Major Adaptive Radiations

Many taxonomic groups have diversified prolifically early in their history after the evolution of some novel characteristic that opened a new **adaptive zone,** a term for a new way of life that presents many previously unexploited opportunities. For example, the development of wings enabled insects to enter an adaptive zone with many new resources, and adaptive radiation produced hundreds of thousands of variations on the basic insect body plan.

The boundary between the Precambrian era and the Paleozoic era is marked by a large increase in the diversity of sea animals, a radiation mentioned earlier. The oldest animals are creatures found in late Precambrian rocks about 700 million years old (see TABLE 23.1). These animals, known from fossilized imprints that have been discovered at several sites around the world, were shell-less invertebrates with body plans quite different from those of their Paleozoic successors. All the animal phyla that exist today evolved during a span of less than 10 million years during the middle Cambrian, the first period of the Paleozoic era. One key evolutionary novelty behind this remarkable diversification may have been the origin of shells and skeletons in a few phyla. This innovation opened a new adaptive zone by making many new complex designs possible and by rewriting the rules for predator-prey relations. More generally, the evolution of genes controlling development may have increased the range of morphological complexity and diversity that was possible. We will evaluate the Cambrian explosion more critically in Chapter 29.

There are probably empty adaptive zones even today. An empty adaptive zone can be exploited only if the appropriate evolutionary novelties arise. For example, flying insects existed at least 100 million years before the flying reptiles and birds that ate the insects evolved. Conversely, an evolutionary novelty cannot enable organisms to take advantage of adaptive zones that do not exist or are already occupied. Mammals, with the many unique features characteristic of their class, existed at least 75 million years before their first major adaptive radiation. The rise in mammalian diversity during the early Cenozoic era may have been associated with the ecological void left by the extinction of the dinosaurs. New adaptive radiations have often followed mass extinctions that swept away old tenants of adaptive zones.

Examples of Mass Extinctions

A species may become extinct because its habitat has been destroyed or because the environment has changed in a direction unfavorable to the species. If ocean temperatures fall by a few degrees, many species that are otherwise beautifully adapted will perish. Even if physical factors in the environment are stable, the biological factors may change; the environment in which a species lives includes the other organisms that live there, and evolutionary change in one species is likely to have some impact on other species in the community. For example, the evolution of shells by some Cambrian animals may have contributed to the extinction of some shell-less forms.

Thus, extinction is inevitable in a changing world. The average rate of extinction has been between 2.0 and 4.6 families per million years (each family may include many species). However, there have been crises in the history of life when global environmental changes have been so rapid and disruptive that a majority of species were swept away, perhaps rather indiscriminately. During periods of mass extinctions, the rate of destruction escalates to as high as 19.3 families per million years. Mass extinctions are known primarily from the decimation of hard-bodied animals of shallow seas, the organisms for which the fossil record is most complete. Of the dozen or so mass extinctions chronicled in the fossil record, two have received the most attention: The Permian extinctions, which occurred about 250 million years ago, and the Cretaceous extinctions of about 65 million years ago (FIGURE 23.10).

The Permian extinctions, which define the boundary between the Paleozoic and Mesozoic eras, claimed over 90% of the species of marine animals about 250 million years ago. Terrestrial life also crashed; for example, 8 out of 27 orders of Permian insects did not survive into the

FIGURE 23.10
Mass extinctions in the history of life. The fossil record profiles mass extinctions during many periods of geological time. Only two are labeled here. The Permian extinctions claimed more than 90% of species on land and in the seas. The Cretaceous extinctions wiped out more than half of all species, including all dinosaurs. The extinctions seem less catastrophic in this figure because we are tracking changes in the number of families, not species. The percentage of species that disappear during mass extinctions is much larger than the percentage of families that disappear. This is because most families consist of many species, and a family is counted as extinct only if all of its species are extinct. Notice that biological diversity has always rebounded in the aftermath of mass extinctions, causing the biological make-overs that define the boundaries of the geological periods and eras.

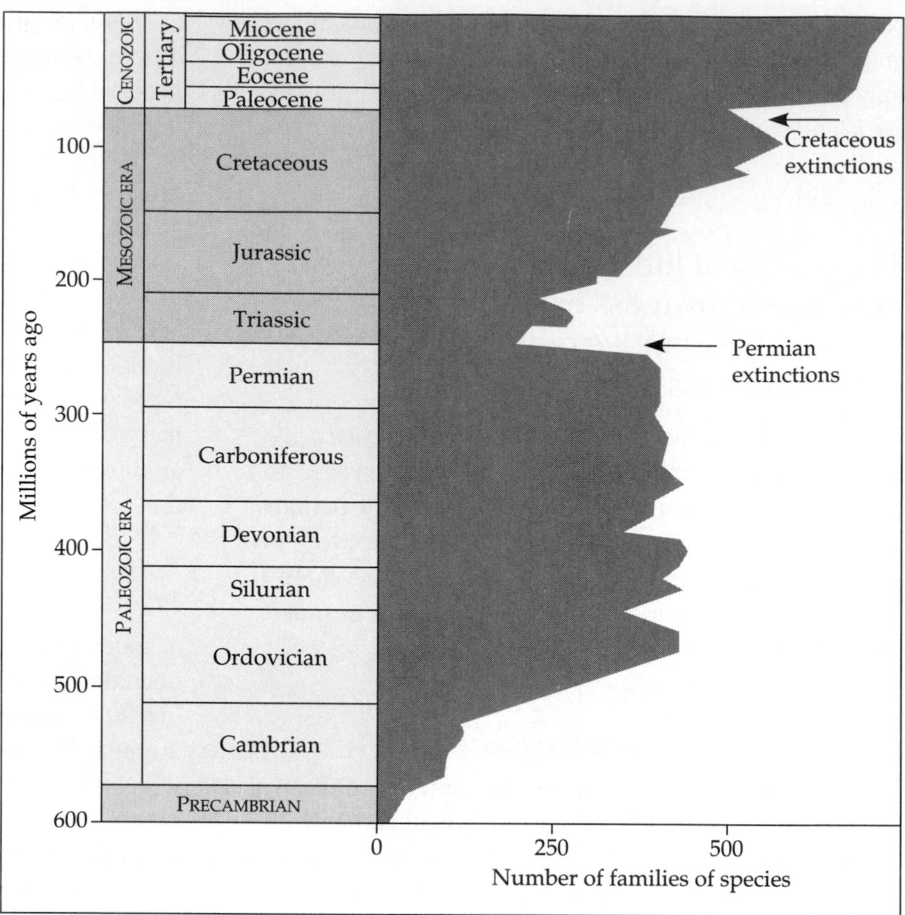

Triassic, the next geological period. The mass extinction occurred in less than 5 million years—possibly much less—an instant in the context of geological time. Several factors may have combined to cause radical environmental change during the late Permian. That was about the time the continents merged to form Pangaea, which disturbed many marine and terrestrial habitats and altered climate. There were also massive volcanic eruptions in what is now Siberia, creating the most extreme episode of vulcanism of the past half-billion years. Atmospheric debris from the eruption of Mount Pinatubo in the Philippines in 1991 had a measurable effect on global climate for many months, blocking enough sunlight to cool Earth by 0.25°C. In comparison, the Siberian volcanoes of the late Permian spewed a million times more lava and ash, which may have altered temperatures enough to contribute to the Permian extinctions.

Another emphatic punctuation, the Cretaceous extinctions of 65 million years ago, delineates the boundary between the Mesozoic and Cenozoic eras. That debacle doomed more than half the marine species and exterminated many families of terrestrial plants and animals, including the dinosaurs. The climate became cooler at that time, and shallow seas receded from continental lowlands. Large volcanic eruptions in what is now India may have contributed to the cooling by releasing material into the atmosphere that blocked sunlight. However, most debate about the causes of the Cretaceous extinctions centers on the **impact hypothesis,** which posits that an asteroid or comet struck Earth and caused mass extinction. Separating Mesozoic from Cenozoic sediments is a thin layer of clay enriched in iridium, an element very rare on Earth, but common in meteorites and other extraterrestrial debris that occasionally falls to Earth. Walter and Luis Alvarez and their colleagues at the University of California, Berkeley, studied the anomalous clay and suggested it is fallout from a huge cloud of dust that billowed into the atmosphere when an asteroid hit Earth. The Alvarezes proposed that the great cloud would have blocked light and disturbed climate severely for several months.

The impact hypothesis really has two parts: that such a collision occurred and that the event caused the Cretaceous extinctions. Much evidence in addition to the iridium layer supports the first part of the impact hypothesis—that a large comet or smallish asteroid crashed into Earth 65 million years ago. Earth is pocked with enough craters to tell us that many large objects have fallen to the planet in the past. Recent research has focused on the Chicxulub crater, a 65-million-year-old scar located beneath sediments on the Yucatan coast of Mexico. Researchers first measured the diameter of the crater as 180 km, about the right size to be caused by an asteroid with a diameter of about 10 km. Collisions of that size probably occur about once every 100 million years. Now there is evidence that the Chicxulub crater is twice as large. In 1993, a team of scientists discovered what seems to be an outer ring of the crater with a diameter of 300 km. An impact of that size is a once-in-a-billion-years event, a very rare catastrophe in the history of Earth.

Critical evaluation of the impact hypothesis now focuses on the second claim—that the collision caused the Cretaceous extinctions. Advocates of the hypothesis argue that the size of the impact was large enough to darken the Earth for years, not months, and that the reduction of photosynthesis would be long enough for food chains to collapse. Some of the minerals in the dust cloud would also have caused severe acid precipitation (see Chapters 2 and 49). And now there is some evidence from sediments at the upper Cretaceous boundary that global fires raged, perhaps ignited by the meteor impact, with smoke further aggravating the atmospheric effects of the impact.

On the other side of the debate over the extinctions are researchers who point out that the coincidence of a meteor impact within a period of mass extinctions does not link the two events as cause and effect. Many paleontologists and geologists believe that changes in climate due to continental drift, increased vulcanism, and other processes on Earth are sufficient to account for mass extinctions without seeking extraterrestrial causes. Researchers are now trying to determine just how sudden and uniform the Cretaceous extinctions were on the scale of geological time. Disappearance of groups as diverse as microscopic marine plankton and dinosaurs all in a relatively small span of time would back the impact hypothesis. A more gradual decline, with the rate varying for different groups of organisms, would favor hypotheses emphasizing terrestrial causes. Hypotheses about the Cretaceous extinctions, however, are not mutually exclusive. One possibility is that the meteor impact was the final, sudden blow in an environmental assault on late Cretaceous life that included more gradual processes.

Whatever the causes, mass extinctions affect biological diversity profoundly. But there is a creative side to the destruction. The species that manage to survive these crises, by their adaptive qualities or by sheer luck, become the stock for new radiations that fill many of the adaptive zones vacated by the extinctions. The world might be a very different place today if a few families of dinosaurs had escaped the Cretaceous extinctions. Or imagine if *Purgatorius*, the one known primate that lived in the Cretaceous period, had not survived.

* * *

So far in this chapter, we have examined some of the geological and biological mechanisms that help us understand the history of macroevolution as told by the fossil record. Paleontology is closely related to the field of systematics, which is concerned with biological diversity, past and present. The fossil record helps systematists determine evolutionary relationships among species, but, as we will see in the next section, evidence from comparing modern organisms is also important.

Systematics connects biological diversity to phylogeny

The evolutionary history of a species or group of related species is called **phylogeny** (Gr. *phylon*, "tribe," and *genesis*, "origin"). These genealogies are traditionally diagrammed as phylogenetic trees that trace putative evolutionary relationships (FIGURE 23.11).

Reconstructing phylogenetic history is part of the scope of **systematics,** the study of biological diversity in an evolutionary context. The diversity of contemporary life reflects past episodes of speciation and macroevolution. In its search for evolutionary relationships among diverse organisms, systematics encompasses taxonomy, the identification and classification of species.

Taxonomy and Hierarchical Classification

The system of taxonomy developed by Linnaeus in the eighteenth century had two main features (see Chapter 20). First, it assigned to each species a two-part Latin name, or **binomial.** The first word of the name is the **genus** (plural, **genera**) to which the species belongs. The second word is the **specific epithet** (name) of the species. For example, the scientific name for the domestic cat is *Felis silvestris.* A given genus may include several related species, each with its own specific name. For instance, the lynx, *Felis lynx,* belongs to the same genus as the domestic cat. Common names—such as cat, black bear, and mountain lilac—often work well in informal communication, but when biologists publish their research, they define the organisms they have studied with scientific names to avoid ambiguity. Many of the scientific names still in use date back to Linnaeus, who assigned binomials to over 11,000 species of plants and animals.

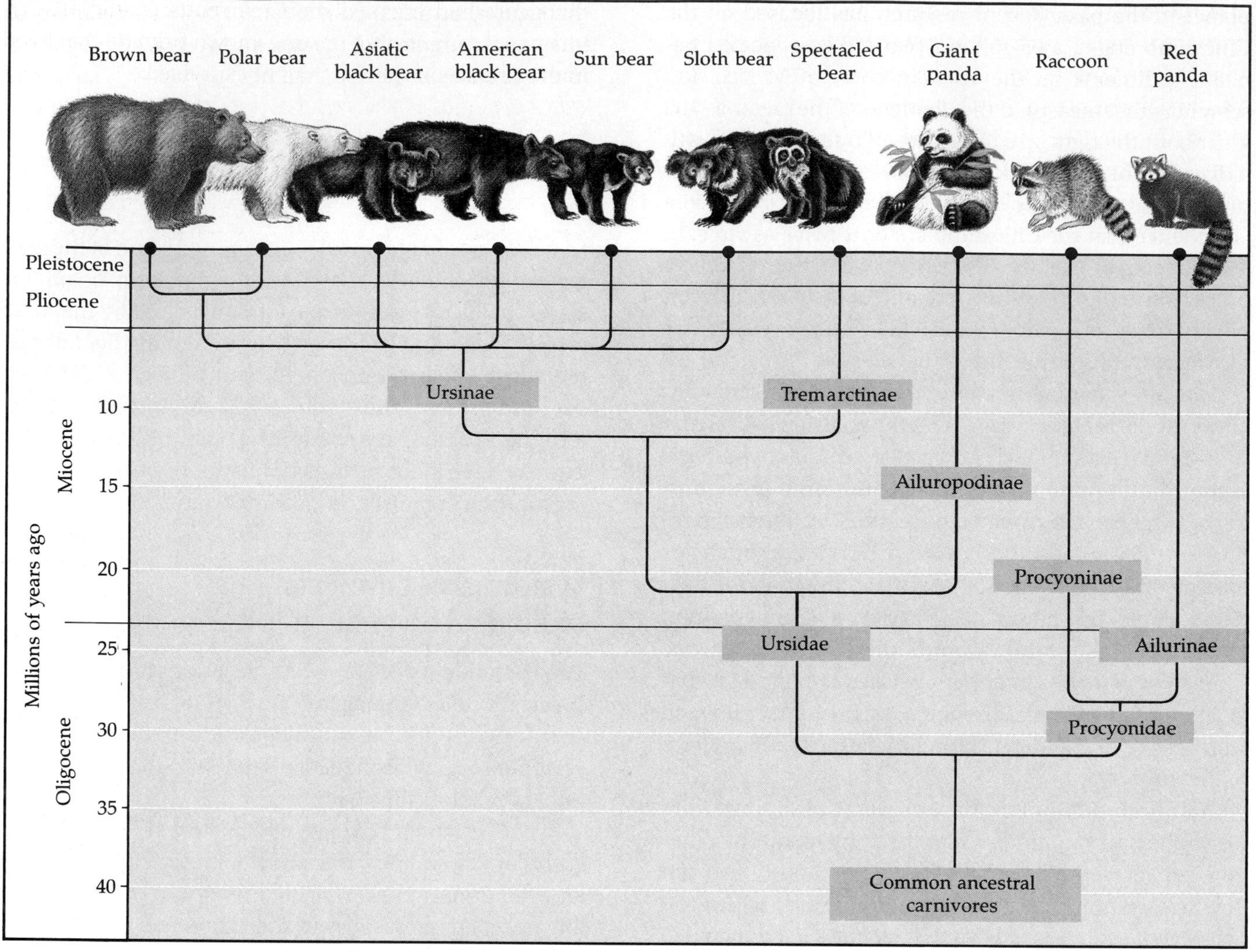

FIGURE 23.11

A phylogenetic tree. Phylogenetic trees are genealogies of probable evolutionary relationships among species and higher taxonomic groups. In this tree of species of bears and raccoons, the two main branches are labeled with the names of the families (e.g., Ursidae), and the subbranches represent subfamilies (e.g., Ursinae). This tree also shows how long ago each evolutionary divergence occurred. Whenever possible, systematists use the fossil record and comparative anatomy to help construct phylogenetic trees, but they also apply other methods. For example, this tree is based mainly on comparisons of the species' DNA and proteins. As a hypothesis of past history based on available data, a phylogenetic tree, like all hypotheses, makes predictions that can be tested by further study. The tree is refined if new information does not fit the existing hypothesis.

The second major contribution Linnaeus made to taxonomy was adopting a filing system for grouping species into a hierarchy of increasingly general categories. The first step in grouping species is built into binomial nomenclature. Species that are very similar, such as the lynx and house cat, are placed in the same genus. Grouping species is natural for us, at least in concept. We lump together several trees we know as oaks and distinguish them from several other species of trees we call maples. Indeed, oaks and maples belong to separate genera. The Linnaean system formalizes this grouping of species into genera and extends the scheme to progressively broader categories of classification, some of which have been added since the time of Linnaeus.

Taxonomists place related genera in the same **family,** group families into **orders,** orders into **classes,** classes into **phyla** (singular, **phylum**), and phyla into **kingdoms.** For example, the genus *Felis* is lumped with various species of the genus *Panthera* (lion, tiger, leopard, and jaguar) in the family Felidae, the cat family. This family belongs to the order Carnivora, which also includes the Canidae (dog family), Ursidae (bear family), and a few other related families. The order Carnivora is grouped with many other orders in the class Mammalia, the mammals. And the class Mammalia is one of several belonging to the phylum Chordata in the kingdom Animalia. Each taxonomic level is more comprehensive than the one below. All members of the family Felidae also belong to the order Carnivora and class Mammalia, but not all mammals are cats. Classifying a species by phylum, class, and so on is analogous to a postal worker sorting mail, first by ZIP codes and then by streets and house numbers. Appendix Two gives taxonomic classification of the major groups of organisms discussed in this book down to the level of class.

As a component of systematics, taxonomy has two main objectives. The first is to sort out closely related organisms and assign them to species, describing the diagnostic characteristics that distinguish the species from one another. Related to this function is the naming of newly discovered species. In the Linnaean tradition, the name is a binomial, with the name of the genus to which the species belongs followed by the specific name, or epithet. The second major objective of taxonomy is classification, the arranging of species into the broader taxonomic categories, from genera to kingdoms. In some cases, there are intermediate categories, such as subfamilies (a category between families and genera). The named taxonomic unit at any level is called a **taxon** (plural, **taxa**). For example, *Pinus* is a taxon at the genus level, the generic name for the various species of pine trees. Mammalia, a taxon at the class level, includes all the many orders of mammals. Only the genus name and specific epithet are italicized, and all taxa at the genus level or higher (broader) are capitalized. International committees establish rules of nomenclature, which are somewhat different for animals, plants, and bacteria. In TABLE 23.2, the house cat and the common buttercup are placed in their appropriate taxa as examples of classification.

TABLE 23.2

Classification of the Domestic Cat and Common Buttercup

CATEGORY	DOMESTIC CAT	COMMON BUTTERCUP
Kingdom	Animalia (animals)	Plantae (plants)
Phylum (animals) or division (plants)	Chordata (chordates)	Anthophyta (flowering plants)
Subphylum	Vertebrata (vertebrates)	—
Class	Mammalia (mammals)	Dicotyledones (dicots)
Order	Carnivora (carnivores)	Ranunculales
Family	Felidae (cats)	Ranunculaceae (crowfoot family)
Genus	*Felis*	*Ranunculus*
Specific epithet	*silvestris*	*acris*

The Relationship Between Classification and Phylogeny

There is more to systematics than taxonomy, the naming and classification of species. Ever since Darwin, systematics has had a goal beyond simple organization: to have classification reflect the evolutionary affinities of species. Each successive group in the taxonomic hierarchy should represent finer and finer branching of phylogenetic trees. A taxon is said to be **monophyletic** if a single ancestor gave rise to all species in that taxon and to no species placed in any other taxon (FIGURE 23.12). A taxon is **polyphyletic** if its members are derived from two or more ancestral forms not common to all members. A **paraphyletic** taxon excludes species that share a common ancestor that gave rise to the species included in the taxon (see FIGURE 23.16a). The ideal in systematics is for each taxon to be a monophyletic grouping, creating a classification that reflects the evolutionary history of organisms. As we will see, achieving this ideal is often easier said than done.

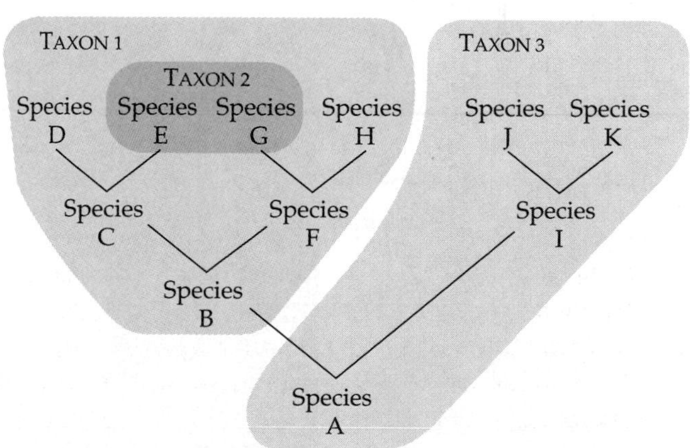

FIGURE 23.12
Grouping species into higher taxa. Taxon 1, with seven species (B–H), qualifies as a monophyletic grouping, the taxonomic ideal. The taxon includes all descendant species along with their immediate common ancestor (species B). However, taxon 2, a subgrouping within taxon 1, is polyphyletic. Species E and species G are derived from two different immediate ancestors (species C versus species F). Taxon 3 is paraphyletic. It includes species A without incorporating all other descendants of that ancestor. Systematists generally value monophyletic classifications over polyphyletic or paraphyletic versions.

Sorting Homology from Analogy

To a large extent, systematics is a comparative science. A systematist classifies species into higher taxa based on the extent of similarities in morphology and other characteristics. Recall from Chapter 20 that likeness attributed to shared ancestry is called **homology.** The forelimbs of mammals are homologous; that is, the similarity in the intricate skeleton that supports the limbs has a genealogical basis (see FIGURE 20.10).

There is a joker in this game of making evolutionary connections by evaluating similarity: Not all likeness is inherited from a common ancestor. Species from different evolutionary branches may come to resemble one another if they have similar ecological roles and natural selection has shaped analogous adaptations. This is called **convergent evolution,** and similarity due to convergence is termed **analogy,** not homology (FIGURE 23.13). The wings of insects and those of birds, for example, are analogous flight equipment that evolved independently and are built from entirely different structures. Convergent evolution has also produced the analogous resemblances between certain Australian marsupials and placental look-alikes that have evolved independently on other continents—the Tasmanian devil and the coyote, for example.

In comparing species to reconstruct evolutionary history, we must build phylogenetic trees on the basis of homologous similarities alone. As a general rule, the greater the number of homologous parts between two species, the more closely the species are related, and this should be reflected in their classification. This guideline is simpler in principle than it is in practice. Adaptation can obscure homologies, and convergence can create misleading analogies. As we saw in Chapter 20, comparing the embryonic development of the features in question can often expose homology that is not apparent in the mature structures.

There is another clue to identifying homology and sorting it from analogy: The more complex two similar structures are, the less likely it is they have evolved independently. Consider the skulls of a human and a chimpanzee, for example. The skulls are not single bones, but a fusion of many, and the chimp skull and human skull match almost perfectly, bone for bone. It is highly improbable that such complex structures matching in so many details could have separate origins. The genes required to build these skulls must have been inherited from a common ancestor.

FIGURE 23.13
Convergent evolution and analogous structures. The ocotillo of southwestern North America (left) looks remarkably similar to the allauidia (right) found in Madagascar. The plants are not closely related and owe their resemblance to analogous adaptations that evolved independently in response to similar environmental pressures.

Molecular biology provides powerful new tools for systematics

The comparison of information-rich macromolecules—proteins and DNA—has become a powerful addition to the other comparative methods systematists use to measure evolutionary relationships between species. For example, molecular data were used to build the phylogenetic tree of bears in FIGURE 23.11. Sequences of nucleotides in DNA are inherited, and they program corresponding sequences of amino acids in proteins. Molecular comparisons go right to the heart of evolutionary relationships.

Protein Comparison

The primary structures of proteins are genetically determined (see Chapter 5). Therefore, a close match in the amino acid sequences of two proteins from different species indicates that the genes for those proteins evolved from a common gene present in a shared ancestor. The degree of similarity is evidence of the extent of common genealogy.

One advantage of this molecular tool of systematics is that it is objective and quantitative. A second advantage is that it can be used to assess relationships between groups of organisms that are so phylogenetically distant that they share very few morphological similarities. For example, the amino acid sequence of cytochrome *c*, an ancient protein common to all aerobic organisms, has been determined for a wide variety of species ranging from bacteria to complex plants and animals. The sequences for humans and chimpanzees match perfectly for all 104 positions along the polypeptide chain, and the cytochromes of both species differ from the version found in the rhesus monkey by just one amino acid. All three species belong to the same mammalian order: Primates. Comparing these proteins to the forms found in nonprimates, we find that the differences increase as the species become more phylogenetically distant. For instance, human cytochrome *c* differs from that of a dog by 13 amino acids, from that of a rattlesnake by 20 amino acids, and from that of a tuna by 31 amino acids. Phylogenetic trees based on cytochrome *c* are consistent with evidence from comparative anatomy and the fossil record.

DNA Comparison

Comparing the genes or genomes of two species is the most direct measure of common inheritance from shared ancestors. Comparisons can be made by three methods: DNA-DNA hybridization, restriction mapping, and DNA sequencing.

Whole genomes can be compared by **DNA-DNA hybridization,** which measures the extent of hydrogen bonding between single-stranded DNA obtained from two sources. How tightly the DNA of one species can bind to the DNA of the other depends on the degree of similarity, as base pairing between complementary sequences holds the two strands together (see the Methods Box on p. 474). Evolutionary trees constructed by this technique generally agree with the phylogeny deduced by other methods such as comparative morphology, but DNA-DNA hybridization has the potential to settle some old debates. For example, mammalogists once disagreed about whether the giant panda is a true bear or a member of the raccoon family; the evidence from DNA-DNA hybridization groups the giant panda with the bears, but places the red panda in the raccoon family (see FIGURE 23.11).

Although DNA-DNA hybridization can estimate the overall similarity of two genomes, it does not give precise information about the matchup in specific nucleotide sequences of the DNA. An alternative approach is **restriction mapping** of DNA. This method employs the same restriction enzymes used in recombinant DNA technology (see Chapter 19). Each type of restriction enzyme recognizes a specific sequence of a few nucleotides and cleaves DNA wherever such sequences are found in the genome. The DNA fragments obtained after treatment with a restriction enzyme can be separated by electrophoresis and compared to restriction fragments derived from the DNA of another species. Two samples of DNA with similar maps for the locations of restriction sites will produce similar collections of fragments. In contrast, two genomes that have diverged extensively since their last common ancestor will have a very different distribution of restriction sites, and the DNA will not match closely in the sizes of restriction fragments.

Because so many fragments are obtained from the nuclear genome, restriction mapping is more practical for comparing smaller segments of DNA, usually a few thousand nucleotides long. Several laboratories are using restriction maps to compare mitochondrial DNA (mtDNA), which is relatively small. There is the added benefit that mtDNA changes by mutation about ten times faster than does the nuclear genome, which makes it possible to sort out phylogenetic relationships between very closely related species or even between different populations of the same species. In one anthropological application of mtDNA comparison, a research team measured the relationships between different groups of Native Americans. Their results corroborate evidence from the study of languages that the Pima of Arizona, the Maya of Mexico, and the Yanomami of Venezuela are closely related, probably descending from the first of three waves of immigrants to cross the Bering Land Bridge from Asia to the Americas during the glaciation of the late Pleistocene epoch.

The most precise and powerful method for comparing DNA from two species is **DNA sequencing**—that is, actually determining the nucleotide sequences of entire DNA segments. The application of polymerase chain reaction (PCR; see Chapter 19) to clone traces of DNA, coupled with automated sequencing, has made the collection of DNA sequence data much simpler and faster. Comparing nucleotide sequences between corresponding DNA segments from two species tells us exactly how much divergence there has been in the evolution of two genes derived from the same ancestral gene. As the technology for the rapid sequencing of DNA continues to improve, DNA sequencing will become an increasingly powerful tool in systematics. A related approach is the sequencing of ribosomal RNA (rRNA), gene products found in all organisms. Because DNA coding for rRNA changes slowly relative to most other DNA, differences in rRNA sequences can be used to trace some of the earliest branching in the tree of life. We will see in Chapter 25 that comparisons of rRNA sequences have been especially useful in sorting out the phylogenetic relationships among the bacteria.

Analysis of Fossilized DNA

With the help of PCR, systematists have been able to extend nucleotide sequence analysis to DNA traces recovered from some types of fossils that retain organic material (see FIGURE 23.1d). The first success came in 1990, when researchers used PCR to amplify DNA extracted from magnolia leaves found in Idaho in sediments that

METHODS: DNA-DNA HYBRIDIZATION AS A TOOL IN SYSTEMATICS

In Chapter 19, you learned about the basic method of nucleic acid hybridization in the context of probes used to detect the presence of certain DNA sequences. Systematists have adapted DNA hybridization as a tool to measure evolutionary relatedness between species. In this example, we use DNA hybridization to compare how closely related the fruit fly *Drosophila melanogaster* is to two other *Drosophila species, D. simulans* and *D. funebris.* ① We extract DNA from the three species and shear it into pieces a few hundred base pairs in length. ② We then heat the three DNA samples to break hydrogen bonds and separate the complementary strands. ③ Next we mix single-stranded DNA from *D. simulans* and *D. funebris* with DNA from *D. melanogaster,* cooling the DNA to allow hydrogen bonds to reform. How firmly the DNA strands from one species bind to DNA from another species depends on the degree of similarity between base sequences. Double-stranded DNA reconstituted from strands derived from a single species serves as a control, a perfect match against which hybrid DNA can be compared. ④ The extent of hydrogen bonding between strands of the hybrid DNA is measured by gradually increasing temperature until the strands separate. The more closely related the DNA, the greater the number of hydrogen bonds, and the higher the temperature must be to pry the strands apart. The data from this example, shown in the graph, support the hypothesis that *D. melanogaster* is more closely related to *D. simulans* than it is to *D. funebris.*

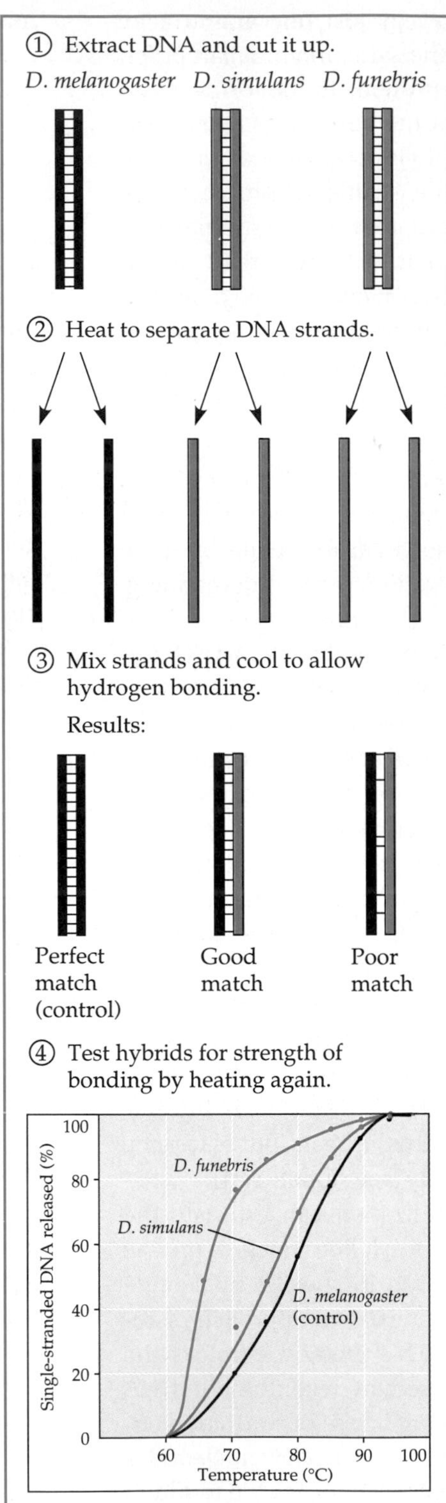

are 17 million years old. The scientists were able to compare a short piece of this ancient DNA to homologous DNA from modern magnolias. Since then, research teams have sequenced DNA fragments from many other fossils, including a 40,000-year-old frozen mammoth and a 40-million-year-old insect fossilized in amber. In 1994, Svante Pääbo and his colleagues at the University of Munich reported the first analysis of DNA preserved in a frozen 5000-year-old Stone Age man discovered by hikers in the Tyrolean Alps (FIGURE 23.14).

Also in 1994, Mary Schweitzer, a graduate student, found DNA traces in the bone marrow of a dinosaur, a 65-million-year-old fossil of *Tyrannosaurus rex* in Montana. Analysis is under way to make sure the DNA trace is not contamination from a fungus or some other organism. If the DNA *is* of dinosaur origin, that does not mean that the scenario of recreating a live dinosaur from ancient DNA in the movie *Jurassic Park* is feasible. Pieces of DNA mined from fossils represent tiny fractions of an organism's whole genome. Even if we had the whole genome of a dinosaur or some other ancient creature, we do not yet understand the developmental steps that translate a library of genetic instructions into an organism. While it is unlikely that we are close to recreating life from the past, it is certain that molecular biology, which has already revolutionized so many fields, will also make its mark in paleontology, systematics, and anthropology.

FIGURE 23.14
Tyrolean Ice Man, a source of prehistoric human DNA. Hikers discovered the frozen 5000-year-old Stone Age man in a melting glacier in the Alps in 1991. Some skepticism that the man was a fraudulently transplanted South American mummy persisted until 1994, when researchers analyzed mitochondrial DNA recovered from the fossil. The DNA closely matches that of northern Europeans, not Native Americans. Continuing analysis of the Stone Age man's DNA may provide more clues about his place in human evolution.

Molecular Clocks

Proteins evolve at different rates, but for a given type of protein—cytochrome *c*, for instance—the rate of evolution seems to be quite constant with time. If homologous proteins are compared for taxa that are known to have diverged from common ancestors during certain periods in the past, the number of amino acid substitutions is proportional to the time that has elapsed since the lineages branched apart. The homologous proteins of bats and dolphins are much more alike than those of sharks and tuna, which is consistent with the fossil evidence that sharks and tuna have been on separate evolutionary lines much longer than bats and dolphins. In this case, molecular divergence has kept better track of the time than have superficial changes in body form.

As a clock to date branch points in phylogenetic trees, DNA comparisons are even more promising than protein comparisons. As with protein clocks, the DNA clock may be relatively accurate; the dating of phylogenetic branchings based on nucleotide substitutions in DNA generally approximate the dates determined from the fossil record. In many cases, the difference in DNA between two taxa is more closely correlated with how long they have been on separate evolutionary branches than is the degree of morphological difference between the taxa.

Molecular clocks are calibrated by graphing the number of amino acid or nucleotide differences against the times for a series of evolutionary branch points known from the fossil record. The graph can then be used to estimate the time of divergence for species when there is no clear fossil evidence for their time of origin from other forms.

The validity of molecular clocks in evolutionary biology rests on the assumption that mutation rates for the genes (and hence proteins) of interest are relatively constant. That assumption seems to be reasonably solid for groups of species that are closely related. For more distantly related groups, differences in generation time and metabolic rates are two of the factors that affect mutation rates and make biological clocks less reliable. Within a closely related group of species, the consistent rate of DNA divergence implies that there is a significant background of neutral mutations that gradually changes the whole genome more than the specific genetic changes associated with adaptation. Evolutionary biologists disagree about the extent of neutral variations (see Chapter 21). Many biologists doubt that neutral evolution is prevalent, so they also question the credibility of molecular clocks as tools for *absolute* dating of the origin of taxa. There is less skepticism about the value of molecular clocks for determining the *relative* sequence of branch points in phylogeny. The modern systematist evaluates any available molecular data along with all other evidence in order to reconstruct phylogeny.

Cladistics highlights the phylogenetic significance of systematics

Phylogenetic trees have two significant structural features. One feature is the location of branch points along the tree, symbolizing the relative time of origin of different taxa. The second is the extent of divergence between branches, representing how different two taxa have become since branching from a common ancestor. If classification is to be based on evolutionary history, which property of phylogenetic trees should be given the greatest weight when grouping species into taxa? This question has divided systematics into three schools of thought: phenetics, cladistics, and classical evolutionary systematics. Of these three approaches, we will see that cladistics places greatest emphasis on the sequence of phylogenetic branching.

Phenetics

Endeavoring to make classification less subjective, **phenetics** (Gr. *phainein*, "to appear"; the term *phenotype* is derived from this same root) makes no phylogenetic assumptions and decides taxonomic affinities entirely on the basis of measurable similarities and differences. As many anatomical characteristics (known as characters) as possible are compared, with no attempt to sort homology from analogy. Pheneticists contend that if enough phenotypic characters are examined, the contribution of analogy to overall similarity will be swamped by the degree of homology. Critics of phenetics argue that overall phenotypic similarity is not a reliable index of phylogenetic proximity. Although a strictly phenetic approach has few proponents today, the methods used in phenetics, especially the emphasis on multiple quantitative comparisons with the help of computers, has had an important impact on systematics. Such "number crunching" is especially useful for analyzing DNA sequence data and other molecular comparisons between species.

Cladistics

Clades (Gr. *clados*, "branch") are evolutionary branches. **Cladistics** classifies organisms according to the order in time that branches arise along a **cladogram,** a dichotomous tree that branches repeatedly. Cladistics, also called phylogenetic systematics, does not consider the degree of evolutionary divergence between the organisms represented in the cladogram. Each branch point in a cladogram is defined by novel homologies unique to the various species on that branch.

In a simplified example, let's apply this cladistic concept to five vertebrates: a lizard, a horse, a seal, a lion, and a domestic cat (FIGURE 23.15). Each species has a mixture of primitive characters that already existed in the common ancestor, along with characters that have evolved more recently. The sharing of primitive characters tells us nothing about the pattern of evolutionary branching from a common ancestor. For example, we cannot use the presence of five separate toes to divide these vertebrate species among evolutionary branches. According to fossil evidence, the remote ancestor common to all species in our list was five-toed, and hence this homology is a *shared primitive character.* Seals and horses apparently lost the trait independently, whereas the other species retained the trait. We must base our cladogram on **synapomorphies,** which are *shared derived charac-*

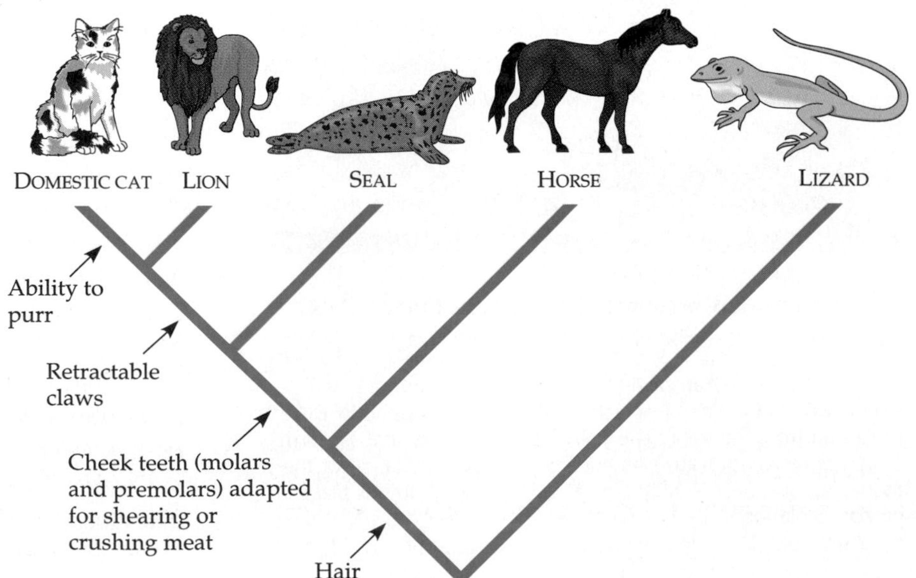

FIGURE 23.15
A simplified cladogram. Each branch point is defined by synapomorphies, derived homologies unique to the lineage that arises at that point. The synapomorphies noted here are not the only ones at each branch. Only the sequence of branching is represented in a cladogram, with no consideration of the degree of divergence between branches. The branch points (forks) represent the most recent ancestor common to all species beyond that point. For example, the lion and cat share a common ancestor that lived more recently than did the ancestor that also gave rise to the evolutionary lineage leading to the seal. However, this does not mean that the seal itself evolved before lions. You will see many cladograms in the next unit of chapters, as we trace the evolution of biological diversity.

ters. These are homologies that evolved in an ancestor common to all species on *one* branch of a fork in the cladogram, but not common to the other branch. Hair and mammary glands are two of the synapomorphies that define a branch point with the lion, cat, seal, and horse on one limb and the lizard on the other. Now we must determine the branching sequence along the mammalian branch. The lion, cat, and seal share many skeletal and dental modifications not present in the horse, and these are among the synapomorphies that define the next branch point in the cladogram. The lion and cat branch from the lineage leading to seals at a later point defined by retractable claws and a number of other synapomorphic characters. FIGURE 23.16 is a more detailed example of how to build a cladogram.

Cladistic systematics produces some taxonomic surprises. For example, the branch point between birds and crocodiles is more recent than the branch point between crocodiles and the other reptiles; that is, birds and crocodiles have synapomorphies not present in snakes and lizards. Indeed, the fossil record supports the hypothesis that birds and crocodiles both evolved during the Mesozoic era from a reptilian ancestor different from the ancestor of snakes and lizards. In the *strictly* cladistic view, there are no such taxa as the class Aves (birds) and the class Reptilia, as we conventionally know them, because birds intrude in the cladogram of the set of animals we call reptiles. (To use terms defined earlier in the chapter, a class Reptilia that excludes birds is paraphyletic, not monophyletic.) Birds seem so superficially different because of the extensive morphological remodeling associated with flight that has occurred since birds branched from their reptilian ancestors.

Classical Evolutionary Systematics

Classical evolutionary systematics, termed *classical* because this approach predates phenetics and cladistics, attempts to balance the dual criteria of extent of divergence and branching sequence. In cases where this leads to a conflict, a subjective judgment is made about which type of information should be given higher priority. For example, while classical systematists acknowledge that crocodiles share a more recent common ancestor with birds than with lizards, they combine lizards and crocodiles in a taxon that excludes birds because the ability to fly was an evolutionary breakthrough that placed birds in a major new adaptive zone. This resulted in adaptive divergence of birds so extensive that classical systematists assign birds to their own class (Aves).

By using evolutionary criteria for classification—the sequence of branching, the extent of divergence, or some combination of these features of evolution—systematists are helping trace the history of life. It is a goal that Darwin set in *The Origin of Species,* where he wrote: "Our classifications will come to be, as far as they can be so made, genealogies." Darwin defined modern systematics, as the Darwinian theme of descent with modification has shaped the entire science of biology. Evolutionary theory has itself evolved, and it is by reviewing some of these developments that we conclude Unit Four.

Is a new evolutionary synthesis necessary?

Evolutionary biology has not had a quiet moment since Darwin published *The Origin of Species* more than 135 years ago. The closest thing to a consensus has been the modern synthesis, the paradigm that has dominated evolutionary theory for the past 50 years (see Chapter 21). Called a synthesis because the ideas were fashioned from several disciplines, including paleontology, biogeography, systematics, and population genetics, it has continued to absorb the discoveries of such new fields as molecular biology. The modern synthesis reaffirmed the Darwinian view of life and updated it by applying the principles of genetics. The paradigm is distinctly gradualist in its view that large-scale evolutionary changes are the accumulations of many minute changes occurring over vast spans of time. Microevolution, the changes in allele frequencies in populations, is extrapolated to explain most macroevolution.

In the classical version of the modern synthesis, natural selection is the major cause of evolution at all levels. Populations adapt by natural selection, new species arise when isolated populations diverge as different adaptations evolve, and continued divergence due to natural selection differentiates the higher taxa. The modern synthesis recognizes, and in fact first described, how genetic drift can cause rapid, nonadaptive evolution. But the major emphases of the synthesis are gradualism and natural selection.

A number of evolutionary biologists dissent from the view that the evolution recorded in the fossil record can be explained by extrapolating the processes of microevolution. The debate is partly about the pace of evolution. Many transitions in the fossil record are punctuational, not gradual. Gradualists argue that apparent abruptness partly derives from the imperfection of the fossil record and partly is a semantic issue clouded by the vastness of geological time. Do we call an evolutionary episode that required 10,000 years "sudden" or "gradual"? Punctuationalists counter that the imperfection of the fossil record is not enough to account for the rarity of transitional forms if speciation and the origin of higher taxa were primarily gradual extensions of microevolution. The debate is not just about the tempo of evolution, but also about the degree to which microevolution compounded over time is sufficient to explain macroevolution.

	Archaeopteryx	Allosaurus	Plateosaurus	Stegosaurus	Parasaurolophus	Pachycephalosaurus	Triceratops
	+	+	+	+	+	+	+
	–	–	–	+	+	+	+
	–	–	–	–	+	+	+
	–	–	–	–	–	+	+
	+	+	+	–	–	–	–
	+	+	–	–	–	–	–

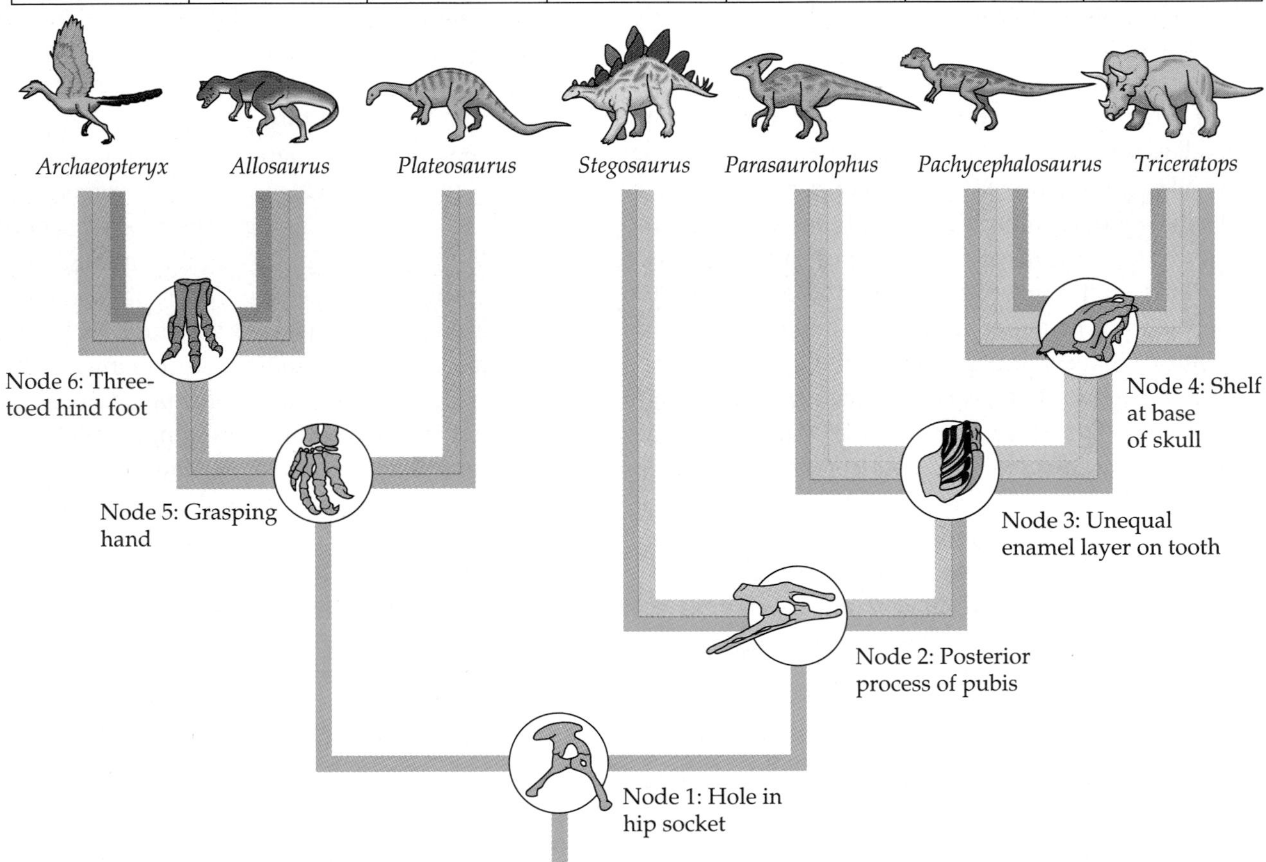

FIGURE 23.16
Building a cladogram of dinosaurs. The top part of this illustration tabulates the presence (+) or absence (–) of key synapomorphies (shared derived characters) among seven dinosaurs. The cladogram below the table is the simplest tree that is consistent with the comparative data of the table. Each node in the cladogram represents an ancestor common to all species beyond that node. And each node is defined by a unique synapomorphy present in all dinosaurs beyond that branch point. These anatomical features are color coded so that you can relate the nodes of the cladogram to the data in the table. For example, the key synapomorphy at node 1 is a hole in the hip socket, which is present in *all* dinosaurs, but in no other four-legged vertebrates. This adaptation is associated with the erect posture of dinosaurs. Node 2, at the base of one of the two major branches of the dinosaur cladogram, is defined by a backward-pointing extension of the pubis, part of the pelvic skeleton. Similarly, each of the other nodes in the cladogram is marked by a labeled synapomorphy. The branches of this cladogram are also color coded so that you can trace each synapomorphy through the tree. Notice, for instance, that *Pachycephalosaurus* and *Triceratops* match in more synapomorphies (4 of the 6 in the table) than does any other combination of two species. This suggests a relatively close evolutionary relationship. Thus, once systematists build such a cladogram, they can use it to group species into a hierarchy of monophyletic taxa.

Some researchers favor a hierarchical theory, with different mechanisms being most important at different levels of evolution. In this view, natural selection is the key to adaptive evolution of a population but is not usually the most important factor in speciation; it plays even less of a role at the level of macroevolution. Most new species begin as small populations isolated from their parent populations by either geographical barriers or genetic accidents, such as chromosomal mutations. The small, isolated population can evolve relatively rapidly, its divergence from the parent population due at least as much to genetic drift as to selection. Chance may produce a new species that is reproductively isolated even before selection has fashioned new adaptations. Chance also figures prominently in macroevolution. Continental drift and mass extinctions have probably had at least as much effect on the history of biological diversity as gradual adaptation caused by selection operating on gene pools at the population level. Among those who see evolution in this dicey context, **historical contingency,** the occurrence of unforeseen events, has become a popular concept.

The importance of natural selection is not under fire; the various factions agree that natural selection is the mechanism of adaptation and should therefore be the centerpiece of evolutionary biology. Selection fine-tunes a population to its environment with generation-to-generation changes in the gene pool that are adaptive. When a new species or higher taxon comes into being, it is natural selection that refines unique adaptations. Although the events that lead to speciation and episodes of macroevolution may have more to do with contingency than with adaptation, new species only persist long enough to be entered into the fossil record if they have adapted to their environment through natural selection.

Perhaps those who have challenged the orthodoxy of the modern synthesis are quibbling, as John Maynard Smith suggests in the interview that opened this unit. The synthesis has never claimed that evolution is always smooth and gradual, or that processes other than changes in gene pools due to selection are unimportant. The questions are not as much about the nature of evolutionary mechanisms as about their relative importance. The modern synthesis may not need radical surgery, only a face-lift.

Vigorous debate about how life evolves is a healthy sign; evolutionary biology is a robust science that will not wallow in dogma. The debates will continue as long as we are curious about our origins and our relationship with the rest of the living world.

REVIEW OF KEY CONCEPTS (with page numbers and key figures)

- The fossil record documents macroevolution (pp. 455–457, FIGURE 23.1)
  - The fossil record provides the historical archives biologists use to study the history of life.
- Paleontologists use a variety of methods to date fossils (pp. 457–459, TABLE 23.1)
  - Sedimentary strata reveal the relative ages of fossils in successive geological periods.
  - The absolute ages of fossils in years can be determined by radiometric dating and other methods. Absolute dating of sedimentary strata has defined the ages for the different geological eras and periods, each of which corresponds to a major transition in the composition of fossil species. The chronology of geological periods and eras makes up the geological time scale.
- What are the major questions about macroevolution? *an overview* (p. 460)
  - Macroevolution is the history of life above the species level. It involves the origin of evolutionary novelties, the study of evolutionary trends, and global episodes of major adaptive radiations and mass extinctions.
- Some evolutionary novelties are modified versions of older structures (p. 460)
  - One mechanism that makes evolutionary novelty possible is preadaptation, the gradual modification of an existing structure for a new function.
- Genes that control development play a major role in evolutionary novelty (pp. 460–462, FIGURE 23.2)
  - Changes in the temporal or spatial details of development are important in the evolution of novel characteristics.
- Recognizing trends in the fossil record does not mean that macroevolution is goal-oriented (pp. 462–465, FIGURE 23.7)
  - Species selection may cause evolutionary trends. According to this hypothesis, species with certain characteristics endure longer and speciate more frequently than species with other characteristics.
- Macroevolution has a biogeographical basis in continental drift (pp. 465–467, FIGURE 23.9)
  - Continental drift has had a significant impact on the history of life by causing major geographical rearrangements affecting biogeography and evolution. The formation of the supercontinent Pangaea during the late Paleozoic era and its subsequent breakup during the early Mesozoic era explain many puzzling cases of biogeography.
- The history of life is punctuated by mass extinctions followed by adaptive radiations of the survivors (pp. 467–469, FIGURE 23.10)
  - Evolutionary history has not been a series of smooth gradations. Long, relatively stable periods have been interrupted by intervals of extensive species turnover—mass extinctions followed by grand episodes of adaptive radiation.

- Systematics connects biological diversity to phylogeny (pp. 469–472, FIGURE 23.12)
  - Systematics, the study of biological diversity in an evolutionary context, includes taxonomy, the identification and classification of species.
  - Phylogenetic relationships are decided on the basis of homology, structural similarity due to common ancestry.
- Molecular biology provides powerful new tools for modern systematics (pp. 472–476)
  - Molecular systematics is the ultimate level for determining homology. Evolutionary relationships can be revealed by comparing amino acid sequences of proteins and nucleotide sequences of DNA. Molecular evolution may occur at a rate consistent enough to function as a crude clock for determining the relative sequence of branch points in phylogeny.
- Cladistics highlights the phylogenetic significance of systematics (pp. 476–478, FIGURES 23.15, 23.16)
  - Phylogeny has two aspects: the relative time of origin for different taxa and the extent of divergence from a common ancestor at each branch point.
  - Phenetics ignores the timing of branch points and classifies organisms solely on the basis of similar characteristics.
  - Cladistics ignores overall similarity and bases classification on the timing of branch points as defined by synapomorphies, or shared derived characters.
  - Classical evolutionary systematics considers overall similarity and branching sequence.
- Is a new evolutionary synthesis necessary? (p. 478)
  - The modern synthesis combines a variety of disciplines to explain evolution. It emphasizes the Darwinian concepts of gradualism and natural selection.
  - Some researchers question the degree to which macroevolution is the cumulative product of microevolution due to natural selection. In this view, chance episodes play an important role in speciation and may be the most important factors in macroevolution.
  - In the final analysis, the debate among evolutionary biologists is more about the relative importance of evolutionary mechanisms than about their nature. Natural selection, as the mechanism of adaptation, remains central to evolutionary biology.

SELF-QUIZ

1. A paleontologist estimates that when a particular rock formed, it contained 12 mg of the radioactive isotope potassium-40. The rock now contains 3 mg of potassium-40. The half-life of potassium-40 is 1.3 billion years. About how old is the rock?
   a. 0.4 billion years
   b. 0.3 billion years
   c. 1.3 billion years
   d. 2.6 billion years
   e. 5.2 billion years

2. If humans and pandas belong to the same class, then they must also belong to the same
   a. order
   b. phylum
   c. family
   d. genus
   e. species

3. In the case of comparing birds to other vertebrates, having four appendages is (explain your answer)
   a. a shared primitive character
   b. a synapomorphy
   c. a character useful for distinguishing the birds from other vertebrates
   d. an example of analogy rather than homology
   e. a character useful for sorting the avian (bird) class into orders

4. Advocates of the theory known as species selection propose that most evolutionary trends result from
   a. the tendency for natural selection to perfect adaptations
   b. stepwise progression of an unbranched lineage, with each step furthering the evolutionary trend
   c. phyletic transformation of a single species (anagenesis)
   d. preadaptation of species for possible changes in the environment
   e. differences between species in their longevity and/or rates of speciation

5. The greatest adaptive radiation of the animal kingdom occurred during the
   a. early Precambrian era
   b. late Precambrian era
   c. early Paleozoic era
   d. early Mesozoic era
   e. early Cenozoic era

6. The DNA from two species is compared by the method of restriction mapping. Extensive similarity between the species in the collection of DNA fragments from treatment with a restriction enzyme indicates that
   a. the genes being compared have the same functions
   b. most sites recognized by the restriction enzyme have equivalent locations in the DNA samples from the two species
   c. the two species normally possess the same restriction enzyme
   d. the DNA fragments that match in size between the two species have identical base sequences
   e. the genomes of the same species are about the same size in their total amount of DNA

7. Extensive adaptive radiations have usually followed in the wake of mass extinctions mainly because
   a. many adaptive zones are vacated
   b. conditions of the physical environment are usually at their most favorable after some crisis has passed
   c. the survivors have superior adaptations that enable them to move into many environments when conditions improve after the extinction episode
   d. preadaptation assures that survivors will radiate to give rise to many new species
   e. given a stable environment, biological diversity tends to increase

8. Which of the following would be most useful for constructing a cladogram showing the taxonomic relationships among several fish species? (explain your answer)
   a. several analogous characteristics shared by all the fishes
   b. a single homologous characteristic shared by all the fishes
   c. the total degree of morphological similarity among various fish species
   d. several characteristics thought to have evolved after different fish diverged from one another
   e. a single characteristic that is different in all the fishes

Coyote

Red Wolf

Gray Wolf

9. The evolutionary transformation of a fish's primitive lung into a swim bladder (float) is an example of
   a. convergent evolution
   b. divergent evolution
   c. preadaptation
   d. adaptive radiation
   e. paedomorphosis

10. The differences between the modern synthesis and a more hierarchical view of evolution include all the following *except*
    a. gradualism versus punctuated equilibrium
    b. natural selection versus chance as central to adaptation
    c. the relative importance of microevolution in macroevolution
    d. phyletic transitions (anagenesis) versus species selection
    e. the relative role of chance in speciation and biological diversity

CHALLENGE QUESTIONS

1. In the "DNA clock," some nucleotide changes cause amino acid substitutions in the encoded protein (nonsynonymous changes), and others do not (synonymous changes). In a comparison of rodent and human genes, rodents were found to accumulate synonymous changes 2.0 times faster than humans and nonsynonymous substitutions 1.3 times as fast. What factors could explain this difference? How do such data complicate the use of molecular clocks in absolute dating?

2. Imagine that Pangaea were re-formed today. What might be some of the environmental effects, and how might these changes influence life on Earth?

3. Refer to the cladogram in FIGURE 23.12. Group the eleven species into a hierarchy of five higher taxa, each of which is monophyletic.

SCIENCE, TECHNOLOGY, AND SOCIETY

1. In 1992, a Montana state legislator introduced a bill making it a felony for anyone except selected scientists to collect fossils on federal lands. Argue in favor of or against this bill. (For more information about the bill and the issues see *Newsweek,* January 11, 1993, p. 8.)

2. It is not clear whether the North American red wolf is a hybrid between the coyote and the gray wolf or a separate species (see photos above). The solution to this taxonomic puzzle has become the center of a debate about whether or not the red wolf qualifies for protection under the Endangered Species Act. If more detailed DNA studies convincingly demonstrate that the red wolf is indeed a hybrid between the coyote and gray wolf, should efforts to protect and breed the red wolf be terminated? Why or why not?

3. Experts estimate that human activities cause the extinction of hundreds of species every year. The natural rate of extinction is thought to be a few species per year. As we continue to alter the global environment, especially by cutting down tropical rain forests, the resulting extinction will probably rival that at the end of the Cretaceous period. Most biologists are alarmed at this prospect. What are some reasons for their concern? Consider that life has endured numerous mass extinctions and has always bounced back. How is the present mass extinction different from previous extinctions? Why? What might be the consequences for the surviving species? For humans?

FURTHER READING

Bak, P., H. Flyvbjerg, and K. Sneppen. "Can We Model Darwin?" *New Scientist,* March 12, 1994. Are mass extinctions predictable?

Cohen, J. "Will Molecular Data Set the Stage for a Synthesis?" *Science,* February 11, 1994. The impact of molecular biology on evolutionary thought.

Gaffney, E. S., L. Dingus, and M. K. Smith. "Why Cladistics?" *Natural History,* June 1995. Reassessing the phylogeny of dinosaurs.

Gibbons, A. "Possible Dino DNA Find Is Greeted with Skepticism." *Science,* November 18, 1994. Contamination is a challenging problem in the study of DNA from fossils.

Gould, S. J. "The Evolution of Life on the Earth." *Scientific American,* October 1994. The role of contingency in evolution.

Gould, S. J."We Are All Monkey's Uncles." *Natural History,* June 1992. Cladists and human classification.

Handt, O., et al. "Molecular Genetic Analyses of the Tyrolean Ice Man." *Science,* June 17, 1994. Sounds like a *National Enquirer* headline, but it's for real!

Hecht, J. "The Geological Timescale." *New Scientist,* May 20, 1995. How to date fossils.

Hellemans, A. "Who Profits from Ecological Disaster?" *Science,* October 7, 1944. How does life rebound after mass extinctions?

Monastersky, R. "Impact Wars." *Science News,* March 5, 1994. Debate about the Cretaceous extinctions.

Pääbo, S. "Ancient DNA." *Scientific American,* November 1993. Analyzing DNA from fossils.

Waters, T. "Greetings from Pangaea." *Discover,* February 1992. Environments were extreme on the supercontinent.

Wray, G. A. "Punctuated Evolution of Embryos." *Science,* February 24, 1995. The importance of developmental regulation in evolution.

UNIT FIVE

THE EVOLUTIONARY HISTORY OF BIOLOGICAL DIVERSITY

AN INTERVIEW WITH

EDWARD O. WILSON

By learning about tiny animals, Edward O. Wilson has become one of the giants of twentieth-century science. Many biologists are specialists who choose experimental organisms to fit questions in a particular research field. Edward Wilson's style is different: He has followed a group of organisms, the ants, into many fields of biology by studying their diversity, behavior, evolution, and ecology. Dr. Wilson has also helped synthesize entirely new disciplines, including sociobiology, which seeks evolutionary explanations for the social behavior of animals. Perhaps Dr. Wilson's greatest contribution is his influence in restoring the study of whole organisms to the center of biology. In the past few years, he has built on that influence to become a leading activist for the conservation of biological diversity.

Dr. Wilson has received most of the major national and international scientific prizes, including the National Medal of Science, the International Prize for Biology, the Eminent Ecologist Award, and the Crafoord Prize, generally considered the equivalent of the Nobel Prize for ecologists and evolutionary biologists. Wilson's students at Harvard have honored him with both of the outstanding teacher awards available, and he was recently named University Professor, one of only 15 on the Harvard faculty. Professor Wilson has also shared his view of life with general audiences in several books, two of which won the Pulitzer Prize. In one of his most recent books, the autobiographical Naturalist *(Island Press, 1994), Dr. Wilson explains his lifelong interest in ants: "Most children have a bug period, and I never grew out of mine." That was evident when I met Professor Wilson in his Harvard office, which is cohabited by thousands of leaf-cutter ants marching between plastic boxes over a bowed tree branch that functions as an arched bridge.*

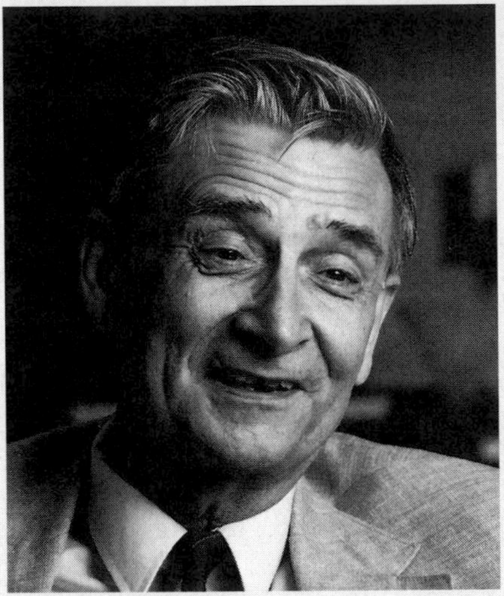

In *Naturalist* you write that images from your youth created a "gravitational force" that pulled your research career and still defines you as a scientist. What were some of those formative experiences?

I had the good fortune of growing up in the beautiful environment of the Gulf Coast states, which remained relatively unexplored in natural history when I was a boy. Many species of frogs, salamanders, and fishes (not to mention insects) were still unknown to science. At a very early age I felt the excitement of an explorer-naturalist. I was able to have that experience before I reached high school age, and I had my career set. It has consisted substantially of trying to repeat these early experiences through one cycle after another of research.

Once you embraced nature and became an explorer-naturalist, how did your interest become focused on insects?

When I was seven years old, I lost the sight in my right eye in an accident. I therefore grew up with vision in only one eye, but fortunately it was relatively acute, particularly for objects at a short distance. Since I was predestined to be a naturalist and biologist, my disability directed me toward insects, for which you do not need great distance vision, and I made the most of it.

Why did you specialize in ants?

I include ants among what I like to call the little things that run the Earth. They, and many other groups of abundant insects, are extremely important in maintaining the balance of nature in the land environment as predators, scavengers, feeders on vegetation, and soil workers. Worldwide, they are more important than earthworms in the role of soil workers. Ants are essential to our existence. We tend to overlook them because they are hard to see—the ant is only about one-millionth the size of the human being—but they are there in vast numbers. I have made an estimate, or educated guess I suppose, that at any given time there are somewhere between a million million and 10 million million (10^{12}–10^{13}) ants alive. Remarkably, that immense legion adds up to approximately the same total mass as humanity.

You established yourself early in your research career as an ant taxonomist. What is it that you find so satisfying about taxonomy, and why do you think taxonomy is still of such fundamental importance to biology?

I believe that the taxonomist, more than any other specialist, is able daily to look upon the face of creation, by which I mean the immense variety of life. The taxonomist doesn't just identify and classify, but serves as the steward of the group on which he works; whether it is deep-sea fishes or ants or orchids or any of hundreds of other groups. The taxonomist, as his expertise broadens and deepens, becomes interested in the marvelous intricacies of anatomy and physiology, and, in the case of animals, behavior. He is a leading authority on that group overall. This sense of mastering a part of the creation creates constant excitement, as do the vast unknown phenomena that await the taxonomist upon further investigation in the field and within the laboratory. As a taxonomist, I know that I will be able to make a discovery of some kind or another—a new species, a new idea about relationships, a new anatomical structure—almost every day. When I go out in the field, the flow of discoveries increases. You know, in most disciplines of science the research stays inside, most activities of biology included. The number of important discoveries per investigator per year is dropping off steadily. In most fields, teams and great expense are needed. In particle physics, the teams often consist of a hundred or more people working together. In molecular biology, commonly a half-dozen or more. To be able to work alone, or at most with a single collaborator, with a steady flow of discoveries—that is exceptionally rewarding!

From your work as an ant taxonomist, your interests extended to general evolutionary theory. What influence drove that trend in your research?

I always wanted to do more than just natural history. I think that there is in me a need to put order into whatever I encounter in life. Someone once remarked that among scientists as well as nonscientists, there appear to be two classes of people: those who upon discovering disorder wish to convert it into order, and those who discovering order wish to reduce it to disorder. There are successful scientists in both genres in evolutionary biology. I have a passion for creating order. But beyond that, I have the ambition to create new theoretical systems in evolutionary biology. Taxonomy and natural history provide a tremendous background of information for creating new theory.

In *Naturalist,* you write that "unlike experimental biologists, evolutionary biologists well versed in natural history already have an abundance of answers from which to choose. What they need most are the right questions." Can you explain?

I believe that the best evolutionary biologists are naturalists. The physical scientists and molecular biologists like to work from questions towards solutions on a single track. These questions take them deeper and deeper through molecular levels. But a naturalist tries to learn all that he can about the group of organisms or about a particular ecosystem. All forms of information regarded by the naturalist have value.

I was a typical naturalist throughout my career. I compiled information of all kinds about ants, and I had access to information from others' work. Patterns began to emerge in which generalities could be drawn. Evolutionary biology is full of patterns, rules, and trends. What then becomes important is: "What do these patterns, rules, and trends reveal?" In order to extend this mode of investigation, the naturalist has to go to the answers already before him and ask the right questions about the evolutionary process, about the history of the world. This has been by far the most

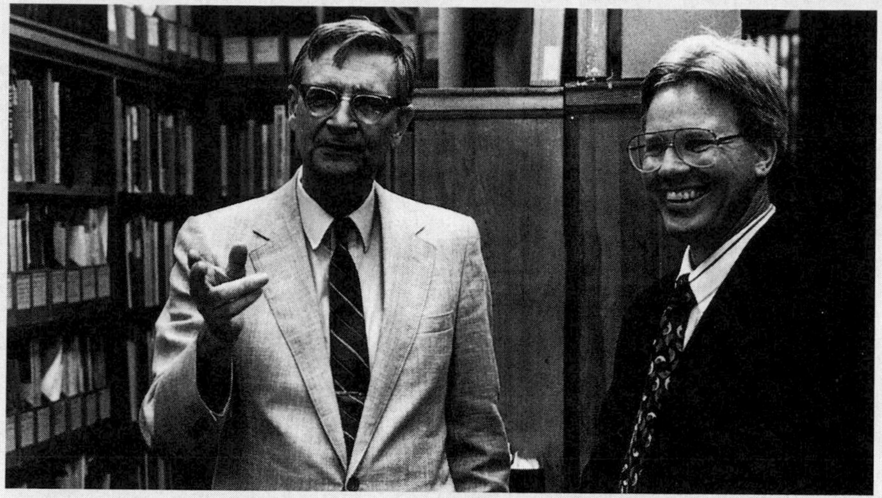

important procedure in the history of evolutionary biology. For example, it is the process that Darwin followed when he developed the theory of evolution by natural selection, which has dominated evolutionary biology since his time.

When you speak about patterns and rules and trends and evolution, that all sounds very mathematical, and yet you have held yourself up as an example of a scientist who has made it in spite of "math anxiety." What encouraging words do you have for science students who are not math wizards?

I think I can give some good advice to young, would-be biologists. Some mathematics is important in pursuing a career in evolutionary biology—at the very least, models entailing probability theory and statistics. I am not a gifted mathematician. I discovered this the hard way, through years of study and lack of notable success. My greatest success in quantitative theory came when I deliberately teamed up with first-class mathematicians who then found my knowledge in natural history equally valuable. That is a fine way to do evolutionary biology—a mathematician teaming with an evolutionary biologist.

Furthermore, in certain areas of evolutionary biology, mathematical ability of a high level is not necessary. I have a rule that I follow: for every level of mathematical ability the biologist possesses, there exists a field within biology still poorly enough worked so that that level of ability is sufficient to do first-class theoretical work. Therefore, the advice I give to young biologists is, do not in your career rush to where most of the scientists are already gathered. Look for the less popular subjects where you will have the greatest chance to innovate.

And we can find some of these opportunities in natural history and evolutionary biology?

Modern biology consists of two major fronts of advance. One of them addresses the physical and chemical basis of life's operation and the development of organisms. The other addresses the behavior and the living together of organisms, as studied in behavioral biology and ecology, increasingly with a new emphasis on biodiversity. These latter areas—behavioral biology and ecology, or evolutionary biology for short—are attaining new importance to society. They also present some of the most intriguing unsolved problems of science. The main route into these fields is by way of natural history.

Your work on the natural history of ants catalyzed your interest in animal behavior, especially social behavior. What led you to write *Sociobiology,* which synthesized a new field?

There is a common feeling that I had some grand vision of producing an overarching theory and worked toward it in my career, but that isn't true at all. I started with ants and natural history. I followed a rule of research strategy while working with ants that goes as follows: For every group of organisms there is a set of problems with solutions for which that group is ideally suited. So having become fascinated with ants, I became increasingly fascinated with social behavior. At the same time, I was also gaining an interest in population biology because I saw this was a broad road into advanced theory of evolution. I saw at this point that the way to synthesize what we knew about ants and other social insects was to treat colonies as populations and to apply all the techniques and ideas of population biology to the colony of insects. The result was a book about social insects in 1971. I pulled the study of sociobiology, as I now called it, from natural history and population biology.

After finishing the book, *The Insect Society,* I decided that this approach, which proved to be successful with insects, should also apply to vertebrate animals, including birds. So I set out to extend all that I had done to the vertebrates. I wrote the book *Sociobiology, the New Synthesis* primarily to cover the social insects and vertebrate animals. I then saw that I could not leave out the most familiar vertebrate animal, *Homo sapiens.* I didn't intend to stir up the hornet's nest of controversy that in fact the book did create. I included two chapters in *Sociobiology* on human beings primarily for completeness. I admit that as I wrote the chapters I realized that many of the ideas coming from evolutionary biology and the current study of animal societies would find relevance in the social clime.
I expected the book would have an impact, but I didn't expect it would have as controversial a result in the social science community as it did.

And with the benefit of retrospect, why do you think *Sociobiology* did trigger such strong reactions, not only from some social scientists but also from some biologists?

Certain biologists had strong political convictions. You must remember that *Sociobiology* came out in the middle of the 1970s, a time when most scholars in the social sciences believed not only that heredity has no importance in human social behavior, but that it is dangerous to speak of it because it might imply that human destiny is fixed, and that there is nothing we can do about social ills. This was a primary reason for resistance both from social scientists, who had already settled on a sociocultural explanation, and from some biologists. Since the 1970s, the evidence for genetic influence, certainly not genetic determination in a rigid sense but strong genetic influence on human social behavior, has grown very substantially. It is now perhaps a mainstream of thinking.

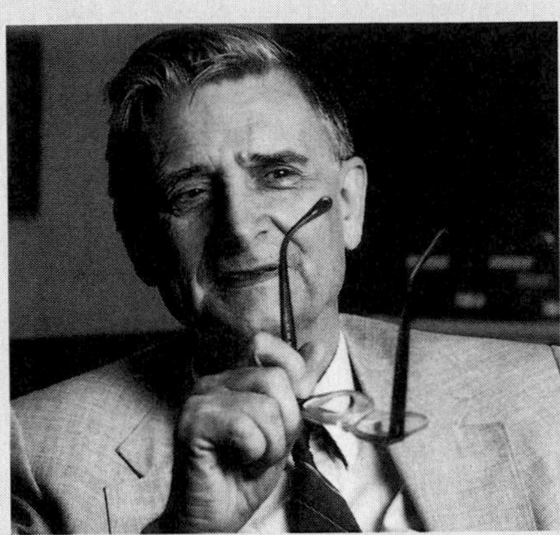

In writing about sociobiology, you once claimed that some day biology must serve as part of the foundation of the social sciences. Has that started to happen as you predicted?

Yes, I believe that sociobiology has already had a profound influence in some social sciences. For example, it has spawned a whole new and quite successful field in psychology called evolutionary psychology, which deals with biological components of human behavior. It has had considerable influence in anthropology, especially in the developing field of comparative studies of nonhuman primate species in relation to human social organization. It has had some influence in economics in the consideration of human choice behavior. And in legal theory, sociobiology has begun to influence thinking about matters such as adoption rights.

Is this all part of a generally increasing awareness about the relationship of science to society?

Science is no longer just a fun thing, like landing on the moon or discovering a new species of bird. It is vital—and people know it. They see science as a major part of modern ethics and legislative action. They also see the environment as something that they have got to know about.

You helped attract attention to biodiversity. What does the term mean?

Biologists define biodiversity in the broadest sense as meaning all of the variety of life—from the different genes at the same chromosome position within populations, up through different species of organisms, on up to different aggregations of species in ecosystems. The emphasis in this definition is on levels of biological organization. It is very important to study each one in turn and to understand fully how they are related to each other: the genes, the species, and the ecosystems.

Given your lifelong interest in biological diversity, how do you account for the timing of your relatively recent emergence as a major activist for conserving biodiversity?

I was a bit slow in becoming an activist. I was aware of the dangers to biodiversity even during my student days in the late 1940s and early 1950s, especially when I began fieldwork in Cuba and other tropical countries where the destruction was well advanced. But I held off becoming an activist until finally in the late 1970s. More precise information on the rate of the deforestation was becoming available then. And I and others were able to use the model of island biogeography that Robert McArthur and I developed to make at least a crude estimate of the rate of species extinction caused by habitat destruction in the case of rain forests. That did it for me. I realized that I should not leave conservation advocacy to others. I and other scientists should become more involved. The issue was just too important.

In one of your books, *The Diversity of Life,* you articulate an environmental ethic. What is it, and how is it derived?

Let me give you the brief description of my own ethic, and that is to make every effort to save all species. This will not be possible in every case everywhere in the world. We are destined to lose a substantial fraction, 10% or more, of the world's biodiversity, no matter what we do. But we should never knowingly allow a species to go extinct if appropriate measures can save it. That, in essence, is the biodiversity ethic.

How many species are known, and what other estimates are there as to how many actually exist?

Roughly 1.5 million species of organisms have been described and given formal scientific names to date. Beyond that, biologists do not know with certainty to the nearest order of magnitude how many species actually exist. Most would agree that there are at least 10 million alive on Earth today, the largest numbers of which are insects and other arthropods and microorganisms like bacteria. But estimates have been made closer to 100 million, and I tend to be one of the radicals who believes that it probably approaches that number. One of the most fascinating problems of evolutionary biology, therefore, is exactly how many—or even approximately how many—kinds of organisms there are on Earth. The second great unsolved problem of evolutionary biology is: Why?

Is a near-complete catalog of species a realistic goal?

It is widely believed among biologists that biodiversity is so great, and so little is known about it, that it is impossible to study at all. But the fact is that a total global survey could be conducted by as few as 25,000 systematists devoting a lifetime career to the subject. Even if we stored this information the old-fashioned way, in books on library shelves, by devoting each page in a book to one species, the complete cataloguing of 100 million species would still occupy only 6 kilometers of library shelving—about as much as is in a medium-sized public library.

You once said, "The key to taking measure of biodiversity lies in a downward adjustment of scale." Can you explain what you mean?

The smaller the organism, the less we know about it. Probably fewer than one-tenth of the species of insects on Earth have been given a scientific name. It is obvious that it is far easier to find and study a bird or a mammal than a tiny insect. The percentage of identified bacterial species is probably much smaller yet. There may be up to 5000 species of bacteria in a single gram of forest soil, almost all of which are unknown to science. Bacteria are in the world of the very small, so they are the greatest challenge of biodiversity study.

You have said that the study and cataloguing of biodiversity is a great responsibility. Why is it so important?

Only by creating a complete catalogue of biodiversity, with information on the traits and the geographic distribution of all the species, can we fully understand the function of ecosystems and locate the most endangered species.

You have also called biodiversity our most valuable but least appreciated resource. What is the value of biodiversity, and why is it worth conserving?

The value of biodiversity is immense to the extent of being immeasurable. In future years it will be the source of countless new pharmaceuticals, genes for disease resistance, petroleum substitutes, and other products vital to human survival. The great majority of species have not even been considered with reference to the many contributions they can make to science and economic wealth.

One reason we should study and maintain biodiversity is the great benefit that it can provide. Once lost, the species and its library of genetic information, with potential importance to the ecosystem, is gone forever and cannot be retrieved.

How does the current extinction rate compare to historical rates?

Using data from the fossil records, I have estimated that species are becoming extinct today at a rate of 1000 to 10,000 times faster than was the case before humanity evolved.

We are doing more damage than asteroids!

We are. The human species is our own home-grown asteroid.

Some conservation organizations now embrace what you call a "new environmentalism." How are approaches to conservation biology changing, particularly in light of economic impact in developing countries?

I introduced the term "new environmentalism" to denote the movement that began in the late 1970s and early 1980s to combine conservation efforts with local economic development. Conservationists realize that it is quite possible to combine the preservation of natural environments with the use of those natural environments in ways that actually contribute to the economic growth of local regions. Therefore, a lot of the efforts of conservation organizations are now devoted to the twin goals of saving species and habitats as always, but also of devising ways to make that compatible with economic development.

You've written about a haunting dream you have. What happens in that dream?

Well, everyone has anxiety dreams, and mine takes a peculiar naturalist turn. I find myself in the dream on an island in the South Pacific. That's where some of my main research efforts were in the early part of my career. In the dream I realize that I have not gotten around to exploring the ends of the island, but my plane is due to leave later that day. The dream consists of looking for natural habitats; however, as I go searching in the dream, I can never find them. What looks like the edge of a forest turns into a hedgerow planted by people. When I wake up from that particular anxiety dream, I usually feel not only anxious but also guilty and disappointed. You can take that as a kind of metaphor of what the modern conservationists feel—that we never, never do enough.

If there is any encouraging word in the biodiversity crisis, it is biophilia, a term you coined. What does it mean?

Biophilia was the offspring of sociobiology. I realized in the early 1980s that there was growing evidence for a genetic predisposition for many kinds of human behavior. I and others believed that there was such a thing as a human preference for certain types of natural habitats and an inherent tendency to affiliate with other forms of life. Since then more evidence has accumulated, and the idea of biophilia has become a popular subject in science as well as in conservation circles. The reason the phenomenon is so important if it does exist (and I believe it exists) is that it will provide a powerful base for an environmental ethic. In my opinion, we need to value biodiversity for the contribution made to the satisfaction of human nature and not just for the physical welfare it provides us.

At the end of *Naturalist,* you distill your philosophy down to three truths. What are they?

Quite simply they are: that the human species is a product of biological evolution; that human beings arose in an arena of natural environments and biodiversity, and that therefore natural environments are a precious part of human heritage; and finally, that neither philosophy nor religion can ever make much sense unless they take the first two points into consideration.

CHAPTER 24

EARLY EARTH AND THE ORIGIN OF LIFE

KEY CONCEPTS

- Life on Earth originated between 3.5 and 4.0 billion years ago
- The first cells may have originated by chemical evolution on a young Earth: *an overview*
- Abiotic synthesis of organic monomers is a testable hypothesis: *science as a process*
- Laboratory simulations of early Earth conditions have produced organic polymers
- Protobionts can form by self-assembly
- RNA was probably the first genetic material
- The origin of hereditary information made Darwinian evolution possible
- Debate about the origin of life abounds
- Arranging the diversity of life into kingdoms is a work in progress

Life is a continuum extending from the earliest organisms through the various phylogenetic branches to the great variety of forms alive today. In Unit Five, we will survey the diversity of contemporary life and trace the evolution of this diversity over 3.5 billion years of history.

One recurrent theme in these chapters is the association between biological and geological history. Geological events that alter environments change the course of biological evolution. The formation and subsequent breakup of the supercontinent Pangaea, for instance, affected the diversity of life tremendously (see Chapter 23). Conversely, life has changed the planet it inhabits. For example, the evolution of photosynthetic organisms that released oxygen into the air completely altered Earth's atmosphere. (This early photosynthetic life included the cyanobacteria present in the algal mats in the foreground of the painting of early Earth that opens this chapter.) Much more recently, the emergence of Homo sapiens *has changed the land, water, and air on a scale and at a rate unprecedented for a single species. The histories of Earth and its life are inseparable.*

These chapters also emphasize key junctures in evolution that have punctuated the history of biological diversity. Earth history and biological history have been episodic, marked by what were in essence revolutions that opened many new ways of life.

Historical study of any sort is an inexact discipline, dependent on the preservation, reliability, and interpretation of past records. The fossil record of past life is generally less and less complete the farther into the past we delve. Fortunately, each organism alive today carries traces of its evolutionary history in its molecules, metabolism, and anatomy. As we saw in Unit Four, such traces are clues to the past that augment the fossil record, much as similarities and differences among extant cultures help social scientists understand historical relationships among the cultures. Still, the evolutionary episodes of greatest antiquity are generally the most obscure. This chapter is the most speculative of the unit, for its main subject is the origin of life on a young Earth, and no fossil record of that seminal episode exists. We begin with the oldest known fossils, in order to certify the antiquity of life on Earth. Then we will examine hypotheses about how natural processes on the youthful planet could have created life. The last section of the chapter introduces the various kingdoms of life, as a prelude to the survey of biological diversity in Chapters 25 through 30.

Life on Earth originated between 3.5 and 4.0 billion years ago

The metaphor of an evolutionary tree implies that the history of life chronicles an increasing diversity of organisms all descended from the primitive creatures that were the first living things. The apparent absence of fossils of ancestral organisms in Precambrian rocks was incongruous with Darwin's view that complex life evolved from simpler forms, and he wrote in *The Origin of Species:* "To the question why we do not find rich fossiliferous deposits belonging to these assumed earliest periods prior to the Cambrian system, I can give no satisfactory answer. . . . The case at present must remain inexplicable, and may be truly urged as a valid argument against the views here entertained." Some of Darwin's adversaries seized the cue and declared the beginning of the Cambrian period as the time of Genesis and all creation. Only in the past few decades has the discovery of older fossils filled in the Precambrian blank. The Cambrian fauna was preceded by a less diverse collection of animals dating back 700 million years. Before then there was a succession of microorganisms spanning nearly 3 billion years (FIGURE 24.1). For most of that time, only prokaryotes inhabited Earth. One would guess from the relatively simple structure of the prokaryotic cell (compared with the eukaryotic cell) that the earliest organisms were bacteria, and the fossil record now supports that presumption.

Prokaryotes originated within a few hundred million years after Earth's crust cooled and solidified (FIGURE 24.2). Though geologists have discovered minerals that crystallized about 4.1 billion years ago and sedimentary rocks that date back 3.8 billion years, no fossils that old have yet been found. However, fossils that appear to be prokaryotes about the size of bacteria *have* been discovered in southern Africa in a rock formation called the Fig Tree Chert, which is 3.4 billion years old. Evidence of even more ancient prokaryotic life has been found in rocks called stromatolites (Gr. *stroma,* "bed," and *lithos,* "rock"). **Stromatolites** are banded domes of sediment strikingly similar to the layered mats constructed by colonies of bacteria and cyanobacteria living today in very salty marshes. The layers are sediments that stick to the jellylike coats of the motile microbes, which continually migrate out of one layer of sediment and form a new one above, producing the banded pattern (FIGURE 24.3; compare the ancient mats depicted on page 486). Fossils resembling spherical and filamentous prokaryotes have been found in stromatolites that are 3.5 billion years old in western Australia and southern Africa. For now, these are the oldest evidence of life. However, the western

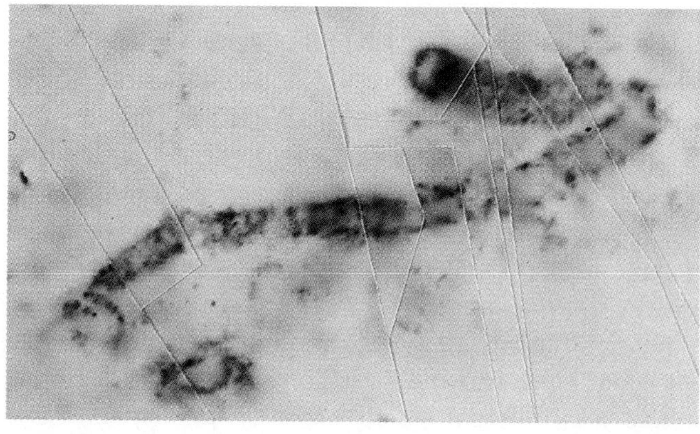

FIGURE 24.1
An early prokaryote. Precambrian fossils escaped notice until about 35 years ago, partly because they are microscopic. This filamentous prokaryote, about 3.5 billion years old, was collected in Western Australia (LM).

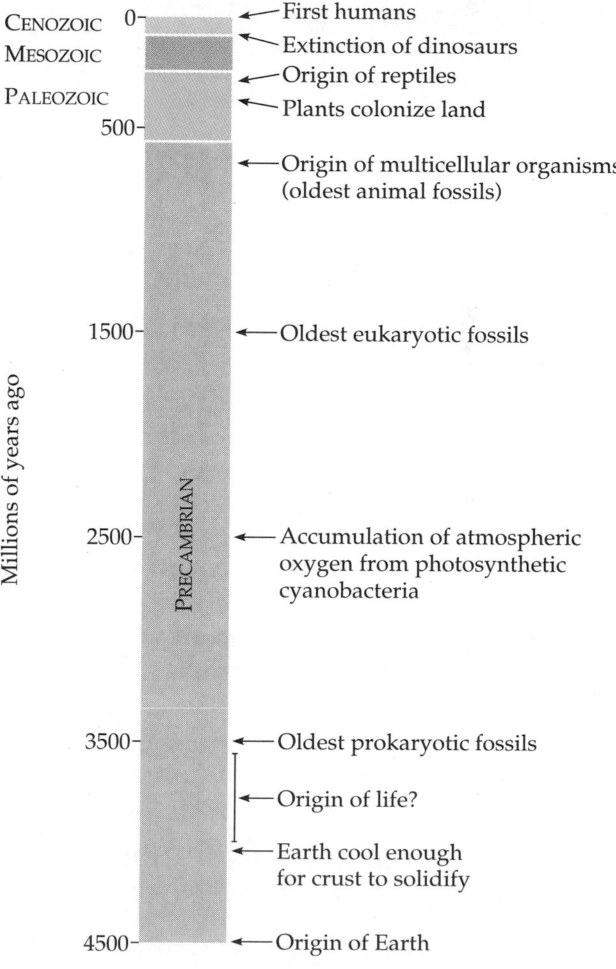

FIGURE 24.2
Some major episodes in the history of life.

(a)

(b)

(c)

FIGURE 24.3
Bacterial mats and stromatolites. The mats are sedimentary structures produced by colonies of bacteria and cyanobacteria that live, uncropped by predators, in environments inhospitable to most other life. (**a**) Lynn Margulis and Kenneth Nealson, who study the history of life, are collecting bacterial mats in a Baja California lagoon. (**b**) The bands, seen in this section of a mat, are layers of sediment that adhere to the sticky prokaryotes, which produce the succession of layers by migrating. (**c**) Fossilized mats known as stromatolites resemble the layered structures formed by contemporary bacterial colonies. This stromatolite is a western Australian specimen about 3.5 billion years old. Microfossils, such as the one in FIGURE 24.1, are present in many stromatolites.

Australia fossils appear to be those of photosynthetic organisms, perhaps oxygen producers. If so, it is likely that life had been evolving long before these organisms lived, possibly as early as 4 billion years ago. To put this in perspective, recall that Earth is about 4.6 billion years old. On the vast scale of geological time, life originated relatively early.

The first cells may have originated by chemical evolution on a young Earth: *an overview*

The question of how life began is more specifically about the genesis of prokaryotes. Sometime between about 4.0 billion years ago, when Earth's crust began to solidify, and 3.5 billion years ago, when the planet was inhabited by bacteria advanced enough to build stromatolites, the first organisms came into being. What was their origin? Most biologists subscribe to the hypothesis that life developed on Earth from nonliving materials that became ordered into molecular aggregates that were eventually capable of self-replication and metabolism. Life cannot arise by spontaneous generation from inanimate material today, as far as we know, but conditions were very different when Earth was only a billion years old. The atmosphere was different (there was little atmospheric O_2, for instance), and lightning, volcanic activity, meteorite bombardment, and ultraviolet radiation were all more intense than what we experience today (see Chapter 2). In that ancient environment, the origin of life was evidently possible, and it is likely that at least the early stages of biological inception were inevitable. However, debate abounds about what occurred during these early stages.

According to one hypothetical scenario, the first organisms were products of a chemical evolution in four stages: (1) the abiotic (nonliving) synthesis and accumulation of small organic molecules, or monomers, such as amino acids and nucleotides; (2) the joining of these monomers into polymers, including proteins and nucleic acids; (3) the aggregation of abiotically produced molecules into droplets, called protobionts, that had chemical characteristics different from their surroundings; and (4) the origin of heredity (which may have been under way even before the "droplet" stage). In laboratory experiments, it is possible to test the plausibility of these stages of chemical evolution.

Abiotic synthesis of organic monomers is a testable hypothesis: *science as a process*

In the 1920s, A. I. Oparin of Russia and J. B. S. Haldane of Great Britain independently postulated that conditions on the primitive Earth favored chemical reactions that synthesized organic compounds from inorganic precursors present in the early atmosphere and seas. This cannot happen in the modern world, Oparin and Haldane reasoned, because the present atmosphere is rich in oxygen produced by photosynthetic life. The oxidizing atmosphere of today is not conducive to the spontaneous synthesis of complex molecules because the oxygen attacks chemical bonds, extracting electrons. Before oxygen-producing photosynthesis, Earth had a much less oxidizing atmosphere, derived mainly from volcanic vapors. Such a reducing (electron-adding) atmosphere would have enhanced the joining of simple molecules to form more complex ones. Even with a reducing atmosphere, making organic molecules would require considerable energy, which was probably provided by lightning and the intense UV radiation that penetrated the primitive atmosphere. The modern atmosphere has a layer of ozone produced from oxygen, and this ozone shield screens out most UV radiation. There is also evidence that young suns emit more UV radiation than older suns. Oparin and Haldane envisioned an ancient world with the necessary chemical conditions and energy resources for the abiotic synthesis of organic molecules.

In 1953, Stanley Miller and Harold Urey tested the Oparin-Haldane hypothesis by creating, in the laboratory, conditions comparable to those of the early Earth. Their apparatus produced a variety of amino acids and other organic compounds found in living organisms today (FIGURE 24.4; also see FIGURE 4.1).

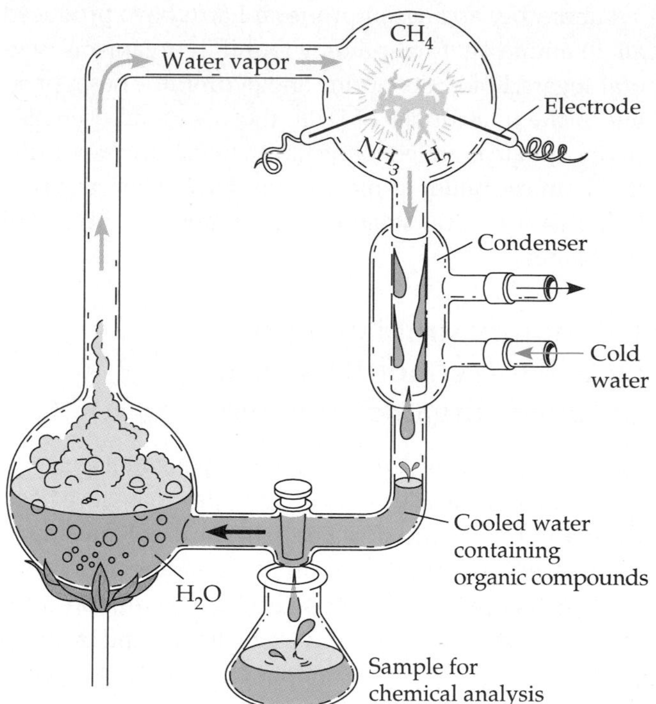

FIGURE 24.4
Abiotic synthesis of organic molecules in a model system. Stanley Miller and Harold Urey used an apparatus similar to this one to simulate chemical dynamics on the primitive Earth. A warmed flask of water simulated the primeval sea. The "atmosphere" consisted of H_2O, H_2, CH_4, and NH_3. Sparks were discharged in the synthetic atmosphere to mimic lightning. A condenser cooled the atmosphere, raining water and any dissolved compounds back to the miniature sea. As material circulated through the apparatus, the solution in the flask changed from clear to murky brown. After one week, Miller and Urey analyzed the contents of the solution and found a variety of organic compounds, including some of the amino acids that make up the proteins of organisms.

The atmosphere in the Miller-Urey model was made up of H_2O, H_2, CH_4 (methane), and NH_3 (ammonia), the gases that researchers in the 1950s believed prevailed in the ancient world. This atmosphere was probably more strongly reducing than the actual atmosphere of the early Earth. The vapors of modern volcanoes include CO, CO_2, and N_2, and it is likely that these gases were abundant in the ancient atmosphere produced from volcanoes. Traces of O_2 may even have been present, formed from reactions among other gases as they baked under powerful UV radiation. Many laboratories have repeated the Miller-Urey experiment using a variety of recipes for the atmosphere, including a mixture having a very low concentration of O_2. Abiotic synthesis of organic compounds occurred in these modified models, although yields were generally less than in the original experiment. The most important characteristic of the early atmosphere seems to be the rarity of the strong oxidizing agent O_2.

Laboratory analogs of primeval Earth have produced all 20 amino acids commonly found in organisms, several sugars, lipids, the purine and pyrimidine bases present in the nucleotides of DNA and RNA, and even ATP (if phosphate is added to the flask). Before there was life, its chemical building blocks may have been accumulating as a natural stage in the chemical evolution of the planet.

Laboratory simulations of early Earth conditions have produced organic polymers

Abiotic synthesis of more complex organic molecules by the joining together of smaller ones also may have been inevitable on the primitive Earth. Organic polymers such as proteins are chains of similar building blocks, or monomers. They are synthesized by dehydration reactions (condensation) that remove hydrogen and hydroxyl (—OH) groups from the monomers, forming a water molecule as a by-product of each new linkage in the polymer (see Chapter 5). In the living cell, specific enzymes catalyze the reactions. Abiotic synthesis of polymers would have had to occur without the help of these efficient enzymes, and the dilute concentrations of the monomers dissolved in an excess of water would not favor spontaneous condensation reactions that form more water. Polymerization *does* occur in laboratory experiments when dilute solutions of organic monomers are dripped onto hot sand, clay, or rock. This process vaporizes water and concentrates the monomers on the substratum. Using this method, Sidney Fox of the University of Miami has made what he calls proteinoids, which are polypeptides produced by abiotic means. Perhaps waves or rain splashed dilute solutions of organic monomers onto fresh lava or other hot rocks on the early Earth and then rinsed proteinoids and other polymers back into the water.

Clay, even cool clay, may have been especially important as a substratum for the polymerization reactions prerequisite to life. Clay concentrates amino acids and other organic monomers from dilute solutions because the monomers bind to charged sites on the clay particles. At some of the binding sites, metal atoms, such as iron and zinc, function as catalysts facilitating the reactions that link monomers. Clay, having many of these binding sites, could have functioned as a lattice that brought monomers close together and then assisted in joining them into polymers. As an alternative to clay, iron pyrite (fool's gold), which consists of iron and sulfur, has been proposed as the substratum of organic synthesis. This hypothesis is advocated by Günter Wachtershäuser of Germany, who points out properties of pyrite that could have catalyzed abiotic synthesis of organic polymers. Pyrite provides a charged surface, and the formation of this mineral from iron and sulfur yields electrons that could support bonding between organic molecules to form more complex products.

Protobionts can form by self-assembly

The properties of life emerge from an interaction of molecules organized into higher levels of order. Living cells may have been preceded by **protobionts,** aggregates of abiotically produced molecules. Protobionts are not capable of precise reproduction, but maintain an internal chemical environment different from their surroundings and exhibit some of the properties associated with life, including metabolism and excitability.

Laboratory experiments demonstrate that protobionts could have formed spontaneously from abiotically produced organic compounds. When mixed with cool water, proteinoids self-assemble into tiny droplets called microspheres (FIGURE 24.5a). Coated by a protein membrane that is selectively permeable, the microspheres undergo osmotic swelling or shrinking when placed in solutions of different salt concentrations. Some microspheres also store energy in the form of a membrane potential, a voltage across the surface. The protobionts can discharge the voltage in nervelike fashion; such excitability is characteristic of all life (which is not to say that microspheres are alive, only that they display *some* of the properties of life). Droplets of another type, called liposomes, form spontaneously when the organic ingredients include certain lipids. These lipids organize into a molecular bilayer at the surface of the droplet, much like the lipid bilayer of cell membranes. The liposomes behave dynamically, sometimes growing by engulfing smaller liposomes and then splitting, other times "giving birth" to smaller liposomes (FIGURE 24.5b). Still another type of protobiont, called a coacervate, is a droplet that self-assembles when a solution of polypeptides, nucleic acids, and polysaccharides is shaken. If enzymes are included among the ingredients, they are incorporated into the coacervates. The coacervates are then able to absorb substrates from their surroundings and release the products of the reactions catalyzed by the enzymes.

Unlike some laboratory models, protobionts that formed in the ancient seas would not have possessed refined enzymes, which are made in cells according to inherited instructions. Some molecules produced abiotically, however, do have weak catalytic capacities, and there could well have been protobionts that modified the substances they took in across their membranes by a rudimentary metabolism.

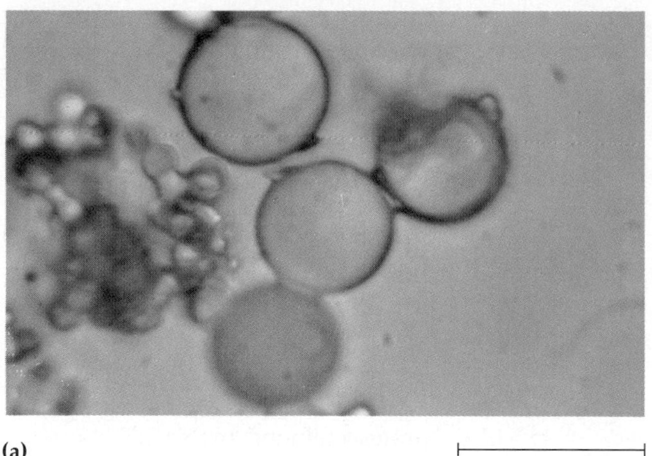

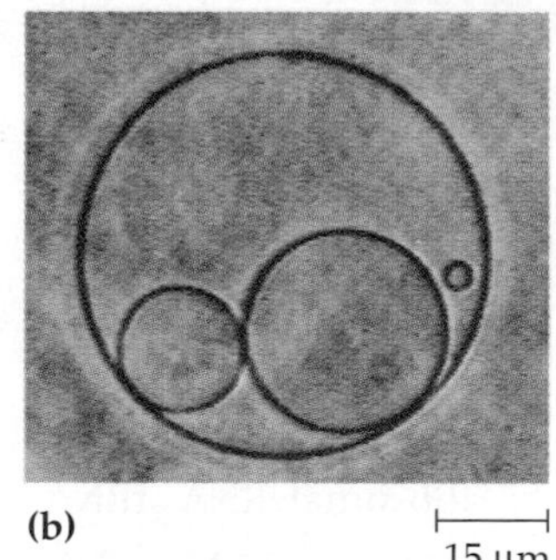

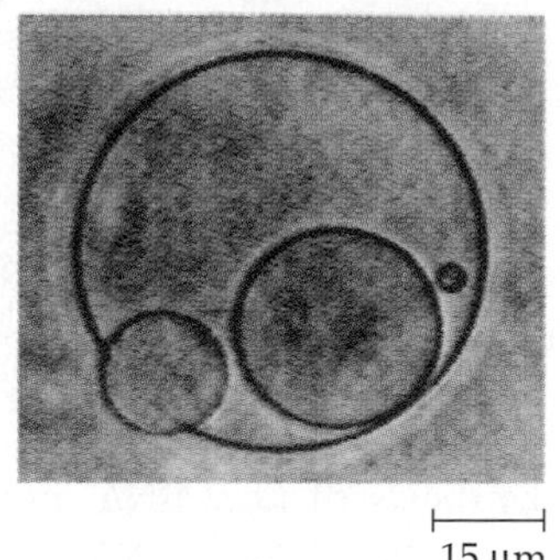

FIGURE 24.5
Laboratory versions of protobionts, aggregates of organic molecules with some biological properties. (**a**) These microspheres are made by cooling solutions of proteinoids, polypeptides created abiotically from amino acids polymerized on hot surfaces. Microspheres grow by absorbing free proteinoids until they reach an unstable size, when they split to form daughter microspheres. Of course, this division lacks the precision of cellular reproduction (LM). (**b**) In an aqueous environment, certain kinds of lipids self-assemble to form liposomes (left). One type of liposome made by scientists at Emory University in 1994 gives rise to smaller liposomes (right) (LMs).

RNA was probably the first genetic material

Imagine a tidepool, pond, or moist clay on the primeval Earth with a suspension of protobionts varying in chemical composition, permeability, and catalytic capabilities. Those droplets most stable and best able to accumulate organic molecules from the environment would grow and split, distributing their chemical components to the "baby" droplets. Other droplets would fall apart or fail to grow and divide. In this way, the environment may have selected in favor of some molecular aggregates and against others. But competition among the various protobionts could not lead to long-range improvement because there was no way to perpetuate success. As prolific droplets grew, split, grew, and split again, their unique catalysts and other functional molecules would become increasingly diluted. The chemical aggregates that were the forerunners of cells could not build on the past and evolve until the development of some mechanism for replicating their characteristics—some mechanism of heredity. Genetic information would make it possible for molecular aggregates to pass along not just samples of key molecules but also instructions for making more of those molecules.

A cell stores its genetic information as DNA, transcribes the information into RNA, and then translates the messages into specific enzymes and other proteins (discussed in Chapters 15 and 16). Instructions are transmitted by the replication of DNA when a cell divides. The DNA ⟶ RNA ⟶ protein axis of cellular control employs intricate machinery that could not have evolved all at once but must have emerged bit by bit as improvements to much simpler processes. Even before DNA, some primitive mechanism may have existed for aligning amino acids along strands of RNA that could replicate themselves. According to this hypothesis, the first genes were not DNA molecules but short strands of RNA that began self-replicating in the prebiotic world.

Some scientists studying the origin of life envision an "RNA world" and are testing the hypothesis of RNA self-replication. Short polymers of ribonucleotides have been produced abiotically in test-tube experiments. If RNA is added to a test-tube solution containing monomers for making more RNA, sequences about five to ten nucleotides long are copied from the template according to the base-pairing rules. If zinc is added as a catalyst, sequences up to 40 nucleotides long are copied with less than 1% error.

In the 1980s, Thomas Cech and his co-workers at the University of Colorado, Boulder, revolutionized thinking about the evolution of life when they discovered that RNA molecules are important catalysts in modern cells. This disproved the long-held view that only proteins (enzymes) serve as biological catalysts. Cech and others found that modern cells use RNA catalysts, called **ribozymes,** to do such things as remove introns from RNA (see Chapter 16). Ribozymes also help catalyze the

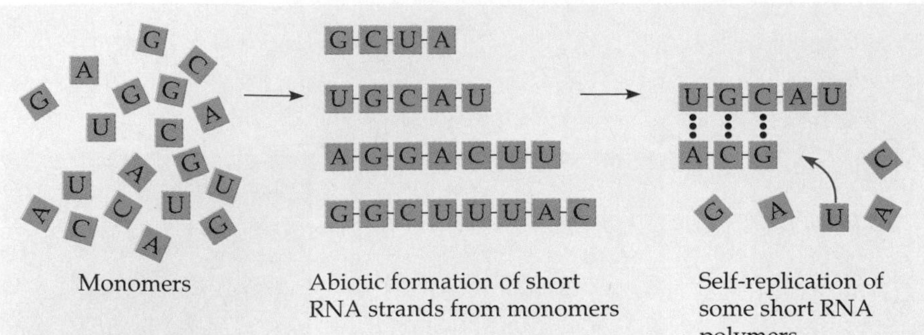

FIGURE 24.6
Abiotic replication of RNA. According to this hypothesis, the first genes were RNA molecules that polymerized abiotically and replicated themselves autocatalytically while bound to clay surfaces. The letters A, G, C, and U symbolize the four RNA nucleotide bases, which pair specifically in AU and GC combinations (see Chapter 16).

synthesis of new RNA, notably ribosomal RNA, tRNA, and mRNA. Thus, RNA is autocatalytic, and in the prebiotic world, long before there were enzymes (proteins) or DNA, RNA molecules may have been fully capable of self-replication (FIGURE 24.6).

Natural selection on the molecular level works on diverse populations of replicating autocatalytic RNA molecules. Unlike double-stranded DNA, which takes the form of a uniform helix, single-stranded RNA molecules assume a variety of specific three-dimensional shapes mandated by their nucleotide sequences. Each sequence folds into a unique conformation, reinforced by hydrogen bonding between regions of the strand having complementary sequences of bases. The molecule thus has a genotype (its nucleotide sequence) and a phenotype (its conformation, which interacts with surrounding molecules in specific ways). In a particular environment, RNA molecules of certain base sequences are more stable and replicate faster with fewer errors than other sequences. Beginning with a diversity of RNA molecules that must compete for monomers to replicate, the sequence best suited to the temperature, salt concentration, and other features of the surrounding solution and having the greatest autocatalytic activity will prevail. Its descendants will be not a single RNA species but a family of closely related sequences (because of copying errors). Selection screens mutations in the original sequence, and occasionally a copying error results in a molecule that folds into a shape even more stable or more adept at self-replication than the ancestral sequence. Natural selection has been observed operating on RNA populations in test tubes, and it probably happened in prebiotic times as well.

The rudiments of RNA-directed protein synthesis may have been in the weak binding of specific amino acids to bases along RNA molecules, which functioned as simple templates holding a few amino acids together long enough for them to be linked. (Indeed, this is one function of rRNA in modern ribosomes.) If RNA happened to synthesize a short polypeptide that in turn behaved as an enzyme helping the RNA molecule to replicate, then the early chemical dynamics included molecular cooperation as well as competition (FIGURE 24.7a).

Thus, the first steps toward the replication and translation of genetic information may have been taken by molecular evolution even before RNA and polypeptides became packaged within membranes. Once primitive genes and their products became confined to membrane-enclosed compartments, the protobionts could evolve as units (FIGURE 24.7b). Molecular cooperation could be refined because components that interacted in ways favorable to the success of the protobiont as a whole were concentrated together in a microscopic volume rather than being spread throughout a puddle or film on the surface of clay. Suppose, for example, that an RNA molecule ordered amino acids into a primitive enzyme that extracted energy from an organic fuel taken up from the surroundings and made the energy available for other reactions within the protobiont, including replication of RNA. Natural selection could favor such a gene only if its product were kept close by, rather than being shared with competing RNA sequences in its environment.

The origin of hereditary information made Darwinian evolution possible

In this scenario, we have built a hypothetical antecedent of the cell by incorporating genetic information into an aggregate of molecules that selectively accumulates monomers from its surroundings and uses enzymes programmed by genes to make polymers and carry out other chemical reactions. The protobiont grows and splits, distributing copies of its genes to offspring. Even if only one such protobiont arose initially by the abiotic processes that have been described, its descendants would vary because of mutations, errors in the copying of RNA. Evolution in the true Darwinian sense—differential reproductive success of varying individuals—presumably accumulated many refinements to primitive metabolism and inheritance. One trend apparently led to DNA becoming the hereditary material. Initially, RNA could have

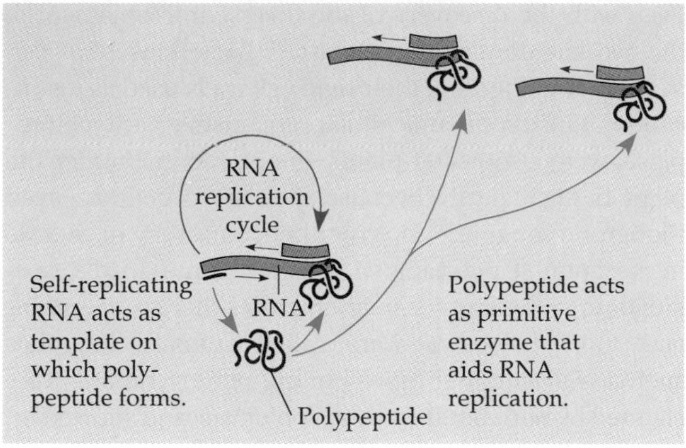

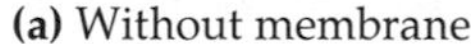

(a) Without membrane

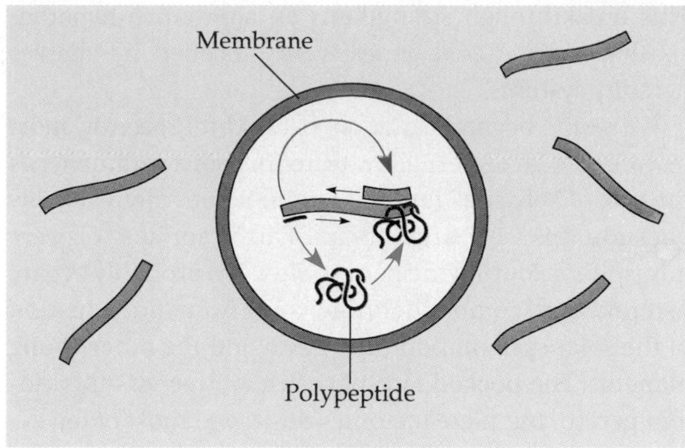

(b) Within membrane-enclosed compartment

FIGURE 24.7

One hypothesis for the beginnings of molecular cooperation. (**a**) An RNA strand enhances its own replication if it orders amino acids into a polypeptide that in turn functions as an enzyme helping the RNA strand replicate. In such cooperation may have been the rudiments for the translation of genetic information into protein structure. The benefits of this protein, however, would have been shared by competing RNA molecules not yet segregated within protobionts. (**b**) Once genes were in membrane-enclosed compartments, they would have benefited exclusively from their protein products.

provided the template on which DNA nucleotides were assembled. DNA is a much more stable repository for genetic information than RNA, and once DNA appeared, RNA molecules would have begun to take on their modern roles as intermediates in the translation of genetic programs. The RNA world gave way to a DNA world.

Debate about the origin of life abounds

Laboratory simulations cannot prove that the kind of chemical evolution that has been described here actually created life on the primitive Earth, but only that some of the key steps *could* have happened. The origin of life remains a matter of scientific speculation, and there are alternative views of how several key processes occurred.

Some researchers question whether abiotic synthesis of organic monomers was necessary as a first step in the origin of life. It is possible that at least some organic compounds reached the early Earth from space. This idea, called *panspermia,* holds that hundreds of thousands of meteorites and comets hitting the early Earth brought with them organic molecules formed by abiotic reactions in outer space. Extraterrestrial organic compounds, including amino acids, have been found in modern meteorites, and it seems likely that these bodies could have seeded the early Earth with organic compounds. And biochemists recently demonstrated that organic molecules extracted from a meteorite produce small vesicles when mixed with water. Both panspermia and chemical evolution could have contributed to the pool of organic molecules that formed the earliest life, but most scientists who study the origin of life believe that extraterrestrial sources made only a minor contribution.

Some biologists interested in the origin of life challenge the idea of an "RNA world." They point out that even under optimal test-tube conditions, it is difficult to make RNA without enzymes. Even short RNA strands may be too complicated to be good candidates as the first self-replicating molecules. In 1991, Julius Rebek, Jr., and his colleagues at the Massachusetts Institute of Technology synthesized a simple organic molecule that acts as a template to produce copies of itself (FIGURE 24.8).

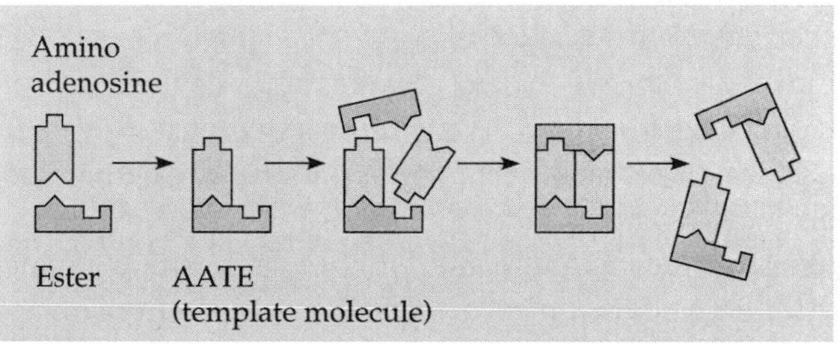

FIGURE 24.8

A synthetic self-replicating molecule. Were RNA or DNA genes preceded by simpler heredity systems? The laboratory creation of a relatively simple organic molecule that makes copies of itself is influencing ideas on the origin of life. The molecule is aminoadenosine triacid ester (AATE), which consists of two components, amino adenosine and an ester. From a pool of these two building blocks, AATE can catalyze the synthesis of additional AATE by acting as a template. No one is arguing that the first genetic material resembled AATE, but this research does demonstrate that molecules much smaller than nucleic acids can replicate.

This breakthrough strengthens an alternative hypothesis that nucleic acid genes were preceded by simpler heredity systems.

Where life began is another issue. Until recently, most researchers favored shallow water or moist sediments as the most likely sites for life's origin. Some scientists now question this view, arguing that Earth's surface was very inhospitable during the period when life probably began. Asteroids and comets, debris left over from the formation of the solar system, pounded Earth and the other young planets. (The pocked face of the moon records this violent period, but plate tectonics destroyed most of the evidence on Earth.) Some scientists speculate that incipient life could not survive this cosmic assault—unless, that is, it began on the less exposed sea floor. In the late 1970s, marine explorers discovered deep-sea vents where hot water and minerals gush through gaps in Earth's crust. Perhaps earlier vents supplied the energy and chemical precursors for the origin of protobionts.

Debate about the origin of life abounds, and we have sampled only a few of the issues. Whatever way prebiotic chemicals accumulated, polymerized, and eventually reproduced, the leap from an aggregate of molecules that reproduces to even the simplest prokaryotic cell is immense and must have been taken in many smaller evolutionary steps. The point at which we stop calling membrane-enclosed compartments that metabolize and replicate their genetic programs protobionts and begin calling them living cells is as fuzzy as our definitions of life. We do know that prokaryotes were already flourishing at least 3.5 billion years ago, and that all kingdoms of life descended from those ancient prokaryotes.

Arranging the diversity of life into kingdoms is a work in progress

In Chapter 23, we looked at systematics as the study of biological diversity in an evolutionary context. Now that we have gone backward in time to the very origin of life on Earth, systematics is once again relevant as we attempt to reconstruct evolutionary relationships among the immense diversity of forms that arose from those early organisms.

Systematists have traditionally considered the kingdom to be the highest—the most inclusive—taxonomic category. We grow up with the bias that there are only two kingdoms of life—plants and animals—because we live in a macroscopic, terrestrial realm where we rarely encounter organisms that do not fit neatly into a plant-animal dichotomy. The two-kingdom scheme also had a long tradition in formal taxonomy; Linnaeus divided all known forms of life between plant and animal kingdoms. Even with the discovery of the diverse microbial world, the two-kingdom system persisted. Bacteria were placed in the plant kingdom, their rigid cell walls used as justification. Eukaryotic unicellular organisms with chloroplasts were also called plants. Fungi, too, fell under the plant banner, partly because they are sedentary, even though no fungi are photosynthetic and they have little in common structurally with green plants. In the two-kingdom system, unicellular creatures that move and ingest food—protozoa—were called animals. Microbes such as *Euglena* that move but are photosynthetic were claimed by both botanists and zoologists, and showed up in the taxonomies of plant *and* animal kingdoms. Schemes with additional kingdoms were proposed, but none became popular with the majority of biologists until Robert H. Whittaker of Cornell University argued effectively for a five-kingdom system in 1969. Whitaker designated these five kingdoms as Monera, Protista, Plantae, Fungi, and Animalia (FIGURE 24.9).

The five-kingdom system recognizes the two fundamentally different types of cell, prokaryotic and eukaryotic, and sets the prokaryotes apart from all eukaryotes by placing them in their own kingdom, Monera. The prokaryotes are bacteria, including cyanobacteria, formerly called blue-green algae.

Organisms of the other four kingdoms all consist of cells organized on the eukaryotic plan (see Chapter 7). The kingdoms Plantae, Fungi, and Animalia are multicellular eukaryotes, each kingdom defined by characteristics of structure and life cycle discussed in upcoming chapters. Plants, fungi, and animals also generally differ in their modes of nutrition (the criterion originally used

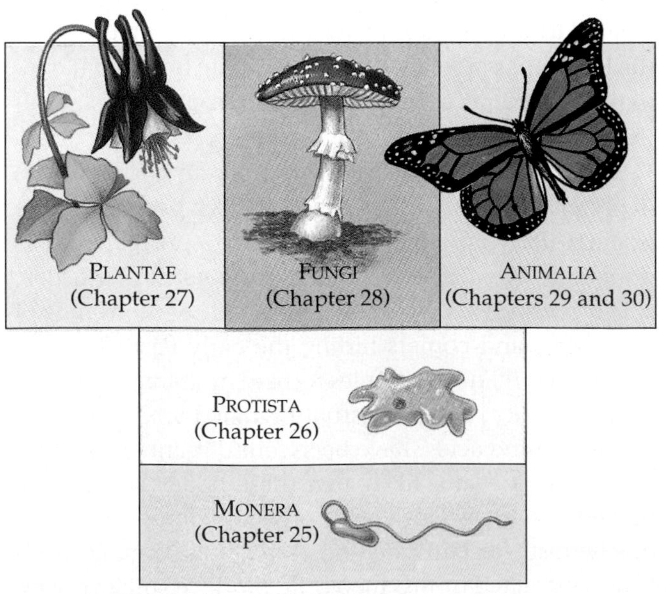

FIGURE 24.9
The traditional five-kingdom system.

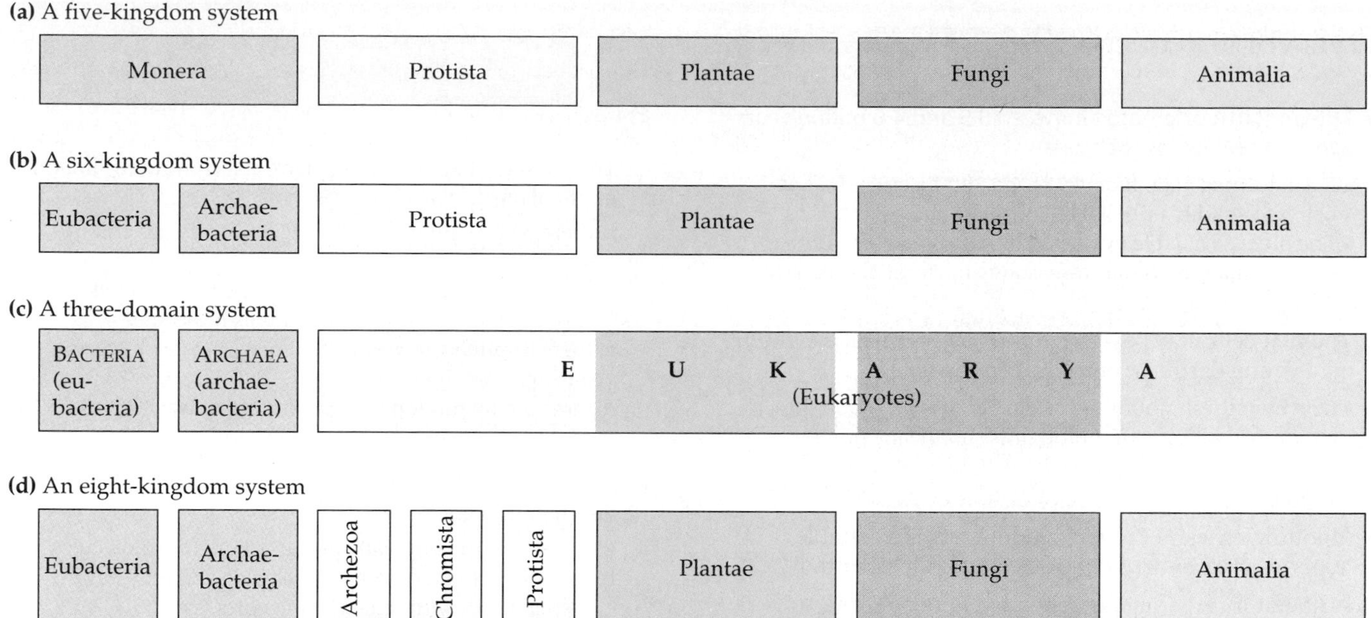

FIGURE 24.10

The five-kingdom system compared with three alternative schemes of classification. (**a**) The five-kingdom system is color-coded here to match FIGURE 24.9. This color coding is continued throughout this figure to make it easier for you to compare alternatives to the five-kingdom scheme. (**b**) This six-kingdom alternative divides the prokaryotes into two kingdoms. This modification is based on molecular evidence for an early evolutionary divergence between eubacteria (most bacteria) and archaebacteria, an ancient lineage of prokaryotes with many unique characteristics (discussed in Chapter 25). (**c**) This scheme assigns even more significance to the ancient evolutionary split between eubacteria and archaebacteria by using a superkingdom taxon called the domain. The phylogenetic rationale for this solution is discussed in Chapter 25. The domain Eukarya consists of the four kingdoms of eukaryotic organisms. (**d**) This eight-kingdom system is one example of a "kingdom-splitting" solution to the taxonomic problems posed by certain distinct lineages of microorganisms. In addition to two separate prokaryotic kingdoms, this system also splits the protists into three kingdoms; the rationale for this is discussed in Chapter 26. What is important for you to understand for now is that new information has reopened issues of biological diversity at the highest taxonomic levels.

by Whittaker). Plants are autotrophic in nutritional mode, making their food by photosynthesis. Fungi are heterotrophic organisms that are absorptive in nutritional mode. Most fungi are decomposers that live embedded in their food source, secreting digestive enzymes and absorbing the small organic molecules that are the products of digestion. Animals live mostly by ingesting food and digesting it within specialized cavities.

We are left with the kingdom Protista (also known as Protoctista). In the five-kingdom system, Protista became a grab-bag containing all eukaryotes that did not fit the definitions of plants, fungi, or animals. Most protists are unicellular forms, but the kingdom also includes relatively simple multicellular organisms that are believed to be direct descendants of unicellular protists.

Keep in mind that the five-kingdom system is not a natural fact, but a human construct. It is one attempt to order the diversity of life into a scheme that is useful and, hopefully, phylogenetically reasonable. During the past few years, systematists using comparisons of nucleic acids and proteins to probe the relationships between different groups of organisms have been pointing out problems with the traditional five-kingdom scheme. FIGURE 24.10 compares three of the alternative classification systems that have emerged from these challenges to the five-kingdom system. We will consider some of the arguments for these alternative taxonomic constructs in the next several chapters. Defining the kingdoms of life is a work in progress, an evolving view of biodiversity that reflects our increased understanding of the characteristics and evolutionary histories of different organisms.

From this discussion of biological kingdoms, we turn to prokaryotes, the first forms of life and the *only* ones for at least 2 billion years. They remain tremendously important on modern Earth. In the next chapter, we will address the diversity and history of prokaryotic life.

REVIEW OF KEY CONCEPTS (with page numbers and key figures)

- Life on Earth originated between 3.5 and 4.0 billion years ago (pp. 487–488, FIGURE 24.2)
  - Earth formed 4.6 billion years ago, but its crust did not begin to solidify until 4.0 billion years ago.
  - The oldest available evidence for life appears in stromatolites containing fossils resembling bacteria dating back 3.5 billion years.
- The first cells may have originated by chemical evolution on a young Earth: *an overview* (pp. 488–489)
  - One hypothesis about the origin of life is based on the chemical evolution of protobionts, abiotically produced molecular droplets with distinctive chemical characteristics.
- Abiotic synthesis of organic monomers is a testable hypothesis: *science as a process* (pp. 489–490; FIGURE 24.4)
  - Laboratory experiments performed under conditions simulating those of the primitive Earth have produced diverse organic molecules from inorganic precursors.
- Laboratory simulations of early Earth conditions have produced organic polymers (p. 490)
  - Small organic molecules polymerize when they are concentrated on hot sand, rock, or clay.
- Protobionts can form by self-assembly (p. 490, FIGURE 24.5)
  - Organic molecules synthesized in the laboratory have spontaneously assembled into a variety of droplets—microspheres, liposomes, and coacervates—with some of the properties associated with life.
- RNA was probably the first genetic material (pp. 491–492, FIGURE 24.6)
  - The first genes may have been abiotically produced RNA, whose base sequence served as a template for both alignment of amino acids in polypeptide synthesis and alignment of complementary nucleotide bases in a primitive form of self-replication.
- The origin of hereditary information made Darwinian evolution possible (pp. 492–493, FIGURE 24.7)
  - Once genetic information became incorporated inside membrane-enclosed compartments, protobionts would have acquired heritability and the ability to evolve as units.
- Debate about the origin of life abounds (pp. 493–494)
  - Researchers continue to debate how the stepwise origin of life actually occurred. Abiotic synthesis of organic molecules versus import via meteorites, origins in shallow water versus deep-sea vents, and RNA genes versus simpler self-replicating molecules are just three of the issues.
- Arranging the diversity of life into kingdoms is a work in progress (pp. 494–495, FIGURE 24.10)
  - The traditional five-kingdom system classifies organisms as Monera (prokaryotes), Protista (relatively simple eukaryotes), Plantae, Fungi, and Animalia.
  - New information is leading to taxonomic alternatives to the five-kingdom system.

SELF-QUIZ

1. The *main* explanation for the lack of a continuing abiotic origin of life on Earth today is that
   a. there is not sufficient lightning to provide an energy source
   b. our oxidizing atmosphere is not conducive to the spontaneous formation of complex molecules
   c. there is much less visible light reaching Earth to serve as an energy source
   d. there are no molten surfaces on which weak solutions of organic molecules would polymerize
   e. all habitable places are already filled
2. Stromatolites are
   a. aggregates of abiotically produced organic molecules
   b. meteorites that contain amino acids and may have seeded Earth with organic molecules
   c. layers of clay that may have facilitated the polymerization of abiotically produced monomers
   d. a group of ancient eukaryotes
   e. banded domes of sediment that contain the oldest known fossils
3. Which of the following was *not* simulated in the experimental setup designed by Miller and Urey?
   a. an energy source
   b. a reducing atmosphere
   c. the effects of alternating day and night
   d. evaporation of the ocean's waters and subsequent rains
   e. an aqueous environment
4. Clays and other fine mineral particles may have contributed all of the following to abiotic origins of organic polymers *except*
   a. a charged surface holding reactants close together
   b. catalysis
   c. enzymatic action on reactants
   d. a supply of electrons
   e. the ability to concentrate reactants from the surrounding solution
5. The formation of microspheres, liposomes, and coacervates all require
   a. RNA
   b. a membrane potential
   c. self-assembly
   d. phospholipids
   e. primitive genes
6. Competition among various protobionts would have led to evolutionary improvement only when
   a. they were able to catalyze chemical reactions
   b. some kind of heredity mechanism developed
   c. they were able to grow and reproduce
   d. the protobionts acquired selectively permeable membranes
   e. DNA first appeared

7. Which of the following represents a probable order of evolution of prelife components on Earth?
   a. protobionts before mitosis
   b. ozone in the atmosphere followed by a reducing atmosphere
   c. amino acids and sugars after the corresponding polymers
   d. DNA before RNA
   e. eukaryotic cell structure before photosynthesis

8. One current debate raises the issue that, rather than beginning in shallow pools, life could have begun
   a. as plate tectonics changed the surface of our planet
   b. near thermal vents on the floor of the ocean
   c. from viruses
   d. in Northern Africa
   e. as chunks that broke off from the moon bombarded Earth

9. Which of the following steps has *not* yet been accomplished by scientists studying the origin of life?
   a. abiotic synthesis of small RNA polymers
   b. abiotic synthesis of polypeptides
   c. the formation of molecular aggregates with selectively permeable membranes
   d. the formation of protobionts that use DNA to direct the polymerization of amino acids
   e. abiotic synthesis of organic monomers

10. Current debates about the number and boundaries of the kingdoms of life center *mainly* on which groups of organisms?
    a. plants and animals
    b. plants and fungi
    c. prokaryotes and relatively simple eukaryotes
    d. fungi and animals
    e. bacteria and the most complex eukaryotes

CHALLENGE QUESTIONS

1. Describe the minimum structural, metabolic, and genetic equipment of a postprotobiont that you would consider to be a true primitive cell.

2. Carl Woese and his colleagues at the University of Illinois compared the ribosomal RNAs of primitive prokaryotes called archaebacteria with rRNAs from other organisms. They found that archaebacteria are all very similar to each other, but different from both other bacteria and eukaryotes. Other characteristics of archaebacteria are actually more like eukaryotes than other bacteria. What problems does this pose for the current version of the five-kingdom classification scheme?

3. Billions of years ago, life apparently arose from inorganic chemicals, but all life today apparently arises only by the reproduction of preexisting life. If life could come from nonlife on the ancient Earth, why do you think this does not continue to happen today?

SCIENCE, TECHNOLOGY, AND SOCIETY

1. There is no reason to believe that the processes that led to the origin of life are unique to Earth. Many scientists think that life may be present on many other planets throughout the universe. During the early 1990s, astronomers first detected planets circling stars beyond our solar system. In 1994, a research team found evidence for the amino acid glycine in the spectrum of light emitted from a star-forming region hundreds of light years from our solar system. If this discovery can be confirmed, it suggests that the abiotic synthesis of organic molecules that made life possible on Earth may have occurred elsewhere in the galaxy. In what ways do you think our perspective would be changed by convincing evidence of extraterrestrial life?

2. As we learn more about possible ways that life might have originated on Earth, we approach the possibility of experimentally producing simple, self-replicating, cell-like structures. Would such structures be "alive"? Should this research be encouraged or discouraged? Why?

FURTHER READING

Aldhous, P. "New Ingredient for the Primeval Soup." *New Scientist,* February 25, 1995. One of the abiotically produced organic molecules may have been a versatile catalyst.

Cohen, J. "Getting All Turned Around Over the Origins of Life on Earth." *Science,* March 3, 1995. Given that many organic monomers occur in "left- and right-handed" forms (see Chapter 4), why did life generally become locked into specific isomeric forms?

Day, S. "Hot Bacteria and Other Ancestors." *New Scientist,* April 9, 1994. The study of certain microorganisms is helping biologists understand the origin and early diversification of life.

Emsley, J. "Babies Born to Artificial Cells." *New Scientist,* April 9, 1994. Some lifelike properties of laboratory-made protobionts.

Freedman, D. H. "The Handmade Cell." *Discover,* August 1992. Will researchers be able to build a cell from scratch?

Holmes, B. "Still Life in Mouldy Bread." *New Scientist,* March 26, 1994. On the trail of enzymes that may help us understand the transition from an RNA to a DNA world.

Orgel, L. E. "The Origin of Life on Earth." *Scientific American,* October 1994. Emphasizes the origin of genetic information.

Rebek, J. "Synthetic Self-Replicating Molecules." *Scientific American,* July 1994. The possibility of molecular reproduction before nucleic acids.

Travis, J. "Hints of First Amino Acid Outside Solar System." *Science,* June 17, 1994. Abiotic synthesis of organic compounds may not be rare in the Milky Way.

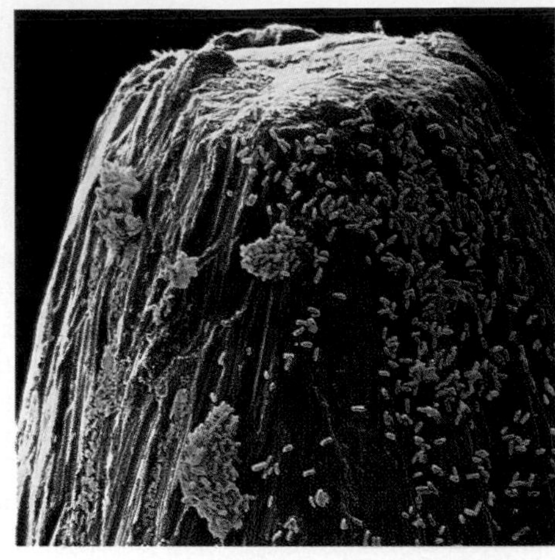

CHAPTER 25

PROKARYOTES AND THE ORIGINS OF METABOLIC DIVERSITY

KEY CONCEPTS

- They're (almost) everywhere! *an overview of prokaryotic life*
- Archaea and Bacteria are the two main branches of prokaryotic evolution
- The success of prokaryotic life is based on diverse adaptations of form and function
- All major types of nutrition and metabolism evolved among prokaryotes
- The evolution of prokaryotic metabolism was both cause and effect of changing environments on Earth
- Molecular systematics is leading to a phylogenetic classification of prokaryotes
- Prokaryotes continue to have an enormous ecological impact

The history of prokaryotic life is a success story spanning at least 3.5 billion years. Prokaryotes were the earliest organisms, and they lived and evolved all alone on Earth for 2 billion years. They have continued to adapt and flourish on an evolving Earth, and in turn they have helped to change the Earth. In this chapter, you will become more familiar with prokaryotes by studying their form and physiology, their origins and evolution, their diversity, and their ecological significance.

They're (almost) everywhere! *an overview of prokaryotic life*

In terms of metabolic impact and numbers, prokaryotes still dominate the biosphere, outnumbering all eukaryotes combined. More prokaryotes inhabit a handful of dirt or the human mouth or skin than the total number of people who have ever lived. Prokaryotes are not only the most numerous organisms by far but also the most pervasive. Wherever we find life of any kind, prokaryotes are among the organisms present. Prokaryotic species also thrive in habitats too hot, too cold, too salty, too acidic, or too alkaline for any eukaryote. Incomparably bountiful and omnipresent, prokaryotes have endured and expanded through their billions of years of descent from the first cells that were the beginnings of all life.

Most prokaryotic cells are relatively small. The colorized scanning electron micrograph on this page demonstrates the size of prokaryotes in relation to a pinpoint. In contrast to eukaryotic cells, prokaryotes lack membrane-enclosed organelles. Although most prokaryotes have cell walls, they differ in molecular composition and construction from the cell walls of plants, walled protists, and fungi. Compared to eukaryotes, bacterias also have smaller, simpler genomes, and they differ from eukaryotes in some of the details of genetic replication, expression (protein synthesis), and recombination.

Although prokaryotes are individually microscopic, their collective impact on Earth and all of life is gigantic. We rarely notice these ubiquitous microbes because they are usually invisible to the unaided eye. Illness caused by infectious prokaryotes occasionally reminds us that these tiny organisms exist, but prokaryotic life is no rogues' gallery. Only a minority of prokaryotes cause disease in humans or any other organisms. The great majority of prokaryotic species are essential to all life on Earth. For example, certain prokaryotes decompose matter from dead organisms and return vital chemical elements to the environment in the form of inorganic compounds required by plants, which in turn are consumed by animals. If for some reason all prokaryotes were suddenly to perish, the chemical cycles that sustain life would halt, and all other forms of life would also be doomed. In contrast, prokaryotic life would undoubtedly persist in the absence of eukaryotes, as it once did for a long time.

Prokaryotes often live in close associations among themselves and with eukaryotes in what are called symbiotic relationships. In the most historically important case of such symbiosis, mitochondria and chloroplasts evolved from prokaryotes that became residents within larger host cells. Thus, animals, plants, fungi, and protists probably evolved from symbiotic associations of ancestral prokaryotes. (We will examine this theory of eukaryotic origins further in Chapter 26.)

Modern prokaryotes are diverse in structure and physiology. About 4000 prokaryotic species are known, and estimates of actual prokaryotic diversity range from about 400,000 to 4 million species. As E. O. Wilson put it in the interview that precedes this unit, a true sense of biodiversity requires a "downward adjustment of scale."

Archaea and Bacteria are the two main branches of prokaryotic evolution

In the traditional five-kingdom system of classification, prokaryotes make up the kingdom Monera, and the four eukaryotic kingdoms are Protista, Plantae, Fungi, and Animalia (see FIGURE 24.9). This scheme emphasizes the structural dichotomy between prokaryotic and eukaryotic cells. In the past decade, however, systematists have questioned whether a separate kingdom incorporating all prokaryotes is consistent with evolutionary history. By comparing ribosomal RNA and other genetic products among extant species, researchers have identified two major branches of prokaryotic evolution. The common names for these two groups are **archaebacteria** and **eubacteria.** The term archaebacteria refers to the antiquity of the group's origin from the earliest cells (Gr. *archaio,* "ancient"). Most species of archaebacteria inhabit extreme environments, such as hot springs and salt ponds. Few, if any, other modern organisms can survive in these environments, which may resemble habitats on the early Earth. Most prokaryotes, however, are eubacteria. They differ from archaebacteria in many key structural, biochemical, and physiological characteristics, differences that will be highlighted later in the chapter.

Archaebacteria and eubacteria diverged so early in the history of life that many researchers, led by Carl Woese of the University of Illinois, first proposed a six-kingdom system: two prokaryotic kingdoms along with the four eukaryotic kingdoms (see FIGURE 24.10). (This division of prokaryotes into two kingdoms is also incorporated into an eight-kingdom system you will learn about in Chapter 26.) Many systematists also favor organizing the diversity of life into three **domains,** a taxonomic level higher than kingdom:

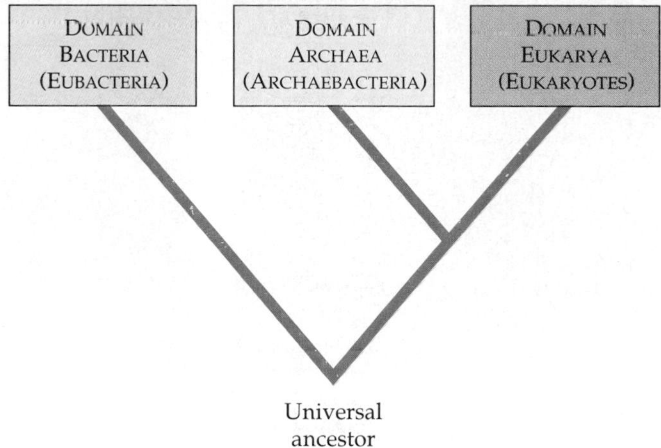

Notice that prokaryotes account for two of the domains: the **domain Archaea** and the **domain Bacteria.** These are the archaebacteria and eubacteria (or simply bacteria), respectively, and we will continue to use these common names throughout the chapter. All eukaryotes are placed in the third domain, Eukarya (or Eucarya). Notice also that the simple cladogram above incorporates the hypothesis that Eukarya and Archaea share a common ancestor that lived more recently than the ancestor common to Archaea and Bacteria. In fact, evidence from molecular systematics supports this hypothesis that archaebacteria are more closely related to eukaryotes than they are to eubacteria. If this view is correct, then a single prokaryotic kingdom (Monera) is at odds with phylogeny.

Taxonomic problems aside, archaebacteria and eubacteria are both structurally organized at the prokaryotic level, which is the rationale for combining them in this chapter. The distinction between archaebacteria and eubacteria will become more apparent after we examine some of the structural, genetic, and metabolic adaptations that contribute to the pervasiveness of prokaryotes on Earth.

The success of prokaryotic life is based on diverse adaptations of form and function

Morphological Diversity of Prokaryotes

Most prokaryotes are unicellular. However, some species tend to aggregate reversibly in two-celled to several-celled groups. Others have the form of true colonies, which are permanent aggregates of identical cells. And some bacterial species even exhibit a simple multicellular organization in which there is a division of labor between two or more specialized types of cells.

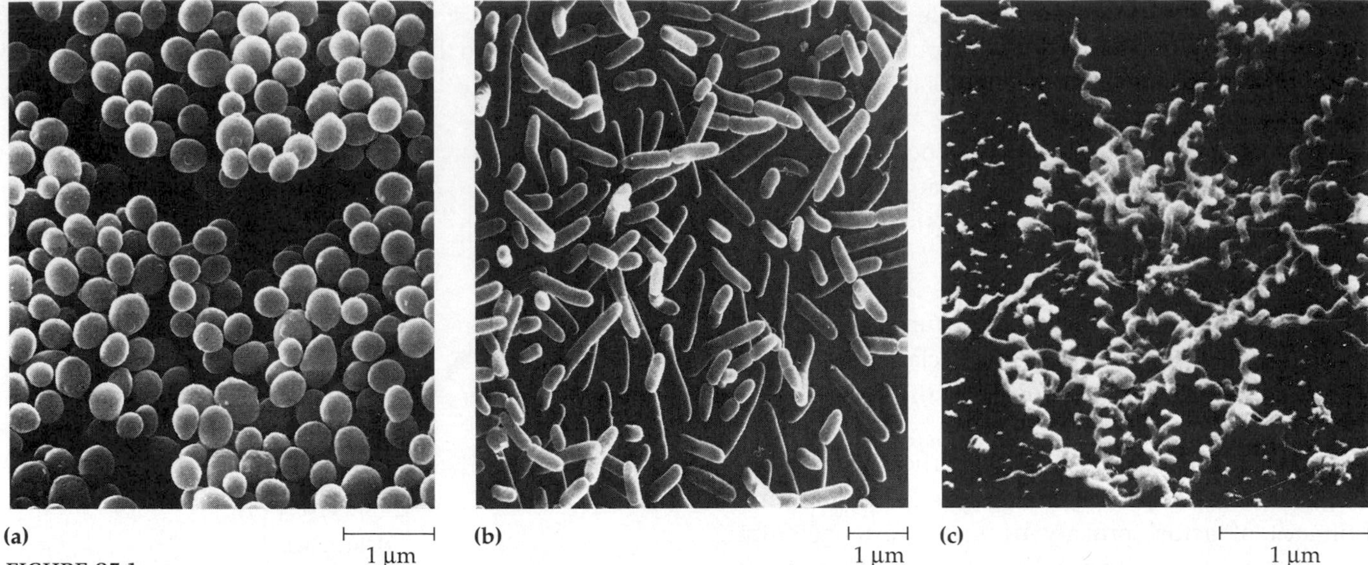

FIGURE 25.1

The most common shapes of prokaryotes. (**a**) Cocci (singular, coccus), or spherical prokaryotes, occur singly or in pairs (diplococci), in chains of many cells (streptococci), and in clusters resembling bunches of grapes (staphylococci). (**b**) Rod-shaped prokaryotes, or bacilli (singular, bacillus), are most commonly solitary, but there are also forms with the rods arranged in chains. (**c**) Helical prokaryotes include the spirilla and the corkscrew-shaped spirochetes. (All SEMs.)

There is a diversity of cell shapes among prokaryotes, the three most common being spheres (cocci), rods (bacilli), and helices (including the bacteria known as spirilla and spirochetes). An important step in identifying bacteria is determining their shapes by microscopic examination (FIGURE 25.1).

Most prokaryotic cells have diameters in the range of 1–5 µm, compared to 10–100 µm for the majority of eukaryotic cells. There are, however, notable exceptions to this size disparity. The largest prokaryotic cell discovered so far is a rod-shaped species that measures about a half-millimeter in length, dwarfing most eukaryotic cells (FIGURE 25.2).

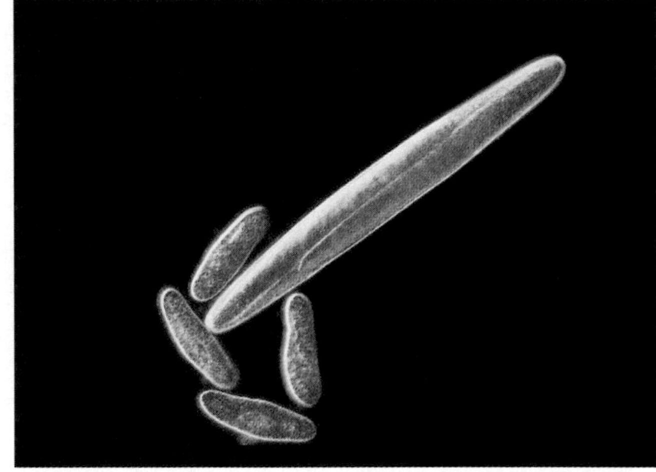

FIGURE 25.2

The largest known prokaryote. In this light micrograph, *Epulopiscium fishelsoni,* about a half-millimeter long, dwarfs four eukaryotes (the protist *Paramecium*). The prokaryotic giant lives as a symbiont in the gut of surgeonfish.

The Cell Surface of Prokaryotes

Nearly all prokaryotes have cell walls external to their plasma membranes. The wall maintains the shape of the cell, affords physical protection, and prevents the cell from bursting in a hypotonic environment (see Chapter 8). Like other walled cells, however, prokaryotes plasmolyze and may die in a hypertonic medium, which is why heavily salted meat can be kept so long without being spoiled by bacteria.

The presence of a cell wall is one reason bacteria were grouped with plants in the old two-kingdom system. But the walls of prokaryotes and plants are analogous rather than homologous; they have a completely different molecular composition. Instead of cellulose, the staple of plant walls, most walls of eubacteria contain a unique material called **peptidoglycan,** which consists of polymers of modified sugars cross-linked by short polypeptides that vary from species to species (the walls of archaebacteria lack peptidoglycan). The effect is a single, giant, molecular network enclosing and protecting the cell. External to this fabric are other substances that also differ from species to species.

One of the most valuable tools for identifying eubacteria (bacteria) is the **Gram stain,** which can be used to separate many eubacteria into two groups based on a difference in their cell walls. **Gram-positive** bacteria

METHODS: THE GRAM STAIN

This method, named for Hans Christian Gram, a Danish physician who developed the technique in the late 1800s, distinguishes between two different kinds of bacterial cell walls. Bacteria are stained with a violet dye and iodine, rinsed in alcohol, and then stained again with a red dye. The structure of the cell wall determines the staining response. Gram-positive bacteria (top) have cell walls with a large amount of peptidoglycan that traps the violet dye (LM). Gram-negative bacteria (bottom) have less peptidoglycan, which is located in a periplasmic gel between the plasma membrane and an outer membrane. The violet dye used in the Gram stain is easily rinsed from gram-negative bacteria, but the cells retain the red dye (LM). (The colors in the diagrams do not represent the stains.)

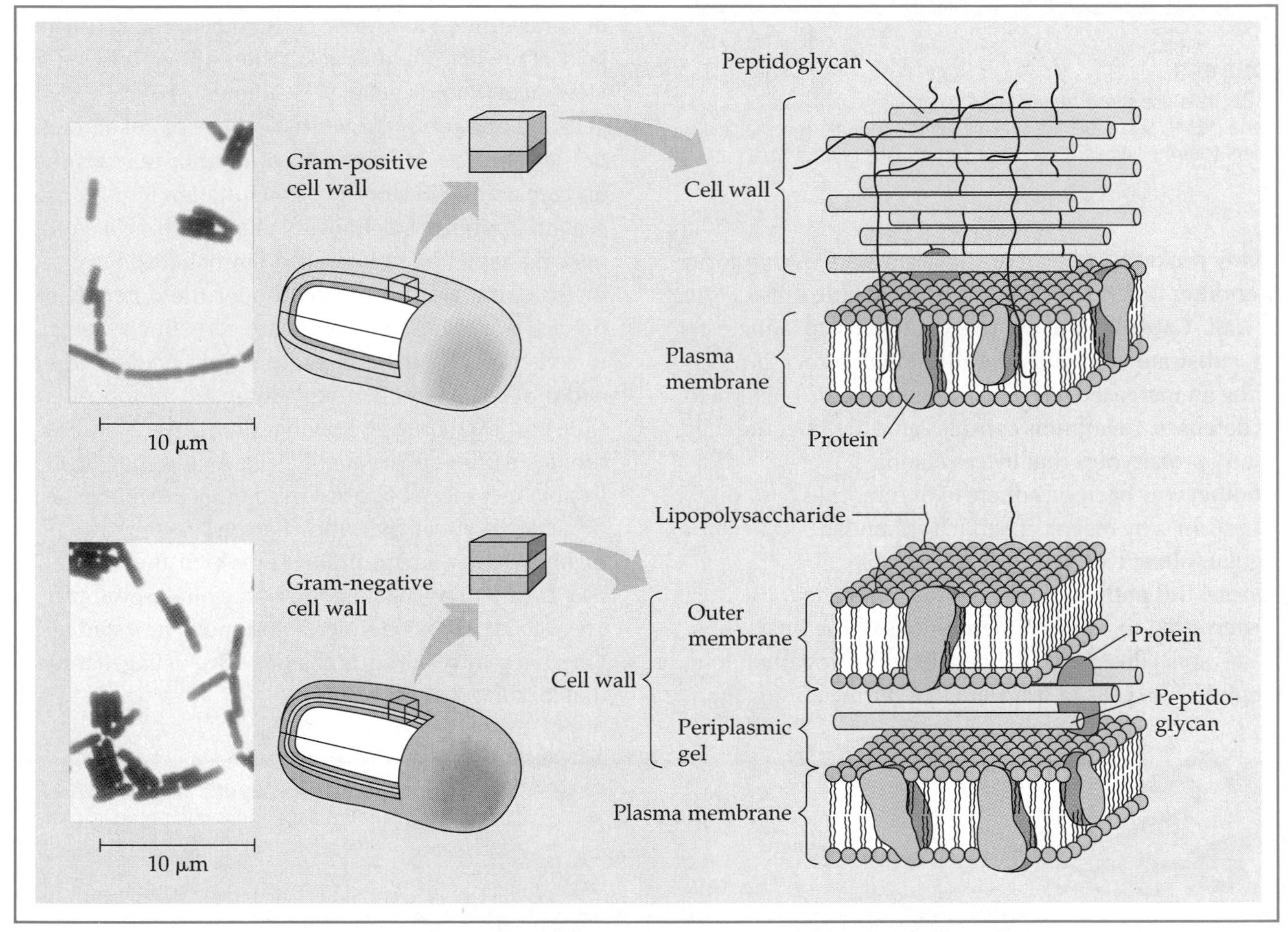

have simpler walls, with a relatively large amount of peptidoglycan. The walls of **gram-negative** bacteria have less peptidoglycan and are more complex in structure. An outer membrane on the gram-negative cell wall contains lipopolysaccharides, carbohydrates bonded to lipids (see the Methods Box).

Among pathogenic, or disease-causing, bacteria, gram-negative species are generally more threatening than gram-positive species. The lipopolysaccharides on the walls of gram-negative bacteria are often toxic, and the outer membrane helps protect the pathogens against the defenses of their hosts. Furthermore, gram-negative bacteria are commonly more resistant than gram-positive species to antibiotics because the outer membrane impedes entry of the drugs.

Many antibiotics, including penicillins, inhibit the synthesis of cross-links in peptidoglycan and prevent the formation of a functional wall, particularly in gram-positive species. These drugs are like selective bullets that cripple many species of infectious bacteria without adversely affecting humans and other eukaryotes, which do not make peptidoglycan.

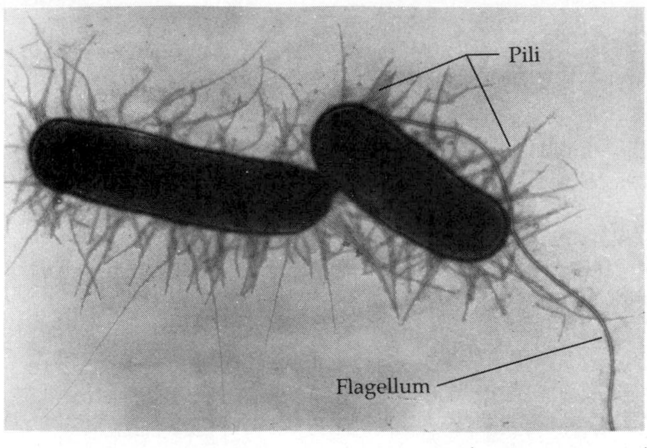

FIGURE 25.3
Pili. Bacteria use these appendages to attach to surfaces or other bacteria (TEM). Some pili are specialized for conjugation, holding partners together while DNA is transferred (see FIGURE 17.11).

Many prokaryotes secrete sticky substances that form still another protective layer called a **capsule** outside the cell wall. Capsules enable the organisms to adhere to their substrate and provide additional protection, including an increased resistance of pathogenic bacteria to host defenses. Gelatinous capsules glue together the cells of many prokaryotes that live as colonies.

Another way bacteria adhere to one another or to some substratum is by means of surface appendages called **pili** (singular, **pilus;** FIGURE 25.3). For example, *Neisseria gonorrhoeae,* the pathogen that causes gonorrhea, uses pili to fasten itself to mucous membranes of the host. Some pili are specialized for holding bacteria together long enough for the cells to transfer DNA during conjugation.

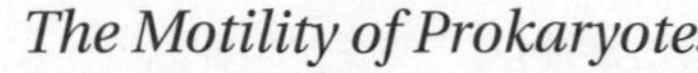

The Motility of Prokaryotes

About half of all prokaryotic species are capable of directed movement. Before examining the ways that prokaryotes move, let's consider what it would be like for a microbe to move through water at the scale of a few micrometers. Relative to our experience, a solution of water, ions, and other solutes would seem as thick as molasses. Other environments, such as the mucous linings of lungs, would seem even thicker, like pudding or mud.

Motile prokaryotes use one of three mechanisms to move. The most common motive force is by means of flagella, which may be scattered over the entire cell surface or concentrated at one or both ends of the cell. The flagella of prokaryotes and eukaryotes differ entirely. (To review eukaryotic flagella, see Chapter 7.) Prokaryotic flagella are one-tenth the width of those of eukaryotes, are not covered by an extension of the plasma membrane, and are unique in structure and function (FIGURE 25.4). A second motility mechanism characterizes a group of helical-shaped bacteria called spirochetes. Several filaments spiral around the cell under the outer sheath of the cell wall. These filaments are very much like flagella in structure. Their basal motors are attached at either end of the cell, and filaments attached at opposite ends slide past each other, behaving somewhat like the microtubules within eukaryotic flagella. When they do so, the flexible cell moves like a corkscrew. This mechanism is particularly effective in moving spirochetes through the highly viscous environments in which they sometimes live. In the third mechanism of motility, some prokaryotes secrete slimy chemicals and move by a gliding motion that may result from the presence of flagellar motors that lack flagellar filaments.

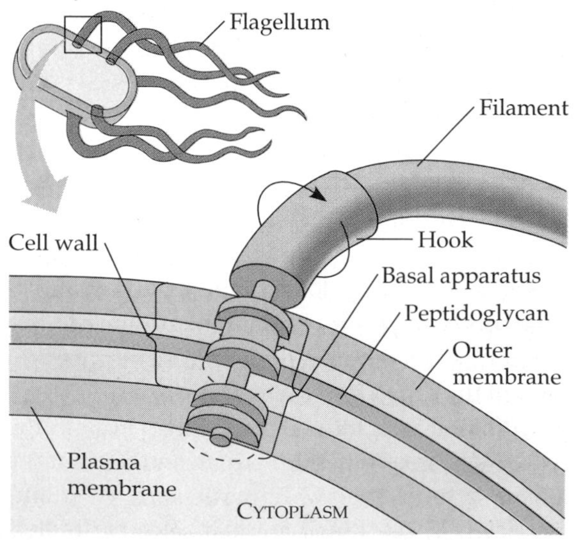

FIGURE 25.4
How prokaryotic flagella work. The entire flagellum is composed of protein and arranged in three basic parts. Chains of a globular protein, flagellin, wound in a tight spiral produce a filament with a relatively rigid, helical form. This filament is attached to another protein that forms a curved hook, which is, in turn, inserted into a basal apparatus composed of about 35 different proteins. This apparatus consists of a system of rings that sit in the various cell wall layers. (The TEM and drawing are characteristic of gram-negative bacteria. The arrangement of rings is different in gram-positive bacteria.) With the basal apparatus functioning as a motor, the filament rotates; the prokaryotic flagellum is the only known wheel in the living world. The basal motor is powered by the diffusion of protons (H^+) into the cell after they have been pumped outward across the plasma membrane at the expense of ATP.

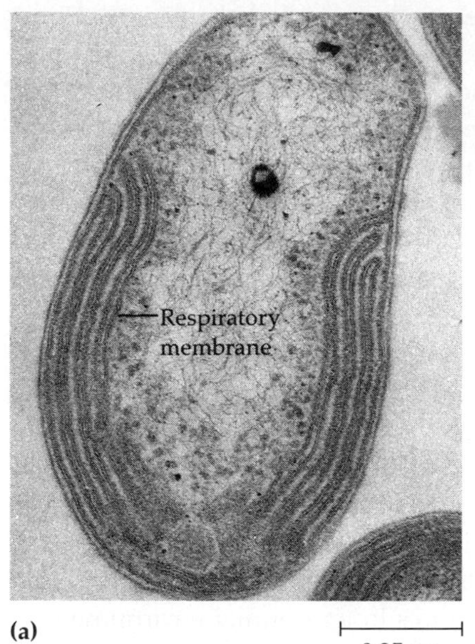

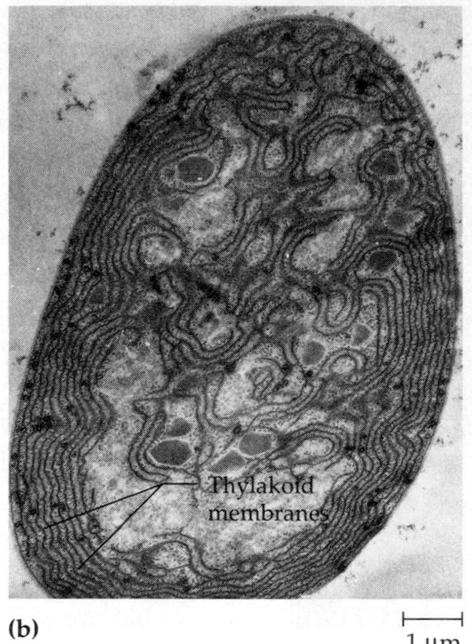

FIGURE 25.5
Specialized membranes of prokaryotes. (**a**) These infoldings of the plasma membrane, reminiscent of the cristae of mitochondria, function in cellular respiration of some aerobic prokaryotes (TEM). (**b**) Prokaryotes called cyanobacteria have thylakoid membranes, much like those in chloroplasts, that function in photosynthesis (TEM).

In an environment that is fairly uniform, flagellated prokaryotes wander randomly. In a heterogeneous environment, however, many prokaryotes are capable of **taxis,** movement toward or away from a stimulus (Gr. *taxis,* "to arrange"). With chemotaxis, for example, prokaryotes respond to chemical stimuli, perhaps moving toward food or oxygen (a positive chemotaxis) or away from some toxic substance (a negative chemotaxis). Several kinds of receptor molecules that detect specific substances are located on the surfaces of chemotactic prokaryotes. Motile prokaryotes that are photosynthetic generally display a positive phototaxis, a behavior that keeps them in the light. There are even prokaryotes with tiny magnets that help distinguish up from down and cause the cells to migrate toward the nutrient-rich sediments at the bottoms of ponds and shallow seas.

Internal Membranous Organization

Prokaryotic cells lack the extensive compartmentalization by internal membranes characteristic of eukaryotes. However, various prokaryotes do have a variety of specialized membranes that perform many of their metabolic functions. These membranes are usually invaginated regions of the plasma membrane (FIGURE 25.5).

Prokaryotic Genomes

On average, prokaryotes have only about one-thousandth as much DNA as a eukaryotic cell. Recall that prokaryotes are named for their lack of true nuclei enclosed by membranes (see FIGURE 7.4). In most prokaryotic cells, the DNA is concentrated as a snarl of fibers in a **nucleoid region** that stains less dense than the surrounding cytoplasm in electron micrographs. The mass of fibers is actually the bacterial chromosome, one double-stranded DNA molecule in the form of a ring. The DNA has very little protein associated with it. The term *genophore* is sometimes used for the prokaryotic chromosome to distinguish it from eukaryotic chromosomes, which have a very different structure. The eukaryotic genome consists of linear DNA molecules packaged along with proteins into a number of chromosomes characteristic of the species.

In addition to its one major chromosome, the prokaryotic cell may also have much smaller rings of DNA called plasmids, most consisting of only a few genes. In most environments, bacteria can survive without their plasmids because all essential functions are programmed by the chromosome. However, plasmids endow the cell with genes for resistance to antibiotics, for the metabolism of unusual nutrients not present in the normal environment, and for other special contingencies. Plasmids replicate independently of the main chromosome, and many can be readily transferred between partners when bacteria conjugate (see Chapter 17).

Although the broad outlines for DNA replication and the translation of genetic messages into proteins are alike for eukaryotes and prokaryotes, some of the details differ. For example, the bacterial ribosome is slightly smaller than the eukaryotic version and differs in its protein and RNA content. The disparity is great enough that selective antibiotics, including tetracycline and chloramphenicol, bind to the ribosomes of bacteria and block protein synthesis, while not inhibiting eukaryotic ribosomes.

Growth, Reproduction, and Gene Exchange

Neither mitosis nor meiosis occurs among prokaryotes; this is another fundamental difference between prokaryotic and eukaryotic life. Prokaryotes reproduce only

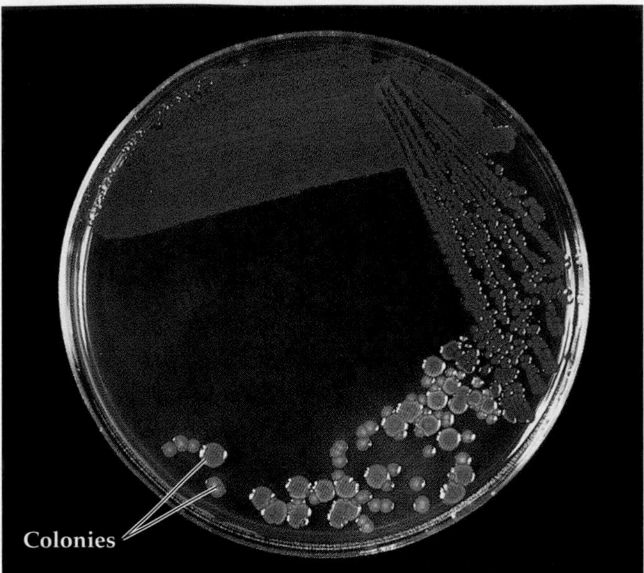

FIGURE 25.6
Cultured bacterial colonies. Bacteria are grown in the laboratory by culturing them in petri dishes or test tubes containing liquid or solid media of known composition. The media are sterilized to ensure that no unwanted microbes will grow, then a sample of bacteria, sometimes just a single cell, is introduced. The plates or tubes are incubated at an appropriate temperature. When the bacteria are grown on solid media, colonies are usually large enough to be visible to the unaided eye after a day or two. The size, shape, texture, and color of a colony provide clues to identification of the bacteria, as do the nutrients and physical conditions required for growth. Shown here are the colonies of several species of bacteria. Microscopic examination of bacteria from the colony is another step in identification.

asexually by the mode of cell division called **binary fission,** synthesizing DNA almost continuously. (Binary fission is described in Chapter 11.) A single bacterium in a favorable environment will give rise by repeated divisions to a colony of offspring (FIGURE 25.6). The word *growth* as applied to bacteria actually refers more to the multiplication of cells and population growth than to the enlargement of individual cells. The conditions for optimal growth—temperature, pH, salt concentrations, nutrient sources, and so on—vary according to species. Refrigeration retards food spoilage because most bacteria and other microorganisms grow only very slowly at such low temperatures.

The resistance of some bacterial cells to environmental destruction is impressive. Some bacteria form resistant cells called **endospores** (see the micrograph of *Bacillus* in TABLE 25.3, pp. 510–511). The original cell replicates its chromosome, and one copy becomes surrounded by a durable wall. The outer cell disintegrates, but the endospore it contained survives all sorts of trauma, including lack of nutrients and water, extreme heat or cold, and most poisons. Boiling water is not hot enough to kill most endospores in a relatively short length of time. Home canners and the food-canning industry, therefore, must take extra precautions to kill endospores of dangerous bacteria. To sterilize media, glassware, and utensils in the laboratory, microbiologists use an appliance called an autoclave, a pressure cooker that kills even endospores by heating to temperatures higher than 120°C. In less hostile environments, endospores may remain dormant for centuries. They will hydrate and revive to the vegetative (colony-producing) state in hospitable environments.

In an environment without limiting resources, bacterial growth is effectively geometric: One cell divides to form 2, which divide again to produce a total of 4 cells, then 8, 16, and so on, the numbers in a colony doubling with each generation. Most bacteria have generation times in the range of 1 to 3 hours, but some species can double every 20 minutes in an optimal environment. If the latter growth rate were sustained, a single cell would give rise to a colony weighing 1 million kg in just 24 hours. However, bacterial growth both in the laboratory and in nature is usually checked at some point when the cells exhaust some nutrient, or the colony poisons itself with an accumulation of metabolic wastes.

In most natural environments, prokaryotes must compete for space and nutrients. A general feature of many microorganisms (including certain species of prokaryotes, protists, and fungi) is the release of **antibiotics,** chemicals that inhibit the growth of other microorganisms. Humans have discovered some of these compounds and use them to combat pathogenic bacteria.

The sexual cycle of meiosis and syngamy (the union of haploid nuclei; see Chapter 12), so important as a source of genetic variation in eukaryotes, does not occur in the reproduction of prokaryotes. However, as you learned in Chapter 17, there are three mechanisms of genetic recombination in prokaryotes: **transformation,** in which genes are taken up from the surrounding environment, allowing for considerable genetic transfer in prokaryotes; **conjugation,** in which genes are transferred directly from one prokaryote to another; and **transduction,** in which genes are transferred between prokaryotes by viruses. These processes, however, involve the unilateral passage of a variable amount of DNA—nothing like the meiotic sex of eukaryotes, in which two parents each contribute homologous genomes to a zygote. Mutation is the major source of genetic variation in prokaryotes. Because generation times are measured in minutes or hours, a favorable mutation can be rapidly propagated to a large number of offspring.

A short generation span enables prokaryotic populations to adapt very rapidly to environmental change, as natural selection screens new mutations and novel genomes resulting from recombination. This adaptive evolution is important to the continuing success of prokaryotes, as it was when prokaryotic life began to diversify billions of years ago. We now examine the most profound result of that adaptive radiation of prokaryotes, the evolution of diverse modes of nutrition and metabolism.

All major types of nutrition and metabolism evolved among prokaryotes

Metabolic diversity is greater among prokaryotes than among all eukaryotes combined. Every type of nutrition observed in eukaryotes is represented among prokaryotes, plus some nutritional modes unique to prokaryotic organisms.

Major Modes of Nutrition

Nutrition refers here to how an organism obtains two resources for synthesizing organic compounds: energy and a source of carbon. Species that use light energy are termed *phototrophs. Chemotrophs* obtain their energy from chemicals taken from the environment. If an organism needs only the inorganic compound CO_2 as a carbon source, it is called an *autotroph. Heterotrophs* require at least one organic nutrient—glucose, for instance—as a source of carbon for making other organic compounds. We can combine the phototroph-versus-chemotroph (energy source) and autotroph-versus-heterotroph (carbon source) criteria to classify prokaryotes based on four major modes of nutrition:

1. **Photoautotrophs** are photosynthetic organisms that harness light energy to drive the synthesis of organic compounds from carbon dioxide. The specialized metabolic machinery of these organisms includes internal membranes with light-harvesting pigment systems (see Chapter 10 and FIGURE 25.5b). Among the diverse groups of photosynthetic prokaryotes are the cyanobacteria. All photosynthetic eukaryotes—plants and certain protists—also fit this nutritional category.
2. **Chemoautotrophs** need only CO_2 as a carbon source, but instead of using light for energy, these bacteria obtain energy by oxidizing inorganic substances. Chemical energy is extracted from hydrogen sulfide (H_2S), ammonia (NH_3), ferrous ions (Fe^{2+}), or some other chemical, depending on the species. This mode of nutrition is unique to certain prokaryotes. For instance, archaebacteria of the genus *Sulfobolus* oxidize sulfur.
3. **Photoheterotrophs** can use light to generate ATP but must obtain their carbon in organic form. This mode of nutrition is restricted to certain prokaryotes.
4. **Chemoheterotrophs** must consume organic molecules for both energy and carbon. This nutritional mode is found widely among prokaryotes, protists, fungi, animals, and even some plants.

TABLE 25.1 reviews the four major modes of nutrition.

Nutritional Diversity Among Chemoheterotrophs

The majority of prokaryotes are chemoheterotrophs. This category includes **saprobes,** decomposers that absorb their nutrients from dead organic matter, and **parasites,** which absorb their nutrients from the body fluids of living hosts.

The specific organic nutrients needed for growth vary extensively among chemoheterotrophic prokaryotes. Some species are very exacting in their requirements; for example, bacteria of the genus *Lactobacillus* will grow well only in a medium containing all 20 amino acids, several vitamins, and other organic compounds. Among species less fastidious in their nutritional needs, *E. coli*

TABLE 25.1

Major Nutritional Modes

MODE OF NUTRITION	ENERGY SOURCE	CARBON SOURCE
Autotroph		
Photoautotroph	Light	CO_2
Chemoautotroph	Inorganic chemicals	CO_2
Heterotroph		
Photoheterotroph	Light	Organic compounds
Chemoheterotroph	Organic compounds	Organic compounds

can grow on a medium containing glucose as the only organic ingredient, and the metabolism of the organism is so versatile that many other compounds can substitute for glucose as the sole organic nutrient.

There is such a diversity of chemoheterotrophs that almost any organic molecule can serve as food for at least some species. For example, some bacteria are capable of metabolizing petroleum; they are used to clean up oil spills. Those few classes of synthetic organic compounds, including some kinds of plastics, that cannot be broken down by any chemoheterotrophs are said to be nonbiodegradable.

Nitrogen Metabolism

Nitrogen metabolism is another facet of nutritional diversity among prokaryotes. Nitrogen is an essential component of proteins and nucleic acids. While animals, plants, and other eukaryotes are limited in the forms of nitrogen they can use, diverse prokaryotes are able to metabolize most nitrogenous compounds.

Key steps in the cycling of nitrogen through ecosystems are performed only by bacteria. (For an overview of the nitrogen cycle, see FIGURE 49.11.) Some chemoautotrophic bacteria, such as *Nitrosomonas,* convert NH_3 to NO_2^-. Other bacteria, such as a few species of *Pseudomonas,* "denitrify" NO_2^- or NO_3^- to atmospheric N_2 gas. And diverse species of prokaryotes, including some cyanobacteria, are able to use atmospheric nitrogen directly as a source of nitrogen. In this process, called **nitrogen fixation,** bacteria convert atmospheric N_2 to NH_3 (ammonia). Nitrogen fixation, unique to certain prokaryotes, is the only biological mechanism that makes atmospheric nitrogen available to organisms for incorporation into organic compounds. In terms of nutrition, nitrogen-fixing cyanobacteria are the most self-sufficient of all organisms. They are photoautotrophs that require only light energy, CO_2, N_2, water, and some minerals in order to grow.

Metabolic Relationships to Oxygen

Another metabolic variation among prokaryotes is in the effect that oxygen has on growth (see Chapter 9). **Obligate aerobes** use oxygen for cellular respiration and cannot grow without it. **Facultative anaerobes** will use oxygen if it is present but can also grow by fermentation in an anaerobic environment. **Obligate anaerobes** cannot use oxygen and are poisoned by it. Some obligate anaerobes live exclusively by fermentation; other species extract chemical energy by anaerobic respiration, in which inorganic molecules other than O_2 accept electrons at the "downhill" end of electron transport chains.

Now that we have surveyed variation in nutrition and metabolism among prokaryotes, let's trace the evolutionary roots of this metabolic diversity.

The evolution of prokaryotic metabolism was both cause and effect of changing environments on Earth

All forms of nutrition and nearly all metabolic pathways evolved among prokaryotes before eukaryotes arose. As early prokaryotes evolved, they were met with constantly changing physical and biological environments. In response to these changes, new metabolic capabilities evolved that, in turn, changed the environment faced by the next community of prokaryotes. All the major metabolic capabilities seen among contemporary prokaryotes probably evolved in the first billion years of life. Reasonable hypotheses about the early history of prokaryotes and the origins of metabolic diversity are now possible. The scenario described here is just one hypothetical sequence based on inferences from molecular systematics, from comparisons of energy metabolism among extant prokaryotes, and from geological evidence about conditions on the early Earth.

The Origin of Glycolysis

The first prokaryotes, which originated at least 3.5 billion years ago, were probably chemoheterotrophs that absorbed free organic compounds generated in the primordial seas by abiotic synthesis (see Chapter 24). ATP was probably among those nutrients. The universal role of ATP as an energy currency in all modern organisms implies that prokaryotes became fixed on its use very early. As the bacteria began to deplete the supply of free ATP, natural selection would have favored cells with enzymes that could regenerate ATP from ADP using energy extracted from other organic nutrients that were still available. The result may have been the step-by-step evolution of glycolysis, a metabolic pathway that breaks organic molecules down to simpler waste products and uses the energy to generate ATP by substrate phosphorylation (see Chapter 9). Glycolysis is the only metabolic pathway common to nearly all modern organisms, suggesting great antiquity.

Further, glycolysis does not require O_2, which was rare in Earth's ancient atmosphere. Fermentation, in which electrons extracted from nutrients during glycolysis are transferred to organic recipients, became a way of life on the anaerobic Earth. Certain archaebacteria and other obligate anaerobes that live today by fermentation deep in the soil or in stagnant swamps have forms of nutrition that may resemble those of the original prokaryotes.

The Origin of Electron Transport Chains and Chemiosmosis

The chemiosmotic mechanism of ATP synthesis is common to all three domains of life—Archaea, Bacteria, and Eukarya—implying a relatively early origin. Recall from Chapter 9 that chemiosmosis uses electron transfers along a chain of membrane proteins to pump hydrogen ions, and then taps the H^+ gradient to power ATP synthesis. Transmembrane proton pumps may have functioned originally to expel hydrogen ions that accumulated when fermentation produced organic acids as waste products (FIGURE 25.7, Stage 1). However, the cell would have to spend a large portion of its ATP to regulate internal pH by driving proton pumps. The first electron transport chains may have saved ATP by coupling the oxidation of organic acids to the transport of H^+ out of the cell (FIGURE 25.7, Stage 2). Finally, in some prokaryotes, electron transport systems efficient enough to extrude more H^+ than necessary for regulating pH evolved. These cells could then use the inward gradient of the H^+ to reverse the proton pump, which now generated ATP rather than consuming it (FIGURE 25.7, Stage 3). This type of energy metabolism, called anaerobic respiration, persists in some modern prokaryotes. For instance, in waterlogged, anaerobic soils, some species of the bacterial genus *Pseudomonas* can pass electrons down transport chains from organic substrates to NO_3^- (instead of O_2, the electron acceptor in aerobic respiration).

Stage 1. Proton pumps, driven by ATP, are used to regulate cellular pH by extruding hydrogen ions.

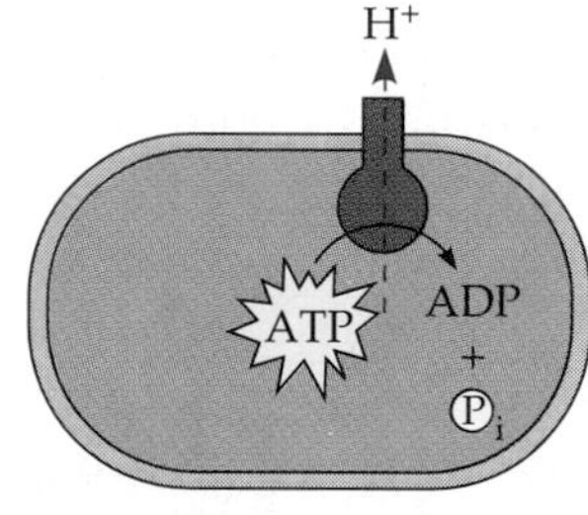

Stage 2. Electron transport chains take over the function of pH regulation by using the oxidation of organic acids to drive proton pumps.

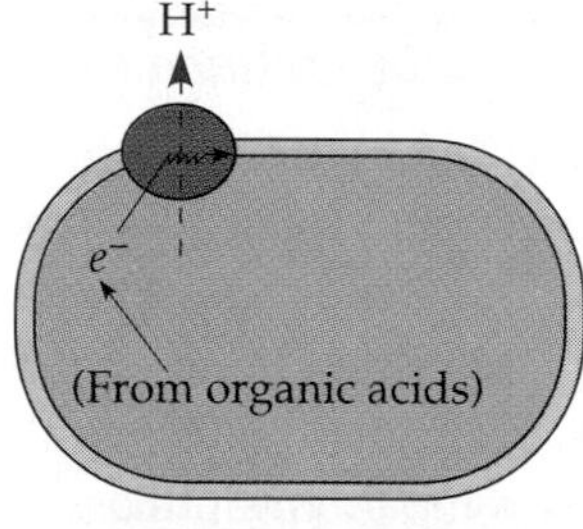

Stage 3. Electron transport chains become efficient enough to generate a gradient of hydrogen ions that can be used to drive ATP synthesis.

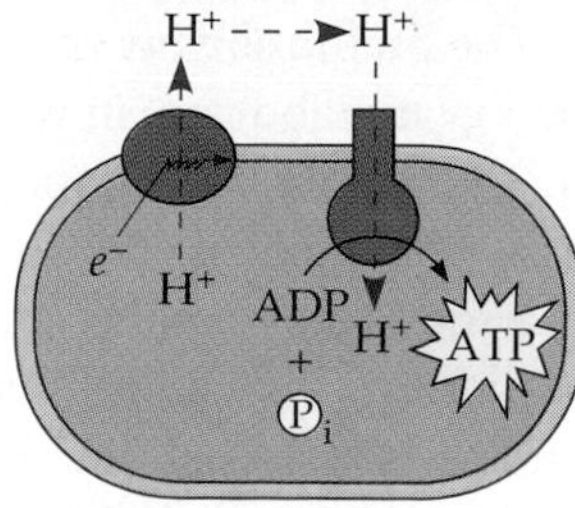

FIGURE 25.7
A hypothetical sequence for the evolution of electron transport chains.

The Origin of Photosynthesis

Life confronted its first energy crisis when the supply of free ATP dwindled. It faced its second when the fermenting prokaryotes consumed organic nutrients faster than the compounds could be replaced by abiotic synthesis. An organism that could make its own organic molecules from inorganic resources would have had a tremendous advantage.

In the earliest prokaryotes, light-absorbing pigments may have been used to absorb excess light energy (particularly ultraviolet) that was harmful to cells growing in surface communities. Later, these energized pigments were coupled with electron transport systems to drive ATP synthesis. In modern archaebacteria called extreme halophiles, a pigment that captures light energy, known as bacteriorhodopsin, is built into the plasma membrane. (This molecule is structurally related to visual pigments in the retina of the eye.) Bacteriorhodopsin absorbs light and uses the energy to pump hydrogen ions out of the cell. The gradient of hydrogen ions then drives the synthesis of ATP. This is the simplest mechanism of photophosphorylation known. Researchers are studying the halophiles as model systems of solar energy conversion.

Other prokaryotes had pigments and photosystems that used light to drive electrons from hydrogen sulfide (H_2S) to $NADP^+$, generating reducing power that could be used to fix CO_2 (see Chapter 10). They probably coopted components of electron transport chains that had previously functioned in anaerobic respiration and continued to use the chains to power ATP synthesis, as well as to provide reducing power (NADPH). The modern organisms with nutrition most like the early photosynthetic prokaryotes are believed to be the green sulfur bacteria and purple sulfur bacteria. They owe their color to bacteriochlorophyll, which functions instead of chlorophyll *a* as their main photosynthetic pigment. Because these bacteria split H_2S instead of H_2O as a source of electrons, they produce no O_2.

Cyanobacteria, the Oxygen Revolution, and the Origins of Cellular Respiration

Some of the photosynthetic eubacteria eventually had the metabolic machinery to use plentiful H_2O instead of

H_2S or other compounds as a source of electrons and hydrogen for reducing CO_2. These were the first **cyanobacteria,** formerly known as blue-green algae (FIGURE 25.8). Capable of making organic compounds from water and CO_2, they flourished and changed the world by releasing O_2 as a by-product of their photosynthesis.

Cyanobacteria evolved between 2.5 and 3.4 billion years ago, living along with other prokaryotes in colonies that built the stromatolites that have been found all over the world (see FIGURE 24.3). Some marine sediments that are 2.5 billion years old are banded iron formations, red layers rich in iron oxide and valuable as a source of iron ore today. The sediments probably formed over a period when abundant cyanobacteria released oxygen that reacted with dissolved iron ions, which precipitated as iron oxide. This reaction would have prevented any accumulation of free O_2 for perhaps a few hundred million years until precipitation exhausted the dissolved iron. Only then would the seas become saturated with O_2, which began gassing out and accumulating in the atmosphere. Beginning about 2 billion years ago, terrestrial rocks rich in iron were rusted red by oxidation with atmospheric O_2.

The gradual change to a more oxidizing atmosphere created a crisis for Precambrian prokaryotes, because oxygen attacks the bonds of organic molecules. The corrosive atmosphere probably caused the extinction of many prokaryotes unable to cope. Other species survived in habitats that remained anaerobic, where we find their descendants living today as obligate anaerobes. The evolution of antioxidant mechanisms enabled other prokaryotes to tolerate the rising oxygen levels. Among photosynthetic prokaryotes, some species went a step further than mere oxygen tolerance to actually using the oxidizing power of O_2 to pull electrons from organic molecules down existing transport chains. Thus, aerobic respiration may have originated as modifications of electron transport chains co-opted from photosynthesis. Among photoheterotrophs, the purple nonsulfur bacteria still use an electron transport system that is a hybrid of photosynthetic and respiratory equipment. Several other bacterial lineages gave up photosynthesis and reverted to chemoheterotrophic nutrition, their electron transport chains adapted to function exclusively in aerobic respiration.

FIGURE 25.8

A bloom of cyanobacteria. Resembling thick, green paint, a population of *Microcystis* (inset, LM) colors the shoreline of Balgavies Loch in Scotland. Although there are several groups of photosynthetic prokaryotes, cyanobacteria are the only ones that split water and release O_2 in their light-harvesting reactions. Earth's relationship with life changed when O_2 produced by early cyanobacteria began to accumulate in the atmosphere more than two billion years ago.

Now that we have surveyed the structural and metabolic adaptations of prokaryotes, we have the evolutionary context we need for a closer look at the diverse groups of archaebacteria and eubacteria.

Molecular systematics is leading to a phylogenetic classification of prokaryotes

Researchers' first hint of the early archaebacteria/eubacteria split was the correlation of each of the two prokaryotic domains with unique **signature sequences,** taxon-specific base sequences at comparable locations in ribosomal RNA or other nucleic acids. TABLE 25.2 highlights other ways in which archaebacteria differ from eubacteria. The third domain, Eukarya, is included in the table to reinforce the point that archaebacteria have at least as much in common with eukaryotes as they do with eubacteria. However, the archaebacteria also have many unique traits, as should be expected of a taxon that has followed a separate evolutionary path for so long.

Domain Archaea (Archaebacteria)

You learned earlier in this chapter that most archaebacteria inhabit the more extreme environments of Earth. Biologists who study the prokaryotic life of such places have identified three main groups of archaebacteria: the methanogens, the extreme halophiles, and the extreme thermophiles.

The **methanogens** are named for their unique form of energy metabolism, in which H_2 is used to reduce CO_2 to methane (CH_4). Methanogens are among the strictest of anaerobes, poisoned by oxygen. They live in swamps and marshes where other microbes have consumed all the oxygen; the methane that bubbles out at these sites is known as marsh gas. Methanogens are also important decomposers employed for sewage treatment. Some farmers have experimented with the use of these microbes to convert garbage and dung to methane, a valuable fuel. Other species of methanogens inhabit the anaerobic en-

TABLE 25.2

A Comparison of the Three Domains of Life

	DOMAIN		
CHARACTERISTIC	Bacteria	Archaea	Eukarya
Nuclear envelope	Absent	Absent	Present
Membrane-enclosed organelles	Absent	Absent	Present
Peptidoglycan in cell wall	Present	Absent	Absent
Membrane lipids	Unbranched hydrocarbons	Some branched hydrocarbons	Unbranched hydrocarbons
RNA polymerase	One kind	Several kinds	Several kinds
Initiator amino acid for start of protein synthesis	Formyl-methionine	Methionine	Methionine
Introns (noncoding parts of genes)	Absent	Present in some genes	Present
Antibiotic sensitivity	Growth inhibited by streptomycin and chloramphenicol	Not inhibited by these antibiotics	Not inhibited by these antibiotics

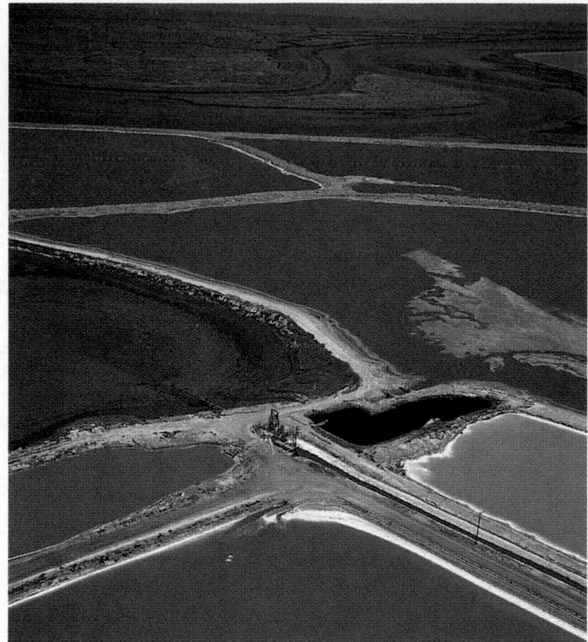

FIGURE 25.9
Extreme halophiles. These archaebacteria live in extremely saline waters. The colors of these seawater evaporating ponds at the edge of San Francisco Bay result from a dense growth of extreme halophiles that thrive in the ponds when the water reaches a salinity of 15% to 20%. (Before evaporation, the salinity of seawater is about 3%.) The ponds are used for commercial salt production; the halophilic bacteria are harmless.

vironment within the guts of animals, playing an important role in the nutrition of cattle, termites, and other herbivores that subsist mainly on a diet of cellulose.

The **extreme halophiles** (Gr. *halo,* "salt," and *philos,* "lover") live in such saline places as the Great Salt Lake and the Dead Sea. Some species merely tolerate salinity, whereas others actually require an environment ten times saltier than seawater to grow (FIGURE 25.9). Colonies of halophiles form a purple-red scum that owes its color to bacteriorhodopsin (see p. 507).

As their name implies, the **extreme thermophiles** thrive in hot environments. The optimal conditions for these archaebacteria are temperatures of 60°C to 80°C. *Sulfolobus* inhabits hot sulfur springs in Yellowstone National Park, obtaining its energy by oxidizing sulfur. Another sulfur-metabolizing thermophile lives in the 105°C water near deep-sea hydrothermal vents. Comparisons of key proteins have convinced James Lake of the University of California, Los Angeles, that the extreme thermophiles are the prokaryotes most closely related to eukaryotes. He highlights this evolutionary significance by calling the extreme thermophiles *eocytes,* meaning "dawn cells."

Domain Bacteria (Eubacteria)

Eubacteria account for most prokaryotes, with every major mode of nutrition and metabolism represented among the thousands of known species. The eubacteria diversified so long ago that evolutionary ties between the various taxonomic groups were, until recently, hazy. Molecular systematics offers the most powerful tool for tracing prokaryotic evolution, and researchers can now propose taxonomic subdivisions of eubacteria that are phylogenetically reasonable. Most prokaryotic systematists recognize about a dozen eubacterial groups. TABLE 25.3 features five of these groups. FIGURE 25.10 (p. 512), an expanded version of the tree on page 499, will help you review the evolutionary relationships of these eubacteria to the archaebacteria and eukaryotes. You should study the table and figure before going on to the chapter's last section, which highlights the continuing importance of prokaryotic life.

Five Major Phylogenetic Groups of Eubacteria (based mainly on comparisons of signature sequences in ribosomal RNA)

GROUP	CHARACTERISTICS	EXAMPLE
Proteobacteria	The most diverse group of bacteria, with three main subgroups:	
	1. Purple bacteria: photoautotrophic or photoheterotrophic organisms with bacteriochlorophylls built into invaginations of the plasma membrane; extract electrons from molecules other than H_2O, such as H_2S, and thus release no oxygen (the yellow globules in the cells at the right consist of sulfur produced as a waste product of H_2S-splitting); most species are obligate anaerobes; found in the sediments of ponds, lakes, and mudflats; many species are flagellated. (LM)	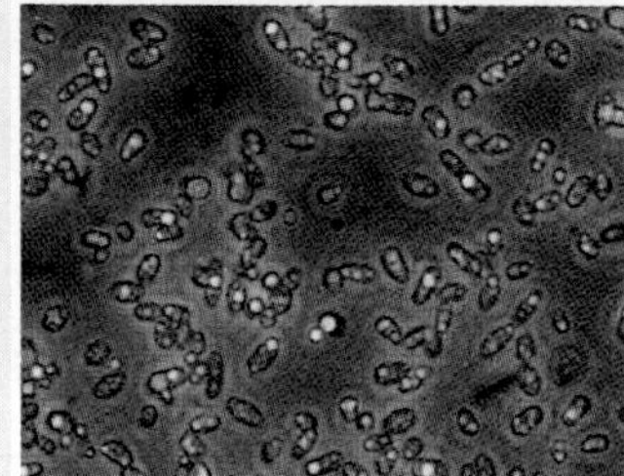 *Chromatium* 1 µm
	2. Chemoautotrophic proteobacteria: includes free-living and symbiotic species; many play key roles in the nitrogen cycles of ecosystems, including nitrogen fixation (conversion of atmospheric N_2 to nitrogenous minerals that plants can use); for example, the genus *Rhizobium* lives symbiotically in root nodules of peas and other legumes, contributing to the nutrition of those plants (you can see these bacteria within the vesicles of a root nodule cell in the photograph, at the arrows). (TEM)	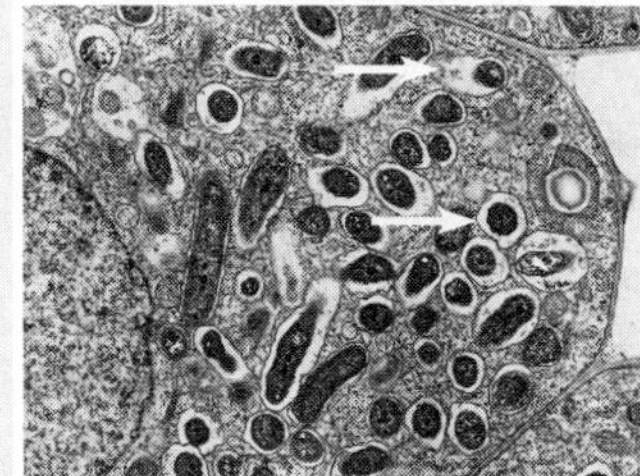 *Rhizobium* 2.5 µm
	3. Chemoheterotrophic proteobacteria: includes the enteric bacteria, which inhabit the intestinal tracts of animals; most enterics are rod-shaped facultative anaerobes; many, such as *E. coli,* are usually harmless; others are generally pathogenic, including *Salmonella,* one of the micro-organisms that causes food poisoning. (SEM)	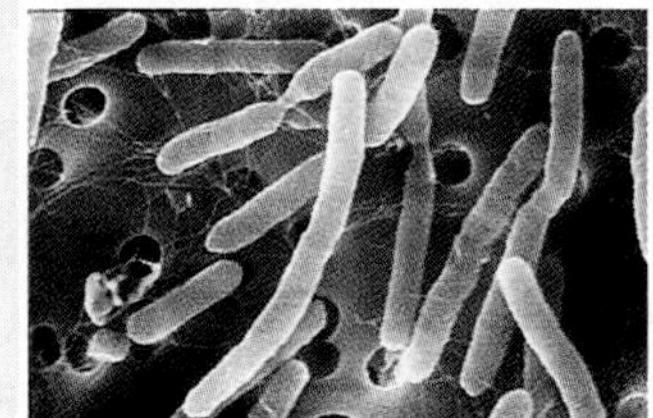 *Salmonella* 2.5 µm
Gram-positive eubacteria	Most *are* gram-positive, but the name of the group is misleading because some species are actually gram-negative and are grouped in this taxon because molecular systematics indicates a close relationship to the gram-positive bacteria. The group includes some photosynthetic members, but most species are chemoheterotrophs; many, including *Clostridium* and *Bacillus,* form endospores (boxed area in the photograph) that are resistant to harsh conditions. (TEM)	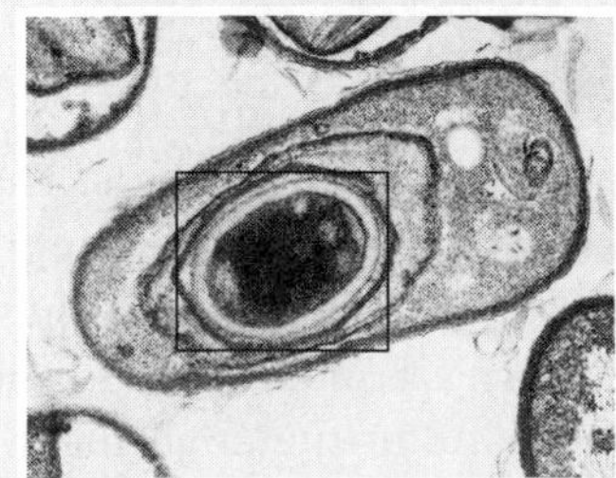 *Bacillus* 1 µm

TABLE 25.3 (continued)

GROUP	CHARACTERISTICS	EXAMPLE
	Among those gram-positive bacteria that do not form spores are the mycoplasmas, the smallest of all known cells with diameters of only 0.10–0.25 μm; the only eubacteria that lack cell walls, they are common in soil, and some are pathogenic in animals (e.g., *Mycoplasma pneumoniae* causes "walking pneumonia" in humans). (SEM)	*Mycoplasma* 2.5 μm
	The gram-positive group also includes the actinomycetes, soil bacteria that form branching colonies resembling fungi; many actinomycetes, including *Streptomyces,* are important commercial sources of antibiotics. (LM)	*Streptomyces* 2.5 μm
Cyanobacteria	Photoautotrophs with plantlike photosynthesis; have chlorophyll *a* and use two photosystems to split water, yielding O_2 as a by-product; most inhabit fresh water, but there are also marine species and symbionts that live along with fungi as lichens; cell walls are often thick and gelatinous; flagella absent, motile forms glide; among cyanobacteria are single-celled forms, colonial species, and truly multicellular organisms with a division of labor between specialized cells (boxed area in the photograph highlights a heterocyst, a cell specialized for nitrogen fixation). (LM)	*Anabaena* 50 μm
Spirochetes	Helical cells, sometimes very long (up to 0.25 mm, but too thin to be resolved without a microscope); internal flagellar filaments function in corkscrewlike movements; chemoheterotrophs, including both free-living species and pathogens such as *Treponema pallidum* (the cause of syphilis) and *Borrelia burgdorferi* (the cause of Lyme disease). (TEM)	*Leptospiru* 0.5 μm
Chlamydias	Obligate intracellular parasites of animals (see arrow); obtain all their ATP from host cells; gram-negative cell walls, but unusual among eubacteria in lacking peptidoglycan; *Chlamydia trachomatis* is the most common cause of blindness in the world and also causes the most common form of sexually transmitted disease (nongonococcal urethritis) in the United States. (TEM)	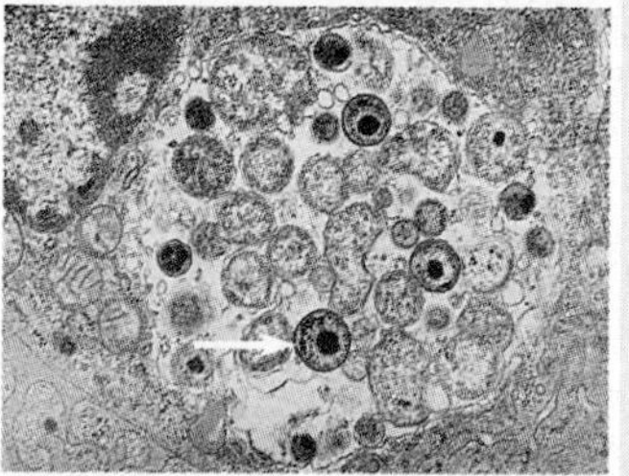 *Chlamydia* 2.5 μm

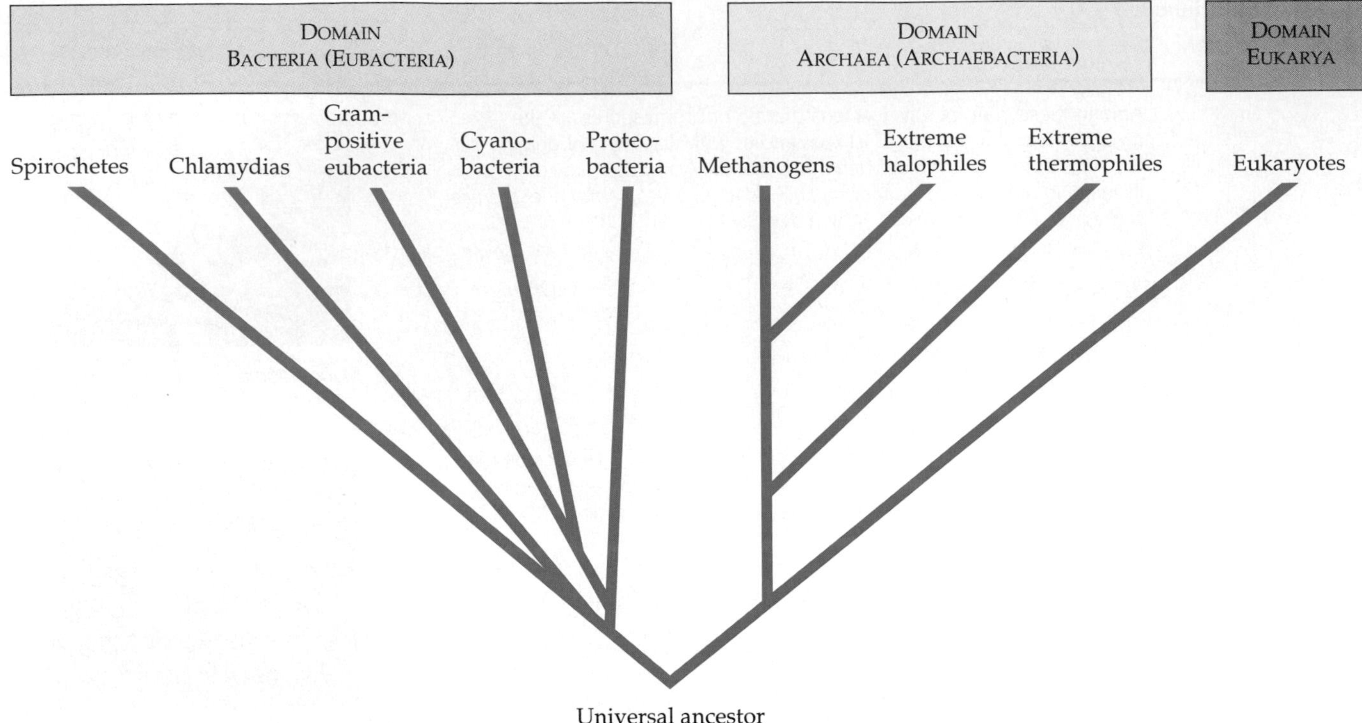

FIGURE 25.10
Evolutionary relationships of the prokaryotes: a review. This tentative phylogeny, like all such trees, is a hypothesis about evolutionary history. This particular hypothesis is based mainly on molecular systematics, particularly comparisons of rRNA signature sequences. (Of the approximately twelve eubacterial groups, only those featured in TABLE 25.3 are included in this simplified tree.)

Prokaryotes continue to have an enormous ecological impact

One of the themes of this unit is changing life on a changing planet—the interactions of geological history and evolving biological forms. Organisms as pervasive, abundant, and diverse as the prokaryotes have a tremendous impact on Earth and all its inhabitants. Here we consider some of these relationships.

Prokaryotes and Chemical Cycles

Not too long ago, in geological terms, the atoms of the organic molecules in our bodies were parts of the inorganic compounds of soil, air, and water, as they will be again. Ongoing life depends on the recycling of chemical elements between the biological and physical components of ecosystems. Prokaryotes are indispensable links in these chemical cycles (to be discussed in detail in Chapter 49). If it were not for such **decomposers,** carbon, nitrogen, and other elements essential to life would become locked in the organic molecules of corpses and feces.

Prokaryotes also mediate the return of elements from the nonbiological sources in the environment (air, inorganic soil, and water). Autotrophic prokaryotes fix CO_2, supporting food chains through which organic nutrients pass from the prokaryotes to prokaryote-eaters, then on to secondary consumers. Because of their many unique metabolic capabilities, prokaryotes are the only organisms able to transform nonbiological molecules containing elements such as iron, sulfur, nitrogen, and hydrogen. Cyanobacteria not only synthesize food and restore oxygen to the atmosphere but fix nitrogen, stocking the soil and water with nitrogenous compounds that other organisms can use to make proteins. And when plants and the animals that eat them die, it is soil prokaryotes that keep the cycle running by returning the nitrogen to the atmosphere. All life on Earth depends on prokaryotes and their unparalleled metabolic diversity.

Symbiotic Bacteria

Prokaryotes rarely function singly in the environment. More often they interact in groups, often consisting of different prokaryotic species with complementary metabolism. They also often depend on close relationships with eukaryotic organisms. **Symbiosis,** (Gr. for "living together") is the term used to describe ecological relationships between organisms of different species that are in direct contact. The organisms involved are known as **symbionts.** If one of the symbionts is much larger than the other, the larger is also termed the **host.** There are three categories of symbiotic relationships: mutualism, commensalism, and parasitism. In **mutualism,** both symbionts benefit. In **commensalism,** one symbiont receives benefits while neither harming nor helping the other in any significant way. In **parasitism,** one symbiont, called a **parasite** in this case, benefits at the expense of the host.

Symbiosis among prokaryotes is undoubtedly common but poorly understood because until recently, research has been restricted to pure cultures of single species. Symbiosis likely played a major role in the evolution of the prokaryotes and also in the origin of the early eukaryotes (as you will see in Chapter 26). Regarding symbiosis with modern eukaryotes, prokaryotic life is represented extensively in all three types of symbiosis. For example, plants of the legume family (peas, beans, alfalfa, and others) have lumps on their roots called nodules, which are home to mutualistic bacteria that fix nitrogen used by the host (see *Rhizobium* in TABLE 25.3). The plant reciprocates with a steady supply of sugar and other organic nutrients. The bacteria inhabiting the inner and outer surfaces of the human body consist mostly of commensal species, but some species are mutual symbionts. For instance, fermenting bacteria living in the vagina produce acids that maintain a pH between 4.0 and 4.5, suppressing the growth of yeast and other potentially harmful microorganisms. Humans also benefit from the metabolic products of *E. coli,* an inhabitant of the intestine. Among parasitic bacteria are those classified as pathogens because they cause disease.

Bacteria and Disease

Bacteria are almost everywhere, and exposure to pathogenic ones is a certainty. Most of us are well most of the time because our defenses check the growth of harmful bacteria and other pathogens to which we are exposed. Occasionally, the balance shifts in favor of a pathogen, and we become ill. To be pathogenic, a parasite must invade the host, resist internal defenses well enough to begin growing, then harm the host in some way. Approximately half of all human disease is caused by bacteria.

Bacteria need not be exotic intruders in order to be pathogenic. Some pathogens are **opportunistic,** meaning they are normal residents of the human body that inflict illness only when defenses have been weakened by such factors as poor nutrition or a recent bout with the flu. For example, *Streptococcus pneumoniae* lives in the throats of most healthy people, but this opportunist can multiply and cause pneumonia when the host's defenses are down.

Louis Pasteur, Joseph Lister, and other scientists began linking disease to pathogenic microbes in the late 1800s. The first to actually connect certain diseases to specific bacteria was Robert Koch, a German physician who identified the bacteria responsible for anthrax and tuberculosis. His methods established four criteria, now called **Koch's postulates,** that are still the guidelines for medical microbiology. To substantiate a specific pathogen as the cause of a disease, the researcher must (1) find the same pathogen in each diseased individual investigated, (2) isolate the pathogen from a diseased subject and grow the microbe in a pure culture, (3) induce the disease in experimental animals by transferring the pathogen from the culture, and (4) isolate the same pathogen from the experimental animals after the disease develops. The postulates can be applied for most pathogens, but prudent exceptions must be made for some cases. For example, no one has yet been able to culture the bacterium that causes syphilis *(Treponema pallidum)* on artificial media, but the volume of circumstantial evidence associating the organism with the disease leaves no doubt in this case.

How do pathogenic bacteria actually produce symptoms of disease? Some bacteria disrupt the physiology of the host by their actual growth and invasion of tissues. For example, two actinomycetes that induce disease by growing into tissues are the species that cause tuberculosis and leprosy.

Pathogenic bacteria more commonly cause illness by producing toxins. These poisons are of two types: exotoxins and endotoxins. **Exotoxins** are proteins secreted by the bacterial cell. Exotoxins can produce symptoms even without the bacteria actually being present. For example, when *Clostridium botulinum* grows anaerobically in poorly canned foods, one of the by-products of its fermentation is an exotoxin that causes the potentially fatal disease botulism. Exotoxins are among the most potent poisons known; one gram of botulism toxin would be sufficient to kill a million humans. Another exotoxin-producing species is the enteric bacterium *Vibrio cholerae,* which can infect the lower intestine of humans and cause cholera, a dangerous disease characterized by severe diarrhea. Resulting from the consumption of water contaminated with human feces, cholera is epidemic among refugees of the recent civil war in Rwanda.

Even *E. coli* can be an exotoxin-releasing culprit. Traveler's diarrhea results from toxins released by alien strains of this intenstinal inhabitant.

In contrast to exotoxins, **endotoxins** are not secreted by the pathogens but are instead components of the outer membranes of certain gram-negative bacteria. Endotoxins induce the same general symptoms—fever and aches—regardless of the bacterial species, whereas exotoxins elicit specific symptoms. Examples of endotoxin-producing bacteria include certain enteric bacteria, such as nearly all members of the genus *Salmonella,* which are not normally present in healthy animals. *Salmonella typhi* causes typhoid fever, and several other species of *Salmonella,* some of which are commonly found in poultry, cause food poisoning.

With the discovery in the nineteenth century that "germs" cause disease, public health officials took steps to upgrade hygiene. Sanitation measures played a significant role in reducing infant mortality and extended life expectancy dramatically in developed countries. In the past few decades, medical technology has increased its success at combating bacterial disease with a variety of antibiotics. More than half of our antibiotics (including streptomycin, neomycin, erythromycin, aureomycin, and tetracycline) come from soil bacteria of the genus *Streptomyces* (an actinomycete; see TABLE 25.3). In the wild, these compounds prevent encroachment by other microbes. Pharmaceutical companies culture various species of *Streptomyces* to produce these antibiotics in commercial amounts.

Bacterial disease has certainly not been conquered, but its decline over the past century, probably due more to public health policies than to "wonder drugs," has so far been the greatest achievement of biomedical research and its application. However, the rapid evolution of antibiotic-resistant strains of pathogenic bacteria is a serious health threat aggravated by imprudent and excessive antibiotic use.

Putting Prokaryotes to Work

Humans have learned many ways of exploiting the diverse metabolic capabilities of prokaryotes, both for scientific research and for practical purposes. Much of what we know about metabolism and molecular biology has been learned in laboratories using bacteria as relatively simple model systems. In fact, *E. coli,* the "white rat" of so many research labs, is the best understood of all organisms. Methanogens are important decomposers utilized for sewage treatment (FIGURE 25.11). Soil bacteria called pseudomonads (proteobacteria) decompose pesticides and other synthetic compounds. (They can also be pests.

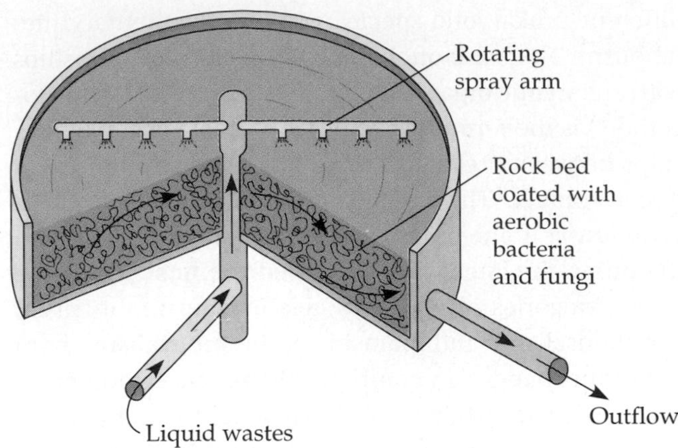

FIGURE 25.11
Bacterial decomposers in sewage treatment. In modern sewage treatment, bacteria are instrumental in decomposing both solid and liquid matter. After liquids and solids are separated, the solid matter, called sludge, is added gradually to a culture of anaerobic bacteria. The bacteria convert the organic matter in the sludge to material than can be used as landfill or fertilizer, after chemical sterilization. Liquid wastes are treated separately from the sludge. In a trickling filter system like the one shown here, a horizontal spray arm rotates slowly, spraying liquid wastes through the air onto a thick bed of rocks, the filter. Aerobic bacteria and fungi growing on the rocks decompose much of the organic matter in the waste. Outflow from the rock bed is chemically sterilized and then released, usually into a river or ocean.

Their ability to feed on unusual carbon sources enables them to invade hot tubs, drug solutions, and even antiseptic solutions meant to prevent bacterial growth.) The chemical industry grows immense cultures of bacteria that produce acetone, butanol, and several other products. Pharmaceutical companies culture bacteria that make vitamins and antibiotics. The food industry uses bacteria to convert milk to yogurt and various kinds of cheese. And recombinant DNA techniques promise a new era in the commercial importance of prokaryotes (see Chapter 19).

* * *

In this chapter, we have surveyed the prokaryotes and traced their history. On an ancient Earth inhabited only by prokaryotes, all the diverse forms of nutrition and metabolism evolved. Most subsequent evolutionary breakthroughs were structural rather than metabolic. The most significant development was the origin of eukaryotic cells from prokaryotic ancestors, a juncture in the history of life we will explore in the next chapter.

REVIEW OF KEY CONCEPTS (with page numbers and key figures)

- They're (almost) everywhere! *an overview of prokaryotic life* (pp. 498–499)
  - Prokaryotes were the first organisms, and persist today as the most numerous and pervasive of all living things.
- Archaea and Bacteria are the two main branches of prokaryotic evolution (p. 499, FIGURE 25.10)
  - Prokaryotes account for two of the three domains (superkingdoms) of life.
  - Molecular systematics suggests that archaebacteria (domain Archaea) may be more closely related to eukaryotes (domain Eukarya) than to eubacteria (domain Bacteria).
- The success of prokaryotic life is based on diverse adaptations of form and function (pp. 499–505, FIGURE 25.1)
  - Prokaryotes are generally single-celled organisms, although some occur as aggregates, colonies, or simple multicellular forms.
  - The three most common prokaryotic shapes are spherical (cocci), rod-shaped (bacilli), and helical forms.
  - Nearly all prokaryotes have external cell walls, which protect and shape the cell and prevent osmotic bursting. Cell walls of eubacteria typically contain the polymer peptidoglycan. Gram-positive and gram-negative bacteria differ in the structure of their walls and other surface layers.
  - Many species secrete sticky substances that form capsules. Some have surface appendages called pili outside the cell wall. Both structures help the cells adhere to one another, and some pili are specialized for conjugation.
  - Motile bacteria propel themselves by flagella, use flagella-like filaments positioned inside the cell wall (spirochetes), or glide on slime secretions.
  - Prokaryotic cells are not compartmentalized by endomembranes. However, invaginations of the plasma membrane may provide internal membrane surface for specialized functions.
  - The prokaryotic genome consists of a single circular DNA molecule in a nucleoid region unbounded by a membrane. Many species also possess smaller separate rings of DNA called plasmids, which code for special metabolic pathways and resistance to antibiotics.
  - Bacteria reproduce asexually by binary fission.
  - Genetic variation occurs in prokaryotes through mutation and gene transfer by transformation, conjugation, or viral transduction.
- All major types of nutrition and metabolism evolved among prokaryotes (pp. 505–506, TABLE 25.1)
  - Prokaryotes are the most metabolically diverse organisms on Earth. Photoautotrophs use light energy and chemoautotrophs use inorganic substances to synthesize their organic compounds from carbon dioxide. Photoheterotrophs require organic molecules for metabolic processes and synthesize ATP using light energy. Most bacteria are chemoheterotrophs, which require organic molecules as a source of both energy and organic carbon.
  - Several groups of bacteria metabolize nitrogen compounds unavailable to other organisms. By doing so, these prokaryotes play critical roles in the cycling of nitrogen in the environment.
  - The ability or inability to survive in the presence of oxygen also reflects variation in metabolism. Obligate aerobes require oxygen, obligate anaerobes are poisoned by it, and facultative anaerobes can survive with or without oxygen.
- The evolution of prokaryotic metabolism was both cause and effect of changing environments on Earth (pp. 506–508, FIGURE 25.7)
  - The first prokaryotes were likely chemoheterotrophs that absorbed free organic compounds. Glycolysis evolved early as a mechanism of regenerating ATP.
  - Electron transport chains could have evolved from transmembrane pumps that originally served to regulate pH.
  - Early photosynthetic prokaryotes used pigments and light-powered photosystems to fix carbon dioxide. The first cyanobacteria began making organic compounds from water and carbon dioxide, releasing free oxygen as a by-product. This drastically changed Earth's ancient atmosphere and affected subsequent biological evolution.
- Molecular systematics is leading to a phylogenetic classification of prokaryotes (pp. 508–511, TABLES 25.2, 25.3)
  - Comparisons of selected macromolecules suggest an early split of the prokaryotes into the archaebacteria, which live in extreme environments reminiscent of conditions on the primordial Earth, and all other prokaryotes, the eubacteria.
  - Methanogens, extreme halophiles, and extreme thermophiles are the three subgroups of archaebacteria.
  - For a review of the eubacteria, see TABLE 25.3.
- Prokaryotes continue to have an enormous ecological impact (pp. 512–514)
  - Prokaryotes, along with fungi, are decomposers that recycle chemical elements in ecosystems.
  - Some prokaryotes live with other species in symbiotic relationships of mutualism, commensalism, or parasitism.
  - Some parasitic prokaryotes are pathogenic, causing disease in the host by invading tissues or poisoning with endotoxins or exotoxins.
  - Bacteria have been put to work in laboratories, sewage treatment plants, and the food and drug industry. One especially exciting development has been the use of prokaryotes in recombinant DNA technology.

SELF-QUIZ

1. Home canners pressure-cook vegetables as a precaution primarily against
   a. mycoplasmas
   b. endospore-forming bacteria
   c. enteric bacteria
   d. pseudomonads
   e. actinomycetes

2. Photoautotrophs use
   a. light as an energy source and can use water or hydrogen sulfide as a source of electrons for producing organic compounds
   b. light as an energy source and oxygen as an electron source
   c. inorganic substances for energy and CO_2 as a carbon source
   d. light to generate ATP but need organic molecules for a carbon source
   e. light as an energy source and CO_2 to reduce organic nutrients

3. Which of the following statements about the two domains of prokaryotes is *not* true?
   a. The lipid composition of the plasma membrane found in archaebacteria is different from that of eubacteria.
   b. The archaebacteria and eubacteria probably diverged very early in evolutionary history.
   c. Both archaebacteria and eubacteria have cell walls, but those of archaebacteria lack peptidoglycan.
   d. Of the two groups, eubacteria are more closely related to domain Eukarya.
   e. Eubacteria include the cyanobacteria.

4. A prokaryotic genome is different from a eukaryotic genome in that
   a. it has only one-half as much DNA as does a typical eukaryotic genome
   b. it consists of a single-stranded DNA molecule
   c. it has less protein associated with its DNA and is not enclosed in a nuclear envelope
   d. it is made of ribosomes that are smaller and chemically distinct
   e. it consists of RNA rather than DNA

5. The first prokaryotes were probably
   a. cyanobacteria
   b. chemoheterotrophs that used abiotically made organic compounds
   c. anaerobic photosynthetic organisms
   d. mycoplasmas
   e. parasitic bacteria

6. Banded iron formations in marine sediments indicate that
   a. early cyanobacteria were probably producing oxygen during that time period
   b. the early atmosphere was very reducing
   c. aerobic bacteria were the dominant life form in the seas at that time
   d. the pH of the early seas was quite low because of early prokaryotic excretion of organic acids from fermentation
   e. mats of bacterial colonies were forming stromatolites

7. Which of the following statements about prokaryotic flagella is true?
   a. The flagella are composed of several microtubules of protein surrounded by the plasma membrane.
   b. All bacteria have flagella.
   c. There is always only one flagellum per cell.
   d. Motion is produced by lashing back and forth, like a whip.
   e. Motion is produced by rotation of a semirigid, helical filament.

8. Penicillins function as antibiotics mainly by inhibiting the ability of some bacteria to
   a. form spores
   b. replicate DNA
   c. synthesize normal cell walls
   d. produce functional ribosomes
   e. synthesize ATP

9. Plantlike photosynthesis that releases oxygen occurs in the
   a. cyanobacteria
   b. chlamydias
   c. archaebacteria
   d. actinomycetes
   e. chemoautotrophic bacteria

10. According to the cladogram in FIGURE 25.10, cyanobacteria share their most recent noncyanobacterial ancestor with the
    a. eukaryotes
    b. proteobacteria
    c. gram-positive bacteria
    d. methanogens
    e. chlamydias

CHALLENGE QUESTIONS

1. Nitrogen-fixing bacteria work either alone or in close conjunction with plants. If you were a scientist investigating the biochemistry of nitrogen fixation, would you choose a solitary or symbiotic species for study? Explain your answer.
2. Lynn Margulis of the University of Massachusetts has suggested that we could probe for extraterrestrial life by simply determining the mixture of gases in the atmospheres of planets. If you were to conduct such research, what would you look for? Why?
3. Under suitable conditions, some bacteria are capable of reproducing by binary fission once every half-hour. Suppose you placed a single bacterium in a culture medium and it (and its descendants) proceeded to reproduce at this rate. How many bacteria would be present after 2 hours? After a total of 12 hours? (This is easier to figure out with a calculator.) What determines how long this kind of increase can continue?
4. What are the arguments for and against the combining of all prokaryotes in the kingdom Monera of the traditional five-kingdom system of classification?

SCIENCE, TECHNOLOGY, AND SOCIETY

1. Despite improvements in sanitation and antibiotics, people continue to be plagued by bacterial diseases, especially in the developing countries. In the United States, health officials have recently become concerned by a resurgence of tuberculosis (TB), a bacterial lung disease spread by airborne droplets. In particular, they are worrying about the current TB epidemic caused by bacteria resistant to standard antibiotics. Drugs can alleviate TB symptoms in a few weeks, but it takes much longer to halt the infection; a patient is likely to discontinue treatment while bacteria are still present. Why can bacteria quickly reinfect a patient if they are not wiped out? How might this result in the evolution of drug-resistant bacteria? How might urban poverty, homelessness, AIDS, and drug abuse contribute to the rise in the incidence of TB?

2. Many archaebacteria live in extreme habitats, including temperatures near 100°C, very high pressures, saturated salt solutions, and extremes of pH. Since genes from any organism can theoretically be transferred to any other organism, what is the importance of knowing more about the archaebacteria? What industrial and environmental uses can you think of for archaebacterial genes?

3. Many local newspapers publish a weekly list of restaurants that have been cited by inspectors for poor sanitation. Locate such a report and highlight the cases that are likely associated with potential food contamination by pathogenic bacteria.

FURTHER READING

Daviss, B. "Power Lunch." *Discover,* March 1995. Using bacteria to generate electricity.

Day, S. "Hot Bacteria and Other Ancestors." *New Scientist,* April 9, 1994. The evolutionary significance of heat-loving bacteria.

Kantor, F. S. "Disarming Lyme Disease." *Scientific American,* September 1994. Progress in fighting the tick-carried spirochete.

Lipkin, R. "Enzyme Helps Microorganism Thrive in Heat." *Science News,* March 11, 1995. Possible applications for the enzymes of archaebacteria.

Maynard Smith, J. "Bacteria Break the Antibiotic Bank." *Natural History,* June 1994. Newly evolved drug-resistant pathogens pose a serious health threat.

Moffat, A. "Microbial Mining Boosts the Environment, Bottom Line." *Science,* May 6, 1994. Using prokaryotes to extract precious metals.

Nowak, R. "Flesh-Eating Bacteria: Not New, But Still Worrisome." *Science,* June 17, 1994. An opportunistic pathogen that made headlines.

Pennisi, E. "Static Evolution." *Science News,* March 12, 1994. How much has the form of prokaryotic species changed in the past two billion years?

Portera, C. "From Bacteria: A New Weapon Against Fungal Infection." *Science,* July 29, 1994. A newly discovered class of antibiotics.

Tortora, G. J., B. R. Funke, and C. L. Case. *Microbiology: An Introduction,* 5th ed. Redwood City, CA: Benjamin/Cummings, 1995. A general text.

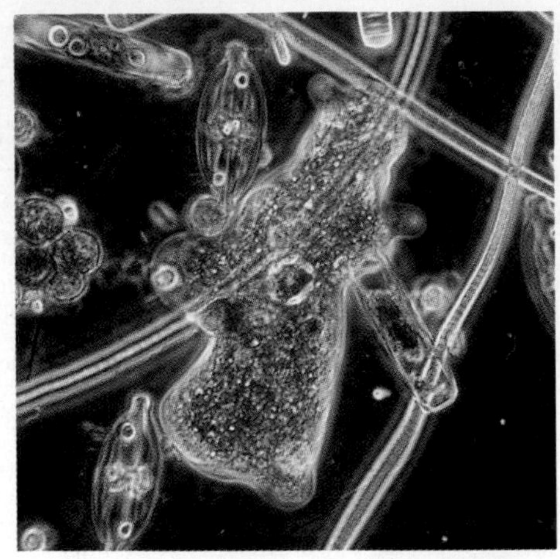

CHAPTER 26

THE ORIGINS OF EUKARYOTIC DIVERSITY

KEY CONCEPTS

- Eukaryotes originated by symbiosis among prokaryotes
- Archezoans provide clues to the early evolution of eukaryotes
- The diversity of protists represents different "experiments" in the evolution of eukaryotic organization
- Protistan taxonomy is in a state of flux
- Diverse modes of locomotion and feeding evolved among protozoa
- Funguslike protists have morphological adaptations and life cycles that enhance their ecological role as decomposers
- Eukaryotic algae are key producers in most aquatic ecosystems
- Systematists continue to refine their hypotheses about eukaryotic phylogeny
- Multicellularity originated independently many times

"No more pleasant sight has met my eye than this of so many thousands of living creatures in one small drop of water," wrote Anton von Leeuwenhoek after his discovery of the microbial world more than three centuries ago. It is a world every biology student rediscovers by peering through a microscope into a droplet of pond water filled with diverse creatures we now call protists (see the photograph on this page). Most of the 60,000 known species of extant protists are unicellular, but we also find colonial forms and even some multicellular organisms with tissues arranged in relatively simple body plans.

Protists are eukaryotic, and thus even the simplest are much more complex than the prokaryotes. The first eukaryotes to evolve from prokaryotic ancestors were probably unicellular and would therefore be classified as protists. The very word implies great antiquity (Gr. protos, *"first"). The primal eukaryotes were not only the predecessors of the great variety of modern protists but were also ancestral to plants, fungi, and animals, the eukaryotic organisms most familiar to us. Two of the most significant chapters in the history of life—the origin of the eukaryotic cell and the subsequent emergence of multicellular eukaryotes—unfolded during the evolution of protists.*

This chapter traces the origins of eukaryotic cells; examines the diversity, evolution, and ecology of protists; and considers the evolutionary origins of multicellular organization. Along the way, we will see why some systematists advocate replacing the five-kingdom system with alternative taxonomic schemes that split into three kingdoms the diverse organisms traditionally grouped in the kingdom Protista.

Eukaryotes originated by symbiosis among prokaryotes

The many differences between prokaryotic and eukaryotic cells represent a distinction much greater than that between the cells of plants and animals. During the genesis of protists, the cellular structures and processes unique to eukaryotes arose: a membrane-enclosed nucleus, mitochondria, chloroplasts, the endomembrane system, the cytoskeleton, 9 + 2 flagella, multiple chromosomes consisting of linear DNA molecules compactly arranged with proteins, diploid stages in life cycles, mitosis, meiosis, and sex. Among the most fundamental questions in biology is how the complex eukaryotic cell evolved from much simpler prokaryotic cells.

The small size and relatively simple construction of a prokaryotic cell have many advantages but also impose limits on the number of different metabolic activities that can be handled at one time. The relatively small size of the prokaryotic genome limits the number of genes coding for the enzymes that control these activities. This is not to say that prokaryotes are less successful than eukaryotes. Bacteria have been evolving and adapting since the dawn of life, and they are the most widespread organisms even today. But in at least some prokaryotic groups, natural selection favored increasing complexity—higher levels of organization with emergent properties. One trend was the evolution of multicellular prokaryotes, such as the filaments of some cyanobacteria, where different cell types are specialized for different functions. A second trend was the evolution of complex bacterial communities, where each species benefited from the metabolic specialties of other species. A third trend was the compartmentalization of different functions within single cells, an evolutionary solution that produced the first eukaryotes.

How did compartmental organization of the eukaryotic cell evolve from the simpler prokaryotic condition? The endomembrane system of eukaryotic cells—the nuclear envelope, endoplasmic reticulum, Golgi aparatus, and related structures—may have evolved from specialized invaginations (infoldings) of the prokaryotic plasma membrane (FIGURE 26.1a). Another process, called endosymbiosis, led to mitochondria, chloroplasts, and perhaps some of the other features of eukaryotic cells.

According to the **endosymbiotic theory** of the origin of eukaryotes, the forerunners of eukaryotic cells were symbiotic consortiums of prokaryotic cells, with certain species living within larger prokaryotes (The term *endosymbiont* is used for a cell that lives within another cell, termed the host cell.) Developed most extensively by Lynn Margulis of the University of Massachusetts, the endosymbiotic theory focuses on the origins of chloroplasts and mitochondria (FIGURE 26.1b). Chloroplasts are the descendants of photosynthetic prokaryotes, probably cyanobacteria, that became endosymbionts within larger cells. The proposed ancestors of mitochondria were endosymbiotic bacteria that were aerobic heterotrophs. Perhaps they first gained entry to the larger cell as undigested prey or internal parasites. By whatever means the relationships began, it is not hard to imagine the symbiosis eventually becoming mutually beneficial. A heterotrophic host could derive nourishment from photosynthetic endosymbionts. And in a world that was becoming increasingly aerobic, a cell that was itself an anaerobe would have benefited from aerobic endosymbionts that turned the oxygen to advantage. As host and endosymbionts became more interdependent, the conglomerate of prokaryotes would gradually be integrated into a single organism, its parts inseparable.

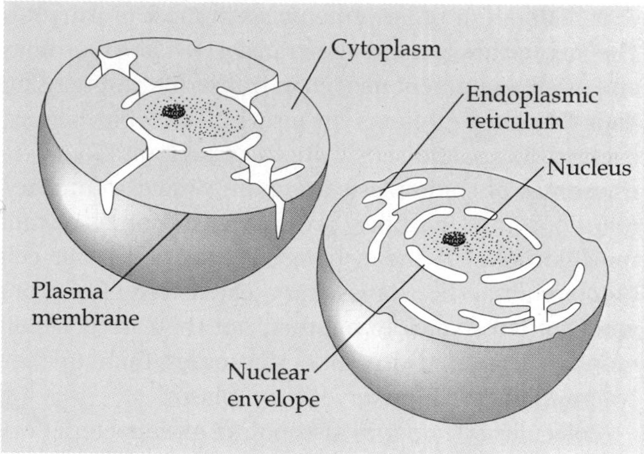

(a) Invagination of the plasma membrane

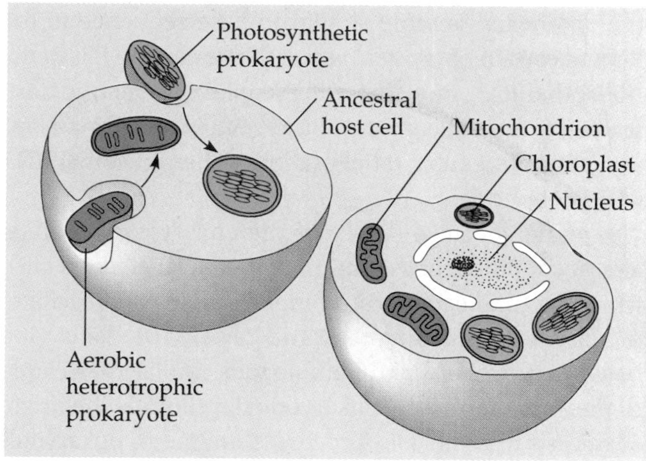

(b) Endosymbiosis

FIGURE 26.1

Two processes that may have contributed to eukaryotic origins. **(a)** The role of membrane invagination is speculative. **(b)** We can be more certain of the contribution of endosymbiosis, for which there is extensive evidence.

The feasibility of an endosymbiotic origin of chloroplasts and mitochondria rests partially on the existence of endosymbiotic relationships in the modern world. Another line of evidence is the similarity between eubacteria and the chloroplasts and mitochondria of eukaryotes. Chloroplasts and mitochondria are the appropriate size to be descendants of eubacteria. The inner membranes of chloroplasts and mitochondria, perhaps derived from the membranes of endosymbiotic prokaryotes, have several enzymes and transport systems that resemble those found on the plasma membranes of modern prokaryotes. Mitochondria and chloroplasts reproduce by a splitting process reminiscent of binary fission in bacteria. Chloroplasts and mitochondria contain DNA in the form of circular molecules not associated with histones or other proteins, as in most prokaryotes. The organelles contain the transfer RNAs, ribosomes, and other equipment needed to transcribe and translate their DNA into proteins. In terms of size, biochemical characteristics, and sensitivity to certain antibiotics the ribosomes of chloroplasts are more similar to prokaryotic ribosomes than they are to the ribosomes outside the chloroplast in the cytoplasm of the eukaryotic cell. Mitochondrial ribosomes vary extensively from one group of eukaryotes to another, but they are generally more similar to prokaryotic ribosomes than to their counterparts in the eukaryotic cytoplasm.

Molecular systematics also points to eubacterial origins for chloroplasts and mitochondria. Ribosomal RNA of chloroplasts, which is transcribed from genes within the organelles, is more similar in base sequence to the RNA of certain photosynthetic eubacteria than it is to the ribosomal RNA in eukaryotic cytoplasm, which is transcribed from nuclear DNA. Base-sequence comparisons also support a eubacterial origin for the ribosomal RNA of mitochondria.

A comprehensive theory of the eukaryotic cell must also account for the evolution of 9 + 2 flagella and cilia, which are *analogous,* not *homologous,* to the flagella of prokaryotes. (See Chapter 23 to review the distinction between analogous and homologous similarity.) Related to the evolution of the eukaryotic flagellum is the origin of mitosis and meiosis, processes unique to eukaryotes that also employ microtubules. Mitosis made it possible to reproduce the large genomes of the eukaryotic nucleus, and the closely related mechanics of meiosis became an essential process in eukaryotic sex. Among eukaryotes, sexual life histories are the most varied among the protists. More generally, the diversity of protistan life may reflect the evolutionary "experimentation" that occurred among the earliest eukaryotes.

Archezoans provide clues to the early evolution of eukaryotes

Giardia lamblia is a flagellated unicellular eukaryote that infects the human intestine, causing abdominal cramps and severe diarrhea. The parasite is transmitted mainly in water contaminated with human feces. Evolutionary biologists are interested in this protist for a different reason: Cell structure and molecular systematics point to *Giardia* as a modern representative of an ancient lineage that branched from the eukaryotic tree very early, perhaps more than two billion years ago.

Giardia belongs to a group of single-celled eukaryotes called diplomonads, one of a few phyla grouped together as **archezoa,** a name that accentuates antiquity (Gr. *arkhaios,* "ancient"). These organisms lack mitochondria and plastids, and their cytoskeletons are relatively simple compared to other eukaryotes. In some characteristics the ribosomes of archezoa are more like those of prokaryotes than those of other eukaryotes. Sequencing of ribosomal RNA provides evidence that archezoans are the eukaryotic organisms most closely related to prokaryotes. In fact, the archezoans are so distinct from other eukaryotes that some systematists advocate placing them in their own kingdom.

Giardia and other diplomonads are characterized by two separate haploid nuclei, giving the cells a facelike appearance when viewed with a microscope (FIGURE 26.2). Given the antiquity of the archezoan lineage, perhaps dual nuclei are a vestige of an early stage in the evolution of eukaryotes. Prokaryotes have haploid genomes (one set of genes), and some researchers speculate that the first eukaryotes had a single haploid nucleus bounded by a membranous envelope. Most modern eukaryotes have diploid stages in their life cycles, with a fusion of haploid nuclei forming the diploid nucleus with two sets of chromosomes. Diplomonads, with their dual haploid nuclei, may represent a "missing link" in the evolution of diploidy in eukaryotes.

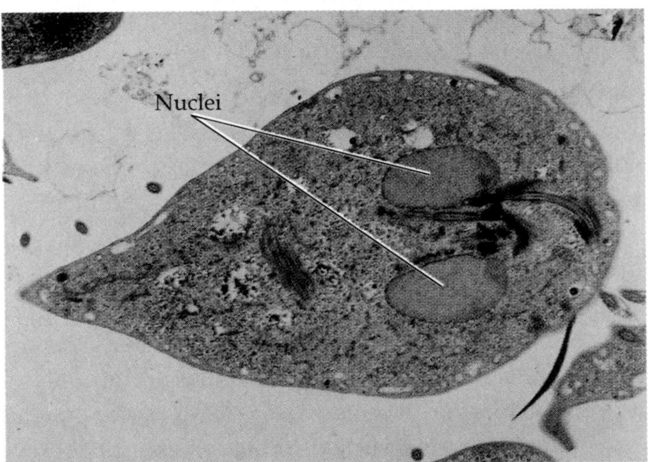

FIGURE 26.2
***Giardia:* a key to eukaryotic history?** Molecular systematics places *Giardia* and other archezoans as descendants of a lineage that originated very early during the prokaryote-eukaryote transition more than 2 billion years ago. Ultrastructural evidence is consistent with this hypothesis. Archezoans lack mitochondria and chloroplasts and have a very simple cytoskeleton. The facelike appearance of *Giardia* is typical of an archezoan phylum distinguished by two haploid nuclei. Some researchers speculate that this condition is a remnant of an early stage in the evolution of diploidy in the life cycles of eukaryotes—an evolutionary stage after the origin of a nuclear envelope and a doubling of the genome, but before the fusion of haploid nuclei to form diploid nuclei with two chromosome sets (colorized TEM).

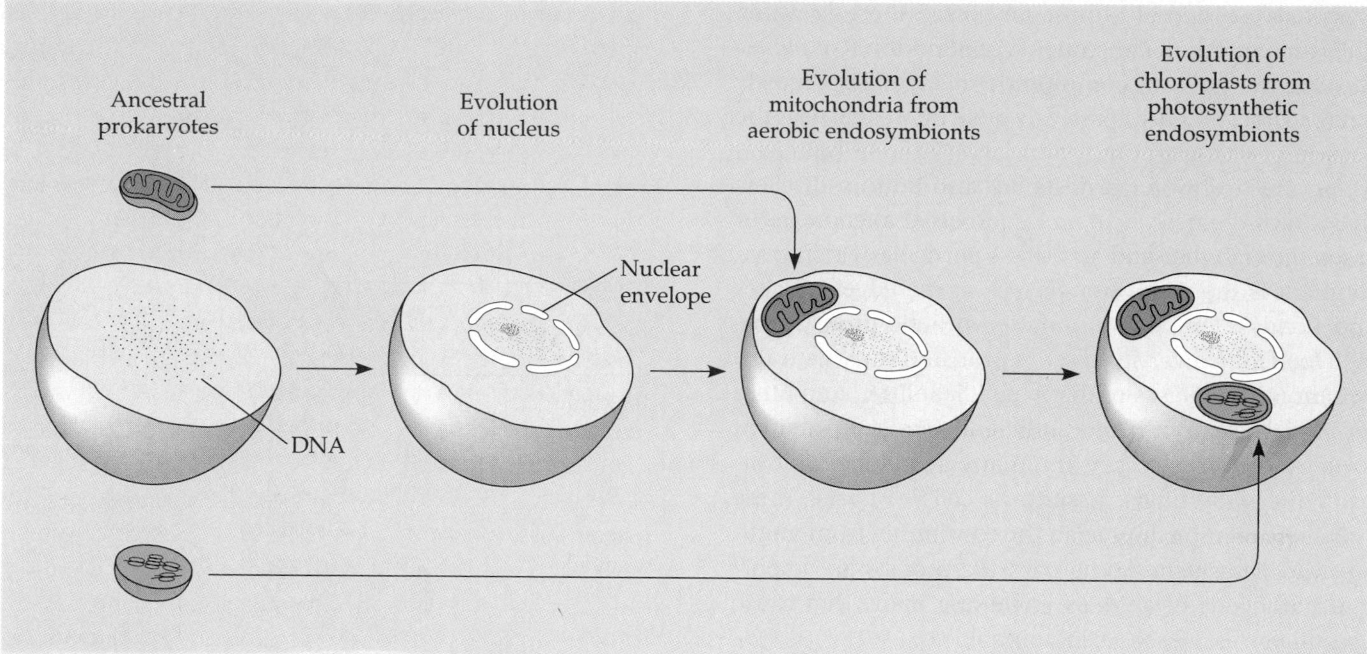

FIGURE 26.3
A review of eukaryotic origins by serial endosymbiosis. Evidence that archezoans, which lack mitochondria and chloroplasts, evolved very early during eukaryotic history suggests that a membrane-enclosed nucleus was in place before mitochondria and plastids. Almost all other eukaryotes, whether heterotrophic or autotrophic, have mitochondria. This is consistent with other evidence that mitochondria predated chloroplasts in the serial episodes of symbiosis that led to eukaryotic diversity.

Of course, *Giardia* and other diplomonads are not ancient organisms; they are as modern as humans. But each branch of the tree of life evolves in the context of historical constraints, adapting the equipment that was in place in its ancestors. If diplomonads are a side branch that diverged from the eukaryotic tree before the processes of nuclear fusion and meiosis evolved, then their dual nuclei may be a clue to the past. And the absence of mitochondria in diplomonads and other archezoans is consistent with an origin predating the endosymbiotic event that gave rise to mitochondria in a world that was becoming increasingly aerobic.

FIGURE 26.3 summarizes a model for the origin of eukaryotes by serial endosymbiosis (a sequence of endosymbiotic events). As an evolutionary mechanism contributing to the origin of eukaryotes, serial endosymbiosis is very different from the mechanisms of evolution you learned about in Unit Four of this book. In those chapters we studied how a lineage of organisms changes over time (anagenesis), and how lineages sometimes split into two or more divergent lineages of organisms (cladogenesis). In the case of endosymbiosis, a *merger* of evolutionary lineages gave rise to a new form of life.

The diversity of protists represents different "experiments" in the evolution of eukaryotic organization

At least a billion years before the origin of plants, fungi, and animals, there were protists, the earliest eukaryotic descendants of prokaryotes. The oldest putative fossils of protists are Precambrian remnants known as **acritarchs** (Gr., "of uncertain origin"). Some are the right size and structure to be the ruptured coats of cysts similar to those made by certain protists today. In 1992, paleobiologists discovered the oldest acritarchs known so far in Michigan rocks that are about 2.1 billion years old. Over the next billion years, adaptive radiation produced a diversity of protists. These evolutionary "experiments" were variations of structure and function possible for eukaryotic cells.

Considering that modern protists descended from these divergent branches near the base of the eukaryotic tree, it is not surprising that they vary in morphology and lifestyle more than any other group of organisms. In fact, protists vary so extensively in cellular anatomy, ecological roles, and life histories that few general characteristics can be cited without exceptions.

Protists are found almost anywhere there is water. They are important constituents of **plankton** (Gr. *planktos*, "wandering"), the communities of organisms, mostly microscopic, that drift passively or swim weakly near the surface of oceans, ponds, and lakes. Among both seawater and freshwater protists are also bottom-dwellers that attach themselves to rocks and other anchorages or creep through the sand and silt. A particularly rich environment is the still-water surface at the edges of lakes and ponds. Here, floating photosynthetic protists provide food for a diversity of other protists. Protists are also common inhabitants of damp soil, leaf litter, and other terrestrial habitats that are sufficiently moist. In addition to free-living protists are the many symbionts that inhabit the body fluids, tissues, or cells of hosts. These symbiotic relationships span the continuum from mutualism to parasitism. Some parasitic protists are important pathogens of animals, including many that cause potentially fatal diseases in humans.

Nearly all protists are aerobic in their metabolism, using mitochondria for cellular respiration (a few lack mitochondria and either live in anaerobic environments or contain mutualistic respiring bacteria). As a group, protists are the most nutritionally diverse of all eukaryotes. Some protists are photoautotrophs with chloroplasts, some are heterotrophs that absorb organic molecules or ingest larger food particles, and still others, called mixotrophs, combine photosynthesis *and* heterotrophic nutrition (an example is *Euglena*, shown in FIGURE 26.4). It is convenient (though phylogenetically inaccurate) to divide this nutritional diversity into three categories: photosynthetic (plantlike) protists—**algae** (singular, **alga**); ingestive (animal-like) protists—**protozoa** (singular, **protozoan**); and absorptive (funguslike) protists (these have no other general name).

Most protists have flagella or cilia at some time in their life cycles. As mentioned previously, it is important to understand that prokaryotic and eukaryotic versions of flagella are not homologous structures. Bacterial flagella are attached to the cell surface (see FIGURE 25.4). In contrast, eukaryotic flagella and cilia are extensions of the cytoplasm, with bundles of microtubules covered by the plasma membrane (see FIGURE 7.24). Eukaryotic cilia and flagella have the same basic ultrastructure (the 9 + 2 arrangement of microtubules), but cilia are shorter and more numerous. They move a cell with their rhythmic power strokes, analogous to the oars of a boat (see FIGURE 7.23b).

Although all protists are eukaryotic, nuclear organization and cell division are perplexingly varied among members of the kingdom. Mitosis occurs in most phyla of protists, but there are many variations in the process unknown in any other kingdom. Reproduction and life histories also vary extensively among the protists. All protists can reproduce asexually. Some forms are exclusively asexual; others can also reproduce sexually or at least use the sexual processes of meiosis and **syngamy** (the union of two gametes) to shuffle genes between two individuals that then go on to reproduce asexually. In Chapter 12, you learned about three basic types of sexual life cycles that

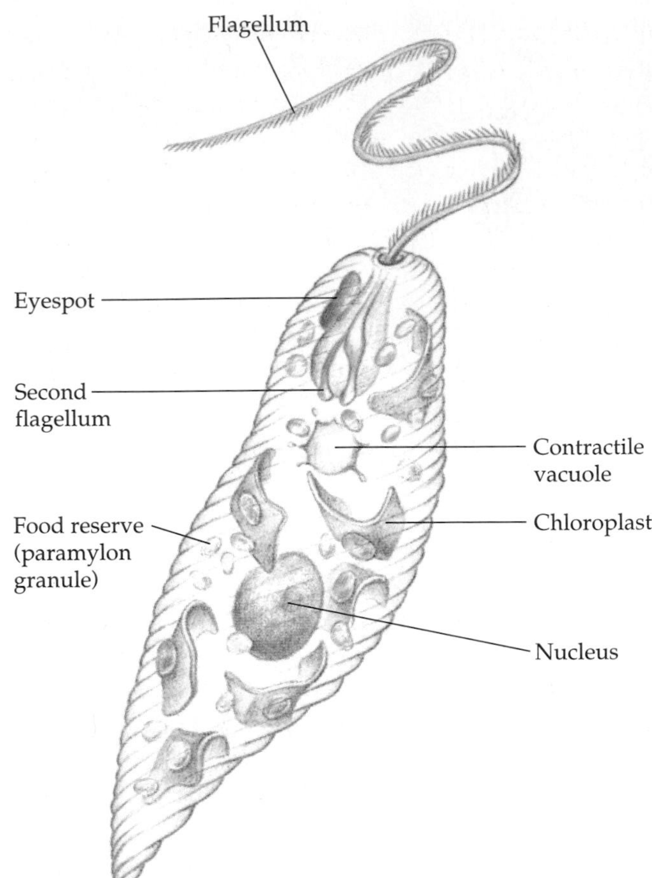

FIGURE 26.4
***Euglena:* an example of eukaryotic complexity.** This green microbe is one of the most common inhabitants of murky pond water. The drawing illustrates the greater structural complexity of eukaryotic microorganisms compared to prokaryotes. *Euglena* uses its long, "tinseled" flagellum to pull itself through the water. The eyespot at the base of the flagellum functions as a pigment shield. Depending on the position of the organism, the eyespot allows light from only a certain direction to strike a light detector, the swelling at the base of the long flagellum. The result is movement of the organism toward light of appropriate intensity, an important adaptation that enhances photosynthesis. *Euglena* lacks a cell wall but has strong, flexible plates made of protein beneath its plasma membrane. The contractile vacuole functions as a "bilge pump," expelling excess water that enters the cell by osmosis from the hypotonic environment. Surplus food made in the cell's chloroplasts is stored in granules as a polysaccharide called paramylon. If placed in the dark, *Euglena* can live as a heterotroph by absorbing organic nutrients from the environment. Some related species lack chloroplasts and ingest food by phagocytosis.

differ in the timing of meiosis and syngamy (see FIGURE 12.4). All three types are represented among protists, along with some variations that do not quite fit any of the three basic life cycle patterns. At some point in the life history of many protists, resistant cells called **cysts** that can survive harsh conditions are formed.

Because most protists are unicellular, they are justifiably considered to be the simplest eukaryotic organisms. But at the *cellular* level, many protists are exceedingly complex—the most elaborate of all cells. We should expect this of organisms that must carry out within the boundaries of single cells all the basic functions performed by the collective of specialized cells that makes up the bodies of plants and animals. Each unicellular protist is not at all analogous to a single cell from a human or other multicellular organism, but is itself an organism as complete as any whole plant or animal.

The more complex a structure, the more structural variation possible. The origin of eukaryotic cells was a major breakthrough in the history of biological diversity. Some of the eukaryotic "experiments" led to plants, animals, and fungi through divergent lineages of protistan ancestors. We see the products of other evolutionary experiments in the dazzling diversity of modern protists.

Protistan taxonomy is in a state of flux

Classification schemes and the phylogeny they are meant to reflect are always tentative—hypotheses about evolutionary history based on the available evidence. When Whittaker popularized the five-kingdom system of classification in 1969, he assigned unicellular eukaryotes to their own kingdom, the kingdom Protista. The trend during the 1970s and 1980s was to expand the boundaries of the kingdom Protista to include some phyla of multicellular organisms classified in earlier versions of the five-kingdom system as either plants or fungi. These taxonomic transfers were based mainly on comparisons of cell ultrastructure and details of life cycles. For example, the available evidence suggested that seaweeds are more closely related to certain unicellular algae (photosynthetic protists) than to plants. Multicellularity apparently evolved several times during protistan history, giving rise not only to plants, fungi, and animals, but also to multicellular organisms such as seaweeds that lack the distinctive traits defining those kingdoms. In its expanded form, the kingdom Protista also encompassed phyla of funguslike organisms, such as the forms known as slime molds and water molds, which may have their closest relatives among the protozoa. (Slime molds are only funguslike in the sense that a whale is fishlike; the resemblance is due to convergent evolution of morphological adaptations, not to common ancestry.) Because the term *protist* had come to connote unicellular forms, some advocates of the expanded version of the kingdom recommended changing the name to Protoctista. Whatever the name, the tendency was to treat this kingdom as the taxonomic home for all eukaryotes that did not fit comfortably into the definitions of plants, fungi, or animals.

During the past decade, molecular systematics, especially comparisons of ribosomal RNA, has been reshaping protistan taxonomy. This research is stimulating three main trends in the systematics and taxonomy of eukaryotic life:

1. Reassessment of the number and membership of protistan phyla (for example, all unicellular organisms with pseudopodia were once grouped in a phylum called Sarcodina, but most current classifications split these protists into several phyla).
2. Arrangement of the phyla into a cladogram based largely on what molecular methods and comparisons of cell structure reveal about evolutionary relationships of protists.
3. Reevaluation of the five-kingdom system and debate about whether additional kingdoms should be added to accommodate some forms formerly classified as protists (the proposal for a kingdom Archezoa is an example).

This last trend is perhaps the most crucial, for it challenges our view of biological diversity at the highest taxonomic levels. Does a kingdom Protista still make sense? It is a question we will defer until the end of the chapter, after we have studied some of the groups of organisms traditionally called protists. You have already met representatives of two of these phyla, *Euglena* (phylum Euglenophyta) and *Giardia* (phylum Diplomonada). In the next three sections, we will survey about a dozen more phyla. This is only a sample of the 30 to 40 phyla now recognized by most taxonomists. In order to emphasize the diverse ecological roles of protists, this survey groups the phyla into the nutritional categories of protozoa, funguslike protists, and algae.

Diverse modes of locomotion and feeding evolved among protozoa

Protozoa means "first animals," a misnomer in the context of the five-kingdom system or any other current classification scheme. The term persists to refer informally to protists that live primarily by ingesting food, an animal-like mode of nutrition (TABLE 26.1). These heterotrophs actively seek and consume bacteria, other protists, and **detritus** (dead organic matter). There are also symbiotic protozoa, including some parasites that cause human

TABLE 26.1

A Partial List of Protozoa (Animal-like Protists; Nutrition Mainly by Ingestion)

PHYLUM	BRIEF DESCRIPTION
Rhizopoda	Naked and shelled amoebas, with broad pseudopodia for motility and feeding
Antinopoda	Occupy planktonic habitats; usually spherically symmetrical; feed with axopodia, slender, radiating pseudopodia supported by internal microtubules; radiozoans possess internal siliceous skeletons, while heliozoans do not
Foraminifera	"Forams": feed and move with slender, interconnected pseudopodia exuding from spirally arranged calcareous compartments
Apicomplexa	Formerly called sporozoans; parasitic with complex life cycles in animal hosts
Zoomastigophora	Zooflagellates: use flagella for motility and feeding; mostly unicellular, but some colonial
Ciliophora	Ciliates: cilia used for motility and feeding; mostly unicellular, with a few sessile, colonial species

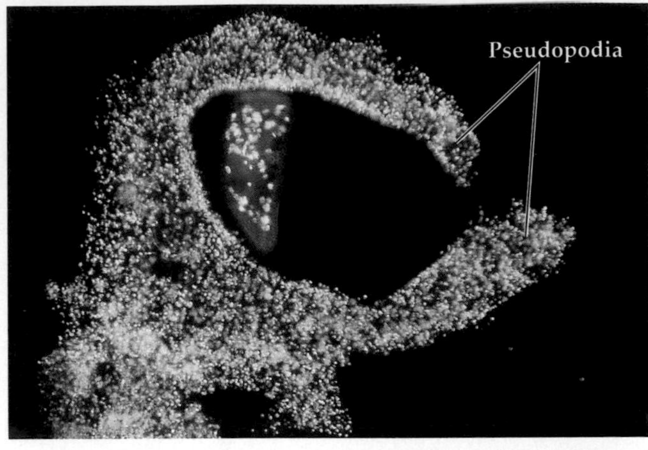

FIGURE 26.5
Rhizopoda. Amoebas use pseudopodia to move and feed. This light micrograph shows the pseudopodia of an amoeba engulfing a ciliated protozoan. The food, engulfed by the process of phagocytosis (see Chapter 8), will be packaged in a food vacuole where digestion will occur.

diseases. The protozoa are subdivided into phyla partly on the basis of how they feed and move. In this section, we examine six of the protozoan phyla. As systematists continue to reevaluate protistan taxonomy, they are splitting some of these phyla, combining others, and proposing different names to reflect these changes. What is important to take from this survey is not so much the formal taxonomy of protozoa as the diversity of the locomotory and feeding mechanisms.

Rhizopoda (Amoebas)

Members of the phylum Rhizopoda, the **amoebas** and their relatives, are all unicellular. Some of them have shells. No stages in their life histories are flagellated. Instead, amoebas use cellular extensions called **pseudopodia** to move and to feed (FIGURE 26.5). You have probably observed this mode of motility in *Amoeba proteus* in the laboratory. Pseudopodia may bulge from virtually anywhere on the cell surface. When an amoeba moves, it extends a pseudopodium and anchors its tip, and then more cytoplasm streams into the pseudopodium (*Rhizopoda* means "rootlike feet"). The cytoskeleton, consisting of microtubules and microfilaments, functions in amoeboid movement (see FIGURE 7.27). Pseudopodial activity may appear chaotic, but in fact amoebas show taxis as they creep slowly toward a food source.

Meiosis and sex are not known to occur in this phylum. The organisms reproduce asexually by various mechanisms of cell division. Spindle fibers form, but the typical stages of mitosis are not apparent in most amoebas. In many genera, for instance, the nuclear envelope persists during cell division.

Amoebas inhabit both freshwater and marine environments and are also abundant in soils. The majority of amoebas are free-living, but some are important parasites, including *Entamoeba histolytica,* which causes amoebic dysentery in humans. These organisms spread via contaminated drinking water, food, or eating utensils.

Actinopoda (Heliozoans and Radiozoans)

Actinopoda means "ray feet," a reference to the slender pseudopodia called axopodia that radiate from the beautiful protists that compose the phylum (FIGURE 26.6). Each axopodium is reinforced by a bundle of microtubules, covered by a thin layer of cytoplasm. The projections place an extensive area of cellular surface in contact with the surrounding water, help the organisms float, and function in feeding. Smaller protists and other microorganisms stick to the axopodia and are phagocytized by the thin layer of cytoplasm. Cytoplasmic streaming then carries the engulfed prey down to the main part of the cell.

Actinopoda are components of plankton. Most **heliozoans** ("sun animals") live in fresh water, whereas **radiozoans** are primarily marine. Radiozoa have delicate shells, most commonly made of silica, the material of

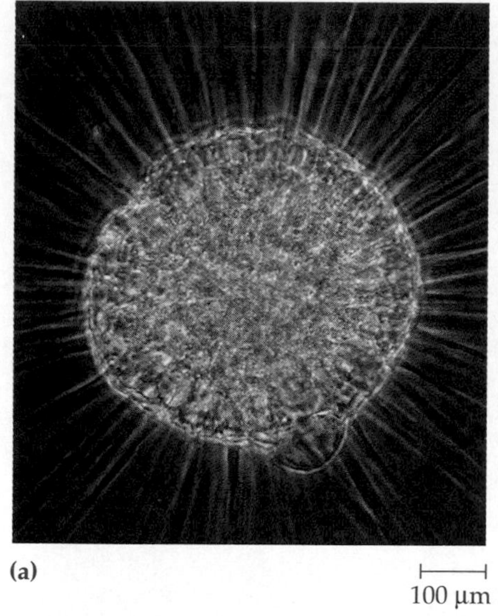

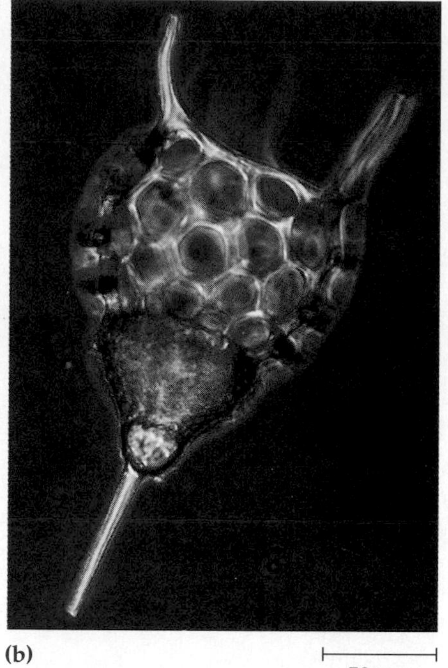

FIGURE 26.6
Actinopoda. (**a**) Heliozoans are mainly freshwater protists with stiff axopodia used for feeding. (**b**) Radiozoans are mostly marine forms with glassy shells, different in shape for each species. (Both LMs).

glass. After these organisms die, their shells settle to the seafloor, where they have accumulated as an ooze that is hundreds of meters thick in some locations.

Foraminifera (Forams)

Foraminifera, or **forams,** are exclusively marine. The majority of members of this phylum live in the sand or attach themselves to rocks and algae, but some families are also abundant in plankton. The phylum is named for the porous shells of its members (L. *foramen,* "little hole," and *ferre,* "to bear"). The shells are generally multichambered and consist of organic material hardened with calcium carbonate. Strands of cytoplasm extend through the pores, functioning in swimming, shell formation, and feeding. Many forams also derive nourishment from the photosynthesis of symbiotic algae that live beneath the shells.

Ninety percent of all identified species of foraminifera are fossils. The shells of fossilized forams are important components of marine sediments, including sedimentary rocks that are now land formations, such as the chalky white cliffs of Dover (FIGURE 26.7). Foram fossils are excellent markers for correlating the vintages of sedimentary rocks in different parts of the world.

FIGURE 26.7
Foraminifera. (**a**) The calcium carbonate shells of these protists have left an excellent fossil record. This living foram has a snail-like shell. The largest forams grow to diameters of several centimeters (LM). (**b**) The white cliffs of Dover, England, are sedimentary deposits that owe their rich calcium carbonate content to the fossil shells of forams and other protists, especially shelled unicellular algae called coccolithophorids.

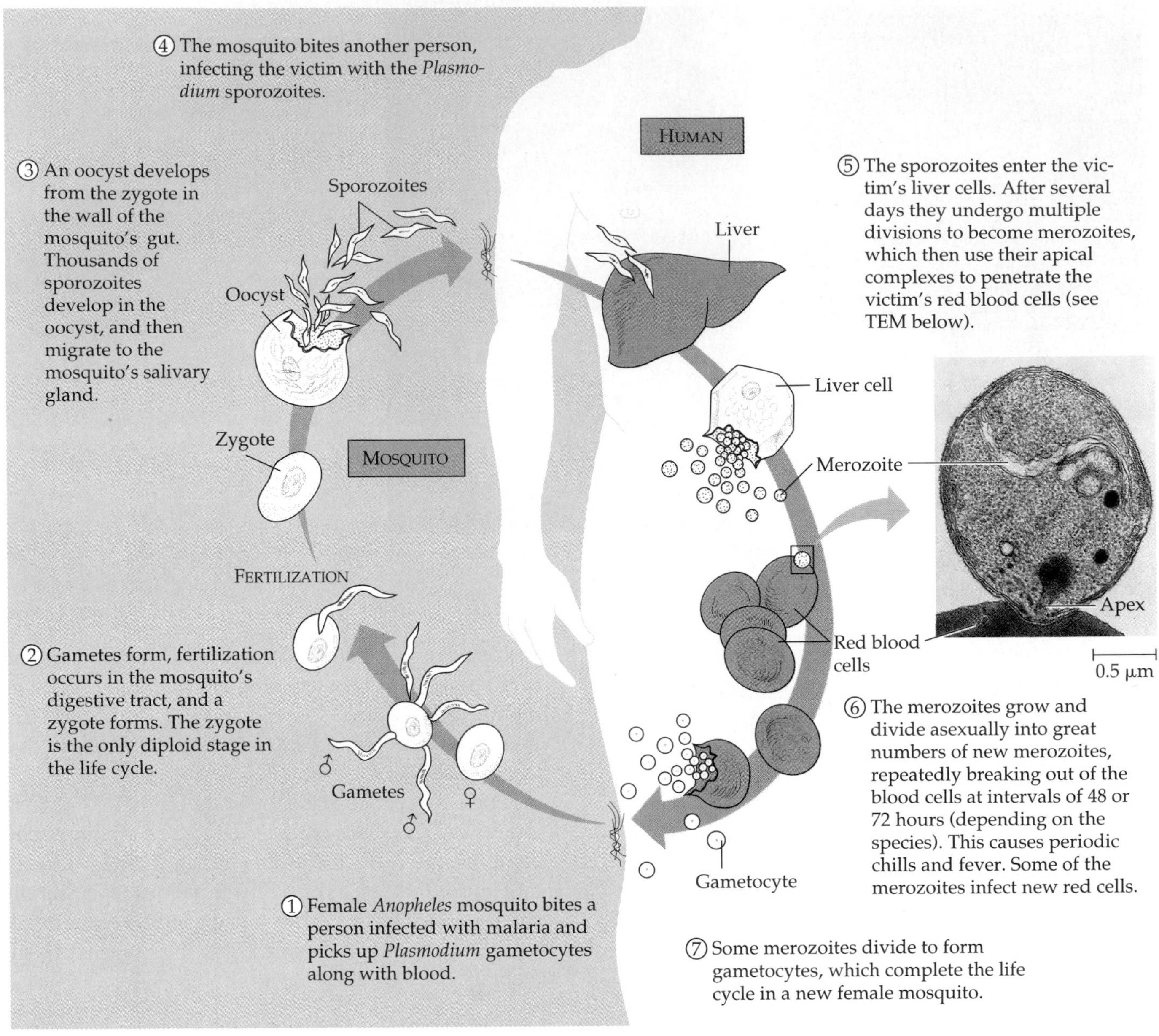

FIGURE 26.8
The life history of *Plasmodium*, the apicomplexan that causes malaria.

Apicomplexa (Sporozoans)

All members of the phylum Apicomplexa, which were formerly called sporozoans, are parasites of animals. Some cause serious human diseases. The parasites disseminate as tiny infectious cells called **sporozoites.** As seen with the electron microscope, one end (the apex) of the sporozoite cell contains a complex of organelles specialized for penetrating host cells and tissues, thus the phylum name Apicomplexa. Most apicomplexans have intricate life cycles with both sexual and asexual stages, and these cycles often require two or more different host species for completion. An example is *Plasmodium,* the parasite that causes malaria (FIGURE 26.8). The incidence of malaria was greatly diminished in the 1960s by the use of insecticides that reduced populations of *Anopheles* mosquitoes, which spread the disease, and by drugs that killed the parasites in humans. However, the multiplication of resistant varieties of both the mosquitoes and *Plasmodium* species have caused a resurgence of the disease. Each year, about 300 million people are infected in the tropics, and up to 2 million die from the disease.

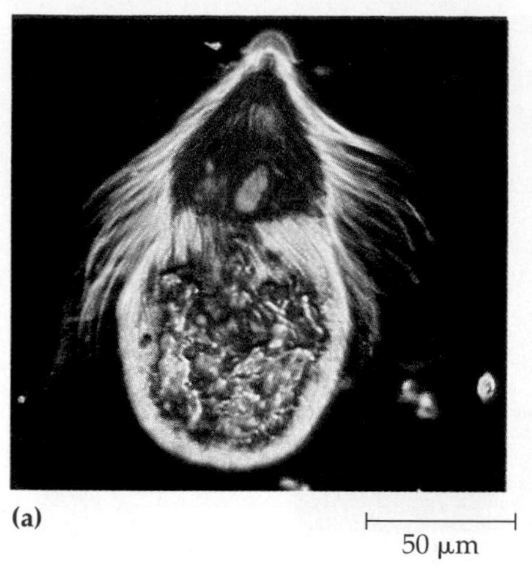
(a) 50 μm

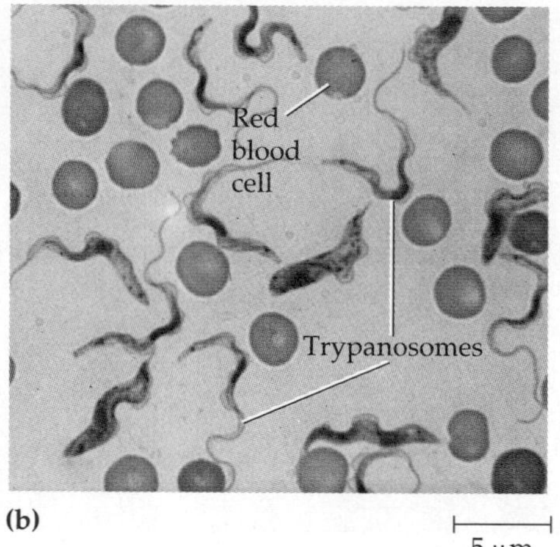

(b) 5 μm

FIGURE 26.9
Zoomastigophora (zooflagellates).
(**a**) *Trichonympha* is one of several symbiotic flagellates inhabiting the gut of termites. The wider end of the cell contains wood particles that are being digested.
(**b**) *Trypanosoma,* seen here in human blood, is the flagellate that causes African sleeping sickness. The molecular composition of these pathogens' coats changes frequently, preventing immunity from developing in hosts. (Both LMs.)

Considerable research on possible malarial vaccines has been carried out, with little success. *Plasmodium* is an extremely evasive parasite. It spends most of its time inside human liver and blood cells, thus hiding from the immune system. The problem is compounded by changes in the surface proteins of *Plasmodium*—the parasite changes the "face" that it shows to the infected person's immune system. Colombian scientists recently developed a promising vaccine based on a "cocktail" of synthetic proteins mimicking serveral *Plasmodium* membrane proteins.

Zoomastigophora (Zooflagellates)

Phylum Zoomastigophora is named for the whiplike flagella these protozoa use to propel themselves (Gr. *mastix,* "whip"). Also called **zooflagellates,** these heterotrophs absorb organic molecules from the surrounding medium or engulf prey by phagocytosis. Most live as solitary cells, but some form colonies of cells. There are both free-living and symbiotic zooflagellates. Living within the gut of a termite, for instance, are symbiotic flagellates that digest cellulose in the wood eaten by the host (FIGURE 26.9a). At the opposite end of the spectrum of symbiotic relationships are parasitic flagellates, some of which are pathogenic to humans. For example, species of *Trypanosoma* cause African sleeping sickness, which is spread by the bite of the tsetse fly (FIGURE 26.9b). Molecular systematists have recently confirmed that zooflagellates are closely related to the group of flagellated protists that includes the photosynthetic *Euglena* (see FIGURE 26.4).

Ciliophora (Ciliates)

The diverse protists of the phylum Ciliophora are characterized by their use of cilia to move and feed. Most members of Ciliophora, or **ciliates,** live as solitary cells in fresh water. In contrast to most flagella, cilia are relatively short and beat synchronously. They are associated with a submembrane system of microtubules that may coordinate the movement of the thousands of cilia.

The organization of the cytoskeleton and other components in the cell's cortex (outer layer of cytoplasm) also provides the information that orders cilia into a specific pattern as the cell grows. When researchers use microsurgery to remove a piece of cortex and then reinsert the piece with its polarity relative to the ends of the cell reversed, this also reverses the pattern of cilia that develops in that location, even after many generations of the cell division. In this developmental mechanism, called *cytotaxis,* it is cytoplasmic organization, not genes (at least not directly), that provides heritable information.

Some ciliates are completely covered by rows of cilia, whereas others have their cilia clustered into fewer rows or tufts. The specific arrangements adapt the ciliates for their diverse lifestyles. Some species, for instance, scurry about on leglike structures constructed from many cilia bonded together. Other forms, such as *Stentor,* have rows of tightly packed cilia that function collectively as locomotor membranelles. Ciliates are among the most complex of all cells (FIGURE 26.10).

A unique feature of ciliate genetics is the presence of two types of nuclei, a large macronucleus and usually several tiny micronuclei. The macronucleus has fifty or more copies of the genome. The genes are not distributed in typical chromosomes but are instead packaged into a much larger number of small units, each with hundreds of copies of just a few genes. The macronucleus controls the everyday functions of the cell by synthesizing RNA and is also necessary for asexual reproduction. Ciliates generally reproduce by binary fission, during which the macronucleus elongates and splits, rather than undergoing mitotic division. The micronuclei, of which species

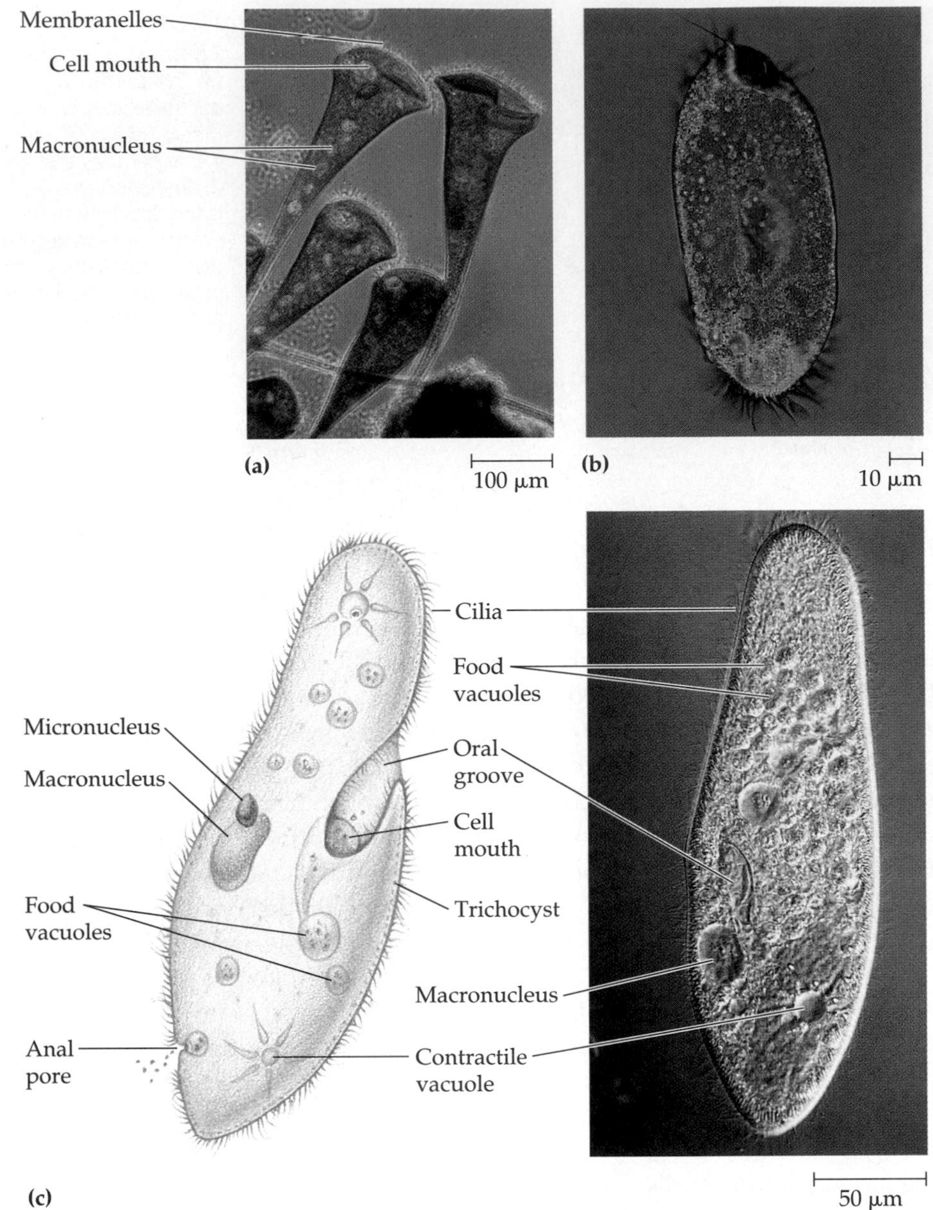

FIGURE 26.10

Ciliophora. (**a**) The beautiful freshwater ciliate *Stentor* moves by individual cilia along the cell sides and by rows of finlike membranelles that spiral around the broad, anterior end of the protozoan. *Stentor* often attaches its narrow (posterior) end to debris and, by beating the anterior membranelles, causes a whirlpool-like current that moves food into the cell mouth. The macronucleus can be seen as light beaded strands running the length of these cells. (**b**) *Stylonychia* belongs to a group of ciliates that often have no individual cilia at all. Cilia on the cell mouth side of the cell are united into leglike cirri. These ciliates scurry around on sediments, feeding among bits of organic debris. (**c**) *Paramecium,* an example of ciliate complexity, is covered by thousands of individual cilia. Associated with each cilium are trichocysts, bulblike organelles that discharge sticky, proteinaceous threads. Although these threads are released in the presence of predators, they have little defensive effect and are thought to function mainly in stabilizing the cell during feeding. Other genera have toxic trichocysts. *Paramecium* feeds mainly on bacteria. Rows of cilia along a funnel-shaped oral groove move food down into the cell mouth, where the food is engulfed by phagocytosis. The food vacuoles combine with lysosomes and, as the food is digested, the vacuoles follow a looping path, first to the anterior end and finally to the posterior end. The undigested contents are released when the vacuoles fuse with a specialized region of the plasma membrane that functions as an anal pore. *Paramecium,* like other freshwater protists, constantly takes in water by osmosis from the hypotonic environment. Bladderlike contractile vacuoles accumulate the excess water from radial canals and periodically expel it through the plasma membrane by contractions of the surrounding cytoplasm (see Chapter 8). (All LMs.)

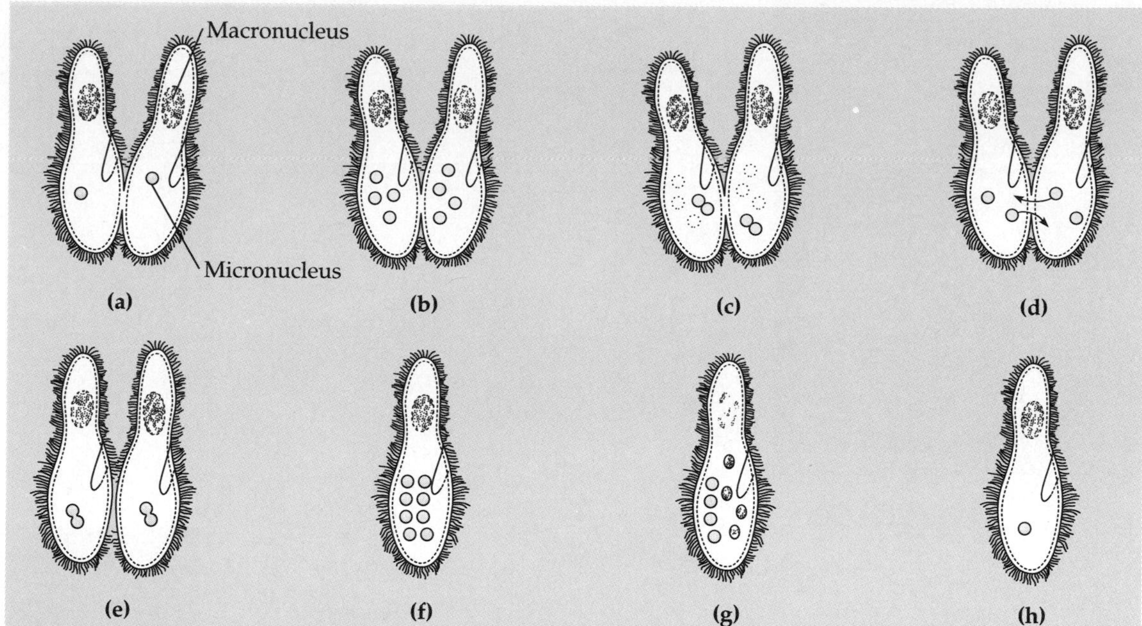

FIGURE 26.11

Conjugation and genetic recombination in *Paramecium caudatum.*
(**a**) Two individuals of compatible mating strains align side by side and partially fuse. In each cell, all but one diploid micronucleus disintegrate. (**b**) The micronucleus that remains undergoes meiosis to produce four haploid micronuclei. (**c**) One of these divides by mitosis, and the other three disintegrate. (**d**) Mates then swap one micronucleus. (**e**) Syngamy occurs when the micronucleus a cell acquired from its partner fuses with its remaining micronucleus, forming a fresh diploid nucleus with a mixture of chromosomes derived from the two individuals. The partners separate. (**f**) From now on, only one partner is shown. In each individual, the newly constituted micronucleus divides repeatedly by mitosis, resulting in eight identical micronuclei. (**g**) Subsequently, in each cell the original macronucleus disintegrates. Four micronuclei develop into new macronuclei by repeated replication of the DNA without nuclear division. The other four remain as micronuclei. (**h**) After two cell divisions (without nuclear divisions), the new macronuclei and micronuclei are parceled out into four new individuals for each original conjugating cell. (Only one individual is shown here.) Note that all of these eight individuals ultimately resulting from one conjugation are identical genetically. However, they represent a genetic makeup different from either of the two original cells that conjugated.

of *Paramecium* have from one to as many as 80, do not function in growth, maintenance, and asexual reproduction of the cell but are required for sexual processes that generate genetic variation. The sexual shuffling of genes occurs during the process known as **conjugation** (FIGURE 26.11). Notice in the diagram that in ciliates, sexual mechanisms of meiosis and syngamy are processes separate from reproduction.

Funguslike protists have morphological adaptations and life cycles that enhance their ecological role as decomposers

Slime molds and water molds resemble fungi in appearance and lifestyle, but the similarities are the result of convergence. In their cellular organization, reproduction, and life cycles, slime molds depart from the true fungi and probably have their closest relatives among the amoeboid protists. Although water molds lack chloroplasts, molecular comparisons suggest they are related to certain algae. The superficial similarity of the funguslike protists to true fungi is partly due to convergent evolution of filamentous body structure, a morphological adaptation that increases exposure to the environment and enhances the ecological role of these organisms as decomposers. The complex life cycles of funguslike protists are adaptations that contribute to survival in changing habitats and facilitate dispersal to new food sources.

Myxomycota (Plasmodial Slime Molds)

The phylum Myxomycota consists of **plasmodial slime molds,** which are more attractive than their name implies. Many species are brightly pigmented, usually yellow or orange, but slime molds are not photosynthetic; all are heterotrophs. The feeding stage of the life cycle is an

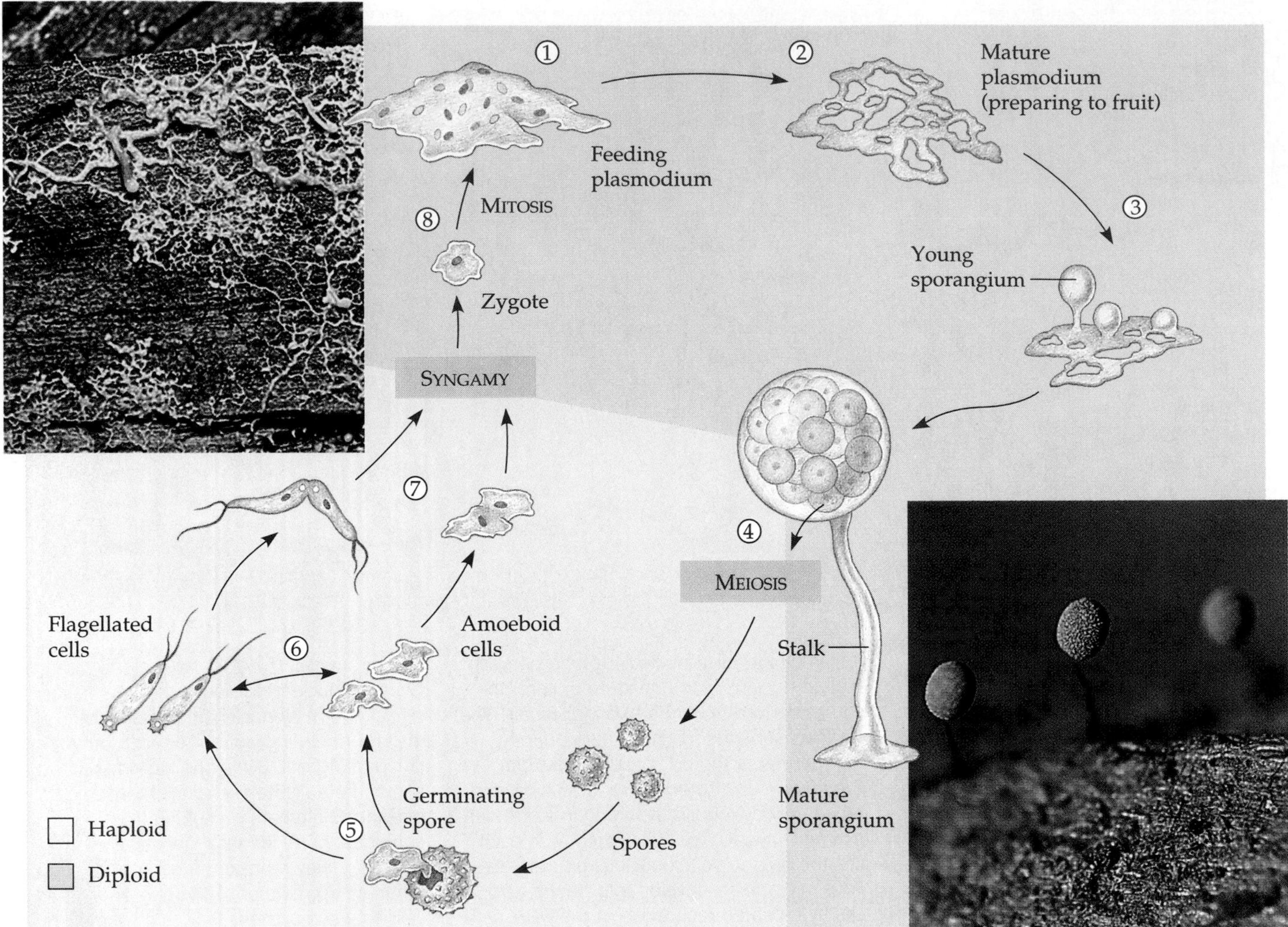

FIGURE 26.12

The life cycle of a plasmodial slime mold. ① The feeding stage is a multinucleate (coenocytic) plasmodium that lives on organic refuse. ② The plasmodium often takes a weblike form, an adaptation that increases the surface area that contacts food, water, and oxygen (left inset). ③ The plasmodium rounds into a mound and erects stalked fruiting bodies called sporangia when conditions become harsh (right inset, LM). ④ Within the bulbous tips of the sporangia, meiosis produces haploid spores. ⑤ The resistant spores germinate to become active haploid cells when conditions are again favorable. ⑥ These cells are either amoeboid or flagellated, the two forms readily reverting from one to the other. ⑦ These cells unite in like pairs (flagellated with flagellated and ameboid with ameboid) to form diploid zygotes. ⑧ Repeated division of the nucleus of the zygote by mitosis, without cytoplasmic division, forms a feeding plasmodium and completes the life cycle.

Notice that the style and color coding of this life cycle conform to the conventions introduced in Chapter 12. This color coding for haploid and diploid stages is used throughout the book to help you compare the life cycles of diverse organisms.

amoeboid mass called a **plasmodium,** which may grow to a diameter of several centimeters (FIGURE 26.12). Large as it is, the plasmodium is not multicellular; it is known as a coenocytic mass, a multinucleate continuum of cytoplasm undivided by membranes or walls. In most species, the nuclei of the plasmodium are diploid and divisions are synchronous, with each of thousands of nuclei going through each phase of mitosis at the same time. Because of this characteristic, plasmodial slime molds have been used to study the molecular details of mitosis. Within the fine channels of the plasmodium, cytoplasm streams first one way, then the other, in pulsing flows that are beautiful to watch through a microscope. The cytoplasmic streaming apparently helps distribute nutrients and oxygen. The plasmodium engulfs food particles by phagocytosis as it grows by extending pseudopodia through moist soil, leaf mulch, or rotting logs. If the habitat of a slime mold begins to dry up or there is no food left, the plasmodium ceases growth and differentiates into a stage of the life cycle that functions in sexual reproduction.

Acrasiomycota (Cellular Slime Molds)

Members of Acrasiomycota, the **cellular slime molds,** pose a semantic question about what it means to be an

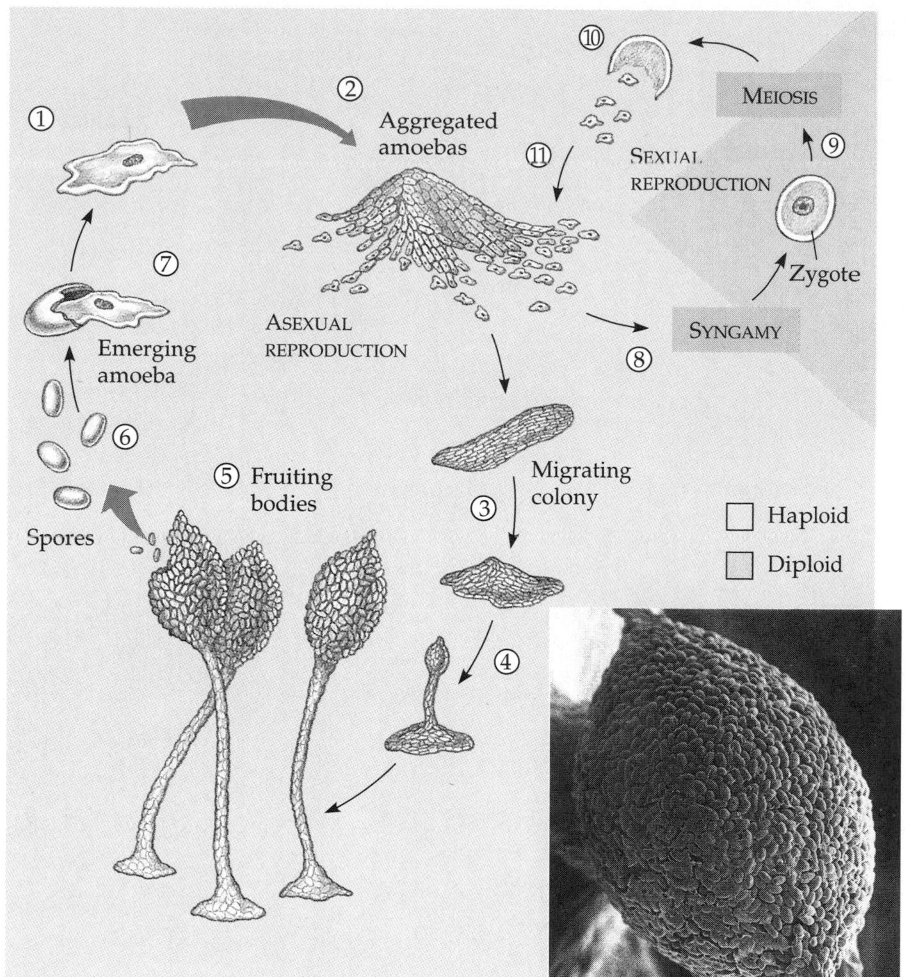

FIGURE 26.13
The life cycle of a cellular slime mold (*Dictyostelium*). ① The feeding stage of the life cycle consists of solitary cells that engulf bacteria while creeping by amoeboid movement through damp compost. ② When food is depleted, the amoeboid cells migrate toward an aggregation center, where hundreds of the cells congregate in response to a chemical attractant they secrete. ③ The sluglike colony of amoeboid cells may migrate as a unit for a while before ④ settling down and developing stalked fruiting bodies that function in asexual reproduction. ⑤ As each fruiting body forms, some cells dry up and form a supportive stalk, while others continue to crawl up over the dried cells, amass, then turn into spores. Thus, a cluster of somewhat resistant spores forms at the tip of each fruiting body (inset, SEM). ⑥ After the spores are released and exposed to a favorable environment, ⑦ amoeboid cells emerge from their protective coats and begin feeding, completing the asexual portion of the life cycle. ⑧ In the sexual phase of *Dictyostelium*, a pair of haploid amoebas fuse to form a zygote, the only diploid stage in the life cycle. ⑨ A zygote becomes a giant cell by consuming surrounding haploid amoebas. The giant cell then becomes surrounded by a resistant wall. ⑩ The encysted giant cell undergoes meiosis, followed by several mitotic divisions. ⑪ New haploid amoebas are released when the cyst ruptures.

individual organism. The feeding stage of the life cycle consists of solitary cells that function individually. When there is no more food, the cells form an aggregate that functions as a unit (FIGURE 26.13). Although the mass of cells resembles a plasmodial slime mold, the important distinction is that the cells of a cellular slime mold maintain their identity and remain separated by their membranes.

In addition to not being coenocytes, cellular slime molds differ from plasmodial slime molds in other ways. Cellular slime molds are haploid organisms, whereas the diploid condition predominates in the life cycles of most plasmodial slime molds (compare FIGURES 26.12 and 26.13). Cellular slime molds have fruiting bodies that function in asexual reproduction. Also, most cellular slime molds have no flagellated stages.

Oomycota (Water Molds)

This phylum includes water molds, white rusts, and downy mildews. These organisms consist of coenocytic hyphae (fine, branching filaments), an adaptation analogous to the morphology of true fungi. However, water molds and their relatives have cell walls most commonly made of cellulose, while the walls of true fungi are made of another polysaccharide, chitin. The diploid condition prevails in the life cycles of most members of Oomycota but is reduced in true fungi. Biflagellated cells occur in the life cycles of oomycotes, while almost all true fungi lack flagella.

Oomycota means "egg fungi," a reference to the mode of sexual reproduction in water molds. A relatively large egg cell is fertilized by a smaller "sperm nucleus," forming a resistant zygote (FIGURE 26.14).

Most water molds are decomposers that grow as cottony masses on dead algae and animals, mainly in fresh water. They are important decomposers in aquatic ecosystems. There are also parasitic water molds, such as those that grow on the skin and gills of fish in ponds or aquariums, but they usually attack only injured tissue. White rusts and downy mildews are close relatives of water molds but generally live on land as parasites of plants. They are dispersed primarily by windblown

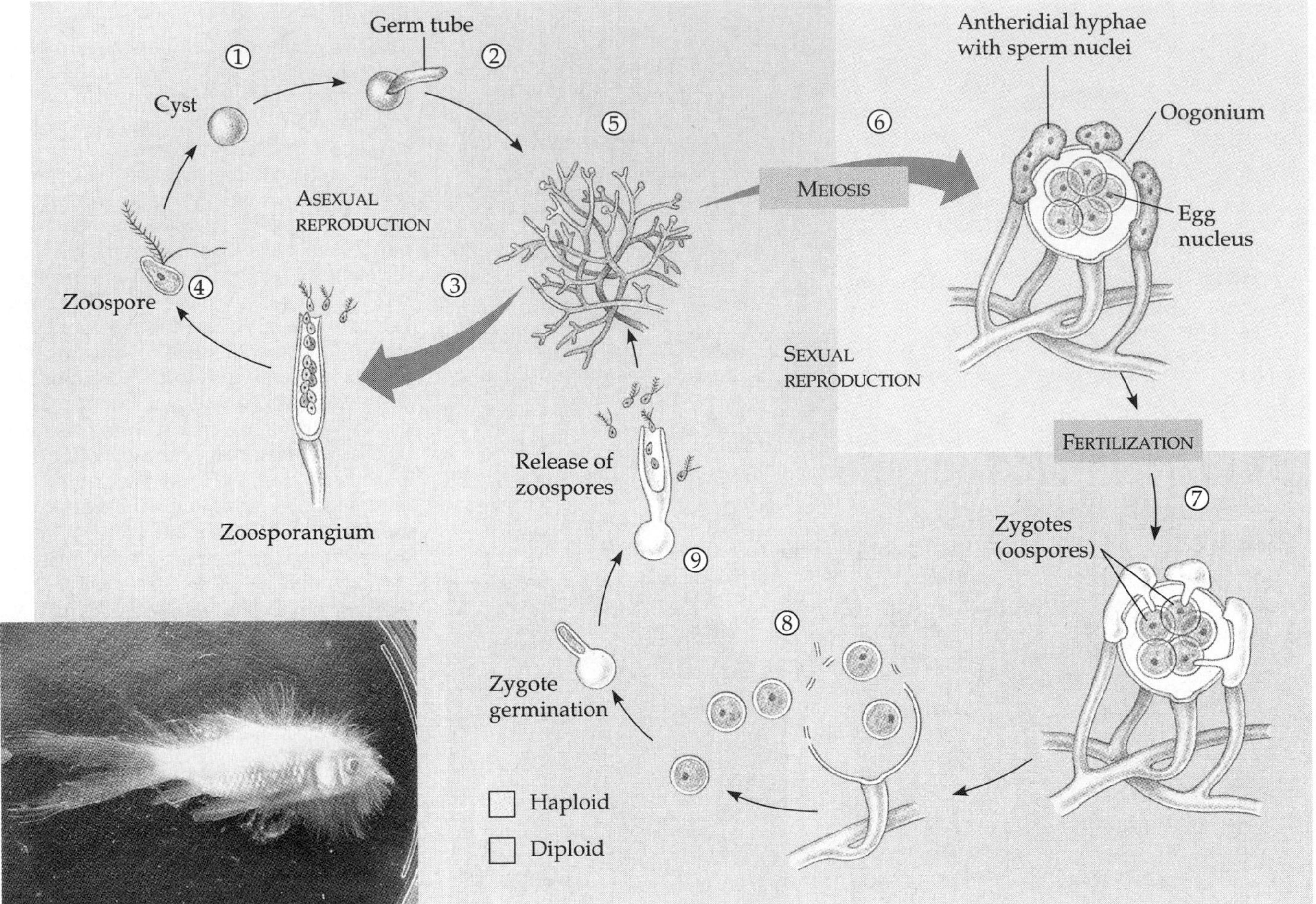

FIGURE 26.14

The life cycle of a water mold. Water molds commonly help decompose dead insects, fish, and other animals. ① Encysted zoospores land on a substrate and ② germinate, growing into a tufted body of coenocytic hyphae. ③ Within several days, the ends of the hyphae form tubular zoosporangia, ④ each of which produces about thirty biflagellated zoospores asexually (the hyphal mass on the goldfish in the inset is in the early zoosporangial stage). ⑤ Several days later, the organism begins to form sexual structures. ⑥ Meiosis produces eggs within structures called oogonia. On separate branches of the same or different individuals, meiosis produces several haploid "sperm nuclei" contained within compartments called antheridial hyphae. ⑦ These hyphae grow like hooks around the oogonium and deposit their nuclei through fertilization tubes that lead to the eggs. The resulting zygotes (oospores) may develop resistant walls but are also protected within the walls of the old oogonia. ⑧ After a period of dormancy, during which the oogonium wall usually disintegrates, ⑨ the oospores germinate and form short hyphae tipped by zoosporangia, and the cycle is completed.

spores, but they also form flagellated zoospores at some point during their life cycles. Some of the most devastating plant pathogens are members of the Oomycota, including the downy mildew that threatened the French vineyards in the 1870s and the species that causes late potato blight, which contributed to the Irish famine in the nineteenth century.

Eukaryotic algae are key producers in most aquatic ecosystems

The phyla grouped here as eukaryotic algae consist mainly of photosynthetic organisms, although some phyla include heterotrophic or mixotrophic members. The term *alga* refers to relatively simple aquatic organisms that are photoautotrophs. Except for the prokaryotic cyanobacteria (formerly called blue-green algae), the organisms generically called algae are eukaryotes that most systematists classified in kingdom Protista in the five-kingdom system.

Algae are extremely important ecologically, accounting for about half the photosynthetic production of organic material on a global scale. As freshwater and marine phytoplankton and intertidal seaweeds, algae are the bases of aquatic food webs, supporting an enormous abundance and diversity of animals.

All algae have chlorophyll *a*, the same "primary" pigment found in plants. But algae differ considerably in

TABLE 26.2

Characteristics of Some Phyla of Eukaryotic Algae

PHYLUM	APPROX. NUMBER OF SPECIES	PREDOMINANT COLOR (PHOTOSYNTHETIC PIGMENTS)	CARBOHYDRATE FOOD RESERVE	NUMBER AND POSITION OF FLAGELLA	CELL WALL COMPONENTS	HABITAT
Euglenophyta (*Euglena* and its relatives)	800	Green (chlorophyll *a*, chlorophyll *b*, carotenoids, xanthophylls)	Paramylon (a β1–3 glucose polymer)	1 to 3, apical	No cell wall; submembrane protein	Mostly fresh water
Dinoflagellata (dinoflagellates)	1,100	Brown (chlorophyll *a*, chlorophyll *c*, carotenoids, xanthophylls)	Starch (an α1–4, branched glucose polymer)	1 lateral, 1 posterior	Submembrane cellulose	Marine and fresh water
Bacillariophyta (diatoms)	10,000	Olive brown (chlorophyll *a*, chlorophyll *c*, carotenoids, xanthophylls)	Leucosin (a β1–3 glucose polymer)	1 in sperm only	Hydrated silica in organic matrix	Fresh water and marine
Chrysophyta (golden algae)	850	Golden olive (chlorophyll *a*, often chlorophyll *c*, carotenoids, xanthophylls)	Laminarin (a β1–3 glucose polymer)	1 or 2, apical	Pectic compounds with siliceous material	Mostly fresh water
Phaeophyta (brown algae)	1,500	Olive brown (chlorophyll *a*, chlorophyll *c*, carotenoids, xanthophylls)	Laminarin (a β1–3 glucose polymer)	2 lateral in sperm only	Cellulose matrix with other polysaccharides	Almost all marine; flourish in cold ocean waters
Rhodophyta (red algae)	4,000	Red to black (chlorophyll *a*, carotenoids, phycobilins, chlorophyll *d* in some)	Floridean starch (a glycogenlike α1-4 branched glucose polymer)	None	Cellulose matrix with other polysaccharides	Mostly marine, but some fresh water; many species tropical
Chlorophyta (green algae)	7,000	Green (chlorophyll *a*, chlorophyll *b*, carotenoids)	Plant starch	2 or more, apical or subapical	Cellulose	Mostly fresh water, but some marine

their accessory pigments, pigments that trap wavelengths of light to which chlorophyll *a* is not as sensitive. These pigments include other forms of chlorophyll (greenish), carotenoids (yellow-orange), xanthophylls (brownish), and phycobilins (red and blue varieties). The mixture of pigments in the chloroplasts lends characteristic colors to related algae. Many of the scientific and common names of algal phyla are based on these colors (the Chlorophyta, or green algae, for example). Study of these pigment mixes has helped establish taxonomic affinities among the algae. Additional clues have come from chloroplast structure; the chemistry of cell walls (a few algal phyla lack walls); the number, type, and position of flagella; and the form of food stored by the cells (TABLE 26.2). The following survey of selected algal phyla will give you a sense of the diversity of forms and life cycles that has evolved.

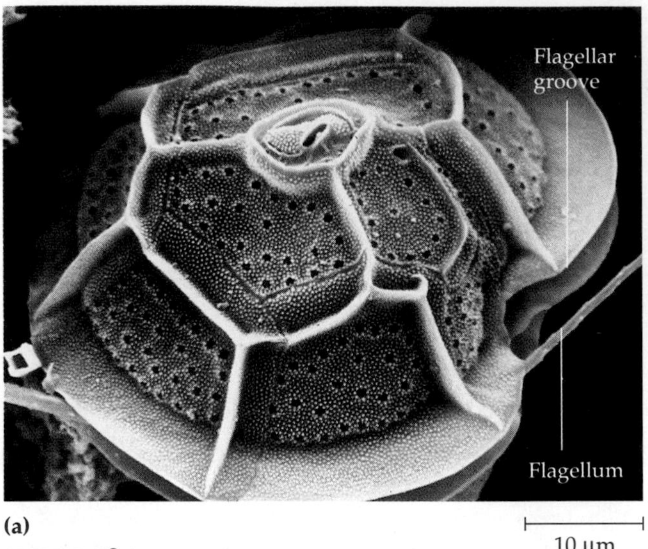

(a)

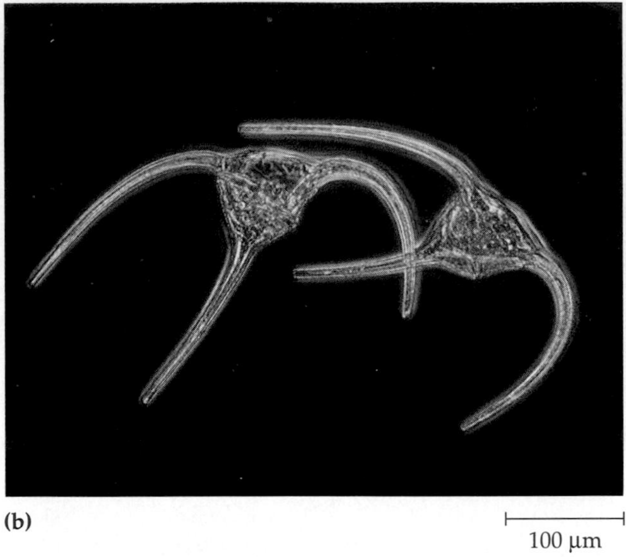

(b)

FIGURE 26.15
Dinoflagellates. These unicellular algae are characterized by a pair of flagella in perpendicular grooves. Beating of the flagella causes the cell to spin as it swims. Each species has a distinctively shaped internal wall. (**a**) *Gonyaulax tamarensis* (SEM). (**b**) *Ceratium* (LM).

Dinoflagellata (Dinoflagellates)

Dinoflagellates are abundant components of the vast aquatic pastures of phytoplankton, microscopic algae floating near the surface of the sea that provide the foundation of most marine food chains (FIGURE 26.15).

Dinoflagellate blooms, episodes of explosive population growth, cause the red tides that occur occasionally in warm coastal waters. These blooms are brownish-red because of predominant xanthophylls in the chloroplasts. When suspension-feeding shellfish such as oysters feed on these blooms, they concentrate the algae along with toxic compounds released by the dinoflagellate cells. These toxins are extremely dangerous to humans, who collect and eat the invertebrates. As a consequence, collecting shellfish during red tides is often regulated to prevent the widespread occurrence of "paralytic shellfish poisoning."

Of the several thousand known species, most dinoflagellates are unicellular, but there are some colonial forms. Each dinoflagellate species has a characteristic shape reinforced by internal plates of cellulose. The beating of two flagella in perpendicular grooves in this "armor" produces a spinning movement for which these organisms are named (Gr. *dinos,* "whirling"). The structure of the dinoflagellate nucleus and its division during asexual reproduction are unusual (see FIGURE 11.11).

Some dinoflagellates live as mutualistic symbionts of animals called cnidarians that build coral reefs; the photosynthetic output of these dinoflagellates is the main food source for reef communities. Other dinoflagellates lack chloroplasts and live as parasites within marine animals. There are even carnivorous species. The existence of both photosynthetic and heterotrophic forms closely related enough to be grouped in the same phylum reinforces the point made earlier that the terms *protozoa* and *algae,* although useful in an ecological context, have no basis in phylogeny. In fact, dinoflagellates are probably more closely related to the protozoa called zooflagellates than they are to any of the other phyla grouped here as algae.

Bacillariophyta (Diatoms)

The members of the phylum Bacillariophyta, or **diatoms,** are yellow or brown in color. They are closely related to two other algal phyla with brown plastids, the golden algae (Chrysophyta) and brown algae (Phaeophyta).

Diatom cells have unique glasslike walls consisting of hydrated silica embedded in an organic matrix (FIGURE 26.16). Each wall is in two parts that overlap like a shoe box and lid. Many diatoms are capable of a gliding movement caused by chemicals secreted out of slits in their cell walls.

Most of the year, diatoms reproduce asexually by mitotic cell divisions, with each daughter cell receiving half of the cell wall of its parent and regenerating a new second half. Cysts are formed by some species as resistant stages. Sexual stages are not common and involve the

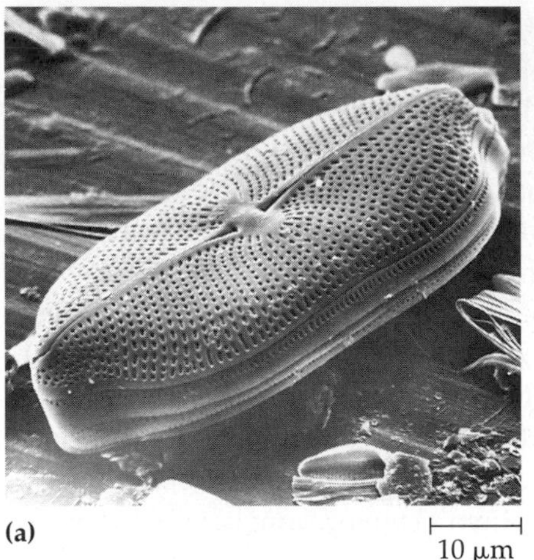

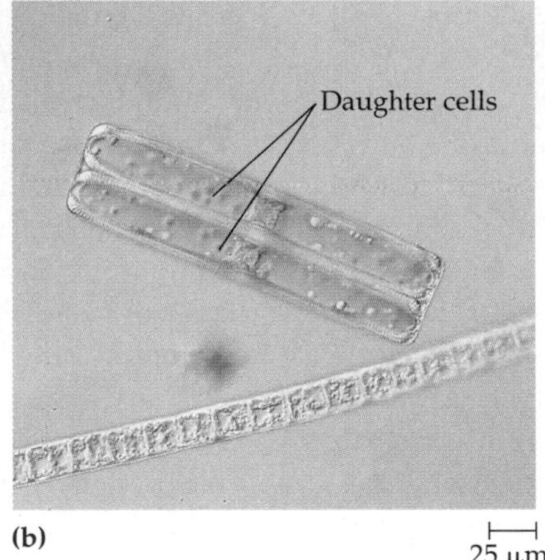

FIGURE 26.16

Diatoms. (**a**) The glasslike shells consist of two halves that fit like the bottom and lid of a shoe box. Tiny pores in the ornate shells allow for the exchange of gases and other substances between the cell and its environment. The shape of the shell and its pattern of pores are used for classifying diatoms. This species is *Navicula monilifera* (SEM). (**b**) In this side view of a species of *Pinnularia,* the cell has just divided by mitosis. Each daughter cell keeps half of the parent cell's wall and builds a new complementary half (LM).

formation of eggs and sperm (each with a single flagellum) by meiosis. The sperm are the only flagellated cells in this phylum.

Both freshwater and marine plankton are rich in diatoms; a bucket of water scooped from the surface of the sea may have millions of these microscopic algae. Planktonic species store food reserves in the form of oils, which also provide a buoyancy that counteracts the relatively heavy weight of their walls and keep the diatoms near the water surface and sunlight.

Massive accumulations of fossilized diatom walls are major constituents of the sediments known as diatomaceous earth, which is mined for its quality as a filtering medium and for many other uses.

Chrysophyta (Golden Algae)

Golden algae (Gr. *chrysos,* "golden") are named for their color, which results from yellow and brown carotenoid and xanthophyll accessory pigments. Their cells are typically biflagellated, with both flagella attached near one end of the cell. Golden algae live among freshwater plankton. Most species are colonial (FIGURE 26.17). In ponds and lakes that freeze in winter or dry up in summer, golden algae survive by forming resistant cysts, from which active cells emerge when conditions are favorable. Microfossils resembling the ruptured cysts of chrysophytes and other algae are among the acritarchs found in Precambrian rocks.

Phaeophyta (Brown Algae)

The largest and most complex algae are members of the phylum Phaeophyta (Gr. *phaios,* "dusky," "brown"), or **brown algae.** All are multicellular and most are marine, including the largest seaweeds. Brown algae are especially common along temperate coasts, where the water is cool. They owe their characteristic brown or olive color to accessory pigments in the chloroplasts. The chloroplast structure and pigment composition of brown algae are homologous to the photosynthetic equipment of golden algae and diatoms. The following subsections will consider the brown algae as an example of the complex morphology and life cycles that evolved among the giant algae called seaweeds.

Evolutionary Adaptations of Seaweeds. The term *seaweed* applies generally to large marine algae. In addition to the seaweeds classified as Phaeophyta (brown algae), two other algal phyla—Rhodophyta (red algae) and Chlorophyta (green algae)—also include seaweeds.

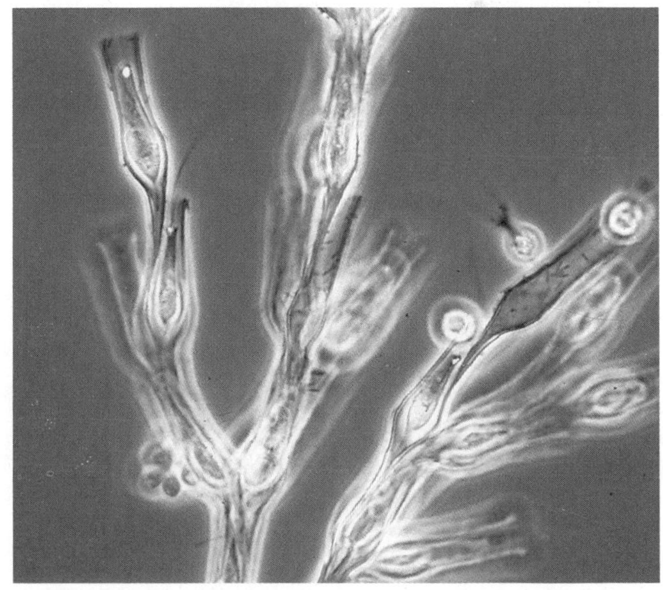

FIGURE 26.17

A golden alga. *Dinobryon,* a freshwater organism, is one of many colonial forms of this algal phylum (LM).

This discussion of seaweed adaptations will include all three of these phyla, though the emphasis will be on brown algae.

Seaweeds, along with many animals and other heterotrophs these algae support, inhabit the intertidal and subtidal zones of coastal waters. The intertidal zone presents unique challenges to life. At times it is violently active, churned by waves and wind. Two times each day, at low tide, intertidal seaweeds are exposed to the drying atmosphere and unfiltered rays of the sun. Twice each day, at high tide, the same seaweeds are covered by up to 5 m of water. Seaweeds have unique anatomical and biochemical adaptations that enable them to survive and thrive in this rough-and-tumble environment.

Seaweeds have the most complex multicellular anatomy of all algae. Some even have differentiated tissues and organs that resemble those we find in plants. The similarities, however, are analogous, not homologous. Seaweeds are more closely related to unicellular members of their phyla than to any true plant, and their anatomical complexity evolved independently. The term **thallus** (plural, **thalli;** Gr. *thallos,* "sprout") is used for a seaweed body that is plantlike but lacks true roots, stems, and leaves. A typical seaweed thallus consists of a rootlike **holdfast,** which anchors the alga, and a stemlike **stipe,** which supports leaflike **blades** (FIGURE 26.18). The blades provide most of the surface for photosynthesis. Some brown algae are equipped with floats, which keep the blades near the water surface. Beyond the intertidal zone in deeper waters live the giant seaweeds known as kelps (FIGURE 26.19). The stipes of these brown algae may be as long as 100 m.

In addition to these structural adaptations of thalli, some seaweeds are also endowed with biochemical adaptations to intertidal and subtidal conditions. For example, their cell walls are composed of cellulose and gel-forming polysaccharides, accounting for the slimy and rubbery feel of many seaweeds. These substances help cushion the thalli against the agitation of the waves and also help prevent the thalli from drying during low tides. Some red algae in the lower intertidal and subtidal zones incorporate large amounts of calcium carbonate in the cell walls, making them virtually unpalatable to marine invertebrate grazers.

Human Uses of Seaweeds. Coastal people, particularly in Asia, harvest seaweeds for food. For example, in Japan and Korea, the brown alga *Laminaria* is used in soups (Japanese "kombu") and the red alga *Porphyra* is used to wrap sushi (Japanese "nori"). Marine algae are rich in iodine and other essential minerals, but much of their organic material consists of unusual polysaccharides that humans cannot digest, which prevents seaweeds from becoming staple foods. They are used mostly for their rich tastes and unusual textures. The gel-forming substances in their cell walls (algin in brown

Blade
Stipe
Holdfast

FIGURE 26.18
Seaweeds: adapted to life at the ocean's margins. The sea palm, *Postelsia,* lives on rocks in the crashing surf along the northwest coast of the United States and Canada. Its thallus is well adapted to maintaining a firm foothold in this extreme environment. *Postelsia* is a brown alga (phylum Phaeophyta).

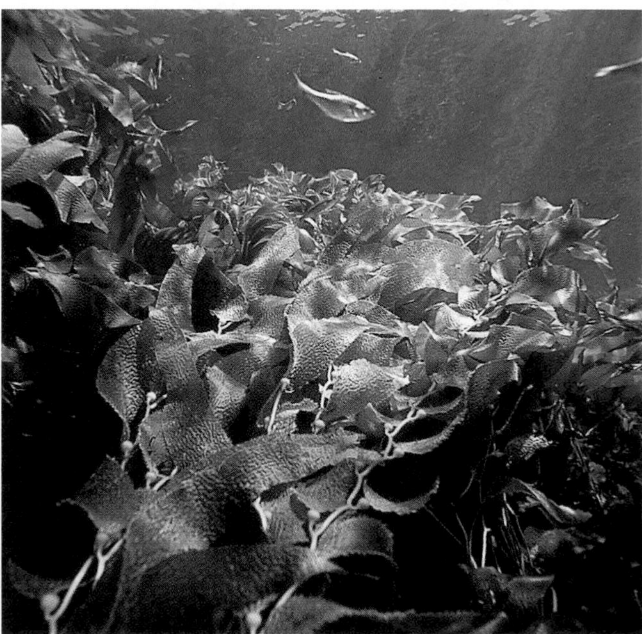

FIGURE 26.19
A kelp forest. The great kelp beds of temperate coastal waters provide habitat and food for a variety of organisms, including many fish caught by humans. The kelps, brown algae (Phaeophyta), are prodigiously productive. This alga, *Macrocystis,* common along the U.S. Pacific coast, grows to a length of more than 60 m in a single season, the fastest linear growth of any organism. Kelp is a renewable resource reaped by special boats that cut and collect the tops of the algae.

algae, agar and carageenan in red algae) are extracted in commercial operations. These substances are widely used in the manufacture of thickeners for such processed foods as puddings and salad dressing, and as lubricants in oil drilling. Agar is also used as the gel-forming base for microbiological culture media.

Alternation of Generations in the Life Cycles of Certain Algae. A variety of life cycles has evolved in the phylum Phaeophyta. In the most complex cycles, there is an **alternation of generations,** the alternation of multicellular haploid forms and multicellular diploid forms in a life history. (Notice that haploid and diploid conditions alternate in *all* sexual life cycles—human gametes, for example, are a haploid stage—but the term "alternation of generations" is reserved for life cycles that include haploid and diploid stages that are both multicellular organisms.) Alternation of generations has also evolved in certain groups of red algae and green algae. As we will see in the next chapter, alternation of generations is also a key feature in the life cycles of all plants.

We can examine the brown alga *Laminaria* as an example of a complex life cycle with an alternation of generations (FIGURE 26.20). The diploid individual is called the **sporophyte** because it produces reproductive cells called spores. The haploid individual is called the **gametophyte,** named for its production of gametes. Notice in FIGURE 26.20 that these two generations alternate—that is, they take turns producing one another. Spores released from the sporophyte develop into gametophytes,

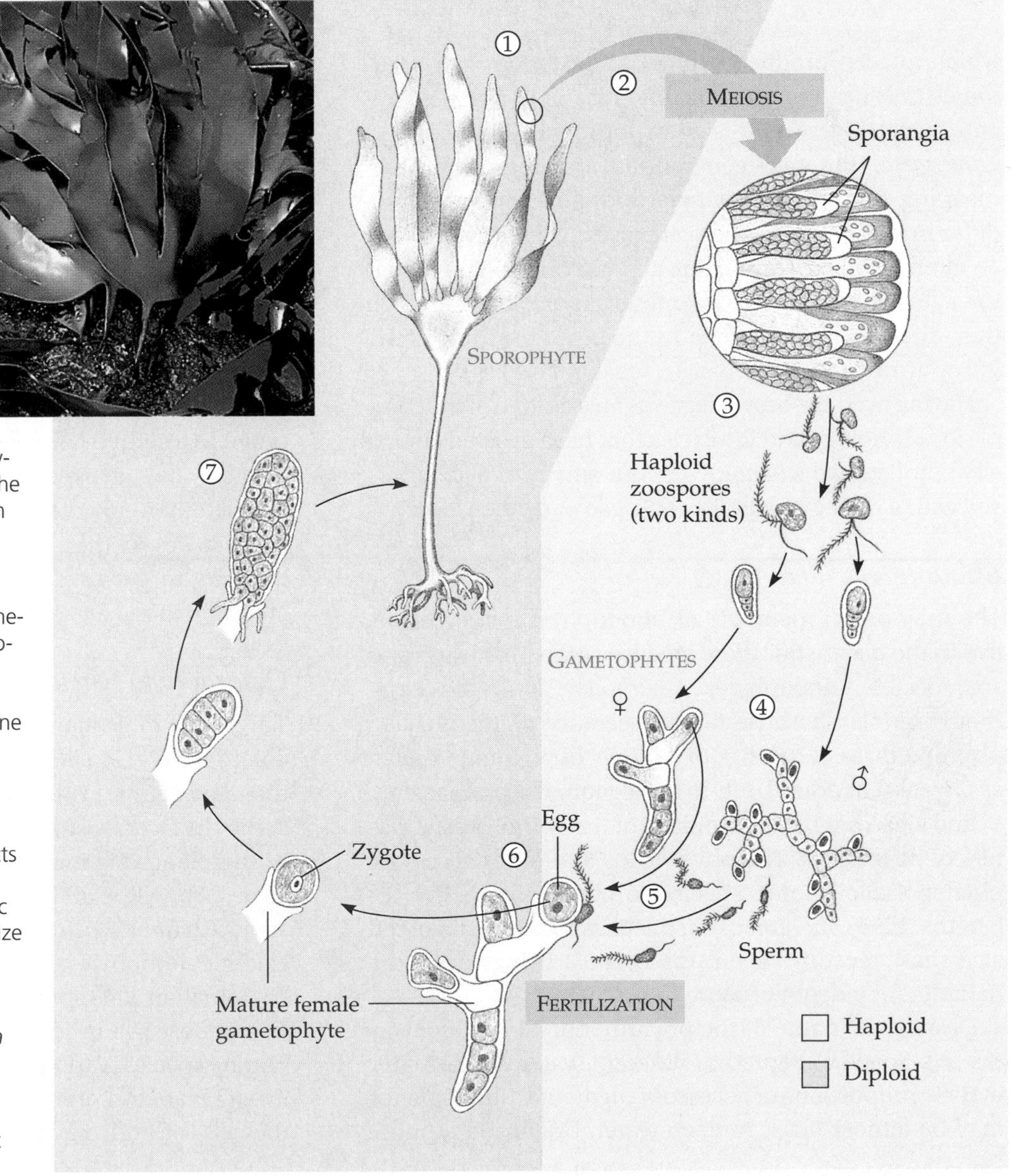

FIGURE 26.20
The life cycle of *Laminaria:* An example of heteromorphic alternation of generations. ① The sporophytes of this seaweed are usually found in water just below the line of lowest tides, attached to rocks with branching holdfasts. ② In early spring, at the end of the main growing season, cells on the surface of the blade develop into sporangia, which ③ produce two types of zoospores by meiosis. ④ One type grows into male gametophytes and the other into female gametophytes. The gametophytes look nothing like the sporophytes, being short, branched filaments that grow on the surface of subtidal rocks, often entangled in one another. ⑤ Male gametophytes release sperm and female gametophytes produce eggs, which remain attached to the gametophyte. Eggs secrete a chemical signal that attracts sperm of the same species, thereby increasing the probability of gametic union in the ocean. ⑥ Sperm fertilize the eggs and ⑦ the zygotes grow into new sporophytes, starting life attached to the remains of the old female gametophyte. The *Laminaria* life cycle is an example of a heteromorphic alternation of generations, in which sporophyte and gametophyte forms are noticeably different in appearance.

(a)

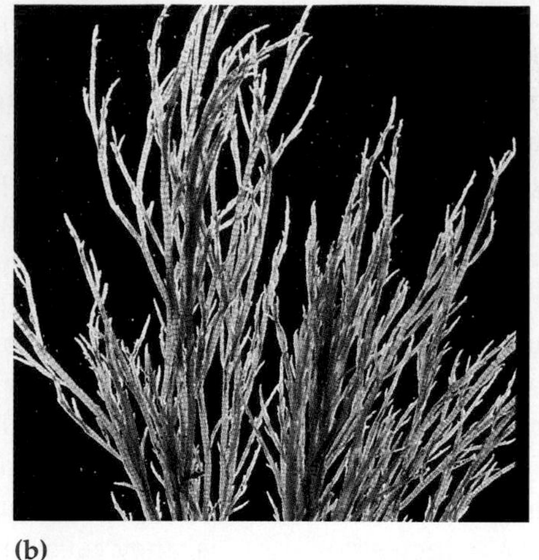

(b)

(c)

FIGURE 26.21

Red algae. (**a**) Dulce *(Palmaria),* an edible species with a "leafy" form. (**b**) *Polysiphonia,* a filamentous red alga. **(c)** This species is among the coralline algae that are members of the biological communities called coral reefs. Calcium carbonate hardens the cell walls of coralline algae.

which in turn produce gametes. The union of two gametes (fertilization, or syngamy) results in a diploid zygote, which gives rise to a new sporophyte. In the case of *Laminaria,* the two generations are **heteromorphic,** meaning that the sporophyte and gametophyte forms differ in morphology. There are also algal life cycles with an alternation of **isomorphic** generations, meaning that the sporophytes and gametophytes look alike, although they differ in chromosome number. An example is the life cycle of *Ulva,* a green alga (see FIGURE 26.25).

Having used the brown algae as an example of the complex morphology and life cycles that have evolved among algal protists, we will complete our survey of algal diversity with a closer look at the red algae and green algae.

Rhodophyta (Red Algae)

The majority of members of Rhodophyta, or **red algae,** live in the ocean, but there are also some freshwater and soil species. Rhodophytes (Gr. *rhodos,* "red") are commonly reddish because of an accessory pigment called phycoerythrin. It belongs to a family of pigments known as phycobilins, found only in red algae and cyanobacteria.

Red algae are most abundant in the warm coastal waters of the tropics. The phycobilins and other accessory pigments allow some species to absorb filtered wavelengths (blues and greens) in deep water. A species of red alga has recently been discovered living near the Bahamas at a depth of more than 260 m.

Despite the name of the phylum, not all rhodophytes are red. Species adapted to different water depths differ in their proportions of accessory pigments. Rhodophytes may be almost black in deep water, bright red at more moderate depths, and greenish in very shallow water, owing to less phycoerythrin masking the green of the chlorophyll. Some tropical species lack pigmentation altogether and function heterotrophically as parasites on other red algae.

Most red algae are multicellular, and the largest share the designation "seaweeds" with the brown algae, although none of the reds are as big as the giant browns (kelps). The thalli of many red algae are filamentous, often branched and interwoven in delicate lacy patterns (FIGURE 26.21). The base of the thallus is usually differentiated as a simple holdfast.

Life cycles are especially diverse among the red algae. Unlike other eukaryotic algae, red algae have no flagellated stages in their life cycles. Gametes rely on water currents to get together. Alternation of generations is common in red algae.

Chlorophyta (Green Algae)

This phylum is named for its members' grass-green chloroplasts (Gr. *chloros,* "green"), which are much like those of plants in ultrastructure and pigment composition. In fact, most botanists believe that the ancestors of the plant kingdom were related to a group of green algae. We will examine the evidence for this hypothesis in Chapter 27; for now, let's survey the diversity of extant chlorophytes.

More than 7000 species of **green algae** have been identified. Most live in fresh water, but there are also many marine species. Various species of unicellular green algae live as plankton, inhabit damp soil or snow, or occupy the cells or body cavities of protozoa and invertebrates

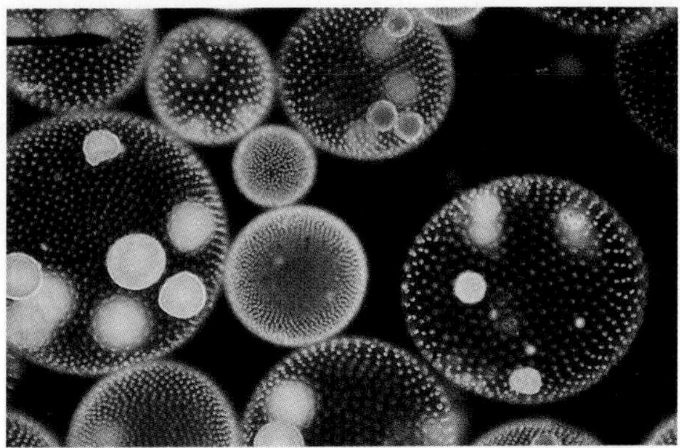

(a)

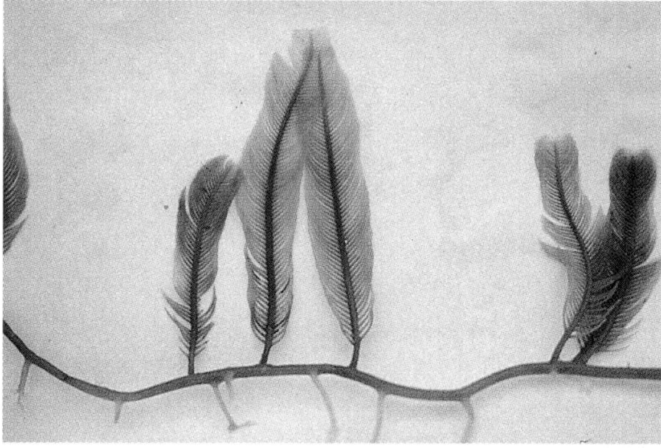

(b)

(c)

FIGURE 26.22

Colonial and multicellular green algae. (**a**) Species of *Volvox* are colonial chlorophytes that inhabit fresh water. The colony is a hollow ball, with its wall composed of hundreds or thousands of biflagellated cells embedded in a gelatinous matrix. The cells are usually connected by strands of cytoplasm; if isolated, these cells cannot reproduce. The large colonies seen here will eventually release the small "daughter" colonies within them (LM). (**b**) Species of *Bryopsis* are found in the marine intertidal zone. The branched filaments of these green algae lack cross-walls and thus are multinucleate. In effect, the thallus is one huge "supercell." (**c**) *Ulva,* or sea lettuce, is an edible seaweed with a thallus differentiated into leaflike blades and a rootlike holdfast that anchors the alga against turbulent waves and tides. The thallus is truly multicellular, consisting of specialized cells organized into tissues.

as photosynthetic symbionts that contribute to the food supply of the hosts (see FIGURE 29.10d). Chlorophytes are also among the algae that live symbiotically with fungi in the mutualistic collectives known as **lichens** (see Chapter 28).

The simplest chlorophytes are biflagellated unicells such as *Chlamydomonas,* which resemble the gametes of more complex green algae. In addition to unicellular chlorophytes, there are colonial species, many of them filamentous forms that contribute to the stringy masses known as pond scum. There are even some truly multicellular chlorophytes, with bodies large and complex enough that marine species qualify as seaweeds along with the large brown algae and red algae.

Three separate evolutionary trends have probably produced the diverse forms of colonial and multicellular chlorophytes from unicellular, flagellated ancestors. Larger size and greater complexity have evolved by: (1) the formation of colonies of individual cells, as seen in species of *Volvox* (FIGURE 26.22a); (2) the repeated division of nuclei with no cytoplasmic division, as seen in multinucleate filaments of *Bryopsis* (FIGURE 26.22b); and (3) the formation of true multicellular forms, as in *Ulva* (FIGURE 26.22c).

Most green algae have complex life histories, with both sexual and asexual reproductive stages. Nearly all reproduce sexually by way of biflagellated gametes having cup-shaped chloroplasts. The exceptions are the conjugating algae, such as *Spirogyra* (FIGURE 26.23), which produce amoeboid gametes.

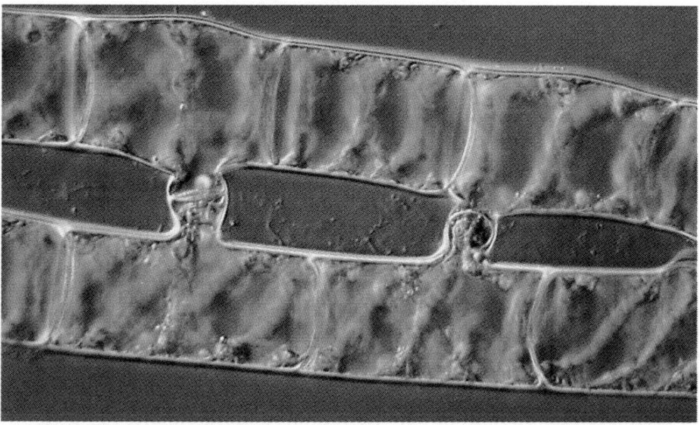

50 μm

FIGURE 26.23

***Spirogyra,* a conjugating alga.** The cells of *Spirogyra* are arranged in long, unbranched filaments. Each cell contains one or more large, spiral-shaped chloroplasts. The filaments grow in length and reproduce asexually by fragmentation as the cells divide by mitosis. During sexual reproduction, as seen here, cells of adjacent mating strands join by conjugation tubes (LM).

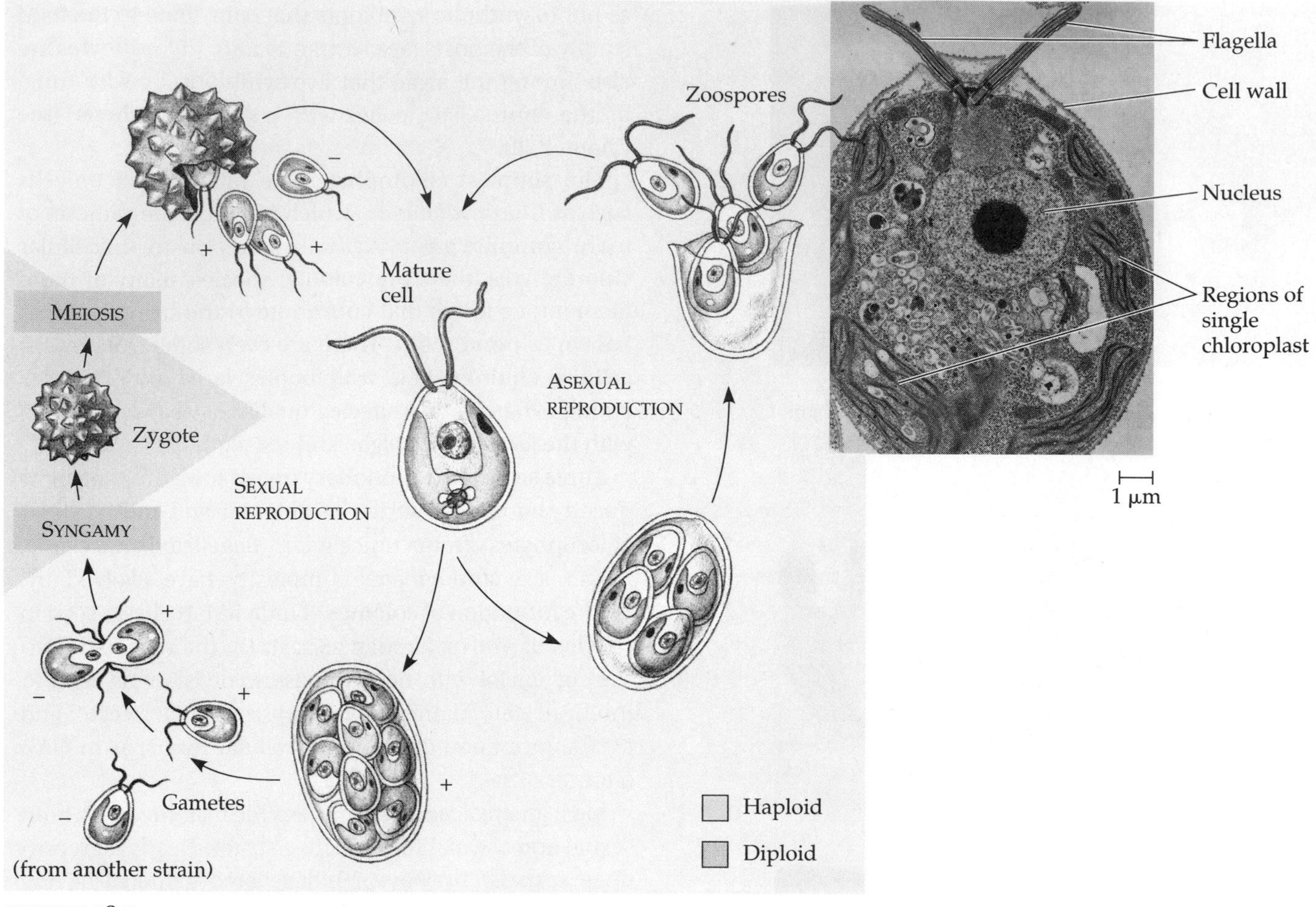

FIGURE 26.24

The life cycle of *Chlamydomonas.* This green alga exhibits sexual as well as asexual reproduction. The gametes are morphologically identical, a condition known as isogamy. When fertilization involves isogametes, rather than distinct sperm and ovum, it is usually referred to by the general term syngamy. The inset shows a vegetative cell prior to reproduction (TEM).

Let's examine the life cycle of *Chlamydomonas,* a unicellular chlorophyte (FIGURE 26.24). The mature organism is a single haploid cell. When it reproduces asexually, the cell resorbs its flagella and then divides twice by mitosis, forming four cells (more in some species). These daughter cells develop flagella and cell walls and then emerge as swimming zoospores from the wall of the parent cell, which had enclosed them. The zoospores grow into mature haploid cells, completing the asexual life cycle. A shortage of nutrients, drying of the pond, or some other stress triggers sexual reproduction. Within the wall of the parent cell, mitosis produces many haploid gametes. After their release, gametes from opposite mating strains (designated + and −) pair off and cling together by the tips of their flagella. The gametes are morphologically indistinguishable, and their fusion is known as **isogamy,** which literally means a "marriage of equals." The gametes fuse slowly, forming a diploid zygote, which secretes a durable coat that protects the cell against harsh conditions. When the zygote breaks dormancy, meiosis produces four haploid individuals (two of each mating type) that emerge from the coat and grow into mature cells, completing the sexual life cycle.

Many of the features of *Chlamydomonas* sex are believed to be primitive, meaning that they evolved early in the chlorophyte lineage. Based on this basic life cycle, we can identify many refinements that have evolved among chlorophytes. Some green algae, for instance, produce gametes that differ morphologically from vegetative cells, and in some species the male and female gametes differ in size, known as **anisogamy,** or morphology. Many species exhibit a type of anisogamy called **oogamy,** a flagellated sperm fertilizing a nonmotile egg. Another refinement is an alternation of generations in some multicellular species, with alternate haploid and diploid individuals producing each other in turn. An example is

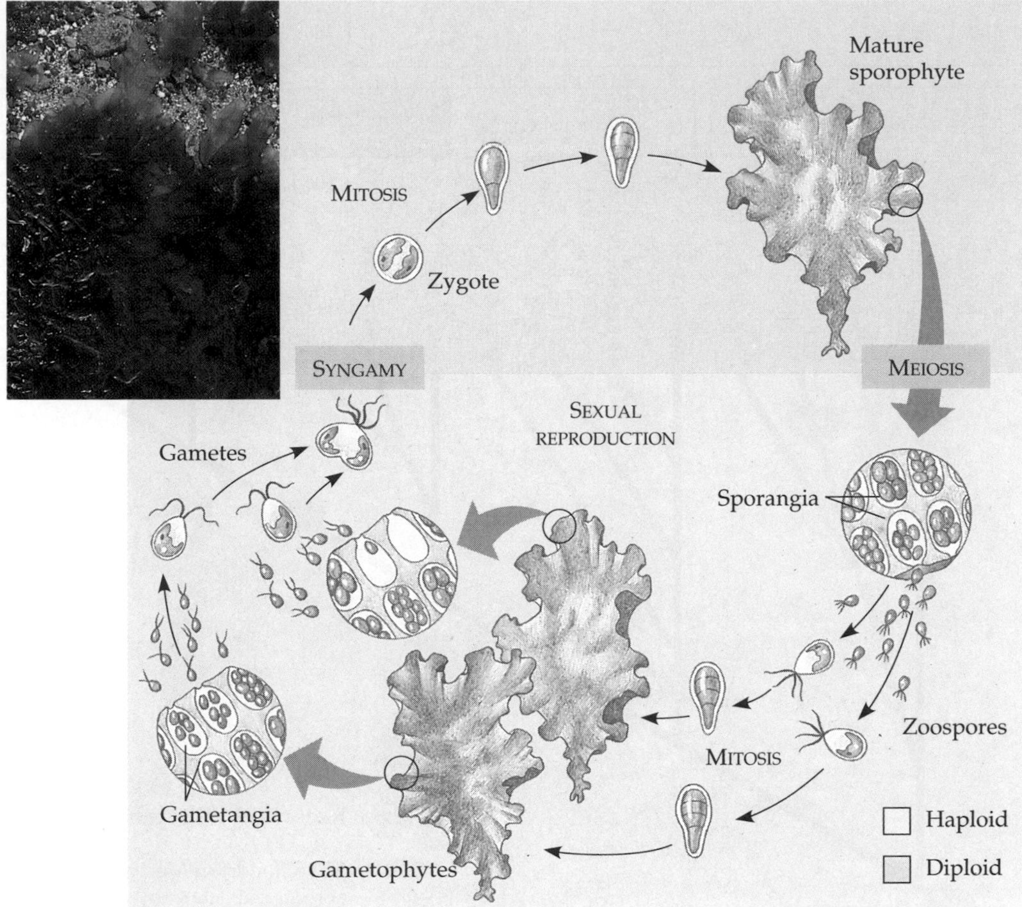

FIGURE 26.25

The life cycle of *Ulva:* an example of isomorphic alternation of generations. The haploid, sexual generation (gametophyte) produces the diploid, asexual generation (sporophyte), and vice versa. Gametophytes produce gametes, which form zygotes by syngamy. Sporophytes produce reproductive cells called spores, which develop directly into gametophytes. In the case of *Ulva,* the two generations are isomorphic, or identical in appearance. Compare *Ulva's* life cycle to the heteromorphic life history of *Laminaria* in FIGURE 26.20.

the life cycle of *Ulva,* in which the sporophyte (diploid thallus) and gametophyte (haploid thallus) are morphologically identical, or isomorphic (FIGURE 26.25).

Now that we have surveyed the protozoa, funguslike protists, and algae, let's return to the systematic issue of how these phyla fit into a phylogenetic classification of eukaryotes.

Systematists continue to refine their hypotheses about eukaryotic phylogeny

The most expansive version of the kingdom Protista within the five-kingdom system includes all the groups of organisms you learned about in this chapter. Many systematists who work on the puzzle of eukaryotic relationships contend that such a classification scheme is obsolete in light of new evidence from fossils, cell ultrastructure, and molecular comparisons. The debates are partly about subdivision of the eukaryotes into phyla: how many phyla, their boundaries, and what to name them. But at a higher taxonomic level, many of these researchers argue that the five-kingdom system does not reflect phylogeny as we understand it at present and should therefore be abandoned.

FIGURE 26.26 synthesizes a hypothetical phylogeny of eukaryotes based on molecular systematics and other evidence. The tree relates some of the phyla you learned about in this chapter to the origin of eukaryotes by serial endosymbiosis among prokaryotic ancestors and to the

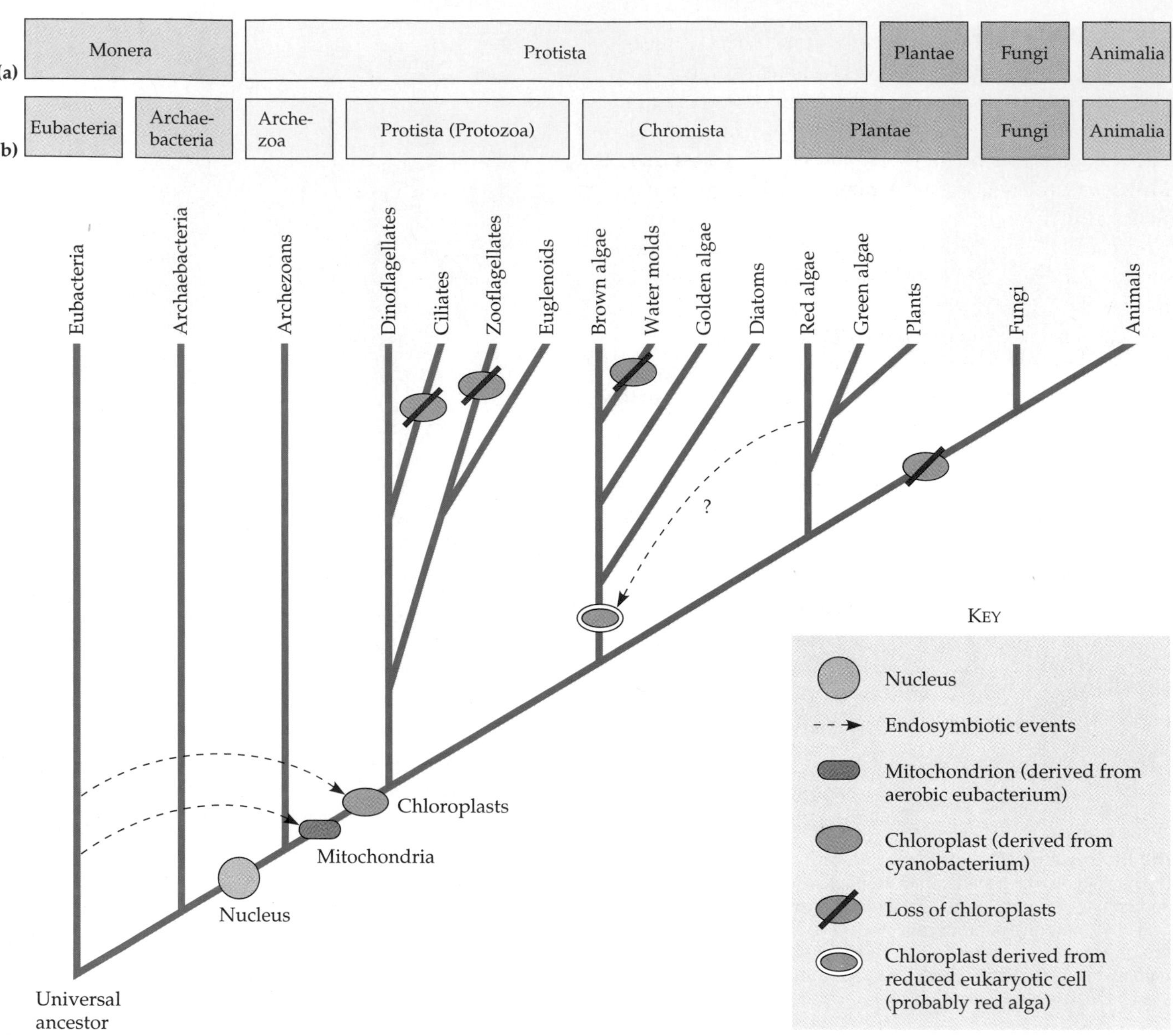

FIGURE 26.26

The origins of eukaryotic diversity: a hypothetical phylogeny. The diagram compares how biological diversity is interpreted by (**a**) the five-kingdom system and (**b**) one alternative classification, an eight-kingdom taxonomic scheme.

later diversification of eukaryotes leading to plants, fungi, and animals. Notice that chloroplasts probably evolved from cyanobacterial endosymbionts relatively early in eukaryotic history, but several lineages of heterotrophic eukaryotes lost their chloroplasts during their subsequent evolution.

The top of FIGURE 26.26 compares the five-kingdom system to one of the alternative classification schemes, an eight-kingdom system. The eight-kingdom classification incorporates the split of prokaryotes into separate kingdoms for archaebacteria and eubacteria, as discussed in Chapter 25. (Remember from Chapters 24 and 25 that some systematists emphasize this split even more by advocating Archaea, Bacteria, and Eukarya as three domains, or superkingdoms; see FIGURE 24.10.) The eight-kingdom system also recognizes the early (premitochondrial) branching of archezoa (such as *Giardia*) from the eukaryotic tree by assigning them to their own kingdom, **kingdom Archezoa.**

The eight-kingdom system retains many of the eukaryotic phyla you learned about in this chapter, including the protozoa and some of the algae, in the kingdom Protista (although some researchers suggest changing the name of this less inclusive version of the kingdom to Protozoa). But notice that the eight-kingdom alternative pulls the brown algae and some related phyla out of the kingdom Protista into a separate **kingdom Chromista.** Chromistans are distinguished by unusual chloroplasts that have two additional membranes outside the usual chloroplast envelope, a small amount of cytoplasm, and a vestigial nucleus. There is increasing evidence that these chloroplasts descended from endosymbionts that were eukaryotic cells, probably red algae. (The chloroplasts of red algae also began as endosymbionts, cyanobacteria living in a host cell. Thus, chromistan chloroplasts may be derived from endosymbionts containing endosymbionts!) Molecular systematics also places water molds (Oomycota) on the chromistan branch, though they lost their chloroplasts during their evolution.

In addition to adding kingdoms, the eight-kingdom system also moves the green algae and red algae from the kingdom Protista to the plant kingdom. The evolutionary ties of red algae are still uncertain, but there is little doubt about a close relationship between green algae and plants. In the next chapter, you will learn about the evidence that the organisms we traditionally call plants evolved from a particular group of green algae.

It is too early to predict whether there will be a consensus among biologists to replace the five-kingdom system with the eight-kingdom scheme or some other alternative classification system. The actual diversity of life, of course, is unaffected by taxonomic systems, which are constructed by human minds in our attempts to understand life on Earth. But if the goal is a classification that reflects phylogeny, then debates about how to arrange biological diversity into kingdoms are not simply semantic issues, but rather dialogues about how to interpret the history of life.

We complete this chapter with another important development in eukaryotic history, the origin of multicellularity.

Multicellularity originated independently many times

The origin of eukaryotic organization ignited an explosion of biological diversification. Put simply, more variations are possible for complex structures than for simpler ones. Unicellular protists, which are organized on the complex eukaryotic plan, are much more diverse in morphology than the simpler prokaryotes. Protists in which multicellular bodies evolved broke through another threshold in structural organization and became the stock for new waves of adaptive radiations. Multicellularity evolved several times among early eukaryotes, with the diverse multicellular algae as examples of the results. The evolution of multicellularity also gave rise to plants, fungi, and animals.

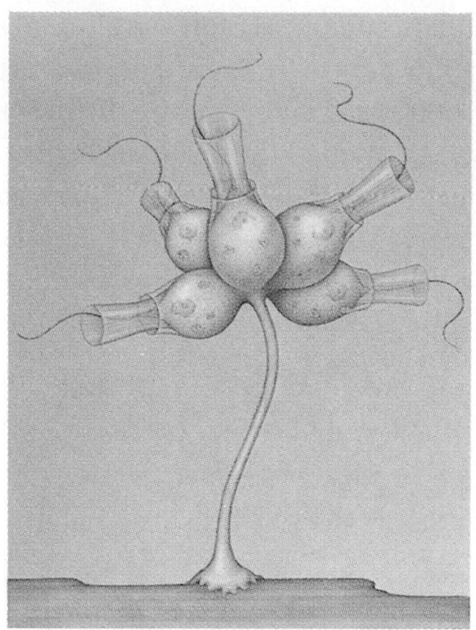

FIGURE 26.27
The colonial connection between unicellular and multicellular life. The organism represented in the drawing is a choanoflagellate, a colonial protist that many zoologists believe is related to the ancestor of animals. More generally, colonial organisms of diverse kinds may have been intermediates in the many origins of multicellular life.

The most widely held view is that links between multicellular organisms and their unicellular ancestors were colonies, or loose aggregates of interconnected cells (FIGURE 26.27). Multicellular algae, plants, fungi, and animals probably descended from several lineages of colonial protists that formed by amalgamations of individual cells. The evolution of multicellularity from colonial aggregates involved increasing cellular specialization and division of labor. Initially, in colonial aggregates ancestral to multicellular algae, plants, and animals, all cells may have been motile with flagella. As cells in the colony tended to become increasingly interdependent, some of the cells may have lost their flagella and become more proficient in performing functions other than locomotion.

Another early form of division of labor probably involved the distinction of sex cells (gametes) from somatic (nonreproductive) cells. We see this type of specialization and intercellular cooperation today in several colonial organisms, such as the green alga *Volvox* (see FIGURE 26.22a). Gametes are specialized for reproduction, and they depend on somatic cells while developing. Evolution of the extensive division of labor required to perform all the nonreproductive functions in multicellular organisms involved many additional steps in somatic cell specialization. In seaweeds, for example, there is extensive division of labor among the different tissues that make up the thalli. Multicellular forms of life more complex than these algae did not appear until about 700 million years ago, during the twilight of the Precambrian era. A variety of animal fossils has been found in late Precambrian strata, and many new forms evolved after the Paleozoic era dawned with the Cambrian period, about 550 million years ago. Seaweeds and other complex algae were also abundant in Cambrian oceans and lakes. The land, however, was barren. About 400 million years ago, certain green algae living along the edges of lakes gave rise to primitive plants. In the next chapter, we will trace the long evolutionary trek of plants onto land.

REVIEW OF KEY CONCEPTS (with page numbers and key figures)

- Eukaryotes originated by symbiosis among prokaryotes (pp. 518–520, FIGURE 26.3)
  - Eukaryotic cells arose as a result of prokaryotes taking up residence inside other prokaryotes. Chloroplasts and mitochondria are thus descendants of cyanobacteria and aerobic, heterotrophic symbionts, respectively.
  - The endomembrane system of eukaryotes may have evolved from specialized infoldings of the host cell's membrane.
- Archezoans provide clues to the early evolution of eukaryotes (pp. 520–521, FIGURE 26.2)
  - Archezoans, which lack mitochondria and have other characteristics that systematists consider to be primitive, make up the most ancient eukaryotic lineage.
- The diversity of protists represents different "experiments" in the evolution of eukaryotic organization (pp. 521–523)
  - Most protists are unicellular, but colonial and simple multicellular forms also exist.
  - Protists are found wherever there is water, living as plankton, submerged bottom-dwellers, or inhabitants of moist soil or body fluids of other organisms.
  - Of all eukaryotes, protists are the most nutritionally diverse. Photoautotrophs, heterotrophs, and mixotrophs are all represented.
- Protistan taxonomy is in a state of flux (p. 523)
  - Protists exhibit the most diverse spectrum of structure and life cycles of all known organisms.
  - Taxonomic debate persists about how to classify the protists.
- Diverse modes of locomotion and feeding evolved among protozoa (pp. 523–529, TABLE 26.1)
  - Rhizopods, unicellular amoebas and their relatives, move by cellular extensions called pseudopodia.
  - Actinopods are protozoa with slender, raylike axopodia that help them float and feed. Heliozoan and radiozoan species are components of plankton in freshwater and marine environments, respectively.
  - The marine forams have porous shells, through which strands of cytoplasm extend.
  - Apicomplexans are parasitic protozoa with complex life cycles characterized by both sexual and asexual stages that often require two or more host species.
  - Zoomastigotes are flagellated heterotrophs.
  - Ciliates use cilia to move and feed. Ciliates are among the most complex cells.
- Funguslike protists have morphological adaptations and life cycles that enhance their ecological role as decomposers (pp. 529–532, FIGURES 26.12–26.14)
  - The myxomycotes are the plasmodial slime molds. These diploid organisms feed by means of a coenocytic amoeboid plasmodium capable of differentiating into sexually reproducing sporangia when moisture or food is scarce.
  - Acrasiomycotes, the cellular slime molds, include a group of haploid organisms that lead unicellular lives until food is depleted. They then aggregate into a multicellular amoeboid mass that erects asexual fruiting bodies.
  - Oomycotes, the water molds, have cell walls of cellulose and biflagellated stages in their life cycles. The phylum also includes white rusts and downy mildews, some of which are serious plant pathogens.
- Eukaryotic algae are key producers in most aquatic ecosystems (pp. 532–541, TABLE 26.2)
  - Algae are aquatic, photosynthetic organisms. Classification characteristics include variations in chloroplast structure, accessory pigments, cell walls, flagella, and forms of food storage.
  - Dinoflagellates are unicellular algae abundant in marine plankton. They move in a spinning motion by the beating of flagella.
  - Bacillariophytes are the diatoms, primarily unicellular organisms with unique glasslike walls of silica.
  - Chrysophytes are the biflagellated freshwater golden algae, named for the color of their carotenoid pigments.
  - Phaeophytes are multicellular, primarily marine, brown algae, including kelps. Most species show some type of alternation of generations.
  - Seaweeds include thallus-forming, marine species among the brown, red, and green algae. They are well adapted to life along the turbulent margins of the oceans.
  - Rhodophytes, the red algae, possess the red accessory pigment phycoerythrin. Red algae are multicellular, often lacy forms that reproduce sexually, usually as part of an alternation of generations.

- Chlorophytes, the green algae, are the likely ancestors of the plant kingdom. Diverging evolutionary pathways have generated an assortment of unicellular, colonial, multinucleate, and multicellular species that live in a variety of environments. Complicated life histories include both sexual and asexual reproductive stages, including examples of alternation of generations.

- Systematists continue to refine their hypotheses about eukaryotic phylogeny (pp. 541–543, FIGURE 26.26)
  - As systematists continue to unravel eukaryotic phylogeny, some are proposing alternatives to the five-kingdom system.
- Multicellularity originated independently many times (pp. 543–544, FIGURE 26.27)
  - Multicellularity evolved many times in the kingdom Protista apparently by the aggregation of individual cells into colonial forms, in which cellular specialization and division of labor developed.

SELF-QUIZ

1. Which of the following is the most accurate general description of the kingdom Protista as it exists in the five-kingdom system?
   a. eukaryotic unicellular organisms that may be photosynthetic or heterotrophic
   b. eukaryotic, heterotrophic and/or photosynthetic, unicellular or simple multicellular organisms that are different enough from multicellular plants, fungi, or animals to be placed in this diverse group
   c. eukaryotic plankton, which may be flagellated at some point in their life cycle and which reproduce asexually
   d. eukaryotic photosynthetic or heterotrophic organisms that inhabit moist environments, form resistant cysts, and reproduce with flagellated gametes
   e. relatively simple versions of plants, animals, and fungi

2. Which of the following protozoa are *incorrectly* paired with their description?
   a. rhizopods—naked and shelled amoebas
   b. actinopods—planktonic with slender, raylike axopodia
   c. forams—flagellated algae, free-living or symbiotic
   d. apicomplexans—parasites with complex life cycles
   e. ciliates—complex, unicellular organisms with macronucleus and micronuclei

3. Which of the following algae are *incorrectly* paired with their description?
   a. dinoflagellates—marine plankton, whirling, spinning movement, characteristic wall
   b. chrysophytes—golden algae, xanthophylls predominant, flagellated, freshwater plankton
   c. bacillariophytes—diatoms, two-piece shells of silica
   d. phaeophytes—multicellular brown algae, seaweeds
   e. rhodophytes—cause red tides, xanthophylls predominant

4. In the eight-kingdom system of classification, the kingdom Chromista is distinguished partly by the presence of
   a. chloroplasts derived from eukaryotic endosymbionts
   b. chloroplasts derived from cyanobacteria
   c. mitochondria
   d. chlorophyll *a*
   e. heterotrophic nutrition

5. Unlike plasmodial slime molds, cellular slime molds
   a. exhibit phagocytosis
   b. form fruiting bodies
   c. have more than one nucleus per cell
   d. are haploid organisms except for the giant-celled zygote
   e. can move as an amoeboid mass

6. Acritarchs are
   a. symbiotic associations between algae and fungi
   b. multicellular eukaryotes
   c. the oldest putative eukaryotic fossils
   d. fossils that resemble the ruptured coats of cyanobacteria
   e. diatom shells

7. Which of the following is an *incorrect* statement about the possible endosymbiotic origins of chloroplasts and mitochondria?
   a. They are the appropriate size to be descendants of bacteria.
   b. They contain their own genome and produce all their own proteins.
   c. They contain circular DNA molecules not associated with histones.
   d. Their membranes have enzymes and transport systems that resemble those found in the plasma membranes of prokaryotes.
   e. Their ribosomes are more similar to those of eubacteria than to those of eukaryotes.

8. The organism that caused the Irish potato famine is
   a. an actinopod
   b. a ciliate
   c. an oomycote
   d. a plasmodial slime mold
   e. a cellular slime mold

9. A student collected a flask of pond water and placed it near a window. A brownish-green scum collected on the side of the flask facing the light. When the flask was turned around, the scum moved to the side facing the window. Which of the following phyla are most likely represented in this pond scum?
   a. Zoomastigophora and Ciliophora
   b. Chrysophyta and Euglenophyta
   c. Chlorophyta and Phaeophyta
   d. Euglenophyta and Acrasiomycota
   e. Phaeophyta and Bacillariophyta

10. Which of the following groups probably represents the earliest branch in the evolution of eukaryotes?
   a. archaebacteria
   b. archezoa
   c. fungi
   d. amoebas
   e. diatoms

CHALLENGE QUESTIONS

1. Different genes are expressed at different stages of the life cycle of the malaria-causing apicomplexan *Plasmodium,* which causes different proteins to appear on the outer coat of the infecting cells. Sporozoites are injected by mosquitoes and travel in the blood to liver cells, where they continue their life cycle. Researchers discovered that the sporozoites produce protein coats that are sloughed off and continuously replaced. Host antibodies can attack these proteins by specific complementary binding. How do the continual sloughing and replacing of the protein coat work as an adaptive mechanism to prevent immune destruction of the sporozoite before it gets inside the liver cell, where it is protected from bloodborne antibodies?

2. Of all photosynthetic prokaryotes, why are cyanobacteria the most reasonable candidates for the origin of chloroplasts? How could a molecular systematist test the hypothesis that cyanobacteria are the prokaryotes most closely related to chloroplasts?

3. Write a paragraph contrasting the five- and eight-kingdom systems in their treatment of the eukaryotic phyla discussed in this chapter.

SCIENCE, TECHNOLOGY, AND SOCIETY

1. The burning of fossil fuels is increasing the amount of carbon dioxide in the atmosphere, and many experts think this will intensify the greenhouse effect and lead to a warming of the global climate. The photosynthesis of diatoms and other microscopic algae in the oceans uses enormous quantities of carbon dioxide. These algae need minute quantities of iron, and some researchers suspect that a shortage of iron may limit photosynthesis, especially in the Antarctic Ocean. One oceanographer suggests that one way to slow global warming might be to fertilize the ocean with iron. This would stimulate the growth of diatom populations, which would remove carbon dioxide from the air. The scientist estimates that a single supertanker of iron dust, spread over a wide enough area, could reduce CO_2 levels significantly. Do you think this would be worth a try? Why or why not? Do you see any reasons to be cautious? Why?

2. Because the malarial parasite is able to evade the immune system, it has been difficult to develop a malaria vaccine (see Challenge Question 1). An additional problem is that fewer scientists are engaged in malaria research, and less money is spent on it, than on diseases such as cystic fibrosis, which affects far fewer people than does malaria. What are the possible reasons for this imbalance?

FURTHER READING

Anderson, D. M. "Red Tides." *Scientific American,* August 1994. How dinoflagellate blooms cause human suffering.

Fenchel, T., and B. J. Finlay. "The Evolution of Life Without Oxygen." *American Scientist,* January/February 1994. Clues to the origins of eukaryotes.

Jacobs, W. P. "Caulerpa." *Scientific American,* December 1994. This tropical alga may be the largest single-celled organism.

Knoll, A. "The Early Evolution of Eukaryotes: A Geological Perspective." *Science,* May 1, 1992. What can Precambrian rocks tell us about the origin of eukaryotic cells?

Lewin, R. "Bacteria Rule OK?" *New Scientist,* June 3, 1995. Genetic evidence for the endosymbiotic origins of eukaryotes from bacteria.

Nussenzweig, R. S., and C. A. Long. "Malaria Vaccines: Multiple Targets." *Science,* September 2, 1994. Some encouraging news.

Richardson, S. "Mergers and Acquisitions." *Discover,* December 1994. Did the endosymbiotic origin of eukaryotic organelles include endosymbionts that were themselves eukaryotes?

Saffo, M. B. "New Light on Seaweeds." *BioScience,* October 1987. The scientific method reasserts itself, as a century-old hypothesis explaining vertical distribution of marine algae turns out to be too simple.

CHAPTER 27

PLANTS AND THE COLONIZATION OF LAND

KEY CONCEPTS

- Structural and reproductive adaptations made the colonization of land possible: *an overview of plant evolution*
- Plants probably evolved from green algae called charophytes
- Bryophytes are embryophytes that generally lack vascular tissue and require environmental water to reproduce
- The origin of vascular tissue was an evolutionary breakthrough in the colonization of land
- Ferns and other seedless plants dominated the Carboniferous "coal forests"
- Reproductive adaptations catalyzed the success of the seed plants
- Gymnosperms began to dominate landscapes as climates became drier at the end of the Paleozoic era
- The evolution of flowers and fruits contributed to the radiation of angiosperms
- Plant diversity is a nonrenewable resource

It is difficult to picture the land barren, totally uninhabited by life of any form. But that is how we must imagine Earth for almost the first 90% of the time life has existed. Life was cradled in the seas and ponds, and there it evolved in confinement for 3 billion years. Paleobiologists have recently discovered fossils of cyanobacteria that may have coated moist soil about 1.2 billion years ago, but the long evolutionary pilgrimage of more complex organisms onto land did not begin until about 460 million years ago. Plants led the way, followed by herbivorous animals and their predators. The terrestrial communities founded by green plants transformed the biosphere. Consider, for example, that humans would not exist had it not been for the chain of evolutionary events that began when certain descendants of green algae first colonized land.

The evolutionary history of the plant kingdom is a story of adaptation to changing terrestrial conditions. That is the historical context in which this chapter surveys the current diversity of plants and traces their origins.

Structural and reproductive adaptations made the colonization of land possible: *an overview of plant evolution*

General Characteristics of Plants

All plants, as defined in this book, are multicellular eukaryotes that are photosynthetic autotrophs. However, not all organisms with these characteristics are plants; such characteristics also apply to some algae, including the giant brown algae we have classified as protists or chromists (see Chapter 26).

Plants as we are defining them are nearly all terrestrial organisms, although some plants have returned secondarily to water during their evolution. Living on land poses very different problems from living in the water. As plants have adapted to the terrestrial environment, complex bodies with extensive specialization of cells for different functions have evolved. Aerial parts of most plants, such as stems and leaves, are coated with a waxy **cuticle** that helps prevent desiccation, a major problem on land. Gas exchange cannot occur across the waxy surfaces, but carbon dioxide and oxygen diffuse between the interior of leaves and the surrounding air through microscopic pores on the leaf's surface called **stomata** (singular, **stoma**). Besides their special adaptations for terrestrial life, land plants share many features with their progenitors, the green algae. For example, their photosynthetic cells contain chloroplasts having the pigments chlorophyll *a*, chlorophyll *b*, and a variety of yellow and orange carotenoids. Plant cells also have walls, and the staple material of their walls is cellulose. Carbohydrate is stored in the form of starch, generally in chloroplasts and other plastids.

The Embryophyte Condition

The move onto land paralleled a new mode of reproduction. In contrast to the reproductive style of algae, gametes now had to be dispersed in a nonaquatic environment, and embryos, like mature body structures, had to be protected against desiccation.

Nearly all plants reproduce sexually, and most are also capable of asexual propagation. Plants produce their gametes within **gametangia,** organs having protective jackets of sterile (nonreproductive) cells that prevent the delicate gametes from drying out during their development (FIGURE 27.1). The egg is fertilized within the female gametangium, where the zygote develops into an embryo that is retained and nourished for some time within the jacket of protective cells. In contrast, developing algae are not retained as embryos within a parent. This difference is so fundamental that plants are sometimes referred to as **embryophytes,** a term that emphasizes a key adaptation that contributed to survival on land.

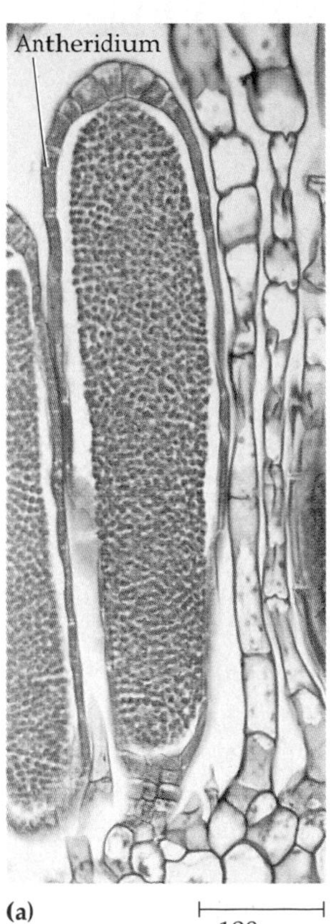

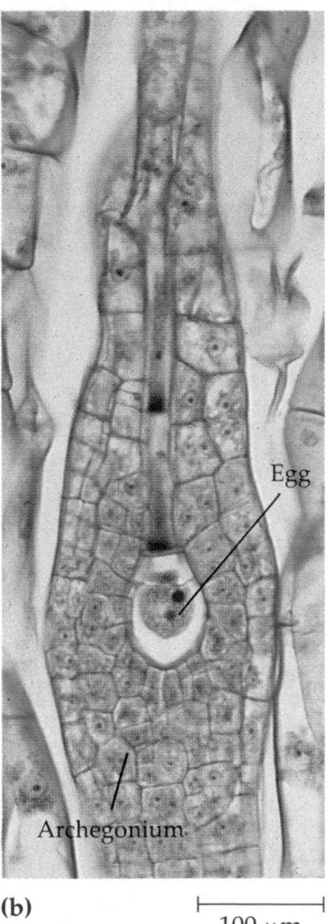

FIGURE 27.1

Gametangia and protected embryos as terrestrial adaptations. Gametes and zygotes of plants develop within gametangia, moist chambers protected by a coat of sterile cells. Shown here are the gametangia of a moss. (**a**) Male gametangia are called antheridia; (**b**) eggs are fertilized within the female gametangia, called archegonia. In the embryophyte condition that helps define plants, the zygote is retained within the archegonium, where it develops into an embryo that is protected and nourished by the parent plant. (Both LMs.)

Alternation of Generations: A Review

In the life cycles of all plants, an alternation of generations occurs, in which haploid gametophytes and diploid sporophytes take turns producing one another (FIGURE 27.2; also see Chapters 12 and 26). In the life cycles of all extant plants, the sporophyte and gametophyte generations differ in morphology; that is, they are heteromorphic. In all plants except the mosses and their relatives, the diploid sporophyte is the larger and more noticeable individual. The physiology of plant reproduction is covered in Chapter 34; in this chapter, we look in some detail at representative plant life cycles. It is valuable to understand these life cycles for two reasons. First, they clarify one of the main trends in plant evolution, toward reduction of the haploid generation and domi-

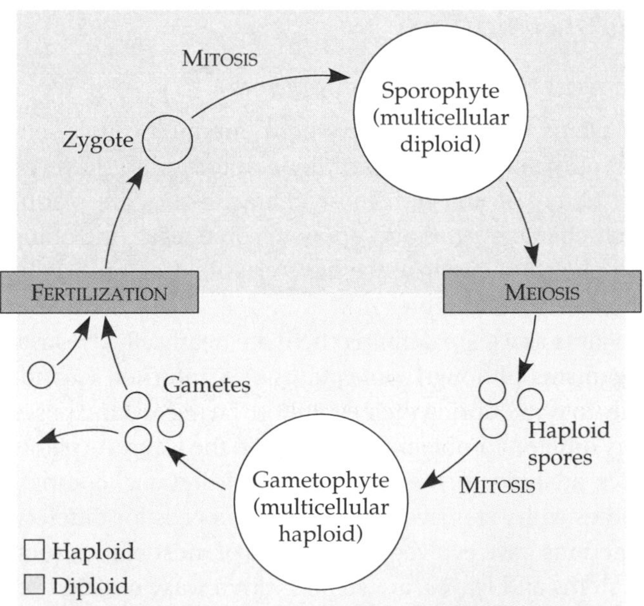

FIGURE 27.2

Alternation of generations: a generalized scheme. The life cycles of all plants include a gametophyte (haploid generation) and a sporophyte (diploid generation). The two generations alternate, each producing the other. The two plant forms are named for the type of reproductive cells they produce: Gametophytes form gametes by mitosis; sporophytes produce spores by meiosis. Spores develop directly into organisms. Gametes *cannot* develop directly into organisms but unite, sperm and egg, to form a zygote. The zygote, in turn, gives rise to an organism, the new sporophyte.

nance of the diploid. Second, in many cases, features of the life cycle, such as the replacement of flagellated sperm by pollen, are key evolutionary adaptations to terrestrial environments.

Some Highlights of Plant Phylogeny

The fossil record chronicles four major periods of plant evolution, which are also evident in the diversity of contemporary plants (FIGURE 27.3). Each period was an adaptive radiation that followed the evolution of structures that opened new adaptive zones on the land.

The first period of evolution was associated with the origin of plants from aquatic ancestors, green algae, during the late-Ordovician period of the Paleozoic era, about 460 million years ago. The first terrestrial adaptations included a cuticle and jacketed gametangia that protected gametes and embryos. **Vascular tissue,** consisting of cells joined into tubes that transport water and nutrients throughout the plant body, also evolved relatively early in plant history. Most mosses lack vascular tissue, and hence they are sometimes categorized as nonvascular plants. However, water-conducting tubes *are* present in some mosses. Researchers have not yet resolved whether this vascular tissue in mosses is analogous (separate evolutionary development) or homologous (common evolutionary development) to the water-conducting tissue of other plants. Evidence from molecular systematics does suggest an early split between mosses and the lineages leading to ancient ferns and other vascular plants.

The second major period of plant evolution was the diversification of vascular plants during the early Devonian period, about 400 million years ago. The earliest vascular plants lacked seeds, a condition still represented by ferns and a few other groups of seedless vascular plants.

The third major period of plant evolution began with the origin of the seed, a structure that advanced the colonization of land by further protecting plant embryos from desiccation and other hazards. A seed consists of an embryo packaged along with a store of food within a protective covering. The first vascular plants with seeds arose

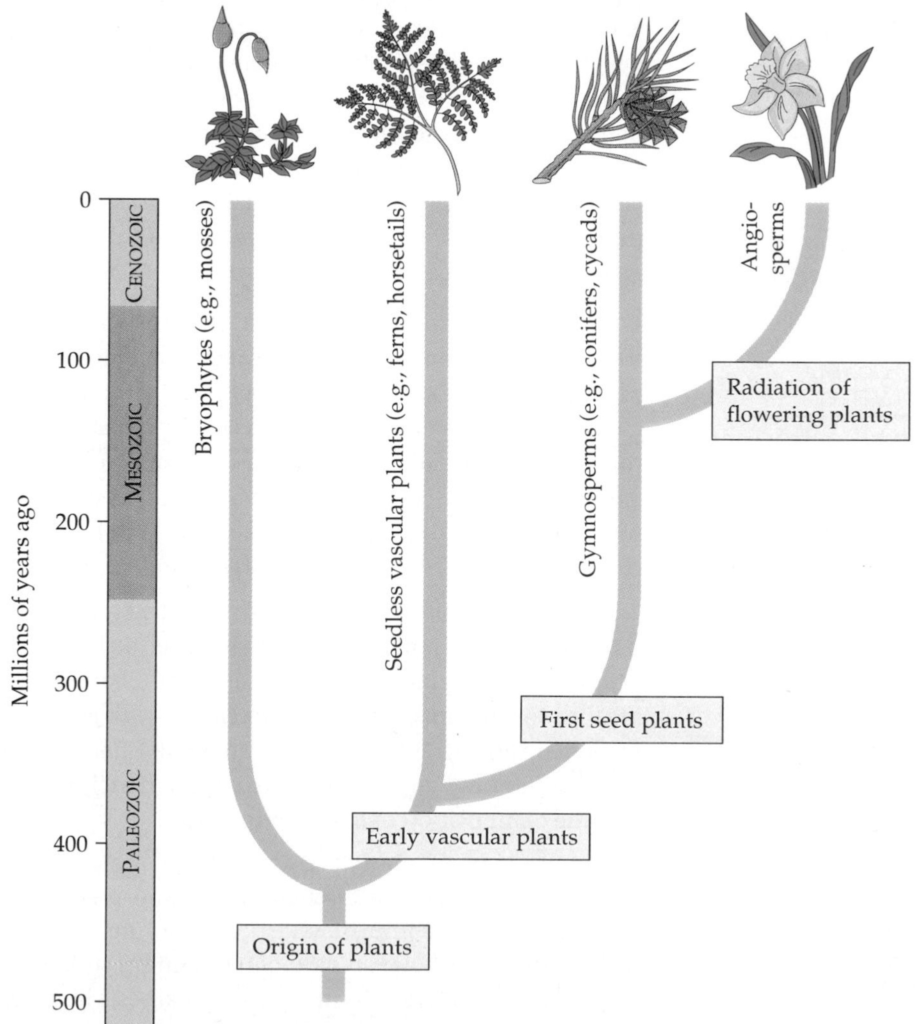

FIGURE 27.3
Some highlights of plant evolution. The origins of plants from green algae, the adaptive radiation of early vascular plants, the emergence of seed plants, and the origin and diversification of flowering plants (angiosperms) are four important chapters in the history of the plant kingdom. Modern representatives of major plant lineages are illustrated at the top of this evolutionary tree.

about 360 million years ago, near the end of the Devonian period. Their seeds were not enclosed in any specialized chambers. Early seed plants gave rise to many types of **gymnosperms** (Gr. *gymnos,* "naked," and *sperma,* "seed"), including the conifers, which are the pines and other plants with cones. Gymnosperms coexisted with ferns and other seedless plants in great forests that dominated the landscape for more than 200 million years.

The fourth major episode in the evolutionary history of plants was the emergence of flowering plants during the early Cretaceous period in the Mesozoic era, about 130 million years ago. The flower is a complex reproductive structure that bears seeds within protective chambers called ovaries, which contrasts with the bearing of naked seeds by gymnosperms. The great majority of contemporary plants are flowering plants, or **angiosperms** (Gr. *angion,* "container," referring to the ovary, and *sperma,* "seed").

Classification of Plants

Plant biologists use the term **division** for the major plant groups within the plant kingdom. This taxonomic category corresponds to phylum, the highest unit of classification within the animal kingdom. Divisions, like phyla, are further subdivided into classes, orders, families, and genera.

The classification scheme used in this book recognizes twelve divisions within the kingdom Plantae. In addition to listing the twelve divisions, TABLE 27.1 associates them with the stages of evolution discussed in the preceding section. Although these broader groupings do not have the status of formal taxonomic categories, they help fit the current diversity of plants into the historical context of a long evolutionary journey onto land.

TABLE 27.1

A Classification of Plants

	COMMON NAME	APPROXIMATE NUMBER OF EXTANT SPECIES
Nonvascular Plants*		
Division Bryophyta	Mosses	10,000
Division Hepatophyta	Liverworts	6,500
Division Anthocerophyta	Hornworts	100
Vascular Plants		
Seedless Plants		
Division Psilophyta	Whiskferns	10–13
Division Lycophyta	Club mosses	1,000
Division Sphenophyta	Horsetails	15
Division Pterophyta	Ferns	12,000
Seed Plants		
Gymnosperms		
Division Coniferophyta	Conifers	550
Division Cycadophyta	Cycads	100
Division Ginkgophyta	Ginkgo	1
Division Gnetophyta	Gnetae	70
Angiosperms		
Division Anthophyta	Flowering plants	235,000

*Use of the term *nonvascular* for mosses must be qualified: Water-conducting tissue is present in some species.

Plants probably evolved from green algae called charophytes

For decades, systematists have recognized that green algae are the photosynthetic protists most closely related to plants. (In fact, some systematists now advocate inclusion of green algae in the plant kingdom; see FIGURE 26.26.) The evidence for this kinship includes similarities in cell wall composition and in the structure and pigmentation of chloroplasts. Because there is a great diversity of green algae, recent research has focused on which group of these aquatic organisms represents the plant kingdom's closest algal relatives. Abundant evidence now points to the green algae called **charophytes** (FIGURE 27.4). By comparing cell ultrastructure, biochemistry, and hereditary information (DNA and its RNA and protein products), researchers have turned up homologies between these charophytes and plants, including:

1. *Homologous chloroplasts.* Of all photosynthetic protists, only the green algae match plants in having chlorophyll *b* and beta-carotene as accessory pigments. The chloroplasts of green algae are also similar to plant chloroplasts in having their thylakoid membranes stacked as grana. When molecular systematists compared the chloroplast DNA of various green algae with that of plants, the closest match was between charophytes and plants.

2. *Biochemical similarity.* Cellulose is a structural component in the cell walls of most green algae, a characteristic shared with plants. Among green algae, charophytes are the most plantlike in wall composition, with cellulose making up 20% to 26% of the wall material. Charophytes are also the only algae with peroxisomes matching the enzyme composition of plant peroxisomes.

3. *Similarity in the mechanisms of mitosis and cytokinesis.* During cell division in charophytes and plants, the nuclear envelope completely disperses during late prophase, and the mitotic spindle persists until cytokinesis begins. In charophytes, as in plants, cytokinesis involves the cooperation of microtubules, actin microfilaments, and vesicles in the formation of a cell plate.

FIGURE 27.4

Charophytes, closest algal relatives of the plant kingdom (**a**) *Chara braunii* is a pond organism popular with researchers because of its giant cells (the cylinder between consecutive whorls of "branches" is a single cell). For example, physiologists studying the role of electricity in membrane transport impale these giant cells with electrodes to measure voltage changes. (**b**) *Coleochaete orbicularis* is a charophyte that retains zygotes on the parent after fertilization. Nonreproductive cells grow around the zygotes (inset), perhaps transporting nutrients that enable the zygotes to enlarge before they undergo meiosis.

4. *Similarity in sperm ultrastructure.* In the details of sperm ultrastructure, charophytes are more similar to certain plants than to other green algae.

5. *Genetic relationship.* Molecular systematists have examined certain nuclear genes and ribosomal RNA in charophytes and plants, and the data agree with other evidence placing charophytes as the closest relatives of plants.

It is important to understand that *modern* charophytes, such as those in FIGURE 27.4, are *not* the ancestors of plants. The available evidence, however, supports the hypothesis that modern charophytes and plants both evolved from a common ancestor that would probably be classified as a charophyte. Researchers may learn more about how plants work by studying the physiology of their simpler relatives, the charophytes.

The Origin of Alternation of Generations in Plants

In Chapter 26, you learned that the life cycles of certain brown algae, red algae, and green algae exhibit alternation of haploid and diploid generations. This phenomenon apparently evolved independently many times in algal history. However, alternation of generations does *not* occur among modern charophytes. This implies that alternation of generations had a separate origin in plants; that is, this feature of life history is analogous, not homologous, to the alternation of generations observed in various groups of algae.

How did alternation of generations evolve in the ancestors of plants? We can find clues in certain modern charophytes, including the genus *Coleochaete* (see FIGURE 27.4b). The thallus (body) of *Coleochaete* is haploid. Its mode of sexual reproduction is very unusual compared to that of other algae. Most algae release their gametes to the surrounding water, where fertilization takes place. In contrast, the parental thallus of *Coleochaete* retains the eggs, and after fertilization occurs, the zygotes remain attached to the parent. Nonreproductive cells of the thallus grow around each zygote, which then enlarges, perhaps nourished by the surrounding cells. The grown zygote then undergoes meiosis, releasing haploid swimming spores that develop into new individuals.

Notice that the only diploid stage in the life cycle of *Coleochaete* is the zygote; there is no alternation of *multicellular* diploid and haploid generations. But imagine an ancestral charophyte in which meiosis was delayed until after the zygote first divided by mitosis to give rise to a mass of diploid cells still attached to the haploid parent. Such a life cycle would fit the definition of alternation of generations. In this case, the rudimentary sporophyte (the clump of diploid cells) would be dependent on the gametophyte (the haploid parent). If specialized

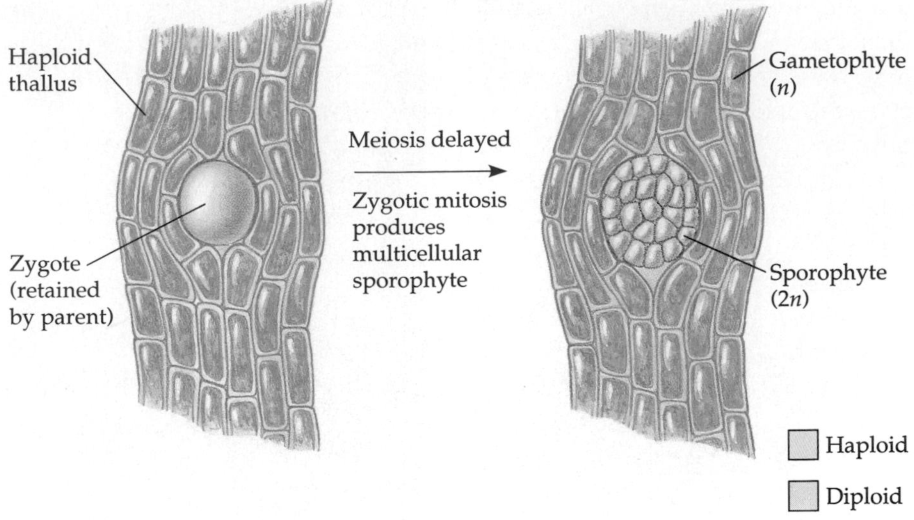

FIGURE 27.5
Delayed meiosis, a hypothetical mechanism for the origin of alternation of generations in the ancestor of plants. According to the hypothesis illustrated here, alternation of generations evolved in the ancestor of plants by the postponement of meiosis until after mitotic division of the zygote produced a mass of diploid cells, the sporophyte. Subsequent meiosis by multiple diploid cells would have increased the number of haploid offspring (gametophytes) possible from each sexual union of sperm and ovum.

cells of the gametophyte formed protective layers around the tiny sporophyte, such a hypothetical ancestor would also qualify as a primitive embryophyte (FIGURE 27.5).

What would be the advantage of delaying meiosis and forming a mass of diploid cells? If the zygote undergoes meiosis directly, each fertilization event results in only a few haploid spores. But *mitotic* division of the zygote to form a sporophyte amplifies the sexual product, with meiosis by the many diploid cells producing a large number of haploid spores. This may have been an important adaptation for maximizing the output of sexual reproduction in environments where a shortage of water decreased the probability of swimming sperm fertilizing eggs.

Adaptations to Shallow Water as Preadaptations for Living on Land

Many species of modern charophytes are found in shallow water around the edges of ponds and lakes. Some of the ancient charophytes that lived about the time land was first colonized may have inhabited shallow water habitats subject to occasional drying. The transition from the Ordovician period to the Silurian period, about 440 million years ago, was a time of mass extinction. Repeated periods of glaciation and other climatic factors changed patterns of rainfall and drought, and contributed to seasonal and longer-term changes in the water levels of lakes and ponds. Among the algae living on the fringes of such bodies of water, natural selection would have favored those that could survive through periods when they were not submerged. Waxy cuticles and the protection of developing gametes and embryos within jacketed organs on the parent are examples of adaptations to living in shallow water that would also prove useful on land.

At least one lineage of organisms that evolved from Ordovician algae accumulated adaptations enabling the organisms to live permanently above the water line. The evolutionary novelties of these first plants opened an adaptive zone that had never before been occupied. The new frontier was spacious, the bright sunlight was unfiltered by water and algae, the soil was rich in minerals, and, at least at first, there were no hervibores on land.

As we survey the diversity of modern plants, remember that the past is the key to the present. Continue to think about the problems and opportunities facing organisms that began living on land.

Bryophytes are embryophytes that generally lack vascular tissue and require environmental water to reproduce

Until recently, the nonvascular plants—mosses, liverworts, and hornworts—were grouped together in a single division, Bryophyta (Gr. *bryon,* "moss"). This book adopts the current view that mosses, liverworts, and hornworts deserve separate divisions because they are probably not closely related. In this taxonomy, only mosses belong to the formal division Bryophyta. However, most plant biologists still use the common name bryophyte to refer to liverworts and hornworts as well as mosses. This informal usage is appropriate, because these three plant groups share some key characteristics.

Bryophytes display two adaptations that first made the move onto land possible. They are covered by a waxy cuticle that helps the body retain water, and their gametes develop within gametangia (see FIGURE 27.1). The male gametangium, known as an **antheridium,** produces flagellated sperm. In each female gametangium, or **archegonium,** one egg (ovum) is produced. The egg is fertilized within the archegonium, and the zygote develops into an embryo within the protective jacket of the female organ. The retention of the zygote and the sporophyte that develops from the zygote is a refined version of the process illustrated in FIGURE 27.5. Thus, bryophyte adaptations to terrestrial conditions include the embryophyte condition.

Even with their cuticles and protected embryos, bryophytes are not totally liberated from their ancestral aquatic habitat. First of all, these plants need water to reproduce, for their sperm, like those of most green algae, are flagellated and must swim from the antheridium to the archegonium to fertilize the egg. For many bryophyte species, a film of rainwater or dew is sufficient for fertilization to occur. In addition, most bryophytes have no vascular tissue to carry water from the soil to the aerial parts of the plant (the exceptions, as mentioned earlier, are certain mosses with elongated water-conducting cells). As water moves over the surface of most bryophytes, they must imbibe it like sponges and distribute it throughout the plant by the relatively slow processes of diffusion, capillary action, and cytoplasmic streaming. This mode of hydration helps explain why damp, shady places are the most common habitats of bryophytes.

Bryophytes lack the woody tissue required to support tall plants on land. Although they may sprawl horizontally as mats over a large surface, bryophytes always have a low profile (FIGURE 27.6). Most are only 1–2 cm in height, and even the largest are usually less than 20 cm tall.

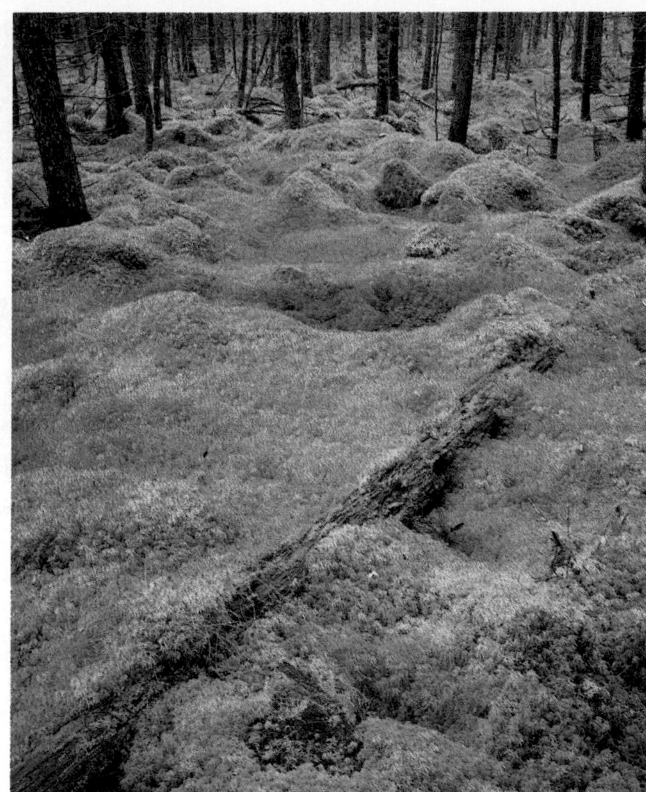

FIGURE 27.6
A moss bog. Lacking rigid supporting tissue, bryophytes are low-profile plants most common in damp habitats. The matlike plants of this green carpet are gametophytes, the dominant generation in the life cycle of bryophytes.

Mosses (Division Bryophyta)

The most familiar bryophytes are **mosses.** A mat of moss actually consists of many plants growing in a tight pack, helping to hold one another up. The mat has a spongy quality that enables it to absorb and retain water. Each plant of the mat grips the substratum with elongate cells or cellular filaments called rhizoids. Most photosynthesis occurs in the upper part of the plant, which has many small stemlike and leaflike appendages. The "stems," "leaves," and "roots" (rhizoids) of a moss, however, are not homologous to these structures in vascular plants.

In the life cycle of a moss, we see a specific example of an alternation of haploid and diploid generations (FIGURE 27.7). The diploid sporophyte produces haploid spores via meiosis in a structure called a **sporangium;** the spores go on to form new gametophytes. The haploid gametophyte is the dominant generation in mosses and other bryophytes. The sporophyte is generally smaller and shorter lived, and it depends on the gametophyte for water and nutrients. This contrasts with the life cycles of vascular plants, where the diploid sporophyte is the dominant generation.

Liverworts (Division Hepatophyta)

Liverworts are even less conspicuous plants than mosses. The bodies of some are divided into lobes, giving an appearance that must have reminded someone of the lobed liver of an animal (*wort* means "herb").

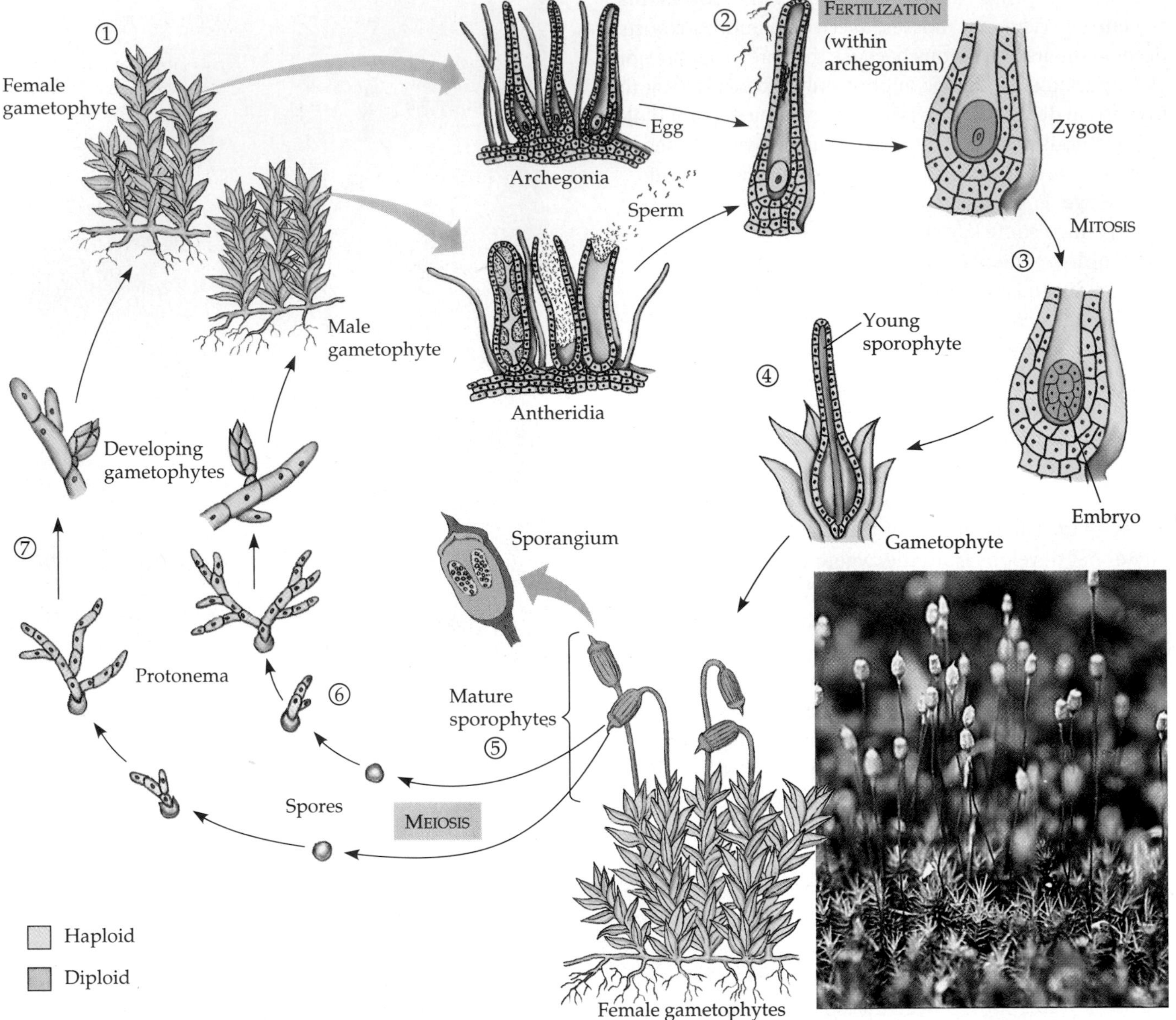

FIGURE 27.7

The life cycle of a moss. The gametophyte is the prevalent generation in the bryophyte life cycle, a characteristic that contrasts with other plants. ① Most species of moss have separate male and female gametophytes, which have antheridia and archegonia, respectively. ② After a sperm swims through a film of moisture to an archegonium and fertilizes the egg, ③ the diploid zygote divides by mitosis and develops into an embryonic sporophyte within the archegonium. ④ During the next stage of its development, the sporophyte grows a long stalk that emerges from the archegonium, but the base of the sporophyte remains attached to the female gametophyte. ⑤ At the tip of the stalk is a sporangium, a capsule in which meiosis occurs and haploid spores develop. When the sporangium bursts, the spores scatter. ⑥ A spore germinates by mitotic division, forming a small, green, threadlike protonema resembling a green alga. ⑦ The haploid protonema continues to grow and differentiates into a new gametophyte, completing the life cycle.

The life cycle of a liverwort is much like that of a moss. Within the sporangia of some liverworts are coil-shaped cells that spring out of the capsule when it opens, helping to disperse the spores. Liverworts can also reproduce asexually from little bundles of cells called gemmae, which are bounced out of cups on the surface of the gametophyte by raindrops (FIGURE 27.8).

Hornworts (Division Anthocerophyta)

Hornworts resemble liverworts but are distinguished by their sporophytes, which are elongated capsules that grow like horns from the matlike gametophyte. The photosynthetic cells of hornworts each have a single large chloroplast rather than the many smaller ones more typical of most plants.

Gemma cup

FIGURE 27.8
Liverworts. The gemmae cups function in asexual reproduction. The tiny plantlets (gemmae) within the cups are dispersed by the impact of raindrops.

The three divisions of bryophytes—mosses, liverworts, and hornworts—have had a long success dating back at least 400 million years, and there are more than 16,000 species today. Bryophytes are elegantly adapted to a limited range of terrestrial habitats, and probably never dominated much of the landscape. The vascular plants have additional terrestrial adaptations that enabled them to claim much more territory.

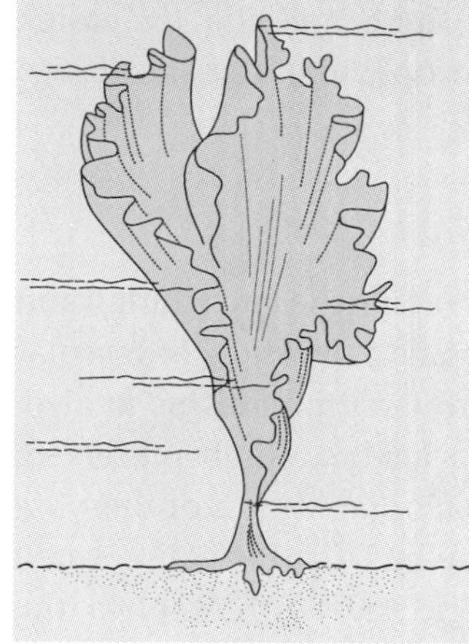
Medium (water) supportive
Whole alga has direct access to environmental water and minerals
Photosynthesis occurs in most cells
Availability of light often limits photosynthesis

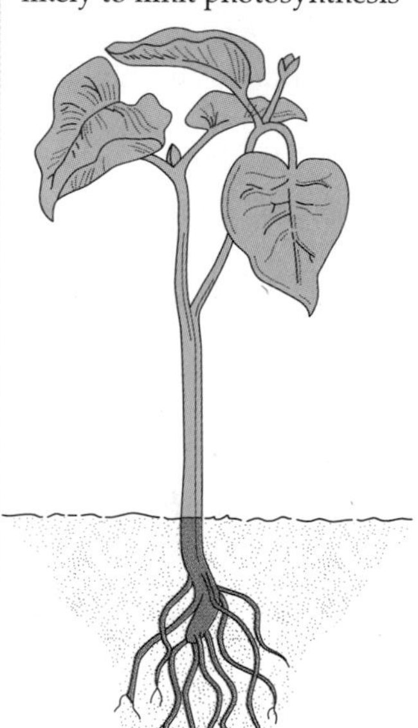
Medium (air) nonsupportive
Aerial parts of plant not in direct contact with water and minerals; tend to lose water to air
Photosynthesis confined to aerial parts of plant
Availability of light less likely to limit photosynthesis

FIGURE 27.9
A comparison of conditions faced by algae and plants.

The origin of vascular tissue was an evolutionary breakthrough in the colonization of land

During their long evolution from aquatic ancestors, vascular plants accumulated many terrestrial adaptations in addition to cuticles and jacketed sex organs. The colonization of land entailed solutions to a new set of problems that aquatic algae did not face (FIGURE 27.9).

The resources a land plant needs to live are spatially separated: The soil provides water and minerals, but there is no light underground for photosynthesis. The body of a vascular plant is differentiated into a subterranean root system that absorbs water and minerals and an aerial shoot system of stems and leaves that makes food.

Regional specialization of the plant body solved one problem but presented new ones. Roots anchor the plant, but for the shoot system to stand up straight in the air, it must have support. This is not a problem in the water: Huge seaweeds need no skeletons because they are buoyed by the surrounding water. An important terrestrial adaptation of vascular plants is lignin, a hard material embedded in the cellulose matrix of the walls of cells that function in support. Turgor pressure (see Chapter 8) contributes to the support of small plants, but it is the skeleton of lignified walls that holds up a tree or any other large vascular plant.

With increasing specialization of the root system and shoot system came the new problem of transporting vital materials between the distant organs. Water and minerals must be conducted upward from the roots to the leaves. Sugar and other organic products of photosynthesis must be distributed from leaves to the roots. These problems are solved by a vascular system that is continuous throughout the plant. The two conducting tissues of the vascular system are **xylem** and **phloem.** Tube-shaped cells in the xylem carry water and minerals up from the roots. These water-conducting cells are actually dead; only their walls remain to provide a system of microscopic water pipes. The walls are generally lignified, and thus xylem functions in support as well as water transport. Phloem is a living tissue with food-conducting cells arranged into tubes that distribute sugar, amino acids, and other organic nutrients throughout the plant. (Transport in plants is discussed further in Chapter 32.)

In some groups of vascular plants, additional adaptations to living on land evolved, including the seed, the replacement of flagellated sperm with pollen as a means of delivering gametes outside water, and the increasing dominance of the diploid sporophyte in the alternation of generations.

The Earliest Vascular Plants

Encased in the sedimentary strata of the late Silurian and early Devonian periods are fossils of a variety of vascular plants, among the oldest terrestrial organisms known. Many of these fossilized plants are beautifully preserved, right down to the microscopic organization of their tissues. The oldest is *Cooksonia,* which has been discovered in Silurian rocks in both Europe and North America (the two continents were probably joined during the Silurian period). It was a simple plant with dichotomous (repeated "Y") branching. Some of the stems terminated in bulbous sporangia (FIGURE 27.10). *Cooksonia* was followed by a diversity of early Devonian species as vascular plants became geographically widespread.

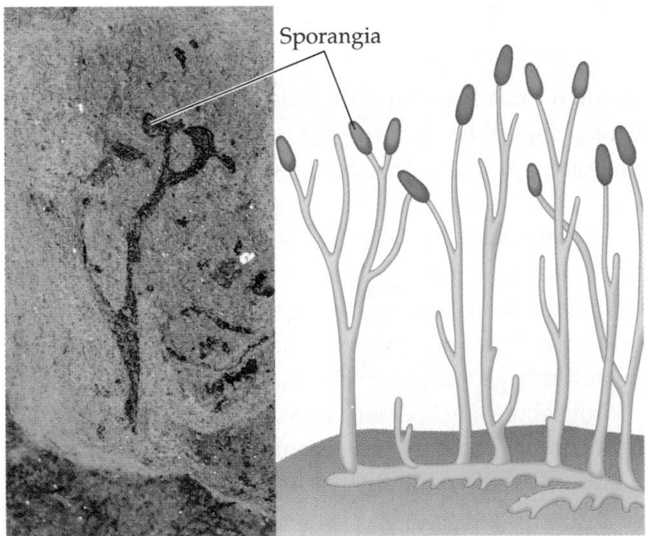

FIGURE 27.10
Cooksonia, a vascular plant of the Silurian. The dichotomous branching and terminal sporangia characteristic of *Cooksonia* are evident in the photograph of the fossil on the left. True roots and leaves were absent. The plant was anchored by a rhizome, a horizontal stem. *Cooksonia* grew in dense stands around marshes. The largest species was about 50 cm tall.

Ferns and other seedless plants dominated the Carboniferous "coal forests"

The earliest vascular plants were seedless, and they dominated the forest landscapes of the Carboniferous period. Although less widespread and diverse than during the Carboniferous, four divisions of seedless plants are represented among the modern flora. As you read about these seedless vascular plants and compare them to the bryophytes, notice an important evolutionary trend in life cycles: In vascular plants, the sporophyte is the dominant generation. This contrasts to the bryophytes, in which the gametophyte is the dominant generation.

Division Psilophyta

This division of relatively simple plants has only two genera. The better known is *Psilotum,* which is widespread in the tropics and subtropics. It is known in the United States by the common name whiskfern (although it is not a true fern). In the diploid sporophyte generation, *Psilotum* has dichotomous branching reminiscent of some of the early vascular plants (FIGURE 27.11). True roots and leaves are absent. The subterranean part of the plant consists of a rhizome (horizontal stem) covered with tiny hairs called rhizoids. The upright stems bear emergences, which, unlike true leaves, lack vascular tissue.

Lycopods (Division Lycophyta)

The extant **lycopods,** of the division Lycophyta, are relicts of a far more eminent past. Lycopods first evolved during the Devonian period and became a major part of the landscape during the Carboniferous period, which began about 340 million years ago and lasted until 280 million years ago. By that time, the division Lycophyta split into two evolutionary lines. One group evolved into woody trees that had diameters as large as 2 m and heights of more than 40 m. A second line of lycopods remained small and herbaceous (nonwoody). The giant lycopods thrived in the Carboniferous swamps for millions of years, but became extinct when the swamps began to dry up at the end of that geological period. The small lycopods survived, and they are represented today by about a thousand species, most belonging to the genera *Lycopodium* and *Selaginella.* Common names for these plants are club mosses or ground pines, though they are neither mosses nor pines.

Many species of *Lycopodium* are tropical plants that grow on trees as **epiphytes**—plants that use another organism as a substratum but are not parasites. Other species of *Lycopodium* grow close to the ground on forest

FIGURE 27.11
The sporophyte of *Psilotum*. The scales on these dichotomously branched stems are not true leaves; they lack vascular tissue. The knobs along the stems are sporangia, which release haploid spores that germinate in the soil. Tiny subterranean gametophytes lack chlorophyll and depend for their food on symbiotic soil fungi that decompose organic matter. Flagellated sperm swim through the moist soil from antheridia to archegonia of the gametophytes. The zygote develops into an embryo within the archegonium, and soon a young sporophyte emerges from the gametophyte, which then dies.

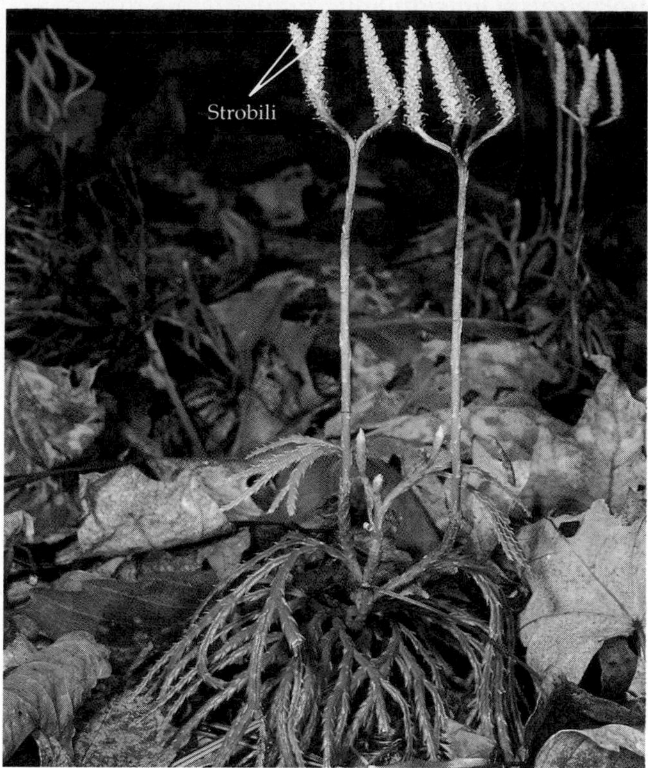

FIGURE 27.12
***Lycopodium,* a club moss.** Club mosses are common inhabitants of forest floors in the northeastern United States. The small plant has a horizontal rhizome that gives rise to roots and vertical branches and has true leaves containing strands of vascular tissue. The sporangia of *Lycopodium* are borne by specialized leaves called sporophylls. In some species, such as the one shown here, the sporophylls are clustered at the tips of branches into club-shaped structures called strobili (hence the common name club mosses).

floors in temperate regions, including the northeastern United States.

The club moss in FIGURE 27.12 is the sporophyte, the diploid generation. The sporangia of *Lycopodium* are borne on **sporophylls,** leaves specialized for reproduction. After their discharge, the spores develop into inconspicuous gametophytes that may live underground for ten years or longer. These tiny haploid plants are, like the gametophytes of *Psilotum,* nonphotosynthetic, and are nurtured by symbiotic fungi. Each gametophyte develops archegonia with eggs and antheridia that make flagellated sperm. After a swimming sperm fertilizes an egg, the diploid zygote gives rise to a new sporophyte.

Lycopodium makes a single type of spore, which develops into a bisexual gametophyte having both female and male sex organs (archegonia and antheridia); it is thus said to be **homosporous.** In **heterosporous** plants, such as the lycopod genus *Selaginella,* the sporophyte makes two kinds of spores. **Megaspores** develop into female gametophytes bearing archegonia, and **microspores** become male gametophytes with antheridia. The gametophytes of heterosporous plants are unisexual, either female or male. The diagrams below will help you compare the homosporous and heterosporous conditions:

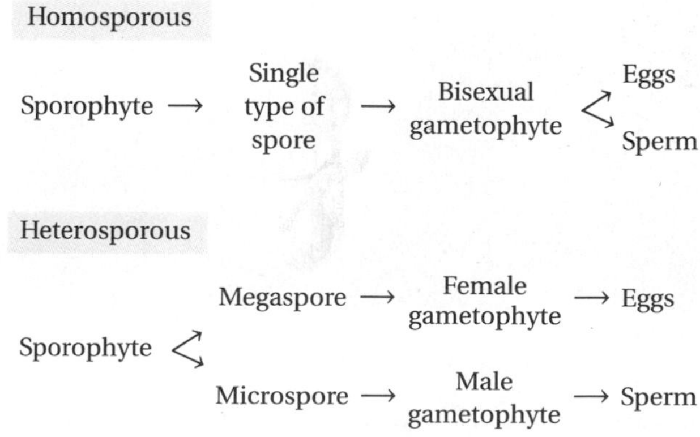

We will encounter the homosporous and heterosporous conditions again as we continue our survey of vascular plants.

Horsetails (Division Sphenophyta)

Sphenophyta, whose members are commonly called **horsetails,** is another ancient lineage of seedless plants dating back to the Devonian radiation of early vascular plants. The group reached its zenith during the Carboniferous period, when many species grew as tall as 15 m. All that survives of this division of plants are about 15 species of a single genus, *Equisetum. Equisetum* is widely distributed but is most common in the Northern Hemisphere, generally in damp locations such as stream banks (FIGURE 27.13).

The conspicuous horsetail plant is the sporophyte generation. Meiosis occurs in the sporangia, and haploid spores are released. The gametophytes that develop from these spores are only a few millimeters long, but they are photosynthetic and free-living (not dependent on the sporophyte for food). Horsetails are homosporous; the single type of spore gives rise to a bisexual gametophyte with both antheridia and archegonia. Flagellated sperm fertilize eggs in the archegonia, and young sporophytes later emerge.

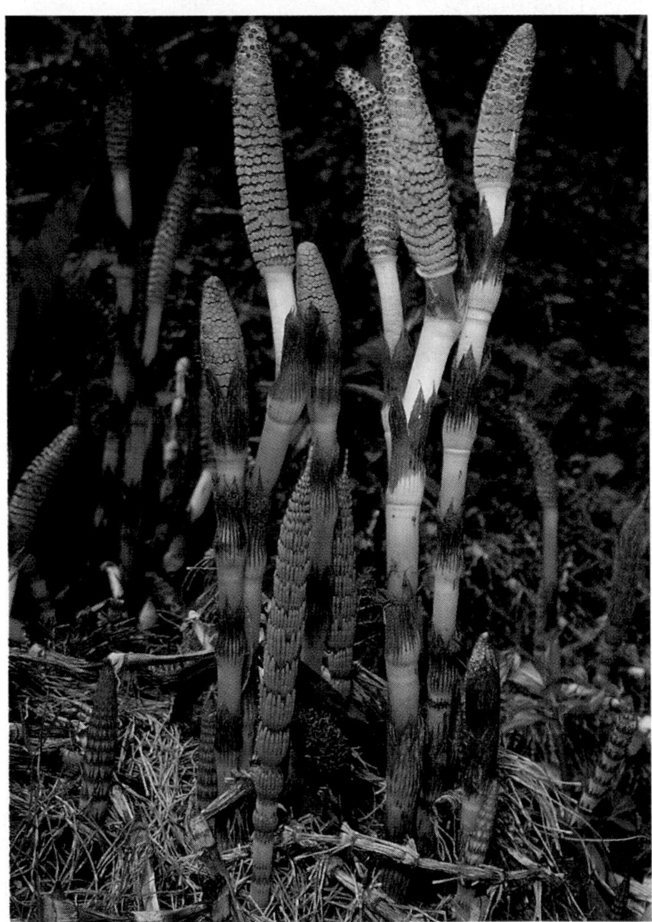

FIGURE 27.13
***Equisetum* (horsetail).** *Equisetum* has an underground rhizome from which vertical stems arise. The straight, hollow stems are jointed, and whorls of small leaves or branches emerge at the joints. At the tips of some stems of *Equisetum* are conelike structures bearing sporangia. The epidermis, the outer layer of cells, is embedded with silica, which gives the plants an abrasive texture. Before scouring pads, people used the abrasive stems of horsetails to scrub pots and pans, which is why these plants are also known as scouring rushes.

Ferns (Division Pterophyta)

From their Devonian beginning, **ferns** (division Pterophyta) radiated into the many species that stood alongside tree lycopods and horsetails in the great forests of the Carboniferous period. Of all seedless plants, ferns are by far the most extensively represented in the modern floras (FIGURE 27.14). More than 12,000 species of ferns live today. They are most diverse in the tropics, but a variety of species is also found in temperate forests.

The leaves of ferns are generally much larger than those of lycopods and probably evolved in a different way. The origin of leaves is currently a subject of much research. The small leaves of lycopods probably evolved as emergences from the stem that contained a single strand of vascular tissue. Leaves with this origin are called microphylls. Each leaf of a fern, termed a megaphyll, has a branched system of veins. Megaphylls probably evolved by the formation of webbing between many separate branches growing close together.

Most ferns have leaves, commonly called fronds, that are compound, meaning each leaf is divided into several leaflets. The frond grows as its coiled tip, the fiddlehead, unfurls. The leaves may sprout directly from a prostrate stem, as they do in brackens and sword ferns. Large tropical tree ferns, by contrast, have upright stems many meters tall.

FIGURE 27.14
Ferns. An ostrich fern *(Matteuccia)* growing on a forest floor in New York. The fiddleheads in the inset are young fronds ready to unfurl.

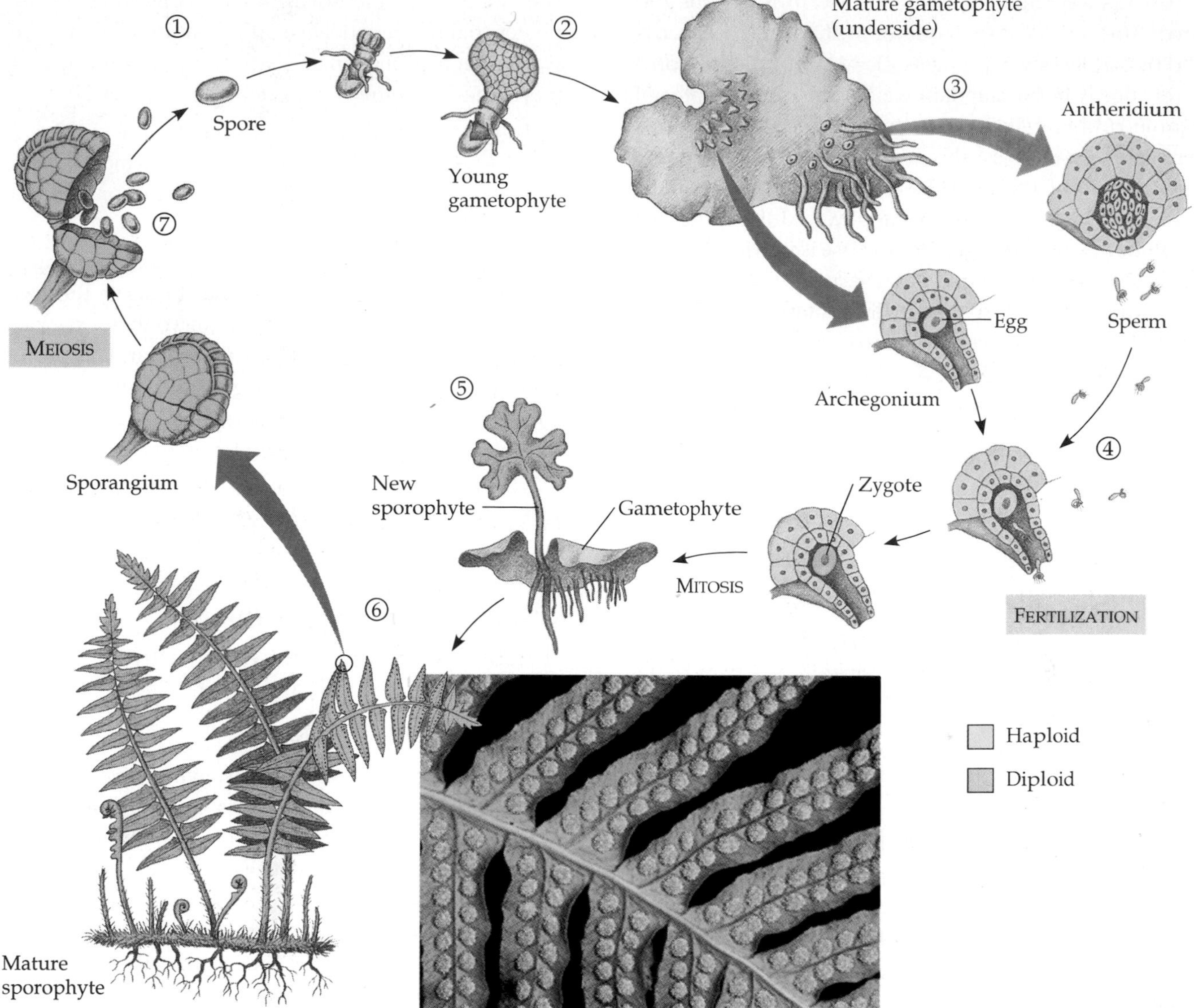

FIGURE 27.15

The life cycle of a fern. ① After a fern spore settles in a favorable place, ② it develops into a small, heart-shaped gametophyte that sustains itself by photosynthesis. ③ Most ferns are homosporous. Each gametophyte has both male and female sex organs, but the archegonia and antheridia usually mature at different times, assuring cross-fertilization between gametophytes. ④ Fern sperm, like those of club mosses and horsetails, use flagella to swim through moisture from antheridia to eggs in the archegonia and then fertilize the egg. A sex attractant secreted by archegonia helps direct the sperm. ⑤ A fertilized egg develops into a new sporophyte, and the young plant grows out from an archegonium of its parent, the gametophyte. ⑥ The spots on the underside of reproductive leaves (sporophylls) are called sori (photograph). Each is a cluster of sporangia, ⑦ which release the spores that give rise to gametophytes.

FIGURE 27.15 illustrates the life cycle of a fern. The leafy fern plant familiar to us is the sporophyte generation. Some of the leaves are specialized sporophylls with sporangia on their undersides. The sporangia of many ferns are arranged in clusters called sori and are equipped with springlike devices that catapult spores several meters. Once airborne, spores can be blown by the wind far from their origin. With their swimming sperm and fragile gametophytes, most ferns are restricted to relatively damp habitats.

The Coal Forests

The four divisions of plants that we have just surveyed represent the extant lineages of seedless vascular plants that formed vast forests during the Carboniferous period (about 300 to 350 million years ago). Seedless plants of the Carboniferous forests left not only living relicts but also fossilized fuel in the form of coal.

Coal formed during several geological periods, but the most extensive beds of coal are found in strata deposited during the Carboniferous, a time when most of the continents were flooded by shallow swamps. Europe and North America, near the equator at that time, were covered by tropical swamp forests. Dead plants did not completely decay in the stagnant waters, and great depths of organic rubble called peat accumulated. The swamps were later covered by the sea, and marine sediments piled on top of the peat. Heat and pressure gradually converted the peat to coal. Coal powered the Industrial Revolution, and a resurgence in its use is inevitable as we continue to deplete oil and gas reserves.

Growing along with the seedless plants in the Carboniferous swamps were primitive seed plants. These gymnosperms were not the dominant plants at that time, but they rose to prominence after the swamps began to dry up at the end of the Carboniferous period.

Reproductive adaptations catalyzed the success of the seed plants

Three life cycle modifications contributed to the success of seed plants as terrestrial organisms:

1. *The gametophytes of seed plants became even more reduced than in ferns and other seedless plants.* Rather than developing in the soil as an independent generation, the minute gametophytes of seed plants are protected from desiccation by being retained within the moist reproductive tissue of the sporophyte generation. Notice that this evolutionary trend reverses the gametophyte-sporophyte relationship as it probably originated in the ancestor of plants (see FIGURE 27.5). Some plant biologists speculate that the shift toward diploidy in land plants was related to the harmful impact of the sun's ionizing radiation, which causes mutations. This damaging radiation is more intense on land than in aquatic habitats because of the light-filtering properties of water. Of the two generations of land plants—gametophyte and sporophyte—the diploid form (sporophyte) may cope better with mutagenic radiation. A diploid organism homozygous for a particular essential allele has a "spare tire" in the sense that one copy of the allele may be sufficient for survival if the other is damaged. According to this hypothesis, the increasing prevalence of sporophytes during the evolution of vascular plants can be interpreted as another adaptation to terrestrial conditions.

Why has the gametophyte generation not been completely eliminated from the plant life cycle? One speculation is that small, relatively simple gametophytes provide a mechanism for "screening" alleles, including new mutations, with a minimum investment of the parent plant's resources. Since the gametophyte is haploid, no alleles can "hide" from the environment. Only those gametophytes that are genetically competent in the existing environment have a chance of surviving and producing gametes that combine to start new sporophytes.

2. *Pollination replaced swimming as the mechanism for delivering sperm to eggs.* Pollen, which contains sperm cells, is disseminated by the wind or by insects and other animals.

3. *The seed evolved.* Instead of the zygote developing into an embryonic sporophyte that fends for itself, the zygote of a seed plant develops into an embryo that is packaged along with a food supply within a seed coat. This protects the dormant embryo from drought, cold, and other harsh conditions. Seeds also function in overland dispersal; they may be carried far from their parents by wind, water, or animals. In seed plants, the seed has replaced the spore as the stage in the life cycle that disperses offspring.

Gymnosperms began to dominate landscapes as climates became drier at the end of the Paleozoic era

Of the two groups of seed plants, gymnosperms appear much earlier than angiosperms in the fossil record. Gymnosperms lack the enclosed chambers (ovaries) in which angiosperm seeds develop.

There are four divisions of gymnosperms (FIGURE 27.16). Three are relatively small divisions: Cycadophyta, Ginkgophyta, and Gnetophyta. The cycads (division Cycadophyta) resemble palms but are not true palms, which are flowering plants. Being gymnosperms, cycads bear naked seeds on the scales of cones. The ginkgos (division Ginkgophyta) have fanlike leaves that turn gold and are deciduous in autumn, an unusual trait for a gymnosperm. Division Gnetophyta consists of three genera that are probably not closely related. One, *Weltwitschia*, is shown in FIGURE 27.16c. Plants of the second genus, *Gnetum*, grow in the tropics as trees or vines. *Ephedra* (Mormon tea), the third genus of Gnetophyta, is a shrub in the American deserts. By far the largest of the four gymnosperm divisions is Coniferophyta, the conifers.

Conifers (Division Coniferophyta)

The term **conifer** (L. *conus*, "cone," and *ferre*, "to carry") comes from the reproductive structure of these plants, the cone. Pines, firs, spruce, larches, yews, junipers, cedars, cypresses, and redwoods all belong to this division of gymnosperms. Most are large trees. Although there are only about 550 species, conifers dominate

(a)

(b)

(c)

(d)

FIGURE 27.16

The four divisions of gymnosperms. (**a**) **Cycadophyta.** Cycads, such as this *Cycas revoluta,* resemble palms and in fact are sometimes called sago palms. (**b**) **Ginkgophyta.** The ginkgo, also known as the maidenhair tree, is a popular ornamental tree in cities because it can survive air pollution and other environmental insults. (**c**) **Gnetophyta.** *Welwitschia,* shown here, lives only in the deserts of southwestern Africa. Its straplike leaves are the largest known leaves. (**d**) **Coniferophyta.** Pines, firs, and redwoods are among the cone-bearing plants, or conifers. This giant sequoia *(Sequoiadendron giganteum),* the General Grant tree in California's Kings Canyon National Park, is over 80 m tall.

vast forested regions of the Northern Hemisphere, where the growing season is relatively short because of latitude or altitude.

Nearly all conifers are evergreens, meaning they retain leaves throughout the year. Even during winter, a limited amount of photosynthesis occurs on sunny days. And when spring comes, conifers already have fully developed leaves that can take advantage of the sunnier days.

The needle-shaped leaves of pines and firs are adapted to dry conditions. A thick cuticle covers the leaf, and the stomata are located in pits, further reducing water loss. The conifer needle, despite its shape, is a megaphyll, as are the leaves of all seed plants (see p. 558).

We get most of our lumber and paper pulp from the wood of conifers. What we call wood is actually an accumulation of lignified xylem tissue, which gives the tree structural support.

Coniferous trees are among the tallest, largest, and oldest living organisms on Earth. Redwoods, found only in a narrow coastal strip of Northern California, grow to heights of more than 110 m; only certain eucalyptus trees in Australia are taller. The largest (most massive) organisms alive are the giant sequoias, relatives of redwoods that grow in the Sierra Nevada mountains of California (see FIGURE 27.16d). One, known as the General Sherman tree, has a trunk with a circumference of 26 m and weighs more than the combined weight of a dozen space shuttles. Bristlecone pines, another species of California conifer, are among the oldest organisms alive. One bristlecone, named Methuselah, is more than 4600 years old; it was a young tree when humans invented writing.

The Life History of a Pine

The pine tree, a representative conifer, is a sporophyte, with its sporangia located on cones. The gametophyte generation develops from haploid spores that are retained within the sporangia. Conifers are heterosporous; male and female gametophytes develop from different types of spores produced by separate cones. Each tree usually has both types of cones. Small pollen cones produce small microspores that develop into the male gametophytes. Larger, more complex ovulate cones usually develop on separate branches of the tree and make larger megaspores that develop into female gametophytes (FIGURE 27.17). From the time young cones appear on the tree, it takes nearly three years for a complicated series of events to produce mature seeds. The scales of the ovulate cone then separate, and the winged seeds travel on the wind. A seed that lands in a habitable place germinates, its embryo emerging as a pine seedling. TABLE 27.2 puts the complicated life cycle of a conifer into perspective by highlighting some important ways it compares to the life cycles of other plants.

The History of Gymnosperms

Gymnosperms probably descended from a group of Devonian plants called progymnosperms. They were originally seedless plants, but by the end of the Devonian period, seeds had evolved. Adaptive radiation during the Carboniferous and early Permian produced the various divisions of gymnosperms.

In the history of life, the Permian period was one of great crises. Formation of the supercontinent Pangaea (see Chapter 23) may have been one reason that continental interiors became warmer and drier as the Permian progressed. The flora and fauna of Earth changed dramatically, as many groups of organisms disappeared and others emerged as their successors. The changeover was most pronounced in the seas, but terrestrial life was affected as well. In the animal kingdom, amphibians decreased in diversity and were replaced by reptiles, which were better adapted to the arid conditions. Similarly, the lycopods, horsetails, and ferns that dominated the Carboniferous swamps were largely replaced by conifers and cycads, which were more suited to the drier climate. The world and its life had changed so markedly that geologists use the end of the Permian period, about 248 million years ago, as the boundary between the Paleozoic and Mesozoic eras. (This boundary was originally defined by the changeover in marine fossils.) The Mesozoic is sometimes referred to as the age of dinosaurs, when giant reptiles were supported by a vegetation consisting mostly of conifers and great palmlike cycads. When the climate changed again at the end of the Mesozoic, becoming cooler, the dinosaurs became extinct. Some of the gymnosperms, particularly conifers, persisted, however, and are still an important part of Earth's flora.

TABLE 27.2

A Comparison of Reproduction for Some Major Plant Groups

GROUP	DOMINANT STAGE OF LIFE CYCLE	HOMOSPOROUS OR HETEROSPOROUS	MECHANISM FOR COMBINING GAMETES
Mosses	Gametophyte	Heterosporous (most species)	Flagellated sperm swims through film of water to egg
Ferns	Sporophyte	Homosporous	Flagellated sperm
Conifers	Sporophyte	Heterosporous	Sperm transported in windblown pollen
Flowering plants	Sporophyte	Heterosporous	Pollen transferred by wind or animals

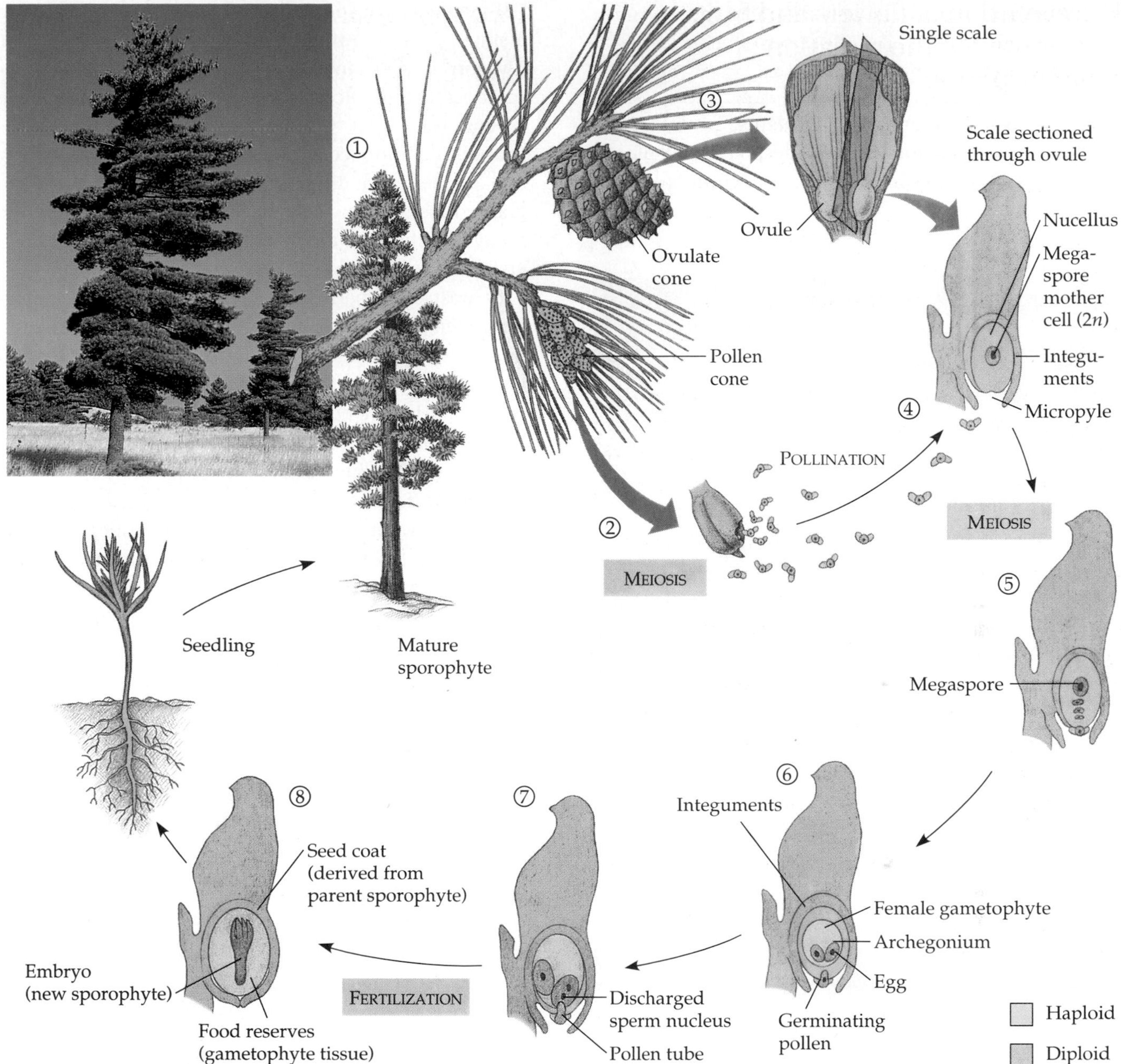

FIGURE 27.17

The life cycle of a pine. ① Trees (sporophytes) of most species bear both pollen cones and ovulate cones. ② A pollen cone contains hundreds of sporangia held in small reproductive leaves. Cells in the sporangia undergo meiosis, giving rise to haploid microspores that develop into pollen grains (immature male gametophytes). ③ An ovulate cone consists of many scales, each with two ovules. Each ovule includes a sporangium, called the nucellus, enclosed in protective integuments with a single opening, the micropyle. ④ During pollination, windblown pollen falls on the ovulate cone and is drawn into the ovule through the micropyle. The pollen grain germinates in the ovule, forming a pollen tube that begins to digest its way through the nucellus. Fertilization usually occurs more than a year after pollination. During that year, ⑤ a megaspore mother cell in the nucellus undergoes meiosis to produce four haploid cells. One of these cells survives as a megaspore, which divides repeatedly, giving rise to the immature female gametophyte. ⑥ Two or three archegonia, each with an egg, then develop within the gametophyte. ⑦ By the time eggs are ready to be fertilized, two sperm cells have developed in the male gametophyte (pollen grain) and the pollen tube has grown through the nucellus to the female gametophyte. Fertilization occurs when one of the sperm nuclei, injected into an egg cell by the pollen tube, unites with the egg nucleus. All the eggs in an ovule may be fertilized, but usually only one zygote develops into an embryo. ⑧ The pine embryo, or the new sporophyte, has a rudimentary root and several embryonic leaves. A food supply, consisting of the female gametophyte, surrounds and nourishes the embryo. The ovule has developed into a pine seed, which consists of an embryo (new sporophyte), its food supply (derived from gametophyte tissue), and a surrounding seed coat derived from the parent tree (parent sporophyte).

The evolution of flowers and fruits contributed to the radiation of angiosperms

Today, angiosperms, or flowering plants, are by far the most diverse and geographically widespread of all plants. About 235,000 species are now known, compared with 721 gymnosperm species. All angiosperms are placed in a single division, Anthophyta (Gr. *antho,* "flower"). The division is split into two classes: Monocotyledones (monocots) and Dicotyledones (dicots), which differ in several ways that will be described in Chapter 31. Examples of monocots are lilies, orchids, yuccas, palms, and grasses, including lawn grasses, sugar cane, and grain crops (corn, wheat, rice, and others). Among the many dicot families are roses, peas, buttercups, sunflowers, oaks, and maples (FIGURE 27.18).

In most angiosperms, insects and other animals transfer pollen to female sex organs, which makes pollination less random than the wind-dependent pollination of gymnosperms. (Some flowering plants are wind-pollinated, but we do not know whether this condition is primitive or whether it evolved secondarily from ancestors that were pollinated by animals.)

Vascular tissue also became more refined during angiosperm evolution. The cells that conduct water in conifers are **tracheids,** believed to be a relatively early type of xylem cell (FIGURE 27.19). The tracheid is an elongated, tapered cell that functions in both mechanical support and movement of water up the plant. In most angiosperms, shorter, wider cells called **vessel elements** evolved from tracheids. Vessel elements are arranged end to end, forming continuous tubes that are more specialized than tracheids for transporting water but less specialized for support. The xylem of angiosperms is reinforced by a second cell type, the **fiber,** which also evolved

(a)

(b)

FIGURE 27.18

Two classes of angiosperms. Venation, or pattern of veins, is one characteristic that differs between the two classes. (**a**) Monocots, such as these pink lady's slipper orchids, generally have leaves with parallel veins. (**b**) Dicots, such as this blue violet, usually have leaves with netlike veins.

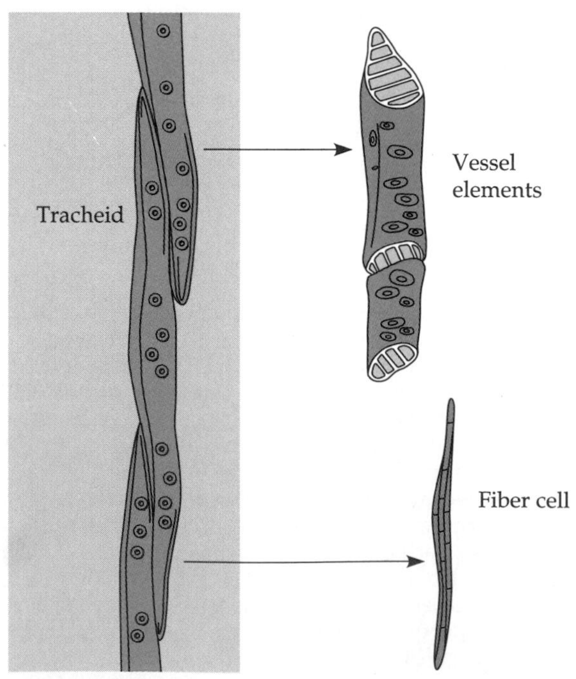

FIGURE 27.19

The evolution of xylem cells. In angiosperm evolution, tracheids gave rise to vessel elements specialized for conducting water and fiber cells specialized for support.

from the tracheid. With their thick lignified walls, the xylem fibers are specialized for support. Fiber cells evolved in conifers, but vessel elements did not.

The refinements in vascular tissue and other structural advances surely contributed to the success of angiosperms, but the greatest factor in the rise of angiosperms was probably the evolution of the flower, a remarkable apparatus that enhances the efficiency of reproduction by attracting and rewarding pollen-carrying animals.

The Flower

The **flower** is the reproductive structure of an angiosperm. A flower is a compressed shoot with four whorls of modified leaves (FIGURE 27.20). Starting at the bottom of the flower are the **sepals,** which are usually green. They enclose the flower before it opens (think of a rosebud). Above the sepals are the **petals,** brightly colored in most flowers. They aid in attracting insects and other pollinators. Flowers that are wind-pollinated, such as those of many grasses, are generally drab in color. The sepals and petals are sterile floral parts not directly involved in reproduction. Within the ring of petals are the reproductive organs, **stamens** and **carpels.** A stamen consists of a stalk called the **filament** and a terminal sac, the **anther,** where pollen is produced. At the tip of the carpel is a sticky **stigma** that receives pollen. A **style** leads to the **ovary** at the base of the carpel. Protected within the ovary are the ovules, which develop into seeds after fertilization. Recall that the enclosure of seeds within the ovary is one of the features that distinguishes angiosperms from gymnosperms. The carpel probably evolved from a seed-bearing leaf that became rolled into a tube.

Botanists recognize four evolutionary trends in the flower structure of various angiosperm lineages:

1. The number of floral parts has become reduced.
2. Floral parts have become fused. For example, some flowers have compound carpels formed by the fusion of several carpels. The general term *pistil* is sometimes used for a single carpel or several fused carpels.
3. Symmetry has changed from radial, in which any cut down its central axis will divide the flower into two equal halves, to bilateral, in which the flower has distinct left and right halves.
4. The ovary has dropped to a position below the petals and sepals, where the ovules are better protected.

With modification in floral structure, many angiosperms are specialists at using specific animals for pollination, as we will see later in the chapter.

The Fruit

A **fruit** is a mature ovary. As seeds develop after fertilization, the wall of the ovary thickens. A pea pod is an example of a fruit, with seeds (mature ovules, the peas) encased in the ripened ovary (the pod). Fruits protect dormant seeds and aid in their dispersal (FIGURE 27.21).

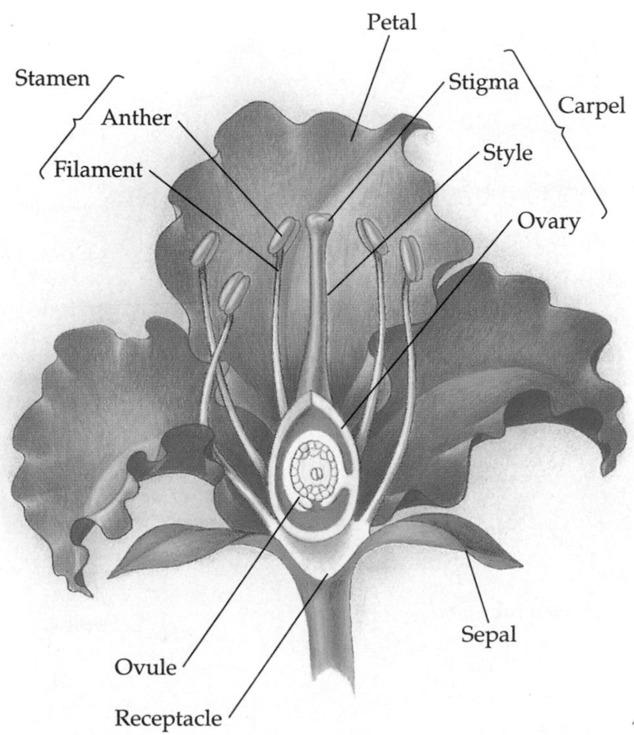

FIGURE 27.20
The structure of a flower.

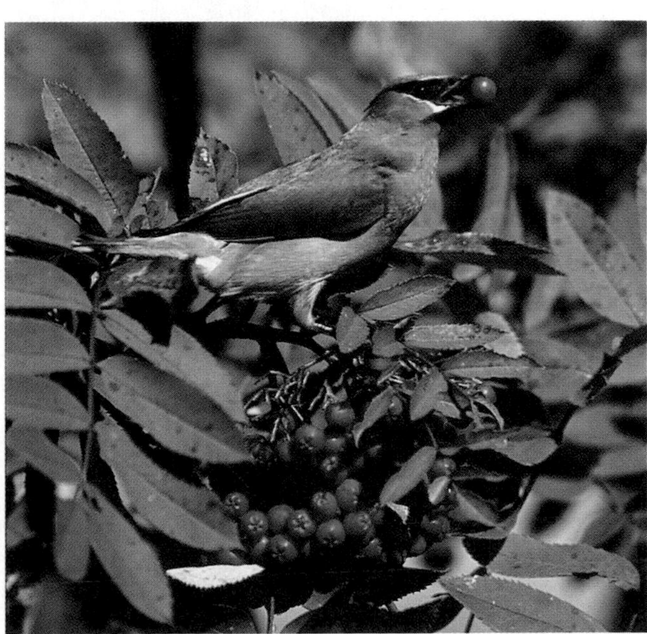

FIGURE 27.21
Fruit as an adaptation for seed dispersal. After pollination and fertilization, the ovules of flowers develop into seeds and the ovaries develop into fruits, such as these mountain ash berries. Edible fruits are injested by animals, most commonly mammals or birds, such as this cedar waxwing. Seed coats usually prevent digestion of the seeds, which the animal may deposit in its feces some distance from the parent plant.

Various modifications in fruits help disperse seeds. Some flowering plants, such as dandelions and maples, have seeds within fruits acting as kites or propellers that aid in dispersal by wind. Most angiosperms use animals to carry seeds. Some of these plants have fruits modified as burrs that cling to animal fur (or the clothes of humans). Other angiosperms produce edible fruits. When it eats the fruit, the animal digests the fleshy part, but the tough seeds usually pass unharmed through the digestive tract. Mammals and birds may deposit seeds, along with a fertilizer supply, miles from where the fruit was eaten. Interactions with animals that tote seeds and pollen has helped angiosperms become the most successful plants on Earth.

The Life Cycle of an Angiosperm

Angiosperms are heterosporous. The flower of the sporophyte produces microspores that form male gametophytes and megaspores that produce female gametophytes (FIGURE 27.22). The immature male gametophytes are **pollen grains,** which develop within the anthers of stamens. Each pollen grain has two haploid cells. **Ovules,** which develop in the ovary, contain the female gametophyte, an **embryo sac** with eight haploid nuclei in

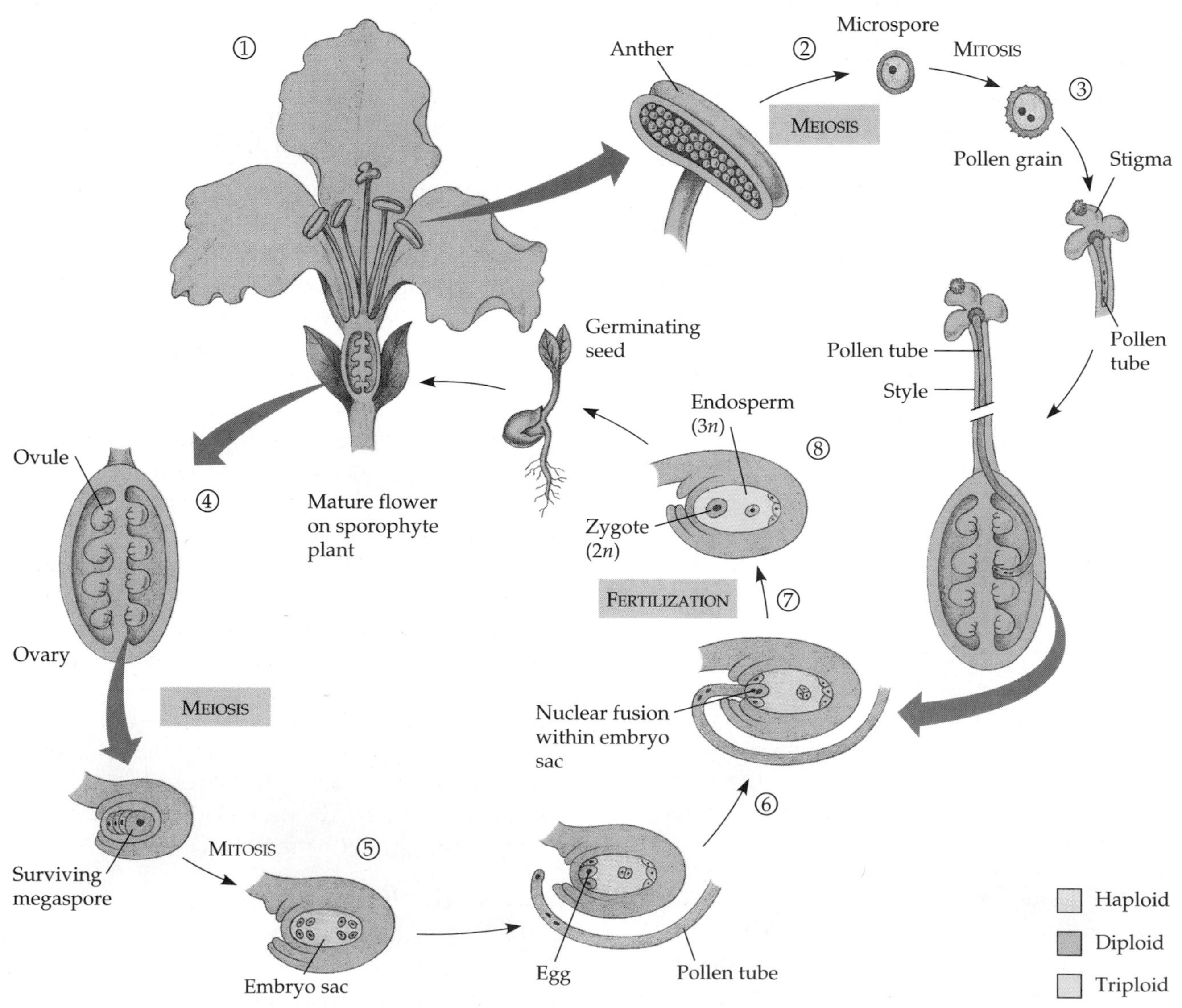

FIGURE 27.22

The life cycle of an angiosperm. ① The flower of the sporophyte produces ② microspores that form ③ male gametophytes (pollen) and ④ megaspores that form ⑤ female gametophytes (embryo sacs) within ovules. ⑥ Pollination brings the gametophytes together in the ovary. ⑦ Fertilization occurs, and ⑧ zygotes develop into sporophyte embryos that are packaged along with food into seeds.

seven cells (a large central cell has two haploid nuclei). One of the cells is the egg. (The development of pollen and the embryo sac will be described in more detail in Chapter 34.)

After its release from the anther, the pollen is carried to the sticky stigma at the tip of a carpel. Although some flowers self-pollinate, most have mechanisms that ensure **cross-pollination,** the transfer of pollen from flowers of one plant to flowers of another plant of the same species. For example, stamens and carpels of a single flower may mature at different times, or the organs may be so arranged within the flower that self-pollination is unlikely.

The pollen grain germinates after it adheres to the stigma of a carpel. Now a mature male gametophyte, the pollen grain extends a tube that grows down the style of the carpel. After it reaches the ovary, the pollen tube penetrates through a pore in the integuments of the ovule and discharges two sperm cells into the embryo sac. One sperm nucleus unites with the egg, forming a diploid zygote. The other sperm nucleus fuses with two nuclei in the center cell of the embryo sac. This central cell now has a triploid ($3n$) nucleus. The pollen of conifers, remember, also releases two sperm nuclei, but one disintegrates. In contrast, both sperm nuclei of angiosperm pollen fertilize cells in the embryo sac. This phenomenon, known as **double fertilization,** is characteristic of angiosperms. (Researchers have recently discovered that double fertilization also occurs in *Ephedra,* a member of division Gnetophyta, the gymnosperm most closely related to angiosperms.)

After double fertilization, the ovule matures into a seed. The zygote develops into a sporophyte embryo with a rudimentary root and one or two seed leaves, the **cotyledons** (monocots have one seed leaf and dicots have two). The triploid nucleus in the center of the embryo sac divides repeatedly, giving rise to a triploid tissue called **endosperm,** rich in starch and other food reserves. Monocot seeds such as corn store most of their food in the endosperm. Beans and many other dicots transfer most of the nutrients from the endosperm to the developing cotyledons.

What is the function of double fertilization? According to one hypothesis, double fertilization synchronizes the development of food storage in the seed with development of the embryo. If a particular flower is not pollinated or sperm cells are not discharged into the embryo sacs, fertilization does not occur and neither embryo nor endosperm forms. Perhaps the requirement for double fertilization prevents flowering plants from squandering nutrients on infertile ovules.

The seed is a mature ovule, consisting of the embryo, endosperm, and a seed coat derived from the integuments (outer layers of the ovule). In a suitable environment, the seed germinates. The coat ruptures and the embryo emerges as a seedling, using the food stored in the endosperm and cotyledons.

The Rise of Angiosperms

Darwin called the origin of the angiosperms an "abominable mystery." The mystery endures. The problem is the relatively sudden appearance of angiosperms in the fossil record. The oldest fossils that are widely accepted as angiosperms are found in rocks of the early Cretaceous period, about 120 million years old. They are sparsely represented among a much greater abundance of ferns and gymnosperms. By the end of the Cretaceous, 65 million years ago, the angiosperms had radiated and become the dominant plants on Earth, as they are today.

Some paleobotanists believe the relative suddenness of the appearance of angiosperms during the early Cretaceous is an artifact of an imperfect fossil record. Angiosperms may have originated somewhat earlier in the highlands or some other location where fossilization was unlikely, and their sudden appearance may be the result of their spread to locations where fossils form more frequently. (In fact, paleobotanists have recently discovered what may be angiosperm pollen in Triassic rocks about 200 million years old.) Another view, in the spirit of the evolutionary theory known as punctuated equilibrium (see Chapter 23), holds that angiosperms actually did evolve and radiate rather abruptly (in geological time) from a gymnosperm ancestor.

Whatever and whenever their origin, the rise to prominence of angiosperms during the Cretaceous is amply documented in the fossil record. The end of the Cretaceous was another crisis period when many old groups of organisms were replaced by new ones. Cooler climates may have contributed to the changeover. Again, the frequency of extinctions was greatest in the seas, but significant changes in terrestrial fauna and flora also occurred. The dinosaurs disappeared, as did many of the cycads and conifers that had thrived during the Mesozoic era. They were replaced by mammals and flowering plants. The change in fossils during the late Cretaceous is so extreme that geologists use the end of that period as the boundary between the Mesozoic and Cenozoic eras.

Relationships Between Angiosperms and Animals

Ever since they followed plants onto the land, animals have influenced the evolution of terrestrial plants, and vice versa. The fact that animals must eat affects the natural selection of both animals and plants. For instance, with animals crawling and foraging for food on the forest floor, natural selection must have favored plants that kept their spores and gametophytes up in the treetops, rather than dropping these critical structures to hungry animals on the ground. This, in turn, may have been a selection factor in the evolution of flying insects. On the other hand, as plants with flowers and fruits evolved, some herbivores became beneficial to the plants by carrying the pollen and seeds of plants they used as food. Certain animals became specialists at these tasks, feeding on specific plants. Natural selection reinforced these interactions, for they improved the reproductive success of both partners. The plant got pollinated and the animal got fed. The mutual evolutionary influence between two species is termed **coevolution.** (This definition will be refined in Chapter 48.)

Coevolution of angiosperms and their pollinators is partly responsible for the diversity of flowers. Some flowers are pollinated by a specific animal, such as a particular type of bee, beetle, bird, or bat (FIGURE 27.23). These exclusive relationships ensure that the plant's pollen will not be wasted by being carried to the flower of a different species. At the same time, the pollinator has a monopoly on a food source.

In most cases, relationships between plants and their pollinators are less specific than in the extreme coevolution between one plant species and one animal species. For example, the flowers of a particular plant species may be adapted for attracting insects rather than birds, but

(a)

(b)

(c)

FIGURE 27.23
Relationships between angiosperms and their pollinators. (**a**) This scotch broom has a tripping mechanism that dusts pollen onto the back of a visiting bee. (**b**) Some flowers have nectaries at the bottom of long tubes. This butterfly *(Heliconius erato)* has a slender proboscis (coiled in this photograph) that reaches the nectary. (**c**) Wahlberg's epauletted bat *(Epomophorus wahlbergi)* feeds on the baobab flower. Its body collects and distributes the plant's pollen. The baobab, like many plants that depend on bats, has blossoms that bloom at night. They are light-colored, large, and scented, and are therefore easily found by nocturnal feeders.

many different insect species may serve as pollinators. Conversely, a single animal species—a honeybee species, for example—may pollinate many different plant species. But even in these less specific relationships, flower color, fragrance, and structure usually reflect specialization for a particular *group* of pollinators, such as various species of bees or hummingbirds. Flowers pollinated by bees, for instance, often have nectar guides, markings that help direct the bee as it taxies to the nectaries, glands that secrete the sugary solution bees eat. On its way to or from the nectar, the bee becomes dusted with pollen (FIGURE 27.23a). The markings on some bee-pollinated flowers are invisible to humans but are apparently vivid to bees, whose eyes are sensitive to ultraviolet light. Flowers pollinated by birds are usually red, to which bird eyes are especially sensitive.

The shape of the flower may also specify a particular group of pollinators. Flowers pollinated by hummingbirds, for instance, have their nectaries located deep in a floral tube where only the long, thin tongues of hummingbirds are likely to reach.

Relationships between angiosperms and animals are also evident in the edible fruits of angiosperms. Fruits that are not yet ripe are usually green, hard, and distasteful (at least to humans). This helps the plant retain its fruit until the seeds are mature and ready for dispersal. As it ripens, the fruit becomes softer and its sugar content increases. Many fruits also become fragrant and brightly colored, advertising their ripeness to animals. One of the most common colors for ripe fruit is red, which insects cannot see very well. Thus, most of the fruit is saved for birds and mammals, animals large enough to disperse the seeds (see FIGURE 27.21). Again, we see that one of the keys to angiosperm success has been interaction with animals.

Angiosperms and Agriculture

Flowering plants provide nearly all our food. All of our fruit and vegetable crops are angiosperms. Corn, rice, wheat, and the other grains are grass fruits. The endosperm of the grain seeds is the main food source for most of the people of the world and their domesticated animals. We also grow angiosperms for fiber, medications, perfumes, and decoration.

Like other animals, early humans probably collected wild seeds and fruits. Agriculture was gradually invented as humans began sowing seeds and cultivating plants to have a more dependable food source. As they domesticated certain plants, humans began to intervene in plant evolution by selective breeding designed to improve the quantity and quality of the foods the crops produced. We have developed a very special relationship with the plants we cultivate. We water them and fertilize them, try to protect them from insects, and plant their seeds. Many of these plants are so genetically removed from their origins that they probably could not survive in the wild. Agriculture is a unique case of an evolutionary relationship between plants and animals.

Plant diversity is a nonrenewable resource

The exploding human population and its demand for space and natural resources are extinguishing plant species at an unprecedented rate. The problem is especially critical in the tropics, where more than half the human population lives and population growth is fastest. Tropical rain forests are being destroyed at a frightening pace. The most common cause of this destruction is slash-and-burn clearing of the forest for agricultural use. Fifty million acres, an area about the size of the state of Washington, are cleared each year, a rate that would completely eliminate Earth's tropical forests within 25 years. As the forest disappears, so do thousands of plant species. Extinction is irrevocable; plant diversity is a nonrenewable resource. Insects and other rainforest animals that depend on these plants are also vanishing. In all, researchers estimate that the destruction of habitat in the rain forest and other ecosystems is claiming hundreds of species per year. The toll is greatest in the tropics, because that is where most species live; but the environmental assault is a generically human tendency. Europeans eliminated most of their forests centuries ago, and in North America, habitat destruction is endangering many species (FIGURE 27.24).

FIGURE 27.24
Forest mismanagement. This old-growth forest in Oregon was clear-cut to harvest lumber.

Many people are ethically troubled about contributing to the extinction of living forms. But there are also selfish reasons to be concerned about the loss of plant diversity. We depend on plants for thousands of products, including food, building materials, and medicines. So far, we have explored the potential uses of only a tiny fraction of the 250,000 known plant species. For example, almost all our food is based on the cultivation of only about two dozen species. More than 120 prescription drugs are extracted from plants. However, researchers have investigated fewer than 5000 plant species as potential sources of medicine. Pharmaceutical companies were led to most of these species by local people who use the plants in preparing their traditional medicines.

The tropical rain forest may be a medicine chest of healing plants that could be extinct before we even know they exist. This is only one reason to value what is left of plant diversity and to search for ways to slow the loss. The solutions we propose must be economically realistic. If the goal is only profit for the short term, then we will continue to slash and burn until the forests are gone. If, however, we begin to see rain forests and other ecosystems as living treasures that can regenerate only slowly, we may learn to harvest their products at sustainable rates. This will only work if the local people who are the custodians of plant diversity in their region receive a fair share of profits from the development of medicine and other plant products. What else can we do to preserve plant diversity? Few questions are as important.

* * *

In this chapter, we have tracked 460 million years of plant evolution, from the descendants of green algae that moved onto shore to the flowering plants that now dominate most landscapes. We have seen once again that the organisms of today are best understood by their evolutionary history. This chapter has also reinforced the theme of the connectedness of biology and geology. Transitions such as changes in climate certainly influenced plant evolution, but plants have also changed the course of geology by altering soil and transforming the land in many other ways. The next chapter considers the history and environmental impact of fungi, another kingdom of organisms that spread onto land with the plants.

REVIEW OF KEY CONCEPTS (with page numbers and key figures)

- Structural and reproductive adaptations made the colonization of land possible: *an overview of plant evolution* (pp. 547–550, FIGURE 27.3)
  - All plants are photosynthetic multicellular eukaryotes. Stomata and the cuticle of stems and leaves are two important terrestrial adaptations.
  - As plants colonized the land, jacketed sex organs, called gametangia, evolved and protected gametes and embryos from desiccation.
  - All modern plants show a heteromorphic alternation of generations, with distinctive haploid gametophyte and diploid sporophyte forms.
  - Important innovations in plant structure catalyzed four major periods of plant evolution. First, terrestrial adaptations enabled plants to make the transition from water to land about 460 million years ago. The second major period occurred with the emergence of vascular tissue in certain plants. Third, the origin of the seed about 360 million years ago enabled embryos to leave the parent plant. The fourth major episode occurred 130 million years ago with the evolution of the flower, a specialized reproductive structure that produces seeds enclosed within an ovary.
- Plants probably evolved from green algae called charophytes (pp. 550–552, FIGURE 27.4)
  - Molecular systematics, comparative biochemistry, and the analysis of cell ultrastructure place charophytes as the closest algal relatives of plants.
  - In descending from the ancestral charophyte, plants may have acquired alternation of generations by a delay in meiosis of a zygote retained by the haploid parent (gametophyte). Mitosis by the zygote would have produced a sporophyte.
- Bryophytes are embryophytes that generally lack vascular tissue and require environmental water to reproduce (pp. 552–555, FIGURE 27.7)
  - The mostly nonvascular bryophytes include three plant divisions: Bryophyta (mosses), Hepatophyta (liverworts), and Anthocerophyta (hornworts). Bryophytes have a waxy cuticle and gametangia that protect the gametes and the embryo on land, but they still require a moist habitat for fertilization and for imbibing water in the absence of vascular tissue. Lack of woody tissue dictates their short stature.
  - The haploid gametophyte is the dominant stage in the life cycle of bryophytes.
- The origin of vascular tissue was an evolutionary breakthrough in the colonization of land (pp. 555–556, FIGURE 27.9)
  - In vascular plants, distant organs are interconnected by the vascular tissue, with water and minerals transported by xylem and organic nutrients by phloem.
  - Over evolutionary time, there was a trend toward increasing dominance of the diploid sporophyte in the life cycles of vascular plants.

- Ferns and other seedless plants dominated the Carboniferous "coal forests" (pp. 556–560, FIGURE 27.15)
  - Division Psilophyta is a small group with relatively simple structures to which the widespread genus *Psilotum* (whisk-ferns) belongs.
  - Division Lycophyta consists of the lycopods, or club mosses—small, herbaceous survivors of a dominant ancient division that once included large treelike forms.
  - The horsetails belong to a single surviving genus of the division Sphenophyta.
  - Division Pterophyta consists of the ferns, the most species-rich group of living seedless plants. The fronds of the sporophyte generation form sporangia that produce spores germinating into small gametophytes.
- Reproductive adaptations catalyzed the success of the seed plants (p. 560)
  - The success of seed plants on land may be attributed to three developments: further reduction of the gametophyte and its retention within the sporophyte; replacement of swimming sperm with pollination; and development of the seed, which functions in protection and dispersal of embryos.
- Gymnosperms began to dominate landscapes as climates became drier at the end of the Paleozoic era (pp. 560–563, FIGURE 27.17)
  - Coniferophyta is the largest of the four gymnosperm divisions. Almost all conifers are evergreens with needle-shaped leaves.
  - The cone is the distinguishing characteristic of conifers. In the pine, two different types of cones produce male and female gametophytes on the sporophyte tree. Pollen grains released as immature male gametophytes land on female cones housing immature female gametophytes inside complex ovules. After a period of gametophyte maturation, fertilization occurs. The zygote develops into an embryo, which is packaged into a winged seed that disperses by the wind.
  - Three smaller divisions of gymnosperms are the Cycadophyta (cycads), Ginkgophyta (ginkgo), and Gnetophyta.
- The evolution of flowers and fruits contributed to the radiation of angiosperms (pp. 564–569, FIGURE 27.22)
  - Angiosperms, or flowering plants, belong to the division Anthophyta, which contains the most diverse and widespread members of the plant kingdom. Angiosperms are divided into monocots and dicots.
  - The origin of refined xylem tissue for transport and support contributed to the rise of the angiosperms. But the greatest contribution to their success was probably the evolution of the flower, which greatly improved reproductive efficiency.
  - The flower is a reproductive structure housing stamens and carpels within sterile sepals and petals.
  - Fruits form from the ripened ovaries of flowers. They protect dormant seeds and are modified in various ways that aid in seed dispersal.
  - Pollen grains are immature male gametophytes that form in the anthers of stamens and germinate on the sticky stigma of the carpel. A pollen tube grows down to the ovary, where one sperm cell fertilizes the egg and another cell combines with two female haploid cells to make a triploid endosperm that functions in food storage. This double fertilization produces a seed containing the embryonic sporophyte surrounded by the endosperm and seed coat.
  - Interactions between plants and animals has contributed to the adaptive radiation of angiosperms.
  - A special case of an evolutionary relationship between plants and animals is the human invention of agriculture.
- Plant diversity is a nonrenewable resource (pp. 569–570)
  - Destruction of rain forests and other ecosystems is eliminating thousands of plants.

SELF-QUIZ

1. Which characteristic of plants is *absent* in their closest relatives, the charophytes?
   a. chloroplyll *b*
   b. cellulose in cell walls
   c. alternation of generations
   d. sexual reproduction
   e. formation of a cell plate during cytokinesis

2. All bryophytes (mosses, liverworts, and hornworts) share certain characteristics. These are
   a. reproductive cells in protective chambers and a waxy cuticle
   b. a waxy cuticle, true leaves, and reproductive cells in protective chambers
   c. vascular tissues, true leaves, and a waxy cuticle
   d. reproductive cells in protective chambers and vascular tissues
   e. vascular tissues and a waxy cuticle

3. Which of the following is *not* common to all divisions of vascular plants?
   a. the development of seeds
   b. alternation of generations
   c. dominance of the diploid generation
   d. xylem and phloem
   e. the addition of lignin to cell walls

4. A heterosporous plant is one that
   a. produces a gametophyte that bears both sex organs
   b. produces microspores and megaspores in separate sporangia, giving rise to separate male and female gametophytes
   c. is a seedless vascular plant
   d. produces two kinds of spores, one asexually by mitosis and one type by meiosis
   e. reproduces only sexually

5. During the Carboniferous period, the dominant plants, which later formed the great coal beds, were mainly
   a. the giant lycopods, horsetails, and ferns
   b. the conifers
   c. the angiosperms
   d. charophytes
   e. the bryophytes that dominated early swamps

6. The male gametophyte of an angiosperm is the
   a. anther
   b. embryo sac
   c. microspore
   d. germinated pollen grain
   e. ovule

7. A fruit is most commonly
   a. a mature ovary
   b. a thickened style
   c. an enlarged ovule
   d. a modified root
   e. a mature female gametophyte

8. Important terrestrial adaptations that evolved *exclusively* in seed plants include all of the following *except*
   a. pollination by wind or animal instead of fertilization by swimming sperm
   b. transport of water through vascular tissue
   c. retention of the gametophyte plant within the sporophyte
   d. dispersal of new plants by seeds
   e. protection and nourishment of the embryo within the seed

9. A land plant produces flagellated sperm and the dominant generation is diploid. The plant is most likely a
   a. fern
   b. moss
   c. conifer
   d. charophyte
   e. dicot

10. Plant diversity is greatest in
    a. tropical forests
    b. deserts
    c. salt marshes
    d. the temperate forests of Europe
    e. farmlands

CHALLENGE QUESTIONS

1. The history of life has been punctuated by several mass extinctions. The impact of a meteorite may have wiped out the dinosaurs and many forms of marine life at the end of the Cretaceous period. Fossils indicate that plants were much less severely affected by this and other mass extinctions. What adaptations may have enabled plants to withstand these disasters better than animals?

2. Contrast a seed plant to an alga in terms of adaptations for life on land versus life in the water.

3. In the cladogram below, discuss at least one derived character (synapomorphy; see Chapter 23) that defines the branch points indicated by the arrows.

GREEN ALGAE MOSSES FERNS GYMNOSPERMS ANGIOSPERMS

SCIENCE, TECHNOLOGY, AND SOCIETY

1. Why are tropical rain forests being destroyed at such an alarming rate? What kinds of social, technological, and economic factors are responsible? Most forests in the northern developed countries have already been cut. (In the United States, less than 2% of the forest outside Alaska remains undisturbed.) Do the developed nations have a right to ask the southern developing nations to slow or stop the destruction of their forests? Defend your answer. What kinds of benefits, incentives, or programs might slow the assault on the rain forest?

2. Plants have coexisted with parasitic organisms such as certain bacteria and fungi for millions of years. Defensive adaptations include many plant chemicals that inhibit the growth of these pathogens. How might these chemicals be useful to humans?

FURTHER READING

Cox, P. A., and M. J. Balick. "The Ethnobotanical Approach to Drug Discovery." *Scientific American,* June 1994. What we can learn about medicinal plants from other cultures.

Diamond, J. "How to Tame a Wild Plant." *Discover,* September 1994. How we domesticated food plants.

Graham, L. E. *Origin of Land Plants.* New York: Wiley, 1993. What can plant biologists learn by studying the algal relatives of plants?

Holden, C. "Ancient Trees Down Under." *Science,* January 20, 1995. Botanists discover a "living fossil" near Sydney, Australia, a conifer species thought to be extinct for 50 million years.

Lewington, A. *Plants for People.* New York: Oxford University Press, 1990. The many uses of plant products.

Palmer, D. "First Flowers Emerge from Triassic Mud." *New Scientist,* January 29, 1994. Paleobotanists may push back the origin of angiosperms by 100 million years.

Retallack, G. J. "Permian-Triassic Life on Land." *Science,* January 6, 1995. How the forests changed after a mass extinction.

Stewart, D. "Green Giants." *Discover,* April 1990. Why redwoods and sequoias, the largest of all organisms, face extinction.

CHAPTER 28

FUNGI

KEY CONCEPTS

- Structural and life history adaptations equip fungi for an absorptive mode of nutrition
- The three major divisions of fungi differ in details of reproduction
- Molds, yeasts, lichens, and mycorrhizae represent unique lifestyles that evolved independently in all fungal divisions
- Fungi have a tremendous ecological impact
- Fungi and animals probably evolved from a common protistan ancestor

The words fungus *and* mold *may evoke some unpleasant images. Fungi rot timbers, attack plants, spoil food, and afflict humans with athlete's foot and worse maladies. However, ecosystems would collapse without fungi to decompose dead organisms, fallen leaves, feces, and other organic materials, thus recycling vital chemical elements back to the environment in forms other organisms can assimilate. Virtually all plants depend on symbiotic fungi that help their roots absorb minerals and water from the soil. In addition to these ecological roles, fungi have been used by humans in various ways for centuries. We eat some fungi (mushrooms, for instance), culture fungi to produce antibiotics and other drugs, add them to dough to make bread rise, and use them to ferment beer and wine. Many fungi are also quite beautiful, as Canada recognized in issuing the stamps shown on this page. Whatever our initial subjective perceptions may be, fungi are fascinating as objects of study. They are a form of life so distinctive that they have been accorded their own taxonomic kingdom.*

This chapter is an introduction to fungi. We will characterize the members of the kingdom Fungi, survey their diversity, discuss their ecological and commercial impact, and consider hypotheses about their evolutionary origin. As we did with the plant kingdom, we will look in some detail at life cycles, mainly for what they tell us about the phylogeny and evolutionary adaptations of fungi.

Structural and life history adaptations equip fungi for an absorptive mode of nutrition

Fungi are eukaryotes, and nearly all are multicellular. Fungi were once grouped with plants in the two-kingdom system. Now it is clear that fungi are not primitive or degenerate plants lacking chlorophyll, but are unique organisms that generally differ from other eukaryotes in nutritional mode, structural organization, and growth and reproduction.

Nutrition and Habitats

All fungi are heterotrophs. In contrast to animals—heterotrophs that, for the most part, ingest food—fungi acquire their nutrients by **absorption.** In this mode of nutrition, small organic molecules are absorbed from the surrounding medium. A fungus digests food outside its body by secreting powerful hydrolytic enzymes into the food. The enzymes decompose complex molecules to the simpler compounds that the fungus can absorb and use.

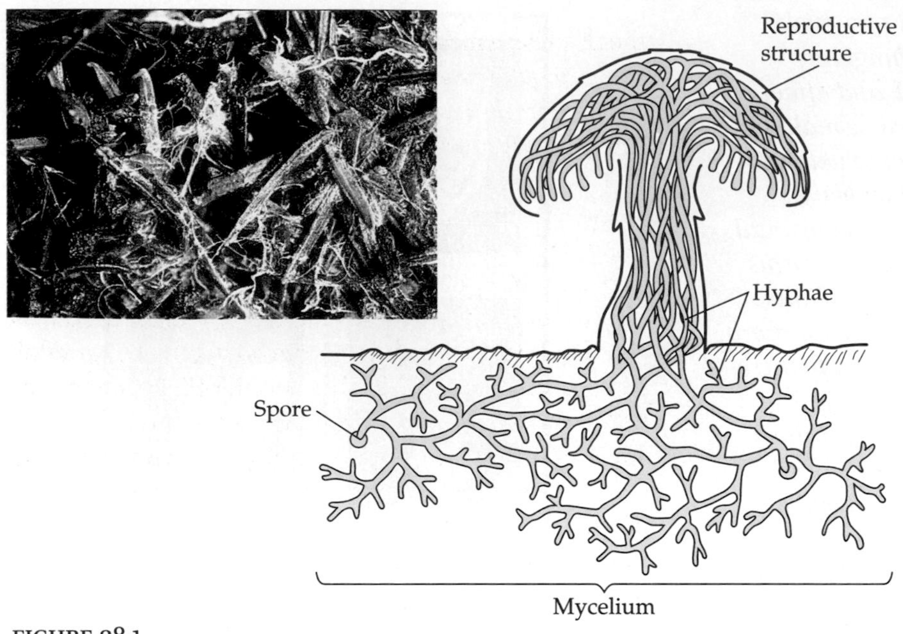

FIGURE 28.1

A fungal mycelium. The drawing illustrates the relationship of the thin hyphae that make up a mycelium to the visible structure we recognize as a mushroom. The mushroom functions in reproduction by producing tiny cells called spores. (Other kinds of fungi have different types of reproductive structures.) At left, the vegetative part of a mycelium is shown decomposing conifer needles, while at right, the mushrooms (*Mycena*) that grow up from this mycelium each fall are shown. A fungal mycelium begins with the germination of a fungal spore in a suitable habitat.

Their absorptive mode of nutrition specializes fungi as saprobes, parasites, or mutualistic symbionts. *Saprobic fungi* absorb nutrients from nonliving organic material, such as fallen logs, animal corpses, or the wastes of live organisms. In the process of this saprobic nutrition, fungi decompose the organic material. *Parasitic fungi* absorb nutrients from the cells of living hosts. Some of these fungi, such as certain species infecting human lungs, are pathogenic. *Mutualistic fungi* also absorb nutrients from another organism, but they reciprocate with functions beneficial to their partners in some way, such as aiding a plant in the uptake of minerals from the soil.

Fungi inhabit diverse environments and are associated symbiotically with many organisms. Although most common in terrestrial habitats, some fungi inhabit aquatic environments where they are associated with both marine and freshwater organisms and their remains. Lichens, symbiotic associations of fungi and algae, are widespread, and are found in some of the most inhospitable habitats on Earth: cold, dry deserts on Antarctica and alpine and arctic tundras. Other symbiotic fungi live inside the healthy tissues of plants, and still other species form cellulose-consuming mutualisms with insects, including ants and termites.

Structure

The vegetative (nutritionally active) bodies of most fungi are usually hidden, being diffusely organized around and within the tissues of their food sources. Except in yeasts, these bodies are constructed of basic building units called **hyphae** (singular, **hypha**) (FIGURE 28.1). Hyphae are minute threads composed of tubular walls surrounding plasma membranes and cytoplasm. The cytoplasm contains the usual eukaryotic organelles. The hyphae form an interwoven mat called a **mycelium,** the "feeding" network of a fungus. Fungal mycelia can be huge, although they usually escape our notice—because they are subterranean, for instance. Researchers in Washington state recently reported a mycelium of a single individual of the fungus *Armillaria ostoyae* that was spread through 6.5 km^2 of soil. This fungus is probably thousands of years old and hundreds of tons in weight, qualifying it among Earth's oldest and largest organisms.

Most fungal hyphae are divided into cells by cross-walls, or **septa** (singular, **septum**). The septa generally have pores large enough to allow ribosomes, mitochondria, and even nuclei to flow from cell to cell (FIGURE 28.2a). The cell walls of fungi differ from the cellulose walls of plants. Most fungi build their walls mainly of **chitin,** a strong but flexible nitrogen-containing polysaccharide similar to the chitin found in the external skeletons of insects and other arthropods. Some fungi are aseptate; that is, their hyphae are not divided into cells by cross-walls. Know as **coenocytic** fungi, they consist of a continuous cytoplasmic mass with hundreds or thousands of nuclei (FIGURE 28.2b). The coenocytic con-

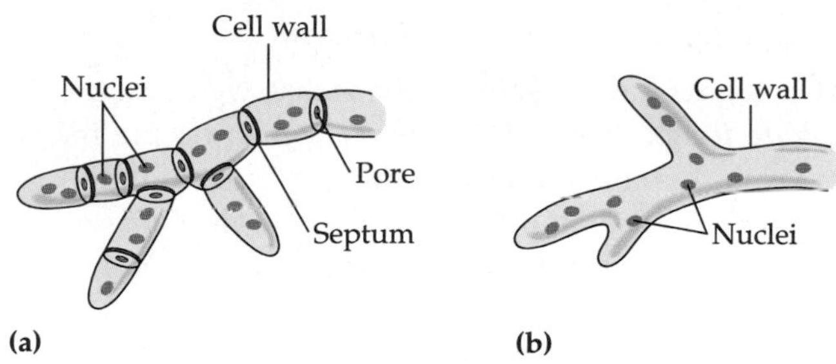

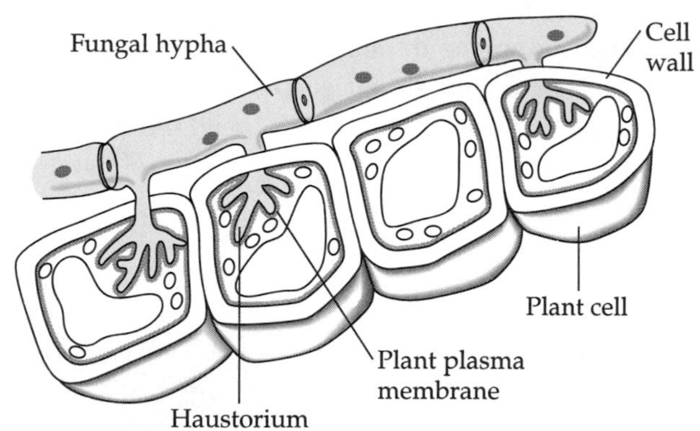

FIGURE 28.2

Characteristics of fungal hyphae. (**a**) Septate hyphae. (**b**) Coenocytic hyphae. (**c**) Specialized hyphae called haustoria parasitize the host cell from *outside,* separated from the host cell's cytoplasm by the plasma membrane of the plant cell (gray). (The effect is like sinking your fingers into an underinflated balloon.)

dition results from the repeated division of nuclei without cytoplasmic division.

The correlation between structure and function is a fundamental theme of biology. The filamentous structure of the mycelium provides an extensive surface area that suits the absorptive nutrition of fungi: 10 cm^3 of rich organic soil may contain as much as 1 km of hyphae. If these hyphae were 10 μm in diameter, about 314 cm^2 of fungal surface area would interface with that 10 cm^3 of soil. Parasitic fungi usually have some of their hyphae modified as **haustoria,** nutrient-absorbing hyphal tips that penetrate the tissues of the host (FIGURE 28.2c).

Growth and Reproduction

A fungal mycelium grows rapidly, adding as much as a kilometer of hyphae each day as it branches within a food source. Such fast growth is possible because proteins and other materials synthesized by the entire mycelium are channeled by cytoplasmic streaming to the tips of the extending hyphae. The fungus concentrates its energy and resources on adding hyphal length rather than girth, another growth pattern adapted to the absorptive lifestyle. Fungi are nonmotile organisms; they cannot run, swim, or fly in search of food or mates. But the mycelium makes up for the lack of mobility by swiftly extending the tips of its hyphae into new territory.

The chromosomes and nuclei of fungi are relatively small, and the nuclei divide in a manner different from that of most other eukaryotes. From prophase to anaphase of mitosis, the nuclear envelope remains intact around an internal spindle. Following anaphase, the nuclear envelope pinches in two, and the spindle disappears.

Fungi reproduce by releasing spores that are produced either sexually or asexually. Fungal spores come in all shapes and sizes. Usually they are unicellular, but multicellular spores also occur. They are produced in, or from, specialized hyphal compartments. When conditions are habitable and stable, fungi generally clone themselves by producing enormous numbers of spores asexually. Carried by wind or water, the spores germinate if they land in a moist place where there is an appropriate substratum, or surface. Spores thus function in dispersal and account for the wide geographical distribution of many species of fungi. The airborne spores of fungi have been found more than 160 km (100 mi) above Earth.

For many fungi, as for protists, sex is a contingency mode of reproduction that occurs when there has been some change in the environment. Compared to asexual reproduction, sexual reproduction results in greater genetic diversity among offspring. As raw material for natural selection, this individual variation among offspring may contribute to adaptation in changing environments.

The nuclei of fungal hyphae and spores are haploid, except for transient diploid stages that form during sexual life cycles. However, some mycelia may become genetically heterogeneous through the fusion of two hyphae that have genetically different nuclei. In some cases, the different nuclei stay in separate parts of the same mycelium, which is then a mosaic in terms of genotype and phenotype. In other cases, the different nuclei mingle and may even exchange chromosomes and genes in a process similar to crossing over.

A special case of genetic heterogeneity occurs during the sexual cycle of fungi. Syngamy, the sexual union of cells from two individuals, occurs in two stages that are separated in time. These two stages of syngamy are called **plasmogamy** (the fusion of cytoplasm) and **karyogamy** (the fusion of nuclei). After plasmogamy, the nuclei from each parent pair up but do not fuse, forming a **dikaryon** ("two nuclei"). The pairs of nuclei may coexist and divide in tandem in a dikaryotic cell or mycelium for months or years. This condition has some of the advantages of diploidy; one haploid genome may be able to compensate for harmful mutations in the other nucleus, and vice versa. Finally, as a step in sexual reproduction, the nuclei fuse to form a diploid cell that undergoes immediate meiosis.

The three major divisions of fungi differ in details of reproduction

More than 100,000 species of fungi are now known, and mycologists describe another thousand or so each year. The taxonomic scheme used in this chapter classifies the species into three divisions. Use of the botanical term *division* instead of *phylum* is a vestige of the two-kingdom system, when fungi were grouped with plants. The three divisions differ in the structures involved in plasmogamy, the length of time spent as a dikaryon, and the location of karyogamy. In fact, the three divisions are named after the sexual cells in which karyogamy occurs.

Zygote Fungi (Division Zygomycota)

Mycologists have described about 600 zygomycetes, or **zygote fungi.** (The suffix *-mycete,* which occurs many times in this chapter, means "fungus.") These fungi are mostly terrestrial and live in soil or on decaying plant and animal material. One group of major importance forms **mycorrhizae,** mutualistic associations with the roots of plants (see FIGURE 28.13). Zygomycete hyphae are coenocytic, with septa found only where reproductive cells are formed. The name of this division comes from zygosporangia, resistant structures formed during sexual reproduction (FIGURE 28.3).

A common zygomycete is black bread mold, *Rhizopus*

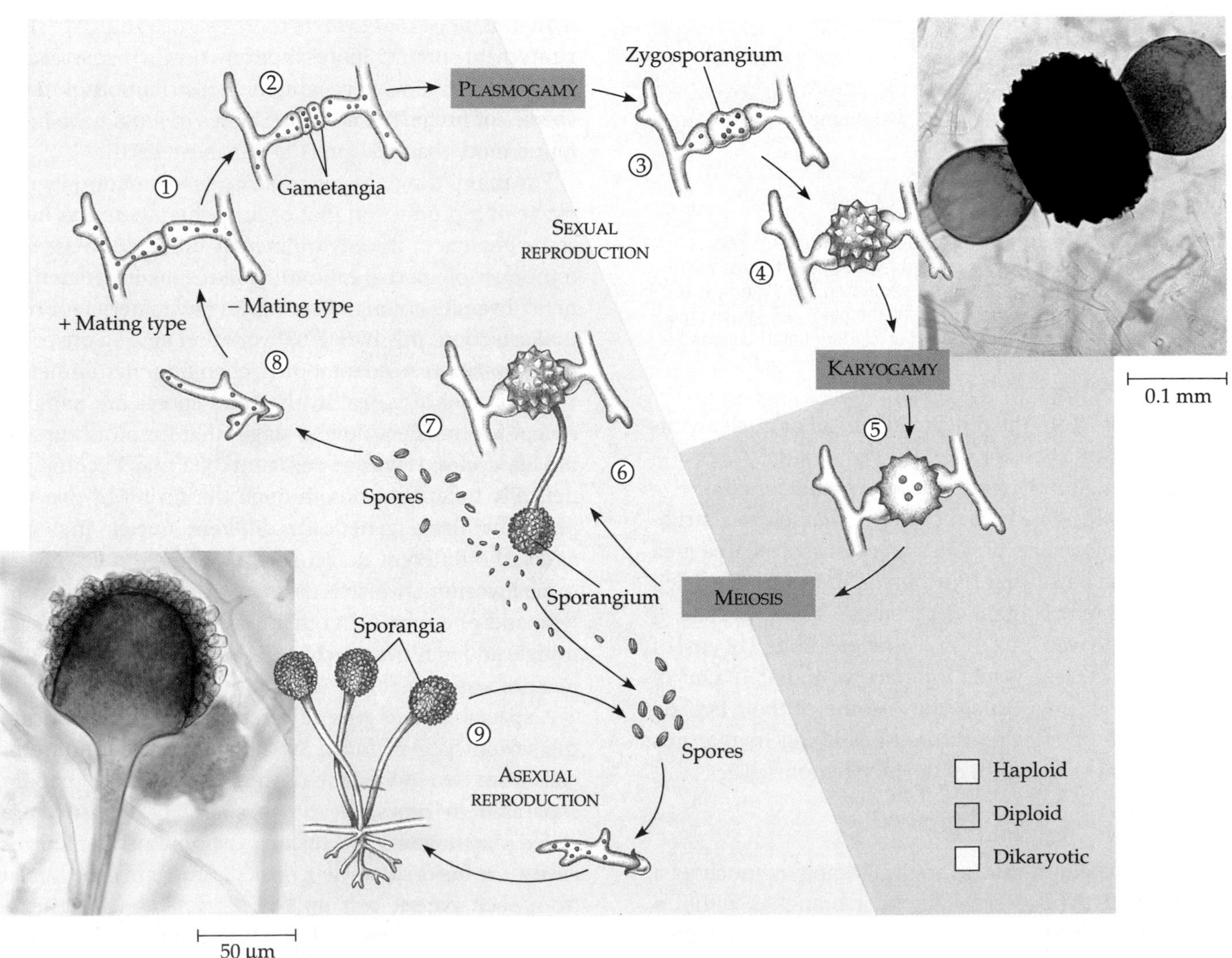

FIGURE 28.3
The life cycle of the zygomycete *Rhizopus.* ① Neighboring mycelia of opposite mating types (designated + and –) ② form hyphal extensions called gametangia, each walled off around several haploid nuclei by a septum. ③ The gametangia undergo plasmogamy and the haploid nuclei pair off, forming a dikaryotic zygosporangium. ④ This cell develops a rough, thick-walled coating (right LM) that can resist dry conditions and other harsh environments for months. ⑤ When conditions are favorable again, karyogamy between paired nuclei occurs, followed rapidly by meiosis. ⑥ The zygosporangium then breaks dormancy, germinating into a short sporangium that ⑦ disperses the genetically diverse, haploid spores. ⑧ These spores germinate and grow into new mycelia. ⑨ *Rhizopus* can also reproduce asexually by forming sporangia (left LM).

stolonifer, still an occasional household pest despite the addition of preservatives to most processed foods. Horizontal hyphae spread out over the food, penetrate it, and absorb nutrients into a rapidly expanding mycelium. In the asexual phase, bulbous black sporangia develop at the tips of upright hyphae. Within each sporangium, hundreds of haploid spores develop and are dispersed through the air. Spores that happen to land on moist food germinate, growing into new mycelia. If environmental conditions deteriorate (for instance, if all the food is used up) and mycelia of opposite mating types are present, this species of *Rhizopus* reproduces sexually (see FIGURE 28.3). The zygosporangia formed are resistant to freezing and desiccation and are metabolically inactive. When conditions improve, the zygosporangia release haploid spores that recolonize the new substrate.

Air currents are not a very precise way to disperse spores, but *Rhizopus* releases the tiny cells in great numbers. Though they drift aimlessly, enough land in hospitable places. Some zygomycetes, however, can actually aim their spores. One is *Pilobolus,* a fungus that decomposes animal dung. *Pilobolus* bends its sporangium-bearing hyphae toward light, a direction where grass is likely to be growing. The whole sporangium is then shot off on an explosive squirt of cytoplasm out the end of the hypha, sometimes carrying the sporangium as far as 2 m. This adaptation disperses the spores away from the mass of dung and onto surrounding grass, which will be eaten by a herbivore such as a cow. The asexual life cycle is completed when the animal scatters the spores in feces.

Sac Fungi (Division Ascomycota)

Over 60,000 species of ascomycetes, or **sac fungi,** have been described from a wide variety of habitats. They range in size and complexity from unicellular yeasts to minute leaf-spot fungi to elaborate cup fungi and truffles (FIGURE 28.4). Ascomycetes include some of the most devastating plant pathogens, which will be discussed later in the chapter. However, many are important saprobes, particularly of plant material. About half the

(a)

(b)

(c)

FIGURE 28.4

Ascomycota. Ascomycetes range from yeasts to such gastronomical delicacies as truffles and morels. (**a**) Many ascomycetes form ascocarps shaped like tiny flasks. In the carbon fungus, *Hypoxylon multiforme,* these ascocarps appear in clusters on the wood the mycelium decomposes. (**b**) Truffles are ascocarps that fruit underground and emit strong odors, which attract animals that eat the fungi and disperse the ascospores. *Tuber melanosporum* is highly prized for its flavor by gourmet cooks, who pay over $600 a pound for this truffle. (**c**) *Morchella esculenta,* the succulent morel, is often found under trees in orchards. Many, if not all, morels produce mycorrhizae.

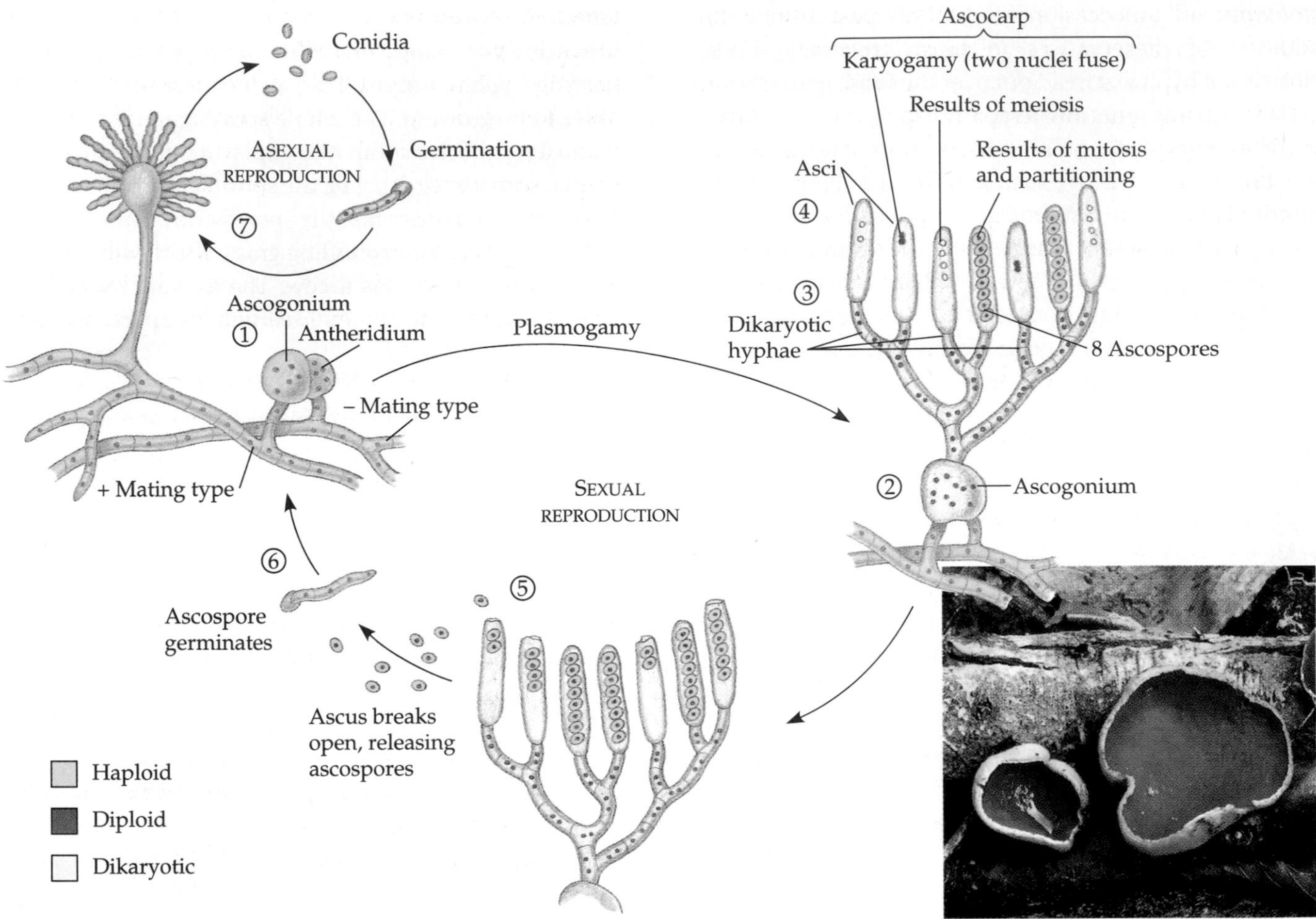

FIGURE 28.5
The life cycle of an ascomycete.
① Haploid mycelia of opposite mating types become intertwined. One acts as a "female," producing a structure called an ascogonium, which receives many haploid nuclei from the antheridium of the "male." ② The ascogonium then has a pool of nuclei from both parents, but karyogamy does not occur at this time. ③ The ascogonium gives rise to dikaryotic hyphae that are incorporated into an ascocarp, the cup of a cup fungus. The photograph features ascocarps of the scarlet cup (*Sarcoscypha coccinea*) growing on dead twigs in early spring. ④ The tips of the ascocarps' dikaryotic hyphae are partitioned into asci. Karyogamy occurs within these asci, and the diploid nucleus divides by meiosis, yielding four haploid nuclei. Each of these haploid nuclei divides once by mitosis, and the ascus now contains eight nuclei. Cell walls develop around these nuclei to form ascospores. ⑤ When mature, all ascospores in an ascus are dispersed at once out the end of the ascus. A collapsing ascus jars neighboring asci and causes them to fire their spores. The chain reaction releases a visible cloud of spores with an audible hiss. ⑥ Germinating ascospores give rise to new haploid mycelia. ⑦ Ascomycetes can also reproduce asexually by producing airborne spores called conidia.

ascomycete species live with algae in the symbiotic associations called lichens. Some ascomycetes, including truffles and morels, form mycorrhizae with plants. Others live in leaves on the surface of mesophyll cells, where the fungi apparently help protect these plant tissues from insects by releasing toxic compounds. Marine ascomycetes are the major nonbacterial saprobes on plant and seaweed material in saltwater habitats.

The defining feature of the Ascomycota is the production of sexual spores in saclike **asci** (singular, **ascus**) (FIGURE 28.5). Unlike the zygote fungi, most sac fungi bear their sexual stages in macroscopic fruiting bodies, or **ascocarps** (see FIGURE 28.4a). Before developing ascocarps, ascomycetes reproduce by producing enormous numbers of asexual spores, which are often dispersed by wind. The asexual spores are produced at the ends of hyphae, often in long chains or clusters. These spores are not formed inside sporangia, as in the Zygomycota. Such naked spores are called **conidia,** from the Greek for "dust."

Compared to zygomycetes, ascomycetes are characterized by a more extensive dikaryotic stage, which is associated with the formation of ascocarps (see FIGURE 28.5). Plasmogamy gives rise to the dikaryotic hyphae, and the cells at the tips of these hyphae become the asci. Within the asci, karyogamy combines the two parental genomes, and then meiosis forms genetically varied ascospores. In many asci, eight ascospores are lined up in a row in the order in which they formed from a single zygote. This arrangement provides geneticists with a unique opportunity to study genetic recombination. Genetic differences between mycelia grown from ascospores taken from one ascus reflect crossing over and independent assortment of chromosomes during meiosis.

Club Fungi (Division Basidiomycota)

Approximately 25,000 fungi, including mushrooms, shelf fungi, puffballs, and rusts, are classified in the division Basidiomycota (FIGURE 28.6). The name derives from the **basidium** (L., "little pedestal"), a transient diploid stage in the organism's life cycle. The clublike shape of the basidium also gives rise to the common name **club fungus.**

Basidiomycetes are important decomposers of wood and other plant material. The division also includes mycorrhiza-forming mutualists and plant parasites. Of all fungi, the saprobic basidiomycetes are best at decomposing the complex polymer lignin, an abundant component of wood. Many shelf fungi (FIGURE 28.6b) live as parasites on the wood of weak or damaged trees and function later as saprobes after the trees die. Half of the mushroom-forming club fungi are saprobic and the other half form mycorrhizae. There are very few strictly parasitic mushrooms. However, two groups of basidiomycetes, the rusts and smuts, include particularly obnoxious plant parasites.

The life cycle of a club fungus usually includes a long-lived dikaryotic mycelium (FIGURE 28.7). Periodically, in response to environmental stimuli, this mycelium reproduces sexually by producing elaborate fruiting bodies called **basidiocarps.** The numerous basidia of a basidiocarp are the sources of sexual spores. Asexual reproduction in basidiomycetes is much less common than in ascomycetes, but also occurs as conidia.

A mushroom is an example of a basidiocarp. The cap of the mushroom supports and protects a large surface area of basidia on gills; each common, store-bought mushroom has a gill surface area of about 200 cm^2. Such a mushroom may release a billion basidiospores, which drop beneath the cap and are blown away. A giant puffball may be over 1 m in diameter and produces trillions of spores, which are released when the basidiocarp breaks apart from animal activity, wind, or rain. Basidiomycetes called stinkhorns produce basidiospores in a slimy, stinky mass on upper surfaces of the basidiocarp,

(a)

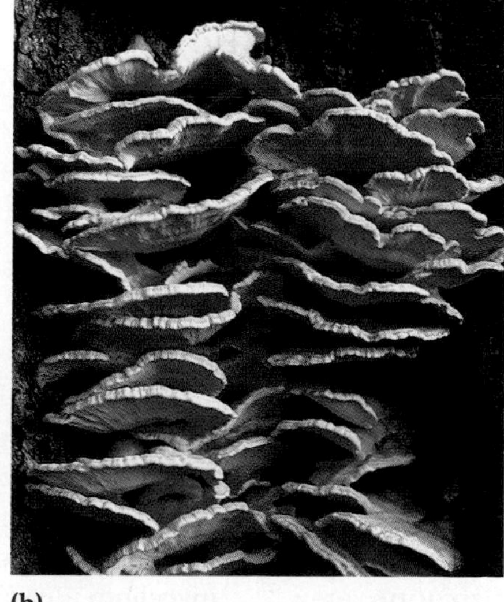
(b)

(c)

FIGURE 28.6

Basidiomycota. These photographs showcase diverse basidiocarps, the fruiting bodies that produce sexual spores. (**a**) A miniature waxy cap (*Hygrophorus*) is mycorrhizal with oaks. (**b**) A shelf fungus growing on the trunk of a tree. Shelf fungi are important decomposers of wood. (**c**) The common puffball (*Lycoperdon gemmatum*) is a saprobe on buried plant litter. When raindrops hit the basidiocarp, basidiospores are expelled.

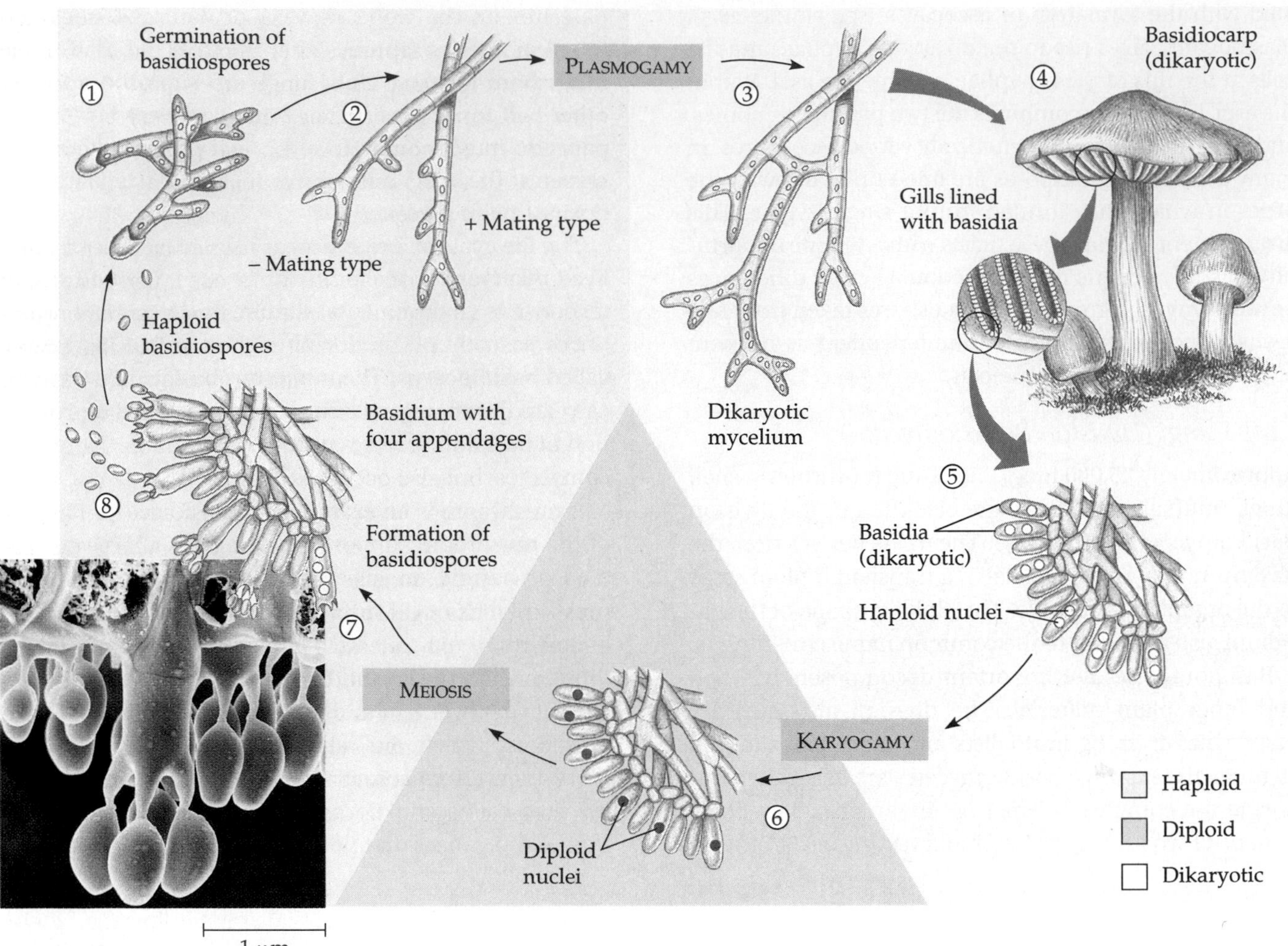

FIGURE 28.7

The life cycle of a mushroom-forming basidiomycete. ① Haploid basidiospores germinate in a suitable environment and grow into short-lived haploid mycelia. ② Undifferentiated hyphae from two haploid mycelia of opposite mating type undergo plasmogamy, ③ creating a dikaryotic mycelium that grows faster than, and ultimately crowds out, the parent haploid mycelium. The mycelium of the mushroom illustrated here (*Cortinarius*) forms mycorrhizae with trees. Environmental cues such as rain, temperature changes, and, for mycorrhizal species, seasonal changes in the plant host, ④ induce the dikaryotic mycelium to form compact masses that develop into mushrooms. Cytoplasm streaming in from the mycelium and from the attached mycorrhizae swells the hyphae of mushrooms, causing them to "pop up" overnight. The dikaryons of basidiomycetes are long-lived, generally producing a new crop of basidiocarps (mushrooms, in this case) each year. ⑤ Karyogamy occurs in the terminal dikaryotic cells that line the surfaces of the gills (SEM at left). ⑥ Each cell swells to form a diploid basidium, which rapidly undergoes meiosis and yields four haploid nuclei. ⑦ The basidium then grows four appendages, and one haploid nucleus enters each appendage and develops into a basidiospore. ⑧ When mature, the basidiospores are propelled slightly (by electrostatic forces) into the spaces between the gills. After the spores drop below the cap, they are dispersed by the wind.

which resembles rotting meat in odor and appearance. Flies and other carrion-eating insects attracted to stinkhorns disperse the sticky spores.

A ring of mushrooms, popularly called a fairy ring, may appear on a lawn overnight. These mushrooms are saprobes on dead plant material, not parasites on living grass. Although the grass in the center of the ring is normal, after a few days the grass near the ring is stunted and the grass just outside the garland of mushrooms is especially lush. As the underground mycelium grows outward, its center portion and the mushrooms above it die because the mycelium has consumed all the available nutrients. The grass beneath the mushrooms is stunted because it cannot compete for minerals with the active mycelium. But fungal enzymes are carried in the soil solution ahead of the advancing mycelium, and the grass there is fertilized by the minerals that become available. The fairy ring slowly increases in diameter as the mycelium advances at a rate of about 30 cm per year. Some giant fairy rings may be centuries old.

Molds, yeasts, lichens, and mycorrhizae represent unique lifestyles that evolved independently in all fungal divisions

Certain ways of living that involve both morphological and ecological specialization have evolved independently among the three divisions of the kingdom Fungi. These lifestyles allow the fungi to take advantage of unusual habitats. Humans have learned to exploit the abilities of these fungi for a variety of commercial purposes. This section discusses four fungal forms with unique ways of life: molds, yeasts, lichens, and mycorrhizae.

Molds

Mention of fungi may bring the ubiquitous molds to mind. A **mold** is a rapidly growing, asexually reproducing fungus (FIGURE 28.8). The mycelia of these fungi grow as saprobes or parasites on a great variety of substrates. You are already familiar with one example, bread mold (*Rhizopus*). Molds may go through a series of different reproductive stages. Early in life, a mold produces asexual spores. The term *mold* applies only to these asexual stages. Later, the same fungus *may* reproduce sexually, producing zygosporangia, ascocarps, or basidiocarps.

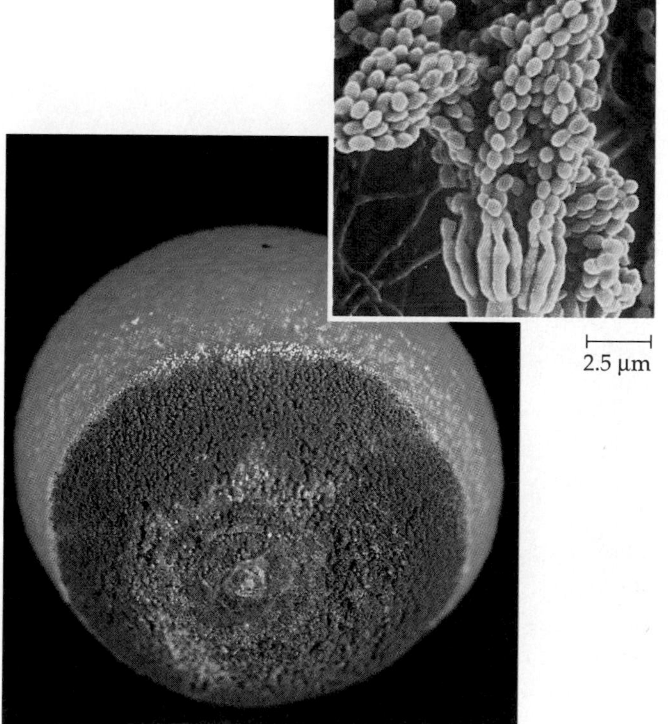

FIGURE 28.8
A moldy orange. "Blue mold" is usually caused by saprobic species of *Penicillium,* an ascomycete. This mold reproduces asexually by producing chains of conidia on hyphae called conidiophores (right, SEM).

There are also molds that cannot be classified as zygomycetes, ascomycetes, or basidiomycetes, because they have no known sexual stages. Those molds are classified as Deuteromycota, or **imperfect fungi** (from the botanical use of the term *perfect* to refer to the sexual stages of life cycles). Imperfect fungi, like many other molds, reproduce asexually by producing conidia on specialized hyphae called conidiophores. Among the more unusual imperfect fungi are predatory fungi in soil that trap, kill, and consume small animals, especially roundworms, or nematodes (FIGURE 28.9). This provides additional nitrogen-containing compounds, which are in short supply in the wood these molds decompose.

Humans have found many commercial uses for molds. Some are sources of antibiotics; pharmaceutical companies grow these molds in large liquid cultures, then extract the antibiotics. Penicillin is produced by some species of *Penicillium,* an ascomycete. Other *Penicillium* species are important fermenters on the surfaces of blue cheese, Brie, and Camembert and produce the unusual colors and flavors we associate with each type of cheese. Roquefort is goat milk cheese that has been incubated in certain caves in the Roquefort region of France, where wild *Penicillium roquefortii* is allowed to "infect" the cheese naturally.

Yeasts

Yeasts are unicellular fungi that inhabit liquid or moist habitats, including plant sap and animal tissues. Yeasts reproduce asexually, by simple cell division or by the pinching of small "bud cells" off a parent cell. Some yeasts reproduce sexually, by forming asci or basidia, and

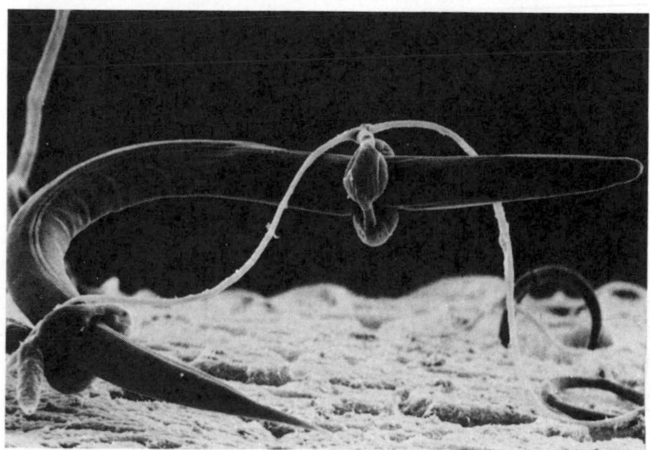

FIGURE 28.9
***Arthrobotrys,* a predatory deuteromycete.** Portions of the hyphae are modified as hoops that constrict around nematodes (roundworms) in less than a second, when an unsuspecting worm rubs the inside of the hoop. The fungus then penetrates its prey with hyphae and digests the inner tissues (SEM).

FIGURE 28.10

Effect of nutrient supply on yeast growth form. (**a**) When food is plentiful, baker's yeast (*Saccharomyoes cerevisiae*) grows as single cells, reproducing asexually when new cells bud from parental cells. (**b**) When starved for key nutrients, the cells elongate and stay linked together to form a mycelium. This change in form may help the nonmotile organism grow into a region where nutrients are more plentiful. (Both SEMs.)

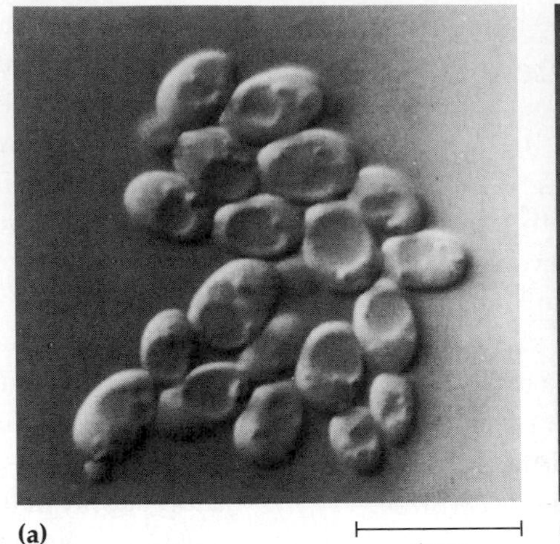

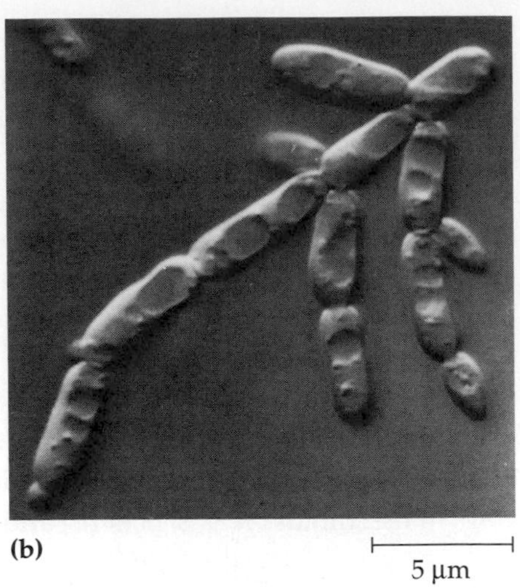

are then classified as Ascomycota or Basidiomycota. Others are placed in the division Deuteromycota because no sexual stages are known. Some fungi can grow as either single cells (yeasts) or as a filamentous mycelium, depending on the availability of nutrients (FIGURE 28.10).

Humans have used yeasts to raise bread and ferment alcoholic beverages for thousands of years. Only relatively recently have the yeasts involved been separated into pure culture for more controlled human use. The yeast *Saccharomyces cerevisiae,* an ascomycete, is the most important of all domesticated fungi (see FIGURE 28.10). The tiny yeast cells, available as many strains of baker's yeast and brewer's yeast, are very active metabolically. The cells release small bubbles of CO_2 that leaven dough. Cultured anaerobically in breweries and wineries, *Saccharomyces* ferments sugars to alcohol. Researchers use *Saccharomyces* to study the molecular genetics of eukaryotes because these microbes are easy to culture and manipulate (see Chapter 19).

Some yeasts cause problems for humans. A pink yeast, *Rhodotorula,* grows on shower curtains and other moist surfaces in our homes. Another yeast is *Candida,* one of the normal inhabitants of moist human epithelial tissue, such as the vaginal lining. Certain circumstances can cause *Candida* to become pathogenic by growing too rapidly and releasing harmful substances. This can occur, for example, with an environmental change, such as a pH change, or when the immune system of the human host is compromised—by AIDS, for instance.

Lichens

At a distance, lichens are often mistaken for mosses or other simple plants growing on rocks, rotting logs, trees, and roofs (FIGURE 28.11). In fact, lichens are *not* mosses or any other kind of plant, nor are they even individual organisms. A **lichen** is a symbiotic association of millions of photosynthetic microorganisms tangled in a mesh of fungal hyphae. The fungal component is most commonly an ascomycete, but several basidiomycete lichens are known. The photosynthetic partners are usually unicellular or filamentous green algae or cyanobacteria. The merger of fungus and alga is so complete that lichens are actually given genus and species names, as though they were single organisms. The scientific name of the lichen is the name of the fungus. Over 25,000 species have been described.

FIGURE 28.11

Lichens. Lichens representing three growth forms inhabit the bark of this maple branch. *Parmelia* (lower left) is foliose (having a flattened, leaflike appearance. *Ramalina* (upper right) is fruticose (shrub-like). Elsewhere, several crustose (paint smearlike) species in the genera *Lecanora* and *Bacidia* are forming disk-shaped ascocarps. The minute orange lichen is *Xanthoria,* a foliose form.

Lichens vary considerably in the details of their architecture and physiology, but there are some key similarities. The fungus usually gives the lichen its overall shape and structure, and tissues formed by hyphae account for most of the lichen's mass. The algal component usually occurs in an inner layer below the lichen surface. In most cases that have been examined, each partner provides things the other could not obtain on its own. The alga always provides the fungus with food. Cyanobacteria in lichens fix nitrogen (see Chapter 25) and provide organic nitrogen. The fungus provides the alga with a suitable physical environment for growth. Lichens absorb most of the minerals they need either from air in the form of dust or from rain. The physical arrangement of hyphae retains water and minerals, allows for gas exchange, and protects the algae (FIGURE 28.12). The fungi produce unique organic compounds with several functions. Fungal pigments help shade the algae from intense sunlight. Some fungal compounds are toxic and prevent lichens from being eaten by consumers. The fungi also secrete acids, which aid in the uptake of minerals.

The fungi of many lichens reproduce sexually by forming ascocarps or, rarely, basidiocarps. Lichen algae reproduce independently of the fungus by asexual cell division. As might be expected of "dual organisms," asexual reproduction as symbiotic units also occurs commonly, either by fragmentation of the parental lichen or by the formation of specialized structures called **soredia** (see FIGURE 28.12). Soredia are small clusters of hyphae with embedded algae.

Whether the lichen symbiosis is mutualistic or parasitic is still debated. Lichens are able to live in environments where neither fungi nor algae could live alone. While the fungal components do not grow alone in the wild, some lichen algae also occur as free-living organisms. The fungi and algae of some lichens have been experimentally separated and cultured. Such cultures look like free-living molds and algae and lose certain physiological behaviors typical of lichens. For instance, the cultured fungi do not produce lichen compounds. Cultured lichen algae do not "leak" carbohydrate from their cells as they do in lichens. Most evidence points to a mutualistic type of symbiosis. In some lichens, however, the fungus invades the algal cells with haustoria and may kill some of them, though not as fast as the alga replenishes its numbers by reproduction. Citing this apparent harm to the algae, some lichenologists refer to lichen symbiosis as "controlled parasitism" rather than mutualism.

Lichens are important pioneers on newly cleared rock and soil surfaces, such as burned forests and volcanic flows. Physical penetration of the outer crystals of rocks and chemical attack of rock by lichen acids help break down the rock and establish soil-trapping lichens. This process makes it possible for a succession of plants to grow. Nitrogen-fixing lichens also add organic nitrogen to some ecosystems.

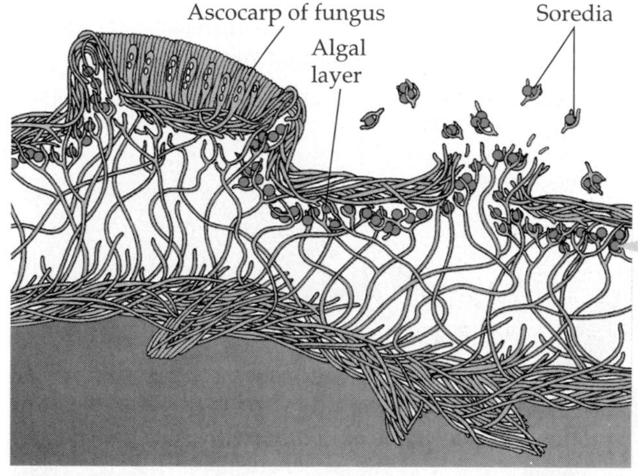

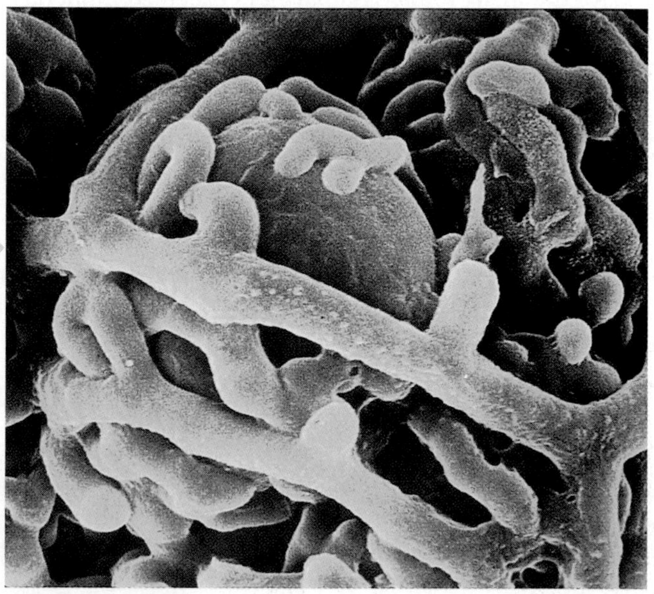

FIGURE 28.12

Anatomy of a lichen. The upper and lower surfaces are protective layers of tightly packed fungal hyphae. Just beneath the upper surface are the algae, enmeshed in a net of hyphae (SEM, right). The middle of a lichen generally consists of loosely woven hyphae of the fungus. Reproductive structures generally form on the upper surface. A sexual ascocarp of the fungus and several asexual soredia that disperse both fungal and algal components are shown here.

Some lichens tolerate severe cold. In the arctic tundra, great herds of caribou and reindeer graze on carpets of reindeer lichen at times of the year when other foods are unavailable. Lichens can also survive desiccation. When it is foggy or rainy, lichens may absorb more than ten times their weight in water. Photosynthesis begins almost immediately when the water content reaches 65% to 75%. In dry air, lichens rapidly dehydrate, and photosynthesis stops. Thus, in arid climates lichens grow very slowly, often in spurts of less than a millimeter per year. Some lichens are thousands of years old, rivaling the oldest plants as the elder organisms on Earth.

As tough as lichens are, many do not stand up very well to air pollution. Their passive mode of mineral uptake from rain and moist air makes them particularly sensitive to sulfur dioxide and other aerial poisons. The death of sensitive lichens and an increase in hardier species in an area is an early warning that air quality is deteriorating.

Mycorrhizae

Physiological cousins of lichens are mycorrhizae, mutualistic associations of plant roots and fungi. The word *mycorrhizae* means "fungus roots," referring to the structures formed by both root cells and hyphae from the associated fungus. The anatomy of this symbiosis varies, depending on the type of fungus (FIGURE 28.13). The extensions of the fungal mycelium from the hyphae forming the mycorrhizae greatly increase the absorptive surface of the plant roots. The partners exchange minerals accumulated from the soil by the fungus for organic nutrients synthesized by the plant.

Mycorrhizae are enormously important in natural ecosystems and agriculture. Over 95% of all vascular plants have mycorrhizae. The fungi involved are permanent associates with their hosts and periodically form fruiting bodies (structures for sexual reproduction). Basidiomycota, Ascomycota, and Zygomycota all have members that form mycorrhizae. In fact, half of all species of mushroom-forming basidiomycetes live as mycorrhizae with oak, birch, and pine trees. The mushrooms that sprout around the bases of these trees are the surface evidence of the underground symbiotic relationship between plants and fungi. (The structure and physiology of mycorrhizae will be described in more detail in Chapter 33.)

Fungi have a tremendous ecological impact

Fungi as Decomposers

Fungi and bacteria are the principal decomposers that keep ecosystems stocked with the inorganic nutrients essential for plant growth. Without decomposers, carbon, nitrogen, and other elements would accumulate in corpses and organic wastes, no longer available as raw material for new generations of life. Imagine what would become of a forest if its decomposers rested for even a few years. Leaves, logs, feces, and dead animals would pile up on the forest floor. Plants and the animals they feed would starve because elements taken from the soil would not be returned. Gradually, the forest would die. This hypothetical forest is a symbol of the fate that would befall all ecosystems without decomposers. In reality, an endless recycling of chemical elements occurs between organisms and their nonliving surroundings (see Chapter 49).

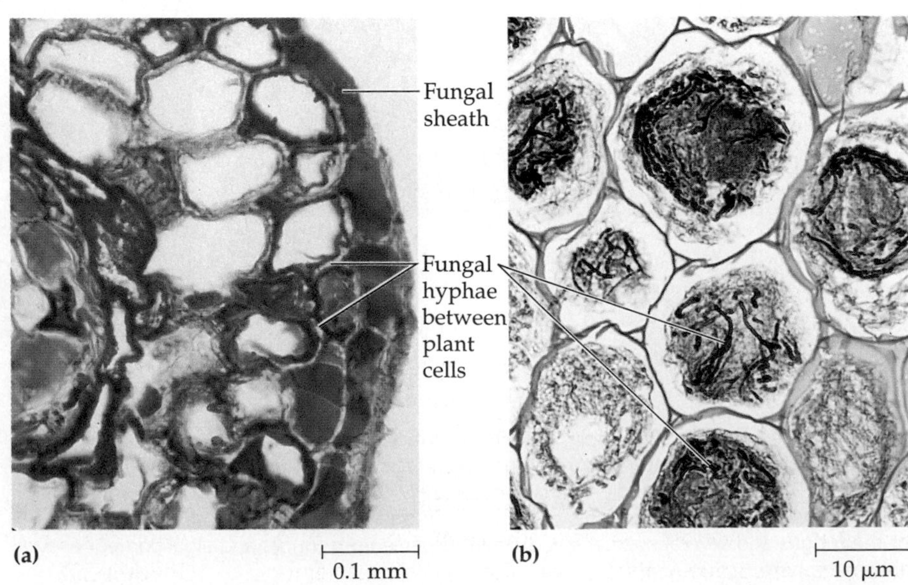

FIGURE 28.13
Mycorrhizae. (**a**) Ectomycorrhizae (sheathing mycorrhizae). The fungal mycelium forms a sheath around this pine root, with some hyphae extending between the plant cells into the root interior. The fungi of ectomycorrhizae are usually mushroom-forming basidiomycetes. (**b**) Endomycorrhizae. Certain zygomycetes establish these mutualistic associations with roots, in this case the roots of an orchid. Fungal hyphae that penetrate the root form branched haustoria that exchange materials with the plant. These endomycorrhizae do not form dense sheaths around the root. (Both LMs.)

Fungi are well adapted as decomposers of plant material. Their invasive hyphae enter the tissues and cells of dead organic matter and hydrolyze polymers, including the cellulose and lignin of plant cell walls. A succession of fungi, in concert with bacteria and, in some environments, invertebrate animals, are responsible for the complete breakdown of plant litter. The air is so loaded with fungal spores that as soon as a leaf falls or an insect dies, it is covered with spores and is soon infiltrated by saprobic hyphae.

Fungi as Spoilers

We may applaud fungi that decompose forest litter or dung, but it is a different story when molds attack our fruit or our shower curtains. A wood-digesting saprobe does not distinguish between a fallen oak limb and the oak planks of a boat. During the Revolutionary War, the British lost more ships to fungal rot than to enemy attack. Soldiers stationed in the tropics during World War II watched their tents, clothing, boots, and binoculars be destroyed by molds. Some fungi can even decompose certain plastics.

Between 10% and 50% of the world's fruit harvest is lost each year to fungal attack. Researchers at Ohio State University recently discovered that ethylene, a plant hormone that causes fruit to ripen, also stimulates fungal spores on the fruit surface to germinate. This timing mechanism enables fungi to invade when fruit is most vulnerable and nutritious.

In addition to their rotting action, some molds also taint food by producing poisons. For example, some species of the mold *Aspergillus* contaminate improperly stored grain by secreting compounds called aflatoxins, which are carcinogenic.

Pathogenic Fungi

Some fungi are pathogens. Among the diseases that fungi cause in humans are athlete's foot, vaginal yeast infections, and lung infections that may be fatal.

Plants are particularly susceptible to fungal diseases. For example, the ascomycete that causes Dutch elm disease has drastically changed the landscape of the northeastern United States. The fungus was accidentally introduced to the United States on logs that were sent from Europe to help pay World War I debts. Carried from tree to tree by bark beetles, the fungus is on its way to completely eliminating the American elm. Another ascomycete has almost eliminated the native American chestnut.

Some of the fungi that attack food crops are toxic to humans. One ascomycete forms purple structures called ergots on rye. If diseased rye is inadvertently milled into flour and consumed, poisons from the ergots cause gangrene, nervous spasms, burning sensations, hallucinations, and temporary insanity. One epidemic in about 944 A.D. killed more than 40,000 people. One of the hallucinogens that has been isolated from ergots is lysergic acid, the raw material from which LSD is made. On the other hand, toxins extracted from fungi often have medicinal uses when administered in weak doses. For example, an ergot compound is helpful in treating high blood pressure and stopping maternal bleeding after childbirth.

Edible Fungi

In the food webs of most ecosystems, fungi are consumed as food by a variety of animals, including humans. Most of us have eaten mushrooms (basidiomycetes), but in the United States we usually restrict our consumption to one species of *Agaricus*, which is cultivated commercially on compost in the dark. In many countries, however, people eat a variety of cultivated and wild mushrooms. There are no simple rules to help the novice distinguish poisonous from edible mushrooms. Only experts in mushroom identification can safely collect wild mushrooms for eating.

The fungi most prized by gourmets are truffles, the underground ascocarps of mycelia that are mycorrhizal on the roots of trees. Their complex flavor is variously described as nutty, musky, cheesy, or all three. The fruiting bodies (ascocarps) release strong odors that attract mammal and insect consumers that excavate the truffles and disperse their spores. In some cases, the odors mimic sex attractants of certain mammals. Truffle hunters traditionally used pigs to locate their prizes, although dogs are now more commonly used, because they have the nose for the scent without the fondness for the flavor.

Fungi and animals probably evolved from a common protistan ancestor

Systematists are making rapid progress in sorting out the phylogenetic relationships between fungi and other eukaryotes. The link between fungi and protists may be a group of organisms called **chytrids.** Until recently, some systematists emphasized the absence of flagellated cells as a membership requirement for the kingdom Fungi. By that criterion, chytrids were excluded and placed instead in the kingdom Protista because they form flagellated spores called zoospores. However, chytrids and fungi share many characteristics, including an absorptive mode of nutrition. Like fungi, chytrids have cell walls of chitin, and most species form hyphae. Fungi and chytrids also have some key enzymes and metabolic pathways in common that are not found in the so-called funguslike protists (slime molds and water molds; see Chapter 26).

In the past few years, molecular systematists comparing the sequences of proteins and nucleic acids have added strong support for combining the chytrids with the fungi as a monophyletic branch of the eukaryotic tree. In this taxonomic view, the chytrids make up a division (phylum), Chytridiomycota, within the kingdom Fungi. The molecular evidence also supports the hypothesis that chytrids are the most primitive fungi, meaning that they belong to the lineage that diverged earliest in the phylogeny of fungi. A reasonable extension of this hypothesis is that fungi evolved from protists that had flagella, a feature retained by the chytrids (FIGURE 28.14).

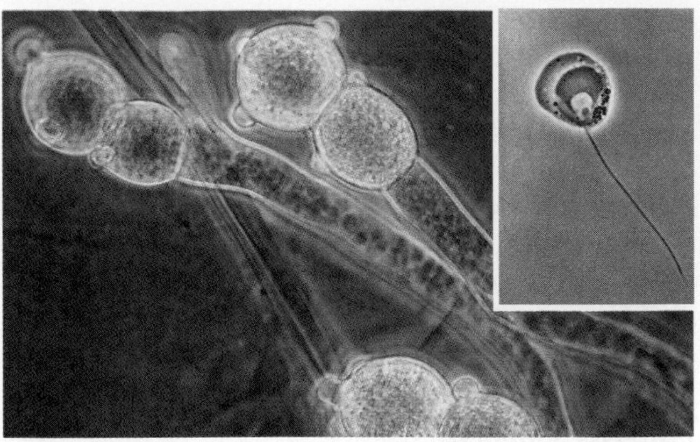

FIGURE 28.14
Chytridiomycota. The branched hyphae of *Allomyces* expose a large surface to the surrounding medium, from which the organism absorbs nutrients (LM). Chytrids are the only fungi with a flagellated stage, the zoophore in the inset (TEM).

Animals probably evolved from flagellated protists too. There is compelling evidence that animals and fungi diverged from a common protistan ancestor. Comparisons of several proteins and ribosomal RNA indicate that fungi are more closely related to animals than either of these kingdoms is to plants. Based on the data of molecular systematists, the most likely protistan ancestor common to fungi and animals was a choanoflagellate (see FIGURE 26.27).

Perhaps the diversity of fungi we observe today had its phylogenetic origin in adaptive radiation when life began to colonize land. If the first fungi were aquatic flagellated organisms, then the three major fungal divisions we surveyed in this chapter—Zygomycota, Ascomycota, and Basidiomycota—apparently lost their flagellated stages early in their evolutionary history. Many of the differences between these divisions center on contrasting solutions to the problem of reproducing and dispersing on land. These fungi may have diverged during the transition from aquatic to terrestrial habitats. The oldest undisputed fossils of fungi are about 440 million years old, about the time plants began to colonize land. Fossils of the first vascular plants dating back to the late Silurian period have petrified mycorrhizae. From their inception, terrestrial communities were apparently dependent on fungi as decomposers and symbionts.

REVIEW OF KEY CONCEPTS (with page numbers and key figures)

- Structural and life history adaptations equip fungi for an absorptive mode of nutrition (pp. 573–575, FIGURE 28.2)
  - The fungi are a eukaryotic, primarily multicellular group. Most have cell walls made of chitin.
  - All fungi are heterotrophs, acquiring their nutrients by absorption. There are saprobic decomposers, parasitic species, and mutualistic forms.
  - The fungal vegetative bodies consist of mycelia, netlike collections of branched hyphae adapted for absorption. Parasitic fungi penetrate their hosts with specialized hyphae called haustoria.
  - Although aseptate (coenocytic) forms occur, most fungi have their hyphae partitioned into cells by septa, with pores allowing cell-to-cell continuity.
  - Fungi reproduce by dispersing enormous numbers of spores.
  - Fungal life cycles vary by division but generally start with haploid mycelia. Early reproduction is usually asexual and later reproduction is sexual.
  - The sexual cycle involves cell fusion (plasmogamy) and nuclear fusion (karyogamy), with an intervening dikaryotic stage (with two haploid nuclei). The diploid phase is short-lived and rapidly undergoes meiosis to produce haploid spores.
- The three major divisions of fungi differ in details of reproduction (pp. 576–580; FIGURES 28.3, 28.5, 28.7)
  - The zygote fungi of the division Zygomycota, fungi that live in soil or decaying organic matter, include the familiar black bread mold. Asexual spores develop in aerial sporangia. The division is named for its sexually produced zygosporangia, which are dikaryotic structures capable of persisting through unfavorable conditions.
  - The sac fungi of the division Ascomycota include plant parasites, fungal components of lichens, and saprobes. Asexual reproduction by conidia is common. Sexual reproduction involves the formation of spores in sacs, or asci, at the ends of dikaryotic hyphae, usually in ascocarps.

- The club fungi of the division Basidiomycota include mushrooms, shelf fungi, puffballs, and rusts. Mycelia of club fungi can last years as dikaryons. Sexual reproduction involves the formation of spores on club-shaped basidia at the ends of dikaryotic hyphae in fruiting bodies, such as mushrooms.

- Molds, yeasts, lichens, and mycorrhizae represent unique lifestyles that evolved independently in all fungal divisions (pp. 581–584; FIGURES 28.8, 28.10–28.12)
  - Molds are either asexual stages of fungi classified as zygomycetes, ascomycetes, or basidiomycetes, or conidia-producing fungi not known to reproduce sexually (deuteromycetes, or imperfect fungi). Molds are important in the commercial production of antibiotics, such as penicillin.
  - Yeasts are unicellular fungi adapted to life in liquids such as plant saps. They may be classified as ascomycetes, basidiomycetes, or deuteromycetes.
  - Lichens are such highly integrated symbiotic associations of algae and fungi that they are classified as single organisms. Lichens are rugged organisms that set the stage for plant growth by slowly breaking down bare rocks.
  - Mycorrhizae are mutualistic associations of fungi with the roots of vascular plants. The plant is more efficiently able to absorb minerals and water with the aid of the fungus, which is provided with its carbohydrate needs in return.
- Fungi have a tremendous ecological impact (pp. 584–585)
  - Without fungi and bacteria as decomposers, biological communities would be deprived of the essential recycling of chemical elements between the biological and nonbiological world. Fungi are particularly important decomposers of wood.
  - Fungi also decompose food and other useful objects.
  - Some fungi cause disease, harming humans with a variety of ills. Plants are especially vulnerable to fungal infections.
  - Many fungi are food for humans and other animals.
- Fungi and animals probably evolved from a common protistan ancestor (pp. 585–586, FIGURE 28.14)
  - The link between fungi and protists may be chytrids, which are fungi that retain a flagellated condition. Molecular evidence supports the hypothesis that fungi and animals diverged from a choanoflagellate.

SELF-QUIZ

1. *All* fungi share which one of the following characteristics?
   a. symbiotic
   b. heterotrophic
   c. flagellated
   d. pathogenic
   e. saprobic

2. In the sexual life cycles of fungi, __________ usually occurs shortly after __________.
   a. syngamy; karyogamy
   b. meiosis; fusion of haploid nuclei
   c. asexual reproduction; sexual reproduction
   d. spore production; formation of a dikaryotic mycelium
   e. syngamy; meiosis

3. Which of the following cells or structures are associated with *asexual* reproduction in fungi?
   a. ascospores
   b. basidiospores
   c. conidia
   d. zygosporangia
   e. ascogonia

4. Which of the following lists three lifestyles that have evolved independently in the three divisions of fungi?
   a. sac fungi, club fungi, molds
   b. zygote fungi, yeasts, mycorrhizae
   c. yeasts, molds, sac fungi
   d. lichens, molds, yeasts
   e. sac fungi, zygote fungi, club fungi

5. Fungi are classified into three major divisions based *mainly* on differences in
   a. mode of nutrition
   b. type of motility
   c. reproductive adaptations
   d. composition of the cell wall
   e. the presence or absence of a dikaryotic stage

6. Sporangia on erect hyphae that produce asexual spores are characteristic of
   a. Ascomycota
   b. Basidiomycota
   c. Deuteromycota
   d. Zygomycota
   e. lichens

7. Which of the following statements is true about Basidiomycota?
   a. hyphae fuse to give rise to a dikaryotic mycelium
   b. spores line up in a sac after they are formed by meiosis
   c. the vast majority of spores are formed asexually
   d. basidiomycetes are the most important commercial sources of antibiotics
   e. basidiomycetes have no known sexual stages

8. Mycorrhizae are
   a. asexual reproductive structures formed by lichens
   b. thin hyphae that grow directly into host tissues
   c. the mycelium that forms fairy rings
   d. compact, dikaryotic hyphae that form a basidiocarp
   e. mutualistic associations between plant roots and fungi

9. The photosynthetic symbiont of a lichen is most commonly a(n)
   a. moss
   b. green alga
   c. red alga
   d. ascomycete
   e. small vascular plant

10. The closest relatives of fungi are probably
    a. animals
    b. vascular plants
    c. mosses
    d. brown algae
    e. slime molds

CHALLENGE QUESTIONS

1. In what way might the use of a wide-spectrum fungicide that kills all fungal species have a harmful effect on vegetation, such as a forest?

2. Zygomycetes fuse dikaryotic nuclei to form zygotes, only to restore the haploid state again by meiosis before the growth of new mycelia. What does this formation of a transient diploid stage accomplish?

3. Many fungi produce antibiotics, such as penicillin, that are valuable in medicine. Of what possible value are antibiotics to the fungi that produce them? Similarly, fungi often produce compounds with unpleasant tastes and odors as they digest their food. What might be the value of these chemicals to the fungi? To other organisms that might eat the decomposing food? How might the production of antibiotics and odors have evolved?

SCIENCE, TECHNOLOGY, AND SOCIETY

1. Many fungi and lichens in the United States are threatened with extinction due to the destruction of habitats. Should they come under legal protection by their inclusion on lists of endangered species? Defend your answer.

2. American chestnut trees once made up more than 25% of the hardwood forests of the eastern United States. These trees were wiped out by a fungus accidentally introduced on imported Asian chestnuts, which are not affected. More recently, a fungus has killed large numbers of dogwood trees from New York to Georgia; some experts suspect the parasite was accidentally introduced from elsewhere. Why are plants particularly vulnerable to fungi imported from other regions? What kinds of human activities might contribute to the spread of plant diseases? Do you think the introduction of plant pathogens like chestnut blight are more or less likely to occur in the future? Why?

FURTHER READING

Barron, G. "Jekel-Hyde Mushrooms." *Natural History,* March 1992. How some wood-decomposing fungi supplement their diets by trapping nematodes.

Bradley, D. "How Ripening Fruit Invite Fungal Attack." *New Scientist,* August 27, 1994.

Brune, J. "Porky's Defeat." *Discover,* January 1991. A new electronic instrument is putting pigs and dogs out of business as truffle rooters.

Cherfas, J. "Disappearing Mushrooms: Another Mass Extinction?" *Science,* December 6, 1991. Fungi are vanishing from Europe, causing concern about forest ecology.

Jaenike, J. "Behind-the-Scenes Role of Parasites." *Natural History,* June 1994. Mushrooms are key junctures in some food webs.

Kendrick, B. *The Fifth Kingdom,* 2nd ed. Newburyport, MA: Focus Information Group, 1992. A lively introduction to the fungi.

Kiester, E. "Prophets of Gloom." *Discover,* November 1991. Lichen death means more bad news about air quality.

Lewis, R. "A New Place for Fungi?" *BioScience,* June 1994. The case for an animal-fungus link in evolution.

Newhouse, J. R. "Chestnut Blight." *Scientific American,"* July 1990. A new parasite may help control the blight.

Radetsky, P. "The Yeast Within." *Discover,* March 1994. Surprising new discoveries about how yeasts develop.

Sternberg, S. "The Emerging Fungal Threat," *Science,* December 9, 1994. Pathogenic fungi are especially deadly to AIDS patients and others with compromised immune systems.

Wood, W. "Deadly Fungus Is Kind to Hearts," *New Scientist,* February 11, 1995. Pharmaceuticals from fungi.

CHAPTER 29

INVERTEBRATES AND THE ORIGIN OF ANIMAL DIVERSITY

KEY CONCEPTS

- What is an animal?
- Comparative anatomy and embryology provide clues to animal phylogeny: *an overview of animal diversity*
- Sponges are sessile animals lacking true tissues
- Cnidarians and ctenophores are radiate, diploblastic animals with gastrovascular cavities
- Flatworms and other acoelomates are bilateral, triploblastic animals lacking body cavities
- Rotifers, nematodes, and other pseudocoelomates have complete digestive tracts and blood vascular systems
- Mollusks and annelids are among the major variations on the protostome body plan
- The protostome phylum Arthropoda is the most successful group of animals ever to live
- The deuterostome lineage includes echinoderms and chordates
- The Cambrian explosion produced all the major animal body plans

Animal life began in Precambrian seas with the evolution of multicellular forms that lived by eating other organisms. This new way of life allowed the exploitation of previously untapped resources and adaptive zones and led to an evolutionary radiation of diverse forms. Early animals populated the seas, fresh water, and eventually the land. The dazzling diversity of animal life on Earth today is the result of over half a billion years of evolution from those ancestral forms.

More than a million extant species of animals are known, and at least as many more will probably be identified by future generations of biologists. Based largely on anatomical and embryological criteria, animals are grouped into about 35 phyla, the exact number depending on the views of different systematists. Most of these phyla consist mainly of aquatic animals. Living as we do on land, our sense of animal diversity is biased in favor of vertebrates, the animals with backbones, which are well represented in terrestrial environments. But vertebrates make up one subphylum within the phylum Chordata, less than 5% of all animal species. If we were to sample the animals inhabiting a tidepool (as in the photograph on this page) or a coral reef, we would find ourselves in the realm of ***invertebrates,*** *the animals without backbones.*

The diversity of invertebrates is the main subject of this chapter. The chapter also describes general characteristics of animals, examines hypotheses about the origin and early radiation of animals, and discusses possible relationships among the phyla.

What is an animal?

Constructing a good definition of animals is not as easy as it might first appear. There are exceptions to nearly every criterion for distinguishing an animal from other life forms. However, when taken together, the following characteristics of animals will serve our purposes.

1. Animals are multicellular, heterotrophic eukaryotes. In contrast to the autotrophic nutrition of plants, animals must take into their bodies preformed organic molecules; they cannot construct them from inorganic chemicals. Most animals do this by **ingestion**—eating other organisms or organic material that is decomposing, called detritus.

2. Animals typically store their carbohydrate reserves as glycogen, whereas plants store theirs as starch.

3. Animal cells lack the cell walls that characterize plant cells. Also, animals have unique types of intercellular

junctions: tight junctions, desmosomes, and gap junctions (see FIGURE 7.31).

4. Also unique among animals are two types of tissues responsible for impulse conduction and movement: nervous tissue and muscle tissue.

5. A few key features of life history also distinguish animals. Most animals reproduce sexually, with the diploid stage usually dominating the life cycle. In most species, a small flagellated sperm fertilizes a larger, nonmotile egg to form a diploid zygote (oogamy; see Chapter 26). The zygote then undergoes **cleavage,** a succession of mitotic cell divisions. During the development of most animals, cleavage leads to the formation of a multicellular stage called a **blastula,** which often takes the form of a hollow ball. Following the blastula stage is the process of **gastrulation,** during which embryonic tissues of the adult body parts are produced. Some animals develop directly through transient stages of maturation into adults, but the life cycles of many animals include larval stages. The **larva** is a free-living, sexually immature form. It is morphologically distinct from the adult stage, usually eats different food, and may even have a different habitat than the adult, as in the case of a frog tadpole. Animal larvae eventually undergo **metamorphosis,** a resurgence of development that transforms the animal into an adult.

Animals inhabit nearly all environments of the biosphere. The seas, where the first animals probably arose, are still home to the greatest number of animal phyla. The freshwater fauna is extensive, but not nearly as rich in diversity as the marine fauna.

Terrestrial habitats pose special problems for animals, as they do for plants (see Chapter 27), and few animal phyla have made successful evolutionary treks onto land. Earthworms (phylum Annelida) and land snails (phylum Mollusca) are generally confined to moist soil and vegetation. Only the vertebrates and arthropods, including insects and spiders, are represented by a great diversity of animal species adapted to various terrestrial environments.

Comparative anatomy and embryology provide clues to animal phylogeny: *an overview of animal diversity*

The origin of most animal phyla and major body plans took place in Precambrian and early Cambrian times. On the scale of geological time, animals diversified so rapidly that it is difficult from the fossil record to sort out the sequence of branching in animal phylogeny. Thus, when reconstructing the evolutionary history of animal phyla, systematists depend largely on clues from the comparative anatomy and embryology of living forms. Molecular systematics is now providing additional clues.

Although lively debate continues, most systematists now agree that the animal kingdom is monophyletic; that is, if we could trace the main branches of animal evolution backward to the Precambrian, they would converge on a single protistan ancestor. That ancestor was probably a choanoflagellate (see FIGURE 26.27). By the early Cambrian period, about a half-billion years ago, all animal phyla had evolved from the first animals of the late Precambrian. FIGURE 29.1 is an evolutionary tree depicting one set of ideas about animal phylogeny. The circled numbers in the figure highlight four key evolutionary branch points, which are also numbered in the following discussion.

① *The Parazoa-Eumetazoa Split*

Sponges (phylum Porifera) have unique development and an anatomical simplicity that separates them from all other animal phyla. They lack true tissues and are called **parazoa** (meaning "beside the animals"). Other animal phyla are called **eumetazoa.**

② *The Radiata-Bilateria Split*

The eumetazoa have been divided into two major branches, partly on the basis of body symmetry. Hydras, jellyfishes, and their relatives have **radial symmetry** (FIGURE 29.2a, p. 592) and are collectively called the **Radiata.** A radial animal has a top and bottom, or an oral (mouth) and an aboral side, but no head end and rear end and no left and right. The other major branch of eumetazoan evolution led to animals with **bilateral** (two-sided) **symmetry** (FIGURE 29.26b). A bilateral animal has not only a top, or **dorsal** side, and bottom, or **ventral** side, but also a head, **anterior,** end and tail, or **posterior,** end and a left and right side. Animals of this evolutionary branch are collectively called the **Bilateria.**

Associated with bilateral symmetry is **cephalization,** an evolutionary trend toward the concentration of sensory equipment on the anterior end, the end of a traveling animal that is usually first to encounter food, danger, and other stimuli. A head end is an adaptation for movement, such as crawling, burrowing, or swimming. The symmetry of an animal generally fits its lifestyle. Many radial animals are sessile forms (attached to a substratum) or plankton (drifting or weakly swimming aquatic forms). Their symmetry equips them to meet the environment equally well from all sides. More active animals are generally bilateral. These two fundamentally different kinds of symmetry probably arose very early in the history of animal life.

Body symmetry alone is not a foolproof criterion for assigning an animal phylum to a particular evolutionary

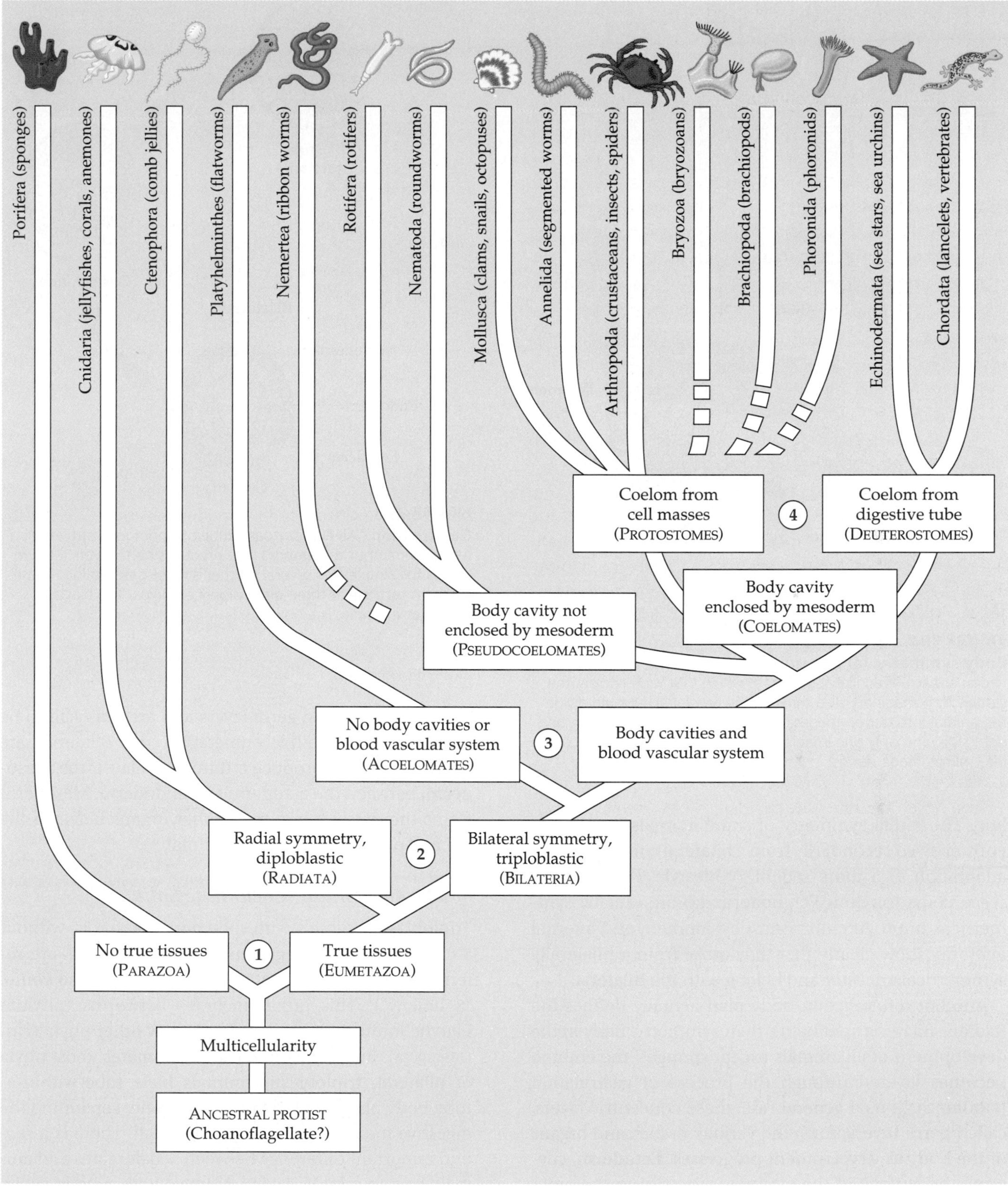

FIGURE 29.1

A hypothetical phylogeny of animals. The circled numbers key four main branch points to the discussion in the text. In our survey of the animal phyla, miniature versions of this tree will appear as orientation diagrams to help you keep sight of evolutionary relationships. (Broken branches indicate relationships that are particularly uncertain.)

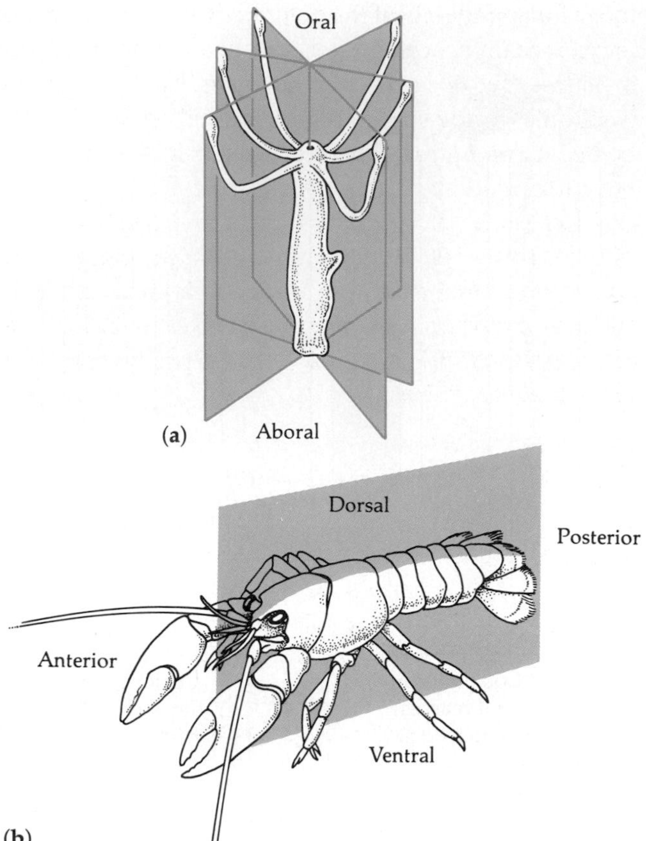

FIGURE 29.2
Body symmetry. (**a**) The parts of a radial animal, such as a hydra, are arranged like spokes of a wheel that radiate from the center. Any imaginary slice through the central axis would divide the animal into mirror images.(**b**) A bilateral animal has a left and right side, and only one imaginary cut would divide the animal into mirror-image halves.

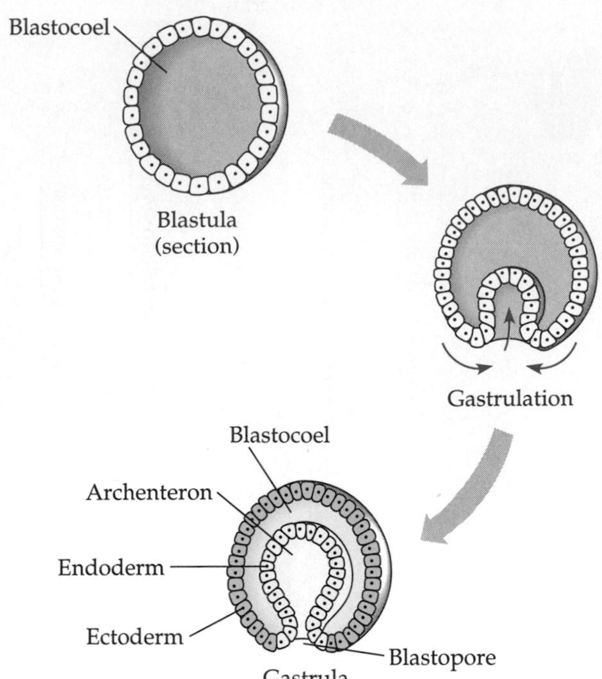

FIGURE 29.3
Gastrulation. All eumetazoan embryos undergo gastrulation, a rearrangement of the embryo that produces germ layers. Shown here is the simplest of several mechanisms of gastrulation. These are cross sections of three-dimensional embryos. The blastula stage, for example, is a hollow ball.

line. The radial symmetry of some animals has apparently evolved secondarily from a bilateral condition as an adaptation to a more sedentary lifestyle. For example, sea urchins (phylum Echinodermata) are radially symmetrical, but their embryonic development and internal anatomy show clearly that they arose from a bilaterally symmetrical ancestor and belong with the Bilateria.

Another difference in body plan actually defines the Radiata-Bilateria split better than symmetry. Early in the development of all animals except sponges, the embryo becomes layered through the process of gastrulation (FIGURE 29.3). As a general rule, these concentric layers, called **germ layers,** form the various tissues and organs of the body as development progresses. **Ectoderm,** covering the surface of the embryo, gives rise to the outer covering of the animal and, in some phyla, to the central nervous system. **Endoderm,** the innermost germ layer, lines the rudimentary gut, or **archenteron,** and gives rise to the lining of the digestive tract and its outpocketings, such as the liver and lungs of vertebrates. Cnidarians and ctenophores, the animals that make up the Radiata, produce only these two germ layers and are thus said to be **diploblastic.** All other eumetazoa, the Bilateria, are **triploblastic** and produce a third germ layer, the **mesoderm,** between the ectoderm and endoderm. Mesoderm forms the muscles and most other organs between the gut and outer covering of the animal.

③ *The Acoelomate-Coelomate Split*

Triploblastic animals with solid bodies—that is, without a cavity between the gut and outer body wall—are referred to as the **acoelomates** (Gr. *a,* "without," and *koilos,* "a hollow"). This group includes flatworms (phylum Platyhelminthes) and animals of a few other phyla (FIGURE 29.4a). In contrast to these acoelomates, most phyla of bilateral, triploblastic animals have tube-within-a-tube body plans, with a fluid-filled cavity separating the digestive tract from the outer body wall. There is a second important difference between acoelomates and animals having a body cavity: Animals with a body cavity also have some form of blood vascular system—blood that is circulated through a network of spaces or vessels; acoelomates lack a blood vascular system.

Among animals with a body cavity, there are differences in how the cavity develops. If the cavity is not completely lined by tissue derived from mesoderm, it is

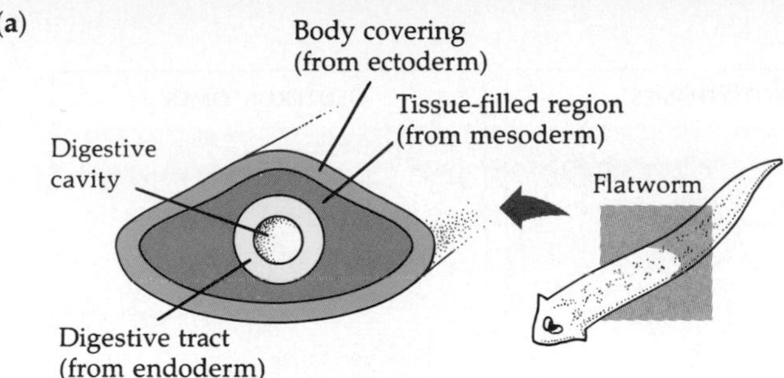

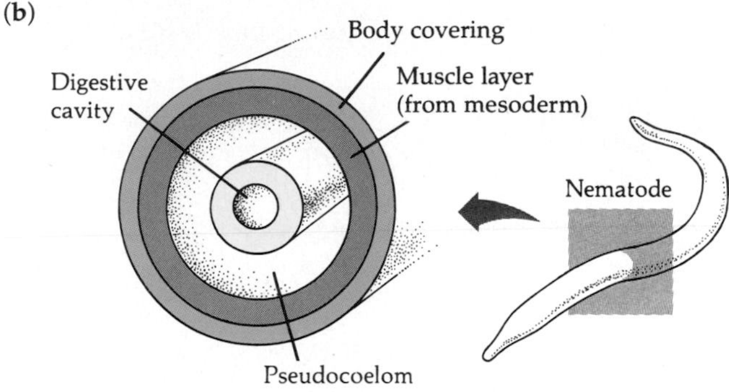

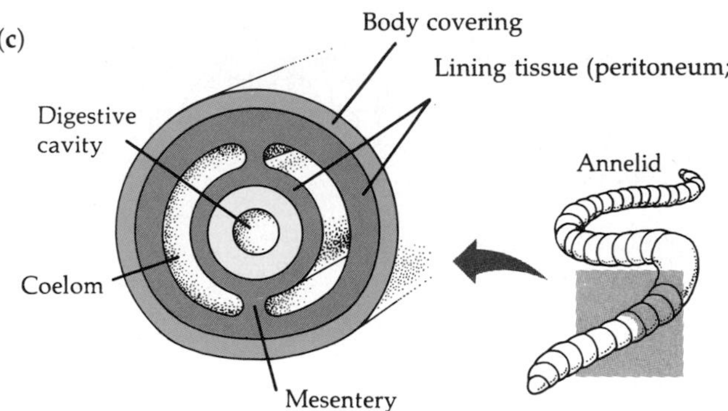

Figure 29.5

FIGURE 29.4
Body plans of the bilateria. The various organ systems of the animal develop from the three germ layers that form in the embryo. (**a**) Acoelomates lack a body cavity between the gut and outer body wall. (**b**) Pseudocoelomates have a body cavity only partially lined by mesodermally derived tissue. (**c**) Coelomates have a true coelom, a body cavity completely lined by mesodermally derived tissue.

termed a **pseudocoelom.** Animals with this body plan, such as rotifers (phylum Rotifera), roundworms (phylum Nematoda), and a few other phyla, are called **pseudocoelomates** (FIGURE 29.4b). **Coelomates** are animals with a true **coelom,** a fluid-filled body cavity completely lined by tissue derived from mesoderm. The inner and outer layers of tissue that surround the cavity connect dorsally and ventrally to form mesenteries, which suspend the internal organs (FIGURE 29.4c).

A body cavity has many functions. Its fluid cushions the suspended organs, helping to prevent internal injury. The cavity also enables the internal organs to grow and move independently of the outer body wall. If it were not for your coelom, every beat of your heart or ripple of your intestine could deform your body surface, and exercise would distort the shapes of the internal organs. In soft-bodied coelomates such as earthworms, the noncompressible fluid of the body cavity is under pressure and functions as a hydrostatic skeleton against which muscles can work. Though they may have originated as adaptations for burrowing by soft-bodied animals, coeloms evolved independently at least twice, as depicted on the tree in FIGURE 29.1, in the protostomes and in the deuterostomes.

④ *The Protostome-Deuterostome Split*

The coelomate phyla can be divided into two distinct evolutionary lines. Mollusks, annelids, arthropods, and several other phyla represent one of these lines and are collectively called **protostomes.** Echinoderms, chordates, and some other phyla, collectively called **deuterostomes,** represent the other line. Protostomes and deuterostomes are distinguished by several fundamental differences in their development.

Cleavage. Differences between animals of the two coelomate lineages are evident as early as the cleavage divisions that transform the zygote into a ball of cells. Many protostomes undergo **spiral cleavage,** in which planes of cell division are diagonal to the vertical axis of the embryo. As seen in the eight-cell stage resulting from spiral cleavage, small cells lie in the grooves between larger, underlying cells (FIGURE 29.5a). Furthermore, the so-called **determinate cleavage** of some protostomes rigidly casts the developmental fate of each embryonic cell very early. A cell isolated at the four-cell stage from a protostome, such as a snail, forms an inviable embryo that lacks parts.

In contrast to the protostome pattern, the zygote of many deuterostomes undergoes **radial cleavage.** Here, the cleavage planes are either parallel or perpendicular to the vertical axis of the egg; as seen in the eight-cell stage, the cells are aligned, one directly above the other. Most deuterostomes are further characterized by **indeterminate cleavage,** meaning that each cell produced by early cleavage divisions retains the capacity to develop into a complete embryo. If the cells of a sea star embryo, for example, are separated at the four-cell stage, each will go on to form a normal larva. It is the indeterminate cleavage of the human zygote that makes identical twins possible.

Blastopore Fate. Another difference between protostomes and deuterostomes is apparent later in development. In gastrulation, the rudimentary gut of an embryo forms as a blind pouch (the archenteron), which has a single opening to the outside known as the **blastopore**

FIGURE 29.5
A comparison of early development in protostomes and deuterostomes. (**a**) Protostomes have spiral, determinate cleavage; deuterostomes have radial, indeterminate cleavage. (**b**) The blastopore of the gastrula forms the mouth in protostomes; the mouth forms from a secondary opening in deuterostomes. (**c**) Coelom formation also begins in the gastrula stage. In the schizocoelous development of protostomes, the coelom forms from splits in the mesoderm. In the enterocoelous development of deuterostomes, the coelom forms from mesodermal outpocketings of the archenteron (blue = ectoderm, yellow = endoderm, red = mesoderm).

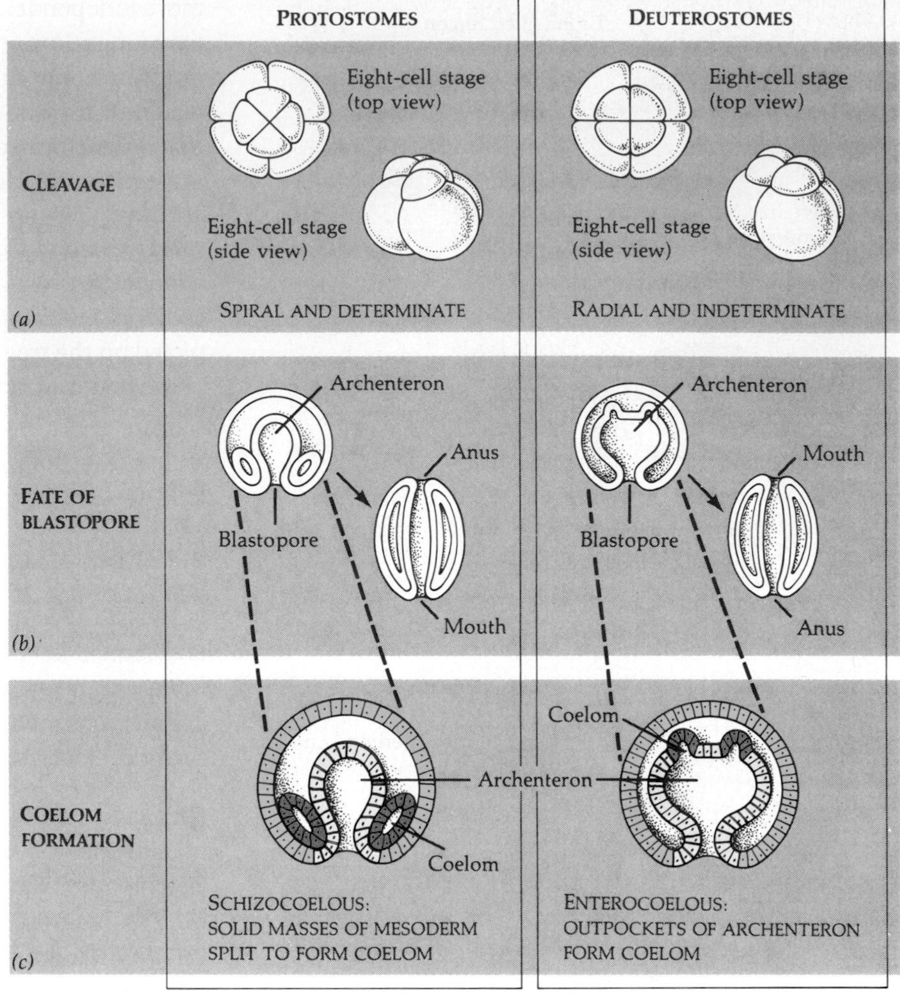

(FIGURE 29.5b). A second opening forms later at the opposite end of the archenteron to produce a digestive tube with a mouth and anus. The mouth of a typical protostome develops from the first opening, the blastopore, and it is for this characteristic that the protostome line is named (Gr. *protos*, "first," and stoma, "mouth"). By contrast, the mouth of a deuterostome (Gr. *deuteros*, "second") is derived from a secondary opening, and the blastopore usually forms the anus, not the mouth.

Coelom Formation. A third fundamental difference between protostomes and deuterostomes is in the development of the coelom (FIGURE 29.5c). As the archenteron forms in a protostome, initially solid masses of mesoderm split to form the coeloms; this is called **schizocoelous** development (Gr. *schizo*, "split"). Development of the body cavities of deuterostomes is termed **enterocoelous:** The mesoderm buds from the wall of the archenteron and hollows to become the coelomic cavities.

Keeping in mind that any phylogenetic drawing is a hypothesis, FIGURE 29.1 will serve as our road map as we survey some of the most important phyla of invertebrate animals in the following pages. Although we consider some aspects of function, such as reproduction and nutrition, as they characterize particular groups, we will leave the main discussion of animal physiology for Unit Seven.

Sponges are sessile animals lacking true tissues

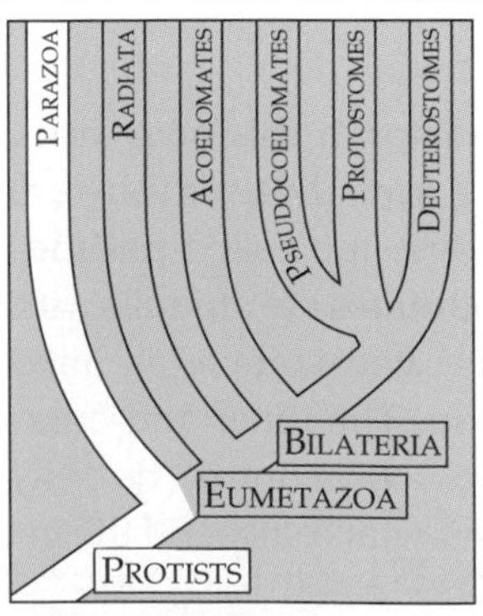

Phylum Porifera

Sponges (phylum Porifera) are sessile animals that appear so sedate to the human eye that the ancient Greeks believed them to be plants (FIGURE 29.6). Sponges range in height from about 1 cm to 2 m. Of the 9000 or so species of sponges, only about 100 live in fresh water; the rest are marine. The body of a simple sponge resembles a sac perforated with holes (*Porifera* means "pore bearers"). Water is drawn through the pores

FIGURE 29.6
A sponge. Sessile animals without specilized organs and tissues, sponges filter food from water pumped through their porous bodies. A sponge must filter about 1 ton of water to grow 1 ounce. The diverse species of sponges vary in shape and color, some brightly pigmented by symbiotic algae. This is an azure vase sponge, *Callyspongia plicifera.*

into a central cavity, the **spongocoel,** then flows out of the sponge through a larger opening called the **osculum** (FIGURE 29.7). More complex sponges have folded body walls and branched spongocoels.

Sponges are suspension-feeders (also known as filter-feeders), which are animals that collect food particles from water passed through some type of food-trapping equipment. Sponges trap food from the water circulated through the porous body. Lining the inside of the body are flagellated **choanocytes,** or collar cells (for the membranous collar around the base of the flagellum). The flagella generate a water current, and the collars trap food particles and ingest them by phagocytosis. The similarity between choanocytes and choanoflagellates (the protistan cells in FIGURE 26.27) supports the hypothesis that sponges evolved from colonial choanoflagellates.

The body wall of a sponge consists of two layers of cells separated by a gelatinous region called the **mesohyl.** Wandering through the mesohyl are cells called **amoebocytes,** named for their use of pseudopodia. Amoebocytes have many functions. They take up food from the choanocytes, digest it, and carry nutrients to other cells. Amoebocytes also form tough skeletal fibers within the mesohyl. In some groups of sponges, these fibers are sharp spicules made from calcium carbonate or silica; other sponges produce more flexible fibers composed of a protein called spongin. Variation in the chemical composition of the skeleton is one way systematists group sponges into the different classes of the phylum Porifera.

Most sponges are **hermaphrodites** (Gr. *Hermes,* the god, and *Aphrodite,* the goddess), meaning that each individual functions as both male and female in sexual reproduction by producing sperm *and* eggs. Gametes arise from choanocytes or amoebocytes. Eggs reside in the mesohyl, but sperm cells are released into the spongocoel and are carried out of the sponge by the water current. Cross-fertilization results from some of the sperm being drawn into neighboring individuals. Fertilization occurs in the mesohyl, where the zygotes develop into larvae covered with flagella. The swimming larvae disperse after their release into the spongocoel and their exit through the osculum. A tiny fraction of the larvae survive to settle on a suitable substratum and begin the sessile existence characteristic of sponges. Upon settling, the larva "turns inside out" during metamorphosis, and the flagellated cells then face the interior as choanocytes. Sponges are capable of extensive regeneration, the replacement of lost parts. They use regeneration not only for repair but also to reproduce asexually from fragments broken off a parent sponge.

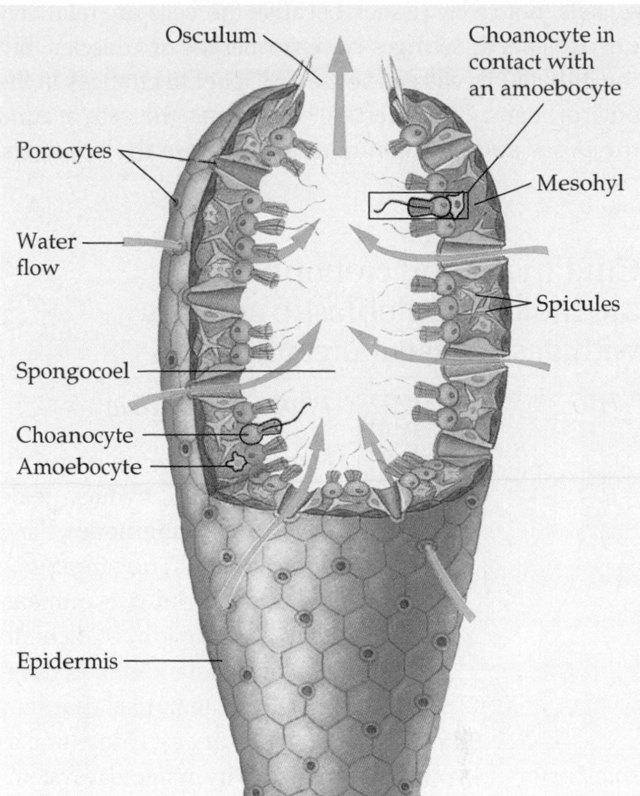

FIGURE 29.7
Anatomy of a sponge. The wall of this simple sponge has two layers of cells separated by a gelatinous matrix, the mesohyl ("middle matter"). The outer layer consists of tightly packed epidermal cells. The incurrent pores are channels through porocytes, cells shaped like elongated donuts that span the body wall. The spongocoel is lined mainly by choanocytes, each with a flagellum ringed by a collar of fingerlike projections that trap food particles. Mobile amoebocytes produce materials for skeletal fibers (spicules) and also transport nutrients from the choanocytes to other cells of the body.

Sponges are among the least complex of all animals. They lack organs, and the cell layers are loose federations of cells, not really tissues because the cells are relatively unspecialized. Sponges have no nerves or muscles, but the individual cells can sense and react to changes in the environment. Under certain conditions, the cells around the pores and osculum contract to close the openings.

Cnidarians and ctenophores are radiate, diploblastic animals with gastrovascular cavities

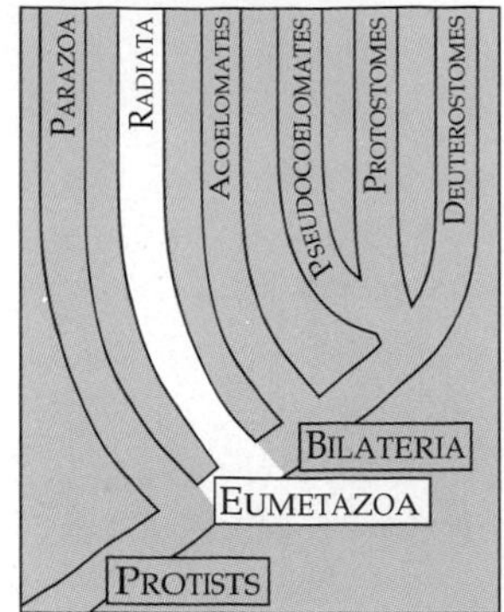

Phylum Cnidaria

Members of the phylum Cnidaria include hydras, jellyfishes, sea anemones, and coral animals. They are among the most primitive eumetazoa. Their diploblastic condition and radial symmetry suggest to systematists that they evolved very early in eumetazoan history, a hypothesis supported by molecular systematics. The absence of mesoderm has limited these animals to a relatively simple body construction. In spite of these constraints, however, the cnidarians are a diverse group with over 10,000 living species, most of which are marine.

The basic body plan of a cnidarian is a sac with a central digestive compartment, the **gastrovascular cavity.** A single opening to this cavity functions as both mouth and anus. This basic body plan has two variations: the sessile polyp and the floating medusa (FIGURE 29.8). **Polyps** are cylindrical forms that adhere to the substratum by the aboral end of the body and extend their tentacles, waiting for prey. Examples of the polyp form are hydras and sea anemones. A **medusa** is a flattened, mouth-down version of the polyp. It moves freely in the water by a combination of passive drifting and weak contractions of the bell-shaped body. The animals we generally call jellyfishes are medusas. The tentacles of the jellyfish dangle from the oral surface, which points downward. Some cnidarians exist only as polyps, others only as medusas, and still others pass sequentially through both medusa and polyp stages in their life cycles. This dimorphic (dual body form) life history is a unique feature of the cnidarians.

Cnidarians are carnivores that use tentacles arranged in a ring around the mouth to capture prey and push the food into the gastrovascular cavity, where digestion begins. The undigested remains are egested through the mouth/anus. The tentacles are armed with batteries of **cnidocytes,** unique cells that function in defense and the capture of prey (FIGURE 29.9).

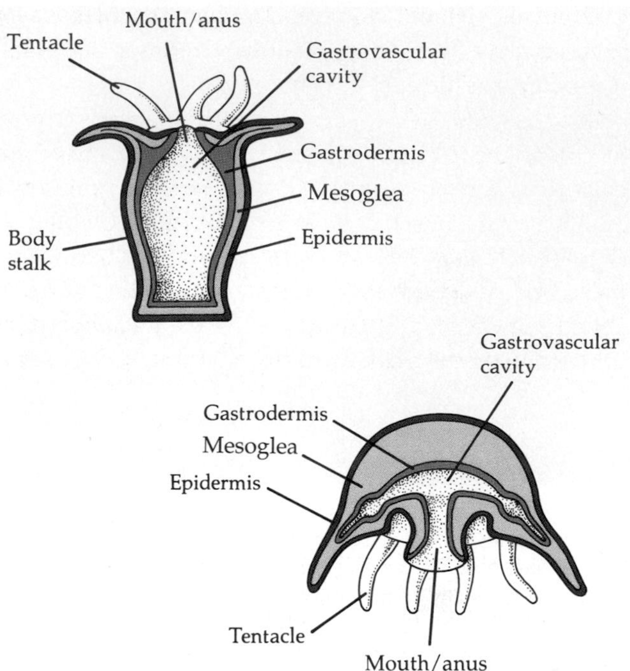

FIGURE 29.8
Polyp and medusa forms of cnidarians. The body wall of a polyp (top) or medusa (bottom) has two layers of cells, an outer layer of epidermis specialized for protection and an inner layer of gastrodermis for digestion. After the animal ingests food, the gastrodermis secretes digestive enzymes into the gastrovascular cavity. The gastrodermal cells engulf small pieces of the partially digested food by phagocytosis, and digestion is completed within the cells in food vacuoles. Flagella on the gastrodermal cells keep the contents of the gastrovascular cavity agitated and help distribute nutrients. Sandwiched between the epidermis and gastrodermis is a gelatinous layer of mesoglea. In many medusas, the mesoglea is thick and jellylike—thus their name, jellyfish.

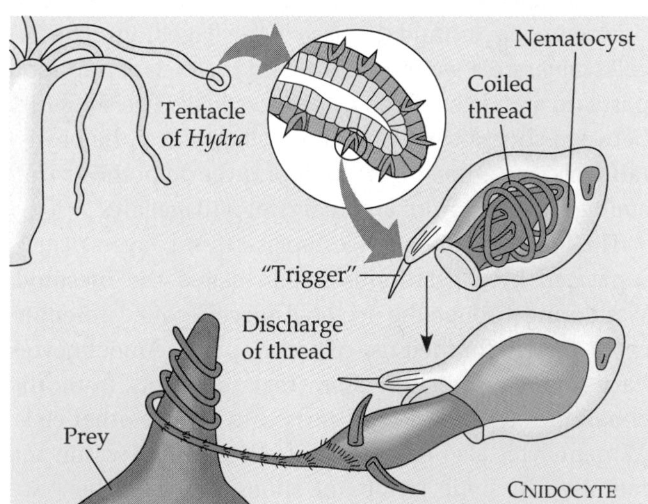

FIGURE 29.9
Cnidocytes of cnidarians. Each cnidocyte contains a stinging capsule, the nematocyst, which itself contains an inverted thread-like weapon. When a "trigger" is stimulated by touch or by certain chemicals, the thread shoots out from the nematocyst. Some of the threads are long and entangle the appendages of small animals that bump into the tentacles. Other threads puncture the prey and inject a poison that paralyzes the victim.

Cnidocytes, which contain stinging capsules called **nematocysts,** give the phylum Cnidaria its name (Gr. *cnide,* "nettle").

Muscles and nerves occur in their simplest forms in cnidarians. Cells of the epidermis (outer layer) and gastrodermis (inner layer) have bundles of microfilaments arranged into contractile fibers (see Chapter 7). True muscle tissue does not appear in diploblastic animals, since it is derived from mesoderm. The gastrovascular cavity acts as a hydrostatic skeleton against which the contractile cells can work. When the animal closes its mouth, the volume of the cavity is fixed, and contraction of selected cells causes the animal to change shape. The slow movements are coordinated by a nerve net. Cnidarians have no brain, and cnidarian behavior seems to be completely rigid; no one has yet trained a jellyfish. The noncentralized nerve net is associated with simple sensory receptors that are distributed radially around the body. Thus, the animal can detect and respond to stimuli equally from all directions.

The phylum Cnidaria is divided into three major classes: Hydrozoa, Scyphozoa, and Anthozoa (FIGURE 29.10).

(a) (b) (c) (d)

FIGURE 29.10

Representatives of the cnidarian classes. (**a**) Polyps of a colonial species belonging to the class Hydrozoa. (**b**) Jellyfish of the class Scyphozoa. The medusa is the conspicuous stage of the scyphozoan life cycle. The largest scyphozoan species have tentacles over 30 m long dangling from umbrellas 2 to 3 m in diameter. (**c**) Sea anemones and other members of the class Anthozoa exist only as polyps. (**d**) A colony of coral polyps (class Anthozoa). Many corals harbor symbiotic algae that contribute to the food supply of the polyps. Coral reefs, which provide habitats for an enormous variety of invertebrates and fishes, are restricted to warm, shallow seas. This is a star coral.

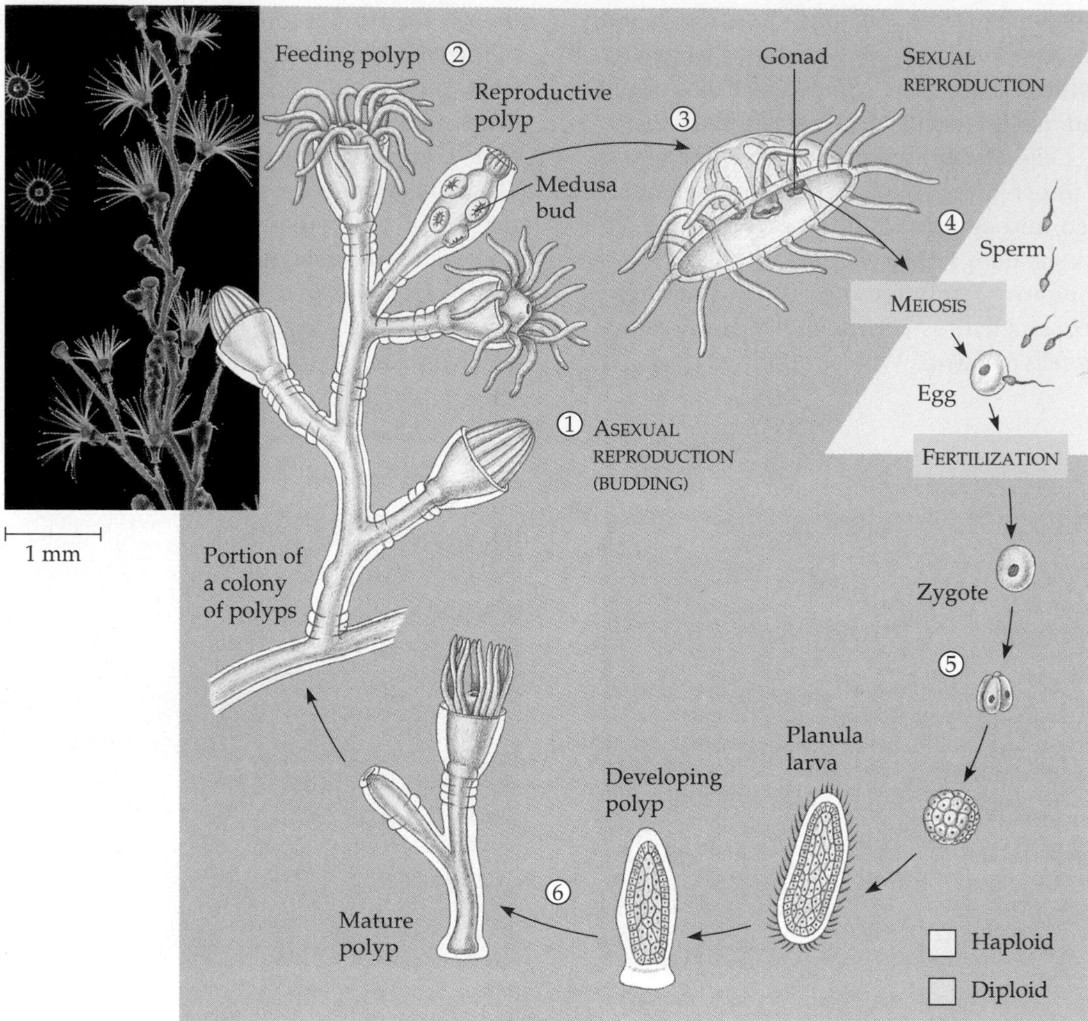

FIGURE 29.11

The life cycle of the hydrozoan *Obelia.* ① The polyp reproduces asexually by budding to form a colony of interconnected polyps (inset, LM). ② Some polyps, equipped with tentacles, are specialized for feeding. ③ Other polyps, specialized for reproduction, lack tentacles and produce tiny medusas by asexual budding. ④ The medusas swim off, grow, and reproduce sexually. ⑤ The zygote develops into a solid, ciliated larva called the planula. ⑥ The planula eventually settles and develops into a new polyp. The polyp stage is asexual, the medusa stage is sexual, and these two stages alternate, one producing the other. But do not confuse this with the alternation of generations that occurs in the plant kingdom. Both polyp and medusa are diploid organisms, whereas one of the plant generations is haploid.

Class Hydrozoa. Most hydrozoans alternate polyp and medusa forms, as in the life cycle of *Obelia* (FIGURE 29.11). The polyp stage, a colony of interconnected polyps in the case of *Obelia,* is more conspicuous than the medusae. *Hydra,* one of the few cnidarian genera found in fresh water, is an unusual member of the class Hydrozoa in that it exists only in the polyp form. When environmental conditions are favorable, *Hydra* reproduces asexually by budding, the formation of outgrowths that pinch off from the parent to live independently (see FIGURE 12.1). When environmental conditions deteriorate, *Hydra* reproduces sexually, forming resistant zygotes that remain dormant until conditions improve.

Class Scyphozoa. In the class Scyphozoa, it is generally the medusa that prevails in the life cycle. The medusas of most species live among the plankton as jellyfishes. Most coastal scyphozoans go through a small polyp stage during their life cycle, but jellyfishes that live in the open ocean have generally eliminated the sessile polyp.

Class Anthozoa. Sea anemones and corals belong to the class Anthozoa ("flower animals"). They occur only as polyps. Coral animals live as solitary or colonial forms and secrete hard external skeletons of calcium carbonate. Each polyp generation builds on the skeletal remains of earlier generations to construct "rocks" with shapes

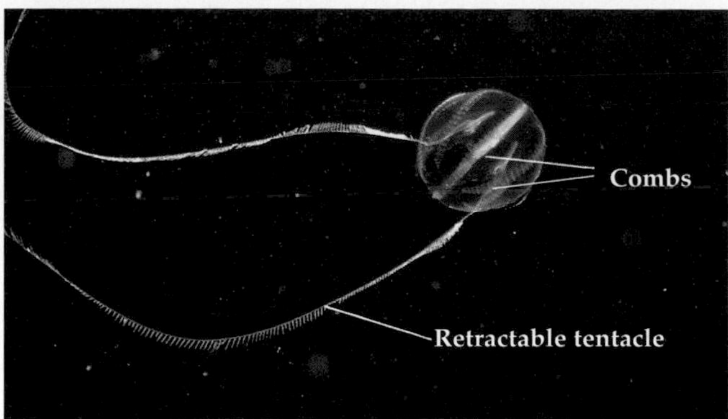

FIGURE 29.12

A ctenophore, or comb jelly. This planktonic marine animal is named for its eight combs of cilia, used for locomotion. The retractable tentacles capture food. Ctenophores and cnidarians are radiate, diploblastic animals, but many zoologists believe these groups have little else in common.

characteristic of the species. It is these skeletons that we call coral.

Phylum Ctenophora

Comb jellies, the common name for members of the phylum Ctenophora, superficially resemble small cnidarian medusas. However, the relationship between ctenophores and cnidarians is uncertain. There are only about 100 species of comb jellies, all of which are marine. The transparent animals range in diameter from about 1 to 10 cm (FIGURE 29.12). Most are spherical or ovoid, but there are elongate and ribbonlike forms up to 1 m long. Ctenophores ("comb-bearers") are named for their eight rows of comblike plates composed of fused cilia. They are the largest animals to use cilia for locomotion. An aboral sensory organ functions in orientation, and nerves running from the sensory organ to the combs of cilia coordinate movement. Most comb jellies have a pair of long, retractable tentacles that function in the capture of food.

Flatworms and other acoelomates are bilateral, triploblastic animals lacking body cavities

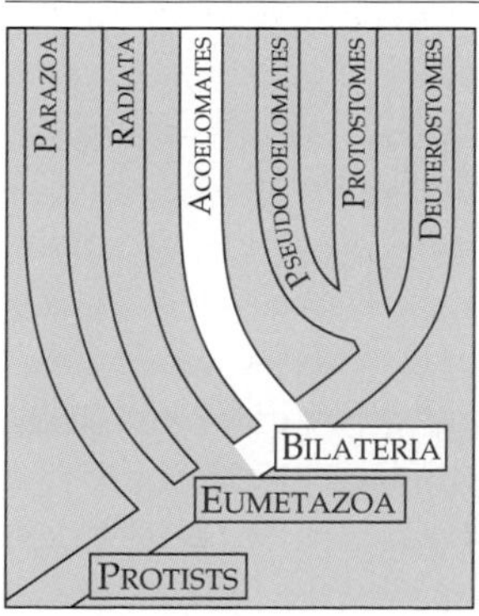

Phylum Platyhelminthes

There are about 20,000 species of flatworms (phylum Platyhelminthes) living in marine, freshwater, and damp terrestrial habitats. In addition to many free-living forms, flatworms include the flukes and the tapeworms, which are exclusively parasitic on other animals. Flatworms are so named because their bodies are generally flattened dorsoventrally (the word *platyhelminth* means "flat worm"). They range in size from nearly microscopic free-living species to certain tapeworms over 20 m long.

Flatworms represent several important evolutionary developments compared to the cnidarians. They are bilaterally symmetrical with unidirectional movement and moderate cephalization. Furthermore, they produce embryonic mesoderm, a third germ layer that contributes to the development of more complex organs and organ systems, and true muscle tissue. However, the gut of a typical flatworm, like that of a cnidarian, is a gastrovascular cavity with only one opening. Another clue that flatworms evolved early in the history of bilateria is the absence of a body cavity between the gut and outer body wall.

Flatworms are divided into four classes: Turbellaria (mostly free-living flatworms), Trematoda and Monogenea (flukes), and Cestoda (tapeworms).

Class Turbellaria. Turbellarians are nearly all free-living (nonparasitic) and mostly marine (FIGURE 29.13), but members of the genus *Dugesia,* commonly known as planarians, abound in ponds and streams. **Planarians** are carnivores that prey on smaller animals or feed on carrion (dead animals). The anatomy of a planarian is illustrated in FIGURE 29.14.

Planarians and other flatworms lack organs specialized for gas exchange and circulation. The flat shape of the body places all cells close to the surrounding water, and fine branching of the gastrovascular cavity distributes food throughout the animal. Nitrogenous waste in the form of ammonia diffuses directly from the cells into the surrounding water. Flatworms also have a relatively simple excretory apparatus that functions mainly to maintain osmotic balance between the animal and its surroundings. This system consists of ciliated cells called

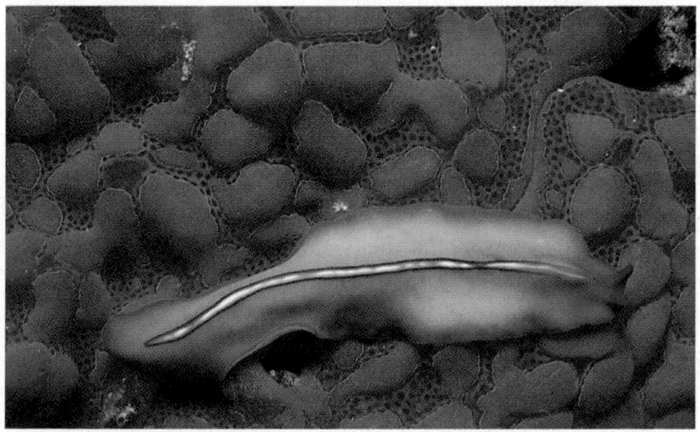

FIGURE 29.13

A flatworm. Class Turbellaria consists mainly of free-living marine flatworms, such as this colorful species.

flame cells that waft fluid through branched ducts opening to the outside (see FIGURE 40.6). The evolution of osmoregulatory structures was a major factor in allowing some turbellarians to invade freshwater and even moist terrestrial environments.

Planarians move by using cilia on the ventral epidermis, gliding along a film of mucus they secrete. Some turbellarians occasionally also use their muscles to swim through water with an undulating motion.

A planarian has a head (is cephalized) with a pair of eyespots that detect light and lateral flaps that function mainly for smell. The planarian nervous system is more complex and centralized than the nerve nets of cnidarians. Planarians can learn to modify their responses to stimuli.

Planarians can reproduce asexually through regeneration. The parent constricts in the middle, and each half regenerates the missing end. Sexual reproduction also occurs. Although planarians are hermaphrodites, copulating mates cross-fertilize.

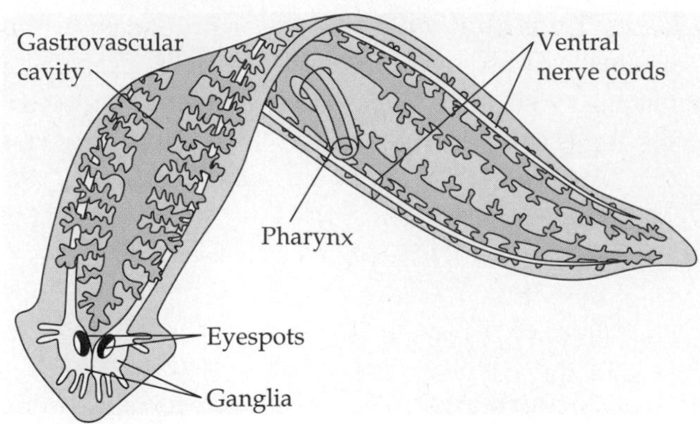

FIGURE 29.14
Anatomy of a planarian. The mouth is at the tip of a muscular pharynx that extends from the middle of the ventral side of the animal. Digestive juices are spilled onto prey, and the pharynx sucks small pieces of food into the gastrovascular cavity, where digestion continues. Digestion is completed within the cells lining the gastrovascular cavity, which has three branches, each with fine sub-branches that provide an extensive surface area. Undigested wastes are egested through the mouth. Located at the anterior end of the worm, near the main sources of sensory input, is a pair of ganglia, dense clusters of nerve cells. From the ganglia, a pair of ventral nerve cords runs the length of the body.

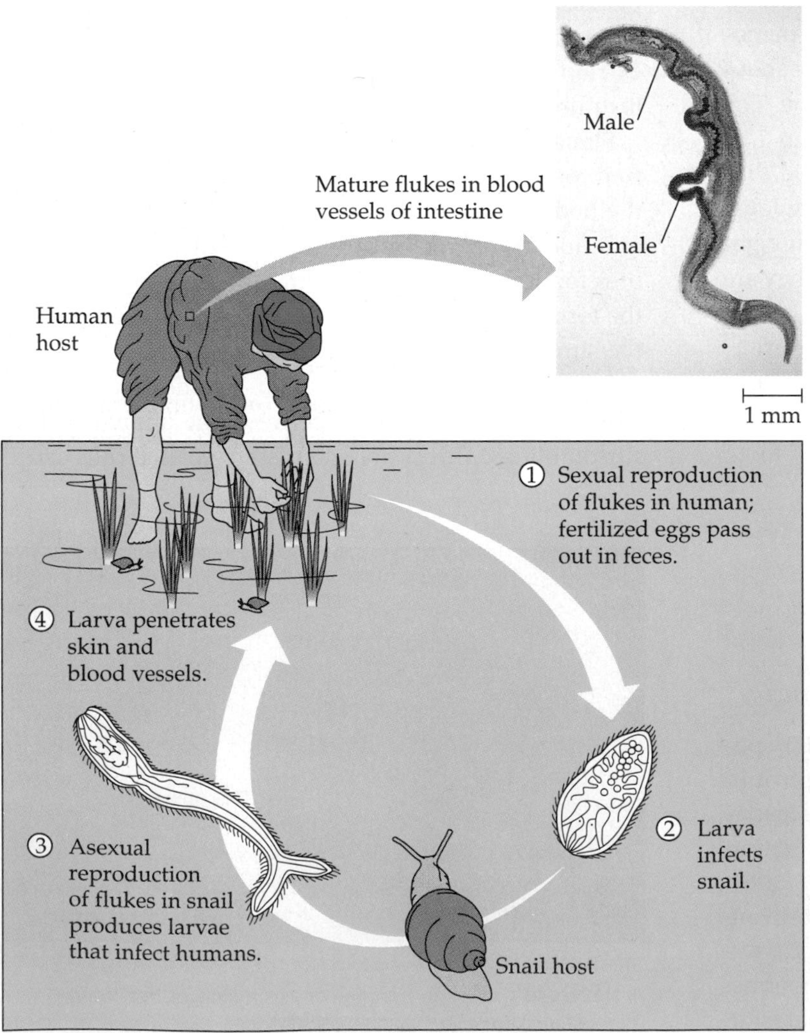

FIGURE 29.15
The life history of a blood fluke *(Schistosoma mansoni).* ① Blood flukes reproduce sexually in the human host. A female fluke fits into a groove running the length of the larger male's body (inset, LM). The fertilized eggs exit the host in feces, and ② develop in water into ciliated larvae. These larvae can enter snails, the intermediate hosts in the life cycle of blood flukes. ③ Asexual reproduction within a snail results in another type of motile larva, which escapes from the host. ④ People working in irrigated fields contaminated with human feces may be exposed to these larvae, which can penetrate the skin of the host.

Classes Trematoda and Monogenea. Flukes live as parasites in or on other animals. Many have suckers for attaching to internal organs of the host, and a tough covering helps protect the parasite. Reproductive organs nearly fill the interior of a mature fluke.

Flukes generally have life cycles with an alternation of sexual and asexual stages. Many species require an intermediate host where larvae develop before infecting the final host, where the adult fluke lives. For example, flukes that parasitize humans spend parts of their life histories in snails (FIGURE 29.15). The 200 million people around the world who are infected with blood flukes (*Schistosoma*) suffer body pains, anemia, and dysentery.

Class Cestoda. Tapeworms, of the class Cestoda, are parasitic flatworms. The adults live mostly in vertebrates, including humans. The tapeworm head, or scolex, is armed with suckers and often menacing hooks that lock the worm to the intestinal lining of the host (FIGURE 29.16). Posterior to the scolex is a long ribbon of units called proglottids, which are little more than sacs of sex organs. The tapeworm lacks a digestive system; its food is predigested by the host. Mature proglottids, loaded with thousands of eggs, are released from the posterior end of the worm and leave the body with feces. In one type of cycle, human feces contaminate the food or water of intermediate hosts, such as pigs or cattle, and the tapeworm eggs develop into larvae that encyst in muscles of these animals. Humans acquire the larvae by eating undercooked meat contaminated with cysts, and the worms develop into mature adults within the human. Large tapeworms, which may be 20 m or more in length, can cause intestinal blockage and can rob enough nutrients from the human host to cause nutritional deficiencies.

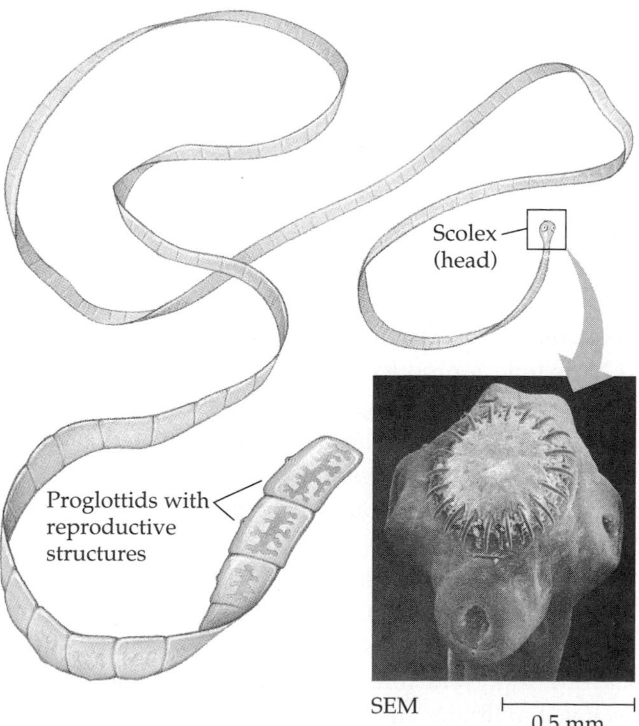

FIGURE 29.16
Anatomy of a tapeworm.

Phylum Nemertea

Members of the phylum Nemertea are called ribbon worms or proboscis worms (FIGURE 29.17). Their phylogenetic position is presently being debated, although molecular systematics supports anatomical evidence that they are related to the protostome lineage. A ribbon worm's body is structurally acoelomate, like that of the flatworm, but it contains a small fluid-filled sac that is used to hydraulically operate an extensible proboscis by which the worm captures prey. Zoologists do not yet agree on whether this sac is a true coelom. Thus, the position of Nemertea in the phylogenetic tree of FIGURE 29.1 is uncertain, but the phylum is discussed here to highlight comparison with Platyhelminthes.

Proboscis worms range in length from less than 1 mm to more than 30 m. Nearly all of the 900 or so members of this phylum are marine, but a few species inhabit fresh water or damp soil. Some are active swimmers, and others burrow in the sand.

Proboscis worms probably evolved from flatworms. The two phyla have similar excretory, sensory, and nervous systems. But, in addition to the unique proboscis apparatus, two anatomical features not found in flatworms have evolved in the phylum Nemertea: a simple blood vascular system (circulatory system) and a **complete digestive tract**—that is, a digestive tube with a separate mouth and anus. The circulatory system consists of vessels through which blood flows, and some species have red blood cells containing a form of hemoglobin,

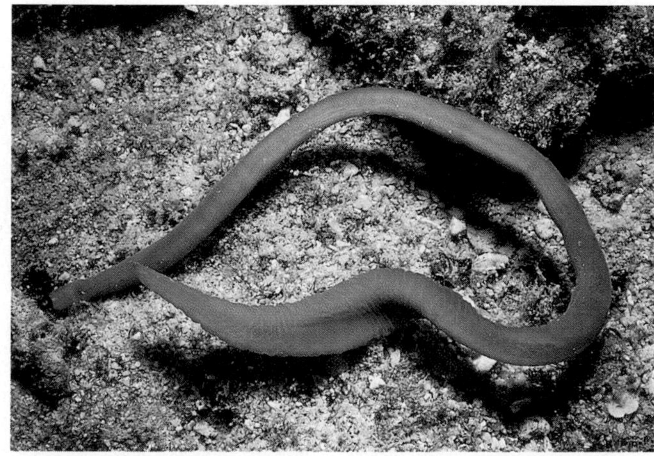

FIGURE 29.17
A ribbon worm, phylum Nemertea.

which transports oxygen. Proboscis worms have no heart, but muscles squeeze the vessels to propel the blood. A digestive tube and a blood vascular system are features nemerteans share with animals that follow in this survey of the invertebrates.

Rotifers, nematodes, and other pseudocoelomates have complete digestive tracts and blood vascular systems

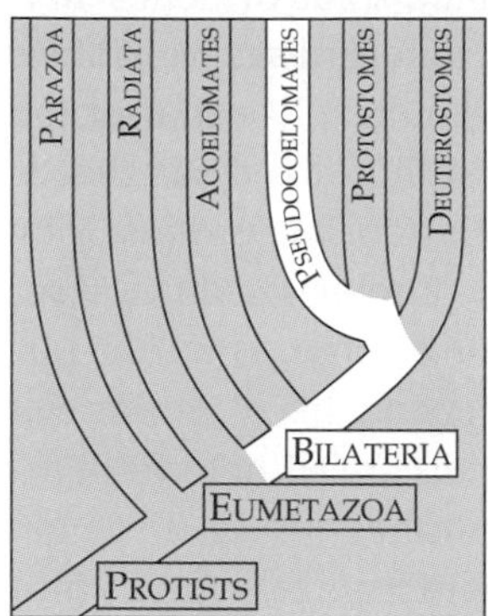

A pseudocoelomate body plan (see FIGURE 29.4b) has evolved in several phyla of small animals. Their evolutionary relationships with other groups and among themselves are still unclear, but they are probably more closely related to protostomes than to deuterostomes. In fact, the pseudocoelomate condition probably arose independently several times. We discuss here only two of these phyla: Rotifera and Nematoda.

Phylum Rotifera

Rotifers (about 1800 species) are tiny animals that mainly inhabit fresh water, although some live in the sea or in damp soil. Ranging in size from only about 0.05 to 2.0 mm, smaller than many protozoa, rotifers are nevertheless truly multicellular and have a complete digestive tract and other specialized organ systems (FIGURE 29.18). Internal organs lie within the pseudocoelom. The fluid that fills the pseudocoelom serves as a hydrostatic skeleton and as a medium for the internal transport of nutrients and wastes in these tiny animals. Movement of a rotifer's body distributes the fluid within the pseudocoelom, and thus this body cavity and its fluid functions as a blood vascular system.

The word *rotifer,* derived from Latin, means "wheel-bearer," a reference to the crown of cilia that draws a vortex of water into the mouth. Posterior to the mouth is a jawlike organ that grinds food, mostly microorganisms suspended in the water.

Rotifer reproduction is unusual. Some species consist only of females that produce more females from unfertilized eggs, a type of reproduction called **parthenogenesis.** Other species produce two types of eggs that develop by parthenogenesis, one type forming females and the other type developing into degenerate males that cannot even feed themselves. The males survive long enough to produce sperm that fertilize eggs, forming resistant zygotes that can survive when a pond dries up. When conditions are favorable again, the zygotes break dormancy and develop into a new female generation that then reproduces by parthenogenesis until conditions become unfavorable again.

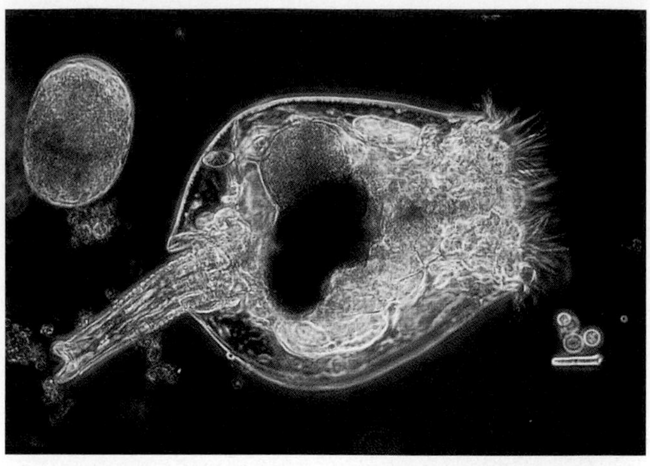

FIGURE 29.18
A rotifer. These pseudocoelomates, smaller than many protozoa, are much more anatomically complex than flatworms (LM).

Phylum Nematoda

Nematodes are cylindrical pseudocoelomate worms with tapered ends. They are called roundworms and are among the most numerous of all animals in both species and individuals (FIGURE 29.19). Roundworms are found in most aquatic habitats, in wet soil, in the moist tissues of plants, and in the body fluids and tissues of animals. About 80,000 species of nematodes are known, and perhaps ten times that number actually exist. Roundworms range from less than 1 mm to more than 1 m in length. They have a complete digestive tract, and the pseudocoelom with its fluid serves as a blood vascular system that transports nutrients throughout the body. Roundworms have a tough but transparent covering, the cuticle. The muscles of nematodes are all longitudinal, and their contraction produces a thrashing motion.

Nematode reproduction is usually sexual. The sexes are separate in most species, females generally being larger than males. Fertilization is internal, and a female may deposit 100,000 or more fertilized eggs per day. The zygotes of most species are resistant cells capable of surviving harsh conditions.

Great numbers of nematodes live in moist soil and in decomposing organic matter on the bottoms of lakes and oceans. These extremely numerous free-living worms play an important role in decomposition and nutrient cycling, but little is known about most species.

Phylum Nematoda also includes many important agricultural pests that attack the roots of plants. Other

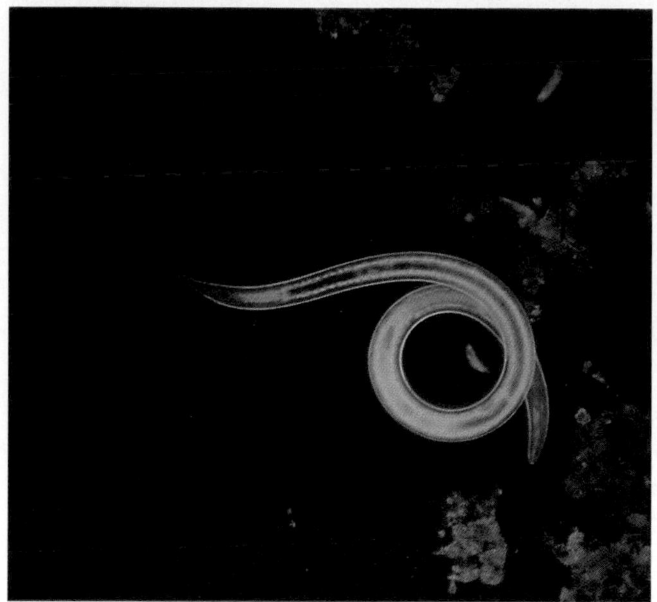

(a) 100 µm

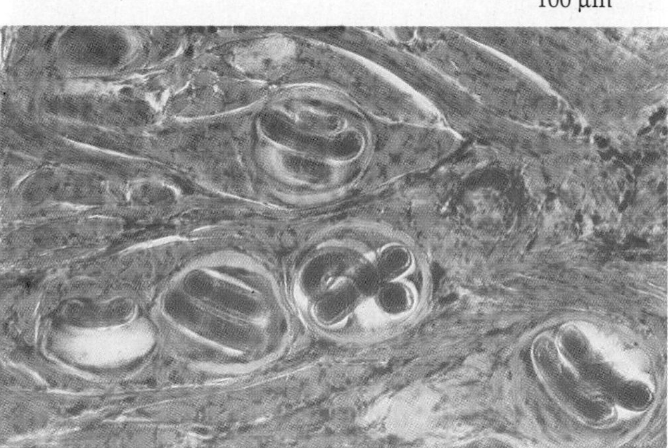

(b) 50 µm

FIGURE 29.19

Nematodes. (**a**) A free-living nematode. Some of the internal organs of this freshwater worm are visible through the transparent skin. (**b**) Juveniles of *Trichinella spiralis* encysted in human muscle tissue. This parasitic nematode causes trichinosis, characterized by severe nausea and sometimes death when large numbers of the juveniles penetrate heart muscle. (Both LMs.)

species of roundworms parasitize animals. Humans host at least 50 nematode species, including various pinworms and hookworms. One notorious nematode is *Trichinella spiralis,* the worm that causes trichinosis (FIGURE 29.19b). Humans acquire this nematode by eating undercooked infected pork or other meat with juvenile worms encysted in the muscle tissue. Within the human intestine, the juveniles develop into sexually mature adults. Females burrow into the intestinal muscles and produce more juveniles, which bore through the body or travel in lymph vessels to encyst in other organs, including skeletal muscles.

Mollusks and annelids are among the major variations on the protostome body plan

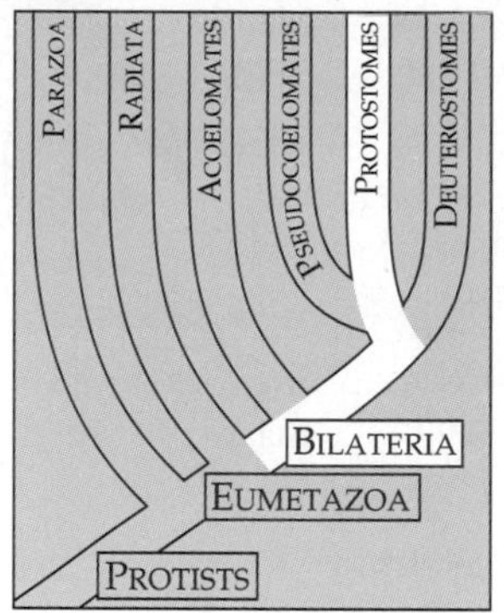

The protostome lineage of coelomate animals gave rise to the mollusks, to the annelids and several other phyla of worms with a coelom that functions as a hydrostatic skeleton for burrowing, and to the arthropods, the most diverse and widespread of all animal phyla. (Arthropods are covered in a separate section.) Before taking a closer look at these animals, you might want to review the key protostome characteristics in FIGURE 29.5.

Phylum Mollusca

Snails and slugs, oysters and clams, and octopuses and squids are all mollusks. In all, the phylum Mollusca has more than 50,000 known species. Most mollusks are marine, though some inhabit fresh water, and there are snails and slugs that live on land. Mollusks are soft-bodied animals (L. *molluscus,* "soft"), but most are protected by a hard shell made of calcium carbonate. Squids and octopuses have reduced shells that have been internalized, or they have lost their shells completely during their evolution.

Despite their apparent differences, all mollusks have a similar body plan (FIGURE 29.20). The body has three main parts: a muscular **foot,** usually used for movement; a **visceral mass,** containing most of the internal organs; and a **mantle,** a heavy fold of tissue that drapes over the visceral mass and may secrete a shell. The mantle also extends beyond the visceral mass, producing a water-filled chamber, the **mantle cavity,** which houses the gills, anus, and excretory pores. Many mollusks feed by using a straplike rasping organ called a **radula** to scrape up food. Most mollusks have separate sexes, with gonads (ovaries or testes) located in the visceral mass. Many snails, however, are hermaphrodites.

Zoologists debate the relationship of mollusks to other coelomate protostomes. The life cycle of many marine mollusks includes a ciliated larva called the **trochophore,** also characteristic of marine annelids (segmented worms) and some other protostomes. But mollusks lack the one trait that most defines an annelid heritage—true segmentation. The mollusk *Neopilina* (class Monoplacophora) has some of its internal organs repeated, but this condition may have evolved secondarily from an ancestral mollusk with unrepeated organs. Many zoologists

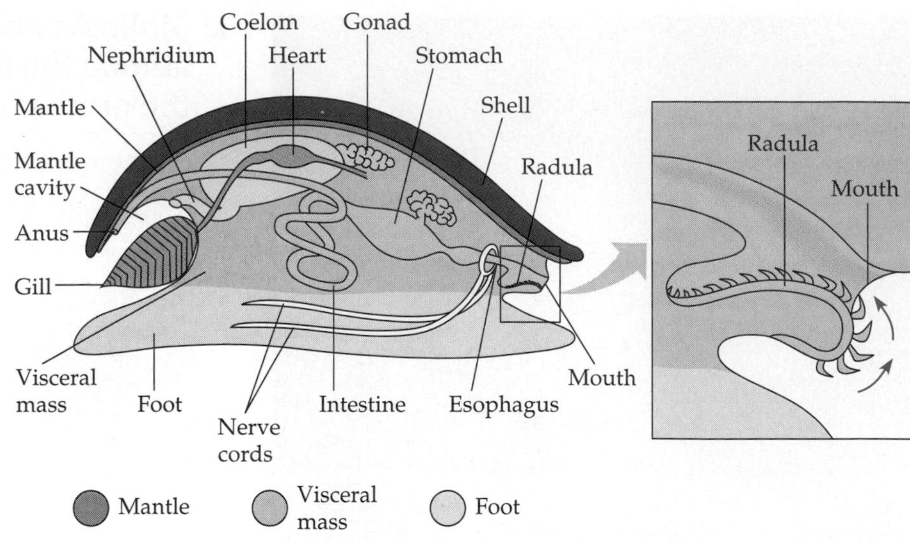

FIGURE 29.20

The basic body plan of mollusks. Three hallmarks of the phylum are the mantle, visceral mass, and foot. A mantle cavity houses gills in many species. The long digestive tract is coiled in the visceral mass. Most mollusks have an open circulatory system with a dorsal heart that pumps circulatory fluid (hemolymph) through arteries into sinuses (body spaces) bathing the organs. Excretory organs called nephridia remove metabolic wastes from the hemolymph. The nervous system consists of a nerve ring around the esophagus, from which nerve cords extend. The enlargement shows the mouth region with the radula, a rasplike feeding organ present in many mollusks. The radula is a belt of backward-curved teeth that extends from the mouth and slides back and forth, scraping and scooping like a backhoe.

doubt that mollusks evolved from an annelidlike ancestor and favor the hypothesis that they arose earlier on the protostome line, before the evolution of segmentation.

The basic body plan of mollusks has evolved in various ways in the different classes of the phylum. Of the eight classes, we examine four here: Polyplacophora (chitons), Gastropoda (snails and slugs), Bivalvia (clams, oysters, and other bivalves), and Cephalopoda (squids, octopuses, and nautiluses).

Class Polyplacophora. Chitons are marine animals with oval shapes and shells divided into eight dorsal plates (the body itself, however, is unsegmented). You can find chitons clinging to rocks along the shore during low tide (FIGURE 29.21). Try to dislodge a chiton by hand, and you will be surprised at how tenaciously its foot, acting as a suction cup, grips the rock. Using this muscular foot in the same fashion as a snail does, a chiton can creep slowly over the rock surface. Chitons are grazers that use their radulas to cut and ingest algae.

FIGURE 29.21

A chiton. Clinging tenaciously to rocks in the intertidal zone, this chiton (class Polyplacophora) displays the eight-plate shell characteristic of this class of mollusks.

Class Gastropoda. The largest of the molluscan classes, Gastropoda has more than 40,000 species. Most gastropods are marine, but there are also many freshwater species. Garden snails and slugs have adapted to land.

The most distinctive characteristic of the class Gastropoda is an embryonic process known as **torsion.** During embryonic development, one side of the visceral mass grows faster than the other. The uneven growth causes the visceral mass to rotate up to 180 degrees, so that the anus and mantle cavity are placed above the head in the adult (FIGURE 29.22). Some zoologists speculate that the advantage of torsion is to place the visceral mass and heavy shell more centrally over the snail's body.

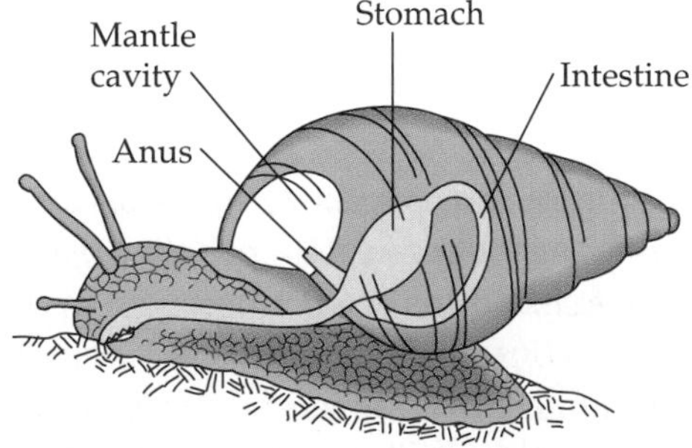

FIGURE 29.22

The results of torsion in a gastropod. Because of torsion (twisting of the visceral mass) during embryonic development, the digestive tract is coiled and the anus is near the mouth at the head end of the animal. After torsion, some of the organs that were bilateral subsequently atrophy on one side of the body. Torsion should not be confused with the formation of a coiled shell, which is an independent process.

Most gastropods are protected by single, spiraled shells into which the animals can retreat when threatened

(a)

(b)

FIGURE 29.23

Gastropods. (**a**) Shell collectors find delight in the variety of Gastropoda, one of the most diverse animal classes. (**b**) Nudibranchs, or sea slugs, have eliminated the shell during their evolution.

(FIGURE 29.23). The shell is often conical, but abalones and limpets have somewhat flattened shells. Slugs and nudibranchs (sea slugs) have lost their shells during their evolution. Many gastropods have distinct heads with eyes at the tips of tentacles. Gastropods inch along literally at a snail's pace by a rippling motion of the elongated foot. Most gastropods use their radula to graze on algae or plant material. Several groups, however, are predators, and the radula is modified for boring holes in the shells of other mollusks or for tearing apart tough animal tissue. In one group, the cone snails, the teeth of the radula form separate poison darts, which penetrate prey.

Gastropods are among the few invertebrate groups to have successfully populated the land. Terrestrial snails lack the gills typical of most aquatic gastropods, and instead, the vascularized lining of the mantle cavity functions as a lung to exchange respiratory gases with the air.

FIGURE 29.24

A bivalve. This scallop has many eyes peering out between the two halves of the hinged shell.

Class Bivalvia. The mollusks of the class Bivalvia include many species of clams, oysters, mussels, and scallops. Bivalves have shells divided into two halves (FIGURE 29.24). The two parts of the shell are hinged at the mid-dorsal line, and powerful adductor muscles draw the two halves tightly together to protect the soft-bodied animal. When the shell is open, the bivalve may extend its hatchet-shaped foot for digging or anchoring.

The mantle cavity of a bivalve contains gills that are used for feeding as well as gas exchange (FIGURE 29.25). Most bivalves are suspension-feeders. They trap fine

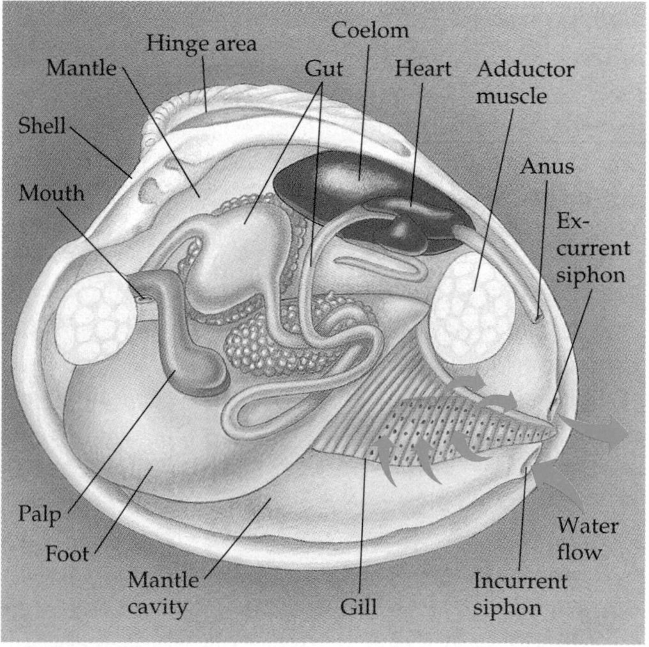

FIGURE 29.25

Anatomy of a clam. The left half of the bivalve shell has been removed. Food particles suspended in water that enters through the incurrent siphon are collected by the gills and passed via cilia and elongated flaps called palps to the mouth.

food particles in mucus that coats the gills, and then use cilia to convey the particles to the mouth. Water flows into the mantle cavity through an incurrent siphon, passes over the gills, and then exits the mantle cavity through an excurrent siphon. Bivalves have no distinct head, and the radula has been lost.

Being suspension-feeders, most bivalves lead rather sedentary lives. Sessile mussels secrete strong threads that tether to rocks, docks, boats, and the shells of other animals. Clams can pull themselves into the sand or mud, using the muscular foot for an anchor. In addition to digging, scallops can also skitter along the seafloor by flapping their shells, rather like the mechanical false teeth sold in novelty shops.

Class Cephalopoda. Unlike the sluggish gastropods and the sedentary bivalves, cephalopods are built for speed, an adaptation that fits their carnivorous diet (FIGURE 29.26). Squids and octopuses use beaklike jaws to bite their prey; they then inject poison to immobilize the victim. The mouth is at the center of several long tentacles. A mantle covers the visceral mass, but the shell is usually either reduced and internal (squids) or missing altogether (octopuses). One shelled cephalopod, the chambered nautilus, survives today (FIGURE 29.26c).

A squid darts about, usually backward, by drawing water into its mantle cavity and then firing a jet stream of water through the excurrent siphon that points anteriorly. The animal steers by pointing the siphon in different directions. The foot of a cephalopod has become modified into this muscular siphon and probably other parts of the tentacles and head. (*Cephalopod* means "head foot.") Most species of squid are less than 75 cm long, but there are also giant squids, the largest of all invertebrates. The biggest specimen on record was 17 m long (including the tentacles) and weighed about 2 tons.

Rather than swimming as squids do in the open seas, most octopuses live on the seafloor, where they creep and scurry about in search of crabs and other food.

Cephalopods are the only mollusks with a **closed circulatory system,** a type of blood vascular system in which blood is always contained in vessels. Cephalopods also have a well-developed nervous system with a complex brain. The ability to learn and behave in a complex manner is probably more critical to fast-moving predators than to sedentary animals such as clams. Squids and octopuses also have well-developed sense organs.

The ancestors of octopuses and squids were probably shelled mollusks that took up a predaceous lifestyle, the loss of the shell occurring in later evolution. Shelled cephalopods called **ammonites,** many of them very large, were the dominant invertebrate predators of the seas for hundreds of millions of years until their disappearance during the mass extinctions at the end of the Cretaceous period.

The Lophophorate Animals

In subdividing coelomate animals into protostomes and deuterostomes, three phyla—Phoronida, Bryozoa, and Brachiopoda—have been the subject of a good deal of controversy. These phyla are collectively called the **lophophorate animals** after the most distinctive structure they share, the lophophore (FIGURE 29.27). The **lophophore** is a horseshoe-shaped or circular fold of the body wall bearing ciliated tentacles that surround the mouth at the anterior end of the animal. The anus lies outside the whorl of tentacles. The cilia draw water toward the mouth between the tentacles, which help trap food particles for these suspension-feeders. The common occurrence of this complex apparatus in the lophophorate animals suggests that the three phyla are related. Other similarities,

(a)

(b)

(c)

FIGURE 29.26

Cephalopods. (**a**) Squids, such as this California market squid *(Loligo opalescens),* are speedy carnivores with beaklike jaws and well-developed eyes. (**b**) Octopuses are believed to be among the most intelligent invertebrates. (**c**) Chambered nautiluses are the only cephalopods alive today with external shells.

(a)

(b)

FIGURE 29.27
The lophophorate animals. The most distinctive characteristic of these phyla is the lophophore, an organ that functions in suspension-feeding. (**a**) Bryozoans, such as this common sea mat *(Membranipora membranacea),* are colonial lophophorates, often with hard exoskeletons. (**b**) Brachiopods are lophophorates with a hinged shell. The two parts of the shell are dorsal and ventral, in contrast to the lateral shells of bivalve mollusks.

such as a U-shaped digestive tract and the absence of a distinct head, are adaptations to a sessile existence that also may have evolved convergently.

In their embryonic development, the lophophorate animals as a group most closely resemble deuterostomes. However, in the phoronids, the mouth arises from the embryonic blastopore, a feature associated with protostomes. Because of this and some other inconsistencies, the evolutionary position of the lophophorates remains uncertain (hence the broken branches of the phylogenetic tree of FIGURE 29.1). Molecular systematics, however, places the lophophorate phyla closer to the protostomes than the deuterostomes.

Phoronids are tube-dwelling marine worms ranging from 1 mm to 50 cm in length. Some live buried in the sand within tubes made of chitin, extending their lophophore from the opening of the tube and withdrawing it into the tube when threatened. There are only about 15 species of phoronid worms in two genera.

Bryozoans are tiny colonial animals that superficially resemble mosses. (*Bryozoa* means "moss animals.") In most species, the colony is encased in a hard exoskeleton with pores through which the lophophores of the animals extend (FIGURE 29.27a). Of the 5000 species of bryozoans, most live in the sea where they are among the most widespread and numerous sessile animals. Several species are important reef builders.

Brachiopods, or lamp shells, superficially resemble clams and other bivalve mollusks, but the two halves of the brachiopod shell are dorsal and ventral to the animal rather than lateral, as in clams (FIGURE 29.27b). A brachiopod lives attached to its substratum by a stalk, opening its shell slightly to allow water to flow between the shells and the lophophore. All brachiopods are marine. The living brachiopods are remnants of a much richer past; only about 330 extant species are known, but there are 30,000 species of Paleozoic and Mesozoic fossils. A tie to the past is *Lingula,* a living brachiopod genus that has changed little in 400 million years.

Phylum Annelida

Annelid worms have segmented bodies that give them a ringed appearance. (*Annelida* means "little rings.") There are about 15,000 annelid species, ranging in length from less than 1 mm to the 3-m length of a giant Australian earthworm. Annelids live in the sea, most freshwater habitats, and damp soil. We can describe the anatomy of annelids in terms of a well-known member of the phylum, the earthworm (FIGURE 29.28).

The coelom of the earthworm is partitioned by septa, but the digestive tract, longitudinal blood vessels, and nerve cords penetrate the septa and run the length of the animal (the major vessels have segmental branches). The digestive system has several specialized regions: the pharynx, the esophagus, the crop, the gizzard, and the intestine. The closed circulatory system consists of a network of vessels containing blood with oxygen-carrying hemoglobin. Dorsal and ventral vessels are connected by segmental pairs of vessels. The dorsal vessel and five pairs of vessels that circle the esophagus of an earthworm are muscular and pump blood through the circulatory system. Tiny blood vessels are abundant in the skin, which the earthworm uses as its respiratory organ to obtain oxygen.

In each segment of the worm is a pair of excretory tubes called **metanephridia** with ciliated funnels, called nephrostomes, that remove wastes from the blood and coelomic fluid. The metanephridia lead to exterior pores, through which the metabolic wastes are discharged.

A brainlike pair of cerebral ganglia lies above and in front of the pharynx. A ring of nerves around the pharynx

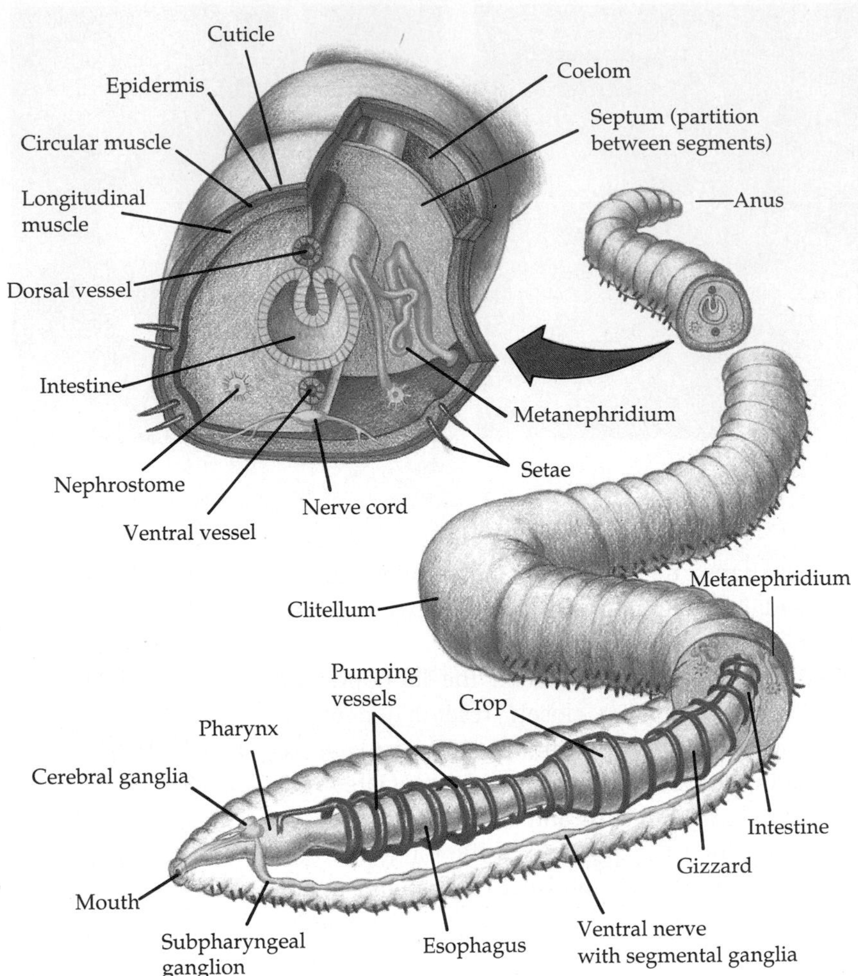

FIGURE 29.28
Anatomy of an earthworm. Annelids are segmented both externally and internally. Many of the internal structures are repeated, segment by segment. Externally, each segment has four pairs of setae, bristles that provide traction for burrowing. Earthworms and many other annelids creep along or burrow by coordinating two sets of muscles, one longitudinal and the other circular (see FIGURE 45.22). These muscles work against the noncompressible coelomic fluid, a hydrostatic skeleton. The muscles can alter the shape of each segment individually because the coelom is divided into separate compartments. When the circular muscles of a segment contract, that segment becomes thinner and elongates. Contraction of the longitudinal muscles causes the segment to shorten and thicken. The worm probes forward as alternating contractions of circular and longitudinal muscles progress along the segments like waves. The coelom and segmentation may have first functioned as adaptations for this type of movement.

connects to a subpharyngeal ganglion, from which a fused pair of nerve cords run posteriorly. All along these ventral nerve cords are segmental ganglia, also fused.

Earthworms are hermaphrodites, but they cross-fertilize. Two earthworms mate by aligning themselves in such a way that they exchange sperm, and then they separate. The received sperm cells are stored temporarily while a special organ, the clitellum, secretes a mucous cocoon. The cocoon slides along the worm, picking up the eggs and then the stored sperm. The cocoon then slips off the worm's head and resides in the soil while the embryos develop. Some earthworms can also reproduce asexually by fragmentation followed by regeneration.

Some aquatic annelids swim in pursuit of food, but most are bottom-dwellers that burrow in the sand and silt; earthworms, of course, are burrowers.

The phylum Annelida is divided into three classes: Oligochaeta (earthworms and their relatives), Polychaeta (polychaetes), and Hirudinea (leeches).

Class Oligochaeta. This class of segmented worms includes the earthworms and a variety of aquatic species. An earthworm eats its way through the soil, extracting nutrients as the soil passes through the digestive tube. Undigested material, mixed with mucus secreted into the digestive tract, is egested as castings through the anus. Farmers value earthworms because the animals till the earth, and the castings improve the texture of the soil. Darwin estimated that 1 acre of British farmland had about 50,000 earthworms that produced 18 tons of castings per year.

Class Polychaeta. Each segment of a polychaete ("many setae," for the bristles on each segment) has a pair of paddlelike or ridgelike structures called parapodia ("almost feet"). In many polychaetes, these richly vascularized parapodia function as gills that provide an extended area of skin for gas exchange (FIGURE 29.29a). Parapodia are also instrumental in locomotion. Each parapodium has several setae made of the polysaccharide chitin.

Most polychaetes are marine. A few adult forms drift and swim among the plankton, many crawl on or burrow in the seafloor, and many others live in tubes, which the worms make by mixing mucus with bits of sand and broken shells. The tube-dwellers include the brightly colored fanworms, which trap microscopic food particles in feathery tentacles that extend from the opening of the tube (FIGURE 29.29b).

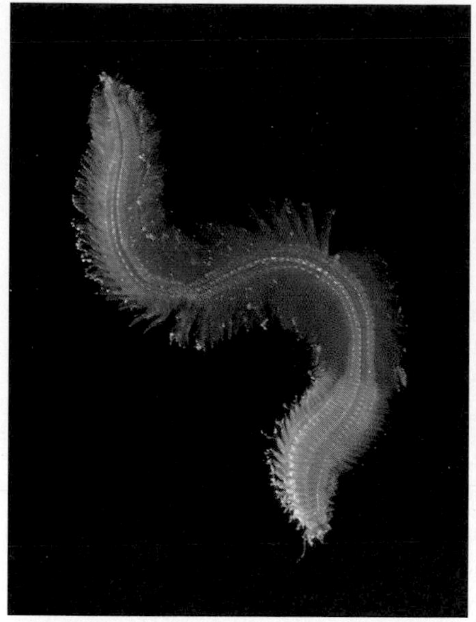

(a)

(b)

(c)

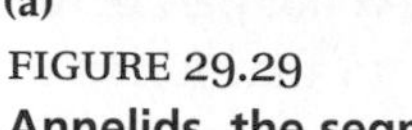

FIGURE 29.29

Annelids, the segmented worms. (**a**) Most annelids of the class Polychaeta are marine worms. Each segment has a pair of lateral flaps that function in movement and as gills for the exchange of respiratory gases with the surrounding water. (**b**) Fanworms (class Polychaeta) are tube-dwellers that use their feathery headdresses for gas exchange and to extract suspended food particles from the seawater. This species is known as a Christmas-tree worm. (**c**) Leeches (class Hirudinea) are free-living carnivores or parasites that suck blood from other animals. This species inhabits the Malaysian rain forest.

Class Hirudinea. The majority of leeches inhabit fresh water, but there are also land leeches that move through moist vegetation. Many leeches feed on other small invertebrates, but others are blood-sucking parasites that feed by attaching temporarily to other animals, including humans (FIGURE 29.29c). Leeches range in length from about 1 to 30 cm. Some parasitic species use bladelike jaws to slit the skin of the host, whereas others secrete enzymes that digest a hole through the skin. The host is usually oblivious to this attack because the leech secretes an anesthetic. After making the incision, the leech secretes another chemical, hirudin, which keeps the blood of the host from coagulating. The parasite then sucks as much blood as it can hold, often more than ten times its own weight. After this gorging, a leech can last for months without another meal. Until this century, leeches were frequently used by physicians for bloodletting. Leeches are still used for treating bruised tissues and for stimulating the circulation of blood to fingers or toes that have been sewn back to hands or feet after accidents.

* * *

Before leaving annelids, let's highlight two evolutionary innovations that are well developed in this phylum: the coelom and segmentation. The evolutionary significance of the coelom cannot be overemphasized. In addition to providing a hydrostatic skeleton and allowing new and diverse methods of locomotion, it has other advantages. It provides body space for storage and for complex organ development; it serves as a cushion that protects internal structures; and it allows a functional separation of the action of body wall muscles from those of internal organs, such as the muscles of the gut.

Segmentation set the foundation for specialization of body regions; groups of segments became modified for different functions. This regional specialization is seen to some degree in annelids, but it reached its zenith among the arthropods.

The protostome phylum Arthropoda is the most successful group of animals ever to live

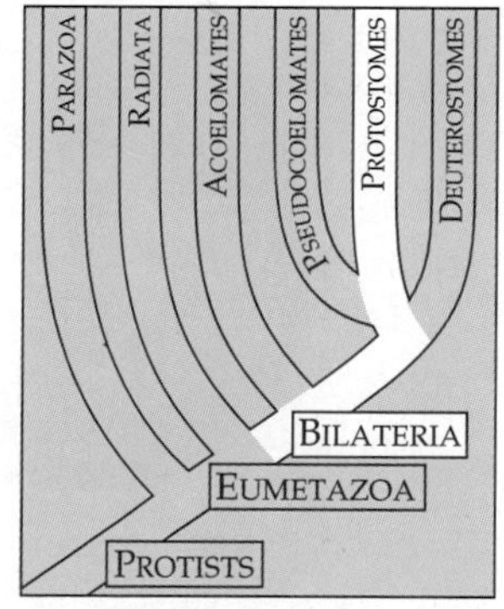

It is estimated that the arthropod population of the world, including crustaceans, spiders, and insects, numbers about 1 billion billion (10^{18}) individuals. Nearly 1 million arthropod species have been described, mostly insects. In fact, two out of every three organisms known are arthropods, and the phylum is represented in nearly all habitats of the biosphere. On the criteria of species diversity, distribution, and sheer numbers, Arthropoda must be regarded as the most successful phylum of animals ever to live.

General Characteristics of Arthropods

The diversity and success of arthropods are largely related to their segmentation, hard exoskeleton, and jointed appendages. (*Arthropoda* means "jointed feet.") Groups of segments and their appendages have become specialized for a great variety of functions. This evolutionary flexibility resulted not only in great diversification but also in an efficient body plan by the division of labor among regions. For example, the appendages are variously modified for walking, feeding, sensory reception, copulation, and defense. FIGURE 29.30 illustrates the diverse appendages and other arthropod characteristics of a lobster.

The body of an arthropod is completely covered by the **cuticle,** an **exoskeleton** (external skeleton) constructed from layers of protein and chitin. The cuticle can be modified into thick, hard armor over some parts of the body or be paper-thin and flexible in other locations, such as the joints. The exoskeleton protects the animal and provides points of attachment for the muscles that move the appendages. The skeleton of arthropods is both strong and relatively impermeable to water. As we will see, both of these qualities were largely responsible for the move onto land by various arthropod groups. The rigid exoskeleton also posed some evolutionary problems. For example, in order to grow, an arthropod must occasionally shed its old exoskeleton and secrete a larger one. This process, called **molting,** is energetically expensive and leaves the animal temporarily vulnerable to predators and other dangers.

Arthropods tune in to their environment with well-developed sensory organs, including eyes, olfactory receptors for smell, and antennae for touch and smell. Cephalization is extensive, with most sensory organs concentrated at the anterior end of the animal.

Arthropods have **open circulatory systems** in which fluid called hemolymph is propelled by a heart through short arteries and then into spaces called sinuses surrounding the tissues and organs. (The term *blood* is reserved for fluid in a closed circulatory system.) Hemolymph reenters the arthropod heart through pores that are usually equipped with valves. The body sinuses are collectively called the hemocoel, which is not part of the coelom. In most arthropods, the coelom that forms in the embryo becomes much reduced as development progresses, and the hemocoel becomes the main body cavity in adults. Although this condition resembles the open circulatory system of mollusks, the two probably arose independently.

A variety of organs specialized for gas exchange have evolved in arthropods. These organs must allow the diffusion of respiratory gases in spite of the exoskeleton. Most aquatic species have gills with thin feathery exten-

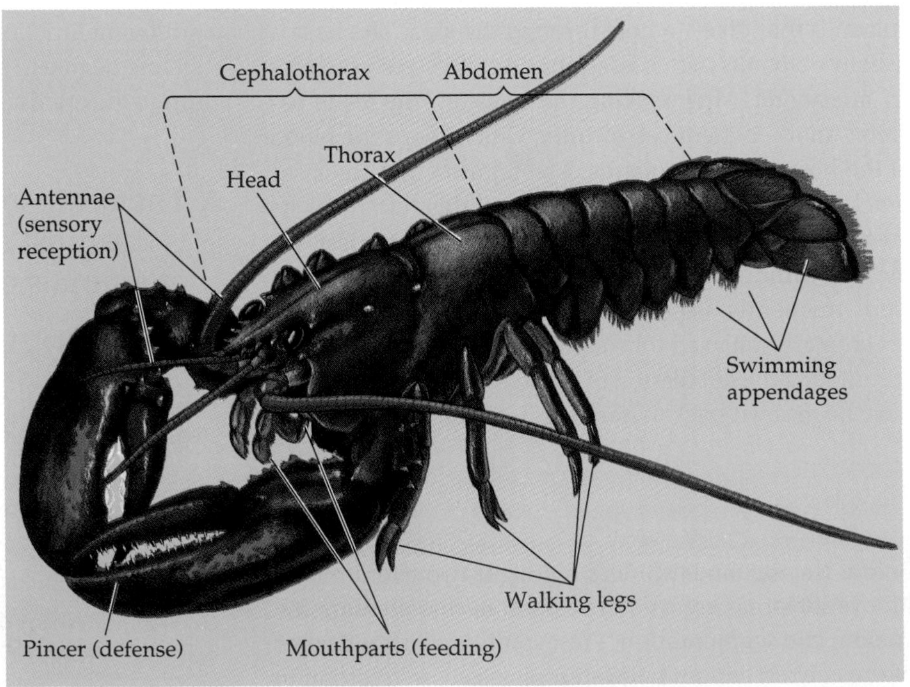

FIGURE 29.30
External anatomy of a lobster, an arthropod. In this dorsal view of the animal, many of the characteristic features of arthropods are apparent, including the jointed exoskeleton, sensory antennae and eyes, and multiple appendages modified for different functions. Arthropods are segmented, but this characteristic is obvious only in the abdominal portion of the lobster.

sions that place an extensive surface area in contact with the surrounding water. Terrestrial arthropods generally have internal surfaces specialized for gas exchange. Most insects, for instance, have tracheal systems, branched air ducts leading into the interior from pores in the cuticle.

Arthropod Phylogeny and Classification

Arthropods are segmented animals, and most biologists think that arthropods either evolved from annelids or shared a common ancestor. Some believe early arthropods resembled onychophorans, animals that appear like walking worms (FIGURE 29.31). Contemporary onychophorans do differ from arthropods in general by having unjointed appendages. However, Cambrian fossils of jointed-legged animals that resemble segmented worms add to the evidence of an evolutionary link between Annelida and Arthropoda.

In counterpoint, some systematists using comparisons of ribosomal RNA and other macromolecules argue that onychophorans are true arthropods, not "missing links," and that annelids and arthropods are not so closely related. Their interpretations support an alternative hypothesis that segmentation evolved independently in these two phyla, the last common ancestor of annelids and arthropods being an unsegmented protostome (perhaps an early mollusk).

Although debate continues about the origin of arthropods, most zoologists agree that in its subsequent evolution, Arthropoda diverged into four subphyla:

Subphylum	*Representatives*
Trilobitomorpha	Trilobites (extinct)
Cheliceriformes	Spiders, ticks, scorpions, sea spiders, eurypterids (extinct)
Uniramia	Insects, centipedes, millipedes
Crustacea	Crabs, lobsters, shrimps, barnacles, and other crustaceans

FIGURE 29.31
***Peripatus* (an onychophoran), the walking worm.** *Peripatus* is distinctly segmented and has excretory organs, musculature, and certain other features that are annelidlike. *Peripatus* resembles arthropods in its respiratory and circulatory systems, its cuticle made of chitin, and its jaws modified from appendages. Most animal systematists assign *Peripatus* and a few related genera to a separate phylum, Onychophora. However, recent analysis by molecular systematists argues in favor of placing these animals within the Arthropoda.

Trilobites

Among the early arthropods were the **trilobites** (FIGURE 29.32). They were common denizens of the shallow seas throughout the Paleozoic era but disappeared with the great Permian extinctions that closed that era, about 250 million years ago. Trilobites had pronounced segmentation, but their appendages showed little variation from segment to segment. As arthropods continued to evolve, the segments tended to fuse and become fewer in number and the appendages became specialized for a variety of functions. (Compare the trilobite in FIGURE 29.32 with the lobster in FIGURE 29.30.)

Spiders and Other Chelicerates

The trilobites were outlasted by the **eurypterids,** or sea scorpions. These marine predators, up to 3 m long, were chelicerates. The body of a chelicerate is divided into an anterior cephalothorax and a posterior abdomen. The appendages are more specialized than those of trilobites, and the most anterior appendages are modified as either pincers or fangs. **Chelicerates** (Gr. *cheilos,* "lips," and *cheir,* "arm") are named for these feeding appendages,

FIGURE 29.32
A fossil arthropod. Trilobites were prevalent arthropods throughout the Paleozoic era. About 4000 trilobite species have been described from fossils.

the **chelicerae.** Most of the marine chelicerates, including the eurypterids, are extinct; one survivor is the horseshoe crab (FIGURE 29.33).

The bulk of modern chelicerates are found on land in the form of the **class Arachnida,** which includes scorpions, spiders, ticks, and mites (FIGURE 29.34). Arachnids have a cephalothorax with six pairs of appendages: the chelicerae, a pair of appendages called pedipalps that usually function in sensing or feeding, and four pairs of walking legs (FIGURE 29.35). Spiders use their fanglike chelicerae, equipped with poison glands, to attack prey. As the chelicerae masticate (chew) the prey, the spider spills digestive juices onto the torn tissues. The food softens, and the spider sucks up the liquid meal.

In most spiders, gas exchange is carried out by **book lungs,** stacked plates contained in an internal chamber. The extensive surface area of these respiratory organs is

FIGURE 29.33
Horseshoe crabs. These "living fossils," which have changed little in hundreds of millions of years, have survived from a rich diversity of chelicerates that once filled the seas. Horseshoe crabs are common on the Atlantic and Gulf coasts of the United States.

FIGURE 29.34
Arachnids. (**a**) Scorpions, which hunt by night, were among the first terrestrial carnivores, preying on other arthropods that fed on the early land plants. The pedipalps of scorpions are pincers specialized for defense and the capture of food. The tip of the tail bears a poisonous sting. (**b**) This magnified house-dust mite is a ubiquitous scavenger in human dwellings (colorized SEM). Unlike some mites that carry disease-causing bacteria, dust mites are harmless except to people who are allergic to them. (**c**) Many arthropods parasitize other arthropods. In this light micrograph, you can see parasitic mites inhabiting the tracheae (air tubes) of a honeybee (an insect).

(a)

(b) 100 μm

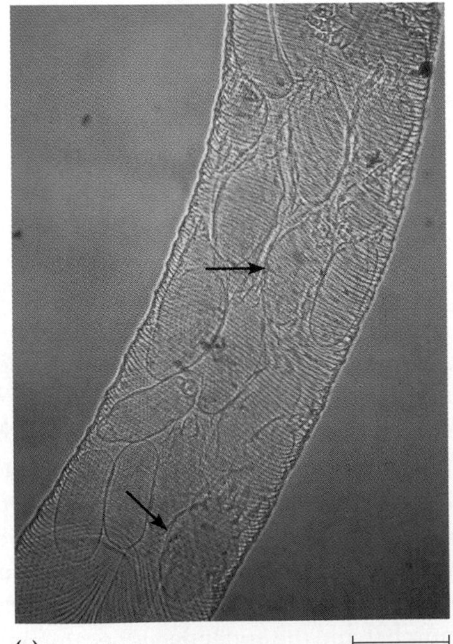

(c) 0.5 μm

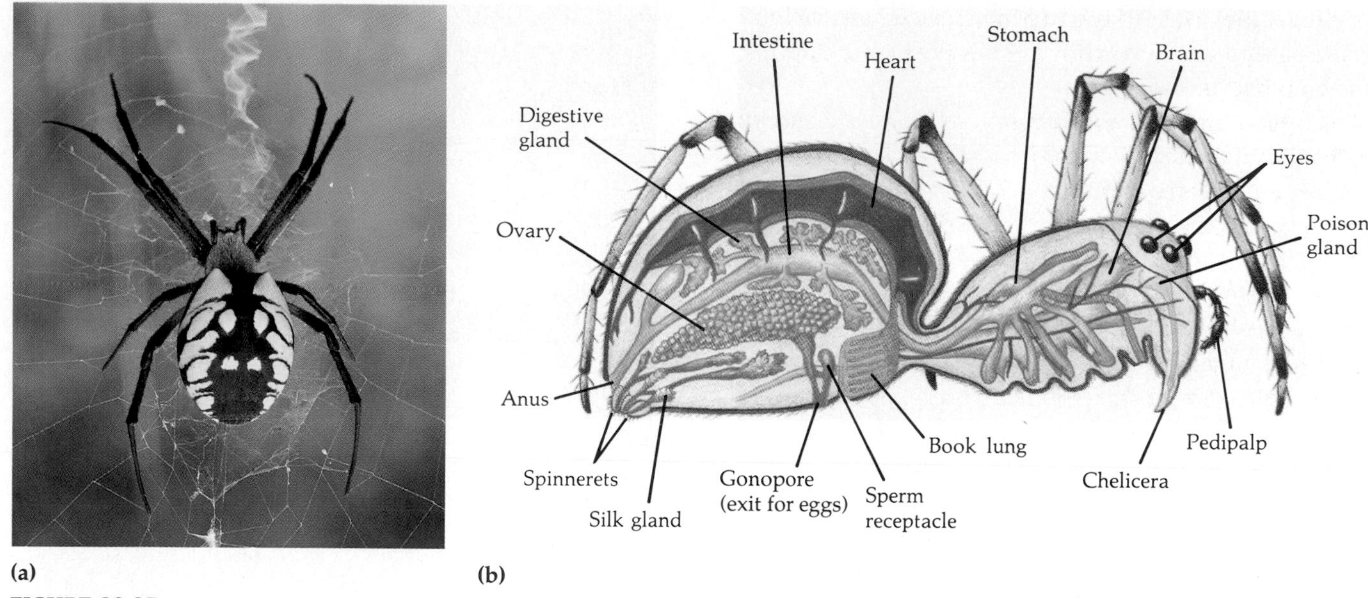

(a) (b)

FIGURE 29.35

Spiders (class Arachnida). (**a**) Spiders are generally most active during the daytime, when they hunt for prey or trap insects in webs. (**b**) Anatomy of a spider.

a stuctural adaptation that enhances the exchange of O_2 and CO_2 between the hemolymph and air.

A unique adaptation of many spiders is the ability to catch flying insects by stringing webs of silk, a protein produced as a liquid by special abdominal glands. The silk is spun by organs called spinnerets into fibers that solidify. Each spider engineers a style of web characteristic of its species and constructs the web perfectly on the first try. This complex behavior is apparently inherited. Besides building their webs from silk, various spiders use these fibers in other ways: as droplines for rapid escape, as cloth that covers eggs, and even as "gift wrapping" for food that certain male spiders offer females during courtship.

Comparing Chelicerates to Uniramians and Crustaceans

Aside from the chelicerates, another major line of arthropod evolution produced the uniramians and crustaceans. Rather than having clawlike chelicerae, these arthropods have jawlike **mandibles.** They are also distinguished from chelicerates in having one or two pairs of sensory **antennae** and usually a pair of **compound eyes** (multifaceted eyes with many separate focusing elements). Chelicerates lack antennae, and most have simple eyes (eyes with a single lens).

Uniramians have one pair of antennae and uniramous (unbranched) appendages; **crustaceans** have two pairs of antennae and typically biramous (branched) appendages. Crustaceans are primarily aquatic and are believed to have evolved in the ocean. Uniramians, on the other hand, are believed to have evolved on land.

The move onto land by chelicerates and uniramians was made possible in part by the exoskeleton. When the exoskeleton first evolved in the seas, its main functions were probably protection and anchorage for muscles, but it eventually helped certain arthropods live on land by solving the problems of water loss and support. The arthropod cuticle is relatively impermeable to water, helping prevent desiccation. The firm exoskeleton also solved the problem of support when arthropods left the buoyancy of water. Uniramians and chelicerates both spread onto land during the early Devonian period, following the colonization by plants. The oldest fossil evidence of terrestrial animals is burrows of millipedelike arthropods about 450 million years old. Fossilized arachnids almost as old have also been found.

Insects and Other Uniramians

Millipedes, centipedes, and insects are the three classes of uniramians.

Millipedes (**class Diplopoda**) are wormlike, with a large number of walking legs (two pairs per segment), though fewer than the thousand their name implies (FIGURE 29.36a). They eat decaying leaves and other plant matter. Millipedes were probably among the earliest animals on land, living on mosses and primitive vascular plants.

(a)

(b)

FIGURE 29.36
Diplopods (millipedes) and chilopods (centipedes). (**a**) Millipedes feed on decaying plant matter. (**b**) The house centipede (*Scutigera coleoptrata*), a fast-moving carnivore, feeds on insects, including cockroaches, and other small invertebrates.

Centipedes (**class Chilopoda**) are terrestrial carnivores. The head has a pair of antennae and three pairs of appendages modified as mouthparts, including the jawlike mandibles. Each segment of the trunk region has one pair of walking legs (FIGURE 29.36b). Centipedes use poison claws on the anteriormost trunk segment to paralyze prey and to defend themselves.

In species diversity, insects (**class Insecta**) outnumber all other forms of life combined. They live in almost every terrestrial habitat and in fresh water, and flying insects fill the air. Insects are rare, though not absent, in the seas, where crustaceans are the dominant arthropods. Class Insecta is divided into about 26 orders, some of which are described in TABLE 29.1 (pp. 616–617). **Entomology,** the study of insects, is a vast field with many subspecialties, including physiology, ecology, and taxonomy. Here we can only examine the general characteristics of this class of animals.

The oldest insect fossils date back to the Devonian period, which began about 400 million years ago. However, when flight evolved during the Carboniferous and Permian periods, it spurred an explosion in insect variety. A fossil record of diverse insect mouth parts indicates that specialized feeding on gymnosperms and other Carboniferous plants also contributed to the adaptive radiation of insects. A widely held hypothesis is that the greatest diversification of insects paralleled the evolutionary radiation of flowering plants during the Cretaceous period about 65 million years ago. This view is challenged by new research suggesting that the major diversification of insects preceded the angiosperm radiation. Thus, during the coevolution of flowering plants and the herbivorous insects that pollinated them, insect diversity may have been more a cause of angiosperm radiation than an effect.

Flight is obviously one key to the great success of insects. A flyer can escape predators, find food and mates, and disperse to new habitats much faster than animals that must crawl about on the ground. Many insects have one or two pairs of wings that emerge from the dorsal side of the thorax (FIGURE 29.37). Because the wings are extensions of the cuticle and not true appendages, insects can fly without sacrificing any walking legs. (By contrast, the flying vertebrates—birds and bats—have one of their two pairs of walking legs modified for wings and are generally quite clumsy on the ground.)

Insect wings may have first evolved as cuticular extensions that helped the insect body absorb heat, only later becoming organs for flight. Other views suggest that wings allowed the animals to glide from vegetation to the ground, or even served as gills in aquatic insects. Still another hypothesis is that insect wings functioned for swimming before they functioned for flight; insects called stone flies still use their small wings to skim across the surfaces of ponds and streams.

Dragonflies, with two coordinated pairs of wings, were among the first insects to fly (see FIGURE 29.37). Several insect orders that evolved later than dragonflies have modified flight equipment. Bees and wasps, for instance, hook their wings together and move them as a single pair. Butterflies get a similar result by overlapping their anterior and posterior wings. In beetles, the posterior wings are used for flight, while the anterior ones are modified as covers that protect the flight wings when the beetle is on the ground or burrowing.

The internal anatomy of an insect includes several complex organ systems (FIGURE 29.38). The digestive tract is a tube pinched into several regions, each with its own function in the breakdown of food and the absorption of nutrients. Like other arthropods, an insect has an open circulatory system, with a heart pumping hemolymph through the sinuses of the hemocoel. Metabolic wastes are removed from the hemolymph by unique excretory organs called **Malpighian tubules,** which are outpocketings of the gut. Gas exchange in insects is accomplished by a **tracheal system** of branched, chitin-lined tubes that infiltrate the body and carry oxygen directly to cells. The tracheal system opens to the outside of the body through spiracles, pores that can open or close to regulate air flow and limit water loss.

The insect nervous system consists of a pair of ventral nerve cords with several segmental ganglia. The two cords meet in the head, where the ganglia of several anterior segments are fused into a dorsal brain close to the antennae, eyes, and other sense organs concentrated on the head. Insects are capable of complex behavior, though this behavior seems to be largely innate. Even the intricate social behavior of some bees and ants is apparently inherited.

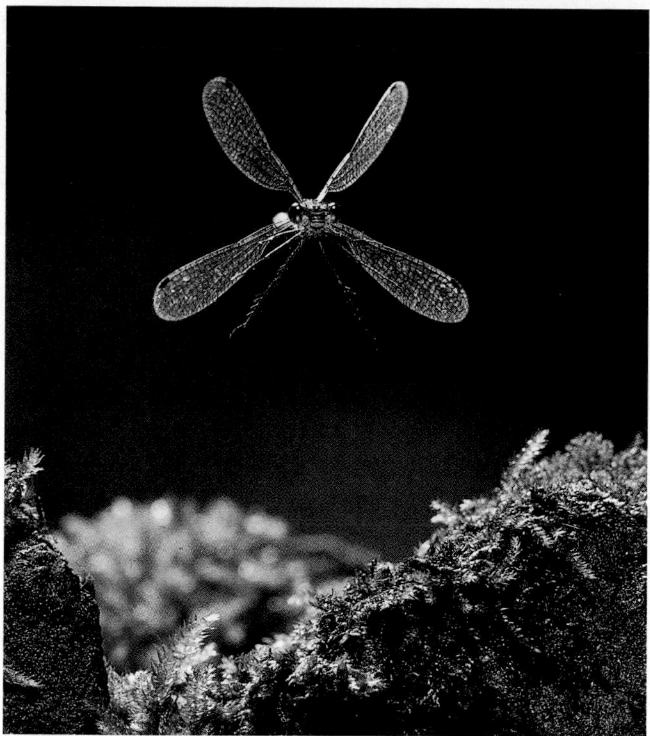

FIGURE 29.37
Insect flight. Insect wings, such as this dragonfly's, are not modified appendages but extensions of the cuticle. Some insects beat their wings at speeds of several hundred cycles per second by using muscles to warp the shape of the entire cuticle covering the thorax. As the wings flap, they change angles, producing lift on both the up and down strokes.

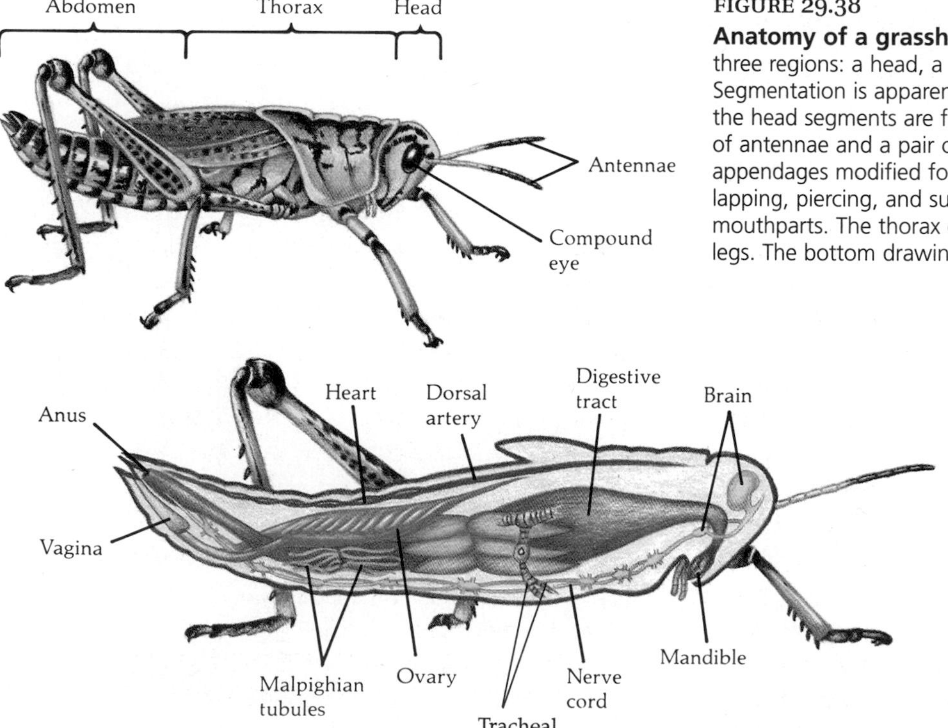

FIGURE 29.38
Anatomy of a grasshopper, an insect. The insect body has three regions: a head, a thorax, and an abdomen (top drawing). Segmentation is apparent along the thorax and abdomen, but the head segments are fused. On the insect head are one pair of antennae and a pair of compound eyes. Several pairs of appendages modified for chewing (as in grasshoppers) or for lapping, piercing, and sucking (in certain other insects) form the mouthparts. The thorax of the insect bears three pairs of walking legs. The bottom drawing identifies major internal organs.

TABLE 29.1

Some Major Orders of Insects

ORDER	APPROXIMATE NUMBER OF SPECIES	MAIN CHARACTERISTICS	EXAMPLES	
Anoplura	2,400	Wingless; sucking mouthparts; small with flattened body, reduced eyes; legs with clawlike tarsi for clinging to skin; incomplete metamorphosis; very host-specific.	Sucking lice	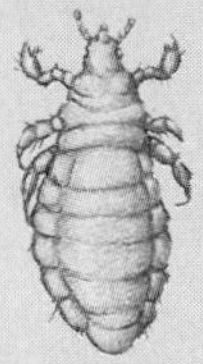 Human body louse
Coleoptera	500,000	Two pairs of horny, membranous wings; heavy, armored exoskeleton; biting and chewing mouthparts; complete metamorphosis.	Beetles, weevils	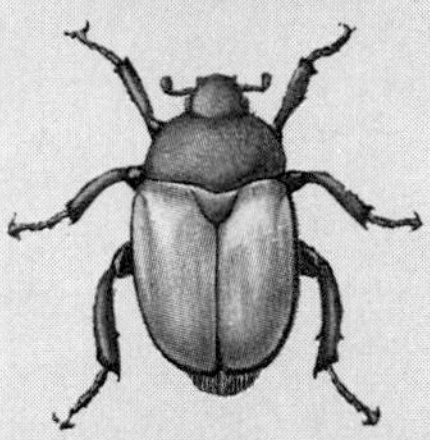 Japanese beetle
Dermaptera	1,000	Two pairs of leathery, membranous wings; biting mouthparts; large pincers in males; incomplete metamorphosis.	Earwigs	 Earwig
Diptera	80,000	One pair of wings and halteres (balancing organs); sucking, piercing, lapping mouthparts; complete metamorphosis.	Flies, mosquitoes	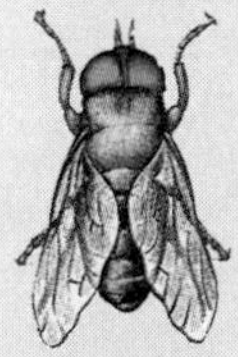 Horsefly
Hemiptera	55,000	Two pairs of horny, membranous wings; piercing, sucking mouthparts; incomplete metamorphosis.	True bugs; assassin bug, bedbug, chinch bug	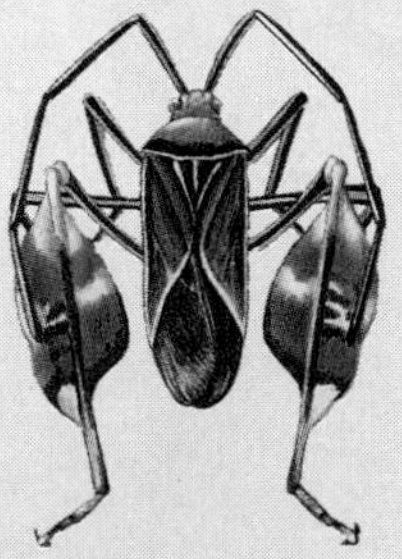 Leaf-footed bug
Hymenoptera	90,000	Two pairs of membranous wings; head mobile; well-developed eyes; chewing and sucking mouthparts; stinging; complete metamorphosis; many species social.	Ants, bees, wasps	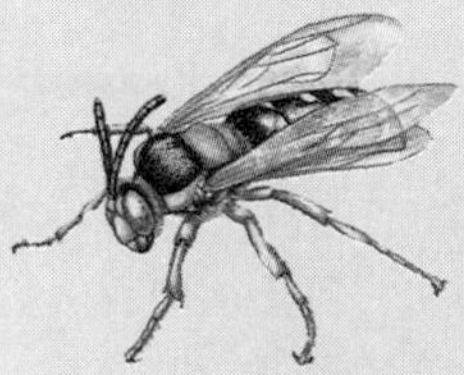 Cicada-killer wasp

TABLE 29.1

Some Major Orders of Insects *(Continued)*

ORDER	APPROXIMATE NUMBER OF SPECIES	MAIN CHARACTERISTICS	EXAMPLES	
Isoptera	2,000	Two pairs of wings, but some stages are wingless; chewing mouthparts; social; division of labor for reproduction, work, defense; incomplete metamorphosis.	Termites	Termite
Lepidoptera	140,000	Two pairs of wings; hairy bodies; long coiled tongue for sucking; complete metamorphosis.	Butterflies, moths	Swallowtail butterfly
Odonata	5,000	Two pairs of wings; biting mouthparts; incomplete metamorphosis.	Damselflies, dragonflies	Dragonfly
Orthoptera	30,000	Two pairs of horny, membranous wings; biting and chewing mouthparts in adults; incomplete metamorphosis.	Crickets roaches, grasshoppers, mantids	Katydid
Siphonaptera	1,200	Small, wingless, laterally compressed; piercing and sucking mouthparts; jumping legs; complete metamorphosis.	Fleas	Flea
Trichoptera	7,000	Two pairs of hairy wings; lapping mouthparts; complete metamorphosis; aquatic larvae build movable cases of sand and gravel bound together by secreted silk.	Caddisflies	Caddisfly

Many insects undergo metamorphosis in their development. In the **incomplete metamorphosis** of grasshoppers and some other orders, the young resemble adults but are smaller and have different body proportions. The animal goes through a series of molts, each time looking more like an adult, until it reaches full size. Insects with **complete metamorphosis** have larval stages, known by such names as maggot, grub, or caterpillar, that look entirely different from the adult stage (FIGURE 29.39). The main job of the larva is to eat and grow. The primary function of the adult is to find a mate and reproduce. Mates come together and recognize each other as members of the same species by advertising with bright colors (butterflies), sound (crickets), or odors (moths). After mating, a female lays her eggs on an appropriate food source where the larvae can begin eating as soon as they hatch.

Reproduction in insects is usually sexual, with separate male and female animals. Fertilization is generally internal. In most species, sperm cells are deposited directly into the female's vagina at the time of copulation, though in some species the male deposits a sperm packet outside the female, and the female picks it up. An internal structure in the female called the spermatheca stores the sperm, usually enough to fertilize more than one batch of eggs. Though most insect species produce a multitude of eggs, some flies produce live offspring, usually one at a time. Many insects mate only once in a lifetime.

Animals so numerous, diverse, and widespread as insects are bound to affect the lives of all other terrestrial organisms, including humans. On the one hand, we depend on insects to pollinate many of our crops and orchards. On the other hand, insects are carriers for many diseases, including malaria and African sleeping sickness. Furthermore, insects compete with humans for food. In parts of Africa, for instance, insects claim about 75% of the crops. Trying to minimize their losses, farmers

FIGURE 29.39
Metamorphosis of a butterfly. The larva (caterpillar) spends its time eating and growing, molting as it grows. After several molts, the larva encases itself in a cocoon and becomes a pupa. Within the pupa, the larval tissues are broken down, and the adult is built by the division and differentiation of cells that were quiescent in the larva. Finally, the adult emerges from the cocoon. Fluid is pumped into veins of the wings and then withdrawn, leaving the hardened veins as struts supporting the wings. Now the insect flies off and reproduces, deriving much of its nourishment from the calories stored by the feeding larva.

in the United States spend billions of dollars each year on pesticides, spraying crops with massive doses of some of the deadliest poisons ever invented. Try as they may, not even humans have challenged the preeminence of insects and their arthropod kin. Or, as Cornell University's Thomas Eisner puts it: "Bugs are not going to inherit the Earth. They own it now. So we might as well make peace with the landlord."

Crustaceans

While arachnids and insects thrived on land, crustaceans, for the most part, remained in the seas and ponds, where they are now represented by about 40,000 species (FIGURE 29.40).

The multiple appendages of crustaceans are extensively specialized. Lobsters and crayfish, for instance, have a toolkit of 19 pairs of appendages (see FIGURE 29.30). Crustaceans are the only arthropods with two pairs of antennae. Three or more pairs of appendages are modified as mouthparts, including the hard mandibles. Walking legs are present on the thorax, and, unlike insects, crustaceans have appendages on the abdomen. A lost appendage can be regenerated.

Small crustaceans exchange gases across thin areas of the cuticle, but larger forms have gills. The circulatory system is open, with a heart pumping hemolymph through arteries into sinuses that bathe the organs. Crustaceans excrete nitrogenous wastes by diffusion through thin areas of the cuticle, but a pair of glands regulates the salt balance of the hemolymph.

Sexes are separate in most crustaceans. In the case of the lobster, the male uses a specialized pair of appendages to transfer sperm to the reproductive pore of the female during copulation. Most aquatic crustaceans go through one or more swimming larval stages.

Lobsters, crayfish, crabs, and shrimp are all relatively large crustaceans called decapods. The exoskeleton, or cuticle, is hardened by calcium carbonate; the portion that covers the dorsal side of the cephalothorax forms a shield called the carapace. Most decapods are marine. Crayfish, however, live in fresh water, and some tropical crabs live on land.

The isopods are mostly small marine crustaceans, but this group also includes sow bugs and pill bugs, familiar land animals. These terrestrial crustaceans live mostly in moist soil and other damp places.

Another group of small crustaceans, the copepods, are among the most numerous of all animals. They are important members of the plankton communities that are the foundation consumers of marine and freshwater food chains.

In addition to copepods, plankton includes larvae of many larger crustaceans. Another group of shrimplike crustaceans called krill are planktonic organisms that provide a major food source for many species of whales (FIGURE 29.40b).

Barnacles are sessile crustaceans with parts of their cuticles hardened into shells by calcium carbonate. They feed by using their appendages to strain food from the water (FIGURE 29.40c).

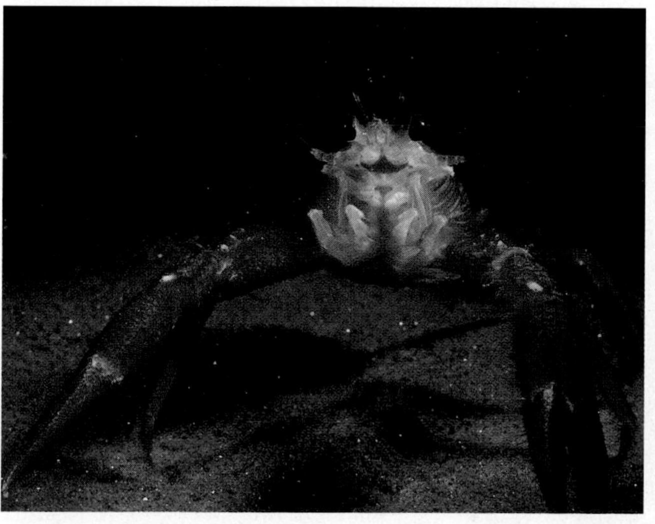

(a)

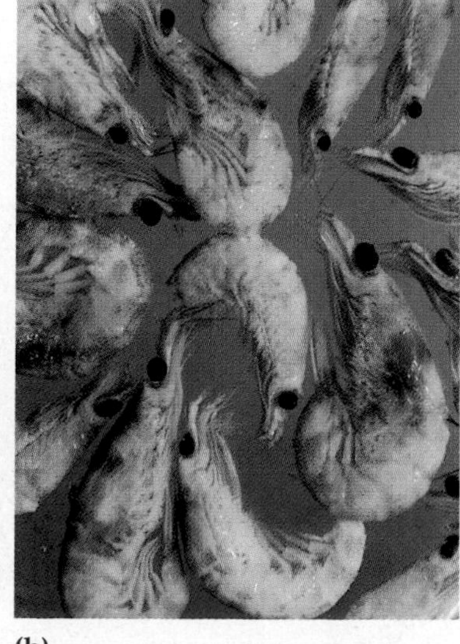

(b)

(c)

FIGURE 29.40

Crustaceans. (**a**) Crabs, lobsters, crayfish, and shrimp are the most familiar crustaceans. This is a red crab. (**b**) Planktonic crustaceans known as krill are consumed by the ton by whales and other large suspension-feeders. (**c**) Barnacles are sessile crustaceans with a shell (exoskeleton) hardened by calcium carbonate. The jointed appendages projecting from the shell capture small plankton and organic particles suspended in the water.

The deuterostome lineage includes echinoderms and chordates

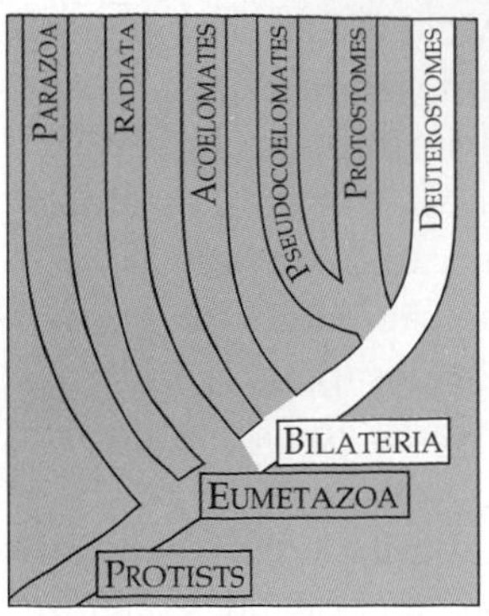

At first glance, sea stars and other echinoderms may seem to have little in common with the phylum Chordata, which includes the vertebrates: fishes, amphibians, reptiles, birds, and mammals. These animals, however, share features characteristic of deuterostomes: radial cleavage, development of the coelom from the archenteron, and formation of the mouth at the end of the embryo opposite the blastopore (see FIGURE 29.5).

Phylum Echinodermata

Sea stars and most other **echinoderms** (Gr. *echin,* "spiny," and *derma,* "skin") are sessile or slow-moving animals with radial symmetry. The internal and external parts of the animal radiate from the center, often as five spokes. A thin skin covers an endoskeleton of hard calcareous plates. Most echinoderms are prickly, from skeletal bumps and spines having various functions. Unique to echinoderms is the **water vascular system,** a network of hydraulic canals branching into extensions called **tube feet** that function in locomotion, feeding, and gas exchange.

Sexual reproduction of echinoderms usually involves separate male and female individuals that release their gametes into the seawater. The radial adults develop by metamorphosis from bilateral larvae. The early embryology of echinoderms clearly aligns them with the deuterostomes.

The 7000 or so echinoderms, all marine, are divided into six classes: Asteroidea (sea stars), Ophiuroidea (brittle stars), Echinoidea (sea urchins and sand dollars), Crinoidea (sea lilies), Holothuroidea (sea cucumbers), and Concentricycloidea (sea daisies). The sea daisies, discovered only recently, are little creatures that live on waterlogged wood in the deep sea.

Class Asteroidea. Sea stars have five arms (sometimes more) radiating from a central disk (FIGURE 29.41). The undersurfaces of the arms bear tube feet, each of which can act like a suction disk. By a complex set of hydraulic and muscle actions, the suction can be created or released. The sea star coordinates its tube feet to adhere firmly to rocks or to creep along slowly as the tube feet extend, grip, contract, release, extend, and grip again (FIGURE 29.42a). Sea stars also use their tube feet to grasp prey, such as clams and oysters (see FIGURE 29.42b). The arms of the sea star embrace the closed bivalve, hanging on tightly by the tube feet. The sea star then turns its stomach inside-out, everting it through its mouth and between the shells of the bivalve. The digestive tract of the sea star secretes juices that begin digesting the soft body of the mollusk within its shell.

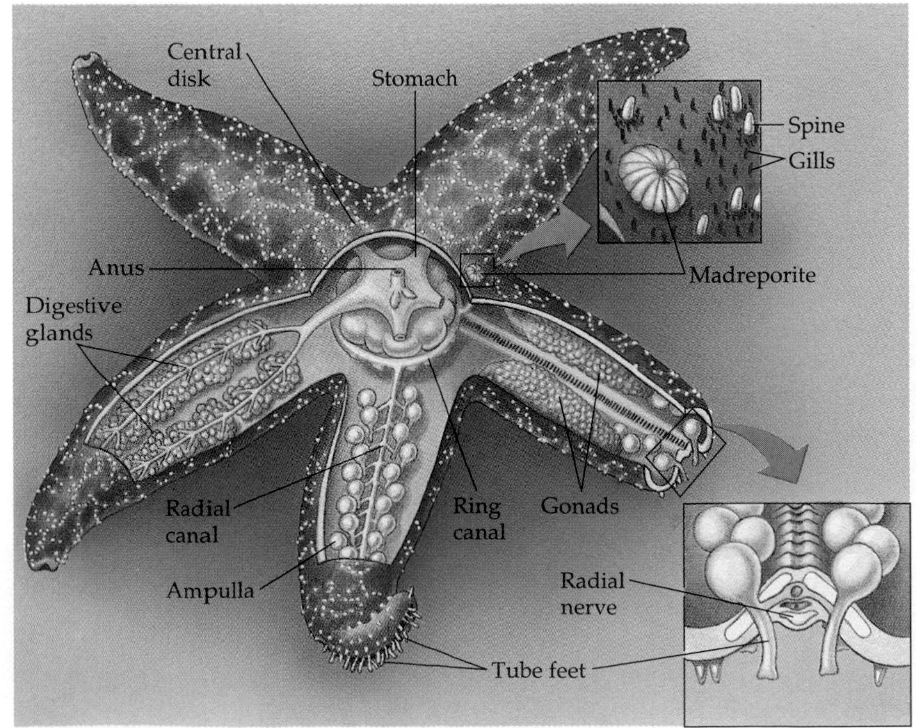

FIGURE 29.41
Anatomy of a sea star. The surface of a sea star is covered by spines that help defend against animals and by small gills for gas exchange. Internal organs are suspended by mesenteries in a well-developed coelom. A short digestive tract runs from the mouth on the bottom of the central disk to the anus on the top of the disk. Digestive glands secrete digestive juices and aid in the absorption and storage of nutrients. The central disk has a nerve ring and nerve cords radiating from the ring into the arms. The water vascular system consists of a ring canal in the central disk and five radial canals, each running the length of an arm in a groove. The system connects to the outside by way of the madreporite. Branching from each radial canal are hundreds of tube feet filled with fluid continuous with the rest of the water vascular system. Attached to each tube foot, and instrumental in its functioning, is a water bulb called the ampulla.

(a) (b) (c) (d) (e) (f)

FIGURE 29.42

Echinoderms. Exclusively marine, the mostly sedentary or slow-moving echinoderms are characterized by radial symmetry, bony endoskeletons, and hydraulic tube feet that function in movement, feeding, and, in some species, gas exchange. (**a**) A sea star (class Asteroidea) on coral. (**b**) Sea star eating a clam. (**c**) A brittle star (class Ophiuroidea). (**d**) A sea urchin (class Echinoidea). (**e**) A sea lily (class Crinoidea). (**f**) A sea cucumber (class Holothuroidea).

Sea stars and some other echinoderms are capable of regeneration. Sea stars can regrow lost arms, but the process is very slow. However, they cannot generally regrow an entire body from a single arm.

Class Ophiuroidea. Brittle stars have distinct central disks, and the arms are long and flexible (FIGURE 29.42c). Their tube feet lack suckers, and they move by serpentine lashing of the arms. Feeding mechanisms vary among the different species.

Class Echinoidea. Sea urchins and sand dollars have no arms, but they do have five rows of tube feet that function in slow movement (FIGURE 29.42d). Sea urchins also use muscles for pivoting their long spines, which aids in moving. The mouth of an urchin is ringed by complex jawlike structures adapted for eating seaweeds and other food. Sea urchins are roughly spherical in shape, whereas sand dollars are flattened and disk-shaped.

Class Crinoidea. Some sea lilies live attached to the substratum by stalks; others crawl about by using their long, flexible arms, which they also use in suspension-feeding (FIGURE 29.42e). The arms circle the mouth, which is directed upward, away from the substratum. Crinoidea is an ancient class that has been very conservative in its evolution; fossilized sea lilies some 500 million years old could pass for contemporary members of the class.

Class Holothuroidea. On casual inspection, sea cucumbers do not look much like other echinoderms (FIGURE 29.42f). They lack spines, and the hard endoskeleton is much reduced. Sea cucumbers are elongated in the oral-aboral axis, giving them the shape for which they are named and further disguising their relationship to sea stars and sea urchins. Closer examination, however, reveals five rows of tube feet, part of the water vascular system found only in echinoderms. Some of the tube feet around the mouth are developed as feeding tentacles.

Phylum Chordata

This phylum, our own, consists of two subphyla of invertebrate animals plus the subphylum Vertebrata, the animals with backbones. Grouping the chordates with echinoderms as deuterostomes on the basis of similarities in early embryonic development is not meant to imply that one phylum evolved from the other. Chordates and echinoderms have existed as distinct phyla for at least half a billion years; if the developmental similarities stem from shared ancestry, then the evolutionary paths of the two phyla must have diverged very early. We will trace the phylogeny of chordates in Chapter 30, focusing on the history of vertebrates. This chapter concludes by considering some more general questions about early diversification of the animal kingdom and summarizing the evolutionary history of the groups we have discussed.

The Cambrian explosion produced all the major animal body plans

The animal kingdom probably originated from a colonial protist related to choanoflagellates. FIGURE 29.43 diagrams one hypothesis for how such an ancestor may have evolved into simple animals with specialized cells arranged in two or more layers. Whatever way animal life began, we know that the diversification that produced the many phyla occurred relatively rapidly on the vast scale of geological time—an evolutionary episode known as the Cambrian explosion. In a span of only 5 to 10 million

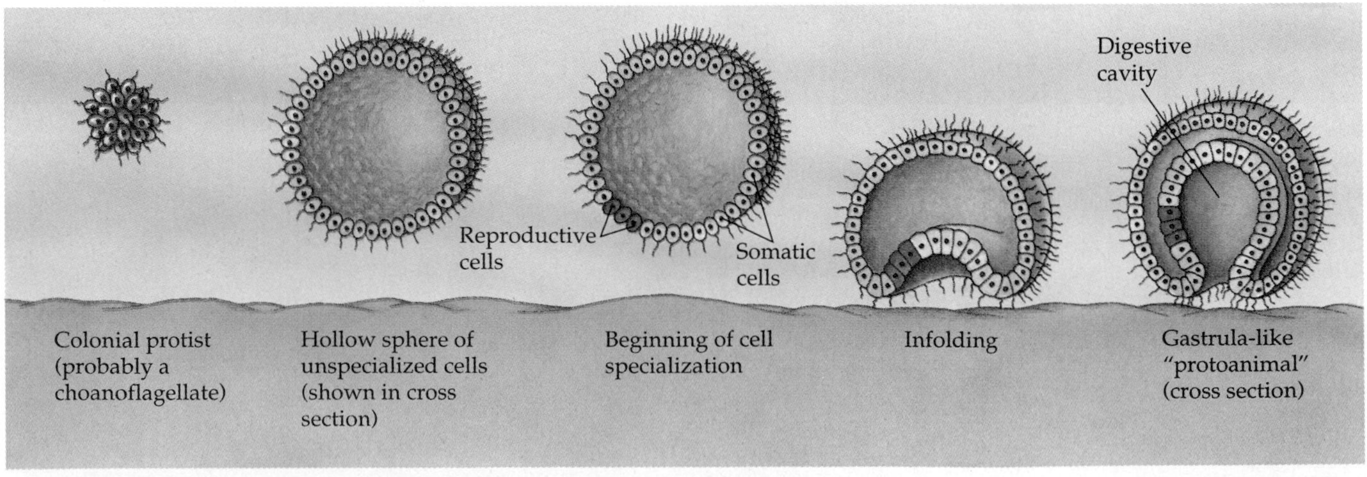

FIGURE 29.43
One hypothesis for the origin of animals from a flagellated protist.

years at the beginning of the Cambrian period 545 million years ago, all the major animal body plans we see today evolved. Of course, new species and even whole new taxonomic classes of organisms appeared later. Class Mammalia, for example, did not evolve until about 220 million years ago. However, mammals represent a variation on the basic chordate plan that originated during the Cambrian explosion.

The diverse animals of the early Cambrian were preceded by a much less diverse fauna dating back about 700 million years. Paleontologists have named this last period of the Precambrian era the **Ediacaran period,** for the Ediacara Hills of Australia where fossils of Precambrian animals were first discovered. Paleontologists have since found similar animals of the same vintage on other continents. Most of the Ediacaran fossils appear to represent cnidarians, but fossilized burrows probably left by worms suggest that bilateral animals also lived before the dawn of the Cambrian. Was the Ediacaran fauna the spark of biological diversity that ignited the Cambrian explosion? Or was the Ediacaran fauna a separate, earlier evolutionary "experiment," its lineages ending in extinction before a second radiation of animals from ancestral protists produced the Cambrian explosion? Paleontologists are still debating the phylogenetic connections between the animals on opposite sides of the Cambrian boundary. But that debate does not detract from the evolutionary drama of the Cambrian explosion, a burst of morphological origins unparalleled in the history of life.

The Burgess Shale in British Columbia, Canada, is the most famous fossil bed documenting the diversity of Cambrian animals. Two other fossil sites, one in Greenland and the other in the Yunnan region of China, predate the Burgess Shale by more than 10 million years. These sites push our view of animal diversification back to within 5 million years of the beginning of the Cambrian. Even so close to the start of the Cambrian explosion, a remarkable diversity of phyla had evolved.

There are at least two contrasting interpretations of the Burgess shale fossils, which are mostly bizarre-looking in the context of the marine invertebrates we know today (FIGURE 29.44). Some researchers argue that the Cambrian explosion produced dozens of extinct phyla in addition to all the modern ones, and that some of the forms of the Burgess shale represent these extinct "experiments" in animal diversity. In this view, mass extinction at the end of the Cambrian culled this stock of animal phyla to the 35 or so that exist today. Other researchers believe that most of the Cambrian fossils, as strange as they may appear to us, are simply ancient variations within the taxonomic boundaries of phyla still represented in the modern fauna. Indeed, the number of exclusively Cambrian phyla seems to be dropping as the fossils are studied more closely and are classified in extant phyla.

What caused the Cambrian explosion? And why haven't any major new body plans (phyla) evolved since? Hypotheses abound, and they are not necessarily mutually exclusive; that is, a combination of factors may have contributed to the explosion and the lack of subsequent major diversification. One hypothesis is that the Cambrian explosion was simply an adaptive radiation made

FIGURE 29.44
A sample of some of the animals that evolved during the Cambrian explosion. This drawing is based on fossils collected from the Burgess Shale in British Columbia, Canada. Taxonomic debate about the Cambrian fauna centers on whether it includes a large number of extinct phyla or consists mainly of variations on the anatomical themes of modern phyla.

possible by the origin of the first animals, which diversified as they adapted to various ecological niches that were previously unoccupied. A related idea emphasizes the emergence of predator-prey relationships, which triggered diverse evolutionary adaptations, including various kinds of shells and different modes of locomotion. Some researchers argue that a major environmental change set the stage for the Cambrian explosion. For example, perhaps atmospheric oxygen finally reached a high enough concentration during the Cambrian to support the more active metabolism required for feeding and other activities by mobile animals.

Other scientists favor internal causes—changes in the organisms themselves—as the catalysts of the Cambrian explosion. For example, the origin of mesoderm, a third tissue layer that makes the development of more complex anatomy possible, may have been one of the evolutionary breakthroughs behind the origin of diverse body plans. But researchers are now focusing their attention on certain genes that control pattern formation during animal development, determining such features as segmentation and the placement of appendages and other structures (see Chapters 18 and 43). At least some of these genes are common to diverse animal phyla, but variation in their expression during embryonic development results in some of the morphological differences that distinguish the phyla. Perhaps this developmental plasticity was partly responsible for the relatively rapid origin of diverse types of animals during the Cambrian. Some researchers who favor this hypothesis also speculate that the phyla became locked into developmental patterns that constrained morphological evolution enough that no additional phyla could evolve after the Cambrian explosion. Animal evolution, of course, did not come to a halt; variations in pattern formation continue to allow more subtle variation, leading to speciation and the origin of taxa below the phylum level.

Continuing research will help test these multiple hypotheses. But as the Cambrian explosion becomes less mysterious, it will seem no less wonderful. In the last half-billion years, animal evolution has mainly generated new variations on old "designs." In this chapter, we have traced the evolution of these major body plans as a way to make phylogenetic sense of some of the living animal phyla. The next chapter takes a closer look at the evolutionary history of one of those phyla, Chordata, the animal group that includes humans and all other vertebrates.

REVIEW OF KEY CONCEPTS (with page numbers and key figures)

- What is an animal? (pp. 589–590)
  - Animals are multicellular eukaryotes distinguished by a type of heterotrophy called ingestion.
  - Animal life began with the Precambrian evolution of multicellular marine forms that ate other organisms.
  - Subsequent evolution generated about 35 phyla of animals. About 95% of animal species are invertebrates (lacking backbones).
  - Animal cells lack walls and store carbohydrate reserves as glycogen. In most animals, cells are successively organized into tissues, organs, and organ systems.
  - Animal reproduction is primarily sexual, and the life cycle is dominated by the diploid stage, gametes generally being the only haploid cells. Asexual budding or reproduction by regeneration occurs in some species.
  - In sexual reproduction, fertilization of an egg by a flagellated sperm initiates cleavage of the zygote and the formation of a hollow ball of cells called the blastula.
  - Many animals go through metamorphosis, a second stage of development that transforms a sexually immature larva into a morphologically distinct sexual adult.
  - Muscles and nerves, which control active behavior, are unique to animals.
- Comparative anatomy and embryology provide clues to animal phylogeny: *an overview of animal diversity* (pp. 590–594, FIGURE 29.1)
  - Systematists recognize four major branchings in the tree of animals. The first distinguishes the Parazoa (sponges), which lack true tissues, from all other animals (the Eumetazoa), which have tissues.
  - In the second dichotomy, Eumetazoa diverged early into two major branches, the Radiata and the Bilateria. Members of the branch Radiata are jellyfishes and their relatives, sedentary and planktonic forms with radial symmetry. Members of the branch Bilateria are characterized by bilateral symmetry and cephalization. Another important difference is that radial animals are diploblastic (two-layered), while bilateral animals are triplobastic (three-layered: ectoderm, mesoderm, and endoderm).
  - The third major branch point splits Bilateria into animals lacking body cavities and blood vascular systems (acoelomates) and animals having those anatomical features. In pseudocoelomates, the cavity is incompletely lined by embryonic mesoderm. Coelomates have a true coelom, a body cavity completely lined by mesoderm.
  - In the fourth major branching, based on features of embryonic development, coelomate phyla are divided into two main groups: protostomes, including the annelids, mollusks, and arthropods; and deuterostomes, including echinoderms and chordates.

- Sponges are sessile animals lacking true tissues (pp. 594–596, FIGURE 29.7)
  - The least complex animals are the sponges, which lack tissues and organs, such as muscles and nerves.
  - Sponges filter-feed by drawing water through pores.
- Cnidarians and ctenophores are radiate, diploblastic animals with gastrovascular cavities (pp. 596–599, FIGURE 29.8)
  - Phylum Cnidaria consists of primarily marine carnivores possessing tentacles armed with stinging cnidocytes that aid in defense and the capture of prey. The simple, radial body exists as a sessile polyp or a floating medusa (or both stages, in some cases). Two cell layers are organized around a central gastrovascular cavity with a single opening for both mouth and anus.
  - Phylum Cnidaria is divided into three classes. Class Hydrozoa usually alternates polyp and medusa forms, although the polyp is more conspicuous. Jellyfishes belong to the class Scyphozoa, in which the medusa is the prevalent form of the life cycle. Class Anthozoa contains the sea anemones and corals, which occur only as polyps.
  - Phylum Ctenophora, the comb jellies, are transparent animals whose locomotion depends on eight rows of cilia.
- Flatworms and other acoelomates are bilateral, triploblastic animals lacking body cavities (pp. 599–602, FIGURE 29.14)
  - Phylum Platyhelminthes, the flatworms, are the simplest members of Bilateria. Ribbonlike animals with a single opening to their gastrovascular cavities, the phylum is divided into four classes. Class Turbellaria is made up of mostly free-living, primarily marine species. Classes Trematoda and Monogenea, the flukes, live as parasites in animals. Class Cestoda consists of parasitic tapeworms.
  - The proboscis worms of the phylum Nemertea are named for the retractable tube they use for defense and prey capture. This group has a simple circulatory system and a complete digestive tract with both mouth and anus.
- Rotifers, nematodes, and other pseudocoelomates have complete digestive tracts and blood vascular systems (pp. 602–603)
  - Phylum Rotifera is made up of tiny animals possessing a crown of cilia that draws food into the mouth.
  - The roundworms of the phylum Nematoda are among the most numerous animals in both species and individuals. The phylum includes both free-living species and parasites.
- Mollusks and annelids are among the major variations on the protostome body plan (pp. 603–609, FIGURES 29.20, 29.28)
  - Phylum Mollusca includes a diverse spectrum of soft-bodied species possessing various modifications of a muscular foot, a visceral mass, and an overlying mantle. In many species the mantle secretes a shell of calcium carbonate.
  - Four of the eight classes of the phylum Mollusca are as follows. Class Polyplacophora is made up of the chitons, oval-shaped marine animals encased in an armor of dorsal plates. The largest molluscan class is Gastropoda, the snails and their relatives. Most gastropods are protected by single, spiraled shells, although unshelled slugs also occur. Embryonic torsion of the body is a distinctive characteristic. The clams and their relatives of the class Bivalvia have hinged shells divided into two halves. Bivalves are sedentary, headless suspension-feeders that use gills for both gas exchange and feeding. Class Cephalopoda includes squids and octopuses, carnivores with beaklike jaws surrounded by tentacles of the modified foot. The shell is either reduced or absent in most genera, consistent with their active lifestyles.
  - The lophophorate animals, the phyla Phoronida, Bryozoa, and Brachiopoda, are grouped together on the basis of their lophophores, horseshoe-shaped, suspension-feeding organs bearing ciliated tentacles. Their protostome-deuterostome affinity is uncertain, although molecular systematics places the phyla closer to protostomes.
  - Phylum Annelida is a group characterized by body segmentation. Their progressive, wavelike locomotion results from alternating contractions of circular and longitudinal muscles against a fluid-filled, compartmentalized coelom. The phylum is divided into three classes. Class Oligochaeta includes earthworms and various aquatic species. Representatives of the class Polychaeta possess vascularized, paddlelike parapodia that function as gills and aid in locomotion. Class Hirudinea consists of the leeches.
- The protostome phylum Arthropoda is the most successful group of animals ever to live (pp. 609–619, FIGURE 29.30)
  - Phylum Arthropoda has more known species than all other phyla combined. The arthropods have bodies with jointed appendages and an exoskeleton that must be shed during molting to allow growth.
  - Subphylum Chelicerata consists of arthropods with pincer or fanglike feeding appendages. A separate line of evolution produced arthropods with jawlike mandibles. Subphylum Uniramia contains the insects, centipedes, and millipedes, which have one pair of antennae and unbranched (uniramous) appendages. Subphylum Crustacea contains primarily aquatic organisms with two pairs of antennae and branched appendages. The five principal classes of Arthropoda follow.
  - Class Arachnida includes spiders, ticks, scorpions, and mites, extant chelicerates. Spiders have two pairs of feeding appendages and four pairs of walking legs.
  - The vegetarian millipedes (class Diplopoda) have two walking legs per segment, whereas the carnivorous centipedes (class Chilopoda) have one pair per segment and are armed with poison claws.
  - In terms of species diversity, members of the class Insecta outnumber all other forms of life combined. Part of their success is due to the evolution of flight. The insect head, which bears one pair of antennae, a pair of compound eyes, and variously modified mouthparts, is connected to a segmented thorax and abdomen. The thorax has three pairs of walking legs and often one or two pairs of wings. In many species, young go through complete or incomplete metamorphosis.
  - Lobsters and crayfish are members of the class Crustacea. Among a number of multiple appendages are two pairs of antennae and appendages modified for pinching, chewing, locomotion, and copulation.

- The deuterostome lineage includes echinoderms and chordates (pp. 620–622, FIGURE 29.41)
  - Sea stars and their relatives make up six classes of the marine phylum Echinodermata, radially symmetrical animals with a unique water vascular system ending in tube feet used for locomotion and feeding. A thin, bumpy, or spiny skin covers a calcareous endoskeleton.
  - Phylum Chordata, discussed in detail in Chapter 30, is made up of two subphyla of invertebrates plus a vertebrate subphylum, to which humans belong.
- The Cambrian explosion produced all the major animal body plans (pp. 622–624, FIGURE 29.44)
  - Animals probably evolved from colonial flagellates (choanoflagellates).
  - The oldest known fauna consisted of soft-bodied animals that lived during the Ediacaran period. During the ensuing Cambrian period, a much more diverse fauna evolved, which included many species with hard shells and skeletons, and all the modern phyla. All the basic designs observed in modern animals arose during the Cambrian radiation.

SELF-QUIZ

1. The distinction between the branches Parazoa and Eumetazoa is based mainly on the absence versus the presence of
   a. body cavities
   b. a complete digestive tract
   c. true tissues
   d. a blood vascular system
   e. mesoderm

2. As a group, acoelomates are characterized by
   a. gastrovascular cavities
   b. the absence of mesoderm
   c. deuterostome development
   d. a coelom that is not completely lined with mesoderm
   e. a solid body without a cavity surrounding internal organs

3. Which of the following is *not* descriptive of deuterostomes?
   a. radial cleavage
   b. includes humans
   c. formation of the coelom from outpocketings of archenteron
   d. development of a blastopore into a mouth
   e. echinoderms and chordates

4. The branches Radiata and Bilateria of the eumetazoa both exhibit
   a. cephalization
   b. bilateral symmetry of larval forms
   c. dominance of diploid stage in life cycle
   d. a complete digestive tract with separate mouth and anus
   e. three germ layers in embryonic development

5. Bilateral symmetry in the animal kingdom is best correlated with
   a. an ability to sense equally in all directions
   b. the presence of a skeleton
   c. motility and active predation and escape
   d. development of a true coelom
   e. adaptation to terrestrial environments

6. A land snail, a clam, and an octopus all share
   a. a mantle
   b. a radula
   c. gills
   d. embryonic torsion
   e. distinct cephalization

7. Which of the following is *not* a characteristic of the phylum Annelida?
   a. hydrostatic skeleton
   b. segmentation
   c. metanephridia
   d. pseudocoelom
   e. closed circulatory system

8. Which of the following is *not* true of the chelicerates?
   a. they have antennae
   b. their body is divided into a cephalothorax and an abdomen
   c. the horseshoe crab is one surviving marine member
   d. they include ticks, scorpions, and spiders
   e. their anterior appendages are modified as pincers or fangs

9. Which of the following combinations of phylum and description is *incorrect*?
   a. Echinodermata—branch Bilateria, coelom from archenteron
   b. Nematoda—roundworms, pseudocoelomate
   c. Cnidaria—radial symmetry, polyp and medusa body forms
   d. Platyhelminthes—flatworms, gastrovascular cavity, acoelomate
   e. Porifera—gastrovascular cavity, mouth from blastopore

10. Which of the following subdivisions of the animal kingdom encompasses all the others in the list?
   a. protostomes
   b. bilateria
   c. pseudocoelomates
   d. coelomates
   e. deuterostomes

CHALLENGE QUESTIONS

1. Compare and contrast the body plans of a planarian (a flatworm) and an earthworm.

2. A marine biologist has dredged up a previously unknown animal species from the seafloor. Describe some of the important characteristics the biologist should look for to determine the animal phylum to which the creature should be assigned.

3. Lynn Margulis of the University of Massachusetts has suggested that observing an explosion of animal diversity in Cambrian strata is like viewing Earth from a satellite over a long period of time and noticing the emergence of cities only after they are large enough to be evident at that distance. What do you think Dr. Margulis was saying about the Cambrian "explosion"?

4. Write a paragraph defining "animal."

5. Compare and contrast spiders and insects.

SCIENCE, TECHNOLOGY, AND SOCIETY

1. Under what circumstances should we regard insects as pests? Develop arguments either for or against the use of pesticides in specific situations.

2. An irrigation project has allowed farmers in an African country to grow more food. In the past, crops were planted only after spring rains. Now fields can be watered year-round. Improved crop yield has had an unexpected cost: an increase in the incidence of schistosomiasis, or blood fluke disease (see FIGURE 29.15). Why do you think the irrigation project increased the incidence of schistosomiasis? Suggest three methods that could be tried to prevent people from becoming infected. What ecological, economic, and social factors must you take into account in implementing your plans?

3. Coral reefs harbor a greater diversity of animals than any other environment in the sea. Australia's Great Barrier Reef has been protected as a marine reserve and is a mecca for scientists and nature enthusiasts. In other parts of the world, such as Indonesia and the Philippines, coral reefs are in danger. Many reefs have been depleted of fish, and runoff from the shore has covered coral heads with sediment. Nearly all the changes in the reefs can be traced back to human activities. What kinds of activities do you think might be contributing to the decline of the reefs? What are some reasons to be concerned about this decline? What might be done on a local and global scale to halt the decline?

4. Give examples from at least three phyla of animals that humans use for food.

FURTHER READING

Briggs, D. "Giant Predators from the Cambrian of China." *Science,* May 27, 1994. Jawed invertebrates 2 meters long were part of the Cambrian seascape.

Brusca, R. G., and G. J. Brusca. *Invertebrates.* Sunderland, MA: Sinauer Associates, 1990. An evolutionary approach to the animal kingdom.

Craig, C. L. "Webs of Deceit." *Natural History,* March 1995. Some spiderwebs actually *attract* prey.

Gould, S. J. "Of Tongue Worms, Velvet Worms, and Water Bears." *Natural History,* January 1995. The Cambrian fauna is a taxonomic puzzle.

Gould, S. J. *Wonderful Life: The Burgess Shale and the Nature of History.* New York: Norton, 1989. A best-seller about contingency, evolution, and the history of animals.

Levinton, J. "The Big Bang of Animal Evolution." *Scientific American,* November 1992. What evolutionary mechanisms made the Cambrian explosion possible?

Marden, J. H. "Flying Lessons from a Flightless Insect." *Natural History,* February 1995. Did insects use their wings to swim before they could fly?

Wilson, E. O. "Empire of the Ants." *Discover,* March 1990. The Pulitzer Prize–winning biologist writes about his favorite animals.

CHAPTER 30

THE VERTEBRATE GENEALOGY

KEY CONCEPTS

- Vertebrates belong to the phylum Chordata
- Invertebrate chordates provide clues to the origin of vertebrates
- The evolution of vertebrate characteristics is associated with increased size and activity
- Vertebrate diversity and phylogeny: *an overview*
- Agnathans are jawless vertebrates
- Placoderms were armored fishes with jaws and paired fins
- Sharks and their relatives have adaptations for powerful swimming
- Bony fishes are the most abundant and diverse vertebrates
- Amphibians are the oldest class of tetrapods
- The evolution of the amniotic egg expanded the success of vertebrates on land
- A reptilian heritage is evident in all amniotes
- Birds began as flying reptiles
- Mammals diversified extensively in the wake of the Cretaceous extinctions
- Primate evolution provides a context for understanding human origins
- Humanity is one very young twig on the vertebrate tree

Most of us are curious about our genealogies. On the personal level, we wonder about our family ancestry. As biology students, we are interested in retracing human ancestry within the broader scope of the evolutionary history of the entire animal kingdom. The questions we must ask are: What were our ancestors like? How are we related to other animals? What are our closest relatives? In this chapter, we trace the evolution of the vertebrates, the group that includes humans and their closest relatives. Mammals, birds, reptiles, amphibians, and the various classes of fishes are all classified as ***vertebrates.*** *They share many features unique to the vertebrates, including the cranium and backbone, a series of vertebrae for which the group is named. You can see these vertebrate hallmarks in the photograph of a snake skeleton that opens this chapter. Our first step in tracking the vertebrate genealogy is to determine where vertebrates fit in the animal kingdom.*

Vertebrates belong to the phylum Chordata

The vertebrates make up one subphylum within the phylum Chordata. **Chordates** also include two subphyla of invertebrates, the urochordates and the cephalochordates.

Chordate Characteristics

Based on certain similarities in early embryonic development, chordates are grouped as deuterostomes along with the echinoderms (see Chapter 29). Although chordates vary widely in appearance, they are distinguished as a phylum by the presence of four anatomical structures that appear at some point during the animal's lifetime, often only during the embryo stage (FIGURE 30.1).

1. *Notochord.* All chordate embryos have a **notochord,** which is a longitudinal, flexible rod located between the gut and the nerve cord. The notochord is composed of large, fluid-filled cells encased in fairly stiff, fibrous tissue. It extends through most of the length of the animal as a relatively simple skeleton. Chordates are named for this structure. In some invertebrate chordates and primitive vertebrates, the notochord persists to support the adult. However, in most vertebrates a more complex, jointed skeleton develops and the adult retains only remnants of the embryonic notochord—as gelatinous material of the disks between the vertebrae of humans, for example.

2. *Dorsal, hollow nerve cord.* The nerve cord of a chordate embryo develops from a plate of ectoderm that rolls into a tube located dorsal to the notochord. The result is

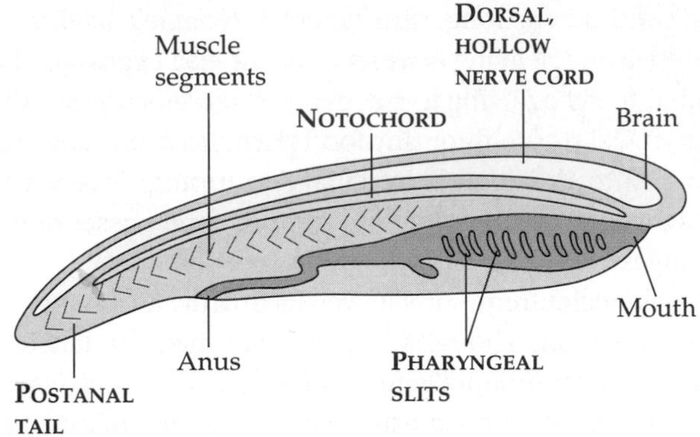

FIGURE 30.1
Chordate characteristics. All chordates possess the four trademarks of the phylum: a notochord; a dorsal, hollow nerve cord; pharyngeal slits; and a postanal tail.

a dorsal, hollow nerve cord unique to chordates. Other animal phyla have solid nerve cords, usually ventrally located. The nerve cord of a chordate embryo develops into the central nervous system: the brain and spinal cord.

3. *Pharyngeal slits.* The digestive tube of chordates extends from the mouth to the anus. The region just posterior to the mouth is the pharynx, which opens to the outside of the animal through several pairs of slits. These pharyngeal slits allow water that enters the mouth to exit without continuing through the entire digestive tract. The pharyngeal slits function as suspension-feeding devices in many invertebrate chordates and have become modified for gas exchange and other functions during vertebrate evolution.

4. *Muscular postanal tail.* Most chordates have a tail extending beyond the anus. By contrast, most nonchordates have a digestive tract that extends nearly the whole length of the body. The chordate tail contains skeletal elements and muscles and provides much of the propulsive force in many aquatic species.

Invertebrate chordates provide clues to the origin of vertebrates

We can examine the two subphyla of invertebrate chordates, Urochordata and Cephalochordata, to see the chordate body plan in its most "stripped-down" versions—without the additional features that evolved in vertebrates. But the study of urochordates and cephalochordates also provides clues to the origin of vertebrates.

Subphylum Urochordata

Urochordates are commonly called **tunicates.** Most tunicates are sessile marine animals that adhere to rocks, docks, and boats (FIGURE 30.2a). Others are planktonic. Some species are colonial. Seawater enters the animal through an incurrent siphon, passes through the pharyngeal slits into a chamber called the atrium, and

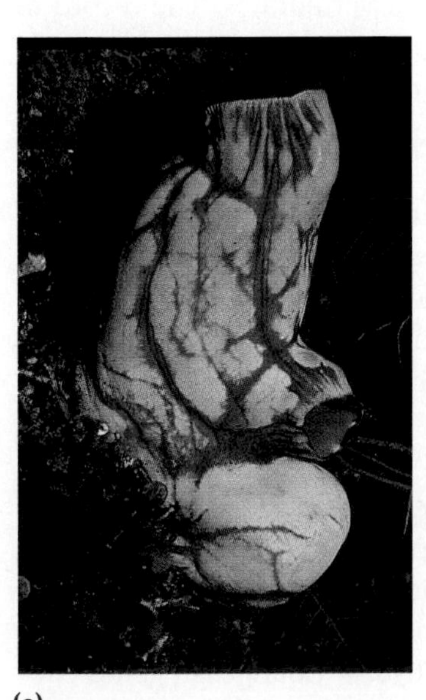
(a)

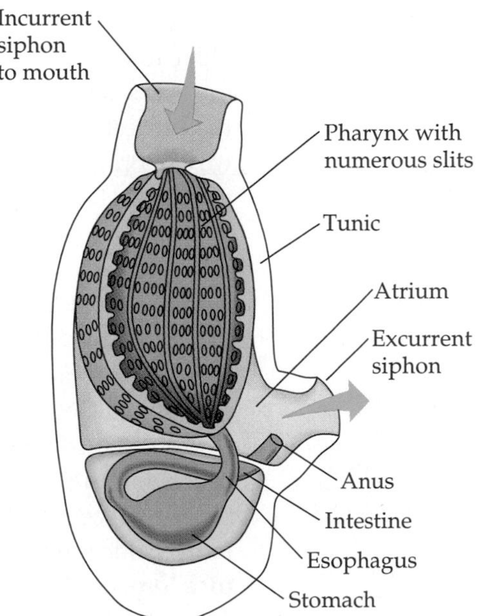

(b)

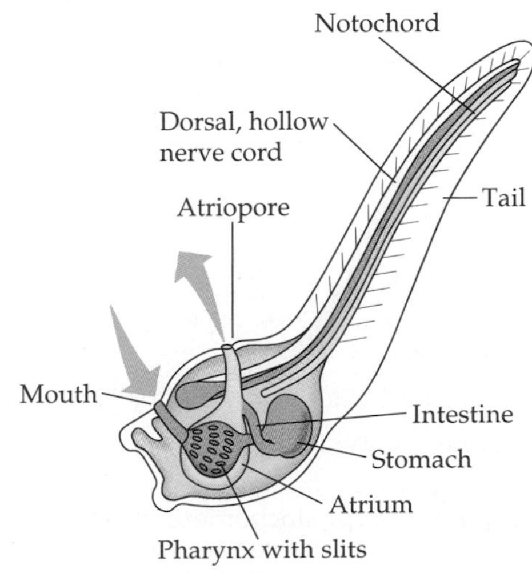

(c)

FIGURE 30.2
Subphylum Urochordata: a tunicate. (**a**) An adult tunicate, or sea squirt, is a sessile animal commonly arranged in a U shape. (**b**) In the adult, prominent pharyngeal slits function in suspension-feeding, but the other chordate characteristics are not obvious. (**c**) In the tunicate larva, which is a free-swimming suspension-feeder, the chordate characteristics are evident. Some urochordates lack the sessile adult stage, existing as free-swimming, larva-like forms throughout life.

exits through an excurrent siphon, or atriopore (FIGURE 30.2b). The food filtered from this water current by a mucous net is passed by cilia into the intestine. The anus empties into the excurrent siphon. The entire animal is cloaked in a tunic made of a celluloselike carbohydrate. Because they shoot a jet of water through the excurrent siphon when molested, tunicates are also called sea squirts.

The adult tunicate scarcely resembles a chordate. It displays no trace of a notochord, nor is there a nerve cord or tail. Only the pharyngeal slits suggest a link to other chordates. But all four chordate trademarks are manifest in the larval form of some groups of tunicates (FIGURE 30.2c). The larva swims until it attaches by its head to a surface and undergoes metamorphosis, during which most of its chordate characteristics disappear.

Subphylum Cephalochordata

Known as **lancelets** because of their bladelike shape, **cephalochordates** closely resemble the idealized chordate in FIGURE 30.1. The notochord; dorsal, hollow nerve cord; numerous gill slits; and postanal tail all persist into the adult stage (FIGURE 30.3). A tiny marine animal only a few centimeters long, the lancelet wriggles backward into the sand, leaving only its anterior end exposed. The animal feeds by using a mucous net secreted across the pharyngeal slits to filter tiny food particles from seawater drawn into the mouth by ciliary pumping. The water exits through the slits, and the trapped food passes down the digestive tube.

The lancelet frequently leaves its burrow and swims to a new location. Though a feeble swimmer, the lancelet displays, in rudimentary form, the mechanism of swimming that fishes use. Coordinated contraction of muscles serially arranged like rows of chevrons («««) along the sides of the notochord flexes the notochord from side to side in a sinusoidal (∽) pattern. This serial musculature is evidence of the lancelet's segmentation. The muscle segments develop from blocks of mesoderm called **somites** arranged along each side of the notochord of a chordate embryo. Chordates are segmented animals (FIGURE 30.4), but this anatomical feature evolved independently of segmentation in annelids and arthropods.

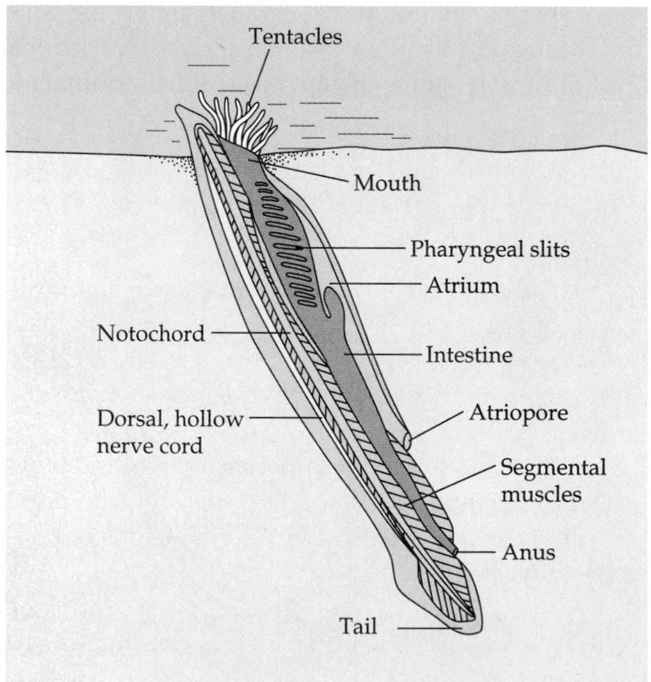

FIGURE 30.3
Subphylum Cephalochordata: the lancelet *Branchiostoma*. This small invertebrate displays all four chordate characteristics. The pharyngeal slits are used for suspension-feeding. Water passes into the pharynx and through slits into the atrium, a chamber that vents to the outside via the atriopore. Food particles trapped by a mucous net are swept by cilia into the digestive tract. The muscle segments produce the sinusoidal swimming of these animals.

The Relationship Between Invertebrate Chordates and Vertebrates

Phylum Chordata, like most animal phyla, makes its first appearance in the fossil record in Cambrian rocks. Paleontologists have found fossilized invertebrates resembling cephalochordates in the Burgess Shale of British Columbia, Canada (see FIGURE 29.44). Those fossils are about 545 million years old, about 50 million years older than the oldest known vertebrates. The record of the rocks is too incomplete for us to retrace the origin of the earliest vertebrates from invertebrate ancestors, but we can propose reasonable hypotheses about this evolution based on comparative anatomy and embryology.

Most zoologists think that the ancestor of vertebrates was a suspension-feeder similar to a cephalochordate, with all four of the fundamental chordate characteristics. Recent research by molecular systematists supports the hypothesis that cephalochordates are the vertebrates' closest relatives.

Both cephalochordates and vertebrates may have evolved from a common sessile ancestor by **paedogenesis,** the precocious development of sexual maturity in a larva. Notice that a cephalochordate appears more akin to a urochordate larva than to an adult urochordate (compare FIGURES 30.2 and 30.3). Changes in genes controlling development can alter the timing of developmental events, such as the maturation of gonads. Perhaps such a change occurred in the ancestor of cephalochordates and vertebrates, causing the gonads to mature in

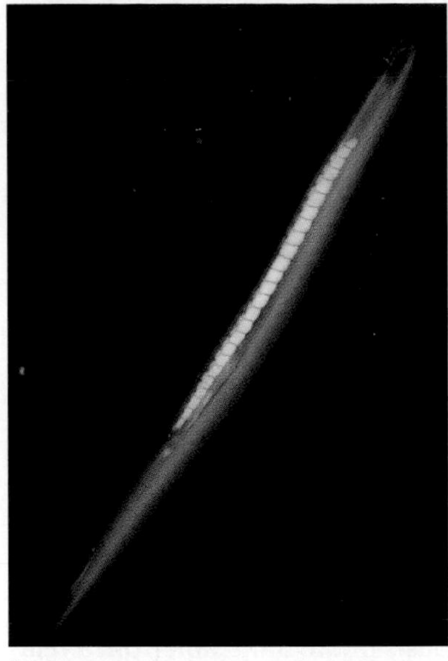

(a)

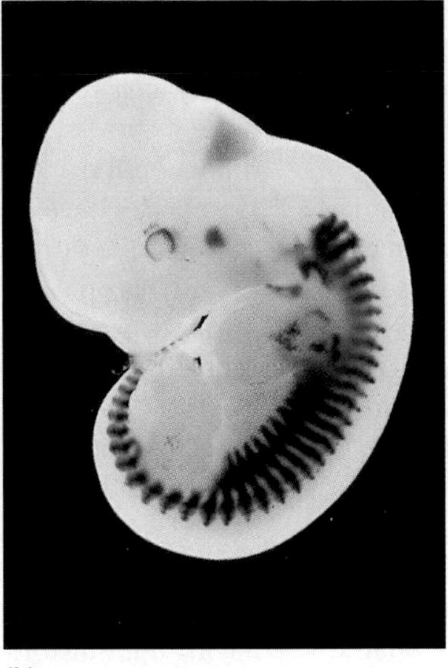

(b)

(c)

FIGURE 30.4

Chordate segmentation. Segmentation evolved independently in the phyla Arthropoda and Chordata. Here are three examples of the chordate version. (**a**) The segmented muscles of cephalochordates and vertebrates develop from somites, the blocks of tissue visible in this lancelet. (**b**) Somites are also visible in this mouse embryo. (**c**) Much of the segmental anatomy so apparent in vertebrate embryos is disguised in the adult. Evidence of segmentation persists in the serial arrangement of vertebrae and in patterns of certain muscle groups, such as the abdominal muscles.

swimming larvae before the onset of metamorphosis to the sessile adult form. If reproducing larvae were very successful, natural selection may have reinforced paedogenesis and eliminated metamorphosis.

Although cephalochordates and vertebrates probably evolved from a common chordate ancestor, they diverged about a half-billion years ago and differ in many important ways. Most of those differences can be interpreted as vertebrate adaptations to the problems of larger size and a more mobile lifestyle.

The evolution of vertebrate characteristics is associated with increased size and activity

Vertebrates retain the primitive chordate characteristics while adding other specializations, shared derived features (synapomorphies; see Chapter 23) that distinguish the subphylum from invertebrate chordates. The evolution of these unique vertebrate structures was probably associated with increasing size and more active foraging for food. These vertebrate adaptations include cephalization, a skeleton that includes a cranium and vertebral column, and anatomical equipment that supports the active metabolism required for a more energetic lifestyle.

1. *Compared to cephalochordates, vertebrates are much more cephalized.* Recall from Chapter 29 that cephalization is the evolutionary concentration of neural and sensory equipment on the head (anterior) end, the part of a mobile animal that comes into contact with most environmental stimuli first. Your eyes, ears, nose, and brain (the enlarged anterior end of the dorsal, hollow nerve cord) are all anatomical reminders of the evolution of navigation equipment in vertebrates.

2. *The cranium and vertebral column replaced the notochord as the main axis of the vertebrate body.* The notochord is a sufficient skeleton for a lancelet, which swims only occasionally and spends most of its time filter-feeding from its burrow in the sand. In contrast, the early vertebrates were stronger swimmers that actively searched for food. The axial skeleton of vertebrates helped make larger size and stronger, faster movement possible. The vertebral column provides physical support and also functions as a strong, jointed strut that

gives leverage to the segmental swimming muscles. The cranium protects the brain. Most vertebrates also have ribs, which anchor muscles and protect internal organs, and an appendicular skeleton supporting two pairs of appendages (fins or limbs).

The vertebrate endoskeleton may be made of hard bone or more flexible cartilage, or some combination of these two materials. Although the skeleton consists mostly of a nonliving matrix, the living cells of the skeleton secrete and maintain the materials of the matrix. The living endoskeleton of a vertebrate can grow with the animal, unlike the nonliving exoskeleton of arthropods.

A group of embryonic cells called the **neural crest,** unique to vertebrates, contributes to the formation of certain skeletal elements and many other structures that distinguish vertebrates from other chordates. The dorsal, hollow nerve cord found in all chordates develops when the edges of an ectodermal plate on the surface of the embryo roll together to form the neural tube. In vertebrates, the neural crest forms near the dorsal margins of this closing tube (FIGURE 30.5). Cells of the neural crest then migrate to various targets in the embryo and participate in the formation of diverse vertebrate structures, including some of the bones and cartilage of the head.

3. *Vertebrate adaptations support the greater metabolic demands of increased size and activity.* When vertebrates move in pursuit of food or when escaping predators, they regenerate their ATP supply mainly by cellular respiration, which consumes oxygen. Adaptations of the vertebrate respiratory and circulatory systems support busy mitochondria in muscle cells and other active tissues. Vertebrates have a closed circulatory system, with a ventral, chambered heart that pumps blood through arteries to microscopic vessels called capillaries that branch throughout every tissue in the body. The blood is oxygenated as it passes through capillaries in gills or lungs.

A vigorous lifestyle also requires a relatively large supply of organic fuel. Vertebrate adaptations for feeding, digestion, and nutrient absorption help support active behavior. For example, muscles in the walls of the digestive tract propel food from organ to organ along the tract. These are all examples of vertebrate form and function having their historical basis in the transition from a relatively sedentary lifestyle to a more active one.

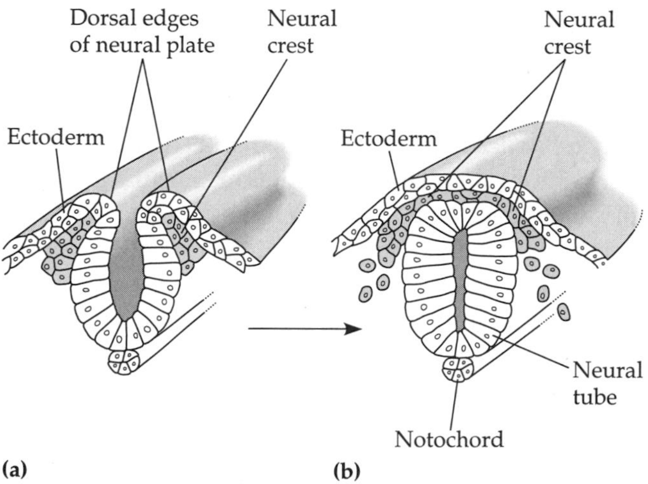

FIGURE 30.5
The neural crest, embryonic source of many vertebrate synapomorphies. (**a**) The neural crest consists of bilateral bands of cells near the margins of the embryonic folds that meet to form the dorsal, hollow nerve cord (neural tube). (**b**) Cells from the neural crest migrate to distant sites in the embryo, where they give rise to some of the anatomical structures unique to vertebrates, including some of the bones and cartilage of the skull.

Vertebrate diversity and phylogeny: *an overview*

The taxonomic scheme adopted for this textbook recognizes seven extant classes of the subphylum Vertebrata (TABLE 30.1, p. 634). Three of these are commonly called fishes: Agnatha (jawless vertebrates), Chondrichthyes (cartilaginous fishes: sharks and rays), and Osteichthyes (bony fishes). The other four classes—Amphibia (frogs and salamanders), Reptilia (reptiles), Aves (birds), and Mammalia (mammals)—are collectively called **tetrapods** (Gr. *tetra,* "four," and *pod,* "foot") because most animals in these classes have two pairs of limbs that support them on land. Reptiles, birds, and mammals have additional terrestrial adaptations that distinguish them from amphibians. One of these is the **amniotic egg,** a shelled, water-retaining egg. The amniotic egg functions as a "self-contained pond" that enables these vertebrates to complete their life cycles on land. Although most mammals don't lay eggs, they retain other key features of the amniotic condition. In recognition of this important evolutionary breakthrough, reptiles, birds, and mammals are collectively called **amniotes.**

In the following pages, we will survey the diversity of organisms in each vertebrate class and highlight some of the major evolutionary trends among the vertebrates. The survey also includes a brief description of placoderms, an extinct class of early jawed vertebrates. As a prelude to learning about the various groups of vertebrates, examine the evolutionary tree in FIGURE 30.6. This tree represents one set of hypotheses about the relationships among the classes of the subphylum Vertebrata.

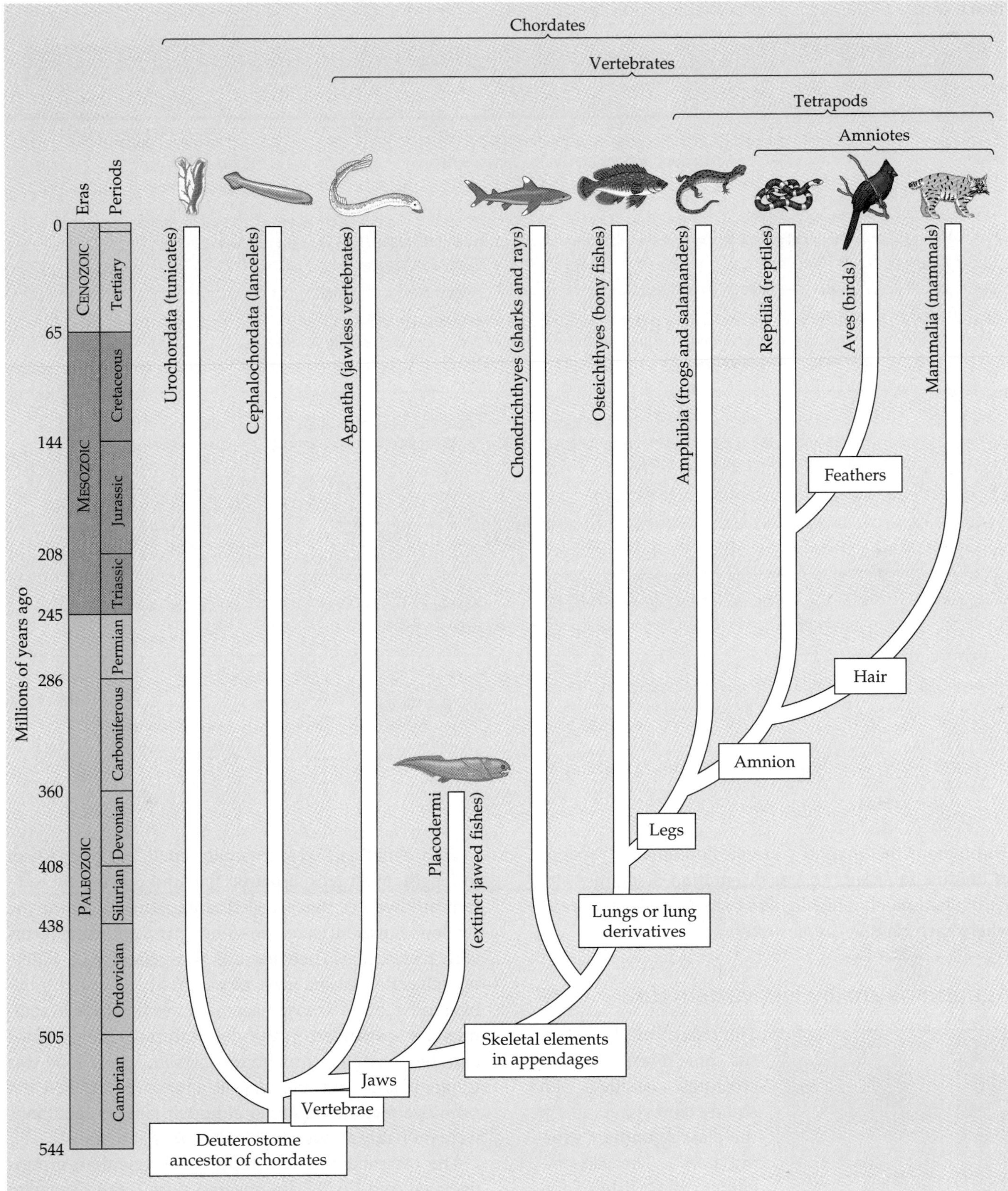

FIGURE 30.6

A hypothetical phylogeny of the chordates. The boxed characteristic at the base of a branch is an example of the derived features (synapomorphies) that are shared by the vertebrate classes beyond that branch point.

TABLE 30.1

The Extant Vertebrate Classes		
CLASS	MAIN CHARACTERISTICS	EXAMPLES
Agnatha	Jawless vertebrates: cartilaginous skeleton; notochord persists throughout life; marine and freshwater; living species lack paired appendages.	Lampreys, hagfishes
Chondrichthyes	Cartilaginous fishes: cartilaginous skeleton; jaws; notochord replaced by vertebrae in adult; paired appendages; respiration through gills; internal fertilization, may lay eggs or bear live young; acute senses, including lateral line system.	Sharks, skates, rays, chimaeras
Osteichthyes	Bony fishes: bony skeletons and jaws; most species have external fertilization and lay large numbers of eggs; respiration mainly through gills; many have a swim bladder; marine and freshwater.	Bass, trout, perch, tuna
Amphibia	Appendages adapted for moving on land (tetrapod condition); aquatic larval stage metamorphosing into terrestrial adult (many species); may lay eggs or bear live young; respiration through lungs and/or skin.	Salamanders, newts, frogs, toads, caecilians
Reptilia	Terrestrial tetrapods with scaly skin: respiration via lungs; lay amniotic shelled eggs or bear live young.	Snakes, lizards, turtles, crocodiles
Aves	Tetrapods with feathers: forelimbs modified as wings; respiration through lungs; endothermic; internal fertilization; shelled amniotic eggs; acute vision.	Owls, sparrows, penguins, eagles
Mammalia	Tetrapods with young nourished from mammary glands of females; hair; diaphragm that ventilates lungs; endothermic; amniotic sac, but most bear live young.	Monotremes (such as platypuses); marsupials (such as kangaroos); placentals (such as rodents)

Throughout the chapter you will find smaller versions of the tree in FIGURE 30.6 as orientation diagrams, with particular branches highlighted to help you keep track of where each class fits in the vertebrate genealogy.

Agnathans are jawless vertebrates

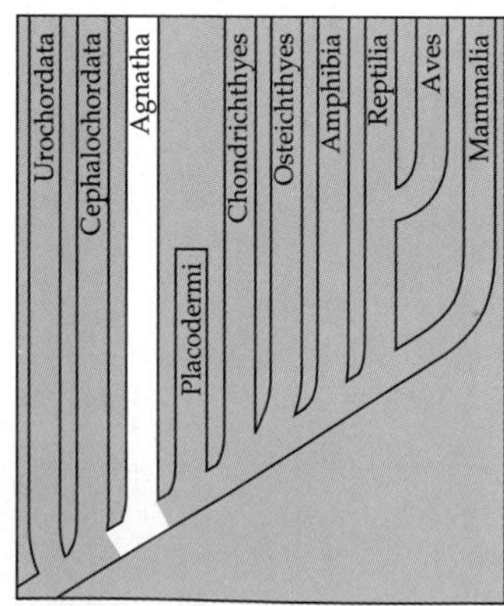

The oldest vertebrate fossils are diverse jawless creatures classified with some extant vertebrates in the **class Agnatha** ("without jaws"). The class includes extinct fishlike animals called **ostracoderms** that were encased in an armor of bony plates. Traces of these agnathans are found in Cambrian strata, but most date back to the Ordovician and Silurian periods, about 400 to 500 million years ago.

Early agnathans were generally small, less than 50 cm in length. Most lacked paired fins and apparently were bottom-dwellers that wiggled along streambeds or the seafloor, but there were also some active, midwater forms with paired fins. Their mouths were circular or slitlike openings that lacked jaws. Most agnathans were probably mud-suckers or suspension-feeders that took in sediments or suspended organic debris through their mouths and then passed it through the gill slits, where food was trapped. Thus, the pharyngeal apparatus retained the primitive feeding function, although gills in agnathans were probably also the major sites of gas exchange.

The ostracoderms and most other agnathan groups declined and finally disappeared during the Devonian period, but about 60 species of jawless vertebrates are alive today in the form of lampreys and hagfishes (FIGURE 30.7). In common with many extinct agnathans, lampreys and hagfishes lack paired appendages. In contrast to many ostracoderms, however, they have no external armor. The eel-shaped sea lamprey feeds by clamping its round mouth onto the flank of a live fish, using a rasping

FIGURE 30.7

A sea lamprey, a jawless vertebrate (class Agnatha). Acting as both a predator and a parasite, a lamprey uses this rasping mouth (inset) to bore a hole in the side of a fish, living on the blood and other tissues of its host.

tongue to penetrate the skin of its prey, and ingesting the prey's blood. Sea lampreys live as larvae for years in freshwater streams and then migrate to the sea or lakes as they mature into adults. The larva is a suspension-feeder that looks very much like the lancelet, a cephalochordate. Some species of lampreys feed only as larvae. Following several years in streams, they attain sexual maturity, reproduce, and die within a few days.

Hagfishes resemble lampreys, but they are mainly scavengers rather than blood-suckers or suspension-feeders, and their mouthparts are not adapted for rasping. Some species feed on sick or dead fish, whereas other hagfishes eat marine worms. Hagfishes lack a larval stage and live entirely in salt water.

Placoderms were armored fishes with jaws and paired fins

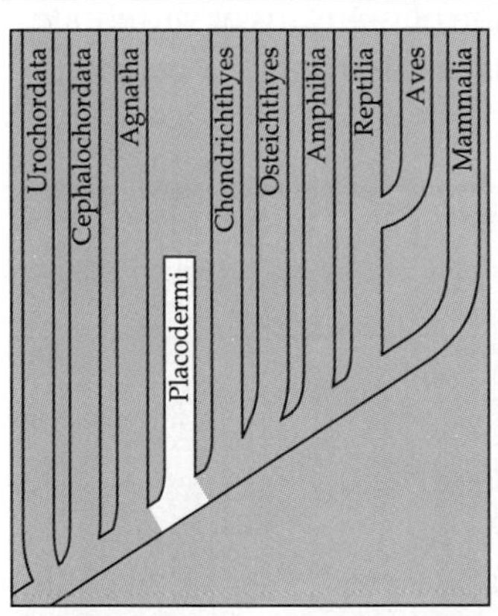

During the late Silurian and early Devonian periods, the agnathans were largely replaced by armored fishes called **placoderms.** The largest were more than 10 m long, but most were less than 1 m long. In contrast to many ostracoderms, placoderms had paired fins, which greatly enhanced their swimming ability. Placoderms also had jaws, making their mouth more than just a fixed orifice for rasping or scooping sediments. With paired fins and hinged jaws, many species were active predators, capable of chasing prey and biting off chunks of flesh. Thus, these two modifications of the early vertebrate body plan allowed the diversification of both lifestyles and nutrient sources.

The hinged jaws of vertebrates evolved by modification of the skeletal rods that had previously supported the anterior pharyngeal (gill) slits (FIGURE 30.8). The remaining gill slits, no longer required for suspension-feeding, remained as the major sites of respiratory gas exchange with the external environment.

The origin of vertebrate jaws from these skeletal parts was a major adaptive event, illustrating a general feature of evolutionary change: New adaptations usually evolve by the modification of existing structures. Hinged jaws also evolved in arthropods, but these had a totally different origin than vertebrate jaws (see Chapter 29). Arthropod jaws are modified appendages that work from side to side rather than up and down like vertebrate jaws. As a mechanism of adaptation, evolution is limited by the raw material with which it must work; evolution is generally more of a remodeling process than a creative one.

The Devonian period (about 350 to 400 million years ago) is known as the age of fishes, when the placoderms and another group of jawed fishes called acanthodians (placed in a separate class) radiated and many new forms evolved in both fresh and salt water. Placoderms and acanthodians dwindled and disappeared almost com-

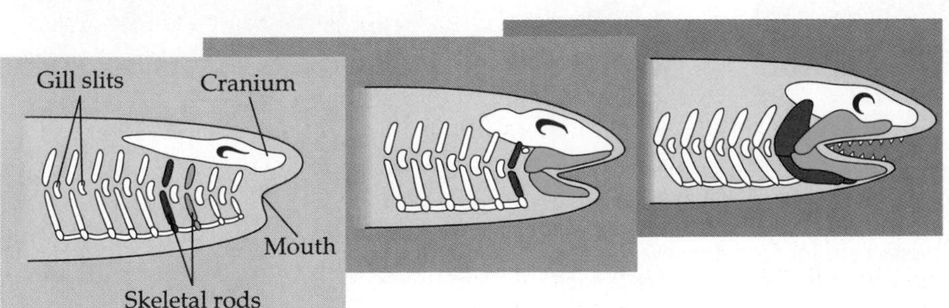

FIGURE 30.8

The evolution of vertebrate jaws. The skeleton of the jaws and their supports evolved from two pairs of skeletal rods located between gill slits that were near the mouth. Pairs of rods anterior to those that formed the jaws were either lost or incorporated into the jaws.

pletely by the beginning of the Carboniferous period, 350 million years ago. During the Devonian or earlier, however, ancestors of placoderms and acanthodians may have also given rise to sharks (class Chondrichthyes). Another group of large, jawed predators—the bony fishes (class Osteichthyes)—diverged from the common ancestor some 425 to 450 million years ago. Sharks and bony fishes are the reigning vertebrates over the watery two-thirds of the surface of Earth.

Sharks and their relatives have adaptations for powerful swimming

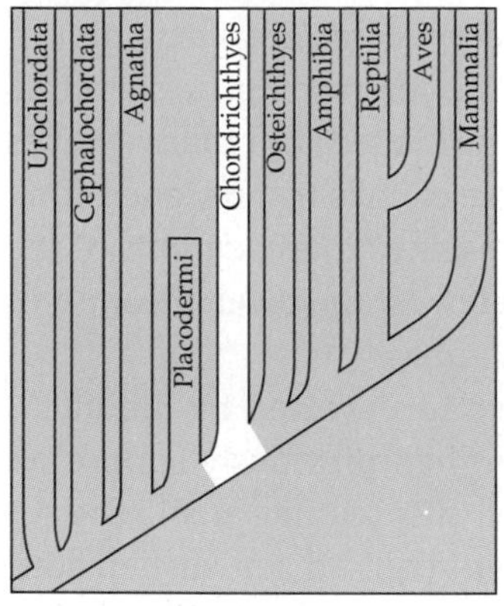

The vertebrates of **class Chondrichthyes,** sharks and their relatives, are called cartilaginous fishes because they have relatively flexible skeletons made of cartilage rather than bone. There are about 750 extant species in this class. Jaws and paired fins, which first appeared in the placoderms, are well developed in the cartilaginous fishes. The largest and most diverse subclass consists of the sharks and rays (FIGURE 30.9). A second subclass is composed of a few dozen species of unusual fishes called chimaeras.

The cartilaginous skeleton of these fishes is a derived characteristic, not a primitive one; that is, the ancestors of Chondrichthyes had bony skeletons, and the cartilaginous skeleton characteristic of the class evolved secondarily. During the development of most vertebrates, the skeleton is first cartilaginous and then becomes bony as the matrix is hardened with calcium phosphate (ossification). Apparently, some modification in the developmental process of cartilaginous fishes prevents or slows ossification. The powerful swimming muscles of a shark flex the cartilaginous skeleton, which is more elastic and lighter than a skeleton of bone.

Most sharks have streamlined bodies and are swift swimmers, but they do not maneuver very well. Their stiff caudal (tail) fin helps propel the animal. The dorsal fins are used largely as stabilizers, and the paired pectoral (fore) and pelvic (hind) fins provide lift in the water. Although a shark gains additional buoyancy by storing a large amount of oil in its huge liver, the animal is still more dense than water, and it sinks if it stops swimming. Continual swimming also ensures that water will flow into the mouth and out through the gills, where gas exchange occurs. However, some sharks and many skates and rays spend a good deal of time resting on the seafloor. When doing so, these fishes use muscles of the jaws and pharynx to pump water over the gills.

The largest sharks and rays are suspension-feeders that feed on plankton. The whale shark, for example, strains 1 million L of water per hour to obtain sufficient food to support its massive body. Most sharks, however, are carnivores that swallow their prey whole or use their powerful jaws and sharp teeth to tear flesh from animals too large to swallow in one piece. Shark teeth probably evolved from the jagged scales that cover the abrasive skin. The digestive tract of many sharks is proportionately shorter than the digestive tube of many other vertebrates. Within the shark intestine is a spiral valve, a corkscrew-

(a)

(b)

FIGURE 30.9

Cartilaginous fishes (class Chondrichthyes). (**a**) Fast swimmers with acute senses and powerful jaws, sharks, such as this blacktip reef shark, are well adapted to their predatory way of life. Sharks have paired pectoral and pelvic fins. (**b**) Most rays are flattened bottom-dwellers that crush mollusks and crustaceans for food. However, some species cruise in open water, such as this manta ray whose gaping mouth scoops food.

shaped ridge that increases surface area and prolongs the passage of food along the short digestive tract.

Acute senses are adaptations that go along with the active, carnivorous lifestyle of sharks. Sharks have sharp vision but cannot distinguish colors. The nostrils of sharks, like those of most fishes, open into dead-end cups. They can be used only for olfaction (smelling), not for breathing. Along with eyes and nostrils, the shark head also has a pair of regions in the skin that can detect electrical fields generated by the muscle contractions of nearby fish and other animals. Running the length of each flank of the shark is the **lateral line system,** a row of microscopic organs sensitive to changes in the surrounding water pressure, enabling the shark to detect minor vibrations. Sharks can also hear by sensing percussions with a pair of auditory organs. Sharks and other fishes have no eardrums, structures that terrestrial vertebrates use to transmit sound waves traveling through air into the ear toward the auditory organs. Sound reaches the shark through water, and the animal's entire body transmits the sound to the hearing organs of the inner ear.

Shark eggs are fertilized internally. The male has a pair of claspers on its pelvic fins that transfer sperm into the reproductive tract of the female. Some species of sharks are **oviparous;** they lay eggs that hatch outside the mother's body. These sharks release their eggs after encasing them in protective coats. Other species are **ovoviviparous;** they retain the fertilized eggs in the oviduct. Nourished by the egg yolk, the embryos develop into young that are born after hatching within the uterus. A few species are **viviparous;** the young develop within the uterus, nourished prior to birth by nutrients received from the mother's blood through a placenta. The reproductive tract of the shark empties along with the excretory system and digestive tract into the **cloaca,** a common chamber that expels through a single vent. All other vertebrates except placental mammals also have cloacas.

Although rays are closely related to sharks, they have adopted a very different lifestyle. Most rays are flattened bottom-dwellers that feed by using their jaws to crush mollusks and crustaceans. A ray's pectoral fins are greatly enlarged and used to propel the animal through the water. The tail of many rays is whiplike and, in some species, bears venomous barbs that function in defense.

Bony fishes are the most abundant and diverse vertebrates

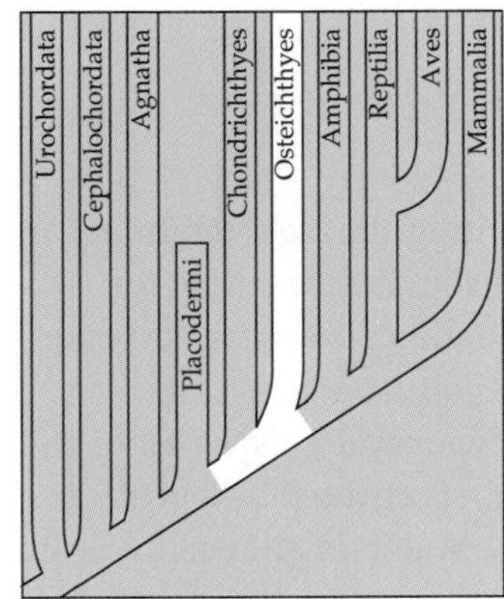

Of all vertebrate classes, bony fishes, of the **class Osteichthyes,** are the most numerous, both in individuals and in species (about 30,000). Fishes of this class are abundant in the seas and in nearly every freshwater habitat. Bony fishes range in size from about 1 cm to more than 6 m long. Most of the fishes familiar to us belong to class Osteichthyes (FIGURE 30.10).

In contrast to the cartilaginous fishes, the skeleton of most bony fishes is reinforced by a hard matrix of calcium phosphate. The skin is often covered by flattened bony scales that differ in structure from the toothlike scales of sharks. Glands in the skin of a bony fish secrete a mucus that gives the animal its characteristic sliminess,

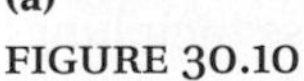

(a)

(b)

FIGURE 30.10

Bony fishes (class Osteichthyes). These two species belong to the largest subclass of bony fishes, the ray-finned fishes. (**a**) A longnose gar. (**b**) A perch.

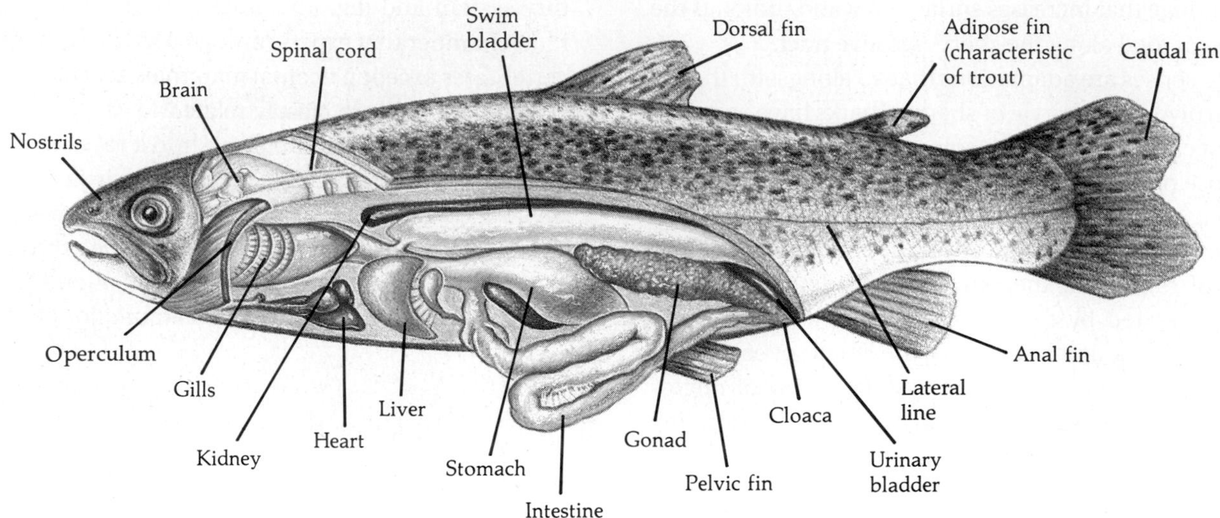

FIGURE 30.11
Anatomy of a trout, a representative bony fish.

an adaptation that reduces drag during swimming. In common with sharks, bony fishes have a lateral line system clearly evident as a row of tiny pits in the skin on either side of the body (FIGURE 30.11).

Bony fishes breathe by drawing water over four or five pairs of gills located in chambers covered by a protective flap called the **operculum.** Water is drawn into the mouth, through the pharynx, and out between the gills by movement of the operculum and contraction of muscles surrounding the gill chambers. This process enables a bony fish to breathe while stationary.

Another adaptation of most bony fishes not found in sharks is the **swim bladder,** an air sac that helps control the buoyancy of the fish. The transfer of gases between the swim bladder and the blood varies the inflation of the bladder and adjusts the density of the fish. Thus, many bony fishes, in contrast to most sharks, can conserve energy by remaining almost motionless.

Bony fishes are generally maneuverable swimmers, their flexible fins better for steering and propulsion than the stiffer fins of sharks. The fastest bony fishes, which can swim in short bursts of up to 80 km/hr, have the same basic body shape as a shark. In fact, this body shape, termed fusiform (tapering on both ends), is common to all fast fishes and aquatic mammals such as seals and whales. Water is about a thousand times more dense than air, and thus the slightest bump that causes drag is even more impeding to a fish than to a bird. Regardless of their different origins, we should expect speedy fishes and marine mammals to have similar streamlined shapes, because the laws of hydrodynamics are universal. This is an example of convergent evolution.

Details about the reproduction of bony fishes vary extensively. Most species are oviparous, reproducing by external fertilization after the female sheds large numbers of small eggs. However, internal fertilization and birthing characterize other species. Some bony fishes display complex mating rituals (see Chapter 50).

Both cartilaginous and bony fishes diversified extensively during the Devonian and Carboniferous periods, but whereas sharks arose in the sea, bony fishes probably originated in fresh water. The swim bladder was modified from simple lungs that had been used to augment the gills for gas exchange, perhaps in stagnant swamps with low oxygen content. By the end of the Devonian, the two subclasses of bony fishes that exist today had already diverged. They are the ray-finned fishes (subclass Actinopterygii; Gr. *aktin,* "ray," and *pteryg,* "wing" or "fin") and the fleshy-finned fishes (sub-class Sarcopterygii; Gr. *sarkodes,* "fleshy").

Nearly all the families of fishes familiar to us are **ray-finned fishes.** The various species of bass, trout, perch, tuna, and herring are examples. The fins, supported mainly by long flexible rays, have become modified for maneuvering, defense, and other functions. Ray-finned fishes spread from fresh water to the seas during their long history. (Adaptations that solve the osmotic problems of this move to salt water will be discussed in Chapter 40.) Numerous species of ray-finned fishes returned to fresh water at some point in their evolution. Some of these, including salmon and sea-run trout, replay their evolutionary round trip from fresh water to seawater back to fresh water during their life cycle.

In contrast to the ray-fins, most **fleshy-finned fishes** remained in fresh water and continued to use their lungs to aid the gills in breathing. Three orders of fleshy-finned fishes evolved: lungfishes, coelocanths, and rhipidistians. Together, coelocanths and rhipidistians are referred to as **lobe-finned fishes.** Lobe-fins and some lungfishes had fleshy, muscular pectoral and pelvic fins that were supported by extensions of the bony skeleton. Many were large, apparently bottom-dwelling forms that may have used their paired, fleshy fins as aids to "walking" on the substrate under water. Some may also have been able to waddle occasionally on land.

Three genera of lungfishes live today in the Southern Hemisphere. They generally inhabit stagnant ponds and swamps, surfacing to gulp air into lungs connected to the pharynx of the digestive tract. When ponds shrink during the dry season, lungfishes can burrow in the mud and aestivate (wait in a state of torpor).

Lobe-finned fishes are represented today by only one known species, a coelocanth (*Latimeria*). Although most Devonian lobe-fins were probably freshwater animals with lungs, the extant coelocanth belongs to a lineage that entered the seas at some point in its evolution (FIGURE 30.12). The other order of lobe-finned fishes, the rhipidistians, is extinct. Few of us encounter lungfishes or lobe-fins, and today these animals are far less numerous than the ray-fins. However, the lobe-finned fishes of the Devonian are of great importance in the vertebrate geneology; the ancestor of amphibians was probably one of the Devonian rhipidistians (FIGURE 30.13).

FIGURE 30.12
***Latimeria,* the only extant lobe-finned fish.** This large coelocanth lives in the deep sea off Madagascar.

CHONDRICHTHYES
OSTEICHTHYES
TETRAPODS
PLACODERMI (extinct)
Ray-finned fishes
Lungfishes
Coelocanths
Rhipidistians (extinct)
Lobe-finned fishes
Fleshy-finned fishes
Ancestral jawed fishes of the Devonian period

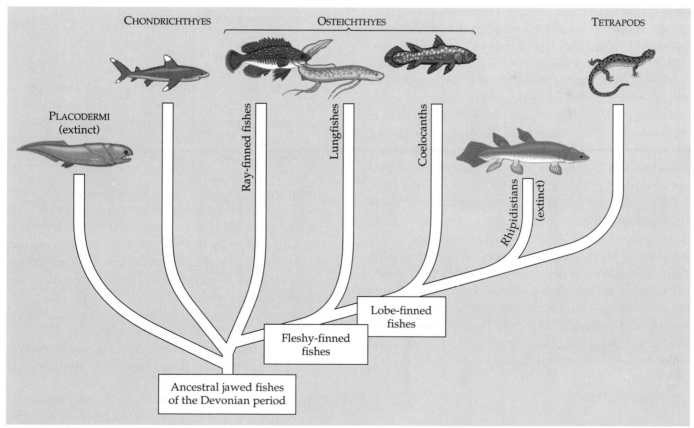

FIGURE 30.13
The Devonian radiation of fishes.

Amphibians are the oldest class of tetrapods

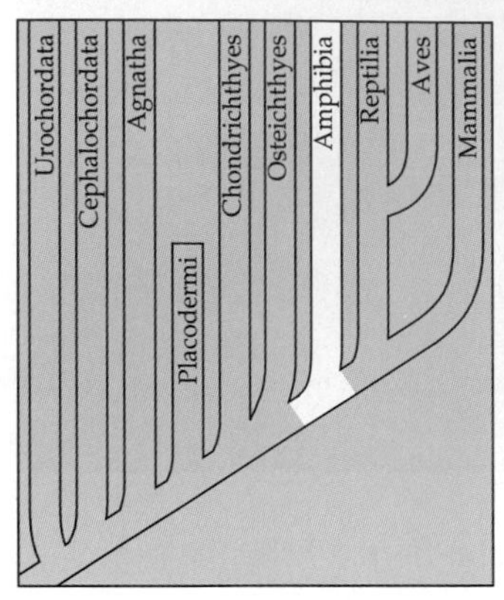

The first vertebrates on land were members of the **class Amphibia.** Today the class is represented by a total of about 4000 species of frogs, salamanders, and caecilians (limbless creatures that burrow in tropical forests and freshwater lakes).

Early Amphibians

As is the case today, parts of the Devonian world were subject to cycles of drought, followed by heavy rainfall and then drought again. In such areas, some of the freshwater fishes apparently adapted to the unreliable conditions, and lobe-finned species with lungs prevailed. The lobe-fins are of particular interest because the skeletal structure of their fins suggests that these paired appendages could have assisted in movement on land. Some of the fossil lobe-fins, including a rhipidistian named *Eusthenopteron,* exhibited many other anatomical similarities to the earliest amphibians, suggesting a relationship to the ancestor of tetrapods (FIGURE 30.14a).

The oldest amphibian fossils date back to late Devonian times, about 365 million years ago (FIGURE 30.14b). New adaptive zones opened to these first vertebrates on land, offering relative safety from carnivorous fishes and a cornucopia of food previously unexploited by backboned animals. From the start, amphibians were predators that ate insects and other invertebrates that preceded them onto land. Adaptive radiation of the earliest tetrapods gave rise to a diversity of new forms. Many Carboniferous species superficially resembled reptiles. Some reached 4 m in length. Because amphibians were the only vertebrates on land in late Devonian and early Carboniferous times, the age of amphibians is an appropriate name for the Carboniferous period. Amphibians began to decline during the late Carboniferous. As the Mesozoic era dawned with the Triassic period, about 230 million years ago, most survivors of the amphibian lineage resembled modern species.

Modern Amphibians

There are three extant orders of amphibians (FIGURE 30.15): Urodela ("tailed ones"—salamanders); Anura ("tail-less ones"—frogs, including toads); and Apoda ("legless ones"—caecilians).

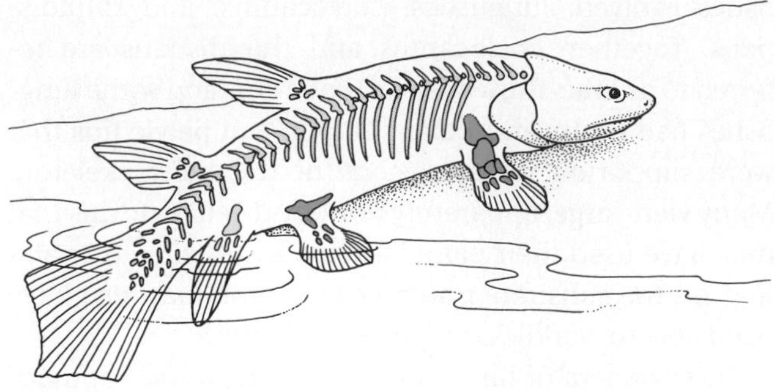

(**a**) Lobe-finned fish

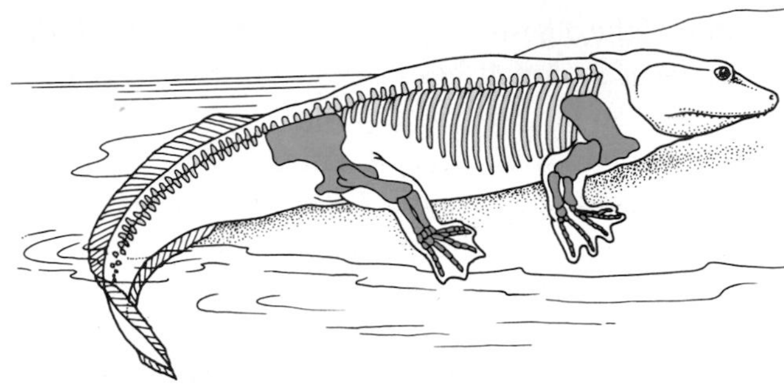

(**b**) Early amphibian

FIGURE 30.14

The origin of tetrapods. (**a**) The ancestor of tetrapods was probably a lobe-finned fish, one of the rhipidistians having muscular fins with extensions of the skeleton, which provided some support for the animal on land. (**b**) This reconstruction, based on Devonian fossils, depicts an early amphibian.

There are only about 400 species of **urodeles.** Some are entirely aquatic, but others live on land as adults or throughout life. Most salamanders that live on land walk with a side-to-side bending of the body that may resemble the swagger of the early tetrapods. Aquatic salamanders swim sinusoidally or walk along the bottom of streams or ponds.

Anurans, numbering nearly 3500 species, are more specialized than urodeles for moving on land. Adult frogs use their powerful hind legs to hop along the terrain. A frog nabs insects by flicking out its long sticky tongue, which is attached to the front of the mouth. Frogs display a great variety of adaptations that help them avoid being eaten by larger predators. In common with other amphibians, many exhibit color patterns that camouflage. The skin glands of frogs secrete distasteful, or even poisonous, mucus. Many poisonous species have bright coloration that apparently warns predators, who associate the coloration with danger (see Chapter 48).

Apodans, the caecilians (about 150 species), are legless and nearly blind, and superficially resemble earthworms. Caecilians inhabit tropical areas where most species burrow in moist forest soil; a few South American apodans live in freshwater ponds and streams.

(a)

(b)

(c)

FIGURE 30.15

Amphibian orders. (**a**) Urodeles (salamanders) retain their tails as adults. Some are entirely aquatic, but others live on land. This is a northern red salamander. (**b**) Anurans, such as this blue poison arrow frog, lack tails as adults. Poison arrow frogs inhabit tropical forests; their skin glands secrete deadly nerve toxins used by Central and South American natives to coat arrow tips. (**c**) Apodans, such as this Sri Lankan caecilian, are legless, mainly burrowing amphibians.

Amphibian means "two lives," a reference to the metamorphosis of many frogs (FIGURE 30.16). The tadpole, the larval stage of a frog, is usually an aquatic herbivore with internal gills, a lateral line system resembling that of fishes, and a long finned tail. The tadpole lacks legs and swims by undulating like its fishlike ancestors. During the metamorphosis that leads to the "second life," legs develop, and the gills and lateral line system disappear. The young tetrapod with air-breathing lungs, a pair of external eardrums, and a digestive system capable of digesting animal protein crawls onto shore and begins life as a terrestrial hunter. In spite of the name amphibian, however, many members of the class, including some frogs, do not go through the aquatic tadpole stage, and many do

(a)

(b)

(c)

FIGURE 30.16

The "dual life" of a frog *(Rana temporaria).* (**a**) The male grasps the female, stimulating her to release eggs. The eggs are laid and fertilized in water. They have a jelly coat but lack shells and would desiccate in air. (**b**) The tadpole is an aquatic herbivore with a fishlike tail and internal gills. (**c**) During metamorphosis, the gills and tail are resorbed, and walking legs develop.

not live a dualistic—aquatic and terrestrial—life. There are strictly aquatic and strictly terrestrial species in all three extant orders. Moreover, salamander and caecilian larvae look much like adults, and typically both the larvae and the adults are carnivorous. Paedogenesis is common among some groups of salamanders; the mudpuppy *(Necturus),* for instance, retains gills and other larval features when sexually mature.

Most amphibians maintain close ties with water and are most abundant in damp habitats such as swamps and rain forests. Even those frogs that are adapted to drier habitats spend much of their time in burrows or under moist leaves, where the humidity is high. Adult amphibians either have small, rather inefficient lungs or lack lungs altogether. Most species rely heavily on their skin to carry out gas exchange with the environment, and amphibians that live on land must cope with the problem of keeping their skin moist, a requirement for gas exchange. Many species also use moist surfaces of the mouth for gas exchange.

The frog egg has no shell, and it dehydrates quickly in dry air. Fertilization is external in most species, with the male grasping the female and spilling his sperm over the eggs as the female sheds them (see FIGURE 30.16a). Amphibians generally lay their eggs in ponds or swamps or at least in moist environments. Some species lay vast numbers of eggs, and mortality is high. Desert frogs often breed explosively in temporary pools. In contrast are species that display various types of parental care and that lay relatively few eggs. Depending on the species, either males or females may incubate eggs on their back, in the mouth, or even in the stomach. Certain tropical tree frogs stir their egg masses into moist foamy nests that resist drying. There are also live-bearing amphibians (ovoviviparous and even some viviparous species) that retain the eggs in the female reproductive tract, where embryos can develop without drying out.

Many amphibians exhibit complex and diverse social behavior, especially during their breeding seasons. Frogs are usually quiet creatures, but many species fill the air with their mating calls during the breeding season. Males may vocalize to defend breeding territory or to attract females. In some terrestrial species, migrations to specific breeding sites may involve vocal communication, celestial navigation, or chemical signaling.

The evolution of the amniotic egg expanded the success of vertebrates on land

The evolution of reptiles from an amphibian ancestor involved many adaptations that we can interpret as specializations for terrestrial living. The amniotic egg, a reproductive adaptation that enabled terrestrial vertebrates to complete their life cycles on land and sever their last ties with their aquatic origins, was a particularly important breakthrough. (Seeds played a similar role in the evolution of land plants, as you learned in Chapter 27.) Amphibians, remember, must deposit their shell-less eggs in water. In contrast, a shell prevents desiccation of the egg, which can therefore be laid in a dry place. Specialized membranes within the egg function in gas exchange, waste storage, and the transfer of stored nutrients to the embryo. These are called **extraembryonic membranes** because they are not part of the body of the developing animal, although they develop from tissue layers that grow out from the embryo. The amniotic egg is named for one of these membranes, the amnion, which encloses a compartment of amniotic fluid that bathes the embryo and acts as a hydraulic shock absorber (FIGURE 30.17).

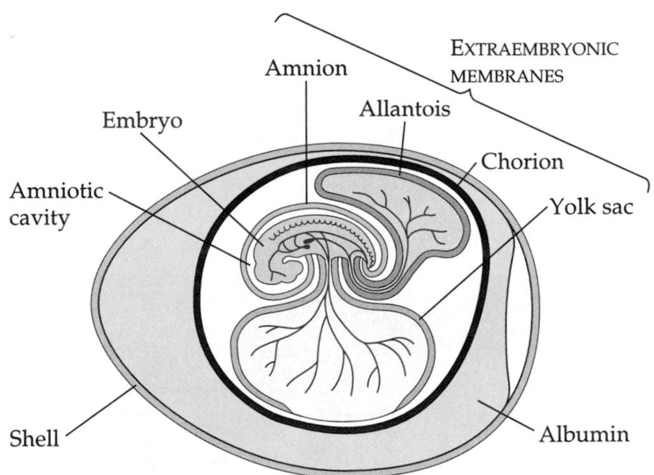

FIGURE 30.17
The amniotic egg. The embryos of reptiles, birds, and mammals form four extraembryonic membranes: the amnion, yolk sac, allantois, and chorion. The amnion protects the embryo in a fluid-filled cavity that prevents dehydration and cushions mechanical shocks. The yolk sac expands over the yolk mass, a stockpile of nutrients stored in the egg. Blood vessels in the yolk sac membrane transport nutrients from the yolk into the embryo. Other nutrients are stored in the albumin (the "egg white"). The allantois functions as a disposal sac for certain metabolic wastes produced by the embryo. The membrane of the allantois also functions with the fourth membrane, the chorion, as a respiratory organ that exchanges gases between the embryo and the surrounding air. Oxygen and carbon dioxide diffuse freely across the egg's shell. However, the shell is waterproof, an adaptation that prevents the embryo and the fluid compartments around it from drying out on land. This drawing represents the extraembryonic membranes in the shelled egg of a reptile or bird. Placental mammals retain the four extraembryonic membranes characteristic of all amniotes, but the embryos develop without shells in the reproductive tract of the mother.

Birds and mammals both evolved from reptiles. The amniotic eggs of birds, like those of reptiles, are shelled. Placental mammals have dispensed with the shell; instead, the embryo implants in the wall of the uterus and obtains its nutrients from the mother. However, the extraembryonic membranes characteristic of the amniotic egg are retained as important structures in the development of mammalian embryos. Together, the reptiles, birds, and mammals make up a monophyletic group, the amniotes. However, we will follow the tradition of dividing the amniotes into three classes: Reptilia, Aves, and Mammalia.

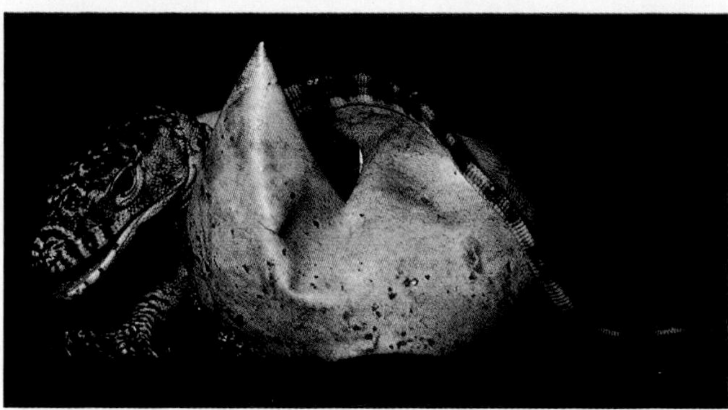

FIGURE 30.18

A hatching reptile. This Komodo dragon is breaking out of a parchmentlike shell, a common type of shell among reptiles. The shells of some reptiles are harder because of an abundance of calcium carbonate.

A reptilian heritage is evident in all amniotes

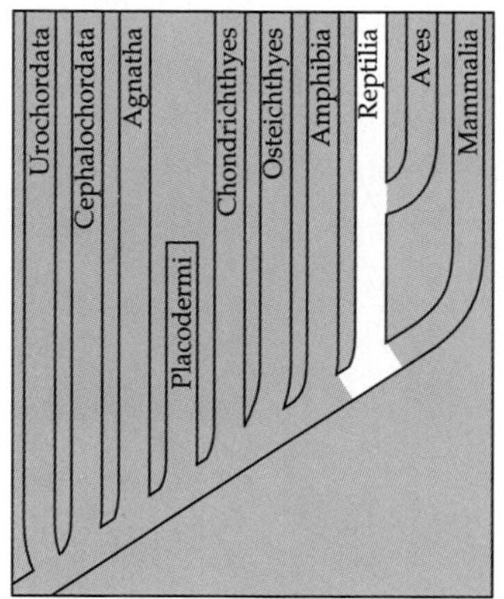

Class Reptilia, a diverse group with many extinct lineages, is represented today by about 7000 species of lizards, snakes, turtles, and crocodilians. Reptiles have several adaptations for terrestrial living not generally found in amphibians. The evolutionary history of reptiles provides a context for understanding not only modern reptiles, but also birds and mammals, the other two classes of amniotes.

Reptilian Characteristics

Scales containing the protein keratin waterproof the skin of a reptile, helping prevent dehydration in dry air. Keratinized skin is the vertebrate analogue of the chitinized cuticle of insects and the waxy cuticle of land plants. Because they cannot breathe through their dry skin, most reptiles obtain all their oxygen with lungs. Many turtles also use the moist surfaces of their cloaca for gas exchange.

Most reptiles lay shelled amniotic eggs on land (FIGURE 30.18). Fertilization in reptiles must occur internally, before the shell is secreted as the egg passes through the reproductive tract of the female. Class Reptilia also includes some species of viviparous ("live-bearing") snakes and lizards. In these reptiles, the extraembryonic membranes form a placenta that enables the embryo to obtain nutrients from its mother.

Reptiles are sometimes labeled "cold-blooded" animals because they do not use their metabolism extensively to control body temperature. But reptiles do regulate body temperature by using behavioral adaptations. For example, many lizards can regulate their internal temperature by basking in the sun when the air is cool and seeking shade when the air is too warm. Because they absorb external heat rather than generating much of their own, reptiles are said to be **ectotherms,** a term more appropriate than cold-blooded. (The control of body temperature will be discussed in more detail in Chapter 40.) By heating directly with solar energy rather than through the metabolic breakdown of food, a reptile can survive on less than 10% of the calories required by a mammal of equivalent size. Having relatively modest food requirements and being adapted to arid conditions, many reptiles thrive in deserts.

During the Mesozoic era, reptiles were far more widespread, numerous, and diverse than they are today. Let's look briefly at this period, known as the age of reptiles.

The Age of Reptiles

The Origin and Early Evolutionary Radiation of Reptiles. The oldest reptilian fossils are found in rocks from the upper Carboniferous period; they are about 300 million years old. Their ancestor was among the Devonian amphibians. In two great waves of adaptive radiation, reptiles became the dominant terrestrial vertebrates in a dynasty that lasted more than 200 million years.

The first major reptilian radiation occurred by the dawn of the Permian, the last period of the Paleozoic era, giving rise to two main evolutionary branches:

1. The **synapsids.** This branch included a diversity of mammal-like reptiles called therapsids, including the probable ancestor of true mammals.
2. The **sauropsids.** This branch gave rise to all modern amniotes *except* the mammals. Relatively early in their history, the sauropsids split into two subbranches:
   a. The **anapsids.** Turtles are the only survivors of this reptilian group.

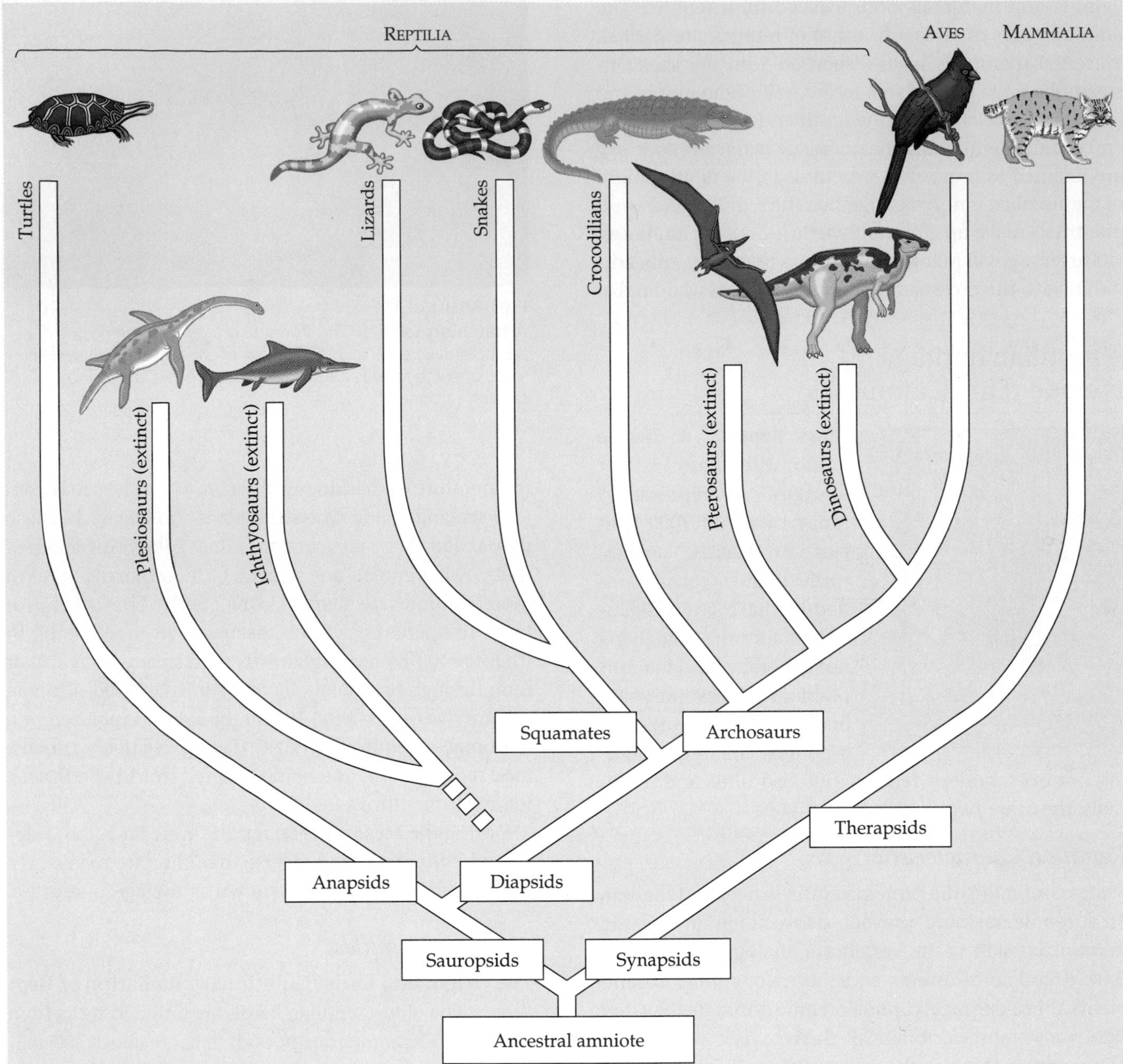

FIGURE 30.19

Phylogeny of the amniotes. The radiation of amniotes during the early Mesozoic era gave rise to a diversity of terrestrial forms, including the dinosaurs. Various reptiles also returned to aquatic habitats (plesiosaurs and ichthyosaurs) and took to the air (pterosaurs). The birds are an extant group of flying "reptiles," although this book follows the taxonomic tradition of placing birds in their own class (Aves). Another class of amniotes, the mammals, evolved from the synapsids, a reptilian branch that diverged from the other reptiles very early. Notice that as a taxonomic cluster, the amniotes are monophyletic; modern reptiles, birds, and mammals all descended from a common ancestor. In Chapter 23, you learned that a monophyletic taxon includes a common ancestor *and all of its descendants*. Thus, the class Reptilia is *not* a monophyletic taxon because it excludes birds and mammals. Zoologists recognize the cladistic inconsistency, but most still find it convenient to classify reptiles, birds, and mammals as three classes of amniotic vertebrates.

b. The **diapsids.** Lizards, snakes, and crocodilians are the extant diapsids classified as reptiles. The dinosaurs and some other groups of extinct reptiles were also diapsids. The closest living relatives of dinosaurs are birds.

FIGURE 30.19 will help you review the evolutionary relationships of amniotes.

Dinosaurs and Their Relatives. The second great reptilian radiation was under way by late Triassic times (a little more than 200 million years ago), and was marked mainly by the origin and diversification of two groups of reptiles: the dinosaurs, which lived on land, and the pterosaurs, or flying reptiles (see FIGURE 30.19). These groups were the dominant vertebrates on Earth for millions of years. Pterosaurs had wings formed from a membrane of skin stretched from the body wall, along the forelimb, to the tip of an elongated finger. Stiff fibers provided support for the skin of the wing. Dinosaurs, an extremely diverse group varying in body shape, size, and habitat, included the largest animals ever to inhabit land. Fossils of gigantic dinosaurs that were 45 m long have recently been unearthed in New Mexico and Utah.

Contradicting the long-standing view that dinosaurs were slow, sluggish creatures, there is increasing evidence that many dinosaurs were agile, fast-moving, and, in some species, social. Paleontologists have also discovered signs of parental care among dinosaurs (FIGURE 30.20). There is continuing debate about whether dinosaurs were **endothermic,** capable of keeping the body warm through metabolism. Some anatomical evidence supports this hypothesis, but many experts remain skeptical. The Mesozoic climate was relatively warm and consistent, and behavioral adaptations such as basking may have been sufficient for maintaining a suitable body temperature, especially for the land-dwelling dinosaurs. Also, large dinosaurs had low surface-to-volume ratios that reduced the effects of daily fluctuations in air temperature on the internal temperature of the animal.

The Cretaceous Crisis. During the Cretaceous, the last period of the Mesozoic era, the climate became cooler and more variable. This was a period of mass extinctions. A quarter of the families of marine invertebrates disappeared, and the Cretaceous crisis was the final curtain for the dinosaurs. Nearly all of them were gone by about the close of the Mesozoic era, some 65 million years ago. (Fossils of some dinosaurs that survived into the early Cenozoic have recently been discovered.) There is evidence that the decline of the dinosaurs occurred over a period of 5 to 10 million years. But even this long a span is relatively sudden, on the scale of geological time, for a group so diverse and dominant to become extinct. (Disappearance of the dinosaurs, one of the most engaging mysteries in the history of life, is discussed in Chapter 23.) The age of reptiles ended when the dinosaurs disappeared, but a few reptilian groups are still successful today.

FIGURE 30.20
Dinosaur social behavior and parental care. This artist's recreation of duck-billed dinosaurs is based on 80-million-year-old fossils discovered in Montana. Dinosaurs of this species were relatively large, about 7 m in length. They used their bills to browse on plants. Fossilized footprints indicate that duck-billed dinosaurs traveled in large social groups, or herds. Fossilized nests contain both eggs and offspring up to a few months old, evidence that these dinosaurs took care of their young. In fact, the genus name for this dinosaur, *Maiasaura,* means "good mother lizard."

Modern Reptiles

The three largest and most diverse extant orders of reptiles are **Chelonia** (turtles), **Squamata** (lizards and snakes), and **Crocodilia** (alligators and crocodiles).

Turtles evolved among the anapsids during the Mesozoic era and have scarcely changed since (FIGURE 30.21a). The usually hard shell, an adaptation that protects against predators, has certainly contributed to this long success. Those turtles that returned to water during their evolution crawl ashore to lay their eggs.

Lizards, which descended from diapsids, are by far the most numerous and diverse reptiles alive today (FIGURE 30.21b). Most are relatively small; perhaps they were able to survive the Cretaceous "crunch" by nesting in

(a)

(b)

FIGURE 30.21

Extant reptiles. (**a**) Turtles have changed little since their origin early in the Mesozoic era. This species is a yellow-bellied turtle. (**b**) Lizards, such as this Australian frillneck, are the most numerous and diverse of the extant reptiles. A frillneck only spreads its frill when threatened, a response that probably discourages many predators. (**c**) Snakes may have evolved from lizards that adapted to a burrowing lifestyle. This is an annulated boa. (**d**) Crocodiles and alligators are the reptiles most closely related to dinosaurs. An alligator is shown here.

(c)

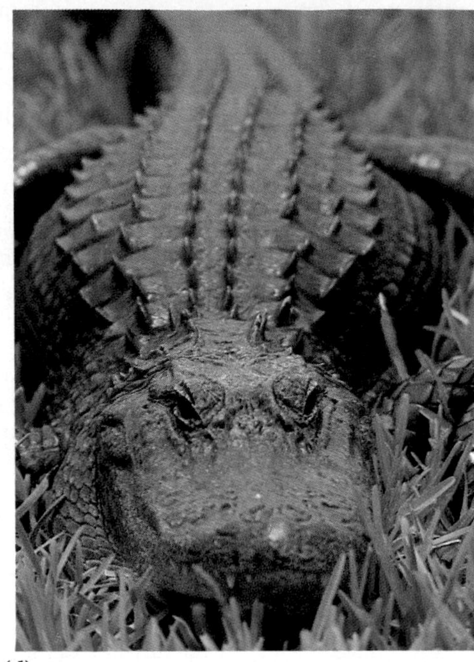

(d)

crevices and decreasing their activity during cold periods, a practice many modern lizards use.

Snakes are apparently descendants of lizards that adopted a burrowing lifestyle (FIGURE 30.21c). Today, most snakes live above the ground, but they have retained the limbless condition. Vestigial pelvic and limb bones in primitive snakes such as boas, however, are evidence that snakes evolved from reptiles with legs.

Snakes are carnivorous, and a number of adaptations aid them in hunting prey. They have acute chemical sensors, and though they lack eardrums, snakes are sensitive to ground vibrations, which helps them detect the movements of prey. Heat-detecting organs between the eyes and nostrils of pit vipers, including rattlesnakes, are sensitive to minute temperature changes, enabling these night hunters to locate warm animals. Poisonous snakes inject their toxin through a pair of sharp hollow or grooved teeth. The flicking tongue is not poisonous but helps fan odors toward olfactory organs on the roof of the mouth. Loosely articulated jaws enable most snakes to swallow prey larger than the diameter of the snake itself.

Crocodiles and alligators, also descendants of diapsids, are among the largest living reptiles (some turtles are heavier) (FIGURE 30.21d). They spend most of their time in water, breathing air through their upturned nostrils. Crocodiles and alligators are confined to the warm regions of Africa, China, Indonesia, India, Australia, South America, and the southeastern United States, where alligators are making a strong comeback after spending years on the endangered species list. Among the modern animals generally classified as reptiles, crocodilians are the most closely related to the dinosaurs. However, the modern animals that share the most recent ancestor with the dinosaurs are the birds.

Birds began as flying reptiles

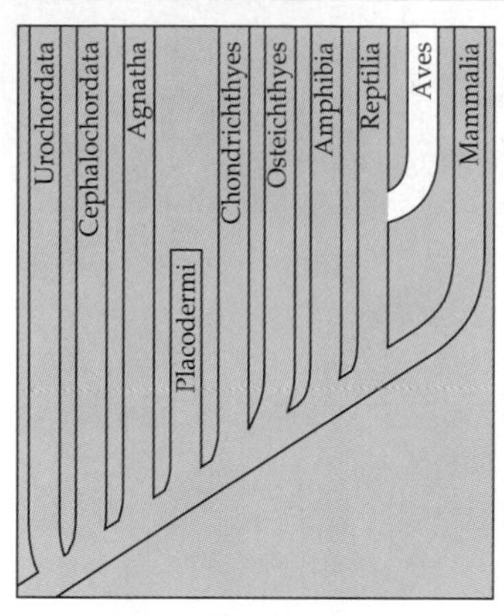

Class Aves (birds) evolved during the great reptilian radiation of the Mesozoic era (see FIGURE 30.19). Amniotic eggs and scales on the legs are just two of the reptilian remnants we see in birds. But birds look quite different from lizards and other modern reptiles because of their feathered wings and other distinctive flight equipment (FIGURE 30.22).

Characteristics of Birds

Almost every part of a typical bird's anatomy is modified in some way that enhances flight. The bones have an internal structure that is honeycombed, making them strong but light (see FIGURE 30.23). The skeleton of a frigate bird, for instance, has a wingspan of more than 2 m but weighs only about 113 g (4 oz). Another adaptation reducing the weight of birds is the absence of some organs. Females, for instance, have only one ovary. Also, modern birds are toothless, an adaptation that trims the weight of the head. Food is not chewed in the mouth but ground in the gizzard, a digestive organ near the stomach. (Crocodiles also have gizzards, as did some dinosaurs.) The bird's bill, made of keratin, has proven to be very adaptable during avian evolution, taking on a great variety of shapes suitable for different diets.

Flying requires a great expenditure of energy from an active metabolism. Birds are endothermic; they use their own metabolic heat to maintain a warm, constant body temperature. Feathers and a layer of fat provide insulation that enables birds to retain their metabolically generated heat. An efficient circulatory system with a four-chambered heart, which segregates oxygenated blood from oxygen-poor blood, supports the high metabolic rate of the bird's cells. The efficient lungs have tiny tubes leading to and from elastic air sacs that help dissipate heat and reduce the density of the body.

For safe flight, senses, especially vision, must be acute. Birds have excellent eyes, perhaps the best of all vertebrates. The visual areas of the brain are well developed, as are the motor areas; flight also requires excellent coordination.

With brains proportionately larger than those of reptiles and amphibians, birds generally display very complex behavior. Avian behavior is particularly intricate during breeding season, when birds engage in elaborate rituals of courtship. Since eggs are shelled when laid, fertilization must be internal. The act of copulation is somewhat awkward because the male of most bird species has no penis. He must climb atop the female's back and then twist her tail so the mates' vents, the openings to their cloacas, can come together. After eggs are laid, the avian embryo must be kept warm through brooding by the mother, father, or both, depending on the species.

A bird's most obvious adaptation for flight is its wings. Bird wings are airfoils that illustrate the same principles of aerodynamics as the wings of an airplane. Providing power for flight, birds flap their wings by contractions of large pectoral (breast) muscles anchored to a keel on the sternum (breastbone). Some birds, such as hawks, have wings adapted for soaring on air currents and flap their wings only occasionally; other birds, including hummingbirds, must flap continuously to stay aloft. In either case, it is the shape and arrangement of the feathers that form the wing into an airfoil.

In being both extremely light and strong, feathers are among the most remarkable of vertebrate adaptations (FIGURE 30.23). Feathers are made of keratin, the same protein that forms our hair and fingernails and the scales of reptiles. Feathers may have functioned first as insulation during the evolution of endothermy, only later being co-opted as flight equipment. Besides providing support and wing shape, feathers can be manipulated to control air movements around the wing.

The evolution of flight required radical alteration in body form, but flight provides many benefits. It enables aerial reconnaissance, which enhances hunting and scavenging, and allows birds to exploit flying insects as an abundant, highly nutritious food resource. Flight also provides ready escape from earthbound predators and

FIGURE 30.22
Flight, avian style. The special flight adaptations associated with birds, such as this bald eagle, are superimposed on ancestral reptilian traits. Birds can reach great speeds and cover enormous distances. The fastest bird is the swift, which can fly 170 km/hr. The bird that travels farthest in its annual migration is the arctic tern, which flies round-trip between the North Pole and South Pole each year.

FIGURE 30.23
Form fits function: the avian wing and feather. A wing is supported by a remodeled version of the tetrapod forelimb. The honeycombed anatomy of the bones (photo, right) is a flight adaptation that reduces weight. A feather consists of a central hollow shaft, from which radiate the vanes. The vanes are made up of barbs, in turn bearing even smaller branches called barbules. Birds have contour feathers and downy feathers. Contour feathers are the stiff ones that contribute to the aerodynamic shapes of the wings and body. Their barbules have hooks that cling to barbules on neighboring barbs. When a bird preens, it runs the length of a feather through its bill, engaging the hooks and uniting the barbs into a precisely shaped vane. Downy feathers lack hooks, and the free-form arrangement of barbs produces a fluffiness that provides excellent insulation because of the trapped air.

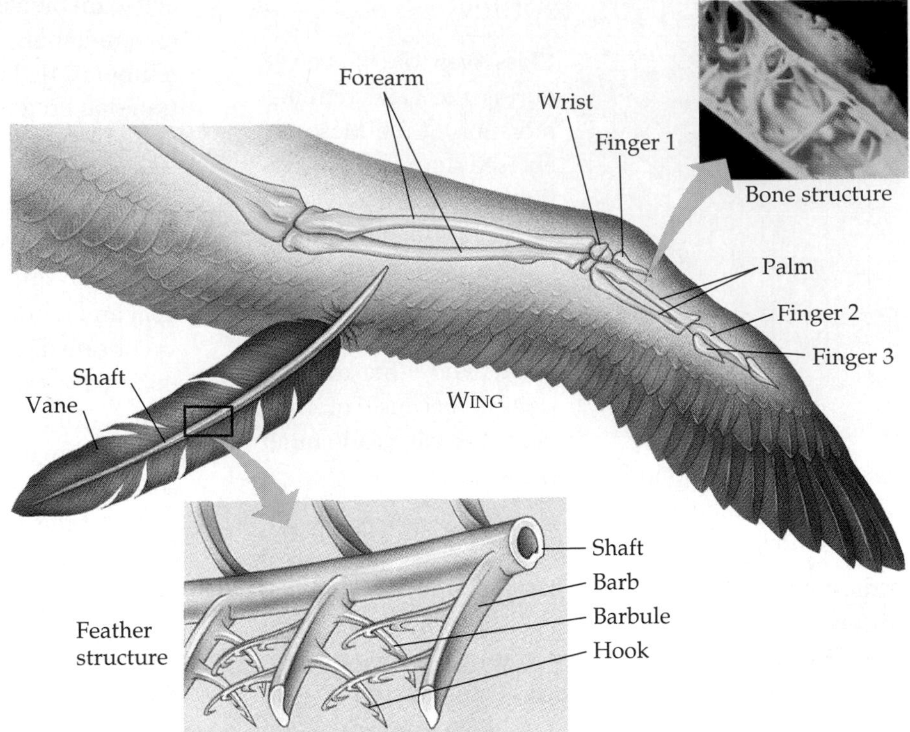

enables some birds to migrate great distances to utilize different food resources and seasonal breeding areas.

The Origin of Birds

For many zoologists, the presence of feathers is enough to classify an animal as a bird. And certainly, if we want to trace the ancestry of birds, we must search for the oldest fossils with feathers. Fossils of an ancient bird named *Archaeopteryx lithographica* have been found in Bavarian limestone in Germany, dating back some 150 million years, into the Jurassic period. Unlike modern birds, but similar to reptiles, *Archaeopteryx* had clawed forelimbs, teeth, and a long tail containing vertebrae (FIGURE 30.24). Indeed, if it were not for the preservation of its feathers, *Archaeopteryx* would be regarded as a member of a diverse group of extinct reptiles called theropods. In common with birds, many ancient reptiles, including some theropods, built nests, cared for their young, and may have been endothermic. In fact, some zoologists advocate a strictly cladistic taxonomy in which modern birds and reptiles are classified in the same vertebrate class.

Archaeopteryx is not considered the ancestor of modern birds, and paleontologists place it on a side branch of the avian lineage. Nonetheless, *Archaeopteryx* probably was derived from ancestral forms that also gave rise to modern birds. Its skeletal anatomy indicates that it was a weak flyer, perhaps mainly a tree-dwelling glider, and this may have been the earliest mode of flying in the bird lineage. Many questions about the origin of birds remain unanswered.

Modern Birds

There are about 8600 extant species of birds classified in about 28 orders. Flight ability is typical of birds, but there are several flightless species, including the ostrich, kiwi, and emu. Flightless birds are collectively called **ratites**

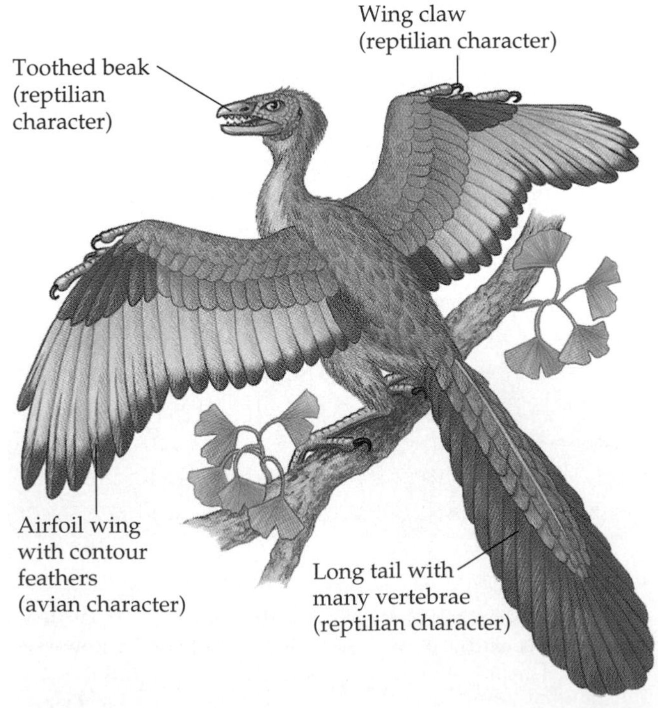

FIGURE 30.24
Archaeopteryx, a Jurassic bird-reptile.

(a)

(b)

(c)

(d)

FIGURE 30.25

Carinate birds. (**a**) In common with many species, the harlequin duck, which inhabits mountain streams and coastal areas of the Pacific Northwest, exhibits pronounced color differences between the sexes. (**b**) Most birds have specific courtship and mating behavior. The western grebe virtually runs on the water surface while courting. (**c**) Hummingbirds can hover while they feed on nectar because they have short, rigid wings that can turn like propellers in virtually any direction. (**d**) The western tanager is a member of the order Passeriformes. Passeriforms are called perching birds because the toes of their feet can lock around a tree branch, enabling the bird to rest in place for long periods of time.

(L., "flat-bottomed") because their breastbone lacks a keel. Ratite birds also lack the large breast muscles that attach to the sternal keel and provide flight power in flying birds.

In contrast to ratites, other birds are called **carinates** because they have a carina, or sternal keel, supporting their large breast muscles. The demands of flight have rendered the body form of carinate birds similar to one another, yet experienced bird watchers can distinguish many species by their body profile. Carinate birds also exhibit great variety in feather colors, bill and foot shape, behavior, and flying style (FIGURE 30.25). Among the most unusual carinate birds are the penguins, which do not fly, but use their powerful breast muscles in swimming. Nearly 60% of the living bird species belong in one carinate order called the **passeriforms,** or perching birds. These are the familiar jays, swallows, sparrows, warblers, and many others.

Mammals diversified extensively in the wake of the Cretaceous extinctions

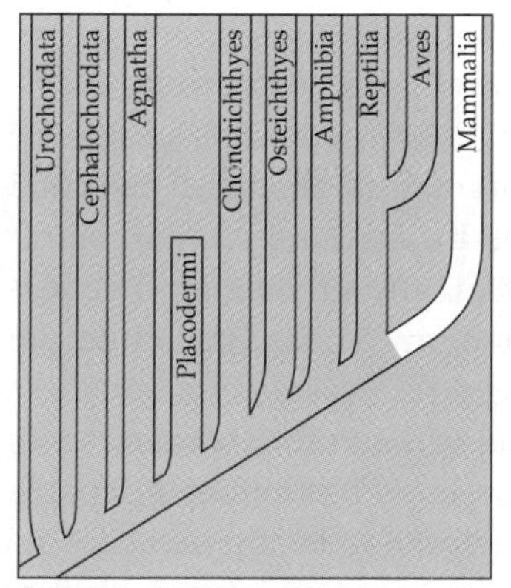

The extinction of the dinosaurs at the close of the Mesozoic era opened many adaptive zones, and mammals underwent an extensive adaptive radiation that filled the void. There are about 4500 species of mammals on Earth today. As mammals ourselves, we naturally have a special interest in this class of vertebrates. Let's examine some of the features we share with all other mammals.

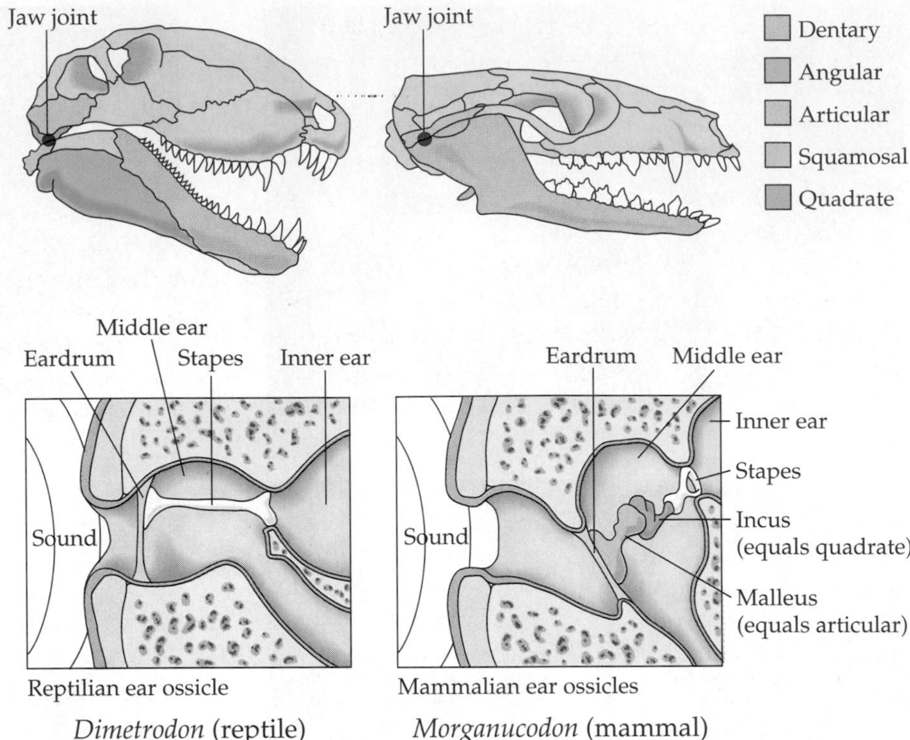

FIGURE 30.26
The evolution of the mammalian jaw and ear bones. The reptile *Dimetrodon* (left) was an early synapsid, the lineage that eventually gave rise to the mammals. *Morganucodon* (right) was one of the first mammals. Notice that the lower jaw of the reptile is composed of several fused bones (top left). In the reptile, two small jawbones, the quadrate and articular, form part of the joint. The mammalian lower jaw is reduced to a single bone, the dentary, and the location of the jaw joint has shifted (top right). During the evolutionary remodeling of the mammalian skull, these jawbones became incorporated into the middle ear (bottom) as two of the three bones that transmit sound from the eardrum to the inner ear. These ear bones enhance hearing by amplifying sound, and the single-bone lower jaw of mammals is stronger than the multibone jaw of reptiles. The steps in this evolutionary remodeling are evident in a succession of fossils of mammal-like reptiles.

Mammalian Characteristics

Vertebrates of **class Mammalia** have hair, a characteristic as diagnostic as the feathers of birds. Hair, like feathers, is made of keratin, although zoologists are uncertain about its evolutionary origin. Also like feathers, hair insulates, helping the animal maintain a warm and constant body temperature. Mammals are endothermic, and their active metabolism is supported by an efficient respiratory system that uses a sheet of muscle called the diaphragm to help ventilate the lungs. The four-chambered heart of a mammal prevents the mixing of oxygen-rich blood with oxygen-poor blood.

Mammary glands that produce milk are as distinctively mammalian as hair. All mammalian mothers nourish their babies with milk, a balanced diet rich in fats, sugars, proteins, minerals, and vitamins.

Most mammals are born rather than hatched. Fertilization is internal, and the embryo develops inside the uterus of the female's reproductive tract. In placental mammals, the lining of the mother's uterus and extraembryonic membranes arising from the embryo collectively form a **placenta,** where nutrients diffuse into the embryo's blood.

Mammals have larger brains than other vertebrates of equivalent size, and they seem to be the most capable learners. The relatively long duration of parental care extends the time for offspring to learn important survival skills by observing their parents.

Differentiation of teeth is another important mammalian trait. Whereas the teeth of reptiles are generally conical and uniform in size, the teeth of mammals come in a variety of sizes and shapes adapted for chewing many kinds of foods. Our own dentition, for example, includes a mixture of knifelike teeth modified for shearing (incisors) and grinding teeth specialized for crushing (molars). Another part of the feeding apparatus, the jaw, was also remodeled during the evolution of mammals from reptiles, and two of the jawbones were incorporated into the mammalian inner ear (FIGURE 30.26).

The Evolution of Mammals

Mammals evolved from reptilian stock even earlier than birds. The oldest fossils believed to be mammalian date back 220 million years, into the Triassic period. The ancestor of mammals was among the mammal-like reptiles known as **therapsids,** part of the synapsid branch of reptilian phylogeny (see FIGURE 30.19). The therapsids disappeared during the reign of the dinosaurs, but their mammalian offshoots coexisted with the dinosaurs throughout the Mesozoic era. Most mesozoic mammals were very small—about the size of shrews—and most probably ate insects. A variety of evidence, such as the size of the eye sockets, suggests that these tiny mammals were nocturnal.

Geologists mark the end of the Cretaceous period as the boundary between the Mesozoic and Cenozoic eras because of the massive changeover in fossils. Mass extinctions and new radiations transformed life in the seas and on the land. Flowering plants became abundant during the Cretaceous and replaced gymnosperms as the dominant land plants in most locations. (Some scientists speculate that this change in vegetation contributed to the decline of the dinosaurs, which may have depended on gymnosperms for food.) As the Cenozoic era dawned, mammals were in the midst of a great adaptive radiation. That diversity is represented today in the three major groups: monotremes (egg-laying mammals), marsupials (mammals with pouches), and placental mammals.

Monotremes

Monotremes—the platypuses and the echidnas (spiny anteaters)—are the only living mammals that lay eggs (FIGURE 30.27a). The egg, which is reptilian in structure and development, contains enough yolk to nourish the developing embryo. Monotremes have hair and produce milk, two of the most important trademarks of Mammalia. On the belly of a monotreme mother are specialized glands that secrete milk. After hatching, the baby sucks the milk from the fur of the mother, who has no nipples. The mixture of ancestral reptilian characters and derived characters of mammals suggests that monotremes descended from a very early branch in the mammalian genealogy. Today, monotremes are found only in Australia and New Guinea.

Marsupials

Opossums, kangaroos, and koalas are examples of **marsupials,** mammals that complete their embryonic development in a maternal pouch called a marsupium (FIGURE 30.27b). The egg contains a moderate amount of yolk that feeds the embryo as it begins its development within the mother's reproductive tract. Marsupials are born very early in their development. A red kangaroo, for instance, is about the size of a honeybee at its birth, just 33 days after fertilization. Its hind legs are merely buds, but the forelimbs are strong enough for the offspring to crawl from the exit of the reproductive tract to the mother's pouch, a journey lasting a few minutes. In the pouch, the newborn fixes its mouth to a teat and completes its development while nursing.

In Australia, marsupials have radiated and filled niches occupied by placental mammals in other parts of the world. Convergent evolution has resulted in a diversity of marsupials that resemble their placental counterparts occupying similar ecological roles. For example, Australia has marsupial "rats" that are very similar in appearance and lifestyle to placental rats.

The opossums of North and South America are the only extant marsupials outside the Australian region, though South America had a diverse marsupial fauna throughout the Tertiary period. The distribution of modern and fossil

(a) A monotreme

(b) A marsupial

(c) A placental mammal

FIGURE 30.27
The three major subclasses of mammals. (**a**) Monotremes, such as this echidna, are the only mammals that lay eggs (inset). Monotremes have hair and milk, but no nipples. (**b**) The young of marsupials, such as this kangaroo, are born very early in their development and finish their growth while nursing from a teat in their mother's pouch. (**c**) In placental mammals, such as this zebra, young develop within the uterus of the mother. There they are nurtured by the flow of maternal blood through the dense capillary network of the placenta, the reddish portion of the afterbirth clinging to this newborn zebra.

TABLE 30.2

Major Orders of Placental Mammals

ORDER	MAIN CHARACTERISTICS	EXAMPLES	
Artiodactyla	Possess hooves with an even number of toes on each foot; herbivorous.	Sheep, pigs, cattle, deer, giraffes	Bighorn sheep
Carnivora	Carnivorous; possess sharp, pointed canine teeth and molars for shearing.	Dogs, wolves, bears, cats, weasels, otters, seals, walruses	Coyote
Cetacea	Marine forms with fish-shaped bodies, paddlelike forelimbs and no hind limbs; thick layer of insulating blubber.	Whales, dolphins, porpoises	Pacific white-sided porpoise
Chiroptera	Adapted for flying; possess a broad skinfold that extends from elongated fingers to body and legs.	Bats	Frog-eating bat
Edentata	Have reduced or no teeth.	Sloths, anteaters, armadillos	Tamandua
Insectivora	Insect-eating mammals	Moles, shrews, hedgehogs	Star-nosed mole

marsupials begins to make sense in the context of plate tectonics and continental drift. According to recent fossil evidence, marsupials probably originated in what is now North America, spreading southward when the land masses were still joined. After the breakup of Pangaea, South America and Australia became island continents, and their marsupials diversified in isolation from the placental mammals that began an adaptive radiation on the northern continents. Australia has not been in contact with another continent since early in the Cenozoic era, about 65 million years ago. The South American fauna, meanwhile, has not remained cloistered. Placental mammals reached South America throughout the Cenozoic. The most important migrations occurred about 12 million years ago and then again about 3 million years ago, when North and South America joined at the Pana-

TABLE 30.2

Major Orders of Placental Mammals, continued			
ORDER	MAIN CHARACTERISTICS	EXAMPLES	
Lagomorpha	Possess chisel-like incisors, hind legs longer than forelegs and adapted for jumping.	Rabbits, hares, pikas	Jackrabbit
Perissodactyla	Possess hooves with an odd number of toes on each foot; herbivorous.	Horses, zebras, tapirs, rhinoceroses	Indian rhinoceros
Primates	Opposable thumb; forward-facing eyes; well-developed cerebral cortex; omnivorous.	Lemurs, monkeys apes, humans	Golden lion tamarin
Proboscidea	Have a long, muscular trunk; thick, loose skin; upper incisors elongated as tusks.	Elephants	African elephant
Rodentia	Possess chisel-like, continuously growing incisor teeth.	Squirrels, beavers rats, porcupines, mice	Red squirrel
Sirenia	Aquatic herbivores; possess finlike forelimbs and no hind limbs.	Sea cows (manatees)	Manatee

manian isthmus. Extensive two-way traffic of animals took place over the land bridge. The biogeography of mammals is another example of the interplay between biological and geological evolution.

Placental Mammals

Placental mammals complete their embryonic development within the uterus, joined to the mother by the placenta (FIGURE 30.27C). Adaptive radiation during late Cretaceous and early Tertiary periods (about 70 to 45 million years ago) produced the orders of placental mammals we recognize today (TABLE 30.2). The phylogenetic relationship between marsupials and placental mammals is somewhat obscure, but scientists believe they are more closely related to each other than either is to the monotremes. Fossil evidence indicates that placentals and marsupials may have diverged from a common ancestor about 80 to 100 million years ago.

The evolutionary relationships among the many orders of placental mammals are, for the most part, also unsettled. Most mammalogists now favor a tentative genealogy that recognizes at least four main evolutionary lines of placental mammals.

One branch of placental mammals consists of the orders Chiroptera (bats) and Insectivora (shrews), which resemble early mammals. Bats, whose forelimbs are modified as wings, probably evolved from insectivores that fed on flying insects. In addition to the insect-eaters, some bat species feed on fruit, whereas others bite mammals and lap up their blood. Most bats are nocturnal.

The second branch started with a lineage of medium-sized herbivores that underwent a spectacular adaptive radiation during the Tertiary period, eventually giving rise to such modern orders as Lagomorpha (rabbits and their relatives); Perissodactyla (odd-toed ungulates, including horses and rhinoceroses; ungulates walk on the tips of toes); Artiodactyla (even-toed ungulates, including deer and swine); Sirenia (sea cows); Proboscidea (elephants); and Cetacea (porpoises and whales).

A third branch in the evolution of placental mammals produced the order Carnivora, which includes cats, dogs, raccoons, skunks, and the pinnipeds (seals, sea lions, and walruses). Carnivores probably first appeared during the early Cenozoic era. Seals and their relatives apparently evolved from middle Cenozoic carnivores that became adapted for swimming.

A fourth and very extensive adaptive radiation of placental mammals produced the primate-rodent complex. Order Rodentia includes rats, squirrels, and beavers, and the order Primates includes monkeys, apes, and humans.

Primate evolution provides a context for understanding human origins

We have now tracked the vertebrate genealogy to the mammalian order that includes *Homo sapiens* and its closest kin. We are primates. To learn what that means, we must trace our ancestry back to the trees, where some of our most treasured traits had their beginnings as adaptations to an arboreal (tree-dwelling) existence.

Evolutionary Trends in Primates

The first primates were probably small arboreal mammals. Dental structure suggests that they descended from insectivores late in the Cretaceous period. A fossil species called *Purgatorius unio,* found in Montana in beds of the Cretaceous/Tertiary boundary, is considered by many authorities to be the oldest known primate.

Thus, by the end of the Mesozoic era, 65 million years ago, our order was already defined by characteristics that had been shaped, through natural selection, by the demands of living in the trees. For example, primates have limber shoulder joints, which make it possible to brachiate (swing from one hold to another). The dexterous hands of primates can hang onto branches and manipulate food. Claws have been replaced by nails in many species, and the fingers are very sensitive. The eyes of primates are close together on the front of the face. The overlapping fields of vision of the two eyes enhance depth perception, an obvious advantage when brachiating. Excellent eye-hand coordination is also important for arboreal maneuvering. Parental care is essential for young animals in the trees. Mammals devote more energy to caring for their young than most other vertebrates, and primates are among the most attentive parents of all mammals. Most primates have single births and nurture their offspring for a long time. Though humans do not live in trees, we retain in modified form many of the traits that evolved there.

Modern Primates

The two suborders of the Primates are the Prosimii and Anthropoidea. The **prosimians** ("premonkeys") probably resemble early arboreal primates (FIGURE 30.28). The lemurs of Madagascar and the lorises, pottos, and tar-

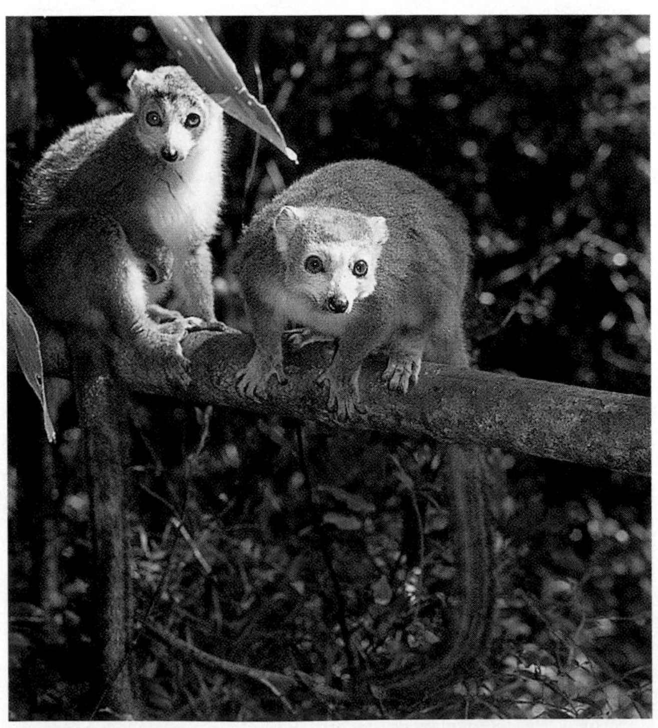

FIGURE 30.28
A crowned lemur, a prosimian.

siers that live in tropical Africa and southern Asia are examples of prosimians. The **anthropoids** include monkeys, apes, and humans.

The evolution of anthropoids from prosimians is a topic of active debate. Paleontologists divide prosimian fossils into two groups: one ancestral to tarsiers; the other ancestral to lemurs, lorises, and pottos (FIGURE 30.29). This split occurred at least 50 million years ago. Until recently, the choice was between these two branches as the lineage that also gave rise to anthropoids. However, paleontologists have recently discovered fossils in Asia and Africa that may be more similar to anthropoids than are either of the other two groups of prosimian fossils. The age of these fossils is still in dispute, but their discoverers date them back at least 50 million years. These discoveries raise the possibility that the prosimians had already split into at least three lineages by 50 million years ago, with the anthropoid ancestry already differentiated from the two branches leading to modern prosimians. Thus, the current debate about anthropoid origins can be phrased this way: Are anthropoids more closely related to tarsiers, to lemurs and lorises, or to some third group of prosimians lacking extant representatives? More fossils will help answer this question.

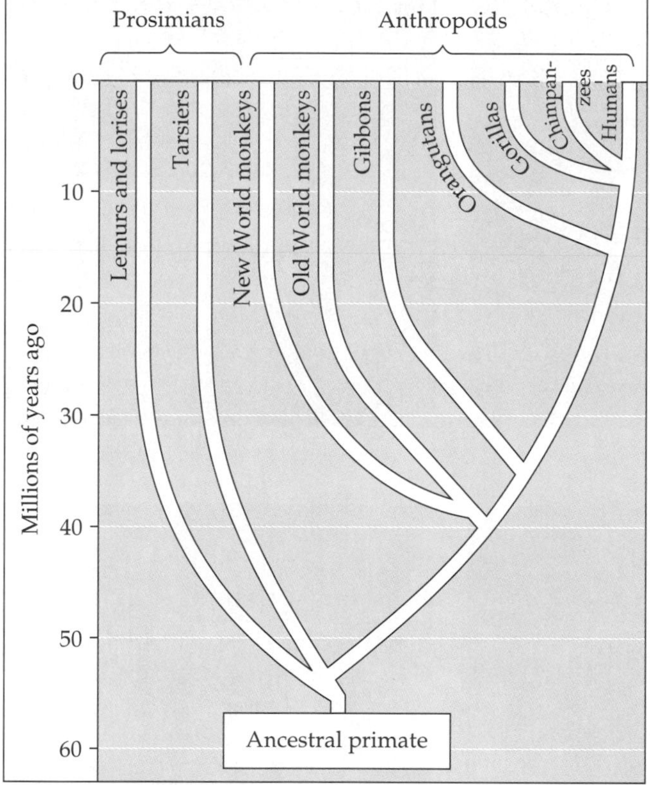

FIGURE 30.29
A hypothetical phylogeny of primates.

Fossils of monkeylike primates indicate that anthropoids were already established in Africa and Asia by 40 million years ago. Africa and South America had already drifted apart, and it is not clear whether ancestors of New World monkeys reached South America by rafting on logs or other debris from Africa or by migration southward from North America. What is certain is that New World monkeys and Old World monkeys have been evolving along separate pathways for many millions of years (FIGURE 30.30). All New World monkeys are arboreal, whereas Old World monkeys include ground-dwelling as well as arboreal species. Most monkeys of both groups are

(a)

(b)

FIGURE 30.30
New World monkeys and Old World monkeys compared. (**a**) New World monkeys, such as spider monkeys, squirrel monkeys, and capuchins, have prehensile tails and nostrils that open to the sides. This is a squirrel monkey. (**b**) Old World monkeys lack prehensile tails, and their nostrils open downward. The tough seat pad is unique to the Old World group, which includes macaques, mandrills, baboons, and rhesus monkeys. These are pig-tailed macaques.

FIGURE 30.31

Apes. (**a**) Gibbons have long arms and are among the most acrobatic of all primates. These Asian primates are also the only monogamous apes. (**b**) The orangutan is a shy and solitary ape that lives in the rain forests of Sumatra and Borneo. Orangutans spend most of their time in trees, but they do venture onto the forest floor occasionally. (**c**) The gorilla is the largest ape, with some males almost 2 m tall and weighing about 200 kg. These herbivores are confined to Africa, where they usually live in small groups of about ten to twenty individuals. (**d**) The chimpanzee lives in tropical Africa. Chimpanzees feed and sleep in trees, but also spend a great deal of time on the ground. Chimpanzees are intelligent, communicative, and social. Some molecular evidence indicates that chimpanzees are more closely related to humans than they are to other apes.

(a) (b) (c) (d)

diurnal (active during the day) and usually live in bands held together by social behavior.

In addition to monkeys, the anthropoid suborder also includes the four genera of apes, shown in FIGURE 30.31: *Hylobates* (gibbons), *Pongo* (orangutans), *Gorilla* (gorillas), and *Pan* (chimpanzees). Modern apes are confined exclusively to tropical regions of the Old World. With the exception of gibbons, modern apes are larger than monkeys, with relatively long arms and short legs and no tails. Although all the apes are capable of brachiation, only gibbons and orangutans preserve a primarily arboreal existence. Social organization varies among the genera of apes; gorillas and chimpanzees are highly social. Apes have larger brains than monkeys, and their behavior is consequently more adaptable.

Humanity is one very young twig on the vertebrate tree

In the continuum of life spanning over 3.5 billion years, humans and apes have shared ancestry for all but the last few million years. **Paleoanthropology,** the study of human origins and evolution, focuses on this tiny fraction of geological time during which humans and chimpanzees diverged from a common ancestor.

Some Common Misconceptions

Paleoanthropology has a checkered history. Until about 20 years ago, researchers often gave new names to fossil forms that were undoubtedly the same species as fossils found by competing scientists. Elaborate theories have often been based on a few teeth or a fragment of jawbone. During the early part of this century, baseless speculations spawned many misconceptions about human evolution that still persist in the minds of much of the general population, long after these myths have been replaced by fossil discoveries.

Let's first dispose of the myth that our ancestors were chimpanzees or any other modern apes. Chimpanzees and humans represent two divergent branches of the anthropoid tree that evolved from a common, less-specialized ancestor.

Another misconception envisions human evolution as a ladder with a series of steps leading directly from an ancestral anthropoid to *Homo sapiens.* This is often illustrated as a parade of fossil hominids (members of the human family) becoming progressively more modern as they march across the page. If human evolution is a parade, then it is a disorderly one, with many splinter groups having traveled down dead ends. At times in hominid history, several different human species coexisted (FIGURE 30.32). Human phylogeny is more like a multibranched bush than a ladder, our species being the tip of the only twig that still lives. If the theory of punctuated equilibrium applies to humans, most change occurred as new hominid species came into existence, not by phyletic (anagenic) change within an unbranched hominid lineage.

One more myth we must bury is the notion that various human characteristics, such as upright posture and an enlarged brain, evolved in unison. A popular image is of early humans as half-stooped, half-witted cave-dwellers. Different features evolved at different rates—a phenomenon known as **mosaic evolution**—with erect posture, or bipedalism, leading the way. Our pedigree includes ancestors who walked upright but had a brain as small as an ape's.

After dismissing some of the folklore on human evolution, however, we must admit that many questions about our ancestry remain.

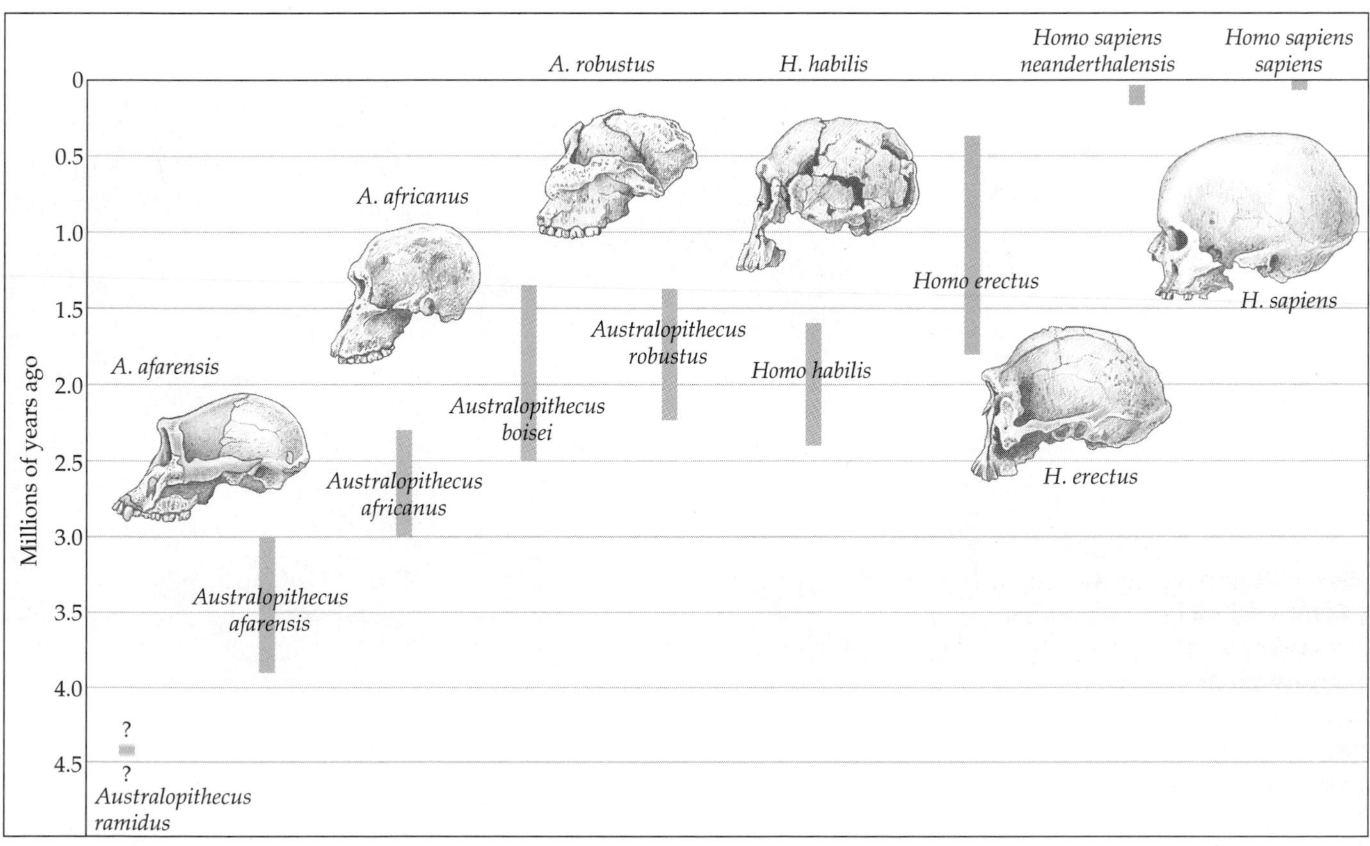

FIGURE 30.32

A timeline of some hominid species. Notice that there have been times in the history of human evolution when two or more hominids coexisted.

Early Anthropoids

The oldest known fossils of apes have been assigned to the genus *Aegyptopithecus*, the "dawn ape." This anthropoid was a cat-sized tree-dweller that lived about 35 million years ago. During the Miocene epoch, which began about 25 million years ago, descendants of the first apes diversified and spread into Eurasia. About 20 million years ago, the Indian plate collided with Asia and thrust up the Himalayan range. The climate became drier and the forests of what is now Africa and Asia contracted, isolating these two regions of anthropoid evolution. Among the African anthropoids was the common ancestor of humans and chimpanzees. Based on the fossil record and comparisons of DNA between humans and chimpanzees, most anthropologists now agree that humans and apes diverged from this common ancestor only about 6 to 8 million years ago.

Australopithecines: The First Humans

In 1924, British anthropologist Raymond Dart announced that a fossilized skull discovered in a South African quarry was the remains of an early human. He named his "ape-man" *Australopithecus africanus* ("southern ape of Africa"). With the discovery of more fossils, it became clear that *Australopithecus* was a legitimate hominid that walked fully erect and had humanlike hands and teeth. However, the brain of *Australopithecus* was only about one-third the size of that of a modern human. Various species of *Australopithecus* existed for over 3 million years, beginning at least 4.4 million years ago.

In 1974, in the Afar region of Ethiopia, paleoanthropologists discovered an *Australopithecus* skeleton that was 40% complete (FIGURE 30.33). "Lucy," as the fossil was named, was petite—only about 1 m tall, with a head about the size of a softball. The skeleton is 3.18 million years old. Lucy and similar fossils have been considered sufficiently different from *Australopithecus africanus* to be named a separate species, *Australopithecus afarensis* (for the Afar region). The skeletons indicate that *A. afarensis* walked on two legs. Fossils discovered in the early 1990s extend the longevity of *A. afarensis* as a species to a span of about 1 million years.

In 1994, paleoanthropologists pushed the fossil record of hominids back a half-million years. At a site only 75 km from where Lucy was discovered, researchers have been unearthing bone fragments from a species of hominid that lived about 4.4 million years ago. Clearly different from *A. afarensis*, this oldest known hominid has been named *Australopithecus ramidus* (which means "root" in the language of the Afar people). A skull fragment indicat-

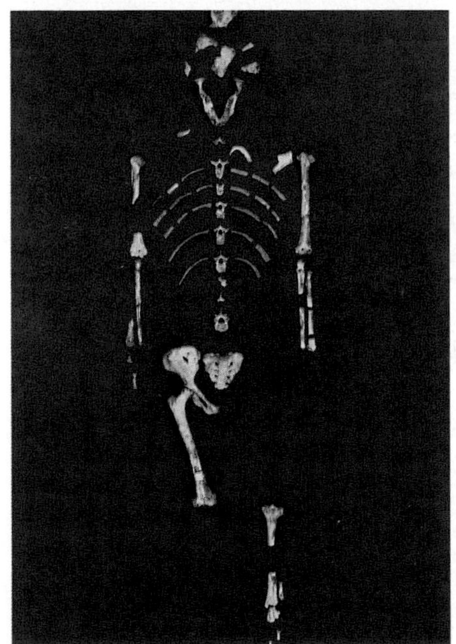

(a)

(b)

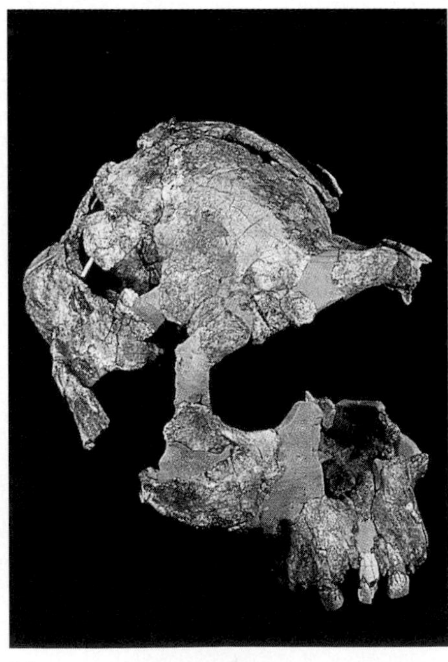

(c)

FIGURE 30.33

The antiquity of upright posture. (**a**) Lucy, a 3.18-million-year-old skeleton, represents the hominid species *Australopithecus afarensis*. Fragments of the pelvis and skull put *A. afarensis* on two feet. (**b**) The Laetoli footprints, over 3.5 million years old, confirm that upright posture evolved quite early in hominid history. (**c**) This skull, discovered in 1993, is among the *A. afarensis* fossils that now span almost a million years, from 3.9 to 3.0 million years ago.

ing that the head balanced on top of the spinal column is among the evidence that upright posture (bipedalism) had already evolved in *A. ramidus.* Skeletons of forest-dwelling animals were found along with the bones, challenging the view that bipedalism evolved when humans began living on the African savanna (grasslands dotted with trees). With the discovery of *A. ramidus,* the fossil record of humanity is creeping closer to the ape-human split estimated to have occurred 6 to 8 million years ago.

The placement of *A. ramidus* and *A. afarensis* on the human bush is stirring much debate. One issue is whether *A. ramidus* is the ancestor of *A. afarensis* or an evolutionary side-branch that faded into extinction. A similar question centers on the evolutionary fate of *A. afarensis.* For the million years that it is represented in the fossil record, *A. afarensis* changed little. Then, beginning about 3 million years ago, an adaptive radiation began that produced several new hominid species, including *A. africanus* and several heavier-boned *Australopithecus* species. This period of speciation also produced *Homo habilis,* the first member of our genus, which debuts in the fossil record about 2.5 million years ago. Was *A. afarensis* ancestral to any or all of these diverse hominids? Or was Lucy a member of a species that branched early from an unknown ancestor that also gave rise to *Homo?*

Whatever the outcome of debates about the phylogeny of early hominids, one fact of our history is clear from the fossil record of the Australopithecines: hominids walked upright for two million years with no substantial enlargement of the brain. Perhaps bipedalism freed the hands for gathering food or caring for babies; the making of tools came much later. Evolutionary biologist Stephen Jay Gould puts it this way: "Mankind stood up first and got smart later."

Homo habilis

Enlargement of the human brain is first evident in fossils dating back to the latter part of australopithecine times, about 2.5 million years ago. Skulls have been found with brain capacities of about 650 cubic centimeters (cc), compared to 500 cc for *A. africanus.* Simple stone tools are sometimes found with the larger-brained fossils. Most paleoanthropologists believe the advances are great enough to place these fossils in the genus *Homo,* naming it *Homo habilis* ("handy man"). After walking upright for at least two million years, hominids were finally beginning to use their brains and hands to fashion tools.

The hominid radiation that included *Homo habilis* was part of a more general burst of speciation among African animals about 2.5 million years ago. Africa's climate became drier during that period, and the transformation of the fauna was associated with adaptation to savannas, which were replacing forests. *Homo habilis* coexisted for a million years with the smaller-brained *Austraopithecus,* including *Australopithecus robustus,* named for its stocky features and heavy skull. Early *Homo* and australopithecines may not have competed directly, although all may have scavenged for food, hunted animals, and gathered fruits and vegetables. According to one hypothesis of human origins, the late australopithecines and *Homo habilis* were distinct lines of hominids, neither evolving from the other. If this scenario is correct, then *A. africanus* was an evolutionary dead end, but *H. habilis* was on the path to modern humans, leading first to *Homo erectus,* which later gave rise to *Homo sapiens.*

Homo erectus *and Descendants*

The first hominid to migrate out of Africa into Asia and Europe was *Homo erectus* ("upright man"). The fossils known as Java Man and Beijing Man are examples of this species. *H. erectus* lived from about 1.8 million years ago until 300,000 years ago. Fossils covering that entire range are found in Africa, where *H. erectus* continued to live contemporaneously with *H. erectus* populations on other continents. We need not picture this migration as a mad dash for new territory, or even a casual stroll. If *H. erectus* simply expanded its range from Africa by about a mile per year, it would take only about 15,000 years to reach Java and other parts of Asia and Europe. The gradual spread may have been associated with a change in diet to include a larger proportion of meat. In general, animals that hunt require more geographical territory than animals that feed exclusively on vegetation.

Homo erectus was taller than *H. habilis* and had a larger brain capacity. During the 1.5 million years the species existed, the *H. erectus* brain increased to as large as 1200 cc, a brain capacity that overlaps the normal range for modern humans.

The intelligence that evolved during the African origins of *H. habilis* enabled these humans to survive in the colder climates of the north, once migration began. *Homo erectus* resided in huts or caves, built fires, clothed themselves in animal skins, and designed stone tools that were more refined than the tools of *Homo habilis.* In anatomical and physiological adaptations, *H. erectus* was poorly equipped for life outside the tropics but made up for the deficiencies with intelligence and social cooperation.

Some African, Asian, European, and Australasian (Indonesia, New Guinea, and Australia) populations of *H. erectus* gave rise to regionally diverse descendants that had even larger brains. Among these descendants of *H. erectus* were the Neanderthals, who lived in Europe, the Middle East, and parts of Asia from about 130,000 years ago to about 35,000 years ago. (They are named Neanderthals because their fossils were first found in the

Neander Valley of Germany.) Compared with us, Neanderthals had slightly heavier brow ridges and less pronounced chins, but their brains, on average, were slightly larger than ours. Neanderthals were skilled toolmakers, and they participated in burials and other rituals that required abstract thought. Much current research on Neanderthal skulls addresses an intriguing question: Did Neanderthals have the anatomical equipment necessary for speech?

The oldest known post–*H. erectus* fossils, dating back over 300,000 years, are found in Africa. Many paleoanthropologists group these African fossils along with Neanderthals and various Asian and Australasian fossils as the earliest forms of our species, *Homo sapiens*. These regionally diverse descendants of *H. erectus* are sometimes referred to as "archaic *Homo sapiens*."

The Emergence of Homo sapiens: *Out of Africa . . . But When?*

What was the fate of Neanderthals and their contemporaries who populated different parts of the world? In the view of some anthropologists, these archaic *Homo sapiens* gave rise to fully modern humans. According to this model, called the *multiregional model,* modern humans evolved in parallel in different parts of the world. If this view is correct, then the geographic diversity of humans originated relatively early, when *Homo erectus* spread from Africa into the other continents between 1 and 2 million years ago. The model accounts for the great genetic similarity of all modern people by pointing out that occasional interbreeding among neighboring populations has always provided corridors for gene flow throughout the entire geographical range of humanity.

Based on alternative interpretations of the fossil record, some paleoanthropologists, including Christopher Stringer of University College, London, began to question the multiregional model during the past decade. The debate has focused partly on the relationship between Neanderthals and the modern humans of Europe and the Middle East. Perhaps the most famous fossils of modern *Homo sapiens*—skulls and other bones that look essentially like those of today's humans—are the remains of Cro-Magnons. Named for the French cave in which these fossils were first discovered, Cro-Magnons date back about 35,000 years. But the oldest fully modern fossils of *Homo sapiens,* about 100,000 years old, are found in Africa; similar fossils almost as ancient have also been discovered in caves in Israel. The Israeli fossils were found not far from other caves containing Neanderthal-like fossils ranging in vintage from about 120,000 to 60,000 years old. These two human types apparently coexisted in this region for at least 40,000 years, from the time modern forms appeared 100,000 years ago. And, according to Stringer and his supporters, the persistence of distinct Neanderthal and modern forms during their time of coexistence means that these two human types did not interbreed. If this interpretation of the Israeli fossils is correct, then the Neanderthals of that region could not have been the ancestors of the modern humans that also lived there. In fact, based on the Israeli bones and other fossil data, Stringer postulates that Neanderthals and the other so-called archaic *Homo sapiens* outside of Africa were evolutionary dead ends.

Thus, Stringer is among the anthropologists who contend that modern humans evolved from *Homo erectus* first in Africa, then migrated to other continents, replacing Neanderthals and the other regional descendants of *H. erectus*. This model, called the "out of Africa" or *monogenesis model,* contrasts sharply with the multiregional model. According to the monogenesis model, modern humanity did not emerge in many different parts of the world, but only in Africa, from which modern humans dispersed relatively recently. If this view is correct, then the geographical diversification of modern humans occurred within just the past 100,000 years (compared to at least 1 million years for the multiregional model). To Stringer, the strongest argument for an exclusively African genesis for modern humans is that only African fossils chronicle the complete transition from archaic *Homo sapiens* to modern humans.

In the late 1980s, Rebecca Cann and other geneticists in the laboratory of the late Allan Wilson at the University of California, Berkeley, made news with research results that seemed to support the monogenesis model of human origins. Instead of digging in the field for bones, Wilson's group probed in the DNA of living humans for clues about our ancestry. Specifically, these scientists compared the mitochondrial DNA (mtDNA) of a multiethnic sample of more than 100 people representing four continents. The greater the difference between the mtDNA of two people, the longer ago that mtDNA diverged from a common source. Using computers to analyze their data, the geneticists traced the source of all human mtDNA back to Africa. The surprise was the calculation that the divergence of mtDNA from this common source began just 200,000 years ago, much too late to represent the dispersal of *Homo erectus*. Advocates of the monogenesis model cheered the genetic evidence for a second, much later dispersal of Africans in the form of modern humans.

In 1992, several researchers challenged the Berkeley group's interpretation of the mtDNA data. At issue are the methods used to construct cladograms from the mtDNA comparisons and the reliability of changes in mtDNA as a molecular clock for timing branch points in these trees (see Chapter 23). The criticism has encouraged proponents of the multiregional model, who continue to argue that multiregional evolution of modern humans fits the fossil evidence better than does African monogenesis. These scientists interpret certain fossils in various parts of the world as links between the regional versions of archaic *Homo sapiens* and the current people native to those continents.

The debate is being reheated by comparisons of nuclear DNA, and new methods for using the mtDNA data to trace the relationships of populations are again strengthening the case for a spread of modern humans out of Africa. For example, researchers are finding greater genetic diversity within African populations south of the Sahara than in other parts of the world. This result is predicted by the hypothesis that modern humans originated in southern Africa and therefore have the longest history of genetic diversification in that region. If populations in other regions began by later migration, the founder effect and genetic drift (see Chapter 21) could account for the smaller genetic diversity.

The debate about the origin of modern humans continues (FIGURE 30.34). We can rephrase the problem this way: Are all contemporary people more closely related to one another than any population native to a particular continent is related to the archaic *Homo sapiens* that lived in that part of the world? The multiregional and monogenesis models predict opposite answers to this question. The ability to test these predictions may actually be within reach. Molecular systematists are now trying to recover DNA samples from fossilized skulls of Neanderthals and other archaic *H. sapiens*. The next step would be to use polymerase chain reaction (PCR) to amplify the DNA traces, producing enough for comparison to the DNA of living humans. Molecular biology and paleontology are converging on the question of human origins. *Wherever, whenever,* and *however* modern humanity evolved, it eventually transformed environments in all parts of the world.

Cultural Evolution: A New Force in the History of Life

An erect stance was the most radical anatomical change in our evolution; it required major remodeling of the foot, pelvis, and vertebral column. Enlargement of the brain was a secondary alteration made possible by prolonging the growth period of the skull and its contents. The brains

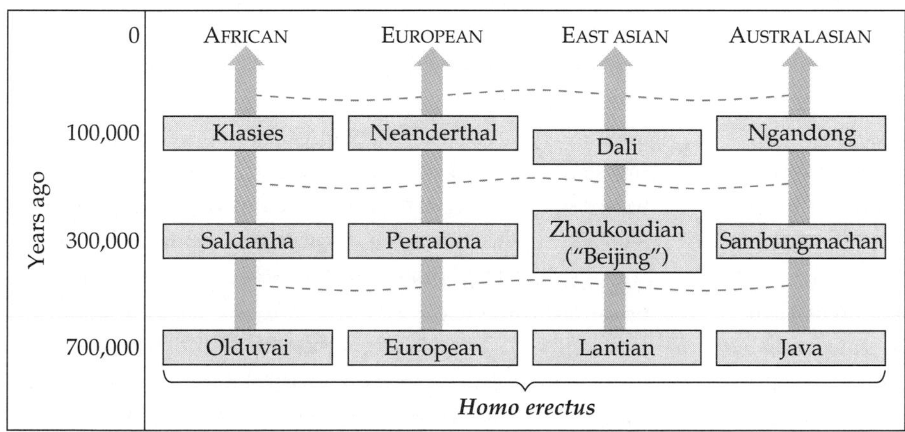

(a) Multiregional model

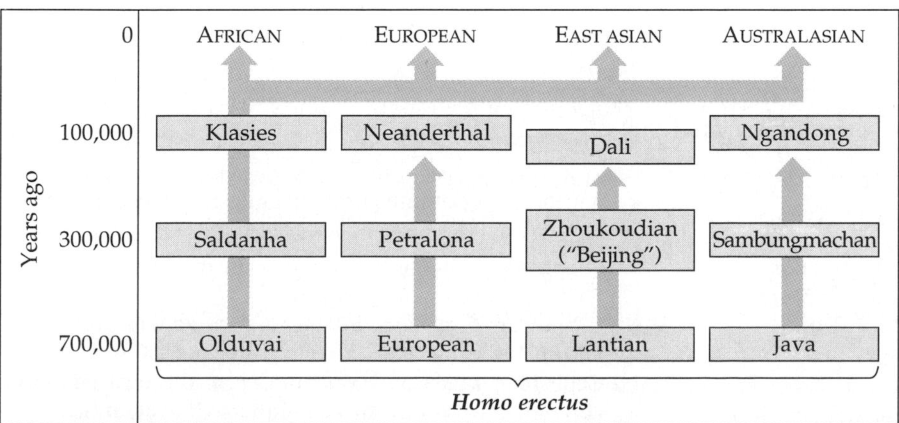

(b) Monogenesis model

FIGURE 30.34
Two models for the origin of modern humans. (**a**) According to the multiregional model, modern humans evolved in many parts of the world from regional descendants of *Homo erectus,* who dispersed from Africa between 1 and 2 million years ago. The boxed names indicate various fossils associated with each region. The dashed lines symbolize interbreeding and gene flow between different populations. (**b**) According to the monogenesis model, only the African descendants of *Homo erectus* gave rise to modern humans. All other regional descendants of *H. erectus,* including Neanderthals, became extinct without contributing to the gene pool of modern humanity. Advocates of the monogenesis model argue that modern humans began spreading from Africa just 100,000 years ago, giving rise to all the diverse populations of contemporary humans.

of nonprimate mammalian fetuses grow rapidly, but the growth usually slows down and stops not long after birth. The primate brain continues to grow after birth, and the period of growth is longer for a human than for any other primate. The extended period of human development also lengthens the time parents care for their offspring, which contributes to the child's ability to benefit from the experiences of earlier generations. This is the basis of culture—the transmission of accumulated knowledge over the generations. The major means of this transmission is language, written and spoken.

Cultural evolution is continuous, but there have been three major stages. The first stage began with nomads who hunted and gathered food on the African grasslands 2 million years ago. They made tools, organized communal activities, and divided labor. The second stage came with the development of agriculture in Africa, Eurasia, and the Americas about 10,000 to 15,000 years ago. Along with agriculture came permanent settlements and the first cities. The third major stage in our cultural evolution was the Industrial Revolution, which began in the eighteenth century. Since then, new technology has escalated exponentially; a single generation spanned the flight of the Wright brothers and Neil Armstrong's walk on the moon. Through all this cultural evolution, from simple hunter-gatherers to high-tech societies, we have not changed biologically in any significant way. We are probably no more intelligent than our forebears in Africa and in Eurasian caves. The same toolmaker who chipped away at stones now fashions microchips. The know-how to build skyscrapers, computers, and spaceships is stored not in our genes but in the cumulative product of hundreds of generations of human experience, passed along by parents, teachers, and books.

Evolution of the human brain may have been anatomically simpler than acquiring an upright stance, but the consequences of cerebral expansion have been enormous. Cultural evolution made *Homo sapiens* a new force in the history of life—a species that could defy its physical limitations and shortcut biological evolution. We do not have to wait to adapt to an environment through natural selection; we simply change the environment to meet our needs. We are the most numerous and widespread of all large animals, and everywhere we go, we bring environmental change. There is nothing new about environmental change. As we have seen throughout this unit, the history of life is the story of biological evolution on a changing planet. But it is unlikely that change has ever been as rapid as in the age of humans. Cultural evolution outpaces biological evolution by orders of magnitude. We may be changing the world faster than many species can adapt; the rate of extinctions in this century is fifty times greater than the average for the past 100,000 years.

This rapid rate of extinction is mainly a result of habitat destruction and chemical pollution, both functions of human cultural changes and overpopulation. Feeding, clothing, and housing the nearly 6 billion people who now exist impose an enormous strain on Earth's capacity to sustain life. If all these people suddenly assumed the high standard of living enjoyed by many people in developed nations, it is likely that Earth's support systems would be overwhelmed. Already, for example, current rates of fossil-fuel consumption, mainly by developed nations, are so great that waste carbon dioxide may be causing the temperature of the atmosphere to increase enough to alter world climates. Today, not just individual species, but entire ecosystems, the global atmosphere, and the oceans, are seriously threatened. Tropical rain forests, which play a vital role in moderating global weather, are being cut down at a startling rate. Scientists have hardly begun to study these ecosystems, and many species in them may become extinct before they are even discovered.

Of the many crises in the history of life, the impact of one species, *Homo sapiens*, is the latest and potentially the most devastating.

REVIEW OF KEY CONCEPTS (with page numbers and key figures)

- Vertebrates belong to the phylum Chordata (pp. 628–629, FIGURE 30.1)
  - Chordates are a diverse phylum of deuterostomes, classified together by virtue of shared structures: a notochord; a dorsal, hollow nerve cord; pharyngeal slits; and a postanal tail.
- Invertebrate chordates provide clues to the origin of vertebrates (pp. 629–631; FIGURES 30.2, 30.3)
  - Subphylum Urochordata includes the marine, suspension-feeding tunicates, or sea squirts.
  - Subphylum Cephalochordata is an invertebrate group best known for the lancelet, an exemplary chordate.
  - Cephalochordates and vertebrates probably evolved from a common ancestor by the elaboration of larval forms.

- The evolution of vertebrate characteristics is associated with increased size and activity (pp. 631–632)
  - One hallmark of the vertebrates is cephalization and the development of highly specialized brains.
  - Vertebrates possess a cranium and a column of vertebrae that encloses the central nervous system (the brain and spinal cord).
  - Vertebrates have anatomical and physiological adaptations that support an active metabolism.
  - A mass of embryonic cells called the neural crest gives rise to many of the uniquely vertebrate characteristics.
- Vertebrate diversity and phylogeny: *an overview* (pp. 632–634, FIGURE 30.6)
  - Extant vertebrates include three classes of fishes and four classes of tetrapods.
- Agnathans are jawless vertebrates (pp. 634–635, FIGURE 30.7)
  - The oldest vertebrate fossils are jawless agnathans. Today, the class Agnatha is represented only by lampreys and hagfishes.
- Placoderms were armored fishes with jaws and paired fins (pp. 635–636, FIGURE 30.8)
  - Placoderms, now extinct, were Devonian fishes with hinged, biting jaws that evolved from the skeletal supports of gills.
- Sharks and their relatives have adaptations for powerful swimming (pp. 636–637, FIGURE 30.9)
  - Sharks, rays, and chimaeras—the class Chondrichthyes—have paired fins, cartilaginous skeletons, and biting jaws characteristic of the group.
- Bony fishes are the most abundant and diverse vertebrates (pp. 637–639, FIGURE 30.13)
  - Osteichthyes is the most species-rich vertebrate class. The bony fishes have skeletons reinforced by calcium phosphate.
  - Bony fishes can adjust their density and thus control their buoyancy by means of a swim bladder.
  - Most fish familiar to us are ray-fins. Supported by long, flexible rays, their fins are modified for various functions.
  - Most fleshy-finned fishes remained in fresh water and used lungs to aid their gills in breathing. Devonian lobe-finned fishes gave rise to tetrapods.
- Amphibians are the oldest class of tetrapods (pp. 640–642, FIGURE 30.15)
  - The first terrestrial vertebrates, amphibians radiated during early Carboniferous times.
  - The lifestyles and life cycles of modern amphibians attest to their aquatic heritage. A moist skin complements the lungs in gas exchange. Most species lay their unshelled eggs in wet environments. Most frogs and their relatives undergo metamorphosis of an aquatic larval stage into a terrestrial adult.
  - The extant amphibian orders are the urodeles (salamanders), anurans (frogs), and apodans (legless, burrowing amphibians).
- The evolution of the amniotic egg expanded the success of vertebrates on land (pp. 642–643, FIGURE 30.17)
  - The amniotic egg is a shelled egg with extraembryonic membranes and fluids that protect and hydrate the embryo. This adaptation, characteristic of reptiles, birds, and mammals, made it possible for vertebrates to reproduce on land.
- A reptilian heritage is evident in all amniotes (pp. 643–646, FIGURE 30.19)
  - Reptiles, a diverse group represented today by lizards, snakes, turtles, and crocodilians, have numerous terrestrial adaptations not present in amphibians.
  - Reptiles have lungs, are covered with waterproof scales, and have a shelled amniotic egg that permits development in a dry environment.
  - The first major reptilian radiation, during the Permian, produced three major branches: the anapsids, represented today only by turtles; the diapsids, which gave rise to lizards, snakes, crocodilians, dinosaurs, and birds; and the synapsids, which gave rise to mammals.
  - The second great reptilian radiation produced flying pterosaurs and terrestrial dinosaurs from Triassic diapsids.
  - The major reptilian orders that survived the Cretaceous crisis are the Chelonia, Squamata, and Crocodilia.
  - Chelonians, the turtles, are hard-shelled reptiles.
  - Order Squamata includes lizards and snakes. Lizards are the most numerous and diverse reptiles today. Ancestral burrowing species gave rise to snakes.
  - Crocodilians, the crocodiles and alligators, are close relatives of the dinosaurs.
- Birds began as flying reptiles (pp. 647–649, FIGURES 30.23–30.25)
  - The amniotic eggs and scaled legs of birds give testimony to their reptilian heritage; birds and dinosaurs descended from a common ancestor.
  - Almost every part of avian anatomy enhances flight. A body of low density is the result of honeycombed bones and the absence of certain bilateral organs.
  - A heart that completely separates oxygen-poor from oxygen-rich blood and a highly efficient lung and air sac system make possible the active, endothermic metabolism of birds.
  - Vision has reached its peak development in birds. Birds also have relatively large brains and display complex behavior.
  - Only birds have feathers, which shape wings into airfoils that provide both propulsion and lift.
- Mammals diversified extensively in the wake of the Cretaceous extinctions (pp. 649–654, FIGURE 30.27)
  - Hair and mammary glands are the two diagnostic characteristics of mammals. Hair helps insulate the body.
  - An active endothermic metabolism is supported by a four-chambered heart and respiratory system ventilated by a muscular diaphragm.
  - Of all vertebrates, mammals are endowed with the largest brains and the greatest capability of learning.
  - Small, insect-eating mammals arose from the therapsids (a group of synapsids) and coexisted with the dinosaurs throughout the Mesozoic era. Extinction of the dinosaurs reopened many adaptive zones and catalyzed an extensive mammalian radiation.

- Monotremes are egg-laying mammals, today represented only by the platypuses and echidnas found in Australia and New Guinea.
- Marsupials include opossums, kangaroos, and koalas, animals whose young complete their embryonic development inside a maternal pouch, the marsupium. Marsupials isolated in Australia show convergent evolution with placental mammals in other parts of the world.
- The most widespread and diverse modern mammals are the placentals, a group whose young complete their embryonic development attached to a placenta inside the mother's uterus.

- Primate evolution provides a context for understanding human origins (pp. 654–656, FIGURE 30.29)
  - The first primates were probably small arboreal animals, which descended from insectivores in the late Cretaceous period. Life in trees demanded limber shoulder joints; dexterous, sensitive hands and fingers; close-set eyes with overlapping fields of vision for depth perception; excellent eye-hand coordination; and extended parental care—all of which extant primates retain.
  - Two subgroups of modern primates are the prosimians—lemurs and their relatives—and the anthropoids. Anthropoids diverged early into New World and Old World monkeys. Modern apes—gibbons, orangutans, gorillas, and chimpanzees—are confined to the Old World.

- Humanity is one very young twig on the vertebrate tree (pp. 656–662, FIGURE 30.32)
  - The first humanlike fossil to be discovered was *Australopithecus africanus*, a hominid with a small brain that walked erect. Subsequent discovery of *A. afarensis* and *A. ramidus* confirmed the early evolution of upright posture in the hominids.
  - Enlargement of the brain and the appearance of simple stone tools occurred about 2.5 million years ago, as indicated by fossils of *Homo habilis*. This species coexisted with the smaller-brained *Australopithecus* and led to the evolution of *Homo erectus*.
  - *Homo erectus* was the first hominid to venture out of the tropics and into colder climates.
  - *Homo erectus* gave rise to regionally diverse versions of archaic *Homo sapiens*, including Neanderthals.
  - According to the multiregional model, modern humans evolved in several locations from Neanderthals and other archaic *Homo sapiens*. In contrast, the monogenesis model views all but the African archaic *Homo sapiens* as evolutionary dead ends. According to this model, a relatively recent dispersal (100,000 years ago) of modern Africans gave rise to today's human diversity.
  - Enlargement of the human brain and its extended period of development, which requires long periods of parental care, gave rise to language and the far-reaching consequences of culture. The first stage of cultural evolution, which began in Africa with wandering hunter-gatherers, progressed to the development of agriculture about 15,000 years ago. The third stage of cultural evolution was the Industrial Revolution, which today continues as accelerating technological change. The exploding human population now threatens Earth's ecosystems.

SELF-QUIZ

1. Vertebrates and sea stars may seem as different as two animal groups can be, yet they share
   a. the same type of body symmetry as adults
   b. a high degree of cephalization
   c. certain developmental features, including the type of coelom formation
   d. an endoskeleton that includes a cranium
   e. the presence of a notochord; a dorsal, hollow nerve cord; and pharyngeal slits.

2. Which of the following groups is entirely extinct?
   a. cephalochordates
   b. agnathans
   c. placoderms
   d. fleshy-finned fishes
   e. ratite birds

3. The amniotic egg first evolved in
   a. bony fishes
   b. amphibians
   c. reptiles
   d. birds
   e. mammals

4. Mammals and birds share all of the following characteristics *except*
   a. endothermy
   b. descent from reptiles
   c. a four-chambered heart
   d. teeth specialized for diverse diets
   e. the ability of some species to fly

5. If you were to observe a monkey in a zoo, which characteristic would indicate a New World origin for that monkey species?
   a. distinct "seat pads"
   b. eyes close together on the front of the skull
   c. use of the tail to hang from a tree limb
   d. occasional bipedal walking
   e. downward orientation of the nostrils

6. Only an animal species with a diaphragm can be expected to have
   a. hair
   b. feathers
   c. scales
   d. lungs
   e. moist skin

7. Unlike placental mammals, both monotremes and marsupials
   a. lack nipples
   b. have some embryonic development outside the mother's uterus
   c. lay eggs
   d. are found in Australia and Africa
   e. include only insectivores and herbivores

8. Which of the following is *not* thought to be ancestral to humans?
   a. a reptile
   b. a bony fish
   c. a primate
   d. an amphibian
   e. a bird

9. As humans diverged from other primates, which of the following appeared first?
   a. the development of culture
   b. language
   c. an erect stance
   d. toolmaking
   e. an enlarged brain

10. The multiregional and monogenesis hypotheses for the origin of modern humans agree that
   a. *Homo erectus* had an African origin
   b. modern *Homo sapiens* originated only in Africa
   c. Neanderthals are the ancestors of modern humans in Europe
   d. Australopithecines migrated out of Africa
   e. North America had the first population of modern humans

CHALLENGE QUESTIONS

1. Compare and contrast bats and birds, two flying vertebrates, in as many ways as you can.

2. Throughout chordate evolution, many existing structures have been co-opted for new functions. Describe several such cases of preadaptation.

3. Evidence supports the hypothesis that alligators, dinosaurs, mammals, and birds are more closely related to each other than any of these groups is to turtles. If we seek to have taxonomic groups that reflect evolutionary relationships—all members of a group having a common ancestor and the group including all the descendants of that common ancestor—how should we revise the classification of vertebrates?

4. Write a short essay comparing and contrasting vertebrates with invertebrate chordates.

SCIENCE, TECHNOLOGY, AND SOCIETY

1. There are 90 million acres in the U.S. National Wildlife Refuge System, representing most of the habitats within the United States and protecting thousands of animal species. Many refuges permit logging, grazing, farming, mineral exploration, and recreational activities. Each year 400,000 animals are legally taken from refuges by hunters and trappers. Pressure is mounting to permit oil and gas exploration in the Arctic National Wildlife Refuge of Alaska. What level of human activity do you think should be permitted on wildlife refuges? Should they be open to "consumptive" uses, such as logging, farming, and hunting? Or should refuges be sanctuaries where animals are touched as little as possible by human activities? Give reasons for your answers.

2. A lot of money and attention have been directed toward saving a few endangered animals, such as the black-footed ferret and the California condor. (Millions of dollars have been spent in efforts to maintain and breed these two species—more than has been spent on all endangered invertebrates, reptiles, and plants combined.) Condors and ferrets have recently been returned to the wild, but it is questionable whether they can withstand the human encroachments that endangered them in the first place. Should we work so hard to save these animals? Why or why not? Would it be better to use our resources to save more species that have a better chance of survival? What are some biological arguments for focusing preservation efforts on whole ecosystems instead of individual endangered species?

FURTHER READING

Blaustein, A. R., and D. B. Wake. "The Puzzle of Declining Amphibian Populations." *Scientific American,* April 1995. Is it habitat destruction or damage to the ozone shield in the atmosphere that's causing loss of amphibians?

Bunney, S. "Most Ancient Human Came from Afar." *New Scientist,* October 1, 1994. The discovery of the oldest known hominid.

Carroll, R. L. *Vertebrate Paleontology and Evolution.* New York: W. H. Freeman, 1987. An authoritative yet accessible text.

Clotles, J. "Rhinos and Lions and Bears (Oh My!)" *Natural History,* May 1995. Recently discovered cave art created by very artistic humans 20,000 years ago.

Coppens, Y. "The East Side Story: The Origin of Humankind." *Scientific American,* May 1994. Africa's Rift Valley holds the secrets of human origins.

Dean, D., and E. Delson. "*Homo* at the Gates of Europe." *Nature,* February 9, 1995. New evidence in the debate about when *Homo erectus* spread from Africa.

Forey, P., and P. Janvier. "Evolution of the Early Vertebates." *American Scientist,* November/December 1994. The Devonian radiation of fishes.

Gee, H. "Return of the Amphioxus." *Nature,* August 18, 1994. The case for cephalochordates as our closest invertebrate relatives.

Gibbons, A. "Neanderthal Language Debate: Tongues Wag Anew." *Science,* April 3, 1992. Could Neanderthals speak?

Gould, S. J. "Lucy on the Earth in Stasis." *Natural History,* September 1994. Evidence that punctuated equilibrium applies to human evolution.

Griffith, R. W. "The Life of the First Vertebrates." *BioScience,* June, 1994. Deducing the characteristics of the ancestral Cambrian vertebrates.

Luck-Baker, A. "Taking the Temperature of *T. rex.*" *New Scientist,* July 23, 1994. Were dinosaurs endotherms?

McClanahan, L. L., R. Ruibal, and V. H. Shoemaker. "Frogs and Toads in Deserts." *Scientific American,* March 1994. Adaptations to extreme environments.

Rice, J. A. "The Marvelous Mammalian Parade." *Natural History,* April 1994. A series of beautifully illustrated articles about the evolution of our vertebrate class.

Shell, E. R. "Flesh and Bone." *Discover,* December 1991. What role did women play in early human evolution?

Szabo, M. "Australia's Marsupials—Going, Going, Gone?" *New Scientist,* January 28, 1995. Another serious case of the biodiversity crisis.

"The Thunder Lizards." *Natural History,* December 1991. Several articles about how dinosaurs lived.

UNIT SIX

PLANTS: FORM AND FUNCTION

AN INTERVIEW WITH ADRIENNE CLARKE

Australia is home to an extraordinary number of world leaders in botanical and agricultural research, and Adrienne Clarke is one of the most prominent of these Australian plant biologists. In fact, as chairman of the CSIRO board, the national agency for science and technology, Dr. Clarke is one of Australia's most influential scientists. I traveled to Australia to interview Dr. Clarke in her office at the University of Melbourne, where she is Professor of Botany. Professor Clarke arrived for the interview from a meeting with top Australian government officials, so it seemed timely for me to begin by asking her about CSIRO and its national importance.

What is CSIRO, and what is its mission?

CSIRO stands for the Commonwealth Scientific and Industrial Research Organization. It is the main government body that does research for the benefit of Australia. We do scientific research related to the main economic sectors of the country, primarily mining, energy, and agriculture. We also do a lot of research on environmental problems. As our economy is changing from the traditional base of agriculture and mining, manufacturing is becoming more important, so we are putting more emphasis into some aspects of manufacturing technology and information technology.

What are your responsibilities as chairman of the CSIRO board?

As the chairman of the board, my responsibility is to report to government on the research activities of the organization in relation to the governing act of Parliament and a set of ministerial guidelines. The board is drawn from the community and like most boards its role is to set policy, to hire (and fire) the chief executive (through the minister), to monitor the progress of the organization, and to make sure that the financial reporting and the achievements of the organization are consistent with the government's goals.

Is your work as chairman of the CSIRO board politically sensitive?

Well, because we are a government-funded organization to a large extent—70% of our budget comes from the government—and we report to the government, there is clearly a political component to my role.

Would you say that scientists and politicians approach problems differently?

Yes, I think they do. They are under different pressures and are from different backgrounds. The scientists, very comfortable with acknowledging that they do not know something, can say, "Well, I don't know what the answers are," and then set about systematically acquiring the knowledge to solve the problem. On the other hand, it is often difficult for politicians to admit an ignorance. They are often under enormous time pressures to come up with some decision, and they are subject to bombardment from various interest groups that are often competing. I think politicians and scientists have to understand each other's culture and try to come to grips with working productively together.

In addition to your position with CSIRO, you also direct the Plant Cell Biology Research Centre at the University of Melbourne. What are the research objectives of this group?

The Plant Cell Biology Research Centre is one of the original special research centers that was set up by the Australian government. In this center we study two related fundamental questions in biology: how plants recognize and resist fungal infection, and how plants fertilize themselves or other plants. If you look at flowers, you will see that many have the female part, the pistil, very closely pressed to the male part, the anther. If the flowers were left to their own devices, the sperm in the pollen from a particular plant would fertilize the female in that same plant. This would lead to inbreeding and inbreeding depression, so very early in evolution plants devised a way of ensuring outcrossing that would make the female recognize and reject the male from the same plant—a mechanism of self-incompatibility. How the plant can distinguish self from nonself is a very fundamental question in plant biology. Not only does it tell us something about how plants achieve fertilization, it also tells us more broadly about how plant cells recognize each other.

In animals, cells are recognized as being self or nonself (foreign) by the immune system. Humans have a circulating blood system with white blood cells that can recognize and reject foreign material. We basically understand how recognition of foreign materials happens in the animal system, but we have very little idea at all of how it happens in plants. By understanding how a plant recognizes self and nonself pollen, we will get some insight into how plants recognize and reject other foreign invaders such as fungi and other pathogenic organisms.

What is the genetic basis for self-incompatibility? How does a plant distinguish self and nonself?

We don't know the full answer yet, but working with plants in the tomato family (such as tomato, tobacco, and potato), we found a single genetic locus with multiple alleles that controls self-incompatibility. This gene—the *S* gene (for self-incompatibility)—directs the secretion of a glycoprotein (a sugar joined to a protein) into the cells of the pistil. This is the pistil tissue through which the pollen tubes must grow to reach the ovules and fertilize eggs. As the pollen tubes grow on their way down to the ovary, they are bathed with the glycoprotein. We isolated the glycoprotein, cloned the gene, and found, very interestingly, that it is a ribonuclease—that is, it has an enzymatic function to degrade RNA. We believe that it acts as a toxin to self pollen. What happens is that RNAase is taken up from the female tissues into the male pollen tube. If it is a self-incompatible situation—if the pollen and pistil have matching alleles for the *S* gene—we can see that the ribosomal RNA of the pollen is degraded. Pollen, unlike most other tissues and cells, is not able to make more ribosomal RNA. When pollen is produced, it has all the ribosomal RNA it is ever going to have. Once it's gone, the pollen can no longer produce proteins, and growth of the pollen tube grinds to a halt.

The missing link in all of this, the part we don't understand yet, is the nature of the product of the *S* gene in pollen and what controls the specificity. How is it that the RNAase from the pistil tissue destroys RNA in self pollen but not in non-self pollen? It is a very hot topic, and there are many labs around the world working on it now.

What impact is biotechnology having on agriculture?

A particularly important impact of biotechnology would be in the area of disease and insect resistance in plants. There is huge pressure in this area to reduce the use of chemicals. There is a lot of interest in using biotechnology to engineer plants to produce natural toxins, for instance, so that the use of pesticides and other chemicals can be reduced.

What influenced you to become a scientist?

I grew up on a farm, and I think the rural experience influences a lot of people to become scientists. In the bush, we're observers of the natural world from an early age. I went to a small girls' school, and we didn't have much in the way of science teaching. However, I had one very inspirational teacher who taught me math and physics. She also did a wonderful thing. She took a band of teenage girls, including myself, on a trip to the Barrier Reef, and there I saw a whole new world under the water. I was astounded at the diversity of the forms and the colors. I also found I was good at science. I did well, got scholarships, and wanted to go on.

And what attracted you to your research specialty of plant cell biology?

I started in the field of plant carbohydrate chemistry, and the initial attraction was the people I knew I would be working with. For me, the environment is as important as the work itself. When you start on a scientific project, it is usually going to last for several years, and you want to feel very comfortable with the people you are working with. First of all, they have to have a professional skill that you admire, and second, they have to be nice people. I think you have to feel comfortable, you have to feel happy with what you do.

You continue to be a strong voice for the value of teamwork in science. Why do you think this is so important?

I think that science has now, in many areas, become so complex. The biggest problems you tackle, which are the ones that will make an impact, require different skills and different technologies. You have to build a team because no one person can have all those technologies under their belt. For instance, we might have a physical chemist working on the flow properties of a gum thickener for the food industry. That physical chemist will also need to know what this material is made of, so there's a need for an analytical chemist. And if the material comes from a plant, then a plant biologist should be part of the team. There are all sorts of examples of those situations. You need multidisciplinary teams to tackle big problems.

Why do you think women are still underrepresented in scientific careers?

As I see it, it is often the conflict between biological destiny and a career pathway that doesn't allow for people to be out of the work force at some time. I struggled with this. When I came back to Australia after several years in America in 1963, I was pregnant with my third child, and I said, "So, I will be a mother now." I cleaned out the cupboard and got all these research papers together and threw them in the fire. As I saw the flames leaping, I changed my mind. Luckily, I was able to save the papers in time.

One of the things that we have done at the University of Melbourne is to ensure that women who are talented and enthusiastic in their careers may have part-time work or time out to have a baby.

What advice do you have for students who are considering careers in science?

My advice is that if you do the basic training in science, you can then branch off to do all sorts of things. For young women considering futures in science, it's important to understand that it is possible, with some difficulty, to have a family and have a satisfying career too. This will become less difficult as our social structure continues to change and men take more of a role in household and child-rearing duties. There are also tremendous opportunities opening up now for people with combinations of skills—science with accounting, science with law, science and engineering, and a whole range of combinations in science and technology. I would say that if women are interested in science careers, they should do it. As our structures are changing, it is very much easier both to pursue a career and to raise a family than it was back when I was doing it.

CHAPTER 31

PLANT STRUCTURE AND GROWTH

KEY CONCEPTS

- Plant biology reflects the major themes in the study of life
- A plant's root and shoot systems are evolutionary adaptations to living on land
- The many types of plant cells are organized into three major tissue systems
- Meristems generate cells for new organs throughout the lifetime of a plant: *an overview of plant growth*
- Apical meristems extend roots and shoots: *a closer look at primary growth*
- Lateral meristems add girth to stems and roots: *a closer look at secondary growth*

Plants are the pillars of most terrestrial ecosystems. Their photosynthesis supports their own growth and maintenance, and it feeds, directly or indirectly, an ecosystem's various consumers, including animals. Although most humans now live far from natural ecosystems and farms, our dependence on plants is evident in our lumber, fabrics, paper, medicines, and—most importantly—our food. The study of plants began when early humans learned to distinguish edible plants from poisonous ones and began to make things from wood and other plant products. Modern plant biology still has its pragmatic side—in research aimed at improving crop productivity, for example—but the pure fun of discovery is what motivates most plant scientists. Plant biology, perhaps the oldest branch of science, is driven by a combination of curiosity and need—curiosity about how plants work and a need to apply this knowledge judiciously to feed, clothe, and house a burgeoning human population.

Plant biology reflects the major themes in the study of life

Plant biology is in the midst of a renaissance. New methods coupled with clever choices of experimental organisms have catalyzed a research explosion. For example, many scientists interested in the genetic control of plant development are focusing their research on *Arabidopsis thaliana,* a little weed that belongs to the mustard family. *Arabidopsis* is small enough to be grown in test tubes, and its short generation span of about six weeks makes it an excellent model for genetic studies. Plant biologists are also attracted to *Arabidopsis* by its tiny genome; the amount of DNA per cell ranks among the least of all known plants. Efforts to associate specific genetic functions with certain regions of DNA are simplified, and it is likely that researchers will map the entire *Arabidopsis* genome within this decade. Already, plant biologists have pinpointed some of the genes that control the development of flowers and have learned the functions of these genes (FIGURE 31.1). This research is just one example of what seems to be a major thrust of modern plant biology: relating processes that occur at the molecular and cellular levels to what we observe at the level of the whole plant. Once again, we see the theme of a hierarchy of structural levels, with emergent properties arising from the ordered arrangement and interactions of component parts.

The other biological themes of this book will also guide our study of plants. The structure and functions of plants

(a)

(b)

(c)

FIGURE 31.1

***Arabidopsis,* a little weed that is helping plant biologists answer big questions about the development of plant form.** (**a**) Plant biologists have made this tiny member of the mustard family one of the organisms of choice for the experimental study of how genes control plant development. Three characteristics that make *Arabidopsis* such a convenient research model are small size, a short life cycle, and a relatively small genome. Numerous laboratories are now mapping *Arabidopsis's* genome and identifying the functions of many of its genes. With the help of this model system, plant biologists hope to discover how genes and environment interact to transform a zygote into a plant. (**b**) Much of this research has centered on flower development. *Arabidopsis* normally has four whorls of flower parts: sepals (Se), petals (Pe), stamens (St), and carpels (Ca). (**c**) Researchers have identified several mutations that cause abnormal flowers to develop. This flower, for example, has an extra whorl of petals in place of stamens and an internal flower where normal plants have carpels.

are shaped by interactions with the environment on two time scales. In their long evolutionary journey from water onto land, plants became adapted by natural selection to the specific problems posed by terrestrial environments. The evolution of tissues, such as the woody tissue of trees, that can support plants against the pull of gravity is just one example. Over the short term, individual plants exhibit structural and physiological responses to environmental stimuli. As an example of how individual plants, far more than individual animals, adapt *structurally* to their environment, look at how wind has affected the growth of branches of the trees illustrated in FIGURE 31.2. Plants, like animals, also have many evolutionary adaptations in the form of *physiological* responses to short-term change. For example, the stomata of many plants—the pores of the leaf surface that function in gas exchange and transpiration—close during the hottest time of day. This response helps the plant conserve water. As we analyze the evolutionary adaptations of plants and the responses of individual plants to their environment, our main tool will be a related theme: the correlation between structure and function. We will learn, for example, how the opening and closing of stomata are consequences of the structure of the guard cells that border these pores.

In studying the structure and function of plants, comparisons with the animal kingdom are inevitable and sometimes useful. Plants and animals confront many of the same problems, which they may solve in different ways. For example, an elephant and a redwood tree support their enormous weight on land by means of very different kinds of "skeletons": bones in the elephant and wood in the tree. In some cases, plant and animal solutions to a common problem seem similar. For example, large organisms require internal transport systems to carry water and other substances between body parts, and networks of tubes have evolved in both plants and animals. But on closer examination, we will see that any anatomical similarities between plants and animals are superficial. These similar adaptations are, in the lexicon of biology, analogous, not homologous (see Chapter 23). The multicellular organization of plants evolved independently from that of animals, from unicellular ancestors with an entirely different mode of nutrition. The autotrophic/heterotrophic distinction has placed plants and animals on separate evolutionary paths. Setting the

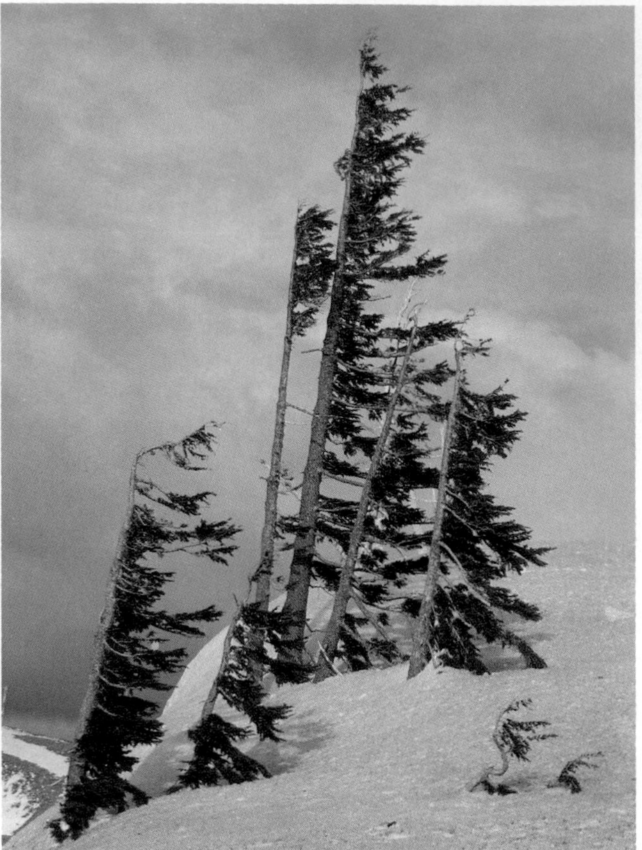

FIGURE 31.2
The effect of the environment on plant form. The "flagging" of these firs growing on a windy ridge on Mt. Hood, Oregon, resulted partly from the mechanical disturbance of the prevailing winds inhibiting limb growth on the windward sides of the trees. This growth response reduces the number of limbs that are broken during strong winds. Although environment affects the growth and development of all organisms, environmental impact on plant form is particularly impressive. In contrast, animal form is much less plastic; wind, for example, does not alter the number, size, or placement of human limbs.

stage for our survey of plant biology, this chapter introduces the general body plan of flowering plants, beginning with external structure. We will see that the architecture of plants is dynamic, continuously shaped by the plants' genetically directed growth patterns and their responses to the environment.

A plant's root and shoot systems are evolutionary adaptations to living on land

Plant biologists study plant architecture on two levels: morphology and anatomy. **Plant morphology** (Gr. *morphe,* "form") is the study of the external structure of plants. The placement of leaves along a stem is an example of morphology. **Plant anatomy** is concerned with internal structure—for example, the arrangement of cells and tissues within a leaf. This section features the morphology of flowering plants, or angiosperms. (The structure of algae, mosses, ferns, and gymnosperms was covered in Unit Five, along with the evolutionary relationships of these plant groups to the angiosperms.)

Angiosperms are characterized by flowers and fruits, evolutionary adaptations that function in reproduction and the dispersal of seeds. With about 275,000 known species, angiosperms are by far the most diverse and widespread group of plants. Taxonomists split the angiosperms into two classes: **monocots,** named for their single cotyledon (seed leaf), and **dicots,** which have two cotyledons. Monocots and dicots have several other structural differences as well (FIGURE 31.3).

The basic morphology of plants reflects their evolutionary history as terrestrial organisms. The algal ancestors of plants were bathed in a solution of water and minerals, including bicarbonate, the source of CO_2 for photosynthesis in aquatic habitats. In contrast, the resources a terrestrial plant needs are divided between the soil and air, and the plant must inhabit these two very different environments at the same time. Soil provides water and minerals, but air is the main source of CO_2, and light does not penetrate far into the soil. The evolutionary solution to this separation of resources was differentiation of the plant body into two main systems: a subterranean **root system** and an aerial **shoot system** consisting of stems, leaves, and flowers (FIGURE 31.4). Neither system can live without the other. Lacking chloroplasts and living in the dark, roots would starve without sugar and other organic nutrients imported from the photosynthetic tissues of the shoot system. Conversely, the shoot system depends on water and minerals absorbed from the soil by roots. Vascular tissues, continuous throughout the plant, transport materials between roots and shoots. Each vein has two types of vascular tissue: **xylem,** which conveys water and dissolved minerals upward from roots into the shoots; and **phloem,** which transports food made in the leaves to the roots and to nonphotosynthetic parts of the shoot system. As we take closer looks at the morphology of roots and shoots, try to view these systems from the evolutionary perspective of adaptation to living on land.

The Root System

Roots anchor the plant in the soil, absorb minerals and water, conduct water and nutrients, and store food. The structure of roots is well adapted to these functions.

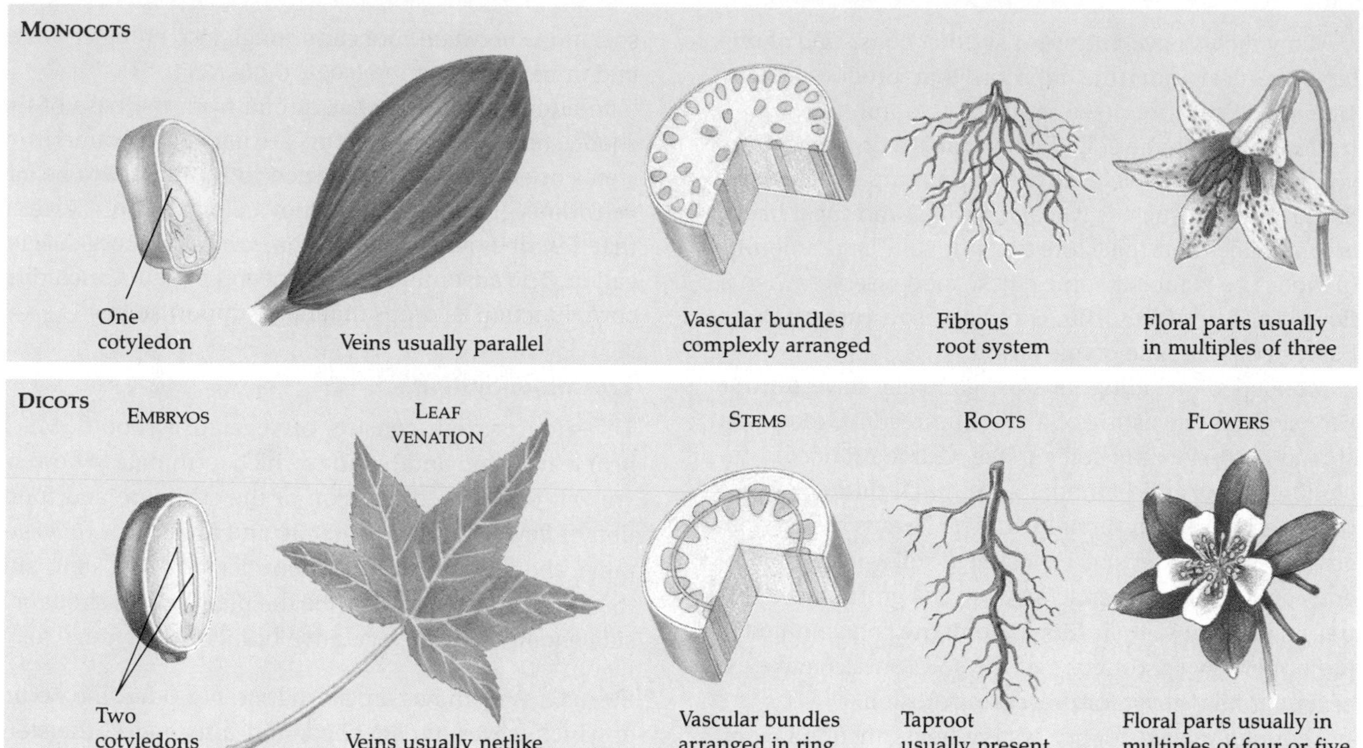

FIGURE 31.3

A comparison of monocots and dicots. These classes of angiosperms are named for the number of cotyledons, or seed leaves, present in the seed of the plant. Monocots include orchids, bamboos, palms, lilies, and yuccas, as well as the grasses, such as wheat, corn, and rice. A few examples of dicots are roses, beans, sunflowers, maples, and oaks.

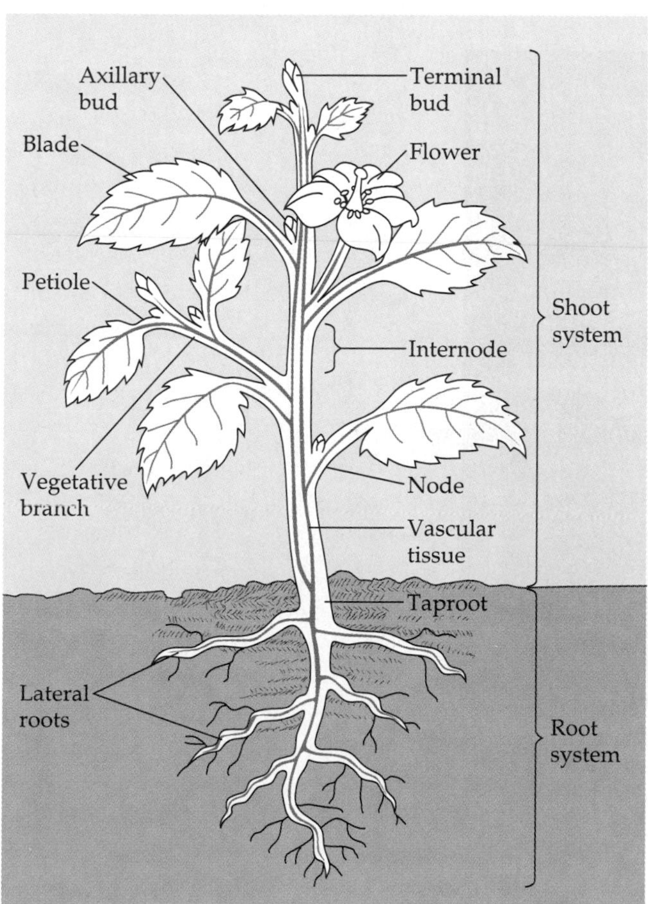

FIGURE 31.4

Morphology of a flowering plant: an overview. The plant body is divided into a root system and a shoot system, connected by vascular tissue that is continuous throughout the plant. The root system of this dicot consists of a taproot and several lateral roots. Shoots consist of stems, leaves, and flowers. The blade, the expanded portion of a leaf, is attached to a stem by a petiole. Nodes, the regions of a stem where leaves attach, are separated by internodes. At a shoot's tip is the terminal bud, the main growing point of the shoot. Axillary buds are located in the upper angles of leaves. Most of these axillary buds are dormant, but they have the potential to develop into vegetative (leaf-bearing) branches or into flowers.

Many dicots have a **taproot** system, consisting of one large, vertical root (the taproot) that produces many smaller lateral roots (see FIGURES 31.3 and 31.4). Penetrating deep into the soil, the taproot is a firm anchor, as you know if you have ever tried to pull up a dandelion. Some taproots, such as carrots, turnips, and sugar beets, are modified roots that store exceptionally large amounts of food. The plant consumes these food reserves when it flowers and produces fruit. For this reason, root crops are harvested before the plants flower.

Monocots, including grasses, generally have **fibrous root** systems consisting of a mat of threadlike roots that spread out below the soil surface. (Large monocots, including palms and bamboo, have much thicker roots—ropelike rather than threadlike.) The fibrous root system gives the plant extensive exposure to soil water and minerals and anchors it tenaciously to the ground (see FIGURE 31.3). Because their root systems are concentrated in the upper few centimeters of the soil, grasses make excellent ground cover for preventing erosion.

Although the entire root system helps anchor a plant, most absorption of water and minerals in both monocots and dicots occurs near the root tips, where vast numbers of tiny **root hairs** increase the surface area of the root tremendously (FIGURE 31.5). Mycorrhizae, symbiotic associations between roots and fungi, also enhance water and mineral absorption (see FIGURE 28.13).

In addition to roots that extend from the base of the shoot, some plants have roots arising aboveground from stems or even from leaves. Such roots are said to be **adventitious** (L. *adventicius,* "not belonging to"), a term that describes any plant part that grows in an unusual location. The adventitious roots of some plants, including corn, function as props that help support stems.

The Shoot System

The shoot system consists of vegetative shoots, which bear leaves, and floral shoots, which terminate in flowers. We will postpone discussion of the structure and function of flowers until Chapter 34 and focus here on vegetative shoots. A vegetative shoot consists of a stem and the attached leaves; it may be the plant's main shoot or a side shoot, called a vegetative branch (see FIGURE 31.4).

Stems. A stem has an alternation of **nodes,** the points at which leaves are attached, and **internodes,** the stem segments between nodes (see FIGURE 31.4). In the angle formed by each leaf and the stem is an **axillary bud,** which is an embryonic side shoot. Most axillary buds are dormant; growth is usually concentrated at the apex (tip)

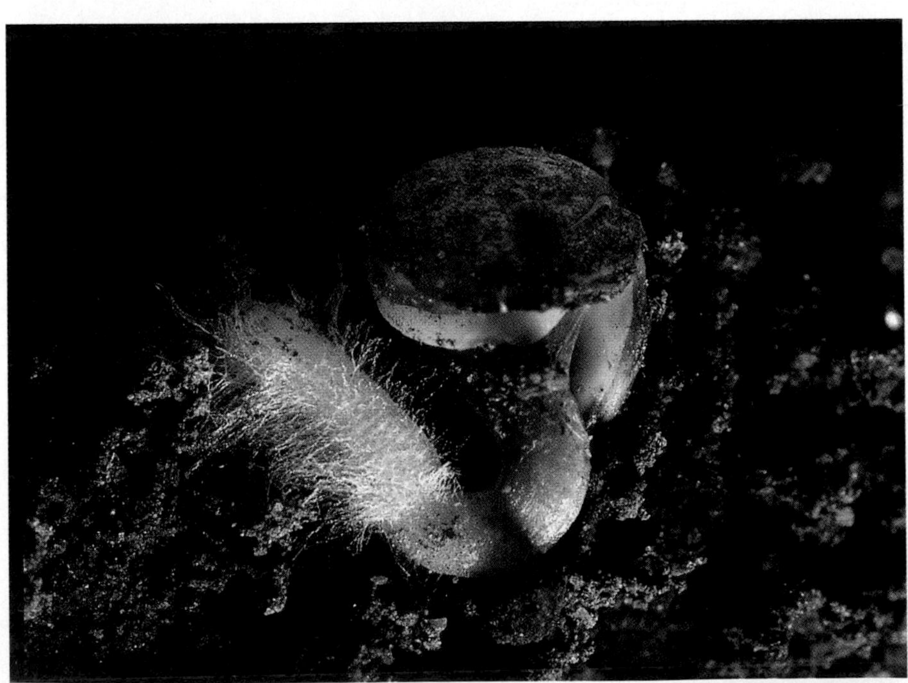

FIGURE 31.5
Root hairs of a radish seedling. Growing by the thousands just behind the tip of each root, the hairs cling tightly to soil particles and increase the surface area for the absorption of water and minerals by the roots.

of a shoot, where there is a **terminal bud** with developing leaves and a compact series of nodes and internodes. The presence of the terminal bud is partly responsible for inhibiting the growth of axillary buds, a phenomenon called **apical dominance.** By concentrating resources on growing taller, apical dominance is an evolutionary adaptation that increases the plant's exposure to light, especially in a location with dense vegetation. However, branching is also important for increasing the exposure of the shoot system to the environment, and under certain conditions, axillary buds begin growing. Each bud has the dual potential to give rise to either a reproductive shoot bearing flowers or a vegetative branch complete with its own terminal bud, leaves, and axillary buds. In some cases, the growth of axillary buds can be stimulated by removing the terminal bud. This is the rationale for pruning trees and shrubs and "pinching back" houseplants to make them bushy.

Modified stems with diverse functions have evolved in many plants and are often mistaken for roots (FIGURE 31.6). Stolons are horizontal stems that grow along the surface of the ground. The "runners" of strawberry plants are examples. Rhizomes, such as those of irises, are horizontal stems that grow underground. Some rhizomes end in enlarged tubers where food is stored, as in white potatoes. Bulbs, such as those of onions, are vertical, underground shoots with fleshy leaf bases modified for food storage.

Leaves. Leaves are the main photosynthetic organs of most plants, although green stems also perform photosynthesis. Leaves vary extensively in form, but they generally consist of a flattened **blade** and a stalk, the **petiole,** which joins the leaf to a node of the stem (see FIGURE 31.4). Grasses and many other monocots lack petioles; instead, the base of the leaf forms a sheath that envelopes the stem. Some monocots, including palm trees, do have petioles.

The leaves of monocots and dicots differ in how their major veins are arranged (see FIGURE 31.3). Most monocots have parallel major veins that run the length of the leaf blade. In contrast, dicot leaves generally have a multibranched network of major veins. All leaves have numerous minor cross-veins. Vascular arrangement, leaf shape, and leaf placement on the stem are among the characteristics used by plant taxonomists to help identify or classify plants (FIGURE 31.7). Although most leaves are specialized for photosynthesis, some plants have leaves that have become adapted by evolution for other functions (FIGURE 31.8).

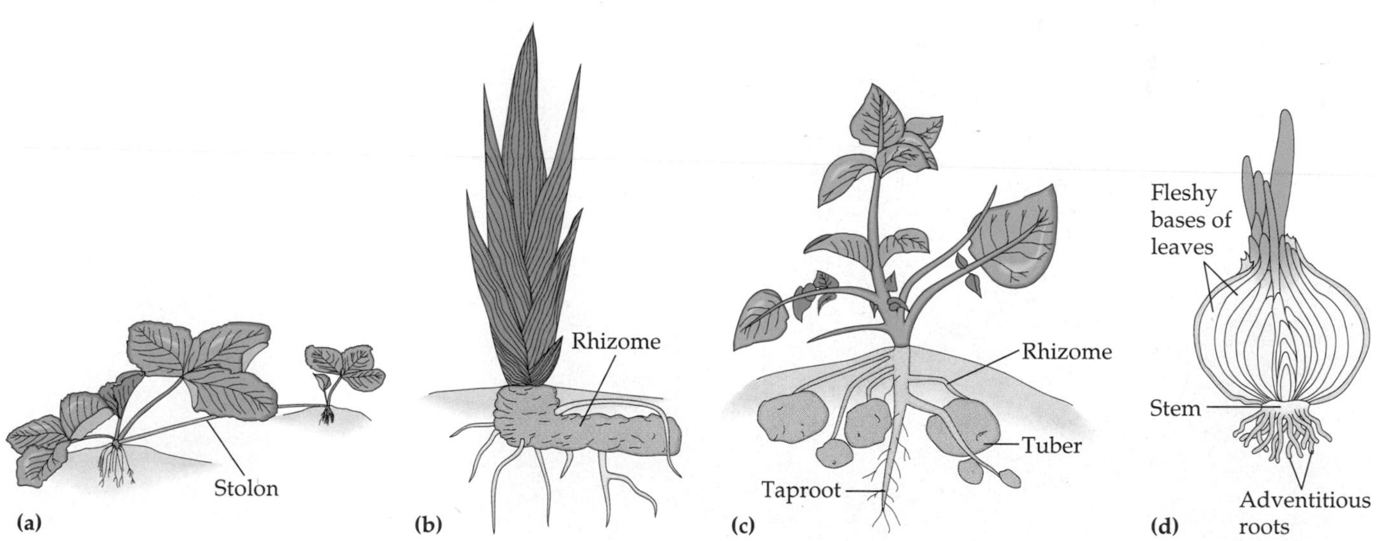

FIGURE 31.6
Modified stems. (**a**) Stolons, shown here on a strawberry plant, grow on the surface of the ground. (**b**) Rhizomes, like the one on this iris plant, are horizontal stems that grow underground. (**c**) Tubers are swollen ends of rhizomes specialized for storing food. The "eyes" arranged in a spiral pattern around the potato are clusters of buds that mark the nodes. (**d**) Bulbs are vertical, underground shoots consisting mostly of the swollen bases of leaves that store food. You can see the many layers of modified leaves attached to the short stem by slicing an onion bulb lengthwise.

FIGURE 31.7

A survey of leaf morphology.
(**a**) Leaves are arranged on the stem in a variety of patterns. If each node has a pair of leaves 180° apart, the leaves are said to be opposite. The leaf placement is alternate when each node has a single leaf and the leaves of adjacent nodes point in different directions. If a node has three or more leaves attached, the arrangement is termed whorled. (**b**) A leaf is said to be simple if it has a single, undivided blade. If the blade is divided into several leaflets, then the leaf is compound. (You can distinguish a compound leaf from a stem with several closely spaced simple leaves by examining the locations of axillary buds. There is no bud at the base of a leaflet, but there is an axillary bud where the petiole of the compound leaf joins the stem.) (**c–e**) Leaves also vary in shape, in the contour of their margins, and in their pattern of veins.

(a)

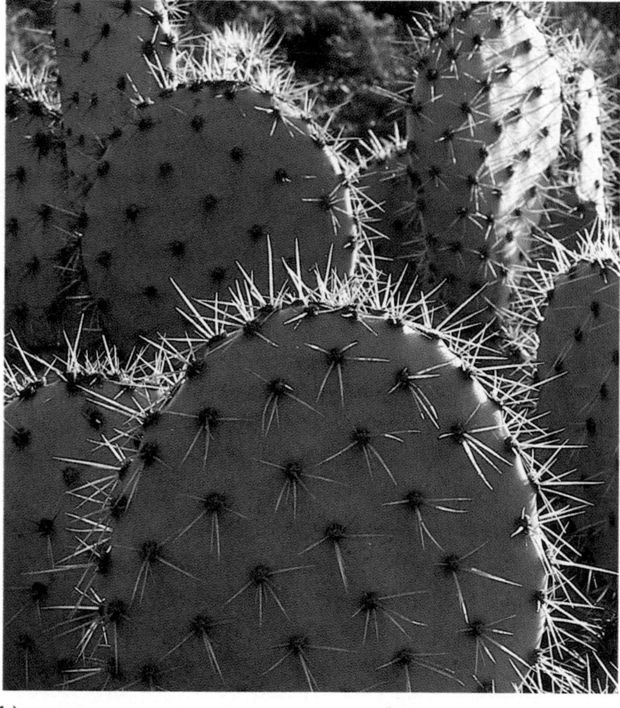

(b)

(c)

(d)

FIGURE 31.8

Modified leaves. (**a**) The tendrils used by this pea plant to cling to supports are modified leaflets. (**b**) The spines of cacti, such as this prickly pear, are actually leaves, and photosynthesis is carried out mainly by the fleshy green stems. (**c**) Most succulents, such as ice plant, have leaves modified for storing water. (**d**) In many plants, brightly colored leaves help attract pollinators to the flower. The red "petals" of the poinsettia are actually leaves that surround a group of flowers.

So far we have examined the structural organization of the whole plant as we see it with the unaided eye. We can now dissect the plant and explore its microscopic organization.

The many types of plant cells are organized into three major tissue systems

In this section, an introduction to plant anatomy, you will learn how the structural specializations of plant cells enable them to perform certain functions. You will also learn how these cells are organized into three main tissue systems and how these tissues are arranged in roots, stems, and leaves. FIGURE 31.9 will help you review the general structure of plant cells before you proceed to the following survey of specific cell types.

Types of Plant Cells

What distinguishes a multicellular organism from a colony of cells is a division of labor among cells differing in structure and function. As you consider each major

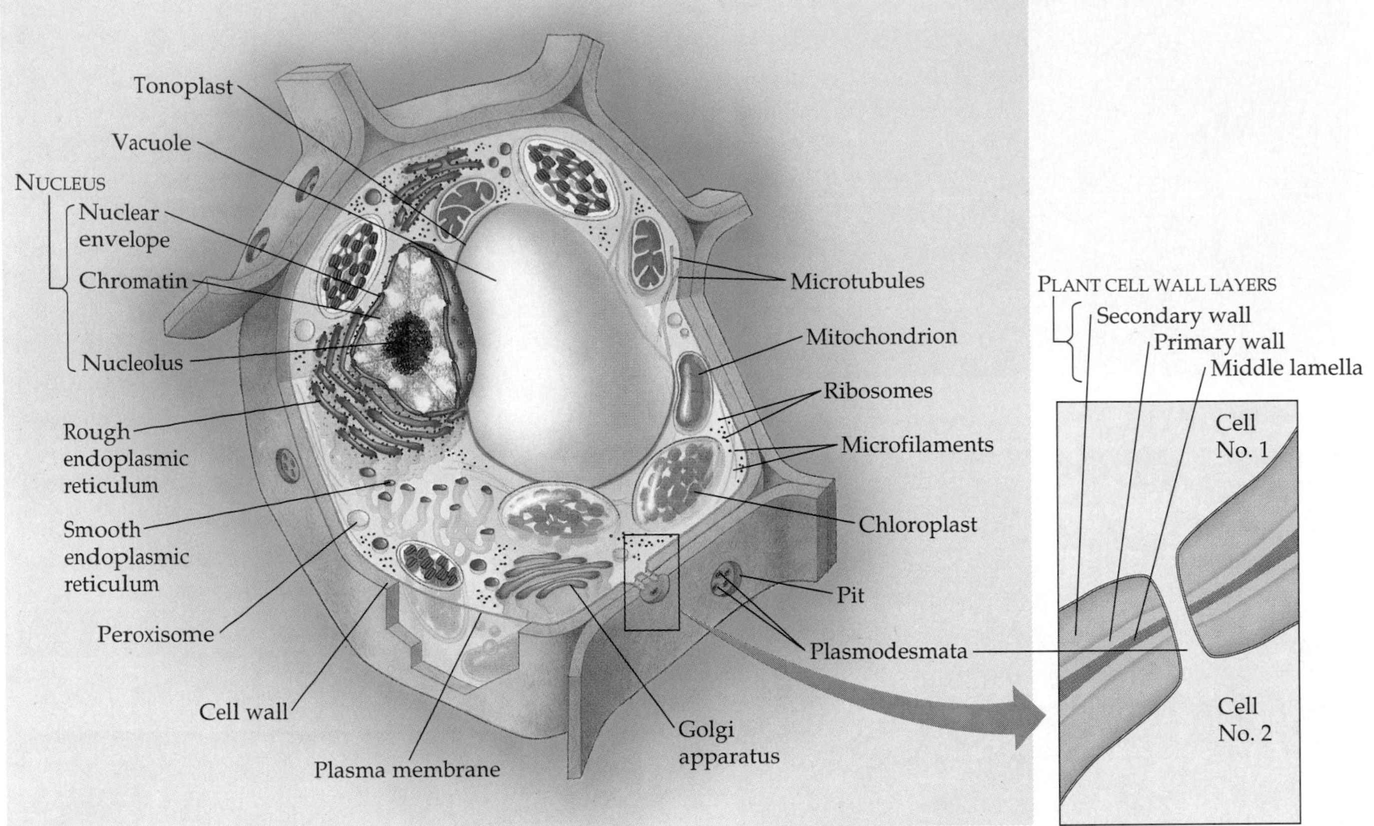

FIGURE 31.9
Plant cell structure: a review. A plant cell consists of a protoplast enclosed in a cell wall. The protoplast—the whole cell, excluding the cell wall—is bounded by the plasma membrane. Outside the plasma membrane is the primary cell wall and in some plants a secondary cell wall. Between the primary walls of adjacent plant cells is the middle lamella, a sticky layer that cements the cells together. The protoplasts of neighboring cells are generally connected by plasmodesmata, cytoplasmic channels that pass through pores in the walls. The plasmodesmata may be concentrated in areas called pits, where the distance between adjacent protoplasts is narrowed. When mature, most living plant cells have a large central vacuole that occupies as much as 90% of the volume of the protoplast. A membrane called the tonoplast separates the contents of the vacuole from the thin layer of cytoplasm, in which the mitochondria, plastids, and other organelles are located. Within the vacuole is the cell sap, a complex aqueous solution that helps the vacuole play an important role in maintaining the turgor, or firmness, of the cell.

type of plant cell, notice the structural adaptations that make specific functions possible. In some cases, we will find distinguishing characteristics within the **protoplast,** the contents of the cell exclusive of the cell wall. For example, only the protoplasts of photosynthetic cells contain chloroplasts. But also notice that modifications of cell walls are important in how the specialized cells of a plant function.

Parenchyma Cells. Because they are the least specialized of all plant cells, **parenchyma cells** are often depicted as "typical" plant cells (FIGURE 31.10a). Mature parenchyma cells have primary walls that are relatively thin and flexible. Most parenchyma cells lack secondary walls. The protoplast generally has a large central vacuole.

Parenchyma cells perform most of the metabolic functions of the plant, synthesizing and storing various organic products. For example, photosynthesis occurs within the chloroplasts of mesophyll cells, the parenchyma cells in the leaf. Some parenchyma cells in stems and roots have colorless plastids that store starch. The fleshy tissue of most fruit is composed mostly of parenchyma cells.

Developing plant cells of all types usually have the generalized structure of parenchyma cells before specializing further in structure and function. Mature parenchyma cells do not generally undergo cell division, but most of them retain the ability to divide and differentiate into other types of plant cells under special conditions—during the repair and replacement of organs after injury

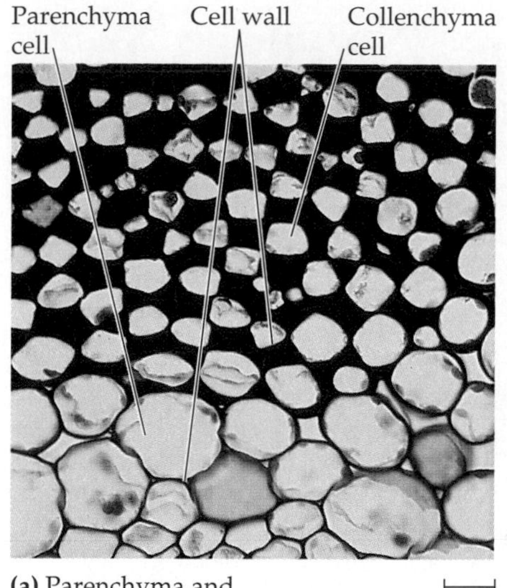

(a) Parenchyma and collenchyma
100 μm

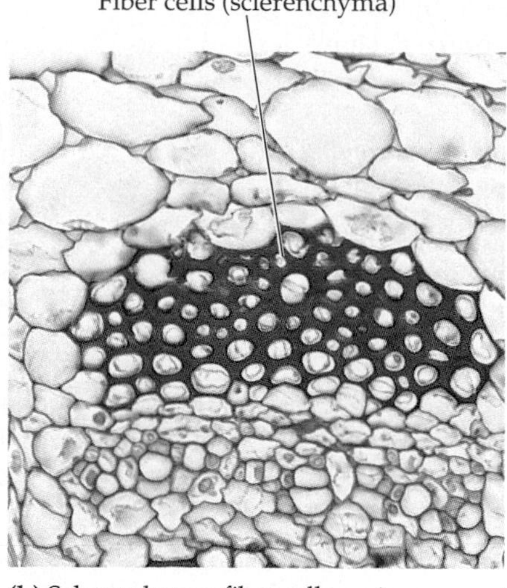

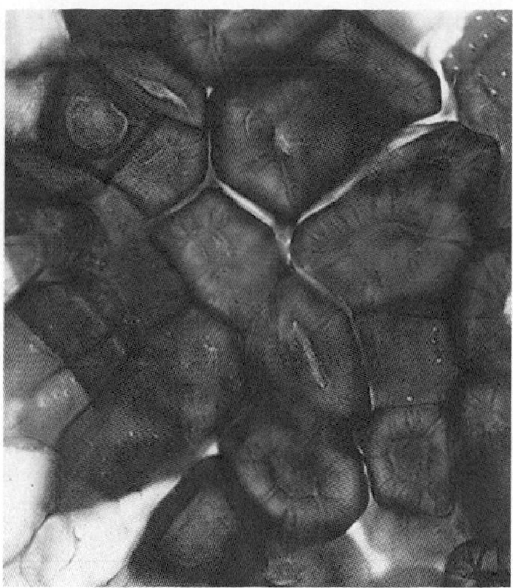

(b) Sclerenchyma: fiber cells (left) and sclereids or stone cells (right)
50 μm
10 μm

FIGURE 31.10

Types of plant cells. (**a**) Parenchyma cells are relatively unspecialized, with thin, flexible primary walls. These cells carry on most of the plant's metabolic functions. Collenchyma cells have unevenly thickened primary walls and provide support to parts of the plant that are still growing. (**b**) Sclerenchyma cells, specialized for support, have secondary walls hardened with lignin and may be dead (lacking protoplasts) at functional maturity. The fiber cells in the left micrograph are elongated sclerenchyma cells. Sclereids (right) are irregularly shaped sclerenchyma cells with very thick, lignified secondary walls. (**c**) The water-conducting cells of xylem include tapered tracheids (left) and vessel elements arranged end to end, forming vessels (right). Both cell types have secondary walls and are dead at functional maturity. In gymnosperms, tracheids have the dual functions of water transport and structural support. In most angiosperms, both vessel elements and tracheids conduct water, and support is provided mainly by fiber cells. (**d**) The food-conducting cells of phloem are sieve-tube members, which are arranged end to end with porous walls (sieve plates) between them. (In the transverse section on the right, two sieve-tube members, including the labeled one, are sectioned through sieve plates.) The cells are living at functional maturity, but lack nuclei. Alongside each sieve-tube member is a nucleated companion cell. (All LMs.)

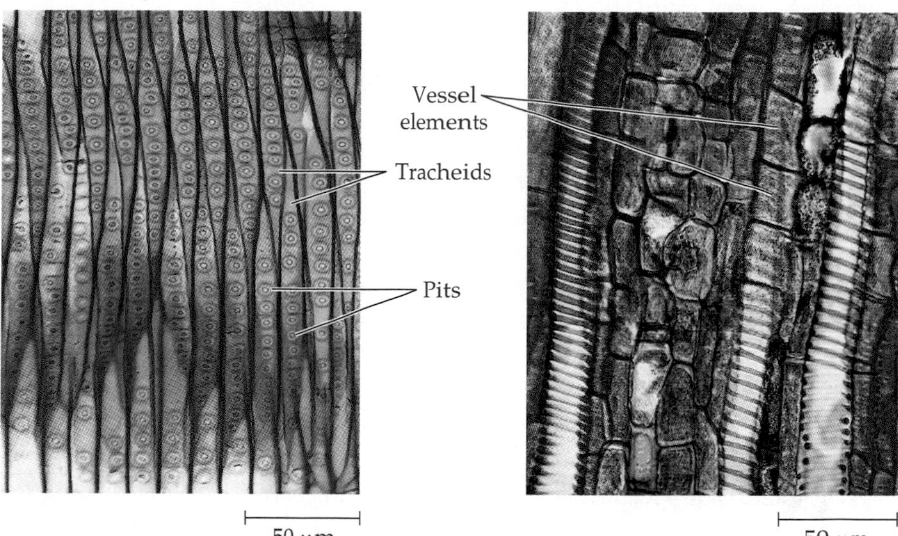

(c) Xylem showing tracheids (left) and vessel elements (right) in longitudinal sections

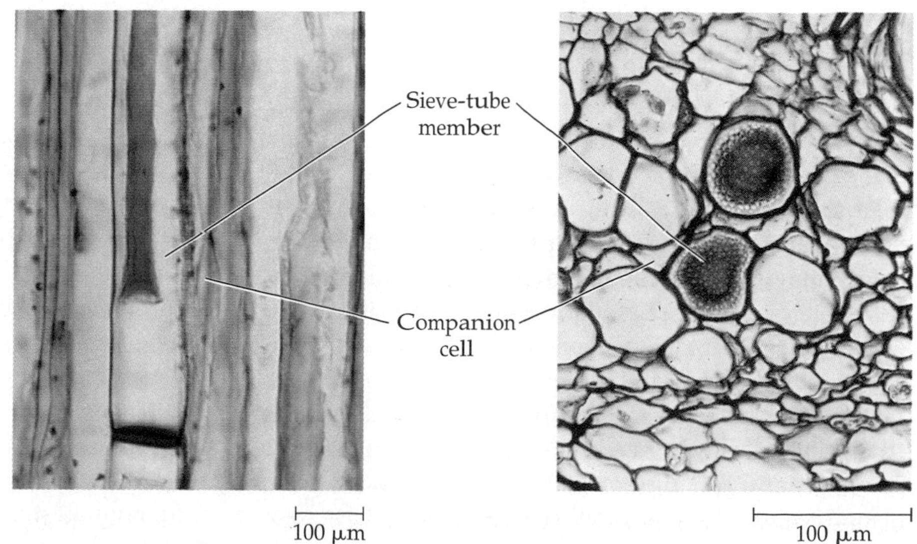

(d) Phloem showing sieve-tube members and companion cells in longitudinal (left) and transverse (right) sections

to the plant, for instance. It is even possible in the laboratory to regenerate an entire plant from a single parenchyma cell.

Collenchyma Cells. Compared to parenchyma cells, **collenchyma cells** have thicker primary walls, though the walls are unevenly thickened (FIGURE 31.10a). Grouped in strands or cylinders, collenchyma cells help support young parts of the plant. Young stems, for instance, often have a cylinder of collenchyma just below their surface. Because they lack secondary walls and the hardening agent lignin is absent in their primary walls, collenchyma cells provide support without restraining growth. Unlike sclerenchyma cells, which we discuss next, mature, functioning collenchyma cells are living and elongate with the stems and leaves they support.

Sclerenchyma Cells. Also functioning as supporting elements in the plant, but with thick secondary walls strengthened by lignin, **sclerenchyma cells** are much more rigid than collenchyma cells. Mature sclerenchyma cells cannot elongate, and they occur in regions of the plant that have stopped growing in length. So specialized are sclerenchyma cells for support that many lack protoplasts at functional maturity, the stage in a cell's development when it is fully specialized for its function. Thus, at functional maturity a sclerenchyma cell may actually be dead, its rigid wall serving as scaffolding to support the plant.

The two forms of sclerenchyma cells are **fibers** and **sclereids** (FIGURE 31.10b). Long, slender, and tapered, fibers usually occur in bundles. Some plant fibers are used commercially, such as hemp fibers for making rope and flax fibers for weaving into linen. Sclereids are shorter than fibers and irregular in shape. Nutshells and seed coats owe their hardness to sclereids, and sclereids scattered among the soft parenchyma tissue give the pear fruit its gritty texture.

Tracheids and Vessel Elements: Water-Conducting Cells. The water-conducting elements of xylem are elongated cells of two types: **tracheids** and **vessel elements** (FIGURE 31.10c). Both types of cells are dead at functional maturity, but they produce secondary walls before the protoplast dies. In parts of the plant that are still elongating, the secondary walls are deposited unevenly in spiral or ring patterns that enable them to stretch like springs as the cell grows. Like the wire that reinforces the wall of a garden hose, these wall thickenings strengthen the water-conducting cells of the plant. Tracheids and vessel elements that form in parts of the plant that are no longer elongating usually have secondary walls with **pits,** thinner regions where only primary walls are present. A tracheid or vessel element completes its differentiation when its protoplast disintegrates, leaving behind a nonliving conduit through which water can flow (FIGURE 31.11).

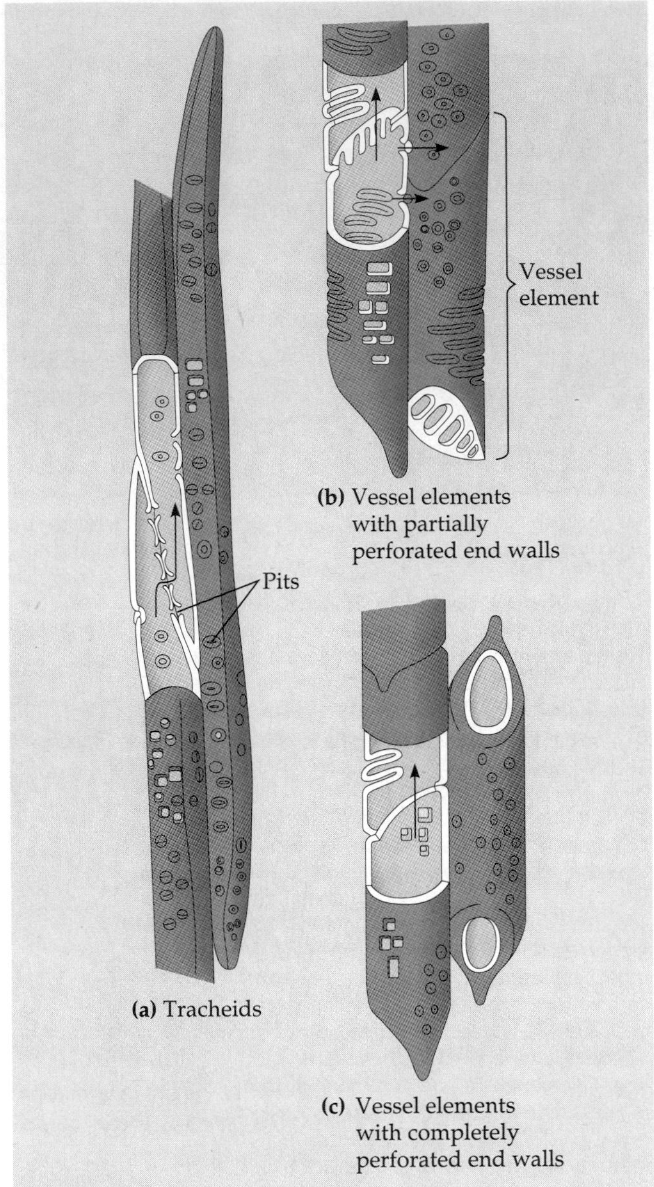

FIGURE 31.11
Water-conducting cells of xylem. Arrows indicate the flow of water. (**a**) Tracheids are spindle-shaped cells with pits through which water flows from cell to cell. (**b**) Vessel elements are individual cells linked together end to end, forming long tubes, or xylem vessels. Water streams from element to element through perforated end walls. Water can also migrate laterally between neighboring vessels through pits. (**c**) Resistance to water flow in some xylem vessels is lowered by the complete perforation of walls between the vessel elements.

Tracheids are long, thin cells with tapered ends. Water moves from cell to cell mainly through pits. Because their secondary walls are hardened with lignin, tracheids function in support as well as water transport.

Vessel elements are generally wider, shorter, thinner-walled, and less tapered than tracheids. Vessel elements are aligned end to end, forming long micropipes, the **xylem vessels.** The end walls of vessel elements are perforated, enabling water to flow freely through xylem vessels.

Sieve-Tube Members: Food-Conducting Cells. Sucrose, other organic compounds, and some mineral ions are transported within the phloem of a plant through tubes formed by chains of cells called **sieve-tube members** (see FIGURE 31.10d). In contrast to the water-conducting cells of xylem, sieve-tube members are alive at functional maturity, although their protoplasts lack such organelles as the nucleus, ribosomes, and a distinct vacuole. In angiosperms, the end walls between sieve-tube members, called **sieve plates,** have pores that presumably facilitate the flow of fluid from cell to cell along the sieve tube.

Alongside each sieve-tube member is at least one **companion cell,** which is connected to the sieve-tube member by numerous plasmodesmata. The nucleus and ribosomes of the companion cell may serve not only that cell but also the adjacent sieve-tube member, which has no nucleus or ribosomes of its own. In some plants, companion cells also help load sugar produced in the mesophyll into the sieve-tube members of leaves.

The Three Tissue Systems of a Plant

The cells of a plant are organized into three tissue systems: the dermal, vascular, and ground tissue systems. Each tissue system is continuous throughout the plant body, although the specific characteristics of the tissues and their spatial relationships to one another vary in different organs of the plant (FIGURE 31.12). Here, we survey the three tissue systems as they occur in a young, nonwoody plant.

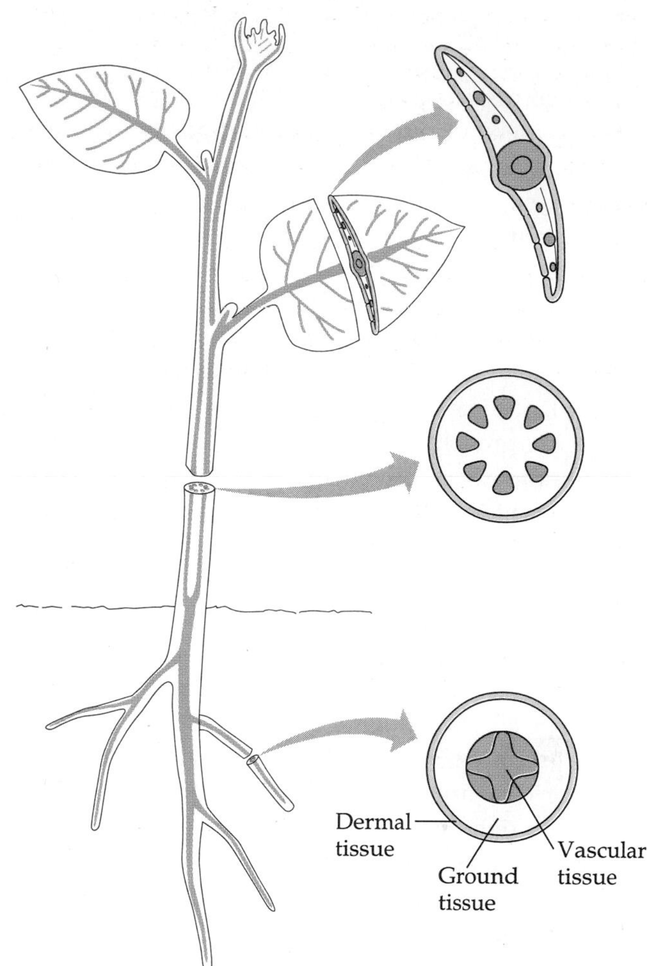

FIGURE 31.12
The three tissue systems. The dermal tissue system, or epidermis, is a single layer of cells that covers the entire body of a young plant. The vascular tissue system is also continuous throughout the plant, but it is arranged differently in each organ. The ground tissue system is located between the dermal tissue and vascular tissue in each organ.

The **dermal tissue system,** or **epidermis,** is generally a single layer of tightly packed cells that covers and protects all young parts of the plant—the "skin" of the plant. In addition to the general function of protection, the epidermis has more specialized characteristics consistent with the function of the particular organ it covers. For example, the root hairs so important in the absorption of water and minerals are extensions of epidermal cells near the tips of roots. The epidermis of leaves and most stems secretes a waxy coating called the **cuticle** that helps the aerial parts of the plant retain water, an important adaptation to living on land.

The continuum of xylem and phloem throughout the plant forms the **vascular tissue system,** which functions in transport and support. The specific organization of vascular tissue in stems and roots is discussed in the next section.

The **ground tissue system** makes up the bulk of a young plant, filling the space between the dermal and vascular tissue systems. Ground tissue is predominantly parenchyma, but collenchyma and sclerenchyma are also commonly present. Among the diverse functions of ground tissue are photosynthesis, storage, and support.

Learning how a plant grows will help you understand how the tissue systems are organized in the different plant organs. We focus on plant growth in the next three sections.

Meristems generate cells for new organs throughout the lifetime of a plant: *an overview of plant growth*

From season to season and from year to year, the growth of plants alters our surroundings—yards, campuses, parks, vacant lots, woods, and the other landscapes in our communities. The growth of a plant from a seed is a fascinating transformation. The early stages of this growth—germination of the seed and emergence of the seedling—are among the topics of Chapter 34. Here, we will learn how plants continue to grow after their shoot and root systems are established.

Most plants continue to grow as long as they live, a condition known as indeterminate growth. Most animals, in contrast, are characterized by determinate growth; that is, they cease growing after reaching a certain size. While whole plants usually show indeterminate growth, certain plant organs, such as leaves and flowers, exhibit determinate growth.

Indeterminate growth does not imply immortality. Most plants probably have lifespans that are genetically programmed; such plants have a fixed longevity even when grown in constant, favorable conditions. Other plants have lifespans that are environmentally determined; if the plants are grown under controlled temperature and light conditions and are protected from disease, they may live much longer than they typically do in natural environments. Plants known as **annuals** complete their life cycle—from germination through flowering and seed production to death—in a single year or less. Many wildflowers are annuals, as are the most important food crops, including the cereal grains and legumes. A plant is called a **biennial** if its life generally spans two years. Flowering usually occurs during the second year, after a year of vegetative growth. Beets and carrots are biennials, but we rarely leave them in the ground long enough to see them flower. Plants that live many years, including trees, shrubs, and some grasses, are known as **perennials.** Some of the buffalo grass of the North American plains is believed to have been growing for 10,000 years from seeds that sprouted at the close of the last ice age.

Plants have the capacity for indeterminate growth because they have perpetually embryonic tissues called **meristems** in their regions of growth. Meristematic cells are unspecialized, and they divide to generate additional cells. Some of the products of this division remain in the meristematic region to produce still more cells, while others become specialized and are incorporated into the tissues and organs of the growing plant. Cells that remain as wellsprings of new cells in the meristem are called initials; those that are displaced from the meristem and are then destined to specialize within a developing tissue are called derivatives.

The pattern of plant growth depends on the locations of the meristems (FIGURE 31.13). **Apical meristems,** located at the tips of roots and in the buds of shoots, supply cells for the plant to grow in length. This elongation, called **primary growth,** enables roots to ramify throughout the soil and shoots to increase their exposure to light and carbon dioxide. In herbaceous (nonwoody) plants, only primary growth occurs. In woody plants, however, there is also **secondary growth,** a progressive thickening of the roots and shoots formed earlier by primary growth. Secondary growth is the product of **lateral meristems,** cylinders of dividing cells extending along the length of roots and shoots. These lateral meristems replace the epidermis with a secondary dermal tissue that is thicker and tougher, and they also add layers of vascular tissue. Wood is the secondary xylem that accumulates over the years.

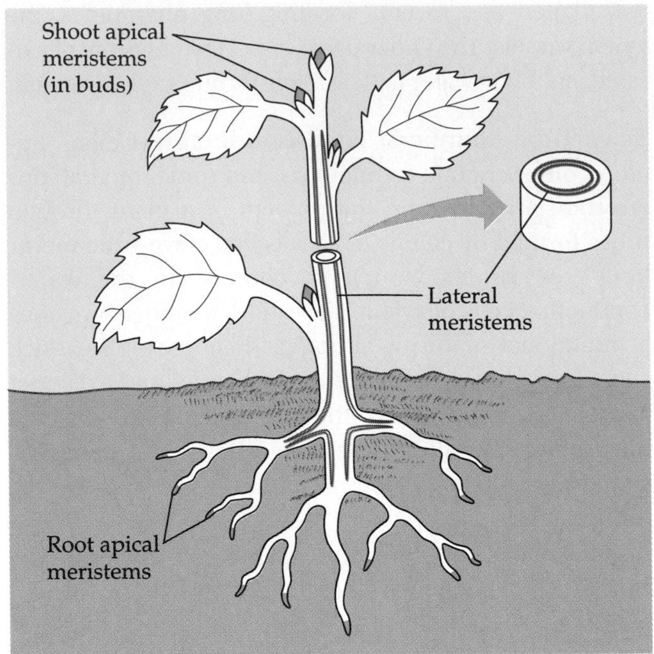

FIGURE 31.13
Locations of major meristems: An overview of plant growth. Meristems are self-renewing populations of cells that divide and provide cells for plant growth. Apical meristems, located near the tips of roots and shoots, are responsible for primary growth, or growth in length. Woody plants also have lateral meristems that function in secondary growth, which adds girth to roots and shoots.

In woody plants, primary and secondary growth occur at the same time, but in different locations. Primary growth is restricted to the youngest parts of the plant—the tips of roots and shoots, where the apical meristems are located. The lateral meristems develop in slightly older regions of the roots and shoots, some distance away from the tips. There, secondary growth adds girth to the organs. The oldest region of a root or shoot—the base of a tree branch, for example—has the greatest accumulation of secondary tissues formed by the lateral meristems. Each growing season, primary growth produces young extensions of roots and shoots, while secondary growth thickens and strengthens the older parts of the plant. Closer study of primary and secondary growth in the next two sections will help you understand the morphology and anatomy of plants.

Apical meristems extend roots and shoots: *a closer look at primary growth*

Primary growth produces what is called the **primary plant body,** which consists of the three tissue systems: dermal, vascular, and ground tissues (see FIGURE 31.12). A herbaceous plant and the youngest parts of a woody plant represent the primary plant body. Although apical meristems are responsible for the extension of both roots and shoots, there are important differences in the primary growth of these two kinds of organs.

Primary Growth of Roots

Primary growth pushes roots through the soil. The root tip is covered by a thimblelike **root cap,** which protects the delicate meristem as the root elongates through the abrasive soil. The cap also secretes a polysaccharide slime that lubricates the soil around the growing root tip. Growth in length is concentrated near the root's tip, where three zones of cells at successive stages of primary growth are located. From the root tip upward, they are the zone of cell division, the zone of elongation, and the zone of maturation. These regions grade together, with no sharp boundaries (FIGURE 31.14).

The **zone of cell division** includes the apical meristem and its derivatives, called primary meristems. The apical meristem, at the heart of the zone of cell division, produces the cells of the primary meristems and also replaces cells of the root cap that are sloughed off. Near the center of the apical meristem is the **quiescent center,** a population of cells that divide much more slowly than the other meristematic cells. Cells of the quiescent center are relatively resistant to damage from radiation and toxic chemicals, and they may function as reserves that can be recruited to restore the meristem if it is somehow damaged. In experiments where part of the apical meristem is removed, cells of the quiescent center become more mitotically active and produce a new meristem. Just above the apical meristem, the products of its cell division form three concentric cylinders of cells that continue to divide for some time. These are the primary

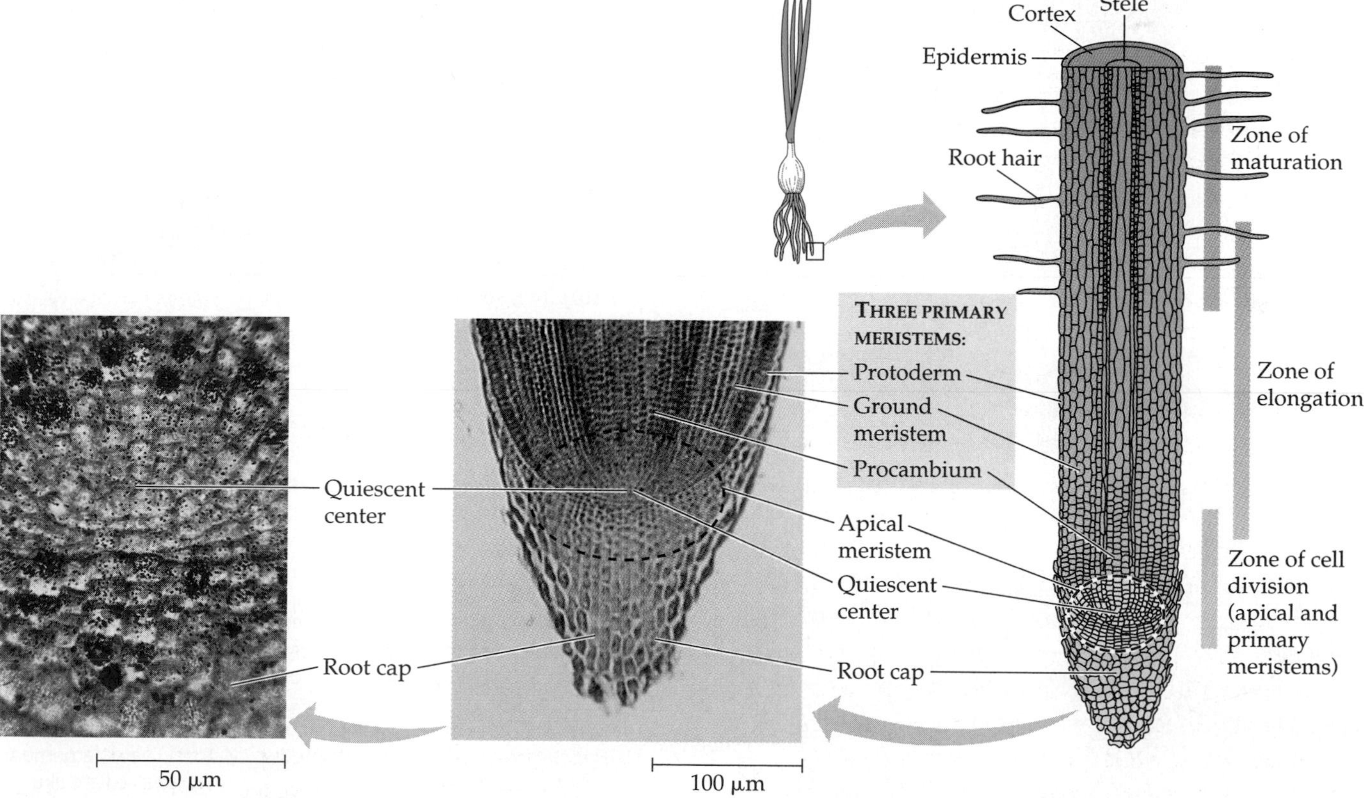

FIGURE 31.14

Primary growth of a root. Mitosis is concentrated in the zone of cell division, where the apical meristem and its products, the three primary meristems, are located. The apical meristem also maintains the root cap by generating new cells that replace those that are sloughed off (right LM). If the apical meristem is damaged, its quiescent center is activated and restores the meristem by means of cell division. Most lengthening of the root is concentrated in the zone of elongation. Cells become functionally mature in the zone of maturation. The zones of the root grade into one another without sharp boundaries. The far left micrograph of the apical meristem shows "hot spots" of cell division. This root was grown in a solution containing radioactive nucleotides, which are incorporated into the DNA of dividing cells. With the technique of autoradiography (see p. 30), the radioactive DNA exposes photographic film placed over the sectioned root, producing the black dots in the micrograph.

meristems—the **protoderm, procambium,** and **ground meristem**—which will produce the three primary tissue systems of the root: dermal, vascular, and ground tissues.

The zone of cell division blends into the **zone of elongation.** Here the cells elongate to more than ten times their original length. Although the meristem provides the new cells for growth, the elongation of cells is mainly responsible for pushing the root tip, including the meristem, ahead. The meristem sustains growth by continuously adding cells to the youngest end of the zone of elongation. Even before they finish elongating, the cells of the root begin to specialize in structure and function where the zone of elongation grades into the **zone of maturation.** In this latter region of the root, the three tissue systems produced by primary growth complete their differentiation.

Primary Tissues of Roots. The three primary meristems give rise to the three primary tissues of roots, shown in FIGURE 31.15. The protoderm, the outermost primary meristem, gives rise to the epidermis, a single layer of cells covering the root. Water and minerals that enter

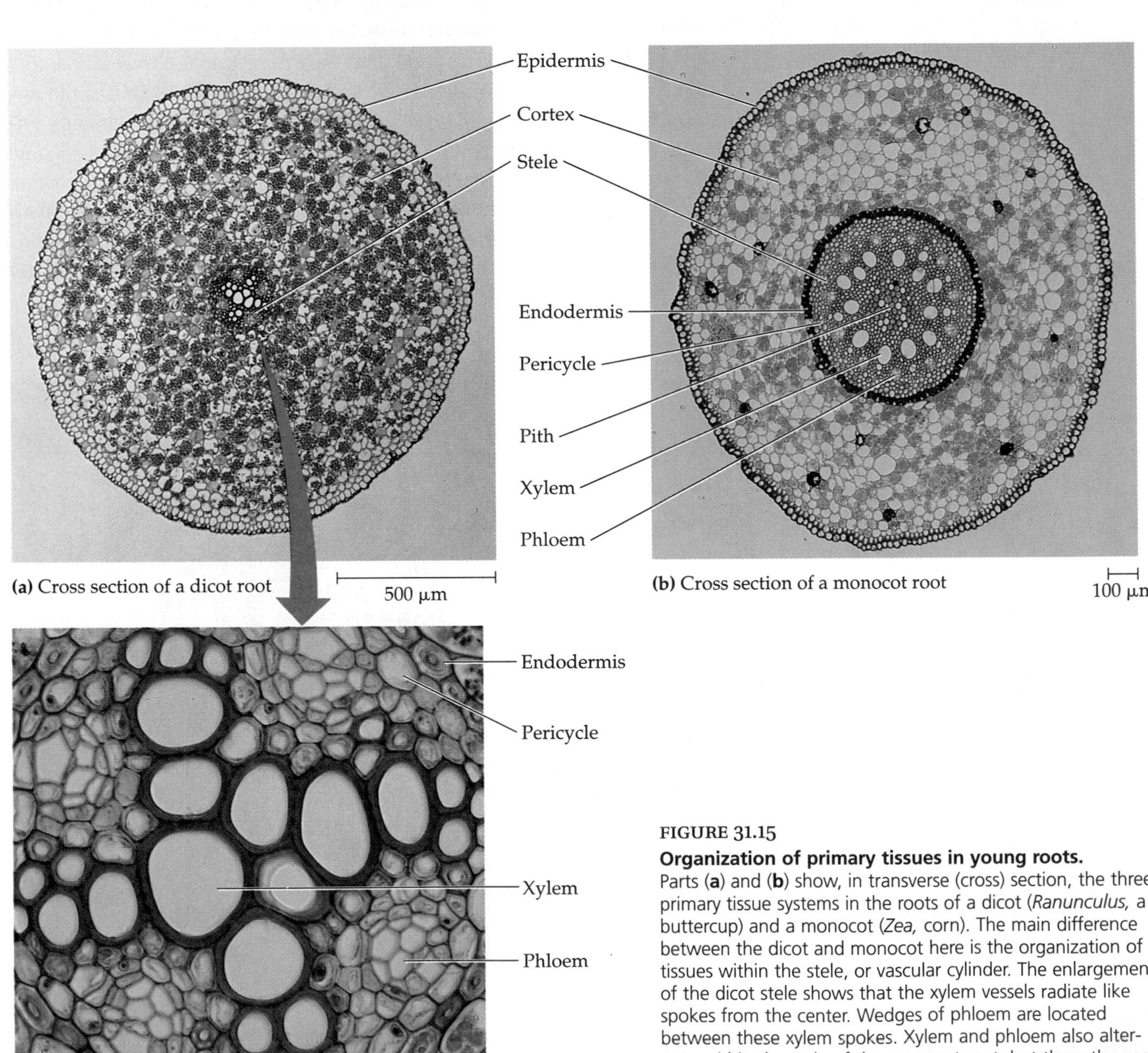

FIGURE 31.15
Organization of primary tissues in young roots.
Parts (**a**) and (**b**) show, in transverse (cross) section, the three primary tissue systems in the roots of a dicot (*Ranunculus,* a buttercup) and a monocot (*Zea,* corn). The main difference between the dicot and monocot here is the organization of tissues within the stele, or vascular cylinder. The enlargement of the dicot stele shows that the xylem vessels radiate like spokes from the center. Wedges of phloem are located between these xylem spokes. Xylem and phloem also alternate within the stele of the monocot root, but there the vascular tissues surround a core of parenchyma cells called the pith. In both dicots and monocots, the stele is circled by the endodermis, the innermost region of cells of the cortex. Just inside the endodermis is the pericycle, a layer of cells with the potential to divide and give rise to lateral roots. (All LMs.)

the plant from the soil must cross the epidermis. The root hairs enhance this process by greatly increasing the surface area of epidermal cells.

The procambium gives rise to a central vascular cylinder, or **stele,** where xylem and phloem develop. The specific arrangement of the two vascular tissues varies. In most dicots, the xylem cells radiate from the center of the stele in two or more spokes, with phloem developing in the wedges between the spokes. The stele of a monocot generally has a central core of parenchyma cells, often called the **pith,** which is ringed by vascular tissue with an alternating pattern of xylem and phloem.

Between the protoderm and procambium is the ground meristem, which gives rise to the ground tissue system. The ground tissue, which is mostly parenchyma, fills the **cortex,** the region of the root between the stele and epidermis. Ground tissue cells of the root store food, and their plasma membranes are active in the uptake of minerals that enter the root with the soil solution. The innermost layer of the cortex is the **endodermis,** a cylinder one cell thick that forms the boundary between the cortex and the stele. The endodermis functions as a selective barrier that regulates the passage of substances from the soil solution into the vascular tissue of the stele.

An established root may sprout **lateral roots,** which arise from the outermost layer of the stele, the **pericycle** (FIGURE 31.16). Just inside the endodermis, the pericycle is a layer of cells that may become meristematic and begin dividing again. Originating as a clump of cells formed by mitosis in the pericycle, a lateral root elongates and pushes through the cortex until it emerges from the primary root. The stele of the lateral root retains its connection with the stele of the primary root, making the vascular tissue continuous throughout the root system.

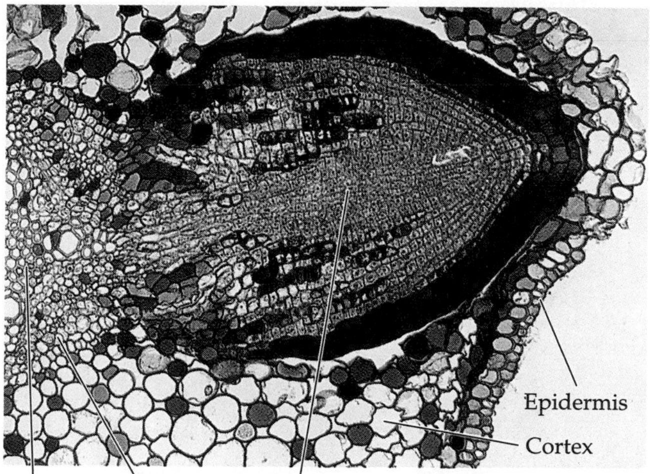

Epidermis
Cortex
Stele Pericycle Lateral root
50 μm

FIGURE 31.16

The formation of lateral roots. In this transverse section of a willow root, a lateral root emerges from the pericycle, the outermost layer of the stele (LM).

Primary Growth of Shoots

The apical meristem of a shoot is a dome-shaped mass of dividing cells at the tip of the terminal bud (FIGURE 31.17). As in the root, the apical meristem of the shoot tip gives rise to the primary meristems—protoderm, procambium, and ground meristem—which will differentiate into the three tissue systems. Leaves arise as leaf primordia on the flanks of the apical meristem. Axillary buds develop from islands of meristematic cells left by the apical meristem at the bases of the leaf primordia.

Within a bud, nodes, with their leaf primordia, are crowded close together, because internodes are very short. Most of the actual elongation of the shoot occurs by the growth of slightly older internodes below the shoot apex. This growth is due to both cell division and cell elongation within the internode. In some plants, including grasses, internodes continue to elongate all along the length of the shoot over a prolonged period.

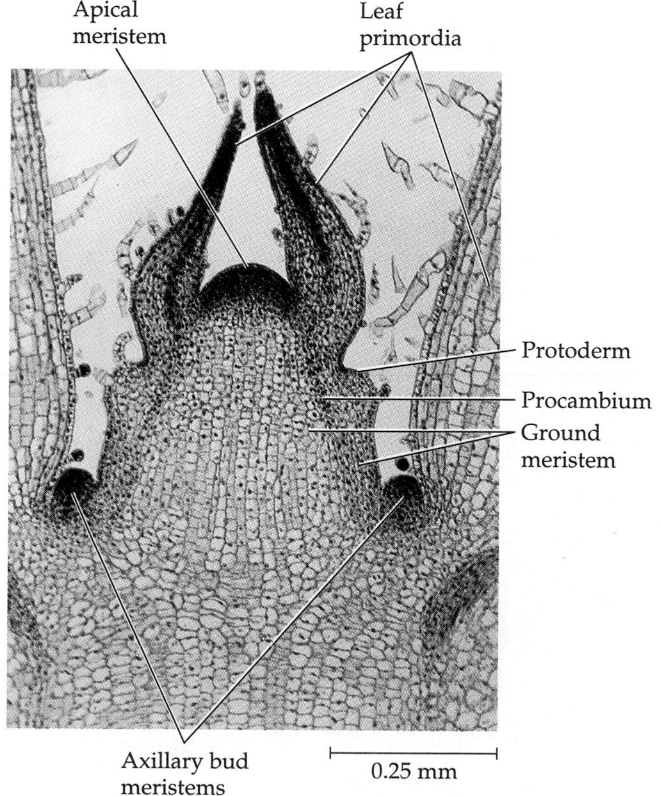

FIGURE 31.17

The terminal bud and primary growth of a shoot. Leaf primordia arise from the flanks of the apical dome. The apical meristem gives rise to protoderm, procambium, and ground meristem, which in turn develop into the three tissue systems. This is a longitudinal section of the shoot tip of *Coleus* (LM).

This is possible because these plants have meristematic regions, called intercalary meristems, at the base of each internode.

Axillary buds have the potential to form branches of the shoot system at some later time (see FIGURE 31.4). Thus, there is an important difference in how roots and shoots form lateral organs. Lateral roots originate from deep within a main root as outgrowths from the pericycle (see FIGURE 31.16). In contrast, branches of the shoot system originate from axillary buds, located at the surface of a main shoot. Only by extending from the stele can a lateral root be connected to the plant's vascular system. The vascular tissue of a stem, however, is near the surface, and branches can develop with connections to the vascular tissue without having to originate from deep within the main shoot.

Primary Tissues of Stems. Vascular tissue runs the length of a stem in several strands called **vascular bundles** (FIGURE 31.18). This arrangement contrasts with the root, where the vascular tissue forms a single stele consisting of the entire united set of vascular bundles (see FIGURE 31.15). At the transition zone where the shoot grades into the root, the vascular bundles converge to join the root stele.

Each vascular bundle of the stem is surrounded by ground tissue. In most dicots, the vascular bundles are arranged in a ring, with pith to the inside of the ring and cortex external to the ring. Both pith and cortex are part of the ground tissue system. The vascular bundles have their xylem facing the pith and their phloem facing the cortex side. The pith and cortex are connected by thin rays of ground tissue between the vascular bundles. In the stems of most monocots, the vascular bundles are scattered throughout the ground tissue rather than being arranged in a ring. The ground tissue of the stem is mostly parenchyma, but many stems are strengthened by collenchyma located just beneath the epidermis.

The protoderm of the terminal bud gives rise to the epidermis, which covers stems and leaves as part of the continuous dermal tissue system.

Tissue Organization of Leaves. The leaf is cloaked by its epidermis, with cells tightly interlocked like pieces of a puzzle (FIGURE 31.19). This epidermis, like our own skin, is a first line of defense against physical damage and

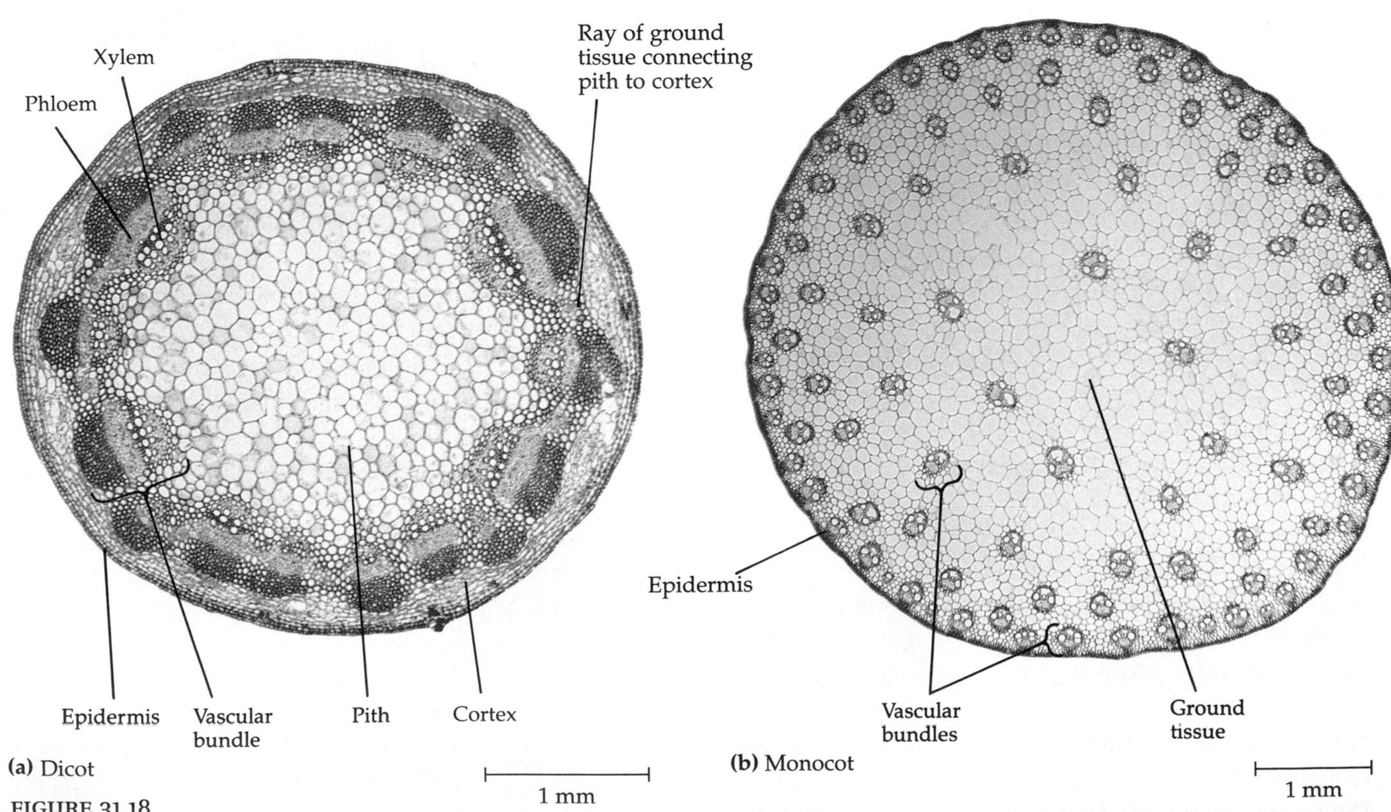

FIGURE 31.18
Organization of primary tissues in young stems. (**a**) A dicot stem (sunflower) with vascular bundles arranged in a ring. The ground tissue system consists of an outer cortex and an inner pith surrounded by vascular bundles. (**b**) A monocot stem (corn) with vascular bundles arranged in a complex manner throughout the ground tissue. (Both LMs.)

FIGURE 31.19

Leaf anatomy. (**a**) This cutaway drawing of a leaf illustrates the organization of the three tissue systems: dermal tissue (epidermis), vascular tissue, and ground tissue (mesophyll, consisting of palisade parenchyma and spongy mesophyll). (**b**) This surface view of a *Tradescantia* leaf shows the cells of the epidermis and the stomata with their guard cells (LM). The lower leaf surface generally has more stomata than the upper surface, an adaptation that helps reduce water loss by transpiration. (**c**) Palisade and spongy regions of mesophyll are present within the leaf of a lilac, a dicot (LM).

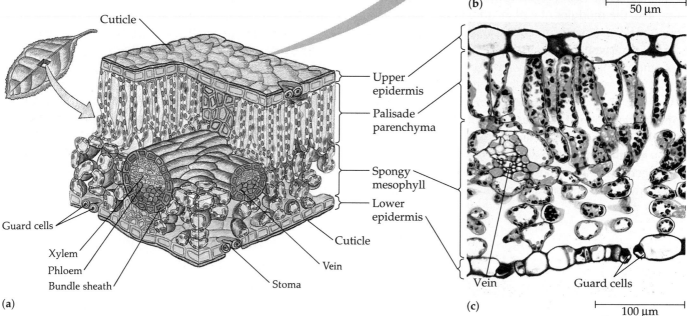

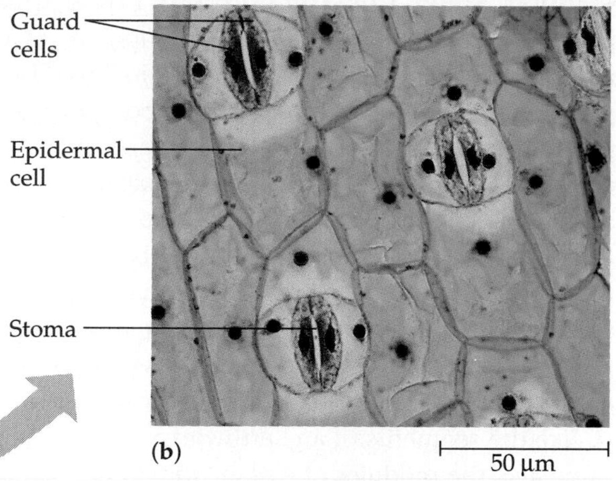

pathogenic organisms. Also, the waxy cuticle of the epidermis is a barrier to the loss of water from the plant. The epidermal barrier is interrupted only by the **stomata,** tiny pores flanked by specialized epidermal cells called **guard cells.** Each stoma is actually a gap between a pair of guard cells. The stomata allow gas exchange between the surrounding air and the photosynthetic cells inside the leaf. Stomata are also the major avenues for the loss of water from the plant by evaporation, a process called **transpiration.**

The ground tissue of a leaf, sandwiched between the upper and lower epidermis, is the **mesophyll** (Gr. *mesos,* "middle," and *phyll,* "leaf"). It consists mainly of parenchyma cells equipped with chloroplasts and specialized for photosynthesis. The leaves of many dicots have two distinct regions of mesophyll. On the upper half of the leaf are one or more layers of palisade parenchyma, made up of cells that are columnar in shape. Below the palisade region is the spongy mesophyll, which gets its name from the labyrinth of air spaces through which carbon dioxide and oxygen circulate around the irregularly shaped cells and up to the palisade region. The air spaces are particularly large in the vicinity of stomata, where gas exchange with the outside air occurs. In most plants, stomata are more numerous on the bottom surface of a leaf than on top. This adaptation minimizes water loss, which occurs more rapidly through stomata on the sunny upper side of a leaf. Once again, it helps to view the functional structure of plants in the evolutionary context of adaptation to land.

The vascular tissue of a leaf is continuous with the xylem and phloem of the stem. Leaf traces, which are branches from vascular bundles in the stem, pass through petioles and into leaves. Within a leaf, veins subdivide repeatedly and branch throughout the mesophyll. This brings xylem and phloem into close contact with the photosynthetic tissue, which obtains water and minerals from the xylem and loads its sugars and other organic products into the phloem for shipment to other parts of the plant. The vascular infrastructure also functions as a skeleton that supports the ground tissue (mesophyll) of the leaf.

Modular Shoot Construction and Phase Changes During Development. Serial development of nodes and internodes within the shoot apex, followed by elongation of the internodes, produces a shoot having a modular construction—a series of segments, each consisting of a stem, one or more leaves, and an axillary bud associated with each leaf (FIGURE 31.20). The development of this modular morphology should not be confused with the development of the segmented anatomy of certain animals such as earthworms. In animal development, the rudiments of all organs form in the embryo. Plants, in contrast, add organs at their tips for as long as they live. Unlike the segments of an earthworm, which are all the same age, the modules of a plant vary in age in proportion to their distance from an apical meristem.

From what you have learned so far about the shoot apex and primary growth, it would seem as if the meristem lays down a series of identical modules for as long as the shoot lives. In fact, the apical meristem can change from one developmental phase to another during its history. One of these phase changes is a gradual transition from a juvenile vegetative (leaf-producing) state to a mature vegetative state. Modification of leaf morphology is usually the most obvious sign of this phase change: The leaves of juvenile versus mature shoot modules differ in shape and other features. Once the meristem has laid down juvenile nodes and internodes, they retain that status even as the shoot continues to elongate and the meristem eventually changes to the mature phase. If axillary buds give rise to branches, those shoots reflect the developmental phase of the main shoot modules from which they arise. The juvenile-to-mature phase transition is another case where it is misleading to compare plant and animal development. In an animal, this transition occurs at the level of the entire organism. In plants, phase changes during the history of apical meristems can result in juvenile and mature regions coexisting along the axis of a shoot.

In some cases, a shoot apex undergoes a second phase transition from a mature vegetative state to a reproductive (flower-producing) state. Unlike vegetative growth, which is self-renewing, the production of a flower by an apical meristem terminates primary growth of that shoot tip; the apical meristem is consumed in the production of the flower's organs.

In Chapter 34, we will study flower development in more detail, and in Chapter 35, we will examine how this phase change from vegetative growth of a shoot to the reproductive growth of flowering is controlled.

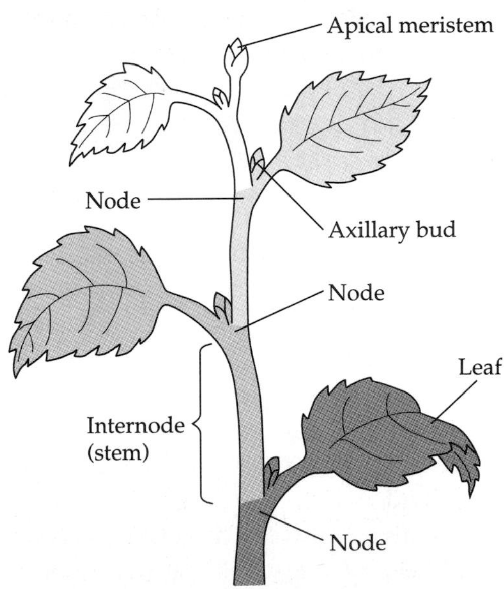

FIGURE 31.20
Modular construction of a shoot. Primary growth lays down a series of segments (different colors in this drawing), each consisting of a node with one or more leaves, an axillary bud in the axil of each leaf, and an internode. Miniature modules develop within the apical meristem and then grow, pushing the apex onward, where it forms the next module, and so on. This serial addition of segments at the growing end of the shoot contrasts with the development of segmentation in certain animals, in which all the segments form at about the same time in the embryo. (Imagine if our arms and legs were different ages, like the appendages of a plant!)

Lateral meristems add girth to stems and roots: *a closer look at secondary growth*

Most vascular plants undergo secondary growth, increasing in girth as well as length. The **secondary plant body** consists of the tissues produced during this secondary growth in diameter. Two lateral meristems function in secondary growth: the **vascular cambium,** which produces secondary xylem and phloem; and the **cork cambium,** which produces a tough, thick covering for stems and roots that replaces the epidermis. Secondary growth occurs in all gymnosperms. Among angiosperms, secondary growth takes place in most dicot species but is rare in monocots.

Secondary Growth of Stems

Vascular Cambium. The vascular cambium forms from parenchyma cells that develop the capacity to divide; that is, the cells become meristematic. This transition to meristematic activity takes place in a layer between the primary xylem and primary phloem of each vascular bundle and in the rays of ground tissue between the bundles (FIGURE 31.21). The cambium within the vascular bundle is called fasicular cambium; the portion of the cambium in the rays between vascular bundles is

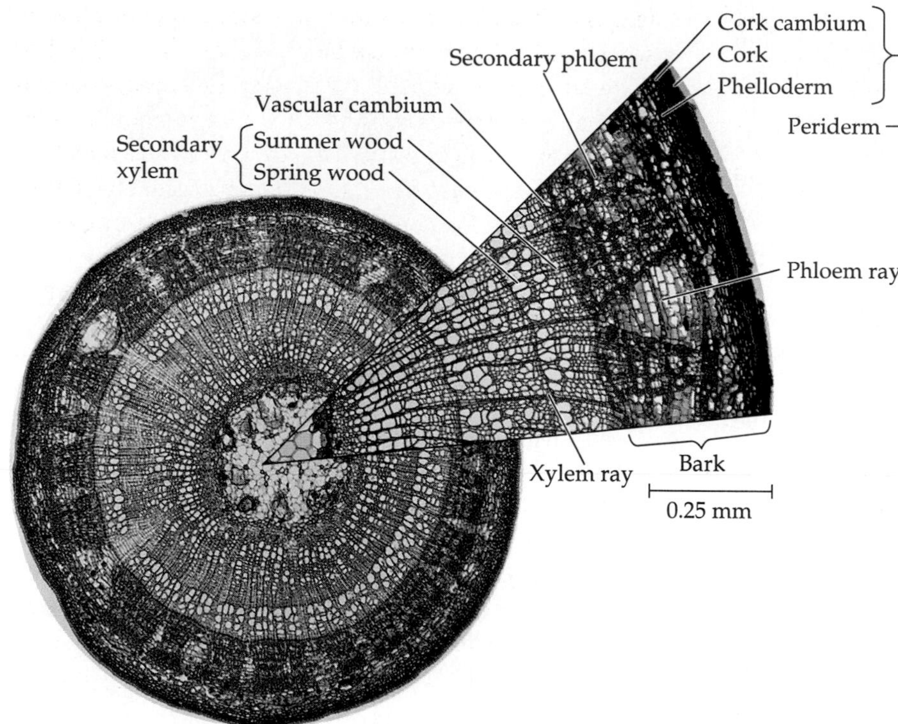

FIGURE 31.21

Anatomy of a woody dicot stem. A few years of secondary growth are apparent as growth rings in this transverse section of a stem from *Tilia,* the American linden (basswood). At the boundary of one season's growth to the next, notice the spring wood and summer wood. Vascular cambium produces the secondary growth. Secondary xylem and secondary phloem are derived from fusiform initials, cambium cells that develop within the vascular bundles. The cambium cells that develop in the intervening rays of ground tissue are called ray initials; they give rise to the xylem and phloem rays, structures that function in radial transport and storage. A second lateral meristem, the cork cambium, produces cork cells to its outside and phelloderm cells to its inside. Together, the cork cambium and its derivatives, cork and phelloderm, make up the periderm, which is the protective covering of the secondary plant body. The bark consists of all tissues external to the vascular cambium. (Both LMs.)

called interfasicular cambium. Together, the meristematic bands in these fasicular and interfasicular regions give rise to the vascular cambium as a continuous cylinder of dividing cells surrounding the primary xylem and pith of the stem.

The meristematic cells of the interfasicular cambium are called **ray initials.** They produce radial files of parenchyma cells known as **xylem rays** and **phloem rays,** which function as living avenues for the lateral transport of water and nutrients and in the storage of starch and other reserves. The cambium cells within the vascular bundles (fasicular cambium) are the **fusiform initials,** a name that refers to the shape of these cells, which have tapered (fusiform) ends and are elongated along the axis of the stem. Fusiform initials produce new vascular tissue, forming secondary xylem to the inside of the vascular cambium and secondary phloem to the outside (FIGURE 31.22).

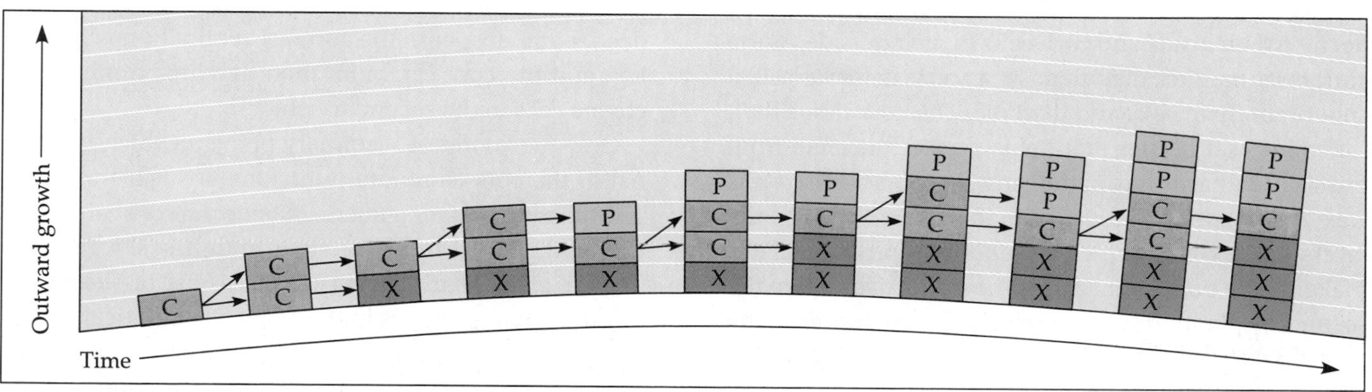

FIGURE 31.22

Production of secondary xylem and phloem by the vascular cambium. This diagram traces the file of cells that develops from the meristematic activity of a single fusiform initial of the vascular cambium as viewed in transverse section. The cambium cell (C) gives rise to xylem (X) on the inside and phloem (P) on the outside. Each time an initial divides, one daughter cell retains its status as an initial, and the other, the derivative, differentiates into a xylem or phloem cell. As layers of xylem are added, the position of the cambium becomes more distant from the center of the stem.

As secondary growth continues over the years, layer upon layer of secondary xylem accumulates, producing what we call wood. Wood consists mainly of tracheids, vessel elements (in angiosperms), and fibers. These cells, dead at functional maturity, have thick, lignified walls that give wood its hardness and strength. In temperate regions of the world, secondary growth in perennial plants is interrupted each year when the vascular cambium becomes dormant during winter. When secondary growth resumes in the spring, the first tracheids and vessel elements to develop usually have relatively large diameters and thin walls compared to the secondary xylem produced later in the summer. Thus, it is usually possible to distinguish spring wood from summer wood (see FIGURE 31.21). The annual growth rings that are evident in cross sections of most tree trunks in temperate regions result from this yearly activity of the vascular cambium: cambium dormancy, spring wood production, and summer wood production. The boundary between one year's growth and the next is usually quite conspicuous, sometimes allowing us to estimate the age of a tree by counting its annual rings.

The secondary phloem, external to the vascular cambium, does not accumulate as extensively over the years as the secondary xylem does. As a tree grows in girth, the older (outermost) secondary phloem, and all tissues external to it, develop into bark, which eventually splits and sloughs off the tree trunk.

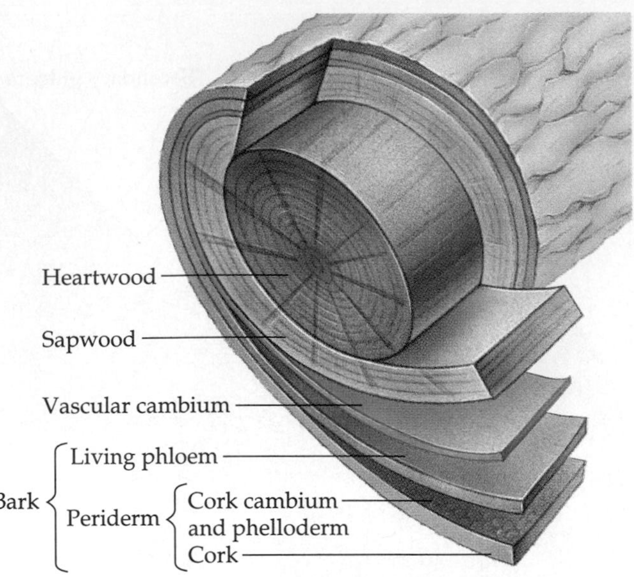

FIGURE 31.23
Anatomy of a tree trunk. Beginning at the center of the tree and tracing outward, we can distinguish several zones. Heartwood and sapwood both consist of secondary xylem. Heartwood is older and no longer functions in water transport; the lignified walls of its dead cells form a central column that supports the tree. This wood owes its rich color to resins and other compounds that clog the cell cavities and help protect the core of the tree from fungi and insects. Sapwood is so named because its secondary xylem cells still function in the upward transport of water and minerals (xylem sap). Since each new layer of secondary xylem has a larger circumference, secondary growth enables the xylem to transport more sap each year, providing water and minerals to an increasing number of leaves.

Cork Cambium. During secondary growth, the epidermis produced by primary growth splits, dries, and falls off the stem. It is replaced by new protective tissues produced by the cork cambium, a cylinder of meristematic tissue that first forms in the outer cortex of the stem (see FIGURE 31.21). As initials in the cork cambium divide, they give rise to parenchyma cells called **phelloderm** to their inside and **cork cells** to their outside. As the cork cells mature, they deposit a waxy material called suberin in their walls and then die. The cork tissue then functions as a barrier that helps protect the stem from physical damage and pathogens. And because cork is waxy, it impedes water loss from the stems. Together, the layers of cork plus the cork cambium and phelloderm make up the **periderm.** This is the protective coat of the secondary plant body that replaces the epidermis of the primary body. The term **bark,** more inclusive than periderm, refers to all tissues external to the vascular cambium. Thus, in an outward direction, bark consists of phloem, phelloderm, cork cambium, and cork. Put another way, bark is phloem plus periderm (FIGURE 31.23).

Unlike the vascular cambium, which grows in diameter, the original cork cambium is a cylinder of fixed size. After a few weeks of cork production, the cork cambium loses its meristematic activity, and its remaining cells differentiate into cork. Expansion of the stem splits the original periderm. How is it renewed to keep pace with continued secondary growth? New cork cambium forms deeper and deeper in the cortex. Eventually, no cortex is left, and the cork cambium then develops from parenchyma cells in the secondary phloem.

Only the youngest secondary phloem, which is internal to the cork cambium, functions in sugar transport. The older secondary phloem, outside the cork cambium, dies and helps protect the stem until it is sloughed off as part of the bark during later seasons of secondary growth. Spongy regions in the bark called **lenticels** make it possible for living cells within the trunk to exchange gases with the outside air for cellular respiration.

The result of many years of secondary stem growth can be seen by examining an old tree trunk in cross section (see FIGURE 31.23). FIGURE 31.24 will help you review the relationships among the primary and secondary tissues of a woody plant.

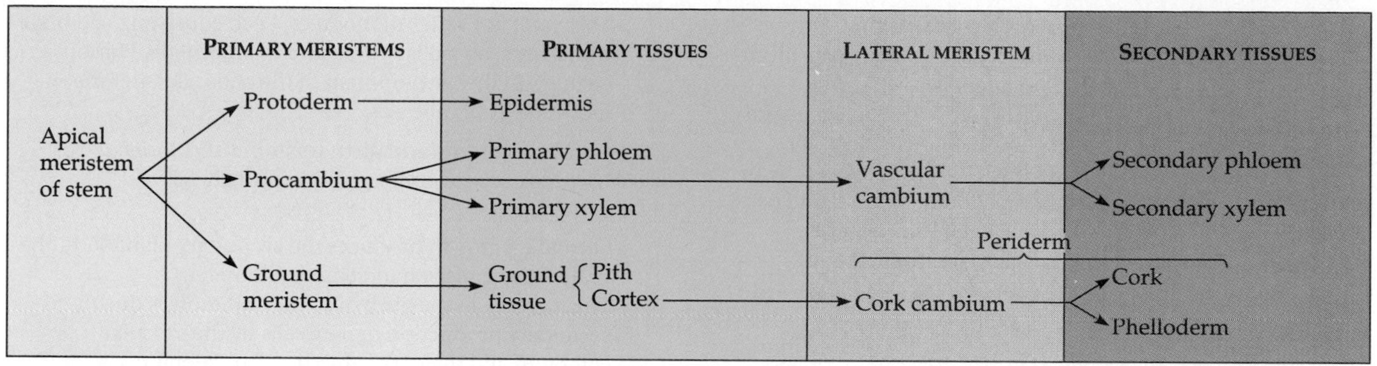

FIGURE 31.24
A summary of primary and secondary growth in a woody stem.

Secondary Growth of Roots

The two lateral meristems, vascular cambium and cork cambium, also develop and produce secondary growth in roots. The vascular cambium forms within the stele and produces secondary xylem to its inside and secondary phloem to its outside. As the stele grows in diameter, the cortex and epidermis are split and shed. A cork cambium forms from the pericycle of the stele and produces the periderm, which becomes the secondary dermal tissue. Unlike the primary epidermis of a younger root, periderm is impermeable to water. Therefore, it is only the youngest roots, those representing the primary plant body, that absorb water and minerals from the soil. Older roots, with secondary growth, function mainly to anchor the plant and to transport water and solutes between the younger roots and the shoot system.

Over the years, the root becomes more woody, and annual rings are usually evident in the secondary xylem. The tissues external to the vascular cambium form a thick, tough bark. After extensive secondary growth, old stems and old roots are quite similar.

*   *   *

In dissecting the plant to examine its parts, as we have done in this chapter, we must remember that the whole plant functions as an integrated organism. In the following chapters, you will learn more about how materials are transported within the plant, how plants obtain nutrients, how plants reproduce and develop, and how the various functions of the plant are coordinated. Your understanding of the working plant will be enhanced by remembering that structure fits function and that interactions with the environment affect the anatomy and physiology of plants.

REVIEW OF KEY CONCEPTS (with page numbers and key figures)

- Plant biology reflects the major themes in the study of life (pp. 668–670)
  - Plants are the main producers in most terrestrial ecosystems.
  - A plant is adapted to living on land through evolution and through individual response to the environment.
- A plant's root and shoot systems are evolutionary adaptations to living on land (pp. 670–675, FIGURE 31.4)
  - Based on differences in anatomy, the angiosperms can be divided into two classes: monocots and dicots.
  - Differentiation of the plant body into an underground root system and an aerial shoot system is an adaptation to terrestrial life.
  - Vascular tissues integrate the parts of the plant body. Water and minerals move up from the roots in the xylem; sugar travels to nonphotosynthetic parts in the phloem.
  - The structure of roots is adapted to anchor the plant, absorb and conduct water and minerals, and store food. Tiny root hairs near the root tips enhance absorption.
  - The shoot system consists of stems, leaves, and flowers.
  - Leaves are attached by their petioles to the nodes of stems. Axillary buds that are stimulated to grow may become flowers or vegetative branches. Stolons, rhizomes, and bulbs are modified stems.
- The many types of plant cells are organized into three major tissue systems (pp. 675–679, FIGURES 31.10, 31.12)
  - Parenchyma cells are the least specialized plant cells, peforming general metabolic and storage functions. They retain the ability to divide and differentiate into other cell types under certain conditions.
  - Collenchyma cells support young parts of the plant shoot without restraining growth.

- Sclerenchyma cells, fibers and sclereids, are supportive cells with thick, lignified secondary walls. Many lack protoplasts; thus, at maturity they are unable to elongate.
- Water-conducting xylem tissue is composed of elongated tracheid and vessel element cells that are dead at functional maturity. Tracheids are long, thin, tapered cells with lignified secondary walls that function in support and permit water flow through pits. The wider, shorter, and thinner-walled vessel elements have perforated ends through which water flows freely.
- Sieve-tube members are living cells that form phloem tubes for the transport of sucrose and other organic nutrients. Each sieve-tube member is connected to one or more companion cells by plasmodesmata.
- Plant tissues are arranged into three continuous systems. The dermal tissue system, or epidermis, is an external layer of tightly packed cells that functions in protection. The vascular tissue system, consisting of xylem and phloem, provides transport and support. The predominantly parenchymous ground tissue system functions in organic synthesis, storage, and support.

■ Meristems generate cells for new organs throughout the lifetime of a plant: *an overview of plant growth* (pp. 679–680, FIGURE 31.13)

- Because they possess permanently embryonic meristems, plants, unlike animals, show indeterminate growth.
- Apical meristems at root tips and shoot buds initiate primary growth (growth in length) and the formation of the three tissue systems. Lateral meristems are responsible for secondary growth (growth in thickness).

■ Apical meristems extend roots and shoots: *a closer look at primary growth* (pp. 681–686, FIGURES 31.14–31.19)

- Primary growth produces the primary plant body, which consists of the three tissue systems.
- Root tips, protected by the root caps, grow and develop by the activities of cells in the zones of cell division, elongation, and maturation.
- Just behind the apical meristem in the zone of cell division are the three primary meristems of the root. The protoderm gives rise to the epidermis, the procambium forms the central vascular stele, and the ground meristem produces the ground tissue of the cortex. Subsequent lateral roots arise from the pericycle of the stele.
- The elongation of shoots comes from the dome-shaped apical meristem at the top of the terminal bud. Leaf primordia arise from the sides of the apical dome, and axillary buds arise from residual islands of meristematic cells at the bases of leaf primordia.
- In contrast to the single stele of the root, the vascular tissue of stems runs in vascular bundles surrounded by ground tissue in characteristic patterns that differ between monocots and dicots.
- Leaves are covered with a waxy epidermis. Pairs of guard cells flank openings called stomata, through which gas exchange and transpiration occur. Between the upper and lower epidermis, the ground tissue, or mesophyll, consists mainly of parenchyma cells equipped with chloroplasts for photosynthesis. A strand of vascular tissue called the leaf trace connects the veins of the leaf with the vascular tissue of the stem.
- The shoot is a series of modules, each consisting of a node with leaves, an axillary bud, and an internode. Phase changes in the development of the shoot tip alter the morphology of modules.

■ Lateral meristems add girth to stems and roots: *a closer look at secondary growth* (pp. 686–689, FIGURE 31.24)

- Secondary growth produces the secondary plant body, the tissues that cause an increase in diameter.
- The increase in the girth of stems and roots is due to secondary production of new cells by the vascular cambium and the cork cambium, two lateral meristems.
- The vascular cambium, a continuous cylinder of meristematic cells, produces secondary xylem internally and secondary phloem externally.
- The cork cambium, a meristematic cylinder in the outer cortex of the stem, produces waxy cork cells externally and phelloderm (a parenchyma) internally. The cork cambium, phelloderm, and cork make up the periderm, which replaces the epidermis that sloughs off during secondary growth. Secondary phloem gives rise to new cork cambium after the original cortex is shed. Bark consists of phloem plus periderm.
- In roots, the vascular cambium arises between the xylem and phloem of the stele and functions similarly to that in stems. Cork cambium, produced from the pericycle of the stele, forms the periderm that replaces cortex and epidermis.

SELF-QUIZ

1. Which structure is *incorrectly* paired with its tissue system?
   a. root hair—dermal tissue
   b. mesophyll—ground tissue
   c. guard cell—dermal tissue
   d. companion cell—ground tissue
   e. tracheid—vascular tissue

2. The lateral roots of a young dicot originate from the
   a. pericycle of the taproot
   b. endodermis of fibrous roots
   c. meristematic cells of the protoderm
   d. vascular cambium
   e. root cortex

3. A sieve-tube member would likely lose its nucleus in which zone of growth in a root?
   a. zone of cell division
   b. zone of elongation
   c. zone of maturation
   d. root cap
   e. quiescent center

4. Vessel elements of the primary plant body originate from the
   a. protoderm
   b. procambium
   c. ground meristem
   d. xylem rays
   e. cork cambium

5. Which of the following is *not* a correctly stated difference between monocots and dicots?
   a. parallel veins in monocots; branching, netlike venation in dicot leaves
   b. vascular bundles scattered in monocot stems; central vascular stele in dicot stems
   c. flower parts in threes in monocots; flower parts in multiples of four or five in dicots
   d. usually only primary growth in monocots; secondary growth in many dicots
   e. one cotyledon in monocots; two cotyledons in dicots

6. Ivy (*Hedera helix*) undergoes a gradual change from a juvenile vegetative state to a mature vegetative state. This results in mature leaves on upper branches having a different shape than juvenile leaves on lower branches. If the phase of the axillary buds is fixed, then the lateral buds of lower branches can develop to form ____________ branches, and the lateral buds of upper branches can develop to form ____________ branches.
   a. only juvenile; only mature
   b. only mature; only juvenile
   c. juvenile or mature; only juvenile
   d. only mature; juvenile or mature
   e. juvenile or mature; only mature

7. Wood consists mostly of
   a. bark
   b. periderm
   c. secondary xylem
   d. secondary phloem
   e. cork

8. Which of the following is not part of an older tree's bark?
   a. cork
   b. cork cambium
   c. lenticels
   d. secondary xylem
   e. secondary phloem

9. Each module of a primary plant body's shoot system consists of a
   a. stem, leaf, and axillary bud
   b. stem, axillary bud, and apical bud
   c. leaf, flower, and stem
   d. stem, flower, and axillary bud
   e. node, internode, and apical bud

10. Which of the following cell types or structures is *incorrectly* paired with its meristematic origin?
   a. epidermis—protoderm
   b. stele—procambium
   c. cortex—ground meristem
   d. secondary phloem—cork cambium
   e. three primary meristems—apical meristem

CHALLENGE QUESTIONS

1. If you were to live for the next several decades in a treehouse built on the large, lower branches of a tree, would you gain much altitude as the tree grew? Explain your answer.
2. Starting at the surface of a tree trunk and working to the center, describe the structure and function of the tissue layers.
3. Describe some important differences in the growth and development of plants and animals.
4. Choose three specialized types of plant cells and describe how their structures are adapted for their specific functions.

SCIENCE, TECHNOLOGY, AND SOCIETY

1. Make a list of the plants and plant products you use in a typical day. How do you use these various plant products? Do you think the number of plants and plant products used in everyday life has increased or decreased in the last century? Do you think the number is likely to increase or decrease in the future? Why?
2. On your next trip to the grocery store, take a notepad and list the types of produce (fruits and vegetables) in one column and the parts of plants they represent in a parallel column.

FURTHER READING

Bolz, D. M. "A World of Leaves: Familiar Forms and Surprising Twists." *Smithsonian*, April 1985. A delightful article on the adaptations of leaves, featuring exquisite photographs.

Dale, J. "How Do Leaves Grow?" *BioScience*, June 1992. Using new techniques of molecular and cell biology, researchers are answering long-standing questions about plant structure and growth.

Galston, A. W. *Life Processes of Plants*. New York: W. H. Freeman, 1994. Plant structure and physiology, with an emphasis on interactions with light.

Gillis, A.M. "Using a Mousy, Little Flower to Understand Flamboyant Ones." *BioScience*, May 1995. The value of *Arabidopsis* as a model for developmental genetics.

Kaplan, D. R., and W. Hagemann. "The Relationship of Cell and Organism in Vascular Plants: Are Cells the Building Blocks of Plant Form?" *BioScience*, November 1991. Important differences in how plants and animals are organized.

Meyorowitz, E. M. "The Genetics of Flower Development." *Scientific American*, November 1994. The value of mutations in research.

Moore, R., W. D. Clark, and K. R. Stern. *Botany*. Dubuque, IA: W. C. Brown, 1995. An excellent introduction to plant biology.

Poethig, R. S. "Phase Change and the Regulation of Short Morphogenesis in Plants." *Science*, November 16, 1990. The molecular and cellular basis of shoot development.

CHAPTER 32

TRANSPORT IN PLANTS

KEY CONCEPTS

- The traffic of water and solutes occurs on cellular, organ, and whole-plant levels: *an overview of transport in plants*
- Roots absorb water and minerals from soil
- The ascent of xylem sap depends mainly on transpiration and the physical properties of water
- Guard cells mediate the transpiration-photosynthesis compromise
- A bulk-flow mechanism translocates phloem sap from sugar sources to sugar sinks

The algal ancestors of plants were completely immersed in water and dissolved minerals, and none of their cells was far from these ingredients. The evolutionary journey onto land involved the differentiation of the plant body into roots, which absorb water and minerals from soil, and shoots, which are exposed to light and atmospheric CO_2. This body plan enables plants to survive in an environment where chemical resources are divided between two media, soil and air. But the morphological solution to a dual environment posed a new problem: the need to transport materials between roots and shoots, sometimes over long distances. For example, the leaves of the aspen trees in the photograph that opens this chapter are more than 100 m away from the roots. These remote organs are bridged by vascular tissues that transport sap throughout the plant body. Water and minerals absorbed by roots are drawn upward in the xylem to shoots. Sugar produced by photosynthesis is exported from leaves to other organs via the phloem. The whole plant depends on this commerce between organs to integrate the activities of its specialized parts. The mechanisms responsible for this internal transport are the subjects of this chapter.

The traffic of water and solutes occurs on cellular, organ, and whole-plant levels: *an overview of transport in plants*

Transport in plants occurs on three levels: (1) the uptake and release of water and solutes by individual cells, such as the absorption of water and minerals from the soil by cells of a root; (2) short-distance transport of substances from cell to cell at the level of tissues and organs, such as the loading of sugar from photosynthetic cells of a leaf into the sieve tubes of phloem; and (3) long-distance transport of sap within xylem and phloem at the level of the whole plant. FIGURE 32.1 provides an overview of these transport functions in plants.

Transport at the Cellular Level

The transport of solutes and water across biological membranes was covered in detail in Chapter 8. Here, we reexamine a few of these transport processes in the specific context of plant cells.

A Review of the Active and Passive Transport of Solutes. The selective permeability of a plant cell's plasma membrane controls the movement of solutes between the cell and the extracellular solution. Recall from Chapter 8 that solutes tend to diffuse down their gradi-

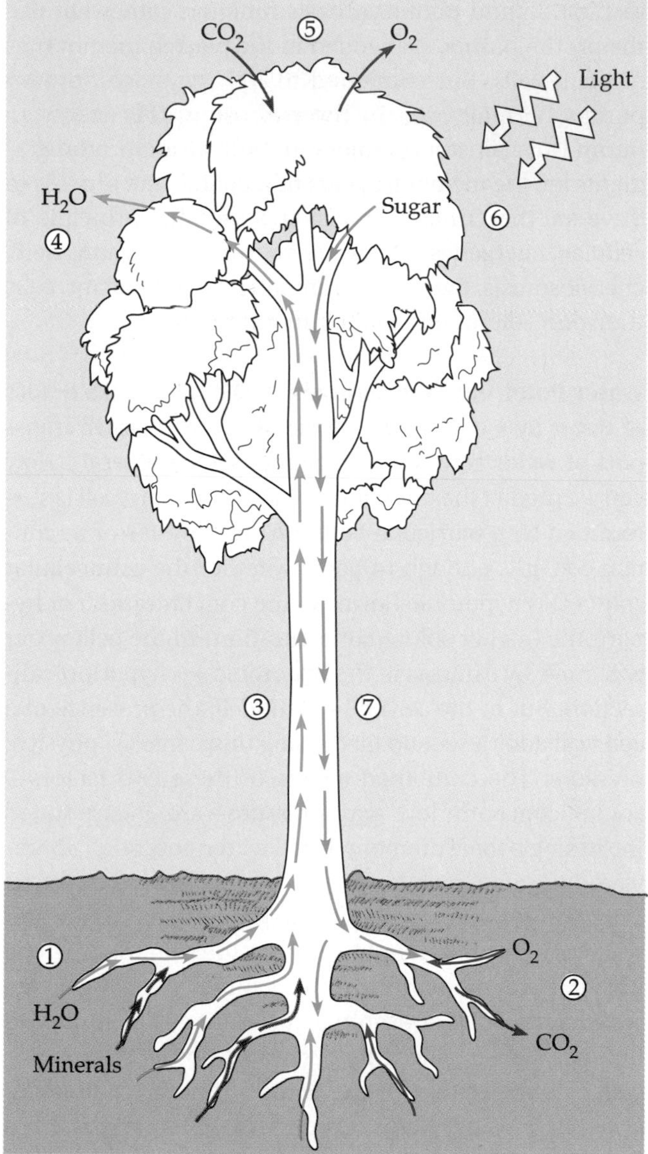

FIGURE 32.1

An overview of transport in plants. ① Roots absorb water and dissolved minerals from the soil. ② Roots also exchange gases with the air spaces of soil, taking in O_2 and discharging CO_2. This gas exchange supports the cellular respiration of root cells. ③ Water and minerals are transported upward as xylem sap within xylem, from the roots into the shoot system. ④ Transpiration, the evaporation of water from leaves (mostly through stomata), creates a force within leaves that pulls xylem sap upward. ⑤ Leaves also exchange gases through stomata, taking in the CO_2 that provides carbon for photosynthesis and expelling O_2. ⑥ Sugar is produced by photosynthesis in the leaves and ⑦ is transported within phloem in a solution called phloem sap to roots and other nonphotosynthetic parts of the plant.

ents, and when this occurs across a membrane, the process is termed passive transport (*passive,* because it happens without the direct expenditure of metabolic energy by the cell). Most solutes, however, diffuse very slowly across a membrane unless they can pass through **transport proteins** embedded in the membrane. Some of these function as specific **carrier proteins,** which facilitate diffusion by binding selectively to a solute on one side of the membrane and releasing the solute on the opposite side. Transfer of the solute across the membrane involves a conformational (shape) change by the carrier protein. Other transport proteins function as **selective channels.** Unlike carriers, which physically transport solutes, channels are simply selective passageways across the membrane. For example, the membranes of most plant cells have potassium channels that allow potassium ions (K^+) to pass, but not similar ions, such as sodium (Na^+). Some channels are gated; that is, certain environmental stimuli can cause the channels to open or close. For instance, we will see later in the chapter how the regulation of K^+ gates in the membranes of guard cells functions in the opening and closing of stomata.

Recall also that active transport is the pumping of solutes across membranes against their gradients. It is termed *active* because the cell must expend metabolic energy, usually in the form of ATP, to transport a solute "uphill"—that is, counter to the direction in which the solute diffuses. One important active transporter in plant cells is the **proton pump,** which hydrolyzes ATP and uses the released energy to pump hydrogen ions (H^+) out of the cell. This results in a proton gradient, with the H^+ concentration higher outside the cell than inside the cell. The gradient is a form of stored energy, because the H^+ ions tend to diffuse "downhill," back into the cell. And because the proton pump moves positive charge, in the form of H^+, out of the cell, the pump also generates a membrane potential. Membrane potential is a voltage, a separation of opposite charges across a membrane. Proton pumping makes the inside of a plant cell negative in charge relative to the outside. This voltage is called a membrane *potential* because the charge separation is a form of potential (stored) energy that can be harnessed to perform cellular work.

Plant cells use energy stored in the proton gradient and the membrane potential to drive the transport of many different solutes (FIGURE 32.2). Consider, for example, one mechanism root cells use to absorb potassium from the soil solution. Because potassium ions are positively charged, and the inside of the cell is negatively charged compared to the outside, the membrane potential helps drive K^+ into the cell. Because K^+ is diffusing down its electrochemical gradient (see Chapter 8), accumulation of the ion by this mechanism represents passive transport. But it is the active transport of H^+ that maintains the membrane potential and makes it possible for the cell to accumulate K^+. In other cases, energy stored by H^+ pumping can actually be used to drive the transport of solutes *against* their electrochemical gradients. For example, many negatively charged minerals,

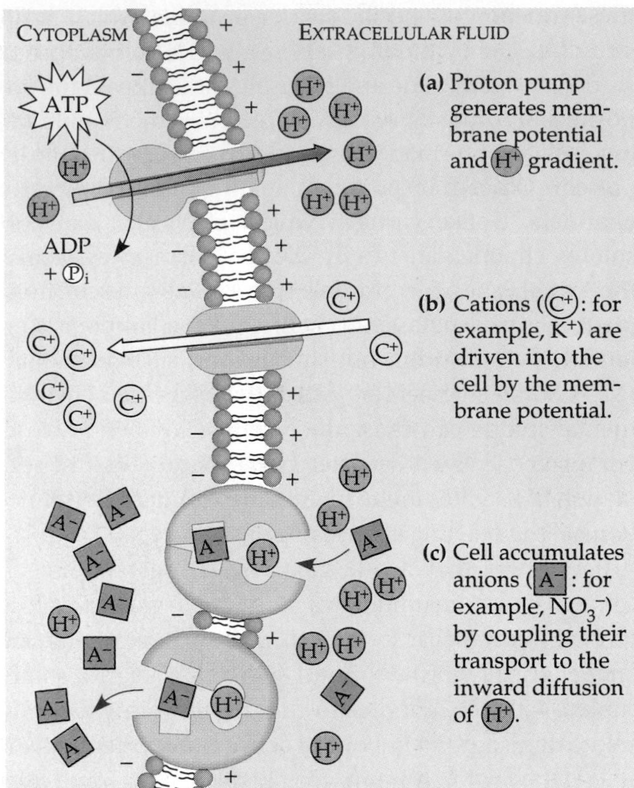
CYTOPLASM
EXTRACELLULAR FLUID
ATP
ADP
+ Ⓟi
(a) Proton pump generates membrane potential and H+ gradient.
(b) Cations (C+: for example, K+) are driven into the cell by the membrane potential.
(c) Cell accumulates anions (A−: for example, NO3−) by coupling their transport to the inward diffusion of H+.

FIGURE 32.2

A chemiosmotic model of solute transport in plant cells. (**a**) Much of the solute transport across the plasma membrane of a plant cell is driven indirectly by proton pumps that maintain a voltage (membrane potential) and an H^+ gradient across the membrane. (**b**) The membrane potential drives cations such as K^+ into the cell. (**c**) The cell can tap the energy stored in the H^+ gradient to accumulate anions such as NO_3^- *against* their electrochemical gradients (active transport). A transport protein couples accumulation of the anion to the inward diffusion of H^+. These transport mechanisms are part of a more general chemiosmotic theory, which emphasizes the versatile role of transmembrane proton gradients in the bioenergetics of cells.

such as nitrate (NO_3^-), enter root cells through carriers that also allow H^+ to reenter the cells. This mechanism is called **cotransport**. A transport protein couples the downhill passage of one solute (H^+) to the uphill passage of another (NO_3^-, in this case). This coattail effect is also responsible for the uptake of the sugar sucrose by plant cells. A membrane protein cotransports sucrose with H^+, which moves down its gradient through the protein.

The role of proton pumps in the transport processes of plant cells is a specific application of the general mechanism called **chemiosmosis** (see Chapter 9). The key feature of chemiosmosis is a transmembrane proton gradient, which links energy-releasing processes to energy-consuming processes in cells. For example, you learned in Chapters 9 and 10 that mitochondria and chloroplasts use proton gradients generated by electron transport chains to drive ATP synthesis. The ATP synthases that couple H^+ diffusion to ATP synthesis during cellular respiration and photosynthesis function somewhat like the proton pumps embedded in the plasma membranes of plant cells. But compared to ATP synthases, proton pumps normally run in reverse, using ATP energy to pump H^+ against its gradient. In both cases, proton gradients are the metabolic gears that enable one process to drive another. Chemiosmosis is a unifying principle of cellular energetics. Relevant to our discussion here, chemiosmosis figures prominently in how plant cells transport solutes across their membranes.

Water Potential and Osmosis. The net uptake or loss of water by a cell occurs by **osmosis,** the passive transport of water across a membrane (see Chapter 8). How can we predict the direction of osmosis when a cell is surrounded by a particular solution? In the case of an animal cell, it is enough to know whether the extracellular solution is hypotonic (lower solute concentration) or hypertonic (higher solute concentration) to the cell; water will move by osmosis in the hypotonic ⟶ hypertonic direction. But in the case of a plant cell, the presence of a cell wall adds a second factor affecting osmosis: physical pressure. The combined effects of these two factors—solute concentration and pressure—are incorporated into a single measurement called **water potential,** abbreviated by the Greek letter psi (ψ). The most important thing for you to learn about water potential is that water will move across a membrane from the solution with the higher water potential to the solution with the lower water potential. For example, if a plant cell is immersed in a solution having a higher water potential than the cell, osmotic uptake of the water will cause the cell to swell. By moving, water can perform work (expanding a cell, for instance). The *potential* in water potential refers to this potential energy, the capacity to perform work when water moves from a region of higher ψ to a region of lower ψ.

Plant biologists measure ψ in units of pressure called **megapascals** (abbreviated MPa). An MPa is equal to about 10 atmospheres of pressure. (An atmosphere is the pressure exerted at sea level by an imaginary column of air—about 1 kg of pressure per cm^2.) A couple of nonbiological examples will give you some idea of the magnitude of a megapascal: A car tire is usually inflated to a pressure of about 0.2 MPa; the water pressure in home plumbing is about 0.25 MPa.

Let's see how solute concentration and pressure affect water potential. For purposes of comparison, the water potential of pure water in a container open to the atmosphere is defined as zero megapascals ($\psi = 0$ MPa). The addition of solutes lowers the water potential. And since ψ is standardized as 0 MPa for pure water, any solution at atmospheric pressure has a negative water potential due to the presence of solutes. For instance, a 0.1 molar *(M)* solution of any solute has a water potential of –0.23 MPa. If this solution is separated from pure water by a selectively permeable membrane, water will move by osmosis into the solution, from the region of higher ψ (0 MPa) to the region of lower ψ (–0.23 MPa). So far, this is just another way of saying that the water is moving in the hypotonic ⟶ hypertonic direction. But we have not yet factored in the influence of physical pressure on ψ.

In contrast to the inverse relationship of ψ to solute concentration, water potential is directly proportional to pressure; increasing pressure raises ψ. Remember that ψ measures the relative tendency for water to leave one location in favor of another. Physical pressure—pressing the plunger of a syringe filled with water, for example—causes water to escape via any available exits. If a solution is separated from pure water by a selectively permeable membrane, external pressure on the solution can counter its tendency to take up water due to the presence of solutes. In fact, even greater pressure will force water across the membrane from the solution to the compartment containing pure water. It is also possible to create a negative pressure, or **tension,** on water or solutions. For example, if you pull up on the plunger of a syringe, the negative pressure within the syringe draws a solution through the needle.

The movement of water due to a pressure difference between two locations is called **bulk flow.** It is usually much faster than simple diffusion, the process responsible for water movement due to a difference of solute concentration.

The combined effects of pressure and solute concentration on water potential are incorporated into the following equation:

$$\psi = \psi_P + \psi_S$$

where ψ_P is the pressure potential (physical pressure on a solution) and ψ_S is the solute potential (which is proportional to the solute concentraton of a solution; ψ_S is also called osmotic potential). Pressure on a solution (ψ_P) can be either a positive number or a negative number (tension, a negative pressure). In contrast, a solution's solute potential (ψ_S) is always a negative number, and the greater the solute concentration, the more negative the value of ψ_S. This makes sense if you think of ψ_S as the effect solutes have on a solution's overall water potential (ψ), which is always to lower ψ below the water potential of pure water, defined as 0 MPa.

Let's see how this equation works, with the help of FIGURE 32.3. A 0.1 *M* solution has a ψ_S of –0.23 MPa. Thus, in the absence of a physical pressure ($\psi_P = 0$), water potential, as stated earlier, is –0.23 MPa for a 0.1 *M* solution:

$$\psi = \psi_P + \psi_S = 0 + (-0.23) = -0.23$$

If we apply a physical pressure of +0.23 MPa to this solution, we raise its water potential from a negative value

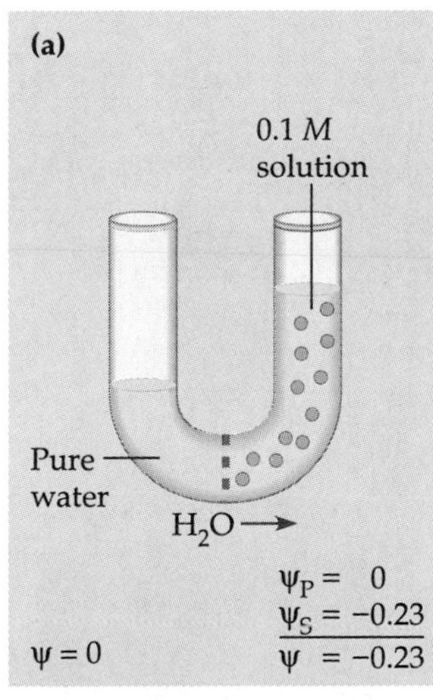

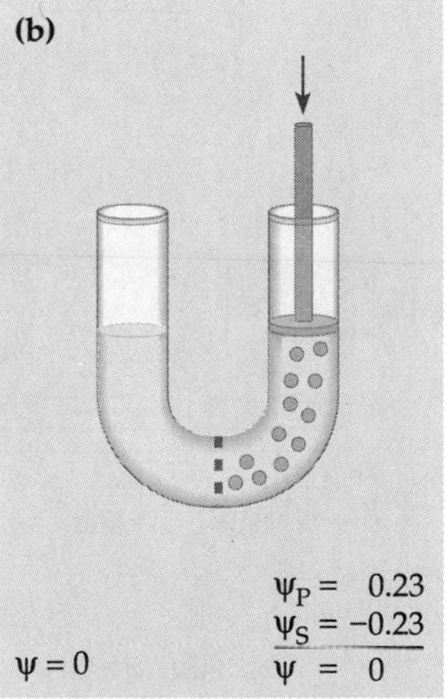

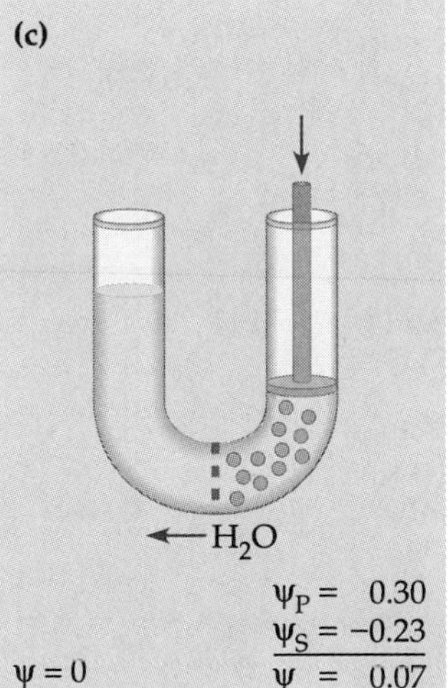

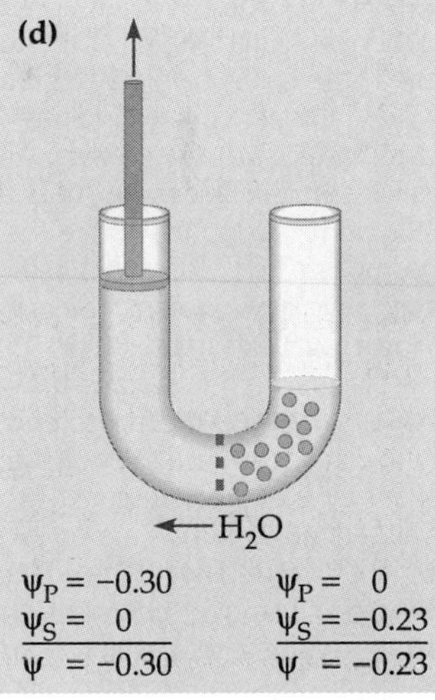

FIGURE 32.3

Water potential and water movement: a mechanical model. Water moves across a selectively permeable membrane from where water potential is higher to where it is lower. The water potential (ψ) of pure water at atmospheric pressure is 0 MPa. The addition of solutes reduces water potential (to a negative value); application of physical pressure increases water potential. If we know the values of pressure potential (ψ_P) and solute potential (or osmotic potential, ψ_S), we can calculate water potential: $\psi = \psi_P + \psi_S$. **(a)** In this U-shaped apparatus, a selectively permeable membrane separates pure water from a 0.1 molar *(M)* solution containing a particular solute that cannot pass freely across the membrane. Water will move by osmosis into the solution, increasing its volume. (The values for ψ and ψ_S are given for *initial* conditions, *before* any net movement of water.) **(b)** If we use a piston to apply just enough physical pressure to offset the solute potential and increase the water potential of the solution to 0, there will be no net movement of water across the membrane. **(c)** If physical pressure exceeds the opposing effect of solute potential, we can force water from the solution into the reservoir of pure water. **(d)** A negative pressure, or tension, reduces water potential.

to 0 ($\psi = 0.23 - 0.23$). If this pressurized solution is separated from water by a selectively permeable membrane, there will be no net water flow between the two compartments. If we increase ψ_P to +0.3 MPa, then the solution has a water potential of +0.07 MPa ($\psi = 0.3 - 0.23$), and this solution will lose water by bulk flow to a compartment containing pure water. In dissecting water potential to see these opposing effects of pressure and solutes, it is important to remember the key point: Water will move across a membrane in the direction of lower (more negative) water potential.

Let's now apply what we have learned about water potential to the uptake and loss of water by plant cells. First, imagine a flaccid cell (that is, $\psi_P = 0$) bathed in a solution of higher solute concentration than the cell itself (FIGURE 32.4a). Since the external solution has the lower (more negative) water potential, water will leave the cell by osmosis, and the cell will plasmolyze, or shrink and pull away from its wall. Now let's place the same flaccid cell in pure water ($\psi = 0$; FIGURE 32.4b). The cell has a lower water potential because of the presence of solutes, and water enters the cell by osmosis. The cell begins to swell and push against the wall, producing a **turgor pressure.** The partially elastic wall pushes back against the turgid cell. When this wall pressure is great enough to offset the tendency for water to enter because of the solutes in the cell, then ψ_P and ψ_S are equal in magnitude, and thus $\psi = 0$. This matches the water potential of the extracellular environment—in this example, 0 MPa. A dynamic equilibrium has been reached, and there is no further net movement of water, although an equal exchange of water across the membrane continues.

The Role of the Tonoplast. Although our emphasis on transport at the cellular level has been on traffic across the plasma membrane, the **tonoplast,** the membrane of the central vacuole, is another important site of regulation. Transport proteins embedded in the tonoplast control the movement of solutes between the cytosol and the vacuole. For example, the tonoplast has proton pumps that expel H^+ from the cytoplasm into the vacuole. This augments the ability of the proton pumps of the plasma membrane to maintain a low cytosolic concentration of H^+. Proton pumping at the tonoplast also generates a membrane potential across that membrane, with the vacuole side of the membrane positive in charge relative to

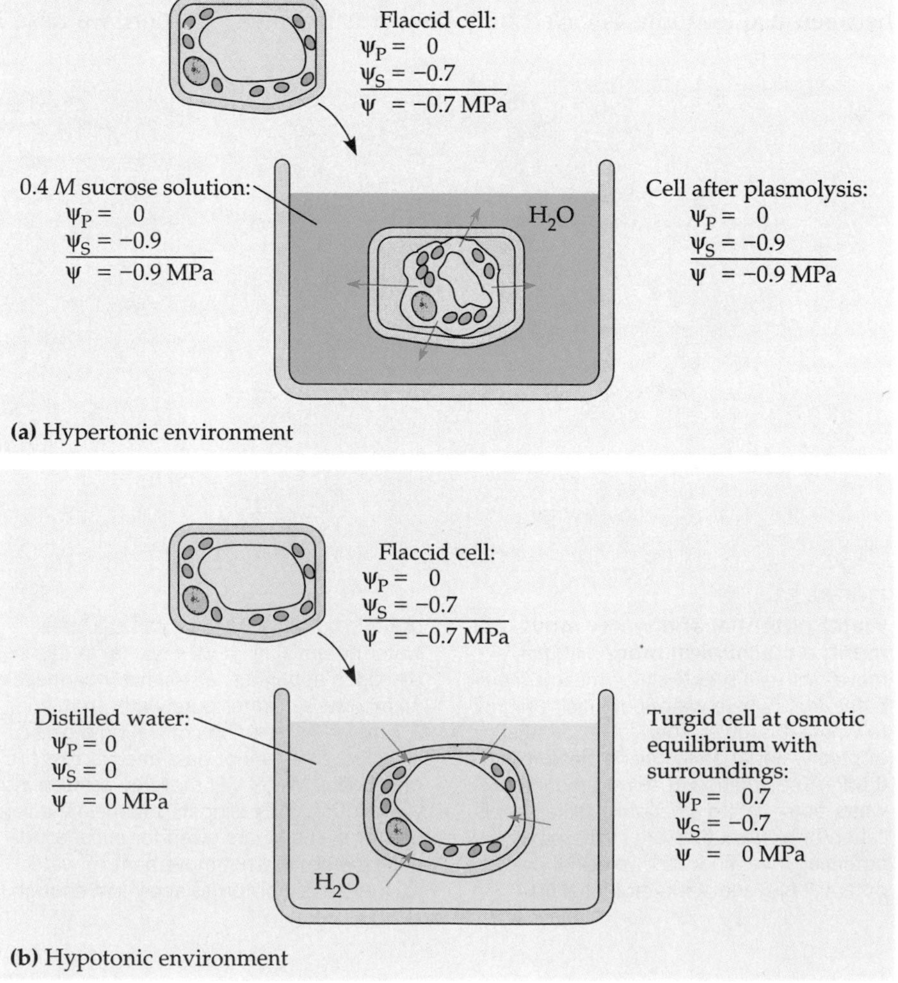

FIGURE 32.4
Water relations of plant cells. In these two experiments, flaccid cells are transferred from an isotonic environment into hypertonic and hypotonic environments. (Flaccid cells are in contact with their walls, but lack turgor pressure.) (**a**) In a hypertonic environment, the cell initially has a greater water potential than its surroundings. The cell loses water and plasmolyzes. After plasmolysis is complete, the water potentials of the cell and its surroundings are the same. Plasmolysis kills most plant cells. (**b**) In a hypotonic environment, the cell initially has a lower water potential than its surroundings. There is a net uptake of water by osmosis, causing the cell to become turgid. When this tendency for water to enter is offset by the back pressure of the elastic wall, water potentials are equal for the cell and its surroundings. (The volume change of the cell is exaggerated in this diagram. The osmotic uptake of a relatively small amount of water does not actually increase the volume of the cell much. This explains why the solute potential, ψ_S, does not change by a significant amount when a cell becomes more turgid.) Turgid (firm) plant cells contribute support to nonwoody parts of a plant. You can see the consequences of turgor loss in a wilted house plant.

the cytosolic side. By mechanisms similar to those we saw at the plasma membrane, the proton gradient and voltage difference across the tonoplast drive the transport of several solutes between the vacuole and the cytosol.

Short-Distance (Lateral) Transport at the Level of Tissues and Organs

How do water and solutes move from one location to another within plant tissues and organs? For example, what mechanisms transport water and minerals absorbed by a root from the outer cells to the inner cells of the root? Such short-distance transport is sometimes called lateral transport because its usual direction is along the radial axis of plant organs, rather than up and down along the length of the plant.

Three routes are available for lateral transport. By the first route, substances move out of one cell, across the cell wall, and into the neighboring cell, which may then pass the substances along to the next cell in the pathway by the same mechanism (FIGURE 32.5a). This route requires repeated crossings of plasma membranes, as the solutes exit one cell and enter the next.

The second route, via the **symplast,** the continuum of cytoplasm within a plant tissue, requires only one crossing of a plasma membrane (FIGURE 32.5b). After entering one cell, solutes and water can then move from cell to cell via **plasmodesmata,** the cytoplasmic channels that connect plant cells through pores in cell walls.

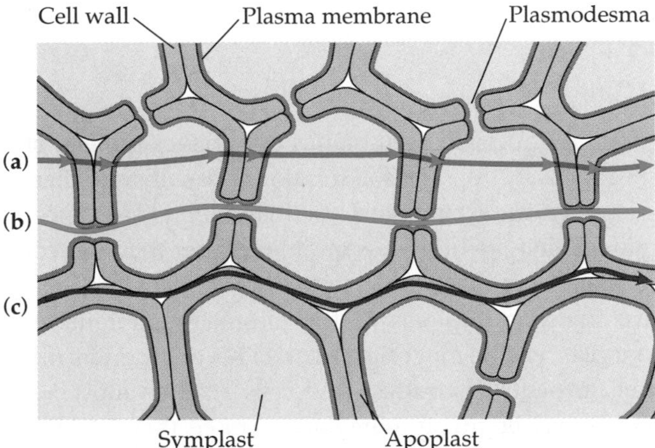

FIGURE 32.5
Three routes for lateral transport in a plant tissue or organ. (**a**) Solutes and water can move across an organ by the repeated crossing of the plasma membranes and walls of the cells along the pathway. (**b**) After entering one cell, substances can move across an organ via the symplast, the cytoplasmic compartment made continuous by the presence of plasmodesmata. (**c**) Water and solutes can also travel across a tissue or organ via the apoplast, the extracellular continuum formed by the cell walls. In this diagram, substances seem confined to one of the three routes; in fact, substances may transfer from one route to another during their commute across an organ.

The third route for lateral transport within a plant tissue or organ is along the **apoplast,** the extracellular pathway consisting of cell walls (FIGURE 32.5c). Before ever entering a cell, water and solutes can move from one location to another within a root or other organ along the byways provided by the continuum of cell walls.

During the course of their transit across a plant organ, solutes and water can change pathways. For example, minerals absorbed by a root may move inward along the apoplast for some distance, and then switch to the symplastic route after being taken up by a root cell. Diffusion due to concentration differences and bulk flow due to pressure differences are the main mechanisms determining the direction of short-distance transport within plant tissues and organs.

Long-Distance Transport at the Whole-Plant Level

Diffusion is much too slow to function in long-distance transport within a plant—the transport of water and minerals from roots to leaves, for example. Water and solutes move through xylem vessels and sieve tubes by bulk flow, the movement of a fluid driven by pressure. In phloem, for example, hydrostatic pressure is generated at one end of a sieve tube, and this forces sap to the opposite end of the tube. In xylem, it is actually tension, a negative pressure, that drives long-distance transport. Transpiration, the evaporation of water from a leaf, reduces pressure in the leaf xylem. This creates a tension that pulls xylem sap upward from the roots.

Now that we have an overview of the basic mechanisms of transport at the cellular, tissue, and whole-plant levels, we are ready to take a closer look at how these mechanisms work together in the overall transport functions that enable a plant to survive on land. For example, bulk flow due to a pressure difference is the mechanism of long-distance transport of phloem sap, but it is active transport of sugar at the cellular level that maintains this pressure difference. The four transport functions we will examine in more detail are the absorption of water and minerals by roots, the ascent of xylem sap, the control of transpiration, and the transport of organic nutrients within phloem.

Roots absorb water and minerals from soil

Water and mineral salts enter the plant through the epidermis of roots, cross the root cortex, pass into the stele, and then flow up xylem vessels to the shoot system. In this section, we focus on the soil ⟶ epidermis ⟶ root cortex ⟶ xylem segments of this transport pathway. Use FIGURE 32.6 to reinforce the text discussion.

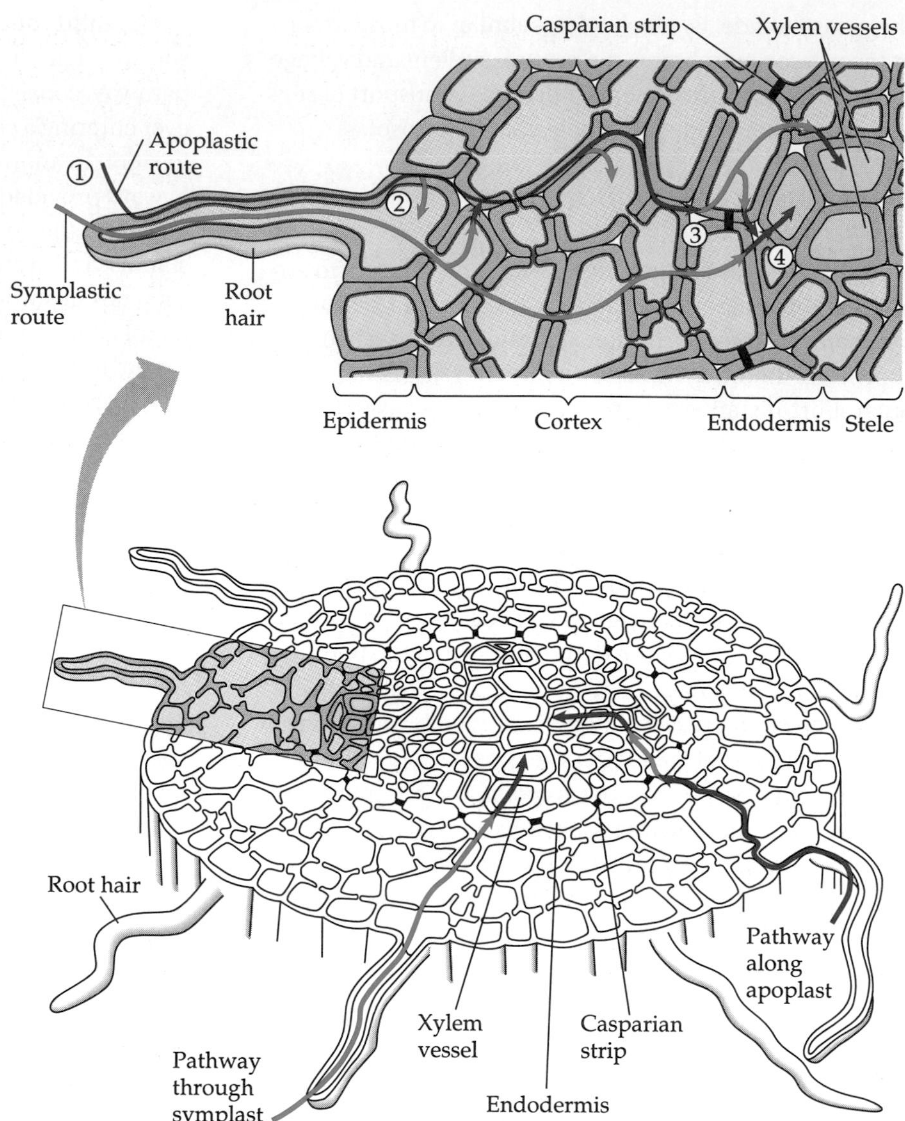

FIGURE 32.6
Lateral transport of minerals and water in roots. Minerals are absorbed with the soil solution by the root surface, especially by root hairs. The water and minerals then move across the root cortex to the vascular cylinder by a combination of the apoplastic and symplastic routes (see FIGURE 32.5). ① The uptake of soil solution by the hydrophilic walls of the epidermis provides access to the apoplast, and water and minerals can soak into the cortex along this matrix of cell walls. Minerals and water that cross the plasma membranes of root hairs enter the symplast. ② As soil solution moves along the apoplast, some water and minerals are transported into cells of the epidermis and cortex and then move inward via the symplast. ③ Water and minerals that move all the way to the endodermis along cell walls cannot continue into the stele via the apoplastic route. Within the wall of each endodermal cell is a belt of waxy material (black band) that blocks the passage of water and dissolved minerals. This barrier to apoplastic transport is called the Casparian strip. Only minerals that are already in the symplast or enter that pathway by crossing the plasma membrane of an endodermal cell can detour around the Casparian strip and pass into the stele. Thus, the transport of minerals into the stele is selective; only those minerals that are admitted into cells by membranes gain access to the vascular tissue. ④ Endodermal cells, and parenchyma cells within the stele, discharge water and minerals into their walls, which, as part of the apoplast, are continuous with the xylem vessels. Water and minerals absorbed from soil are now ready for upward transport into the shoot system.

Most absorption of water and minerals occurs near root tips, where the epidermis is permeable to water and where root hairs are located. Root hairs, extensions of epidermal cells, account for most of the surface area of roots. (Most plants also have mycorrhizae, symbiotic associations of roots and fungi, which increase the surface area for water and mineral absorption.) Soil particles, which are usually coated with water and dissolved minerals, adhere tightly to the hairs. The soil solution flows into the hydrophilic walls of epidermal cells and passes freely along the apoplast into the root cortex. This exposes all the parenchyma cells of the cortex to soil solution, providing a much greater surface area of membrane for the uptake of water and minerals into cytoplasm than the surface area of the epidermis alone. As the soil solution moves along cell walls, some of the water and solutes are taken up by cells of the epidermis and cortex, switching pathways from the apoplast to the symplast. It is this crossing of plasma membranes that makes mineral absorption selective. The soil solution is usually very dilute, and roots can accumulate essential minerals to concentrations that are hundreds of times higher than the concentrations of these minerals in soil. For example, selective transport proteins of the plasma membrane and tonoplast enable root cells to extract K^+, an essential mineral nutrient. In contrast, the cells exclude most Na^+, which may be much more concentrated than K^+ in the soil solution.

The **endodermis,** the innermost layer of cells in the root cortex, surrounds the stele and functions as a last checkpoint for the selective passage of minerals from the cortex into the vascular tissue. Minerals already in the symplast when they reach the endodermis continue right through the plasmodesmata of the endodermal cells and pass into the stele. These minerals were already screened by the selective membrane they had to cross to enter the symplast in the cortex. Those minerals that reach the endodermis via the apoplast find a dead end in that route, which blocks their passage into the stele. In the wall of each endodermal cell is a belt made of suberin, a waxy material that is impervious to water and dissolved minerals. This ring of wax, called the **Casparian strip**, is tangential to the cylindrical outline of the stele. Thus, water and minerals cannot cross the endodermis and enter vascular tissue via the apoplast. The only way past this barrier is for the water and minerals to cross the plasma membrane of an endodermal cell and to enter the stele via the symplast. The endodermis, with its Casparian strip, ensures that no minerals can reach the vascular tissue of the root without crossing membranes. And you know that transport across membranes is selective. If minerals do not enter cells in the cortex, they must enter endodermal cells or be excluded from the vascular tissue. The structure of the endodermis and its strategic location in the root fit its function as sentry of the cortex-stele border, a function that contributes to the ability of roots to preferentially transport certain minerals from the soil into the xylem. The Casparian strip also prevents the backflow of water and minerals from the stele into the cortex.

The last segment in the soil ⟶ xylem pathway is the passage of water and minerals into the tracheids and vessel elements of the xylem. These water-conducting cells lack protoplasts, and thus the lumens of the cells, as well as their walls, are part of the apoplast. But you have just learned that water and minerals enter the stele from the cytoplasm of endodermal cells. Thus, the entry of water and minerals into the xylem requires their transfer from the symplast to the apoplast. Endodermal cells and parenchyma cells within the stele discharge minerals into their walls. Both diffusion and active transport are probably involved in this transfer of solutes from symplast to apoplast, and the water and minerals are now free to enter the tracheids and xylem vessels. The water and mineral nutrients we have tracked from the soil to root xylem can now be transported upward as xylem sap to the shoot system.

The ascent of xylem sap depends mainly on transpiration and the physical properties of water

Xylem sap flows upward in vessels at rates of 15 m per hour or faster. Veins branch throughout each leaf, placing xylem vessels close to every cell. Leaves depend on this efficient delivery system for their supply of water. Plants lose an astonishing amount of water by **transpiration,** the evaporation of water from leaves and other aerial parts of the plant. An average-sized maple tree, for instance, loses more than 200 L of water per hour during the summer. Unless the transpired water is replaced by water transported up from the roots in xylem, leaves wilt and eventually die. The flow of xylem sap upward in the plant also brings mineral nutrients to the shoot system.

Xylem sap rises against gravity, without the help of any mechanical pump, to reach heights of more than 100 m in the tallest trees. Is the sap *pushed* upward from the roots, or is it *pulled* upward by the leaves? Let's evaluate the relative contributions of these two possible mechanisms.

Pushing Xylem Sap: Root Pressure

At night, when transpiration is very low or zero, the root cells are still expending energy to pump mineral ions into the xylem. The endodermis surrounding the stele of the root helps prevent the leakage of these ions back out of the stele. The accumulation of minerals in the stele lowers its water potential, and water flows in, generating a positive pressure that forces fluid up the xylem. This upward push of xylem sap in some plants is called **root pressure.**

Root pressure causes **guttation,** the exudation of water droplets that can be seen in the morning on tips of grass blades or the leaf margins of some small, herbaceous (nonwoody) dicots. During the night, when the rate of transpiration is low, the roots of some plants keep accumulating minerals, and root pressure pushes xylem sap into the shoot system. More water enters leaves than is transpired, and the excess is forced as guttation through specialized structures called hydathodes, which function as escape valves.

In most plants, root pressure is not the major mechanism driving the ascent of xylem sap. At most, root pressure can force water upward only a few meters, and many plants, including some of the tallest trees, generate no root pressure at all. Even in most small plants that display guttation, root pressure cannot keep pace with transpiration after sunrise. For the most part, xylem sap is not pushed from below by root pressure but pulled upward by the leaves themselves.

Pulling Xylem Sap: The Transpiration-Cohesion-Tension Mechanism

If we want to move material upward, we can either push it from below or pull it from above. It is not as obvious that something as seemingly fluid as water could be pulled up a pipe. Nevertheless, that is what happens in the xylem vessels of plants. As we investigate this mechanism of transport, we will see that transpiration provides the pull, and the cohesion of water due to hydrogen bonding transmits the upward pull along the entire length of the xylem to the roots.

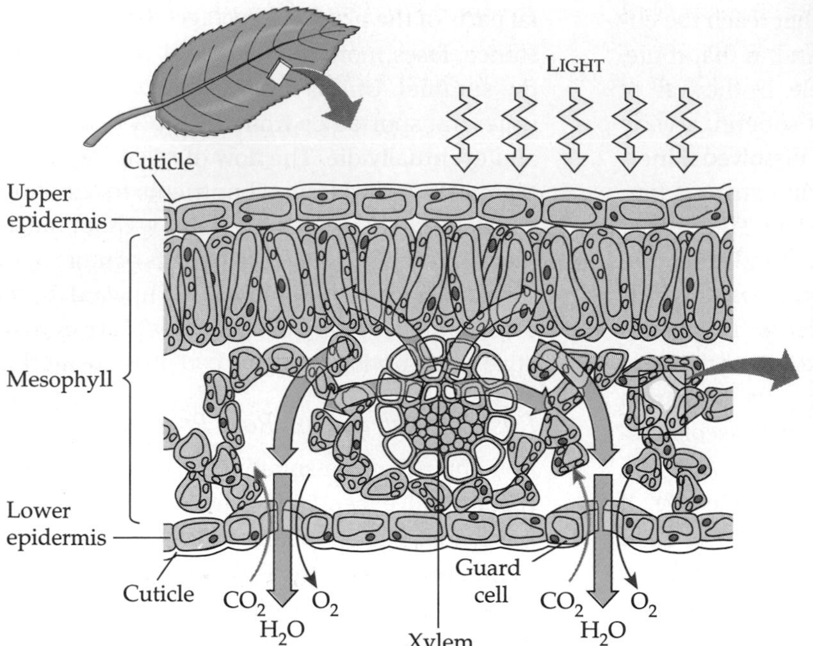

Radius of curvature (μm)	Hydrostatic pressure (MPa)
a = 1.00	a = −0.15
b = 0.10	b = −1.50
c = 0.01	c = −15.00

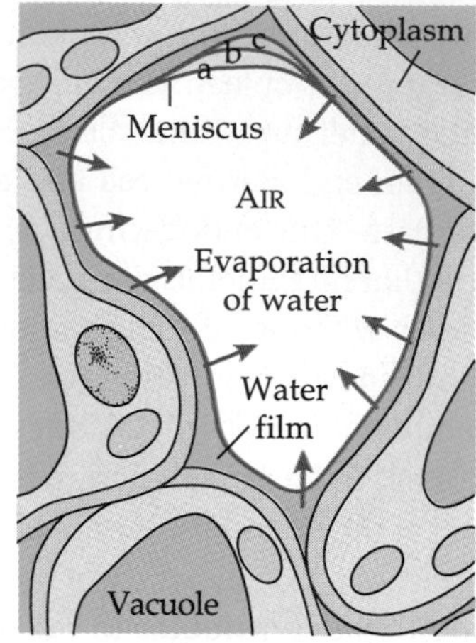

FIGURE 32.7

The generation of transpirational pull in a leaf. Water vapor diffuses from the moist air spaces of the leaf to the drier air outside via stomata. (Stomata are also avenues for the exchange of CO_2 and O_2 between the photosynthetic tissue and the atmosphere.) Evaporation from the water film coating the mesophyll cells maintains the high humidity of the air spaces. This loss of water causes the water film to form menisci, which become more and more concave as the rate of transpiration (evaporative water loss from the leaf) increases. A meniscus has a tension that is inversely proportional to the radius of the curved water surface. Thus, as the water film recedes and its menisci become more concave, the tension of the water film increases. Tension is a negative pressure—a force that pulls water from locations where hydrostatic pressure is greater. The negative pressure (tension) of water lining the air spaces of the leaf is the physical basis of transpirational pull, which draws water out of xylem and through the mesophyll tissue to the surfaces near stomata. The cohesion of water due to hydrogen bonding enables transpiration to pull water up the narrow xylem vessels and tracheids without these columns of water breaking apart. In fact, transpirational pull, with the help of water's cohesion, is transmitted all the way from leaves to roots. The bulk flow of water to the top of a tree is solar-powered, because it is the absorption of sunlight by leaves that causes the evaporation responsible for transpirational pull.

Transpirational Pull. Stomata, the microscopic pores on the surface of a leaf, lead to a maze of internal air spaces that expose the mesophyll cells to the carbon dioxide they need for photosynthesis (FIGURE 32.7). The air in these spaces is saturated with water vapor because it is in contact with the moist walls of the cells. On most days, the air is drier outside the leaf; that is, it has a lower water concentration than the air inside the leaf. Therefore, gaseous water, diffusing down its concentration gradient, exits the leaf via the stomata. It is this loss of water from the leaf that we call transpiration.

How is transpiration translated into a pulling force for the movement of water in a plant? The mechanism depends on the generation of a negative pressure (tension) in the leaf due to the unique physical properties of water. Evaporation from the thin film of water that coats the mesophyll cells replaces the water vapor that is lost from the leaf's air spaces by transpiration. As water evaporates, the remaining film of liquid water retreats into the pores of the cell walls, attracted by adhesion to the hydrophilic walls (see FIGURE 32.7). At the same time, cohesive forces in the water resist an increase in the surface area of the film (a surface tension effect; see Chapter 3). The combination of the two forces acting on the water—adhesion to the wall and surface tension—causes the surface of the water film to form a meniscus, or concave shape. In a sense, the water is being "pulled on" by the adhesive and cohesive forces. Thus, the water film at the surface of leaf cells has a negative pressure, a pressure less than atmospheric pressure. And the more concave the menisci (plural), the more negative the pressure of the water film. This negative pressure, or tension, is the pulling force that draws water out of the leaf xylem, through the mesophyll, and toward the cells and surface film bordering the air spaces near stomata. Transit across the mesophyll occurs via both symplastic and apoplastic routes. This water flow fits with what you learned earlier about water potential. In the equation for water potential, a tension (negative pressure) *lowers* the potential. And since water moves from where its potential is higher to where it is lower, mesophyll cells will lose water to the surface film lining air spaces, which in turn loses water by transpiration. The water lost via the stomata is replaced by water that is pulled out of the leaf xylem.

Cohesion and Adhesion of Water. The transpirational pull on xylem sap is transmitted all the way from the leaves to the root tips and even into the soil solution. The cohesion of water due to hydrogen bonding makes it possible to pull a column of sap from above without the water separating. Water molecules exiting the xylem in the leaf tug on adjacent molecules, and this pull is relayed, molecule by molecule, down the entire column of water in the xylem. Also helping to fight gravity is the strong adhesion of water molecules (again by hydrogen bonds) to the hydrophilic walls of the xylem cells. The very small diameter of tracheids and vessel elements contributes to the importance of this factor in overcoming the downward force of gravity.

The upward pull on the cohesive sap creates tension within the xylem. Pressure will cause an elastic pipe to swell, but tension will pull the walls of the pipe inward. (You can actually measure a decrease in the diameter of a tree trunk on a warm day when transpirational pull puts the xylem under tension.) The rings of secondary walls prevent xylem vessels from collapsing, much as the wire rings maintain the shape of a vacuum hose. Transpirational pull puts the xylem under tension all the way down to the root tips, even in the tallest trees. This tension lowers water potential in the root xylem to such an extent that water flows passively from the soil, across the root cortex, and into the stele.

Transpirational pull can extend down to the roots only through an unbroken chain of water molecules. Cavitation, the formation of a water vapor pocket in a xylem vessel, such as when xylem sap freezes in winter, breaks the chain. Small plants can use root pressure to refill xylem vessels in spring, but in trees, root pressure cannot push water to the top, so a vessel with a water vapor pocket can never function as a water pipe again. However, the transpiration stream can detour around the water vapor pocket through pits between adjacent xylem vessels, and secondary growth adds a layer of new xylem vessels each year. In some angiosperm trees, including oaks and elms, the youngest, outermost growth ring of xylem transports most of the water. The older xylem functions to support the tree (see Chapter 31).

Review: The Long-Distance Transport of Water in Plants by Solar-Powered Bulk Flow

The transpiration-cohesion-tension mechanism that transports xylem sap against gravity is an excellent example of how physical principles apply to biological problems. Long-distance transport of water from roots to leaves occurs by bulk flow, the movement of fluid driven by a pressure difference at opposite ends of a conduit. In a plant, the conduits are xylem vessels or chains of tracheids. The pressure difference is generated at the leaf end by transpirational pull, which lowers pressure (increases tension) at the "upstream" end of the xylem. On a smaller scale, gradients of water potential drive the movement of water from cell to cell within root and leaf tissue. Differences in solute concentration and pressure both contribute to this microscopic transport. In contrast, bulk flow, the mechanism for long-distance transport up xylem vessels, depends only on pressure. And the plant expends none of its own metabolic energy to lift water up to the leaves by bulk flow. The absorption of sunlight drives transpiration by causing water to evaporate from the moist walls of mesophyll and maintaining a high humidity in the air spaces within a leaf. Thus, the ascent of xylem sap is ultimately solar-powered.

Guard cells mediate the transpiration-photosynthesis compromise

A leaf may transpire more than its weight in water each day. Leaves are kept from wilting by a transpiration stream in xylem vessels that flows as fast as 75 cm per minute, about the speed of the tip of a second hand sweeping around a wall clock. The tremendous requirement for water by a plant is part of the cost of making food by photosynthesis. Guard cells, by controlling the size of stomata, help balance the need to conserve water with the requirement for photosynthesis.

The Photosynthesis-Transpiration Compromise

To make food, a plant must spread its leaves to the sun and obtain CO_2 from the air. Carbon dioxide diffuses into the leaf, and oxygen produced as a by-product of photosynthesis diffuses out of the leaf, through the stomata (see FIGURE 32.7). The stomata lead to the honeycomb of air spaces through which CO_2 diffuses to the photosynthetic cells of the mesophyll. The internal surface area of the leaf may be 10 to 30 times greater than the external surface area we see when we look at the leaf. This structural feature of leaves amplifies photosynthesis by increasing exposure to CO_2, but at the same time it increases the surface area for the evaporation of water, which exits the plant freely through open stomata. About 90% of the water a plant loses escapes through stomata, though these pores account for only 1% to 2% of the external leaf surface. The waxy cuticle limits water loss through the remaining surface of the leaf. Recall from Chapter 31 that the stomata of most plants are more concentrated on the lower surface of leaves, reducing transpiration because the bottom surface receives less sunlight than the top surface.

One way to evaluate how efficiently a plant uses water is to determine its transpiration-to-photosynthesis ratio, the amount of water lost per gram of CO_2 assimilated into organic material by photosynthesis. A common ratio is 600:1, meaning the plant transpires 600 g of water for each gram of CO_2 that becomes incorporated into carbohydrate. However, corn and other plants that assimilate atmospheric CO_2 by the C_4 pathway have transpiration-to-photosynthesis ratios of 300:1 or less. With the same concentration of CO_2 within the air spaces of the leaf, C_4 plants can assimilate that CO_2 at a greater rate than C_3 plants can. Since water loss is the trade-off for enabling CO_2 to diffuse into the leaf, the photosynthetic return for each gram of water sacrificed is greater for plants that can assimilate CO_2 at greater rates when stomata are partially closed.

In addition to supplying water to leaves, the transpiration stream assists in the transfer of minerals and other substances from the roots to the shoot and leaves. Transpiration also results in evaporative cooling, which can lower the temperature of a leaf by as much as 10–15°C compared with the surrounding air. This prevents the leaf from reaching temperatures that could inhibit the enzymes that catalyze the photosynthetic reactions, as well as other enzymes involved in the leaf's metabolic processes. Desert succulents, which have low rates of transpiration, can tolerate high leaf temperatures; in this case, the loss of water due to transpiration is a greater threat than overheating.

As long as leaves can pull water from the soil fast enough to replace what they lose, transpiration is no problem. When transpiration exceeds the delivery of water by xylem, as when the soil begins to dry out, the leaves begin to wilt as their cells lose turgor pressure. The potential rate of transpiration will be greatest on a day that is sunny, warm, dry, and windy, because these are the environmental factors that increase the evaporation of water. Plants are not helpless against the elements, however, for they are capable of adjusting to their environment. In the photosynthesis-transpiration compromise, mechanisms that regulate the size of the stomatal openings strike a balance.

How Stomata Open and Close

Each stoma is flanked by a pair of guard cells, which are kidney-shaped in dicots and dumbbell-shaped in monocots. The guard cells are suspended by their epidermal neighbors over an air chamber, which leads to the leaf's maze of air spaces.

Guard cells control the diameter of the stoma by changing shape, thereby widening or narrowing the gap between the two cells (FIGURE 32.8a). When guard cells take in water by osmosis, they become more turgid and swell. In most dicots, the cell walls of guard cells are not uniformly thick, and the cellulose microfibrils are oriented in a radial manner. These structural adaptations cause the guard cells to buckle outward when they are turgid, increasing the size of the gap between the cells. When the cells lose water and become flaccid, they sag together and close the space between them. This basic mechanism also applies to the stomata of monocots.

The changes in turgor pressure that open and close stomata result primarily from the reversible uptake and loss of potassium ions (K^+) by the guard cells. Stomata open when guard cells actively accumulate K^+ from neighboring epidermal cells (FIGURE 32.8b). This uptake of solute causes the water potential to become more negative within the guard cells, and the cells become more turgid as water enters by osmosis. Most of the K^+ and water are stored in the vacuole, and thus the tonoplast also plays a role. The cell's increase in positive charge due to the influx of K^+ is offset by the uptake of chloride (Cl^-), by the pumping out of the cell of hydrogen ions released from organic acids, and by the negative charges of these organic acids after they lose their hydrogen ions. Stomatal closing results from an exodus of K^+ from guard cells, which leads to an osmotic loss of water.

The K^+ fluxes across the guard cell membrane are probably coupled to the generation of membrane potentials by proton pumps. Stomatal opening correlates with active transport of H^+ out of the guard cell. The resulting voltage (membrane potential) drives K^+ into the cell through specific membrane channels (see FIGURE 32.2). Plant physiologists are using a method called patch clamping to study regulation of the guard cell's proton pumps and K^+ channels (see the Methods Box, p. 704).

In general, stomata are open during the day and closed at night. This prevents the plant from needlessly losing water when it is too dark for photosynthesis. At least three cues contribute to stomatal opening at dawn. First, light itself stimulates guard cells to accumulate potassium and become turgid. This response is triggered by the illumination of a blue-light receptor in a guard cell, perhaps built into the plasma membrane. Activation of these blue-light receptors stimulates the activity of

CELLS TURGID/STOMA OPEN

CELLS FLACCID/STOMA CLOSED

Radially oriented cellulose microfibrils

Cell wall

Vacuole

Guard cell

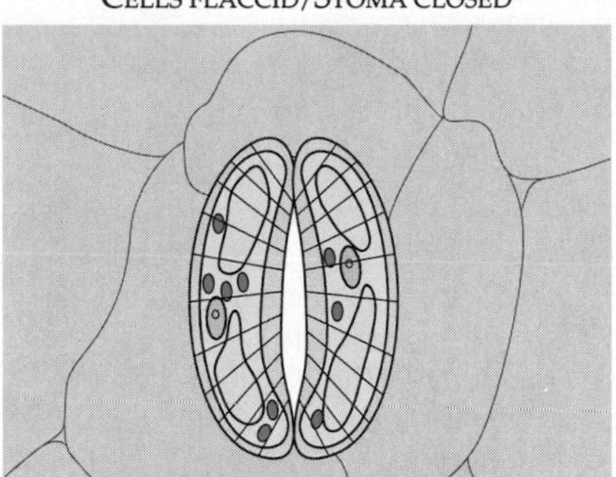

(a) Changes in guard cell shape and stomatal opening and closing (surface view)

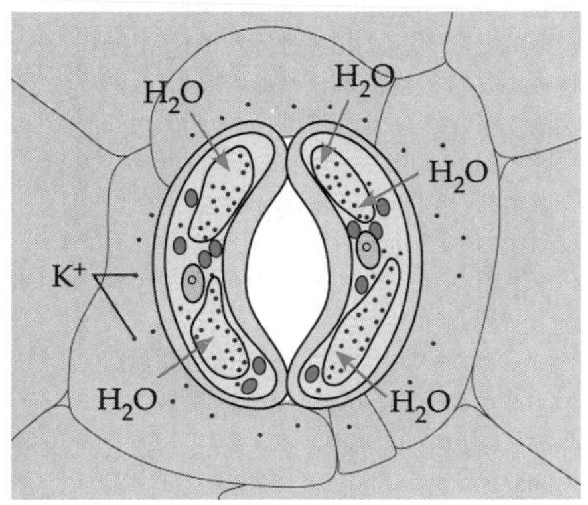

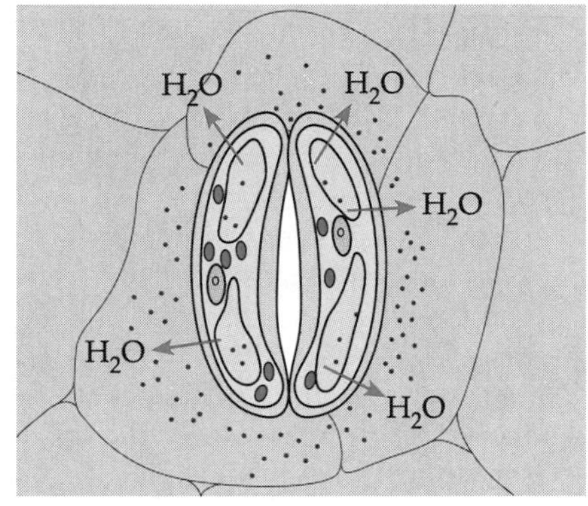

(b) Role of potassium in stomatal opening and closing

FIGURE 32.8

The control of stomatal opening and closing. (**a**) Guard cells of a dicot are illustrated in their turgid (stoma open) and flaccid (stoma closed) states. The radial microfibrils in the walls and the uneven thickness of the walls cause the pair of guard cells to buckle outward when turgid. Guard cells respond to a complex set of signals, including environmental factors and cues within the plant itself. The uptake or loss of water causes the guard cells to change their shape and widen or narrow the gap between them, a response that affects the rate of photosynthesis and controls transpiration. (**b**) The transport of K^+ (potassium ions) across the plasma membrane and tonoplast causes the turgor changes of guard cells. Stomata open when guard cells accumulate potassium (red dots), which lowers the cells' water potential and causes them to take up water by osmosis. The cells become turgid. An exodus of K^+ from guard cells causes the stomatal closure.

ATP-powered proton pumps in the plasma membrane of the guard cells, which, in turn, promotes the uptake of K^+ (see the Methods Box). Light may also stimulate stomatal opening by driving photosynthesis in guard cell chloroplasts, making ATP available for the active transport of H^+ ions. (Guard cells are the only epidermal cells equipped with chloroplasts.) A second factor causing stomata to open is depletion of CO_2 within air spaces of the leaf, which occurs when photosynthesis begins in the mesophyll. A plant can be tricked into opening its stomata at night by placing it in a chamber devoid of CO_2. A third cue in stomatal opening is an internal clock located in the guard cells. Even if you keep a plant in a dark closet, stomata will continue their daily rhythm of opening and closing. All eukaryotic organisms have internal clocks that somehow keep track of time and regulate cyclical processes. Cycles that have intervals of approximately 24 hours are called **circadian rhythms.** You will learn more about circadian rhythms and the biological clocks that control them in Chapter 35.

Environmental stress of various kinds can cause stomata to close during the daytime. When the plant is suffering a water deficiency, guard cells may lose turgor. In addition, a hormone called abscisic acid, which is produced in the mesophyll cells in response to water deficiency, signals guard cells to close stomata. This response reduces further wilting, but it also slows down photosynthesis; this is one reason droughts reduce crop yields. High temperatures also induce stomatal closure, probably by stimulating cellular respiration and increasing CO_2 concentration within the air spaces of the leaf. High temperature and excessive transpiration may combine to cause stomata to close briefly during midday. Thus, guard cells arbitrate the photosynthesis-transpiration compromise on a moment-to-moment basis by integrating a variety of internal and external stimuli.

Evolutionary Adaptations That Reduce Transpiration

Plants adapted to arid climates are called xerophytes. They have various leaf modifications that reduce the rate of transpiration. Many xerophytes have small, thick leaves, an adaptation that limits water loss by reducing surface area relative to volume. A thick cuticle gives some of these leaves a leathery consistency. The stomata are concentrated on the lower leaf surface, and they are often located in depressions that shelter the pores from the dry wind. During the driest months, some desert plants shed

METHODS: PATCH-CLAMP RECORDING

In 1991, Erwin Neher and Bert Sakmann of Germany's Max Planck Institute shared the Nobel Prize in physiology or medicine for their invention of a technique called patch clamping. This method enables researchers to record the traffic of ions through specific channels in membranes. Patch clamping makes it possible to isolate a very tiny "patch" of membrane and study ion movement through a single type of channel—a K^+ or H^+ channel, for instance.

To apply the patch-clamp method to plant cells, it is first necessary to remove the cell walls. This is usually accomplished by using enzymes to digest the walls, resulting in naked (wall-less) cells, or protoplasts. FIGURE a is a light micrograph of guard cell protoplasts from tobacco (*Nicotiana*) leaves.

The tool for making a patch is a micropipette with a tip only about 1 μm in diameter. A larger pipette is used to hold the cell in place (FIGURE b). Slight suction draws a membrane "blister" into the opening of the micropipette, and the rim of the micropipette seals to the membrane. This partitions a small patch from the rest of the plasma membrane. The micropipette, hooked up to appropriate equipment, now functions as an electrode to record ion fluxes across the tiny patch of membrane.

Several options for studying ion transport across this patch are now available. With the whole cell still attached to the micropipette by the tight seal around the membrane patch, it is possible to record ion transport between the cytoplasm and the artificial solution that fills the micropipette, as shown in FIGURE c. An alternative is to pull the micropipette away from the cell, separating the membrane patch from the rest of the cell, as shown in FIGURE d. This enables the researcher to control the composition of solutions on *both* sides of the membrane patch. Because the patch is so small, and because the ions in the bathing solutions are known, it is possible to record the transport of a single kind of ion through selective channels or pumps in the membrane. FIGURE e shows such a recording for H^+ transport across a membrane patch of a guard cell. Passage of H^+ across the membrane represents an electrical current that can be measured. The graph indicates current due to H^+ flux during a recording that lasts several minutes. This particular experiment measures the effect of blue light, which increases H^+ current across the membrane.

Along with other evidence, patch-clamp studies support the hypothesis that blue light and other stimuli regulate guard cells by affecting the proton (H^+) pumps of the membrane.

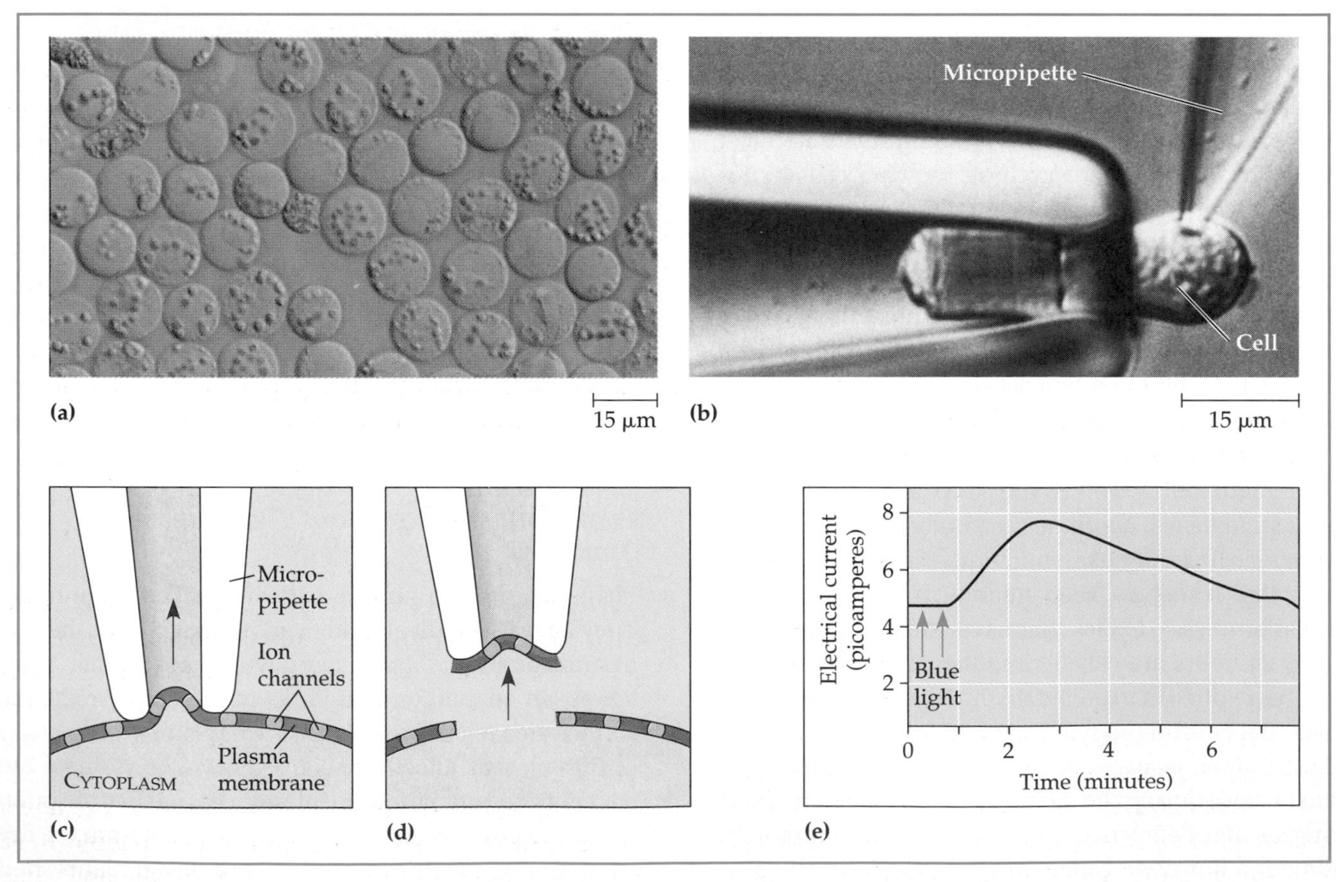

their leaves. Others, such as cacti, subsist on water the plant stores in its fleshy stems during the rainy season. (These modified stems are the photosynthetic organs of cacti; the spines are modified leaves.)

One of the most elegant adaptations to an arid habitat is found in succulent plants of the family Crassulaceae and representatives of many other plant families. These plants assimilate their CO_2 by an alternative photosynthetic pathway known as CAM, for crassulacean acid metabolism (see Chapter 10). Mesophyll cells in a CAM plant have enzymes that can incorporate CO_2 into organic acids during the night. During the daytime, the organic acids are broken down to release CO_2 in the same cells, and sugars are synthesized by the conventional (C_3) photosynthetic pathway. Since the leaf takes in its CO_2 at night, the stomata can close during the day when transpiration would be most severe. The circadian rhythm for stomatal opening in CAM plants is out of phase with other plants by about 12 hours. Stomatal behavior provides examples of both short-term physiological adjustments and evolutionary adaptations.

A bulk-flow mechanism translocates phloem sap from sugar sources to sugar sinks

Xylem sap generally flows in the wrong direction to function in exporting sugar from leaves to other parts of the plant. A second vascular tissue, the phloem, transports the organic products of photosynthesis throughout the plant. This transport of food in the plant is called **translocation.** In angiosperms, the specialized cells of phloem that function in translocation are the sieve-tube members, which are arranged end to end to form long sieve tubes (see FIGURE 31.10d). Between the members are sieve plates, porous cross walls that allow the flow of sap along the sieve tube.

Phloem sap is an aqueous solution that differs markedly in composition from xylem sap. By far the prevalent solute in phloem sap is sugar, primarily the disaccharide sucrose. The sucrose concentration may be as high as 30% by weight, giving the sap a syrupy thickness. Phloem sap may also contain minerals, amino acids, and hormones in transit from one part of the plant to another.

Source-to-Sink Transport

In contrast to the unidirectional transport of xylem sap, from roots to leaves, the direction that phloem sap travels is variable. The one generalization that holds is that sieve tubes carry food from a sugar source to a sugar sink. A **sugar source** is a plant organ in which sugar is being produced by either photosynthesis or the breakdown of starch. Leaves are usually sources. A **sugar sink** is an organ that consumes or stores sugar. Growing roots, shoot tips, and fruits are sugar sinks supplied by phloem, as are nongreen stems and trunks. A storage organ, such as a tuber or bulb, may be either a source or a sink depending on the season. When the storage organ is stockpiling carbohydrates during the summer, it is a sugar sink. After breaking dormancy in the early spring, however, the storage organ becomes a source as its starch is broken down to sugar, which is carried away in the phloem to the growing buds of the shoot system.

Other solutes may be transported to sinks along with sugar. For example, minerals that reach leaves in xylem may later be transferred in the phloem to developing fruit.

A sugar sink usually receives its sugar from the sources nearest to it. The upper leaves on a branch may send sugar to the growing shoot tip, whereas the lower leaves export sugar to roots. A growing fruit requires so much food that it may monopolize the sugar sources all around it. One sieve tube in a vascular bundle may carry phloem sap in one direction, while sap in a different tube in the same bundle flows in the opposite direction. For each sieve tube, the direction of transport depends only on the locations of the source and sink connected by that tube, and the direction may change with the season or developmental stage of the plant.

Phloem Loading and Unloading

Sugar produced in the mesophyll cells of a leaf must be loaded into sieve-tube members before it can be exported to sugar sinks. In some species, including certain plants of the mint and squash families, sucrose may move all the way from mesophyll to sieve-tube members via the symplast, passing from cell to cell through plasmodesmata. In other species, sucrose reaches sieve-tube members by a combination of symplastic and apoplastic pathways (FIGURE 32.9a). For example, in corn leaves, sucrose diffuses through the symplast from mesophyll cells into small veins. Much of the sugar then moves out of the cells into the apoplast in the vicinity of sieve-tube members and companion cells. This sucrose is accumulated from the apoplast (walls) by the sieve-tube members and their companion cells, which pass the sugar into the sieve-tube members through plasmodesmata linking the cells. In some plants, companion cells have numerous ingrowths of their walls, an adaptation that increases the cells' surface area and enhances the transfer of solutes between apoplast and symplast. Such modified cells are called **transfer cells.**

In corn and many other plants, sieve-tube members accumulate sucrose to concentrations two to three times higher than concentrations in mesophyll, and thus phloem loading requires active transport. Proton pumps

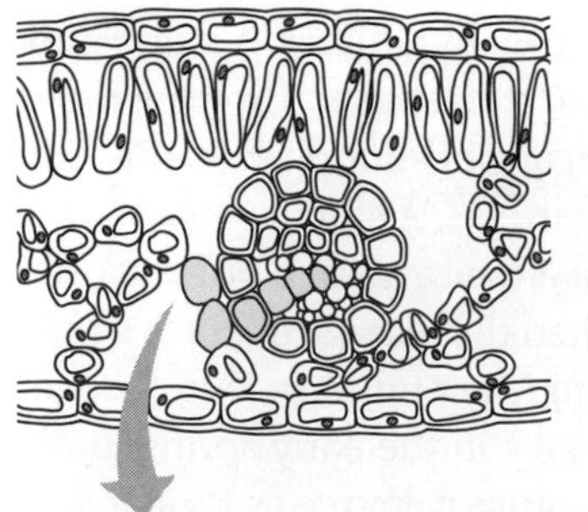

FIGURE 32.9

Loading of sucrose into phloem. (**a**) Sucrose manufactured in mesophyll cells can travel via the symplast (blue arrows) to sieve-tube members. In some species, sucrose exits (magenta arrow) the symplast near sieve tubes and is actively accumulated from the apoplast by sieve-tube members and their companion cells. Some companion cells are modified as transfer cells, which have wall ingrowths that increase the surface area for the transport of solutes. (**b**) A chemiosmotic mechanism is responsible for the active transport of sucrose into companion cells and sieve-tube members. Proton pumps (ATPase) generate an H^+ gradient, which drives sucrose accumulation with the help of a membrane protein (symport) that couples sucrose transport to the diffusion of H^+ back into the cell.

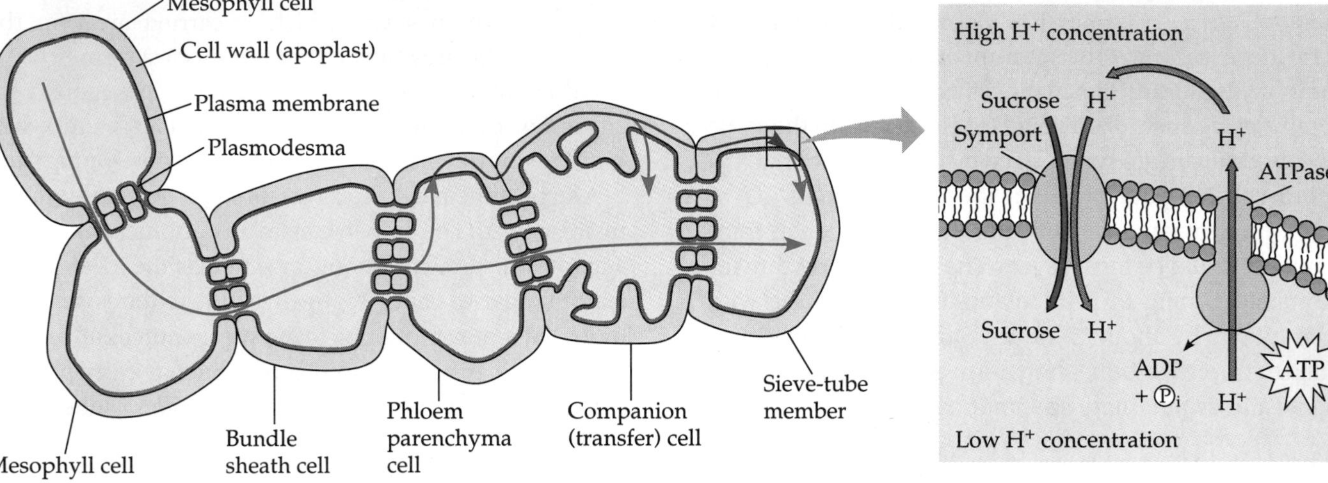

do the work that enables the cells to accumulate sucrose (FIGURE 32.9b). The ATP-driven pumps move H^+ out of the cell and store energy in the form of the difference in H^+ concentration across the plasma membrane. Another membrane protein uses this energy source to cotransport sucrose into the cell along with returning hydrogen ions.

Downstream, at the sink end of a sieve tube, phloem unloads its sucrose. Pathways and mechanisms of sugar movement in sink organs vary, depending on species. Both symplastic and apoplastic pathways may be involved. In some plants, sucrose may be unloaded from phloem by active transport. In other species, diffusion is sufficient to move sucrose from phloem to the surrounding cells of the sink organ.

Pressure Flow (Bulk Flow) of Phloem Sap

Phloem sap flows from source to sink at rates as great as 1 m per hour, which is much too fast to be accounted for by either diffusion or cytoplasmic streaming. Phloem sap moves by bulk flow, which is driven by pressure (thus the synonym, pressure flow). Phloem loading results in a high solute concentration at the source end of a sieve tube, which lowers the water potential and causes water to flow into the tube (FIGURE 32.10). Hydrostatic pressure develops within the sieve tube, and the pressure is greatest at the source end of the tube. At the sink end, the pressure is relieved by the loss of water, owing to water potential being lowered outside the sieve tube by the exodus of sucrose. (In cells of some sink organs, the metabolic breakdown of sugar or the storage of sugar in starch maintains a low sucrose concentration, which favors the continuing exit of sugar from sieve tubes.) The building of pressure at one end of the tube and reduction of that pressure at the opposite end cause water to flow from source to sink, carrying the sugar along. Water is recycled back from sink to source by xylem vessels.

The pressure-flow model explains why phloem sap always flows from a sugar source to a sugar sink, regardless of their locations in the plant. Researchers have devised several kinds of experiments to test the model. FIGURE 32.11 illustrates an innovative experiment that takes advantage of natural phloem probes: aphids that feed on phloem sap. The case for pressure flow as the mechanism of translocation in angiosperms is convincing. It is not yet known, however, if this model applies to gymnosperms.

In our study of how sugar moves in plants, we have seen examples of plant transport on all three levels: at the cellular level of transport across plasma membranes (sucrose accumulation by active transport in phloem cells); at the short-distance level of lateral transport within organs (sucrose migration from mesophyll to phloem via the symplast and apoplast); and at the long-distance level of transport between organs (bulk flow within sieve tubes).

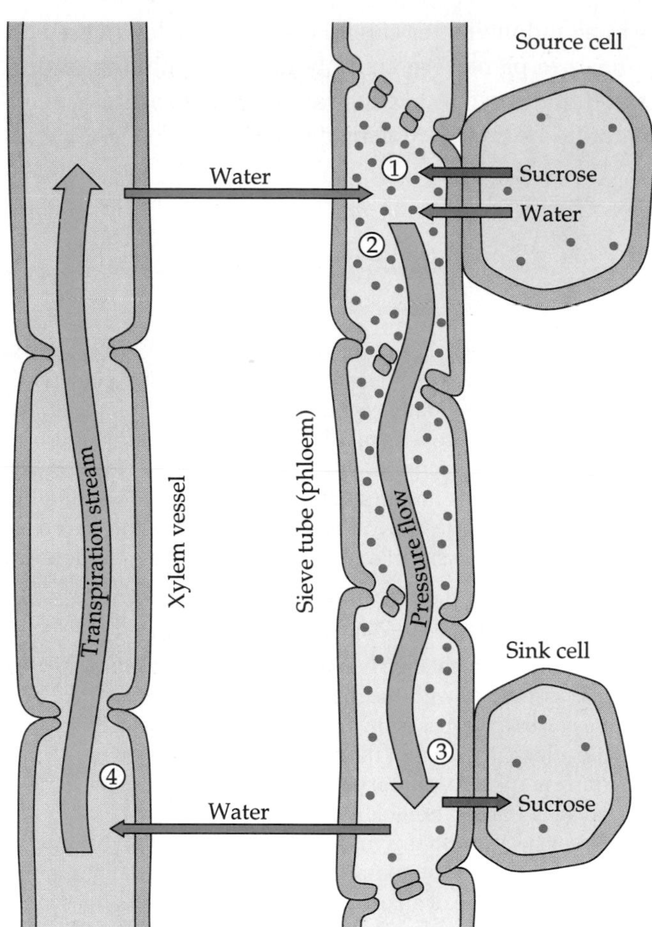

FIGURE 32.10

Pressure flow in a sieve tube. ① Loading of sugar into the tube at the source reduces the water potential inside sieve-tube members. This causes the sieve tubes to take up water from surrounding tissues by osmosis. ② This absorption of water generates a hydrostatic pressure that forces the sap to flow along the tube. ③ The gradient of pressure in the tube is reinforced by the unloading of sugar and the consequent loss of water from the sieve tube at the sink. Sugar does not accumulate in sink cells because it is either consumed in metabolism or converted to storage compounds such as starch. ④ Xylem recycles water from sink to source.

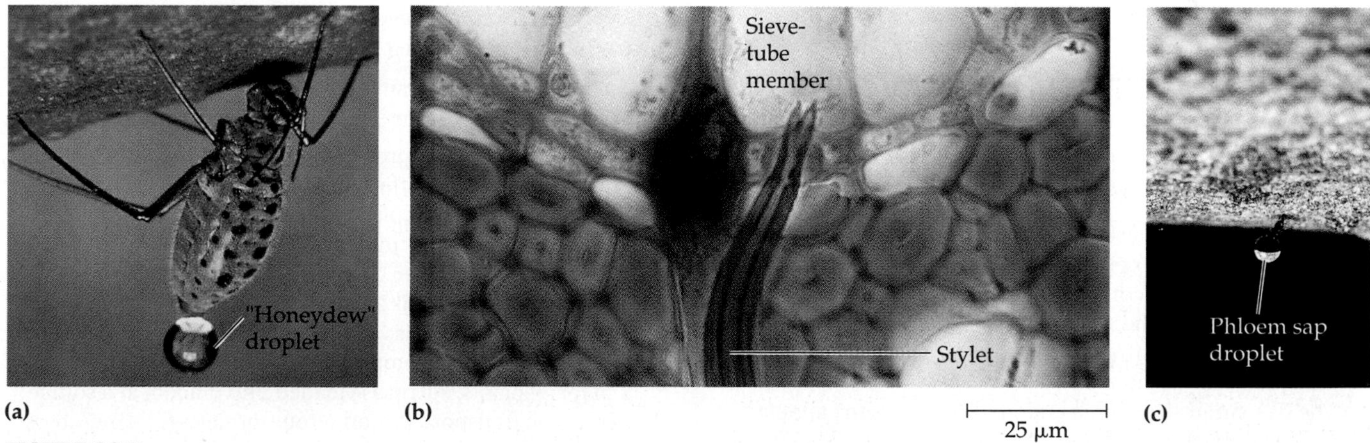

FIGURE 32.11

Tapping phloem sap with the help of an aphid. (**a**) The "honeydew" droplet exuded from the anus of this aphid consists of phloem sap minus some nutrients absorbed by the insect. (**b**) The aphid inserts a modified mouthpart called a stylet into the plant and probes until the tip of this hypodermiclike organ penetrates a single sieve-tube member (LM). The pressure within the sieve tube force-feeds the aphid, swelling it to several times its original size. (**c**) While the aphid is feeding, it can be anesthetized and severed from its stylet, which then serves the researcher as a miniature tap that exudes phloem sap for hours. The closer the stylet is to a sugar source, the faster the sap will flow out and the greater its sugar concentration. These results are predicted by the hypothesis that pressure is generated at the source end by the pumping of sugar into sieve tubes.

Plant physiologists still have much to learn about the mechanisms of transport in the elaborate vascular system of xylem and phloem. William Harvey, the great seventeenth-century physiologist, speculated that plants and animals have similar circulatory systems. The idea was abandoned after careful dissection failed to turn up a heart in plants. We are only now beginning to understand how the plant keeps sap flowing through its veins without the help of moving parts.

REVIEW OF KEY CONCEPTS (with page numbers and key figures)

- The traffic of water and solutes occurs on cellular, organ, and whole-plant levels: *an overview of transport in plants* (pp. 692–697, FIGURE 32.1)
  - Transport occurs at the level of individual cells, at the short-distance level of lateral transport within plant organs, and at the long-distance level of sap flow within xylem and phloem. Different mechanisms operate at these different levels of transport.
  - At the cellular level, solutes move across membranes by both passive transport (diffusion) and active transport. A chemiosmotic mechanism drives most active transport: Proton pumps store energy in the form of an H^+ gradient across the membrane, and specific transport proteins couple the diffusion of H^+ to the movement of other solutes.
  - Differences of water potential (ψ) drive the osmotic movement of water into and out of plant cells. Solutes lower water potential, and pressure increases water potential. When a plant cell is placed in a hypotonic environment, the lower ψ in the cell due to its higher solute concentration causes the osmotic uptake of water. Eventually, pressure exerted by the elastic wall equilibrates ψ of the cell and its surroundings, and water entry and exit are now balanced for the turgid cell.
  - The lateral transport of solutes and water in plant organs can occur via the symplast (cytoplasmic continuum) or apoplast (continuum of cell walls), or by some combination of these pathways.
  - Long-distance transport of sap in xylem and phloem occurs by bulk flow, the pressure-driven movement of a fluid.
- Roots absorb water and minerals from soil (pp. 697–699, FIGURE 32.6)
  - Water and dissolved solutes gain access to the roots through epidermal cells and their root hairs. The hydrophilic cell walls, in contact with those of the root cortex, provide access to the apoplastic pathway across the cortex. Minerals extracted from the apoplast by cortical cells enter the symplastic pathway.
  - The Casparian strip of the endodermis blocks the apoplastic entry of water and minerals to the stele. Of minerals that reach the Casparian strip via the apoplast, only those that can cross the plasma membranes of endodermal cells gain access to the xylem. Thus, the endodermis helps regulate the mineral composition of the xylem sap.
- The ascent of xylem sap depends mainly on transpiration and the physical properties of water (pp. 699–701, FIGURE 32.7)
  - The upward flow of xylem sap supplies minerals to the shoots and replaces water lost by transpiration.
  - The transpiration-cohesion-tension mechanism transports xylem sap. Evaporative water loss during transpiration generates surface tension of the water film coating mesophyll cells. This tension, or negative pressure, causes water to move by bulk flow out of the xylem vessels. The cohesion of water due to hydrogen bonding relays the transpirational pull on xylem sap all the way down to the roots.
- Guard cells mediate the transpiration-photosynthesis compromise (pp. 701–705, FIGURE 32.8)
  - Plants strike a balance between gas exchange and water loss associated with the presence of stomata.
  - Because water loss is the trade-off for allowing carbon dioxide to diffuse into the leaf, plants with lower transpiration-to-photosynthesis ratios are generally at an advantage in arid habitats.
  - Plants balance photosynthesis and transpiration by regulating the size of stomatal openings through changes in the turgor pressure of guard cells. These physical changes are due to fluxes in potassium ions.
  - Guard cells usually open at dawn because of carbon dioxide depletion, an inherent circadian rhythm, and ion movements triggered by light-detecting pigments.
  - Xerophytic plants in dry habitats have leaves with morphological or physiological adaptations that reduce transpiration.
  - CAM plants have a reverse stomatal rhythm; they conserve water by opening their stomata at night.
- A bulk-flow mechanism translocates phloem sap from sugar sources to sugar sinks (pp. 705–708, FIGURE 32.10)
  - Translocation is the process of transporting photosynthetically produced food throughout the body of the plant in the phloem. In most plants, sucrose is the main solute translocated in phloem.
  - Sieve tubes of phloem carry food from a sugar source to a sugar sink. A source is an organ that produces sugar by either photosynthesis or the breakdown of starch, whereas a sink consumes or stores sugar.
  - In many plants, sucrose is loaded into phloem at a source by active transport. Proton pumps of sieve-tube members and companion cells drive sucrose loading by chemiosmotic coupling.
  - Pressure moves phloem sap along sieve tubes by bulk flow. The pressure is generated at the source end of a sieve tube by the accumulation of sucrose and the resulting osmotic uptake of water.

SELF-QUIZ

1. Which of the following would *not* contribute to water uptake by a plant cell?
   a. an increase in the water potential (ψ) of the surrounding solution
   b. a decrease in pressure on the cell exerted by the wall
   c. the uptake of solutes by the cell
   d. a decrease in ψ of the cytoplasm
   e. an increase in tension on the surrounding solution

2. In which of the following processes is osmosis *least* involved?
   a. the long-distance transport of xylem sap
   b. the swelling of guard cells
   c. water uptake by cells immersed in a hypotonic solution
   d. root pressure
   e. water movement between neighboring cells of the root cortex

3. Stomata open when guard cells
   a. sense an increase in CO_2 in the air spaces of the leaf
   b. flop open because of a decrease in turgor pressure
   c. become more turgid because of an influx of K^+, followed by the osmotic entry of water
   d. reverse their circadian rhythm
   e. accumulate water by active transport

4. Which of the following is *not* part of the transpiration-cohesion-tension mechanism for the ascent of xylem sap?
   a. the evaporation of water from the mesophyll cells, which initiates a pull of water molecules from neighboring cells and eventually from the xylem
   b. the transfer of transpirational pull from one water molecule to the next due to the cohesion caused by hydrogen bonds
   c. the hydrophilic walls of the narrow tracheids and xylem vessels that help maintain the column of water against the force of gravity
   d. active pumping of water into the xylem of roots
   e. the reduction of water potential in the surface film of mesophyll due to increased surface tension

5. Both ____________ and ____________ are sugar sinks.
   a. a growing root; a developing fruit
   b. a photosynthesizing leaf; a growing root
   c. a growing shoot tip; a tuber where starch is being broken down
   d. a photosynthesizing leaf; a developing fruit
   e. a photosynthesizing leaf; a tuber where starch is being broken down

6. Which of the following does *not* appear to involve active transport?
   a. the movement of mineral nutrients from the apoplast to the symplast
   b. the movement of sugar from mesophyll into sieve-tube members in corn
   c. the movement of sugar from one sieve-tube member to the next
   d. K^+ uptake by guard cells during stomatal opening
   e. the movement of mineral nutrients into cells of the root cortex

7. The movement of sap from a sugar source to a sugar sink
   a. occurs through the apoplast of sieve-tube members
   b. may translocate sugars from the breakdown of stored starch up to developing shoots
   c. is similar to the flow of xylem sap in depending on tension, or negative pressure
   d. depends on the active pumping of water into sieve tubes at the source end
   e. results mainly from diffusion

8. The productivity of a crop begins to decline when leaves begin to wilt mainly because (explain your answer)
   a. the chlorophyll of wilting leaves decomposes
   b. flaccid mesophyll cells are incapable of photosynthesis
   c. stomata close, preventing CO_2 from entering the leaf
   d. photolysis, the water-splitting step of photosynthesis, cannot occur when there is a water deficiency
   e. an accumulation of CO_2 in the leaf inhibits the enzymes required for photosynthesis

9. Imagine cutting a live twig from a tree and examining the cut surface of the twig with a magnifying glass. You locate the vascular tissue and observe a growing droplet of fluid exuding from the cut surface. This fluid is probably (explain your answer)
   a. phloem sap
   b. xylem sap
   c. guttation fluid
   d. fluid of the transpiration stream
   e. cell sap from the broken vacuoles of cells

10. Which structure or compartment is *not* part of the plant's apoplast?
    a. the lumen of a xylem vessel
    b. the lumen of a sieve tube
    c. the cell wall of a mesophyll cell
    d. the cell walls of transfer cells
    e. the cell walls of root hairs

CHALLENGE QUESTIONS

1. A botanist used a device called a dendrometer to measure slight changes in the diameter of a tree trunk caused by water movement in the xylem. The device made simultaneous measurements at different heights, graphed below. Explain the daily pattern of variation in diameter. Do these data suggest that water is "pushed" or "pulled" up the trunk? Explain.

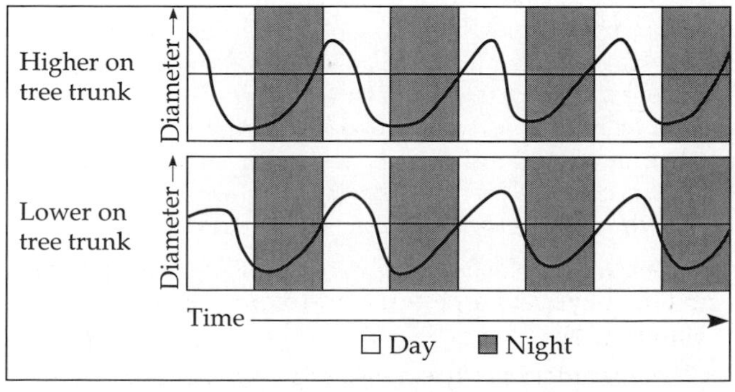

2. Wind ventilates the leaf surface, increases transpiration, and replenishes the supply of carbon dioxide for photosynthesis. Stomata close when the carbon dioxide concentration inside the leaf increases. How do you think stomatal response to carbon dioxide is related to wind speed, and what might be the adaptive significance of this relationship?

3. Describe the environmental conditions that would minimize the transpiration-to-photosynthesis ratio for a C_3 plant.

4. A tip for making cut flowers last longer without wilting is to cut off the ends of the stems under water and then transfer the flowers to a vase full of water while drops of water are still present on the cut ends of the stems. Explain why this works.

SCIENCE, TECHNOLOGY, AND SOCIETY

1. In a reverse osmosis unit, pressure is applied to a solution, forcing water through a selectively permeable membrane but leaving solute molecules behind. Why is this called reverse osmosis? Relate reverse osmosis to the water potential concept discussed in this chapter. What might be some possible commercial applications of reverse osmosis?

2. Many scientists believe that increased carbon dioxide in the atmosphere may result in warming of the global climate in coming decades. This could have a serious impact on agriculture. Imagine that you are in charge of a project to use selective breeding and recombinant DNA techniques to "engineer" crop plants that can grow under hotter, drier conditions. What traits would you try to build into the "ideal" corn or wheat plant?

FURTHER READING

Heinrich, B. "Nutcracker Sweets." *Natural History,* February 1991. How red squirrels obtain maple syrup from the sap flowing within trees.

Neher, E., and B. Sakmann. "The Patch Clamp Technique." *Scientific American,* March 1992. The Nobel Prize–winning developers of this method explain how it can be used to study membrane transport.

Salisbury, F. B., and C. W. Ross. *Plant Physiology,* 4th ed. Belmont, CA: Wadsworth, 1992. A rigorous text with excellent chapters on transport.

Szabo, M. "Plant Sugars Move in Mysterious Ways." *New Scientist,* December 4, 1993. Agricultural applications of understanding source-sink transport in plants.

Taiz, L., and E. Zeiger. *Plant Physiology.* Redwood City, CA: Benjamin/Cummings, 1991. Chapters 3–7 of this text cover transport on a detailed but accessible level.

CHAPTER 33

PLANT NUTRITION

KEY CONCEPTS

- Plants require at least seventeen essential nutrients
- The symptoms of a mineral deficiency depend on the function and mobility of the element
- Soil characteristics are key environmental factors in terrestrial ecosystems
- Soil conservation is one step toward sustainable agriculture
- The metabolism of soil bacteria makes nitrogen available to plants
- Improving the protein yield of crops is a major goal of agricultural research
- Predation and symbiosis are evolutionary adaptations that enhance plant nutrition

Every organism is an open system connected to its environment by a continuous exchange of energy and materials. In the energy flow and chemical cycling that keep an ecosystem alive, plants and other photosynthetic autotrophs perform the key step of transforming inorganic compounds into organic ones. Autotrophic does not mean autonomous, however. Plants need sunlight as the energy source for photosynthesis. But to synthesize organic matter, plants also require raw materials in the form of inorganic substances: carbon dioxide, water, and a variety of minerals present as inorganic ions in the soil. With its ramifying root system and shoot system (see the photograph of a bean seedling on this page), a plant is extensively networked with its environment—the soil and air, which are the reservoirs of the plant's inorganic nutrients. In this chapter, you will learn more about the nutritional requirements of plants and examine some of the structural and physiological adaptations for plant nutrition that have evolved.

Plants require at least seventeen essential nutrients

The Chemical Composition of Plants

Watch a large plant grow from a tiny seed, and you cannot help wondering where all the mass comes from. Aristotle thought soil provided the substance for plant growth because plants seemed to spring from the ground. Leaves, he believed, functioned only to shade the developing fruit. In the seventeenth century, a Belgian physician named Jean-Baptiste van Helmont performed an experiment to test the hypothesis that plants grew by absorbing soil. He planted a willow seedling in a pot that contained 90.9 kg of soil. After five years, the willow had grown into a tree weighing 76.8 kg, but only 0.06 kg of soil had disappeared from the pot. Van Helmont concluded that the willow had grown mainly from the water he had added regularly. A century later, Stephen Hales, an English physiologist, postulated that plants were nourished mostly by air.

As it turns out, none of the early ideas about plant nutrition is entirely incorrect. Plants *do* extract minerals that are essential nutrients from the soil, but these minerals make only a small contribution to the overall mass of the plant, as van Helmont discovered. About 80% to 85% of a herbaceous (nonwoody) plant is water, and plants grow mainly by accumulating water in the central vacuoles of their cells. Furthermore, water can truly be considered a nutrient because it supplies most of the hydrogen and some of the oxygen that are incorporated into organic compounds by photosynthesis. However, only a small fraction of the water that enters a plant contributes atoms to organic molecules. Generally, more than 90% of the water absorbed by plants is lost by transpiration, and most of the water that is retained by the plant actually functions as a solvent, makes cell elongation possible, and serves to maintain the form of soft tissue by keeping cells turgid. By weight, the bulk of the organic material of a plant is derived not from water, but from the CO_2 that is assimilated from the atmosphere (FIGURE 33.1).

We can measure water content by comparing the weight of plant material before and after it is dried. We can then analyze the chemical composition of the dry residue. Organic substances account for about 95% of the dry weight, with inorganic minerals making up the remaining 5%. Most of the organic material is carbohydrate, including the cellulose of cell walls. Thus, carbon, oxygen, and hydrogen, the ingredients of carbohydrates, are the most abundant elements in the dry weight of a plant. Because some organic molecules contain nitrogen, sulfur, or phosphorus, these elements are also relatively abundant in plants.

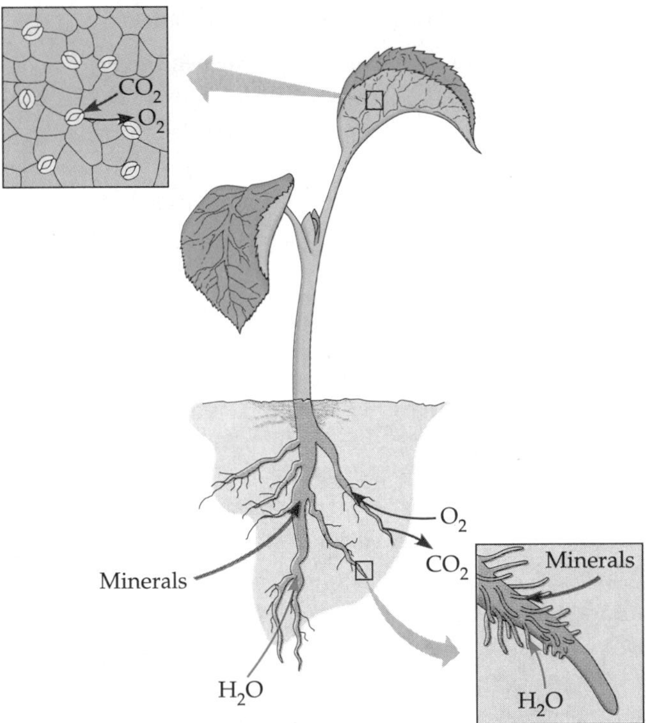

FIGURE 33.1
The uptake of nutrients by a plant: an overview. Roots absorb water and minerals from the soil, with root hairs greatly increasing the area of the epidermal surface that functions in this absorption. Carbon dioxide, the source of carbon for photosynthesis, diffuses into leaves from the surrounding air through stomata. (Plants also need O_2 for cellular respiration, although the plant is a net producer of O_2.) From these inorganic nutrients, the plant can produce all of its own organic material.

More than 50 chemical elements have been identified among the inorganic substances present in plants, but it is unlikely that all these elements are essential. Roots are able to absorb minerals somewhat selectively, enabling the plant to accumulate essential elements that may be present in the soil in very minute quantities. To a certain extent, however, the minerals in a plant reflect the composition of the soil in which the plant is growing. Plants growing on mine tailings, for instance, may contain gold or silver. Studying the chemical composition of plants provides clues about their nutritional requirements, but we must distinguish elements that are essential from those that are merely present in the plant.

Essential Nutrients

A particular chemical element is considered to be an **essential nutrient** if it is required for a plant to grow from a seed and complete the life cycle, producing another generation of seeds. A method known as hydroponic culture can be used to determine which of the mineral elements are actually essential nutrients (see the Methods Box, p. 714). Such studies have helped identify 17 elements that are essential nutrients in all plants and a few other elements that are essential to certain groups of plants (TABLE 33.1). Most research has involved crop plants; little is known about the specific nutritional needs of uncultivated plants, even some of the most commercially important conifers.

Elements required by plants in relatively large amounts are called **macronutrients.** There are nine macronutrients in all, including the six major ingredients of organic compounds: carbon, oxygen, hydrogen, nitrogen, sulfur, and phosphorus. The other three macronutrients are calcium, potassium, and magnesium. (TABLE 33.1 lists some of their functions).

Elements that plants need in very small amounts are called **micronutrients.** The eight micronutrients are iron, chlorine, copper, manganese, zinc, molybdenum, boron, and nickel. These elements function in the plant mainly as cofactors of enzymatic reactions (see Chapter 6). Iron, for example, is a metallic component of cytochromes, the proteins that function in the electron transport chains of chloroplasts and mitochondria. It is because micronutrients generally play catalytic roles that plants need only minute quantities of these elements. The requirement for molybdenum, for example, is so modest that there is only one atom of this rare element for every 16 million atoms of hydrogen in dried plant material. Yet a deficiency of molybdenum or any other micronutrient can weaken or kill a plant.

TABLE 33.1

Essential Nutrients in Plants

ELEMENT	FORM AVAILABLE TO PLANTS	MAJOR FUNCTIONS
Macronutrients		
Carbon	CO_2	Major component of plant's organic compounds.
Oxygen	CO_2	Major component of plant's organic compounds.
Hydrogen	H_2O	Major component of plant's organic compounds.
Nitrogen	NO_3^-, NH_4^+	Component of nucleic acids, proteins, hormones, and coenzymes.
Sulfur	SO_4^{2-}	Component of proteins, coenzymes.
Phosphorus	$H_2PO_4^-$, HPO_4^{2-}	Component of nucleic acids, phospholipids, ATP, several coenzymes.
Potassium	K^+	Cofactor functional in protein synthesis; major solute functioning in water balance; operation of stomata.
Calcium	Ca^{2+}	Important in formation and stability of cell walls; maintenance of membrane structure and permeability; activates some enzymes; regulates many responses of cells to stimuli.
Magnesium	Mg^{2+}	Component of chlorophyll; activates many enzymes.
Micronutrients		
Chlorine	Cl^-	Required for water-splitting step of photosynthesis; functions in water balance.
Iron	Fe^{3+}, Fe^{2+}	Component of cytochromes; activates some enzymes.
Boron	$H_2BO_3^-$	Cofactor in chlorophyll synthesis; may be involved in carbohydrate transport and nucleic acid synthesis.
Manganese	Mn^{2+}	Active in formation of amino acids; activates some enzymes.
Zinc	Zn^{2+}	Active in formation of chlorophyll; activates some enzymes.
Copper	Cu^+, Cu^{2+}	Component of many redox and lignin-biosynthetic enzymes.
Molybdenum	MoO_4^{2-}	Essential for nitrogen fixation; cofactor functional in nitrate reduction.
Nickel	Ni^{2+}	Cofactor for an enzyme functioning in nitrogen metabolism.

The symptoms of a mineral deficiency depend on the function and mobility of the element

The symptoms of a mineral deficiency depend partly on the function of that nutrient in the plant. For example, a deficiency of magnesium, an ingredient of chlorophyll, causes yellowing of the leaves, or chlorosis (FIGURE 33.2). In some cases, the relationship between a mineral deficiency and its symptoms is less direct. For instance, iron deficiency can cause chlorosis even though chlorophyll contains no iron, because this metal is required as a cofactor in one of the steps of chlorophyll synthesis.

Mineral deficiency symptoms depend not only on the role of the nutrient in the plant but also on its mobility within the plant. If a nutrient moves about freely from one part of the plant to another, symptoms of a deficiency will show up first in older organs. This is because young, growing tissues have more "drawing power" than old tissues for nutrients that are in short supply. A plant starved for magnesium, for example, will show signs of chlorosis first in its older leaves. Magnesium, which is relatively mobile in the plant, is shunted preferentially to young leaves. In contrast, a deficiency of a nutrient that is relatively immobile within a plant will affect young parts of the plant first. Older tissues may have adequate amounts of the mineral, which they are able to retain during periods of short supply. A deficiency of iron, which does not move freely in the plant, will cause yellowing of young leaves before any effect on older leaves is visible.

The symptoms of a mineral deficiency are often distinctive enough for a plant physiologist or farmer to diagnose its cause. One way to confirm the diagnosis of a specific deficiency is to analyze the mineral content of the plant and soil. Deficiencies of nitrogen, potassium, and phosphorus are the most common problems. Shortages of micronutrients are less common and tend to be geographically localized because of differences in soil

METHODS: HYDROPONIC CULTURE TO IDENTIFY ESSENTIAL NUTRIENTS

In the technique called hydroponic culture, a researcher bathes the roots of plants in solutions of various minerals dissolved in known concentrations. Aerating the water provides the roots with oxygen for cellular respiration. A particular mineral, such as potassium, can be omitted from the culture medium to test whether it is essential to the plants.

If the element deleted from the mineral solution is an essential nutrient, then the incomplete medium will cause plants to become abnormal in appearance compared with controls grown on a complete mineral medium. The most common symptoms of a mineral deficiency are stunted growth and discolored leaves.

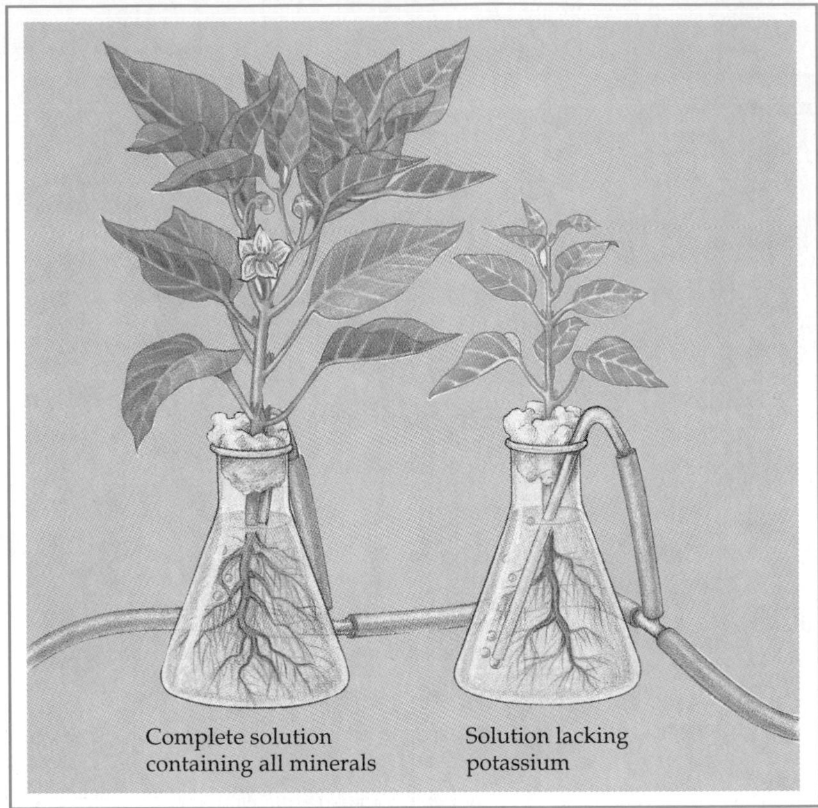

(a) (b)

FIGURE 33.2

Magnesium deficiency in tomato. Compare the control plant (**a**) to the experimental plant (**b**), which shows signs of magnesium deficiency. Yellowing of the leaves (chlorosis) is the result of an inability to synthesize chlorophyll, which contains magnesium.

FIGURE 33.3
Hydroponic farming. In this apparatus, a nutrient solution flows over the roots of lettuce growing on a slat. Perhaps astronauts living in a space station will one day grow their vegetables hydroponically, but because of the expense, it is unlikely that this type of farming will soon relieve hunger here on Earth.

composition. The amount of a micronutrient needed to correct a deficiency is usually quite small. For example, a zinc deficiency in fruit trees can usually be cured by hammering a few zinc nails into each tree trunk. Moderation is important because overdoses of some micronutrients can be toxic to plants.

One way to ensure optimal mineral nutrition is to grow plants hydroponically on nutrient solutions that can be precisely regulated (FIGURE 33.3). Hydroponics is currently practiced commercially, but only on a limited scale because the requirements for equipment and labor make hydroponic farming relatively expensive compared with growing crops in soil.

Soil characteristics are key environmental factors in terrestrial ecosystems

The texture and chemical composition of soil are major factors determining what kinds of plants can grow well in a particular location, be it a natural ecosystem or an agricultural region. (Climate, of course, is another important factor.) Plants that grow naturally in a certain type of soil are adapted to its mineral content and texture and are able to absorb water and extract essential nutrients from that soil. In interacting with the soil that supports their growth, plants, in turn, affect the soil, as we will soon see. The soil-plant interface is a critical component of the nutrient cycles that sustain terrestrial ecosystems.

Texture and Composition of Soils

Soil has its origin in the weathering of solid rock. Water that seeps into crevices and freezes in winter fractures the rock, and acids dissolved in the water also help to break down the rock. Once organisms are able to invade the rock, they accelerate the decomposition. Lichens, fungi, bacteria, mosses, and plant roots all secrete acids, and the expansion of roots growing in fissures cracks rocks and pebbles. The eventual result of all this activity is **topsoil,** a mixture of decomposed rock of varying texture, living organisms, and humus, a residue of partially decayed organic material. The topsoil and other distinct soil layers, or **horizons,** are often visible in vertical profile where there is a roadcut (FIGURE 33.4).

The texture of topsoil depends on the size of its particles, which are classified in a range from coarse sand to microscopic clay particles. The most fertile soils are usually **loams,** made up of a mixture of sand, silt, and clay.

FIGURE 33.4
Soil horizons. This roadcut exposes a vertical profile of three soil layers, or horizons. The A horizon is the topsoil, a mixture of decomposed rock of varying textures, living organisms, and decaying organic matter. Topsoil is subject to extensive physical and chemical weathering. The B horizon contains much less organic matter than the A horizon and is less weathered. Clay particles and minerals leached from the A horizon by water may accumulate in the B horizon. The C horizon serves as the parent material for the upper layers of soil. This horizon is mainly composed of partially weathered rock.

Loamy soils have enough fine particles to provide a large surface area for retaining minerals and water, which adhere to the particles. But loams also have enough coarse particles to provide air spaces containing oxygen that can be used by roots for cellular respiration. If soil does not drain adequately, roots suffocate because the air spaces are replaced by water; the roots may also be attacked by molds favored by the soaked soil. These are common hazards for houseplants that are overwatered in pots that do not drain. Some plants, however, are adapted to waterlogged soil. For example, mangroves and many other plants that inhabit swamps and marshes have some of their roots modified as hollow tubes that grow upward and function as snorkels, bringing down oxygen from the air. Variation in the degree of drainage required by plants is one of the many ways soil conditions help determine what type of vegetation prevails in a particular location.

Soil is home to an astonishing number and variety of organisms. A teaspoon of soil has about five billion bacteria that cohabit with various fungi, algae and other protists, insects, earthworms, nematodes, and the roots of plants. The activities of all these organisms affect the physical and chemical properties of the soil. Earthworms, for instance, aerate the soil by their burrowing and add mucus that holds fine soil particles together. The metabolism of bacteria alters the mineral composition of the soil. Plant roots extract water and minerals but also affect soil pH and reinforce the soil against erosion.

Humus is the decomposing organic material formed by the action of bacteria and fungi on dead organisms, feces, fallen leaves, and other organic refuse. Humus prevents clay from packing together and builds a crumbly soil that retains water but is still porous enough for the adequate aeration of roots. Humus is also a reservoir of mineral nutrients that are returned gradually to the soil as microorganisms decompose the organic matter.

The Availability of Soil Water and Minerals

After a heavy rainfall, water drains away from the larger spaces of the soil, but smaller spaces retain water because of its attraction for the soil particles, which have electrically charged surfaces. Some of this water adheres so tightly to the hydrophilic soil particles that it cannot be extracted by plants. In the tiniest spaces of the soil, however, is a film of water bound less tightly to the particles; this is the water generally available to plants (FIGURE 33.5a). It is not pure water, but a soil solution containing dissolved minerals. Roots absorb this soil solution.

Many minerals in soil, especially those that are positively charged, such as potassium (K^+), calcium (Ca^{2+}), and magnesium (Mg^{2+}), adhere by electrical attraction to the negatively charged surfaces of clay particles. The presence of clay in a soil helps prevent the leaching

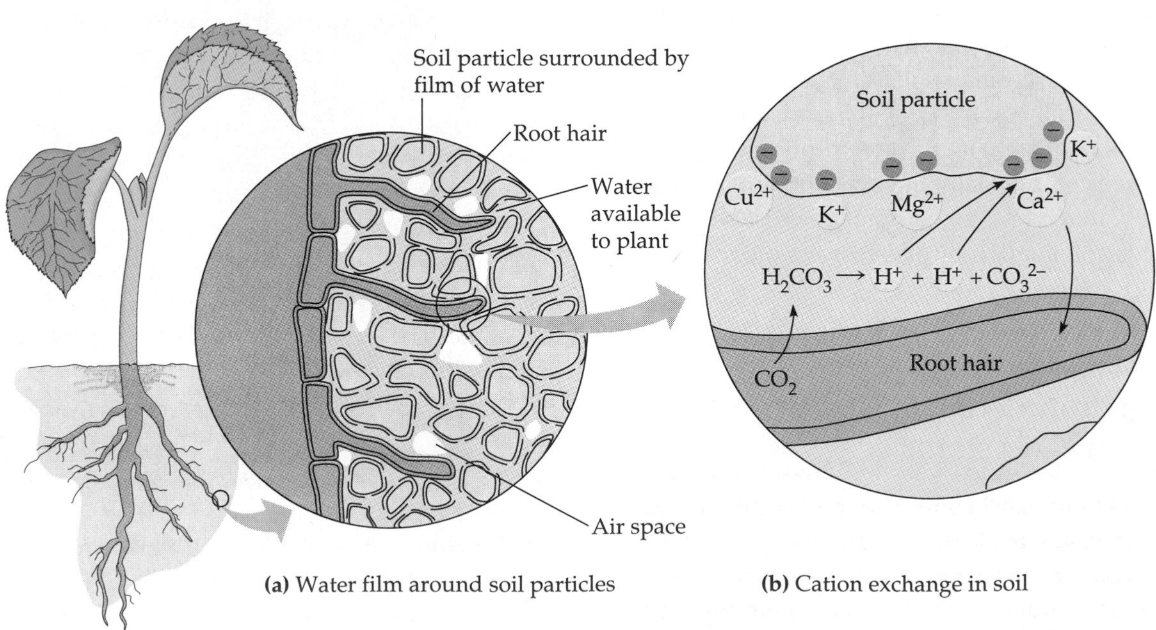

FIGURE 33.5
The availability of soil water and minerals. (**a**) A plant cannot extract all the water in the soil because some of it is tightly held by hydrophilic soil particles. Water retained in tiny spaces in the soil, bound less tightly to soil particles, can be imbibed by the root—especially by the epidermis, which includes the root hairs. (**b**) Hydrogen ions in the soil solution help make certain nutrients available to plants by displacing positively charged minerals (cations) that were bound tightly to the surface of fine soil particles. Plants contribute to the pool of H^+ in the soil in the following way: Cellular respiration in roots releases CO_2 to the soil solution, where the CO_2 reacts with water to form carbonic acid (H_2CO_3). Dissociation of this acid adds hydrogen ions to the soil.

(draining away) of mineral nutrients during heavy rain or irrigation because the finely divided particles provide so much surface area for binding minerals. Minerals that are negatively charged, such as nitrate (NO_3^-), phosphate ($H_2PO_4^-$), and sulfate (SO_4^{2-}), are usually not bound tightly to soil particles and thus tend to leach away more quickly. On the other hand, clay particles must release their bound minerals to the soil solution in order for roots to absorb the nutrients. Positively charged minerals are made available to the plant when hydrogen ions in the soil displace the mineral ions from the clay particles. This process, called **cation exchange,** is stimulated by the roots themselves, which release acids that add hydrogen ions to the soil solution (FIGURE 33.5b).

As roots absorb water and minerals, the supply of these resources is depleted in the immediate surroundings of the existing roots. However, the rapid growth of roots, with surface areas expanded by root hairs, brings the plant into contact with a larger and larger volume of soil and thus a new supply of nutrients.

Soil conservation is one step toward sustainable agriculture

It may take centuries for a soil to become fertile through decomposition and the accumulation of organic material, but human mismanagement can destroy that fertility within a few years. Good soil conservation, on the other hand, can preserve fertility, resulting in sustained agricultural productivity.

To understand soil conservation, we must begin with the premise that agriculture is unnatural. In forests, grasslands, and other natural ecosystems, mineral nutrients are usually recycled by the decomposition of dead organic material in the soil. In contrast, when farmers harvest a crop, or a gardener uses a grass catcher on the lawn mower, essential elements are diverted from the chemical cycles going on in that location. In general, agriculture depletes the mineral content of the soil. To grow a ton of wheat grain, the soil gives up 18.2 kg of nitrogen, 3.6 kg of phosphorus, and 4.1 kg of potassium. Each year the fertility of the soil diminishes, unless fertilizers are applied to replace the lost minerals. Many crops also use far more water than the natural vegetation that once grew on that land, forcing farmers to irrigate. Prudent fertilization, thoughtful irrigation, and the prevention of erosion are three of the most important goals of soil conservation.

Fertilizers

Prehistoric farmers may have started fertilizing their fields after noticing that grass grew faster and greener where animals had defecated. The Romans used manure to fertilize their crops, and Native Americans buried fish along with seeds when they planted corn. In developed nations today, most farmers use commercially produced fertilizers containing minerals that are either mined or prepared by industrial processes. These fertilizers are usually enriched in nitrogen, phosphorus, and potassium, the three mineral elements that are most commonly deficient in farm soils.

Manure, fishmeal, and compost are referred to as "organic" fertilizers because they are of biological origin and contain organic material that is in the process of decomposing. However, before the elements in compost can be of any use to plants, the organic material must be decomposed to the inorganic nutrients that roots can absorb. In the end, the minerals a plant extracts from the soil are in the same form whether they came from organic fertilizer or from a chemical factory. Compost releases minerals gradually, however, whereas the minerals in commercial fertilizers are available immediately, but may not be retained by the soil for long. Excess minerals not taken up by the plants are wasted because they may be rapidly leached from the soil by rainwater or irrigation. To make matters worse, this mineral runoff may enter the groundwater and eventually pollute streams and lakes. Agricultural researchers are attempting to develop ways to reduce the use of fertilizers while maintaining crop yields.

To fertilize judiciously, the farmer must pay close attention to the pH of the soil. Acidity not only affects cation exchange but also influences the chemical form of all minerals. Even though an essential element may be abundant in the soil, plants may be starving for that element because it is bound too tightly to clay or it is in a chemical form the plant cannot absorb. Managing the pH of soil is touchy; a change in hydrogen ion concentration may make one mineral more available to the plant while causing another mineral to become less available. At pH 8, for instance, the plant can absorb calcium, but iron is almost completely unavailable. The pH of the soil should be matched to the specific mineral needs of the crop. If the soil is too alkaline, sulfate can be added to lower the pH. Soil that is too acidic can be adjusted by liming (adding calcium carbonate or calcium hydroxide).

Irrigation

Even more than mineral deficiencies, the availability of water most often limits the growth of plants. Irrigation can transform a desert into a garden, but farming in arid regions is a huge drain on water resources. Many of the rivers in the southwestern United States have been reduced to trickles by the diversion of water for irrigation. (Quenching the thirst of growing cities adds to the problem.) Another problem is that irrigation in an arid region

FIGURE 33.6
Irrigation. (**a**) Flood irrigation. After a field is flooded, much of the water evaporates, leaving the salts that were dissolved in the irrigation water behind to accumulate in the soil. (**b**) Drip irrigation. Instead of flooding the field or filling ditches with water, perforated pipes drip the water slowly into the soil close to the plant roots. Drip irrigation reduces the loss of water from evaporation and drainage.

(a)

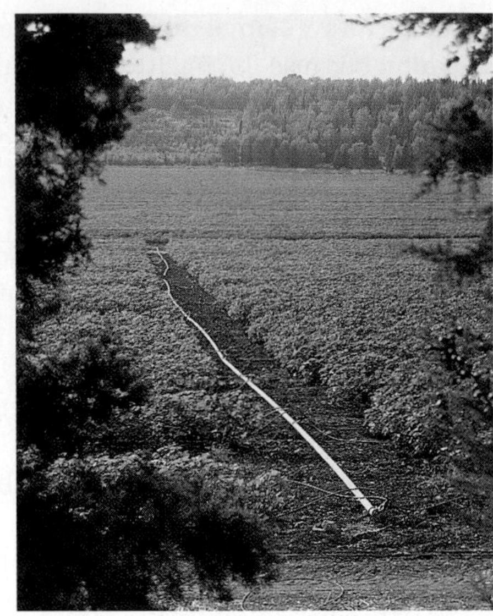

(b)

can gradually make the soil so salty that it becomes completely infertile (FIGURE 33.6a). Salts dissolved in the irrigation water accumulate in the soil as the water evaporates. Eventually, the salt makes the soil solution hypertonic to root cells, which then *lose* water instead of absorbing it. As the world population continues to grow, more and more acres of arid land will have to be cultivated. New methods of irrigation may reduce the risks of running out of water or losing farmland to salinization (salt accumulation). For instance, drip irrigation is now used as an alternative to flooding fields for many of the crops and orchards in Israel and the western United States (see FIGURE 33.6b). In another approach to solving some of the problems of dryland farming, plant breeders are working to develop varieties of plants that require less water or that can tolerate more salinity.

Erosion

Topsoil from thousands of acres of farmland is lost to water and wind erosion each year in the United States alone. Certain precautions can help reduce these losses. Rows of trees dividing fields make effective windbreaks, and terracing a hillside can prevent the topsoil from washing away in a heavy rain. Such crops as alfalfa and wheat provide good ground cover and protect the soil better than corn and other crops that are usually planted in rows.

If managed properly, soil is a renewable resource on which farmers can grow food for generations to come. The goal is **sustainable agriculture,** a commitment embracing a variety of farming methods that are conservation-minded, environmentally safe, and profitable (FIGURE 33.7).

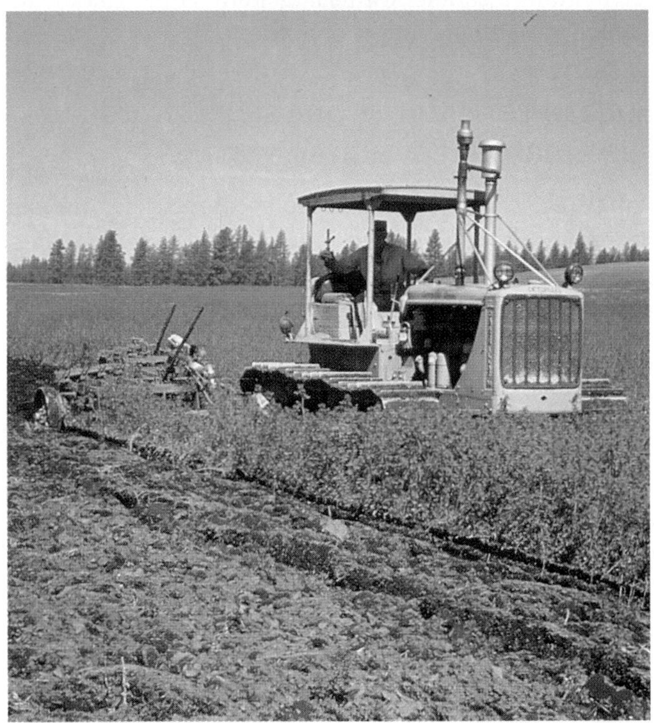

FIGURE 33.7
Sustainable agriculture and "green manure." The "green manure" being mulched into the soil of this Washington state farm is sweet clover. One out of every three years, clover is planted and the crop is plowed under. This improves the physical structure and nutrient content of the soil for growing wheat and other crops during the other two years of the crop-rotation cycle. Substituting green manure for chemical additives, using beneficial insects that feed on crop pests as an alternative to insecticides, and choosing crops that are suited to the amount of rainfall and other climatic conditions of the locale are examples of actions aimed at developing sustainable agriculture.

The metabolism of soil bacteria makes nitrogen available to plants

Of all mineral elements, nitrogen is the one that most often limits the growth of plants and the yields of crops. Plants require nitrogen as an ingredient of proteins, nucleic acids, and other important organic molecules. It is ironic that plants sometimes suffer nitrogen deficiencies, for the atmosphere is nearly 80% nitrogen. This atmospheric nitrogen, however, is gaseous N_2, and plants cannot use nitrogen in that form. For plants to absorb nitrogen, it must first be converted to ammonium (NH_4^+) or nitrate (NO_3^-). In contrast to other minerals, the NH_4^+ and NO_3^- in soil are not derived from the breakdown of parent rock. Over the short term, the main source of nitrogenous minerals is the decomposition of humus by microbes, including ammonifying bacteria (FIGURE 33.8). In this way, nitrogen present in organic compounds, such as proteins, is repackaged in inorganic compounds that can be recycled when they are absorbed as minerals by roots. However, nitrogen is lost from this local cycle when soil microbes called denitrifying bacteria convert NO_3^- to N_2, which diffuses from the soil to the atmosphere. Still other bacteria, called **nitrogen-fixing bacteria,** restock nitrogenous minerals in the soil by converting N_2 to NH_3 (ammonia), a metabolic process called **nitrogen fixation.** The complex cycling of nitrogen in ecosystems will be traced in detail in Chapter 49. Here, we focus on nitrogen fixation and the other steps that lead directly to nitrogen assimilation by plants.

Nitrogen Fixation

All of Earth's life depends on nitrogen fixation, an exclusive function of certain prokaryotes. Soil is populated by several species of free-living bacteria that are among the nitrogen-fixing prokaryotes. The conversion of atmospheric nitrogen (N_2) to ammonia (NH_3) is a complicated, multistep process, but we can simplify nitrogen fixation by just indicating the reactants and products:

$$N_2 + 8\,e^- + 8\,H^+ + 16\,ATP \longrightarrow 2\,NH_3 + H_2 + 16\,ADP + 16\,Ⓟ_i$$

One enzyme complex, called **nitrogenase,** catalyzes the entire reaction sequence, which reduces N_2 to NH_3 by adding electrons along with hydrogen ions. Notice that nitrogen fixation is very expensive in terms of metabolic energy, costing the bacteria 8 ATP molecules for each ammonia molecule synthesized. Nitrogen-fixing bacteria are most abundant in soils rich in organic material, which provides fuel for cellular respiration.

In the soil solution, ammonia picks up another hydrogen ion to form ammonium (NH_4^+), which plants can absorb. However, plants acquire their nitrogen mainly in

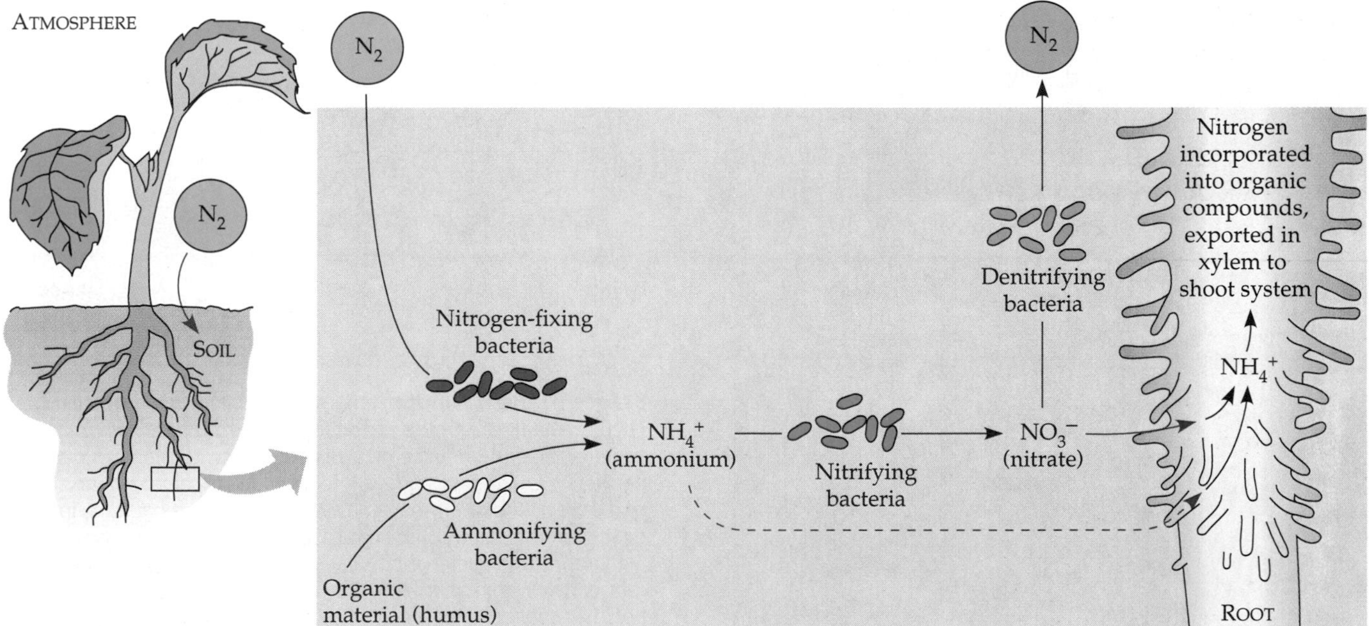

FIGURE 33.8
The role of soil bacteria in the nitrogen nutrition of plants. Soil bacteria form ammonium by fixing atmospheric N_2 (nitrogen-fixing bacteria) and by decomposing organic material (ammonifying bacteria). Although plants absorb some ammonium from the soil, they absorb mainly nitrate, which is produced from ammonium by nitrifying bacteria. The form of nitrogen transported up to the shoot system depends on the plant species. Most often, nitrogen from nitrate is incorporated into organic compounds such as amino acids in roots.

the form of nitrate (NO_3^-), which is produced in the soil by nitrifying bacteria that oxidize ammonium (see FIGURE 33.8). After nitrate is absorbed by roots, the nitrogen is usually incorporated into organic compounds. Most plant species export nitrogen from roots to shoots, via the xylem, in the form of amino acids and other organic compounds.

Symbiotic Nitrogen Fixation

Plants of the legume family, including peas, beans, soybeans, peanuts, alfalfa, and clover, have a built-in source of fixed nitrogen (FIGURE 33.9). Their roots have swellings called **nodules** composed of plant cells that contain nitrogen-fixing bacteria of the genus *Rhizobium* ("root living"). Inside the nodule, the *Rhizobium* assume a form called **bacteroids,** which are contained within vesicles formed by the root cell. Each legume is associated with a particular species of *Rhizobium.* FIGURE 33.10 describes the steps in the development of root nodules.

The symbiotic relationship between a legume and its nitrogen-fixing bacteria is mutualistic, with both partners benefiting. The bacteria supply the legume with fixed nitrogen, and the plant provides the bacteria with carbohydrates and other organic compounds. The exquisite coevolution of partners is evident in their cooperative synthesis of a molecule named leghemoglobin, with

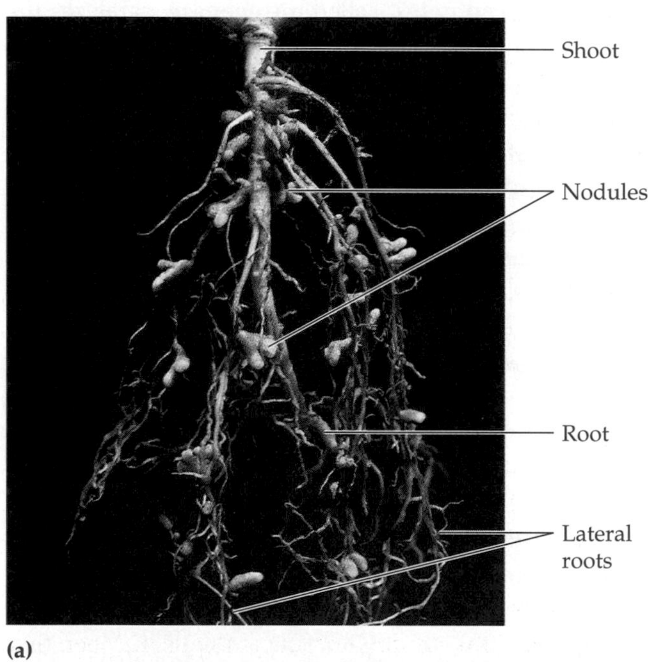

(a)

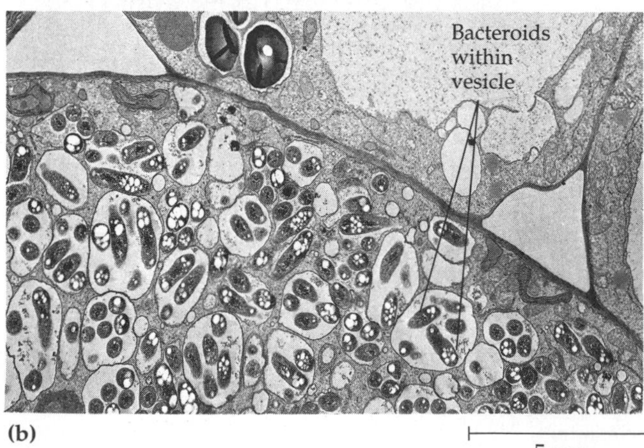

(b)

FIGURE 33.9

Root nodules on a legume. (**a**) The root nodules of this clover plant contain symbiotic bacteria that fix nitrogen and obtain photosynthetic products supplied from the plant. (**b**) In this transmission electron micrograph, a cell from a root nodule of soybean is filled with bacteroids. The adjoining cell remains uninfected.

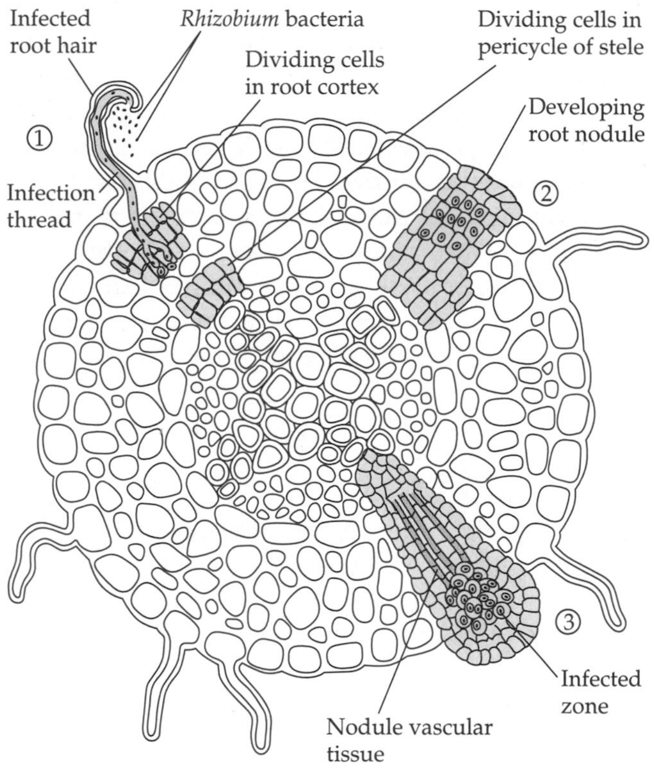

FIGURE 33.10

Stages in the development of a soybean root nodule. The coordinated activities of the legume and the *Rhizobium* bacteria depend on a chemical dialogue between the symbiotic partners. ① The first stage is infection. Roots secrete chemicals that attract *Rhizobium* bacteria living nearby. The bacteria, in turn, emit chemical signals that stimulate root hairs to elongate and curl around the bacterial population. The bacteria penetrate the root cortex within an infection thread. At the same time, the root begins responding to infection with a division of cells in the cortex and in the pericycle of the stele. Vesicles containing the bacteria, now known as bacteroids, bud into cortical cells from the tips of the branching infection thread. ② Growth continues in the affected regions of cortex and pericycle, and these two masses of dividing cells fuse, forming the nodule. ③ The nodule continues to grow, and vascular tissue connecting the nodule to the xylem and phloem of the stele develops. This vascular tissue supplies the nodule with sugar and other organic substrates for metabolism and carries nitrogenous compounds produced in the nodule into the stele for distribution to the rest of the plant.

the plant and the bacteria each making part of the molecule. Leghemoglobin is an iron-containing protein that, like the hemoglobin of human red blood cells, binds reversibly to oxygen (*leg-* is for legume). Leghemoglobin releases oxygen for the intense respiration required to produce ATP for nitrogen fixation. More importantly however, leghemoglobin keeps the concentration of free oxygen low in root nodules, an essential function because the enzyme nitrogenase is inhibited by oxygen.

Most of the ammonium produced by symbiotic nitrogen fixation is used by the nodules to make amino acids, which are then transported to the shoot and leaves via the xylem. When conditions are favorable, root nodules fix so much nitrogen that they actually secrete the excess ammonium, which increases the fertility of the soil for nonlegumes. This is one reason for crop rotation. One year a nonlegume such as corn is planted, and the following year alfalfa or some other legume is planted to restore the concentration of fixed nitrogen in the soil. Instead of being harvested, the legume crop is often plowed under so that it will decompose as "green manure" and add even more fixed nitrogen to the soil (see FIGURE 33.7). To ensure that the legume encounters its specific *Rhizobium,* the seeds are soaked in a culture of the bacteria or dusted with bacterial spores before sowing (FIGURE 33.11).

A few groups of nonlegumes, including alders and certain tropical grasses, host nitrogen-fixing bacteria of the actinomycete group (see Chapter 25). Rice, a crop of great commercial importance, benefits indirectly from symbiotic nitrogen fixation. Rice farmers culture a water fern called *Azolla* in their paddies. The fern has symbiotic cyanobacteria that fix nitrogen and increase the fertility of the rice paddy. The growing rice shades and kills the *Azolla,* and decomposition of this organic material adds more nitrogenous minerals to the paddy.

Improving the protein yield of crops is a major goal of agricultural research

The ability of plants to incorporate fixed nitrogen into proteins and other organic substances has a major impact on human welfare; the most common form of malnutrition in humans is protein deficiency. Either by choice or by economic necessity, the majority of people in the world have a predominantly vegetarian diet, and thus, particularly in developing countries, depend mainly on plants for protein. Unfortunately, many plants have a low protein content, and the proteins that are present may be deficient in one or more of the amino acids that humans need from their diet. Improving the quality and quantity of proteins in crops is a major goal of agricultural research.

Plant breeding has resulted in new varieties of corn, wheat, and rice that are enriched in protein. However, many of these "super" varieties have an extraordinary demand for nitrogen, which is usually supplied in the form of commercial fertilizer. The industrial production of ammonia and nitrate from atmospheric nitrogen is, like biological nitrogen fixation, very expensive in energy costs. A chemical factory making fertilizer consumes large quantities of fossil fuels. Generally, the countries that most need high-protein crops are the ones least able to afford to pay the fuel bill. New catalysts based on the mechanism by which nitrogenase fixes nitrogen may make commercial fertilizer production less costly in the future. Biochemists determined the structure of *Rhizobium* nitrogenase in 1992, providing a model for chemical engineers to design catalysts by imitating nature.

Improving the productivity of symbiotic nitrogen fixation is another strategy with the potential to increase protein yields of crops. Normally, the accumulation of fixed

(a)

(b)

FIGURE 33.11
Agricultural application of *Rhizobium* infection. (**a**) This Rwandan farmer is mixing legume seeds with the appropriate species of *Rhizobium* in order to inoculate the seeds with the nitrogen-fixing bacteria. (**b**) A field experiment compares plants grown from the inoculated seeds (left) with uninfected plants (right), whose yellow color is evidence of nitrogen deficiency.

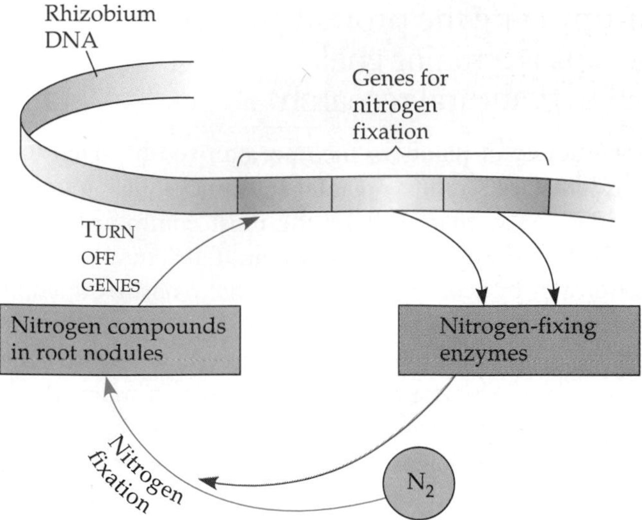

FIGURE 33.12
The regulation of nitrogen fixation in *Rhizobium*. This feedback circuit is one factor that limits the amount of fixed nitrogen the root nodules add to soil and also limits the protein yield of legume crops.

nitrogen in the nodules of legumes switches off the bacterial genes that code for nitrogenase and other enzymes involved in nitrogen fixation (FIGURE 33.12). Microbiologists have isolated mutant strains of *Rhizobium* that keep on making these enzymes even after fixed nitrogen accumulates. Farmers may someday grow plants infected with the mutant bacteria, which would increase the protein content of the legume crop and also add more fixed nitrogen to the soil. Geneticists are also working to improve the efficiency of symbiotic nitrogen fixation. The total food yield of a legume crop is often relatively low because so much of the carbohydrate produced in photosynthesis is consumed as energy for nitrogen fixation. Selecting for legume and *Rhizobium* varieties that can fix nitrogen at a lower cost in photosynthetic energy could be a boon to human nutrition.

In the past few years, researchers in China and Australia have been able to induce root nodule formation in nonlegumes, including rice and wheat. The problem these workers had to solve was eliminating the requirement for specific binding of symbiotic bacteria to the root surface, which is normally the first step in the infection of a legume by *Rhizobium*. The solution is to treat the roots of nonlegumes with chemicals that damage the cell walls of the epidermis. This enables nitrogen-fixing bacteria to infect the cortex and trigger nodule formation. In some of the experiments with wheat, the nodules fix nitrogen. More research will reveal how much nitrogen is actually fixed, and how much of this fixed nitrogen is assimilated into the proteins of the host plant.

Genetic engineering promises additional improvements in the protein yields of crops. Molecular biologists have already succeeded in transferring some of the genes required for nitrogen fixation from *Rhizobium* into other bacteria, and it may be possible through genetic engineering to create varieties of *Rhizobium* that can infect nonlegumes. Transplanting the genes for nitrogen fixation directly into the genomes of plants is also a possibility, using the plasmids of bacteria that cause a plant tumor called crown gall as vehicles for the gene transfer (see Chapter 19). This research has a long way to go from the laboratory to the field, but the prospect of wheat, potatoes, and other nonlegumes fixing their own nitrogen is a strong incentive for the work to continue.

Predation and symbiosis are evolutionary adaptations that enhance plant nutrition

Symbiotic nitrogen fixation underscores the relationship between plants and their environment, which includes the other organisms that interact with plants. We conclude this chapter by exploring three other types of plant adaptations that enhance nutrition through interactions with other organisms. Some plants parasitize other plants, others prey on animals, and most use fungi to absorb minerals. The correlation between structure and function is manifest in the specialized anatomical equipment of these plants.

Parasitic Plants

The mistletoe we find tacked above doorways during the holiday season lives in nature as a parasite on oaks and other trees (FIGURE 33.13a). Mistletoe is photosynthetic, but it supplements its nutrition by using projections called haustoria to siphon xylem sap from the vascular tissue of the host tree. Some parasitic plants, such as dodder (FIGURE 33.13b and c), do not perform photosynthesis, drawing all their nutrients from other plants.

Plants called epiphytes (Gr. *epi*, "upon," and *phyton*, "plant") are sometimes mistaken for parasites. An epiphyte is a plant that nourishes itself but grows on the surface of another plant, usually on the branches or trunks of trees. An epiphyte is anchored to its living substratum, but it absorbs water and minerals mostly from rain that falls on the leaves. Examples of epiphytes are staghorn ferns, some mosses, Spanish moss (actually an angiosperm), and many species of bromeliads and orchids.

Carnivorous Plants

Living in acid bogs and other habitats where soil conditions are poor, especially in nitrogen, are plants that fortify themselves by occasionally feeding on animals.

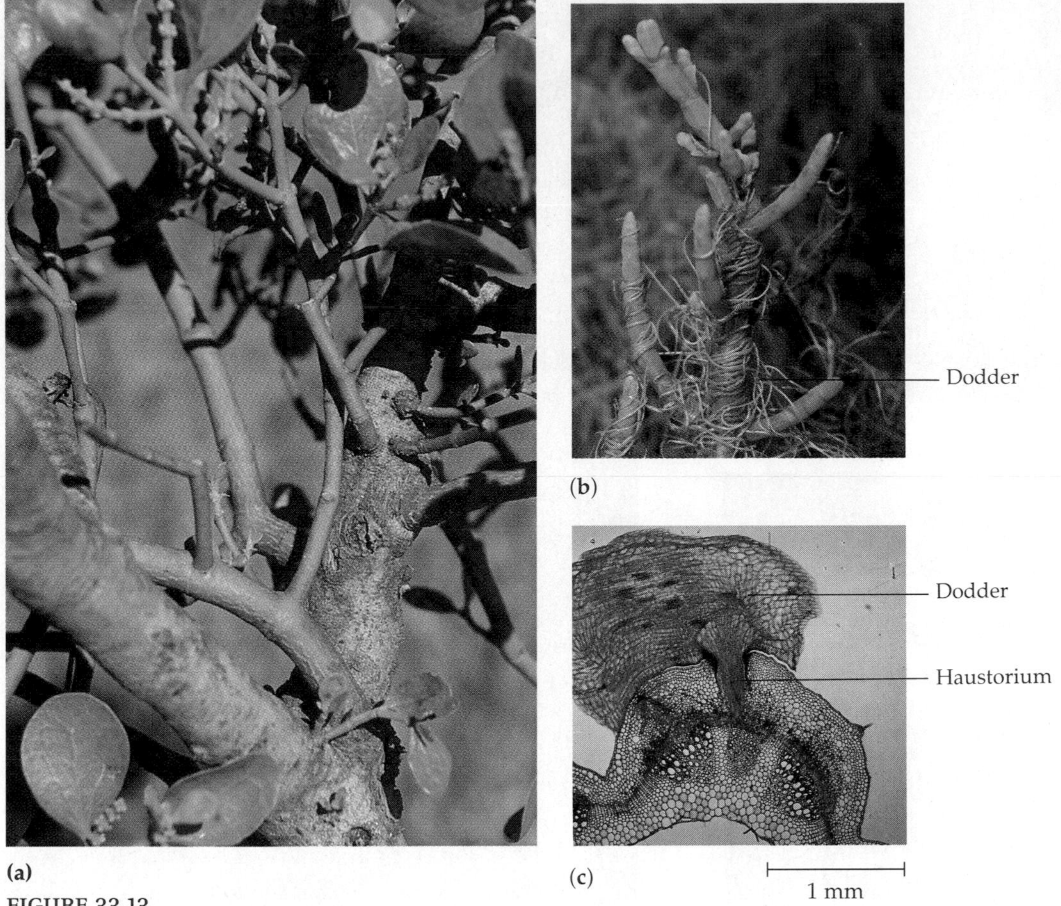

FIGURE 33.13
Parasitic plants. (**a**) Mistletoe growing on an oak. (**b**) Dodder growing on a California pickleweed. (**c**) In this transverse section of a host stem supporting dodder, a haustorium (modified root) of the parasite can be seen tapping the host plant's vascular tissue for water and mineral nutrients (LM).

These carnivorous plants make their own carbohydrates from photosynthesis, but they obtain some of their nitrogen and minerals by killing and digesting insects. Various kinds of insect traps have evolved by the modification of leaves (FIGURE 33.14). The traps are usually equipped with glands that secrete digestive juices.

Mycorrhizae

Most plants have roots modified as **mycorrhizae** (Gr. *mykos,* "fungus," and *rhiza,* "root"), which are actually mutualistic associations between the roots and fungi. The fungal partners of mycorrhizae are discussed in Chapter 28. These fungi secrete growth factors that stimulate roots to grow and branch. As the roots develop into mycorrhizae, the fungus either sheathes the root and extends hyphae among the cortex cells or invades the root cells themselves (see FIGURE 28.13). The other ends of the fungal mycelium function to provide the root with a greatly increased surface area through which the root gains minerals, especially phosphate, and water. The fungi are more efficient at absorbing these materials and secrete acid that increases the solubility of some minerals. Minerals taken up by the fungus are transferred to the plant, and photosynthetic products from the plant nourish the fungus. The fungi also protect the plant against certain soil pathogens.

Almost all plants are capable of forming mycorrhizae if they are exposed to the appropriate species of fungi. In most natural ecosystems, these fungi are present in soil, and seedlings develop mycorrhizae. But if seeds are collected in one environment and planted in foreign soil, the plants may show signs of malnutrition resulting from the absence of the plants' mycorrhizal partners. Researchers observe similar results in experiments where soil fungi are poisoned (FIGURE 33.15). Farmers and foresters are already applying the lessons of such research. For example, inoculating pine seeds with spores of mycorrhizal fungi promotes the formation of mycorrhizae by the seedlings. Pine seedlings so infected grow more vigorously than those trees without the fungal association.

(a)

(b)

(c)

FIGURE 33.14

Carnivorous plants. (**a**) The Venus flytrap is a modified leaf with two lobes that close together rapidly enough to capture an insect. An insect that enters the trap touches sensory hairs, initiating an electrical impulse that triggers closure of the trap. Movement of the trap is essentially a very rapid growth response, with cells in the outer region of each lobe accumulating water and enlarging. This changes the shape of the lobes, bringing their margins together. Glands in the trap then secrete digestive enzymes, and nutrients are later absorbed by the modified leaf. In spite of its name, the flytrap catches more ants and grasshoppers than it does flies. (**b**) Sundews capture their prey in a sticky secretion that makes the trap like flypaper. Hairs on the trap bend over the insect and restrain it, and the entire leaf cups around the prey. The hairs then secrete digestive enzymes. (**c**) Pitcher plants use a pitfall to capture insects. Insects slip into a long funnel containing water in the bottom. After the insect drowns, it is digested by enzymes secreted into the water.

FIGURE 33.15

An experimental test of the benefits of mycorrhizae. The experimental soybean plant on the left lacks mycorrhizae because it is growing in soil that has been treated with a fungicide (poison that kills fungi). The plant's growth is stunted, probably due to a phosphorus deficiency. The control plant on the right has mycorrhizae, which enhance the uptake of phosphate and other minerals.

Fossil evidence indicates that the earliest plants on land had mycorrhizae. This symbiosis of plants and fungi enabled roots to extract from soil enough minerals to supply the entire plant and thus may have contributed to the colonization of land. To understand plant nutrition and life in general, it is essential to view organisms in their ecological contexts. Each organism represents a functional unit adapted through evolution to the physical and biological features of its environment. Look for examples of how this theme applies to plant nutrition as you review the concepts you learned in this chapter.

REVIEW OF KEY CONCEPTS (with page numbers and key figures)

- Plants require at least seventeen essential nutrients (pp. 711–712, TABLE 33.1)
  - As photosynthetic autotrophs, plants produce all their own organic compounds but require inorganic nutrients in the form of carbon dioxide, water, and minerals.
  - As a source of hydrogen in photosynthesis, water is a plant nutrient.
  - Atmospheric carbon dioxide is incorporated into a plant's organic material, most of which is carbohydrate.
  - Minerals (inorganic ions) are selectively absorbed from the soil by roots.
  - Plants require nine elements, the macronutrients, in fairly large amounts.
  - Eight micronutrients function mainly as cofactors in enzymatic reactions.
- The symptoms of a mineral deficiency depend on the function and mobility of the element (pp. 713–715)
  - Mineral deficiencies reflect the composition of the soil and cause an array of symptoms that depend on the function and mobility of the nutrient in the plant.
- Soil characteristics are key environmental factors in terrestrial ecosystems (pp. 715–717, FIGURE 33.5)
  - Soil texture depends on the size of particles in topsoil. The most fertile soils are generally loams, which contain both fine particles that retain adequate water and minerals and coarse particles that provide adequate drainage.
  - A variety of living organisms inhabit the soil. Some of these organisms are involved in producing humus, decomposing organic matter that improves the texture and mineral content of soil.
  - Fine particles of clay in soil are negatively charged and attract water and cations. By releasing acids, roots obtain these solutes in a process called cation exchange.
- Soil conservation is one step toward sustainable agriculture (pp. 717–718)
  - Unlike natural ecosystems, agriculture depletes the mineral content of soil, taxes water reserves, and encourages erosion. Proper soil conservation is essential to avoid squandering a precious resource.
- The metabolism of soil bacteria makes nitrogen available to plants (pp. 719–721, FIGURE 33.8)
  - Plant growth and crop yields are often limited by nitrogen, an essential ingredient of proteins and nucleic acids.
  - Although the atmosphere is rich in nitrogen, plants require the assistance of bacteria living in soil to supply them with the forms of nitrogen they need. Nitrogen-fixing bacteria possess nitrogenase, an enzyme that converts atmospheric nitrogen to ammonia. The ammonia is then converted in the soil to nitrate or ammonium, which is absorbed by the plant.
  - The roots of legumes have nodular swellings that house nitrogen-fixing bacteria, which have coevolved with the plants into a mutualistic symbiotic relationship.
- Improving the protein yield of crops is a major goal of agricultural research (pp. 721–722)
  - Agricultural research in developing protein-rich plant breeds and enhancing nitrogen-fixing capabilities may result in practical benefits.
- Predation and symbiosis are evolutionary adaptations that enhance plant nutrition (pp. 722–725)
  - Parasitic plants either supplement their photosynthetic nutrition or give up photosynthesis entirely by tapping into the vascular tissues of host plants.
  - Carnivorous plants obtain nitrogen and minerals by killing and digesting insects.
  - Mycorrhizae, mutualistic associations between roots and fungi, help the plant by enhancing mineral nutrition, water absorption, and resistance to pathogens.

SELF-QUIZ

1. Most of the mass of organic material of a plant comes from
   a. water
   b. carbon dioxide
   c. soil minerals
   d. atmospheric oxygen
   e. nitrogen

2. Micronutrients are needed in very small amounts because
   a. most of them are mobile in the plant
   b. most function as cofactors of enzymes
   c. most are supplied in large enough quantities in seeds
   d. they play only a minor role in the health of the plant
   e. only the growing regions of the plants require them

3. Which of the following is a correct statement about nitrogen fixation?
   a. plants convert atmospheric nitrogen to ammonia
   b. ammonia is converted to N_2, which is the form of nitrogen most easily absorbed by plants
   c. bacteria housed in mycorrhizae are capable of producing ammonium
   d. mutant strains of *Rhizobium* are able to secrete excess proteins into the soil
   e. the enzyme nitrogenase reduces N_2 to form ammonia

4. Two groups of tomatoes were grown under laboratory conditions, one with humus added to the soil and one a control without the humus. The leaves of the plants grown without humus were yellowish (less green) than those of the plants growing in humus-enriched soil. The best explanation for this difference is that
   a. the healthy plants used the food in the decomposing leaves of the humus for energy to make chlorophyll
   b. the humus made the soil more loosely packed, so the plants' roots would grow with less resistance
   c. the humus contained minerals such as magnesium and iron needed for the synthesis of chlorophyll
   d. the heat released by the decomposing leaves of the humus caused more rapid growth and chlorophyll synthesis
   e. the plants absorbed chlorophyll from the humus

5. We would expect the greatest difference in size and general appearance between two groups of plants of the same species, one group with mycorrhizae and one without, in an environment
   a. that is shaded
   b. that has soil with poor drainage
   c. that has hot summers and cold winters
   d. in which the soil is relatively deficient in minerals
   e. that is near a body of water such as a pond or a river

6. Which of the following nutrients is *incorrectly* paired with its function in a plant? (Consult TABLE 33.1.)
   a. calcium—formation of cell walls
   b. magnesium—constituent of chlorophyll
   c. iron—component of chlorophyll
   d. potassium—important in osmotic regulation
   e. phosphorus—component of nucleic acids

7. Most of the water taken up by the plant is
   a. split during photosynthesis as a source of electrons and hydrogen
   b. lost by transpiration through stomata
   c. absorbed by cells during their elongation
   d. returned to the soil by osmosis from the roots
   e. incorporated directly into organic material

8. A mineral deficiency is likely to affect older leaves more than younger leaves if
   a. the mineral is a micronutrient
   b. the mineral is very mobile within the plant
   c. the mineral is required for chlorophyll synthesis
   d. the deficiency persists for a long time
   e. the older leaves are in direct sunlight

9. Carnivorous adaptations of plants mainly compensate for soil that has a relatively low content of
   a. potassium
   b. nitrogen
   c. calcium
   d. water
   e. phosphate

10. Based on our retrospective view, the most reasonable conclusion from van Helmont's famous experiment on the growth of a willow tree is that
   a. the tree increased in mass mainly by producing its own matter
   b. the increase in the mass of the tree could not be accounted for by the consumption of soil
   c. most of the increase in the mass of the tree was due to the uptake of O_2
   d. soil simply provides physical support for the tree without providing nutrients
   e. trees do not require water to grow

CHALLENGE QUESTIONS

1. Explain why an acre of corn actually yields more total protein than an acre of soybeans. (*Note:* Recall that nitrogen-fixation requires large amounts of metabolic energy in the form of ATP.)

2. Researchers at the University of Sydney artificially induced root nodules on the roots of a wheat seedling and treated the roots with a chemical that softens cell walls. This enabled *Azospirillum*, a genus of nitrogen-fixing bacteria that normally live free in the soil, to infect the root cortex of seedlings. The nodules fix nitrogen. Design an experiment that uses radioactive N_2 to determine whether the nitrogen fixed in these nodules is incorporated into leaf proteins.

SCIENCE, TECHNOLOGY, AND SOCIETY

1. Based on what you have learned in this chapter, identify some of the problems the world faces for future food production, and outline some possible solutions. What economic and environmental costs are associated with each of these solutions?

2. About 10% of U.S. cropland is irrigated. Agriculture is by far the biggest user of water in arid western states, including Colorado, Arizona, and California. The populations of these states are growing, and there is an ongoing conflict between cities and farm regions over water. To ensure water supplies for urban growth, cities are purchasing water rights from farmers. This is often the least expensive way for a city to obtain more water, and it is possible for some farmers to make more money selling water than growing crops. Discuss the possible consequences of this trend. Is this the best way to allocate water for all concerned? Why or why not?

FURTHER READING

Albert, V., S. Williams, and M. Chase. "Carnivorous Plants: Phylogeny and Structural Evolution." *Science,* September 11, 1992. Convergent evolution has produced insect-catching species in several plant families.

Aldhous, P. "Ecologists Draft Plan to Dig in the Dirt." *Science,* September 9, 1994. Most soil organisms have not yet been identified.

Baskin, Y. "Forests in the Gas." *Discover,* October 1994. How will a global increase in atmospheric CO_2 affect plant nutrition?

Beardsley, T. "A Nitrogen Fix for Wheat." *Scientific American,* March 1991. Artificially induced root nodules on nonlegumes.

Holmes, B. "Can Sustainable Farming Win the Battle of the Bottom Line?" *Science,* June 25, 1993. The relationship between environmental goals and economic realities.

Nap, J.-P., and T. Bisseling. "Developmental Biology of a Plant-Prokaryote Symbiosis: The Legume Root Nodule." *Science,* November 16, 1990.

Stone, J. "Little Bog Man." *Discover,* February 1990. The profile of a high school student who is one of the world's experts on carnivorous plants.

It has been said that an oak is an acorn's way of making more acorns. Indeed, in a Darwinian view of life, the fitness of an organism is measured only by its ability to replace itself with healthy, fertile offspring. Consider the century plant (Agave) *in the photograph on this page. It lives for decades without flowering, and then one spring, the century plant grows a floral stalk as tall as a telephone pole. That season the plant produces seeds and then withers and dies, its food reserves, minerals, and water spent in the formation of its massive bloom. Although not all flowering plants are as completely consumed as the century plant in leaving offspring, most of their other functions can be interpreted, in the broadest Darwinian sense, as mechanisms contributing to propagation.*

Modifications in reproduction were key adaptations that enabled plants descended from aquatic ancestors to spread into a variety of terrestrial habitats. During the life cycles of most algae, flagellated sperm swim to eggs, and the offspring develop in the water without protection. In the conifers and angiosperms, the plants currently most widespread on land, pollen carried by wind or animals has replaced flagellated sperm as a means of bringing gametes together. The zygotes then develop into embryos protected within seeds. These two developments, pollen and seeds, are among the most important adaptations of plants to life on land. Many plants can also reproduce without sex by mechanisms that promote the propagation of successful individuals in specific environments.

This chapter focuses on the reproduction and development of flowering plants. (The life cycles of other plant and algal groups were covered in Chapters 26 and 27.) After comparing the sexual and asexual modes of reproduction, we will examine some of the cellular mechanisms responsible for plant development. The complex interactions of hormones and environmental cues that control these events will be discussed in Chapter 35.

CHAPTER 34

PLANT REPRODUCTION AND DEVELOPMENT

KEY CONCEPTS

- Sporophyte and gametophyte generations alternate in the life cycles of plants: *an overview*
- Male and female gametophytes develop within anthers and ovaries, respectively
- Pollination brings female and male gametophytes together
- The ovule develops into a seed containing a sporophyte embryo and a supply of nutrients
- The ovary develops into a fruit adapted for seed dispersal
- Evolutionary adaptations in the process of germination increase the probability that seedlings will survive
- Many plants can clone themselves by asexual reproduction
- Vegetative reproduction of plants is common in agriculture
- Sexual and asexual reproduction are complementary in the life histories of many plants: *a review*
- Growth, morphogenesis, and differentiation produce the plant body: *an overview of developmental mechanisms in plants*
- The cytoskeleton guides the geometry of cell division and expansion: *a closer look*
- Cellular differentiation depends on the control of gene expression: *a closer look*
- Mechanisms of pattern formation determine the location and tissue organization of plant organs: *a closer look*

Sporophyte and gametophyte generations alternate in the life cycles of plants: *an overview*

In Chapter 27, you learned about the life cycle of a flowering plant from an evolutionary perspective. The life cycles of angiosperms and other plants are characterized by an **alternation of generations,** in which haploid (n) and diploid $(2n)$ generations take turns producing each other (see FIGURE 27.2). The diploid plant, called the **sporophyte,** produces haploid spores by meiosis. A spore

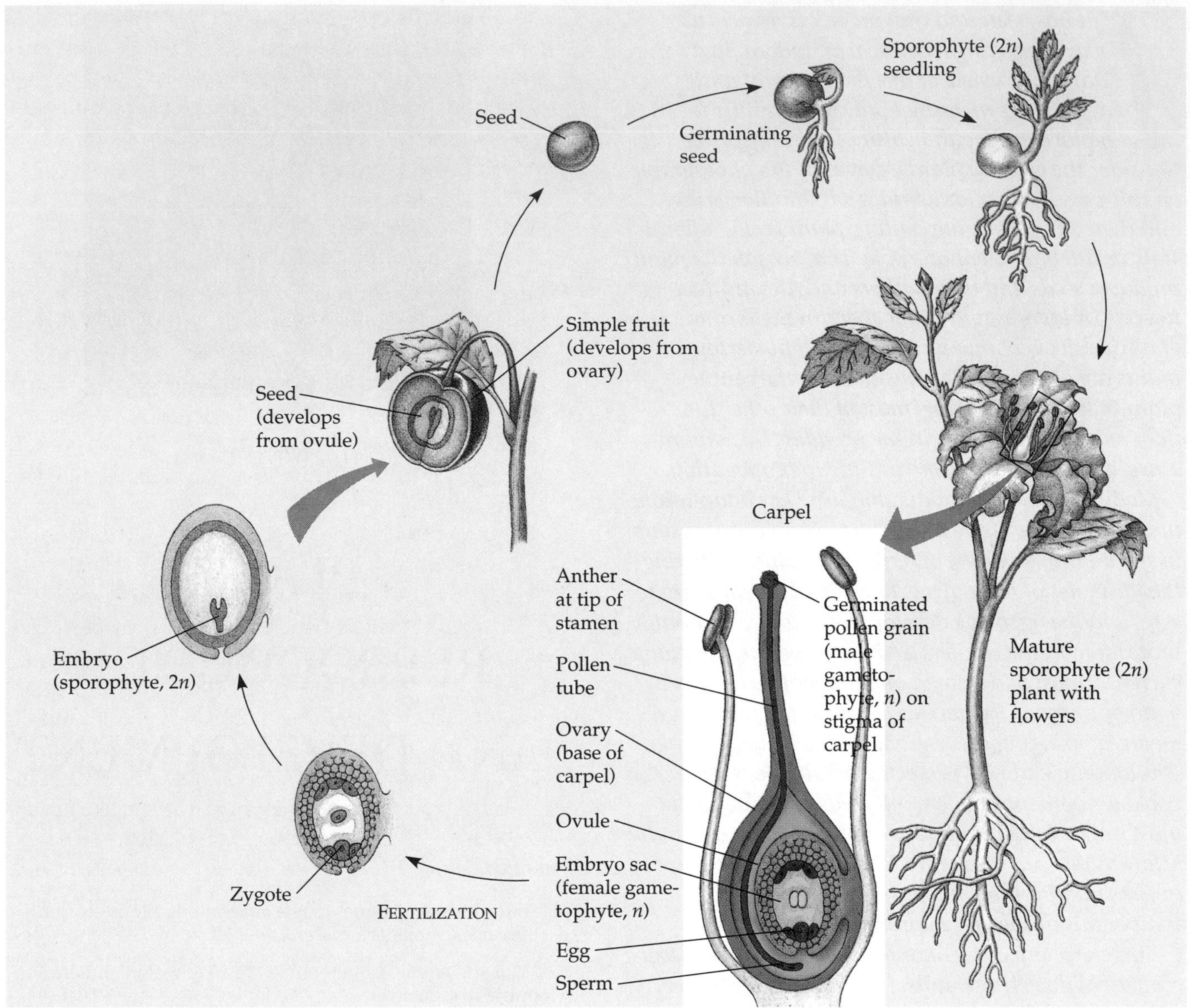

FIGURE 34.1

The angiosperm life cycle: an overview. Within the ovary of a flower, the egg of an ovule is fertilized by a sperm cell released from a pollen tube. The egg is part of the embryo sac, the female gametophyte, and the sperm-bearing pollen grain is the male gametophyte. After fertilization, the ovule matures into a seed containing the embryo, and the ovary develops into a fruit, which aids in dispersal of the seed. In a suitable habitat, the seed germinates, its embryo developing into a seedling.

divides by mitosis, giving rise to a multicellular male or female **gametophyte,** the haploid generation. Mitosis in the gametophytes produces gametes—sperm and eggs. Fertilization results in diploid zygotes, which divide by mitosis and form new sporophytes. FIGURE 34.1 follows the main stages of the angiosperm life cycle. In angiosperms, the sporophyte is the dominant generation in the sense that it is the conspicuous plant we see. Gametophytes became reduced during evolution to tiny structures totally contained within and dependent upon their sporophyte parents.

The reproductive structures of angiosperm sporophytes are flowers. Flowers evolved from compressed shoots with four whorls of modified leaves separated by very short internodes. These four floral organs, in sequence from the outside to the inside of the flower, are the **sepals, petals, stamens,** and **carpels.** Their structure and function are reviewed in FIGURE 34.2. The stamens and carpels of flowers contain the sporangia, the chambers where male and female gametophytes, respectively, develop. The male gametophytes are sperm-containing pollen grains, which form within the chambers of anthers at the tips of stamens. The female gametophytes are egg-containing structures called embryo sacs. Embryo sacs develop inside structures called **ovules,** which are enclosed by the ovaries (the bases of carpels). Thus, stamens and carpels are the reproductive organs of flowers, while sepals and petals are nonreproductive organs.

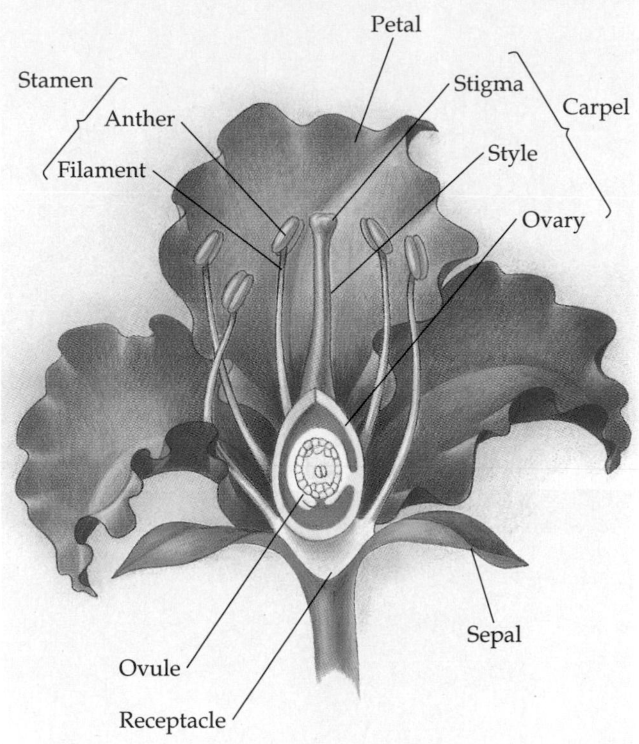

FIGURE 34.2

A review of an idealized flower. Sepals, petals, stamens, and carpels are arranged in four whorls, all attached to the receptacle at the end of a modified stem. Usually green, the sepals of most flowers have retained a more leaflike appearance than the other floral parts. Sepals enclose and protect a floral bud before it opens. Petals, generally more brightly colored than sepals, advertise the flower to insects and other pollinators. Stamens and carpels are the reproductive parts of flowers. Each stamen consists of a stalk called the filament and a terminal structure called the anther. Within the anther are chambers where the pollen grains (male gametophytes) develop. A carpel has a slender neck, the style, leading to an ovary located at the base of the carpel. Developing within the ovary are one or more ovules, in which egg-containing embryo sacs (female gametophytes) develop. This flower has a single carpel, but many plants have multiple carpels that are fused, forming an ovary with two or more chambers where ovules develop. At the tip of the carpel is a sticky stigma, which serves as a landing platform for pollen brought from other flowers by wind or animals.

Pollination occurs when pollen grains released from anthers and carried by wind or animals land on the sticky stigmas at the tips of carpels (though not necessarily on the same flower or plant). Pollen tubes grow down the carpels and discharge sperm into embryo sacs, resulting in the fertilization of eggs. Each zygote gives rise to an embryo, and as the embryo grows, the ovule develops into a seed. The entire ovary, meanwhile, develops into a fruit containing one or more seeds, depending on the species. Fruits, carried by wind or by animals, help disperse seeds some distance from their source plants. If deposited in sufficiently moist soil, seeds germinate; that is, their embryos start growing into seedlings, a new generation of flowering sporophytes. The next gametophyte generation arises within the anthers and carpels of the new sporophytes, and the complex sexual life cycle of flowering plants continues.

More About Flowers

Numerous floral variations evolved during the 130 million years of angiosperm history (see Chapter 27). In certain flowers, one or more of the four basic floral organs—sepals, petals, stamens, and carpels—has been eliminated. Plant biologists distinguish between **complete flowers,** those having all four organs, and **incomplete flowers,** those lacking one or more of the four floral parts. For example, most grasses have incomplete flowers lacking petals.

A flower equipped with both stamens and carpels is termed a **perfect flower,** even if it is incomplete because it lacks sepals or petals. **Imperfect flowers** are incomplete flowers missing either stamens or carpels. These unisexual flowers are called staminate or carpellate, depending on which set of reproductive organs is present. If staminate and carpellate flowers are located on the same individual plant, then that plant species is said to be **monoecious** (Gr., "one house"). Corn is an example. The "ears" are derived from clusters of carpellate flowers. The tassels of a corn plant consist of staminate flowers. In contrast, a **dioecious** ("two houses") species has staminate flowers and carpellate flowers on separate plants, analogous to the presence of testes and ovaries on separate male and female animals. Date palms are dioecious. Because dates develop only on the carpellate (female) palms, commercial date growers plant mostly carpellate individuals. A few males (staminate plants) provide pollen enough for hundreds of females.

In addition to these differences based on the presence of floral organs, flowers have many variations in size, shape, and color (FIGURE 34.3). Much of this diversity represents adaptations of flowers to different pollinators. Indeed, the presence of animals in the environment was a key factor in angiosperm evolution (see Chapter 27).

* * *

The following sections describe in more detail the development of pollen and ovules, then trace how pollination leads to fertilization and the development of seeds and fruits. As you read these sections, refer frequently to the appropriate figures, which illustrate the different processes and the relationships among them. Keep in mind, however, that there are many variations in details in these processes, depending on species.

(a) (b) (c) (d) (e) (f)

FIGURE 34.3

A few examples of floral diversity. (**a**) This lily flower is complete, meaning that sepals, petals, stamens, and carpels are all present. In the case of the lily, the sepals and petals look the same. (**b**) Lupines are examples of plants with inflorescences, clusters of flowers. (**c**) This sunflower represents a plant family characterized by composite flowers. What appears to be a single flower is actually a collection of hundreds. The central disk consists of tiny complete flowers. What appear to be petals ringing the central disk are actually imperfect flowers called ray flowers. (**d**) The diverse shapes, colors, and odors of flowers also reflect adaptations to different modes of pollination. For example, *Hibiscus* is pollinated by hummingbirds, which are attracted to the red color. (The yellow anthers look like they are attached to the carpel, but stamen filaments are actually fused as a tube that ensheaths the style of the carpel.) As the hummingbird probes deep in the flower for nectar, its feathers become dusted with pollen, while pollen carried from other *Hibiscus* flowers is transferred to the sticky stigma. (**e**) Corn is a monoecious species with inflorescences of staminate (male) and carpellate (female) flowers on the same individual plant. The staminate inflorescences are the tassels at the tip of a plant. An "ear" of corn is a collection of kernels (one-seeded fruits) that develops from an inflorescence of fertilized carpellate flowers. (**f**) *Sagittaria* is dioecious, its staminate (left) and carpellate (right) flowers on separate plants.

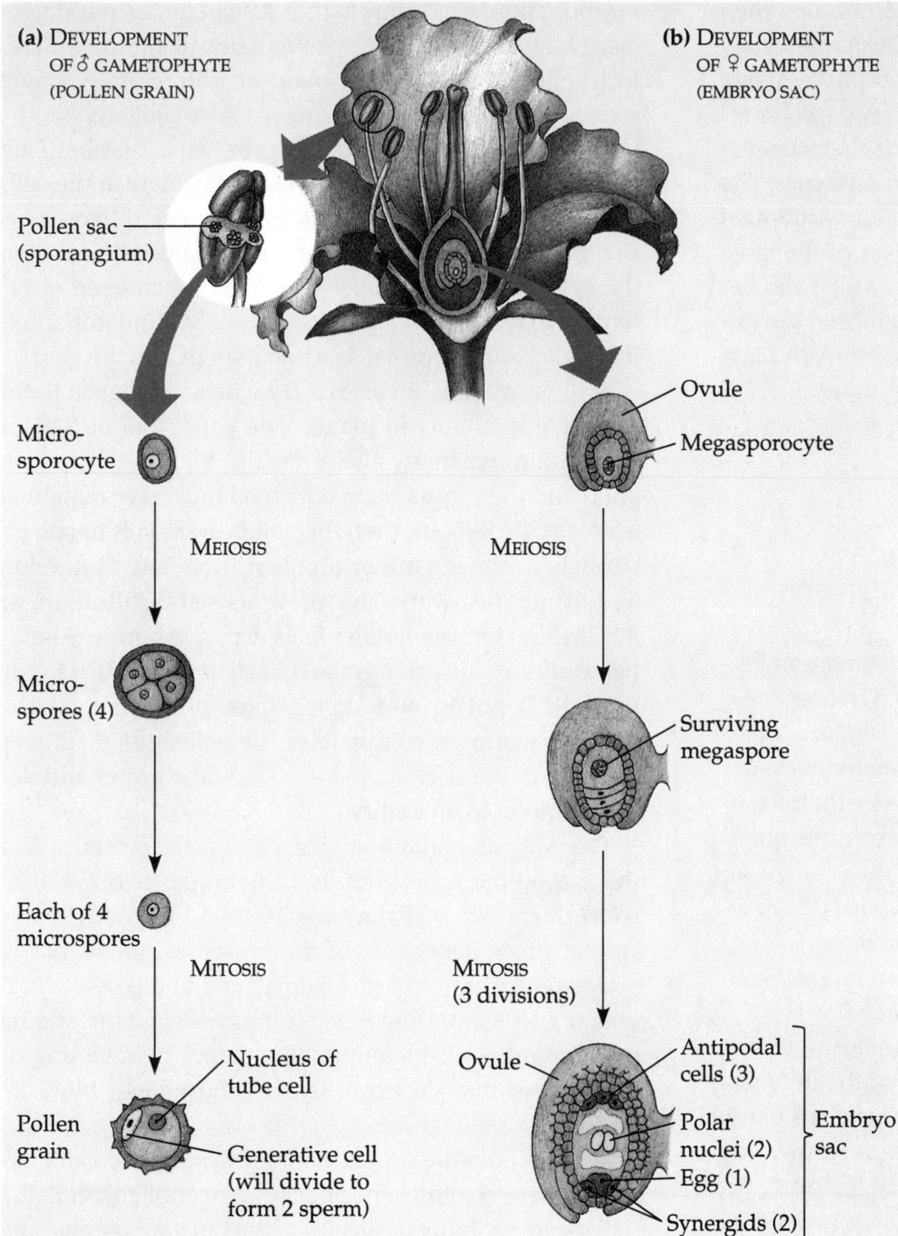

FIGURE 34.4
The development of angiosperm gametophytes (pollen and embryo sacs). (**a**) Pollen grains develop within the sporangia (pollen sacs) of anthers at the tips of stamens. Within each sac are numerous microsporocytes, the diploid cells that give rise to pollen. Meiosis forms four haploid microspores from each microsporocyte. Each microspore then undergoes a mitotic division, giving rise to a pollen grain, an immature male gametophyte consisting of a generative cell and a tube cell. The pollen grain has a thick, tough wall. This pollen becomes a *mature* male gametophyte when the generative cell divides to form two sperm. This usually occurs after a pollen grain lands on the stigma of a carpel and the pollen tube begins to grow (see FIGURE 34.5). (**b**) The embryo sac (female gametophyte) develops within an ovule, itself enclosed by the ovary at the base of a carpel. A diploid cell, the megasporocyte, divides by meiosis and gives rise to four cells, but only one of these survives as the megaspore. (This contrasts with pollen formation, in which all four products of meiosis go on to form gametophytes.) Three mitotic divisions of the megaspore form the embryo sac, a multicellular female gametophyte. At one end of the embryo sac are the egg and two synergid cells. Three antipodal cells are located at the opposite end. The large central cell has two nuclei called polar nuclei. The ovule now consists of the embryo sac along with the surrounding integuments (protective tissues).

Male and female gametophytes develop within anthers and ovaries, respectively

Pollen Development

Within the sporangia (pollen sacs) of an anther, diploid cells called microsporocytes undergo meiosis, each forming four haploid **microspores** (FIGURE 34.4a). Each microspore eventually divides once by mitosis and produces two cells, a generative cell and a tube cell. The two-cell structure is encased in a thick wall that becomes sculptured into an elaborate pattern unique to the particular plant species. Together, the two cells and their wall constitute a pollen grain, or immature male gametophyte.

Pollen grains are extremely durable. Their tough coats are chemically different from other plant cell walls and are resistant to biodegradation. Fossilized pollen has provided many important clues to the evolutionary history of flowering plants, a research approach you will learn more about in the interview with Margaret Davis that precedes Unit Eight.

Ovule Development

Ovules, each containing a sporangium, form within the chambers of the ovary. One cell in the sporangium of each ovule, the megasporocyte, grows and then goes through meiosis, producing four haploid **megaspores** (FIGURE 34.4b). In most angiosperms, only one of these survives. This megaspore continues to grow, and its nucleus divides by mitosis three times, resulting in one large cell with eight haploid nuclei. Membranes then partition this mass into a multicellular structure called the **embryo sac,** which is the female gametophyte. At one end of the embryo sac are three cells: the egg cell, or female gamete, and two cells called synergids that flank the egg cell. At the opposite end are three antipodal cells. The other two nuclei, called polar nuclei, are not partitioned into separate cells but share the cytoplasm of the large central cell of the embryo sac. The ovule, which will develop into the seed, now consists of the embryo sac (female gametophyte) and the integuments, protective layers of sporophyte tissue around the embryo sac.

Pollination brings female and male gametophytes together

Pollination

For the egg to be fertilized, the male and female gametophytes must meet and unite their gametes. The first step is **pollination,** the placing of pollen onto the stigma of a carpel. Some plants, including grasses and many trees, use wind as a pollinating agent. They compensate for the randomness of this dispersal by releasing enormous quantities of the tiny grains. At certain times of the year, the air is loaded with pollen, as anyone plagued with pollen allergies can attest. Many angiosperms, however, do not rely on the aimless wind to carry pollen but interact with animals that transfer pollen directly between flowers.

Some flowers self-pollinate, but the majority of angiosperms have mechanisms that make it difficult or impossible for a flower to pollinate itself. The various barriers that prevent self-pollination contribute to genetic variety by ensuring that sperm and eggs come from different parents. Dioecious plants, of course, cannot self-pollinate because they are unisexual, being either staminate or carpellate. In some plants with perfect flowers, the stamens and carpels mature at different times. Many flowers that are pollinated by animals are structurally arranged in such a way that it is unlikely the pollinator could transfer pollen from anthers to the stigma of the same flower. Other flowers are **self-incompatible;** if a pollen grain from an anther happens to land on a stigma of a flower on the same plant, a biochemical block prevents the pollen from completing its development and fertilizing an egg.

The Molecular Basis of Self-Incompatibility

In the interview that precedes this unit of chapters, Adrienne Clarke discusses her work on the genetic and molecular mechanisms of self-incompatibility in plant reproduction. Clarke and other researchers are trying to learn how the carpels of certain plants recognize and reject pollen from the same plant or from closely related individuals. This plant response is analogous to the immune response of animals, in the sense that both are based on the ability of organisms to distinguish the cells of "self" from those of "nonself." The key difference is that the animal immune system rejects nonself, as when the system mounts a defense against a pathogen or attempts to reject a transplanted organ. Self-incompatibility in plants, by contrast, is a rejection of self.

A single genetic locus, the S-locus, is responsible for self-incompatibility in plants. The gene pool of a plant population has many alleles for the S-locus. If a pollen grain and the stigma upon which it lands have matching alleles at the S-locus, then the pollen grain fails to adhere strongly to the stigma or a pollen tube fails to develop and invade the ovary. This prevents self-fertilization, or fertilization between individuals with a common S-locus (a genetic equality that is most likely if plants are closely related). If pollen and stigma have different S-alleles, then the mating is compatible: The pollen grain adheres strongly to the stigma, and a pollen tube grows and deposits sperm in an embryo sac.

The S-locus actually codes for several proteins, but these multiple transcription units are so tightly linked (close together) that they are inherited as though they were a single gene. One of the protein products of the S-locus is a ribonuclease, an enzyme that digests RNA. If pollen and stigma have matching S-alleles, the stigma cells mount a self-incompatibility attack by releasing ribonuclease that destroys RNA in the pollen, blocking growth of a pollen tube.

But how does the stigma distinguish self pollen and target its attack enzymes specifically at that pollen? Studying self-incompatibility in species of the mustard family, June Nasrallah and her colleagues at Cornell University are tracing the recognition system to other protein products of the S-locus. One of those proteins is secreted into the cell walls of stigma cells, where it may function as a specific receptor for molecules secreted by pollen of matching S-type. Another product of the S-locus is a regulatory protein called a kinase, which is probably built into the plasma membranes of stigma cells. Nasrallah is designing some of her current experiments to test the hypothesis that the receptor protein in the cell wall signals the presence of molecules from self pollen, and the signal is relayed by the kinase across the plasma membrane to the cytoplasm of the stigma cell. This signal pathway, according to the hypothesis, triggers the cell's attack on the pollen.

As Clarke, Nasrallah, and other researchers unravel the

molecular basis of self-incompatibility in plants, some of their discoveries may lead to agricultural applications. Plant breeders sometimes hybridize between different varieties of a crop plant to combine the best traits of the varieties and counter the loss of vigor that can result from excessive inbreeding (see Chapter 13). Many of the plants important in agricultural are self-compatible. To maximize the number of hybrid seeds, breeders currently must prevent self-fertilization by laboriously removing the anthers from the parent plants that provide the seeds. Eventually, it may be possible instead to impose the molecular mechanisms of self-incompatibility on crop species that are normally self-compatible.

Double Fertilization

Once the stigma identifies a pollen grain as compatible, the pollen grain produces a tube that extends down between the cells of the style toward the ovary (FIGURE 34.5). The generative cell divides by mitosis and forms two sperm, the male gametes. The pollen grain, now with a tube containing two sperm, constitutes the mature male gametophyte. Directed by a chemical attractant, usually calcium, the tip of the pollen tube enters the ovary, probes through the micropyle (an opening through the integuments), and discharges its two sperm within the embryo sac. One sperm fertilizes the egg to form the zygote. The other combines with the two polar nuclei to form a triploid ($3n$) nucleus in the center of the large central cell of the embryo sac. This large cell will give rise to the **endosperm,** a food-storing tissue. The union of two sperm cells with two cells of the embryo sac is termed **double fertilization.** Double fertilization ensures that the endosperm will develop only in ovules where the egg has been fertilized, thereby preventing angiosperms from squandering nutrients. After double fertilization, each ovule develops into a seed, and the ovary develops into a fruit enclosing the seed (or seeds, depending on the species).

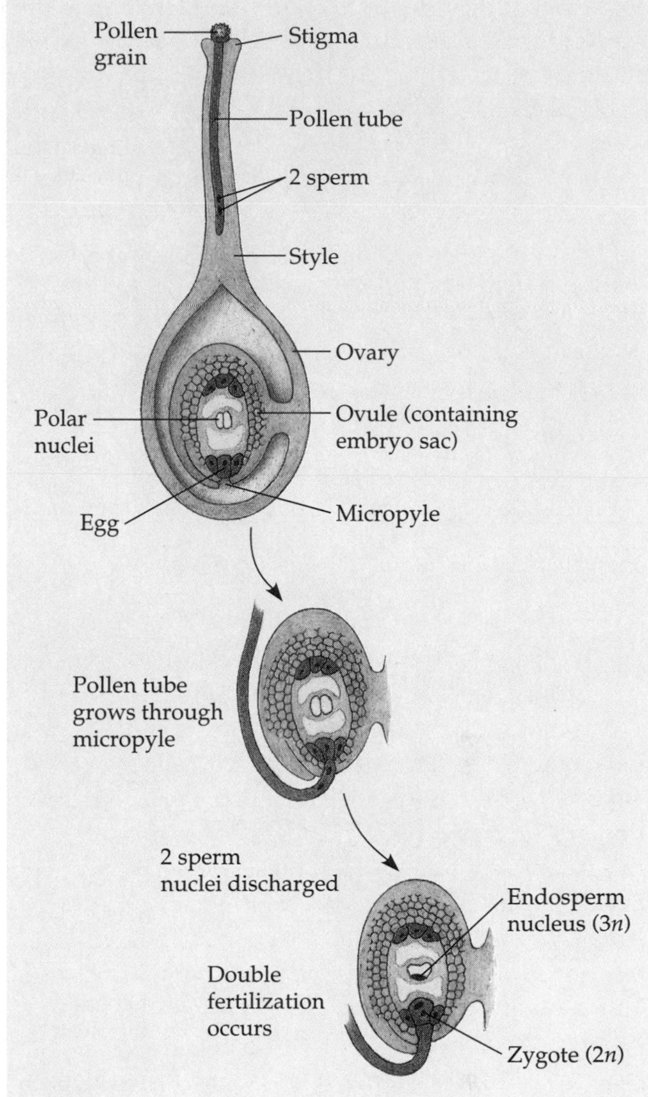

FIGURE 34.5
Growth of the pollen tube and double fertilization in angiosperms. After a pollen grain is carried by wind or an animal to the stigma, a long pollen tube begins growing down the style toward the ovary. The tube discharges two sperm into the embryo sac of an ovule. One sperm fertilizes the egg, forming the zygote. The other combines with the two polar nuclei of the embryo sac's large central cell and forms a triploid cell that will develop into a nutritive tissue called endosperm.

The ovule develops into a seed containing a sporophyte embryo and a supply of nutrients

In this section, we continue our closer look at the angiosperm life cycle by tracing the transition from ovule to seed.

Endosperm Development

Endosperm development usually begins before embryo development. After double fertilization, the triploid nucleus of the ovule's central cell divides, forming a multinucleate "supercell" having a milky consistency. This mass, the endosperm, becomes multicellular and more solid when cytokinesis forms membranes and walls between the nuclei.

The endosperm is rich in nutrients, which it provides to the developing embryo. In most monocots, endosperm also stocks nutrients that can be used by the seedling after germination. In many dicots, the food reserves of the endosperm are exported to the cotyledons before the seed completes its development, and consequently the mature seed lacks endosperm.

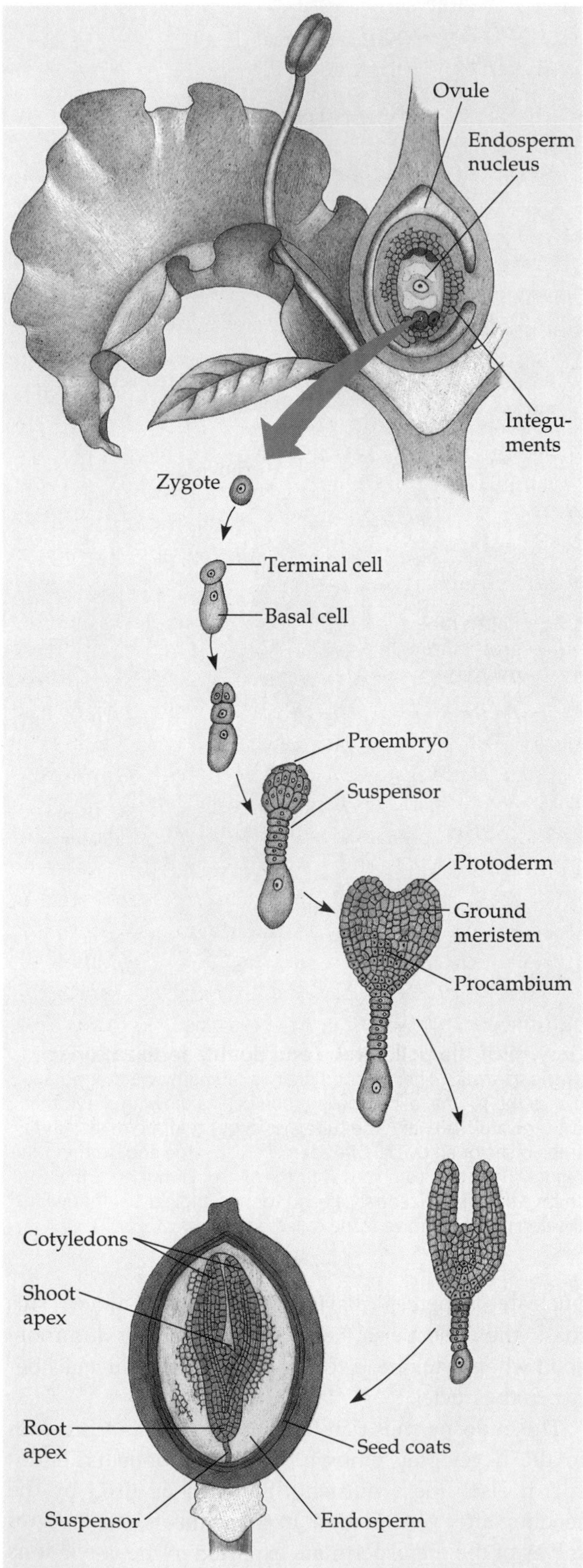

FIGURE 34.6
The development of a dicot plant embryo. By the time the ovule becomes a mature seed, the zygote has given rise to an embryonic plant with rudimentary organs.

Embryo Development (Embryogenesis)

The first mitotic division of the zygote is transverse, splitting the fertilized egg into a basal cell and a terminal cell (FIGURE 34.6). The terminal cell eventually gives rise to most of the embryo. The basal cell continues to divide transversely, producing a thread of cells called the suspensor, which will anchor the embryo to the ovule integuments and transfer nutrients to it from the parent plant. Meanwhile, the terminal cell divides several times and forms a spherical proembryo attached to the suspensor. The cotyledons, or seed leaves, begin to form as bumps on the proembryo. A dicot, with its two cotyledons, is heart-shaped at this stage. Only one cotyledon develops in monocots.

Soon after the rudimentary cotyledons appear, the embryo elongates. Cradled between the cotyledons is the apical meristem of the embryonic shoot. At the opposite end of the embryo's axis, where the suspensor attaches, is the apex of the embryonic root, also with a meristem. (In some species, the basal cell gives rise to part of the root meristem.) After the seed germinates, the apical meristems at the tips of shoot and root will sustain primary growth as long as the plant lives. The three primary meristems—protoderm, ground meristem, and procambium—are also present in the embryo. Thus, embryogenesis establishes two features of plant form: the root-shoot axis, with meristems at opposite ends; and a radial pattern of protoderm, ground meristem, and procambium, set to give rise to the three tissue systems (dermal, ground, and vascular tissues).

Structure of the Mature Seed

During the last stages of its maturation, the seed dehydrates until its water content is only about 5% to 15% of its weight. By now, the embryo has ceased growing and will remain quiescent until the seed germinates. The embryo is surrounded by its enlarged cotyledons, by endosperm, or by both. The embryo and its food supply are enclosed by a **seed coat** formed from the integuments of the ovule, the progenitor of the seed.

We can take a closer look at one type of dicot seed by splitting open the seed of a common bean (FIGURE 34.7a). At this stage, the embryo is an elongate structure, the embryonic axis, attached to fleshy cotyledons. Below the point at which the cotyledons are attached, the embryonic axis is called the **hypocotyl** (Gr. *hypo,* "under"). The hypocotyl terminates in the **radicle,** or embryonic root. The portion of the embryonic axis above the cotyledons is the **epicotyl** (Gr. *epi,* "on" or "over"). At its tip is the plumule, consisting of the shoot tip with a pair of miniature leaves.

The cotyledons of the common bean are fleshy before the seed germinates because they absorbed food from

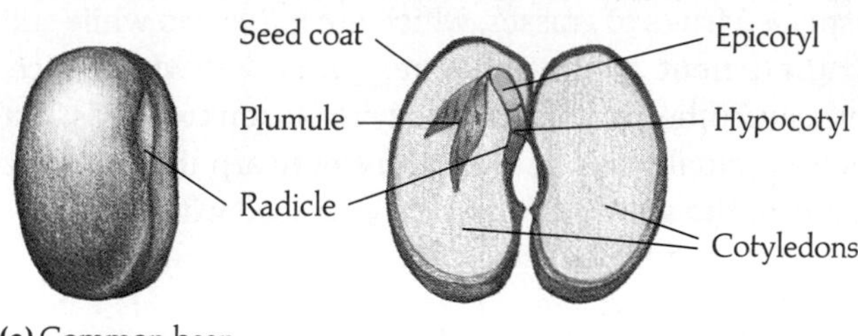

(a) Common bean

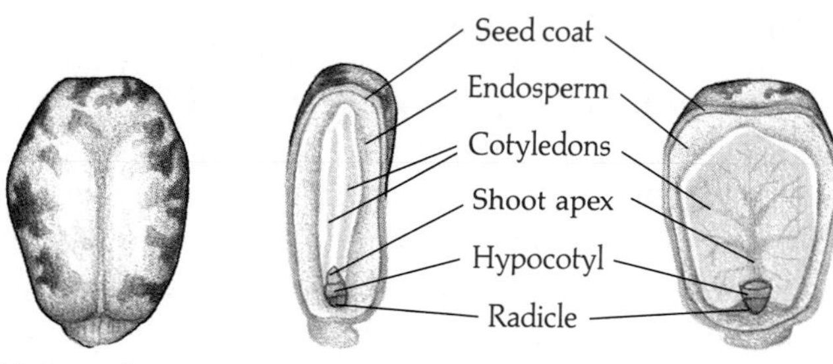

(b) Castor bean

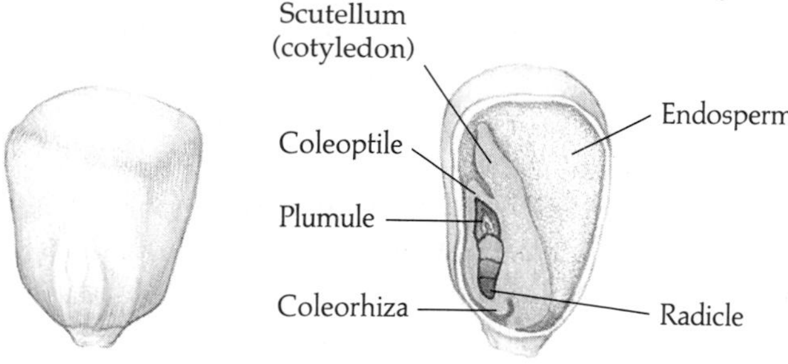

(c) Corn

FIGURE 34.7
Seed structure. (**a**) The fleshy cotyledons of the common garden bean store food that was absorbed from the endosperm when the seed developed. (**b**) The castor bean has membranous cotyledons that will absorb food from the endosperm when the seed germinates. (**c**) Corn, a monocot, has only one cotyledon (the scutellum). The rudimentary shoot is sheathed in a coleoptile.

the endosperm when the seed developed. However, the seeds of some dicots, such as castor beans, retain their food supply in the endosperm and have cotyledons that are very thin (FIGURE 34.7b). The cotyledons will absorb nutrients from the endosperm and transfer them to the embryo when the seed germinates.

The seed of a monocot, such as corn, has a single cotyledon (FIGURE 34.7c), also called the **scutellum.** This seed leaf is very thin but has a large surface area, the better to absorb nutrients from the endosperm during germination. The embryo is enclosed by a sheath consisting of a **coleorhiza,** which covers the root, and a **coleoptile,** which cloaks the embryonic shoot.

The ovary develops into a fruit adapted for seed dispersal

While the seeds develop from ovules, the ovary of the flower develops into a **fruit,** which protects the enclosed seeds and aids in their dispersal by wind or animals. In some angiosperms, other floral parts also contribute to what we call a fruit in grocery store vernacular. The fleshy part of an apple, for instance, is derived mainly from the fusion of flower parts located at the base of the flower, and only the core is a true fruit—a ripened ovary.

The fruit begins to develop after pollination triggers hormonal changes that cause the ovary to grow tremendously (FIGURE 34.8). The wall of the ovary becomes the **pericarp,** the thickened wall of the fruit. As the ovary grows, the other parts of the flower wither away in many plants. This transformation of the flower, called fruit set, parallels the development of the seeds. If a flower has not been pollinated, fruit usually does not set, and the entire flower withers and falls away.

Fruits are classified into several types, depending on their developmental origin (TABLE 34.1). A fruit derived from a single ovary is called a **simple fruit.** A simple fruit may be fleshy, such as a cherry, or dry, such as a soybean pod. An **aggregate fruit,** such as a blackberry, results from a single flower that has several separate carpels. A **multiple fruit,** such as a pineapple, develops from an inflorescence, a group of separate flowers tightly clustered together. When the walls of the many ovaries start to thicken, they fuse together and become incorporated into one fruit.

The fruit usually ripens about the time the seeds it contains are completing their development. For a dry fruit such as a soybean pod, ripening is little more than senescence (aging) of the fruit tissues, which allows the fruit to open and release the seeds. The ripening of a fleshy fruit

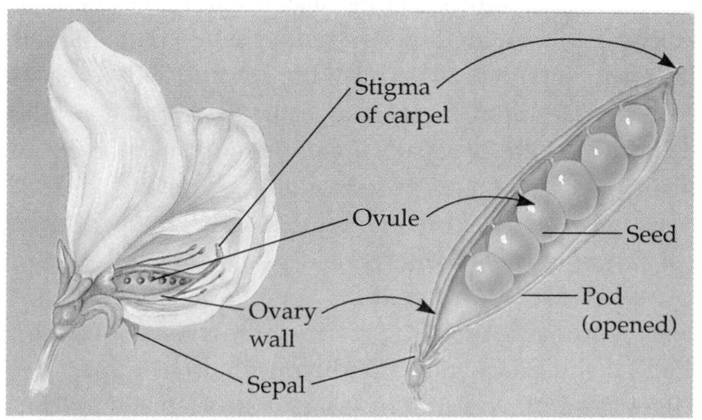

FIGURE 34.8
Relationship between a pea flower and a fruit (pea pod).

TABLE 34.1

Classification of Fleshy Fruits

TYPE OF FRUIT	FLORAL ORIGIN	EXAMPLE
Simple	Single ovary of one flower	Cherry
Aggregate	Many ovaries of one flower	Blackberry (unripe)
Multiple	Many ovaries of many flowers	Pineapple

is more elaborate, its steps guided by the complex interactions of hormones. In this case, ripening results in an edible fruit that serves as an enticement to the animals that help spread seeds. The "pulp" of the fruit becomes softer as a result of enzymes digesting components of the cell walls. There is usually a color change from green to some other color such as red, orange, or yellow. The fruit becomes sweeter as organic acids or starch molecules are converted to sugar, which may reach a concentration of as much as 20% in a ripe fruit.

By selectively breeding plants, humans have capitalized on the production of edible fruits. The apples, oranges, and other fruits in grocery stores are amplified versions of much smaller natural varieties of fleshy fruits. However, the staple foods for humans are the dry, wind-dispersed fruits of grasses, which are harvested while still on the parent plant. The cereal grains of wheat, rice, corn, and other grasses are easily mistaken for seeds, but each is actually a fruit with a dry pericarp that adheres tightly to the seed coats of the single seed within.

Evolutionary adaptations in the process of germination increase the probability that seedlings will survive

To many people, the germination of a seed symbolizes the beginning of life, but in fact the seed already contains a miniature plant, complete with an embryonic root and shoot. At germination, the plant does not begin life but rather resumes the growth and development that was temporarily suspended when the seed matured and its embryo became quiescent. Some seeds germinate as soon as they are in a suitable environment. Other seeds are dormant and will not germinate, even if sown in a favorable place, until a specific environmental cue causes them to break dormancy.

Seed Dormancy

Evolution of the seed was one of the most important factors in the adaptation of plants to the special problems of living and reproducing on land. Terrestrial habitats are generally less stable than lakes and seas, with fluctuating environmental conditions such as temperature and water availability. Seed dormancy increases the chances that germination will occur at a time and place most advantageous to the seedling. Seeds of desert plants, for instance, germinate only after a substantial rainfall. If they were to germinate after a modest drizzle, the soil might soon be too dry to support the seedlings. Where natural fires are common, many seeds require intense heat to break dormancy; seedlings are therefore most abundant after fire has cleared away competing vegetation. Where winters are harsh, seeds may require extended exposure to cold; seeds sown during summer or fall do not germinate until the following spring. This assures a long growth season before the *next* winter. Very small seeds, such as those of some lettuce varieties, require light for germination and will break dormancy only if they are buried shallow enough for the seedlings to poke through the soil surface. Some seeds have coats that must be weakened by chemical attack as they pass through an animal's digestive tract and thus are likely to be carried some distance before germinating.

The length of time a dormant seed remains viable and capable of germinating varies from a few days to decades or even longer, depending on the species and environ-

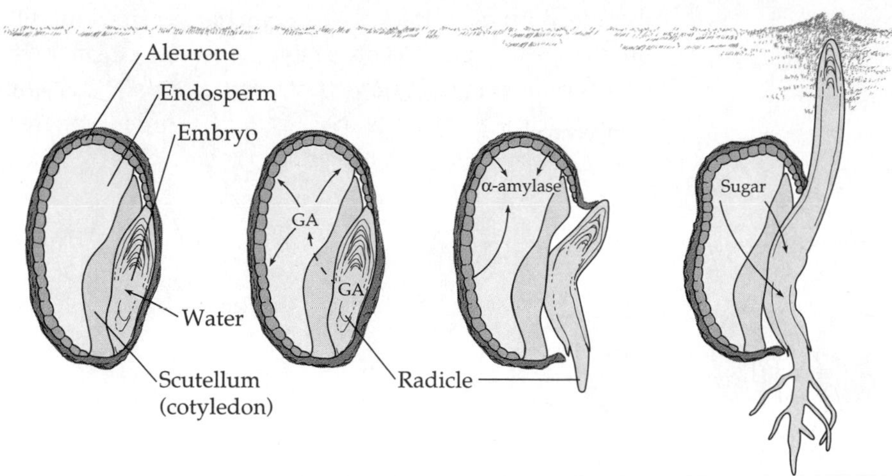

FIGURE 34.9
Mobilization of nutrients during the germination of a cereal. After the seed imbibes water, the embryo releases hormones called gibberellins (GA) as signals to the aleurone. The aleurone responds by synthesizing and secreting α-amylase and other digestive enzymes that hydrolyze stored foods in the endosperm, producing small, soluble molecules. Sugars and other nutrients absorbed from the endosperm by the scutellum (cotyledon) are consumed during growth of the embryo into a seedling.

mental conditions. Most seeds are durable enough to last a year or two until conditions are favorable for germinating. Thus, the soil has a pool of ungerminated seeds that may have accumulated for several years. This is one reason vegetation can come back so rapidly after a fire, drought, flood, or some other disruption.

From Seed to Seedling

The first step in the germination of many seeds is **imbibition,** the absorption of water due to the low water potential of the dry seed. Hydration causes the seed to expand and rupture its coat and also triggers metabolic changes in the embryo that cause it to resume growth. Enzymes begin digesting the storage materials of the endosperm or cotyledons, and the nutrients are transferred to the growing regions of the embryo. This mobilization of food reserves has been studied most extensively in the grains of barley and other cereals, so we will use a cereal as an example of the process (FIGURE 34.9). Soon after water is imbibed, the aleurone, the thin outer layer of the endosperm, begins making α-amylase and other enzymes that digest the starch stored in the endosperm. (A similar enzyme in our saliva helps us digest bread and other foods made from the starchy endosperm of ungerminated cereal grains.) If the embryo is dissected out of the seed before water is added, no α-amylase is produced, suggesting that the embryo sends some kind of messenger to the aleurone to initiate enzyme production. This chemical signal has been identified as a class of plant hormones called gibberellins (discussed in more detail in Chapter 35).

The first organ to emerge from the germinating seed is the radicle, the embryonic root. Next, the shoot tip must break through the soil surface. In garden beans and many other dicots, a hook forms in the hypocotyl, and growth pushes the hook aboveground. Stimulated by light, the hypocotyl straightens, raising the cotyledons and epicotyl. Thus, the delicate shoot apex and bulky cotyledons are pulled aboveground, rather than being pushed tip-first through the abrasive soil. The epicotyl now spreads its first foliage leaves, which expand, become green, and begin making food by photosynthesis. The cotyledons shrivel and fall away from the seedling, their food reserves having been consumed by the germinating embryo.

Light seems to be the main cue that tells the seedling it has broken ground. The hypocotyl of a common garden bean will continue to elongate and push its hook upward until it is out of darkness. Only when the seedling senses light will the hook straighten and the epicotyl begin to elongate. We can trick a bean seedling into behaving as though it is still buried by germinating the seed in darkness. The unilluminated seedling extends an exaggerated hypocotyl with a hook at its tip, and the foliage leaves fail to turn green. After it exhausts its food reserves, the spindly seedling stops growing and dies.

Peas, although in the same family as beans, have a different style of germinating. A hook forms in the epicotyl rather than the hypocotyl, and the shoot tip is lifted gently out of the soil by elongation of the epicotyl and straightening of the hook. Pea cotyledons, unlike those of beans, remain behind in the ground.

Monocots, such as corn, use yet a different method for breaking ground when they germinate. The coleoptile, the sheath enclosing and protecting the embryonic shoot, pushes upward through the soil and into the air. The shoot tip then grows straight up through the tunnel provided by the tubular coleoptile (FIGURE 34.10).

Germination of a plant seed, like the birth or hatching of an animal, is a critical stage in the life cycle. The tough seed gives rise to a fragile seedling that will be exposed to predators, parasites, wind, and other hazards. In the wild, only a small fraction of seedlings endure long enough to become parents themselves. Production of enormous

FIGURE 34.10

Seed germination. The radicle, the root of the embryo, emerges from the seed first. Then the shoot breaks the soil surface by one of the following mechanisms. (**a**) In beans, straightening of a hook in the hypocotyl pulls the shoot and cotyledons from the soil. (**b**) In peas, the hook is above the cotyledons on the epicotyl, and the cotyledons remain in the ground. (**c**) In corn and other monocots, the shoot grows straight up through the tube of the coleoptile.

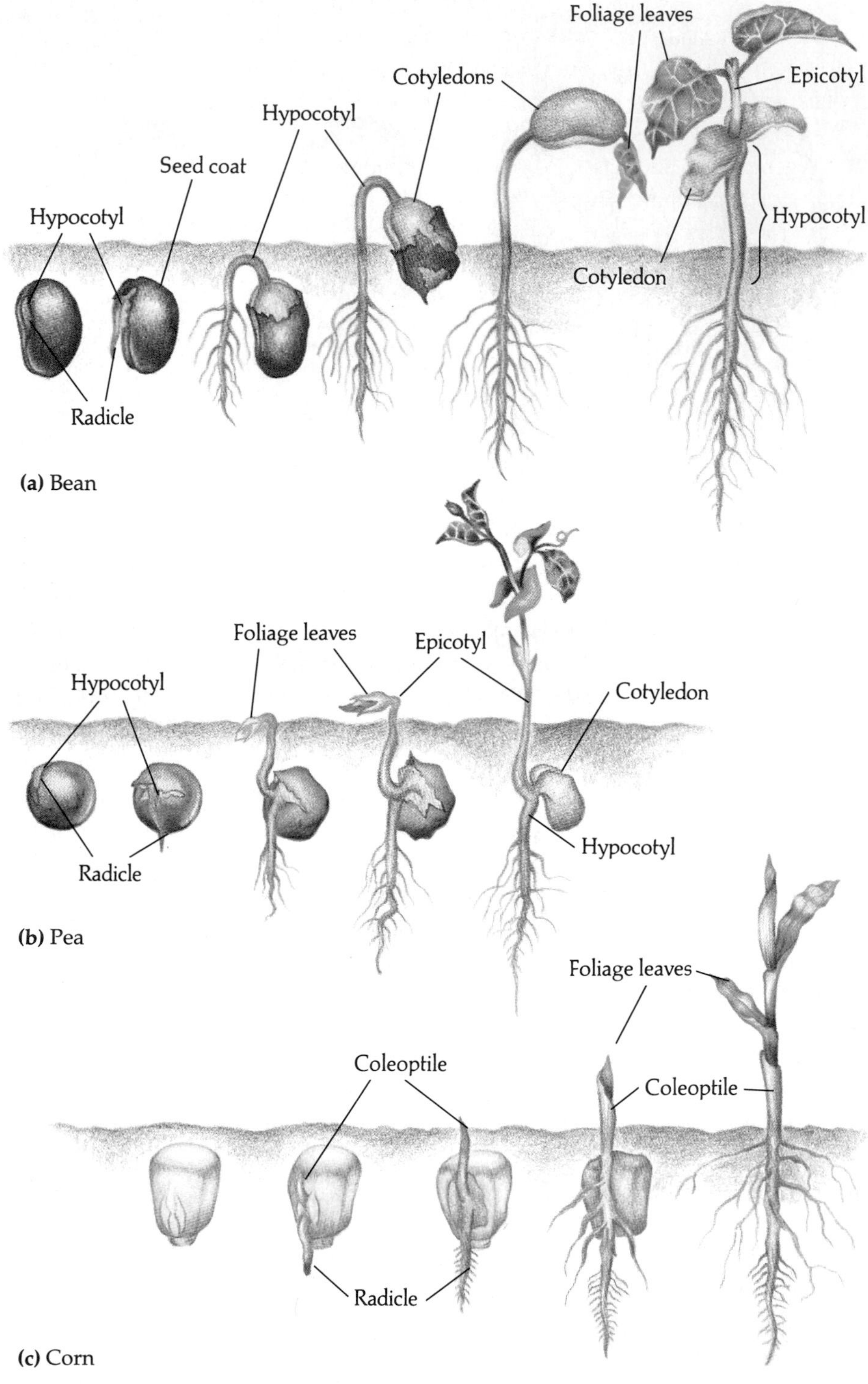

numbers of seeds and fruits compensates for the odds against individual survival and gives natural selection ample material to screen for the most successful genetic combinations. However, this is a very expensive means of reproduction in terms of the resources consumed in flowering and fruiting. Asexual reproduction, generally simpler and less hazardous for offspring than sexual reproduction, is an alternative means of plant propagation.

Many plants can clone themselves by asexual reproduction

Imagine some of your fingers separating from your body, taking up life on their own, and eventually developing into entire copies of yourself. This would be an example of asexual reproduction, offspring derived from a single parent without genetic recombination. The result would be a clone, a population of asexually produced, genetically identical organisms. Some animals *can* reproduce asexually (though not humans, of course). And many plant species do clone themselves by asexual reproduction, also called **vegetative reproduction.**

Vegetative reproduction is an extension of the capacity of plants for indeterminate growth. Plants, remember, have meristematic tissues of dividing, undifferentiated cells that can sustain or renew growth indefinitely. In addition, parenchyma cells throughout the plant can divide and differentiate into the various types of specialized cells, enabling plants to regenerate lost parts. Detached fragments of some plants can develop into whole offspring; a severed stem, for instance, may develop adventitious roots that reestablish the plant. **Fragmentation,** the separation of a parent plant into parts that re-form whole plants, is one of the most common modes of vegetative reproduction (FIGURE 34.11a). A variation of this process occurs in some species of dicots, in which the root system of a single parent gives rise to many adventitious shoots that become separate shoot systems. The result is a clone formed by asexual reproduction from one parent (FIGURE 34.11b). Such asexual propagation has produced the oldest of all known plant clones, a ring of creosote bushes in the Mojave Desert of California, believed to be at least 12,000 years old.

An entirely different mechanism of asexual reproduction has evolved in dandelions and some other plants, which produce seeds without their flowers being fertilized. A diploid cell in the ovule gives rise to the embryo, and the ovules mature into seeds, which in the dandelion are dispersed by windblown fruit. Thus, though these plants clone themselves by an asexual process, they also have the advantage of seed dispersal, an adaptation usually associated with sexual reproduction of plants. This asexual production of seeds is called **apomixis.**

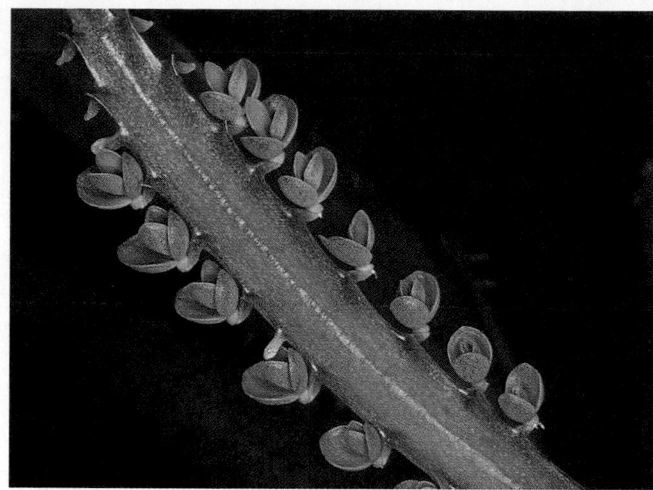

(a)

(b)

FIGURE 34.11
Natural mechanisms of vegetative reproduction.
(**a**) *Kalanchoe* is known as the maternity plant because of the numerous plantlets it produces along its leaf margins. The asexually produced plantlets fragment from their parent and become independent plants. (**b**) Some aspen groves, such as those shown here, are actually clones of thousands of trees descended by asexual reproduction from the root system of one parent. Notice that genetic differences among the clones result in different timing for the development of fall color and the loss of leaves.

Vegetative reproduction of plants is common in agriculture

With the objective of improving crops, orchards, and ornamental plants, humans have devised various methods for propagating plants by vegetative reproduction. Most of these methods are based on the ability of plants to form adventitious roots or shoots.

Clones from Cuttings

Most houseplants, woody ornamentals, and orchard trees are asexually reproduced from plant fragments called cuttings. In some cases, shoot or stem cuttings are used. At the cut end of the shoot, a mass of dividing, undifferentiated cells called a **callus** forms, and then adventitious roots develop from the callus. Some plants, including African violets, can be propagated from single leaves rather than stems. For still other plants, cuttings are taken from specialized storage stems. For example, a potato can be cut up into several pieces, each with a vegetative bud, or "eye," that regenerates a whole plant.

In a modification of vegetative reproduction from cuttings, a twig or bud from one plant can be grafted onto a plant of a closely related species or a different variety of the same species. Grafting makes it possible to combine the best qualities of different species or varieties into a single plant. The graft is usually done when the plant is young. The plant that provides the root system is called the **stock;** the twig grafted onto the stock is referred to as the **scion.** For example, scions from French varieties of vines that produce superior wine grapes are grafted onto root stock of American varieties, which are more resistant to certain diseases. The quality of the fruit, determined by the genes of the scion, is not diminished by the genetic makeup of the stock. In some cases of grafting, however, the stock can alter the characteristics of the shoot system that develops from the scion. For example, dwarf fruit trees are made by grafting normal twigs onto dwarf stock varieties that retard the vegetative growth of the shoot system. Since seeds are produced by the part of the plant derived from the scion, they would give rise to plants of the scion species if planted.

Test-Tube Cloning and Related Techniques

In a new kind of botanical alchemy, test-tube methods are now being used to create and clone novel plant varieties. It is possible to grow whole plants by culturing small explants (pieces of tissue cut from the parent), or even single parenchyma cells, on an artificial medium containing nutrients and hormones (FIGURE 34.12). The cultured cells divide and form an undifferentiated callus. When the hormonal balance is manipulated in the culture medium, the callus can sprout shoots and roots with fully differentiated cells. The test-tube plantlets can then be transferred to soil, where they continue their growth. A single plant can be cloned into thousands of copies by subdividing calluses as they grow. This method is used for propagating orchids and also for cloning pine trees that deposit wood at unusually fast rates.

Cultured explants of certain plants give rise to embryo-like structures that are smaller and simpler than plantlets. They are called somatic embryos because they are derived asexually from somatic (nonreproductive) cells. Plant biotechnologists have learned how to make artificial seeds by packaging somatic embryos along with nutrients within a polysaccharide gel.

Plant tissue culture also facilitates genetic engineering in plants. Most techniques for the introduction of foreign genes into plants require the use of small pieces of plant tissue or single plant cells as the starting material. Test-tube culture makes it possible to regenerate genetically altered plants from a single plant cell into which the for-

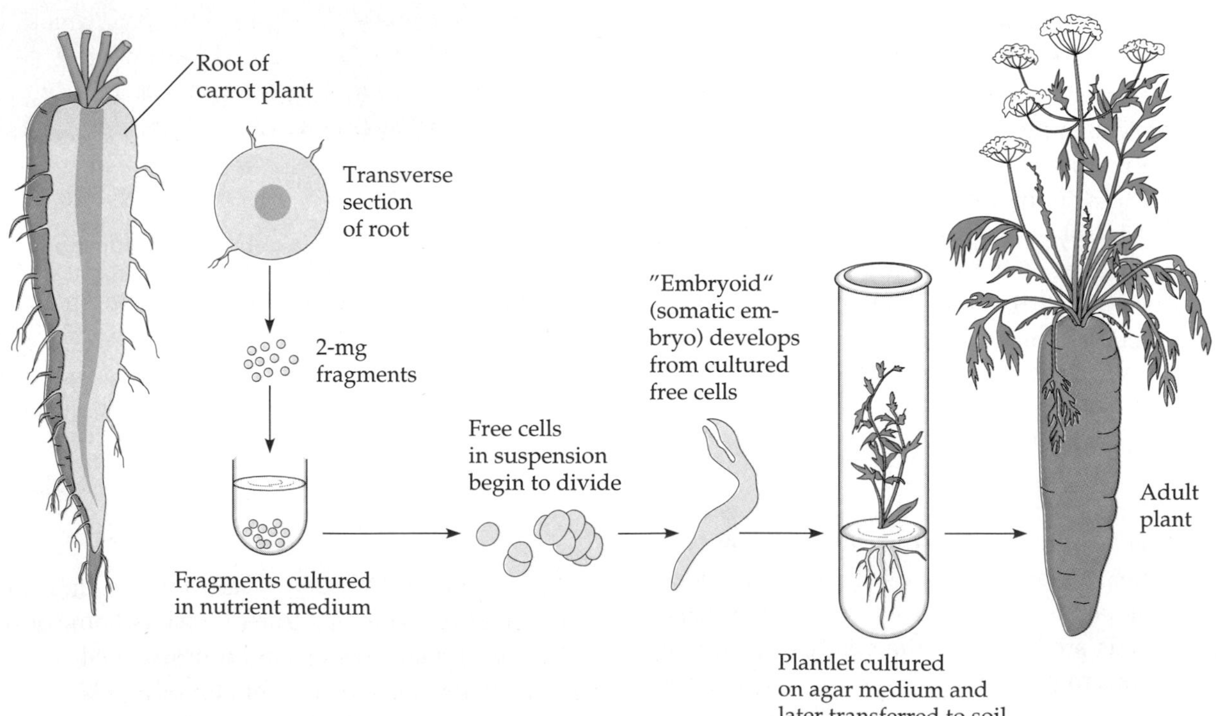

FIGURE 34.12
Test-tube cloning of carrots. In classic experiments conducted during the 1950s, F. C. Steward and his students at Cornell University demonstrated that whole plants could be regenerated from somatic (nonreproductive) cells dissected from a carrot. The products are genetic duplicates of the parent plant. Today, there are many agricultural applications of this basic method of vegetative reproduction. For example, test-tube cloning is used to reproduce the qualities of "super" fruit trees and supply the copies to growers by the thousands.

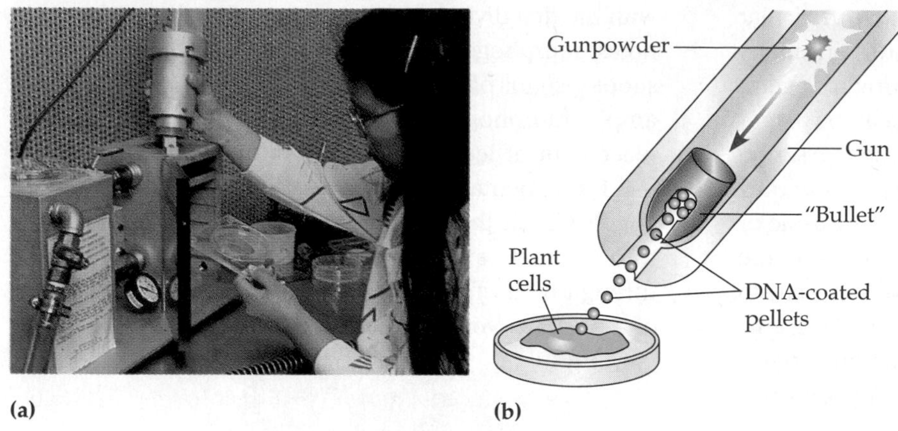

FIGURE 34.13
A DNA gun. (**a**) This researcher is preparing to use a modified .22-caliber gun to shoot foreign DNA into cultured plant cells. (**b**) The gun fires a plastic bullet loaded with tiny metallic pellets coated with DNA. (A different type of DNA gun uses a burst of gas rather than an explosion to propel pellets.) When a plate stops the bullet shell at the end of the gun, the pellets continue toward the cellular targets. (This drawing represents the pellets as much larger than they actually are, compared to the size of the gun and petri dish.) The projectiles penetrate cell walls and membranes, introducing foreign DNA to the nuclei of some cells. A cell that integrates this DNA into its genome can be cultured to produce a plantlet, and it is then possible to clone the transgenic plant.

eign DNA has been incorporated. For example, researchers have used recombinant DNA technology to transfer a gene for bean protein into cultured cells from a sunflower plant. The experiment improved the protein quality of sunflower seeds harvested from the transgenic plants. Firing DNA-coated pellets from a gun is one method researchers use to insert foreign DNA into plant cells (FIGURE 34.13). (The potential agricultural impact of genetic engineering is discussed in more detail in Chapter 19.)

A technique known as **protoplast fusion** is being coupled with tissue culture methods to actually invent new plant varieties that can be cloned. Protoplasts are plant cells that have had their cell walls removed. Before they are cultured, the protoplasts can be screened for mutations that may improve the agricultural value of the plant. It is also possible in some cases to fuse two protoplasts from different plant species that would otherwise be sexually incompatible, and then culture the hybrid protoplasts. Each of the many protoplasts can regenerate a wall and eventually form a hybrid plantlet. One success of this method has been a hybrid between a potato and a wild relative called black nightshade. The nightshade is resistant to an herbicide that is commonly used to kill weeds. The hybrids are also resistant, and this makes it possible to "weed" a potato field with the herbicide without killing the potato plants.

Benefits and Risks of Monoculture

By conscious effort, genetic variability in many crops has been virtually eliminated. For grains and other crops grown from seed, plant breeders have selected for varieties that self-pollinate. Whenever practical, vegetative reproduction is used to clone exceptional plants. Genetic uniformity ensures that all plants in a field grow at the same rate, fruit on all the trees in an orchard ripens in unison, and yields at harvest time are dependable. There is no doubt that **monoculture,** the cultivation of large areas of land with a single plant variety, has helped farmers feed human populations. But modern farms are very fragile ecosystems: Where there is little genetic variability, there is also little adaptability.

A clone is a kind of superorganism; in terms of natural selection, it is one genetic individual. What is good for one plant is good for all, and what is bad for one threatens the entire clone. For example, monoculture was a major factor in the nineteenth-century Irish famine, which was caused by potato blight (a water mold). Spanish explorers brought potatoes to Europe from South America in the late sixteenth century. Europeans used vegetative reproduction to grow the "new" food, which became the staple of the Irish diet. When the blight hit almost two centuries later, the potato plants, which were genetically quite uniform, were all susceptible to the parasite. Plant scientists fear that some plant disease could again devastate thousands of acres of an important monoculture. Plant breeders have responded by maintaining "gene banks," where they store seeds of many plant varieties that can be used to breed new hybrids.

Sexual and asexual reproduction are complementary in the life histories of many plants: *a review*

The dilemma of monoculture provides us with insight about sexual and asexual reproduction in the wild. Many plants are capable of both modes of reproduction, and each offers advantages in certain situations. Sex generates variation in a population, an asset in an environment where evolving pathogens and other variables affect survival and reproductive success. An additional benefit of sexual reproduction in plants is the seed, which can disperse to new locations and can also wait to grow until hostile environmental conditions have improved.

On the other hand, a plant well suited to a particular environment can use asexual reproduction to clone many copies of itself rapidly. Moreover, the offspring of vegetative reproduction, usually mature fragments of the parent plant, are not as frail as the seedlings produced by sexual reproduction. A sprawling clone of prairie grass may cover an area so thoroughly that seedlings of the same or other species have little chance of competing. But in the soil is a pool of seeds, waiting in the wings for some cue to germinate. After a fire, drought, or some other disturbance clears patches of the turf, seedlings can finally get a foothold when conditions improve. The seedlings are unequal in their traits, for their genotypes are products of the sexual recombination of genes. A new competition ensues, in which certain plants excel and reproduce themselves asexually. Both modes of reproduction, sexual and asexual, have had featured roles in the evolutionary adaptation of plant populations to their environments.

Growth, morphogenesis, and differentiation produce the plant body: *an overview of developmental mechanisms in plants*

Reproduction and development are closely related topics. Whether a plant arises from a sexually produced zygote or by vegetative reproduction, its transformation into a whole new individual depends on mechanisms that shape organs such as leaves and roots and generate specific patterns of specialized cells and tissues within those organs. **Development** is the sum of all of the changes that progressively elaborate an organism's body. These last four sections of the chapter present what researchers are learning about the cellular mechanisms responsible for plant development.

A plant zygote is a single cell that bears no resemblance to the organism it becomes. Three overlapping developmental processes transform the fertilized egg into a plant: growth, morphogenesis, and cellular differentiation.

Growth, an irreversible increase in size, results from cell division and cell enlargement. By a series of mitotic divisions, the zygote gives rise to the multicellular embryo within a seed. After germination, mitosis resumes, concentrated mostly in the apical meristems near the tips of roots and shoots. But it is the enlargement of these newly made cells that accounts for most of the actual increase in the size of a plant.

If development were simply a matter of growth, then the zygote would give rise to an expanding ball of cells. In reality, growth is accompanied by **morphogenesis,** the development of form. The embryo encased in a seed has cotyledons and a rudimentary root and shoot, products of morphogenetic mechanisms that begin operating with the first division of the zygote. After the seed germinates, morphogenesis continues to shape the root and shoot systems of the growing plant (FIGURE 34.14). For example, morphogenesis at the shoot tip establishes the placement of leaves and other morphological features.

A fundamental difference between plants and animals is reflected in their morphogenesis. Most animals *move* through their environments; plants, in contrast, *grow* through their environments. The indeterminate growth of a plant, as you have learned, is a function of regions

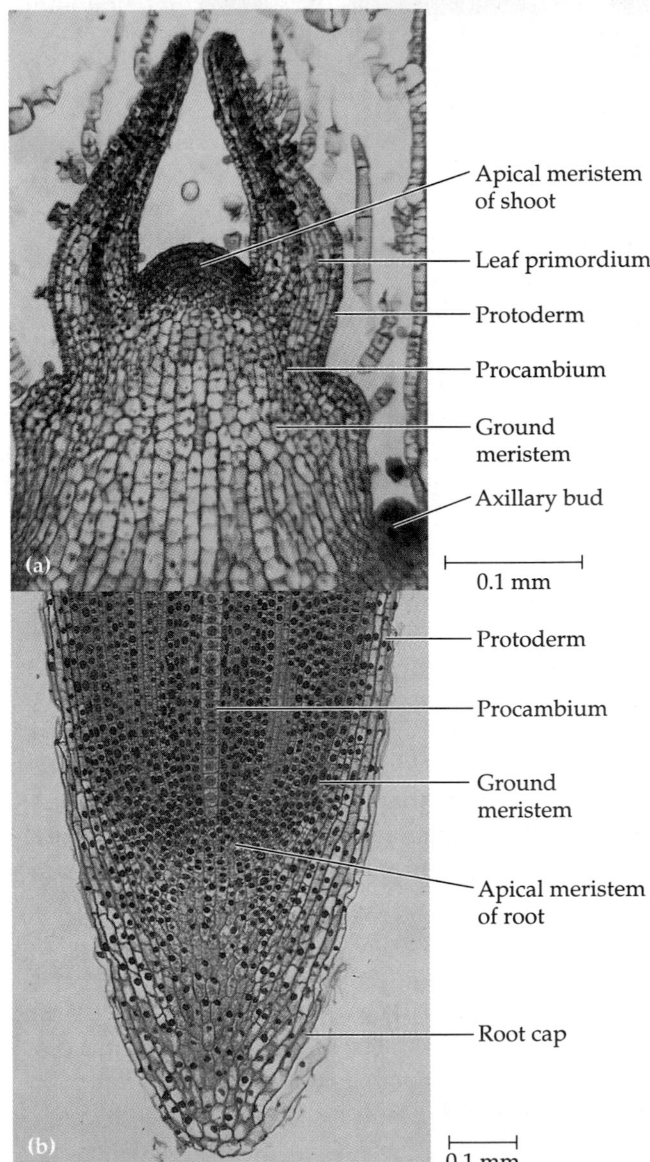

FIGURE 34.14
Lifelong morphogenesis in plants. In contrast to animals, which cease growing and change shape little after reaching maturity, plants exhibit indeterminate growth and persistent morphogenesis for as long as they live. For elaboration of the primary plant body, the centers of ongoing morphogenesis are at a plant's tips, in the regions of the apical meristems of (**a**) shoots and (**b**) roots. Morphogenesis is especially evident at the shoot tip, where the meristem gives rise to a succession of modules (see Chapter 31), each consisting of a leaf-bearing node, internode, and axillary bud. (Both LMs.)

that remain embryonic for the life of the plant at its shoot and root tips. These regions are centers not only of continuing growth, but of continuing morphogenesis as well.

Shape alone, of course, does not enable a plant organ to perform its various functions. Each organ—a leaf, for example—has a diversity of cell types specialized for certain functions and fixed in certain locations. For example, the guard cells that border stomata differ markedly in structure and function from the surrounding cells of the epidermis. Another example is the development of xylem and phloem from the vascular cambium during secondary growth. Recall from Chapter 31 that the cambium gives rise to xylem on the inside and phloem on the outside. This acquisition of a cell's specific structural and functional features is called **cellular differentiation** (see Chapter 18).

Although we have dissected plant development into growth, morphogenesis, and differentiation, it is important to realize that these processes occur in concert as the plant develops. Their integration will become apparent as we take a closer look at the cellular mechanisms of plant development.

The cytoskeleton guides the geometry of cell division and expansion: *a closer look*

Reexamine FIGURE 34.14, and you can see evidence of a basic principle of plant morphology: The shape of a plant organ depends mostly on the spatial orientations of cell divisions and cell expansions. In the case of the root tip, for example, notice that the young cells issue from the apical meristem in files. This is because most of the cell divisions in this region are oriented in a transverse plane, perpendicular to the long axis of the root. When the new cells begin growing, they mainly elongate; that is, most of their expansion parallels the long axis of the root.

Here we can see another important difference between plant and animal development. Animal morphogenesis also involves oriented cell division and growth, but the migration of cells plays a major role as well. In contrast, plant cells cannot move about individually within a developing organ, because they are immobilized by cell walls that are cemented to neighboring cells. This means that oriented cell division and expansion are the chief mechanisms of plant morphogenesis. Plant biologists have learned that the cytoskeleton controls these processes.

Orienting the Plane of Cell Division

The plane in which a cell will divide is determined during late interphase (G_2 of the cell cycle; see Chapter 11). The first sign of this spatial orientation is a rearrangement of the cytoskeleton. Microtubules in the cortex (outer cytoplasm) of the cell become concentrated into a ring called the **preprophase band** (FIGURE 34.15). The band disappears before metaphase, but it has already set the future plane of cell division. The "imprint" consists of an ordered array of actin microfilaments that remain after the microtubules of the preprophase band disperse. These microfilaments hold the nucleus in a fixed orientation until the spindle forms, and later they direct movement of the vesicles that produce the cell plate. (To review the role of the cell plate in the division of plant cells, see Chapter 11.) When the cell finally divides, the walls separating the daughter cells form in the plane defined earlier by the preprophase band. Researchers are now investigating what controls the placement of the preprophase band.

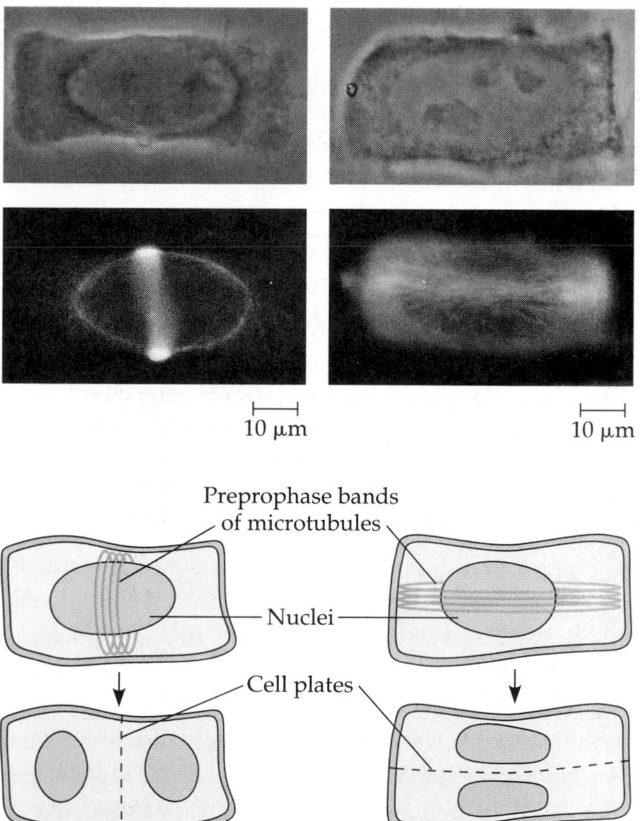

FIGURE 34.15
The preprophase band and the plane of cell division. The band is a ring of microtubules that forms just inside the plasma membrane during late interphase. Its location predicts the future plane of cell division. Although the left and right cells are similar in shape, they will divide in different planes. Each cell is represented by two light micrographs, one unstained (top) and the other stained (bottom) with a fluorescent dye that binds specifically to microtubules. The microtubules form a "halo" (preprophase band) around the nucleus in the cortex of the cytoplasm.

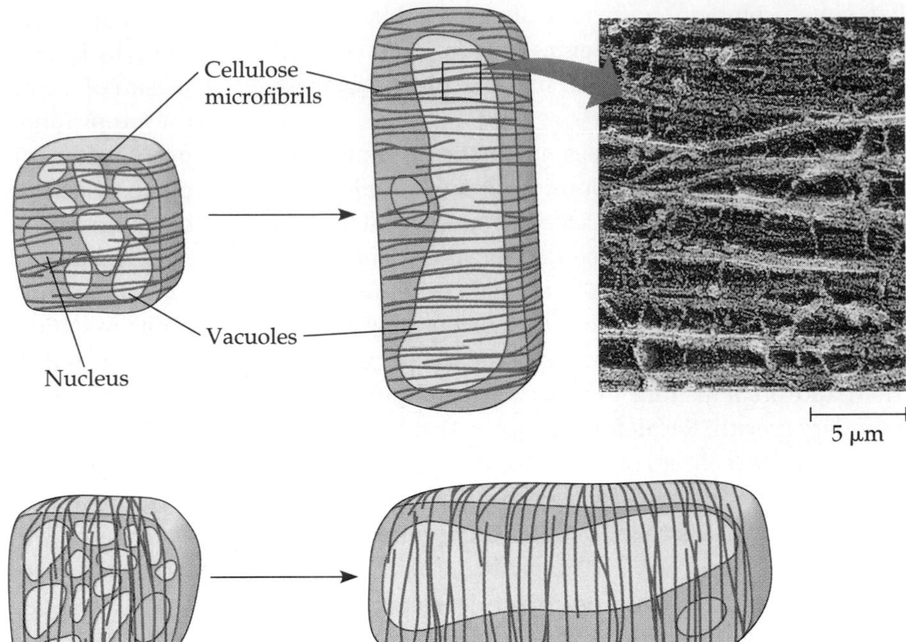

FIGURE 34.16
The orientation of plant cell expansion. Growing plant cells expand mainly in the plane perpendicular to the orientation of cellulose microfibrils in the wall. The microfibrils are embedded in a matrix of other (noncellulose) polysaccharides, some of which form the cross-links visible in the micrograph (TEM). Weakening of these cross-links reduces restraint on the turgid cell, which can then take up more water and expand. Small vacuoles, which accumulate most of this water, coalesce and form the cell's central vacuole.

Orienting the Direction of Cell Expansion

The shape of a plant organ is the outcome of the oriented growth of its cells, and we will see how the cytoskeleton controls this differential enlargement. But first we must learn a little more about *how* plant cells grow.

In a region of active growth, such as the zone of elongation in roots, cells can expand to a volume as much as fifty times their original size. This expansion occurs when the cell wall yields to the turgor pressure of the cell. Acid secreted by the cell causes chemical changes that weaken the cross-links between cellulose microfibrils in the wall. The cell is hypertonic to the surrounding solution, and with its wall loosened, it can take up additional water by osmosis and expand. Growth continues until the wall again becomes knit tightly enough to offset the cell's turgor pressure. Once again, it is useful to highlight a difference between plants and animals. Animal cells grow mainly by synthesizing protein-rich cytoplasm, a metabolically expensive process. Growing plant cells also produce additional organic material in their cytoplasm, but the uptake of water typically accounts for about 90% of a plant cell's expansion. Most of this water is packaged in the large vacuole, which forms by the coalescence of numerous smaller vacuoles as a cell grows. A plant can grow rapidly and economically because a small amount of cytoplasm can go a long way. Bamboo shoots, for instance, may elongate more than 2 m per week. Rapid extension of shoots and roots increases the exposure to light and soil, an important evolutionary adaptation to the immobile lifestyle of plants.

Plant cells rarely expand equally in all directions. For example, cells near the root tip may elongate to twenty times their original length, with relatively little increase in width. The orientation of cellulose microfibrils in the innermost layers of the cell wall causes this differential growth (FIGURE 34.16). The microfibrils cannot stretch much, so the cell expands mainly in the direction perpendicular to the "grain" of the microfibrils. Microfibrils are synthesized by a complex of enzymes built into the plasma membrane (FIGURE 34.17). The pattern of microfibrils in the wall mirrors the orientation of microtubules located just across the plasma membrane in the cortex of the cell. According to one hypothesis, the microtubules confine the flow of the cellulose-producing enzymes to a specific direction along the membrane. This specifies the alignment of microfibrils in the wall, which in turn determines the direction of cell expansion. As with the role of the cytoskeleton in the plane of cell division, an important question remains: What regulates the orientation of microtubules in the cell's cortex? Many researchers are searching for answers.

Cellular differentiation depends on the control of gene expression: *a closer look*

It is remarkable that cells as diverse as guard cells, sieve-tube members (phloem), and xylem vessel elements all descend from a common cell, the zygote. Cellular differentiation, the progressive development of a cell's specialized structure and function, is one of the most engaging research problems in modern biology.

Differentiation reflects the synthesis of different proteins in different types of cells. For example, a developing

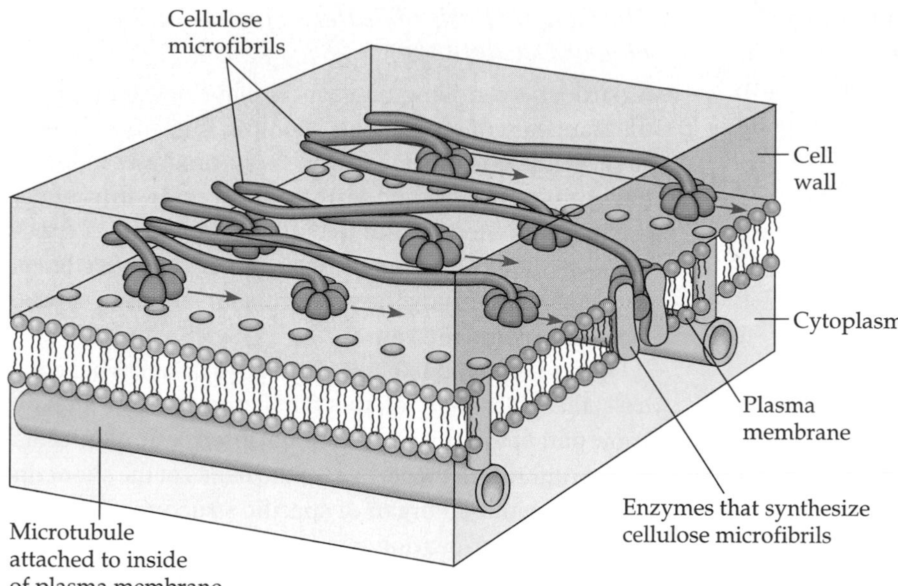

FIGURE 34.17
A hypothetical mechanism for how microtubules orient cellulose microfibrils. Cellulose microfibrils are synthesized at the cell surface by complexes of enzymes that can move in the plane of the plasma membrane. According to one hypothesis, microtubules form "banks" that confine the movement of the enzymes to channels of specified direction. Each enzyme complex advances along one of these channels as the microfibril it extends becomes locked in place by cross-linking to other microfibrils.

xylem cell makes the enzymes that produce lignin, which hardens the cell wall. In contrast, guard cells lack the enzymes for lignin synthesis and have flexible walls. This difference in enzyme content contributes to a structural difference correlated with the contrasting functions of these two cell types. Xylem vessels function in support (as well as transport), and their hard cell walls fit this function. Guard cells, on the other hand, control the size of stomata by changing their shape, a function that requires flexible walls. In its final stage of differentiation, a xylem cell unleashes hydrolytic enzymes that destroy the protoplast, leaving only the wall. No such metabolic episode occurs during the differentiation of a guard cell.

What makes differentiation so fascinating is that the cells of a developing organism synthesize different proteins and diverge in structure and function even though they share a common genome. The cloning of whole plants from somatic cells supports the conclusion that the genome of a differentiated cell remains intact (see FIGURE 34.12). If a mature cell removed from a root or leaf can "dedifferentiate" in tissue culture and give rise to the diverse cell types of a plant, then it must possess all the genes necessary to make these many kinds of cells. This means that cellular differentiation depends, to a large extent, on the control of gene expression—the regulation of transcription and translation leading to specific proteins. Cells with the same genomes follow different developmental pathways because they selectively express certain genes at specific times during their differentiation. A guard cell has the genes that program the self-destruction of a xylem protoplast, but it does not express those genes. A xylem vessel element *does* express them, but only at a specific time in its differentiation, after the cell has elongated and produced its secondary wall. Researchers are beginning to unravel the molecular mechanisms that switch specific genes on and off at critical times in a cell's development (see Chapters 18 and 43).

Mechanisms of pattern formation determine the location and tissue organization of plant organs: *a closer look*

Cellular differentiation has a spatial component: Each plant organ has a characteristic pattern of tissues and cell types within those tissues. On a larger scale, spatial organization is also apparent in the placement of a plant's organs—in the arrangement of a flower's parts, for example. The development of specific structures in specific locations is called **pattern formation.** It is a key element in the overall development of an organism's form.

Positional Information

Pattern formation depends on **positional information,** signals of some kind that indicate each cell's location within an embryonic structure, such as a shoot tip. By affecting localized rates and planes of cell division and expansion, these signals would cause organs such as leaf primordia to emerge as "bumps" at certain places on the plant. Within a developing organ, each cell would continue to detect positional information and respond by differentiating into a particular type of cell.

Developmental biologists are accumulating evidence that gradients of specific molecules, generally proteins, provide positional information. Perhaps, for example, a substance diffusing from a shoot's apical meristem "informs" the cells below of their distance from the shoot tip. We can imagine that the cells gauge their radial positions within the developing organ by detecting a second chemical signal that emanates from the outermost cells. The gradients of these two substances would be sufficient for each cell to "get a fix" on its position relative to the long axis and radial axis of the rudimentary organ. This idea of diffusible chemical signals is just one of several alternative hypotheses that developmental biologists are testing to learn how an embryonic cell detects its location.

Clonal Analysis of the Shoot Apex

In the process of shaping a rudimentary organ, patterns of cell division and cell expansion also affect the differentiation of cells by placing them in specific locations relative to other cells. Thus, positional information underlies all the processes of development: growth, morphogenesis, and differentiation. One approach to studying the relationships among these processes is clonal analysis, in which the cell lineages (clones) derived from each cell in an apical meristem are mapped as organs develop. Researchers can do this by using radiation or chemicals to induce somatic mutations that alter chromosome number or otherwise tag a cell in some way that distinguishes it from its neighbors in the shoot tip. The lineage of cells derived by mitosis from the mutant meristematic cell will also be "marked."

One of the important questions that clonal analysis can address is: How early is the developmental fate of a cell determined by its position in an embryonic structure? To some extent, the developmental fates of cells in the shoot apex are predictable. For example, almost all the cells derived from division of the outermost meristematic cells end up as part of the dermal tissue of leaves and stems. But it is not possible to pinpoint precisely which cells of the meristem will give rise to specific organs and tissues. Apparently random changes in rates and planes of cell division can reorganize the meristem. For example, the outermost cells *usually* divide in a plane perpendicular to the surface of the shoot tip, resulting in the addition of cells to the surface layer. But occasionally a cell at the surface divides in a plane parallel to this meristematic layer, placing one daughter cell beneath the surface among cells derived from different lineages. Thus, the cells of the meristem are not dedicated early to forming specific organs and tissues. Put another way, a cell's developmental fate is not determined by its membership in a lineage derived from a particular meristematic cell. Rather, it is the cell's *final* position in an emerging organ that determines what kind of cell it will become, presumably as a result of positional information.

The Genetic Basis of Pattern Formation in Flower Development

A particularly striking passage in plant development is the transition of a vegetative shoot tip into a floral meristem. While it is vegetative, the shoot tip grows indeterminately, reiterating node-internode-node-internode, module after module. But if it is converted to floral function, the shoot tip becomes determinate. Its meristem is consumed in forming the primordia for whorls of sepals, petals, stamens, and carpels.

Once a shoot meristem is induced to flower, what controls the placement of the floral organs? You already know part of the answer. Positional information commits each primordium that arises on the flanks of the shoot tip to develop into an organ of specific structure and function—an anther-bearing stamen, for example. In the past few years, plant biologists have identified some of the genes that are regulated by positional information and function in this development of floral pattern. Mutations in these genes, called **organ-identity genes,** substitute one type of floral organ where another would normally form. For example, a particular mutation may cause an extra whorl of sepals to develop in a flower where there ought to be petals. The implication is that the wild-type alleles for these organ-identity genes are responsible for the development of normal floral pattern.

Many plant biologists interested in genes that control development have focused on *Arabidopsis thaliana* as their experimental organism. Recall from Chapter 31 that this wild mustard is a small plant with a rapid life cycle, characteristics that make it suitable to culture in the laboratory (see FIGURE 31.1). *Arabidopsis* also has a relatively small genome, a feature that makes the search for specific genes more manageable. Researchers have identified several organ-identity genes that affect flower development in *Arabidopsis* and have used recombinant DNA methods to clone a few of those genes. Very similar organ-identity genes have also been found in snapdragon *(Antirrhinum majus),* which is not closely related to *Arabidopsis.* This suggests that evolution has been very conservative with genes that control the development of an angiosperm's basic body plan.

Organ-identity genes code for regulatory proteins called transcription factors (see Chapter 18). These proteins help control the expression of *other* genes by binding to DNA at specific sites in the genome and affecting rates of RNA synthesis (transcription). Positional information determines which organ-identity genes are expressed in a particular floral-organ primordium. The resulting transcription factors probably induce the expression of those genes responsible for building an organ of specific structure and function. FIGURE 34.18 illustrates a hypothesis for

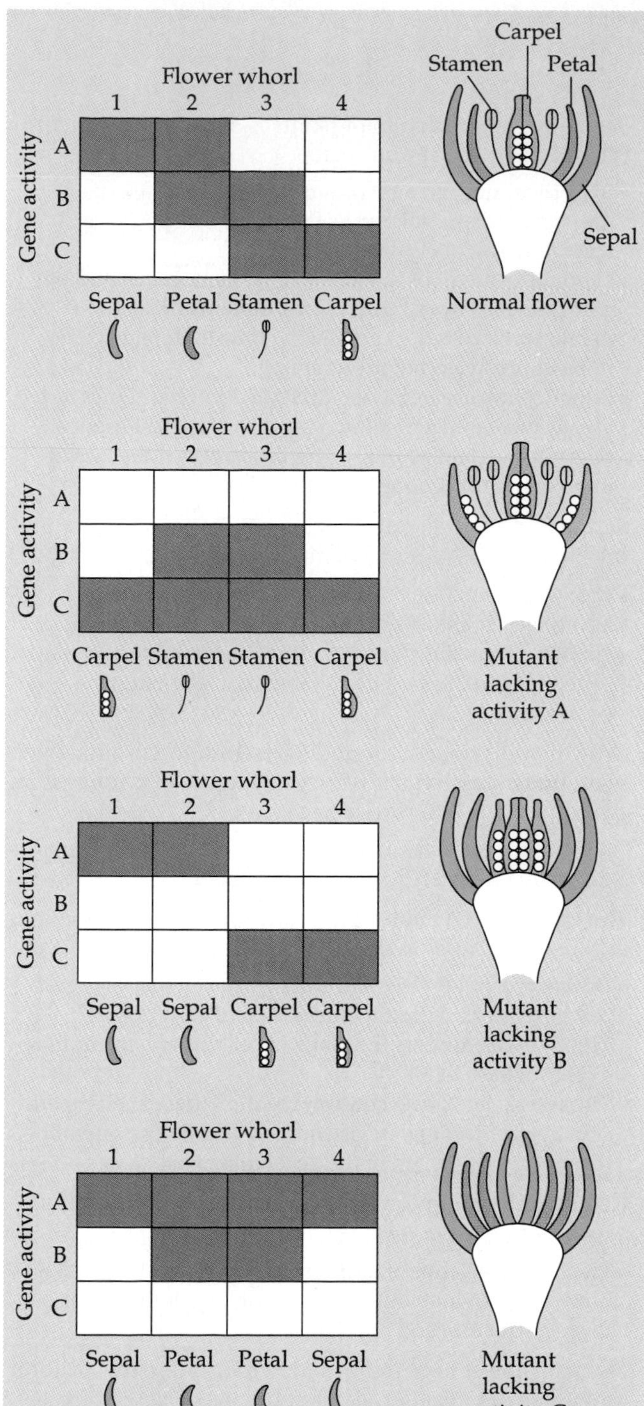

FIGURE 34.18

Organ-identity genes and pattern formation in flower development. Working with floral mutants in *Arabidopsis,* researchers are identifying the genetic activities that determine whether specific regions in a primordial flower develop into sepals, petals, stamens, or carpels. Compare the normal flower at the top to the three organ-identity mutants. The tables illustrate a hypothesis that accounts for the anomalies with a spatial pattern in the expression of three genes, designated here as gene activities A, B, and C. The shaded boxes indicate which genes are expressed in different locations in flower development. Notice that there are four spatial zones of gene activity, corresponding to the regions that develop into the four floral organs in normal flowers. In the most basal (bottom) zone of a floral primordium, only the A gene is expressed, a condition that causes "bumps" of cells there to develop into sepals. In the next zone, both A and B genes are active, causing the development of petals. Stamens develop in the next zone, where the B and C genes are both expressed. In the innermost zone, only the C gene is active, specifying the development of carpels. In examining the patterns of gene activity for the three mutants, notice that the expression of genes A and C are complementary; one or the other must be expressed in each zone, but never both. If a mutation knocks out A function, then C expression expands into all four zones. Conversely, a mutant lacking C activity (bottom mutant in this figure) expresses the A gene in all four zones. With these genetic rules of flower structure, you can relate the gene activity tables on the left to the resulting floral morphologies on the right.

how three regulatory products of organ-identity genes are responsible for normal flower development. By constructing such hypotheses and designing experiments to test them, researchers are tracing the genetic basis of pattern formation. More generally, with the help of excellent experimental systems such as *Arabidopsis* and the powerful methods of modern cell biology, plant biologists are making rapid progress toward understanding how a single cell becomes a plant.

* * *

Throughout our study of plant reproduction and development in this chapter, the role of the environment has been underemphasized. In fact, as you learned in Chapter 31, environmental factors such as light and wind have an enormous impact on plant form and function. In the next chapter, you will learn how plants tune their morphology and physiology to the outside world.

REVIEW OF KEY CONCEPTS (with page numbers and key figures)

- Sporophyte and gametophyte generations alternate in the life cycles of plants: *an overview* (pp. 727–730, FIGURE 34.1)
  - The flower is an angiosperm structure of modified leaves specialized for reproduction.
  - In alternation of generations in angiosperms, the dominant stage is the diploid sporophyte. Spores develop inside the flower into tiny, haploid gametophytes: the male pollen grain and the female embryo sac.
- Male and female gametophytes develop within anthers and ovaries, respectively (pp. 731–732, FIGURE 34.4)
  - Pollen develops from microspores inside the sporangia of the anther.
  - Within an ovule, a haploid megaspore divides by mitosis and forms the embryo sac, the female gametophyte.
- Pollination brings female and male gametophytes together (pp. 732–733, FIGURE 34.5)
  - Fertilization is preceded by pollination, the placing of pollen on the stigma of the carpel.
  - In some plant species, self-incompatibility results from the rejection of pollen having an S-allele that matches that of the stigma.
  - The pollen grain produces a pollen tube that extends down the style toward the embryo sac. Two sperm are released and effect a double fertilization, resulting in a diploid zygote and a triploid endosperm.
- The ovule develops into a seed containing a sporophyte embryo and a supply of nutrients (pp. 733–735, FIGURE 34.6)
  - The zygote gives rise to an embryo with apical meristems and one or two cotyledons.
  - Mitotic division of the triploid endosperm gives rise to a multicellular, nutrient-rich mass that feeds the developing embryo and later (in some plants) the young seedling.
- The ovary develops into a fruit adapted for seed dispersal (pp. 735–736, TABLE 34.1)
  - A fruit is a mature ovary that protects the enclosed seeds and aids in their dispersal via wind or animals.
- Evolutionary adaptations in the process of germination increase the probability that seedlings will survive (pp. 736–738, FIGURE 34.10)
  - Germination begins when seeds imbibe water. This expands the seed, rupturing its coat, and triggers metabolic changes that cause the embryo to resume growth.
  - The embryonic root, or radicle, is the first structure to emerge from the germinating seed. Next, the embryonic shoot breaks through the soil surface.
- Many plants can clone themselves by asexual reproduction (pp. 738–739)
  - Asexual reproduction (cloning) is the production of genetically identical offspring from a single parent.
  - The fragmentation of a parent plant into parts that re-form whole plants demonstrates the versatility and latent potential of meristematic and parenchymal tissues.
- Vegetative reproduction of plants is common in agriculture (pp. 739–741, FIGURE 34.12)
  - In horticulture and agriculture, plants can be asexually propagated from isolated leaves, pieces of specialized storage stems, or cuttings (shoots).
  - Laboratory methods can clone large numbers of desired plant varieties by culturing small explants or single parenchyma cells. It is possible to transfer foreign genes into cultured cells before cloning.
  - Although the use of monoculture of desirable clones has greatly increased agricultural productivity, the absence of genetic variability may prove disastrous in the face of uncertain future conditions.
- Sexual and asexual reproduction are complementary in the life histories of many plants: *a review* (pp. 741–742)
  - Both sexual and asexual modes of reproduction offer advantages in different situations. Asexual reproduction enables successful clones to spread. Sexual reproduction generates genetic variation that makes adaptation possible.
- Growth, morphogenesis, and differentiation produce the plant body: *an overview of developmental mechanisms in plants* (pp. 742–743; FIGURE 34.14)
  - Researchers are beginning to unravel the cellular mechanisms that transform a zygote into a plant.
- The cytoskeleton guides the geometry of cell division and expansion: *a closer look* (pp. 743–744; FIGURES 34.15, 34.16)
  - During growth of a plant, the planes of cell divisions and cell expansions determine the shape of each organ.
  - The cytoskeleton sets the plane of cell division by forming a preprophase band.
  - The cytoskeleton also controls the direction of cell expansion by determining the orientation of cellulose microfibrils that are deposited in the developing wall.
- Cellular differentiation depends on the control of gene expression: *a closer look* (pp. 744–745)
  - The basic challenge of cellular differentiation is to explain how cells with matching genomes diverge into cells of diverse structure and function.
- Mechanisms of pattern formation determine the location and tissue organization of plant organs: *a closer look* (pp. 745–747, FIGURE 34.18)
  - Pattern formation, the emergence of organs and tissues in specific locations, depends on the ability of developing cells to detect and respond to positional information.
  - Clonal analysis of shoot tips suggests that a cell's developmental fate is determined by its final location within a primordial organ.
  - By studying organ-identity genes that cause floral organs to develop in the wrong locations, plant biologists are investigating the genetic basis of pattern formation.

SELF-QUIZ

1. Which of the following would definitely be an imperfect flower? A flower that
   a. is also incomplete
   b. lacks sepals
   c. is found on a monoecious plant
   d. is staminate
   e. cannot self-pollinate

2. Germinated pollen grain is to ______ as ______ is to female gametophyte.
   a. male gametophyte—embryo sac
   b. embryo sac—ovule
   c. ovule—sporophyte
   d. anther—seed
   e. petal—sepal

3. A seed develops from
   a. an ovum
   b. a pollen grain
   c. an ovule
   d. an ovary
   e. an embryo

4. A fruit is a
   a. mature ovary
   b. mature ovule
   c. seed plus its integuments
   d. fused carpel
   e. enlarged embryo sac

5. Which of these conditions is needed by almost all seeds to break dormancy?
   a. exposure to light
   b. imbibition
   c. abrasion of the seed coat
   d. exposure to cold temperatures
   e. covering of fertile soil

6. Which of the following structures is unique to the seed of a monocot?
   a. coleoptile
   b. radicle
   c. seed coat
   d. endosperm
   e. cotyledon

7. Which of the following does *not* appear to have a major effect on plant morphology?
   a. cell division
   b. environmental factors
   c. cell migration
   d. cell expansion
   e. selective expression of genes

8. The direction in which a plant cell will expand seems to a large extent to depend on
   a. the size and position of the central vacuole
   b. a ring of actin microfilaments in the cortex of the cell
   c. the location of the preprophase band
   d. the distribution of solutes in the cytoplasm
   e. the orientation of microtubules just inside the plasma membrane

9. The type of mature cell that a particular embryonic plant cell will become appears to be determined mainly by
   a. the selective loss of genes
   b. the cell's final position in a developing organ
   c. the cell's pattern of migration
   d. the cell's age
   e. the cell's particular meristematic lineage

10. Based on the hypothesis in FIGURE 34.18, predict the floral morphology of a *double* mutant lacking activity of both gene B and gene C. Explain your answer.
    a. carpel–petal–petal–carpel
    b. petal–petal–petal–petal
    c. sepal–sepal–sepal–sepal
    d. sepal–carpel–carpel–sepal
    e. carpel–carpel–carpel–carpel

CHALLENGE QUESTIONS

1. Explain some advantages of propagating houseplants by asexual reproduction.
2. Imagine you are a member of a team of plant biologists who have isolated and purified a substance that might function in pattern formation. It is secreted only by the tips of floral meristems and diffuses down the flanks of the shoot tip. Your team has been able to grow isolated bits of floral meristem in cell culture, treat them with the substance, and in this way induce them to form floral parts. What floral parts would you expect to see develop at relatively high and low concentrations of the substance if its gradient provides positional information?

SCIENCE, TECHNOLOGY, AND SOCIETY

1. Seed companies gather seeds from all over the world and use them to develop new kinds of food and medicinal plants through crossbreeding and biotechnology. The greatest diversity of potentially useful plants is in the tropics. The developing tropical countries lack the resources for conserving plants and developing new crops, but they resent large corporations exploiting their plant resources and selling seeds back to them. What would you suggest to resolve this conflict, allowing both the developing nations and the seed companies to benefit from finding, preserving, and breeding useful plants?
2. As our understanding of how organ-identity genes control the development of floral organs progresses, it will be possible to engineer crop plants to produce a greater or lesser number of specific flower parts. Given that seeds and fruits provide much of our food, discuss how certain floral mutations could possibly increase the food supply.

FURTHER READING

Dale, J. "How Do Leaves Grow?" *BioScience*, June 1992. The application of cell and molecular biology in the study of plant development.

Fosket, D. E. *Plant Growth and Development*. San Diego, CA: Academic Press, 1994. A new textbook appropriate for undergraduates.

Goldberg, R. B., G. dePaiva, and R. Yadegari. "Plant Embryogenesis: Zygote to Seed." *Science*, October 28, 1994. Progress in plant development research.

Grant, M. C. "The Trembling Giant." *Discover*, October 1993. Is an aspen grove the most massive organism?

Mestel, R. "Not Doing It, Plant-Style." *Discover*, January 1995. What prevents certain plants from fertilizing themselves?

Meyerowitz, E. M. "The Genetics of Flower Development." *Scientific American*, November 1994. Progress in our understanding of pattern formation.

Stix, G. "A Recombinant Feast." *Scientific American*, March 1995. The first bioengineered crops reach the market.

Vaughan, D. A., and L. A. Sitch. "Gene Flow from the Jungle to Farmers." *BioScience*, January 1991. The problems and methods of maintaining genetic diversity in crop plants.

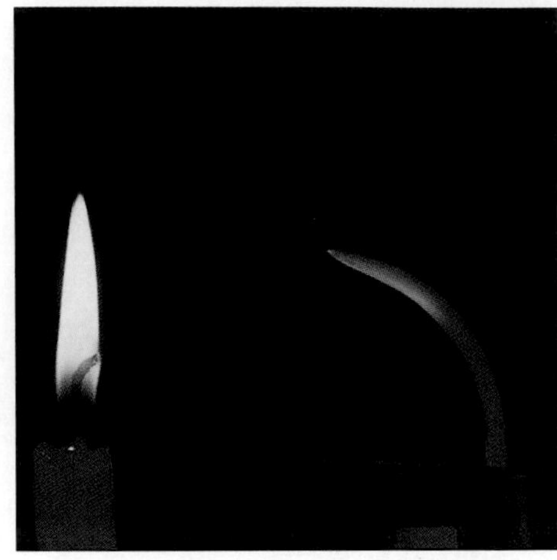

CHAPTER 35

CONTROL SYSTEMS IN PLANTS

KEY CONCEPTS

- Research on how plants grow toward light led to the discovery of plant hormones: *science as a process*
- Plant hormones help coordinate growth, development, and responses to environmental stimuli
- Tropisms orient the growth of plant organs toward or away from stimuli
- Turgor movements are relatively rapid, reversible plant responses
- Biological clocks control circadian rhythms in plants and other eukaryotes
- Photoperiodism synchronizes many plant responses to changes of season
- Phytochrome functions as a photoreceptor in many plant responses to light and photoperiod
- Control systems enable plants to cope with environmental stress
- Signal-transduction pathways mediate the responses of plant cells to environmental and hormonal stimuli

At every stage in the life of a plant, sensitivity to the environment and coordination of responses are evident. One part of a plant can send signals to other parts. For example, the terminal bud at the apex of a shoot is able to suppress the growth of axillary buds that may be many meters away. Plants keep track of the time of day and also the time of year. They can sense gravity and the direction of light and respond to these stimuli in ways that seem to us to be completely appropriate (the grass seedling in the photograph on this page is growing toward light). It is tempting to explain these responses in terms of such human qualities as desire, need, wisdom, and decisiveness, or to imagine that a plant behaves a certain way to accomplish a particular result. Science, however, searches for a mechanism—a link between cause and effect. A houseplant orients its leaves toward a window not because it is trying to find more light, but because cells on the darker side of stems and petioles grow faster than cells on the brighter side. This mechanism causes the stems and petioles to curve toward the light. Natural selection has favored mechanisms of plant response that enhance reproductive success, but this implies no purposeful planning on the part of the plant.

As we examine the control mechanisms at work in plants, keep in mind the theme that response to the environment occurs on two time scales: Individual plants and other organisms respond adaptively to what goes on around them, but these control systems are themselves adaptations that evolved over countless generations of plants interacting with their environments.

For the most part, plants and animals respond to environmental stimuli by very different means. Animals, being mobile, respond mainly by behavioral mechanisms, moving toward positive stimuli and away from negative stimuli. Rooted to one location for life, a plant generally responds to environmental cues by adjusting its pattern of growth and development. Because the program for development of the plant remains somewhat plastic, plants of the same species vary in body form much more than animals of the same species. All lions have four limbs and approximately the same body proportions, but oak trees are less regular in their number of limbs and their shapes.

In this chapter, you will learn how a plant's morphology and physiology are constantly tuned to the surroundings by complex interactions of environmental factors and internal signals. Some of the most important signals are chemical messengers called hormones.

Research on how plants grow toward light led to the discovery of plant hormones: *science as a process*

The word *hormone* is derived from a Greek verb meaning "to excite." Found in all multicellular organisms, **hormones** are chemical signals that coordinate the parts of the organism. As first defined by animal physiologists, a hormone is a compound that is produced by one part of the body and is then transported to other parts of the body, where it triggers responses in target cells and tissues. Another important characteristic of hormones is that only minute concentrations of these chemical messengers are required to induce substantial change in an organism.

The concept of chemical messengers in plants emerged from a series of classic experiments on how stems respond to light. A houseplant on a windowsill grows toward light. If you rotate the plant, it will soon reorient its growth until its leaves again face the window. The growth of a shoot toward light is called positive **phototropism** (growth away from light is negative phototropism). In a forest or other natural ecosystem where plants may be crowded, phototropism directs growing seedlings toward the sunlight that powers photosynthesis. What is the mechanism for this adaptive response? Much of what is known about phototropism has been learned from studies of grass seedlings, particularly oats. The shoot of a grass seedling is enclosed in a sheath called the coleoptile, which grows straight upward if the seedling is kept in the dark or if it is illuminated uniformly from all sides. If the growing coleoptile is illuminated from one side, it will curve toward the light (see the photograph that opens the chapter). This response results from a differential growth of cells on opposite sides of the coleoptile; the cells on the darker side elongate faster than cells on the brighter side (FIGURE 35.1).

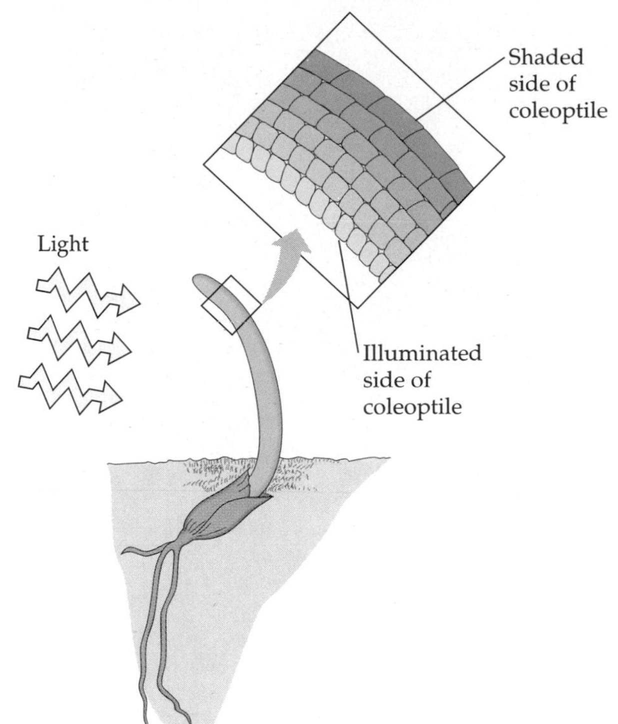

FIGURE 35.1
Phototropism. The coleoptile of an oat seedling grows toward light because cells on the darker side of the organ elongate faster than cells on the brighter side.

Some of the earliest experiments on phototropism were conducted in the late nineteenth century by Charles Darwin and his son, Francis. They observed that a grass seedling could bend toward light only if the tip of the coleoptile was present (FIGURE 35.2). If the tip was removed, the coleoptile would not curve. The seedling would also fail to grow toward light if the tip was covered with an opaque cap; neither a transparent cap over the tip nor an opaque shield placed farther down the coleoptile prevented the phototropic response. It was the tip of the coleoptile, the Darwins concluded, that was responsible for sensing light. However, the actual growth response, the curvature of the coleoptile, occurred some distance below the tip. Charles and Francis Darwin proposed the hypothesis that some signal was transmitted downward from the tip to the elongating region of the coleoptile. A few decades later, Peter Boysen-Jensen of Denmark tested this hypothesis and demonstrated that the signal was a mobile substance of some kind. He separated the tip from the remainder of the coleoptile by a block of gelatin, which would prevent cellular contact but allow chemicals to pass. These seedlings behaved normally, bending toward light. However, if the tip was segregated from the lower coleoptile by an impermeable barrier of mica, no phototropic response occurred.

In 1926, F. W. Went, a young plant physiologist in Holland, extracted the chemical messenger for phototropism by modifying the experiments of Boysen-Jensen (FIGURE 35.3). Went removed the coleoptile tip and placed it on a block of agar, a gelatinous material. The chemical messenger from the tip, Went reasoned, should diffuse into the agar, and the agar block should then be able to substitute for the coleoptile tip. Went placed the agar blocks on decapitated coleoptiles that were kept in the dark. A block that was centered on top of the coleoptile caused the stem to grow straight upward. However, if the block was placed off center, then the coleoptile began to bend away from the side with the agar block, as though growing toward light. Went concluded that the agar block

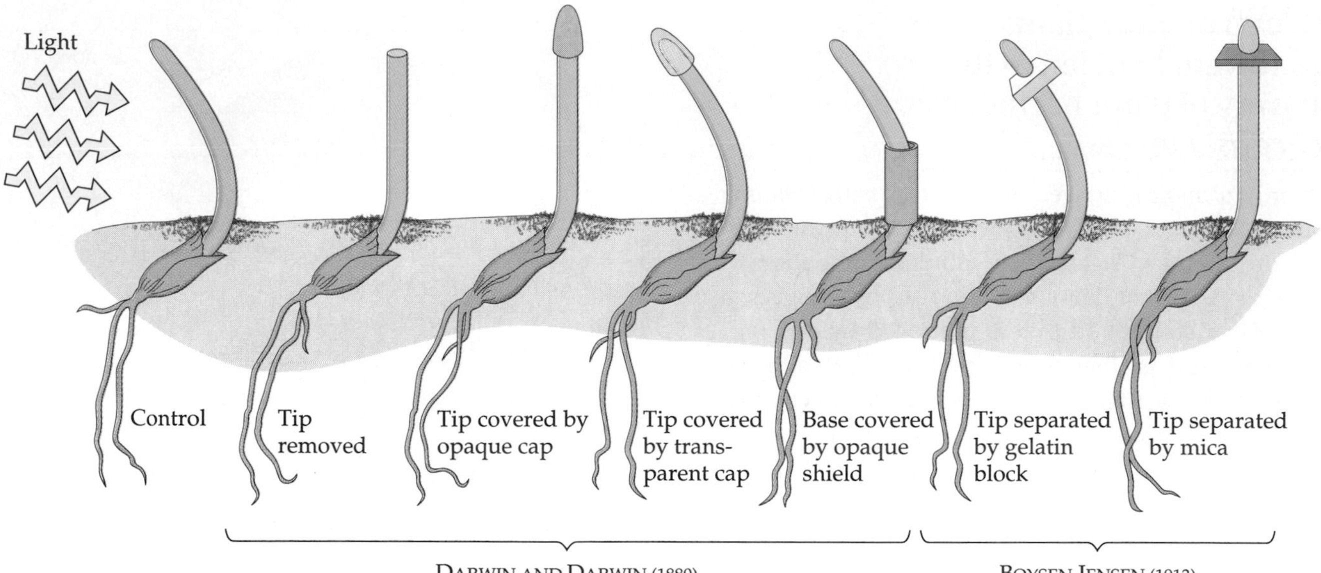

FIGURE 35.2
Early experiments on phototropism. Only the tip of the coleoptile can sense the direction of light, but the bending response occurs some distance below the tip. A signal of some kind must travel downward from the tip. The signal can pass through a permeable barrier (gelatin block) but not through a solid barrier (mica), suggesting that the signal for phototropism is a mobile chemical.

contained a chemical produced in the coleoptile tip, that this chemical stimulated growth as it passed down the coleoptile, and that a coleoptile curved toward light because of a higher concentration of the growth-promoting chemical on the darker side of the coleoptile. For this chemical messenger, or hormone, Went chose the name auxin (Gr. *auxein*, "to increase"). Auxin was later purified and its structure determined by Kenneth Thimann and his colleagues at the California Institute of Technology. Other plant hormones were subsequently discovered.

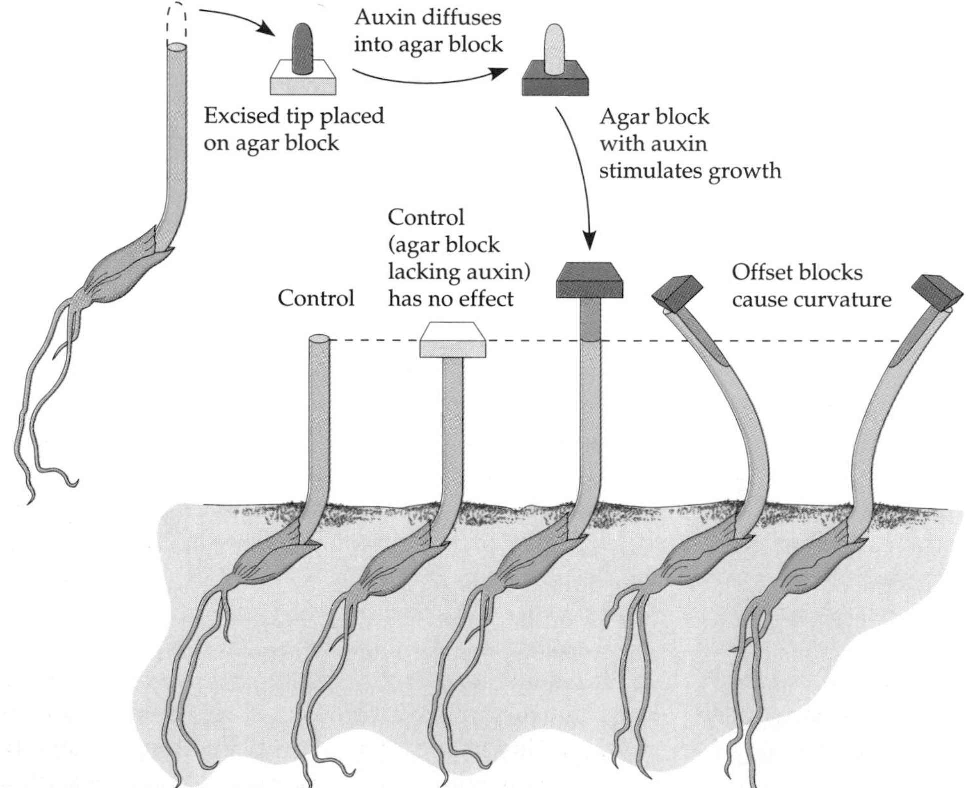

FIGURE 35.3
The Went experiments. Some chemical (indicated by pink) that can pass into an agar block from a coleoptile tip stimulates elongation of the coleoptile when the block is substituted for a tip. If the block is placed off-center on the top of a decapitated coleoptile kept in the dark, the organ bends as if responding to illumination from one side. The chemical is the hormone auxin, which stimulates elongation of cells in the shoot.

TABLE 35.1

An Overview of Plant Hormones

HORMONE	WHERE PRODUCED OR FOUND IN PLANT	MAJOR FUNCTIONS
Auxin (such as IAA) 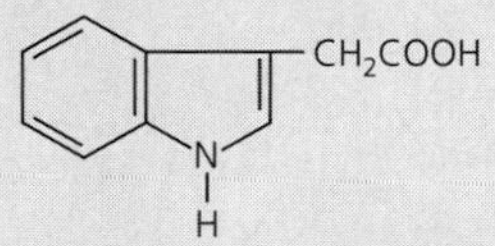	Embryo of seed, meristems of apical buds, young leaves	Stimulates stem elongation root growth, differentiation and branching, development of fruit; apical dominance; phototropism and gravitopism.
Cytokinins (such as zeatin) 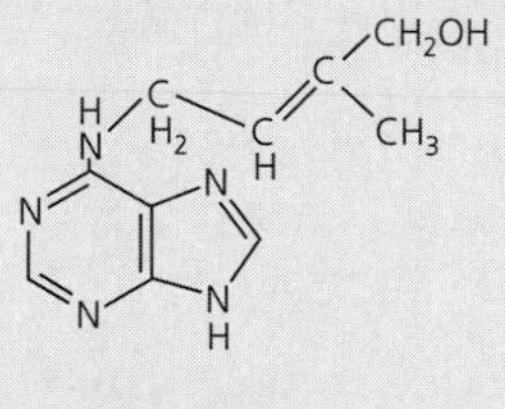	Synthesized in roots and transported to other organs	Affect root growth and differentiation; stimulate cell division and growth, germination, and flowering; delay senescence.
Gibberelins (such as GA_3) 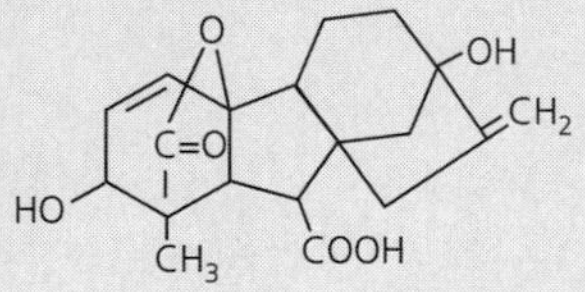	Meristems of apical buds and roots, young leaves, embryo	Promote seed and bud germination, stem elongation, leaf growth; stimulate flowering and development of fruit; affect root growth and differentiation.
Abscisic acid 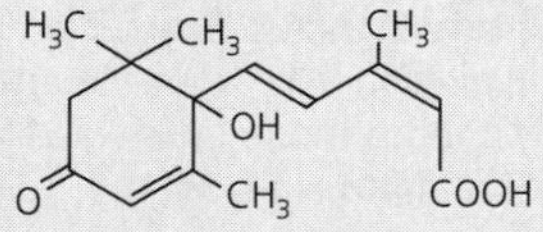	Leaves, stems, green fruit	Inhibits growth; closes stomata during water stress; counteracts breaking of dormancy.
Ethylene 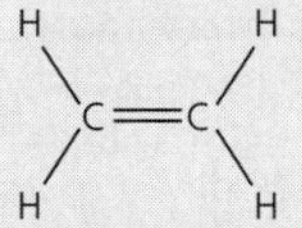	Tissues of ripening fruits, nodes of stems, senescent leaves and flowers	Promotes fruit ripening; opposes some auxin effects; promotes or inhibits growth and development of roots, leaves, flowers, depending on species.

Plant hormones help coordinate growth, development, and responses to environmental stimuli

TABLE 35.1 previews five major classes of plant hormones: auxin, cytokinins (actually a class of related chemicals), gibberellins (also a class of similar chemicals), abscisic acid, and ethylene. In general, these hormones control plant growth and development by affecting the division, elongation, and differentiation of cells. Some hormones also mediate shorter-term physiological responses of plants to environmental stimuli.

Each hormone has a multiplicity of effects, depending on its site of action, the developmental stage of the plant, and the concentration of the hormone. Notice in TABLE 35.1 that all the plant hormones are relatively small molecules. Their transport from cell to cell often involves passage across cell walls, a pathway that blocks the movement of large molecules.

Plant hormones are produced in very small concentrations, but a minute amount of hormone can have a profound effect on the growth and development of a plant organ. This implies that the hormonal signal must be amplified in some way. A hormone may act by altering the expression of genes, by affecting the activity of existing enzymes, or by changing properties of membranes. Any of these actions could redirect the metabolism and development of a cell responding to a small number of hormone molecules.

Reaction to a hormone usually depends not so much on the absolute amount of that hormone as on its relative concentration compared with other hormones. It is hormonal balance, rather than hormones acting in isolation, that may control the growth and development of the plant. These interactions will become apparent in the following survey of hormone function.

Auxin

The term **auxin** is actually used to describe any chemical substance that promotes the elongation of coleoptiles. The natural auxin that has been extracted from plants is a compound named indoleacetic acid, or IAA. In addition to this natural auxin, several other compounds, including some synthetic ones, have auxin activity. Throughout this chapter, however, the name auxin is used specifically to refer to IAA. Although auxin affects several aspects of plant development, one of its most important functions is to stimulate the elongation of cells in young developing shoots.

Auxin and Cell Elongation. The apical meristem of a shoot is a major site of auxin synthesis. As auxin from the shoot apex moves down to the region of cell elongation (see Chapter 31), the hormone stimulates growth of the cells. Auxin has this effect only over a certain concentration range, from about 10^{-8} to 10^{-3} *M*. At higher concentrations, auxin may inhibit cell elongation. This is probably due to a high level of auxin inducing the synthesis of another hormone, ethylene, which generally acts as an inhibitor of plant growth due to cell elongation.

The speed at which auxin is transported down the stem from the shoot apex is about 10 mm per hour—much too fast for diffusion, although slower than translocation in phloem. Auxin seems to be transported directly through parenchyma tissue, from one cell to the next. It moves only from shoot tip to base, not in the reverse direction. This unidirectional transport of auxin is called polar transport. Polar transport has nothing to do with gravity, for auxin travels upward in experiments where a stem or coleoptile segment is placed upside down. Polar auxin transport requires energy. FIGURE 35.4 illustrates how proton pumps, driven by ATP, couple metabolic energy to auxin transport. The mechanism of polar auxin transport is another example of cellular work driven by chemiosmosis, the harnessing of H^+ gradients generated by proton pumps. Notice another important feature of the

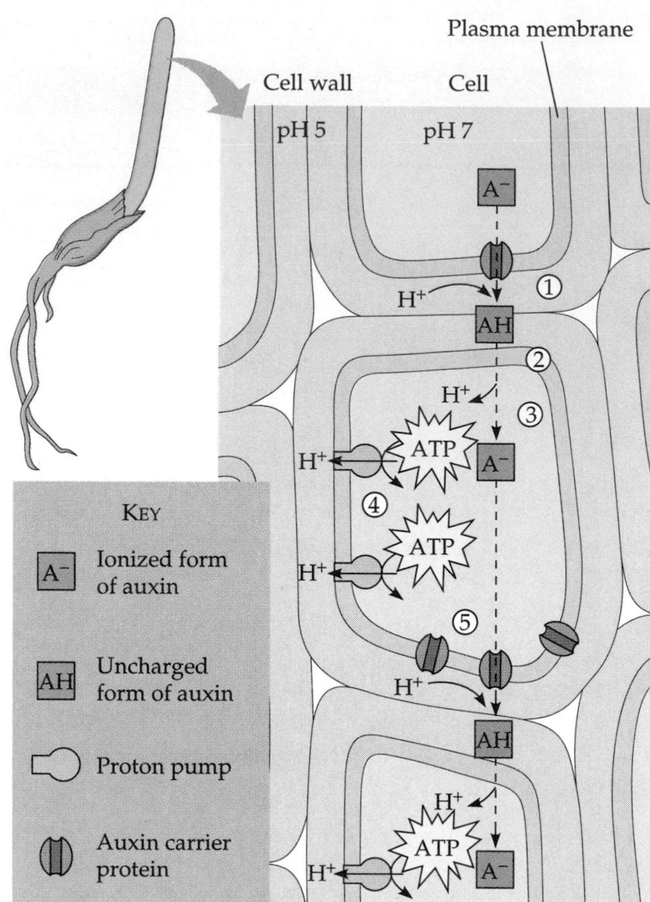

FIGURE 35.4
Polar auxin transport: a chemiosmotic model. In growing shoots, auxin is transported unidirectionally, from the apex down the shoot. Along this pathway, the hormone enters a cell at the apical end, exits at the basal end, diffuses across the wall, and enters the apical end of the next cell. ① When auxin encounters the acidic environment of the wall, the molecule picks up a hydrogen ion to become electrically neutral. ② As a relatively small, neutral molecule, auxin passes across the plasma membrane. ③ Once inside a cell, the pH 7 environment causes auxin to ionize. This temporarily traps the hormone within the cell, because the plasma membrane is less permeable to ions than to neutral molecules the same size. ④ ATP-driven proton pumps maintain the pH difference between the inside and outside of the cell. ⑤ Auxin can only exit the cell at the basal end, where specific carrier proteins are built into the membrane. The proton pumps contribute to this auxin efflux by generating a membrane potential (voltage) across the membrane, which favors the transport of anions out of the cell.

model: Auxin exits each cell along a polar pathway via specific carrier proteins that are restricted to the basal end of the cell.

Proton pumps located at the plasma membrane also play a role in the growth response of cells to auxin. In a shoot's region of elongation, auxin stimulates the proton pumps, an action that lowers the pH in the wall (FIGURE 35.5). This acidification of the wall causes cross-links between cellulose microfibrils to break, loosening the

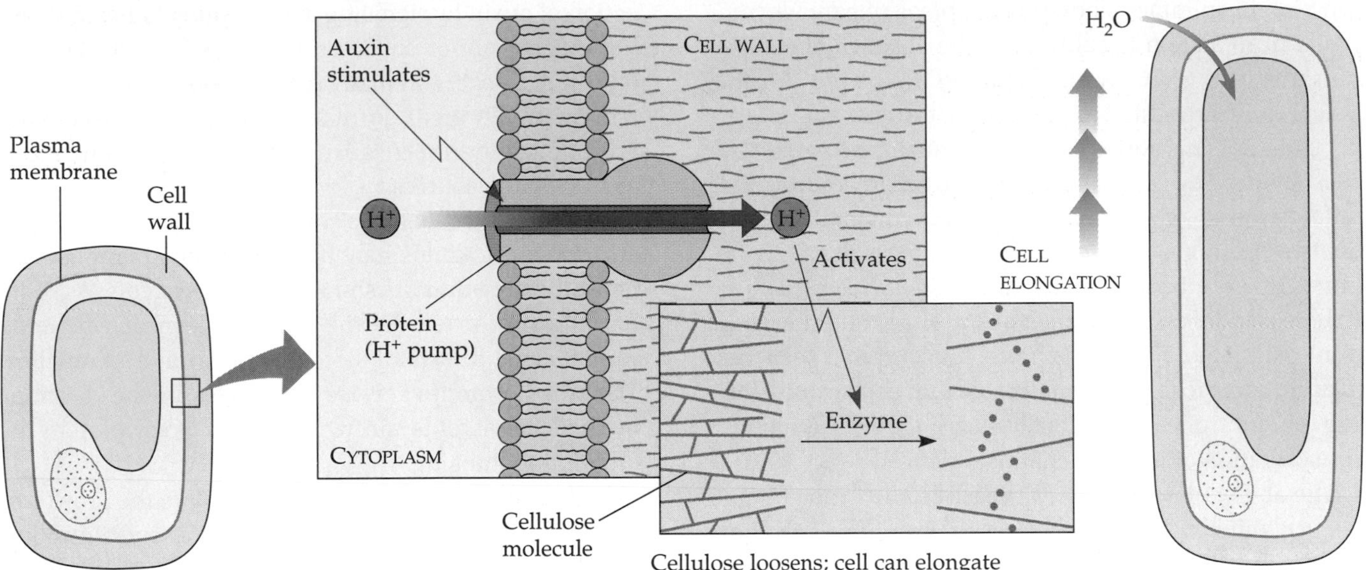

FIGURE 35.5
Cell elongation in response to auxin: the acid growth hypothesis.

fabric of the wall. With its wall now more plastic, the cell is free to take up additional water by osmosis and elongate. Experimental evidence supports this *acid growth hypothesis.* The mechanism works relatively fast. If auxin is applied to young stems, it induces wall loosening and cell elongation within 20 minutes. For sustained growth after this initial spurt, however, cells must make more cytoplasm and wall material. Auxin also stimulates this longer-term growth response. Later in the chapter, we will investigate how cells actually detect the presence of auxin and other signals and how these stimuli lead to specific responses.

Other Effects of Auxin. In addition to stimulating cell elongation for primary growth, auxin affects secondary growth by inducing cell division in the vascular cambium and by influencing the differentiation of secondary xylem. Auxin also promotes the formation of adventitious roots at the cut base of a stem, an effect employed in horticulture by dipping cuttings in rooting media containing synthetic auxins. Developing seeds also synthesize auxin, which promotes the growth of fruit in many plants. Synthetic auxins sprayed on tomato vines induce fruit development without a need for pollination. This makes it possible to grow seedless tomatoes by substituting for the auxin that would normally be synthesized by seeds.

One of the most widely used herbicides (weed killers) is 2,4-D, a synthetic auxin that disrupts the normal balance of plant growth. Dicots are more sensitive than monocots to this herbicide, perhaps because dicots absorb 2,4-D more readily. Thus, 2,4-D can be used to selectively remove dandelions and other broad-leaf weeds from a lawn or grain field.

Cytokinins

Cytokinins were discovered during trial-and-error attempts to find chemical additives that would enhance the growth and development of plant cells in tissue culture. In the 1940s, Johannes van Overbeek, working at the Cold Spring Harbor Laboratory in New York, found he could stimulate the growth of plant embryos by adding coconut milk, the liquid endosperm of the giant coconut seed, to his culture medium. A decade later, Folke Skoog and Carlos O. Miller, at the University of Wisconsin, induced tobacco cells being grown in culture to divide by adding degraded samples of DNA. The active ingredients of both experimental additives turned out to be modified forms of adenine, one of the components of nucleic acids. These growth regulators were named cytokinins because they stimulate cytokinesis, or cell division. Of the variety of cytokinins that occur naturally in plants, the most common is zeatin, so named because it was first discovered in corn (*Zea mays*). Several synthetic compounds with cytokinin activity have also been produced. As you read about a few of the functions of cytokinins, notice that these hormones are complemented or countered by other hormones, especially auxin.

Control of Cell Division and Differentiation. Cytokinins are produced in actively growing tissues, particularly in roots, embryos, and fruits. Cytokinins produced in the root reach their target tissues by moving up the plant in the xylem sap. Acting in concert with auxin, cytokinins stimulate cell division and influence the pathway of differentiation.

The effects of cytokinins on cells growing in tissue culture provide clues about how this class of hormones may function in an intact plant. When a piece of parenchyma tissue from a stem is cultured in the absence of cytokinins, the cells grow very large but do not divide. If cytokinins alone are added to the culture, they have no effect. If cytokinins are added along with auxin, however, the cells divide. The ratio of cytokinin to auxin controls the differentiation of the cells. When the concentration of the two hormones is about equal, the mass of cells continues to grow, but it remains an undifferentiated callus. If there is more cytokinin than auxin, shoot buds develop from the callus. If auxin is more concentrated than cytokinin, roots form. It is remarkable that gene expression can be controlled so extensively by manipulating the concentration of just two chemical signals.

The ability of cytokinins to trigger cell division and influence differentiation could result from the fact that these hormones stimulate RNA and protein synthesis. These newly synthesized proteins may be involved in cell division.

Control of Apical Dominance. We can see another interaction of cytokinins and auxin in the control of apical dominance, the ability of the terminal bud to suppress the development of axillary buds. In this case, the two hormones are antagonistic. Auxin transported down the shoot from the terminal bud restrains axillary buds from growing, causing a shoot to lengthen at the expense of lateral branching. If the terminal bud is removed, the plant may become bushier (FIGURE 35.6). However, cytokinins entering the shoot system from roots counter the action of auxin by signaling axillary buds to begin growing. Auxin cannot suppress the growth of these buds once it has begun. In many plant species, lower buds on a shoot usually break dormancy before buds closer to the apex, reflecting the relative distance away from auxin and cytokinin sources.

The check-and-balance control of lateral branching by auxin and cytokinins may be one way the plant coordinates the growth of its shoot and root systems. As roots become more extensive, the increased level of cytokinins would signal the shoot system to form more branches. The two hormones reverse their roles in the development of lateral roots; auxin stimulates and cytokinins inhibit root branching.

Both auxin and cytokinins may regulate the growth of axillary buds indirectly by changing the concentration of still another hormone, ethylene. The levels of different nutrients in a bud may also affect its response to these hormones.

Cytokinins as Anti-Aging Hormones. Cytokinins can retard the aging of some plant organs, perhaps by inhibiting protein breakdown, by stimulating RNA and protein synthesis, and by mobilizing nutrients from surrounding tissues. If leaves removed from a plant are dipped in a cytokinin solution, they stay green much longer than they otherwise would. It is probable that cytokinins also slow the deterioration of leaves on intact plants. Because of this anti-aging effect, florists use cytokinin sprays to keep cut flowers fresh. Cytokinin sprays can also prolong the

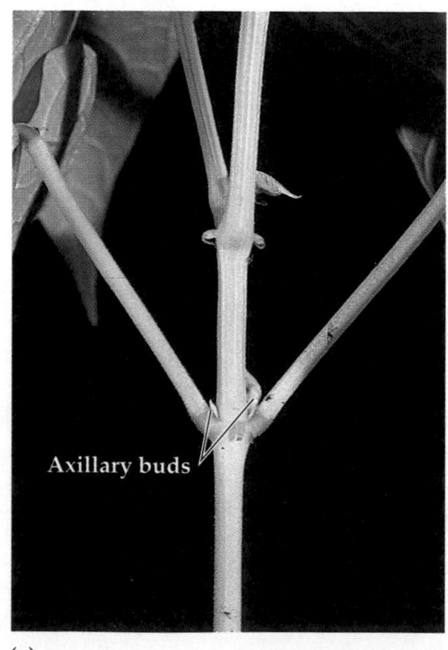

(a)

(b)

FIGURE 35.6
Apical dominance. (**a**) Auxin from the apical bud inhibits the growth of axillary buds. This favors elongation of the shoot's main axis over lateral branching. Cytokinins, which are transported upward from roots, counter auxin, stimulating the growth of axillary buds. This explains why, in most plants, axillary buds near the shoot tip are less likely to grow than those closer to the roots. (**b**) Removal of the apical bud from the same plant enabled lateral branches to grow. Unchecked by auxin, cytokinin stimulates the growth of axillary buds, which produce lateral branches of the shoot system.

shelf life of fruits and vegetables after harvest, although this latter application has not yet been approved by the U.S. Food and Drug Administration (FDA).

Gibberellins

A century ago, rice farmers in Asia noticed some exceptionally tall seedlings growing in their paddies. Before these rice seedlings could mature and flower, they grew so tall and spindly that they toppled over. In Japan, this aberration in growth pattern became known as *bakanae*, or "foolish seedling disease." In 1926, E. Kurosawa, a Japanese scientist, discovered that the disease was caused by a fungus of the genus *Gibberella* (FIGURE 35.7). By the 1930s, Japanese scientists had determined that the fungus produced hyperelongation of rice stems by secreting a chemical, which was given the name **gibberellin.** Western scientists finally learned of gibberellin after World War II. In the past 30 years, scientists have identified more than 70 different gibberellins, many of them occurring naturally in plants. All the gibberellins are subtle variations on a common molecular theme, but some forms are much more active than others in the plant. Foolish rice seedlings, it seems, suffer from an overdose of growth regulators normally found in plants in lower concentrations. Gibberellins have a variety of effects in plants.

FIGURE 35.7
"Foolish seedling disease" in rice. The spindly rice plants on the right are infected with the fungus *Gibberella*. The pathogen secretes gibberellins, a growth stimulant that uninfected plants (left) also produce, but in smaller quantity.

Stem Elongation. Roots and young leaves are major sites of gibberellin production. Gibberellins stimulate growth in both the leaves and the stem, but they have little effect on root growth. In stems, gibberellins stimulate cell elongation *and* cell division. In a growing stem, gibberellins and auxin must be acting simultaneously in some synergistic manner we do not yet understand.

Enhancement of stem elongation by gibberellins can be seen by applying the hormones to certain dwarf varieties of plants. For instance, dwarf pea plants (including the variety Mendel studied; see Chapter 13) grow to normal height if treated with gibberellins. The extent of the growth of dwarf plants is generally correlated with the concentration of the added hormone (see the Methods Box). If gibberellins are applied to plants of normal size, there is often no response. Apparently, these plants are

METHODS: BIOASSAY

A bioassay is a technique that determines the concentration of a chemical by measuring the response of living material. In this example, a plant's quantitative response to a gibberellin is used to measure the concentration of the hormone. A sample of unknown concentration is applied to dwarf pea plants, and after a certain amount of growth time, their height is compared with dwarfs treated with a range of known gibberellin concentrations. Using the degree of coleoptile curvature in phototropism to determine auxin concentration is another example of a bioassay.

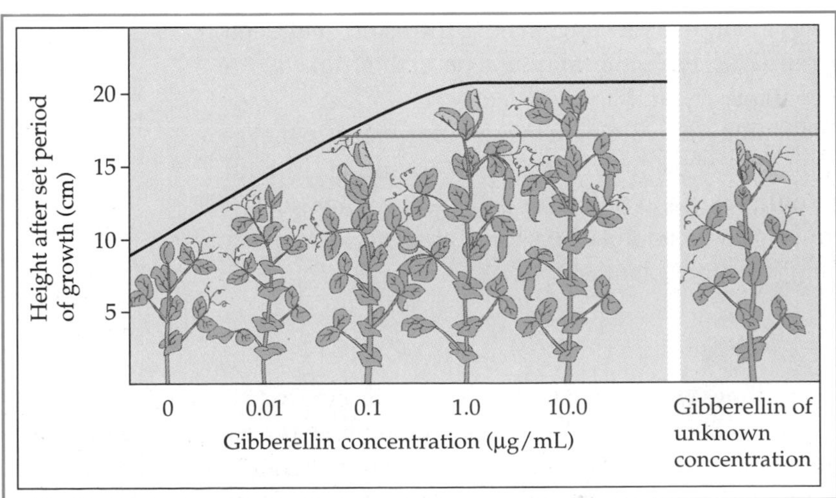

already producing their own optimal dose of the hormone. Gibberellins may be at low levels or absent in dwarf varieties of some species, or the target cells may be less responsive to the hormones.

A specific case in which gibberellins cause rapid elongation of stems is bolting, the growth of a floral stalk. In their nonflowering stage, some plants develop a rosette form; that is, they are low to the ground with very short internodes. The plant switches to reproductive growth when a surge of gibberellins induces stems to elongate rapidly, which elevates flowers developing from buds at the tips of the stems.

Fruit Growth. Fruit development is another case in which we can observe dual control by auxin and gibberellins. In some plants, both hormones must be present for fruit to set. The most important commercial application of gibberellins is in the spraying of Thompson seedless grapes. The hormones cause the grapes to grow larger and farther apart.

Germination. Many seeds have a high concentration of gibberellins, particularly in the embryo. After water is imbibed, the release of gibberellins from the embryo signals the seeds to break dormancy and germinate. Some seeds that require special environmental conditions to germinate, such as exposure to light or cold temperatures, will break dormancy if they are treated with a gibberellin solution. In nature, gibberellins in the seed are probably the link between the environmental cue and the metabolic processes that renew the growth of the embryo.

Gibberellins support the growth of cereal seedlings by stimulating the synthesis of digestive enzymes such as α-amylase that mobilize stored nutrients (see FIGURE 34.9). Even before these enzymes appear, gibberellin stimulates the synthesis of messenger RNA coding for α-amylase. Here is a case in which a hormone controls development by affecting gene expression, although the hormone molecule may not be acting directly on the genome.

Gibberellins also function to break dormancy in the resumption of growth by apical buds in spring. In both seed dormancy and bud dormancy, gibberellins act antagonistically to another hormone, abscisic acid, which generally inhibits plant growth.

Abscisic Acid

The hormones we have studied so far—auxin, cytokinins, and gibberellins—usually stimulate plant growth. By contrast, there are times in the life of a plant when it is adaptive to slow down growth and assume a dormant state. The hormone **abscisic acid (ABA),** produced in the terminal bud, slows growth and directs leaf primordia to develop into the scales that will protect the dormant bud during winter. The hormone also inhibits cell division in the vascular cambium. Thus, ABA helps prepare the plant for winter by suspending both primary and secondary growth. Abscisic acid was named when it was believed that the hormone caused the abscission of leaves from deciduous trees during autumn (L. *ab,* "loss," and *caedere,* "cut"). However, no clear-cut role for ABA in abscission has been demonstrated.

Another stage in the life of a plant when it is advantageous to suspend growth is the onset of seed dormancy, and again it may be abscisic acid that acts as the growth inhibitor. The seed will germinate when ABA is overcome by its inactivation or removal or by the increased activity of gibberellins. The seeds of some desert plants break dormancy when a heavy rain washes ABA out of the seed. Other seeds require light or some other stimulus to trigger the degradation of abscisic acid. In most cases, the ratio of ABA to gibberellins determines whether the seed will remain dormant or germinate. Similarly, the dormancy of terminal buds is controlled more by a balance of hormones than by their absolute concentrations. In apple trees, for instance, the concentration of ABA is actually higher in growing buds than in dormant buds, but an excess of gibberellins overpowers the inhibitory hormone.

In addition to its role as a growth inhibitor, abscisic acid acts as a "stress" hormone, helping the plant cope with adverse conditions. For example, when a plant begins to wilt, ABA accumulates in leaves and causes stomata to close, reducing transpiration and preventing further water loss. In one variety of tomato that is deficient in ABA and suffers chronic wilting, the experimental addition of abscisic acid closes stomata by triggering potassium loss from the guard cells (see Chapter 32). In this response, we see how a small amount of hormone has its impact amplified by acting on a membrane.

Ethylene

Early in this century, citrus was ripened by "curing" the fruit in sheds equipped with kerosene stoves. Fruit growers believed it was the heat that ripened the fruit, but newer, cleaner-burning stoves did not work. Plant physiologists learned later that ripening in the sheds was actually due to **ethylene,** a gaseous by-product of kerosene combustion. Researchers later demonstrated that plants produce their own ethylene as a hormone, and that this hormone elicits a variety of responses in addition to fruit ripening. Unique among plant hormones because it is a gas, ethylene diffuses through the plant in the air spaces between cells. Ethylene can also move in the cytosol, traveling from cell to cell in the symplast and in the phloem. In some cases, ethylene acts to inhibit cell elongation. Many of the inhibitory effects once attributed to auxin are now believed to be the result of ethylene synthesis induced by a high concentration of auxin. It is probably ethylene, for example, that inhibits root elongation and the development of axillary buds in the presence of an excess of auxin. In addition to its role as a growth inhibitor, ethylene is also associated with a variety of aging processes in plants.

Senescence in Plants. Aging, or **senescence,** is a progression of irreversible change that eventually leads to death. A normal part of plant development, senescence may occur at the level of individual cells, entire organs, or the whole plant. Xylem vessel elements and cork cells age and die before assuming their specialized functions. Autumn leaves and withering flower petals are examples of senescent organs. Plants that are annuals age and die soon after flowering. Ethylene probably has important functions in all these cases of senescence, but the aging processes where the effects of the hormone have been studied most extensively are fruit ripening and leaf abscission.

Fruit Ripening. Several changes in structure and metabolism accompany the ripening of an ovary into a fruit. Some of these changes, including the degradation of cell walls, which softens the fruit, and the decrease in chlorophyll content, which causes loss of greenness, can be regarded as aging processes. Ethylene initiates or hastens these deteriorative changes and also causes some ripened fruits to drop from the plant.

A chain reaction occurs during ripening, as ethylene triggers senescence, and the aging cells then release more ethylene. Because ethylene is a gas, the signal to ripen even spreads from fruit to fruit: One bad apple really does spoil the lot. If you pick or buy green fruit, you may be able to speed ripening by storing the fruit in a plastic bag, so that ethylene gas will accumulate. On a commercial scale, many kinds of fruit are ripened in huge storage containers into which ethylene gas is piped—a modern variation on the old curing shed. In other cases, measures are taken to retard ripening caused by natural ethylene. Apples, for instance, are stored in bins flushed with carbon dioxide. Circulating the air prevents ethylene from accumulating, and carbon dioxide somehow inhibits the action of whatever ethylene has not been flushed away. Stored in this way, apples picked in autumn can be shipped to grocery stores the following summer. Recently, a team of molecular biologists devised a method to manipulate the expression of one of the genes required for ethylene synthesis (FIGURE 35.8).

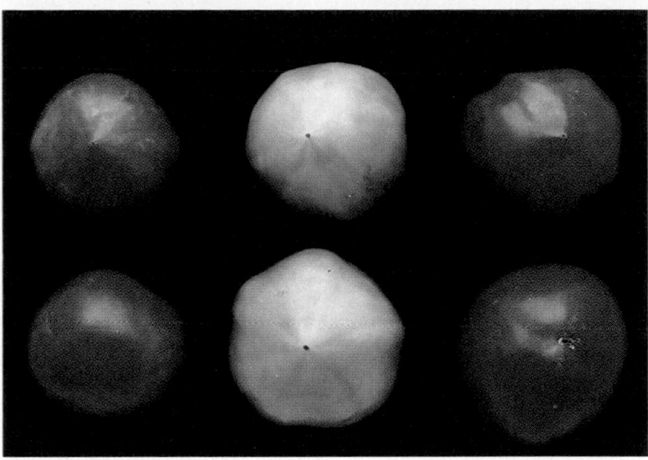

FIGURE 35.8
The control of fruit ripening by antisense RNA. The tomato fruits on the left have ripened naturally as a result of their production of the hormone ethylene. Molecular biologists suppressed ripening of the tomatoes in the center by adding an antisense RNA, which blocks transcription of one of the genes required for ethylene synthesis (see Chapter 19). The fruit can then be cued to ripen on demand by adding ethylene gas (tomatoes on the right). As such methods are refined, they may reduce spoilage of fruits and vegetables, a problem that currently ruins almost half the produce harvested in the United States.

Leaf Abscission. The loss of leaves each autumn is an adaptation that keeps deciduous trees from desiccating during winter when the roots cannot absorb water from the frozen ground. Before leaves are abscised, many of their essential elements are shunted to storage tissues in the stem. These nutrients are recycled back to developing leaves the following spring. The autumn leaf stops making new chlorophyll and loses its greenness. The fall colors are a combination of new pigments made during autumn and pigments that were already present in the leaf but concealed by the dark green chlorophyll.

When an autumn leaf falls, the breaking point is an abscission layer located near the base of the petiole (FIGURE 35.9). The small parenchyma cells of this layer have very thin walls, and there are no fiber cells around the vascular tissue. The abscission layer is further weakened when enzymes hydrolyze polysaccharides in the cell walls. Finally, the weight of the leaf, with the help of wind, causes a separation within the abscission layer. Even before the leaf falls, a layer of cork forms a protective scar on the twig's side of the abscission layer, preventing pathogens from invading the plant.

A change in the balance of ethylene and auxin controls abscission. An aging leaf produces less and less auxin, and this drop in concentration makes the cells of the abscission layer more sensitive to ethylene. This shift in hormonal balance is self-reinforcing, as cells in the abscission layer begin producing additional ethylene, which inhibits the leaf's synthesis of auxin. As the influence of ethylene on the abscission layer prevails, the cells produce enzymes that digest the cellulose and other components of cell walls.

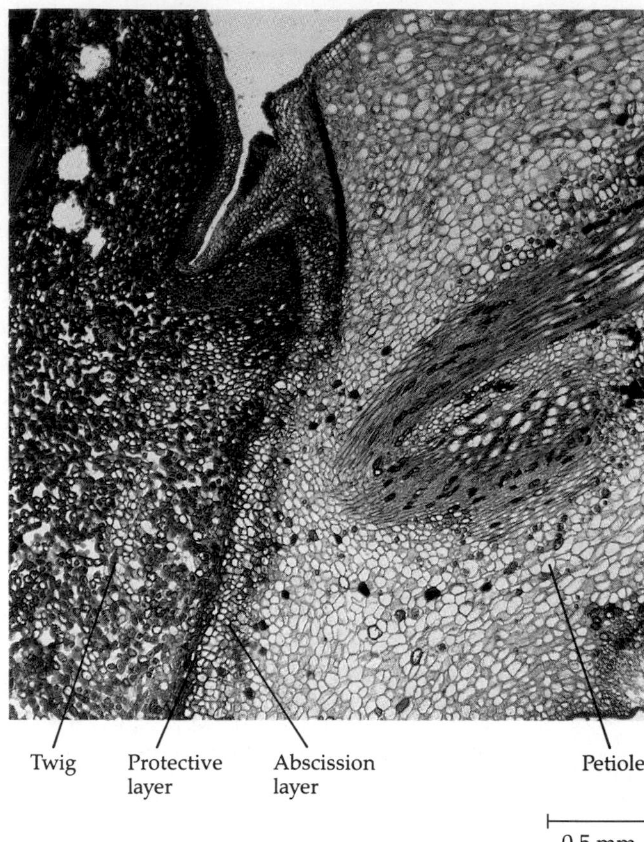

FIGURE 35.9
Abscission of a maple leaf. Abscission is controlled by a change in the balance of ethylene and auxin. The abscission layer can be seen here as a vertical band at the base of the petiole. After the leaf falls, a protective layer of cork becomes the leaf scar that helps prevent pathogens from invading the plant (LM).

The environmental stimuli for leaf abscission are the shortening days and cooler temperatures of autumn. This relationship between outside stimuli and the hormonal signals and responses within a plant brings us to an important juncture in our study of control systems. How do plants actually detect changes in their surroundings? Plant biologists are approaching this problem by studying plant movements. To the casual observer, most plants do not appear to be very dynamic, but time-lapse photography reveals they are capable of quite precise movements. The two types of plant movements are tropisms and turgor movements.

Tropisms orient the growth of plant organs toward or away from stimuli

The environment has a great influence on a plant's shape. **Tropisms** are growth responses that result in curvatures of whole plant organs toward or away from stimuli (Gr. *tropos,* "turn"). The mechanism for a tropism is a differential rate of elongation of cells on opposite sides of the organ. Three of the stimuli that induce tropisms, and a consequent change of body shape, are light (phototropism), gravity (gravitropism), and touch (thigmotropism).

Phototropism

As we have seen, the classical hypothesis for what causes grass coleoptiles to grow toward light maintains that cells on the darker side of a stem elongate faster than cells on the brighter side because of an asymmetrical distribution of auxin moving down from the shoot tip. There is less support for this hypothesis from studies of phototropism by organs other than grass coleoptiles. For example, there is no evidence that unilateral light causes an asymmetrical distribution of auxin in the stems of sunflowers, radishes, and other dicots. There *is,* however, an asymmetrical distribution of certain substances that may act as growth inhibitors, with these substances more concentrated on the lighted side of a stem.

Whether phototropism results from auxin stimulating cell elongation on the darker side of a stem, or some other chemical messenger inhibiting elongation on the lighter side, most researchers agree that the shoot tip is the site of the photoreception that triggers the growth response. The photoreceptors are pigment molecules that are most sensitive to blue light. The most likely candidate is a yellow pigment related to the vitamin riboflavin. The same receptor may be involved in stomatal opening and some other responses of plants to light.

Gravitropism

Place a seedling on its side, and it will adjust its growth so that the shoot bends upward and the root curves downward. In their responses to gravity, or **gravitropism,** roots display positive gravitropism and shoots exhibit negative gravitropism. Gravitropism functions as soon as a seed germinates, ensuring that the root grows into the soil and the shoot reaches sunlight regardless of how the seed happens to be oriented when it lands.

Plants may tell up from down by the settling of **statoliths,** specialized plastids containing dense starch grains, to the low points of cells (FIGURE 35.10). In roots, statoliths are located in certain cells of the root cap. According to one hypothesis, the aggregation of statoliths at the low points of these cells triggers the redistribution of calcium, which in turn causes lateral transport of auxin within the root. The calcium and auxin accumulate on the lower side of the root's zone of elongation. Because these chemicals are dissolved, they do not respond to gravity but must be actively transported to one side of the root. At high concentration, auxin inhibits cell elongation, an effect that slows growth on the lower side of the root. The more rapid elongation of cells on the upper side causes the root to curve as it grows. This tropism continues until the root is growing straight down.

Some researchers are challenging the hypothesis that "falling" statoliths trigger gravitropism. According to calculations by Randy Wayne of Cornell University, the impact of starch grains on the bottom of a cell does not release enough energy to be the mechanism for gravitational detection. Wayne also points out that many plants lacking starch grains still distinguish up from down. He studies gravitropism in *Chara,* a green alga closely related to plants. Proteins attach the protoplast of each cell to the inside of the cell wall. According to Wayne's hypothesis, the downward settling of the entire protoplast, which stretches protein tethers at the top of the cell and compresses those at the bottom, gives the cell its sense of up and down. As one test of this hypothesis, Wayne placed *Chara* in a solution more dense than the alga's cytoplasm. The protoplasts floated upward instead of falling downward, and the growth pattern of the alga was also upside down. Wayne and other plant physiologists are now trying to determine whether this mechanism of gravitational detection is also at work in vascular plants.

Thigmotropism

Most vines and other climbing plants have tendrils that coil around supports (see FIGURE 31.8a). These grasping organs usually grow straight until they touch something; the contact stimulates a coiling response caused by differential growth of cells on opposite sides of the tendril.

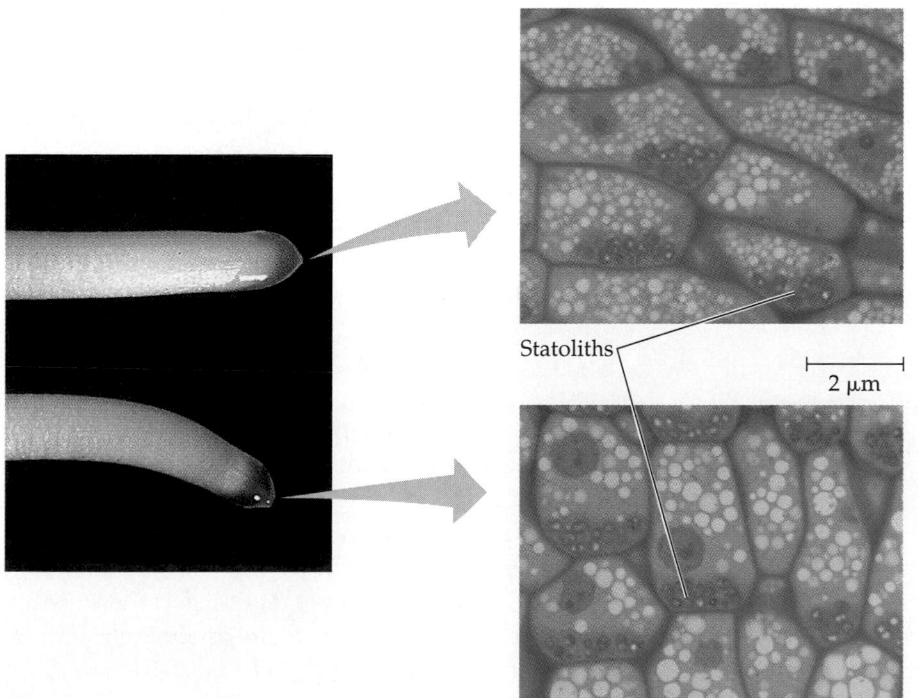

FIGURE 35.10
The statolith hypothesis and gravitropism. These corn roots were placed on their sides and photographed before (top) and 1.5 hours after the gravitropic response. The light micrographs show the locations of statoliths, modified plastids, within cells of the root cap. The settling of statoliths to the low points of these cells may be the gravity-sensing step that leads to the redistribution of auxin and the differential rates of elongation by cells on opposite sides of the root. According to an alternative hypothesis, it is the downward settling of the entire protoplast that enables the cell to detect gravity.

This directional growth in response to touch is called thigmotropism (Gr. *thigma*, "touch").

Mechanical stimulation can also cause a much more general response. One experiment demonstrated that rubbing stems with a stick a few times results in plants that are shorter and thicker than controls that are not as manipulated. In the wild, wind can cause a similar stunting of growth, enabling the plant to hold its ground against strong gusts. A tree growing on a windy mountain ridge, for instance, will usually have a shorter, stockier trunk than a tree of the same species growing in a more sheltered location. This developmental response to mechanical perturbation is referred to as **thigmomorphogenesis.** It usually results from an increased production of ethylene in response to chronic mechanical stimulation.

Turgor movements are relatively rapid, reversible plant responses

In addition to the relatively long-lasting changes in body shape resulting from tropisms, plants are also capable of reversible movements caused by changes in the turgor pressure of specialized cells in response to stimuli.

Rapid Leaf Movements

When the compound leaf of the sensitive plant *Mimosa* is touched, it collapses and its leaflets fold together (FIGURE 35.11). This response, which takes only a second or two, results from a rapid loss of turgor by cells within pulvini, specialized motor organs located at the joints of the leaf. The motor cells suddenly become flaccid after stimulation because they lose potassium, which causes water to leave the cells by osmosis. It takes about ten minutes for the cells to regain their turgor and restore the natural form of the leaf. The function of the sensitive plant's behavior invites speculation. Perhaps by folding its leaves and reducing its surface area when jostled by strong winds, the plant conserves water. Or perhaps because the collapse of the leaves exposes thorns on the stem, the rapid response of the sensitive plant discourages herbivores.

A remarkable feature of rapid leaf movements is the transmission of the stimulus through the plant. If one leaf on a sensitive plant is touched with a hot needle, first that leaf collapses, then the adjacent leaf responds, then the next leaf along the stem, and so on until all the plant's leaves are drooping. From the point of stimulation, the message that produces this response travels wavelike through the plant at a speed of about a centimeter per second. Chemical messengers probably have a role in this transmission, but an electrical impulse can also be detected by attaching electrodes to the plant. These impulses, called **action potentials,** resemble nervous messages in animals, though the action potentials of plants are thousands of times slower than those of animals. Action potentials, which have been discovered in many species of algae and plants, may be widely used as a form of internal communication. Another example is the Venus flytrap, in which action potentials are transmitted from sensory hairs in the trap to the cells that respond by closing the trap.

Sleep Movements

Bean plants and many other members of the legume family lower their leaves in the evening and raise them to a horizontal position in the morning (FIGURE 35.12). These **sleep movements** are powered by daily changes in the turgor pressure of motor cells in pulvini similar to those of the sensitive plant. When leaves are horizontal, cells on one side of the pulvinus are turgid, whereas cells on the opposite side are flaccid. This is reversed at night when the leaves close to their "sleeping" position. Paralleling the opposing changes in volume of the motor cells is a massive migration of potassium ions from one side of the

(a)

(b)

FIGURE 35.11
Rapid turgor movements by the sensitive plant (*Mimosa pudica*). (**a**) In the unstimulated plant, leaflets are spread apart. (**b**) Within a second or two of being touched, the leaflets have folded together.

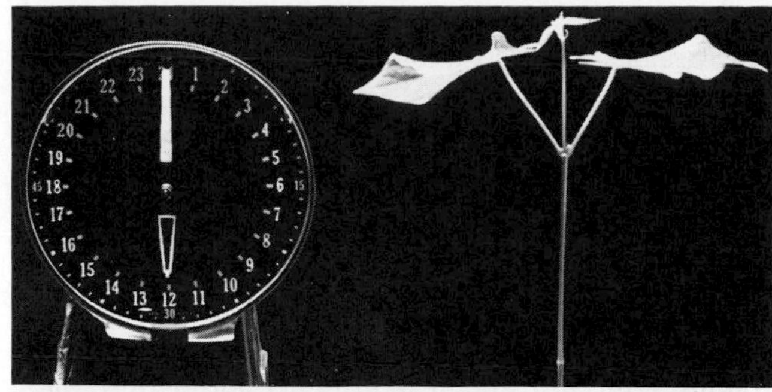

FIGURE 35.12
Sleep movements of a bean plant. Leaf position at noon and leaf position at midnight. The movements are caused by reversible changes in the turgor pressure of cells on opposing sides of the pulvini, swollen regions at the base of the petiole.

pulvinus to the other. Apparently, the potassium is an osmotic agent that leads to the reversible uptake and loss of water by the motor cells. In this respect, the mechanism of sleep movements is similar to stomatal opening and closing. Sleep movements are only one example of the many responses that depend on the ability of plants to keep track of time.

Biological clocks control circadian rhythms in plants and other eukaryotes

Your pulse, blood pressure, temperature, rate of cell division, blood cell count, alertness, urine composition, metabolic rate, sex drive, and responsiveness to medications all fluctuate with the time of day. Some insects are more vulnerable to insecticides in the afternoon than in the morning. Certain fungi produce spores for several hours the same time each day. Unicellular algae that glow in the dark switch on their bioluminescence like clockwork. Plants also display rhythmic behavior; the sleep movements of legumes and the opening and closing of stomata are examples. All these rhythmic phenomena and many others are controlled by biological clocks, internal oscillators that keep accurate time. Biological clocks seem to be ubiquitous features of eukaryotic organisms, and our first evidence for biological rhythms came from studies of plants.

A physiological cycle with a frequency of about 24 hours is called a **circadian rhythm** (L. *circa,* "approximately," and *dies,* "day"). Are these rhythms truly prompted by an internal clock, or are they merely daily responses to some environmental cycle, such as the rotation of Earth (night and day)? Circadian rhythms persist even when the organism is sheltered from environmental cues. A bean plant, for example, will continue its sleep movements even if kept in constant light or constant darkness; the leaves are not simply responding to sunrise and sunset. Organisms, including humans, continue their rhythms when placed in the deepest mine shafts or when orbited in satellites. All research thus far indicates that the oscillator for circadian rhythms is endogenous (internal). This clock, however, is entrained (set) to a period of precisely 24 hours by daily signals from the environment. If an organism is kept in a constant environment, circadian rhythms deviate from a 24-hour period (a "period" is the duration of one cycle). These free-running periods, as they are called, vary from about 21 to 27 hours, depending on the particular rhythmic response. The sleep movements of bean plants, for instance, have a period of 26 hours when kept under the free-running conditions of constant darkness.

Deviation of the free-running period from exactly 24 hours does not mean that biological clocks drift erratically. The clocks are still keeping perfect time, but they are not synchronized with the outside world. The light-dark cycle resulting from Earth's rotation is the most common factor that entrains biological clocks. If the timing of these cues changes, it takes a few days for the clock to be reset. Thus, a plant kept for several days in the dark will be out of phase with plants growing in a normal environment where the sun rises and sets each day. The same thing happens when we cross several time zones in an airplane; when we reach our destination, the clocks on the wall are not synchronized with our internal clocks. All eukaryotes are probably prone to jet lag.

Most biologists now agree that organisms possess built-in clocks, but the nature of the internal oscillator is still unknown. Where is the clock, and how does it work? In attempting to answer these questions, we must be careful to differentiate between the oscillator and the rhythmic processes it controls. The sweeping leaves of sleep movements are the "hands" of the biological clock, but these movements are not the essence of the clockwork itself. If the leaves of a bean plant are restrained for several hours so they cannot move, they will, on release, rush to the position appropriate for the time of day. We can interfere with a biological rhythm, but the clock goes right on ticking off the time. Most scientists who study circadian rhythms place the clock at the cellular level, either in membranes or in the machinery for protein synthesis. For example, in 1993, researchers at the University of Connecticut demonstrated that the sleep movements of a legume species are correlated with the rhythmic opening and closing of potassium channels in the membranes of the motor cells. This traces the cause of circadian leaf movements to the subcellular level, but only changes the question from "What times the rhythmic movement of leaves?" to "What times the rhythm of potassium gates?"

Photoperiodism synchronizes many plant responses to changes of season

Seasonal events are important in the life cycles of most plants. Seed germination, flowering, and the onset and breaking of bud dormancy are examples of stages in plant development that usually occur at specific times of the year. The environmental stimulus plants most often use to detect the time of year is the photoperiod, the relative lengths of night and day. A physiological response to day length, such as flowering, is called **photoperiodism.**

Photoperiodic Control of Flowering

One of the earliest clues to how plants detect the progression of seasons came from an unusual variety of tobacco studied by W. W. Garner and H. A. Allard in 1920. This variety, named Maryland Mammoth, grew exceptionally tall but failed to flower during summer when normal tobacco plants flowered. Maryland Mammoth finally bloomed in a greenhouse in December. After trying to induce earlier flowering by varying temperature, moisture, and mineral nutrition, Garner and Allard learned that it was the shortening days of winter that stimulated Maryland Mammoth to flower. If the plants were kept in light-tight boxes so that lamps could be used to manipulate durations of "day" and "night," flowering would occur only if the day length were 14 hours or shorter. The Maryland Mammoths did not flower during summer because, at Maryland's latitude, the days were too long during that season.

Garner and Allard termed Maryland Mammoth a **short-day plant,** because it apparently required a light period *shorter* than a critical length to flower. Chrysanthemums, poinsettias, and some soybean varieties are a few of the other short-day plants, which generally flower in late summer, fall, or winter. Another group of plants dependent on photoperiod will flower only when the light period is *longer* than a certain number of hours. These **long-day plants** generally flower in late spring or early summer. Spinach, for example, flowers when days are 14 hours or longer. Radish, lettuce, iris, and many cereal varieties are also long-day plants. Flowering in a third group, day-neutral plants, is unaffected by photoperiod. Tomatoes, rice, and dandelions are examples of **day-neutral plants** that flower when they reach a certain stage of maturity, regardless of day length at that time.

Critical Night Length. In the 1940s, researchers discovered that night length, not day length, actually controls flowering and other responses to photoperiod. Many of these scientists worked with cocklebur, a short-day plant that flowers only when days are 16 hours or less in length (and nights are at least 8 hours long). If the daytime portion of the photoperiod is broken by a brief exposure to darkness, there is no effect on flowering. However, if the nighttime part of the photoperiod is interrupted by even a few minutes of dim light, the plants will not flower (FIGURE 35.13). Cocklebur requires at least 8 hours of *continuous* darkness to flower. Short-day plants are really long-night plants, but the older term is embedded firmly in the jargon of plant physiology. Long-day plants are actually short-night plants; grown on photoperiods of long nights that would not normally induce flowering, long-day plants will flower if the period of continuous darkness is shortened by a few minutes of light at any point.

Thus, photoperiodic responses depend on a critical night length. Short-day plants will flower if the duration of night is *longer* than the critical length (8 hours for cocklebur); long-day plants will flower when the night is *shorter* than the critical length. The floriculture (flower-growing) industry has applied this knowledge to produce flowers out of season. Chrysanthemums, for instance, are short-day plants that normally bloom in fall, but their blooming can be stalled until Mother's Day in May by punctuating each long night with a flash of light, thus turning one long night into two short nights.

Notice that we distinguish long-day from short-day plants *not* by an absolute night length but by whether the critical night length sets a maximum (long-day plants) or minimum (short-day plants) number of hours of darkness required for flowering.

Some plants bloom after a single exposure to the photoperiod required for flowering. Other species need several successive days of the appropriate photoperiod. Still other plants will respond to photoperiod only if they have been previously exposed to some other environmental stimulus, such as a period of cold temperatures. Winter wheat, for example, will not flower unless it has been exposed to several weeks of temperatures below 10°C. This requirement for pretreatment with cold before flowering is called vernalization. Several weeks after winter wheat is vernalized, a photoperiod with long days (short nights) induces flowering.

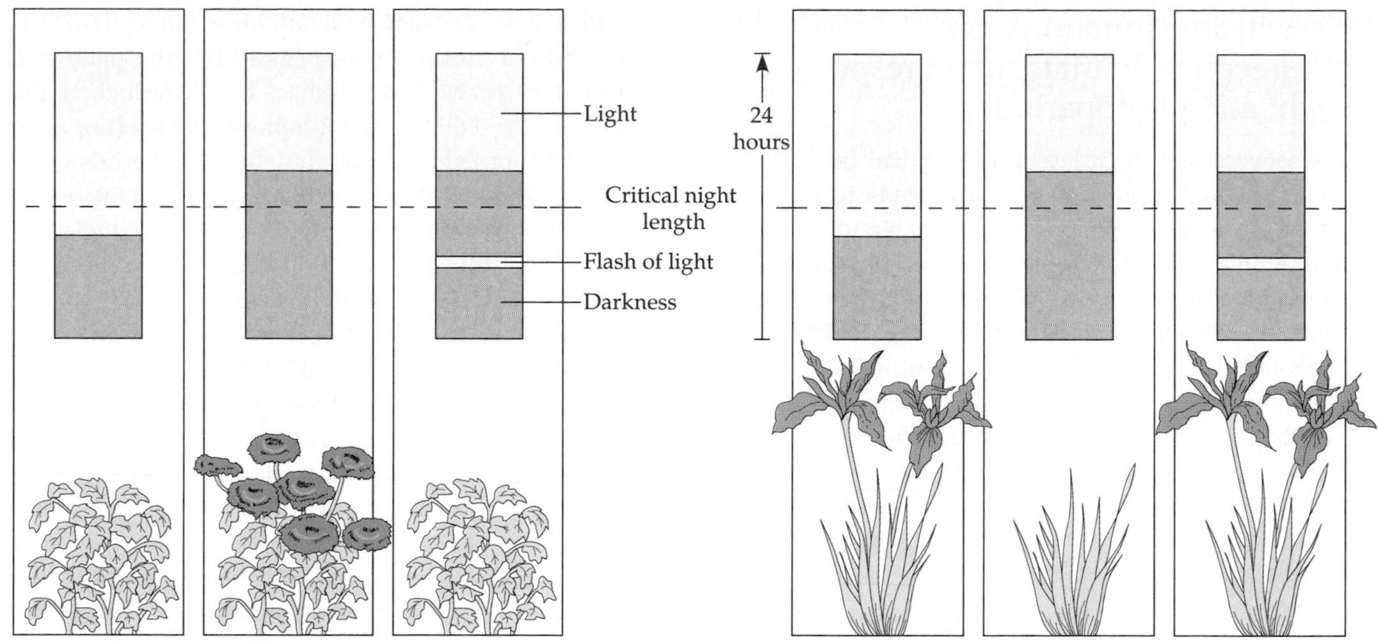

FIGURE 35.13
Photoperiodic control of flowering. A short-day (long-night) plant flowers when night exceeds a critical dark period. A flash of light interrupting the dark period prevents flowering. A long-day (short-night) plant flowers only if the night is shorter than a critical dark period. The night can be artificially shortened with a flash of light.

Evidence for a Flowering Hormone. Buds produce flowers, but the photoperiod is detected by leaves. To induce a short-day plant or long-day plant to flower, it is enough in many species to expose a single leaf to the appropriate photoperiod. Indeed, if only one leaf is left attached to the plant, photoperiod is detected and floral buds are induced. If all leaves are removed, however, the plant is blind to photoperiod. Apparently, some message to flower is transported from leaves to buds. Most plant physiologists believe this message is a hormone (FIGURE 35.14). The signal to flower that travels from leaves to buds appears to be the same for short-day and long-day plants, although the two groups of plants differ in the photoperiodic conditions required for leaves to send this signal.

The evidence for a flowering hormone is compelling, but researchers have not yet identified the chemical messenger. It is possible that the flowering impulse is not a single chemical but a specific mixture of several substances, including hormones.

FIGURE 35.14
Experimental evidence for a hormone that induces flowering. If a plant that has been induced to flower by photoperiod is grafted to a plant that has not been induced, both plants flower, indicating the transmission of a flower-inducing substance. This works in some cases even if one is a short-day plant and the other is a long-day plant.

Phytochrome functions as a photoreceptor in many plant responses to light and photoperiod

The discovery that night length is a critical factor controlling seasonal responses of plants leads to another question: How does a plant measure the length of darkness in a photoperiod? A pigment named **phytochrome** is part of the answer. It was discovered as a result of studies on how different colors of light affect flowering, seed germination, and other responses to photoperiod.

Red light, a wavelength of 660 nanometers (nm), is the most effective light in interrupting night length. A short-day plant kept under conditions of critical night length fails to flower if a brief exposure to red light breaks the dark period. Conversely, a flash of red light during the dark period will induce a long-day plant to flower even if the total night length exceeds the critical number of hours (long-day plants, remember, require nights shorter than a critical length). The effect of the red flash is to shorten the plant's perception of night length.

The shortening of night length by red light can be negated by a subsequent flash of light having a wavelength of about 730 nm. This wavelength is in the far-red part of the spectrum, just barely visible to humans. If red light (R) during the dark period is followed by far-red light (FR), the plant perceives no interruption in night length. A short-day plant will not flower if a night of critical length is broken by an R flash, but the plant *will* flower if it receives two flashes, first R and then FR. Reversing this sequence to FR-R prevents flowering. Each wavelength of light cancels the effect of the other. No matter how many flashes of light are given, the wavelength of only the last one affects the plant's measurement of night length. Thus, a succession of flashes with the sequence R-FR-R-FR-R prevents short-day plants from flowering, but flowering occurs if the sequence is R-FR-R-FR-R-FR. As expected, the opposite behavior occurs in long-day plants (FIGURE 35.15).

The photoreceptor responsible for the reversible effects of red and far-red light is phytochrome, a protein containing a chromophore (the light-absorbing portion). Phytochrome alternates between two forms that differ only slightly in structure, but one absorbs red light and the other absorbs far-red light. The two variations of phytochrome—P_r (red absorbing) and P_{fr} (far-red absorbing)—are said to be photoreversible:

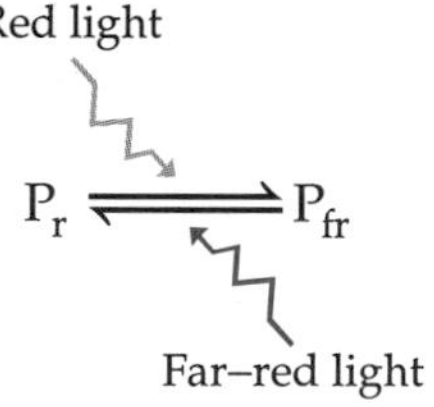

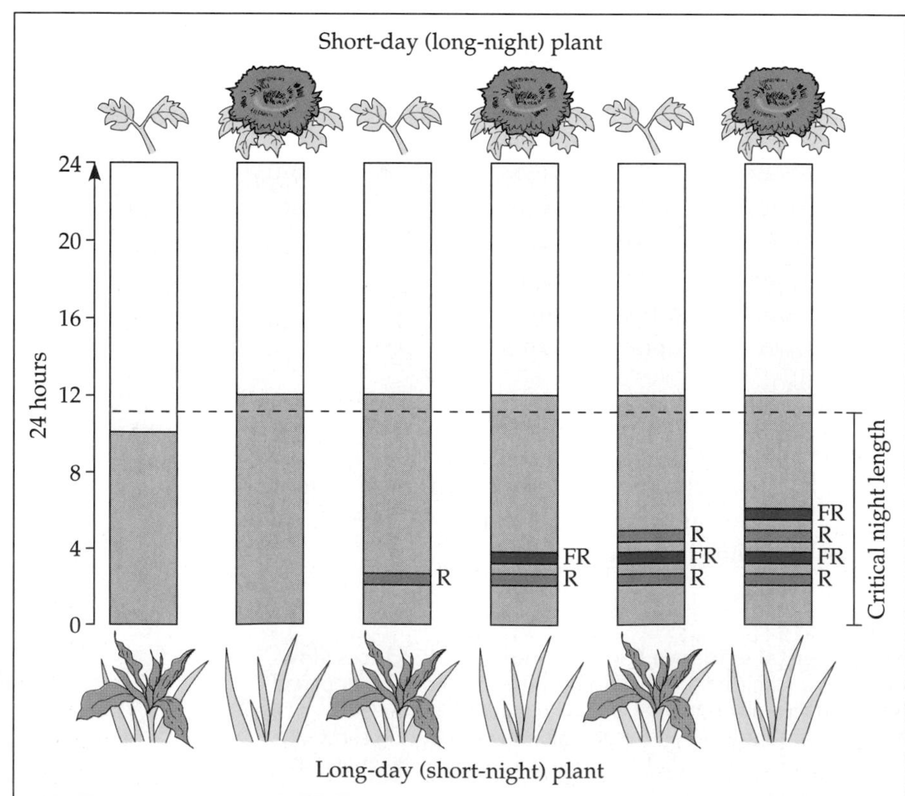

FIGURE 35.15
Reversible effects of red and far-red light on photoperiodic response.
A flash of red light shortens the dark period. A subsequent flash of far-red light cancels the effect of the red flash.

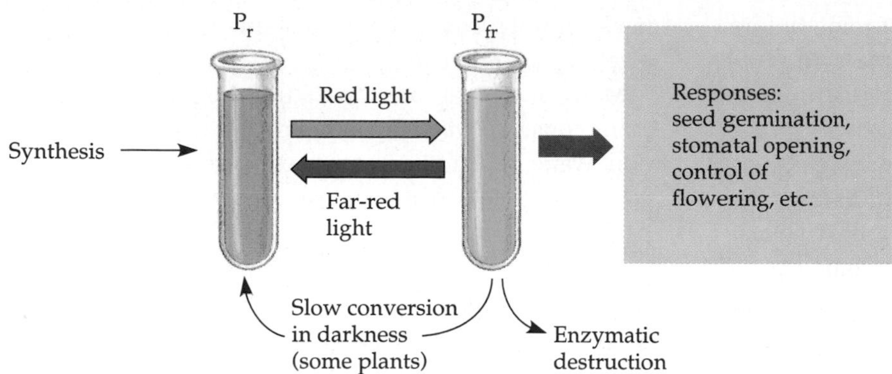

FIGURE 35.16

Phytochrome: a molecular switching mechanism. The test tubes in this diagram contain solutions of the two photoreversible forms of phytochrome. Absorption of red light causes the bluish P_r to change to the blue-greenish P_{fr}. Far-red light reverses this conversion. In most cases, it is the P_{fr} form of the pigment that switches on physiological responses in the plant.

This $P_r \rightleftharpoons P_{fr}$ interconversion acts as a switching mechanism that controls various events in the life of the plant (FIGURE 35.16).

The Ecological Significance of Phytochrome as a Photoreceptor

Light is a key environmental variable in the life of a plant. In the photoperiodic control of flowering and many other plant responses to illumination, phytochrome functions as a photodetector that tells the plant whether light is present. Plants synthesize phytochrome as P_r, and if they are kept in the dark, the pigment remains in that form. Then, if the phytochrome is illuminated with sunlight, some of the P_r is converted to P_{fr} because the pigment is exposed to red light (along with all the other wavelengths in sunlight) for the first time. This appearance of P_{fr} is one way plants detect sunlight. P_{fr} is the form of phytochrome that triggers many of a plant's developmental responses to light, such as the germination of seeds that require light to break dormancy.

The phytochrome system also provides the plant with information about the *quality* of light. Sunlight includes both red and far-red radiation. Thus, during the day, the $P_r \rightleftharpoons P_{fr}$ photoreversion reaches a dynamic equilibrium, with the ratio of the two phytochrome forms indicating the relative amounts of red and far-red light. This sensing mechanism enables plants to adapt to changes in light conditions. Consider, for example, a "shade-avoiding" tree that requires relatively high light intensity. If this tree becomes shaded by other trees in a forest, the phytochrome ratio shifts in favor of P_r because the forest canopy screens out more red light than far-red light. This cue induces the tree to use most of its resources to grow taller.

Plant cells actually have very little phytochrome, and yet photoreception by the pigment can have an enormous effect on the whole plant. This implies that the photoconversion of phytochrome from P_r to P_{fr} is a signal that becomes amplified in the steps that lead to the plant's responses to light. Two mechanisms by which photoconversion of relatively few phytochrome molecules could have a very large effect are by altering membrane permeability or by affecting gene expression. There is evidence for both mechanisms among phytochrome-controlled responses. In the sleep movements of legumes, for example (FIGURE 35.12), photoconversion of phytochrome triggers the potassium fluxes across membranes that cause osmotic swelling and shrinking of cells in the pulvini. An example of phytochrome affecting gene expression is the light-induced synthesis of starch-digesting α-amylase required for the germination of some seeds.

Other photoreceptors complement phytochrome in tuning a plant's growth and development to its environment. For example, we have already learned about a blue-absorbing pigment in shoot tips that functions as the receptor in phototropism.

The Interaction of Phytochrome and the Biological Clock in Photoperiodism

In darkness, the phytochrome ratio shifts gradually in favor of the P_r form. This is partly due to turnover in the overall phytochrome pool. The pigment is synthesized in the P_r form, and degradative enzymes destroy more P_{fr} than P_r (see FIGURE 35.16). In addition, in some plant species, P_{fr} present at sundown slowly converts to P_r by a biochemical mechanism. In darkness, there is no means for the accumulating P_r to be reconverted to P_{fr}. Then at sunrise, the P_{fr} level suddenly increases again by rapid photochemical conversion from P_r. Phytochrome conversion marks the beginning and end of the dark segment of a photoperiod. Does the plant use these signals to measure night length, the environmental variable that controls flowering and other responses to photoperiod? Perhaps the gradual conversion of P_{fr} to P_r in the dark is a chemical hourglass that gauges the passage of night. However, the conversion is usually completed within a few hours after sunset. If the plant used the disappearance of P_{fr} to measure the length of the dark period, it would lose track of time during the middle of the night.

Night length is measured not by phytochrome but by the biological clock. The role of phytochrome in photoperiodism may be to synchronize the clock to the environment by telling it when the sun sets and rises.

If the photoperiodic requirement for flowering is met, the clock triggers some sort of alarm that causes leaves to send a flowering stimulus (perhaps a hormone) to buds. Night length is measured very accurately; some short-day plants will not flower if night is even one minute shorter than the critical length. Some plant species always flower on the same day each year. According to the hypothesis described here, plants tell the season of year by using the clock, apparently entrained with the help of phytochrome, to keep track of photoperiod.

Control systems enable plants to cope with environmental stress

Environmental fluctuations challenge plants every day of their lives. Occasionally, factors in the environment change severely enough to put plants under stress. We'll define stress here as an environmental condition that can have an adverse effect on a plant's growth, reproduction, and survival. This definition narrows our survey of how plants cope with stress by excluding evolutionary adaptations that enable certain plants to thrive in environments that are stressful to most other plants. For example, some plants, called halophytes ("salt plants"), have special anatomical and physiological adaptations that enable them to grow best in salty soil. The leaves of some halophytes, for instance, are equipped with salt glands, desalination machines that pump salt out of the plants. As a result, a saline environment is not an environmental stress for halophytes; plants lacking such specialized equipment would be stressed by the same environment. Within the range of salinity the plants can survive, how do control systems minimize the damage? How do plants cope with water deficiency, oxygen deprivation, excessive heat, or an environment that turns stressfully cold? And what mechanisms help plants survive stressful biological factors such as herbivores and pathogens? These are the questions to keep in mind as we examine some plant responses to environmental stress.

Responses to Water Deficit

On any bright, warm, dry day, a plant may be stressed by a water deficiency because it is losing water by transpiration faster than the water can be restored by uptake from the soil. Prolonged drought can stress crops and the plants of natural ecosystems for weeks or months. Severe water deficit, of course, will kill a plant, as you may know from experience with neglected houseplants. But plants have control systems that enable them to cope with less extreme water deficits.

Many of a plant's responses to water deficit help the plant conserve water by reducing the rate of transpiration. Water deficit in a leaf causes guard cells to lose turgor, a simple control mechanism that slows transpiration by closing stomata (see Chapter 32). Water deficit also stimulates increased synthesis and release of abscisic acid from mesophyll cells in the leaf, and this hormone helps keep stomata closed by acting on guard cell membranes. Leaves respond to water deficit in several other ways. Because cell expansion is a turgor-dependent process, a water deficit will inhibit the growth (expansion) of young leaves. This response minimizes the transpirational loss of water by slowing the increase in leaf surface. When the leaves of many grasses and other plants wilt from a water deficit, they roll into a shape that reduces transpiration by exposing less leaf surface to the sun. While all of these responses of leaves help the plant conserve water, they also reduce photosynthesis. This is one reason a drought diminishes crop yield.

Root growth also responds to water deficit. During a drought, the soil usually dries from the surface down. This inhibits the growth of shallow roots, partly because cells cannot maintain the turgor required for elongation. Deeper roots surrounded by soil that is still moist continue to grow. Thus, the root system proliferates in a way that maximizes exposure to soil water.

Responses to Oxygen Deprivation

An overwatered houseplant may suffocate because the soil lacks the air spaces that provide oxygen for cellular respiration in the roots. Some plants are structurally adapted to very wet habitats. For example, the submerged roots of trees called mangroves, which inhabit coastal marshes, are continuous with aereal roots that provide access to oxygen. But how do plants less specialized for aquatic environments cope with oxygen deprivation in waterlogged soils? One structural change is the formation of air tubes that provide oxygen to submerged roots (FIGURE 35.17).

Responses to Salt Stress

An excess of sodium chloride or other salts in the soil threatens plants for two reasons. First, by lowering the water potential of the soil solution, salt can cause a water deficit in plants even though the soil has plenty of water. This is because in an environment with a water potential more negative than that of the root tissue, roots will *lose* water rather than absorb it (see Chapter 32). The second problem with saline soil is that sodium and certain other ions are toxic to plants when their concentration is rela-

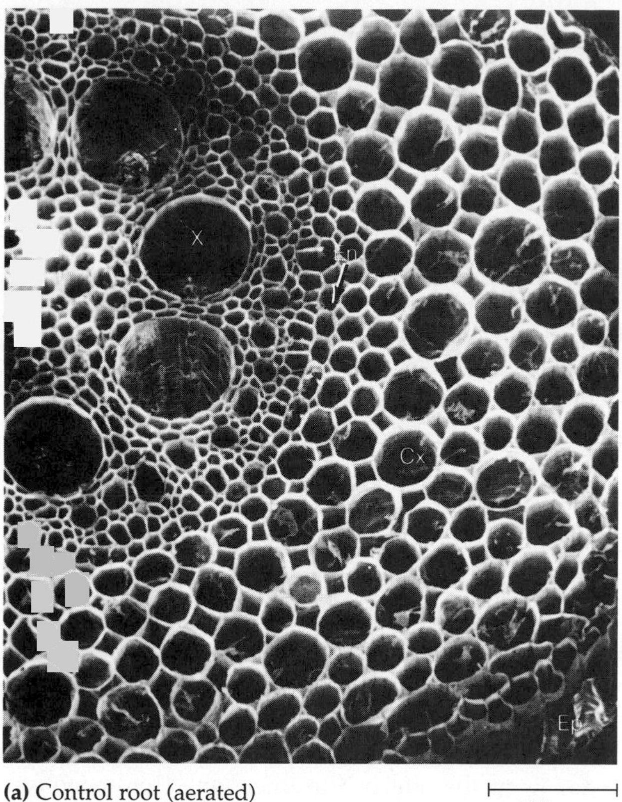

(a) Control root (aerated)

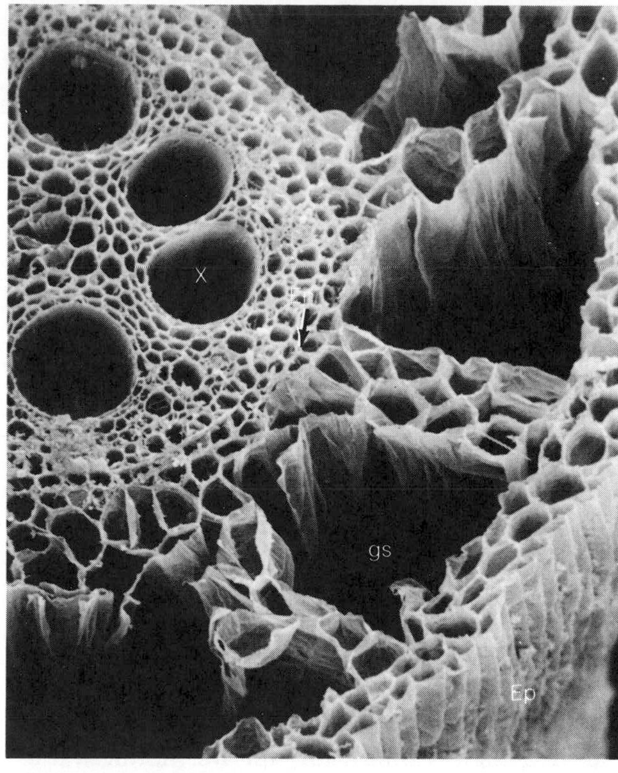

(b) Experimental root (nonaerated)

FIGURE 35.17
An anatomical response of corn roots to oxygen deprivation. (**a**) A transverse section of a control root grown in aerated hydroponic medium. (**b**) An experimental root grown in a nonaerated hydroponic medium. Oxygen deprivation stimulates the production of the hormone ethylene, which causes some of the cells in the root cortex to age and die. Enzymatic destruction of the cell walls creates air tubes that function as "snorkels," providing oxygen to the submerged roots. Key to labels: X = xylem, Cx = root cortex, Ep = epidermis, gs = gas (air)-filled air tubes. (Both SEMs.)

tively high. The selectively permeable membranes of root cells impede the uptake of most harmful ions, but this only aggravates the problem of acquiring water from soil that is rich in solutes. Many plants can respond to moderate soil salinity by producing *compatible solutes,* organic compounds that keep the water potential of cells more negative than the soil solution without admitting toxic quantities of salt. However, except for the specially equipped halophytes, plants cannot survive salt stress for long. You learned in Chapter 33 that flood irrigation in arid regions compromises the long-term productivity of farmland as evaporation of the irrigation water concentrates salts in the soil.

Responses to Heat Stress

Excessive heat can harm and eventually kill a plant by denaturing its enzymes and damaging its metabolism in other ways. One function of transpiration is evaporative cooling. On a warm day, for example, the temperature of a leaf may be 3°–10°C below ambient air temperature. Of course, hot, dry weather also tends to cause water deficiency in many plants; the closing of stomata in response to this stress conserves water but sacrifices evaporative cooling. This dilemma is one reason very hot, dry days take such a toll on most plants.

Most plants have a backup response that enables them to survive heat stress. Above a certain temperature—about 40°C for most plants that inhabit temperate regions—plant cells begin synthesizing relatively large quantities of special proteins called **heat-shock proteins.** Researchers have also discovered this response in heat-shocked animals and microorganisms. Some heat-shock proteins are identical to chaperone proteins, which function in unstressed cells as temporary scaffolds that help other proteins fold into their functional conformations (see Chapter 5). As heat-shock proteins, perhaps these molecules embrace enzymes and other proteins and help prevent denaturation.

Responses to Cold Stress

One problem plants face when the temperature of the environment falls is a change in the fluidity of cell membranes. Recall from Chapter 8 that a biological membrane is a fluid mosaic with proteins and lipids drifting laterally in the plane of the membrane. When a membrane cools below a critical point, it loses its fluidity as the lipids become locked into crystalline structures. This alters solute transport across the membrane and also adversely affects the functions of membrane proteins. Plants respond to cold stress by altering the lipid composition of their membranes. For example, membrane lipids increase in their proportion of unsaturated fatty acids, which have shapes that help keep membranes fluid at lower temperatures by impeding crystal formation (see FIGURE 8.3). Such molecular modification of the membrane requires from several hours to days, which is one reason rapid chilling is generally more stressful to plants than a more gradual drop in air temperature.

Freezing is a more severe version of cold stress. At subfreezing temperatures, ice crystals begin to form in most plants. If the ice is confined to cell walls and intercellular spaces, the plant will probably survive. However, if ice begins to form within protoplasts, the sharp crystals perforate membranes and organelles, killing the cells. Oaks, maples, roses, rhododendrons, and other woody plants native to regions where winters are cold have special adaptations that enable them to cope with freezing stress. For example, changes in the solute composition of live cells allows the cytosol to supercool without ice forming, although ice crystals may form in the cell walls.

Responses to Herbivores

Herbivory—animals eating plants—is a stress that plants face in any ecosystem. Plants counter excessive herbivory with both physical defenses, such as thorns, and chemical defenses, such as the production of distasteful or toxic compounds. For example, some plants produce an unusual amino acid called canavanine, named for one of its sources, the jackbean (*Canavalia ensiformis*). Canavanine resembles arginine, one of the twenty amino acids organisms incorporate into their proteins. If an insect eats a plant containing canavanine, the molecule is incorporated into the insect's proteins in place of arginine. Because canavanine is different enough from arginine to adversely affect the conformation and hence the function of proteins, the insect dies. You will learn more about how plants defend against herbivores in Chapter 48.

Defense Against Pathogens: Systemic Acquired Resistance

Plants, like animals, benefit from a surface barrier as a first line of defense against pathogenic microorganisms. For a plant, this "skin" is the epidermis of the primary plant body and the periderm of the secondary plant body. This first defense system, however, is not impenetrable. Viruses, bacteria, and the spores and hyphae of fungi can enter the plant through injuries or through natural openings in the epidermis, such as stomata. Once a pathogen invades, the plant mounts a chemical attack that functions as a second line of defense. The infected plant produces a variety of compounds called **phytoalexins,** antibiotics that destroy microorganisms or inhibit their growth.

Researchers are beginning to unravel the steps that enable a plant to sense and respond to pathogens. During infection, molecules belonging to the pathogens and compounds released from the injured plant tissue function as "alarm substances." These molecules induce rapid responses at the site of the infection, such as a cross-linking of molecules in the cell wall. This sets up a local barricade that slows the spread of pathogens to other parts of the plant. The alarm substances also cause cells in the infected tissue to synthesize and release a hormone that is transported throughout the plant. By "spreading the news" of infection, the hormone triggers phytoalexin production and other defense measures in cells far from the site of the infection. This response, called **systemic acquired resistance (SAR),** helps protect uninfected tissue should the pathogen spread from its original point of invasion.

Scientists have recently identified **salicylic acid** as the hormone responsible for activating SAR. A modified form of this compound, acetylsalicylic acid, is the active ingredient in aspirin. Centuries before aspirin was sold as a pain reliever, some cultures had learned that chewing the bark of a willow tree (*Salix*) would lessen the pain of a toothache or headache. With the discovery of systemic acquired resistance, biologists have finally learned one function of salicylic acid in plants. Aspirin turns out to be a natural medicine in the plants that produce it, but with effects entirely different from the medicinal effects in humans who consume the drug.

* * *

We complete our study of control systems in plants by investigating how hormonal and environmental signals are relayed within cells from the points of reception to the cellular components that actually respond.

Signal-transduction pathways mediate the responses of plant cells to environmental and hormonal stimuli

In 1990, Stanford University researchers studying *Arabidopsis* growth made a serendipitous discovery. They sprayed plants with hormones to see if there was any effect on gene expression. Indeed, the treatment increased transcription of five specific genes. But when control plants were sprayed with water, activity of the same five genes increased as well. In fact, merely touching the plants a couple of times a day had the same effect. It was mechanical stimulation, not the hormones or water, that induced transcription of these specific genes. Touching the plants also affected morphology. Mechanically stimulated plants did not grow as tall as control plants that were unstimulated (FIGURE 35.18). Recall that such an effect of touch on plant development is called thigmomorphogenesis. It is a reasonable hypothesis that the genes induced by touching the plants code for proteins that function in the growth response. Somehow, the touch stimulus must be transduced into an intracellular signal that is relayed to nuclei, altering gene expression. Such a mechanism linking stimulus to response in cells is called a **signal-transduction pathway.**

FIGURE 35.18
Altering gene expression by touch in *Arabidopsis.* The shorter plant on the left was touched twice a day. Compare it to the unmolested plant on the right, which grew much taller. Researchers have discovered that this developmental response to touch is associated with the induction of five specific genes. A signal-transduction pathway is activated by mechanical stimulation of cells and relays the message to respond to the genome in the cell nuclei.

All the hormones and environmental stimuli you have learned about in this chapter act on plant cells via signal-transduction pathways. Plant biologists are just beginning to work out the steps of these pathways, but it is already apparent that the basic mechanisms resemble pathways that transduce signals in animal cells (see Chapter 41). We started this chapter by contrasting how whole plants and animals cope with environmental change. But when we trace their behavior to the cellular level, the similarities are remarkable.

We can dissect a signal-transduction pathway into three main steps: reception, transduction, and induction. *Reception* is the cell's detection of an environmental signal or hormone. In the case of responses to light, for example, the reception step is the absorption of light of a specific wavelength by a pigment within the cell. Photoconversion of phytochrome is this type of reception. For responses to hormones, the reception step is the binding of the hormone to a specific receptor, usually a protein molecule (FIGURE 35.19a). If the receptor is embedded in the plasma membrane, the hormone can trigger changes in the cell without even entering. In other cases, the hormone receptor is dissolved in the cytosol or associated with a particular organelle within the cell. A specific hormone can only affect those types of cells that have receptors for that hormone. These are the **target cells** for that hormone. Even if other cells are exposed to the hormone, they do not respond, because there is no means of detecting the stimulus. This may be one reason, for example, that cells in a shoot's zone of elongation are especially sensitive to the growth-promoting action of auxin. Thus, the reception step is one key to the specificity of responses to hormones.

The *transduction* step in a signal-transduction pathway amplifies the stimulus and converts it to a chemical form capable of activating the cell's responses. The critical link in transduction is a **second messenger,** some substance that increases in concentration within a cell that is stimulated by the first messenger—a hormone, for example. The binding of a hormone to its specific receptor evokes the second message. Even if the receptor is a membrane protein at the cell surface, the hormonal signal can be relayed to intracellular sites by the second messenger. For example, binding of the hormone may activate the receptor or associated membrane proteins to catalyze a chemical reaction that activates the second messenger on the cytoplasmic side of the membrane. Evidence is accumulating that calcium ions (Ca^{2+}) function as the second messenger in many plant responses. In these cases, a hormone or environmental stimulus triggers an increase in cytoplasmic Ca^{2+}. The signal may release the ion from

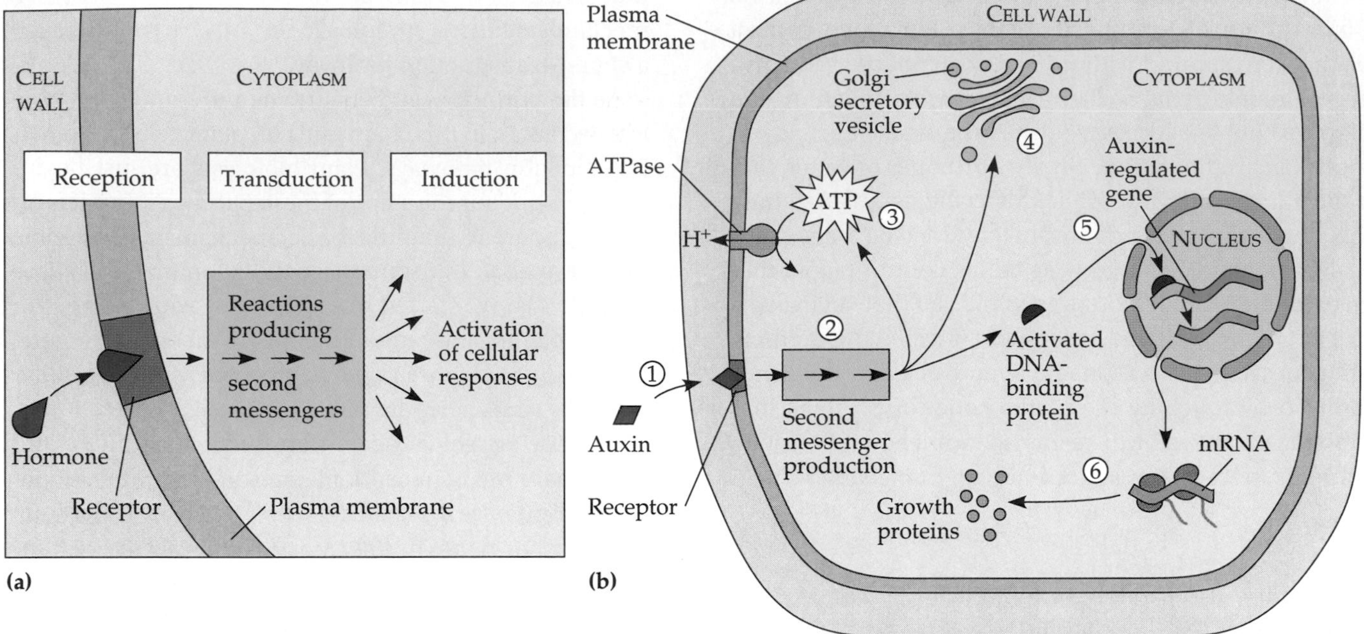

FIGURE 35.19

Signal-transduction pathways in plant cells. (**a**) A general model. A hormone binding to a specific receptor stimulates the cell to produce second messengers. A second messenger triggers the cell's various responses to the original signal. In this diagram, the receptor is on the surface of the target cell. In other cases, hormones enter cells and bind to specific receptors inside. Environmental stimuli can also initiate signal-transduction pathways. For example, phytochrome conversion is the first step in the transduction pathways that lead to a cell's responses to red light. (**b**) A specific example: a hypothetical mechanism for auxin's stimulation of cell elongation. ① The hormone binds to an auxin receptor, and ② this signal is transduced into second messengers within the cell, triggering various reponses. ③ Proton pumps are activated, and secretion of acid loosens the wall, enabling the cell to elongate. ④ The Golgi is stimulated to discharge vesicles containing materials that maintain the thickness of the cell wall. ⑤ The signal-transduction pathway also activates DNA-binding proteins that induce the transcription of specific genes. ⑥ This leads to the production of proteins required for sustained growth of the cell.

organelles, or it may open Ca^{2+} channels in the plasma membrane, facilitating a rush of extracellular Ca^{2+} into the cell. The second messenger, Ca^{2+}, then binds to a specific protein called **calmodulin.** In a chain reaction, the calcium-calmodulin complex activates other target molecules within the cell. This is probably one way red light acts on phytochrome-containing cells. Absorption of red light converts P_r to P_{fr}. The P_{fr} then acts on the membrane to raise cytoplasmic Ca^{2+} concentration. The Ca^{2+} combines with calmodulin, which triggers a cascade of protein activations that are manifest as the phytochrome-mediated responses we observe.

Notice two important properties of the transduction step in signal pathways. First, signal transduction amplifies a stimulus. Binding of a single hormone molecule to a receptor can give rise to a large population of second messengers, which can activate an even larger number of proteins and other cellular molecules. Second, signal transduction contributes to the specificity of response. Two cell types that both have receptors for a particular hormone may respond differently, because the cells contain different target proteins that key on the second messengers. Although we have used the calcium burst as an example, there are other second messengers (see Chapter 41).

In the third main step of a signal-transduction pathway, the *induction* step, the amplified signal induces the cell's specific responses to the stimulus. Some of these responses are relatively rapid. For example, abscisic acid can cause stomata to close within minutes via a signal-transduction pathway that induces K^+ efflux from guard cells. Another example is the auxin-induced acidification of cell walls that causes cell elongation (FIGURE 35.19b). Other responses take longer, usually because they require changes in gene expression. You learned earlier in the chapter, for instance, that auxin stimulates longer-term elongation of cells by inducing specific genes that code for proteins functioning in cell wall synthesis. And you have also learned that touch, acting on the genome through a signal-transduction pathway, alters the morphology of plants.

* * *

Botanists investigating control systems are getting to the heart of how a plant adapts to its environment. These scientists, along with thousands of other plant biologists working on other problems and millions of students experimenting with plants, are all extending a centuries-old tradition of curiosity about the form and function of the organisms that feed the biosphere.

REVIEW OF KEY CONCEPTS (with page numbers and key figures)

- Reseach on how plants grow toward light led to the discovery of plant hormones: *science as a process* (pp. 751–752; FIGURES 35.2, 35.3)
  - Hormones are traveling chemical messengers that coordinate functions between distant parts of an organism.
  - Experiments on phototropism led to identification of the first plant hormone, auxin.
- Plant hormones help coordinate growth, development, and responses to environmental stimuli (pp. 753–760, TABLE 35.1)
  - Five known classes of hormones control plant growth and development by multiple effects on the division, elongation, and differentiation of cells. The site of action, the developmental stage of the plant, the concentration of the hormone, and the presence of other hormones all affect reaction to a hormone.
  - Produced primarily in the apical meristem of the shoot, auxin stimulates cell elongation in different target tissues.
  - The acid-growth hypothesis maintains that auxin causes the stimulation of proton pumps that acidify the cell wall region. This breaks cross-links between cellulose microfibrils in the cell wall, enabling the cell to take up additional water and elongate.
  - Auxin also affects secondary growth and differentiation, initiates adventitious root formation, and promotes fruit growth.
  - Cytokinins stimulate cell division.
  - Actively growing tissues, such as roots, embryos, and fruits, are rich in cytokinins, which, in concert with auxin, stimulate cell division and influence differentiation. Subtle changes in the cytokinin-to-auxin ratio have specific effects on plant development.
  - Cytokinins and auxin also work together to control apical dominance in a complex interaction.
  - Gibberellins produced in roots and young leaves stimulate growth in leaves and stems. In conjunction with auxin, gibberellins increase elongation in stems.
  - Gibberellins and auxin also stimulate fruit development.
  - Gibberellins support germination by stimulating the synthesis of key enzymes involved in the mobilization of seed storage material.
  - Abscisic acid slows plant growth and favors the dormant state by inducing the development of bud scales, inhibiting cell division in the vascular cambium, and suspending growth in buds and seeds. Abscisic acid is also a stress hormone that helps plants cope with adverse conditions.
  - Ethylene, a gaseous hormone, diffuses through the plant in air spaces. It inhibits root growth and the development of axillary buds in the presence of high auxin concentrations. Ethylene also stimulates fruit ripening and induces several aspects of senescence in plant cells and organs.
  - The mechanics of leaf abscission involve decreased auxin and increased ethylene production.
- Tropisms orient the growth of plant organs toward or away from stimuli (pp. 760–762, FIGURE 35.10)
  - Tropisms are growth responses that result in the curvature of entire plant organs toward or away from stimuli.
  - Phototropism enhances photosynthesis by bending shoots toward light.
  - Gravitropism may be mediated by statoliths.
  - Thigmotropism leads to an adaptive coiling of tendrils on touching a support. Thickening of stems in response to chronic, strong winds is an example of thigmomorphogenesis.
- Turgor movements are relatively rapid, reversible plant responses (pp. 762–763)
  - Turgor movements are responses to stimuli caused by changes in turgor pressure of specialized cells. Sleep movements of legumes are examples.
- Biological clocks control circadian rhythms in plants and other eukaryotes (pp. 763–764, FIGURE 35.12)
  - Circadian rhythms are physiological cycles that have a frequency of about 24 hours. The absence of environmental cues leads to free-running periods, in which the rhythms may deviate by a few hours from a 24-hour period but are still precise within their own cycles. The light-dark cycle probably entrains biological clocks.
- Photoperiodism synchronizes many plant responses to changes of season (pp. 764–765, FIGURE 35.13)
  - Photoperiodism, the response to relative lengths of night and day, helps regulate stages of plant development.
  - Photoperiodic control of flowering actually depends on a critical night length. This sets a minimum (short-day plants) or maximum (long-day plants) number of hours of darkness required for flowering. Flowering in day-neutral plants is unaffected by photoperiod.
- Phytochrome functions as a photoreceptor in many plant responses to light and photoperiod (pp. 766–768, FIGURE 35.15)
  - Phytochrome, which exists in two photoreversible states, is one factor that signals sunrise and sunset. Actual night length is measured by the biological clock.
  - The P_R:P_{fr} ratio also measures light quality, triggering such responses as rapid elongation of shoots that are shaded.
- Control systems enable plants to cope with environmental stress (pp. 768–770)
  - Stomatal closing and other responses that conserve water help plants survive water deficits.
  - In waterlogged soils, some roots develop air tubes as a response to oxygen deprivation.
  - The production of compatible solutes is one way plants respond to salt stress.
  - Plants produce heat-shock proteins that help protect other proteins during heat stress.
  - Plant cells adjust the lipid composition of their membranes in response to cold stress.
  - Most plants produce physical and chemical defenses in response to herbivory.
  - Salicylic acid stimulates systemic acquired resistance (SAR), which helps a plant defend against pathogens.

- Signal-transduction pathways mediate the responses of plant cells to environmental and hormonal stimuli (pp. 770–772, FIGURE 35.19)
  - Hormones stimulate cells by binding to specific receptors (reception).
  - Reception triggers the production of second messengers within the cell (transduction).
  - Second messengers activate a chain of events that mobilize the cell's specific response (induction).

SELF-QUIZ

1. Which of the following plant hormones is incorrectly paired with its function?
   a. auxin—promotes stem growth through cell elongation
   b. cytokinin—initiates senescence
   c. gibberellin—stimulates seed and bud germination
   d. abscisic acid—promotes seed and bud dormancy
   e. ethylene—inhibits cell elongation

2. Spraying some plants with a combination of auxin and gibberellins
   a. promotes fruit growth
   b. kills broad-leaf dicot plants
   c. prevents senescence
   d. promotes fruit ripening
   e. is used to treat dwarfism in plants

3. Buds and sprouts often form on tree stumps. Which of the following hormones would you expect to stimulate their formation?
   a. auxin
   b. cytokinins
   c. abscisic acid
   d. ethylene
   e. gibberellins

4. Which of the following is not part of the acid-growth hypothesis?
   a. Auxin stimulates proton pumps in cell membranes.
   b. Lowered pH results in the breakage of cross-links between cellulose microfibrils.
   c. The wall fabric becomes looser (more elastic).
   d. Auxin-activated proton pumps stimulate cell division in meristems.
   e. The turgor pressure of the cell exceeds the restraining pressure of the loosened cell wall, and the cell takes up water and elongates.

5. The signal for flowering could be released prematurely in a long-day plant experimentally exposed to flashes of
   a. far-red light during the night
   b. red light during the night
   c. red light followed by far-red light during the night
   d. far-red light during the day
   e. red light during the day

6. The phytochrome system helps set the biological clock by indicating to a plant that light is present when
   a. P_r is rapidly converted to P_{fr}
   b. P_{fr} is slowly converted to P_r
   c. P_r and P_{fr} are equal in concentration
   d. red light is absorbed by P_{fr}
   e. photosynthetic production of ATP powers phytochrome conversion

7. If a long-day plant has a critical night length of 9 hours, which of the following 24-hour cycles would prevent flowering?
   a. 16 hours light/8 hours dark
   b. 14 hours light/10 hours dark
   c. 15.5 hours light/8.5 hours dark
   d. 4 hours light/8 hours dark/4 hours light/8 hours dark
   e. 8 hours light/8 hours dark/light flash/8 hours dark

8. The probable role of salicylic acid in systemic acquired resistance of plants is to
   a. destroy pathogens directly.
   b. activate plant defenses throughout the plant before infection spreads.
   c. close stomata, thus preventing the entry of pathogens.
   d. activate heat-shock proteins.
   e. sacrifice infected tissues by hydrolyzing cells.

9. Auxin triggers the acidification of cell walls that results in rapid growth, followed by sustained, long-term cell elongation. What best explains how auxin brings about this two-stage growth response?
   a. Auxin binds to different receptors in different cells.
   b. Different concentrations of auxin have different effects.
   c. Auxin causes a second messenger to activate both proton pumps and certain genes.
   d. The dual effects are due to two different auxins.
   e. Auxin's effects are modified by other antagonistic hormones.

10. In signal-transduction pathways, hormones generally function as
    a. first messengers
    b. second messengers
    c. phytoalexins
    d. enzymes
    e. receptors on membranes

CHALLENGE QUESTIONS

1. Discuss how day length, phytochrome, the biological clock, gibberellins, and abscisic acid may interrelate in the germination of seeds planted just below the soil surface.
2. Explain how it is possible that a short-day plant and a long-day plant growing in the same location could flower on the same day of the year.
3. Write a short paragraph explaining how pruning results in fruit trees having more branches.
4. Dandelion flowers open each morning and close each evening. Describe an experiment that would determine whether this daily activity is controlled by an internal biological clock or simply by the presence or absence of light.
5. There are some species of bamboo that flower only when they are 120 years old. What mechanisms can you suggest to explain how these plants "count" years? How would you test these hypotheses? What are possible advantages of this reproductive "strategy" relative to seed predation?

SCIENCE, TECHNOLOGY, AND SOCIETY

1. Imagine the following scenario: A plant scientist discovers a synthetic chemical that mimics the effects of a plant hormone. The chemical can be sprayed on apples before harvest to prevent flaking of the natural wax that is formed on the skin. This makes the apples shinier and gives them a deeper red color. What kinds of questions do you think should be answered before farmers start using this chemical on apples?

2. Certain herbicides disrupt growth by mimicking the action of plant hormones. The weed killer 2,4-D, for example, is a synthetic auxin. Recombinant DNA techniques can be used to alter crop plants to make them more resistant to herbicides, so that the herbicides can be used to "weed" crops without harming them. What are some arguments for and against this technology?

3. Based on your study of this chapter, write a short essay explaining at least three examples of how knowledge about the control systems of plants is applied in agriculture or horticulture.

FURTHER READING

Evans, M. L., R. Moore, and K.-H. Hasenstein. "How Roots Respond to Gravity." *Scientific American*, December 1986.

Jones, A. M. "Surprising Signals in Plant Cells." *Science*, January 14, 1994. How aspirin and other hormones trigger plant responses.

Moffat, A. S. "Mapping the Sequence of Disease Resistance." *Science*, September 23, 1994. Molecular biologists are identifying the genes that help protect plants from pathogens.

Page, T. L. "Time Is the Essence: Molecular Analysis of the Biological Clock." *Science*, March 18, 1994. How do cells keep track of time?

Taiz, L., and E. Zeiger. Plant Physiology. Redwood City, CA: Benjamin/Cummings, 1991. Chapters 16–21.

Vogel, S. "When Leaves Save the Tree." *Natural History*, September 1993. How leaf responses help prevent wind damage to trees.

Wayne, R. "Excitability in Plant Cells." *American Scientist*, March/April 1993. Nervelike impulses in plants.

UNIT SEVEN

ANIMALS: FORM AND FUNCTION

AN INTERVIEW WITH

PATRICIA CHURCHLAND

Nervous systems distinguish animals from all other kingdoms of life. And certain properties of the human brain distinguish our species from all other animals. The human brain is, after all, the only known collection of matter that tries to understand itself. To most biologists, the brain and the mind are one and the same; understand how the brain is organized and how it works, and we'll understand such mindful functions as abstract thought and feelings. Some philosophers are less comfortable with this mechanistic view of the mind, finding Descartes' concept of a mind-body duality more attractive. Patricia Churchland has taken center stage in this debate about the human mind.

Dr. Churchland is a professor of philosophy at the University of California, San Diego, and an adjunct professor of neuroscience at the Salk Institute. Boundaries between the humanities and the sciences dissolve as Professor Churchland attempts to synthesize a philosophy of the mind based on what neuroscience is learning about the brain. Her seminal book in this new synthesis is Neurophilosophy: Toward a Unified Science of the Mind-Brain *(MIT Press, 1986). Biology is a multidisciplinary adventure that integrates the natural sciences and connects to the humanities and social sciences. Neurophilosopher Patricia Churchland is helping make those connections.*

What led you, as a philosopher, to neuroscience?

The questions I was interested in as a philosophy graduate student were really questions about the human mind, about the nature of learning and perception, about what it is for something to be conscious, about the difference between the actions we call voluntary and actions we call involuntary—the free-will problem. As time went on, it became increasingly clear to me that these were really questions about the brain and that I needed to know the nuts and bolts of the brain, what neurons were and how they talked to each other. The more neuroscience I knew, the more it seemed to me that we really had the key to understanding the nature of the mind via neuroscience. That is not to say that we had a key that you could use independently of psychology — behavioral descriptions — but that it was a crucial element.

I was always unconvinced by arguments that in addition to the brain, there is a nonphysical soul, and it's the soul that makes decisions, the soul that feels and thinks. If you're unconvinced by that, then the nature of the brain and its organization have to be relevant in understanding these fundamental questions that philosophers are interested in.

How did you then begin to learn neuroscience?

Well, I knew some basic biology, basic cell physiology and biochemistry, but I realized that I needed to know brain anatomy. By this time, I was a faculty member at the University of Manitoba. So I phoned the head of the anatomy department at the University of Manitoba Medical School and explained my interests. He said, "Why don't you just come and take the basic neuroscience course with the medical students and do as much or as little as you want." I did all the anatomy and the basic physiology, but that wasn't enough. Then I got associated with a lab and learned the basic techniques, such as recording from single cells.

Was it uncomfortable for you, coming from a philosophy background, to challenge the tradition of mind-body duality?

Not for me personally, because I don't think that distinction of mind and body, or brain, ever seemed terribly plausible. I was always part of that tradition that says that complexity is not predictable from looking at the constituents, but put them together in certain ways, and you get these really extraordinary properties. The mental properties for perception, for knowledge, for learning, for memory all had to come out of the complexity of the organization of matter.

If we view the major mind functions as emergent properties of the brain organization, what do we need to understand about neural complexity to explain these mental functions?

I think the major area we're missing is at the level of neural networks. Assuming that neurons form themselves into micronets, and micronets interact with larger units, or macronets, it's really the properties at that larger level that we don't yet understand. With the advent of artificial neural nets we can design on computers, we are beginning to get at least a conceptual framework for making that bridge between the individual neurons on the one hand and the systems on the other. We have a long way to go.

Take the case of categorization, for example. Some regions of the brain seem to be specialized for categories of natural things and others for manmade things. And within the regions specialized for manmade things, there is further categorization—for example, tools that you use with your hands, and other kinds of things like automobiles. We don't know how brains do this regionalization, so we don't know how to make nets that can help us understand it. Another key question about brains is how they get things done in time. How can an eagle intercept its prey? How can you catch an outfield fly? How you get the timing right is a major issue for a lot of neural network theory now. What we desperately need is more understanding of neuroanatomy at the network level.

Modeling neural nets on a computer is one thing, but is the brain itself a type of computer?

It's useful to think of the brain as a kind of computer because that allows you a framework for thinking about how individual neurons interact to achieve a certain effect. For example, if what you are trying to do is focus on a given object and your head is turning, thinking of the brain as a computer gives you a way of understanding the interaction between the neurons in the vestibular organ, which function in the sensation of movement, and the neurons that control eye movement. But unlike a desktop computer, the very elements in the brain that process perception are also elements that store information about what you perceive. It isn't that the memory is in one place and the processing for perception is in another place. The brain is a very different kind of computer, but I find the computer to be a useful metaphor.

Do you think knowledge of the brain can enable us to build better computers or different kinds of computers?

Yes, I really do. I think that there is likely to be big technological payoff as we understand in more detail how neuronets function and how they solve problems. That is, how do they manage to be so flexible? How, with so few neurons, can a bee solve problems that are really quite complicated? In general, I think the flexibility and the capacity to generalize that we see with brains and don't see in the very brittle architecture of artificial computers are things we will understand as we get more of the story of the brain.

Do you think it's possible to build machines that think?

I think in principle it is. Since we are machines that think, and evolution built us, then yes. Whether using the kinds of components that we are now using we can mimic the brain depends on the nature of the problem. If what we want to do is get the timing right, we may have to mimic what neurons actually do down to the level of membrane proteins.

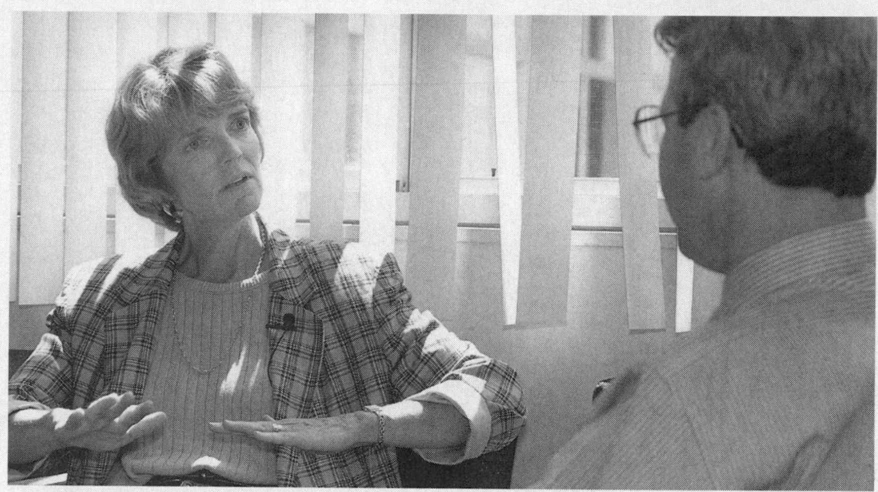

All of which depends on learning more about how the brain itself works. Are we getting closer to understanding consciousness, memory, or emotions?

We have made progress on a number of major questions—certainly on the learning and memory front, and on the interconnectedness of reason and emotion. Researchers have studied the idea that emotion and feelings—the ones we actually call gut feelings, the feeling of uneasiness or that something is wrong — help bias us in ways that allow us to do the kinds of calculations we normally call pure reason. The main point is that in situations requiring a certain amount of analysis, you need not just frontal cortex but input from the limbic structures that are cued, in turn, from the viscera, the skin, and so forth. That is very revealing about the way we actually work. It shows us that for really heavy intelligence stuff—determining what, in the general sense, is relevant and what isn't—you've got to have the emotions involved. It may mean that there is something right about the strategy that people invoke when they say it's important to be in touch with your feelings. Say you come into a room and you sense something is not right. I certainly wouldn't say that it is a completely reliable clue, but it's something worth relying on for telling you to look closer, to look harder.

Is that part of what we mean by being conscious?

Some people have the idea that before you can study consciousness, you need a nice, precise, clean definition of what consciousness is. I want to resist that. Clean, nice, precise definitions are what you get after you've done the science and you've got a nice theory. Before you have a good theory, often what you have to do is go with good examples, with phenomena where you've got quite a lot of agreement. Doctors and most lay people are pretty good at telling whether somebody is in a coma or vegetative state or conscious. We're pretty good at telling when somebody is in deep sleep and when they're awake—and that's a *big* difference in consciousness. When you're in deep sleep, you're not aware of things, even though you may do some things that are somewhat intelligent. For example, you move a lot in deep sleep and you navigate around the bed. Even if you're in a relatively novel bed and there are other people in the bed, you generally manage to navigate. And we can tell when someone is in REM sleep, which seems to be different from both of those other states. So what we'd like to know is: what happens in the brain when somebody is in deep sleep, and when awake, and in REM sleep?

Why *do* we sleep?

There's been interesting recent work on the anatomy and physiology of sleep. But why we need to sleep is really puzzling. I think sleep plays a critical role; that really important things are happening seems obvious to me, but what they are, we don't know.

But we do a lot of it, and we like it. People, by and large, like to sleep. There must be a significant reason why the brain does it. Some people think that it has to do with restoration of basic neurotransmitters, and that it's fundamentally a metabolic issue. That's quite possible. The startling thing that people discovered many years ago, however, is that your neurons are not quiet in deep sleep; they're going like blazes. There's very little difference between neural activity when you're in deep sleep and overall neural activity when you are conscious. Both deep sleep and REM sleep are quite puzzling because we incur a survival risk in both. We're terribly vulnerable. Evolution has put a good deal into this strategy—but exactly why, it's hard to say.

Do you see potential medical applications of neurophilosophy?

Insofar as I see neurophilosophy as a part of the general discipline of neuroscience and insofar as I see neuroscience as having a major medical benefit, then yes. There are still a number of serious medical problems we haven't quite gotten a grip on. One is schizophrenia, another is Alzheimer's, and another is Parkinson's. In all of those cases we have palliatives for treating people to make those diseases slightly less awful, but we are a long way from enabling a person to have a normal life and a normal death. When you realize that about 1% of the population have schizophrenia, and something like 15% over age 65 suffer from Alzheimer's, and something comparable for Parkinson's, it is really important to understand these diseases. I think the answers will come out of further understanding of basic neuroscience.

Unifying brain and mind may seem like a reasonable objective to most biologists. Why are some of your philosophy colleagues uncomfortable with this synthesis you call neurophilosophy?

Part of it is that science discovers that the reality behind the appearances is quite different from what we thought. Aristotle thought, for pretty good reasons, that an object is not going to keep moving unless you keep a force applied to it. Giving that up based on the Newtonian framework means accepting a very different picture of the nature of kinetics. And similarly, thinking that things come into being in a Darwinian fashion rather than at the moment of creation is, for some people, very counterintuitive. I think some people worry that the next step is that science is going to say, "You aren't what you thought you were." I think lots of us find that hard. It's one thing to have a counterintuitive theory of motion but I don't want a counterintuitive theory of myself; there can't be a counterintuitive theory of me because I'm the best authority for how I work. Also, I think some people think that if science is applied to humans, then the dignity of humans is at stake. I actually take a very different view, that some of the very damaging superstitions we have about humans may actually be replaced by much more caring, more humanitarian hypotheses or approaches, in just the way that it is really much more humanitarian to have a pharmacological way of treating a schizophrenic than to put that person on a dung pile in order to chase the demons out. As in other places in our universe, I think that a scientific understanding of the mind will actually promote humanitarian values rather than detract from them.

Maybe the other thing that worries some philosophers is that these questions about the mind have been their property for a long time. It's a turf thing. But I think the coming generation of graduate students and young faculty in philosophy find it obvious that scientific data on the brain are relevant to our understanding of the mind. In the meantime, the turf thing continues. A leading critic of neurophilosophy makes an argument something like this: Sure the brain is probably all there is, but you will never explain consciousness, the painfulness of pain, and so forth in terms of neurons, in terms of ions passing back and forth across membranes, etc. My response is, that's an argument from ignorance. That's an argument from, "I can't imagine." So what if you can't imagine it? That's a fact about you. That's not a fact about what we can and can't discover. I am unimpressed by that argument.

As neurophilosophy brings the brain and mind closer, how will this change how we view ourselves as a species?

Maybe in a general way, one could say that we are learning increasingly how much similarity there is between us and other animals. And that many aspects of our behavior are rooted in our evolutionary past via evolution of the brain. The brain has many of the aspects it has because we had to survive. We shouldn't find this alarming or sad; I think we should revel in it. But at the same time, there are things that make us different. Just as there are things chimpanzees can do that we can't do, there are things we can do that they can't do. It would be nice to know what the difference is. One hypothesis is that what makes our intelligence possible is allowing the nervous system to mature at different rates. In our case, immaturity in the brain is extended for quite a long time. And things the developing brain learned earlier can be useful in teaching the brain later. Perhaps these developmental delays enable us to do more complicated things. So it's not that we're rational, and other animals are not; it's a quantitative difference partly related to the developmental timing that prolongs brain development. That understanding might even have the effect of allowing people to have more regard for other species and less of the attitude that we are the greatest, so we get to squash all the others. Evolution doesn't start from scratch. It has to be gradual, to some degree. A small difference in some aspect of an organism can make the thing as a whole look like it's got vastly different properties. Our brains are so similar to chimpanzee brains, but some relatively small difference in developmental timing magnifies into very different properties. My guess is that the study of brain development and child psychology are where the really big action is going to be in neuroscience in the next few decades.

CHAPTER 36

AN INTRODUCTION TO ANIMAL STRUCTURE AND FUNCTION

KEY CONCEPTS

- The functions of animal tissues and organs are correlated with their structures
- Bioenergetics is fundamental to all animal functions
- An animal's size and shape affect its interactions with the external environment
- Homeostatic mechanisms regulate an animal's internal environment

The chapters in this unit examine the correlation between structure and function in a variety of animals and selected protists. Our main goal is to see how species of diverse evolutionary history and varying complexity solve problems common to all. For instance, how do organisms as different as an amoeba, a clam, a bumblebee, and a human obtain energy from the environment? And how do they obtain O_2 for cellular respiration while disposing of the waste gas CO_2?

Animals provide vivid examples of biology's core theme: the capacity to adjust to the environment over the long term by adaptation due to natural selection. A bumblebee like the one in the photograph here has a long, thin tongue, an adaptation enabling it to probe deep into flowers for nectar. Another adaptation for feeding, portions of the bee's hind legs are fringed with long, curved hairs, forming a food basket that collects plant pollen (the yellowish lump on this bee's hind leg).

A foraging bumblebee illustrates another major theme of this unit: the capacity of organisms to adjust to the environment over the short term by physiological responses. Many insects are inactive when it is cold, but bumblebees can forage for nectar and pollen when air temperatures are as low as 5°C. On cool mornings in the early spring when the first flowers are out, a bumblebee's muscles will contract rapidly, much like yours do when you shiver. Heat produced by the contracting muscles warms the bee to over 30°C, the temperature at which its wings can beat fast enough to fly. This unusual ability enables the bumblebee to forage when it is too cold for most other insects to compete with them for resources.

Searching for food, generating body heat, responding to external stimuli, and all other animal activities require fuel in the form of chemical energy. Being warm enough to fly in cold weather depends on a bumblebee's ability to obtain enough fuel to at least match the rate at which its body cells use chemical energy. An appreciation of bioenergetics as it applies to animals—how animals obtain, process, and use chemical energy—is key to understanding animal physiology, and will integrate our comparative study of animals throughout this unit.

This introductory chapter has three main objectives: to illustrate the hierarchy of structural order characterizing animals, to emphasize the importance of energetics in animal life, and to examine how animal body forms affect their interactions with the environment. We begin with a look at the basic body structure of animals.

The functions of animal tissues and organs are correlated with their structures

All life is characterized by hierarchical levels of organization. The cell holds a special place in the hierarchy of life because it is the lowest level of organization that can live as an organism. Protozoa, for example, have specialized organelles that perform particular jobs, enabling them to digest food, move about, sense environmental change, excrete waste products, and reproduce—all within the confines of a single cell. Protozoa represent the cellular level of organization, the simplest level possible for an organism. Multicellular organisms, including animals, have specialized cells grouped into tissues, the next higher level of structure and function. In most animals, various tissues are combined into functional units called organs, and the organs cooperate in teams called organ systems. For example, our own digestive system consists of a stomach, a small intestine and a large intestine, a gallbladder, and several other organs, each one a composite of different types of tissues. As we explore the hierarchy of animal organization, pay special attention to how the structure of a tissue or organ fits its function.

Animal Tissues

Tissues are groups of cells with a common structure and function. A tissue may be held together by a sticky extracellular matrix that coats the cells (see Chapter 7) or that weaves them together in a fabric of fibers. Indeed, the term *tissue* is from a Latin word meaning "weave."

We can classify tissues into four main categories: epithelial tissue, connective tissue, nervous tissue, and muscle tissue. These are present to some extent in all but the simplest animals; the following survey emphasizes the tissues as they appear in vertebrates.

Epithelial Tissue. Occuring in sheets of tightly packed cells, **epithelial tissue** covers the outside of the body and lines organs and cavities within the body (FIGURE 36.1). The cells of an epithelium are closely joined, with little material between them. In many epithelia, the cells are

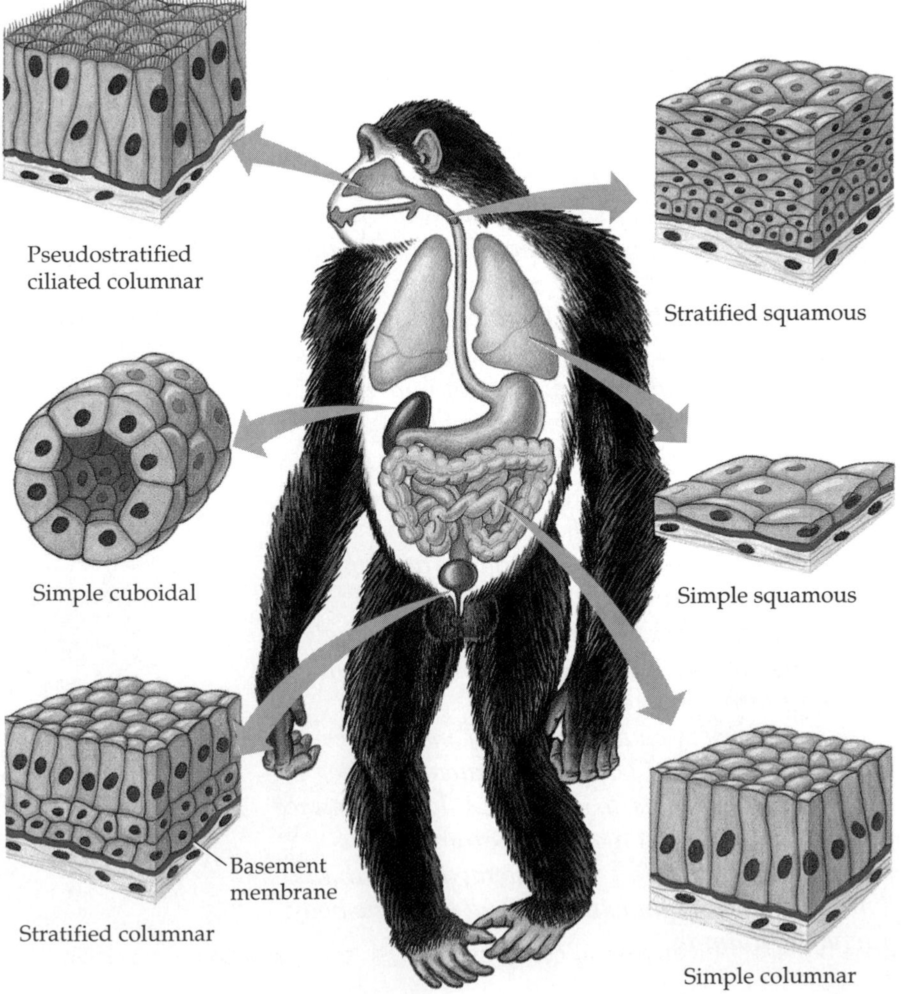

FIGURE 36.1
The structure and function of epithelial tissue. The structure of an epithelium fits its function. For instance, simple squamous epithelia, which are relatively leaky, are specialized for the exchange of materials by diffusion. These epithelia line the blood vessels and air sacs of the lungs. Stratified squamous epithelia regenerate rapidly by cell division near the basement membrane. The new cells are pushed to the free surface as replacements for cells that are continually sloughed off. This type of epithelium is commonly found on surfaces subject to abrasion, such as the outer skin and linings of the esophagus, anus, and vagina. Columnar epithelia, having cells with relatively large cytoplasmic volumes, are often located where secretion or the active absorption of substances is an important function. For example, the intestines are lined by columnar epithelia that secrete digestive juices and absorb nutrients. Cuboidal cells specialized for secretion make up the epithelia of kidney tubules and many glands, including the thyroid gland and salivary glands.

riveted together by tight junctions like those described in Chapter 7 (see FIGURE 7.31). This tight packing is consistent with the function of an epithelium as a barrier protecting against mechanical injury, invading microorganisms, and fluid loss. The free surface of the epithelium is exposed to air or fluid, whereas the cells at the base of the barrier are attached to a **basement membrane,** a dense mat of extracellular matrix. Cell biologists are finding that the basement membrane performs many different functions, such as helping organize sequential events in cellular metabolism, filtering wastes from the blood in the kidney, and providing routes of migration for cells during development.

Two criteria for classifying epithelia are the number of cell layers and the shape of the cells on the free surface. A **simple epithelium** has a single layer of cells, whereas a **stratified epithelium** has multiple tiers of cells. A pseudostratified epithelium is single-layered, but it appears stratified because the cells vary in length. The shape of the cells that are at the free surface of an epithelium may be **cuboidal** (like dice), **columnar** (like bricks on end), or **squamous** (like flat floor tiles). Combining the features of cell shape and number of layers, we get such terms as *simple cuboidal epithelium* and *stratified squamous epithelium.*

As well as protecting the organs they line, some epithelia are specialized for absorbing or secreting chemical solutions. For example, the epithelial cells that line the lumen (cavity) of the digestive and respiratory tracts are called **mucous membrane** because they secrete a slimy solution called mucus that lubricates the surface and keeps it moist. The mucous membrane lining the small intestine also releases digestive enzymes and absorbs nutrients. The free epithelial surfaces of some mucous membranes have beating cilia that move the film of mucus along the surface. For example, the ciliated epithelium of our respiratory tubes helps keep our lungs clean by trapping dust and other particles and sweeping them back up the trachea (windpipe).

Connective Tissue. Connective tissue functions mainly to bind and support other tissues (FIGURE 36.2). In

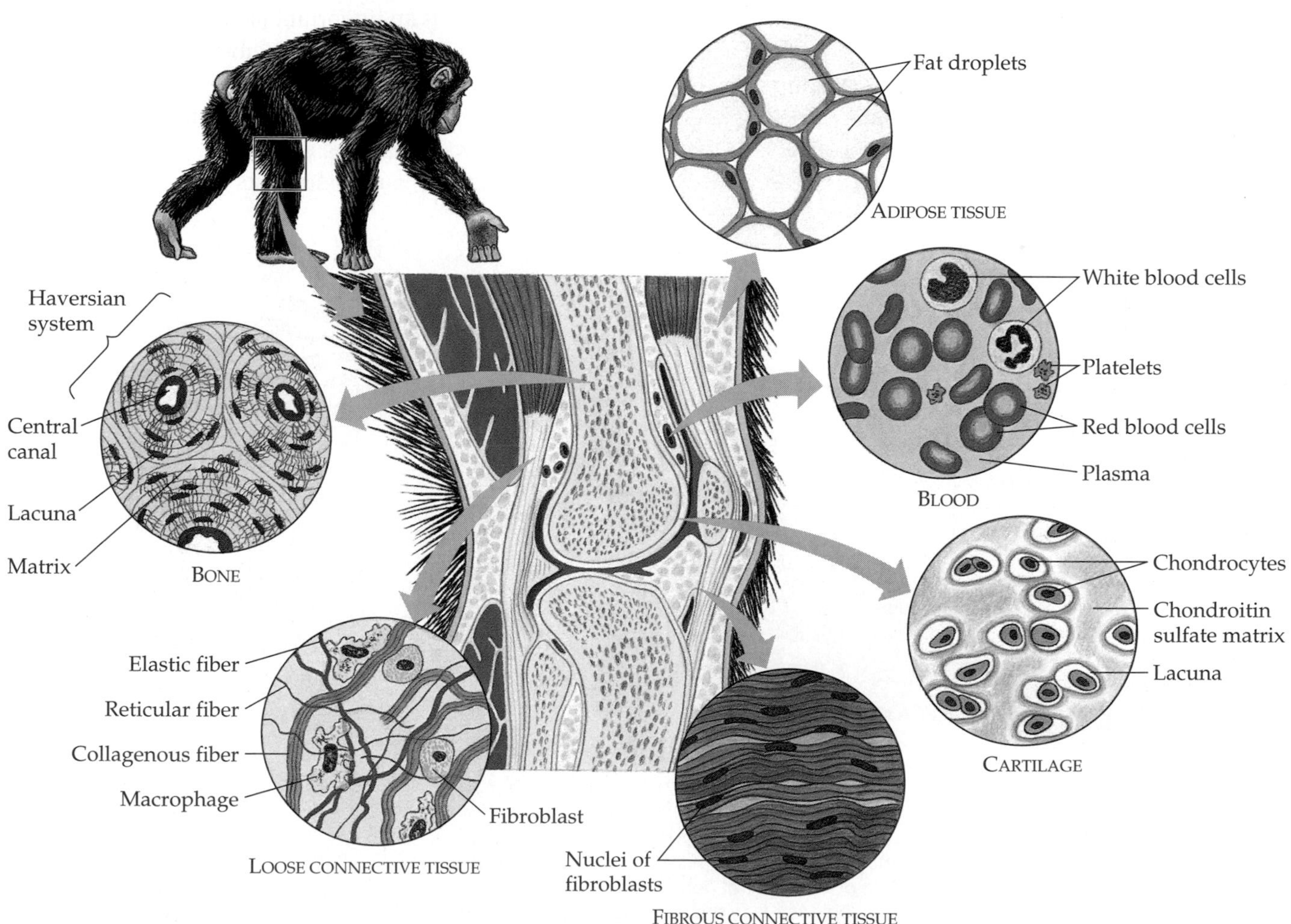

FIGURE 36.2
Some representative types of connective tissue. The area pictured is the region around the knee joint.

contrast to epithelia with their tightly packed cells, connective tissues have a sparse population of cells scattered through an extracellular matrix. The matrix generally consists of a web of fibers embedded in a homogeneous ground substance that may be liquid, jellylike, or solid. In most cases, the substances of the matrix are secreted by the cells of the connective tissue. The major types of connective tissue in vertebrates are loose connective tissue, adipose tissue, fibrous connective tissue, cartilage, bone, and blood. Each has a structure correlated with its specialized functions.

The most widespread connective tissue in the vertebrate body is **loose connective tissue.** It binds epithelia to underlying tissues and functions as packing material, holding organs in place. This type of connective tissue gets its name from the loose weave of its fibers. These fibers, which are made of protein, are of three kinds: collagenous fibers, elastic fibers, and reticular fibers. **Collagenous fibers** are made of collagen, perhaps the most abundant protein in the animal kingdom (FIGURE 36.3). Collagenous fibers have great tensile strength, which means that they do not tear easily when pulled lengthwise. If you pinch and pull some skin on the back of your hand, it is mainly collagen that keeps the flesh from tearing away from the bone. **Elastic fibers** are long threads made of a protein called elastin. As their name implies, these fibers are elastic, unlike collagenous fibers, which resist stretching. Elastic fibers endow loose connective tissue with a resilience that complements the tensile strength of collagenous fibers. When you pinch the back of your hand and then let go, the rubbery elastic fibers quickly restore your skin to its original shape. **Reticular fibers** are very thin and branched. Composed of collagen and continuous with collagenous fibers, they form a tightly woven fabric that joins connective tissue to adjacent tissues.

Among the cells scattered in the fibrous mesh of loose connective tissue, two types predominate: fibroblasts and macrophages. **Fibroblasts** secrete the protein ingredients of the extracellular fibers. **Macrophages** are amoeboid cells that roam the maze of fibers, engulfing bacteria and the debris of dead cells by phagocytosis (see Chapter 8). They are weapons in an elaborate arsenal of defense you will learn more about in Chapter 39.

Adipose tissue is a specialized form of loose connective tissue that stores fat in adipose cells distributed throughout its matrix. Adipose tissue pads and insulates the body and stores fuel molecules. Each adipose cell contains a large fat droplet that swells when fat is stored and shrinks when fat is used by the body as fuel. Although heredity is an important factor in obesity, there is also some evidence that the number of fat cells in our connective tissue is determined partly by the amount of fat we stored when we were babies. This might be discouraging news for dieting adults who were overweight at an early age, because it suggests that they have more

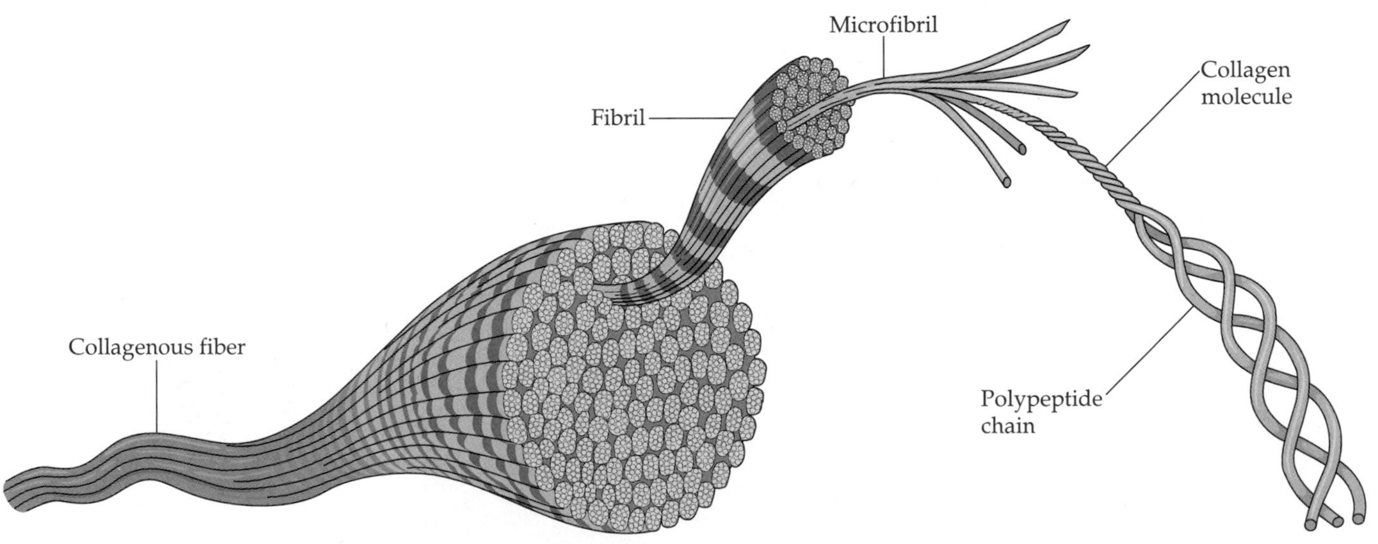

FIGURE 36.3

The structure of collagen. A collagenous fiber is a nonelastic, ropelike bundle of many fibrils, each of which in turn is a bundle of many microfibrils. A microfibril consists of helically coiled collagen molecules, each consisting of three helical polypeptide chains. The arrangement of collagen molecules makes the fibrils appear striped through an electron microscope.

fat cells than people who may have put on a few pounds later in life.

Fibrous connective tissue is dense, owing to its enrichment in collagenous fibers. The fibers are organized into parallel bundles, an arrangement that maximizes tensile strength. We find this type of connective tissue in **tendons,** which attach muscles to bones, and in **ligaments,** which join bones together at joints.

Cartilage has an abundance of collagenous fibers embedded in a rubbery matrix called chondroitin sulfate, a protein-carbohydrate complex. The chondroitin sulfate and collagen are secreted by **chondrocytes,** cells confined to scattered spaces called lacunae in the matrix. The composite of collagenous fibers and chondroitin sulfate makes cartilage a strong yet somewhat flexible support material. The skeleton of a shark is made of cartilage. Other vertebrates, including humans, have cartilaginous skeletons during the embryo stage, but most of that skeleton is replaced by bone as the embryo matures. We nevertheless retain cartilage as flexible support in certain locations, such as the nose, the ears, the rings that reinforce the windpipe, the discs that act as cushions between our vertebrae, and the caps on the ends of some bones.

The skeleton that supports the bodies of most vertebrates is made of **bone,** a mineralized connective tissue. Bone-forming cells called **osteoblasts** deposit a matrix of collagen, but they also release calcium phosphate, which hardens within the matrix into the mineral hydroxyapatite. The combination of hard mineral and flexible collagen makes bone harder than cartilage without being brittle. The microscopic structure of hard mammalian bone consists of repeating units called **Haversian systems.** Each system has concentric layers of the mineralized matrix, which are deposited around a central canal containing blood vessels and nerves that service the bone. Once osteoblasts become trapped in their own secretions, they are called osteocytes. The osteocytes are located in lacunae, spaces surrounded by the hard matrix and connected to each other by long, thin extensions. In long bones, such as the femur (shank) of your thigh, only the hard outer region is hard compact bone built from Haversian systems. The interior is a spongy bone tissue honeycombed with spaces filled with bone marrow. Blood cells are manufactured in red bone marrow located near the ends of long bones. (Bones and skeletons will be examined in more detail in Chapter 45.)

Although **blood** functions differently from other connective tissues, it does meet the criterion of having an extensive extracellular matrix. In this case, the matrix is a liquid called plasma, consisting of water, salts, and a variety of dissolved proteins. Suspended in the plasma are two classes of blood cells, erythrocytes (red blood cells) and leukocytes (white blood cells), and cell fragments called platelets. Red cells carry oxygen; white cells function in defense against viruses, bacteria, and other invaders; and platelets are involved in the clotting of blood. The composition and functions of blood will be discussed in detail in Chapters 38 and 39.

Nervous Tissue. **Nervous tissue** senses stimuli and transmits signals from one part of the animal to another. The functional unit of nervous tissue is the **neuron,** or nerve cell, which is uniquely specialized to transmit signals called nerve impulses (FIGURE 36.4). It consists of a cell body and two or more extensions called dendrites and axons, which may be as long as a meter in humans. Dendrites transmit impulses toward the cell body, and axons transmit impulses away from the cell body. We postpone a detailed discussion of the structure and function of neurons until Chapter 44.

Muscle Tissue. **Muscle tissue** is composed of long, excitable cells capable of considerable contraction.

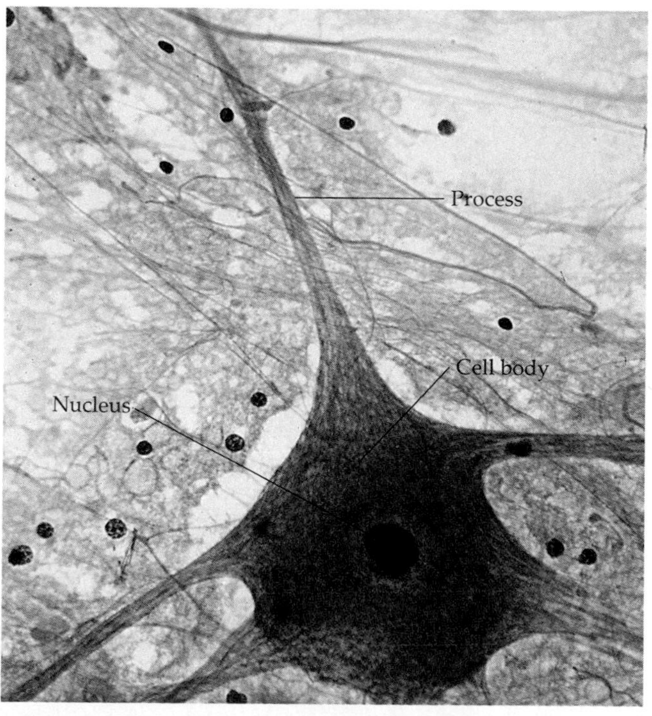

FIGURE 36.4
The basic structure of a neuron. This nerve cell from the spinal cord has a large cell body with multiple processes that transmit electrical impulses (LM).

Arranged in parallel within the cytoplasm of muscle cells are large numbers of microfilaments made of the contractile proteins actin and myosin. Consistent with a high priority on movement, muscle is the most abundant tissue in most animals, and muscle contraction accounts for much of the energy-consuming cellular work in an active animal.

In the vertebrate body, there are three types of muscle tissue: skeletal muscle, cardiac muscle, and smooth muscle (FIGURE 36.5). Attached to bones by tendons, **skeletal muscle** is generally responsible for the voluntary movements of the body. Adults have a fixed number of muscle cells; weight lifting and other methods of building muscle do not increase the number of cells but simply enlarge those already present. Skeletal muscle is also called **striated muscle** because the arrangement of overlapping filaments gives the cells a striped (striated) appearance under the microscope.

Cardiac muscle forms the contractile wall of the heart. It is striated like skeletal muscle, but cardiac cells are branched, and the ends of the cells are joined by structures called intercalated discs, which relay signals from cell to cell during a heartbeat.

Smooth muscle, so named because it lacks cross-striations, is found in the walls of the digestive tract, bladder, arteries, and other internal organs. The cells are spindle-shaped. They contract more slowly than skeletal muscles, but they can remain contracted longer.

Skeletal and smooth muscles are controlled by different kinds of nerves. Skeletal muscles are often called voluntary because an animal can generally contract them at will. Smooth muscles are often called involuntary; they are not generally subject to conscious control. You can decide to raise your hand, but you are usually unaware when smooth muscles churn your stomach or constrict your arteries. You will learn more about the control and contraction of muscles in Chapter 45.

Organs and Organ Systems

In all but the simplest animals (sponges and some cnidarians), different tissues are organized into the specialized centers of function called **organs.** In some organs, the tissues are arranged in layers. The vertebrate stomach, for example, has four major layers (FIGURE 36.6). The lumen is lined by a thick epithelium that secretes mucus and digestive juices. Outside this layer is a zone of connective tissue surrounded by a thick layer of smooth muscle. The entire stomach is encapsulated by another layer of connective tissue. A layered organization is also apparent in the skin of the body.

Many of the organs of vertebrates are suspended by sheets of connective tissue called **mesenteries** into body

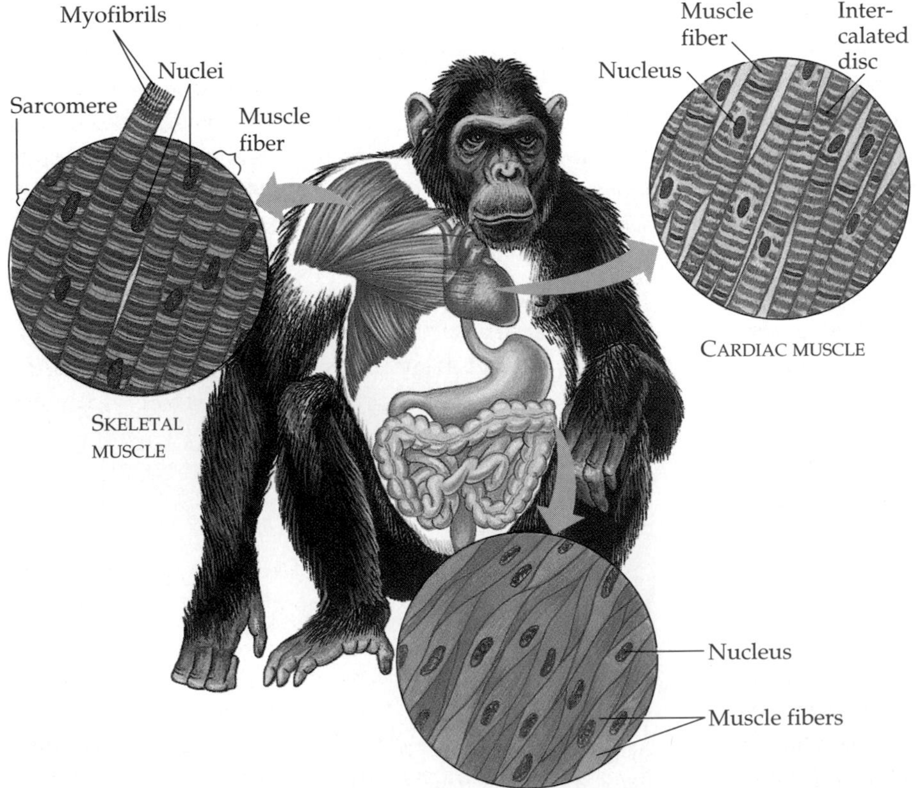

FIGURE 36.5

Types of vertebrate muscle. Skeletal muscle consists of bundles of long cells called fibers; each fiber is a bundle of strands called myofibrils. The myofibrils are a linear array of sarcomeres, the basic contractile units of the muscle. Skeletal muscle is said to be striated because the alignment of sarcomere subunits in adjacent myofibrils forms light and dark bands. Cardiac muscle, also striated, has contractile properties similar to those of skeletal muscle. Unlike skeletal muscle, however, cardiac muscle fibers branch and interconnect via intercalated discs, which help synchronize the heartbeat. Smooth muscle consists of spindle-shaped cells lacking cross-striations.

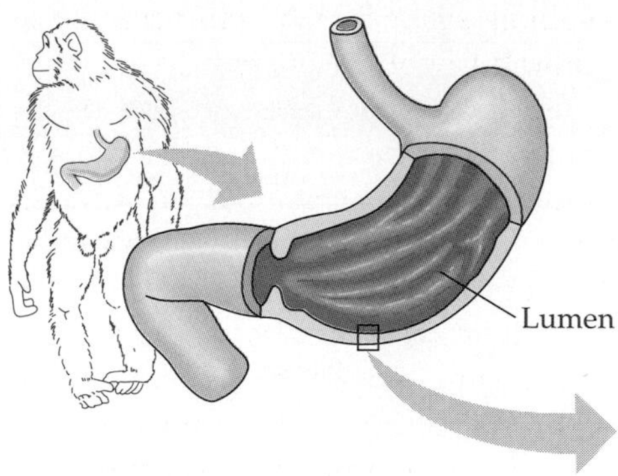

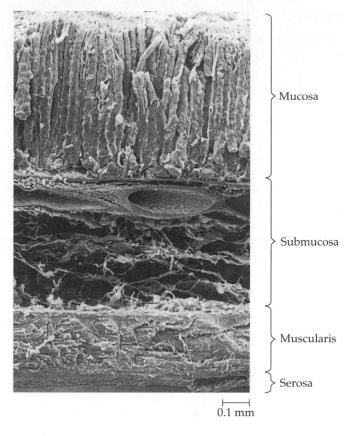

FIGURE 36.6
Tissue layers of the stomach, a digestive organ. Each organ of the digestive system has a wall with four main tissue layers: the mucosa, submucosa, muscularis, and serosa. The mucosa is an epithelial layer that lines the lumen. The submucosa is a matrix of connective tissue that contains blood vessels and nerves. The muscularis has an inner layer of circular muscles and an outer layer of longitudinal muscles. External to the muscularis is the serosa, a thin layer of connective tissue and epithelial tissue. (SEM from *Tissues and Organs: A Text-Atlas of Scanning Electron Microscopy* by Richard G. Kessel and Randy H. Kardon. W. H. Freeman and Company. Copyright © 1979.)

cavities moistened or filled with fluid. Mammals have a **thoracic cavity** housing the lungs and heart and separated from a lower **abdominal cavity** by a sheet of muscle called the diaphragm.

There is a level of organization still higher than organs. **Organ systems,** each consisting of several organs, carry out the major functions of vertebrates and members of most invertebrate phyla (TABLE 36.1). Each organ system has specific functions, but the efforts of all systems must be coordinated for the animal to survive. For instance, nutrients absorbed from the digestive tract are distributed throughout the body by the circulatory system. But the heart that pumps blood through the circulatory system depends on nutrients absorbed by the digestive tract and also on oxygen obtained from the air or water by the respiratory system. The organism, whether a protozoan or an assembly of organ systems, is a living whole greater than the sum of its parts.

Bioenergetics is fundamental to all animal functions

One of the defining features of life is the exchange of energy with the environment. As heterotrophs, animals obtain chemical energy in food, which contains organic molecules synthesized by other organisms. Food is digested by enzymatic hydrolysis, and energy-containing molecules are absorbed by body cells. The cells harvest chemical energy from some of the molecules, generating ATP by the catabolic processes of cellular respiration and fermentation (see Chapter 9). The chemical energy of ATP powers cellular work, enabling organ systems to perform functions that keep an animal alive. An animal constantly exchanges energy with its environment as cellular work generates heat, which the animal loses to its surroundings. Chemical energy remaining after the needs of staying alive are met can be used in biosynthesis; new tissues and reproductive products develop as cells construct their own macromolecules using chemical energy and carbon skeletons obtained from food molecules (FIGURE 36.7, p. 787).

Metabolic Rate

We can learn a lot about an animal's adaptations to its environment by studying bioenergetics. How much of the total energy an animal obtains from food does it need just to stay alive? What are the energy costs for an insect, a bird, or a mammal to walk, run, or swim from one place to another? Physiologists obtain answers to such questions by measuring the rates at which animals use chemical energy. The total amount of energy an animal uses in

TABLE 36.1

Organ Systems: Their Main Components and Functions in Mammals

ORGAN SYSTEM	MAIN COMPONENTS	MAIN FUNCTIONS
Digestive	Mouth, pharynx, esophagus, stomach, intestines, liver, pancreas, anus	Food processing (ingestion, digestion, absorption, elimination)
Circulatory	Heart, blood vessels, blood	Internal distribution of materials
Respiratory	Lungs, trachea, other breathing tubes	Gas exchange (uptake of oxygen; disposal of carbon dioxide)
Immune and Lymphatic	Bone marrow, lymph nodes, thymus, spleen, lymph vessels, white blood cells	Body defense (fighting infections and cancer)
Excretory	Kidneys, ureters, urinary bladder, urethra	Disposal of metabolic wastes; regulation of osmotic balance of blood
Endocrine	Pituitary, thyroid, pancreas, other hormone-secreting glands	Coordination of body activities (e.g., digestion, metabolism)
Reproductive	Ovaries, testes, and associated organs	Reproduction
Nervous	Brain, spinal cord, nerves, sensory organs	Coordination of body activities; detection of stimuli and formulation of responses to them
Integumentary	Skin and its derivatives (e.g., hair, claws, skin glands)	Protection against mechanical injury, infection, drying out
Skeletal	Skeleton (bones, ligaments, cartilage)	Body support, protection of internal organs
Muscular	Skeletal muscles	Movement, locomotion

a unit of time is called its **metabolic rate.** Energy is measured in **calories (cal)** or **kilocalories (kcal).** (A kilocalorie is 1000 calories. The term *Calorie,* with a capital C, as used by many nutritionists, is actually a kilocalorie.)

Metabolic rate can be determined in several ways. Because heat results from the use of chemical energy, metabolic rate can be measured by monitoring an animal's heat loss in a unit of time. An animal is placed in a calorimeter, a closed, insulated chamber equipped with a device that records heat production in calories or kilocalories. Calorimeters are used most often with birds and small mammals, such as mice, which have high metabolic rates. They are cumbersome with large animals and less precise with small animals having low metabolic rates. Another way to measure metabolic rate is to determine the amount of oxygen consumed by an animal's cellular respiration (see the Methods Box, p. 788).

Every animal has a range of metabolic rates. Minimal rates support the basic functions that maintain life, such as breathing and the beating of the heart. Maximal metabolic rates occur during peak activity, such as all-out running or high-speed swimming. Between these extremes, many factors can influence an animal's metabolic rate, including its age, sex, size, body temperature, the temperature of its surroundings, the quality and quantity of its food, its activity level, amount of available oxygen, hormonal balance, and the time of day. Birds, humans, and many insects are mostly active and have their high-

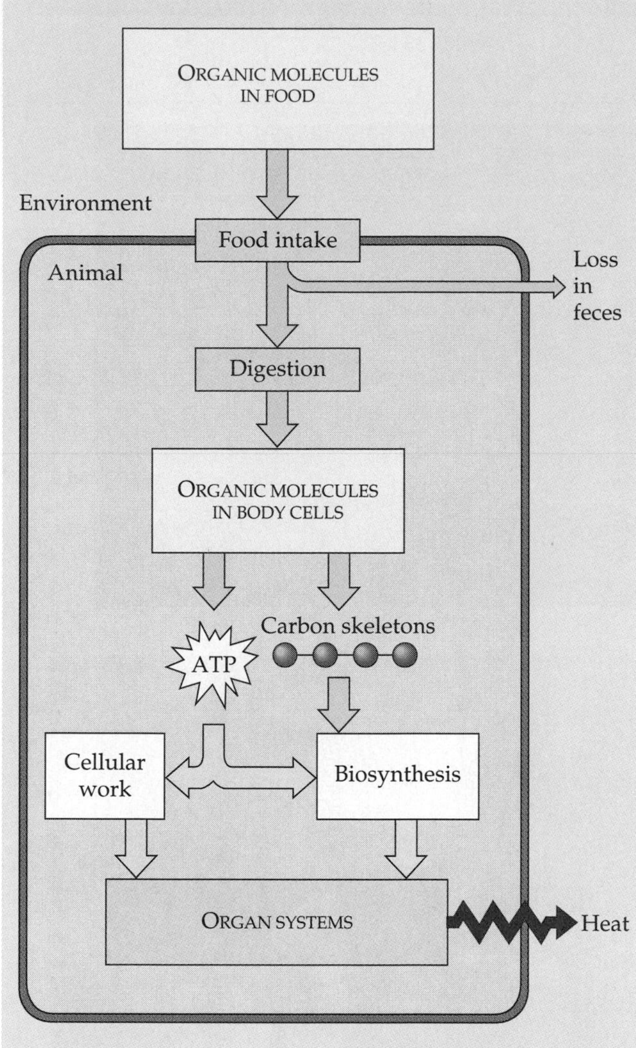

FIGURE 36.7
Bioenergetics of an animal: an overview. Animals derive chemical energy from the environment in the organic molecules in food. Hydrolytic enzymes break digestible food down into small molecules that body cells can absorb. ATP that cells produce by oxidizing the small molecules fuels the cellular work underlying the life-sustaining activities of organ systems (see TABLE 36.1). Body cells also use ATP and carbon skeletons from digested food molecules for biosynthesis—energy-storing processes that construct macromolecules, forming new tissues for growth and repair of damaged ones and developing the components necessary for reproduction. Biosynthesis thereby keeps organ systems functioning. A portion of the energy in food passes back to the environment in the undigestible material in feces. In addition, all cellular work, including that performed in biosynthesis and by the digestive system in processing food, involves a loss of energy to the environment as heat.

est metabolic rates during daylight hours. By contrast, bats, mice, and many other mammals generally are active and have their highest metabolic rates at night or during the hours of dawn and dusk.

Birds and mammals are mainly endothermic, meaning their bodies are warmed by heat generated by metabolism, and their body temperature must be maintained at a certain level to sustain life. By contrast, most fishes, reptiles, amphibians, and invertebrates are ectothermic, meaning they absorb most of their body heat from the external environment. The number of kilocalories an endotherm requires to sustain minimal life functions is generally higher than that of an ectotherm, because of the energy cost of heating (and cooling) an endothermic body. (Thermoregulation is discussed in detail in Chapter 40.)

The metabolic rate of an endotherm at rest, on an empty stomach (since digestion requires cellular work), and experiencing no stress is called the **basal metabolic rate (BMR).** The BMR for humans—the number of kilocalories we "burn" lying motionless—averages about 1600 to 1800 kcal per day for adult males and about 1300 to 1500 kcal per day for adult females. These BMRs are about equivalent to the daily energy consumption of a 100-watt light bulb.

Ectotherms are energetically quite different from endotherms. Their body temperature changes with the temperature of their surroundings, and so does their metabolic rate. Unlike BMRs, which can be determined within a range of environmental temperatures, the minimal metabolic rate of an ectotherm must be determined at a specific temperature. The metabolic rate of a resting, fasting, nonstressed ectotherm is called its **standard metabolic rate (SMR).**

Activity generally has an enormous effect on metabolic rate, and any activity, even a person working quietly at a desk or an insect extending its wings, consumes energy beyond the BMR or SMR. Metabolic rates measured when animals are active give a better idea of the energy costs of everyday life. Metabolic rates are highest during intense physical exercise, with maximal rates about five to ten times greater than the BMR or SMR.

Body Size and Metabolic Rate

One of animal biology's most intriguing and as yet unanswered questions has to do with the relationship between body size and metabolic rate. By monitoring the metabolic rates of hundreds of species of birds and mammals, physiologists have determined that the amount of energy it takes to maintain each gram of body weight is inversely related to body size. Each gram of a mouse, for instance,

METHODS: MEASURING METABOLIC RATE

An animal's metabolic rate is the number of kilocalories of energy consumed per unit of time. For an animal that is respiring aerobically, a convenient way to measure metabolic rate is to determine the amount of oxygen the animal consumes in a unit of time. The rate of oxygen consumption can be monitored by enclosing the animal in a chamber called a respirometer.

The top photograph shows a ghost crab in a respirometer. Temperature is held constant in the chamber, with air of known O_2 concentration flowing through. The crab's metabolic rate is calculated from the amount of O_2 entering and leaving the respirometer. For every liter of O_2 consumed by the crab, respiration liberates about 4.83 kcal of energy from food molecules. If, for example, the crab used 0.05 L of oxygen per hour, then its metabolic rate is 0.2415 kcal/h (0.05 L/h $\times$ 4.83 kcal/L).

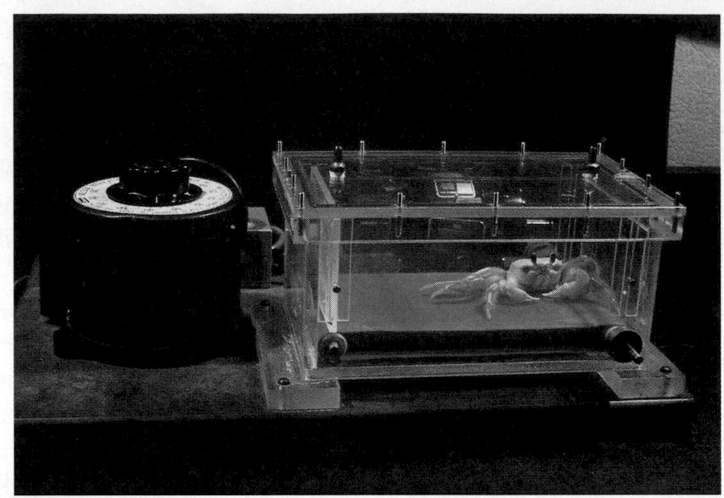

This ghost crab is on a treadmill, running at a constant speed as measurements are made. This setup provides estimates of an animal's metabolic rate during activity. Similarly, in the bottom photograph, the metabolic rate of a man fitted with a plastic breathing mask is being monitored while he works out on a stationary bike.

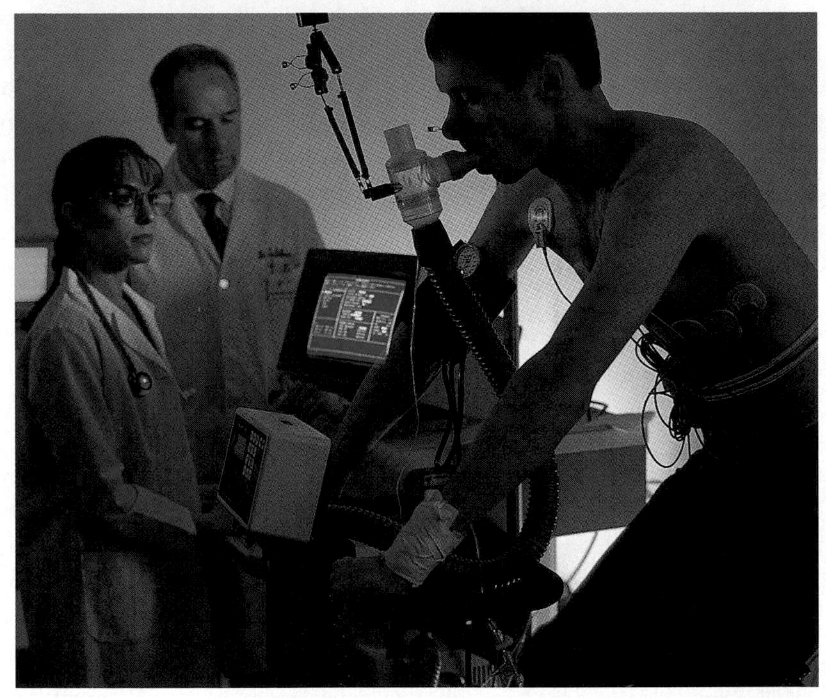

consumes about ten times more calories than a gram of an elephant (even though the whole elephant uses far more calories than the whole mouse). The higher metabolic rate of a smaller animal's body tissues demands a proportionately greater rate of delivery of oxygen to the tissues. And correlated with its higher metabolic rate, the smaller animal also has a higher breathing rate, blood volume, and heart rate (pulse).

What causes the inverse relationship between metabolic rate and size? Perhaps you would hypothesize that it is a matter of geometry; the smaller the animal, the greater its surface area relative to its volume. The greater the surface-to-volume ratio, the greater the loss or absorption of heat from the surroundings and the greater the energy cost of maintaining a stable body temperature. This may explain the relationship between body size and metabolic rate for endotherms. However, physiologists have found that the inverse relationship between body size and metabolic rate applies to ectotherms as well as endotherms. The search for more general causes underlying the inverse relationship between body size and metabolic rate continues.

An animal's size and shape affect its interactions with the external environment

An animal's size and shape, features that biologists often call body plans or designs, are fundamental aspects of form and function that significantly affect the way an animal interacts with its environment. Use of the terms *plan* and *design* in no way implies that animal forms are products of a conscious invention. The body plan or design of an animal results from a pattern of development programmed by the genome, itself the product of millions of years of evolution due to natural selection.

Body Size, Proportions, and Posture

An engineer designing a bridge or tall building must take into account the effects of changes in size, or scale. An increase in size from a small-scale model to the real thing has a significant effect on building design. Physical laws dictate that the strength of a building support depends on its cross-sectional area, which increases as the square of its diameter. In sharp contrast, the strain on the supports depends on the building's weight, which increases as the cube of its height or other linear dimension. These principles apply to animals as well as buildings, and just as an engineer designs building supports to accommodate size-weight relationships, so must an animal's body design account for the greater demand for support that comes with increasing size.

A large animal, such as an elephant, has very different body proportions than a small animal, such as a mouse. Imagine a mouse, with its very slender legs, scaled up to elephant size. If the imaginary animal kept its mouselike body proportions, its legs would probably collapse under its weight. Applying engineering principles, we might predict that the real size of an elephant's leg bones is directly proportional to the strain imposed by its body weight. On the contrary, however, researchers have recently found that there is no clear relationship between leg bone size and body weight, or between body weight and the strain imposed on the legs. At least in mammals and birds, the most important design feature in supporting body weight seems to be posture—specifically, the position of the legs relative to the main body. For example, the legs of an elephant and other large and medium-sized mammals are held in a more upright position than those of small mammals (FIGURE 36.8). Large mammals also run with their legs more nearly extended, thereby reducing strain, compared to small mammals, which bend their legs when running and tend to crouch when standing. The importance of posture in bearing the load of the body emphasizes once again the pivotal role of bioenergetics in animal physiology: posture is partly a function of the work (contraction) of muscle cells, powered by chemical energy.

(a)

(b)

FIGURE 36.8

The relationship between body posture and size. (**a**) A coyote stands and runs more upright, with its legs straighter than one of its favorite prey species, (**b**) the deer mouse. The coyote's erect posture, like that of most larger mammals, supports more body weight than the crouched posture of the mouse. As a consequence, the coyote is less agile; it cannot turn as abruptly or stop and start as quickly as a mouse, which often escapes being eaten because of its more maneuverable body design.

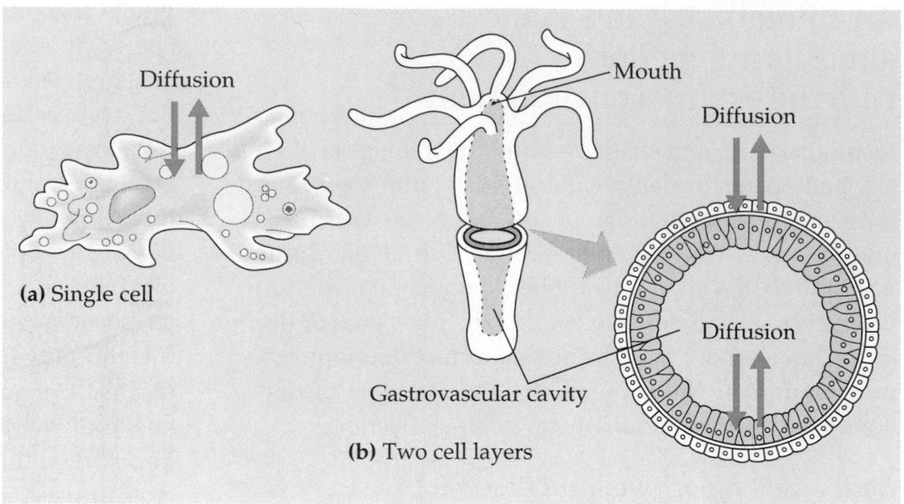

FIGURE 36.9
Contact with the environment. (**a**) In a unicellular organism, such as this amoeba, the entire surface area contacts the environment. Because of its small size, the cell has a large surface area relative to its volume through which to exchange materials with the external world. (**b**) Each cell in the bilayered *Hydra* directly contacts the environment and exchanges materials with it.

Body Plans and Exchange with the Environment

An animal's size and shape have a direct effect on how the animal exchanges energy and materials with its surroundings. As a requirement to maintain the fluid integrity of the plasma membranes of its cells, an animal body must be arranged so all of its living cells are bathed in an aqueous medium. Exchange with the environment occurs as dissolved substances diffuse across the plasma membranes between the cells and their aqueous surroundings. Because it is so small, a protozoan living in water has a sufficient surface area of plasma membrane to service its entire volume of cytoplasm (FIGURE 36.9a). Geometry imposes limits on the size that is practical for a single cell. A large cell has less surface area relative to its volume than a smaller cell of the same shape. This is one reason nearly all cells are microscopic. A multicellular animal is composed of microscopic cells, each with its own plasma membrane that functions as a loading and unloading platform for a modest volume of cytoplasm. But this only works if all the cells of the animal have access to a suitable aqueous environment. The freshwater cnidarian *Hydra,* built on the sac plan, has a body wall only two cell layers thick (FIGURE 36.9b). Because its body cavity opens to the exterior, both outer and inner layers of cells are bathed in water. A flat body shape is another way to maximize exposure to the surrounding medium. For instance, tapeworms may be several meters long, but they are very thin, and thus most cells are bathed in the intestinal fluid of the worm's vertebrate host.

Two-layered sacs and flat shapes are designs that put a large surface area in contact with the environment, but these simple forms do not allow much complexity in internal organization. Most animals are bulkier, with outer surfaces that are relatively small compared with the animal's volume. As an extreme comparison, the surface-to-volume ratio of a whale is millions of times smaller than that of a protozoan, yet every cell in the whale must be bathed in fluid and have access to oxygen, nutrients, and other resources.

Most complex animals have internal surfaces specialized for exchange with the environment (FIGURE 36.10). Extensive folding or branching gives these moist internal membranes expansive surface areas. Materials are shuttled between all these exchange surfaces by the circulatory system.

Although the logistical problems of exchange with the environment are much more complicated for an animal with a compact form than for one with all its cells exposed directly to its surroundings, a compact form has some distinct benefits. Because the animal's external surface need not be bathed in water, it is possible for the animal to live on land. Also, because the immediate environment for the cells is the internal body fluid, the animal can control the quality of the solution that bathes its cells.

Homeostatic mechanisms regulate an animal's internal environment

Over a century ago, French physiologist Claude Bernard made the distinction between the external environment surrounding an animal and the internal environment in which the cells of the animal actually live. The internal environment of vertebrates is called the **interstitial fluid.** This fluid, which fills the spaces between our cells, exchanges nutrients and wastes with blood contained in microscopic vessels called capillaries. Bernard also recognized the power of many animals to maintain rela-

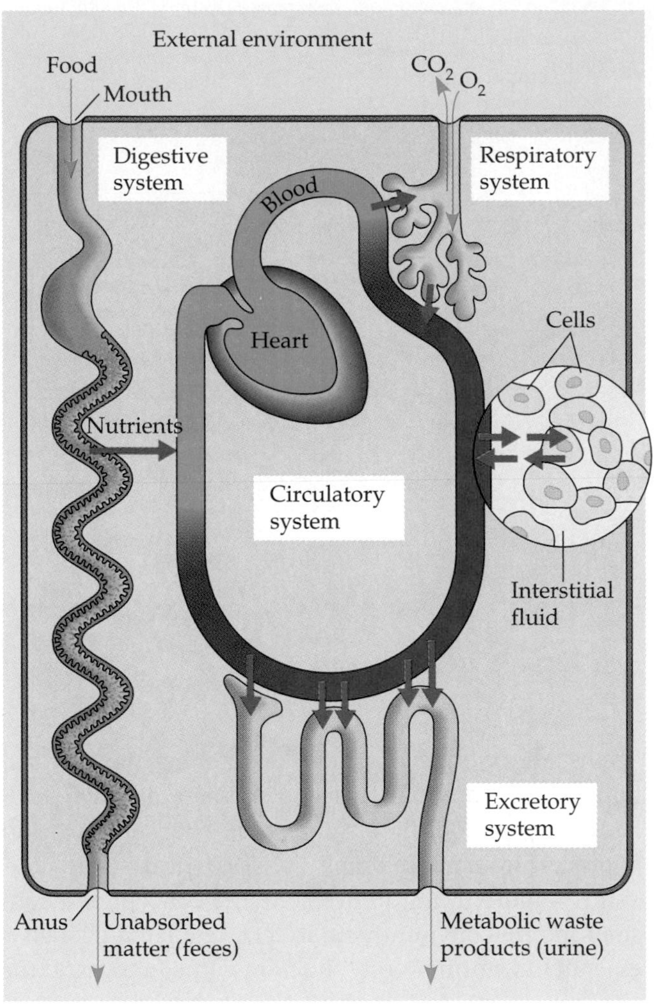

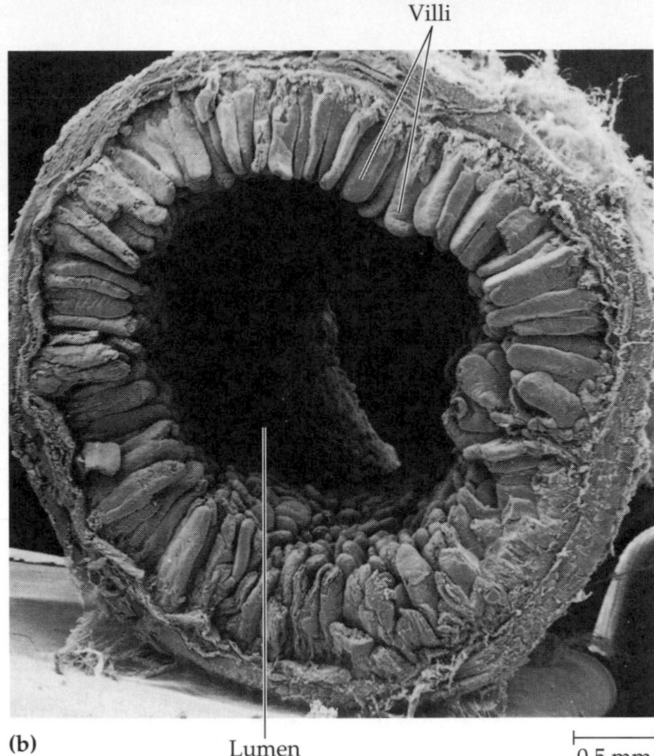

FIGURE 36.10
Internal surfaces of complex animals. (**a**) Most animals have specialized surfaces for exchanging materials with the environment. Usually internal, but connected to the environment via openings on the body surface, these exchange surfaces are branched or folded, and therefore large in area. A circulatory system transports substances among the various internal surfaces and other parts of the body. (**b**) The lining of the mammalian small intestine is an example of a large surface specialized for absorption. The intestinal lining has numerous projections called villi, giving the organ a very large surface area that absorbs nutrients. (SEM from *Tissues and Organs: A Text-Atlas of Scanning Electron Microscopy* by Richard G. Kessel and Randy H. Kardon. W. H. Freeman and Company. Copyright © 1979.)

tively constant conditions in their internal environment, even when the external environment changes. A pond-dwelling *Hydra* is powerless to affect the temperature of the fluid that soaks its cells, but the human body can maintain its "internal pond" at a more-or-less constant temperature of about 37°C. Our bodies also can control the pH of our blood and interstitial fluid to within a tenth of a pH unit of 7.4, and regulate the amount of sugar in our blood so that it does not fluctuate for long from a concentration of 0.1%. There are times, of course, during the development of an animal when major changes in the internal environment are programmed to occur. For example, the balance of hormones in human blood is altered radically during puberty. Still, the stability of the internal environment is remarkable.

Today, Bernard's "constant internal milieu" is incorporated into the concept of **homeostasis,** which means "steady state." One of the main objectives of modern physiology is to learn how animals maintain homeostasis. Any homeostatic control system has three functional components: a receptor, a control center, and an effector. The *receptor* detects a change in some variable of the animal's internal environment, such as change in body temperature. The *control center* processes information it receives from the receptor and directs an appropriate response by the *effector.* As a nonliving example of how these components interact to maintain homeostasis, consider how the temperature of a room is controlled. In this case, the control center, called a thermostat, also contains the receptor (a thermometer). When room temperature falls below a **set point,** say 20°C, the thermostat switches on the heater (the effector). When the thermometer detects a temperature above the set point, the thermostat switches the heater off. This type of control circuit is called **negative feedback,** because a change in the variable being monitored triggers a response that

FIGURE 36.11

Negative feedback as a mechanism of homeostasis. The thermostatic control of room temperature illustrates the principles of homeostatic control in an animal. A receptor (thermometer) detects a change in a variable (room temperature). A control center processes the change and directs an effector (heater) to make an appropriate change. The mechanism is called negative feedback because a change in a variable triggers an effector to counteract the change.

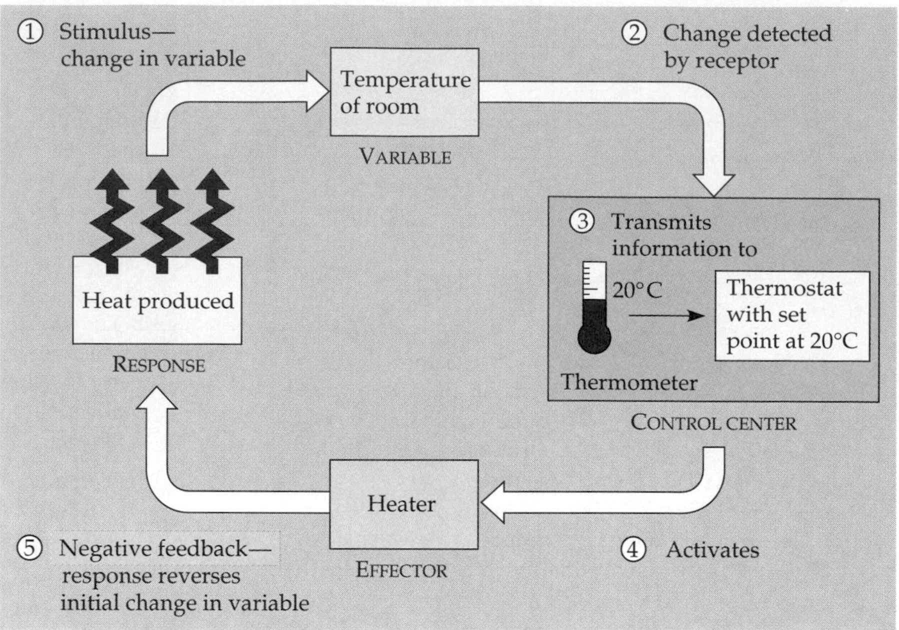

counteracts the initial fluctuation (FIGURE 36.11). Owing to a lag time between reception and response, the variable drifts slightly above and below the set point, but the fluctuations are moderate. Negative-feedback mechanisms prevent small changes from becoming too large. Most known homeostatic mechanisms in animals operate on this principle of negative feedback.

Our own body temperature is kept close to a set point of 37°C by the cooperation of several negative-feedback circuits that regulate energy exchange with the environment. One of these involves sweating as a means to dispose of metabolic heat and cool the body. A part of the brain called the hypothalamus monitors the temperature of the blood. If this thermostat detects a rise in body temperature above the set point, it sends nerve impulses directing sweat glands to increase their production of sweat, thereby lowering body temperature by evaporative cooling (see Chapter 3). When body temperature drops below the set point, the thermostat in the hypothalamus stops sending the signals to the glands, and the body retains more of the heat produced by metabolism. We will see several examples of negative feedback in the chapters that follow.

In contrast to negative feedback, **positive feedback** involves a change in some variable that triggers mechanisms that amplify rather than reverse the change. During childbirth, for instance, the pressure of the baby's head against sensors near the opening of the uterus stimulates uterine contractions, which cause greater pressure against the uterine opening, heightening the contractions, which causes still greater pressure. Positive feedback brings childbirth to completion, a very different sort of process from maintaining a physiological steady state.

It is important not to overstate the concept of a constant internal environment. In fact, regulated change is essential to normal body functions. In some cases, the changes are cyclical, such as the changes in hormone levels responsible for the menstrual cycle in women (see Chapter 42). In other cases, a regulated change is a reaction to some contingency. For example, the human body reacts to certain infections by raising the set point for temperature to a slightly higher level, and the resulting fever helps fight the infection. Over the short term, homeostatic mechanisms keep body temperature close to a set point, whatever it is at that particular time. But over the longer term, homeostasis allows regulated change in the body's internal environment, illustrating adaptive response as one of the themes of organismal biology.

*   *   *

Having examined some general principles of animal body form and function, we are now ready to compare how diverse animals perform such activities as digestion, circulation, gas exchange, excretion of wastes, reproduction, and coordination.

REVIEW OF KEY CONCEPTS (with page numbers and key figures)

- The functions of animal tissues and organs are correlated with their structures (pp. 780–785, TABLE 36.1)
  - The organization of an animal's body emerges from the grouping of specialized cells into tissues, tissues into organs, and organs into organ systems.
  - Epithelial tissue covers the outside of the body and lines internal organs and cavities. Consistent with their barrier function, epithelial cells are tightly packed. They rest on a dense mat of extracellular matrix, the basement membrane, which may function in metabolism, development, and excretion. Epithelia are described according to the number of cell layers (simple or stratified) and the shape of the surface cells (cuboidal, columnar, or squamous).
  - Some epithelia are specialized for absorption and secretion. The mucus secreted by the mucous membranes lining the digestive and respiratory tracts lubricates and moistens these surfaces.
  - Connective tissues bind and support other tissues. Unlike epithelia, the cells of connective tissue are scattered through a nonliving extracellular matrix of fibers and ground substance.
  - Loose connective tissue, the body's binding and packing material, consists of fibroblasts and macrophages interspersed among collagenous, elastic, and reticular fibers. Adipose (fat) tissue is a specialized type of loose connective tissue. Fibrous connective tissue, found in tendons and ligaments, is made of dense, parallel bundles of collagenous fibers. Cartilage, bone, and blood are also connective tissues. Cartilage is a strong yet flexible support material consisting of collagenous fibers and a rubbery ground substance secreted by chondrocytes. The hard substance of bone is secreted by cells called osteoblasts that become embedded in repeating units called Haversian systems.
  - Neurons, the functional units of nervous tissue, are composed of a cell body with extending dendrites and axons that transmit electrical signals called impulses.
  - Muscle tissue is composed of long, excitable cells containing parallel microfilaments of contractile proteins. The muscle tissue of vertebrates consists of skeletal, cardiac, and smooth muscles, which differ in shape, striation, and nervous control.
  - In most animals, tissues are organized into organs, functional units with complex structures. Many organs of vertebrates are suspended by mesenteries inside fluid-filled body cavities. Each organ system, or group of organs, has specific functions, and the activities of all systems are coordinated in the functioning of the whole animal.
- Bioenergetics is fundamental to all animal functions (pp. 785–788, FIGURE 36.7)
  - Animals are heterotrophs that obtain chemical energy by eating and digesting food produced by other organisms. The functions of an animal's cells, tissues, organs, and organ systems depend on cellular work powered by chemical energy in ATP.
  - An animal's metabolic rate is the total amount of energy used in a unit of time. Measured by the amount of heat an animal gives off or by how much oxygen it consumes, metabolic rates range from minimal levels for supporting life to maximal levels for performing peak activities. Minimal metabolic rates for birds and mammals, which maintain a fairly constant body temperature using metabolic heat, are generally higher than those of most fishes, reptiles, amphibians, and invertebrates, whose body temperature changes with that of their surroundings. An animal's metabolic rate per gram is inversely related to its body size, a relationship that is not yet fully understood.
- An animal's size and shape affect its interactions with the external environment (pp. 789–790, FIGURE 36.10)
  - Natural laws govern the relationship between an animal's body weight and the strength of its support structures. Posture often reduces the strain on support structures produced by a larger, heavier body.
  - Each cell of a multicellular animal must have access to an aqueous environment. Simple two-layered sacs and flat shapes maximize exposure to the surrounding medium. More complex, compact body plans have highly folded, moist internal surfaces specialized for exchanging materials with the environment. A circulatory system carries these materials throughout an animal's body.
- Homeostatic mechanisms regulate an animal's internal environment (pp. 790–792, FIGURE 36.11)
  - The internal environment surrounding the cells making up an animal's body is usually very different from the external environment surrounding the entire animal. In addition, the internal environment is carefully controlled and regulated by the process of homeostasis.
  - Homeostatic mechanisms usually involve negative feedback. These mechanisms also enable regulated change, as they react to occasional shifts in the body's set points for variables, such as temperature.

SELF-QUIZ

1. We would describe the layered, flat, scaly epithelium of human skin as
   a. simple squamous
   b. stratified squamous
   c. stratified columnar
   d. pseudostratified cuboidal
   e. ciliated columnar

2. Which of the following structures or substances is *incorrectly* paired with a tissue?
   a. Haversian system—bone
   b. platelets—blood
   c. fibroblasts—skeletal muscle
   d. chondroitin sulfate—cartilage
   e. basement membrane—epithelium

3. Which structures contribute most to the tensile strength of loose connective tissues?
   a. elastic fibers
   b. myofibrils
   c. collagenous fibers
   d. chondroitin sulfate matrix
   e. reticular fibers

4. The involuntary muscles that cause the wavelike contractions pushing food along our intestine are
   a. striated muscles
   b. cardiac muscles
   c. skeletal muscles
   d. smooth muscles
   e. intercalated muscles

5. Which of the following is *not* considered to be a tissue?
   a. cartilage
   b. mucous membrane lining the stomach
   c. blood
   d. brain
   e. cardiac muscle

6. The membranes that suspend vertebrate organs in the body cavity are called
   a. smooth muscle
   b. loose connective tissue
   c. coelomic membranes
   d. ligaments
   e. mesenteries

7. Compared to a smaller cell, a larger cell of the same shape has
   a. less surface area
   b. less surface area per unit of volume
   c. the same surface-to-volume ratio
   d. a lesser average distance between its mitochondria and the external source of oxygen
   e. a smaller cytoplasm-to-nucleus ratio

8. Which of the following vertebrate organ systems does *not* open directly to the external environment?
   a. digestive system
   b. circulatory system
   c. excretory system
   d. respiratory system
   e. reproductive system

9. Most of our cells are surrounded by
   a. blood
   b. a fluid equivalent to seawater in salt composition
   c. interstitial fluid
   d. pure water
   e. air

10. Which of the following physiological responses is an example of *positive* feedback?
   a. An increase in the concentration of glucose in the blood stimulates the pancreas to secrete insulin, a hormone that lowers blood glucose concentration.
   b. A high concentration of carbon dioxide in the blood causes deeper, more rapid breathing, which expels carbon dioxide.
   c. Stimulation of a nerve cell causes sodium ions to leak into the cell, and the sodium influx triggers the inward leaking of even more sodium.
   d. The body's production of red blood cells, which transport oxygen from the lungs to other organs, is stimulated by a low concentration of oxygen.
   e. The pituitary gland secretes a hormone called TSH, which stimulates the thyroid gland to secrete another hormone called thyroxine; a high concentration of thyroxine suppresses the pituitary's secretion of TSH.

CHALLENGE QUESTIONS

1. Red blood cells pick up oxygen as they travel through the capillaries (microscopic blood vessels) of the lungs, and then deposit the oxygen as they pass through capillaries in other organs. Considering this function, suggest an advantage for our blood being populated by enormous numbers of very small red blood cells rather than fewer large cells. (Assume that the *total* volume of red blood cells is the same in both cases.)
2. Choose three vertebrate tissues, and describe how their structures fit their functions.
3. Suggest your own hypothesis to explain the reverse relationship between body size and metabolic rate per gram of tissue. How could you test your hypothesis?
4. Identify each of the tissues shown on the next page (LMs). Be as specific as possible.

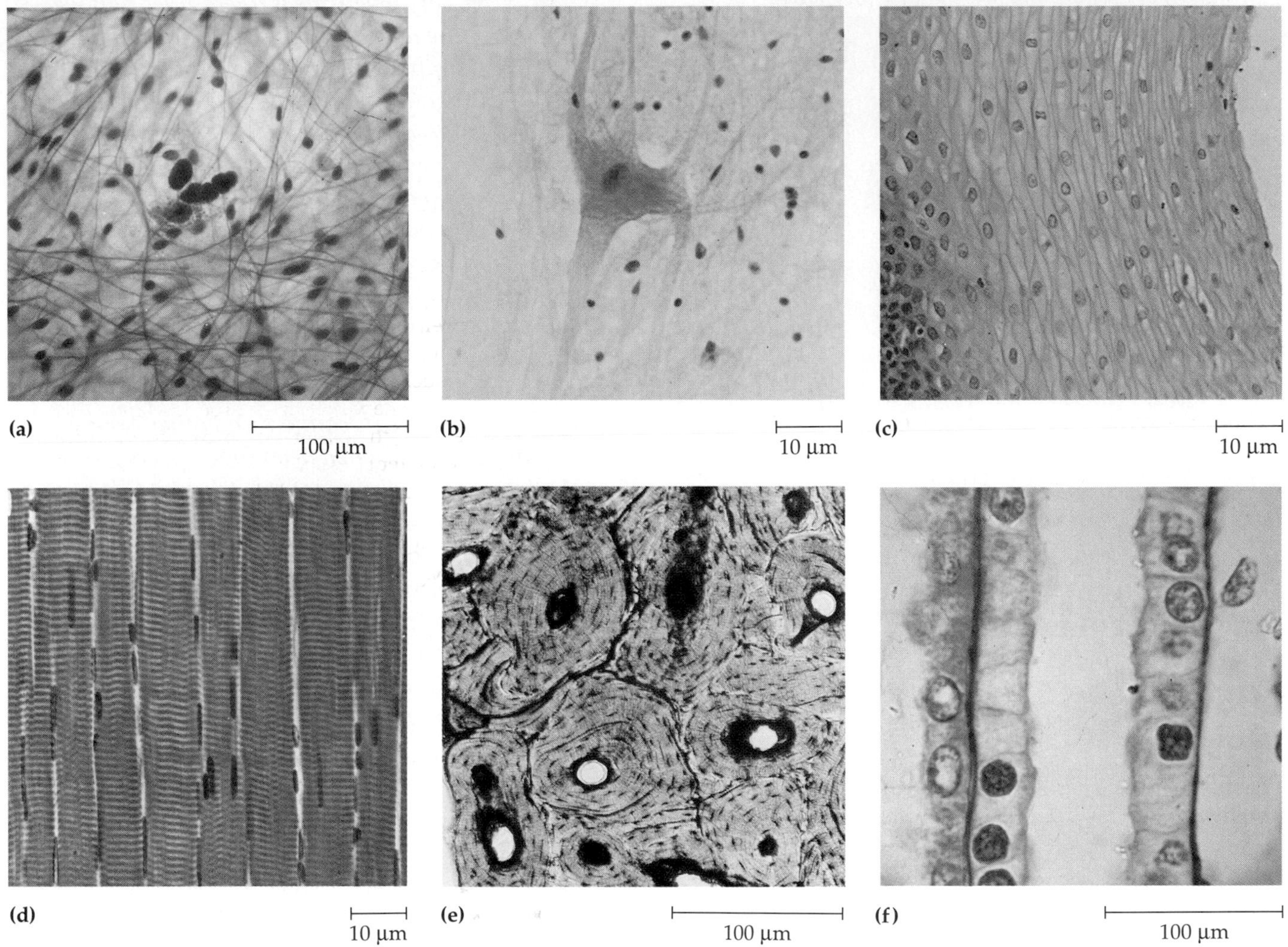

SCIENCE, TECHNOLOGY, AND SOCIETY

Medical researchers are investigating the possibilities of artificial substitutes for various human tissues. Examples are a liquid that could serve as "artificial blood" and a fabric that could temporarily serve as artificial skin for victims of serious burns. In what other situations might artificial blood or skin be useful? What characteristics would these substitutes need in order to function effectively in the body? Why do real tissues work better? Why not use the real things if they work better? Can you think of other artificial tissues that might be useful? Can you anticipate problems in developing and applying them?

FURTHER READING

Amato, I. "Heeding the Call of the Wild." *Science,* August 30, 1991. Materials scientists study the structure and function of skeletons and other animal products.

Diamond, J. "Building to Code." *Discover,* May 1993. An exploration of the safety factors built into animal bodies by natural selection.

Houston, C. S. "Mountain Sickness" *Scientific American,* October 1992. High altitude interferes with homeostasis.

Marieb, E. *Human Anatomy and Physiology,* 3rd ed. Redwood City, CA: Benjamin/Cummings, 1995. A basic text, beautifully illustrated.

McMasters, J. H. "The Flight of the Bumblebee and Related Myths of Entomological Engineering." *American Scientist,* March/April 1989. A thoughtful essay on the relationship between engineering and natural designs.

Schmidt-Nielsen, K. *Animal Physiology: Adaptation and Environment,* 4th ed. New York: Cambridge University Press, 1990. A basic text emphasizing form and function, adaptations, and homeostasis.

Schmidt-Nielsen, K. "How Are Control Systems Controlled?" *American Scientist,* January/February 1994. An exploration of unanswered questions about how organ systems and behavioral regulatory mechanisms are controlled.

Sochurek, H., and P. Miller. "Medicine's New Vision." *National Geographic,* January, 1987. Remarkable methods for photographing the human interior.

Stone, G. N. "Hot-Blooded Bees." *Natural History,* July 1993. Temperature regulation by insects as an adaptation to cool northern climates.

CHAPTER 37

ANIMAL NUTRITION

KEY CONCEPTS

- Diets and feeding mechanisms vary extensively among animals
- Ingestion, digestion, absorption, and elimination are the four main stages of food processing
- Digestion occurs in food vacuoles, gastrovascular cavities, and alimentary canals
- A tour of the mammalian digestive system
- Vertebrate digestive systems exhibit many evolutionary adaptations associated with diet
- An adequate diet provides fuel, carbon skeletons for biosynthesis, and essential nutrients

Every mealtime is a reminder that we are animals, dependent on a regular supply of food derived from other organisms. Like all heterotrophs, animals are unable to live on inorganic nutrients alone and rely on organic compounds in their food for energy and raw materials for growth and repair. The ability of an animal to feed itself figures prominently in reproductive success, and natural selection has produced many fascinating nutritional adaptations during the long evolution of the animal kingdom. The snowshoe hare shown in the photograph on this page, is adapted for life in northern forests. It often eats conifers and other woody plants in the winter when deep snow covers grasses and other nonwoody vegetation. Adapted to a diet of plants alone, hares and rabbits have a large intestinal pouch housing bacteria that digest cellulose. In this chapter, we compare and contrast the diverse mechanisms by which animals obtain and process their food. We will take a closer look at the digestive adaptations of vertebrates, after a brief introduction to feeding and digestion and a tour of the mammalian digestive system.

Diets and feeding mechanisms vary extensively among animals

Most animals ingest other organisms, dead or alive, whole or by the piece. (Exceptions are certain parasitic animals, such as tapeworms, which absorb organic molecules directly across their outer body surface.) Diets vary. **Herbivores,** including gorillas, cows, hares, many snails, and sponges, eat autotrophs (plants, algae, and autotrophic bacteria). **Carnivores,** such as sharks, hawks, spiders, and snakes, eat other animals. **Omnivores** consume both animals and autotrophs. Cockroaches, crows, raccoons, and humans, who evolved as hunters and gatherers, are examples of omnivores.

The terms *herbivore, carnivore,* and *omnivore* represent the kinds of food an animal usually eats and the adaptations enabling it to obtain and process that food. For any particular animal or species, the diet may not conform exactly to these categories. For example, cows and deer, which are herbivores, may occasionally eat small animals or bird eggs, along with grass and other plants. Most carnivores obtain some nutrients from plant materials that remain in the digestive tract of the prey they eat.

The mechanisms by which animals obtain their food also vary. Many aquatic animals are **suspension-feeders** that sift small food particles from the water. Clams and

FIGURE 37.1
Suspension-feeding: baleen whales. The gray whale and other baleen whales use comblike plates suspended from the upper jaw to sift small invertebrates from enormous volumes of water. The whale opens its mouth and fills an expandable oral pouch with water, then closes its mouth and contracts the pouch. This forces water out of the mouth through the baleen, leaving a mouthful of trapped food.

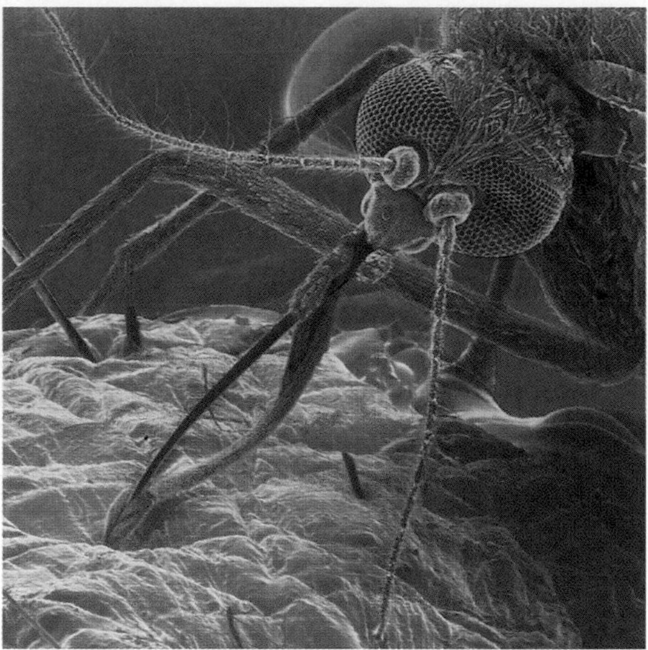

FIGURE 37.3
Fluid-feeding: a mosquito. This parasite has impaled the skin of its human host with a hollow needlelike mouthpart and is filling its gut with a blood meal (colorized SEM).

oysters, for example, use their gills to trap tiny morsels, which are then swept along with a film of mucus to the mouth by beating cilia. Baleen whales, the largest animals ever to live, are also suspension-feeders. They swim with their mouths agape, straining millions of small animals from huge volumes of water forced through screenlike plates attached to their jaws (FIGURE 37.1).

Substrate-feeders live in or on their food source, eating their way through the food. Examples are leaf miners, which are larvae of various insects that tunnel through the interior of leaves (FIGURE 37.2). Earthworms are also substrate-feeders or, more specifically, **deposit-feeders.** Eating their way through the dirt, earthworms salvage partially decayed organic material consumed along with soil.

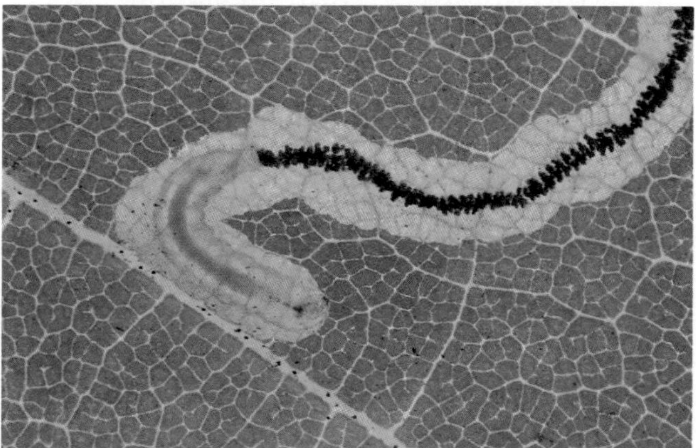

FIGURE 37.2
Substrate-feeding: a leaf miner. Eating its way through the soft mesophyll of an oak leaf, this caterpillar, the larva of a moth, leaves a trail of dark feces in its wake.

Fluid-feeders make their living by sucking nutrient-rich fluids from a living host (FIGURE 37.3). Aphids, for example, tap the phloem sap of plants, and leeches and mosquitoes suck blood from animals. Because these particular fluid-feeders harm their hosts, they are considered parasites. Hummingbirds and bees, on the other hand, benefit the host plants, transferring pollen as they move from flower to flower to obtain nectar.

Most animals, rather than filtering food from water, eating their way through the substrate, or sucking fluids, are **bulk-feeders** that eat relatively large pieces of food (FIGURE 37.4). Their adaptations include such diverse utensils as tentacles, pincers, claws, poisonous fangs, and jaws and teeth that kill their prey or tear off pieces of meat or vegetation.

Ingestion, digestion, absorption, and elimination are the four main stages of food processing

The act of eating, called **ingestion,** is the first stage of food processing. **Digestion,** the second stage, is the process of breaking food down into molecules small enough for the

FIGURE 37.4

Bulk-feeding: a python. Most animals ingest relatively large pieces of food, though rarely as large compared to the size of the animal as in the case illustrated here. In this amazing scene, a rock python is beginning to ingest a gazelle.

body to absorb. The bulk of the organic material in food consists of proteins, fats, and carbohydrates in the form of starch and other polysaccharides. Although these macromolecules are suitable raw materials, animals cannot use them directly, for two reasons. First, macromolecules are too large to pass through membranes and enter the cells of the animal. Second, the macromolecules that make up an animal are not identical to those of its food. In building their macromolecules, however, all organisms use common monomers. For instance, soybeans, cattle, and people all assemble their proteins from the same 20 amino acids (see Chapter 5). Digestion cleaves macromolecules into their component monomers, which the animal then uses to make its own molecules. Polysaccharides and disaccharides are split into simple sugars, fats are digested to glycerol and fatty acids, proteins are broken down to amino acids, and nucleic acids are cleaved into nucleotides.

Recall from Chapter 5 that when a cell makes a macromolecule by linking together monomers, it does so by removing a molecule of water for each new covalent bond formed. Digestion reverses this process by breaking bonds with the enzymatic addition of water. This splitting process is called **enzymatic hydrolysis.** Hydrolytic enzymes catalyze the digestion of each of the classes of macromolecules found in food. This chemical digestion is usually preceded by mechanical fragmentation of the food—by chewing, for instance. Breaking food into smaller pieces increases the surface area exposed to digestive juices containing hydrolytic enzymes.

An animal must digest its food in some type of specialized compartment where hydrolytic enzymes can attack the food molecules without damaging the animal's own cells. The last two stages of food processing occur after the food is digested. In the third stage, **absorption,** the animal's cells take up (absorb) small molecules such as amino acids and simple sugars from the digestive compartment. Finally, **elimination** occurs, as undigested material passes out of the digestive compartment.

Digestion occurs in food vacuoles, gastrovascular cavities, and alimentary canals

Intracellular Digestion in Food Vacuoles

The simplest digestive compartments are food vacuoles, organelles where a single cell can digest its food without the hydrolytic enzymes mixing with the cell's own cytoplasm. Protozoa digest their meals in food vacuoles, usually after engulfing the food by endocytosis (phagocytosis or pinocytosis; see Chapter 8). The food vacuoles fuse with lysosomes, which are organelles containing hydrolytic enzymes. This mixes the food with the enzymes, allowing digestion to occur safely within a compartment that is enclosed by a membrane. This digestive mechanism is termed **intracellular digestion** (FIGURE 37.5). Sponges are unusual among animals in that they digest

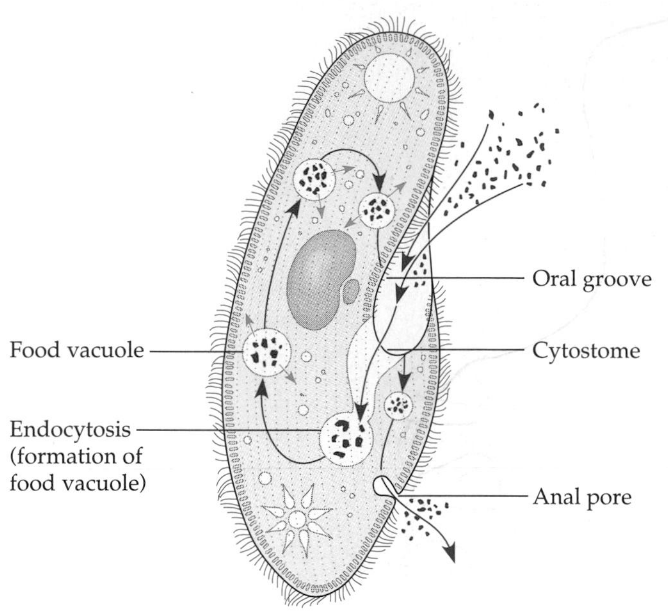

FIGURE 37.5
Intracellular digestion in *Paramecium*. *Paramecium* has a specialized feeding structure called the oral groove, which leads to the cell's "mouth" (cytostome). Cilia that line the groove draw water and suspended food particles, mostly bacteria, toward the mouth, where the food is packaged by endocytosis into a food vacuole that functions as a miniature digestive compartment. Cytoplasmic streaming carries food vacuoles around the cell, while hydrolytic enzymes are secreted into the vacuoles. As molecules in the food are digested, the nutrients (sugars, amino acids, and other small molecules, represented by blue arrows) are transported across the membrane of the vacuole into the cytoplasm. Later, the vacuole fuses with an anal pore, a specialized region of the plasma membrane where undigested material can be eliminated by exocytosis.

their food entirely by the intracellular mechanism (FIGURE 37.6). In most animals, at least some hydrolysis occurs by **extracellular digestion** within compartments that are continuous, via passages, with the outside of the animal's body.

Digestion in Gastrovascular Cavities

Many animals with relatively simple body plans have digestive sacs with single openings. These pouches, called **gastrovascular cavities,** function in both digestion and distribution of nutrients throughout the body (hence, the *vascular* part of the term). The cnidarian *Hydra* provides a good example of how a gastrovascular cavity works. *Hydra* is a carnivore that stings its prey with specialized organelles called nematocysts. It then uses its tentacles to stuff the food through its mouth into the gastrovascular cavity. With food in the cavity, specialized cells of the gastrodermis, the tissue layer that lines the cavity, secrete

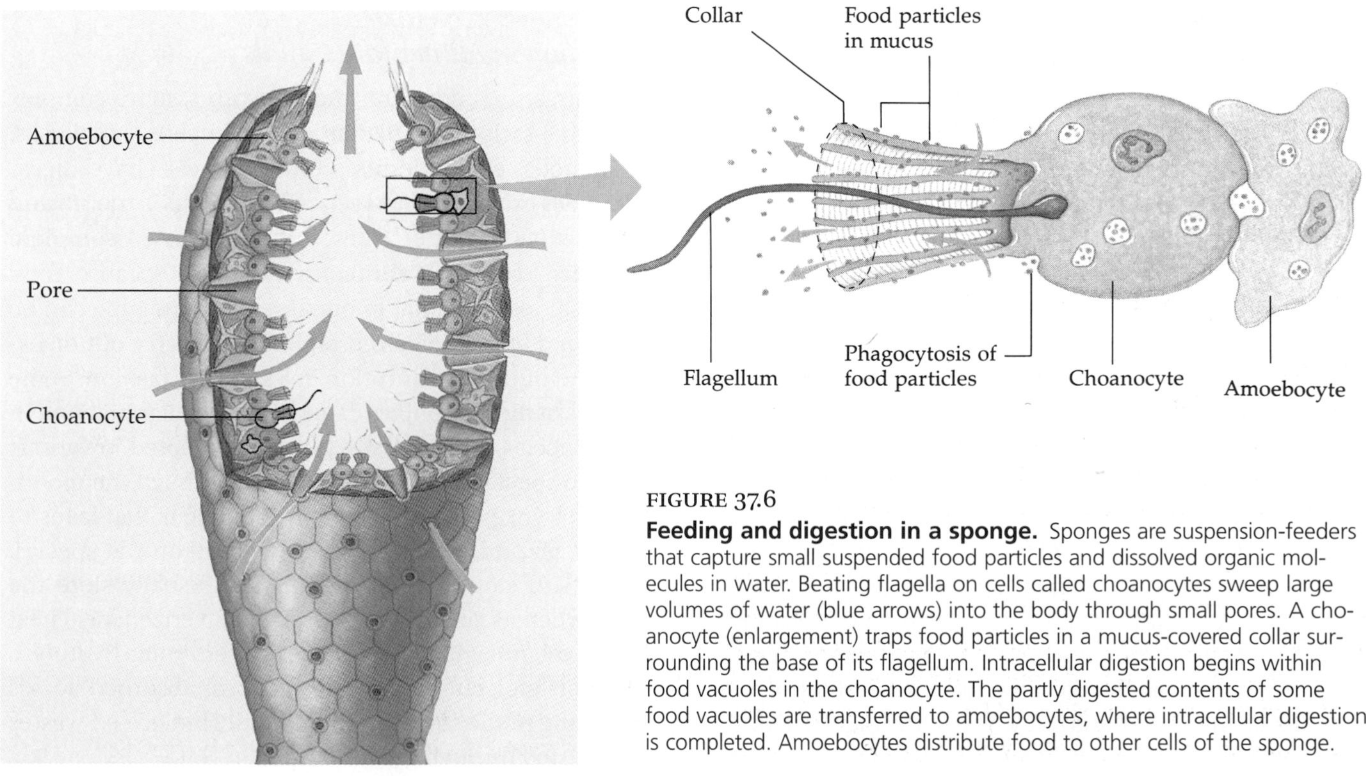

FIGURE 37.6
Feeding and digestion in a sponge. Sponges are suspension-feeders that capture small suspended food particles and dissolved organic molecules in water. Beating flagella on cells called choanocytes sweep large volumes of water (blue arrows) into the body through small pores. A choanocyte (enlargement) traps food particles in a mucus-covered collar surrounding the base of its flagellum. Intracellular digestion begins within food vacuoles in the choanocyte. The partly digested contents of some food vacuoles are transferred to amoebocytes, where intracellular digestion is completed. Amoebocytes distribute food to other cells of the sponge.

FIGURE 37.7

Gastrovascular cavities. (**a**) *Hydra.* The outer epidermis has protective and sensory functions, whereas the inner gastrodermis is specialized for digestion. The mesoglea, a jellylike layer, separates the two layers of cells. Enzymes released from gland cells into the gastrovascular cavity initiate digestion, which is completed intracellularly after small food particles are taken into the nutritive cells of the gastrodermis by phagocytosis. Undigested waste is egested through the mouth, the single opening of the gastrovascular cavity. (**b**) A planarian has a gastrovascular cavity with extensive branching, which increases the surface area available for digestion and functions in the transport of nutrients to all parts of the animal.

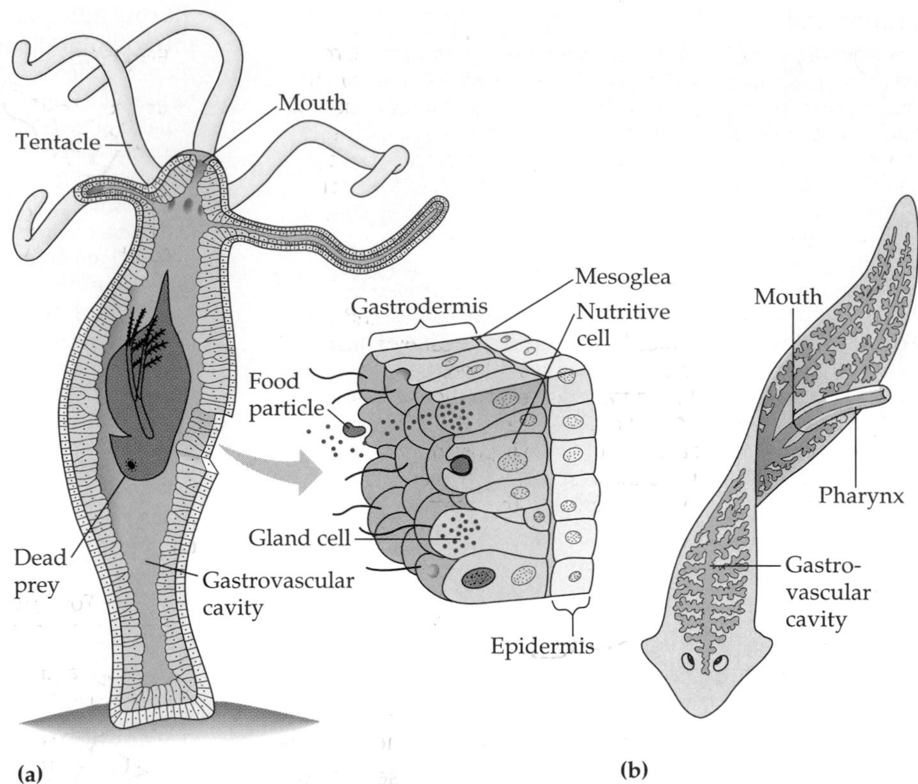

digestive enzymes that break the soft tissues of the prey into tiny pieces (FIGURE 37.7a). Beating flagella on the gastrodermal cells keep the small food particles from settling and distribute them throughout the cavity. Nutritive cells of the gastrodermis then take the food particles in by phagocytosis, and food molecules are hydrolyzed within food vacuoles. Extracellular digestion within the gastrovascular cavity initiates breakdown of the food, but most of the actual hydrolysis of macromolecules is accomplished by intracellular digestion. After *Hydra* has digested its meal, undigested materials remaining in the gastrovascular cavity, such as the exoskeletons of small crustaceans, are eliminated through the single opening, which functions in the dual role of mouth and anus.

If the gastrodermal cells are capable of phagocytosis and digestion, what is the advantage of having an extracellular cavity? Phagocytosis is limited to the ingestion of microscopic food. Extracellular cavities for digestion are adaptations that enable animals to devour larger prey.

Most flatworms are carnivores, eating small invertebrates and dead animal matter. Like *Hydra,* they have gastrovascular cavities with single openings (FIGURE 37.7b). Planarians have a pharynx, a muscular tube that can be everted ventrally through the mouth. The pharynx penetrates prey and releases a digestive juice, beginning digestion outside the planarian's body. The partly digested food is sucked into the gastrovascular cavity, which has three main branches and many secondary ones, increasing the surface area of the cavity extensively. Digestion continues within the gastrovascular cavity and is completed within cells that take up food particles by phagocytosis.

Digestion in Alimentary Canals

In contrast to cnidarians and flatworms, more complex animals—including nematodes, annelids, mollusks, arthropods, echinoderms, and chordates—have digestive tubes extending between two openings, a mouth and an anus (FIGURE 37.8). These tubes are called **complete digestive tracts** or **alimentary canals.** Because food moves along the canal in one direction, the tube can be organized into specialized regions that carry out digestion and nutrient absorption in a stepwise fashion. Some regions of the alimentary canal have a similar function in most species, although the tubes are adapted in various ways to specific diets. Food ingested through the mouth and pharynx passes through an esophagus that leads to a crop, gizzard, or stomach, depending on the species. Crops and stomachs are organs that usually store the food, whereas gizzards grind it. The pulverized meal next enters the intestine, where digestive enzymes hydrolyze the food molecules, and nutrients are absorbed across the lining of the tube into the blood. Undigested wastes are egested through the **anus.**

FIGURE 37.8

Alimentary canals. (**a**) The digestive tract of an earthworm consists of five specialized regions (organs) between the mouth and anus. A muscular pharynx sucks food in through the mouth. Food passes through the esophagus and is stored and moistened in the crop. The muscular gizzard, which contains small bits of sand and gravel, pulverizes the food. Digestion and absorption occur in the intestine, which has a dorsal fold, the typhlosole, that increases the surface area for nutrient absorption. Undigested material is expelled through the anus. (**b**) A grasshopper has several digestive chambers grouped into three main regions: a foregut, with an esophagus and crop; a midgut; and a hindgut. Food is moistened and stored in the crop, but most digestion occurs in the midgut. Gastric ceca, pouches extending from the midgut, absorb nutrients and transfer them to the grasshopper's hemolymph (blood). (**c**) A bird has three separate chambers—the crop, stomach, and gizzard—where food is pulverized and churned before passing into the intestine. Most chemical digestion and absorption of nutrients occur in the intestine.

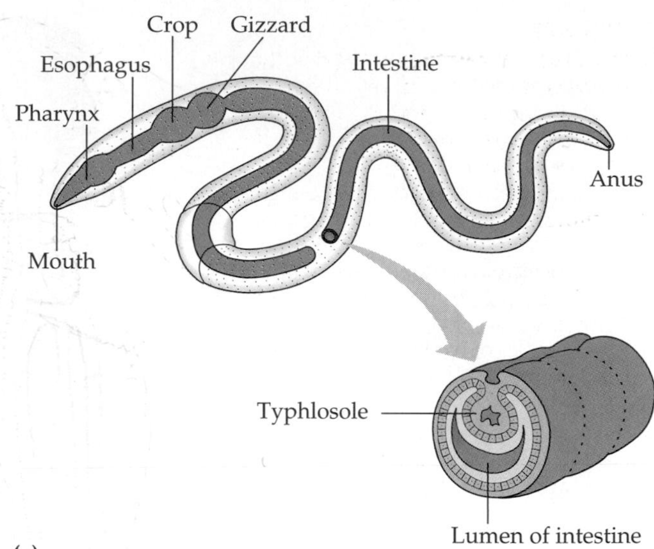

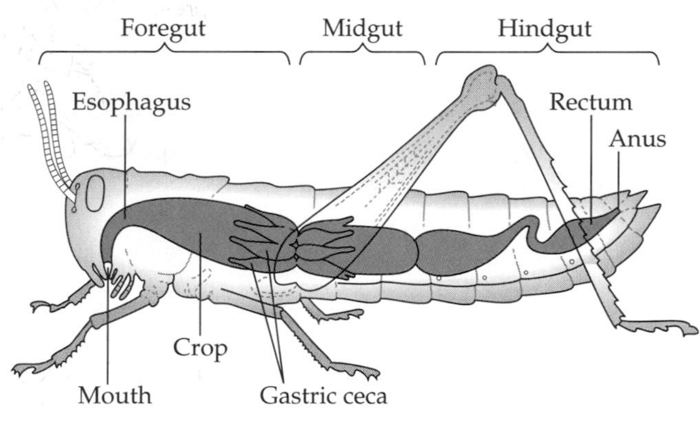

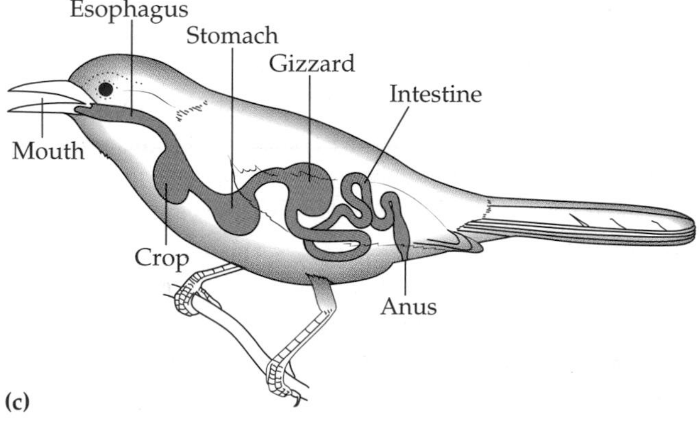

A tour of the mammalian digestive system

The mammalian digestive system consists of the alimentary canal and various accessory glands that secrete digestive juices into the canal through ducts. From the lower esophagus through the large intestine, the mammalian digestive tract has a four-layered wall (see FIGURE 36.6). The lumen (cavity) is lined by a mucous membrane, the mucosa. Next comes a layer made up mostly of connective tissue, followed by a layer of smooth muscle. The outermost layer is a sheath of connective tissue attached to the membrane of the body cavity. **Peristalsis,** rhythmic waves of contraction by the smooth muscles, pushes the food along the tract. At some of the junctions between specialized segments of the tube, the muscular layer is modified into ringlike valves called **sphincters,** which close off the tube like drawstrings, regulating the passage of material between chambers of the canal.

The accessory glands of the mammalian digestive system are three pairs of **salivary glands,** the **pancreas,** and the **liver** with its storage organ, the **gallbladder.** The accessory glands originate in the embryo as outpocketings of the developing alimentary canal.

Using the human as an example, we now follow a meal through the alimentary canal, seeing in more detail what happens to the food in each of the processing stations along the way (FIGURE 37.9).

The Oral Cavity

Physical and chemical digestion of food both begin in the mouth. During chewing, teeth of various shapes cut, smash, and grind food, making the food easier to swallow and increasing its surface area. The presence of food in the **oral cavity** triggers a nervous reflex that causes the salivary glands to deliver saliva through ducts to the oral cavity. Even before food is actually in the mouth, salivation may occur in anticipation because of learned associations between eating and the time of day, cooking odors, or other stimuli.

In humans, more than a liter of saliva is secreted into the oral cavity each day. Dissolved in the saliva is a slippery glycoprotein (carbohydrate-protein complex) called mucin, which protects the soft lining of the mouth from abrasion and lubricates the food for easier swallowing.

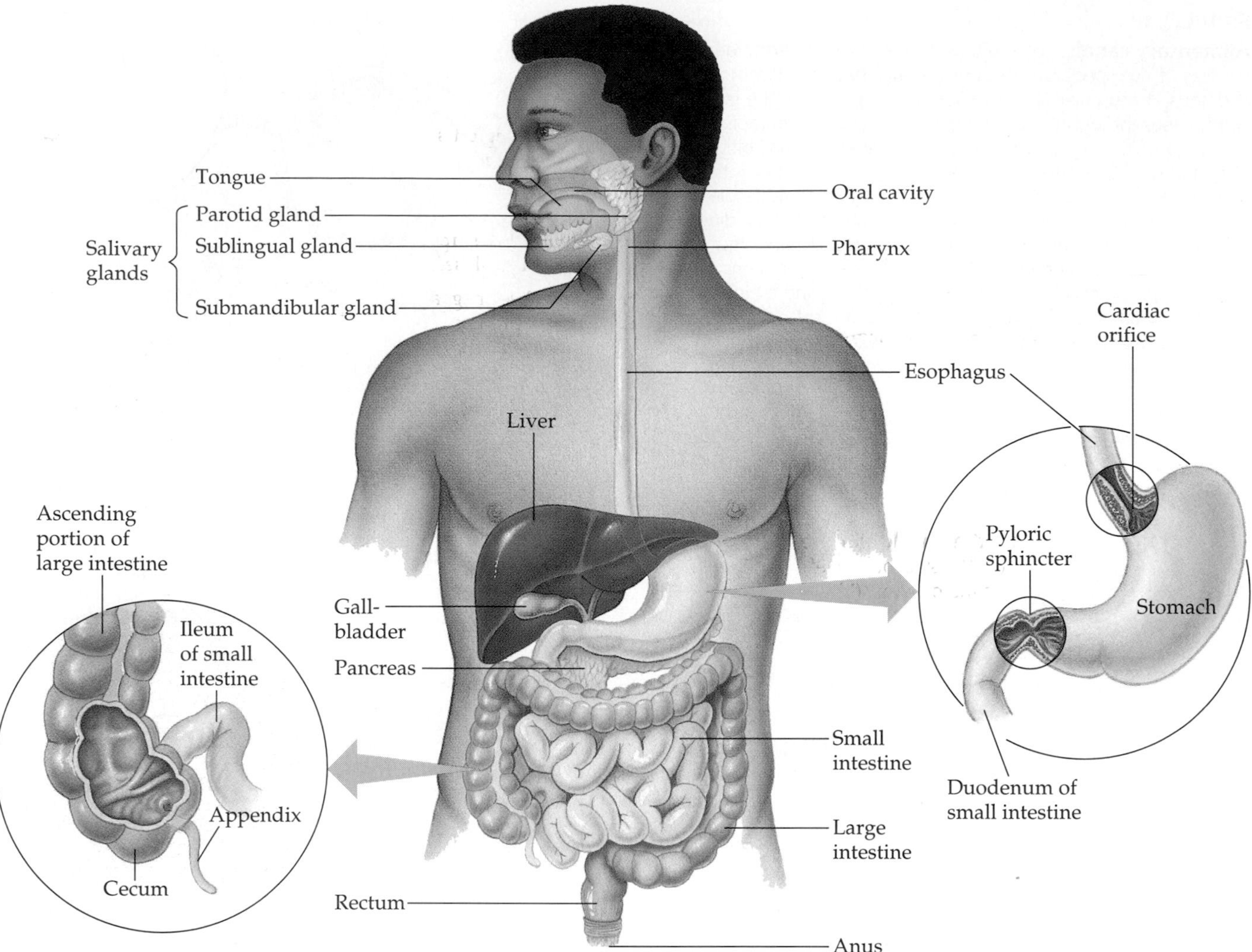

FIGURE 37.9
The human digestive system. After chewing and swallowing, it takes only 5 to 10 seconds for food to pass down the esophagus and into the stomach, where it spends 2 to 6 hours being partially digested. Final digestion and nutrient absorption occur in the small intestine over a period of 5 to 6 hours. In 12 to 24 hours, any undigested material passes through the large intestine, and feces are expelled through the anus.

Saliva contains buffers that help prevent tooth decay by neutralizing acid in the mouth. Also, antibacterial agents in saliva kill many of the bacteria that enter the mouth with food. Finally, **salivary amylase,** a digestive enzyme that hydrolyzes the glucose polymers starch (from plants) and glycogen (from animals), is present in saliva. The main products of this enzyme's action are smaller polysaccharides and the double sugar (disaccharide) maltose.

The tongue tastes food, manipulates it during chewing, and helps shape the food into a ball called a **bolus.** During swallowing, the tongue pushes a bolus to the very back of the oral cavity and into the pharynx.

The Pharynx

The region we call our throat is the **pharynx,** an intersection that leads to both the esophagus and the windpipe (trachea). When we swallow, the top of the windpipe moves so that its opening, the glottis, is blocked by a cartilaginous flap, the **epiglottis** (FIGURE 37.10a, and b). You can see this motion in the bobbing of the "Adam's apple" during swallowing. Closing the opening of the windpipe guards the respiratory system against the entry of food or fluids during swallowing. The swallowing mechanism normally ensures that a bolus will be guided into the entrance of the esophagus.

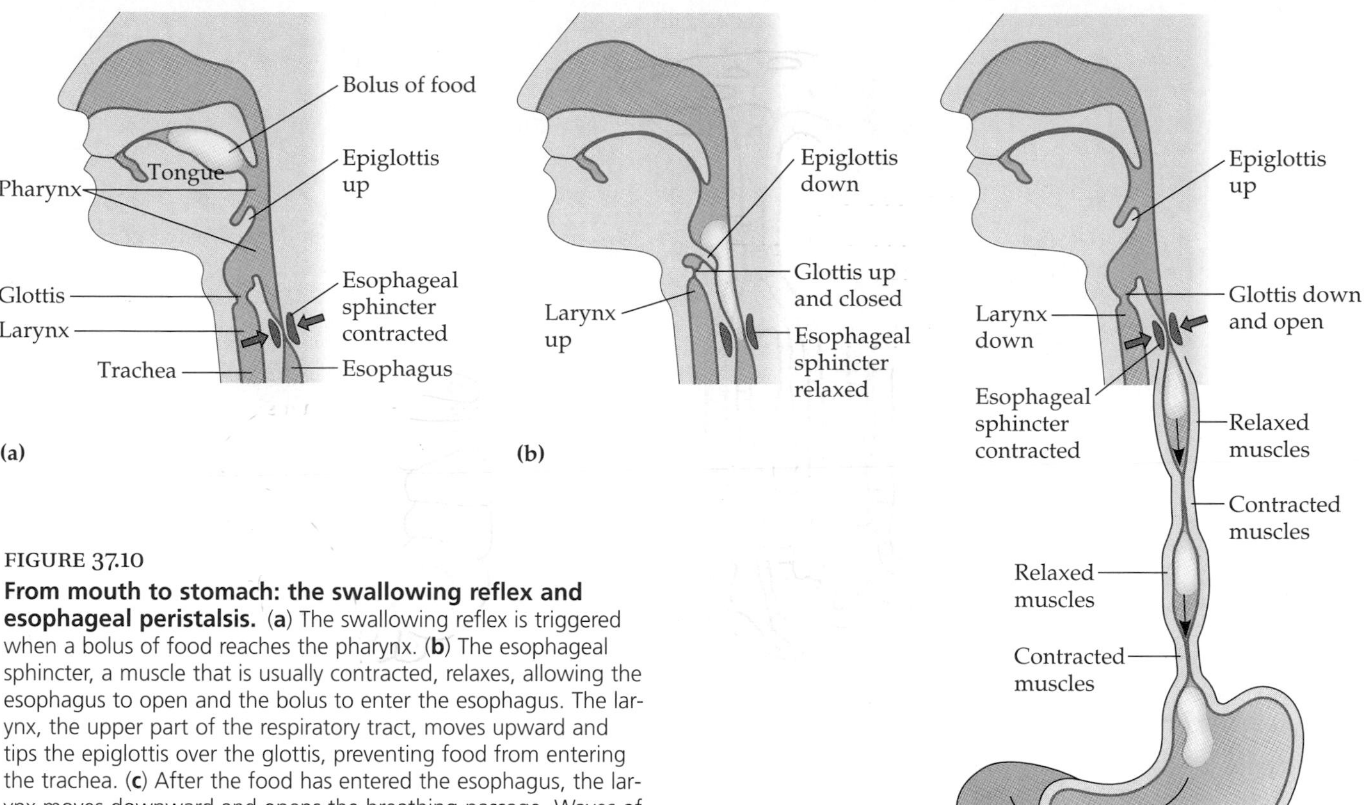

FIGURE 37.10
From mouth to stomach: the swallowing reflex and esophageal peristalsis. (**a**) The swallowing reflex is triggered when a bolus of food reaches the pharynx. (**b**) The esophageal sphincter, a muscle that is usually contracted, relaxes, allowing the esophagus to open and the bolus to enter the esophagus. The larynx, the upper part of the respiratory tract, moves upward and tips the epiglottis over the glottis, preventing food from entering the trachea. (**c**) After the food has entered the esophagus, the larynx moves downward and opens the breathing passage. Waves of muscular contraction (peristalsis) move the bolus down the esophagus to the stomach.

The Esophagus

The **esophagus** conducts food from the pharynx down to the stomach. Peristalsis squeezes a bolus along the narrow esophagus (FIGURE 37.10C). Only the muscles at the very top of the esophagus are striated (voluntary). Thus, the act of swallowing is initiated voluntarily, but then the involuntary waves of contraction by the smooth muscles take over. Salivary amylase continues to hydrolyze starch and glycogen as the bolus passes through the esophagus.

The Stomach

The **stomach** is located on the left side of the abdominal cavity, just below the diaphragm. Because this large organ can store an entire meal, we do not need to eat constantly. With a very elastic wall and accordionlike folds, the stomach can stretch to accommodate about 2 L of food and fluid.

The epithelium that lines the lumen of the stomach secretes **gastric juice,** a digestive fluid that mixes with the food. With its high concentration of hydrochloric acid, the gastric juice has a pH of about 2—acidic enough to dissolve iron nails. One function of the acid is to disrupt the extracellular matrix that binds cells together in meat and plant material. The acid also kills most bacteria that are swallowed with the food.

Also present in gastric juice is **pepsin,** an enzyme that hydrolyzes proteins. Hydrolysis is incomplete, however, because pepsin can only break peptide bonds adjacent to specific amino acids. Pepsin's action cleaves proteins into smaller polypeptides. Pepsin is one of the few enzymes that works best in a strongly acidic environment. Indeed, the low pH of gastric juice denatures the proteins in food, increasing exposure of their peptide bonds to the pepsin. Even the salivary amylase that is swallowed ceases to work soon after reaching the stomach and is digested along with the food proteins.

What prevents pepsin from destroying cells of the stomach wall that produce this hydrolytic enzyme? Pepsin is synthesized and secreted in an inactive form called **pepsinogen.** Hydrochloric acid in the gastric juice converts pepsinogen to active pepsin by removing a short segment of the protein's polypeptide chain, an alteration that exposes the active site of pepsin. Because the acid and pepsinogen are secreted by different kinds of cells, the two ingredients do not mix until their release into the lumen of the stomach. Once some of the pepsinogen is activated by acid, a chain reaction occurs because pepsin itself can activate additional molecules of pepsinogen.

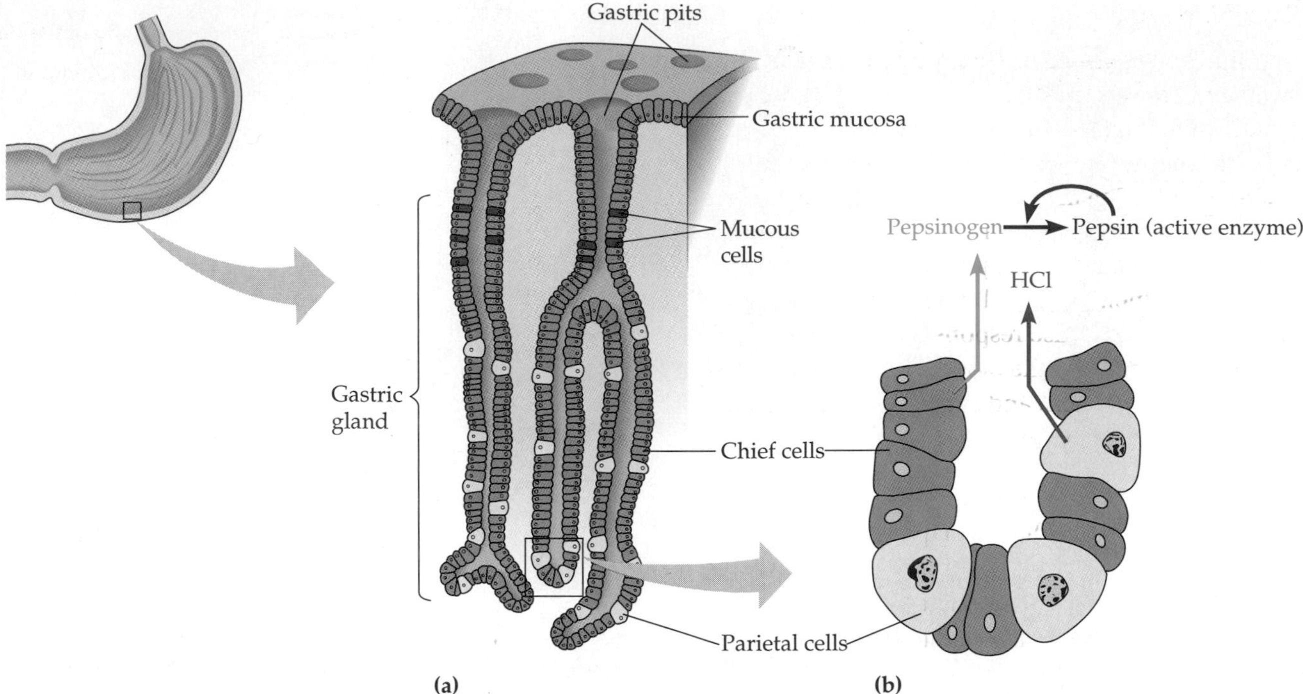

FIGURE 37.11

The secretion of gastric juice in the stomach. (**a**) The mucosa that lines the stomach consists of simple columnar epithelium organized into tubular gastric glands. Gastric pits lead into the gastric glands, which have three types of secretory cells. The cells that line the openings of the glands secrete mucus that helps protect the stomach from digesting itself. Deeper in the gland are parietal cells, which secrete hydrochloric acid, and chief cells, which secrete pepsinogen. (**b**) Within the gastric glands, hydrochloric acid converts the inactive pepsinogen to pepsin, which activates more pepsin—an example of positive feedback.

This domino effect is an example of positive feedback (FIGURE 37.11). We will see other cases where protein-digesting enzymes are secreted in inactive forms, generally called **zymogens.**

A coating of mucus secreted by the epithelial cells helps protect the stomach lining from being digested by the pepsin and acid in gastric juice. Still, the epithelium is constantly eroded, and mitosis generates enough cells to completely replace the stomach lining every three days. Lesions in the stomach lining, called gastric ulcers, are caused mainly by bacteria (*Helicobacter pylori*), but they may worsen if pepsin and acid destroy the lining faster than it can regenerate.

Gastric secretion is controlled by a combination of nervous impulses and hormones. When we see, smell, or taste food, impulses from the brain to the stomach initiate the secretion of gastric juice. Then certain substances in the food stimulate the stomach wall to release a hormone called **gastrin** into the circulatory system. As the gastrin gradually recirculates in the bloodstream back to the stomach wall, the hormone stimulates further secretion of gastric juice. Thus, an initial burst of gastric secretion at mealtime is followed by a sustained secretion that continues to add gastric juice to the food for some time. If the pH of the stomach contents becomes too low, the acid will inhibit the release of gastrin, decreasing the secretion of gastric juice; this is an example of negative feedback. Each day, the stomach wall secretes about 3 L of gastric juice.

About every 20 seconds, the stomach contents are mixed by the churning action of smooth muscles. You may feel hunger pangs when your empty stomach churns. (Sensations of hunger are also associated with centers in the brain that monitor the nutritional status of the blood.) As a result of mixing and enzyme action, what begins in the stomach as a recently swallowed meal soon becomes a nutrient broth known as **acid chyme.**

Much of the time, the stomach is closed off at either end (see FIGURE 37.9). The opening from the esophagus to the stomach, the cardiac orifice, normally dilates only when a bolus driven by peristalsis arrives. The occasional backflow of acid chyme from the stomach into the lower end of the esophagus causes heartburn. (If backflow is a persistent problem, an ulcer may develop in the esophagus.) The cardiac orifice opens intermittently with each wave of peristalsis that delivers a bolus. At the opening from the stomach to the small intestine is the **pyloric sphincter,** which helps regulate the passage of chyme into the intestine. A squirt at a time, it takes about 2 to 6 hours after a meal for the stomach to empty.

The Small Intestine

Although limited digestion of starch takes place in the oral cavity and partial digestion of proteins by pepsin in the stomach, most enzymatic hydrolysis of the macromolecules in food occurs in the **small intestine.** With a length of more than 6 m in humans, the small intestine is the longest section of the alimentary canal (its name is based on its small diameter, compared with the diameter of the large intestine). The small intestine is the major organ of digestion and is also responsible for the absorption of most nutrients into the blood. Let's focus first on the main organs, secretions, and events in digestion and then take a look at absorption.

Digestion. The pancreas, liver, and gallbladder, as well as the small intestine itself, participate in digestion. The pancreas produces several hydrolytic enzymes and an alkaline solution rich in bicarbonate. The bicarbonate acts as a buffer, offsetting the acidity of chyme from the stomach.

The liver performs a wide variety of important functions in the body, including the production of **bile,** a mixture of substances that is stored in the gallbladder until needed. Bile contains no digestive enzymes, but it does contain bile salts, which aid in the digestion and absorption of fats. Bile also contains pigments that are by-products of red blood cell destruction in the liver; these bile pigments are eliminated from the body with the feces.

The first 25 cm or so of the small intestine is called the **duodenum.** It is here that acid chyme seeping in from the stomach mixes with digestive juices from the pancreas, liver, gallbladder, and gland cells of the intestinal wall itself.

At least four regulatory hormones help ensure that digestive secretions are present only when needed (FIGURE 37.12). We have already seen that gastrin is released from the stomach lining in response to the presence of food. The acidic pH of the chyme that enters the duodenum stimulates the intestinal wall to release a second hormone, **secretin.** This hormone signals the pancreas to release bicarbonate, which neutralizes the acid chyme. A third hormone, **cholecystokinin (CCK),** produced by cells in the lining of the duodenum, causes the gallbladder to contract and release bile into the small intestine. CCK also triggers the release of pancreatic enzymes. The chyme, particularly if rich in fats, also causes the duodenum to release a fourth hormone, **enterogastrone,** which inhibits peristalsis in the stomach, thereby slowing down the entry of food into the small intestine.

Let's now follow the action of enzymes from the pancreas and intestinal wall in digesting macromolecules. TABLE 37.1, p. 806, summarizes the actions of the digestive enzymes of the entire digestive system.

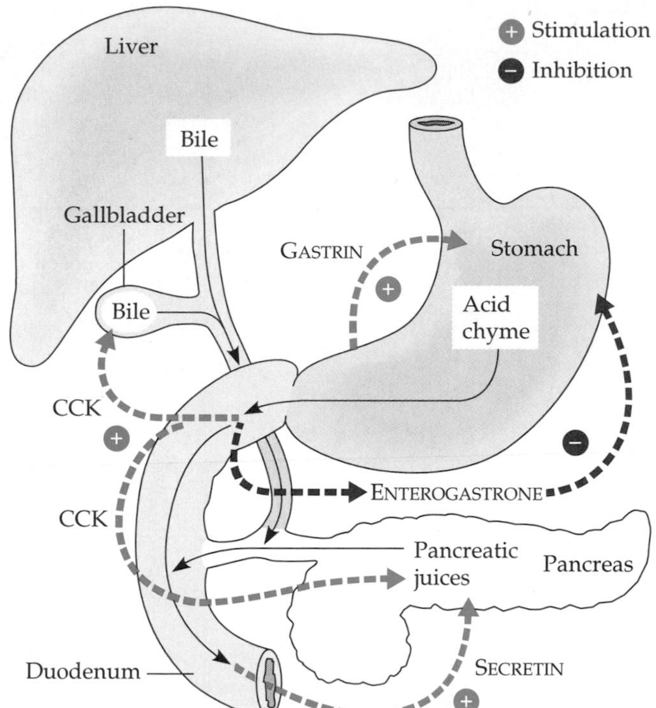

FIGURE 37.12
Hormonal control of digestion. Gastrin from the stomach stimulates the production of gastric juices. Most digestion of food occurs in the duodenum, and secretion of three hormones by this region of the small intestine helps regulate digestion. Secretin stimulates the pancreas to release sodium bicarbonate, which neutralizes acid chyme from the stomach. The presence of amino acids or fatty acids in the duodenum triggers the release of cholecystokinin (CCK), which stimulates the release of digestive enzymes by the pancreas and bile by the gallbladder. Enterogastrone inhibits peristalsis and acid secretion by the stomach, thereby slowing digestion when acid chyme rich in fats (which require additional digestion time) enters the duodenum.

The digestion of the carbohydrates starch and glycogen begun by salivary amylase in the oral cavity continues in the small intestine. Pancreatic amylases hydrolyze starch, glycogen, and smaller polysaccharides into disaccharides, including maltose. The enzyme maltase completes the digestion of maltose, splitting it into two molecules of the simple sugar glucose. Maltase is one of a family of disaccharidases, each one specific for the hydrolysis of a different disaccharide. Sucrase, for instance, hydrolyzes table sugar (sucrose), and lactase digests milk sugar (lactose). (In general, adults have much less lactase than children.) The disaccharidases are built into the membranes and extracellular matrix covering the intestinal epithelium (the brush border in TABLE 37.1). Thus, the terminal steps in carbohydrate digestion occur at the site of sugar absorption.

Protein digestion in the small intestine involves completion of the work begun by pepsin in the stomach.

TABLE 37.1

Enzymatic Digestion in the Human Digestive System

	ORAL CAVITY, PHARYNX, ESOPHAGUS	STOMACH	SMALL INTESTINE: Lumen	SMALL INTESTINE: Brush border
Carbohydrate digestion	Polysaccharides (starch, glycogen) → SALIVARY AMYLASE → Smaller polysaccharides, maltose		Polysaccharides → PANCREATIC AMYLASES → Maltose and other disaccharides	→ DISACCHARIDASES → Monosaccharides
Protein digestion		Proteins → PEPSIN → Small polypeptides	Polypeptides → TRYPSIN, CHYMOTRYPSIN → Smaller polypeptides → AMINOPEPTIDASE, CARBOXYPEPTIDASE → Amino acids	Small peptides → DIPEPTIDASES → Amino acids
Nucleic acid digestion			DNA, RNA → NUCLEASES → Nucleotides	→ NUCLEOTIDASES → Nucleosides → NUCLEOSIDASES → Nitrogenous bases, sugars, phosphates
Fat digestion			Fat globules → BILE SALTS → Fat droplets (emulsified) → LIPASE → Glycerol, fatty acids, glycerides	

Enzymes in the duodenum dismantle polypeptides into their component amino acids or into small peptides (fragments only two or three amino acids long). **Trypsin** and **chymotrypsin** are specific for peptide bonds adjacent to certain amino acids, and thus, like pepsin, break large polypeptides into shorter chains. **Carboxypeptidase** splits off one amino acid at a time, beginning at the end of the polypeptide that has a free carboxyl group. **Aminopeptidase** works in the opposite direction. Either aminopeptidase or carboxypeptidase alone could completely digest a protein. But teamwork among these enzymes and the trypsin and chymotrypsin that attack the interior of the protein speeds up hydrolysis tremendously. Other enzymes called **dipeptidases,** attached to the intestinal lining, further hasten digestion by splitting small peptides.

The protein-digesting enzymes, including trypsin, chymotrypsin, and carboxypeptidase, are secreted as inactive zymogens by the pancreas. An intestinal enzyme called **enteropeptidase** triggers activation of these enzymes within the lumen of the small intestine (FIGURE 37.13).

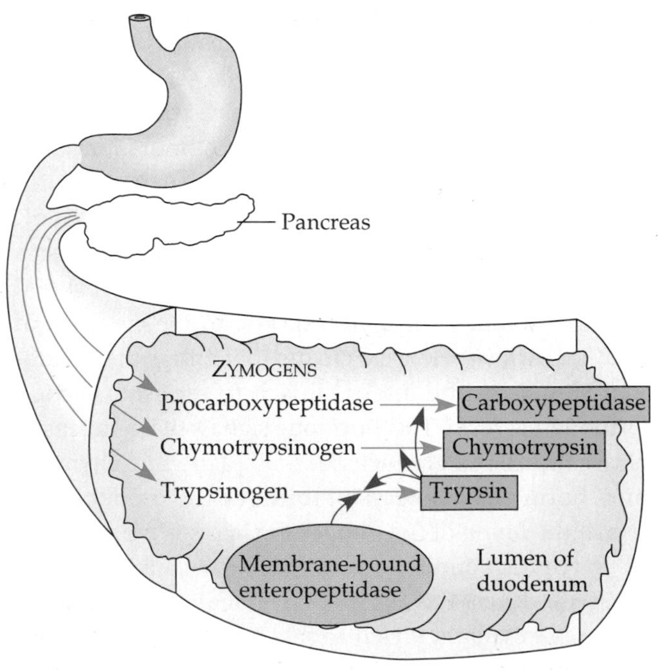

FIGURE 37.13
Activation of zymogens in the small intestine. The pancreas secretes protein-digesting enzymes in an inactive (zymogen) form into the lumen of the duodenum. An enzyme called enteropeptidase, which is bound to the intestinal epithelium, begins the activation of zymogens by converting trypsinogen to trypsin. Trypsin then activates procarboxypeptidase and chymotrypsinogen.

The digestion of nucleic acids involves a hydrolytic assault similar to that mounted on proteins. A team of enzymes called **nucleases** hydrolyzes DNA and RNA in food into their component nucleotides. Other hydrolytic enzymes then break nucleotides down further into

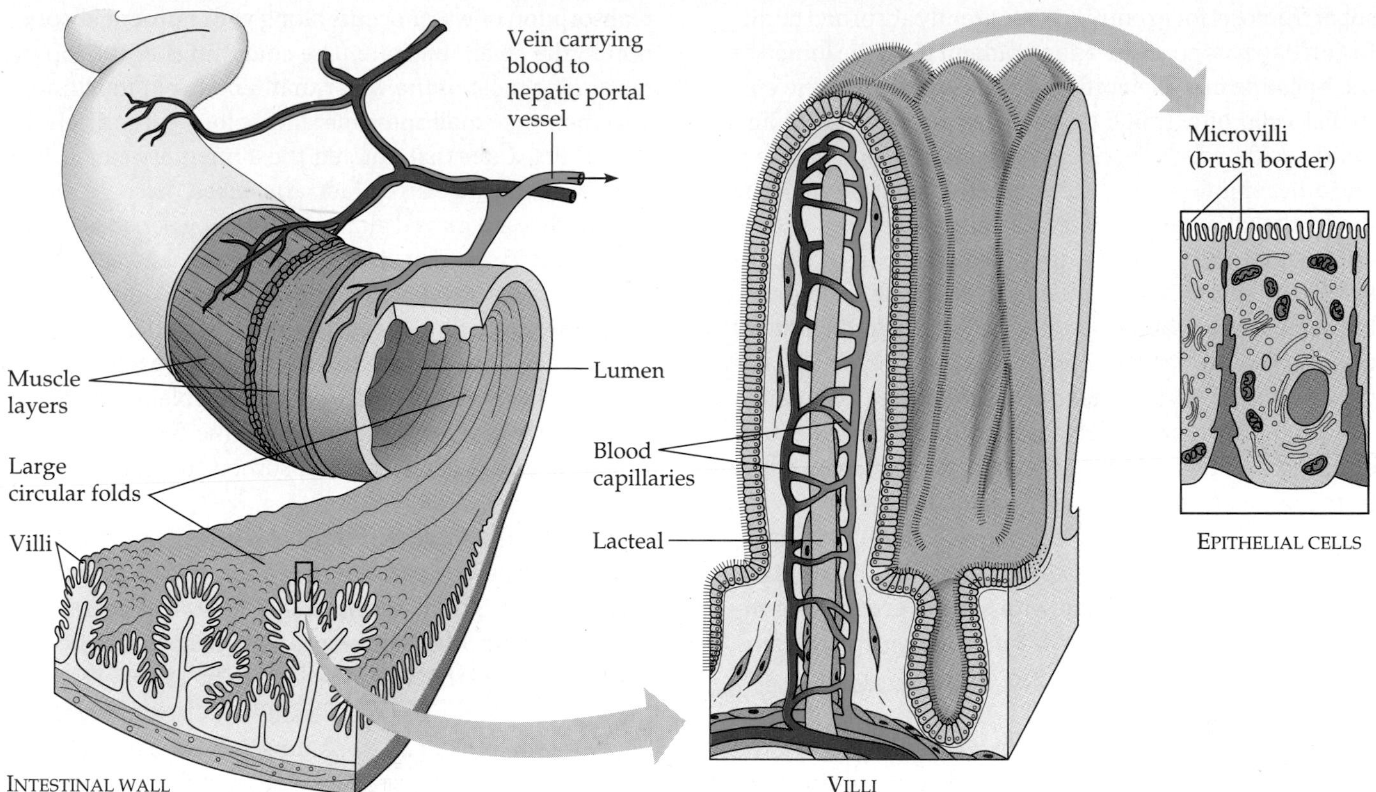

FIGURE 37.14

The structure of the small intestine. Large folds of epithelium line the small intestine. Villi project outward from the folds. Each villus has a small lymphatic vessel called a lacteal and a network of blood capillaries surrounded by a layer of epithelial cells. The epithelial cells have microscopic projections, microvilli, that extend into the intestinal lumen as a "brush border." Microfilaments within the microvilli wiggle the tiny appendages in the nutrient broth of the intestine's lumen. Veins distribute blood with absorbed nutrients to the hepatic portal vessel, which conveys it directly to the liver.

nucleosides (see Chapter 5), nitrogenous bases, sugars, and phosphates.

Nearly all the fat in a meal reaches the small intestine completely undigested. Hydrolysis of fats is a special problem, because fat molecules are insoluble in water. Bile salts secreted into the duodenum coat tiny fat droplets and keep them from coalescing, a process called **emulsification.** Because the droplets are small, there is a large surface area of fat exposed to **lipase,** an enzyme that hydrolyzes the fat molecules.

Thus, the macromolecules from food are completely hydrolyzed to their component monomers as peristalsis moves the mixture of chyme and digestive juices along the small intestine. Most digestion is completed early in this journey, while the chyme is still in the duodenum. The remaining regions of the small intestine, the **jejunum** and **ileum,** function mainly in the absorption of nutrients.

Absorption. To enter the body, nutrients that accumulate in the lumen when food is digested must cross the lining of the digestive tract. A limited number of nutrients are absorbed in the stomach and large intestine, but most absorption occurs in the small intestine.

The lining of the small intestine has a huge surface area of about 300 m^2, roughly the size of a tennis court. Large circular folds in the lining bear fingerlike projections called **villi,** and each of the epithelial cells of a villus has many microscopic appendages called **microvilli,** which are exposed to the lumen of the intestine. Commonly called a **brush border** for its bristlelike appearance, the huge microvillar surface is an adaptation well suited to the task of absorbing nutrients (FIGURE 37.14).

Only two single layers of epithelial cells separate nutrients in the lumen of the intestine from the bloodstream. Penetrating the core of each villus is a net of microscopic blood vessels (capillaries) and a small vessel of the lymphatic system called a **lacteal.** (In addition to their circulatory system that carries blood, vertebrates have an auxiliary system of vessels—the lymphatic system—which carries a clear fluid called lymph; discussed in Chapter 38.) Nutrients are absorbed across the epithelium and then across the unicellular wall of the capillaries or lacteals. In some cases, the transport is passive. The simple sugar fructose, for example, is apparently absorbed by diffusion down its concentration gradient from the lumen of the intestine into the epithelial cells, then out of the epithelial cells into capillaries. Other nutrients, including amino acids, small peptides, vitamins, glucose, and several other simple sugars, are pumped against gradients by the epithelial membranes. The absorption of some nutrients seems to be coupled to the active transport of sodium across the membranes of the epithelial cells. The membrane pumps sodium out of the cell and into the lumen, and the passive reentry of the ions is harnessed by cotransport to drive the uptake of nutrients (see Chapter 8).

Amino acids and sugars pass through the epithelium, enter capillaries, and are carried away from the intestine by the bloodstream. After glycerol and fatty acids are absorbed by epithelial cells, they are recombined within those cells to form fats again. The fats are then mixed with cholesterol and coated with special proteins, forming small globules called **chylomicrons,** which are transported by exocytosis out of the epithelial cell and into a lacteal.

The capillaries and veins that drain nutrients away from the villi all converge into a single circulatory channel, the **hepatic portal vessel,** which leads directly to the liver. The rate of flow in this large vessel, about 1 L per minute, ensures that the liver, which has the metabolic versatility to interconvert various organic molecules, has first access to nutrients absorbed after a meal is digested. The blood that leaves the liver may have a very different balance of nutrients from the blood that entered via the hepatic portal vessel. For example, blood exiting the liver usually has a glucose concentration of very close to 0.1%, regardless of the carbohydrate content of a meal. From the liver, blood travels to the heart, which pumps the blood and the nutrients it contains to all parts of the body.

The Large Intestine

The **large intestine,** or **colon,** is connected to the small intestine at a T-shaped junction, where a sphincter acts as a valve controlling the movement of material. One arm of the T is a pouch called the **cecum** (see FIGURE 37.9). Compared to many other mammals, humans have a relatively small cecum with a fingerlike extension, the **appendix,** which is dispensable. (Lymphoid tissue in the appendix makes a minor contribution to body defense.) The main branch of the human colon is shaped like an upside-down U about 1.5 m long.

A major function of the colon is to reabsorb water that has entered the alimentary canal as the solvent of the various digestive juices. Altogether, about 7 L of fluid are secreted into the lumen of the digestive tract each day. Most reabsorption of water occurs along with nutrient absorption in the small intestine. The colon finishes the job by reclaiming most of the water that remains in the lumen. Together, the small intestine and colon reabsorb about 90% of the water that entered the alimentary canal. The wastes of the digestive tract, the **feces,** become more solid as they are moved along the colon by peristalsis. The movement is sluggish, and it takes about 12 to 24 hours for material to travel the length of the organ. If the lining of the colon is irritated—by a viral or bacterial infection, for instance—less water than normal may be reabsorbed, resulting in diarrhea. The opposite problem, constipation, occurs when peristalsis moves the feces along too slowly. An excess of water is reabsorbed, and the feces become compacted.

Living in the large intestine is a rich flora of mostly harmless bacteria. One of the common inhabitants of the human colon is *Escherichia coli,* a favorite research organism of molecular biologists. Intestinal bacteria live on organic material that would otherwise be eliminated with the feces. As by-products of their metabolism, many of the bacteria generate gases, including methane and hydrogen sulfide, and some produce vitamins, which are absorbed by the host; in fact, intestinal bacteria are probably the main source of vitamin K for humans. As much as 40% of the mass of human feces consists of intestinal bacteria; the presence of *E. coli* in lakes and streams is an indication that the water is contaminated by untreated sewage.

Feces also contain cellulose and other undigested ingredients of food. Although cellulose fibers have no caloric value to humans, their presence in the diet helps move food along the digestive tract. Feces may also contain an abundance of salts. For instance, when iron and calcium concentrations in the blood get too high, the colon lining excretes salts of these elements into the lumen, and they are eliminated in the feces.

The terminal portion of the colon is called the **rectum,** where feces are stored until they can be eliminated. Between the rectum and the anus are two sphincters, one involuntary and the other voluntary. Once or more each day, strong contractions of the colon create an urge to defecate.

Vertebrate digestive systems exhibit many evolutionary adaptations associated with diet

The digestive systems of mammals and other vertebrates are variations on a common plan, but there are many intriguing adaptations, often associated with the animal's diet. We will examine just a few.

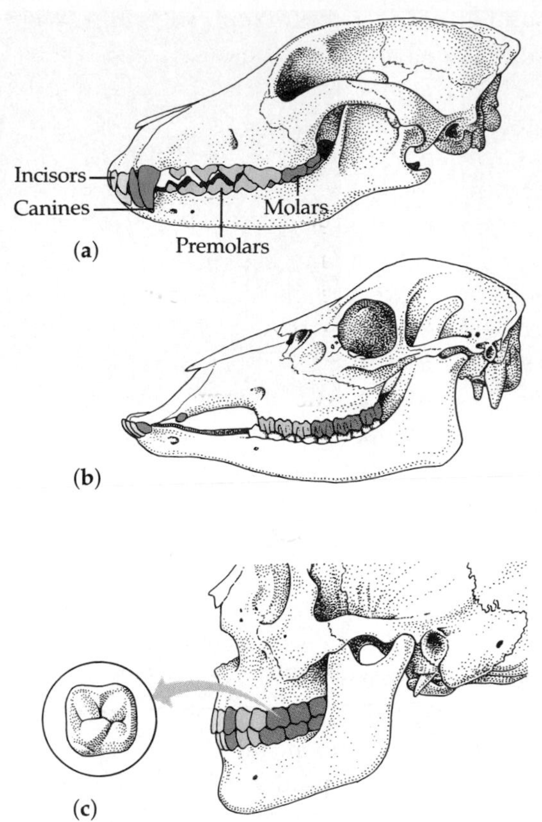

FIGURE 37.15
Dentition and diet. (**a**) Carnivores, such as members of the dog and cat families, generally have pointed incisors and canines that can be used to kill prey and rip away pieces of flesh. The jagged premolars and molars are modified for crushing and shredding. (**b**) In contrast, herbivorous mammals, such as horses and cows, usually have teeth with broad, ridged surfaces that work like millstones for grinding tough plant material. The incisors and canines are generally modified for biting off pieces of vegetation. (**c**) Humans, being omnivores adapted for eating both vegetation and meat, have a relatively unspecialized dentition. The permanent (adult) set of teeth is 32 in number. Beginning at the midline of the upper and lower jaw are two bladelike incisors for biting, a pointed canine for tearing, two premolars for grinding, and three molars for crushing.

Dentition, an animal's assortment of teeth, is one example of anatomical variation reflecting diet, especially in mammals. Compare the dentition of herbivores, carnivores, and omnivores in FIGURE 37.15. Nonmammalian vertebrates generally have less specialized dentition, but there are interesting exceptions. For example, poisonous snakes, such as rattlesnakes, have fangs, modified teeth that inject venom into prey. Some fangs are hollow, like syringes, while others drip the poison along grooves on the surfaces of the teeth. Snakes in general have another important anatomical adaptation associated with feeding: The lower jaw is loosely hinged to the skull by an elastic ligament that permits the mouth and throat to open *very* wide for swallowing large prey (once again, witness the astonishing episode recorded in FIGURE 37.4).

The length of the vertebrate digestive system is also correlated with diet. In general, herbivores and omnivores have longer alimentary canals relative to their body size than carnivores (FIGURE 37.16). Vegetation is more difficult to digest than meat because it contains cell walls. A longer tract furnishes more time for digestion and more surface area for the absorption of nutrients, which are usually less concentrated in vegetation than in meat. A model case is the frog, which changes its diet upon metamorphosis. Algae-eating tadpoles (frog larvae) have a coiled intestine that is very long relative to their size. During metamorphosis, the rest of the body grows more than the intestine, leaving the carnivorous adult frogs with a shorter intestine relative to their size.

In some vertebrates, the *functional* length of the alimentary canal is actually longer than superficial appearance reveals. For example, most sharks are carnivores, with a relatively short intestine lacking the "switchbacks" that lengthen the human intestine. However, the lining of a shark's intestine has folds that form a spiral valve, a corkscrewlike apparatus that increases surface area and provides a much longer route than if food were to move through the tube in a straight line.

In addition to being very long, the alimentary canals of many herbivorous mammals have special fermentation chambers where symbiotic bacteria and protozoa live. These microorganisms have enzymes that can digest cellulose, which the animal itself cannot digest. The microorganisms not only digest the cellulose to simple sugars but also convert the sugar to a variety of nutrients essential to the animal. Many herbivores, including horses, house symbiotic microorganisms in a large cecum, the pouch where the small and large intestines connect. The

FIGURE 37.16

The digestive tracts of a carnivore (coyote) and a herbivore (koala) compared. Although these two mammals are about the same size, the koala's intestines are much longer, an adaptation that enhances processing of fibrous, protein-poor eucalyptus leaves from which it obtains virtually all its food and water. Extensive chewing chops the leaves into very small pieces, increasing exposure of the food to digestive juices. The koala's cecum—at 2 m, the longest of any animal of equivalent size—functions as a fermentation chamber where symbiotic bacteria convert the shredded leaves into a more nutritious diet. The shorter length of the coyote's digestive tract is sufficient for digesting meat and absorbing nutrients from this diet.

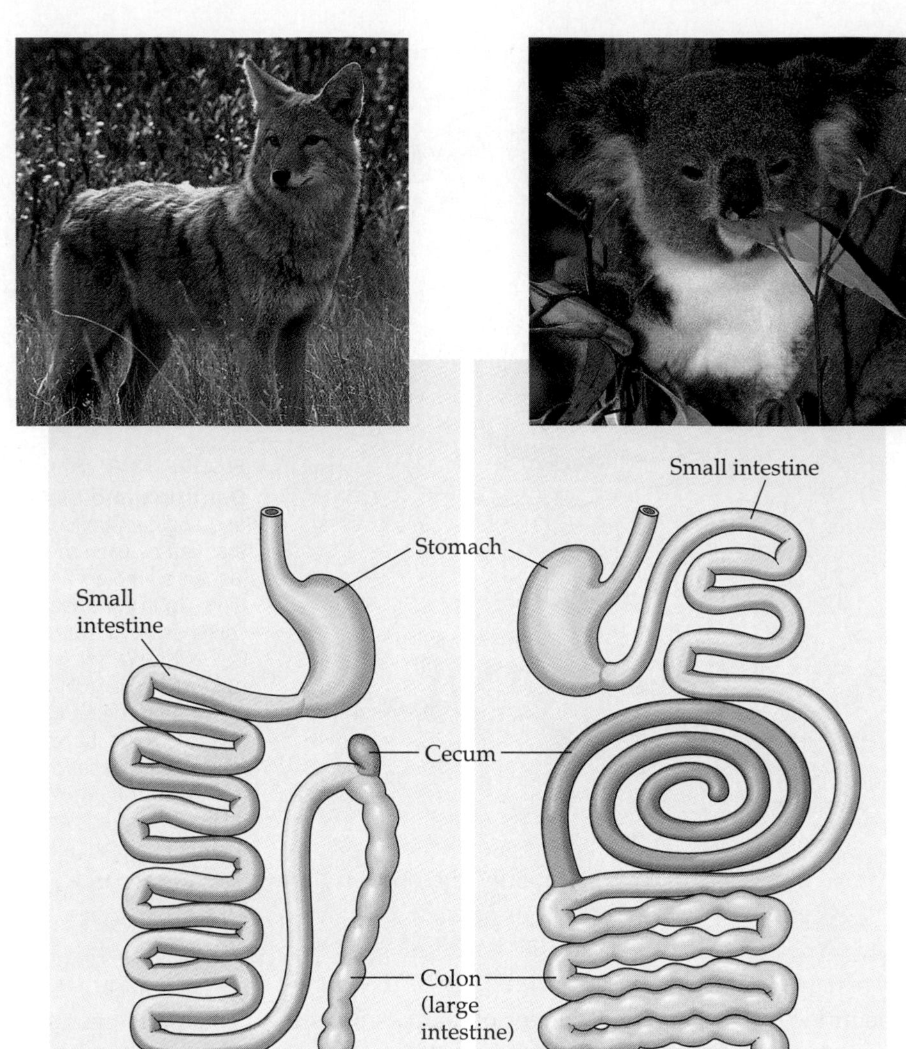

symbiotic bacteria of rabbits and some rodents live in the large intestine as well as in the cecum. Since most nutrients are absorbed in the small intestine, however, potentially nourishing by-products of fermentation by bacteria in the large intestine are initially lost with the feces. Rabbits and rodents obtain these nutrients by eating some of their feces and passing the food through the alimentary canal a second time. (The familiar rabbit "pellets," which are not reingested, are the feces eliminated after food has passed through the digestive tract the second time.) The koala, an Australian marsupial, also has an enlarged cecum, where symbiotic bacteria ferment finely shredded eucalyptus leaves (see FIGURE 37.16). The most elaborate adaptations for a herbivorous diet have evolved in the **ruminants,** which include cattle and sheep (FIGURE 37.17).

Having sampled some of the many evolutionary adaptations of digestive systems, we now shift our attention to the nutritional needs that must be satisfied by feeding and digestion.

An adequate diet provides fuel, carbon skeletons for biosynthesis, and essential nutrients

A nutritionally adequate diet satisfies three needs: fuel (chemical energy) for all the cellular work of the body; the organic raw materials animals use in biosynthesis (carbon skeletons to make many of their own molecules); and essential nutrients, substances the animal cannot make for itself from *any* raw material and that must be obtained in food in prefabricated form.

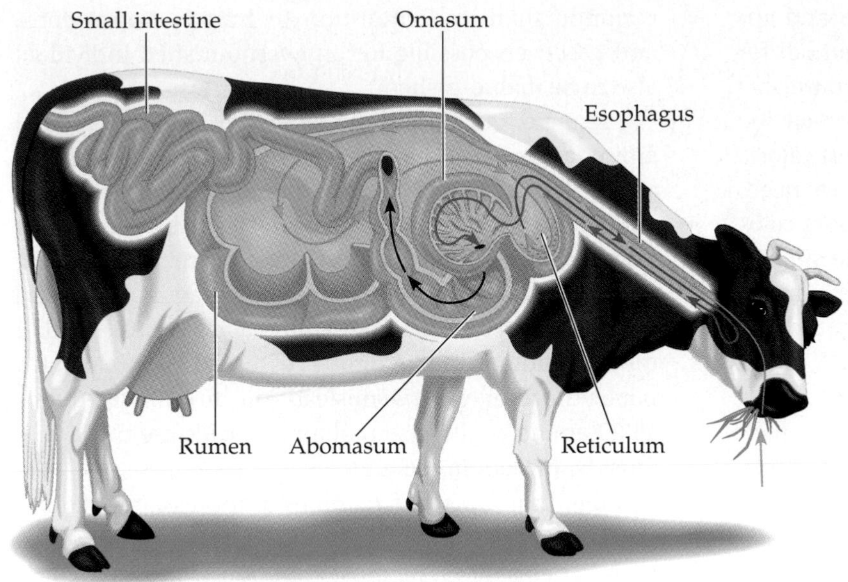

FIGURE 37.17
Ruminant digestion. The stomach of a ruminant has four chambers, three of which were probably modified from the lower end of the esophagus. When the cow first chews and swallows a mouthful of grass, boluses (green arrows) enter both the rumen and the reticulum, where symbiotic bacteria go to work on the cellulose-rich meal. As by-products of their metabolism, the bacteria secrete fatty acids. The cow periodically regurgitates and rechews the cud (red arrow), which further breaks down the fibers, making them more accessible to further bacterial action. The cow then reswallows the cud (blue arrow), which moves to the omasum, where water is removed. The cud, containing great numbers of bacteria, finally passes to the abomasum for digestion by the cow's own enzymes (black arrow). Because of the bacterial action, the diet from which a ruminant actually absorbs its nutrients is much richer than the grass the animal originally ate. In fact, a ruminant eating grass or hay obtains many of its nutrients by digesting the symbiotic bacteria, which reproduce rapidly enough in the rumen to maintain a stable population.

Food as Fuel

Animals obtain the chemical energy that powers the work of their body cells from the oxidation of complex organic molecules (see Chapter 9)—the carbohydrates, fats, and proteins that make up the bulk of the diet of most animals. The monomers of any of these substances can be used as fuel for the generation of ATP by cellular respiration, though priority is usually given to carbohydrates and fats. (Only when these substances are in short supply is protein used as the major fuel.) Fats are especially rich in energy; the oxidation of 1 g of fat liberates about 9.5 kcal, twice the energy liberated by 1 g of carbohydrate or protein.

As we saw in Chapter 36, animals have basal energy requirements that must be met to maintain the metabolic functions for sustaining life. When an animal takes in more calories than it consumes to meet its energy requirements, the excess calories are stored. The liver and muscles store energy in the form of glycogen. A human can store enough glycogen to supply about a day's worth of basal metabolism. If glycogen stores are full and caloric intake still exceeds caloric expenditure, the excess food is stored primarily in adipose tissue in the form of fat. This happens even if the diet contains little fat, because the liver converts excess carbohydrates or proteins into fat, which is then distributed to adipose tissue for storage.

Between meals or when the diet is deficient in calories, fuel is taken out of storage. The body generally expends glycogen stores in the liver first, and then draws on muscle glycogen and fat. Most of us have enough fat to sustain us in calories for several weeks. An animal can also consume its own proteins as a source of fuel, but this rarely occurs, unless the animal is starving or stressed in some other way for a prolonged period.

An **undernourished** person or other animal is one whose diet is deficient in calories. When starvation for calories persists and the body begins breaking down its own proteins for fuel, muscles begin to decrease in size, and even the brain can become protein-deficient. If the undernourished person survives, some of the damage is irreversible. Even a diet of a single staple such as rice or corn provides calories; thus, undernourishment is generally common only where drought, war, or some other crisis has severely disrupted the food supply. Another alarming cause of undernourishment is anorexia nervosa, an eating disorder associated with a compulsive aversion to body fat.

In the United States and other developed countries, overnourishment, or obesity, is a far more common problem than undernourishment. Obesity increases the risk of heart attack, diabetes, and several other disorders. Imagine the extra strain on your heart if you were to carry a 50-lb suitcase around with you all day. How heavy we can be before we are considered unhealthy is a disputed matter. Charts of "ideal weights" for people of various heights have been revised downward, upward, and then downward again over the past few years, depending on the current consensus of researchers. The billions of dollars spent each year on diet books and diet aids indicate that a substantial portion of the population would like to lose weight because of a concern with health and appearance. Unfortunately, there are no magic methods for slimming, and some of the fad diets—"eat all you want as long as it's carbohydrate," for instance—can actually be dangerous. Effective weight control is a matter of caloric bookkeeping. If we take in more calories than we need, the excess will be stored as fat. If we expend more calories than we take in, we balance the deficit by consuming body fat. Weight is stable when caloric demand is matched by caloric supply from the diet. To lose weight, we must eat less and exercise more.

Food for Biosynthesis

As heterotrophs, animals cannot make organic molecules from raw materials that are entirely inorganic. To synthesize the molecules it needs to grow and replenish itself, an animal must obtain organic precursors from its food. Given a source of organic carbon (such as sugar) and a source of organic nitrogen (such as amino acids from the digestion of protein), the animal can fabricate a great variety of organic molecules by using enzymes to rearrange the molecular skeletons of the precursors acquired from food. For instance, a single type of amino acid can supply nitrogen for the synthesis of several other types of amino acids that may not be present in the food. Also, we have seen that animals can synthesize fats from carbohydrates. In vertebrates, the conversion of nutrients from one type of organic molecule to another occurs mainly in the liver.

So we see that although animals depend on food as a source of organic carbon and nitrogen, they exhibit considerable versatility in the biosynthesis of their own organic molecules. However, there are some substances the animal cannot fabricate from any precursors.

Essential Nutrients

In addition to providing fuel and carbon skeletons for biosynthesis, an animal's diet must also supply certain substances in preassembled form. Chemicals an animal requires but cannot make are called **essential nutrients.** They vary from species to species, depending on the biosynthetic capabilities of the animal. A particular molecule can be an essential nutrient for one animal and yet be nonessential in the diet of another animal, one that can produce the molecule for itself from some precursor.

An animal whose diet is missing one or more essential nutrients is said to be **malnourished** (recall that *undernourished* refers to caloric deficiency). For example, cattle and other herbivores may suffer mineral deficiencies if they graze on plants grown in soil that itself is deficient in key minerals (FIGURE 37.18). Malnutrition is much more common than undernutrition in human populations, and it is even possible for an overnourished individual also to be malnourished.

There are four classes of essential nutrients: essential amino acids, essential fatty acids, vitamins, and minerals. Of the 20 amino acids required to make proteins, about half can be synthesized by most animals, as long as their diet includes organic nitrogen. The remaining ones, the **essential amino acids,** must be obtained from food in prefabricated form. Eight amino acids are essential in the adult human diet (a ninth is essential for infants). Do not let the term *essential* mislead you; an animal requires all 20 amino acids. Essential amino acids are those that must be present in the *diet.*

A diet that lacks one or more of the essential amino acids results in a form of malnutrition known generally as protein deficiency. The most common type of malnutrition among humans, protein deficiency is concentrated in geographical regions where there is a great gap between food supply and population size. The victims are

FIGURE 37.18
Obtaining essential nutrients. A giraffe, an herbivore of East Africa, chews on old bones from a dead mammal. Bones contain calcium phosphate, and osteophagia ("bone-eating") is common among herbivores living where soils and plants are deficient in phosphorus, a mineral nutrient required to make ATP, nucleic acids, phospholipids, and bones.

usually children, who, if they survive infancy, are likely to be retarded in physical, and, perhaps, mental development. In Africa, the syndrome is named **kwashiorkor,** meaning "rejected one," a reference to the onset of impaired physical development when a child is weaned from its mother's milk and placed on a starchy diet after a sibling is born. The problem of protein deficiency in some developing countries has been compounded by a trend away from breast-feeding altogether, an unfortunate dietary change that has been catalyzed by an aggressive marketing campaign by companies that make baby formula. Impoverished mothers often "stretch" the expensive formula by diluting it with water to such an extent that the protein content is insufficient. Also, uninformed consumers pay too little attention to hygiene when mixing and storing formula. In contrast, breast-feeding not only provides a balanced meal but also transfers temporary immunity to several infectious diseases from mother to child.

There is, of course, an economic component to protein deficiency. The most reliable sources of essential amino acids are meat and animal by-products (for example, eggs and cheese), which are relatively expensive. The proteins in animal products are complete, which means that they provide all the essential amino acids in their proper proportions. Most plant proteins, on the other hand, are incomplete, being deficient in one or more essential amino acids. Zein, for example, the principal protein in corn, is deficient in the amino acid lysine. People forced by economic necessity to obtain nearly all their calories from corn would show symptoms of protein deficiency, as would those who eat only rice, wheat, potatoes, or any other single staple. A vegetarian, however, can obtain sufficient quantities of all essential amino acids. The key is to eat a combination of plant foods that complement one another; two plants with incomplete proteins may not be deficient in the same amino acids. Beans, for example, supply the lysine that is missing in corn; and whereas beans are deficient in methionine, this amino acid is present in corn. Thus, a meal of beans and corn can provide all the essential amino acids (FIGURE 37.19). The combination of vegetables must be consumed at the same meal. The body cannot store amino acids, so a deficiency of a single essential amino acid retards protein synthesis and limits the use of other amino acids. Most cultures have, by trial and error, developed balanced diets that prevent protein deficiency.

Animals can synthesize most of the fatty acids they need. The **essential fatty acids,** the ones they cannot make, are certain unsaturated fatty acids (fatty acids having double bonds; see Chapter 5). In humans, for example, linoleic acid must be present in the diet in prefabricated form. This essential fatty acid is required to make some of the phospholipids found in membranes. Most diets furnish ample quantities of essential fatty acids, and thus deficiencies are rare.

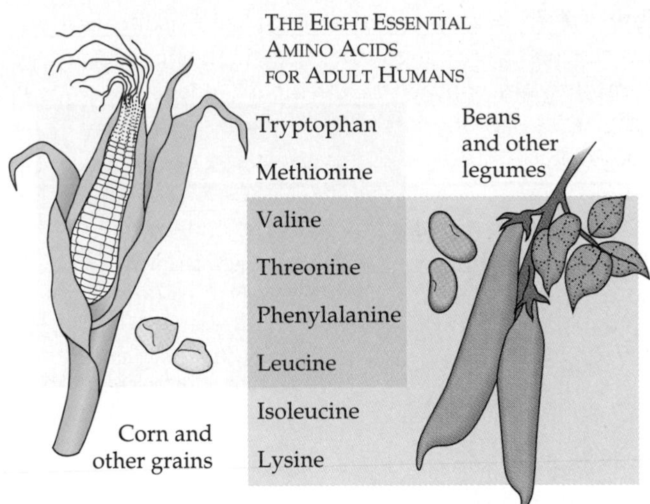

FIGURE 37.19
Essential amino acids from a vegetarian diet. Corn has little isoleucine and lysine. Beans have ample isoleucine and lysine, but little tryptophan and methionine. An adult human can obtain all eight essential amino acids by eating a meal of corn and beans.

Vitamins are organic molecules required in the diet in amounts that are quite small compared with the relatively large quantities of essential amino acids and fatty acids animals need (TABLE 37.2). Tiny amounts of vitamins may suffice, from about 0.01 to 100 mg per day, depending on the vitamin.

Although requirements for vitamins are modest, these molecules are absolutely essential in a nutritionally adequate diet. Deficiencies can cause severe problems. Indeed, the first vitamin to be identified, thiamine (vitamin B_1), was discovered as a result of the search for the cause of a disease called beriberi. Its symptoms include loss of appetite, fatigue, and nervous disorders. Beriberi was first described when it struck soldiers and prisoners in the Dutch East Indies during the nineteenth century. The dietary staple for these men was polished rice, which had the hulls (seed coats and other outer layers) removed to increase storage life. Not only the men who ate this diet developed beriberi, but also the chickens that ate the table scraps. It was found that beriberi could be prevented in both the men and the chickens by supplementing their diets with unpolished rice. Later, the active ingredient of rice hulls was isolated. Since it belongs to the chemical family known as amines, the compound was named "vitamine" (a vital amine). The *e* was later dropped, and the term has persisted, even though many of the vitamins subsequently discovered are not amines.

TABLE 37.2

Vitamin Requirements of Humans

VITAMIN	MAJOR DIETARY SOURCES	FUNCTIONS IN THE BODY	POSSIBLE SYMPTOMS OF DEFICIENCY OR EXTREME EXCESS
Water-Soluble Vitamins			
Vitamin B_1 (thiamine)	Pork, legumes, peanuts, whole grains	Coenzyme used in removing CO_2 from organic compounds	Beriberi (nerve disorders, emaciation, anemia)
Vitamin B_2 (riboflavin)	Dairy products, meats, enriched grains, vegetables	Component of coenzymes FAD and FMN	Skin lesions such as cracks at corners of mouth
Niacin	Nuts, meats, grains	Component of coenzymes NAD^+ and $NADP^+$	Skin and gastrointestinal lesions, nervous disorders Flushing of face and hands, liver damage
Vitamin B_6 (pyridoxine)	Meats, vegetables, whole grains	Coenzyme used in amino acid metabolism	Irritability, convulsions, muscular twitching, anemia Unstable gait, numb feet, poor coordination
Pantothenic acid	Most foods: meats, dairy products, whole grains, etc.	Component of coenzyme A	Fatigue, numbness, tingling of hands and feet
Folic acid (folacin)	Green vegetables, oranges, nuts, legumes, whole grains (also made by colon bacteria)	Coenzyme in nucleic acid and amino acid metabolism	Anemia, gastrointestinal problems May mask deficiency of vitamin B_{12}
Vitamin B_{12}	Meats, eggs, dairy products	Coenzyme in nucleic acid metabolism; needed for maturation of red blood cells	Anemia; nervous system disorders
Biotin	Legumes, other vegetables, meats	Coenzyme in synthesis of fat, glycogen, and amino acids	Scaly skin inflammation; neuromuscular disorders
Vitamin C (ascorbic acid)	Fruits and vegetables, especially citrus fruits, broccoli, cabbage, tomatoes, green peppers	Used in collagen synthesis (e.g., for bone, cartilage, gums); antioxidant; aids in detoxification; improves iron absorption	Scurvy (degeneration of skin, teeth, blood vessels), weakness, delayed wound healing, impaired immunity Gastrointestinal upset
Fat-Soluble Vitamins			
Vitamin A (retinol)	Provitamin A (beta-carotene) in deep green and orange vegetables and fruits; retinol in dairy products	Component of visual pigments; needed for maintenance of epithelial tissues; antioxidant; helps prevent damage to lipids of cell membranes	Vision problems; dry, scaling skin Headache, irritability, vomiting, hair loss, blurred vision, liver and bone damage
Vitamin D	Dairy products, egg yolk (also made in human skin in presence of sunlight)	Aids in absorption and use of calcium and phosphorus; promotes bone growth	Rickets (bone deformities) in children; bone softening in adults Brain, cardiovascular, and kidney damage
Vitamin E (tocopherol)	Vegetable oils, nuts, seeds	Antioxidant; helps prevent damage to lipids of cell membranes	None well documented in humans; possibly anemia
Vitamin K (phylloquinone)	Green vegetables, tea (also made by colon bacteria)	Important in blood clotting	Defective blood clotting Liver damage and anemia

TABLE 37.3

Mineral Requirements of Humans

MINERAL	DIETARY SOURCES	SOME MAJOR FUNCTIONS IN THE BODY	SYMPTOMS OF DEFICIENCY*
Calcium (Ca)	Dairy products, dark green vegetables, legumes	Bone and tooth formation, blood clotting, nerve and muscle function	Retarded growth, possibly loss of bone mass
Phosphorus (P)	Dairy products, meats, grains	Bone and tooth formation, acid-base balance, nucleotide synthesis	Weakness, loss of minerals from bone, calcium loss
Sulfur (S)	Proteins from many sources	Component of certain amino acids	Symptoms of protein deficiency
Potassium (K)	Meats, dairy products, many fruits and vegetables, grains	Acid-base balance, water balance, nerve function	Muscular weakness, paralysis, nausea, heart failure
Chlorine (Cl)	Table salt	Acid-base balance, formation of gastric juice	Muscle cramps, reduced appetite
Sodium (Na)	Table salt	Acid-base balance, water balance, nerve function	Muscle cramps, reduced appetite
Magnesium (Mg)	Whole grains, green leafy vegetables	Cofactor; ATP bioenergetics	Nervous system disturbances
Iron (Fe)	Meats, eggs, legumes, whole grains, green leafy vegetables	Component of hemoglobin and of electron-carriers in energy metabolism; enzyme cofactor	Iron-deficiency anemia, weakness, impaired immunity
Fluorine (F)	Drinking water, tea, seafood	Maintenance of tooth (and probably bone) structure	Higher frequency of tooth decay
Zinc (Zn)	Meats, seafood, grains	Component of certain digestive enzymes and other proteins	Growth failure, scaly skin inflammation, reproductive failure, impaired immunity
Copper (Cu)	Seafood, nuts, legumes, organ meats	Enzyme cofactor in iron metabolism, melanin synthesis, electron transport	Anemia, bone and cardiovascular changes
Manganese (Mn)	Nuts, grains, vegetables, fruits, tea	Enzyme cofactor	Abnormal bone and cartilage
Iodine (I)	Seafood, dairy products, iodized salt	Component of thyroid hormones	Goiter (enlarged thyroid)
Cobalt (Co)	Meats and dairy products	Component of vitamin B_{12}	None, except as B_{12} deficiency
Selenium (Se)	Seafood, meats, whole grains	Enzyme cofactor; antioxidant functioning in close association with vitamin E	Muscle pain; possibly heart muscle deterioration
Chromium (Cr)	Brewer's yeast, liver, seafood, meats, some vegetables	Involved in glucose and energy metabolism	Impaired glucose metabolism
Molybdenum (Mo)	Legumes, grains, some vegetables	Enzyme cofactor	Disorder in excretion of nitrogen-containing compounds

*All of these minerals are also harmful when consumed in excess.

So far, 13 vitamins essential to humans have been identified (see TABLE 37.2). The compounds are grouped into two categories: water-soluble vitamins and fat-soluble vitamins. Water-soluble vitamins include the B complex, which consists of several compounds that generally function as coenzymes in key metabolic processes. Vitamin C (ascorbic acid) is also water-soluble. Ascorbic acid is required for the production of connective tissue. Excesses of water-soluble vitamins are excreted in urine, and moderate overdoses of these vitamins are probably harmless.

The fat-soluble vitamins are A, D, E, and K. Vitamin A is incorporated into visual pigments of the eye. Vitamin D aids in calcium absorption and bone formation. The function of vitamin E is not yet fully understood, but it seems to protect the phospholipids in membranes from oxidation. Vitamin K is required for blood clotting. Excesses of fat-soluble vitamins are not excreted but are deposited in body fat, so overdoses may result in an accumulation of these compounds to toxic levels.

The subject of vitamin dosage has aroused heated debate. Some believe it is sufficient to meet recommended daily allowances (RDAs), the nutrient intake recommendations for healthy people, deliberately set to be higher than most people's actual needs. Others argue that RDAs are set too low for some vitamins and that we should be thinking in terms of *optimal* requirements. Research is far from complete, and debate continues, especially over optimal doses of vitamins C and E. All that can be said with any certainty at this time is that people who eat a balanced diet are not likely to develop symptoms of vitamin deficiency.

Research involving experimental animals is of limited value in assessing the vitamin needs of humans. A compound that is a vitamin for one species is not a vitamin for another species that can synthesize the compound for itself. For example, ascorbic acid is a vitamin for humans, other primates, guinea pigs, and some birds and snakes, but not for most other animals.

Minerals are inorganic nutrients, usually required in very small amounts, from less than 1 mg to about 2500 mg per day, depending on the mineral (TABLE 37.3, p. 815). As with vitamins, mineral requirements vary with animal species. Humans and other vertebrates require relatively large quantities of calcium and phosphorus for the construction and maintenance of bone. Calcium is also necessary for the normal functioning of nerves and muscles, and phosphorus is also an ingredient of ATP and nucleic acids. Iron is a component of the cytochromes that function in cellular respiration (see Chapter 9) and of hemoglobin, the oxygen-binding protein of red blood cells. Magnesium, iron, zinc, copper, manganese, selenium, and molybdenum are cofactors built into the structure of certain enzymes; magnesium, for example, is present in enzymes that split ATP. Vertebrates need iodine to make thyroid hormones, which regulate metabolic rate. Sodium, potassium, and chlorine are important in nerve function and also have a major influence on the osmotic balance between cells and the interstitial fluid. Herbivores, such as deer, cattle, and sheep, often seem to crave salt; vegetation generally contains a very low concentration of sodium chloride. Salt licks, either natural or placed on the range by ranchers, attract large numbers of herbivores (FIGURE 37.20). Most humans, on the other hand, ingest far more salt than they need. In the United States, the average person consumes enough salt to provide about twenty times the required amount of sodium.

FIGURE 37.20
A natural salt lick. Many herbivores cannot obtain enough sodium and chloride from plants and supplement their diet by visiting natural salt licks, as this female bighorn sheep and her lamb are doing.

In summary, a nutritionally adequate diet must supply enough calories to satisfy energy needs, raw materials for biosynthesis, and ample quantities of the essential nutrients.

* * *

Although the focus of this chapter has been on nutrition and digestion, nerves, blood vessels, and hormones have all been part of the discussion. No organ system functions alone. The animal is an integrated whole, a concert of organ systems working together. In the next chapter, you will learn about the parts played by the circulatory and respiratory systems.

REVIEW OF KEY CONCEPTS (with page numbers and key figures)

- Diets and feeding mechanisms vary extensively among animals (pp. 796–797)
  - All animals are heterotrophs and must obtain their nutrients by consuming organic molecules.
  - Animals may obtain nutrients by suspension-, substrate-, deposit-, fluid-, or bulk-feeding.
- Ingestion, digestion, absorption, and elimination are the four main stages of food processing (pp. 797–798)
  - Food processing in animals involves ingestion (the act of eating), digestion (enzymatic breakdown of the macromolecules of food into their monomers), absorption (the uptake of nutrients by body cells), and elimination (the passage of undigested materials out of the body in feces).
  - Digestion occurs inside compartments separated from living cytoplasm.
- Digestion occurs in food vacuoles, gastrovascular cavities, and alimentary canals (pp. 798–800; FIGURES 37.7, 37.8)
  - In intracellular digestion, food particles are engulfed by endocytosis and digested within food vacuoles.
  - Most animals use extracellular digestion, with enzymatic hydrolysis occurring outside cells in a gastrovascular cavity or alimentary canal.
  - The gastrovascular cavities of cnidarians and flatworms have a single opening through which food enters and undigested wastes pass.
  - Most animals have digestive tracts, or alimentary canals, that move food through a one-way tube with specialized regions for digestion and absorption.
- A tour of the mammalian digestive system (pp. 801–808, FIGURE 37.9, TABLE 37.1)
  - The mammalian digestive tract has a four-layered wall over most of its length. The smooth muscle layer propels food along the tract by peristalsis and regulates its passage through strategic points by means of sphincters.
  - Mammals have accessory glands that add digestive secretions to the tract through ducts. These are the salivary glands, pancreas, and liver.
  - Digestion begins in the oral cavity, where teeth chew food into smaller particles that are exposed to salivary amylase. Saliva also contains buffers, antibacterial agents, and mucin for lubricating the food.
  - The pharynx is the intersection leading to the trachea and the esophagus. The epiglottis usually prevents food from entering the trachea.
  - The esophagus conducts food from the pharynx to the stomach by involuntary peristaltic waves.
  - The stomach stores food and secretes gastric juice, which converts a meal to acid chyme. Gastric juice includes hydrochloric acid and the enzyme pepsin. Nerve impulses and the hormone gastrin regulate gastric motility and secretion.
  - Most digestion and virtually all absorption occur in the small intestine, the longest segment of the alimentary canal.
  - The pancreas and gallbladder, which stores bile secreted by the liver, empty by ducts into the duodenum, the first part of the small intestine. Regulatory hormones, such as secretin and cholecystokinin, regulate the activities of the pancreas and gallbladder.
  - Carbohydrate digestion, begun in the mouth, continues in the duodenum in the presence of pancreatic amylase in the lumen and disaccharidases built into the plasma membranes of intestinal epithelial cells.
  - Trypsin, chymotrypsin, carboxypeptidase, and aminopeptidase, pancreatic enzymes that work together to hydrolyze polypeptides, are secreted into the small intestine as inactive zymogens that are subsequently activated by enteropeptidase, an intestinal enzyme. Dipeptidases on the intestinal cells complete protein digestion.
  - Teams of hydrolytic enzymes (nucleases, nucleotidases, and nucleosidases) in the small intestine digest the nucleic acids DNA and RNA into their component nitrogenous bases, sugars, and phosphates.
  - Fats are broken up into smaller droplets during emulsification by bile salts in the intestinal lumen, thereby allowing maximum exposure to the enzyme lipase.
  - Most digestion is completed in the duodenum, and the remaining regions of the small intestine, the jejunum and ileum, are involved mainly in absorption.
  - The structure of the small intestine fits its absorptive function. The large folds of the lining have fingerlike villi, whose cells have microscopic microvilli, all greatly increasing the surface area. In the core of each villus is a network of capillaries and a lacteal that take up and distribute absorbed nutrients. Nutrients are absorbed by passive diffusion or by active transport.
  - Nutrient-laden blood from the villi is conveyed through the large hepatic portal vessel to the liver, which regulates the nutrient content of the blood.
  - The colon aids the small intestine in reabsorbing water and houses bacteria, some of which synthesize vitamin K. Feces pass through the rectum and out the anus.
- Vertebrate digestive systems exhibit many evolutionary adaptations associated with diet (pp. 808–810; FIGURES 37.15, 37.16)
  - A mammal's dentition is generally correlated with its diet.
  - Herbivores generally have longer alimentary canals, reflecting the longer time needed to digest vegetation. Many herbivorous mammals have special fermentation chambers in the stomach, cecum, or intestines, where symbiotic microorganisms digest cellulose.
- An adequate diet provides fuel, carbon skeletons for biosynthesis, and essential nutrients (pp. 810–816; TABLES 37.2, 37.3)
  - Carbohydrates and fats are most often used as fuel. Monomers of carbohydrates, proteins, fats, and nucleic acids are used in biosynthesis.
  - Animals store excess calories as glycogen in the liver and muscles and as fat in adipose tissue. Undernourished animals have diets deficient in calories.

- Essential nutrients must be supplied in preassembled form because the body lacks the machinery for their biosynthesis. Malnourished animals are missing one or more of the essential nutrients.
- Essential amino acids are those an animal cannot make from nitrogen-containing precursors.
- Animals can synthesize most essential fatty acids; there are a few essential unsaturated fatty acids.
- Vitamins are organic molecules, many of which serve as coenzymes or parts of coenzymes; they are required in small amounts.
- Minerals are inorganic nutrients that are required in varying amounts, from about 1 to 2500 mg per day. Many are cofactors of enzymes. Some are important in bone and tooth formation, muscle activity, cellular bioenergetics, and iron metabolism.

SELF-QUIZ

1. In function, a paramecium's food vacuole is most analogous to the human
   a. mouth c. esophagus e. anus
   b. small intestine d. liver

2. Which of the following animals lack alimentary canals (complete digestive systems)?
   a. earthworms c. insects e. birds
   b. jellyfish d. fishes

3. Our oral cavity, with its dentition, is most functionally analogous to an earthworm's
   a. intestine c. gizzard e. anus
   b. pharynx d. stomach

4. Which of the following enzymes has the lowest pH optimum? (Explain your answer.)
   a. salivary amylase d. pancreatic amylase
   b. trypsin e. pancreatic lipase
   c. pepsin

5. After surgical removal of an infected gallbladder, a person must be especially careful to restrict his or her dietary intake of
   a. starch c. sugar e. water
   b. protein d. fat

6. Trypsinogen, a pancreatic zymogen secreted into the duodenum, can be activated by
   a. chymotrypsin c. secretin e. pepsin
   b. enterogastrone d. trypsin

7. Pointed molars and premolars would most likely be part of the dentition of a
   a. human c. lion e. hawk
   b. cow d. rabbit

8. What do the typhlosole of an earthworm, the spiral valve of a shark, and the villi of a mammal all have in common?
   a. all are adaptations for the efficient digestion and absorption of meat
   b. they are all adaptations of the stomach
   c. they are all microscopic structures
   d. they all increase the absorptive surface area of intestinal epithelium
   e. they are all homologous structures

9. If you were to sprint 100 m a few hours after lunch, which stored fuel would you probably tap?
   a. muscle proteins d. fats stored in adipose tissue
   b. muscle glycogen e. blood proteins
   c. fat stored in the liver

10. Which of the following is *not* an essential nutrient in humans?
    a. vitamin A d. calcium
    b. lysine e. iron
    c. glucose

CHALLENGE QUESTIONS

1. Trace a bacon, lettuce, and tomato sandwich through the human alimentary canal, describing what happens to the food in each region of the tract.

2. Some essential mineral nutrients, such as selenium and molybdenum, are required in the human diet in minuscule amounts—thousandths of a milligram. What kinds of experiments and observations would help demonstrate that a mineral is required by humans?

SCIENCE, TECHNOLOGY, AND SOCIETY

1. According to recent surveys, 75% of Americans 18 to 35 years old think they are fat. At any given time, 30 million women and 18 million men are on diets. Only 25% are *actually* medically overweight. What do you think causes so many people who are not overweight to be concerned about their weight? Other data show that more than twice as many high school girls as boys see themselves as overweight, and 45% of underweight women see themselves as fat. Why do you think there is a difference between men and women regarding their perception of their weight?

2. Famine plagues certain parts of the world today. Some people argue that distributing food more equitably to the various countries would reduce starvation, at least for a while. Others counter that it is erroneous and ultimately harmful to perceive starvation as a global problem, because the causes and long-term solutions are usually regional. Evaluate the biological, political, and ethical aspects of this debate.

FURTHER READING

Alper, J. "Ulcers as an Infectious Disease." *Science*, April 9, 1993. The evidence for bacteria as the cause of gastric ulcers and a discussion of potential treatments.

Christian, J. L., and J. L. Greger. *Nutrition for Living*, 3rd ed. Redwood City, CA: Benjamin/Cummings, 1991. A detailed discussion of nutrition, related to life circumstances.

Diamond, J. "The Athlete's Dilemma." *Discover*, August 1991. Have marathoners and other endurance athletes reached the limits of what metabolism can support?

Diamond, J. "Dining with the Snakes." *Discover*, April 1994. How pythons and other large constrictors obtain and digest their food.

Grajal, A., and S. D. Strahl. "A Bird with the Guts to Eat Leaves." *Natural History*, August 1991. Only one species of bird, the hoatzin, an herbivore found in tropical South America, has a foregut in which bacteria digest plant material in the manner of ruminant mammals.

Sanderson, S., and R. Wasserug. "Suspension Feeding Vertebrates." *Scientific American*, March 1990. This article emphasizes suspension-feeding whales.

Scrimshaw, N. "Iron Deficiency." *Scientific American*, October 1991. One of the most common nutritional deficiencies in the world.

Willett, W. C. "Diet and Health: What Should We Eat?" *Science*, April 22, 1994. A critical review of human nutrition and the relationship between diets and major diseases; includes recommendations based on current research.

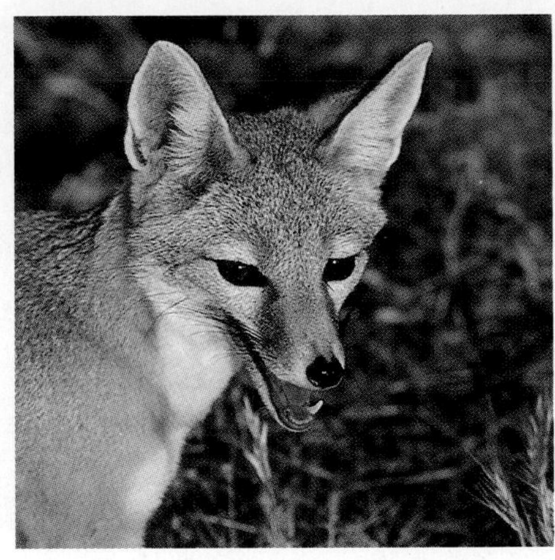

CHAPTER 38

CIRCULATION AND GAS EXCHANGE

KEY CONCEPTS

- Transport systems functionally connect body cells with the organs of exchange: *an overview*
- Most invertebrates have a gastrovascular cavity or a circulatory system for internal transport
- Diverse adaptations of a cardiovascular system have evolved in vertebrates
- Rhythmic pumping of the mammalian heart drives blood through pulmonary and systemic circuits
- The lymphatic system returns fluid to the blood and aids in body defense
- Blood is a connective tissue with cells suspended in plasma
- Cardiovascular diseases are the leading cause of death in the United States and many other developed nations
- Gas exchange supplies oxygen for cellular respiration and disposes of carbon dioxide: *an overview*
- Gills are respiratory adaptations of most aquatic animals
- Tracheae are respiratory adaptations of insects
- Lungs are the respiratory adaptations of most terrestrial vertebrates

Every organism must exchange materials and energy with its environment, and as we have seen, this commerce ultimately occurs at the cellular level. Cells live in aqueous surroundings, with the resources they need, such as nutrients and oxygen, passing into the cytoplasm across the plasma membrane, and metabolic wastes, such as carbon dioxide, passing out. Substances crossing the plasma membrane are dissolved in the water bathing the cell. For a protozoan living in an aquatic habitat, the environment is the surrounding pond water or seawater, and chemical exchange is accomplished simply by diffusion or active transport across the plasma membrane. Because a protozoan is small, its external surface is sufficient to service the entire volume of the organism. This also works for some animals, such as Hydra *and many sponges, whose body plans expose virtually every cell to the surroundings (see* Figure 36.9*).*

The exposed tongue of the kit fox in the photograph that opens this chapter presents a sizable moist surface area to the outside environment, and much of the animal's excess body heat passes out across its tongue and mouth surfaces. However, little or no oxygen or nutrients diffuse inward this way. Foxes and most other animals have organ systems specialized for exchanging materials with the environment, and many have an internal transport system that conveys fluid (blood or interstitial fluid) throughout the body.

In this chapter, you will learn about mechanisms of internal transport in animals. You will also learn about one of the most important cases of chemical transfer between animals and their environment: the exchange of the gases oxygen and carbon dioxide, which is essential to cellular respiration and bioenergetics.

Transport systems functionally connect body cells with the organs of exchange: *an overview*

The time it takes for a substance to diffuse from one place to another is proportional to the square of the distance the chemical will travel. For example, if it takes 1 second for a given quantity of glucose to diffuse 100 μm, it will take 100 seconds for the same quantity to diffuse 1 mm. Clearly, diffusion alone is not adequate for transporting chemicals over macroscopic distances in animals—for example, for moving glucose from the digestive tract and oxygen from the lungs to the brain of a mammal.

The circulatory system solves the problem by ensuring that no substance has to diffuse far to enter or leave a

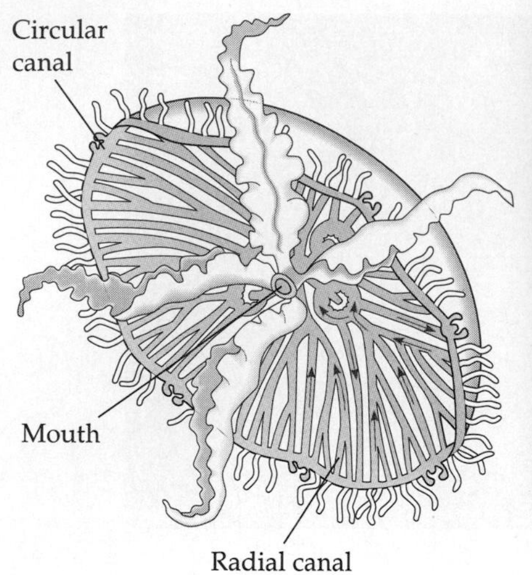

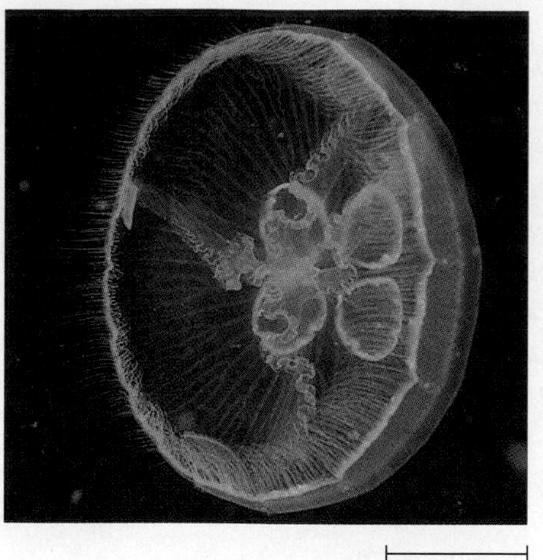

FIGURE 38.1
Internal transport in the cnidarian *Aurelia.* The mouth leads to an elaborate gastrovascular cavity (shown in gold) that has branches radiating to and from a circular canal. Ciliated cells lining the canals circulate fluid in the directions indicated by the arrows. The animal is viewed here from its underside (oral surface).

cell. By transporting fluid throughout the body, it functionally connects the aqueous environment of the body cells to the organs specialized for exchanging gases, absorbing nutrients, and disposing of wastes. In the lungs of a mammal, for example, oxygen from inhaled air diffuses across a thin epithelium and into the blood, while carbon dioxide diffuses in the opposite direction. The circulatory system then carries the oxygen-rich blood to all parts of the body. As the blood streams through the tissues within microscopic vessels called capillaries, chemicals are transported between the blood and the interstitial fluid that directly bathes the cells.

Our focus in this chapter is the functional relationship between the circulatory system and gas exchange organs. However, we will also see that by transporting blood throughout the body, the circulatory system plays a key role in homeostasis (see Chapter 36). Having an internal "pond"—blood and the interstitial fluid it services—makes it possible to control the chemical and physical properties of the immediate surroundings of the cells. The circulatory system facilitates this control by conveying fluid between the body cells and organs such as the liver and kidneys, which regulate the fluid's contents of nutrients and wastes. We will examine the role of the circulatory and waste-disposal systems in homeostasis in detail in Chapter 40.

Most invertebrates have a gastrovascular cavity or a circulatory system for internal transport

Gastrovascular Cavities

The body plan of *Hydra* and other cnidarians makes any specialized system for internal transport unnecessary. A body wall only two cells thick encloses a central gastrovascular cavity, which serves the dual functions of digestion and distribution of substances throughout the body (see FIGURE 37.7a). The fluid inside the cavity is continuous with the water outside through a single opening; thus, both inner and outer layers of tissue are bathed by fluid. Thin strands of the gastrovascular cavity extend into the tentacles of *Hydra,* and some cnidarians have even more elaborate gastrovascular cavities (FIGURE 38.1). Since digestion begins in the cavity, only the cells of the inner layer have direct access to nutrients, but the nutrients have only a short distance to diffuse to the cells of the outer layer.

Planarians and other flatworms also have gastrovascular cavities that exchange materials with the environment through a single opening (see FIGURE 37.7b). The flat shape of the body and the branching of the gastrovascular cavity throughout the animal ensure that all cells are bathed by a suitable medium.

Open and Closed Circulatory Systems

A gastrovascular cavity is inadequate for internal transport within animals having many layers of cells, especially if the animals live out of water. In insects, other arthropods, and most mollusks, blood bathes the internal organs directly. This arrangement is called an **open circulatory system** (FIGURE 38.2a). There is no distinction between blood and interstitial fluid, and the general body fluid is more correctly termed **hemolymph.** Chemical exchange between the fluid and body cells occurs as one or more hearts pump the hemolymph into an interconnected system of **sinuses,** which are spaces surrounding the organs. Hemolymph is circulated by body movements that squeeze the sinuses and by the contraction of the heart. In grasshoppers and other arthropods, for example, the heart is an elongated tube located dorsally. When the heart contracts, it pumps hemolymph

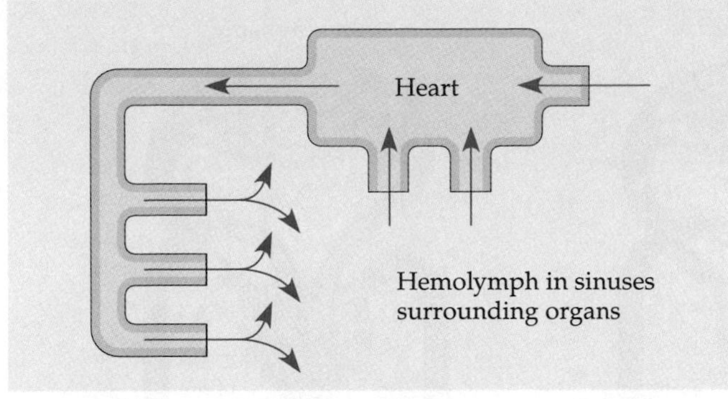

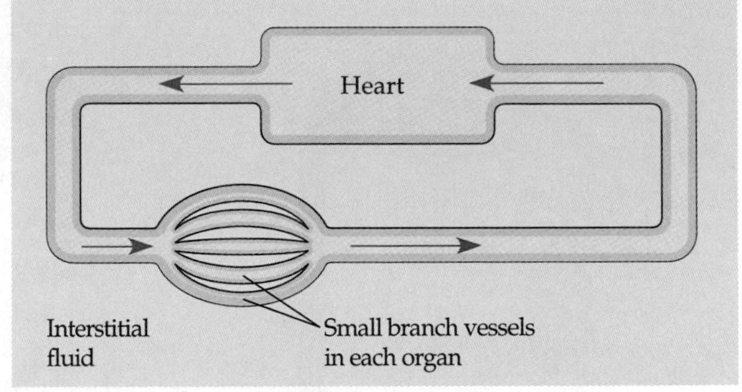

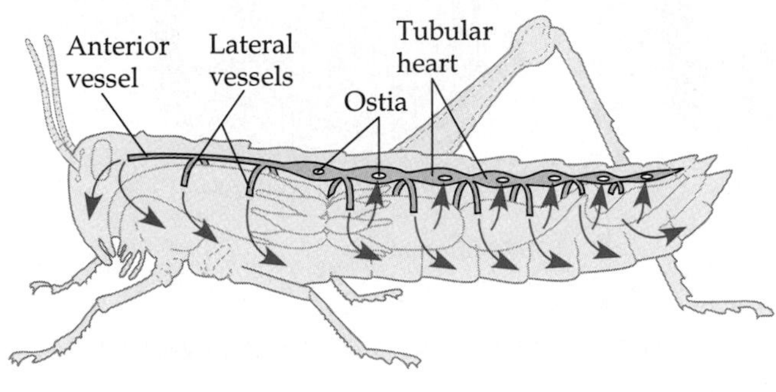

(a)

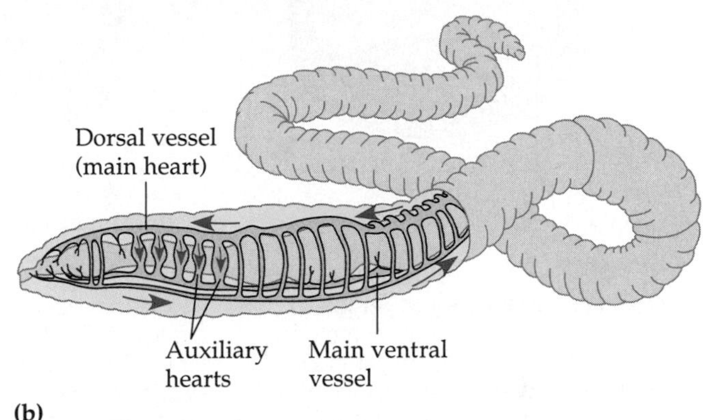

(b)

FIGURE 38.2

Open and closed circulatory systems. (**a**) In an open circulatory system, blood and interstitial fluid are the same, and this fluid is called hemolymph. The heart pumps hemolymph through vessels into sinuses, where materials are exchanged between the hemolymph and cells. Hemolymph returns to the heart through ostia. In a grasshopper, a dorsal tubular heart pumps hemolymph forward through an anterior vessel and to the sides of the animal through a series of lateral vessels. Squeezed rearward through sinuses by body movements, the hemolymph reenters the heart through ostia, which are equipped with valves that close when the heart contracts. (**b**) A closed circulatory system contains blood within vessels, distinct from the interstitial fluid. In an earthworm, three major vessels, one dorsal and two ventral, branch into smaller vessels that supply blood to the various organs. The dorsal vessel functions as the main heart, pumping blood forward by peristalsis. Near the worm's anterior end, five pairs of vessels loop around the digestive tract, connecting the dorsal and the main ventral vessel. The paired vessels function as auxiliary hearts, propelling blood ventrally. Blood flows rearward in the ventral vessels.

out into sinuses. When the heart relaxes, it draws hemolymph in through pores called ostia.

In a **closed circulatory system,** blood is confined to vessels and is distinct from the interstitial fluid (FIGURE 38.2b). The heart (or hearts) pumps blood into large vessels that branch into smaller ones coursing through the organs. The blood exchanges materials with the interstitial fluid bathing the cells. Earthworms, squids, octopuses, and vertebrates have closed circulatory systems.

In general, blood percolates through the sinuses of an open circulatory system more slowly than it flows through the vessels of a closed system. Because most animals depend on their circulatory system to transport oxygen for cellular respiration, we might expect to find open systems only in animals that move sluggishly. Yet flying insects, among the most active of all animals, have open circulatory systems. Insects, however, do not use blood to carry oxygen long distances. Oxygen infiltrates the insect body through air ducts called tracheae, which we will discuss in detail later in the chapter.

It is important to avoid thinking of animals with open circulatory systems as disadvantaged in any way. Open and closed systems are alternative solutions to the problem of moving fluids within large multicellular bodies, and both are highly effective adaptations.

Diverse adaptations of a closed cardiovascular system have evolved in vertebrates

Internal transport is accomplished in humans and other vertebrates by a closed circulatory system, also called the cardiovascular system. The components of the **cardiovascular system** are the heart, blood vessels, and blood. The heart has one **atrium** or two **atria** (plural), the chambers that receive blood returning to the heart, and one or two **ventricles,** the chambers that pump blood out of the heart. Arteries, veins, and capillaries are the three main

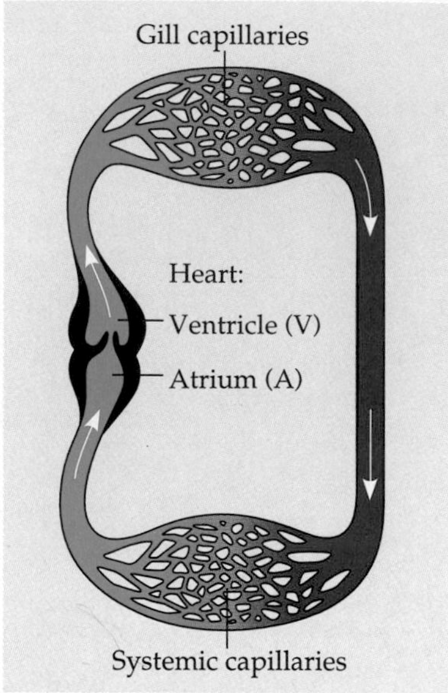

(a) Fish

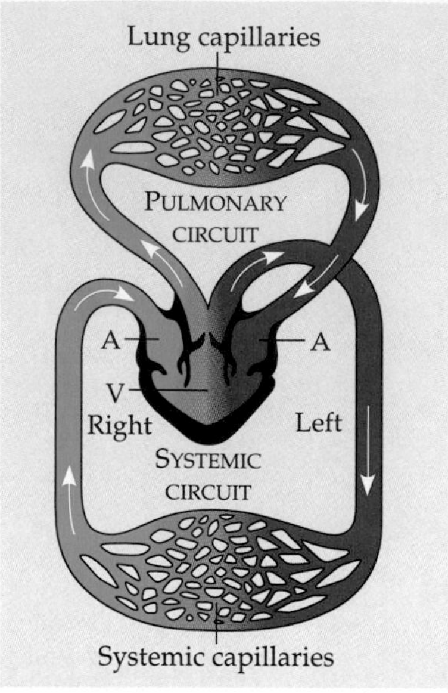

(b) Amphibian

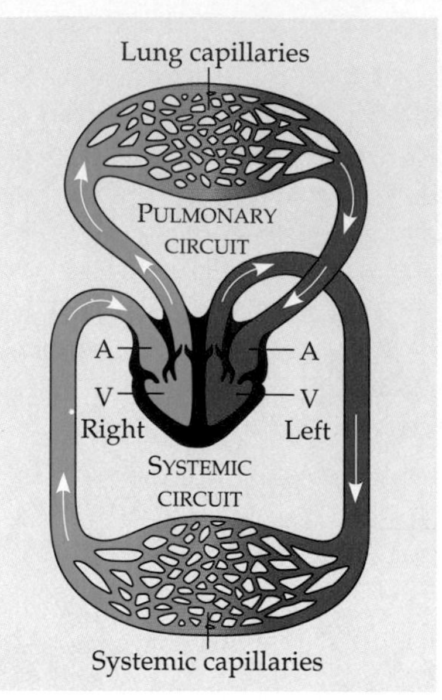

(c) Mammal

FIGURE 38.3
Generalized circulatory schemes of vertebrates. Red symbolizes oxygen-rich blood, and blue represents oxygen-poor blood. (**a**) Fishes have a two-chambered heart and a single circuit of blood flow, with the atrium of the heart receiving oxygen-poor blood from veins and the ventricle pumping blood to the gills. (**b**) Amphibians have a three-chambered heart and two circuits of blood flow: pulmonary and systemic. This double circulation delivers blood to systemic organs under high pressure. In the single ventricle, there is some mixing of oxygen-rich with oxygen-poor blood. (It is customary to illustrate cardiovascular systems with the right side of the heart on the left and the left side on the right, as if the body is facing you from the page.) (**c**) Mammals have a four-chambered heart and double circulation. Within the heart, oxygen-rich blood is kept completely segregated from oxygen-poor blood.

kinds of blood vessels, which in the human body have been estimated to extend a total distance of 100,000 km. **Arteries** carry blood away from the heart to organs throughout the body. Within these organs, arteries branch into **arterioles,** small vessels that convey blood to the capillaries. **Capillaries** are microscopic vessels with very thin, porous walls. Networks of these vessels, called **capillary beds,** infiltrate each tissue. It is across the thin walls of capillaries that chemicals are exchanged between the blood and the interstitial fluid surrounding the cells. At their "downstream" end, capillaries converge into **venules,** and venules converge into veins. **Veins** return blood to the heart. Notice that arteries and veins are distinguished by the *direction* in which they carry blood, not by the quality of the blood they contain. Not all arteries carry oxygen-rich blood, and not all veins carry blood with a low concentration of oxygen. But all arteries do carry blood from the heart *toward* capillaries, and only veins return blood to the heart *from* capillaries. We will now examine the routes of blood flow in different classes of vertebrates.

Various adaptations of the general circulatory scheme just described have evolved among vertebrates.

A fish has a heart with two main chambers, one atrium and one ventricle (FIGURE 38.3a). Blood pumped from the ventricle travels first to the gills, where the blood picks up oxygen and disposes of carbon dioxide across capillary walls. The gill capillaries converge into a vessel that carries the oxygen-rich blood to capillary beds in all other parts of the body. Blood then returns in veins to the atrium of the heart. Notice that in a fish, blood must pass through *two* capillary beds during each circuit, one in the gills and a second one, called systemic capillaries, in an organ other than the gills. When blood flows through a capillary bed, blood pressure, the hydrostatic pressure that pushes blood through vessels, drops substantially (for reasons that will be explained shortly). Therefore, oxygen-rich blood leaving the gills flows to other organs in the fish quite slowly, but the process is aided by the movements of the body during swimming.

Frogs and many other amphibians have a three-chambered heart, with two atria and one ventricle (FIGURE 38.3b). The ventricle pumps blood into a forked artery that directs the blood through two circuits: the pulmonary circuit and the systemic circuit. The **pulmonary circuit** leads to the lungs and skin, where the blood picks up oxygen as it flows through capillaries. The oxygen-rich blood returns to the left atrium of the heart, then most of it is pumped into the systemic circuit. The **systemic circuit** carries blood to all organs except the lungs and then returns the blood to the right atrium via the veins. This scheme, called **double circulation,** ensures a vigorous flow of blood to the brain, muscles, and other organs because the blood is pumped a second time after it loses pressure in the capillary beds of the lungs. This is distinctly different from the single circulation in the fish, where blood flows directly from the respiratory organs (gills) to other organs under reduced pressure.

In the single ventricle of the frog, there is some mixing of oxygen-rich blood that has returned from the lungs with oxygen-poor blood that has returned from the rest of the body. However, a ridge within the ventricle diverts most of the oxygen-rich blood from the left atrium into the systemic circuit and most of the oxygen-poor blood from the right atrium into the pulmonary circuit. In reptiles, there is even less mixing of oxygen-rich with oxygen-poor blood. Although the reptilian heart is three-chambered, the single ventricle is partially divided. In one order of reptiles, the crocodilians, the ventricle is completely divided into right and left chambers.

The four-chambered heart of a bird or mammal has two atria and two completely separated ventricles (FIGURE 38.3c). There is double circulation, as in amphibians and reptiles, but the heart keeps oxygen-rich blood fully segregated from oxygen-poor blood. The left side of the heart handles only oxygen-rich blood, and the right side receives and pumps only oxygen-poor blood. Delivery of oxygen to all parts of the body is enhanced because there is no mixing of the oxygenated and deoxygenated blood, and double circulation restores pressure after blood has passed through the lung capillaries. As endotherms, which use heat released from metabolism to warm the body, birds and mammals require more oxygen per gram of body weight than other vertebrates of equal size. Birds and mammals descended from different reptilian ancestors, and their four-chambered hearts evolved independently—an example of convergent evolution. A more detailed diagram of blood flow through the mammalian circulatory system is shown in FIGURE 38.4.

We will now see how the mammalian heart and circulatory system actually work, using the human as our example.

Rhythmic pumping of the mammalian heart drives blood through pulmonary and systemic circuits

The Heart: General Form and Function

The human heart is a cone-shaped organ about the size of a clenched fist, located just beneath the breastbone (sternum). It is enclosed in a sac with a two-layered wall. A lubricating fluid fills the space between the two walls, enabling them to slide past each other as the heart pulsates. The heart itself is mostly cardiac muscle tissue (see Chapter 36). Its two atria have relatively thin walls and function as collection chambers for blood returning to the heart, pumping blood only the short distance to the ventricles. The ventricles have thicker walls and are much more powerful than the atria—especially the left ventricle, which must pump blood out to all body organs through the systemic circuit (see FIGURE 38.4b).

The chambers of the heart alternately contract, pumping blood, and relax, filling with blood, in a rhythmic cycle. One complete sequence of filling and pumping is called the **cardiac cycle.** A contraction phase of the cycle is called **systole,** and a relaxation phase is called **diastole** (FIGURE 38.5).

Four valves in the heart, each consisting of flaps of connective tissue, prevent a backflow of blood when the ventricles contract (see FIGURE 38.4b). Between each atrium and ventricle is an **atrioventricular (AV) valve.** The AV valves are anchored by strong fibers that prevent them from turning inside out. Strong force generated by the powerful contraction of the ventricles during systole closes the AV valves, keeping blood from flowing back into the atria. **Semilunar valves** are located at the two exits of the heart, where the aorta leaves the left ventricle and the pulmonary artery leaves the right ventricle. The blood is pumped out into the arteries through the semilunar valves, which are forced open by ventricular contraction. The elastic walls of the arteries expand and then snap back. Along with relaxation of the ventricles, recoil of the arteries closes the semilunar valves, which prevents blood from flowing back into the ventricles.

The heart sounds we can hear with a stethoscope are caused by the vigorous closing of the valves. (You can even hear them without a stethoscope by pressing your ear against the sternum of a friend.) The sound pattern is "lub-dup, lub-dup, lub-dup." The first heart sound ("lub") is created by the recoil of blood against the closed AV valves. The second sound ("dup") is the recoil of blood against the semilunar valves.

A defect in one or more of the valves causes a condition known as a heart murmur, which may be detectable as a hissing sound when a stream of blood squirts backward through a valve. Some people are born with heart murmurs, while others have their valves damaged by infection (from rheumatic fever, for instance). Most heart murmurs do not reduce the efficiency of blood flow enough to warrant surgery. More serious murmurs may be corrected by replacing the damaged valves with artificial ones or with human valves taken from a cadaver.

The number of times the heart beats each minute is called the **heart rate.** You can measure your heart rate by taking your **pulse,** which is the rhythmic stretching of the arteries caused by the pressure of blood driven by

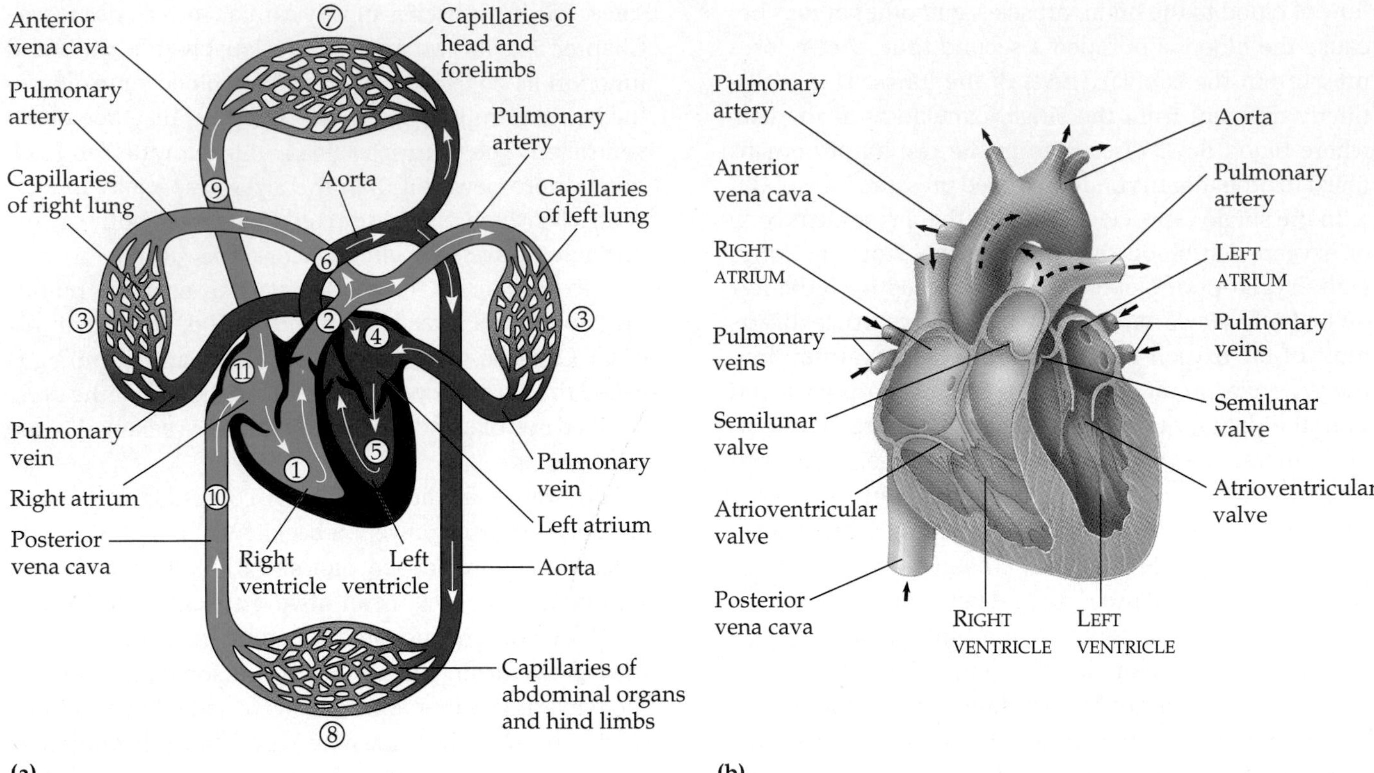

FIGURE 38.4
The mammalian cardiovascular system: an overview. (**a**) The numbers in this diagram trace the flow of blood through the heart, main blood vessels, and capillary beds. Beginning with the pulmonary (lung) circuit, ① the right ventricle pumps blood to the lungs via ② the pulmonary arteries. As the blood flows through ③ capillary beds in the right and left lungs, it loads up on oxygen and unloads carbon dioxide. Oxygen-rich blood returns from the lungs via the pulmonary veins to ④ the left atrium of the heart. The left atrium pumps the oxygen-rich blood into ⑤ the left ventricle, which in turn pumps it out to body tissues through the systemic circuit. Blood leaves the left ventricle via ⑥ the aorta, which arches over the heart, conveying blood to arteries leading throughout the body. The first branches from the aorta are the coronary arteries (not shown), which supply blood to the heart muscle itself. Then come branches leading to capillary beds ⑦ in the head and arms (or forelimbs). The aorta continues in a posterior direction, supplying oxygen-rich blood to arteries leading to ⑧ capillary beds in the abdominal organs and legs (or hind limbs). Within each organ, arteries branch into arterioles, which in turn branch into capillaries where the blood gives up much of its oxygen and picks up the carbon dioxide produced by cellular respiration. Capillaries rejoin to form venules, which lead to veins. Oxygen-poor blood from the head, neck, and forelimbs is channeled into a large vein called ⑨ the anterior vena cava. Another large vein called ⑩ the posterior vena cava drains blood from the trunk and hind limbs. The two venae cavae empty their blood into ⑪ the right atrium. The right atrium pumps the blood into the right ventricle. (**b**) In this detailed view of the structure of the heart, notice the valves, which prevent backflow of blood within the heart, and the relative thickness of the walls of the heart chambers. The atria, which pump blood only into the ventricles, have thinner walls than the ventricles.

the powerful contractions of the ventricles. On average, a healthy young adult at rest has a heart rate of about 60 beats per minute. Individuals who exercise regularly often have a slower resting pulse than those who are less fit. Your own heart rate will vary, depending on your level of activity, and recent studies indicate that every few seconds, a normal heart rate may change by as many as 20 beats/min, and in a single day it may range from 40 beats/min to 180 beats/min. Researchers are currently testing the hypothesis that too much regularity in the heart rhythm—rather than irregularity, as formerly thought—is a sign of developing heart problems. Computer analyses indicate that weeks before serious problems develop, the heart begins to lose some of what appears to be its normal irregularity.

In comparing different mammals, we see an inverse relationship between size and heart rate. An elephant, for instance, has a heart rate of only 25 beats/min, while the heart of a tiny shrew races at about 600 beats/min. To understand the significance of this difference, we need to revisit the concept of bioenergetics. Recall from Chapter 36 that metabolic rate (hence, oxygen consumption) per gram of tissue is proportionately greater for smaller mammals than for larger ones. A rapid heart rate is one adaptation that enhances the delivery of oxygen for cellular respiration.

The volume of blood per minute that the left ventricle pumps into the systemic circuit is called **cardiac output.** This volume depends on two factors: heart rate (pulse) and **stroke volume,** the amount of blood pumped by

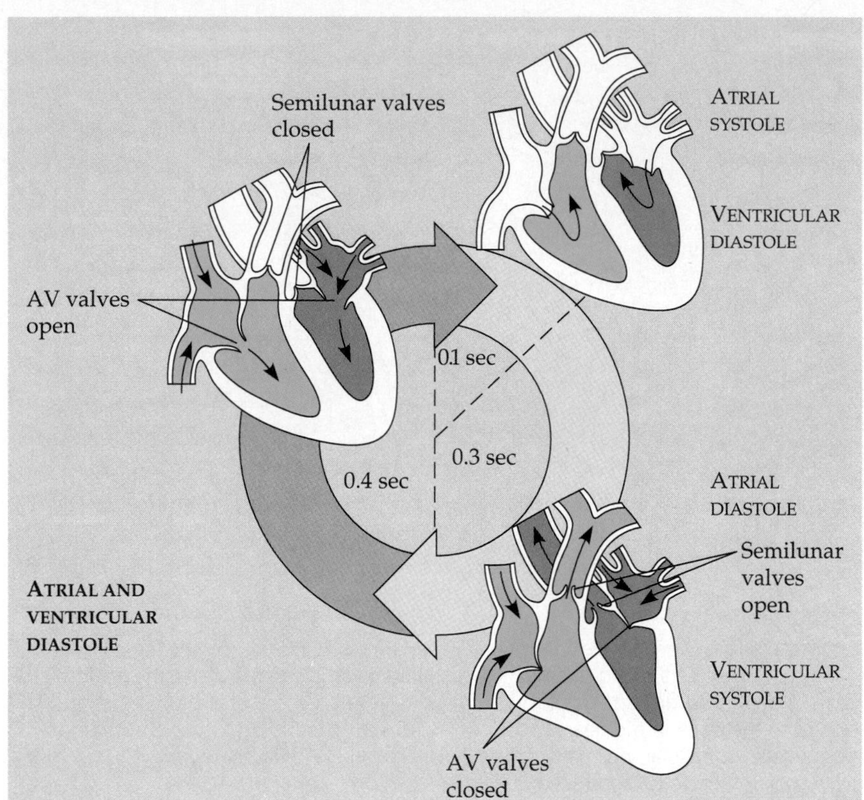

FIGURE 38.5
The cardiac cycle. The cardiac muscles of the heart contract (systole) and relax (diastole) in a rhythmic cycle. In an average adult human at rest (with a pulse of about 65 to 75 beats per minute), one complete cardiac cycle takes about 0.8 second. During a relaxation phase (atria and ventricles in diastole) lasting about 0.4 sec, blood returning from the large veins flows into the atria and ventricles. A brief period (about 0.1 sec) of atrial systole then forces all the blood out of the atria into the ventricles. During the remaining 0.3 sec of the cardiac cycle, ventricular systole pumps blood into the large arteries. Note that seven-eighths of the time—all but the first 0.1 sec of the cardiac cycle—the atria are relaxed and filling with blood returning in the veins.

the left ventricle each time it contracts. The average stroke volume for a human is about 75 mL per beat. A person with this stroke volume and a resting pulse of 70 beats/min has a cardiac output of 5.25 L/min. This is about equivalent to the total volume of blood in the human body. Cardiac output can increase about fivefold during heavy exercise.

Control of the Heart

Control of the rhythmic beating of the heart is shown in FIGURE 38.6. The cells of vertebrate cardiac muscle are self-excitable; they are said to be autorhythmic, meaning they can contract without any signal from the nervous system. Individual cardiac muscle cells removed from a heart and viewed with a microscope can be seen to pulsate, but at irregular intervals. A region of the heart called the **sinoatrial (SA) node,** or **pacemaker,** maintains the heart's pumping rhythm by setting the rate at which all cardiac muscle cells contract. Composed of specialized muscle tissue, the SA node is located in the wall of the right atrium, near the point where the superior vena cava enters the heart.

The SA node generates electrical impulses much like those produced by nerve cells. Because cardiac muscle cells are electrically coupled (by the intercalated discs between adjacent cells; see FIGURE 36.4), impulses from the SA node spread rapidly through the walls of the atria, making them contract in unison. The impulses also pass to another region of specialized muscle tissue, a relay point called the **atrioventricular (AV) node,** in the wall between the right atrium and right ventricle. Here the impulses are delayed for about 0.1 sec, which ensures that the atria will contract first and empty completely before the ventricles contract. After this delay, specialized muscle fibers conduct the signal to contract throughout the walls of the ventricle. The impulses that travel through cardiac muscle during the heart cycle produce electrical currents that are conducted through body fluids to the body surface, where the currents can be detected by electrodes placed on the skin and recorded as an **electrocardiogram (EKG** or **ECG).**

The SA node sets the tempo for the entire heart, but this pacemaker itself is influenced by a variety of cues. Two sets of nerves oppose each other in adjusting heart rate; one set speeds up the pacemaker, and the other set slows it down. At any given time, heart rate is a compromise regulated by the opposing actions of these two sets of nerves. The pacemaker is also controlled by hormones secreted into the blood by glands. For example, epinephrine, the "fight-or-flight" hormone from the adrenal glands, increases heart rate (see Chapter 41). Body temperature is another factor that affects the pacemaker. A temperature increase of only 1°C raises the heart rate by about 10 beats/min. This is the reason your pulse increases substantially when you have a fever. The heart

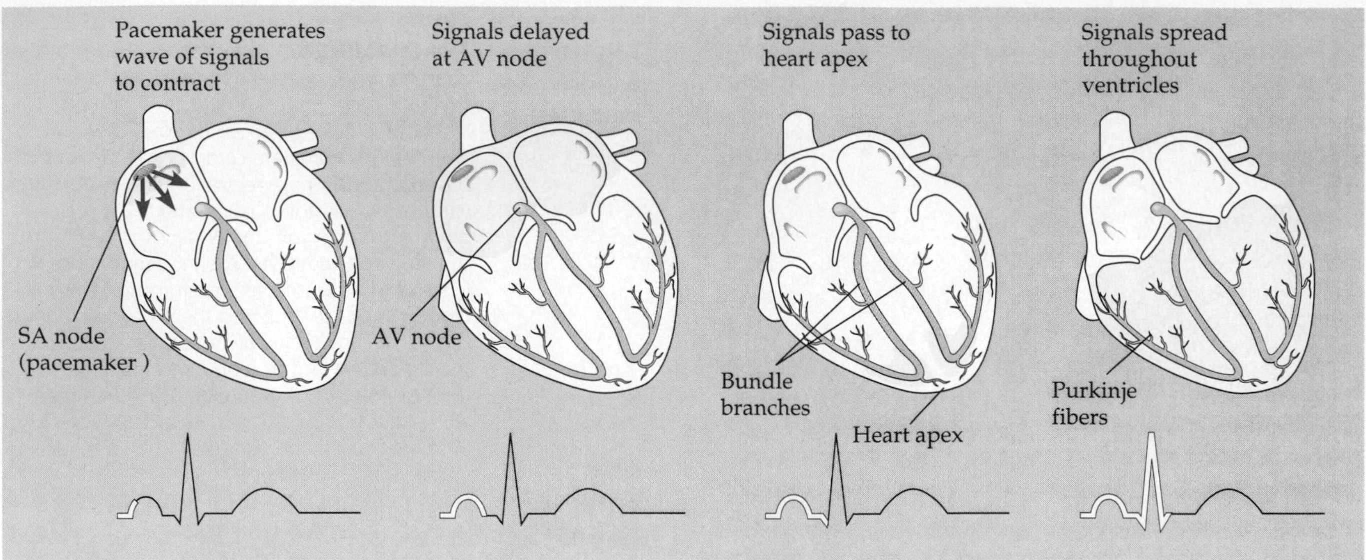

FIGURE 38.6

The control of heart rhythm. The SA node, or pacemaker, sets the tempo of the heartbeat by generating a wave of electrical signals (gold) that spread through both atria, making them contract simultaneously. The signals to contract are delayed at the AV node for about 0.1 sec, during which the atria empty into the ventricles. Specialized muscle fibers called bundle branches and Purkinje fibers then conduct the signals to the apex of the heart and throughout the ventricular wall. The signals trigger a wave of powerful contractions of both ventricles from the apex toward the atria, driving blood into the large arteries. Gold color in the graphs at the bottom indicates the components of an electrocardiogram corresponding to the sequence of electrical events in the heart.

rate also increases with exercise, an adaptation that enables the circulatory system to provide the additional oxygen needed by muscles hard at work.

Blood Vessel Structure

The wall of an artery or vein has three layers (FIGURE 38.7). On the outside is a zone of connective tissue with elastic fibers that enable the vessel to stretch and recoil. The middle layer consists of smooth muscle and more elastic fibers. The middle and outer layers are thicker in the arteries, which must be stronger and more elastic than veins. (The walls of major arteries are so thick that they must themselves be supplied by blood vessels.) Blood vessels are lined by **endothelium,** a simple squamous epithelium (see Chapter 36). The endothelium, along with its basement membrane, forms the inner layer of a blood vessel. Capillaries lack the outer layers, and their very thin walls consist only of endothelium and its basement membrane.

Blood Flow Velocity

Blood flows through the vessels of the circulatory system at an uneven speed. For instance, blood travels over a thousand times faster in the aorta (about 30 cm/sec on average) than in the capillaries (about 0.026 cm/sec). To understand why the blood decelerates, we need to consider the *law of continuity,* a rule that governs the flow of fluids through pipes. If a pipe changes diameter over its length, a fluid will stream through narrower segments of the pipe faster than it flows through wider segments. The *volume* of flow per second must be constant through the entire pipe, so the fluid must flow faster as the cross-sectional area of the pipe narrows. For instance, compare the velocity of water squirted by a hose with and without a nozzle.

Based on the law of continuity, it may at first seem that blood should travel faster through capillaries than through arteries, because the diameter of capillaries is much smaller. However, it is the *total* cross-sectional area of the pipes delivering the fluid that determines flow rate. Although an individual capillary is very narrow, each artery conveys blood to such an enormous number of capillaries that the *total* diameter of the vessels is actually much greater in capillary beds than in any other part of the circulatory system. For this reason, the blood slows down substantially as it enters the arterioles from arteries and flows slowest in the capillary beds. Capillaries are the only vessels with walls thin enough to permit the transfer of substances between the blood and interstitial fluid, and the slower flow of blood through these tiny vessels enhances this chemical exchange. As blood leaves the capillary beds and passes to the venules and veins, it speeds up again, a result of the reduction in total cross-sectional area (FIGURE 38.8).

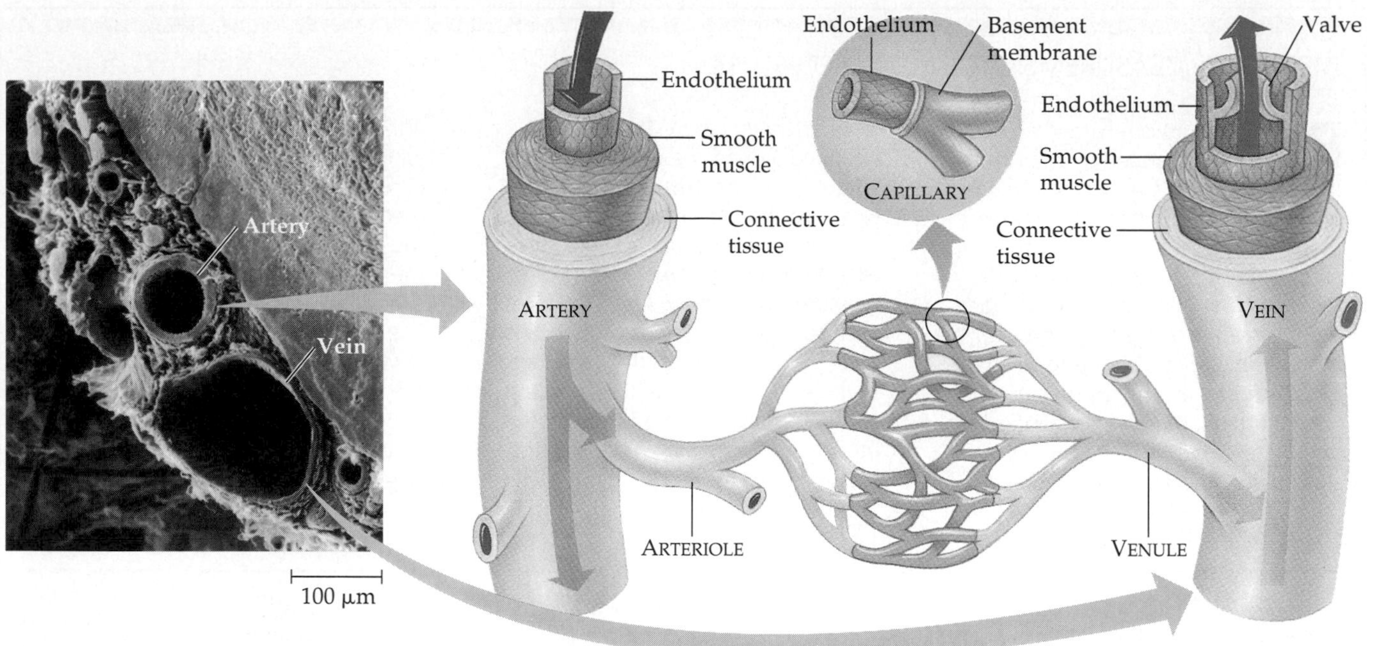

FIGURE 38.7

The structure of blood vessels. In the micrograph, an artery can be seen next to a thinner-walled vein. The wall of an artery or a vein has three layers: an inner layer of endothelium, a middle layer of smooth muscle with elastic fibers, and an outer layer of connective tissue also with elastic fibers. Capillaries have much smaller diameters than arteries and veins, and their walls consist of only a single layer of endothelium and a basement membrane. (SEM from *Tissues and Organs: A Text-Atlas of Scanning Electron Microscopy* by Richard G. Kessel and Randy H. Kardon. W. H. Freeman and Company. Copyright © 1979.)

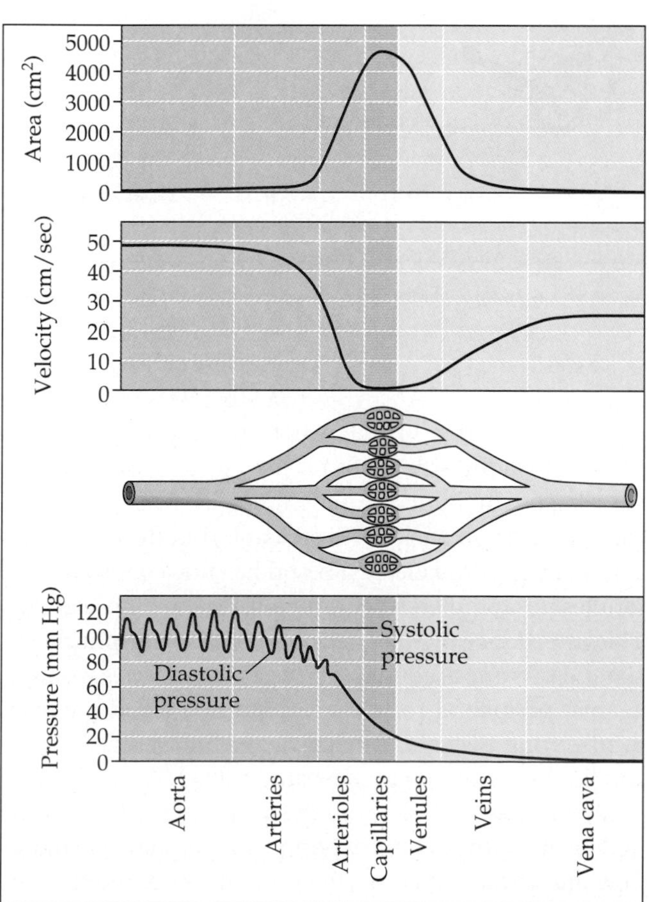

FIGURE 38.8

The interrelationship of blood flow velocity, cross-sectional area of blood vessels, and blood pressure. Blood flow velocity decreases markedly in the arterioles and is slowest in the capillaries, owing to an increase in total cross-sectional area (smallest in the arteries, greater in the arterioles, and greatest in the capillaries). Blood pressure, the main force driving blood from the heart to the capillaries, is highest in the arteries. Peaks in blood pressure corresponding to ventricular systole alternate with lower pressures corresponding to diastole. Resistance to flow through the arterioles and capillaries, due to contact of the blood with a greater surface area of endothelium, reduces blood pressure and eliminates the pressure peaks.

METHODS: MEASUREMENT OF BLOOD PRESSURE

Blood pressure is recorded as two numbers separated by a slash; the higher number is the systolic pressure, and the lower number is the diastolic pressure. A typical blood pressure reading for a 20-year-old is 120/70 (FIGURE a). The units for these numbers are millimeters of mercury (mm Hg); a blood pressure of 120 is a force that can support a column of mercury 120 mm high. A sphygmomanometer, an inflatable cuff attached to a pressure gauge, measures blood pressure in an artery (FIGURE b). The cuff is wrapped around the upper arm and inflated until the pressure closes the artery, so that no blood flows past the cuff. When this occurs, the pressure exerted by the cuff exceeds the blood pressure in the artery. A stethoscope is used to listen for sounds of blood flow below the cuff. If the artery is closed, there is no pulse below the cuff. The cuff is gradually deflated until blood begins to flow into the forearm, and sounds from blood pulsing into the artery below the cuff can be heard with the stethoscope (FIGURE c). This occurs when the blood pressure is greater than the pressure exerted by the cuff. The pressure at this point is the systolic pressure, the high pressure exerted by the ventricles contracting. The cuff is loosened further until blood flows freely through the artery, and the sounds below the cuff disappear. The pressure at this point is the diastolic pressure remaining in the artery when the heart is relaxed (FIGURE d).

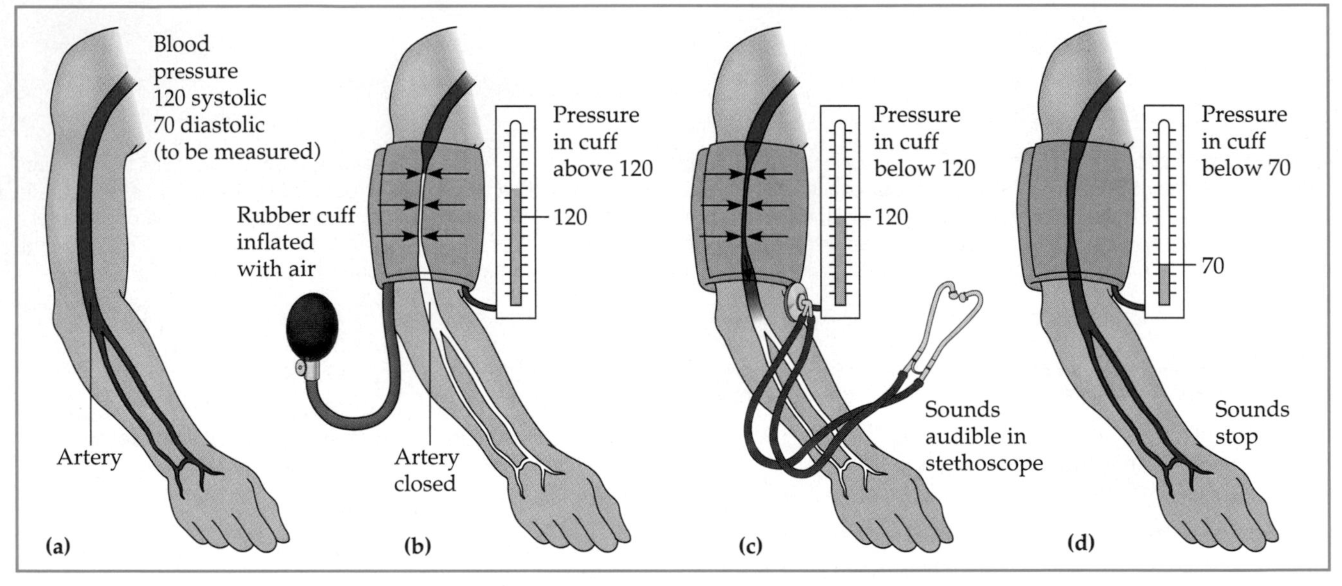

Blood Pressure

Fluids exert a force called hydrostatic pressure against surfaces they contact, and it is that pressure that drives fluids through pipes. The hydrostatic force that blood exerts against the wall of a vessel is called **blood pressure** (see the Methods Box). This pressure is much greater in arteries than in veins and is greatest in arteries when the heart contracts during ventricular systole (see FIGURE 38.5). Blood pressure is the main force that propels blood from the heart through the arteries and arterioles to the capillary beds.

When you take your pulse by placing your fingers on your wrist, you can actually feel an artery bulge with each heartbeat. The surge of pressure is partly due to the narrow openings of arterioles impeding the exit of blood from the arteries. Thus, when the heart contracts, blood enters the arteries faster than it can leave, and the vessels stretch from the pressure. The elastic walls of the arteries snap back during diastole, but the heart contracts again before enough blood has flowed into the arterioles to completely relieve pressure in the arteries. This impedance by the arterioles is called **peripheral resistance.** As a consequence of the elastic arteries working against peripheral resistance, there is a blood pressure even during diastole, driving blood into arterioles and capillaries continuously (see FIGURE 38.8).

Blood pressure is determined partly by cardiac output and partly by the degree of peripheral resistance to blood flow due to the arterioles, the bottlenecks of the circulatory system. Contraction of smooth muscles in the walls

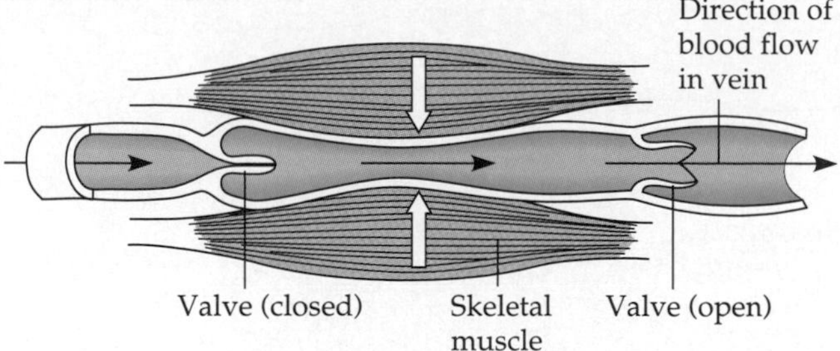

FIGURE 38.9
Blood flow in veins. Contracting muscles squeeze the veins, in which flaps of tissue act as one-way valves that keep blood moving only toward the heart. If we sit or stand too long, the lack of muscular activity causes our feet to swell with stranded fluid unable to return to the heart.

of the arterioles constricts the tiny vessels, increases resistance, and therefore increases blood pressure in the arteries. When the muscles relax, the arterioles dilate, and pressure in the arteries falls. Nerve impulses, hormones, and other signals control these muscles. Stress, both physical and emotional, can raise blood pressure by triggering nervous and hormonal responses that constrict the blood vessels.

By the time blood reaches the veins, its pressure is not affected much by the heart. This is because the blood encounters so much resistance as it passes through the millions of tiny arterioles and capillaries that the force from the pumping heart can no longer propel the blood in the veins. How, then, does blood return to the heart, especially when it must travel from the lower extremities against gravity? The answer is that veins are sandwiched between muscles, and whenever we move, our skeletal muscles pinch our veins and squeeze blood through the vessels (FIGURE 38.9). Within large veins are flaps of tissue that function as one-way valves, allowing the blood to flow only in the direction of the heart. Muscular activity during exercise, as well as breathing, increase the rate at which blood returns to the heart. When we inhale, the change in pressure within the thoracic (chest) cavity causes the venae cavae and other large veins near the heart to expand and fill.

Blood Flow Through Capillary Beds

At any given time, only about 5% to 10% of the body's capillaries have blood flowing through them. Because each tissue has so many capillaries, however, every part of the body is supplied with blood at all times. Capillaries in the brain, heart, kidneys, and liver usually carry a full load of blood, but in many other sites, the blood supply varies from time to time.

Two mechanisms, both dependent on smooth muscles controlled by nerve signals and hormones, regulate the distribution of blood in capillary beds. In one mechanism, contraction of the smooth muscle layer in the wall of an arteriole constricts the arteriole, decreasing blood flow through it to a capillary bed. When the muscle layer relaxes, the arteriole dilates, allowing blood to enter the capillaries. In the other mechanism, rings of smooth muscle, called precapillary sphincters because they are located at the entrance to capillary beds, control the flow of blood between arterioles and venules (FIGURE 38.10).

The blood supply to capillary beds varies locally as blood is diverted from one destination to another. After a meal, for instance, arterioles in the wall of the digestive tract dilate, capillary sphincters open, and the digestive

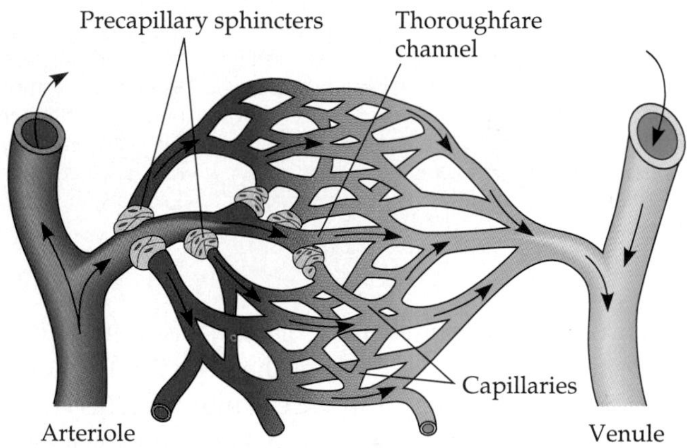

(a) Sphincters relaxed

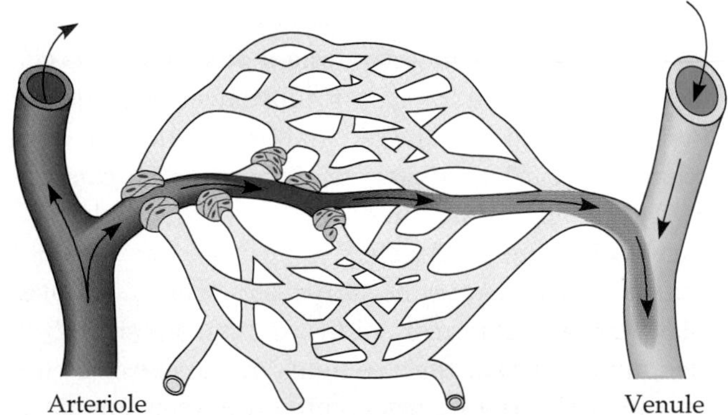

(b) Sphincters contracted

FIGURE 38.10
Blood flow in capillary beds. Precapillary sphincters regulate the passage of blood into many capillary beds. Some blood flows directly from arterioles to venules through capillaries called thoroughfare channels, which are always open. (**a**) When the precapillary sphincters are relaxed, capillaries branching from the thoroughfare channel are open, and blood flows into the capillary bed. (**b**) When the sphincters are contracted, the capillaries are closed, and blood flow through the capillary bed is shut down.

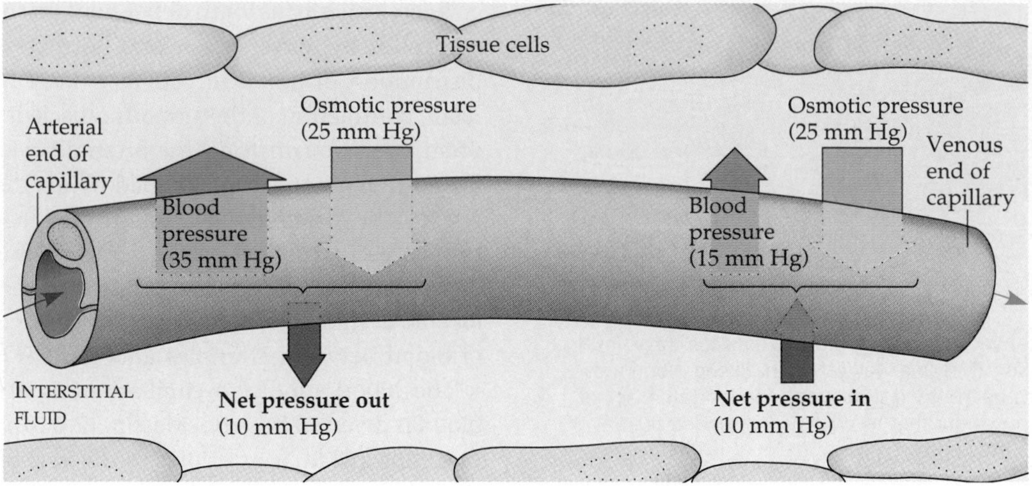

FIGURE 38.11

The movement of fluid between capillaries and the interstitial fluid. Fluid flows out of a capillary at the upstream end near an arteriole and reenters a capillary downstream near a venule. The direction of fluid movement at any point along the capillary depends on the difference between two opposing forces: hydrostatic pressure (blood pressure) and osmotic pressure. Blood pressure tends to force fluid out of the capillary. Osmotic pressure is the tendency for water to enter the capillary because of the relatively high solute concentration of the blood. At the arterial end of the capillary, the blood pressure forcing fluid outward exceeds the osmotic pressure drawing water inward, resulting in a net movement of fluid out of the capillary. Because the endothelium is selectively permeable, it filters the fluid so that water and small solutes leave the capillary, while most proteins remain behind in the blood. As blood continues along the capillary, the blood pressure decreases as a result of resistance and loss of fluid volume. At the venous end of the capillary, the tendency for fluid to exit because of blood pressure is overpowered by the tendency for fluid to enter because of osmotic pressure.

tract receives a larger share of blood. During strenuous exercise, blood is diverted from the digestive tract and supplied more generously to skeletal muscles. This is one reason heavy exercise immediately after a big meal may cause indigestion.

Capillary Exchange

Now we come to the most important business of the circulatory system: the transfer of substances between the blood and the interstitial fluid that bathes the cells. This exchange takes place across the thin walls of the capillaries. The capillary wall, remember, is an endothelium, a single layer of flattened cells.

Some substances may be carried across an endothelial cell in vesicles that form by endocytosis on one side of the cell and then release their contents by exocytosis on the opposite side; others simply diffuse between the blood and the interstitial fluid. Small molecules, such as oxygen and carbon dioxide, diffuse down their concentration gradients across the membranes of the endothelial cells. Diffusion can also occur through the clefts between adjoining cells. However, transport through these clefts occurs mainly by bulk flow, the movement of fluid due to pressure. Hydrostatic pressure (blood pressure) within the capillary pushes fluid (water and small solutes such as sugars, salts, oxygen, and urea) through the capillary clefts. Blood cells and proteins dissolved in the blood are too large to pass readily through the endothelium and remain in the capillaries.

Fluid flows out of a capillary at the upstream end near an arteriole, but reenters downstream near a venule (FIGURE 38.11). About 85% of the fluid that leaves the blood at the arterial end of a capillary bed reenters from the interstitial fluid at the venous end, and the remaining 15% of the fluid lost from capillaries is eventually returned to the blood by the vessels of the lymphatic system.

The lymphatic system returns fluid to the blood and aids in body defense

So much blood passes through the capillaries that the cumulative loss of fluid adds up to about 4 L per day. There is also some leakage of blood proteins, even though the capillary wall is not very permeable to these large molecules. The lost fluid and proteins return to the blood via the **lymphatic system** (see FIGURE 39.4). Fluid enters this system by diffusing into tiny lymph capillaries that are intermingled among capillaries of the cardiovascular system. Once inside the lymphatic system, the fluid is called **lymph;** its composition is about the same as that of interstitial fluid. The lymphatic system drains into the circulatory system near the shoulders (see FIGURE 39.4).

Whenever interstitial fluid accumulates rather than being returned to the blood by the lymphatic system, tissues and body cavities become swollen, a condition known as edema. One cause of edema is severe dietary protein deficiency. When starved for amino acids, the body consumes its own blood proteins. This reduces the osmotic pressure of the blood, causing interstitial fluid to accumulate in body tissues rather than being drawn back into capillaries. A child suffering from protein deficiency may have a swollen belly because of all the fluid that collects in the body cavity.

Lymph vessels, like veins, have valves that prevent the backflow of fluid toward the capillaries. Like veins, lymph vessels depend mainly on the movement of skeletal muscles to squeeze fluid along. Rhythmic contractions of the vessel walls also help draw fluid into lymphatic capillaries.

Along a lymph vessel are specialized swellings called **lymph nodes.** By filtering the lymph and attacking viruses and bacteria, lymph nodes play an important role in the body's defense. Inside a lymph node is a honeycomb of connective tissue whose spaces are filled by white blood cells specialized for defense. When the body is fighting an infection, these cells multiply rapidly, and the lymph nodes become swollen and tender (which is why your doctor checks for swollen nodes in your neck).

The lymphatic system, then, helps defend the body against infection and maintains the fluid level and protein concentration of the blood. In addition, lymph capillaries called lacteals penetrate the villi of the small intestine; the lacteals absorb fats and transport them from the digestive tract to the circulatory system (see Chapter 37).

Blood is a connective tissue with cells suspended in plasma

We now shift our focus from the structure and function of blood vessels to the composition of the blood itself. Vertebrate blood is a type of connective tissue consisting of several kinds of cells suspended in a liquid matrix called **plasma.** The average human body contains about 4 to 6 L of blood. If a blood sample is taken, the cells can be separated from the plasma by spinning the whole blood in a centrifuge. (An anticoagulant must be added to prevent the blood from clotting.) The cellular elements (cells and cell fragments), which occupy about 45% of the volume of blood, settle to the bottom of the centrifuge tube, forming a dense red pellet. Above this pellet is the transparent, straw-colored plasma (FIGURE 38.12).

Plasma

Blood plasma is about 90% water. Among a variety of solutes dissolved in the water are inorganic salts, sometimes referred to as blood electrolytes, present in the plasma in the form of dissolved ions. The combined concentration of these ions is important in maintaining osmotic balance of the blood. Some of the ions also help buffer the blood, which has a pH of 7.4 in humans. And the ability of muscles and nerves to function normally depends on the concentration of key ions in the interstitial fluid, which reflects their concentration in plasma. The kidney maintains plasma electrolytes at precise concentrations, an example of homeostasis.

Another important class of solutes is the plasma proteins, which have a number of functions. Collectively, they act as buffers against pH changes, help maintain the osmotic balance between the blood and interstitial fluid, and contribute to the blood's viscosity (thickness). The various types of plasma proteins also have specific functions. Some serve as escorts for lipids, which are insoluble in water and can travel in blood only when bound to proteins. Another class of proteins, the immunoglobulins, are the antibodies that help combat viruses and other foreign agents that invade the body (see Chapter 39). And some of the plasma proteins, called fibrinogens, are clotting factors that help plug leaks when blood vessels are injured. Blood plasma that has had these clotting factors removed is called serum.

Plasma also contains various substances in transit from one part of the body to another, including nutrients, metabolic waste products, respiratory gases, and hormones. Blood plasma and interstitial fluid are similar in composition, except that plasma has a much higher protein concentration than interstitial fluid (capillary walls, remember, are not very permeable to proteins).

Cellular Elements

Dispersed throughout blood plasma are two classes of cells: red blood cells, which transport oxygen, and white blood cells, which function in defense. A third cellular element, platelets are pieces of cells that are involved in blood clotting (see FIGURE 38.12).

Red blood cells, or **erythrocytes,** are by far the most numerous blood cells. Each cubic millimeter of human blood contains 5 to 6 million red cells, and there are about 25 trillion of these tiny cells in the body's 5 L of blood.

The structure of the red blood cell is another excellent example of structure fitting function. A human erythrocyte is a biconcave disk, thinner in the center than at its edges. Mammalian erythrocytes lack nuclei, an unusual characteristic for living cells (the other vertebrate classes

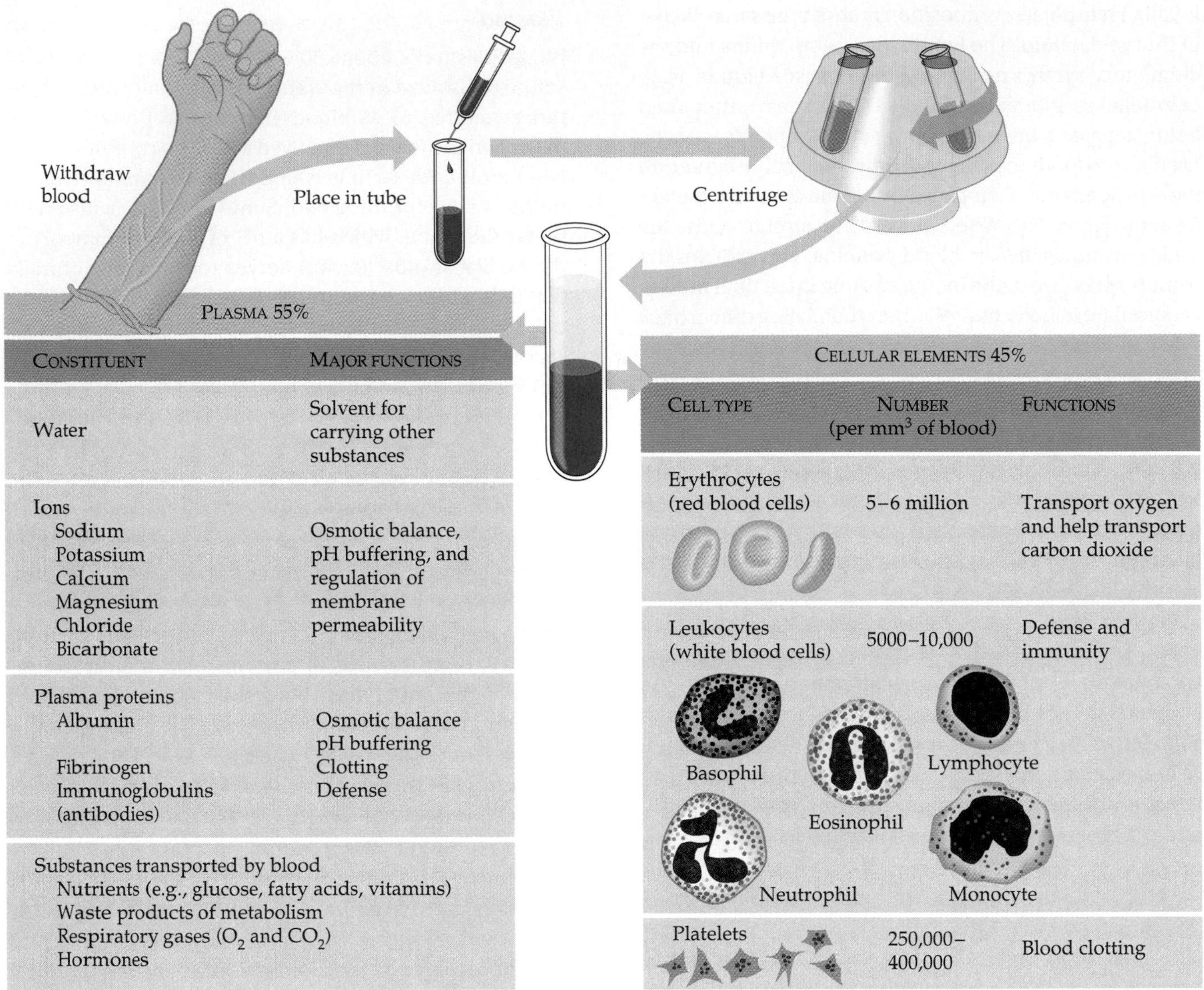

PLASMA 55%	
CONSTITUENT	MAJOR FUNCTIONS
Water	Solvent for carrying other substances
Ions Sodium Potassium Calcium Magnesium Chloride Bicarbonate	Osmotic balance, pH buffering, and regulation of membrane permeability
Plasma proteins Albumin Fibrinogen Immunoglobulins (antibodies)	 Osmotic balance pH buffering Clotting Defense
Substances transported by blood Nutrients (e.g., glucose, fatty acids, vitamins) Waste products of metabolism Respiratory gases (O_2 and CO_2) Hormones	

CELLULAR ELEMENTS 45%		
CELL TYPE	NUMBER (per mm^3 of blood)	FUNCTIONS
Erythrocytes (red blood cells)	5–6 million	Transport oxygen and help transport carbon dioxide
Leukocytes (white blood cells)	5000–10,000	Defense and immunity
Platelets	250,000–400,000	Blood clotting

FIGURE 38.12
The composition of mammalian blood.

have nucleated erythrocytes). Moreover, all red blood cells lack mitochondria and generate their ATP exclusively by anaerobic metabolism. The major function of erythrocytes is to carry oxygen, and they would not be very efficient if their own metabolism were aerobic and consumed some of the oxygen they carry. The small size of erythrocytes also suits their function. For oxygen to be transported, it must diffuse across the plasma membranes of the red blood cells. The smaller the cells, the greater the total area of plasma membrane in a given volume of blood. The biconcave shape of the erythrocyte also adds to its surface area.

As small as a red cell is, it contains about 250 million molecules of **hemoglobin,** an oxygen-carrying protein containing iron (see Chapter 5). As red cells pass through the capillary beds of lungs, gills, or other respiratory organs, oxygen diffuses into the erythrocytes and hemoglobin binds the oxygen. This process is reversed in the capillaries of the systemic circuit, with the hemoglobin unloading its cargo of oxygen.

Red blood cell production, which occurs in the red marrow of bone, is controlled by a negative-feedback mechanism that is sensitive to the amount of oxygen reaching the tissues via the blood. If the tissues are not receiving enough oxygen, the kidney converts a plasma protein to a hormone called **erythropoietin,** which stimulates production of erythrocytes in the bone marrow. If blood is delivering more oxygen than the tissues can use, the level of erythropoietin is reduced, and erythrocyte production slows.

There are five major types of **white blood cells,** or **leukocytes:** monocytes, neutrophils, basophils, eosinophils, and lymphocytes. Their collective function is to fight infections in various ways. For example, monocytes and neutrophils are phagocytes, which engulf and digest bacteria and debris from our own dead cells. As we will see in Chapter 39, lymphocytes become specialized as B cells and T cells, which produce the immune response against foreign substances. Normally, a cubic millimeter of human blood has about 5,000 to 10,000 leukocytes, most of which are in transit. White blood cells spend most of their time outside the circulatory system, patrolling through interstitial fluid and the lymphatic system, where most of the battles against pathogens are waged. The number of leukocytes increases temporarily whenever the body is fighting an infection.

The third cellular element of blood, **platelets** are fragments of cells about 2 to 3 μm in diameter. They have no nuclei and originate as pinched-off cytoplasmic fragments of large cells in the bone marrow. Platelets then enter the blood and function in the important process of blood clotting.

The Formation of Blood Cells and Platelets

The cellular elements of the blood (erythrocytes, leukocytes, and platelets) wear out and are replaced constantly throughout a person's life. Erythrocytes, for example, usually circulate for only about 3 to 4 months and then are destroyed by phagocytic cells located mainly in the liver and spleen. Enzymes digest the old cell's macromolecules, and biosynthetic processes construct new ones using many of the monomers, such as amino acids, obtained from old blood cells, as well as new materials and energy derived from food. Many of the iron atoms derived from the hemoglobin in old red blood cells are built into new hemoglobin molecules.

Erythrocytes, leukocytes, and platelets all develop from a common source, a single population of cells called **pluripotent stem cells** in the red marrow of bones, particularly the ribs, vertebrae, breastbone, and pelvis. ("Pluripotent" means these cells have the potential to differentiate into any type of blood cell or into cells that produce platelets.) The population of pluripotent stem cells arises in the early embryo. From then on, it renews itself by mitosis while replenishing the blood with cellular elements (FIGURE 38.13). An example of homeostasis, the number of new blood cells that differentiate from stem cells equals the number of old cells that die in a given period of time.

Recently, researchers have been able to isolate pluripotent stem cells (by means of monoclonal antibody techniques, discussed in Chapter 39) and grow these cells in laboratory cultures. Purified pluripotent stem cells may soon provide an effective treatment for a number of human diseases, such as leukemia. A person with

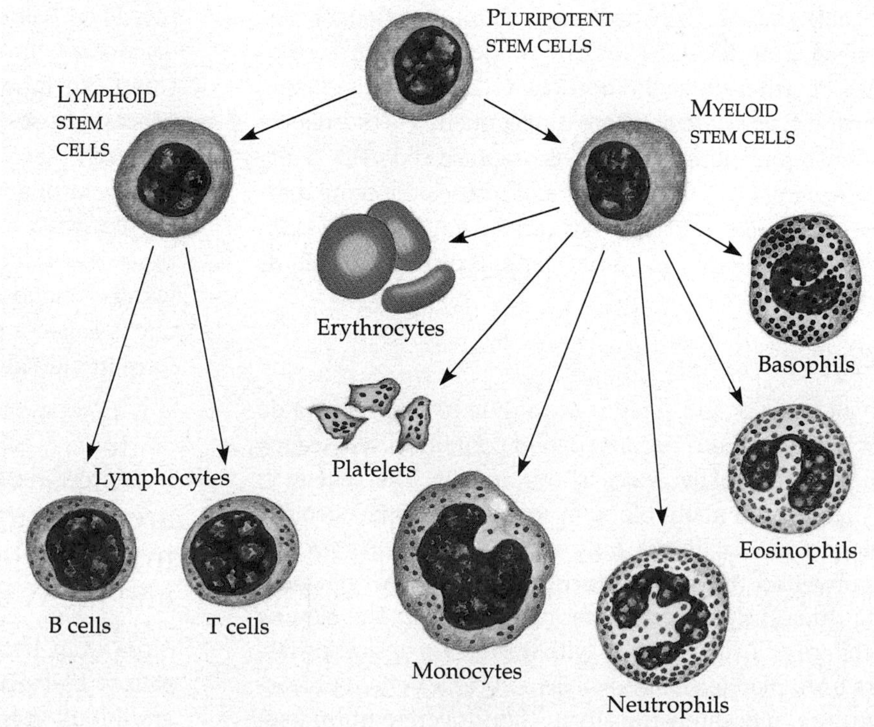

FIGURE 38.13
Differentiation of blood cells. All blood cells develop from a common source, a population of pluripotent stem cells in red bone marrow. Some of these cells differentiate into lymphoid stem cells, which then develop into B cells and T cells, two types of lymphocytes that function in the immune response (see Chapter 39). All other blood cells differentiate from myeloid stem cells, also derived from the population of pluripotent stem cells.

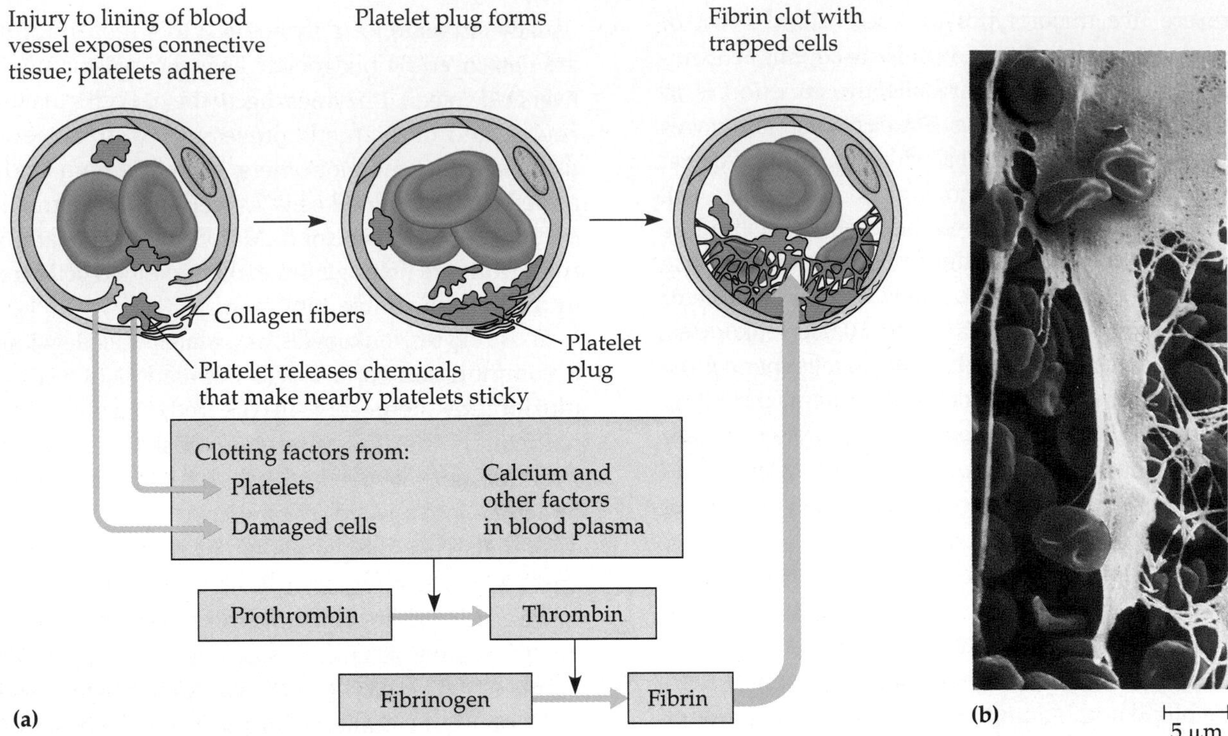

FIGURE 38.14

Blood clotting. (**a**) The clotting process begins when the endothelium of a vessel is damaged and connective tissue in the vessel wall is exposed to blood. Platelets adhere to collagen fibers in the connective tissue and release a substance that makes nearby platelets sticky. The platelets form a plug that provides emergency protection against blood loss. This seal is reinforced by a clot of fibrin when vessel damage is more severe. Clotting factors released from the clumped platelets or damaged cells mix with clotting factors in the plasma forming an activator, which converts a plasma protein called prothrombin to its active form, thrombin. Calcium and vitamin K are among the plasma factors required for this step. Thrombin itself is an enzyme that catalyzes the final step of the clotting process, the conversion of fibrinogen to fibrin. Thus, in a cascade of reactions, injury activates prothrombin, which then activates fibrinogen. The threads of fibrin become interwoven into a patch. (**b**) Red blood cells trapped in a clot of fibrin (colorized SEM).

leukemia has an unusually high number of leukocytes. Derived from a cancerous line of stem cells, the leukocytes obstruct normal blood cell formation in the bone marrow. One strategy for treating leukemia is to isolate a patient's pluripotent stem cells, replace the cell's faulty genes using DNA technology, and restock the bone marrow with the genetically engineered stem cells. The strategy may also work for other genetic disorders, such as sickle-cell disease.

Blood Clotting

We all get cuts and scrapes from time to time, yet we do not bleed to death because blood contains a self-sealing material that plugs leaks in our vessels. The sealant is always present in our blood in an inactive form called **fibrinogen.** A clot forms only when this plasma protein is converted to its active form, **fibrin,** which aggregates into threads that form the fabric of the clot. The clotting mechanism usually begins with the release of clotting factors from platelets and involves a complex chain of reactions that ultimately transforms fibrinogen to fibrin (FIGURE 38.14). More than a dozen clotting factors have been discovered, and the mechanism is still not fully understood. An inherited defect in any step of the clotting process causes **hemophilia,** a disease characterized by excessive bleeding from even minor cuts and bruises.

Anticlotting factors in the blood normally prevent spontaneous clotting in the absence of injury. Sometimes, however, platelets clump and fibrin coagulates within a blood vessel, blocking the flow of blood. Such a clot is called a **thrombus.** These clots are more likely to form in individuals with cardiovascular disease.

Cardiovascular diseases are the leading cause of death in the United States and many other developed nations

More than half of all deaths in the United States are caused by **cardiovascular disease,** diseases of the heart and blood vessels. Most often, the final blow from cardiovascular disease is either a heart attack or a stroke. A **heart attack** is the death of cardiac muscle tissue resulting from prolonged blockage of one or more coronary arteries, the vessels that supply oxygen-rich blood to the heart. A **stroke** is the death of nervous tissue in the brain, usually resulting from blockage of arteries in the head. Heart attacks and strokes are frequently associated with a thrombus that clogs an artery. A thrombus that causes a heart attack or stroke may form in a coronary artery or an artery in the brain, or it may develop elsewhere in the circulatory system and reach the heart or brain via the bloodstream. Such a moving clot is called an embolus. The embolus is swept along until it becomes lodged in an artery too small for the clot to pass. Cardiac or brain tissue downstream from the obstruction may die. If the damage in the heart interrupts the conduction of electrical impulses through the cardiac muscle, the heart rate may change drastically or the heart may stop beating altogether. Still, the person may survive if heartbeat is restored by cardiopulmonary resuscitation (CPR) or some other emergency procedure within a few minutes of the attack. The effects of a stroke and the individual's chance of survival depend on the extent and location of the damaged tissue.

The suddenness of a heart attack or stroke belies the fact that the arteries of most victims had become gradually impaired by a chronic cardiovascular disease known as **atherosclerosis.** During the course of this disease, growths called plaques develop on the inner walls of the arteries, narrowing their bore (FIGURE 38.15). A plaque forms at a site where the smooth muscle layer of an artery thickens abnormally and becomes infiltrated with fibrous connective tissue and lipids such as cholesterol. In some cases, the plaques even become hardened by calcium deposits, resulting in a form of atherosclerosis called **arteriosclerosis,** commonly known as hardening of the arteries. An embolus is more likely to become trapped in a vessel that has been narrowed by plaques. Furthermore, plaques are common sites of thrombus formation. Healthy arteries have smooth linings. The rougher lining of an artery affected by atherosclerosis seems to encourage the adhesion of platelets, which triggers the clotting process.

As atherosclerosis progresses, arteries become more and more clogged by plaque, and the threat of heart attack or stroke becomes much greater. Sometimes, there are warnings. For example, if a coronary artery is partially blocked by atherosclerosis, a person may feel occasional chest pains, a condition known as angina pectoris. The pain is a signal that part of the heart is not receiving a sufficient supply of oxygen, and it is most likely to occur when the heart is laboring hard because of physical or emotional stress. However, many people with atherosclerosis are completely unaware of their disease until catastrophe strikes.

Hypertension (high blood pressure) promotes atherosclerosis and increases the risk of heart attack and stroke. Conversely, atherosclerosis tends to increase blood pressure by narrowing the bore of the vessels and reducing their elasticity. According to one hypothesis, chronic punishment to the lining of arteries by hypertension damages the endothelium and initiates plaque formation. Acting alone or in lethal combination, hypertension and atherosclerosis, the two most common cardiovascular diseases, lead to the majority of deaths in the United States and other developed nations. Hypertension is sometimes called the "silent killer" because a person with the disease may experience no symptoms until a stroke or heart attack occurs. Fortunately, hypertension is simple to diagnose and can usually be controlled by medication, diet, exercise, or a combination of these. A diastolic pressure above 90 may be cause for concern,

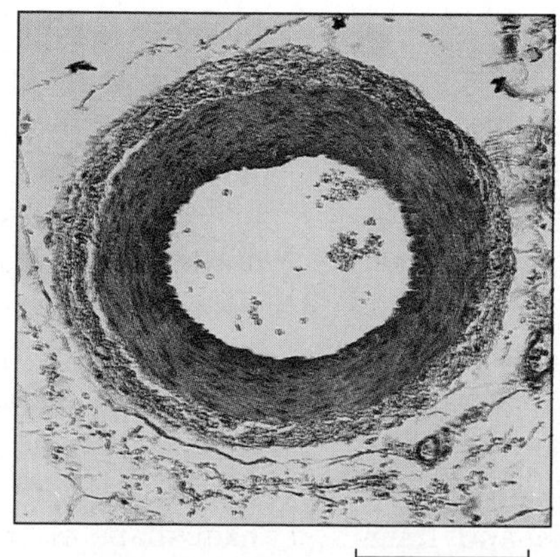

FIGURE 38.15
Atherosclerosis. These light micrographs contrast a normal artery (left) with one partially closed by an atherosclerotic plaque (right). Plaques consist mostly of fibrous connective tissue and smooth muscle cells infiltrated with lipids.

and living with extreme hypertension—say, 200/120—is courting disaster.

To some extent, the tendency for hypertension and atherosclerosis is inherited, making certain people more predisposed than others to cardiovascular disease. We cannot do much about our genes, but the health of our cardiovascular system is not completely out of our hands. Smoking, lack of exercise, obesity, and a diet rich in animal fats and cholesterol are among the factors that have been correlated with an increased risk of cardiovascular disease.

An abnormally high concentration of cholesterol in blood plasma is one of the most important correlates of potential atherosclerosis. Cholesterol travels in the blood mainly in the form of **low-density lipoproteins (LDLs),** plasma particles consisting of thousands of cholesterol molecules and other lipids bound to a protein. In contrast to LDLs, another form of cholesterol carriers, called **high-density lipoproteins (HDLs),** actually *reduce* the depositing of cholesterol in arterial plaques. Many researchers now believe that the ratio of LDLs to HDLs is more reliable than total plasma cholesterol as an indicator of impending cardiovascular disease. Exercise tends to increase HDL concentration, while smoking has the opposite effect on the LDL-to-HDL ratio.

Let's end the discussion of cardiovascular disease with some good news: Over the past 20 years, the death rate from cardiovascular disease in the United States has declined by more than 25%. So far, heart transplants, bypass surgery, and other radical methods for treating heart disease have not made a statistically significant contribution to this decline. On the other hand, diagnosis and treatment of hypertension may be preventing a large number of heart attacks and strokes, and improved methods of intensive care for cardiovascular patients may be increasing the chances of surviving heart attacks and strokes once they occur. Also, many Americans are now more conscious of their health; as a group, we are smoking less, exercising more, and watching our diets. Education is potent medicine.

Gas exchange supplies oxygen for cellular respiration and disposes of carbon dioxide: *an overview*

A major function of circulatory systems is to transport oxygen and carbon dioxide between respiratory organs and other parts of the body. We now focus on the actual exchange of these gases between animals and their environments.

Gas Exchange and the Respiratory Medium

Animals require a continuous supply of oxygen (O_2) for cellular respiration (see Chapter 9), and they must expel carbon dioxide (CO_2), the waste product of this process. Gas exchange supports the metabolic process of cellular respiration by supplying oxygen and removing carbon dioxide.

Earth's main reservoir of oxygen is the atmosphere, which is about 21% O_2. Oceans, lakes, and other bodies of water also contain oxygen in the form of dissolved O_2. The source of oxygen, called the **respiratory medium,** is air for a terrestrial animal and water for an aquatic one. The part of an animal where oxygen from the environment diffuses into living cells and carbon dioxide diffuses out is called the **respiratory surface.** All living cells must be bathed in water to maintain their plasma membranes. Thus, the respiratory surfaces of terrestrial as well as aquatic animals are moist, and oxygen and carbon dioxide diffuse across them after first dissolving in water. In addition, an animal's respiratory surfaces must be large enough to provide O_2 and expel CO_2 for the entire body.

Respiratory Surfaces: Form and Function

A variety of solutions to the problem of providing a large enough respiratory surface have evolved, depending mainly on the size of the organism and whether it lives in water or on land. Gas exchange occurs over the entire surface area of protozoa and other unicellular organisms. Similarly, for some animals, such as sponges, cnidarians, and flatworms, the plasma membrane of virtually every cell in the body contacts the outside environment, providing ample respiratory surface. In many animals, however, the bulk of the body does not have direct access to the respiratory medium, and the respiratory surface is a single layer of cells, a moist epithelium separating the respiratory medium from the blood or capillaries, which transport gases to and from the rest of the body (FIGURE 38.16).

Some animals use their entire outer skin as a respiratory organ. An earthworm, for example, has moist skin and exchanges gases by diffusion across its general body surface. Just below the skin is a dense net of capillaries. Because the respiratory surface must be moist, earthworms and many other skin breathers, including some amphibians, must live in water or damp places.

Most animals that use their moist skin as a respiratory organ are relatively small and have a long, thin (wormlike) shape or flat shape with a high ratio of surface to volume. For most other animals, the general body sur-

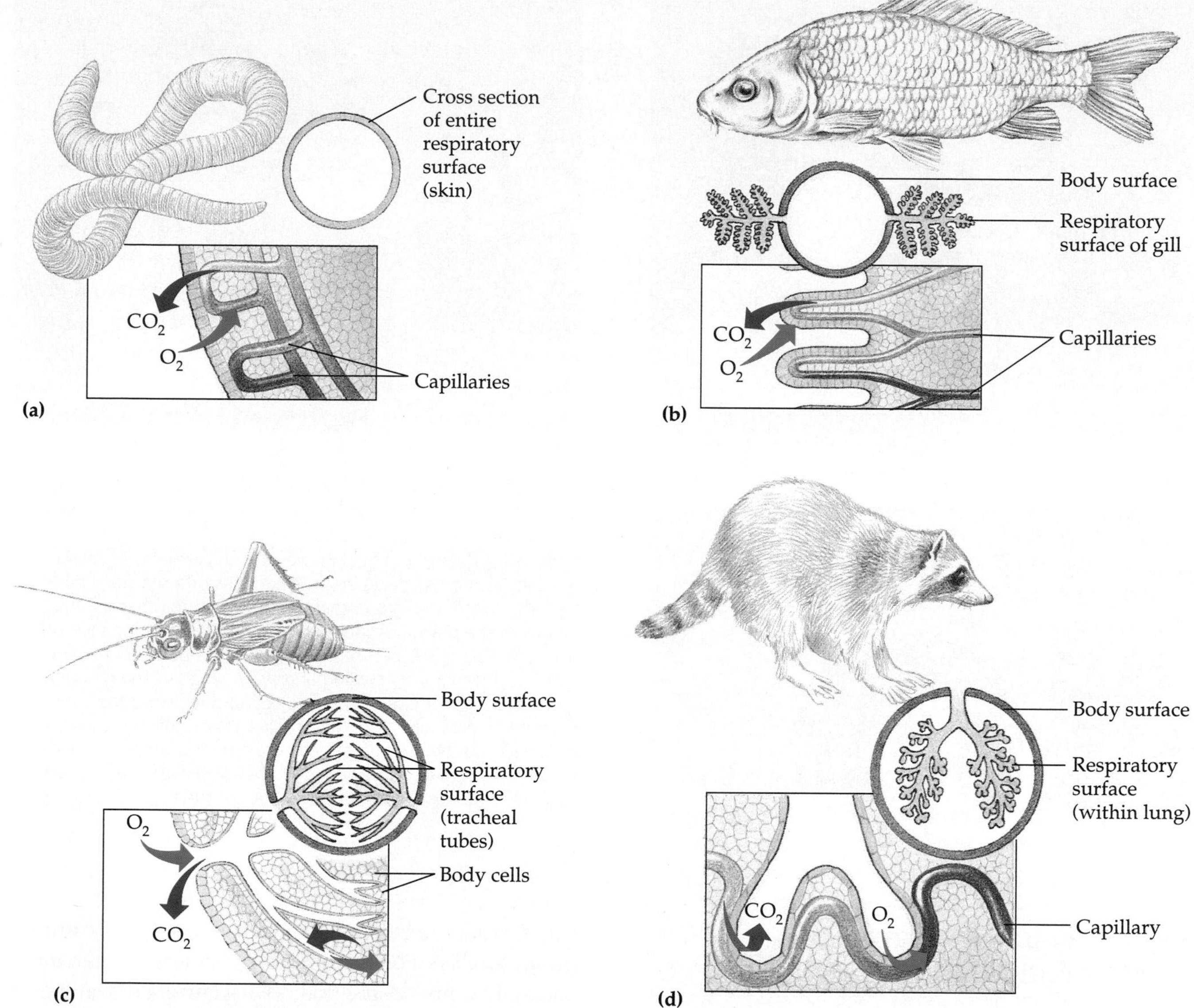

FIGURE 38.16
Respiratory organs. (**a**) Some small animals, such as earthworms, use their entire moist outer skin as a respiratory organ. (**b**) Fishes, some salamanders, and many other aquatic animals exchange gases through specialized respiratory organs called gills. (**c**) Insects have an extensive system of internal tubes called tracheae that channel air directly to body cells. (**d**) Most terrestrial vertebrates exchange gases across the lining of the lungs, internal organs containing respiratory surfaces supplied with blood.

face lacks sufficient area to exchange gases for the whole body. The solution is a region of the body surface that is extensively folded or branched, thereby enlarging the area of the respiratory surface for gas exchange. The expanded respiratory surface of most aquatic animals is external and bathed by fresh water. These extensions of the body surface are called gills. Gills are generally unsuitable for an animal living on land, because an expansive surface of wet membrane exposed to air would lose too much water to evaporation, and because the gills would collapse as their fine filaments, no longer supported by water, would cling together. Most terrestrial animals have their respiratory surfaces within the body, opening to the atmosphere only through narrow tubes. The lungs of terrestrial vertebrates and the tracheae of insects are two variations of this plan. We will now examine more closely the three most common respiratory organs: gills, tracheae, and lungs.

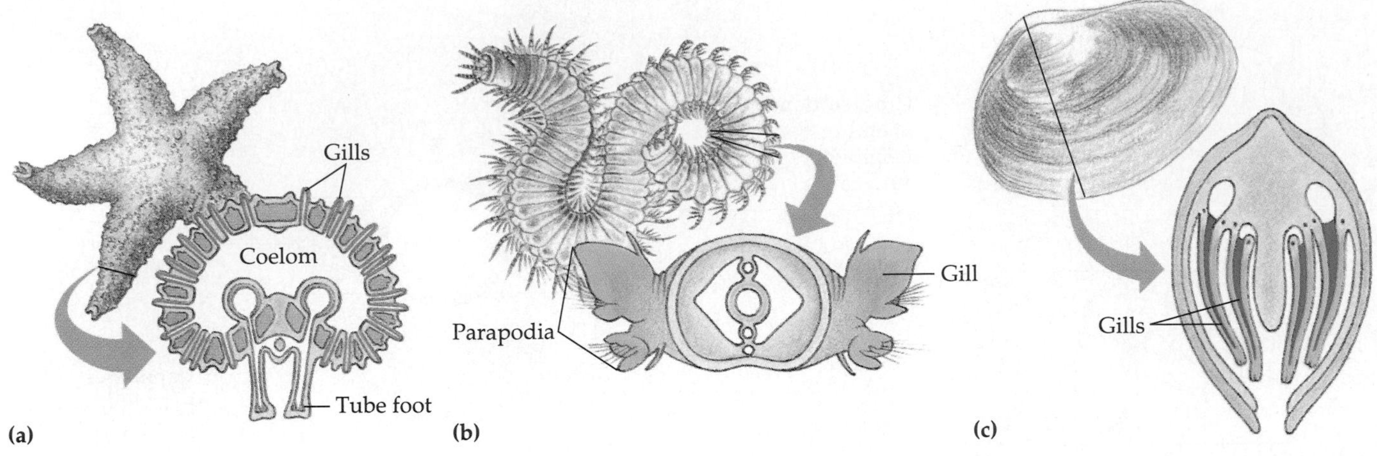

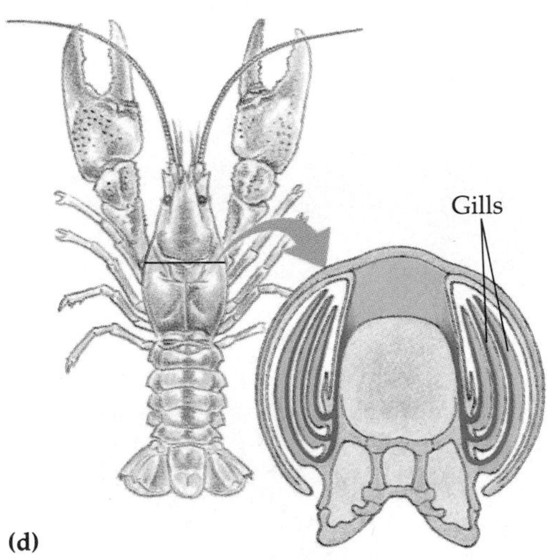

FIGURE 38.17
Invertebrate gills. (**a**) The gills of a sea star are simple tubular projections of the skin. The hollow core of each gill is an extension of the coelom (body cavity). Gas exchange occurs across the moist gill surfaces, and fluid in the coelom circulates in and out of the gills, aiding gas transport. The surfaces of a sea star's tube feet also function in gas exchange. (**b**) Many polychaetes, marine worms of the phylum Annelida, have a pair of gills for each body segment. Each gill is part of a parapodium, a body appendage that also functions in crawling and swimming. (**c**) The gills of a clam are long, flattened plates projecting downward from the main body mass inside the hard shell. Cilia on the gills circulate water around the gill surfaces. (**d**) Crayfish and other crustaceans have long, feathery gills covered by the exoskeleton. Specialized body appendages drive water over the gill surfaces.

Gills are respiratory adaptations of most aquatic animals

Gills are outfoldings of the body surface specialized for gas exchange (FIGURE 38.17). In some invertebrates, such as sea stars (phylum Echinodermata), the gills have a simple shape and are distributed over much of the body. Many segmented worms have flaplike gills that extend from each segment of the body or long feathery gills clustered at the head or tail. The gills of clams, crayfish, and many other animals are restricted to a local body region, and the total surface area of the gills is much greater than that of the rest of the body.

As a respiratory medium, water has both advantages and disadvantages. There is no problem keeping the cell membranes of the respiratory surface moist, since the gills are completely surrounded by the aqueous environment in which the animal lives. But the oxygen concentration in water is much lower than in air; and the warmer and saltier the water, the less dissolved oxygen it holds. Thus, gills must be very efficient to obtain enough oxygen from water. One process that helps is **ventilation,** a term that refers to any method of increasing flow of the respiratory medium (air or water) over the respiratory surface (lungs or gills). For example, crayfish and lobsters have paddlelike appendages that drive a current of water over the gills. The gills of a bony fish are ventilated continuously by a current of water that enters the mouth, passes through slits in the pharynx, flows over the gills, and exits at the back of the operculum (gill cover) (FIGURE 38.18). Ventilation brings a fresh supply of oxygen and removes carbon dioxide expelled by the gills. Because water is much denser and contains much less oxygen per unit of volume than air, a fish must expend a considerable amount of energy to ventilate its gills.

The arrangement of capillaries in the gills of a fish also enhances gas exchange. Blood flows in the opposite direction to which water passes over the gills. This pattern makes it possible for oxygen to be transferred to the blood by a very efficient process called **countercurrent exchange** (FIGURES 38.18 and 38.19). As blood flows through the capillary, it becomes more and more loaded with oxygen, but it simultaneously encounters water that is more and more concentrated in oxygen because the water is just beginning its passage over the gills. This means that along the entire length of the capillary, there

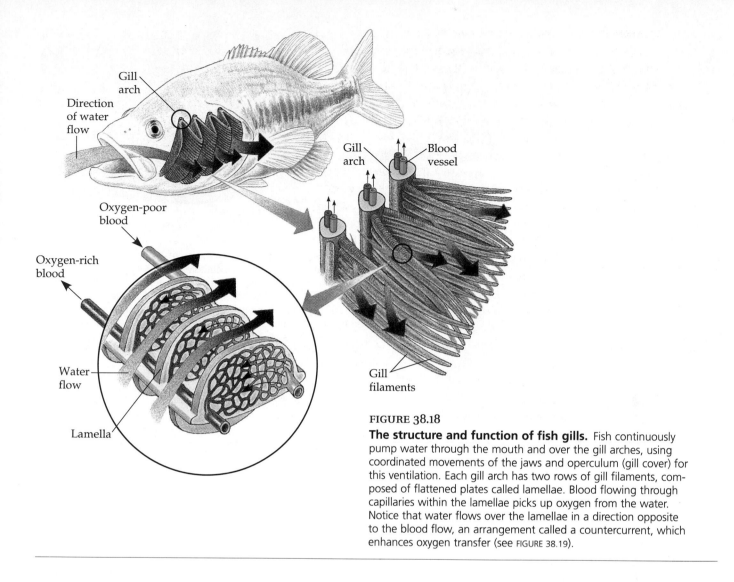

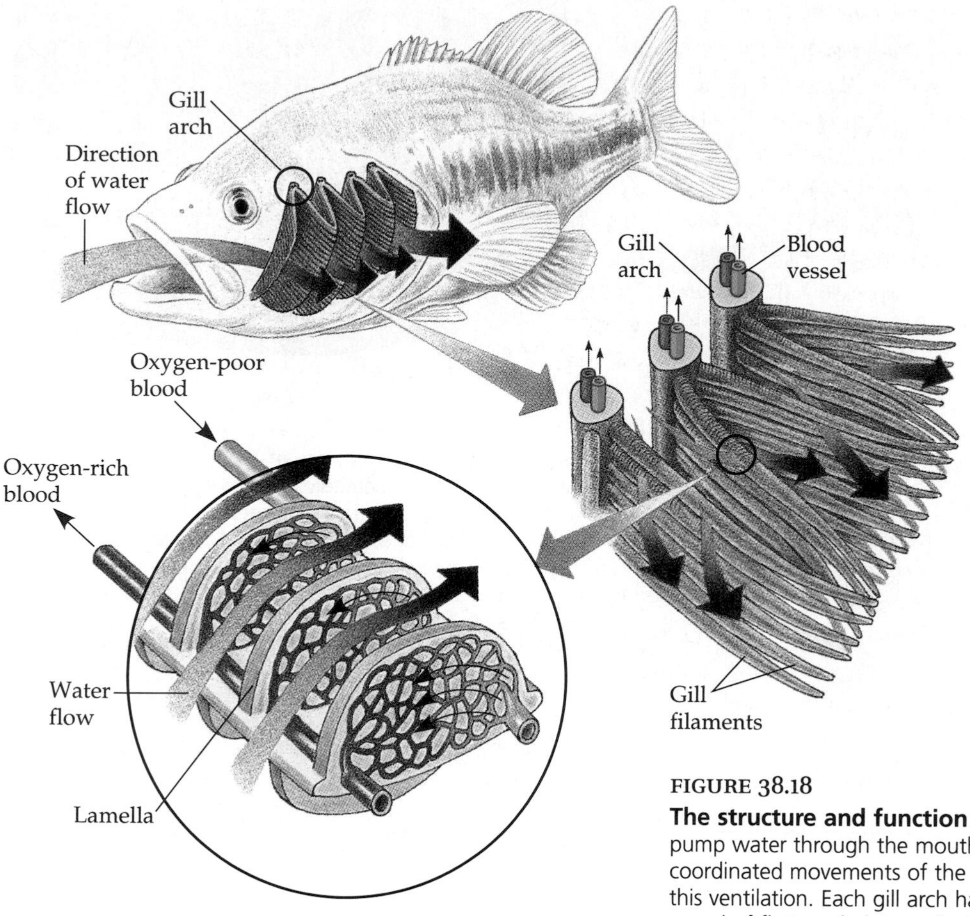

Gill arch
Direction of water flow
Gill arch
Blood vessel
Oxygen-poor blood
Oxygen-rich blood
Water flow
Lamella
Gill filaments

FIGURE 38.18

The structure and function of fish gills. Fish continuously pump water through the mouth and over the gill arches, using coordinated movements of the jaws and operculum (gill cover) for this ventilation. Each gill arch has two rows of gill filaments, composed of flattened plates called lamellae. Blood flowing through capillaries within the lamellae picks up oxygen from the water. Notice that water flows over the lamellae in a direction opposite to the blood flow, an arrangement called a countercurrent, which enhances oxygen transfer (see FIGURE 38.19).

FIGURE 38.19

Countercurrent exchange. The arrangement of blood vessels in fish gills is an adaptation that maximizes oxygen transfer from the water to the blood. (**a**) The direction of blood flow through capillaries in the lamellae of a gill filament is opposite that of water flowing over the lamellae. This countercurrent flow maintains a concentration gradient down which O_2 diffuses from the water into the blood over the whole length of a capillary. As the blood flows through a lamellar capillary, it becomes more and more loaded with oxygen because it continuously passes by water that has given up less of its O_2. Because of countercurrent exchange, a lamella can extract over 80% of the O_2 dissolved in the water passing over it. (**b**) By contrast, if blood flowed through capillaries of the lamellae in the same direction as water flowing over the lamellae (concurrent flow), gills could pick up at most 50% of the oxygen dissolved in the water. As O_2 diffused from the water into the blood, the concentration gradient would become less and less steep and finally cease to exist when the same amount of O_2 was dissolved in both the blood and the water. Fish gills exhibit countercurrent, not concurrent, flow.

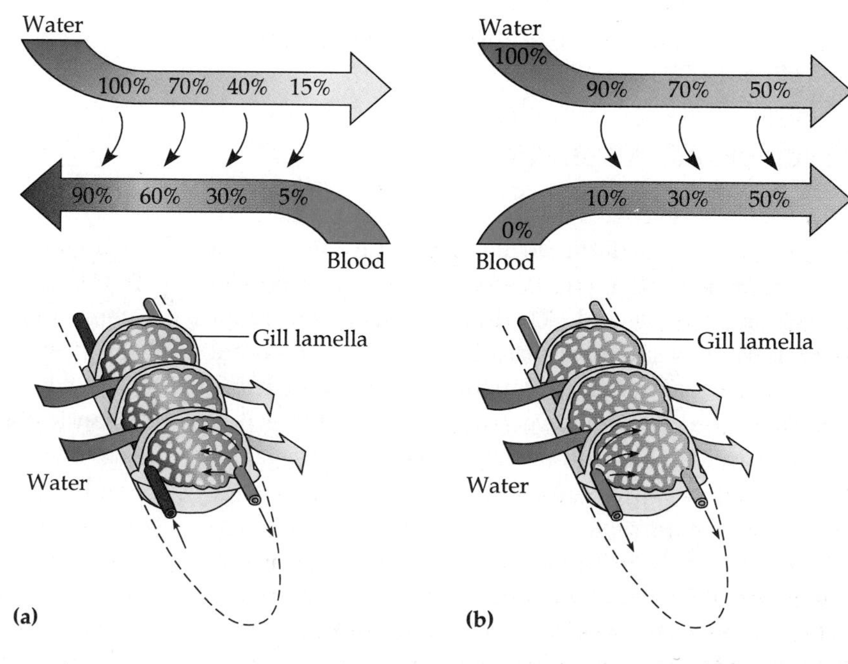

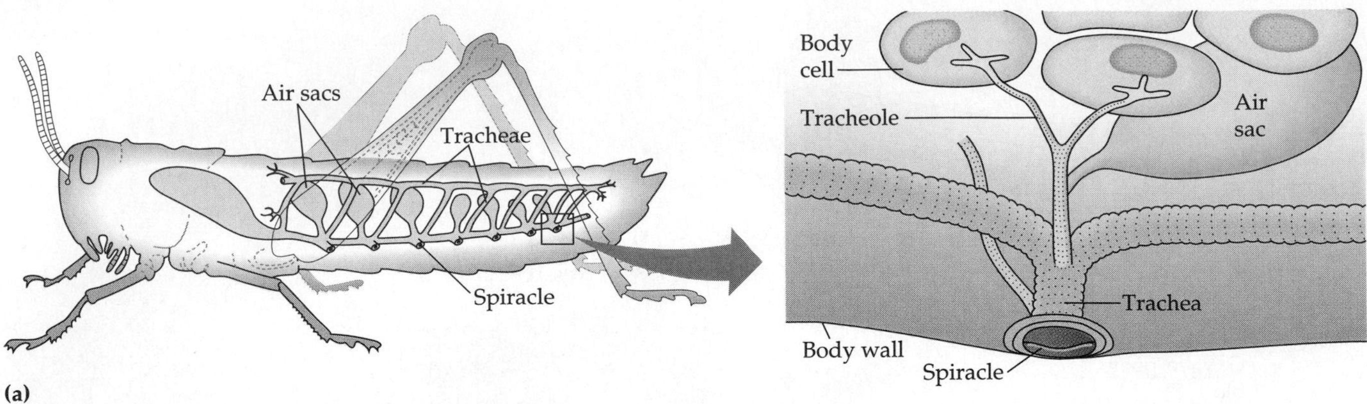

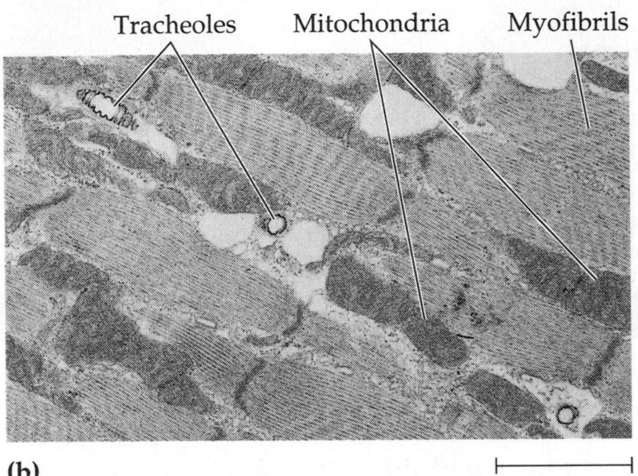

FIGURE 38.20

Tracheal systems. (**a**) The respiratory system of an insect consists of internal tubes, or tracheae, that branch repeatedly and deliver air directly to body cells. Rings of chitin reinforce the tracheae, keeping them from collapsing. Enlarged portions of tracheae form air sacs near organs that require a large supply of oxygen. Air enters the tracheae through openings called spiracles on the insect's body surface and passes into smaller tubes called tracheoles. The tracheoles terminate on the plasma membranes of individual cells. The amount of liquid in the tracheole endings determines the amount of air that contacts the cells. When the O_2 requirement increases, some liquid is withdrawn, increasing the surface area of air in contact with cells. (**b**) This micrograph shows cross sections of tracheoles in a tiny piece of insect flight muscle (TEM). Each of the numerous mitochondria in the muscle cells lies within about 5 μm of one or more tracheoles, and gas exchange occurs by diffusion across this short distance.

is a diffusion gradient favoring the transfer of oxygen from the water to the blood. So efficient is this countercurrent exchange mechanism that the gill can remove more than 80% of the oxygen dissolved in the water passing over the respiratory surface. The mechanism of countercurrent exchange is also important in temperature regulation and several other physiological processes, as we will see in Chapter 40.

Tracheae are respiratory adaptations of insects

As a respiratory medium, air has many advantages, not the least of which is a much higher concentration of oxygen. Also, since O_2 and CO_2 diffuse much faster in air than in water, respiratory surfaces exposed to air do not have to be ventilated as thoroughly as gills. As the respiratory surface removes oxygen from the air and expels carbon dioxide, diffusion rapidly brings more oxygen to the respiratory surface and carries the carbon dioxide away. When a terrestrial animal does ventilate, less energy is expended because air is much easier to move than water. But offsetting these advantages of air as a respiratory medium is a problem: The respiratory surface, which must be large and moist, continuously loses water to the air by evaporation. The solution is a respiratory surface folded into the body. One variation of this plan is the tracheal system of insects.

Tracheae are air tubes that branch throughout the insect body (FIGURE 38.20). The finest branches extend to the surface of nearly every cell, where gas is exchanged by diffusion across the moist epithelium that lines the terminal ends of the tracheal system. With virtually all body cells exposed to the respiratory medium, the open circulatory system of insects is not involved in transporting oxygen and carbon dioxide.

For a small insect, diffusion alone brings enough O_2 from the air into the tracheal system and removes enough CO_2 to support cellular respiration. Larger insects with higher energy demands ventilate their tracheal systems with rhythmic body movements that compress and expand the air tubes like bellows. An insect in flight has a high metabolic rate, consuming 10 to 100 times more O_2 than it does at rest. In many flying insects, alternating contraction and relaxation of the flight muscles compress and expand the body, rapidly pumping air through the tracheal system. Also supporting the high metabolic rate, the flight muscle cells are packed with mitochondria, and the tracheal tubes supply each of these ATP-generating organelles with ample oxygen (FIGURE 38.20b). Thus, we see a direct relationship between adaptations of tracheal systems and the theme of bioenergetics.

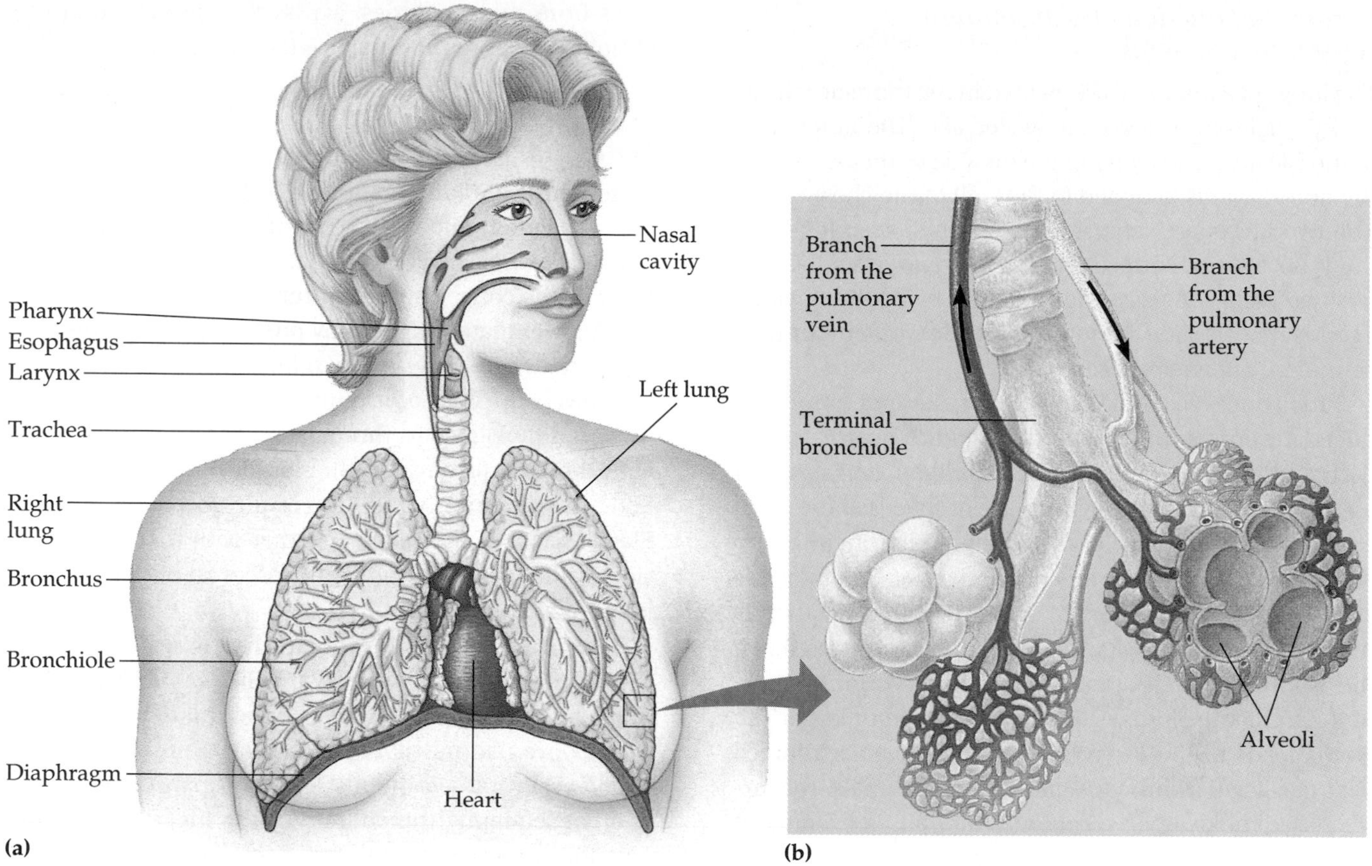

Nasal cavity
Pharynx
Esophagus
Larynx
Trachea
Left lung
Right lung
Bronchus
Bronchiole
Diaphragm
Heart
(a)

Branch from the pulmonary vein
Branch from the pulmonary artery
Terminal bronchiole
Alveoli
(b)

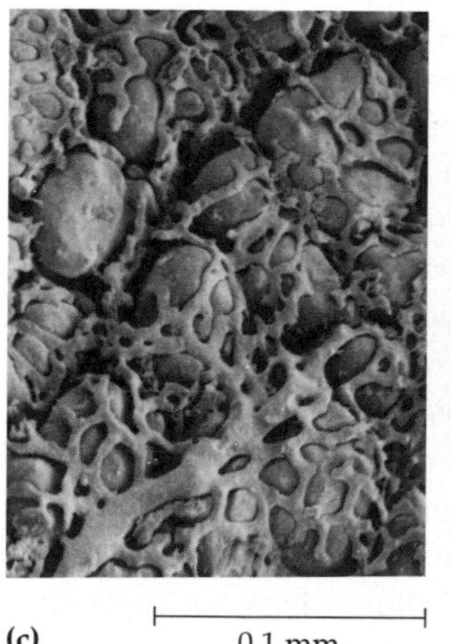

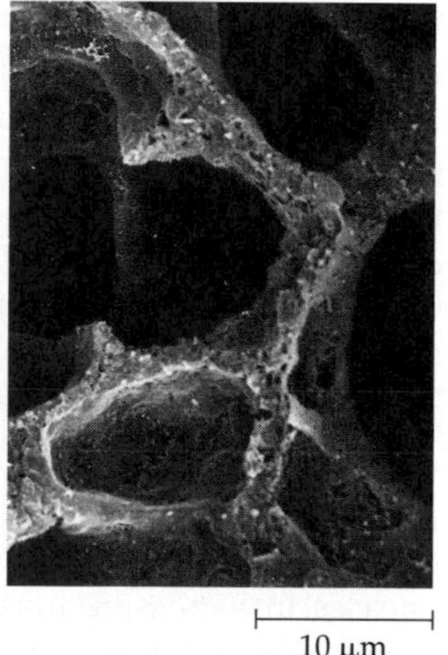

(c) 0.1 mm 10 μm

FIGURE 38.21

The mammalian respiratory system. (**a**) From the nasal cavity and pharynx, inhaled air passes through the larynx and down the trachea and bronchi to the tiniest bronchioles, which end in lobed, microscopic air sacs, the alveoli. (**b**) A thin, moist epithelium lining the inner surfaces of the alveoli forms the respiratory surface. Branches of the pulmonary artery (see FIGURE 38.4) convey oxygen-poor blood to the alveoli, while branches of the pulmonary vein transport oxygen-rich blood from the alveoli back to the heart. (**c**) The left micrograph shows the dense capillary bed that envelops the alveoli (SEM, from *Tissues and Organs: A Text-Atlas of Scanning Electron Microscopy,* by Richard G. Kessel and Randy H. Kardon. W. H. Freeman and Company. Copyright © 1979). The right micrograph (SEM) is a cutaway view of alveoli at higher magnification.

Lungs are the respiratory adaptations of most terrestrial vertebrates

In contrast to the respiratory channels that branch throughout the insect body, **lungs** are restricted to one location. Because the respiratory surface of a lung is not in direct contact with all other parts of the body, the gap must be bridged by the circulatory system, which transports oxygen from the lungs to the rest of the body. Lungs have a dense net of capillaries located just beneath the epithelium that forms the respiratory surface. This solution to the problem of gas exchange has evolved as an internal mantle in land snails, as book lungs in spiders, and as lungs in terrestrial vertebrates: reptiles, birds, mammals, and many amphibians. The lungs of most frogs are balloonlike, and do not provide a large respiratory surface area, but frogs also obtain some of their oxygen by diffusion across their skin. In contrast, the lungs of mammals have a spongy texture and are honeycombed with a moist epithelium that serves as the respiratory surface (FIGURE 38.21). The total surface area of the epithelium (about 100 m^2 in humans) is sufficient to carry out gas exchange for the whole body.

Form and Function of Mammalian Respiratory Systems

The lungs of a mammal are located in the thoracic (chest) cavity, enclosed by a double-walled sac. The inner layer of the sac adheres tightly to the outside of the lungs, and the outer layer is attached to the wall of the chest cavity. The two layers are separated by a thin space filled with fluid. Because of surface tension, the two layers behave like two plates of glass stuck together by a film of water. The layers can slide smoothly past each other, but they cannot be pulled apart easily.

Air is conveyed to the lungs by a system of branching ducts. Air enters this system through the nostrils and is then filtered by hairs, warmed, humidified, and sampled for odors as it flows through a maze of spaces in the nasal cavity. The nasal cavity leads to the pharynx, an intersection where the paths for air and food cross. When food is swallowed, the opening of the windpipe (glottis) is pushed against the epiglottis, and food is diverted down the esophagus to the stomach (see FIGURE 37.10). The rest of the time, the glottis is open, and we can breathe. The glottis leads to the **larynx,** which has a wall reinforced with cartilage. In humans and other mammals, the larynx is adapted as a voicebox. When air is exhaled, it rushes by a pair of **vocal cords** in the larynx, and sounds are produced when voluntary muscles in the voicebox are tensed, stretching the cords so they vibrate. High-pitched sounds result when the cords are stretched tight and vibrate fast; low-pitched sounds come from less tense cords vibrating slowly.

From the larynx, air passes into the **trachea,** or windpipe. Rings of cartilage (shaped like the letter C) maintain the shape of the trachea, much as metal rings keep the hose of a vacuum cleaner from collapsing. The trachea forks into two **bronchi** (singular, **bronchus**), one leading to each lung. Within the lung, the bronchus branches repeatedly into finer and finer tubes called **bronchioles.** The entire system of air ducts has the appearance of an inverted tree, the trunk being the trachea. The epithelium lining the major branches of this respiratory tree is covered by cilia and a thin film of mucus. The mucus traps dust, pollen, and other contaminants, and the beating cilia move the mucus upward to the pharynx, where it can be swallowed into the esophagus. This process helps cleanse the respiratory system.

At their tips, the tiniest bronchioles dead-end as a cluster of air sacs called **alveoli** (singular, **alveolus**). The thin epithelium of the millions of alveoli in the lung serves as the respiratory surface. Oxygen in the air conveyed to the alveoli by the respiratory tree dissolves in the moist film and diffuses across the epithelium and into a web of capillaries that surrounds each alveolus. Carbon dioxide diffuses from the capillaries, across the epithelium of the alveolus, and into the air space (FIGURE 38.21b and c).

Ventilating the Lungs

Vertebrates ventilate their lungs by **breathing,** the alternate inhalation and exhalation of air. Ventilation maintains a maximal oxygen concentration and minimal carbon dioxide concentration within the alveoli.

A frog ventilates its lungs by **positive pressure breathing,** actually pushing air down its windpipe. The animal first lowers the floor of its mouth, enlarging the oral cavity and drawing air in through the open nostrils. Then, with the nostrils and mouth closed, the frog raises the floor of its mouth, which forces air down the trachea. Elastic recoil of the lungs, together with their compression by the muscular body wall, forces air back out of the lungs during exhalation.

In contrast, mammals ventilate their lungs by **negative pressure breathing,** which works like a suction pump, pulling air down into the lungs rather than pushing it in from above (FIGURE 38.22). This is accomplished by the action of several sets of muscles that expand the rib cage and consequently the volume of the lungs. During inhalation, contraction of the rib muscles expands the rib cage by pulling the ribs upward. Movement of the lungs is coupled to movement of the rib cage by the surface tension of fluid in the thin space between the two layers of the sac enclosing the lungs; when the rib cage expands, so do the lungs. Because this movement increases the lung volume, air pressure within the alveoli is reduced to less than atmospheric pressure. Because air always flows from a region of higher pressure to a region of lower pressure, air rushes through the nostrils and down the respiratory tree to the alveoli. When the rib muscles relax, the lung volume is reduced, and the increase in air pressure within the alveoli forces air up the respiratory tree and out through the nostrils. The action of the rib muscles in increasing lung volume is most important during vigorous exercise. However, it accounts for less than one-third of the increase in volume during shallow breathing, when a mammal is at rest.

Most of the increase in lung volume during shallow inhalation results from the action of the diaphragm rather than the rib muscles. The **diaphragm** is a sheet of muscle that forms the bottom wall of the thoracic cavity. During inhalation, the contraction of the diaphragm causes it to descend like a piston enlarging the thoracic cavity and lowering the pressure in the lungs. Air is exhaled when the diaphragm relaxes and moves upward, decreasing the volume of the lungs.

The volume of air an animal inhales and exhales with each breath is called **tidal volume.** It averages about

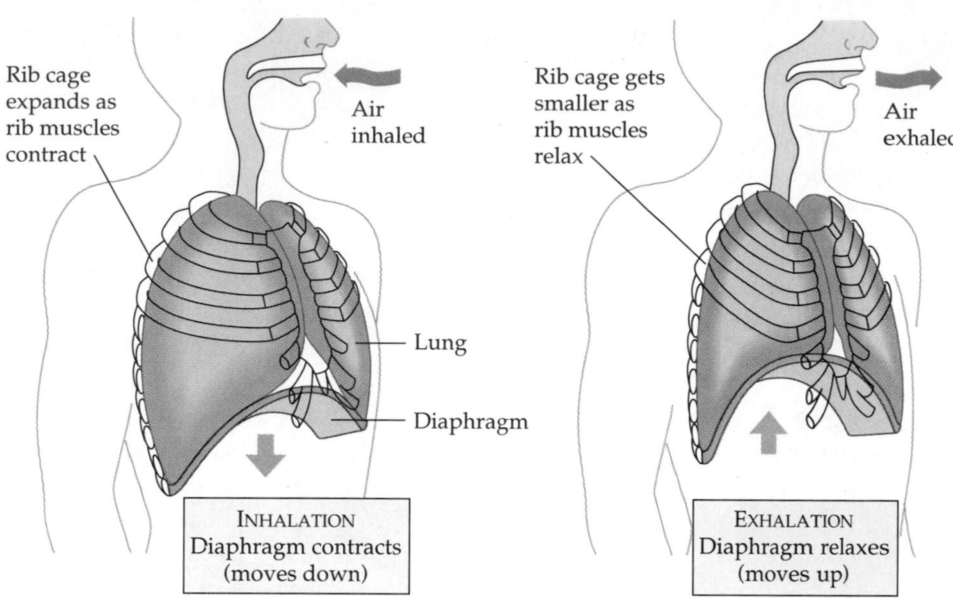

FIGURE 38.22
Negative pressure breathing. A mammal breathes by changing the air pressure within its lungs relative to the pressure of the outside atmosphere. During inhalation, the rib muscles and diaphragm contract. The volume of the thoracic cavity and lungs increases as the diaphragm moves down and the rib cage expands. Air pressure in the lungs falls below that of the atmosphere, and air rushes into the lungs. Exhalation occurs when the rib muscles and diaphragm relax, restoring the thoracic cavity to its smaller volume.

500 mL in humans. The maximum volume of air that can be inhaled and exhaled during forced breathing is called **vital capacity,** which averages about 3400 mL and 4800 mL for college-age females and males, respectively. Among other factors, vital capacity depends on the resilience of the lungs. The lungs actually hold more air than the vital capacity, but since it is impossible to completely collapse the alveoli, a **residual volume** of air remains in the lungs even after we forcefully blow out as much air as we can. As lungs lose their resilience as a result of aging or disease (such as emphysema), residual volume increases at the expense of vital capacity.

Ventilation is much more complex in birds than in mammals (FIGURE 38.23). Besides lungs, birds have eight or nine air sacs that penetrate the abdomen, neck, and even the wings. The air sacs do not function directly in gas exchange, but act as bellows that keep air flowing through the lungs. The air sacs also trim the density of the bird, an important adaptation for flight (see Chapter 30). The entire system—lungs and air sacs—is ventilated when the bird inhales and exhales. Air flows through the interconnected system in a circuit that passes through the lungs in one direction only, regardless of whether the bird is inhaling or exhaling. Alveoli, which are dead ends, would not be suitable in such a system. Instead, the lungs of a bird have tiny channels called **parabronchi,** through which air can flow continuously in one direction.

The Control of Breathing

We can hold our breath voluntarily a short while or consciously breathe faster and deeper, but most of the time, automatic mechanisms regulate our breathing. Automatic control ensures that the work of the respiratory system is coordinated with that of the cardiovascular system.

Our **breathing control centers** are located in two regions of the brain, the medulla oblongata and the pons (FIGURE 38.24). Aided by the control center in the pons, the medulla's center sets the basic breathing rhythm. When we take deep breaths, a negative-feedback mechanism prevents our lungs from overexpanding; stretch sensors in the lung tissue send nerve impulses back to the medulla, inhibiting its breathing control center.

The medulla's control center also helps maintain homeostasis by monitoring the CO_2 level of the blood and regulating the amount of CO_2 our alveoli dispose of when we exhale. Its main cues about CO_2 come from slight changes in the pH of the blood and tissue fluid (cerebrospinal fluid) bathing the brain. Carbon dioxide reacts with water to form carbonic acid, which lowers the pH. When the medulla's control center registers a slight drop in pH (increase in CO_2) of the cerebrospinal fluid or the blood, it increases the depth and rate of breathing, and the excess CO_2 is eliminated in exhaled air. This occurs when we exercise.

The concentration of O_2 in the blood usually has little effect on the breathing control centers. However, when the O_2 level is severely depressed—at very high altitudes for instance—O_2 sensors in the aorta and carotid arteries in the neck send alarm signals to the breathing control centers, and the centers respond by increasing the breathing rate. A rise in carbon dioxide concentration is usually a good indication of a drop in oxygen concentration, because CO_2 is produced by the same process that

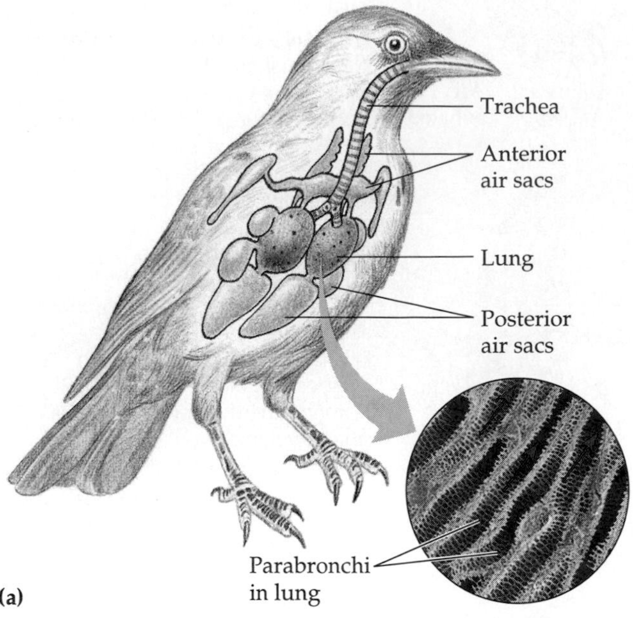

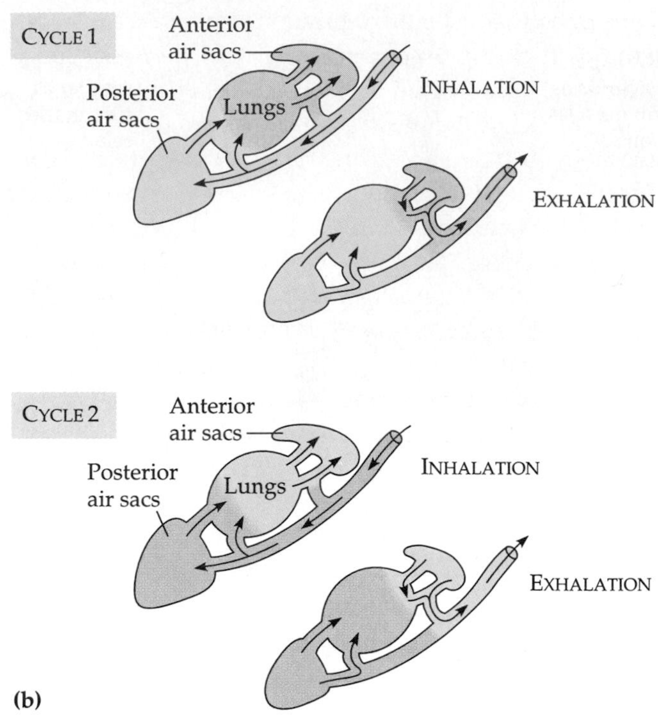

FIGURE 38.23

The avian respiratory system. (**a**) A bird has air sacs in addition to lungs. Contraction of the air sacs ventilates the lungs, forcing air in one direction (red arrows) through tiny parallel tubes called parabronchi (inset, SEM). Gas exchange occurs across the walls of the parabronchi. (**b**) This sequence of diagrams traces one "breath" of fresh air (blue) through the respiratory system of a bird. Two cycles of inhalation and exhalation are required for the air to pass all the way through the system and out of the bird. During the first inhalation, most of the air bypasses the lungs and enters the posterior air sacs. That air passes through the lungs during exhalation and the next inhalation, ending up in the anterior air sacs. At the same time, the posterior sacs draw in another breath of fresh air (purple).

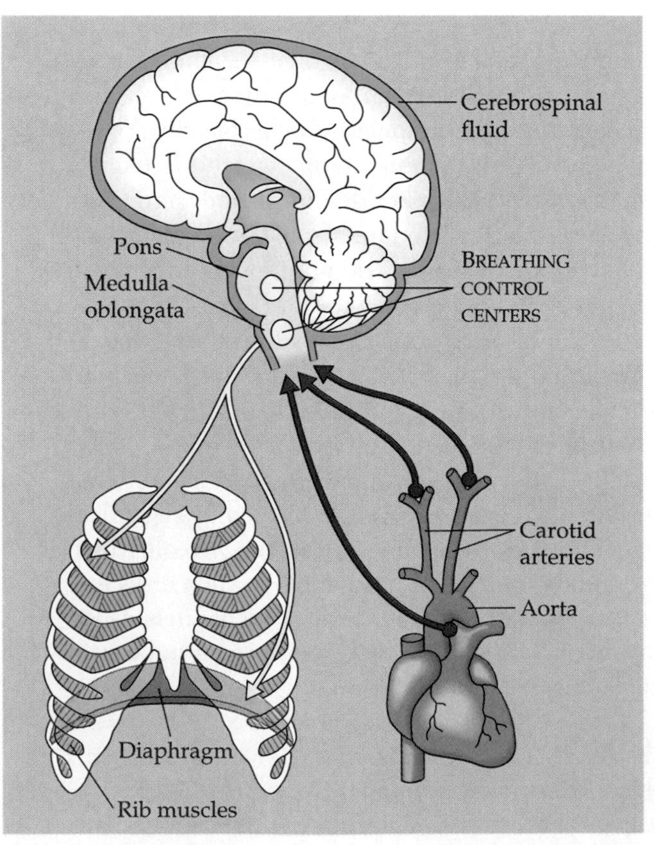

FIGURE 38.24

Automatic control of breathing. Making us inhale, nerves from a breathing control center in the medulla oblongata of the brain (yellow) send impulses to the diaphragm and rib muscles, stimulating them to contract. When we are at rest, these nerves send out impulses that result in about 10 to 14 inhalations per minute. Between inhalations, the muscles relax, and we exhale. A control center in the pons modulates the basic rhythm set by the medulla, smoothing out the transitions between inhalations and exhalations. The medulla's control center also helps regulate the CO_2 level of the blood. Sensors in the medulla itself detect changes in the pH (a measure of CO_2) of the blood and cerebrospinal fluid bathing the surface of the brain. Other sensors in the walls of the aorta and carotid arteries in the neck detect changes in blood pH and send nerve impulses to the medulla (red arrows). In response, the medulla's breathing control center alters the rate and depth of breathing, increasing it to dispose of excess CO_2 or slowing it if CO_2 levels are depressed. The sensors in the aorta and carotid arteries also detect changes in O_2 levels in the blood and signal the medulla to increase the breathing rate when levels become very low.

consumes O_2—cellular respiration. It is possible, however, to trick the breathing center by hyperventilating. Excessively deep, rapid breathing purges the blood of so much CO_2 that the breathing center temporarily ceases to send impulses to the rib muscles and diaphragm. Breathing stops until the CO_2 level increases enough to switch the breathing center back on.

The breathing center, then, responds to a variety of nervous and chemical signals, adjusting the rate and depth of breathing to meet the changing demands of the body. Control of breathing is only effective, however, if it is coordinated with control of the circulatory system. During exercise, for instance, cardiac output is matched to the increased breathing rate, which enhances O_2 supply and CO_2 removal as blood flows through the lungs.

Loading and Unloading of Oxygen and Carbon Dioxide

To understand how gases are exchanged at various locations around the body, recall that substances diffuse down their gradients. For a gas, whether present in air or dissolved in water, diffusion depends on differences in a quantity called **partial pressure.** At sea level, the atmosphere exerts a total pressure of 760 mm Hg. This is a downward force equivalent to that exerted by a column of mercury 760 mm high. Since the atmosphere is 21% oxygen (by volume), the partial pressure of oxygen is 0.21×760, or about 160 mm Hg. (This is the portion of atmospheric pressure contributed by oxygen, hence the term *partial pressure.*) The partial pressure of carbon dioxide at sea level is only 0.23 mm Hg. These partial pressures are symbolized by P_{O_2} and P_{CO_2}. When water is exposed to air, the amount of any gas that dissolves in the water is proportional to its partial pressure in the air and its solubility in water. An equilibrium is eventually reached when the gas molecules enter and leave the solution at the same rate. At this point, the gas is said to have the same partial pressure in the solution as it does in the air. Thus, the P_{O_2} in a glass of water exposed to air is 160 mm Hg, and the P_{CO_2} is 0.23 mm Hg. A gas will always diffuse from a region of higher partial pressure to a region of lower partial pressure.

Blood arriving at a lung via the pulmonary artery has a lower P_{O_2} and a higher P_{CO_2} than the air in the alveoli (FIGURE 38.25). As blood enters the capillary beds around the alveoli, carbon dioxide diffuses from the blood to the air within the alveoli. Oxygen in the air dissolves in the fluid that coats the epithelium and diffuses across the surface and into a capillary. By the time the blood leaves the lungs in the pulmonary veins, its P_{O_2} has been raised and its P_{CO_2} has been lowered. After returning to the heart, this blood is pumped through the systemic circuit. In the systemic capillaries, gradients of partial pressure favor the diffusion of oxygen out of the blood and carbon dioxide into the blood. This is because cellular respiration rapidly depletes the oxygen content of interstitial fluid and adds carbon dioxide to the fluid (again, by diffusion). After the blood unloads oxygen and loads carbon dioxide, it is returned to the heart by systemic veins. The blood is then pumped to the lungs again, where it exchanges gases with air in the alveoli.

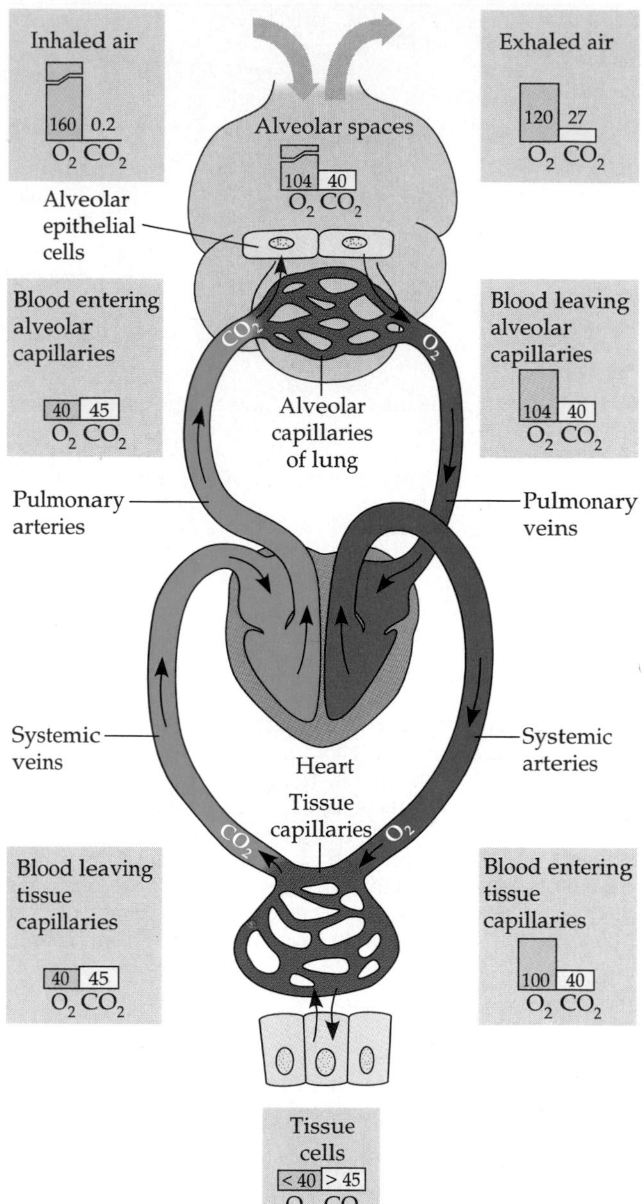

FIGURE 38.25

Loading and unloading of respiratory gases. Gases diffuse from a region of higher partial pressure to a region of lower partial pressure. The colored bars indicate the partial pressures of O_2 (P_{O_2}) and CO_2 (P_{CO_2}) that account for gas exchange in the capillary beds of the lungs and tissues. The numbers in the bars are the actual partial pressures of the gases in mm Hg.

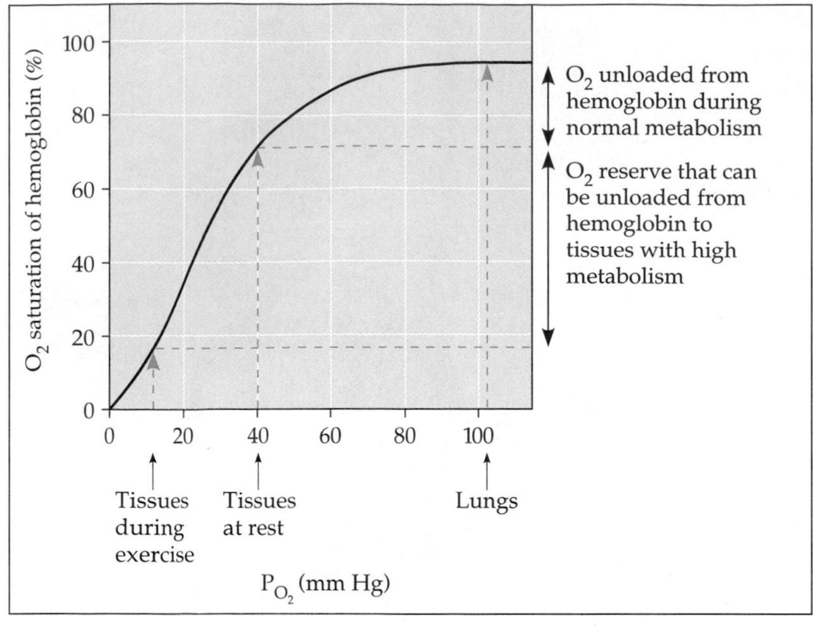

(a)

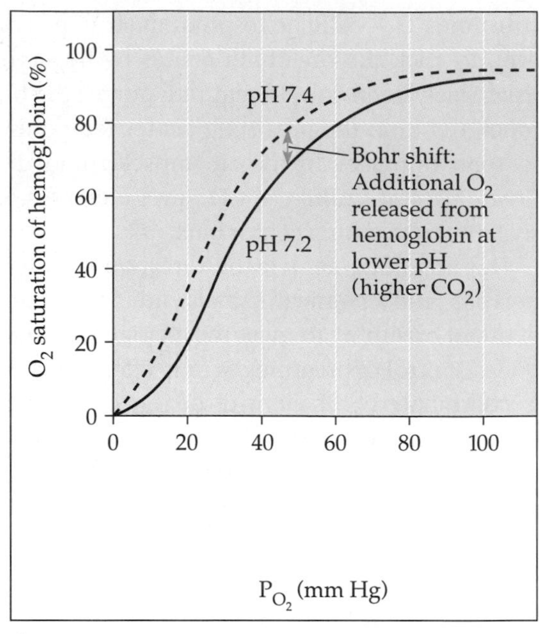

(b)

FIGURE 38.26
Oxygen dissociation curves for hemoglobin. (**a**) This is the dissociation curve for hemoglobin at 37°C and pH 7.4. The curve shows the relative amounts of oxygen bound to hemoglobin when the pigment is exposed to solutions varying in their partial pressure of dissolved oxygen. At a P_{O_2} of 100 mm Hg, typical in the lungs, hemoglobin is about 98% saturated with oxygen. At a P_{O_2} of 40 mm Hg, common in the vicinity of tissues, hemoglobin is only about 70% saturated; that is, it gives up about 28% of its oxygen. Hemoglobin can release its reserve of O_2 to tissues that are metabolically very active—muscle tissue during exercise. (**b**) Hydrogen ions affect the conformation of hemoglobin, and thus a drop in pH shifts the oxygen dissociation curve toward the right. Notice that at an equivalent P_{O_2}, say 40 mm Hg, hemoglobin gives up more oxygen at pH 7.2 than it does at pH 7.4, the normal pH of human blood. This occurs in very active tissues because the CO_2 produced by respiration reacts with water to form carbonic acid, thus lowering the pH. Hemoglobin then releases more O_2, which supports a high level of cellular respiration.

Respiratory Pigments and Oxygen Transport

Because oxygen is not very soluble in water, very little is transported in blood in the form of dissolved O_2. In most animals, oxygen is carried by **respiratory pigments** in the blood. These pigments are proteins that owe their color to metal atoms built into the molecules. Several oxygen-carrying proteins are found in the blood of various invertebrates. One of these, called **hemocyanin,** has copper as its oxygen-binding component, coloring the blood blue. Common in arthropods and many mollusks, hemocyanin is dissolved in plasma rather than being confined to cells.

The respiratory pigment of almost all vertebrates is hemoglobin, contained in red blood cells. (Presumably, one advantage of packaging the hemoglobin in cells is that the blood can have a high concentration of this protein without increasing the osmotic pressure of the plasma.) Hemoglobin consists of four subunits, each with a cofactor called a heme group that has an iron atom at its center. It is the iron that actually binds to oxygen; thus, each hemoglobin molecule can carry four molecules of O_2 (see FIGURE 5.23b). To function as an oxygen vehicle, hemoglobin must bind the gas reversibly, loading oxygen in the lungs or gills and unloading it in other parts of the body. In this loading and unloading, there is cooperation among the four subunits of the hemoglobin molecule. (See Chapter 6 to review the concept of cooperativity in allosteric proteins.) The binding of oxygen to one subunit induces the remaining subunits to change their shape slightly so that their affinity for oxygen increases. The hesitant loading of the first O_2 molecule results in the rapid loading of three more. And when one subunit unloads its oxygen, the other three quickly follow the lead as a conformational change lowers their affinity for oxygen.

This mechanism of cooperative oxygen binding and release is evident in the **dissociation curve** for hemoglobin (FIGURE 38.26). Over the range of oxygen partial pressures (P_{O_2}) where the dissociation curve has a steep slope, even a slight change in P_{O_2} will cause hemoglobin to load or unload a substantial amount of oxygen. Notice that the steep part of the curve corresponds to the range of oxygen partial pressures found in body tissues. When cells in a particular location begin working harder—during exercise, for instance—P_{O_2} dips in the vicinity, as the

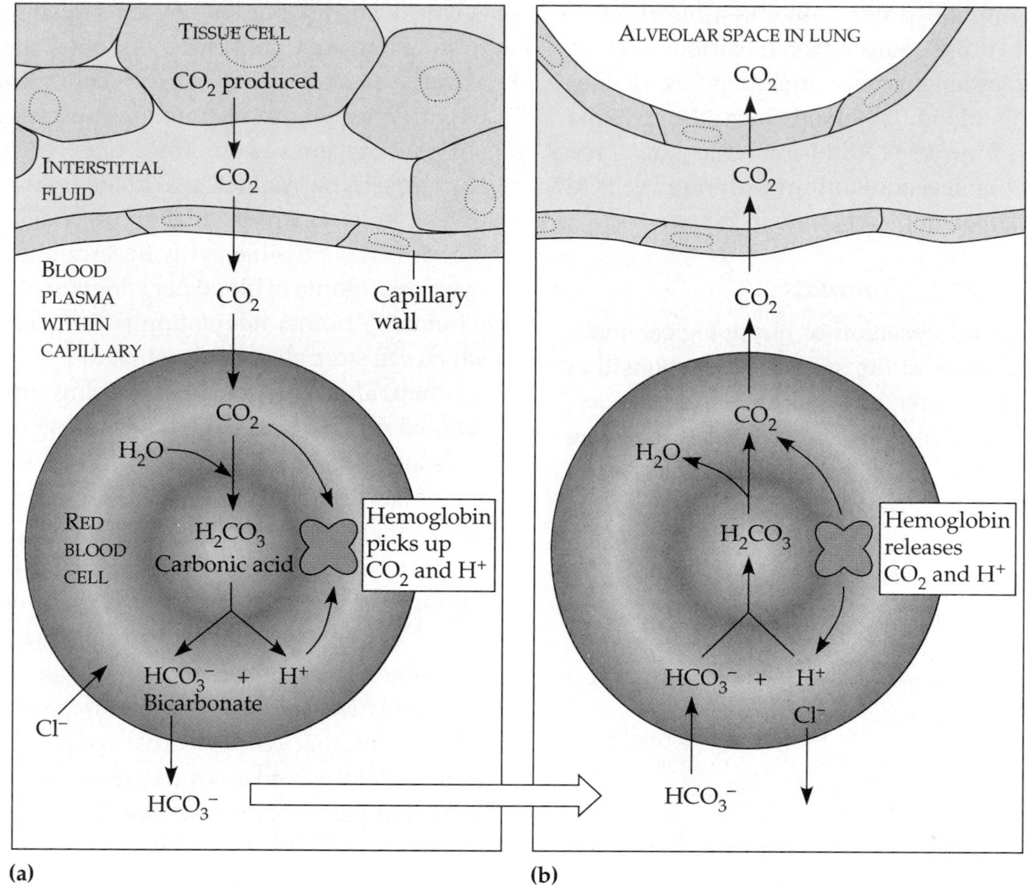

FIGURE 38.27

Carbon dioxide transport in the blood. (**a**) Carbon dioxide produced by body tissues diffuses into the interstitial fluid and into the plasma. Less than 10% remains in the plasma as dissolved CO_2.The rest diffuses into the erythrocytes, where some is picked up and transported by hemoglobin. Most of the CO_2 reacts with H_2O in the erythrocyte to form carbonic acid. (Red blood cells contain the enzyme carbonic anhydrase, which catalyzes this reaction.) Carbonic acid dissociates into a bicarbonate ion and a hydrogen ion (H^+). Most of the bicarbonate ions diffuse into the plasma and are carried in this form to the lungs (white arrow). Hemoglobin binds most of the H^+ ions from carbonic acid, preventing them from acidifying the blood. The reversibility of the carbonic acid–bicarbonate conversion also helps buffer the blood, releasing or removing H^+, depending on pH. (**b**) The processes that occur in the tissue capillaries are reversed in the lungs. CO_2 is formed from carbonic acid and unloaded from hemoglobin. It diffuses out of the blood, into the interstitial fluid, and into the alveolar space, from which it is expelled during inhalation.

O_2 is consumed in cellular respiration. Because of the effect of subunit cooperativity, a slight drop in P_{O_2} is enough to cause a relatively large increase in the amount of oxygen the blood unloads.

As with all proteins, hemoglobin's conformation is sensitive to a variety of environmental factors. For example, a drop in pH lowers the affinity of hemoglobin for O_2, an effect called the *Bohr shift* (FIGURE 38.26b). Because CO_2 reacts with water to form carbonic acid, an active tissue will lower the pH of its surroundings and induce hemoglobin to give up more of its oxygen, which can be used for cellular respiration. Hemoglobin is a remarkable molecule, its structure well suited for its role of transporting oxygen from regions of supply to regions of demand.

Carbon Dioxide Transport

In addition to its role in oxygen transport, hemoglobin also helps the blood transport carbon dioxide and assists in buffering the blood, that is, preventing harmful changes in pH (FIGURE 38.27).

Only about 7% of the carbon dioxide released by respiring cells is transported as dissolved CO_2 in blood plasma. Another 23% binds to the multiple amino groups of hemoglobin. Most carbon dioxide, about 70%, is transported in the blood in the form of bicarbonate ions. Carbon dioxide expelled by respiring cells diffuses into the blood plasma and then into the red blood cells, where the CO_2 is converted to bicarbonate. Carbon dioxide first reacts with water to form carbonic acid, which then dissociates into a hydrogen ion and a bicarbonate ion. Most of the hydrogen ions attach to various sites on hemoglobin and other proteins and therefore do not change the pH of blood. As blood flows through the lungs, the process is reversed. Diffusion of CO_2 out of the blood shifts the chemical equilibrium within red cells in favor of the conversion of bicarbonate to CO_2.

Adaptations of Diving Mammals

We can further our appreciation of physiological diversity by examining some of the special adaptations that make it possible for air-breathing mammals, such as certain species of seals, whales, and dolphins, to make long underwater dives. One diving mammal that has been studied extensively is the Weddell seal, a large antarctic predator (adults weigh about 400 kg). Weddell seals catch large cod and other deep-sea fish by plunging to depths of 200 to 500 m during dives that typically last about 20 minutes. They occasionally submerge for more than an hour, perhaps to escape predators or explore new routes beneath the ice (FIGURE 38.28).

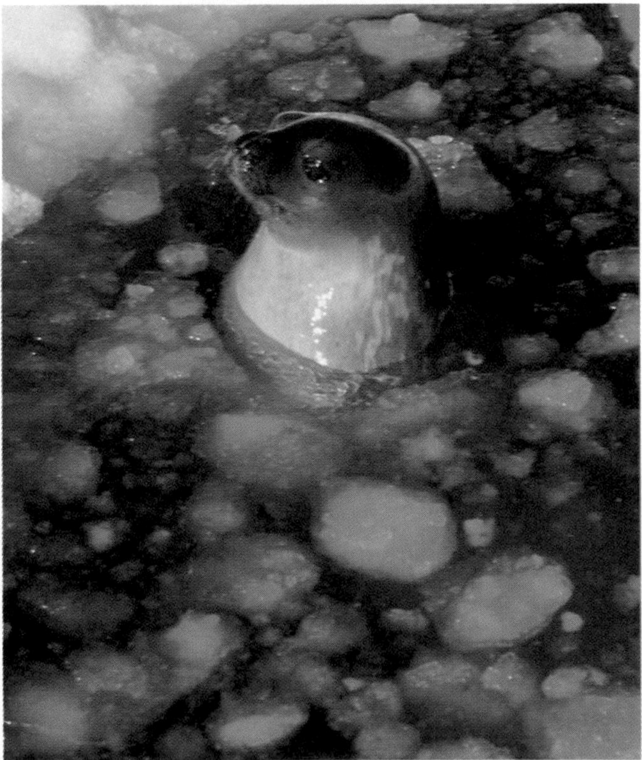

FIGURE 38.28
The Weddell seal (*Leptonychotes weddelli*), a diving mammal. Adaptations of the circulatory and respiratory systems enable these antarctic seals to travel underwater for more than an hour.

One diving adaptation of the Weddell seal is its ability to store oxygen. Compared to humans, the seal contains about twice as much oxygen per kilogram of body weight, mostly in the blood and muscles. About 36% of our total oxygen is in our lungs, and 51% is in our blood. In contrast, the Weddell seal holds only about 5% of its oxygen in its relatively small lungs, stocking 70% in the blood. This is possible partly because the seal has about twice the volume of blood per kilogram of body weight as a human. Another adaptation is the seal's huge spleen, which can store about 24 L of blood. The spleen probably contracts after a dive begins, fortifying the blood with additional erythrocytes loaded with oxygen. Diving mammals also have a higher concentration than most other mammals of an oxygen-storing protein called **myoglobin** in their muscles. The Weddell seal can stow about 25% of its oxygen in muscle, compared to only 13% in humans. Thus, the Weddell seal owes part of its diving virtuosity to blood and muscles enriched with oxygen.

Diving mammals not only begin an underwater trip with a relatively large reservoir of oxygen, they also have adaptations that conserve that oxygen. A diving reflex slows the pulse, and an overall reduction in oxygen consumption parallels the lower cardiac output. Regulatory mechanisms affecting peripheral resistance route most blood to the brain, spinal cord, eyes, adrenal glands, and placenta (in pregnant seals). Blood supply to the muscles is restricted, and it is shut off altogether during the longest dives. Thus, during dives of more than about 20 minutes, the muscles deplete the oxygen stored in their myoglobin and then derive their ATP from anaerobic instead of aerobic respiration (see Chapter 9).

The unusual abilities of the Weddell seal and other diving mammals to power their bodies during long dives showcase a central theme in our study of organisms—the response to environmental pressures over the short term by physiological adjustments and over the long term by natural selection

* * *

We have seen throughout this chapter that the circulatory and respiratory systems cooperate extensively. The main function of the respiratory system—to supply O_2 and dispose of CO_2 in support of cellular bioenergetics—depends on the transport work of the circulatory system. Still another activity of the body that depends on the circulatory system is defense, which we will examine in the next chapter.

REVIEW OF KEY CONCEPTS (with page numbers and key figures)

- Transport systems functionally connect body cells with the organs of exchange: *an overview* (pp. 819–820)
  - The exchange of oxygen, carbon dioxide, nutrients, and metabolic wastes between an organism's cells and the environment occurs across fluid-bathed membranes.
  - Because most of their cells are too far from the outside environment to be serviced by diffusion or active transport, many animals have a special system that transports chemicals within the body.
  - By transporting fluids (blood or interstitial fluid) throughout the body, transport systems provide a lifeline between the aqueous environment of living cells and the organs, such as the lungs, that exchange chemicals with the outside environment.
  - The transport system also functions in homeostasis, enabling other organ systems to regulate the chemical and physical properties of the cellular environment.
- Most invertebrates have a gastrovascular cavity or a circulatory system for internal transport (pp. 820–821; FIGURES 38.1, 38.2)
  - Cnidarians and flatworms have gastrovascular cavities that function in circulation as well as digestion.
  - Arthropods and most mollusks have open circulatory systems, in which tissues are bathed directly in hemolymph pumped by a heart into sinuses.
  - Annelids and some mollusks have closed circulatory systems, with blood confined to vessels, some of which pulsate and function as hearts.
- Diverse adaptations of a closed cardiovascular system have evolved in vertebrates (pp. 821–823, FIGURE 38.3)
  - In vertebrates, blood flows in a closed cardiovascular system consisting of blood vessels and a two- to four-chambered heart. The heart has one atrium or two atria, which receive blood from veins, and one or two ventricles, which pump blood into arteries. Arteries branch into arterioles, which convey blood to capillaries, the sites of chemical exchange between blood and interstitial fluid. Capillaries rejoin into venules that converge into veins.
  - In fishes, the heart has a single atrium and a single ventricle that pumps blood to gills for oxygenation; the blood then travels to other capillary beds of the body before returning to the heart.
  - Amphibians and most reptiles have a three-chambered heart in which the single ventricle pumps blood to both lungs and body in the pulmonary and systemic circuits. These two circuits return blood to separate atria. This double circulation repumps blood returning from the capillary beds of the respiratory organ, thus ensuring a strong flow of blood to the rest of the body.
  - Birds and mammals, both endotherms, have four-chambered hearts that keep oxygen-rich and oxygen-poor blood completely separated.
- Rhythmic pumping of the mammalian heart drives blood through pulmonary and systemic circuits (pp. 823–830, FIGURE 38.4)
  - The cardiac cycle consists of periods of contraction, called systole, and periods of relaxation, called diastole.
  - Heart valves dictate a one-way flow of blood through the heart.
  - Together with the stroke volume, heart rate (pulse) determines cardiac output, the volume of blood pumped into the systemic circulation per minute.
  - The intrinsic contraction of the cardiac muscle is coordinated by a conduction system originating in the sinoatrial (SA) node (pacemaker) of the right atrium. The pacemaker initiates a wave of contraction that spreads to both atria, hesitates momentarily at the atrioventricular (AV) node, and then progresses to both ventricles. The pacemaker is itself influenced by nerves, hormones, and body temperature, and by atrial volume changes during exercise.
  - All blood vessels, including capillaries, are lined by a single layer of endothelium. Arteries and veins have two additional outer layers composed of characteristic proportions of smooth muscle, elastic fibers, and connective tissue.
  - The velocity of blood flow varies in the circulatory system, being slowest in the capillary beds as a result of the high resistance and large total cross-sectional area of the arterioles and capillaries. This slower flow enhances the exchange of substances between the blood and interstitial fluid.
  - Blood pressure is determined by cardiac output and peripheral resistance due to variable constriction of the arterioles.
  - Muscular activity and pressure changes during breathing propel blood back to the heart in veins equipped with one-way valves.
  - The steady supply of blood to different organs is determined by variable constriction of arterioles and capillary sphincters.
  - Capillary exchange is the ultimate function of the circulatory system. Substances traverse the endothelium in endocytotic-exocytotic vesicles, by diffusion, or are dissolved in fluids forced out by blood pressure at the arterial end of the capillary.
- The lymphatic system returns fluid to the blood and aids in body defense (pp. 830–831)
  - Fluid reenters the circulation directly at the venous end of the capillary or indirectly through the lymphatic system.
  - Lymph nodes help fight against infection.
- Blood is a connective tissue with cells suspended in plasma (pp. 831–834, FIGURE 38.12)
  - Whole blood consists of cellular elements (cells and pieces of cells) suspended in a liquid matrix called plasma.
  - Plasma is a complex aqueous solution of inorganic electrolytes, proteins, nutrients, metabolic waste products, respiratory gases, and hormones.
  - Plasma proteins influence blood pH, osmotic pressure, and viscosity, and function in lipid transport, immunity (antibodies), and blood clotting (fibrinogens).

- Red blood cells, or erythrocytes, transport oxygen, a function reflected in their small size, biconcave shape, anaerobic metabolism, and hemoglobin content. Pluripotent stem cells in red bone marrow give rise to all types of blood cells.
- Five types of white blood cells, or leukocytes, function in defense by phagocytosing bacteria and debris or by producing antibodies.
- Platelets are fragments of cells produced in the bone marrow that function in blood clotting, a cascade of complex reactions that converts the plasma fibrinogen to fibrin.

- Cardiovascular diseases are the leading cause of death in the United States and many other developed nations (pp. 834–836)
  - A leading cause of death in the United States is cardiovascular disease, a deterioration of the heart and blood vessels. Gradual plaque buildup during atherosclerosis or arteriosclerosis narrows the diameter of blood vessels and may be associated with vessel blockage and consequent heart attack or stroke.

- Gas exchange supplies oxygen for cellular respiration and disposes of carbon dioxide: *an overview* (pp. 836–837, FIGURE 38.16)
  - Metabolism demands a constant supply of oxygen and removal of carbon dioxide. Animals require large, moist respiratory surfaces for the adequate diffusion of respiratory gases between their cells and the respiratory medium, either air or water.

- Gills are respiratory adaptations of most aquatic animals (pp. 838–840, FIGURE 38.18)
  - Gills are outfoldings of the body surface specialized for gas exchange. The efficiency of gas exchange in some gills, including those of fishes, is increased by ventilation and countercurrent flow of blood and water.

- Tracheae are respiratory adaptations of insects (p. 840, FIGURE 38.20)
  - The tracheae of insects are tiny branching tubes that penetrate the body, bringing oxygen directly to cells.

- Lungs are the respiratory adaptations of most terrestrial vertebrates (pp. 841–848, FIGURE 38.21)
  - Most terrestrial vertebrates, land snails, and spiders have internal lungs restricted to one location.
  - Mammalian lungs are enclosed in a double-walled sac whose layers adhere to one another, to the lungs, and to the chest cavity. Air inhaled through the nostrils passes through the pharynx into the trachea, bronchi, bronchioles, and dead-end alveoli, where gas exchange occurs.
  - Lungs must be ventilated by breathing. Frogs ventilate by positive pressure, pumping air into their lungs. Mammals ventilate with negative pressure by contracting and relaxing rib muscles and the diaphragm, which changes the volume and hence the pressure of the thoracic cavity and lungs relative to the atmosphere.
  - Tidal volume, the amount of air normally inhaled or exhaled, is less than the vital capacity of the lungs. Even after forceful exhalation, a residual volume of air remains in the alveoli.
  - Birds have one-way ventilation of the lungs, made possible by a system of air sacs, which also decrease density and dissipate heat, important adaptations for flight.
  - Breathing is regulated automatically by breathing control centers in the medulla oblongata and pons of the brain. Impulses from the medulla set the basic breathing rhythm. Sensors detect blood pH changes reflecting CO_2 concentrations and O_2 levels in the blood, and the medulla adjusts the rate and depth of breathing to match the metabolic demands of the body.
  - Oxygen and carbon dioxide diffuse from where their partial pressures are higher to where they are lower.
  - Respiratory pigments increase the amount of oxygen that blood can carry. Arthropods and many mollusks use copper-containing hemocyanin, whereas vertebrates and some invertebrates use iron-containing hemoglobin.
  - Hemoglobin molecules have four iron-containing subunits, each one capable of binding a molecule of oxygen. The binding of the first O_2 molecule increases the affinity of the other subunits for oxygen. The dissociation curve for hemoglobin is steepest in the range of O_2 concentrations found in body tissues. A drop in pH reduces hemoglobin's affinity for oxygen, thus providing more oxygen to active tissues.
  - Most carbon dioxide generated during metabolism is transported in the form of bicarbonate ions, which result from the dissociation of carbonic acid formed in the red blood cells from the chemical union of carbon dioxide and water. Hydrogen ions from the dissociation are bound to hemoglobin and other proteins, serving to buffer the blood. The entire process is reversed when blood enters the lungs, allowing free carbon dioxide to diffuse into the environment.
  - Diving mammals have a larger volume of blood, a higher concentration of myoglobin in their muscles, and a diving reflex that slows heart rate and reroutes blood flow away from muscles and noncritical organs.

SELF-QUIZ

1. Which of the following respiratory systems is not closely associated with a blood supply?
   a. vertebrate lungs
   b. fish gills
   c. tracheal systems of insects
   d. the outer skin of an earthworm
   e. the parapodia of a polychaete worm

2. Blood returning to the mammalian heart in a pulmonary vein will drain first into the
   a. vena cava
   b. left atrium
   c. right atrium
   d. left ventricle
   e. right ventricle

3. Pulse is a direct measure of
   a. blood pressure
   b. stroke volume
   c. cardiac output
   d. heart rate
   e. breathing rate

4. The respiratory medium flows unidirectionally over the respiratory surface in the gas exchange systems of
   a. mammals
   b. frogs
   c. birds
   d. insects
   e. sea stars

5. In negative pressure breathing, inhalation results from
   a. forcing air from the throat down into the lungs
   b. contracting the diaphragm
   c. relaxing the muscles of the rib cage
   d. using muscles of the lungs to expand the alveoli
   e. contracting the abdominal muscles

6. The maximum volume of air you can forcefully exhale after taking the deepest possible breath is called
   a. tidal volume
   b. residual volume
   c. vital capacity
   d. total respiratory volume
   e. alveolar volume

7. A decrease in the pH of human blood caused by exercise would
   a. decrease breathing rate
   b. increase heart rate
   c. decrease the amount of O_2 unloaded from hemoglobin
   d. decrease cardiac output
   e. decrease CO_2 binding to hemoglobin

8. Compared to the interstitial fluid that bathes active muscle cells, blood reaching that tissue in arteries has a
   a. higher P_{O_2}
   b. higher P_{CO_2}
   c. greater bicarbonate concentration
   d. lower pH
   e. greater osmotic pressure

9. Which of the following reactions prevails in red blood cells traveling through pulmonary capillaries? (Hb = hemoglobin)
   a. $Hb + 4\ O_2 \longrightarrow Hb(O_2)_4$
   b. $Hb(O_2)_4 \longrightarrow Hb + 4\ O_2$
   c. $CO_2 + H_2O \longrightarrow H_2CO_3$
   d. $H_2CO_3 \longrightarrow H^+ + HCO_3^-$
   e. $Hb + 4\ CO_2 \longrightarrow Hb(CO_2)_4$

10. Compared to a human, a diving mammal of equal size has
    a. less blood, an adaptation that helps conserve oxygen
    b. larger lungs
    c. a larger spleen
    d. less oxygen stored in muscles
    e. less oxygen stored in blood

CHALLENGE QUESTIONS

1. Because they support their weight against gravity and repeatedly accelerate limbs from standing starts, terrestrial vertebrates consume more energy in locomotion than do fishes swimming through water. In other words, it takes more calories per gram of animal to move 1 m on land than it does to move 1 m in water (assuming, of course, that the animal is in its natural habitat, either land or water). How does this disparity fit in with the evolution of the vertebrate cardiovascular system?

2. The hemoglobin of a human fetus differs from adult hemoglobin. Compare the dissociation curves of the two hemoglobins in the graph below, and then explain the physiological significance of the difference.

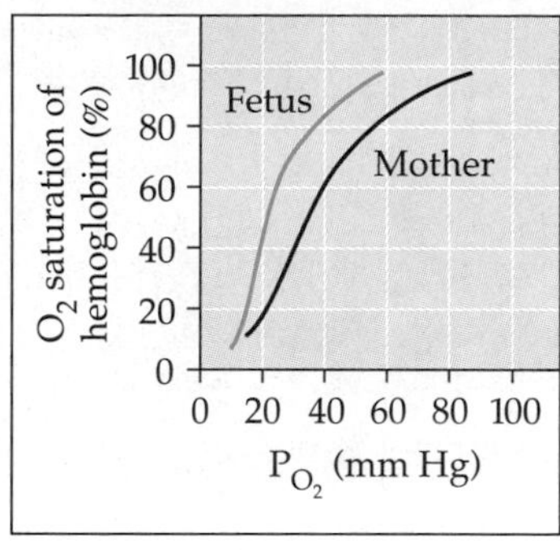

3. Describe some adaptations you think would be helpful for a mammal living at very high altitude, where the air is "thin" (relatively low P_{O_2}).

SCIENCE, TECHNOLOGY, AND SOCIETY

1. The incidence of cardiovascular disease is much lower in the Mediterranean countries (Spain, Italy, Greece, etc.) than in North America. Suggest several alternative hypotheses that might explain this difference. How could you test your hypotheses?

2. Hundreds of studies have linked smoking with cardiovascular and lung disease. According to most health authorities, smoking is the leading cause of preventable, premature death in the United States. The government bans television advertising of cigarettes and requires the tobacco industry to place health warnings on packages and in print ads. Antismoking and health groups have proposed that cigarette advertising be banned entirely. What are some arguments in favor of a total ban on cigarette advertising? What are arguments in opposition? Do you favor or oppose such a ban? Why?

FURTHER READING

Bartecchi, C. E., T. D. Mackenzie, and R. W. Schrier. "The Global Tobacco Epidemic." *Scientific American,* May 1995. How marketing is spreading a cardiovascular/respiratory risk.

Chien, K. R. "Molecular Advances in Cardiovascular Biology." *Science,* May 14, 1993. A review of molecular biology's current and potential applications in cardiovascular medicine.

Diamond, J. "The Saltshaker's Curse." *Natural History,* October 1991. Explores the genetic basis for the unusually high incidence of hypertension in African Americans.

Harken, A. H. "Surgical Treatment of Cardiac Arrhythmias." *Scientific American,* July 1993. A new method for correcting lethally fast heartbeats.

Houston, C. S. "Mountain Sickness." *Scientific American,* October 1992. How high altitude interferes with homeostasis.

Lawn, R. "Lipoprotein (a) in Heart Disease." *Scientific American,* June 1992. A cholesterol-binding protein that raises the risk of cardiovascular disease.

Pennisi, E. "Thicker Than Water." *Science News,* October 5, 1992. The hydrodynamics of blood flow.

Radetsky, P. "The Mother of All Blood Cells." *Discover,* March 1995. Potential application of research on pluripotent stem cells.

Zapol, W. M. "Diving Adaptations of the Weddell Seal." *Scientific American,* June 1987. Lab and field studies of a living diving machine.

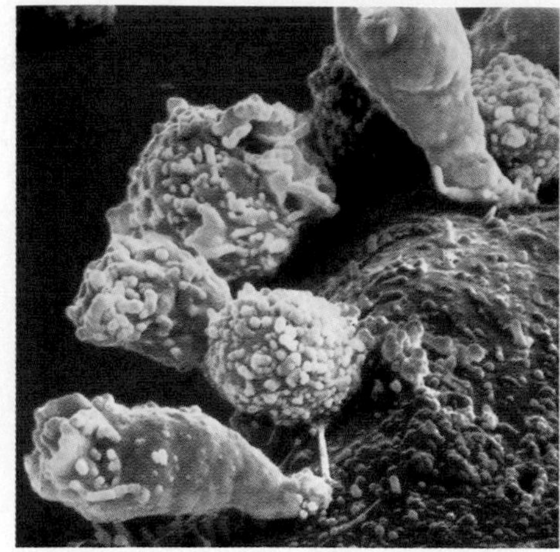

CHAPTER 39

THE BODY'S DEFENSES

KEY CONCEPTS

- Nonspecific mechanisms provide general barriers to infection
- The immune system defends the body against specific invaders: *an overview*
- Clonal selection of lymphocytes is the cellular basis for immunological specificity and diversity
- Memory cells function in secondary immune responses
- Molecular markers on cell surfaces function in self/nonself recognition
- In the humoral response, B cells defend against pathogens in body fluids by generating specific antibodies
- In the cell-mediated response, T cells defend against intracellular pathogens
- Complement proteins participate in both nonspecific and specific defenses
- The immune system's capacity to distinguish self from nonself is critical in blood transfusion and transplantation
- Abnormal immune function leads to disease states
- Invertebrates exhibit a rudimentary immune system

An animal must defend itself against unwelcome intruders: the many potentially dangerous viruses, bacteria, and other pathogens it encounters in the air and in food and water. It must also deal with abnormal cells that periodically appear in the body and, if not eliminated, may develop into cancer. Two cooperative defense systems that counter these threats have evolved. One of these systems is nonspecific in nature—that is, it does not distinguish one infectious agent from another. This nonspecific system includes two lines of defense, which an invader encounters in sequence. The first line of defense is external: It consists of the epithelial tissues that cover and line our bodies (skin and mucous membranes) and the secretions these tissues produce. The second line of nonspecific defense is internal: It is triggered by chemical signals and uses antimicrobial proteins and phagocytic cells that indiscriminately attack any invader that penetrates the body's outer barriers. In some cases, inflammation is a sign that this second line of defense has been deployed.

The other major defense system, the immune system, constitutes a third line of defense, which comes into play simultaneously with the second line of nonspecific defense. However, the immune system responds specifically to the particular type of invader. This immune response includes the production of specific defensive proteins called antibodies. It also involves the participation of several different types of cells that are derived from the white blood cells called lymphocytes (see Chapter 38). The photograph that opens this chapter (a colorized SEM) shows specialized lymphocytes (green) attacking a cancer cell. This chapter examines how an animal's nonspecific and specific defenses work together to protect the body (Figure 39.1). *Our main focus is on the defense mechanisms of vertebrates, which include a highly developed immune system.*

Nonspecific mechanisms provide general barriers to infection

An invading microbe must penetrate an external barrier formed by the skin and mucous membranes, which cover the surface and line the openings of an animal's body (see Chapter 36). If it succeeds in doing so, the pathogen encounters the second line of nonspecific defense: interacting mechanisms that include phagocytosis, antimicrobial proteins, and the inflammatory response.

FIGURE 39.1
An overview of the body's defenses.

Nonspecific defense mechanisms		Specific defense mechanisms (immune system)
First line of defense	Second line of defense	Third line of defense
• Skin • Mucous membranes and their secretions	• Phagocytic white blood cells • Antimicrobial proteins • The inflammatory response	• Lymphocytes • Antibodies

The Skin and Mucous Membranes

The intact skin is a barrier that cannot normally be penetrated by bacteria or viruses, although minute abrasions may allow their passage. Likewise, the mucous membranes that line the digestive, respiratory, and genitourinary tracts bar the entry of potentially harmful microbes. Beyond their role as a physical barrier, the skin and mucous membranes counter pathogens with chemical defenses. In humans, for example, secretions from oil and sweat glands give the skin a pH ranging from 3 to 5, which is acidic enough to discourage many microorganisms from colonizing there. (Bacteria that make up the normal flora of the skin are adapted to its acidic, relatively dry environment.) Saliva, tears, and mucous secretions bathe the surface of exposed epithelia, washing away many potential invaders. In addition, these secretions contain various antimicrobial proteins. One of these protective proteins is **lysozyme,** an enzyme that digests the cell walls of many bacteria and destroys many microbes entering the upper respiratory system and the openings around the eyes.

Mucus, the viscous fluid secreted by cells of the mucous membranes, also traps particles that contact it. Microbes entering the upper respiratory system are often caught in the mucus and are then swallowed or expelled. Lining the trachea are specialized epithelial cells equipped with cilia that sweep out microbes and other particles trapped by the mucus, preventing them from entering the lungs (FIGURE 39.2). Microbes present in food or trapped in swallowed mucus from the upper respiratory system must pass through the highly acidic gastric juice produced by the stomach lining, which destroys most of the microbes before they can enter the intestinal tract.

Phagocytic White Cells and Natural Killer Cells

The body's internal mechanisms of nonspecific defense depend mainly on **phagocytosis,** the ingestion of invading particles by certain types of white blood cells (see FIGURE 8.17a). The phagocytic cells called **neutrophils** comprise about 60% to 70% of all white blood cells. Attracted by chemical signals, neutrophils can leave the blood and enter infected tissue by amoeboid movement, destroying microbes there. (This migration toward the source of a chemical attractant is called *chemotaxis.)* However, neutrophils tend to self-destruct as they destroy foreign invaders, and their average life is only a few days.

Monocytes, although constituting only about 5% of the white blood cells, provide an even more effective phagocytic defense. After maturing, monocytes circulate in the

FIGURE 39.2
First-line respiratory defenses. An important part of the body's first line of defense is found in the upper part of the respiratory system. Specialized cells (orange) produce mucus that traps microbes before they can enter the lungs. The mucous membrane is also equipped with ciliated cells, which are colored yellow here (colorized SEM). Synchronized beating of the cilia expels mucus and the trapped microbes.

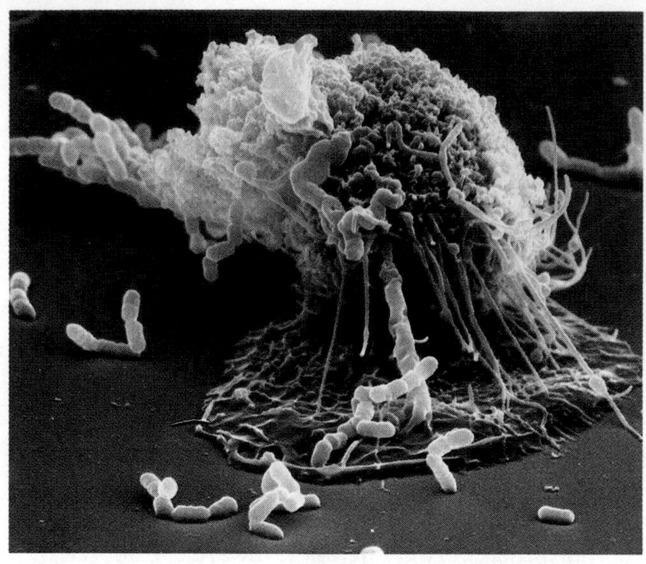

FIGURE 39.3
Phagocytosis by a macrophage. This micrograph shows fibril-like pseudopodia of a macrophage attaching to rod-shaped bacteria, which will be ingested and destroyed (colorized SEM).

blood for a few hours, then migrate into tissues, enlarging and developing into **macrophages** ("big eaters"). Macrophages, the largest phagocytic cells, are especially effective, long-living phagocytes. These amoeboid cells extend pseudopodia that pull in microbes (FIGURE 39.3), which are then destroyed by digestive enzymes and reactive forms of oxygen within the macrophages. However, mechanisms for evading phagocytosis have evolved in some microbes. Certain bacteria, for example, have special capsules to which the macrophage cannot attach. Others have developed a resistance to the lytic enzymes of the phagocyte and can even reproduce within macrophages. The presence or absence of such an adaptation often explains why one microbe is a pathogen and a similar microbe is not.

Some macrophages reside permanently in organs and connective tissues. In the lungs, for example, they are the alveolar macrophages; in the liver, they are called Kupffer's cells. Although fixed in place, they are located where infectious agents circulating in the blood and lymph will contact them. Fixed macrophages are especially numerous in the lymph nodes and in the spleen, key organs of the lymphatic system (FIGURE 39.4). Other macrophages migrate through all tissues of the body.

About 1.5% of the white cells are **eosinophils,** which have only limited phagocytic activity but contain destructive enzymes within cytoplasmic granules. Their main contribution to defense is against larger parasitic invaders, such as worms. Eosinophils position themselves against the external wall of a worm and discharge the destructive enzymes from their granules.

Nonspecific defense of the body also includes **natural killer cells.** They do not attack microorganisms directly, but rather destroy the body's own infected cells, especially cells harboring viruses, which can reproduce only within host cells. The natural killers also assault aberrant cells that could form tumors. The mode of destruction is not phagocytosis but an attack on the membrane of the target cell, which causes that cell to lyse (break open).

Antimicrobial Proteins

A variety of proteins function in nonspecific defense either by attacking microorganisms directly or by impeding their reproduction. You have already learned about lysozyme, an antimicrobial protein present in tears, saliva, and mucous secretions. Complement proteins and interferons are important antimicrobial proteins in the tissues and the blood. The **complement system** is a group of at least 20 proteins, named for its cooperation with (complementation of) other defense mechanisms. The complement proteins act together in a cascade of activation steps that culminates with lysis of invading microbes. Some components of the complement system also function in chemotaxis as attractants for the recruitment of phagocytes to sites of infection. Since the complement system is an essential part of both the nonspecific and the specific defense systems, we will look at how it functions after our introduction to both systems.

First identified in 1957, **interferons** were discovered by researchers to be substances that virus-infected cells produce, helping other cells interfere with, or resist infection by, the virus. There are three known types of interferons: alpha, beta, and gamma. Interferons are secreted by an infected cell as an early, nonspecific defense before specific antibodies appear. Interferons cannot save the infected cell, but they diffuse to neighboring cells, where they stimulate the production of other proteins that inhibit viral replication in those cells. Thus, an infected cell can help protect uninfected cells. The defense is not virus-specific; interferon produced in response to one viral strain confers resistance to unrelated viruses. Interferons are most effective in controlling short-term infections, such as colds and influenza. In addition to its role as an antiviral agent, interferon-gamma activates phagocytes, enhancing their ability to ingest and kill microorganisms. Interferons are now mass-produced by recombinant DNA technology and are being tested clinically for the treatment of viral infections and cancer (see Chapter 19).

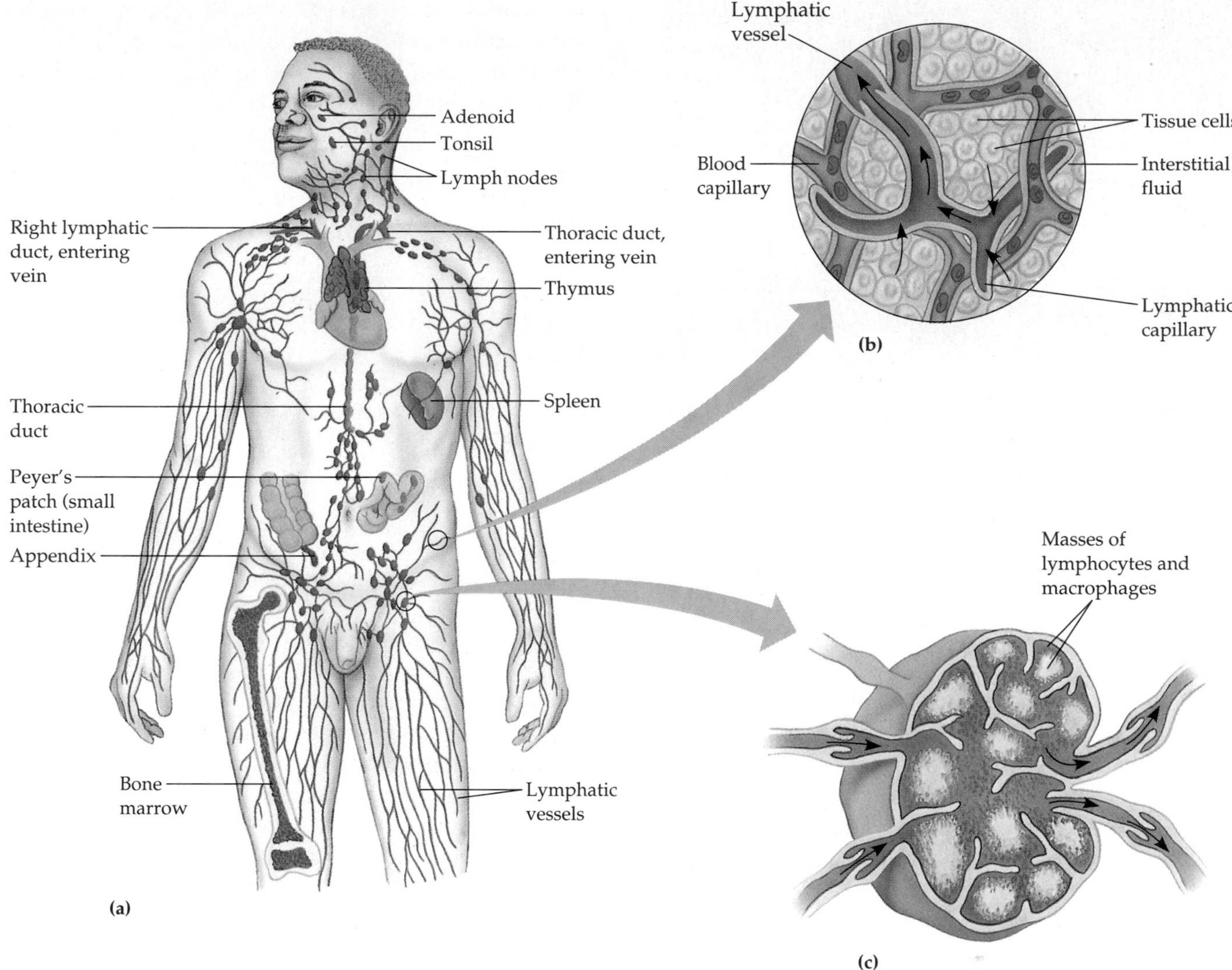

FIGURE 39.4

The human lymphatic system. (**a**) The lymphatic system returns fluid from the interstitial spaces to the circulatory system. In addition to the lymphatic vessels, the system includes various satellite organs that are important in the body's defense system, including the spleen, adenoids, tonsils, appendix, Peyer's patches, and numerous lymph nodes. Also depicted are the bone marrow and the thymus, sites of white blood cell development. (**b**) Some of the interstitial fluid that bathes tissues is taken up by the lymphatic capillaries. The fluid, now called lymph, flows through the system of vessels, eventually returning to the blood circulatory system near the shoulders where the right lymphatic duct and the thoracic duct drain into veins, such as the subclavian vein. (**c**) Along the way, lymph must pass through numerous lymph nodes, where any pathogens present in lymph encounter macrophages and lymphocytes, another class of white blood cells with defensive functions.

The Inflammatory Response

Damage to tissue by a physical injury, such as a cut, or by the entry of microorganisms, triggers an **inflammatory response** (FIGURE 39.5). Small blood vessels in the vicinity of the injury dilate (vasodilation), increasing the blood supply to the injured area, causing the characteristic redness and heat (inflammation means "setting on fire"). The dilated blood vessels also become more permeable, and fluids move from the blood into neighboring tissues, causing the edema (swelling) associated with inflammation.

The processes that constitute the inflammatory response are initiated by chemical signals. One of these signals is **histamine,** which is contained in circulating white cells called **basophils** and in cells called **mast cells** found in connective tissue. Injury to these cells causes the release of histamine, which triggers local vasodilation and makes capillaries in the vicinity leakier. White blood cells and damaged tissue cells also discharge prostaglandins (see Chapter 41) and other substances that promote blood flow to the site of injury. Vasodilation and the increase in blood vessel permeability also deliver clotting

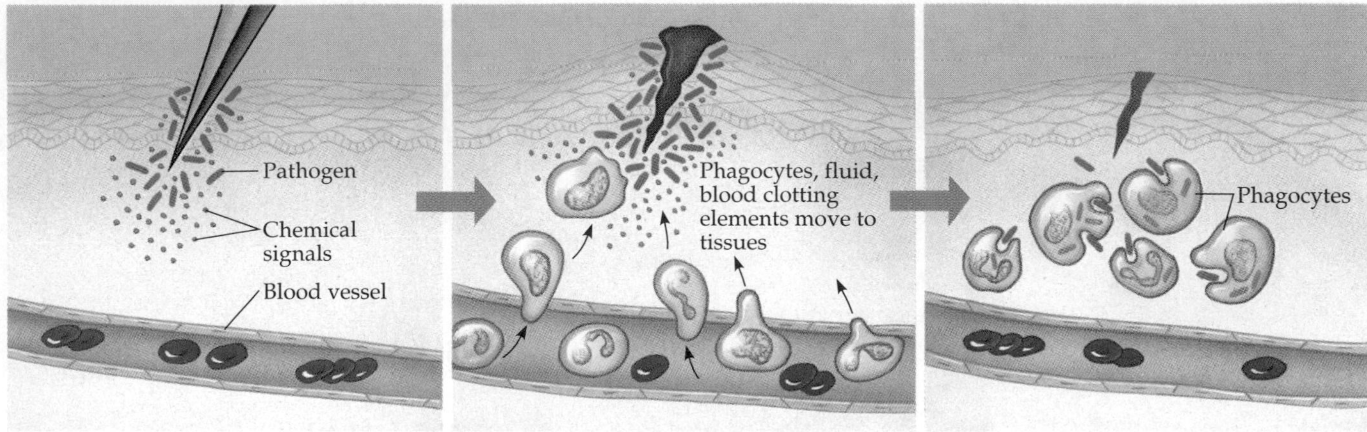
Pathogen
Chemical signals
Blood vessel
Phagocytes, fluid, blood clotting elements move to tissues
Phagocytes
① Tissue injury; release of chemical signals (histamine, prostaglandins).
② Vasodilation (increased blood flow); increased vessel permeability; phagocyte migration.
③ Phagocytes (macrophages and neutrophils) consume pathogens and cell debris; tissue heals.

FIGURE 39.5
A simplified view of the inflammatory response. ① The localized response is triggered when cells of tissue injured by bacteria or physical damage release chemical signals such as histamine and prostaglandins. ② These signals induce increased permeability of the capillaries and blood flow to the affected area. Also released are certain chemicals that attract phagocytic cells and lymphocytes. ③ After their arrival at the site of injury, the phagocytes consume pathogens and cell debris, and the tissue heals.

elements into the injured area. Blood clotting marks the beginning of the repair process and helps block the spread of pathogenic microbes to other parts of the body.

Probably the most important element of the inflammatory response is phagocyte migration from the blood into the injured tissues, which usually begins within an hour after injury. The increased flow of blood to the injured area and the leakage from the capillaries enhance the migration of phagocytic cells into the interstitial fluid. In this chemotaxis, several chemical mediators, including some of the proteins of the complement system, attract the phagocytes to the damaged tissue. Neutrophils arrive at the site of the injury first, followed by monocytes that develop into macrophages. Macrophages not only devour pathogens, but also clean up the remains of tissue cells and neutrophils that have self-destructed after eliminating many microorganisms. The pus that often accumulates at the site of an infection consists mostly of dead cells and the fluid that leaked from the capillaries during the inflammatory response. If pus remains in place, it is gradually absorbed by the body over a period of a few days.

The inflammatory responses described so far are localized, as in infections caused by a splinter. But the body's reaction to an infection may also be systemic—that is, more widespread. For example, injured cells emit molecules that stimulate the release of more neutrophils from bone marrow—a call for reinforcements. The number of white cells in the blood may increase severalfold within a few hours after the inflammatory reaction begins. This dramatic increase in leukocyte (white blood cell) count often indicates a severe infection, such as meningitis or appendicitis. Another systemic response to infection is fever. Toxins produced by pathogens may trigger the fever, but certain white blood cells also release molecules called pyrogens, which set the body's thermostat at a higher temperature. A very high fever may be dangerous, but a moderate fever contributes to defense. It inhibits the growth of some microorganisms, perhaps in part by decreasing the amount of iron available to them. Fever may also facilitate phagocytosis and, by speeding up body reactions, may speed the repair of tissues.

Let's review the body's nonspecific defense systems: The first line of defense, the skin and mucous membranes, prevents most pathogens from entering the body; the second line of defense uses phagocytes, natural killer cells, antimicrobial proteins, and inflammatory responses to defend against pathogens that have managed to enter the body. These two lines of defense are termed nonspecific because they do not target specific pathogens, such as a certain bacterial species.

The immune system defends the body against specific invaders: *an overview*

Key Features of the Immune System

While microbes are being assaulted by the inflammatory response, antimicrobial agents, and phagocytes, they inevitably come into contact with cells of the immune system, the body's third line of defense. The immune system develops a specific response against each type of foreign microbe, toxin, or transplanted tissue. Four key features characterize the immune system: specificity, diversity, memory, and self/nonself recognition.

Specificity. The immune system has the ability to recognize and eliminate particular microorganisms and foreign molecules—a certain strain of flu virus, for example. A foreign substance that elicits this immune response is called an **antigen.** The immune system responds to an antigen by activating specialized lymphocytes and producing specific proteins called **antibodies.** (*Antigen* is a contraction of *anti*body-*gen*erating.) Antigens that trigger an immune response include molecules belonging to viruses, bacteria, fungi, protozoa, and parasitic worms. Antigenic molecules also mark the surfaces of such foreign materials as pollen, insect venom, and transplanted tissue, such as skin or organs. Each antigen has a unique molecular shape and stimulates the production of the very type of antibody that defends against that specific antigen. Thus, in contrast to the nonspecific defenses, each response of the immune system targets a specific invader, distinguishing it from other foreign molecules that may be very similar.

Diversity. The immune system has the ability to respond to millions of kinds of invaders, each recognized by its antigenic markers. This diversity of response is possible because the immune system is equipped with an enormous variety of lymphocyte populations, each population bearing receptors for a particular antigen. Among antibody-producing lymphocytes, for example, each population is stimulated by a specific antigen and responds by synthesizing and secreting the appropriate type of antibody.

Memory. The immune system has the ability to "remember" antigens it has encountered and to react to them more promptly and effectively on subsequent exposures. This characteristic, called **acquired immunity,** was recognized about 2400 years ago by Thucydides of Athens, who described how those sick and dying during an epidemic of plague were cared for by those who had recovered, "for no one was ever attacked a second time." This concept of immunity is familiar to all of us: If we had chickenpox as a child, we are unlikely to get it again.

Self/Nonself Recognition. The immune system distinguishes the body's own molecules from foreign molecules (antigens). Failure of self/nonself recognition can lead to autoimmune disorders, in which the immune system destroys the body's own tissues. You will learn about the mechanism of self/nonself recognition and autoimmune disorders later in the chapter.

Active Versus Passive Acquired Immunity

Immunity conferred by recovering from an infectious disease such as chickenpox is called **active immunity** because it depends on the response of a person's own immune system. In this case, the active immunity is *naturally* acquired. Active immunity can also be acquired *artificially,* by vaccination. Vaccines may be inactivated bacterial toxins, killed microorganisms, or living but weakened microorganisms. These agents can no longer cause disease, but they retain the ability to act as antigens and stimulate an immune response. A vaccinated person who encounters the actual pathogen will have the same quick defensive reaction based on immunological memory as a person who has had the disease.

Antibodies can also be transferred from one individual to another, providing **passive immunity.** This occurs naturally when a pregnant woman's body passes some of her antibodies across the placenta to the fetus. A newborn's immune system is not fully operative, but if the mother is immune to chickenpox, for example, the infant will also be temporarily immune to this disease. Certain antibodies are also passed from the mother to her nursing infant in breast milk, especially in the colostrum, or first secretions. Passive immunity persists for only a few weeks or months, however, and the infant's own immune system must take over the defense of the body. Passive immunity can also be transferred artificially by introducing antibodies from an animal or human who is already immune to the disease. For example, rabies is treated in humans by injecting antibodies from people who have been vaccinated against rabies. This produces an immediate immunity, which is important because rabies progresses rapidly and the response to vaccination would take too long. The injected antibodies last only a few weeks, but by then the person's own immune system has produced antibodies against the rabies virus.

Humoral Immunity and Cell-Mediated Immunity

The immune system can actually mount two different types of responses to antigens: a humoral response and a cell-mediated response. **Humoral immunity** results in the production of antibodies, which are secreted by certain lymphocytes and circulate as soluble proteins in blood plasma and lymph, fluids that were long ago called humors. Around the turn of the century, researchers transferred such fluids from one animal to another and found that this process transferred immunity. However, the researchers also found that immunity to some conditions could be passed along only if lymphocytes were transferred. This second type of immunity, which depends on the direct action of cells (certain types of lymphocytes) rather than antibodies, became known as **cell-mediated immunity.**

Thus, the immune system is made up of two branches. The circulating antibodies of the humoral branch defend mainly against toxins, free bacteria, and viruses present in body fluids. In contrast, lymphocytes of the cell-mediated branch are active against bacteria and viruses inside the host's cells and against fungi, protozoa, and worms. Cell-mediated immunity is also involved in attacks on transplanted tissue and cancer cells, both of which are perceived as "nonself."

Cells of the Immune System

The vertebrate body is populated by two main classes of lymphocytes: **B cells** (B lymphocytes), which carry out the humoral immune response, and **T cells** (T lymphocytes), which function mainly in the cell-mediated immune response. (Later in the chapter, we will see that one subclass of T cells also participates in humoral immunity by stimulating B cells.) Lymphocytes, like all blood cells, originate from pluripotent stem cells in the bone marrow (see FIGURE 38.13) or, in the fetus, mainly in the liver. Initially, all lymphocytes are alike, but they later differentiate into T cells or B cells, depending on where they continue their maturation (FIGURE 39.6). Lymphocytes that migrate from the bone marrow to the thymus, a gland in the upper region of the chest, develop into T cells (T for thymus). Lymphocytes that remain in the bone marrow and continue their maturation there become B cells. (The B actually stands for the bursa of Fabricius, an organ unique to birds where B cells mature; these lymphocytes were first discovered in chickens. It may help you remember the site of maturation for these cells, however, if you equate B with bone marrow.)

Mature B cells and T cells are most concentrated in lymph nodes, the spleen, and other lymphatic organs, where the lymphocytes are most likely to encounter antigens (see FIGURE 39.4). Both B cells and T cells are equipped with specific **antigen receptors** on their plasma membranes. Antigen receptors on a B cell are actually membrane-bound antibody molecules specific for a certain antigen. The antigen receptors of T cells are different from antibodies, but these **T cell receptors** recognize antigens as specifically as antibodies do. Thus, the specificity and diversity of the immune system depend on receptors on each B cell and T cell that enable that lymphocyte to identify and respond to a particular antigen.

When antigens bind to specific receptors on the surface of a lymphocyte, the lymphocyte is activated to divide and differentiate, giving rise to a population of **effector cells,** the cells that actually defend the body in an immune response. In the case of humoral responses, B cells activated by antigen binding give rise to effector cells called **plasma cells,** which secrete antibodies that help eliminate that particular antigen. The effectors of cell-mediated responses include two populations of T cells. **Cytotoxic T cells (T_C)** kill infected cells and cancer cells. **Helper T cells (T_H)** secrete protein factors called **cytokines,** molecules that are secreted by one cell as a regulator of neighboring cells. Cytokines help regulate B cells and T cells and therefore play a pivotal role in both humoral and cell-mediated responses. After we take a closer look at the key features of the immune system—specificity, diversity, memory, and self/nonself recognition—we will examine the humoral and cell-mediated responses in more detail.

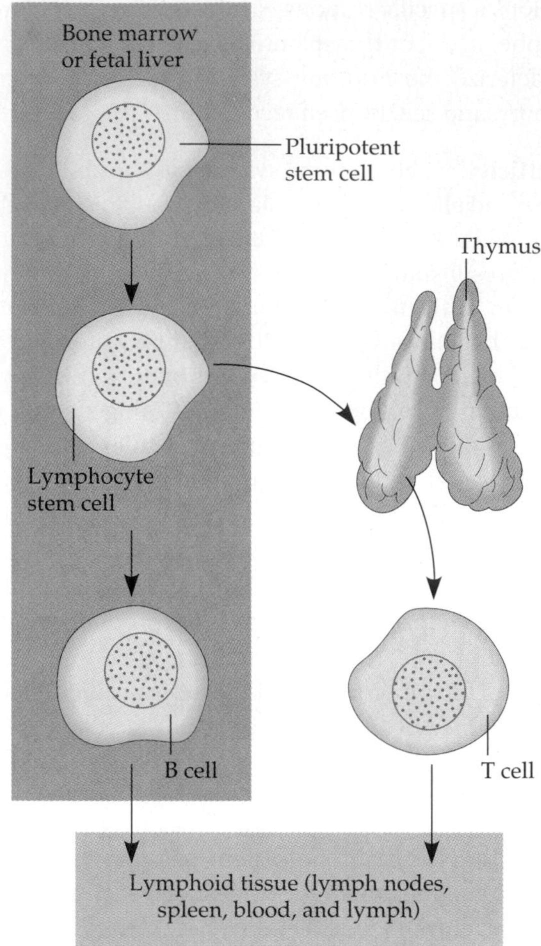

FIGURE 39.6
The development of lymphocytes. Like other blood cells, lymphocytes differentiate from multipotent stem cells in bone marrow (see FIGURE 38.13). Lymphocytes that continue their maturation in bone marrow develop into B cells, while lymphocytes that move to the thymus and complete their maturation there differentiate into T cells. Both classes of lymphocytes populate the lymph nodes, spleen, and other lymphatic organs shown in FIGURE 39.4. The B and T cells are now ready for their functions in humoral immunity and cell-mediated immunity, respectively.

Clonal selection of lymphocytes is the cellular basis for immunological specificity and diversity

We can imagine two possible mechanisms by which an animal's immune system could respond specifically to the millions of potential antigens. In one scenario, each lymphocyte is plastic in its response, adapting to a particular antigen and tailoring its action to match whatever antigen happens to be present at that time—by changing the type of antibody it secretes, for example. The evidence supports the alternate view: Each lymphocyte recognizes and responds to only one antigen, and the ability of the immune system to defend against an almost unlimited variety of antigens depends on the enormous diversity of antigen-specific lymphocytes. (The genetic basis for this immunological diversity was discussed in Chapter 18.)

Each lymphocyte's specificity for an antigenic target is rigidly predetermined during embryonic development, before an encounter with that antigen ever takes place. The mark of this specificity is the antigen receptor the lymphocyte bears on its surface. The lymphocyte may or may not ever come into contact with the corresponding antigen. If that antigen *does* enter the body and binds to receptors on the specific lymphocytes, then those (and only those) lymphocytes are activated to mount an immune response. The selected cells proliferate by cell division and develop into a large number of identical effector cells—a *clone* of cells that combats the very antigen that provoked the response. For example, plasma cells that develop from an activated B cell all secrete the same type of antibody that functioned as the antigen receptor on the original B cell that first encountered the antigen. This antigen-specific selection and cloning of lymphocytes is called **clonal selection** (FIGURE 39.7). The concept of clonal selection is so fundamental to understanding immunity that it is worth restating: Each antigen, by binding to specific receptors, selectively activates a tiny fraction of cells from the body's diverse pool of lymphocytes; this relatively small number of selected cells gives rise to a clone of millions of effector cells, all dedicated to eliminating the specific antigen that stimulated the humoral or cell-mediated immune response.

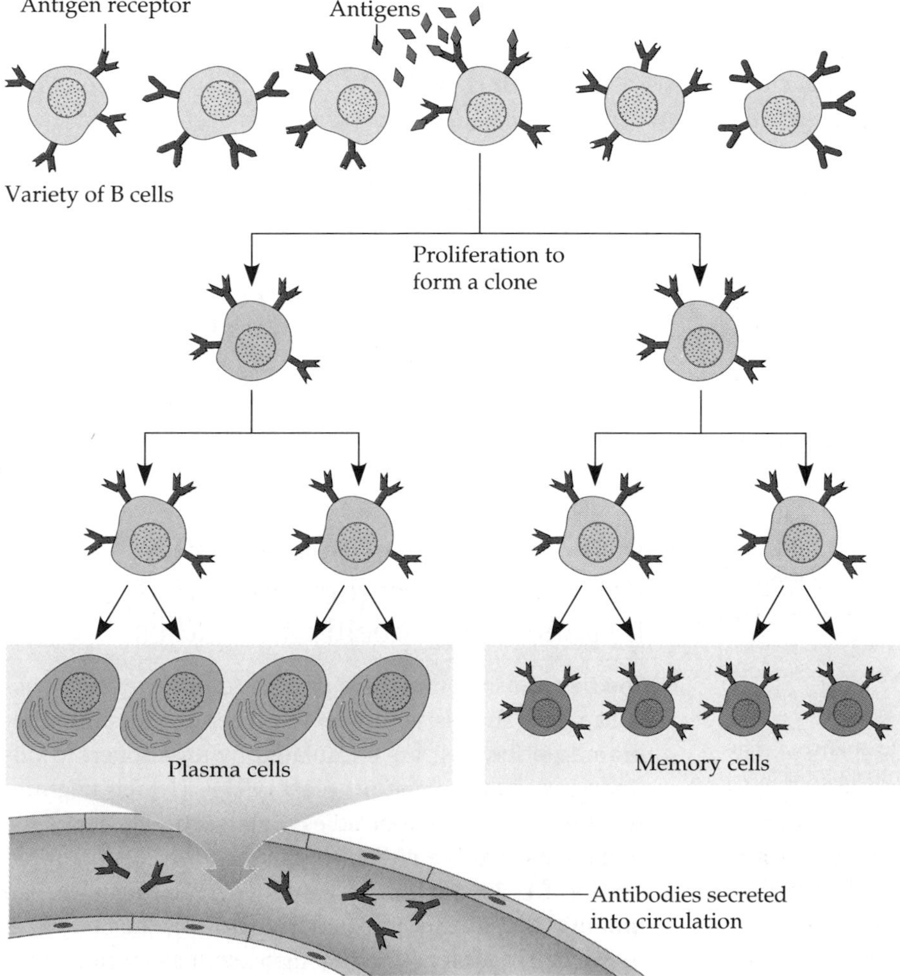

FIGURE 39.7
Clonal selection. The B cells and T cells of the body collectively recognize an almost infinite number of antigens, but each cell recognizes only one type of antigen. (Notice the variation in the shapes of antigen receptors on the six B cells at the top of this illustration.) When an antigen binds to a B or T cell, that cell proliferates, forming a clone of effector cells with the same specificity. Here, the antigen selects a particular B cell and stimulates it to reproduce and give rise to a population of identical effector cells called plasma cells. The plasma cells secrete antibodies specific for the antigen. Notice that relatively long-lived memory cells for the antigen are also formed. These cells can respond rapidly upon subsequent exposure to the same antigen.

Memory cells function in secondary immune responses

The selective proliferation of lymphocytes to form clones of effector cells upon first exposure to an antigen constitutes the **primary immune response.** Between initial exposure to an antigen and maximum production of effector cells, there is a lag period of 5 to 10 days. During this lag period, the lymphocytes selected by the antigen are differentiating into effector T cells and antibody-producing plasma cells. If the body is exposed to the same antigen at some later time, the response is faster (only 3 to 5 days) and more prolonged than the primary response. This is the **secondary immune response** (FIGURE 39.8). The antibodies produced at this time are also more effective in binding to the antigen than those produced during the primary response.

The immune system's ability to recognize an antigen as previously encountered is called immunological memory. This ability is based on long-lived **memory cells,** which are produced along with the relatively short-lived effector cells of the primary immune response (see FIGURE 39.7). During the primary response, these memory cells are not active. However, they survive for long periods and proliferate rapidly when exposed again to the same antigen that caused their formation. The secondary immune response gives rise to a new clone of memory cells, as well as to new effector cells. By this mechanism, childhood exposure to diseases such as chickenpox usually confers immunity for a lifetime.

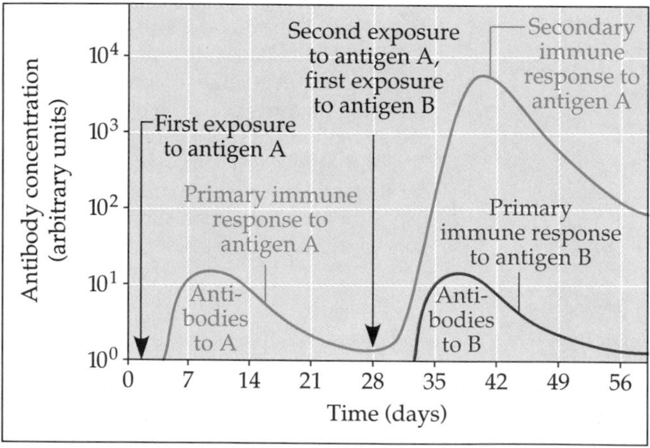

FIGURE 39.8
Immunological memory. An initial exposure to antigen A stimulates a primary immune response, with the eventual production of antibodies against that antigen. Notice the time lag and a relatively small response. A second exposure to antigen A at day 28 produces a faster and greater secondary immune response. If antigen B were also injected at day 28, the reaction to that antigen would be a primary response, not a secondary one. This experiment demonstrates that the secondary response, which is due to the presence of long-lived memory cells, is specific.

Molecular markers on cell surfaces function in self/nonself recognition

You have already learned that a key feature of the immune system is its ability to distinguish self from nonself, and that antigen receptors on the surfaces of lymphocytes are responsible for detecting foreign molecules that enter the body. Normally, there are no lymphocytes that are reactive against the body's *own* molecules. **Self-tolerance** begins to develop as T and B lymphocytes bearing antigen receptors mature in the thymus and bone marrow, and continues to develop even as the cells migrate to lymphoid tissues. Any lymphocytes with receptors for molecules present in the body are destroyed or are rendered nonfunctional, leaving only lymphocytes that are reactive against foreign molecules.

Among the important "self markers," the native molecules tolerated by an individual's immune system, are a collection of molecules encoded by a family of genes called the **major histocompatibility complex (MHC).** In humans, the MHC is also called the HLA (human leukocyte antigen) group. These genes encode glycoproteins (proteins with carbohydrate attached) embedded in the plasma membranes of cells. Because there are at least 20 MHC genes and as many as 100 alleles for each gene, it is virtually impossible for any two people, except identical twins, to have matching sets of MHC markers on their cells. Thus, the major histocompatibility complex is a biochemical fingerprint unique to each individual.

Two main classes of MHC molecules mark cells as "self." **Class I MHC** molecules are located on all nucleated cells—almost every cell of the body. **Class II MHC** molecules are restricted to a few specialized cell types of the body's defense system, including macrophages, B cells, and activated T cells. Class II MHC molecules play an important role in interactions between cells of the immune system. You will learn more about the functions of MHC as we now examine the humoral and cell-mediated immune responses in more detail.

In the humoral response, B cells defend against pathogens in body fluids by generating specific antibodies

You have learned that the humoral immune response occurs when B cells with specific receptors (membrane-bound antibodies) are stimulated by an antigen to differentiate into a clone of plasma cells, which begin to secrete antibodies. Antibodies are most effective against pathogens that are circulating in the blood and lymph, such as free bacteria and viruses. You have also learned that this selective activation of B cells arms the body with long-lived memory cells that function in a secondary hu-

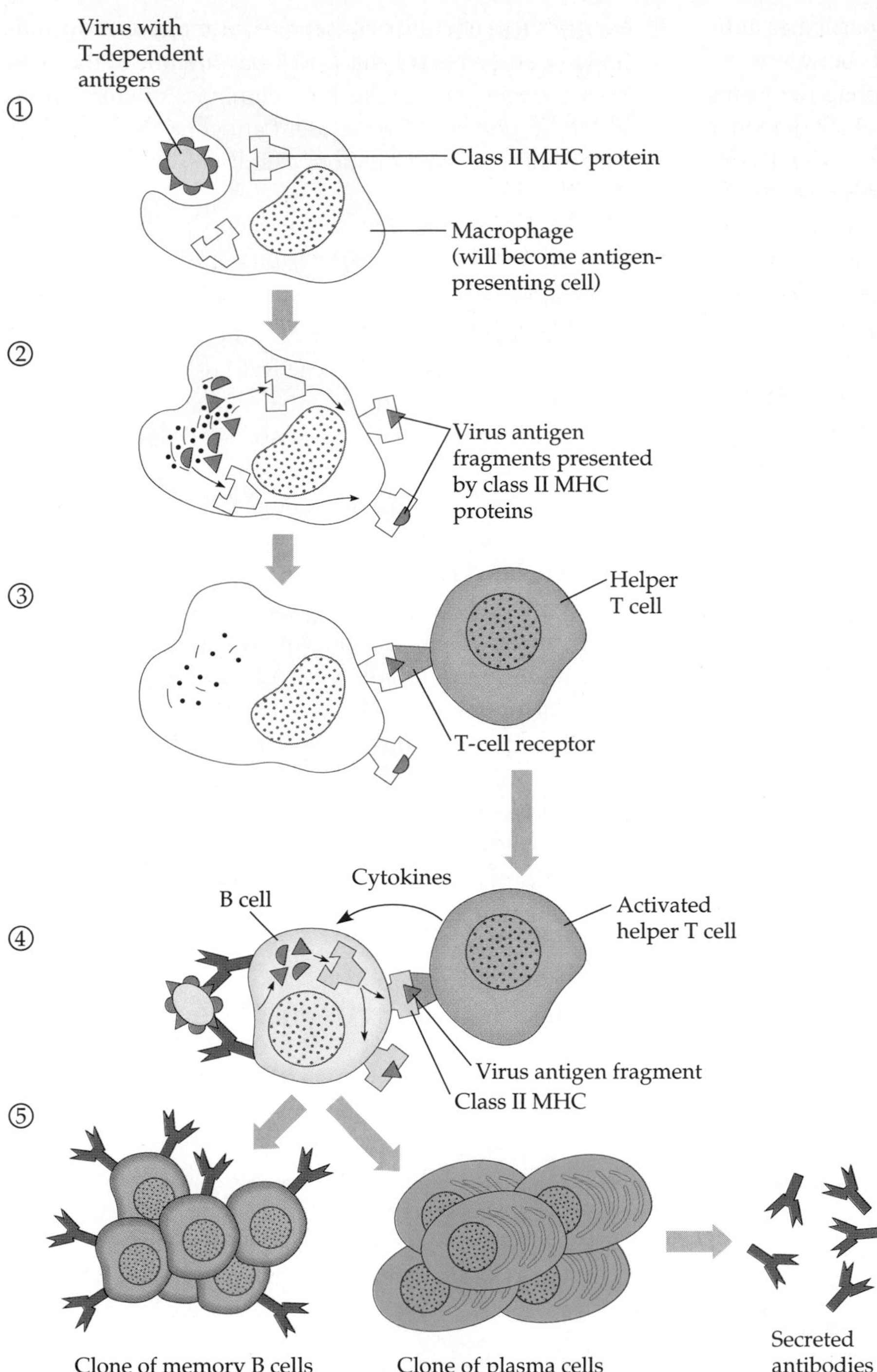

FIGURE 39.9
Response to a T-dependent antigen. ① A macrophage ingests a pathogen and will become an antigen-presenting cell (APC). ② Antigen fragments from the partially digested pathogen form complexes with class II MHC proteins. These self/nonself complexes are then transported to the cell surface, where they are presented to other cells of the immune system. ③ A helper T cell with a receptor specific for the presented antigen interacts with the macrophage by binding to the MHC-antigen complex. ④ The activated helper T cell then interacts with a B cell that has taken in antigens by endocytosis and displays a fragment of the antigen along with class II MHC proteins. The helper T cell secretes chemical signals called cytokines that stimulate the B cell. ⑤ This activates the B cell to divide repeatedly and differentiate into memory cells and plasma cells. Plasma cells are the antibody-secreting effector cells of humoral immunity. An antigen that triggers this type of immune response is termed a T-dependent antigen because of the requirement for participation by helper T cells.

moral response. In the following discussion, you will see that the humoral response involves other cells as well, including T cells. What distinguishes this branch of the immune system is its production of antibodies, which are not involved in cell-mediated defense. Notice, however, that the two branches often operate together, so that our distinctions sometimes break down.

The Activation of B Cells

In most cases, the selective activation of a B cell to form a clone of plasma cells and memory cells is a two-step process. One step, discussed earlier in the chapter, is the binding of antigen to specific receptors on the surface of the B cell. The other step involves two other types of cells: macrophages and helper T cells. As you can see in FIGURE 39.9, after a macrophage engulfs pathogens by phagocytosis, fragments of the partially digested antigen molecules are bound by class II MHC molecules. The two molecules are transported to the plasma membrane and displayed on the surface of the macrophage, which is now known as an **antigen-presenting cell** (**APC**). A helper T cell's specific receptors recognize this self/nonself combination of MHC and a particular antigen fragment. Antigen-specific contact between the T cell and the antigen-presenting macrophage activates the T cell, which proliferates and forms a clone of helper T cells keyed to the specific antigen. The helper T cells then secrete cytokines, which selectively stimulate B cells that have already encountered that particular antigen. A helper T cell contacts a B cell in the same way it contacts a macrophage displaying the antigen. When antigen binds to antigen receptors on a B cell, the cell takes in a few of these foreign molecules by endocytosis. The B cell then displays antigen fragments bound to class II MHC markers on the cell surface. The helper T cell's receptor recognizes and binds to this antigen-MHC complex.

Thus, macrophages and B cells both act as APCs in their interactions with helper T cells, but there is an important difference: Each macrophage can display a wide variety of antigens, from the many different pathogens it has taken in by phagocytosis; but each B cell, being specific, can bind to and subsequently display only the antigens of a particular pathogen. Helper T cells are also antigen-specific and can be activated only by those macrophages presenting the appropriate antigen in combination with a class II MHC molecule. In this way, the nonspecific defense afforded by macrophages also enhances specific defense by selectively priming T cells. The activated T cells then stimulate the appropriate B cells to mount a humoral immune response against a certain antigen.

T-Dependent and T-Independent Antigens

Antigens that evoke the response you have just learned about are known as **T-dependent antigens** because they cannot stimulate antibody production without the involvement of T cells (see FIGURE 39.9). Most antigens are T-dependent. However, certain types of antigens, called **T-independent antigens,** trigger humoral immune responses without involving macrophages or T cells. These antigenic molecules are usually long chains of repeating units, such as polysaccharides or proteins with many similar polypeptide subunits. Such antigens are found in bacterial capsules and bacterial flagella. Apparently, the numerous subunits of such antigens bind simultaneously to a number of the antigen receptors on the B-cell surface. This provides enough stimulus to the B cells without the assistance of T cells. However, the humoral response (antibody production) to T-independent antigens is generally much weaker than the response to T-dependent antigens. Furthermore, no memory cells are generated in T-independent responses.

Once activated by T-dependent or T-independent antigens, a B cell gives rise to a clone of plasma cells, and each of these effector cells secretes as many as 2000 antibodies per second for the 4- to 5-day lifetime of the cells. These specific antibodies help eliminate the foreign invader. Response to T-dependent antigens also generates a clone of long-lived memory cells.

The Molecular Basis of Antigen-Antibody Specificity

Most antigens are proteins or large polysaccharides. These molecules are often outer components of the coats of viruses, the capsules and cell walls of bacteria, and the surface molecules of many other types of cells. Foreign molecules associated with transplanted tissues and organs or with blood cells from other individuals or species can also incite an immune response. The surfaces of foreign substances such as pollen also include antigens.

Antibodies do not generally recognize the antigen as a whole molecule. Rather, they identify a localized region on the surface of an antigen called an antigenic determinant, or **epitope** (FIGURE 39.10). A single antigen such as a bacterial protein may have several effective epitopes, stimulating several different B cells to make distinct antibodies against it. And different parts of the bacterial cell may possess different antigens. Thus, the immune system responds to a particular species of bacterial cell by producing many antibodies specific for the various epitopes marking the antigens of the bacterial cell wall, capsule, and flagella.

Antibodies constitute a class of proteins called **immunoglobulins (Igs).** Every antibody molecule has at

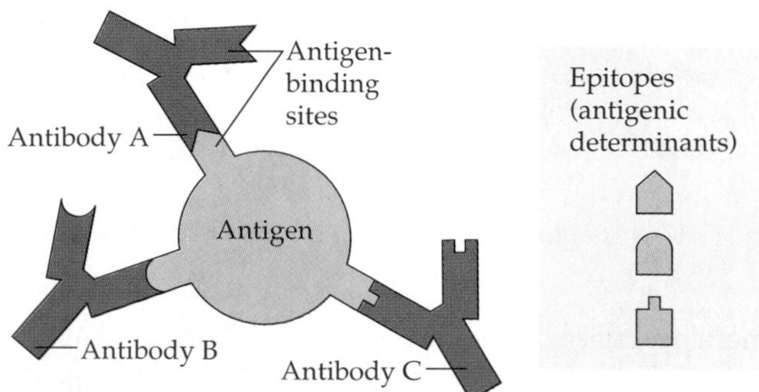

FIGURE 39.10
Epitopes (antigenic determinants). Antibodies bind to epitopes on the surface of an antigen. In this example, specific antibodies of three types react with different epitopes on the same large antigen molecule.

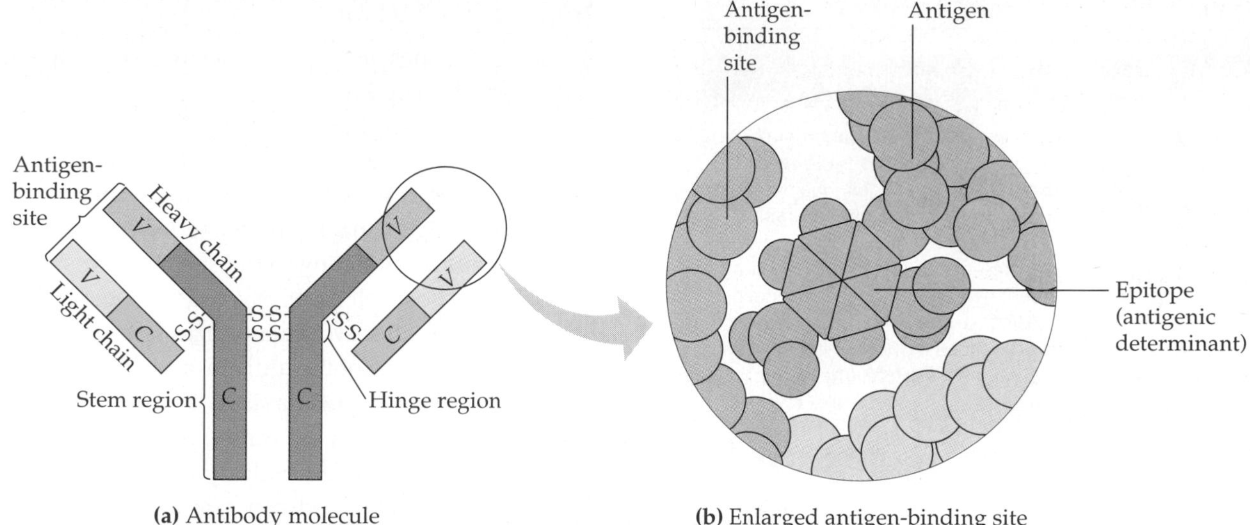

(a) Antibody molecule

(b) Enlarged antigen-binding site

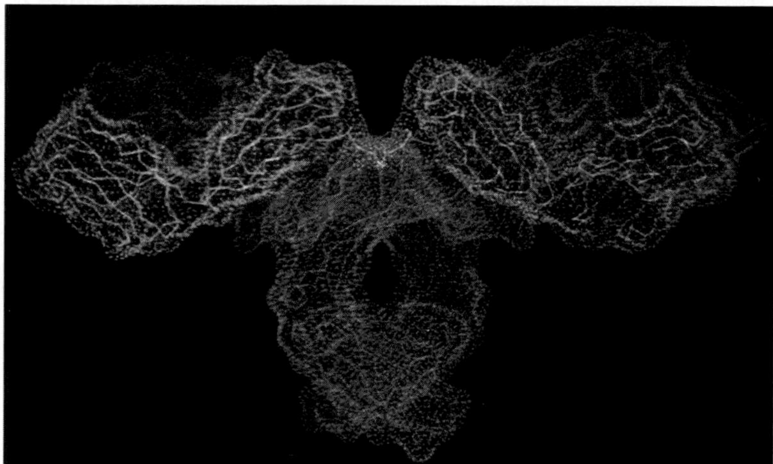

(c) Model of an antibody molecule

FIGURE 39.11

The structure of a typical antibody molecule. (**a**) The Y-shaped molecule (monomer) is composed of two light and two heavy chains linked together by disulfide bridges (S-S). Most of the molecule is the same for all antibodies of the same immunoglobulin class and is termed the constant region *(C).* The amino acid sequences of the variable regions *(V),* which make up the two identical antigen-binding sites, are different in each specific type of antibody, giving these sites specific shapes that fit certain antigenic epitopes. The stem of the Y binds to the surface of a cell in some cases. (**b**) This enlargement shows an antigenic epitope bound to an antigen-binding site. (**c**) A computer graphic image of an antibody molecule.

least two identical sites that bind to the epitope that provoked its production. A typical antibody is diagrammed in FIGURE 39.11a. It has four polypeptide chains joined to form a Y-shaped molecule: two identical light chains and two identical heavy chains. Both heavy and light chains have *constant* regions. These regions are called constant because their sequences of amino acids vary little among the antibodies that perform a particular type of defense, despite a wide variation in antigen specificity. At the tips of the Y-shaped molecule's two arms are the *variable* regions of the heavy and light chains, so named because their amino acid sequences vary extensively from antibody to antibody (see FIGURE 18.10). These tips of the Y function as the antigen-binding sites. The dimensions and contours of the binding sites are determined by the unique amino acid sequences of the variable regions of the heavy and light chains. The association between an antigen-binding site and an epitope resembles that between an enzyme and its substrate: Several weak bonds form between contiguous chemical groups on the respective molecules (FIGURE 39.11b).

The antigen-binding site is responsible for an antibody's recognition function—its ability to identify a specific epitope of an antigen. The tail of the antibody molecule, consisting of the constant regions of the polypeptide chains, is responsible for the antibody's effector function, the mechanism by which it inactivates or helps destroy an antigenic invader.

There are five types of constant regions, and these determine the five major classes of mammalian immunoglobulins: IgM, IgG, IgA, IgD, and IgE. The shapes and functions of these basic antibody types are summarized in TABLE 39.1. Each class is characterized by a type of constant region that enables the antibody molecules to perform certain defense functions. For example, IgA can be transported across epithelia, and it is present in saliva, sweat, and tears. Within each Ig class is an enormous variety of specific antibodies with unique antigen-binding sites.

TABLE 39.1

The Five Classes of Immunoglobulins	
 IgM (pentamer)	IgMs are the first circulating antibodies to appear in response to an initial exposure to an antigen. Their concentration in the blood declines rapidly. This is diagnostically useful because the presence of IgM usually indicates a current infection by the pathogen causing its formation. IgM consists of five Y-shaped monomers arranged in a pentamer structure. The numerous antigen-binding sites make it very effective in agglutinating antigens and in reactions involving complement. IgM is too large to cross the placenta and does not confer maternal immunity.
 IgG (monomer)	IgG is the most abundant of the circulating antibodies. It readily crosses the walls of blood vessels and enters tissue fluids. IgG also crosses the placenta and confers passive immunity from the mother to the fetus. IgG protects against bacteria, viruses, and toxins circulating in the blood and lymph, and triggers action of the complement system.
 IgA (dimer)	IgA is produced primarily in the form of two Y-shaped monomers (a dimer) by cells abundant in mucous membranes. The main function of IgA is to prevent the attachment of viruses and bacteria to epithelial surfaces. IgA is also found in many body secretions, such as saliva, perspiration, and tears. Its presence in colostrum (the first milk of a nursing mammal) helps protect the infant from gastrointestinal infections.
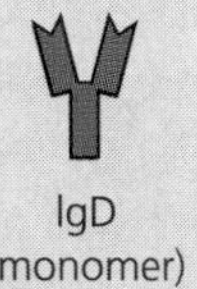 IgD (monomer)	IgD antibodies do not activate the complement system and cannot cross the placenta. They are mostly found on the surfaces of B cells, probably functioning as an antigen receptor required for initiating the differentiation of B cells into plasma cells and memory B cells.
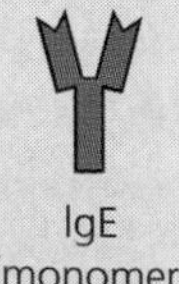 IgE (monomer)	IgE antibodies are slightly larger than IgG molecules and represent only a very small fraction of the total antibodies in the blood. The tail regions attach to receptors on mast cells and basophils and, when triggered by an antigen, cause the cells to release histamine and other chemicals that cause an allergic reaction.

How Antibodies Work

An antibody does not usually destroy an antigenic invader directly. The binding of antibodies to antigens to form an antigen-antibody complex is the basis of several effector mechanisms (FIGURE 39.12). The simplest of these is neutralization, in which the antibody blocks certain sites on an antigen, making it ineffective. Antibodies neutralize a virus by attaching to the sites the virus must use to bind to its host cell. Similarly, coating a bacterial toxin with antibodies effectively neutralizes it. Eventually, phagocytic cells dispose of the antigen-antibody complex.

Another effector mechanism is the agglutination (clumping) of bacteria by antibodies. Agglutination is possible because each antibody molecule has at least two antigen-binding sites and can cross-link adjacent antigens. The bacterial clumps are easier for phagocytic cells to engulf than are single bacteria. A similar mechanism is precipitation, the cross-linking of soluble antigen molecules (rather than cells) to form immobile precipitates that are captured by phagocytes.

One of the most important effector mechanisms of the humoral response is the activation of the complement system by antigen-antibody complexes. Recall that the complement system is a group of proteins that acts cooperatively with elements of the nonspecific and specific defense systems. Antibodies often combine with complement proteins, activating the complement proteins to produce lesions in a foreign cell's membrane that result in lysis (bursting) of the cell. We will examine this process in more detail later.

Monoclonal Antibody Technology

Until the late 1970s, the only source of antibodies for research or the treatment of disease was the blood of immunized animals. Antibodies obtained from the blood of an immunized animal are called *polyclonal antibodies* because they arise from many different B-cell clones, each specific for a particular epitope of the introduced antigen. Indeed, any normal immune response is polyclonal. The animal's blood also contains antibodies specific for other antigens to which it has been recently exposed. It is therefore difficult to isolate the desired antibodies from this mixture. The problem was solved by the work of Cesar Milstein and Georges Kohler who, in 1975, developed a method for making **monoclonal antibodies.** The term *monoclonal* means that all the cells producing such antibodies are descendants of a single cell; thus, they all produce identical antibody molecules. Monoclonal antibody technology makes it possible to produce commercial quantities of a specific antibody relatively inexpensively. Many new diagnostic methods for

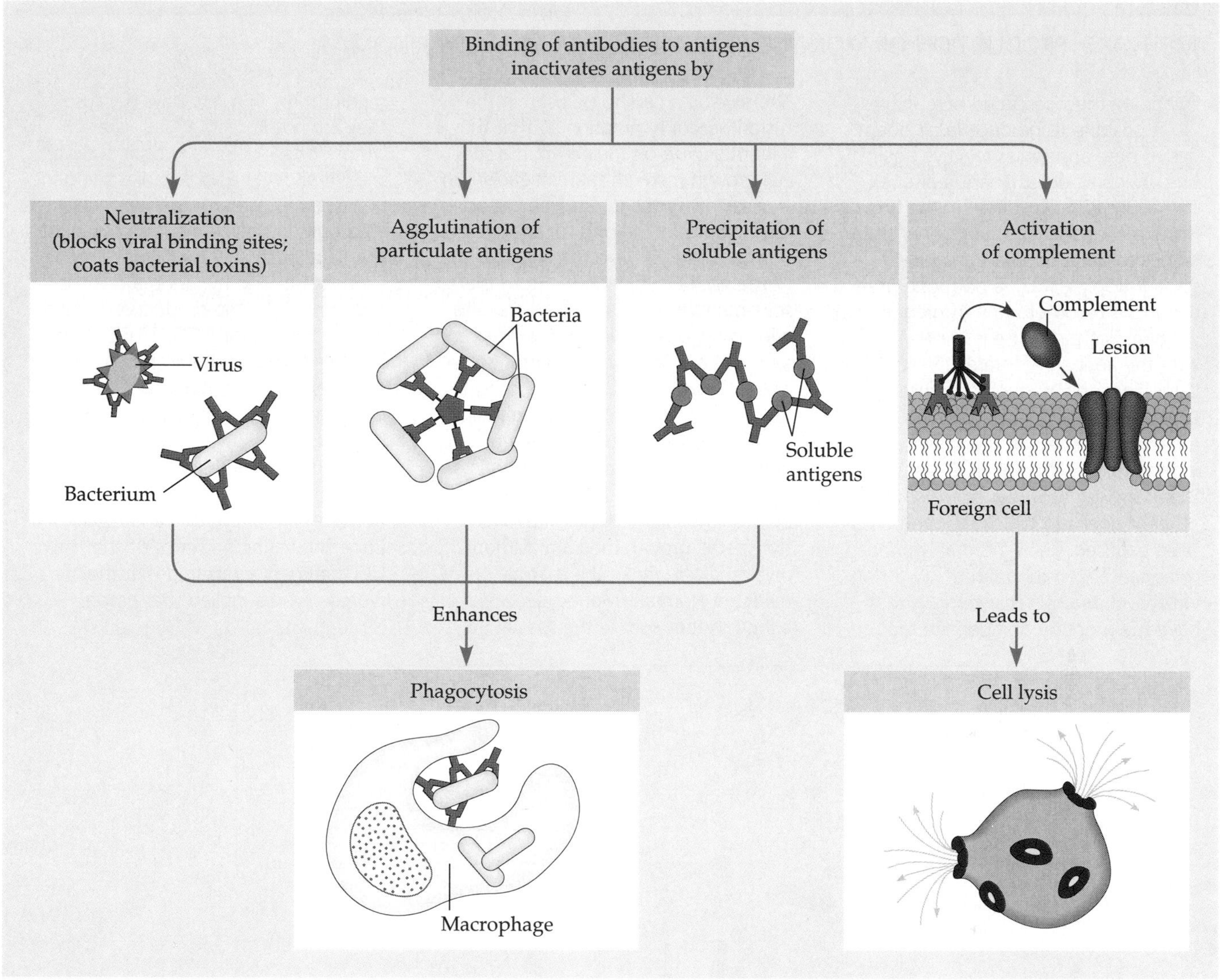

FIGURE 39.12

Effector mechanisms of humoral immunity. The binding of antibodies to antigens tags foreign cells and molecules for destruction by phagocytes or the complement system of proteins.

detecting pathogenic microbes in clinical samples depend on monoclonal antibodies. Monoclonal antibodies are also the basis of the over-the-counter pregnancy tests that detect a hormone excreted only in the urine of a pregnant woman.

Monoclonal antibodies will have increasing application as therapeutic agents—for example, to combat microbial toxins in the bloodstream. There is also the prospect of using monoclonal antibodies to treat cancer. The basic idea is to produce monoclonal antibodies against a patient's cancer cells and attach a toxin to the antibodies. The antibody-toxin complex, called an immunotoxin, acts as a "magic bullet" to selectively find and destroy the cancer cells when injected into that same patient.

The technology that makes monoclonal antibodies possible is the fusion of two cells to form a hybrid cell called a **hybridoma**. The two types of cells from which a hybridoma is made have distinct abilities, neither of which can be exploited alone. One type of cell is derived from a cancerous cell known as a myeloma. These cancer cells, unlike normal cells, can be cultured indefinitely. The other type of cell is a normal antibody-producing plasma cell, obtained from the spleen of an animal immunized with the antigen of interest. This cell can be cultured for only a few generations. Fusing the myeloma cell with the plasma cell to form a hybridoma combines key qualities of the two cells; the hybridoma produces a single type of antibody and can be cultured indefinitely to manufacture that antibody on a large scale (see the Methods Box, p. 866).

METHODS: PRODUCTION OF MONOCLONAL ANTIBODIES WITH HYBRIDOMAS

The method described here makes it possible to produce large quantities of pure antibodies reactive against one antigenic determinant (epitope). The application illustrated in the figure is the preparation of monoclonal antibodies to detect human chorionic gonadotropin (HCG), a hormone present in the blood or urine of pregnant women. First, a mouse is immunized with the antigen of interest, here HCG. Later, once the immune response has developed, the mouse's spleen is removed. The cells of the spleen include many different HCG-reactive B cells as well as B cells specific for other antigens to which the mouse has been exposed. These normal lymphocytes can be "immortalized" by fusion with myeloma cells, cancer cells that have the property of indefinite growth. The myeloma cells to be used in the fusion harbor a mutation so that they cannot survive on their own in a special growth medium that is deficient in a necessary nutrient. This is an important feature that allows for the selection of cells that have fused properly.

The normal lymphocytes from the spleen and the mutant myeloma cells are mixed together in conditions that favor fusion. Some of the lymphocytes and myeloma cells fuse to become hybridomas, while other cells remain unfused or are fused with "like partners" (myeloma-myeloma, lymphocyte-lymphocyte). All the cells are then placed in a culture dish containing the special growth medium. Mutant myeloma cells die in the special medium. Normal lymphocytes, although they survive the special medium, die in a few days because they are not immortal. Only the lymphocyte-myeloma hybrids survive; the lymphocyte DNA supplies the gene necessary for growth in the special medium, and myeloma DNA supplies the gene necessary for immortality.

Each surviving hybridoma clone is then tested for the production of antibodies specific for HCG. Clones that test positive are cultured for large-scale production of the antibody. The monoclonal antibody specific for HCG now becomes a sensitive diagnostic tool that is used to detect HCG in a blood or urine sample. A positive reaction between the antibodies and blood or urine shows the presence of the HCG antigen, indicating that the woman who supplied the sample is pregnant.

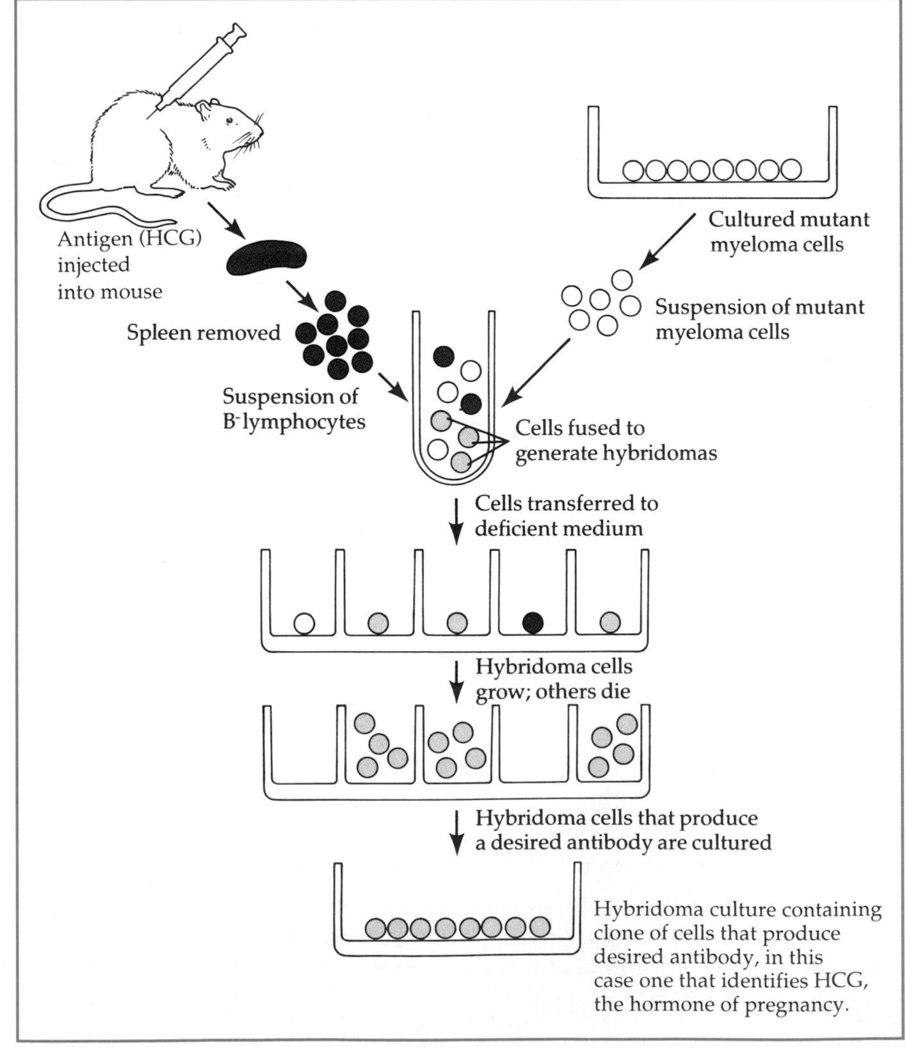
Antigen (HCG) injected into mouse
Spleen removed
Suspension of B-lymphocytes
Cultured mutant myeloma cells
Suspension of mutant myeloma cells
Cells fused to generate hybridomas
Cells transferred to deficient medium
Hybridoma cells grow; others die
Hybridoma cells that produce a desired antibody are cultured
Hybridoma culture containing clone of cells that produce desired antibody, in this case one that identifies HCG, the hormone of pregnancy.

In the cell-mediated response, T cells defend against intracellular pathogens

Recall that the body's specific response to antigens is a *dual* defense, consisting of humoral and cell-mediated immunity. Many pathogens, including all viruses, are obligate intracellular parasites that can reproduce only within the cells of the host. The humoral immune response helps the defense network identify and destroy extracellular pathogens, but it is cell-meditated immunity that battles pathogens that have already entered cells. The key components of cell-mediated immunity are helper T cells (T_H) and cytotoxic T cells (T_C). Recall that these lymphocytes complete their maturation in the thymus and migrate to lymphoid organs such as the lymph nodes and the spleen.

The Activation of T Cells

In contrast to B cells, T cells cannot detect free antigens present in body fluids. T cells respond only to antigenic epitopes displayed on the surfaces of the body's own cells. These bound antigens are recognized by specific T-cell receptors embedded in the plasma membranes of T cells. We have already seen one example, the binding of helper T cells to antigens displayed on the surfaces of antigen-presenting macrophages and B cells in the activation of humoral immunity (see FIGURE 39.9). Recall from that example that the receptor of a T_H cell actually recognizes a combination of an antigen fragment and one of the body's self markers, class II MHC, on the surface of a macrophage or B cell. In contrast, the receptor of a T_C cell recognizes a combination of an antigen fragment and a *class I* MHC molecule, a self marker found on every nucleated cell of the body. In both cases, the MHC protein that presents the antigen is shaped something like a hammock that nestles the antigen. An MHC molecule can associate with a variety of antigens. However, each combination of MHC and antigen forms a unique complex that is recognized by specific T cells.

The interaction between a T_H cell and an antigen-presenting cell (APC) is greatly enhanced by the presence of a T-cell surface molecule called CD4. Present on most helper T cells, CD4 has an affinity for a part of the class II (MHC) molecule. The CD4–class II MHC interaction helps keep the T_H and the APC engaged while antigen-specific contact is going on (FIGURE 39.13). Similarly, a cytotoxic T cell bears surface molecules called CD8, which help in the interaction with class I MHC molecules.

The MHC-antigen complex is like a red flag to T cells, calling them into action against cells infected by the pathogen represented by that particular antigen. The situation stimulates T cells that have the appropriate receptors to proliferate and form clones of activated T_H or T_C cells specialized for fighting that particular pathogen.

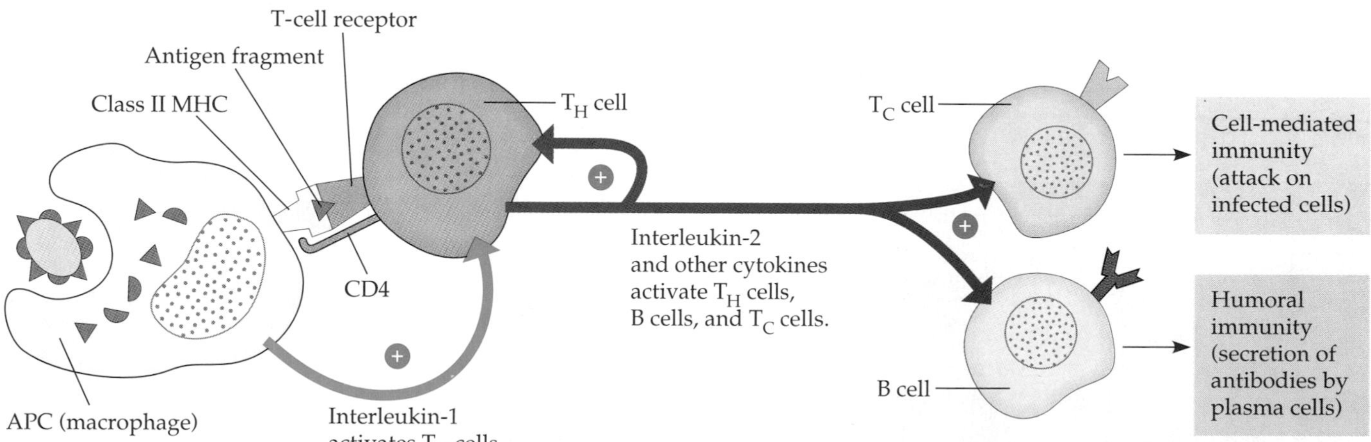

FIGURE 39.13
The central role of helper T cells. Helper T cells mobilize both humoral and cell-mediated branches of the immune response. The receptor of the T_H cell recognizes the class II MHC–antigen complex displayed on the surface of an antigen-presenting cell, usually a macrophage. The interaction of the two cells is enhanced by CD4, a T_H molecule that bonds to the class II MHC protein on an APC. The macrophage secretes interleukin-1, a cytokine that contributes to the activation of the T_H cells. An activated T cell grows and divides, producing a clone of T_H cells, all with receptors keyed to MHC markers in combination with the specific antigen that triggered the response. The T_H cells then secrete the cytokine interleukin-2, which amplifies the cell-mediated response by stimulating the proliferation and activity of other T_H cells, all specific for the same antigenic determinant. Interleukin-2 and other cytokines also help activate B cells, which function in humoral immunity, and T_C cells, which function in the cell-mediated immune response.

We have already examined one function of helper T cells: their role in activating B cells to secrete antibodies against T-dependent antigens during humoral responses. T_H cells also activate other types of T cells to mount cell-mediated responses to antigens.

The ability of helper T cells to stimulate other lymphocytes depends on cytokines. As a macrophage engulfs and presents antigen, it is stimulated to secrete a cytokine called **interleukin-1** (see FIGURE 39.13). This signals the T_H cell to release another cytokine known as **interleukin-2.** In an example of positive feedback, interleukin-2 stimulates T_H cells to grow and divide more rapidly, increasing the supply of both T_H cells and interleukin-2. The interleukin-2 and other cytokines released by this growing population of T_H cells also help activate B cells, thus stimulating the humoral response against a specific antigen. And cytokines from T_H cells help arm the cell-mediated response by stimulating T_C cells.

How Cytotoxic T Cells Work

Cytotoxic T cells kill host cells infected by viruses or other intracellular pathogens. Such infected cells display antigens complexed with class I MHC molecules. Notice an important difference in the functions of the receptors on T_H and T_C cells: The receptors on helper T cells enable those lymphocytes to stimulate immunity by binding specifically to antigen-presenting cells marked by *class II* MHC molecules, such as macrophages and B cells. In contrast, cytotoxic T cells, by recognizing specific antigens in association with *class I* MHC markers, can bind to any cell of the body infected by that particular antigenic invader. When a T_C cell binds to an infected cell, it releases *perforin,* a protein that forms an open lesion in the infected cell's membrane. The target cell loses cytoplasm through the lesion, leading to cell lysis (FIGURE 39.14a). While this destroys the host cell, it also deprives the pathogen of a place to reproduce and exposes

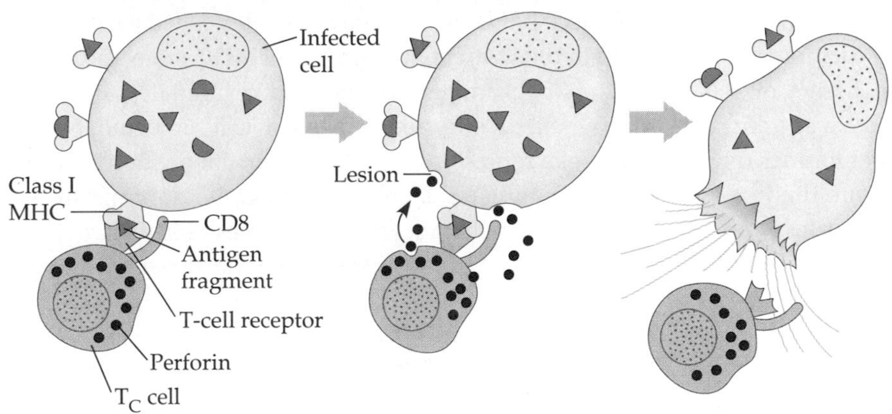

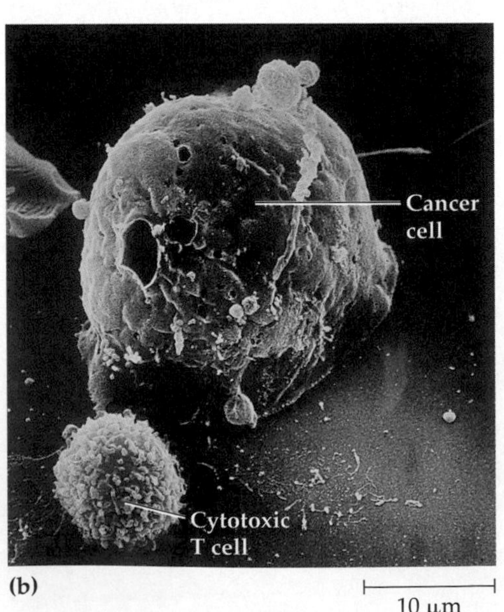

FIGURE 39.14
Cytotoxic T cells. (**a**) The T cell receptor of a cytotoxic T cell recognizes the class I MHC–antigen complex on the surface of an infected cell or cancer cell. This interaction is enhanced by CD8, which binds to the class I MHC protein. The activated T_C cell discharges the protein perforin, which lyses the infected or cancer cell. (**b**) The T_C cell (smaller cell) has lysed a cancer cell (SEM).

the pathogen to circulating antibodies. This cell-mediated action is an essential feature of the body's overall defense system, because antibodies cannot attack pathogens that have already invaded host cells. After destroying an infected cell, the T_C cell continues to live and can kill many other cells.

Cytotoxic T cells also function in defense against cancer. Cancer cells arise periodically in the body and, because they carry distinctive molecular markers not found on normal cells, are identified as nonself by the immune system. T_C cells target the cancer cells and lyse them (FIGURE 39.14b). Cancers are more likely to become established in individuals with defective immune systems and elderly people with declining immunity.

Although T_C cells and natural killer cells attack target cells by similar mechanisms, there is an important

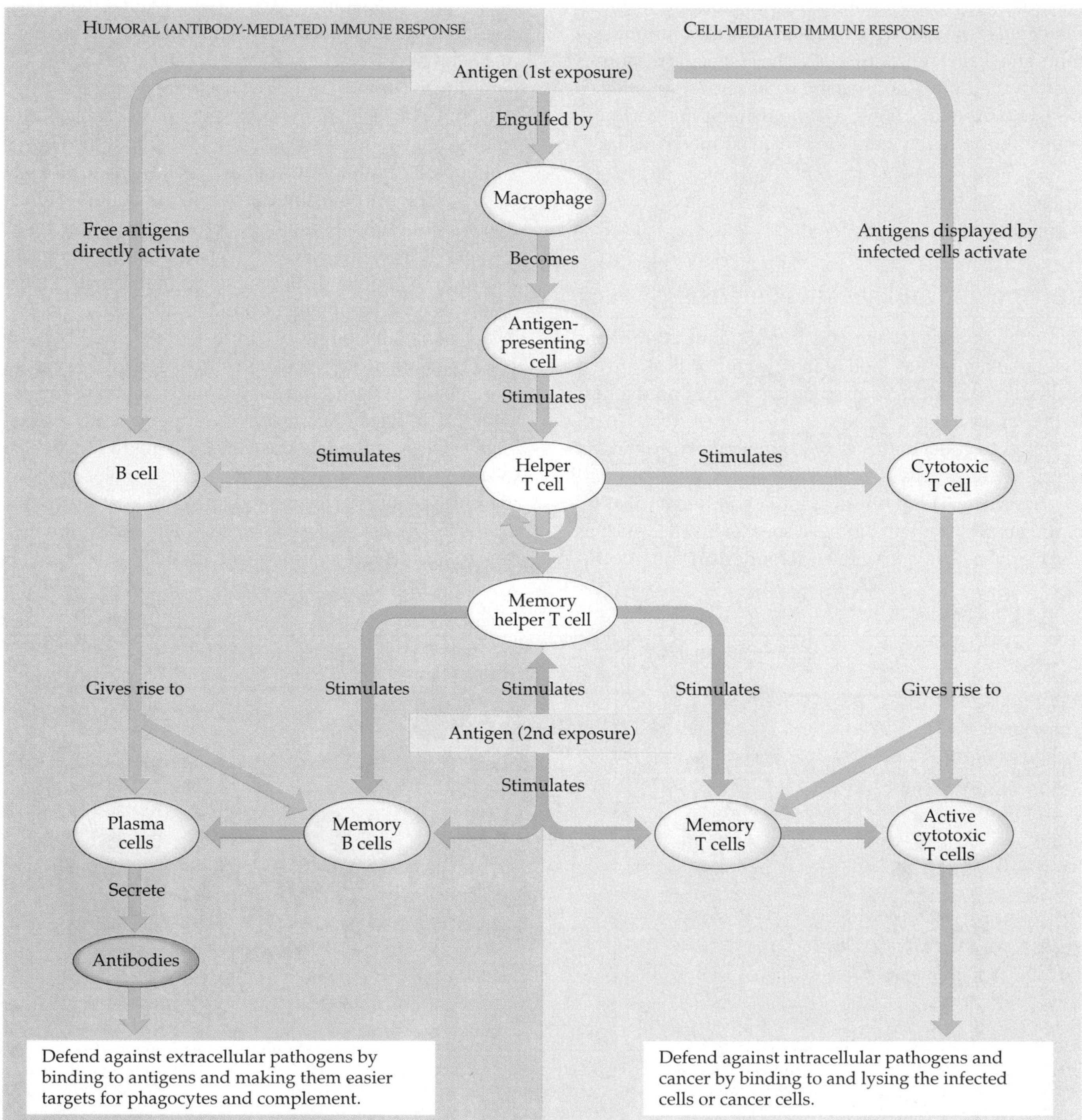

FIGURE 39.15

A summary of the immune responses. In this simplified flow chart, green arrows track the primary response, and blue arrows track the secondary response.

difference between the roles of these cells in defense: Natural killer cells, part of the body's *nonspecific* defense, do not respond to specific antigens (see p. 854).

A third type of T lymphocyte, called a **suppressor T cell** (**T_S**) is not well understood. (Some immunologists believe that T_S cells are actually a type of T_H cell rather than a separate subclass of T cells.) T_S cells probably function in turning off the immune response when an antigen is no longer present (hence their name).

FIGURE 39.15 (p. 869) summarizes the humoral and cell-mediated responses, the two branches of the immune system's attack on foreign invaders. These specific responses overlap the body's nonspecific defenses. As we will see next, the proteins known as complement are partly responsible for this cooperation of defense mechanisms.

Complement proteins participate in both nonspecific and specific defenses

The complement system consists of about 20 proteins that circulate in the blood in inactive form. They are activated in cascade fashion, each one activating the next in the series.

The activation pathway of complement proteins in specific defense was identified first and is called the *classical pathway.* This pathway is initiated when antibodies of the immune system bind to a specific invader, such as a bacterial cell, targeting the cell for destruction. A complement protein then bridges the gap between two adjacent antibody molecules. This association of antibodies and complement activates complement proteins, leading to the formation of a *membrane attack complex* (FIGURE 39.16). The complex produces a small lesion in the pathogen's membrane, resulting in lysis of the cell. This is similar to the way T_C cells kill infected host cells, but the target cell in this case is the invader itself.

The *alternative pathway* of complement action does not require cooperation with antibodies, and the targets are therefore not specific. This alternative complement pathway can also generate a membrane attack complex. Substances found in many bacteria, yeasts, viruses, virus-infected cells, and protozoan parasites can activate complement to form such a complex without the help of antibodies. Another outcome of the alternative pathway is complement's contribution to inflammation. By binding to histamine-containing cells, some complement proteins trigger the release of histamine, the chemical alarm to the body's defense systems that signals a local injury. Several complement proteins also attract phagocytes to the infected site.

Complement also cooperates with phagocytes in the destruction of pathogens. In a process called *opsonization* (L., "to relish"), complement proteins attach to foreign cells and stimulate phagocytes to ingest those cells. (Coating the pathogens with antibody molecules can have the same effect.) In another example of teamwork in the body's defense systems, complement, antibodies, and phagocytes function together in a phenomenon

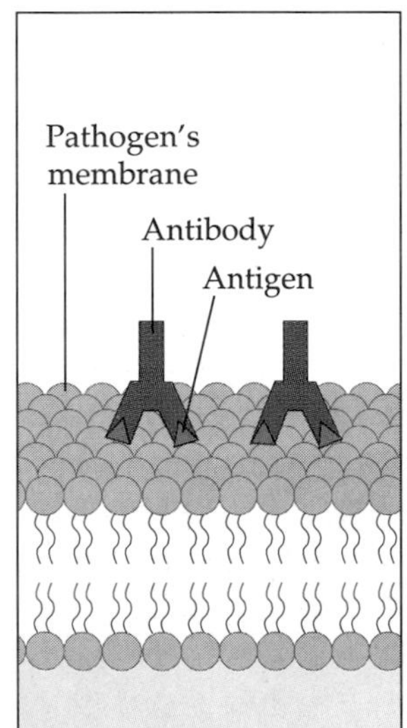

① Antibody molecules attach to antigens on pathogen's plasma membrane.

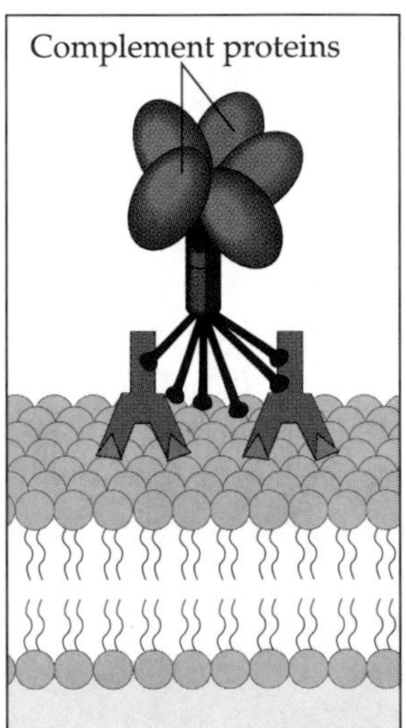

② Complement proteins attach to pair of antibodies.

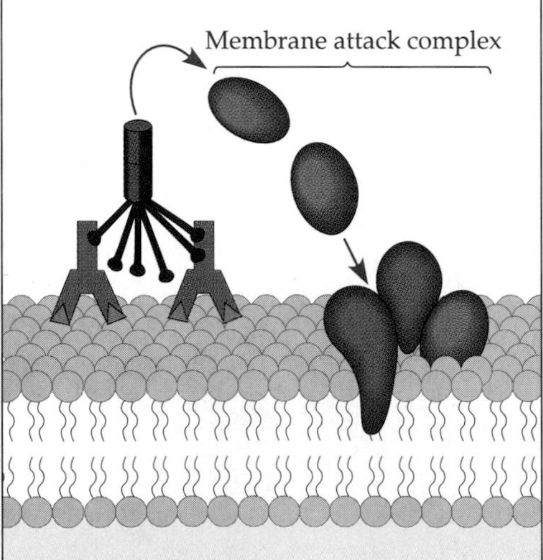

③ Activated complement proteins attach to pathogen's membrane in step-by-step sequence, forming a membrane attack complex.

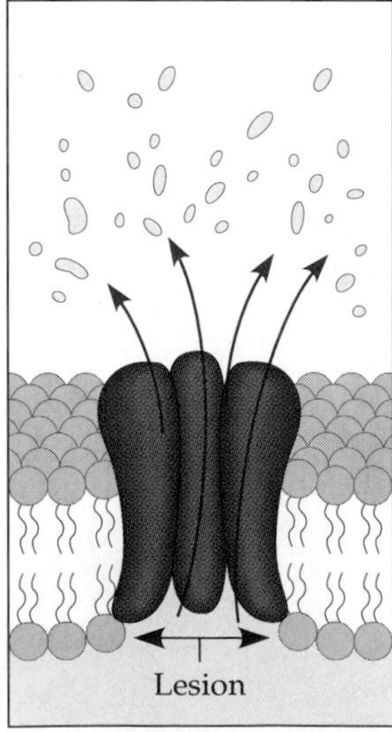

④ Complement proteins lyse target membrane, resulting in lesion and death of pathogen.

FIGURE 39.16
The classical complement pathway, resulting in lysis of a target cell.

called *immune adherence.* A microbe coated with both antibodies and complement proteins adheres to surfaces such as the walls of blood vessels, making the pathogens easier prey for phagocytic cells circulating in the blood.

Now that we have examined how the body's defenses work, the next two sections will survey some important applications.

The immune system's capacity to distinguish self from nonself is critical in blood transfusion and transplantation

In addition to distinguishing between an animal's own cells and pathogens such as bacteria and viruses, the immune system wages war against cells from other individuals of the same species. For example, a skin graft from one person placed on another person (not an identical twin) will look healthy for a day or two, but after that it will be destroyed by the immune system. It is interesting that a mother does not reject a fetus as a foreign body. Apparently, the structure of the placenta, the link between mother and fetus, is the key to this acceptance (see FIGURE 42.17); the mother rejects fetal tissue placed elsewhere in her body.

In this section, you will learn about the potential problems associated with blood transfusions and organ transplants, two important applications of the immune system's response to nonself.

Blood Groups

The genetic basis for **ABO blood groups** was covered in Chapter 13. An individual with type A blood has A antigens on the surface of his or her red blood cells. The A molecule is referred to as an antigen here because it may be identified as foreign if placed in the body of another person; the type A antigen is *not* antigenic to its owner. And an individual with type A blood will not produce antibodies against A antigens (see FIGURE 13.11). However, a person with type B blood has antibodies against A antigens (anti-A antibodies). Therefore, if type A blood is introduced into the bloodstream of a type B person, the transfused red blood cells will be destroyed by complement-mediated lysis (see FIGURE 39.16). The reverse situation, transfusion of type B cells into a type A person, would have a similarly disastrous result. Type O individuals are known as universal donors because their blood cells carry neither A nor B antigens. Type O blood cells are not subject to destruction by anti-A or anti-B antibodies. Type O individuals can receive only type O blood, however, because they have both anti-A and anti-B antibodies. Type AB individuals are universal recipients; because their red cells have both antigens, they have neither anti-A nor anti-B antibodies. However, type AB people can donate blood only to other type AB individuals.

You may find it odd that antibodies to blood group antigens already exist in the body prior to a blood transfusion. Indeed, this is a unique case; these antibodies arise in response to normal bacterial flora in the body and cross-react with (also bind to) blood group antigens.

Antibodies against the A and B antigens are generally of the IgM class, and therefore do not cross the placenta to harm a developing fetus with a blood type different from its mother's (see TABLE 39.1). However, another red blood cell antigen, the **Rh factor,** is notorious in cases where antibodies produced by a pregnant woman react with the red blood cells of her developing fetus. The situation arises if the mother is Rh-negative (lacks the Rh factor) but the fetus is Rh-positive, having inherited the factor from the father. The mother develops antibodies against the Rh factor when small amounts of fetal blood cross the placenta and come in contact with her lymphocytes, usually late in pregnancy or during delivery of the baby. Typically, the mother's response to this first exposure is mild and without medical consequences for the baby. The real danger occurs in subsequent pregnancies, when the mother's immune response against the Rh factor has already been primed and her antibodies, now IgG, can cross the placenta and destroy the red blood cells of an Rh-positive fetus. To prevent this, the mother may be injected with anti-Rh antibodies after delivering her first Rh-positive baby. The antibodies destroy the Rh-positive red cells that have entered the mother's circulation before her own immune system has been stimulated by the Rh antigen.

Tissue Grafts and Organ Transplants

The major histocompatibility complex (MHC), the molecular fingerprint unique to each self, is responsible for stimulating the rejection of tissue grafts and organ transplants. Foreign MHC molecules are antigenic, causing cytotoxic T cells to mount a cell-mediated response against the donated tissue or organ. To minimize this rejection, attempts are made to match the MHC of the organ donor and recipient as closely as possible. In the absence of an identical twin, siblings or parents usually provide the closest match. Various drugs are necessary to suppress the immune response when most transplants are performed. The complication with this strategy is that the organ recipient is more vulnerable to infection when drugs compromise the immune system. Such drugs as cyclosporine and FK 506 have the advantage of selectively suppressing T-cell activation without crippling humoral immunity.

Note that the body's disastrous reaction to an incompatible blood transfusion or a transplanted organ is not a disorder of the immune system, but a normal action taken by a healthy immune system exposed to foreign antigens. However, like any complex system, the immune apparatus does malfunction in some individuals, resulting in a variety of serious disorders.

Abnormal immune function leads to disease states

Autoimmune Diseases

Sometimes, for reasons immunologists are only beginning to understand, the immune system goes awry and turns against components of the body, leading to a variety of autoimmune diseases. In systemic lupus erythematosus, for example, people develop immune reactions against components of their own cells, especially nucleic acids released by the normal breakdown of skin and other tissues. Rheumatoid arthritis is a crippling autoimmune disease in which inflammation damages the cartilage and bone of joints. In insulin-dependent diabetes, an autoimmune reaction seems to cause the destruction of insulin-producing cells of the pancreas. Rheumatic fever is an autoimmune condition that was once a major killer of young adults and is now making a reappearance in certain areas. Antibodies produced in response to streptococcal infections (such as strep throat) react with heart muscle tissue in some people, damaging the heart valves. Repeated episodes of streptococcal infection result in more antibodies and more heart damage. In Graves' disease, antibodies attach to receptors on the thyroid gland that normally react with a thyroid-stimulating hormone produced by the pituitary. The abnormal binding of antibodies to these receptors stimulates the thyroid gland to produce thyroid hormones in excessive amounts. People suffering from Graves' disease often show bulging eyes and a greatly enlarged thyroid gland.

Allergy

Allergies are hypersensitivities of the body's defense system to certain environmental antigens called allergens. One hypothetical explanation for allergies is that they are evolutionary remnants of the immune system's response to parasitic worms. The defense mechanism that combats worms is similar to the allergic response that causes such disorders as hay fever or asthma.

The most common allergies involve antibodies of the IgE class (see TABLE 39.1). For example, hay fever and other allergies caused by pollen occur when plasma cells secrete IgE specific for these allergens. The IgE antibodies attach by their tails to mast cells, noncirculating cells

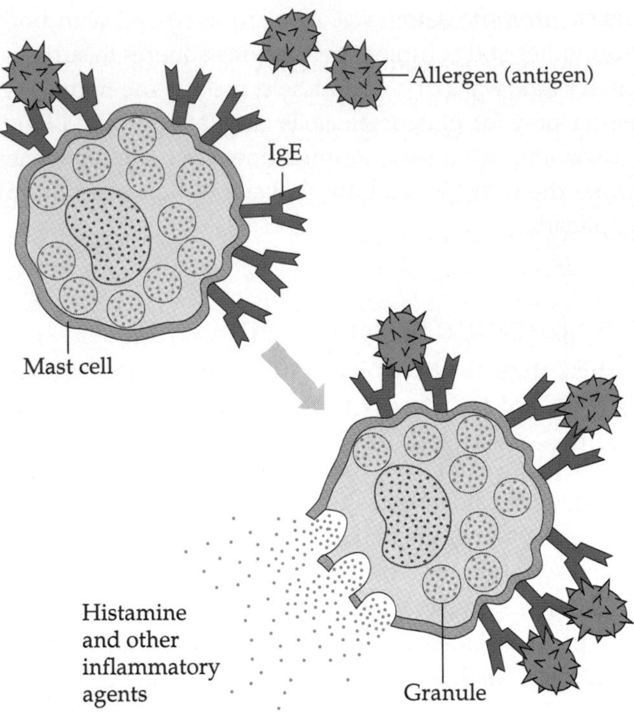

FIGURE 39.17
Mast cells and the allergic response. Upon first exposure to an allergen, plasma cells secrete IgE specific for the allergen. Some of these antibodies are attached by their tails to a mast cell. When, upon second exposure, the allergen binds to IgE already on the mast cell, it triggers the degranulation of the cell. Degranulation releases histamine, leading to most of the symptoms of the allergy.

found in connective tissue. In this way, a susceptible person becomes sensitized to the specific pollen antigen. Later, when a pollen grain binds to the IgE and bridges the space between two adjacent IgE monomers, the mast cell responds with a rapid reaction called degranulation, releasing histamine and other inflammatory agents (FIGURE 39.17). Recall that histamine causes dilation and increased permeability of small blood vessels. In the case of an allergy, response to histamines includes such symptoms as sneezing, a runny nose, and smooth muscle contractions that often result in breathing difficulty. Antihistamines are drugs that interfere with the action of histamine.

The most serious type of acute allergic response is anaphylactic shock, a life-threatening reaction to injected or ingested antigens. Hypersensitivity to wasp or bee stings is an example. Anaphylactic shock occurs when mast cell degranulation triggers an abrupt dilation of peripheral blood vessels and causes a precipitous drop in blood pressure. Death may occur within a few minutes. People who are extremely allergic to certain foods, like peanuts or fish, have died from eating tiny amounts of these foods. Some individuals with such severe hypersensitivities carry syringes containing the hormone epinephrine, which counteracts the allergic response.

Immunodeficiency

Certain individuals are inherently deficient in either humoral or cell-mediated immune defenses. Both branches of the immune system fail to function in a congenital disorder known as *severe combined immunodeficiency* (*SCID*). For people with this genetic disease, long-term survival requires a successful bone marrow transplant that will continue to supply functional lymphocytes. One risk of such transplants is that the transplanted cells will identify their new hosts as foreign and mount an attack against the marrow recipient (graft-versus-host disease). One type of SCID, caused by deficiency of the enzyme adenosine deaminase (ADA), has been treated with some success by gene therapy (see p. 386). The treatment eliminates the danger of graft-versus-host disease because the individual's own cells are genetically engineered and reintroduced to the patient.

Immunodeficiency is not always an inborn condition; in some cases, individuals acquire immunodeficiency later in life. For example, certain cancers depress the immune system, especially Hodgkin's disease, which damages the lymphatic system and makes the patient susceptible to many infections. The viral disease AIDS, discussed in the next section, causes very severe depression of the immune system. Immunodeficiency also results from deliberate suppression of the immune system with drugs in order to minimize the possibility of rejection of transplanted organs.

Nearly 2000 years ago, the Greek physician Galen recorded that people suffering from depression were more likely than others to develop cancer. In fact, there is growing evidence that physical and emotional stress can compromise immunity. Hormones secreted by the adrenal glands during stress affect the numbers of white blood cells and may suppress the immune system in other ways.

There is also evidence of direct links between the nervous system and the immune system. A network of nerve fibers penetrates deep into lymphoid tissue, including the thymus. Also, receptors for chemical signals secreted by nerve cells have been discovered on the surfaces of lymphocytes. Some signals secreted by nerve cells when we are relaxed and happy may actually enhance immunity. These and other observations (and speculations) have led physiologists to take a serious look at how general health and state of mind affect immunity. In one study, students were examined just after a vacation and again during final exams. Their immune systems were impaired in various ways during exam week; for example, interferon levels were lower. Most of us have been recipients of the admonition, "Don't get run down; it'll lower your resistance, and you'll get sick." This is good advice, but difficult to heed in a hectic collegiate environment.

Acquired Immunodeficiency Syndrome (AIDS)

In 1981, health care workers in the United States became aware of an increased frequency of Kaposi's sarcoma, a cancer of the skin and blood vessels, and of *Pneumocystis* pneumonia, a respiratory infection caused by a protozoan. Both of these diseases are extremely rare in the general population but occur more frequently in severely immunosuppressed individuals. This observation led to the recognition of an immune system disorder that was named **acquired immunodeficiency syndrome,** or **AIDS.** By 1983, virologists working in France and the United States had identified a causative agent, a virus now known as the **human immunodeficiency virus,** or **HIV** (FIGURE 39.18; also see FIGURE 17.7). With a mortality rate close to 100%, this may be the most lethal pathogen ever encountered. HIV probably arose by evolution from another virus in central Africa and may have caused unrecognized cases of AIDS there for many years. The virus has been identified in preserved blood samples from as early as 1959 in African nations and in England.

HIV infects cells that bear surface CD4 molecules. Recall that CD4 on helper T cells assists in the interaction

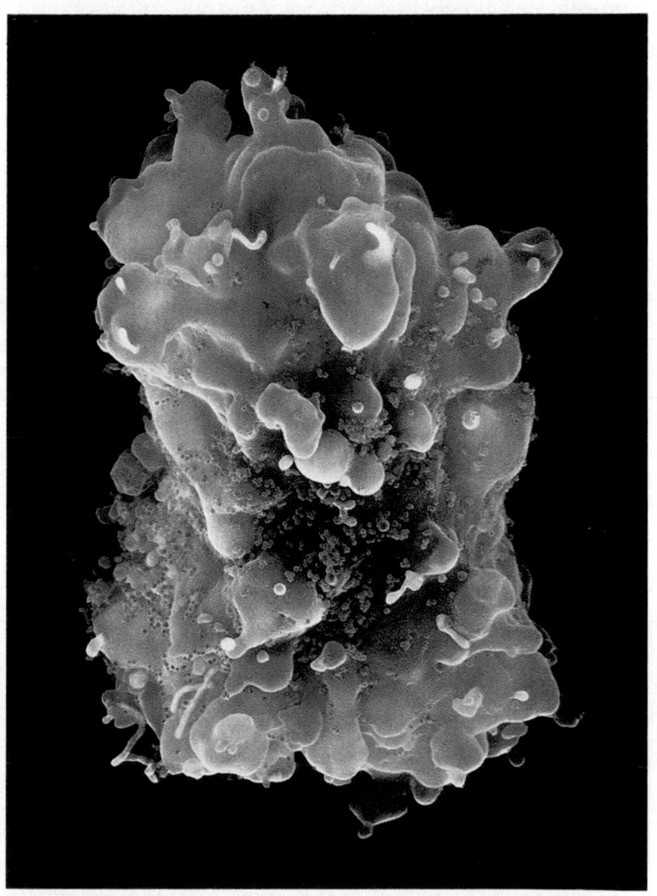

FIGURE 39.18
A T cell infected with HIV. The viruses (blue) bud continuously from the surface of the T cell (orange; colorized SEM). The cell will die, but only after it produces many copies of its killer.

between a T_H cell and an antigen-presenting cell. HIV, however, uses CD4 as its receptor. The infection process begins when a glycoprotein on the HIV envelope binds to CD4. The virus then enters the cell and begins to replicate. Newly formed viruses bud from the host cell, circulate, and infect other cells. The infected cells may produce new viruses for an extended time or may be killed quickly, either by the virus or by the response of the immune system. Because helper T cells are the main target of the virus, we can understand the devastating effects the infection has on the immune system. If these T_H cells, with their central role in both humoral and cell-mediated responses, are depleted, the immune system is essentially paralyzed. In addition to T_H cells, other cells, including some macrophages and a few subclasses of B cells that have CD4, as well as a few cell types that lack CD4, may also be infected.

HIV may be sustained in an infected individual for many years as a provirus assimilated into the genomes of host cells. The provirus is invisible to the immune system. The ability of HIV to exist as a provirus is one reason anti-HIV antibodies fail to eradicate the disease. Probably more important, however, are the extremely rapid mutational changes that occur in the virus as it replicates. Indeed, most new HIV viruses in an infected individual probably differ in at least one small way from the HIV versions that were originally acquired from another individual. The immune system responds effectively against HIV infection at first and may continue to fight HIV in body fluids for many years, but the defense system is eventually overwhelmed by the accumulation of variants with new and different antigens. In FIGURE 39.19, notice that the number of viruses gradually increases as the helper T-cell population (and hence the body's protection) decreases. When the damage to the immune system reaches a certain point, cell-mediated immunity collapses, and so-called *opportunistic diseases* such as Kaposi's sarcoma and *Pneumocystis* pneumonia can become established. AIDS is the name given to this late stage of HIV infection. It is defined by a blood concentration of T cells below a specified level and by the appearance of the characteristic opportunistic infections and cancer.

The time required for an HIV infection to progress to full-blown AIDS varies, but it averages about ten years. During most of this time, the patient exhibits only moderate hints of illness, such as swollen lymph glands and, later, yeast infections of the mucous membranes. Infants infected by their mothers *in utero* often progress to AIDS much more rapidly.

People who have been exposed to HIV have circulating antibodies against the virus, and detection of these antibodies is the most common method for identifying infected individuals. This is the test used to screen blood donations, for example. Thus, **HIV-positive** means that

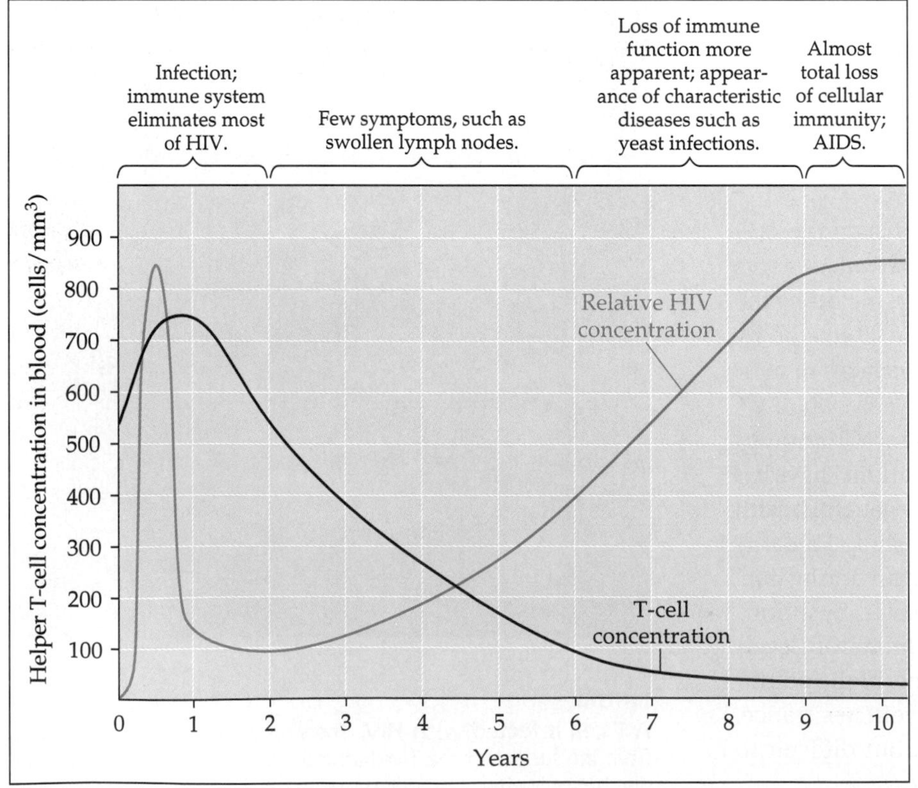

FIGURE 39.19

The stages of HIV infection. HIV concentration increases rapidly after the initial infection but is then almost eliminated by the immune response. However, some viruses survive and replicate slowly over time. The immune system continues to combat the viruses during this period, which may last for years, but the helper T-cell concentration gradually decreases. AIDS is the last stage of the process.

a person is infected with the virus, while AIDS refers to the set of symptoms that are eventually produced by that infection.

Transmission of HIV requires the transfer of body fluids, such as blood or semen, containing free viruses or infected cells. Unprotected sex (without a condom) among male homosexuals and transmission between intravenous drug users via unsterilized needles account for most of the AIDS cases reported thus far in the United States and Europe. However, the transmission of AIDS among heterosexuals is rapidly increasing as a result of unprotected sex with infected partners. In Africa and Asia, transmission has been primarily by heterosexual sex, especially where there is a high incidence of other sexually transmitted diseases that result in genital lesions. These sores facilitate the transmission of HIV, as the skin barrier (first line of defense) is breached and HIV-susceptible cells, such as macrophages and T cells, are attracted to the area by the inflammatory response.

HIV is not transmitted by casual contact or even kissing. Breast milk, however, has transmitted the disease from mothers to nursing infants. Blood transfusions have been almost eliminated as a route of transmission in developed countries by the test for anti-HIV antibodies. This type of test will never guarantee a completely safe blood supply, however, because a person may be infected by the virus for several weeks or months before the antibodies become detectable.

At this time, AIDS is incurable. Some antiviral drugs, such as AZT, ddC, and ddI, which inhibit the viral enzyme reverse transcriptase (see Chapter 17), may extend the lives of patients, but they do not completely eliminate the virus. Many drugs are useful for treating the opportunistic infections common in AIDS patients and can prolong life, but again, these drugs do not cure AIDS. Research to develop a vaccine is very active, but the antigenic variability of the virus poses a major problem. For the present, the best approach for slowing the spread of AIDS seems to be to educate people about the practices that transmit HIV, such as unprotected sex and sharing needles. Condoms do not completely eliminate the risk of transmitting the virus but they do reduce it. Anyone who has unprotected sex—vaginal, oral, or anal—with a partner who may have had unprotected sex with another individual during the past 15 years risks exposure to HIV.

Invertebrates exhibit a rudimentary immune system

This chapter has emphasized the nonspecific and specific defenses that occur in vertebrates, because relatively little is known about how invertebrates react against pathogens that have penetrated the skin and other outer barriers. However, experiments have established that one fundamental facet of defense, the ability to distinguish self from nonself, is well developed in invertebrates. For example, if the cells of two sponges of the same species are mixed, the cells from each individual sort themselves out and aggregate, excluding cells from the other individual. In many invertebrates, amoeboid cells called coelomocytes identify and destroy foreign materials.

Studies of tissue grafting in earthworms have shown a memory response in the defense systems of these annelids. When a portion of body wall is grafted from one worm to another, the recipient's coelomocytes attack the foreign tissue. If donor and recipient are from the same population, then the grafted tissue survives for about eight months before it is completely rejected. However, if a worm receives a graft from a donor taken from a distant location (different population), the graft is rejected in just two weeks. If a second graft from that same donor to the same recipient is attempted, it takes coelomocytes less than a week to eliminate the foreign tissue. Additional research on the defense systems of invertebrates may help biologists understand how the vertebrate immune system evolved.

* * *

The immune response is one of the many adaptive processes that enable animals to adjust to the adversities of the environment. The next chapter describes several other processes that help maintain favorable conditions within animals as they cope with varying external environments.

REVIEW OF KEY CONCEPTS (with page numbers and key figures)

- Nonspecific mechanisms provide general barriers to infection (pp. 852–856, FIGURE 39.5)
  - The first line of nonspecific defense is the intact skin and mucous membranes, including the physical barrier formed by these tissues and the microbe-attacking enzyme lysozyme; mucus; ciliated cells lining the upper respiratory system; and gastric juices.
  - The second line of nonspecific defense is primarily dependent upon neutrophils and macrophages, phagocytic white cells in the blood and lymph. Natural killer cells also participate in nonspecific defense.
  - The most important antimicrobial proteins in the blood and tissues are interferons and the complement system. Interferons are proteins secreted by virus-infected cells that inhibit neighboring cells from making new viruses. Complement proteins are involved in nonspecific and specific defense.
  - A local inflammatory response is triggered by tissue damage. Injured cells release histamine, a chemical signal that dilates blood vessels and increases capillary permeability, allowing large numbers of phagocytic white blood cells to enter the interstitial fluid.
- The immune system defends the body against specific invaders: *an overview* (pp. 856–858; FIGURE 39.15)
  - The immune system recognizes foreign microbes, toxins, or transplanted tissues as not belonging to itself and develops an immune response to inactivate or destroy the specific type of invader.
  - A foreign substance that elicits an immune response is called an antigen. The immune system responds to antigens by activating specialized lymphocytes and producing specific proteins called antibodies.
  - Immunity is the result of the immune system's enhanced response to a previously encountered pathogen. Active immunity can be acquired by exposure to an actual disease or to a vaccine that simulates a disease. Passive immunity can be acquired by administering antibodies formed in others, or it can be passed from mother to child via the placenta and milk.
  - The immune system has two main functional branches. Humoral immunity is based on circulation of antibodies in the blood and lymph, and defends against free viruses, bacteria, and other extracellular threats. Cell-mediated immunity defends against intracellular pathogens by destroying infected cells. Cell-mediated immunity also reacts against transplanted tissue and cancer cells.
  - Lymphocytes, which originate in bone marrow, are the main cells of the immune system. B lymphocytes mature in the bone marrow and function in humoral immunity. T lymphocytes mature in the thymus and function mainly in cell-mediated immunity.
- Clonal selection of lymphocytes is the cellular basis of immunological specificity and diversity (p. 859, FIGURE 39.7)
  - Clonal selection occurs when a lymphocyte is activated by the binding of an antigen to its specific antigen receptor and proliferates to produce a clone of infection-fighting effector cells, all specific for that particular antigen.
- Memory cells function in secondary immune responses (p. 860, FIGURE 39.8)
  - Proliferation of a specific lymphocyte clone (the primary immune response) also produces the long-lived memory cells responsible for the enhanced response (secondary immune response) to future exposure to the antigen.
- Molecular markers on cell surfaces function in self/nonself recognition (p. 860)
  - Self-tolerance develops as lymphocytes bearing receptors for native molecules are destroyed or are rendered nonresponsive. Self cells are marked by molecules of the major histocompatibility complex (MHC).
- In the humoral response, B cells defend against pathogens in body fluids by generating specific antibodies (pp. 860–866; FIGURES 39.9, 39.12)
  - In humoral immunity, the production of plasma cells often depends on the cooperation of macrophages functioning as antigen-presenting cells (APCs) and specialized T cells called helper T cells. The APC displays antigenic fragments along with class II MHC markers. The helper T cell functions as a liaison between the APC and a B cell, which is activated in the process. The B cell then differentiates into memory cells and plasma cells. Plasma cells secrete enormous numbers of specific antibodies.
  - Most antigens are proteins or large polysaccharides; they may have many epitopes (antigenic determinants) on their surfaces.
  - Antibodies constitute a class of proteins called immunoglobulins (Ig). The variable region of the Ig molecule binds specifically to a certain antigenic determinant.
  - The constant region of the Ig molecule is the same for all antibodies of the same class. There are five major Ig classes: IgG, IgM, IgA, IgD, and IgE.
  - An antibody does not usually destroy an antigen directly but targets it for elimination by complement or phagocytes.
  - Monoclonal antibody production is a method for making pure antibodies on an industrial scale for use in diagnostic tests, treatment, and research.
- In the cell-mediated response, T cells defend against intracellular pathogens (pp. 867–870; FIGURES 39.13, 39.14)
  - Cell-mediated immunity is mainly a function of two types of T cells: helper T cells (T_H) and cytotoxic T cells (T_C).
  - A T cell is activated when its receptor binds to a specific MHC-antigen complex.
  - Cell secretions called cytokines enable T_H cells to stimulate both humoral and cell-mediated immunity, activating B cells and cytotoxic T cells.
  - Cytotoxic T cells kill infected or cancerous cells with perforin, which opens a lesion in the target cell.
  - Suppressor T cells (T_S) may turn off the immune response when antigen is no longer present.

- **Complement proteins participate in both nonspecific and specific defenses** (pp. 870–871, FIGURE 39.16)

  Complement, a group of blood proteins, can lyse a target cell by combining with antibodies (the classical pathway). In the alternative pathway, complement functions as an antimicrobial defense without the help of antibodies.

- **The immune system's capacity to distinguish self from nonself is critical in blood tranfusion and transplantation** (pp. 871–872)

  Some of the antigens on red blood cells determine whether a person has type A, B, AB, or O blood. Antibodies to foreign blood types already exist in the body. If incompatible blood is transfused, the transfused cells are killed by antibodies and complement proteins. The Rh factor, another red blood cell antigen, may create difficulties when an Rh-negative mother carries successive Rh-positive fetuses. Because the immune system rejects transplanted tissue and organs as nonself, immunosuppressive drugs are usually used after a transplant.

- **Abnormal immune function leads to disease states** (pp. 872–875)

  Sometimes the immune system goes awry and turns against itself, leading to autoimmune diseases such as rheumatoid arthritis and insulin-dependent diabetes. In allergies such as hay fever, histamine is released from mast cells. This release (degranulation) is triggered by an allergen, such as pollen. Some people are naturally deficient in humoral or cell-mediated immune defenses, or both. Acquired immunodeficiency syndrome (AIDS) is caused by the destruction of CD4-bearing T cells and other cells by HIV, the human immunodeficiency virus, over a period of years. AIDS, the final stage of this process, is marked by a low level of helper T cells and opportunistic diseases characteristic of a deficient cell-mediated immune response.

- **Invertebrates exhibit a rudimentary immune system** (p. 875)

  Invertebrates have the ability to distinguish between self and nonself. Amoeboid cells called coelomocytes can identify and destroy foreign substances. Experiments with earthworms show that their defense systems form a memory response to tissue grafts.

SELF-QUIZ

1. Which of the following molecules is *incorrectly* paired with its source?
   a. lysozyme—saliva
   b. histamine—injured cells
   c. interferons—virus-infected cells
   d. immunoglobulins—neutrophils
   e. interleukin-2—helper T cell

2. Epitopes (antigenic determinants) bind to which portions of an antibody?
   a. variable regions
   b. constant regions
   c. only light chains
   d. only heavy chains
   e. the stem region

3. Which of the following is *not* characteristic of the early stages of a localized inflammatory response?
   a. increased permeability of capillaries
   b. attack by cytotoxic T cells
   c. release of clotting proteins
   d. release of histamine
   e. dilation of blood vessels

4. The major difference between humoral immunity and cell-mediated immunity is that
   a. humoral immunity is nonspecific, whereas cell-mediated immunity is specific for particular antigens
   b. only humoral immunity is a function of lymphocytes
   c. humoral immunity cannot function independently; it is always activated by cell-mediated immunity
   d. humoral immunity acts against free-floating antigens, whereas cell-mediated immunity works against pathogens that have entered body cells
   e. only humoral immunity displays immunological memory

5. Monoclonal antibodies are
   a. produced by clones formed from memory cells
   b. used to produce large quantities of interferon
   c. produced by cultures of hybridoma cells
   d. produced by clones of T cells fused with tumor cells
   e. produced by recombinant DNA methods

6. In which of the following does active immunity *not* result?
   a. A child receives a vaccination consisting of inactivated polio virus.
   b. An adult becomes sick with hepatitis B and does not get the disease when exposed to it again in the future.
   c. A baby born to a woman who once had chickenpox has temporary resistance to this disease.
   d. Dogs are inoculated against a particular canine viral disease by being given part of the protein coat of that virus.
   e. A health care worker accidentally pricks her finger with a needle recently used to draw blood from someone with tuberculosis (TB), has a small reaction, and some years later tests positive for the antibody against the TB pathogen.

7. Which of the following cells is *incorrectly* paired with its function?
   a. plasma cell—produces antibodies
   b. helper T cell—lyses foreign cells
   c. memory cell—rapidly proliferates into clones of effector cells when it encounters antigen
   d. macrophage—engulfs bacteria and viruses
   e. cytotoxic T cell—releases perforin that lyses infected cells

8. Which blood transfusion would agglutinate blood?
   a. A donor ⟶ A recipient
   b. A donor ⟶ O recipient
   c. A donor ⟶ AB recipient
   d. O donor ⟶ A recipient
   e. O donor ⟶ AB recipient

9. HIV compromises the immune system mainly by infecting
   a. cytotoxic T cells
   b. helper T cells
   c. suppressor T cells
   d. plasma cells
   e. B cells

10. The body produces antibodies complementary to foreign antigens. The process by which the body comes up with the correct antibodies to fight a given disease is most like which of the following?
    a. going to a tailor and having a suit made to fit you
    b. ordering the lunch special at a restaurant without looking at the menu
    c. going to a shoe store and trying on shoes until you find a pair that fits
    d. picking the first video that you haven't already seen
    e. selecting a lottery winner by means of a random drawing

CHALLENGE QUESTIONS

1. Assuming that immunological memory is intact, how can you explain people getting colds or flu year after year?
2. Explain why the passive immunity transferred from mother to child is only temporary.
3. Write a paragraph highlighting the central role of helper T cells in humoral and cell-mediated immunity.
4. Compare and contrast nonspecific defense against pathogens with the specific defense system (immune system).

SCIENCE, TECHNOLOGY, AND SOCIETY

1. Concern is increasing over the rising rate of HIV infection among teenagers. Schools in some large cities have instituted programs to make condoms available to students, along with advice and counseling about safer sex. These plans have divided school boards and communities. Some citizens and some church groups are opposed to giving condoms to students, because it might appear to encourage their sexual activity. Many school and public health officials view the situation differently. Former New York City Schools Chancellor Joseph Fernandez put it this way: "This is not an issue of morality. It is a matter of life and death." All agree that the spread of AIDS is a serious problem. The heart of the controversy seems to be whether the schools should take such a direct role in this area of student life. What do you think the schools' role should be?
2. New immunosuppressant drugs have allowed researchers to attempt transplants of tissues and organs from animals to humans. In 1992, the first baboon-to-human liver transplant was performed. The patient was a 35-year-old man whose liver had been destroyed by hepatitis B. A human transplant was not possible because the virus would have attacked the new liver. Scientists believe that the human hepatitis B virus does not infect baboons. The hospital where the operation took place was picketed by animal rights protesters opposed to such transplants, carrying signs saying "Baboons and Humans: Both Victims" and "Animals Are Not Ours to Experiment On," and chanting "Animals are not spare parts!" Do you think sacrificing the life of an animal is justified in this situation? Why or why not? How do you feel about the routine use of animals as sources of organs for humans? What other scientific and medical uses of animals are justified or unjustified, in your opinion? Why?

FURTHER READING

Aldhous, P. "HIV's War of Attrition." *New Scientist,* May 13, 1995. New evidence that HIV has no latency period; the immune system battles the virus throughout the long infection.

Beardsley, T. "Better Than a Cure." *Scientific American,* January 1995. Why vaccination is the most cost-effective strategy in medicine.

Johnson, H., J. Russell, and C. Pontzer. "Superantigens in Human Disease." *Scientific American,* April 1992. What causes anaphylactic shock?

"Life, Death, and The Immune System." *Scientific American,* September 1993. A special issue devoted to the immune system.

Oliwenstein, L. "The Bug That Can Say No." *Discover,* April 1992. Immunity in cockroaches is providing clues about the evolution of immune systems.

Robbins, A., and P. Freeman. "Obstacles to Developing Vaccines for the Third World." *Scientific American,* November 1988. Who will pay for immunizing the children of developing countries?

"Vaccines: Frontiers in Medicine." *Science,* September 2, 1994. A special section devoted to recent advances in the development of a number of important vaccines.

CHAPTER 40

CONTROLLING THE INTERNAL ENVIRONMENT

KEY CONCEPTS

- Homeostatic mechanisms protect an animal's internal environment from harmful fluctuations: *an overview*
- Cells require a balance between water uptake and loss
- Osmoregulation depends on transport epithelia
- Tubular systems function in osmoregulation and excretion in many invertebrates
- The kidneys of most vertebrates are compact organs with many excretory tubules
- The kidney's transport epithelia regulate the composition of blood
- The water-conserving ability of the mammalian kidney is a key terrestrial adaptation
- Diverse adaptations of the vertebrate kidney have evolved in different habitats
- An animal's nitrogenous wastes are correlated with its phylogeny and habitat
- Thermoregulation maintains body temperature within a range conducive to metabolism
- Ectotherms derive body heat mainly from their surroundings and endotherms derive it mainly from metabolism
- Thermoregulation involves physiological and behavioral adjustments
- Comparative physiology reveals diverse mechanisms of thermoregulation
- Regulatory systems interact in the maintenance of homeostasis

Most animals can survive fluctuations in the external environment that are more extreme than any of their individual cells could tolerate. During the course of a day, a human may be exposed to substantial changes in outside temperatures but will die if the internal body temperature goes more than a few degrees above or below an average of about 37°C. The spring peeper, a North American tree frog (shown in the photograph here), often spends an entire winter with much of its body frozen. Most of its blood and interstitial fluid turns to ice, but its cells and a film of water surrounding them are kept from freezing by specialized proteins and a high level of glucose throughout the frog's body. Without these protectants, ice crystals would rupture the cell membranes, and the animal would die.

Wintering spring peepers illustrate several themes central to our study of animals. The theme of bioenergetics is evident in that being frozen and inactive enables a frog to go through the entire winter without eating; its metabolic rate is so low its cells survive on the energy resources they have when the animal freezes. Illustrating the theme of short-term physiological adjustment to environmental change, a spring peeper's organ systems respond quickly to low temperature. For example, the liver secretes freeze-protectant chemicals into the animal's blood as soon as the frog's skin starts freezing. Finally, the mechanisms that maintain homeostasis, keeping the frog's internal environment within ranges that cells can tolerate, are adaptations that have evolved in populations facing certain environmental problems.

This chapter concentrates on homeostasis, focusing on ***osmoregulation,*** *how animals regulate solute balance and the gain and loss of water;* ***excretion,*** *how they get rid of the nitrogen-containing waste products of metabolism such as urea; and* ***thermoregulation,*** *how they maintain internal temperature within a tolerable range.*

Homeostatic mechanisms protect an animal's internal environment from harmful fluctuations: *an overview*

In most animals, the majority of cells are not in direct contact with the external environment but are bathed by an internal body fluid. Insects and other animals with an open circulatory system have an internal "pond" composed of hemolymph, which bathes all body cells (see Chapter 38). In vertebrates and other animals with a closed circulatory system, the internal pond is interstitial fluid serviced by blood. (Note that animals with a closed

FIGURE 40.1

Homeostasis. Represented here by a gray rectangle, an animal has homeostatic mechanisms (purple box) that regulate the fluid environment bathing its cells. Conditions of the external environment (blue), such as the salt concentration in the water where the animal lives, may fluctuate widely. By contrast, conditions of the internal environment vary within only a narrow range. Regulating the internal environment as outside forces tend to change it, homeostatic mechanisms maintain conditions within a range in which the animal's metabolic processes can occur.

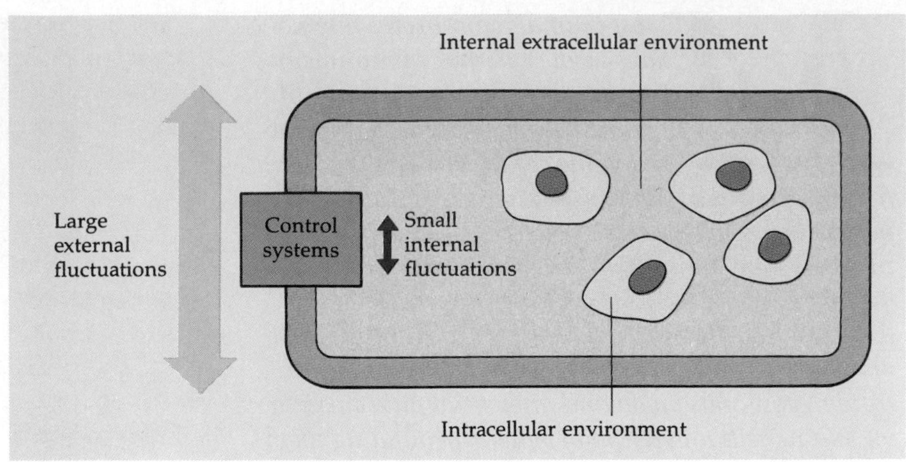

circulatory system actually have three internal fluid compartments: an intracellular compartment, consisting of the cytosol of cells; and two extracellular compartments, the blood plasma and interstitial fluid.) Homeostatic mechanisms temper changes in an animal's body fluids, cushioning them from the potentially harmful impact of fluctuations in the external environment (FIGURE 40.1).

Cells require a balance between water uptake and loss

Think about the ways water enters and leaves your body. We acquire most of our water in our food and drink and a smaller amount by oxidative metabolism. ("Metabolic water" is produced by cellular respiration when electrons and hydrogen are added to oxygen; see Chapter 9.) We lose water by urinating and defecating, and by evaporative loss due to sweating and breathing. For aquatic animals, evaporation is unimportant, but these animals experience the uptake and loss of water across the body surface by osmosis. Even if an animal is protected by a covering that impedes water loss or gain, specialized epithelia that must be exposed to the environment in order to exchange gases (such as gills, lungs, and tracheae) cannot be waterproof.

Whether an animal inhabits land, fresh water, or salt water, or moves back and forth between these environments, one general problem occurs: The cells of the animal cannot survive a net water gain or loss. Water continuously enters and leaves an animal cell across the plasma membrane; however, uptake and loss must balance. Animal cells swell and burst if there is a net uptake of water or shrivel and die if there is a net loss of water. Even an ice-bound spring peeper, whose blood and body fluids are mostly frozen, has its cells bathed in a film of interstitial fluid that is osmotically balanced with the cytosol. Glucose and proteins in the interstitial fluid ensure the balance.

Recall from Chapter 8 that osmosis, a special case of diffusion, is the movement of water across a selectively permeable membrane. It occurs whenever two solutions separated by the membrane differ in total solute concentration, or **osmolarity** (total solute concentration expressed as molarity, or moles of solute per liter of solution; see Chapter 3). The unit of measurement for osmolarity used in this chapter is milliosmoles per liter (mosm/L). This unit is equivalent to a total solute concentration of 10^{-3} *M*. For example, the osmolarity of human blood is about 300 mosm/L, while seawater commonly has an osmolarity of about 1000 mosm/L. Two solutions are said to be isotonic if they are equal in osmolarity. There is no *net* osmosis between isotonic solutions. When two solutions differ in osmolarity, the one with the greater concentration of solutes is referred to as hypertonic and the more dilute solution as hypotonic. Water will flow by osmosis across a membrane from a hypotonic solution to a hypertonic one.

Osmoconformers and Osmoregulators

There are two basic solutions to the problem of balancing water gain with water loss. One solution for a marine animal is to be isotonic with its saltwater environment. Such animals, which do not actively adjust their internal osmolarity, are known as **osmoconformers.** By contrast, animals whose body fluids are not isotonic with the outside environment, called **osmoregulators,** must either discharge excess water if they live in a hypotonic environment or continuously take in water to offset osmotic loss if they inhabit a hypertonic environment. A net movement of water occurs only in an osmotic gradient (from a region of lower osmolarity to a region of higher osmolarity), and osmoregulators must expend energy to maintain osmotic gradients, to move water either in or out. They do so by manipulating solute concentrations in their body fluids.

The ability to osmoregulate enables animals to live, for example, in fresh water, where the osmolarity is too low to support cellular life, and on land, where water is usually in short supply. However, osmoregulation is energetically costly, depending mainly on how different an animal's osmolarity is from its surroundings and on how much membrane-transport work is required to actively transport solutes. For example, osmoregulation accounts for nearly 5% of the resting metabolic rate of many marine and freshwater bony fishes. For the brine shrimp, a small crustacean that lives in the Great Salt Lake in Utah and other extremely salty environments, the cost is as high as 30% of the resting metabolic rate.

Most animals, whether osmoconformers or osmoregulators, cannot tolerate substantial changes in external osmolarity. Such animals are said to be **stenohaline** (Gr. *stenos,* "narrow"; haline refers to salt). However, some animals, called **euryhaline** animals (Gr. *eurys,* "broad"), do survive radical fluctuations of osmolarity in their surroundings. They either conform to the changes or regulate their internal osmolarity within a narrow range even as the external osmolarity changes. One example of a euryhaline animal, a bony fish called the tilapia, a native of Africa grown widely for human food, can adjust to any salt concentrations between fresh water and twice that of seawater.

All freshwater animals and many marine animals are osmoregulators, maintaining an internal osmolarity that differs from the surrounding water. Humans and other terrestrial animals, also osmoregulators, must compensate for water loss. With these general approaches to water balance in mind, we can now survey some specific examples of osmoregulatory adaptations to various environments.

Maintaining Water Balance in Different Environments

Marine Animals. Animals first evolved in the sea, which remains the most common environment for the majority of phyla. Most marine invertebrates are osmoconformers; however, even these animals differ from seawater in their concentrations of specific salts. The difference is usually slight, but in some cases it is substantial. Thus, an animal that conforms to the osmolarity of its surroundings may still regulate its internal composition of ions.

Among the vertebrates, the hagfishes (jawless members of the class Agnatha) are isotonic with the surrounding seawater, but most marine vertebrates osmoregulate. Sharks and most other cartilaginous fishes (class Chondrichthyes) maintain internal salt concentrations lower than that of seawater. Their kidneys excrete some salts, and a salt-excreting organ called the rectal gland excretes sodium chloride out of the body through the anus. Researchers have learned much about membrane function by studying the active transport of ions by shark rectal gland cells.

Despite its relatively low salt concentration, a marine shark is slightly hypertonic to seawater. It does not drink water, and the water that enters its body by osmosis is disposed of in **urine,** the waste fluid formed by the excretory organs, the **kidneys.** If you have ever eaten shark meat, you know it should be soaked in fresh water before cooking. Soaking washes out urea, a waste product of nitrogen metabolism. A large amount of urea dissolved in the body fluids accounts for a shark's being slightly hypertonic to seawater. Sharks also produce and retain another organic compound, trimethylamine oxide (TMAO), which protects proteins from being damaged by the urea.

Bony fishes (class Osteichthyes) evolved from ancestors that entered freshwater habitats. In their subsequent evolution, many groups of bony fishes became marine but internally remained more similar to fresh water in osmolarity. Marine bony fishes constantly lose water by osmosis to their hypertonic surroundings (FIGURE 40.2a) They compensate by drinking large amounts of seawater, pumping out excess salts, and excreting urine in relatively small amounts.

Freshwater Animals. The osmoregulatory problems of freshwater animals are opposite those of marine animals. Freshwater animals are constantly taking in water by osmosis because the osmolarity of their internal fluids is much higher than that of their surroundings. Freshwater protozoa such as *Amoeba* and *Paramecium* have contractile vacuoles that pump out excess water (see FIGURE 26.10c). Many freshwater animals, including fishes, bail out water by excreting large amounts of very dilute urine and regaining lost salts in their food or by active uptake from their surroundings (FIGURE 40.2b).

Salmon and other fishes that migrate between seawater and fresh water are euryhaline. While in the ocean, salmon drink seawater and excrete excess salt from the gills, osmoregulating like other marine fishes. With their migration to fresh water, salmon cease drinking, and their gills start taking up salt from the dilute environment, as do the gills of other freshwater fishes.

Life in Temporary Waters. Dehydration dooms most animals, but some aquatic invertebrates living in temporary ponds and films of water around soil particles can lose almost all their body water and survive in a dormant state when their habitats dry up. This remarkable adaptation is called **anhydrobiosis** ("life without water"), or cryptobiosis ("hidden life"). Among the most striking examples are the tardigrades, or water bears, tiny

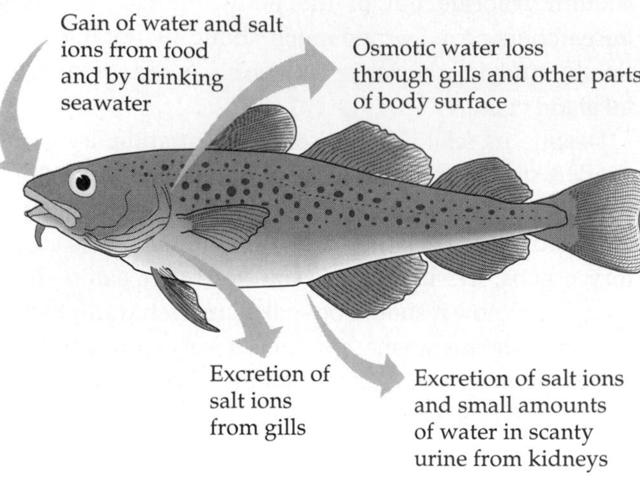

(a) Osmoregulation in a saltwater fish

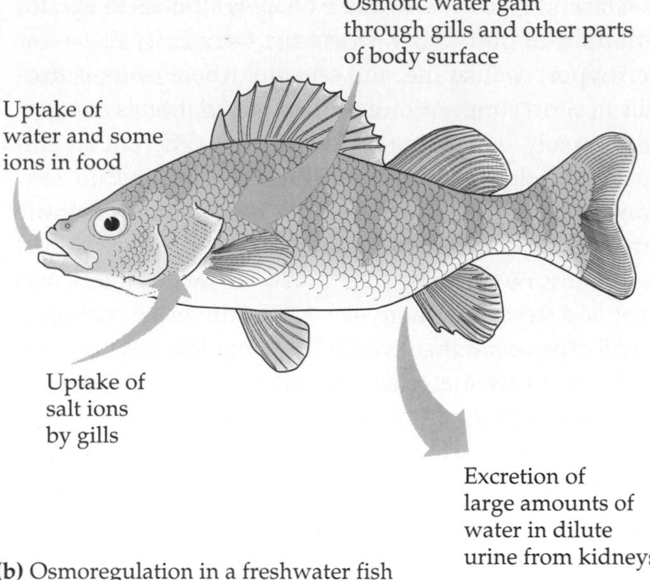

(b) Osmoregulation in a freshwater fish

FIGURE 40.2

Osmoregulation in marine and freshwater bony fishes: a comparison. (**a**) A marine fish, such as this cod, is hypotonic to the surrounding seawater and therefore constantly loses water by osmosis. The fish drinks large amounts of seawater; its gills and general body surface dispose of sodium chloride (special cells called chloride cells actively transport Cl^- ions out, and Na^+ ions follow passively); and its kidneys dispose of excess calcium (Ca^{2+}), magnesium (Mg^{2+}), and sulfate (SO_4^{2-}) ions while excreting only small amounts of water. (**b**) Facing the opposite situation, a freshwater fish, such as this perch, constantly gains water because it is hypertonic to its surroundings. The water uptake is balanced by the excretion of large amounts of urine that is hypotonic to the body fluids. Salts lost in the urine are replenished by foods and by uptake across the gills; chloride cells in the gills actively transport Cl^- ions in.

invertebrates less than 1 mm long (FIGURE 40.3). In their active, hydrated state, these animals contain about 85% water by weight but can dehydrate to less than 2% water and survive in an inactive state, dry as dust, for a decade or more. Just add water, and within minutes the rehydrated tardigrades are moving about and feeding.

A dehydrated animal and a frozen one (such as the wintering frog discussed earlier) face similar problems. They both must have adaptations that keep cell membranes intact. Researchers are just beginning to learn how tardigrades survive drying out, but studies of roundworms (class Nematoda) that are anhydrobiotic show that dehydrated individuals contain a large amount of sugars, especially a disaccharide called trehalose. Consisting of two glucose units, trehalose seems to protect the cells by replacing the water associated with membranes and proteins. Many insects that survive freezing in the winter also utilize trehalose as a membrane protectant.

Terrestrial Animals. Few truly terrestrial animals are capable of anhydrobiosis. Humans, for example, die if they lose about 12% of their body water. The threat of desiccation is perhaps the most important problem confronting terrestrial life, both plants and animals. The

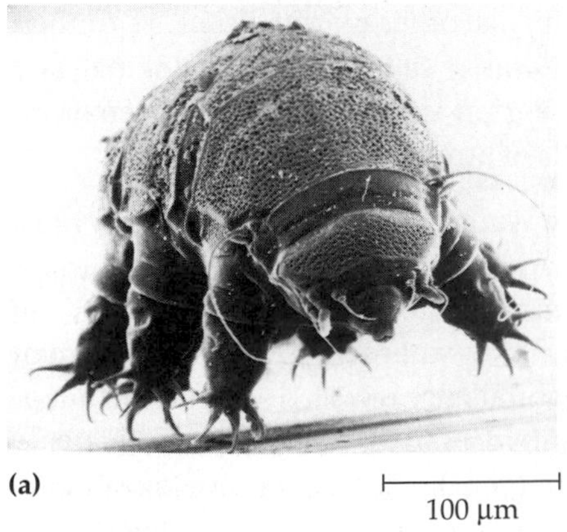

(a)

(b)

FIGURE 40.3

Anhydrobiosis. (**a**) Tardigrades (water bears) inhabit temporary ponds and films of water in soil and on moist plants. Even a moss-covered roof provides a suitable habitat. A water bear is active, feeding on plant juices, when its habitat is wet. (**b**) When it dries up, the animal can lose more than 95% of its body water and survive in a dehydrated, inactive state for decades (SEMs). (The dehydrated tardigrade on the right is actually smaller than the one on the left, as you would expect; the scale of the right photograph is about twice that of the left.)

severity of this problem may be one reason only two groups of animals, arthropods and vertebrates, have colonized the land with great success. (Although other phyla have some representatives on land, most of their species are aquatic.)

What *are* some of the evolutionary adaptations that have made it possible for animals, which consist mostly of water, to survive on land? Much as a waxy cuticle contributes to the success of plants on land, most terrestrial animals are covered by relatively impervious surfaces that help prevent dehydration. Examples are the waxy layers of the exoskeletons of insects, the shells of land snails, and the multiple layers of dead, keratinized skin cells covering most terrestrial vertebrates. Still, most terrestrial animals lose a considerable amount of water that must be replenished by drinking and eating moist foods. Behavioral adaptations, such as nervous and hormonal mechanisms that control thirst, are important osmoregulatory mechanisms in land-dwelling animals (discussed later in the chapter). Many terrestrial animals, especially in deserts, are nocturnal, another important behavioral adaptation that reduces dehydration. The kidneys and other excretory organs of terrestrial animals often exhibit adaptations that help conserve water (also discussed later in the chapter). Some mammals are so well adapted to minimizing water loss that they can survive in deserts without drinking. For example, kangaroo rats lose so little water that they can recover 90% of the loss by using metabolic water (FIGURE 40.4).

(a)

		Water Balance in a Kangaroo Rat	Water Balance in a Human
Water gain (mL/day)	Ingested in liquid	0	1500 (60%)
	Ingested in food	6.0 (10%)	750 (30%)
	Derived from metabolism	54.0 (90%)	250 (10%)
		60.0 (100%)	2500 (100%)
Water loss (mL/day)	Evaporation	43.9 (73%)	900 (36%)
	Urine	13.5 (23%)	1500 (60%)
	Feces	2.6 (4%)	100 (4%)
		60.0 (100%)	2500 (100%)

Source: Kangaroo rat data from Schmidt-Nielsen. *Animal Physiology: Adaptation and Environment,* 4th ed. Cambridge: Cambridge University Press, 1990, p. 339.

(b)

FIGURE 40.4
Water balance in two terrestrial mammals. (**a**) Kangaroo rats, which live in American southwestern deserts, eat mostly dry seeds and do not drink water. (**b**) A kangaroo rat loses water mainly by evaporation during gas exchange and gains water mainly from cellular metabolism (water produced when hydrogen and oxygen are combined during aerobic respiration). In contrast, a human loses a large amount of water in urine and regains it mostly in food and drink.

Osmoregulation depends on transport epithelia

Although the problems of water balance in environments as diverse as salt water, fresh water, and land are very different, the solutions in osmoregulators have a common theme: the regulation of solute movement (and hence water movement, which follows solutes by osmosis) between the animals' internal fluids and the external environment. Specialized epithelia, called transport epithelia, regulate the solute movements.

A **transport epithelium** is usually a single sheet of cells facing the external environment or some channel that leads to the exterior through an opening on the body surface. The cells of the epithelium are joined by impermeable tight junctions (see FIGURE 7.34a), forming a continuous barrier at the tissue-environment boundary. This configuration, another example of how form fits function, ensures that any solute passing between the extracellular fluid and the environment must pass through the selectively permeable membranes of cells. The molecular composition of the epithelium's plasma membrane determines the specific osmoregulatory functions. Transport epithelia vary in their passive permeabilities to water and salts, and in the number, type, and orientation of membrane proteins responsible for active transport. For example, differences in membrane structure and function account for the transport epithelia of the gills of marine fishes pumping salt outward, while the gills of freshwater fishes pump salt inward.

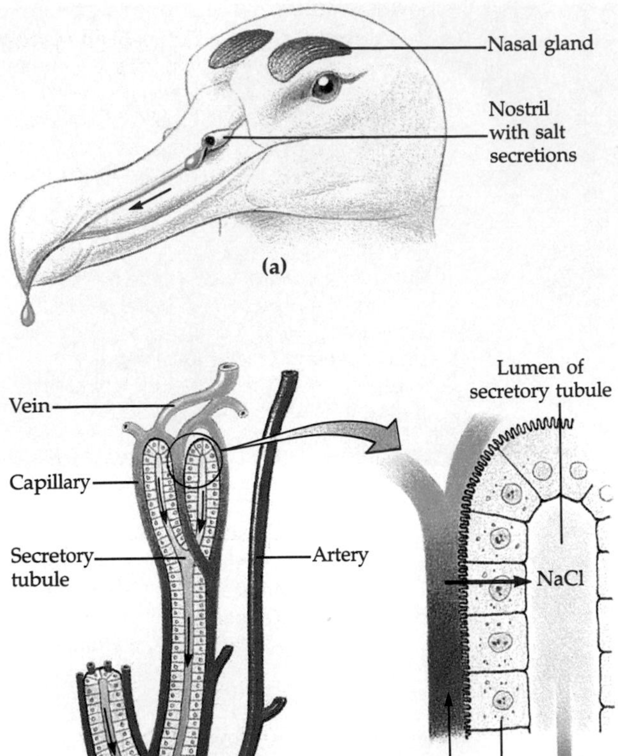

FIGURE 40.5

Salt-excreting glands in birds. (**a**) Many marine birds, such as this albatross, can drink seawater because they have a pair of salt-excreting nasal glands. The glands empty via a duct into the nostrils, and the salty solution runs along a ridge to the tip of the beak or is exhaled in a fine mist from the nostrils. (**b**) This drawing shows one of several thousand secretory tubules in a salt-excreting gland. Each tubule is lined by a transport epithelium, surrounded by capillaries, and drains into a central duct. (**c**) The secretory cells of the transport epithelium pump salt from the blood into the tubules. Notice that blood flows counter to the flow of salt secretion. By maintaining a concentration gradient of salt in the tubule (graded blue shading), this countercurrent system enhances salt transfer from the blood to the lumen of the tubule (see Chapter 38).

One of the most efficient transport epithelia is found in marine birds that spend months or years at sea and obtain water from the ocean (FIGURE 40.5). They have specialized salt-excreting glands whose transport epithelia are dedicated exclusively to osmoregulation—maintaining salt and water balance. In other cases, transport epithelia function in the excretion of nitrogenous wastes as well as osmoregulation. In the excretory systems of most animals, transport epithelia are arranged into tubular networks with extensive surface areas, as in the vertebrate kidney and in various invertebrate excretory systems.

Tubular systems function in osmoregulation and excretion in many invertebrates

Protonephridia: The Flame-Bulb System of Flatworms

Flatworms (phylum Platyhelminthes) have simple tubular excretory systems called protonephridia. A **protonephridium** is a network of closed tubules lacking internal openings (FIGURE 40.6). The tubules branch throughout the body, and the smallest branches are capped by a cellular unit called a flame bulb. Interstitial fluid bathing the tissues of the animal passes through the flame bulb and enters the tubule system. The flame bulb has a tuft of cilia projecting into the tubule, and the beating of these cilia propels fluid along the tubule, away from the flame bulb. (The beating cilia look like a flickering flame, for which these cellular structures are named.) In planaria, tributaries of the tubular system drain into excretory ducts that empty into the external environment through numerous openings called nephridiopores.

The excreted fluid is very dilute in the case of freshwater flatworms, helping balance the osmotic uptake of water from the hypotonic environment. The cellular mechanisms for this osmoregulation are unknown, however. It is likely that the lining of the tubules is a transport epithelium specialized for reabsorbing certain salts before the fluid exits the body. The flame-bulb systems of freshwater flatworms function mainly in osmoregulation; most metabolic wastes diffuse out from the body surface or are excreted into the gastrovascular cavity and eliminated through the mouth (see Chapter 36). However, in some parasitic flatworms, which are isotonic to the surrounding fluids of their host organisms, protonephridia function mainly in excretion, disposing of nitrogenous wastes. This difference in function illustrates how anatomical equipment common to a group of organisms can be adapted in diverse ways by evolution in different environments. Protonephridia are also found in rotifers, some annelids, the larvae of mollusks, and lancelets, which are invertebrate chordates. (See Chapters 29 and 30 to review these animal phyla.)

Metanephridia of Earthworms

In contrast to protonephrida, another type of tubular excretory system, the **metanephridium,** has internal openings that collect body fluids. Metanephridia are found in most annelids, including earthworms (FIGURE 40.7). Each segment of a worm has a pair of metanephridia, which are tubules immersed in coelomic fluid and enveloped by a network of capillaries. (Recall that earthworms have a closed circulatory system, meaning that blood is re-

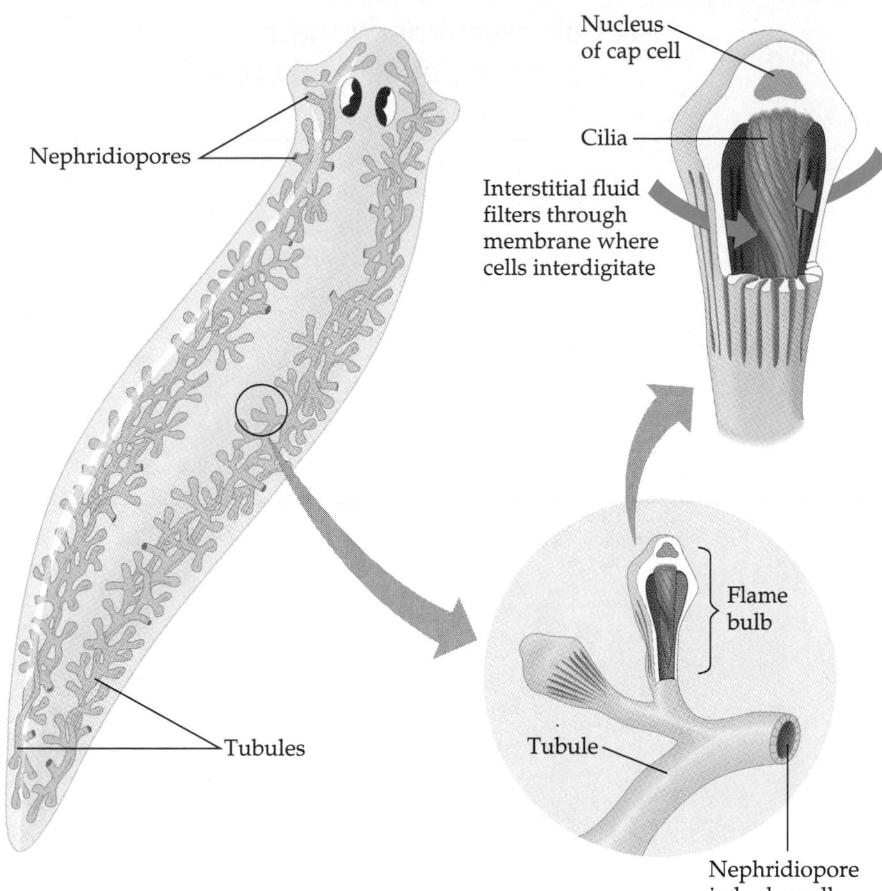

FIGURE 40.6
Protonephridia: the flame-bulb system of a planarian. Protonephridia are branching internal tubules that function mainly in osmoregulation. A single cell caps the internal end of each tubule and interlocks with a tubule cell, forming a flame bulb. Interstitial fluid filters into the lumen of the tubule across the interdigitating membranes of the cap cells and tubule cells. Cilia on the cap cell keep the fluid moving through the tubules. The planarian's tubules produce dilute urine, which passes out through small openings called nephridiopores. Thus, this freshwater flatworm balances the osmotic uptake of water from its hypotonic environment.

stricted to vessels.) The internal opening of a metanephridium is surrounded by a ciliated funnel, the nephrostome, that collects coelomic fluid.

An earthworm's metanephridia have excretory and osmoregulatory functions. As the fluid moves along the tubule, the transport epithelium bordering the lumen pumps essential salts out of the tubule, and the salts are reabsorbed into the blood circulating through the capillaries. The urine that exits through the nephridiopore contains nitrogenous wastes and is hypotonic to the body fluids. By excreting this dilute urine in amounts up to 60% of the body weight of the worm per day, the metanephridia offset the continuous osmosis taking place across the skin of the animal from the damp soil. (The skin is moist and permeable and functions as a respiratory organ.)

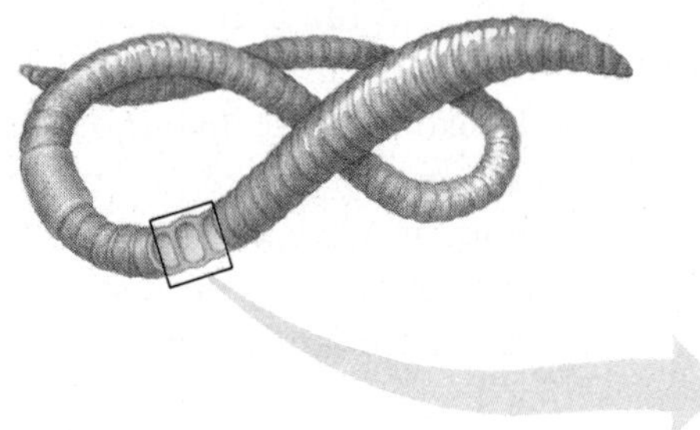

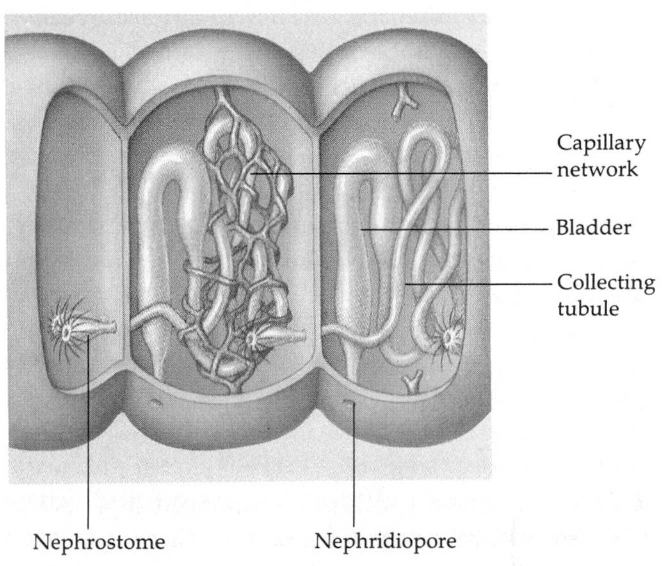

FIGURE 40.7
Metanephridia of an earthworm. Each segment of the worm contains a pair of metanephridia, which collect coelomic fluid from the adjacent anterior segment. Fluid enters the nephrostome and passes through the coiled collecting tubule, which includes a storage bladder that opens to the outside through the nephridiopore. Nitrogenous wastes remain in the fluid, but certain salts are pumped back into the blood. Balancing osmotic uptake of water through the skin, an earthworm's urine is very dilute.

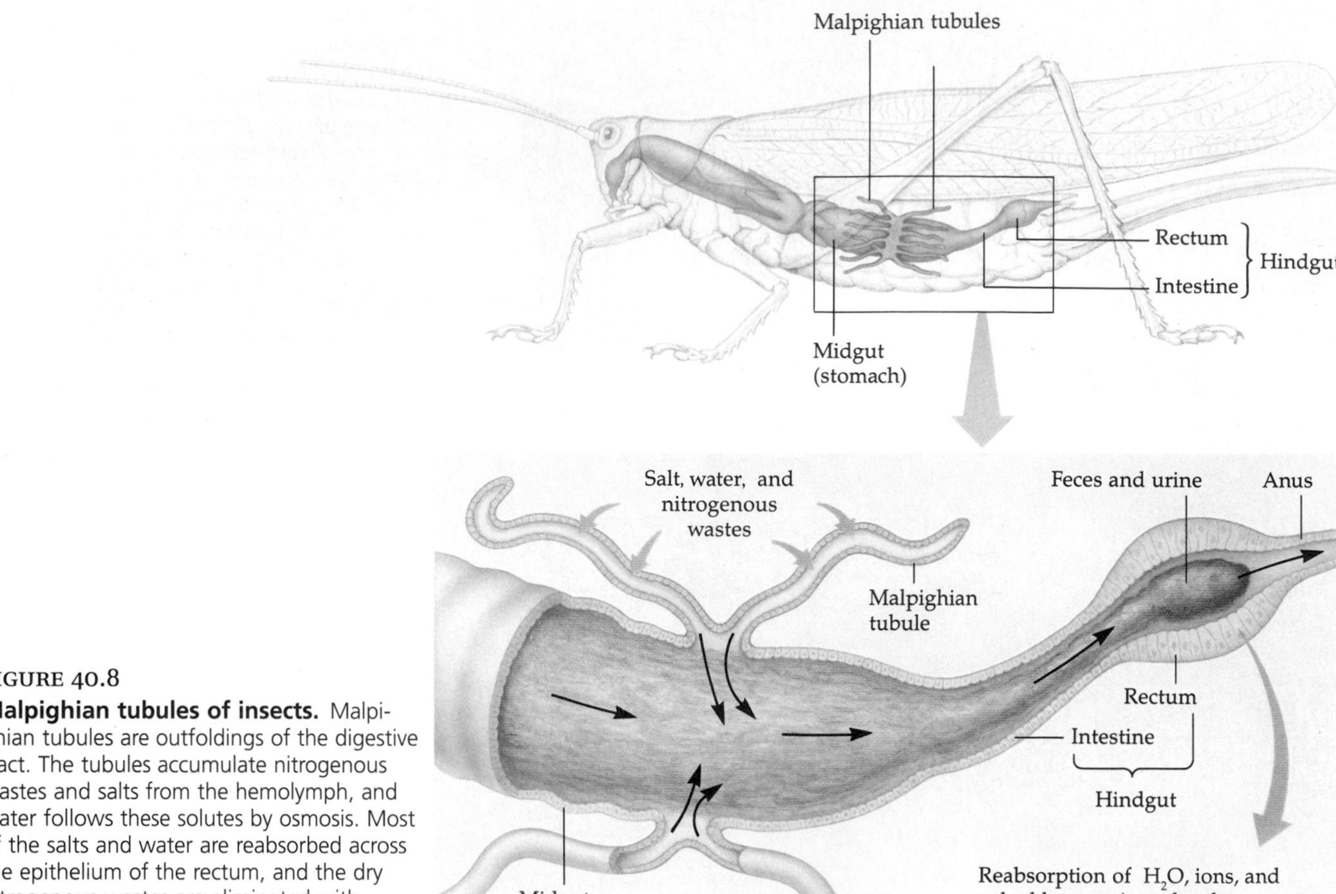

FIGURE 40.8
Malpighian tubules of insects. Malpighian tubules are outfoldings of the digestive tract. The tubules accumulate nitrogenous wastes and salts from the hemolymph, and water follows these solutes by osmosis. Most of the salts and water are reabsorbed across the epithelium of the rectum, and the dry nitrogenous wastes are eliminated with the feces.

Malpighian Tubules of Insects

Insects and other terrestrial arthropods have open circulatory systems, with tissues bathed directly in hemolymph contained in sinuses. Their excretory organs, called **Malpighian tubules,** remove nitrogenous wastes from the hemolymph and also function in osmoregulation (FIGURE 40.8). These organs open into the digestive tract at the juncture of the midgut and hindgut. The tubules, which dead-end at the tips away from the digestive tract, are immersed in hemolymph. The transport epithelium that lines a tubule pumps certain solutes, including salts and nitrogenous wastes, from the hemolymph into the lumen of the tubule. The fluid within the tubule then passes through the hindgut into the rectum. The epithelium of the rectum pumps most of the salt back into the hemolymph, and water follows the salts by osmosis. The nitrogenous wastes are eliminated as nearly dry matter along with the feces. The insect excretory system is one adaptation that has contributed to the tremendous success of these animals on land, where conserving water is essential.

The kidneys of most vertebrates are compact organs with many excretory tubules

Vertebrates evolved from a group of invertebrate chordates (see Chapter 30), and the excretory structures of their ancestors probably were segmentally arranged throughout the body like the metanephridia of earthworms. Indeed, hagfishes, which are among the most primitive living vertebrates, have kidneys with segmentally arranged excretory tubules. By contrast, the kidneys of other vertebrates are compact organs containing numerous tubules that are not segmentally arranged. A dense network of capillaries intimately associated with the tubules is also part of the kidney. In vertebrates that osmoregulate, the tubules function in both excretion and osmoregulation.

The kidneys, the blood vessels that serve them, and the structures that carry urine formed in the kidneys out of the body are the components of the vertebrate excretory system (FIGURE 40.9). We will focus first on the

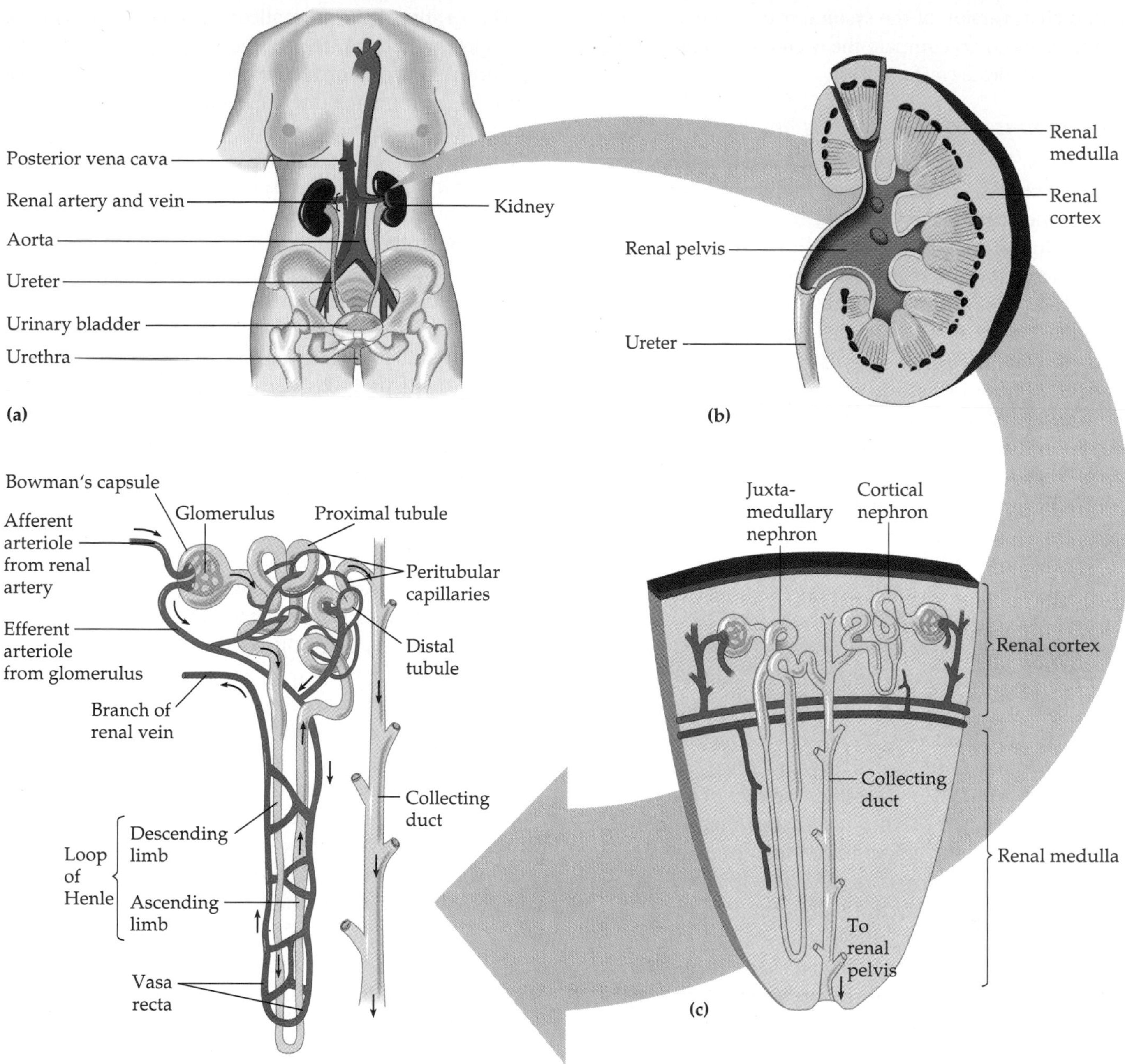

FIGURE 40.9

The human excretory system at four size scales. (**a**) The kidneys produce urine and regulate the composition of the blood. Urine is conveyed to the urinary bladder via the ureter and to the outside via the urethra. Branches of the aorta, the renal arteries, convey blood to the kidneys; renal veins drain blood from the kidneys into the posterior vena cava. (**b**) Urine is formed in two distinct regions of the kidney: the outer renal cortex and inner renal medulla. It then drains into a central chamber, the renal pelvis, and into the ureters. (**c**) Excretory tubules (nephrons and collecting ducts) and associated blood vessels pack the cortex and medulla. The human kidney has about a million nephrons, representing about 80 km of tubules. Cortical nephrons are restricted mainly to the renal cortex. Juxtamedullary nephrons have a long, hairpinlike portion that extends into the renal medulla. Several nephrons empty into each collecting duct, which drains into the renal pelvis. (**d**) Each nephron consists of a glomerulus, or capillary cluster, surrounded by Bowman's capsule; a proximal tubule; a loop of Henle; and a distal tubule. Blood enters the glomerulus via an afferent arteriole and leaves via an efferent arteriole, which conveys it to peritubular capillaries surrounding the proximal and distal tubules, and to the vasa recta, capillaries surrounding the loop of Henle. The nephrons, collecting duct, and associated blood vessels produce urine from a filtrate (water and small solutes) forced into Bowman's capsule from the glomerulus. As the filtrate travels from Bowman's capsule to the collecting duct, its chemical makeup is changed as substances pass via the interstitial fluid between the nephron and the surrounding capillaries. Filtrate processing continues in the collecting duct. The flow of blood in the vasa recta is opposite that of the filtrate in the loop of Henle (arrows).

mammalian version of the system, using humans as our example, and then compare the excretory systems of the various vertebrate classes.

The Mammalian Excretory System

In humans, the kidneys are a pair of bean-shaped organs about 10 cm long (FIGURE 40.9a). Blood enters each kidney via the **renal artery** and leaves each kidney via the **renal vein.** Although the kidneys account for less than 1% of the weight of the human body, they receive about 20% of the blood pumped with each heartbeat. Urine exits the kidney through a duct called the **ureter.** The ureters of both kidneys drain into a common **urinary bladder.** During urination, urine leaves the body from the urinary bladder through a tube called the **urethra,** which empties near the vagina in females or through the penis in males. Sphincter muscles near the junction of the urethra and the bladder control urination.

The Nephron and Associated Structures

The kidney has two distinct regions, an outer **renal cortex** and an inner **renal medulla** (FIGURE 40.9b). Packing both regions are microscopic excretory tubules called nephrons and collecting ducts, associated with tiny blood vessels (FIGURE 40.9c). The **nephron,** which is the functional unit of the vertebrate kidney, consists of a single long tubule and a ball of capillaries called the **glomerulus** (FIGURE 40.9d). The blind end of the tubule forms a cup-shaped swelling, called **Bowman's capsule,** which surrounds the glomerulus. Blood pressure forces water, urea (a nitrogenous waste), salts, and other small solutes from the blood in the glomerulus into the lumen of Bowman's capsule. Fluid in the lumen of the nephron is called the **filtrate.** From Bowman's capsule, the filtrate passes successively through three regions of the nephron: the **proximal tubule;** the **loop of Henle,** a hairpin turn with a descending limb and an ascending limb; and the **distal tubule.** The distal tubule empties into a **collecting duct,** which receives filtrate from many other nephrons. The many collecting ducts of the kidney empty into the renal pelvis.

In the human kidney, about 80% of the nephrons, the **cortical nephrons,** have reduced loops of Henle and are almost entirely confined to the renal cortex. The other 20% of the nephrons, the **juxtamedullary nephrons,** have well-developed loops that extend into the renal medulla (see FIGURE 40.9c). Only mammals and birds have juxtamedullary nephrons; the nephrons of other vertebrates lack loops of Henle. As we will soon see, the juxtamedullary nephrons play a key role in the ability of mammals to excrete urine that is hypertonic to body fluids, an adaptation that conserves water.

The nephron and the collecting duct are lined by a transport epithelium that processes the filtrate to form the urine. From about 1100 to 2000 L of blood that flows through the human kidneys each day, the nephrons and collecting ducts process about 180 L of filtrate, but the kidneys excrete only about 1.5 L of urine. The rest of the filtrate, including about 99% of the water, is reabsorbed into the blood.

Each nephron is supplied with blood by an **afferent arteriole,** a branch of the renal artery that subdivides into the capillaries of the glomerulus. The capillaries converge as they leave the glomerulus, forming an **efferent arteriole.** This vessel subdivides again into a second network of capillaries, the **peritubular capillaries.** These capillaries intermingle with the proximal and distal tubules of the nephron. Additional capillaries extend downward to form the **vasa recta,** the capillary system that serves the loop of Henle. The vasa recta is also a loop, with a descending vessel and an ascending vessel conveying blood in opposite directions.

Although the excretory tubules and their surrounding capillaries are closely associated, they do not exchange materials directly across their walls. The tubules and capillaries are immersed in interstitial fluid, through which various substances pass back and forth between the plasma within capillaries and the filtrate within the nephron tubules.

Keeping in mind the correlation between structure and function, let's now investigate how the complex organization of the excretory tubules and their blood vessels explains how the kidneys produce urine and regulate the composition of the blood.

The kidney's transport epithelia regulate the composition of blood

Production of Urine from a Blood Filtrate

The transport epithelia of nephrons and collecting ducts regulate the composition of blood by a combination of three processes that transfer material between the tubules and the capillaries that serve them: filtration, secretion, and reabsorption.

Filtration of the Blood. Blood pressure forces fluid from the capillaries of the glomerulus across the epithelium of the Bowman's capsule into the lumen of the nephron tubule. The porous capillaries, along with specialized cells of the capsule called **podocytes,** function as a filter, being permeable to water and small solutes but not to blood cells or large molecules such as plasma proteins (FIGURE 40.10). **Filtration** is nonselective with regard to small molecules; any substance small enough to be forced through the capillary wall and between the podocytes by blood pressure enters the lumen of the nephron tubule. Thus, at this point, the filtrate contains solutes such as salts, glucose, and vitamins; nitrogenous wastes such as urea; and other small molecules—a mixture that mirrors the concentrations of these substances in blood plasma.

Secretion. As filtrate travels through the nephron tubule, it is joined by substances that are transported across the tubule epithelium from the surrounding interstitial fluid. Because small molecules pass freely from the plasma within capillaries into the interstitial fluid, the net effect of renal **secretion** is the addition of plasma solutes to the filtrate within the tubule. The proximal and distal tubules are the most common sites of secretion. Unlike filtration, which is nonselective, secretion is a very selective process involving both passive and active transport.

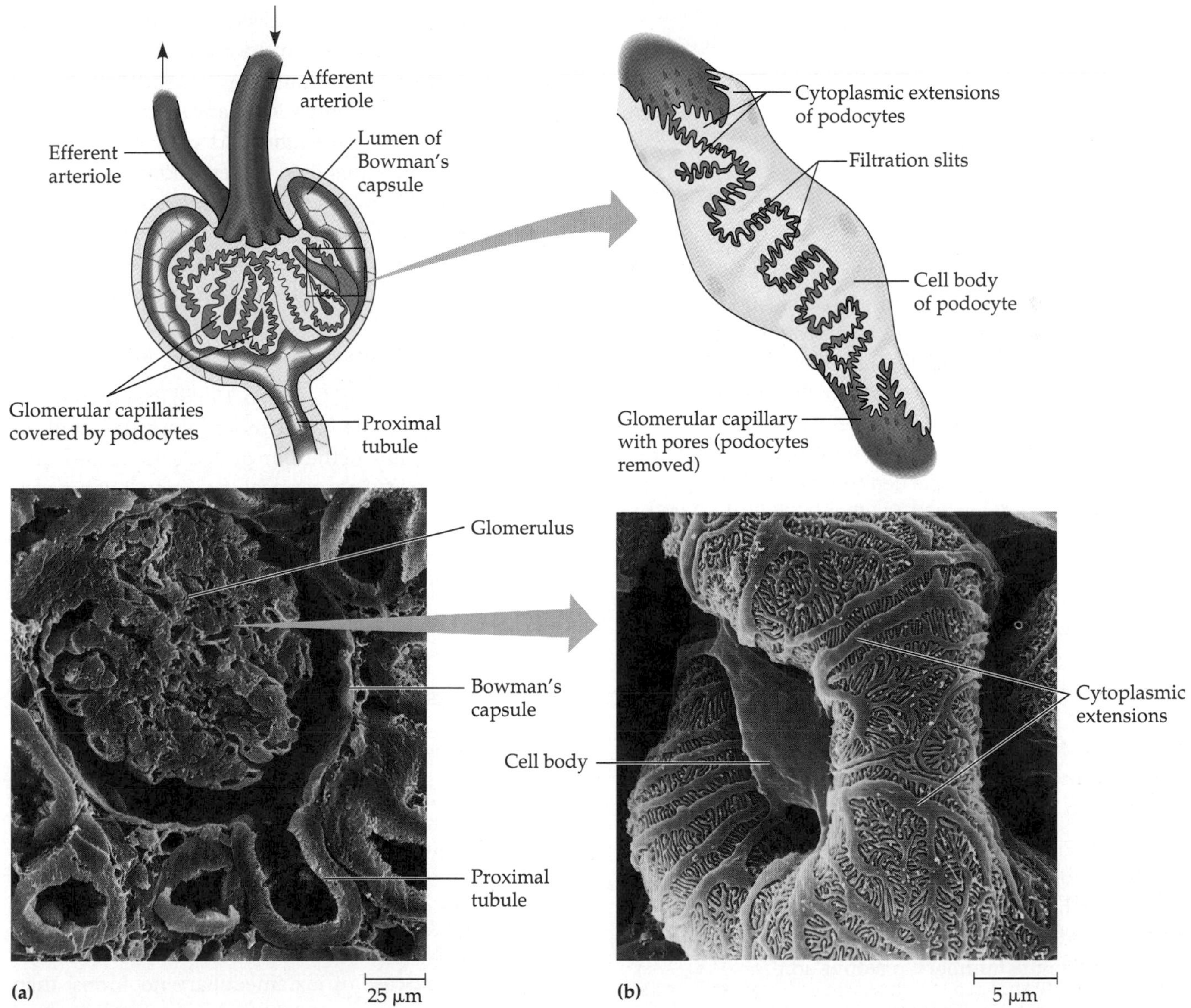

FIGURE 40.10

The glomerular filtration apparatus. (**a**) Bowman's capsule has a simple squamous epithelium making up the outer wall of the cup and an inner layer of podocytes surrounding the capillaries of the glomerulus. The glomerulus is supplied with blood by an afferent arteriole and drained by an efferent arteriole. (SEM from *Tissues and Organs: A Text-Atlas of Scanning Electron Microscopy* by Richard G. Kessel and Randy H. Kardon. W. H. Freeman and Company. Copyright © 1979.) (**b**) Cytoplasmic extensions of neighboring podocytes interdigitate, forming slits. Along with the numerous pores of the capillaries themselves, the podocyte slits function in filtering the blood, allowing blood pressure to force water and small solutes into the lumen of the capsule but keeping blood cells and macromolecules such as plasma proteins in the blood (SEM).

For example, the controlled secretion of hydrogen ions from the interstitial fluid into the nephron tubule is important in maintaining a constant pH for the body fluids.

Reabsorption. Because filtration is nonselective, it is important that small molecules essential to the body be returned to the interstitial fluid and blood plasma. This selective transport of substances across the epithelium of the excretory tubule from the filtrate to the interstitial fluid is called **reabsorption.** The proximal and distal tubules and the loop of Henle all contribute to reabsorption, as does the collecting duct. Nearly all the sugar, vitamins, and other organic nutrients present in the initial filtrate are reabsorbed. Most of the water of the filtrate is also reabsorbed in the kidneys of mammals and birds. Together, selective reabsorption and secretion control the concentrations of various salts in body fluids. These key functions of the nephron and collecting ducts modify the composition of the filtrate, increasing the concentrations of some substances and decreasing the concentrations of others in the urine that is finally excreted.

The overall effect of filtration, secretion, and reabsorption is analogous to cleaning out a drawer (blood) by first removing all the small articles (filtration), returning useful items to the drawer (reabsorption), adding additional useless household items to the refuse pile (secretion), and then discarding all the unwanted objects (excretion). These main functions of the nephron and collecting duct are central to homeostasis, for they enable the kidney to clear the blood of metabolic wastes and respond to imbalances in body fluids by excreting more or less of a particular ion. To better understand how the nephron and collecting duct work, we need to examine the specialized functions of the transport epithelium forming their walls.

Transport Properties of the Nephron and Collecting Duct

The filtrate formed by glomerular filtration is essentially identical to blood plasma in overall osmolarity and in the concentrations of small solutes. In this section, we focus on how the filtrate becomes urine as it flows through the nephron and collecting duct. The circled numbers below correspond to the numbers in FIGURE 40.11.

① *Proximal tubule.* The transport epithelium of the proximal tubule alters the volume and composition of filtrate substantially, by both reabsorption and secretion. For example, it produces and secretes ammonia. It also helps maintain a constant pH in body fluids by the controlled secretion of hydrogen ions and by reabsorbing about 90% of the important buffer bicarbonate (HCO_3^-) from the filtrate. Drugs and other poisons that have been processed in the liver are secreted into the filtrate by the epithelium of the proximal tubule. They pass from the peritubular capillaries into the interstitial fluid, and then across the epithelium of the tubule into the lumen. On the other hand, nutrients, including glucose and amino acids, are actively transported from the filtrate to the interstitial fluid, and then into the blood within the peritubular capillaries. Without this reabsorption, these nutrients would be lost with the urine. Potassium (K^+) is also reabsorbed.

One of the most important functions of the proximal tubule is the reabsorption of NaCl (salt) and water. Salt in the filtrate diffuses into the cells of the transport epithelium, and the membranes of the cells actively transport Na^+ out of the cells and into the interstitial fluid. This transfer of positive charge is balanced by the passive transport of Cl^- out of the tubule. As salt moves from the filtrate to the interstitial fluid, water follows passively by osmosis. The side of the epithelium facing the exterior of the tubule has a much smaller surface area than the side facing the lumen, which minimizes the leakage of salt and water back into the tubule. Instead, the salt and water now diffuse from the interstitial fluid into the peritubular capillaries.

② *Descending limb of the loop of Henle.* Reabsorption of water continues as the filtrate moves along the tubule to the descending limb of the loop of Henle. Here, the transport epithelium is freely permeable to water but not very permeable to salt and other small solutes. For water to move out of the tubule by osmosis, the interstitial fluid bathing the tubule must be hypertonic to the filtrate. The osmolarity of the interstitial fluid does in fact increase gradually, becoming progressively greater from the outer cortex to the inner medulla of the kidney. (The mechanism that maintains this gradient will be discussed shortly.) Thus, filtrate moving downward from the cortex to the medulla within the descending limb of the loop of Henle continues to lose water to interstitial fluid of greater and greater osmolarity. At the same time, the NaCl concentration of the filtrate increases as water departs by osmosis.

③ *Ascending limb of the loop of Henle.* The filtrate reaches the tip of the loop, located deep in the renal medulla in the case of juxtamedullary nephrons, then moves to the cortex again within the ascending limb of the loop. In contrast to the descending limb, the transport epithelium of the ascending limb is permeable to salt but not permeable to water. The ascending limb actually has two specialized regions: a thin segment near the loop tip and a thick segment leading to the distal tubule. As filtrate ascends in the thin segment, NaCl, which became concentrated in the descending limb, dif-

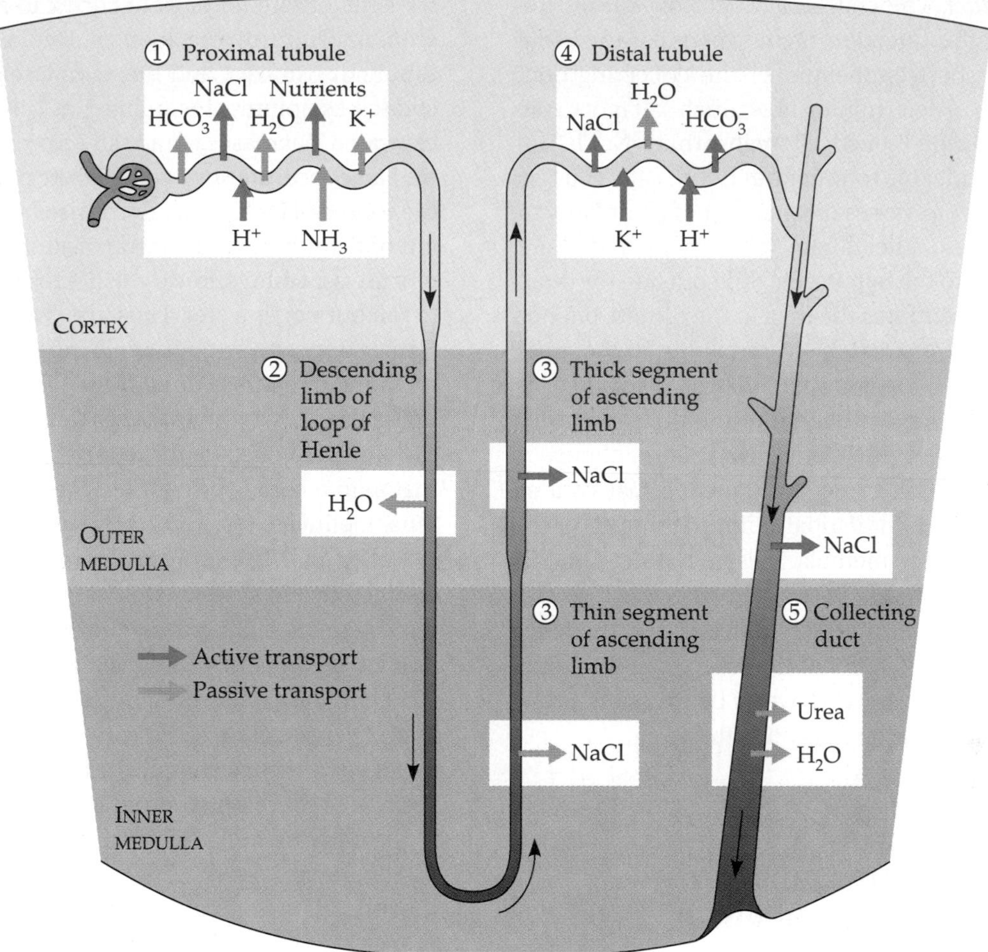

FIGURE 40.11

The nephron and collecting duct: regional functions of the transport epithelium. In this diagram, purple arrows indicate passive transport and red arrows symbolize active transport. ① The proximal tubule plays an important role in homeostasis by its controlled secretion and reabsorption of several substances. For example, about two-thirds of the NaCl and water filtered from blood into the nephron tubule are reabsorbed across the epithelium of the proximal tubule. This region also functions in the reabsorption of nutrients and in controlling pH by the secretion of H^+ and the reabsorption of HCO_3^-. ② The descending limb of the loop of Henle is permeable to water but not to salt. Osmotic loss of water from the filtrate, as the descending limb penetrates the renal medulla, concentrates NaCl in the filtrate. ③ The ascending limb of the loop of Henle consists of a thin segment and a thick segment. Both have epithelia that are virtually impermeable to water. The thin segment is permeable to NaCl, and the salt that was concentrated in the filtrate within the descending limb now diffuses out of the ascending limb, contributing to a high interstitial osmolarity in the inner medulla of the kidney. The thick segment continues the transfer of salt from the filtrate to the interstitial fluid, but now the transport is active. ④ The distal tubule is another important region of controlled secretion and reabsorption. For example, it helps regulate blood pH by the reabsorption of bicarbonate (HCO_3^-), a buffer. The distal tubule also functions in K^+ and Na^+ homeostasis. ⑤ The specialized epithelium of the collecting duct is permeable to water but not to salt. The duct carries the filtrate toward the renal medulla for a second time, and the filtrate becomes more and more concentrated as water is lost to the interstitial fluid. The bottom portion of the collecting duct is permeable to urea, and leakage of this solute into the interstitial fluid contributes to the high osmolarity of the medulla.

fuses out of the tubule into the interstitial fluid. This loss of salt contributes to the high osmolarity of the interstitial fluid in the medulla. The exodus of salt from the filtrate continues in the thick segment of the ascending limb, but here the transport epithelium actively transports NaCl into the interstitial fluid. By losing salt without giving up water, the filtrate becomes progressively more dilute as it moves up to the cortex again in the ascending limb of the loop.

④ *Distal tubule.* The distal tubule is another important site of selective secretion and absorption. For example, the distal tubule plays a key role in regulating the K^+ and NaCl concentration of body fluids by varying the amount of the K^+ that is secreted into the filtrate and the amount of NaCl that is reabsorbed from the filtrate. Like the proximal tubule, the distal tubule also contributes to pH regulation, by the controlled secretion of H^+ and by the reabsorption of bicarbonate (HCO_3^-).

⑤ *Collecting duct.* The collecting duct now carries the filtrate back in the direction of the medulla and renal pelvis. The transport epithelium of the collecting duct plays a large role in determining how much salt is actually excreted in the urine by actively reabsorbing NaCl. The epithelium is permeable to water but not to salt. Thus, as the collecting duct traverses the gradient of osmolarity in the interstitial fluid, the filtrate loses more and more water by osmosis to the hypertonic fluid outside the duct. Loss of water concentrates the urea in the filtrate, but not all of this urea is immediately passed along to the renal pelvis in the urine. At the bottom of the collecting duct, in the inner medulla, the epithelium of the duct is permeable to urea. Because of the high concentration of urea in the filtrate at this point, some of the urea diffuses out of the duct and into the interstitial fluid, bathing the portions of nephrons in the medulla. This interstitial urea is a major solute contributing, along with the NaCl, to the high osmolarity of the interstitial fluid in the medulla. And it is this high osmolarity of the interstitial fluid that enables the kidney to conserve water by excreting urine that is hypertonic to the general body fluids.

The water-conserving ability of the mammalian kidney is a key terrestrial adaptation

The osmolarity of human blood is about 300 mosm/L, but the kidney can excrete urine up to four times as concentrated—about 1200 mosm/L. The cooperative action of the loop of Henle and the collecting duct maintain the gradient of osmolarity in the interstitial tissue of the kidney that makes it possible to concentrate the urine. The two solutes responsible for this osmolarity gradient are NaCl, which is deposited in the renal medulla by the loop of Henle, and urea, which leaks across the epithelium of the collecting duct in the inner medulla.

Conservation of Water by Two Solute Gradients

To better understand the physiology of the mammalian kidney as a water-conserving organ, let's retrace the flow of filtrate through the excretory tubule, this time focusing on how the juxtamedullary nephrons maintain an osmolarity gradient in the kidney and use that gradient to excrete a hypertonic urine (FIGURE 40.12). Filtrate passing from Bowman's capsule to the proximal tubule has an osmolarity of about 300 mosm/L, the same as blood. As the filtrate flows through the proximal tubule, located in the renal cortex, a large amount of water *and* salt is reabsorbed; thus, the volume of filtrate decreases substantially at this stage, but the osmolarity remains about the same.

As the filtrate flows from cortex to medulla in the descending limb of the loop of Henle, water leaves the tubule by osmosis, and the osmolarity of the filtrate increases as solutes, including NaCl, become more concentrated. Increasing gradually from cortex to medulla, the salt concentration of the filtrate peaks at the elbow of the loop of Henle. This maximizes the diffusion of salt out of the tubule as the filtrate rounds the curve and enters the ascending limb, which, remember, is permeable to salt but not to water. Thus, the two limbs of the loop of Henle cooperate in maintaining the gradient of osmolarity in the interstitial fluid of the kidney. The descending limb produces a progressively saltier filtrate, and the ascending limb exploits this concentration of NaCl to help maintain a high osmolarity in the interstitial fluid of the renal medulla.

Notice that the loop has some of the qualities of a countercurrent system, similar in principle to the countercurrent mechanism that maximizes oxygen absorption by the gills of fishes (see FIGURES 38.18 and 38.19). Although the two limbs of the loop of Henle are not in direct physical contact, they are close enough together to affect each other's chemical exchanges with a common interstitial fluid. The loop of Henle can concentrate salt in the inner medulla only because traffic in the descending limb counters the osmolarity gradient produced by the ascending limb in the interstitial fluid.

What prevents the capillaries of the renal medulla from dissipating the osmolarity gradient by carrying away the NaCl that leaks from the ascending limb into the interstitial fluid? Notice in FIGURE 40.9d that the vasa recta is also a countercurrent system, with descending and ascending vessels carrying blood in opposite directions through the kidney's osmolarity gradient. As the descending vessel conveys blood toward the inner medulla, water is lost from the blood and NaCl diffuses into the blood. These fluxes are simply reversed as blood flows back toward the cortex in the ascending vessel, with water reentering the blood and salt diffusing out of the blood. Thus, the vasa recta can supply nutrients and other important substances carried by blood without interfering with the osmolarity gradient that makes it possible for the kidney to excrete a hypertonic urine.

By the time the filtrate reaches the distal tubule, it is *not* hypertonic to body fluids at all but is actually hypotonic. This is because the thick segment of the ascending limb of the loop of Henle actively pumps NaCl out of the tubule, making the filtrate more and more dilute. Now the filtrate descends once again toward the medulla, this time in the collecting duct, which, remember, is permeable to water but not to salt. Flowing from cortex to medulla, the filtrate loses water by osmosis as it encounters interstitial fluid of increasing osmolarity. Urea is thus

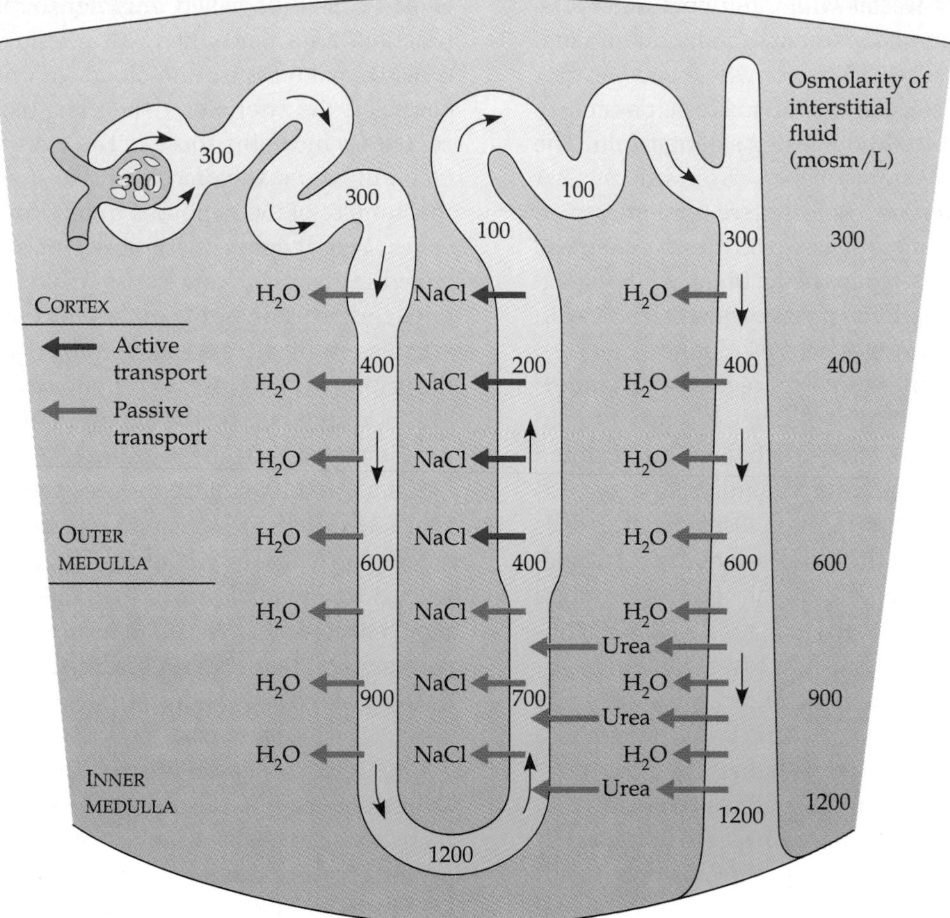

FIGURE 40.12

How the human kidney concentrates urine: the two-solute model. From the cortex to the inner medulla, the interstitial fluid of the kidney increases in osmolarity from about 300 to 1200 mosm/L. Two solutes contribute to this gradient of osmolarity: NaCl and urea. The loop of Henle maintains the interstitial gradient of NaCl. The filtrate concentration of this salt increases by the loss of water from the descending limb, then the ascending limb leaks the salt into the interstitial fluid. Additional salt is actively transported out of the thick segment of the ascending limb. The second solute, urea, is added to the interstitial fluid of the medulla by diffusion out of the collecting duct (urea remaining in the collecting duct is excreted). Urea reenters the tubule by diffusion into the ascending limb of the loop of Henle. The filtrate makes a total of three trips between the cortex and medulla: first down, then up, and then down one more time in the collecting duct. As the filtrate flows in the collecting duct past interstitial fluid of increasing osmolarity, more and more water moves out of the duct by osmosis, thereby concentrating the solutes, including urea, that are left behind in the filtrate. Under conditions in which the kidney conserves as much water as possible, urine can reach an osmolarity of about 1200 mosm/L, considerably hypertonic to blood (about 300 mosm/L). This ability to excrete nitrogenous wastes with a minimal loss of water is a key terrestrial adaptation of mammals.

concentrated in the filtrate. Some of it leaks out of the lower portion of the collecting duct, thereby making a contribution to the high interstitial osmolarity of the inner medulla. (This urea is recycled by its diffusion into the loop of Henle, but continual urea leakage from the collecting duct maintains a high interstitial concentration of this solute.) The urea remaining in the collecting duct is excreted with a minimal loss of water from the body because osmosis causes the filtrate in the collecting duct to equal the osmolarity of the interstitial fluid, which can be as high as 1200 mosm/L in the inner medulla. Notice that urine, at its most concentrated, is actually *isotonic* to the interstitial fluid of the inner medulla; however, it is *hypertonic* to blood and interstitial fluid elsewhere in the body. The juxtamedullary nephron, with its urine-concentrating features, is a key adaptation to terrestrial life, enabling mammals to get rid of nitrogenous wastes without squandering water.

Regulation of Kidney Function by Feedback Circuits

Although it is true that the kidneys *can* excrete hypertonic urine, it is not always beneficial for them to do so. Nevertheless, if you are dehydrated and water is unavailable, the kidneys can excrete a small volume of hypertonic urine as concentrated as 1200 mosm/L, making it possible to discharge wastes with a minimal water loss. But if you have consumed an excessive amount of fluid, the kidneys can actually excrete a large volume of hypotonic urine as dilute as 70 mosm/L, making it possible to eliminate a lot of water without losing essential salts. The kidney is a versatile osmoregulatory organ, where water and salt reabsorption are subject to a combination of nervous and hormonal controls. (Hormones, chemical signals between various organs of the body, are discussed in detail in Chapter 41. Here, we are concerned only with the effects of a few hormones on the kidneys.)

One hormone important in osmoregulation is **antidiuretic hormone (ADH)** (FIGURE 40.13a). It is produced in a part of the brain called the hypothalamus and stored and released from the pituitary gland, which is positioned just below the hypothalamus. Osmoreceptor cells in the hypothalamus monitor the osmolarity of blood, stimulating the release of additional ADH when blood osmolarity rises above a set point of 300 mosm/L (in humans). Excessive water losses due to sweating or diarrhea are examples of crises that could cause an increase in blood osmolarity. More ADH is then discharged into the bloodstream and reaches the kidney. The main targets of ADH are the distal tubules and collecting ducts of the kidney, where the hormone increases the permeability of the epithelium to water. This amplifies water reabsorption, which helps prevent further deviation of blood osmolarity from the set point. By negative feedback, the subsiding osmolarity of the blood reduces the activity of osmoreceptor cells in the hypothalamus, and less ADH is secreted. Only the intake of additional water in food and drink can bring osmolarity all the way back down to 300 mosm/L. When very little ADH is released, as would occur after a large volume of water has lowered the blood osmolarity, the kidneys would absorb little water, resulting in an increased discharge of dilute urine. (Increased urination is called diuresis, and it is because ADH opposes this state that it is called *anti*diuretic hormone.) Alcohol can disturb water balance by inhibiting the release of ADH, causing excessive loss of water in the urine and dehydrating the body. Some of the symptoms of a hangover may be due to this dehydration. Normally, however, blood osmolarity, ADH release, and water reabsorption in the kidney are all linked in a feedback loop that contributes to homeostasis.

A second mechanism that regulates kidney function involves a specialized tissue called the **juxtaglomerular apparatus (JGA),** located in the vicinity of the afferent arteriole, which supplies blood to the glomerulus (FIGURE 40.13b). When the blood pressure or blood volume in the afferent arteriole drops (sometimes as a result of reduced salt intake), the enzyme renin initiates chemical reactions that convert a plasma protein called angiotensinogen to a peptide called **angiotensin II.** Angiotensin II functions as a hormone, with multiple effects that increase blood pressure and blood volume. For example, it constricts the arterioles, decreasing blood flow to many capillaries including those of the kidney, thereby raising blood pressure. Angiotensin II also stimulates the proximal tubules of the nephrons to reabsorb more NaCl and water. This reduces the amount of salt and water excreted in the urine and consequently raises blood volume and pressure. Yet another effect of angiotensin II is stimulation of the adrenal glands, organs located atop the kidneys, to release a hormone called **aldosterone.** This hormone acts on the nephrons' distal tubules, making them reabsorb more sodium (Na^+) and water and increasing blood volume and pressure. In summary, the **renin-angiotensin-aldosterone system (RAAS)** is part of a complex feedback circuit that functions in homeostasis. A drop in blood pressure and blood volume triggers renin release from the JGA. In turn, the rise in blood pressure and volume resulting from the various actions of angiotensin II and aldosterone reduce the release of renin (see FIGURE 40.13b).

It may seem that the functions of ADH and the RAAS are redundant, but this is not the case. It is true that both increase water reabsorption, but they counter different osmoregulatory problems. The release of ADH is a response to an increase in the osmolarity of the blood, as when the body is dehydrated from an inadequate intake of water, for instance. But imagine a situation that causes an excessive loss of salt and body fluids—an injury, for example, or severe diarrhea. This reduces the blood's volume without increasing its osmolarity. The RAAS would save the day by increasing water and Na^+ reabsorption in response to the drop in blood volume caused by fluid loss. Normally, ADH and the RAAS are partners in homeostasis; ADH alone would lower blood Na^+ concentration by stimulating water reabsorption in the kidney, but the RAAS helps maintain balance by stimulating Na^+ reabsorption.

Still another hormone, a peptide called **atrial natriuretic factor (ANF),** opposes the RAAS. The walls of the atria of the heart release ANF in response to an increase in blood volume and pressure. ANF inhibits the release of renin from the JGA, inhibits NaCl reabsorption by the collecting ducts, and reduces aldosterone release from the adrenal glands. These actions lower blood volume and pressure. Thus, ADH, the RAAS, and ANF provide an elaborate system of checks and balances that regulate the kidney's ability to control the osmolarity, salt concentration, volume, and pressure of blood.

Having considered the mammalian kidney and its regulation in detail, we can now compare the structures and functions of kidneys in other vertebrate classes.

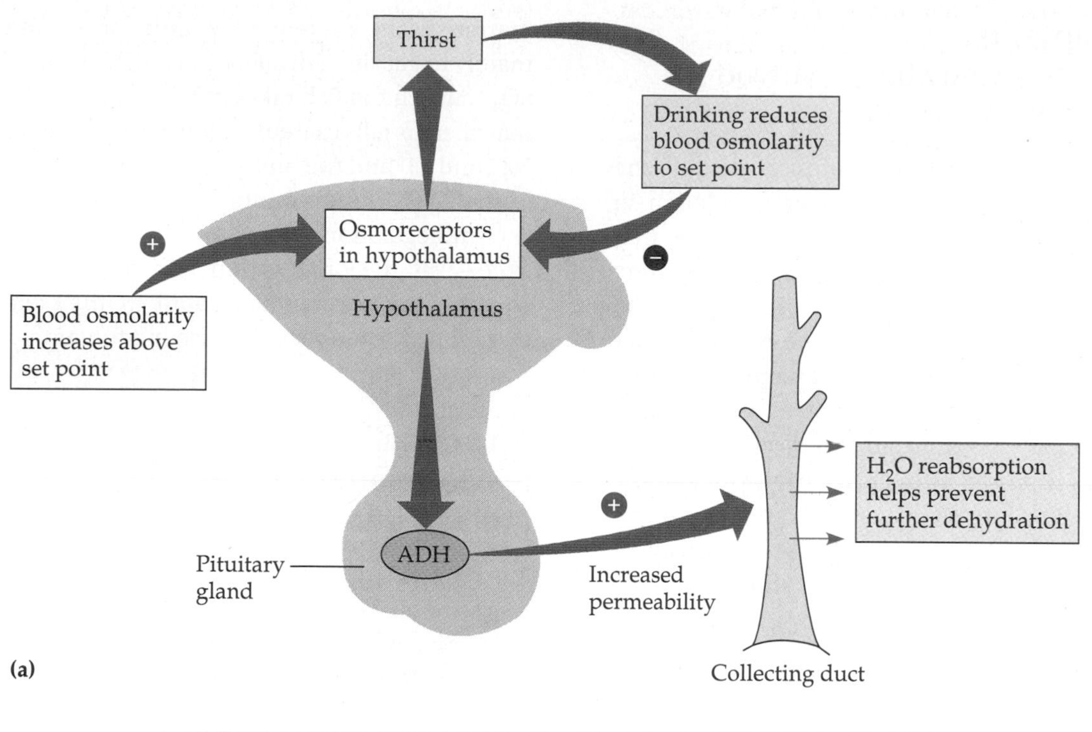

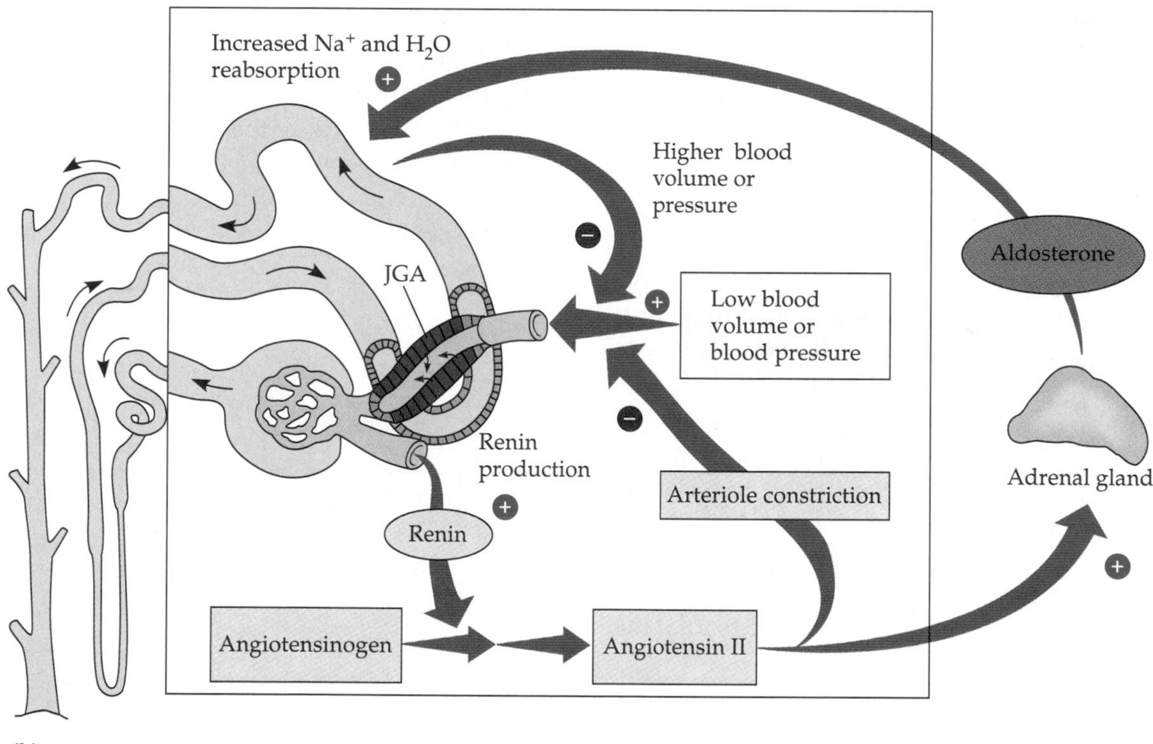

FIGURE 40.13

Hormonal control of the kidney by negative feedback circuits. (**a**) Antidiuretic hormone (ADH), produced in the hypothalamus of the brain and secreted into the bloodstream from the pituitary gland, enhances fluid retention by making the kidneys reclaim more water. The release of ADH is triggered when osmoreceptor cells in the hypothalamus detect an increase in the osmolarity of the blood. In this situation, the osmoreceptor cells also promote thirst. Drinking reduces the osmolarity of the blood, which inhibits the secretion of ADH, thereby completing the feedback circuit. (**b**) The renin-angiotensin-aldosterone system (RAAS) centers on the juxtaglomerular apparatus (JGA). The JGA responds to a decrease in blood pressure or blood volume by releasing the enzyme renin into the bloodstream (small black arrows). In the blood, renin initiates the conversion of angiotensinogen to angiotensin II. Angiotensin II increases blood pressure by causing arterioles to constrict. It also increases blood volume in two ways: by signaling the proximal tubules of the nephrons to reabsorb more NaCl and water (not illustrated) and by stimulating the adrenal glands to release aldosterone, a hormone that makes the distal tubules reabsorb more Na^+ and water. This leads to an increase in blood volume and pressure, completing the feedback circuit by suppressing the release of renin.

Diverse adaptations of the vertebrate kidney have evolved in different habitats

Variations in nephron structure and function equip the kidneys of different vertebrates for osmoregulation in their various habitats. We have seen, for instance, that nephrons of the mammalian kidney can concentrate urine and conserve water. Mammals that excrete the most hypertonic urine, such as kangaroo rats and other mammals adapted to the desert, have exceptionally long loops of Henle. Long loops maintain steep osmotic gradients in the kidney, resulting in urine becoming very concentrated as it passes from cortex to medulla in the collecting ducts. In contrast, beavers, which spend much of their time in fresh water and rarely face problems of dehydration, have nephrons with very short loops, resulting in dilute urine.

Birds, like mammals, have kidneys with juxtamedullary nephrons that specialize in conserving water. However, bird nephrons have much shorter loops of Henle than mammalian nephrons. Bird kidneys, therefore, cannot concentrate urine to the osmolarities achieved by mammalian kidneys.

The kidneys of reptiles, having only cortical nephrons, produce urine that is, at best, isotonic to body fluids. However, the epithelium of the cloaca (see Chapter 30) helps conserve fluid by reabsorbing some of the water present in urine and feces. Also, most terrestrial reptiles excrete nitrogenous wastes in an insoluble form known as uric acid, which helps conserve water because it does not contribute to the osmolarity of the urine. (This adaptation is discussed in more detail in the next section.)

In contrast to mammals and birds, a freshwater fish has the problem of excreting excess water because the animal is hypertonic to its surroundings. Instead of conserving water, the nephrons use cilia to sweep a large volume of very dilute urine from the body. Freshwater fishes conserve salts by efficient reabsorption of ions from the filtrate in the nephrons.

Amphibian kidneys function much like those of freshwater fishes. When in fresh water, the skin of the frog accumulates certain salts from the water by active transport, and the kidneys excrete dilute urine. On land, where dehydration is the most pressing problem of osmoregulation, frogs conserve body fluid by reabsorbing water across the epithelium of the urinary bladder.

Bony fishes that live in seawater, being hypotonic to their surroundings, have the opposite problem of their freshwater relatives. In many species, nephrons lack glomeruli and Bowman's capsules, and concentrated urine is formed by secreting ions into the excretory tubules. Thus, as mentioned previously, the kidneys of marine fishes excrete very little urine and function mainly to get rid of divalent ions such as Ca^{2+}, Mg^{2+}, and SO_4^{2-}, which the fish takes in by its incessant drinking of seawater. Its gills excrete mainly monovalent ions such as Na^+ and Cl^- and the bulk of its nitrogenous wastes in the form of NH_4^+ (ammonium).

Osmoregulation—the control of salt and water balance—was the original function of the kidney. In the course of evolutionary development, the excretion of nitrogenous wastes became a second function.

An animal's nitrogenous wastes are correlated with its phylogeny and habitat

Metabolism produces toxic by-products. Perhaps the most troublesome is the nitrogen-containing waste from the metabolism of proteins and nucleic acids. Nitrogen is removed from these nutrients when they are broken down for energy, or when they are converted to carbohydrates or fats. The nitrogenous waste product is ammonia, a small and very toxic molecule. Excreting ammonia directly is a bioenergetically efficient way to dispose of waste because there is essentially no metabolic cost involved. However, many animals first convert ammonia to compounds such as urea or uric acid, which are much less toxic but require energy in the form of ATP to produce.

NH_3

Ammonia

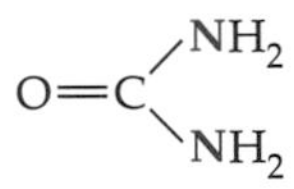

$O{=}C(NH_2)_2$

Urea

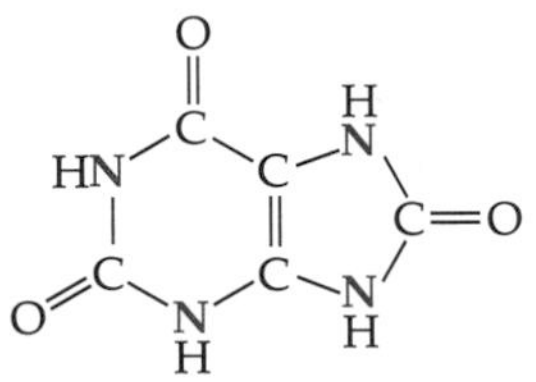

Uric acid

We will see that the form of nitrogenous wastes an animal excretes depends on both the animal's evolutionary history and its habitat (FIGURE 40.14).

Ammonia

Most aquatic animals excrete nitrogenous wastes as **ammonia.** Ammonia molecules are very soluble in water, so they easily permeate membranes. In soft-bodied invertebrates, ammonia diffuses across the whole body surface into the surrounding water. In fishes, most of the ammo-

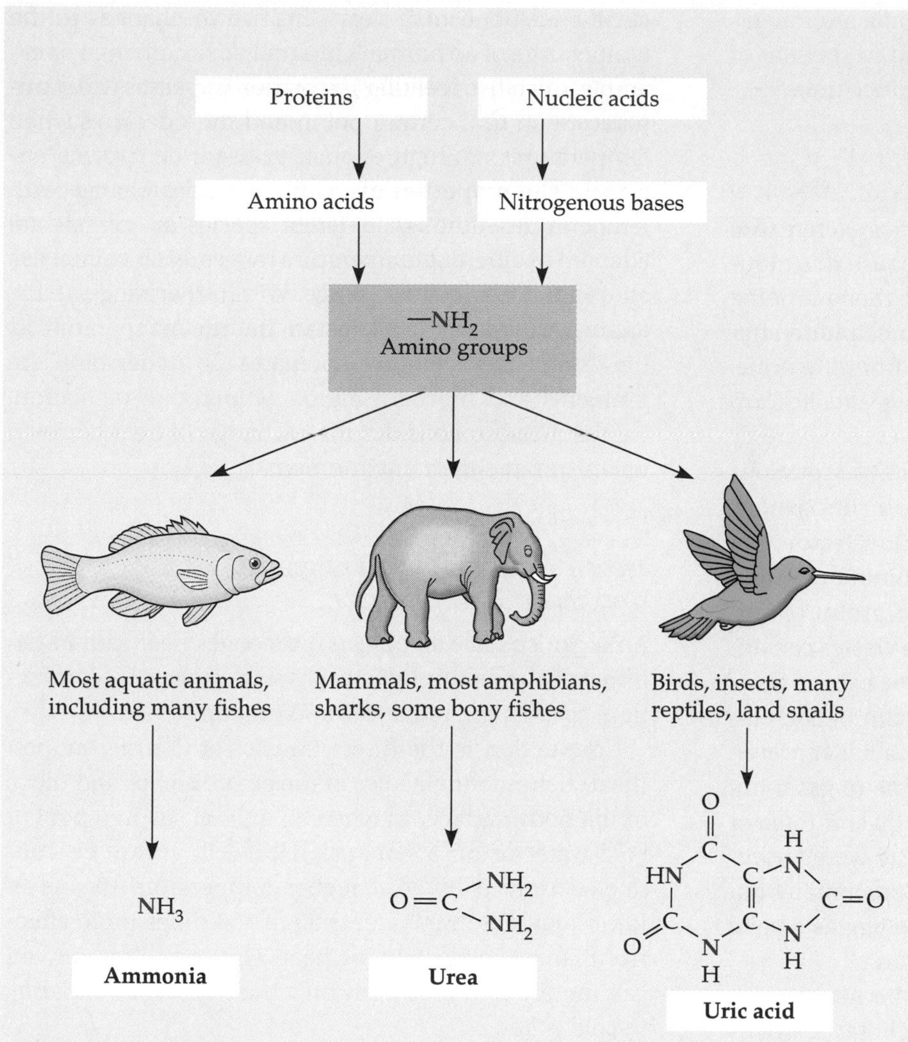

FIGURE 40.14
Nitrogenous wastes. Ammonia is a toxic by-product of the metabolic removal of nitrogen (deamination) from proteins and nucleic acids. Most aquatic animals get rid of ammonia by excreting it from body fluids. Most terrestrial animals convert the ammonia to urea or uric acid, which conserves water because these less toxic wastes can be transported in the body and hence excreted in more concentrated form.

nia is lost as ammonium ions (NH_4^+) across the epithelium of the gills, with kidneys playing only a minor role in the excretion of nitrogenous wastes. In freshwater fishes, the epithelium of the gills takes up Na^+ from the water in exchange for NH_4^+, which helps maintain Na^+ concentrations much higher than the Na^+ concentration in the surrounding water.

Urea

Although it works in water, ammonia excretion is unsuitable for disposing of nitrogenous wastes on land. Ammonia is so toxic that it can only be transported in an animal and excreted in a very dilute solution, and terrestrial animals simply cannot dispose of it quickly enough. Instead, mammals and most adult amphibians excrete **urea.** (Many marine fishes and turtles, which have the problem of conserving water in their hypertonic environment, also excrete urea.) This substance can be tolerated in a much more concentrated form because it is about 100,000 times less toxic than ammonia. Urea excretion enables an animal to sacrifice less water while disposing of its nitrogenous wastes, an important adaptation for living on land.

Urea is produced in the liver by a metabolic cycle that combines ammonia with carbon dioxide. The circulatory system carries urea to the kidneys. As we have seen, not all urea is excreted immediately in mammals; some of it is retained in the kidneys, where it contributes to osmoregulation by helping maintain the osmolarity gradient that functions in water reabsorption. Recall that sharks also produce urea and retain a relatively high concentration of it in the blood, which helps balance the osmolarity of body fluids with the surrounding seawater.

Amphibians that undergo metamorphosis generally switch from excreting ammonia to excreting urea during the transformation from an aquatic larva, the tadpole, to the terrestrial adult. This biochemical modification, however, is not always coupled with metamorphosis. Frogs that remain aquatic, such as the South African clawed toad (*Xenopus*), continue excreting ammonia after metamorphosis. But if these animals are forced to stay out of water for several weeks, they begin to produce urea.

Uric Acid

Land snails, insects, birds, and many reptiles excrete **uric acid** as the major nitrogenous waste. Because it is thousands of times less soluble in water than either ammonia or urea, uric acid can be excreted as a precipitate after nearly all the water has been reabsorbed from the urine. In birds and reptiles, urine is eliminated in pastelike form along with feces.

Uric acid and urea represent two different adaptations that enable terrestrial animals to excrete nitrogenous wastes with a minimal loss of water. One factor that seems to have been important in determining which of these alternatives evolved in a particular group of animals is the mode of reproduction. Soluble wastes can diffuse out of a shell-less amphibian egg or be carried away by the mother's blood in the case of a mammalian embryo. The vertebrates that excrete uric acid, however, produce shelled eggs, which are permeable to gases but not to liquids. If an embryo released mostly ammonia or urea within a shelled egg, the soluble waste would accumulate to toxic concentrations. Uric acid precipitates out of solution and can be stored within the egg as a solid that is left behind when the animal hatches.

In grouping vertebrates according to the nitrogenous wastes they excrete, the boundaries are not drawn strictly along phylogenetic lines but depend also on habitat. Among reptiles, for example, lizards, snakes, and terrestrial turtles excrete mainly uric acid; crocodiles excrete ammonia in addition to uric acid; and aquatic turtles excrete both urea and ammonia. Some turtles can modify their nitrogenous wastes when their environment changes. For example, a tortoise that usually produces urea can shift to uric acid production when the temperature increases and water becomes less available. This is another example of how response to the environment occurs on two levels: Evolution determines the limits of physiological responses for a species, but individual organisms make adjustments within these evolutionary constraints. This principle also applies to the regulation of body temperature, as we will see next.

Thermoregulation maintains body temperature within a range conducive to metabolism

Thermoregulation is the maintenance of body temperature within a range that enables cells to function efficiently. Metabolism is very sensitive to changes in the temperature of an animal's internal environment. For example, the rate of cellular respiration increases with temperature up to a certain point, and then declines when temperatures are high enough to begin denaturing enzymes. The properties of membranes also change with temperature. Although different species of animals are adapted to different temperature ranges, each animal has an optimal temperature range. Within that range, many animals can maintain a constant internal temperature as the external temperature fluctuates. To understand the problems and mechanisms of temperature regulation, we first need to consider the exchange of heat between organisms and their environment.

Heat Transfer Between Organisms and Their Surroundings

An organism, like all objects, exchanges heat with its external environment by four physical processes: conduction, convection, radiation, and evaporation.

Conduction is the direct transfer of thermal motion (heat) between molecules of the environment and those of the body surface, as when an animal sits in a pool of cold water or on a hot rock. Heat will always be conducted from a body of higher temperature to one of lower temperature. Water is 50 to 100 times more effective than air in conducting heat. This is one reason you can rapidly cool your body on a hot day just by standing in cold water.

Convection is the transfer of heat by the movement of air or liquid past the surface of a body, as when a breeze contributes to heat loss from the surface of an animal with dry skin. Convection also contributes to the comfort a fan brings a human on a hot, still day, but most of this effect is due to evaporative cooling. On the other hand, a wind-chill factor compounds the harshness of cold winter temperatures.

Radiation is the emission of electromagnetic waves produced by all objects warmer than absolute zero, including an animal's body and the sun. Radiation can transfer heat between objects that are not in direct contact, as when an animal absorbs heat radiating from the sun. Researchers have recently discovered a specific adaptation for exploiting solar radiation in polar bears. The fur of these animals is actually clear, not white. Each hair functions somewhat like an optical fiber that transmits ultraviolet radiation to the black skin, where the energy is absorbed and converted to body heat.

Evaporation is the loss of heat from the surface of a liquid that is losing some of its molecules as gas. Evaporation of water from an animal has a significant cooling effect on the animal's surface.

If you were to sit at rest in still air at a comfortable temperature cooler than your body (for example, an air temperature of 23°C), conduction would account for only about 1% of your heat loss, convection for about 40%, radiation for another 50%, and evaporation for about 9%. Convection and evaporation are the most variable causes of heat loss. A breeze of just 15 km/hr will increase total heat loss substantially by increasing convection fivefold. Evaporative cooling is increased greatly by the production of sweat. However, evaporation can only occur if the surrounding air is not saturated with water molecules (that is, the relative humidity is less than 100%). This is the biological basis for the common complaint, "The heat is not as bad as the humidity."

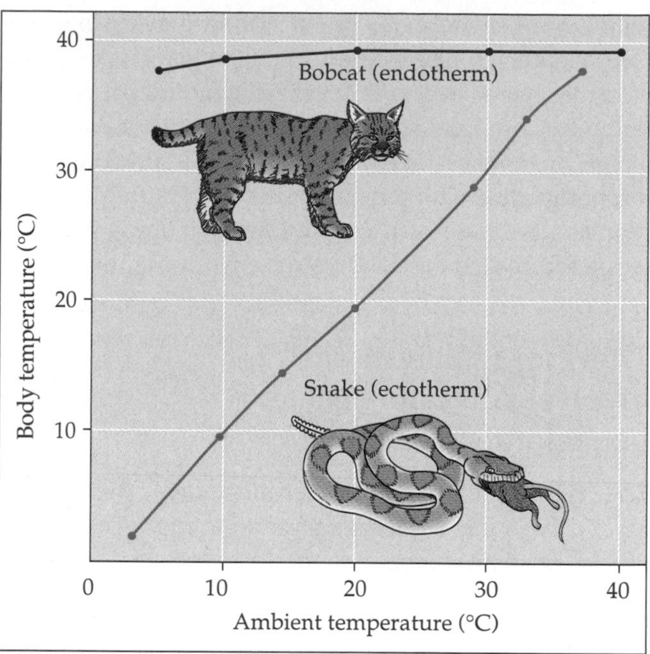

FIGURE 40.15
The relationship between body temperature and ambient (environmental) temperature in an ectotherm and an endotherm. An ectotherm (such as a snake) obtains most of its body heat from the surroundings. An endotherm (such as a bobcat) derives its body heat mainly from metabolism and uses metabolic energy for heating *and* cooling mechanisms that keep body temperature relatively constant. (Modified from P. T. Marshall and G. M. Hughes, *Physiology of Mammals and Other Vertebrates,* 2nd ed. Cambridge: Cambridge University Press, 1980.)

Ectotherms derive body heat mainly from their surroundings and endotherms derive it mainly from metabolism

One way to classify the thermal characteristics of animals is to emphasize the major source of body heat. An **ectotherm** warms its body mainly by absorbing heat from its surroundings. The amount of heat it derives from its own metabolism is usually negligible. Most invertebrates, fishes, amphibians, and reptiles are ectotherms. In contrast, an **endotherm** derives most or all of its body heat from its own metabolism. Mammals, birds, some fishes, and numerous insects are endotherms. Many endotherms maintain a consistent internal temperature even as the temperature of their surroundings fluctuates (FIGURE 40.15). However, a constant body temperature does not distinguish endotherms from ectotherms. For example, many ectothermic marine fishes and invertebrates inhabit water with such stable temperatures that their body temperature varies less than that of humans and other endotherms. Also, the terms *cold-blooded* and *warm-blooded* are misleading. Many lizards, which are ectotherms, have higher body temperatures when active than mammals.

Note again that the terms *ectotherm* and *endotherm* are not based on body temperature but rather on the main source of body heat. However, even this distinction of environmental versus metabolic heat sources is not absolute. Most ectothermic insects and some fishes are actually partial endotherms, retaining metabolic heat to warm only certain parts of their body, such as the thorax, where the muscles used in locomotion are located; and birds and mammals, which are endotherms, may add body heat by basking in the sun.

Endothermy solves certain problems of living on land. It enables terrestrial animals to maintain a constant body temperature in the face of environmental temperature fluctuations that are generally more severe than those an aquatic animal confronts. In general, endothermic vertebrates—birds and mammals—are warmer than their surroundings, but these animals also have mechanisms for cooling the body in a hot environment. A consistently warm body temperature requires active metabolism, but conversely, a warm body temperature contributes to the high levels of aerobic metabolism (cellular respiration) required for the endurance of intense physical activity. This is one reason endotherms can generally endure vigorous activity longer than ectotherms. Being endothermic is liberating, but it is also energetically expensive, especially in a cold environment. For example, at 20°C, a human at rest has a metabolic rate of 1300 to 1800 kcal per day (see Chapter 36). In contrast, a resting ectotherm of similar weight, such as an American alligator, has a metabolic rate of only about 60 kcal per day at 20°C. Thus, endotherms generally consume much more food (as measured in kilocalories) than ectotherms of equivalent size.

The bioenergetic connections between body temperature, active aerobic metabolism, and mobility were important in the evolution of endothermy; moving on land requires considerably more effort than moving in water (see Chapter 45). The efficient circulatory and respiratory systems of birds and mammals can be thought of as adaptations accompanying the evolution of endothermy and a high metabolic rate. This is not to say that ectothermy is incompatible with terrestrial success. Amphibians and reptiles have their own adaptations for coping with the temperature changes of terrestrial environments.

Thermoregulation involves physiological and behavioral adjustments

Both ectothermic and endothermic animals thermoregulate using some combination of up to four general categories of adaptations.

1. *Adjusting the rate of heat exchange between the animal and its surroundings.* Body insulation, such as hair, feathers, and fat located just beneath the skin, reduce an animal's heat loss. Other mechanisms that regulate heat exchange usually involve adaptations of the circulatory system. For example, many endotherms and some ectotherms can alter the amount of blood flowing to their skin. Increased blood flow usually results from **vasodilation,** an increase in the diameter of superficial blood vessels (those near the body surface). Nerve signals usually cause the muscles of the blood vessel walls to relax, and more blood flows through the vessels. When this occurs, more heat is transferred to the environment by conduction, convection, and radiation. The reverse adjustment, **vasoconstriction,** reduces blood flow and heat loss by decreasing the diameter of superficial vessels.

Another type of adaptation that alters heat exchange is a special arrangement of arteries and veins called a **countercurrent heat exchanger** (FIGURE 40.16). Countercurrent heat exchange is important in controlling heat loss in many endothermic animals. For example, marine mammals and many birds face the problem of losing large amounts of heat from their extremities. Arteries carrying warm blood down the legs of a bird or the flippers of a dolphin or whale are in close contact with veins conveying blood in the opposite direction, back toward the trunk of the body. This countercurrent arrangement facilitates

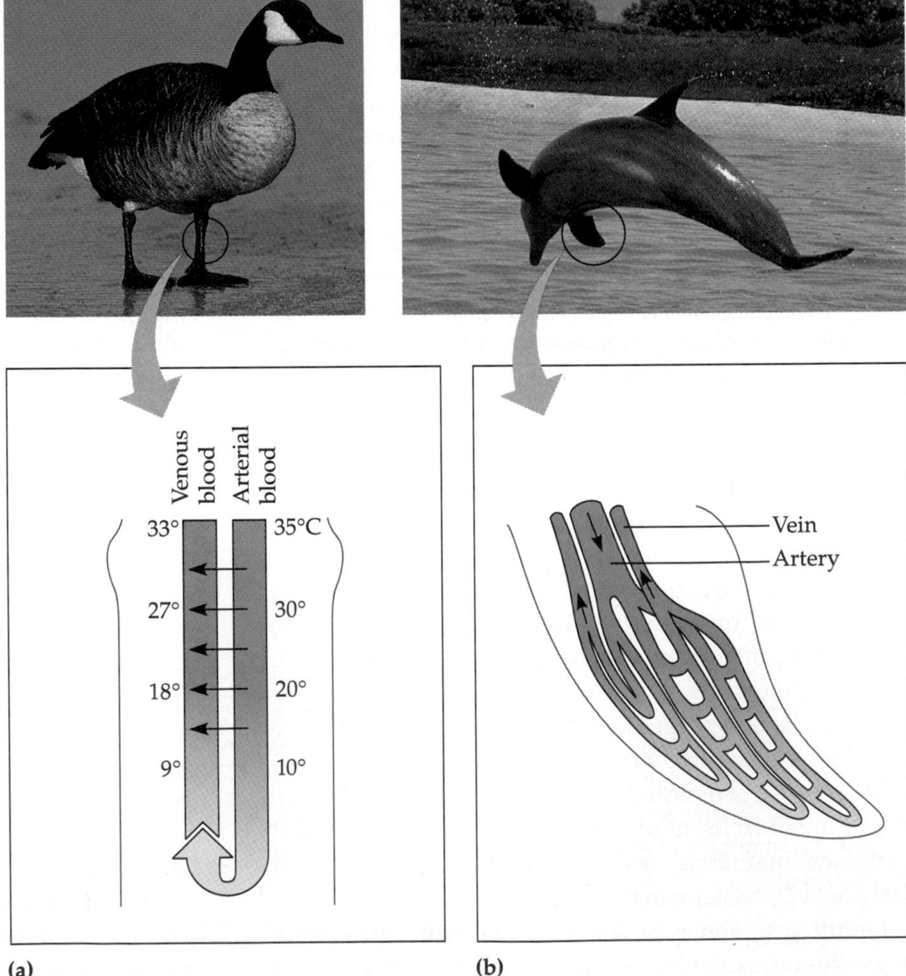

FIGURE 40.16
Countercurrent heat exchange.
(**a**) Many birds, such as this Canada goose, possess countercurrent systems in their legs that reduce heat loss. The arteries carrying blood down the legs contact the veins that return blood to the body core. In a cold environment, arterial blood transfers heat to venous blood returning to the body (black arrows). The countercurrent flow facilitates heat exchange by setting up a thermal gradient between arterial and venous blood over the entire length of the vessels. The body's core temperature is further protected by the constriction of surface veins. (**b**) In the flippers of marine mammals, such as this Pacific bottlenose dolphin, each artery is surrounded by several veins in a countercurrent arrangement that allows efficient heat exchange between venous and arterial blood.

(a)

(b)

FIGURE 40.17
Behavioral adaptations for thermoregulation. (**a**) Bathing in cool water brings immediate relief from the heat and continues to cool the surface for some time by evaporation. (**b**) Basking is an important mechanism for warming the body by radiation, convection, and conduction (from warm rocks, for example), especially in ectotherms such as these marine iguanas inhabiting the Galapagos Islands.

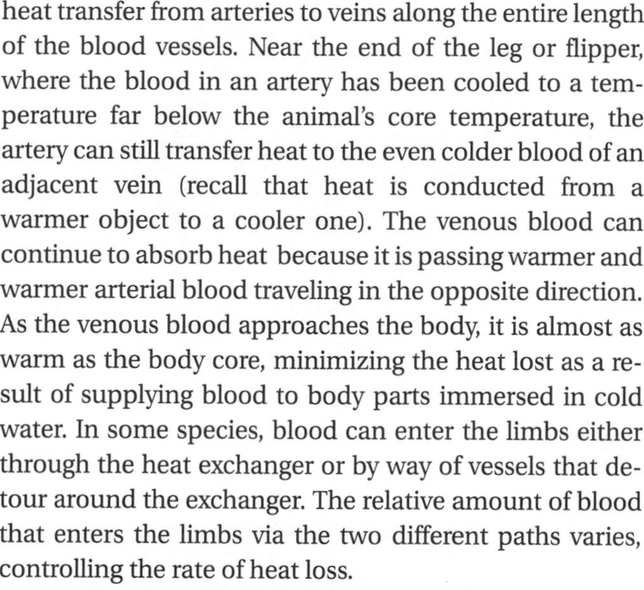

heat transfer from arteries to veins along the entire length of the blood vessels. Near the end of the leg or flipper, where the blood in an artery has been cooled to a temperature far below the animal's core temperature, the artery can still transfer heat to the even colder blood of an adjacent vein (recall that heat is conducted from a warmer object to a cooler one). The venous blood can continue to absorb heat because it is passing warmer and warmer arterial blood traveling in the opposite direction. As the venous blood approaches the body, it is almost as warm as the body core, minimizing the heat lost as a result of supplying blood to body parts immersed in cold water. In some species, blood can enter the limbs either through the heat exchanger or by way of vessels that detour around the exchanger. The relative amount of blood that enters the limbs via the two different paths varies, controlling the rate of heat loss.

2. *Cooling by evaporative heat loss.* Terrestrial endotherms and ectotherms lose water in breathing and across their skin. If the humidity of the air is low enough, the water will evaporate and the animal will lose heat by evaporative cooling. Evaporation from the respiratory system can be increased by panting. Evaporative cooling via the skin can be increased by such means as sweating in mammals.

3. *Behavioral responses.* Many animals can increase or decrease body heat loss by relocating. They will bask in the sun or on warm rocks in winter; find cool, damp areas or burrow in summer; or even migrate to a more suitable climate (FIGURE 40.17).

4. *Changing the rate of metabolic heat production.* This fourth category of thermoregulatory adaptation applies only to endotherms, particularly mammals and birds. By means we will discuss later, many species of mammals and birds can double or triple their metabolic heat production when exposed to cold.

Comparative physiology reveals diverse mechanisms of thermoregulation

Invertebrates

Most invertebrates have very little control over their body temperature, but some do adjust temperature by behavioral or physiological mechanisms. The desert locust, for example, must reach a certain temperature to become active. It orients in a direction that maximizes the absorption of sunlight.

Some species of large flying insects, such as bees and large moths, can generate internal heat and are endothermic (see the introduction to Chapter 36). They are able to "warm up" before taking off by contracting all of the flight muscles in synchrony, so that only slight movements of the wings occur but large amounts of heat are produced. This higher temperature of the flight muscles enables the insects to sustain the intense activity required for flight on cold days and at night (FIGURE 40.18a). Endotherms such as bumblebees, honeybees, and certain moths called noctuids that survive and fly during cold winter months have a countercurrent heat exchanger that helps maintain a high temperature in the thorax. For example, the heat exchanger keeps the flight muscles of some noctuid species at about 30°C, even on cold, snowy nights (FIGURE 40.18b).

Honeybees use an additional mechanism that depends on social organization to increase body temperature. In cold weather, they increase their movements and huddle together, thereby retaining heat. They maintain a relatively constant temperature by changing the density of the huddling. Individuals move from the cooler outer edges of the cluster to the warmer center and back again, thus circulating and distributing the heat. Honeybees also control the temperature of their hive by transporting water to it in hot weather and fanning with their wings, which promotes evaporation and convection. A honeybee colony uses many of the mechanisms of thermoregulation also seen in single, larger organisms.

Amphibians and Reptiles

The optimal temperature range for amphibians varies substantially with the species. For example, closely related species of salamanders have average body temperatures ranging 7°–25°C. Amphibians produce very little heat, and most lose heat rapidly by evaporation from their body surfaces, making it difficult to control body temperature. However, behavioral adaptations enable them to maintain body temperature within a satisfactory range most of the time, by moving to a location where solar heat is available or into water, for instance. When the surroundings are too warm, the animals seek cooler microenvironments, such as shaded areas. Some amphibians, including bullfrogs, can vary the amount of mucus they secrete from their surface, a physiological response that regulates evaporative cooling.

Reptiles are generally ectotherms with relatively low metabolic rates that contribute little to normal body temperature. Reptiles warm themselves mainly by behavioral adaptations. They seek warm places, orienting themselves toward heat sources to increase heat uptake and expanding the body surface exposed to a heat source. Reptiles do not simply maximize heat uptake, however; they may behave in such a way as to truly regulate their temperature within a range. If a sunny spot is too warm, for instance, a lizard may sit alternately in the sun and in the shade, or turn in another direction, thereby reduc-

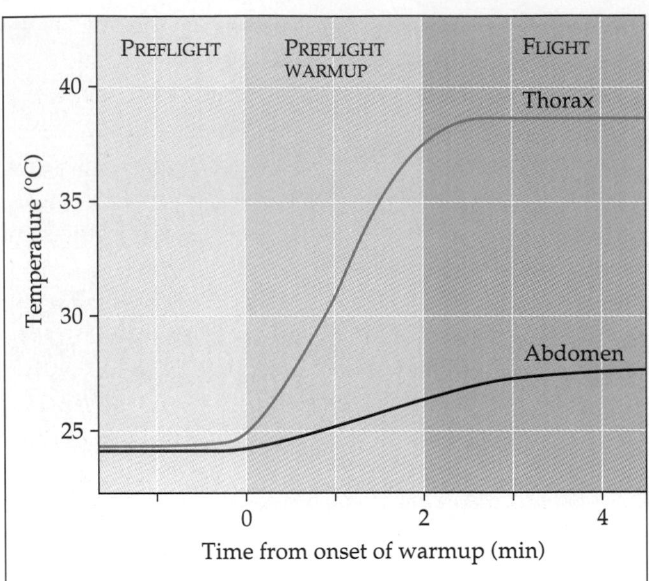

(a) Preflight warmup in the sphinx moth

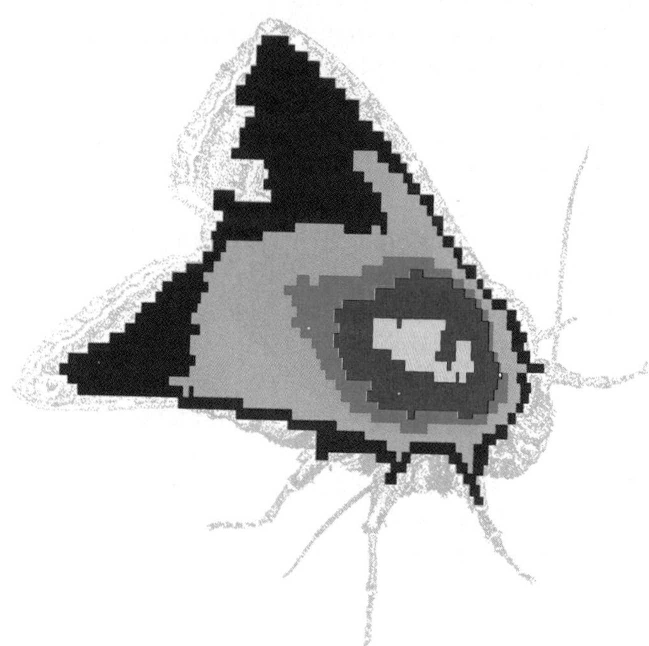

(b) Internal temperature in the winter moth

FIGURE 40.18

Thermoregulation in moths. (**a**) The sphinx moth *(Manduca sexta)* is one of the many insect species using a shiveringlike mechanism for preflight warmup of its thoracic flight muscles. Once the moth takes off, metabolic activity of the flight muscles maintains a high thoracic temperature. (**b**) Various endothermic adaptations, including a countercurrent heat exchanger in the thorax, help keep the flight muscles of moths that are active in winter warmed to a temperature of 30°C, even though the external temperature may be subfreezing. An infrared map superimposed over an image of a winter moth portrays the distribution heat immediately after a flight. Yellow, located here in the thorax region, indicates the highest temperature. Moving outward from the thorax, the variously colored zones correspond to regions of increasingly cooler body temperatures.

ing the surface area exposed to the sun. By seeking favorable microclimates within the environment, many reptiles maintain body temperatures that are quite stable.

Some reptiles also have physiological adaptations that regulate heat loss. For example, in the Galapagos iguana, body heat is conserved by superficial vasoconstriction, routing more blood to the central core of the body when the animal is swimming in the cold ocean. A few reptiles are endothermic for brief periods of time. For instance, when incubating eggs, female pythons increase their metabolic rate by shivering, generating enough heat to maintain their body temperature 5°–7°C above the surrounding air. Researchers continue to debate whether certain groups of dinosaurs were endothermic (see Chapter 30).

Fishes

The body temperature of most fishes is usually within 1°–2°C of the surrounding water temperature. Metabolic heat generated by swimming muscles is lost to the surrounding water when blood passes through the gills, and the large dorsal aorta conveys the blood directly inward from the gills, cooling the body core. Endothermic fishes include several large, active species such as the bluefin tuna, swordfish, and the great white shark. Their swimming muscles produce enough metabolic heat to elevate temperatures at the body core, and adaptations of the circulatory system retain the heat. Large arteries convey most of the cold blood from the gills to tissues just under the skin (FIGURE 40.19). Branches deliver blood to the deep muscles, where the small vessels are arranged into a

FIGURE 40.19

Thermoregulation in large, active fishes. (**a**) The bluefin tuna maintains internal temperatures much higher than the surrounding water. (Colors indicate swimming muscles cut in transverse section.) The temperatures shown were recorded for a tuna in 19°C water. (**b**) Like the bluefin tuna, the great white shark has a countercurrent heat exchanger in its swimming muscles that reduces the loss of metabolic heat. All fishes lose some heat to surrounding water when their blood passes through the gills. However, in contrast to ectothermic fishes, endothermic species have a relatively small-diameter dorsal aorta, which carries little blood from the gills to the core of the body. Instead, most of the blood from the gills is conveyed via large arteries just under the skin, keeping cool blood away from the body core. As shown in the enlargement, a network of capillaries carrying cool blood inward from the skin arteries is paralleled by venous capillaries carrying warm blood outward from the inner body. This countercurrent flow retains heat in the muscles.

21°
25°
23°
27°
29°
31.4°
31°
Body cavity
(a)

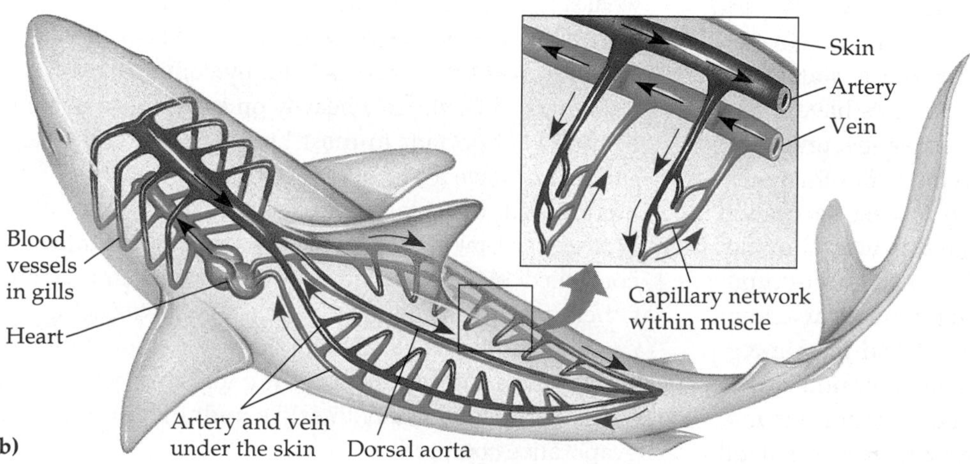

countercurrent heat exchanger. Endothermy enhances vigorous activity by keeping the swimming muscles several degrees warmer than the tissues near the animal's surface, which is about the same temperature as the surrounding water. A current hypothesis is that endothermy in fishes evolved as an adaptation to foraging in cold seas.

Mammals and Birds

Mammals and birds generally maintain high body temperatures within a narrow range of about 36°–38°C for most mammals and about 40°–42°C for most birds. In contrast to endothermic insects and fishes, which regulate their body temperature by controlling heat loss, birds and mammals regulate the rate of metabolic heat production, balancing it with the rate at which they gain or lose heat from their surroundings. The rate of heat production can be increased in one of two ways: by the increased contraction of muscles (by moving or by shivering) or by the action of hormones that increase the metabolic rate and the production of heat instead of ATP. The hormonal triggering of heat production is called **nonshivering thermogenesis.** Occurring in numerous mammals and a few birds, nonshivering thermogenesis takes place throughout the body, and some mammals have a tissue called **brown fat** in the neck and between the shoulders that is specialized for rapid heat production.

Mammals and birds have an array of mechanisms that regulate heat exchange with their environment. Vasodilation and vasoconstriction effect heat exchange and may also contribute to regional temperature differences within the animal. For example, on a cool day a human's temperature may be several degrees lower in the arms and legs than in the trunk, where most vital organs are located. The insulating power of a layer of fur or feathers depends on how much still air the layer traps. Accordingly, most land mammals and birds react to cold by raising their fur or feathers, thereby trapping a thicker layer of still air. Humans rely more on a layer of fat just beneath the skin as insulation against heat loss (FIGURE 40.20). Goose bumps are a vestige of hair-raising left over from our furry ancestors.

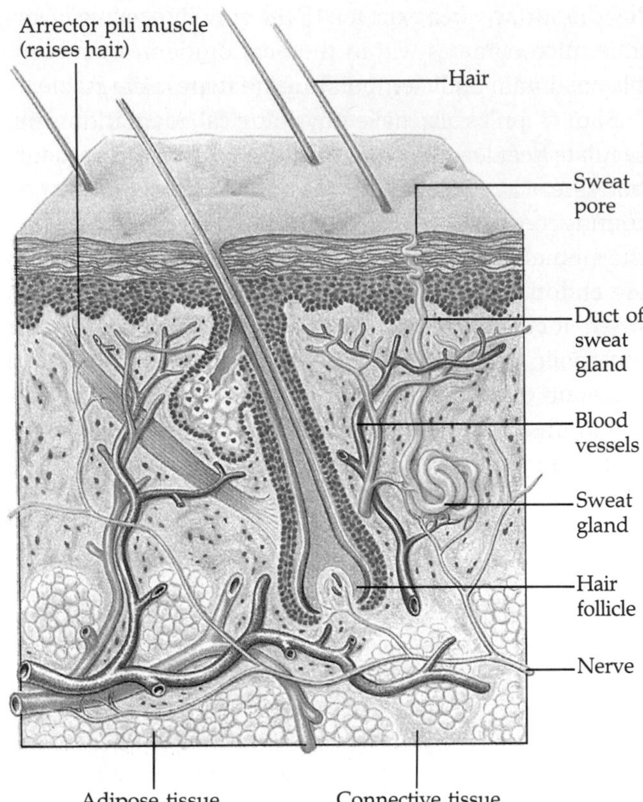

FIGURE 40.20
Skin as an organ of thermoregulation. Fat (adipose tissue) and hair help insulate mammals. Heat loss to the environment can be regulated by the constriction and dilation of superficial blood vessels and by the erection and compaction of fur. Sweat glands, under nervous control, function in evaporative cooling.

Hair loses much of its insulating power in water, and marine mammals such as whales and seals have a very thick layer of insulating fat called blubber, just under the skin. All marine mammals live in water colder than their body temperature, and many spend at least part of the year in nearly freezing arctic or antarctic water. Although the loss of heat to water occurs 50 to 100 times more rapidly than heat loss to air, marine mammals maintain body temperatures of about 36°–38°C and have metabolic rates close to those of land mammals of similar size. This suggests that adaptations that conserve heat in marine mammals are more effective than those of land mammals. The flippers or tail of a whale or seal lack insulating blubber, but countercurrent heat exchangers effectively reduce heat loss in these extremities, as they do in the legs of many birds (see FIGURE 40.16).

Many mammals and birds live where endothermy requires behavorial and physiological adaptations that cool the body as well as keep it warm. For example, when a marine mammal moves into warm seas, as do many whales to reproduce, it disposes of excess metabolic heat by vasodilation, enhanced by large numbers of blood vessels in the outer layer of its skin. By contrast, terrestrial mammals and birds rely heavily on evaporative cooling. Panting is important in most birds and mammals, and some birds have a vascularized pouch in the floor of the mouth that they can flutter, increasing evaporation. Many terrestrial mammals have sweat glands, which are controlled by the nervous system (see FIGURE 40.20). Other mechanisms that promote evaporative cooling include spreading saliva on body surfaces, an adaptation of some kangaroos and rodents for combating severe heat stress. Some bats use both saliva and urine to enhance evaporative cooling.

Feedback Mechanisms in Thermoregulation

The regulation of body temperature in humans is an example of a complex homeostatic system facilitated by feedback mechanisms. Nerve cells that control thermoregulation, as well as those that control many other aspects of homeostasis, are concentrated in the hypothalamus (discussed in detail in Chapter 44). The hypothalamus serves as a thermostat, responding to changes in body temperature above and below a set point (actually above or below a normal range) by activating mechanisms that promote heat loss or gain (FIGURE 40.21). Nerve cells that sense body temperatures are located in the skin, the hypothalamus itself, and some other parts of the nervous system. Some of these are warm receptors that signal the thermostat in the hypothalamus when the temperature of the skin or blood increases. Others are cold receptors that signal the thermostat when the temperature decreases. Responding to body temperature below the normal range, the thermostat inhibits heat-loss mechanisms and activates heat-saving ones such as the vasoconstriction of superficial vessels and the erection of fur, while stimulating heat-generating mechanisms (shivering and nonshivering thermogenesis). In response to elevated body temperature, the thermostat shuts down heat-saving mechanisms and promotes body cooling by vasodilation, sweating, or panting.

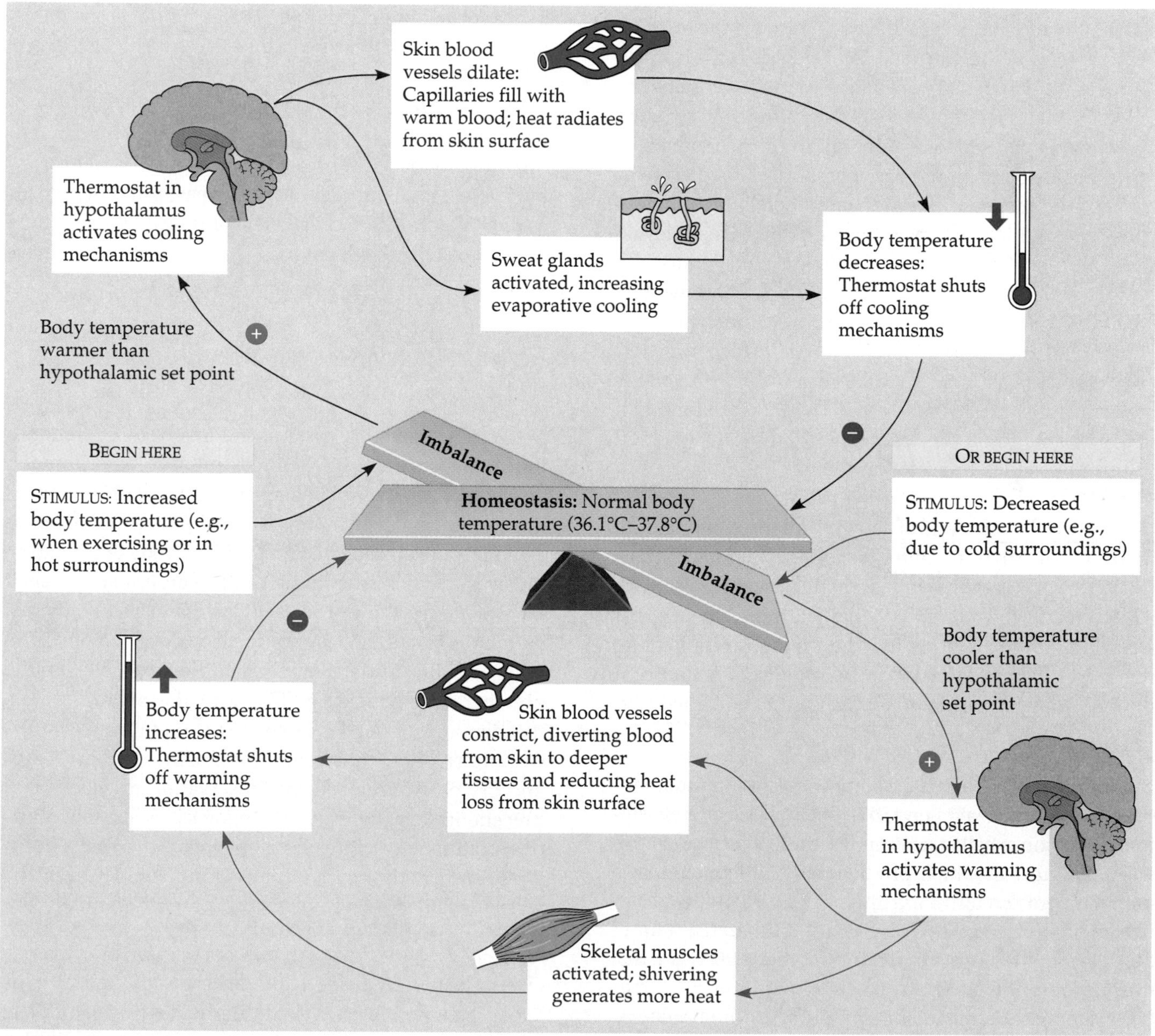

FIGURE 40.21
The thermostat function of the hypothalamus and feedback mechanisms in human thermoregulation.

Torpor: Conserving Energy During Environmental Extremes

Torpor is an alternative physiological state in which metabolism decreases and the heart and respiratory system slow down. Many endotherms enter a state of torpor in which their body temperature declines. In effect, their body's thermostat is turned down, thereby conserving energy when food supplies are low and environmental temperatures are extreme. **Hibernation** is long-term torpor during which the body temperature is lowered as an adaptation to winter cold and food scarcity. **Aestivation,** or summer torpor, is characterized by slow metabolism and inactivity. It enables an animal to survive long periods of high temperatures and scarce water supplies. Hibernation and aestivation are often triggered by seasonal changes in the length of daylight. As the days shorten, some animals will eat huge quantities of food before hibernating. Ground squirrels, for instance, will more than double their weight in a month of gorging.

Many small mammals and birds exhibit a daily period of torpor that seems to be adapted to their feeding patterns. For instance, most bats and shrews feed at night and go into torpor when they are inactive during daylight hours. Chickadees and hummingbirds feed during the day and often undergo torpor on cold nights; the body temperature of chickadees that spend winters in cold, northern forests may drop as much as 10°C at night. All endotherms that show a daily torpor are relatively small; when active, they have a high metabolic rate and thus a very high rate of energy consumption. During hours when they cannot feed, their daily period of torpor enables them to survive on energy stored in their tissues.

An animal's daily cycle of activity and torpor appears to be a built-in rhythm controlled by the biological clock (see Chapter 35). Even if food is made available to a shrew all day, it still goes through its daily torpor. The need for sleep in humans and the slight drop in body temperature that accompanies it may be a remnant of a more pronounced daily torpor in our early mammalian ancestors.

Temperature Range Adjustments

Many animals can adjust to a new range of environmental temperatures over a period of days or weeks, a physiological response called **acclimatization.** Seasonal change is one context in which physiological adjustments to a new temperature range are important. The bullhead catfish, for instance, can survive water temperatures up to 36°C in summer, but a temperature above 28°C is lethal during winter.

Physiological adjustment to a new external temperature range is multifaceted. It may involve changes in the mechanisms that control an animal's internal temperature. Acclimatization may also involve cellular adjustments. For example, cells may increase the production of certain enzymes, helping compensate for the lowered activity of each enzyme molecule at temperatures that are not optimal. In other cases, cells produce variants of enzymes that have the same function but different temperature optima. Membranes may also change in the proportions of saturated and unsaturated lipids they contain. This response helps keep membranes fluid at different temperatures (see Chapter 8).

Cells can often make rapid adjustments to temperature changes. For example, the cells of tree frogs respond to cold shock within hours. Mammalian cells grown in laboratory cultures respond to a marked increase in temperature and to other forms of severe stress, such as toxins, rapid pH changes, and viral infections, by accumulating special molecules called **stress-induced proteins,** including **heat-shock proteins.** When "shocked" by a rapid change in temperature from 37°C to about 43°C, cultured mammalian cells begin synthesizing heat-shock proteins within minutes. These molecules help maintain the integrity of other proteins that would be denatured by severe heat. Found in bacteria, yeasts, and plant cells as well as in animal cells, stress-induced proteins help prevent cell death when challenged with severe changes in the external environment.

■ Regulatory systems interact in the maintenance of homeostasis

Numerous regulatory systems are involved in maintaining homeostasis in an animal's internal environment, and one of the major challenges in animal physiology is unraveling how regulatory systems interact and are controlled. For example, the regulation of body temperature involves mechanisms that also have an impact on such parameters of the internal environment as osmolarity, metabolic rate, blood pressure, tissue oxygenation, and body weight. Under some conditions, usually at the physical extremes compatible with the organism's life, the demands of one system might come into conflict with those of other systems. For instance, in very warm and dry environments, the conservation of water takes precedence over evaporative heat loss. As a result, many desert animals tolerate occasional hyperthermia (abnormally high body temperature). Normally, however, the various regulatory systems act together to maintain homeostasis in the internal environment.

Roles of the Liver in Homeostasis

Our discussion of homeostasis would be incomplete without mentioning the liver, the vertebrate body's largest and most functionally diverse organ. Liver func-

TABLE 40.1

Major Functions of the Liver

FUNCTIONS	MAJOR EFFECTS ON HOMEOSTASIS	CHAPTER(S)
Synthesis		
Nitrogenous wastes (ammonia, urea, uric acid) from amino acids	Assist kidney in waste disposal	40
Plasma proteins (e.g., prothrombin, fibrinogen, albumin)	Blood clotting; maintain osmotic balance of blood	38
Growth factors	Help regulate growth and development	11, 41
Bile	Emulsifies fats in small intestine	37
Lipids, cholesterol, lipoproteins	Help regulate blood chemistry; store energy; help maintain cell membranes	5, 37, 38, 8
Freeze-protectant chemicals	Maintain cell membrane integrity	40
Storage		
Fat-soluble vitamins	Help maintain skeleton; chemistry of vision	37, 45
Iron	Oxygenation of tissues (constituent of hemoglobin)	5, 38
Glycogen	Bioenergetics (energy reserves)	5, 9
Conversion		
Excess glucose in blood to glycogen Lactic acid to glycogen Stored glycogen to glucose	Bioenergetics (energy storage and use)	5, 9, 37, 38, 41
Recycling		
Contents of old red blood cells (e.g., iron and other constituents of hemoglobin)	Oxygenation of tissues	5, 38
Detoxification		
Many harmful chemicals (e.g., alcohol, barbiturate drugs)	Assist kidney in toxin disposal	40

tions are pivotal to homeostasis and involve interaction with most of the body's organ systems (TABLE 40.1). For example, the liver supports the excretory role of the kidneys by detoxifying many chemical poisons and synthesizing ammonia, urea, or uric acid, using nitrogen from amino acids. Many of the liver's activities are important in bioenergetics. Liver cells take up glucose from the blood, store excess amounts as glycogen, then convert glycogen back to glucose, releasing glucose to the blood in response to a demand for fuel. With the liver performing the essential tasks, homeostatic mechanisms regulated by feedback circuits maintain tight control on the glucose content of the blood. The feedback circuits themselves involve nervous communication and hormones. We address the hormonal control of homeostasis in the next chapter.

REVIEW OF KEY CONCEPTS (with page numbers and key figures)

- Homeostatic mechanisms protect an animal's internal environment from harmful fluctuations: *an overview* (pp. 879–880, FIGURE 40.1)
  - Homeostasis enables animals to survive large fluctuations in their external environment by tempering changes in body fluids.
  - Physiological adjustments, usually involving feedback mechanisms, regulate the hemolymph or interstitial fluid that bathes all body cells.
- Cells require a balance between water uptake and loss (pp. 880–883, FIGURE 40.2)
  - Water uptake must balance water loss, requiring various mechanisms of osmoregulation in different environments. Osmoconformers are isotonic with their aqueous surroundings and do not regulate their osmolarity. Osmoregulators control water uptake and loss in a hypertonic or hypotonic environment.
  - Stenohaline animals cannot tolerate marked osmotic changes. Euryhaline animals tolerate large osmotic changes in their environments, often by altering their osmoregulatory mechanisms.
  - Most marine invertebrates are osmoconformers, whereas most marine vertebrates osmoregulate. Sharks have an osmolarity slightly higher than seawater because they retain urea. Marine bony fishes lose water to their hypertonic environment and must compensate by drinking large quantities of seawater. Marine vertebrates excrete excess salt through rectal glands, gills, salt-excreting glands, or kidneys.
  - Freshwater organisms constantly take in water from their hypotonic environment. Protozoa pump out excess water with contractile vacuoles, and freshwater animals excrete copious amounts of dilute urine. Salt loss is replaced by eating or by ion uptake by gills.
  - Terrestrial animals combat desiccation by drinking and eating food with high water content and through the nervous and hormonal control of thirst, behavioral adaptations, and water-conserving excretory organs.
- Osmoregulation depends on transport epithelia (pp. 883–884)
  - Osmoregulation is usually accomplished by the transport of salt across a transport epithelium, followed by the osmotic flow of water.
- Tubular systems function in osmoregulation and excretion in many invertebrates (pp. 884–886, FIGURES 40.6–40.8)
  - Extracellular fluid is filtered into the protonephridia of the flame-bulb system in flatworms. These blind-ended tubules excrete a dilute fluid and also function in osmoregulation.
  - Each segment of an earthworm has a pair of open-ended, excretory tubules enveloped by capillaries. These metanephridia collect coelomic fluid, the transport epithelia pump out salts for reabsorption, and dilute urine is excreted through nephridiopores.
  - In insects, Malpighian tubules function in osmoregulation and removal of nitrogenous wastes from the hemolymph. Insects produce a relatively dry waste matter, an important adaptation to terrestrial life.
- The kidneys of most vertebrates are compact organs with many excretory tubules (pp. 886–888; FIGURE 40.9)
  - The excretory tubules of vertebrates are arranged into compact organs, the kidneys, which, along with associated blood vessels and excretory ducts, constitute the excretory system.
  - Excretory tubules, consisting of nephrons and collecting ducts, and associated blood vessels pack the kidney. Each nephron consists of a Bowman's capsule, which surrounds a ball of capillaries called the glomerulus; a proximal tubule; a loop of Henle; and a distal tubule. Each collecting duct obtains fluid from several nephrons and passes urine into the kidney's central receptacle, the renal pelvis. A ureter conveys urine from the renal pelvis to the urinary bladder.
  - Blood supply to the nephron is through an afferent arteriole, which divides into the capillaries of the glomerulus. An efferent arteriole carries blood away from Bowman's capsule and subdivides into the peritubular capillaries embracing the proximal and distal tubules. The vasa recta is a system of capillaries that services the loop of Henle.
- The kidney's transport epithelia regulate the composition of blood (pp. 888–892; FIGURE 40.11)
  - Nephrons control the composition of the blood by filtration, secretion, and reabsorption.
  - During filtration, blood pressure nonselectively filters water and small solutes (the blood filtrate) from the glomerulus into the lumen of the nephron tubule.
  - Additional substances destined for excretion are directly secreted from the interstitial fluid into the tubule by active and passive transport.
  - Filtered substances that must be returned to the blood, such as vital nutrients and water, are reabsorbed from the filtrate at various points along the nephron.
  - Most of the salt and water filtered from the blood is reabsorbed by the proximal tubule. In addition, ammonia, drugs, and hydrogen ions (for the control of body pH) are selectively secreted into the filtrate; glucose and amino acids are actively transported out of the filtrate; and potassium is reabsorbed.
  - The descending limb of the loop of Henle is permeable to water but not to salt; water moves by osmosis into the hypertonic interstitial fluid. Salt diffuses out of the concentrated filtrate as it moves through the salt-permeable ascending limb of the loop of Henle.
  - The distal tubule is specialized for selective secretion and reabsorption, playing a key role in regulating potassium concentration and blood pH.

- The water-conserving ability of the mammalian kidney is a key terrestrial adaptation (pp. 892–895, FIGURES 40.12 and 40.13)
  - The collecting duct, which is permeable to water but not to salt, carries the filtrate through the osmolarity gradient of the medulla, and more water exits by osmosis.
  - Urea also diffuses out of the tubule, joining salt in forming the osmotic gradient that enables the kidney to produce urine that is hypertonic to the blood.
  - The osmolarity of the filtrate varies as it travels through a nephron. The differing permeabilities to water, salt, and urea in regions of the tubule, combined with the active transport of salt, produce two solute (NaCl and urea) osmolarity gradients that result in the production of a hypertonic urine.
  - The osmolarity of the urine can vary widely, depending on the hydration needs of the body, because it is regulated by nervous and hormonal control of water and salt reabsorption in the kidneys. Antidiuretic hormone (ADH), released in response to a rise in blood osmolarity signaled by osmoreceptor cells in the hypothalamus, increases water reabsorption by the tubule.
  - The juxtaglomerular apparatus (JGA) responds to decreased blood pressure or blood volume by the release of renin, which triggers the formation of angiotensin II. This blood peptide causes arterioles to constrict and the adrenal glands to release aldosterone, a hormone that stimulates the reabsorption of Na^+ and the passive flow of water from the filtrate. Both ADH and the renin-angiotensin-aldosterone system (RAAS) result in the reabsorption of water and more concentrated urine.
  - Atrial natriuretic factor (ANF), released by the atria of the heart in response to increased blood pressure, inhibits the release of renin and counters the RAAS.
- Diverse adaptations of the vertebrate kidney have evolved in different habitats (p. 896)
  - The original function of the vertebrate kidney was osmoregulation. The form and function of nephrons in the various vertebrate classes are related primarily to the requirements for osmoregulation in the animal's habitat. The excretion of nitrogenous wastes became a secondary function of the kidney in the course of vertebrate evolution.
- An animal's nitrogenous wastes are correlated with its phylogeny and habitat (pp. 896–898, FIGURE 40.14)
  - The metabolism of proteins and nucleic acids generates ammonia, a toxic waste product that is excreted in one of three forms, depending on the habitat and evolutionary history of the animal.
  - Most aquatic animals excrete ammonia, a highly toxic but very soluble molecule that easily passes across the body surface or gill epithelia into the surrounding water.
  - The liver of mammals and most adult amphibians converts ammonia to the less toxic urea, which is carried by the circulatory system to the kidneys and excreted in concentrated form with a minimal loss of water.
  - Uric acid is an insoluble precipitate excreted in the paste-like urine of land snails, insects, birds, and many reptiles.
  - The mode of reproduction of terrestrial animals is related to the form of their nitrogenous wastes. Differences in waste product within phyla are related to habitat, and some organisms can shift the form of their nitrogenous wastes, depending on environmental conditions.
- Thermoregulation maintains body temperature within a range conducive to metabolism (pp. 898–899)
  - The maintenance of body temperature within a range that enables cells to function efficiently involves heat transfer between the organism and the external environment. Heat exchange involves the physical processes of conduction, convection, radiation, and evaporation.
- Ectotherms derive body heat mainly from their surroundings and endotherms derive it mainly from metabolism (pp. 899–900, FIGURE 40.15)
  - Endothermy enables terrestrial animals to maintain a relatively uniform body temperature and a high level of aerobic metabolism, which facilitate movement on land and active swimming in cold oceans.
- Thermoregulation involves physiological and behavioral adjustments (pp. 900–901, FIGURE 40.21)
  - Ectotherms and endotherms regulate body temperature by adjusting the rate of heat exchange with their surroundings, by evaporative cooling, and by various behavioral responses. Birds and mammals can also change the rate of metabolic heat production.
  - Adaptations that alter the rate of heat exchange include body insulation, vasodilation, vasoconstriction, and countercurrent heat exchangers.
  - Adaptations that affect evaporative cooling include panting, sweating, and bathing.
- Comparative physiology reveals diverse mechanisms of thermoregulation (pp. 901–906)
  - Many large flying insects generate metabolic heat by muscle contractions, and many have countercurrent heat exchangers that retain it. Honeybees regulate body temperature by social behavior.
  - Reptiles and amphibians maintain internal temperatures within tolerable ranges mainly by various behavioral adaptations. Some lizards use vasoconstriction and vasodilation to regulate heat exchange with their surroundings. Some female snakes can generate metabolic heat and raise their body temperature by shivering when incubating eggs.
  - Although the body temperature of most fishes matches the environment, some large, active species maintain a higher temperature in their swimming muscles with a countercurrent heat exchanger.
  - Mammals and birds can adjust their rate of metabolic heat production by shivering and nonshivering thermogenesis.
  - Marine mammals maintain their high body temperatures in cold water by a thick layer of insulating blubber and countercurrent heat exchange between arterial and venous blood.
  - Birds may thermoregulate by panting, increasing evaporation from a vascularized pouch in the mouth, and by passing blood going to the legs through a countercurrent heat exchanger.

- In humans, thermoregulatory areas of the hypothalamus serve as the body's thermostat, receiving nerve signals from warm and cold receptors and responding by initiating either cooling or warming processes.
- Torpor, including hibernation and aestivation, is a physiological state characterized by a decrease in metabolic, heart, and respiratory rates. This state enables the animal to temporarily withstand varying periods of unfavorable temperatures or the absence of food and water.
- Physiological adjustment to a new range of environmental temperatures occurs gradually, often seasonally, and may involve changes in an animal's thermostatic control mechanisms and various responses at the cellular level.

- Regulatory systems interact in the maintenance of homeostasis (pp. 906–907, TABLE 40.1)
  - Homeostasis is a dynamic response to the external environment involving the cooperative interaction of numerous regulatory systems.
  - The vertebrate liver performs diverse functions vital to homeostasis. Feedback circuits involving nervous communication and hormones integrate homeostatic mechanisms.

SELF-QUIZ

1. *Unlike* an earthworm's metanephridia, a mammalian nephron
   a. is intimately associated with a capillary network
   b. forms urine by changing the composition of fluid inside the tubule
   c. functions in both osmoregulation and the excretion of nitrogenous wastes
   d. filters blood instead of coelomic fluid
   e. has a transport epithelium

2. The majority of water and salt filtered into Bowman's capsule is reabsorbed by
   a. the transport epithelia of the proximal tubule
   b. diffusion from the descending limb of the loop of Henle into the hypertonic interstitial fluid of the medulla
   c. active transport across the transport epithelium of the thick upper segment of the ascending limb of the loop of Henle
   d. selective secretion and diffusion across the distal tubule
   e. diffusion from the collecting duct into the increasing osmotic gradient of the renal medulla

3. The high osmolarity of the renal medulla is maintained by all the following *except*
   a. diffusion of salt from the ascending limb of the loop of Henle
   b. active transport of salt from the upper region of the ascending limb
   c. the spatial arrangement of juxtamedullary nephrons
   d. diffusion of urea from the collecting duct
   e. diffusion of salt from the descending limb of the loop of Henle

4. Ammonia (or ammonium) is excreted by most
   a. bony fishes
   b. organisms that produce shelled eggs
   c. adult amphibians
   d. land snails
   e. insects

5. Nonshivering thermogenesis is
   a. a behavioral adaptation for absorbing heat in ectotherms
   b. a hormone-triggered rise in metabolic rate, often associated with brown fat
   c. a countercurrent heat exchange of blood going to the limbs
   d. a heat-producing method of large, active fishes
   e. muscle contractions that have been identified in winter moths

6. Physiological adjustments, or acclimatization, by an ectotherm to cooler seasonal temperatures may include
   a. an increase in metabolic rate
   b. aestivation
   c. changes in the lipid components of cell membranes
   d. increased vasodilation and countercurrent heat exchange
   e. hibernation

7. Malpighian tubules are excretory organs found in
   a. vertebrates
   b. insects
   c. flatworms
   d. annelids
   e. jellyfish

8. Which process in the nephron is *least* selective?
   a. secretion
   b. reabsorption
   c. transport across the epithelium of a collecting duct
   d. filtration
   e. salt pumping by the loop of Henle

9. The key difference between an ectotherm and an endotherm is that
   a. ectotherms generate energy mainly from fermentation; endotherms mainly from cellular respiration
   b. ectotherms are mostly aquatic animals; endotherms are mostly terrestrial
   c. ectotherms warm their bodies mainly by absorbing environmental heat; endotherms mainly use metabolic heat to warm their bodies
   d. ectotherms are "cold-blooded" animals with body temperatures that cannot reach the high body temperatures of endotherms
   e. ectotherms are all invertebrates; endotherms are all vertebrates

10. The vertebrate liver functions in all of the following regulatory processes *except*
   a. osmoregulation by variable excretion of salts
   b. maintenance of blood sugar concentration
   c. detoxification of harmful substances
   d. production of nitrogenous wastes
   e. caloric storage in the form of glycogen

CHALLENGE QUESTIONS

1. Compare and contrast the osmoregulatory problems and adaptations of a marine bony fish with a freshwater bony fish.

2. A large part of the terrestrial success of arthropods and vertebrates is attributable to their osmoregulatory capabilities. Compare and contrast the Malpighian tubule with the nephron in regard to anatomy, relationship to circulation, and physiological mechanisms for conserving body water.

3. Write an essay describing the diverse thermoregulatory mechanisms of a human.

SCIENCE, TECHNOLOGY, AND SOCIETY

1. The kidneys remove many drugs from the blood, and these substances show up in the urine. Some employers require a urine drug test at the time of hiring and at intervals during the term of employment. An employee who fails a drug test can lose his or her job. What are some arguments for and against drug testing of individuals in certain occupations?

2. Kidneys were the first organs to be successfully transplanted. A donor can live a normal life with a single kidney, making it possible for individuals to donate a kidney to an ailing relative or even an unrelated individual with a similar tissue type. In some countries, poor people *sell* kidneys to transplant recipients through organ brokers. What are some of the ethical issues associated with this organ commerce?

FURTHER READING

Diamond, J. "Pearl Harbor and the Emperor's Physiologists." *Natural History,* December 1991. How our childhood locale affects our ability to tolerate hot climates.

Heinrich, B. "Some Like It Cold." *Natural History,* February 1994. Thermoregulatory adaptations in winter moths.

Marieb, E. N. *Human Anatomy and Physiology,* 3rd ed. Redwood City, CA: Benjamin/Cummings, 1995. Chapter 26 describes the human excretory system.

McClanahan, L. L., R. Ruibal, and V. H. Shoemaker. "Frogs and Toads in Deserts." *Scientific American,* March 1994. Osmoregulation and thermoregulation in amphibians adapted to extreme environments.

McKenzie, A. "Seeking the Mechanisms of Hibernation." *BioScience,* June 1990. The role of the brain in a fascinating adaptation.

Smith, H. W. *From Fish to Philosopher.* Boston: Little, Brown, 1953. Vertebrate evolution as revealed by kidney structure and function; a classic.

Vickers-Rich, P., and T. H. Rich. "Australia's Polar Dinosaurs." *Scientific American,* July 1993. How did dinosaurs living near the South Pole survive the cold, dark climate?

CHAPTER 41

CHEMICAL SIGNALS IN ANIMALS

KEY CONCEPTS

- A variety of chemical signals coordinates body functions: *an overview*
- Hormone binding to specific receptors triggers signaling mechanisms at the cellular level
- Many chemical signals are relayed and amplified by second messengers and protein kinases
- Invertebrate control systems often integrate endocrine and nervous system functions
- The hypothalamus and pituitary integrate many functions of the vertebrate endocrine system
- The vertebrate endocrine system coordinates homeostasis and regulates growth, development, and reproduction
- The endocrine system and the nervous system are structurally, chemically, and functionally related

People offer hormones as an explanation for the howling of alley cats and the moodiness of teenagers. Over a million diabetics in the United States take the hormone insulin, and other hormones are used in cosmetics intended to keep the skin smooth or are added to livestock feed to fatten cows. Hormones are chemical signals that communicate regulatory messages within an animal. They affect bioenergetics and metabolism, growth, development, behavior, and homeostasis. The monarch butterfly in the photograph that opens this chapter has just emerged from the silvery cocoon attached to the twig above it. In becoming an adult, the butterfly has undergone a complete change of body form, a process regulated by hormones. Internal communication involving those hormones makes it possible for different parts of the insect's body to develop in concert and to function in homeostasis.

Two structurally and functionally overlapping systems of internal communication that have evolved in animals are the ***nervous system*** *and the* ***endocrine system.*** *The nervous system, which we will study in Chapters 44 and 45, conveys high-speed signals along specialized cells called neurons. These rapid messages function in such activities as the movement of body parts in response to sudden environmental changes—jerking one's hand away from a flame, for example. Slower means of communication regulate other biological processes, such as the maturation of a butterfly. Different parts of the body must be informed how fast to grow and when to develop the characteristics that distinguish male and female or the juvenile and adult of a species. In some cases, neighboring cells communicate by chemical messages that simply diffuse through the interstitial fluid from one cell to another. In many cases, however, messages must travel over greater distances to reach their targets. The endocrine system is a major source of such dispatches in the form of hormones. This chapter surveys the hormonal control of animal form and function.*

A variety of chemical signals coordinates body functions: *an overview*

All animals exhibit some form of coordination by chemical signals (FIGURE 41.1). The hormones of the endocrine system convey information between organs within the body, while other types of chemical messengers perform other functions. For example, pheromones are chemical signals between different individuals, as in mate attraction. Still other messengers act only between cells on a localized scale. The most familiar of these local regulators

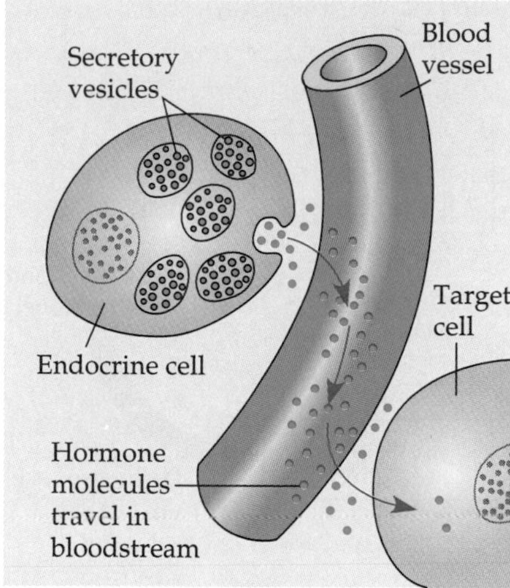

HORMONE FROM AN ENDOCRINE GLAND

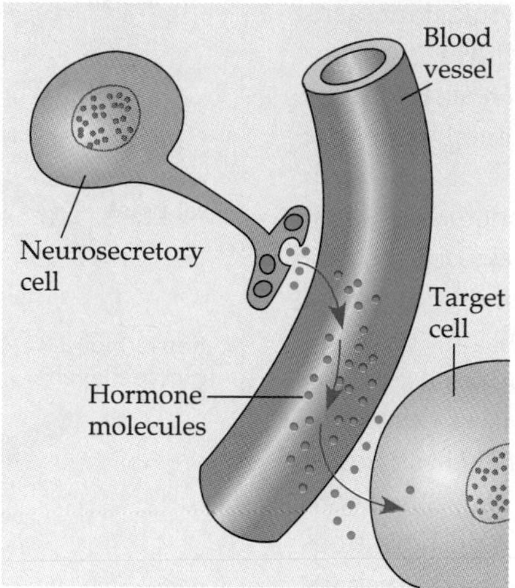

HORMONE FROM A NEUROSECRETORY CELL

(a) Two forms of long-distance signaling

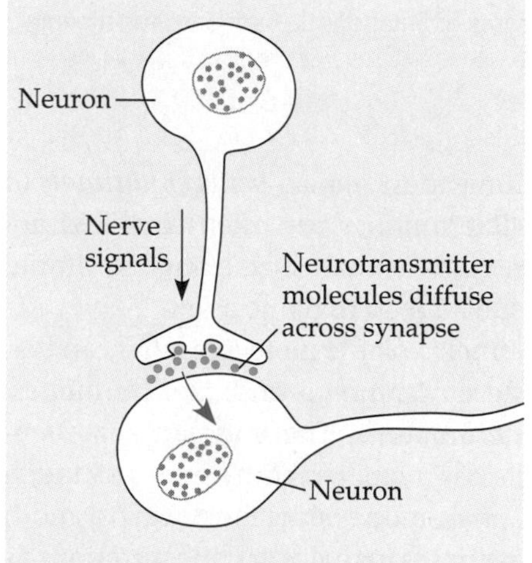

SYNAPTIC SIGNALING

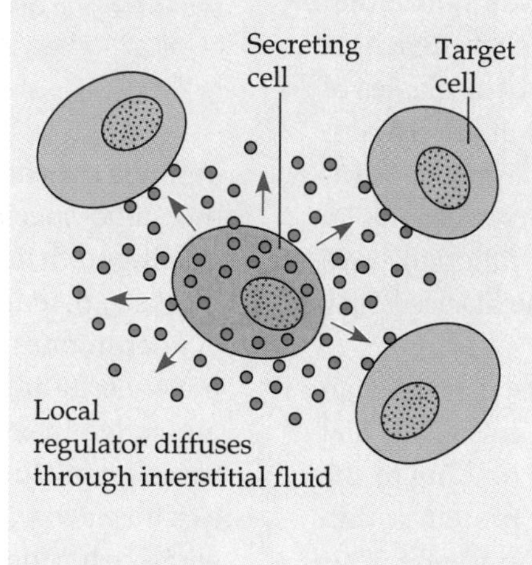

PARACRINE SIGNALING

(b) Two forms of local signaling

FIGURE 41.1
An overview of chemical signaling. (**a**) The endocrine system transmits chemical signals (hormones) between organs in the body. Specialized endocrine cells secrete hormones into body fluids (often the bloodstream, as shown here). Hormones may contact virtually all body cells, but only specific target cells have receptors for the hormone, which enable them to respond to the chemical signal. Endocrine cells are often grouped into endocrine glands. Specialized cells of the nervous system called neurosecretory cells also synthesize and secrete hormones into the bloodstream. (**b**) There are two mechanisms of local chemical signaling. In synaptic signaling, a nerve cell (neuron) releases neurotransmitter molecules into a synapse, the narrow space between the neuron and a single target cell, often another neuron. In paracrine signaling, a secreting cell affects nearby target cells by discharging local regulators into the interstitial fluid.

are the neurotransmitters that carry information between cells of the nervous system. Current research on chemical signals is blurring some of the lines between categories of signals. For instance, some chemicals can function as hormones, local regulators, pheromones, and even as regulators within individual cells. With this in mind, let's take a look at the general classes of chemical signals.

Hormones

An animal **hormone** (Gr. *hormon,* "excite") is a chemical signal that is secreted into body fluids, most often into the blood, by specialized cells called endocrine cells or by specialized nerve cells called neurosecretory cells. **Neurosecretory cells** are neurons that receive signals from other nerve cells and respond by releasing hormones into body fluids or into a storage organ from which hormones are released at a later time. Although a hormone may reach all parts of the body, only certain types of cells, the **target cells,** are equipped to respond. Thus, a given hormone traveling in the bloodstream elicits specific responses—a change in metabolism, for example—from selected target cells, while other cell types ignore that particular hormone. Hormones are very potent regulators; they are effective in minute amounts. Even a slight change in a hormone's concentration can have a significant impact on the body.

Endocrine cells are generally assembled into organs called endocrine glands. A gland is a secretory organ, and animal glands can be classified as exocrine or endocrine. Exocrine glands produce a variety of substances, such as sweat, mucus, and digestive enzymes, and convey their products to the appropriate locations by means of ducts. In contrast, **endocrine glands** are ductless glands. They produce hormones and secrete these chemical messengers into body fluids. Many organs perform both endocrine and exocrine functions. The pancreas, for example, secretes at least two hormones directly into the circulatory system. However, endocrine cells make up only 1% to 2% of the total weight of the pancreas. The rest of the organ is exocrine tissue that produces bicarbonate ions and digestive enzymes that are carried to the small intestine through a duct (see Chapter 37).

There are more than 50 known hormones in the human body, and the endocrine systems of other animals are similarly elaborate. In terms of their chemical structure, these hormones can be grouped into two general classes: lipid, or steroid, hormones; and hormones derived from amino acids. **Steroid hormones,** which include the sex hormones, are lipid molecules the body makes from cholesterol (see FIGURE 5.14). Hormones derived from amino acids include amine hormones, modified versions of single amino acids; peptide hormones, short chains of amino acids (as few as three); protein hormones; and glycoproteins. (For examples of all classes of hormones, see TABLE 41.1, p. 923). Notice that our classification of hormones as steroids or amino acid derivatives groups the hormones according to similarities in chemical structure without considering function. Two hormones belonging to the same chemical class may have completely unrelated functions.

Each hormone molecule has a specific shape that can be recognized by that hormone's target cells. The first step in a hormone's action is the specific binding of the chemical signal to a hormone receptor, a protein within the target cell or built into its plasma membrane. This meeting of hormone and receptor triggers the target cell's responses to the hormonal signal. Cells are unresponsive to a particular hormone if they lack the appropriate receptors.

Our overview of hormones includes another general principle: The effects of a hormone are often countered by an antagonistic (opposing) signal, often another hormone. For example, insulin, a hormone secreted by the pancreas, acts to decrease glucose concentration in the blood, while insulin's antagonist, the hormone glucagon, increases blood glucose level (FIGURE 41.2). Feedback mechanisms that adjust the balance of these two hormones maintain blood sugar at a concentration very close to a set point. We will see that antagonistic hormones function in many other examples of homeostasis.

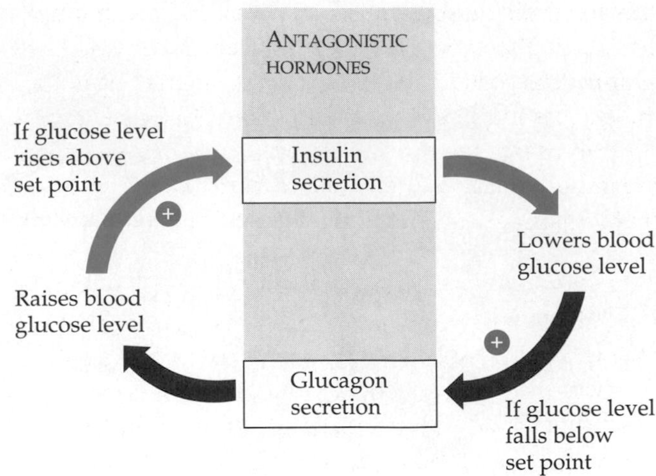

FIGURE 41.2
Antagonistic hormones and homeostasis: an example. Feedback mechanisms in which blood glucose concentration controls the secretion of insulin and glucagon maintain glucose levels close to the set point. Later in the chapter, you will learn more about this and other cases of antagonistic hormones functioning in homeostasis.

Pheromones

Pheromones are chemical signals that function much like hormones, with one important exception: Instead of coordinating the parts of a single animal's body, pheromones are communication signals *between animals* of the same species. Pheromones are often classified according to their functions as mate attractants, territorial markers, or alarm substances, to name a few.

Pheromones are small, volatile molecules that can disperse easily into the environment and, like hormones, are active in minute amounts. The mate attractants of some female insects can be detected by males as much as a mile away. The pheromone of the female gypsy moth elicits behavioral responses in males at concentrations as low as 1 molecule of pheromone in 10^{17} molecules of other gases in the air.

Compared to most animals, humans do not have well-developed olfactory senses, but it is interesting to speculate whether we use pheromones to communicate. Some indirect evidence suggests we do—for example, cases of women living together for several months, as in dormitories, convents, or prisons, who begin to have synchronous menstrual cycles. The role of pheromones in social behavior is considered in more detail in Chapter 50.

Local Regulators

Chemical messengers that affect target cells adjacent to or near their point of secretion function in local regulation (see FIGURE 41.1b). The neurotransmitters are a familiar group of local regulators. These substances carry information from one neuron to another or from a neuron to a muscle, gland, or other target cell. In *synaptic signaling,* a neuron dispatches its transmitter into a synapse, a junction with a single target cell. Thus, neurotransmission is the most direct form of chemical communication. Less direct, but still very localized, is *paracrine signaling*. This occurs when a cell dispatches regulatory substances into the surrounding interstitial fluid, affecting only nearby target cells. Examples of paracrine signals are histamine and interleukins, which are among the local regulators that coordinate the body's defenses (see Chapter 39).

Many types of cells produce gases, such as nitric oxide (NO), that have multiple functions as local regulators. Highly reactive and potentially toxic, NO usually affects its targets in a few seconds and then breaks down. Discoveries made in the 1980s confirmed that NO and another potentially toxic gas, carbon monoxide (CO), can act as both neurotransmitters and paracrine signals. We will examine their role as neurotransmitters in Chapter 44. As a paracrine signal, NO released by endothelial cells in blood vessels makes the adjacent smooth muscles relax, dilating the vessel walls. And NO secreted by white blood cells kills certain cancer cells and bacteria in body fluids. Two other groups of local regulators are growth factors and prostaglandins.

Growth factors are peptides and proteins that regulate the behavior of cells in growing and developing tissues. They must be present in the extracellular environment for many types of cells to grow, divide, and develop normally. Growth factors are generally named for the first function discovered for them, and this can be misleading because each growth factor can affect several kinds of cells and have a variety of functions. For instance, a protein called nerve growth factor (NGF) that speeds the rate of development of certain embryonic nerve cells also affects developing white blood cells and several other kinds of cells such as fibroblasts.

The action of growth factors has been studied mainly in cell cultures, but a variety of experiments show that growth factors work within the animal body as well as in cultures. For instance, injecting one called epidermal growth factor (EGF) into fetal mice accelerates epidermal development. And a group of peptides known as insulinlike growth factors (IGFs), produced by the liver, are essential to skeletal development. It is likely that the interaction of numerous growth factors regulates cell behavior in developing tissues and organs of animals.

Research on how growth factors actually produce their effects on target cells is moving very rapidly. It is known that growth factors bind with specific proteins on the outer surface of plasma membranes, and the binding is the first step in a signal-transduction pathway leading to a change in gene expression in the target cell nucleus. We discussed the basic mechanism of signal transduction in Chapter 8 and examined specific examples in plants in Chapter 35. We will study signal transduction in animals in detail when we look at hormone action in the next section.

Prostaglandins (PGs) are modified fatty acids, often derived from lipids of the plasma membrane. Released from most types of cells into the interstitial fluid, prostaglandins function as local regulators affecting nearby cells in various ways.

About 16 prostaglandins have been discovered. Very subtle differences in their molecular structures can result in profound differences in how these signals affect target cells. For example, prostaglandin E (PGE) and prostaglandin F (PGF) have opposite effects on the smooth muscle cells in the walls of blood vessels serving the lungs. PGE causes the muscles to relax, which dilates the blood vessels and promotes oxygenation of the blood. PGF signals the muscles to contract, which constricts the vessels and reduces blood flow through the lungs. Thus, these two chemical signals are antagonistic, and shifts in their relative concentrations contribute to moment-to-moment adjustments an animal makes to cope with changing circumstances. The use of antagonistic signals as counterbalances is a common regulatory mechanism in chemical and nervous coordination of the body.

Some of the best-known actions of prostaglandins are on the female reproductive system. For example, prostaglandins secreted by cells of the placenta cause chemical changes in the nearby muscles of the uterus, making them more excitable and thereby helping to induce labor during childbirth.

Prostaglandins also function as local regulators in the defense mechanisms of vertebrates. Various prostaglandins help induce fever and inflammation and also intensify the sensation of pain (which can be thought of as contributing to the body's defense by sounding an alarm that something harmful is going on). Aspirin may reduce these symptoms of injury or infection by inhibiting the synthesis of prostaglandins.

So far, biologists have probably recognized only a fraction of the diverse functions of prostaglandins as local regulators of animal tissues. Scientists continue to discover new hormones and other chemical signals and are learning more about their actions. In fact, a central question in endocrinology is: How do hormones actually trigger the changes they do?

Hormone binding to specific receptors triggers signaling mechanisms at the cellular level

A set of general principles seems to govern the activity of the hormones we know about. First, hormones can act at very low concentrations. Second, a given hormone can affect different target cells within an animal differently, or it may affect different species differently. A striking example is thyroxine, a hormone of the thyroid gland. In humans and other vertebrates, it is responsible for metabolic regulation. But thyroxine also plays diverse roles in animal development. In a specific case, thyroxine triggers the development of an adult frog from the tadpole stage, stimulating resorption of the tadpole's tail and other morphological changes during metamorphosis. The diversity of responses of target cells to hormones is seemingly endless and depends on the number and affinity of receptor proteins within the target cells.

In sharp contrast to the varied responses it can elicit, a hormone seems to trigger changes in target cells by one of only two general mechanisms. Some hormones, especially the steroids, actually enter their target cells. But most nonsteroid hormones attach to the cell surface and influence activity within the cell through second messengers of signal-transduction pathways. As we saw in Chapter 8, a **signal-transduction pathway** is a sequence of steps that enables a chemical signal—in this case, a hormone—to produce an intracellular change. Both mechanisms of hormone action involve the binding of receptor proteins to the hormones, and it is the hormone-receptor complex that actually triggers the effects of the hormone.

Steroid Hormones and Gene Expression

The general signaling mechanism for steroid hormones was discovered by studying two vertebrate hormones, estrogen and progesterone. In most mammals, including humans, these hormones are necessary for the normal development and function of the female reproductive system. In the early 1960s, researchers demonstrated that cells in the reproductive tract of female rats accumulate estrogen. The hormone was found within the nuclei of these cells, but not in the cells of such tissues as the spleen, which do not respond to estrogen. Progesterone also enters the nuclei of target cells. Such observations led to the hypothesis that cells sensitive to a steroid hormone contain receptor molecules that bind specifically to that hormone. It is now known that the steroid receptors are specialized proteins within the cells.

The specificity of receptor proteins is a key to understanding steroid hormone action. According to the model illustrated in FIGURE 41.3, a steroid hormone that crosses the membrane of its target cell binds to a receptor protein in the cytoplasm. The hormone-receptor complex then has the proper conformation to bind to certain sites on DNA. It is transported into the nucleus and attaches to certain regulatory sites along regions of the target cell's genome. The binding of the hormone-receptor complex to these sites can either induce or suppress the expression of specific genes (see Chapter 18). For example, estrogen induces certain cells in the reproductive system of a female bird to synthesize large quantities of ovalbumin, the main storage protein stockpiled in egg white. Different proteins are synthesized in the liver cells of the same animal in response to estrogen.

How can two types of cells respond differently to a common chemical signal (estrogen, in this case)? It is likely that the estrogen receptor is the same protein in both target cells, but the hormone-receptor complex probably binds with regulatory sites controlling different genes in the two kinds of cells.

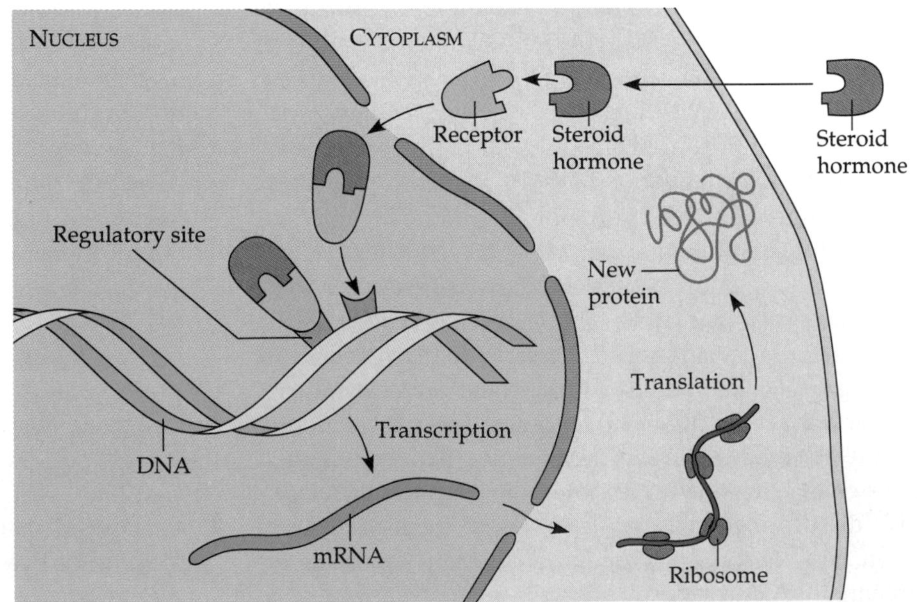

FIGURE 41.3
Steroid hormone action. In this model, the lipid-soluble steroid passes through the plasma membrane and binds to a receptor protein present only in target cells. The hormone-receptor complex then enters the nucleus and binds to specific regulatory sites, stimulating the expression of specific genes.

Peptide Hormones and Signal Transduction

A dramatic example of hormone action is illustrated when a frog's skin turns darker or lighter, an adaptation that helps camouflage the frog as lighting changes. This color change is controlled by a peptide hormone called melanocyte-stimulating hormone (MSH), secreted by the pituitary gland at the base of the brain. Melanocytes are specialized skin cells that contain the dark brown pigment melanin in cytoplasmic organelles called melanosomes. The frog's skin appears light when the melanosomes are clustered tightly around the cell nucleus and darker when the melanosomes spread throughout the cell. When MSH is added to the interstitial fluid surrounding the pigment-containing cells, the melanosomes disperse. However, direct microinjection of MSH into individual melanocytes does not induce melanosome dispersion. Why is it that some hormones will not act directly within a target cell?

Peptide hormones and most other hormones derived from amino acids are unable to pass through the plasma membrane of their target cells, but they still influence cellular activity. Unlike steroids, these hormones bind to specific sites on receptor proteins embedded in the cell's plasma membrane. The receptors are components of signal-transduction pathways that convert the hormone's extracellular signal to intracellular signals that change the behavior of the target cells.

Many chemical signals are relayed and amplified by second messengers and protein kinases

Our current understanding of how hormones act via signal-transduction pathways has advanced from the pioneering work of Earl W. Sutherland, whose research led to a Nobel Prize in 1971. Sutherland and his colleagues at Vanderbilt University were investigating how the hormone epinephrine stimulates breakdown (depolymerization) of the storage polysaccharide glycogen within liver cells and muscle cells. Glycogen depolymerization releases the sugar glucose-1-phosphate, increasing the energy supply for cells. Thus, one effect of epinephrine, secreted from the adrenal gland during times of physical or mental stress, is the mobilization of fuel reserves. Sutherland's research team discovered that epinephrine stimulates glycogen breakdown by activating a cytoplasmic enzyme called glycogen phosphorylase. However, if epinephrine was added to a test-tube mixture containing the phosphorylase and its substrate, glycogen, no depolymerization occurred. Epinephrine could activate glycogen phosphorylase only if the hormone was added to the extracellular solution bathing intact cells with plasma membranes. Thus, the search began for a second messenger that transmits the signal from the plasma membrane to the metabolic machinery in the cytoplasm.

Cyclic AMP

Sutherland found that the binding of epinephrine to the plasma membrane of a liver cell elevates the cytoplasmic concentration of a compound called cyclic adenosine monophosphate, abbreviated **cyclic AMP** or **cAMP** (FIGURE 41.4). The enzyme **adenylyl cyclase** converts ATP to cAMP in response to a hormonal signal—in this case, epinephrine. Adenylyl cyclase, a membrane protein with its active site facing the cytoplasm, is the effector in the signal-transduction pathway. The enzyme is idle until activated by the binding of epinephrine to the hormone's receptor protein. Thus, the first messenger, the hormone, triggers the synthesis of cAMP, which functions as a

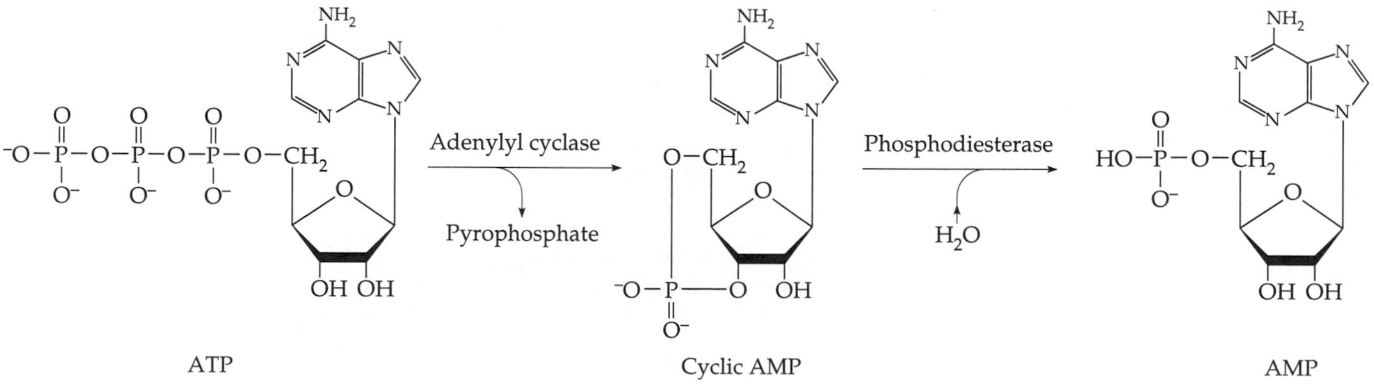

ATP — Adenylyl cyclase ($\rightarrow$ Pyrophosphate) — Cyclic AMP — Phosphodiesterase (H_2O) — AMP

FIGURE 41.4
Cyclic AMP, a second messenger. Binding of a peptide hormone to its specific receptor on the surface of a target cell causes adenylyl cyclase, an enzyme embedded in the plasma membrane, to convert ATP to cyclic AMP (cAMP). The cAMP functions as a second messenger that relays the signal from the membrane to the metabolic machinery of the cytoplasm. When the first message, the hormonal signal, ceases, the second message is also shut down by the action of phosphodiesterase, an enzyme that converts cAMP to inactive AMP.

second messenger that relays the signal to the cytoplasm. It is not epinephrine itself but its cytoplasmic emissary, cAMP, that activates glycogen depolymerization within the cell. When the extracellular concentration of the hormone declines, this reduces the second message. Hormones bind strongly but reversibly to their receptors, and less epinephrine means that there are, at any particular instant, fewer hormone-receptor complexes to activate adenylyl cyclase. Also, the second messenger does not persist for long in the absence of the first message, because another enzyme converts the cAMP to an inactive product. Another surge of epinephrine will again boost the cytoplasmic concentration of the second messenger. Epinephrine is only one of many hormones derived from amino acids that signal target cells by using cAMP as a second messenger.

Subsequent research has revealed additional players in the signal-transduction pathways of hormones. Although the hormone receptor and effector enzyme are both built into the plasma membrane, they do not interact directly. A third class of membrane proteins functions as a relay. One type of relay, called **G proteins,** binds to GTP (guanosine triphosphate), an energy carrier closely related to ATP. Docking of the hormone with the receptor activates a G protein, making it bind GTP, and in turn activating adenylyl cyclase to produce cAMP (FIGURE 41.5).

A G protein provides a versatile relay between a variety of hormone receptors and adenylyl cyclase. The same G protein that stimulates cAMP production in response to epinephrine also transduces several other hormonal signals. For example, MSH, the hormone that triggers pigment dispersion in the melanocytes of a frog, works via the receptor–G-protein–adenylyl-cyclase linkage. The receptors, however, are unique for each hormone.

A second type of G protein inhibits adenylyl cyclase. Binding of an inhibitory hormone to its receptor causes this G protein to lower the activity of adenylyl cyclase, thus reducing the cytoplasmic concentration of cAMP. These opposing effects of two G proteins in regulating adenylyl cyclase enable the cell to fine-tune its metabolism in response to slight changes in the proportions of antagonistic hormones.

Signal Amplification and Specificity

Having examined how membrane proteins transduce a hormonal signal to the second message, let's next consider how cAMP actually causes a particular metabolic response by the cell. In the case of glucose mobilization

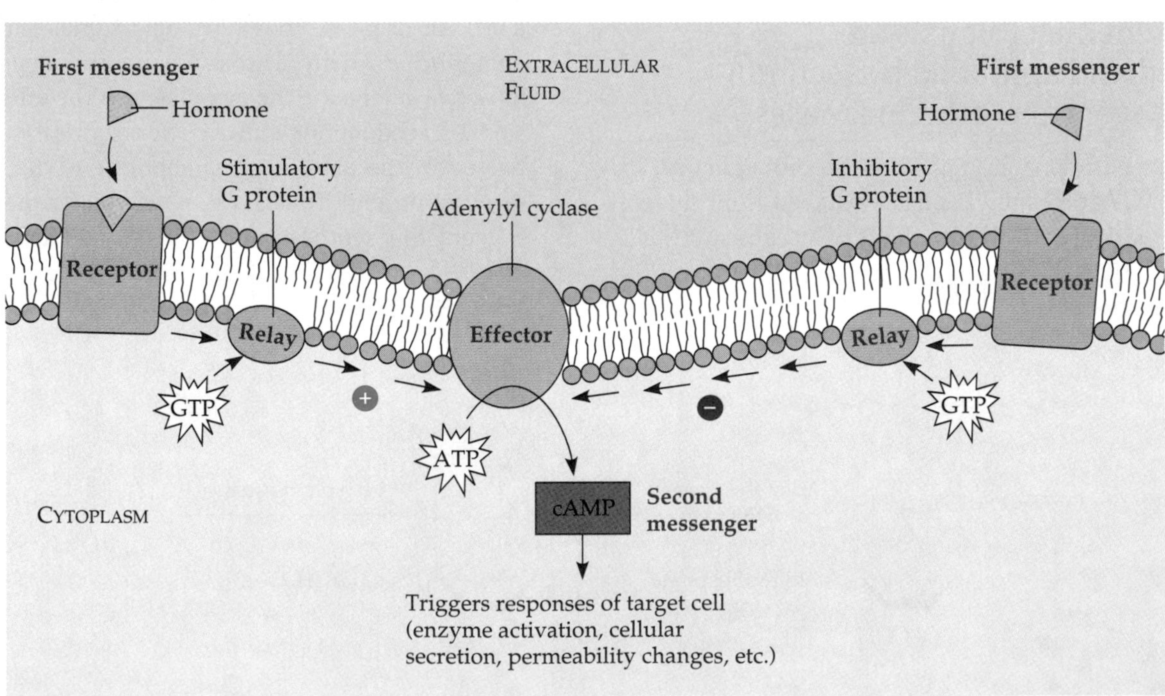

FIGURE 41.5
Signal-transduction pathways involving cAMP as a second messenger. A hormone, the first messenger, binds with, and activates, a specific receptor protein in the plasma membrane of a target cell. The hormone-receptor complex then binds with a molecular relay, a G protein, also in the membrane. In turn, the G protein binds with GTP, thereby stimulating an effector, the enzyme adenylyl cyclase. The enzyme converts ATP to cyclic AMP (the second messenger). Some G proteins stimulate the activity of adenylyl cyclase, thereby increasing the concentration of cAMP in the cytoplasm and the hormone's effect on the target cell. Inhibitory G proteins have opposite effects.

in response to epinephrine, we can specifically ask how cAMP stimulates the activity of glycogen phosphorylase, the enzyme that catalyzes the depolymerization of glycogen. The stimulation is indirect, using two intermediaries (FIGURE 41.6). First, cAMP activates an enzyme called cAMP-dependent protein kinase. A **protein kinase** is an enzyme that catalyzes the transfer of a phosphate group from ATP to proteins, the process of phosphorylation. Depending on the type of protein that receives the phosphate group, phosphorylation either increases or decreases the activity of the protein. In the case of liver cells responding to epinephrine, the cAMP-dependent protein kinase stimulates *another* kinase enzyme, phosphorylase kinase. This second kinase adds a phosphate group to glycogen phosphorylase, the enzyme that actually depolymerizes glycogen. Thus, the metabolic response to epinephrine involves an **enzyme cascade,** where each step activates an enzyme that in turn activates the next enzyme in the series: cAMP starts the cascade within the cytoplasm by activating protein kinase, which activates phosphorylase kinase, which activates glycogen phosphorylase, which mobilizes glucose by hydrolyzing glycogen.

One advantage of this elaborate enzyme cascade is that it amplifies response to the hormone. With each catalytic step in the cascade, the number of activated products is much greater than in the preceding step. A relatively small number of adenylyl cyclase molecules can produce a large number of cAMP molecules, which activate even more molecules of protein kinase, which activate still more molecules of phosphorylase kinase, and so on. As a result of this amplification, a very few molecules of epinephrine binding to receptors on the surface of a liver or muscle cell can quickly result in the release of millions of glucose molecules from glycogen.

Such an indirect mechanism of hormone action offers another advantage. An animal can use a single basic mechanism to mediate the responses of diverse cell types to many different hormones. We have already seen that two different hormones can sound the same second message via a common linkage of G proteins to adenylyl cyclase. The second messenger, cAMP, alters the activity of various cytoplasmic proteins by activating a cAMP-dependent protein kinase. Given this generalized action of hormones, how can we account for specificity in the responses of various target cells to hormonal signals?

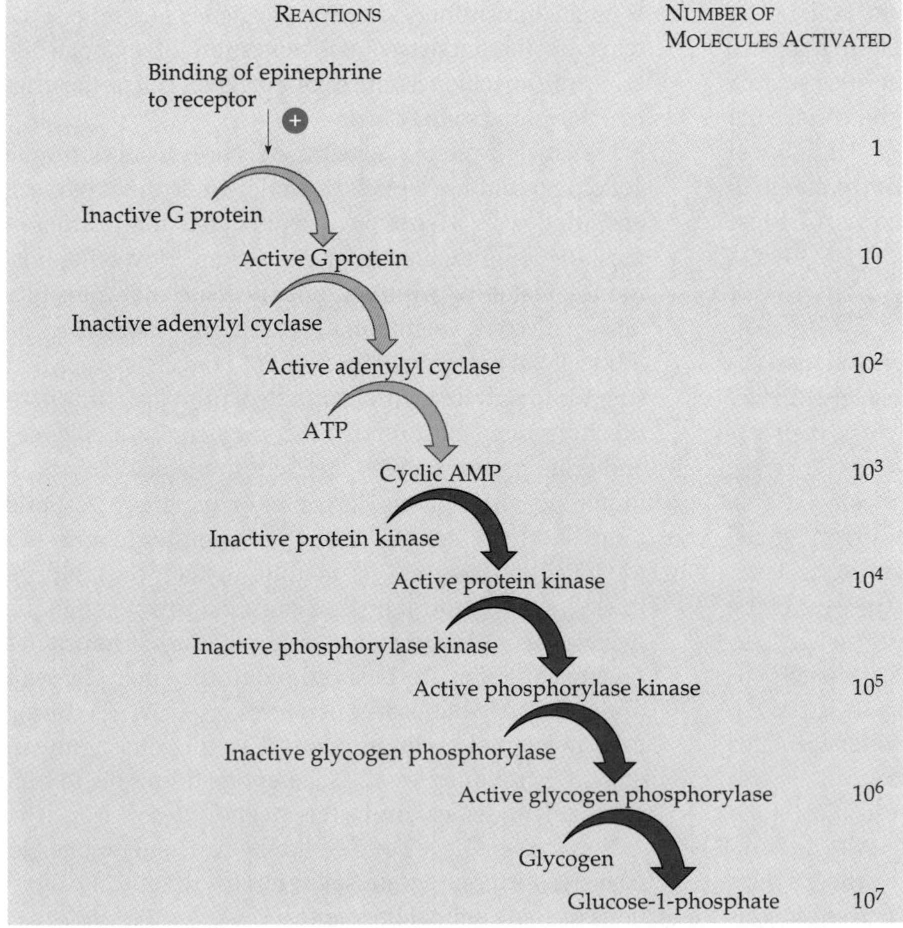

FIGURE 41.6
How an enzyme cascade amplifies response to a hormone. In this system, the hormone epinephrine acts through cyclic AMP to activate a succession of enzymes, leading to extensive depolymerization of glycogen. The number of activated molecules at each step is actually much larger than shown here. With each step, the hormonal signal is amplified, because each enzyme molecule can activate many molecules of its substrate, the next protein in the cascade.

First, a particular type of cell is only tuned-in to certain hormones—those that fit the receptors found on the surface of that cell type. Second, cAMP-dependent protein kinases vary somewhat in structure and function from tissue to tissue. Third, the steps in the enzyme cascade that follow the activation of protein kinase differ markedly from one cell type to another. Different kinds of specialized cells contain different proteins capable of being phosphorylated by protein kinase. In a liver cell responding to epinephrine, protein kinase activates phosphorylase kinase in the enzyme cascade that leads to glycogen breakdown. In a frog melanocyte responding to MSH, protein kinase apparently activates proteins of the cytoskeleton that function in pigment dispersion. In other cases, protein kinase phosphorylates membrane proteins and alters the transport of substances across the membrane. As it differentiates, each type of cell acquires its individual responses to peptide hormones by producing a unique set of proteins regulated by cAMP-dependent protein kinases. In this way, the same second messenger, cAMP, regulates different things in different cells.

Researchers have recently discovered another important function of cAMP, that of inhibiting or stimulating key enzyme components of the signal-transduction pathways of many growth factors. This functional linkage between cAMP as a participant (second messenger) in many signal-transduction pathways and as a regulator of others is an example of the complex interplay among chemical signals in cells. Studies of the molecular interplay in growth factor regulation may give us a better understanding of cancers in which abnormal signal-transduction pathways lead to uncontrolled cell division.

Cytoplasmic Ca^{2+} and Signal Transduction

In contrast to those using cAMP, many chemical messengers in animals, including neurotransmitters, growth factors, and some hormones, induce responses in their target cells via signal-transduction pathways that increase the cytoplasmic concentration of calcium ions (Ca^{2+}). Binding of the messenger to a specific receptor on the cell surface results in the uptake of Ca^{2+} from extracellular fluid or in the release of Ca^{2+} from a reservoir within the cell, usually the endoplasmic reticulum. The Ca^{2+} increase in the cytoplasm alters the activities of specific enzymes, either directly or by first binding to a protein called **calmodulin,** which then binds to other proteins and changes their activities. Given these discoveries, calcium was once believed to be a second messenger in the responses of cells to extracellular signals. However, additional research has shown that calcium is actually a *third* messenger. The second messenger, which functions between the hormonal signal and the rise in cytoplasmic Ca^{2+} concentration, is a compound called **inositol trisphosphate (IP_3).**

Inositol trisphosphate is derived from a particular type of phospholipid present in the plasma membrane (FIGURE 41.7). The production of IP_3 involves a signal-transduction pathway analogous to the mechanism that transduces a hormonal signal into the production of cAMP. A hormone binds to its specific receptor in the plasma membrane, and the hormone-receptor complex activates a relay, a G protein that also resides in the membrane. This G protein, different from those that function in the cAMP system, stimulates a membrane enzyme called phospholipase C. This enzyme then cleaves a phospholipid into two products, IP_3 and another compound called diacylglycerol. Both products may function as second messengers. The diacylglycerol activates another membrane enzyme called protein kinase C, which triggers some of the cytoplasmic responses to the hormone by phosphorylating specific proteins (similar to the action of cAMP-dependent protein kinase, but protein kinase C activates a different set of cytoplasmic proteins). The IP_3 released from the plasma membrane may cause the cell to take up Ca^{2+} from the surrounding fluid or interact with the membrane of the endoplasmic reticulum, causing Ca^{2+} to leak from the ER. The signal is finally transmitted to various enzymes and other proteins by the increased concentration of cytoplasmic Ca^{2+}, and specific responses of the target cell to the original hormonal signal occur.

Let's review the two general mechanisms of hormone action. Steroid hormones generally enter the target cell and bind with a specific receptor, and the hormone-receptor complex alters gene expression. Most other hormones, including peptides, operate via signal-transduction pathways, regulating cytoplasmic enzymes via intracellular messengers, often cAMP or Ca^{2+}. Notice that steroids mainly affect the *synthesis* of proteins, while peptide hormones most often affect the *activity* of enzymes and other proteins already present in the cell. This is an important distinction. Target cells generally respond more slowly to steroids than to peptide hormones, but the changes induced by steroids usually last longer. Indeed, steroid hormones are commonly involved in the development of tissues. For example, steroid hormones secreted by a woman's ovary stimulate the gradual growth of the uterine wall over a period of a few weeks during each menstrual cycle. In contrast to this time frame, it takes only minutes for MSH, a peptide hormone, to trigger pigment dispersion in the melanocytes of frog skin.

Now that we have considered the two general mechanisms of hormone action, let's examine the specific functions of some animal hormones.

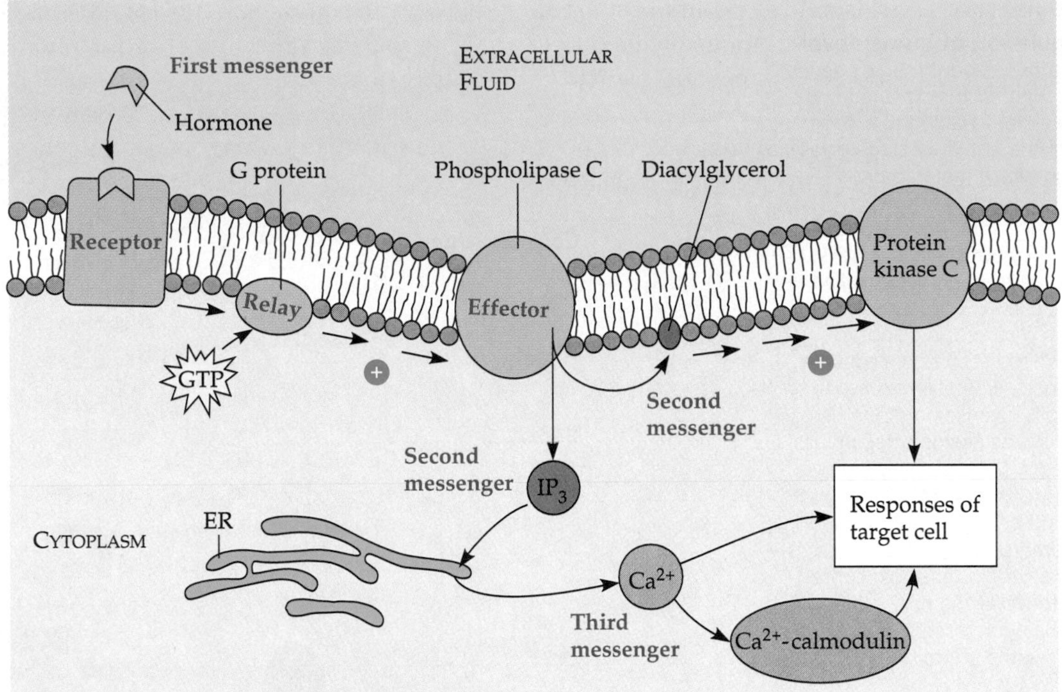

FIGURE 41.7
Ca^{2+} as an intracellular messenger. The first messenger, the hormone, binds to its specific receptor, activating a G protein that in turn stimulates phospholipase C, an effector enzyme that resides in the plasma membrane. Phospholipase C cleaves a membrane phospholipid into two products—inositol trisphosphate (IP_3) and diacylglycerol—that function as second messengers. Diacylglycerol stimulates still another membrane enzyme, protein kinase C, which triggers various responses of the target cell by phosphorylating specific proteins. The IP_3 released from the membrane in response to the hormonal signal triggers the release of Ca^{2+} into the cytoplasm. The Ca^{2+} increase regulates the activity of many cellular proteins, either by itself or by first binding to the protein calmodulin.

Invertebrate control systems often integrate endocrine and nervous system functions

Although diverse invertebrate hormones function in homeostasis—by regulating water balance, for instance—the hormones that have been most extensively studied function in reproduction and development. In *Hydra,* for example, one hormone stimulates growth and budding (asexual reproduction) but prevents sexual reproduction. In more complex invertebrates, the endocrine and nervous systems are generally integrated in the control of reproduction and development. One well-studied example of nerve and hormone interaction in controlling both reproductive physiology and behavior involves the hormone that regulates egg laying in the mollusk *Aplysia.* This peptide hormone, secreted by specialized neurons, stimulates the laying of thousands of eggs and also inhibits feeding and locomotion, activities that interfere with reproduction.

All groups of arthropods have extensive endocrine systems. Crustaceans, for example, have hormones for growth and reproduction, water balance, movement of pigments in the integument and in the eyes, and the regulation of metabolism. Insects have been the most thoroughly studied arthropods. Having exoskeletons that cannot stretch, insects appear to grow in spurts, shedding the old and secreting a new exoskeleton with each molt. Further, most insects acquire their adult characteristics in a single, terminal molt. By studying these events, scientists have identified how three important hormones interact to control insect development (FIGURE 41.8).

Insect molting is triggered by the steroid hormone **ecdysone,** secreted from a pair of prothoracic glands just behind the head. Besides stimulating the molt, ecdysone also favors the development of adult characteristics, as in the change from a caterpillar to a butterfly. Ecdysone has the same mechanism of action as vertebrate steroid hormones, stimulating the transcription of specific genes. Ecdysone production is itself controlled by a second hormone, called **brain hormone (BH).** A peptide from the brain, this hormone promotes development by stimulating the prothoracic glands to secrete ecdysone.

Brain hormone and ecdysone are balanced by **juvenile hormone (JH),** the third hormone in this system. JH is secreted by a pair of small glands just behind the brain, the

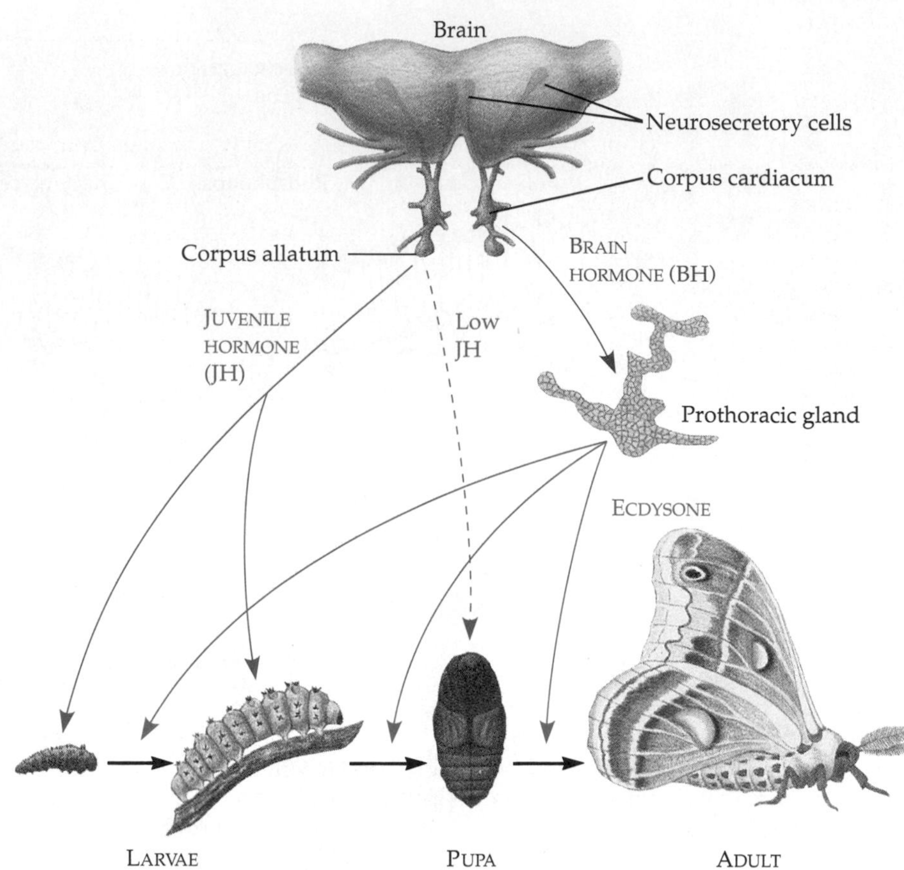

FIGURE 41.8
Hormonal regulation of insect development. Most insects go through a series of larval stages, with each molt (shedding of the old exoskeleton) leading to a larger larva. Molting of the last larval stage gives rise to a pupa, in which metamorphosis produces the adult form of the insect. Neurosecretory cells in the brain produce brain hormone (BH), but the hormone is stored and released from an organ called the corpus cardiacum. BH signals its main target organ, the prothoracic gland, to produce the hormone ecdysone. Ecdysone secretion is episodic, with each release stimulating a molt. Juvenile hormone (JH), secreted by the corpus allatum, determines the result of the molt. At relatively high concentrations of JH, ecdysone-stimulated molting produces another larval stage. Thus, JH suppresses metamorphosis. But when levels of JH fall below a certain threshold concentration, a pupa forms at the next ecdysone-induced molt. The adult insect emerges from the pupa (see the photo on p. 912).

corpora allata. Juvenile hormone promotes the retention of larval characteristics. In the presence of a relatively high concentration of juvenile hormone, ecdysone can still stimulate molting, but the product is a larger larva. Only when the level of juvenile hormone wanes can ecdysone-induced molting produce a developmental stage called a pupa. Within the pupa, metamorphosis replaces larval anatomy with the insect's adult form. (Synthetic versions of JH are now being used as insecticides to prevent insects from maturing into reproducing adults.)

In all these invertebrate examples, we see the importance of the nervous system to hormone activity, and we will see these interactions repeatedly as we survey the vertebrate endocrine system.

The hypothalamus and pituitary integrate many functions of the vertebrate endocrine system

Of the numerous hormones regulating body functions of vertebrates, some affect only one or a few tissues. Others, such as the sex hormones, which promote male and female characteristics, affect most of the tissues of the body. Some hormones, called **tropic hormones,** have other endocrine glands as their targets and are particularly important to our understanding of chemical coordination. In studying hormonal regulation of the vertebrate body in the following sections, you may find FIGURE 41.9 (p. 924), showing where the major endocrine glands are located in the human body, and TABLE 41.1, summarizing the functions of the major vertebrate hormones, especially helpful. We begin our discussion of the vertebrate endocrine system by looking at the major control axis of the system.

Structural and Functional Relationships of the Hypothalamus and the Pituitary Gland

The **hypothalamus** plays an important role in integrating the vertebrate endocrine and nervous systems. This region of the lower brain receives information from nerves throughout the body and from other parts of the brain, then initiates endocrine signals appropriate to environmental conditions. In many vertebrates, for example, the brain passes sensory information about seasonal changes and the availability of a mate to the hypothalamus by means of nerve signals; the hypothalamus then triggers the release of reproductive hormones required for breeding.

TABLE 41.1

Major Vertebrate Endocrine Glands and Some of Their Hormones

GLAND	HORMONE	CHEMICAL CLASS	REPRESENTATIVE ACTIONS	REGULATED BY
Hypothalamus	Hormones released by the posterior pituitary and hormones that regulate the anterior pituitary (see below)			
Pituitary gland				
Posterior pituitary (releases hormones made by hypothalamus)	Oxytocin	Peptide	Stimulates contraction of uterus and mammary gland cells.	Nervous system
	Antidiuretic hormone (ADH)	Peptide	Promotes retention of water by kidneys.	Water/salt balance
Anterior pituitary	Growth hormone (GH)	Protein	Stimulates growth (especially bones) and metabolic functions.	Hypothalamic hormones
	Prolactin (PRL)	Protein	Stimulates milk production and secretion.	Hypothalamic hormones
	Follicle-stimulating hormone (FSH)	Glycoprotein	Stimulates production of ova and sperm.	Hypothalamic hormones
	Luteinizing hormone (LH)	Glycoprotein	Stimulates ovaries and testes.	Hypothalamic hormones
	Thyroid-stimulating hormone (TSH)	Glycoprotein	Stimulates thyroid gland.	Thyroxine in blood; hypothalamic hormones
	Adrenocorticotropic hormone (ACTH)	Peptide	Stimulates adrenal cortex to secrete glucocorticoids.	Glucocorticoids; hypothalamic hormones
Thyroid gland	Triiodothyronine (T_3) and thyroxine (T_4)	Amine	Stimulate and maintain metabolic processes.	TSH
	Calcitonin	Peptide	Lowers blood calcium level.	Calcium in blood
Parathyroid glands	Parathyroid hormone (PTH)	Peptide	Raises blood calcium level.	Calcium in blood
Pancreas	Insulin	Protein	Lowers blood glucose level.	Glucose in blood
	Glucagon	Protein	Raises blood glucose level.	Glucose in blood
Adrenal glands				
Adrenal medulla	Epinephrine and norepinephrine	Amine	Raise blood glucose level; increase metabolic activities; constrict certain blood vessels.	Nervous system
Adrenal cortex	Glucocorticoids	Steroid	Raise blood glucose level.	ACTH
	Mineralocorticoids	Steroid	Promote reabsorption of Na^+ and excretion of K^+ in kidneys.	K^+ in blood
Gonads				
Testes	Androgens	Steroid	Support sperm formation; development and maintenance of male secondary sex characteristics.	FSH and LH
Ovaries	Estrogens	Steroid	Stimulate uterine lining growth; development and maintenance of female secondary sex characteristics.	FSH and LH
	Progesterone	Steroid	Promotes uterine lining growth.	FSH and LH
Pineal gland	Melatonin	Amine	Involved in biological rhythms.	Light/dark cycles
Thymus	Thymosin	Peptide	Stimulates T lymphocytes.	Not known

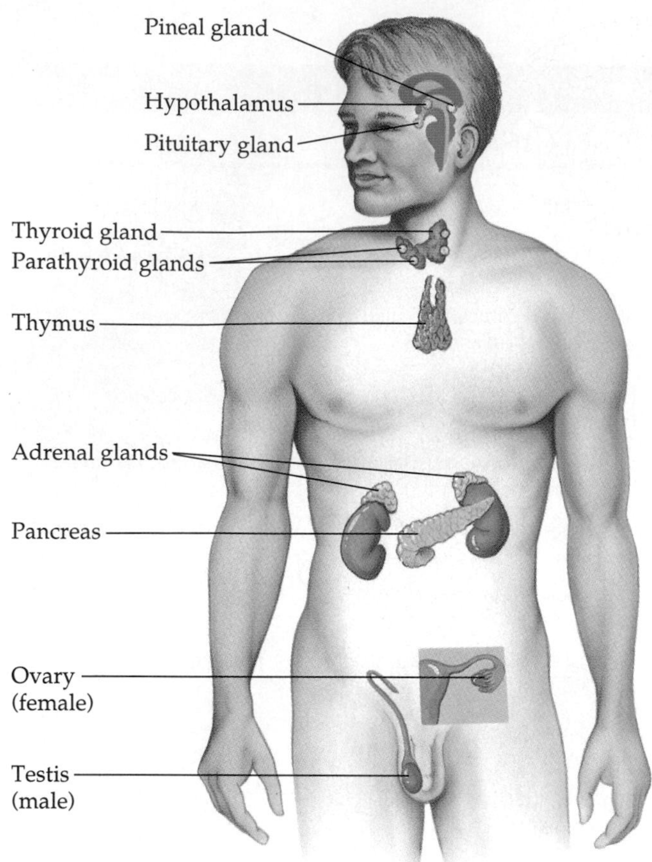

FIGURE 41.9
Human endocrine glands surveyed in this chapter. (The endocrine functions of the stomach, small intestine, heart, and kidneys were covered in Chapters 37, 38, and 40. The liver secretes a number of growth factors.)

The hormone-releasing cells in the hypothalamus are two sets of neurosecretory cells whose secretions are stored in or regulate the activity of the **pituitary gland,** a small organ with multiple endocrine functions, located at the base of the hypothalamus. The pituitary gland was formerly called the "master gland" because so many of its hormones regulate other endocrine functions. However, the pituitary itself obeys hormonal orders from the hypothalamus. The pituitary has two discrete parts, each with a different function (FIGURE 41.10). The **posterior pituitary,** or **neurohypophysis,** is an extension of the brain that stores and secretes two peptide hormones that are made by a set of neurosecretory cells in the hypothalamus. The two hormones released from the posterior lobe act directly on muscles of the uterus and on the kidneys, rather than affecting other endocrine glands.

Unlike the posterior pituitary, the **anterior pituitary,** or **adenohypophysis,** is composed of non-nervous endocrine tissue and synthesizes its own hormones. A set of neurosecretory cells in the hypothalamus exerts control over the anterior pituitary by secreting two kinds of hormones into the blood. **Releasing hormones** make the anterior pituitary secrete its hormones. **Inhibiting hormones** from the hypothalamus make the anterior pituitary stop secreting hormones (see FIGURE 41.10).

Posterior Pituitary Hormones

The posterior pituitary stores and releases the hypothalamic peptide hormones oxytocin and antidiuretic hormone (ADH). These hormones are synthesized in the hypothalamus and secreted into the posterior pituitary. Oxytocin induces contraction of the uterine muscles during childbirth and causes the mammary glands to eject milk during nursing. ADH acts on the kidneys, increasing water retention and thus decreasing urine volume.

ADH is part of an elaborate feedback scheme that helps regulate the osmolarity of the blood. This mechanism was introduced in Chapter 40, but it is worth reviewing here to illustrate how hormones contribute to homeostasis and how negative feedback controls hormone levels. Blood osmolarity is monitored by a group of nerve cells that function as osmoreceptors in the hypothalamus. When the osmolarity of the plasma increases, these cells shrink slightly (due to osmosis) and transmit nerve impulses to certain neurosecretory cells in the hypothalamus. These cells respond by releasing ADH from their tips (located in the posterior pituitary) into the general circulation. When ADH reaches the kidneys, it binds to receptors on the surface of the cells lining the collecting ducts. This binding activates a signal-transduction pathway that increases the water permeability of the collecting duct. Water exits the collecting ducts and enters nearby capillaries, helping prevent a further increase in blood osmolarity above the set point. The brain's osmoreceptors also stimulate a thirst drive, and drinking water brings blood osmolarity back down to the set point. Thus, the hormonal reaction to high blood osmolarity is augmented by a behavioral response controlled by the nervous system. As the more dilute blood arrives at the brain, the hypothalamus responds to the reduction in osmolarity by slowing the release of ADH and diminishing the sensation of thirst. The effects of these hormonal and behavioral responses—increased water reabsorption in the kidneys and drinking, respectively—prevent overcompensation by shutting off further secretion of the hormone and quenching thirst. We see once again how negative feedback helps maintain homeostasis. This example also highlights the central role of the hypothalamus as a member of both the endocrine system and the nervous system.

FIGURE 41.10

Hormones of the hypothalamus and pituitary glands. The pituitary gland, located at the base of the brain and surrounded by bone, consists of the posterior pituitary (neurohypophysis) and the anterior pituitary (adenohypophysis). (**a**) **The posterior pituitary**. Neurosecretory cells in the hypothalamus synthesize antidiuretic hormone (ADH) and oxytocin, peptide hormones that are transported down the axons to the posterior pituitary, where they are stored. The posterior pituitary releases the hormones into the blood, where they circulate and bind to target cells in the kidneys (ADH) and mammary glands and uterus (oxytocin). (**b**) **The anterior pituitary.** Endocrine cells in the anterior pituitary manufacture a number of hormones and secrete them into the circulation, but the release of these hormones is controlled by the hypothalamus. Neurosecretory cells in the hypothalamus secrete releasing hormones and inhibiting hormones into a capillary network located above the stalk of the pituitary. Blood containing the hormones travels through short portal vessels and into a second capillary network within the anterior pituitary.

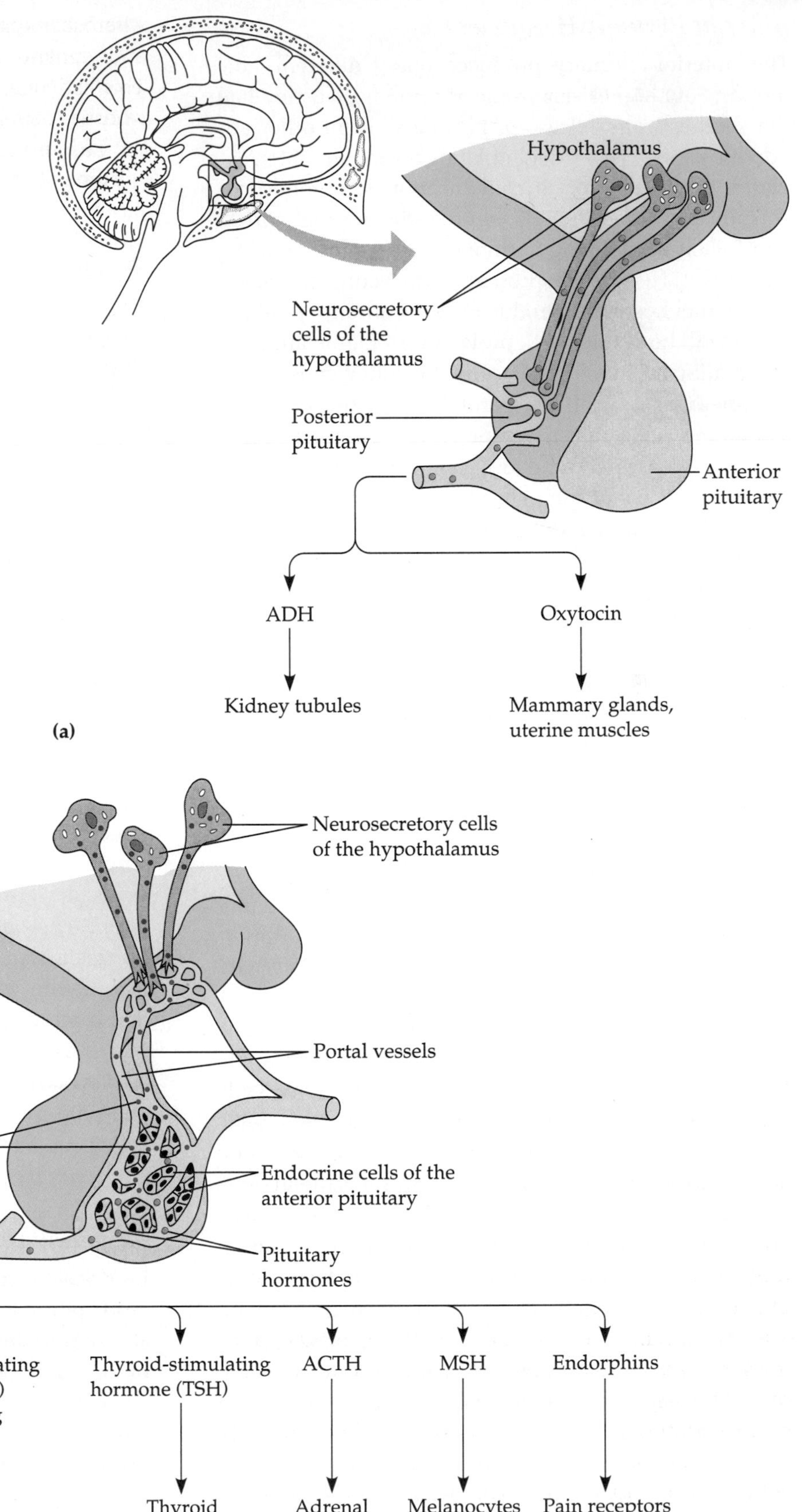

Anterior Pituitary Hormones

The anterior pituitary produces many different hormones. Four of these are tropic hormones that stimulate the synthesis and release of hormones from other endocrine glands. Thyroid-stimulating hormone (TSH) regulates the release of thyroid hormones; adrenocorticotropic hormone (ACTH) controls the adrenal cortex, and follicle-stimulating hormone (FSH) and luteinizing hormone (LH) govern reproduction by acting on the gonads. Other hormones produced by the anterior pituitary are growth hormone (GH), prolactin (PRL), melanocyte-stimulating hormone (MSH), and the endorphins.

Growth hormone (GH), a protein consisting of almost 200 amino acids, affects a wide variety of target tissues and has both direct and tropic effects. GH promotes growth directly and stimulates the production of growth factors. For example, the ability of GH to stimulate the growth of bones and cartilage is partly due to the hormone's signaling the liver to produce **insulinlike growth factors (IGFs),** which circulate in blood plasma and directly stimulate bone and cartilage growth. (This endocrine response to growth hormone qualifies GH as a tropic hormone. And IGF secretion by the liver qualifies that organ as an endocrine gland, among its many other functions; see TABLE 40.1.) In the absence of GH, skeletal growth of an immature animal will stop. If GH is injected into an animal that has been deprived of its own GH, growth will be partially restored.

Several human growth disorders are related to abnormal GH production. Excessive production of GH during development can lead to gigantism, while excessive GH production during adulthood results in the abnormal growth of bones in the hands, feet, and head, a condition known as acromegaly. Deficient GH production in childhood can lead to hypopituitary dwarfism. Children with GH deficiency have been treated successfully with human growth hormones isolated from cadaver pituitaries. However, the supply falls short of demand, and growth hormones from most other animals are ineffective. One of the most dramatic achievements of genetic engineering has been the production of GH by bacteria with genes for human GH spliced into their genomes (see Chapter 19). The product is now being used to treat children with hypopituitary dwarfism. Some athletes take GH (legally or illegally) to build muscles.

Prolactin (PRL) is a protein so similar to GH that it is believed they are encoded in genes that evolved from the same ancestral gene. The physiological roles of these two hormones, however, are different. Prolactin's most remarkable characteristic is the great diversity of effects it produces in different vertebrate species. For example, PRL stimulates mammary gland growth and milk synthesis in mammals; regulates fat metabolism and reproduction in birds; delays metamorphosis in amphibians, where it may also function as a larval growth hormone; and regulates salt and water balance in freshwater fish. This list suggests that prolactin is an ancient hormone whose functions have diversified during the evolution of the various vertebrate classes.

Three of the tropic hormones secreted by the anterior pituitary are closely related chemically. **Follicle-stimulating hormone (FSH), luteinizing hormone (LH),** and **thyroid-stimulating hormone (TSH)** are all similar glycoproteins, protein molecules with carbohydrate attached to them. FSH and LH are also called **gonadotropins** because they stimulate the activities of the male and female gonads, the testes and ovaries. TSH stimulates the production of hormones by the thyroid gland.

The remaining hormones from the anterior pituitary all come from a single parent molecule called pro-opiomelanocortin. This large protein is cleaved into several short fragments inside the pituitary cells. At least three of these fragments are active peptide hormones. **Adrenocorticotropic hormone (ACTH)** stimulates the production and secretion of steroid hormones by the adrenal cortex. As described earlier, **melanocyte-stimulating hormone (MSH)** regulates the activity of pigment-containing cells in the skin of some vertebrates. Very small amounts of MSH are secreted by the human pituitary, but the functions of this hormone in humans are uncertain. The other derivatives of pro-opiomelanocortin are a class of hormones called **endorphins.** These molecules are also produced by certain neurons in the brain. They are sometimes called the body's natural opiates because they inhibit the perception of pain. In fact, heroin and other opiate drugs mimic endorphins and bind to the same receptors in the brain. Some researchers speculate that the so-called runner's high results partly from the release of endorphins when stress and pain in the body reach critical levels. (Endorphins are discussed further in Chapter 44.)

Hypothalamic Hormones

Let's now return to the hypothalamus to see how the anterior pituitary is controlled. The hypothalamic signals called releasing and inhibiting hormones are produced by neurosecretory cells and released into capillaries in a region at the base of the hypothalamus (see FIGURE 41.10b). The capillaries drain into short portal vessels that subdivide into a second capillary bed within the anterior pituitary. (Recall a similar portal circulation in the liver and kidneys; see Chapters 37 and 40, respectively.) In this way, the hypothalamic hormones have direct access to the gland they control. Every anterior pituitary hormone is controlled by at least one releasing hormone, and some have both a releasing hormone and an inhibiting hormone.

The vertebrate endocrine system coordinates homeostasis and regulates growth, development, and reproduction

Now that we have examined the hypothalamus and pituitary as the main control axis of the endocrine system, let's survey how some other major endocrine glands function in the control of homeostasis, growth, development, and reproduction.

The Thyroid Gland

In humans and other mammals, the **thyroid gland** consists of two lobes located on the ventral surface of the trachea (see FIGURE 41.9). In other vertebrates, the halves of the gland may be more separated on the two sides of the pharynx. The thyroid gland produces two very similar hormones derived from the amino acid tyrosine: **triiodothyronine (T_3),** which contains three iodine atoms, and tetraiodothyronine, or **thyroxine (T_4),** which contains four iodine atoms:

TRIIODOTHYRONINE (T_3)

THYROXINE (T_4)

In mammals, T_3 is usually the more active of the two hormones, although both have the same effects on their target cells.

The thyroid gland plays a crucial role in vertebrate development and maturation. A striking example is the thyroid's control of metamorphosis of a tadpole into a frog, which involves massive reorganization of many different tissues. The thyroid is equally important in human development. An inherited condition of thyroid deficiency known as cretinism results in markedly retarded skeletal growth and poor mental development. These defects can often be overcome, at least partially, if treatment with thyroid hormones is begun early in life. Studies with nonhuman animals have shown that thyroid hormones are required for the normal functioning of bone-forming cells and for the branching of nerve cells during embryonic development of the brain.

The thyroid gland also plays a vital role in homeostasis. In adult mammals, for instance, thyroid hormones help maintain normal blood pressure, heart rate, muscle tone, digestion, and reproductive functions. Throughout the body, T_3 and T_4 are important in bioenergetics, generally increasing the rate of oxygen consumption and cellular metabolism. Too much or too little of these hormones in the blood can result in serious metabolic disorders. For example, in humans, an excessive secretion of thyroid hormones, known as hyperthyroidism, produces such symptoms as high body temperature, profuse sweating, weight loss, irritability, and high blood pressure. The opposite condition, hypothyroidism, can cause cretinism in infants and produce symptoms such as weight gain, lethargy, and intolerance to cold in adults. Another condition associated with a shortage of thyroid hormones is an enlargement of the thyroid called goiter, often caused by a deficiency of iodine in the diet.

The secretion of thyroid hormones is controlled by the hypothalamus and pituitary in a complex negative feedback system (FIGURE 41.11). The mammalian thyroid gland also contains endocrine cells that secrete **calcitonin.** This peptide lowers calcium levels in the blood as part of calcium homeostasis, described in the next section.

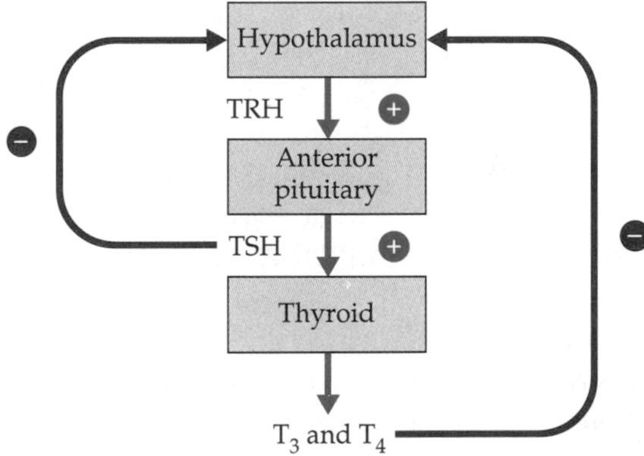

FIGURE 41.11

Feedback control loops regulating the secretion of thyroid hormones. The hypothalamus secretes TRH (TSH-releasing hormone), which stimulates the anterior pituitary to secrete TSH (thyroid-stimulating hormone). When TSH binds to specific receptors in the thyroid gland, a signal-transduction pathway involving cAMP as a second messenger triggers the synthesis and release of the thyroid hormones T_3 and T_4. The system is balanced by negative feedback loops (red arrows); high levels of T_3, T_4, and TSH in the blood inhibit TRH secretion by the hypothalamus. There is also evidence that additional feedback loops are involved; for example, high levels of thyroid hormones may inhibit the anterior pituitary and the thyroid gland itself, and high levels of TSH may inhibit TSH secretion by the anterior pituitary. The hypothalamus–anterior pituitary–thyroid feedback system explains why iodine deficiencies lead to goiter. In the absence of sufficient iodine, the thyroid gland cannot synthesize adequate amounts of T_3 and T_4. Consequently, the pituitary continues to secrete TSH, leading to an enlargement of the thyroid.

The Parathyroid Glands

The four **parathyroid glands,** embedded in the surface of the thyroid, function in the homeostasis of calcium ions. They secrete **parathyroid hormone (PTH),** which raises blood levels of calcium and thus has an effect opposite that of the thyroid hormone calcitonin. Parathyroid hormone elevates blood Ca^{2+} by stimulating Ca^{2+} reabsorption in the kidneys, and by inducing specialized bone cells called osteoclasts to decompose the mineralized matrix of bone and release Ca^{2+} to the blood. Calcitonin has just the opposite effects on the kidneys and bone, thus decreasing blood Ca^{2+}. Vitamin D, synthesized in the skin and converted to its active form in many tissues, is essential to PTH function, so it is also required for complete calcium balance. A lack of PTH causes blood levels of calcium to drop dramatically, leading to convulsive contractions of the skeletal muscles. If unchecked, this condition, known as tetany, is fatal. The control of blood calcium level is an example of how homeostasis is often maintained by the balancing of two antagonistic hormones—in this case, PTH and calcitonin (FIGURE 41.12).

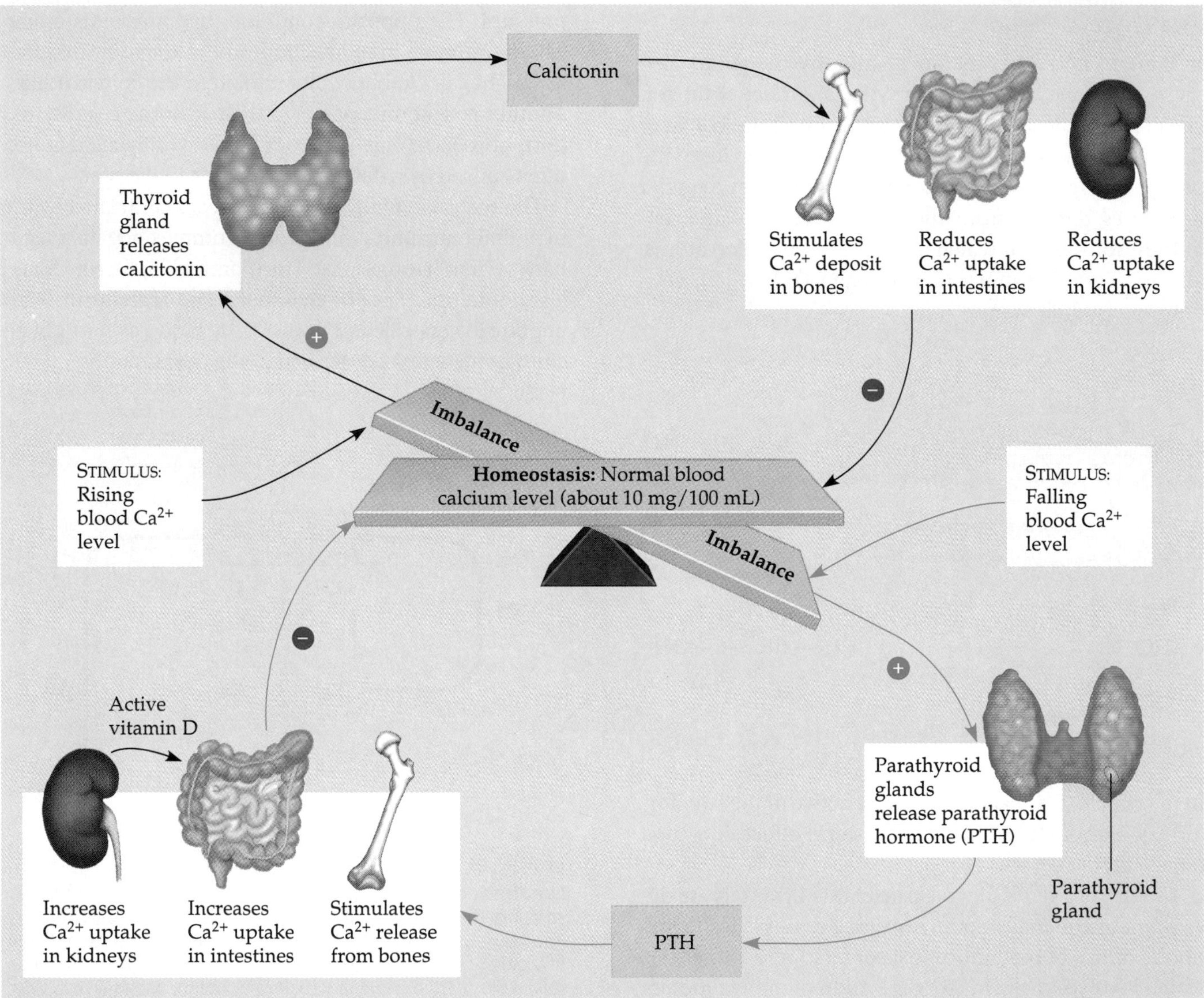

FIGURE 41.12

Hormonal control of calcium homeostasis in mammals. A negative feedback system involving two antagonistic hormones, calcitonin and parathyroid hormone (PTH), maintains the concentration of calcium in blood within a very narrow range of about 10 mg/100 mL. A rise in blood Ca^{2+} induces the thyroid gland to secrete calcitonin, which lowers the Ca^{2+} concentration by increasing bone deposition, reducing Ca^{2+} uptake in the intestines, and reducing reabsorption in the kidneys. These effects are reversed by PTH, which is secreted from the parathyroid glands when the concentration of blood Ca^{2+} falls below the set point. Blood calcium levels begin to increase as target cells in the kidneys, intestines, and bone respond to PTH. Blood Ca^{2+} will only rise so far before the thyroid counters by secreting more calcitonin. In classic feedback fashion, these two hormones balance each other's effects, thereby minimizing fluctuations in the concentration of blood Ca^{2+}, an ion essential to the normal functioning of all cells. Synthesized in an inactive form by skin exposed to sunlight, vitamin D plays an important role in calcium homeostasis. Vitamin D is carried in the blood and converted to its active form in many tissues such as the liver and kidneys. The active form enables PTH to increase Ca^{2+} uptake by the intestines.

The Pancreas

The **pancreas** consists mostly of exocrine tissue that produces digestive enzymes and exports them to the small intestine via the pancreatic duct (see Chapter 37). Scattered among this exocrine tissue are the **islets of Langerhans,** clusters of endocrine cells. Each islet has a population of **alpha cells,** which secrete the peptide hormone **glucagon,** and a population of **beta cells,** which secrete the hormone **insulin.**

Insulin and glucagon are antagonistic hormones that regulate the concentration of glucose in the blood (see FIGURE 41.2). This is a critical bioenergetic and homeostatic function, because glucose is a major fuel for cellular respiration and a key source of carbon skeletons for the synthesis of other organic compounds. Metabolic balance depends on the maintenance of blood glucose at a concentration near a set point, which is about 90 mg/100 mL in humans (FIGURE 41.13). When blood glucose exceeds this level, insulin is released and acts to lower the glucose concentration. When blood glucose drops below the set point, glucagon increases glucose concentration. By negative feedback, blood glucose concentration determines

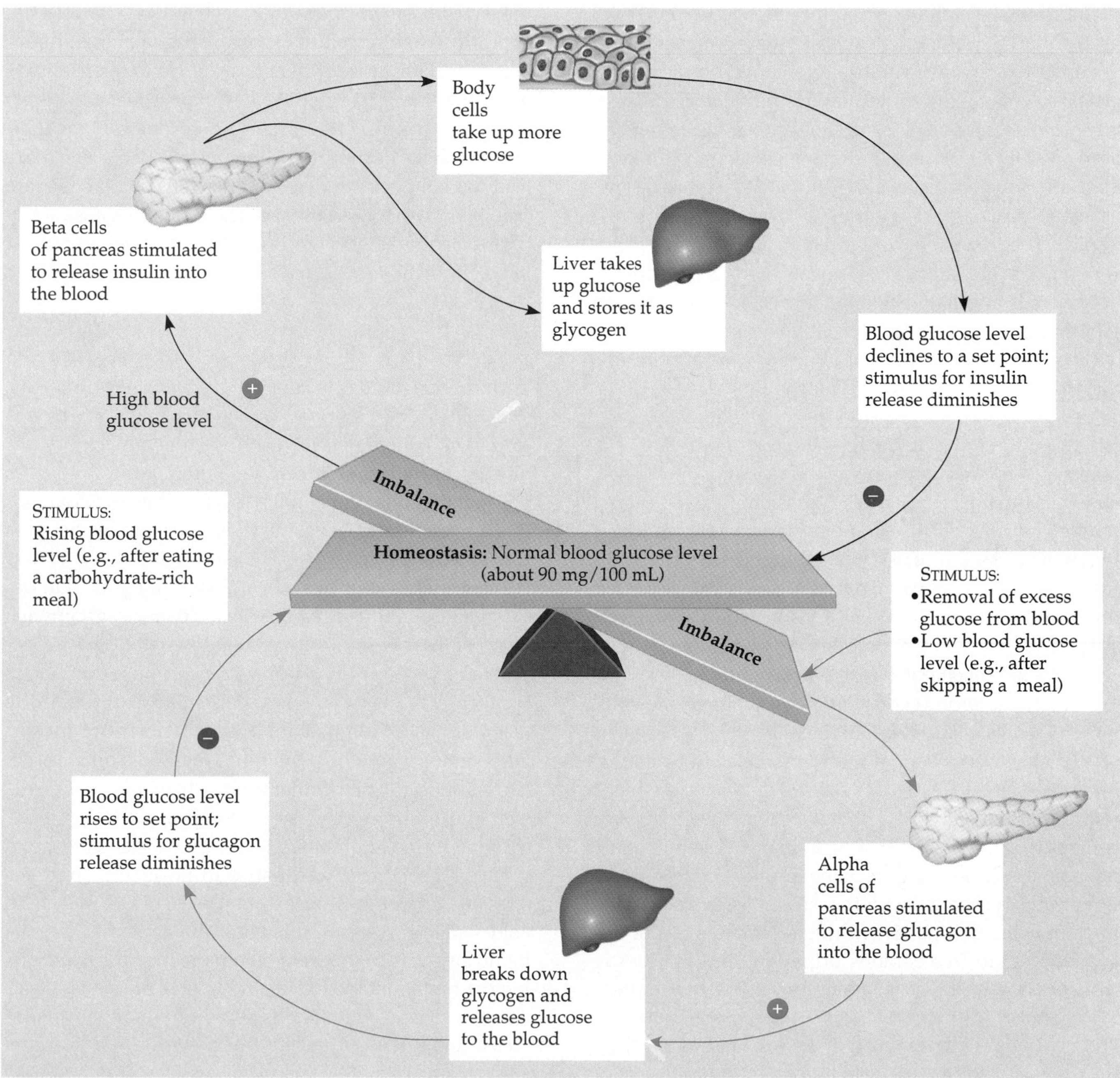

FIGURE 41.13
Glucose homeostasis maintained by insulin and glucagon. A rise in blood glucose above the set point of about 90 mg/100 mL in humans stimulates the pancreas to secrete insulin, which triggers its target cells to take up the excess glucose from the blood. Once the excess is removed or when blood glucose concentration dips below the set point, the pancreas responds by secreting glucagon, which acts on the liver to raise the blood glucose level. (See FIGURE 41.2 for an overview.)

the relative amounts of insulin and glucagon secreted by the islet cells.

Insulin and glucagon both influence blood glucose concentration by multiple mechanisms. Insulin lowers blood glucose levels by stimulating virtually all body cells except those of the brain to take up glucose from the blood. (Brain cells are unusual in being able to take up glucose without insulin; as a result, the brain has access to circulating fuel molecules almost all the time.) Insulin also decreases blood glucose by slowing glycogen breakdown in the liver and inhibiting the conversion of amino acids and fatty acids to sugar.

The liver, skeletal muscles, and adipose tissues store large amounts of fuel molecules and are especially important in bioenergetics. The liver and muscles store sugar as glycogen, whereas adipose tissue cells convert sugars to fats. The liver is a key fuel-processing center because only liver cells are sensitive to glucagon. Normally, glucagon starts having an effect before blood glucose levels even drop below the set point. In fact, as soon as excess glucose is cleared from the blood, glucagon signals the liver cells to increase glycogen hydrolysis, convert amino acids and fatty acids to glucose, and start slowly releasing glucose back into the circulation.

The antagonistic effects of glucagon and insulin are vital to glucose homeostasis, a mechanism that precisely manages both fuel storage and fuel use by body cells. We will revisit the topic of fuel management and see that it involves additional hormones when we study the adrenal glands in the next section. The liver's ability to perform its vital roles in glucose homeostasis results from the metabolic versatility of its cells and its access to absorbed nutrients via the hepatic portal vessel, which carries blood directly from the small intestine to the liver (see Chapter 37).

When the mechanisms of glucose homeostasis go awry, there are serious consequences. Diabetes mellitus, perhaps the best known endocrine disorder, is caused by a deficiency of insulin or a loss of response to insulin in target tissues. The result is high blood glucose—so high, in fact, that the diabetic's kidneys excrete glucose, which explains why the presence of sugar in urine is one test for diabetes. As more glucose concentrates in the urine, more water is excreted with it, resulting in excessive volumes of urine and persistent thirst. (*Diabetes*, from the Greek, refers to this copious urination, and *mellitus* is from the Greek word for "honey," referring to the presence of sugar in the urine.) Because glucose is unavailable as a major fuel source for diabetics, fat must serve as the main substrate for cellular respiration. In severe cases of diabetes, acidic metabolites formed during fat breakdown accumulate in the blood, threatening life by lowering blood pH.

There are actually two major forms of diabetes with very different causes. **Type I diabetes mellitus** (insulin-dependent diabetes) is an autoimmune disorder, in which the immune system mounts an attack on the cells of the pancreas. (Chapter 39 discusses possible causes of autoimmune reactions.) This usually occurs rather suddenly during childhood, destroying the person's ability to produce insulin. Treatment consists of insulin injections, which are usually taken several times daily. Until recently, insulin for injections was extracted from animal pancreases, but genetic engineering has provided a relatively inexpensive source of human insulin by inserting the gene for the hormone into bacteria (see Chapter 19). **Type II diabetes mellitus** (non–insulin-dependent diabetes) is characterized either by a deficiency of insulin or, more commonly, by reduced responsiveness in target cells due to some change in insulin receptors. Type II diabetes usually occurs after about age 40, becoming more likely with increasing age. More than 90% of diabetics are type II, and many can manage their blood glucose solely by exercise and dietary control. Heredity is a major factor in type II diabetes.

The Adrenal Glands

The **adrenal glands** are adjacent to the kidneys. In mammals, each adrenal gland is actually made up of two glands with different cell types, functions, and embryonic origins: the **adrenal cortex,** or outer portion, and the **adrenal medulla,** or central part of the gland. Nonmammalian vertebrates have quite different arrangements of the same tissues.

What makes your heart beat faster and your skin develop goose bumps when you sense danger or approach a stressful situation, like speaking in public? These reactions are part of the fight-or-flight response stimulated by two hormones of the adrenal medulla, **epinephrine** (also known as adrenaline) and **norepinephrine** (noradrenaline). These hormones are members of a class of compounds, the **catecholamines,** which are synthesized from the amino acid tyrosine (FIGURE 41.14).

Epinephrine, norepinephrine, and other catecholamines are secreted in response to positive or negative stress—everything from extreme pleasure to increased cold to life-threatening danger. Their release into the blood gives the body a rapid bioenergetic boost, increasing the basal metabolic rate and having dramatic effects on several targets. Epinephrine and norepinephrine increase the rate of glycogen breakdown in the liver and skeletal muscles and glucose release into the blood by liver cells. They also stimulate the release of fatty acids from fat cells. The fatty acids may be used by cells for energy. In addition to increasing the availability of energy

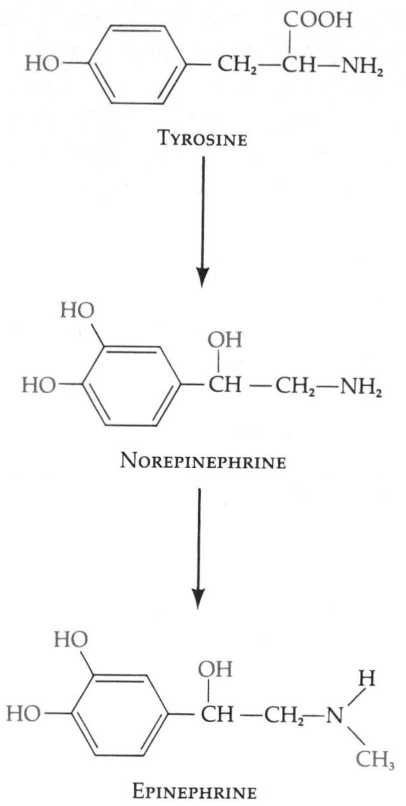

TYROSINE

NOREPINEPHRINE

EPINEPHRINE

FIGURE 41.14
The synthesis of catecholamine hormones. Cells in the adrenal medulla synthesize the catecholamines norepinephrine and epinephrine from the amino acid tyrosine. Norepinephrine is made by removing a carboxyl group (red) and adding two hydroxyl groups (green). Epinephrine is made from norepinephrine by adding a methyl group (blue).

sources, epinephrine and norepinephrine have profound effects on the cardiovascular and respiratory systems. For example, they increase both the rate and the stroke volume of the heartbeat and dilate the bronchioles in the lungs, effects that increase the rate of oxygen delivery to body cells. (This is why doctors prescribe epinephrine as a heart stimulant and to open breathing tubes during asthma attacks.) The catecholamines also cause smooth muscles of some blood vessels to contract and muscles of other vessels to relax, with an overall effect of shunting blood away from the skin, digestive organs, and kidneys, while increasing the blood supply to the heart, brain, and skeletal muscles.

What causes the release of catecholamines during the response to stress? The adrenal medulla is under the control of nerve cells from the sympathetic division of the autonomic nervous system (see Chapter 44). When nerve cells are excited by some form of stressful stimulus, they release the neurotransmitter acetylcholine in the adrenal medulla. Acetylcholine combines with receptors on the cells, stimulating the release of epinephrine. Norepinephrine is released independently of epinephrine. Its functions are similar to those of epinephrine, but its primary role is in sustaining blood pressure, while epinephrine generally has a stronger effect on heart and metabolic rates. Norepinephrine also functions as an important neurotransmitter in the nervous system, as we will see in Chapter 44.

The adrenal cortex, like the adrenal medulla, reacts to stress. But it responds to endocrine signals rather than to nervous input. Stressful stimuli cause the hypothalamus to secrete a releasing hormone that stimulates the anterior pituitary to release the tropic hormone ACTH. When it reaches its target via the bloodstream, ACTH stimulates cells of the adrenal cortex to synthesize and secrete a family of steroids called **corticosteroids.** In another case of negative feedback, elevated levels of corticosteroids in the blood suppress the secretion of ACTH.

Many corticosteroids have been isolated from the adrenal cortex. The two main types in humans are the **glucocorticoids,** such as cortisol, and the **mineralocorticoids,** such as aldosterone (FIGURE 41.15).

The primary effect of glucocorticoids is on bioenergetics, specifically on glucose metabolism. Augmenting the fuel-mobilizing effects of glucagon from the pancreas, glucocorticoids promote the synthesis of glucose from noncarbohydrate sources, such as proteins, making more glucose available as fuel. Glucocorticoids act on skeletal muscle, causing a breakdown of muscle proteins. The resulting carbon skeletons are transported to the liver and kidney, where they are converted to glucose and released into the blood. The synthesis of glucose from muscle proteins is a homeostatic mechanism, providing circulating fuel when body activities require more than the liver can mobilize from its glycogen stores. It can also be part of a broader role of the glucocorticoids, that of helping the body withstand long-term environmental challenge.

Abnormally high doses of glucocorticoids administered as medication suppress certain components of the body's immune system—for example, the inflammatory reaction that occurs at the site of an infection. Glucocorticoids are used to treat diseases in which excessive inflammation is a problem. Cortisone, for instance, was once thought to be a miracle drug that could cure serious inflammatory conditions such as arthritis. It has become clear, however, that long-term use of the corticosteroids can result in increased susceptibility to infection and disease because of their immunosuppressive effects.

Mineralocorticoids have their major effects on salt and water balance. Aldosterone, for example, stimulates cells in the kidney to reabsorb sodium ions and water from the filtrate, raising blood pressure and volume. Aldosterone (of the renin-angiotensin-aldosterone system, or RAAS), ADH from the posterior pituitary, and

FIGURE 41.15
Steroid hormones from the adrenal cortex and gonads. Cortisol (a glucocorticoid) and aldosterone (a mineralocorticoid), both made in the adrenal cortex, are structurally similar to the sex hormones testosterone (an androgen), estradiol (an estrogen), and progesterone (a progestin). The precursor for the synthesis of all steroid hormones is cholesterol (see FIGURE 5.14). Most of the androgens (male hormones) that circulate in the blood are made by the testes, and most of the estrogens and progestins (female hormones) are produced by the ovaries; however, small amounts of both types of sex hormones are also made by the adrenal cortex.

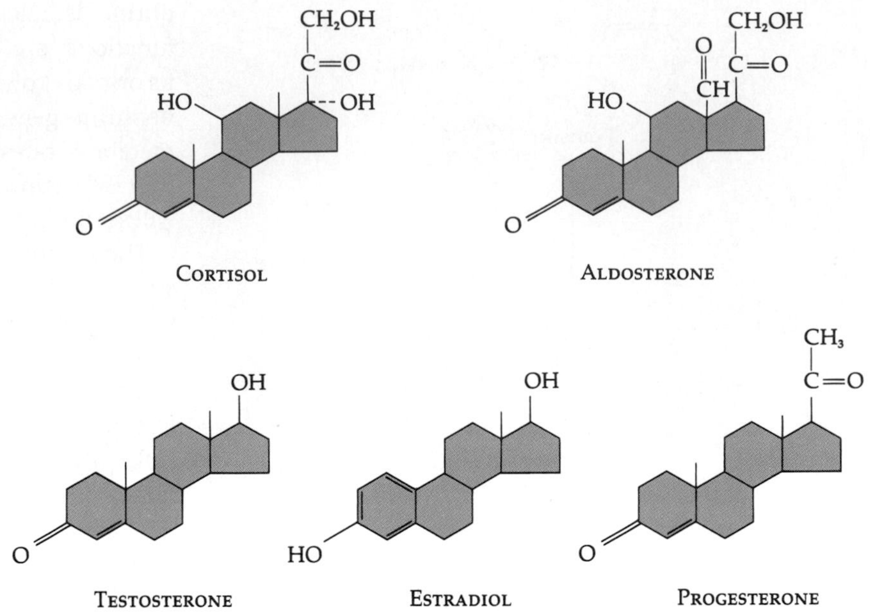

atrial natriuretic factor (ANF) from the heart form a regulatory complex with multiple feedback loops that underly the kidney's ability to maintain the ion and water homeostasis of the blood (see Chapter 40). Aldosterone secretion is largely regulated by the RAAS itself, in response to changes in plasma ion concentration. However, when an individual is under severe stress, the hypothalamus tends to secrete more releasing hormones that increase the rate of secretion of ACTH by the anterior pituitary. The rise in ACTH in the blood then increases the rate of secretion of aldosterone by the adrenal cortex.

Evidence is mounting that both glucocorticoids and mineralocorticoids help maintain homeostasis when the body experiences stress over an extended period of time. FIGURE 41.16 compares the long-term stress-induced actions of corticosteroids with the short-term ones of epinephrine and norepinephrine.

A third group of corticosteroids are sex hormones, mainly androgens (male hormones) similar to testosterone produced by the testes and small amounts of estrogens and progesterone (female hormones). There is evidence that adrenal androgens account for the sex drive in adult females, but otherwise the physiological roles of the adrenal sex hormones are not well understood. We will discuss the sex hormones from the gonads in the next section.

The Gonads

Steroids produced in the testes of males and ovaries of females affect growth and development and also regulate reproductive cycles and behaviors. There are three major categories of gonadal steroids: androgens, estrogens, and progestins (see FIGURE 41.15). All three types are found in both males and females but in different proportions.

The testes primarily synthesize **androgens,** the main such hormone being **testosterone.** In general, androgens stimulate the development and maintenance of the male reproductive system. Androgens produced early in the development of an embryo determine that the fetus will develop as a male rather than a female. At puberty, high concentrations of androgens are responsible for the development of human male secondary sex characteristics, such as male patterns of hair growth and a low voice.

Estrogens, the most important of which is estradiol, have a parallel role in the maintenance of the female reproductive system and the development of female secondary sex characteristics. In mammals, **progestins,** which include progesterone, are primarily involved with preparing and maintaining the uterus, which supports the growth and development of an embryo.

The synthesis of both estrogens and androgens is controlled by gonadotropins, FSH and LH, from the anterior pituitary gland. FSH and LH secretion is controlled by one releasing hormone from the hypothalamus, GnRH (gonadotropin-releasing hormone). We will examine the complex feedback relationships that regulate the secretion of gonadal steroids in detail in Chapter 42.

Other Endocrine Organs

Many organs with primarily nonendocrine functions also secrete hormones, several of which we have encountered in earlier chapters. The digestive tract, for example, is the source of at least eight hormones, including gastrin and secretin (see Chapter 37). Erythropoietin, from the kid-

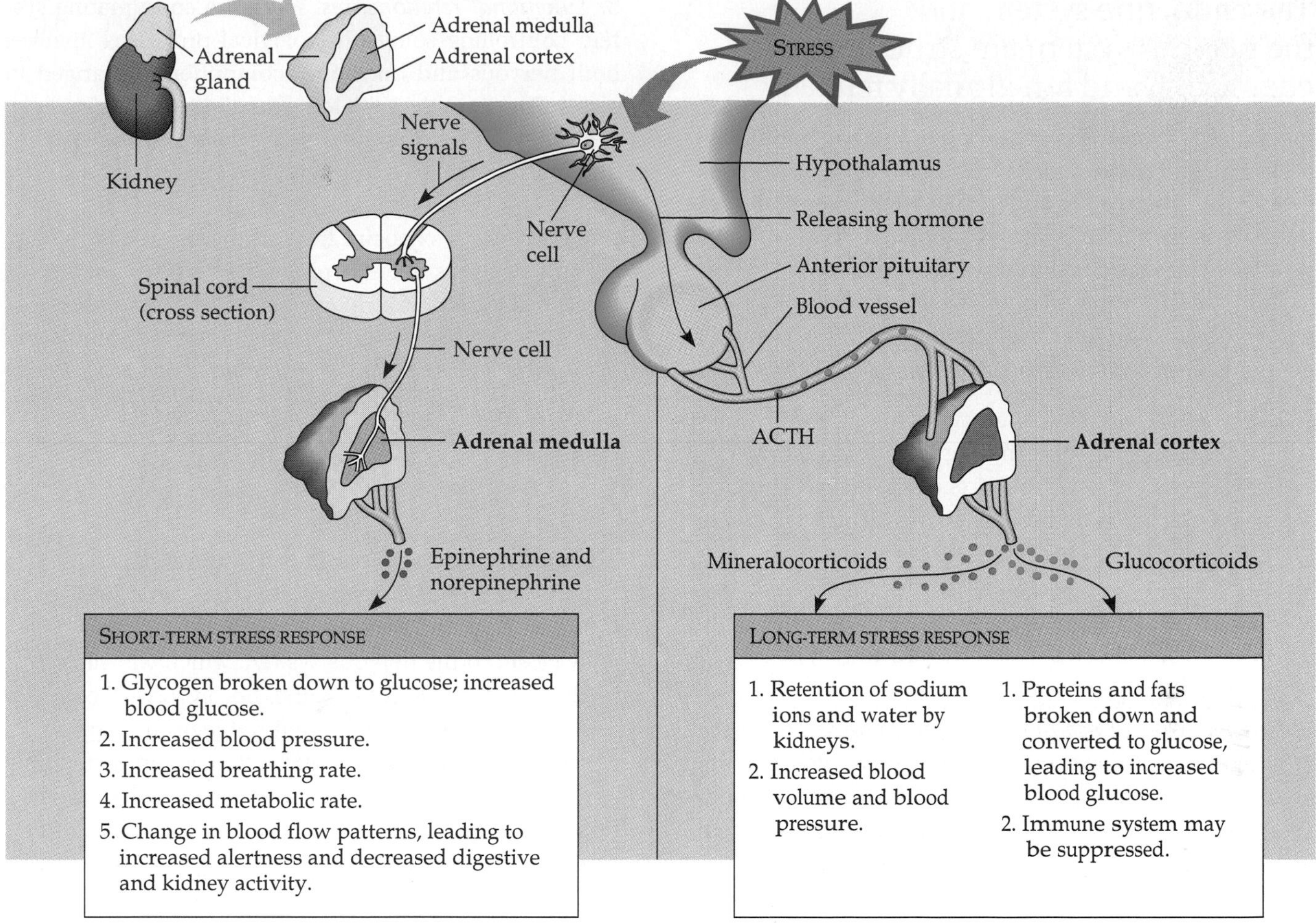

FIGURE 41.16
Stress and the adrenal gland. Stressful stimuli cause the hypothalamus to activate the adrenal medulla via nerve impulses and the adrenal cortex via hormonal signals. The adrenal medulla mediates short-term responses to stress by secreting the catecholamine hormones epinephrine and norepinephrine. The adrenal cortex controls more prolonged responses by secreting steroid hormones.

ney, stimulates red blood cell production (see Chapter 38). Two other endocrine organs that deserve some attention are the pineal gland and the thymus.

The **pineal gland** is a small mass of tissue near the center of the mammalian brain (closer to the brain surface in some other vertebrates). Although we know considerably more about the pineal than we did when Descartes described it as the seat of the soul, it still remains largely a mystery. The pineal secretes the hormone **melatonin,** a modified amino acid. The pineal contains light-sensitive cells or has nervous connections from the eyes, and melatonin regulates functions related to light and to seasons marked by changes in day length. For example, melatonin, like MSH, affects skin pigmentation in many vertebrates. Most of the pineal's functions, though, are related to biological rhythms associated with reproduction. Since melatonin is secreted at night, the amount secreted depends on the length of the night. In winter, for example, the days are short and the nights are long, so more melatonin is secreted. Thus, melatonin production is a link between a biological clock and daily or seasonal activities, such as reproduction. However, the precise role of melatonin in mediating rhythms is not yet clear.

It was not until the 1960s that the role of the **thymus** in the immune system was discovered. This gland, which lies under the breastbone in humans, is quite large during childhood. At puberty, when the immune system is well established, the thymus begins to decline, and by adulthood it is largely replaced by adipose and fibrous tissue. Though much reduced in size, the thymus continues to function throughout life. It secretes several messengers, including **thymosin,** that stimulate the development and differentiation of T lymphocytes after they leave the thymus (see Chapter 39). The subject of an intense research effort, thymosin may inhibit certain cancers by enhancing cellular immunity.

The endocrine system and the nervous system are structurally, chemically, and functionally related

There have been many examples in this chapter of interactions between the endocrine system and the nervous system. Indeed, we have seen that the two systems are often inseparable and may function as a single unit. We can synthesize much of our understanding of the chemical communication and coordination in animal bodies by examining three types of relationships between the endocrine system and the nervous system.

1. *Structural relationships.* Many endocrine glands are made of nervous tissue. The vertebrate hypothalamus and posterior pituitary, and parts of the insect brain, are examples of nerve tissues that secrete hormones into the blood. Other endocrine glands that are not nervous tissue in their present form have evolved from the nervous system. The adrenal medulla is derived from the same cells that produce certain ganglia (clusters of nerve cell bodies outside the central nervous system).

2. *Chemical relationships.* Several vertebrate hormones are used as signals by the nervous system as well as by the endocrine system. Norepinephrine, for example, functions in the body both as an adrenal hormone and as a neurotransmitter in the nervous system.

3. *Functional relationships.* First, the coordinating system controlling some physiological processes involves both nervous and hormonal components arranged in series. For example, milk letdown, the release of milk by a mother during nursing, is controlled by a neuroendocrine reflex: Suckling stimulates sensory cells in the nipples, and nervous signals to the hypothalamus trigger the release of oxytocin from the posterior pituitary. Second, each system affects the output of the other. We have seen several examples of how the nervous system controls the endocrine glands, including the stimulation of the adrenal medulla. But the endocrine system also affects both the development of the nervous system and its output—behavior.

* * *

Thus, we have seen that animals have two related coordinating systems. We have studied the endocrine system in some detail, and that discussion has drawn us increasingly closer to the nervous system, which we will examine in Chapters 44 and 45. First, however, we'll consider one of the most fundamental subjects in biology, in which the role of the endocrine system is central not only to the survival of the individual but also to the propagation of the species. In Chapters 42 and 43, we explore reproduction and development.

REVIEW OF KEY CONCEPTS (with page numbers and key figures)

- A variety of chemical signals coordinates body functions: *an overview* (pp. 912–915, FIGURE 41.1)
  - The endocrine system regulates the activity of target organs at distant sites by means of chemical messengers called hormones.
  - A hormone is a molecule secreted by endocrine cells or neurosecretory cells that travels in body fluids to a target cell, where it binds with specific receptors and elicits a response. Antagonistic (opposing) hormones function in many cases of homeostasis.
  - Pheromones act as communication signals between different individuals of the same species.
  - Local regulators, such as neurotransmitters, growth factors, and prostaglandins, affect target cells in the immediate vicinity of their secretion.
- Hormone binding to specific receptors triggers signaling mechanisms at the cellular level (pp. 915–917, FIGURE 41.3)
  - Steroid hormones penetrate the plasma membrane of target cells and bind to specific protein receptors. Hormone-receptor complexes then enter the nucleus, bind to an acceptor protein on a chromosome, and initiate transcription.
  - Nonsteroid hormones, which cannot pass through the cell membrane, bind to specific receptors on the plasma membrane. Through signal transduction involving second messengers such as cyclic AMP and inositol triphosphate, these hormones trigger a cascade of metabolic reactions within the cells.
- Many chemical signals are relayed and amplified by second messengers and protein kinases (pp. 917–920, FIGURES 41.5–41.7)
  - In signal transduction involving cAMP, the binding of a hormone (the first messenger) to a surface receptor activates a G protein (a relay), which in turn activates the membrane protein adenylyl cyclase (an effector) to produce cAMP. The cAMP then activates cAMP-dependent protein kinase, an enzyme that phosphorylates other proteins.
  - Inositol trisphosphate (IP_3) serves as a second messenger for neurotransmitters, growth factors, and some hormones. The binding of a hormone to its receptor activates a G protein that stimulates a membrane enzyme to cleave a membrane phospholipid into IP_3 and diacylglycerol. Diacylglycerol activates a protein kinase, and IP_3 causes the uptake of Ca^{2+} and its release from storage in the endoplasmic reticulum. Ca^{2+}, acting alone or bound to calmodulin, alters the activities of certain enzymes.

- Invertebrate control systems often integrate endocrine and nervous system functions (pp. 921–922, FIGURE 41.8)
  - Diverse hormones regulate different aspects of homeostasis in invertebrates. Hormones controlling reproduction and development have been extensively researched.
  - Arthropods have well-developed endocrine systems. In insects, molting and development are controlled by an interplay between ecdysone and juvenile hormone (JH).
- The hypothalamus and pituitary integrate many functions of the vertebrate endocrine system (pp. 922–926, FIGURES 41.9–41.11)
  - Neurosecretory cells of the hypothalamus integrate endocrine and neural function by influencing the pituitary gland. Under the direction of releasing and inhibiting hormones from the hypothalamus, the anterior pituitary produces several tropic hormones that act on other endocrine glands. The posterior pituitary is an extension of the brain that stores and releases two peptide hormones produced by neurosecretory cells in the hypothalamus.
  - The posterior pituitary is a repository for oxytocin and antidiuretic hormone (ADH). Oxytocin induces uterine contractions and milk ejection, and ADH enhances water reabsorption in the kidneys.
  - The anterior pituitary produces an array of hormones, including thyroid-stimulating hormone (TSH), follicle-stimulating hormone (FSH), luteinizing hormone (LH), growth hormone (GH), prolactin (PRL), adrenocorticotropic hormone (ACTH), melanocyte-stimulating hormone (MSH), and the endorphins.
  - The chemically related tropic hormones TSH and the gonadotropins (FSH and LH) stimulate the thyroid gland and the gonads, respectively, to produce their hormones.
  - GH promotes growth directly and stimulates the production of growth factors.
  - Prolactin, named for its stimulation of lactation in mammals, has diverse effects in different vertebrates.
  - ACTH has a tropic effect on the adrenal cortex. MSH influences skin pigmentation in some vertebrates. Endorphins, the brain's natural opiates, inhibit the perception of pain.
  - Special portal vessels convey hypothalamic hormones to the anterior pituitary. These hormones stimulate or inhibit the secretion of specific hormones from the anterior pituitary.
- The vertebrate endocrine system coordinates homeostasis and regulates growth, development, and reproduction (pp. 927–933, TABLE 41.1)
  - The thyroid gland produces iodine-containing hormones (T_3 and T_4) that stimulate metabolism and influence development and maturation in vertebrates. The thyroid also secretes calcitonin, which lowers calcium levels in the blood.
  - The parathyroid glands raise plasma calcium levels by secreting parathyroid hormone (PTH). PTH works with calcitonin to affect calcium homeostasis by actions on the intestines, bone, and kidneys.
  - The endocrine portion of the pancreas consists of islet cells that secret insulin and glucagon. High blood glucose levels stimulate the release of insulin, which increases the cellular uptake of glucose, promotes the formation and storage of glycogen in the liver, and stimulates protein synthesis and fat storage. Low blood glucose levels trigger glucagon release, which increases blood glucose by stimulating the conversion of glycogen to glucose in the liver and increasing the breakdown of fat and protein. Type I diabetes mellitus is an autoimmune disorder resulting in a lack of insulin. Type II diabetes is usually caused by the loss of responsiveness of target cells to insulin.
  - The adrenal gland consists of an outer cortex and an inner medulla. The adrenal medulla releases epinephrine and norepinephrine in response to stress-activated impulses from the nervous system. These hormones mediate various fight-or-flight responses. The adrenal cortex releases corticosteroids, including sex hormones, glucocorticoids, and mineralocorticoids. Glucocorticoids influence glucose metabolism and the immune system; mineralocorticoids affect salt and water balance.
  - The gonads—testes and ovaries—produce varying proportions of androgens, estrogens, and progestins, steroid hormones that affect growth, development, physical sexual differentiation, and reproductive cycles and behaviors.
  - The pineal gland secretes melatonin, which influences skin pigmentation, biological rhythms, and reproduction in various vertebrates.
  - The thymus secretes thymosin and other chemical messengers that stimulate the development and differentiation of T lymphocytes.
- The endocrine system and the nervous system are structurally, chemically, and functionally related (p. 934)
  - The endocrine and nervous systems often function inseparably in maintaining homeostasis, development, and reproduction. Neurosecretory cells of the nervous system secrete many hormones, and some endocrine glands have evolved from nervous tissue. Several hormones, such as epinephrine, are used as signals by both the endocrine and the nervous system. And many body functions are regulated by both systems.

SELF-QUIZ

1. Which of the following is *not* an accurate statement about hormones?
   a. Hormones are chemical messengers that travel to target cells through the circulatory system.
   b. Hormones often regulate homeostasis through antagonistic functions.
   c. Hormones of the same chemical class usually have the same general function.
   d. Hormones are secreted by specialized cells usually located in endocrine glands.
   e. Hormones are often regulated through feedback loops.

2. A major difference in the mechanism of action between steroid and peptide hormones is that
   a. steroid hormones mainly affect the synthesis of proteins, whereas peptide hormones mainly affect the activity of proteins already in the cell
   b. target cells react more rapidly to steroid hormones than they do to peptide hormones
   c. steroid hormones enter the nucleus, whereas peptide hormones stay in the cytoplasm
   d. steroid hormones bind to a receptor protein, whereas peptide hormones bind to G protein
   e. steroid hormones affect metabolism, whereas peptide hormones affect membrane permeability

3. Which of the following accurately represents the sequence of components in a cellular response to a peptide hormone?
   a. hormone binding to adenylyl cyclase—G protein—protein kinase—phosphorylation of enzymes
   b. hormone binding to receptor—G protein—transcription factor—protein kinase
   c. hormone binding to cAMP—G protein—cAMP-dependent protein kinase—adenylyl cyclase
   d. hormone binding to G protein—adenylyl cyclase—protein kinase—phosphorylation of proteins
   e. hormone binding to receptor—G protein—adenylyl cyclase—protein kinase
4. Growth factors are local regulators that
   a. are produced by the anterior pituitary
   b. are modified fatty acids that stimulate bone and cartilage growth
   c. are found on the surface of cancer cells and stimulate abnormal cell division
   d. are proteins that bind to surface receptors and stimulate target cell growth and development
   e. include histamines and interleukins and are necessary for cellular differentiation
5. Which of the following hormones is *incorrectly* paired with its action?
   a. oxytocin—stimulates uterine contractions during childbirth
   b. thyroxine—stimulates metabolic processes
   c. insulin—stimulates glycogen breakdown in the liver
   d. ACTH—stimulates the release of glucocorticoids by the adrenal cortex
   e. melatonin—affects biological rhythms, seasonal reproduction
6. An example of antagonistic hormones controlling homeostasis is
   a. thyroxine and parathyroid hormone in calcium balance
   b. insulin and glucagon in glucose metabolism
   c. progestins and estrogens in sexual differentiation
   d. epinephrine and norepinephrine in fight-or-flight responses
   e. oxytocin and prolactin in milk production
7. Which of the following human conditions is *incorrectly* paired with a hormone?
   a. acromegaly—growth hormone
   b. diabetes—insulin
   c. cretinism—thyroid hormone
   d. tetany—PTH
   e. hypopituitary dwarfism—ACTH
8. A portal vessel carries blood from the hypothalamus directly to the
   a. thyroid
   b. pineal gland
   c. anterior pituitary
   d. posterior pituitary
   e. thymus
9. A second messenger derived from membrane lipids is
   a. cyclic AMP
   b. calmodulin
   c. inositol triphosphate
   d. protein kinase
   e. calcium
10. The main target organs for tropic hormones are
   a. muscles
   b. blood vessels
   c. endocrine glands
   d. kidneys
   e. nerves

CHALLENGE QUESTIONS

1. A woman with a hypothyroid condition is treated with thyroxine. How is this medication likely to affect levels of TSH and TSH-releasing hormone (TRH)?
2. Describe three specific examples of the interaction between the nervous system and the endocrine system.
3. Write a paragraph explaining how the same hormone can have one effect on one type of target cell, a different effect on another type of target cell, and no effect on a third type of target cell.

SCIENCE, TECHNOLOGY, AND SOCIETY

Growth hormone (GH) produced by DNA technology has enabled hundreds of children who suffer from hypopituitary dwarfism to grow normally and reach a stature within the normal range. Now that the hormone is readily available and relatively inexpensive, many parents who are concerned that their children are not growing fast enough want to use GH to make them grow faster and taller. There can be potentially harmful effects, such as a reduction in body fat and an increase in muscle mass. And no one yet knows if GH injections will have seriously harmful long-term effects in individuals who do not have a hypopituitary condition. Do you think GH therapy should be regulated? In your opinion, what criteria should determine the cases in which GH treatments or other hormone therapies are appropriate?

FURTHER READING

Lienhard, G. E., J. W. Slot, D. E. James, and M. M. Mueckler. "How Cells Absorb Glucose." *Scientific American,* January 1992. Insulin may act on the membrane proteins that transport glucose.

Linder, E., and A. Gilman. "G Proteins." *Scientific American,* July 1992. Signal-transduction pathways in hormone action.

Pacchioli, D. "Potent Aromas." *Discover,* November 1991. The role of pheromones in the reproduction of mice.

Richardson, S. "The Brain-Boosting Sex Hormone." *Discover,* April 1994. The female sex hormone estrogen may help prevent Alzheimer's disease.

Snyder, S., and D. Bredt. "Biological Roles of Nitric Oxide." *Scientific American,* May 1992. Research on a chemical signal with multiple functions.

Snyder, S. H. "The Molecular Basis of Communication Between Cells." *Scientific American,* October 1985. An excellent article emphasizing the relationship between the endocrine and nervous systems.

Weiss, R. "Heightened Concern Over Growth Hormone." *Science News,* December 8, 1990. The benefits, risks, and ethics of a new product of genetic engineering.

Weiss, R. "Promising Protein for Parkinson's." *Science,* May 21, 1993. A newly discovered growth factor may prevent the loss of dopamine-producing neurons.

Weissmann, G. "Aspirin." *Scientific American,* January 1991. This painkiller probably suppresses the synthesis of prostaglandins.

CHAPTER 42

ANIMAL REPRODUCTION

KEY CONCEPTS

- Both asexual and sexual reproduction occur in the animal kingdom
- In sexual reproduction, gametes unite in the external environment or within the female
- Diverse reproductive systems have evolved in the animal kingdom
- Human reproduction involves intricate anatomy and complex behavior
- Spermatogenesis and oogenesis both involve meiosis but differ in three significant ways
- A complex interplay of hormones regulates reproduction
- Embryonic and fetal development occur during pregnancy in humans and other placental mammals
- Contraception prevents pregnancy
- New technologies offer help for reproductive problems

The many aspects of animal form and function we have studied so far can be viewed, in the broadest context, as adaptations contributing to reproductive success. Individuals are transient. A population transcends finite lifespans only by reproduction, the creation of new individuals from existing ones. The two earthworms in the photograph on this page are mating. Unless disturbed, they will remain above the ground and joined like this for several hours. Each worm produces both sperm and eggs, and each donates and receives sperm during mating. In a few weeks, sexual reproduction will be completed when many new individuals hatch from the fertilized eggs resulting from this union.

Animal reproduction is the subject of this chapter. We will first compare the diverse reproductive mechanisms that have evolved in the animal kingdom, and then examine the details of mammalian, particularly human, reproduction.

Both asexual and sexual reproduction occur in the animal kingdom

There are two general modes of animal reproduction. **Asexual reproduction** is the creation of offspring whose genes all come from one parent without the fusion of egg and sperm. In most cases, asexual reproduction relies entirely on mitotic cell division. **Sexual reproduction** in animals is the creation of offspring by the fusion of two haploid **gametes** (sperm and egg) to form a diploid **zygote.** Gametes are formed by meiosis, and sexual reproduction usually involves two parents, both contributing genes to the offspring (see Chapter 12).

Asexual Reproduction

Many invertebrates can reproduce asexually by **fission,** the separation of a parent into two or more individuals of approximately equal size (FIGURE 42.1). Also common among invertebrates, **budding** involves new individuals splitting off from existing ones. For example, in certain cnidarians and tunicates, new individuals grow out from the body of a parent (see FIGURE 12.1). The offspring may either detach from the parent or remain joined, eventually forming extensive colonies. Stony corals, which may be more than 1 m across, are cnidarian colonies of several thousand connected individuals. In another form of asexual reproduction, some invertebrates release specialized groups of cells that can grow into new individuals. For example, the **gemmules** of sponges are formed when cells of several types migrate together within the sponge and become surrounded by a protective coat.

FIGURE 42.1
Two from one: asexual reproduction of a sea anemone (*Anthopleura elegantissima*). The individual in the center of this photograph is undergoing fission, a type of asexual reproduction. Two smaller individuals will soon form as the parent divides approximately in half. The offspring will be genetic copies of the parent.

Yet another type of asexual reproduction is **fragmentation,** the breaking of the body into several pieces, some or all of which develop into complete adults. For an animal to reproduce this way, fragmentation must be accompanied by **regeneration,** the regrowth of lost body parts. Reproduction by fragmentation and regeneration occurs in many sponges, cnidarians, polychaete annelids, and tunicates. Many animals can also replace lost appendages by regeneration—most sea stars can grow new arms when injured, for example—but this is not reproduction because new individuals are not created.

Asexual reproduction has several potential advantages. For instance, it enables animals that live in isolation to produce offspring without locating mates. It may also allow many offspring to be produced in a short amount of time, which is ideal for colonizing a habitat rapidly. Theoretically, asexual reproduction is most advantageous in stable, favorable environments because it perpetuates successful genotypes precisely.

Sexual Reproduction

In sexual reproduction, the female gamete, the **ovum** (unfertilized egg), is usually a relatively large and nonmotile cell. The male gamete, the **spermatozoon,** is generally a small, flagellated cell (though arthropods produce nonmotile sperm that must be placed near the female reproductive tract).

Sexual reproduction increases genetic variability among offspring by generating unique combinations of genes inherited from two parents (see Chapters 12 and 13). Many biologists believe that sexual reproduction, by producing offspring having varying phenotypes, may enhance the reproductive success of parents when pathogens or other environmental factors change relatively rapidly.

Reproductive Cycles and Patterns

Most animals show definite cycles in reproductive activity, often related to changing seasons. The periodic nature of reproduction allows animals to conserve resources and reproduce when more energy is available than is needed for maintenance and when environmental conditions favor the survival of offspring. Ewes (female sheep), for example, have 15-day reproductive cycles and ovulate at the midpoint of each cycle. But these cycles occur only during fall and early winter, so lambs are born in the spring. Even animals that live in apparently stable habitats, such as the tropics or the ocean, generally reproduce only at certain times of the year. Reproductive cycles are controlled by a combination of hormonal and environmental cues, the latter including such factors as seasonal temperature, rainfall, day length, and lunar cycles.

Animals may reproduce asexually or sexually exclusively, or they may alternate between the two modes. In aphids, rotifers, and the freshwater crustacean *Daphnia,* each female can produce eggs of two types, depending on environmental conditions such as the time of year. One type of egg is fertilized, but the other type develops by **parthenogenesis,** a process in which the egg develops without being fertilized. The adults produced by parthenogenesis are often haploid, and their cells do not undergo meiosis in forming new eggs. In the case of *Daphnia,* the switch from sexual to asexual reproduction is often related to season. Asexual reproduction occurs under favorable conditions and sexual reproduction during times of environmental stress.

Parthenogenesis has a role in the social organization of certain species of bees, wasps, and ants. Male honeybees, or drones, are produced parthenogenetically, whereas females, both sterile workers and reproductive females (queens), develop from fertilized eggs.

Among vertebrates, several genera of fishes, amphibians, and lizards reproduce exclusively by a complex form of parthenogenesis that involves the doubling of chromosomes after meiosis to create diploid "zygotes." For example, there are about 15 species of whiptail lizards (genus *Cnemidophorus*) that reproduce exclusively by parthenogenesis. There are no males in these species, but the lizards imitate courtship and mating behavior typical of sexual species of the same genus. During the breeding season, one female of each mating pair mimics a male (FIGURE 42.2a). The roles change two or three times during the season, female behavior occurring when the level of the female sex hormone estrogen is high prior to ovulation (the release of eggs) and male behavior occurring after ovulation when the level of estrogen drops (FIGURE 42.2b). In fact, ovulation is more likely to occur if one individual is mounted by another during the critical time of the hormone cycle; isolated lizards lay fewer eggs than those that go through the motions of sex. Apparently, these parthenogenetic lizards, which evolved from species having two sexes, still require certain sexual stimuli for maximum reproductive success.

Sexual reproduction presents a special problem for sessile or burrowing animals, such as barnacles and earthworms, or for parasites, such as tapeworms, which may have difficulty encountering a member of the opposite sex. One solution to this problem is **hermaphroditism,** in which each individual has both male and female reproductive systems (the term is derived from Hermes and Aphrodite, names of a Greek god and goddess). Although some hermaphrodites fertilize themselves, most must mate with another member of the same species. When this occurs, each animal serves as both male and female, donating and receiving sperm, as we saw for earthworms. Each individual encountered is a potential mate, resulting in twice as many offspring than if only one individual's eggs were fertilized.

Another remarkable reproductive pattern is **sequential hermaphroditism,** in which an individual reverses its sex during its lifetime. In some species, the sequential hermaphrodite is **protogynous** (female first), while other species are **protandrous** (male first). In various species of reef fishes called wrasses, sex reversal is associated with age and size. For example, the Caribbean bluehead wrasse is a protogynous species in which only the largest (usually the oldest) individuals change from female to male (FIGURE 42.3). These fish live as harems consisting of a single

(a)

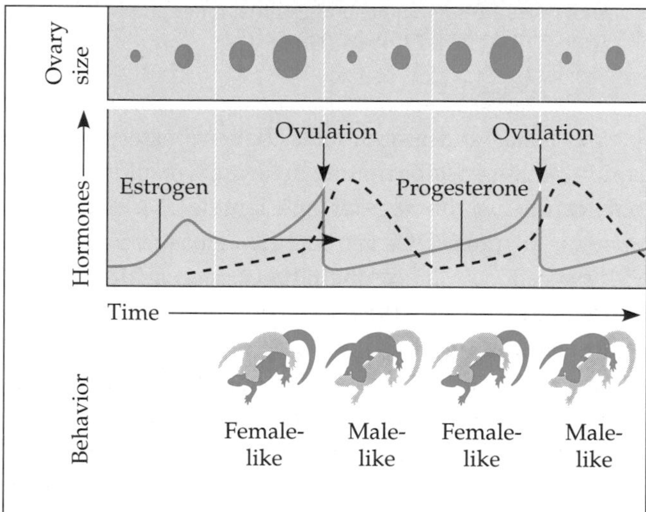

(b)

FIGURE 42.2
Sexual behavior in parthenogenetic lizards. The desert-grassland whiptail lizard (*Cnemidophorus uniparens*) is an all-female species. These reptiles reproduce by parthenogenesis; eggs undergo a chromosome doubling after meiosis and develop into lizards without being fertilized. However, ovulation is enhanced by courtship and mating rituals that imitate the behavior of closely related species that reproduce sexually. (**a**) Both lizards in this photograph are *C. uniparens* females. The one on top is playing the role of a male. Every two or three weeks during the breeding season, individuals switch sex roles. (**b**) The sexual behavior of *C. uniparens* is correlated with the cycle of ovulation mediated by sex hormones. Before ovulation, an individual behaves like a female. She has relatively large ovaries and her behavior is correlated with high blood levels of estrogen. After ovulation, small ovaries, an abrupt drop in estrogen, and a rapid increase in circulating progesterone are correlated with an individual behaving like a male.

FIGURE 42.3
Sex reversal in a sequential hermaphrodite. In many species of reef fishes called wrasses, sex is not fixed but can change at some time in the animal's life. Sex reversal is often correlated with size. In this scene, a male Caribbean bluehead wrasse and two smaller females are feeding on a sea urchin. All wrasses of this species are born females, but the oldest, largest individuals change sex and complete their lives as males.

male and several females. If the male dies or is removed in experiments, the largest female in the harem changes sex and becomes the new male. Within a week, the transformed individual is producing sperm instead of eggs. In this species, the male defends the harem against intruders, and thus larger size may give a greater reproductive advantage to males than it does to females. In contrast, there are protandrous animals that change from male to female when size increases. In such cases, greater size may increase the reproductive success of females more than it does males. For example, the production of huge numbers of gametes is an important asset for sedentary animals, such as oysters, that release their gametes into the surrounding water. Egg cells are generally much larger than sperm cells, so females produce fewer gametes than males. Larger females tend to produce more eggs than smaller ones, and species of oysters that are sequential hermaphrodites are generally protandrous.

The diverse reproductive cycles and patterns we observe in the animal kingdom are adaptations that have evolved by natural selection. We will see many other examples as we survey the various mechanisms of sexual reproduction.

FIGURE 42.4
The release of eggs and external fertilization. Many amphibians shed gametes into the environment, and fertilization occurs outside the female's body. In most species, behavioral adaptations ensure that a male is present when the female releases eggs. Here, a female toad, clasped by a male (on top), has just released a mass of eggs. The male released sperm (not shown) at the same time, and external fertilization has already occurred in the surrounding water.

In sexual reproduction, gametes unite in the external environment or within the female

The mechanisms of **fertilization,** the union of sperm and egg, play an important part in sexual reproduction. The two major patterns of fertilization that have evolved are external and internal fertilization, each having specific environmental and behavioral requirements. In **external fertilization,** eggs are shed by the female and fertilized by the male in the environment (FIGURE 42.4). **Internal fertilization** occurs when sperm are deposited in (or nearby) the female reproductive tract, and egg and sperm unite within her body.

Because external fertilization requires an environment where an egg can develop without desiccation or heat stress, it occurs almost exclusively in moist habitats. Many aquatic invertebrates simply shed their eggs and sperm into the surroundings, and fertilization occurs without the parents actually making physical contact. Timing is crucial to ensure that mature sperm encounter ripe eggs. Environmental cues such as temperature or day length may cause all the individuals of a population to release gametes at once, or pheromones from one individual releasing gametes may trigger gamete release in others.

Most fishes and amphibians that use external fertilization exhibit specific mating behaviors, resulting in one male fertilizing the eggs of one female. Courtship behavior is a mutual trigger for the release of gametes, with two effects: The probability of successful fertilization is increased, and the choice of mates may be somewhat selective.

Internal fertilization requires cooperative behavior, leading to copulation. In some cases, uncharacteristic sexual behavior is eliminated by natural selection in a direct manner; for example, female spiders will eat males if specific reproductive signals are not followed during mating. (Other examples of sexual selection and mating behavior will be discussed in Chapter 50.) Internal fertilization also requires sophisticated reproductive systems. Copulatory organs for the delivery of sperm and receptacles for its storage and transport to the eggs must be present.

Protection of the Embryo

External fertilization usually results in enormous numbers of zygotes, but the proportion that survive and develop further is often quite small. Internal fertilization usually produces fewer zygotes, but this may be offset by greater protection of the embryos and parental care of the young. Major types of protection include resistant eggshells, development of the embryo within the reproductive tract of the female parent, and parental care of the eggs and offspring.

Many species of terrestrial animals produce eggs that can withstand harsh environments. Birds, reptiles, and monotremes have amniote eggs with calcium and protein shells that resist water loss and physical damage. By contrast, the eggs of fishes and amphibians have only a gelatinous coat.

Rather than secreting a protective shell around the egg, many animals retain the embryo, which develops within the female reproductive tract. Among mammals, marsupials such as kangaroos and opossums retain their embryos for a short period in the uterus; the embryos then crawl out and complete fetal development attached to a mammary gland in the mother's pouch. The embryos of placental mammals develop entirely within the uterus, being nourished by the mother's blood supply through a special organ, the placenta (see Chapter 30).

When a kangaroo crawls out of its mother's pouch for the first time, or a human is born, it still is not capable of independent existence. We are familiar with adult birds feeding their young and mammals nursing their offspring, but parental care is much more widespread than we might suspect, and it takes a variety of unusual forms. In one species of tropical frog, for instance, the male carries the tadpoles in his stomach until they metamorphose and hop out as young frogs. There are also many cases of parental care among invertebrates (FIGURE 42.5).

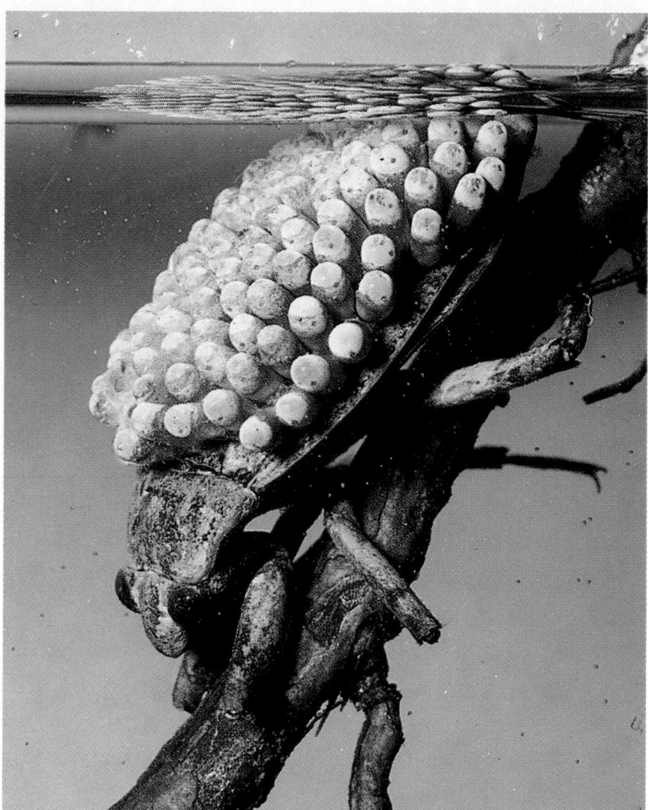

FIGURE 42.5
Parental care in an invertebrate. Compared to many insects and other arthropods, giant water bugs produce relatively few offspring, but parental protection enhances the survival of those offspring. Fertilization is internal, and the female glues her fertilized eggs to the back of the male (shown here). Whereas the males of most insect species provide no parental care for their offspring, the male giant water bug carries them for days, frequently fanning water over them, which helps keep the eggs moist, aerated, and free of parasites.

Diverse reproductive systems have evolved in the animal kingdom

To reproduce sexually, animals must have systems that produce and deliver gametes to the gametes of the opposite sex. These reproductive systems are varied. The least complex systems do not even contain distinct **gonads,** the organs that produce gametes in most animals. The most complex reproductive systems contain many sets of accessory tubes and glands that carry and protect the gametes and developing embryos. If we can make one generalization, it is that the complexity of the reproductive system is not entirely related to the phylogeny of the animal; the reproductive systems of parasitic flatworms, for example, are among the most complex in the animal kingdom.

Invertebrate Reproductive Systems

Diverse reproductive systems have evolved among invertebrates. Among the least complex systems are those of polychaete annelids. Most polychaetes have separate sexes but do not have distinct gonads; rather, the eggs and sperm develop from undifferentiated cells lining the coelom. As the gametes mature, they are released from the body wall and fill the coelom. Depending on the species, mature gametes may be shed through the excretory openings, or the swelling mass of eggs may split the body open, killing the parent and spilling the eggs into the environment.

Insects have separate sexes with complex reproductive systems. In the male, sperm develop in a pair of testes and are conveyed along a coiled duct to two seminal vesicles, where they are stored. During mating, sperm are ejaculated into the female reproductive system. In the female, eggs develop in a pair of ovaries and are conveyed through ducts to the vagina, where fertilization occurs. In many species, the female reproductive system includes a

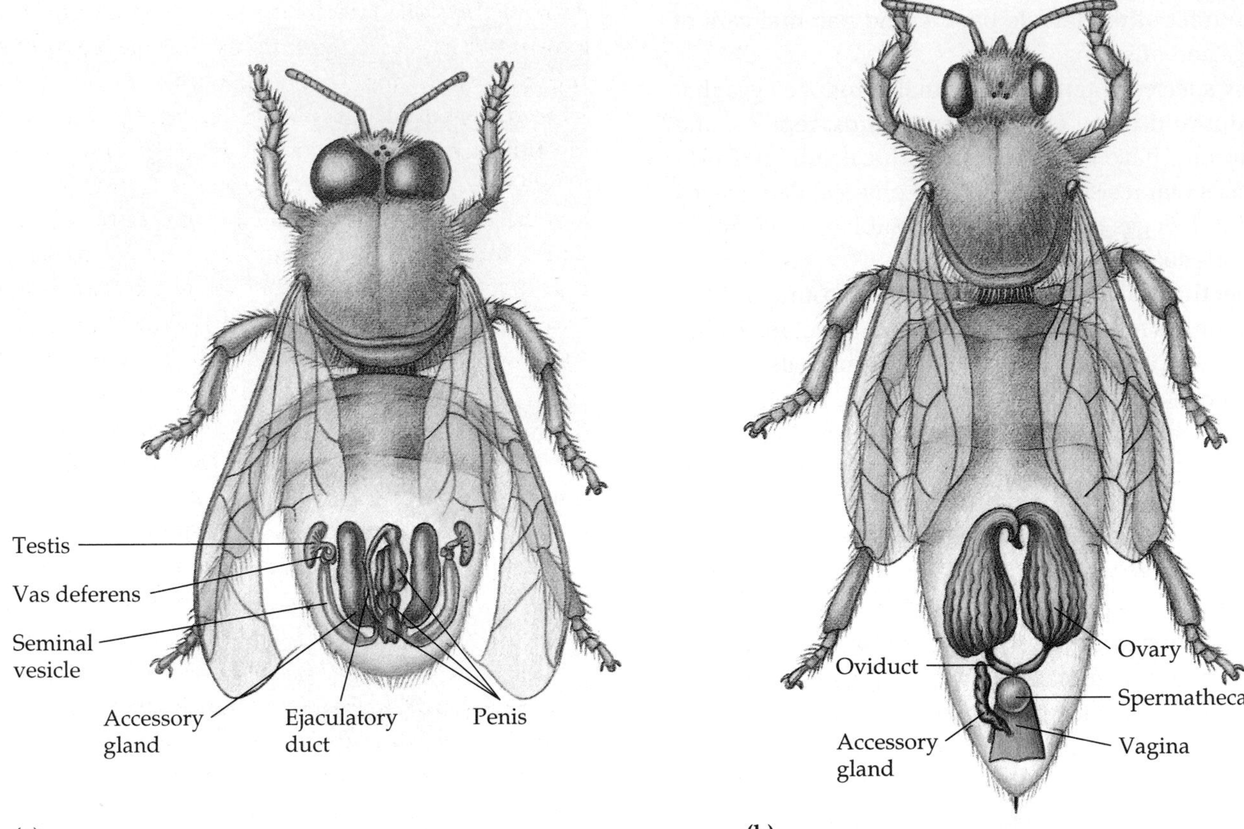

FIGURE 42.6
Insect reproductive anatomy. (**a**) A male honeybee. Sperm form in the testes, pass through the sperm duct (vas deferens), and are stored in the seminal vesicle. The male ejaculates sperm along with fluid from the accessory glands. Males of some species of insects and other arthropods have appendages called claspers that grasp the female during copulation. (**b**) A female honeybee. Eggs develop in the ovaries, pass through the oviducts, and into the vagina. A pair of accessory glands (only one is shown) add protective secretions to the eggs in the vagina. After mating, sperm are stored in the spermatheca, a sac connected to the vagina by a short duct.

spermatheca, a sac in which sperm may be stored for a year or more (FIGURE 42.6).

Most flatworms (phylum Platyhelminthes) are hermaphroditic (FIGURE 42.7). In some species, the female reproductive system includes yolk and shell glands, as well as a uterus where eggs are fertilized and development begins. The male system includes a copulatory apparatus sometimes called a penis. Copulation in flatworms is usually mutual, each partner inseminating the other. The mechanisms of insemination range from insertion of the penis into the vagina to hypodermic impregnation, in which the penis injects sperm into the body tissues and sperm migrate to the female reproductive tract.

Vertebrate Reproductive Systems

The basic plan of all vertebrate reproductive systems is quite similar, but there are some important variations on this common theme. In many nonmammalian vertebrates, the digestive, excretory, and reproductive systems have a common opening to the outside, the **cloaca,** which was probably present in the ancestors of all vertebrates (see Chapter 30). By contrast, most mammals lack a cloaca and have a separate opening for the digestive tract, and most female mammals have separate openings for the excretory and reproductive systems as well. The uterus of most vertebrates is partly or completely divided into two chambers. However, in humans and other mammals that produce only a few young at a time, as well as in birds and snakes, the uterus is a single structure. Male reproductive systems differ mainly in the copulatory organs. Nonmammalian vertebrates do not have well-developed penises and may just evert the cloaca to ejaculate.

For the remainder of this chapter, we will focus on mammalian reproduction, using humans as an example. The reproductive anatomy of other mammals is similar.

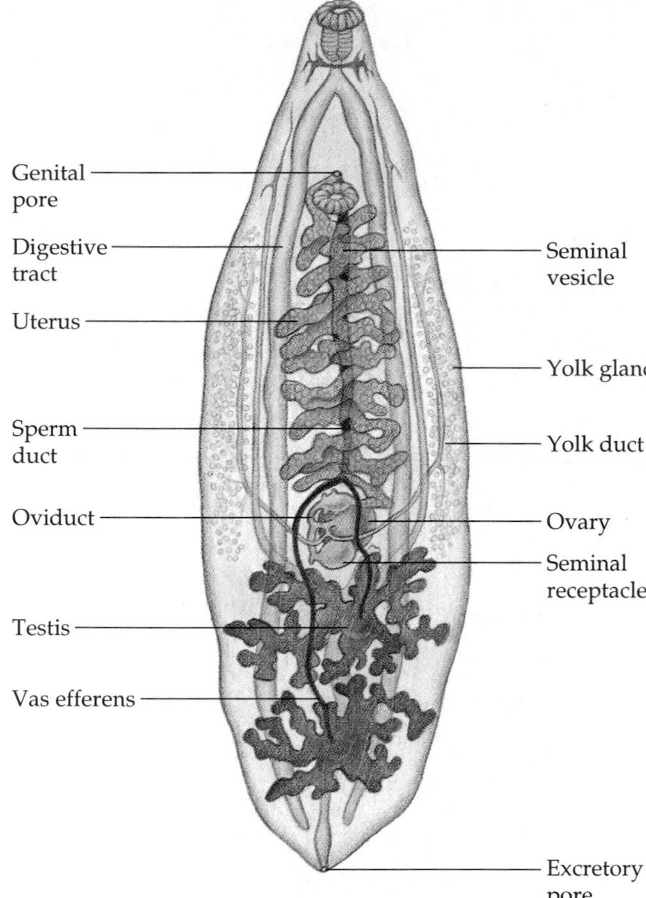

FIGURE 42.7
Reproductive anatomy of a parasitic flatworm. Hermaphroditic individuals have complex male and female reproductive systems. Both systems open to the outside through the genital pore. Sperm produced in the testes pass through a pair of ducts (vasa efferentia) into a single sperm duct (vas deferens) and are stored in the seminal vesicle. During mating, sperm are ejaculated into the female (usually another individual) and then move through the uterus to the seminal receptacle. Eggs from the ovary pass into the oviduct, where they are fertilized by sperm from the seminal receptacle and coated with yolk and tough shell material secreted by the yolk glands. From the oviduct, the fertilized, shelled eggs pass into a long, coiled uterus from which they are shed through the genital pore.

Human reproduction involves intricate anatomy and complex behavior

Reproductive Anatomy of the Human Male

The human reproductive system is frequently described as two sets of organs: the internal reproductive organs and the external **genitalia**. The male genitalia are the scrotum and penis. The male internal reproductive organs consist of gonads that produce gametes (sperm cells) and hormones, accessory glands that secrete products essential to sperm movement, and a set of ducts that carry the sperm and glandular secretions (FIGURE 42.8).

The male gonads, or **testes** (singular, **testis**), consist of many highly coiled tubes surrounded by several layers of connective tissue. These tubes are the **seminiferous tubules,** where sperm form. The **interstitial cells** scattered between the seminiferous tubules produce testosterone and other androgens, the male sex hormones.

Sperm production cannot occur at normal body temperatures in most mammals, but the testes of humans and many other mammals are held outside the abdominal cavity in a fold of skin called the **scrotum.** The temperature in a scrotum is about 2°C below that in the abdominal cavity. The testes develop high in the abdominal cavity and descend into the scrotum just before birth. In some mammals (not humans), the testes are drawn back into the abdominal cavity between breeding seasons. Whales and bats are exceptional in retaining the testes within the abdominal cavity permanently.

From the seminiferous tubules of a testis, the sperm pass into the coiled tubules of the **epididymis,** which stores sperm and is the site where they begin to gain motility and the ability to fertilize. During **ejaculation,** the sperm are propelled from the epididymis through the muscular **vas deferens.** These two ducts (one from each epididymis) run from the scrotum around and behind the urinary bladder, where each joins a duct from the seminal vesicle, forming a short **ejaculatory duct.** The ejaculatory ducts open into the **urethra,** the tube that drains both the excretory system and the reproductive system. The urethra runs through the penis and opens to the outside at the tip of the penis.

Three sets of accessory glands add secretions to the **semen,** the fluid that is ejaculated. A pair of **seminal vesicles** contributes about 60% of the total volume of the semen. The fluid from the seminal vesicles is thick and clear, containing mucus, amino acids, and large amounts of fructose (sugar), which provides energy for the sperm. The seminal vesicles also secrete prostaglandins (see Chapter 41). Once in the female reproductive tract, prostaglandins stimulate contractions of the uterine muscles, which help move the semen up the uterus. Proteins in the seminal fluid cause the semen to coagulate after it is deposited in the female reproductive tract, making it easier for uterine contractions to move the semen.

The **prostate gland** is the largest of the semen-secreting glands. It secretes its products directly into the urethra through several small ducts. Prostatic fluid is thin and milky, contains several enzymes, and is quite alkaline. It balances the acidity of any residual urine in the urethra and the natural acidity of the vagina, and helps activate sperm. The prostate gland is the source of some of the most common medical problems of men over age 40. Benign (noncancerous) enlargement of the prostate occurs in more than half of all men in this age group.

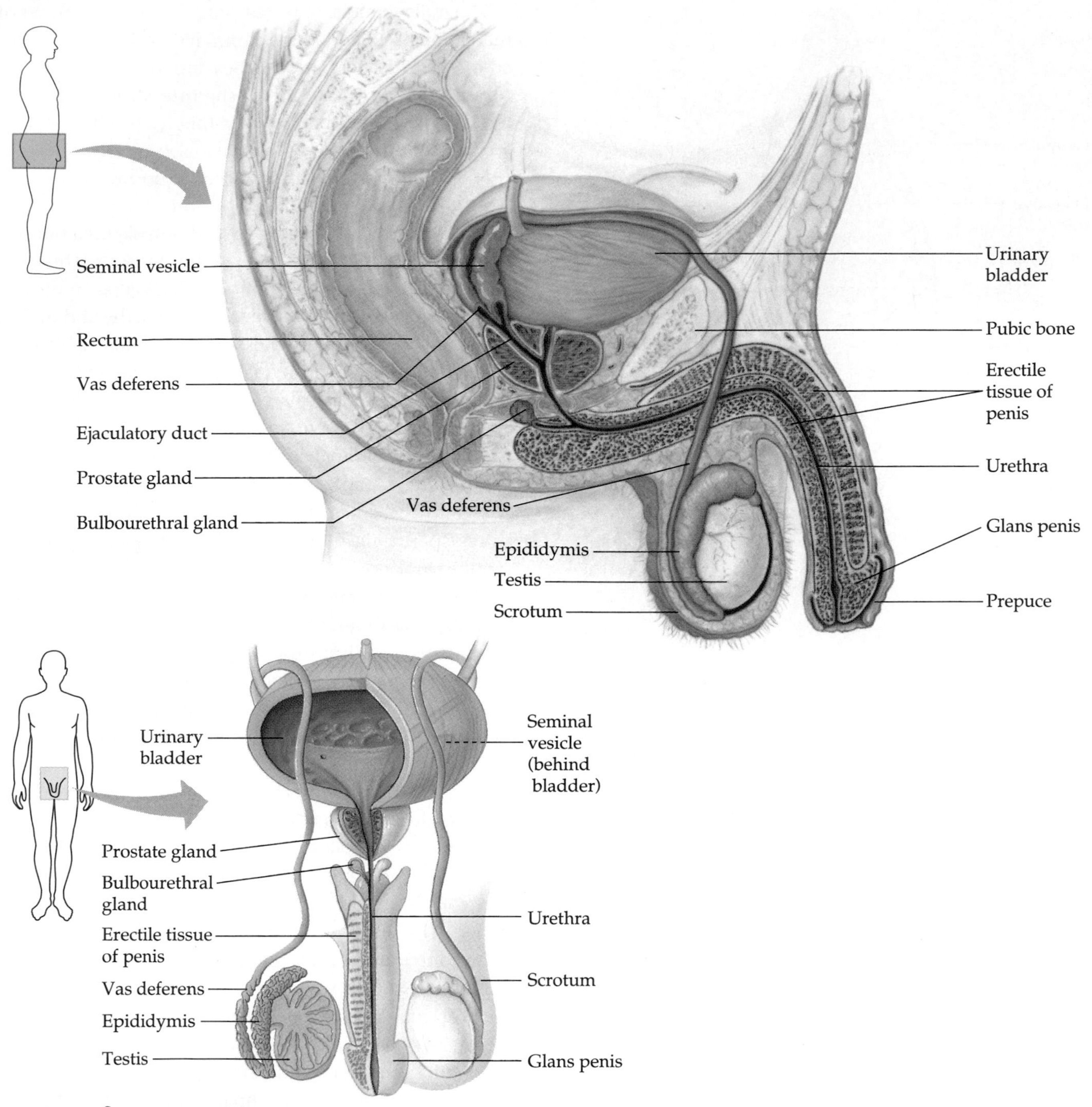

FIGURE 42.8
Reproductive anatomy of the human male.

The **bulbourethral glands** are a pair of small glands along the urethra below the prostate. Their function is still in question, but before ejaculation they secrete a clear viscous fluid that may neutralize any remaining acidic urine in the urethra. Bulbourethral fluid also carries some sperm released before ejaculation, which is one reason for the high failure rate of the withdrawal method of birth control.

The human **penis** is composed of three cylinders of spongy erectile tissue derived from modified veins and capillaries. During sexual arousal, the erectile tissue fills with blood from the arteries. As it fills, the increasing pressure seals off the veins that drain the penis, causing it to engorge with blood. The resulting erection is essential to insertion of the penis into the vagina. Rodents, raccoons, walruses, and several other mammals also possess a **baculum,** a bone that is contained in, and helps stiffen, the penis.

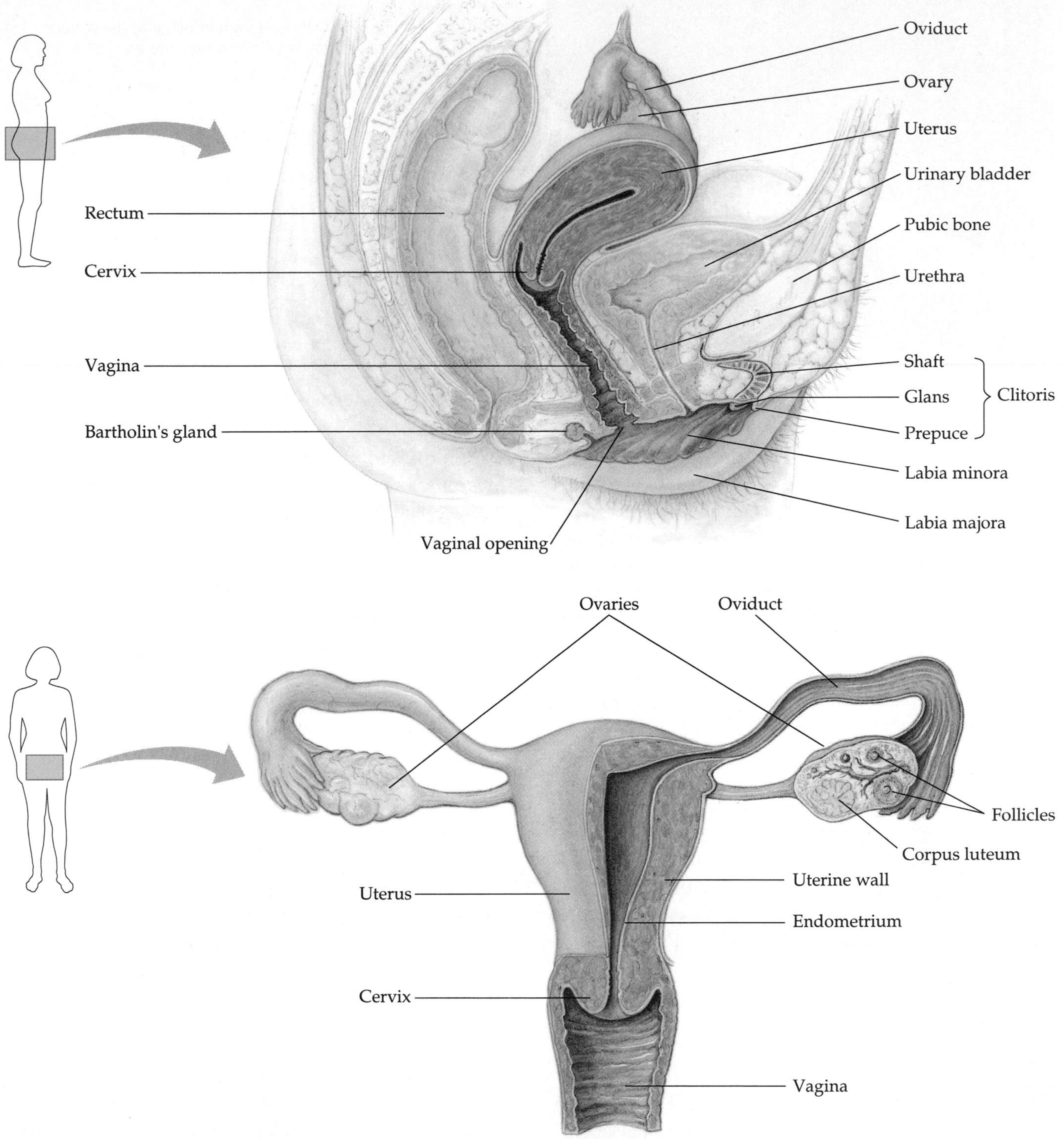

FIGURE 42.9
Reproductive anatomy of the human female.

The main shaft of the penis is covered by relatively thick skin. The head, or **glans penis,** has a much thinner covering and is consequently more sensitive to stimulation. The human glans is covered by a fold of skin called the foreskin, or **prepuce,** which may be removed by circumcision. Circumcision, which arose from religious traditions, has no verifiable basis in health or hygiene.

Reproductive Anatomy of the Human Female

The female external genitalia are the clitoris and two sets of labia surrounding the clitoris and vaginal opening. The internal reproductive organs consist of a pair of gonads and a system of ducts and chambers to conduct the gametes and house the embryo and fetus (FIGURE 42.9).

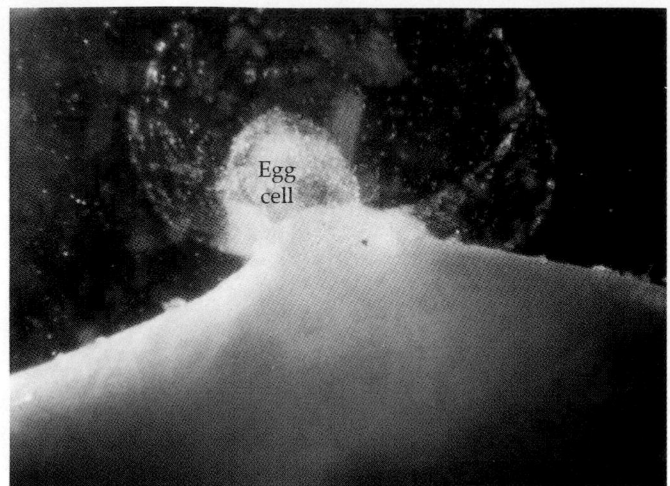

FIGURE 42.10
Ovulation. An egg cell is released from a follicle at the surface of the ovary. The orangish mass below the ejected egg cell is part of a mammalian ovary.

The female gonads, the **ovaries,** lie in the abdominal cavity, flanking, and attached by a mesentery to, the uterus. Each ovary is enclosed in a tough protective capsule and contains many follicles. A **follicle** consists of one egg cell surrounded by one or more layers of follicle cells, which nourish and protect the developing egg cell. All of the 400,000 follicles a woman will ever have are formed before birth. Of these, only several hundred will release egg cells during the woman's reproductive years. Starting at puberty and continuing until menopause, usually one follicle matures and releases its egg cell during each menstrual cycle. The cells of the follicle also produce the primary female sex hormones, the estrogens. The egg cell is expelled from the follicle in the process of **ovulation** (FIGURE 42.10). The remaining follicular tissue then grows within the ovary to form a solid mass called the **corpus luteum.** The corpus luteum secretes additional estrogen and progesterone, the hormone that maintains the uterine lining during pregnancy. If the egg cell is not fertilized, the corpus luteum disintegrates, and a new follicle matures during the next cycle.

The female reproductive system is not completely closed, and the egg cell is released into the abdominal cavity near the opening of the **oviduct,** or fallopian tube. The oviduct has a funnel-like opening, and cilia on the inner epithelium lining the duct help collect the egg cell by drawing fluid from the body cavity into the duct. The cilia also convey the egg cell down the duct to the **uterus,** also known as the womb. The uterus is a thick, muscular organ that can expand during pregnancy to accommodate a 4-kg fetus. The inner lining of the uterus, the **endometrium,** is richly supplied with blood vessels.

The neck of the uterus is the **cervix,** which opens into the vagina. The **vagina** is a thin-walled chamber that forms the birth canal through which the baby is born; it is also the repository for sperm during copulation.

At birth, and usually until sexual intercourse or vigorous physical activity ruptures it, a vascularized membrane called the **hymen** partly covers the vaginal opening in humans. The vaginal opening and the separate urethral opening are located within a region called the **vestibule,** bordered by a pair of slender skin folds, the **labia minora.** A pair of thick, fatty ridges, the **labia majora,** encloses and protects the labia minora and vestibule. At the front edge of the vestibule, the **clitoris** consists of a short shaft supporting a rounded glans, or head, covered by a small hood of skin, the prepuce. During sexual arousal, the clitoris, vagina, and labia minora all engorge with blood and enlarge. The clitoris consists largely of erectile tissue. Richly supplied with nerve endings, it is one of the most sensitive points of sexual stimulation. During sexual arousal, **Bartholin's glands,** located near the vaginal opening, secrete mucus into the vestibule, keeping it lubricated and facilitating intercourse.

The **mammary gland,** or breast, is another structure important to mammalian reproduction, although it is not part of the reproductive tract itself. Within the gland, small sacs of epithelial tissue secrete milk, which drains into a series of ducts, opening at the nipple. Fatty (adipose) tissue forms the main mass of the mammary gland of a nonlactating mammal. The absence of estrogen in males prevents the development of both the secretory apparatus and the fat deposits, so male breasts remain small and the nipples are not connected to the ducts.

Human Sexual Response

Many vertebrates and invertebrates have elaborate and complex mating behaviors, but these are usually stereotyped interactions that involve specific sequences of reciprocal behaviors (see Chapter 50). The hallmark of human sexuality is the diversity of stimuli and responses.

Behind this variable sexual behavior, however, is a common physiological pattern, often called the sexual response cycle. As is true of reproductive anatomy, the sexual responses of males and females have similarities as well as differences.

Two types of physiological reactions predominate in both sexes: **vasocongestion,** the filling of a tissue with blood caused by increased blood flow through the arteries of that tissue, and **myotonia,** increased muscle tension. Both skeletal and smooth muscle may show sustained or rhythmic contractions, including those associated with orgasm.

The sexual response cycle can be divided into four phases: excitement, plateau, orgasm, and resolution. An important function of the excitement phase is preparation of the vagina and penis for **coitus** (sexual intercourse). During this phase, vasocongestion is particularly evident in erection of the penis and clitoris; enlargement of the testes, labia, and breasts; and vaginal lubrication. Myotonia may occur, resulting in nipple erection or tension of the arms and legs.

The plateau phase continues these responses. In females, the outer third of the vagina becomes vasocongested, while the inner two-thirds becomes slightly expanded. This change, coupled with the elevation of the uterus, forms a depression that receives sperm at the back of the vagina. Reactions in nonreproductive organs continue as breathing increases and heart rate rises, sometimes to 150 beats per minute—not in response to the physical effort of sexual activity, but as an involuntary response to stimulation of the autonomic nervous system (see Chapter 44).

Orgasm is characterized by rhythmic, involuntary contractions of the reproductive structures in both sexes. Male orgasm has two stages. Emission is the contraction of the glands and ducts of the reproductive tract, which forces semen into the urethra. Expulsion, or ejaculation, occurs when the urethra contracts and the semen is expelled. During female orgasm, the uterus and outer vagina contract, but the inner two-thirds of the vagina do not. Orgasm is the shortest phase of the sexual response cycle, usually lasting only a few seconds. In members of both sexes, contractions occur at about 0.8-sec intervals and may involve the anal sphincter and several abdominal muscles.

The resolution phase completes the cycle and reverses the responses of the earlier stages. Vasocongested organs return to their normal size and color, and muscles relax. Most of the changes during resolution are completed in 5 minutes. Loss of penile and clitoral erection, however, may take longer. An initial loss of erection, or detumescence, is rapid in both sexes, but a return of the organs to their nonaroused size may take as long as an hour.

Spermatogenesis and oogenesis both involve meiosis but differ in three significant ways

Spermatogenesis, the production of mature sperm cells, is a continuous and prolific process in the adult male. Each ejaculation of a human male contains 250 to 400 million sperm cells, and males can ejaculate daily with little loss of fertilizing capacity.

The structure of a sperm cell fits its function. In most species, a thick head containing the haploid nucleus is tipped with a special body, the **acrosome,** which contains enzymes that help the sperm penetrate the egg. Behind the head, the sperm cell contains large numbers of mitochondria (or a single large one, in some species) that provide ATP for movement of the tail, which is a flagellum (FIGURE 42.11). Mammalian sperm shape is quite variable, with a head resembling a slender comma, an oval form as in the human sperm, or nearly spherical. Spermatogenesis occurs in the seminiferous tubules of the testes (FIGURE 42.12).

Oogenesis, the development of ova (mature, unfertilized egg cells), differs from spermatogenesis in three important ways (FIGURE 42.13, p. 949). First, during the meiotic divisions of oogenesis, cytokinesis is unequal, with almost all the cytoplasm monopolized by a single daughter cell, the secondary oocyte. This large cell can go on to form the ovum, and the other products of meiosis, smaller cells called polar bodies, degenerate. This contrasts with spermatogenesis, when all four products of meiosis I and II develop into mature sperm (compare FIGURES 42.12 and 42.13). Second, while the cells from which sperm develop continue to divide by mitosis throughout the male's reproductive years, this is not the case for oogenesis in the female. At birth, an ovary already contains all the cells it will ever have that will develop into eggs.

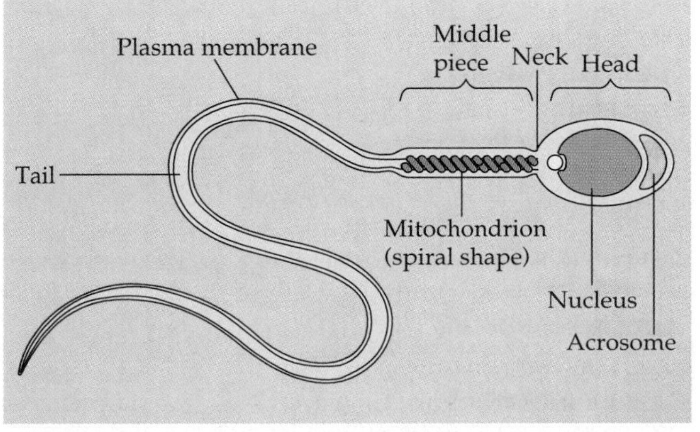

FIGURE 42.11
Structure of a human sperm cell.

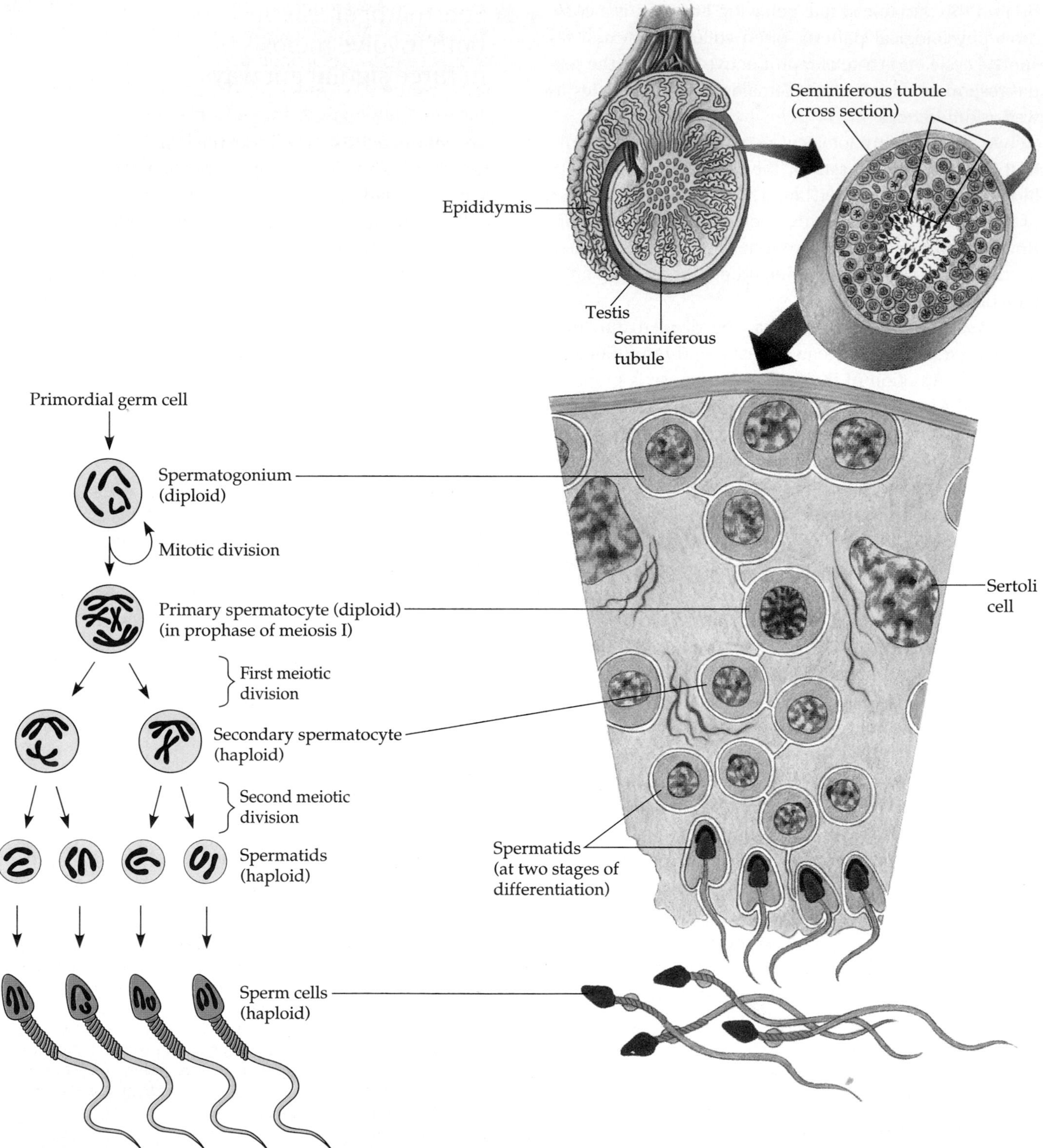

FIGURE 42.12

Spermatogenesis. These drawings correlate the meiotic stages in sperm development (left) with the structure of seminiferous tubules. Primordial germ cells of the embryonic testes differentiate into spermatogonia, the diploid cells that are the precursors of sperm. Located near the outer wall of the seminiferous tubules, spermatogonia undergo repeated mitoses, which produce large populations of potential sperm. In a mature male, about 3 million spermatogonia per day differentiate into primary spermatocytes. The chromosome number is reduced by half as the primary spermatocytes undergo the first division of meiosis. This diagram simplifies by representing the diploid number ($2n$) as only 4 (the actual $2n$ number in humans is 46). Notice that the secondary spermatocytes each have only two chromosomes (the haploid number), and these chromosomes are still duplicated, each consisting of two identical chromatids. The second meiotic division produces four spermatids, each with two single chromosomes. Spermatids then differentiate into mature spermatozoa, or sperm cells. This involves association of the developing sperm with large Sertoli cells, which transfer nutrients to the spermatids. During spermatogenesis, the developing sperm are gradually pushed toward the center of the seminiferous tubule and make their way to the epididymis, where they acquire motility. This process, from spermatogonia to motile sperm, takes 65 to 75 days in the human male.

Third, oogenesis has long "resting" periods, in contrast to spermatogenesis, which produces mature sperm from precursor cells in an uninterrupted sequence.

Between birth and puberty, the egg cells (primary oocytes) enlarge, and the follicles around them grow. The primary oocytes replicate their DNA and enter prophase I of meiosis and do not change again unless reactivated by hormones. Beginning at puberty, FSH (follicle-stimulating hormone) periodically stimulates a follicle to begin growing again and induces its primary oocyte to complete the first meiotic division. Meiosis then stops again with the secondary oocyte, released during ovulation, not undergoing the second meiotic division. In humans, penetration of the egg cell by the sperm triggers the second meiotic division, and only then is oogenesis actually complete.

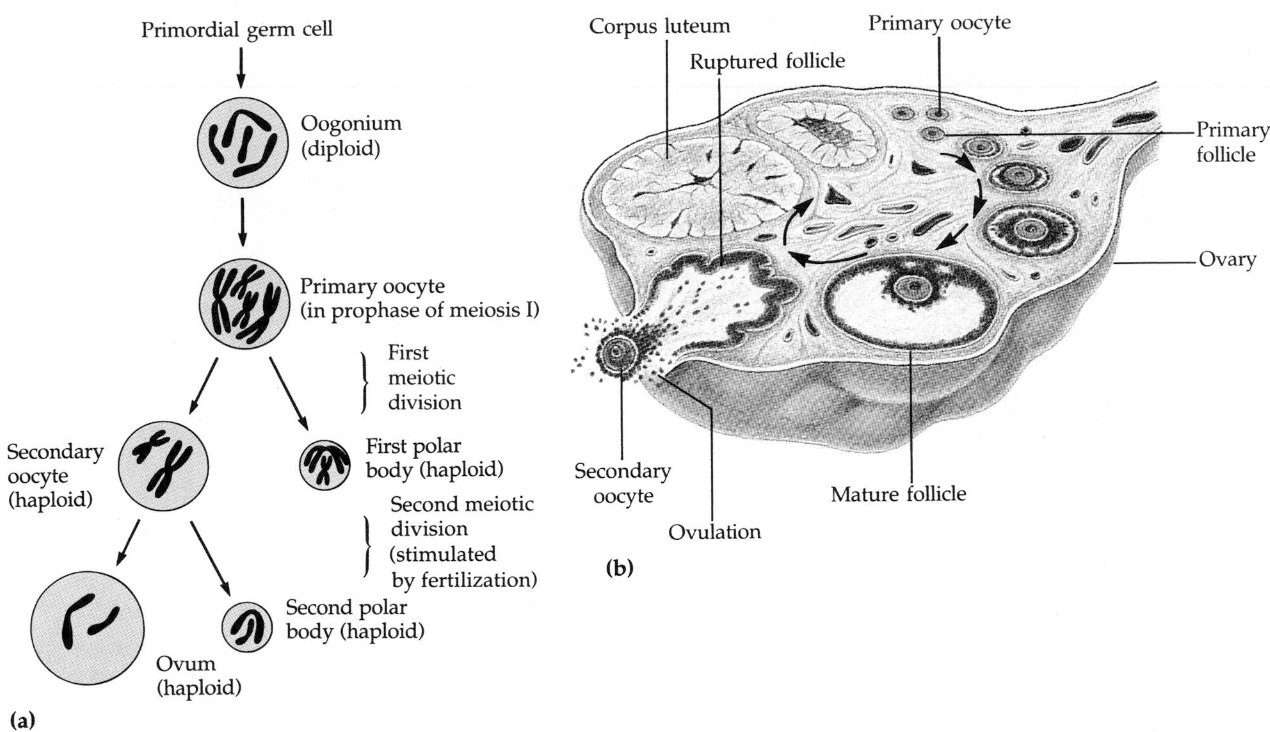

FIGURE 42.13

Oogenesis. (**a**) The production of ova begins with mitosis of the primordial germ cells in the embryo, producing diploid oogonia ($2n = 4$, in this simplified diagram). Each oogonium develops into a primary oocyte, which is also diploid. Starting at puberty, a single primary oocyte usually completes meiosis I each month. The meiotic divisions in oogenesis involve unequal cytokinesis. The first meiotic division produces a large cell, the secondary oocyte, and a much smaller polar body. The second meiotic division, which produces the ovum and another small polar body, occurs only if a sperm cell penetrates the secondary oocyte. After meiosis is completed and the second polar body separates from the ovum, the haploid nuclei of the sperm and the mature ovum fuse in the actual process of fertilization. (**b**) This cutaway view of an ovary illustrates the series of developmental stages of an ovarian follicle that accompany oogenesis: growth of the follicle, ovulation, and formation and disintegration of the corpus luteum (disintegration results if fertilization does not occur). For convenience, the stages are presented as a cycle (arrows), although they occur at different times and are never actually present simultaneously within the ovary. In a real ovary, each follicle stays in one place as it goes through the sequential stages.

A complex interplay of hormones regulates reproduction

The Male Pattern

In the male, the principal sex hormones are the androgens, of which testosterone is the most important. Androgens, steroid hormones produced mainly by the interstitial cells of the testes, are directly responsible for the primary and secondary sex characteristics of the male. Primary sex characteristics are those associated with the reproductive system: development of the vasa deferentia and other ducts, development of the external genitalia, and sperm production. Secondary sex characteristics are features that are not directly related to the reproductive system, including deepening of the voice, distribution of facial and pubic hair, and muscle growth (androgens stimulate protein synthesis). Androgens are also potent determinants of behavior in mammals and other vertebrates. In addition to specific sexual behaviors and sex drive, androgens increase general aggressiveness and are responsible for such actions as singing in birds and calling by frogs. Hormones from the anterior pituitary and hypothalamus control both androgen secretion and sperm production by the testes (FIGURE 42.14).

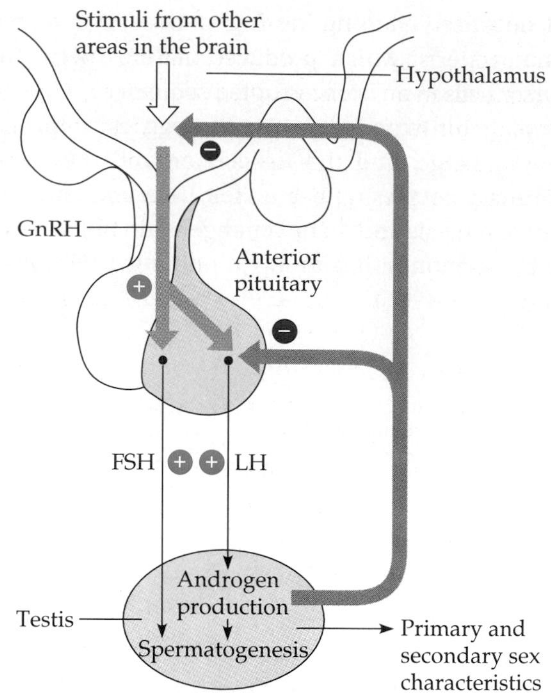

FIGURE 42.14
Hormonal control of the testes. The pituitary secretes two gonadotropic hormones with different effects on the testes. Luteinizing hormone (LH) stimulates androgen production by the interstitial cells. Follicle-stimulating hormone (FSH) acts on the seminiferous tubules to increase spermatogenesis. Since androgens are also required for sperm production, LH stimulates spermatogenesis indirectly. LH and FSH are in turn regulated by a single hormone from the hypothalamus, gonadotropin-releasing hormone (GnRH). How GnRH controls the release of two different hormones at different times is still unknown. LH, FSH, and GnRH concentrations in the blood are regulated by negative feedback by androgens. GnRH is also controlled by negative feedback from the two pituitary gonadotropins. In human males, these feedback loops keep the hormones at relatively constant levels, but in many other mammalian species, seasonal cycles in hormone concentration regulate breeding patterns.

The Female Pattern

In the female, the pattern of hormone secretion and the reproductive events they regulate are cyclic, very different from the male pattern. Whereas males produce sperm continuously, females release only one egg or a few eggs at one time during each cycle. Control of the female cycle is quite complex.

Two different types of cycles occur in female mammals. Humans and many other primates have **menstrual cycles,** whereas other mammals have **estrous cycles.** In both cases, ovulation occurs at a time in the cycle after the endometrium has started to thicken and develop a rich blood supply, which prepares the uterus for the possible implantation of an embryo. One difference between the two types of cycles involves the fate of the uterine lining if pregnancy does not occur. In menstrual cycles, the endometrium is shed from the uterus through the cervix and vagina in a bleeding called **menstruation.** In estrous cycles, the endometrium is reabsorbed by the uterus, and no extensive bleeding occurs.

Other major distinctions include more pronounced behavioral changes during estrous cycles than during menstrual cycles and stronger effects of season and climate on estrous cycles. Whereas human females may be receptive to sexual activity throughout their cycles, most mammals will copulate only during the period surrounding ovulation. This period of sexual activity, called **estrus** (L. *oestrus,* "frenzy," "passion"), is the only time vaginal changes permit mating. Estrus is sometimes called heat, and indeed the female's body temperature increases slightly. The length and frequency of reproductive cycles vary widely among mammals. The human menstrual cycle averages 28 days; the estrous cycle of the rat is only 5 days. Bears and dogs have one cycle per year, but elephants cycle several times per year.

Let's examine the reproductive cycle of the human female in more detail as a case study of how a complex function is coordinated by hormones. Only about 30% of women have cycle lengths within a day or two of the statistical average of 28 days. Cycles vary from one woman to another, ranging from about 20 to 40 days. In some women the cycles are usually very regular, but in other individuals the timing varies from cycle to cycle.

The term *menstrual cycle* refers specifically to the changes that occur in the uterus (FIGURE 42.15). By convention, the first day of a woman's menstrual period, the first day of menstruation, is designated day 1 of the cycle. The **menstrual flow phase** of the cycle, during which menstrual bleeding (loss of most of the endometrium) occurs, usually persists for a few days (FIGURE 42.15d). Then the thin remaining endometrium begins to regenerate and thicken for a week or two, during what is called the **proliferative phase** of the menstrual cycle. During the next phase, the **secretory phase,** usually about two weeks in duration, the endometrium continues to thicken, becomes more vascularized, and develops glands that secrete a fluid rich in glycogen. If an embryo has not implanted in the uterine lining by the end of the secretory phase, a new menstrual flow commences, marking day 1 of the next cycle.

Paralleling the menstrual cycle is an **ovarian cycle** (FIGURE 42.15c). It begins with the **follicular phase,** during which several follicles in the ovary begin to grow. The egg cell enlarges, and the coat of follicle cells becomes multilayered. Of the several follicles that start to grow, only one usually continues to enlarge and mature, while the others disintegrate. The maturing follicle develops an internal fluid-filled cavity and grows very large, forming a bulge near the surface of the ovary. The follicular phase ends with ovulation and the **ovulatory phase,** when the follicle and adjacent wall of the ovary rupture, releasing the egg cell. The follicular tissue that remains in the ovary after ovulation is transformed into the corpus luteum, an endocrine tissue that secretes female hormones during the **luteal phase** of the ovarian cycle. The next cycle begins with a new growth of follicles.

Hormones coordinate the menstrual and ovarian cycles in such a way that growth of the follicle and ovulation are synchronized with preparation of the uterine lining for possible implantation of an embryo. Five hormones participate in an elaborate scheme involving both positive and negative feedback. These hormones are gonadotropin-releasing hormone (GnRH), secreted by the hypothalamus; follicle-stimulating hormone (FSH) and luteinizing hormone (LH), the two gonadotropins secreted by the anterior pituitary; and estrogens (a family of closely related hormones) and progesterone, the female sex hormones secreted by the ovaries. The levels of the pituitary and ovarian hormones in blood plasma are traced in FIGURE 42.15a and b, along with the ovarian and menstrual cycles. As you read the following discussion, refer to the figure as a guide to understanding how the hormones regulate the female reproductive system.

During the follicular phase of the ovarian cycle, the pituitary secretes small amounts of FSH and LH in response to stimulation by GnRH from the hypothalamus. At this time, the cells of immature ovarian follicles have receptors for FSH but not for LH. The FSH stimulates follicle growth, and the cells of these growing follicles secrete estrogen. Notice in FIGURE 42.15b that there is a slow rise in the amount of estrogen secreted during most of the follicular phase. This small increase in estrogen inhibits secretion of the pituitary hormones, keeping the levels of FSH and LH relatively low during most of the follicular phase. These hormonal relationships change radically and rather abruptly when the rate of estrogen secretion by the growing follicle begins to rise steeply. Whereas a slow rise of estrogen inhibits the secretion of pituitary gonadotropins, a high estrogen concentration has the opposite effect and *stimulates* the secretion of gonadotropins by acting on the hypothalamus to increase its output of GnRH. You can see this response in FIGURE 42.15a as a steep incline of FSH and LH levels that follows closely behind the increase in estrogen concentration. The effect is greater for LH because the high estrogen concentration, in addition to stimulating GnRH secretion, also increases the sensitivity of LH-releasing mechanisms in the pituitary to the hypothalamic signal (GnRH). By now, the follicles have receptors for LH and can respond to this hormonal cue. In an example of positive feedback, the increase in LH concentration caused by increased estrogen secretion from the growing follicle induces final maturation of the follicle, and ovulation occurs about a day after the LH surge.

Following ovulation, LH stimulates the transformation of the follicular tissue left behind in the ovary to form the corpus luteum, a glandular structure. (It is for this "luteinizing" function that LH is named.) Under continued stimulation by LH during the luteal phase of the ovarian cycle, the corpus luteum secretes estrogen and a second steroid hormone, progesterone. The corpus luteum usually reaches its maximum development about 8 to 10 days after ovulation. As the progesterone and estrogen levels rise, the combination of these hormones exerts negative feedback on the hypothalamus and pituitary, inhibiting the secretion of LH and FSH. When the LH concentration plummets, the corpus luteum, which requires LH in order to function, begins to disintegrate. Consequently, estrogen and progesterone concentrations decline sharply near the end of the luteal phase. The dropping levels of ovarian hormones liberate the hypothalamus and pituitary from the inhibitory effects of these hormones. The pituitary then begins to secrete enough FSH to stimulate the growth of new follicles in the ovary, initiating the follicular phase of the next ovarian cycle.

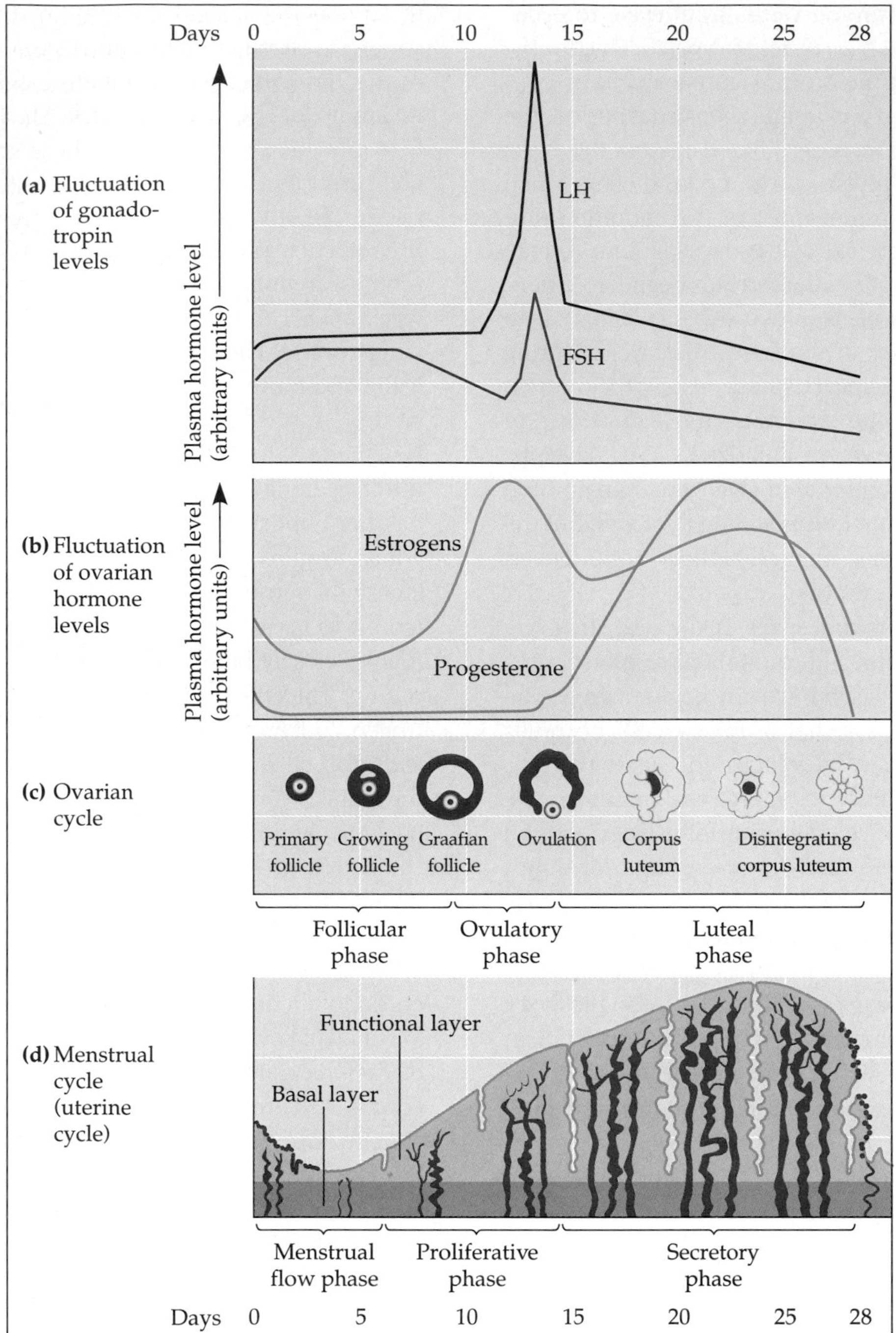

FIGURE 42.15

The reproductive cycle of the human female. Hormones coordinate the ovarian and menstrual cycles, preparing the uterine lining (endometrium) for implantation of an embryo even before ovulation. (**a**) Changes in LH and FSH levels. (**b**) Changes in the levels of estrogens and progesterone. (**c**) The ovarian cycle consists of a follicular phase, during which follicles grow and secrete increasing amounts of estrogens; an ovulatory phase, when ovulation occurs; and a luteal phase, during which the corpus luteum formed from the follicular tissue after ovulation secretes estrogens and progesterone. Length of the follicular phase varies among women, and in some women from one cycle to the next. The luteal phase usually lasts 13 to 15 days, regardless of the overall length of the cycle. (**d**) The menstrual cycle consists of a menstrual flow phase, a proliferative phase, and a secretory phase. Menstruation, the shedding of the endometrium, occurs during the menstrual flow phase. The first day of flow marks day 1 of the menstrual cycle. During the proliferative phase, estrogens from the growing follicle stimulate the endometrium to thicken and become increasingly vascularized. During the secretory phase, the endometrium continues to thicken, its arteries enlarge, and endometrial glands grow. These endometrial changes require estrogens and progesterone, secreted by the corpus luteum after ovulation. Thus, the secretory phase of the menstrual cycle parallels the luteal phase of the ovarian cycle. Disintegration of the corpus luteum at the end of the luteal phase reduces the amount of estrogens and progesterone available to the endometrium, so it is shed. The first day of menstruation, marking the beginning of the next cycle, usually occurs 13 to 15 days after ovulation. In the event of pregnancy, additional mechanisms maintain high levels of estrogens and progesterone, preventing loss of the endometrium.

How is the ovarian cycle synchronized with the menstrual cycle? Estrogen, secreted in increasing amounts by growing follicles, is a hormonal signal to the uterus, causing the endometrium to thicken. Thus, the follicular phase of the ovarian cycle is coordinated with the proliferative phase of the menstrual cycle. *Before* ovulation, the uterus is already being prepared for a possible embryo. *After* ovulation, estrogen and progesterone secreted by the corpus luteum stimulate continued development and maintenance of the endometrium, including an enlargement of arteries supplying blood to the uterine lining and growth of endometrial glands that secrete a nutrient fluid that can sustain an early embryo before it actually implants in the uterine lining. Thus, the luteal phase of the ovarian cycle is coordinated with the secretory phase of the menstrual cycle. The rapid drop in the level of ovarian hormones when the corpus luteum disintegrates causes spasms of arteries in the uterine lining that deprive the endometrium of blood. Disintegration of the endometrium results in menstruation and the beginning of a new menstrual cycle. In the meantime, ovarian follicles that will stimulate renewed thickening of the endometrium are just beginning to grow. Cycle after cycle, the maturation and release of egg cells from the ovary is integrated with changes in the uterus, the organ that must accommodate an embryo if the egg cell is fertilized. In the absence of pregnancy, a new cycle begins. We will soon see that there are "override" mechanisms that prevent disintegration of the endometrium in the event of pregnancy.

In addition to their role in coordinating reproductive cycles, estrogens are also responsible for the secondary sex characteristics of the female. The hormones induce deposition of fat in the breasts and hips, increase water retention, affect calcium metabolism, stimulate breast development, and mediate female sexual behavior.

Hormones and Sexual Maturation

Mammals cannot reproduce until they have undergone substantial growth and development after birth. For example, a human male can achieve an erection at birth but has no sperm to ejaculate. In humans, the onset of reproductive ability, or puberty, is a gradual process that usually begins about two years earlier in females than in males. Between the ages of 8 and 14, depending on the individual, the hypothalamus begins secreting increasing amounts of GnRH, leading to higher levels of FSH followed by increased LH. These gonadotropins trigger maturation of the reproductive system and development of the secondary sex characteristics (by increasing secretion of sex hormones from the gonads). The first indication of puberty is a growth spurt, followed by first menstruation at about age 12 or 13 in girls or first ejaculation of viable sperm at age 13 or 14 in boys. The age of puberty is quite variable, and recent analysis of historical data suggests that there has been little or no change in the average age throughout modern times.

Embryonic and fetal development occur during pregnancy in humans and other placental mammals

From Conception to Birth

In placental mammals, **pregnancy,** or **gestation,** is the condition of carrying one or more **embryos,** new developing individuals, in the uterus. Pregnancy is preceded by **conception,** the fertilization of the egg by a sperm cell, and continues until the birth of the offspring. Human pregnancy averages 266 days (38 weeks) from conception, or 40 weeks from the start of the last menstrual cycle. Duration of pregnancy in other species correlates with body size and the extent of development of the young at birth. Many rodents (mice and rats) have gestation periods of about 21 days, whereas those of dogs are closer to 60 days. In cows, gestation averages 270 days (almost the same as humans); in giraffes, it is about 420 days; and in elephants, gestation is more than 600 days.

Human gestation can be divided for convenience of study into three **trimesters** of about 3 months each. The first trimester is the time of most radical change for both the mother and the baby. Fertilization occurs in the oviduct. About 24 hours later, the resulting zygote begins dividing, a process called **cleavage** (FIGURE 42.16). Cleavage continues, with the embryo becoming a ball of cells by the time it reaches the uterus about 3 to 4 days after fertilization. By about 1 week after fertilization, cleavage has produced a hollow ball of cells, an embryonic stage called the **blastocyst.** In a process that takes about 5 more days, the blastocyst implants into the endometrium. Differentiation of body structures now begins in earnest. (Embryonic development will be described in detail in Chapter 43.) During implantation, the blastocyst bores into the endometrium, which responds by growing over the blastocyst. The embryo obtains nutrients directly from the endometrium during the first 2 to 4 weeks of development. Meanwhile, tissues grow out from the developing embryo and mingle with the endometrium to form the **placenta**. This disk-shaped organ, containing embryonic and maternal blood vessels, grows to about the size of a dinner plate and weighs somewhat less than 1 kg. Diffusion of material between maternal and embryonic circulations provides nutrients, exchanges respiratory gases, and disposes of metabolic wastes for the embryo. Blood from the embryo travels to the placenta

FIGURE 42.16

Formation of the zygote and early postfertilization events. (**a**) Fertilization occurs soon after a secondary oocyte enters the oviduct. (**b**) Cleavage, or cell division, begins in the oviduct and continues as the developing embryo moves down the oviduct toward the uterus, propelled by peristalsis and cilia. (**c**) By the time the embryo reaches the uterus, cleavage has transformed the embryo into a ball of cells. Cleavage continues while the embryo floats freely in the uterus for several days, nourished by fluid secreted by endometrial glands. (**d**) About 7 days after conception, the embryo is a hollow ball stage called the blastocyst, which implants into the endometrium.

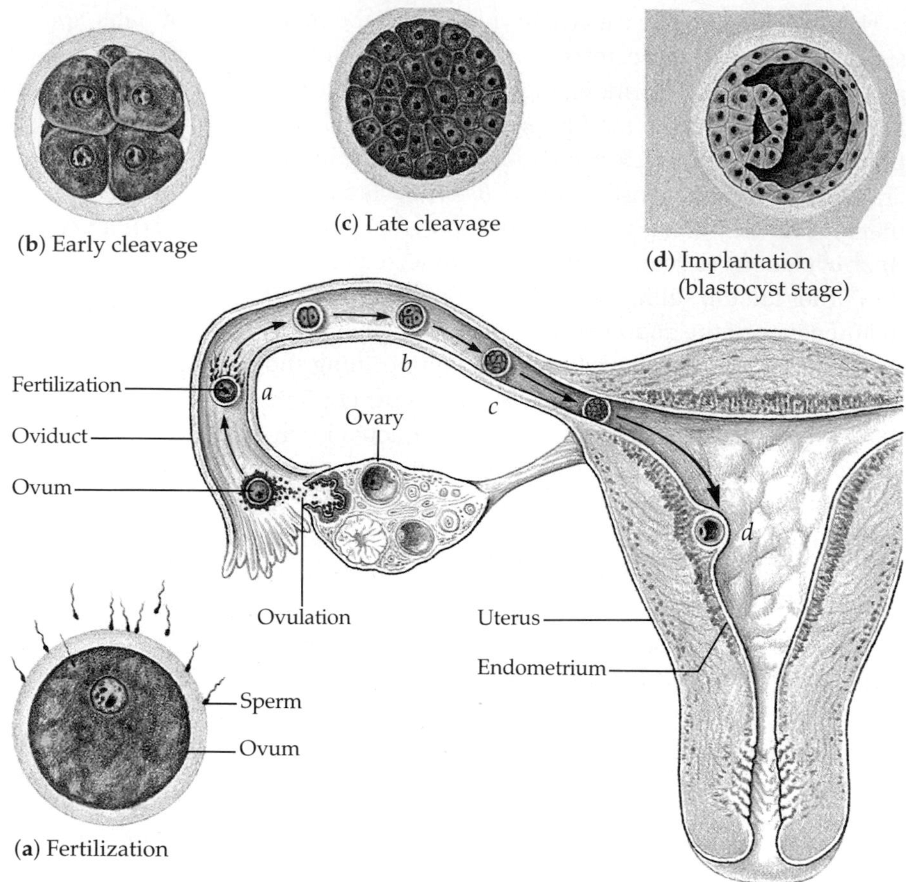

through arteries of the umbilical cord and returns via the umbilical vein, passing through the liver of the embryo (FIGURE 42.17).

The first trimester is also the main period of **organogenesis,** the development of the body organs (FIGURE 42.18). The heart begins beating by the fourth week and can be detected with a stethoscope by the end of the first trimester. By the end of the eighth week, all the major structures of the adult are present in rudimentary form. At this point, the embryo is called a **fetus.** Although well differentiated, the fetus is only 5 cm long by the end of the first trimester. Because of its rapid organogenesis, the embryo is most sensitive during the first trimester to such threats as radiation and drugs that can cause birth defects.

The first trimester is also a time of rapid change for the mother. The embryo secretes hormones that signal its presence and control the mother's reproductive system. One embryonic hormone, **human chorionic gonadotropin (HCG),** acts like pituitary LH to maintain progesterone and estrogen secretion by the corpus luteum through the first trimester. In the absence of this hormonal override, the decline in maternal LH due to inhibition of the pituitary by progesterone would result in menstruation and spontaneous abortion of the embryo. Levels of HCG in the maternal blood are so high that some is excreted in the urine, where it can be detected in pregnancy tests (see the Methods Box on p. 866). High levels of progesterone initiate changes in the pregnant woman's reproductive system, including increased mucus in the cervix that forms a protective plug, growth of the maternal part of the placenta, enlargement of the uterus, and (by negative feedback on the hypothalamus and pituitary) cessation of ovulation and menstrual cycling. The breasts also enlarge rapidly and are often quite tender.

During the second trimester, the fetus grows rapidly to about 30 cm and is very active. The mother may feel movements during the early part of the second trimester, and fetal activity may be visible through the abdominal wall by the middle of this time period. Hormone levels stabilize as HCG declines, the corpus luteum deteriorates, and the placenta secretes its own progesterone, which maintains the pregnancy. During the second trimester, the uterus will grow enough for the pregnancy to become obvious.

The third and final trimester is one of rapid growth of the fetus to about 3–3.5 kg in weight and 50 cm in length. Fetal activity may decrease as the fetus fills the available

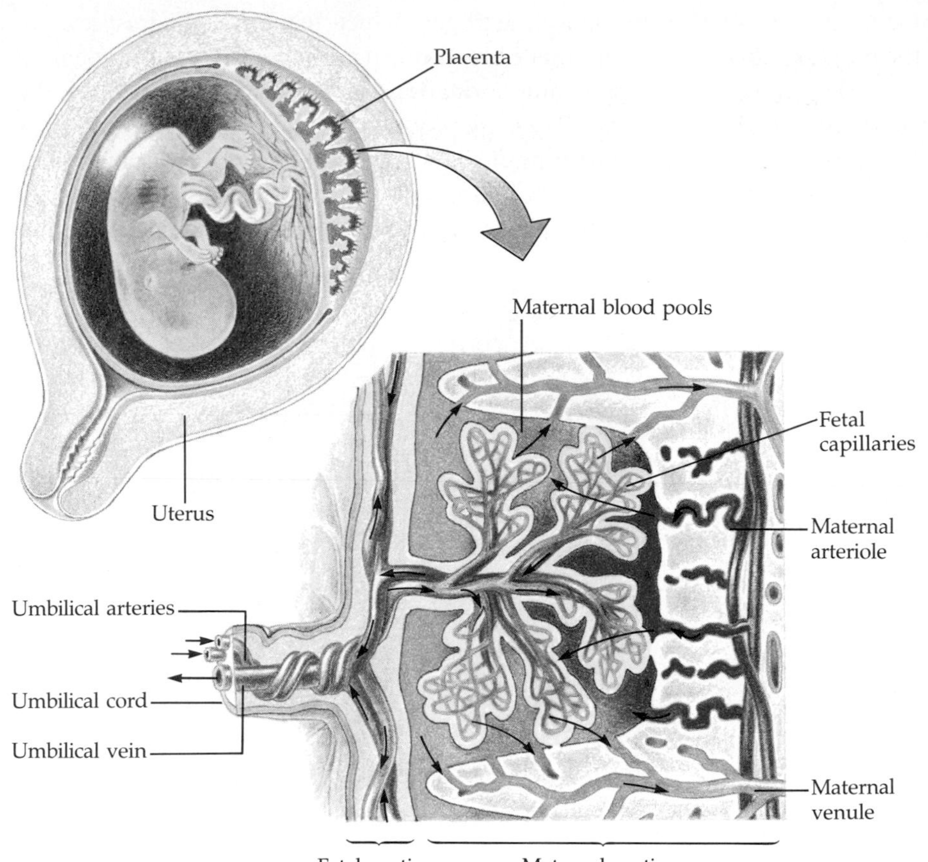

FIGURE 42.17
Placental circulation. From the fourth week of development until birth, the placenta, a combination of maternal and embryonic tissues, transports nutrients, respiratory gases, and wastes to and from the embryo or fetus. Maternal blood enters the placenta in arterioles, flows through blood pools in the endometrium, and leaves via venules. Embryonic or fetal blood, which remains in vessels, enters the placenta through arteries and passes through capillaries in fingerlike chorionic villi, where oxygen and nutrients are acquired. As indicated in the drawing, the fetal or embryonic capillaries and villi project into the maternal portion of the placenta. Embryonic blood leaves the placenta through veins leading back to the embryo. Materials are exchanged by diffusion, active transport, and selective absorption between the embryonic capillary bed and the maternal blood pools.

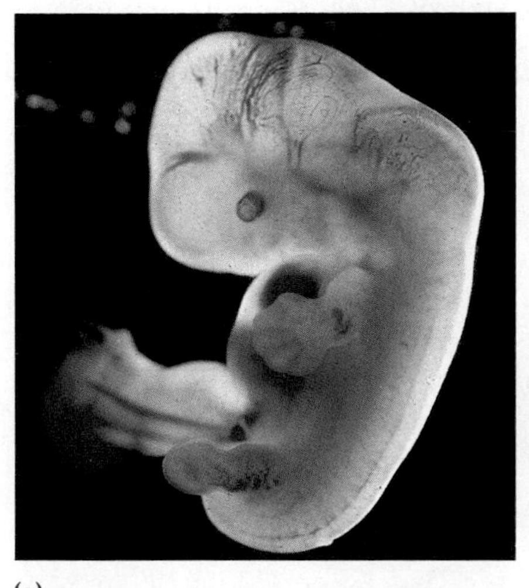
(a)

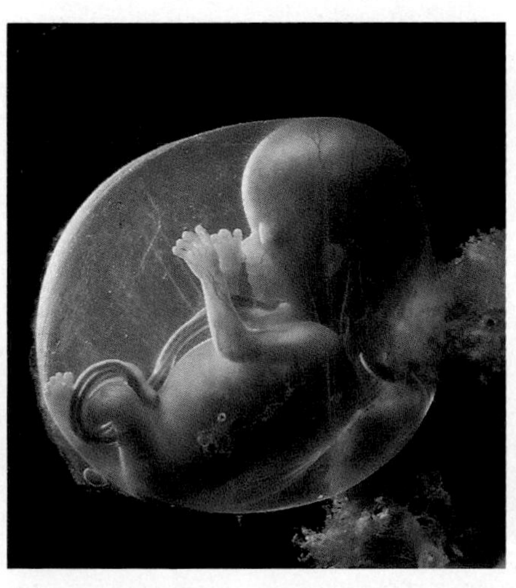
(b)

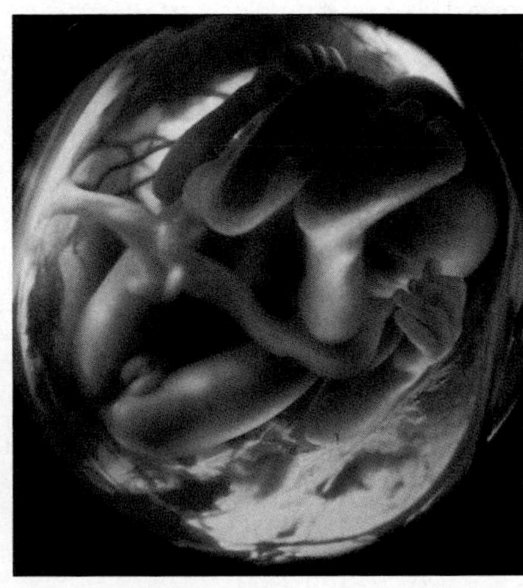
(c)

FIGURE 42.18
Human fetal development. (**a**) At 5 weeks, limb buds, eyes, the heart, liver, and rudiments of all other organs have started to develop in the embryo, which is only about 1 cm long. (**b**) Growth and development of the offspring, now called a fetus, continue during the second trimester. This fetus is 14 weeks old and about 6 cm long. (**c**) The fetus in this photograph is 20 weeks old. By the end of the second trimester (at 24 weeks), the fetus grows to about 30 cm in length.

space within the embryonic membranes. As the fetus grows and the uterus expands around it, the mother's abdominal organs become compressed and displaced, leading to frequent urination, digestive blockages, and strain in the back muscles. A complex interplay of the hormones estrogen and oxytocin and local regulators (prostaglandins) induces and regulates labor (FIGURE 42.19a). Estrogen, which reaches its highest level in the mother's blood during the last weeks of pregnancy, triggers the formation of oxytocin receptors on the uterus. Oxytocin, produced by the fetus and the mother's posterior pituitary, stimulates powerful contractions by the smooth muscles of the uterus. Oxytocin also stimulates the placenta to secrete prostaglandins, which en-

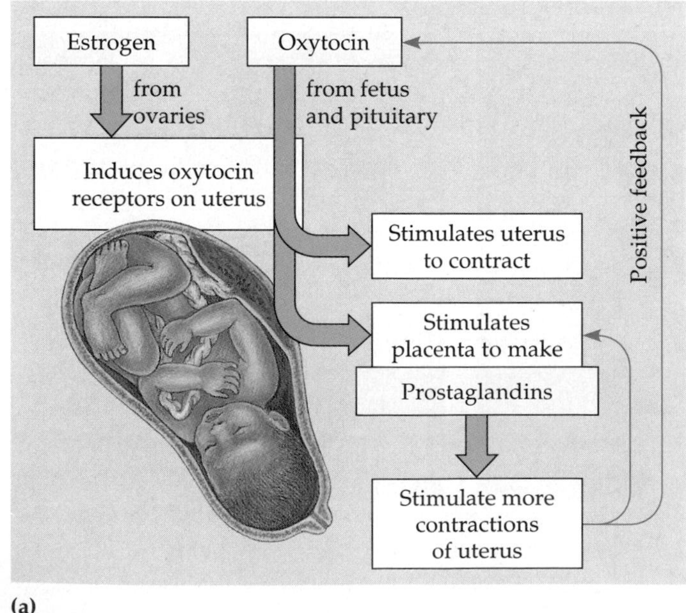

(a)

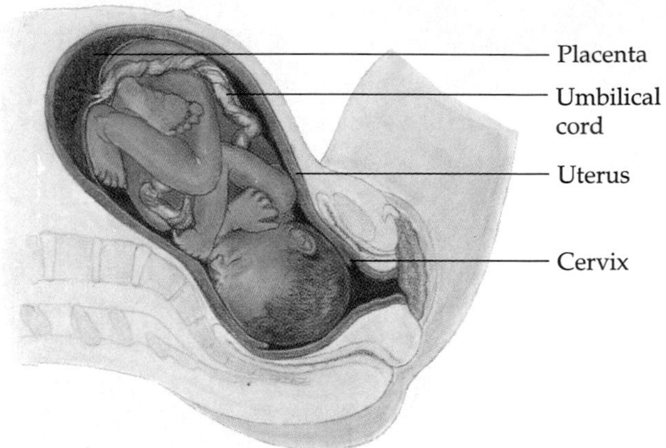

Dilation of the cervix

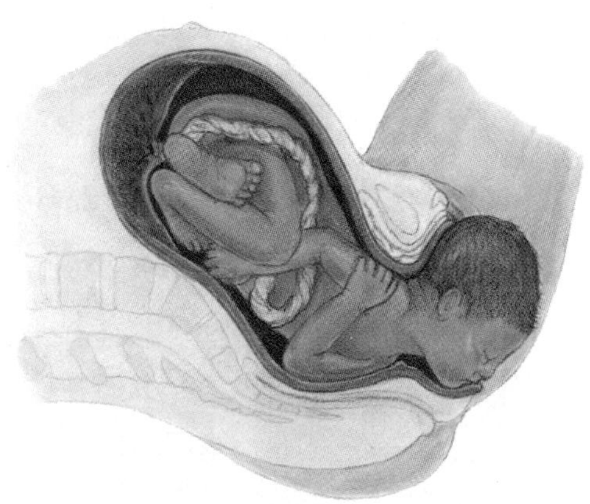

Expulsion: delivery of the infant

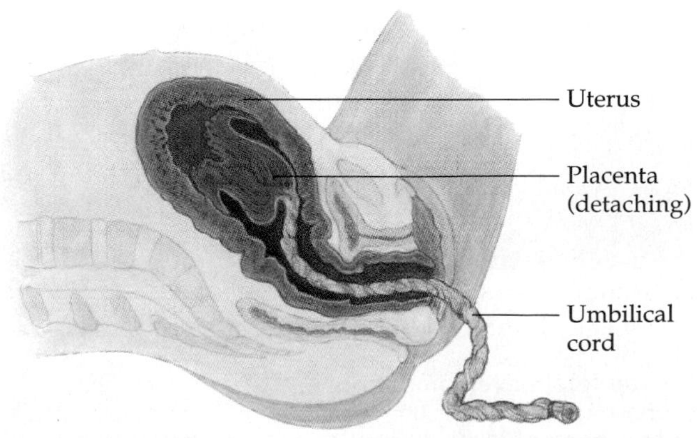

Delivery of the placenta

(b)

FIGURE 42.19
The birth of a human baby. (**a**) Hormonal induction of labor. (**b**) The three stages of labor.

hance the contractions. In turn, the physical and emotional stresses associated with the contractions stimulate the release of more oxytocin and prostaglandins, a positive feedback system that underlies the three stages of labor (FIGURE 42.19b).

Birth, or **parturition,** occurs through a series of strong, rhythmic uterine contractions, commonly known as **labor.** The first stage is the opening up and thinning of the cervix, ending with complete dilation. The second stage is expulsion, or delivery, of the baby. Continuous strong contractions force the fetus down and out of the uterus and vagina. The umbilical cord is cut and clamped at this time. The final stage of labor is delivery of the placenta, which normally follows the baby.

Lactation is an aspect of postnatal care unique to mammals. After birth, decreasing levels of progesterone free the anterior pituitary from negative feedback and allow prolactin secretion. Prolactin stimulates milk production after a delay of 2 or 3 days. The release of milk from the mammary glands is controlled by oxytocin, described in Chapter 41.

Reproductive Immunology

Pregnancy is an immunological enigma. Half of the embryo's genes are inherited from the father, and thus many of the chemical markers present on the surface of the embryo will be foreign to the mother. Why, then, does the mother not reject the embryo as a foreign body as she would repel a tissue or organ graft bearing antigens from another person? Reproductive immunologists are only beginning to solve this puzzle.

Part of the answer is the presence of a physical barrier. A protective layer called the trophoblast prevents the embryo from actually contacting maternal tissue. But the trophoblast develops along with the embryo from the cells of the blastocyst, and this protective barrier, which penetrates the endometrium, may also be foreign to the mother. According to one hypothesis, the trophoblast does *not* develop paternal markers and thus does not trigger an immune response by the mother. However, researchers have discovered paternal antigens on portions of the trophoblast. There is evidence that the trophoblast produces a chemical signal that induces the development of a special type of white blood cell in the uterus that prevents other white cells from mounting an attack on the foreign tissue. This suppressor cell may work by secreting a substance that blocks the action of interleukin-2, the cytokine required for a normal immune response (see Chapter 39). One hypothesis suggests that this local dampening of the immune response occurs, paradoxically, only after nearby white blood cells have first identified the trophoblast as foreign tissue and have taken the first steps of the immune response. If this immunological alarm is not intense enough—that is, if the father's cellular markers are too similar to the mother's—then no suppressor cells are produced.

Some researchers speculate that if the initial immune response is too weak to trigger suppression, the continued attack on the foreign tissue, though weak, may lead to spontaneous abortion of the embryo. According to this view, failure to suppress the immune response in the uterus may account for many cases of women who have multiple miscarriages for no other apparent reason. There has been some success treating frequent miscarriers by sensitizing the woman's immune system to her mate's antigens through immunization—that is, by injecting appropriate chemical markers into the mother prior to pregnancy. The subsequent response to the foreign embryological tissue is intensified to a level that activates the suppressor mechanism. Some critics of this interpretation argue that the psychological support women receive during this experimental treatment counts for more than the immunotherapy itself. Only more research will resolve the interesting and important questions about how a woman's immune system tolerates a 9-month relationship with a large foreign organism.

Contraception prevents pregnancy

Contraception literally means "against taking in"—in this case, the taking in of a child. The term has come to mean preventing pregnancy through one of several methods. Some prevent the release of mature eggs and sperm from gonads, others prevent fertilization by keeping sperm and eggs apart, and still others prevent implantation of an embryo or abort the embryo (FIGURE 42.20). The following brief introduction to the biology of these methods makes no pretense of being a contraception manual. For more complete information, you should consult a respected text on human sexuality.

Fertilization can be prevented by abstinence from sexual intercourse or by any of several barriers that keep live sperm from contacting the egg. Temporary abstinence, often called the **rhythm method** of birth control, or **natural family planning,** depends on refraining from intercourse when conception is most likely. Because the egg can survive in the oviduct for 24 to 48 hours and sperm for up to 72 hours, a couple practicing temporary abstinence should not engage in intercourse during the few days before and after ovulation. The most effective methods for timing ovulation combine several indicators, including changes in cervical mucus and body temperature during the menstrual cycle. Thus, natural family planning requires that the couple be knowledgeable about these physiological signs. A pregnancy rate of 10% to 20% is

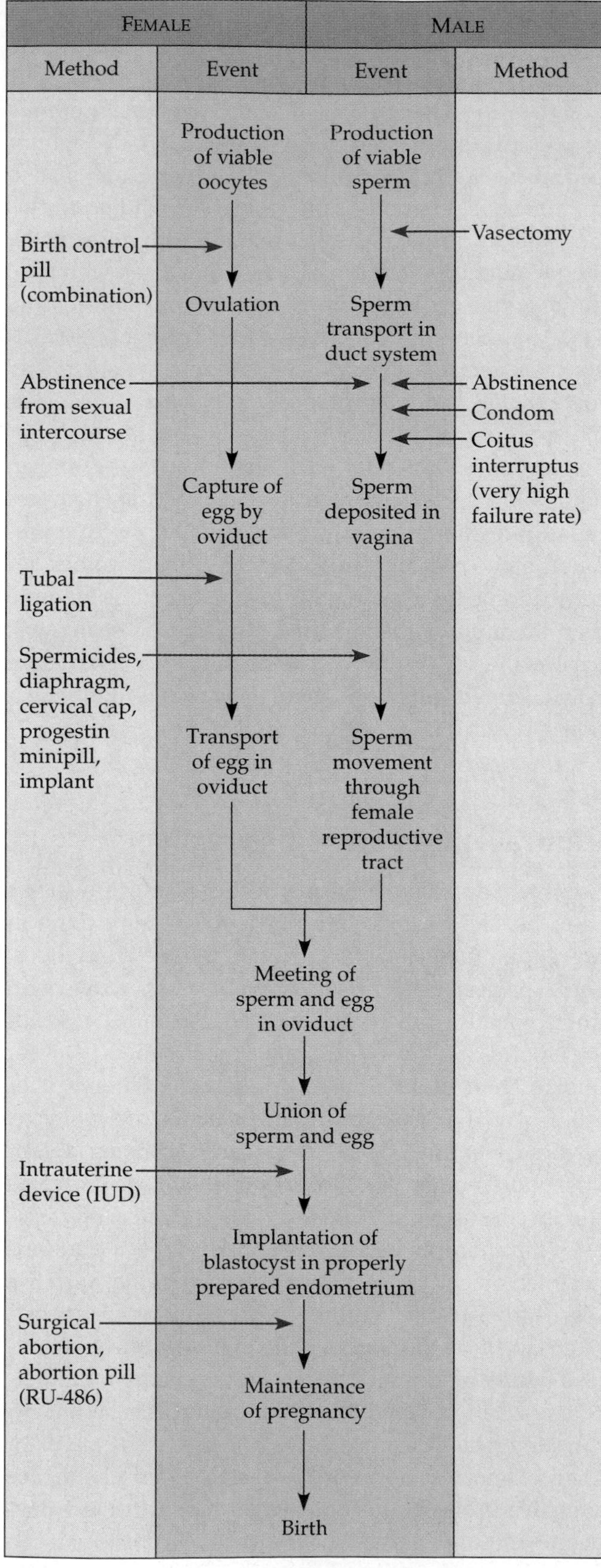

FIGURE 42.20
Mechanisms of some contraceptive methods. The magenta arrows indicate the stages where contraception works.

typical for couples practicing natural family planning. (Pregnancy rate is the number of women who become pregnant during a year out of every 100 women using a particular family-planning method, expressed as a percentage.) Some couples use the natural family planning method to *increase* the probability of conception, so the rate of unplanned pregnancies among couples skilled with this method is actually less than 10% to 20%.

The several **barrier methods** of contraception that block the sperm from meeting the egg have pregnancy rates of less than 10%. For the male, the **condom** is a thin, natural membrane or latex rubber sheath that fits over the penis to collect the semen. For the female, the diaphragm is a dome-shaped rubber cap fitted into the upper portion of the vagina before intercourse. Both of these methods are more effective when used in conjunction with a spermicidal (sperm-killing) foam or jelly. More recently introduced barriers include the cervical cap, which fits tightly around the opening of the cervix, held in place for a prolonged period by suction, and the new female condom.

The intrauterine device (IUD) probably prevents implantation of the blastocyst in the uterus by irritating the endometrium, but its precise mechanism of preventing pregnancy is unknown. IUDs are small, usually plastic, devices in a variety of shapes that fit into the uterine cavity. IUDs have a low pregnancy rate but cause harmful effects in a small percentage of women. Problems of persistent vaginal bleeding, uterine infection, perforation of the uterus, tubal pregnancy (implantation of the embryo in the oviduct), and spontaneous expulsion of the devices have been reported and have led to many lawsuits against IUD manufacturers. New IUD products have recently become available.

As a method of preventing fertilization, coitus interruptus, or withdrawal (removal of the penis from the vagina before ejaculation), is unreliable. Sperm may be present in secretions that precede ejaculation, and a lapse in timing or willpower can result in late withdrawal.

Besides complete abstinence from sexual intercourse, the methods that prevent the release of gametes are the most effective means of birth control. Chemical contraception—birth control pills—have pregnancy rates of less than 1%, and sterilization is nearly 100% effective. The most commonly used birth control pills are combinations of a synthetic estrogen and a synthetic progestin (progesteronelike hormone). These two hormones act by negative feedback to stop the release of GnRH by the hypothalamus and FSH (an estrogen effect) and LH (a progestin effect) by the pituitary. By blocking LH release, the progestin prevents ovulation. As a backup measure, the estrogen inhibits FSH secretion so that no follicles develop. A second type of birth control pill, called the minipill, contains only progestin. The minipill prevents fertilization mainly by altering a woman's cervical mucus so that it blocks sperm from entering the uterus. In 1990, the FDA approved a version of the progestin-only minipill that is implanted under the skin. Steadily releasing a tiny amount of progestin into the blood, the implant produces effective birth control for about 5 years.

Birth control pills have been the center of much debate, particularly because of long-term harmful effects of the estrogens. No solid evidence exists for cancers caused by the pill, but cardiovascular problems are a major concern. Birth control pills have been implicated in abnormal blood clotting, atherosclerosis, and heart attacks. Smoking while using chemical contraception increases the risk of mortality tenfold or more. Although the pill places women at risk for these diseases, it eliminates the dangers of pregnancy; women on birth control pills have mortality rates about one-half those of pregnant women.

Sterilization is the permanent prevention of gamete release. **Tubal ligation** in women usually involves cauterizing or tying off (ligating) a section of the oviducts to prevent eggs from traveling into the uterus. **Vasectomy** in men is the cutting of each vas deferens to prevent sperm from entering the urethra. Both male and female sterilization are relatively safe and free from harmful effects. Both are also difficult to reverse, so the procedures should be considered permanent.

Abortion is the termination of a pregnancy in progress. Spontaneous abortion, or miscarriage, is very common; it occurs in as many as one-third of all pregnancies, often before the woman is even aware she is pregnant. In addition, about 1.5 million women in the United States annually choose abortions performed by physicians. A drug called RU-486, developed in France, enables a woman to terminate pregnancy nonsurgically within the first few weeks. An analog of progesterone, RU-486 blocks progesterone receptors in the uterus, thus preventing progesterone from maintaining pregnancy. The use of RU-486 in the United States has been delayed by the controversy over abortion.

Of all contraceptives for sexually active individuals, latex condoms are the only ones that offer some protection against sexually transmitted diseases, including AIDS. This protection is, however, not absolute.

New technologies offer help for reproductive problems

Recent scientific and technological advances have made it possible to deal with problems of reproduction in striking ways. For example, it is now possible to diagnose many genetic diseases and congenital (present at birth) disorders while the fetus is in the uterus. Noninvasive procedures use high-frequency sound waves, or **ultrasound imaging,** to detect fetal condition (FIGURE 42.21). Amniocentesis and chorionic villus sampling are more invasive techniques (see FIGURE 13.18). In **amniocentesis,** a long needle is inserted into the amnion (the fluid-containing sac surrounding the fetus), and a sample of amniotic fluid is withdrawn. Fetal cells in the fluid are cultured for 2 to 4 weeks, and the cultured cells can then be analyzed for genetic disorders and chromosomal problems, such as Down syndrome (see Chapter 14). **Chorionic villus sampling (CVS)** is a more recent technique in which a small sample of tissue is removed for genetic and metabolic analysis from the fetal portion of the placenta (the chorion). This test carries a greater risk than amniocentesis (5% to 20% versus 1% spontaneous abortions following testing), but results may be obtained in a matter of hours rather than weeks, and CVS can be performed earlier in the pregnancy.

Ultrasound, amniocentesis, and CVS all pose important ethical questions. To date, essentially all detectable disorders remain untreatable in the uterus, and many cannot be corrected even after birth. Parents may be faced with difficult decisions about whether to terminate a pregnancy or cope with a child who may have profound defects and a short life expectancy. These are complex issues that demand careful, informed thought and competent counseling.

Another breakthrough in reproductive technology is **in vitro fertilization.** First accomplished in 1978 in England,

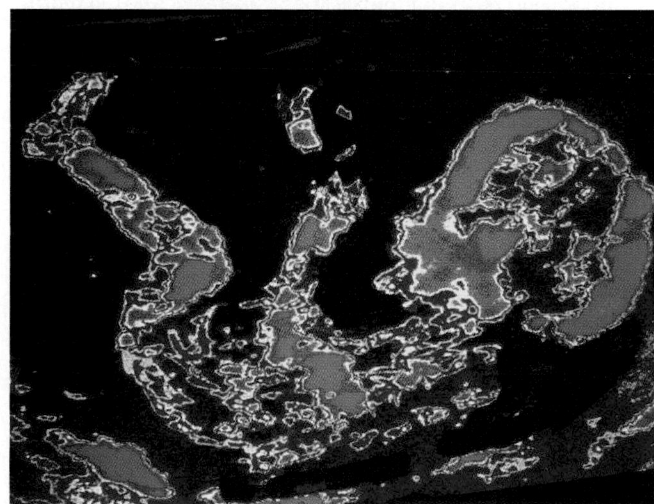

FIGURE 42.21
Ultrasound imaging. This color-enhanced image shows a fetus in the uterus at about 18 weeks. The image is produced on a computer screen when high-frequency sounds from an ultrasound scanner held against a pregnant woman's abdomen bounce off the fetus.

this procedure is now performed in major medical centers throughout the world. Women whose oviducts are blocked can have ova surgically removed following hormonal stimulation of their follicles. The ova are then fertilized in culture dishes in a laboratory. After about 2½ days, when the embryo has reached the eight-cell stage, it is placed in the uterus and allowed to implant. In vitro fertilization is a difficult and costly procedure. At this time, only about one out of six attempts is successful, at a cost of $5000 or more per attempt. This success rate may seem low, but in fact it is probably not different from the pregnancy rate resulting from insemination by sexual intercourse. Embryos can be frozen for later use if the first attempt is unsuccessful. Multiple embryos are often placed in the uterus to increase the chances of a successful pregnancy. However, the goal of researchers is to perfect in vitro techniques so that only a single embryo is transferred, because transferring multiple embryos tends to increase the chance of miscarriage, prematurity, and multiple births. A couple may choose to have oocytes obtained from another woman fertilized with the male partner's sperm, although more serious ethical issues must be resolved when donors are used. Several thousand children have thus far been conceived by in vitro fertilization, and there is no evidence of any abnormalities associated with this procedure.

One area of reproductive research finally receiving much attention is male contraception. Male chemical contraceptives have proved quite elusive. Testosterone will block the release of pituitary gonadotropins, but testosterone itself stimulates spermatogenesis. Estrogens are effective, but they inhibit sex drive and can be feminizing. The best prospects so far are for analogs of GnRH, which are potent inhibitors of spermatogenesis. Other treatments under study include progestins and gossypol, a compound extracted from cotton seeds.

* * *

In this chapter, we have considered the structural and physiological bases of animal reproduction. The next chapter focuses on the mechanics of development that transform a zygote into an animal form, and other topics in animal development.

REVIEW OF KEY CONCEPTS (with page numbers and key figures)

- Both asexual and sexual reproduction occur in the animal kingdom (pp. 937–940)
  - Asexual reproduction produces offspring whose genes all come from a single parent. Fission, budding, and fragmentation with regeneration are mechanisms of asexual reproduction in various invertebrates.
  - Sexual reproduction requires the fusion of male and female gametes to form a diploid zygote. The production of offspring with varying genotypes and phenotypes may enhance reproductive success in fluctuating environments.
  - Animals may reproduce exclusively sexually or asexually, or they may alternate between the two, depending on environmental conditions. Variations on these two modes are made possible through parthenogenesis, hermaphroditism, and sequential hermaphroditism.
  - Reproductive cycles are controlled by hormones and environmental cues, such as changes in temperature, rainfall, day length, and seasonal lunar cycles.
- In sexual reproduction, gametes unite in the external environment or within the female (pp. 940–941)
  - External fertilization requires critical timing, mediated by environmental cues, pheromones, and/or courtship behavior. External fertilization is most common in aquatic or moist habitats, where the zygote can develop without desiccation and heat stress.
  - Internal fertilization requires important behavioral interactions between male and female animals, as well as compatible copulatory organs. Although internal fertilization usually results in fewer zygotes than external fertilization, it is accompanied by greater protection of embryos and parental care of the young.
- Diverse reproductive systems have evolved in the animal kingdom (pp. 941–942, FIGURES 42.6, 42.7)
  - Invertebrate reproductive systems range from the simple production of gametes by undifferentiated cells in the body cavity to complex assemblages of male and female gonads with accessory tubes and glands that carry and protect gametes and developing embryos. The reproductive systems of insects, which have separate sexes, and flatworms, which are hermaphroditic, are among the most complex in the animal kingdom.
  - Vertebrate reproductive systems are similar, with nonmammals having a common opening from the digestive, excretory, and reproductive systems and most mammals having a separate opening for the digestive tract.
- Human reproduction involves intricate anatomy and complex behavior (pp. 943–947, FIGURES 42.8, 42.9)
  - Human male reproductive anatomy consists of internal organs and external genitalia, the scrotum and penis. The

gonads, or testes, reside in the cool environment of the scrotum. They possess endocrine interstitial cells surrounding sperm-forming seminiferous tubules that successively lead into the epididymis, vas deferens, ejaculatory duct, and urethra, which exits at the tip of the penis. Accessory glands add secretions to the semen.

- Human female reproductive anatomy consists internally of two ovaries and oviducts, the uterus, and the vagina. External genitalia include the vestibule containing separate openings of the vagina and urethra, the labia minora bordering the vestibule, the labia majora, and the clitoris. Bartholin's glands secrete lubricating mucus into the vestibule.
- The ovaries are stocked with follicles containing diploid primary oocytes formed before birth. Beginning at puberty, one or more follicles mature during each menstrual cycle. The oocyte contained in a maturing follicle undergoes the first meiotic division, and a secondary oocyte, which is haploid, is expelled from the surface of the ovary during ovulation. After ovulation, the remaining tissue of the follicle forms a corpus luteum that secretes progesterone and estrogen for a variable duration, depending on whether or not pregnancy occurs.
- The oviduct draws the secondary oocyte into its open end and transports it to the uterus by ciliary action. The uterus opens through the cervix into the muscular vagina, which serves as a sperm receptacle and birth canal.
- Although separate from the reproductive system, the mammary gland, or breast, evolved in association with parental care.
- Common patterns of sexual response underlie apparent differences in the rich diversity of human sexuality. Both males and females experience the erection of certain body tissues due to vasocongestion and myotonia, which culminate in orgasm.

- Spermatogenesis and oogenesis both involve meiosis but differ in three significant ways (pp. 947–949, FIGURES 42.12, 42.13)
  - Cytokinesis is unequal in oogenesis, producing one large ovum.
  - Production of sperm is continuous, although, in humans, the number of future egg cells is set at birth.
  - Spermatogenesis is an uninterrupted sequence, but there are long delays in oogenesis.

- A complex interplay of hormones regulates reproduction (pp. 950–953, FIGURES 42.14, 42.15)
  - Androgens from the testes cause the development of primary and secondary sex characteristics in the male. Androgen secretion and sperm production are both controlled by hypothalamic and pituitary hormones.
  - Female hormones are secreted in a rhythmic fashion reflected in the menstrual or estrous cycle.
  - In both types of female cycles, the endometrium thickens in preparation for possible implantation. The menstrual cycle, however, is punctuated by endometrial bleeding and lacks the clear-cut period of sexual receptivity limited to the heat period of the estrous cycle.
  - The human menstrual cycle consists of the menstrual flow phase, proliferative phase, and secretory phase. The ovarian cycle includes the follicular and luteal phases.
  - The female reproductive cycle is orchestrated by cyclic secretion of GnRH from the hypothalamus and FSH and LH from the anterior pituitary. The developing follicle produces estrogen, and the corpus luteum secretes progesterone and estrogen. Positive and negative feedback produce the changing levels of these five hormones, which coordinate the menstrual and ovarian cycles.
  - Puberty is a gradual, hormonally regulated process that involves the development of secondary sex characteristics and the onset of reproductive capability.

- Embryonic and fetal development occur during pregnancy in humans and other placental mammals (pp. 953–957, FIGURES 42.16, 42.17)
  - Human pregnancy can be divided into three trimesters. Organogenesis is completed by eight weeks.
  - Birth, or parturition, results from strong, rhythmic uterine contractions that bring about the three stages of labor: dilation of the cervix, expulsion of the baby, and delivery of the placenta. Positive feedback involving the hormones estrogen and oxytocin, and prostaglandins, regulate labor.
  - The ability of a pregnant woman to accept her "foreign" fetus may be due to the suppression of the immune response in her uterus.

- Contraception prevents pregnancy (pp. 957–959, FIGURE 42.20)
  - Contraceptive methods include preventing the release of mature gametes from the gonads, preventing gamete union in the female tract, and preventing implantation of the zygote.

- New technologies offer help for reproductive problems (pp. 959–960)
  - Current technological methods of detecting fetal condition include ultrasound imaging, amniocentesis, and chorionic villus sampling. Current technology also provides in vitro fertilization, and ongoing research promises to yield new developments in contraception.

SELF-QUIZ

1. Which of the following characterizes parthenogenesis?
   a. an individual may change its sex during its lifetime
   b. specialized groups of cells may be released and grow into new individuals
   c. an organism is first a male and then a female
   d. an egg develops without being fertilized
   e. both members of a mating pair have male and female reproductive organs

2. Which of the following structures is *incorrectly* paired with its function?
   a. gonads—gamete-producing organs
   b. spermatheca—sperm-transferring organ found in male insects
   c. cloaca—common opening for reproductive, excretory, and digestive systems
   d. baculum—bone that stiffens the penis, found in some mammals
   e. endometrium—lining of the uterus, forming the maternal part of the placenta

3. Which of the following male and female structures are *least* alike in function?
   a. seminiferous tubules—vagina
   b. interstitial cells of testes—follicle cells
   c. testes—ovaries
   d. spermatogonia—oogonia
   e. vas deferens—oviduct
4. A difference between estrous and menstrual cycles is that
   a. nonmammalian vertebrates have estrous cycles, whereas mammals have menstrual cycles
   b. the endometrial lining is shed in menstrual cycles but reabsorbed in estrous cycles
   c. estrous cycles occur more frequently than menstrual cycles
   d. estrous cycles are not controlled by hormones
   e. ovulation occurs before the endometrium thickens in estrous cycles
5. Peaks of LH and FSH production occur during
   a. the flow phase of the menstrual cycle
   b. the follicular phase of the ovarian cycle
   c. the period surrounding ovulation
   d. the end of the luteal phase of the ovarian cycle
   e. the secretory phase of the menstrual cycle
6. The *direct* function of GnRH is to
   a. stimulate production of estrogen and progesterone
   b. initiate ovulation
   c. inhibit secretion of pituitary hormones
   d. stimulate secretion of LH and FSH
   e. initiate the flow phase of the menstrual cycle
7. During human gestation, organogenesis occurs
   a. in the first trimester
   b. in the second trimester
   c. in the third trimester
   d. while the embryo is in the oviduct
   e. during the blastocyst stage
8. Except for abstinence from sexual intercourse, the contraceptive method most effective in reducing the chance of contracting AIDS and other sexually transmitted diseases is the
   a. birth control pill
   b. diaphragm
   c. latex condom
   d. IUD
   e. RU-486 pill
9. Fertilization of human eggs most often takes place in the
   a. vagina
   b. ovary
   c. uterus
   d. oviduct (fallopian tube)
   e. vas deferens
10. In mammalian males, the excretory and reproductive systems share the
   a. testes
   b. urethra
   c. ureter
   d. vas deferens
   e. prostate

CHALLENGE QUESTIONS

1. Describe how sexual and asexual reproduction differ in mechanism and result.
2. Explain why menstruation and ovulation do not occur during pregnancy.
3. Compare and contrast oogenesis and spermatogenesis.

SCIENCE, TECHNOLOGY, AND SOCIETY

1. New techniques for sorting sperm, combined with in vitro fertilization, make it possible for a couple to choose their baby's sex. Would you want to do this if you had the chance? Why or why not? What potential problems can you foresee if this procedure becomes widely available?
2. Because of a uterine tumor, Julie is unable to carry a fetus to term. Eggs were surgically removed from her ovaries and fertilized in a glass dish with her husband Ron's sperm. One of the fertilized eggs was implanted in the uterus of a young woman named Michelle, whom Julie and Ron had hired as a surrogate mother. When the baby was born 9 months later, Michelle decided she wanted to keep it, and she refused to accept payment. Julie and Ron sued to gain custody of the baby. Who do you think should get the baby? Who is the "real" mother? What are your criteria for making a decision?
3. New technology has made it possible for doctors to save a small percentage of babies born 16 weeks prematurely. A baby born this early weighs just over a pound and faces months of treatment in an intensive care nursery. The cost for care may be hundreds of thousands of dollars per infant. Many of the surviving infants have mental and physical disabilities. Some people wonder whether such a huge technological and financial investment should be devoted to such a small number of babies. They feel that the resources might better be directed at providing prenatal care that could prevent many premature births. What is your opinion in this controversy? Defend your position.

FURTHER READING

Buss, D. M. "The Strategies of Human Mating." *American Scientist,* May/June 1994. An analysis of human mate selection and its evolutionary underpinning.

Crews, D. "Animal Sexuality." *Scientific American,* January 1994. A comparison of mechanisms determining gender in several different species, including all-female lizards.

Crooks, R., and K. Baur. *Our Sexuality,* 5th ed. Redwood City, CA: Benjamin/Cummings, 1993.

Duellman, W. "Reproductive Strategies of Frogs." *Scientific American,* July 1992. Fascinating reproductive diversity.

Fackelmann, K. A. "Cloning Human Embryos." *Science News,* February 5, 1994. A critical review of cloning experiments, what cloning really means, and the ethical questions associated with in vitro fertilization.

Freedman, D. "The Aggressive Egg." *Discover,* June 1992. Does the human egg attract sperm to it?

Richardson, S. "*Guinness Book* Gametes." *Discover,* March 1995. Fruit flies, very small animals, have the largest known animal sperm. How does this contribute to reproductive success?

Ulmann, A., G. Teutsch, and D. Philibert. "RU-486." *Scientific American,* June 1990. A history of the development of the pill that induces abortion.

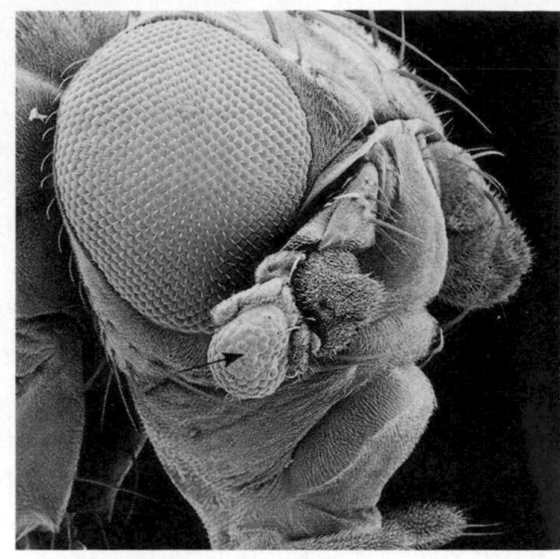

CHAPTER 43

ANIMAL DEVELOPMENT

KEY CONCEPTS

- From egg to organism, an animal's form develops gradually: *the concept of epigenesis*
- Embryonic development involves cell division, differentiation, and morphogenesis
- Fertilization activates the egg and brings together the nuclei of sperm and egg
- Cleavage partitions the zygote into many smaller cells
- Gastrulation rearranges the blastula to form a three-layered embryo with a primitive gut
- Organogenesis forms the organs of the animal body from the three embryonic germ layers
- Amniote embryos develop in a fluid-filled sac within a shell or uterus
- The developmental fate of cells depends on cytoplasmic environment, location, and cell-cell interactions
- Pattern formation in *Drosophila* is controlled by a hierarchy of gene activations
- Comparisons of genes that control development reveal homology in animals as diverse as flies and mammals

It is difficult to imagine that each of us began life as a single cell about the size of the period at the end of this sentence. Less than a month after conception, our brains were taking form and our developing hearts had already begun to pulsate. It took a total of only about nine months—the length of a school year—to be transfigured from zygote to newborn human.

By combining molecular genetics with classical approaches to embryology, developmental biologists are now poised to answer many of the basic questions about how a single fertilized egg cell gives rise to a specific animal form built of thousands or billions of differentiated cells organized into specialized tissues and organs. Molecular biologist Sidney Brenner has wryly commented that developmental biology is about "how to make a mouse," and indeed, mice are favorite experimental models of researchers studying mammalian development.

*The fruit fly (*Drosophila*) is another important research model because much of its well-known genetic makeup has direct application to development studies. In a striking example, Swiss researchers demontrated in 1995 that a particular gene functions as a master switch that triggers the development of an eye in* Drosophila*. The scanning electron micrograph on this page shows the head of an abnormal fly with extra eyes on its antennae (arrow). Expression of the master gene for eye development in an abnormal location caused the extra eyes. The same gene also triggers eye development in mice and other mammals. In fact, developmental biologists are discovering remarkable similarities in the genetic, molecular, and cellular mechanisms that shape diverse animals.*

Embryonic development, from a fertilized egg to a new adult animal, is the main topic of this chapter, but it is important to realize that animals actually develop throughout their lifetime. As adults, they undergo progressive changes in body form and function, regenerate lost body parts, undergo wound healing, and age. Many of the same basic cellular mechanisms that transform an embryo underlie these postembryonic changes.

From egg to organism, an animal's form develops gradually: *the concept of epigenesis*

The question of how an egg becomes an animal has been asked for centuries. As recently as the eighteenth century, the prevailing view was that the egg or sperm contains a preformed, miniature embryo that simply grows during its development. This idea of **preformation** came to include the notion that the embryo must contain all its descendants: a series of successively smaller embryos within embryos, like Russian nesting dolls. One theologian proposed that Eve, in the Garden of Eden, stored all future humanity within her.

The competing theory of embryology was an idea called **epigenesis,** originally proposed 2000 years earlier by Aristotle, that the form of an embryo emerges gradually from a relatively formless egg. As microscopy improved during the nineteenth century, biologists could see that embryos took shape in a series of progressive steps, and epigenesis displaced preformation as the favored explanation among embryologists.

Modern biology, of course, has completely discarded the idea of a tiny person living in an egg or sperm cell. But when interpreted in broader terms, the concept of preformation may have some merit. Although an embryo's form emerges gradually as it develops from a fertilized egg, something *was,* in effect, preformed in the zygote. An organism's development is largely determined by the genome of the zygote and the organization of the cytoplasm of the egg cell. Messenger RNA, proteins, and other components are heterogeneously distributed in the unfertilized egg, and this has a profound impact on the development of the future embryo in most animal species. After fertilization, division of the zygote partitions the cytoplasm in such a way that nuclei of different embryonic cells are exposed to different cytoplasmic environments. This sets the stage for the expression of different genes in different cells. As the embryo develops, inherited traits emerge by mechanisms that selectively control gene expression (see Chapter 18). The timely communication of instructions, telling cells precisely what to do when, is essential, and signal-transduction mechanisms often convey developmental messages (see Chapters 8 and 41). A strong research effort focuses on working out the functional details of cell-cell signaling and how it affects the main processes in embryonic development.

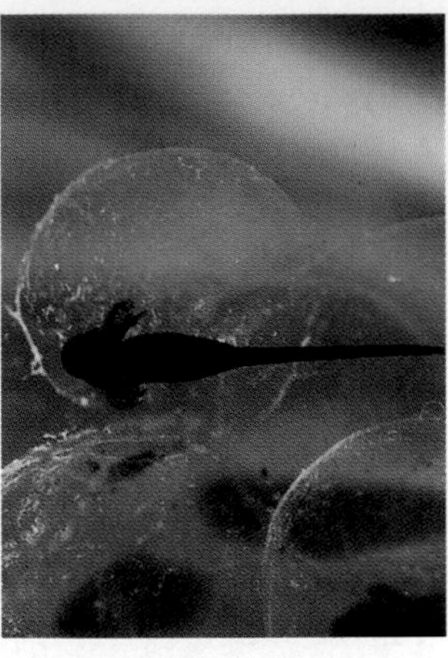

FIGURE 43.1
From egg to animal: what a difference a week makes. It took just one week for cell division, differentiation, and morphogenesis to transform this fertilized egg (left) into a hatching tadpole (right).

Embryonic development involves cell division, differentiation, and morphogenesis

Three interdependent processes in embryonic development are cell division, differentiation, and morphogenesis (FIGURE 43.1). Through a succession of mitotic divisions, the zygote gives rise to a large number of cells. Cell division alone, however, would produce a great ball of identical cells, which is nothing like an animal. During embryonic development, cells not only increase in number, they also undergo **differentiation,** becoming the diverse specialized cells that are organized into the tissues and organs of the animal. The physical processes that actually give shape to the animal and its organs are called **morphogenesis,** which literally means "creation of form." Movements of cells and tissues are necessary to convert the cell mass of the early embryo into the characteristic three-dimensional form of the larval or juvenile stage of the species. Such movements and rearrangements are components of morphogenesis. As we will see, cell division and differentiation also take part in morphogenesis, as does the programmed death of certain cells.

The importance of precise regulation of the developmental processes is evident in human disorders that result from developmental mechanisms going awry. For example, the condition known as cleft palate, in which the upper part of the mouth cavity fails to close completely, is a defect of morphogenesis. The failure of normal cell death in morphogenesis can result in such defects as webbed fingers and toes. When we consider how many things *can* go wrong during cell division, differentiation, and morphogenesis, it is all the more remarkable that most animals develop normally.

Regulated cell division, differentiation, and morphogenesis form the heart of our inquiry into how animals develop. In the first half of the chapter, we will view the role of these processes during the early stages of embryonic development, when the basic body plan of an animal takes form. In the second half of the chapter, we will take a closer look at the cellular mechanisms that control development.

Fertilization activates the egg and brings together the nuclei of sperm and egg

The gametes, sperm and egg that unite during fertilization, are both highly specialized cell types produced by a complex series of developmental events in the testes and ovaries of the parents (see Chapter 42). One of the functions of fertilization is to combine haploid sets of chromosomes from two individuals into a single diploid cell, the zygote. Its other function is activation of the egg: Contact of the sperm with the egg's surface initiates metabolic reactions within the egg that trigger the onset of embryonic development.

Fertilization has been studied most extensively by combining the gametes of sea urchins in the laboratory. Although the details of fertilization vary with different animal groups, sea urchins (phylum Echinodermata) provide a good general model for the important events of fertilization.

The Acrosomal Reaction

The eggs of sea urchins are fertilized externally after the animals release their gametes into the surrounding seawater. When a sperm cell is exposed to molecules from the slowly dissolving jelly coat that surrounds an egg, a vesicle at the tip of the sperm called the acrosome discharges its contents by exocytosis (FIGURE 43.2). This **acrosomal reaction** releases hydrolytic enzymes that enable an elongating structure called the acrosomal process to penetrate the jelly coat of the egg. The tip of the acrosomal process is coated with a protein that adheres to specific receptor molecules located on the vitelline layer just external to the plasma membrane of the egg. In sea urchins, "lock-and-key" recognition of molecules ensures that eggs will be fertilized only by sperm of the same animal species. This is especially important when fertilization occurs externally in water, where gametes of other species are likely to be present.

The acrosomal reaction leads to the fusion of sperm and egg plasma membranes and the entry of a single

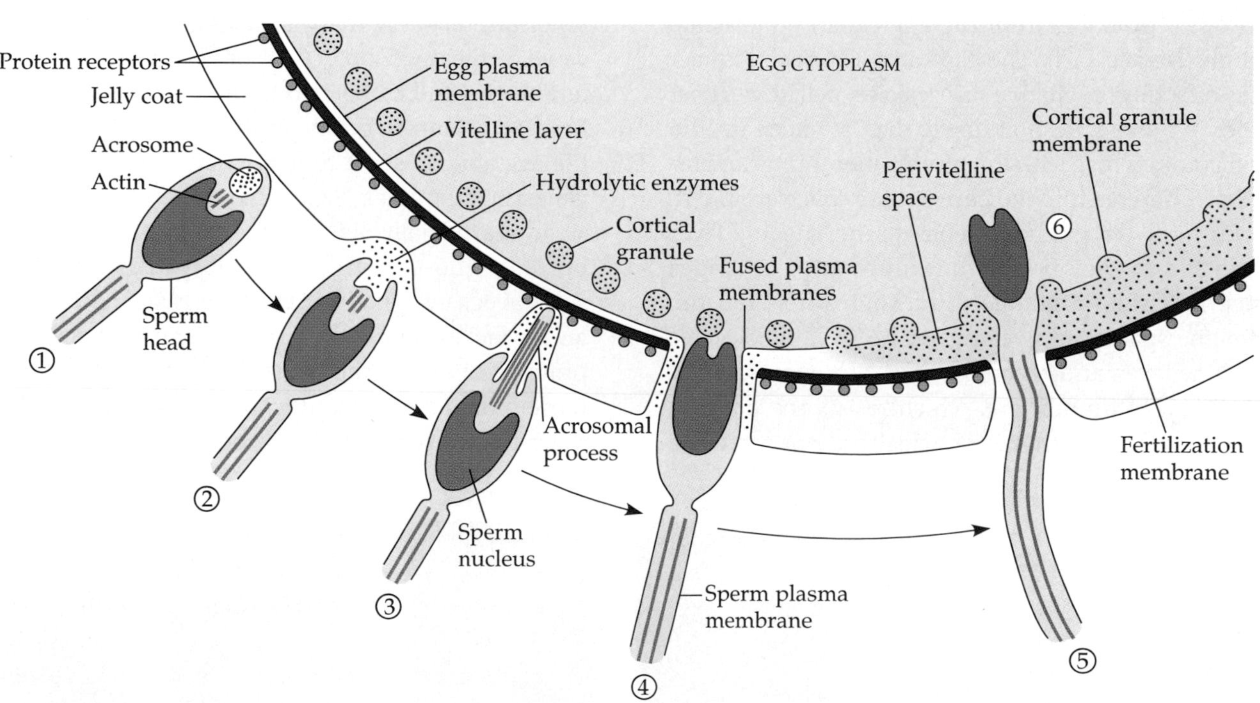

FIGURE 43.2
The acrosomal and cortical reactions during sea urchin fertilization. The events following contact of a single sperm and egg lead to the fusion of gamete nuclei and ensure that only one sperm nucleus enters the cytoplasm of the egg. ① The sperm cell contacts the egg's jelly coat. ② The acrosomal reaction begins with the release of hydrolytic enzymes from the acrosome in the sperm head. The enzymes excavate a hole in the jelly coat, while the polymerization of actin, a protein, creates an extension from the sperm head, the acrosomal process. ③ The acrosomal process extends through the jelly coat and binds to the egg's vitelline layer. Enzymes on the acrosomal process probably digest a hole in the vitelline layer, and the acrosomal process contacts the egg's plasma membrane. ④ As a result of the acrosomal reaction, the plasma membranes of the sperm and egg fuse, and ⑤ the sperm nucleus enters the cytoplasm of the egg. Fusion of the gamete membranes triggers both an electrical change in the plasma membrane of the egg and the cortical reaction, blocking entry by other sperm. ⑥ In the cortical reaction, cortical granules in the egg fuse with the plasma membrane and discharge enzymes and other macromolecules that raise and harden the vitelline layer, forming a sperm-proof fertilization membrane.

sperm nucleus into the cytoplasm of the egg. Fusion of the membranes causes a nervelike electrical response by the egg's plasma membrane. Ion channels open, allowing sodium ions to flow into the egg cell and change the membrane potential, the voltage across the membrane (see Chapter 8). This membrane depolarization, as the electrical response is called, is common among animal species. Occurring within about 1 to 3 seconds after a sperm cell binds to the vitelline layer, the depolarization is also called the **fast block to polyspermy** because it prevents more than one sperm cell from fusing with the egg's plasma membrane. Without the block, multiple fertilizations could result in an aberrant chromosome count and abnormal mitosis.

The Cortical Reaction

Another major effect of the fusion of egg and sperm plasma membranes is the **cortical reaction,** a series of changes in the outer zone (cortex) of the egg cytoplasm (see FIGURE 43.2, step 6). In sea urchins and many other species, the sperm-egg fusion activates a signal transduction pathway involving a G protein relay in the egg (see Chapter 41). The G protein triggers the release of calcium (Ca^{2+}), probably from the egg cell's endoplasmic reticulum (FIGURE 43.3). The Ca^{2+} acts as a second messenger, effecting a change in vesicles called **cortical granules** in the egg. Responding to the Ca^{2+} increase, the cortical granules fuse with the plasma membrane and release their contents into the perivitelline space around it. Enzymes from the granules separate the vitelline layer from the plasma membrane while mucopolysaccharides produce an osmotic gradient, drawing water into the perivitelline space and swelling it. The swelling elevates the vitelline layer, and other enzymes harden it, forming the **fertilization membrane,** which resists the entry of additional sperm. By this time, usually about a minute after sperm and egg fuse, the voltage across the plasma membrane has returned to normal, and the fast block to polyspermy no longer functions. But the fertilization membrane, along with other changes in the egg's surface, functions as a **slow block to polyspermy.**

Activation of the Egg

The sharp rise in the egg's cytoplasmic concentration of Ca^{2+} not only triggers the cortical reaction but also incites metabolic changes within the egg cell. The unfertilized egg has a very slow metabolism, but within a few minutes of fertilization, the rates of cellular respiration and protein synthesis increase substantially. With these rapid changes, the egg cell is said to be activated. In sea urchins and many other species, the rise in Ca^{2+} (the second messenger) triggers a loss of hydrogen ions from the egg, and the egg cytosol becomes slightly alkaline. This pH change seems to be indirectly responsible for the metabolic responses of the egg to fertilization.

The binding and fusion of sperm are triggers for egg activation; however, sperm cells do not contribute any materials required for activation to proceed. Indeed, the unfertilized eggs of many species can be artificially activated by the injection of Ca^{2+} or by a variety of mildly injurious treatments, such as temperature shock. This artificial activation switches on the metabolic responses of the egg and causes it to begin developing by parthenogenesis (without fertilization by a sperm). It is even possible to artificially activate an egg that has had its own nucleus removed. (Of course, embryonic development of such an egg terminates at a very early stage.) The fact that an egg lacking a nucleus can begin making new kinds of proteins upon activation means that mRNA coding for these proteins must be stockpiled in an inactive form in the cytoplasm of the unfertilized egg.

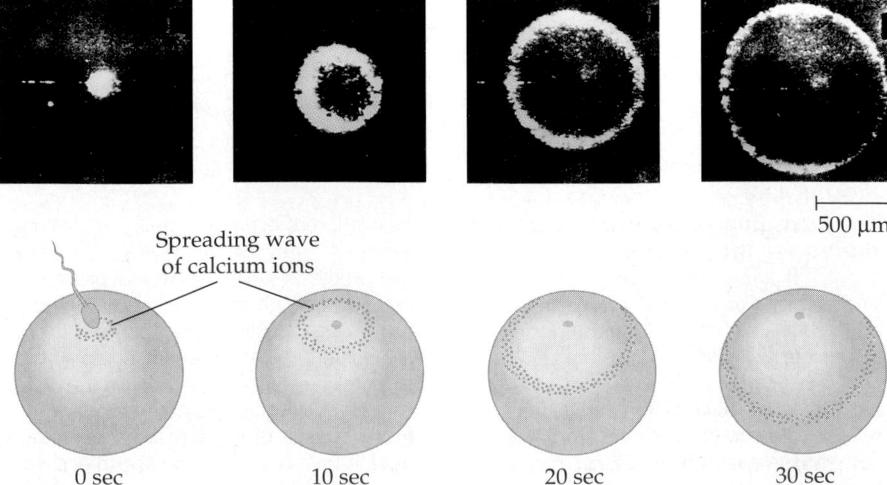

FIGURE 43.3
A wave of Ca^{2+} release during the cortical reaction.
A fluorescent dye that glows when it binds free Ca^{2+} was used in this experiment to track the cortical reaction from the point of sperm contact (0 sec) during the fertilization of a fish egg (LM). The spreading wave of Ca^{2+} ions, released into the cytosol from some intracellular reservoir (probably the endoplasmic reticulum), functions as a second messenger (sperm contact being the first) that helps activate metabolic changes in the egg.

While the activated egg gears up its metabolism, the nucleus of the sperm cell within the egg starts to swell. After about 20 minutes, the sperm nucleus merges with the egg nucleus, creating the diploid nucleus of the zygote. DNA synthesis begins, and the first cell division occurs (in the case of sea urchins and some frogs) in about 90 minutes. The events of fertilization in sea urchins are summarized in FIGURE 43.4.

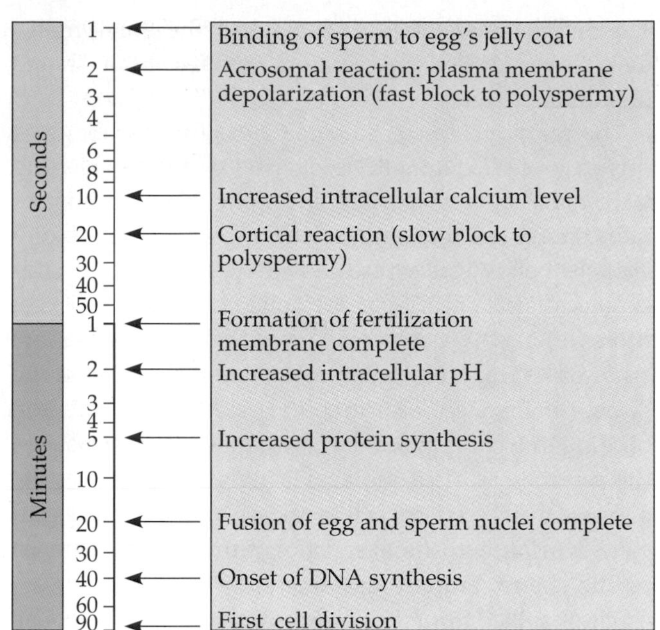

FIGURE 43.4
Timeline for the fertilization of sea urchin eggs. Notice that the scale is logarithmic. The process begins when a sperm cell contacts the jelly coat of an egg (top of chart).

Fertilization in Mammals

In Chapter 42, you learned how in vitro fertilization of human eggs is enabling some couples with fertility problems to have children. Test-tube fertilization has also made it possible for developmental biologists to study the process of fertilization in mammals. Many of the events turn out to be similar to what has been observed in sea urchins, but there are also important differences.

In contrast to the external fertilization of sea urchins and most other marine invertebrates, fertilization in terrestrial animals, including mammals, is generally internal (FIGURE 43.5). Secretions in the mammalian female reproductive tract alter certain molecules on the surface of sperm cells that have been deposited during the male's ejaculation and also increase the motility of the sperm.

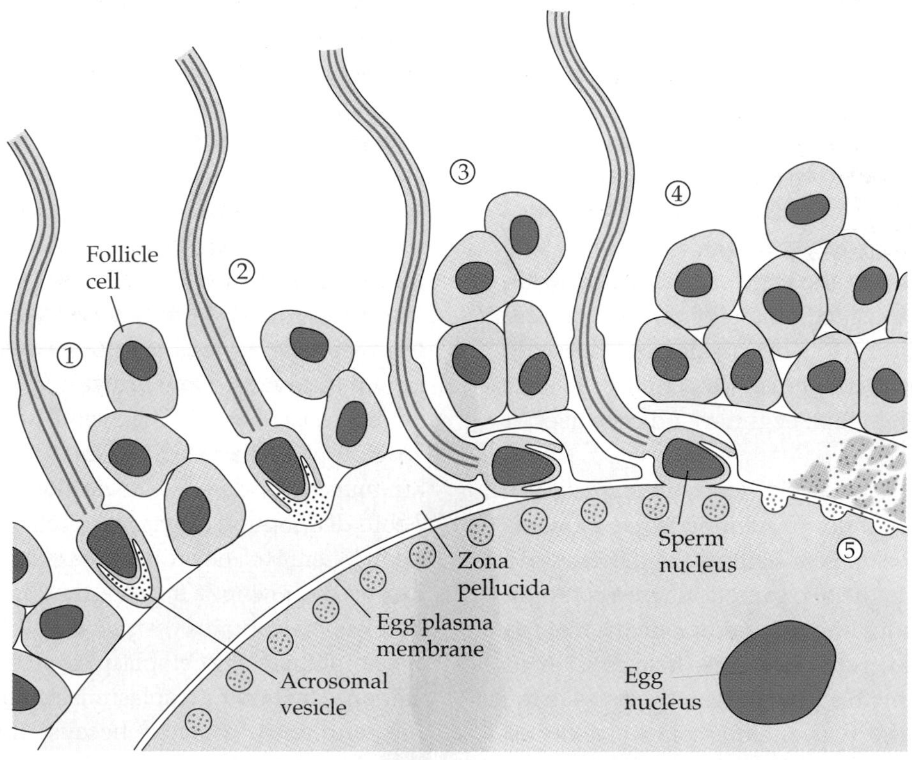

FIGURE 43.5
Fertilization in mammals. ① The sperm migrates through the coat of follicle cells and binds to a receptor molecule in the zona pellucida of the egg. ② This binding induces the acrosomal reaction, in which the sperm releases digestive enzymes into the zona pellucida. ③ With the help of these hydrolytic enzymes, the sperm reaches the egg, and a membrane protein of the sperm binds to a receptor on the egg membrane. ④ The plasma membranes fuse, making it possible for the sperm cell to enter the egg. ⑤ Enzymes released during the egg's cortical reaction harden the zona pellucida, which now functions as a block to polyspermy.

This enhancement of sperm function in the female reproductive tract, called capacitation, requires about 6 hours in humans.

The mammalian egg (actually a secondary oocyte at this stage; see Chapter 42) is cloaked by follicle cells that were released with the egg during ovulation. A capacitated sperm cell must migrate through this layer of follicle cells before it reaches the **zona pellucida,** the extracellular matrix of the egg. The zona pellucida consists of three different glycoproteins forming filaments that are cross-linked in a three-dimensional network. One of the glycoproteins, ZP3, also functions as a sperm receptor, binding to a complementary molecule on the surface of the sperm head. This molecular interaction induces the acrosome of the sperm cell to release its contents by exocytosis in an acrosomal reaction similar to that of sea urchin sperm. Protein-digesting enzymes and other hydrolases spilled from the acrosome enable the sperm cell to penetrate the zona pellucida and reach the plasma membrane of the egg. The acrosomal reaction also exposes a protein in the sperm membrane that binds and fuses with the egg membrane.

Binding of a sperm cell to the egg triggers depolarization of the egg membrane, which functions as a fast block to polyspermy, as in sea urchin fertilization. A cortical reaction, in which granules in the cortex of the egg release their contents to the outside of the cell via exocytosis, also occurs. In contrast to sea urchins, however, the cortical reaction does not raise a fertilization membrane. Instead, enzymes released from the cortical granules catalyze a hardening of the zona pellucida, which now functions as the slow block to polyspermy.

Fingerlike extensions of the egg cell, called microvilli, take the whole sperm cell, tail and all, into the egg. The basal body of the sperm's flagellum divides and forms the centrioles of the zygote, which will function in cell division; unfertilized mammalian eggs have no centrioles of their own.

In contrast to sea urchin fertilization, the haploid nuclei of sperm and egg do not fuse immediately in mammals. Instead, the envelopes of both nuclei disperse and the chromosomes from the two gametes share a common spindle apparatus during the first mitotic division of the zygote. Thus, it is not until after this first division as diploid nuclei form in the two daughter cells that the chromosomes from the two parents come together in common nuclei to form the genome of the offspring.

Cleavage partitions the zygote into many smaller cells

In the early stages of development following fertilization, three embryonic processes establish an animal's basic body plan. First, a special type of cell division called cleavage creates a multicellular embryo, the blastula, from the zygote. The second process, gastrulation, produces a three-layered embryo called the gastrula. The third process, called organogenesis, generates rudimentary organs from which adult structures grow.

Cleavage is a succession of rapid cell divisions that follow fertilization (FIGURE 43.6). During cleavage, the cells undergo the S (DNA synthesis) and M (mitosis) phases of the cell cycle but often virtually skip the G_1 and G_2 phases (see Chapter 11). Gene transcription is virtually shut down during cleavage, perhaps because of rapid DNA replication, and the embryo does not grow during this period of development. Cleavage simply partitions the cytoplasm of one large cell, the zygote, into many smaller cells called **blastomeres,** each with its own nucleus. Thus, different regions of cytoplasm present in the original undivided egg cell end up in separate blastomeres. And because the regions may contain different cytoplasmic components, the partitioning sets the stage for later developmental events.

The eggs of most animals have a definite polarity, and the planes of division during cleavage follow a specific pattern relative to poles of the zygote. The polarity is defined by concentration gradients of cytoplasmic components in the egg, such as mRNA, proteins, and **yolk** (nutrients stored in the egg). In many frogs and other animals, the concentration gradient of yolk is a key factor in determining polarity and influencing the pattern of cleavage. Yolk is most concentrated at one pole of the egg, called the **vegetal pole,** while the opposite pole, the **animal pole,** has the lowest concentration of yolk. The animal pole is also the site where the polar bodies of meiosis are budded from the cell (see Chapter 41), and in most animals, it marks the point where the most anterior part of the embryo will form.

The hemispheres of the zygote are named for their respective poles. In the eggs of many frogs, the hemispheres are marked by different coloration due to the heterogeneous distribution of cytoplasmic substances. The animal hemisphere has melanin granules embedded in the outer cytoplasm, giving it a deep gray hue, while the vegetal hemisphere contains the yellow yolk. A rearrangement of the amphibian egg cytoplasm occurs at the time of fertilization. The outer cytoplasm is rotated toward the point of sperm entry, probably because the centriole brought into the egg by the sperm cell reorganizes the cytoskeleton. The rotation exposes a light-colored region of cytoplasm, a narrow band called the **gray crescent.** Located near the equator of the egg on the side opposite the point of sperm entry, the gray crescent is an important early marker of the egg's polarity, as we will see later.

Because yolk tends to impede cell division, cleavage of

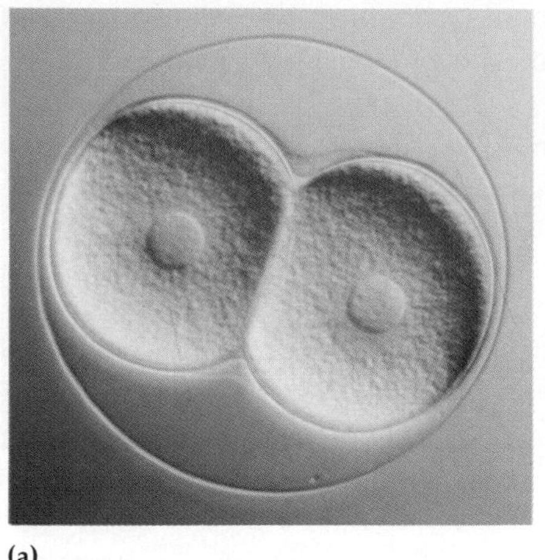

(a)

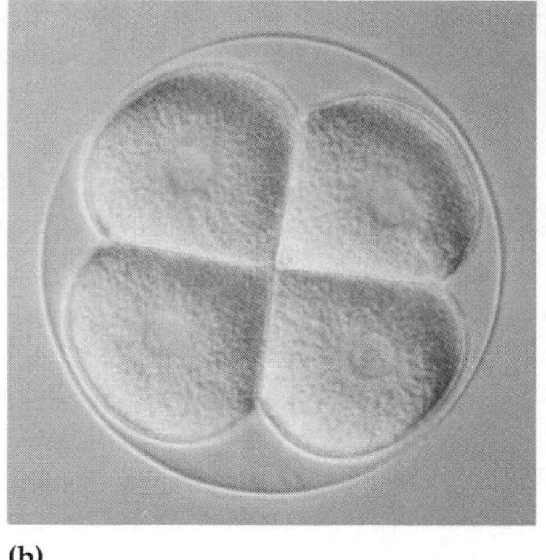

(b)

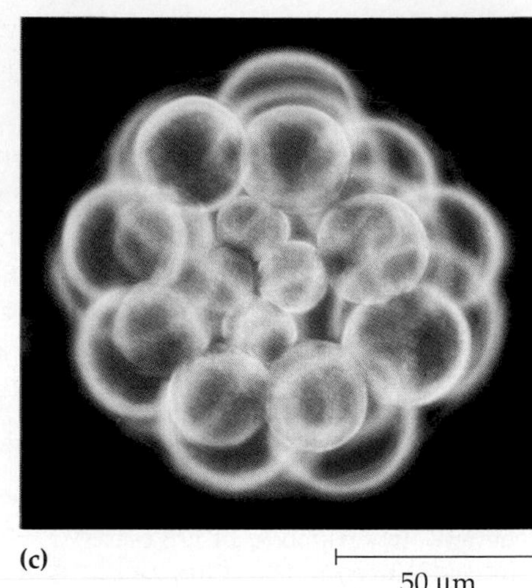

(c)

FIGURE 43.6
Cleavage in an echinoderm (sea urchin) embryo. Cleavage is a series of mitotic divisions and cytokineses that transform the zygote, a single large cell, into a ball of much smaller cells, called blastomeres. These light micrographs of living sea urchin embryos illustrate three stages of cleavage. (**a**) The two-cell stage, following the first cleavage division, occurs about 45 to 90 minutes after fertilization (notice that the fertilization membrane is still present). (**b**) The second cleavage division produces the four-cell stage. (**c**) In a few hours, repeated cleavage divisions have formed a multicellular ball. The embryo is still retained within the fertilization membrane, from which the motile larva that develops from the embryo will eventually hatch.

the frog zygote occurs more rapidly in the animal hemisphere than in the vegetal hemisphere, resulting in an embryo with different-sized cells. In contrast to frog eggs, those of sea urchins and many other animals have less yolk but still have an animal-vegetal axis, owing to concentration gradients of other substances. Without the restraint imposed by yolk, their cleavage divisions all occur at about the same rate, producing blastomeres of virtually equal size.

In both sea urchins and frogs, the first two cleavage divisions are polar (vertical), resulting in four cells that each extend from animal to vegetal pole. The third division is equatorial (horizontal), producing an eight-celled embryo with two tiers of four cells each (FIGURE 43.7a). The general pattern up to this point is the same in both types of embryos. In fact, echinoderms, chordates, and the other animal phyla grouped as deuterostomes share many features of early embryonic development. These similarities distinguish the deuterostomes from the protostomes, the evolutionary branch that includes the mollusks, annelids, and arthropods (see FIGURE 29.5). For example, cleavage in deuterostomes is radial, meaning that the upper (animal) tier of four cells is aligned directly over the lower (vegetal) tier at the eight-cell stage. In contrast, most protostomes exhibit spiral cleavage, in which the cells of the upper tier sit in the grooves between the cells of the lower tier.

Continued cleavage produces a solid ball of cells known as the **morula,** Latin for "mulberry," in reference to the lobed surface of the embryo at this stage (FIGURE 43.7b). A fluid-filled cavity called the **blastocoel** forms within the morula, creating a hollow ball stage of development called the **blastula** (FIGURE 43.7c). In sea urchins, the blastocoel is centrally located in the blastula. However, in frogs, because of unequal cell division, the blastocoel is located in the animal hemisphere (FIGURE 43.7d).

Yolk is most plentiful and has its most pronounced effect on cleavage in the eggs of birds, reptiles, many fishes, and insects. In birds, for example, the part of the egg we commonly call the yolk is actually the egg cell (ovum), swollen with yolk nutrients. This enormous cell is surrounded by a protein-rich solution (the egg white) that will provide additional nutrients for the growing embryo. Cleavage of the fertilized egg is restricted to a small disc of yolk-free cytoplasm at the animal pole of the egg cell. This incomplete division of a yolk-rich egg is known as **meroblastic cleavage.** It contrasts with **holoblastic cleavage,** the term for the complete division of eggs having little yolk (as in sea urchins) or a moderate amount of yolk (as in frogs).

The yolk-rich eggs of insects, such as *Drosophila,* undergo a unique type of meroblastic cleavage (FIGURE 43.8). After fertilization, the zygote's nucleus is situated within a mass of yolk. Cleavage begins with the

FIGURE 43.7

Cleavage in a frog embryo. Yolk, concentrated near the vegetal pole of the egg, impedes the formation of cleavage furrows. (**a**) After two equal polar divisions, the third cleavage division is perpendicular to the polar axis but is displaced toward the animal pole by yolk. Thus, in this eight-celled embryo viewed from the animal pole, the four blastomeres near the animal pole are smaller than the four blastomeres near the vegetal pole. (**b**) As cleavage continues, the cells near the animal pole divide more frequently than the yolk-laden cells near the vegetal pole. A frog embryo consisting of 16 to 64 cells is called a morula (this is a sideview). (**c**) An embryo of 128 cells is a blastula. (**d**) As shown in this diagram of a cross section, the frog blastula contains a blastocoel surrounded by several layers of cells and situated in the animal hemisphere. (a, b, and c are SEMS.)

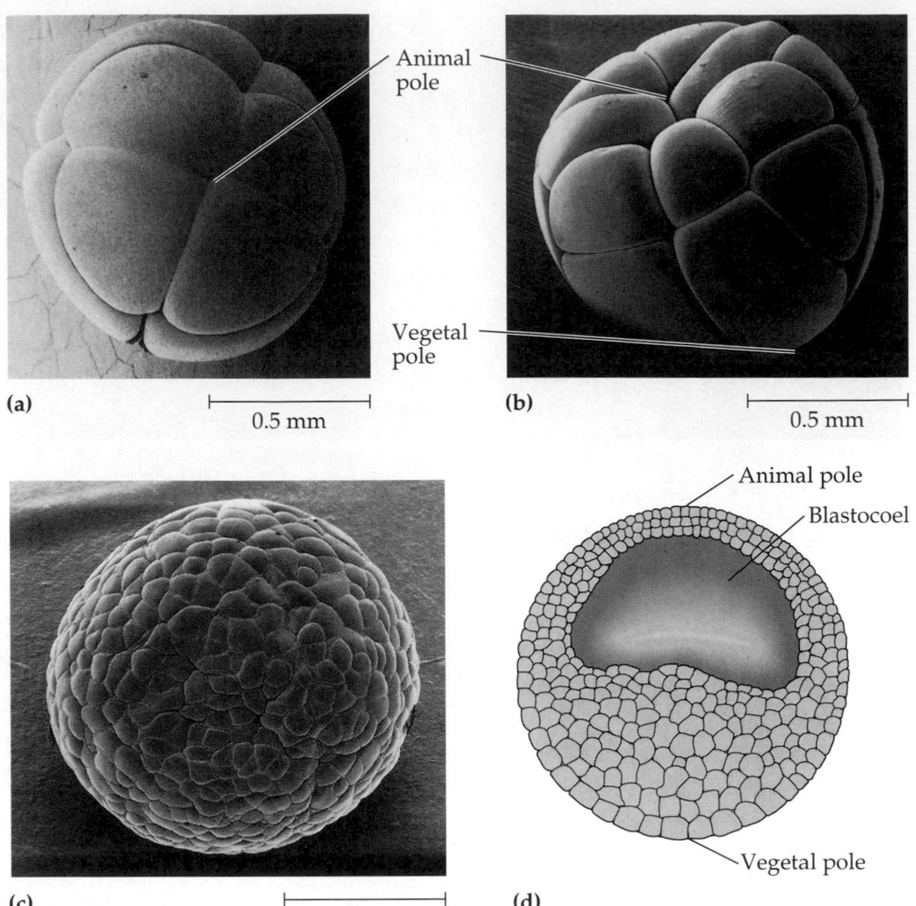

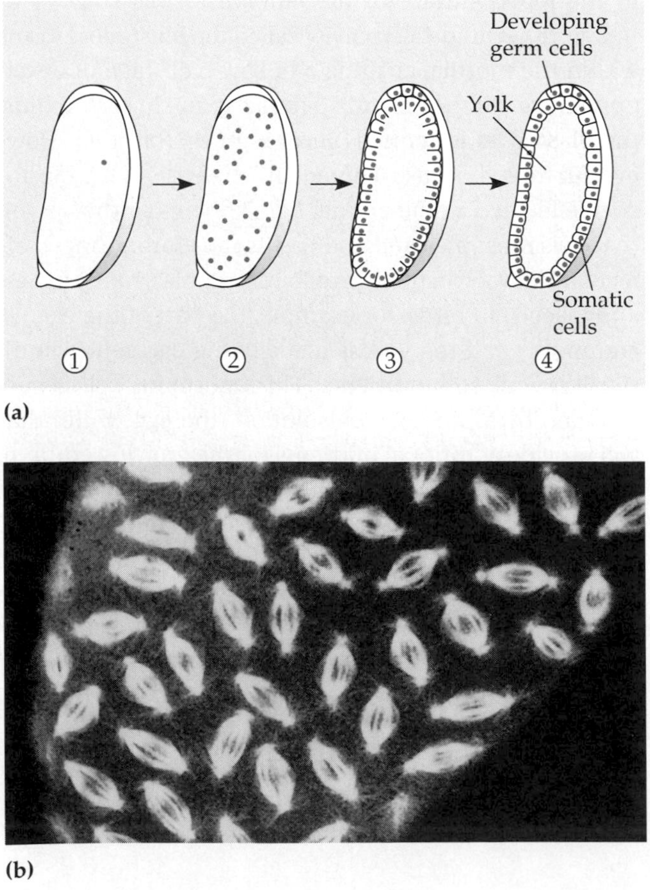

Developing germ cells

Yolk

Somatic cells

① ② ③ ④

(a)

(b)

FIGURE 43.8

Cleavage in *Drosophila*. (**a**) A series of mitotic divisions unaccompanied by cytokineses first transforms ① the zygote to ② a syncytium with many nuclei scattered through a mass of yolk. ③ The nuclei move to the surface of the embryo, and membranes develop around them. ④ The fruit fly blastula consists of a central mass of yolk surrounded by a single layer of cells. By this time, a cluster of germ cells, which will become the new individual's gametes, is distinguished from the somatic cells, which will form all other parts of the fly. (**b**) Many of the nuclei in this fruit fly embryo (a syncytial stage) stained with fluorescent dye are undergoing mitosis (LM).

nucleus undergoing mitotic divisions that are not accompanied by cytokinesis. These mitotic divisions produce several hundred nuclei, which migrate to the outer edge of the egg. At the surface, each nucleus is surrounded by its own domain of cytoskeletal proteins but is not set off from other nuclei by membranes. The insect embryo at this stage is an example of a **syncytium,** a cell containing many nuclei all within a common cytoplasm. After several more rounds of mitosis, plasma membranes form around the nuclei, and the embryo, now a blastula, consists of a single layer of about 6000 cells surrounding a mass of yolk.

Gastrulation rearranges the blastula to form a three-layered embryo with a primitive gut

The morphogenetic process called **gastrulation** is a dramatic rearrangement of the cells of the blastula. Gastrulation differs in detail from one animal group to another, but a common set of cellular changes drives this spatial rearrangement of embryos. These general cellular mechanisms are changes in cell motility, changes in cell shape, and changes in cellular adhesion to other cells and to molecules of the extracellular matrix (see Chapter 7). The essential result of gastrulation is that some of the cells at or near the surface of the blastula move to a new, more interior location. This transforms the blastula into a three-layered embryo called the **gastrula.**

The three layers produced by gastrulation are embryonic tissues called **ectoderm, endoderm,** and **mesoderm,** also collectively termed the embryonic germ layers. The ectoderm forms the outer layer (skin) of the gastrula. The endoderm lines the embryonic digestive tract. And the mesoderm partly fills the space between the ectoderm and the endoderm. Eventually, these three cell layers develop into all the parts of the adult animal. For instance, our nervous system and the outer layer (epidermis) of our skin come from ectoderm; the innermost lining of our digestive tract and associated organs, such as the liver and pancreas, arise from endoderm; and most other organs and tissues, such as the kidney, heart, muscles, and the inner layer of our skin (dermis), develop from mesoderm.

Let's examine gastrulation in a sea urchin embryo (FIGURE 43.9). The wall of the sea urchin blastula consists of a single layer of cells. Gastrulation begins at the vegetal pole, where the cells flatten slightly to form a plate that buckles inward, a process called **invagination.** Other cells near the plate detach from the blastula wall and enter the blastocoel as migratory cells called mesenchyme cells. The buckled vegetal plate then undergoes extensive rearrangement of its cells, a process that transforms the

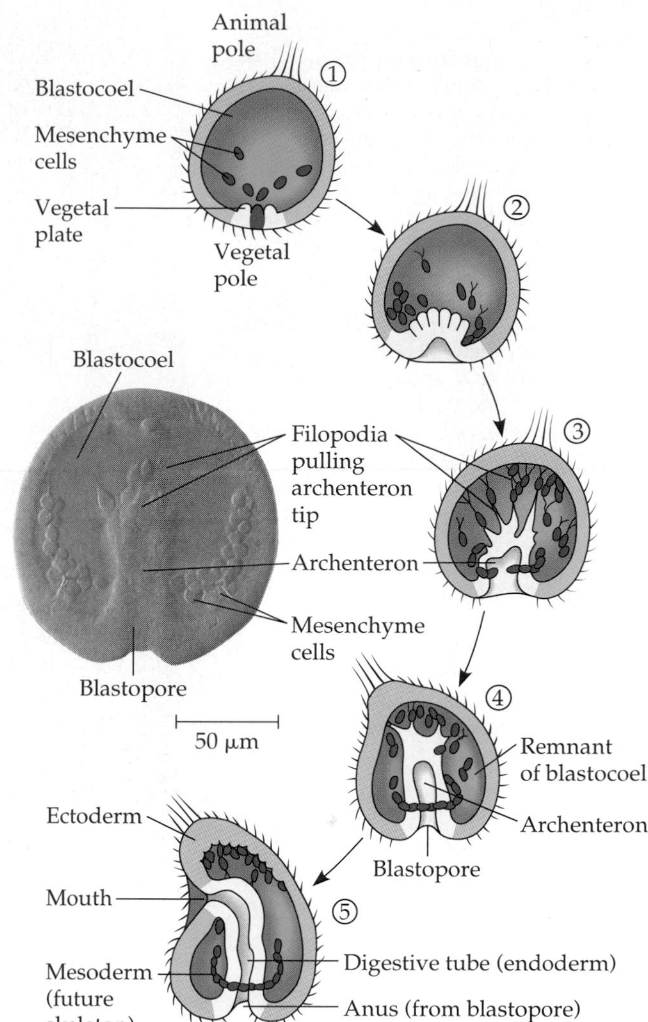

FIGURE 43.9
Sea urchin gastrulation. ① Formed by cleavage, the blastula consists of a single layer of ciliated cells surrounding the blastocoel. Gastrulation begins with the formation of the vegetal plate at the vegetal pole. Blue = future ectoderm; yellow = future endoderm; red = future mesoderm. Mesenchyme cells (future mesoderm) detach from the vegetal plate and migrate into the blastocoel. ② The vegetal plate in this early gastrula invaginates (buckles inward). Mesenchyme cells begin to form extensions (filopodia). ③ Endoderm cells form the archenteron (future digestive tube). Mesenchyme cells form filopodial connections between the tip of the archenteron and the ectoderm cells of the blastocoel wall (inset, LM). ④ Contraction of the filopodia in a late gastrula drags the archenteron the rest of the way across the blastocoel, where the endoderm of the archenteron will fuse with ectoderm of the blastocoel wall. ⑤ Gastrulation is complete. The gastrula has a functional digestive tube formed from the endoderm of the archenteron, with a mouth and an anus. Ectoderm forms the embryo's ciliated skin. Some of the mesenchyme cells of the mesoderm have secreted minerals that will form a simple internal skeleton.

FIGURE 43.10

Gastrulation in a frog embryo. ① The blastocoel of the frog blastula is off-center and surrounded by a wall that is more than one cell thick. At this stage, the colors indicate regions of the blastula that will become the embryo's three germ layers. ② Gastrulation begins when a small tuck, the dorsal lip of the blastopore, appears on one side of the blastula. The tuck is formed by cells burrowing inward from the surface. Additional cells that will become endoderm and mesoderm then roll inward over the dorsal lip (involution) and move away from the blastopore into the interior of the gastrula. Meanwhile, cells of the animal pole, which will form ectoderm, spread over the embryo's outer surface. ③ Externally, the lip of the blastopore starts becoming circular. Internally, the three germ layers start forming as cells continue migrating inward. The advancing endoderm, mesoderm, and the archenteron, lined by endoderm, are filling the space occupied by the blastocoel. ④ Late in gastrulation, the circular blastopore surrounds a plug of yolk cells (the yolk plug) and the three germ layers are in place, ready for organogenesis.

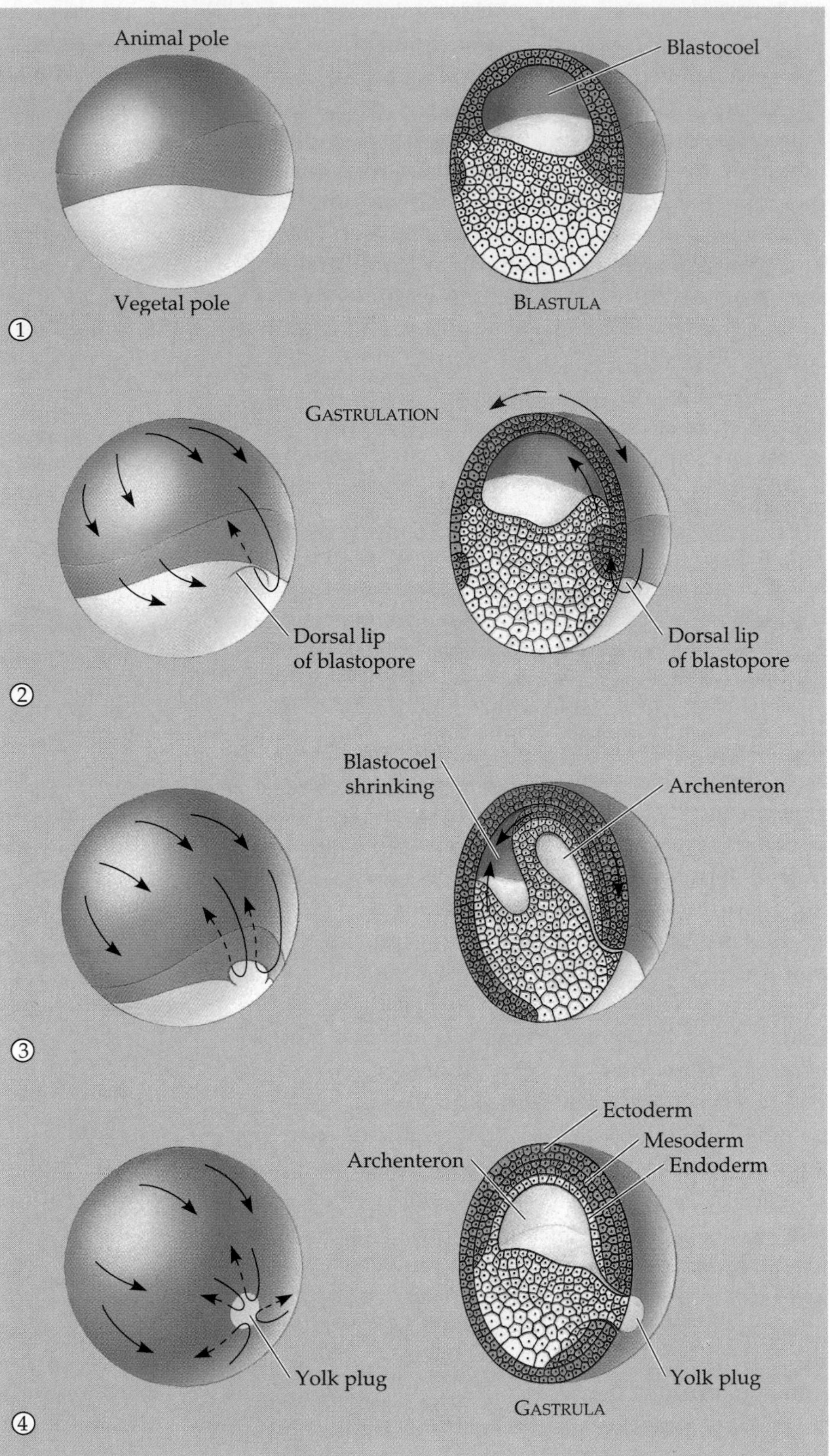

shallow invagination into a deeper, narrower pouch called the **archenteron,** or primitive gut. This opening of the archenteron, which will become the anus, is called the **blastopore.** A second opening forms at the other end of the archenteron, which is then a developing digestive tube with a mouth and anus. Gastrulation has produced an embryo with a primitive gut and three germ layers: ectoderm (color-coded blue throughout this chapter), endoderm (yellow), and mesoderm (red). Thus, the triploblastic (three-layered) body plan characteristic of most animal phyla is established very early in development (see Chapter 29). In the sea urchin, the gastrula eventually develops into a ciliated larva that drifts near the ocean surface as plankton, feeding on bacteria and unicellular algae.

Gastrulation during frog development also produces a three-layered embryo with an archenteron, as shown in FIGURE 43.10, to the left. The mechanics of gastrulation are much more complicated in a frog, however, because of the large, yolk-laden cells of the vegetal hemisphere, and because the wall of the blastula is more than one cell thick in most species. The first sign of gastrulation is a small crease on one side of the blastula where the blastopore will eventually be located. This invagination is produced by a cluster of cells burrowing inward. An external tuck produced by the invagination becomes the upper edge, or **dorsal lip,** of the blastopore. The dorsal lip of the blastopore forms where the gray crescent was located in the zygote. Gastrulation continues with cells on the surface of the embryo rolling over the edge of the dorsal lip into the interior of the embryo, a process called **involution.** Once inside, these cells migrate away from the blastopore along the roof of the blastocoel. Involution continues, with migrating internal cells becoming organized into layered mesoderm and endoderm, and the archenteron forming within the endoderm. Eventually, the complex cell movements of gastrulation produce a three-layered embryo. As the process is completed, the lip of the blastopore encircles a **yolk plug** consisting of large, food-laden cells from the vegetal pole of the embryo. Except for the yolk plug, cells remaining on the surface make up the ectoderm, surrounding the layers of mesoderm and endoderm. With the three germ layers in place, gastrulation is complete, and the embryo's organs begin to form.

Organogenesis forms the organs of the animal body from the three embryonic germ layers

Various regions of the three germ layers develop into the rudiments of organs during the process of **organogenesis** (TABLE 43.1). Three kinds of morphogenetic changes—folds, splits, and dense clustering (condensation) of cells—that occur within the layered embryonic tissues are the first evidence of organ building. The organs that begin to take shape first in the embryos of frogs and other chordates are the neural tube and notochord, the skeletal rod characteristic of all chordate embryos (see Chapter 30).

FIGURE 43.11 shows early organogenesis as it occurs in a frog. The **notochord** is formed from condensation of dorsal mesoderm just above the archenteron, and the neural tube originates as a plate of dorsal ectoderm just above the developing notochord. The neural plate soon undergoes folding, actually rolling itself into the **neural tube,** which will become the central nervous system—the brain and spinal cord. These organs are hollow in chordates because of this mechanism of development. The notochord elongates and stretches the embryo along its anterior-posterior axis. Later, the notochord will function as a core around which mesodermal cells gather and form the vertebrae.

Other condensations occur in strips of mesoderm lateral to the notochord, which separate into blocks called **somites.** The somites are arranged serially on both sides along the length of the notochord (FIGURE 43.11C). Cells from the somites not only give rise to the vertebrae of the backbone, they also form the muscles associated with the axial skeleton. This serial origin of the axial skeleton and muscles reinforces the point made in Chapter 30 that chordates are basically segmented animals, although the segmentation becomes less obvious later in development. (There are signs of segmentation even in the adult, as in the series of vertebrae in a human or the segments of chevron-shaped muscles in a fish.) Lateral to the somites, the mesoderm splits into two layers that form the lining of the body cavity, or coelom.

As organogenesis progresses, morphogenesis and cellular differentiation, which we will discuss later in the chapter, continue to refine the organs that arise from the three embryonic germ layers. Review TABLE 43.1, which lists the embryonic sources of the major organs and tissues in frogs and other vertebrates. Unique to vertebrate embryos, a band of cells called the neural crest develops along the border where the neural tube pinches off from the ectoderm (see Chapter 30). Cells of the neural crest later migrate to various parts of the embryo, forming pigment cells of the skin, some of the bones and muscles of the skull, the teeth, the medulla of the adrenal glands, and peripheral components of the nervous system, such as sensory and sympathetic ganglia.

Embryonic development of the frog leads to a larval stage, the tadpole (see FIGURE 43.1), which hatches from the jelly coat that originally cloaked the egg. Later, metamorphosis will transform the frog from the aquatic, herbivorous tadpole to the terrestrial, carnivorous adult.

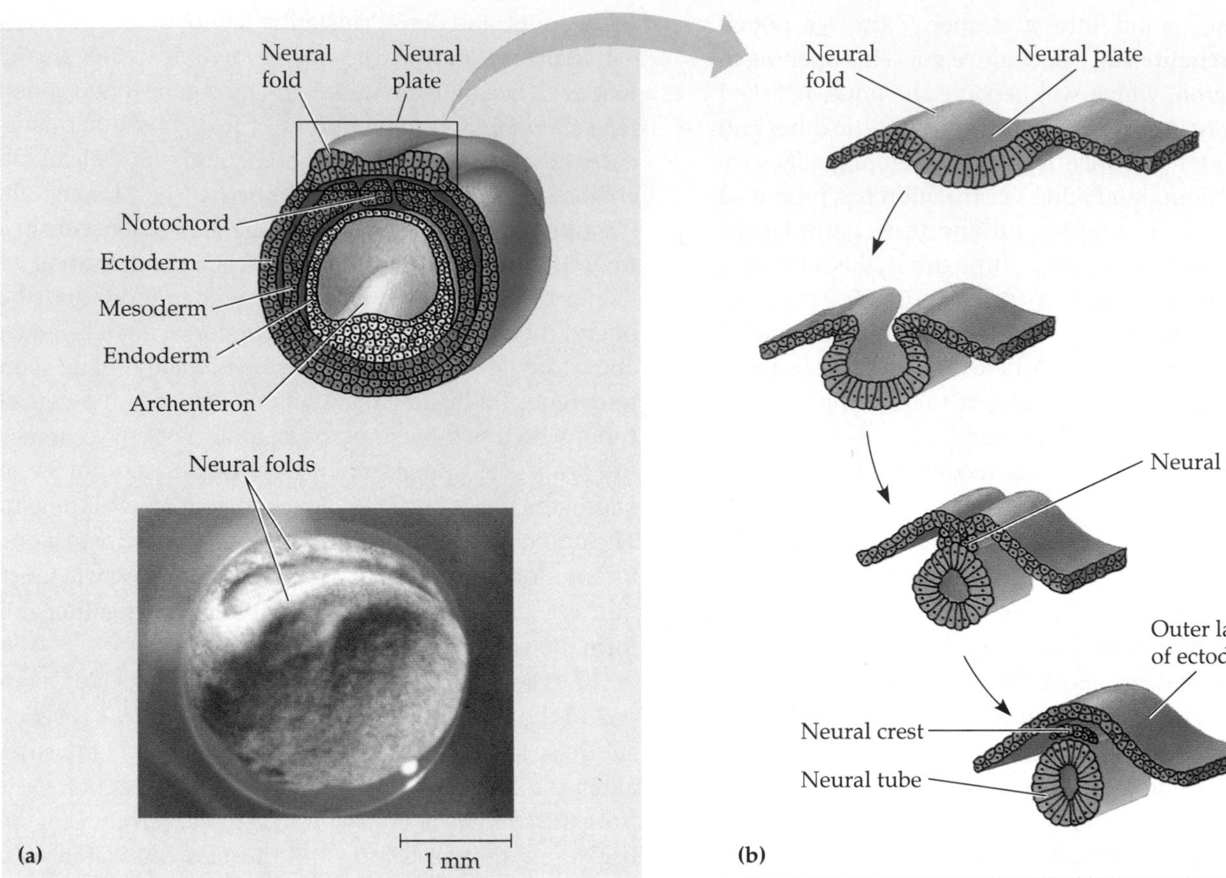

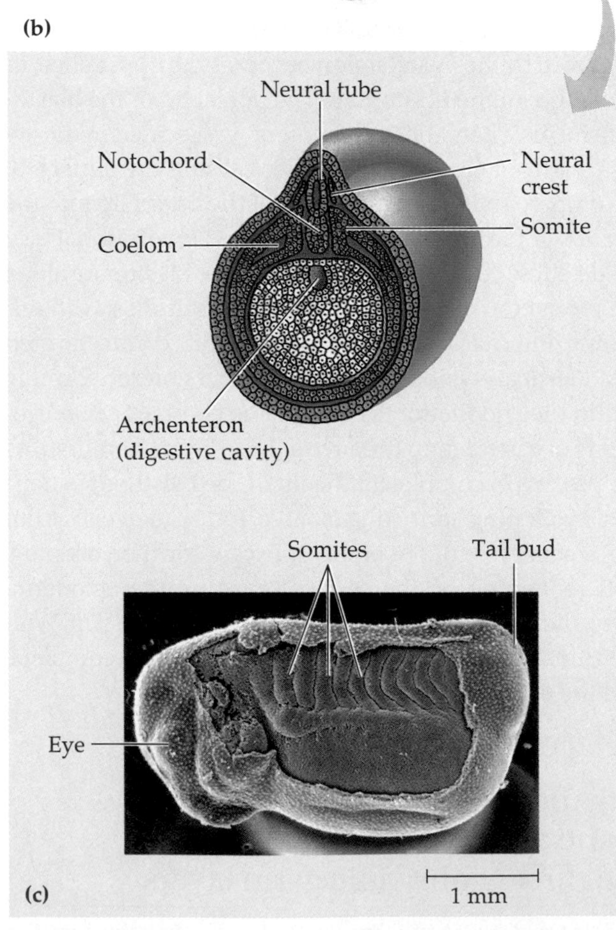

FIGURE 43.11

Organogenesis in a frog embryo. (**a**) A cross section of a frog embryo at the beginning of organogenesis shows the three germ layers and the rudiments of the notochord and neural plate. The notochord has developed from dorsal mesoderm, and the dorsal ectoderm has thickened, forming the neural plate. Two pronounced ridges, the neural folds, form the lateral edges of the neural plate. The light micrograph shows a whole frog embryo at the neural plate stage of development. (**b**) These diagrams illustrate how the neural plate rolls into a hollow cylinder, the neural tube. Tissue at the meeting margins of the tube separate from the tube as the neural crest, a source of migrating cells that eventually form many structures, including bones and muscles of the skull, skin pigment cells, adrenal medulla glands, and peripheral ganglia of the nervous system. (**c**) In this cross section, an embryo with a completed neural tube has somites flanking the notochord. Formed from mesoderm, the somites will give rise to segmental structures such as vertebrae and serially arranged skeletal muscles. The lateral mesoderm has begun to separate into the two tissue layers that line the coelom. In the scanning electron micrograph of a whole embryo at the tail-bud stage (side view), part of the ectoderm has been removed to reveal the somites.

TABLE 43.1

Derivatives of the Three Embryonic Germ Layers in Vertebrates

GERM LAYER	ORGANS AND TISSUES IN THE ADULT
Ectoderm	Epidermis of skin and its derivatives (e.g., skin glands, nails); epithelial lining of mouth and rectum; sense receptors in epidermis; cornea and lens of eye; nervous system; adrenal medulla; tooth enamel; epithelium of pineal and pituitary glands.
Endoderm	Epithelial lining of digestive tract (except mouth and rectum); epithelial lining of respiratory system; liver; pancreas; thyroid; parathyroids; thymus; lining of urethra, urinary bladder, and reproductive system.
Mesoderm	Notochord; skeletal system; muscular system; circulatory and lymphatic systems; excretory system; reproductive system (except germ cells, which differentiate during cleavage); dermis of skin; lining of body cavity; adrenal cortex.

Amniote embryos develop in a fluid-filled sac within a shell or uterus

All vertebrate embryos require an aqueous environment for development. In the case of fish and amphibians, the egg is laid in the surrounding sea or pond and needs no special water-filled enclosure. The vertebrate move onto land required solving the problem of reproduction in dry environments, and two major solutions evolved: the shelled egg of reptiles and birds and the uterus of placental mammals. Within the shell or uterus, the embryos of birds, reptiles, and mammals are surrounded by a fluid-filled sac formed by a membrane called the amnion. Vertebrates of these three classes are therefore called **amniotes** (see Chapter 30). We have already examined the embryology of a vertebrate that lacks an amnion, the frog. For comparison, we will now study the early development of two amniotes, a bird and a mammal.

Avian Development

After fertilization, a bird egg undergoes meroblastic cleavage in which cell division occurs only in a small region of yolk-free cytoplasm atop the large mass of yolk. The early cleavage divisions produce a cap of cells called the **blastodisc,** which rests on the undivided yolk. The blastomeres then sort into upper and lower layers, the epiblast and hypoblast (FIGURE 43.12). The cavity between these two layers is the avian version of the blastocoel. And

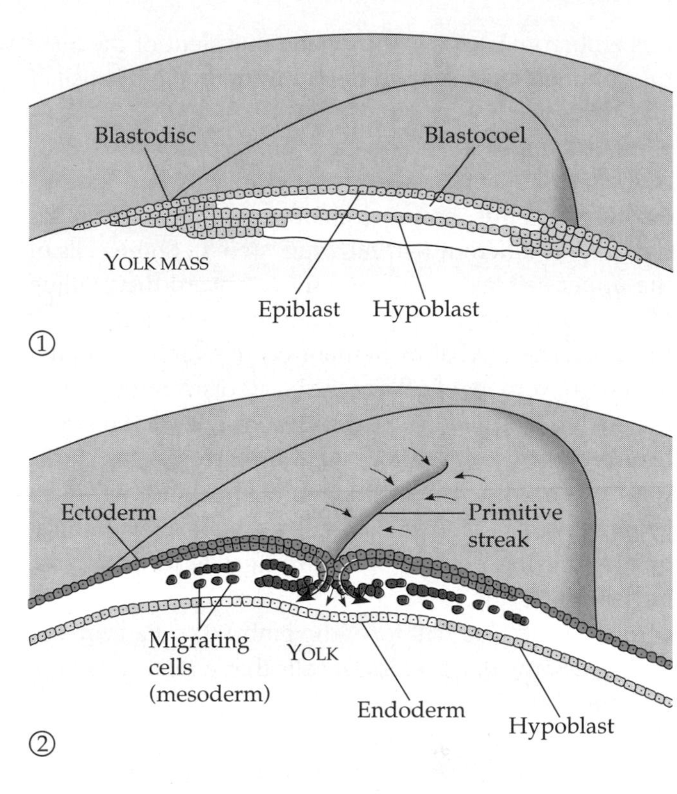

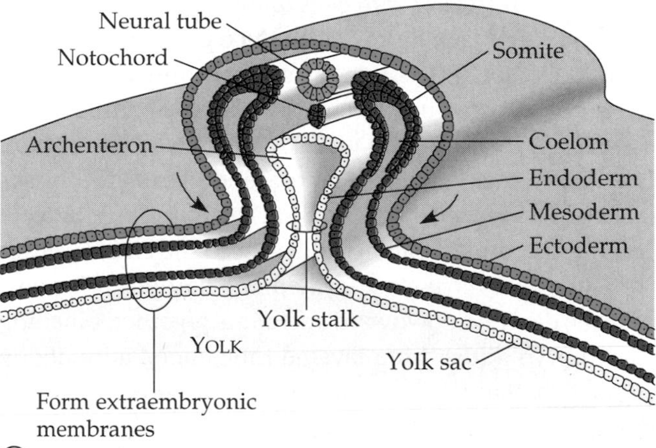

FIGURE 43.12

Cleavage and gastrulation in a chick embryo. ① For the eggs of birds and reptiles, which store a very large amount of yolk, cleavage is meroblastic, or incomplete. Cell division is restricted to a small cap of cytoplasm at the animal pole. Cleavage produces a blastodisc that rests on the large, undivided mass of yolk. The blastodisc becomes arranged into two layers (epiblast and hypoblast) that bound the blastocoel, forming the avian version of a blastula. ② During gastrulation, some cells of the epiblast migrate (arrows) into the interior of the embryo through the primitive streak, shown here in transverse section. Some of these cells move laterally to form mesoderm, while others migrate downward to form the endoderm. ③ The rudimentary gut (archenteron) is formed when lateral folds pinch the embryo from the yolk. About midway along its length, the embryo will remain attached to the yolk by the yolk stalk formed mainly by hypoblast cells. The three germ layers and hypoblast cells also contribute to the system of extraembryonic membranes that support further development.

this embryonic stage is the avian equivalent of the blastula, although its form is different from the hollow ball of an early frog embryo.

Gastrulation, as in the frog embryo, involves cells moving from the surface of the embryo to an interior location. In birds, however, the route of this cell migration is very different (FIGURE 43.12, step 2). Some cells of the upper cell layer (epiblast) move toward the midline of the blastodisc, then detach and move inward toward the yolk. The medial movement on the surface and the inward movement of cells at the blastodisc's midline produce a groove called the **primitive streak.** As the streak lengthens over the surface of the blastodisc, it marks what will become the bird's anterior-posterior axis. The primitive streak is functionally equivalent to the dorsal lip of a frog blastopore, but it is arranged like a linear tuck rather than a ring.

All the cells that will form the embryo come from the epiblast. Some of the epiblast cells that pass through the primitive streak move laterally into the blastocoel, producing the mesoderm. Other epiblast cells, which will produce the endoderm, migrate through the streak and downward, mingling with cells of the hypoblast. The epiblast cells that remain on the surface give rise to the ectoderm. Although the hypoblast contributes no cells to the embryo, it seems to help direct the formation of the primitive streak and is required for normal development. Eventually segregating from the endoderm, the hypoblast cells form portions of a sac surrounding the yolk and a stalk connecting the yolk mass to the embryo. After the three germ layers are formed, the borders of the embryonic disc fold downward and come together, pinching the embryo into a three-layered tube joined at midbody to the yolk (FIGURE 43.12, step 3). Neural tube formation, development of the notochord and somites, and other events in organogenesis occur much as in the frog embryo (FIGURE 43.13).

Notice in FIGURE 43.12, step 3, that only part of each germ layer contributes to the embryo itself. The tissue layers that form external to the embryo develop into four **extraembryonic membranes** that support further embryonic development within the egg. These four membranes are the **yolk sac,** the **amnion,** the **chorion,** and the **allantois** (FIGURE 43.14).

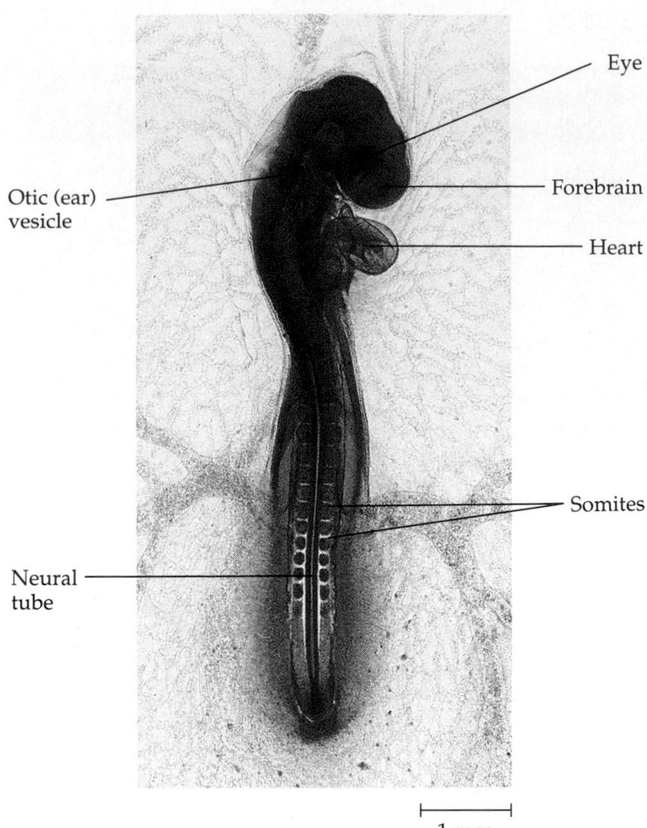

FIGURE 43.13
Organogenesis in a chick embryo. Rudiments of most major organs have already formed in this chick, which is about 56 hours old (LM).

Mammalian Development

In most mammalian species, eggs are fertilized in the oviduct, and the earliest stages of development occur while the embryo completes its journey down the oviduct to the uterus (see Chapter 42). In contrast to the large, yolky eggs of birds and reptiles, the egg of a placental mammal is quite small, storing little in the way of food reserves. Cleavage of the mammalian zygote is therefore holoblastic, but gastrulation and early organogenesis follow a pattern similar to that of birds and reptiles. (Recall from Chapter 30 that mammals descended from reptilian stock during the early Mesozoic era.)

Cleavage is relatively slow in mammals. In the case of humans, the first division is complete about 36 hours after fertilization, the second division at about 60 hours, and the third division at about 72 hours. The mammalian zygote has no apparent polarity. The cleavage planes seem to be randomly oriented, and the blastomeres are equal in size.

Further development of the human embryo is shown in FIGURE 43.15. By about 7 days after fertilization, the embryo has over 100 cells arranged around a central cavity. This is the embryonic stage known as the **blastocyst.** Protruding into one end of the blastocyst cavity is a cluster of cells called the **inner cell mass,** which will subsequently develop into the embryo proper and some of the extraembryonic membranes. The outer epithelium surrounding the cavity is the **trophoblast,** which, along with mesodermal tissue, will form the fetal portion of the placenta.

The embryo reaches the uterus by the blastocyst stage and begins to implant soon thereafter. The trophoblast secretes enzymes that enable the blastocyst to penetrate

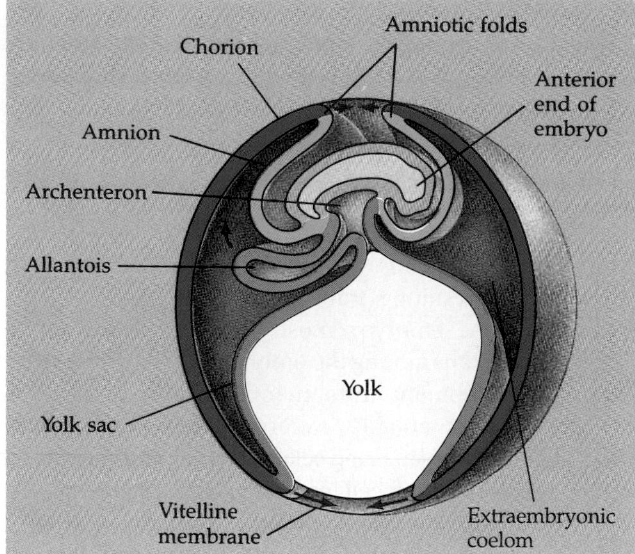

FIGURE 43.14

The development of extraembryonic membranes in a chick. Each of the four membranes, which provide support services for the embryo, develops from epithelial sheets external to the embryo proper. The yolk sac expands over the surface of the yolk mass. Cells of the yolk sac will digest yolk, and blood vessels that develop within the membrane will carry nutrients into the embryo. Lateral folds of extraembryonic tissue extend over the top of the embryo and fuse to form two more membranes, the amnion and the chorion, which are separated by an extraembryonic extension of the coelom. The amnion encloses the embryo in a fluid-filled amniotic sac, protecting the embryo from desiccation and cushioning it against mechanical shocks. The fourth membrane, the allantois, originates as an outpocketing of the embryo's hindgut. The allantois is a sac that extends into the extraembryonic coelom separating the chorion and amnion. It functions as a disposal sac for uric acid, the insoluble nitrogenous waste of the embryo. As the allantois continues to expand, it presses the chorion against the vitelline membrane, the inner lining of the eggshell. Together, the allantois and chorion form a respiratory organ that serves the embryo. Blood vessels that form in the epithelium of the allantois transport oxygen to the embryonic chick. The extraembryonic membranes of reptiles and birds are adaptations associated with the special problems of development on land.

the uterine lining. Bathed in blood spilled from eroded capillaries in the endometrium (uterine lining), the trophoblast thickens and extends fingerlike projections into the surrounding maternal tissue. The placenta eventually forms from this proliferated trophoblast and the region of endometrium it invades (see FIGURE 42.17). About the time the blastocyst implants in the uterus, the inner cell mass forms a flat disc with an upper layer of cells, the epiblast, and a lower layer, the hypoblast. These layers are homologous to those of birds, and as in birds, the embryo develops entirely from epiblast cells, while the hypoblast cells form the yolk sac. Gastrulation occurs by the inward movement of cells from the upper layer through a primitive streak to form mesoderm and endoderm, just as it does in the chick.

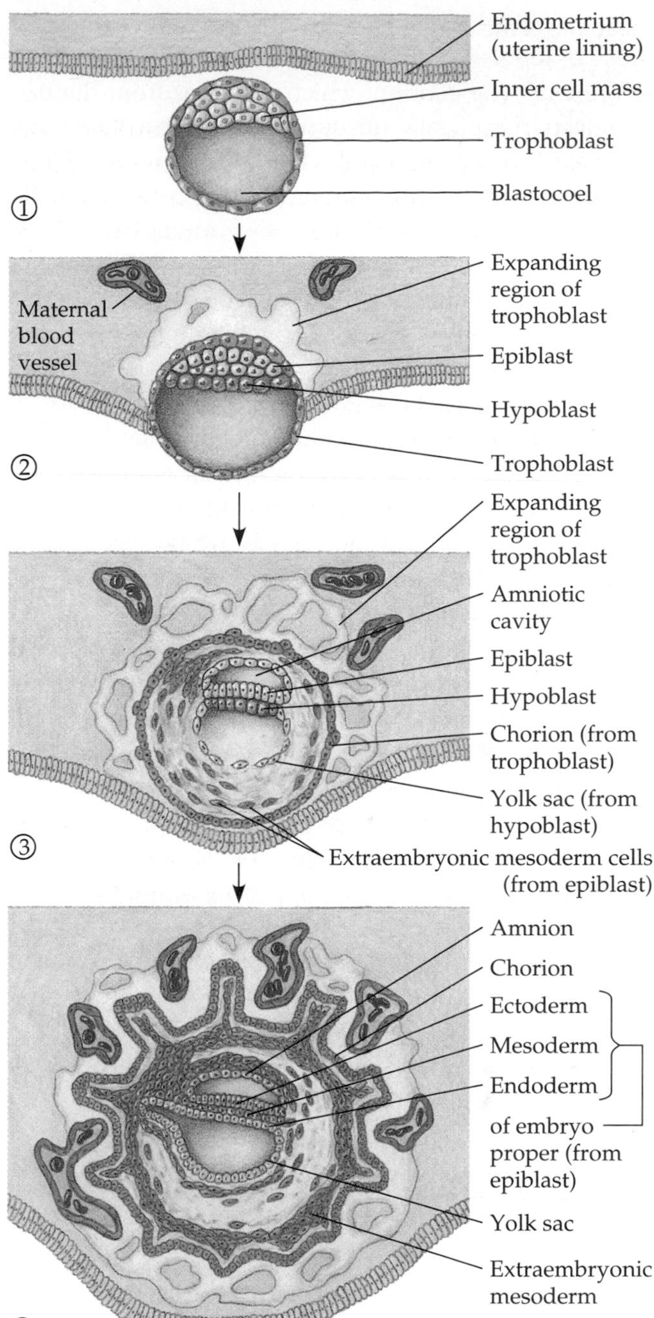

FIGURE 43.15

Early development of a human embryo and its extraembryonic membranes. This series of drawings illustrates four stages in transverse section. ① Cleavage produces a blastocyst, consisting of a trophoblast surrounding a blastocoel and an inner cell mass. The blastocyst implants in the uterine lining. ② Coincident with implantation, the inner cell mass forms an epiblast cell layer, which will develop into the three germ layers of the embryo, and a hypoblast, which will form the yolk sac. ③ By this stage, the trophoblast has begun to form the chorion and continues to expand into the endometrium. The epiblast has begun to form the amnion, surrounding a fluid-filled cavity. Mesodermal cells that will become part of the placenta are also derived from the epiblast. ④ Gastrulation by the inward movement of epiblast cells has produced a three-layered embryo surrounded by proliferating extraembryonic mesoderm.

Four extraembryonic membranes homologous to those of reptiles and birds form during mammalian development. The chorion, which develops from the trophoblast, completely surrounds the embryo and the other extraembryonic membranes. The amnion begins as a dome above the proliferating epiblast and eventually encloses the embryo in a fluid-filled amniotic cavity. (The fluid from this cavity is the "water" expelled from the vagina of the mother when the amnion breaks just prior to childbirth.) Below the developing embryo proper, the yolk sac encloses a fluid-filled cavity. Although this cavity contains no yolk, the membrane that surrounds it is given the same name as the homologous membrane in birds and reptiles. The yolk sac membrane of mammals is a site of early formation of blood cells, which later migrate into the embryo proper. The fourth extraembryonic membrane, the allantois, develops as an outpocketing of the embryo's rudimentary gut, as it does in the chick. The allantois is incorporated into the umbilical cord, where it forms blood vessels that transport oxygen and nutrients from the placenta to the embryo and rid the embryo of carbon dioxide and nitrogenous wastes. Thus, the extraembryonic membranes of shelled eggs, where embryos are nourished with yolk, were conserved as mammals diverged from reptiles in the course of evolution, but with modifications adapted to development within the reproductive tract of the mother.

Organogenesis begins with the formation of the neural tube, notochord, and somites. By the end of the first trimester of human development, rudiments of all the major organs have developed from the three germ layers, as summarized in TABLE 43.1.

The developmental fate of cells depends on cytoplasmic environment, location, and cell-cell interactions

Having traced the sequence of changes that transform a fertilized egg into a multicellular, three-layered embryo with developing organs, we now focus on the underlying mechanisms of development. Coupled with the morphogenetic mechanisms that give an animal and its organs their characteristic shapes, development also requires the timely differentiation of many kinds of cells in specific locations. These diverse cell types generally share a common genome, but differ in their patterns of gene expression. Two general principles integrate our current knowledge of the genetic and cellular mechanisms that underly embryonic development:

1. *The heterogeneous organization of cytoplasm in the unfertilized egg leads to regional differences in the early embryo of many animal species.* By partitioning the heterogeneous cytoplasm of the egg, cleavage parcels out different mRNAs, proteins, and other molecules to different blastomeres. These local differences in cytoplasmic composition establish body axes and influence the expression of genes that affect the developmental fate of cells in the early embryo.
2. *Cell-cell interactions compound differences initiated by cytoplasmic organization and by the relocation of cells during morphogenesis.* The differences between the early blastomeres resulting from cytoplasmic heterogeneity determine the embryo's most basic features. Beyond that, interactions among the embryonic cells themselves induce development of the many specialized cell types making up a new animal. As morphogenetic movements, such as gastrulation, bring cells from different regions of the embryo together, cell interactions can elicit changes in gene expression that occur only among neighboring cells. This may be accomplished by the transmission of chemical signals, or if the cells are actually in contact, by membrane interactions.

It will help to keep these two principles in mind as we take a closer look at the molecular and cellular mechanisms of animal development.

Polarity and the Basic Body Plan

A bilaterally symmetrical animal has an anterior-posterior axis, a dorsal-ventral axis, and left and right sides (see Chapter 29). Establishing this basic body plan is a first step in morphogenesis and is prerequisite to the development of tissues and organs. In humans and other mammals, the basic polarities do not seem to be established until after cleavage. However, in most species, fundamental instructions (as to the head end, the tail end, and so on) are set down in the unfertilized egg or during early cleavage. In many cases, all three axes of the embryo are defined before the onset of cleavage. For example, in many frogs, the animal-vegetal axis of the egg, defined by concentration gradients of cytoplasmic chemicals, marks the anterior-posterior axis of the embryo. And the gray crescent marks both the dorsal-ventral axis and the left-right axis. The embryo takes shape along these axes. You will learn later how researchers studying *Drosophila* are working out the genetic basis of an animal's spatial organization.

Fate Maps and the Analysis of Cell Lineages

Classic studies performed in the 1920s by German embryologist W. Vogt established that in embryos whose axes are defined early in development, it is often possible to determine which parts of the embryo will be derived from each region of the zygote or blastula. Following the lead of earlier work on marine worms and mollusks, Vogt charted a **fate map,** a type of territorial diagram, for embryos (FIGURE 43.16a). Using nontoxic dyes, he labeled

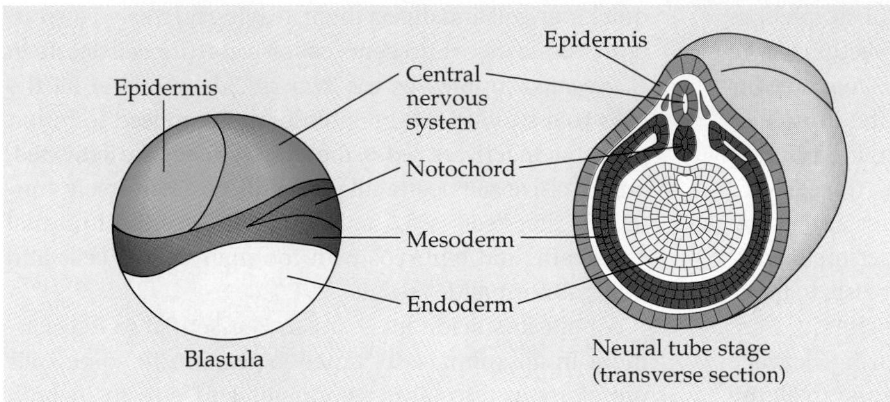

(a) Fate map of a frog embryo

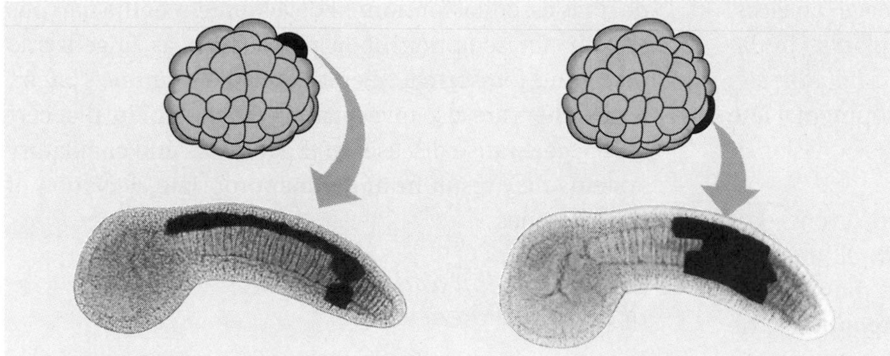

(b) Cell lineage analysis in a tunicate

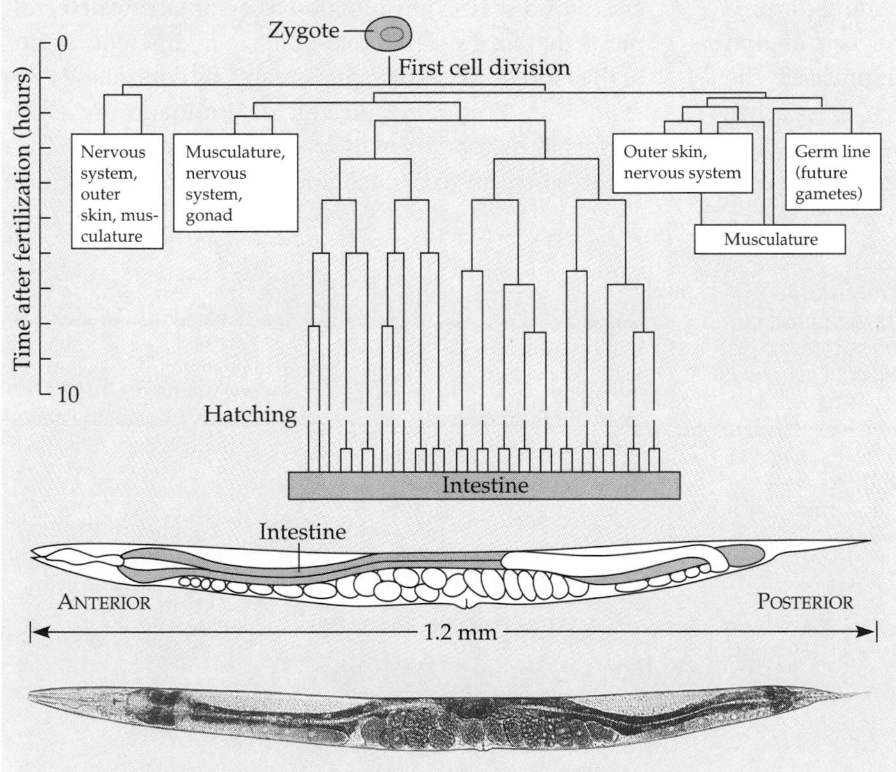

(c) Cell lineages of *Caenorhabditis elegans*

FIGURE 43.16

Examples of fate maps and cell lineage analysis. (**a**) The fates of cells of a frog embryo were determined by marking different regions of the blastula surface with dyes of various colors, then determining the locations of dyed cells in the gastrula or at later stages of development, such as at this neural tube stage. (**b**) The drawings depict 64-cell embryos of a tunicate, an invertebrate chordate. An individual cell can be injected with a dye used as a marker, enabling a researcher to determine which cells in a later embryo are derived from the marked cells. The two light micrographs of larvae contrast the regions that develop from two different blastomeres indicated by the two drawings. (**c**) The nematode *Caenorhabditis elegans* is transparent at all stages of its development, making it possible for researchers to trace the lineage of every cell, from the zygote to the 2000 cells of the adult worm (LM). The diagram shows a detailed lineage only for the intestine, which is derived exclusively from one of the first four cells formed during cleavage.

cells of different regions of the surface of amphibian blastulas with different colors and later sectioned the embryos to see where the colors turned up. Vogt's results were among the earliest indications that the lineage of cells making up the three germ layers created by the complex rearrangements of gastrulation is traceable to cells in the blastula (compare FIGURES 43.16a and 43.10).

New methods and clever choices of experimental organisms have enabled developmental biologists to make even more detailed fate maps. Various techniques are used to mark an individual blastomere during cleavage and then follow the marker as it is distributed to all the mitotic descendants of that cell (FIGURE 43.16b). Fate mapping at this level is called cell lineage analysis. (Chapter 34 discussed lineage analysis of plants.) In the case of one organism, the nematode *Caenorhabditis elegans,* it has been possible to map the developmental fate of every cell, beginning with the first cleavage division of the zygote (FIGURE 43.16c).

Lineage analyses of *C. elegans* accent some events crucial to normal development and growth in all animals. For example, this research has illuminated the developmental role of programmed cell death, or **apoptosis.** The timely suicide of cells occurs exactly 131 times in the course of *C. elegans'* normal development. The genome of *C. elegans* includes two "suicide" genes coding for proteins that are either directly toxic to the cell or that change other metabolites into toxins. At precisely the same points in each new worm's cell lineage, these genes are expressed in cells destined to die (FIGURE 43.17). The cells shrink, their nuclei condense, and neighboring cells quickly engulf and digest them, leaving no trace. The protein product of a third gene, called *ced-9* (for cell death) in *C. elegans,* suppresses the two suicide genes in all the cells that survive. When an embryo is exposed to mutagens that inactivate *ced-9,* the suicide genes are activated, and massive cell death kills the embryo. Conversely, mutations that keep *ced-9* active in all cells prevent normal cell death, and embryos with too many nerve cells and other abnormalities result.

A built-in suicide mechanism is essential to development in all animals. Its timely activation in some cells functions in normal development and growth in both embryos and adults. In vertebrates, programmed cell death is essential for normal development of the nervous system, for sculpting body parts (such as fingers and toes), and for normal operation of the immune system. Researchers are also investigating the possibility that certain degenerative diseases of the nervous and circulatory systems may result from the inappropriate activation of suicide genes.

Determination and Restriction of Cellular Potency

In some animals, substances localized within specific regions of an egg's cytoplasm, and consequently within specific blastomeres, may function as **cytoplasmic determinants** that fix the developmental fate of different regions of the embryo very early, presumably by controlling gene expression. Even if cytoplasmic determinants are asymmetrically distributed in the zygote, the first cleavage may occur along an axis that produces two blastomeres of

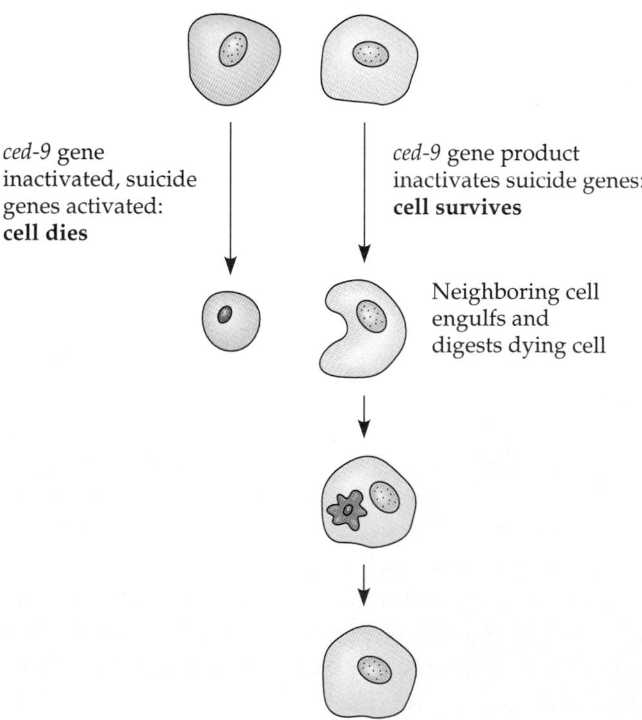
ced-9 gene inactivated, suicide genes activated: **cell dies**
ced-9 gene product inactivates suicide genes: **cell survives**
Neighboring cell engulfs and digests dying cell

FIGURE 43.17
Apoptosis (programmed cell death) in the nematode *C. elegans.* Many cells must die during normal development in all animals. In *C. elegans,* a cell remains alive as long as one of its genes, called *ced-9,* is expressed. The protein product of *ced-9* inhibits two other genes (suicide genes). A cell begins to die when *ced-9* is turned off, enabling the suicide genes to be expressed. Signals from the dying cell turn on other genes in neighboring cells, causing them to engulf and digest the remains, keeping the embryo free of harmful enzymes and metabolites. The mechanism for regulation of *ced-9* is not yet known.

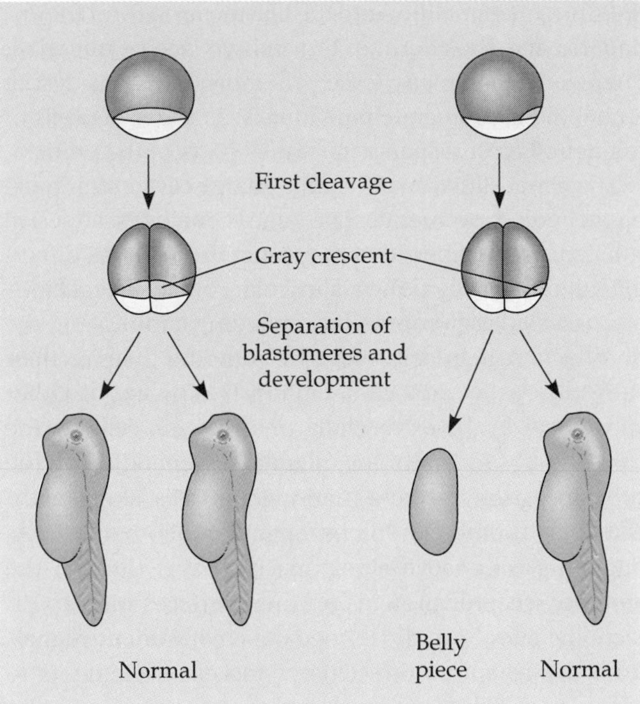

FIGURE 43.18
Cytoplasmic determinants. The first cleavage division of an amphibian zygote normally divides the gray crescent between the two blastomeres, which retain their totipotency after the division (left). However, in experiments where pressure is used to divert the cleavage plane from the crescent (right), only the blastomere obtaining material from the crescent develops normally.

equal developmental potential. In amphibians, for instance, blastomeres experimentally separated at the two-cell stage develop into normal tadpoles. The cells are said to be **totipotent,** meaning they retain the zygote's potential to form all parts of the animal. However, if an amphibian zygote is experimentally manipulated so that the first cleavage plane misses the gray crescent instead of bisecting it, only the blastomere that gets the gray crescent retains the capacity to develop into a normal tadpole (FIGURE 43.18). Thus, a zygote's characteristic pattern of cleavage, along with its distribution of cytoplasmic determinants, affects the destiny of cells in the embryo.

In many species, only the zygote is totipotent. The first cleavage plane divides cytoplasmic determinants in such a way that each blastomere will give rise only to specific parts of the embryo. In sharp contrast, the cells of mammalian embryos remain totipotent until they become arranged into the trophoblast and inner cell mass of the blastocyst. Probably because mammalian eggs lack polarity, the potential destiny of early blastomeres is not narrowed by cytoplasmic determinants. Indeed, up to the eight-cell stage, the blastomeres of a mouse embryo all look alike, and each can form a complete embryo if isolated.

However different the potency of cells of early embryos, the progressive restriction of potency is a general feature of development in all animals. In some species, the cells of early gastrulas retain the capacity to give rise to more than one kind of cell, though they have lost their totipotency. If left alone, the dorsal ectoderm of an early amphibian gastrula will develop into a neural plate above the notochord. If the dorsal ectoderm is experimentally replaced with ectoderm from some other location, the transplanted tissue will form a neural plate. If the same experiment is performed on a late-stage gastrula, however, the transplanted ectoderm will not respond to its new location and will not form a neural plate. In general, the fate of the tissue cells of late gastrulas are fixed. No matter how they are manipulated experimentally, they will give rise to only one type of cell.

Determination is the progressive restriction of a cell's developmental potential. A cell is said to be determined when its developmental fate cannot be reversed by moving it to a different location in the embryo. A determined cell's memory of its developmental commitment is passed on to daughter cells, and they "remember" what they are supposed to become, even if they are moved to a different environment. Determination involves control of the genome by the cytoplasmic environment of the cell. By partitioning the heterogeneous cytoplasm of an egg, cleavage exposes the nuclei of cells to cytoplasmic determinants that may affect which genes will be expressed later, when the cell begins to differentiate.

Morphogenetic Movements

Morphogenetic movements are the changes in cell shape and the cell migrations involved in cleavage, gastrulation, and organogenesis. Changes in the shape of a cell usually involve reorganization of the cytoskeleton. Consider, for example, how the cells of the neural plate become distorted in shape during formation of the neural tube (FIGURE 43.19). First, microtubules lengthen the cells of the plate in the dorsal-ventral axis of the embryo. Then, ordered arrays of microfilaments contract the apical ends of the cells, giving them a wedge shape that deforms the epithelium inward. Similar shape changes are observed for other invaginations (inpocketings) and evaginations (outpocketings) of the tissue layers throughout development.

Amoeboid movement based on extension and contraction is another fundamental process in development. In the gastrulation of some embryos, invagination is initiated by a wedging of cells on the surface of the blastula, but the penetration of cells deeper into the embryo involves the extension of filopodia by cells on the leading edge of the migrating tissue. These amoeboid cells drag

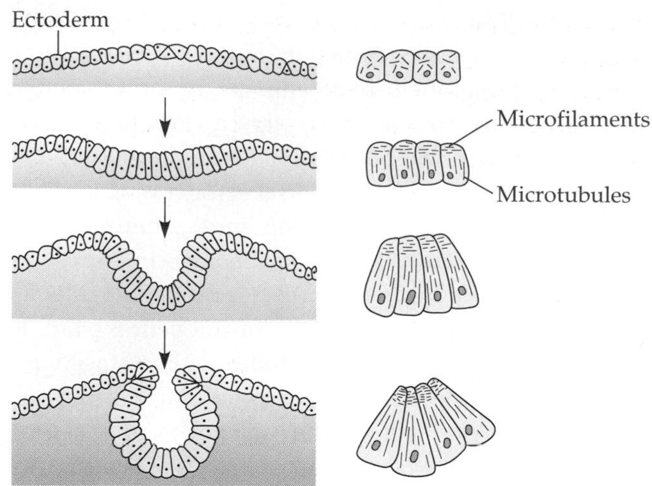

FIGURE 43.19
Change in cellular shape during morphogenesis.
Reorganization of the cytoskeleton is associated with morphogenetic alterations in embryonic tissues. As indicated here, in the formation of the neural tube of vertebrates, microtubules elongate the cells of the neural plate. Microfilaments at the apexes of the cells then contract, deforming the cells into wedge shapes.

the sheet of epithelium into the blastocoel to form the endoderm and mesoderm of the embryo (see FIGURE 43.9). There are also many cases in morphogenesis where amoeboid cells migrate individually, as when the cells of the neural crest disperse to various parts of the embryo.

The extracellular matrix helps guide cells in their morphogenetic movements. The matrix includes adhesive substances and fibers that may function as tracks, directing migrating cells along a particular route. Several kinds of extracellular glycoproteins, including laminin and the fibronectins, help cells move by adhering them to their substratum, the extracellular matrix (FIGURE 43.20). Other substances in the extracellular matrix keep cells on the correct paths by inhibiting migration. Depending on the substances they secrete, nonmigrant cells along pathways may promote or inhibit migration of other cells. As migrating cells move along specific paths through the embryo, receptor proteins on their surfaces pick up directional cues from the immediate environment. Signals from the receptors direct the cytoskeletal elements to propel the cell in the proper direction.

FIGURE 43.20
The extracellular matrix and cell migration. (**a**) Cells from the neural crest migrate along a strip of fibronectin fibrils placed on an artificial substratum (LM). (**b**) Two different fluorescent dyes demonstrate the close relationship between the orientation of fibronectin fibrils in the extracellular matrix (left) and that of contractile microfilaments (right) of the cytoskeleton within two migrating cells (LMs). Notice that the orientations of the intracellular microfilaments and extracellular fibrils correspond. The extracellular fibrils are ordered by the orientation of the cytoskeleton in the cells that secrete the extracellular materials. This is one way that a group of cells can influence the path along which another group of cells migrates during the development of tissues and organs.

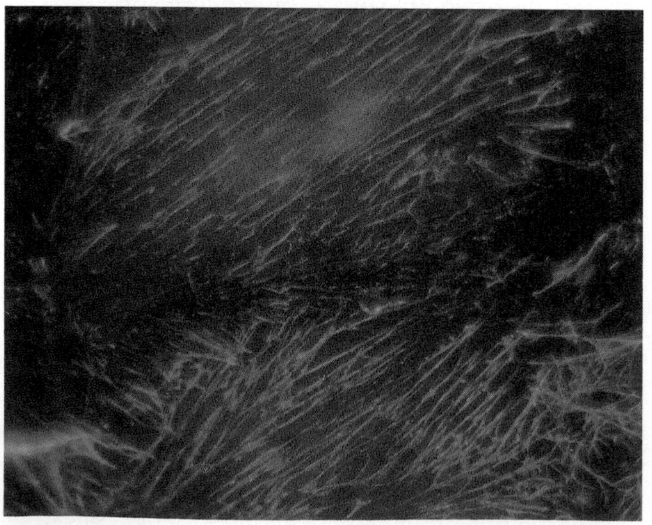

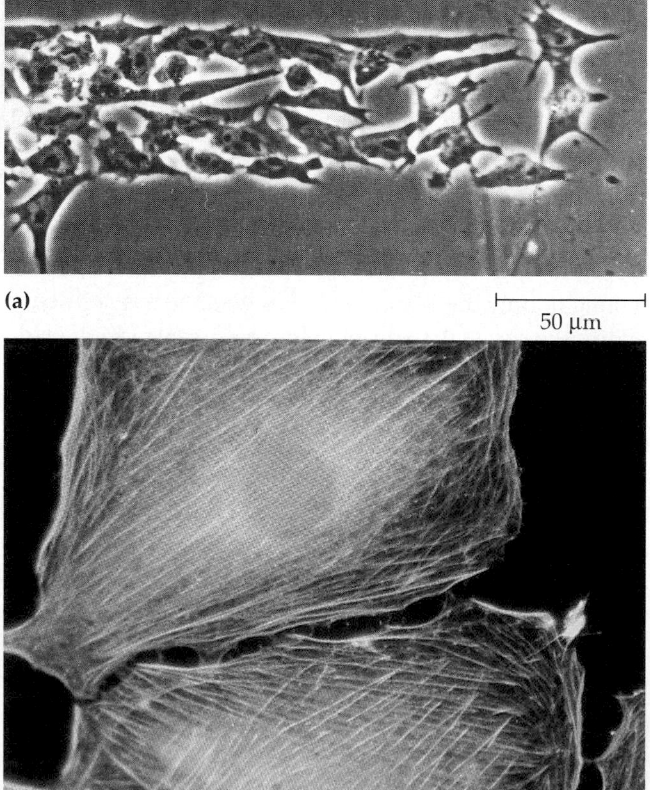

The glycoproteins that adhere migrating cells to the substratum also play a role in holding cells together when migrating cells reach their destinations, and tissues and organs take shape. Also contributing to stable tissue structure and cell migrations are glycoproteins called **cell adhesion molecules (CAMs)** found on the surfaces of cells. CAMs vary in either amount or chemical identity from one type of cell to another, and these differences help regulate morphogenetic movements and tissue building.

Induction

By placing cells in embryonic neighborhoods of differing chemical and physical environments, morphogenetic movements play an important role in cellular determination and differentiation. Interactions among neighboring cells and cell layers are crucial during and after gastrulation in the origin of most organs. The ability of one group of cells to influence the development of an adjacent group of cells is called **induction.** Induction can occur by actual physical contact between two groups of cells or by chemical signals. In either case, the effect is to switch on a set of genes that make the receiving cells differentiate into a specific tissue.

Induction by the rudimentary notochord causes the dorsal ectoderm of the gastrula to thicken into a neural plate. If chordamesoderm, the dorsal mesoderm that forms the notochord, is transplanted from its normal position to a position beneath the ectoderm in some other part of the embryo, a neural plate will form and give rise to a neural tube in an abnormal location. In a series of transplant experiments in the 1920s, German zoologist Hans Spemann and his student Hilde Mangold discovered that the dorsal lip of the blastopore is responsible for setting up the interaction between chordamesoderm and the overlying ectoderm. For example, transplanting a small piece of the dorsal lip to some other part of the surface of an early amphibian gastrula results in dual notochords, which induce the formation of two neural tubes. Spemann referred to the dorsal lip of the blastopore as the *primary organizer* of the embryo because of its role in the early stages of organogenesis.

Developmental biologists have yet to determine the exact nature of Spemann's organizer. However, they have found that certain peptide growth factors seem to induce the cells that move inward through the dorsal lip to form chordamesoderm. The precise action of these extracellular signals remains unclear, and the inducing substance that subsequently causes overlying ectoderm to form the neural plate has yet to be discovered.

The induction of dorsal ectoderm for development into the neural tube is one of many cell-cell interactions that transform the three germ layers into organ systems. Many inductions, such as those responsible for the development of the vertebrate eye, seem to involve a sequence of inductive steps that progressively determine the fate of cells (FIGURE 43.21). In frogs, determination of ectodermal cells that will become the eye lens seems to

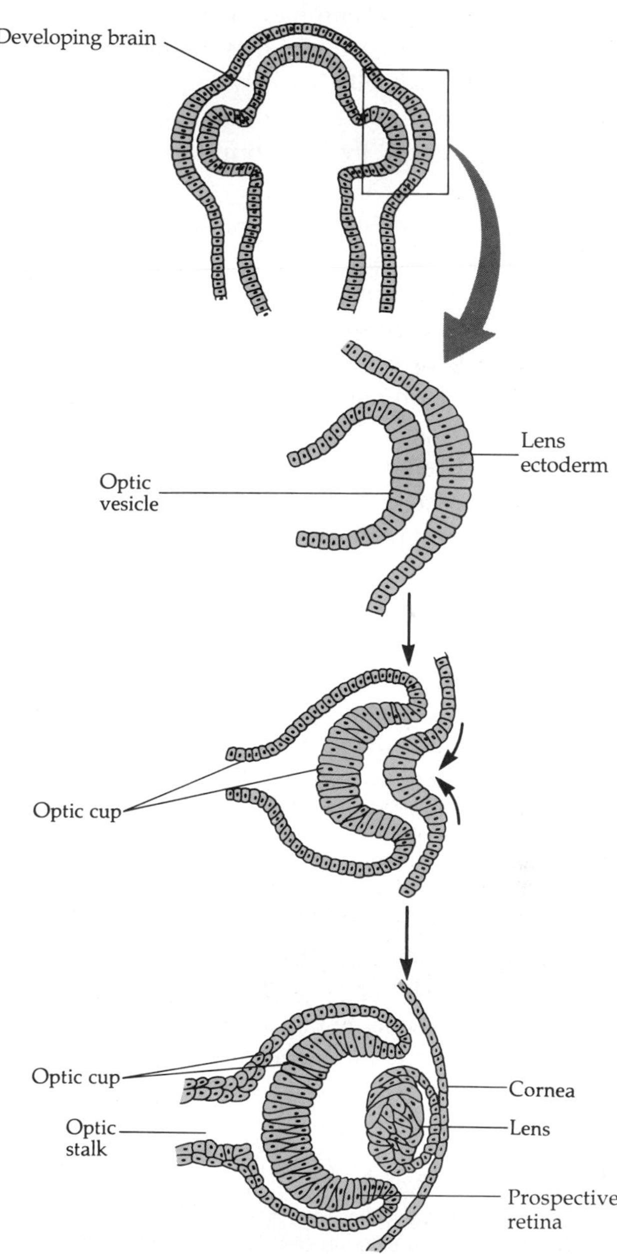

FIGURE 43.21

Induction during eye development. This series of drawings depicts stages in the development of the eye of a frog embryo. The eye forms from two different layers of cells and starts taking shape as the brain develops from the neural tube. An outgrowth of the rudimentary brain, the optic vesicle develops into the optic cup (which will form the retina of the eye) and the optic stalk (which will form the optic nerve). The lens of the eye forms from an invagination of ectodermal cells on the surface of the embryo. By the time the optic vesicle forms, determination of the lens ectoderm is well underway, the lens-forming cells having received a series of inductive signals from other cells in all three germ layers. Signals from the optic cup induce the final determination of lens-forming ectoderm into lens tissue. In turn, the lens cells induce development of the cornea.

begin after the early stages of gastrulation, with the presumptive lens cells first gaining the competence to respond to inductive signals. Late in gastrulation, the presumptive lens cells receive inductive signals from the ectodermal cells that are destined to become the neural plate (probably those that will form the retina of the eye). Additional inductive signals probably come from endodermal cells of the late gastrula, and as the neural plate forms, from mesodermal cells. Inductive signals from the optic cup, an outgrowth of the developing brain, complete the determination of the lens-forming cells.

Differentiation

As the tissues and organs of an embryo take shape, their cells begin to specialize in structure and function during the process known as cellular differentiation. Cellular differentiation is the product of a cell's developmental history extending back to early cleavage; however, a cell is generally determined before it shows signs of differentiation. At the microscopic level, the first signs of specialization are alterations in cellular structure. At the molecular level, differentiation is heralded by the appearance of **tissue-specific proteins,** proteins found only in a certain type of cell.

During differentiation, cells become specialists at making certain proteins. Developing lens cells in vertebrates, for example, synthesize large quantities of crystallins, proteins that aggregate to form transparent fibers that give the lens the ability to transmit and focus light. Because no other vertebrate cell type makes crystallins, these proteins can be used to follow the progress of differentiation by lens cells. The cue for the immature lens cell to turn on its crystallin genes is induction by the still immature retina (see FIGURE 43.21). In response to chemical signals from the rudimentary retinal cells, specific mRNA molecules coding for crystallin are produced by transcription and accumulate in the immature lens cell's cytoplasm. Then synthesis of crystallin begins, and the lens cells devote 80% of their capacity for protein synthesis to making this one type of protein. The cells elongate and flatten as they accommodate the crystallin fibers.

If a lens cell makes crystallin but a blood cell does not, we must conclude that the lens cell expresses a gene that is not expressed in the blood cell. We could account for such differences if cells lost nonessential genes as they differentiated, but most evidence supports the conclusion that nearly all cells of an organism have **genomic equivalence**—that is, they all have the same genes. (There are a few exceptions, as discussed in Chapter 18.) What happens to these genes as a cell begins to differentiate? We can shed some light on this question by asking whether genes are irreversibly inactivated during differentiation.

If genes are permanently turned off during differentiation, we would expect that a normal embryo would not develop if the nucleus of a normal egg or zygote were replaced with a nucleus of a differentiated cell. The pioneering experiments in nuclear transplantation were carried out by American embryologists Robert Briggs and Thomas King during the 1950s and were later extended by British embryologist John Gurdon. These investigators removed or destroyed the nuclei of frog egg cells, then transplanted nuclei from embryonic and tadpole cells of the same species into the enucleated eggs (FIGURE 43.22). The ability of the transplanted nuclei to support normal development turned out to be inversely related to the age of the donor embryos. If the nuclei came from the relatively undifferentiated cells of an early embryo, most of the recipient eggs developed into tadpoles. But with nuclei from the differentiated intestinal cells of a tadpole, fewer than 2% of the eggs developed into normal tadpoles, and most of the embryos failed to make it through even the earliest stages of embryonic development.

Developmental biologists still debate the meaning of

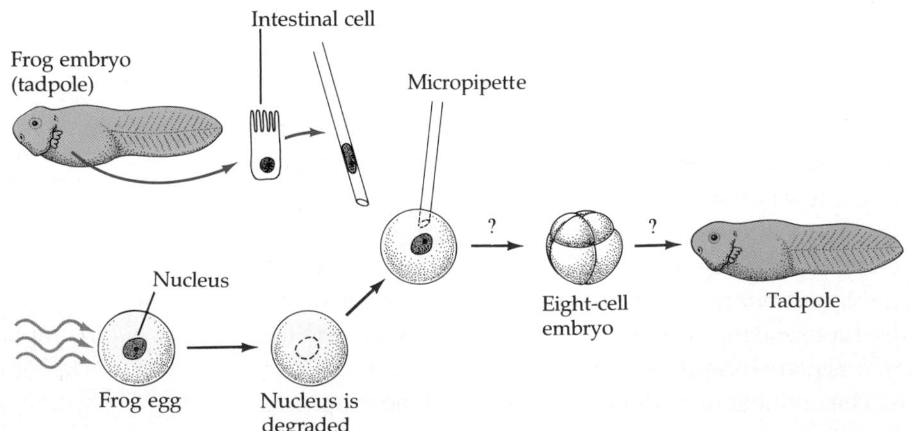

FIGURE 43.22
Nuclear transplantation. After the frog egg nucleus is destroyed by ultraviolet (UV) radiation, a nucleus from a more advanced developmental stage is inserted into the egg to test whether nuclei change irreversibly as cells begin to differentiate. The earlier the developmental stage from which the nucleus comes, the more likely it will support development. Nuclei from very early stages frequently prove to be totipotent, whereas nuclei from late stages (such as a tadpole) rarely are.

these results, but most agree on two conclusions. First, nuclei *do* change in some way as they prepare for differentiation. (Recall from Chapter 18 that genomes are rearranged in various ways.) Second, this change is not always irreversible, implying that the nucleus of a differentiated cell has all the genes required for making all other parts of the organism. The cells of the body differ in structure and function not because they contain different genes, but because they express different portions of a common genome. Possible mechanisms for this tissue-specific gene expression were discussed in Chapter 18.

Pattern Formation

There is more to building a specialized organ than the determination and differentiation of its cells. A human arm and a leg have the same mixture of tissues—muscle, connective tissue, cartilage, and skin—but with different spatial arrangements. And arms and legs normally develop in certain locations relative to each other and other parts of the body. The development of an animal's spatial organization, with organs and tissues all in their characteristic places along the three dimensions of the animal, is called **pattern formation.** The set of molecular cues that control pattern formation, collectively called **positional information,** indicate a cell's location relative to neighboring cells and determine how the cell and its offspring will respond to future molecular signals.

Several developmental biologists have been investigating the role of positional information in pattern formation by investigating the development of chick limbs (wings and legs). Vertebrate limbs begin as undifferentiated limb buds that give rise to their unique forms (FIGURE 43.23). Each bone and muscle and every other limb component has a precise location and orientation relative to three axes: the proximal-distal axis, from the base of the limb to the tip of the digits; the anterior-posterior axis, from the front edge of the limb to the rear edge; and the dorsal-ventral axis, from the top of the limb to the bottom. Embryonic cells within a limb bud must receive positional information indicating location along these three axes.

A region of the chick's limb consisting of mesoderm located just beneath the *apical epidermal ridge,* at the tip of the wing bud, is thought to assign position to the embryonic tissue along the proximal-distal axis as the limb bud grows. Positional information along the dorsal-ventral axis probably reflects the relative distance of developing tissue from the dorsal and ventral ectoderm of the limb bud. In experiments where the ectoderm of a bud is separated from the mesoderm, then replaced with its orientation rotated by 180°, the limb develops with its dorsal-ventral orientation reversed. (This is equivalent to reversing the palm and back of your hand.)

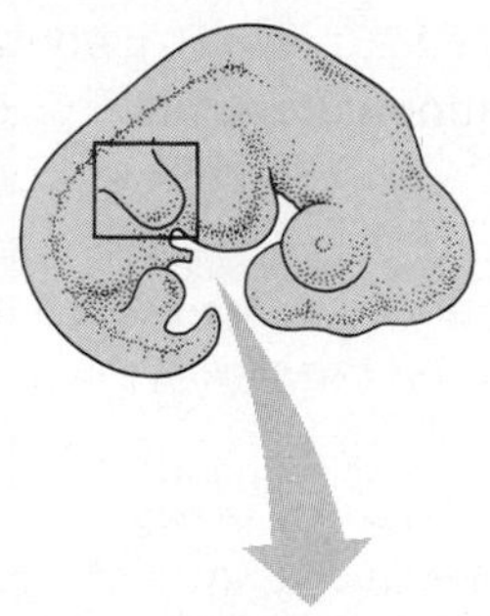

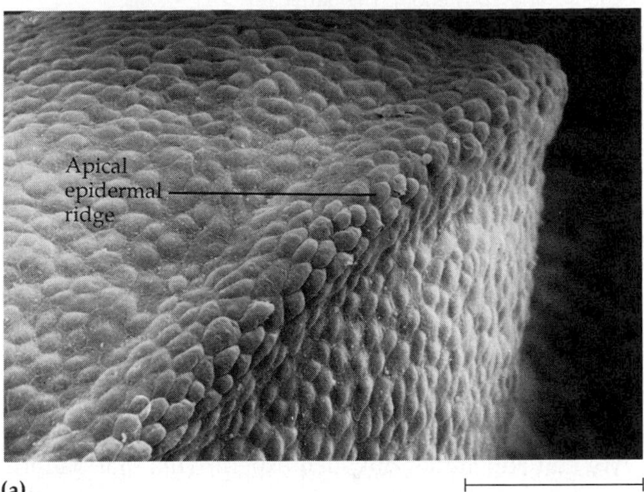

(a)

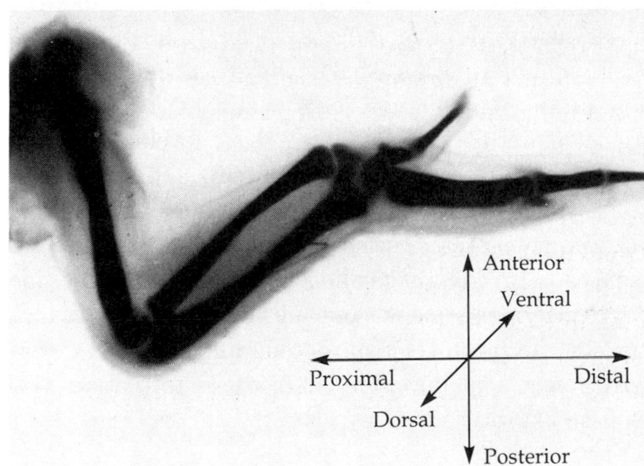

(b)

FIGURE 43.23
Pattern formation during vertebrate limb development. (**a**) Vertebrate limbs develop from rudiments called limb buds (SEM). (**b**) As the bud develops into a limb, such as this wing of a chick embryo, a specific pattern of tissues emerges. This requires each embryonic cell to receive some kind of positional information indicating location along the three axes of the limb.

Another region of the chick's limb bud called the *zone of polarizing activity (ZPA),* located where the posterior side of the bud is attached to the body, apparently assigns position along the anterior-posterior axis to developing tissues. Cells nearest the ZPA give rise to posterior

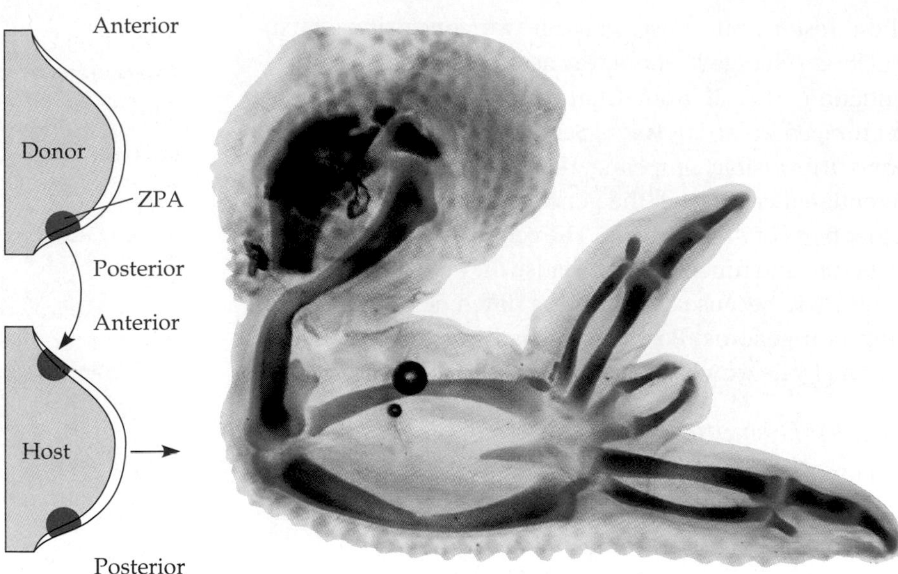

FIGURE 43.24
Altering positional information with grafting experiments. A region called the zone of polarizing activity (ZPA) is the reference point for indicating the position of cells along the anterior-posterior axis of a vertebrate limb bud. In this experiment, a second ZPA is added to the anterior margin of a limb bud by transplanting the tissue from a donor. Cells near the grafted ZPA, like cells near the host bud's own ZPA, apparently receive positional information indicating "posterior." The pattern that emerges in the developing limb is a mirror image, with an arrangement of digits equivalent to two human hands joined together at the thumbs.

structures, such as the digit homologous to our little finger, and cells farthest from the ZPA form anterior structures, such as the avian equivalent of our forefinger and thumb. One of the experiments supporting this hypothesis is shown in FIGURE 43.24.

We can conclude from such experiments that pattern formation requires cells to receive and interpret environmental cues that vary from one location to another. Variation in these cues along three axes would provide the positional information a cell needs to determine its location in the three-dimensional realm of a developing organ. In the case of vertebrate limb development, researchers are closing in on certain polypeptides as positional signals. Regional variation in the production of these polypeptides is itself an example of differential gene expression in different locations of the embryo. Thus, positional signposts such as the apical epidermal ridge and the ZPA are products of positional information at even earlier stages of development. Pattern formation is a sequential process.

Pattern formation in *Drosophila* is controlled by a hierarchy of gene activations

Many developmental biologists have turned to the fruit fly as a research model for the genetic basis of pattern formation in the steps from egg to animal. The importance of the research was recognized in 1995 when the Nobel Prize for physiology or medicine was awarded to Edward Lewis, Eric Weischaus, and Christiane Nuesslein-Volhard for their pioneering research on the developmental genetics of *Drosophila*. In this section, we will focus on experimental manipulation of *Drosophila* as a way of demonstrating how pattern formation works.

Pattern formation has its genetic basis in genes that control the overall body plan of an animal. Fruit flies and other arthropods have a modular construction, an ordered series of segments. Although the embryonic segments look alike, they become anatomically more distinct as development progresses. For example, the first thoracic segment of an adult *Drosophila* bears a pair of legs, while the second thoracic segment has a pair of legs plus a pair of wings, and the third thoracic segment bears a pair of legs plus a pair of balancing organs called halteres.

Drosophila's modular body plan is not preformed in the egg or early embryo, but emerges stepwise (epigenesis). Positional information operating progressively on finer and finer scales first sets the gross form of the animal (head end versus tail end, for example), then cues the origin of a specific number of segments with correct orientations relative to the axes of the embryo, and finally triggers the development of each segment's specific anatomical features. This positional information consists mainly of regional differences in the type or concentrations of regulatory proteins that control gene expression. Positional information leading to this hierarchy of gene activations begins even before the egg has been fertilized.

Egg-Polarity Genes

Female fruit flies that carry certain mutant alleles produce offspring with abnormal polarity, such as two heads, one on each end of the embryo. These genes are examples of **maternal-effect genes,** meaning that the phenotypes they produce in offspring are exclusively products of the mother's contribution to the genomes of embryos, with the father's genotype having no effect on these traits. Specifically, the maternal-effect genes that set the polar axes of the future embryo are called **egg-polarity genes.** One group of these genes determines the

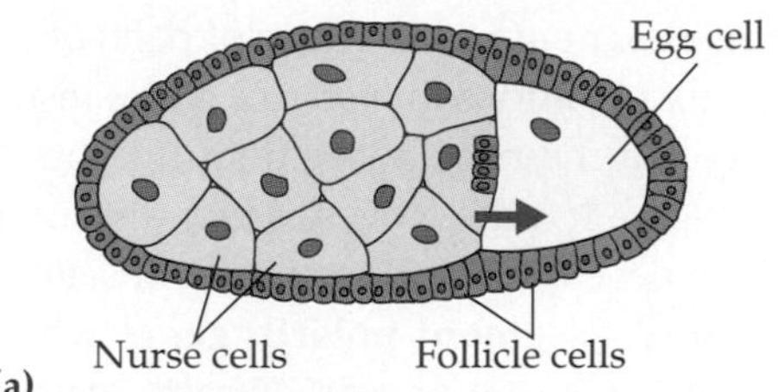

(a)

(b)

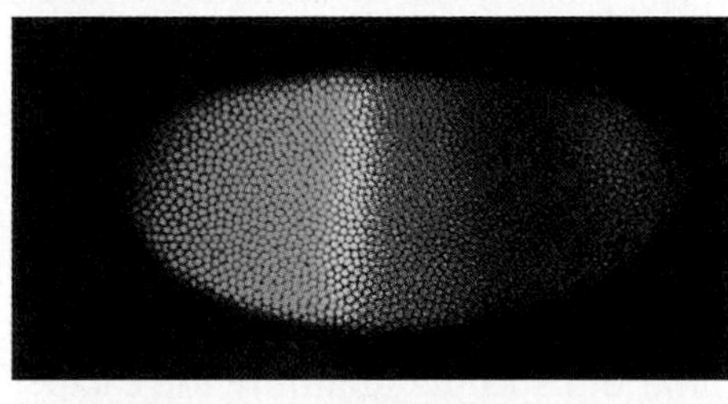

(c)

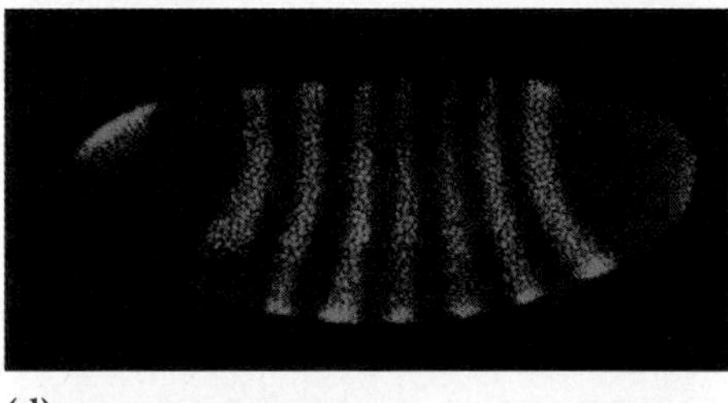

(d)

(e)

FIGURE 43.25
The genetic basis of pattern formation in *Drosophila.*
A hierarchy of gene activations beginning in the maternal genome determines the body plan of fruit flies. Genes in the follicle cells and nurse cells in the female fly's ovary nourish and support the unfertilized egg and produce mRNAs and proteins that regulate its early development. (**a**) Egg-polarity genes in some of the follicle cells and nurse cells transcribe mRNA whose protein products set up the body axes in the new fly. The mRNA (magenta arrow) from one of the egg-polarity genes, named *bicoid,* passes into the egg cell and remains near its entry point, thereby fixing the future anterior end of the animal. (**b**) The *bicoid* mRNA is translated when the egg is laid after fertilization. The resulting bicoid protein diffuses from the anterior end of the zygote, setting up a concentration gradient. This computer-generated image shows the bicoid-protein gradient in a fruit fly zygote (purple = highest concentration; blue = lowest). The site where the protein is most concentrated becomes the head end. Bicoid protein is only one of several morphogens that give the embryo its first coordinates. The bluish area in this computer image contains morphogens that initiate the development of posterior structures. (**c–e**) Soon after fertilization, bicoid protein and other maternal gene products initiate a hierarchy of activity by segmentation genes in the embryo itself. These photographs of developing fly embryos illustrate successive bands of regulatory proteins produced by segmentation genes. These DNA-binding proteins direct the division of the body into segments, characteristic of flies and other arthropods. The colors result from fluorescent dyes on antibodies bound to the protein products of the segmentation genes. (**c**) Gap genes, the first set of segmentation genes turned on, produce these broad bands of gene-regulating proteins, prescribing a coarse subdivision of the embryo. (**d**) Next, localized products of gap genes turn on a second set of segmentation genes, the pair-rule genes, whose protein products produce these narrower bands. The pair-rule genes prescribe further subdivision of the embryo. (**e**) The final directions for subdividing the fly embryo into segments are in place when the pair-rule proteins affect localized expression of various segment-polarity genes, the third set of segmentation genes. Products of these segment-polarity genes account for the narrow bands visible in this computer-enhanced micrograph of a *Drosophila* embryo. Each of the compartments between these protein bands represents an actual body segment of the embryo.

anterior-posterior axis of the embryo, while a second group sets the dorsal-ventral axis.

Egg-polarity genes define the axes of a future embryo while the unfertilized egg is still in the ovary, before it has been laid. Ovarian cells called follicle cells and nurse cells surround each developing egg (FIGURE 43.25a). It is the activity of egg-polarity genes in these support cells that stamps the egg with anterior-posterior and dorsal-ventral polarity. As an example, we will examine one egg-polarity gene called *bicoid,* which is active in nurse cells at the end of the egg that will later give rise to the anterior end of the embryo. (*Bicoid* means "two tails," a reference to the absence of a head end in offspring of a mother that is homozygous for a recessive mutant allele; the wild-type dominant allele specifies the anterior end of the egg.) When this gene is transcribed, the mRNA for the bicoid protein is secreted by the nurse cells into the egg. The mRNA is held at that end of the egg, probably by its attachment to the cytoskeleton. After the egg is fertilized and the zygote is laid, the mRNA is translated. The product, the bicoid protein, diffuses slowly through the cytoplasm while the embryo is still a syncytium, its multiple nuclei sharing an undivided cytoplasm (see FIGURE 43.8).

When cytokinesis forms the blastoderm at the surface of the embryo, the concentration of bicoid protein is highest in cells at the pole of the embryo that will become the anterior end and lowest in cells at what will become the posterior end. Bicoid protein is an example of a **morphogen,** a substance that provides positional information in the form of a gradation in its concentration along an embryonic axis (FIGURE 43.25b). The *bicoid* gene, of course, is only part of the story of how the polarity of the embryo is determined; differential activity of egg-polarity genes in follicle cells all around the unfertilized egg also mark the posterior pole and the dorsal-ventral axis.

Bicoid protein binds to DNA and regulates the expression of certain genes in the cells of the blastoderm. In the hierarchy of positional information, some of the genes activated by bicoid protein and other egg-polarity morphogens code for proteins that set the basic pattern of segmentation in the *Drosophila* embryo.

Segmentation Genes

The gradiants of bicoid protein and other morphogens set up by spatial differences in the activity of egg-polarity genes in turn cause regional differences in the expression of **segmentation genes,** which regulate the origin of segmentation once the embryo's anterior-posterior and dorsal-ventral axes are defined.

Sequential activation of three sets of segmentation genes provides the positional information for increasingly finer details of the animal's modular body plan. First, products of the **gap genes** map out the basic subdivisions along the anterior-posterior axis of the embryo (FIGURE 43.25c). Mutations in these genes cause "gaps" in the animal's segementation. For example, one gap mutation results in an embryo lacking eight segments, from the first thoracic to the fifth abdominal segment. **Pair-rule genes** are the next segmentation genes to act. They define the modular pattern in terms of pairs of segments (FIGURE 43.25d). Mutations in pair-rule genes result in embryos having half the normal segment number because every other segment (odd or even, depending on the mutation) fails to develop. The third group of segmentation genes to act are the **segment-polarity genes,** which determine the anterior-posterior axis of each segment (FIGURE 43.25e). Embryos with mutations in segment-polarity genes have the normal number of segments, but a part of each segment is replaced by a mirror-image repeat of some other part of the segment.

The products of segmentation genes, like those of egg-polarity genes, are generally DNA-binding proteins, transcription factors that activate the next set of genes in the hierarchical scheme of pattern formation. Thus, the products of egg-polarity genes regulate the localized expression of gap genes, which control the regional expression of pair-rule genes, which in turn activate specific segment-polarity genes in different parts of each segment. The boundaries and axes of the segments are now set. In the hierarchy of gene activations responsible for pattern formation, the next genes to be expressed actually determine the specific anatomy for each segment along the embryo.

Homeotic Genes

The modular anatomy of a normal fly requires that structures such as antennae, legs, and wings develop on the appropriate segments. This anatomical identity of the segments is set by regulatory genes called **homeotic genes.** Once the segmentation genes have staked out the basic segmental body plan of the fly, homeotic genes specify the type of appendages and other structures that are specific for each segment. Mutations in various homeotic genes produce flies with such bizarre traits as an extra set of wings or legs growing from the head in place of antennae (FIGURE 43.26). Thus, homeotic muta-

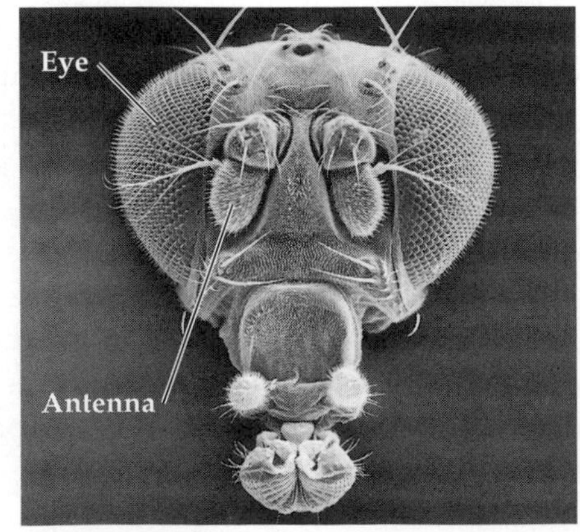

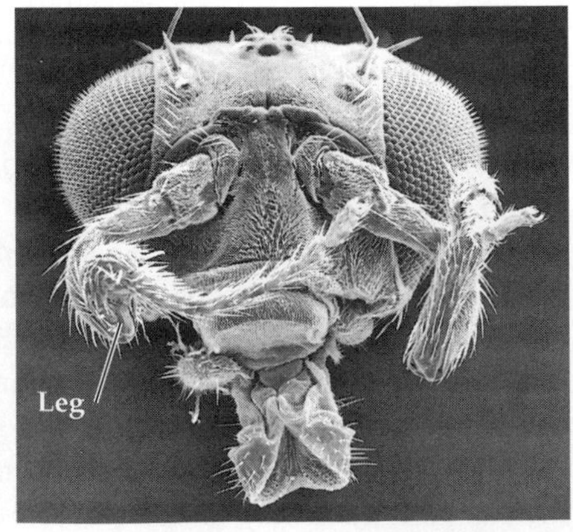

FIGURE 43.26
Homeotic mutations and abnormal pattern formation in *Drosophila.* Homeotic mutations cause a positional replacement of organs from one location in an animal to another. These micrographs contrast the heads of two fruit flies (SEMs). Where small antennae are located in the normal fruit fly (left), one homeotic mutant has legs (right).

tions replace structures characteristic of one part of the animal with structures normally found at some other location. Particular homeotic genes are activated in each segment as a result of the positional imprints left by the earlier activations of segmentation genes.

In the continuing saga of gene activations, the products of homeotic genes are regulatory proteins that bind to DNA. These proteins affect the selective gene expression responsible for continuing development of the embryo's segmental body plan. For example, the protein product of a homeotic gene in a particular thoracic segment of the embryo may select the genes that trigger the actual development of legs. In contrast, a homeotic gene in a certain head segment specifies "antennae go here." A mutation in this latter gene may result in a protein that labels the segment as "thoracic" instead of "head," causing legs to develop in place of antennae.

All the research on pattern formation in fruit flies is helping developmental biologists unravel the genetic basis for epigenesis, the gradual development of animal form. One of the surprising revelations of this research is how conservative evolution has been with the basic genetic mechanisms regulating pattern formation.

Comparisons of genes that control development reveal homology in animals as diverse as flies and mammals

The homeotic genes of *Drosophila* include a 180-nucleotide-long region that is common to all homeotic genes in the fly. This specific DNA sequence within a homeotic gene is called the **homeobox.** The same or a very similar homeobox sequence has been discovered in numerous other eukaryotic organisms, including other insects, the cnidarian *Hydra,* nematodes, mollusks, fish, frogs, birds, and mammals, including humans. Homeotic genes, which prescribe the placement of organs along the animal body, are a particularly striking example of homology in genes that control development (FIGURE 43.27).

In addition to their presence in homeotic genes, homeobox nucleotide sequences are also found within a variety of other genes, most of which are associated in some way with development. When such a gene is translated, the homeobox region specifies a 60-amino-acid-long *homeodomain* that is part of the protein product of the gene. These homeodomains are the portions of the regulatory proteins that bind to DNA in the genome. Other domains of these regulatory proteins are much more variable than the homeodomains and give each regulatory protein its specific function of binding to targets in the genome and activating certain genes. All the

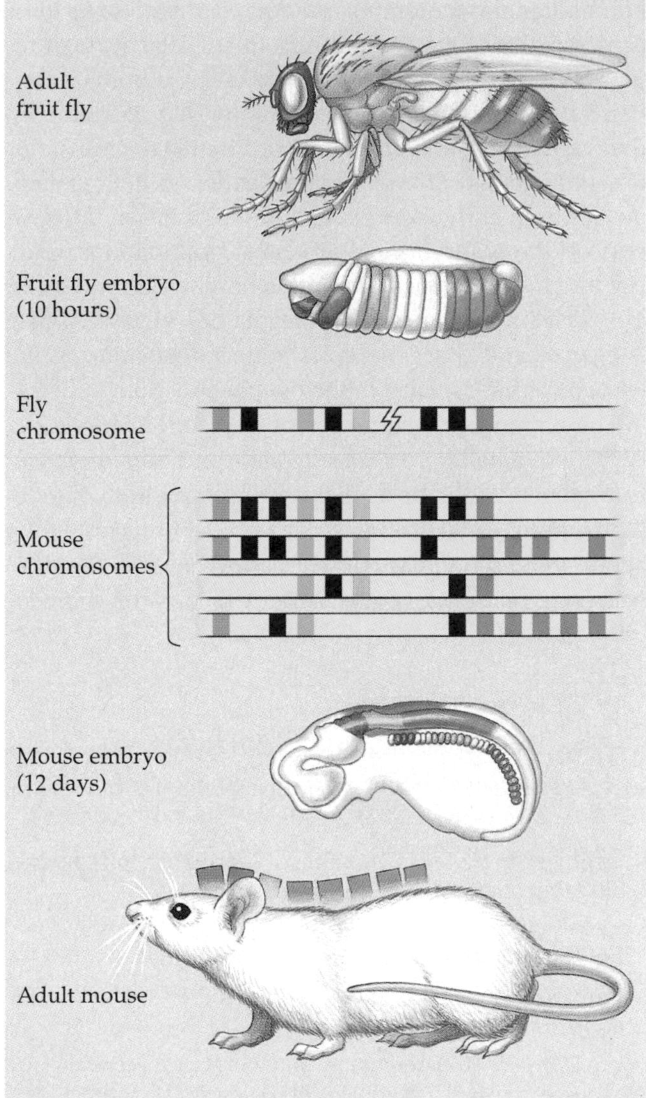

FIGURE 43.27
Homologous genes that affect pattern formation in a fruit fly and a mouse. Genes that control development of the anterior and posterior parts of the body occur in the same linear sequence on chromosomes in *Drosophila* and mice. Each small box represents a homeobox-containing gene. All of these regulatory genes are found on one chromosome in fruit flies. The mouse and other mammals have the same or similar sets of genes on four chromosomes. Genes represented by purple and green boxes code for proteins that regulate pattern formation of anterior body parts; those represented by gray and orange code for proteins regulating the formation of middle and posterior body parts. The color code indicates the parts of the body of a fruit fly embryo and mouse embryo in which these genes are expressed and the adult body regions that result. Colored boxes represent genes with homeoboxes that are essentially identical in flies and mice. The black boxes represent genes with homeoboxes that are similar but not identical between the two animals.

homeodomain-containing proteins examined so far have turned out to be transcription factors that activate or repress the transcription of other genes by binding to enhancer regions of DNA (see Chapter 18). Presumably, they regulate development by coordinating the transcription of batteries of developmental genes, switching them on or off. In early *Drosophila* embryos, cells in different parts of the embryo show different combinations of active and inactive genes containing homeobox sequences. This selective expression of regulatory genes, varying over time and space, controls pattern formation in the embryo, as you learned in the previous section.

The amino acid sequences of homeodomains, while varying somewhat from gene to gene and from organism to organism, have been highly conserved in evolution. Many, perhaps all, animals have homeobox-containing genes organized in virtually the same way as in fruit flies (see FIGURE 43.27). Furthermore, eukaryotic homeodomains in general show some similarity to the DNA-binding domains of prokaryotic regulatory proteins. Researchers are reaching the conclusion that homeoboxes are all derived from a nucleotide sequence that arose very early in the history of life, and that they have continued to function in the regulation of gene expression and development throughout the diversity of life.

* * *

Through a variety of experiments, developmental biologists are begining to learn how the one-dimensional information encoded in the nucleotide sequence of a zygote's DNA is translated into the three-dimensional form of an animal. There are enough unanswered questions, however, to entertain many future generations of scientists.

REVIEW OF KEY CONCEPTS (with page numbers and key figures)

- From egg to organism, an animal's form develops gradually: *the concept of epigenesis* (pp. 963–964)
  - An embryo is not preformed in an egg; it develops by epigenesis, the gradual, gene-directed acquisition of form.
- Embryonic development involves cell division, differentiation, and morphogenesis (p. 964)
  - Cell division, differentiation, and morphogenesis transform a unicellular zygote into a multicellular organism.
- Fertilization activates the egg and brings together the nuclei of sperm and egg (pp. 965–968, FIGURE 43.2)
  - Fertilization both reinstates diploidy and activates the egg to begin a chain of metabolic reactions that triggers the onset of embryonic development.
  - The acrosomal reaction, which occurs when the sperm meets the egg, releases hydrolytic enzymes that digest through material surrounding the egg.
  - Gamete fusion depolarizes the egg cell membrane and sets up a fast block to polyspermy.
  - Sperm-egg fusion also initiates the cortical reaction, involving a signal-transduction pathway in which calcium ions stimulate cortical granules to erect a fertilization membrane that functions as a slow block to polyspermy.
  - In mammalian fertilization, the cortical reaction hardens the zona pellucida as a slow block to polyspermy.
- Cleavage partitions the zygote into many smaller cells (pp. 968–971, FIGURE 43.7)
  - Fertilization is followed by cleavage, a period of rapid cell division without growth, which results in the production of a large number of cells called blastomeres.
  - Holoblastic cleavage, or division of the entire egg, occurs in species whose eggs have little or moderate amounts of yolk. Meroblastic cleavage, incomplete division of the egg, occurs in species with yolk-rich eggs.
  - Cleavage planes usually follow a specific pattern relative to the animal and vegetal poles of the zygote.
  - In many species, cleavage creates a multicellular ball called the blastula, which contains a fluid-filled cavity, the blastocoel.
  - Meroblastic cleavage in insects creates an early syncytial stage that transforms into a blastula comprising a single layer of cells surrounding a mass of yolk.
- Gastrulation rearranges the blastula to form a three-layered embryo with a primitive gut (pp. 971–973, FIGURE 43.10)
  - Gastrulation transforms the blastula into an embryo called the gastrula, which has a rudimentary digestive tube, the archenteron, and three embryonic germ layers: the ectoderm, endoderm, and mesoderm.
- Organogenesis forms the organs of the animal body from the three embryonic germ layers (pp. 973–974, FIGURE 43.11)
  - Early events in organogenesis in vertebrates include formation of the notochord by condensation of dorsal mesoderm, development of the neural tube from folds of the ectodermal neural plate, and formation of the coelom from splits in lateral mesoderm.

- Amniote embryos develop in a fluid-filled sac within a shell or uterus (pp. 975–978, FIGURES 43.12, 43.14)
  - Meroblastic cleavage in the yolk-rich, shelled eggs of birds and reptiles is restricted to a small disc of cytoplasm at the animal pole. A cap of cells called the blastodisc forms and begins gastrulation with the formation of the primitive streak. In addition to the embryo, the three germ layers give rise to the four extraembryonic membranes: the yolk sac, amnion, chorion, and allantois.
  - The eggs of placental mammals are small and store little food, exhibiting a holoblastic cleavage with no apparent polarity. Gastrulation and organogenesis, however, resemble the patterns of birds and reptiles. After fertilization and early cleavage in the oviduct, the blastocyst implants in the uterus. The trophoblast initiates formation of the fetal portion of the placenta, and the embryo proper develops from a single layer of cells, the epiblast, within the blastocyst. Extraembryonic membranes homologous to those of birds and reptiles function in intrauterine development.
- The developmental fate of cells depends on cytoplasmic environment, location, and cell-cell interactions (pp. 978–986, FIGURE 43.16)
  - Embryonic development depends on specific patterns of gene expression, affected by the heterogeneous organization of egg cytoplasm, the relocation of cells during morphogenesis, and interactions among neighboring cells.
  - In most animals except mammals, the polarity of the embryo is established in the unfertilized egg or during early cleavage. Embryos take shape along the axes of polarity.
  - Experimentally derived fate maps of polar embryos have shown that specific regions of the zygote or blastula develop into specific parts of the embryo.
  - The timely death of cells, called apoptosis, is a crucial aspect of development.
  - Progressive restriction of potency, a cell's potential to develop into all parts of an animal, characterizes embryonic development. Embryonic cells are determined when their developmental fate is not reversed by experimental manipulation.
  - Morphogenetic movements—the changes in cell shape and the cell migrations during cleavage, gastrulation, and organogenesis—affect cellular determination by exposing cells to varying chemical and physical environments. Glycoproteins of the extracellular matrix help migrating cells reach specific destinations.
  - Transplant experiments have demonstrated that certain groups of cells can influence the development of adjacent cells by induction.
  - Nuclear transplantation and other experiments reveal the genomic equivalence of most specialized cells. Therefore, differentiation reflects the differential expression of genes selectively activated in a common genome.
  - Cells are ordered into specific three-dimensional positions to produce specific organs and body parts by pattern formation, which involves cells receiving and interpreting positional information that varies with location.
- Pattern formation in *Drosophila* is controlled by a hierarchy of gene activations (pp. 986–989, FIGURE 43.26)
  - A sequence of activations of genes controlling body plan begins before the egg is fertilized and continues with the staking-out of increasingly finer detail in the embryo.
  - Homeotic genes control the types of structures that develop in specific locations of the body.
- Comparisons of genes that control development reveal homology in animals as diverse as flies and mammals (pp. 989–990, FIGURE 43.27)
  - A sequence of DNA nucleotides, called the homeobox, occurs in genes that control development in many different kinds of animals, an indication that these genes arose early in the history of life.

SELF-QUIZ

1. The cortical reaction of sea uchin eggs functions directly in the
   a. formation of a fertilization membrane
   b. production of a fast block to polyspermy
   c. release of hydrolytic enzymes from the sperm cell
   d. generation of a nervelike impulse by the egg cell
   e. fusion of egg and sperm nuclei

2. Which of the following is common to both avian and mammalian development?
   a. holoblastic cleavage
   b. primitive streak
   c. trophoblast
   d. yolk plug
   e. gray crescent

3. The archenteron develops into
   a. the mouth in protostomes
   b. the blastocoel
   c. the endoderm
   d. the lumen of the digestive tract
   e. the placenta

4. In a frog embryo, the blastocoel is
   a. completely obliterated by yolk platelets
   b. lined with endoderm during gastrulation
   c. located primarily in the animal hemisphere
   d. the cavity that becomes the coelom
   e. the cavity that later forms the archenteron

5. Amphibians, unlike reptiles, generally lay their eggs in water or moist places. This difference is related to the absence (amphibians) versus presence (reptiles) of
   a. extraembryonic membranes
   b. yolk
   c. cleavage
   d. gastrulation
   e. development of the brain from ectoderm

6. In an amphibian embryo, a band of cells called the neural crest
   a. rolls up to form the neural tube
   b. develops into the main sections of the brain
   c. produces amoeboid cells that migrate to form teeth, skull bones, and other structures in the embryo
   d. has been shown by experiments to be the organizer region of the developing embryo
   e. induces the formation of the notochord

7. The results of transplantation of frog nuclei into enucleated eggs allow for which conclusion?
   a. Frogs cannot be cloned.
   b. All differentiated cells actually express the same genes.
   c. The later the embryonic stage, the less likely that nuclei from cells at that stage can support the development of an egg into a tadpole.
   d. The nuclei of differentiated cells actually lack some genes found in other cell types.
   e. The differentiated state is unstable.

8. Development of the specific arrangement of appendages along the body of a fruit fly is an example of
   a. pattern formation
   b. induction
   c. cellular differentiation
   d. determination
   e. gastrulation

9. In the early development of an amphibian embryo, the organizer is the
   a. neural tube
   b. notochord
   c. archenteron roof
   d. dorsal lip of the blastopore
   e. dorsal ectoderm

10. Which statement about homeotic genes is *incorrect?*
   a. They code for DNA-binding proteins.
   b. They help control pattern formation.
   c. They include homeobox regions that are homologous in various animals.
   d. They are examples of maternal-effect genes in *Drosophila.*
   e. Mutations in these genes can cause body parts to develop in abnormal locations.

CHALLENGE QUESTIONS

1. Explain how in certain animals (such as *Drosophila*) the organization of the unfertilized egg's cytoplasm affects the form of the adult animal that develops after fertilization.

2. The "snout" of a frog tadpole bears a sucker. A salamander tadpole has a mustache-shaped structure called a balancer in the same area. If ectoderm from the side of a young salamander embryo is transplanted to the snout of a frog embryo, the frog tadpole later has a balancer. If ectoderm is transplanted from the side of a slightly older salamander embryo to the snout of a frog embryo, the frog tadpole ends up with a patch of salamander skin on its snout. Explain the results of this experiment in terms of the mechanisms of development.

3. Compare and contrast frog development with chick development in terms of cleavage, gastrulation, and organogenesis.

SCIENCE, TECHNOLOGY, AND SOCIETY

1. Nerve cells transplanted from aborted fetuses can relieve the symptoms of Parkinson's disease, a brain disorder. Fetal tissue transplants might also be used to treat epilepsy, diabetes, Alzheimer's disease, and spinal cord injuries. Why might tissues from a fetus be particularly useful for replacing diseased or damaged cells? Since 1989, there has been controversy over whether the U.S. government should allow fetal tissues from induced abortions to be used in transplant research. Opponents would allow only tissues from miscarriages to be used. Why would most researchers prefer to use tissues from aborted fetuses? Why do you think some people oppose this? What is your position on this issue, and why?

2. The abortion debate in the United States has provoked thought and created controversy over this question: When does human life begin? What are some of the possible answers? Do you think the continuing study of human embryological development can give a definitive answer and resolve this debate? Why or why not?

FURTHER READING

Barinaga, M. "Focusing on the *Eyeless* Gene." *Science,* March 24, 1995. One gene triggers eye development in animals as diverse as fruit flies and humans.

Barinaga, M. "Frontiers in Biology: Development. Looking into Development's Future." *Science,* October 28, 1994. A preview of major directions in developmental biology's next half-decade.

Caldwell, M. "How Does a Single Cell Become a Whole Body?" *Discover,* November 1992. One of the most important questions in modern biology.

Gilbert, S. F. *Developmental Biology,* 4th ed. Sunderland, MA: Sinauer Associates, 1994. A widely used upper-division text.

Hart, S. "The Drama of Cell Death." *BioScience,* July/August 1994. The importance of programmed cell death in development and immunity.

Kahn, P. "Zebrafish Hit the Big Time." *Science,* May 13, 1994. How developmental biology gained a new research model.

Kessler, D. S., and D. A. Melton. "Vertebrate Embryonic Induction: Mesodermal and Neural Patterning." *Science,* October 28, 1994. A review of progress toward determining the molecular basis of pattern formation in vertebrates.

Lawrence, P. A. *The Making of a Fly.* Oxford: Blackwell Scientific Publications, 1992. A survey of research accomplishments in fruit fly developmental genetics.

McGinnis, W., and M. Kuziora. "The Molecular Architects of Body Design. *Scientific American,* February 1994. Elaboration on the developmental and evolutionary significance of homeobox genes.

Newman, R. A. "Adaptive Plasticity in Amphibian Metamorphosis." *BioScience,* October 1992. How does environment affect an individual animal's development?

Steller, H. "Mechanisms and Genes of Cellular Suicide." *Science,* March 10, 1995. The molecular basis of apoptosis (programmed cell death).

The scanning electron micrograph on this page shows an unusual juxtaposition of the basic components of animal nervous systems and computers. A single nerve cell (a neuron) was removed from a nervous system and grown on the surface of a Motorola 68000 microprocessor. Made of intricate circuit boards like this, computers are capable of complex tasks. However, your own nervous system, made of living neurons, is performing vastly more complex tasks as you read and comprehend these words. The human nervous system is probably the most intricately organized aggregate of matter on Earth. A single cubic centimeter of the human brain may contain several million nerve cells, each of which may communicate with thousands of other neurons in information-processing networks that make the most elaborate computer look primitive. These neural pathways control our every perception and movement and enable us to learn, remember, think, and be conscious of ourselves and our surroundings.*

In many cases, the nervous system and endocrine system cooperate and interact in regulating internal body functions and behavior. In the maintenance of homeostasis, for example, the hypothalamus and other parts of the brain receive and process data about the body's internal environment and send out commands that correct imbalances to other organs via neurons and neurosecretory cells.

Despite their structural and functional linkage (see Chapter 41), the nervous system and endocrine system play somewhat different roles in body coordination. For example, with its incomparable structural complexity, the nervous system can integrate vast amounts of information, such as that required for human thought and speech. Timing can also be a key difference. The endocrine system may take minutes, hours, or even days to act, partly because of the time it takes for hormones to be made and carried in the blood to their target organs. In contrast, the nervous system is a signaling network with branches carrying information directly to and from specific targets. Neurons are specialized for the fast transmission of impulses—as quickly as 100 m/sec (over 200 mph). As a result, information can go from the brain of a human to the hands (or vice versa) in a few milliseconds.

To survive and reproduce, an animal must respond rapidly and appropriately to its environment. The focus of this chapter is how nervous systems mediate these interactions with the environment while cooperating with the endocrine system in maintaining homeostasis. As we will see, a combination of electrical and chemical signals enabling neurons to communicate with one another is the functional basis of nervous systems.

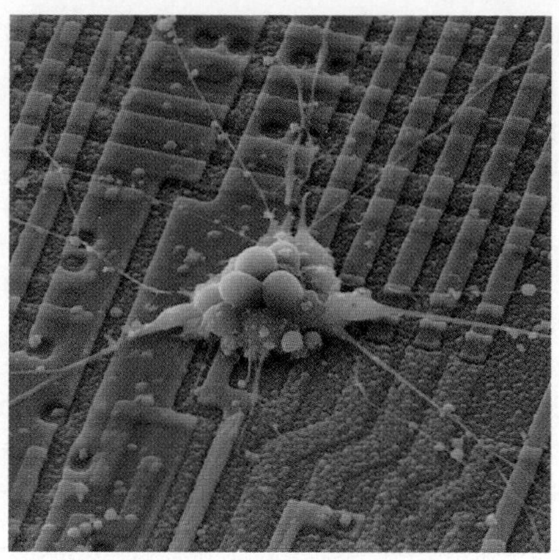

CHAPTER 44

NERVOUS SYSTEMS

KEY CONCEPTS

- Nervous systems perform the three overlapping functions of sensory input, integration, and motor output: *an overview*
- The nervous system is composed of neurons and supporting cells
- Impulses are action potentials, electrical signals propagated along neuronal membranes
- Chemical or electrical communication between cells occurs at synapses
- Invertebrate nervous systems are highly diverse
- The vertebrate nervous system is a hierarchy of structural and functional complexity
- The human brain is a major research frontier

Nervous systems perform the three overlapping functions of sensory input, integration, and motor output: *an overview*

In general, a nervous system has three overlapping functions: sensory input, integration, and motor output (FIGURE 44.1). Input is the conduction of signals from sensory receptors, such as the light-detecting cells in the eyes, to integration centers in the nervous system. Integration is the process by which the information from the environmental stimulation of the sensory receptors is interpreted and associated with appropriate responses of the body. For the most part, integration is carried out in the **central nervous system (CNS),** the brain and spinal cord. Motor output is the conduction of signals from the processing center, the CNS, to **effector cells,** the muscle cells or gland cells that actually carry out the body's responses to stimuli. The signals are conducted by **nerves,** ropelike bundles of extensions of neurons, tightly wrapped in connective tissue. The nerves that communicate motor and sensory signals between the central nervous system and the rest of the body are collectively called the **peripheral nervous system (PNS).** From receptor to effector, information is communicated in a nerve from one neuron to the next by a combination of electrical and chemical signals. We concentrate on communication within nervous systems in this chapter. Chapter 45 connects the nervous system to its inputs and outputs by discussing sensory receptors and the physiology of movement.

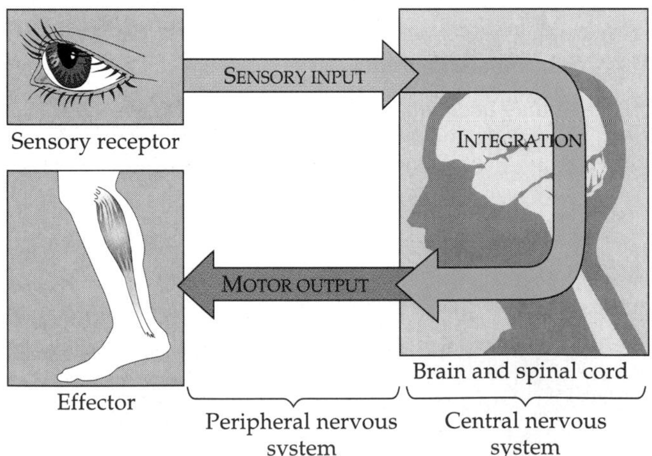

FIGURE 44.1
Overview of a vertebrate nervous system. The brain and spinal cord together form the central nervous system (CNS), which is responsible for the integration of information. A network of nerves forming the peripheral nervous system (PNS) carries information from sensory receptors (sensory input) to the CNS and motor commands from the CNS (motor output) to various target organs, or effectors.

The nervous system is composed of neurons and supporting cells

Two main classes of cells populate the nervous system: neurons and supporting cells. Neurons are the cells that actually conduct messages along the communication pathways of the nervous system. More numerous are the supporting cells, which provide structural reinforcement in the nervous system and also protect, insulate, and generally assist neurons.

Neurons

The **neuron** is the functional unit of the nervous system and is specialized for transmitting signals from one location in the body to another. Although there are many different types of neurons, varying in their structural details depending on their function, most neurons share some common features (FIGURE 44.2). A neuron has a relatively large **cell body** containing the nucleus and a variety of other cellular organelles. The most striking features of neurons are fiberlike extensions, called processes, that increase the distance over which the cells can conduct messages. Neuronal processes are of two general types: **dendrites,** which convey signals from their tips to the rest of the neuron; and **axons,** which conduct messages toward their tips.

The dendrites of many neurons, such as the one in FIGURE 44.2a, are numerous and extensively branched (the name is derived from the Greek *dendron,* "tree"). Thus, dendrites are structural adaptations that increase the surface area of the neuron where it receives inputs from other neurons.

Many neurons have a single axon, which may be very long. In fact, the sciatic nerve in your leg contains axons that extend all the way from the lower part of your spinal cord to muscles of the lower leg and foot, a distance of a meter or more. The axon hillock is the portion of the cell body where the axon originates. As we will see, this is the region where impulses that pass down the axon are usually generated. The axons of many neurons in the vertebrate PNS are enclosed by a chain of supporting cells called **Schwann cells** that form an insulating layer, the **myelin sheath.** The axon may be branched, and each branch may give rise to hundreds or thousands of specialized endings called **synaptic terminals,** which relay signals to other cells by releasing chemical messengers called neurotransmitters (see FIGURE 41.1b). The site of contact between a synaptic terminal and a target cell (either another neuron or an effector cell, such as a muscle cell) is called a **synapse.** Thus, the synapse is the junction where one neuron communicates with another neuron in a neural pathway, or where a neuron communicates with a muscle cell or gland cell.

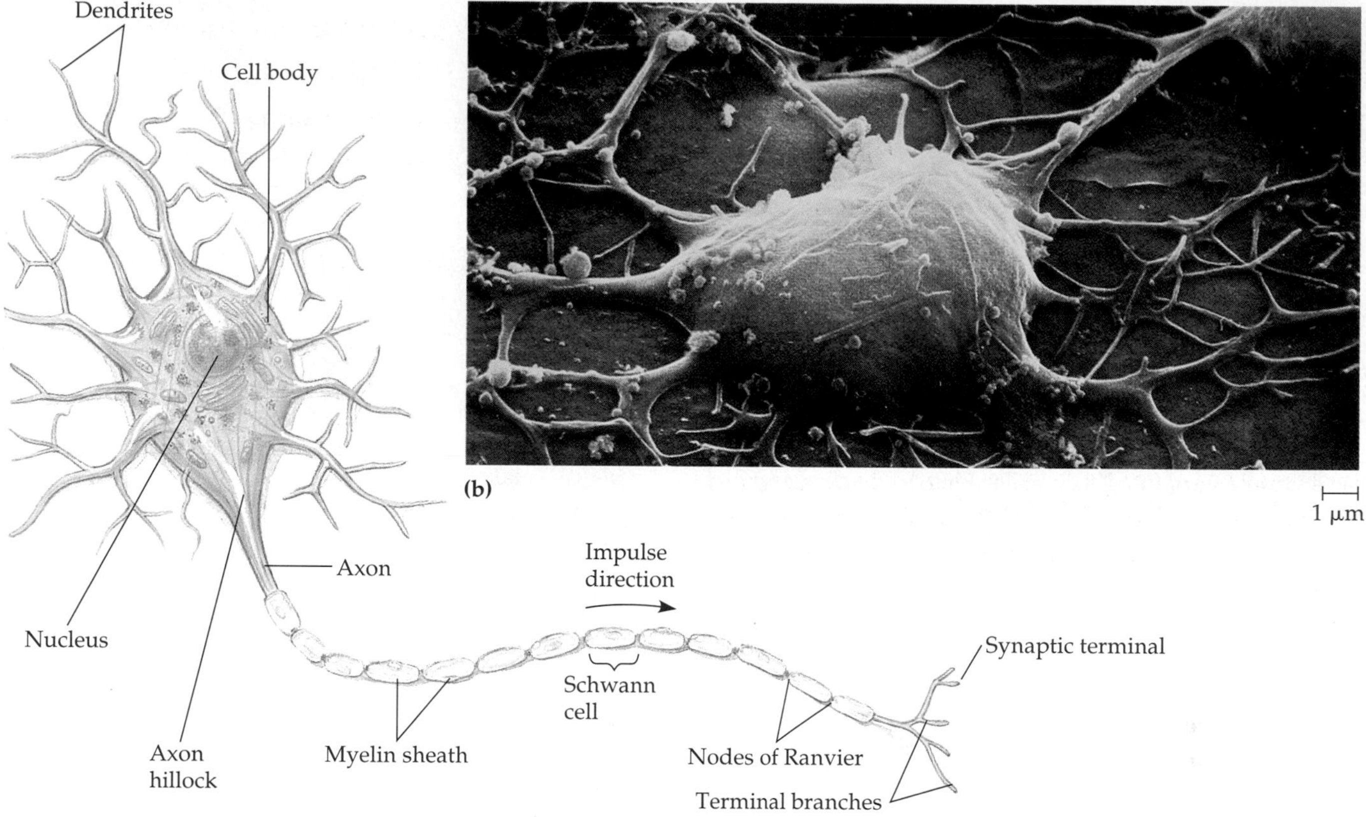

FIGURE 44.2
Structure of a vertebrate neuron. (**a**) The cell body has two types of processes, or extensions: Dendrites generally receive inputs and conduct signals toward the cell body, whereas axons conduct signals away from the cell body. At the end of the axon, terminal branches each bear a synaptic terminal that makes connections with other neurons or target cells. In the PNS, supporting cells called Schwann cells wrap many axons with an insulating myelin sheath. Gaps between successive Schwann cells are called nodes of Ranvier. (**b**) A scanning electron micrograph of a neuron.

There are three major functional classes of nerve cells, corresponding to the three major functions of the nervous system (FIGURE 44.3). **Sensory neurons** communicate information (sensory input) about the external and internal environments from sensory receptors to the central nervous system. Most sensory neurons synapse with interneurons, a second class of nerve cells, located within the CNS. **Interneurons** integrate sensory input and motor output; they make synaptic connections only with other neurons. A third class of nerve cells, **motor neurons** convey impulses (motor output) from the CNS to effector cells. Adapted for different functions, neurons of the three classes differ markedly in shape, and there is also a variety of shapes within each class (FIGURE 44.4).

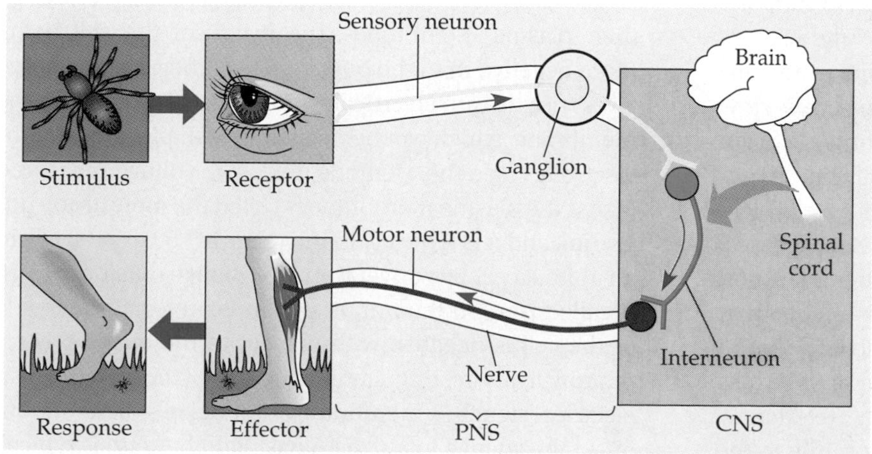

FIGURE 44.3
A neural pathway between a receptor and an effector. A sensory neuron, part of a nerve in the PNS, conveys sensory information to the CNS. The cell bodies of many sensory neurons are located outside the CNS in clusters called ganglia. One or more interneurons in the CNS integrate the information and send appropriate signals to a motor neuron, which in turn conveys appropriate commands to an effector.

FIGURE 44.4
Structural diversity of neurons. These examples illustrate some of the variety in neuron shape. Cell bodies and dendrites are black; axons are red. (**a**) A vertebrate sensory neuron. The short, multibranched dendrites communicate with receptor cells. A single, long axon, which is usually myelinated, conveys signals from the dendrites to synapses with neurons in the CNS. The cell body is connected only to the axon. This configuration is markedly different from that of the neuron in FIGURE 44.2a, which is a vertebrate motor neuron. (**b**) Two types of interneurons found in the mammalian brain. The top one has multiple dendrites and a branched axon; the bottom one has finely branched, meshlike dendrites. (**c**) An invertebrate motor neuron. In contrast to the vertebrate motor neuron in FIGURE 44.2a, the cell body connects only to the dendrites.

Supporting Cells

Although they do not actually conduct nerve impulses, **supporting cells** are essential for the structural integrity of the nervous system and for the normal functioning of neurons. Supporting cells outnumber neurons by tenfold to fiftyfold.

Supporting cells in the central nervous system are called **glial cells** ("glue" cells). There are several types of glial cells in the brain and spinal cord, and as a group they do far more than simply glue neurons together. Astrocytes encircle the capillaries in the brain and contribute to the **blood-brain barrier,** which restricts the passage of most substances into the brain, allowing the extracellular chemical environment of the CNS to be tightly controlled. Recent evidence also suggests that astrocytes actually communicate with one another and with neurons via chemical signals. Other glial cells called oligodendrocytes form insulating myelin sheaths around axons of many CNS neurons. In the PNS, Schwann cells are the supporting cells that form the myelin sheath (see FIGURE 44.2).

Neurons become myelinated in a developing nervous system when Schwann cells or oligodendrocytes grow around axons such that their plasma membranes form concentric layers, somewhat like a jelly roll. The membranes are mostly lipid, which is a poor conductor of electrical currents. Thus, the myelin sheaths provide electrical insulation of the axon, analogous to the rubber insulation covering copper wires. We will see later in this chapter that the myelin sheath also increases the speed of propagation of nerve impulses. In the degenerative disease known as multiple sclerosis, myelin sheaths gradually deteriorate, resulting in a progressive loss of coordination due to the disruption of nerve impulse transmission. Clearly, supporting cells are indispensable partners of neurons in a working nervous system. That being said, let's return to the neuron and its ability to conduct an impulse.

Impulses are action potentials, electrical signals propagated along neuronal membranes

The signal transmitted along the length of a neuron, from a dendrite or cell body to the tip of an axon, is an electrical signal that depends on the flow of ions across the plasma membrane of the cell. In this section, we focus on how an electrical voltage arises in a cell and how a flow of ions *across* the membrane is converted to a signal that travels in a perpendicular direction *along* the neuron.

The Origin of Electrical Membrane Potential

All living cells have an electrical charge difference across their plasma membranes, the inside of the cell being more negative than the outside. This difference in charge gives rise to an electrical voltage gradient across the membrane, which can be measured with ultrafine microelectrodes (see the Methods Box). The voltage measured across the plasma membrane is called the **membrane potential,** and it is typically in the range of –50 to –100 mV in an animal cell. By convention, the voltage outside the cell is called zero, so the minus sign indicates that the inside of the cell is negative with respect to the outside. For a neuron in its resting state (that is, not transmitting an electrical signal), a membrane potential of –70 mV (about 5% of the voltage in a size AA flashlight battery) is typical.

METHODS: MEASURING MEMBRANE POTENTIALS

Due to a difference in the relative concentrations of cations and anions on opposite sides of the plasma membrane, the cytoplasmic side of the membrane is negative in charge compared to the extracellular side. This charge separation is called a membrane potential, or, for an unstimulated neuron, a resting potential.

Electrophysiologists can measure the membrane potential as a voltage by using microelectrodes (shown in FIGURE a) connected to a sensitive voltmeter or oscilloscope. Precise mechanical devices called micromanipulators (moved with the large knobs being manipulated by the scientist in FIGURE b) are used to position one electrode just inside the cell for comparison with a reference electrode located outside the cell. The voltmeter indicates the magnitude of the charge separation across the membrane, typically about –70 mV for an unstimulated neuron. (The minus sign indicates that the inside of the cell is negative with respect to the outside.)

A number of invertebrates, including squids, lobsters, and earthworms, have some unusually large neurons that make excellent model systems for studying nerve impulses. For example, the squid nervous system includes some neurons whose axons have diameters of about 1 mm. These giant axons are relatively easy to impale with microelectrodes. Once the electrodes are in place, they can be used to measure the voltage of the resting potential as well as to record changes in voltage due to ion currents that occur during the transmission of a nerve impulse. Much of the pioneering research on membrane potentials and on the nature of nerve signals was performed using squid giant axons.

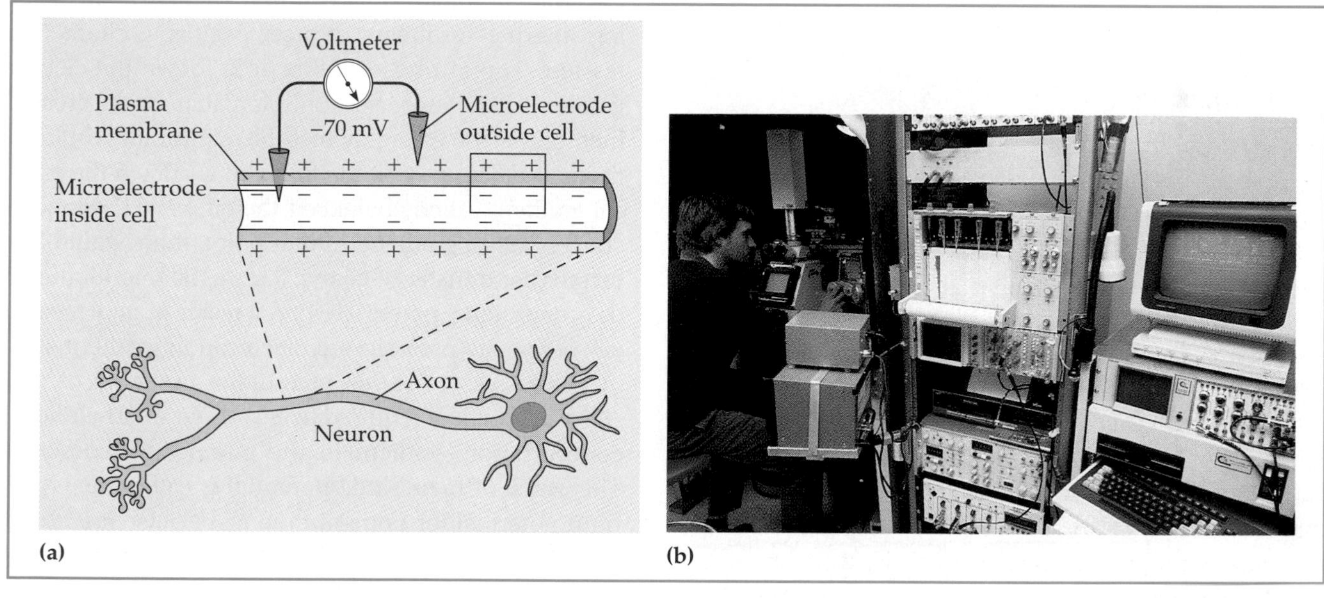

The membrane potential arises from two things: differences in the ionic composition of the intracellular and extracellular fluids, and the selective permeability of the plasma membrane, the barrier between those two fluids (FIGURE 44.5). Intracellular and extracellular fluids both contain various kinds of dissolved substances, including a variety of electrically charged substances (ions). Inside a cell, the principal cation (positively charged ion) is potassium (K^+), although there is also some sodium (Na^+). Outside a cell, the situation is reversed, with Na^+ the principal cation and K^+ having a much lower concentration. Inside a cell, the principal anions are proteins, amino acids, sulfate, phosphate, and other negatively charged ions that we can group and symbolize by A^-; chloride (Cl^-) is also present but in a relatively low concentration. Outside a cell, Cl^- is the main anion, with other anions present but less important in the context of membrane potentials.

Recall from Chapter 8 that the plasma membrane is a phospholipid bilayer with associated membrane proteins. Ions, being electrically charged, cannot dissolve in lipid and thus cannot directly diffuse across the lipid of the plasma membrane. In order to cross the membrane, ions must either be carried by transport proteins or move through ion channels, which are aqueous pores made up of specific transmembrane protein molecules. There are many different kinds of selective ion channels; some channels allow only sodium ions to cross, others allow

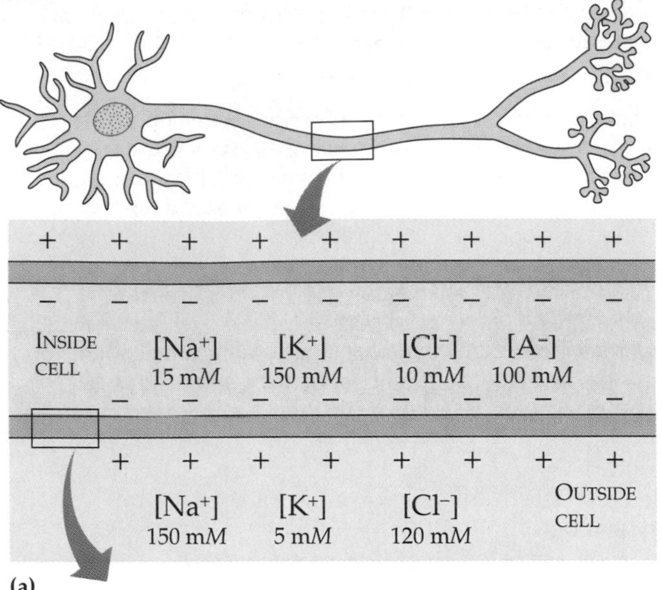

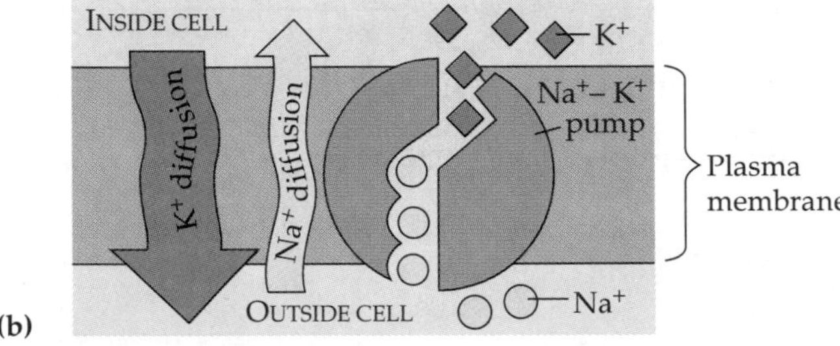

FIGURE 44.5

The basis of the membrane potential. (**a**) Intracellular and extracellular fluids have different ionic compositions. Shown here are the approximate concentrations for a mammalian cell (in millimoles per liter, abbreviated m*M*) of sodium ($[Na^+]$), potassium ($[K^+]$), chloride ($[Cl^-]$), and ($[A^-]$), anions that remain inside the cell. (Brackets indicate concentration of the enclosed substances; e.g., $[Na^+]$ stands for sodium ion concentration.) K^+ diffuses out of the cell down its concentration gradient, but the A^- anions cannot follow, so the interior of the cell develops a net negative charge. (**b**) There is a steady diffusion of K^+ out of the cell (orange arrow) and steady diffusion of Na^+ into the cell (gold arrow); the thickness of the arrows indicates the relative permeability of the membrane to K^+ and Na^+. Over time, diffusion would cause the ionic gradients shown in part a to dissipate. Dissipation is prevented by the sodium-potassium pump, which uses ATP to actively transport Na^+ out of the cell and K^+ into the cell.

only K^+, and still others only Cl^-. Depending on how many ion channels of each kind are present in the plasma membrane of a cell, it is possible for the membrane to have very different permeabilities to each of the different ions in the intracellular and extracellular fluids. Cells usually have much greater permeability to K^+ than to Na^+, suggesting that the membrane has many more potassium channels than sodium channels; in a resting neuron, for instance, potassium permeability is about fiftyfold higher than sodium permeability. Because the internal anions (A^-) are primarily large organic molecules (proteins and amino acids), they cannot cross the membrane and thus are a pool of internal negative charge that remains in the cell.

How does the distribution of ions shown in FIGURE 44.5 give rise to a membrane potential? Consider the case of potassium ions. There is a large concentration gradient for diffusion of K^+ out of the cell, and the membrane has a high permeability to potassium. Thus, there will be a net flux of K^+ out of the cell (efflux) driven by the concentration gradient. But as K^+ exits, it transfers positive charge from the inside to the outside of the cell. Because A^- stays in the cell, the inside of the cell becomes progressively more negative with respect to the outside. As positive charge is lost while negative charge is trapped within, an electrical gradient builds up across the membrane. In essence this *electrical* gradient competes with the effect of the K^+ *concentration* gradient: The increasing internal negativity attracts positively charged potassium, supporting an influx of K^+ down the electrical gradient. If K^+ were the only ion that could cross the membrane, the voltage across the membrane would continue to build up until the influx of K^+ down the electrical gradient exactly balanced the efflux of K^+ down the concentration gradient. At that point, there would be no further net transfer of charge across the membrane, and the membrane potential would reach a stable, resting value. For the potassium concentration gradient shown in FIGURE 44.5, a stable membrane potential of about –85 mV would be required to exactly counterbalance the concentration gradient in the manner just described. This value of membrane potential is called the equilibrium potential for potassium ions, because it is the potential at which there will be no net movement of K^+ across the membrane (in other words, potassium is at equilibrium).

Potassium, however, is not the only ion to which the plasma membrane is permeable. Although the membrane is much less permeable to Na^+ than to K^+, the permeability to Na^+ is not zero. For sodium, both the concentration gradient ($[Na^+]$ greater outside cell) and the electrical gradient (negative inside cell) tend to move sodium ions into the cell. The resulting trickle of positive charge into the cell, carried by Na^+, makes the actual value of the membrane potential somewhat more positive than the –85 mV that would be expected if the membrane were permeable only to potassium ions. This explains why the membrane potential of a resting neuron is typically about –70 mV rather than about –85 mV.

Over time, the steady influx of sodium would cause a progressive increase in internal sodium concentration. Also, because the influx of Na^+ makes the cell interior less negative than the −85 mV required to balance the potassium concentration gradient, there would be a steady efflux of potassium and a progressive decline in internal K^+ concentration. In other words, if the situation were left unchecked, the concentration gradients for Na^+ and K^+ shown in FIGURE 44.5 would gradually dissipate. This is prevented by a different type of plasma membrane protein also found in abundance in neurons: the sodium-potassium pump (see Chapter 8). This protein uses energy from ATP to drive the active transport of sodium back out of the cell, against both the concentration and the electrical gradients for sodium. At the same time, the pump moves potassium into the cell, thus restoring the concentration gradient for this ion as well. In essence, the cell uses metabolic energy, in the form of ATP, to maintain the ionic gradients across the membrane that give rise to the steady-state membrane potential.

Membrane Potential Changes and the Action Potential

All cells have a membrane potential; however, only certain kinds of cells, including neurons and muscle cells, have the ability to generate changes in their membrane potentials. Collectively these cells are called **excitable cells.** The membrane potential of an excitable cell in a resting (unexcited) state is called the **resting potential,** and as we are about to see, a change in the resting potential may result in an active electrical impulse.

Neurons have special ion channels, called **gated ion channels,** that allow the cell to change its membrane potential in response to stimuli the cell receives. In the case of a sensory neuron, the stimulus may come from the organism's environment (for example, light in the case of photoreceptors in the eye, or vibrations in the air in the case of receptors in the ear). In the case of an interneuron, the stimulus will ordinarily be produced by the activation of other neurons that provide inputs to the cell. The effect of the stimulus on the neuron depends on the type of gated ion channel that is opened by the stimulus. If the stimulus opens a potassium channel, an increased efflux of potassium will occur, and the membrane potential will become more negative. Such an increase in the electrical gradient across the membrane is called a **hyperpolarization** (FIGURE 44.6a). If the channel opened

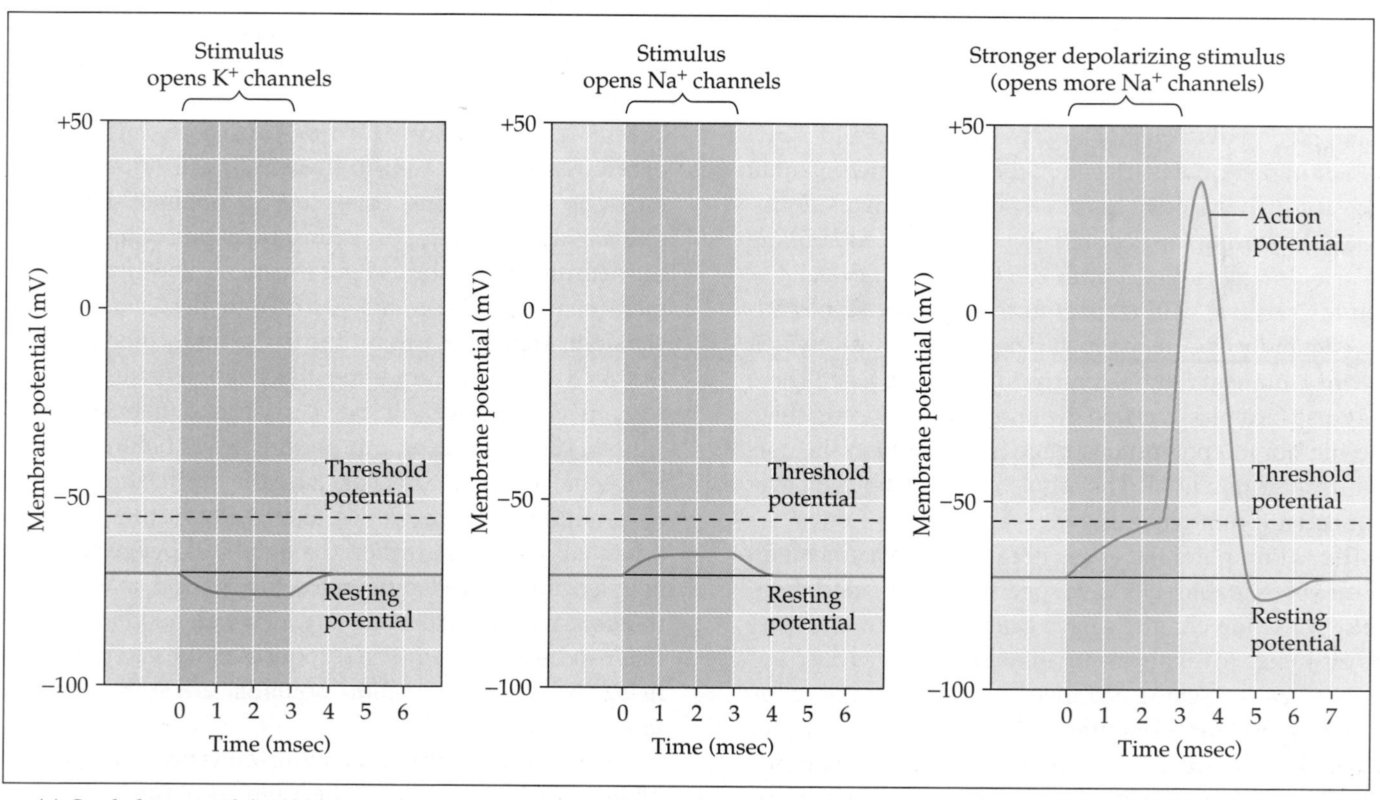

(a) Graded potential: hyperpolarization **(b)** Graded potential: depolarization **(c)** Action potential

FIGURE 44.6

Graded potentials and the action potential in a neuron. Neurons are stimulated by environmental changes that alter the cell's membrane potential. (**a**) Stimuli that open potassium channels hyperpolarize the neuron. (**b**) Stimuli that open sodium channels depolarize the neuron. (**c**) A depolarizing stimulus of sufficient strength will change the membrane potential to a critical level called the threshold potential. This triggers an action potential, or nerve impulse, which, unlike a graded potential, is an all-or-none event; the size of the action potential is not affected by the strength of the stimulus that triggered it.

by the stimulus is a sodium channel, an increased influx of sodium will occur, and the membrane potential will become less negative. Such a reduction in the electrical gradient is called a **depolarization** (FIGURE 44.6b). Voltage changes produced by stimulation of this type are called **graded potentials** because the magnitude of change (either hyperpolarization or depolarization) depends on the strength of the stimulus: A larger stimulus will open more channels and produce a larger change in permeability.

In an excitable cell, such as a neuron, the response to a depolarizing stimulus is graded with stimulus intensity only up to a particular level of depolarization, called the **threshold potential.** If a depolarization reaches the threshold, a different type of response, called an **action potential,** will be triggered (FIGURE 44.6c). The threshold potential in a neuron is typically about 15 to 20 mV more positive than the resting potential (that is, at a potential of −50 to −55 mV). Hyperpolarizing stimuli do not produce action potentials; in fact, hyperpolarization makes it less likely that an action potential will be triggered by making it more difficult for a depolarizing stimulus to reach threshold.

The action potential is the nerve impulse. It is a nongraded *all-or-none event,* meaning that the magnitude of the action potential is independent of the strength of the depolarizing stimulus that produced it, provided the depolarization is sufficiently large to reach threshold. Once an action potential is triggered, the membrane potential goes through a stereotypical sequence of changes (FIGURE 44.7). During the depolarizing phase, the membrane polarity briefly reverses, with the interior of the cell becoming positive with respect to the outside. This is followed rapidly by a steep repolarizing phase, during which the membrane potential returns to its resting level. There may also be a phase, called the undershoot, during which the membrane potential is more negative than the normal resting potential. The whole event is typically over within a few milliseconds.

The action potential arises because the plasma membranes of excitable cells have special **voltage-gated channels.** These ion channels have gates that open and close in response to changes in membrane potential. Two types of voltage-gated channels contribute to the action potential: potassium channels and sodium channels (see FIGURE 44.7). Each potassium channel has a single gate that is voltage-sensitive; it is closed when resting and *opens slowly* in response to depolarization. By contrast, each sodium channel has two voltage-sensitive gates: an activation gate that is closed when resting and responds to depolarization by *opening rapidly,* and an inactivation gate that is open when resting and responds to depolarization by *closing slowly.* In the membrane's resting state, the inactivation gate is open but the activation gate is closed, so the channel does not allow Na^+ to enter the neuron. Upon depolarization, the activation gate opens quickly, causing an influx of Na^+, which depolarizes the membrane further, opening more voltage-gated sodium channels and causing still more depolarization. This inherently explosive process, an example of positive feedback, continues until all the sodium channels at the stimulated site of the membrane are open. Depolarization to the threshold potential causes further depolarization (the action potential). At that point, the sodium permeability is about a thousandfold greater than the permeability of the resting membrane. The positive feedback that underlies the rapid depolarizing explains why the action potential, once triggered, is an all-or-none event.

Two factors underlie the rapid repolarizing phase of the action potential as membrane potential is returned to rest. First, the sodium channel inactivation gate, which is slow to respond to changes in voltage, has time to respond to depolarization by closing, returning sodium permeability to its low resting level. Second, potassium channels, whose voltage-sensitive gates respond relatively slowly to depolarization, have had time to open. As a result, during repolarization, K^+ flows rapidly out of the cell, helping restore the internal negativity of the resting neuron (see FIGURE 44.7). The potassium channel gates are also the main cause of the undershoot, or hyperpolarization, which follows the repolarizing phase. Instead of returning immediately to their resting position, these relatively slow-moving gates remain open during the undershoot, allowing potassium to keep flowing out of the neuron. The continued potassium outflow makes the membrane potential more negative (hyperpolarizes it). Notice in FIGURE 44.7 that during the undershoot, both the activation gate and the inactivation gate of the sodium channel are closed. If a second depolarizing stimulus arrives during this period, it will be unable to trigger an action potential because the inactivation gates have not had time to reopen after the preceding action potential. This period when the neuron is insensitive to depolarization is called the **refractory period,** and it sets the limit on the maximum rate at which action potentials can be generated. The responses of the gated channels involved in the action potential are summarized in TABLE 44.1.

If the action potential is an all-or-none event with amplitude (size) unaffected by the intensity of the stimulus, how can the nervous system distinguish strong stimuli from weaker ones that are still sufficient to trigger action potentials? Strong stimuli result in a greater *frequency* of action potentials than weaker stimuli; if a stimulus is intense, the neuron will fire repeatedly, producing action potentials as rapidly as the refractory period will allow.

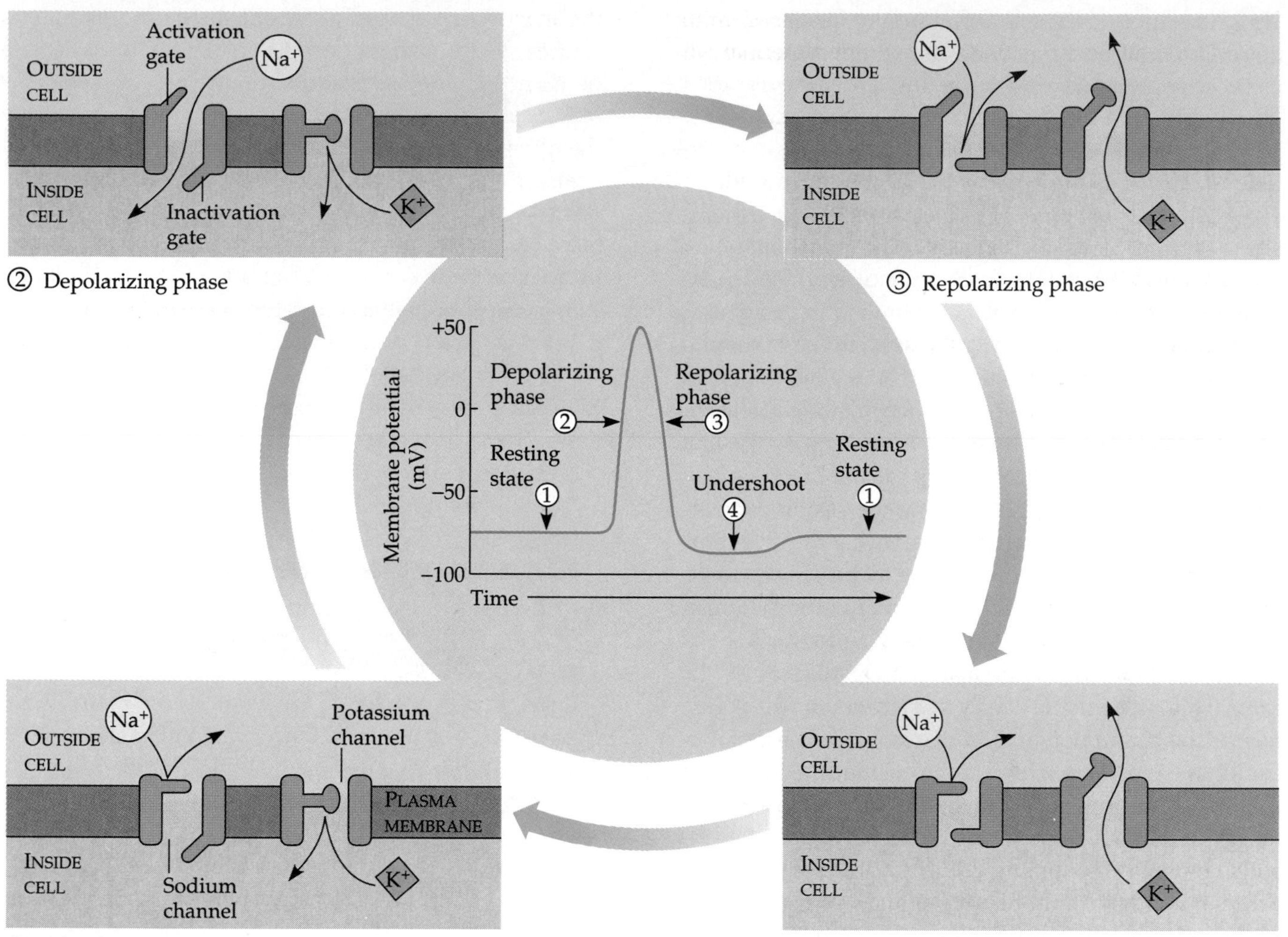

FIGURE 44.7

The role of gated ion channels in the action potential. The circled numbers on the action potential graph correspond to the four diagrams of voltage-gated sodium and potassium channels in a neuron's plasma membrane. Opening and closing of the voltage-sensitive gates underlie a sequence of stereotypical changes in membrane potential that occur when a stimulus triggers an action potential. ① In the resting state, both the sodium channel and the potassium channel are closed, and the membrane's resting potential is maintained. ② During the depolarizing phase, the action potential is generated as activation gates of the sodium channels open, and the potassium channels remain closed. ③ During the repolarizing phase, inactivation gates close the sodium channels, and potassium channels open. ④ During the undershoot, both gates of the sodium channels are closed, but potassium channels remain open because their relatively slow gates have not had time to respond to the repolarization of the membrane. Within another millisecond or two, the resting state is restored, and the system is ready to respond to another stimulus.

TABLE 44.1

Responses of Voltage-Gated Sodium and Potassium Channels to Depolarization

CHANNEL	TYPE OF GATE	RESTING STATE OF GATE	RESPONSE TO DEPOLARIZATION	SPEED OF RESPONSE
Na^+	Activation	Closed	Opens	Fast
Na^+	Inactivation	Open	Closes	Slow
K^+	Activation	Closed	Opens	Slow

Thus, it is the number of action potentials per second, not their amplitude, that codes for stimulus intensity in the nervous system.

Propagation of the Action Potential

An action potential is a localized electrical event, a membrane depolarization at a specific point of stimulation. A neuron is usually stimulated at its dendrites or cell body; for the resulting action potential to function as a signal, it must somehow "travel" along the axon to the other end of the cell. Actually, the action potential does not travel but is regenerated anew in a sequence along the axon. The effect of one action potential is like tipping the first of a row of standing dominoes. Just as the first domino's fall is relayed to the end of the row, the strong depolarization of one action potential assures that the neighboring region of the neuron will be depolarized above threshold, triggering a new action potential at that position, and so on to the end of the axon (FIGURE 44.8).

Once the action potential is propagating along the axon, what prevents Na^+ entry from reexciting the region *behind* the action potential, which would cause depolarization to spread back toward the cell body as well as in the direction of normal propagation? Recall that an action potential is followed by a refractory period, when inactivation gates of sodium channels are closed and an action potential cannot be triggered. A wave of depolarization passing a point along the axon cannot induce another action potential behind it, but only in the forward direction. Thus, the axon is normally a one-way avenue for the conduction of nerve impulses.

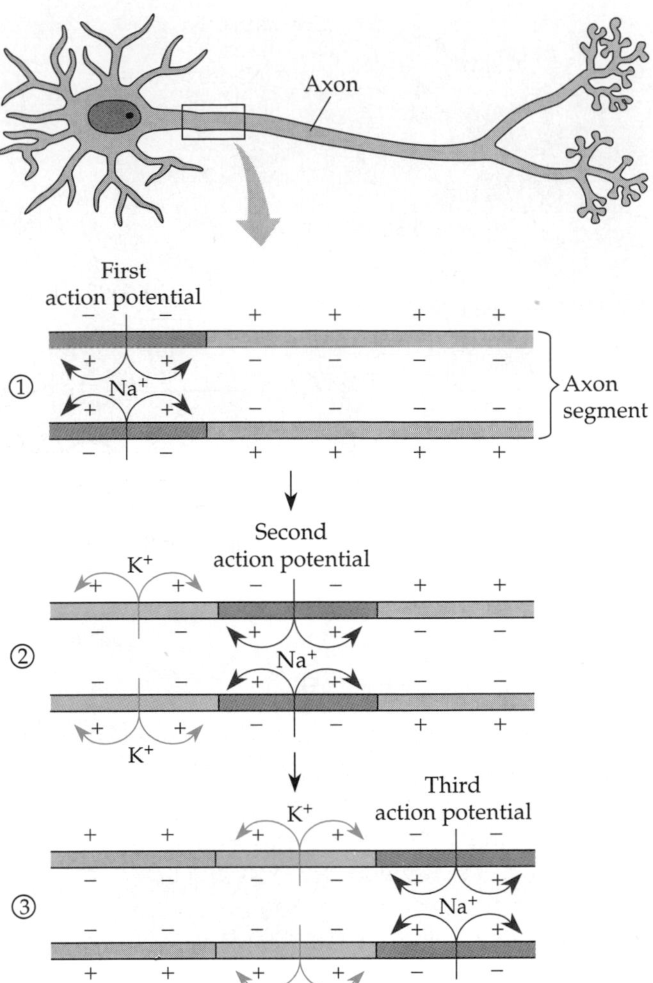

FIGURE 44.8
Propagation of the action potential. The three parts of this figure show the changes that occur in a portion of an axon at three successive times as a nerve signal passes from left to right. At each point along the axon, the voltage-gated channels go through the sequence described in FIGURE 44.7, reproducing the sequence of voltage changes associated with the action potential. Thus, ① an action potential is generated as sodium ions flow inward across the membrane at one location. ② The depolarization of the first action potential has spread to the neighboring region of the membrane, depolarizing it and initiating a second action potential. At the site of the first action potential, the membrane is repolarizing as K^+ flows outward. ③ A third action potential follows in sequence, with repolarization in its wake. In this way, local currents of ions across the plasma membrane give rise to a nerve impulse that passes along the axon.

Action Potential Transmission Speed

One factor that affects the speed at which an action potential propagates along an axon is the diameter of the axon: The larger the axon's diameter, the faster the speed of transmission. This is because resistance to the flow of electrical current is inversely proportional to the cross-sectional area of the "wire" that conducts the current. In a thick axon, the depolarization associated with an action potential at a particular location can effectively reach farther along the interior of the axon and set up a new action potential at a greater distance away than in a thin axon. Transmission speed varies from several centimeters per second in very thin axons to about 100 m/sec in the giant axons of certain invertebrates, including squids and lobsters (see the Methods Box, p. 997). These giant axons function in behavioral responses requiring great speed, such as the backward tail-flip that enables a threatened lobster or crayfish to escape.

A different means of speeding the propagation of action potentials has evolved in vertebrates. Recall that many axons in vertebrate nervous systems are myelinated, coated with insulating layers of membranes deposited by glial cells or Schwann cells. The voltage-gated ion channels that produce the action potential are concentrated in the *nodes of Ranvier,* small gaps between successive Schwann cells along the axon. Also, extra-

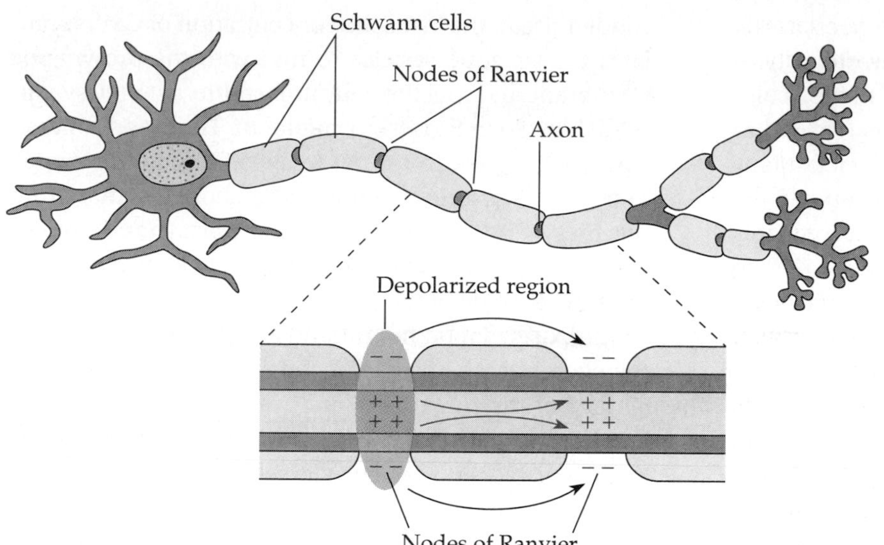

FIGURE 44.9
Saltatory conduction. In a myelinated axon, the depolarization resulting from an action potential at one node of Ranvier spreads along the interior of the axon to the next node (red arrows), triggering an action potential there. The action potential thus jumps from node to node as it propagates along the axon (black arrows).

cellular fluid is in contact with the axon membrane only at the nodes, so that the flow of ions between the inside and outside of the axon can occur only in these regions. For these reasons, the action potential does not propagate in a continuous manner over the length of the axon, but rather "jumps" from node to node, skipping the insulated regions of membrane between the nodes (FIGURE 44.9). This mechanism, called **saltatory conduction** (L. *saltare,* "to leap"), results in faster transmission of the nerve impulse.

We have seen how stimulating the dendrite of a neuron can produce an action potential that is propagated along the axon to the very tip of the neuron. Our next step is to find out how the impulse is transmitted from one neuron to the next in a neural pathway, such as the one illustrated in FIGURE 44.3.

Chemical or electrical communication between cells occurs at synapses

The synapse is a unique junction that controls communication between neurons. Synapses are also found between sensory receptors and sensory neurons, and between motor neurons and the muscle cells they control. Here, we focus on synapses between neurons, which usually conduct signals from an axon's synaptic terminals to dendrites or cell bodies of the next cells in a neural pathway. The transmitting cell is called the **presynaptic cell,** and the receiving cell is called the **postsynaptic cell.** Synapses are of two types: electrical synapses and chemical synapses.

Electrical Synapses

An electrical synapse allows action potentials to spread directly from the presynaptic to the postsynaptic cell. The cells are connected by gap junctions (see FIGURE 7.30c), intercellular channels that allow the local ion currents of an action potential to flow between neurons. The giant axons of lobsters and other crustaceans are connected end to end and coupled by electrical synapses. These make it possible for impulses to travel from neuron to neuron without delay and with no loss of signal strength. Electrical synapses in the central nervous systems of vertebrates synchronize the activity of neurons responsible for some rapid, stereotypical movements. For example, electrical synapses in the brain enable some fishes to flap their tail very rapidly when escaping from predators. However, chemical synapses are much more common than electrical synapses in vertebrates and most invertebrates.

Chemical Synapses

At a chemical synapse, a narrow gap, or **synaptic cleft,** separates the presynaptic cell from the postsynaptic cell. Because of the cleft, the cells are not electrically coupled, and an action potential occurring in the presynaptic cell cannot be transmitted directly to the membrane of the postsynaptic cell. Instead, a series of events converts the electrical signal of the action potential arriving at the synaptic terminal into a chemical signal that travels across the synapse, where it is converted back into an electrical signal in the postsynaptic cell.

The key to understanding the function of a chemical synapse is to examine its structure. Within the cytoplasm of the synaptic terminal, at the tip of the presynaptic axon, are numerous sacs called **synaptic vesicles** (FIGURE 44.10). Each vesicle contains thousands of molecules of a **neurotransmitter,** the substance that is released as an intercellular messenger into the synaptic cleft. Many different neurotransmitters have been discovered in the nervous systems of animals. Most neurons secrete only one kind of neurotransmitter. However, a single neuron may receive chemical signals from a variety of neurons that secrete different neurotransmitters from their synaptic terminals.

A neuron dispatches neurotransmitter molecules into the synapse when an action potential arrives at the synaptic terminal and depolarizes the **presynaptic membrane,** the surface of the synaptic terminal that faces the cleft. Calcium ions play a central role in this conversion of the electrical impulse into a chemical signal. Depolarization of the presynaptic membrane causes Ca^{2+} to rush into the neuron through voltage-gated channels. The sudden rise in the cytosolic concentration of Ca^{2+} stimulates the synaptic vesicles to fuse with the presynaptic membrane and spill the neurotransmitter into the synaptic cleft by exocytosis (see Chapter 8). Thousands of vesicles may respond in unison to a single action potential. The neurotransmitter diffuses the short distance from the presynaptic membrane to the **postsynaptic membrane,** the plasma membrane of the cell body or dendrite on the other side of the synapse.

The postsynaptic membrane is specialized to receive the chemical message. Projecting from the extracellular surface of the membrane are proteins that function as specific receptors for neurotransmitters. The receptors are associated with selective ion channels that open and close, controlling movements of ions across the postsynaptic membrane. A receptor is keyed to a particular type of neurotransmitter, and when it binds to this chemical, the gate of the ion channel opens, allowing specific ions, such as Na^+, K^+, or Cl^-, to cross the membrane. Thus, the

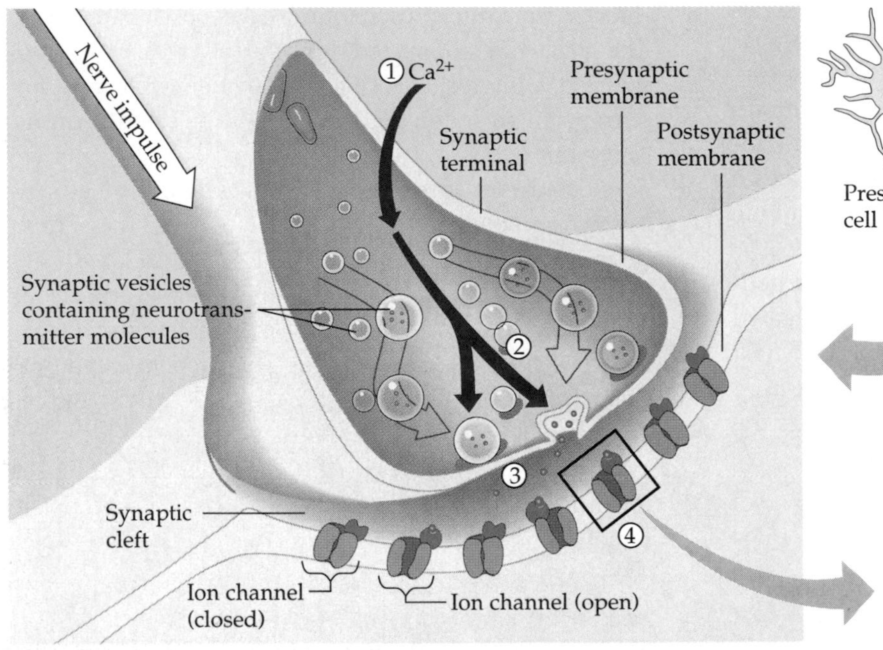

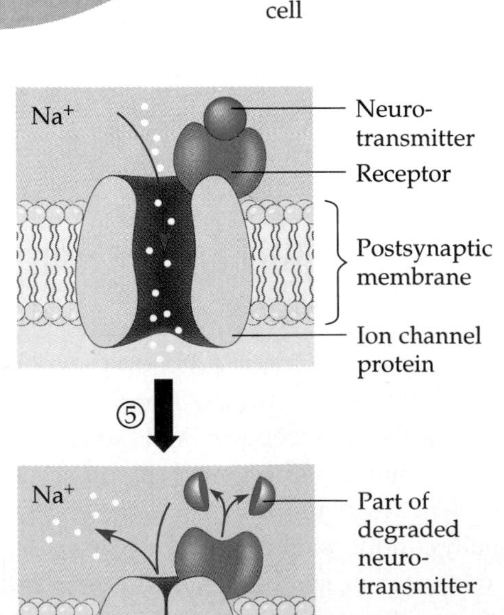

FIGURE 44.10
A chemical synapse. When an action potential depolarizes the membrane of the synaptic terminal, it ① triggers an influx of Ca^{2+} that ② causes synaptic vesicles to fuse with the presynaptic membrane. When the synaptic vesicles fuse with the membrane, ③ they release neurotransmitter molecules into the synaptic cleft. These molecules diffuse across the cleft and bind to receptors on the postsynaptic membrane. ④ The receptors control selective ion channels; the binding of neurotransmitter to its specific receptors opens the ion channels. The resulting ion flux changes the voltage of the postsynaptic membrane, either moving the membrane potential toward the threshold required for an action potential (an excitatory synapse, as illustrated here) or hyperpolarizing the membrane (an inhibitory synapse). ⑤ In either case, the neurotransmitter molecules are quickly degraded by enzymes or are taken up by another neuron, closing the ion channels and terminating the synaptic response.

ion channels of the postsynaptic membrane are chemically gated, in contrast to the voltage-gated channels responsible for the action potential.

The ion movements resulting from the binding of neurotransmitter to its receptors alter the membrane potential of the postsynaptic cell. Depending on the type of receptors and the ion channels they control, neurotransmitters binding to the postsynaptic membrane may either excite the membrane by bringing its voltage closer to the threshold potential or inhibit the postsynaptic cell by hyperpolarizing its membrane. In either case, enzymes quickly break down the neurotransmitter, ensuring that its effect on the postsynaptic cell will be brief and precise, and that the next action potential arriving at the synapse will be transmitted. For example, the neurotransmitter known as acetylcholine is rapidly degraded by cholinesterase, an enzyme present in both the synaptic cleft and the postsynaptic membrane.

Note that one important function of the synapse is that it allows nerve impulses to be transmitted only in a single direction over a neural pathway. Synaptic vesicles are present only in synaptic terminals, and thus only the presynaptic membrane can discharge neurotransmitters. And receptors are restricted to the postsynaptic membrane, ensuring that only this membrane can receive a chemical signal from another neuron.

A postsynaptic membrane is analogous to a tiny circuit board in a computer, receiving and processing information in the form of neurotransmitter molecules. At any particular instant, these living circuit boards integrate multiple positive and negative inputs and either fire an action potential or not. The ability of the nervous system to perform all of its integrating activities and make precise, appropriate responses to stimuli depends on this complex control of postsynaptic membranes. Let's take a more detailed look at the cellular basis of this nervous integration.

Summation: Neural Integration at the Cellular Level

A single neuron may receive information from numerous neighboring neurons via thousands of synapses, some of them excitatory and some of them inhibitory (FIGURE 44.11). Excitatory and inhibitory synapses have

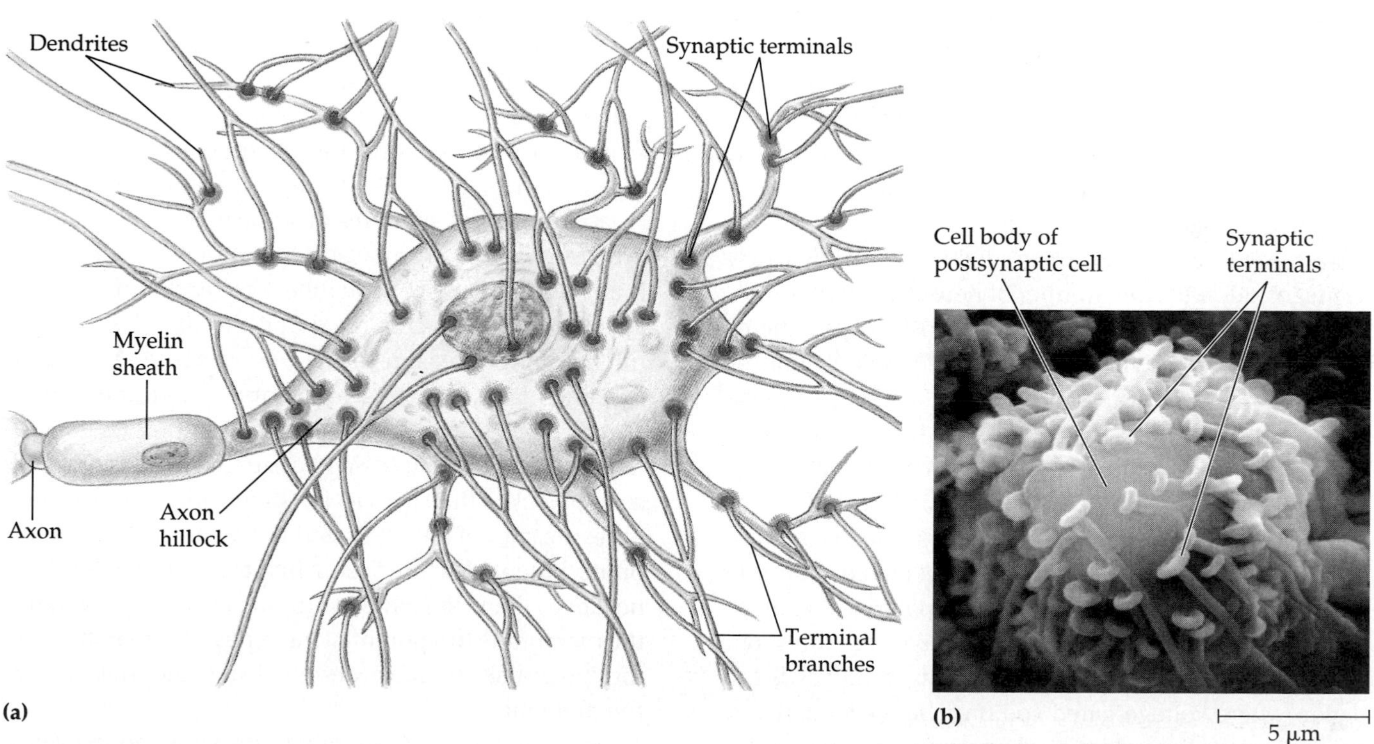

FIGURE 44.11

Integration of multiple synaptic inputs. (**a**) Each neuron, especially in the CNS, is on the receiving end of thousands of synapses, some excitatory (green) and others inhibitory (red). At any instant, an action potential may be generated at the axon hillock if the combined effect of ion currents induced by excitatory and inhibitory synapses depolarizes the membrane to the threshold potential. Synapses close to the axon hillock generally have a stronger impact on the membrane potential than other synapses. This mechanism of integrating multiple positive and negative inputs by adding their individual effects is called summation. (**b**) This micrograph reveals numerous synaptic terminals of presynaptic neurons that communicate with a single postsynaptic cell (SEM).

opposite effects on the membrane potential of the postsynaptic cell.

At an excitatory synapse, neurotransmitter receptors control a type of gated channel that allows Na^+ to enter the cell and K^+ to leave the cell. Because the driving force is greater for Na^+ than for K^+ (remember, voltage and concentration gradient both drive Na^+ into the cell), the effect of opening these channels is a net flow of positive charge into the cell. This depolarizes the cell, moving the membrane potential closer to the threshold voltage and making it more likely that the postsynaptic cell will generate an action potential. Therefore, the electrical change caused by the binding of neurotransmitter to the receptor is called an **excitatory postsynaptic potential,** or **EPSP.**

At an inhibitory synapse, the binding of neurotransmitter molecules to the postsynaptic membrane hyperpolarizes the membrane by opening ion channels that make the membrane more permeable to K^+, which rushes out of the cell; or to Cl^-, which enters the cell because of a large concentration gradient (see FIGURE 44.5); or to both of these ions. These ion fluxes push the membrane potential to a voltage even more negative than the resting potential, making it more difficult for an action potential to be generated. Therefore, the voltage change associated with chemical signaling at an inhibitory synapse is called an **inhibitory postsynaptic potential,** or **IPSP.** Whether a particular neurotransmitter results in an EPSP or an IPSP depends on the type of receptors and gated ion channels on the postsynaptic membrane responding to that neurotransmitter.

Both EPSPs and IPSPs are graded potentials that vary in magnitude with the number of neurotransmitter molecules binding to receptors on the postsynaptic membrane. The change in voltage, either depolarization or hyperpolarization, lasts only a few milliseconds because the neurotransmitters are inactivated by enzymes soon after their release into the synapse. Also, the electrical impact on the postsynaptic cell decreases with distance away from the synapse. For the postsynaptic cell to fire (generate an action potential), the local ion currents due to EPSPs must be strong enough to depolarize the membrane in the region of the axon hillock to the threshold potential, usually about −50 mV. The axon hillock is the region where voltage-gated sodium channels open and generate an action potential when some stimulus has depolarized the membrane to the threshold.

A single EPSP at one synapse, even one close to the axon hillock, is not usually strong enough to trigger an action potential. However, several synaptic terminals acting simultaneously on the same postsynaptic cell, or a smaller number of synaptic terminals discharging neurotransmitters repeatedly in rapid-fire succession, can have a cumulative impact on the membrane potential at the axon hillock. This additive effect of postsynaptic potentials is called **summation** (FIGURE 44.12).

There are two types of summation: temporal summation and spatial summation. In temporal summation, chemical transmissions from one or more synaptic terminals occur so close together in time that each postsynaptic potential affects the membrane before the voltage has returned to the resting potential after the previous stimulation (FIGURE 44.12b). In spatial summation, several different synaptic terminals, usually belonging to different presynaptic neurons, stimulate a postsynaptic cell at the same time and have an additive effect on the membrane potential (FIGURE 44.12c).

By reinforcing one another through temporal or spatial summation, the ion currents associated with several EPSPs can depolarize the membrane at the axon hillock to threshold, causing the neuron to fire. Summation also applies to IPSPs; two or more IPSPs can hyperpolarize the membrane to a voltage more negative than any single release of a neurotransmitter at an inhibitory synapse can achieve. Furthermore, IPSPs and EPSPs counter each other's electrical effects (FIGURE 44.12d).

The axon hillock is the neuron's integrating center, the region where the membrane potential represents the effect of all EPSPs and IPSPs. At any instant, that potential is an average of the depolarization due to summation of all EPSPs and the hyperpolarization due to summation of all IPSPs. (This takes into account the greater impact of synapses at or near the axon hillock.) Whenever EPSPs overpower IPSPs enough for the membrane potential at the axon hillock to reach threshold, an action potential is generated and the impulse is transmitted along the axon to the next synapse. A few milliseconds later, after the refractory period, the neuron may fire again if the sum of all synaptic inputs at that moment is still sufficient to depolarize the membrane at the axon hillock to threshold level. On the other hand, by that time, the sum of all EPSPs and IPSPs may put the membrane potential at the axon hillock at a voltage more negative than the threshold, or even hyperpolarize the membrane to a potential more negative than the resting potential, thereby desensitizing the neuron for the moment.

Action potentials, remember, are all-or-none events. But now we see that the occurrence of these nerve impulses depends on the ability of the neuron to integrate quantitative information in the form of multiple excitatory and inhibitory inputs, each involving the specific binding of a neurotransmitter to a receptor on the postsynaptic membrane.

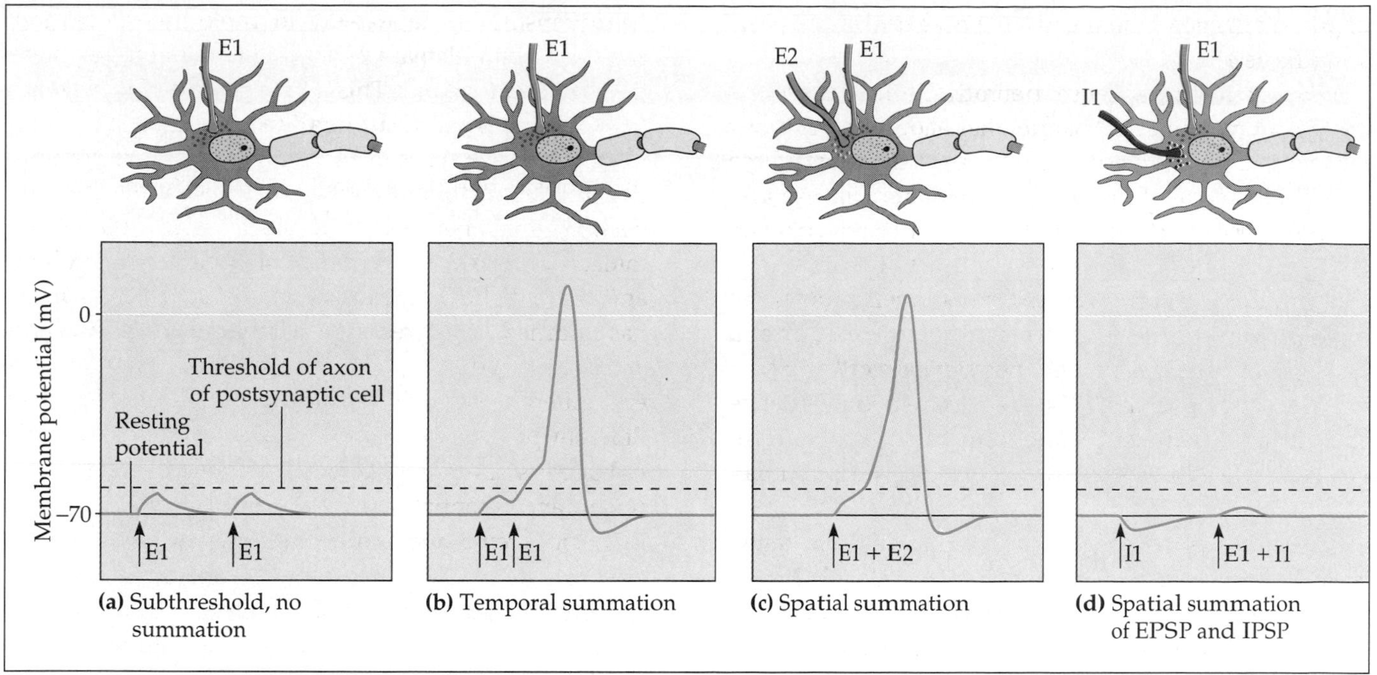

(a) Subthreshold, no summation

(b) Temporal summation

(c) Spatial summation

(d) Spatial summation of EPSP and IPSP

FIGURE 44.12

Summation of postsynaptic potentials. These graphs trace changes in membrane potentials at a postsynaptic neuron's axon hillock. The arrows indicate times when signals trigger changes in membrane potentials at two excitatory synapses (E1 and E2, green) and at one inhibitory synapse (I1, red). Like most single EPSPs, those shown at E1 and E2 cannot depolarize the membrane at the axon hillock all the way to the threshold level, and thus without summation do not trigger an action potential. (**a**) Repeated subthreshold EPSPs that do not overlap in time do not add together and therefore do not depolarize the membrane to threshold. (**b**) Temporal summation occurs when two or more subthreshold EPSPs that overlap in time reinforce each other. Here, a second release of a neurotransmitter at synapse E1 affects the postsynaptic membrane when it is still partially depolarized from a slightly earlier stimulation. The cumulative effect depolarizes the membrane to threshold, triggering an action potential. (**c**) Spatial summation occurs when two or more presynaptic cells release neurotransmitter at the same time, causing a cumulative voltage change greater than the individual EPSPs. Here, the action potential results from spatial summation of EPSPs at synapses E1 and E2, depolarizing the membrane to threshold. (**d**) EPSPs and IPSPs move the membrane potential in opposite directions. An example of spatial summation, an IPSP (at synapse I1) subtracts from the depolarization because of an overlapping EPSP (at synapse E1).

Neurotransmitters and Receptors

Studying the chemistry of the nervous system, identifying neurotransmitters and discovering how they and their receptors work, is one of the most exciting areas of research in biology today. Dozens of different substances, many of them small, nitrogen-containing organic molecules, are known to function as neurotransmitters, and researchers expect to find many more. In general, in order for a particular compound to qualify as a neurotransmitter at a specific type of synapse, it must meet three criteria:

1. The presynaptic cell must contain the compound in synaptic vesicles and discharge the substance when the cell is appropriately stimulated, and the chemical must then affect the membrane potential of the postsynaptic cell.
2. The compound should be able to cause an EPSP or IPSP when experimentally injected into the synapse.
3. The substance must be removed rapidly from the synapse, by either enzymatic degradation or uptake by a cell, allowing the postsynaptic membrane to return to resting potential.

TABLE 44.2 summarizes the major known neurotransmitters. **Acetylcholine** is one of the most common neurotransmitters in both invertebrates and vertebrates. At the vertebrate neuromuscular junction, the synapse between a motor neuron and a skeletal muscle cell, acetylcholine is released from the terminal of the motor axon. Here, it is excitatory, depolarizing the postsynaptic muscle cell. In other cases, acetylcholine is inhibitory. For example, this neurotransmitter slows down the heart rate in vertebrates and mollusks. This versatility of a single neurotransmitter depends on the receptors present on different postsynaptic cells. In some cases, such as

TABLE 44.2

The Major Known Neurotransmitters

NEUROTRANSMITTER	STRUCTURE	FUNCTIONAL CLASS	SECRETION SITES
Acetylcholine	$H_3C-C(=O)-O-CH_2-CH_2-N^+-(CH_3)_3$	Excitatory to vertebrate skeletal muscles; excitatory or inhibitory at other sites	CNS; PNS; vertebrate neuromuscular junction
Biogenic Amines			
Norepinephrine	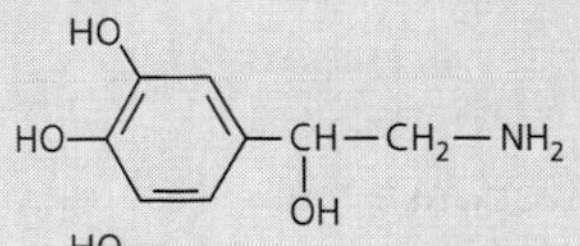 $(HO)_2C_6H_3-CH(OH)-CH_2-NH_2$	Excitatory or inhibitory	CNS; PNS
Dopamine	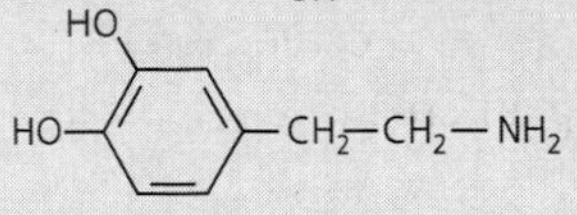 $(HO)_2C_6H_3-CH_2-CH_2-NH_2$	Generally excitatory; may be inhibitory at some sites	CNS; PNS
Serotonin	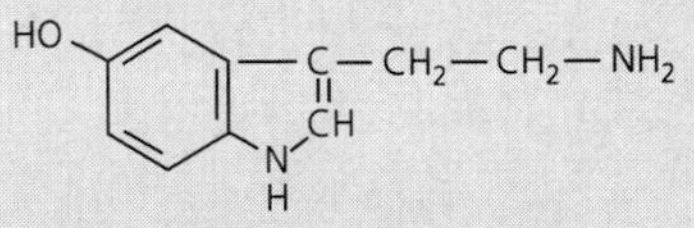 HO-indole-$C-CH_2-CH_2-NH_2$ (ring: $C=CH-NH$)	Generally inhibitory	CNS
Amino Acids			
GABA (gamma aminobutyric acid)	$H_2N-CH_2-CH_2-CH_2-COOH$	Inhibitory	CNS; invertebrate neuromuscular junction
Glycine	H_2N-CH_2-COOH	Inhibitory	CNS
Glutamate	$H_2N-CH(COOH)-CH_2-CH_2-COOH$	Excitatory	CNS; invertebrate neuromuscular junction
Aspartate	$H_2N-CH(COOH)-CH_2-COOH$	Excitatory	CNS
Neuropeptides			
Substance P	Arg Pro Lys Pro Gln Gln Phe Phe Gly Leu Met	Excitatory	CNS; PNS
Met-enkephalin (an endorphin)	Tyr Gly Gly Phe Met	Generally inhibitory	CNS

acetylcholine's effects on skeletal muscle and heart muscle, the receptors are different proteins. In many other cases, the signaling molecule binds to the same receptor but triggers different molecular changes, or signaling mechanisms, in different postsynaptic cells.

The **biogenic amines** are neurotransmitters derived from amino acids. One group, known as catecholamines, is produced from the amino acid tyrosine. This group includes **epinephrine** and **norepinephrine,** which also function as hormones (see Chapter 41), and a closely related compound called **dopamine.** Another biogenic amine, **serotonin,** is synthesized from the amino acid tryptophan. The biogenic amines most commonly function as transmitters within the CNS. However, norepinephrine also functions in a branch of the peripheral nervous system called the autonomic nervous system, which we will examine shortly. Dopamine and serotonin are widespread in the brain and affect sleep, mood, attention, and learning. Imbalances of these neurotransmitters are associated with several disorders; for example, the degenerative illness Parkinson's disease is associated with a lack of dopamine in the brain, and an excess of dopamine is linked to schizophrenia. Some psychoactive drugs, including LSD and mescaline, apparently produce their hallucinatory effects by binding to serotonin and dopamine receptors in the brain.

Four amino acids are known to function as CNS neurotransmitters: **gamma aminobutyric acid (GABA), gly-**

cine, glutamate, and **aspartate.** GABA, believed to be the transmitter at most inhibitory synapses in the brain, produces IPSPs by increasing the chloride permeability of the postsynaptic membrane. The brain has hundreds of times more GABA than any other neurotransmitter.

Several **neuropeptides,** relatively short chains of amino acids, serve as neurotransmitters. A neuropeptide called **substance P** is a key excitatory signal that mediates our perception of pain. The **endorphins** are neuropeptides that function as natural analgesics, decreasing the perception of pain by the CNS. Endorphins were first discovered in the 1970s by neurochemists studying the mechanism of opium addiction. Candace Pert and Solomon Snyder of Johns Hopkins University found specific receptors for the opiates morphine and heroin on neurons in the brain. It seemed odd that humans would have receptors keyed to chemicals from a plant (the opium poppy). Further research showed that, in fact, the drugs bind to these receptors in the brain by mimicking endorphins, the natural painkillers produced in the brain during times of physical or emotional stress, such as the labor of childbirth. In addition to relieving pain, endorphins also decrease urine output (by affecting ADH secretion; see Chapter 41), depress respiration, produce euphoria, and have other emotional effects through specific pathways in the brain. An endorphin is also released from the anterior pituitary gland as a hormone that affects specific regions of the brain. Once again, we see the overlap between nervous and endocrine control.

A neurotransmitter affects the postsynaptic cell by one of two general mechanisms. The amino acid transmitters bind to receptors and alter the permeability of the postsynaptic membrane to specific ions, either depolarizing (EPSP) or hyperpolarizing (IPSP) the membrane. The biogenic amines and the neuropeptides usually have a longer-lasting impact because they affect metabolism within the postsynaptic cell. The binding of these transmitters to their specific receptors on the postsynaptic membrane triggers a signal-transduction pathway, in many cases involving a G protein relay system, an effector enzyme (often adenylyl cyclase), and a second messenger (often cyclic AMP; see Chapter 41).

Considerable research has focused on acetylcholine receptors, which demonstrate that the diversity of response to a neurotransmitter is mainly a function of the type of receptor and its mode of action. For example, the binding of acetylcholine to receptors at synapses in vertebrate heart muscle activates a signal-transduction pathway whose G proteins have two effects on the muscle cells: They inhibit adenylyl cyclase and open K^+ channels in the muscle cell plasma membrane, making it less able to generate an action potential. Both effects reduce the strength and rate of cardiac muscle cell contraction. A second type of acetylcholine receptor functions in vertebrate skeletal muscle cells. It consists of four different peptides that have a direct stimulatory effect on ion channel proteins in the muscle cell plasma membrane.

Gaseous Signals of the Nervous System

In common with many other types of cells, some neurons of the vertebrate PNS and CNS utilize gas molecules, notably nitric oxide (NO) and carbon monoxide (CO), as local regulators. For example, during sexual arousal, certain neurons release NO gas into the erectile tissue of the penis. In response, smooth muscle cells in the blood vessel walls of the erectile tissue dilate and fill with blood, producing an erection. In many other cases, cells release gas molecules in response to other chemical signals. For instance, the neurotransmitter acetylcholine released by neurons into the walls of blood vessels stimulates the endothelial cells of the vessels to synthesize and release NO. In turn, NO signals the neighboring smooth muscle cells to relax, dilating the vessels. The discovery of this mechanism in the late 1980s explained the medicinal action of nitroglycerin, which had been used for a century to treat angina (chest pain associated with reduced blood supply to the heart). Enzymes convert nitroglycerin to NO, which dilates the blood vessels that supply cardiac muscle.

Unlike typical neurotransmitters, NO and other gaseous messengers are not stored in cytoplasmic vesicles; cells synthesize them on demand. They diffuse into neighboring target cells, produce a change, and are broken down—all within a few seconds. In many of its targets, including smooth muscle cells, NO works via a signal-transduction pathway; like many hormones, it stimulates a membrane-bound enzyme to synthesize a second messenger that directly affects cellular metabolism.

Neural Circuits and Clusters

We have now seen how neurons function. In a nervous system, the activity of millions or more of these individual units must be coordinated. Some structural and functional principles above the cellular level apply to all nervous systems.

Neurons are arranged in circuits, groups of nerve cells that feed into and away from one another, carrying information along specific pathways. There are three major patterns of neural circuitry. In a **convergent circuit,** information from several presynaptic neurons comes together at a single postsynaptic neuron. Convergent circuits can bring together information from several sources, such as vision, touch, and hearing—to identify an object in the environment. In a **divergent circuit,** information from a single neuron spreads out to several postsynaptic neurons. This type of circuit can take information from one source, such as an eye, to several parts of the brain. In a **reverberating circuit,** information flows in a circular path, from one neuron to others and then back to its source. Reverberating circuits are thought to play a part in memory storage.

Another key feature of most nervous systems is the clustering of nerve cell bodies into functional groups. In most animals, nerve cell bodies in the PNS are arranged in clusters of similar function called **ganglia** (singular, **ganglion**). Similar functional clusters in the brain are usually called **nuclei** (not to be confused with the nuclei of individual cells). Ganglia and nuclei enable parts of the nervous system to coordinate activities without involving the entire system. This concentration of nervous coordination, called centralization, is an important feature of the evolution of nervous systems in both invertebrates and vertebrates.

Invertebrate nervous systems are highly diverse

While there is remarkable uniformity in how nerve cells function throughout the animal kingdom, there is great diversity in how nervous systems as a whole are organized. The simplest type of nervous system occurs in many cnidarians, such as *Hydra* (FIGURE 44.13a). The cnidarian **nerve net** is a loosely organized system of nerves with little or no central control. Most of the synapses in the nerve net are electrical, impulses are conducted in both directions, and stimulation at any point on the body of a hydra spreads from that site and may cause movements of the entire body. Certain other cnidarians and ctenophores show evidence of nervous system centralization. In jellyfishes, clusters of nerve cells at the margin of the bell (associated with simple sensory structures)

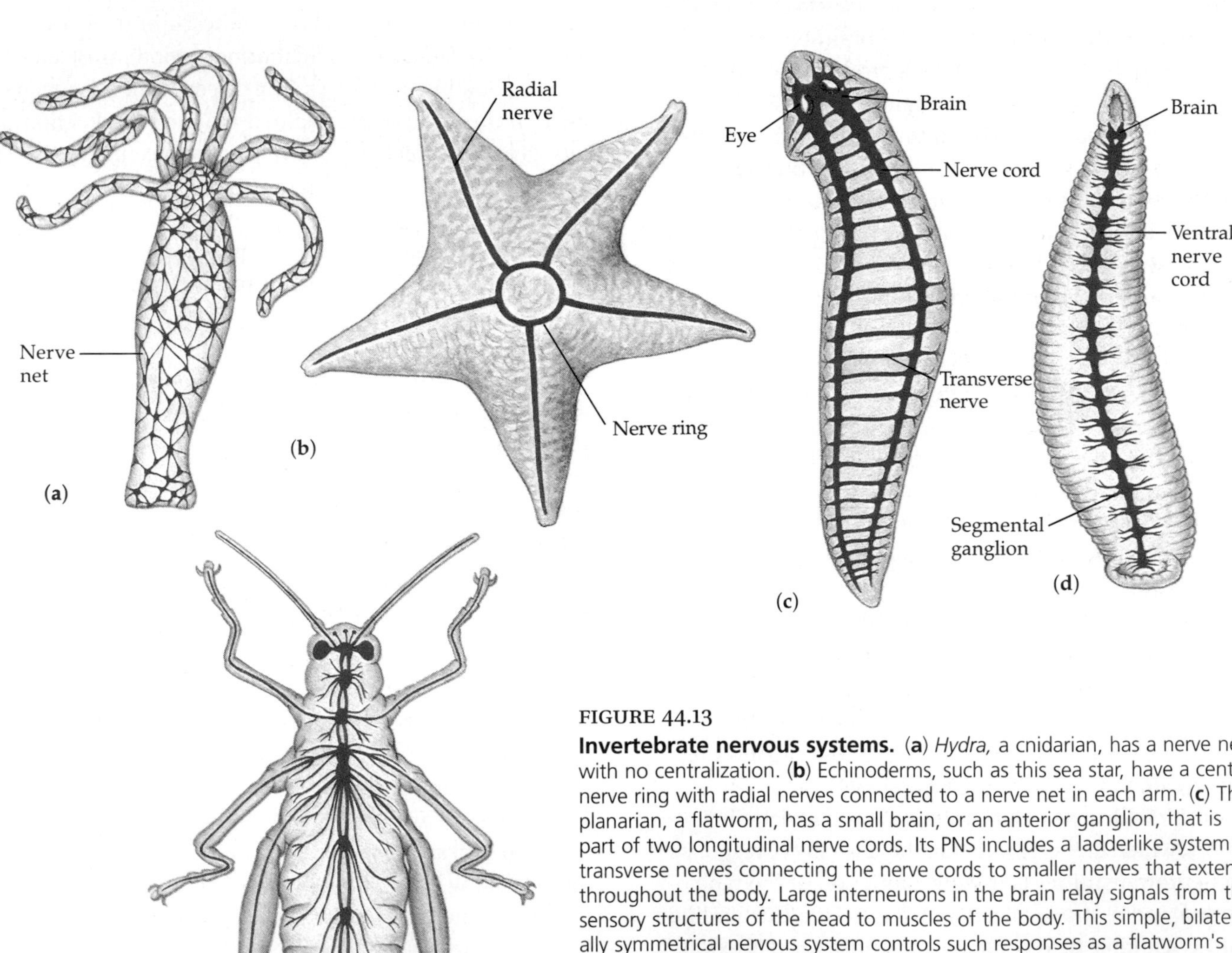

FIGURE 44.13
Invertebrate nervous systems. (**a**) *Hydra,* a cnidarian, has a nerve net with no centralization. (**b**) Echinoderms, such as this sea star, have a central nerve ring with radial nerves connected to a nerve net in each arm. (**c**) The planarian, a flatworm, has a small brain, or an anterior ganglion, that is part of two longitudinal nerve cords. Its PNS includes a ladderlike system of transverse nerves connecting the nerve cords to smaller nerves that extend throughout the body. Large interneurons in the brain relay signals from the sensory structures of the head to muscles of the body. This simple, bilaterally symmetrical nervous system controls such responses as a flatworm's ability to sense bright light and retreat to a darker place, such as the safety of a space beneath a rock. (**d**) Annelids, such as this leech, have a brain with a greater concentration of neurons than that of flatworms. Interneurons in ganglia of the ventral nerve cord coordinate many of the actions of the individual body segments. (**e**) The arthropod nervous system, as shown in this insect, probably evolved from an annelidlike system. With extensive fusion of ganglia in the head and ventral nerve cord, it is less uniformly segmented and more centralized than the annelid nervous system.

and pathways around the bell formulate the neural signals that affect activities such as swimming, which require more complex coordination of the entire body.

The nervous system of echinoderms resembles that of jellyfishes. In the sea star, for example, radial nerves extend through each arm from a central nerve ring around the oral disk (FIGURE 44.13b). Branches of the radial nerves form an interconnected network similar to the cnidarian nerve net. This system coordinates movement regardless of which arm leads.

Cnidarians and echinoderms are radially symmetrical and so are their nervous systems. By contrast, bilaterally symmetrical animals tend to have bilaterally arranged nervous systems. Correlated with bilateral symmetry, they also exhibit some degree of **cephalization,** a concentration of feeding organs and sensory structures at the anterior (head) end, the part of the body most likely to make first contact with food or threatening stimuli (see Chapter 29). Most bilateral animals also have a peripheral and a central nervous system, the latter consisting of a brain in the head and one or more nerve cords. A **nerve cord** is a thick bundle of nerves usually extending longitudinally through the body from the brain. A main thoroughfare of nerve impulses passing between the brain and the PNS, a nerve cord contains nerve cell bodies that integrate sensory information and formulate command signals to effectors. Brains have evolved from anterior enlargements of nerve cords.

Flatworms are among the simplest animals with a clearly defined CNS, composed of a small brain and two or more longitudinal nerve cords (FIGURE 44.13c). Most other invertebrates show a greater degree of centralization of the nervous system. For example, annelids and arthropods have a prominent brain and a ventral nerve cord containing segmentally arranged ganglia (FIGURE 44.13d and e).

Mollusks are good examples of how nervous system complexity correlates with habitat and natural history as well as with phylogeny. Sessile or slow-moving mollusks such as clams have little or no cephalization and only simple sense organs. Their central nervous system consists of widely separated ganglia connected by nerve cords. In contrast, cephalopods have the most sophisticated nervous systems of any invertebrates, rivaling even those of some vertebrates. The large brain of the octopus, accompanied by large, image-forming eyes and rapid conduction along giant axons, correlates well with the active predatory life of the animal. The octopus is capable of learning to discriminate between visual patterns and perform specific tasks in laboratory experiments. Indeed, learning and memory are probably everyday functions of the octopus as it interacts with its natural environment.

Mollusks continue to be key research models in neurobiology. Studies of giant squid axons yielded much of our understanding of action potentials, and snails are especially popular among researchers working in such areas as nerve cell regeneration, neurotransmitters, learning, and memory. For example, a marine snail called the sea hare *(Aplysia)* has large neurons and ganglia that are accessible to manipulation. Its entire nervous system consists of only about 20,000 neurons (compared to about 100 billion in the human brain), yet *Aplysia* displays some simple forms of learning. For instance, it learns to ignore mild touch stimuli that are presented repeatedly, a primitive type of learning called habituation. The combination of simplicity and learning ability makes it an ideal model for studying the neuronal basis of learning.

The vertebrate nervous system is a hierarchy of structural and functional complexity

The vertebrate nervous system is extremely complex, and it is convenient to divide it into a hierarchy of components that differ in function (FIGURE 44.14). The primary division is between the central nervous system, which processes information, and the peripheral nervous

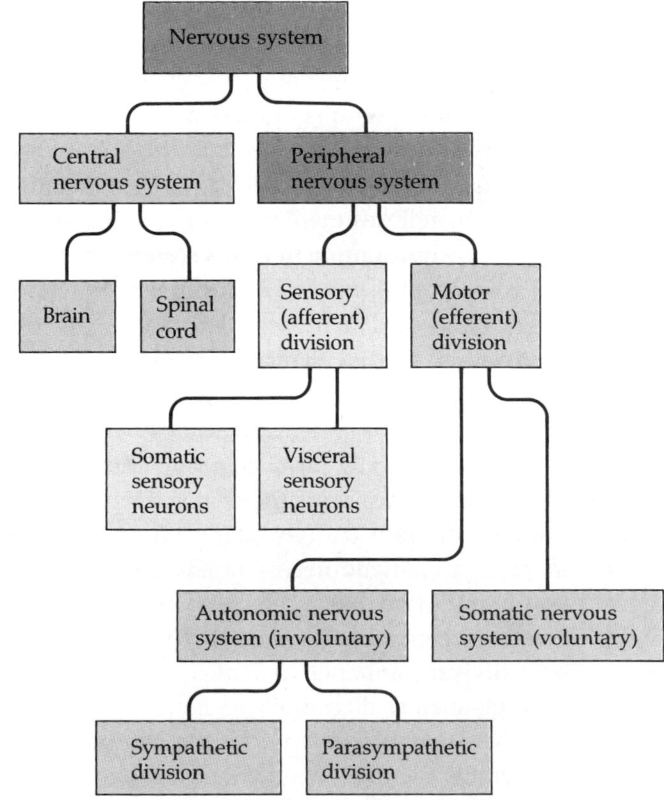

FIGURE 44.14
Functional components of the vertebrate nervous system.

system, which carries information to and from the CNS and the sensory, muscle, and gland cells. We will discuss the peripheral nervous system before going on to the central nervous system and evolution of the vertebrate brain.

The Peripheral Nervous System

Functionally, the vertebrate PNS consists of two main divisions. The **sensory division** is made up of sensory, or afferent, neurons that bring information *to* the CNS from sensory receptors. The **motor division** is composed of efferent neurons that convey signals *from* the CNS to effector cells.

In the human PNS, 12 pairs of cranial nerves originate in the brain and innervate organs of the head and upper body, and 31 pairs of spinal nerves originate in the spinal cord and innervate the entire body. Most of the cranial nerves and all the spinal nerves contain both sensory and motor neurons; a few of the cranial nerves are sensory only (the olfactory and optic nerves, for example).

Nervous systems have two basic functions: controlling responses to the external environment and coordinating the functions of internal organs (that is, maintaining homeostasis). The sensory division of the PNS contributes to both, bringing in stimuli from the external environment and monitoring the status of the internal environment. The motor division has separate components associated with these two functions. The motor neurons of the **somatic nervous system** carry signals to skeletal muscles mainly in response to *external* stimuli. The somatic nervous system is often considered voluntary because it is subject to conscious control, but a substantial proportion of skeletal muscle movement is actually determined by reflexes mediated by the spinal cord or lower brain. The **autonomic nervous system** conveys signals that regulate the *internal* environment by controlling smooth and cardiac muscles and the organs of the gastrointestinal, cardiovascular, excretory, and endocrine systems. This control is generally involuntary.

The autonomic nervous system consists of two subdivisions that are anatomically, physiologically, and chemically distinguishable: the sympathetic division and the parasympathetic division (FIGURE 44.15). When sympathetic and parasympathetic nerves innervate the same organ, they often (but not always) have antagonistic (opposite) effects. In general, signals carried via the **parasympathetic division** enhance activities that gain and conserve energy, such as digestion and slowing the heart rate. In contrast, signals conveyed by the **sympathetic division** generally increase energy consumption and prepare an individual for action by accelerating the heart rate, increasing metabolic rate, and performing related functions.

The somatic and autonomic nervous systems often cooperate in maintaining the all-important balance among organ system functions essential to homeostasis. In response to a drop in temperature, for example, the hypothalamus signals the autonomic nervous system to constrict surface blood vessels, which reduces heat loss; at the same time, the hypothalamus signals the somatic nervous system and causes shivering.

The Central Nervous System

The vertebrate CNS consists of a bilaterally symmetrical group of structures in the brain and spinal cord. The **spinal cord,** which runs down the neck and back inside the vertebral column, or spine, receives information from the skin and muscles and sends out motor commands for movement. The **brain,** at the anterior end of the spinal cord, contains centers for more complex integration of homeostasis, perception, movement, and (in humans, at least) intellect and emotions. The CNS is covered by three protective layers of connective tissue, collectively called the **meninges.**

CNS axons are located in well-defined bundles, or tracts, whose myelin sheaths give them a white appearance. In the brain, this **white matter** is located in the inner region, where pathways extend to the cell bodies in the outer **gray matter.** The situation is reversed in the spinal cord, where the white matter is outside the gray matter.

Representing the dorsal hollow nerve cord of all chordates, the vertebrate CNS contains fluid-filled spaces. The narrow **central canal** of the spinal cord is continuous with fluid-filled spaces, called **ventricles,** in the brain. These cavities are filled with **cerebrospinal fluid,** which is formed in the brain by filtration of the blood. Among the most important functions of cerebrospinal fluid is the absorption of shock, which cushions the brain. Cerebrospinal fluid also performs circulatory functions, bringing nutrients, hormones, and white blood cells to different parts of the brain. Cerebrospinal fluid normally circulates through the ventricles and central canal of the spinal cord and drains back into the veins.

The spinal cord has two principal functions: integrating simple responses to certain kinds of stimuli, and carrying information to and from the brain. Spinal integration usually takes the form of a **reflex,** an unconscious, programmed response to a specific stimulus. The knee-jerk reflex is an example of the simplest type of reflex, an automatic response involving only sensory and motor neurons (FIGURE 44.16a, p. 1014). Most reflexes are more complex, sometimes involving a higher degree of integration by one or more interneurons between the sensory and motor neurons (FIGURE 44.16b). Also, branches

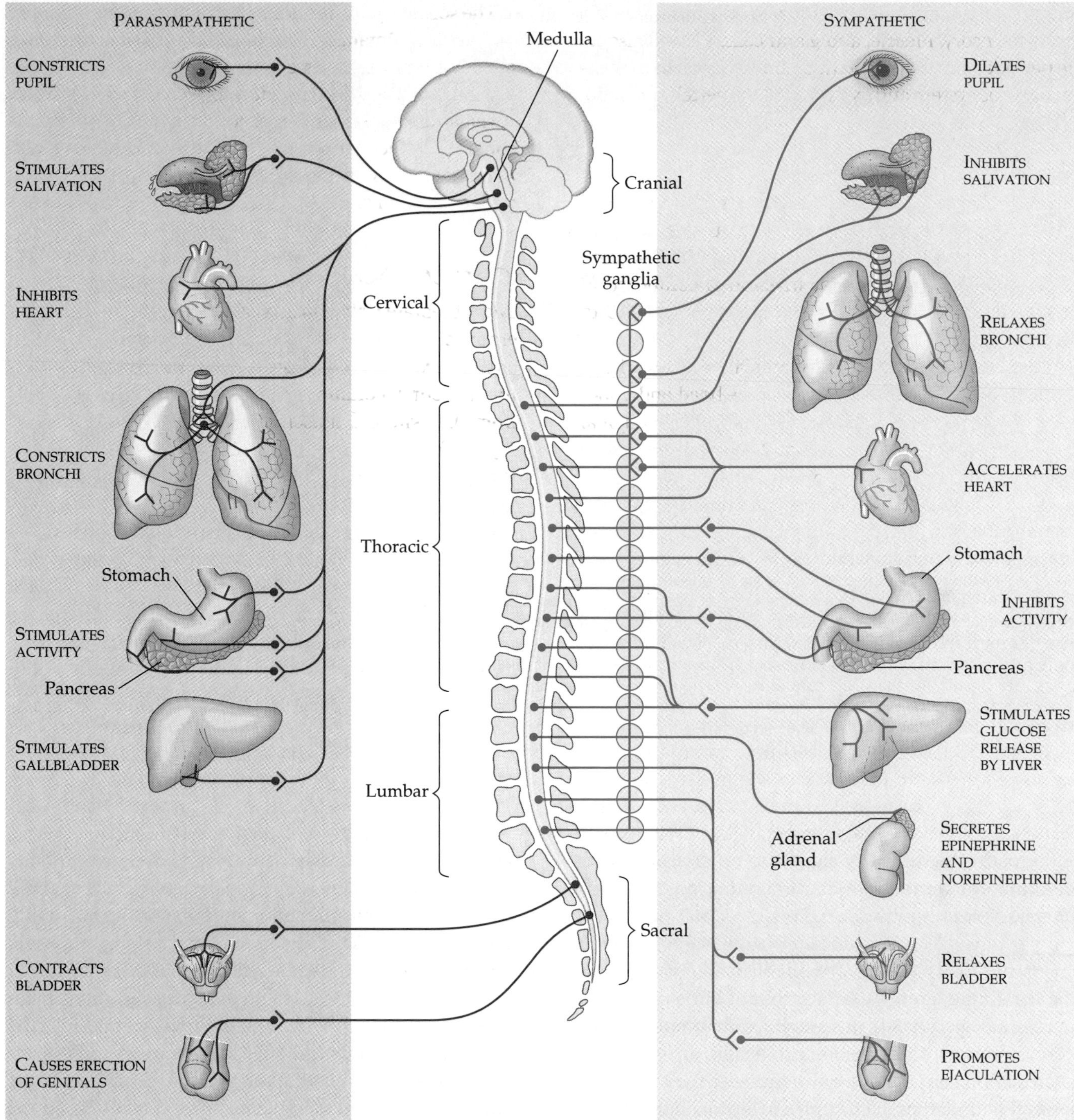

FIGURE 44.15

The autonomic nervous system. Anatomically, the subdivisions of the autonomic nervous system are distinguished by the place of origin of their nerves. Sympathetic nerves emerge from the middle regions of the spinal cord (thoracic and lumbar). Parasympathetic nerves originate from the medulla in the brain and the sacral region of the spinal cord. Most autonomic pathways consist of a chain of two neurons, but the arrangements and positions of the junctions between the presynaptic and postsynaptic cells differ. Most sympathetic nerves have relatively short presynaptic axons that synapse with cell bodies of the second neurons in prominent sympathetic ganglia near the spinal cord; the neurotransmitter released at this synapse is acetylcholine. Long axons of the postsynaptic cells leave the sympathetic ganglia and branch to the target organs. There, the neurotransmitter released from sympathetic synaptic terminals is usually norepinephrine. Presynaptic axons of parasympathetic nerves are typically much longer than presynaptic axons of sympathetic nerves, and synapses with the second neurons usually occur at or near the target organ. Acetylcholine is the neurotransmitter released by parasympathetic neurons, at both the target organ and the synapse between the two neurons in the chain.

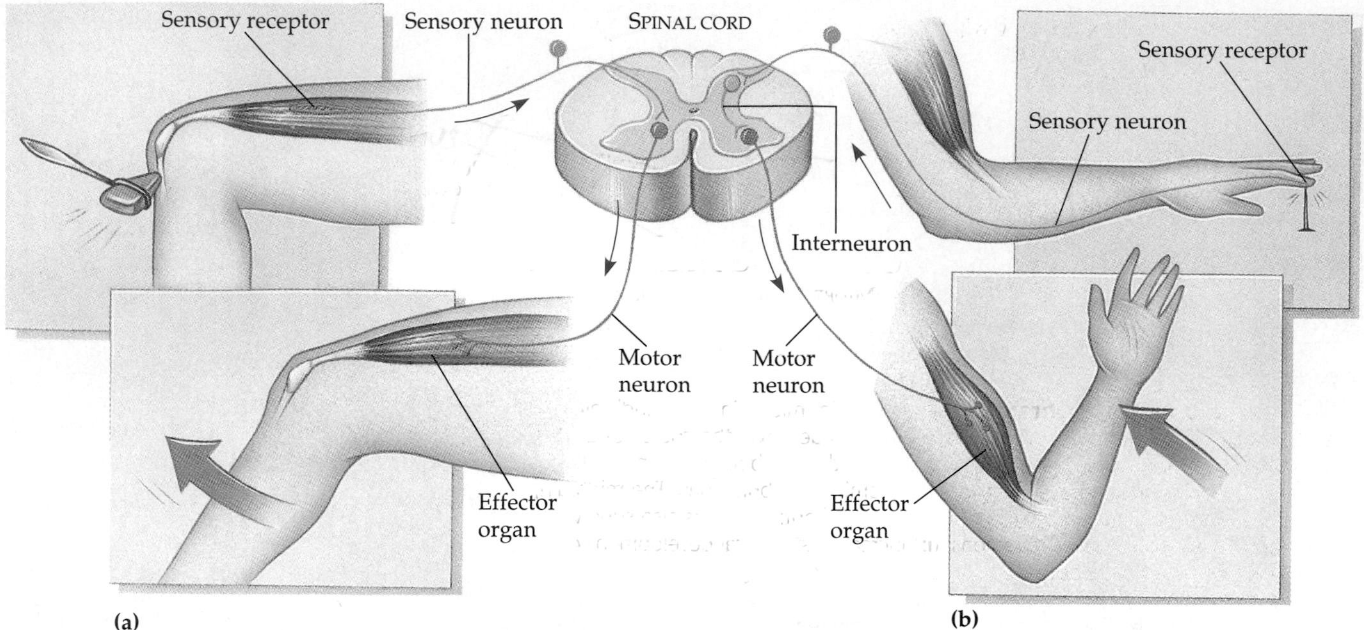

FIGURE 44.16
The spinal cord and spinal reflexes. Gray matter (the butterfly-shaped region) in the center of the spinal cord, shown here in cross section, contains the cell bodies of motor neurons and interneurons. The outer white matter consists of motor and sensory axons. Ganglia outside the spinal cord (dorsal root ganglia) contain cell bodies of the incoming sensory neurons. (**a**) The knee-jerk, or patellar, reflex is initiated when a sensory receptor detects a sudden stretch in the quadriceps muscle in the thigh (here caused by tapping the patellar tendon). A sensory neuron conveys the information (action potentials) to a synapse with a motor neuron in the spinal cord. If the signal is strong enough, an action potential will be generated in the motor neuron, making the quadriceps muscle contract, jerking the lower leg forward. Only two neurons mediate the actual reflex action, but at least one other neuron is also involved. Whenever one motor neuron conveys a signal to contract to the quadriceps muscle, another motor neuron carries an inhibitory signal that prevents a different leg muscle, whose action opposes that of the quadriceps, from contracting. (**b**) Most reflexes involve at least one interneuron coordinating sensory input with motor output.

in the pathway may carry signals to other segments of the spinal cord or to the brain, generating larger-scale or more complex responses.

Evolution of the Vertebrate Brain

The vertebrate brain, especially that of birds and mammals, is a powerful data processor. While keeping all the organ systems running homeostatically, an enormous coordination task, the brain also provides the integrative power that underlies the complex behavior characteristic of all vertebrates.

The vertebrate brain evolved from a set of three bulges at the anterior end of the spinal cord (FIGURE 44.17a). These three ancestral regions—the **forebrain,** the **midbrain,** and the **hindbrain**—are still present in all vertebrates, but they are further subdivided structurally and functionally, providing additional capacity for the integration of complex activities.

Three trends are evident in the evolution of the vertebrate brain. First, the relative size of the brain increases in certain evolutionary lineages. Brain size is a fairly constant function of body weight among fishes, amphibians, and reptiles, but is dramatically larger relative to body size in birds and mammals. A rodent weighing 100 g would have a much larger brain than a 100-g lizard. But the brain of that lizard and the brain of a 100-g fish would be approximately the same size.

A second evolutionary trend is increased compartmentalization of function. In the hindbrain, for example, the region known as the cerebellum becomes a prominent structure for coordinating movements. In the forebrain, one subdivision contains integrating centers called the thalamus and hypothalamus, while another subdivision contains the cerebral cortex, the part of the brain most important in learning and memory. As these specific regions become more complex, the original divisions between the three bulges become blurred. The mammalian midbrain, hindbrain, and lower forebrain are not readily distinguishable in adults, although the three separate bulges are apparent in developing embryos.

The third trend in the evolution of the vertebrate brain is the increasing sophistication and complexity of the forebrain. As amphibians and reptiles made the transition from water to land, the vision and hearing functions

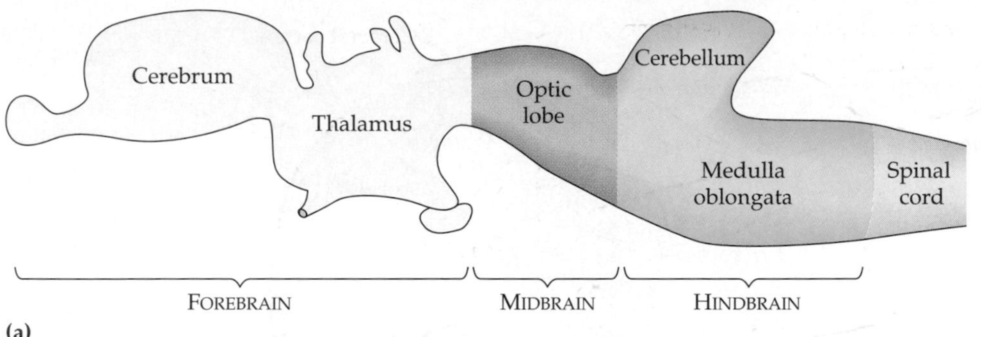

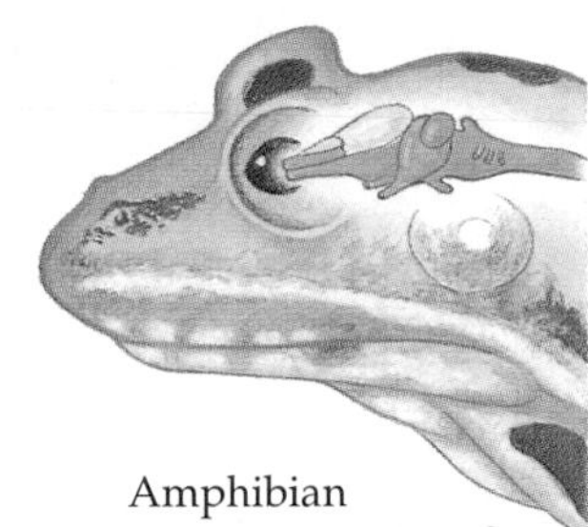

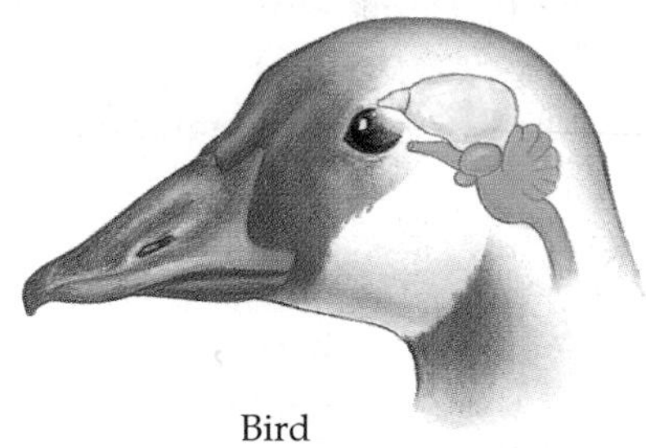

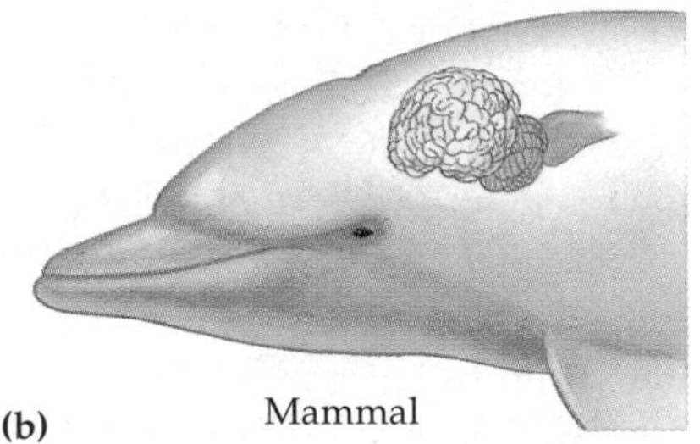

FIGURE 44.17

Evolution of the vertebrate brain. (**a**) The forebrain, midbrain, and hindbrain, the ancestral regions of the vertebrate brain, became subdivided over the course of evolution. The cerebrum and thalamus of the forebrain and the cerebellum and medulla oblongata of the hindbrain are four of the most prominent subdivisions. The midbrain, prominent in fishes, amphibians, and reptiles as a pair of optic lobes, is also subdivided but less dramatically than the other regions. (**b**) Increased structural development correlated with greater integrative power of the forebrain (yellow) is one of three major evolutionary trends in vertebrate brains. The most radical evolutionary change, a general increase in relative size and dominance over other brain regions, occurred in the cerebrum (the expanded portion of the forebrain in these drawings). The cerebrums of birds and mammals develop from different regions of the forebrain, but they share the distinction of being larger relative to other parts of the brain than the cerebrums of other vertebrates. Folds on the surface of the mammalian cerebrum (the cerebral cortex) increase its surface area and thereby its integrative power.

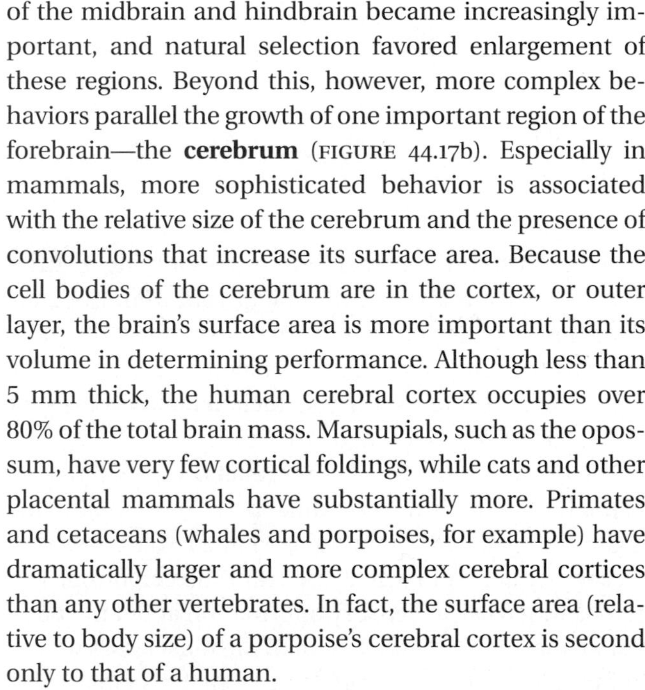

of the midbrain and hindbrain became increasingly important, and natural selection favored enlargement of these regions. Beyond this, however, more complex behaviors parallel the growth of one important region of the forebrain—the **cerebrum** (FIGURE 44.17b). Especially in mammals, more sophisticated behavior is associated with the relative size of the cerebrum and the presence of convolutions that increase its surface area. Because the cell bodies of the cerebrum are in the cortex, or outer layer, the brain's surface area is more important than its volume in determining performance. Although less than 5 mm thick, the human cerebral cortex occupies over 80% of the total brain mass. Marsupials, such as the opossum, have very few cortical foldings, while cats and other placental mammals have substantially more. Primates and cetaceans (whales and porpoises, for example) have dramatically larger and more complex cerebral cortices than any other vertebrates. In fact, the surface area (relative to body size) of a porpoise's cerebral cortex is second only to that of a human.

The human brain is a major research frontier

At 1.35 kg (about 3 lb), the human brain is one of the largest organs in the body. Its soft, almost squishy texture belies the density of its cells and the complexity of its structure and function.

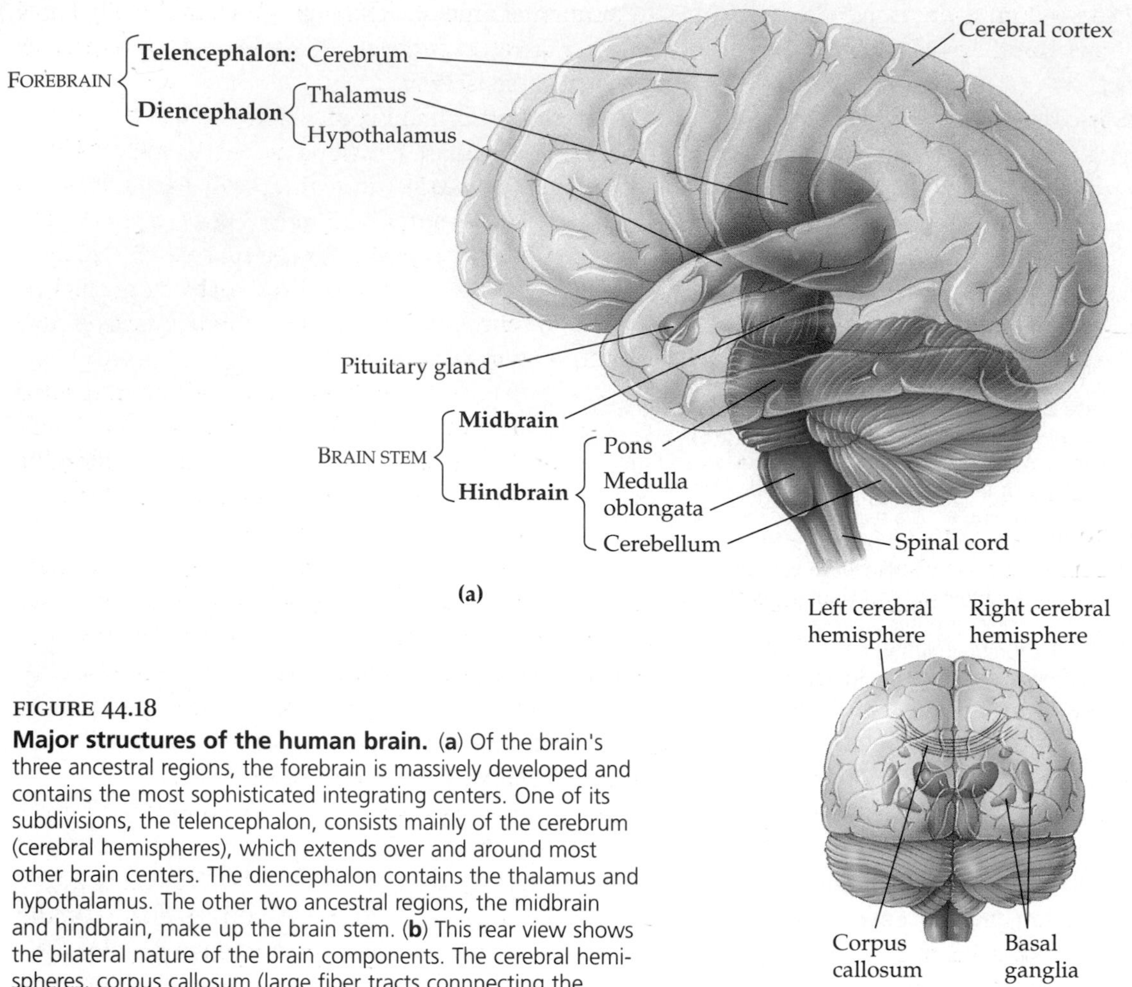

FIGURE 44.18
Major structures of the human brain. (**a**) Of the brain's three ancestral regions, the forebrain is massively developed and contains the most sophisticated integrating centers. One of its subdivisions, the telencephalon, consists mainly of the cerebrum (cerebral hemispheres), which extends over and around most other brain centers. The diencephalon contains the thalamus and hypothalamus. The other two ancestral regions, the midbrain and hindbrain, make up the brain stem. (**b**) This rear view shows the bilateral nature of the brain components. The cerebral hemispheres, corpus callosum (large fiber tracts connnecting the hemispheres), and basal ganglia are parts of the telencephalon.

Anatomy of the Brain

Let's take a closer look at this remarkable organ, starting with the hindbrain and midbrain (FIGURE 44.18). Forming a stalk and caplike swellings atop the spinal cord, these regions are collectively called the **brain stem,** also known as the lower brain. The hindbrain has three parts that function in homeostasis, movement coordination, and data conduction. The lowest part of the brain, the **medulla oblongata** (commonly called the medulla) contains centers that control several visceral (autonomic, homeostatic) functions, including breathing, heart and blood vessel activity, swallowing, vomiting, and digestion. Just above the medulla, the **pons** also participates in some of these activities, having nuclei that regulate the breathing centers in the medulla, for example. All the bundles of axons carrying sensory information to and motor instructions from higher brain regions pass through the hindbrain, making data conduction one of the most important functions of the medulla and pons. The hindbrain also helps coordinate large-scale body movements such as walking. Most of the descending axons carrying instructions about movement to the spinal cord from the midbrain and forebrain cross from one side of the CNS to the other as they pass through the medulla. As a result, the right side of the brain controls much of the movement of the left side of the body, and vice versa.

The primary function of the **cerebellum,** the third part of the hindbrain, is the coordination of movement. This highly convoluted, semidome-shaped outgrowth is tucked behind and partially beneath the cerebrum. The cerebellum receives sensory information about the position of the joints and the length of the muscles, as well as information from the auditory and visual systems. It also receives input from the motor pathways, telling it which actions are being commanded from the cerebrum. The cerebellum uses this information to provide automatic coordination of movements and balance. If one part of the body is moved, the cerebellum will coordinate other parts to ensure smooth action and maintenance of equilibrium. Hand-eye coordination is one example of cerebellar function. If the cerebellum is damaged, the eyes can follow a moving object, but they will not stop at the same place as the object.

The midbrain contains centers for the receipt and integration of several types of sensory information. It also serves as a projection center, sending coded sensory information along neurons to specific regions of the forebrain. Prominent nuclei of the midbrain are the **inferior** and **superior colliculi,** which are part of the auditory and visual systems, respectively. All fibers involved in hearing either terminate in or pass through the inferior colliculi. In nonmammalian vertebrates, the superior colliculi take the form of prominent optic lobes and may be the only visual centers. In mammals, vision is integrated in the forebrain, leaving the superior colliculi to coordinate visual reflexes and perform limited perceptual functions. Some of the nuclei in the midbrain are part of the **reticular formation,** a functional system of neurons localized in the core of the medulla, pons, and midbrain, that regulates states of arousal.

The most sophisticated neural processing occurs in the forebrain. The intricate networks of integrating centers and sensory and motor pathways allow pattern and image formation, as well as associative functions, such as memory, learning, and emotions. Of the two major divisions of the forebrain, the lower **diencephalon** contains two integrating centers, the thalamus and the hypothalamus. The upper **telencephalon** consists of the cerebrum, the most complex integrating center in the CNS.

Within the diencephalon, the most prominent integrating center is the **thalamus,** a major relay station for sensory information on its way to the cerebrum. The thalamus contains many different nuclei, each one dedicated to sensory information of a particular type. Incoming information from all the senses is sorted out in the thalamus and sent on to the appropriate higher brain centers for further interpretation and integration. The thalamus also receives input from the cerebrum and from parts of the brain that regulate emotion and arousal, making it an important station for controlling access to the cerebrum.

Although it weighs only a few grams, the **hypothalamus** is one of the most important sites for the regulation of homeostasis. We have already seen that the hypothalamus is the source of two sets of hormones, posterior pituitary hormones and releasing hormones of the anterior pituitary (see FIGURE 41.10). It contains the body's thermostat, as well as centers for regulating hunger, thirst, and many other basic survival mechanisms. This region also plays a role in sexual response and mating behaviors, the fight-or-flight response, and pleasure. The hypothalamic pleasure centers have been given that name because of the responses seen when they are stimulated in experimental animals, although we cannot really know whether a rat experiences what humans interpret as pleasurable sensations.

Another part of the hypothalamus, called the **suprachiasmatic nucleus,** functions as our biological clock. It maintains our daily biorhythms—for example, when we sleep, when our blood pressure is highest, and when our sex drive peaks. Visual information received via sensory neurons from the eyes enables the clock to remain synchronized with the natural cycles of day length and darkness.

Making up the telencephalon, the cerebrum is divided into right and left **cerebral hemispheres.** Each hemisphere consists of an outer covering of gray matter, the cerebral cortex; internal white matter; and a cluster of nuclei deep within the white matter, the basal ganglia (FIGURE 44.18b). The **basal ganglia** are important centers for motor coordination, acting as switches for impulses from other motor systems. If the basal ganglia are damaged, a person may become passive and immobile because the ganglia no longer allow motor impulses to be sent to the muscles. Degeneration of cells entering the basal ganglia occurs in Parkinson's disease.

The **cerebral cortex** is the largest and most complex part of the human brain, and the part that has changed the most during vertebrate evolution. The highly folded cerebral cortex of humans has a surface area of about 0.5 m^2. Like the rest of the brain, the cerebral cortex is bilaterally symmetrical, and the two hemispheres are connected by a thick band of fibers (cerebral white matter) known as the **corpus callosum.** The surface of each cerebral hemisphere has four discrete lobes, and researchers have identified a number of functional areas within each lobe (FIGURE 44.19).

Two functional cortical areas, motor cortex and somatosensory cortex, form the boundary between the frontal lobe and the parietal lobe. The motor cortex functions mainly in sending commands to skeletal muscles, signaling appropriate responses to sensory stimuli. The somatosensory cortex is a mosaic of regions corresponding to different parts of the body. This functional area receives and partially integrates signals from touch, pain, pressure, and temperature receptors throughout the body. The proportion of somatosensory or motor cortex devoted to a particular part of the body is correlated with the importance of sensory or motor information for that part of the body. For example, more brain surface is committed to sensory and motor communication with the hands than with the entire torso. Impulses transmitted from receptors to specific areas of somatosensory cortex enable us to associate pain, touch, pressure, heat, or cold with specific parts of the body receiving those stimuli. However, the so-called special senses—vision, hearing,

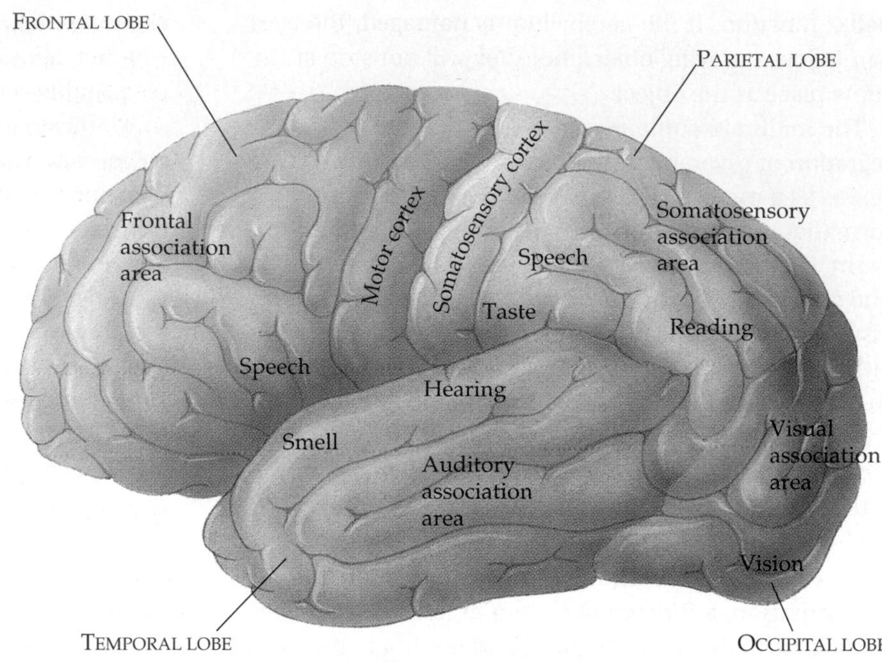

FIGURE 44.19
Functional areas of the cerebral cortex. The surface of each cerebral hemisphere is divided into four lobes. (A fifth lobe, the insula, located within the fold separating the temporal and parietal lobes, is not shown in this drawing of the left hemisphere.) Specialized functions are localized in each lobe. The association areas of the left hemisphere have different functions from those of the right hemisphere.

smell, and taste—are integrated by other cortical regions. Each of these functional regions, as well as the somatosensory cortex, cooperates with an adjacent association area (see FIGURE 44.19). Researchers are just beginning to understand how a complicated interchange of signals among receiving centers and association centers produces our sensory perceptions.

Integration and Higher Brain Functions

Integration of nerve impulses occurs at all levels in the human nervous system. The simplest kinds of integration are the spinal reflexes illustrated in FIGURE 44.16; the most complex integration enables the cerebral cortex to create a work of art or make a scientific discovery. Some aspects of brain activity that are particularly interesting are arousal and sleep, lateralization (hemispheric specialization), language and speech, emotions, and memory.

Arousal and Sleep. As anyone who has sat through a lecture on a warm spring day knows, attentiveness and mental alertness vary from moment to moment. Arousal is a state of awareness of the external world. The counterpart of arousal is sleep, when an individual continues to receive external stimuli but is not conscious of them. The mechanisms of arousal and sleep are fairly well understood, but the question of why we sleep remains a compelling problem. All birds and mammals sleep and show a characteristic sleep-wake cycle, which is probably mediated by the hypothalamus, as it is in humans.

Sleep and wakefulness produce different patterns in the electrical activity of the brain, which can be recorded in an **electroencephalogram,** or **EEG** (FIGURE 44.20). As a general rule, the less mental activity taking place, the more synchronous the brain waves of the EEG. When a healthy person is lying quietly with closed eyes, slow, synchronous *alpha waves* predominate. When the eyes are opened or the person solves a complex problem, faster *beta waves* take over, indicating desynchronization of the parts of the brain.

A sleeping person's EEG reflects the fact that sleep is a dynamic process. In the early stages, *theta waves,* more irregular than beta waves, often predominate. Deeper sleep produces *delta waves,* which are quite slow and highly synchronized. Deep sleep also includes periods when a desynchronized EEG reminiscent of wakefulness occurs. During these periods, called REM (rapid eye movement) sleep, the eyes move actively across the visual field behind closed lids. Most dreaming occurs during REM sleep. Like sleep, dreaming has been ascribed magical or prophetic importance, but its true function remains unknown.

Sleep and arousal are controlled by several centers in the cerebrum and the brain stem. The reticular formation is a vital link in determining states of arousal and consciousness. This group of over 90 separate nuclei extends from the medulla to the thalamus, through which almost all neuron processes reaching the cerebral cortex must pass. The reticular formation is essentially a sensory filter that selects which information reaches the cortex. The more input the cortex receives, the more alert and aware a person is. But arousal is not just a generalized phenomenon; certain stimuli can be ignored while the brain is actively processing other input. Also, specific centers regulate sleep and wakefulness. The pons and medulla con-

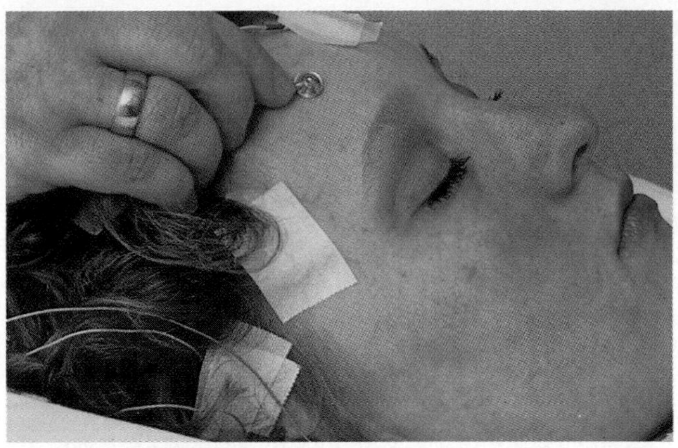

(a)

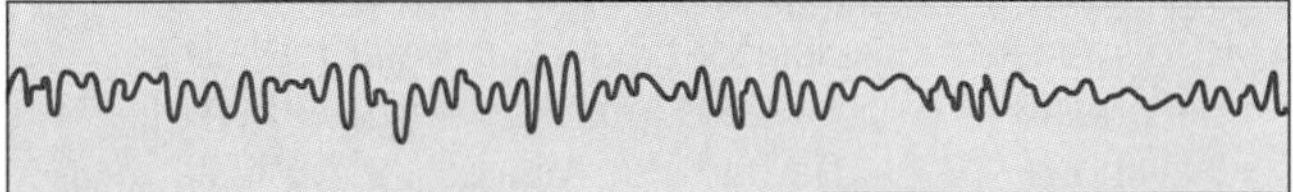

(b)

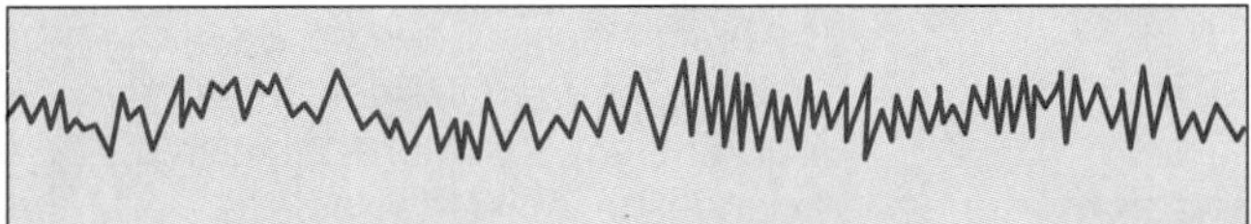

(c)

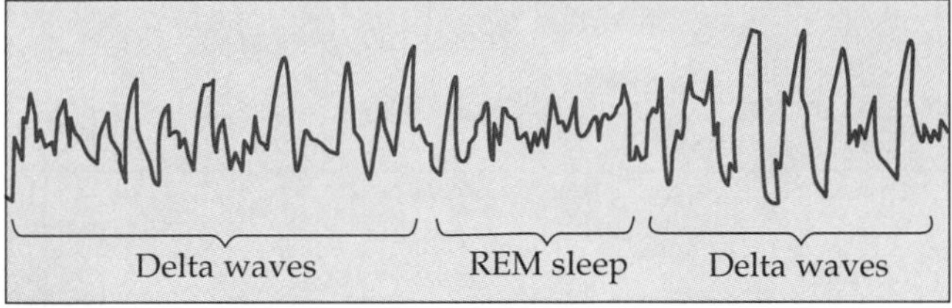

(d)

FIGURE 44.20
Brain waves recorded by an electroencephalogram (EEG). (**a**) Electrical contacts placed on the scalp detect brain waves (electrical activity of the brain). (**b**) When a person is awake but quiet, slow alpha waves dominate the EEG. (**c**) During intense mental activity, rapid, irregular beta waves predominate. (**d**) A sleep cycle usually includes periods of slow, fairly regular delta waves interspersed with periods when brain waves resemble those of the awake state, during REM sleep and dreaming.

tain nuclei that cause sleep when stimulated, and the midbrain has a center that causes arousal. Serotonin may be the neurotransmitter of the sleep-producing centers. Drinking milk before bedtime might induce sleep because milk contains large amounts of tryptophan, the amino acid from which serotonin is synthesized.

Right Brain/Left Brain. The association areas of the cerebral cortex, unlike the motor cortex and somatosensory cortex, are not bilaterally symmetrical; each side of the brain controls different functions. Speech, language, and calculation, for example, are centered in the left hemisphere, while the right hemisphere controls creative ability and spatial perception. Much of what we know about this **lateralization** of the brain comes from the work of Nobel Prize–winner Roger Sperry and his colleagues, who have performed extensive studies of "split-brain" patients. Some forms of epilepsy involve reverberating circuits that send massive electrical discharges back and forth between the right and left hemispheres through the corpus callosum (see FIGURE 44.18b). A now outmoded treatment for this illness was to surgically split the brain by severing the corpus callosum. Surprisingly, this drastic procedure did not seem to affect behavior overtly, but as Sperry's work showed, it did have subtle effects on the patient's brain function.

A person with a severed corpus callosum may appear perfectly normal in most situations, but careful experiments revealed much about lateralization. A subject holding a key in the left hand, with both eyes open, will readily name it as a key. If blindfolded, though, the subject will recognize the key and use it to open a lock (and may be able to describe it) but will be completely unable to name it. The center for speech is in the left hemisphere, but sensory information from the left hand crosses and enters the right side of the brain. Without the corpus callosum to function as a switchboard between the two sides of the brain, knowledge of the size, texture, and function of the object cannot be transferred from the right to the left hemisphere. Thus, sensory input and spoken response are dissociated.

Language and Speech. Language results from some extremely complex interactions among several association areas in the left cerebral hemisphere. For instance, speech and reading association areas in the parietal lobe obtain visual information (e.g., the appearance of words on a page) from the vision centers. The parietal lobe's speech area stores information required for speech content, arranging the words of a learned vocabulary into meaningful speech according to the rules of grammar. Another speech association area in the frontal lobe contains information required for speech production. Told "what to say" by the parietal speech area, it programs the motor cortex to move the tongue, lips, and other speech muscles to articulate the words. You can see these areas "in action" in the PET scans shown in the Methods Box on p. 1020. Damage to either speech area causes very different kinds of aphasia, the inability to speak coherently. Damage to the parietal speech area leads to the production of long strings of words and nonsense syllables. Damage to the frontal speech area leads to a loss of fluency, but some content remains.

METHODS: IMAGING THE NERVOUS SYSTEM

Several imaging techniques are now available to researchers studying normal body functions and physicians diagnosing various disorders, including cancer. These techniques are especially useful in studies of the brain.

One widely used method called **magnetic resonance imaging (MRI)** takes advantage of the behavior of hydrogen atoms in water molecules. The nuclei of hydrogen atoms are usually oriented in random directions. MRI uses powerful magnets to align the nuclei, then knocks them out of alignment with a brief pulse of radio waves. Still under the magnets' influence, the hydrogen nuclei spring back into alignment, giving out faint radio signals of their own, signals detected by the MRI scanner and translated by computer into an image. Soft tissues having relatively high water content, including the nervous system, appear more opaque in the images than dense structures, such as bone, which contain relatively little water. For this reason, MRI is useful for detecting problems in the brain and spinal cord, which are surrounded by bone. FIGURE a is an MRI image of a tumor (artificially colored red) in the human spinal cord.

Another imaging technique called **computed tomography (CT)** produces images of a series of thin X-ray sections through the body. A patient is slowly moved through a CT machine as an X-ray source circles around the body, illuminating successive sections from many angles. A computer then produces high-resolution video images of the sections, which can be studied individually or combined into various three-dimensional views. CT is especially useful for detecting ruptured blood vessels on the brain.

A third imaging technology, widely used in brain research, is called **positron-emission tomography (PET).** PET can reveal the location of a variety of physiological or biochemical processes in the human body. In preparation for a PET scan, water, glucose, or another molecule (depending on the process to be studied) is labeled with a radioactive isotope and injected into a person's bloodstream. (The isotopes are not dangerous because they are used only in small quantities.) The images in FIGURE b are PET scans of the human brain obtained using water labeled with a radioisotope of oxygen that emits particles called positrons. Because radioactive water tends to go to the most active areas of the brain, scientists can use this technique to map those areas used when the brain is performing a particular task. The positrons interact with oppositely charged electrons on the body's atoms, and the resulting radiation is detected by a PET camera. A computer creates the brain maps you see here, which illustrate localized brain activity under four different conditions, all related to language in healthy subjects. Physicians use the PET scanner to evaluate brain and heart disorders and certain types of cancer.

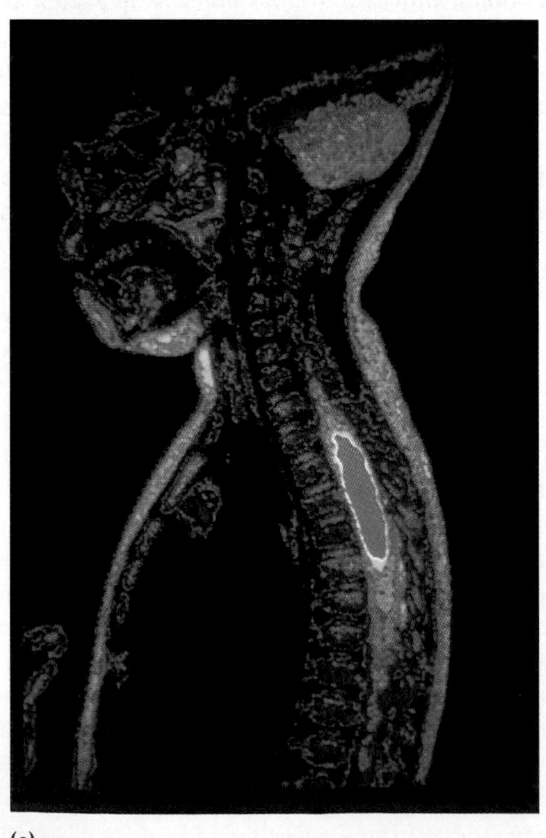

(a)

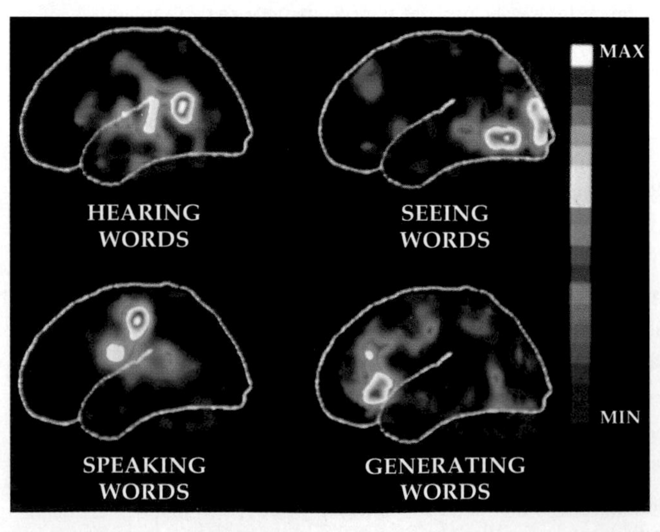

(b)

Emotions. What causes us to laugh, cry, love, and fight has been the subject of much biological and philosophical speculation. Some say our emotions cause facial expressions; others suggest that contracting the facial muscles stimulates the emotional centers of the brain. Some hypotheses propose that emotions result from feedback from the body's organs and muscles to the CNS. Emotions are difficult to study experimentally, because even if an experimental animal seems to *show* emotion, we cannot say conclusively that the animal *feels* the emotion in the same sense that humans do.

Researchers have developed a partial map of the brain regions that generate emotions. The map comes mainly from studies of brain-damaged people and other mammals, but the pace of discovery of how the brain integrates all types of information has been greatly speeded by new imaging technology (see the Methods Box). Much of human emotion depends on a functional group of nuclei and interconnecting axon tracts in the forebrain, called the **limbic system** (FIGURE 44.21). Though still only loosely defined, the limbic system includes parts of the thalamus, hypothalamus, and inner portions of the cerebral cortex, including two nuclei called the **amygdala** and **hippocampus,** which also may function in memory. Its cerebral components are linked to the prefrontal cortex, the part of the cerebral cortex involved in complex learning, reasoning, and personality. The prefrontal cortex makes decisions about the emotional content of these unique human qualities after "consulting" the limbic system. Eliminating this emotional consultation through surgical destruction of the limbic cortex or its connection with the prefrontal cortex, a procedure called frontal lobotomy, has been used as a treatment for severe mental disorders. However, the resulting docility is not necessarily a cure, and such drastic surgery usually impairs other brain functions.

Memory. The manuscript for this chapter was typed on a personal computer that weighs about 11 kg and can retain about 4×10^6 bits (1200 pages) of information. Your brain, a much more powerful data processor, weighs about 1.35 kg and stores billions of bits of information dating back to the beginning of your life.

Memory, which is essential for learning, is the ability to store and retrieve information related to previous experiences. Human memory occurs in two stages. **Short-term memory** reflects an immediate sensory perception of an object or idea and occurs before the image is stored. Short-term memory enables you to dial a phone number after looking it up but without looking at it directly. If you call the number frequently, it becomes stored in **long-term memory** and can be recalled several weeks after you originally looked it up. The transfer of information from short-term to long-term memory is enhanced by rehearsal ("practice makes perfect"), favorable emotional state (we learn best when we are alert and motivated), and the association of new data with data previously learned and stored in long-term memory (it's easier to learn a new card game if you already have "card sense" from playing other games).

In its ability to learn and remember, the human brain apparently distinguishes between facts and skills. When you acquire factual knowledge by memorizing dates, word definitions, the parts of the brain, and other information, this fact memory can be consciously and specifically retrieved from the data bank of your long-term memory. You can even recall visual images, such as the face of a friend. In contrast, skill memory usually involves

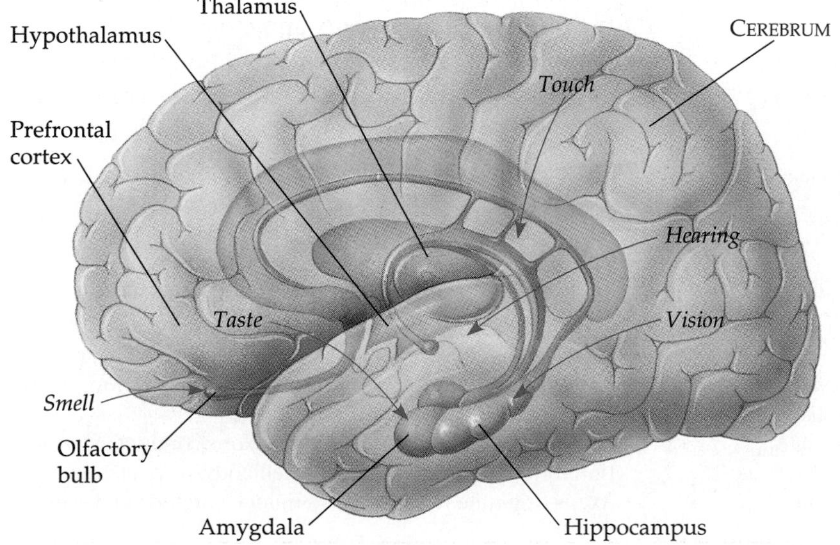

FIGURE 44.21

The limbic system. Portions of the diencephalon (thalamus and hypothalamus) and inner parts of the cerebral cortex, including the amygdala and hippocampus, make up this functional center of human emotions and memory. Signals from the nose enter the brain through the olfactory bulb, which is part of the limbic system. Other sensory information enters the limbic system via other parts of the cerebral cortex (arrows). A higher center of integration in the cerebrum, the prefrontal cortex, apparently consults the limbic system and other brain centers in processing and retrieving memories and may use memories to modify behavior.

motor activities that are learned by repetition without consciously remembering specific information. You perform learned motor skills, such as walking, tying your shoes, riding a bicycle, or writing without consciously recalling the individual steps required to do these tasks correctly. Once a skill memory is learned, it is difficult to unlearn. For example, a person who has played tennis for years with a self-taught, awkward backhand has a much tougher time learning the correct form than a beginner just learning the game. Bad habits, as we know, are difficult to break.

By studying amnesia (memory loss) in humans, experimenting with animals, and using brain imaging techniques, neuroscientists are beginning to map the complex brain pathways involved in memory. For instance, in the case of a telephone number and other fact memories, sensory signals from the eyes go to the vision centers in the brain's occipital lobe, where a visual perception is formed (see FIGURE 44.21). They also pass through the brain's sensory filter, the reticular formation (to determine if attention should be paid to the information), to parts of the hypothalamus and limbic system (to determine whether emotion should be involved), and to higher centers of the forebrain, including the prefrontal cortex (for higher-level integration). The pathway is completed when impulses return to the cortical vision centers where the perception first occurred.

Physiologist Karl Lashley spent several decades of this century searching for what he called the engram, the physical basis of a memory. But because partial damage to one area of the cerebral cortex does not destroy individual memories, he ultimately concluded that there is no highly localized memory trace in the nervous system. Rather, a memory seems to be stored within a certain association area of the cortex with some redundancy.

Neuroscientists are investigating the cellular changes involved in memory and learning. Two limbic nuclei, the amygdala and hippocampus, seem to play key roles (see FIGURE 44.21). The amygdala may act as a type of memory filter, somehow labeling information to be saved by tying it to an event or emotion of the moment. In the hippocampus, a functional change at certain synapses seems to be directly related to memory storage and learning. Called **long-term potentiation (LTP),** this change is an enhanced response by a postsynaptic cell to an action potential. It can result when a presynaptic cell bombards a synapse with a series of brief, repeated action potentials that strongly depolarize the postsynaptic membrane. With LTP established, a single action potential from the presynaptic cell has a much greater effect at the synapse than before. Lasting for a matter of hours, days, or weeks, depending on the number and frequency of repeated action potentials, LTP may be what happens when a memory is being stored or when learning takes place.

LTP is associated with the release of the excitatory neurotransmitter glutamate by the presynaptic cell. Glutamate binds with a specific class of receptors in the postsynaptic membrane, opening gated channels that are highly permeable to calcium ions (Ca^{2+}). In turn, the Ca^{2+} triggers a cascade of intracellular changes that underlie LTP. Contributing further to LTP, the affected postsynaptic cell may signal the presynaptic cell to release more glutamate, an example of positive feedback. One hypothesis being tested is that the postsynaptic cell signals the presynaptic cell, and perhaps synapses on neighboring neurons as well, by secreting the local mediator nitric oxide.

Studies of the chemical and neuronal basis of learning and memory are part of a rapidly expanding research effort in neurobiology. Unraveling more about how networks of neurons in the brain store, retrieve, and use memories, control the internal environment of the body, construct our thoughts and emotions, and make us conscious of ourselves and our surroundings is one of the most challenging and fascinating aspects of modern biology. Neuroscientists such as Patricia Churchland (see the interview on p. 776) are beginning to unify the "mind" and the "brain."

In the next chapter, we examine the relationships of the nervous system to sensory and motor functions: how sense organs gather information about the environment and how skeletal muscles carry out motor commands.

REVIEW OF KEY CONCEPTS (with page numbers and key figures)

- Nervous systems perform the three overlapping functions of sensory input, integration, and motor output: *an overview* (p. 994, FIGURE 44.1)
  - The nervous system's three main functions are sensory input, integration, and motor output to effector cells.
  - The central nervous system (CNS) integrates information, while the interconnecting nerves of the peripheral nervous system (PNS) communicate sensory and motor signals between the CNS and the rest of the body.
- The nervous system is composed of neurons and supporting cells (pp. 994–996, FIGURES 44.2, 44.3)
  - Cells of the nervous system include neurons, which transmit the signals, and supporting cells, which support, insulate, and protect the neurons.
  - A neuron's fiberlike dendrites and axons conduct information toward and away from the cell body, respectively. Axons originate from the axon hillock and terminate in

numerous branches. Synaptic terminals at the ends of axons release neurotransmitter molecules into the synapses, thereby relaying neural signals to the dendrites or cell bodies of other neurons or effectors.

- The CNS consists of the brain and spinal cord. The PNS contains sensory neurons, which transmit information from internal and external environments to the CNS; and motor neurons, which carry information from the CNS to target organs. Interneurons of the CNS integrate sensory input and motor output.
- Glial cells, supporting cells in the CNS, include astrocytes, which line the capillaries in the brain and contribute to the blood-brain barrier; and oligodendrocytes, which wrap and insulate some neurons in a myelin sheath. In the PNS, myelin sheaths are formed by supporting cells called Schwann cells.

- Impulses are action potentials, electrical signals propagated along neuronal membranes (pp. 996–1003, FIGURES 44.7, 44.8)
  - The membrane potential of a nontransmitting neuron is due to the unequal distribution of ions, particularly sodium and potassium, across the plasma membrane; the cytoplasm is more negatively charged than the extracellular fluid. Membrane potential is maintained by differential ion permeabilities and the sodium-potassium pump.
  - A stimulus that affects the membrane's permeability to ions can either depolarize or hyperpolarize the membrane relative to the membrane's resting potential. This local voltage change is called a graded potential, and its magnitude is proportional to the strength of the stimulus.
  - An action potential, or nerve impulse, is a rapid, transient depolarization of the neuron's membrane. A local depolarization to the threshold potential opens voltage-gated sodium channels, and the rapid influx of Na^+ brings the membrane potential to a positive value. The membrane potential is restored to its normal resting value by the delayed opening of voltage-gated K^+ channels and by the closing of the Na^+ channels. A refractory period follows an action potential, corresponding to the period when the voltage-gated Na^+ channels are inactivated.
  - The all-or-none generation of an action potential always creates the same amplitude of voltage change for a given neuron. The frequency of action potentials varies with the intensity of the stimulus.
  - Once an action potential is initiated in an axon, a wave of depolarization propagates a series of action potentials to the end of the axon.
  - The rate of transmission of a nerve impulse is directly related to the diameter of the axon. Saltatory conduction, a mechanism by which action potentials jump between the nodes of Ranvier of myelinated axons, speeds nervous impulses in vertebrates.

- Chemical or electrical communication between cells occurs at synapses (pp. 1003–1010, FIGURE 44.10)
  - Synapses between neurons conduct signals from the axon of a presynaptic cell to a dendrite or cell body of a postsynaptic cell.
  - Electrical synapses directly pass an action potential between two neurons via gap junctions.
  - In a chemical synapse, a depolarization stimulates the fusion of synaptic vesicles with the presynaptic membrane and the release of neurotransmitter molecules into the synaptic cleft. Neurotransmitters bind to receptor proteins associated with particular ion channels on the postsynaptic membrane. The selective opening of chemically sensitive gates either brings the membrane potential closer to threshold (EPSP) or hyperpolarizes the membrane (IPSP). The neurotransmitter is rapidly broken down by enzymes.
  - Whether or not an action potential is created in the postsynaptic cell depends on temporal or spatial summation of EPSPs and IPSPs at the axon hillock.
  - One of the most common invertebrate and vertebrate neurotransmitters is acetylcholine. Other transmitters that have been identified include the biogenic amines (epinephrine, norepinephrine, dopamine, and serotonin); several amino acids; and some neuropeptides, such as the analgesic endorphins. Some neurons also secrete gases, such as nitric oxide, to signal other cells.
  - Groups of neurons may interact and carry information along specific pathways called circuits. Nerve cell bodies are arranged in functional groups usually called ganglia in the PNS and nuclei in the brain.

- Invertebrate nervous systems are highly diverse (pp. 1010–1011, FIGURE 44.13)
  - Invertebrate nervous systems range from the diffuse nerve nets of many cnidarians to the highly centralized nervous systems of the cephalopods, which possess large brains capable of sophisticated learning.
  - The evolution of nervous system centralization was correlated with the evolution of bilateral symmetry.

- The vertebrate nervous system is a hierarchy of structural and functional complexity (pp. 1011–1015, FIGURE 44.14)
  - Functionally, the vertebrate PNS consists of the sensory, or afferent, division, which brings information from sensory receptors to the CNS; and the motor, or efferent, division, which carries signals away from the CNS to effector cells.
  - The motor division consists of the somatic nervous system, which carries signals to skeletal muscles, and the autonomic nervous system, which regulates the primarily automatic, visceral functions of smooth and cardiac muscles.
  - The autonomic nervous system consists of the parasympathetic and sympathetic divisions, which are anatomically, functionally, and chemically distinct and usually antagonistic in effect on target organs.
  - The CNS serves as the integrating link between the sensory and motor subdivisions of the PNS.
  - The spinal cord mediates many reflexes that integrate sensory input with motor output. It also has tracts of neurons that carry information to and from the brain.
  - All vertebrate brains develop and diversify from three regions: the forebrain, the midbrain, and the hindbrain.
  - Evolutionary trends in the vertebrate brain include increases in relative size, in compartmentalization of function, and in complexity of the forebrain, particularly the cerebral cortex.

- The human brain is a major research frontier (pp.1015–1022, FIGURES 44.18, 44.19)
  - In the human brain, the midbrain and hindbrain make up the brain stem.
  - The medulla oblongata and pons of the hindbrain work together to control homeostatic functions and conduct sensory and motor signals between the spinal cord and higher brain centers.
  - The cerebellum of the hindbrain coordinates movement and balance.
  - The midbrain receives, integrates, and projects sensory information to the forebrain.
  - The forebrain is the site of the most sophisticated neural processing, with major integrating centers in the thalamus, hypothalamus, and cerebrum.
  - The thalamus routes neural input to specific areas of the cerebral cortex, the outer gray matter of the cerebrum. The functions of the hypothalamus range from hormone production to the regulation of body temperature, hunger, thirst, sexual response, the flight-or-fight response, and biorhythms.
  - The cerebral cortex contains distinct somatosensory and motor areas, which directly process information; and association areas, which integrate information. Imaging technology enables researchers to identify specific integrating centers in the functioning brain.
  - Sleep and arousal are controlled by several areas in the cerebrum and brain stem, the most important being the reticular formation, which filters the sensory input sent to the cortex.
  - The two cerebral hemispheres control different functions. Speech, language, and analytical ability are centered in the left hemisphere, whereas spatial perception and artistic ability predominate in the right. Nerve tracts of the corpus callosum link the two sides and allow the brain to function as an integrated whole.
  - Human emotions are believed to originate from interactions between the limbic system, a group of nuclei in the diencephalon and inner cerebrum. The limbic system interacts closely with the prefrontal cortex, a higher integrating center of the forebrain.
  - Human memory consists of short-term and long-term memories. The process of learning facts appears to differ from that of learning skills. The hippocampus and amygdala, two components of the limbic system, participate in circular brain pathways involved in fact memory.
  - A functional change at synapses, called long-term potentiation (LTP), may underly memory storage and learning. Resulting from repeated bursts of action potentials, LTP is a heightened sensitivity to a single action potential by a postsynaptic membrane.

SELF-QUIZ

1. Which of the following occurs when a stimulus depolarizes a neuron's membrane?
   a. Na^+ diffuses out of the cell.
   b. The action potential approaches zero.
   c. The membrane potential changes from the resting potential to a voltage closer to the threshold potential.
   d. The depolarization is all or none.
   e. The Na^+-K^+ pump is stimulated.

2. Action potentials are usually propagated in only one direction along an axon because
   a. the nodes of Ranvier conduct only in one direction
   b. the brief refractory period prevents depolarization in the direction from which the impulse came
   c. the axon hillock has a higher membrane potential than the tips of the axon
   d. ions can flow along the axon only in one direction
   e. both sodium and potassium voltage-gated channels open in one direction

3. The depolarization of the presynaptic membrane of an axon *directly* causes
   a. voltage-gated calcium channels in the membrane to open
   b. synaptic vesicles to fuse with the membrane
   c. an action potential in the postsynaptic cell
   d. the opening of chemically sensitive gates that allow neurotransmitter to spill into the synaptic cleft
   e. an EPSP or IPSP in the postsynaptic cell

4. Anesthetics reduce pain by blocking the transmission of nerve impulses. Which of these three chemicals might work as an anesthetic?
   a. a chemical that blocks voltage-gated sodium channels in membranes
   b. a chemical that opens voltage-gated potassium channels
   c. a chemical that blocks neurotransmitter receptors
   d. a, b, and c
   e. only b and c

5. Gray matter is
   a. made up of three protective layers called meninges
   b. located on the outside of the spinal cord
   c. restricted to the brain
   d. populated by cell bodies of neurons
   e. found in the ventricles of the vertebrate brain

6. Which of the following structures or regions is incorrectly paired with its function?
   a. limbic system—screening of information between the spinal cord and the brain; regulates arousal and sleep
   b. medulla oblongata—homeostatic control center
   c. cerebellum—unconscious coordination of movement and balance
   d. corpus callosum—band of fibers connecting left and right cerebral hemispheres
   e. hypothalamus—production of hormones and regulation of temperature, hunger, and thirst

7. Receptor sites for neurotransmitters are located on the
   a. tips of axons
   b. axon membranes in the regions of the nodes of Ranvier
   c. postsynaptic membrane
   d. membranes of synaptic vesicles
   e. presynaptic membrane

8. All the following electrical changes of neurons are graded events *except*
   a. EPSPs
   b. IPSPs
   c. action potentials
   d. depolarizations caused by stimuli
   e. hyperpolarizations caused by stimuli

9. Of the following components of the nervous system, which is the *most inclusive?*
   a. brain
   b. spinal cord
   c. central nervous system
   d. gray matter
   e. neuron

10. A nerve net is characteristic of which animal phylum?
   a. Chordata
   b. Cnidaria
   c. Annelida
   d. Arthropoda
   e. Platyhelminthes

CHALLENGE QUESTIONS

1. From what you know about the action potential, propose one feasible mechanism whereby anesthetics might prevent pain.
2. Describe various ways in which drugs that are stimulants could increase the activity of the nervous system by acting at synapses.
3. Describe the role of calcium in nerve impulses.
4. Emergent properties and the hierarchical organization of living matter are two related themes in biology (see Chapter 1). Write an essay discussing some examples of how these themes apply to functions of the nervous system.

SCIENCE, TECHNOLOGY, AND SOCIETY

1. A woman lay in a coma for nine years after suffering irreversible brain damage in a car accident. Her family asked that she be allowed to die, but legal authorities refused to allow the hospital to withhold life support. A long legal battle went all the way to the U.S. Supreme Court, which ruled in 1990 that a person has a right to refuse medical treatment, but that the state may require clear evidence of the patient's wishes. The woman died twelve days after her feeding tube was removed. Only 10% to 15% of Americans have a living will or other documentation stating their wishes in the event that they become brain dead and are kept alive by artificial life support. What would you want done if you or a member of your family were in this situation? Why? Who should speak for the patient? Is it a physician's job to ask a patient his or her wishes, if that is possible? What constitutes evidence of a patient's wishes?
2. Parkinson's disease is a brain disorder that interferes with motor control, causing slow, jerky movements. It results from a decreased production of the neurotransmitter dopamine. Experiments show that many sufferers can be helped by transplanted tissues obtained from aborted fetuses, a treatment that restores the missing dopamine. In 1992, the U.S. government banned federal funding of fetal implants, fearing that using tissue from aborted fetuses would encourage women to have abortions. It has been suggested that tissue from miscarriages might be used for research. However, researchers doubt there is enough usable tissue from miscarriages for their needs. Do you think a ban on fetal tissue research is justified, or do you think this research should be funded? What are your reasons?

FURTHER READING

Gould, J. L., and C. G. Gould. *The Animal Mind.* New York: W. H. Freeman/ Scientific American Library, 1994. An introduction to the scientific study of cognition in nonhuman animals.

Horgan, J. "Can Science Explain Consciousness?" *Scientific American,* July 1994. A lucid review of current hypotheses and research on how the human brain formulates awareness.

Kandel, E. R., and R. D. Hawkins. "The Biological Basis of Learning and Individuality." *Scientific American,* September 1992. A review of key experiments with mollusks and mammals that led to an understanding of how nervous systems learn.

Kimura, D. "Sex Differences in the Brain." *Scientific American,* September 1992. Evidence of hormonally induced differences in brain development in the sexes.

Lancaster, J. R., Jr. "Nitric Oxide in Cells." *American Scientist,* May/June 1992. A lucid description of the delicate balance the body maintains in using a potentially lethal molecule in diverse physiological roles.

LeDoux, J. E. "Emotion, Memory and the Brain," *Scientific American,* June 1994. Research on the chemical basis of memory in humans and other animals.

Marx, J. "Helping Neurons Find Their Way." *Science,* May 19, 1995. Chemotaxis may direct growing axons to specific targets.

Raichle, M. E. "Visualizing the Mind." *Scientific American,* April 1994. Recent progress in studying neural activity in the thinking brain, using modern imaging techniques.

Travis, J. "Glia: The Brain's Other Cells." *Science,* November 11, 1994. A concise review of current research on supporting cells of the central nervous system.

CHAPTER 45

SENSORY AND MOTOR MECHANISMS

KEY CONCEPTS

- Sensory receptors detect changes in the external and internal environment: *an overview*
- Photoreceptors contain light-absorbing visual pigments: *a closer look*
- Hearing and balance are related in most animals: *a closer look*
- The interacting senses of taste and smell enable animals to detect many different chemicals: *a closer look*
- Movement is a hallmark of animals
- Skeletons support and protect the animal body and are essential to movement
- Muscles move skeletal parts by contracting

In the gathering dusk, a male moth's antennae detect the pheromone of a female moth somewhere upwind. The moth takes to the air, following the scent trail toward the female. Suddenly, vibration sensors in the moth's abdomen signal the presence of ultrasonic chirps of a rapidly approaching bat. The bat's sonar enables the mammal to locate moths and other flying insects for food. Reflexively, the moth's nervous system alters the motor output to its wing muscles, sending the insect into an evasive spiral toward the ground. Although it is probably too late for the moth in the photograph on this page, many moths can escape because they can detect a bat's sonar about 30 m away. The bat has to be within 3 m to sense the moth, but since the bat flies faster, it may still have time to detect, home in, and catch its prey.

The outcome of this interaction depends on the abilities of both predator and prey to sense important environmental stimuli and to produce coordinated movement that is appropriate. Although not all of an animal's moment-by-moment interactions with the environment are as dramatic as such predator-prey struggles, the detection and processing of sensory information and the generation of motor output provide the physiological basis for all animal behavior.

In Chapter 44, we saw how the nervous system transmits and integrates sensory and motor information. We will now examine the input and output of this coordinating system that controls so many aspects of animal behavior. We first consider the sensory receptors that receive information from the environment, and then examine the structure and function of muscles, the motor effectors that bring about movement in response to that information. We will also study skeletons in the context of body movement.

Sensory receptors detect changes in the external and internal environment: *an overview*

Amputees often report pain or numbness in "phantom" limbs that are no longer present. To understand this phenomenon, we must consider the distinction between sensation and perception.

Sensation and Perception

As we saw in Chapter 44, information is transmitted in the nervous system in the form of action potentials. An action potential triggered by light striking the eye is no different from an action potential triggered by air vibrating in the ear, yet we readily distinguish sight from sound. The difference depends on the part of the brain that receives the signal. The air vibrations we call sounds, for example, are converted by the ear into nerve impulses that are received by a particular region of the cerebral cortex. These nerve impulses, conveyed as action potentials along sensory neurons to the brain, are called **sensations.** Once the brain is aware of the sensations, it interprets them, giving us the **perception** of sound. Other kinds of input are sent to other parts of the brain and trigger different perceptions. What matters, then, is where the impulse goes, not what triggers it.

It follows that if neurons from your eyes could be crossed with those from your ears, you might perceive a camera flash as a loud "boom" and a concert as bursts of light. In a less dramatic demonstration of this principle, you have probably experienced light flashes while rubbing your eyes. The rubbing provides enough stimulation to generate action potentials in the sensory neurons of the optic nerves. The stimulus is pressure, but the perception is a spot of light because this is the only perception that can be generated by the part of the brain receiving the sensations. Returning to the example of pain in a phantom limb, severed nerves that carried impulses from the limb of an amputee may remain alive and respond to irritation. They can still transmit sensations to the brain and trigger perception. The resulting pain is just as real as the pain of a person who experiences irritated nerves in an existing arm. But how does the sensory system actually generate sensations?

The General Function of Sensory Receptors

Sensations, and the perceptions they evoke in the brain, begin with excitation of **sensory receptors,** structures that detect changes in the external and internal environment of the animal. Receptors are usually modified neurons or epithelial cells that exist singly or in groups with other cell types within sensory organs, such as the eyes and ears. Sensory receptors are specialized to respond to various stimuli, including heat, light, pressure, and chemicals. All these stimuli represent forms of energy. The general function of receptor cells is to convert the energy of stimuli into changes in membrane potentials and transmit signals to the nervous system. This task can be broken down into five functions common to all receptor cells: reception, transduction, amplification, transmission, and integration.

Reception is the ability of a cell to absorb the energy of a stimulus. Each kind of receptor has a region specifically suited to absorbing a particular type of energy. As we will see, sensory cells in the human eye, for example, have membranes that contain a light-absorbing pigment molecule.

Transduction is the actual conversion of stimulus energy into a change in the membrane potential of a receptor cell. The initial response of the sensory receptor to a stimulus is a change in its membrane permeability, resulting in a graded change in membrane potential called a **receptor potential.** (Recall from Chapter 44 that a graded potential is a change in the voltage across the membrane proportional to the strength of the stimulus.) In some cases, a stimulus such as pressure can stretch the membrane and increase ion flow. In other cases, specific receptor molecules on the membrane of a receptor cell open or close gates to ion channels when the stimulus is present. We will examine specific examples of sensory transduction later in the chapter.

Amplification is the strengthening of stimulus energy that is otherwise too weak to be carried into the nervous system. Amplification of the signal may occur in accessory structures of a complex sense organ, as when sound waves are enhanced by a factor of more than 20 before reaching the receptors of the inner ear. Amplification also may be a part of the transduction process itself. An action potential conducted from the eye to the brain has about 100,000 times as much energy as the few photons of light that triggered it.

Once the energy in the stimulus has been transduced into a receptor potential, **transmission,** or the conduction of impulses to the CNS, can occur. In some instances, such as in the case of "pain cells," the receptor itself is actually a sensory neuron that conducts action potentials to the CNS. Other receptors are separate cells that must transmit chemical signals across synapses to sensory neurons. If the receptor also functions as the sensory neuron, the intensity of the receptor potential will affect the frequency of action potentials that travel as sensations to the CNS. For separate receptor cells, the strength of the stimulus and receptor potential affects the amount of neurotransmitter released by the receptor at its synapse with a sensory neuron, which in turn determines the frequency of action potentials fired by the sensory neuron. Many sensory neurons fire at a low rate spontaneously. Therefore, a stimulus does not really switch the production of action potentials on or off; it modulates their frequency. In this way, the CNS is sensitive not only to the presence or absence of a stimulus but also to changes in stimulus intensity.

Integration, the processing of information, begins as soon as information starts to be received. Signals from receptors are integrated through the summation of graded potentials, as are those within the nervous system. One type of integration by receptor cells is **sensory adaptation,** a decrease in responsiveness during continued stimulation (not to be confused with the term adaptation as used in an evolutionary context). Without sensory adaptation, you would feel every beat of your heart and every bit of clothing on your body. Receptors are selective in the information they send to the CNS, and adaptation reduces the likelihood that a continued stimulus will be transmitted.

Another important aspect of sensory integration is the sensitivity of the receptors. The threshold for firing in receptor cells varies with conditions. For example, the firing thresholds of glucose receptors in the human mouth and in the feet of flies can vary over several orders of magnitude of sugar concentration, as both the general state of nutrition and the amount of sugar in the diet change.

The integration of sensory information occurs at all levels within the nervous system, and the cellular actions just described are only the first steps. Complex receptors such as the eyes have higher levels of integration as signals converge on sensory nerves, and the CNS further processes all incoming signals.

Types of Receptors

The various types of sensory receptors can be classified on the basis of location into two broad groups: **exteroreceptors,** which receive information from the outside world (light, sound, touch), and **interoreceptors,** which provide information about the body's internal environment. Another way of categorizing receptors is in terms of the energy stimulus to which they respond. Based on this criterion, the five types of receptors we will consider are mechanoreceptors, chemoreceptors, electromagnetic receptors, thermoreceptors, and pain receptors.

Mechanoreceptors are stimulated by physical deformation caused by such stimuli as pressure, touch, stretch, motion, and sound—all forms of mechanical energy. Bending or stretching of the plasma membrane of a mechanoreceptor cell increases its permeability to both sodium and potassium ions, resulting in a depolarization (receptor potential).

The human sense of touch relies on mechanoreceptors that are actually modified dendrites of sensory neurons (FIGURE 45.1). Receptors that detect light touch are close to the surface of the skin; they transduce very slight inputs of mechanical energy into receptor potentials. Receptors responding to strong pressure and vibrations in the body are in deep skin layers. Other touch receptors detect hair movement.

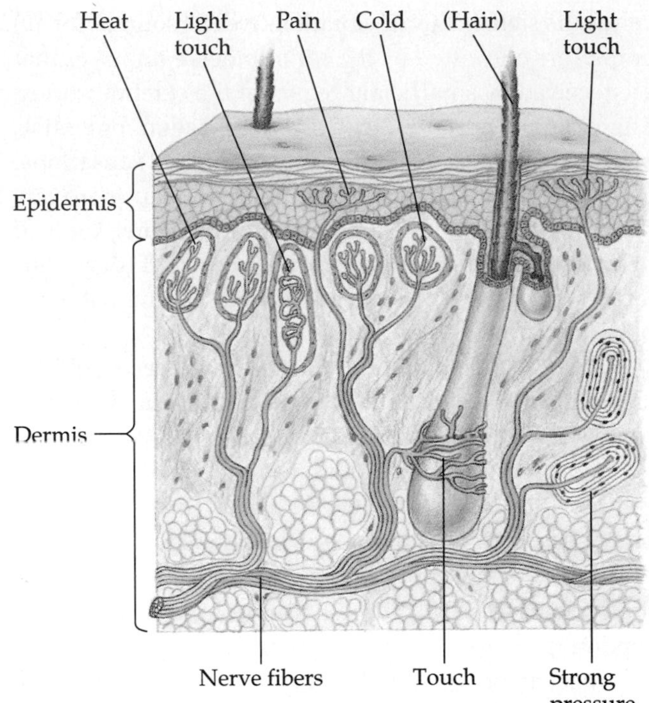

FIGURE 45.1
Sensory receptors in human skin. Each mechanoreceptor is a modified dendrite of a sensory neuron. Most receptors in the inner layer of the skin (dermis) are encapsulated by one or more layers of connective tissue. Those in the outer skin layers (epidermis) and touch receptors wound around the base of hairs are naked dendrites. Touch receptors at the base of the stout whiskers of mammals such as cats and many rodents are extremely sensitive and enable the animal to detect close objects in the dark.

An example of an interoreceptor stimulated by mechanical distortion is the **muscle spindle,** or stretch receptor. This mechanoreceptor monitors the length of skeletal muscles. The muscle spindle contains modified muscle fibers attached to sensory neurons and runs parallel to muscle. When the muscle is stretched, the fibers of the spindle are also stretched, depolarizing the sensory neurons and triggering action potentials that are transmitted back to the spinal cord. It is the activation of the sensory neurons of the muscle spindles that provides the sensory component of the knee-jerk reflex depicted in FIGURE 44.16a.

The **hair cell** is a common type of mechanoreceptor that detects motion. Hair cells are found in the vertebrate ear and in the lateral line organs of fishes and amphibians, where they detect movement relative to the environment (see Chapter 30). The "hairs" are either specialized cilia or microvilli. They project upward from the surface of the hair cell into either an internal compartment, such as the human inner ear, or an external environment, such as a pond. When the cilia or microvilli bend in one direction, they stretch the hair cell membrane and increase its permeability to sodium and potassium ions, leading to an increase in the rate of impulse production in a sensory neuron. When the cilia bend in the opposite direction, ion permeability decreases, reducing the number of action potentials in the sensory neuron. This specificity allows hair cells to respond to the direction of motion as well as to its strength and speed. The role of hair cells in hearing and balance is explored later in the chapter.

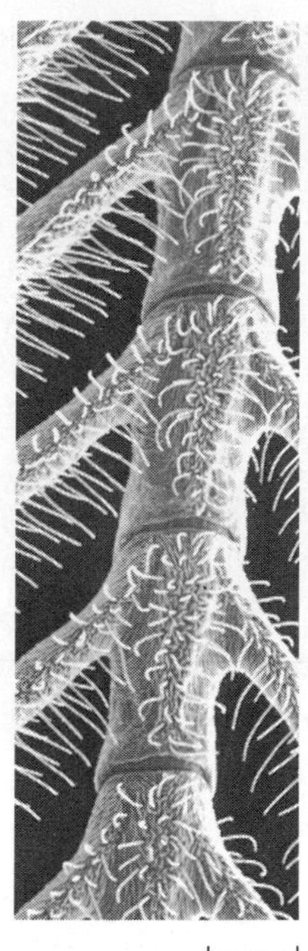

FIGURE 45.2
Chemoreceptors in an insect. The antennae of the male silkworm moth *Bombyx mori* are covered with sensory hairs, visible in the SEM enlargement (right). About 50,000 of these hairs are chemoreceptors that are highly sensitive to the sex pheromone released into the air by the female. Each chemoreceptive hair has thousands of tiny pores that admit air, and each hair contains dendrites of two sensory neurons, one sensitive to bombykol, the other to bombykal, the two components in the pheromone. The male moth begins to respond to the pheromone when as few as 50 or more of his bombykol receptors come into contact with one bombykol molecule per second.

Chemoreceptors include both general receptors that transmit information about the total solute concentration in a solution and specific receptors that respond to individual kinds of molecules. Osmoregulators in the mammalian brain, for example, are general receptors that detect changes in the total solute concentration of the blood and stimulate thirst when osmolarity increases (see Chapter 40). Water receptors in the feet of house flies respond to pure water or to a dilute solution of virtually any substance. Most animals also have receptors specific to important molecules, including glucose, oxygen, carbon dioxide, and amino acids. In all these examples, the stimulus molecule binds to a specific site on the membrane of the receptor cell and initiates changes in membrane permeability. Two other groups of chemoreceptors show intermediate specificity. **Gustatory** (taste) and **olfactory** (smell) **receptors** respond to *categories* of related chemicals. Humans often classify such categories as sweet, sour, salt, or bitter. (Taste and olfaction are discussed in detail later in the chapter.) Two of the most sensitive and specific chemoreceptors known are present in the antennae of the male silkworm moth (FIGURE 45.2). They detect the two chemical components of the female sex pheromone.

Electromagnetic receptors detect various forms of electromagnetic energy, such as visible light, electricity, and magnetism. **Photoreceptors,** which detect the radiation we know as visible light, are often organized into eyes. Snakes have extremely sensitive infrared receptors that detect the body heat of prey standing out against a colder background (FIGURE 45.3a). Some fishes discharge electric currents and use special electroreceptors to locate objects, such as prey, that disturb the electric currents. The platypus, a monotreme mammal, has electroreceptors on its bill that can probably detect electrical fields generated by the muscles of prey, such as crustaceans, frogs, and small fishes. There is also evidence that some animals that home or migrate use the magnetic field lines of Earth to help orient themselves (FIGURE 45.3b). Although the nature of magnetoreceptors is unknown, the ferrous mineral magnetite has been found in the skulls of several animals. (Researchers have

(a)

(b)

FIGURE 45.3
Specialized electromagnetic receptors. (**a**) This rattlesnake and other pit vipers have a pair of infrared receptors, one between each eye and nostril. The organs are sensitive enough to detect the infrared radiation emitted by a warm mouse a meter away. The snake moves its head from side to side until the radiation is detected equally by the two receptors, indicating that the mouse is straight ahead. (**b**) Some migrating animals, such as the beluga whales in this aerial photograph, can apparently sense Earth's magnetic field and use the information, along with other cues, for orientation. The mechanism of the magnetic sense is unknown.

recently found magnetite in human skulls, but there is no evidence that these deposits are associated with a magnetic sense.)

Thermoreceptors, responding to either heat or cold, help regulate body temperature by signaling both surface and body core temperature. There is still debate about the identity of thermoreceptors in the mammalian skin. Possible candidates are two receptors consisting of encapsulated, branched dendrites (see FIGURE 45.1). Many researchers, however, believe that these structures are actually modified pressure receptors and maintain that naked dendrites of certain sensory neurons are the actual thermoreceptors of the skin. There is general agreement that cold and heat receptors in the skin, as well as interothermoreceptors in the anterior hypothalamus of the brain, send information to the body's thermostat, located in the posterior hypothalamus (see Chapter 40).

Pain receptors in humans are a class of naked dendrites in the epidermis of the skin called **nociceptors** (see FIGURE 45.1). Virtually all animals experience pain, although we cannot say what perceptions other animals actually associate with stimulation of their pain receptors. Pain is one of the most important sensations because the stimulus becomes translated into a negative reaction, such as withdrawal from danger. Rare individuals who are born without any pain sensation may die from such conditions as a ruptured appendix because they cannot feel the associated pain and are unaware of the danger.

Different groups of pain receptors respond to excess heat, pressure, or specific classes of chemicals released from damaged or inflamed tissues. Some of the chemicals that trigger pain include histamine and acids. Prostaglandins increase pain by sensitizing the receptors—that is, lowering their threshold. Aspirin reduces pain by inhibiting prostaglandin synthesis. Nociceptive neurons carry impulses to the spinal cord, synapsing with neural pathways leading to the brain.

We have seen that sensory receptors can be classified according to the type of energy stimulus they receive. In many cases, large numbers of a particular type of receptor cell are collected into complex sense organs that detect specific environmental stimuli. We will now examine the structure and function of sense organs responsible for vision, hearing, balance, taste, and smell.

Photoreceptors contain light-absorbing visual pigments: *a closer look*

Photoreceptors range from simple clusters of cells that detect only the direction and intensity of light to complex organs that form images. All photoreceptors contain pigment molecules that absorb light.

Invertebrates

Most invertebrates have photoreceptors. One of the simplest is the **eye cup** of planarians, structures that provide information about light intensity and direction without actually forming an image. Photoreceptor cells are located within a cup formed by a layer of cells containing a screening pigment that blocks light. Light can enter the cup and stimulate the photoreceptors only through an opening on one side where there is no screening pigment (FIGURE 45.4). The mouth of one eye cup faces left and slightly forward, and the mouth of the other cup faces right-forward. Thus, light shining from one side of the planarian can enter only the eye cup on that side. The brain compares the rate of nerve impulses coming from the two eye cups, and the animal turns until the sensations from the two cups are equal and minimal. The result is that the animal moves directly away from the light source and reaches a shaded location beneath a rock or some other object, a behavioral adaptation that helps hide the planarian from predators.

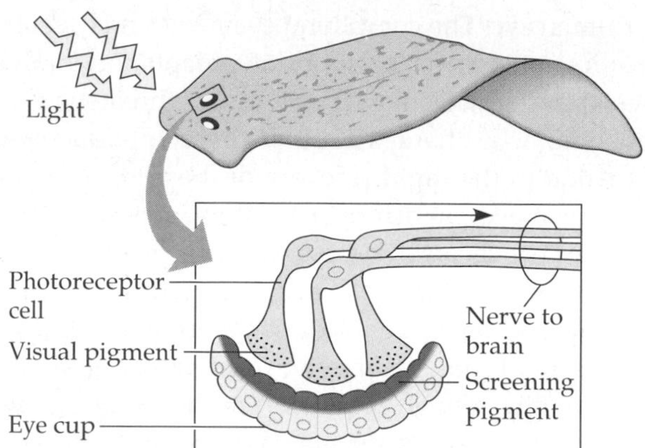

FIGURE 45.4
Eye cups and orientation behavior of a planarian. The head of the flatworm has two eye cups with photoreceptors that send nerve impulses to the brain. Because each eye cup consists of cells containing screening pigment that shades the photoreceptors, light can reach the photoreceptor cells only through the mouth of the cup. The brain directs the body to turn until the sensations from the two cups are equal and minimal, causing the animal to move away from light.

Image-forming eyes of two major types have evolved in invertebrates: the compound eye and the single-lens eye. **Compound eyes** are found in insects and crustaceans (phylum Arthropoda) and some polychaete worms (phylum Annelida). A compound eye consists of up to several thousand light detectors called **ommatidia** (the "facets" of the eye), each with its own cornea and lens (FIGURE 45.5). Each ommatidium registers light from a tiny portion of the visual field. Differences in the intensity of light entering the many ommatidia result in a mosaic image. Although the animal's brain may sharpen the image when it integrates the visual information, the image may still not be as sharp as that produced by

(a)

1 mm

(b)

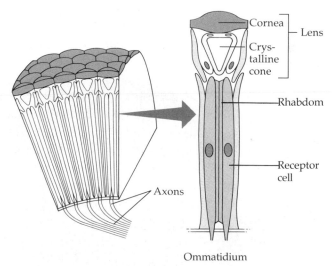

FIGURE 45.5
Compound eyes. (**a**) The faceted eyes on the head of a horsefly, photographed with a stereomicroscope. (**b**) The cornea and crystalline cone of each ommatidium function as a lens that focuses light onto the rhabdom, a stack of pigmented plates on the inside of a circle of receptor cells. The rhabdom traps light and guides it to the receptor cells. The image formed by a compound eye is a mosaic of dots formed by the different intensities of light entering the many ommatidia from different angles.

the human eye. The compound eye is extremely acute at detecting movement, an important adaptation for flying insects and small animals constantly threatened with predation. This characteristic of the compound eye is partly due to the rapid recovery of the photoreceptors. The human eye can distinguish light flashes up to about 50 flashes per second. For this reason, the individual images of a movie, which flash at a faster rate, fuse together to create the perception of smooth motion. The compound eyes of some insects, however, recover from excitation rapidly enough to detect the flickering of a light flashing 330 times per second. Such an insect viewing a movie could easily resolve each frame of the film as a separate still image. Insects also have excellent color vision, and some (including bees) can see into the ultraviolet range of the spectrum, which is invisible to us. In studying animal behavior, we cannot extrapolate our sensory world to other species; different animals have different sensitivities and different brain organization.

The second type of invertebrate eye is the **single-lens eye,** found in some jellyfish, polychaetes, spiders, and many mollusks. It works on a principle similar to that of a camera. The eye of an octopus or squid, for example, has a small opening, the pupil, through which light enters. Analogous to a camera's shutter, an adjustable iris changes the diameter of the pupil; behind the pupil, a single lens focuses light onto the retina, which consists of light-transducing receptor cells. Also similar to a camera's action, muscles move the lens forward or backward to focus images on the retina. The eyes of humans and other vertebrates are also cameralike, but they evolved independently and differ from the single-lens eyes of invertebrates in several details.

Structure and Function of the Vertebrate Eye

The human eye, shown in FIGURE 45.6, is capable of detecting an almost countless variety of colors, forming images of objects miles away, and responding to as little as one photon of light. Remember, however, that it is actually the brain that "sees." Thus, to understand vision, we must begin by learning how the vertebrate eye generates sensations (action potentials), and then follow these signals to the visual centers of the brain, where images are perceived.

The globe of the vertebrate eye, or eyeball, consists of a tough, white outer layer of connective tissue called the **sclera** and a thin, pigmented inner layer called the **choroid.** A delicate layer of epithelial cells forms a mucous membrane, the **conjunctiva,** that covers the outer surface of the sclera and helps keep the eye moist. At the front of the eye, the sclera becomes the transparent **cornea,** which lets light into the eye and acts as a fixed lens. The conjunctiva does not cover the cornea. The anterior choroid forms the donut-shaped **iris,** which gives the eye its color. By changing size, the iris regulates the amount of light entering the **pupil,** the hole in the center of the iris. Just inside the choroid, the **retina** forms the in-

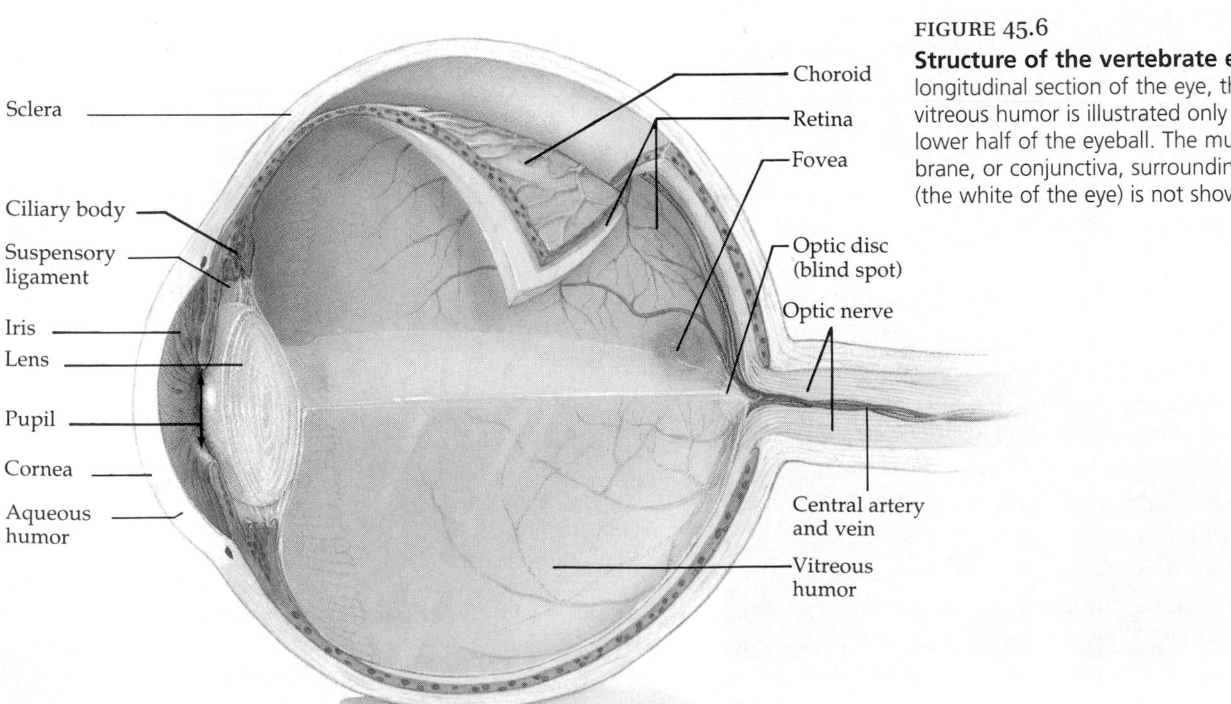

FIGURE 45.6
Structure of the vertebrate eye. In this longitudinal section of the eye, the jellylike vitreous humor is illustrated only in the lower half of the eyeball. The mucous membrane, or conjunctiva, surrounding the sclera (the white of the eye) is not shown.

nermost layer of the eyeball and contains the photoreceptor cells. Information from the photoreceptors leaves the eye at the optic disc, where the optic nerve attaches to the eye. Because there are no photoreceptors in the optic disc, this spot on the lower outside of the retina is a blind spot: Light focused onto that part of the retina is not detected.

The **lens** and **ciliary body** divide the eye into two cavities, one between the lens and the cornea, and a much larger cavity behind the lens within the eyeball itself. The ciliary body constantly produces the clear, watery **aqueous humor** that fills the anterior cavity of the eye. Blockage of the ducts that drain the aqueous humor can produce glaucoma, increased pressure that leads to blindness by compressing the retina. The posterior cavity, filled with the jellylike **vitreous humor,** constitutes most of the volume of the eye. The aqueous and vitreous humors function as liquid lenses that help focus light onto the retina. The lens itself is a transparent protein disc that focuses an image onto the retina. Like squids and octopuses, many fishes focus by moving the lens forward or backward, as in a camera. Humans and other mammals, however, focus by changing the *shape* of the lens. When viewing a distant object, the lens is flat. When focusing on a close object, the lens becomes almost spherical, a change called **accommodation** (FIGURE 45.7).

The human retina contains about 125 million **rod cells** and 6 million **cone cells,** two types of photoreceptors named for their shapes. They account for 70% of all sensory receptors in the body, a fact that underscores the importance of the eyes and visual information in how humans perceive their environment.

Rods and cones have different functions in vision, and the relative numbers of these two photoreceptors in the retina are partly correlated with whether an animal is most active during the day or at night. Rods are more sensitive to light but do not distinguish colors; they enable us to see at night, but only in black and white. Because it takes more light to stimulate cones, cones do not function in night vision. Cones can distinguish colors in daylight. Color vision is found in all vertebrate classes, though not in all species. Most fishes, amphibians, reptiles, and birds have strong color vision, but humans and other primates are among the minority of mammals with this ability. Most mammals are nocturnal, and a maximum number of rods in the retina is an adaptation that gives these animals keen night vision. Cats, usually most active at night, have limited color vision and probably see a pastel world during the day. In the human eye, rods are found in greatest density at the peripheral regions of the retina and are completely absent from the **fovea,** the center of the visual field (see FIGURE 45.6). You cannot see a dim star at night by looking at it directly; if you view it at an angle, however, focusing the starlight onto the regions of the retinas most populated by rods, you will be able to see the star. You achieve your sharpest daylight vision by looking straight at the object of interest because cones are most dense at the fovea, where there are about 150,000 color receptors per mm^2. Some birds have more than a million cones per mm^2, which enables such species as hawks to spot mice and other small prey from high in the sky. In the retina of the eye, as in all biological structures, variations represent evolutionary adaptations.

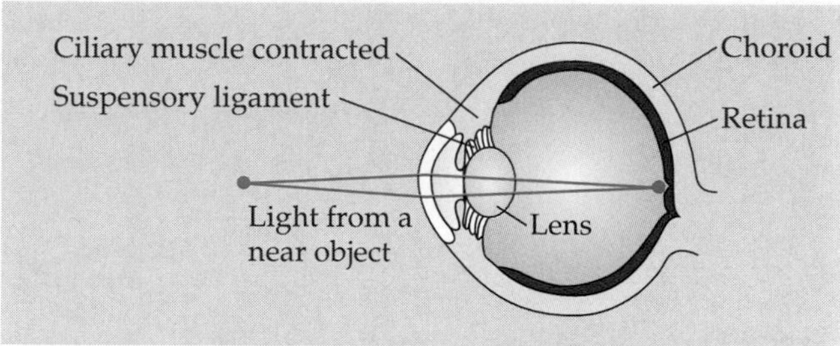

(a) Near vision (accommodation)

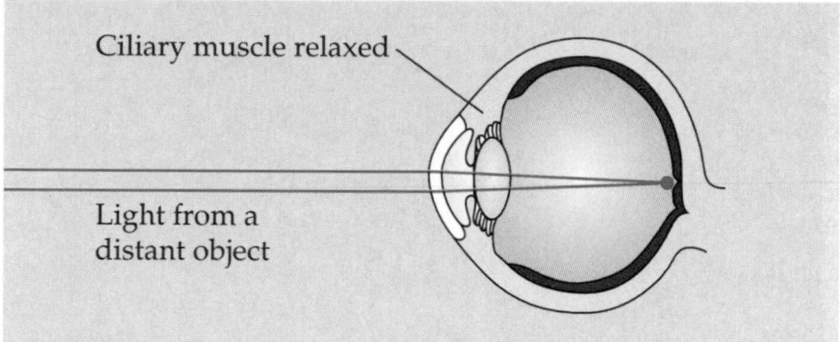

(b) Distance vision

FIGURE 45.7

Focusing in the mammalian eye. The lens bends light and focuses it onto the retina. The thicker the lens, the more sharply the light is bent. The lens is nearly spherical when focusing on near objects and much flatter when focusing on distant objects. Ciliary muscles control the shape of the lens. (**a**) In near vision, the ciliary muscles contract, pulling the border of the choroid layer of the eye toward the lens and causing the suspensory ligaments to relax. With this reduced tension, the elastic lens becomes thicker and rounder, bending light in such a way that near objects can be focused on the retina. This adjustment of the lens for close vision is known as accommodation. (**b**) In distance vision, the ciliary muscles relax, allowing the choroid to expand and put tension on the suspensory ligaments. The lens is pulled into a flatter shape, and the distant object is focused onto the retina.

Signal Transduction in the Eye

When the lens focuses a light image onto the retina, how do the cells of the retina transduce the stimuli into sensations—action potentials that transmit this information about the environment to the brain? Each rod cell or cone cell has an outer segment with a stack of folded membranes, or discs, in which visual pigments are embedded

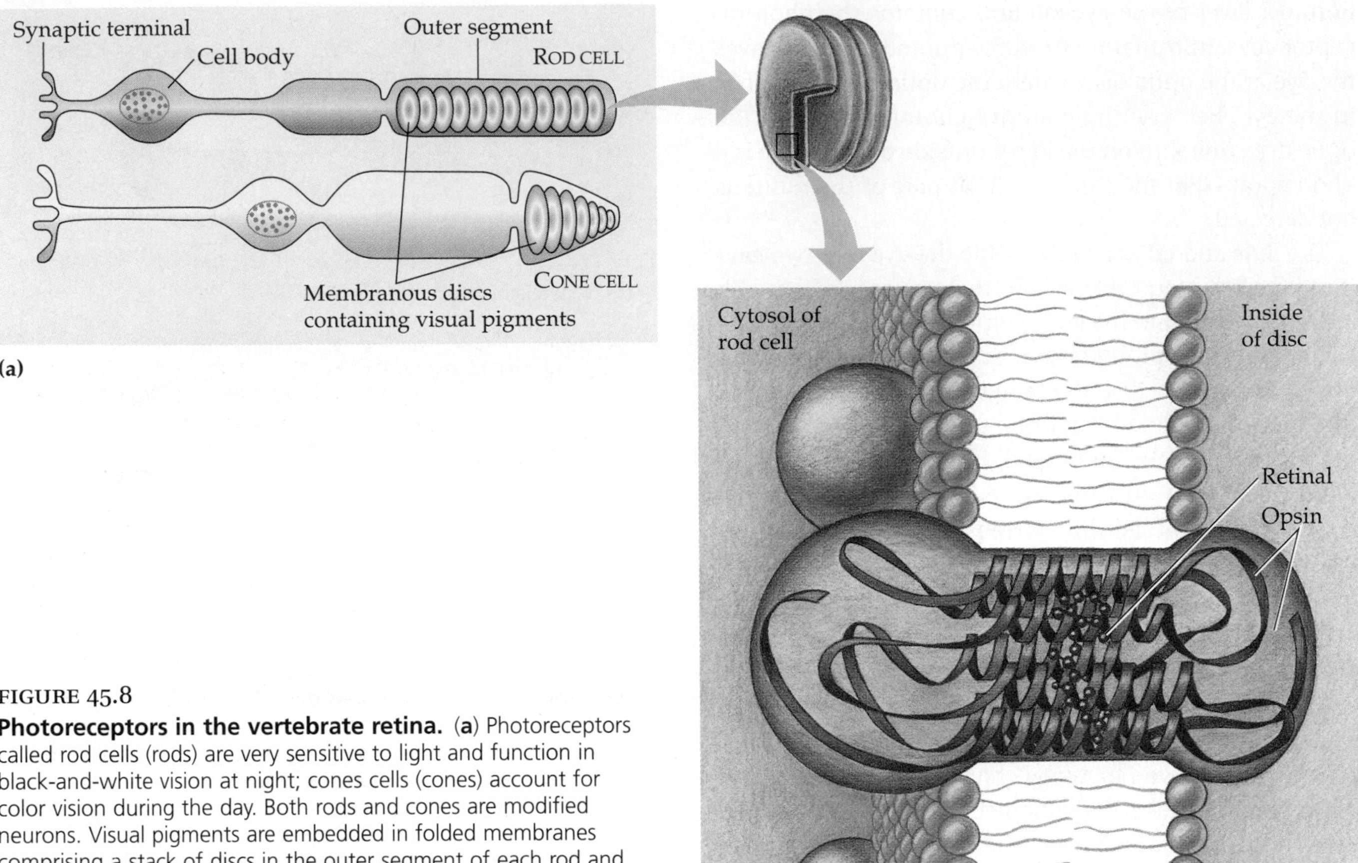

FIGURE 45.8
Photoreceptors in the vertebrate retina. (**a**) Photoreceptors called rod cells (rods) are very sensitive to light and function in black-and-white vision at night; cones cells (cones) account for color vision during the day. Both rods and cones are modified neurons. Visual pigments are embedded in folded membranes comprising a stack of discs in the outer segment of each rod and cone. (**b**) Rhodopsin, the visual pigment in the disc membrane of rods, consists of the light-absorbing molecule retinal bonded to a specific type of membrane protein, an opsin. The opsin has seven regions of alpha helix that span the disc membrane.

(FIGURE 45.8a). The visual pigments consist of a light-absorbing pigment molecule called **retinal** (a derivative of vitamin A) bonded to a membrane protein called an **opsin.** Opsins vary in structure from one type of photoreceptor to another, and the light-absorbing ability of retinal is affected by the specific identity of its opsin partner.

Rods contain their own type of opsin, which, combined with retinal, makes up the visual pigment **rhodopsin** (FIGURE 45.8b). When rhodopsin absorbs light, its retinal component changes shape, triggering a signal-transduction pathway that ultimately results in a receptor potential in the rod cell membrane. Initially, retinal's shape change causes a conformational change in its opsin partner. The altered opsin molecule then activates a relay molecule in the signal-transduction pathway, a G protein called transducin, which is also in the disc membrane. In turn, transducin activates an effector enzyme that chemically alters the second messenger in the rod cell, a nucleotide called cyclic guanosine monophosphate (cGMP).

In the dark, when rhodopsin is inactive, cGMP is bound to sodium ion channels in the rod cell plasma membrane and keeps those channels open. In this state, the rod cell membrane is actually *depolarized* and releases an inhibitory neurotransmitter at its synapses with other neurons in the retina (FIGURE 45.9a). The neurotransmitter inhibits depolarization of the neuron membranes, thus preventing them from developing action potentials. However, when light alters retinal, triggering the rhodopsin signal-transduction pathway, the effector enzyme converts cGMP to GMP, which disengages from the Na^+ channels (FIGURE 45.9b). This closes the channels, decreasing the membrane's permeability to Na^+ and altering the membrane potential. The membrane actually becomes hyperpolarized, and this is the rod cell's receptor potential. In effect, the hyperpolarization slows the rod cell's release of the inhibitory neurotransmitter. Thus, the transduction of light energy into receptor potentials produces a *decrease* in the chemical signal to the cells with which rods synapse, and that decrease is the message that the rods have been stimulated by light.

The light-induced change in retinal, which initiates the light-transducing signal pathway in rod cells, is referred

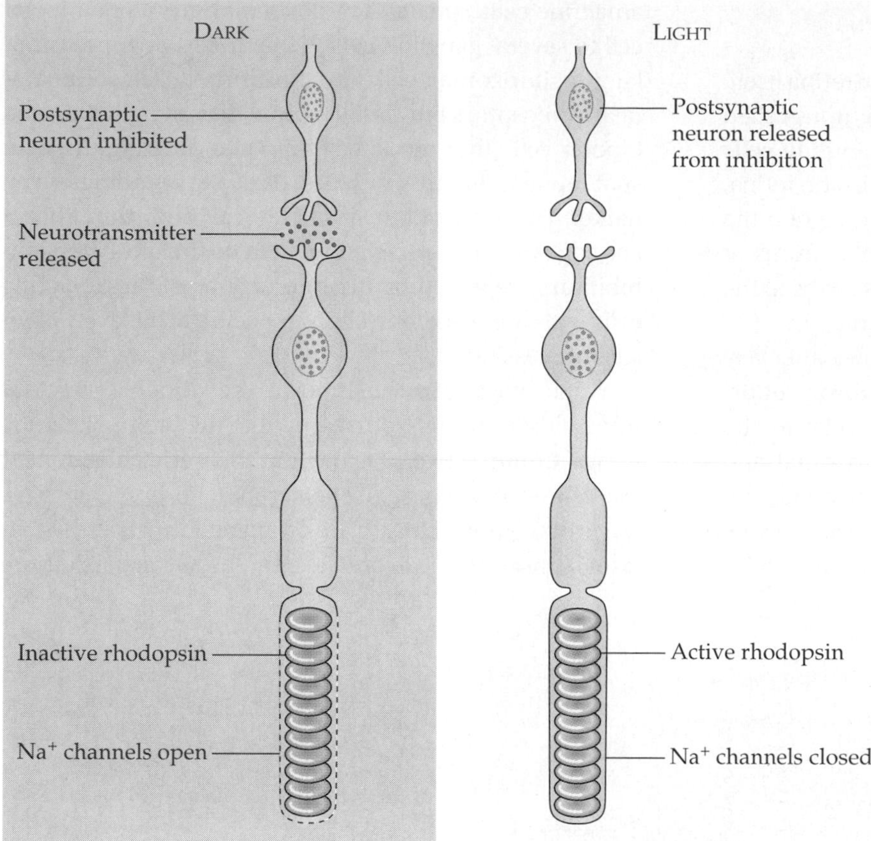

(a) Rod cell depolarized

(b) Rod cell hyperpolarized

CH_3 H_3C CH_3 CH C CH CH CH CH C CH_3 CH HC=O

Light

Enzyme

CH_3 CH_3 H_3C CH_3 CH C CH C HC=O CH CH CH CH

cis form

trans form

(c)

FIGURE 45.9

The effect of light on rod cells and retinal. (**a**) In the dark, rhodopsin is inactive, and the rod cell membrane is highly permeable to sodium, thus depolarized. In this state, the rod cell releases neurotransmitter that inhibits the firing of postsynaptic neurons in the retina. (**b**) In contrast, when light activates rhodopsin, the rod cell membrane becomes less permeable to sodium, and its membrane potential changes (it develops a receptor potential, a hyperpolarization in this case). The synaptic terminals of the rod cell then slow their release of inhibitory neurotransmitter molecules. Freed from inhibition, the postsynaptic neurons can develop action potentials. (**c**) Retinal exists in two forms, isomers of each other. Absorption of light converts the pigment from the *cis* isomer to the *trans* isomer. This change triggers the signal-transduction pathway that converts the light signal into an electrochemical signal, a receptor potential, in the rod cell membrane. When the photoreceptor is no longer stimulated by light, enzymes convert the retinal back to the *cis* form.

to as "bleaching" of rhodopsin. In the dark, enzymes convert the retinal back to its original form, and it recombines with opsin to form rhodopsin (FIGURE 45.9c). Bright light keeps the rhodopsin bleached and rods become unresponsive; cones take over. When you walk from a bright environment into a dark place, such as walking into a movie theater in the afternoon, you are initially almost blind to faint light. There is not enough light to stimulate the cones, and it takes at least a few minutes for the bleached rods to become functional again.

Color vision involves more complex signal processing than the rhodopsin mechanism in rods. Color vision results from the presence of three subclasses of cones in the retina, each with its own type of opsin associated with retinal to form visual pigments collectively called **photopsins.** These photoreceptors are known as red cones, green cones, and blue cones, referring to the colors their kind of photopsin is best at absorbing. The absorption spectra for these pigments overlap, and the brain's perception of intermediate hues depends on the differential stimulation of two or more types of cones. For example, when both red and green cones are stimulated, we may see yellow or orange, depending on which of these two populations of cones is most strongly stimulated. Color blindness, more common in males than females because it is generally inherited as a sex-linked trait (see Chapter 14), is due to a deficiency or absence of one or more types of cones.

Visual Integration

Processing of visual information begins in the retina itself. The axons of rods and cones synapse with neurons called **bipolar cells,** which in turn synapse with **ganglion cells** (FIGURE 45.10). Additional types of neurons in the retina, **horizontal cells** and **amacrine cells,** help integrate the information before it is sent to the brain. The axons of ganglion cells then convey the resulting sensations to the brain as action potentials along the optic nerve.

Signals from the rods and cones may follow either vertical or lateral pathways. In the vertical pathway, information passes directly from the receptor cells to the bipolar cells to the ganglion cells. The horizontal and amacrine cells provide lateral integration of visual signals. Horizontal cells carry signals from one rod or cone to other receptor cells and to several bipolar cells; amacrine cells spread the information from one bipolar cell to several ganglion cells. When a rod or cone stimulates a horizontal cell, the horizontal cell stimulates nearby receptors but inhibits more distant receptors and bipolar cells that are not illuminated, making the light spot appear lighter and the dark surroundings even darker. This integration, called **lateral inhibition,** sharpens edges and enhances contrast in the image. Lateral inhibition is repeated by the interactions of the amacrine cells with the ganglion cells and occurs at all levels of visual processing.

Axons of ganglion cells form the optic nerves that transmit sensations from the eyes to the brain. The optic nerves from the two eyes meet at the **optic chiasm** near the center of the base of the cerebral cortex. The nerve tracts of the optic chiasm are arranged in such a way that visual sensations from both eyes in the left visual field are

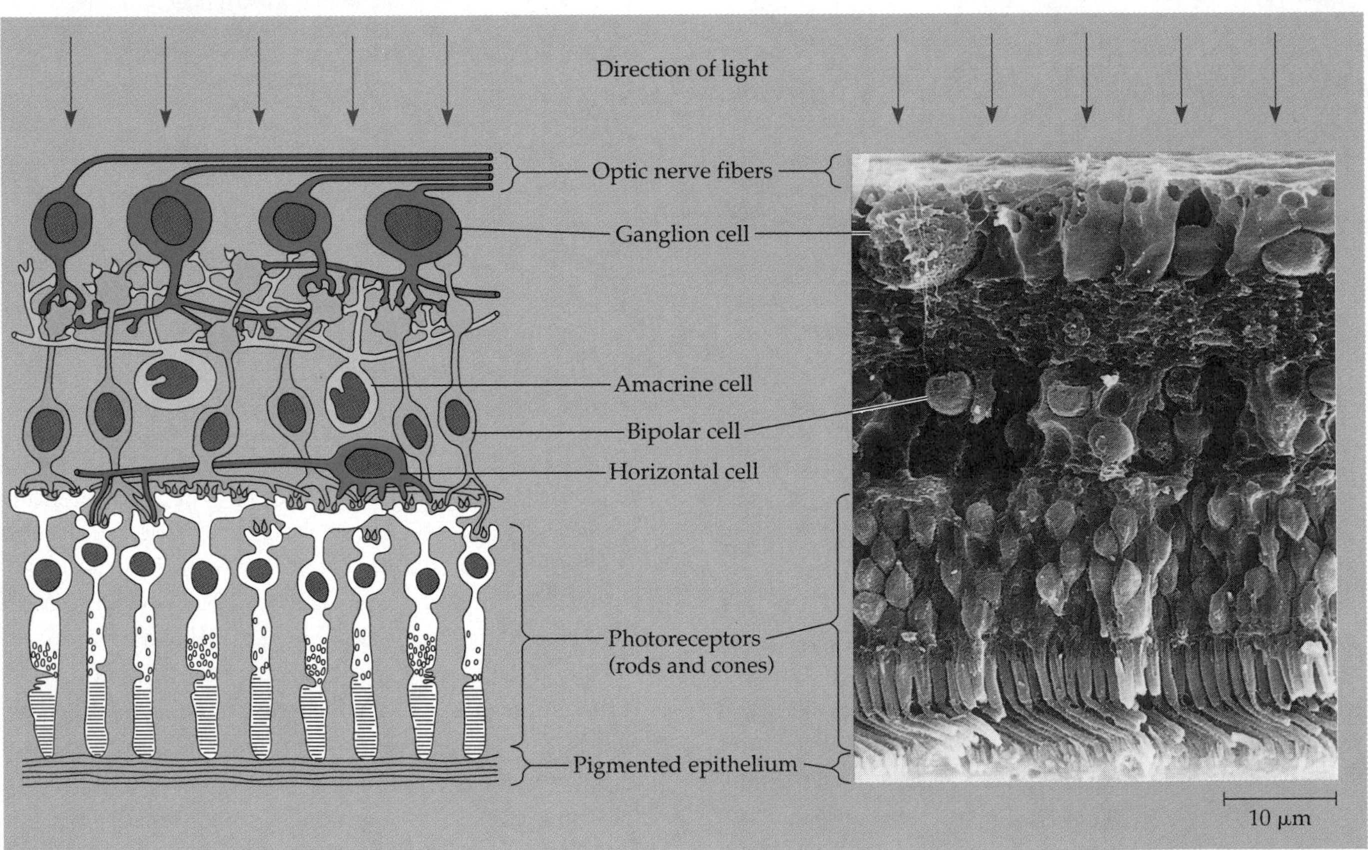

FIGURE 45.10

The vertebrate retina. Light must pass through several relatively transparent layers of cells before reaching the rods and cones. These photoreceptors communicate with ganglion cells via bipolar cells. The axons of the ganglion cells transmit the visual sensations (action potentials) to the brain. There is not a one-to-one relationship between the rods and cones, bipolar cells, and ganglion cells; each bipolar cell receives information from several rods or cones, and each ganglion cell from several bipolar cells. The horizontal and amacrine cells carry information across the retina to integrate the signals. All the rods or cones that feed information to one ganglion cell form the receptive field for that cell. The larger the receptive field (the more rods or cones that supply a ganglion cell), the less sharp the image, because it is less evident exactly where the light struck the retina. The ganglion cells of the fovea have very small receptive fields, so visual acuity is very sharp in this area. The SEM shows the retinal cell layers of a rabbit. (SEM from *Tissues and Organs: A Text-Atlas of Scanning Electron Microscopy* by Richard G. Kessel and Randy H. Kardon. W. H. Freeman and Company. Copyright © 1979.)

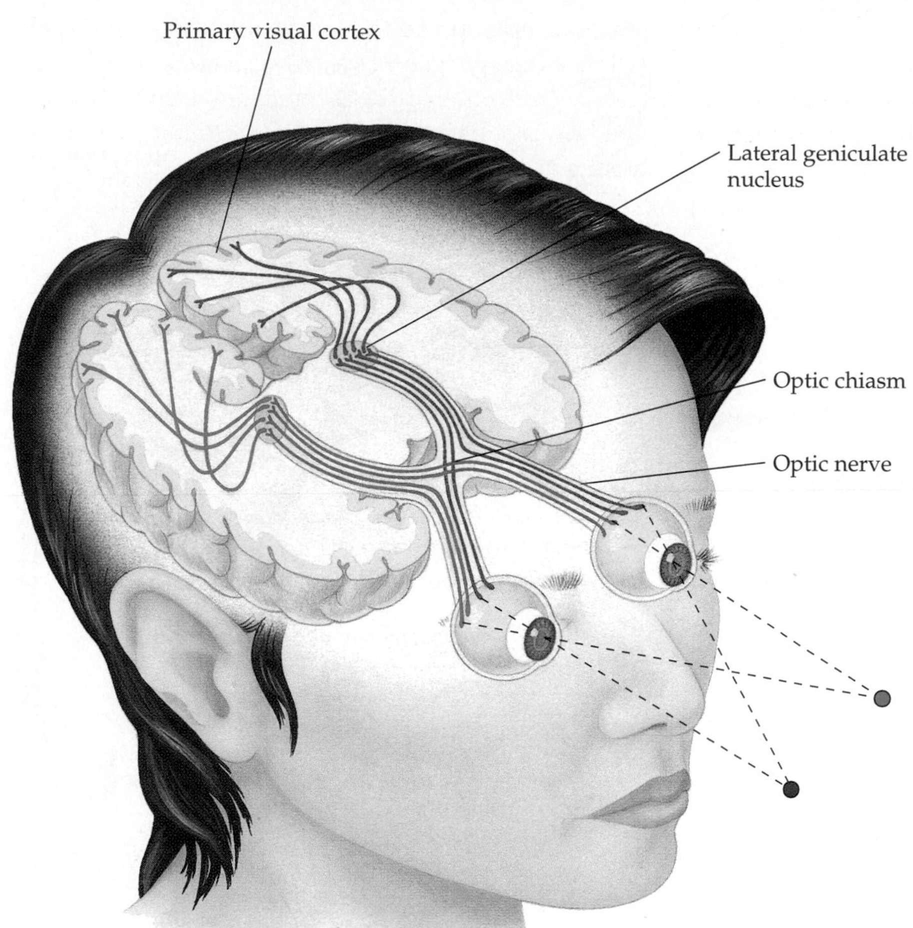

FIGURE 45.11
Neural pathways for vision. Because of the arrangement of neurons in the retinas, optic nerves, and optic chiasm, the right side of the brain receives sensory information about objects in the left visual field (blue), while the left side of the brain receives information from the right visual field (red). Each optic nerve contains about a million axons that synapse with interneurons in the lateral geniculate nuclei. The nuclei relay sensations to the visual cortex, believed to be the first of many brain centers that cooperate in constructing our visual perceptions.

transmitted to the right side of the brain, and visual sensations in the right visual field are transmitted to the left side of the brain (FIGURE 45.11). Most of the ganglion cell axons lead to the **lateral geniculate nuclei** of the thalamus. Neurons of the lateral geniculate nuclei continue back to the **primary visual cortex** in the occipital lobe of the cerebrum. Additional interneurons carry the information to other, more sophisticated visual processing and integrating centers elsewhere in the cortex.

Point-by-point information in the visual field is projected along neurons onto the visual cortex according to its position in the retina, but the information the brain receives is highly distorted. How does the brain convert a complex set of action potentials representing two-dimensional images projected onto our retinas into three-dimensional perceptions of our surroundings? Researchers estimate that fully 30% of the cerebral cortex—hundreds of millions of interneurons in perhaps dozens of integrating centers—take part in processing visual data. Determining how these centers integrate such components of our vision as color, motion, depth, shape, and detail is the focus of an exciting, fast-moving research effort. There is far more to learn than we presently know about how we actually "see."

Hearing and balance are related in most animals: *a closer look*

Both hearing and balance involve mechanoreceptors containing hair cells that produce receptor potentials when the hairs are bent by settling particles or moving fluid. In mammals and most other terrestrial vertebrates, the sensory organs for hearing and balance are located together within the ear. We will discuss humans as an example of mammalian hearing and balance before looking at other vertebrates and invertebrates.

The Human Ear

The human ear contains the organ of hearing and the organ of equilibrium, or balance. Both organs contain hair cells in fluid-filled canals.

The ear itself can be divided into three regions. The **outer ear** consists of the external pinna and the auditory canal, which collect sound waves and channel them to the **tympanic membrane** (eardrum) separating the outer ear from the middle ear. Within the **middle ear,** vibrations are conducted through three ossicles (small bones)—the **malleus** (hammer), **incus** (anvil), and **stapes** (stirrup)—to the inner ear, passing through the

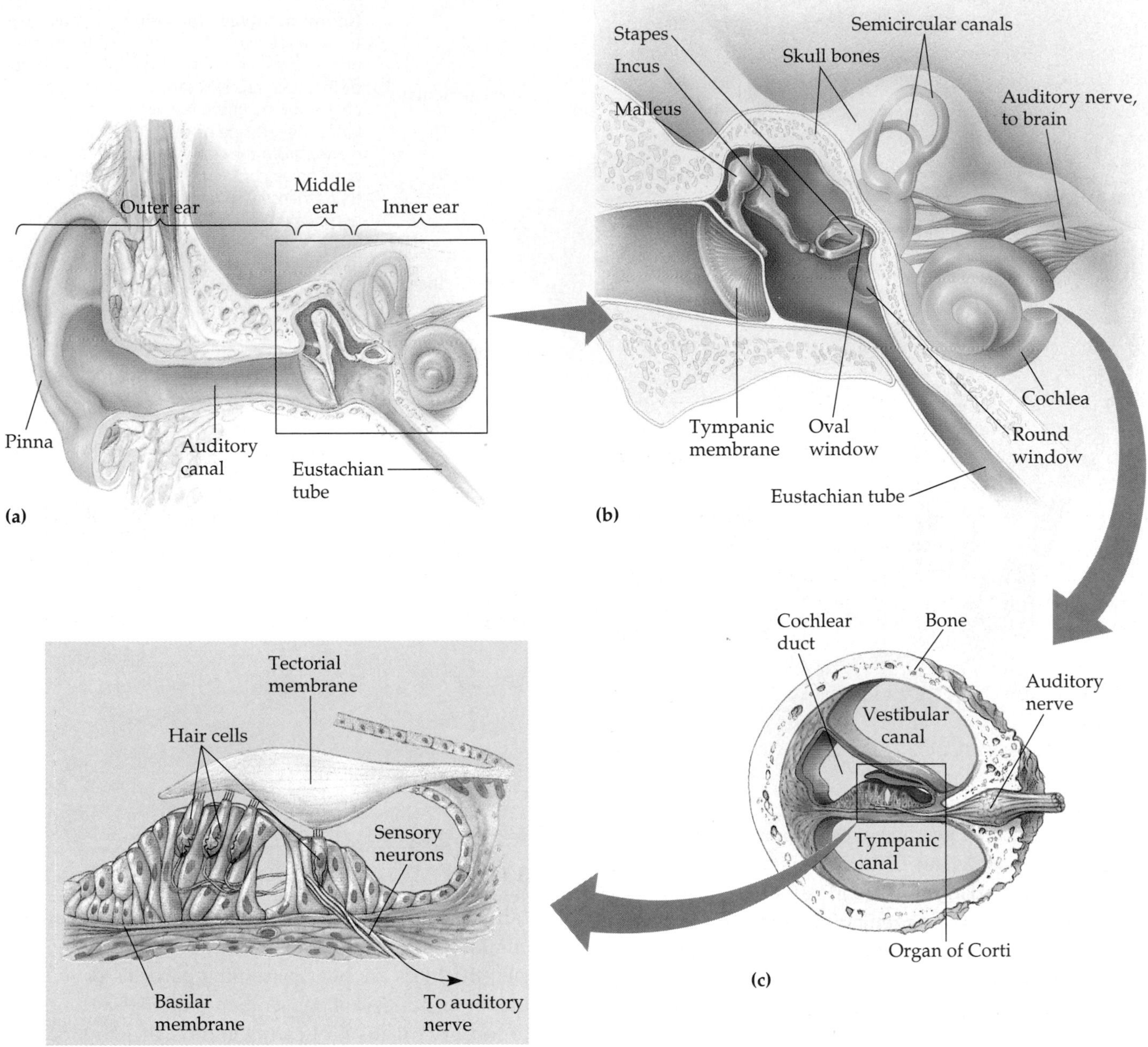

FIGURE 45.12

Structure and function of the human ear. (**a**) Sound waves collected by the outer ear create vibrations in the tympanic membrane that are conducted via three small bones (malleus, incus, and stapes) in the middle ear to the inner ear. (**b**) The inner ear contains the cochlea, a long, coiled tube that contains sound detectors, and the semicircular canals, which control balance. (**c**) A cross-sectional view of the cochlea shows three canals. The vestibular canal and tympanic canal contain the fluid perilymph. Between these two canals is a smaller cochlear duct filled with endolymph. The organ of Corti sits on the basilar membrane, which forms the floor of the cochlear duct. (**d**) The receptor cells—hair cells—are part of the organ of Corti. Suspended over the organ of Corti is the tectorial membrane, to which many of the hairs of the receptor cells are attached. Vibrations of the oval window on the surface of the cochlea create pressure waves in the cochlear fluid. As the basilar membrane vibrates, hair cells repeatedly brush against the tectorial membrane. This stimulus causes the hair cells to depolarize and release neurotransmitter, thereby triggering an action potential in a sensory neuron.

oval window, a membrane beneath the stapes (FIGURE 45.12a and b). The middle ear also opens into the **Eustachian tube,** which connects with the pharynx and equalizes pressure between the middle ear and the atmosphere, enabling you to "pop" your ears when changing altitude, for example. The **inner ear** consists of a labyrinth of channels within a skull bone (the temporal bone). These channels are lined by a membrane and contain fluid that moves in response to sound or movement of the head.

The part of the inner ear involved in hearing is a complex coiled organ known as the **cochlea** (L. for "snail"). The cochlea has two large chambers, an upper vestibular canal and a lower tympanic canal, separated by a smaller cochlear duct (FIGURE 45.12c). The vestibular and tympanic canals contain a fluid called perilymph, and the cochlear duct is filled with a liquid called endolymph. The floor of the cochlear duct, the basilar membrane, bears the **organ of Corti,** which contains the actual receptor cells of the ear, hair cells with hairs projecting into the cochlear duct (FIGURE 45.12d). Many of the hairs are attached to the tectorial membrane, which hangs over the organ of Corti like a shelf.

How is the complex anatomy of the ear correlated with the function of hearing? The ear converts the energy of pressure waves traveling through air into nerve impulses that the brain perceives as sound. Vibrating objects, such as the reverberating strings of a guitar or the vocal cords of a speaking person, create percussion waves in the surrounding air. These waves cause the tympanic membrane to vibrate with the same frequency as the sound. The three bones of the middle ear amplify and transmit the mechanical movements to the oval window, a membrane on the surface of the cochlea. Vibrations of the oval window produce pressure waves in the fluid within the cochlea.

The cochlea transduces the energy of the vibrating fluid into action potentials. The stapes vibrating against the oval window creates a traveling pressure wave in the fluid of the cochlea that passes into the vestibular canal (FIGURE 45.13a). This wave continues around the tip of the cochlea and through the tympanic canal, dissipating as it strikes the **round window.** The pressure waves in the vestibular canal push downward on the cochlear duct and basilar membrane. The basilar membrane vibrates up and down in response to the pressure waves, and its hair cells alternately brush against and are withdrawn from the tectorial membrane. Deflection of the hairs opens ion channels in the plasma membrane of the hair cells, and positive ions (K^+, in this case) enter. The resulting depolarization increases neurotransmitter release from the hair cell and the frequency of action potentials in the sensory neuron with which the hair cell synapses. This neuron carries the sensations to the brain through the auditory nerve.

Sound is detected by increases in the frequency of impulses in the sensory neuron, but how is the quality of that sound determined? Two important aspects of sound are volume and pitch. Volume (loudness) is determined by the amplitude, or height, of the sound wave. The greater the amplitude of a sound, the more vigorous the vibrations of fluid in the cochlea, the greater the bending of the hair cells, and the more action potentials generated in the sensory neurons. **Pitch** is a function of a sound

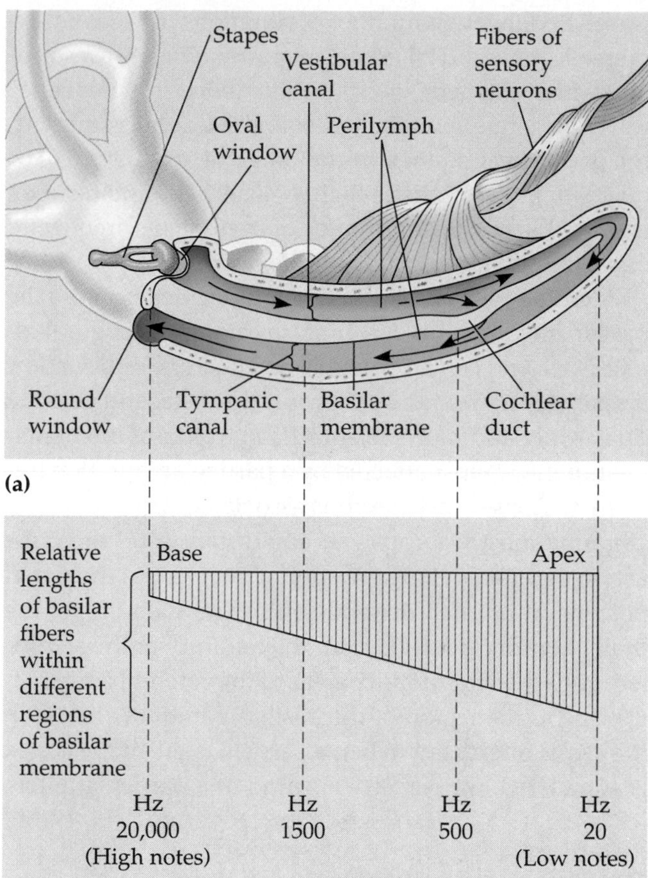

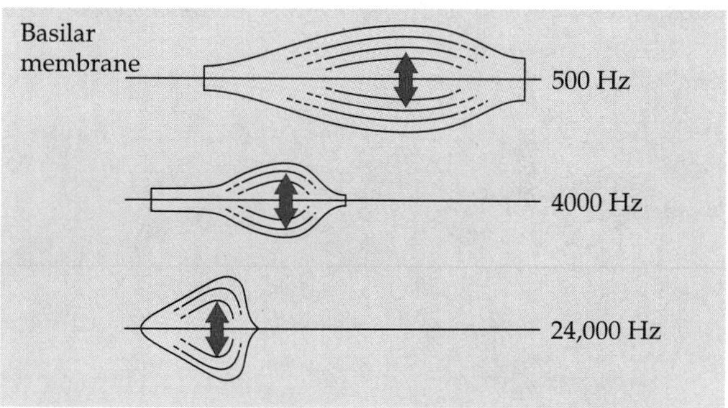

FIGURE 45.13
How the cochlea distinguishes pitch. (**a**) Vibrations of the stapes against the oval window agitate the fluid within the cochlea (uncoiled here), causing pressure waves having a frequency equivalent to the sound waves that entered the ear. The waves (black arrows) pass through the vestibular canal to the apex of the cochlea, then back toward the base of the cochlea via the tympanic canal. The energy causes the cochlear duct, with its basilar membrane and organ of Corti, to vibrate up and down. The bouncing of the basilar membrane stimulates the hair cells within the cochlear duct. (**b**) Fibers span the width of the basilar membrane. Like harp strings, these fibers vary in length, being shorter near the base of the membrane and longer near its apex. The length of the fibers "tunes" specific regions of the basilar membrane to vibrate at specific frequencies. (**c**) Different frequencies of pressure waves in the cochlea cause certain places along the basilar membrane to vibrate, stimulating particular hair cells and sensory neurons. The differential stimulation of hair cells is perceived in the brain as sound of a certain pitch.

wave's frequency, or number of vibrations per second, expressed in hertz (Hz). Short, high-frequency waves produce high-pitched sound, while long, low-frequency waves generate low-pitched sound. Healthy young humans can hear sounds in the range of 20 to 20,000 Hz, dogs can hear sounds as high as 40,000 Hz, and bats can emit and hear clicking sounds of even higher frequency, using this ability to locate objects by sonar.

Pitch can be distinguished by the cochlea because the basilar membrane is not uniform along its length (FIGURE 45.13b and c). The proximal end near the oval window is relatively narrow and stiff, while the distal end near the tip is wider and more flexible. Each region of the basilar membrane is most affected by a particular vibration frequency. The sensory neurons associated with the region vibrating most vigorously at any instant send the most action potentials along the auditory nerve. But the actual perception of pitch depends on neural mapping of the brain. Sensory neurons from the auditory pathway project onto specific auditory areas of the cerebral cortex according to the region of the basilar membrane in which the signal originated. When a particular site of the cortex is stimulated, we perceive a sound of a particular pitch.

Balance and Equilibrium in Humans

The balance apparatus in humans and most other mammals is also centered in the inner ear. Behind the oval window is a vestibule that contains two chambers, the **utricle** and **saccule.** The utricle opens into three **semicircular canals** that complete the apparatus for equilibrium (FIGURE 45.14a).

Sensations related to body position are generated much like sensations of sound in humans and most other mammals. Hair cells in the utricle and saccule respond to changes in head position with respect to gravity and movement in one direction. The hair cells are arranged in clusters, and all the hairs project into a gelatinous material containing many small calcium carbonate particles called otoliths ("ear stones"). Because this material is heavier than the endolymph within the utricle and saccule, gravity is always pulling downward on the hairs of the receptor cells, sending a constant series of action potentials along the sensory neurons of the vestibular branch of the auditory nerve.

Different body angles cause different hair cells and their sensory neurons to be stimulated. When the position of the head changes with respect to gravity (as when the head bends forward), the force on the hair cell changes, and it increases (or decreases) its output of neurotransmitter. The brain interprets the resulting changes in impulse production by the sensory neurons to determine the position of the head. By a similar mechanism, the semicircular canals, arranged in the three spatial planes, detect changes in the rate of rotation or angular movements of the head (FIGURE 45.14b and c).

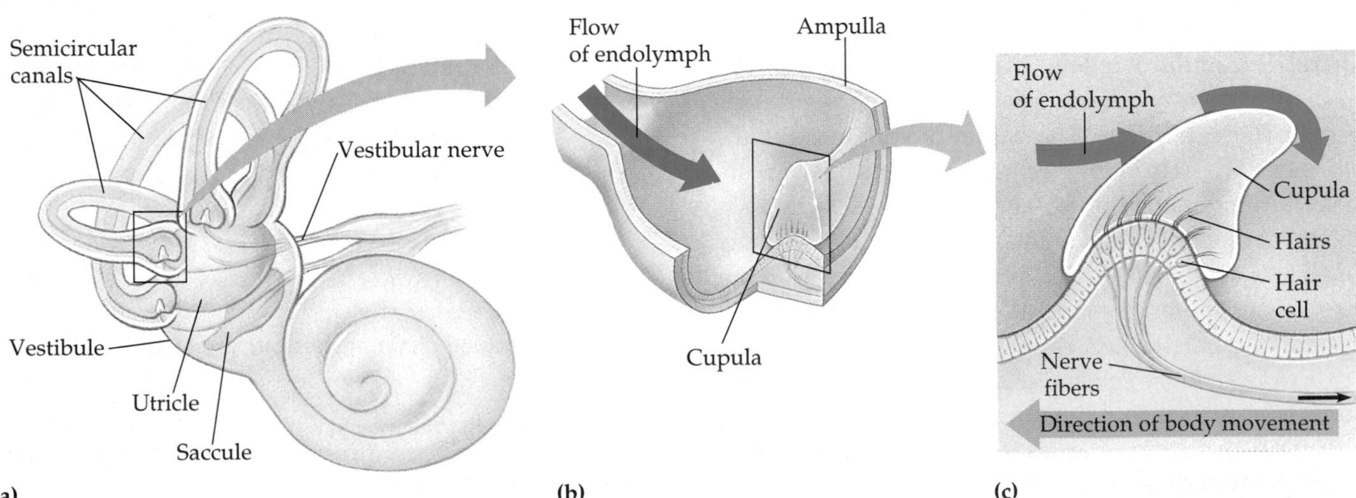

FIGURE 45.14
Organs of balance in the inner ear. (**a**) Three structures of the inner ear—the utricle and saccule, in the vestibule, and the semicircular canals—contain hair cells sensitive to balance and body position. The saccule and utricle tell the brain which way is up and also inform it of the body's fixed position in space or of any linear acceleration associated with movement. Arranged in the three spatial planes, the semicircular canals each have a swelling called an ampulla at their base. (**b**) The ampulla contains a cluster of hair cells with hairs projecting into a gelatinous cap called the cupula. (**c**) When the head changes its rate of rotation, inertia prevents the endolymph in the semicircular canals from moving with the head, so the fluid presses against the cupula, bending the hair cells. The bending increases the frequency of action potentials in the sensory neurons in direct proportion to the amount of rotational acceleration. The mechanism adjusts quickly if rotation continues at a constant speed: The endolymph begins moving with the head, and the pressure on the cupula is reduced. If rotation stops suddenly, however, the fluid continues to flow through the semicircular canals and again stimulates the hair cells. This new stimulus can cause dizziness.

Hearing and Equilibrium in Other Vertebrates

Most fishes and aquatic amphibians have a **lateral line system** along both sides of the body (FIGURE 45.15; also see Chapter 30). The system contains mechanoreceptors that detect movement by a mechanism similar to the function of the inner ear. Water from the animal's surroundings enters the lateral line system through numerous pores and flows along a tube past the mechanoreceptors. The receptor units, called **neuromasts,** resemble the ampullae in our semicircular canals. Each neuromast has a cluster of hair cells, with the sensory hairs embedded in a gelatinous cap, the cupula. As the pressure of moving water bends a cupula, the hair cells transduce the energy into receptor potentials and then into action potentials that are transmitted along a nerve to the brain. This information helps the fish perceive its movement through water or the direction and velocity of water currents flowing over its body. The lateral line system also detects water movements or vibrations generated by other moving objects, including prey and predators.

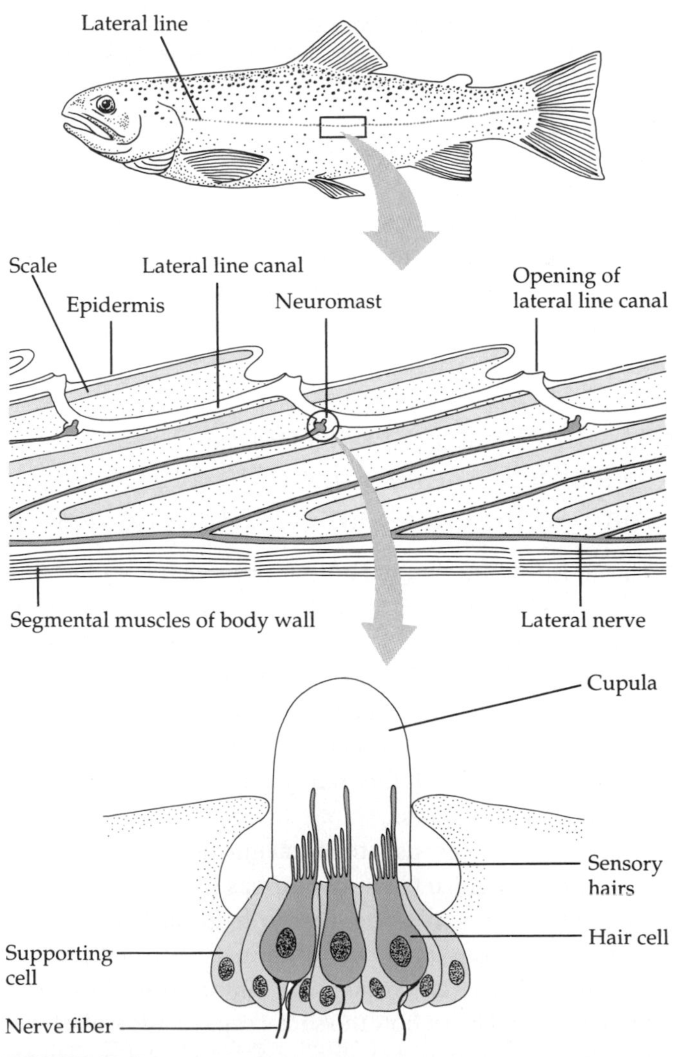

FIGURE 45.15
The lateral line system in a fish. Water flowing through the system bends hair cells. The hair cells transduce the energy into receptor potentials, triggering action potentials, which are conveyed to the brain. The lateral line system enables a fish to monitor water currents, pressure waves produced by moving objects, and low-frequency sounds conducted through water

Like other vertebrates, fishes also have inner ears located near the brain. Along with the lateral line system, the inner ears enable a fish to hear. There is no cochlea, but there are a saccule, a utricle, and semicircular canals, structures homologous to the equilibrium sensors of our own ears. Within these chambers in the inner ear of a fish, sensory hairs are stimulated by the movement of otoliths. Unlike the mammalian hearing apparatus, the ear of a fish has no eardrum and does not open to the outside of the body. Vibrations of the water caused by sound waves are conducted through the skeleton of the head to the inner ears, setting the otoliths in motion and stimulating the hair cells. The air-filled swim bladder (see Chapter 30) also vibrates in response to sound and may contribute to the transfer of sound to the inner ear. Some fishes, including catfishes and minnows, have a series of bones called the Weberian apparatus, which conducts vibrations from the swim bladder to the inner ear. Sound waves also stimulate the hair cells of the lateral line system, but only if the sound is of relatively low frequency. The inner ears extend the hearing of fishes to higher frequencies.

The lateral line system functions only in water. In terrestrial vertebrates, the inner ear has evolved as the main organ of hearing and equilibrium. Some amphibians have a lateral line system as tadpoles, but not as adults living on land. In the ear of a terrestrial frog or toad, sound vibrations traveling in the air are conducted to the inner ear by a tympanic membrane on the body surface and a single middle ear bone. There is recent evidence that the lungs of a frog also vibrate in response to sound and transmit their vibrations to the eardrum via the auditory tube. A small side pocket of the saccule functions as the main hearing organ of the frog, and it is this outgrowth of the saccule that gave rise to the more elaborate cochlea during the evolution of mammals. Birds also have a cochlea, but like amphibians and reptiles, sound is conducted from the tympanic membrane to the inner ear by a single bone, the stapes.

Sensory Organs for Hearing and Equilibrium in Invertebrates

Many invertebrates sense sounds. Hearing structures have been most extensively studied in the arthropods. For example, the body hairs of many insects vibrate in response to sound waves of specific frequencies, depending on the stiffness and length of the hairs. The hairs are commonly tuned to frequencies of sounds produced by other organisms. A male mosquito locates a mate by means of fine hairs on his antennae. The hairs vibrate in a specific way in response to the hum produced by the beating wings of flying females. A tuning fork that vibrates at the same frequency as a female mosquito's wings will also attract males. Some caterpillars, larvae of insects, have vibrating body hairs that detect the buzzing wings of predatory wasps, warning the caterpillars of danger. Many insects also have localized "ears" (FIGURE 45.16). A tympanic membrane (eardrum) is stretched over an internal air chamber. Sound waves vibrate the tympanic membrane, stimulating receptor cells attached to the inside of the membrane and resulting in nerve impulses that are transmitted to the brain. Some moths can hear notes of such high pitch that they detect the sounds bats produce for sonar, and perception of these sounds triggers the moth's escape maneuver, as mentioned earlier.

Most invertebrates have sensory organs called **statocysts** that contain mechanoreceptors and function in their sense of equilibrium (FIGURE 45.17). A common type of statocyst has a layer of hair cells surrounding a chamber containing **statoliths,** which are grains of sand or other dense granules. Gravity causes the statoliths to settle to the low point within the chamber, stimulating hair cells in that location. (This is similar to how the saccule and utricle function in vertebrates, and indeed these structures in the vertebrate inner ear are considered to be specialized types of statocysts.) The statocysts of invertebrates have various locations. For example, many jellyfish have statocysts at the fringe of the "bell," giving the animals an indication of body position. Lobsters and crayfish have statocysts near the bases of their antennules. Crayfish have been tricked into swimming upside down in experiments in which the statoliths were replaced with metal shavings that could be pulled to the upper end of the statocysts with magnets.

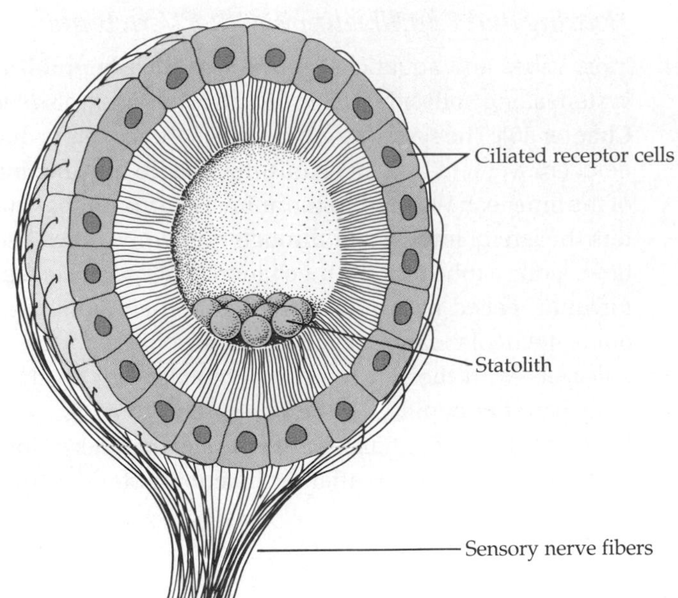

FIGURE 45.17
The statocyst of an invertebrate. The settling of statoliths to the low point within the chamber bends cilia on receptor cells in that location, providing the brain with information about the position of the body.

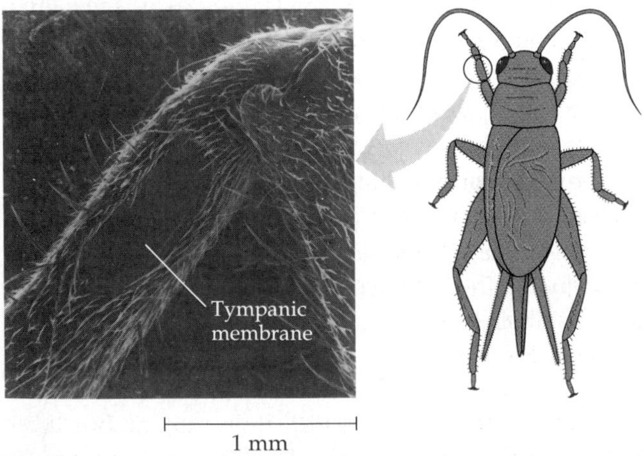

FIGURE 45.16
An insect ear. The tympanic membrane, this one on the front leg of a cricket, vibrates in response to sound waves (SEM). The vibrations stimulate mechanoreceptor cells attached to the inside of the tympanic membrane.

The interacting senses of taste and smell enable animals to detect many different chemicals: *a closer look*

The senses of taste (gustation) and smell (olfaction) depend on chemoreceptors that detect specific chemicals in the environment. In the case of terrestrial animals, taste is the detection of certain chemicals that are present in a solution, and smell is the detection of airborne chemicals. However, these chemical senses are usually closely related, and there really is no distinction in aquatic environments.

Various animals use their chemical senses to find mates, to recognize territory that has been marked with some chemical substance, and to help navigate during migration (FIGURE 45.18). Many animals, especially social insects, also use chemicals to communicate. In Chapter 50, you will learn how the social organization of a beehive or anthill is based on chemical "conversation." In all animals, taste and smell are important in feeding behavior. For example, the cnidarian *Hydra* begins to swallow when chemoreceptors detect the compound glutathione, which is released from prey captured by the hydra's ten-

FIGURE 45.18
Salmon follow their noses home. Various species of Pacific salmon make a one-time round trip from small streams where the fish hatch, to the sea via rivers such as the Columbia, then back to the stream of their origin to spawn. While in the sea, salmon from many river systems school and feed together in the Gulf of Alaska. Sexually mature salmon segregate into groups of common geographic origin and migrate back toward the river from which they emerged as juveniles. During this first stage of the return, they may navigate by the position of the sun. But once a salmon reaches the general region of the river leading to its home stream, a keen sense of smell takes over. The water that flows from each stream into the river carries a unique scent from the types of plants, soil, and other components of that stream. This scent is apparently imprinted in the memory of a young salmon before it migrates to the sea. Years later, on its return journey, a salmon follows these chemical cues at each fork in the river system. Battling white water, sometimes for hundreds of miles, the fish eventually arrives at its place of origin, spawns, and dies.

tacles. If you add glutathione to the water around a hydra in the laboratory, you may observe this feeding behavior for yourself. (In fact, if there are two or more hydras in your container, they may resort to cannibalism.)

The taste receptors of insects are located within sensory hairs called sensillae on the feet and mouthparts. The animals use their sense of taste to select food. A tasting hair contains several chemoreceptor cells, each especially responsive to a particular class of chemical stimuli, such as sugar or salt. By integrating sensations (nerve impulses) from these different receptor cells, the insect's brain can apparently distinguish a very large number of tastes (FIGURE 45.19). Insects can also smell airborne chemicals, using olfactory sensillae, usually located on the antennae (see FIGURE 45.2).

In humans and other mammals, the chemical senses of gustation and olfaction are functionally similar and interrelated. In both cases, a small molecule must dissolve in liquid to reach the receptor cell and stimulate the sensation. That molecule binds to a specific protein in the receptor cell membrane, triggering a depolarization of the membrane and the release of neurotransmitter.

The receptor cells for taste are modified epithelial cells organized into **taste buds** scattered in several areas of the tongue and mouth. Most of the taste buds are on the surface of the tongue or are associated with raised papillae on the tongue. Although we cannot distinguish different types of taste receptors from their structures, we recognize four basic taste perceptions—sweet, sour, salty, and bitter—each detected in a distinct region of the tongue. These basic tastes are associated with specific molecular shapes or charges (the ring structure of glucose for sweetness, for instance, or the positive sodium ion for saltiness) that bind to separate receptor molecules. As with the taste receptors of insects, sensory data transmitted by sensory neurons from taste buds to the mammalian brain represent the differential stimulation of the various classes of receptors. Although each receptor cell is more responsive to a particular type of substance, it can actually be stimulated by a broad range of chemicals. With each taste of food or sip of drink, the brain integrates the differential input from the taste buds, and a complex flavor is perceived.

The olfactory sense of mammals detects certain airborne chemicals. Olfactory receptor cells are neurons that line the upper portion of the nasal cavity and send impulses along their axons directly to the olfactory bulb of the brain (FIGURE 45.20). The receptive ends of the cells contain cilia that extend into the layer of mucus coating the nasal cavity. When an odorous substance diffuses into this region, it binds to specific receptor molecules on the plasma membrane of the olfactory cilia. The binding triggers a signal-transduction pathway involving a G protein relay and, in many cases, the effector enzyme adenylyl cyclase and the second messenger cyclic AMP (see FIGURE 41.5). The second messenger opens Na^+ channels in the olfactory receptor cell membrane, depolarizing it and generating action potentials that go to the brain. Humans can distinguish thousands of different odors, but these are probably based on a few primary odors, analogous to the basic tastes of the gustatory system.

Although the receptors and brain pathways for taste and olfaction are independent, the two senses do interact. Indeed, much of what we call taste is really smell. If

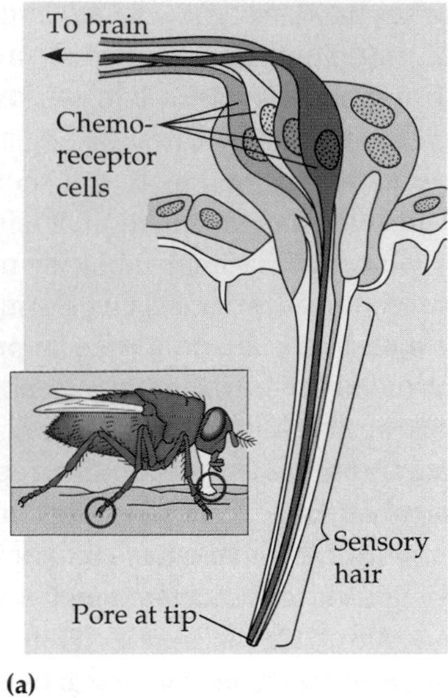

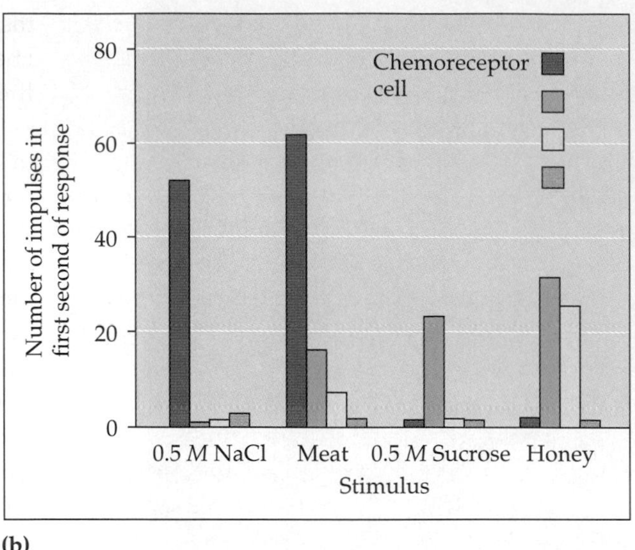

FIGURE 45.19
The mechanism of taste in a blowfly. (**a**) Gustatory sensillae (hairs) on the feet and mouthparts each contain four chemoreceptor cells with dendrites that extend to the pore at the tip of the sensory hair. (**b**) Each chemoreceptor (taste) cell is especially sensitive to a particular class of substance; for example, the receptor colored green here is most responsive to sugars. But this specificity is relative; each cell can respond to some extent to a broad range of chemical stimuli. Thus, any natural food probably stimulates two or more of the receptor cells. The brain apparently integrates the frequencies of impulses arriving along the axons of the four classes of receptor cells and distinguishes a great variety of tastes.

the olfactory system is blocked, as by a head cold, the perception of taste is sharply reduced.

* * *

Throughout our discussions of sensory mechanisms, we have seen many examples of how sensory inputs to the nervous system result in the specific body movements that we observe as animal behavior. The swimming of planarians away from light, the escape behavior of a moth that hears bat sonar, the feeding movements of *Hydra* when it tastes glutathione, and the homing of a salmon that can smell its breeding stream—these are just a few cases that have been mentioned so far. The remainder of the chapter focuses on the motor mechanisms that make these animal responses possible: how animals use their muscles and skeletons to move.

Movement is a hallmark of animals

To catch food, an animal must either move through its environment or move the surrounding water or air past itself. Sessile animals stay put, but they wave tentacles that capture prey or use beating cilia to generate water

FIGURE 45.20
Olfaction in humans. The specific binding of molecules (blue dots) to specific receptor molecules in the plasma membrane of chemoreceptor cells triggers action potentials. Action potentials are conveyed to neurons in the olfactory bulb of the brain by axons of the receptor cells.

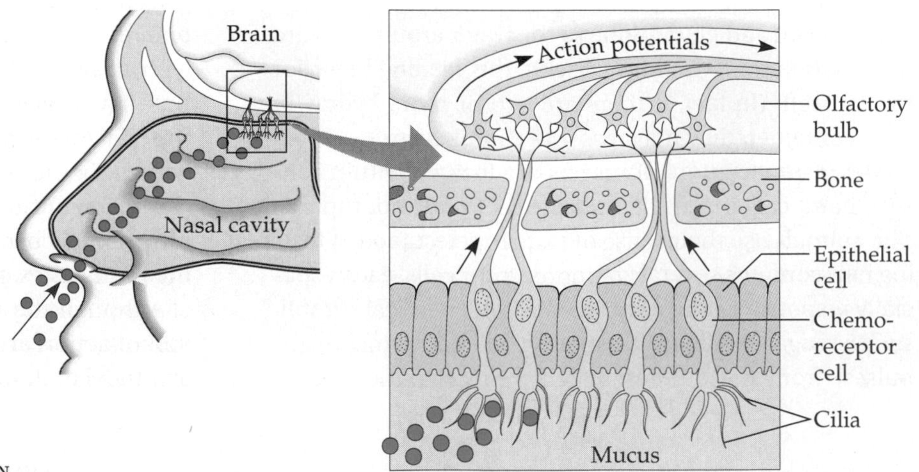

currents that draw and trap small food particles. Most animals, however, are mobile and spend a considerable portion of their time and energy actively searching for food, as well as escaping from danger and looking for mates. Movement is a hallmark of animals.

Locomotion is active travel from place to place. The modes of animal locomotion are diverse. Several animal phyla include species that swim. On land and in the sediments on the floor of the sea and lakes, animals crawl, walk, run, or hop. Adaptations for flight have evolved in only a few animal classes among the insects (phylum Arthropoda) and among the reptiles, birds, and mammals (phylum Chordata).

Whatever the mode of transport, an animal must exert enough force against its environment to overcome friction and gravity. Exerting force requires energy-consuming cellular work. Thus, the study of locomotion returns us to the theme of animal bioenergetics. The energetic cost of transport is different for the various modes of locomotion in different environments (FIGURE 45.21). Because most animals are reasonably buoyant in water, overcoming gravity is less of a problem for swimming animals than for species that move on land or through the air. On the other hand, water is a much denser medium than air, and thus the problem of resistance (friction) is a major one for aquatic animals. A sleek, fusiform (torpedolike) shape is a common adaptation of fast swimmers, and swimming tends to be the most energy-efficient means of locomotion.

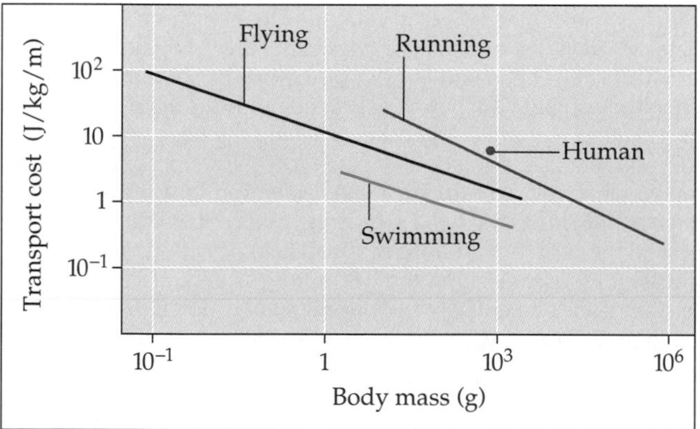

FIGURE 45.21

The cost of transport. This graph compares the transport cost, in joules per kilogram of body weight per meter traveled, for animals specialized for swimming, flying, and running (1 J = 0.24 cal). Notice that both axes are plotted on logarithmic scales. Running animals generally consume more energy per meter traveled than equivalently sized animals specialized for flying or swimming, partly because to run (or walk), an animal must expend energy to overcome inertia. Swimming is the most efficient mode of transport (assuming, of course, that an animal is specialized for swimming). If we were to compare energy consumption per minute rather than per meter, we would find that flying animals use more energy than animals swimming or walking for the same amount of time. Each line on the graph also shows that a larger animal travels more efficiently than a smaller species specialized for the same mode of transport. For example, a horse consumes less energy per kilogram of body weight than a cat running the same distance. (Of course, total energy consumption is greater for the larger animal.)

On land, a walking or running animal must be able to support itself and move against gravity, but, at least at moderate speeds, air poses relatively little resistance. To move in such an environment, strong skeletal support is more important than a streamlined shape. A walking or running animal must also overcome inertia with each step by accelerating a leg from a standing start. A flying animal does not use its skeleton to support itself against the pull of gravity, but gravity poses a major problem in a different way: For an animal to be airborne, the wings must develop enough lift to enable the animal to completely overcome the downward force of gravity. In all these situations, we see adaptations of body shape and specialized appendages representing evolutionary solutions to specific problems of movement.

Underlying the diverse forms of locomotion are fundamental mechanisms common to all animals. At the cellular level, all animal movement is based on one of two basic contractile systems, both of which consume energy to move protein strands against one another. These two systems of cell motility—microtubules and microfilaments—were discussed in Chapter 7. Microtubules are responsible for the beating of cilia and the undulations of flagella. Microfilaments play a major role in amoeboid movement, and they are also the contractile elements of muscle cells. It is the contraction of muscles that concerns us in this chapter, but the work of a muscle in itself cannot translate into movement of the animal. Swimming, crawling, running, and flying all result from muscles working against some type of skeleton.

Skeletons support and protect the animal body and are essential to movement

The three functions of a skeleton are support, protection, and movement. Most land animals would sag from their own weight if they had no skeleton to support them. Even an animal living in water would be a formless mass with no framework to maintain its shape. Many animals have hard skeletons that protect soft tissues. For example, the vertebrate skull protects the brain, and the ribs form a cage around the heart, lungs, and other internal organs. And skeletons aid in movement by giving muscles something firm to work against. There are three main types of skeletons: hydrostatic skeletons, exoskeletons, and endoskeletons.

Hydrostatic Skeletons

A **hydrostatic skeleton** consists of fluid held under pressure in a closed body compartment. This is the main type of skeleton in most cnidarians, flatworms, nematodes, and annelids (see Chapter 29). These animals control their form and movement by using muscles to change the shape of the fluid-filled compartments. Among the cnidarians for example, *Hydra* can elongate by closing its mouth and using contractile cells in the body wall to constrict the central gastrovascular cavity. Because water cannot be compressed very much, decreasing the diameter of the cavity forces it to increase in length. In flatworms (planarians), the interstitial fluid is kept under pressure and functions as the main hydrostatic skeleton. Planarian movement results mainly from muscles in the body wall exerting localized forces against the hydrostatic skeleton. Roundworms (nematodes) hold the fluid in the body cavity (a pseudocoelom; see Chapter 29) at a high pressure, and contractions of longitudinal muscles result in thrashing movements. In earthworms and other annelids, the coelomic fluid functions as a hydrostatic skeleton. The coelomic cavity is divided by septa between the segments of the worm, and thus the animal can change the shape of each segment individually, using both circular and longitudinal muscles. The hydrostatic skeleton enables earthworms and most other annelids to move by **peristalsis,** a type of locomotion produced by rhythmic waves of muscle contractions passing from head to tail (FIGURE 45.22).

Hydrostatic skeletons are well suited for life in aquatic environments. They may cushion internal organs from shocks and provide support for crawling and burrowing. However, a hydrostatic skeleton cannot support the forms of terrestrial locomotion in which an animal's body is held off the ground, such as walking or running.

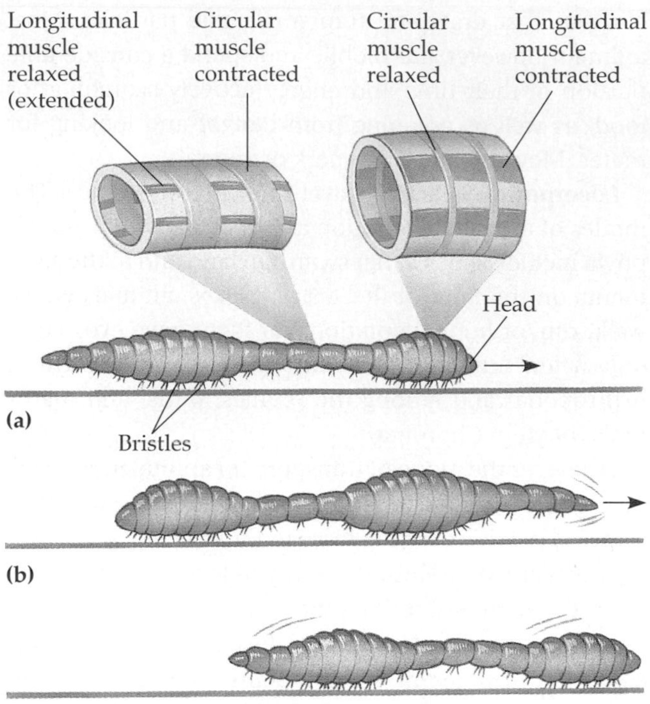

FIGURE 45.22
Peristaltic locomotion in an earthworm. A hydrostatic skeleton, two sets of muscles (one elongating the body, the other shortening it), and bristles holding to the substrate enable an earthworm to crawl over moist ground or burrow through it. Contraction of longitudinal muscles thickens and shortens the worm, while contraction of circular muscles constricts and elongates it. (**a**) In this example, as the worm crawls forward, body segments at its head and in front of the tail are short and thick (longitudinal muscles contracted; circular muscles relaxed) and anchored to the ground by bristles. Behind the head and at the tail, segments are thin and elongated (circular muscles contracted; longitudinals relaxed). (**b**) The head has moved forward because circular muscles in the head segments have contracted. Segments behind the head and in front of the tail are now thick and anchored, thus preventing the worm from slipping backward. (**c**) The head segments are thick again and anchored in their new position. The rear segments have released their hold on the ground and have been pulled forward.

Exoskeletons

An **exoskeleton** is a hard encasement deposited on the surface of an animal. For example, most mollusks are enclosed in calcareous (calcium carbonate) shells secreted by the mantle, a sheetlike extension of the body wall. As the animal grows, it enlarges the diameter of the shell by adding to its outer edge. Clams and other bivalves close their hinged shells using muscles attached to the inside of this exoskeleton.

The jointed exoskeleton typical of arthropods is a **cuticle,** a nonliving coat secreted by the epidermis. Muscles are attached to knobs and plates of the cuticle that extend into the interior of the body. About 30% to 50% of the cuticle consists of **chitin,** a polysaccharide similar to cellulose (see Chapter 5). Fibrils of chitin are embedded in a matrix made of protein, forming a composite material that combines strength and flexibility. Where protection is most important, the cuticle is hardened with organic compounds that cross-link the proteins of the exoskeleton. Some crustaceans, such as lobsters, harden portions of their exoskeletons even more by adding calcium salts. In contrast, at the joints of the legs, where the cuticle must be thin and flexible, there is only a small amount of inorganic salts and little cross-linking of proteins. The exoskeleton of an arthropod must periodically be shed (molted) and replaced by a larger case with each spurt of growth by the animal (see FIGURE 5.9).

Next we will examine endoskeletons, including those of vertebrates (FIGURE 45.23).

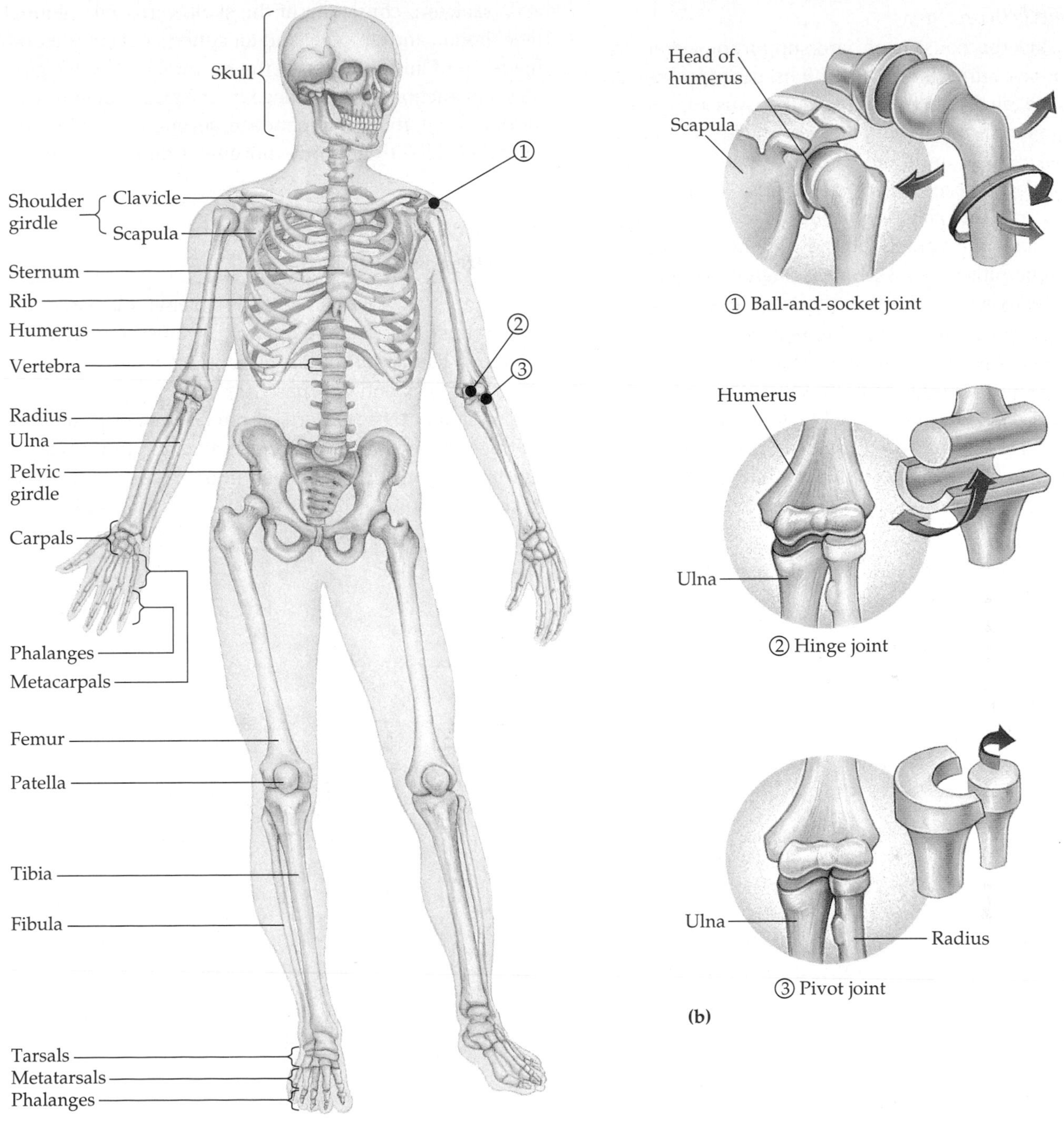

FIGURE 45.23

The human skeleton. (**a**) The axial skeleton (green) provides an axis of support for the upright (bipedal) body and surrounds and protects the brain, spinal cord, lungs, and heart. The appendicular skeleton (gold) supports the arms and legs. Adapted from an ancient model, the human skeleton has many features in common with other vertebrate skeletons. For example, most land vertebrates, though quadrupedal, have the same long bones supporting their arms and legs. Circled numbers are examples of different types of joints, described in part b. (**b**) Joints allow great flexibility in body movement, indicated by arrows. ① Ball-and-socket joints, where the humerus contacts the shoulder girdle and where the femur contacts the pelvic girdle, enable us to rotate our arms and legs and move them in several planes. ② Hinge joints, such as between the humerus and the head of the ulna, restrict movement to a single plane. ③ A pivot joint allows us to rotate the forearm at the elbow. Hinge and pivot joints between bones in our wrists and hands enable us to make precise manipulations.

Endoskeletons

An **endoskeleton** consists of hard supporting elements, such as bones, buried within the soft tissues of an animal. Sponges are reinforced by hard spicules consisting of inorganic material or by softer fibers made of protein. Echinoderms have an endoskeleton of hard plates beneath the skin. These ossicles are composed of magnesium carbonate and calcium carbonate crystals, and the separate plates are usually bound together by protein fibers. Sea urchins have a skeleton of tightly bound ossicles, but the ossicles of sea stars are more loosely bound, allowing the animal to change the shape of its arms.

Chordates have endoskeletons consisting of cartilage, bone, or some combination of these materials (see Chapter 36). The mammalian skeleton is built from more than 200 bones, some fused together and others connected at joints by ligaments that allow freedom of movement. Anatomists divide the vertebrate frame into an axial skeleton, consisting of the skull, vertebral column (backbone), and rib cage; and an appendicular skeleton, made up of limb bones and the pectoral and pelvic girdles that anchor the appendages to the axial skeleton (see FIGURE 45.23). In each appendage, several types of joints provide flexibility for body movement and locomotion.

Muscles move skeletal parts by contracting

As we mentioned earlier, animal movement is based on the contraction of muscles working against some type of skeleton. The action of a muscle is *always* to contract; muscles can extend only passively.

The ability to move parts of the body in opposite directions requires that muscles be attached to the skeleton in antagonistic pairs, each muscle working against the other (FIGURE 45.24). We flex our arm, for instance, by

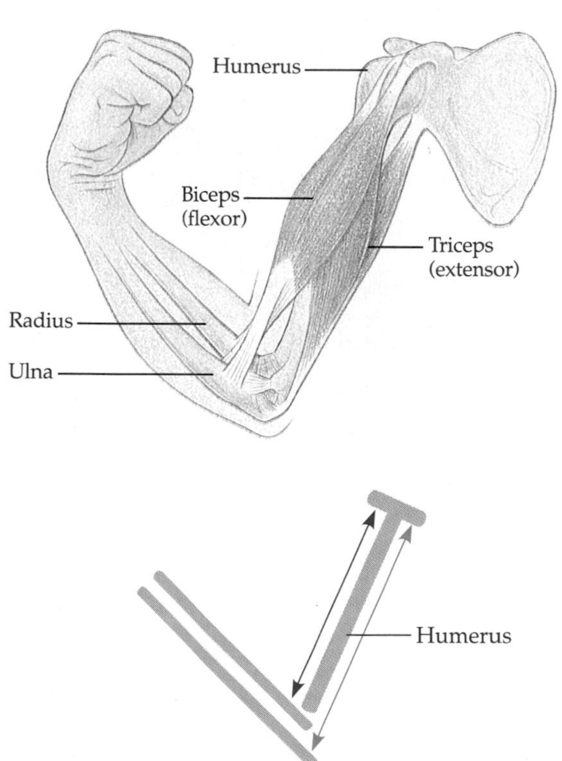

(a)

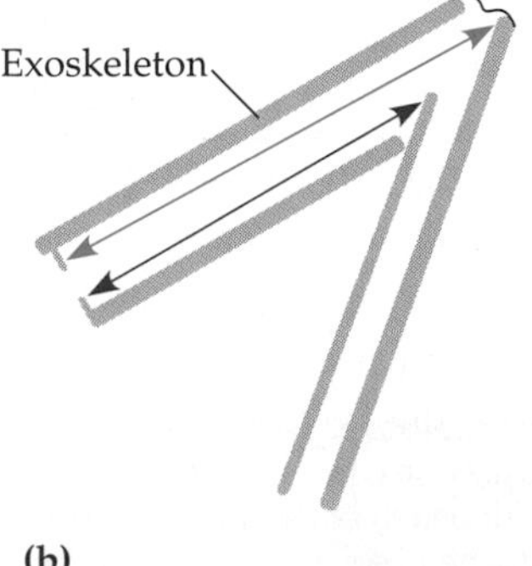

(b)

FIGURE 45.24

The cooperation of muscles and skeletons in movement. Muscles actively contract, but they elongate only when passively stretched. Back-and-forth movement is generally accomplished by antagonistic muscles, each working against the other. This arrangement works with either an endoskeleton or an exoskeleton. (**a**) In humans, contraction of the biceps muscle, represented by red in the bottom diagram, raises (flexes) the forearm. Contraction of the triceps muscle (green) lowers (extends) the forearm. (**b**) Although arthropod muscles are positioned differently and housed within an exoskeleton, the antagonistic action of flexors and extensors is similar to that of a vertebrate. When the flexor muscle (red) in the upper part of a grasshopper's leg contracts, the lower leg is pulled toward the body. In this position the grasshopper is sitting, poised for a jump, as shown here. Alternatively, when the extensor muscle (green) in its upper leg contracts, the leg jerks backward, sending the insect into the air.

contracting the biceps, with the hinged joint of the elbow acting as the fulcrum of a lever. To extend the arm, we relax the biceps while the triceps on the opposite side contracts. But how does a muscle actually contract? As always, the key to function is structure. In this section, we will examine the structure and mechanism of contraction of vertebrate skeletal muscle and then compare this basic pattern with other types of muscle.

Structure and Function of Vertebrate Skeletal Muscle

Vertebrate **skeletal muscle,** which is attached to the bones and responsible for their movement, is characterized by a hierarchy of smaller and smaller parallel units (FIGURE 45.25). A skeletal muscle consists of a bundle of long fibers running the length of the muscle. Each fiber is a single cell with many nuclei, reflecting its formation by the fusion of many embryonic cells. Each fiber is itself a bundle of smaller **myofibrils** arranged longitudinally. The myofibrils, in turn, are composed of two kinds of **myofilaments. Thin filaments** consist of two strands of actin and one strand of regulatory protein coiled around one another, while **thick filaments** are staggered arrays of myosin molecules.

Skeletal muscle is also called striated muscle because the regular arrangement of the myofilaments creates a repeating pattern of light and dark bands. Each repeating unit is a **sarcomere,** the basic functional unit of the muscle. The borders of the sarcomere, the **Z lines,** are lined up in adjacent myofibrils and contribute to the striations visible with a light microscope. The thin filaments are attached to the Z lines and project toward the center of the sarcomere, while the thick filaments are centered in the sarcomere. At rest, the thick and thin filaments do not overlap completely, and the area near the edge of the sarcomere where there are only thin filaments is called the **I band.** The **A band** is the broad region that corresponds to the length of the thick filaments. The thin filaments do not extend completely across the sarcomere, so the **H zone** in the center of the A band contains only thick filaments. This arrangement of thick and thin filaments is the key to how the sarcomere, and hence the whole muscle, contracts.

The Molecular Mechanism of Muscle Contraction

When a muscle contracts, the length of each sarcomere is reduced; that is, the distance from one Z line to the next becomes shorter. In the contracted sarcomere, the A bands do not change in length, but the I bands shorten

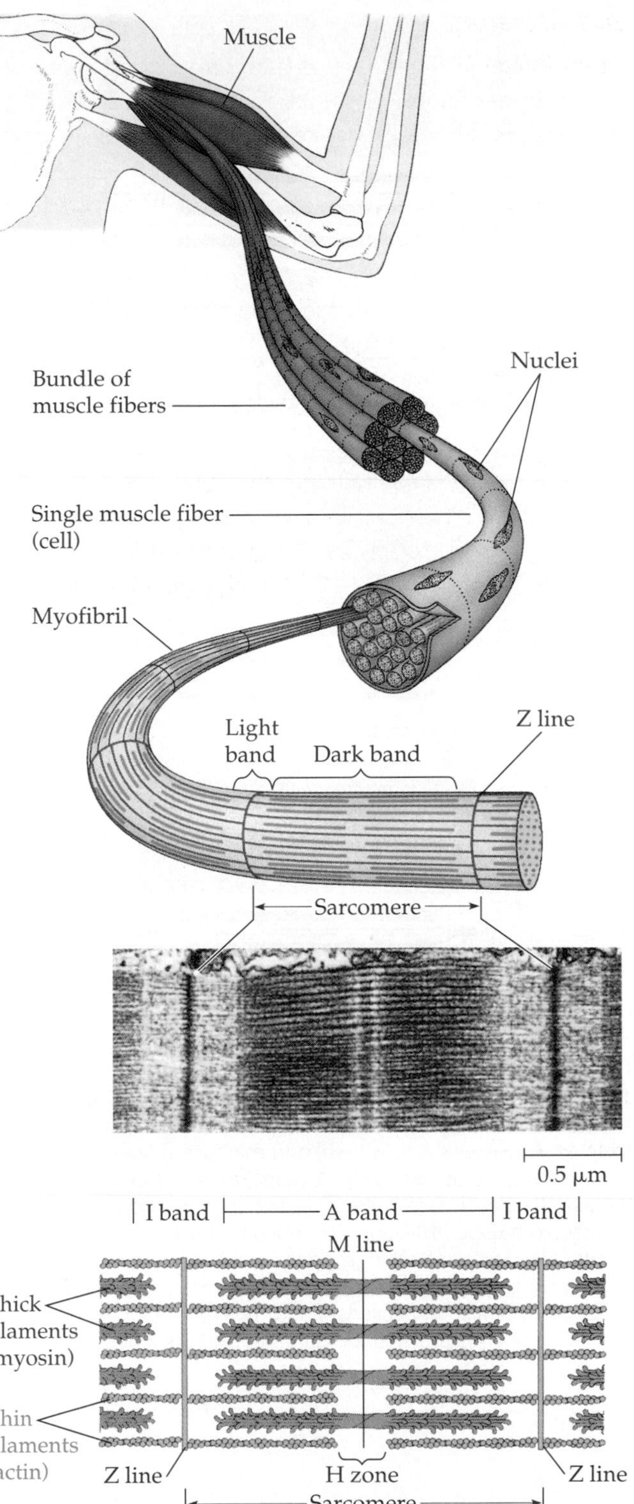

FIGURE 45.25
The structure of skeletal muscle. A muscle consists of bundles of multinucleated muscle fibers (cells), each of which is a bundle of myofibrils. Each myofibril is made of thick and thin filaments aligned in contractile units called sarcomeres. The arrangement of thick and thin filaments appears as alternating light and dark bands when striated muscle is viewed with a microscope, as in the TEM here. As the bottom diagram indicates, only thin filaments occur in the I bands, and the dark zone in the center of each I band, called the Z line, is attached to the thin filaments. The sarcomere is the entire apparatus between two Z lines. The A band includes regions where thick and thin filaments overlap, and a central H zone containing only thick filaments. Connections among the thick filaments form the thin M line within the H zone.

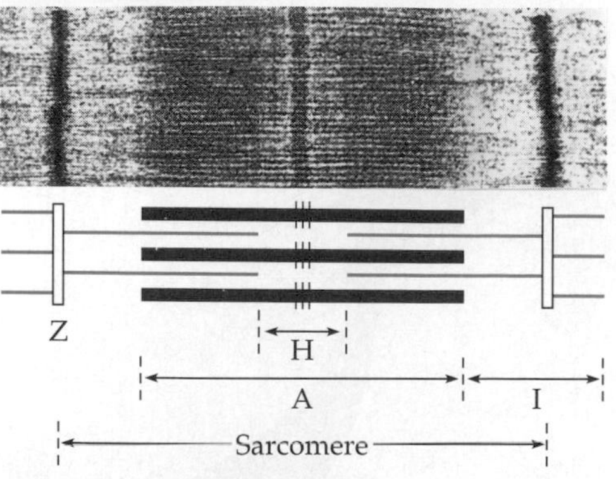

(a) Muscle relaxed (extended)

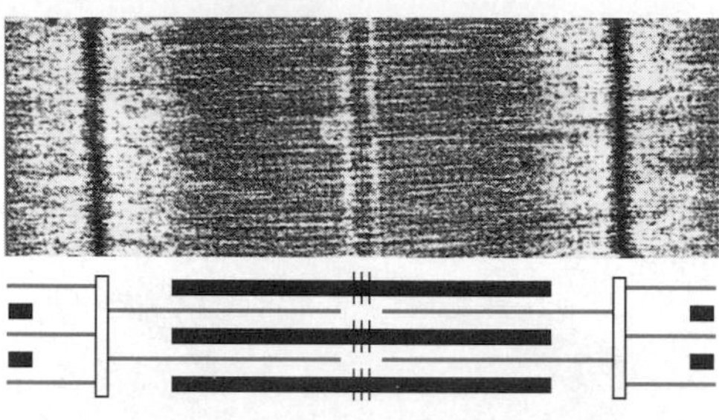
(b) Muscle contracting

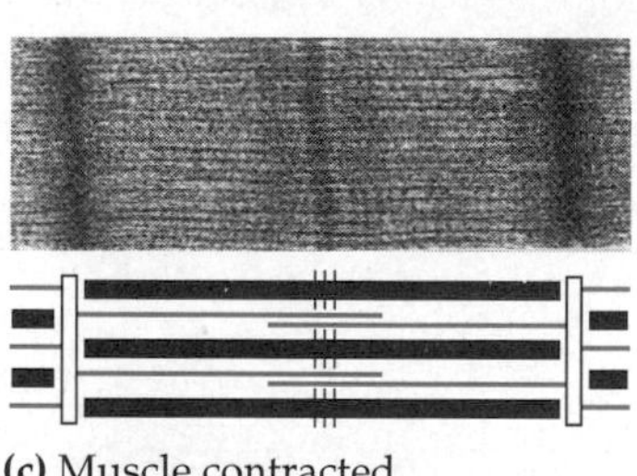
(c) Muscle contracted

FIGURE 45.26
The sliding-filament model of muscle contraction. As seen in these transmission electron micrographs, the lengths of the thick (myosin) filaments (red) and thin (actin) filaments (blue) remain the same as contraction occurs. (**a**) In a relaxed muscle, the length of each sarcomere is greater than in a contracting or contracted muscle. (**b**) During contraction, thick and thin filaments slide past each other, shortening the sarcomere. (**c**) When the muscle is fully contracted, the sarcomere is markedly shortened; the thin filaments overlap, and there is little or no space between the ends of the thick filaments and the Z lines.

and the H zone disappears (FIGURE 45.26). This behavior can be explained by the **sliding-filament model** of muscle contraction. According to this model, neither the thin filaments nor the thick filaments change in length when the muscle contracts; rather, the filaments slide past each other longitudinally, so that the degree of overlap of the thin and thick filaments increases. If the region of overlap increases, both the length occupied only by thin filaments (the I band) and the length occupied only by thick filaments (the H zone) must decrease.

The sliding of the filaments is based on the interaction of the actin and myosin molecules that make up the thin and thick filaments. Myosin consists of a long, fibrous "tail" region, with a globular "head" region sticking off to the side. The tail is where the individual myosin molecules cohere to form the thick filament. The myosin head is the center of bioenergetic reactions that power muscle contractions. It can bind ATP and hydrolyze it into ADP and inorganic phosphate. Some of the energy released by cleaving the ATP is transferred to the myosin, which changes shape to a high-energy configuration (FIGURE 45.27). This energized myosin binds to a specific site on actin, forming a **cross-bridge.** The stored energy is released, and the myosin head relaxes to its low-energy configuration. This relaxation changes the angle of attachment of the myosin head to the fibrous myosin tail; as the myosin bends inward on itself, it exerts tension on the thin filament to which it is bound, pulling the thin filament toward the center of the sarcomere. The bond between the low-energy myosin and actin is broken when a new molecule of ATP binds to the myosin head. In a repeating cycle, the free head can then cleave the new ATP to revert to the high-energy configuration and attach to a new binding site on another actin molecule farther along the thin filament. Each of the approximately 350 heads of a thick filament forms and re-forms about 5 cross-bridges per second, driving filaments past each other.

A muscle cell typically stores only enough ATP for a few contractions. Muscle cells also store glycogen in columns between the myofibrils, but most of the energy needed for repetitive muscle contraction is stored in substances called **phosphagens. Creatine phosphate,** the phosphagen of vertebrates, can supply a phosphate group to ADP to make ATP.

The Control of Muscle Contraction

A skeletal muscle contracts only when stimulated by a motor neuron. When the muscle is at rest, the myosin binding sites on the actin molecules are blocked by the regulatory protein **tropomyosin.** Another set of regulatory proteins, the **troponin complex,** controls the position of tropomyosin on the thin filament (FIGURE 45.28, p. 1052). For a muscle cell to contract, the myosin binding sites on the actin must be uncovered. This occurs when calcium ions bind to troponin, altering the interaction between troponin and tropomyosin. The Ca^{2+} binding causes the whole tropomyosin-troponin complex to change shape and expose the myosin binding sites on actin. When calcium is present, the sliding of thin and thick filaments can occur, and the muscle contracts. When internal calcium concentration falls, the binding sites of actin are covered, and contraction stops.

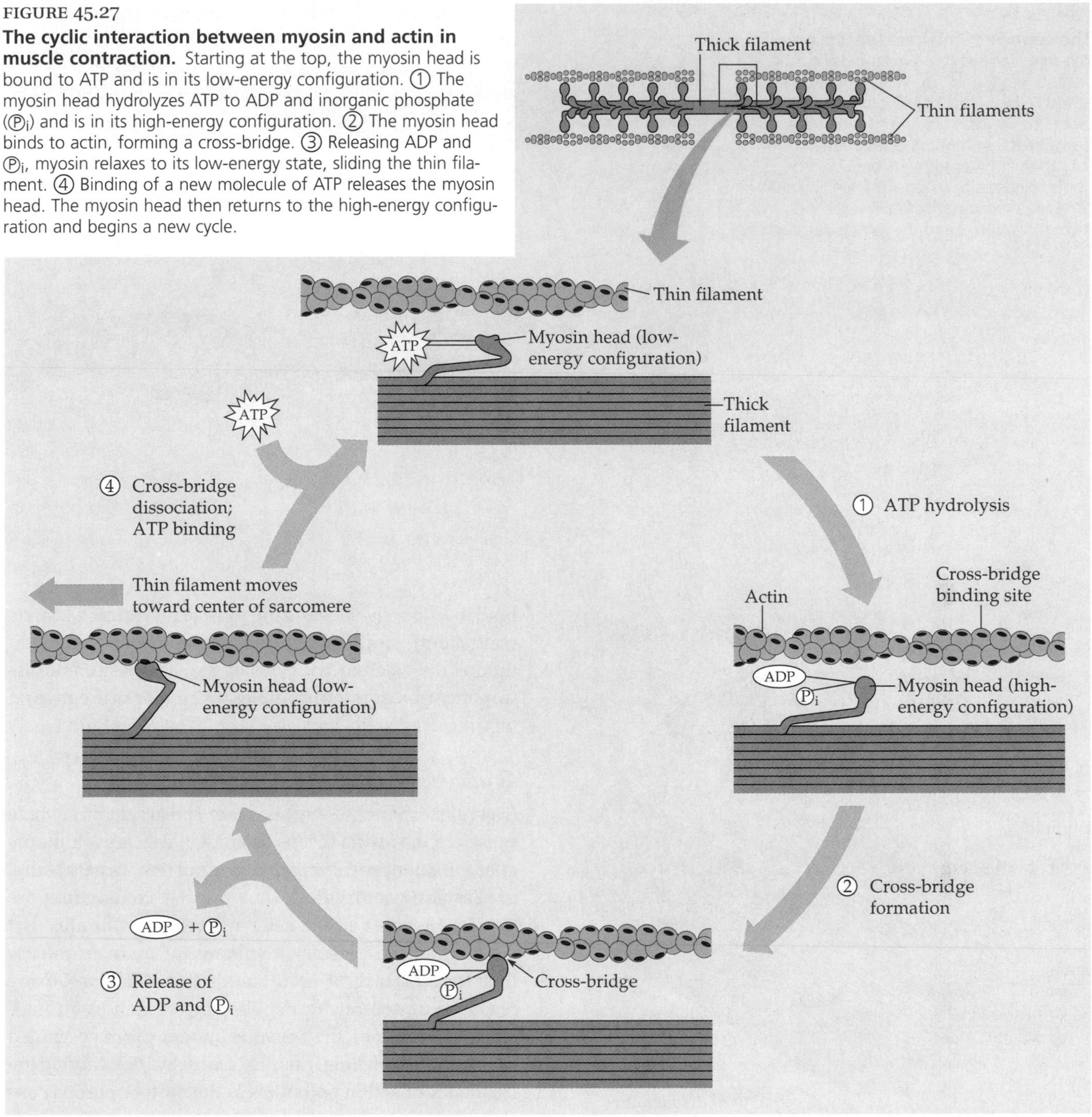

FIGURE 45.27

The cyclic interaction between myosin and actin in muscle contraction. Starting at the top, the myosin head is bound to ATP and is in its low-energy configuration. ① The myosin head hydrolyzes ATP to ADP and inorganic phosphate (Ⓟ$_i$) and is in its high-energy configuration. ② The myosin head binds to actin, forming a cross-bridge. ③ Releasing ADP and Ⓟ$_i$, myosin relaxes to its low-energy state, sliding the thin filament. ④ Binding of a new molecule of ATP releases the myosin head. The myosin head then returns to the high-energy configuration and begins a new cycle.

Calcium concentration in the cytoplasm of the muscle cell is regulated by the **sarcoplasmic reticulum,** a specialized endoplasmic reticulum (FIGURE 45.29). The membrane of the sarcoplasmic reticulum actively transports calcium from the cytoplasm into the interior of the reticulum, which is thus an intracellular storehouse for calcium. The stimulus leading to the contraction of a skeletal muscle cell is an action potential in the motor neuron that makes synaptic connection with the muscle cell. As you learned in Chapter 44, the synaptic terminal of the motor neuron releases acetylcholine at the neuromuscular junction, depolarizing the postsynaptic muscle cell and triggering an action potential in the muscle cell. That action potential is the signal for contraction. The action potential spreads deep into the interior of the muscle cell along infoldings of the plasma membrane called **T (transverse) tubules.** Where the transverse tubules contact the sarcoplasmic reticulum, the action potential changes the permeability of the sarcoplasmic reticulum, causing it to release calcium ions. These calcium ions

FIGURE 45.28

The control of muscle contraction. The thin filament has two strands of actin twisted into a helix. (**a**) When the muscle is at rest, a long, rodlike tropomyosin molecule blocks the myosin binding sites that are instrumental in forming cross-bridges. (**b**) When another protein complex, troponin, binds calcium ions, the binding sites on actin are exposed, cross-bridges with myosin can form, and the muscle contracts.

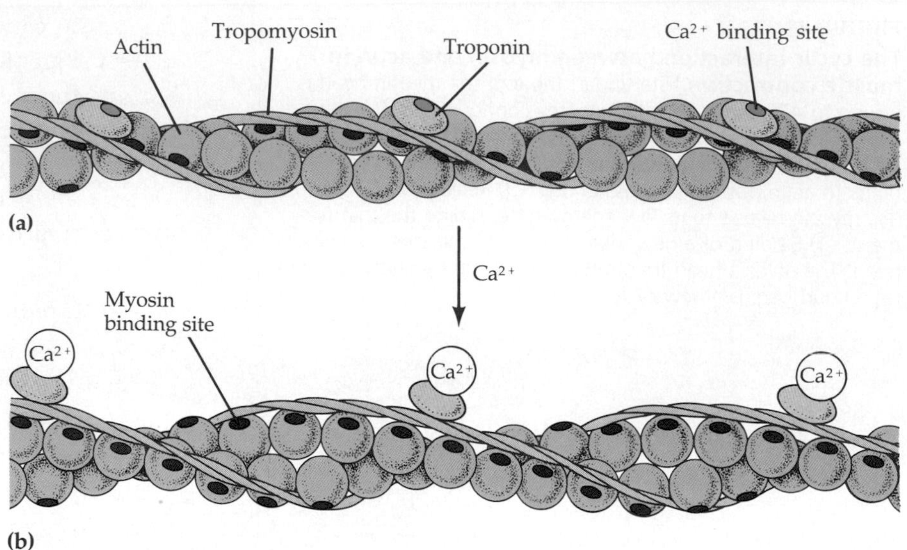

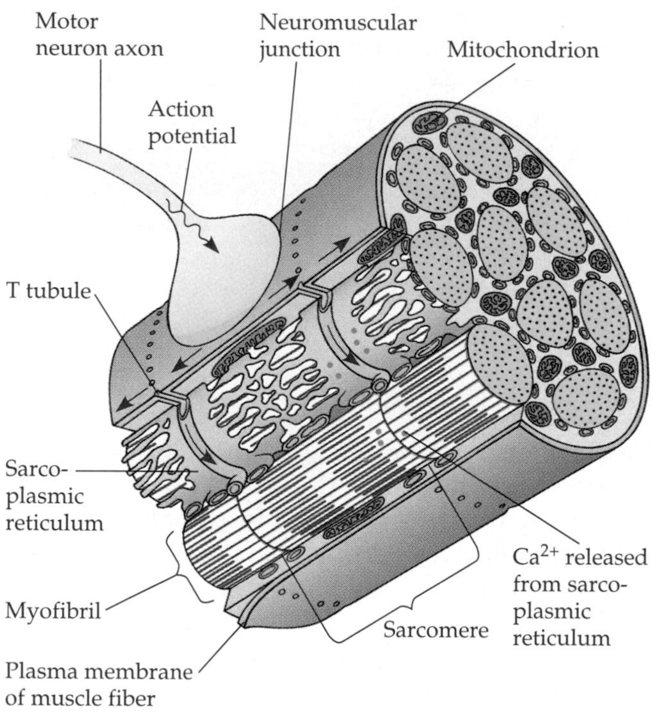

FIGURE 45.29

The role of the muscle fiber's sarcoplasmic reticulum and T tubules in contraction. In response to an action potential arriving at the synaptic terminal of a motor neuron, the neuron membrane releases the neurotransmitter acetylcholine. Diffusing across the neuromuscular junction, acetylcholine depolarizes the plasma membrane of the muscle fiber, and action potentials (black arrows) sweep across the fiber and deep into it along T (transverse) tubules. Within the muscle cell, the action potentials trigger the release of Ca^{2+} (green dots) from the sarcoplasmic reticulum into the cytoplasm. The Ca^{2+} initiates the sliding of filaments by triggering the binding of myosin to actin.

bind to troponin, allowing the muscle to contract. Muscle contraction stops when the sarcoplasmic reticulum pumps the calcium back out of the cytoplasm, and the tropomyosin-troponin complex again blocks the myosin binding sites as the concentration of calcium falls.

Graded Contractions of Whole Muscles

Everyday experience suggests that the action of a whole muscle, such as the biceps, is graded; we can voluntarily alter the extent and strength of contraction. Experimental studies also confirm that whole-muscle contractions are graded. However, at the cellular level, any stimulus that depolarizes the plasma membrane of a single muscle fiber triggers an all-or-none contraction, analogous to the response of neurons to depolarizing stimuli (see Chapter 44). How does the nervous system produce graded contractions of whole muscles? One way is by varying the frequency of action potentials in the motor neurons controlling the muscle. A single action potential will produce an increase in muscle tension lasting about 100 msec or less, a single twitch (FIGURE 45.30). If a second action potential arrives before the response to the first is over, the tension will sum and produce a greater response. If a muscle receives an overlapping series of action potentials, further summation will occur, with the level of tension depending on the rate of stimulation. And if the rate of stimulation is fast enough, the twitches will blur into one smooth and sustained contraction called **tetanus** (not to be confused with the disease of the same name). Motor neurons usually deliver their action potentials in rapid-fire volleys, and the resulting summation of tension

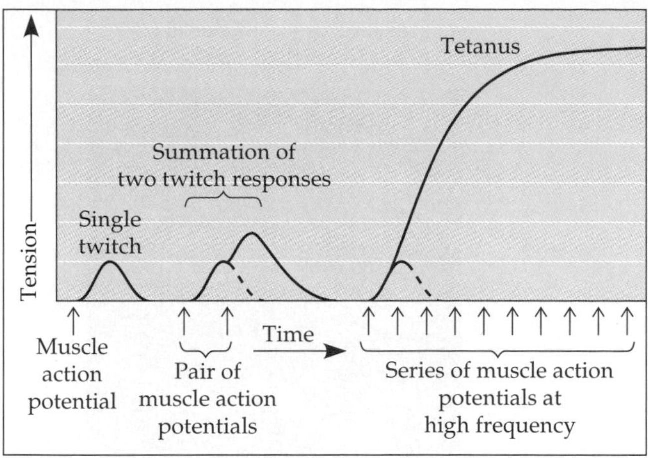

FIGURE 45.30
Temporal summation of muscle cell contractions. This graph compares the tension developed in a muscle in response to a single action potential, a pair of action potentials, and a series of action potentials. The dashed lines show the response that would have resulted if only the first action potential had occurred.

results in smooth contraction typical of tetanus rather than the jerky actions of individual muscle twitches.

The nervous system can also produce graded contraction of a whole muscle by taking advantage of the organization of the muscle cells into motor units. In a vertebrate muscle, each muscle cell is innervated by only one motor neuron, but each branched motor neuron may make synaptic connections with many muscle cells (FIGURE 45.31). There may be hundreds of motor neurons controlling an individual muscle, each with its own pool of muscle fibers scattered throughout the muscle. A **motor unit** consists of a single motor neuron and all the muscle fibers it controls. When the motor neuron fires, all the muscle fibers in the motor unit contract as a group. The strength of the resulting contraction will therefore depend on how many muscle fibers the motor neuron controls. In most muscles, there is wide variation in the number of muscle fibers among motor units; some motor neurons may control only a few muscle cells, while others may control hundreds. The nervous system can thus regulate the strength of contraction in the whole muscle by both determining how many motor units are activated at a given instant and selecting whether large or small motor units are activated. Tension in a muscle can be progressively increased by activating more and more of the motor neurons controlling the muscle, a process called **recruitment** of motor neurons. Depending on the number and size of motor neurons your brain recruits to the task, you can lift a fork, or something much heavier, like your biology textbook.

Some muscles, especially those that hold the body up and maintain posture, are almost always partially contracted. However, prolonged contraction results in muscle fatigue caused by the depletion of ATP, dissipation of the ion gradients required for normal electrical signaling, and the accumulation of lactate (see Chapter 9). In a mechanism that avoids fatigue in postural muscles, the nervous system alternates activation among the various motor units making up the muscle, so that different motor units take turns maintaining the prolonged contraction.

Fast and Slow Muscle Fibers

We have seen that the action potential in a skeletal muscle fiber is only a trigger for the contraction; the actual duration of the contraction is controlled by how long the calcium concentration in the cytoplasm remains elevated. Not all skeletal muscle fibers are identical in this regard. We can identify fast and slow fibers based on the duration of their twitches. **Fast muscle fibers** are used for rapid, powerful contractions. Some, such as the flight muscles of birds, may be able to sustain long periods of repeated contractions without fatiguing. By contrast, **slow muscle fibers,** which can sustain long contractions, are often found in muscles that maintain posture. A slow fiber has less sarcoplasmic reticulum than a fast fiber, so calcium remains in the cytoplasm longer. This causes a twitch in a slow fiber to last about five times longer than in a fast fiber. Slow fibers are also specialized to make use of a steady supply of energy; they have many mitochondria, a rich blood supply, and an oxygen-storing protein called myoglobin. Myoglobin, the brownish-red pigment in the dark meat of poultry and fish, binds oxygen more tightly than hemoglobin, so it can effectively extract oxygen from the blood.

Other Types of Muscle

There are many types of muscles in the animal kingdom, but as noted before, they all share the same fundamental mechanism of contraction: the sliding of actin filaments and myosin filaments past one another. In addition to skeletal muscle, vertebrates have cardiac and smooth muscle (see FIGURE 36.4).

Vertebrate **cardiac muscle** is found in only one place—the heart. Like skeletal muscle, cardiac muscle is striated. The primary differences between skeletal and cardiac muscle are in their electrical and membrane properties. The junctions between cardiac muscle cells contain specialized regions called **intercalated discs,** where gap junctions provide direct electrical coupling among cells. Thus, an action potential generated in one part of the heart will spread to all the cardiac muscle cells, and the whole heart will contract. Skeletal muscle cells will not fire an action potential and contract unless triggered to do so by input from a controlling motor neuron. Cardiac muscle cells, however, can generate action

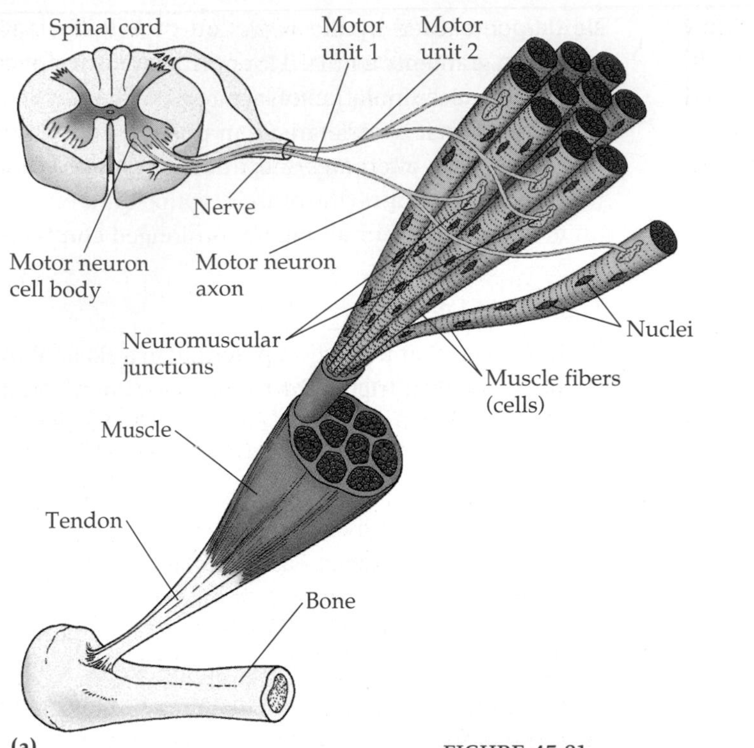

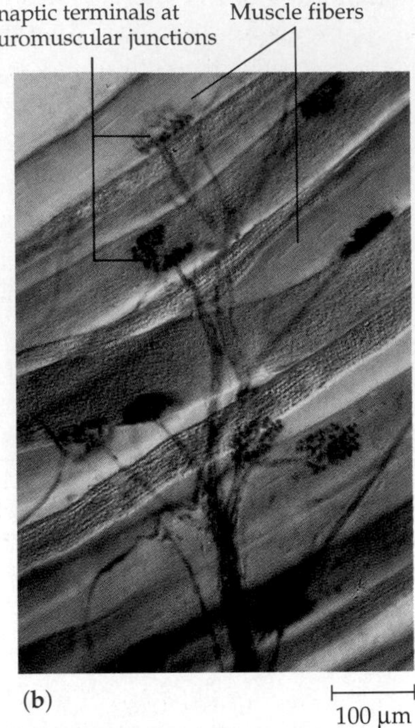

FIGURE 45.31

Motor units in a vertebrate muscle. (**a**) Each muscle fiber (cell) has a single neuromuscular junction, or synaptic connection, with the motor neuron that controls it. However, each motor neuron typically branches and controls several or many muscle fibers. A motor neuron and all the fibers it controls comprise a contractile apparatus called a motor unit. (**b**) You can see several synaptic terminals of a branched motor neuron in this light micrograph.

potentials on their own, without any input from the nervous system. The plasma membrane of a cardiac muscle cell has pacemaker properties that cause rhythmic depolarizations, triggering action potentials and causing single cardiac muscle cells to "beat" even when isolated from the heart and placed in cell culture. (But on the whole-organ level, the heart also has a pacemaker, specialized muscle tissue in the wall of the right atrium that coordinates contractions of cardiac muscle cells throughout the heart; see FIGURE 38.6.) The action potentials of cardiac muscle cells are also different from those of skeletal muscle cells, lasting up to twenty times longer. Whereas the action potential of a skeletal muscle cell serves only as a trigger for contraction and does not control the duration of contraction, in a cardiac cell, the duration of the action potential plays an important role in controlling the duration of contraction.

Smooth muscle lacks the striations of skeletal and cardiac muscle because the actin and myosin filaments are not all regularly arrayed along the length of the cell. Instead, the filaments may have a spiral arrangement within smooth muscle cells. Smooth muscle also contains less myosin than striated muscle, and the myosin is not associated with specific actin strands. Because of the organization, smooth muscle cannot generate nearly as much tension as striated muscle, but it can contract over a much greater range of lengths. Further, smooth muscle has neither a T tubule system nor a well-developed sarcoplasmic reticulum. Calcium ions must enter the cytoplasm via the plasma membrane during an action potential, and the amount reaching the filaments is rather small. Contractions are relatively slow, but there is a greater range of control than in striated muscle. Smooth muscle is found mainly in the walls of hollow organs such as digestive tract organs and blood vessels. Smooth muscle propels substances through the hollow organ by alternately contracting and relaxing.

Invertebrates possess muscle cells similar to vertebrate skeletal and smooth muscle cells. Arthropod skeletal muscles are nearly identical to vertebrate skeletal muscles. However, the flight muscles of insects are capable of independent, rhythmic contraction, so the wings of some insects can actually beat faster than action potentials can arrive from the central nervous system. Another interesting evolutionary adaptation has been discovered in the muscles that hold clam shells closed. The thick filaments of these muscle fibers contain a unique protein called paramyosin that enables the muscles to remain in a fixed state of contraction with a low rate of energy consumption for as long as a month.

Although we have talked about sensory receptors and muscles separately in this chapter, they are the two ends of a single integrated system. An animal's behavior, so fundamental to how the animal interacts with its environment, is the product of a nervous system connecting sensations to responses. Behavior is discussed in Unit Eight, within the broader context of ecology: the study of interactions between organisms and their environment.

REVIEW OF KEY CONCEPTS (with page numbers and key figures)

- Sensory receptors detect changes in the external and internal environment: *an overview* (pp. 1026–1030)
  - Animal behavior depends on the ability of the nervous system to integrate the input from sensory receptors about the internal and external environment and translate the information into appropriate responses in muscles and other effectors.
  - Sensations are action potentials traveling along sensory neurons that are interpreted by different parts of the brain as perceptions.
  - Sensory receptors are usually modified neurons or epithelial cells that detect environmental stimuli and respond with an electrochemical change in their membrane.
  - Reception is the specialized ability of a receptor cell to absorb the energy of a stimulus.
  - Transduction is the conversion of stimulus energy into a change in a membrane potential called a receptor potential.
  - Stimulus energy may be amplified by accessory structures of sense organs or by transduction.
  - Transmission of the receptor potential to the nervous system occurs either as an action potential (when the receptor is a sensory neuron) or as the release of neurotransmitter, which then initiates action potentials in the sensory neuron with which the receptor cell synapses.
  - Integration of information begins at the receptor level by such processes as sensory adaptation and variation in the sensitivity of the receptor. Integration continues in the CNS.
  - Sensory receptors can be classified according to the type of stimulus energy they receive.
  - Mechanoreceptors respond to stimuli such as pressure, touch, stretch, motion, and sound.
  - Chemoreceptors respond either to generalized solute concentrations or to specific molecules.
  - Electromagnetic receptors detect energy in the form of different wavelengths of radiation.
  - Various types of thermoreceptors signal surface and core temperatures of the body.
  - Pain is detected by nociceptors, a group of diverse receptors that respond to excess temperature, pressure, or specific classes of chemicals.
- Photoreceptors contain light-absorbing visual pigments: *a closer look* (pp. 1030–1037, FIGURES 45.4–45.6)
  - The light receptors and visual capabilities of invertebrates vary widely, ranging from the simple light-sensitive eye cup of planarians to the image-forming compound eye of insects and crustaceans and the single-lens eye of some jellyfish, spiders, and many mollusks.
  - The main parts of the vertebrate eye are an outer layer, the sclera, including the transparent cornea; the conjunctiva, a mucous membrane, surrounding all of the sclera except the cornea; the choroid, a pigmented middle layer, which includes the iris, surrounding the pupil; the retina, an inner layer at the back of the eyeball, containing the photoreceptor cells; and the lens, suspended between two chambers in the eye, which focuses light on the retina.
  - Transduction of the light signal occurs in specialized photoreceptors called rods and cones, which contain light-absorbing retinal bonded to specific membrane proteins, collectively called opsins.
  - When rods and cones absorb light, signal-transduction pathways hyperpolarize their membrane, and they release less neurotransmitter. This change in chemical message is transmitted to bipolar cells and then to ganglion cells, whose axons in the optic nerve convey action potentials to the brain. Horizontal cells and amacrine cells in the retina integrate information before it is sent to the brain.
  - Most axons of the optic nerves go to the lateral geniculate nuclei of the thalamus, from which neurons lead to the primary visual cortex. Several integrating centers in the cerebral cortex are active in creating visual perceptions.
- Hearing and balance are related in most animals: *a closer look* (pp. 1037–1042, FIGURES 45.12, 45.15, 45.16)
  - In mammals and most terrestrial vertebrates, the hair cell mechanoreceptors for hearing and balance are located in the inner ear.
  - The outer ear consists of the external pinna and auditory canal. The tympanic membrane (eardrum) transmits sound waves to three small bones of the middle ear, which amplify and transmit the waves through the oval window to the fluid in the coiled cochlea of the inner ear. Pressure waves vibrate the basilar membrane and the attached organ of Corti, which contains receptor hair cells. The bending of the hairs against the tectorial membrane depolarizes the hair cells, triggering action potentials in the auditory nerve to the brain.
  - Volume (loudness) is a function of the amplitude of the sound wave that results in greater bending of the hair cells. Pitch is related to frequency of sound waves. Regions of the basilar membrane vibrate more vigorously at different frequencies and transmit to specific auditory areas of the cerebral cortex.
  - The utricle, saccule, and three semicircular canals in the inner ear function in balance and equilibrium.
  - The detection of water movement in fishes and aquatic amphibians is accomplished by a lateral line system of clustered hair cell receptors.

- Many arthropods have vibrating exoskeletal hairs and localized "ears," consisting of a tympanic membrane and receptor cells, to sense sounds. Some invertebrates detect position in space by means of statocysts.

- The interacting senses of taste and smell enable animals to detect many different chemicals: *a closer look* (pp. 1042–1044, FIGURES 45.19, 45.20)
  - Taste and smell both depend on the stimulation of receptor cells by small, dissolved molecules that bind to proteins on a chemoreceptor membrane.
  - Many insects have sensillae, sensory hairs containing taste receptor neurons, on their feet and mouthparts.
  - In mammals, taste receptors are organized into taste buds that respond to distinct shapes of molecules. Olfactory receptor cells line the upper part of the nasal cavity and send their axons to the olfactory bulb of the brain.

- Movement is a hallmark of animals (pp. 1044–1045, FIGURE 45.21)
  - Locomotion, active travel from place to place, requires an animal to exert enough force against its environment to overcome friction and gravity. Aquatic, aerial, and terrestrial environments pose different problems for locomotion.
  - Animal movement uses energy and is a function of protein strands moving past each other, either in the microtubules of cilia and flagella or by microfilaments in amoeboid movement and muscle contraction.

- Skeletons support and protect the animal body and are essential to movement (pp. 1045–1048, FIGURE 45.23)
  - A hydrostatic skeleton, found in most cnidarians, flatworms, nematodes, and annelids, consists of fluid under pressure in a closed body compartment. Hydrostatic skeletons support peristaltic locomotion, produced by rhythmic waves of muscle contractions passing from the head to the tail of many worms.
  - Exoskeletons, found in most mollusks and arthropods, are hard coverings deposited on the surface of an animal.
  - Endoskeletons, found in sponges, echinoderms, and chordates, are hard supporting elements embedded within the animal's body.

- Muscles move skeletal parts by contracting (pp. 1048–1055, FIGURES 45.25, 45.29)
  - Muscles, often present in antagonistic pairs, contract and pull against the skeleton to provide movement.
  - Vertebrate skeletal muscle consists of a bundle of muscle cells, each of which contains myofibrils composed of thin filaments of actin and thick filaments of myosin.
  - Contraction begins when impulses from a motor neuron are transmitted to the muscle cell membrane through release of acetylcholine at the neuromuscular junction. Action potentials travel to the interior of the cell along the T tubules, stimulating the release of calcium ions from the sarcoplasmic reticulum. The calcium ions bind to the regulatory troponin-tropomyosin complex on the thin filaments, exposing the myosin binding sites on the actin. Cross-bridges form, and bending of the myosin heads pulls the thin filaments toward the center of the sarcomere. The energy to move the myosin heads is provided by ATP, which is hydrolyzed by myosin.
  - A muscle twitch results from a single stimulus. More rapidly delivered signals produce a graded contraction by summation. Tetanus is a state of smooth and sustained contraction, obtained when motor neurons deliver a volley of action potentials.
  - A motor unit consists of a branched motor neuron and the muscle fibers it innervates. Multiple motor unit recruitment results in stronger contractions.
  - Cardiac muscle, found only in the heart, consists of striated, branching cells that are electrically connected by intercalated discs. Cardiac muscle cells can generate action potentials without neural input.
  - In smooth muscle, contractions are slow but can be sustained over long periods of time.

SELF-QUIZ

1. Which of the following receptors is *incorrectly* paired with its category?
   a. hair cell—mechanoreceptor
   b. muscle spindle—mechanoreceptor
   c. taste receptor—chemoreceptor
   d. rod—electromagnetic receptor
   e. gustatory receptor—electromagnetic receptor

2. Some sharks close their eyes just before they bite. Although they cannot see their prey, their bites are on target. Researchers have noted that sharks often misdirect their bites at metal objects, and they can find batteries buried under the sand of an aquarium. This evidence suggests that sharks keep track of their prey during the split second before they bite in the same way that
   a. a rattlesnake finds a mouse in its burrow
   b. a male silkworm moth locates a mate
   c. a bat can find moths in the dark
   d. a platypus locates its prey in a muddy river
   e. a migrating bird finds its way on a cloudy night

3. Which of the following is an *incorrect* statement about the vertebrate eye?
   a. The vitreous humor regulates the amount of light entering the pupil.
   b. The transparent cornea is an extension of the sclera.
   c. The fovea is the center of the visual field and contains only cones.
   d. The ciliary muscle functions in accommodation.
   e. The retina lies just inside the choroid and contains the photoreceptor cells.

4. The transduction of sound waves to action potentials takes place
   a. within the tectorial membrane as it is stimulated by the hair cells
   b. when hair cells are bent against the tectorial membrane, causing them to depolarize and release neurotransmitter molecules that stimulate sensory neurons
   c. as the basilar membrane becomes more permeable to sodium ions and depolarizes, initiating an action potential in a sensory neuron
   d. as the basilar membrane vibrates at different frequencies in response to the varying volume of sounds
   e. within the middle ear as the vibrations are amplified by the malleus, incus, and stapes

5. The utricle and saccule are
   a. involved in lateral integration of visual signals
   b. visual centers in the cortex
   c. part of the apparatus responsible for positional information about the head
   d. semicircular canals that respond to head rotation
   e. organs of balance found in insects and crustaceans

6. When you first walk from a brightly lit area into darkness, which of the following occurs?
   a. The photopsins in your cones become bleached.
   b. Your rod cells become hyperpolarized.
   c. The receptor cells release less neurotransmitter.
   d. Lateral inhibition caused by your horizontal cells ceases.
   e. Your rhodopsin is still dissociated into retinal and opsin, and your rods are temporarily nonfunctional.

7. The role of calcium in muscle contraction is
   a. to break the cross-bridges as a cofactor in the hydrolysis of ATP
   b. to bind with troponin, changing its shape so that the actin filament is exposed
   c. to transmit the action potential across the neuromuscular junction
   d. to spread the action potential through the T tubules
   e. to reestablish the polarization of the plasma membrane following an action potential

8. Tetanus refers to
   a. the partial sustained contraction of major supporting muscles
   b. the all-or-none contraction of a single muscle fiber
   c. a stronger contraction resulting from multiple motor unit summation
   d. the result of wave summation, which produces a smooth and sustained contraction of a muscle
   e. the state of muscle fatigue caused by the depletion of ATP and the accumulation of lactate

9. Which of the following is a *true* statement about cardiac muscle cells?
   a. They lack an orderly arrangement of actin and myosin filaments.
   b. They have less extensive sarcoplasmic reticulum and thus contract more slowly than smooth muscle cells.
   c. They are connected by intercalated discs, through which action potentials spread to all cells in the heart.
   d. They have a resting potential more positive than an action potential threshold.
   e. They only contract when stimulated by neurons.

10. Which of the following changes occurs when a skeletal muscle contracts?
   a. The A bands shorten.
   b. The I bands shorten.
   c. The Z lines slide farther apart.
   d. The thin filaments contract.
   e. The thick filaments contract.

CHALLENGE QUESTIONS

1. Explain the difference between sensation and perception. Give some examples.

2. Compare vision and hearing with respect to receptor type, transduction, amplification, transmission, and integration characteristics.

3. Although skeletal muscles generally fatigue fairly rapidly, clam shell muscles have a unique protein called paramyosin that allows them to sustain contraction for up to a month. From your knowledge of the cellular mechanism of contraction, propose a hypothesis to explain how paramyosin might work. How would you test your hypothesis experimentally?

SCIENCE, TECHNOLOGY, AND SOCIETY

1. Have you ever felt your ears ringing after listening to loud music from a sound system or at a concert? The sound often tops 90 decibels—loud enough to permanently impair hearing. Do you think people are aware of the possible danger from prolonged exposure to loud music? What, if anything, should be done to warn or protect them?

2. You may know an older person who has broken a bone (often a hip) partly because of osteoporosis, a loss of bone density that affects many women after menopause. Researchers think that prevention is the best way to avoid osteoporosis. They recommend exercise and maximum calcium intake during the teenage years and the twenties. Is it realistic to expect young people to view themselves as future senior citizens? How would you recommend that they be encouraged to develop good health habits that might not pay off for 40 or 50 years?

FURTHER READING

Alexander, R. M. *The Human Machine.* New York: Columbia University Press, 1992. A lucid, well-illustrated account of the biomechanical principles underlying human body movement.

Cronin, T. W., N. J. Marshall, and M. F. Land. "The Unique Visual System of the Mantis Shrimp." *American Scientist,* July/August 1994. Structure and function of compound eyes and how they fit the lifestyle of a marine predator.

Freeman, D. H. "In the Realm of the Chemical." *Discover,* June 1993. An essay on the senses of taste and smell and their evolutionary significance in humans.

Grady, D. "The Vision Thing: Mainly in the Brain." *Discover,* June 1993. A survey of current research on how visual perceptions are produced.

Hamill, O. P., and D. W. McBride, Jr. "Mechanoreceptive Membrane Channels." *American Scientist,* January/February 1995. Research on the cellular mechanisms underlying the senses of touch and hearing: membrane channels with gates sensitive to mechanical stimulation.

Konishi, M. "Listening with Two Ears." *Scientific American,* April 1993. Insights on how the brain creates aural perceptions from research on how barn owls locate prey in the dark.

Richardson, S. "The Small Files." *Discover,* August 1995. How the olfactory equipment sorts odors.

Vogel, S. *Life's Devices,* New York: Prentice-Hall, 1991. An award-winning science writer and researcher explores the biomechanics of locomotion.

Zeki, S. *A Vision of the Brain* . Cambridge, MA: Blackwell Scientific, 1993. A top researcher reviews progress in understanding how the brain processes visual information.

Zimmer, C. "Making Senses." *Discover,* June 1993. Current work on surgically implanted electronic devices that may restore hearing and sight.

UNIT EIGHT

ECOLOGY

AN INTERVIEW WITH

MARGARET DAVIS

Margaret Davis is a forest ecologist, but her specialty is the study of forests as they existed thousands of years ago. Dr. Davis travels back in time by analyzing fossil pollen left by the trees that made up the ancient forests. She has used this approach to track tree species as they migrated over the North American landscape when the climate changed during the Ice Ages. This investigation of the past *is now helping ecologists think about the* future, *as they debate how global warming caused by human activities will alter forests and other ecosystems.*

Dr. Davis is a member of the National Academy of Sciences and served as president of the Ecological Society of America in 1988–1989. The Ecological Society also recognized the importance of Dr. Davis's contributions when it selected her for its Eminent Ecologist Award in 1993. In this interview, Dr. Davis, who is Regents Professor of Ecology at the University of Minnesota, tells us about her science and its implications.

Dr. Davis, how did you become interested in science?

I got interested in science as a child because my father was a scientist, a geology professor at Harvard, and he talked a lot about science. My father seemed to really enjoy what he was doing, and that made an impression upon me. It seemed enviable that he had a job that he just really loved and really wanted to do. He never *said* he liked what he was doing; he just acted as if he did.

So did your family encourage you to become a scientist?

No, they didn't. They didn't approve of women being scientists, or even having careers of any sort, actually. However, my

father encouraged it in a way, because he was most comfortable talking to his children about things he was interested in himself. So I think that he did foster my interest in science without meaning to.

Once your interest in science took form, how did it become focused on paleobotany, and more specifically on the study of fossilized pollen?

I suppose I'd always been interested in geology, and I also liked plants and field biology. Paleobotany was a combination of these interests, and I became very intrigued with the fossil record of plants. These weird plants from the past were just as intriguing to me as dinosaurs are now to children. I got interested in ecology as well, and particularly in field botany. I like to go out hiking, especially in the forest. The study of fossil pollen was a way of doing paleobotany and relating it very closely to ecology. That was what the attraction really was. It was ecology with a time dimension.

And what can the study of fossil pollen tell us about the past?

With most kinds of paleobotany, you have real plant parts, so you can tell a lot about the evolution of plants. But you can't tell anything about the quantities of plants. With pollen grains, you have less taxonomic resolution, but there's actually a quantitative relationship between the pollen grains in the sediment and the quantities of trees that were in the surrounding landscape. We're actively investigating this right now in my lab. We don't have all the answers to how you translate the quantities of pollen into quantities of plants. It's really still an active field. Pollen analysis has enabled us to get an idea about paleoclimate and to make a start at reconstructing whole plant communities and placing them on the landscape.

How do you obtain such a pollen history of a particular location?

Sometimes you can see sediment in open section. Actually, this is how the field started out. People were excavating peat bogs in Europe for fuel, and there were many peat bog profiles exposed. People took samples out of them. We don't have very many peat bogs in this country, so we usually work with cores taken through the sediments in a lake. We either go out in the winter and core through a hole in the ice, with the ice serving as a big platform, or we go in the summer with a couple of canoes or inflatable rafts with a platform over them. Through a hole in the middle of the platform, we core down into the sediments. The cores we collect now are big ones, about four inches in diameter. We can use them for radiometric dating and analysis of larger plant parts, and then we also study the pollen fossils up and down the core.

Using pollen as markers for the past distribution and abundance of trees, you've been able to track plant species as they migrated with the advance and retreat of ice sheets during the Ice Ages. What is the mechanism and speed of such tree migration?

Trees spread seeds very effectively. If you think of the leaves around a tree in the fall, there's a little shadow of leaves around the tree. Seeds are disbursed much the same way. When people put out seed traps, they see that the numbers are largest in the vicinity of the tree, and then they fall away quite rapidly. But a few seeds are transported much greater distances. We know this is true because in the fossil record you can find evidence for the existence of a tree on one side of Lake Michigan, and then you begin picking it up on the other side. So obviously seeds have gone all the way across this big lake, which is 100 kilometers across. For example, in the case of hemlock, which sheds its seeds in the winter, it's quite easy to see how this can happen. The cones open when the air is dry, and the seeds then fall out onto the surface of the snow. And if there's a high wind, they're scattered along the surface. So if you walk across a lake, you can see hemlock seeds and birch seeds way out in the middle of the lake on top of the ice. And if the climate is changing at the same time that seeds are dispersed into a new habitat, that habitat might, for the first time, be a place where this tree can become established. And that new colony disperses seeds in all directions, and so the migration continues. Trees can actually migrate this way, at remarkable rates—up to 40 or 50 kilometers in a century. That's pretty fast when you consider that the tree has to shed its seed, and then the seed has to grow up into a new tree and be old enough to produce seed.

Do animals also contribute to tree migration by dispensing seeds?

Some recent studies have been done on the dispersal of seeds by blue jays. They've been observed flying from a forest where they're collecting acorns across a bog to their nests. And then they bury the seeds. That's just the perfect situation for a seed, to actually be planted in the right habitat. Usually, the birds are pretty smart about finding their seed again, but on a big seed year, they bury more than they actually need, so some of those germinate the next spring.

One common image of the Ice Ages is of whole plant communities, such as hardwood forests, migrating southward with the advancing sheets of ice and then back northward with the retreat of the ice. How is your research changing this view?

When we show on a map how different trees have changed their ranges in the past, the maps show clearly that not all trees are changing in synchrony. They are following different migratory paths. At any moment in time, the coincidence of tree ranges is not the same as it is today. By studying fossil pollen, you can find out that different combinations of tree species made up forest communities in the past. During the height of the last glaciation, the forest that existed south of the ice sheet was not similar to the modern boreal forest, it had a different combination of species. People often relate the distribution of modern biomes to different climates. When ecologists think

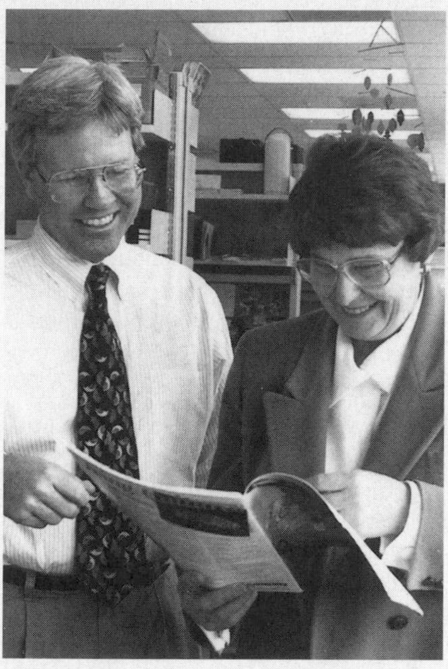

about global climate change in the future, they tend to push these modern biomes into new places. But the fossil record shows clearly that as the climate changes, new plant communities arise. Different combinations of species occur, because the distribution of each species is related to a particular set of climate parameters.

The Ice Ages are examples of past environmental change on a global scale. We now live in a period of global change caused by the human population. Among the concerns are an increasing concentration of atmospheric carbon dioxide and global warming. What is the evidence that global changes are under way?

We don't know for sure whether global warming is under way yet or whether what we're experiencing is just a climate anomaly. We do know that there's a strong relationship between carbon dioxide and climate, because in Antarctica and Greenland there are bubbles of atmosphere preserved in ice sheets, which clearly show that the carbon dioxide concentration was very low when the temperatures were low during the last glacial period. Carbon dioxide levels change from time to time, very much in concert with climate changes. That's some of the best empirical evidence that carbon dioxide levels are correlated with temperature.

What's causing the atmospheric carbon dioxide concentration to increase?

Our use of fossil fuel. There was a lot of carbon dioxide in the atmosphere of Precambrian Earth. After plants began to photosynthesize, they began taking up the carbon dioxide that was in the atmosphere and turning it into organic matter, much of which got buried in rocks. We are now digging up this carbon, which has been buried for a long time, burning it as fossil fuel and turning it into CO_2, which is then entering the atmosphere. There's also a lot of carbon held in living biomass, especially in the tropical forests. And we're very busy cutting down these forests and turning them into carbon dioxide and producing grasslands, which have very low biomass. This is all human participation in what were once natural biogeochemical cycles on the globe.

Why are ecologists and others concerned about increasing atmospheric CO_2 and global warming?

The rate at which we are changing the atmosphere will produce a much more rapid climate change than most of the climate changes in the past. When there's been a very rapid climate change in the past—for example, at the end of the last glacial period—plant species had a hard time keeping up with the temperature change. Now we may be causing climate changes of several degrees in less than a century. That's an order of magnitude faster than most of the climate changes in the past. In the worst-case scenario, you could have a climate that changed so much during the lifetime of a single tree that the tree was only briefly in the best environment for its growth. This would be tremendously disruptive, even in managed ecosystems; in natural ecosystems it would be devastating. There would be widespread extinction of species.

Can you speculate about how this climate change would affect forests?

That's a big question. I think that commercially valuable trees are going to be planted where they can grow, and I think foresters will be very alert to look at long-range predictions and try to plant trees that will grow. So I would expect that commercial forestry will continue. What I'm really worried about is the unmanaged forests, which are the source of a lot of commercially harvested timber. Some of the slow-growing trees that we're used to seeing, like sugar maple and hemlock, are going to lose out, and the landscape will be covered with successional species like aspen and paper birch. We already have a lot of aspen and birch here in Minnesota as a result of logging. I see a great loss in the diversity of forest species.

We have many preserves designed to save certain rare species. Those preserves are going to be hard hit. People will have to try to create communities artificially at more northern latitudes in order to save species. That kind of intervention is often not very successful. Aside from the ethical questions that arise when humans try to manipulate nature at that level of detail, I'm not sure we really *know* how to create an ecosystem similar to an old-growth forest, which takes many centuries to develop. I'm very concerned about this.

How is logging changing the North American landscape?

It's already had a huge effect. Essentially all of eastern North America was clear-cut. In the Great Lakes region, there's only a half of one percent left of the original northern hardwood forest. Actually, some of the cutting was done as recently as 1950. Small remnants of old-growth forest are still being cut here and there because nobody is keeping track of where they are in a systematic way. Individual property owners cut them, and there's no way to protect them. There are just a few preserves under management by the U.S. Forest Service or by the state that are kept as parks. It seems shocking to me that this ecosystem was essentially eliminated as late as the 1950s, when we knew better. People like Aldo Leopold were pointing out that we had an opportunity to save a big enough corridor for species to move north and south, but the corridor wasn't preserved when it could have been saved. And now in the Pacific Northwest, at an earlier stage, there's debate about whether we should save 10% of the forest, and whether it should be in big patches or little patches. I hope people will feel that ecosystems that take a very, very long time to develop should be preserved, and they recognize that you can't recreate them once you've cut them down. Unfortunately, it's clear that logging of old growth will continue because it's economically profitable. But we have a lot of wood that we can cut from plantations and second-growth forests, and that's what we should be concentrating on, rather than going after the old growth.

Logging is built into the economy of some regions. How can you make your point about conserving old-growth forests to people who make their living from timber?

They should learn a lesson from the northern Great Lakes region, which had a big logging boom and then the trees were all gone and there was a depression. I think the Pacific Northwest has to consider what's going to happen when the trees are all gone. How are people going to support themselves? I don't think they're thinking in those terms.

A sustained cutting strategy at a much lower rate would keep a smaller group of people employed forever. The Menominee Indians in Wisconsin, for instance, have resisted contracts that would have allowed logging companies to come in and clear-cut their forests. They're cutting trees themselves, but at a very slow rate that keeps people employed locally. That slow rate of cutting is such that it's very hard to tell the Menominee Reservation forests from old growth.

Beyond the pragmatic arguments for trying to conserve biodiversity in forests and other ecosystems, do you see the human-caused extinction of other species as a moral issue?

People learn in elementary ecology that populations grow at a rapid rate, and then they begin to level off as they reach their environmental limitation. The human population has never behaved this way. It keeps growing as we find more and more habitat where we can manage to support more people. We don't behave like a natural population at all. But there will come a point at which we reach a limitation. We're a dominant species in terms of our numbers and our huge impact on the global ecosystem. But that doesn't give us the right to take away space from other species and to cause their extinction. I feel that human beings, as a very successful and dominant species on Earth, should take responsibility for more vulnerable species. For us to take advantage of our power and walk over other species, causing their extinction, is arrogant, and I believe, immoral.

Do we also have a moral responsibility not to ruin Earth for future generations of our own species?

That's really true. The consequences of destruction can be so slow that one generation causes extinction of a species, and it's a later generation that regrets it. There's no going back to create that species again. Most people aren't thinking about the kind of world their grandchildren are going to inherit. When I was born there were only about two billion people on Earth, and now there are three times that many, increasing at an even more rapid rate. For some reason, concerns about population growth, destruction of habitat, and extinction of species don't seem to be in the general public's eye. I don't quite see why not, but maybe scientists haven't been doing a very good job of making these issues seem important to people.

Do you think biologists have a special responsibility to try to influence the business and political communities on environmental issues?

As an ecologist, you can't escape this responsibility even if you want to. Once you start doing research in an area, before you know it, somebody appears who wants to destroy it. For example, I spent quite a bit of time last year arguing about whether or not snowmobile routes ought to come through an area of old-growth forest. It seemed obvious to me that they shouldn't, but it was not at all obvious to the people who wanted the routes. Most ecologists I know who do long-term studies spend a lot of time just defending the territory where their study is taking place. Even if they initially felt no responsibility at all, they'd end up having to take some. There are certainly a lot of demands on ecologists' time to testify here and there and present these issues in a public forum. Ecologists have a special responsibility to give testimony in any way they can to try to head off environmental disasters.

CHAPTER 46

AN INTRODUCTION TO ECOLOGY: DISTRIBUTION AND ADAPTATIONS OF ORGANISMS

KEY CONCEPTS

- Ecology is the scientific study of the interactions between organisms and their environments
- Basic ecology provides a scientific context for evaluating environmental issues
- Ecological research ranges from the adaptations of organisms to the dynamics of ecosystems
- Climate and other abiotic factors are important determinants of the biosphere's distribution of organisms
- The costs and benefits of homeostasis affect an organism's responses to environmental variation
- The response mechanisms of organisms are related to environmental grain and the time scale of environmental variation
- The geographical distribution of terrestrial biomes is based mainly on regional variations in climate
- Aquatic ecosystems, consisting of freshwater and marine biomes, occupy the largest part of the biosphere

Organisms as open systems that interact continuously with their environments is a theme that has already surfaced many times in this book (see Chapter 1). The scientific study of the interactions between organisms and their environments is called **ecology** *(from the Greek* oikos, *"home," and* logos, *"to study"). This straightforward definition masks an enormously complex and exciting area of biology that is also of increasingly critical practical importance. As the first key concept in this chapter, it is worth restating the definition of ecology so we can look more closely at three key words in our definition: "scientific," "environment," and "interactions."*

Ecology is the scientific study of the interactions between organisms and their environments

As an area of *scientific* study, ecology incorporates the hypothetico-deductive approach, using observations and experiments to test hypothetical explanations of ecological phenomena. As we will see shortly, ecologists often face extraordinary challenges in their research because of the complexity of their questions, the diversity of their subjects, and the large expanses of time and space over which studies must often be conducted. Ecology is also challenging because of its multidisciplinary nature; ecological questions form a continuum with those from other areas of biology, including genetics, evolution, physiology, and behavior, as well as those from other sciences, such as chemistry, physics, geology, and meteorology.

The *environment* includes **abiotic** factors, such as temperature, light, water, and nutrients. Just as important in their effects on organisms are **biotic** factors—all the other organisms that are part of any individual's environment. Other organisms may compete with an individual for food and other resources, prey upon it, or change its physical and chemical environment. As we will see, questions about the relative importance of various environmental factors are frequently at the heart of ecological studies—and the accompanying controversies.

Finally, in our dissection of the definition of ecology, we should highlight the *interaction* between organisms and their environment. Organisms are affected by their environment but, by their very presence and activities, they also change it—often dramatically. Possibly the most dramatic effect organisms have had on their environment occurred some three billion years ago, when cyanobacteria first began utilizing sunlight for energy and gave off oxygen as a by-product of photosynthesis. The aerobic atmosphere that resulted from this change had a profound effect on the entire planet. On a more localized level, trees reduce light levels on the floor of a forest as they grow, sometimes making the environment unsuitable for their own offspring. Throughout our survey of ecology, we will see many more examples of how organisms and their environments affect one another. And we will relate ecological research to concerns about the impact of humans on the environment.

Basic ecology provides a scientific context for evaluating environmental issues

The science of ecology, the study of how organisms interact with their environments, should be distinguished from the same word that is often used informally in popular media to refer to environmental concerns. Nevertheless, we need to understand the often complicated and delicate relationships between organisms and their environments in order to address environmental problems. Much of our current environmental awareness had its beginnings with Rachel Carson's 1962 book *Silent Spring,* which pointed out that the widespread use of pesticides such as DDT was causing declines in many nontarget organisms. Less than a decade later, photographs of the planet taken by Apollo astronauts, such as the one that opens this chapter, contributed to the growing awareness of Earth as a finite home in the vastness of space, rather than an unlimited frontier for human activity. Acid precipitation, localized famine aggravated by land misuse and population growth, the growing list of species extinct or endangered because of habitat destruction, and the poisoning of soil and streams with toxic wastes are just a few of the problems that threaten the home we share with millions of other forms of life. Many influential ecologists, including Margaret Davis, the scientist you met in the preceding interview, are strong advocates for environmental quality, and they recognize their responsibility to educate legislators and the general public about decisions that affect the environment. Usually, any response also involves ethics, economics, and politics, which are considerations beyond our scope here. However, the chapters of this unit will point out many connections between basic ecology and environmental issues.

This chapter introduces ecology by defining its scope. We then examine the important abiotic environmental factors that influence organisms and the ways in which organisms are adapted to those factors. We will also describe some of the large-scale ecological patterns found in various places on Earth.

Ecological research ranges from the adaptations of organisms to the dynamics of ecosystems

The Questions of Ecology

Because there are many levels and types of interactions between organisms and their environments, the questions ecologists address are extremely wide-ranging. Ecology can be divided into four increasingly comprehensive levels of study, from the interactions of individual organisms with the abiotic environment to the dynamics of ecosystems.

Organismal ecology is concerned with the behavioral, physiological, and morphological ways in which individual organisms meet the challenges posed by their abiotic environment. The distribution of organisms is limited by the abiotic conditions they can tolerate.

The next level of organization in ecology is the **population,** a group of individuals in a particular geographic area that belong to the same species. Population ecology concentrates mainly on factors that affect population size and composition.

A **community** consists of all the organisms that inhabit a particular area; it is an assemblage of populations of different species. Questions at this level of analysis involve the ways in which predation, competition, and other interactions among organisms affect community structure and organization.

Ecological study of the **ecosystem** includes all the abiotic factors in addition to the community of species that exists in a certain area. Some critical questions at the ecosystem level concern energy flow and the cycling of chemicals among the various biotic and abiotic components.

Ecology ultimately deals with the highest levels in the hierarchy of biological organization (see Chapter 1). The web of interactions at the heart of ecological phenomena is what makes this branch of biology so engaging and challenging.

Ecology as an Experimental Science

By necessity, humans have always had an absorbing interest in other organisms and their environments. As hunters and gatherers, prehistoric people had to learn where game and edible plants could be found in greatest abundance. Naturalists from Aristotle to Darwin made the process of observing and describing organisms in their natural habitat an end in itself, rather than simply a means to survive. Extraordinary insight can still be gained through this descriptive approach, and thus natural history remains fundamental to ecological science.

Although ecology has a long history as a descriptive

FIGURE 46.1
Ecology as an experimental science. Over the past three decades, ecology has become more of an experimental science. Here, a team of researchers prepares an experiment to test hypotheses about the colonization of islands by insects and other animals. The scientists enclosed a number of small mangrove islands in the Florida Keys with scaffolding that was used to support a plastic tent over the island. The island was then fumigated to destroy its fauna of insects and other small invertebrates. When the tent was removed, the researchers could study recolonization.

science, most modern ecologists are also skilled experimentalists. In spite of the difficulty of conducting experiments that often involve large amounts of time and space, an increasing number of creative ecologists are testing hypotheses in the laboratory and are experimentally manipulating populations and communities in the field. One early example is a classic 1969 study by Daniel Simberloff and E. O. Wilson that involved eliminating all the insects on small islands to study recolonization of the islands from nearby mainland populations (FIGURE 46.1).

Many ecologists devise mathematical models that enable them to simulate large-scale experiments that might be impossible to conduct in the field. In this approach, important variables and their hypothetical relationships are described through mathematical equations. The potential ways in which the variables interact can then be studied. For example, many ecologists, along with climatologists, paleontologists, and other scientists, use sophisticated computer programs to develop models that predict the effects human activities will have on climate, and how climatic changes will affect ecosystems. Of course, such simulations are only as good as the basic information on which the models are based, and obtaining that information still requires extensive fieldwork.

Advances in ecological research have highlighted the complex relationships between organisms and their biotic and abiotic surroundings. Although it makes ecology a dynamic and exciting science, such complexity can sometimes lead to problems. Politicians and lawyers often want specific answers to such questions as: How much forest is needed to save spotted owls? While ecological studies can certainly provide essential information for making policy decisions on habitat preservation, responses to such questions often include further questions: How many owls must be saved? With what certainty must they be saved? How long must they survive in this amount of forest? An important part of the responsibility of ecologists in educating legislators and citizens is communicating the complexity of these decisions.

Ecology and Evolution

The fields of ecology and evolutionary biology are tightly linked. It was the geographical distribution of organisms and their exquisite adaptations to specific environments that provided Charles Darwin with evidence for evolution. An important (although not the only) cause of evolutionary change is the response of organisms to both biotic and abiotic features of their environment. Thus, events that occur in the frame of what is sometimes called *ecological time* translate into effects over the longer scale of *evolutionary time*. For instance, hawks feeding on field mice have an impact on the gene pool of the prey population by curtailing the reproductive success of certain individuals. One long-term effect of such a predator-prey interaction may be a prevalence in the mouse population of fur coloration that camouflages the animals.

An evolutionary theme is inherent in these chapters on ecology: The present distribution and abundance of organisms are products of both long-term evolutionary changes and ongoing interactions with the environment.

Climate and other abiotic factors are important determinants of the biosphere's distribution of organisms

The entire portion of Earth that is inhabited by life is called the **biosphere;** it is the sum of all the planet's communities and ecosystems. The biosphere is a relatively thin layer consisting of seas, lakes, rivers, and streams; the land to a soil depth of a few meters; and the atmosphere to an altitude of a few kilometers.

Ecologists have long recognized striking global and regional patterns in the distribution of organisms within the biosphere (FIGURE 46.2). The term **biomes** refers to the major types of communities and ecosystems that are typical of broad geographic regions—coniferous forests, deserts, and grasslands are some examples. (The worldwide distribution of biomes is shown in FIGURES 46.11 and 46.21.) These global patterns, and more localized variations in the distribution of organisms, mainly reflect

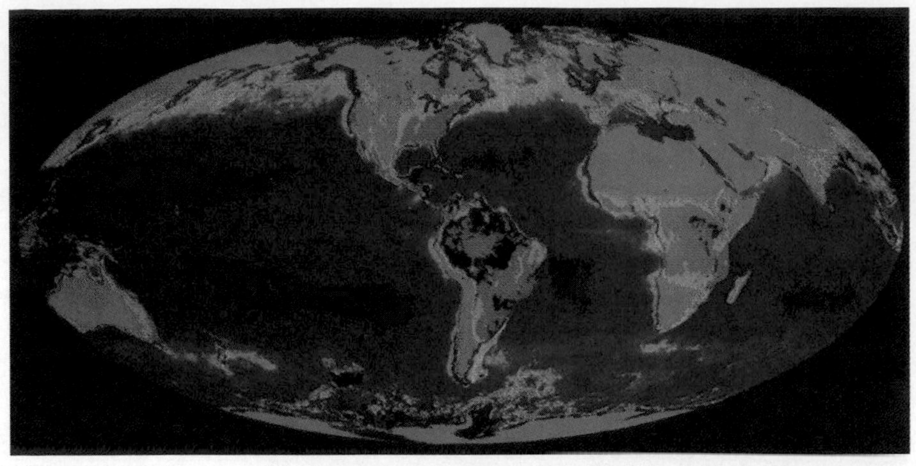

FIGURE 46.2
Patterns of distribution in the biosphere. The regional distribution of life in the biosphere mainly reflects variations in abiotic factors such as temperature and the availability of water. This global patchiness is highlighted in this image of Earth, which is based on data sent from two satellites over an eight-year period. The colors are keyed to the relative abundance of chlorophyll, an indication of the regional densities of life. Dark red spots in the oceans are blooms of phytoplankton. Bright red bands along the coasts correspond to nutrient-rich shallow water where algae abound. Green or black areas on land are dense forests, including the tropical forests of South America and Africa. Orange areas on land are relatively barren regions, such as the Sahara.

regional differences in climate and other abiotic factors in the environment. Almost all organisms ultimately derive their energy from sunlight, and they must tolerate the ranges of temperature, humidity, salinity, and light in their environment. Therefore, before we survey the major biomes, we will analyze how some important abiotic factors such as temperature influence the distribution of organisms. We then consider global, regional, and seasonal variations in climate, an environmental feature that incorporates several abiotic factors, including temperature and rainfall. Throughout this discussion it is important to remember that the physical environment varies in both space and time. Although two regions of Earth may experience different conditions at any given moment, daily and annual fluctuations of abiotic factors sometimes blur or accentuate the distinctions between those regions.

Important Abiotic Factors

Temperature. Environmental temperature is an important factor in the distribution of organisms because of its effect on biological processes and the inability of most organisms to regulate body temperature precisely. Cells may rupture if the water they contain freezes at temperatures below 0°C, and the proteins of most organisms denature at temperatures above 45°C. In addition, few organisms can maintain a sufficiently active metabolism at very low or very high temperatures. Within this range, however, most biochemical reactions and physiological processes occur more rapidly at higher temperature. Extraordinary adaptations enable some organisms to live outside this temperature range. The actual internal temperature of an organism is affected by heat exchange with its environment (see Chapter 40), and most organisms cannot maintain body temperatures more than a few degrees above or below the ambient temperature. As endotherms, mammals and birds are the major exceptions, but even endotherms function best within certain environmental temperature ranges, which vary with the species.

Water. The unique properties of water have effects on organisms and their environments, as you learned in Chapter 3. Water is essential to life, but its availability varies dramatically among habitats. Freshwater and marine organisms live submerged in an aquatic environment, but they face problems of water balance if their intracellular osmolarity does not match that of the surrounding water. Organisms in terrestrial environments encounter a nearly constant threat of desiccation, and their evolution has been shaped by the requirements for obtaining and conserving adequate supplies of water.

Sunlight. Sunlight provides the energy that drives nearly all ecosystems, although only plants and other photosynthetic organisms use this energy source directly. Light intensity is not the most important factor limiting plant growth in many terrestrial environments, but shading by a forest canopy makes competition for light in the understory intense. In aquatic environments, the intensity and quality of light limit the distribution of photosynthetic organisms. Every meter of water selectively absorbs about 45% of the red light and about 2% of the blue light that pass through it. As a result, most photosynthesis in aquatic environments occurs relatively near the surface. However, the photosynthetic organisms themselves absorb some of the light that penetrates, further reducing light levels in the waters below.

Light is also important to the development and behavior of the many plants and animals that are sensitive to photoperiod, the relative lengths of daytime and nighttime. Photoperiod is a more reliable indicator than temperature for cueing seasonal events, such as flowering or migration.

Wind. Wind amplifies the effects of environmental temperature on organisms by increasing heat loss due to evaporation and convection (the wind-chill factor). It also contributes to water loss in organisms by increasing the rate of evaporation in animals and transpiration in plants. In addition, wind can have a substantial effect on the morphology of plants by inhibiting the growth of limbs on the windward side of trees; limbs on the leeward side grow normally, resulting in a "flagged" appearance (see FIGURE 31.2).

Rocks and Soil. The physical structure, pH, and mineral composition of rocks and soil limit the distribution of plants and the animals that feed upon them, thus contributing to the patchiness we see in terrestrial biomes. In streams and rivers, the composition of the substrate can affect water chemistry, which in turn influences the resident plants and animals. In marine environments, the structure of the substrates in the intertidal zone and on seafloors determines the types of organisms that can attach or burrow in those habitats.

Periodic Disturbances. Catastrophic disturbances such as fires, hurricanes, tornadoes, and volcanic eruptions can devastate biological communities. After the disturbance, the area is recolonized by organisms or repopulated by survivors, but the structure of the community undergoes a succession of changes during the rebound. Some disturbances, such as volcanic eruptions, are so infrequent and unpredictable over space and time that organisms have not acquired evolutionary adaptations to them. Fire, on the other hand, although unpredictable over the short term, recurs frequently in some communities, and many plants have adapted to this periodic disturbance. In fact, several communities actually depend on periodic fire to maintain them. We will see examples of this when we look at grasslands and other biomes later in the chapter.

Climate and the Distribution of Organisms

The important abiotic factors just described have a direct effect on the biology of organisms. The first four factors—temperature, water, light, and wind—are the major components of **climate,** the prevailing weather conditions at a locality. We can see the great impact of climate on the distribution of organisms by constructing a climograph, a plot of the temperatures and rainfall in a particular region, often in terms of annual means. For example, FIGURE 46.3 shows a climograph for some of the major North American biomes. Notice that the range of rainfall occurring in coniferous forest regions is similar to that of temperate forest areas, but that temperatures are different. Grasslands are generally drier than either kind of forest, and deserts are drier still.

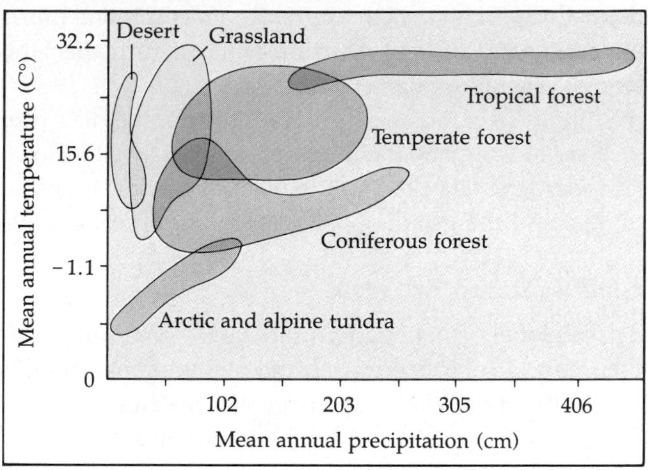

FIGURE 46.3
A climograph for some major North American biomes. The areas plotted here encompass the annual mean temperatures and precipitation occurring in some major North American biomes. The climograph provides only circumstantial evidence, however, that these factors are important in explaining the distribution of the biomes. The areas of overlap, for example, show that these variables alone are not sufficient to explain the observed distribution.

Annual means for temperature and rainfall are reasonably well correlated with the biomes we find in different regions. However, we must always be careful to distinguish a *correlation* between variables from *causation,* a cause-and-effect relationship. Although our climograph provides circumstantial evidence that temperature and rainfall are important to the distributions of biomes, it does not prove that these variables govern their geographic location. Only a detailed analysis of the water and temperature tolerances of individual species could establish the controlling effects of these variables.

We can see in our climograph that factors other than mean temperature and precipitation must also play a role in determining which biomes are found where, as there are regions where biomes overlap. For example, there are areas in North America with a certain temperature and precipitation combination that support a temperate forest; but other areas with the same values for these variables support a coniferous forest; still others, a grassland. How do we explain this variation? First, remember that the climograph is based on annual *means.* Often it is not only the mean climate that is important but also the pattern of climatic variation. For example, some areas may get regular precipitation throughout the year, whereas others with the same annual amount have distinct wet and dry seasons. A similar phenomenon may occur with respect to temperature. Other factors, such as the bedrock in an area, may greatly affect mineral nutrient availability and soil structure, which in turn affect the kind of vegetation that will develop.

With these complex considerations in mind, let's take a closer look at global climate patterns, as well as local and seasonal variations in the physical environment, to understand the geographical distribution of organisms.

Global Climate Patterns

Earth's global climate patterns are largely determined by the input of solar energy and the planet's movement in space. About half the solar energy that reaches the upper layers of the atmosphere is absorbed before it reaches Earth; certain wavelengths of light (including the ultraviolet wavelengths that are damaging to biological systems) are more readily absorbed by oxygen molecules and ozone than are others. Much of the energy that strikes Earth itself is absorbed by land and water (and organisms), although some is reflected back into the atmosphere. The sun's warming effect on the atmosphere, land, and water establishes the temperature variations, cycles of air movement, and evaporation of water that are responsible for dramatic latitudinal variations in climate.

Because solar radiation is most intense when the sun is directly overhead, the shape of the Earth causes latitudinal variation in the intensity of sunlight (FIGURE 46.4). However, the planet is also tilted on its axis by 23.5° relative to its plane of orbit around the sun, and this tilt causes seasonal variation in the intensity of solar radiation (FIGURE 46.5). The angle of incoming solar radiation changes daily everywhere as Earth rotates around the sun, but only the **tropics** (those regions that lie between 23.5° north latitude and 23.5° south latitude) ever receive sunlight from directly overhead. As a result, the tropics experience the greatest annual input and the least sea-

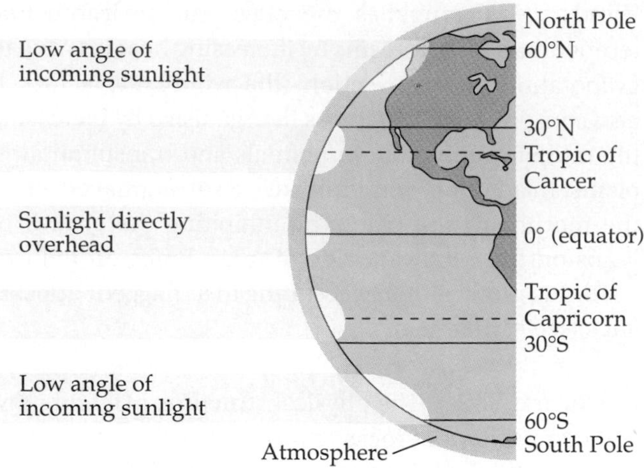

FIGURE 46.4
Solar radiation and latitude. Because sunlight strikes the equator perpendicularly, more heat and light are delivered there per unit of surface area than at higher northern and southern latitudes, where sunlight has a longer path through the atmosphere and strikes the curved surface of Earth at an oblique angle. This uneven distribution of solar radiation creates latitudinal differences in temperature and light intensity and establishes the vertical air currents illustrated in FIGURE 46.6b.

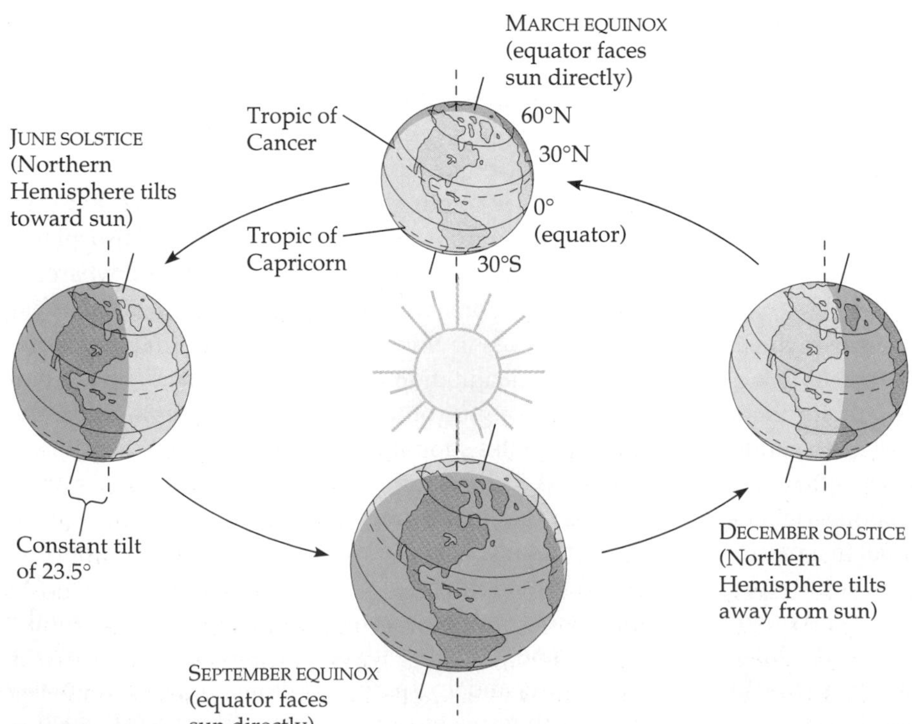

FIGURE 46.5
The cause of the seasons. The permanent tilt of the Earth on its axis causes seasonal variation in temperature and light intensity as the planet revolves around the sun. The December solstice marks the beginning of winter in the Northern Hemisphere, when the North Pole is maximally tilted *away from* the sun. Day length is reduced, and solar radiation arrives at its most oblique angle, producing lower temperatures during short winter days. In contrast, the December solstice marks the beginning of summer in the Southern Hemisphere, when the South Pole is maximally tilted *toward* the sun. Day length increases, and the sun is more nearly overhead, increasing temperatures during the long days of summer. Seasons in the two hemispheres are reversed when the North Pole is tilted toward the sun and the South Pole away from the sun on the June solstice. At the March and September equinoxes, neither pole is tilted toward the sun, and all regions on Earth experience 12 hours of daylight and 12 hours of darkness.

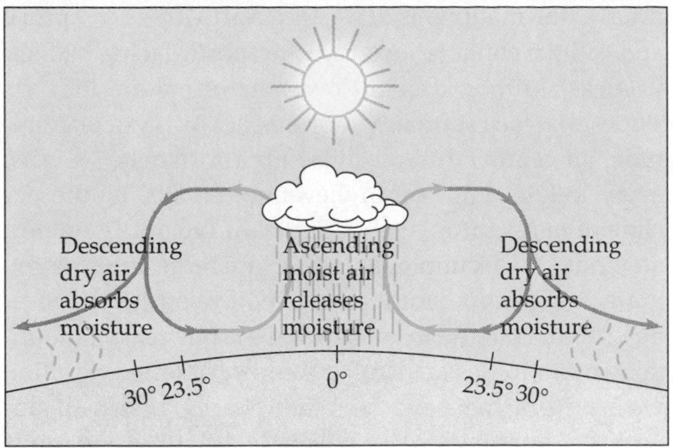

(a) Air circulation and precipitation at the equator

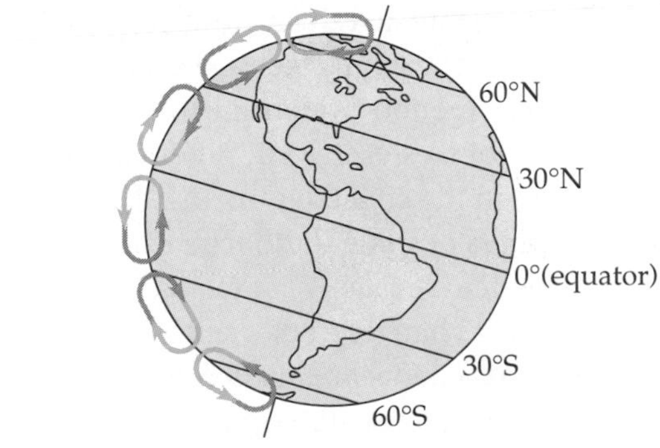

(b) Global air circulation

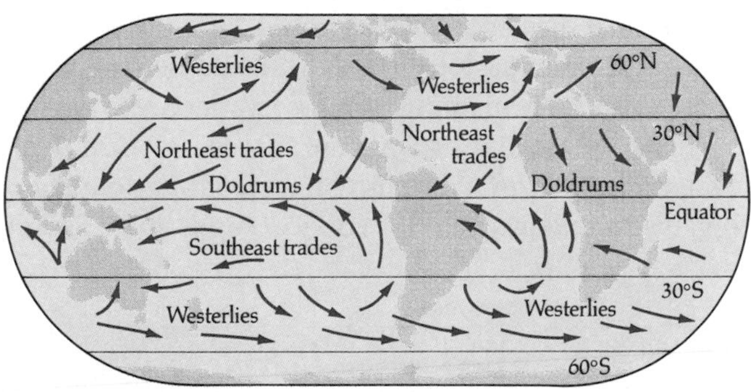

(c) Global wind patterns

FIGURE 46.6

Global air circulation, precipitation, and winds. (**a**) Air masses in the lower atmosphere are warmed by solar radiation and by heat radiating from Earth. Air expands, decreases in density, and rises when it is warmed. Thus, warm air at the equator rises, creating an area of light, shifting winds known as the doldrums. The warm air masses expand as they rise, and, because this expansion distributes their internal heat energy over a larger volume, they also cool as they move upward through the atmosphere. Cool air holds less water vapor than warm air, and the rising air masses drop large amounts of rain in the tropics. The now dry air masses flow toward the two poles at high altitude, cooling further as they spread away from the equator. The density of the air masses increases as they cool, and they descend, absorbing water from the land and creating bands of arid climate around 30° latitude. (**b**) The movement of heated air creates three major air circulation cells on either side of the equator. Within each circulation cell, rising air (blue) releases moisture as precipitation, and descending air (gray) absorbs moisture, creating arid conditions. The first circulation cells, described in part a, are completed when some of the air that descends at 30° latitude flows back toward the equator at low altitude. The remainder of the air flows away from the equator, also at low altitude, initiating midlatitude circulation cells that acquire water and then release it as the air rises around 60° latitude. The third circulation cells carry cool, dry air to the poles where it descends and flows back toward the equator. Air from the different circulation cells merges where the cells meet, constantly mixing the gases in the lower levels of the atmosphere. (**c**) Air flowing in the lower levels of the circulation cells, close to Earth's surface, creates predictable global wind patterns. However, as Earth rotates on its axis, land near the equator moves faster than that at the poles, deflecting the winds from the vertical paths illustrated in part b and creating more easterly and westerly flows. Cooling trade winds blow from east to west in the tropics and subtropics. In contrast, the prevailing westerlies blow from west to east in the temperate zone.

sonal variation in solar radiation of any region on Earth. The seasonality of light and temperature increases steadily toward the poles; polar regions have long, cold winters with periods of continual darkness and short summers with periods of continual light.

Intense solar radiation near the equator initiates a global circulation of air, creating precipitation and winds (FIGURE 46.6). High temperatures in the tropics evaporate water from Earth's surface and cause warm, wet air masses to rise and flow toward the poles. The rising air masses release much of their water content, creating abundant precipitation in tropical regions. Thus, high temperatures, intense sunshine, and ample rainfall are all characteristic of a tropical climate, fostering the growth of lush vegetation in some tropical forests and the development of coral reefs. The high-altitude air masses, now dry, descend toward Earth at latitudes around 30° north and south, absorbing moisture from the land and creating an arid climate conducive to the development of the deserts that are common at these latitudes. Some of the descending air flows toward the poles at low altitude, establishing a midlatitude circulation cell that deposits abundant precipitation (though less than in the tropics) where the air masses again rise and release moisture in the vicinity of 60° latitude. Broad expanses of coniferous forest dominate the landscape at these fairly wet, but generally cool, latitudes. A third circulation cell carries some of the cold and dry rising air to the poles, where it descends and flows back toward the equator, absorbing moisture and creating the comparatively rainless and bitterly cold climates of the Arctic and Antarctic. Although the arctic tundra receives very little annual rainfall, water cannot penetrate the underlying permafrost and accumulates in pools on the shallow topsoil during the short summer.

Local and Seasonal Effects on Climate

Proximity to bodies of water and topographic features such as mountain ranges create a climatic patchiness on a regional scale, and smaller features of the landscape also contribute to local climatic variation. Although global climate patterns help explain the geographical distribution of some major biomes, regional and local variations in climate and soil influence the distributions of some less widely spread communities and individual species.

Ocean currents influence climate along the coasts of continents by heating or cooling overlying air masses, which may then pass across the land. Evaporation from the ocean is also greater than it is over land, and coastal regions are generally moister than inland areas at the same latitude. The rain forests of the Pacific Northwest and the large redwood groves farther south require the cool, misty climate produced by the cold California current that flows southward along the western United States. Similarly, the warm Gulf Stream flowing north out of the Gulf of Mexico and across the North Atlantic tempers the climate on the west coast of the British Isles, making it warmer than the coast of New England, which is actually further south but is cooled by a current flowing south from the coast of Greenland.

As every vacationer knows, oceans and large inland bodies of water generally moderate the climate of nearby terrestrial environments on a daily cycle. During a warm summer day, when the land is hotter than a lake or the ocean, air over the land heats and rises, drawing a cool breeze from the water across the land. At night, by contrast, air over the warmer ocean rises, establishing a circulation that draws cooler air from the land out to sea, replacing it with warmer air from offshore. Proximity to water does not always moderate climate, however. Several regions (including the coast of central and southern California) have a Mediterranean-like climate; in summer, cool, dry ocean breezes are warmed when they contact the land, absorbing moisture and creating hot rainless summers just a few miles inland.

Mountains also have a significant effect on solar radiation, local temperature, and rainfall. South-facing slopes the Northern Hemisphere receive more sunlight than north-facing slopes and are therefore warmer and drier. In the mountains of western North America, spruce and other conifers occupy the north-facing slopes, whereas shrubby, drought-resistant vegetation inhabits slopes that face south. In addition, at any particular latitude, air temperature declines approximately 6°C with every 1000-m increase in elevation, paralleling the decline of temperature with latitude. In the north temperate zone, for example, a 1000-m increase in elevation produces a temperature change equivalent to that over an 880-km increase in latitude. This is one reason mountain communities are similar to those at lower elevation farther from the equator. When warm, moist air approaches a mountain, it rises and cools, releasing moisture on the windward side of the range. On the leeward side of the mountain, cooler, dry air descends, absorbing moisture and producing a rainshadow (FIGURE 46.7). Deserts commonly occur on the leeward sides of mountain ranges, a phenomenon evident in the Great Basin and Mojave Desert of western North America, the Gobi Desert of Asia, and in the small deserts that characterize the southwest corners of some Caribbean islands.

Seasonality generates local environmental variation in addition to the global changes in day length, solar radiation, and temperature described earlier. Because of the changing angle of the sun over the course of the year, the belts of wet and dry air on either side of the equator undergo slight seasonal shifts in latitude that produce marked wet and dry seasons around 20° latitude, where tropical deciduous forests grow. In addition, seasonal changes in wind patterns produce variations in ocean currents, sometimes causing the upwelling of nutrient-rich, cold water from deep ocean layers, thus nourishing organisms that live near the surface. Ponds and lakes are

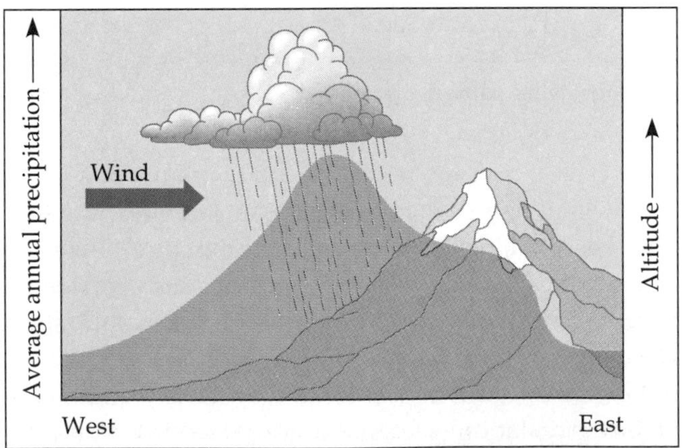

FIGURE 46.7
Rain shadows. As moisture-laden air is deflected upward by the Sierra Nevada mountain range in California, it cools and the moisture condenses, causing heavy precipitation on the western side of the mountain. When the air follows the eastern slope down, it warms and absorbs moisture from the land, creating dry conditions in the Great Basin of the southwestern United States.

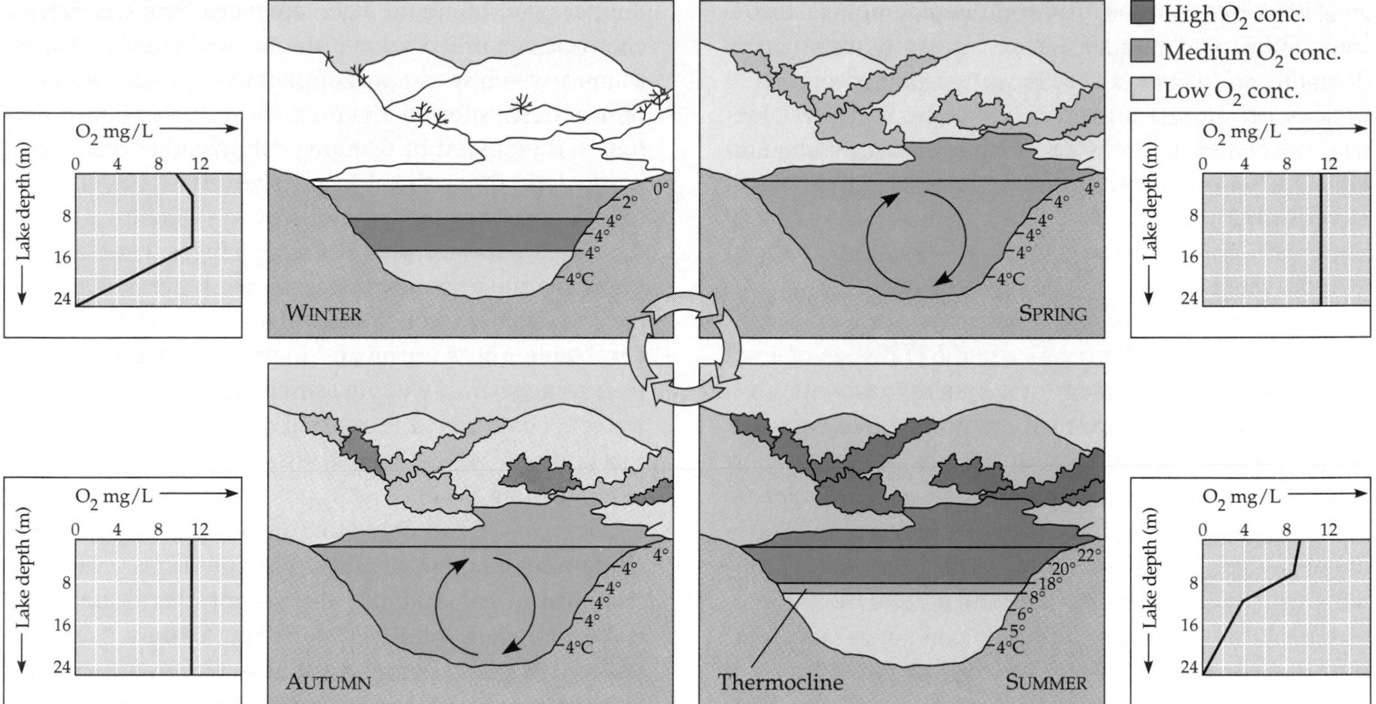

FIGURE 46.8

Lake turnover. The biannual mixing of lake waters occurs because water is most dense at 4°C, and water at that temperature sinks below water that is either warmer or colder. In winter, the coldest water in the lake (0°C) lies just below the surface ice; water is progressively warmer at deeper levels of the lake, typically 4°–5°C at the bottom. In spring, as the sun melts ice, the surface water warms to 4°C and sinks below the cooler layers immediately below, eliminating the thermal stratification established in winter. In the absence of thermal layering, spring winds mix the water to great depth, bringing oxygen to the bottom waters and nutrients to the surface (see graphs). In summer, the lake regains a distinctive thermal profile with warm water at the surface, separated from cold bottom water by a narrow vertical zone of rapid temperature change, called a thermocline. In autumn, as surface water cools rapidly, it sinks below the underlying layers, remixing the lake water until the surface begins to freeze and the winter temperature profile is reestablished.

also extremely sensitive to seasonal temperature changes and undergo a biannual mixing of their waters as a result of changing water-temperature profiles (FIGURE 46.8). This **turnover,** as it is called, brings oxygenated water from the surface of the lake to the bottom and nutrient-rich water from the bottom to the surface in both spring and autumn. These cyclic changes in the abiotic properties of the lake are essential for the survival and growth of organisms at all levels within this ecosystem.

Climate also varies on a very fine scale, called microclimate. For example, ecologists often refer to the microclimate on a forest floor or under a rock. Many features in the environment influence microclimates by casting shade, reducing evaporation from soil, and minimizing the effects of wind. Forest trees frequently moderate the microclimate below. Cleared areas generally experience greater temperature extremes than the forest interior, because of greater solar radiation and wind currents that are established by the rapid heating and cooling of open land; evaporation is generally greater in clearings as well. Low-lying ground is usually wetter than high ground and tends to be occupied by different species of trees within the same forest. If you have ever lifted a log or large stone in the woods, you are well aware that there are organisms (such as salamanders, worms, and some insects) that live in the shelter of this microenvironment, buffered from the extremes of temperature and moisture. Every environment on Earth is similarly characterized by a mosaic of small-scale differences in the abiotic factors that influence the distributions of organisms.

The costs and benefits of homeostasis affect an organism's responses to environmental variation

Ecological communities are composed of populations of individual organisms that are adapted to the physical environments in which they live. Natural selection has produced diverse adaptations to extremes of temperature, light, and other abiotic factors. Some organisms living in the Arctic and Antarctic tolerate air temperatures of –70°C, and other organisms survive desert temperatures higher than 45°C. Aquatic life can be found in water with salt concentration near zero (snowmelt) and in salt lakes several times more saline than seawater. However, no single species can survive the full range of environmental conditions present on Earth, and the geographical distribution of a population or species is partly determined by its ability to tolerate a specific subset of environmental conditions. Various anatomical structures and physiological mechanisms, discussed in Unit Six (for plants) and Unit Seven (for animals), have evolved as adaptations to the constraints of specific environments. In this section, we will briefly survey the general types of adaptations in the context of variation in the physical environment.

The success of an organism at survival and reproduction reflects its overall tolerance to the entire set of environmental variables it confronts. In many cases, the ability to tolerate a particular factor may depend on another factor. For example, many aquatic ectothermic organisms can survive reduced oxygen at low temperatures, but not at high temperatures when their metabolic rates are also high. Coping with a set of environmental problems usually involves imperfect adaptations that represent evolutionary compromises. Panting or sweating, for instance, cools the body on a hot day but can also lead to a water deficit.

Regulators and Conformers

Chapter 36 introduced the term *homeostasis,* the maintenance of a steady-state internal environment in the face of variations in the external environment. Many animals and plants can be described as **regulators** that use behavioral and physiological mechanisms to achieve homeostasis in the face of environmental fluctuations in temperature, moisture, light intensity, and a variety of chemical factors in the environment. For example, Pacific salmon, which spend part of their lives in salt water and part in fresh water, maintain a constant salt concentration in their blood by osmoregulation. Other organisms, particularly those that live in relatively stable environments, are often **conformers,** allowing conditions within their bodies to vary with external changes (FIGURE 46.9a) Many marine invertebrates, such as spider crabs of the genus *Libinia,* live in environments where the salinity is very stable. These organisms do not osmoregulate, and if placed in water of varying salinity, they will lose or gain water to conform to the external environment even when this internal adjustment is extreme enough to cause death (FIGURE 46.9b).

Conforming and regulating represent extremes on a continuum, and few organisms are perfect regulators or conformers. For example, the salmon described above can osmoregulate, but they conform to external temperatures. Even endotherms such as ourselves are not perfect thermoregulators; anyone living in a cold climate has certainly noticed how cool exposed extremities such as hands, nose, and ears become on a cold day.

Many species are conformers under certain environmental conditions but can regulate to some extent under others. Regulation requires the expenditure of energy, and in some environments the cost of regulation may outweigh the benefits of homeostasis. For example, temperature regulation would require the forest-dwelling

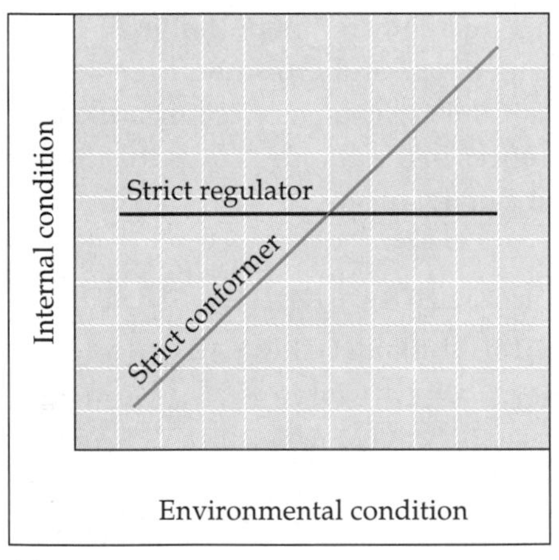

(a)

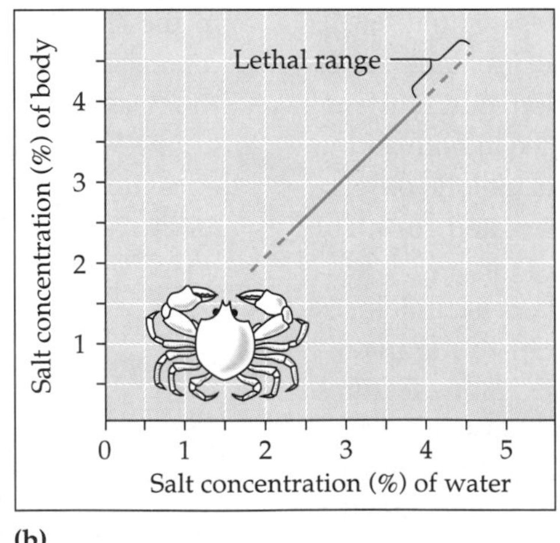

(b)

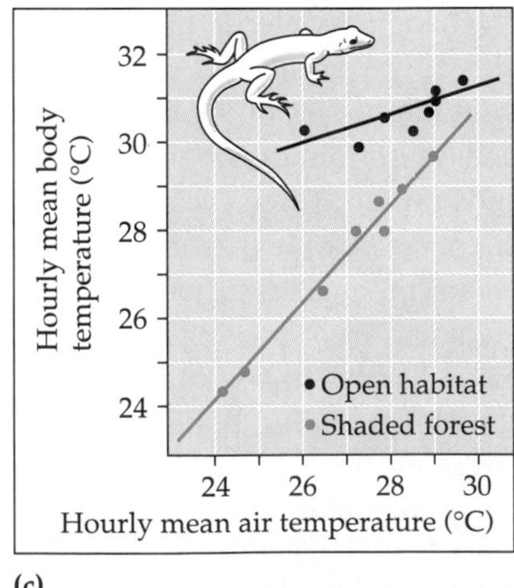

(c)

FIGURE 46.9
Regulators and conformers. (**a**) With respect to an environmental variable, organisms can be described as either regulators, which maintain a nearly constant internal environment over a range of external conditions, or conformers, which allow their internal environment to vary. Although the two "strategies" are idealized in the illustration, most organisms are neither perfect regulators nor perfect conformers. (**b**) Spider crabs *(Libinia)* are osmoconformers with little ability to physiologically regulate internal osmolarity in the narrow range of salinity where they live. If exposed to slightly higher or lower salinity in the laboratory, the crabs continue to conform and soon die. (**c**) Some species change their regulatory mode in different environments. *Anolis cristatellus,* a small lizard on Puerto Rico, behaviorally thermoregulates in open habitats where it can bask in the plentiful patches of sun (black dots), but it is a thermoconformer in shaded forests where basking sites are rarer (blue dots).

lizard *Anolis cristatellus* to travel long distances (and risk capture by a predator) to find an exposed sunny perch. It is therefore likely to survive longer and produce more offspring by allowing its body temperature to conform to that of the forest environment. However, this same species behaviorally thermoregulates in open habitats where it can bask in the plentiful patches of sun (FIGURE 46.9c).

The Principle of Allocation

One concept organismal ecologists have found useful in assessing the responses of organisms to their complex environments is the **principle of allocation.** This principle holds that each organism has a limited amount of energy that can be allocated for obtaining nutrients, escaping from predators, coping with environmental fluctuations (homeostasis), growth, and reproduction. Energy expended for homeostasis is therefore not available for other functions. For example, in grasshoppers, which are moderately active ectotherms, about 30% of assimilated energy remains after the animal's basic maintenance needs are met. This energy can be channeled into growth or reproduction. In contrast, for a very active endotherm, such as a weasel that uses most of its ingested energy to stay warm and active, only 2.5% of assimilated energy remains, and for a wren, only 0.5% remains. For the latter organisms, there must be significant evolutionary advantages to endothermy and higher activity levels that offset the high maintenance costs.

Different priorities in energy allocation are related to the distribution of organisms and their homeostatic mechanisms. Conformers that live in very stable environments, such as the spider crabs in FIGURE 46.9b, might be able to channel more of their energy into growth and reproduction. However, the intolerance of such specialists to environmental change severely restricts their geographical distribution. In contrast, regulators that allocate a larger fraction of their energy to coping with environmental changes may grow and propagate less efficiently, but such organisms are able to survive and reproduce over a wider range of variable environments.

The response mechanisms of organisms are related to environmental grain and the time scale of environmental variation

You have learned that environments vary over both space and time. The importance of environmental variation to a particular organism depends on the spatial and temporal scale of the variation in relation to the organism's size, lifespan, and movement patterns.

Environmental Grain

For a small herbivorous insect living in a field of wildflowers, the differences among the plants may be profound because only some plants are suitable food sources. However, to a grazing mammal such as a horse, all the plants may be more or less the same, and the horse feeds indiscriminately within the field. Ecologists use the concept of **environmental grain** to define the use of spatial variation, or patchiness, by organisms. A *coarse-grained environment* is one in which environmental patches are so large (relative to the size and activity of the organism) that an individual organism can distinguish and choose among patches. The field of wildflowers is coarse-grained for the insect described above because it may spend its entire larval life on one plant, having selected that plant over others.

A *fine-grained environment* is one in which patches are small relative to the size and activities of an organism, and the organism may not even behave as though patches exist. For the horse, the field is a fine-grained environment in which there are no choices to be made. Of course, a large field in which succulent clover grows on one side and spiny thistles on the other represents a coarse-grained environment even to the horse, because it can select which side of the field it prefers.

Temporal variation in the environment can also be coarse-grained or fine-grained, depending on the periodicity of the variation in relation to the lifespan of the organism. A sudden cold snap might have a dramatic effect on the success of an insect such as a mayfly, whose entire adult lifespan may be only a few hours, but have little influence on the activities of a long-lived mammal. Daily variations in environmental factors are usually fine-grained for most organisms, whereas seasonal and longer-term shifts in climate are coarse-grained for even the largest organisms.

Organisms can respond to variations in the environment with a variety of adaptations. Behavioral adaptations are almost instantaneous in their effects and easily reversed, whereas physiological adaptations may be implemented and changed over time scales ranging from seconds to weeks. Morphological adaptations may develop over the lifetimes of individual organisms or between generations. Adaptive genetic changes in populations are slower still, usually evolving over several generations. The appropriate response to environmental change depends partly on the duration of that change.

Behavioral Responses

Behavioral response in the sense of muscular reaction to a stimulus is limited to animals. Behavioral mechanisms are so important to how animals interact with their environments that they will be discussed in detail in

Chapter 50. Let's consider them briefly here from an environmental perspective.

The quickest response of many animals to an unfavorable change in the environment is to move to a new location. Such movement may be fairly localized. For example, lake trout avoid the heat of the upper zone of a lake during summer by moving to deeper water. Many desert animals escape intense heat by burrowing, and they maintain a reasonably constant body temperature while active by shuttling between sun and shade. Some animals are capable of migrating great distances in response to such environmental cues as changes in temperature or changes in photoperiod associated with seasonal transitions. Many migratory birds overwinter in Central and South America, returning to northern latitudes to breed in the summer.

Some animals are able to modify their immediate environment by cooperative social behavior. Honeybees, for instance, can cool the inside of their hive on hot days by the collective beating of their wings. During cold periods, they seal the hive, helping retain the heat generated by their activity inside. Many small mammals huddle within burrows during cold weather, a behavioral mechanism that reduces heat loss by minimizing the total amount of surface the animals expose to the cold air.

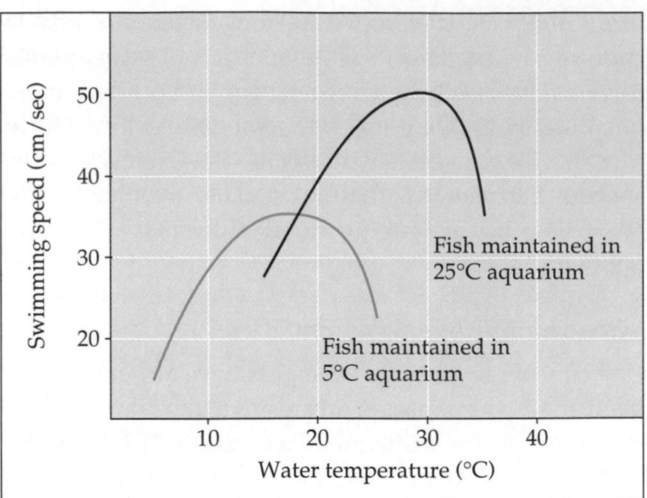

FIGURE 46.10

Tolerance curves and acclimation. Goldfish, like all ectothermic animals, have an optimal temperature for physiological function as well as a range of temperatures they can tolerate. In this example, swimming speed, which varies with temperature, is used as an index of the fish's general well-being. Goldfish normally live at temperatures between 25°and 30°C, but both the optimal temperature and the tolerance range can be shifted somewhat when the fish is gradually acclimated to a water temperature of 5°C. Acclimation is a slow process, and the fish would not survive a sudden, large change in water temperature. Acclimation also involves trade-offs. Notice that the cold-acclimated fish swims faster than the warm-acclimated fish at 15°C but much more slowly at 25°C.

Physiological Responses

As already noted, physiological responses to environmental change are generally slower than behavioral reactions. However, physiological responses that involve relatively small changes in the rates of processes, and that do not require alteration of morphology or biochemical pathways, can occur very rapidly. For example, when you venture outside on a very cold day, blood vessels in your skin may constrict within seconds, a physiological response that minimizes the loss of body heat.

Regulation and homeostasis are the hallmarks of physiological adaptation. However, all organisms, whether regulators or conformers, function most efficiently under certain environmental conditions. We can study an organism's response to changing environmental conditions in the laboratory by varying a single abiotic factor, such as temperature, and measuring some aspect of the organism's performance. The resulting tolerance, or performance, curves are approximately bell-shaped, with peak performance at some optimal condition and the tails of the curve representing the limits of the organism's tolerance to the particular environmental variable. FIGURE 46.10 shows tolerance curves for the effect of temperature on the swimming speed of goldfish. Tolerance limits are important determinants of the geographical distribution of organisms, though biological interactions can prevent a species from occupying a habitat to which it is physiologically adapted (see Chapter 48).

Physiological responses to environmental variation can also include **acclimation,** which involves substantial but reversible changes that shift an organism's tolerance curve in the direction of the environmental change. For example, if you moved from Boston, which is essentially at sea level, to the mile-high city of Denver, one physiological response to the lower oxygen pressure in your new environment would be an increase in the number of your red blood cells. Acclimation is a gradual process, taking days or weeks, and the ability to acclimate is generally related to the range of environmental conditions the species naturally experiences. Species that live in very hot climates, for example, usually do not acclimate to extreme cold.

Morphological Responses

Organisms may react to some change in the environment with responses that alter the form or internal anatomy of the body. In some cases, these responses are examples of acclimation, since they are reversible. Many mammals and birds, for example, grow a heavier coat of fur or feathers in winter; sometimes coat color changes seasonally as well, camouflaging the animal against winter snow and summer vegetation.

Other morphological changes are irreversible over the lifetime of an individual. In many cases, environmental variation can affect growth and differentiation patterns, often leading to remarkable morphological variation within a species. In general, plants are more morphologically plastic than animals; this response helps them compensate for their inability to move from one environmental patch to another. One example is the arrowleaf plant, which can grow on land, rooted in water with its upper leaves emerging above the surface, or completely submerged in water. Leaf morphology varies with the environment in which the leaves grow. Submerged leaves are flexible, bending with the currents, and, lacking a waxy cuticle, are able to absorb mineral nutrients from the surrounding water. Arrowleaf plants growing on land have more extensive root systems, and their leaves are more rigid and covered with a thick cuticle that reduces water loss.

Adaptation Over Evolutionary Time

The various behavioral, physiological, and morphological mechanisms we have examined are responses of individual organisms operating on an ecological time scale. However, it is important to remember that these responses occur within a framework of adaptations fashioned by natural selection acting over evolutionary time. For example, all plants are capable of changing the size of the stomata of their leaves, a physiological response that helps prevent desiccation under environmental conditions when transpiration would exceed delivery of water. In plants living in the desert, the ability to adjust the size of the stomatal openings in response to water stress is superimposed on many other anatomical and physiological adaptations that have accumulated over evolutionary time as these plants have evolved in their arid environments. For instance, some desert plants have their stomata in pits, protected from the hot, dry winds that accelerate transpiration. Also common among desert plants is the CAM pathway of photosynthesis (see Chapter 10), which enables the plants to keep their stomata closed during the daytime.

The distinction between short-term adjustments on the scale of ecological time and adaptation on the scale of evolutionary time begins to blur when we consider that the range of responses of an individual to changes in the environment is itself the product of evolutionary history. For example, when an endotherm such as a mammal uses physiological adjustments to maintain constant body temperature in the face of fluctuations in environmental temperature, it is utilizing mechanisms of homeostasis that are adaptations acquired by natural selection.

As populations adapt to localized environments, their distribution is often constrained. For instance, earthworms are skin breathers that obtain oxygen by diffusion across their moist body surface, a solution to the problem of gas exchange that restricts these animals to damp soils. If pine seeds are blown from the rim of the Grand Canyon to the canyon floor 2000 m below, where conditions are much hotter and drier, the seeds are unlikely to germinate and grow successfully in the new environment. Organisms locked by their adaptations into one type of environment may fail to survive if dispersed to some foreign environment, or they may face extinction if their local environment changes beyond their tolerance limits. On the other hand, the absence of a species in a particular place does not necessarily imply that the species could not survive in that location. Pines would not live even on the rim of the Grand Canyon if they had not dispersed to that location at some time in their evolutionary history. Thus, the existence of a species in a particular place depends on two factors: The species must reach that location, and it must be able to survive and reproduce in that location once it is there. We will evaluate the importance of these factors to the geographical distribution of organisms in Chapter 48. In the next section, we will look at how regional variation in abiotic factors described earlier in the chapter relates to biomes, the generalized ecosystems we recognize by such generic descriptions as "deserts" and "coniferous forests."

The geographical distribution of terrestrial biomes is based mainly on regional variations in climate

Recall that biomes are communities and ecosystems that are typical of broad geographic regions. There is no strict way of defining specific biomes, and different ecologists recognize and organize biomes in different ways. Nevertheless, the patterns in the distributions of organisms within the biosphere are real and have been a major focus of ecological research. All the abiotic factors we covered earlier in the chapter, especially climate, are important in determining why a particular biome is found in a particular area. Because there are latitudinal patterns of climate over Earth's surface (see FIGURES 46.4–46.6), there are also latitudinal patterns of biome distribution. For example, coniferous forests extend in a broad band across North America, Europe, and Asia.

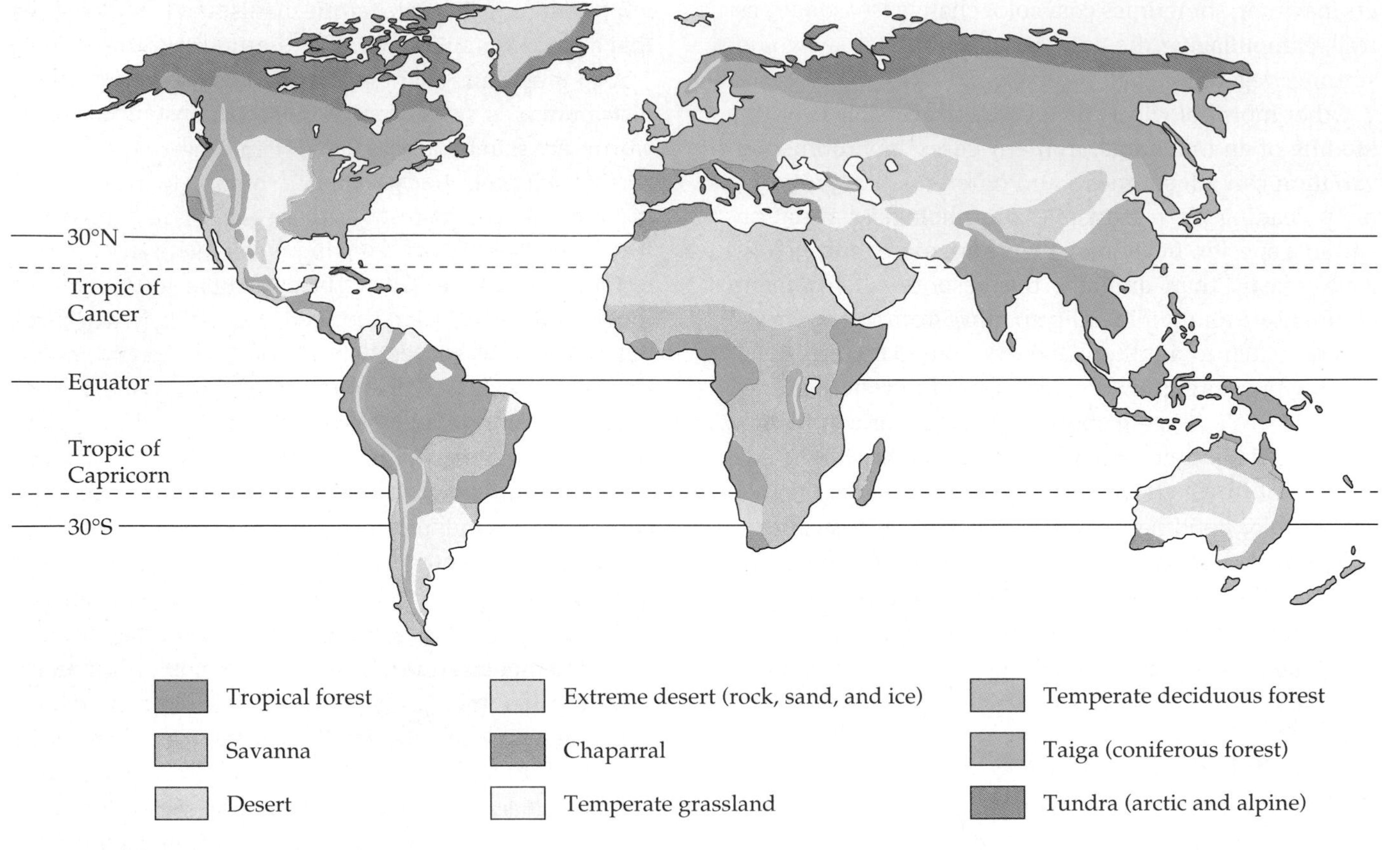

FIGURE 46.11

The distribution of major terrestrial biomes. Although terrestrial biomes are mapped here with sharp boundaries, biomes actually grade into one another, sometimes over relatively large areas. The tropics are the low-latitude regions bordered by the Tropic of Cancer and the Tropic of Capricorn.

The major terrestrial biomes are mapped in FIGURE 46.11. Although we often name biomes for their predominant vegetation, each biome is also characterized by microorganisms, fungi, and animals adapted to that particular environment. Grasslands, for example, are more likely than forests to be populated by large grazing mammals. Biomes usually grade into each other, without sharp boundaries. If the area of intergradation is itself large, it may be recognized as a separate biome, or *ecotone* (see the discussion on savanna later in this section).

As we survey the biomes, it is important to keep in mind that the actual species composition throughout a biome varies from one location to another. In the North American coniferous forest, red spruce is common in the east but does not occur in most other areas, where black spruce and white spruce are abundant. Although the vegetation of African deserts superficially resembles that of North American deserts, the plants are in different families. Such "ecological equivalents" can arise because of convergent evolution (see Chapter 23).

Within a biome there may be extensive patchiness, with several communities represented. Biomes are usually recognized on the basis of the communities that develop as the result of succession (changes in community structure through time), a topic discussed in Chapter 48. Disturbances often allow representatives of earlier successional stages to become reestablished. For example, snowfall may break branches and small trees and cause openings in the coniferous forest, allowing deciduous species, such as aspen and birch, to grow. Most of the eastern United States is classified as temperate deciduous forest, but human activity has eliminated all but a tiny percentage of the undisturbed forest that would otherwise be present. In fact, humans have altered much of Earth's surface, replacing original biomes with urban and agricultural ones (FIGURE 46.12).

Let's now survey the major terrestrial biomes, traveling generally in a direction from the equator to the poles.

Tropical Forest

A variety of **tropical forests** is found within 23.5° latitude of the equator, where the average temperature (around 23°C) and length of daylight (around 12 hours) vary little throughout the year. Rainfall, on the other hand, is quite variable in the tropics, and the amount of precipitation, rather than temperature or photoperiod, is the prime de-

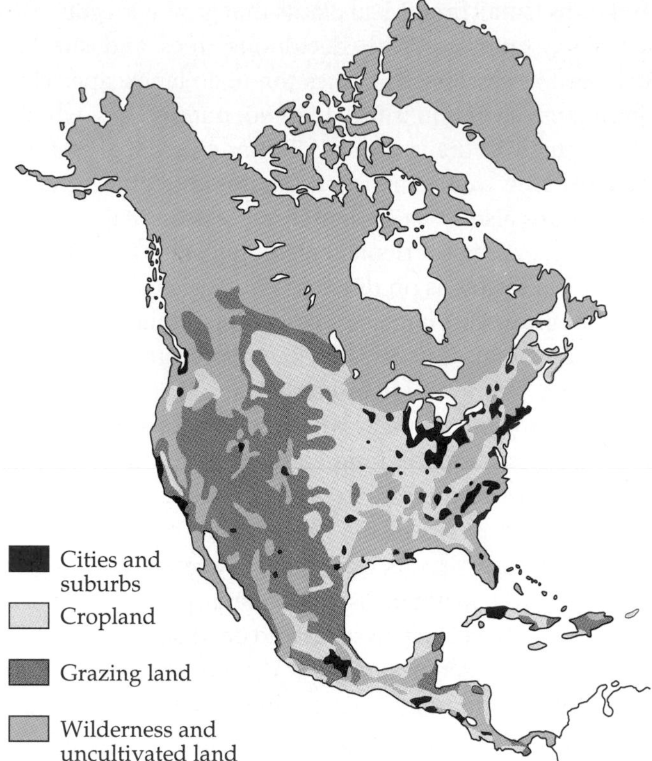

FIGURE 46.12
Urban and agricultural biomes of North America. Many regions on Earth have been disturbed by intense human activity. In urban and agricultural biomes, natural communities have been replaced by housing, industry, cropland, and grazing range. Relatively few undisturbed habitats remain in most regions on the planet.

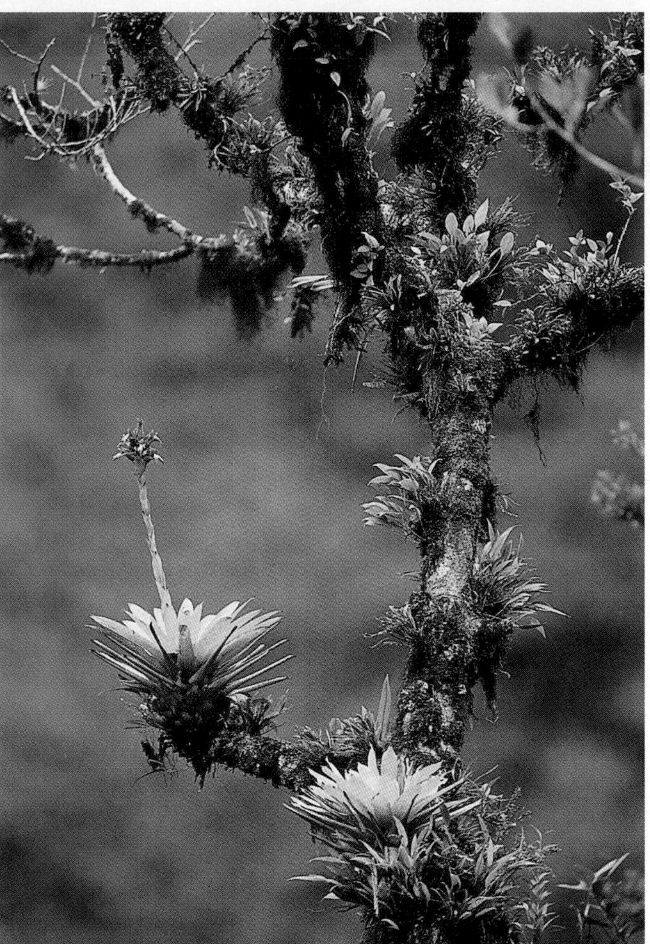

FIGURE 46.13
Tropical rain forest. This photograph was taken about 10 m above the floor of rain forest in Costa Rica. Tropical rain forest often has a closed canopy, with little light reaching the ground below. When an opening does occur, perhaps because of a fallen tree, other trees and large woody vines grow rapidly, competing for light and space as they fill the gap. Many of the trees are covered with epiphytes (plants that grow on other plants rather than in soil), such as orchids and bromeliads.

terminant of the vegetation growing in an area. In lowland areas that have a prolonged dry season and scarce rainfall at any time, **tropical dry forests** predominate. The plants found there are a mixture of thorny shrubs and trees, and succulents. In other areas that have distinct wet and dry seasons, **tropical deciduous forests** are common. Deciduous trees and shrubs drop their leaves during the long dry season (when water lost through transpiration would exceed available supplies) and releaf only during the following heavy rains or monsoons. The luxuriant **tropical rain forest** (FIGURE 46.13) is found in areas near the equator, where rainfall is abundant (greater than 250 cm per year) and the dry season lasts no more than a few months. Though we often equate tropical forest with the lush jungles of tropical rain forest, rain forests include only 25% of forested area in the tropics; 32% is deciduous and 42% is dry.

The tropical rain forest has the greatest diversity of species of all communities, perhaps harboring as many plant and animal species as all other terrestrial biomes combined. As many as 300 species of trees, some of them 50 to 60 m tall, can be found in 1 hectare (10,000 m^2, or about 2.5 acres).

The vegetation in tropical rain forests is divided into five general layers: the trees that emerge above the canopy, the high upper **canopy** or topmost continuous layer of foliage, the low-tree stratum, the shrub understory, and a ground layer of herbaceous plants and ferns. In addition to the vertical stratification, there is also a horizontal mosaic of changing vegetation.

This great plant diversity in the tropical rain forest allows great animal diversity. Animals occupy well-defined feeding groups, from the insectivorous and carnivorous birds and bats that feed above the canopy to the birds, small mammals, and insects that search the ground litter and lower parts of tree trunks for food. Though the forest as a whole is extremely dense, individuals of many plant species are widely scattered and often rely on mutualistic interactions with animals to deliver pollen. Animals are also important in dispersing fruits and seeds.

Human impact on the tropical rain forest is currently a matter of great concern. The rain forests may appear luxurious and thriving, but nutrient-poor soil and low densities of individual populations make them quite fragile. In addition, we are currently logging and converting to other uses millions of hectares of tropical forests. This destruction is proceeding at an alarming rate. More than one-half of Earth's tropical rain forest is already gone, and projections suggest that these communities may disappear entirely by the end of this decade. Although the loss of all types of tropical forests is cause for concern, the loss of biological diversity in rain forests is particularly tragic. There are also selfish reasons for preventing these losses, as we are losing important current and potential sources of medicine and wild genotypes needed to improve agricultural crops and livestock. The large-scale destruction of plants in the tropics may also cause significant changes in world climate, as we will see in Chapter 49.

Savanna

Savanna is grassland with scattered individual trees (FIGURE 46.14). Extensive savanna covers wide tropical and subtropical areas of central South America, central and southern Africa, and parts of Australia. There are generally three distinct seasons in these regions: cool and dry, hot and dry, and warm and wet, in that sequence. Most savanna soils are low in nutrients, due in part to their porosity, which results in the rapid drainage of water. Porous soils have only a thin layer of the rich, partially decomposed organic matter called humus.

Savanna is relatively simple in physical structure but often rich in number of species. Low-growing grasses and forbs (small broad-leaf plants that grow with grasses) are always present, with deciduous trees and shrubs scattered at low density across the open landscape. The dominant vegetation is fire-adapted, but the fires, which are frequent in the savanna, kill many seedlings before they become well established (as do large grazing animals). Fires also function in removing dead plant material and recycling nutrients that support new growth.

Tropical savannas on different continents are home to some of the world's large herbivores, including the giraffe, zebra, antelope, buffalo, and kangaroo. The dominant herbivores, however, are insects, especially ants and termites. Burrowing animals, whose nest sites and shelters are primarily underground, are also common: mice, moles, gophers, ground squirrels, snakes, worms, and arthropods. Animals in the savanna are most apparent during the rainy season; during the dry season, when the aboveground vegetation is sparse, many small animals are dormant or subsist on seeds and dead plant parts, and large mammals often migrate or disperse to other areas.

Savannas, often areas where forest and grassland biomes intergrade, are examples of ecotones. For instance, scattered savanna occurs in North America where the temperate deciduous forest and grasslands merge, in a band running roughly from Minnesota to east Texas. Here the climatic conditions and community features are intermediate between those of forest and grassland.

Humans have had an important impact on savannas for centuries, but in modern times this impact has been severe and often adverse. This is especially apparent in some regions of Africa, where population pressure, intensified grazing, and firewood collection have caused extensive destruction. The loss of vegetation has resulted

FIGURE 46.14
Savanna. This Kenyan savanna is a showcase of large herbivores and their predators. The luxuriant growth of grasses and forbs (small broad-leaf plants) during the rainy season provides a rich food source for animals. However, large grazing mammals must migrate to greener pastures and watering holes during regular periods of seasonal drought.

in less water vapor recirculating to the atmosphere, which in turn has caused climate changes that have devastated the region.

Desert

Deserts (FIGURE 46.15) are the driest of all terrestrial biomes, characterized by low and unpredictable precipitation (less than 30 cm per year). Although some deserts can be very hot (with soil surface temperatures above 60°C during the day), cold deserts also exist; the hot deserts generally experience large daily fluctuations in temperature. Deserts occur in two distinct belts on Earth: between 15° and 35° latitude in both hemispheres. These belts are particularly dry because of global air circulation patterns, which result from descending dry air absorbing available moisture (see FIGURE 46.6). The rain shadows on the lee side of mountain ranges can also cause deserts (see FIGURE 46.7), as can remoteness from oceanic moisture. Hot deserts are found in the southwestern United States, along the west coast of South America, in North Africa, and in the Middle East. Cold deserts are found west of the Rocky Mountains, in eastern Argentina, and in much of central Asia. The driest deserts, where average annual rainfall is less than 2 cm (some years having no rain at all), are the Atacama in Chile, the Sahara in Africa, and parts of central Australia.

The density of desert vegetation is determined largely by the frequency and amount of precipitation. The driest deserts receive too little rainfall to support any perennial vegetation. In less arid regions, the dominant vegetation is sparse, consisting of widely scattered drought-resistant shrubs and cacti or other succulents that store water in their tissues. For example, the "pleated" structure of saguaro cacti enables the plants to expand when they absorb water during wet periods. Periods of rainfall (for example, late winter in the Sonoran Desert of the southwestern United States) are marked by sudden and spectacular blooms of annual plants.

Seed-eating animals, such as ants, birds, and rodents, are common in deserts, feeding on the numerous small seeds produced by the plants. Reptiles, such as lizards and snakes, are important predators of these seed-eaters. Like the desert plants, desert animals are well adapted to scarcity of water and extreme temperatures. Many animals are active only during the cooler months of the year. Others are nocturnal, spending the day in underground burrows where they are shielded from the hot, dry air and intense sunlight. Diurnal animals are often very light in color, thereby reflecting the sunlight. Most desert animals also exhibit remarkable physiological adaptations to an arid environment. Some mice, for example, never drink, deriving all their water from the metabolic breakdown of the seeds they eat. And the development of spadefoot toads from egg to tadpole to frog is completed in less than two weeks in the temporary pools where these amphibians breed.

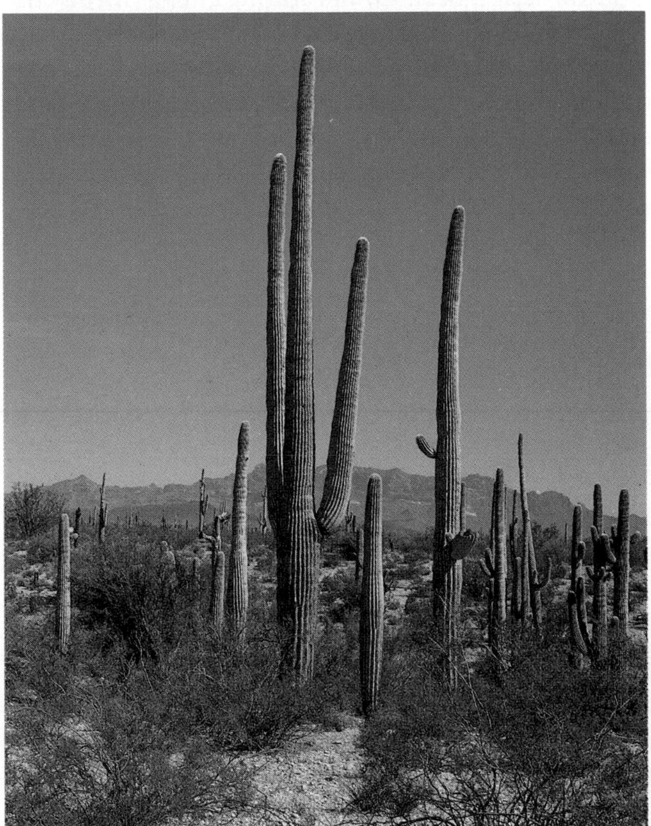

FIGURE 46.15
Desert. The Sonoran Desert of southern Arizona is characterized by giant saguaro cacti and deeply rooted shrubs. Evolutionary adaptations of desert plants include a remarkable array of protective devices, such as spines on cacti and poisons in the leaves of shrubs, that deter feeding by mammals and insects. Many desert plants also rely on CAM photosynthesis, a metabolic adaptation that conserves water in this arid environment (see Chapter 10).

Chaparral

Coastal areas between 30° and 40° latitude are often characterized by mild, rainy winters and long, hot, dry summers. These areas are dominated by **chaparral,** stands of dense, spiny shrubs with tough evergreen leaves (FIGURE 46.16). A combination of environmental stresses in chaparral—aridity, short growing season, low-nutrient soil, and frequent fires—has prevented trees from growing and resulted in the shrubby vegetation. Chaparral is found in the Mediterranean region, and also along coastlines in California, Chile, southwestern Africa, and southwestern Australia. Plants from these regions are unrelated but resemble each other in form and function—for instance, the low-growing eucalyptus shrubs of Australia and the scrub live oak of California. Annual

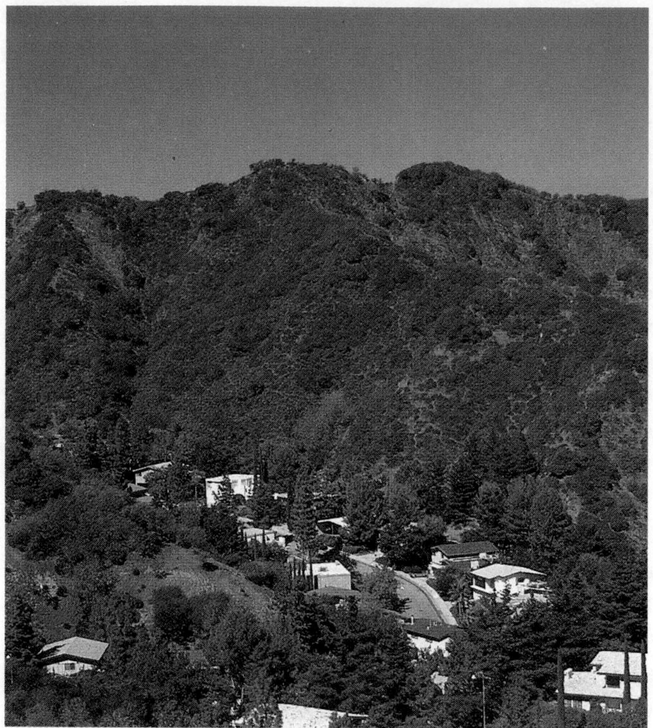

FIGURE 46.16

Chaparral. Plants of the chaparral, such as this California scrubland, are adapted to periodic fires. The dry, woody plants are frequently ignited by lightning and by careless human activities, creating summer and autumn brushfires in the densely populated canyons of southern California and elsewhere. The large storage roots of the plants enable them to resprout more quickly than the houses can be rebuilt.

plants are also common in chaparral regions during winter and early spring, when rainfall is most abundant.

Chaparral is maintained by and adapted to periodic fires. Many of the shrubs store food reserves in their fire-resistant roots, enabling them to resprout quickly and use nutrients released by fires. In addition, many chaparral species produce seeds that will germinate only after a hot fire, whereas other species are clonal, reproducing asexually without complete reliance on seeds.

Animals characteristic of the chaparral are browsers such as deer, fruit-eating birds, and ants and rodents that eat seeds of annual plants, as well as lizards and snakes.

Temperate Grassland

Temperate grasslands share some of the characteristics of tropical savanna, but they are found in regions of relatively cold winter temperatures. Temperate grasslands include the veldts of South Africa, the puszta of Hungary, the pampas of Argentina and Uruguay, the steppes of Russia, and the plains and prairies of central North America (FIGURE 46.17).

The key to the persistence of all grasslands is seasonal drought, occasional fires, and grazing by large mammals, all of which prevent woody shrubs and trees from invading and becoming established. Grassland soils tend to be deep and among the most fertile in the world. An important feature of grasslands is the amount of mulch, or decaying plant material that is deposited each year. Grazing and burning both reduce mulch accumulation; grass-

FIGURE 46.17

Temperate grassland. Temperate grasslands, such as this tallgrass prairie in Kansas, once covered much of central North America. Because grassland soil is both rich in nutrients and deep, these habitats provide fertile land for agriculture. Most grassland in the United States has been converted to farmland (see FIGURE 46.12), and very little natural prairie exists today.

lands that are ungrazed and unburned accumulate a thick layer of mulch that can suppress grass growth and allow the invasion of forbs and woody vegetation. Some mulch is needed, however, to decrease runoff and erosion, stabilize soil temperature, and improve conditions for seed germination; if no mulch accumulates, grassland regresses to weedy plants.

Because they are so conspicuous, we often think of the large vertebrate grazers, such as bison, antelopes, and wild horses, as the most important herbivores in grasslands. However, the most intense grazing occurs underground, where invertebrates consume up to four times as much as all aboveground herbivores.

Humans had their beginnings in the savannas and grasslands of Africa, and have utilized these biomes throughout their history. We have converted the more productive grasslands into monocultures of cereal grains, and the less productive into pasture land, until few natural grasslands remain today.

Temperate Deciduous Forest

Temperate deciduous forests occur throughout midlatitude regions where there is sufficient moisture to support the growth of large trees—most of the eastern United States, most of middle Europe, and part of eastern Asia. These temperate forests are characterized by broad-leaf, deciduous trees (FIGURE 46.18).

Temperatures range from very cold in the winter to hot in the summer (–30°C to 30°C), with a five- to six-month growing season. Precipitation is relatively high and fairly evenly distributed throughout the year, though groundwater may be temporarily unavailable if the soil freezes on very cold winter days. Temperate deciduous forests have a distinct annual rhythm in which trees drop leaves and become dormant in winter, then produce new leaves each spring. Although losing leaves is costly in terms of energy and nutrients, the relatively rich soils in temperate areas provide the nutrients needed for the production of new leaves in the spring. Rates of decomposition are lower in temperate forests than in the tropics, and temperate deciduous forests accumulate a thick layer of leaf litter, which conserves many of the biome's nutrients.

More open than the tropical forest and not as tall, a mature temperate forest has several layers of vegetation, including one or two strata of trees, an understory of shrubs, and low-growing forbs. Species composition varies widely around the world; some of the dominant trees are oak, birch, hickory, beech, and maple species.

Plant diversity is extensive in many temperate deciduous forests, resulting from the high availability of moisture, light, and nutrients. Because of the variety and abundance of food and habitats it offers, the temperate deciduous forest also supports a rich diversity of animal life. The greatest concentration of animals is on and just below the ground layer, where many invertebrates remain in the soil and litter for most of their lives, and many vertebrates, such as mice, shrews, and ground

FIGURE 46.18

Temperate deciduous forest. Dense stands of deciduous trees are trademarks of temperate deciduous forests, such as this one in North Carolina's Great Smoky Mountains National Park. Deciduous forest trees drop their leaves before winter, when temperatures are too low for effective photosynthesis and water lost through transpiration is not easily replaced from frozen soil. Many temperate deciduous forest mammals also enter a dormant winter state called hibernation, and some bird species migrate to warmer climates.

squirrels, burrow for shelter and food. Other mammals and birds move more freely among several strata, but favor one layer over another.

Humans have dramatically altered temperate deciduous forests by logging for building materials and fuel, clearing for agriculture, and introducing exotic pests and diseases; only scattered remnants of the original worldwide forest remain today.

Taiga (Coniferous Forest)

The **taiga,** also known as coniferous or boreal forest, is the largest terrestrial biome on Earth, extending in a broad band across northern North America, Europe, and Asia to the southern border of the arctic tundra (FIGURE 46.19). Taiga is also found at cool high elevations in more temperate latitudes, as in much of the mountainous region of western North America. The taiga is characterized by long, cold winters and short, wet summers that are occasionally warm. There may be considerable precipitation, mostly in the form of snow. Taiga soil is usually thin, nutrient-poor, and acidic. It forms slowly, owing to the low temperatures and the waxy covering of conifer needles, which decompose slowly. Nevertheless, plants grow quickly during the long days of summer (up to 18 hours of daylight) at these high latitudes.

The conifer stands in a particular area typically consist of only one or a few species of spruce, pine, fir, or hemlock, often so dense that little undergrowth is present. Deciduous species such as oak, birch, willow, alder, and aspen occur in particularly wet or disturbed habitats.

The heavy snowfall that may accumulate to several meters each winter has important ecological consequences. By insulating the soil before the coldest temperatures occur, snow prevents the soil from becoming permanently frozen. Mice and other small mammals that would quickly freeze to death above the snow remain active all winter in snow tunnels at ground level, where they continue to forage on old vegetation.

The animal populations in the taiga consist mainly of seed-eaters, such as squirrels, jays, and nutcrackers; herbivores, such as insects that eat leaves and wood; and larger browsers, such as deer, moose, elk, snowshoe hares, beavers, and porcupines. Predators of the taiga include grizzly bears, wolves, lynxes, and wolverines. Many mammals in the taiga have thick winter coats that insulate them against the cold, and some hibernate through the long winter.

Coastal coniferous forests, such as the temperate rain forests and redwood forests of the Pacific Northwest, are similar to taiga, being dominated by dense stands of only one or a few tree species. However, these unique communities are considerably warmer and moister than taiga because of their proximity to the ocean. These forests are being logged at an alarming rate, and old-growth stands of these trees may soon disappear, as Margaret Davis pointed out in this unit's interview.

Tundra

The northernmost limits of plant growth occur in the **arctic tundra,** where plant forms are limited to low shrubby or matlike vegetation (FIGURE 46.20). The arctic tundra encircles the North Pole and extends southward to the coniferous forests of the taiga. Similar communities, called **alpine tundras,** are found on high mountains

FIGURE 46.19
Taiga (coniferous forest). Dense, uniform stands of coniferous trees dominate the taiga, such as this fir forest in Alberta's Banff National Park. Taiga receives heavy snowfall during winter. The conical shape of the conifers prevents much snow from accumulating on and breaking their branches.

FIGURE 46.20
Tundra. Permafrost, bitterly cold temperatures, and high winds are responsible for the absence of trees and other tall plants in this arctic tundra in central Alaska (photographed in autumn). Tundra covers expansive areas of the Arctic, amounting to 20% of Earth's land surface. High winds and cold temperatures create similar plant communities, called alpine tundra, on very high mountaintops at all latitudes, including the tropics.

at altitudes above those where trees can grow. The floras and faunas of the arctic and alpine tundra are generally similar, but there are significant differences in the two environments.

In the arctic tundra, the climate is very cold for most of the year, with little light available during the long winters. Although the upper meter of topsoil may thaw during the summer, the underlying subsoil remains permanently frozen, a condition called **permafrost,** which prevents the roots of plants from growing very deep. The tundra may receive as little precipitation as some deserts, yet the combination of permafrost, low temperatures, and low evaporation leaves the soils continually saturated, further restricting the types of plants that can grow there. This environment supports dwarf perennial shrubs, sedges, grasses, mosses, and lichens. Plant growth and reproduction occur in a rapid burst during the brief summers, marked by nearly continuous daylight.

Alpine tundra occurs at all latitudes, even in the tropics, if the elevation is high enough. Tropical alpine tundra is confined to the very highest mountaintops, where nightly temperatures are usually below freezing.

Animal species diversity is low in both types of tundra. Blackflies, deerflies, and mosquitoes are abundant in the arctic tundra during the short summers. Flies are scarce in the alpine tundra, but beetles, grasshoppers, and butterflies are common. Many of these insects have very short wings, or no wings at all, an adaptation to the constant winds. Because of the extreme environmental conditions, insect development is much slower in the tundra; butterflies often take two years to mature. Herbivorous birds are rare, but a surprising diversity of migratory birds fly great distances each summer to take advantage of the seasonally abundant insects in arctic tundra. The arctic tundra is also home to many herbivorous mammals, such as the large musk oxen and caribou in North America and the reindeer of Europe and Asia, as well as smaller herbivores such as lemmings, which undergo huge cyclical fluctuations in their population densities. Common predators include arctic fox, wolves, and snowy owls, and polar bears near the coast.

Aquatic ecosystems, consisting of freshwater and marine biomes, occupy the largest part of the biosphere

Life first arose in water and evolved there for almost three billion years before plants and animals moved onto land and diversified in terrestrial habitats. Today, the largest part of the biosphere is still occupied by aquatic habitats. Ecologists distinguish between freshwater biomes and marine biomes on the basis of physical and chemical differences that influence the communities occupying these distinctive aquatic habitats. Freshwater biomes, for example, are usually characterized by a salt concentration less than 1%, whereas marine biomes generally have salt concentrations that average 3%. The general theme developed in the last section applies to all of these biomes as well: Where aquatic environments are similar in abiotic factors, similar adaptations and similar communities are usually found. Earth's major freshwater and marine biomes are mapped in FIGURE 46.21.

Freshwater biomes are closely linked to the terrestrial biomes through which they pass or in which they are situated. Streams and rivers are created by the runoff of water from terrestrial habitats, and ponds and lakes form where runoff accumulates in a landlocked basin. The particular characteristics of a freshwater biome are also influenced by the patterns and speed of water flow, and the climate to which the biome is exposed.

Marine biomes are those found in the oceans, which cover nearly 75% of Earth's surface. The evaporation of seawater provides most of the planet's rainfall, and ocean temperatures have a major effect on world climate and wind patterns. In addition, marine algae supply a substantial portion of the world's oxygen and consume huge amounts of atmospheric carbon dioxide.

We will examine two freshwater and five marine biomes, and one aquatic biome—wetlands—that includes both freshwater and marine ecosystems.

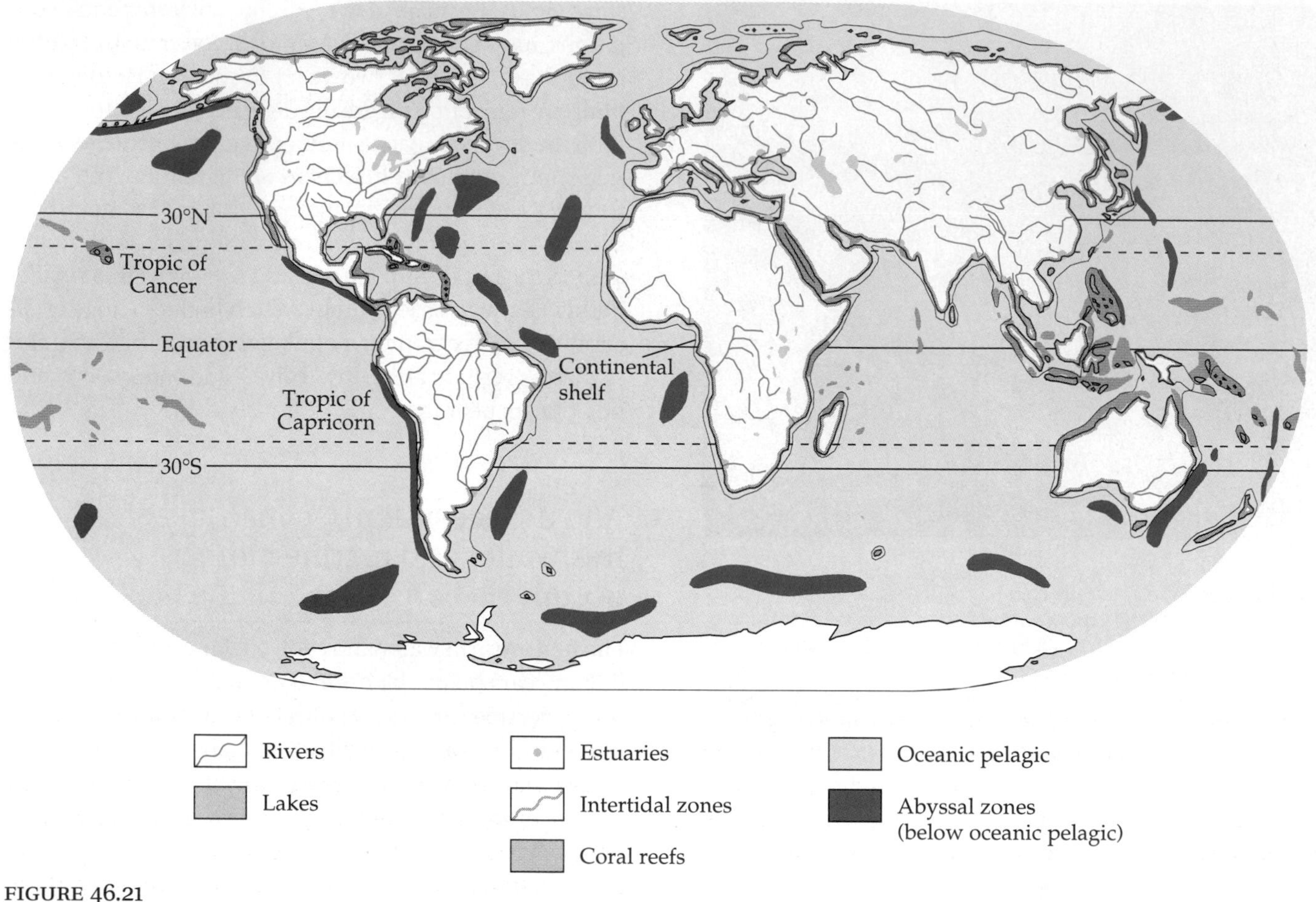

FIGURE 46.21
The distribution of major aquatic biomes. The characteristics of aquatic biomes are determined by the physical and chemical properties of water in different regions. Rivers and streams are created by runoff from the land that collects in channels and flows into standing bodies of water. Lakes form where water accumulates in inland basins, and estuaries are tidal habitats where rivers flow into the sea. A variety of marine biomes exists in the oceans, where salt concentration is usually high. Intertidal communities flourish along coastlines where rising tides periodically inundate the land. Coral reefs develop in the shallow tropical waters of continental shelves. Pelagic communities exist in the open ocean (over continental shelves and in deeper waters). Benthic communities are found in the substrate below all bodies of water. The map also indicates abyssal zones, the habitat of deep benthic communities.

Ponds and Lakes

Standing bodies of fresh water range from a few square meters to thousands of square kilometers in area; small bodies of fresh water are called **ponds** and larger ones **lakes**. Except in the shallowest ponds and lakes, there is usually a significant vertical stratification of important physical and chemical variables. Light is absorbed by both the water itself and the microorganisms in it, so that its intensity decreases rapidly with depth. Ecologists distinguish between the upper **photic zone,** where there is sufficient light for photosynthesis, and the lower **aphotic zone,** where little light penetrates. Water temperature also tends to be stratified, especially during summer in deeper ponds and lakes of temperate zones. Heat energy from sunlight warms the surface waters to whatever depth the sunlight penetrates, but the deeper waters remain quite cold. A narrow vertical zone of rapid temperature change called a **thermocline** separates the uniformly warm upper layer from the uniformly cold bottom water.

Communities of plants and animals are distributed within ponds and lakes according to the depth of the water and its distance from shore. Rooted and floating aquatic plants flourish in the **littoral zone,** the shallow, well-lighted, warm waters close to shore; some have stems and leaves that emerge above the water surface. The littoral community in most lakes is very diverse, including many species of attached algae, especially diatoms, and a variety of grazing snails and suspension-feeding clams, as well as herbivorous and carnivorous insects, crustaceans, fishes, and amphibians. For many of the insects, such as dragonflies and midges, only the egg and larval stages are strictly aquatic. The adults

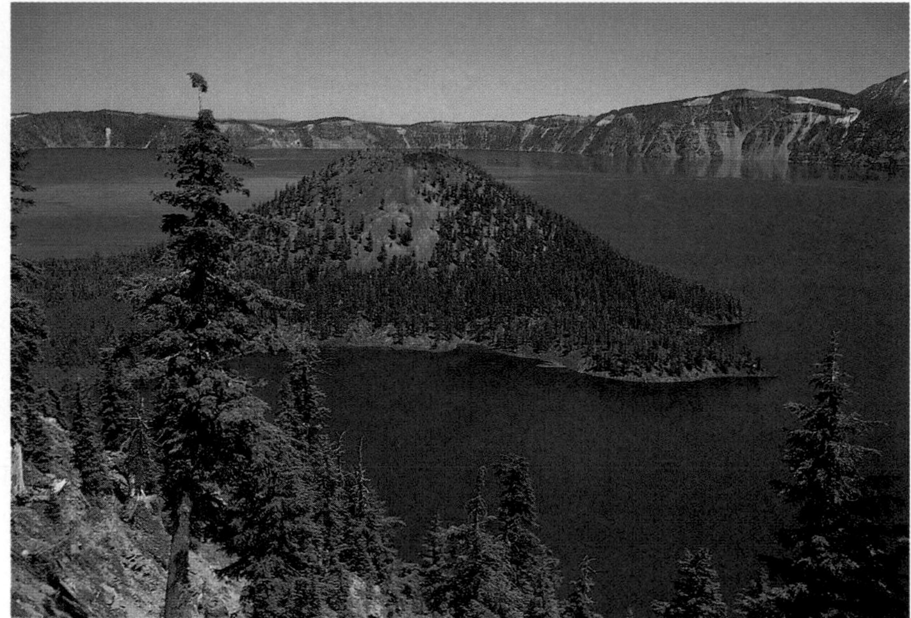

FIGURE 46.22
Ponds and lakes. Oregon's Crater Lake illustrates the pristine quality of a nutrient-poor oligotrophic lake. The shortage of nutrients limits the productivity of phytoplankton in the open water of the limnetic zone; as a result, the water is clear and oxygen-rich, supporting populations of fish and the invertebrates upon which they feed.

emerge from the water as flying insects that complete their life cycle in the air and on land, returning to the water only to lay their eggs. Aquatic and semiaquatic reptiles such as turtles and snakes, waterfowl such as ducks and swans, and some mammals feed on the plants and animals in the littoral zone.

The well-lighted, open surface waters farther from shore, called the **limnetic zone,** are occupied by a variety of phytoplankton, consisting of algae and cyanobacteria. These organisms photosynthesize and reproduce at a high rate during spring and summer. Zooplankton, mostly rotifers and small crustaceans, graze on the phytoplankton. The zooplankton are consumed by many small fish, which in turn become food for larger fish, semiaquatic snakes and turtles, and fish-eating birds.

The small organisms of the limnetic zone are short-lived, and their remains continually sink into the **profundal zone,** the deep, aphotic regions of the pond or lake. Microbes and other organisms in the profundal zone use oxygen for cellular respiration as they decompose this **detritus,** the dead organic material that "rains" down from the limnetic zone. Because the deep waters are colder and thus denser, they do not mix with the surface waters that contact the atmosphere. Decomposers may actually deplete the oxygen supply in the profundal zone by the end of a productive summer, making the deep waters unsuitable for most organisms. Decomposition also releases large quantities of mineral nutrients from the detritus, but these nutrients are trapped in the waters at the bottom of the lake. The biannual turnover that results from changing temperature profiles in temperate lakes brings oxygen to the profundal zone and nutrients to the limnetic zone (see FIGURE 46.8).

Lakes are often classified according to their production of organic matter. **Oligotrophic** lakes are deep and nutrient-poor, and the phytoplankton in the limnetic zone are not very productive (FIGURE 46.22). The water of these lakes is clear, and, because detritus from the limnetic zone is limited, the deep waters contain lots of oxygen all year. **Eutrophic** lakes, in contrast, are usually shallower, and the nutrient content of their water is high. As a result, the phytoplankton are very productive, the waters are murkier than those of an oligotrophic lake, and oxygen supplies may be depleted in the profundal zone in summer. Over long periods of time, oligotrophic lakes may develop into eutrophic lakes as runoff brings in large quantities of mineral nutrients and sediments. Unfortunately, human activities often speed this natural process dramatically. Runoff from fertilized lawns and agricultural fields and the dumping of municipal wastes enrich the lakes with excessive amounts of nitrogen and phosphorus, mineral nutrients that normally limit the growth of phytoplankton and plants. The result of such pollution is often a population explosion of algae, the production of much detritus, and the eventual depletion of oxygen supplies. Such "cultural eutrophication" makes the water unusable and degrades the lake's aesthetic value (see FIGURE 49.15).

Streams and Rivers

Streams and **rivers** are bodies of water moving continuously in one direction. At the headwaters of a stream (perhaps a spring or snowmelt), the water is cold and clear, and it carries little sediment and relatively few mineral nutrients. The channel is usually narrow, with a swift current passing over a rocky substrate. Farther downstream, where numerous tributaries may have joined to form a river, the water is more turbid, carrying substantially more sediment (from the erosion of soil) and nutrients. The channel near the mouth of a river is relatively wide, and the substrate is generally silty from the deposition of sediments over long periods of time.

Many factors influence the flow, the nutrient and oxygen content, and the turbidity of streams and rivers. Shallow water flowing rapidly over a rough bottom causes turbulent flow; where deep water flows slowly over a smooth bottom, pools of water are common; and where deep water flows rapidly over a flat bottom, smooth runs of water are apparent. Nutrient content is largely determined by the terrain and vegetation through which streams and rivers flow. Fallen leaves from dense, overhanging vegetation can add substantial amounts of organic matter, and the erosion of rocks in the underlying streambed can increase the concentration of inorganic nutrients in the flowing water. The turbulent flow of many streams constantly oxygenates the water, whereas the murky, warm waters of large rivers may contain relatively little oxygen. The amount of water in a stream or river varies with rainfall patterns and snowmelt, causing seasonal changes in flow and oxygen content.

Biological communities in streams and rivers are substantially different from those in ponds and lakes. Many fast-flowing streams and rivers do not support large stationary plankton communities because these small organisms are washed away by the flow of water. Instead, photosynthesis by attached algae and rooted plants supports the food chains. However, where dense vegetation on the banks of a narrow stream blocks the sunlight needed for photosynthesis, organic material carried into the stream by runoff provides the most important input of food for consumers. The high concentration of silt near the mouth of large rivers increases the turbidity and can also block the passage of light, reducing photosynthesis.

Because of the dramatic variations in the physical environment described above, the composition of animal communities varies significantly from the headwaters of a stream to the mouth of the river. Upstream, fish such as trout may be present where their requirements for cool temperatures, high oxygen, and clear water are met. In the warmer, murkier waters further downstream, catfish and carp may be abundant. Benthic (bottom-dwelling) communities also change, and many insect species are restricted to the relatively short stretches of a stream or river that provide their specific requirements.

Not surprisingly, stream- and river-dwelling animals exhibit evolutionary adaptations that enable them to resist being carried away by the relentless flow of water. Small animals are typically flat in shape and can attach to rocks temporarily. Many insect species live on the underside of rocks or on their downstream side, thereby exploiting a small habitat that is relatively free of turbulent flow.

The moving water of streams and rivers provides a living for a unique and important group of insects that filter their food from water passing through their mesh nets (FIGURE 46.23). By capturing and processing algal particles, diatoms, and small invertebrates that are suspended in the water, these suspension-feeding insects affect both

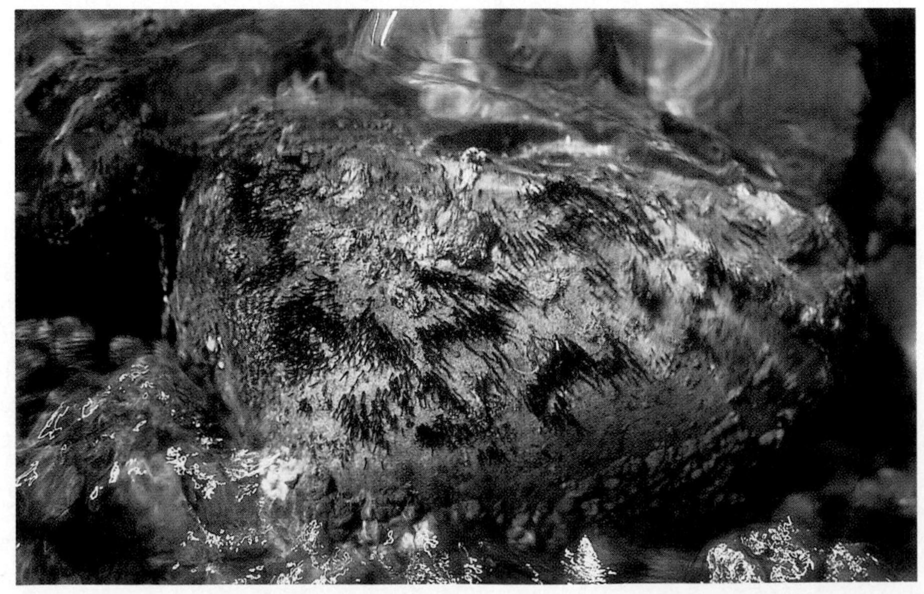

FIGURE 46.23
Rivers. The rock in the photograph is covered with several species of suspension-feeding insects. Caddisfly larvae build silken catchnets that trap particulate foods as the water flows through the net. Blackfly larvae have siphonlike structures on their heads, which serve the same purpose. Suspension-feeding insects, which depend on relatively fast current and can live only in flowing water biomes, have important effects on the amount of suspended nutrients available in streams and rivers.

the quality and the quantity of nutrients available to other members of stream and river communities. Such a lifestyle is not possible in the still water of ponds and lakes.

Many streams and rivers have been affected by pollution from human activities, by stream channelization to speed up water flow, and by dams that hold water. For centuries, humans used streams and rivers as repositories of waste, thinking that these materials would be diluted and carried downstream. While some pollutants are carried far from their source, many settle to the bottom, where they can be taken up by aquatic organisms. Even the pollutants that are carried away contribute to estuary, ocean, and lake pollution. In many cases, dams have completely changed the downstream ecosystems, altering the intensity and volume of water flow and affecting fish and invertebrate populations.

Wetlands

At the most simple level, a **wetland** is an area covered with water that supports aquatic plants (FIGURE 46.24). In fact, wetlands range from periodically flooded regions to soil that is permanently saturated during the growing season. These conditions favor the growth of specially adapted plants called hydrophytes ("water plants"), which can grow in water or in soil that is periodically anaerobic due to the presence of water. Hydrophytes include floating pond lilies and emergent cattails, many sedges, tamarack, and black spruce. Both the hydrology and the vegetation of an area are important determinants of its classification as a wetland—a classification that can be critical when federal, state, and local governments are making preservation decisions based on rigorous, and often conflicting, definitions.

A wide variety of types of wetlands have been recognized, ranging from marshes to swamps to bogs. All these varieties, however, generally form in one of three different topographic situations. Basin wetlands develop in shallow basins, ranging from upland depressions to filled-in lakes and ponds. Riverine wetlands develop along shallow and periodically flooded banks of rivers and streams. Fringe wetlands occur along the coasts of large lakes and seas, where water flows back and forth because of rising lake levels or tidal action. Thus, fringe wetlands include both freshwater and marine biomes. Marine coastal wetlands are closely linked to estuaries, which are considered below. The flow of water through a wetland, as well as the duration, frequency, depth, and season of flooding, determine the types of plants that are present.

Ecologically, wetlands are among the richest of biomes. They contain a diverse community of invertebrates, which support a wide variety of birds. Herbivores from crustaceans to muskrats consume algae, detritus, and plants. In addition to the rich diversity of wildlife that is supported by wetlands, the ecological and economic value of wetlands is much larger than their geographic extent implies; they provide water-storage basins that reduce the intensity of flooding, and they improve water quality by filtering pollutants. In the past, humans have often regarded wetlands as wastelands—sources of mosquitoes, flies, and bad odors—and have destroyed many wetlands, mostly to provide land for agriculture and development. Recently, both governments and private organizations are attempting to protect remaining wetlands through acquisition, economic incentives, and regulation. A great deal of research is underway to determine how wetlands can be created or restored.

FIGURE 46.24
Wetlands. This bog in Great Meadow National Wildlife Refuge (Concord, Massachusetts) is an example of a basin wetland. The soil is waterlogged most of the year, and is often covered by spongy mosses. Heath shrubs and sedges are other common plants in bogs. In some bogs, organic production exceeds the rate of decomposition, and much of the production accumulates as peat. The accumulation of several generations of mosses and peat results in the formation of peatlands.

FIGURE 46.25
Estuaries. This view of an estuary that is part of Chesapeake Bay in Maryland shows the intimate association of river mouths and the marine environment into which they carry water. Unfortunately, the land surrounding Chesapeake Bay is heavily populated and industrialized, and pollution that enters the bay through four major rivers has made it unsuitable for many plant and animal species. What was once a bountiful natural source of seafood and other resources has been degraded and rendered less productive by human activity.

Estuaries

The area where a freshwater stream or river merges with the ocean is called an **estuary;** it is often bordered by extensive coastal wetlands called mudflats and saltmarshes (FIGURE 46.25). Salinity varies spatially within estuaries, from nearly that of fresh water to that of the ocean; it also varies on a daily cycle with the rise and fall of the tides. Nutrients from the river enrich estuarine waters, making estuaries one of the most biologically productive environments on Earth.

Saltmarsh grasses, algae, and phytoplankton are the major producers in estuaries. This environment also supports a variety of worms, oysters, crabs, and many of the fish species that humans consume. Many marine invertebrates and fishes use estuaries as a breeding ground or migrate through them to freshwater habitats upstream. Estuaries are also crucial feeding areas for many semiaquatic vertebrates, particularly waterfowl.

Although estuaries support a wide variety of extremely valuable commercial species, areas around estuaries are also prime locations for commercial and residential developments. In addition, estuaries are unfortunately at the receiving end for pollutants dumped upstream. Very little undisturbed estuary habitat remains, and a large percentage has been totally eliminated by landfill and development. Many states have now—rather belatedly—taken steps to preserve their remaining estuaries.

Marine Zones: An Introduction

Like those in freshwater lakes, marine communities are distributed according to the depth at which they occur and their distance from shore (FIGURE 46.26). There is a photic zone where phytoplankton, zooplankton, and many fish species occur. Below is the aphotic zone. Because water absorbs light so well and the ocean is so deep, most of the ocean volume is virtually devoid of light, except for tiny amounts produced by a few luminescent fishes and invertebrates. The zone where land meets water is called the **intertidal zone.** Beyond the intertidal zone is the **neritic zone,** the shallow regions over the continental shelves. Past the continental shelf is the

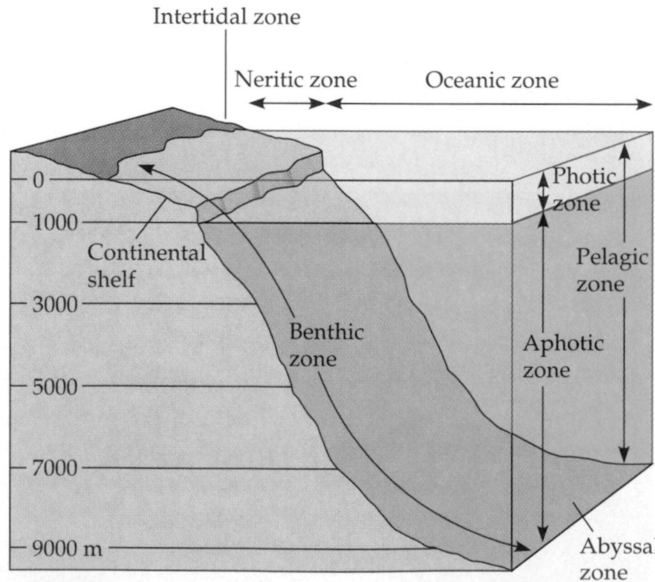

FIGURE 46.26
Marine zones. The marine environment can be classified on the basis of three physical criteria: light penetration (photic and aphotic zones), distance from shore and water depth (intertidal, neritic, and oceanic zones), and open water or bottom (pelagic and benthic zones). The abyssal zone is the benthic region in the very deepest oceans. Ecologists often use two designations, such as the oceanic pelagic zone, to identify the location of a biome.

oceanic zone, reaching very great depths. Open water of any depth is the **pelagic zone,** at the bottom of which is the seafloor, or **benthic zone.**

The Intertidal Zones

An intertidal zone, where the land meets the sea, is alternately submerged and exposed by the daily cycle of tides. Intertidal communities are therefore subject to huge daily variations in the availability of seawater (and the nutrients it carries) and in temperature. Perhaps most significant of all, intertidal organisms are subject to the mechanical forces of wave action, which can dislodge them from the habitat.

The rocky intertidal zone is vertically stratified (FIGURE 46.27). Most of the organisms have structural adaptations that enable them to attach to the hard substrate in this physically tumultuous environment. The uppermost zone, which is submerged only during the highest tides, contains relatively few species of algae, and grazing mollusks and suspension-feeding barnacles that are sometimes eaten by crabs and shorebirds. The middle zone is generally submerged at high tide and exposed at low tide. It is inhabited by a diverse array of algae, sponges, sea anemones, bryophytes, suspension-feeding barnacles and mussels, herbivorous and predatory snails, crabs, sea urchins, sea stars, and small fishes. Tidepool organisms in this zone may experience dramatic increases in salinity as evaporation decreases the volume of water in the pools during low tides. The bottom of the intertidal zone is exposed only during the lowest tides. The low intertidal zone and the neritic subtidal zone just below it house an extraordinary diversity of invertebrates and fish species that live within the dense cover of abundant and productive seaweeds.

On sandy substrates (beaches) or mudflats, the intertidal zone is not as clearly stratified. Wave action constantly moves the particles of mud and sand, and few large algae or plants occupy these habitats. Many animals, such as suspension-feeding worms and clams and predatory crustaceans, bury themselves in sand or mud, feeding when the tides bring sources of food. Other surface-dwelling organisms, such as crabs and shorebirds, are scavengers or predators on these organisms.

Partly because of our strong attraction to the seashore, humans have had a long-term impact on intertidal ecosystems. The recreational use of ocean shores has caused a severe decline in the numbers of many beach-nesting birds and sea turtles. Incoming tides carry in polluted water and old fishing lines and plastic debris that can harm wildlife. The most dramatic intertidal pollutant is probably oil, which harms not only birds and marine mammals but also intertidal algae and invertebrates. The ultimate outcome of oil pollution on intertidal zones is a reduction in species diversity, with increases in the populations of a few oil-resistant species, such as barnacles.

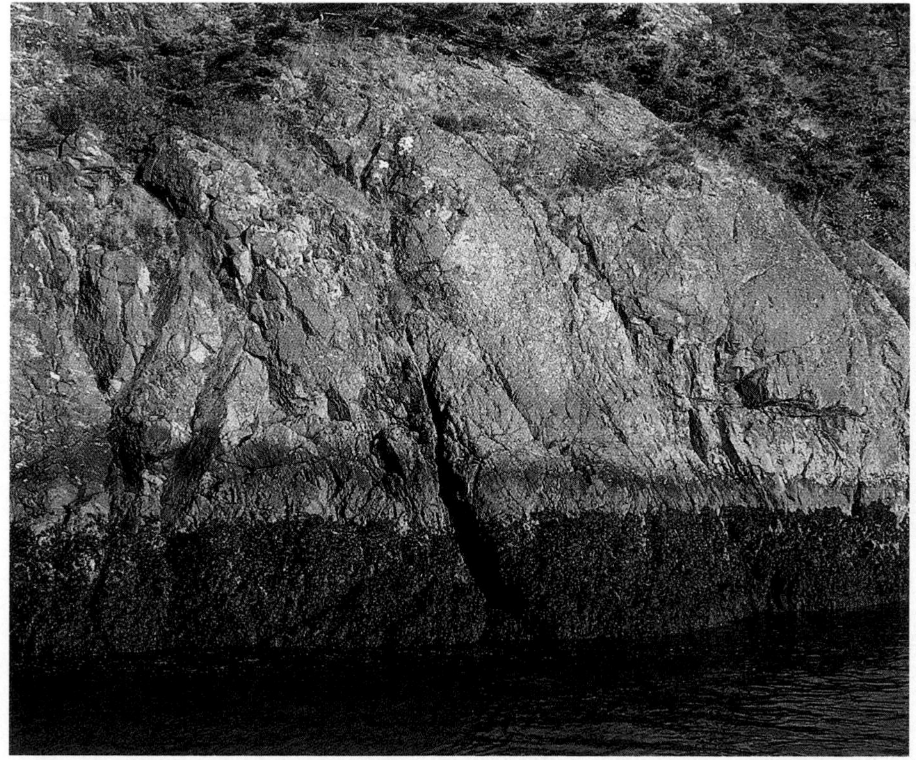

FIGURE 46.27
The intertidal zones. This photograph of the rocky intertidal zones at Washington state's Lopez Island was taken at very low tide to illustrate the vertical zonation of algae and animals. The density of organisms in each of the three major zones is roughly proportional to the percentage of time the zone is submerged. Organisms in the highest zone are frequently exposed to air and sun and have numerous adaptations that prevent dehydration and overheating.

Coral Reefs

In warm tropical waters in the neritic zone, **coral reefs** constitute a conspicuous and distinctive biome. Currents and waves constantly renew nutrient supplies to the reefs, and sunlight penetrates to the ocean floor, allowing photosynthesis.

Coral reefs are dominated by the structure of the coral itself, formed by a diverse group of cnidarians that secrete hard external skeletons made of calcium carbonate. These skeletons vary in shape, forming a substrate upon which other corals, sponges, and algae grow (FIGURE 46.28). Although the coral animals themselves feed on microscopic organisms and particles of organic debris, they are also dependent on the photosynthesis of symbiotic dinoflagellate algae. An immense variety of microorganisms, invertebrates, and fishes live among the coral and algae, making the reefs one of the most diverse and productive biomes on Earth. Prominent herbivores include snails, sea urchins, and fishes, which are in turn consumed by predatory octopuses, sea stars, and carnivorous fishes.

FIGURE 46.28
Coral reefs. This coral reef in Fiji illustrates the diversity of algae and animals that populate these productive tropical biomes. Free-living algae and those symbiotic with the coral animals photosynthesize during the day, but the coral animals themselves extend their polyps and feed on plankton at night. The activities of other invertebrates and fishes also follow a daily cycle, and different components of the fauna are active by day and by night.

Some coral reefs cover enormous expanses of shallow ocean, but this delicate biome is easily degraded by pollution and development, as well as by souvenir hunters who gather the coral skeletons. Corals are also subject to damage from native and introduced predators such as the crown-of-thorns sea star, which has undergone a population explosion in many regions and actually destroyed coral reefs in parts of the western Pacific Ocean. Reef communities are very old and grow very slowly, and they may not be able to withstand continued human encroachment.

The Oceanic Pelagic Biome

Most of the ocean's water lies far from shore in the **oceanic pelagic biome,** constantly mixed by the ever-circulating ocean currents. Nutrient concentrations are typically low in the open ocean because the remains of phytoplankton, zooplankton, and other organisms sink below the photic region into the dark, lower benthic zone, robbing the upper layer of nutrients. Interestingly, nutrient concentration is lowest in tropical waters, despite their high light intensity and warm temperatures. A permanent thermal stratification prevents an exchange of nutrients between the surface and the deep. Temperate oceans are more productive, because, like temperate lakes, they experience a nutrient overturn in the spring and, to a limited extent, in the fall. The recirculation of nutrients from the depths stimulates a surge of spring phytoplankton growth. Although the temperature varies with latitude and depth, pelagic waters are generally cold.

Photosynthetic plankton grow and reproduce rapidly in the photic region of the oceanic biome, representing only the top 100 m of the open ocean. Their activity accounts for less than half the photosynthetic activity on Earth, although oceans cover 70% of Earth's surface. Zooplankton, including protozoans, worms, copepods, shrimplike krill, jellyfishes, and the small larvae of invertebrates and fishes, graze on the phytoplankton. Most plankton exhibit morphological structures like bubble-trapping spines, lipid droplets, gelatinous capsules, and air bladders, which help them stay afloat within the photic zone.

The oceanic pelagic biome also includes free-swimming animals, called nekton, which can move against the currents to locate food. Large squids, fishes, sea turtles, and marine mammals feed on either plankton or each other. Although many of these animals feed in the photic region of the pelagic zone, others live at great depth

FIGURE 46.29

Benthos: a deep-sea vent community. Benthic faunas occupy the ocean bottom from the intertidal zone to the abyssal zone. The species composition of benthos varies dramatically with water depth. Pictured here is a vent community, first discovered at a depth of 2500 m in the late 1970s. These communities are found at spreading centers on the seafloor, where hot magma superheats the water. About a dozen species of bacteria identified near the vents are chemoautotrophic producers that obtain energy by oxidizing H_2S formed by a reaction of the hot water with dissolved sulfate (SO_4^{-2}). Among the animals in these communities are giant tube-dwelling worms (pictured here), some more than 1 m long. They are apparently nourished by chemosynthetic bacteria that live as symbionts within the worms. Many other invertebrates, including arthropods and echinoderms, are also abundant around the vents.

where fish may have enlarged eyes, enabling them to see in the very dim light, or luminescent organs that attract mates and prey. Many pelagic birds, such as petrels, terns, albatrosses, and boobies, catch fishes in the surface waters. A number of marine animals are migratory, following seasonally available food sources or moving between summer breeding grounds and their winter feeding range.

Benthos

The ocean bottom below the neritic and pelagic zones is the benthic zone, occupied by communities of organisms collectively called **benthos.** Like the profundal zone of lakes, nutrients reach the seafloor by "raining down" in the form of detritus from the waters above. Although the benthic zone in shallow, near-coastal waters may receive substantial sunlight, light and temperature decline dramatically with depth. The bottom itself is composed of sand or very fine sediments ("ooze") made up of silt and, in the deep sea, the shells of dead microscopic organisms.

Neritic benthic communities are extremely productive, consisting of bacteria, fungi, seaweeds and filamentous algae, sponges, sea anemones, worms, clams, crustaceans, sea stars, sea urchins, and fishes. Species composition of these communities varies with distance from the shore, water depth, and composition of the bottom. Many organisms live buried in soft substrates and are not apparent to the casual observer.

Deep benthic communities live in the **abyssal zone,** where continuous cold (about 3°C), extremely high water pressure, the near absence of light, and low nutrient concentrations are typical. However, oxygen is usually present in abyssal waters, and a fairly diverse community of invertebrates and fishes occupies this region. Marine scientists have also discovered a unique assemblage of organisms associated with deep-sea hydrothermal vents of volcanic origin in midocean ridges (FIGURE 46.29). In this dark, hot, oxygen-deficient environment, the primary producers are not photosynthesizing organisms but bacteria that are chemoautotrophs (see Chapter 25). These bacteria are consumed by giant polychaete worms, arthropods, echinoderms, and fishes.

* * *

Now that we have surveyed the environmental diversity of the biosphere, we will turn in Chapter 47 to the factors that influence the composition, size, and dispersion of populations.

REVIEW OF KEY CONCEPTS (with page numbers and key figures)

- Ecology is the scientific study of the interactions between organisms and their environments (pp. 1061–1062)
  - The environment includes both abiotic factors, such as temperature, light, water, and nutrients; and biotic factors, other living organisms.
- Basic ecology provides a scientific context for evaluating environmental issues (p. 1062)
  - Although environmental problems have political, economic, and ethical components, it is ecology that provides the scientific basis for understanding these problems.

- Ecological research ranges from the adaptations of organisms to the dynamics of ecosystems (pp. 1062–1063)
  - Ecology spans increasingly comprehensive levels of organization, from the individual organism through populations and communities to the ecosystem and biome.
  - The fields of ecology and evolutionary biology are tightly linked. Ecological interactions affect how organisms evolve, and evolutionary change in turn affects ecological relationships.
- Climate and other abiotic factors are important determinants of the biosphere's distribution of organisms (pp. 1063–1069, FIGURES 46.4–46.6)
  - The biosphere is an environmental mosaic in which several abiotic factors have an important impact on the distribution and abundance of organisms: temperature, water availability and quality, light intensity, wind, soil characteristics, and less predictable disturbances such as fire.
  - Global climates and seasonality are established by the input of solar energy and Earth's rotation around the sun. Differential heating of the atmosphere and Earth's surface produce air circulation cells and latitudinal variation in temperature and precipitation, which in turn account for the geographical distribution of major biomes.
  - Oceans and lakes moderate the climate in coastal localities, and mountains influence temperature and rainfall.
- The costs and benefits of homeostasis affect an organism's responses to environmental variation (pp. 1069–1071, FIGURE 46.9)
  - Although natural selection has produced a wide spectrum of adaptations to the environmental diversity of the biosphere, most species tolerate only a relatively narrow range of environmental variables.
  - Organisms can be described as regulators or conformers, depending on their homeostatic capabilities. However, few organisms are perfect regulators or conformers.
  - According to the principle of allocation, the total amount of energy available to an organism for all of its processes, including response to physical variables, is limited. Natural selection has resulted in different approaches, which are usually related partly to the stability of an organism's environment.
- The response mechanisms of organisms are related to environmental grain and the time scale of environmental variation (pp. 1071–1073)
  - The importance of environmental variation to an organism depends on the spatial and temporal scale of the variation in relation to the organism's size, lifespan, and movement patterns.
  - Animals may respond to the environment behaviorally, by moving or migrating to more favorable locations.
  - Physiological responses often include mechanisms of homeostasis. Tolerance limits and optimal temperatures can shift through reversible physiological adjustments involved in acclimation.
  - Organisms may react to environmental change with responses that alter the body. Many plants exhibit irreversible morphological plasticity, which helps compensate for their inability to move to new locations.
- The geographical distribution of terrestrial biomes is based mainly on regional variations in climate (pp. 1073–1081, FIGURE 46.11)
  - Biomes are communities and ecosystems that are typical of broad geographic regions. Abiotic factors are important in determining why a particular biome is found in a particular area.
  - Tropical forests are found near the equator, where photoperiod and temperature are nearly constant. The tropical rain forest is the most species-rich terrestrial biome.
  - Savanna is a tropical grassland with scattered trees. Precipitation varies greatly between wet and dry seasons.
  - Deserts are arid biomes, with extremes in temperature and very low precipitation.
  - Chaparral consists of scrublands usually found along coastlines and is characterized by mild, rainy winters and long, hot, dry summers.
  - Temperate grasslands occur in relatively cool climates with nutrient-rich, deep soils. Periodic fires and drought inhibit the growth of woody shrubs and trees.
  - Temperate deciduous forests occur in midlatitudes where there is sufficient moisture to support the growth of large, broad-leaf deciduous trees.
  - Taiga consists of the dense coniferous forests. Taiga is characterized by long, cold, snowy winters and short summers.
  - Tundra occurs at the northernmost limits of plant growth and at high altitudes, where plant forms are limited by cold temperature and wind to a low shrubby or matlike morphology.
- Aquatic ecosystems, consisting of freshwater and marine biomes, occupy the largest part of the biosphere (pp. 1081–1089, FIGURE 46.21)
  - Aquatic biomes are often stratified vertically in regard to light, temperature, and community structure. Phytoplankton and zooplankton are the primary food source for the rest of the community.
  - Lakes are classified on the basis of their nutrient content and productivity, with eutrophic lakes being high in nutrients and oligotrophic lakes being nutrient-poor. Many temperate lakes turn over twice a year, mixing oxygen and nutrients that were stratified as a result of the thermocline.
  - Rivers and streams contain freshwater communities that change significantly from the source to the final destination in an ocean or lake. Upstream areas contain organisms associated with clear water, cool temperatures, and rocky substrates. Downstream are organisms that can tolerate murkier water and warmer temperatures.
  - Wetlands range from having periodically flooded to permanently saturated soil. They contain specially adapted plants, called hydrophytes, which can grow in water or soil that is periodically anaerobic. Wetlands can be either freshwater or marine.
  - Marine communities are in the oceans, which comprise nearly 75% of Earth's surface. Oceanic zones can be classified according to degree of light penetration and according to depth.
  - An estuary is the transition zone between a river or stream and the ocean into which it empties. Such areas experience large fluctuations in salinity, but they support an abundance of both aquatic and semiaquatic organisms.

- The rocky intertidal zone is a vertically stratified biome on ocean shorelines that is periodically inundated by seawater. Organisms in the uppermost zone are frequently exposed to air and sun and have adaptations that prevent desiccation and overheating.
- Coral reefs are found in nutrient-rich and warm tropical waters. Skeletons of coral animals form complex structures among which a diversity of invertebrates and fishes live.
- The oceanic pelagic biome includes most of the open ocean where phytoplankton and zooplankton occupy the upper layers.
- Benthic communities subsist largely on detritus that "rains" down from the pelagic zone.

SELF-QUIZ

1. Which statement follows from the principle of allocation?
   a. The number of organisms an area can support is determined by its energy supply.
   b. Physiological adjustments to environmental changes can extend the tolerance limits of organisms.
   c. The total amount of energy available to an organism is partitioned into such processes as reproduction, obtaining nutrients, and coping with the environment.
   d. Organisms that use more energy for growth and reproduction are able to survive in a wider range of variable environments.
   e. Organisms allocate most of their energy for homeostasis.

2. Which statement about tolerance limits is *incorrect?*
   a. They can often be tested experimentally and plotted as a tolerance curve.
   b. They help determine whether organisms can live in particular environments.
   c. They can be extended by acclimation in some cases.
   d. They are likely to be greater in regulators than in conformers.
   e. They are generally greatest for organisms restricted to stable environments.

3. Which of the following biomes is *correctly* paired with the description of its climate?
   a. savanna—cool temperature, precipitation uniform during the year
   b. tundra—long summers, mild winters
   c. temperate deciduous forest—relatively short growing season, mild winters
   d. temperate grasslands—relatively warm winters, most rainfall in summer
   e. tropical forests—nearly constant photoperiod and temperature

4. Which of the following is an *incorrect* comparison between tropical rain forests and temperate grasslands?
   a. They are both highly productive biomes.
   b. They both have rich, deep topsoil.
   c. They are both characterized by rapid decomposition of dead plant material.
   d. Tropical rain forests, but not grasslands, have a high degree of vertical stratification in their vegetation.
   e. Large areas of both biomes have been converted by human use.

5. Which of the following is *correctly* paired with its description?
   a. neritic zone—shallow area over continental shelf
   b. benthic zone—surface water of shallow seas
   c. pelagic zone—seafloor
   d. aphotic zone—zone in which light penetrates
   e. intertidal zone—open water at the edge of the continental shelf

6. In which area are algal blooms most likely to occur?
   a. headwaters of a stream
   b. downstream area of a river
   c. lake or pond
   d. intertidal zone of an ocean
   e. benthic zone of an ocean

7. In general, deserts are located at latitudes where
   a. dry air is descending
   b. moist air is descending
   c. dry air is rising
   d. rising air creates doldrums
   e. air masses are stationary

8. The growing season would generally be shortest in which biome?
   a. tropical rain forest
   b. savanna
   c. taiga
   d. temperate deciduous forest
   e. temperate grassland

9. Imagine some cosmic catastrophe that jolts Earth so that its axis is perpendicular to the line between the sun and Earth. The most predictable effect of this change would be
   a. no more night and day
   b. a big change in the length of the year
   c. a cooling of the equator
   d. a loss of seasonal variations at northern and southern latitudes
   e. the elimination of ocean currents

10. While climbing the Rocky Mountains, one observes transitions in biological communities that are analogous to the changes one encounters
    a. in biomes at different latitudes
    b. at different depths in the ocean
    c. in a community through different seasons
    d. in an ecosystem as it evolves over time
    e. traveling across the United States from east to west

CHALLENGE QUESTIONS

1. In which terrestrial biome is your college or university located? Describe how five local features on your campus influence microclimates.
2. Describe five abiotic factors that might be important to the life of a tree in a temperate deciduous forest. How might the tree adjust to day-to-day changes in these factors?
3. Explain how the following factors change from the source of a river to its mouth: nutrient content, current, sediments, temperature, oxygen content, food sources.

SCIENCE, TECHNOLOGY, AND SOCIETY

1. Near Lawrence, Kansas, there was a rare patch of original North American temperate grasslands that had never been converted to farming. It was home to numerous native grasses, annual plants, and grassland animals. Among the species present were two endangered plants. Environmental activists thought the area should be set aside as a nature preserve, and they started to raise money to save it. In 1990, the owner of the land plowed it, stating that he did not want to be told what he could do with his property. He was within his legal rights, because there are no federal laws protecting endangered species on private land. What issues and values are in conflict in this situation? How do you think such conflicts should be resolved?

2. During the summer of 1988, huge forest fires burned a large portion of Yellowstone National Park. The National Park Service has a natural-burn policy: Fires that start naturally are allowed to burn unless they endanger human settlements. Lightning ignited the Yellowstone fires, so they were allowed to spread and burn themselves out, with firefighters primarily protecting people. This drew a lot of public criticism; the Park Service was accused of letting a national treasure go up in flames. Park Service scientists stuck with the natural-burn policy. Do you think this was the best decision? Support your position.

3. Although chaparral habitats burn regularly, many people in southern California and elsewhere build expensive houses in canyons and on hillsides covered with this vegetation. Should local governments regulate development in such communities? Should the government continue to provide disaster relief to people whose homes burn in these periodic fires so that these people can rebuild in the hills?

FURTHER READING

Abrahamson, W. G., T. G. Whitham, and P. W. Price. "Fads in Ecology." *BioScience,* May 1989. Changing ideas in a dynamic field.

Cunningham, W. P. *Understanding Our Environment: An Introduction.* Dubuque, IA: W. C. Brown, 1994.

Gomez-Pompa, A., and A. Kaus. "Taming the Wilderness Myth." *BioScience,* April 1992. How Western beliefs affect environmental policy.

Heinrich, B. "In Plain Sight." *Natural History,* July 1995. How certain bird species hide their nests on the open tundra.

Holloway, M. "Still Negotiating." *Scientific American,* June 1992. Behind-the-scenes politics that set the stage for the United Nations Earth Summit.

"Managing Planet Earth." *Scientific American,* September 1989. A special issue devoted to the environment.

Pain, S. "Exploring the Riches of the 'Coral Rainforest.'" *New Scientist,* February 26, 1994. Researchers are exploring the biodiversity of coral reefs.

Ray, G., and J. Grassle. "Marine Biology Diversity." *BioScience,* July/August 1991. Why we need a program to conserve marine communities.

Regalado, A. "Listen Up! The World's Oceans May Be Starting to Warm." *Science,* June 9, 1995. What effects will global warming have on marine biomes?

Ricklefs, R. E. *The Economy of Nature,* 3rd ed. New York: W. H. Freeman, 1993.

Robinson, B. H. "Light in the Ocean's Midwaters." *Scientific American,* July 1995. Bioluminescence is a common adaptation in deep-sea biomes.

Every day, we hear about local and global problems that threaten our well-being or provoke disputes between individuals and nations—global warming, toxic waste, conflicts over oil in the Middle East, starvation in Rwanda aggravated by civil war. Contributing to all these apparently unrelated problems is a common factor: the continued increase of the human population in the face of limited resources. The human population explosion is now Earth's most significant biological phenomenon. As the size of our species approaches 6 billion individuals, we require vast amounts of materials and space, including places to live, land to grow our food, and places to dump our waste. (The photo that opens this chapter shows the clearing of land for a shopping center.) Endlessly expanding our presence on Earth, we have devastated the environment for many other species and now threaten to make it unfit for ourselves.

To understand the problem of human population growth on more than a superficial level, we must consider the general principles of population ecology. It is obvious that no population can grow indefinitely. Species other than humans sometimes exhibit population explosions, but their populations inevitably crash. In contrast to these radical booms and busts, many populations are relatively stable over time, with only minor increases or decreases in population size. Population ecology, the subject of this chapter, is concerned with measuring changes in population size and composition, and identifying the causes of these fluctuations.

We can think of populations in a variety of ways. In Chapter 21, our emphasis was on populations as interbreeding groups of individuals of a single species. With a more ecological focus, we can also think of a ***population*** *as individuals of one species that simultaneously occupy the same general area; they rely on the same resources and are influenced by similar environmental factors.*

As we analyze population structure and growth in this chapter, remember this basic theme: The characteristics of a population are shaped by the interactions between individuals and their environments on both ecological and evolutionary time scales, and natural selection can modify these characteristics in a population. Later in this chapter, we will return to our discussion of the human population. But let's begin by examining some of the ways of describing and analyzing populations of any species.

CHAPTER 47

POPULATION ECOLOGY

KEY CONCEPTS

- Two important characteristics of any population are density and the spacing of individuals
- Demography is the study of factors that affect birth and death rates in a population
- The traits that affect an organism's schedule of reproduction and death make up its life history
- A mathematical model for exponential growth describes an idealized population in an unlimited environment
- A logistic model of population growth incorporates the concept of carrying capacity
- Both density-dependent and density-independent factors can affect population growth
- The human population has been growing exponentially for centuries but will not be able to do so indefinitely

Two important characteristics of any population are density and the spacing of individuals

At any given moment, every population has geographical boundaries and a population size (the number of individuals it includes). Ecologists who study population dynamics begin by defining boundaries appropriate to the organisms under study and to the questions being posed. A population's boundaries may be natural ones, such as a specific island in Lake Superior where terns nest, or they may be arbitrarily defined by an investigator, such as the oak trees within a specific county in Minnesota. Regardless of differences in scale, two important characteristics of any population are its density and its dispersion. Population **density** is the number of individuals per unit area or volume—the number of oak trees per km^2 in the Minnesota county, for example. **Dispersion** is the pattern of spacing among individuals within the geographical boundaries of the population.

FIGURE 47.1
An indirect census of a cliff swallow population. The number of birds nesting in this colony can be estimated by counting the number of entrance holes to the mud nests.

Measuring Density

In rare cases, it is possible to determine population size and density by actually counting all individuals within the boundaries of the population. We could count the number of sea stars in a tidepool, for example. Herds of large mammals, such as buffalo or elephants, can sometimes be counted accurately from airplanes. In most cases, however, it is impractical or impossible to count all individuals in a population. Instead, ecologists often use a variety of sampling techniques to estimate densities and total population sizes. For example, they might estimate the number of alligators in the Florida Everglades by counting individuals in a few representative plots of an appropriate size. Such estimates are more accurate when there are more numerous or larger sample plots, and when the habitat is homogeneous.

In some cases, population sizes are estimated not by counts of organisms but by indirect indicators, such as the numbers of nests or burrows (FIGURE 47.1), or signs such as droppings or tracks. Another sampling technique commonly used to estimate wildlife populations is the **mark-recapture method,** described in the Methods Box on p. 1096.

Patterns of Dispersion

A population's **geographical range** is defined as the geographic limits within which it lives. Within the range, local densities may vary substantially because not all areas provide equally suitable habitat, and because individuals exhibit patterns of spacing in relation to other members of the population. The possible patterns vary in a continuum, from **clumped,** if the individuals are aggregated in patches; to **uniform,** if the spacing is even; to **random,** if spacing varies in an unpredictable way (FIGURE 47.2).

Clumping often results from patterns in resource distribution. Plants may be clumped in certain sites where soil conditions and other environmental factors favor germination and growth. For example, the eastern red cedar is often found clumped on limestone outcrops, where soil is less acidic than in nearby areas. Animals often move within the range toward a particular microenvironment that satisfies their requirements. For example, many forest insects and salamanders are clumped under logs where the humidity remains high. Herbivorous animals of a particular species are likely to be most abundant where their food plants are concentrated. Clumping of animals may also be associated with mating or other social behavior. For example, crane flies often swarm in great numbers, a behavior that increases mating chances for these short-lived insects. There may also be "safety in numbers"; fish swimming in large schools, for example, are often less likely to be eaten by predators than fish swimming alone or in small groups (FIGURE 47.2a).

An evenly spaced, or uniform, distribution results from direct interactions between individuals in the population. For example, a tendency toward regular spacing of plants may result from shading and competition for water and minerals; some plants also secrete chemicals that inhibit the germination and growth of nearby individuals that could compete for resources. In animal populations, regular spacing is usually caused by competition for some resource or by social interactions that set

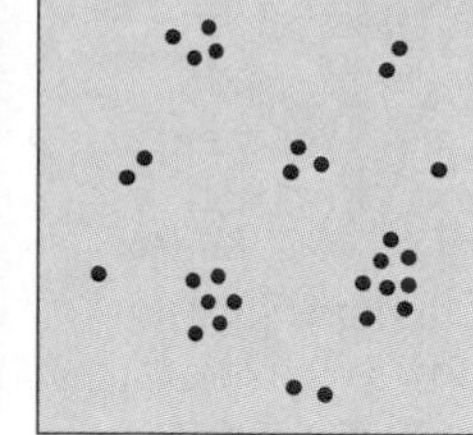

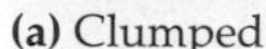

(a) Clumped

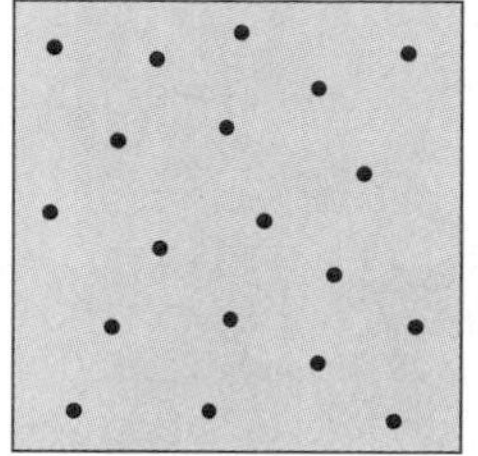

(b) Uniform

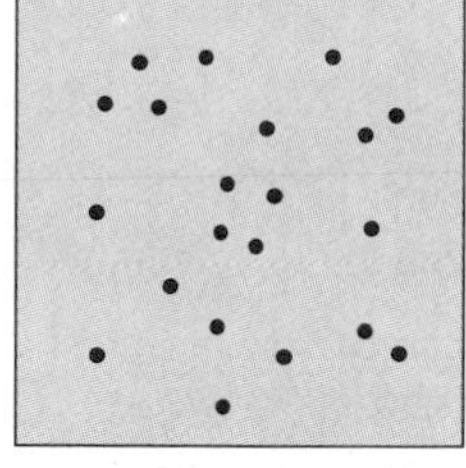

(c) Random

FIGURE 47.2

Patterns of dispersion within a population's geographical range. Individuals within a population frequently exhibit either clumped or uniform distribution within their geographical range, but random distribution is rare. In populations with a clumped distribution, the individuals *within* each clump also show a pattern of dispersion, as do the clumps themselves. (**a**) Butterfly fish, like many fish, are often found clumped in schools. Schooling may increase the hydrodynamic efficiency of swimming, reduce predation risks, and increase feeding efficiency. (**b**) The uniform spacing of these king penguins on South Georgia Island reflects their maintenance of very small breeding territories. (**c**) Trees of the same species are often randomly distributed in tropical rain forests, but this pattern of dispersion is rare in nature.

up individual territories for feeding, breeding, nesting, or resting (FIGURE 47.2b). Territorial behavior is discussed later in this chapter and in Chapter 50 with other concepts of behavioral ecology.

Random spacing occurs in the absence of strong attractions or repulsions among individuals of a population; the position of each individual is independent of other individuals. For example, forest trees are sometimes randomly distributed (FIGURE 47.2c). Overall, however, random patterns are not very common in nature; most populations show at least a tendency toward either clumping or uniform distribution.

We have been looking at local dispersion patterns within populations, but populations within a species also show dispersion patterns, often concentrating in clusters within a species' range. For example, populations of cattails are not evenly distributed throughout the species' range, but are clustered in areas along rivers and lakes and in wetlands. The factors that influence the distribution of a species over its range are the subject of **biogeography.**

METHODS: A MARK-RECAPTURE ESTIMATE OF POPULATION SIZE

Traps are placed within the boundaries of the population being studied, and captured animals are marked with tags, collars, bands, or spots of dye and then released. After a few days or a few weeks, enough time for the marked animals to mix randomly with unmarked members of the population, traps are set again. The proportion of marked to unmarked animals that are captured during the second trapping gives an estimate of the size of the entire population. If there have been no births, deaths, immigration, or emigration, the following simple equation can be used to estimate the population size:

$$N = \frac{\text{Number marked} \times \text{Total catch second time}}{\text{Number of marked recaptures}}$$

For example, suppose that 50 sanderlings are captured in mist net traps, marked with leg bands (FIGURE a), and released. Two weeks later, 100 sanderlings are captured (FIGURE b). If 10 of this second catch are marked birds that have been recaptured, we would estimate that 10% of the total sanderling population is marked. Since 50 birds were originally marked, we would then estimate that the entire population consists of about 500 birds. This method assumes each marked individual has the same probability of being trapped as each unmarked individual. This is not always a safe assumption, however; an animal that has been trapped once, for instance, may be wary of the traps later.

(a)

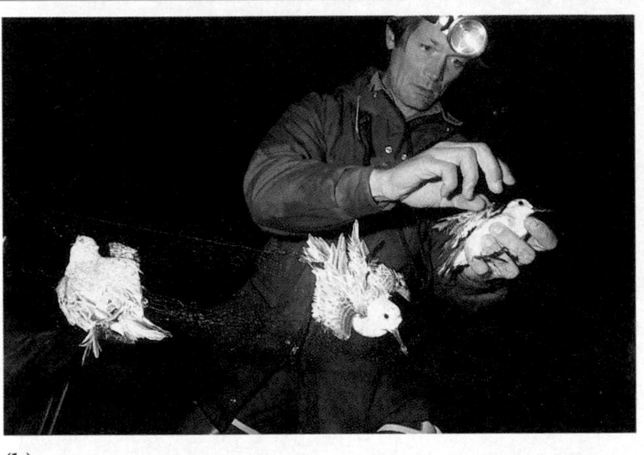
(b)

Demography is the study of factors that affect birth and death rates in a population

Changes in population size reflect the relative rates of processes that add individuals to the population and those that eliminate individuals from it. Additions occur through births (which we will define here to include all forms of reproduction) and immigration, the influx of new individuals from other areas. Opposing these additions are mortality (death) and emigration, the movement of individuals out of a population. Our focus in this chapter is primarily on factors that influence birth rates and death rates.

The study of the vital statistics that affect population size is called **demography.** Birth and death rates usually vary among subgroups within a population, depending in particular on age and sex. It follows that future population size will be determined partly by the existing age structure and sex ratio, two of the most important demographic factors.

Age Structure and Sex Ratio

Many organisms exhibit overlapping generations, or the coexistence of individuals from more than one generation. Only organisms such as annual plants and animals, such as many insects, in which adults all reproduce at about the same time and then die, do not have overlapping generations. The coexistence of generations gives most populations an **age structure,** which is the relative number of individuals of each age. Demographers often use age pyramids to show the age structure of a population (see FIGURE 47.22). A population's structure is very important in determining the rate at which it is growing.

Every age group has a characteristic birth and death rate. **Birth rate,** or **fecundity,** the number of offspring produced during a certain amount of time, is often greatest for individuals of intermediate age. In humans, for example, the birth rate is highest among 20-year-old women. The **death rate** is highest in the first year and, of course, in old age. In many species, juveniles and old individuals are generally more likely to die than individuals of intermediate age, who have the optimum combination of vigor and the ability to find food and avoid predators that comes with maturity. Therefore, a population with a large percentage of individuals of prime reproductive (or slightly younger) age will grow proportionately faster than a population that has an age structure skewed toward older individuals. The implications of such a pattern for human population growth will be discussed later in the chapter.

An important demographic feature related to age structure is **generation time,** the average span between the birth of individuals and the birth of their offspring. In general, generation time is strongly related to body size over a broad range of organisms (FIGURE 47.3). Other factors being equal, a shorter generation time will result in faster population growth (assuming, of course, that the overall birth rate is greater than the death rate). This is simply because the increases in population size attributed to births accumulate more rapidly when individuals reach sexual maturity in a shorter period of time.

The **sex ratio,** the proportion of individuals of each sex, is another important demographic statistic that affects population growth. The number of females is usually directly related to the number of births that can be expected, but the number of males may be less significant because in many species, a single male can mate with several females. In herds of elk, for example, there are fewer males of reproductive age than females, but this has no significant effect on the number of births in the overall population, because each male guards a "harem" of females with which he mates. In many bird species, by contrast, individuals form monogamous pairbonds; any significant reduction in males would be more likely to affect the population's birth rate. Wildlife management often reflects these demographic considerations. For example, deer-hunting regulations are usually more liberal regarding the killing of bucks than does because each buck typically mates with many does.

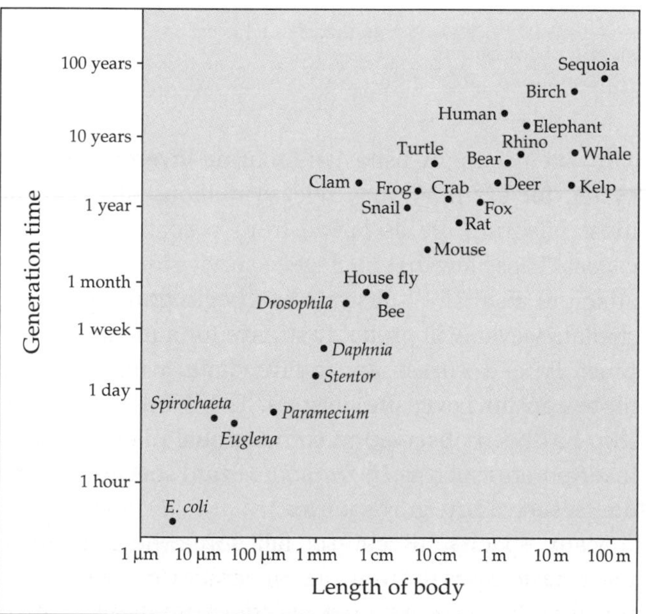

FIGURE 47.3
Generation time and body size. Small organisms generally have short generation times, achieving reproductive maturity quickly. Generation time increases with body size because larger organisms generally take longer to reach the size at which they can reproduce.

Life Tables and Survivorship Curves

About a century ago, when life insurance first became available, insurance companies needed to determine how much longer, on average, an individual of a given age could be expected to live. The result was the development of mortality summaries in what are called, ironically, **life tables.** Population ecologists have adapted this approach for nonhuman populations. While insurance companies are most interested in the chances of survival at a given age, population ecologists often want to know how a population is changing in size, which requires knowing birth rates as well as death rates.

One way to construct a life table is to follow the fate of a **cohort,** a group of individuals of the same age, from birth until all are dead. The table is constructed from the number of individuals that die and the number of offspring born to each member of each age group during the defined time period. Obviously, this approach can be used for only a very limited number of short-lived species. Fortunately, a cross-sectional life table can also be constructed by knowing age-specific mortality and birth rates in a population during a specified period of time.

TABLE 47.1 is a life table for great tits, common and well-studied small birds in Europe. You can gain a great deal of information by looking at age-specific mortality and birth rates. The third, fourth, and fifth columns are different ways of showing how mortality varies with age. The third column shows the proportion of breeding birds in a cohort that are still alive at a given age. The fourth and fifth columns show the number and proportions of birds that died during a given age. For example, we can see that, while fewer than 20% of the original females are alive at age 3, a 3-year-old female has about a 56% chance of surviving her next year. We can also see that mortality rates are relatively constant through the middle

TABLE 47.1

Life Table for Great Tits *(Parus major)* in Scotland*

	AGE	NUMBER ALIVE AT BEGINNING OF YEAR	PROPORTION OF COHORT SURVIVING TO BEGINNING OF YEAR	NUMBER OF DEATHS DURING YEAR	PORPORTION OF COHORT DYING DURING YEAR	NUMBER OF SUCCESSFUL FLEDGLINGS PER INDIVIDUAL
Females	1	1000	1.000	613	0.613	0.359
	2	387	0.387	216	0.558	0.370
	3	171	0.171	95	0.556	0.401
	4	76	0.076	39	0.513	0.518
	5	37	0.037	23	0.622	0.328
	6	14	0.014	10	0.714	0.154
	7	4	0.004	3	0.750	0.000
	8	1	0.001	—	—	0.000
Males	1	1000	1.000	575	0.575	0.326
	2	425	0.425	212	0.499	0.392
	3	213	0.213	104	0.488	0.425
	4	109	0.109	65	0.596	0.580
	5	44	0.044	21	0.477	0.293
	6	23	0.023	15	0.652	0.383
	7	8	0.008	6	0.750	0.643
	8	2	0.002	—	—	0.000

*The original study included more than 1000 birds of each sex, but we have started with 1000 in each cohort to make the table easier to follow.

Source: Adapted from McCleery, R. H., and C. M. Perrins. "Lifetime Reproduction Success of the Great Tit, *Parus major.* In *Reproductive Success: Studies of Individual Variation in Contrasting Breeding Systems,* ed. T. H. Clutton-Brock. Chicago: University of Chicago Press, 1988.

of the birds' lives, but that they are higher in very young and very old individuals. The birth rate, shown in the last column, is highest for 4-year-old females, and lowest for very young and very old females.

A graphic way of representing some of the data in a life table is to draw a **survivorship curve,** a plot of the numbers in a cohort still alive at each age (FIGURE 47.4). Survivorship curves can be classified into three general types. A Type I curve is relatively flat at the start, reflecting low death rates during early and middle life, dropping steeply as death rates increase among older age groups. Humans and many other large mammals that produce relatively few offspring but provide them with good care often exhibit this kind of curve. In contrast, a Type III curve drops sharply at the left of the graph, reflecting very high death rates for the young, but then flattens out as death rates decline for those few individuals that have survived to a certain critical age. This type of curve is usually associated with organisms that produce very large numbers of offspring but provide little or no care, such as many fishes and marine invertebrates. An oyster, for example, may release millions of eggs, but most offspring die as larvae from predation or other causes. Those few that manage to survive long enough to attach to a suitable substrate and begin growing a hard shell, however, will probably survive for a relatively long time. Type II curves are intermediate, with mortality more constant over the lifespan. This kind of survivorship has been observed in some annual plants, various invertebrates such as *Hydra,* some lizard species, and rodents, such as the gray squirrel.

Many species, of course, fall somewhere between these basic types of survivorship or show more complex patterns. In birds, for example, mortality is often high among the youngest individuals (as in a Type III curve) but fairly constant among adults (as in a Type II curve); see the mortality rates in TABLE 47.1. Some invertebrates, such as crabs, may show a "stair-stepped" curve, with brief periods of increased mortality during molts (caused by physiological problems or greater vulnerability to pre-

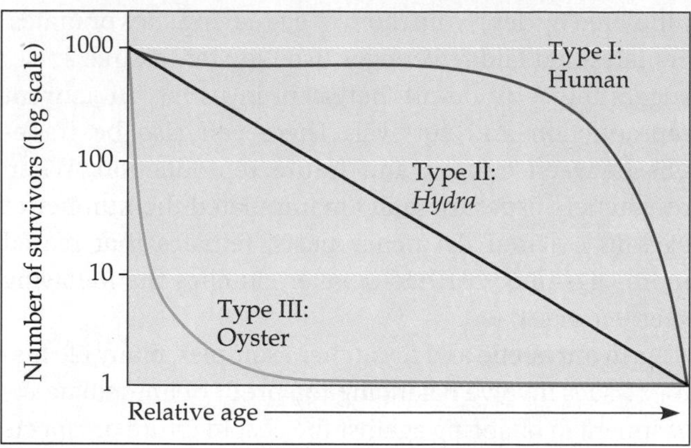

FIGURE 47.4

Survivorship curves. These three curves illustrate idealized patterns of survivorship in different kinds of organisms. As an example of a Type I curve, humans in developed countries experience high survival rates until old age, except during infancy. At the opposite extreme are Type III curves for organisms like oysters, which experience very high mortality as larvae but decreased mortality later in life. Type II survivorship curves are intermediate between the other two types, and result when a constant proportion of individuals die at each age. Notice that the *y* axis is logarithmic and that the *x* axis is on a relative scale, so that species with widely varying lifespans can be compared on the same graph.

dation), followed by periods of lower mortality (when the exoskeleton is hard).

Survivorship is an important factor in the changes in population size over time. Let's now consider some additional characteristics that influence population dynamics.

The traits that affect an organism's schedule of reproduction and death make up its life history

Variation in Life Histories

In Chapter 46, we looked at some of the ways the responses of organisms to environmental variation increase their chances of survival. However, natural selection does not act only on traits that increase survival; organisms that survive a long time but do not reproduce are not very "fit" in the Darwinian sense. In many cases, there are trade-offs between survival and traits such as clutch size (number of offspring per reproductive episode), frequency of reproduction, and investment in parental care. The traits that affect an organism's schedule of reproduction and death make up its **life history.** Of course, a particular life history pattern, like most characteristics of an organism, is the result of natural selection operating over evolutionary time. Life history traits help determine how populations grow.

Because of varying pressures of natural selection, life histories are diverse. Pacific salmon, for example, hatch in the headwaters of a stream, then migrate to open ocean, where they require several years to mature. They eventually return to freshwater streams to spawn, producing millions of small eggs in a single reproductive opportunity, and then they die. In contrast, some lizards produce only a few large eggs during their second year, then repeat the reproductive act annually for several years. The life histories of plants are just as variable. Some species of oaks do not reproduce until the tree is 20 years old, but then produce vast numbers of large seeds each year for a century or more. Annual desert wildflowers generally germinate, grow, produce many small seeds, and then die, all in the span of a month after spring rains. Complicating matters further, important characteristics of life history may vary significantly among populations of a single species, or even among individuals in the same population.

Despite such wide variation in life history traits, there are some patterns in the way in which they vary. Life histories often vary in parallel with environmental factors. Some of the earliest work on life histories resulted from studies in the 1940s by British ornithologist David Lack. Building on previous research, Lack showed that songbirds in the tropics lay fewer eggs than their counterparts at higher latitudes. He suggested that the evolution of clutch size reflected the number of young that parents could successfully feed; in other words, clutch size is an adaptation to food supply. He further suggested that because days are longer during offspring-rearing seasons at high latitudes, temperate birds are able to gather more food than birds in the tropics, where day length is approximately 12 hours all year. Though other researchers have suggested different explanations for this pattern, the pattern of clutch size varying in parallel with latitude is found in other taxa as well; tropical mammals, lizards, and even insects tend to produce fewer eggs than their temperate region counterparts.

Another pattern in life history traits is that they often vary with respect to each other. For example, among birds, fecundity and mortality tend to vary in close association (FIGURE 47.5). At one end of the scale, albatrosses have less than a 5% chance of dying from one breeding season to the next, and only produce, on average, a single surviving offspring every 5 years (about 0.2 offspring per year). At the other extreme are tree sparrows, which have more than a 50% chance of dying from one breeding season to another and produce an average of six fledglings each year. Other traits, such as delayed maturity and high parental investment in each offspring, tend to be correlated with low fecundity and low mortality.

Allocation of Limited Resources

Darwinian fitness is measured not by how many offspring are produced but by how many survive to produce

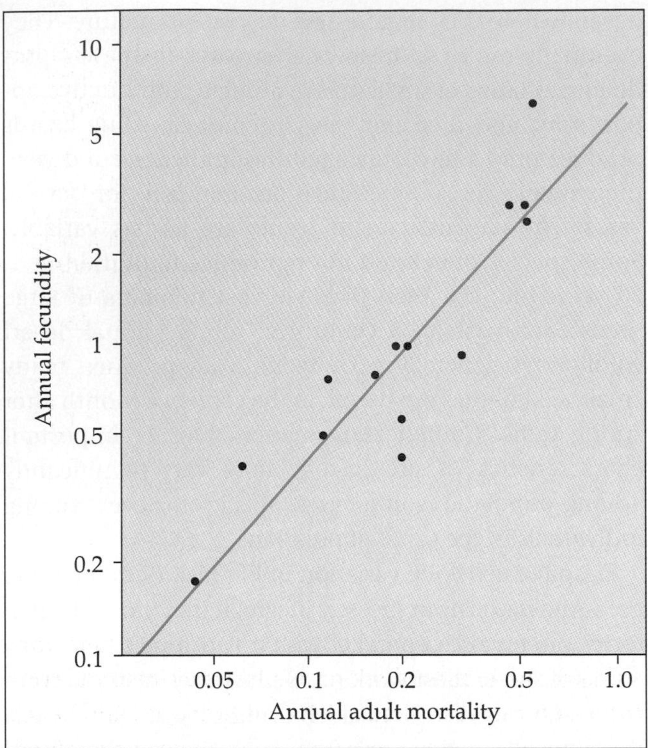

FIGURE 47.5
The relationship between adult mortality and annual fecundity in birds. Birds that have a high probability of dying during any given year usually raise more offspring each year than those with a low probability of dying. The wandering albatross has the lowest fecundity and annual mortality, while the tree sparrow has the highest.

their own offspring: Heritable characteristics of life history that result in the most reproductively successful descendants will become more common within the population. If we were to construct a hypothetical life history that would yield the greatest lifetime reproductive output, we might imagine a population of individuals that begin reproducing at an early age, have large clutch sizes, and reproduce many times in a lifetime. However, natural selection cannot maximize all these variables simultaneously, because organisms have a finite energy budget that mandates trade-offs. For example, the production of many offspring with little chance of survival may result in fewer descendants than the production of a few well-cared-for offspring that can compete vigorously for limited resources in an already dense population.

The life histories we observe in organisms represent a resolution of several conflicting demands. An important part of the study of life histories has been understanding the relationship between limited resources and competing functions: Time, energy, and nutrients that are used for one thing cannot be used for something else. Several experiments have demonstrated such a trade-off. In one study, the fecundity of female bruchid beetles was manipulated by depriving them of egg-laying sites or mates. Females that laid fewer eggs lived longer (FIGURE 47.6a), suggesting a trade-off between investing in current reproduction and survival. There can also be trade-offs between current and future reproduction. When researchers experimentally manipulated the number of eggs in collared flycatcher nests, females that reared more eggs one year had smaller clutches the following year (FIGURE 47.6b).

As in our beetle and flycatcher examples, many life history issues involve balancing the profit of immediate investment in offspring against the cost to future prospects of reproduction. These issues can be phrased in terms of three basic questions: How often should an organism breed? When should it begin to reproduce? How many offspring should it produce during each reproductive episode? The way each population resolves these questions results in the integrated life history patterns we see in nature. We will look more closely at each of these choices, but first it is important to clarify our use of the word "choice." Individual organisms rarely choose when to breed and how many offspring to have. (Humans are an important exception we will consider later in the chapter.) Life history traits are evolutionary outcomes that are reflected in the development and physiology of an organism. Organisms generally breed as soon as they reach sexual maturity, and although the environment may influence when this occurs, the organisms do not consciously choose when to become mature. Similarly, the number of offspring produced during a given reproductive episode is associated with the number of gametes fertilized, not the result of a conscious decision.

Number of Reproductive Episodes Per Lifetime

Some plants and animals invest most of their energy in growth and development, expend this energy in a single large reproductive effort, and then die. Most insects have this type of life history, called **semelparity** (L. *semel,* "once," and *parito,* "to beget"), as do some species of salmon, annual plants, and some perennial plants such as bamboos and century plants. Other organisms produce fewer offspring at a time over a span of many seasons, a life history adaptation called **iteroparity** (L. *itero,* "to repeat"). The relative advantage of each "strategy" can be thought of in terms of a trade-off between fecundity and survival probability. Multiple breeding episodes require that an organism allocate some of its resources to survival. For example, perennial plants invest more in their roots, and also in the formation of freeze-resistant or drought-resistant buds, than do plants that live only a single season. Of course, all the resources not allocated to reproduction are wasted if the organism happens to die before reproducing again.

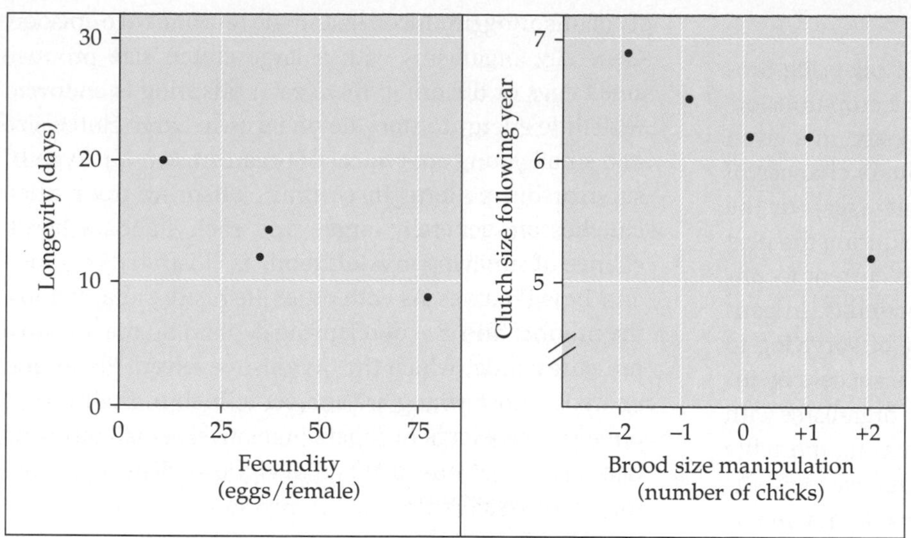

(a) Bruchid beetles

(b) Collared flycatchers

FIGURE 47.6

Effects of current reproductive effort on future reproductive success. (**a**) When the fecundity of female bruchid beetles is manipulated by denying them access to males or egg-laying sites, there is a trade-off between adult longevity and fecundity. (**b**) When the clutch size of collared flycatchers is manipulated by removing or adding one or two eggs, there is a direct trade-off between future and current fecundity. In this study, there was no effect of current fecundity on adult survival.

Under what circumstances do we expect semelparity and iteroparity to evolve? Population ecologists have developed mathematical models to determine the relative payoffs of each, and have shown that the relevant considerations are the probability of survival of both the adult and the immature individual. Semelparity is expected when the cost to parents of staying alive between broods is great, or if there is a large trade-off between fecundity and survival. Iteroparity is expected if individuals survive well once they are established, but immature individuals are unlikely to survive. Thus, in the harsh climates of the desert, most plants live only a single season and put all their energy into a single reproductive effort. In the tropics, where competition and predation make seedling establishment difficult, but where plants can usually live a long time once they are established, there are more iteroparous plants.

Semelparity is rarely found in plants and animals that live for longer than one or two years; once organisms have invested the resources necessary for survival between growing seasons, reproduction every year seems most successful. However, some organisms live for several seasons and then invest all their energy into a single immense reproductive effort—often called "big-bang" reproduction. The agave, or century plant, grows in arid climates with sparse and unpredictable rainfall. Agaves grow vegetatively for several years, then send up a large flowering stalk, produce seeds, and die (FIGURE 47.7). The shallow roots of agaves catch water after rain showers but are dry during droughts. This unpredictable water supply may prevent seed production or seedling establishment for several years at a time. By growing and storing nutrients until an unusually wet year and then putting all their resources into reproduction, the agave's big-bang reproduction is a life history adaptation to severe climate.

Pacific salmon also exhibit big-bang reproduction. During their single reproductive effort, female salmon convert a large portion of their body tissue to eggs, and males to sperm production and mating. The huge cost of migrating upriver to reach their spawning grounds may make it advantageous to make the trip only a single time, and salmon expend so many resources in this effort that it kills them.

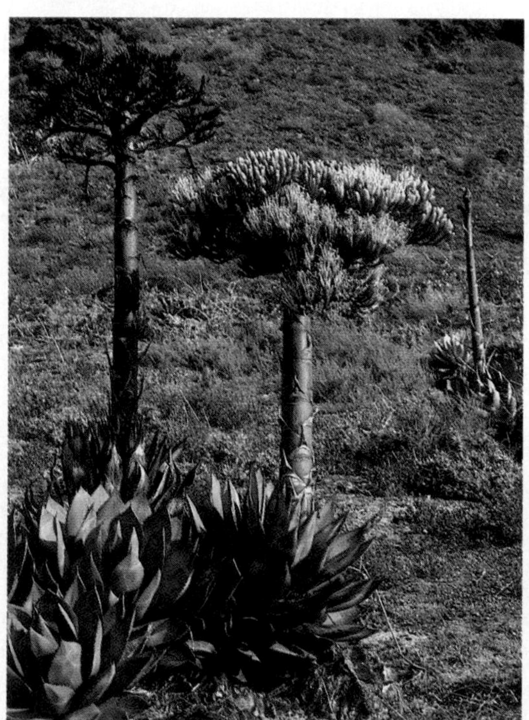

FIGURE 47.7

Big-bang reproduction. Agaves, or century plants, grow without reproducing for several years and then produce a gigantic flowering stalk and many seeds. After this one-time reproductive effort, the plant dies.

Clutch Size

Clutch size, remember, is the number of offspring produced at each reproductive episode. For iteroparous organisms, the total amount of energy invested in a given breeding season could affect an individual's chances of surviving to reproduce again (see FIGURE 47.6a), or the number of offspring it is able to produce during the next season (see FIGURE 47.6b). The trade-off is current fecundity against adult survival and future fecundity. In general, organisms with a lower probability of surviving to another year may maximize reproductive success by investing more in their current effort. Thus, organisms with high losses to predation or high overwintering mortality tend to invest more in a single reproductive episode. Organisms with potentially long lifespans do not generally increase current fecundity enough to jeopardize future reproduction. In some cases, clutch size can vary seasonally within a single population (FIGURE 47.8).

There is also a trade-off between the number and quality of offspring produced in a single reproductive episode. Generally, organisms with a large clutch size produce small eggs or offspring; thus, each offspring is endowed with little energy to start life on its own. Large clutch size and small young are typical of organisms with a Type III survivorship pattern. In contrast, offspring from small clutches are generally larger, and each stands a better chance of surviving to adulthood, as illustrated by Type I and Type II curves. As with other life history adaptations, the number and size of offspring depend on the selective pressures under which the organism evolved. Plants and animals whose young are subject to high mortality rates often produce large numbers of relatively small offspring (FIGURE 47.9a). Thus, plants that colonize disturbed environments usually produce many small seeds, many of which will not get to a suitable environment. Small size might actually benefit such seeds if it enables them to be carried long distances. Birds such as quail and mammals such as rabbits and mice that suffer high predation rates also produce large numbers of small offspring.

In some cases, extra investment on the part of the parent greatly increases the offspring's chance of survival. Oak, walnut, and coconut trees all have large seeds with a store of energy that the seedlings can use to get established (FIGURE 47.9b). In animals, parental investment in offspring does not always end with incubation or gestation. Primates generally have only one or two offspring at a time. Parental care and an extended period of learning in the first several years of life are very important to offspring fitness in these mammals.

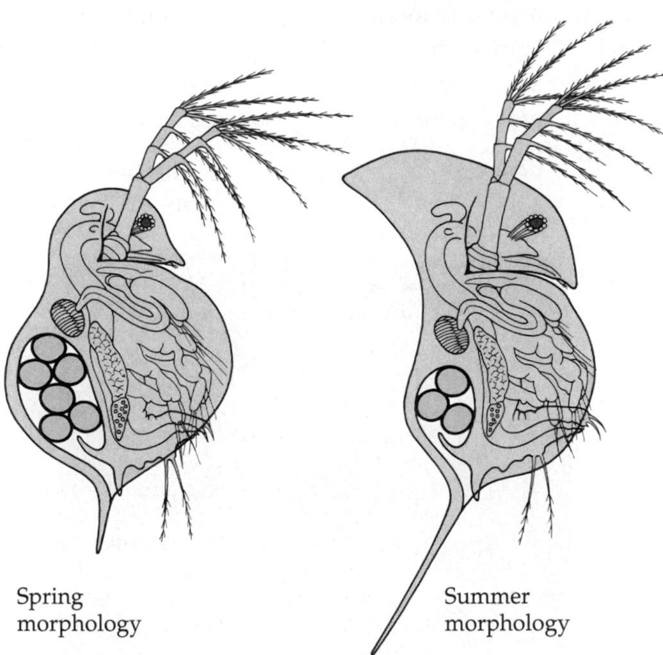

FIGURE 47.8
Seasonal variation in life history caused by predation. The freshwater crustacean *Daphnia retrocurva* shows marked seasonal variation in morphology and clutch size. In spring, the phytoplankton eaten by *Daphnia* are abundant, and predators that eat *Daphnia* are scarce. Under these conditions, individuals develop into a rounder form with a large brood chamber containing six eggs. In summer, however, other plankton that feed on *Daphnia* are abundant. *Daphnia* developing at this time of year have large "helmets" and long tail spines, presumably making them more difficult for predators to consume. These morphological changes use energy reserves and compress the brood chamber so that only half as many eggs can be carried. Natural selection has apparently sacrificed the larger clutch size in favor of a better chance of producing at least some eggs and surviving to reproduce again.

Age at First Reproduction

For organisms that reproduce repeatedly during their lifespan, the timing of first reproduction can have a large effect on the female's lifetime reproductive output. Again, the balance is between current reproduction and survival plus future reproduction. Organisms that delay first reproduction avoid costs that can include courtship, nest building, gamete production, and migrating to breeding areas. In many animals, life experience might reduce some risks of breeding or increase fecundity. Also, many organisms grow throughout their lives, and these size increases often increase fecundity. On the other hand, a female that delays reproduction must invest more energy in maintenance and growth. Although such an adaptation might increase her potential for future reproduction, she may die before producing any offspring at all, thus reducing her fitness to zero.

Mathematical models of these phenomena suggest that if older females produce much larger clutches than younger ones, and if there is a good chance of surviving to an advanced age, then female lifetime reproductive output is maximized when first reproduction is delayed.

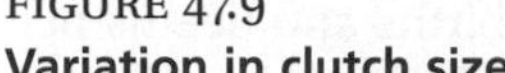

(a) A plant with a large clutch size

(b) A plant with a moderate clutch size

FIGURE 47.9

Variation in clutch size in plants.
(**a**) Most annual plants, such as this dandelion, grow quickly and produce a large number of seeds. Although most of the seeds will not produce mature plants, their large number and ability to disperse to new habitats ensure that at least some will grow and eventually produce seeds themselves. (**b**) Some plants, such as this coconut palm, produce a moderate number of very large seeds. The large endosperm provides nutrients for the embryo (a plant's version of parental care), an adaptation that helps ensure the success of a relatively large fraction of offspring. Animal species exhibit similar trade-offs between clutch size and the amount of nutrients provided to each offspring.

In general, this is what we see in nature. Among birds, for example, age at sexual maturity varies directly with the annual survival rates of adults (FIGURE 47.10). In many cases, birds that live a long time gain experience during a relatively long period as juveniles, which makes them more successful at rearing offspring. Birds that have a low probability of surviving from one year to the next generally start reproducing as soon as possible.

Now that we have analyzed some patterns that underlie diverse life histories, let's examine the effects of these phenomena on the growth of populations and the factors that regulate population size.

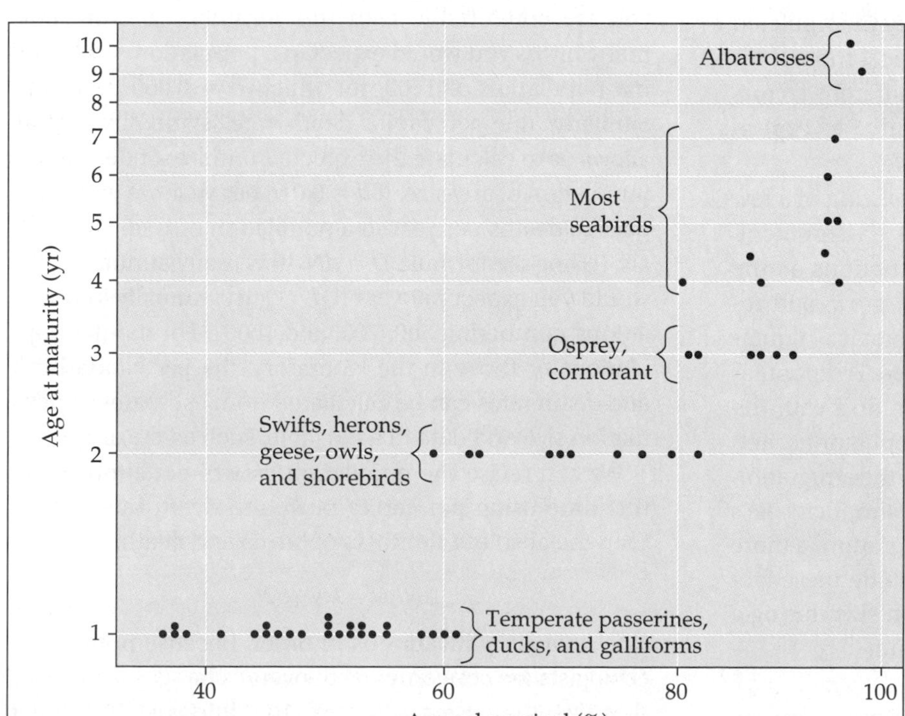

FIGURE 47.10

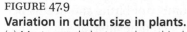

The relationship between age at maturity and annual adult survival in birds. Generally, birds that have a good chance of surviving from one year to the next delay reproduction longer than birds with low survival probabilities. The advantage of delayed reproduction in birds may be experience that makes them more successful at rearing offspring. In some organisms, delayed reproduction might allow larger clutches.

A mathematical model for exponential growth describes an idealized population in an unlimited environment

To begin to understand the potential for population increase, consider a single bacterium that can reproduce by fission every 20 minutes under ideal laboratory conditions. At the end of this time, there would be two bacteria, four after 40 minutes, and so on. If this continued for only a day and a half—a mere 36 hours—there would be bacteria enough to form a layer a foot deep over the entire Earth. At the other life history extreme, elephants may produce only six young in a 100-year lifespan. Darwin calculated that it would take only 750 years for a single mating pair of elephants to produce a population of 19 million. Obviously, indefinite increase does not occur either in the laboratory or in nature. A population that begins at a low level in a favorable environment may increase rapidly for a while, but eventually the numbers must, as a result of limited resources and other factors, stop growing.

As you learned in Chapter 46, the answers to many ecological questions depend on a combination of observation, experimentation, and mathematical modeling. The two major forces affecting population growth—birth rates and death rates—can be measured (or observed) in many populations and used to predict how the populations will change in size over time. Small organisms can be studied in the laboratory to determine how various factors affect their population growth rates, and certain natural populations can be experimentally manipulated to answer the same questions. Mathematical models for testing hypotheses about the effects of different factors on population growth can be an alternative to experiments that would be difficult or impossible.

Imagine a hypothetical population consisting of a few individuals living in an ideal, unlimited environment. Under these conditions, there are no restrictions on the abilities of individuals to harvest energy, grow, and reproduce, aside from the inherent physiological limitations that are the result of their life histories. The population will increase in size with every birth and with the immigration of individuals from other populations, and decrease in size with every death and with the emigration of individuals out of the population. For simplicity, let's ignore the effects of immigration and emigration (a more complex formulation would certainly include these factors). We can define change in population size during a fixed time interval with this verbal equation:

$$\begin{bmatrix}\text{Change in population}\\ \text{size during time interval}\end{bmatrix} = \begin{bmatrix}\text{Births during}\\ \text{time interval}\end{bmatrix} - \begin{bmatrix}\text{Deaths during}\\ \text{time interval}\end{bmatrix}$$

We use mathematical models as a simple way of generalizing the ideas that are expressed in words. If we let N = population size and t = time, we can also specify that ΔN = change in population size and Δt = the time interval (appropriate to the lifespan and generation time of the species) over which we are evaluating population growth. (The Greek letter delta, Δ, indicates change, such as time elapsing.) The verbal equation presented at the bottom of the left column can be rewritten as

$$\Delta N / \Delta t = B - D$$

in which B = the absolute number of births in the population during the time interval and D = the absolute number of deaths.

All populations differ in size, and we want to create a general mathematical model that can be applied to any population. We therefore convert the simple model presented above into one in which births and deaths are expressed as the average number of births and deaths per individual during the specified time interval. Let b = the annual per capita birth rate, or the number of offspring produced per year by an average member of the population. If, for example, a population of 1000 individuals experiences 34 births per year, the per capita birth rate is $^{34}/_{1000}$, or 0.034. If we know the per capita birth and death rates, we can calculate the expected number of births and deaths in a population of any size. For example, if we know that the annual per capita birth rate is 0.034 and the population size is 500, we could use the formula $B = bN$ to calculate the absolute number of births expected in that population: 17 (0.034×500) per year. (To ensure that you are comfortable with this formula, calculate how many births you would expect in a population of 700 and ina population of 1700, for which $b = 0.050$ per year.) Similarly, the per capita death rate, symbolized as d, allows us to calculate the expected number of deaths in a population of any size. If $d = 0.016$ per year, we would expect 16 deaths per year in a population of 1000 individuals. (Using the formula $D = dN$, how many annual deaths would you expect per year if $d = 0.010$ annually in populations numbering 500, 700, and 1700?) For natural populations or those in the laboratory, the per capita birth and death rates can be calculated from estimates of population size and data in a life table, such as TABLE 47.1.

We can revise the population growth equation again, this time using per capita birth and death rates rather than the absolute numbers of births and deaths:

$$\Delta N / \Delta t = bN - dN$$

One final simplification is in order. Because population ecologists are concerned with overall changes in population size, they use r to identify the difference in the per capita birth and death rates ($r = b - d$). This value, the per capita population growth rate, tells whether a population is actually growing (positive value of r) or declining in size (negative value of r). **Zero population growth (ZPG)** occurs when the per capita birth and death rates are equal and r equals 0. Note that births and deaths still occur in the population, but they balance each other exactly. (The relevance of ZPG for the human population and the factors preventing the human population from leveling off are discussed later in this chapter.)

Using the population growth rate, we rewrite the equation as

$$\Delta N / \Delta t = rN$$

Finally, most ecologists use the notation of differential calculus to express population growth in terms of instantaneous growth rates:

$$dN/dt = rN$$

If you have not yet studied calculus, don't be intimidated by the form of the last equation; it is essentially the same as the previous one, except that the time intervals are very small.

We started this section by describing a population living under ideal conditions. In such a situation, the population grows at the fastest rate possible, because all members have access to abundant food and are free to reproduce at their physiological capacity. This maximum population growth rate, called the **intrinsic rate of increase,** is symbolized as r_{max}. Population increase under these conditions is called **exponential population growth:**

$$dN/dt = r_{max}N$$

The size of a population that is growing exponentially increases rapidly, resulting in a J-shaped growth curve when population size is plotted over time (FIGURE 47.11). Although the intrinsic rate of increase is constant as the population grows, the population actually accumulates more new individuals per unit of time when it is large than when it is small—the curve in FIGURE 47.11 gets progressively steeper through time. This occurs because population growth is dependent upon N as well as r_{max}, and larger populations experience more births (and deaths) than small ones growing at the same per capita rate.

We can see in FIGURE 47.11 that a population with a higher intrinsic rate of increase will grow faster than one with a lower rate of increase. The value of r_{max} for a population is influenced by life history features, such as age at the beginning of reproduction, the number of young produced, and how well the young survive. Usually, generation time and r_{max} are inversely related over a broad range of species (FIGURE 47.12).

The J-shaped curve of exponential growth is characteristic of populations that are introduced into a new or unfilled environment, or whose numbers have been drastically reduced by a catastrophic event and are rebounding.

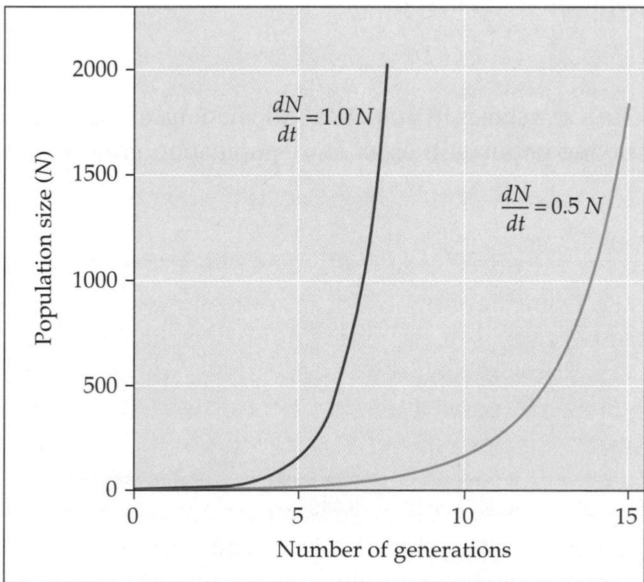

FIGURE 47.11
Population growth predicted by the exponential model. The exponential growth model predicts unlimited population increase under conditions of unlimited resources. This graph compares growth in populations with two different values of r_{max}: 1.0 and 0.5.

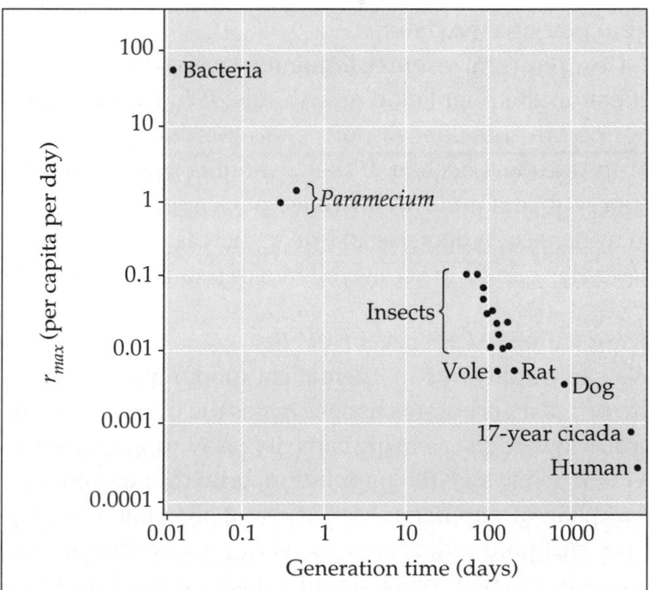

FIGURE 47.12
Generation time and r_{max} (maximum population growth rate). Small organisms that mature quickly tend to have short generation times and high r_{max}. Larger organisms, which take longer to reach sexual maturity, usually have long generation times and low r_{max}. Notice that for comparison here, r_{max} is presented as per capita per day, even for large animals.

A logistic model of population growth incorporates the concept of carrying capacity

The exponential growth model assumes unlimited resources, which is never the case in the real world. Thus, no population—neither bacteria, elephants, nor any other organisms—can grow exponentially indefinitely. As any population grows larger in size, its increased density may influence the ability of individuals to harvest sufficient resources for maintenance, growth, and reproduction. Populations subsist on a finite amount of available resources, and as the population becomes more crowded, each individual has access to an increasingly smaller share. Ultimately, there is a limit to the number of individuals that can occupy a habitat. Ecologists define **carrying capacity** as the maximum stable population size that a particular environment can support over a relatively long period of time. Carrying capacity, symbolized as *K*, is a property of the environment, and it varies over space and time with the abundance of limiting resources. For example, the carrying capacity for songbirds may be high in lush habitats where insects are abundant, but lower in other habitats where food items are less numerous.

Carrying capacity can be determined by factors other than food resources, though energy limitation is perhaps the most common determinant of *K*. Other limiting factors include the availability of specialized nesting sites (as for spotted owls and other hole-nesting organisms), roosting sites (as for some bats), and shelters or refuges from potential predators.

Crowding and resource limitation can have a profound effect on the population growth rate. If individuals cannot obtain sufficient resources to reproduce, per capita birth rates will decline. If they cannot consume enough energy to maintain themselves, per capita death rates may increase. A decrease in *b* or an increase in *d* results in a smaller *r* and a lower overall rate of population growth.

The Logistic Growth Equation

We can modify our mathematical model of population growth to incorporate changes in *r* as the population size grows toward the carrying capacity (as *N* approaches *K*). A model of **logistic population growth** incorporates the effect of population density on *r*, allowing it to vary from r_{max} under ideal conditions to zero as carrying capacity is reached. When a population's size is below the carrying capacity, population growth is rapid according to the logistic model, but as *N* approaches *K*, population growth is slow.

Mathematically, we construct the logistic model by starting with the model of exponential population growth and creating a term that reduces the value of *r* as *N* increases (FIGURE 47.13). If the maximum sustainable population size is *K*, the term $(K - N)$ tells us how many additional individuals the environment can accommodate, and the term $(K - N)/K$ tells us what fraction of *K* is still available for population growth. By multiplying r_{max} by $(K - N)/K$, we reduce the value of *r* as *N* increases:

$$\frac{dN}{dt} = r_{max}N\,\frac{K-N}{K}$$

Thus, the actual growth rate of the population at any population size becomes

$$r_{max}N\,\frac{K-N}{K}$$

TABLE 47.2 shows hypothetical calculations of *r* and *N* at varying population sizes for a population growing ac-

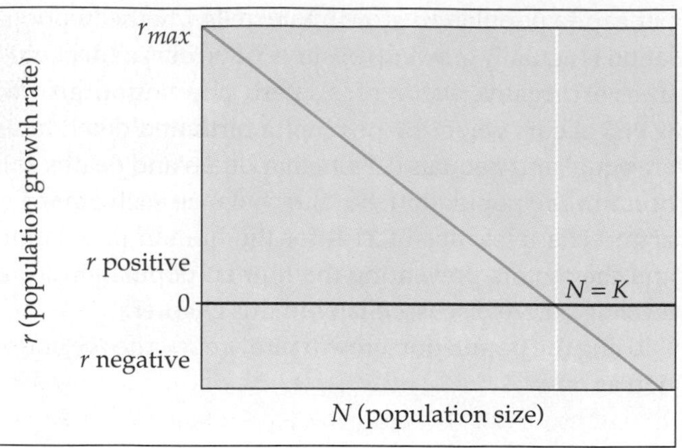

FIGURE 47.13
Reduction of *r* (population growth rate) with increasing *N* (population size). The logistic model of population growth assumes that *r* decreases as *N* increases. When *N* is close to 0, *r* equals r_{max}, and the population grows rapidly. However, as *N* approaches *K*, *r* approaches 0, and population growth slows. If *N* is greater than *K*, *r* is negative, and population size decreases.

TABLE 47.2

Variations in *r* and Δ*N* in a Hypothetical Population Showing Logistic Growth Where *K* Is 1000 and the Intrinsic Rate of Increase Is Constant at 0.05 per Capita per Year*

N	(K – N) /K	r	ΔN
20	0.98	0.049	+ 1
100	0.90	0.045	+ 5
250	0.75	0.038	+ 9
500	0.50	0.025	+13
750	0.25	0.013	+ 9
1000	0.00	0.000	0

*Δ*N* is rounded to the nearest whole number.

cording to the logistic model. Notice that when N is low, the term $(K - N)/K$ is large, and r is close to r_{max}. But when N is large and resources are limiting, the term $(K - N)/K$ will be small, and r is substantially reduced below r_{max}. Zero population growth occurs when the numbers of births and deaths are equal and r equals 0—in this case, when N equals K.

The logistic model of population growth produces a sigmoid (S-shaped) growth curve (FIGURE 47.14) when N is plotted over time. New individuals are added to the population most rapidly at intermediate population sizes when there is not only a breeding population of substantial size, but also lots of available space and other resources in the environment. The population growth rate slows dramatically as N approaches K. Because the rate at which a population grows changes with the density of organisms that are currently in the population, the logistic model is said to be density-dependent.

Notice that we haven't said anything about *how* the growth rate of the population changes as N approaches K. Obviously, either the birth rate (b) must decrease, the death rate (d) must increase, or both. Later in the chapter, we will go into some detail about some of the factors that could affect b and d.

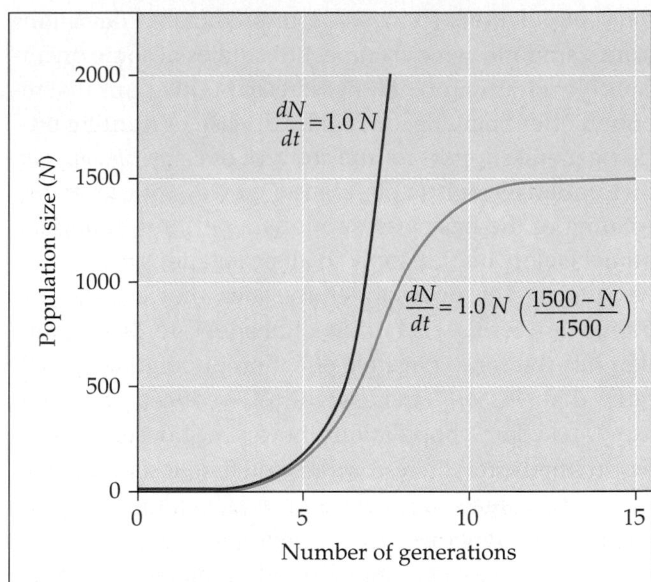

FIGURE 47.14

Population growth predicted by the logistic model. The logistic growth model assumes that there is a maximum population size that the environment can support—the carrying capacity (K). The rate of population growth slows as the population approaches the carrying capacity of the environment. The graph shows logistic growth in a population with an r_{max} of 1.0 and $K = 1500$ individuals. The magenta line illustrates a population growing exponentially with the same r_{max} for comparison.

How Well Does the Logistic Model Fit the Growth of Real Populations?

The growth of laboratory populations of some small animals, such as beetles or crustaceans, and microorganisms, such as paramecia, yeast, and bacteria, fit S-shaped curves fairly well (FIGURE 47.15a and b). However, these experimental populations are grown in a constant environment lacking predators and other species that may compete for resources, idealized conditions that never occur in nature. Even under these laboratory conditions, not all populations stabilize at a clear carrying capacity, and

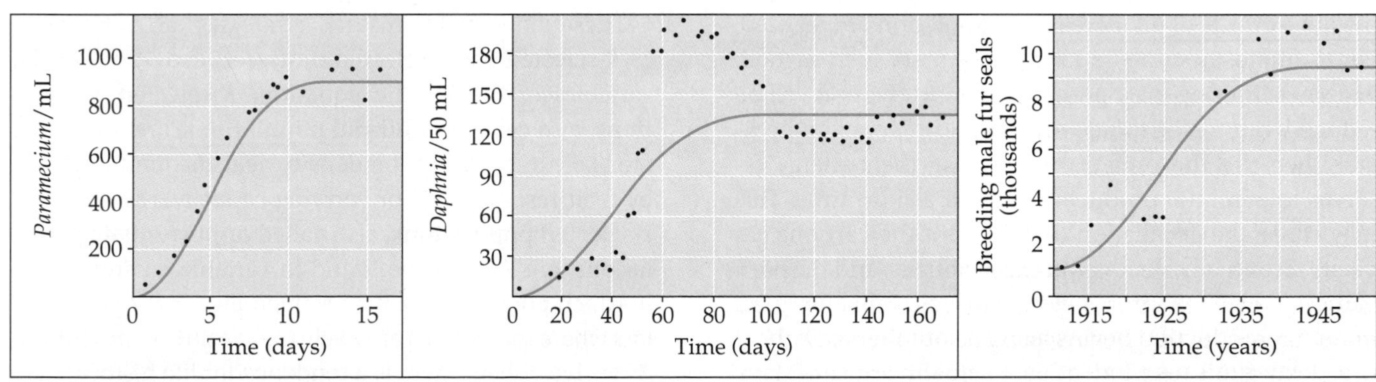

(a) A *Paramecium* population in laboratory culture

(b) A *Daphnia* population in laboratory culture

(c) A fur seal (*Callorhinus ursinus*) population on St. Paul Island, Alaska

FIGURE 47.15

Examples of logistic population growth. In each of these studies, actual data (points on the graph) are compared to an idealized curve (blue) based on the logistic model of population growth. (**a**) The growth of *Paramecium aurelia* in small laboratory cultures closely approximates logistic growth if an experimenter maintains a constant environment by regularly adding food and removing toxic wastes. (**b**) Similarly, the growth of a population of *Daphnia* in a small laboratory culture also shows approximate logistic growth. Notice, however, that this population overshot the carrying capacity of its artificial environment, and then settled back to a relatively stable population size. (**c**) The numbers of male fur seals with "harems" on St. Paul Island, Alaska, were greatly depressed by hunting until 1911. After hunting was banned, the population increased dramatically and now oscillates around an equilibrium number, presumably the island's carrying capacity for this species.

most populations show some unpredictable deviations from a smooth sigmoid curve. But studies of some organisms introduced into new habitats or populations that rebound after being decimated by disease or hunting provide general support for the concept of carrying capacity that underlies logistic population growth (FIGURE 47.15c).

Some of the basic assumptions built into the logistic model clearly do not apply to all populations. For example, the model incorporates the idea that even at low population levels, each individual added to the population has the same negative effect on population growth rate; that is, *any* increase in N reduces the term $(K - N)/K$. Some populations, however, show an *Allee effect* (named after the researcher who first described it), in which individuals may have a more difficult time surviving and reproducing if the population size is too small. For example, a single plant standing alone may suffer from excessive wind but would be protected in a clump of individuals. Some oceanic birds require large numbers at their breeding grounds to provide the necessary social stimulation for reproduction. And conservationists fear that populations of rhinoceri, animals that live solitary lives, may be so small that individuals will not be able to locate mates in the breeding season. In these cases, a greater number of individuals in the population has an enhancing effect, up to a point, on population growth. In addition, when a population is small, there is a greater possibility that chance events will eliminate all individuals, or that inbreeding will lead to a general reduction in fitness.

The logistic model also makes the assumption that populations approach carrying capacity smoothly. In many populations, however, there is a lag time before the negative effects of an increasing population are realized. For example, as some important resource, such as food, becomes limiting for a population, reproduction will be reduced, but the birth rate may not be affected immediately because the organisms may use their energy reserves to continue producing eggs for a short time. This may cause the population to overshoot the carrying capacity. Eventually, deaths will exceed births, and the population may then drop below carrying capacity; even though reproduction begins again as numbers fall, there is a delay until new individuals actually appear. Many populations seem to oscillate about their carrying capacity (see FIGURE 47.19) or to overshoot at least once before attaining a stable size (see FIGURE 47.15b). We will examine some possible reasons for these oscillations later in the chapter.

As you will see in the next section, some populations do not necessarily remain at, or even reach, levels where population density is an important factor. In many insects and other small, quickly reproducing organisms that are sensitive to environmental fluctuations, physical variables such as temperature or moisture reduce the population well before resources become limiting.

Overall, the logistic model is a useful starting point for thinking about how populations grow and for constructing more complex models. Although it fits few, if any, real populations exactly, the logistic model incorporates basic ideas that, with modification, do apply to many populations. And like any good hypothesis, this model has stimulated many experiments and discussions that, whether they support the model or not, lead to a greater understanding of population ecology in general.

Population Growth Models and Life Histories

The logistic model predicts different growth rates for populations under conditions of high and low density, relative to the carrying capacity of the environment. At high densities, each individual has few resources available, and the population is growing slowly, if at all. At low densities, the opposite is true—resources are abundant and the population is growing rapidly. During the late 1960s, population ecologist Martin Cody introduced the concept that different life history adaptations would be favored under these different conditions (TABLE 47.3). He proposed that at high population density, selection favors adaptations that enable organisms to survive and reproduce with few resources. Thus, competitive ability and maximum efficiency of resource utilization are favored in populations that tend to remain at or near their carrying capacity. At low population density, adaptations that promote rapid reproduction, such as increased fecundity and earlier maturity, are selected. High rates of reproduction, regardless of efficiency, are favored in this case.

These different life history "strategies" are referred to as *K*-selected and *r*-selected traits, respectively, after the variables of the logistic equation. ***K*-selected populations,** also called **equilibrial populations,** are those that are likely to be living at a density near the limit imposed by their resources (K, or carrying capacity). By contrast, ***r*-selected populations,** also called **opportunistic populations,** are likely to be found in variable environments in which population densities fluctuate, or in open habitats where individuals are likely to face little competition. As we have seen, there is a tendency for life history traits to vary, often in ways represented in TABLE 47.3, and *r*-selection and *K*-selection occupy an important place in our thinking about life history patterns. However, it has been difficult to demonstrate a direct relationship between population growth rate and specific life history characteristics. Increasingly, ecologists are recognizing that most populations show a mix of the traditional *r*-selected and *K*-selected characteristics; life history evolves in the context of a complex interplay of factors.

TABLE 47.3

Characteristics of Idealized r-Selected (Opportunistic) and K-Selected (Equilibrial) Populations

CHARACTERISTIC	*r*-SELECTED POPULATIONS	*K*-SELECTED POPULATIONS
Homeostatic capability	Limited	Often extensive
Maturation time	Short	Long
Lifespan	Short	Long
Mortality rate	Often high	Usually low
Number of offspring produced per reproductive episode	Many	Few
Number of reproductions per lifetime	Usually one	Often several
Age at first reproduction	Early	Late
Size of offspring or eggs	Small	Large
Parental care	None	Often extensive

Source: Adapted from E. R. Pianka, *Evolutionary Ecology,* 4th ed. New York: Harper & Row, 1987.

Both density-dependent and density-independent factors can affect population growth

The exponential and logistic models predict very different patterns of population growth. The exponential model sets no limit on population increase, whereas the logistic model predicts the regulation of population growth as density increases.

Population ecologists have long debated the most important factors regulating population growth. At one time, one camp emphasized the importance of density-dependent factors in population regulation, and the other emphasized density-independent factors. The current consensus is that the relative importance of these factors differs among species and their specific circumstances, and that often both density-dependent and density-independent factors interact to affect population densities.

Density-Dependent Factors

The major biological implication of the logistic model is that increasing population density reduces the resources available for individual organisms and that resource limitation ultimately limits population growth. Indeed, the logistic model is a model of **intraspecific competition:** the reliance of individuals of the same species on the same limited resources. As the population size increases, the competition becomes more intense, and growth rate (r) declines in proportion to the intensity of competition; the population growth rate is density-dependent. In restricting population growth, **a density-dependent factor** is one that intensifies as the population increases in size. As noted above in our description of the logistic model, density-dependent factors reduce the population growth rate by decreasing reproduction or by increasing mortality in a crowded population. In general, the density-dependent factor that limits a population's growth can be said to determine the carrying capacity, or K, of the environment. Let's look at some examples of how this occurs.

Resource limitation in crowded populations can influence the future size of a population by profoundly reducing reproduction (FIGURE 47.16). Available food supplies

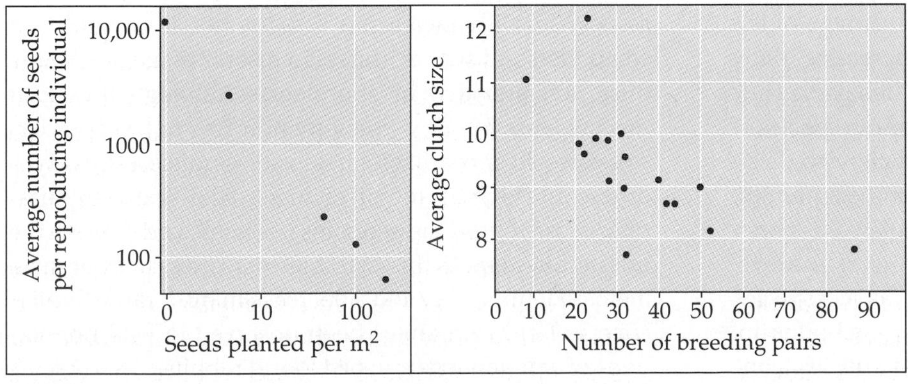

FIGURE 47.16
Decreased fecundity at high population densities. (**a**) The average number of seeds produced by plantain *(Plantago major),* a small herb, decreases with increased sowing density. In this experiment, the germination rate and the proportion of plants that produced any seeds also decrease at higher densities, and mortality rate increases. (**b**) Average clutch size decreases markedly with increasing population density in a woodland population of the great tit *(Parus major).*

often limit the reproductive output of songbirds, for example, and, as bird population density increases in a particular habitat, each female lays fewer eggs. Seed production by plants is similarly affected by crowding. In both of these cases, increasing population density causes decreased nutrient availability, resulting in a lower birth rate. Factors other than nutrients can also limit populations. In many vertebrates and some invertebrates, **territoriality,** the defense of a well-bounded physical space, allows territorial individuals to prevent other individuals from using this space (see FIGURE 47.2b). This behavioral mechanism may reduce intraspecific competition for food and nest sites within the territory, but the space for constructing a territory becomes the resource for which individuals compete. For oceanic birds such as gannets, which nest on rocky islands where they are relatively safe from predators, the limited number of suitable nesting sites allows only a certain number of pairs to nest and reproduce. Up to a certain population size, most birds can find a suitable nest site, but few birds beyond that number breed successfully. Thus, the limiting resource that usually determines *K* for gannets is safe nesting space. As this space is used up, birds that cannot obtain a nesting spot do not reproduce.

Population density also influences the health and survivorship probabilities of plants and animals. Plants grown under crowded conditions tend to be smaller and less robust than those grown at lower density. Small plants are less likely to survive, and those that do survive produce fewer flowers, fruit, and seeds, a phenomenon well recognized by gardeners who thin their flowers and vegetables to produce the best possible yield. Similarly, animals often experience increased mortality at high population densities. In laboratory studies of flour beetles, for example, the percentage of eggs that hatch and survive to adulthood decreases steadily as density increases from moderate to high levels (FIGURE 47.17).

Predation may also be an important density-dependent regulator for some populations if a predator encounters and captures more prey as the population density of the prey increases. This is only a density-dependent effect, however, if a larger percentage of the prey population is taken as its density increases. Many predators, for example, exhibit switching behavior: They begin to concentrate on a particularly common species of prey because it becomes energetically efficient to do so (see the discussion on optimal foraging in Chapter 50). For example, trout may concentrate for a few days on a particular species of insect that is emerging from its aquatic larval stage, then switch as another insect species becomes more abundant. As a prey population builds up, predators may feed preferentially on that species, consuming a higher percentage of individuals; this can cause density-dependent regulation of the prey population.

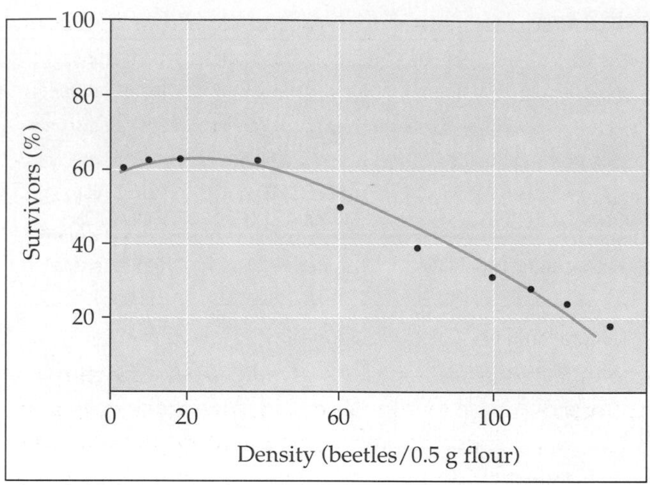

FIGURE 47.17
Decreased vigor and survivorship at high population densities. The percentage of flour beetles *(Tribolium confusum)* surviving from egg stage to adult in a laboratory culture decreases at moderate to high population densities, reducing the numbers of adults in the next generation.

The accumulation of toxic wastes is another component of carrying capacity that can contribute to density-dependent regulation of population size. In laboratory cultures of small microorganisms, for example, metabolic by-products accumulate as the population grows, poisoning the population within this limited, artificial environment. Indeed, ethanol accumulates as a waste product when yeast ferments sugar, and the alcohol content of wine is usually less than 13%, the maximum ethanol concentration that wine-producing yeast cells can tolerate. Some scientists think that waste accumulation could eventually limit human populations.

For some animal species, intrinsic factors, rather than the extrinsic factors just discussed, appear to regulate population size. White-footed mice in a small field enclosure will multiply from a few to a colony of 30 to 40 individuals, but eventually reproduction will decline until the population ceases to grow. This change is clearly associated with increasing density, but it occurs even when food and shelter, the main resources needed by the mice, are provided in abundance. Although the exact mechanisms for this phenomenon are not yet understood, we do know that high densities induce a stress syndrome in which hormonal changes delay sexual maturation, cause reproductive organs to shrink, and depress the immune system. In this case, high densities cause both an increase in mortality and a decrease in birth rates. Similar effects of crowding have been observed in wild populations of woodchucks and Old World rabbits.

We have just considered examples where increased densities cause population growth rates to decline by affecting reproduction, growth, and survivorship in the individuals that make up the populations. Other factors can also affect population growth in ways unrelated to density. In some cases, these density-independent factors can be more important than density-dependent factors. Let's see how.

Density-Independent Factors

The occurrence and severity of **density-independent factors** are unrelated to population size; they affect the same percentage of individuals regardless of population size. The most common and important density-independent factors are related to weather and climate. For example, a freeze in the fall may kill a certain percentage of the insects in a population. Obviously, the timing of the first freeze and just how cold it gets are not affected by the density of the insect population. Some natural populations grow exponentially, not reaching the carrying capacity of their environment, until their numbers are reduced by weather, predators, or another density-independent component of the environment. While no one argues that this is true of many organisms, ecologists debate the relative importance of density-dependent and density-independent factors in regulating populations.

Probably the most often-cited example of density-independent population growth is that of a small Australian insect pest of the genus *Thrips* (FIGURE 47.18). These animals feed on the pollen, leaf, and flower tissues of many plant species; those we will discuss subsist mainly on the pollen of roses and other cultivated plants.

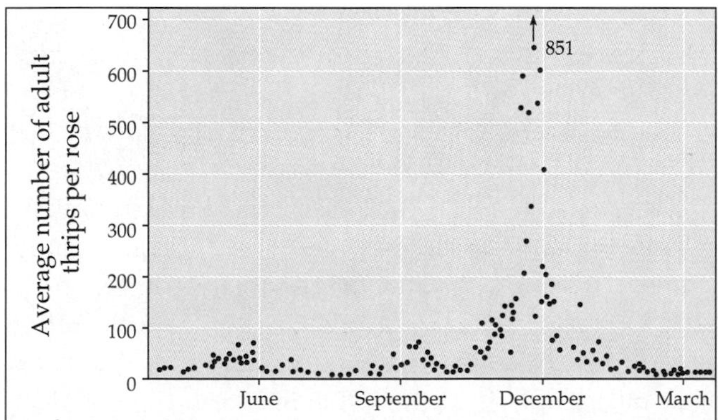

FIGURE 47.18
Density-independent regulation of population size. Populations of the insect genus *Thrips* grow rapidly during spring in the flowers that provide both food and shelter. Before the population reaches carrying capacity, however, numbers are drastically reduced during the dry Australian summer. (In interpreting this graph, remember that summer in the Southern Hemisphere begins in late December.)

The number of flowers available to thrips is correlated with seasonal weather patterns, but there are always some flowers available, and thrips remain active all year. However, during the winter (which begins in June in Australia), cool temperatures lower the development rate and fecundity of the thrips, and many take so long to mature that the flowers in which they live die and fall to the ground, carrying their resident thrips to their deaths. As warm weather arrives in the spring, development rates and fecundity increase, and populations rise to high levels. Population growth is not checked until the heat and dryness of summer cause an increase in adult mortality and the population declines. This occurs well before density-dependent factors become important and before the population shows any evidence of reaching carrying capacity. A few individuals remain in the surviving flowers, and these allow population growth to resume again when favorable conditions return. The population size of many other species, particularly insects and other small organisms, is probably limited at some point primarily by density-independent factors.

Environmental episodes more sporadic than seasonal changes in weather can also affect populations in a density-independent manner. For example, fires and hurricanes may strike often enough in some areas to have a significant impact on some populations. In contrast, volcanic eruptions, though extraordinarily dramatic and destructive, occur so infrequently that they are not very important as general mechanisms of population regulation.

The Interaction of Regulating Factors

Over the long term, many populations remain fairly stable in size and are presumably close to a carrying capacity that is determined by density-dependent factors. Superimposed on this general stability, however, are short-term fluctuations due to density-independent factors. In one case study, researchers monitored the number of European herons, large birds, for 30 years in two different areas of England. The general pattern was the same for both areas: Populations were reasonably stable over the three-decade span, but major declines occurred following unusually cold winters.

In some cases, density-dependent and density-independent factors act together to regulate a population. For example, in very cold, snowy areas, many deer may starve to death during the winter. The severity of this factor is proportional to the season's harshness; colder temperatures increase energy requirements (and therefore the need for food), while deeper snow makes it harder to find food. But the effect is also density-dependent, because each individual in a large population gets a proportionately smaller share of what little food is available.

The relative importance of density-dependent and density-independent controls may also vary seasonally. This complex interaction of factors apparently influences populations of the bobwhite quail. The range of this bird includes southern Wisconsin, where survival is difficult during winter. The number of birds alive at the end of the winter is largely determined by the depth of snow cover, a density-independent factor. If there is little snow, as much as 80% of the population will survive, but with more snow, survivors may decline to only 20%. Whatever happens in winter, however, the population size at the end of summer is quite constant from year to year. This is because surviving adults may benefit from the high mortality of others in a snowy winter by showing a high reproductive rate the following spring, reflecting the per capita abundance of resources. Even the *Thrips* populations described earlier as a classic example of density-independent limitation are probably regulated during part of the year by a density-dependent factor; that is, the number of flowers available in the summer is a limited resource that restricts the number of surviving insects. Most populations are probably regulated by some mix of density-dependent and density-independent factors.

Population Cycles

Some populations of insects, birds, and mammals fluctuate in density with remarkable regularity, more regularity than can be explained by chance alone. Some small herbivores, such as voles and lemmings, tend to have 4-year cycles, while larger herbivores, such as snowshoe hares, muskrat, ruffed grouse, and ptarmigan, have 9- to 10-year cycles. The causes of these cycles undoubtedly vary among species and perhaps even among populations of the same species. We will explore several hypotheses for these cycles, but as yet there seems to be no satisfactory general explanation.

One idea is that crowding regulates cyclical populations, perhaps by affecting the organisms' endocrine systems. As described earlier for white-footed mice, stress resulting from a high population density may alter hormonal balance, which in turn reduces fertility, increases aggressiveness, and induces the mass emigrations that are often portrayed in nature films. However, it is not known whether such changes are a common occurrence in the many species of animals that have population cycles.

Another hypothesis is that population cycles are caused by a time lag in the response to density-dependent factors, creating large fluctuations of population size above and below carrying capacity. As we noted before, extending the logistic model to incorporate time lags in response can lead to fluctuations similar to those observed in nature. Predation could also cause populations to oscillate if increases in prey populations are followed by increases in the number of predators, which in turn become so numerous that they cause the prey densities to decrease. Predator populations would then decrease, allowing prey to increase, and so on. Such a mechanism was once a popular explanation for a correlation in the cycles of the snowshoe hare and its predator, the Canadian lynx (FIGURE 47.19). However, research indicates that the lynx is not the major factor regulating the hare population; in fact, the hares cycle in the same manner in the absence of the lynx. It appears, therefore, that the

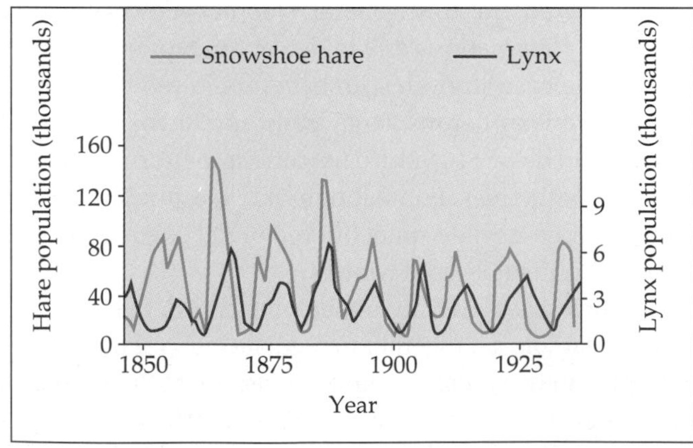

FIGURE 47.19
Population cycles in snowshoe hare and lynx. Oscillations in the population density of hare followed by corresponding changes in lynx density were once interpreted as strong evidence that these populations of prey and predator regulated each other. (Population counts are based on the number of pelts sold by trappers to the Hudson Bay Company.) However, snowshoe hare populations on islands where lynx are absent show similar cycles. The cycles of lynx populations may, indeed, be caused by cyclic fluctuations of the hare, a major food source for lynx. Changing ideas about the hare-lynx interaction emphasize the problem of inferring process from pattern. Most patterns of population growth are likely caused by multiple interacting factors that are difficult to untangle without direct experimentation.

lynx are regulated by their prey, but the hares are not regulated by their predator. The cycling of predator populations in response to prey population cycle is probably common in nature. Evidence for this comes from the fact that predators that feed on small herbivores with 4-year cycles themselves have short population cycles, while predators of larger prey (with longer cycles) have longer cycles themselves.

An alternative explanation for the hare-lynx cycle is that high hare population density causes a deterioration in the quality of their food. Studies indicate that when certain plants are damaged by herbivores, their nutrient content decreases, and they produce increased amounts of defensive chemicals. It is therefore likely that the density of the hare population may be regulated by cyclic changes in the plants they eat—or perhaps by some other resource limitation.

Perhaps the most striking population cycles known are those of periodical cicadas, grasshopper-like insects that complete their life cycles every 13 or 17 years, emerging from the ground at phenomenal densities (as high as 600 per m^2). This long life cycle may be an adaptation that reduces predation; few predators can wait 13 or 17 years for their prey to appear. Cicada populations are locally controlled, however, by a fungus whose spores can remain alive underground for the 13 or 17 years between cicada outbreaks.

The human population has been growing exponentially for centuries but will not be able to do so indefinitely

The exponential growth model of FIGURE 47.11 describes the population explosion of humans. Indeed, it is unlikely that any other population of large animals has ever sustained exponential growth for so long. The explosive growth of our population is the primary cause of severe environmental degradation, and the environmental problems that we now confront cannot be solved without stringent regulation of our numbers.

The human population increased relatively slowly until about 1650, when approximately 500 million people inhabited Earth (FIGURE 47.20). The population doubled to 1 billion within the next two centuries, doubled again to 2 billion between 1850 and 1930, and doubled still again by 1975 to more than 4 billion. The population now numbers almost 6 billion people and increases by about 80 million each year. It takes only 3 years for world population growth to add the population equivalent of another United States. If the present growth rate persists, there will be 8 billion people on Earth by the year 2017.

Human population growth is based on the same general parameters that affect other animal and plant populations: birth rates and death rates. Birth rates increased and death rates decreased when agricultural societies replaced a lifestyle of hunting and gathering about 10,000 years ago. Since the Industrial Revolution, exponential growth has resulted mainly from a drop in death rates, especially infant mortality, even in the least developed countries (FIGURE 47.21). Improved nutrition, better

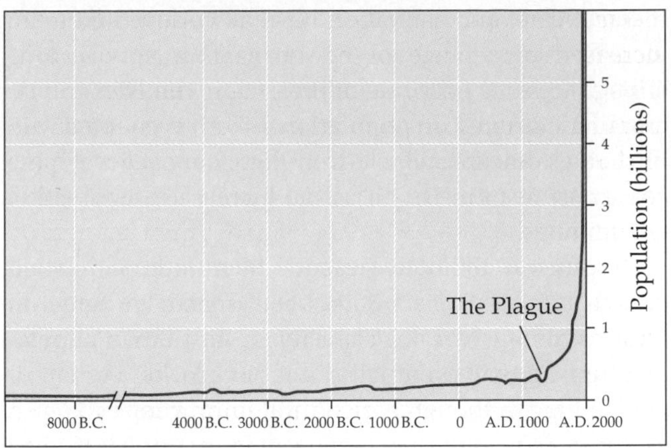

FIGURE 47.20
Human population growth. The human population has grown almost continuously throughout history, but it has skyrocketed since the Industrial Revolution. No other population of organisms has shown such steady population growth, and the human population must eventually either level off or decline. Whether this reduction in population growth will occur because of decreased birth rates or mass deaths is an open question, one that careful population policies can address.

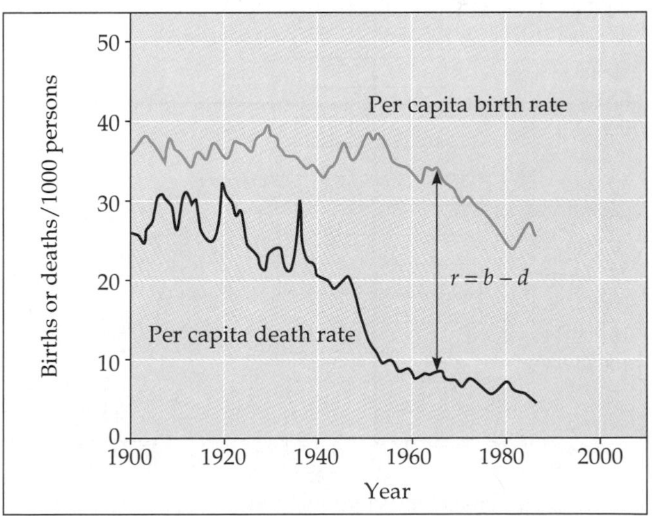

FIGURE 47.21
Changes in birth and death rates in Sri Lanka. Although family planning efforts and better medical care have reduced both per capita birth and death rates in Sri Lanka, the population growth rate (*r*, the difference between the birth and death rates at any point in time) increased after 1940. Population control will occur only after a reduction in the birth rate is greater than a reduction in the death rate.

medical care, and sanitation have all contributed to an increased percentage of newborns that survive long enough to leave offspring of their own. The effect of decreasing mortality on population growth is coupled with birth rates that are still relatively high in most developing countries, resulting in an actual increase in population growth rates.

Despite our ability to measure the human population growth rate, ecologists have been unable to agree on Earth's ultimate carrying capacity for the human population, or what will eventually limit our growth. Perhaps it will be food, although agricultural improvement known as the green revolution has allowed food supplies to keep up with population growth until now. However, we do know that, for energetic reasons, environments can support a larger number of herbivores than carnivores (see Chapter 49). If everyone ate as much meat as the wealthiest people in the world, less than half of the present world population could be fed on current food harvests. Nevertheless, it seems unlikely that people in wealthier countries will abandon the consumption of meat. Perhaps we will eventually be limited by suitable space, like the gannets on ocean islands. Certainly, as our population grows, the conflict over how space will be utilized will intensify, and agricultural land may be developed for housing. There seem to be few limits, however, on how closely humans can be crowded together.

We could also run out of resources other than nutrients and space. Many people are concerned about supplies of resources like some metals and fossil fuels, which are nonrenewable. It is also possible that our population will eventually be limited by the capacity of the environment to absorb the waste disposal and other insults imposed by humans (see Chapter 49). In this case, current human occupants could lower Earth's long-term carrying capacity for future generations.

Current worldwide population growth is actually a mosaic of various rates of growth in different countries. Some developed countries, such as Sweden, are near zero population growth because birth and death rates balance. The human population as a whole, however, continues to grow because birth rates greatly exceed death rates in most nations, particularly in developing countries.

A major factor in the variation of growth rates among countries is the variation in their age structures, an important demographic factor in present and future growth trends (FIGURE 47.22). The relatively uniform age distribution in Sweden, for instance, contributes to that country's stable population size; individuals of reproductive age or younger are not disproportionately represented in the

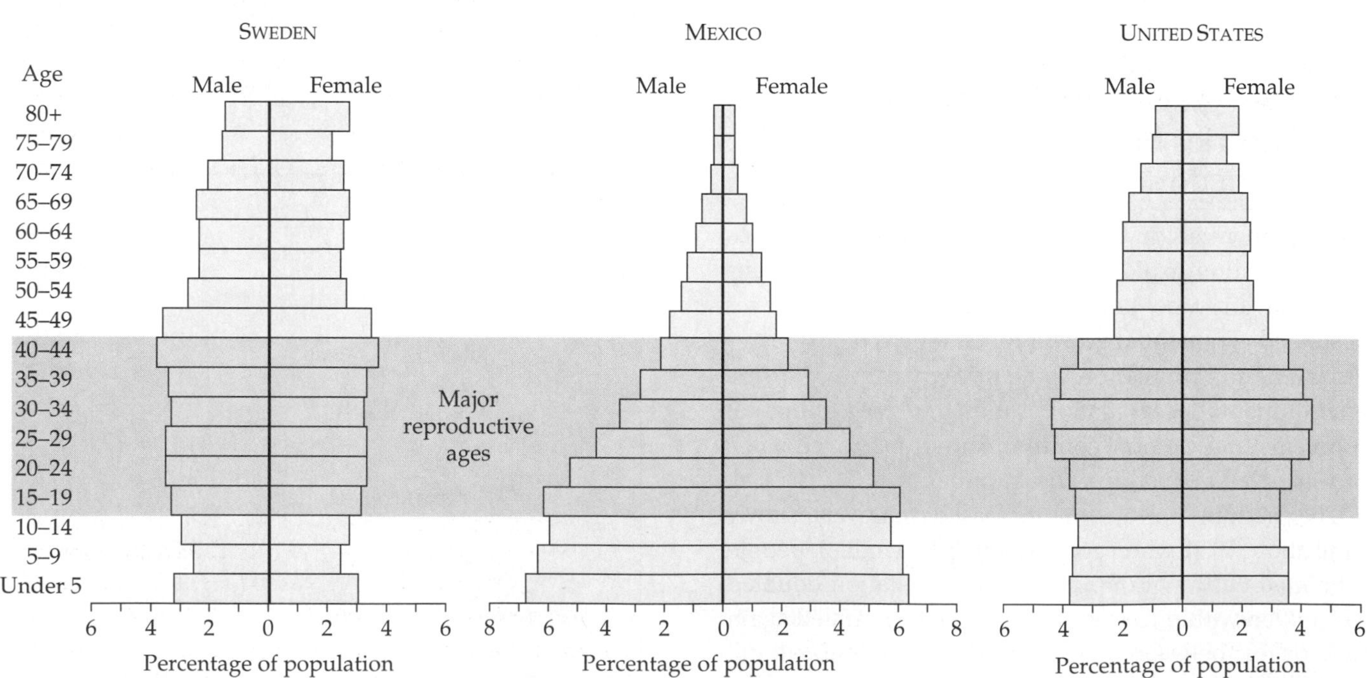

FIGURE 47.22

Age structures of three nations. The proportion of individuals in different age groups (1990 data) has a significant impact on the potential for future population growth. Mexico, for example, has a large fraction of individuals who are young and likely to reproduce in the near future. In contrast, Sweden's population is distributed more evenly through all age classes, with a high percentage of individuals beyond prime reproductive years. The United States has a fairly even age distribution, except for the bulge corresponding to the post–World War II baby boom.

population. In contrast, Mexico has an age structure that is bottom-heavy, skewed toward young individuals who will grow up and sustain the explosive growth with their own reproduction. Notice in FIGURE 47.22 that the age structure for the United States is relatively even except for a bulge that corresponds to the "baby boom" that lasted for about two decades after the end of World War II. Even though couples born during those years are now having an average of fewer than two children, the nation's overall birth rate still exceeds the death rate because there are so many "boomers" of reproductive age.

A unique feature of human population growth is our ability to control it with voluntary contraception and government-sponsored family planning. Reduced family size certainly contributes to successful population control. However, it was evident at the 1994 International Conference on Population and Development in Cairo that there is a great deal of disagreement among world leaders as to how much support should be provided for global family planning efforts. Social change and the rising educational and career aspirations of women in many cultures encourage them to delay marriage and postpone reproduction. Delayed reproduction dramatically decreases population growth rates. You can develop a sense of this phenomenon by imagining two human populations in which women each produce three children but begin reproduction at different ages. In one population, females first give birth at age 15 and in the other at age 30. If we start with a cohort of newborn girls, after 30 years, women in the first population will already begin to have grandchildren, whereas women in the second population will be giving birth to their first children. After 60 years, women in the first population will have a large number of great-great grandchildren (who will themselves begin to reproduce 15 years later), but women in the second population will just begin to see their grandchildren being born.

The problem of defining carrying capacity for humans is confounded by the observation that carrying capacity has changed with human cultural evolution (see Chapter 30). The advent of agricultural and industrial technology has significantly increased K at least twice during human history, and opponents of population control are counting on some new, as yet unidentified technological breakthrough that will allow our population to grow and plateau at some higher level.

Technology has undoubtedly increased Earth's carrying capacity for humans, but no population can continue to grow indefinitely. Exactly what the world's human carrying capacity is and under what circumstances we will approach it are topics of great concern and debate. Ideally, human populations would reach carrying capacity smoothly and then level off. This will occur when birth rates and death rates are equal, and a decrease in birth rate is more desirable than an increase in death rate. If, however, the population fluctuated about K, we would expect periods of increase followed by mass death, as has occurred during plagues, localized famines, and international military conflicts. In any case, the human population must eventually stop growing. Unlike other organisms, we can decide whether zero population growth will be attained through social changes involving individual choice or government intervention, or through increased mortality due to resource limitation and environmental degradation. For better or worse, we have the unique opportunity to decide the fate of our species and the rest of the biosphere.

REVIEW OF KEY CONCEPTS (with page numbers and key figures)

- Two important characteristics of any population are density and the spacing of individuals (pp. 1094–1096, FIGURE 47.2)
  - Density is the number of individuals per unit area or volume.
  - Population ecologists use several methods to measure density, including direct counting and various sampling techniques.
  - Dispersion is the pattern of spacing for individuals and may range from clumped, to uniform, to random, as determined by various environmental or social factors.
- Demography is the study of factors that affect birth and death rates in a population (pp. 1096–1099, FIGURE 47.4)
  - Life tables describe the fate of a cohort of organisms throughout their lives, tabulating mortality rates, the number of survivors from one age to the next, and average reproductive output.
  - Survivorship curves, which plot the number of individuals of a cohort alive at each age, can be classified into three general types, depending on whether mortality is greater among the young or old or constant over all ages.
- The traits that affect an organism's schedule of reproduction and death make up its life history (pp. 1099–1103)
  - Natural selection has led to the evolution of diverse life history "strategies" that maximize lifetime reproductive success.
  - Life history traits represent trade-offs between conflicting demands for limited time, energy, and nutrients. These trade-offs often involve conflicts between current and future reproductive output.

- Semelparous organisms reproduce a single time and then die, while iteroparous organisms reproduce repeatedly over several breeding seasons. When survival between breeding seasons is low or if there is a large trade-off between fecundity and survival, semelparity is favored over iteroparity.
- Organisms allocate resources among offspring in different ways, generally producing either large clutches with small offspring or small clutches with large offspring.
- Some organisms reproduce at a relatively young age, while others delay reproduction. Generally, organisms that have a good chance of surviving to an advanced age, or whose reproductive output increases with age, delay reproduction.

- A mathematical model for exponential growth describes an idealized population in an unlimited environment (pp. 1104–1105, FIGURE 47.11)
  - The exponential growth equation ($dN/dt = r_{max} N$) is a useful way to think about the growth potential of a population in an unlimited environment. This model predicts that the larger a population becomes, the faster it grows. Such exponential growth is never sustained for extended periods of time in any population.

- A logistic model of population growth incorporates the concept of carrying capacity (pp. 1106–1109, FIGURE 47.14)
  - A more realistic model limits growth by incorporating carrying capacity (K), the maximum population size that can be supported by the available resources. The logistic equation ($dN/dt = r_{max} N[K-N]/K$) fits an S-shaped curve in which population growth levels off as size approaches carrying capacity.
  - Few populations fit the logistic model exactly, but many show a generally similar pattern in which the growth rate decreases with increasing density.
  - At densities near carrying capacity (K), selection should favor traits that allow organisms to survive and reproduce with few resources; organisms that tend to live at or near their carrying capacity are called *K*-selected. At low densities, adaptations that promote rapid reproduction (high r) are favored; organisms that are likely to be found in variable environments where their numbers fluctuate, or in open or disturbed habitats, are called *r*-selected.

- Both density-dependent and density-independent factors can affect population growth (pp. 1109–1113)
  - A density-dependent factor is one that intensifies as population density increases, and can eventually stabilize a population near its carrying capacity. Several density-dependent factors—limited resources, increased predation, stress due to crowding, or buildup of toxins—have been shown to cause population growth rates to decline at high densities.
  - Density-independent factors, such as climatic events, reduce population size by a given fraction, regardless of its density. Such factors are often superimposed on density-dependent factors.
  - Defining the importance of specific environmental factors in population regulation is complicated by the overlap of density-dependent and density-independent factors and the difficulty of determining cause-and-effect relationships.
  - Some populations have remarkably regular population cycles that may be a result of the physiological effects of crowding or of time lags in response to density-dependent factors.

- The human population has been growing exponentially for centuries but will not be able to do so indefinitely (pp. 1113–1115, FIGURE 47.20)
  - Human population growth has been sustained by such factors as the Industrial Revolution and improved nutrition, medical care, and sanitation. Different countries have different rates of growth for various environmental, cultural, and historical reasons.
  - The importance of age structure as a demographic factor strongly affecting population growth can be seen in the post–World War II baby boom in the United States and in the bottom-heavy age structure of some countries.
  - The explosive growth of the human population has resulted in severe environmental degradation.
  - Humans are unique in our ability to consciously control our own population growth. Reduced fecundity and delayed reproduction both decrease population growth rates.

SELF-QUIZ

1. A uniform dispersion pattern for a population may indicate that
   a. the population is spreading out and increasing its range
   b. resources are heterogeneously distributed
   c. individuals of the population are competing for some resource, such as water and minerals for plants or nesting sites for animals
   d. there is an absence of strong attractions or repulsions among individuals
   e. the density of the population is low

2. A population that has a relatively low r value will most likely
   a. have large clutch sizes with relatively small offspring
   b. be found in environments that are highly variable
   c. have an early age of first reproduction and a short generation time
   d. produce fewer offspring with more competitive capabilities
   e. be regulated by density-independent factors

3. The term $(K-N)/K$ influences dN/dt such that
   a. the increase in actual population numbers is greatest when N is small
   b. as N approaches K, r_{max} (the intrinsic rate of increase) becomes smaller
   c. when N equals K, population growth is zero
   d. when K is small, the influence of density-dependent factors is smaller
   e. as N approaches K, the birth rate approaches zero

4. A population's carrying capacity is
   a. the number of individuals in that population
   b. reached when the number of deaths exceeds the number of births
   c. inversely related to r_{max}
   d. the population size that can be supported by available resources for that species within the habitat
   e. set at 8 billion for the human population

5. A Type III survivorship curve would be expected in a species in which
   a. mortality occurs at a constant rate over the lifespan
   b. parental care is extensive
   c. a large number of offspring are produced but parental care is minimal
   d. mortality rate is quite low for the young
   e. *K*-selection prevails
6. In a mark-recapture study of a lake trout population, 40 fish were captured, marked, and released. In a second capture, 45 fish were captured; 9 of these were marked. What is the estimated number of individuals in the lake trout population?
   a. 90 c. 360 e. 1800
   b. 200 d. 800
7. The example of the population cycles of the snowshoe hare and its predator, the lynx, illustrates that
   a. predators are the major factor in controlling the size of prey populations and are, in turn, regulated in their numbers by the oscillating supply of prey
   b. the two species must have evolved in close contact with each other because their life histories are intertwined
   c. one should not conclude a cause-and-effect relationship when viewing population patterns without careful observation and experimentation
   d. both populations are controlled by density-independent factors
   e. the hare population is *r*-selected, whereas the lynx population is *K*-selected
8. The current size of the human population is closest to
   a. 2 million d. 6 billion
   b. 3 billion e. 10 billion
   c. 4 billion
9. Consider five human populations that differ demographically *only* in their age structures. The population that will grow the most in the next 30 years is the one with the greatest fraction of people in which age group? Explain your answer.
   a. 10–20 c. 30–40 e. 50–60
   b. 20–30 d. 40–50
10. All these terms are characteristic of the human populations in industrialized countries *except*
   a. relatively small clutch size
   b. several potential reproductions per lifetime
   c. *r*-selected life history
   d. Type I survivorship curve
   e. bottom-heavy age structure

CHALLENGE QUESTION

1. A biologist studied a population of fish called cichlids in an African lake that is 120 hectares (1 hectare = 10,000 m^2) in area. She found that the fish lived only in scattered reed beds that accounted for one-fourth the area of the lake. She caught 185 fish, tagged them, and released them. Two days later, she netted 208 fish and found that 35 of them were tagged. About how many cichlids are in the lake? What is the density of the cichlid population (in this case, in fish per hectare of lake surface)? What is the pattern of dispersion of the fish? How might the dispersion pattern affect your interpretation of their density?
2. Beavers are released in an effort to repopulate a valley where they were trapped to extinction many years ago. Many of their young are killed by predators, but each original pair produces an average of three offspring that survive to maturity. (This means, in effect, 1.5 offspring per parent per generation.) If this rate of growth continues, roughly how many beavers will there be in the sixth generation for each original pair in the first generation? Graph the number of beavers against generation, and compare to the curves in FIGURES 47.11 and 47.14. What kind of growth is seen in the beaver population? What would the growth curve look like if the beavers averaged 1.1 offspring per parent per generation? Do numbers have to double each generation for exponential growth to occur?

SCIENCE, TECHNOLOGY, AND SOCIETY

1. The mountain gorilla, spotted owl, and California condor are all endangered by human activities. In general, are these *r*-selected or *K*-selected species? How might the life histories of these species make them vulnerable to human encroachment? How would you design plans for saving them that reflect their life histories?
2. Many people regard the rapid population growth of developing countries as our most serious environmental problem. Others think that the population growth in developed countries, though smaller, is actually a greater threat to the environment. What kinds of problems result from population growth in (a) developing countries, and (b) the industrialized world? Which do you think is the greater threat, and why?

FURTHER READING

Ackerman, L., et al. "The Successful Animal." *Science 86,* January/February 1986. Cultural and historical aspects of human population growth and control.

Clutton-Brock, T. H., ed. *Reproductive Success: Studies of Individual Variation in Contrasting Breeding Systems.* Chicago: University of Chicago Press, 1988. A collection of 25 longitudinal studies that document life history patterns in animals.

Daily, G. C., and P. R. Ehrlich. "Population, Sustainability, and Earth's Carrying Capacity." *BioScience,* November 1992. The relationship between current population growth and the standard of living for future generations.

Dasgupta, P. S. "Population, Poverty, and the Local Environment." *Scientific American,* February 1995. The value of a child's labor to a family is one of the cultural factors behind the population explosion.

Kates, R. W. "Sustaining Life on the Earth." *Scientific American,* October 1994. The cultural context of environmental concerns.

MacKenzie, D. "Will Tomorrow's Children Starve?" *New Scientist,* September 3, 1994. What will life be like in 2050 when the world has 12 billion people?

Miller, S. K. "Bats Sow Seeds of Rainforest Recovery." *New Scientist,* June 18, 1994. Hurricanes as density-independent regulators of population size.

Peters, R. H. *The Ecological Implications of Body Size.* Cambridge: Cambridge University Press, 1986. Applies the relationship between organismal form and body size to ecological problems.

Pool, R. "Ecologists Flirt with Chaos." *Science,* January 20, 1989. Does chaos theory apply to population ecology?

Savonen, C. "One Salmon, Two Salmon . . . 10,000 Salmon: Counting the Fish in Alaska." *Oceans,* January/February 1985. A population density determination in action.

CHAPTER 48

COMMUNITY ECOLOGY

KEY CONCEPTS

- The interactive and individualistic hypotheses pose alternative explanations of community structure: *science as a process*
- Community interactions can provide strong selection factors in evolution
- Interspecific interactions may have positive, negative, or neutral effects on a population's density: *an overview*
- Predation and parasitism are (+−) interactions: *a closer look*
- Interspecific competitions are (−−) interactions: *a closer look*
- Commensalism and mutualism are (+0) and (++) interactions, respectively: *a closer look*
- A community's structure is defined by the activities and abundance of its diverse organisms
- The factors that structure communitites include competition, predation, and environmental patchiness
- Succession is the sequence of changes in a community after a disturbance
- Biogeography complements community ecology in the analysis of species distribution
- Lessons from community ecology and biogeography can help us preserve biodiversity

On your next walk through a natural field or woodland, or even across campus or through a park, try to observe some of the interactions of the species present. You may see birds using trees as nesting sites, bees pollinating flowers, shelf fungi growing on trees, spiders trapping insects in their webs, ferns growing in shade provided by trees—a sample of the many interactions that exist in any ecological theater. In addition to the physical and chemical factors discussed in Chapter 46, an organism's environment includes other individuals in its population and populations of other species living in the same area. Such an assemblage of species living close enough together for potential interaction is called a ***community.*** *In the photograph that opens this chapter, lion, zebra, vultures, and grasses and other plants are all members of a community in Kenya.*

This chapter examines the diverse kinds of biotic interactions among organisms and addresses the central issue in community ecology: What factors are most significant in structuring a community—in determining its species composition and both the absolute and the relative abundance of species present? The analysis of communities is an active area of ecological research, a field replete with unanswered questions and controversies.

The interactive and individualistic hypotheses pose alternative explanations of community structure: *science as a process*

Why are certain combinations of species found together as members of a community? Two divergent views on this question emerged among ecologists in the 1920s and 1930s, derived primarily from observations of plant distributions. The **individualistic** hypothesis, first enunciated by H. A. Gleason, depicted the community as a chance assemblage of species found in the same area simply because they happen to have similar abiotic requirements—for example, for temperature, rainfall, and soil type. The alternative view, the **interactive hypothesis,** advocated by F. E. Clements, saw the community as an assemblage of closely linked species, locked into association by mandatory biotic interactions that cause the community to function as an integrated unit. Carried to the extreme, Clements' view interpreted communities as a kind of "superorganism." Evidence for the latter view was based on the observation that certain species of plants are consistently found together. For example, deciduous forests in the northeastern United States almost always include certain species of oak, maple, birch, and beech, along with a specific assemblage of shrubs and vines. These two ways of looking at community structure suggest different priorities in studying biological communities. The individualistic hypothesis emphasizes studying single species, while the interactive hypothesis emphasizes entire assemblages of species as the essential units for understanding the interrelationships and distributions of organisms.

We can use the hypothetico-deductive approach to evaluate the individualistic and interactive hypotheses. The two hypotheses of communities make contrasting predictions about how species should be distributed along a gradient of environmental variables, such as moisture or temperature. The individualistic hypothesis predicts that communities should generally lack discrete geographical boundaries because each species has an independent distribution along the environmental gradient. In other words, each species will be distributed according to its tolerance ranges for abiotic factors that vary along the gradient, and communities should change continuously along the gradient with the addition or loss of particular species. According to the interactive hypothesis, in contrast, species should be clustered into discrete communities with distinct boundaries, because the presence or absence of a particular species is largely governed by the presence or absence of other species with which it interacts.

In most actual cases, especially where there are broad regions characterized by gradients of environmental variation, the composition of plant communities does seem to change on a continuum, with each species more or less independently distributed (FIGURE 48.1). Such distributions generally support the view of plant communities as relatively loose associations without discrete boundaries. However, where some key factor in the physical environment changes abruptly, adjacent communities are delineated by correspondingly sharp boundaries. In California, for example, many native plants have been replaced by introduced European grasses, but in some strictly delimited areas where serpentine rocks increase the magnesium content of the soil, native wildflowers persist (FIGURE 48.2). The individualistic hypothesis can account for this type of discontinuity without assuming that certain plant species occur in the same area because they are locked into mandatory community relationships.

The individualistic hypothesis may not apply to the animals in a community, as animals are usually linked more intimately to other organisms. For example, limpkins, long-beaked birds of Florida swamps, feed primarily on one species of snail. The birds' evolutionary adaptations make them extremely efficient in their specialized foraging, but their geographical distribution is restricted by the distribution of their prey. On the other hand, gray squirrels of the eastern United States are not as strongly tied to a particular food. They may be found in a variety of habitats, including forests where pine is abundant, although they are most common in areas of mature hardwoods, such as oak and hickory trees.

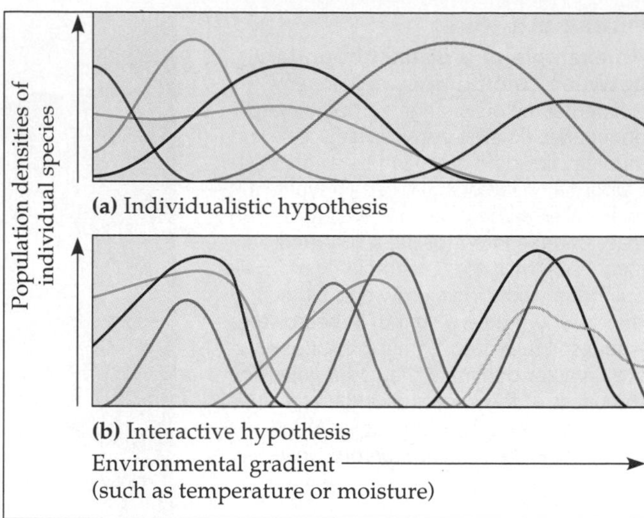

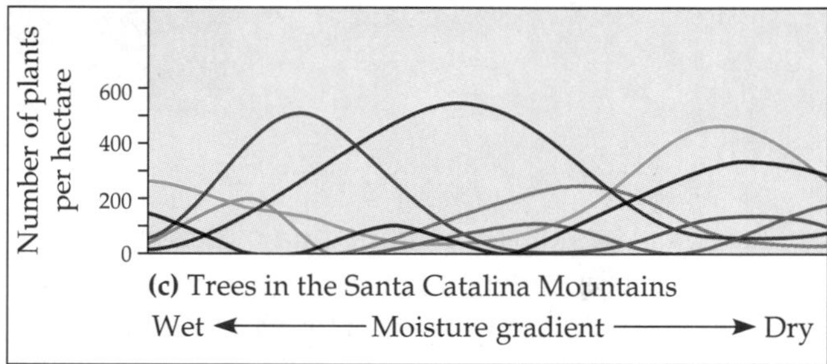

FIGURE 48.1

Testing the individualistic and interactive hypotheses of communities. Ecologist Robert Whittaker tested these two hypotheses by graphing the abundance of different species (*y* axis) along abiotic environmental gradients such as temperature or moisture (*x* axis). Each line on the graphs represents the abundance of one species. (**a**) The individualistic hypothesis poses that species are independently distributed along gradients and that a community is simply the assemblage of species that happen to occupy the same area. (**b**) The interactive hypothesis suggests that communities are discrete groupings of particular species that are closely interdependent and nearly always occur together. (**c**) The distribution of tree species at one elevation in the Santa Catalina Mountains of Arizona supports the individualistic hypothesis. Each tree species has an independent distribution along the gradient, apparently conforming to its tolerance for moisture, and the species that live together at any point along the gradient have similar physical requirements. Because the vegetation changes continuously along the gradient, it is impossible to delimit sharp boundaries for the communities.

Thus, simple generalizations can rarely explain why certain species commonly occur together in communities. Just as we saw in Chapter 47 that population sizes

FIGURE 48.2
An example of a distinct boundary between communities. Where environmental factors change abruptly, sharp boundaries do exist between adjacent communities. In some grassland areas of coastal California, weathering of certain types of rock, called serpentine rock, increases the magnesium content of the soil. Where magnesium-rich soils predominate, native California wildflowers grow and blossom colorfully in the early spring. In nearby areas where soil is derived from sandstone, however, grasses dominate the community. The sharpness of these local boundaries is apparent in this scene from the Jasper Ridge Biological Preserve near Stanford University.

are usually affected by a combination of density-dependent and density-independent factors, we will see that the distributions of most populations in communities are probably affected to some extent by both abiotic gradients and interactions with other species.

Community interactions can provide strong selection factors in evolution

Just as the physical and chemical features of the environment are important factors in adaptation by natural selection, so are interactions between species. In some cases, the adaptation of one species to the presence of another has a relatively obvious evolutionary basis. For example, natural selection has apparently favored peppered moths of coloration that blends in with the living lichens on which the moths sometimes rest, an adaptation that probably makes it harder for predatory birds to spot the moths.

In its broadest sense, **coevolution** is the term used to describe more complex interactions involving reciprocal evolutionary adaptations in two species: A change in one species acts as a selective force on another species, and counteradaptation by the second species, in turn, is a selective force on individuals in the first species. Coevolution has been studied most extensively in predator-prey relationships and in mutualism. One important example of coevolution was discussed in Chapter 27: the interactions between some species of flowering plants and their exclusive pollinators (see FIGURE 27.23).

In many cases where two species appear to have coevolved adaptations, it is difficult to establish that an evolutionary change in one species creates a selective force that induces an evolutionary change in the other. As an example, let's examine a specific relationship that probably illustrates coevolution but also demonstrates the complexities that may confound our attempts to unravel evolutionary history. Passionflower vines of the genus *Passiflora* are protected against most herbivorous insects by their production of toxic compounds in young leaves and shoots. However, the larvae of butterflies of the genus *Heliconius* can tolerate these defensive chemicals. This counteradaptation has enabled *Heliconius* larvae to become specialized feeders on plants that few other insects can eat (FIGURE 48.3).

Survival of the larvae is further enhanced by a behavioral adaptation of the butterflies. The eggs that female *Heliconius* butterflies lay on the leaves of passionflower vines are bright yellow, and other females generally avoid laying eggs on leaves marked by yellow dots. This behavior presumably reduces intraspecific competition among the larvae for food. An infestation of *Heliconius* larvae can devastate a passionflower vine, and these poison-resistant insects are likely to be a strong selection force favoring the evolution of more defenses in the plants.

In some species of *Passiflora,* the leaves have conspicuous yellow spots that look like *Heliconius* eggs, an adaptation that may divert the butterflies to other plants in their search for egg-laying sites. The story is more complicated, however. The yellow "spots" on the passionflower vine are actually nectaries, which attract ants and wasps that prey on *Heliconius* eggs or larvae. There is evidence, too, that the mere presence of ants on a leaf will discourage a *Heliconius* butterfly from laying eggs there. Thus, adaptations that may seem, on superficial

(a)

(b)

(c)

FIGURE 48.3

Evolutionary interactions between a plant and an insect. (**a**) Passionflower vines (*Passiflora*) produce toxic chemicals that help protect leaves from herbivorous insects. A counteradaptation has evolved in the butterfly *Heliconius:* The larvae can feed on the leaves because they have digestive enzymes that break down the toxic compounds. (**b**) The females of some *Heliconius* species avoid laying eggs on passionflower leaves that already carry bright yellow egg clusters deposited by other females. This behavior reduces intraspecific competition on individual leaves. (**c**) These egglike yellow nectaries grow on the leaves of some species of passionflowers. Experiments show that *Heliconius* females avoid laying eggs on leaves with these yellow spots. These nectaries, as well as smaller ones scattered over the leaf, also attract ants and other insects that prey on the butterfly eggs and larvae.

examination, to be coevolutionary responses between just two species may, in fact, result from interactions among many species in the community. It is difficult to sort out the importance of the various selective forces, and the simple idea of coevolution as adaptation-counteradaptation occurring exclusively between two species does not often adequately describe the interactions within communities.

Despite the problems in assessing cause and effect in the evolution of complex ecological relationships, biologists agree that the adaptation of organisms to other species in a community is a fundamental characteristic of life. Put another way, interactions of species in ecological time often translate into adaptations over evolutionary time. We will encounter several examples of such adaptations as we now survey the interspecific interactions of greatest importance in community ecology.

Interspecific interactions may have positive, negative, or neutral effects on a population's density: *an overview*

Although the geographical distributions of many species are largely determined by their adaptations to abiotic environmental factors, all organisms are also influenced by biotic interactions with other individuals in their immediate vicinity. In Chapter 47, we examined one such interaction, intraspecific competition. In this section, we consider **interspecific interactions,** those that occur between populations of different species living together within a community. These interactions may have positive, negative, or neutral effects on one or more of the populations involved.

The possible interactions between any two species living together in a community are summarized in TABLE 48.1. For each interaction, a pair of signs, such as

TABLE 48.1

Interspecific Interactions

INTERACTION	EFFECTS ON POPULATION DENSITY
Predation (+−) (includes parasitism)	The interaction is beneficial to one species and detrimental to the other.
Competition (−−)	The interaction is detrimental to both species.
Commensalism (+0)	One species benefits from the interaction but the other is unaffected.
Mutualism (++)	The interaction is beneficial to both species.

(+−), refers to the effects on population density of the two species involved in the interaction. For example, mutual symbiosis (mutualism) is a (++) interaction, meaning that the density of each species is increased in the presence of the other. Predation is an example of a (+−) interaction, with a positive effect on population density of one species (the predator) and a negative effect on the density of the other population (the prey). The following three sections examine the interspecific interactions of communities in more detail.

Predation and parasitism are (+−) interactions: *a closer look*

The most conspicuous population interactions are those involving **predation,** in which a **predator** eats its **prey.**

Though most of us associate predation with the kinds of interactions illustrated by the lion eating the zebra in the photograph that opens this chapter, there are other important kinds of (+−) interactions. True **parasitism** involves predators that live on or in their hosts, seldom killing them outright. In **parasitoidism,** insects, usually small wasps, lay eggs on living hosts. The larvae then feed within the body of the host, eventually causing its death. **Herbivory** occurs when animals eat plants, and we include it here as a form of predation. (Herbivory can kill an entire organism, as when a mouse eats a seed. However, the kind of herbivory called grazing does not kill the plant and is actually more similar to parasitism than to predation.) While all of the above interactions involve one kind of organism feeding on another, the negative effects in some (+−) interactions are not always due to feeding, as we will see when we discuss brood parasitism.

Predation

Many important feeding adaptations of predators are both obvious and familiar. Most predators have acute senses that enable them to locate and identify potential prey. In addition, many predators have adaptations such as claws, teeth, fangs, stingers, or poison that help catch and subdue, or simply chew, the organisms on which they feed. Rattlesnakes and other pit vipers, for example, locate their prey with special heat-sensing organs located between each eye and nostril, and they kill small birds and mammals by injecting them with toxins through their fangs. Similarly, many herbivorous insects locate appropriate food plants with chemical sensors on their feet, and their mouthparts are adapted to shred tough vegetation. Predators that pursue their prey are generally fast and agile, whereas those that lie in ambush are often camouflaged in their environments.

Through repeated encounters with predators over evolutionary time, various defensive adaptations have evolved in prey species, and we will now survey some of these.

Plant Defenses Against Herbivores. In many cases, herbivores do not consume an entire plant, so their impact on prey density is different from that of a predator that consumes whole prey. Nevertheless, the removal of plant tissue—stems, leaves, bark, and sap—usually affects a plant's fitness and ability to survive, and plants exhibit a variety of defensive adaptations against the animals that might eat them. Many of these defenses are mechanical. For example, thorns may discourage large herbivores such as vertebrates, and some plants have microscopic crystals in their tissues, or hooks or spines on their leaves, that make feeding difficult even for small insects.

Many plants produce chemicals that function in defense by making the vegetation distasteful or harmful to an herbivore (see the interview with Eloy Rodrigeuz on p. 22). These chemicals, classified as *secondary compounds,* are so named because they form as metabolic by-products of major biochemical pathways, such as glycolysis or the Krebs cycle. Some well-known poisons and drugs are secondary compounds that probably function as chemical weapons against herbivores: strychnine, produced by plants of the genus *Strychnos;* morphine, from the opium poppy; nicotine, produced by tobacco; and digitoxin, from the common foxglove. Other secondary compounds that are not toxic to humans but may be distasteful to herbivores are responsible for the familiar flavors of cinnamon, cloves, and peppermint. Some plants even produce secondary products that are analogous to insect hormones and cause abnormal development in some insects that eat them.

The specific defenses of a plant population can act as selective agents that provoke the evolution of counter-adaptations in herbivore populations. The responses of herbivores may nullify the plant defenses and allow subsequent generations of herbivores to eat the descendants of the original plant population. For example, some insects, such as the *Heliconius* larvae depicted in FIGURE 48.3, can absorb or detoxify certain secondary compounds. Some insects even store plant poisons and use them in defense against their own predators; the larvae of monarch butterflies, for instance, store toxins from their host plants (milkweeds) that make them distasteful to some of their predators. Although a plant's defenses may restrict the number of herbivore species that can feed upon it, the herbivores must eat to reproduce, and selection to overcome plant defenses is strong. Hence, no plant defense is likely to provide protection forever.

FIGURE 48.4
A behavioral defense against predators. Many prey species turn the tables and attack their predator. This is common in bird species that nest in groups. Here, two crows are mobbing a barn owl, a predator that often kills and eats eggs and nestlings. Mobbing behavior has been observed in animals as diverse as baboons and ground squirrels.

Animal Defenses Against Predators. Animals can avoid being eaten by using passive defenses, such as hiding, or active ones, such as escaping or defending themselves against predators. Fleeing is perhaps the most direct antipredator response, although it can be energetically very expensive. Many animals flee into a shelter and avoid being caught without expending the energy required for a prolonged flight. Active self-defense is less common, though some large grazing mammals will vigorously defend their young from predators such as lions. Other behavioral defenses include alarm calls, which often bring in many individuals of the prey species that mob the predator. Mobbing can involve either harassment at a safe distance, or direct attack (FIGURE 48.4). Distraction displays direct the attention of the predator away from a vulnerable prey, such as a bird chick, to another potential prey that is more likely to escape, such as the chick's parent.

Many other defenses rely on adaptive coloration, which has evolved repeatedly among animals in a variety of contexts. Camouflage, called **cryptic coloration,** is the quintessential passive defense, making potential prey difficult to spot against its background (FIGURE 48.5). A camouflaged animal need only remain still on an appropriate substrate to avoid detection. The shape of an animal can also help camouflage it; the cryptic shape of the seaweedlike fish in FIGURE 1.14 is an example.

Deceptive markings are another form of adaptive coloration. Large, fake eyes or false heads can apparently deceive predators momentarily, allowing the prey to escape (FIGURE 48.6), or they may induce the predator to strike a nonvital area.

FIGURE 48.5
Camouflage. Cryptic coloration of the canyon tree frog (*Hyla arenicolor*) allows it to "disappear" on a granite background.

Some animals have mechanical or chemical defenses against would-be predators. Most predators are strongly discouraged by the familiar defenses of porcupines and skunks. Some animals, such as poisonous toads and frogs, can synthesize toxins. Others acquire chemical defense passively by accumulating toxins from the plants they eat. The monarch larvae mentioned earlier provide one example of toxin accumulation from plants. Remarkably, the poisons are retained through metamorphosis and are present in adult butterflies. Birds that eat monarch butterflies regurgitate their prey and quickly

FIGURE 48.6
Deceptive coloration. The hindwing markings of the io moth (*Automerisio*) resemble the eyes of a much larger animal. Potential predators may be momentarily startled when the moth moves its forewings, enabling the moth to escape.

FIGURE 48.7
Aposematic (warning) coloration. Many toxic or unpalatable animals are conspicuously colored with black and yellow or red stripes. The fire salamander shown here, for example, can squirt a nerve poison (visible in the photograph as two streams of dots) from glands on its back. Warning coloration probably trains predators quickly to avoid such brightly colored animals.

learn to avoid feeding on that species of insect. This is not a foolproof protection, however; monarchs are still subject to predation by many predatory and parasitoid insects, and even some mammals, such as mice.

Animals with effective chemical defenses are often brightly colored, a warning to predators known as **aposematic coloration** (FIGURE 48.7). This warning coloration seems to be adaptive; there is evidence that predators are more cautious in dealing with bright color patterns in potential prey, perhaps because so many aposematic animals tend to be dangerous prey. In an example of convergent evolution, unpalatable animals in several different taxa have similar patterns of coloration—black with yellow or red stripes characterize unpalatable animals as diverse as yellow-jacket wasps and fire salamanders.

Mimicry. A predator or species of prey may gain a significant advantage through **mimicry,** a phenomenon in which the mimic bears a superficial resemblance to another species, the model. Defensive mimicry in prey often involves aposematic models. While there are many examples of mimicry in which the mimic and the model are related, such as one butterfly mimicking another, mimicry can also cross broad taxonomic lines.

In **Batesian mimicry,** a palatable or harmless species mimics an unpalatable or harmful model. In one intriguing example, the larva of the hawkmoth puffs up its head and thorax when disturbed, looking like the head of a small poisonous snake complete with eyes. The mimicry even involves behavior; the larva weaves its head back and forth and hisses like a snake (FIGURE 48.8). Additional examples of Batesian mimicry are the many harmless snakes that mimic the conspicuous red, white, and black markings of the poisonous coral snake. For Batesian mimicry to be effective, however, models must generally outnumber mimics; otherwise, predators would learn that animals with a particular coloration are good rather than bad to eat.

In **Müllerian mimicry,** two or more unpalatable, aposematically colored species resemble each other. Presumably each species gains an additional advantage, because the pooling of numbers causes predators to learn more quickly to avoid any prey with a particular appearance.

Predators also use mimicry in a variety of ways. For example, some snapping turtles have tongues that resemble a wriggling worm, thus luring small fish; any fish that tries to eat the "bait" is itself quickly consumed as the turtle's strong jaws snap closed.

Parasitism

We are classifying predation and parasitism together here because they are both (+−) interactions in communities. In parasitism, one organism, the **parasite,** derives its nourishment from another organism, its **host,** which is harmed in the process. Organisms that live within host tissues, such as tapeworms and malarial parasites, are called **endoparasites;** others that feed briefly on the external surface of a host, like mosquitoes and aphids, are called **ectoparasites.**

As in other (+−) relationships, natural selection favors parasites that are best able to locate hosts and feed on them. Many endoparasites acquire hosts by passive mechanisms. For example, *Ascaris,* a nematode endoparasite of the human intestine, produces an abundance of eggs that are passed from a host's digestive tract to the external environment. In places without good sanitation, other humans inadvertently ingest the eggs and acquire the parasite themselves. Ectoparasites, on the other hand, often have elaborate host-finding adaptations. Some aquatic leeches, for example, first locate a host by detecting its movement in the water and then confirm its identity on the basis of temperature and chemical cues on the host's skin.

Natural selection has also favored the evolution of defensive capabilities in potential hosts. Some of the secondary plant products that are toxic to herbivores, for instance, are also toxic to such parasites as fungi and

(a)

(b)

FIGURE 48.8
Batesian mimicry. In Batesian mimicry, a palatable species mimics the appearance of a harmful or unpalatable one. When it is disturbed (**a**), the hawkmoth larva resembles (**b**) a snake.

bacteria. In vertebrates, the immune system provides defense against internal parasites. Many parasites, particularly microorganisms, have adapted to particular hosts, often a single species. In such specific interactions, coevolution generally results in a relatively stable relationship that does not kill the host quickly—an excess that would eliminate the parasite, as well.

An example will demonstrate how rapidly natural selection can temper a host-parasite relationship. In the 1940s, Australia was plagued by a population of hundreds of millions of rabbits, which were descended from just 12 pairs imported a century earlier. In 1950, the myxoma virus, which parasitizes rabbits, was introduced in an effort to control the rabbit population. The virus spread rapidly and killed 99.8% of all rabbits infected. However, a second exposure to the virus killed only 90% of the remaining rabbits, and the third infection killed only about 50%. Today, the virus has only a mild effect on the host, and the rabbit population has rebounded. Apparently, viral infection selected for host genotypes that were better able to resist the parasite.

While we usually think of parasites as consuming the host itself, in some cases organisms exploit the behavior of other organisms. The best-known example of this is brood parasitism, which occurs when birds such as cowbirds and European cuckoos lay their eggs in the nests of other species. In many cases, the newly hatched brood parasite ejects other eggs from the nest when it hatches (see FIGURE 50.7a), and the host parents often invest more in feeding the imposter than they do their own offspring (see FIGURE 50.7b). The fitness of the foster parents, or hosts, is lowered by the parasite; thus, this is truly a (+−) interaction. An evolutionary adaptation exhibited by many potential hosts is the ability to detect parasite eggs in their nests and eject them, or, if this is impossible, to start over in their nesting attempt, often abandoning their own eggs at the same time. As a counteradaptation to this host defense, some brood parasite species have an elaborate egg mimicry that tricks the host into accepting the parasite egg as its own.

Interspecific competitions are (− −) interactions: *a closer look*

When two or more species in a community rely on similar limiting resources, they may be subject to **interspecific competition.** Competition is manifested in different ways under natural conditions. Actual fighting over resources is termed **interference competition,** whereas the consumption or use of similar resources is called **exploitative competition.** The density-dependent effects of interspecific competition are similar to those of intraspecific competition, discussed in Chapter 47. As population densities increase, every individual has access to a smaller share of some limiting resource; as a result, mortality rates increase, birth rates decrease, and population growth is curtailed. In interspecific competition, however, the population growth of a species may be limited by the density of competing species as well as by the density of its own population. For example, if several bird species in a forest feed on a limited population of insects, the density of each species may have a negative impact on population growth in the others. Similarly, species may compete for nesting sites, shelters, or any other resource that is in short supply.

The Competitive Exclusion Principle

In the first quarter of this century, two mathematician-biologists, A. J. Lotka and V. Volterra, independently modified the logistic model of population growth (see Chapter 47) to incorporate the effects of interspecific competition. They predicted that two species with similar requirements could not coexist in the same community; one species would inevitably harvest resources and reproduce more efficiently, driving the other to local extinction. Even a slight reproductive advantage would eventually lead to the elimination of the inferior competitor and an increase in the density of the superior one.

In 1934, Russian ecologist G. F. Gause tested this hypothesis with laboratory experiments on the effects of interspecific competition between two closely related species of the protozoan *Paramecium, P. aurelia* and *P. caudatum* (FIGURE 48.9). Grown in separate cultures under constant conditions and with a constant amount of bacteria added every day as food, each population of *Paramecium* grew until it leveled off at what was apparently the carrying capacity. When the two species were cultured together on a constant food supply, however, *P. caudatum,* apparently unable to compete with *P. aurelia,* was driven to extinction in the microcosm of the culture dish. Gause's experiments supported the hypothesis that two species with similar needs for the same limiting resources cannot coexist in the same place. This concept was later termed the **competitive exclusion principle.** Subsequent laboratory experiments on several other species of animals and plants reinforced the principle.

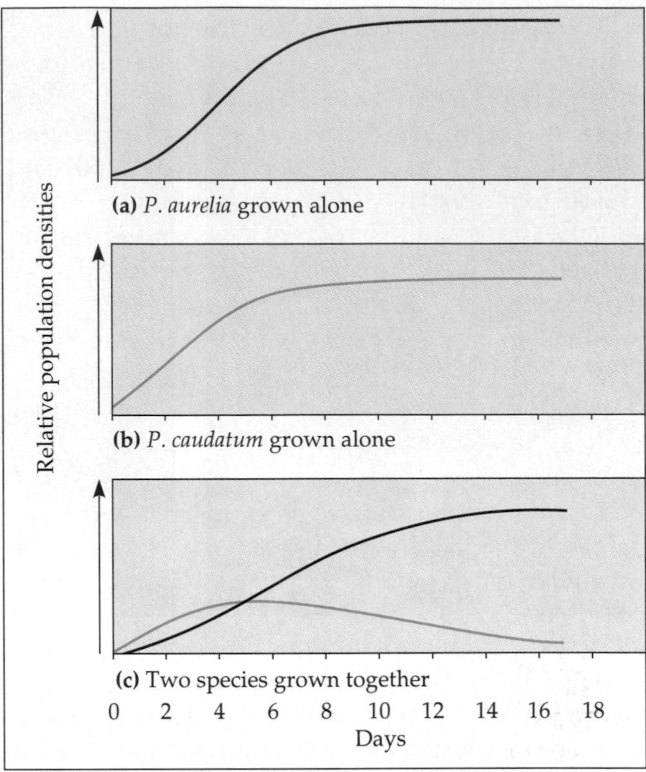

FIGURE 48.9
Competition in laboratory populations of *Paramecium.* In separate laboratory cultures with constant amounts of bacteria added every day for food, populations of the species (**a**) *P. aurelia* and (**b**) *P. caudatum* each grow to carrying capacity. (**c**) When the two species are grown together, *P. aurelia* has a competitive edge in obtaining food, and *P. caudatum* is driven to extinction in the culture.

Although laboratory studies have generally confirmed the predictions of the competitive exclusion principle, natural communities are infinitely more complex than laboratory environments. In order to determine how important competition is in structuring natural populations, ecologists have used both field experiments and indirect evidence from observations of how species use resources in the presence and absence of potential competitors. Before describing evidence about the importance of competition, we must consider how ecologists define and visualize resource use.

Ecological Niches

The concept of the ecological niche is almost inseparable from the concept of interspecific competition, but it is difficult to define rigorously. The **ecological niche** is the sum total of the organism's use of the biotic and abiotic resources in its environment. One way to think of the concept is through a comparison made by ecologist Eugene Odum: If an organism's habitat is its address, the niche is its occupation. Put another way, an organism's niche is its ecological role—how it "fits into" an ecosystem. The niche of a population of tropical tree lizards, for example, consists of, among other variables, the temperature range it tolerates, the size of trees upon which it perches, the time of day in which it is active, and the size and type of insects it eats.

The term **fundamental niche** refers to the set of resources a population is theoretically capable of using under ideal circumstances. In reality, each population is embedded in a web of interactions with populations of other species, and biological constraints, such as competition, predation, or the absence of some usable resources, may force the population to use only a subset of its fundamental niche. The resources a population actually uses are collectively called its **realized niche.**

We can now restate the competitive exclusion principle to say that two species cannot coexist in a community if their niches are identical. However, ecologically similar species can coexist in a community if there are one or more significant differences in their niches.

Evidence for Competition in Nature

One problem with evaluating competition in nature is that if competition is as potent a force as the competitive exclusion principle suggests, we would expect it to be quite rare in natural communities. The reason is simple. There are two possible outcomes to competition between species with the same ecological niche: either the weaker competitor will become extinct, or one of the species will evolve enough to use a different set of resources. In both of these outcomes, there will no longer be competition, which poses an intellectual and practical dilemma for ecologists. It is difficult to demonstrate the existence and importance of a force (competition) that, by its very nature, often cannot operate for long periods of time.

Although we cannot look directly at the evolutionary history of a community, many ecologists stress the importance of past competition, citing several lines of circumstantial evidence suggesting that it has been a major factor in shaping some of the ecological relationships we see today. Thus, in many cases we must study what ecologist Joseph H. Connell has called the "ghost of competition past." We'll look at some examples of this, but first we need to review two terms you first learned in Chapter 22. *Sympatric* populations occur in the same geographic area, and can thus interact, whereas *allopatric* populations occur in different geographic areas.

One line of evidence for past competition is simply the observation that similar species always seem to exhibit some niche differences when they coexist in a community. Patterns of **resource partitioning,** in which sympatric species consume slightly different foods or use other resources in slightly different ways, are well documented, particularly among animals. Several species of small arboreal lizards of the genus *Anolis,* for example, are often sympatric. At one site in the Dominican Republic, seven *Anolis* species live in close proximity to each other, feeding on small arthropods that land within their territories. However, each species uses a characteristic perching site (FIGURE 48.10), and these perch differences presumably minimize competition among the lizard species. Each species also has morphological characteristics, such as body size or leg length, that adapt it to its particular microhabitat, suggesting that natural selection has favored perch site specializations among these sympatric lizards. Similar patterns of resource partitioning are evident in other *Anolis* communities throughout the American tropics. In this case, resource partitioning is the "ghost" that has resulted from past competition among these lizard species.

A second line of circumstantial evidence for the importance of competition comes from comparisons of closely related species whose populations are sometimes sympatric and sometimes allopatric. Although allopatric populations of such species are similar in structure and use

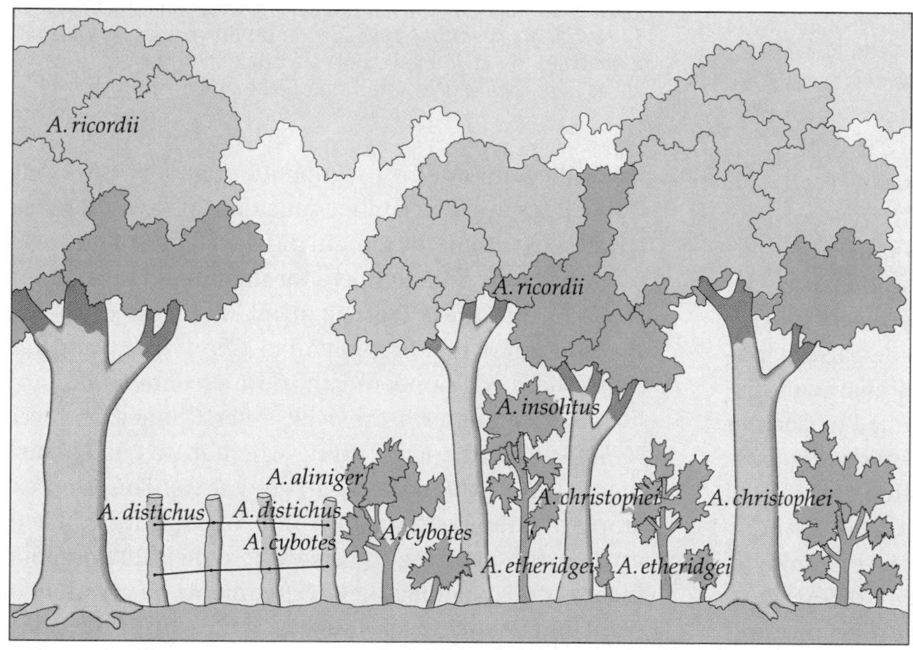

(a)

(b)

(c)

FIGURE 48.10

Resource partitioning in a group of sympatric lizards.

(**a**) Seven species of *Anolis* lizards live in close proximity at La Palma in the Dominican Republic. Each species perches in a characteristic microhabitat, distinguished by the amount of sun it receives and the size of the vegetation. (**b**) *A. distichus,* for example, perches on fenceposts and other sunny surfaces (such as this leaf), whereas (**c**) *A. insolitus* usually perches on shady branches. Such patterns of resource partitioning probably reduce interspecific competition among the members of a community, enabling them to coexist within a small geographic area.

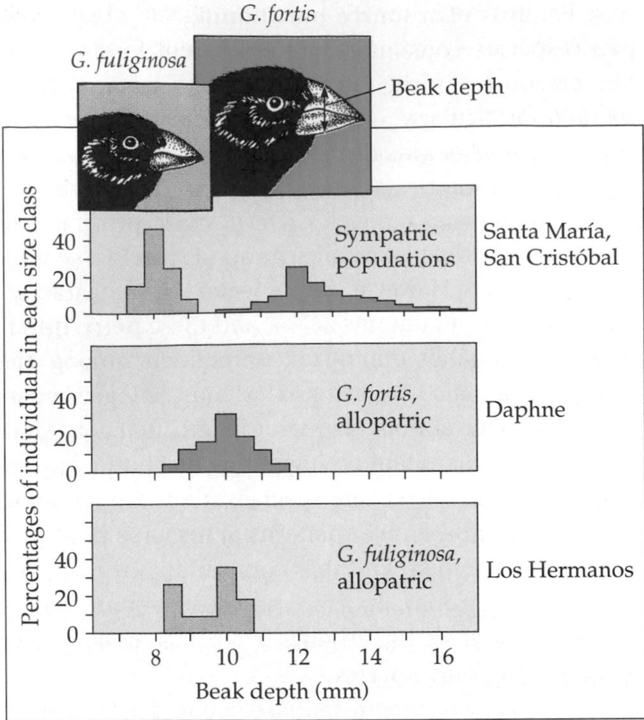

FIGURE 48.11
Character displacement: circumstantial evidence for competition in nature. Although allopatric populations of potential competitors are often similar in morphology and use equivalent resources, sympatric populations may diverge in both characteristics. In this example, two species of Galapagos finches have similar beak morphologies and presumably eat similarly sized seeds where their populations are allopatric on Daphne and Los Hermanos islands. However, where the two species are sympatric on Santa María and San Cristóbal, *Geospiza fuliginosa* has a shallower, smaller beak and *G. fortis* a deeper, larger one. Such evolutionary changes in morphology are thought to reflect resource partitioning. In this case, the two species have adapted to eating different sizes of seeds.

similar resources, sympatric populations often show differences in morphology and the resources they use. The tendency for characteristics to be more divergent in sympatric populations of two species than in allopatric populations of the same two species is called **character displacement.** The Galapagos finches described in Chapter 20 provide a good example of character displacement in beak sizes and, presumably, in the seeds that they can eat most efficiently. Allopatric populations of *Geospiza fuliginosa* and *G. fortis* have beaks of similar size, but on an island where both species occur, a significant difference in beak depth has evolved (FIGURE 48.11). This difference presumably enables the two species to avoid competition by feeding on seeds of different sizes and thus represents the "ghost" that resulted from past competition.

Although examples of resource partitioning and character displacement are compelling, they do not really prove the importance of competition. Controlled field experiments can provide evidence that one species can influence the density and distribution of another species when they are in direct competition. In a classic study, Connell manipulated the densities of two barnacle species that ordinarily grow in different strata of the rocky intertidal zone and compete for the limited attachment space on the rocks (FIGURE 48.12). The heavier-shelled *Balanus* grows more rapidly than *Chthamalus*, and *Balanus* shells edge underneath those of *Chthamalus* and literally pry them off the rock. After Connell removed *Balanus* from the lower strata where it was most common, *Chthamalus* was able to grow there. This is an example of interference competition; one species is able to exclude another from the area where their fundamental niches overlap. Similar field experiments on a variety of plant and animal species suggest that interspecific competition is strong under some circumstances.

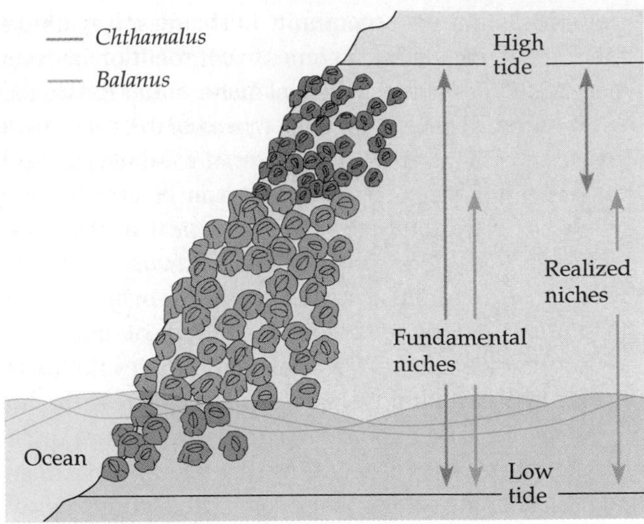

FIGURE 48.12
Experimental evidence for competition in nature. *Balanus balanoides* and *Chthamalus stellatus* are two species of barnacle that grow on rocks along the Scottish coast. These rocks are exposed during low tide. The barnacles have a stratified distribution, with *Balanus* most concentrated on the lower portions of the rocks and *Chthamalus* on the higher portions. The swimming larvae of the barnacles may settle randomly on the rocks and begin to develop into sessile adults, but *Balanus* fails to survive high on the rocks because it is unable to resist desiccation when these areas are exposed to air for several hours during low tides. Its fundamental niche and its realized niche are similar. Although *Chthamalus* is concentrated primarily on the upper strata of rocks, when ecologist Joseph H. Connell removed *Balanus* from the lower strata, the *Chthamalus* population spread into that area. Thus, *Chthamalus could* survive lower on the rocks than where it is generally found, if it were not for competition from *Balanus*. Its realized niche is only a fraction of its fundamental niche.

Commensalism and mutualism are (+0) and (++) interactions, respectively: *a closer look*

You learned in Chapter 25 that **symbiosis** ("living together") is a term that encompasses a variety of interactions in which two species, a **host** and its **symbiont,** maintain a close association. There are three types of symbiotic interactions. In **parasitism,** one organism, the parasite, harms the host; in **commensalism,** one partner benefits without significantly affecting the other; and in **mutualism,** both partners benefit from the relationship. In our survey of interspecific interactions in communities, we included parasitism with predation because both are (+−) interactions in which one organism generally feeds on another. In this section, we complete our survey of community interactions by examining commensal and mutual symbiotic relationships.

Commensalism (+0)

Few absolute examples of commensalism exist because of the unlikelihood that one partner in an ecological interaction will be completely unaffected by the other. "Hitchhiking" species, such as algae that grow on the shells of aquatic turtles or barnacles that attach to whales, are sometimes considered to be commensal. However, the hitchhikers may actually decrease the reproductive success of their hosts slightly by reducing the efficiency of the hosts' movements in their search for food or escape from predators. An association that may be truly commensal is the relationship between cattle egrets and grazing cattle (FIGURE 48.13). The egrets concentrate their predation near grazing cattle, which flush insects and other small animals from the vegetation as they move. Because the birds increase their feeding rates when following cattle, they clearly benefit from the association with cattle. It is difficult to imagine that the cattle either benefit from or are harmed by the relationship in any way. Because commensalism benefits only one of the species involved, any evolutionary change in the relationship is likely to occur only in the beneficiary.

FIGURE 48.13
Commensalism between a bird and a mammal. The cattle egret concentrates its feeding where grazing cattle and other large herbivores, such as this cape buffalo in Tanzania, flush insects from the vegetation. This relationship benefits the bird without apparent harm or benefit to the mammal. Of course, the relationship may help the buffalo in some way not yet discovered.

Mutualism (++)

In contrast to commensal relationships, mutualistic relationships require the evolution of adaptations in both participating species, because changes in either species are likely to affect the survival and reproduction of the other. Many coevolved mutualistic adaptations have been described in earlier chapters: nitrogen fixation by bacteria in the root nodules of legumes; the digestion of cellulose by microorganisms in the gut of termites and ruminant mammals; photosynthesis by unicellular algae in the tissues of corals; the exchange of nutrients in mycorrhizae, the association of fungi and the roots of plants; and specific interactions of certain pollinators and flowering plants. FIGURE 48.14 illustrates yet another interesting mutualistic interaction: the relationship between certain species of acacia and the ants that protect these trees from herbivorous insects.

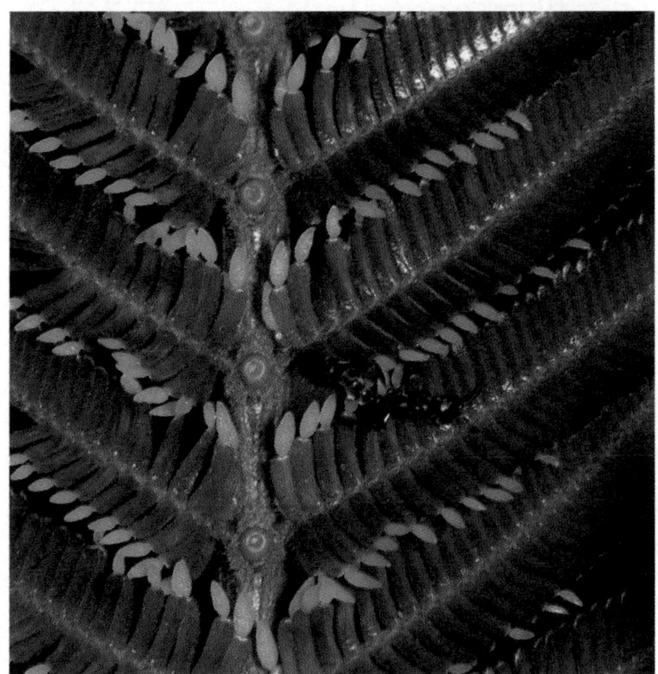

FIGURE 48.14
Mutualism between acacia trees and ants. Certain species of Central and South American acacia trees, called bull's horn acacias, have hollow thorns that house stinging ants of the genus *Pseudomyrmex.* The ants feed on sugar produced by nectaries on the tree and on protein-rich swellings called Beltian bodies (orange in the photograph) that grow at the tips of leaflets. The acacia benefits from housing and feeding a population of pugnacious ants, which attack anything that touches the tree. The ants sting other insects, remove fungal spores and other debris, and clip surrounding vegetation that happens to grow close to the foliage of the acacia. A series of experiments in the 1960s demonstrated the mutualistic nature of the relationship. When the ants on experimental trees were poisoned, the trees died, probably because of damage by herbivores and failure to compete with the surrounding vegetation for light and growing space.

Many mutualistic relationships may have evolved from predator-prey or host-parasite interactions. Certain angiosperm plants, for example, have adaptations that attract animals that function in pollination or seed dispersal; these probably represent counteradaptations to the herbivores' feeding on pollen and seeds. In many cases, pollen is spared when the pollinator is able to consume nectar, and seeds are dispersed by animals that eat the fruit that encloses them. Any plants that could derive some benefit by sacrificing organic materials other than pollen or seeds would increase their reproductive success, and the adaptations for mutualistic interactions would spread through the plant population.

Now that we have examined how different types of interspecific interactions affect the populations within a community, the next step is to take a less reductionist look at community structure.

A community's structure is defined by the activities and abundance of its diverse organisms

Although we understand the dynamics of some interspecific interactions within many communities, these relationships are inextricably tied together into a complex web that itself is much more difficult to study. Some of the first community ecologists were naturalists who listed species in an attempt to show the existence of plant communities with sharply defined boundaries. However, as we saw at the beginning of this chapter, sharp community boundaries rarely occur in nature. This difficulty became more and more apparent, and research began to yield new insights into community structure and processes.

We will focus our discussion of community structure on three features that result from the particular species composition of a community and the biological interactions among the populations that compose it: the feeding relationships that exist among organisms, patterns of abundance and diversity among species, and the ways a community is affected by disturbances.

Feeding Relationships Within Communities

The most basic observations of a natural community will yield information about what the species within the community eat. The various feeding relationships, or **trophic structure** (Gr. *trophe,* "nourishment"), are determined by predator-prey, host-parasite, and plant-herbivore interactions, but studies of trophic structure also identify some interactions that might represent competition for food. The trophic structure of communities has been studied in two ways. Analysis of feeding relationships places species in functional groups with similar trophic position: lumping all plants together as producers, all animals that consume plants—from caterpillars to cattle—as primary consumers (herbivores), and so on. However, as we shall see, lumping species into these functional categories obscures important aspects of community structure.

Another way to look at feeding relationships in communities is with food web analysis, which includes species-level information about the community and emphasizes the myriad connections among community members. This approach recognizes, for example, that not all plants are consumed by all herbivores. Food webs can be very complex, as we will see when we analyze them more thoroughly in Chapter 49. Nevertheless, a reasonably detailed knowledge of feeding relationships can be very useful in understanding community structure.

Species Richness, Relative Abundance, and Diversity

Communities differ dramatically in their **species richness,** the numbers of species they contain. Ecologists also recognize, however, that some communities consist of a few common species and many rare ones, whereas others contain an equivalent number of species that are all about equally common. The **relative abundance** of species within a community has an enormous impact on its general character. These concepts are illustrated in TABLE 48.2, which lists the trees present in a mature woodland. Two trees, yellow poplar and sassafras, make up almost 84% of the entire stand, and four of the ten tree species are represented by only a single individual each. A different community that had the same species richness, but in which the numbers were more evenly divided among the ten species, would seem more diverse. Indeed, the term **species diversity,** as used by ecologists, considers *both* components of diversity: species richness and relative abundance.

We will take a closer look at the importance of species diversity in community structure in the next section. At this point, it is important to consider that the impact of humans on communities usually reduces diversity. We currently use about 60% of Earth's land in one way or another, mostly as cropland, forest, and rangeland. Most crops are grown in monocultures. Forests that are used to produce pulpwood and lumber are often replanted in single-species stands. And the effects of intensive grazing on rangelands often include the removal of native plant species and replacement with a few introduced species.

Disturbances and Community Stability

Disturbances, both natural and human-caused, are common in ecological communities, but communities vary in their responses to disturbance. **Stability** is the tendency of a community to reach and maintain an equilib-

TABLE 48.2

The Trees Present in a Deciduous Forest in West Virginia

SPECIES	NUMBER	PERCENTAGE OF STAND
Yellow poplar (*Liriodendron tulipifera*)	122	44.5
Sassafras (*Sassafras albidum*)	107	39.0
Black cherry (*Prunus serotina*)	12	4.4
Cucumber magnolia (*Magnolia grandiflora*)	11	4.0
Red maple (*Acer rubrum*)	10	3.6
Red oak (*Quercus rubra*)	8	2.9
Butternut (*Juglans cinerea*)	1	.4
Shagbark hickory (*Carya ovata*)	1	.4
American beech (*Fagus grandiflora*)	1	.4
Sugar maple (*Acer saccharum*)	1	.4
	274	100.0

Source: Smith, R. L. *Elements of Ecology,* 3rd ed. New York: HarperCollins, 1992, p. 303.

rium, or relatively constant condition, in the face of disturbance. Forest gaps that fill in with similar tree species provide examples of stability in the face of small disturbances, such as the death of a small group of trees. The same forests might not be stable in the face of a large disturbance—widespread logging, for example. Some communities, such as eucalyptus forests, return very quickly to their original condition and species composition after large-scale disturbances, such as fire. As you can see, stability depends on both the type of community and the nature of the disturbance.

Community resilience, the ability of a community to persist in the face of disturbance, is related to stability but does not imply stability of the individual populations that compose the community. Populations can show wide fluctuations in response to environmental change, but the community itself could persist in its major structural features, and therefore be considered resilient.

As we will see, the subject of community response to disturbances provides another example of the changes in viewpoint that characterize ecology today as a vigorous field of scientific inquiry.

The factors that structure communities include competition, predation, and environmental patchiness

Our analysis of biomes in Chapter 46 identified a number of abiotic factors that determine the nature of the vegetation that grows in various regions of the Earth. But what factors determine the other community characteristics described in the preceding section? Why are some communities diverse and others not? Why are some communities relatively stable, whereas others do not readily recover from disturbances? Careful study in the last few decades has produced insights into the factors that influence the characteristics of particular communities, but you will see that we do not have answers to all of these questions; the most exciting research still lies ahead.

The Role of Competition

In the 1960s and 1970s, many ecologists proposed that competition was a major factor limiting the diversity of species that could occupy a community. This hypothesis was largely based on observations of niche differences and resource partitioning among sympatric species. A given quantity of resources, these ecologists argued, could be partitioned only so finely before the effects of competition would inevitably lead to the extinction of poorer competitors, setting a limit on the number of species that could occur together. Numerous studies, such as the barnacle study described earlier (see FIGURE 48.12), documented the effects of competition in nature, suggesting that it may be a potent force in structuring communities.

Still more evidence for the importance of competition in structuring communities comes from cases where species have been accidentally or purposely introduced into new communities by humans. In many cases, these

FIGURE 48.15
An introduced competitor. The zebra mussel (the smaller of the shelled mollusks in this photo), accidentally introduced to North America from Russia in the mid-1980s, competes with larval fish and other aquatic life for food, and with native clams for space.

introduced species, or **exotic species** have outcompeted native community members and altered community structure. A recent troublesome invader in the United States is the zebra mussel (FIGURE 48.15), which entered the St. Lawrence Seaway in the mid-1980s in ballast water released by a cargo ship that had traveled from the mussels' native Caspian Sea. By the summer of 1993, zebra mussels were found in the Mississippi River as far south as New Orleans. Of the many problems caused by the mussels, those that have provoked the most uproar have resulted from their competition with humans—by clogging reservoir intake pipes, for example. However, zebra mussels also compete with native shellfish for space, and with fish for the plankton used for nourishment. The extent to which this new competitor is altering community structure is yet to be determined.

Despite the evidence we have just examined, interspecific competition does not always lead to competitive exclusion, and competing species may sometimes coexist, albeit at reduced densities. Thus, although we can conclude that competition is probably important in regulating the relative abundance of many species and perhaps the species richness of many communities, its general importance is still being debated.

One general bias in studies of competition is that relatively little research has been conducted on the numerous species of herbivorous insects, organisms that are probably often subject to density-independent population limitation (see Chapter 47). We would not expect competition to be important to the ecology of species that rarely approach their carrying capacities, and few studies have demonstrated significant competition among such species. Herbivore populations in general are rarely regulated by their food supply, and surveys of competition indicate that it is far more prevalent among populations of plants and carnivores.

Many ecologists are still reluctant to embrace competition as the major factor structuring communities because, while it is hard enough to demonstrate that two species are competing in the present, it is even more difficult to assess what has happened in their evolutionary past. Furthermore, competition cannot be important unless population sizes are near carrying capacity and resources are limiting. Let's look at other forces that could also play a major role in shaping ecological communities.

The Role of Predation

In simple laboratory experiments where a single predator species is kept with a single prey species having no refuge, the predator may devour all the prey and then perish from starvation. If this scenario were commonly enacted in nature, predators would always reduce the diversity of species in communities. These effects have been documented in some aquatic systems, where the accidental or purposeful introduction of fish and other aquatic predators has drastically reduced diversity in communities.

Predators do not always reduce diversity. Probably the most important effect of a predator on community structure is to moderate competition among its prey species. Heavy predation can reduce the density of a strong competitor, thereby allowing weaker competitors to persist in the community. Experiments by ecologist Robert Paine in the 1960s were among the first to provide a clear picture of this complex interaction. Paine removed the dominant predator, a sea star of the genus *Pisaster,* from experimental areas within the intertidal zone on the coast of Washington state. The favorite prey of *Pisaster,* the mussel *Mytilus,* then outnumbered many of the other tidepool organisms that compete for attachment space on the rocks. Because *Mytilus* was such a dominant competitor in the experimentally created predator-free environment, the species richness of the community declined from 15 to 8 (FIGURE 48.16). This research and numerous other field experiments have given rise to the concept of a **keystone predator,** a predator that exerts an

(a)

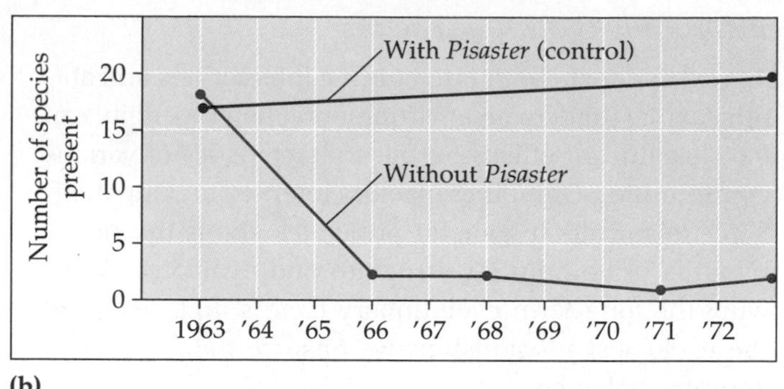

(b)

FIGURE 48.16

A keystone predator. (**a**) The rocky intertidal zone on the coast of Washington state contains a variety of invertebrate and algal species, including *Pisaster ochraceous*. *Pisaster* feeds preferentially on mussels but will also consume other invertebrates. (**b**) When ecologist Robert Paine experimentally removed *Pisaster* from a tidepool in 1963, mussels eventually took over the rock face and eliminated other invertebrates and algae. In a control area from which *Pisaster* was not removed, there was little change in species diversity.

important regulating effect on other species in the community. Keystone predators maintain higher species diversity in a community by reducing the densities of strong competitors, such that competitive exclusion of other species does not occur.

We have seen that both competitive and predatory interactions between members of a community can play roles in determining community diversity. However, in many cases interspecific interactions are not as important as the heterogeneity of the environment and the number of niches available. Next, we take a look at factors that influence environmental heterogeneity.

The Role of Environmental Patchiness

In general, habitats that are more heterogenous can support a more diverse community for the simple reason that they provide more ecological niches. This heterogeneity can be both spatial and temporal. An important aspect of spatial heterogeneity is the vegetation structure found in the community, and, as we saw in our survey of the world's biomes, there is a great deal of diversity in the overall form of vegetation. In turn, the vegetation in a community largely determines the types of animals found there. In general, structurally complex vegetation provides a diversity of microhabitats that can be used by a variety of animal species (see FIGURE 48.10), whereas less complex vegetation does not provide a structural resource that animal populations can partition easily.

Another important type of spatial heterogeneity is environmental patchiness. As described in Chapter 46, all environments are patchy over both space and time. For example, the mineral content of soil varies locally with the chemical composition of the rocks from which it is derived. Similarly, soil moisture varies with topography, such that low-lying areas are generally wetter than high ground. If different species are best adapted to different local conditions, environmental patchiness can increase the community's species diversity by facilitating resource partitioning among potential competitors. The distributions of some low-growing plants in the forests of Indiana, for example, are regulated by their seedlings' tolerance of calcium and organic matter. Because these factors vary locally, wild black cherry occurs in some patches and two species of violets are found in others. If the distribution of these abiotic factors were uniform throughout the forest, one of the species would probably outcompete the others, thereby reducing species diversity in the community. Of course, the importance of patchiness varies with the size and lifespan of the organisms under study. Soil characteristics may have a huge influence on the local distributions of small plants, but they are not likely to have much impact on a large herbivorous mammal that eats the plants.

Temporal use of habitats can also affect community diversity. This is especially pronounced with seasonal changes, with spring flowers giving way to summer and then fall-blooming species, and spring migrant birds being replaced by summer residents, then fall migrants. Over the course of a single day, animals that are active during daylight are replaced by nocturnal animals after dark. These temporal changes can greatly increase total species diversity in a community.

Evaluating Causative Factors

As we have seen, both interspecific interactions and abiotic factors that create environmental heterogeneity can have significant effects on the characteristics of various communities. Often these factors interact, making it difficult to formulate general principles about the determinants of community structure and diversity. Identifying the long-term evolutionary factors that produced the ecological relationships we observe today presents another challenge.

Research on the nesting behavior of two North American bird species that occupy the same trees illustrates the difficulty of identifying causative factors. MacGillivray's warbler nests low in trees; black-headed grosbeaks nest high. At first this might be interpreted as another example of resource partitioning that reduces competition—in this case, for nesting sites. But ecologist Thomas Martin has demonstrated that predation may be the key factor in this segregation of nests; predators locate and feed on the young of these birds more successfully when nests are concentrated in the same area, rather than being more widely distributed in the trees. In other words, the partitioning we see in nest sites reduces predation, and this, rather than a reduction in competition between the bird species, may explain the observed difference.

Thus, careful studies lead to the conclusion that communities are structured by multiple interactions of organisms with their biotic environment and with abiotic factors as well. Which interactions are most important can vary from one type of community to another. Because of the complexity of these networks, there are few, if any, natural communities for which we have a good understanding of all the important relationships or how the relationships evolved.

Community ecology is made even more challenging by the fact that the structure of a community may change, sometimes over relatively short periods of time. Disturbances influence structure, and the activities of the organisms themselves may destabilize an existing structure and favor a new one. The next section examines some of these changes.

Succession is the sequence of changes in a community after a disturbance

Changes in community composition and structure are most apparent after some disturbance—a flood, a fire, the advance and retreat of a glacier, volcanic eruption, overgrazing and other animal activities, or human activity—strips away the existing vegetation. Disturbances change resource availability and create opportunities for new species to become established. The disturbed area may be colonized by a variety of species, which are gradually replaced by others. Such transitions in species composition over ecological time represent a process called **ecological succession.** In traditional views of community ecology, the community passes through a sequence of predictable transitional stages, ultimately achieving a relatively stable state called a **climax community.**

This process is called **primary succession** if it begins in an essentially lifeless area where soil has not yet formed, such as on a new volcanic island, or on the rubble left behind by a retreating glacier. In the case of glaciers, which are still shrinking in places like Glacier Bay, Alaska, the barren ground is first occupied by mosses and lichens, and then by dwarf willows. After about 50 years, alders form dense stands. These eventually give way to Sitka spruce, which are later joined by hemlock, forming the relatively stable spruce-hemlock forest that we recognize as taiga (see Chapter 46). The entire process takes about 200 years (FIGURE 48.17).

Secondary succession occurs where an existing community has been cleared by some disturbance that leaves the soil intact. Often the area begins a return to something like its original state. Old-field succession in the Piedmont region of North Carolina has been studied extensively. If an agricultural field is abandoned in this area, an herbaceous community composed mostly of annual crabgrass develops in the first year. This is followed by other herbaceous plants, and by the third year, pine seedlings invade the field. The pines are eventually replaced by a climax community dominated by oaks and hickories, the vegetation that originally occupied the site before farmers cleared it for planting.

As noted above, succession may seem to produce a final stage called the climax community. Many ecologists once viewed the climax as a common endpoint that communities living under similar environmental conditions inevitably attain—a condition that then persists almost indefinitely. As we will see, however, the notion of a climax community is too simplistic to represent the wealth of variation in nature. Recent research on succession has moved ecologists toward a more pluralistic view of the nature of communities and their development through time.

Causes of Succession

In most cases, a variety of interrelated factors determines the course of succession. Although succession is often presented as a process that involves succeeding assemblages of species replacing each other, it actually takes place among individual species that are competing for available resources. Because resource availability changes over the course of succession, different species compete

(a) (b) (c) (d) (e) (f)

FIGURE 48.17

Succession after the retreat of glaciers. These photographs illustrate the different stages of succession: (**a**) retreating glacier; (**b**) barren landscape after the retreat; (**c**) moss and lichen stage; (**d**) alders and cottonwoods covering the hillsides; (**e**) spruce coming into the alder and cottonwood forest; (**f**) spruce and hemlock forest. Ecologists usually deduce the process of succession by studying a variety of areas that are at different successional stages. These photographs were taken in different places, of course, because the time frame of the changes they illustrate is about 200 years.

better at different stages. Early stages are typically characterized by *r*-selected species that are good colonizers because of their high fecundity and excellent dispersal mechanisms. Many of these may be described as "fugitive" or "weedy" species that do not compete well in established communities, but maintain themselves by constantly colonizing newly disturbed areas before better competitors can become established in the same places.

Tolerance of the abiotic conditions in a barren area also affects the species composition during early successional stages. Many *K*-selected species may colonize an area, but they will not grow in abundance if environmental conditions there are at the extremes of their tolerance limits. Variations in the growth rates and maturation times of colonizing species are also clearly important. For example, if the seeds of two plant species, one an annual herb and the other a tree, colonize a community at the same time, the herbaceous species will have earlier prominence because of its faster growth and shorter generation time. The ecological impact of the tree species may not be realized until the trees are relatively large.

Many of the changes in community structure during succession may be induced by the organisms themselves. Direct biotic interactions may be involved, including *inhibition* of some species by others through exploitative competition, interference competition, or both. The presence of organisms also affects the abiotic environment by modifying local conditions. This may result in *facilitation,* in which the group of organisms representing one successional stage "paves the way" for species typical of the next stage. For example, the alders that are abundant in an intermediate stage of glacial-till succession (FIGURE 48.17d) lower soil pH as their dropped leaves decompose. The change in pH facilitates the entry of spruce and hemlock (FIGURE 48.17f), which require acidic soil. Sometimes the changes that facilitate the development of a later stage actually make the environment unsuitable for the very species responsible for the changes.

Both inhibition and facilitation may be involved throughout the successional process. For example, horseweed is one of the earliest colonizers in the old-field succession described earlier. For a year or two, horseweed may inhibit other species through shading and soil water use. However, as the horseweed and other early species die and decompose, their presence facilitates the entry of later species by adding organic matter to the soil, which aids in holding moisture. The pine trees that dominate a later successional stage require full sunlight; their growth becomes self-inhibitory as the trees shade the ground and prevent their own offspring from growing. At the same time, the pines continue to add organic material to the soil, and the shade they cast keeps the forest floor moist, conditions that facilitate the germination and growth of the hardwoods that follow. At the climax stage, environmental conditions are such that the same species can continue to maintain themselves. For example, the oak-hickory forest that is the climax stage of the old-field succession maintains the moist, shaded environment that allows offspring of these species to grow, while inhibiting most of the species typical of earlier stages of succession.

As noted earlier, some ecologists have challenged the traditional idea of a climax community. There is evidence that what appear to be climax communities may not be stable over very long periods of time. Studies of pollen preserved in lake sediments provide evidence of community composition over thousands or even millions of years. These studies indicate that common tree species sometimes disappear from North American forests for hundreds of years, only to eventually return at a later time (see the interview with Margaret Davis, pp. 1058–1060). The reasons for such changes are unknown. In addition, many communities are routinely disturbed by outside factors during the course of succession. For example, as we saw in Chapter 46, prairie grasslands are maintained by fire. Without fire, some grassland areas would develop into forest, at least in moister areas. We could say that forest is the climax community for such areas, but that would make little sense if the forest community never develops. In this case, periodic fires stabilize the community at a stage that precedes the typical climax state. Let's take a closer look at some of the natural and human disturbances that may set succession in motion or prevent it from occurring.

Natural and Human Disturbance

Disturbances are events that can disrupt communities, changing resource availability and creating opportunities for new species to become established. Variation in the size, frequency, and severity of disturbances affects the magnitude of their impact. Important natural causes of disturbance are fire, drought, wind, and moving water. Many animals are agents of disturbance. Forests that contained large deer populations or overgrazed grasslands both illustrate the impact that herbivores can have on communities. Beavers and gypsy moths are other examples of animals that have created major disturbances in communities throughout the world.

Of all animals, humans have had the greatest impact on communities all over the world. Logging and clearing for urban development, mining, and farming have reduced large tracts of mature hardwood and pine forests to small patches of disconnected woodlots in many parts of the United States and throughout Europe. Similarly, agricultural development has disrupted what were once the vast grasslands of the North American prairie.

After a community is disturbed and then left alone, early stages of secondary succession, which are often dominated by weedy and shrubby vegetation, may persist for many years. This type of vegetation can be found extensively in forests that have been clear-cut, in agricultural fields no longer under cultivation, and in vacant lots and construction sites that are periodically cleared. Much of the United States is now a hodgepodge of early successional growth where mature communities once prevailed, and repeated disturbances often prevent mature communities from developing.

Human disturbance of communities is by no means limited to the United States and Europe, nor is it a recent problem. Tropical rain forests are quickly disappearing as a result of clear-cutting for lumber and pastureland. Centuries of overgrazing and agricultural disturbance have undoubtedly contributed to the current famine in parts of Africa by turning seasonal grasslands into great barren areas.

We tend to think of disturbances as having exclusively negative impacts, but this is not always the case. In many cases, small-scale natural disturbances, such as those that result in patches of different successional stages, can be important to the maintenance of species diversity in a community. Frequent small-scale disturbances can also prevent large-scale disturbances with more negative impacts. The fires that occurred in Yellowstone National Park during the summer of 1988 are an example. Much of this park was dominated by lodgepole pine, a tree that requires the rejuvenating influence of periodic fires. Its cones remain unripened with viable seeds in them until a fire destroys the parent tree, opens the cones, releases the seeds, and prepares a seedbed well-fertilized with ash. The immature trees do not burn easily, but trees that are over 100 years old become increasingly flammable. Fire suppression in this century had prevented small fires that would have resulted in patches of less-flammable early-successional trees, and by 1988, one-third of the Yellowstone forests were 250 to 300 years old. The drought conditions of 1988, combined with the fuel that had accummulated in the forest, resulted in large-scale destruction. In this case, the Yellowstone community demonstrated resilience but not stability. The community has changed; there are larger meadows, stands of seedling lodgepole pines, and thick stands of quaking aspen and herbaceous vegetation (FIGURE 48.18). A hundred years from now, however, the community will probably appear much as it did before the fire.

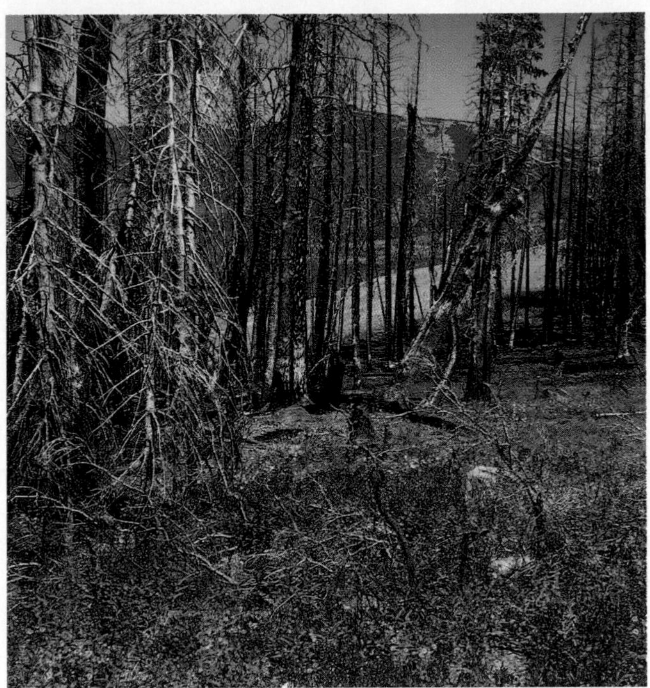

FIGURE 48.18
Secondary succession in the aftermath of the Yellowstone National Park fire of 1988.

Community Equilibrium, Disturbance, and Species Diversity

The traditional concept of ecological succession was that the climax community represents an equilibrium. During the successional process, the *K*-selected species that are stronger competitors replace the *r*-selected early colonizers as population densities increase and the vegetation modifies the site. Once the longer-lived *K*-selected species have become established, the rate at which new species replace old ones slows down. Although there may be some turnover of species in the climax community, the addition of new colonists is balanced by localized extinctions. This equilibrial view of community dynamics emphasizes the importance of interspecific interactions in structuring the community. Predation, competition, and symbiosis become more extensive and varied during succession, making increased diversity possible. According to this model of community development and structure, succession reaches a climax when the web of biotic interactions becomes so intricate that the community is saturated. No additional species can "fit into" the community unless resources become available through the localized extinction of species that are already present.

An opposing view portrays communities as being in a continual state of flux: The identities and numbers of species change during all successional stages, even in the so-called climax stage. This nonequilibrial model of community dynamics emphasizes the importance of less-predictable factors, such as dispersal and disturbance, in the development of community composition and structure. The course of succession may vary, for example, with the identity of the particular species that happen to colonize an area first. Severe disturbances such as fires, hurricanes, and windstorms may prevent a community from ever achieving a state of equilibrium. Proponents of the nonequilibrial view of community development often describe a mature community as an unpredictable mosaic of patches at different successional stages. Local environmental heterogeneity also contributes to the mosaic nature of many communities because different species occupy different habitat patches.

According to the nonequilibrial model, disturbance is a major determinant of community composition and species diversity. When disturbance is severe and frequent, the community may include only good colonizers typical of early stages of succession. If disturbances are mild and rare in a particular location, the late-successional species that are most competitive will make up the community. According to the **intermediate disturbance hypothesis,** species diversity is greatest where disturbances are moderate in both frequency and severity, because organisms typical of different successional stages will be present. Studies of species diversity in tropical rain forests provide some evidence for the intermediate disturbance hypothesis. Scattered throughout these forests are gaps where trees and the vines attached to them have fallen. In these disturbed areas, species from various successional stages coexist within a relatively small space.

Biogeography complements community ecology in the analysis of species distribution

The field of biogeography provides a different approach to an understanding of community properties by analyzing both global and local phenomena, mostly from a historical perspective. **Biogeography** is the study of the past and present distribution of individual species, as well as entire communities.

Traditionally, biogeographers have been concerned with the actual identities of the species that make up particular communities. For example, they show how the present distributions of species reflect their distant evolutionary history, as well as more recent interactions with both biotic and abiotic components of the environment. More than a century ago, Darwin and other naturalists began to recognize biogeographical realms that we can now associate with patterns of continental drift that followed the breakup of Pangaea (FIGURE 48.19). We discussed these historical aspects of biogeography in Chapters 20 and 23. Although this is still an active area of research, some biogeographers have more recently applied the principles of community ecology to the analysis of geographical distribution. The intersection of these fields is fertile ground: Biogeographical analyses

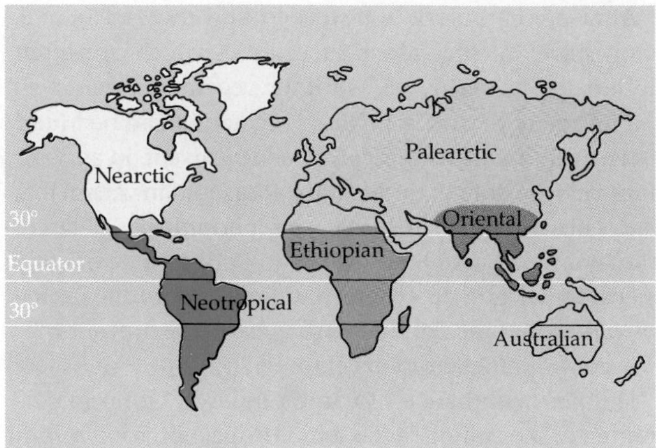

FIGURE 48.19

Biogeographical realms. Continental drift and barriers such as deserts and mountain ranges all contribute to the distinctive floras and faunas found in Earth's major regions. Except for Australia, the realms are not sharply delineated but grade together in zones where taxa from adjacent realms coexist.

have contributed as much to our understanding of community patterns and processes as the study of community ecology has contributed to our knowledge of biogeography. Here, we consider three intersecting topics that have captured the attention of ecologists during the last three decades.

Limits of Species Ranges

Understanding the determinants of a species' geographical range is central to any analysis in community ecology and biogeography. Three general explanations can account for the limitation of a species to a particular range today: (1) The species may never have dispersed beyond its current boundaries; (2) pioneers that did spread beyond the observed range failed to survive; and (3) over evolutionary time, the species has retracted from a once larger range to its present boundaries. Research in paleontology and historical biogeography can identify cases for which the third explanation is valid. For example, we know from fossil evidence that close relatives of living elephants and camels once occupied North America, but local extinctions have caused retractions of the geographical ranges of both groups.

Distinguishing between the first and second explanations is more difficult, because discovering a few colonists that dispersed to a new area but then died is much harder than looking for the proverbial needle in a haystack. However, transplant experiments, in which individuals of a plant or animal species are moved to similar environments outside their range, can provide useful information. Survival of the transplants suggests that the species had not dispersed to suitable locations outside the existing range. Failure of the transplants, in contrast, suggests that the species could not expand its range because of its inability to tolerate abiotic conditions or to compete with resident species. In one simple experiment, researchers transplanted several individuals of the lizard *Anolis cristatellus* from the warm lowlands in Puerto Rico to a cool, forested site at higher elevation. These animals failed to survive for more than a few weeks, presumably because they could not attain sufficiently high body temperatures to capture and digest their food. Therefore, this species does not occupy high-elevation forests because it is not adapted to the physical environment in such habitats.

Global Clines in Species Diversity

Ecologists have long recognized the existence of clines (gradual variation) in species diversity with major geographical gradients. The number of terrestrial bird species in North America, for example, increases steadily from the Arctic to the tropics (FIGURE 48.20). Similar clines

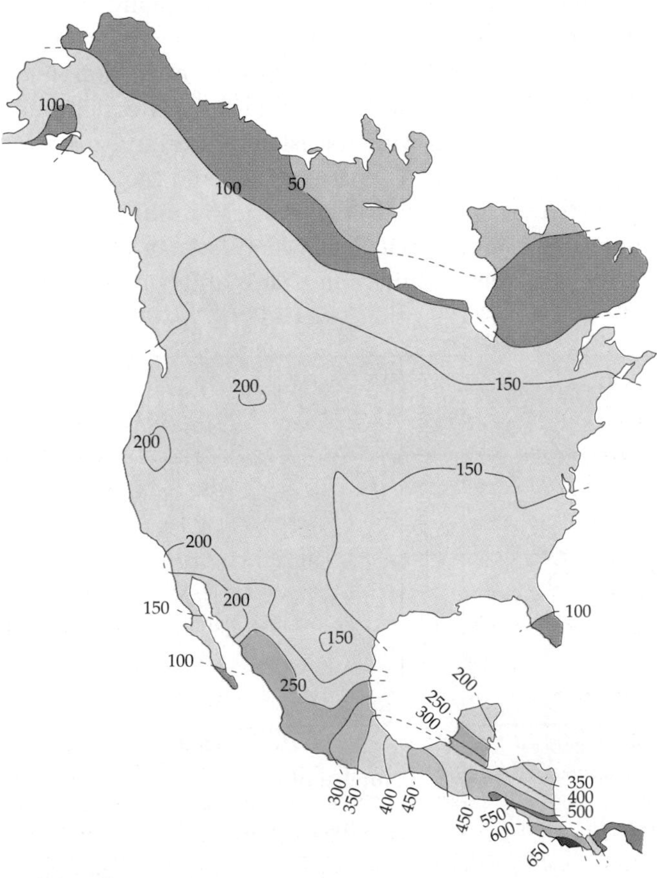

FIGURE 48.20
Species density of North and Central American birds. Biogeographers often plot latitudinal trends in numbers of species in the form of a "topographic" map that illustrates how many species occupy different geographic areas. In this species-density map for North and Central American birds, we can see that fewer than 100 species are found in arctic areas, whereas more than 600 occupy some tropical regions.

have been observed for most other major groups of organisms, including microbes, flowering plants, reptiles, and mammals. Explaining this pattern (and, more generally, the factors that regulate all patterns of diversity in natural communities) has been a goal of many community ecologists and biogeographers during much of this century. Several hypotheses have been proposed, many of which apply factors that regulate species diversity on more local scales. We will briefly consider six of these hypotheses.

1. Some ecologists suggest that tropical communities are very old and rarely experience major natural disturbances. In addition to providing time for a greater variety of plants and animals to evolve, the age of tropical communities has allowed complex population interactions to coevolve and develop to a greater extent than in the temperate zone.
2. Tropical regions may generally experience intermediate levels of disturbance and have greater environmental patchiness, allowing a greater diversity of plant species to form the resource base for diverse communities of animals.
3. The stability and predictability of tropical climates might allow many organisms to specialize on a narrower range of resources. Smaller niches would reduce competition and permit a finer level of resource partitioning among species, which in turn would foster higher species diversity.
4. Increased solar radiation in the tropics increases the photosynthetic activity of plants, which provides an increased resource base for other organisms.
5. The structural complexity of tropical forests creates a great variety of microhabitats that other plants and animals can partition.
6. Diversity is, in a sense, self-propagating because the complex predator-prey and symbiotic interactions in a diverse community prevent any populations from becoming dominant.

Many of these hypotheses can be accepted or rejected for specific groups of organisms or specific communities, but the question of what factors determine latitudinal gradients in species diversity is simply too large to be addressed with simple experiments either in the laboratory or in the field. Although most of the explanations listed above may be true for some organisms and some communities, complex combinations of these factors probably operate under most circumstances.

Another clinal diversity pattern is an increase, with depth, in the number of species in marine benthic faunas. Several ecologists have proposed that, despite the low level of productivity in deep-sea benthic communities, the long-term stability of these habitats has fostered many coevolved relationships among the organisms present. However, no research has yet been able to establish a *causal* relationship between environmental stability and species diversity.

Island Biogeography

Because of their limited size and isolation, islands provide excellent opportunities for studying some of the factors that affect the species diversity of communities. By "islands," we mean not only oceanic islands but also habitat islands on land, such as lakes, mountain peaks separated by lowlands, or natural woodland fragments surrounded by areas disturbed by humans—in other words, any areas surrounded by an environment not suitable for the "island" species. In the 1960s, American ecologists Robert MacArthur and E. O. Wilson developed a general theory of island biogeography to identify the important determinants of species diversity on an island with a given set of physical characteristics (see the interview with Wilson, p. 482). The study of islands may help us understand some of the interactions in more complex systems.

Imagine a newly formed oceanic island some distance from a mainland that will serve as a source of colonizing species. Two factors will determine the number of species that eventually inhabit the island: the rate at which new species immigrate to the island and the rate at which species become extinct on the island. Immigration and extinction rates are, in turn, affected by two important features of the island: its size and its distance from the mainland. Small islands will generally have lower immigration rates, because potential colonizers are less likely to reach a small island. For example, birds blown out to sea by a storm are more likely to land by chance on a larger island than on a small one. Small islands will also have higher extinction rates. They generally contain fewer resources and less diverse habitats for colonizing species to partition, increasing the likelihood of competitive exclusion. Distance from the mainland is also important; for two islands of equal size, a closer island will have a higher immigration rate than one farther away.

The immigration and extinction rates are also affected at any given time by the number of species already present on the island. As the number of species on the island increases, the immigration rate of new species decreases, because any individual reaching the island is less likely to represent a species that is not yet present. At the same time, as more species inhabit an island, extinction rates on the island increase because of the greater likelihood of competitive exclusion.

These relationships are summarized in FIGURE 48.21, where immigration and extinction rates are plotted as a function of the number of species present on the island. The main point of this model is that eventually, an equilibrium will be reached where the rate of species immigration matches the rate of species extinction (FIGURE 48.21a). The number of species at this equilibrium point is correlated with the island's size and distance from the mainland. The theory of island biogeography predicts that the equilibrium species number is greater on a large island than a small island (FIGURE 48.21b), and greater on a near island than a far island (FIGURE 48.21c).

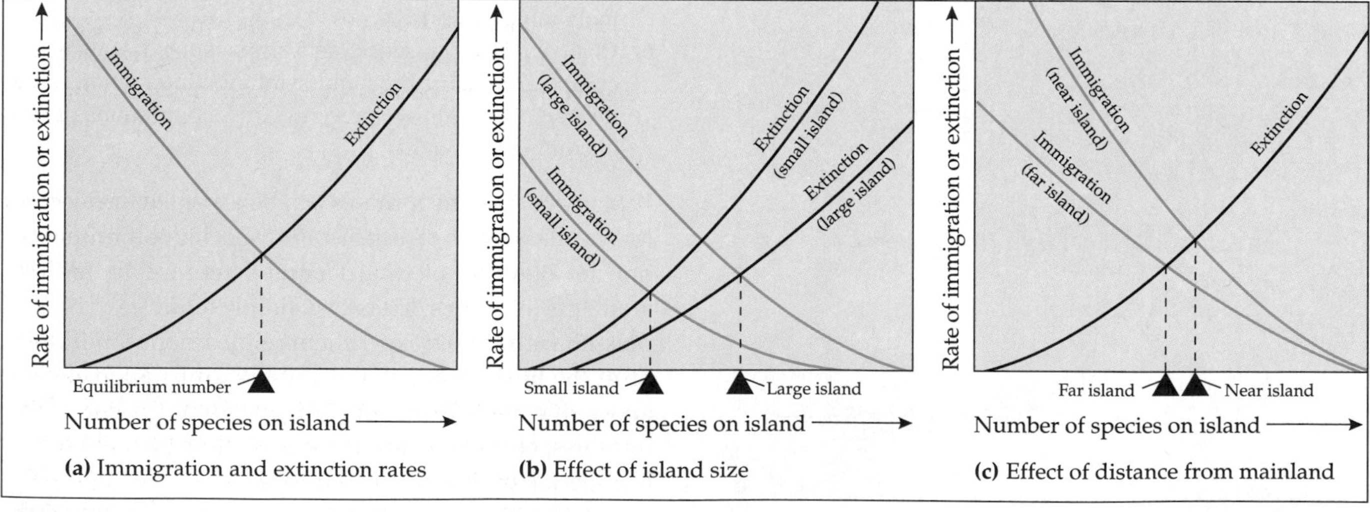

FIGURE 48.21
The theory of island biogeography. (**a**) The equilibrium number (black triangle) of species on an island represents a balance between the immigration of new species to the island and the extinction of species that are already there. (**b**) Large islands will ultimately have a larger equilibrium number of species than small islands because immigration rates are higher and extinction rates are lower on large islands. (**c**) Although extinction rates do not differ with an island's distance from a mainland source of species, near islands will have larger equilibrium numbers of species than far islands, because immigration rates to near islands are higher than those for more distant ones.

(Any equilibrium, of course, is dynamic; immigration and extinction continue, and the exact species composition may change over time.) This theory generally applies over a relatively short period, where colonization is the important process determining species composition. Over a longer period, adaptive evolutionary changes in island species and speciation on the island can affect the species composition and community structure.

Observations and experiments provide evidence that new islands do, indeed, reach an equilibrium in their species richness. For example, within 35 years after a volcanic eruption killed nearly all organisms on the island of Krakatoa in 1883, the diversity of birds that repopulated the island had reached an equilibrium of about 30 species. MacArthur and Wilson's studies of the diversity of amphibians and reptiles on many island chains, including the West Indies, support the prediction that species richness increases with island size (FIGURE 48.22). Species counts also fit the prediction that the number of species decreases with increasing remoteness of the island.

In the late 1960s, Wilson and Daniel Simberloff tested the theory of island biogeography in experiments on small islands of mangroves off the southern tip of Florida (see FIGURE 46.1). Six islands, each about 12 m in diameter, were enclosed in tents and fumigated to kill all resident arthropods. The pesticide used, methyl bromide, decomposes rapidly, and the islands could be recolonized by arthropods from the mainland species pool. Within about a year, species numbers equilibrated on each island, with the fewest species on the island most distant from the coast. Although the equilibrium number for each island was about the same before fumigation and after recolonization, the species composition was different. Chance events—in this case, which species of arthropods happened to disperse to which islands—affected the species composition of the communities.

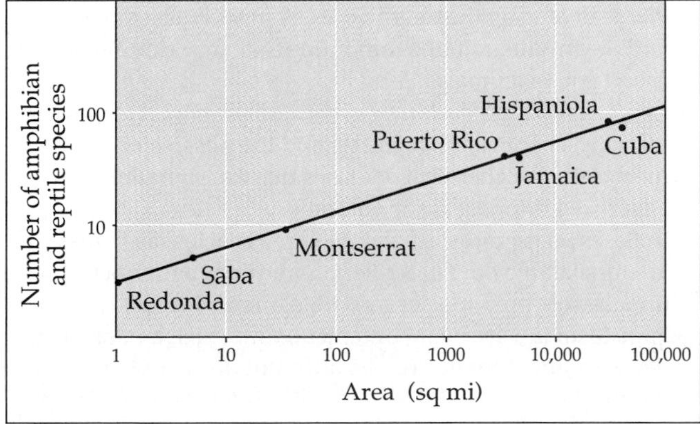

FIGURE 48.22
Species richness and island size. This species-area graph illustrates that the number of amphibian and reptile species found on West Indian islands is closely related to island size. Large islands harbor more species because greater habitat diversity allows greater resource partitioning among the resident species, reducing the likelihood of competitive exclusion.

Lessons from community ecology and biogeography can help us preserve biodiversity

As human encroachment on natural communities continues to advance at a rapid pace, concerned citizens and governments are supporting efforts to conserve as much as possible of the biological diversity that remains on Earth. Global support for conservation stems from our esthetic appreciation for other forms of life, the recognition that products useful to humans may ultimately be derived from species not yet discovered and the understanding that natural environments are necessary to maintain the proper functioning of the biosphere. However, conservation efforts are complicated by continued rapid human population growth and the economic difficulties faced by people in every region of the planet. Strict conservationists might call for an end to all development in wilderness areas, but the political realities suggest that such an approach is not likely to be embraced by governments anywhere.

What lessons from community ecology and biogeography can be applied to conservation so that we may maximize the preservation of biodiversity? Most preservation efforts in the past have focused on preserving endangered species, which is helpful. However, conservation efforts increasingly aim to conserve entire ecosystems. This more sophisticated and realistic approach requires a solid understanding of many aspects of community ecology, as governments and citizen groups attempt to choose the size and location of areas that will be protected from development and pollution. These preserved areas exist (and will continue to exist) as islands in a metaphorical sea of disturbed habitat that is unsuitable for the organisms being protected. Lessons from the study of island biogeography are increasingly being applied in the design of nature preserves; the goal is to understand *in advance* what a preserve with a particular set of characteristics might successfully protect. The species-area relationship portrayed in FIGURE 48.22, for example, indicates that the number of species a preserve can harbor is directly related to the size of the preserve.

In the late 1970s, Thomas Lovejoy, with the support of the World Wildlife Fund and the Brazilian government, used the approach of island biogeography to undertake what has become the largest biological experiment in history. He persuaded landowners who were converting rain forest tracts in the Amazon region into pastureland and farms to leave square patches (islands) of forests ranging from 1 to 1000 hectares in size. The goal of this experiment was to determine the smallest size a rain forest preserve must be in order to sustain the native species. A team of biologists is monitoring diversity in these sites and will continue to do so into the next century. As predicted by the theory of island biogeography, the diversity of the smaller islands is decreasing the most rapidly, but species extinction has been unexpectedly fast, in large part due to the deep penetration of winds, which dry out the forest 100 or more meters from the edge. Many plant and animal species have disappeared from the smaller plots, though a few are actually increasing in numbers. For example, fruit-eating Saki monkeys have left the 10-hectare plots, but howler monkeys, which eat leaves, have been able to find enough food to remain. Colonies of army ants require more than 10 hectares to maintain their worker force, so they quickly disappeared from the small plots, followed by five species of ant birds that survive by following ant swarms and feeding on the insects driven forward by army ant raiding fronts. While there is a mountain of data to be analyzed from this study, we have learned that an Amazon forest chopped into many small fragments loses diversity quickly.

There are many issues to consider in determining the relative benefits of creating one large preserve versus a group of smaller preserves that collectively include the same area as a large one. If the smaller preserves all contain the same representation of species, one large preserve probably would be preferable. However, if the community being conserved includes environmental heterogeneity, such that different species occupy different areas, several small preserves would retain a larger total number of species than one large preserve. In addition, populations within the small preserves might avoid the spread of epidemic diseases that could not cross the unsuitable habitats between them. When ecologists and wildlife managers design a series of small preserves, they are also concerned with their spatial relationship to one another. Reserves that are closely spaced might allow the possibility of recolonization from one preserve to another if a species became locally extinct. Similarly, corridors of suitable habitat that connect the preserves might encourage the spread of protected species among them.

Many questions about the optimal design of nature preserves remain to be answered. Hopefully the Lovejoy study and others like it will teach us how to manage land use in a way that minimizes loss of biodiversity. It is research that highlights the relationships between science, technology, and society—one of the themes of this book.

REVIEW OF KEY CONCEPTS (with page numbers and key figures)

- The interactive and individualistic hypotheses pose alternative explanations of community structure: *science as a process* (pp. 1118–1120, FIGURE 48.1)
  - A community is an assemblage of species living close enough together for potential interaction.
  - The individualistic hypothesis proposes that communities are chance assemblages of independently distributed species that happen to have the same abiotic requirements. The interactive hypothesis states that the species within a community are locked into biotic interactions and that the community functions as an integrated unit.
  - Many plant species seem to be independently distributed; animals, however, are frequently linked to other species.
- Community interactions can provide strong selection factors in evolution (pp. 1120–1121)
  - Coevolution refers to reciprocal interactions between two species that result in a series of adaptations and counteradaptations.
- Interspecific interactions may have positive, negative, or neutral effects on a population's density: *an overview* (pp. 1121–1122, TABLE 48.1)
  - A pair of signs, such as (+−) for predation, symbolizes how an interspecific interaction affects the two species.
- Predation and parasitism are (+−) interactions: *a closer look* (pp. 1122–1125)
  - Predation refers to interactions in which animals feed on other organisms.
  - Plants defend against herbivores by mechanical defenses and the production of compounds that are irritating or even toxic to animals.
  - Animals resist predation by cryptic coloration, deceptive markings, behavioral defenses, and the possession of mechanical or chemical defenses that are sometimes advertised by aposematic coloration.
  - In Batesian mimicry, a palatable prey species resembles an unpalatable one; in Müllerian mimicry, a number of unpalatable prey species resemble one another.
  - Parasitism is a symbiotic interaction in which a parasite derives nourishment from its host, but does not kill it outright.
- Interspecific competitions are (− −) interactions: *a closer look* (pp. 1125–1128, FIGURE 48.12)
  - The competitive exclusion principle states that two species competing for the same limiting resources cannot coexist in the same place.
  - The ecological niche is the sum total of the organism's use of the biotic and abiotic resources of its environment. Closely related species may be able to coexist if there are one or more significant differences in their niches.

- Resource partitioning and character displacement provide indirect evidence for the importance of past competition.
- Field experiments provide direct evidence of competition if the removal of one competitor allows the other to increase in density.

- Commensalism and mutualism are (+0) and (++) interactions, respectively: *a closer look* (pp. 1128–1130)
  - Commensalism refers to symbiotic interactions in which one species benefits and the other is not affected.
  - Mutualism refers to symbiotic interactions in which both species benefit.

- A community's structure is defined by the activities and abundance of its diverse organisms (pp. 1130–1131)
  - The trophic structure of a community refers to all of the feeding relationships in the community.
  - Species diversity within a community refers to both the absolute number of species present and their relative abundance.
  - Stability refers to the tendency of a community to reach and maintain an equilibrium in the face of disturbance.

- The factors that structure communities include competition, predation, and environmental patchiness (pp. 1131–1134)
  - Competition may be important in structuring many communities, reducing the densities of competing species.
  - Keystone predators limit the densities of the most competitive species, thereby increasing species diversity by allowing more species to coexist.
  - Environmental patchiness can also increase species diversity because different species are best adapted to conditions in different patches.
  - Determining the causative factors that structure communities is often difficult because of the complex ways in which populations interact.

- Succession is the sequence of changes in a community after a disturbance (pp. 1134–1138, FIGURE 48.17)
  - Succession involves changes in the species composition of a community over ecological time. Primary succession occurs where no soil previously existed; secondary succession begins in an area where soil remains after a disturbance.
  - Sometimes succession involves inhibition, a phenomenon in which species inhibit the growth of newcomers.
  - Facilitation refers to alterations in the environment by the species of one stage that enable species in the next stage to grow.
  - Disturbances are events that disrupt communities. They have variable effects on communities, depending on the severity and duration of the disturbance.
  - Some ecologists believe that communities are in continual flux, with the identity and numbers of species changing at all stages of succession, even at the climax stage.

- Biogeography complements community ecology in the analysis of species distribution (pp. 1138–1141, FIGURE 48.21)
  - A species would be limited to a given range if it never dispersed beyond that range, if it dispersed but failed to survive in other locations, or if it retracted from a larger range.
  - Species diversity within many groups of organisms increases in a clinal pattern from polar regions to the tropics.
  - A general theory of island biogeography maintains that species richness on an ecological island levels off at some dynamic equilibrium point, where new immigrations are balanced by extinctions. The theory predicts that species richness is directly proportional to size and inversely proportional to distance of the island from the source of colonizers.

- Lessons from community ecology and biogeography can help us preserve biodiversity (pp. 1141–1142)
  - The design of nature preserves is frequently based on the principles established by studies of island biogeography.

SELF QUIZ

1. The concept of trophic structure of a community emphasizes the
   a. prevalent form of vegetation
   b. keystone predator
   c. feeding relationships within a community
   d. effects of coevolution
   e. species richness of the community

2. According to the concept of competitive exclusion,
   a. two species cannot coexist in the same habitat
   b. extinction or emigration are the only possible results of competitive interactions
   c. intraspecific competition results in the success of the best adapted individuals
   d. two species cannot share the same realized niche in a habitat
   e. resource partitioning will allow a species to utilize all the resources of its fundamental niche

3. The effect of a keystone predator within a community may be to
   a. competitively exclude other predators from the community
   b. maintain species diversity by preying on the prey species that is the dominant competitor
   c. increase the relative abundance of the most competitive prey species
   d. encourage the coevolution of predator and prey adaptations
   e. create nonequilibrium in species diversity

4. All the following statements are consistent with the nonequilibrial model of community structure *except*
   a. chance events such as dispersal and disturbance play major roles in determining species diversity
   b. species diversity may be increased by certain types of disturbances
   c. even when a community represents a mature climax stage, species composition and the number of species may continue to change
   d. succession reaches a climax when the intricate web of interactions allows for the addition of new species only when resources are made available by extinction
   e. the course of succession may vary depending on the chance arrival of early colonizers

5. Transplant experiments provide evidence that
   a. transplants always fail
   b. continental drift accounts for the geographical distribution of species
   c. the theory of island biogeography is valid
   d. keystone predators maintain community structure
   e. some species can live outside their normal ranges

6. An example of cryptic coloration is the
   a. green color of a plant
   b. bright markings of a poisonous tropical frog
   c. stripes of a skunk
   d. mottled coloring of moths that rest on lichens
   e. bright colors of an insect-pollinated flower

7. An example of Müllerian mimicry is
   a. a butterfly that resembles a leaf
   b. two poisonous frogs that resemble each other in coloration
   c. a minnow with spots that look like large eyes
   d. a beetle that resembles a scorpion
   e. a carnivorous fish with a wormlike tongue that lures prey

8. Predation and parasitism are similar in that both are
   a. (++) interactions
   b. (+−) interactions
   c. (+0) interactions
   d. (−−) interactions
   e. symbiotic interactions

9. To be certain that two species had coevolved, one would *ideally* need to establish that
   a. the two species originated about the same time
   b. local extinction of one species dooms the other species
   c. each species affects the population density of the other species
   d. one species has adaptations that *specifically* tracked evolutionary change in the other species, and vice versa
   e. the two species are adapted to a common set of environmental conditions

10. According to the theory of island biogeography, species richness would be greatest on an island that is
    a. small and remote
    b. large and remote
    c. large and close to a mainland
    d. small and close to a mainland
    e. environmentally homogeneous

CHALLENGE QUESTIONS

1. An ecologist studying desert plants performed the following experiment. She staked out two identical plots that included a few sagebrush plants and numerous small annual wildflowers. She found the same five wildflower species in similar numbers in both plots. Then she enclosed one of the plots with a fence to keep out kangaroo rats, the most common herbivores in the area. After two years, four species of wildflowers were no longer present in the fenced plot, but one wildflower species had increased dramatically. The control plot had not changed significantly. Using the concepts discussed in the chapter, what do you think happened?

2. In the mountains of California, ecologist Craig Heller found that in most locations, the least chipmunk lived in sagebrush areas and the yellow pine chipmunk lived in higher areas of mixed sagebrush and piñon pines. However, if the yellow pine chipmunk was absent from a location, the least chipmunk lived in both the sagebrush and the sagebrush-piñon pine areas. If the least chipmunk was absent, the yellow pine chipmunk distribution was unchanged. How might you explain this difference in terms of the concepts of community ecology?

3. Write a paragraph contrasting the intermediate disturbance hypothesis with the concept of a stable climax community.

SCIENCE, TECHNOLOGY, AND SOCIETY

1. By 1935, hunting and trapping had eliminated wolves from the United States outside Alaska. Since wolves have been protected as an endangered species, they have moved south from Canada and have become reestablished in the Rocky Mountains and northern Great Lakes. Conservationists who would like to speed up this process have reintroduced wolves into Yellowstone National Park. Local ranchers are opposed to bringing back the wolves because they fear predation on their cattle and sheep. What are some reasons for reestablishing wolves in Yellowstone Park? What effects might the reintroduction of wolves have on the ecological communities in the park? What might be done to mitigate the conflicts between ranchers and wolves?

2. Write a paragraph describing one example of how human activities have reduced species diversity in a biological community within your own local region.

3. Explain how ecologists are applying the theory of island biogeography to the design of nature preserves.

FURTHER READING

Grant, P. R. "Ecological Character Displacement." *Science,* November 4, 1994. Darwin's finches are an enduring model for field research.

Holmes, B. "Noah's New Challenge." *New Scientist,* June 17, 1995. How do we decide which species and communities are most important to preserve?

Jensen, G. C. "To Each His Zone." *Natural History,* July 1995. Interspecific competition among intertidal crabs.

MacArthur, R. H., and E. O. Wilson. *The Theory of Island Biogeography.* Princeton, NJ: Princeton University Press, 1967.

Moore, J. "The Behavior of Parasitized Animals." *BioScience,* February 1995. Some parasites alter the behavior of their hosts in self-serving ways.

Reice, S. R. "Nonequilibrium Determinants of Biological Community Structure." *American Scientist,* September/October 1994. Are communities always recovering from disturbance?

Rennie, J. "Living Together." *Scientific American,* January 1992. The evolution of parasitism.

"The Science and Biodiversity Policy Supplement to *BioScience,*" 1995. A whole issue examining the relationships of biology, culture, politics, and economics in the movement to preserve what is left of biodiversity.

Wilson, E. O. *The Diversity of Life.* New York: Norton, 1992. The evolution and importance of biodiversity.

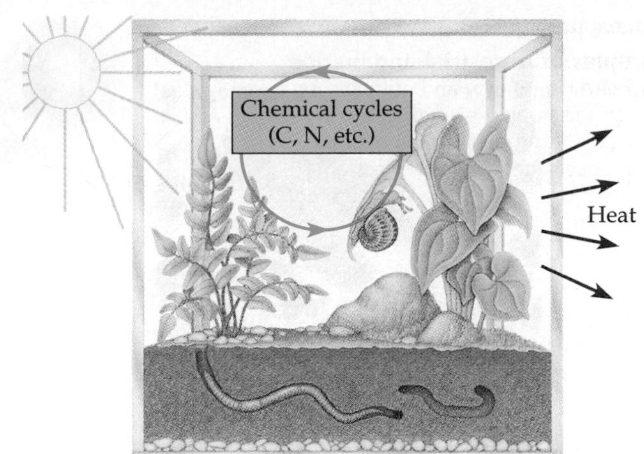

CHAPTER 49

ECOSYSTEMS

KEY CONCEPTS

- Trophic structure determines an ecosystem's routes of energy flow and chemical cycling
- An ecosystem's energy budget depends on primary productivity
- As energy flows through an ecosystem, much is lost at each trophic level
- Matter cycles within and between ecosystems
- A combination of biological and geological processes drives chemical cycles
- Field experiments reveal how vegetation regulates chemical cycling: *science as a process*
- The human population is disrupting chemical cycles throughout the biosphere
- Human activities are altering species distribution and reducing biodiversity
- The Sustainable Biosphere Initiative is reorienting ecological research

*An **ecosystem** consists of all the organisms living in a community as well as all the abiotic factors with which they interact. As with populations and communities, the boundaries of ecosystems are usually not discrete. This unit of study can range from a laboratory microcosm, such as the terrarium illustrated here, to lakes and forests. Indeed, some ecologists regard the entire biosphere as a sort of global ecosystem, a composite of all the local ecosystems on Earth.*

The most inclusive level in the hierarchy of biological organization, an ecosystem involves two processes that cannot be fully described at lower levels: energy flow and chemical cycling. Energy enters most ecosystems in the form of sunlight. It is then converted to chemical energy by autotrophic organisms, passed to heterotrophs in the organic compounds of food, and dissipated in the form of heat. Chemical elements such as carbon and nitrogen are cycled between abiotic and biotic components of the ecosystem. Photosynthetic organisms acquire these elements in inorganic form from the air, soil, and water and assimilate them into organic molecules, some of which are consumed by animals. The elements are returned in inorganic form to the air, soil, and water by the metabolism of plants and animals and by other organisms, such as bacteria and fungi, that break down organic wastes and dead organisms. The movements of energy and matter through ecosystems are related because both occur by the transfer of substances through feeding relationships. However, because energy, unlike matter, cannot be recycled, an ecosystem must be powered by a continuous influx of new energy from an external source (the sun). Thus, energy flows through ecosystems, while matter cycles within them.

This chapter describes the dynamics of energy flow and chemical cycling in ecosystems and considers some of the consequences of human intrusions into these processes.

Trophic structure determines an ecosystem's routes of energy flow and chemical cycling

Each ecosystem has a **trophic structure** of feeding relationships that determines the pathways of energy flow and chemical cycling. Ecologists divide the species in a community or ecosystem into **trophic levels** on the basis of their main source of nutrition. The trophic level that ultimately supports all others consists of autotrophs, or the **primary producers** of the ecosystem. Most producers are

FIGURE 49.1
Examples of terrestrial and marine food chains. Energy and nutrients pass through the trophic levels of an ecosystem when organisms feed on one another. Decomposers (detritivores), important components of every ecosystem, are not shown here.

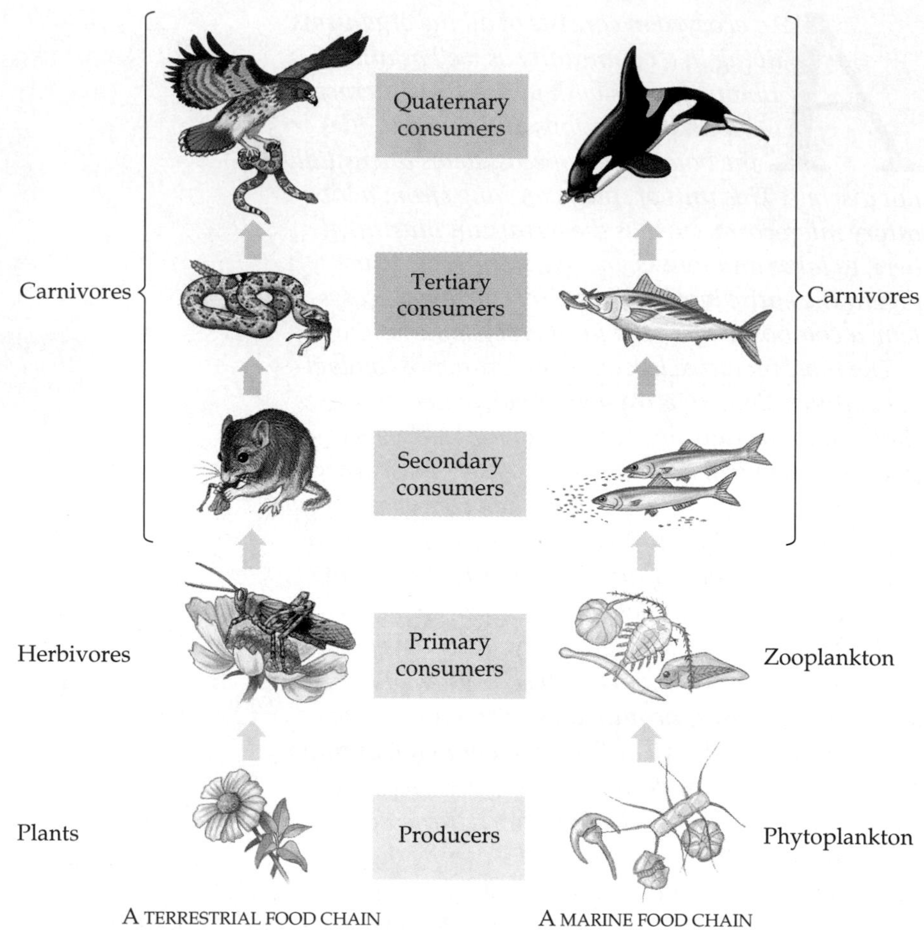

photosynthetic organisms that use light energy to synthesize sugars and other organic compounds, which they then use as fuel for cellular respiration and as building material for growth. All other organisms in an ecosystem are consumers—heterotrophs that directly or indirectly depend on the photosynthetic output of producers. Herbivores, which eat plants or algae, are the **primary consumers.** The next trophic level consists of **secondary consumers,** carnivores that eat herbivores. These carnivores may in turn be eaten by other carnivores that are **tertiary consumers,** and some ecosystems have carnivores of an even higher level. Some consumers, the **decomposers** (also called detritivores), derive their energy from detritus, which is organic waste such as feces or fallen leaves and the remains of dead organisms from the other trophic levels.

The pathway along which food is transferred from trophic level to trophic level, beginning with producers, is known as a **food chain** (FIGURE 49.1). You will learn later in the chapter that the length of food chains is limited by the amount of energy that gets transferred from one level to the next. However, few ecosystems are so simple that they are characterized by a single, unbranched food chain. Several types of primary consumers usually feed on the same plant species, and one species of primary consumer may eat several different plants. Such branching of food chains occurs at the other trophic levels as well. For example, frogs, which are secondary consumers, eat several insect species that may also be eaten by various birds. In addition, some consumers feed at several different trophic levels. An owl, for instance, may eat primary consumers such as field mice and also feed on higher-level consumers such as snakes. Omnivores, including humans, eat producers as well as consumers of different levels. Thus, the feeding relationships in an ecosystem are usually woven into elaborate **food webs** (FIGURE 49.2). The producer ⟶ primary consumer ⟶ secondary consumer chain is thus a simplification of the many permutations that feeding relationships can have.

Let's now reinforce the concept of trophic structure by looking at some examples of producers, consumers, and decomposers in various ecosystems.

Producers

The main producers in most terrestrial ecosystems are plants. In streams, much of the organic material used by consumers is also supplied by terrestrial plants, entering the ecosystem as debris that falls into the water or is washed in by runoff. In the limnetic zone of lakes and in the open ocean, phytoplankton are the most important

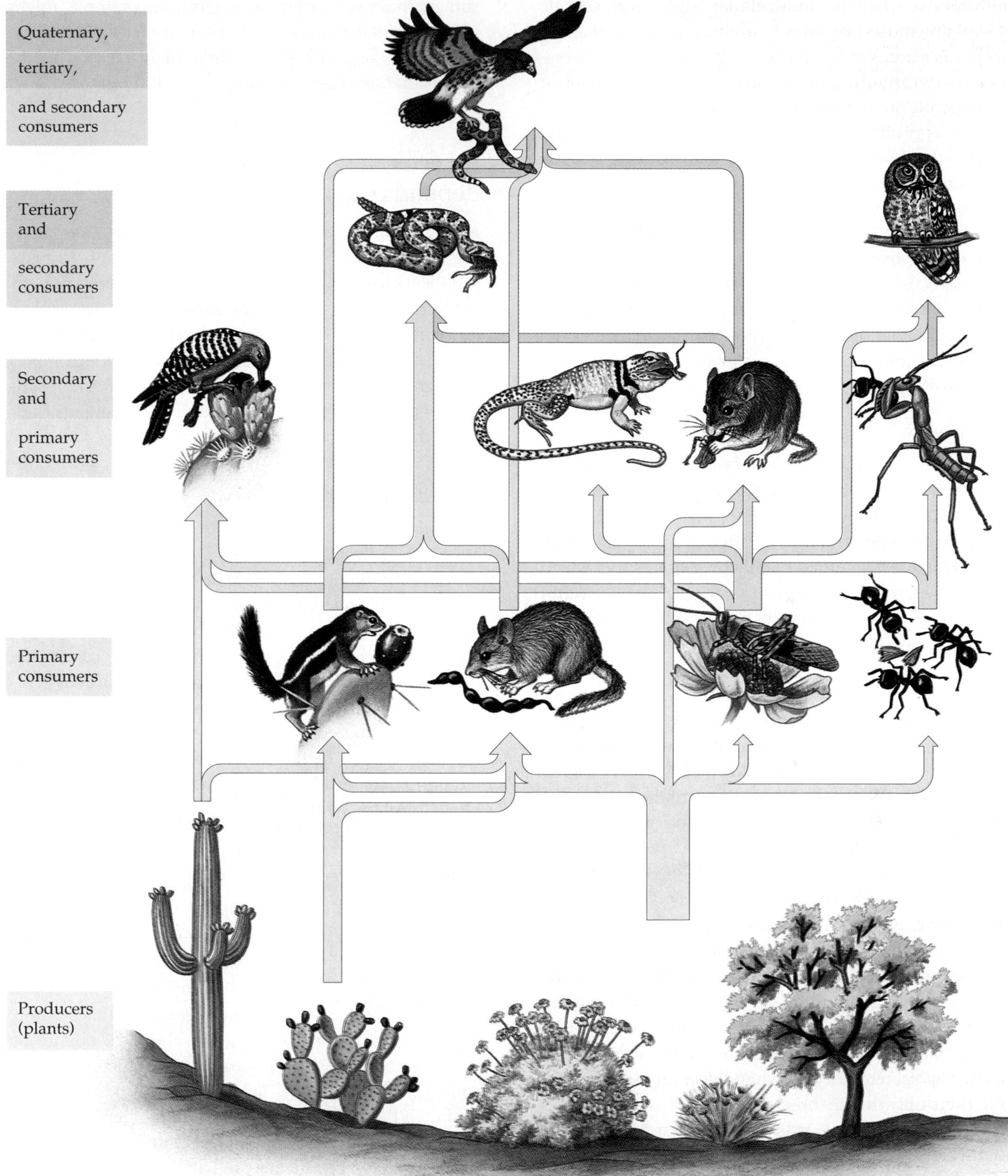

FIGURE 49.2

A food web. This simplified diagram of feeding relationships in the Sonoran Desert in the southwestern United States does not include all species that feed at each trophic level. Decomposers, which feed on the remains of organisms in every trophic level, are also omitted. The arrows are color-coded to indicate the trophic levels at which different species feed.

autotrophs, whereas multicellular algae and aquatic plants are more important producers in the shallow, near-shore areas of both freshwater and marine ecosystems. In the aphotic zone of the deep sea, however, most life depends on photosynthetic production in the photic zone; energy and nutrients rain down from above in the form of dead plankton and other detritus. One notable exception is the community of organisms that live near hot water vents on the deep-sea floor (see FIGURE 46.29). Chemoautotrophic bacteria that derive energy from the oxidation of hydrogen sulfide are the main producers in these ecosystems, which are therefore supported by chemical, rather than solar, energy. Because the bacteria need oxygen derived from photosynthesis to oxidize the hydrogen sulfide, these hot water vent ecosystems are not, however, totally independent of solar energy.

Consumers

The primary consumers, or herbivores, on land are mostly insects, snails, and certain vertebrates, including grazing mammals and the numerous birds and mammals that eat seeds and fruit. In aquatic ecosystems, phytoplankton are consumed mainly by zooplankton, which include heterotrophic protists, various small invertebrates (especially crustaceans and, in the ocean, larval stages of many species that live in the benthos as adults), and some fish.

Examples of secondary consumers in terrestrial ecosystems are spiders, frogs, insect-eating birds, carnivorous mammals, and animal parasites. In aquatic habitats, many fish feed on zooplankton and are in turn fed upon by other fish. In the benthic zone of the seas, algae-eating invertebrates are prey to other invertebrates, such as sea stars.

Decomposers (Detritivores)

The organic material that composes the living organisms in an ecosystem is eventually recycled, broken down and returned to the abiotic environment in forms that can be used by plants. Decomposers, which feed on nonliving organic material, are key to this recycling process. The most important decomposers are bacteria and fungi, which first secrete enzymes that digest organic material and then absorb the breakdown products; some can even digest cellulose. Earthworms and such scavengers as crayfish, cockroaches, and bald eagles are also decomposers, but these animals digest organic material internally after ingesting it. In fact, all heterotrophs, including humans, are decomposers in the sense that they break down organic material and release inorganic products, such as carbon dioxide and ammonia, to the environment. Decomposers often form a major link between primary producers and the secondary and tertiary consumers in an ecosystem. A crayfish, for example, might feed on plant detritus at the bottom of a lake and then be eaten by a bass. In a forest, birds might feed on earthworms that have been feeding on leaf litter in the soil.

An ecosystem's energy budget depends on primary productivity

All organisms require energy for growth, maintenance, reproduction, and, in some species, locomotion. Primary producers use light energy to synthesize energy-rich organic molecules, which can subsequently be broken down to make ATP. Consumers acquire their organic fuels secondhand (or even thirdhand or fourthhand) through food webs. Therefore, the extent of photosynthetic activity sets the spending limit for the energy budget of the entire ecosystem.

The Global Energy Budget

Every day, Earth is bombarded by 10^{22} joules (J) of solar radiation, ($1 \text{ J} = 0.239$ calories). This is the energy equivalent of 100 million atomic bombs the size of the one dropped on Hiroshima. As described in Chapter 46, the intensity of the solar energy striking Earth and its atmosphere varies with latitude, with the tropics receiving the highest input. Most solar radiation is absorbed, scattered, or reflected by the atmosphere in an asymmetrical pattern determined by variations in cloud cover and the quantity of dust in the air over different regions. The amount of solar radiation reaching the globe ultimately limits the photosynthetic output of ecosystems, although photosynthetic productivity is also limited by water, temperature, and nutrient availability.

Much of the solar radiation that reaches the biosphere lands on bare ground and bodies of water that either absorb or reflect the incoming energy. Only a small fraction actually strikes algae and plant leaves, and only some of this is of wavelengths suitable for photosynthesis. Of the visible light that does reach leaves and algae, only about 1% to 2% is converted to chemical energy by photosynthesis, and this photosynthetic efficiency varies with the type of plant, light level, and other factors. Although the fraction of the total incoming solar radiation that is ultimately trapped by photosynthesis is very small, primary producers on Earth collectively create about 170 billion tons of organic material per year—an impressive quantity.

Primary Productivity

The amount of light energy converted to chemical energy (organic compounds) by the autotrophs of an ecosystem during a given time period is called **primary productivity.** Total primary productivity is known as **gross primary productivity (GPP).** Not all of this product is stored as organic material in the growing plants, because the plants use some of the molecules as fuel in their own cellular respiration. Thus, **net primary productivity (NPP)** is equal to the gross primary productivity minus the energy used by the producers for respiration (Rs):

$$\text{NPP} = \text{GPP} - \text{Rs}$$

We may also think of this relationship in terms of the equations for photosynthesis and respiration:

$$6\ CO_2 + 6\ H_2O \underset{\text{Respiration}}{\overset{\text{Photosynthesis}}{\rightleftharpoons}} C_6H_{12}O_6 + 6\ O_2$$

Gross primary productivity results from photosynthesis; net primary productivity is the difference between the yield of photosynthesis and the consumption of organic fuel in respiration.

Net primary productivity is the measurement of interest to us because it represents the storage of chemical energy available to consumers in an ecosystem. Between 50% and 90% of the gross primary productivity of most primary producers remains as net primary productivity after their energetic needs are fulfilled. The NPP-to-GPP ratio is generally smaller for large producers with elaborate nonphotosynthetic structures, such as trees, which support large and metabolically active stem and root systems.

Primary productivity can be expressed in terms of energy per unit area per unit time ($J/m^2/yr$), or as **biomass** (weight) of vegetation added to the ecosystem per unit area per unit time ($g/m^2/yr$). Biomass is usually expressed in terms of the dry weight of organic material because water molecules contain no usable energy, and because the water content of plants varies over short periods of time. An ecosystem's primary productivity should not be confused with the total biomass of plants present at a given time, the **standing crop biomass;** primary productivity is the *rate* at which the vegetation synthesizes *new* biomass. Although a forest has a very large standing crop biomass, its productivity may actually be less than that of some grasslands, which do not accumulate vegetation because animals consume the plants rapidly and because many of the plants are annuals.

Different ecosystems vary considerably in their productivity as well as in their contribution to the total productivity on Earth (FIGURE 49.3). Tropical rain forests are among the most productive terrestrial ecosystems, and because they cover a large portion of the Earth, they contribute a large proportion of the planet's overall productivity. Estuaries and coral reefs also have very high

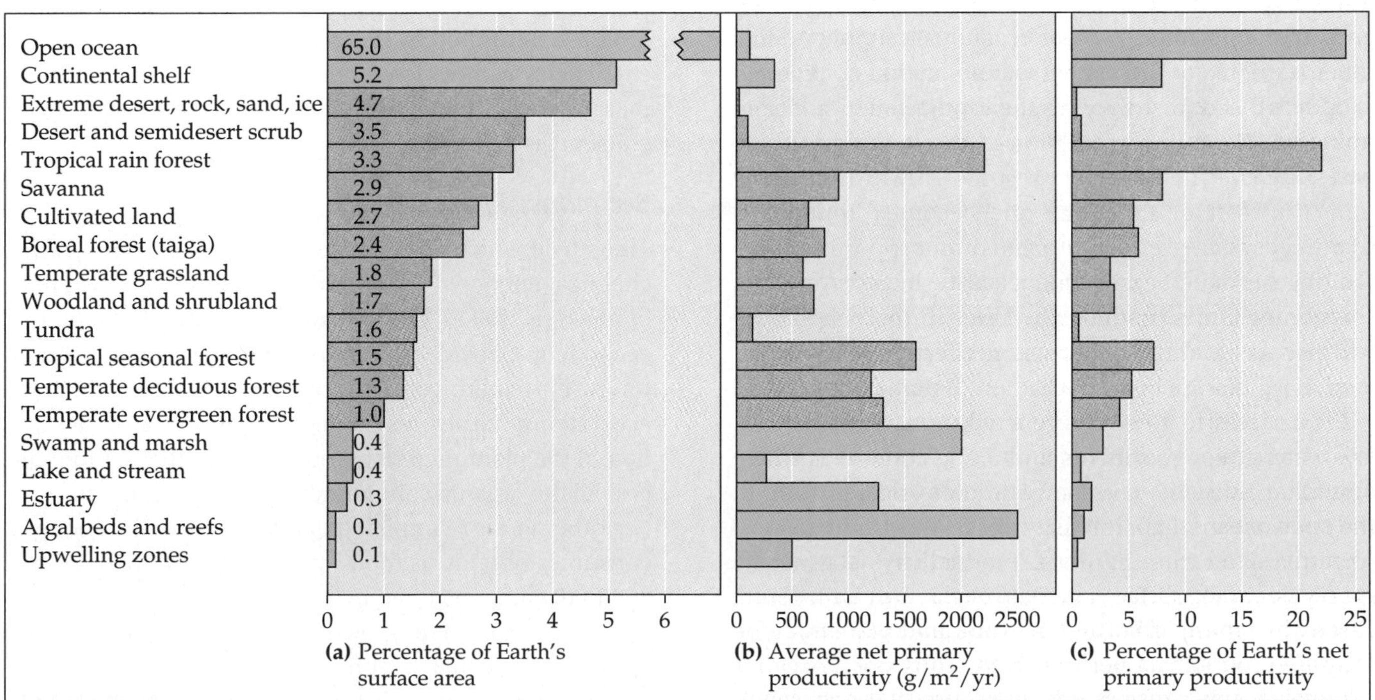

FIGURE 49.3

Productivity of different ecosystems. (**a**) The geographical extent and (**b**) the productivity per unit area of different ecosystems determine their total contribution to (**c**) worldwide primary productivity. Open ocean, for example, contributes a lot to the planet's productivity despite its low productivity per unit area because of its large size, whereas tropical rain forest contributes a lot because of its high productivity. (Aquatic ecosystems are color-coded blue in these histograms; terrestrial ecosystems are brown.)

productivity, but their total contribution to global productivity is relatively small because these ecosystems are not very extensive. The open ocean contributes more primary productivity than any other ecosystem, but this is because of its very large size; productivity per unit area is relatively low. Deserts and tundra also have low productivity.

The factors most important in limiting productivity depend on the type of ecosystem and on seasonal changes in the environment. Productivity in terrestrial ecosystems is generally correlated with precipitation, temperature, and light intensity. Farmers often irrigate their fields, for example, to increase productivity in habitats where the availability of water limits photosynthetic activity; heat and light, as well as water, are provided to plants grown in greenhouses. Productivity generally increases with proximity to the equator because water, heat, and light are more readily available in the tropics.

Inorganic nutrients may also be important in limiting the productivity of many terrestrial ecosystems. Recall from Chapter 33 that plants need a variety of nutrients, some in relatively large quantities and others only in trace amounts—but all are crucial. Primary productivity removes nutrients from a system, sometimes faster than they are returned. At some point, productivity may slow or cease because a specific nutrient is no longer present in sufficient quantity. It is unlikely that all nutrients will be exhausted simultaneously, so that further productivity is limited by the single nutrient—called the **limiting nutrient**—that is no longer present in adequate supply. Adding other nutrients to the system will not stimulate renewed productivity because they are already present in sufficient quantity. However, the addition of the limiting nutrient will stimulate the system to resume growth until some other nutrient or the same one becomes limiting. In many ecosystems, either nitrogen or phosphorus is a key limiting nutrient. Some evidence also suggests that CO_2 sometimes limits productivity. Later in the chapter, we will discuss how human intrusions affecting nutrient balance have disrupted terrestrial and aquatic ecosystems.

Productivity in the seas is generally greatest in the shallow waters near continents and along coral reefs where abundant nutrients and light stimulate algal growth. In the open oceans, light intensity affects the productivity of phytoplankton communities. Productivity is generally greatest near the surface and declines sharply with depth, as light is rapidly absorbed by water and plankton. The primary productivity per unit area of the open ocean is relatively low because inorganic nutrients, especially nitrogen and phosphorus, are in short supply near the surface; at great depth, where nutrients are abundant, there is insufficient light to support photosynthesis. Phytoplankton communities are most productive where upwelling currents bring nitrogen and phosphorus to the surface. This phenomenon is apparent in antarctic seas, which, in spite of the cold water and low light intensity, are actually more productive than most tropical seas. The chemoautotrophic ecosystems near hot water vents are also very productive, but these communities are not widespread, and their overall contribution to marine productivity is small.

In freshwater ecosystems, as in the open ocean, light intensity and its variation with depth appear to be important determinants of productivity. The availability of inorganic nutrients may also limit productivity in freshwater ecosystems as it does in the oceans, but the biannual turnover of lakes mixes the waters, carrying nutrients to the well-illuminated surface layers (see FIGURE 46.08).

As energy flows through an ecosystem, much is lost at each trophic level

As energy flows through an ecosystem, much of it is dissipated before it can be consumed by organisms at the next level. If all of the plants in a prairie were piled into a huge mound, another mound of all the herbivores would be dwarfed beside the plants. But the herbivore mound would be much larger than a mound of secondary consumers. The amount of energy available to each trophic level is determined by net primary productivity and the efficiencies with which food energy is converted to biomass in each link of the food chain. As we will see, these efficiencies are never 100%.

Secondary Productivity

The rate at which an ecosystem's consumers convert the chemical energy of the food they eat into their own new biomass is called the **secondary productivity** of the ecosystem. Consider the transfer of organic matter from producers to herbivores, the primary consumers. In most ecosystems, herbivores manage to eat only a small fraction of the plant material produced, and they cannot digest all the organic compounds that they do ingest.

FIGURE 49.4 is a simplified diagram of how the energy a consumer obtains as food might be partitioned. Of the 200 J (48 cal) consumed by a caterpillar, only about 33 J (one-sixth) is used for growth. The rest is passed as feces or used for cellular respiration. Of course, the energy contained in the feces is not lost from the ecosystem; it can still be consumed by decomposers. However, the energy used for respiration is lost from the ecosystem; thus, while solar radiation is the ultimate source of energy for

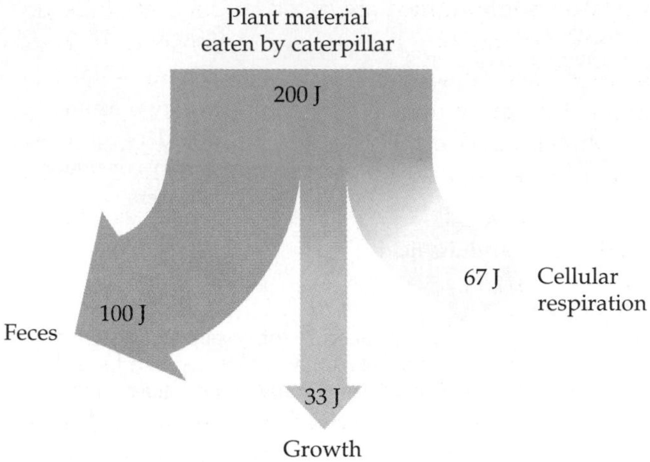

FIGURE 49.4
Energy partitioning within a link of the food chain. Caterpillars digest and absorb only about half of what they eat, passing the rest as feces. Thus, if a caterpillar consumed leaves containing 200 joules of energy, 100 J would be lost in feces (1 J = 0.239 cal). Approximately two-thirds of the absorbed material, or 67 J, would be used in maintenance, as fuel for cellular respiration, which degrades food molecules to inorganic waste products and heat. The remaining 33 J would be converted into caterpillar biomass, and would therefore be available to the next trophic level.

most ecosystems, respiratory heat loss is the ultimate sink. This is why energy is said to flow through, not cycle within, ecosystems. Only the chemical energy stored as growth (or the production of offspring) by herbivores is available as food to secondary consumers. In one sense, our example actually overestimates the conversion of primary productivity into secondary productivity because we did not account for all the plant material that herbivores do not even consume. The fact that natural ecosystems usually look green—they contain large amounts of plant material—indicates that much net primary productivity is not converted over the short term into secondary productivity.

Carnivores are slightly more efficient at converting food into biomass, mainly because meat is more easily digested than vegetation. But in many cases, secondary consumers use more of the energy they assimilate for cellular respiration, which dramatically decreases the amount of chemical energy available to the next trophic level. Endotherms, in particular, devote a large proportion of their assimilated energy to maintaining a high and constant body temperature.

Ecological Efficiency and Ecological Pyramids

Ecological efficiency is the percentage of energy transferred from one trophic level to the next, or the ratio of net productivity at one trophic level to net productivity at the level below. Ecological efficiencies vary greatly among organisms, usually ranging from 5% to 20%. In other words, 80% to 95% of the energy available at one trophic level never transfers to the next. This multiplicative loss of energy from a food chain can be represented diagrammatically by a **pyramid of productivity,** in which the trophic levels are stacked in blocks, with primary producers forming the foundation of the pyramid. The size of each block is proportional to the productivity of each trophic level (per unit of time). Pyramids of productivity are typically quite bottom-heavy because ecological efficiencies are low (FIGURE 49.5).

One important ecological consequence of decreasing energy transfers through a food web can be represented in a **biomass pyramid,** in which each tier represents the standing crop biomass (the total dry weight of all organisms) in a trophic level. Biomass pyramids generally narrow sharply from producers at the base to top-level carnivores at the apex because energy transfers between

FIGURE 49.5
An idealized pyramid of net productivity. In the case illustrated here, 10% of the energy available at each trophic level is converted into new biomass in the trophic level above it. Notice that producers convert only about 1% of the energy in the sunlight available to them into primary productivity. In actual ecosystems, the decline in productivity with the transfer of energy between trophic levels varies with the particular species present; a 10% transfer of energy (ecological efficiency) is a rough average.

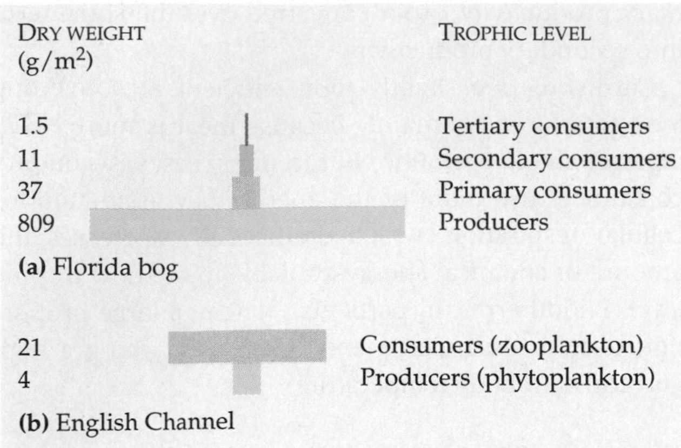

FIGURE 49.6
Pyramids of standing crop biomass. Numbers denote the dry weight (g/m^2) for all organisms at a trophic level. (**a**) Most biomass pyramids show a sharp decrease in biomass at successively higher trophic levels, as illustrated by data from a bog at Silver Springs, Florida. (**b**) However, in some aquatic ecosystems, such as the English Channel, a small standing crop of producers (phytoplankton) supports a larger standing crop of primary consumers (zooplankton). This is because the phytoplankton have a short turnover time; the algae reproduce rapidly and are consumed at a high rate.

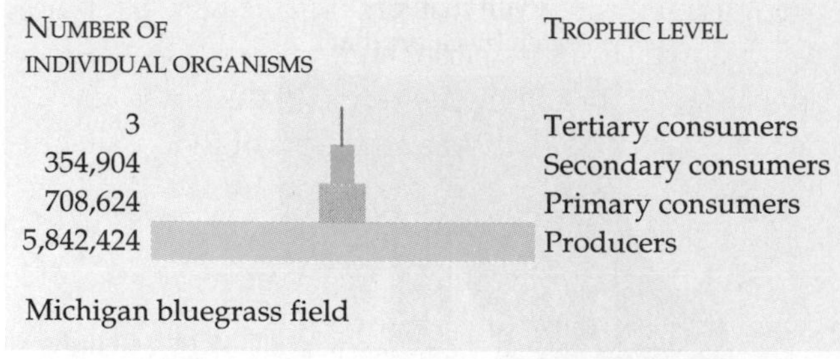

FIGURE 49.7
A pyramid of numbers. Because top-level predators are generally large animals, the small biomass at the top of a food chain is contained in a relatively small number of individuals. In this pyramid of numbers for a bluegrass field in Michigan, only three top carnivores are supported in an ecosystem based on the productivity of nearly 6 million plants.

trophic levels are so inefficient (FIGURE 49.6a). Some aquatic ecosystems, however, have inverted biomass pyramids, with primary consumers outweighing producers. In the waters of the English Channel, for example, the biomass of zooplankton (consumers) is five times the weight of phytoplankton (producers) (FIGURE 49.6b). Such inverted biomass pyramids occur because the zooplankton consume the phytoplankton so quickly that the producers never develop a large population size or standing crop. Instead, the phytoplankton grow, reproduce, and are consumed rapidly. Phytoplankton have a short **turnover time,** or a low standing crop biomass compared to their productivity:

$$\text{Turnover time} = \frac{\text{Standing crop biomass (mg/m}^2\text{)}}{\text{Productivity (mg/m}^2\text{/day)}}$$

Nevertheless, the *productivity* pyramid for this ecosystem is upright, like the one in FIGURE 49.5, because phytoplankton have a higher productivity than zooplankton.

The multiplicative loss of energy from food chains severely limits the overall biomass of top-level carnivores that any ecosystem can support. Only about one-thousandth of the chemical energy fixed by photosynthesis can flow all the way through a food web to a tertiary consumer, such as a hawk or a shark. This partly explains why food webs usually include only three to five trophic levels. There are, for example, no nonhuman predators of adult lions, eagles, and killer whales because their biomass is insufficient to support yet another trophic level.

Because top-level predators tend to be fairly large animals, the limited biomass at the top of an ecological pyramid is concentrated in a relatively small number of individuals. This phenomenon is reflected in a **pyramid of numbers,** in which the size of each block is proportional to the number of individual organisms present in each trophic level (FIGURE 49.7). Populations of top predators are typically very small, and the animals may be widely spaced within their habitats. As a result, predators are highly susceptible to extinction (as well as to the evolutionary consequences of small population size discussed in Chapter 21).

The concept of an energy or biomass pyramid also has implications for the human population. Eating meat is a relatively inefficient way of tapping photosynthetic productivity. A human obtains far more calories by eating grain directly as a primary consumer than by processing that same amount of grain through another trophic level and eating grain-fed beef. Worldwide agriculture could, in fact, successfully feed many more people than it does today if we all consumed only plant material, feeding more efficiently as primary consumers.

Matter cycles within and between ecosystems

Although ecosystems receive an essentially inexhaustible influx of solar energy, chemical elements are available only in limited amounts. (The meteorites that occasionally strike Earth are the only extraterrestrial source of matter, with trivial importance.) Life on Earth therefore depends on the recycling of essential chemical elements. Even while an individual organism is alive, much of its chemical stock is rotated continuously, as nutrients are absorbed and waste products released. Atoms present in the complex molecules of an organism at its time of death are returned as simpler compounds to the atmosphere, water, or soil by the action of decomposers. This decomposition replenishes the pools of inorganic nutrients that plants and other autotrophs use to build new organic matter. Because nutrient circuits involve both biotic and abiotic components of ecosystems, they are also called **biogeochemical cycles.**

A chemical's specific route through a biogeochemical cycle varies with the particular element and the trophic structure of the ecosystem. We can, however, recognize two general categories of biogeochemical cycles. Gaseous forms of carbon, oxygen, sulfur, and nitrogen occur in the atmosphere, and cycles of these elements are essentially global. For example, some of the carbon and oxygen atoms a plant acquires from the air as CO_2 may have been released into the atmosphere by the respiration of an animal in some distant locale. Other elements that are less mobile in the environment, including phosphorus, potassium, calcium, and the trace elements, generally cycle on a more localized scale, at least over the short term. Soil is the main abiotic reservoir of these elements, which are absorbed by plant roots and eventually returned to the soil by decomposers, usually in the same general vicinity.

Before examining the details of some individual cycles, let's look at a general model of nutrient cycling that shows the main reservoirs, or compartments, of elements and the processes that transfer elements between reservoirs (FIGURE 49.8). Most nutrients accumulate in four reservoirs, each of which is defined by two characteristics: whether it contains organic or inorganic materials, and whether or not the materials are directly available for use by organisms. One compartment of organic materials is composed of the living organisms themselves and detritus; these nutrients are available to other organisms when they feed on one another. The second organic compartment includes "fossilized" deposits of once-living organisms (coal, oil, and peat), from which nutrients cannot be assimilated directly. Material moved from the living organic compartment to the fossilized organic compartment long ago, when organisms died and were buried by sedimentation over millions of years to become coal or oil.

Nutrients also occur in two inorganic compartments, one in which they are available for use by organisms and one in which they are not. The available inorganic compartment includes matter (elements and compounds) that is dissolved in water or present in soil or air. Organisms assimilate materials from this compartment directly and return nutrients to it through the fairly rapid processes of respiration, excretion, and decomposition. Elements in the unavailable inorganic compartment are

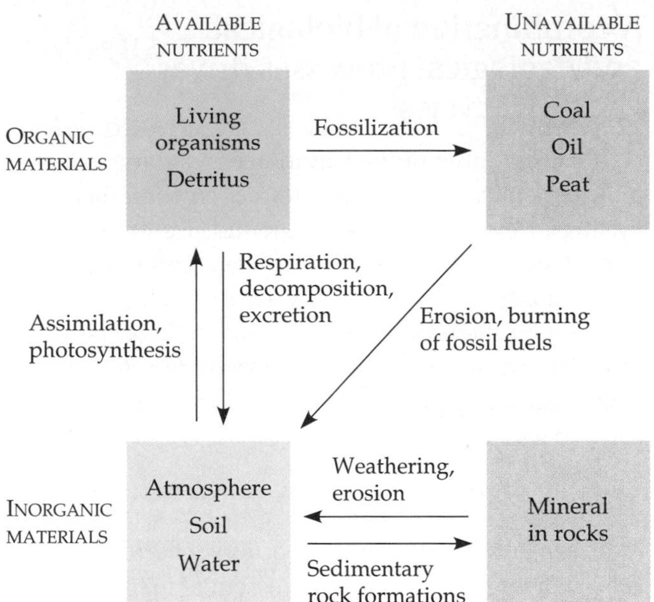

FIGURE 49.8

A general model of nutrient cycling. Most nutrients cycle within the biosphere among four major compartments, or reservoirs. The biological and geological processes that move nutrients from one compartment to another are indicated in this diagram.

tied up in rocks. Although organisms cannot tap into this compartment directly, nutrients slowly become available for use through the action of weathering and erosion. Similarly, unavailable organic materials move into the available inorganic nutrients compartment through erosion or when fossil fuels are burned and their elements are vaporized.

Describing biogeochemical cycles in general theory is much simpler than actually tracing elements through these cycles. Not only are ecosystems exceedingly complex, they usually exchange at least some of their materials with other regions. Even in a pond, which has discrete boundaries, several processes add and remove key nutrients. Minerals dissolved in rainwater or runoff from the neighboring land are added to the pond, as are nutrient-rich pollen, fallen leaves, and other airborne material. And, of course, carbon, oxygen, and nitrogen cycle between the pond and the atmosphere. Birds may feed on fish or the aquatic larvae of insects, which derived their store of nutrients from the pond, and some of those nutrients may then be excreted or eliminated on land far from the pond's drainage area. Keeping track of the inflow and outflow is even more challenging in less clearly delineated terrestrial ecosystems. Nevertheless, ecologists have worked out the general schemes for chemical cycling in several ecosystems, often by adding tiny amounts of radioactive tracers that enable the researchers to follow chemical elements through the various biotic and abiotic constituents of the ecosystems.

A combination of biological and geological processes drives chemical cycles

To illustrate some of the variations and complexities of biogeochemical cycles, we will trace in some detail the cycling of one important compound, water, and three important elements: carbon, nitrogen, and phosphorus. As you study the specific pathways for each cycle, try to understand them in terms of the general compartment model in FIGURE 49.8. And keep in mind the concept that chemical cycling in ecosystems depends on both biological and geological processes.

The Water Cycle

Although only a very small proportion of Earth's water resides in living material, water is essential to living organisms. In addition to water's direct contributions to the fitness of the environment (see Chapter 2), its movement within and between ecosystems also transfers other materials in several biogeochemical cycles. Driven by solar energy, most of the water cycle occurs between the oceans and the atmosphere through evaporation and precipitation (FIGURE 49.9). The amount of water evaporating from oceans exceeds precipitation over oceans, and the excess water vapor is moved by wind to the land. Over land surfaces, precipitation exceeds evaporation and transpiration, the evaporative loss of water from plants. Runoff and groundwater from the land balance the net flow of water vapor from the ocean to the land. The water cycle differs from the other cycles because most water flux through ecosystems occurs by physical, rather than chemical, processes; during evaporation, transpiration, and precipitation, water maintains its form as H_2O. An ecologically (although not quantitatively) important exception is the chemical transformation of water during photosynthesis.

The Carbon Cycle

Carbon is a basic constituent of all organic compounds. Its movement through an ecosystem parallels that of energy more closely than other chemicals; carbohydrates are produced during photosynthesis, and CO_2 is released with energy during respiration. In the carbon cycle, the reciprocal processes of photosynthesis and cellular respiration provide a link between the atmosphere and terrestrial environments (FIGURE 49.10). Plants acquire carbon, in the form of CO_2, from the atmosphere through the stomata of their leaves, and incorporate it into the organic matter of their own biomass through the process of photosynthesis. Some of this organic material then becomes the carbon source for consumers. Respiration by all organisms returns CO_2 to the atmosphere.

Although CO_2 is present in the atmosphere at a relatively low concentration (about 0.03%), carbon recycles at a relatively fast rate because plants have a high demand for this gas. Each year, plants remove about one-seventh of the CO_2 in the atmosphere; this is approximately (but not exactly) balanced by respiration. Some carbon may be diverted from the cycle for longer periods. This happens, for example, when it is accumulated in wood and other durable organic material. Decomposition eventually recycles even this carbon to the atmosphere as CO_2, although fires can oxidize such organic material to CO_2 much faster. Some processes, however, can remove carbon from short-term cycling for millions of years; in

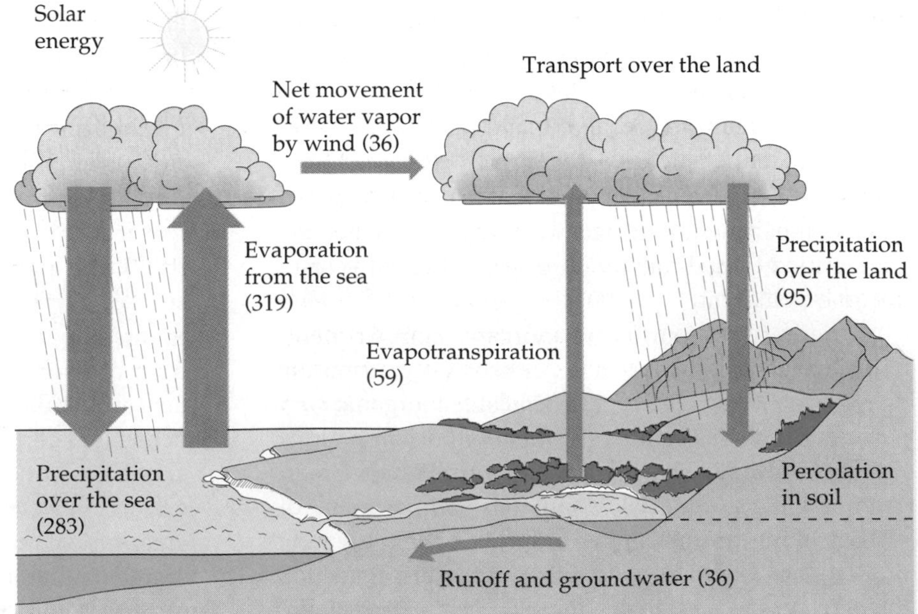

FIGURE 49.9
The water cycle. On a global scale, evaporation exceeds precipitation over the oceans. The result is a net movement of water vapor, carried by winds, from the ocean to the land. The excess of precipitation over evaporation on land results in the formation of surface and groundwater systems that flow back to the sea, completing the major part of the cycle. Over the sea, evaporation forms most water vapor. On land, however, 90% or more of the vaporization is due to plant transpiration. The numbers in this diagram indicate water flow in billion billion (10^{18}) grams per year.

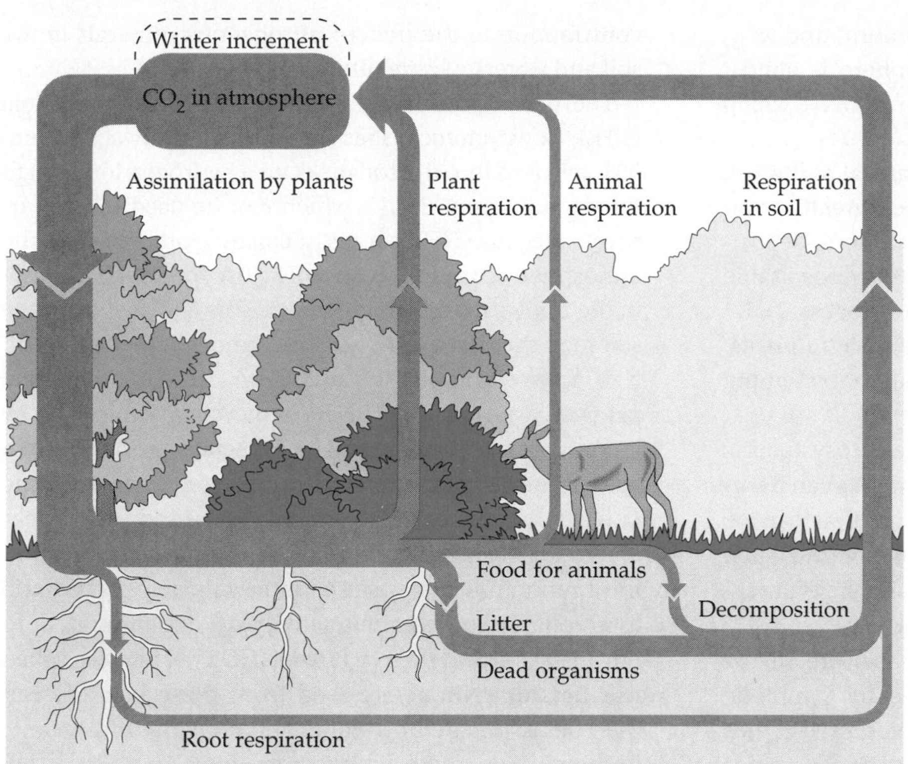

FIGURE 49.10
The carbon cycle. The reciprocal processes of photosynthesis and cellular respiration are responsible for the major transformations and movements of carbon. A seasonal pulse in atmospheric CO_2 is caused by decreased photosynthetic activity during the Northern Hemisphere's winter. On a global scale, the return of CO_2 to the atmosphere by respiration closely balances its removal by photosynthesis. However, the burning of wood and fossil fuels adds more CO_2 to the atmosphere; as a result, the amount of atmospheric CO_2 is steadily increasing. Atmospheric CO_2 also moves into or out of aquatic systems, where it is involved in a dynamic equilibrium with other inorganic forms, including bicarbonates.

some environments, organic litter accumulates much more quickly than decomposers can break it down. Under certain conditions, these deposits eventually form coal and petroleum that become locked away as unavailable organic nutrients.

The amount of CO_2 in the atmosphere varies slightly with the seasons. Concentrations of CO_2 are lowest during the Northern Hemisphere's summer and highest during winter. This seasonal pulse of CO_2 concentration occurs because there is more land in the Northern Hemisphere than in the Southern, and therefore more vegetation. The vegetation has maximal photosynthetic activity during summer, reducing the global amount of atmospheric CO_2. During winter, the vegetation releases more CO_2 by respiration than it uses for photosynthesis, causing a global increase in the gas (see FIGURE 49.17).

Superimposed on this seasonal fluctuation is a continuing increase in the overall concentration of atmospheric CO_2 caused by the combustion of fossil fuels by humans. From a long-term perspective, this can be viewed as a return to the atmosphere of CO_2 that was removed by photosynthesis long ago. But in the millions of years while this material was effectively out of circulation, a new equilibrium developed in the global carbon cycle. Now that balance is being disrupted, with uncertain consequences that we will consider later.

Though the basic processes of respiration and photosynthesis occur in aquatic environments, the cycling of carbon is more complicated because of the interaction of CO_2 with water and limestone. Dissolved carbon dioxide reacts with water to form carbonic acid (H_2CO_3). Carbonic acid in turn reacts with the limestone ($CaCO_3$) that is abundant in many waters, including the ocean, to form bicarbonates and carbonate ions:

$$H_2O + CO_2 \rightleftharpoons H_2CO_3$$

$$H_2CO_3 + CaCO_3 \rightleftharpoons Ca(HCO_3)_2 \rightleftharpoons Ca^{2+} + 2\,HCO_3^-$$

$$\underset{\text{Bicarbonate}}{2\,HCO_3^-} \rightleftharpoons 2\,H^+ + \underset{\text{Carbonate}}{2\,CO_3^{2-}}$$

As CO_2 is used in photosynthesis in aquatic and marine environments, the equilibrium of this reaction series shifts toward the left, converting bicarbonates back to CO_2. Thus, bicarbonates serve as a CO_2 reservoir. Aquatic autotrophs may also use dissolved bicarbonate directly as their source of carbon. Overall, the amount of carbon present in various inorganic forms in the ocean, not including sediments, is about fifty times that available in the atmosphere. Because of these inorganic reactions of CO_2 in water, as well as its uptake by marine phytoplankton, the ocean may serve as an important "buffer" that will absorb some of the CO_2 being added to the atmosphere by the burning of fossil fuels.

The Nitrogen Cycle

Nitrogen is another key chemical in ecosystems. It is found in all amino acids, which make up the proteins of organisms. Nitrogen is available to plants only in the form of two soil minerals: NH_4^+ (ammonium) and NO_3^- (nitrate). Thus, although Earth's atmosphere is almost 80% nitrogen, it is in the form of nitrogen gas (N_2), which is unavailable to plants.

Nitrogen enters ecosystems via two natural pathways, the relative importance of which varies greatly from ecosystem to ecosystem. The first, atmospheric deposition, accounts for approximately 5% to 10% of the usable nitrogen that enters ecosystems. In this process, NH_4^+ and NO_3^-, the two forms of nitrogen available to plants, are added to soil by being dissolved in rain or by settling as part of fine dust or other particulates.

The other pathway for nitrogen to enter ecosystems is via **nitrogen fixation.** Only certain prokaryotes can fix nitrogen—that is, convert N_2 into minerals that can be used to synthesize nitrogenous organic compounds such as amino acids. Indeed, prokaryotes are vital links at several points in the nitrogen cycle (FIGURE 49.11).

Nitrogen is fixed in terrestrial ecosystems by free-living (nonsymbiotic) soil bacteria as well as by symbiotic bacteria (*Rhizobium*) in the root nodules of legumes and certain other plants (see Chapter 33). Some cyanobacteria fix nitrogen in aquatic ecosystems. Organisms that fix nitrogen, of course, are fulfilling their own metabolic requirements, but the excess ammonia they release becomes available to other organisms. In addition to these natural sources of usable nitrogen, industrial fixation of nitrogen for fertilizer now makes a significant contribution to the pool of nitrogenous minerals in the soil and waters of agricultural regions.

The direct product of nitrogen fixation is ammonia (NH_3). However, most soils are at least slightly acidic, and NH_3 released into the soil picks up a hydrogen ion (H^+) to form ammonium, NH_4^+, which can be used directly by plants. Because NH_3 is a gas, it can evaporate back to the atmosphere from soils with a pH close to 7 (such as those in the midwestern United States). This NH_3 lost from the soil may then form NH_4^+ in the atmosphere. As a result NH_4^+ concentrations in rainfall are correlated with soil pH over large regions. This local recycling of nitrogen by atmospheric deposition can be especially pronounced in agricultural areas where both nitrogen fertilizers and lime (a base that decreases soil acidity) are used extensively.

Although plants can use ammonium directly, most of the ammonium in soil is used by certain aerobic bacteria as an energy source; their activity oxidizes ammonia to nitrite (NO_2^-) and then to nitrate (NO_3^-), a process called **nitrification.** Nitrate released from these bacteria can then be assimilated by plants and converted to organic forms, such as amino acids and proteins. Animals can assimilate only organic nitrogen, by eating plants or other animals. Some bacteria can obtain the oxygen they need for metabolism from nitrate rather than from O_2 under anaerobic conditions. As a result of this **denitrification** process, some nitrate is converted back to N_2, returning to the atmosphere.

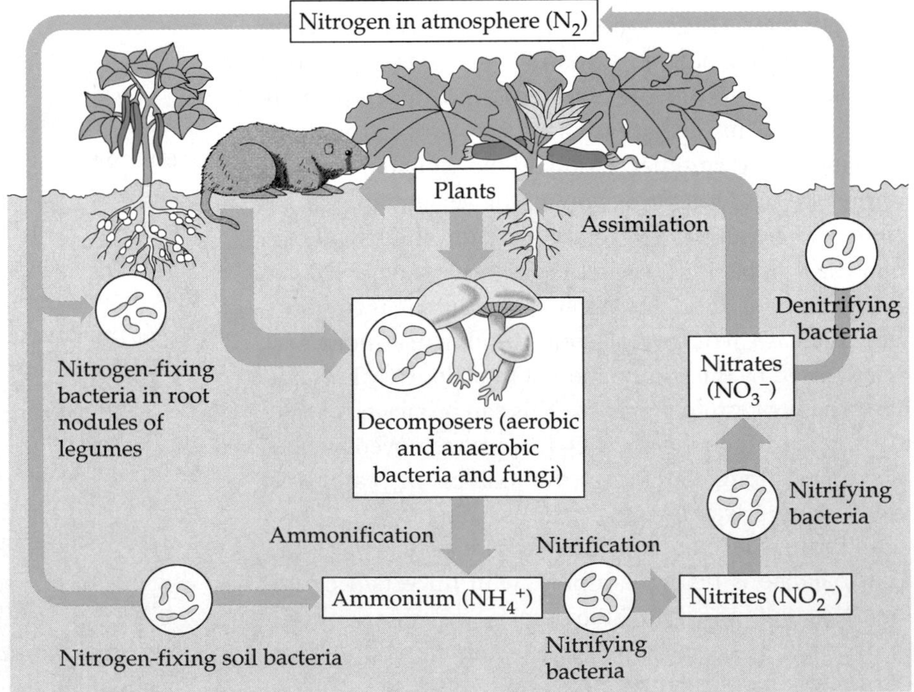

FIGURE 49.11

The nitrogen cycle. Most of the nitrogen cycling through food webs is taken up by plants in the form of nitrate. Most of this, in turn, comes from the nitrification of ammonium that results from the decomposition of organic material. The addition of nitrogen from the atmosphere and its return via denitrification involve relatively small amounts compared to the local recycling that occurs in the soil or water. The widths of the arrows represent the relative amounts of nitrogen that move between each source, but these are extremely variable across ecosystems. Also, in some ecosystems, atmospheric deposition of NH_4^+ and NO_3^- that is dissolved in rain adds nitrogenous minerals to the soil. (The text describes atmospheric deposition, but it is not included in this simplified diagram.)

The decomposition of organic nitrogen back to ammonia, a process called **ammonification,** is carried out by decomposers. This process recycles large amounts of nitrogen to the soil.

Overall, most of the nitrogen cycling in natural systems involves the nitrogenous compounds in soil and water, not atmospheric N_2. Although nitrogen fixation is important in the buildup of a pool of available nitrogen, it contributes only a tiny fraction of the nitrogen assimilated annually by total vegetation. Nevertheless, many common species of plants depend on their association with nitrogen-fixing bacteria to provide this essential nutrient in a form they can assimilate. The amount of N_2 returned to the atmosphere by denitrification is also relatively small. The important point is that although nitrogen exchanges between soil and atmosphere are significant over the long term, the majority of nitrogen in most ecosystems is recycled locally by decomposition and reassimilation.

The Phosphorus Cycle

Organisms require phosphorus as a major constituent of nucleic acids, phospholipids, ATP and other energy shuttles, and as a mineral constituent of bones and teeth.

In some respects the phosphorus cycle is simpler than either the carbon or the nitrogen cycle. Phosphorus cycling does not include movement through the atmosphere because there are no significant phosphorus-containing gases. In addition, phosphorus occurs in only one important inorganic form, phosphate (PO_4^{3+}), which plants absorb and use for organic synthesis. The weathering of rocks gradually adds phosphate to soil (FIGURE 49.12). After producers incorporate phosphorus into biological molecules, it is transferred to consumers in organic form, and added back to the soil by the excretion of phosphate by animals and by the action of decomposers on detritus. Humus and soil particles bind phosphate, so that the recycling of phosphorus tends to be quite localized in ecosystems. However, phosphorus does leach into the water table, gradually draining from terrestrial ecosystems to the sea. Severe erosion can hasten this drain, but in most natural ecosystems, the weathering of rocks can keep pace with the loss of phosphate. Phosphate that reaches the ocean gradually accumulates in sediments and becomes incorporated into rocks that may later be included in terrestrial ecosystems as a result of geological processes that raise the seafloor or lower sea level at a particular location. Thus, most phosphate recycles locally among soil, plants, and consumers on the scale of ecological time, while a parallel sedimentary cycle removes and restores terrestrial phosphorus over geological time. The same general pattern applies to other nutrients that lack atmospheric forms.

Phosphorus probably limits algal productivity in many aquatic habitats, and the addition of phosphorus to

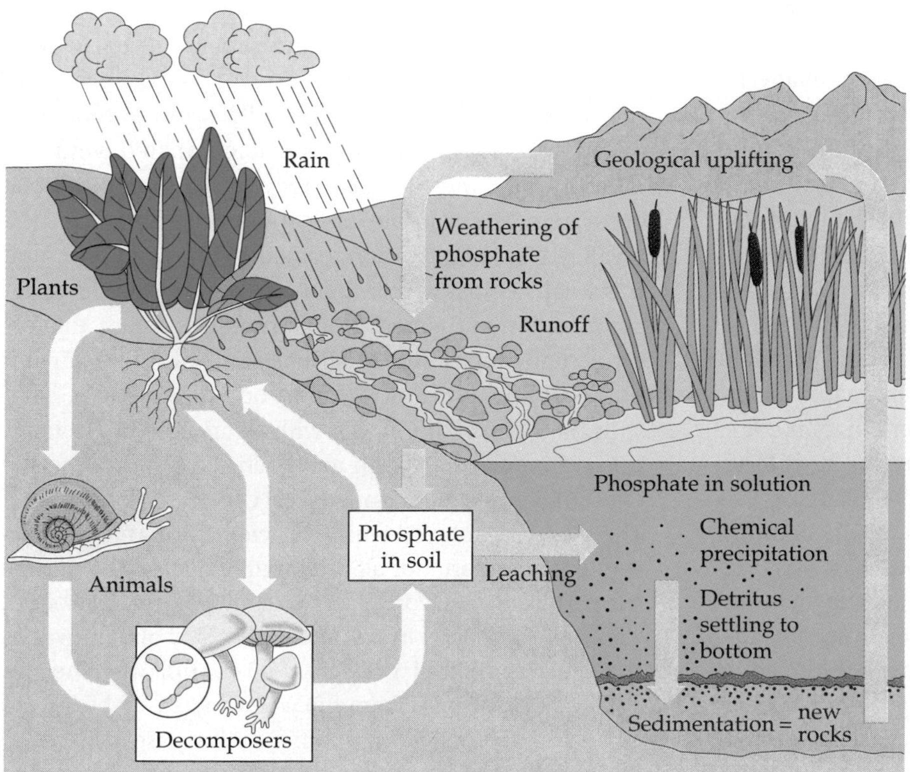

FIGURE 49.12
The phosphorus cycle. Phosphorus, which does not have an atmospheric component, tends to cycle locally (yellow arrows). Exact rates vary in different systems. Generally, small losses from terrestrial systems caused by leaching are balanced by gains from the weathering of rocks. In aquatic systems, as in terrestrial systems, phosphorus is cycled through food webs. Some phosphorus is lost from the ecosystem because of chemical precipitation or through settling of detritus to the bottom, where sedimentation may lock away some of the nutrient before biological processes can reclaim it. On a much longer time scale, this phosphorus may become available to ecosystems again through geological processes such as uplifting (gold arrows). This general pattern applies to many other nutrients, including trace elements.

these ecosystems, in the form of sewage and runoff from fertilized agricultural fields, stimulates production, often with negative consequences. For this reason, many states have banned detergents containing phosphates.

A review of FIGURES 49.10–49.12 at this time will reinforce the concept that chemical cycling in ecosystems involves both geological and biological processes.

Variations in Nutrient Cycling Time

The rates at which nutrients cycle in different ecosystems are extremely variable, mostly due to differences in rates of decomposition. In tropical rain forests, most organic material decomposes in a few months to a few years, while in temperate forests decomposition takes, on average, 4 to 6 years. In the tundra, decomposition can take 50 years, and in aquatic ecosystems, which are often anaerobic, it may occur even more slowly. The temperature and the availability of water and O_2 all affect rates of decomposition, and thus nutrient cycling times. Other factors that can influence nutrient cycling are local soil chemistry and the frequency of fires.

In some parts of a tropical rain forest, key nutrients such as phosphorus occur in the soil at levels far below those typical of a temperate forest. At first this might seem paradoxical because tropical forests generally have very high productivity. The key to this apparent riddle is rapid decomposition in tropical areas because of the warm temperatures and abundant precipitation. In addition, the immense biomass of these forests creates a high demand for nutrients, which are absorbed almost as soon as they become available through the action of decomposers. As a result of rapid decomposition, relatively little organic material accumulates as leaf litter on the floor of tropical rain forests; about 75% of the nutrients in the ecosystem are present in the woody trunks of trees, and about 10% are contained in the soil. The relatively low concentrations of some nutrients in the soil of tropical rain forests result from a fast cycling time, not an overall scantiness of these elements in the ecosystem.

In temperate forests, where decomposition is much slower, the soil may contain 50% of all the organic material in the ecosystem. The nutrients present in temperate forest detritus and soil may remain there for fairly long periods of time before being assimilated by plants.

In aquatic ecosystems, bottom sediments are comparable to the detritus layer in terrestrial ecosystems, except for their very slow decomposition and the fact that algae and aquatic plants usually assimilate nutrients directly from the water. Thus, these sediments often constitute a nutrient sink, and aquatic ecosystems can only be very productive if there is interchange between the bottom layers of water and the surface (see FIGURE 46.8).

Field experiments reveal how vegetation regulates chemical cycling: *science as a process*

Many research groups are conducting **long-term ecological research (LTER)** to track the dynamics of natural ecosystems over relatively long periods of time. In this section, we examine an example of LTER as a capstone to our study of chemical cycling in ecosystems.

Since 1963, a team of scientists has been conducting a long-term study of nutrient cycling in a forest ecosystem under both natural conditions and after vegetation is removed. The study site is the Hubbard Brook Experimental Forest in the White Mountains of New Hampshire. It is a nearly mature deciduous forest with several valleys, each drained by a small creek that is a tributary of Hubbard Brook. Bedrock impenetrable to water is close to the surface of the soil, and each valley constitutes a watershed that can drain only through its creek.

The research team first determined the mineral budget for each of six valleys by measuring the inflow and outflow of several key nutrients. They collected rainfall at several sites to measure the amount of water and dissolved minerals added to the ecosystem. To monitor the loss of water and minerals, they constructed V-shaped, concrete weirs (dams) across the creek at the bottom of each valley (FIGURE 49.13a). About 60% of the water added to the ecosystem as rainfall and snow exits through the stream, and the remaining 40% is lost by transpiration from plants and evaporation from the soil.

Preliminary studies confirmed that internal cycling within a mature terrestrial ecosystem conserves most of the mineral nutrients. Mineral inflow and outflow balanced and were relatively small compared with the quantity of minerals being recycled within the forest ecosystem. For example, only about 0.3% more Ca^{2+} left a valley via its creek than was added by rainwater, and this small net loss was probably replaced by chemical decomposition of the bedrock. During most years, the forest actually registered small net gains of a few mineral nutrients, including nitrogenous ones.

In 1966, one of the valleys, with an area of 15.6 hectares, was completely logged in an experiment that tested the effect of deforestation on nutrient cycling (FIGURE 49.13b). The land was sprayed with herbicides for 3 years to prevent regeneration, but all the original plant material was left in place to decompose. The inflow and outflow of water and minerals in the experimentally altered watershed were compared with those in a control watershed for 3 years. Water runoff from the valley increased by 30% to 40% after deforestation, apparently because there were no trees to absorb and transpire water from the soil. Net

(a)

(b)

FIGURE 49.13

Nutrient recycling in the Hubbard Brook Experimental Forest: an example of long-term ecological research. (**a**) Concrete weirs built across streams at the bottom of watersheds enabled researchers to monitor the outflow of water and nutrients from the ecosystem. Because these small dams were anchored to the impervious bedrock, all water draining from the valleys had to pass through the weirs. (**b**) Some watersheds were completely logged to study the effects of the loss of vegetation on drainage and nutrient cycling. All the logs and rubble from the clear-cutting operation were left in place, and care was taken not to disturb the soil any more than necessary.

losses of minerals from the deforested watershed were huge. The concentration of Ca^{2+} in the creek increased fourfold, for example, and the concentration of K^+ increased by a factor of 15. Most remarkable was the loss of nitrate, which increased in concentration in the creek sixtyfold (FIGURE 49.14). Not only was this vital mineral nutrient drained from the ecosystem, but nitrate in the creek reached a level considered unsafe for drinking water.

This study demonstrated that the amount of nutrients leaving an intact forest ecosystem is controlled by the plants themselves; when plants are not present to retain them, nutrients are lost from the system. These effects are almost immediate, occurring within a few months of deforestation, and continuing as long as plants are absent.

While the Hubbard Brook study was designed to assess the importance of vegetation in regulating ecosystem dynamics, the results provide important insight into the mechanisms by which human activities affect these processes. In the next two sections, we will consider other examples of human intrusion.

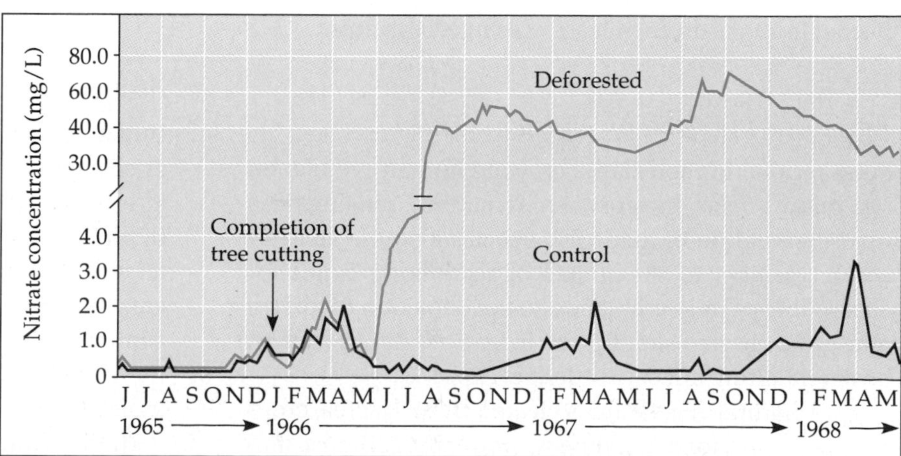

FIGURE 49.14

The loss of nitrate from a deforested watershed in the Hubbard Brook Experimental Forest. The concentration of nitrate in runoff from the deforested watershed was sixty times greater than in a control unlogged watershed.

The human population is disrupting chemical cycles throughout the biosphere

As the human population has grown to an immense size, our activities and technological capabilities have intruded in one way or another into the dynamics of most ecosystems. Even where we have not completely destroyed a natural system, our actions have disrupted the trophic structure, energy flow, and chemical cycling of ecosystems in most areas of the world. The effects are sometimes local or regional, but the ecological impact of humans can be widespread or even global. For example, acid precipitation may be carried by prevailing winds and fall hundreds or thousands of miles from the smokestacks emitting the chemicals that produce it (see Chapter 3).

Human activity often intrudes in nutrient cycles by removing nutrients from one part of the biosphere and adding them to another. This may result in the depletion of key nutrients in one area, excesses in another place, and the disruption of the natural equilibrium in both locations. For example, nutrients in the soil of croplands soon appear in the wastes of humans and livestock, and then appear in streams and lakes through runoff from fields and discharge as sewage. Someone eating a piece of broccoli in Washington, DC, is consuming nutrients that only days before might have been in the soil in California; and a short time later, some of these nutrients will be in the Potomac River on their way to the sea, having passed through an individual's digestive system and the local sewage facilties. Once in aquatic ecosystems, the nutrients may stimulate excessive growth of algae, and the human impact ripples on, from ecosystem to ecosystem.

Humans have intruded on nutrient cycles to such an extent that it is no longer possible to understand any cycle without taking human effects into account. In addition to transporting nutrients from one location to another, we have added entirely new materials, many of them toxic, to ecosystems. Here, we examine a few examples of the impact of humans on chemical cycles in ecosystems.

Agricultural Effects on Nutrient Cycling

Food production for Earth's growing human population has many effects on ecosystem dynamics, ranging from near-elimination of many fish species in some areas by overharvesting, to the spread of toxic compounds for controlling pests in croplands, to the depletion of surface and groundwater supplies by irrigation. Let's focus on how agriculture affects nutrient cycling.

After natural vegetation is cleared from an area, crops may be grown for some time without an additional supplement of nutrients because of the existing reserve of nutrients in the soil. However, in agricultural ecosystems, a substantial fraction of these nutrients is not recycled but is exported from the area in the form of crop biomass. The "free" period for crop production—when there is no need to add nutrients to the soil—varies greatly. When some of the early North American prairie lands were first tilled, for example, good crops could be produced for many years because the large store of organic materials in the soil continued to decompose and provide nutrients. By contrast, some cleared land in the tropics can be farmed for only one or two years because so few of the ecosystems' nutrients are contained in the soil. Eventually, in any area under intensive agriculture, the natural store of nutrients becomes exhausted, and fertilizer must be added. The industrially synthesized fertilizers used extensively today are produced at considerable expense in terms of both money and energy.

Agriculture has a great impact on the nitrogen cycle. Cultivation—breaking up and mixing the soil—increases the rate of decomposition of organic matter, releasing usable nitrogen that is then removed from the ecosystem when crops are harvested. As we saw in the case of Hubbard Brook, ecosystems from which plants are removed lose nitrogen not only because it is removed with the plants themselves, but because without plants to take them up, nitrates continue to be leached from the ecosystem. Industrially synthesized fertilizer is used to make up for the loss of usable nitrogen from agricultural ecosystems; the amount of nitrogen added through the industrial process is now at least as large as that added by all of the nitrogen-fixing bacteria in the world. Nitrogen fertilizers upset the balance between denitrification and nitrogen fixation, causing a release of extra N_2 into the atmosphere. A considerable portion of nitrogen in fertilizers eventually finds its way into aquatic ecosystems, either by leaching out of the soil as nitrates into groundwater or in surface water runoff. Once in aquatic systems, excess nutrients may stimulate algal growth, a problem we will examine next.

Accelerated Eutrophication of Lakes

Under natural conditions, lakes can be either oligotrophic or eutrophic. In an oligotrophic lake, primary productivity is relatively low because the mineral nutrients required by phytoplankton are scarce. In other lakes, basin and watershed characteristics cause the addition of more nutrients that are captured by the primary producers and then continuously recycled through the lake's food webs. Thus, the overall productivity is higher, and the lake is said to be more eutrophic (Gr., "well nourished").

Human intrusion has disrupted freshwater ecosystems by what is termed *cultural eutrophication.* Sewage

FIGURE 49.15
The experimental eutrophication of a lake. The far basin of this lake was separated from the near basin by a plastic curtain and fertilized with inorganic sources of carbon, nitrogen, and phosphorus. Within two months, the fertilized basin was covered with an algal bloom, which appears white in the photograph. The near basin, which was treated with only carbon and nitrogen, remained unchanged. In this case, phosphorus was the key limiting nutrient, and its addition stimulated the explosive growth of algal populations.

and factory wastes; runoff of animal waste from pastures and stockyards; and the leaching of fertilizer from agricultural, recreational, and urban areas has overloaded many streams, rivers, and lakes with inorganic nutrients. This enrichment often results in an explosive increase in the density of photosynthetic organisms (FIGURE 49.15). Shallower areas become weed-choked, making boating and fishing impossible. Large algal blooms become common, possibly resulting in increased oxygen production during the day but reduced oxygen levels at night because of respiration by the excessive algae. As the algae die and organic material accumulates at the lake bottom, decomposers use all the oxygen in the deeper waters. All these effects may make it impossible for some organisms to survive. For example, cultural eutrophication of Lake Erie wiped out commercially important fishes such as blue pike, whitefish, and lake trout by the 1960s. Since then, tighter regulations on the dumping of wastes into the lake have enabled some fish populations to rebound, but many of the native species of fishes and invertebrates have not recovered. And the U.S. Congress recently passed legislation that would loosen the standards for ensuring water quality.

Poisons in Food Chains

Humans produce an immense variety of toxic chemicals, including thousands of synthetics previously unknown in nature, that have been dumped into ecosystems with little regard for the ecological consequences. Many of these poisons cannot be degraded by microorganisms and consequently persist in the environment for years or even decades. In other cases, chemicals released into the environment may be relatively harmless, but they are converted to more toxic products by reaction with other substances or by the metabolism of microorganisms. For example, mercury, a by-product of plastic production, was once routinely expelled into rivers and the sea in an insoluble form. Bacteria in the bottom mud converted the waste to methyl mercury, an extremely toxic soluble compound that then accumulated in the tissues of organisms, including humans who consumed fish from the contaminated waters.

Organisms acquire toxic substances from the environment along with nutrients and water. Some of the poisons are metabolized and excreted, but others accumulate in specific tissues.

An example of a class of industrially synthesized compounds that act in this manner are the chlorinated hydrocarbons, which include many pesticides, such as DDT, and the industrial chemicals called PCBs (polychlorinated biphenols). Current research is implicating many of these compounds and others in endocrine system disruption in a large number of animal species, including humans. One of the reasons these toxins are so harmful is that they become more concentrated in successive trophic levels of a food web, a process called **biological magnification.** Magnification occurs because the biomass at any given trophic level is produced from a much larger biomass ingested from the level below. Thus, top-level carnivores tend to be the organisms most severely affected by toxic compounds that have been released into the environment.

A well-known example of biological magnification involves DDT, which has been used to control insects such as mosquitoes and agricultural pests. DDT persists in the environment and is transported by water to areas far

FIGURE 49.16

Biological magnification of DDT in a food chain. DDT concentration in a Long Island Sound food chain was magnified by a factor of about 10 million, from just 0.000003 parts per million (ppm) as a pollutant in the seawater to a concentration of 25 ppm in a bird at the top of this food pyramid, the fish-eating osprey.

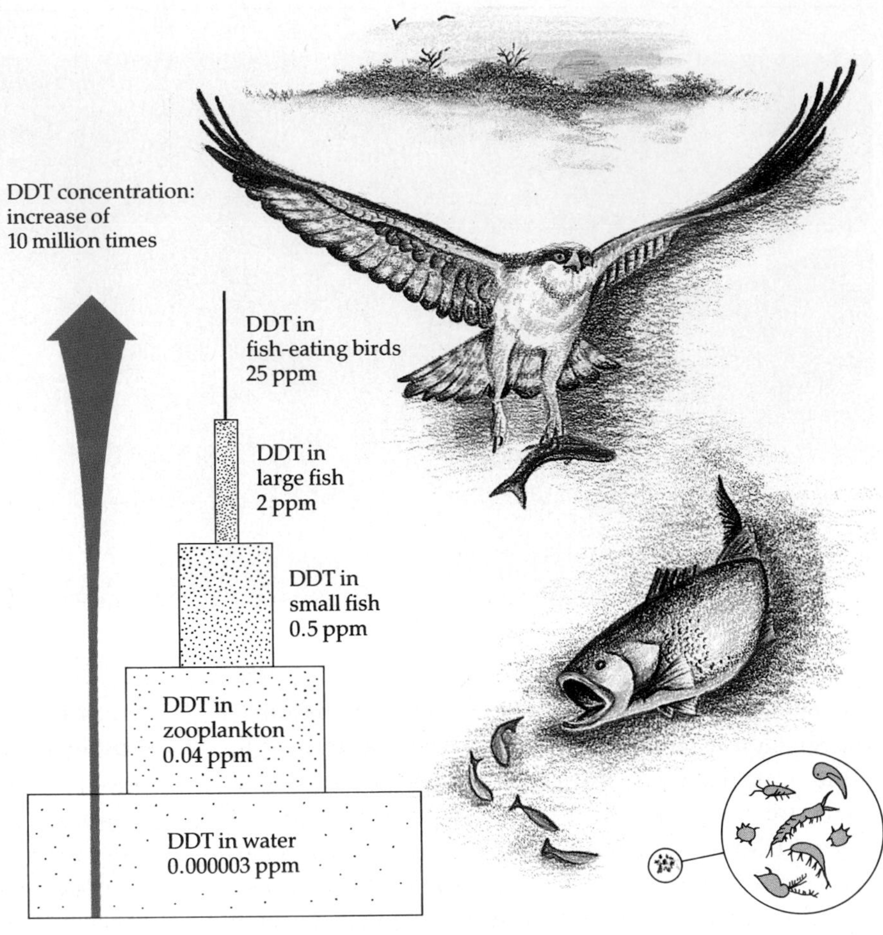

from where it is applied, and it rapidly became a global problem. Because the compound is soluble in lipids, it collects in the fatty tissues of animals, and its concentration is magnified in higher trophic levels (FIGURE 49.16). Traces of DDT have been found in nearly every organism tested; it has even been found in human breast milk throughout the world. One of the first signs that DDT was a serious environmental problem was a decline in the populations of pelicans and eagles, birds that feed at the top of food chains. The accumulation of DDT (and DDE, a product of its partial breakdown) in the tissues of these birds interfered with the deposition of calcium in their eggshells. When these birds tried to incubate their eggs, the weight of the parents broke the affected eggs, resulting in catastrophic declines in their reproduction rates. DDT was banned in the United States in 1971, and a dramatic recovery in populations of the affected bird species followed. The pesticide is still used in other parts of the world, however.

Intrusions in the Atmosphere

Many human activities release a variety of gaseous waste products. We once thought the vastness of the atmosphere could absorb these materials without significant consequences, but we now know that the finite volume of the atmosphere means that human intrusions can cause fundamental changes in its composition and its interactions with the rest of the biosphere. One pressing problem that relates directly to one of the nutrient cycles we examined is the rising level of carbon dioxide in the atmosphere.

Carbon Dioxide Emissions and the Greenhouse Effect

Since the Industrial Revolution, the concentration of CO_2 in the atmosphere has been increasing as a result of the combustion of fossil fuels and burning of enormous quantities of wood removed by deforestation. Various methods have estimated that the average carbon dioxide concentration in the atmosphere before 1850 was about 274 parts per million (ppm). When a monitoring station on Hawaii's Mauna Loa peak began making very accurate measurements in 1958, the CO_2 concentration was 316 ppm (FIGURE 49.17). Today, the concentration of CO_2 in the atmosphere exceeds 357 ppm, an increase of about 13% since the measurements began. If CO_2 emissions continue to increase at the present rate, by the year 2075, the atmospheric concentration of this gas will be double what it was at the start of the Industrial Revolution.

Increased productivity by vegetation is one pre-

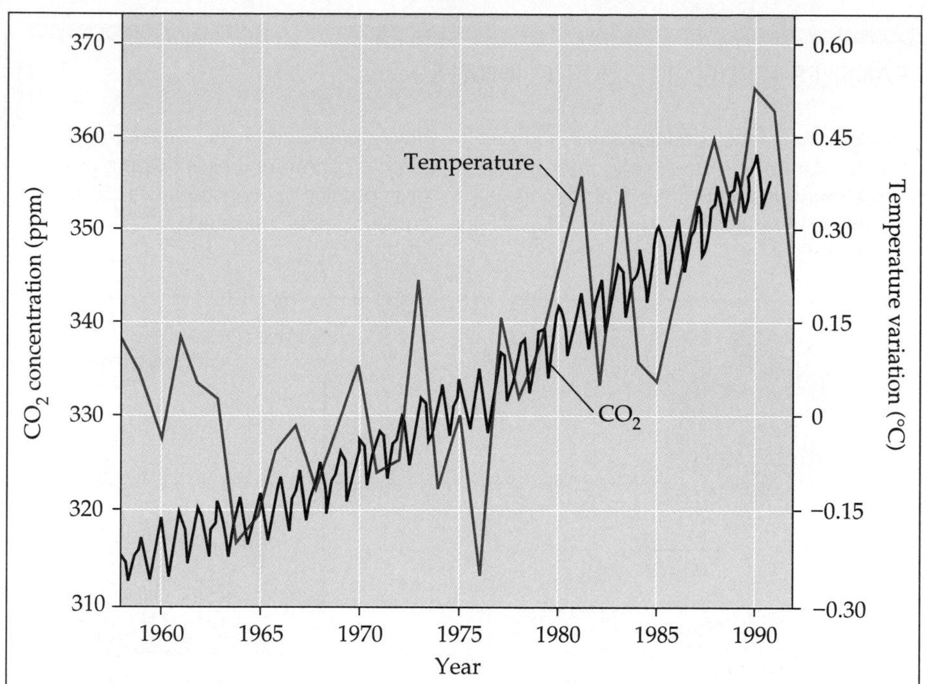

FIGURE 49.17
The increase in atmospheric carbon dioxide and average temperatures from 1958 to 1991. In addition to normal seasonal fluctuations in CO_2 levels (black), this graph shows a steady increase in the total amount of CO_2 in the atmosphere. These measurements are being taken at a relatively remote site in Hawaii, where the air is free from the variable short-term effects that occur near large urban areas. Though average temperatures over the same time period fluctuate a great deal (red), there is a warming trend. Climatologists predict that temperatures could rise up to 4°C over the next 50 years if atmospheric CO_2 levels continue to rise at the current rate.

dictable consequence of increasing CO_2 levels. In fact, when CO_2 concentrations are raised in experimental chambers, such as greenhouses, most plants respond with increased growth. However, because C_3 plants are more limited than C_4 plants by CO_2 availability (see Chapter 10), one effect of increasing CO_2 concentrations on a global scale may be the spread of C_3 species into terrestrial habitats previously favoring C_4 plants. This may have important agricultural implications. For example, corn, a C_4 plant and the most important grain crop in the United States, may be replaced on farms by wheat and soybeans, C_3 crops that will outproduce corn in a CO_2-enriched environment. However, no one can predict the gradual and complex effects that rising CO_2 levels will have on species composition in natural communities.

One factor that complicates predictions about the long-term effects of rising atmospheric CO_2 concentration is its possible influence on Earth's heat budget. Much of the solar radiation that strikes the planet is reflected back toward space. Although CO_2 and water vapor in the atmosphere are transparent to visible light, they intercept and absorb much of the reflected infrared radiation, rereflecting it back toward Earth. This process retains some of the solar heat. If it were not for this **greenhouse effect,** the average air temperature at Earth's surface would be −18°C. The marked increase in atmospheric CO_2 concentrations during the last 150 years concerns ecologists and environmentalists because of its potential effect on global temperature.

While scientists debate how increasing levels of atmospheric CO_2 will affect global temperatures, there is mounting evidence that a doubling of CO_2 concentration, which could occur by the end of the next century, might produce an average temperature increase of 3°–4°C. This evidence is based on mathematical models developed independently by several groups of researchers. Supporting these models is a correlation between CO_2 levels and temperatures in prehistoric times. Climatologists can actually measure CO_2 levels in bubbles trapped in glacial ice at different times in Earth's history. Past temperatures are inferred by several methods, one of which is described in the Methods Box (p. 1164).

An increase of only 1.3°C would make the world warmer than at any time in the past 100,000 years. A worst-case scenario suggests that the warming would be greatest near the poles; the resultant melting of polar ice might raise sea levels by an estimated 100 m, gradually flooding coastal areas 150 or more km inland from the current coastline. New York, Miami, Los Angeles, and many other cities would then be under water. A warming trend would also alter the geographical distribution of precipitation, making major agricultural areas of the central United States much drier. However, the various mathematical models disagree about the details of how climate in each region will be affected. By studying how past periods of global warming and cooling affected plant communities, Margaret Davis and other paleoecologists are using another strategy to help predict the consequences of future temperature changes. Records from pollen cores provide evidence that plant communities change dramatically with changes in temperature. Past climate changes occurred gradually, and plant and animal populations could migrate into areas where abiotic conditions allowed them to survive. However, the rate at which

METHODS: USING FOSSIL POLLEN SAMPLES TO DEDUCE PAST CLIMATES

To deduce what the earth's climate was like in the past, paleoecologists study the ecology of fossil communities. In this case, based on the work of Margaret Davis, the scientist you met in the interview that precedes Chapter 46, the fossils are pollen grains preserved in the sediments at the bottom of lakes and bogs. Sediments are deposited throughout the life of a lake; thus, the pollen at each level is a record of the surrounding vegetation that was present when that sediment was deposited. Each cubic centimeter of lake mud can hold about 100,000 to 200,000 pollen grains, which retain their original shape and ornamentation (FIGURE a), and can usually be classified, at least at the genus level.

Paleoecologists take cores of sediments from beneath lakes and bogs (FIGURE b), usually during the winter, when it is possible to bring coring equipment onto the ice. In the laboratory, the researchers use radiometric dating methods to determine the age of different levels of the core, remove core samples of different ages, and dissolve the matrix of material surrounding the pollen grains (see the Methods Box in Chapter 23, p. 459). By analyzing the pollen, paleoecologists can reconstruct the vegetation that was present around the lake at any time during the period represented by the core.

From hundreds of cores taken throughout eastern North America, paleoecologists have reconstructed the historical distributions of populations of many plant species. The maps in FIGURE c were constructed using data from these cores to infer the movement of spruce populations over the last 18,000 years, as the glaciers that had covered much of the continent retreated. These maps show where high, medium, and low amounts of spruce pollen (represented by dark, medium, and light green, respectively) were found in cores. The receding glacier is shown in white. Using knowledge of climatic conditions that spruce can tolerate, it is possible to infer mean temperatures during the time period covered by the cores. And studying how climatic changes in the past affected communities is helping ecologists predict the consequences of the current global warming.

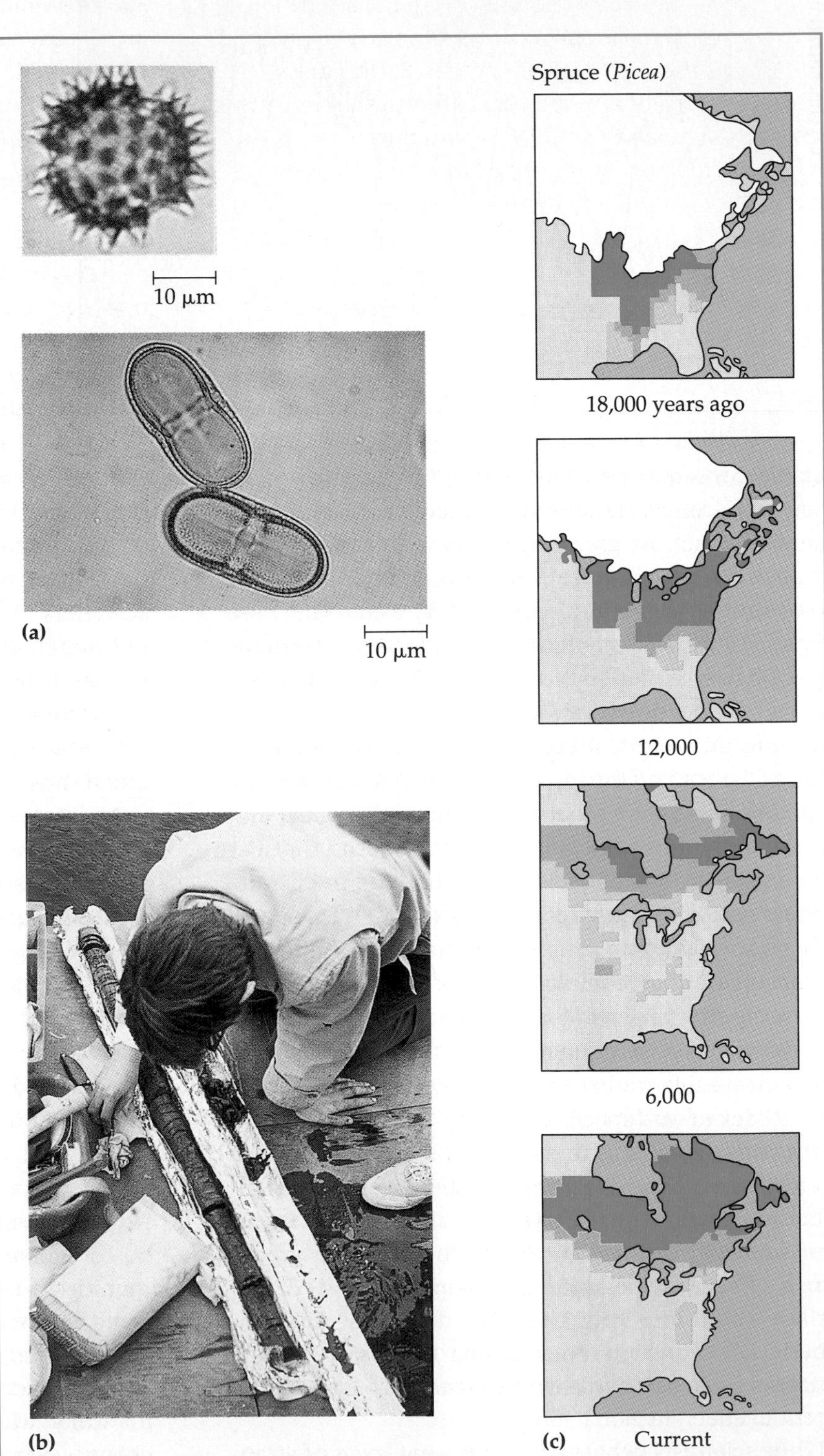

warming is occurring now is much faster than at any time in the past, and many organisms, especially plants that cannot disperse rapidly over very long distances, will probably not be able to survive.

Coal, natural gas, gasoline, wood, and other organic fuels central to modern life cannot be burned without releasing CO_2. The warming of the planet that is now under way as a result of the addition of CO_2 to the atmosphere is a problem of uncertain consequences and no simple solutions. At the 1992 United Nations Earth Summit in Rio de Janeiro, leaders of the world's industrialized countries signed a treaty that provides a framework for stabilizing CO_2 emissions by the end of this century. However, given the importance of combustion to our increasingly industrialized societies, implementation of these guidelines will require concerted international effort and the acceptance of dramatic changes in both personal lifestyles and industrial processes.

Depletion of Atmospheric Ozone

Life on Earth is protected from the damaging effects of ultraviolet (UV) radiation by a protective layer of ozone molecules (O_3) that is present in the lower stratosphere between 17 and 25 km above Earth's surface. Ozone absorbs UV radiation, preventing much of it from contacting organisms in the biosphere. Satellite studies of the atmosphere suggest that the ozone layer has been gradually depleted, or "thinned," since 1975, and that the depletion continues at an increasing rate.

The destruction of atmospheric ozone probably results mainly from the accumulation of chlorofluorocarbons, chemicals used for refrigeration, as propellants in aerosol cans, and in certain manufacturing processes. When the breakdown products from these chemicals rise to the stratosphere, the chlorine they contained reacts with ozone, reducing it to molecular O_2. Subsequent chemical reactions liberate the chlorine, allowing it to react with other ozone molecules in catalytic chain reaction. The effect is most apparent over Antarctica, where cold winter temperatures facilitate these atmospheric reactions. Scientists first described the "ozone hole" over Antarctica in 1985 and have since documented that it is a seasonal phenomenon that grows and shrinks on an annual cycle. However, the magnitude of ozone depletion and the size of the ozone hole have increased steadily in recent years, and the hole sometimes extends as far as the southernmost portions of Australia, New Zealand, and South America. At the more heavily populated middle latitudes, ozone levels have decreased 2% to 10% during the past 20 years.

The consequences of ozone depletion for life on Earth may be quite severe. Some scientists expect increases in both lethal and nonlethal forms of skin cancer and cataracts among humans, as well as unpredictable effects on crops and natural communities, especially the phytoplankton that are responsible for a large proportion of Earth's primary productivity. The danger posed by ozone depletion is so great that many nations have agreed to end the production of chlorofluorocarbons within a decade. Unfortunately, even if all chlorofluorocarbons were banned today, the chlorine molecules that are already in the atmosphere will continue to influence stratospheric ozone levels for at least a century.

Human activities are altering species distribution and reducing biodiversity

As we have seen, the activities and technology of the exploding human population have intruded in one way or another into the functioning of many ecosystems. Even where we have not completely destroyed a natural system, our actions have disrupted the trophic structure, energy flow, and chemical cycling of ecosystems in most areas of the world. Here we address ways that humans are directly affecting the biosphere's distribution and diversity of organisms.

The Introduction of Exotic Species

In Chapter 48, we discussed how ecologists sometimes use transplant experiments to identify why species are restricted to particular geographical ranges. Most transplant experiments have, in fact, been undertaken by people who inadvertently carry hitchhiking seeds or insects with them when they travel throughout the world, or who intentionally introduce foreign plants or animals for agricultural or ornamental purposes. Most transplanted species fail to survive outside their normal range. However, there are numerous examples of viable transplants. In the 1800s, starlings and English sparrows were imported to the northeastern United States from Europe; both species are now abundant in cities and the countryside throughout much of North America, and have replaced native birds in many areas. In fact, if your campus is in an urban area, there is a good chance that the birds you see most frequently as you walk between classes are starlings, English sparrows, and pigeons—all exotic (non-native) species.

Of the exotic species that do succeed in their new environment, many have little effect on the existing ecosystem. For example, pheasants, birds introduced to North America from Asia, have probably had little impact on the distribution or abundance of native species. However, in some cases, exotic species can play an important role in their new communities, usually through predation on native species or competition for resources. In

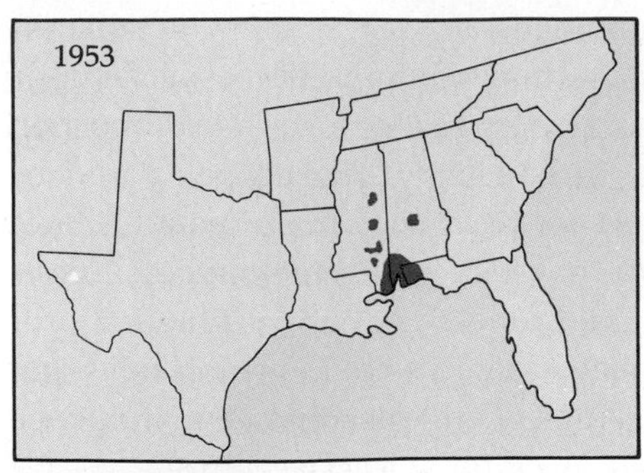

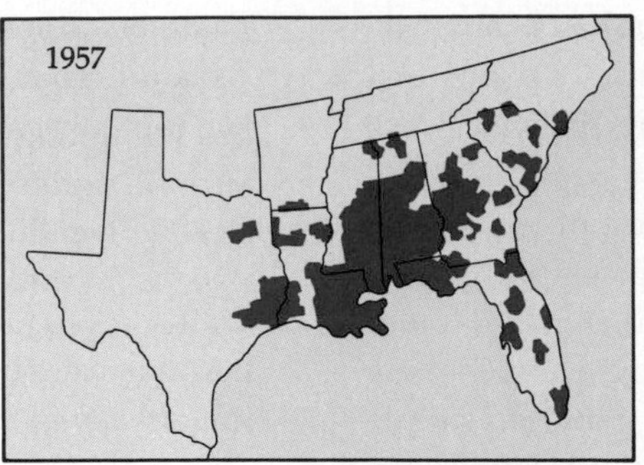

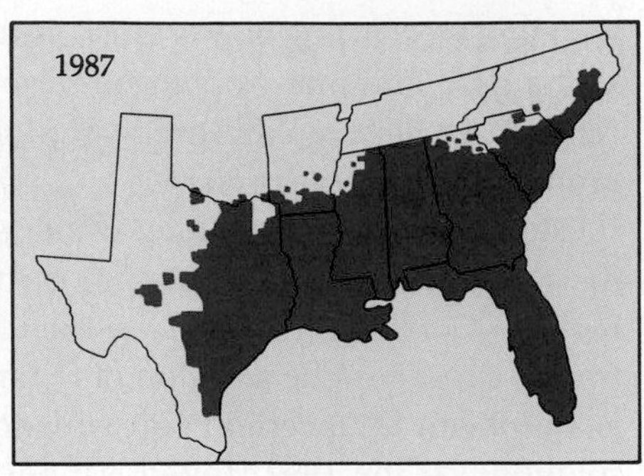

FIGURE 49.18
March of the fire ants. The fire ant (*Solenopsis invicta*) can inflict a beelike sting and tends to attack in large numbers. In the southern United States, more than 70,000 sting victims seek medical help each year. Fire ants were accidentally imported from South America about 45 years ago, probably in the hold of a ship that docked in Mobile, Alabama. The range of the ant has been spreading ever since, as traced by this series of maps.

fact, displacement by introduced species is considered at least partially responsible for 68% of the listings of extinct, endangered, vulnerable, and rare species published by the International Union for the Conservation of Nature and Natural Resources (IUCN). A striking example is the northward march of fire ants, which were accidentally introduced to the southern United States from Brazil in the 1940s (FIGURE 49.18). Not only can fire ants damage crops and inflict painful stings, they are also harmful to natural ecosystems. In their spread across the southeastern United States, they reduced the number of native ant species in one area of Texas from 15 to 5 and killed hatchlings of the threatened brown pelican in wildlife refuges.

Another example that illustrates important ecological concepts is the Nile perch, introduced into Lake Victoria in East Africa by Europeans who wanted to provide additional food and income for the people living in the area. The lake had supported fishing for a variety of endemic fish—mostly cichlids, which feed on detritus and plants. There were no large predators in the ecosystem, and the cichlids lacked adaptations for escaping predators. The introduced perch annihilated the cichlid populations, destroyed the native fishery, and reduced its own food supply to the point that the perch were not abundant enough to be a significant food source for humans. By ignoring basic ecological concepts, those who introduced the perch virtually destroyed the lake's entire fishery. Because energy is lost in each higher trophic level of a food chain, a predator can never be harvested at as high a rate as its prey.

Recently, some ecologists and environmentalists have become concerned about the ecological consequences of releasing genetically engineered organisms into natural and agricultural ecosystems. Advances in gene transfer technology now enable scientists to move genetic material from one species into another. For example, gene transfers are useful for developing herbicide resistance or pest resistance in agricultural crops, and field tests of engineered corn, potatoes, wheat, tomatoes, and many other agricultural and ornamental plants have been under way for several years. Animals can also be genetically engineered.

Potential ecological problems associated with genetic engineering were discussed in Chapter 19, but these concerns are worth reviewing here. Some ecologists worry that the introduced traits might allow crop plants to become weeds or pests if they escape from the fields in which they are grown. Genes for traits such as increased cold tolerance and faster growth have been added to fish. These traits could enable the engineered fish to expand their natural ranges and become more competitive with native species if the individuals carrying them escaped from confined aquaculture areas.

In addition, genetically engineered plants or animals might interbreed with naturally occurring organisms, increasing their ability to invade areas where they might not occur naturally. For example, if a weed species accidentally acquired an introduced gene for herbicide resistance, new mechanisms to control its spread would have to be developed. Similarly, genes for insect resistance in crop plants could act as a selection pressure favoring the coevolution of traits in the insect pests that overcome the plant defenses. The accidental transfer of insect resistance from crop plants to noncrop species could also influence the population dynamics of insects that feed on these plants and on the higher-level consumers that in turn feed on them. There are some cases in which crop plants in the United States have close wild relatives to which introduced genes could potentially be transferred by cross-pollination. However, this problem could be much greater in the tropics, where many of our agricultural plant varieties originated, and where their native progenitors still exist.

It is difficult to predict all the potential consequences of introducing genetically engineered species into the environment. But many ecologists have joined molecular biologists to help develop ways to minimize the chance of negative results.

Habitat Destruction and the Biodiversity Crisis

Human encroachment on natural ecosystems has reached epidemic proportions, as indicated in the map of agricultural and urban biomes in FIGURE 46.12. The clearing of natural ecosystems, which is usually necessary for agricultural, industrial, and residential development, undoubtedly causes the greatest local disruption of natural environments. Clear-cut harvesting of timber also destroys vast tracts of forest. Relatively little undisturbed habitat still exists in many countries; in the United States, for example, only 15% of the original primary forest (most of it in Alaska) and less than 1% of the original tallgrass prairie remain. The statistics on forests are even worse in some other regions, such as Europe, China, and Australia. In recent years, environmentalists have focused attention on the destruction of tropical forests, among the most productive ecosystems on Earth. Some estimates suggest that if tropical forests continue to be cut at the present rate (about 500,000 km^2 worldwide per year), nearly all large tracts of this ecosystem will be eliminated within a decade or two.

Development and logging are certainly not the only human activities that disrupt entire ecosystems. The ecological consequences of war are devastating. During the Vietnam War, for example, the United States used large quantities of chemical defoliants to kill the vegetation in which Viet Cong soldiers concealed themselves (FIGURE 49.19a and b). More recently, as the Persian Gulf War ended in 1991, Iraqi soldiers created other disasters that ravaged the landscape. Huge oil spills polluted marine ecosystems, and the burning of oil wells in Kuwait blackened the skies and left oily deposits on everything nearby (FIGURE 49.19c).

A major concern about the wholesale destruction of any natural habitat, particularly tropical rain forest, is the loss of biodiversity. In fact, the destruction of physical

(a)

(b)

(c)

FIGURE 49.19
Ecological effects of warfare. Military conflicts devastate entire landscapes as well as particular ecosystems. The use of a defoliant (agent orange) during the Vietnam War quickly destroyed vast areas of tropical forest. (**a**) A Vietnamese tropical rain forest before defoliation. (**b**) The same area afterward. One of the chemicals in agent orange is a carcinogen. (**c**) In this satellite view of Kuwait at the end of the 1991 Persian Gulf War, hundreds of burning oil wells are visible as orange dots. The fires blackened the skies, deposited a thick layer of oily soot on surrounding areas, and squandered precious fossil fuel.

habitat has been implicated in 73% of the IUCN's designations of species as extinct, endangered, vulnerable, and rare. The preservation of endangered species is a difficult task because ecologists are uncertain about minimum viable population sizes and the minimum expanses of suitable habitat that need to be protected.

The problem of preserving species is particularly complicated for migratory species that winter in one country and breed in another. Successful conservation efforts for such species require international cooperation and the careful preservation of habitat in both parts of the species' range. Monarch butterflies from eastern North America, for example, migrate from as far north as Canada to a few relatively small overwintering sites in high-altitude fir forests of the Transvolcanic Range of Mexico. Despite the fact that migrants have been extraordinarily abundant in some recent autumns, their migration has been termed an "endangered phenomenon" because of the vulnerability of these sites to human intrusion. In addition, milkweed, the host plant of monarchs, is considered a noxious weed and destroyed in Canada. In the United States, herbicide and insecticide use can affect monarch populations. Habitat preservation in Canada, the U.S., or Mexico alone will not remove the threats to this insect, and the situation is similar for many species of migratory songbirds.

Nobody can accurately estimate the magnitude of the biodiversity crisis, largely because taxonomists have identified and described only a fraction of the species present on Earth. In addition, extinction is an obscure, local process. The death of the last passenger pigeon in captivity was publicized, but this is rare. We are still uncertain whether there are any ivory-billed woodpeckers or Bachman's warblers left in the United States, or Tasmanian wolves in Australia. In order to know that a given species is extinct, we must know its exact distribution and its habits. But the fact that millions of the world's species have not even been given names is a clear indication that we do not know most species well at all.

Some people argue that because the rate of extinctions cannot be clearly documented, there is no strong reason to worry about them at this time. But we are cetain that the extinction of well-known organisms is proceeding at an alarming rate. For example:

- According to the International Council for Bird Preservation, 11% of the 9040 known bird species in the world are endangered. In the past 40 years, population densities of migratory songbirds in the mid-Atlantic United States dropped 50%.
- About 20% of the known freshwater fishes in the world have either become extinct during historical times or are seriously threatened. A recent search for the 266 known species in lowland Malaysia resulted in 122 being observed.
- A survey conducted by the Center for Plant Conservation showed that of the approximately 20,000 known plant species in the United States, 680 are in danger of becoming extinct by the year 2000.

These are only a few of the many documented examples of extinctions. People may still argue that extinction is a natural phenomenon that has been occurring almost since life first evolved. The important distinction is the rate at which extinction is now occurring. Worldwide, species are being lost at a rate about fifty times greater than at any time in the past 100,000 years. In some areas, the picture is even bleaker; in tropical rain forests, ecological models estimate that human activity has increased extinction between 1000 and 10,000 times over the normal "background" rate of one out of every million species per year.

Why should we care about the loss of biodiversity? Perhaps the purest reason is our *biophlilia,* our sense of connection to nature and other forms of life. But in addition to the aesthetic and ethical reasons for preserving biodiversity, there are practical reasons as well. Biodiversity is a crucial natural resource, and species that are threatened could provide crops, fibers, and medicines. In the United States, 25% of all prescriptions dispensed from pharmacies contain substances derived from plants. In the 1970s, researchers discovered that the rosy periwinkle from Madagascar contains alkaloids that inhibit cancer cell growth (FIGURE 49.20). The result of this discovery is remission for most victims of two of the most deadly cancers, Hodgkin's disease and a childhood leu-

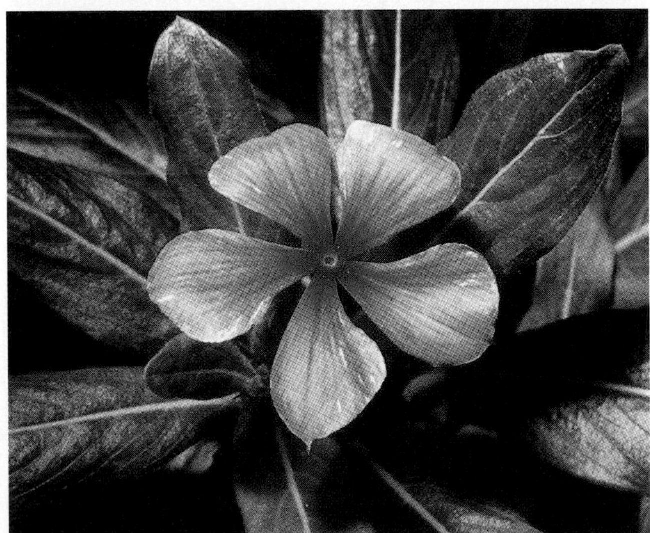

FIGURE 49.20
The rosy periwinkle: a plant that saves lives. Before alkaloids that inhibit cancer cell growth were discovered in the rosy periwinkle 20 years ago, Hodgkin's disease and acute lymphocytic leukemia were two of the deadliest of cancers. Now most victims are cured. This plant is one of hundreds that are used to treat human diseases.

kemia. There are five other species of periwinkles on Madagascar, and one is approaching extinction.

The benefits that individual species provide to humans are often substantial, but when we focus on these, we risk overlooking the fact that humans evolved in the living ecosystems of Earth and that our bodies are finely adjusted to these conditions. While it is possible that we could survive in a world with much less biodiversity, it is not certain. By allowing the extinction of species to continue, we are taking a risk that seems unwise.

The Sustainable Biosphere Initiative is reorienting ecological research

Everyone can and should play a role in reducing the negative impact humans have on ecosystems. But we clearly need a good deal of basic knowledge about the complex interconnections of the biosphere before we can make sensible decisions about how to accomplish this. We must understand how ecosystems work if we are to conserve them. To this end, the Ecological Society of America, the world's largest organization of professional ecologists, has recently endorsed a new research agenda, the **Sustainable Biosphere Initiative.** The goal of this initiative is to define and acquire the basic ecological information necessary for the intelligent and responsible development, management, and conservation of Earth's resources. The research agenda includes studies of global change, including interactions between climate and ecological processes; biological diversity and its role in maintaining ecological processes; and the ways in which the productivity of natural and artificial ecosystems can be sustained. This initiative will require a reorientation of scientific research to meet the goals and the development of new interdisciplinary approaches to study these complex phenomena, as well as a commitment of substantial financial resources to undertake the research. The effort must be far bigger than the Human Genome Project, one of the largest scientific projects that this country has ever undertaken. However, the importance of this research agenda cannot be overstated because life on Earth may ultimately be at stake.

The current state of the biosphere demonstrates clearly that we are treading dangerously on uncharted ecological ground. Perhaps our guide for future activity should be the words of the great American naturalist Aldo Leopold: "If the biota, in the course of aeons, has built something we like but do not understand, then who but a fool would discard seemingly useless parts? To keep every cog and wheel is the first precaution of intelligent tinkering."

REVIEW OF KEY CONCEPTS (with page numbers and key figures)

- Trophic structure determines an ecosystem's routes of energy flow and chemical cycling (pp. 1145–1148, FIGURE 49.2)
  - An ecosystem is the most inclusive level in the hierarchy of biological organization, consisting of the community of organisms in a defined area and the abiotic factors making up the physical environment.
  - Energy flow and chemical cycling are two interrelated processes that occur by the transfer of substances through the feeding levels of ecosystems.
  - Species in an ecosystem are divided into different trophic levels, depending on their main source of nutrition. Primary producers are autotrophic organisms. Heterotrophic organisms are consumers. Herbivores are primary consumers that eat the autotrophs. The secondary and tertiary consumers are carnivores. Decomposers feed on organic wastes and dead organisms from all trophic levels.
  - The pathway along which food from producers is transferred to the various levels of consumers is known as a food chain. However, natural feeding relationships are usually more like webs, because consumers feed at more than one trophic level or on a variety of items at the same level.
- An ecosystem's energy budget depends on primary productivity (pp. 1148–1150, FIGURE 49.3)
  - Ecologists distinguish between two measures of the energy assimilated during photosynthesis: gross primary productivity, the total energy assimilated; and net primary productivity, the accumulation of energy in plant biomass.
  - Only net primary productivity, defined as gross primary productivity minus the energy used by the primary producers for respiration, is available to consumers.
- As energy flows through an ecosystem, much is lost at each trophic level (pp. 1150–1152, FIGURE 49.5)
  - The amount of energy available to each trophic level is determined by the net primary productivity and the efficiencies with which food energy is converted into biomass at each link of the food chain.
  - The percentage of energy transferred from one trophic level to the next is called ecological efficiency; this is generally 5% to 20%.
- Matter cycles within and between ecosystems (pp. 1152–1153, FIGURE 49.8)
  - While nutrients move with energy through the various trophic levels in an ecosystem, they differ from energy in that they can cycle over and over within the ecosystem.

- Nutrient cycles involve both biotic and abiotic components of ecosystems; they are also called biogeochemical cycles.

- A combination of biological and geological processes drives chemical cycles (pp. 1154–1158, FIGURES 49.9–49.12)
  - Water moves in a global cycle that is driven by solar energy.
  - The carbon cycle primarily reflects the reciprocal processes of photosynthesis and cellular respiration.
  - Nitrogen-fixing bacteria convert atmospheric N_2 to nitrogenous minerals that plants can use to synthesize organic compounds, but most of the nitrogen cycling in natural ecosystems involves local cycles between organisms and soil or water.
  - The phosphorus cycle does not involve any atmospheric forms and thus occurs on a more localized scale than the cycles of water, carbon, and nitrogen.
  - The proportion of a nutrient in a particular form and its cycling time in that form vary among ecosystems, mostly because of differences in the rate of decomposition.

- Field experiments reveal how vegetation regulates chemical cycling: *science as a process* (pp. 1158–1159, FIGURE 49.14)
  - Long-term ecological research at Hubbard Brook in New Hampshire showed that complete logging of a mature deciduous forest increased water runoff and caused huge losses of minerals.

- The human population is disrupting chemical cycles throughout the biosphere (pp. 1160–1165)
  - Agricultural practices result in the constant removal of nutrients from ecosystems, so that large supplements are continually required. Considerable amounts of the nutrients in fertilizer move into aquatic ecosystems, where they can stimulate excess algal growth.
  - The dumping of nutrient-rich wastes into aquatic habitats accelerates eutrophication, and the increased biomass depletes O_2.
  - The release of toxic wastes has polluted the environment with harmful substances that often persist for long periods of time. DDT is an example of the many toxic substances that become concentrated along the food chain by biological magnification.
  - Because of the burning of wood and fossil fuels, CO_2 in the atmosphere has been steadily increasing. The ultimate effects are uncertain but may include significant warming and other influences on climate, as well as on the distribution of species and ecosystem productivity.
  - Chlorine-containing pollutants are eroding the ozone layer, which reduces the penetration of UV radiation through the atmosphere.

- Human activities are altering species distribution and reducing biodiversity (pp. 1165–1169)
  - The introduction of exotic species has disrupted many communities and ecosystems, often through predation or competition with native species.
  - Human intrusion into natural ecosystems has destroyed many habitats.
  - Species are becoming extinct at a rate fifty times greater than at any time in the past 100,000 years, largely because of loss of habitat.
  - Loss of biodiversity means that humans are losing important natural resources.

- The Sustainable Biosphere Initiative is reorienting ecological research (p. 1169)
  - Ecologists are embarking on a gigantic new research agenda. The goal of this initiative is to acquire the basic ecological information necessary for the intelligent and responsible development, management, and conservation of Earth's resources.

SELF-QUIZ

1. Which of the following organisms is *incorrectly* paired with its trophic level?
   a. cyanobacteria—primary producer
   b. grasshopper—primary consumer
   c. zooplankton—secondary consumer
   d. eagle—tertiary consumer
   e. fungi—decomposer (detritivore)

2. One of the lessons from a pyramid of productivity is that
   a. only one-half of the energy in one trophic level is passed on to the next level
   b. most of the energy from one trophic level is incorporated into the biomass of the next level
   c. the energy lost as heat or in cellular respiration is 10% of the available energy of each trophic level
   d. ecological efficiency is highest for primary consumers
   e. eating grain-fed beef is an inefficient means of obtaining the energy trapped by photosynthesis

3. The role of decomposers in the nitrogen cycle is to
   a. fix N_2 into ammonia
   b. release ammonia from organic compounds, thus returning it to the soil
   c. denitrify ammonia, thus returning N_2 to the atmosphere
   d. convert ammonia to nitrate, which can then be absorbed by plants
   e. incorporate nitrogen into amino acids and organic compounds

4. The Hubbard Brook Experimental Forest study demonstrated all of the following *except* that
   a. most minerals were recycled within a forest ecosystem
   b. mineral inflow and outflow within a natural watershed were nearly balanced
   c. deforestation resulted in an increase in water runoff
   d. the nitrate concentration in waters draining the deforested area became dangerously high
   e. deforestation causes a large increase in the density of soil bacteria

5. The recent increase in atmospheric CO_2 concentration is mainly a result of an increase in
   a. primary productivity
   b. the biosphere's biomass
   c. the absorption of infrared radiation escaping from Earth
   d. the burning of fossil fuels and wood
   e. cellular respiration by the exploding human population

6. Which of the following is a result of biological magnification?
   a. Top-level predators may be most harmed by toxic environmental chemicals.
   b. DDT has spread throughout every ecosystem and is found in almost every organism.
   c. The greenhouse effect will be most significant at the poles.
   d. Energy is lost at each trophic level of a food chain.
   e. Many nutrients are being removed from agricultural lands and shunted into aquatic ecosystems.

7. Which of these ecosystems has the *lowest* primary productivity per square meter?
   a. a salt marsh
   b. an open ocean
   c. a coral reef
   d. a grassland
   e. a tropical rain forest

8. Quantities of mineral nutrients in soils of tropical rain forests are relatively low because
   a. the standing crop is small
   b. microorganisms that recycle chemicals are not very abundant in tropical soils
   c. the decomposition of organic refuse and reassimilation of chemicals by plants occur rapidly
   d. nutrient cycles occur at a relatively slow rate in tropical soils
   e. the high temperatures destroy the nutrients

9. At present, the most significant cause of dwindling biodiversity is probably
   a. global warming
   b. the deterioraton of the ozone layer
   c. the destruction of habitat
   d. human consumption of native plants for food
   e. biological magnification of DDT

10. Which of the following statements concerning the water cycle is *incorrect?*
   a. There is a net movement of water vapor from oceans to terrestrial environments.
   b. Precipitation exceeds evaporation on land.
   c. Most of the water that evaporates from oceans is returned by runoff from land.
   d. Transpiration makes a significant contribution to evaporative water loss from terrestrial ecosystems.
   e. Evaporation exceeds precipitation over the seas.

CHALLENGE QUESTIONS

1. Imagine you have been chosen as the biologist for the design team for a self-contained space station to be assembled in orbit. It will be stocked with organisms you choose to create an ecosystem that will support you and five other people for two years. Describe the main functions you expect the organisms to perform. List the types of organisms you would select, and explain why you chose them.

2. In Southeast Asia, there's an old saying, "There is only one tiger to a hill." Explain this saying in terms of the energy flow in ecosystems.

3. A carbon atom in CO_2 might be used by a dandelion to make a sugar molecule. The sugar molecule might then be consumed by a grasshopper when it eats the dandelion, and be desposited in the soil in an organic molecule in the grasshopper's feces. A bacterium could release the carbon atom back into the air as CO_2 as it decomposes the soil litter. Continue the story, describing five more places the carbon atom might go, including your own body.

SCIENCE, TECHNOLOGY, AND SOCIETY

1. If we use substantial amounts of crop residues (such as corn stalks) to produce alcohol for use as a fuel instead of plowing those residues back into the soil, how will this affect the carbon cycle? What effects will this practice have on soil structure and function (see Chapter 33)? Do you think this use of crop residues is a good idea? Why or why not?

2. The amount of CO_2 in the atmosphere is increasing, and global temperature has increased over the last century; however, scientists do not agree about the extent to which the two phenomena are related. Most say that greenhouse warming is under way, and we need to take action now to avoid drastic environmental change. Some say it is too soon to tell, and we should gather more data before we act. The latter group dominated the U.S. delegation to the 1992 Earth Summit. Because of U.S. opposition, specific CO_2 targets were eliminated from the global warming agreement. What are the advantages and disadvantages of doing something now to slow global warming? What are the advantages and disadvantages of waiting until more data are available?

2. Some organizations are starting to envision a sustainable society—one in which each generation inherits sufficient natural and economic resources and a relatively stable environment. The Worldwatch Institute, an environmental policy organization, estimates that we must reach sustainability by the year 2030 to avoid economic and environmental disaster. To get there, we must begin shaping a sustainable society during this decade. In what ways is our present system not sustainable? What might we do to work toward sustainability, and what are the major roadblocks to achieving it? How would your life be different in a sustainable society?

FURTHER READING

Brown, L. R. *State of the World.* New York: Norton, 1996. A new edition of this book, prepared by the staff of the Worldwatch Institute, appears each year to provide an up-to-the-minute analysis of human intrusions in ecosystems.

Holloway, M. "Diversity Blues." *Scientific American,* August 1994. The biodiversity crisis extends to oceanic biomes.

Kerr, R. A. "Studies Say—Tentatively—That Greenhouse Warming Is Here." *Science,* June 16, 1995. Evidence linking increasing CO_2 to global warming.

Pimm, S. "Seeds of Our Own Destruction." *New Scientist,* April 8, 1995. The scale of the current mass extinction may be on par with the Cretaceous extinctions.

Repetto, R. "Accounting for Environmental Assets." *Scientific American,* June 1992. Why we should revise our economic indicators to reflect the true cost of damaging ecosystems.

Risser, P. G., J. Lubchenco, and S. A. Levin. "Biological Research Priorities—A Sustainable Biosphere." *BioScience,* October 1991. A brief description of the Sustainable Biosphere Initiative.

Sirica, C. "Taking Stock of Tropical Biodiversity." *Science,* July 29, 1994. The importance of international cooperation in conservation biology.

Sisk, T. D., A. E. Launer, K. R. Switky, P. R. Ehrlich. "Identifying Extinction Threats." *BioScience,* October 1994. A novel approach to identifying countries that are facing exceptionally severe conservation problems.

Volk, T. "The Soil's Breath." *Natural History,* November 1994. An examination of the carbon cycle, focusing on the role of microorganisms in the soil.

Wilson, E. O. "The Diversity of Life." *Discover,* September 1992. Why we should be concerned about the extinction of other species.

CHAPTER 50

BEHAVIOR

KEY CONCEPTS

- Behavior is what an animal does and how it does it
- Behavioral ecology emphasizes evolutionary hypotheses: *science as a process*
- A behavior has both an ultimate and a proximate cause
- Certain stimuli trigger innate behaviors called fixed action patterns
- Learning is experience-based modification of behavior
- Rhythmic behaviors synchronize an animal's activities with daily and seasonal changes in the environment
- Environmental cues guide animal movement
- Behavioral ecologists are using cost/benefit analysis to study foraging behavior
- Sociobiology places social behavior in an evolutionary context
- Competitive social behaviors often represent contests for resources
- Mating behavior relates directly to an animal's fitness
- Social interactions depend on diverse modes of communication
- The concept of inclusive fitness can account for most altruistic behavior
- Human sociobiology connects biology to the humanities and social sciences

Almost everyone has at one time or another listened to an infant lying in its crib, vocalizing an array of coos, babbles, and gurgles. Humans are not alone in this behavior. Thousands of other species of young animals, especially songbirds such as offspring of the brown-headed cowbird in the photograph on this page, engage in this behavior of "talking to oneself." Because the vocalizations are crude, and they occur whether or not anyone is there to listen, it is easy to miss the adaptive nature of this behavior.

Baby sounds apparently provide a mechanism for the young animal to learn to match its vocalizations to inborn or innate sound templates. The process of matching is akin to the human process of learning a new song from a musical score. The score contains the directions for playing, but a musician needs to listen to his or her own practicing in order to get the piece right. Somewhere in the brain there is a similar score: a neural network that dictates the nature of baby sounds. These first sounds of people and songbirds eventually give way to an array of learned sounds. Young birds hear sounds from conspecifics (members of the same species) and then learn to match these other individuals' songs just as they learned to match their innately programmed baby sounds. Communication is the result of genetic cues provided during development that are modified by environmental factors in the life of the individual.

Bird song provides an excellent introduction to the subject of behavior, because biologists are beginning to understand a great deal about its development, functions, and consequences at virtually all levels of biological organization, from molecules to whole organisms to entire populations. In addition to being amenable to modern experimental methods, bird song is attractive to researchers because of several striking parallels between it and human speech, including the one already noted. Bird song has also become a model system for animal behavior research because it demonstrates a very important generalization: Behavior is influenced by both innate and learned factors. The study of bird song has provided guideposts for the study of other complex behaviors that are less well understood. In this chapter, you will learn how biologists study animal behavior and what they are learning about the dual contributions of genes and environment.

Behavior is what an animal does and how it does it

Before we discuss bird song and many other types of behavior in more detail, we need to consider how behavior is defined. A dictionary definition of behavior may read something like "to act, react, or function in a particular way in response to some stimulus." Much of behavior does indeed consist of externally observable muscular activity, the "act" and "react" components of this definition. But a young bird that hears an adult song may show no obviously related muscular activity. Instead, the memory of the song may be stored in the bird's brain. Although this memory may change the brain's functioning, any observable muscular response will come later, perhaps even months later, when the bird begins to learn to match this stored memory of an adult song. Thus, if we think of **behavior** as what an animal does and *how* it does it, this definition will encompass the nonmotor components of behavior as well as an animal's observable actions.

The study of animal behavior is undoubtedly one of the oldest aspects of biology. Tens of thousands of years ago, such knowledge was essential to human survival. By learning the habits of the animals around them, early humans increased their chances of securing a meal and decreased their chances of becoming a meal. Thus, our ancestors' study of animal behavior ultimately enhanced their Darwinian fitness. More generally, our own behaviors and the behaviors of the animals we study have their ultimate basis in evolution.

Behavioral ecology emphasizes evolutionary hypotheses: *science as a process*

While early people increased their *own* fitness by studying animal behavior, today's biologists use the fitness of their animal subjects as a guiding principle for their research. The concept is a simple one: Because natural selection works on the enormous amount of genetic variation generated by mutation and recombination, we expect organisms to possess features that maximize their genetic representation in the next generation. Applied to behavior, this concept means that we expect animals to behave in ways that maximize their fitness. For example, feeding behavior is likely to optimize efficiency, or net energy gain, and the choice of a mate tends to maximize the number of healthy offspring produced.

This expectation of optimal behavior is valid only if genes influence behavior, because if there were no genetic influence, behavior could not be subjected to natural selection and could not evolve. Genes do exert a strong influence on many behaviors, as the lovebird example in FIGURE 50.1 (p. 1174) illustrates. Even learned behaviors typically depend on genes that create a neural system receptive to learning. The research approach based on the expectation that animals increase their Darwinian fitness by optimal behavior is called **behavioral ecology.** This evolutionary approach has dominated the study of animal behavior for almost 20 years. Behavioral ecology is appealing because it leads to testable hypotheses—hypotheses that make predictions that can be checked by experiments and observations. This, of course, is what science is all about.

Suppose you became interested in the observation that many songbirds have a repertoire of song types. Some of these songs sound identical to us but can be easily distinguished when analyzed with a sound spectrograph (FIGURE 50.2, p. 1175). Why has natural selection favored this multisong behavior over the expert vocalization of a single tune? If you followed the practices of behavioral ecology, you could formulate several testable hypotheses, all starting with "A repertoire increases fitness because . . ." (Recall from Chapter 1 that hypotheses are phrased as possible explanations.) You might hypothesize that a repertoire increases fitness because it makes an older, more experienced male more attractive to females. For this hypothesis to be true, it must be the case that: (1) males learn more song types as they get older, so that repertoire size is a reliable indicator of age; and (2) females prefer to mate with males having large repertoires. Thus, your hypothesis makes two clearly testable predictions.

To test the first prediction, you can determine whether there is a correlation between a male's age and the size of his song repertoire. If there is not, your hypothesis will be invalidated, which would be informative although perhaps disappointing. As it turns out, some songbird species show this correlation, while others do not. Next, you can determine whether females are more sexually stimulated by a large song repertoire than by a small one. This can be done by playing tape-recorded male songs to females that have been made especially receptive to male song by the temporary administration of a female hormone. Such females indicate their song preferences by assuming a copulatory posture, even though there is no male around. FIGURE 50.3 (p. 1175) shows another way to assess female responses. All this work may lead to an evolutionary explanation: Bird-song repertoires result in females mating more often with experienced males, who will give their offspring a better chance to survive.

Now suppose you did not use behavioral ecology to guide your research on song repertoires. Without an

FIGURE 50.1

The genetic component of behavior: a case study. Several species of brightly colored African parrots, commonly known as lovebirds, build cup-shaped nests inside tree cavities. Females typically make nests with thin strips of vegetation (or, in the laboratory, paper) that they cut with their beaks. (**a**) In one species, Fischer's lovebird (*Agapornis fischeri*), the bird cuts relatively long strips and carries them back to the nest one at a time in her beak. (**b**) In contrast, the peach-faced lovebird (*A. roseicollis*) cuts shorter strips and usually carries several at a time by tucking them into the feathers of the lower back. Tucking is a fairly complex behavior because the strips must be held just right and pushed in firmly, and the feathers then smoothed over. (**c**) These two species are closely related and have been experimentally interbred. The resultant hybrid females exhibited an intermediate kind of nest-building behavior. The strips cut by the hybrid birds were intermediate in length; even more interesting was the birds' hybrid manner of handling the strips. They usually made some attempt to tuck them into their rear feathers, but in some cases they did not let go after turning and pushing them a short distance. In other cases, the strips were manipulated or inserted improperly or simply dropped. The result was almost a total failure to transport strips by this method. Eventually, the birds learned to transport the strips in their beaks. Even so, they always made at least token tucking attempts. (**d**) After several years, the birds still turned their heads to the rear before flying off with a strip. These observations demonstrate that the phenotypic differences in the behavior of the two species are based on different genotypes. We also see that innate behavior can be modified by experience; the hybrid birds eventually learned to transport the strips.

(a) Nests made with long strips—no tucking behavior

(b) Nests made with short strips—tucking behavior

(c) Hybrid nests made with intermediate-length strips—in first mating season, unsuccessful tucking behavior

(d) In later seasons, only head-turning behavior

approach that generates testable hypotheses and predictions, you would probably record observations of numerous aspects of singing behavior. Although these efforts may produce interesting data, they would not *explain* the behavior. Alternatively, you might hypothesize that repertoires have nothing to do with fitness, but that male birds simply find variety more pleasurable than singing the same boring song over and over again. The method for testing such a hypothesis is not clear, emphasizing again how much more productive it is to use evolutionary principles as a guide to behavioral research.

Evolution is the core theme of biology, and behavioral ecology, with its emphasis on evolutionary explanations, is the focal point of this chapter.

A behavior has both an ultimate and a proximate cause

When we observe a certain behavior, we are apt to ask two types of questions. When our questions are about *why* the animal has the behavior, we are asking about **ultimate causation**—that is, the reason something exists. In the study of animal behavior, questions about ultimate causation are evolutionary questions: Why did natural selection favor this behavior and not a different one? Hypotheses that address a "why" question always take the form of ultimate explanations. In behavioral ecology, such hypotheses propose that the behavior maximized fitness in some particular way.

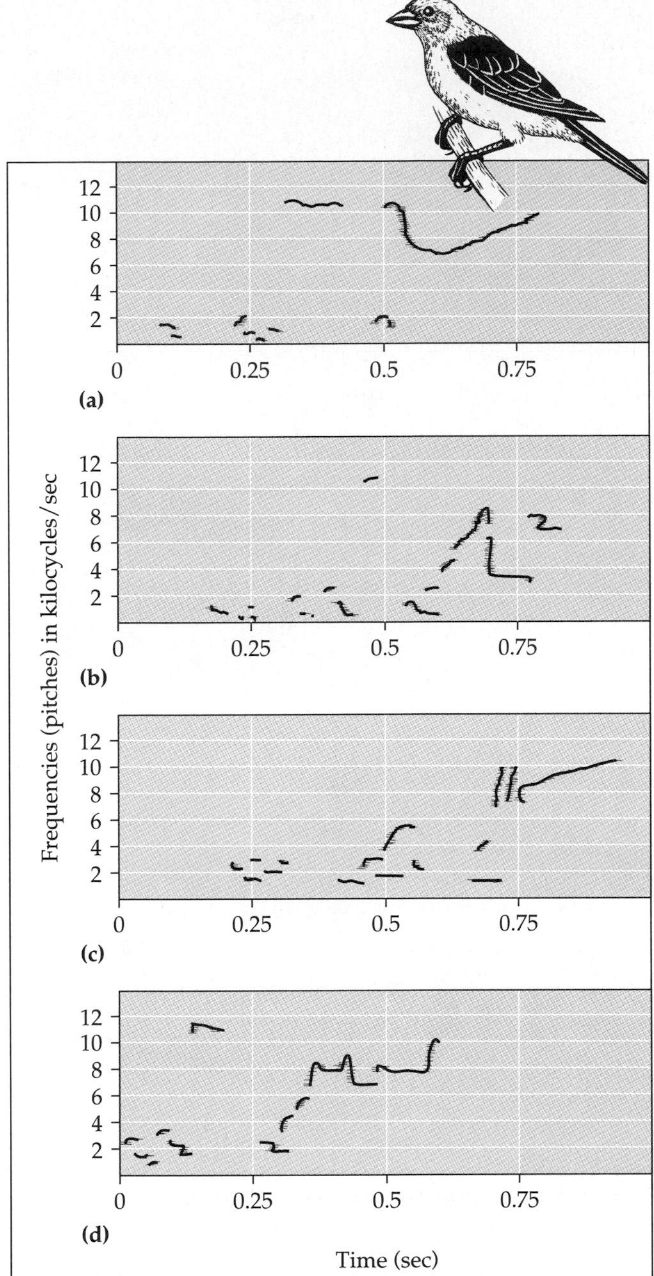

FIGURE 50.2
The repertoire of a songbird. These sonograms, or "voiceprints," show a graph of a sound's frequency (perceived as pitch) versus time. These are four distinct song types of one male brown-headed cowbird. The song on top is quite distinct and easily distinguishable from the other three, which would sound similar to us but probably not to a bird. Individual cowbirds generally have three to six song types, but other species have dozens or even hundreds of types.

In contrast to "why" questions about ultimate causation, we often ask questions about *how* an animal carries out a particular behavior. With "how" questions, we are asking about the immediate cause, or **proximate causation,** of something. In the study of animal behavior, this means that we want to understand the mechanisms underlying the particular behavior. Proximate causes can in-

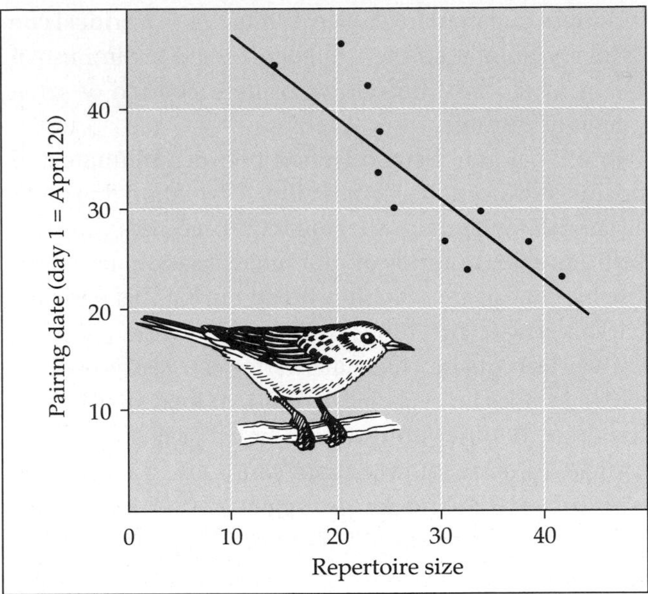

FIGURE 50.3
Female warblers prefer males with large song repertoires. Male sedge warblers in Europe with large repertoires attract females to pair with them earlier in the breeding season than males with small repertoires. Since the first females who choose mates pick males with large repertoires, it is likely that females prefer large repertoires. Males with large repertoires benefit from pairing early because breeding early tends to be more successful than breeding late in the season.

clude the environmental stimulus, if any, responsible for triggering a behavior as well as the internal processes of the animal's neuromuscular, endocrine, and other physiological systems, which provide the proximate mechanisms. Researchers investigating proximate causation must have a keen understanding of a species' natural history and its ways of perceiving the external sensory world. Such an understanding helps researchers place limits on an animal's behavioral options by discarding those that are physically impossible, and it sharpens the focus of hypotheses to be tested. Scientists studying animal behavior use the term *umvelt* to describe the way in which an animal perceives its sensory world and reacts within the constraints of its own physiology.

The study of ultimate versus proximate causation is not an either/or dichotomy; these two levels of causation are related, and most biologists studying behavior are interested in both evolutionary (ultimate) and mechanistic (proximate) causes. Although behavioral ecologists are primarily concerned with the evolutionary significance of a behavior rather than exactly how the behavior is carried out, they need to be aware of proximate causation. By limiting the range of behaviors upon which natural selection can act, proximate mechanisms limit the behavior we expect to see. For example, it would be unreasonable for a behavioral ecologist studying a songbird to propose, at least as the initial hypothesis, that odor is an important cue in mate choice. While this is a critical cue for many animal species, olfaction plays a diminished role in most birds' umvelt, with the exception of some species of vultures.

To drive home the connection between ultimate and proximate causation, consider the observation that bluegill sunfish (and many other animals) breed in spring and early summer. In terms of ultimate causation, a reasonable hypothesis for why they breed during this period is that this is when breeding is most productive or adaptive. The warmth of the water allows for the fast growth of young, as does an especially abundant food supply. Fish that breed at other times would be at a selective disadvantage. In terms of proximate causation, a reasonable hypothesis is that breeding is triggered by the effect of increased day length on a fish's photoreceptors. Bluegills and many other animals can be stimulated to begin breeding by experimentally lengthening their daily exposure to light. This stimulus results in neural and hormonal changes that induce nest building and other reproductive behavior. The increased day length itself has little adaptive significance, but since day length is the most reliable indicator of the time of year, there has been selection for a proximate mechanism that depends on increased day length.

Another example linking the two causes of behavior is the pleasure most people feel when they taste sweet foods. This is a proximate mechanism that increases the likelihood of eating sweet, high-energy foods. Such food was in short supply until the rise of modern agriculture, and the fitness associated with its consumption is probably the ultimate reason for the natural selection of the proximate mechanism we call a sweet tooth. FIGURE 50.4 illustrates yet another example of proximate and ultimate causation in behavior.

Thus, the "how" and "why" questions about animal behavior are related in their evolutionary basis: Proximate mechanisms produce behaviors that ultimately evolved because they increase fitness in some way. This connection will be reinforced as we now take a closer look at innate ("instinctive") components of behavior.

Certain stimuli trigger innate behaviors called fixed action patterns

Before the advent of behavioral ecology, an older discipline called *ethology* dominated the study of animal behavior. Ethology originated in the 1930s with naturalists who tried to understand how a variety of animals behave in their natural habitats. One of the major findings of ethology was that animals can carry out many behaviors without ever having seen them performed. In other

(a)

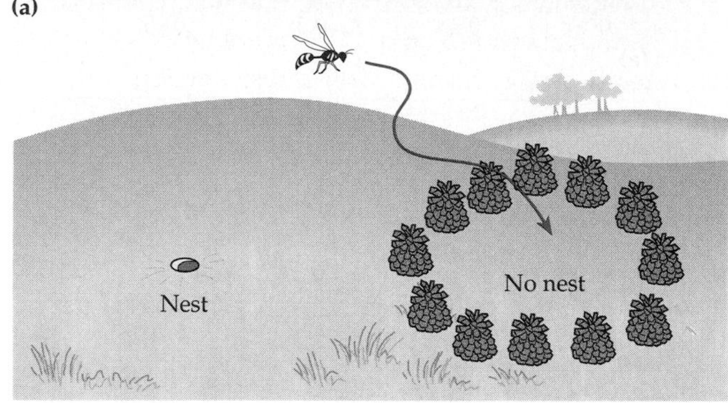

(b)

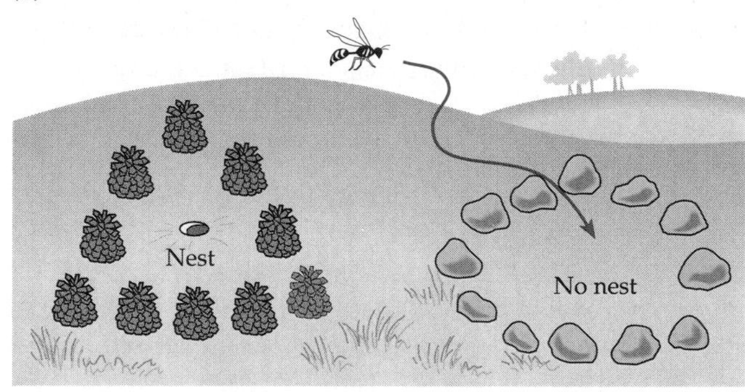

(c)

FIGURE 50.4

Proximate and ultimate causes of the digger wasp's nest-locating behavior. A female digger wasp excavates and cares for four or five separate underground nests. She will fly to each nest daily, bringing food to the single larva in each nest. Biologist Niko Tinbergen designed field experiments to test his hypothesis that the wasp uses visual landmarks to keep track of where her nests are located. (**a**) Tinbergen marked a wasp's nest with a ring of pinecones. (**b**) After the mother wasp visited the nest and flew away, Tinbergen moved the pinecones a few feet to one side of the nest. When the wasp returned, she flew to the center of the pinecone circle instead of the nearby nest. The results of such experiments supported the hypothesis that digger wasps use landmarks to keep track of their nests, and that they can learn new visual cues. (**c**) In a follow-up experiment, Tinbergen restored the pinecones to the actual nest site but arranged them in a triangle instead of a circle. He placed a circle of stones to one side of the nest. The returning wasp flew to the stone ring, a result supporting the hypothesis that the insect was cued by the arrangement of the landmarks rather than the physical objects themselves. Thus, the proximate cause of the wasp's nest-locating behavior is the environmental cue of the landmark arrangement and the response it elicits in the animal. For ultimate causation, a reasonable hypothesis is that this behavior enhances the wasp's fitness by targeting food to her offspring.

words, many behaviors are innately programmed. And while such behaviors seem purposeful because they are clearly beneficial, they are carried out in ways that show the animals are unaware of the significance of their actions. The work that led to these and other findings was initiated primarily by ethologists Konrad Lorenz, Niko Tinbergen (see FIGURE 50.4), and Karl von Frisch, who received a Nobel Prize in 1973 for their discoveries. Their research focused on proximate mechanisms, but with an eye toward the genetic links to behavior and the adaptive nature of behavior, an orientation that behavioral ecologists would later build upon. Also, ethologists' investigations of innate behaviors called fixed action patterns helped transform the study of behavior into a modern hypothesis-oriented part of biology.

Fixed Action Patterns

A **fixed action pattern (FAP)** is a highly stereotypical behavior that is innate. Once an animal initiates a FAP, it usually carries the action pattern to completion even if other stimuli impinge on the animal or the activity becomes inappropriate. A FAP is triggered, or released, by an external sensory stimulus known as a **sign stimulus** or **releaser.** Thus, we can think of a FAP as the innate ability of an animal to detect a certain stimulus associated with the animal's umvelt, combined with an innate behavioral program that is activated by the stimulus to direct some kind of motor activity. In many cases, the sign stimulus is some feature of another species. For example, some moths instantly fold their wings and drop to the ground in response to the ultrasonic signals sent out by predatory bats.

A classic case of releasers in social behavior can be observed in the male three-spined stickleback fish, which attacks other males that invade his territory. The releaser for the attack behavior is the red belly of the intruder. The stickleback will not attack an invading fish lacking a red underside but will readily attack nonfishlike models as long as some red is present (FIGURE 50.5). Tinbergen, who first reported these findings, was inspired to look into the matter by his casual observation that his fish responded aggressively when a red truck passed their tank. As it turns out, red coloration of body parts is a releaser for either aggressive or sexual behavior in many species that have color vision.

Interactions between parents and their offspring often involve FAPs and have been studied extensively in birds. In many bird species, when a parent returns to the nest with food, the newly hatched, blind young respond immediately with begging behavior, raising their heads, opening their mouths (gaping), and cheeping loudly. The releaser for this begging is the impact of the parent landing on the nest. Later, when the young birds' eyes are

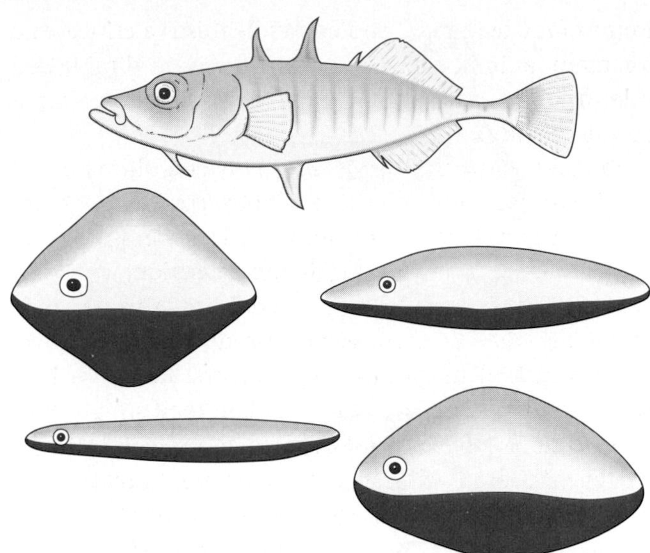

FIGURE 50.5
Sign stimuli used to demonstrate FAPs. Aggression in male three-spined stickleback fish is triggered (released) by a simple cue. The realistic model at the top without a red underside produces no response. All the others produce strong responses because they have the required red underside, a releaser.

more developed, the sight of a parent releases gaping, but the specific stimuli required still are not complex; simple models of the parents work very well. In turn, the releaser for the parent's feeding behavior is remarkably simple—merely a gaping mouth.

A ground-nesting bird such as the graylag goose of Europe sometimes accidentally bumps one of its eggs out of the nest. The sight of an approximately spherical object near the nest is a releaser for the goose's egg-retrieval behavior, in which it rolls the egg back into its nest with side-to-side head motions (FIGURE 50.6). The adaptiveness (ultimate causation) of this behavior is obvious, but it has two features that are very strange from a human perspective. First, if the egg slips away (or is experimentally pulled away) as it is being retrieved, the goose stops her side-to-side head motion but continues the normal retrieval movement with her head extended, as though she were still pulling the egg. Only after she sits down does she "notice" that an egg is still outside the nest, and then she begins another retrieval. If the egg is again removed, the goose goes through the retrieval maneuver again. Second, if an unusual object such as a doorknob or a toy dog is placed near a nest, the goose will retrieve it as well. The toy dog may be discarded after the goose attempts to sit on it, but a smoothly rounded object like a doorknob may be incubated for a long time.

By now, you can see the automatic way animals behave in many situations. If a stickleback processed information like a human, it would realize that the models in FIGURE 50.5 are not real fish, despite their red bellies.

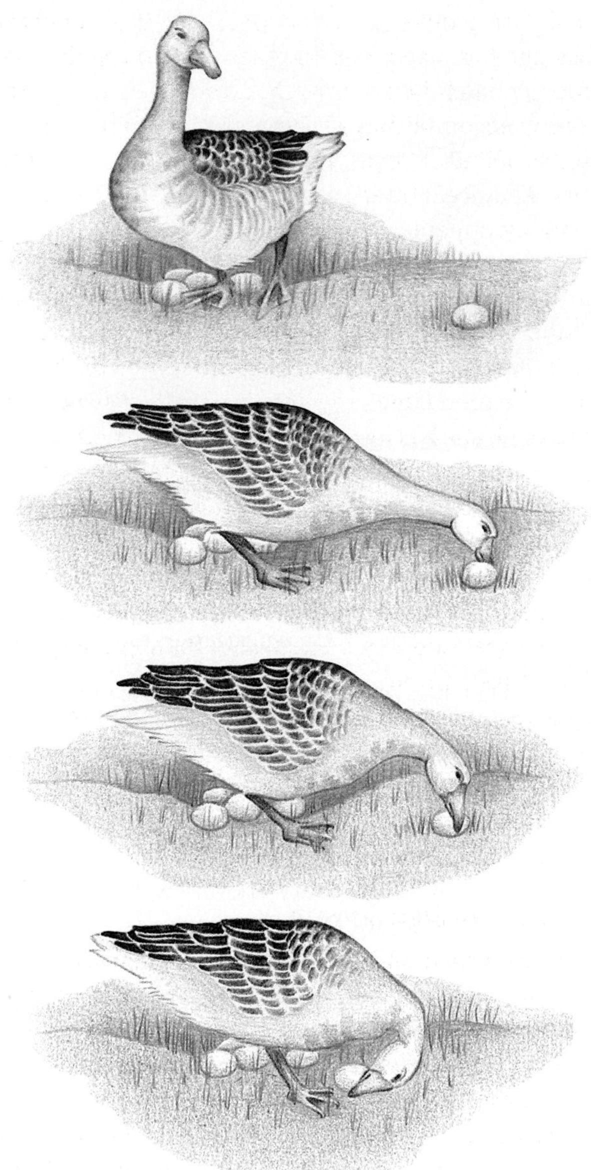

FIGURE 50.6
A graylag goose retrieving an egg. The goose stands up, extends her neck, and retrieves the egg by pulling it slowly with the underside of her bill as she settles back into the nest. The egg is lopsided and tends to roll away, but the goose uses a side-to-side motion of her head to keep the egg on course. She will use this technique for any object that serves as a releaser for retrieval behavior, including a ball or a doorknob.

Humans respond much more to an entire situation than do most other animals, and we generally base our actions on more diverse information. These classic experiments, and many others, indicate that animals use only a limited subset of the available sensory information to elicit behavior and in many instances are not taking a thoughtful or intelligent action. By "intelligent" we mean integrating various kinds of information and then making a decision; in the case of a FAP, the animals are behaving more like robots. FAPs occur in all animals, including humans. Infants grasp strongly with their hands in response to a tactile stimulus. An infant's smile is also a FAP; it is readily induced by simple stimuli such as a sound or a figure consisting of two dark spots on a white circle, a kind of rudimentary representation of a face.

In all these cases, adaptive responses to specific stimuli have been selected for. The variations on experimental releasers that ethologists introduce often present animals with situations that almost never arise naturally. If egglike objects were commonly found near goose nests, natural selection probably would have resulted in a discrimination mechanism that enables geese to identify their own eggs. We know that such mechanisms are possible, because they *have* evolved in many birds. For example, some species can differentiate between their own eggs and those of brood parasites such as cuckoos. These parasites lay their eggs in the nests of another species (host species), and their young often cause the death of the host's own young (FIGURE 50.7a). In terms of the reproductive success (Darwinian fitness) of the host bird, there is a great advantage in recognizing the cuckoo egg as foreign and removing it from the nest. A reed warbler will feed a baby cuckoo once the bird has hatched, but warblers do recognize their own eggs and remove cuckoo eggs that are dissimilar (FIGURE 50.7b). We see here that natural selection can lead to complex behavioral mechanisms when there is a strong selection pressure, as with host species parasitized by cuckoos.

Though natural selection has produced many adaptive responses in animals, these animals will show what seems to us gross incompetence when confronted with novel situations unrelated to past selection pressures, as do geese that retrieve nonegg objects—a situation unlikely to occur in the natural habitat. The ability to confront novel stimuli, learn about them, and adjust behavior as a result of what has been learned is a hallmark of both intelligence and self-awareness. The evolution of intelligence is costly, in both the development of the neural tissue necessary to process the information and its metabolic maintenance. In addition, the evolution of intelligence requires dramatic changes in life history patterns, such as long juvenile phases and high parental investment per offspring. For most species, these costs, measured as reductions in reproductive fitness, far outweigh the costs of an occasional inappropriate use of FAPs, and extensive intelligence has not evolved in many animal groups. We will address this issue in greater detail later in the chapter.

The mechanistic component of behavior based on FAPs is demonstrated most clearly by some animals that seem to have an endless capacity for repeating stereotypical behavior. One example, first described by the nineteenth-century naturalist Jean-Henri Fabre, involves digger wasps. The female wasp builds a nest in the

(a)

(b)

FIGURE 50.7

Brood parasitism. Some species of European cuckoos are brood parasites, meaning that they lay their eggs in the nests of other species. (**a**) Within hours of hatching, an otherwise helpless cuckoo methodically pushes all of the host's eggs out the nest. If the cuckoo hatches after some or all of the nestlings have already hatched, it will evict them as well. There is no opportunity for cuckoo nestlings to learn this behavior from another individual, so it must be innately programmed. (**b**) The host, in this case a reed warbler (left), feeds the brood parasite even though the baby cuckoo looks nothing like a much smaller nestling warbler. The gaping mouth of a nestling, whether true offspring or brood parasite, is the releaser that causes the parent (or foster parent) to feed the baby. Even in those species where some kin discrimination is possible, the gaping mouth of the cuckoo is much larger and more aggressive than that of the host species and thus serves as a supernormal stimulus. Although they do not differentiate among nestlings, reed warblers and many other birds recognize their own eggs and remove cuckoo eggs that are dissimilar. This has resulted in the evolution of cuckoo eggs that are nearly identical to those of many host species.

ground (see FIGURE 50.4) and places in it a paralyzed cricket she has just stung. She then lays an egg that will eventually hatch into a larva that consumes the cricket. The wasp accomplishes all this with a highly stereotypical series of FAPs. She places the cricket a short but rather precise distance (about 2.5 cm) from the nest, enters the nest briefly, apparently for a final inspection, then comes out to get the cricket and carry it into the nest. If an observer moves the cricket a short distance while the wasp is inside, she searches for it after emerging and, after retrieving it, puts it back in its original spot near the entrance. She then enters the nest for another inspection, even though she has just made one, and emerges again a short time later. If the cricket has been moved again, the wasp repeats the cycle—and will continue to repeat it at least forty times, showing no sign of either tiring of the repetition or of circumventing the scientist's game by changing behavior. The wasp can only get past the "inspect-the-nest" step by immediately finding the cricket near the nest. In nature, such behavior may be quite adaptive. Movement indicates that the cricket is not truly paralyzed, and the wasp would have to sting the cricket again and eventually subdue it. A reinspection of the nest by the wasp after a long time spent struggling with the cricket would also be prudent and useful.

The Nature of Sign Stimuli

Ethologists have found from experiments that sign stimuli for FAPs are generally based on one or two simple characteristics. In many cases, the stimulus is the most obvious (or the only) characteristic of a particular situation; for example, the ultrasonic signals of bats are the obvious cue for triggering avoidance behavior in moths. In other cases, it appears that animals have settled on certain characteristics out of an array of possible choices. When an adult herring gull brings food to its chick, it bends its head down and moves its beak, which has a red spot. The chick pecks the beak, stimulating the adult to regurgitate the food. The chick might be cued to peck by a variety of stimuli, including such obvious things as a lump at the end of a rectangle (simulating food at the end of a bill). However, studies have demonstrated that the releaser is a red spot swung horizontally at the end of a long, vertically oriented object. We expect natural selection to have favored cues having a high probability of association with the relevant object or activity, but when there are many possible cues, there is probably some randomness in which one becomes fixed upon as the sign stimulus for a FAP.

Close correlations exist between an animal's sensitivity to general stimuli and the specific sign stimuli to which it responds. For example, frogs have retinal cells that are especially good at detecting movement, and it is the movement of an object that releases the frog's tongue-shooting behavior during feeding. A frog will starve to death if surrounded by dead or motionless flies but will attack one readily if it moves. Sometimes an exaggeration of the relevant stimulus produces a stronger response. Wider gaping by young birds more readily releases the feeding behavior of the adult (see FIGURE 50.7b). Because a young bird gapes more widely in response to increased hunger, the hungriest young are most likely to be fed. This aspect of sign stimuli can be seen in simple experiments in which animals are presented with a *supernormal stimulus*, an artificial stimulus that elicits a stronger response than any naturally occurring stimuli. A graylag goose will attempt to roll a volleyball into her nest instead of her own egg, and an oystercatcher (a shorebird) will attempt to incubate a giant model of her egg instead of the real thing.

Some ethologists have suggested that the use of simple cues to release programmed behavior prevents an animal from wasting time processing or integrating a wide variety of input. Perhaps a better way of interpreting the situation has to do with the limitations of innate behavior and how it evolved. A frog's sensory-neural network for detecting movement is probably much less complex than the apparatus that would be necessary to rapidly distinguish a fly from another object of similar size. In any case, simple cues usually work quite well in an animal's normal sensory world, though not in the experimental world ethologists often create.

Learning is experience-based modification of behavior

Learning, which is formally defined as the modification of behavior in response to specific experiences, often affects even innately programmed behaviors such as FAPs. Before we examine the various ways in which learning occurs, we will review what has been referred to as the nature-versus-nurture controversy.

Nature Versus Nurture

While the early ethologists, who were mostly Europeans, were developing their approach, psychologists in North America were studying animal behavior from a very different perspective. Focusing on learning, they typically studied a few species of captive reared animals, such as laboratory rats. Psychologists were impressed with how effectively they could shape animal behavior with learning, while ethologists were equally impressed with the degree to which animal behavior seemed innate. A clash was inevitable, and it came to be known as the *nature-versus-nurture controversy.* Few scientists ever believed that behavior is either all genetic or all learned. Instead, the debate was mostly over which input—instinct or learning—is of primary importance.

Nearly all biologists today agree that most behavior is a consequence of genetic and environmental influences. Even though an animal may not have to witness a FAP because the basic behavior is innate, learning is still involved. Most FAPs improve with performance, as animals learn to carry them out more efficiently. It might seem that some things, such as the different human languages, are completely learned. It is true that the ability to speak either English or Spanish has no genetic basis, but the ability to learn any language is a function of a complex brain that develops in a particular environmental context under the guidance of a human genome.

Despite the dictum that behavior is some mix of learning and instinct, or of environmental and genetic inputs, it is useful to sort out the relative contributions of genetic and environmental influences of a particular behavior. This analysis can help scientists understand the extent to which the behavior can vary among individuals of a species. We have already identified language differences among people as a completely learned behavioral variation of an innate ability. There are many parallel situations among nonhuman animals, including one involving vocalizations. Many songbirds have different songs in different populations. These "dialects" arise because male birds learn to sing like other males, generally early in life, and this sort of learning is prone to occasional variations that will cause populations to differ. Birds reared in one dialect but moved to another population early in life will learn the new dialect, just as a person born in the northern part of the United States who moves to the deep south early in life will speak like a southerner. Chickadees living in temperate regions learn new song dialects each fall when they shift from living in single-family social groups to large winter flocks. These flock-specific songs help establish a group identity that is useful in defending rare food resources from other conspecific flocks during the winter. As we discuss the various forms of learning, we will see additional examples in which it is productive to analyze a behavior and identify those aspects for which genetic or environmental influences are of primary importance.

Learning Versus Maturation

Some behaviors that are clearly innate, in the sense that one individual does not have to learn them from another, may be performed more quickly or effectively as time goes on. However, this does not necessarily mean that learning has occurred. Behavior may improve because of ongoing developmental changes in neuromuscular systems, a process called **maturation.** We commonly speak of birds "learning" to fly, and you may have seen a hapless fledgling awkwardly fluttering about as if it were practicing. However, young birds have been experimentally reared in restrictive devices so that they could never flap their wings until an age when their normal kin were already flying. Such birds flew immediately and normally when released. Thus, the improvement must have resulted from neuromuscular maturation, not from learning.

Earlier we examined the pecking behavior that young gull chicks direct toward their parents' beaks. Recently hatched chicks are quite indiscriminate and will peck at a wide variety of objects, but chicks a week or two old show their strongest response to accurate models of their parents' beaks. Is this a maturation or a learning process? Experiments involving herring gulls and laughing gulls show that maturation is not involved. A laughing gull chick that has been reared by a herring gull will respond more strongly to its foster parent's beak type than to the beak of its own species. The reverse is true for a herring gull chick cross-fostered by laughing gulls. This is an example of how learning can modify a behavior that is basically instinctive.

Now that we have considered the nature-versus-nurture issue and the importance of maturation, let's survey the major types of learning.

Habituation

Habituation is a very simple type of learning that involves a loss of responsiveness to unimportant stimuli or to stimuli that do not provide appropriate feedback. Examples are widespread. *Hydra* stop contracting if disturbed too often by water currents. Gray squirrels and many other animals recognize alarm calls of conspecifics threatened by a predator, but they eventually stop responding if these calls are not followed by an actual attack (the "cry-wolf" effect). Researchers simulated a similar situation in experiments with baby ducklings. These birds, even as chicks, have a keen sense of vision and are innately frightened by the silhouettes of potential predators such as hawks flying overhead. When artificial hawk-shaped objects are flown over them, the ducklings scurry for cover. Over time, with repeated trials, the ducklings habituate to the stimulus and eventually ignore it altogether. However, if a novel shape of artificial predator is flown overhead, even after many trials of the previous shape, the ducklings will scurry for cover. Their anti-predator response had not been lost but was modified by learning.

Imprinting

Some of the most interesting cases where learning interacts closely with innate behavior involve the phenomenon known as **imprinting,** the only form of learning that attracted much interest from ethologists. You have probably seen young ducks or geese following their mother. Mother-offspring bonding in species with parental care is a critical part of the reproductive cycle. If bonding fails to happen, the parent will not initiate care of the infant. The result is certain death to the offspring and loss of reproductive fitness to the parent. But how do the young know whom—or what—to follow? In his most famous study, Konrad Lorenz divided a clutch of graylag goose eggs, leaving some with the mother and putting the rest in an incubator. The young reared by the mother showed normal behavior, following her about as goslings and eventually growing up to interact and mate with other geese. When the artificially incubated eggs hatched, the geese spent their first few hours with the researcher instead of with their mother. From that day on, they steadfastly followed Lorenz and showed no recognition of their own mother or other adults of their own species (FIGURE 50.8). As adults, the birds continued to prefer the company of Lorenz and other humans over that of their own species, and they sometimes even initiated courtship behavior with humans.

Apparently, graylag geese have no innate sense of "mother" or "I am a goose, you are a goose." Instead, they simply respond to and identify with the first object they

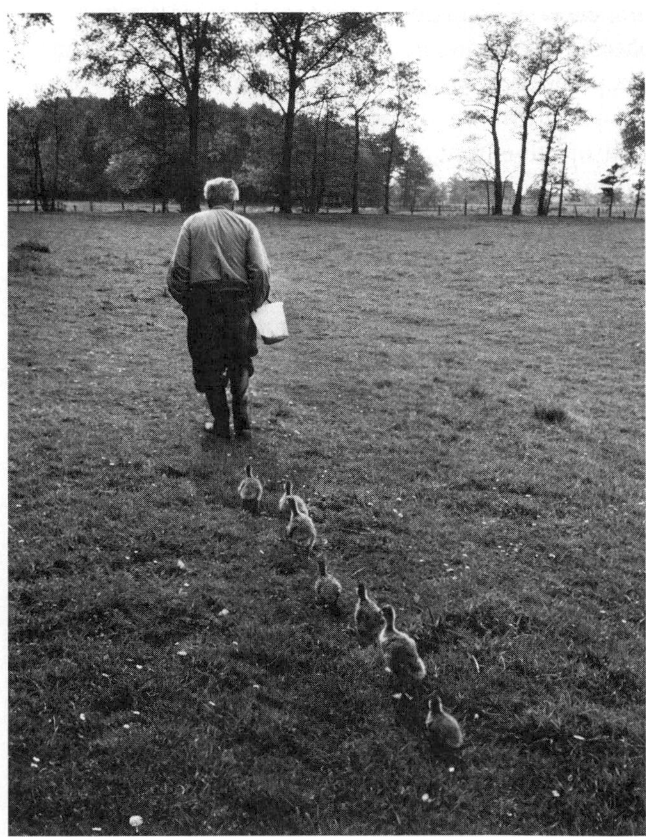

FIGURE 50.8
Imprinting. Konrad Lorenz was "mother" to these imprinted geese.

encounter that has certain simple characteristics. What is innate in these birds is the ability or tendency to respond; the outside world provides the *imprinting stimulus*, something to which the response will be directed. The most important imprinting stimulus in Lorenz's graylag geese was movement of an object away from the young, although the effect was greater if the object emitted some sound. The sound did not have to be that of a goose, however; Lorenz found that a box with a ticking clock in it was readily and permanently accepted as a "mother."

Many other examples of imprinting are now known. Salmon are noteworthy for their long migrations in the open ocean, where they feed, grow, and mature after hatching in freshwater streams. Some species remain at sea for several years. Nonetheless, as demonstrated by the tagging of thousands of young fish, each returning individual goes back to its own home stream to spawn. This involves not only finding the proper major stream to enter at the coast, but also making a number of correct choices among smaller and smaller tributaries as the fish winds its way upstream. While the overall migration pattern of salmon is influenced by a variety of navigational cues, olfactory imprinting plays a crucial role in a salmon's ability to locate the final stream in which the salmon will complete its reproductive cycle. Young fish artificially reared and exposed to a chemical called morpholine can be directed into a stream upon their return if morpholine is introduced into it. Under natural conditions, the fish apparently imprint on the complex bouquet of odors unique to their specific stream, and can recognize those odors and swim toward their source even after being long distances from the stream for many years (see FIGURE 45.18).

Critical Period. One of the two features that distinguish imprinting from other types of learning is that what is learned by imprinting is irreversible. The second feature is a **critical period,** a limited phase in an individual animal's development when learning particular behaviors can take place. Lorenz found, for example, that geese totally isolated from any moving objects during the first two days after hatching, which is the critical period for imprinting on parents, failed to imprint on anything afterward. Imprinting has commonly been thought of as involving very young animals and rather short critical periods. But it is now clear that a similar learning process occurs in older animals, and that the critical period may be of various durations. For example, just as a young bird requires imprinting to "know" its parents, the adults must also imprint to recognize their young. For a day or two after their young hatch, adult herring gulls will accept and even defend a strange chick introduced to their nesting territory. After imprinting, which is probably based largely on individually variable cues such as the call notes of chicks, the adults will kill and eat any strange chick.

By imprinting on their parents, young birds first learn who will care for them, and subsequently learn species identity and the kind of bird they should mate with later in life. Not only does this sexual imprinting occur later, but the critical period lasts longer. For example, in one study involving two closely related species of finches, young males of one species were reared first with members of their own species, then with members of the other species during their several-week-long critical period for sexual identity. Later, when exposed to females of their own species, they mated quite reluctantly. They readily mated with females of the other species, however, even when they had not seen any members of that species for as long as eight years. Identification with the second species had been permanently imprinted. While a critical period and irreversibility characterize imprinting, it is now recognized that these phenomena are not always rigidly fixed. For example, the cross-fostered finches did eventually mate with females of their own species.

Song Development in Birds: A Study of Imprinting. This chapter began with a brief discussion of vocal learning in young birds. In most songbird species, males sing complex vocalizations that have a variety of adaptive functions. They may serve as advertisements for potential mates, as warnings to male competitors, as identification to announce membership within a group, or as signals that initiate group behaviors. Bird songs are a complex interplay of learned and instinctive behaviors that have strong genetic and environmental influences. Some species of birds have rigid song repertoires, while others have repertoires that are more fluid, have regional dialects, and change throughout the individual bird's life. As such, bird song has been both a rewarding and a perplexing subject of study for biologists who examine not only the proximate mechanisms of song, but the ultimate adaptive nature of song as well.

Many studies of song development have focused on common North American species. Thirty years ago, experiments showed that male white-crowned sparrows raised in isolation in soundproof chambers developed abnormal songs that had only a slight resemblance to the normal adult song. This crude and undeveloped song is called the template, a neural "scaffolding" upon which the full adult song is developed. However, if tape recordings of typical adult male white-crowned sparrow songs were played to the experimental birds when they were 10 to 50 days old, they developed a normal song several

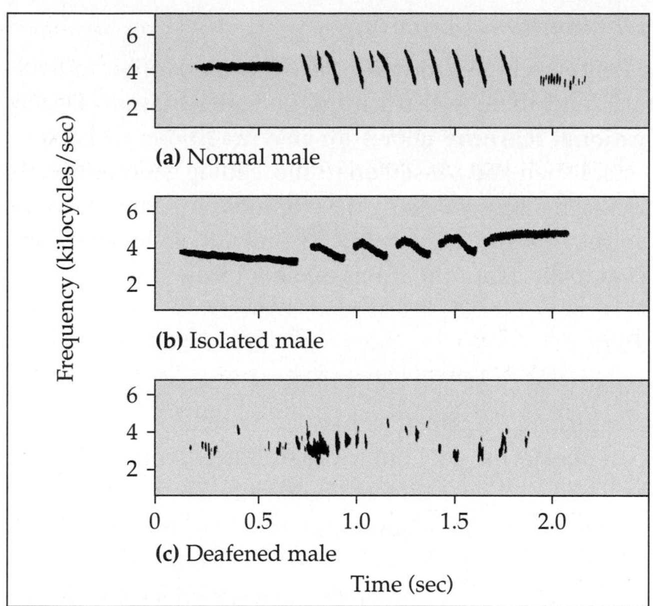

FIGURE 50.9

Sonograms of songs of white-crowned sparrows reared under three conditions. (**a**) Males that heard tape recordings of their species' song before reaching 50 days of age learned to sing a normal song several months later. (**b**) Isolated males were reared in soundproof chambers, which ensured that they never heard their species' song.They sang only the primitive template of the song. (**c**) Males in the last category were deafened *after* they had heard recordings of their species' song, but *before* they ever started to practice singing.

months later. The normal song developed as a result of a bird's comparing its own singing with its memory of the taped song. This was shown by experimentally deafening some birds after exposing them to the taped playbacks: The deaf birds developed songs even simpler and more abnormal than isolated birds that could hear but were never exposed to songs (FIGURE 50.9).

We see that there are two learning processes in the sparrow's song development. First, a bird must acquire a song type by hearing an adult; and second, it must learn to match this song by listening to itself. Young males raised in soundproof chambers and exposed to taped songs of white-crowned sparrows after 50 days of age sang like birds that never heard these songs. In other words, the sparrows were receptive to learning the taped songs only during a critical period, which is why song learning can be considered a type of imprinting. Innate influences were also demonstrated: Young white-crowned sparrows that heard taped songs of other sparrow species during the critical period did not adopt these songs. Apparently, they were genetically predisposed to learn their own species' song.

Additional experiments revealed even more complexity in how birds learn their songs. When isolated white-crowned sparrows more than 50 days old were exposed to live singing adults of another species, they learned the song of the other species. Because a bird can interact socially with it, a live singing adult provides a much stronger and more diverse set of stimuli than a taped song. These strong stimuli can overcome the innately programmed tendency to acquire only a white-crowned sparrow song. Again we see that an innate tendency can be modified by experience and is not necessarily inflexible. Furthermore, the critical period proved to be longer when the stimulus was a live bird than when it was a taped playback.

People also have a critical period for learning vocalizations: It is well known that foreign languages are learned most easily up until the teen years. Of course, this critical period is not rigidly fixed. Adults can learn a new language, but they usually require much more effort and time to become fluent than a child does.

Bird species other than the white-crowned sparrow have different programs for song development. The closely related song sparrow develops a nearly normal song even if reared in total isolation. But males have larger repertoires if they can learn song variants from other birds. Mockingbirds have huge repertoires of at least 150 song types, many of which are accurate mimics of sounds made by other kinds of birds and animals and even telephone rings. The adaptive value of a large repertoire is apparently so great for mockingbirds that genetic programming places few or no restrictions on the range of songs they can acquire. Until recently, researchers assumed that all birds had to listen to themselves in order to develop song. But eastern phoebes experimentally deafened early in life, well before they began to sing, developed normal songs of their species. Thus, there are important exceptions to the song-learning scenario based on extensive study of white-crowned sparrows.

Research with canaries has helped elucidate the relationship between the observed behavior of song learning and the neurological events underlying the process. Canaries are unusual in that they sing variable numbers of songs and develop new songs each year. Thus, they must relearn a new song repertoire prior to each breeding season. Biologist Fernando Nottebohm identified the region of the forebrain responsible for song learning in canaries. He found that this region in male canaries shows dramatic individual variation in size according to the season and the number of different songs in an individual male's repertoire. Brains are largest during the breeding season and for males with the biggest song repertoire. The shrinking of this brain region at the end of the breeding season may be a mechanism for erasing unneeded songs. Subsequent regeneration of neurons in the brain during courtship provides a way for new song learning to take place. From an adaptive perspective, the ability of male canaries to learn new songs more than once in their lifetime must be critically important to their fitness, as this adaptation requires a considerable investment of energy.

The observation of so much variation in the ways birds acquire songs and then deploy them suggests that this is a fairly flexible behavioral system in many groups of birds. The strong connection between song and reproductive fitness in birds has apparently enabled natural selection to modify nearly all aspects of song and its development into the rich diversity that we find in modern birds.

Classical Conditioning

Many animals can learn to associate one stimulus with another, a process known as **associative learning.** A type of associative learning called **classical conditioning** is well known from the work of Ivan Pavlov, a Russian physiologist, around 1900. Pavlov sprayed powdered meat into dogs' mouths, causing them to salivate (primarily a physiological rather than a behavioral response). Just before the spraying, however, he exposed the dogs to a sound such as a ringing bell or clicking metronome. Eventually, the dogs salivated readily in response to the sound alone, which they had learned to associate with the normal stimulus. Similar types of conditioning experiments have been carried out with other animals.

Operant Conditioning

Another type of associative learning that directly affects behavior is **operant conditioning,** also called trial-and-error learning. Here, an animal learns to associate one of its own behaviors with a reward or punishment and then tends to repeat or avoid that behavior. The best known laboratory work involving operant conditioning is that of American psychologist B. F. Skinner. A rat or other animal placed in a "Skinner box" finds and manipulates a lever in the box, usually by accident, and is rewarded by the release of food. The animal quickly learns to associate manipulation of the lever with a food reward. Such learning is the basis for most of the animal training done by humans, in which the trainer typically encourages a behavior by rewarding the animal. Eventually the animal performs the behavior on command, without always receiving a reward.

Operant conditioning is undoubtedly very common in nature. For example, animals quickly learn to associate eating particular food items with good or bad tastes and modify their behavior accordingly. Remember that what tastes good or bad is likely to be affected by natural selection on the basis of a food's value. Thus, as with imprinting, genes influence the outcome of operant conditioning.

Observational Learning

Many vertebrates monitor the behavior of other individuals and thereby learn important information. **Observational learning** allows for new traditions to become established and passed on to succeeding generations. As we have discussed, song development in many birds involves this sort of learning, as individuals listen to and eventually adopt the songs of older birds.

Play

Many mammals and some birds engage in behavior that can best be described as **play.** Such behavior has no apparent external goal but involves movements closely associated with goal-directed behaviors (FIGURE 50.10). For example, playful stalking and attacking of conspecifics occurs in many predator species, such as those in the dog and cat families. Although this behavior does not usually involve painful bites, the animals grab and mouth one another, using movements similar to those used to capture and kill prey. A study of bottlenose dolphins in Australia revealed that young dolphins spend long periods away from their mothers in groups of juveniles engaged in a full range of social and sexual play. Bottlenose dolphins are intelligent and social animals, and social lifestyle is one of the hallmarks of mammalian species that routinely engage in play.

Another common feature of play is that it is potentially dangerous or costly. Baboons sometimes kill and eat ju-

FIGURE 50.10
Play behavior. In the view of behavioral ecologists, the roughhousing behavior of these lion cubs evolved in spite of the energy it consumes and the risks it poses, because play improves individual fitness in some way. Practicing survival behavior such as the capture of prey, experimenting with social roles, and building a healthy body through exercise are three possible benefits of play.

venile vervet monkeys, and they are most successful in doing so when the monkeys are playing in groups away from adults. In a study of caged ibex, a type of wild goat, play resulted in at least five out of fourteen kids sustaining injuries that produced limps. "Horsing around" often produces similar injuries in human kids.

Play obviously consumes energy, and the risks to life and limb result in significant additional costs. What could be the ultimate adaptive basis for such seemingly pointless behavior? The "practice hypothesis" suggests that play is a type of learning that allows animals to perfect behaviors needed in functional circumstances. This hypothesis is supported by the observation that play is most common in young animals. However, movements used in play show little improvement after their first few practices. An equally likely ultimate explanation for play is the "exercise hypothesis," which suggests that play is adaptive because it keeps the muscular and cardiovascular systems in top condition. The exercise hypothesis also predicts that play should be especially common in young animals, because they typically do not have to exert themselves in useful activities while under the protection and care of their parents.

Insight

Insight learning is the ability to perform a correct or appropriate behavior on the first attempt in a situation with which the animal has no prior direct experience. (Some scientists prefer to call this reasoning or innovation, rather than learning.) If a chimpanzee is placed in an area with a banana hung too high above its head to be reached and several boxes on the floor, the chimp can "size up" the situation and stack the boxes, enabling it to reach the food. Notable examples of insight have also been observed in some bird species, especially within the family Corvidae, which includes crows, ravens, and jays. Ravens placed in an experimental situation comparable to that described for chimps were able to devise novel strategies for obtaining food. Most remarkable, however, was the tremendous individual variation in behavior exhibited by the ravens in their attempts to solve the problems imposed on them during these experiments (FIGURE 50.11). In general, insight is best developed in mammals, especially primates, but even in these groups the amount of insight often varies substantially from one situation or species to another.

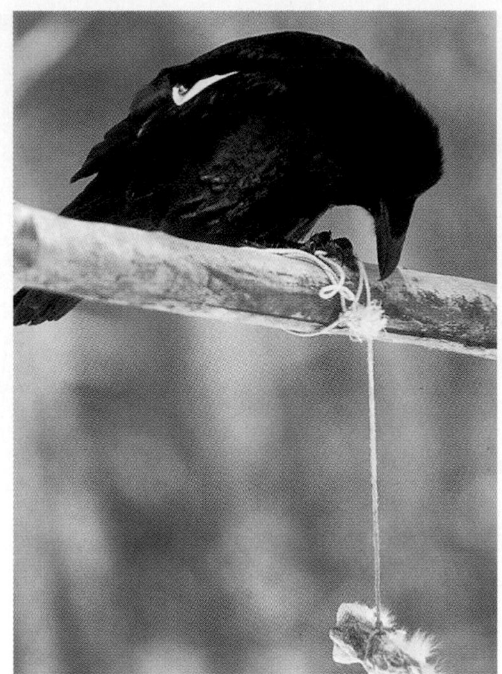

FIGURE 50.11
Insight learning. Behavioral biologist Bernd Heinrich placed ravens in an experimental situation in which they had to solve the problem of retrieving a food reward hanging from a string attached to a wire. The raven shown here solved the problem by using one foot to retrieve the string and the other foot to hold the string as it was being retrieved. Of interest was the tremendous individual variation in insight that Heinrich observed. Some ravens never learned to get at the food, while others solved the problem in different ways.

Animal Cognition

Related to insight is the more general issue of animal cognition. **Cognition** involves an animal's ability to be aware of and make judgments about its environment. A simple but profound question is whether nonhuman animals are cognitive beings. Are other animals consciously *aware* of themselves and of the world around them? Do they "feel" pain or pleasure or sadness as we do? As yet, we have no way of answering these questions directly. We have seen that other animals may sometimes behave like programmed computers, and they certainly do not have the ability to integrate information (to "think") to the same extent as humans do. But is this a matter of degree—a continuum of abilities—or are humans fundamentally different in some behavioral respect?

Donald Griffin of Princeton University is the foremost proponent of the relatively young field of **cognitive ethology.** This discipline views conscious thinking as an inherent and essential part of the behavior of many non-human animals. Griffin argues that if other animals behave in ways we associate with conscious processing in ourselves, perhaps it makes sense to assume that there is the same underlying awareness. In her famous field studies, Jane Goodall has documented cognitive decision making in chimpanzees. Griffin suggests that such abilities may extend to many nonprimate branches of the phylogenetic tree as well. He argues, for example, that cognitive processes are at the heart of such behaviors as the injury-feigning "strategy" of some species of

FIGURE 50.12
Injury-feigning display. This killdeer uses deception to defend her nest against predators or human disturbance. When danger threatens, she leaves the nest, which is usually concealed, and begins an elaborate display as though she were injured with a broken wing. This behavior has the effect of leading the potential predator away from the nest. As the predator gets close, the killdeer simply flies away. She returns to the nest only after the danger has passed. Killdeers show extraordinary individual variation in the form and use of this display. In addition, individual killdeers use multiple variations of the display depending on the type of threat to the nest and whether or not they have seen the threat before. Cognitive ethologists point to versatile behavior such as this to support the hypothesis that nonhuman animals are conscious, thinking beings.

ground-nesting birds (FIGURE 50.12). This view sees cognitive ability arising through the normal process of natural selection, and, like many other major animal functions, forming a phylogenetic continuum that extends far back in evolutionary history. James Gould, also at Princeton, has reported that bees can form and use "mental maps" of their foraging areas. As he notes, this kind of cognitive function is usually considered to be a form of thinking. Many people who have spent a lot of time with pets or wild animals would agree that these animals are not behaving like sophisticated robots.

Because of the difficulty of answering questions about animal cognition with scientific rigor, some researchers assume the most conservative position—that most animals do not "think." There are intermediate positions within the argument, and no behavioral biologist would argue that *all* animals behave in complex and variable ways, but these two polar positions help define the debate. Ultimately, answers to questions about animal cognition may profoundly affect how we interact with other animals—and how we view ourselves, as well.

Now that we have examined the interplay of innate and learned components of behavior, we can apply the basic concepts to a sampling of behavioral adaptations that have evolved in the animal kingdom.

Rhythmic behaviors synchronize an animal's activities with daily and seasonal changes in the environment

Animals exhibit all kinds of regularly repeated behaviors: feeding in the day and sleeping at night (or vice versa), reproducing every spring, migrating in spring and fall, and so on. What determines when a squirrel sleeps and when it awakens? This sort of question has long been the subject of experimentation designed to explain the proximate mechanisms that control behavioral rhythms. The ultimate basis for rhythms is usually quite obvious, as animals typically carry out behaviors when their particular ecological niches can be exploited most safely and profitably. However, as with most questions about behavior, the proximate bases of rhythms are neither as obvious nor as simple as they might seem. We have already discussed circadian (daily) rhythms in some detail in relation to plants (see Chapter 35). Animals, too, are subject to such rhythms, which are regulated by various environmental cues. But what would happen to rhythmic behavior if an animal were placed in an environment with no external cues about time? Put another way, is rhythmic behavior based on *exogenous* (external) timers, *endogenous* (internal) components, or both?

Numerous studies have been performed to assess the relative importance of endogenous and exogenous proximate mechanisms in the rhythmic behavior of many species. These studies show that circadian rhythms usually have a strong endogenous component, referred to as a *biological clock,* but because the endogenous rhythm does not exactly match that of the environment, an exogenous cue is necessary to keep cycles timed to the outside world. The exogenous cue is sometimes called a *zeitgeber* (German for "time giver"). Light is undoubtedly the most common zeitgeber for circadian rhythms. For example, activity of the North American flying squirrel normally begins with the onset of darkness and ends at dawn, which suggests that light is an important exogenous regulator. However, if a squirrel is placed in constant light or constant darkness, the rhythmic activity is not abolished; in fact, it holds up quite well for at least a month or so. The duration of each cycle (one period of activity plus one period of inactivity) deviates slightly from 24 hours. The squirrel's rhythm becomes free-running and eventually loses its synchronization with the environmental cycle (FIGURE 50.13).

(a)

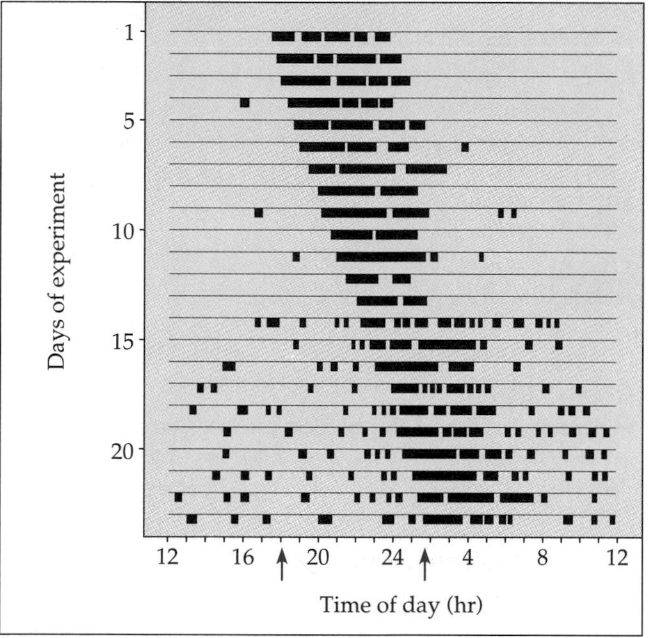

(b)

FIGURE 50.13

Rhythmic activity of a flying squirrel. (**a**) The nocturnal flying squirrel is most active during the first few hours after sunset. In order to trace the activity level of a squirrel kept in constant darkness for 23 days, the animal was placed in a cage with an exercise wheel connected electronically to a chart recorder (which moves graph paper past a pen at a fixed speed). (**b**) Whenever the animal ran, the wheel caused the pen to mark the graph paper, producing the graph shown here. The thicker horizontal lines indicate the squirrel's periods of activity. The pattern displays a definite rhythm, but the period of greatest activity is shifted slightly each day. The free-running cycle for this squirrel was 24 hours, 21 minutes. Thus, after 23 days, its activity cycle was about 8 hours out of synchronization with the actual timing of sunrise and sunset (that is, 23 times 21 min equals 483 min, or 8 hr). The magenta arrows indicate the times when sustained activity began on days 1 and 23 of the experiment. In a normal environment, the daily cues of alternating light and dark periods would adjust the biological clock to a 24-hour cycle. Biological clocks are intrinsic to organisms, but the timing of rhythmic behavior must be adjusted to important events in the environment.

Human circadian rhythms have been studied by placing individuals in comfortable living quarters deep underground, where they could make their own schedules with no external cues of any kind. Under these free-running conditions, the biological clock of humans seems to have a period of about 25 hours, but with much individual variation; like other animals, humans use zeitgebers to adjust their rhythms to 24 hours in the real world.

The role of endogenous timekeepers (biological clocks) in rhythmic behavior that involves longer cycles is even less well understood. In many species, **circannual behaviors,** such as breeding, hibernation, and migration, are based at least in part on physiological and hormonal changes directly linked to exogenous factors such as day length. Little is known about possible endogenous factors, in large part because of practical difficulties in conducting experiments: Animals would have to be maintained for years (instead of days) under constant conditions. However, some long-term studies on ground squirrels indicate that fat deposition, a physiological preparation for hibernation, occurs despite the persistence of a constant daylight cycle. (In nature, day length decreases with the onset of winter, serving as a seasonal cue in a wide variety of organisms.)

Locating the internal mechanisms responsible for behavioral rhythms is a serious challenge for researchers. The problem is magnified because the control mechanism is likely to vary across taxonomic boundaries. A strong candidate for the control of rhythmic behavior is some sort of anatomical structure that serves as a pacemaker. (Do not confuse this general concept with the heart's pacemaker.) In mammals, the pacemaker is located within the hypothalamus, a part of the brain responsible for regulating many basic physiological functions. The suprachiasmatic nucleus (SCN) is a cluster of neurons within the hypothalamus that receives direct nerve signals from the retina. Experiments with hamsters and rats have revealed that the cells of the SCN produced specific proteins in response to changing light-dark cycles. The function of this or any other pacemaker could be the regulation of a variety of physiological processes, such as hormone release, hunger, and heightened sensitivity to external stimuli that motivate specific rhythmic behaviors.

Environmental cues guide animal movement

Animal movements are another aspect of behavior with a long tradition of experimental study. As with rhythms, the emphasis has been on proximate causation, especially the mechanisms animals use to detect and respond to the external cues that guide movements. The ultimate bases for oriented movements are often obvious. Different animals are adapted to different environments and typically have behaviors that bring them to those environments, often at a particular time of year.

Kinesis and Taxis

Animals can use a variety of environmental cues to guide them as they move from one place to another. A **kinesis,** the simplest sort of movement, involves a change in activity rate in response to a stimulus. Sowbugs or woodlice become more active in dry areas and less active in humid ones, a simple behavior that tends to keep these animals in moist environments. Note that a kinesis is actually randomly directed: The animals do not move toward or away from specific conditions, but since they slow down in a favorable environment, they tend to stay there. In contrast, a **taxis** is an oriented movement, more or less automatic, toward or away from some stimulus. For example, housefly larvae are negatively phototactic after feeding, automatically moving away from light; this simple response presumably ensures that the flies remain in an area where they are harder for predators to detect. Trout are positively rheotactic (Gr. *rheos,* "current"); they automatically swim or orient themselves in an upstream direction, which keeps them from being swept away.

Migration Behavior

The regular movement of animals over relatively long distances is called **migration.** Migrating animals generally make one round trip between two regions each year, although there is considerable variation among species. The most notable examples are the migrations of birds, whales, a few butterfly species, and certain oceangoing fish. How is it, for example, that golden plovers find their way over 13,000 kilometers from their arctic breeding grounds to southeastern South America? Even more remarkably, some populations of these birds return to the Hawaiian Islands, a small piece of land in a vast expanse of ocean (FIGURE 50.14).

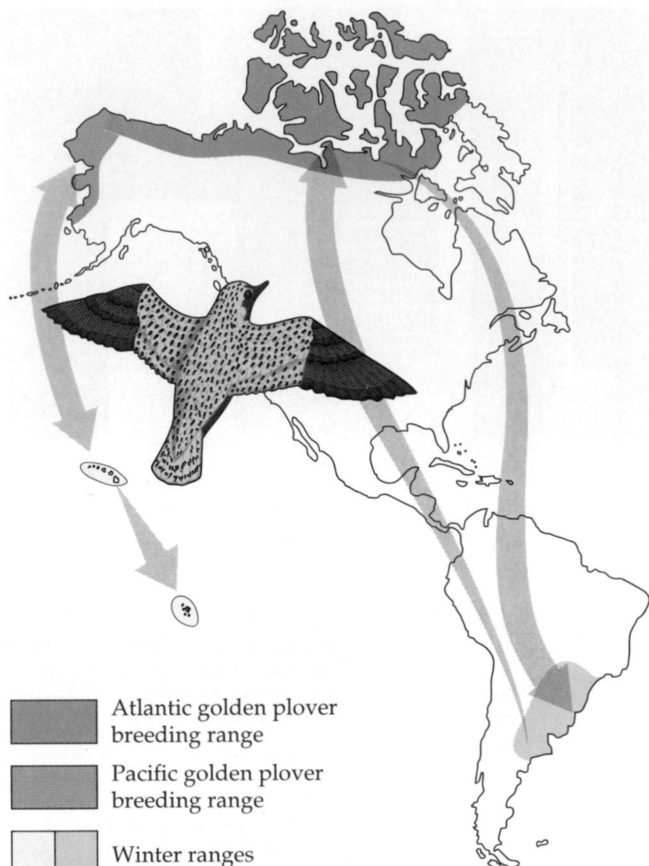

FIGURE 50.14
Migration routes of the golden plover. These birds are able to navigate across vast expanses of ocean to the relatively small Hawaiian and Marquesas islands (yellow).The ground-nesting plovers migrate from warm winter feeding areas to seasonally rich, essentially predator-free breeding grounds in the Arctic during the short arctic summer.

Migrating animals use one of three mechanisms to find their way, or some combination of these three techniques. In *piloting,* an animal moves from one familiar landmark to another until it reaches its destination. Piloting is used mostly for short-distance movements. In *orientation,* an animal can detect compass directions and travels in a particular straight-line path for a certain distance or until it reaches its destination. The most complex process is *navigation,* which involves determining one's present location relative to other locations, in addition to detecting compass direction (orientation). If you were dropped off at an unfamiliar spot and told that your home was directly to the north, you could use a compass and straight-line travel to get home; that is, you could use orientation. But a compass alone would not be adequate if you were not told which way to go; to choose the right direction, you would need to determine where you were in relation to home. You would need a mental picture of your surroundings, a so-called map sense. FIGURE 50.15 reinforces this distinction between orientation and navigation.

What sorts of cues do animals use for orientation and navigation? The answer, surprising at first but now firmly established, is that some species of birds and other animals commonly use the same means of navigation as early sailors: the heavens. The sun (for day-migrating species) and stars (for night migrants) provide excellent, albeit complex, cues to direction.

Orienting by the sun or stellar constellations requires an internal timing device to compensate for the continuous daily movement of celestial objects. Consider what would happen if you started walking one day, orienting yourself by keeping the sun on your left. In the morning you would be heading south, but by evening you would

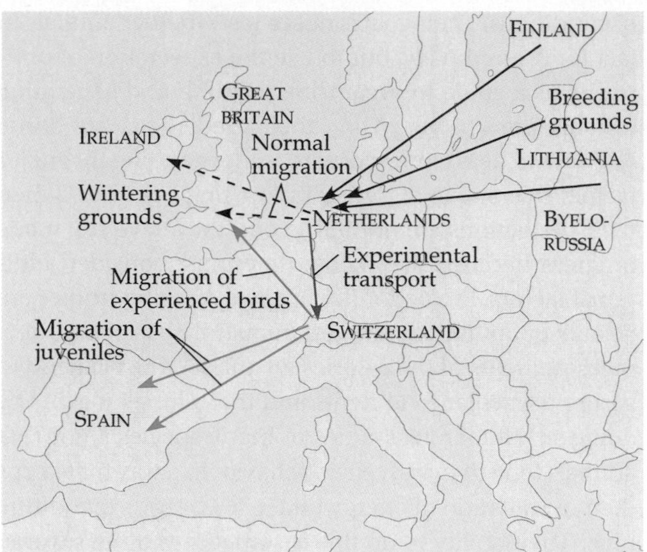

FIGURE 50.15
Orientation versus navigation in juvenile and adult starlings. About 11,000 starlings were captured in the Netherlands on their migration from their breeding grounds in northeastern Europe to their wintering grounds in Great Britain, Ireland, and northern France. After being transported to Switzerland (red arrow) and released, juvenile starlings, which had never made the journey before, continued to fly west and southwest (blue arrows), which brought them to Spain. Adults, all of whom had made the trip at least once before, flew northwest (green arrow), an atypical direction but one that took them to their usual wintering grounds. Members of both age groups were able to detect direction, but only the adults showed true navigation because they had developed a "map sense" and could determine where their original goal was relative to the site to which they were transported.

be heading back north, having made a circle and gotten nowhere. At night, the stars also shift their apparent position as Earth rotates. One night-migrating bird, the indigo bunting, avoids the need for a timing mechanism by the same means as ancient navigators: fixing on the North Star, which moves little. Many migrants, however, use some type of internal clock. For example, if an experimental sun is held in a constant position, starlings change their orientation steadily at a rate of about 15° per hour. This normally compensates for the change in the sun's position as Earth rotates on its axis. How they do it is unknown. The problem is very complex because the apparent location of the sun shifts at a variable rate, being fastest at mid-day. Furthermore, the apparent position of celestial objects changes as the animal moves over its migration route. As with the various internal rhythms discussed earlier, almost nothing is known of the mechanisms that could account for this extraordinary ability.

With some bird species, experimentally obscuring the sky to simulate cloudy weather causes them to flutter about in all directions or to cease migratory behavior. Many species, however, are known to continue migrating quite accurately under clouds or even through fog. There is now good evidence that some birds have the ability to detect and orient themselves to Earth's magnetic field, which varies geographically. The migratory orientation of some species can be experimentally manipulated by changing the magnetic field around them. Very little is known about how birds detect magnetism, but it is intriguing that magnetite, the iron-containing ore once used by sailors as a primitive compass, has been found in the heads of some birds. Magnetite has also been found in the abdomens of bees and in certain bacteria that orient with respect to magnetic field. The role of magnetite in sensing Earth's magnetic field has not been experimentally established. However, magnetic sensing may be a widespread, important orienting mechanism among animals, overlooked until recently because we seem to lack this sense ourselves. Although researchers have recently discovered magnetite in human skulls, there is no other evidence for a magnetic sense in humans.

Behavioral ecologists are using cost/benefit analysis to study foraging behavior

Feeding is obviously a behavior essential to survival and reproductive success, but what determines exactly what an animal eats? Because feeding behavior, or **foraging,** lends itself to experimental analysis at both the proximal and the ultimate level, it has become a favorite research topic among behavioral ecologists.

Animals feed in many ways, using various foraging behaviors that are closely linked to morphological traits. Suspension-feeding, for example, requires behavior different from that required for active predation. Ecological and evolutionary considerations are also extremely important; food habits are a fundamental part of an animal's niche and may be shaped in part by competition with other species. In this section, we will examine foraging from a behavioral perspective and see how behavioral ecologists are using cost/benefit analysis to study the proximate and ultimate causes of diverse foraging "strategies."

Animals of many species can theoretically choose from a large array of potential foods. Some animals tend to be generalists, feeding on a wide variety of items. Gulls feed on material that may be living or dead, aquatic or terrestrial, plant or animal. In contrast, European oyster catchers, wading shorebirds that forage in the intertidal zone, often feed on very specific species of shellfish. In fact, they are so specialized that oyster catchers use individual hunting techniques, learned from their parents, that further restrict the size and location of the shellfish they eat. Specialists usually have morphological and behavioral adaptations that are highly specific to their food, and as a result they are extremely efficient at foraging. Generalists cannot be as efficient at securing any one type of food, but they have the advantage of having other options if a preferred food becomes unavailable.

Most generalists do not choose their food randomly. Often, an animal will concentrate on a particular item when it is abundant, sometimes to the exclusion of other foods. This behavior depends on the animal's developing a **search image,** a set of key characteristics that will lead it to the desired object. We can understand search images from our own experience. If you were looking for a particular package on a kitchen shelf, you would probably scan rapidly, looking for a package of particular size and color rather than reading labels. Eventually, if the favored item becomes scarce relative to others, the animal will develop a new search image. Search images enable an animal to combine efficient short-term specialization with the flexibility of generalization.

The switching described here, and other choices animals make while foraging, have generated enormous interest among behavioral ecologists. Much recent research has emphasized *optimal foraging.* Optimal foraging theory predicts that natural selection will favor animals that choose foraging strategies that maximize the differential between benefits and costs. Benefits are usually considered in terms of energy (calories) gained. However, other optimization criteria, such as specific nutrients, are sometimes more important than energy. Costs or trade-offs associated with foraging consist of the energy needed to locate, catch, and eat food; the risk of being caught by a predator during feeding; and time taken away from other vital activities, such as searching for a mate.

Many trade-offs must be considered in an optimal foraging study. A given food item may be large and therefore contain considerable energy, but if it is farther away than a small item, moving to it will require more energy. In addition, the larger item may require more time to subdue or manipulate before swallowing, time in which other prey could be pursued. The smallmouth bass, for example, readily consumes both minnows and crayfish. The fact that it does not show an overall preference suggests that the trade-offs balance, with minnows being the optimal prey in some situations and crayfish in others. Minnows contain more usable energy per unit weight (the crayfish has a lot of hard-to-digest exoskeleton), but they may require more energy expenditure to pursue. However, even though crayfish may be easier to catch, their large claws and aggressive resistance make them harder to subdue. Trade-offs also include the relative abundance and size of each type of prey.

Behavioral ecologists use these and many other factors to predict how an animal will forage given a particular set of conditions. Their goal is not to test whether animals in fact forage optimally, but to use the expectation of optimality as a guide to organizing research and generating testable predictions. When their predictions are borne out, researchers come closer to understanding the major factors that shape an animal's foraging behavior. When their predictions are not borne out, they have still made progress because they know they must consider additional factors. Predictions in optimal foraging studies are usually quantitative; they are frequently based on direct measurements of the calories an animal must expend to secure particular food items and the calories it gains by doing so. Numerous studies of many species show that animals tend to modify their behavior in a way that keeps their overall ratio of energy intake to energy expenditure high. Their ability to do this is sometimes quite surprising. The smallmouth bass is somehow able to factor in all the relevant variables and forage in a highly efficient manner, switching between minnows and crayfish as conditions change. The proximal mechanisms responsible for this process are not known. They must include innate cues, but experience undoubtedly modulates the behavior.

The bluegill sunfish provides an interesting example of how animals maximize the energy intake-to-expenditure ratio. These animals feed on small crustaceans called *Daphnia,* generally selecting larger prey, which supply the most energy. However, smaller prey will be selected if a larger one is too far away (FIGURE 50.16a). The optimal foraging approach predicts that the proportion of small to large prey eaten will also vary with the overall density of prey. At very low prey densities, bluegill sunfish should exhibit little size selectivity, because all the prey encountered are needed to meet energy requirements. At higher prey densities, it is more efficient to concentrate on larger crustaceans. In actual experiments, bluegill sunfish did become more selective at higher prey densities, though not to the extent that would theoretically maximize efficiency (FIGURE 50.16b). Young bluegill sunfish forage fairly efficiently, but not as close to the optimum as older individuals, who are apparently able to make more complex distinctions. It may be that younger fish judge size and distance less accurately because their vision is not yet completely developed. It is unclear, however, whether the improvement in the older fish is due solely to maturation or whether it also involves learning.

Feeding is so fundamental to an animal's survival, and hence fitness, that the evolutionary emphasis of behavioral ecology is made to order for the study of foraging. But what can evolutionary biology bring to the study of social behavior, the interactions between conspecifics? The remaining six sections of this chapter explore the proximate and ultimate causes of social behaviors.

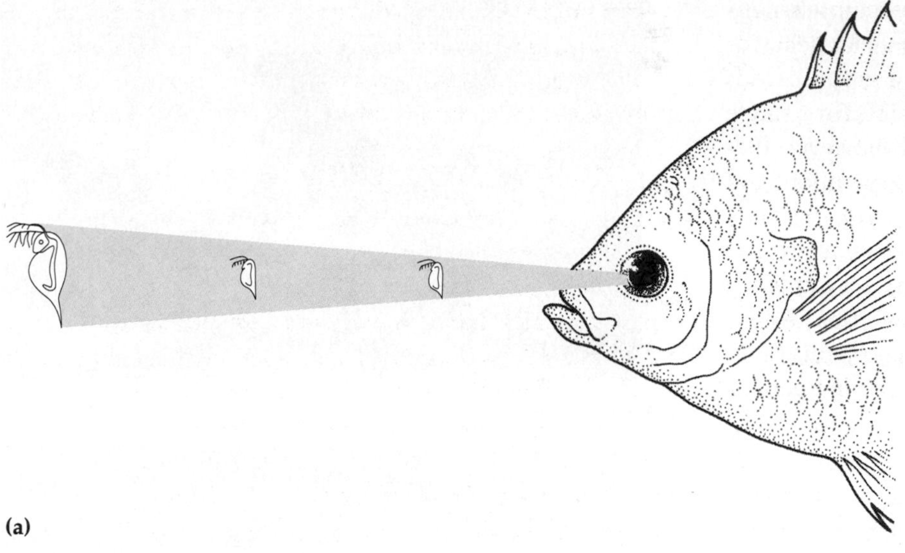

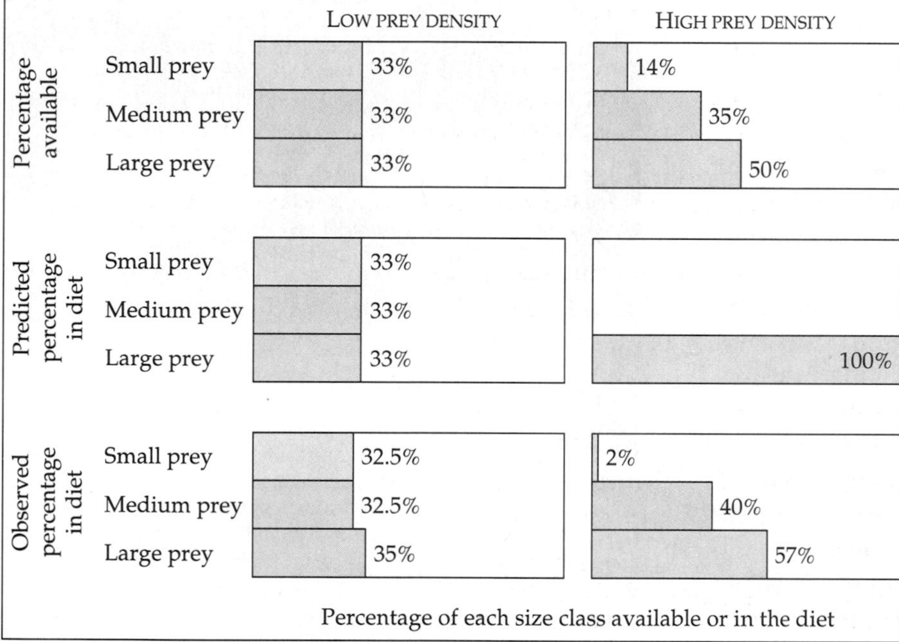

FIGURE 50.16

Feeding by young bluegill sunfish. (**a**) In feeding on *Daphnia* (water fleas), the fish do not feed randomly but tend to select prey based on "apparent size," information about both prey size and distance. Confronted with various potential prey, the animal will pursue the one that looks largest. Small prey (low energy yield) at the middle distance in this example may be ignored. But small prey at the closest distance may be taken with a relatively small energy expenditure. More distant but larger prey will require more energy expenditure but will provide a high yield and thus may be chosen over small prey at middle and even close distances. (**b**) When prey are at low density, calculations based on optimal foraging theory predict that bluegill sunfish will not be selective, but will eat any size prey as it is encountered. At higher prey densities, the ratio of energy intake to energy expended can be maximized by concentrating only on larger prey. In the experiments described here, we see that bluegills did forage nonselectively at lower prey density. At higher prey density, they favored larger prey, though not to the extent predicted.

Sociobiology places social behavior in an evolutionary context

Social behavior, broadly defined, is any kind of interaction between two or more animals, usually of the same species. Though most sexually reproducing species must be social for part of their life cycle in order to reproduce, some species spend most of their lives in close association with conspecifics. Social interactions have long been a research focus for scientists who study behavior. The complexity of behavior increases dramatically when interactions among individuals are considered. Aggression, courtship, cooperation, and even deception are part of the behavioral landscape of social behavior. Social behavior has both costs and benefits to members of those species that interact extensively. The relatively new discipline of **sociobiology** applies evolutionary theory as a foundation for the study and interpretation of social behavior. The development of sociobiology into a coherent method of analysis and interpretation was catalyzed by the work of E. O. Wilson and the 1975 publication of his watershed book *Sociobiology: The New Synthesis* (see the interview on p. 482). The book used the concepts of fitness and the genetic basis of behavior in analyzing the evolution and maintenance of social behavior in animals. The evolutionary emphasis of sociobiology also inspired the more general field of behavioral ecology as it is practiced today.

Because members of a population have a common niche, there is a strong potential for conflict, especially among members of species that normally maintain densities near carrying capacity. Sometimes social behavior seems to involve cooperation, as when a group carries out behavior more efficiently than is possible for a single individual (FIGURE 50.17). Keep in mind, however, that even when behavior requires some cooperation and seems to be mutually beneficial to interacting individuals, as in mating behavior, each participant usually acts in a way that will maximize its benefits, even if this is at a cost to the other participant. In the next section, we examine competitive social interactions, where this "selfish" aspect of behavior is most obvious. Later sections focus on social behaviors involving cooperation.

(a)

(b)

FIGURE 50.17
Cooperative prey capture. (**a**) This pack of African wild dogs is killing a wildebeest much larger than themselves. (**b**) This line of white pelicans is moving toward a school of fish. The cooperative behavior makes it difficult for fish to escape by swimming around the birds. Although all participants benefit from these sorts of cooperation, each individual is behaving in a manner that maximizes its own benefits.

Competitive social behaviors often represent contests for resources

Agonistic Behavior

In **agonistic behavior,** a contest involving both threatening and submissive behavior determines which competitor gains access to some resource, such as food or a mate. Sometimes the encounter involves tests of strength. More commonly, the contestants engage in threat displays that make them look large or fierce, often with exaggerated posturing and vocalizations. Eventually one individual stops threatening and ends with a submissive or appeasement display, in effect surrendering. Much of this behavior includes **ritual,** the use of symbolic activity, so that usually no serious harm is done to either combatant (FIGURE 50.18). Dogs and wolves show aggression by baring their teeth; erecting their ears, tail, and fur; standing upright; and looking directly at their opponent—all of which make the animal appear large and threatening. The eventual loser, on the other hand, sleeks its fur, tucks its tail, and looks away. This appeasement display inhibits any further aggressive activity. The degree to which combat is ritualized depends on the scarcity of the resource and the likelihood that the resource will be available again. For example, male ground squirrels often inflict severe injury upon, or even kill, each other when battling for access to sexually receptive females. In this case, the females for which the ground squirrels are competing are in estrus and receptive to male courtship for only a few hours each year, and thus a male's entire reproductive fitness may depend on his ability to compete against other males that one day.

When animals do inflict injury, natural selection favors a strong tendency to end the contest as soon as the winner is established, because violent combat could injure the victor as well as the defeated. Any future interaction between the same two animals is usually settled much more quickly in favor of the original victor.

Dominance Hierarchies

If several hens who are unfamiliar with one another are put together, they respond by pecking each other and skirmishing about. Eventually, the group establishes a clear "pecking order"—a more or less linear **dominance hierarchy.** Within a group, the alpha (top-ranked) hen controls the behavior of all others, often by mere threats rather than actual pecking. The beta (second-ranked) hen similarly subdues all others except the alpha, and so on down the line to the omega, or lowest, animal. The advantage to the top-ranked bird is obvious because it is assured of access to resources such as food. Even for lower-ranked animals, the system ensures that they do not waste energy or risk harm in futile combat.

Wolves typically function in packs, cooperation being necessary for killing large prey. Within each pack there is a dominance hierarchy among the females, and the top

FIGURE 50.18
Ritual wrestling by rattlesnakes. Rattlesnakes attempt to pin each other to the ground, but they never use their deadly fangs in such combat.

FIGURE 50.19
Territories. Gannets nest virtually only a peck apart and defend their territories by calling and pecking at each other. This is a population of Australian gannets in New Zealand.

female controls the mating of the others. When food is generally abundant, the top female mates and allows others to do so also; when food is scarce, she allows less mating by other females, thereby making more food available for her own young. Although these examples illustrate dominance hierarchies among females, male hierarchies are also common.

Territoriality

A **territory** is an area that an individual defends, usually excluding other members of its own species. Territories are typically used for feeding, mating, rearing young, or combinations of these activities. Usually a territory is fixed in location, its size varying with the species, the territory's function, and the amount of resources available. Song sparrow pairs, for example, may have territories of about 3000 m^2, in which they carry out all activities during the several months of their breeding season. Gannets and other seabirds, in contrast, mate and nest in territories of only a few square meters or less and feed away from their territories (FIGURE 50.19). Bull sea lions defend small territories used only for mating, whereas red squirrels have rather large territories apparently based on feeding patterns. In many species that defend territories only during the breeding season, individuals may form social groups at other times. This is the case with the chickadees mentioned earlier in the chapter. The monogamous breeding pairs that defend small territories in the summer form larger flocks in the winter, enabling the birds to forage more efficiently and benefit from the increased protection from predators that results from membership in a large group.

Note that there is a distinction between a territory and a home range, which is simply the area in which an animal roams about and which is often not defended. In some species, such as breeding song sparrows, territory and home range are the same; but for other species, such as gannets, a territory is considerably smaller than the home range. The distinction cannot always be made clearly. Gray squirrels, for example, typically have home ranges that overlap extensively, but one individual may defend part of the range from competitors.

Territories are established and defended through agonistic behavior, and an individual that has gained a territory is often difficult to dislodge. Why do owners usually win? One explanation in behavioral ecology is that a territory is worth more to an owner than to an intruder because the owner is already familiar with it. Thus, because it has more to gain from a territory, an owner is more likely to escalate a battle than is an intruder. In addition, established territory holders are likely to be older and more experienced at engaging in agonistic interactions.

Ownership of territory is usually continually proclaimed; this is a primary function of most familiar bird songs, as well as the noisy bellowing of sea lions and the chatter of the red squirrel. Other animals may use scent marks or frequent patrols that warn potential invaders

(a)

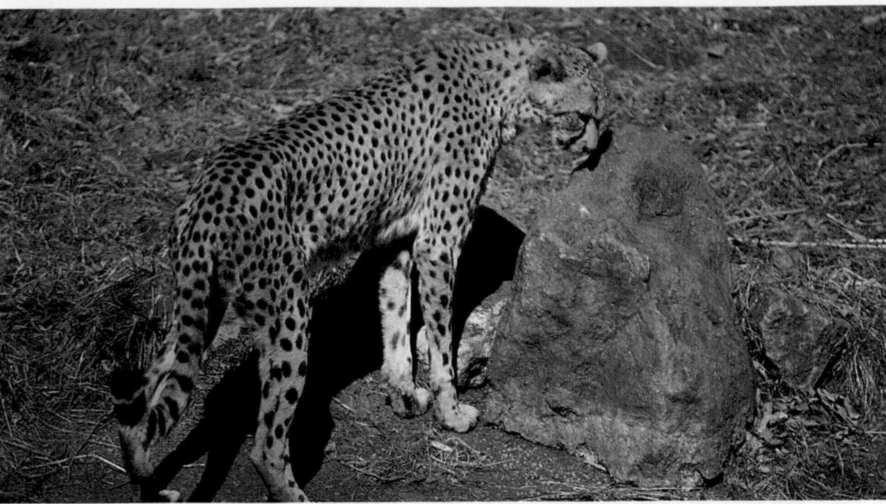

(b)

FIGURE 50.20
Staking out territory with chemical markers. (**a**) This male cheetah, a resident of Africa's Serengeti National Park, is spray-urinating on rocks. The odor will serve as a chemical "No Trespassing" sign to other males. (**b**) Another male cheetah sniffs a marked rock. With their keen olfactory sense, cheetahs can distinguish their own marks from those left by others. These signals help prevent face-to-face meetings that could escalate into violence.

(FIGURE 50.20). Grey wolves, which live in packs on huge territories (hundreds of square kilometers), use multiple strategies to advertise territorial boundaries, including scent marking and howling. Multiple signals help dispel any ambiguities as to the boundaries of a territory, thereby minimizing the risk that one group will accidentally stray into the territory of a rival pack. This is especially important in wolves because actual face-to-face meetings between groups are often violent.

Defense of territory is usually directed only at conspecifics; a white-crowned sparrow may live within a song sparrow's territory because, as we saw in Chapter 48, a different species usually has a different niche and is less likely to be a direct competitor. Another adaptive reason for concentrating defense on conspecifics is that they are likely to mate with a territory holder's mate.

Although dominance hierarchies and territoriality evolved as a result of their advantages to individuals, such systems have important effects at the population level because they tend to stabilize density. If resources were allocated evenly among all members of a population, the "fair share" that each individual received might not be enough to sustain anyone, leading to occasional population crashes. Dominance and territoriality usually mean that at least some individuals receive an adequate amount of a resource. Often, in fact, if a resource such as food becomes scarce, territories expand somewhat. In addition, there are usually individuals low in the hierarchy or lacking territories who are ready to move up or step in if one of the successful individuals dies. The result is relatively stable populations from year to year.

Mating behavior relates directly to an animal's fitness

All aspects of reproductive behavior receive extensive attention from behavioral ecologists. The reason is that researchers can often determine the number of young an individual produces, which comes very close to determining an animal's fitness. The correlation between measurable behaviors and fitness may not be as strong for other aspects of behavioral ecology, such as optimal foraging.

Courtship

Most animals probably do not have any conscious sense of reproduction as an important function in their lives, nor do they have the kind of continuous attraction to members of the opposite sex found in most humans. Often there is a strong tendency for an animal to view any organism of the same species as a threatening competitor to be driven off, if possible. Even within many social species, individuals usually avoid contact with one another. How, then, is mating accomplished? In many animals, potential partners must go through a complex courtship interaction, unique to the species, before mating. This complex behavior often consists of a series of fixed action patterns, each triggered by some action of the other partner and initiating, in turn, the other partner's next required behavior. This sequence of events assures each animal not only that the other is not a threat, but also that the other animal's species, sex, and physiological condition are all correct.

In some species, courtship also plays an important

role in allowing one or both sexes to choose a mate from a number of candidates. When individuals choose mates, females nearly always show greater discrimination than males, because in most species they have more parental investment in each offspring. **Parental investment** is defined as the time and resources an individual must expend to produce offspring. Eggs are usually much larger and much more costly to produce than sperm. Placental mammals have less disparity in gamete size than most other animals, but female placental mammals invest considerable time and energy in carrying the young before birth. In most species, females are very choosy; picking a poor-quality male can be a costly error. By contrast, males of most animal species mate with as many females as possible. Males usually compete with one another for mates, sometimes by trying to impress females. Thus, males of most species perform more intense courtship displays than females; in many species, in fact, it is only males who court. Secondary sex characteristics are usually much more pronounced in males; bright plumage in birds and antlers in deer are examples. Differential reproductive success that results from advantages in attracting or competing for mates, called *sexual selection,* which was discussed in more detail in Chapter 21.

In some animals, competition among individuals of the same sex (usually males) almost entirely determines which animals of that sex will mate. But in other species, individuals, usually females, actively choose among potential mates, based on specific characteristics of the male or resources under his control. This process is called *assessment.* There are two ultimate bases for such choices. First, if the other sex provides parental care, it is advantageous to choose as competent a mate as possible. For example, male common terns (a bird species related to gulls) carry fish and display them to potential mates as part of their mating ritual. Eventually, a male may begin to feed fish to a female. This behavior may be a good proximate indicator of a male's ability to provide food for tern chicks. In some animal species, females prefer males who are capable of the most extreme and energetic courtship displays or who have the most extreme secondary sex characteristics, such as long tails. Perhaps these characteristics are proximate indicators that the males are vigorous and in good health.

The second ultimate basis for mate choice is genetic quality. This is likely to be most important when males provide no parental care and sperm are their only contribution to offspring. In a number of bird and insect species, males display communally in a small area called a **lek.** Females visit the lek and choose among the displaying males. After mating occurs, there is no further contact between females and males. In terms of a female's fitness, it is beneficial for her to mate with a male that has favorable characteristics, since the success of her offspring will depend on both her genes and those of her mate. The proximate basis for this sort of choice is also likely to be a preference for males that court the most vigorously and are adorned with the showiest secondary sex characteristics. An especially important factor in male genetic quality is the ability to counter pathogens and parasites (see FIGURE 21.13).

It is often very difficult to determine which of these bases for female assessment is more important, and in some cases it is even difficult to determine whether differential mating success among males is due to male-male competition, female assessment, or both. One of the best known examples of ritualized courtship is that of the three-spined stickleback fish mentioned earlier (FIGURE 50.21). Although the courtship sequence is based on releasers and FAPs, it does not necessarily proceed smoothly or quickly. Often a female will begin to follow a male and then hesitate. She may cut off the sequence at any point, sometimes because she is simultaneously being courted by other nearby males. The final result is successful mating by some but not necessarily all males. Males provide all the parental care in this species. Therefore, proximate mechanisms of assessment by the female must gauge such relevant characteristics as nest quality and territorial defense by the male.

Ritualized acts, whether of courtship or of agonistic behavior, probably evolved from actions whose meaning was at one time more direct. We can see an interesting example of this process in the balloon flies of the family Empididae. In a few species, males spin oval balloons of silk and carry them while flying in a swarm, which is approached by females seeking mates. After a female chooses a male, she accepts his balloon, and the two fly off to copulate. A clue about how this ritual originated comes from examining the behavior of other species in the family. In some predatory species, the male brings a dead insect for the female to eat while he mates with her; perhaps this keeps her from attacking him. In other species, the dead insect is carried inside a silk balloon like the one just described, a variation that may have evolved because silk was helpful in subduing the insect or because it made the male's gift look larger. In balloon flies, which are not predators but instead eat nectar, the ritual has apparently evolved into merely bringing something associated with food in ancestral species.

Mating Systems

The mating relationship between males and females varies a great deal among species. In many species, mating is **promiscuous,** with no strong pair-bonds or lasting relationships. In species where the mates remain together for a longer period, the relationship may be

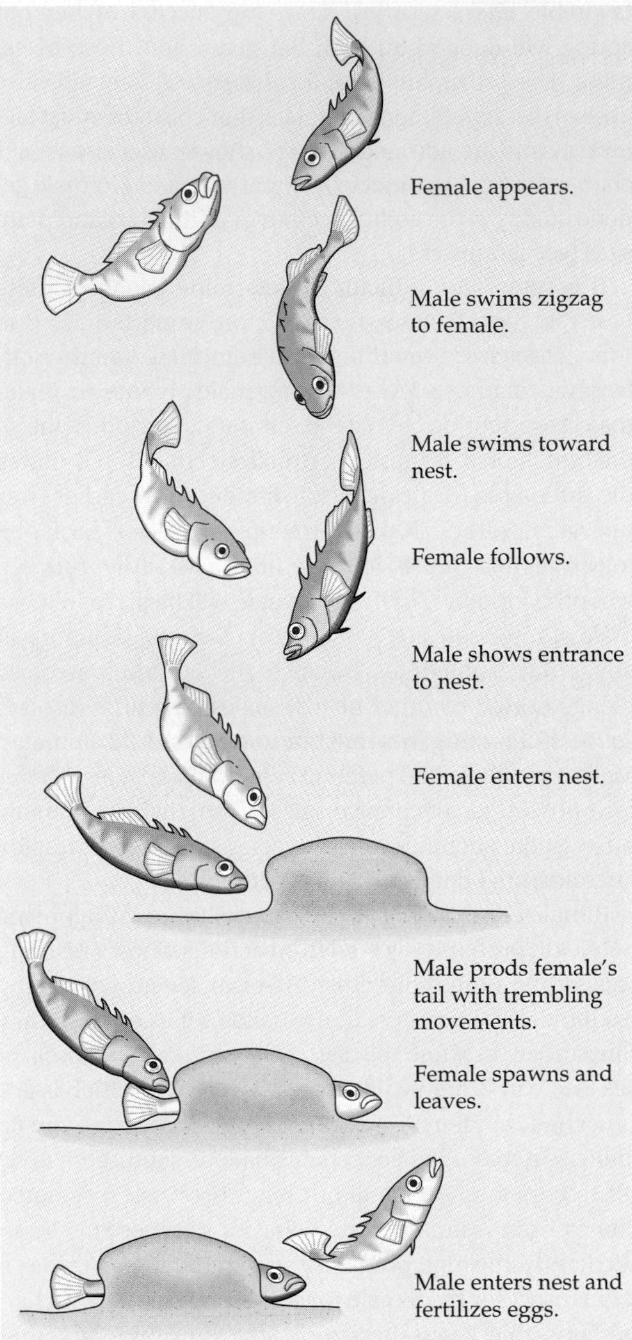
Female appears.
Male swims zigzag to female.
Male swims toward nest.
Female follows.
Male shows entrance to nest.
Female enters nest.
Male prods female's tail with trembling movements.
Female spawns and leaves.
Male enters nest and fertilizes eggs.

FIGURE 50.21
Courtship behavior in the three-spined stickleback. Males are strongly territorial, defending an area in which they have built a nest. If a gravid (egg-carrying) female approaches, her swollen belly releases zigzag swimming in the male. This entices her to swim closer, which in turn stimulates him to swim to the nest and stick his snout inside. This action stimulates the female to wriggle into the tunnel. The male then nuzzles her tail, which stimulates her to spawn, after which she swims out the other end of the nest. The male then enters and deposits sperm on the eggs, after which he immediately and quite aggressively drives the female out of the area. Her lack of a swollen belly after egg laying no longer provides a sign stimulus that inhibits the male's aggression and serves as a releaser for courtship behavior.

monogamous (one male mating with one female) or **polygamous** (an individual of one sex mating with several of the other). Polygamous relationships most often involve a single male and many females, called **polygyny,** which can be explained in terms of a difference in parental investment. However, there are also species in which a single female mates with several males, a relationship called **polyandry.**

The needs of the young are an important ultimate factor in the evolution of mating systems. Most newly hatched birds cannot care for themselves and require a large, continuous food supply that one parent may not be able to provide. In such a system, a male may ultimately leave more viable offspring by helping a single mate than by going off to seek more mates. This may explain why most birds are monogamous. In birds with young that feed and care for themselves almost immediately after hatching, there is less need for parents to stay together. Males of these species can maximize their reproductive success by seeking other mates, and polygyny is relatively common in such birds. In the case of mammals, the lactating female is often the only food source for the young. Males usually play no role, or, if they protect the females and young, they typically take care of many at once in a harem.

Another factor that influences mating systems and parental care is certainty of paternity. Young born or eggs laid by a female definitely contain the female's genes. But even within a normally monogamous relationship, these young may have been fathered by a male other than the female's usual mate. For example, this was the case in a population of red-winged blackbirds studied by researchers who used DNA fingerprinting to determine the paternity of nestlings (see the Methods Box). The certainty of paternity is relatively low in most species with internal fertilization because the acts of mating and birth (or mating and egg laying) are separated over time. This could explain why parental care is exclusively by males in very few bird or mammal species. However, certainty of paternity is much higher when egg laying and mating occur together, as in external fertilization. This may explain why parental care in fishes and amphibians, when it occurs at all, is at least as likely to be by males as by females. Male parental care occurs in only 2 out of 28 (7%) fish and amphibian families with internal fertilization, but in 61 out of 89 (69%) families with external fertilization. In fish, even when parental care is given exclusively by males, the mating system is often polygynous, with several females laying eggs in a nest tended by one male.

It is important to point out again that when behavioral ecologists use terms such as "certainty of paternity," they do not mean that animals are aware of those factors

METHODS: USING MOLECULAR BIOLOGY TO ANSWER BEHAVIORAL QUESTIONS

Molecular biology has had an impact on every field of biology, including behavioral ecology. Here, we examine the specific case of a research project that applied DNA fingerprinting to determine the paternity of nestlings hatched in a population of red-winged blackbirds (FIGURE a). The DNA fingerprinting was based on restriction fragments (RFLPs), as described in Chapter 19.

In most bird species, mating relationships are fixed, typically involving one male and one to several females that are joint inhabitants of the male's territory. By counting the number of young in the nest or nests within a male's territory, it should be possible to compare the relative reproductive success (fitness) of different males. Behavioral ecologists have sometimes used this approach to compare the effects of behavioral variations on the fitness of different birds in a population. But there is a problem with this method of measuring how many offspring each male sires. Although a male generally guards his mate (or mates) during her fertile period, there is evidence that this vigilance is imperfect at preventing the females from mating with other males, and females may occasionally mate with males from neighboring territories.

In the 1970s, biologists at the U.S. Fish and Wildlife Service tried to control populations of red-winged blackbirds, which sometimes threaten agricultural crops. Rather than kill the birds, the biologists performed vasectomies on the males. The researchers studying paternity found, surprisingly, that 50% or more of the eggs in each vasectomized male's territory had been fertilized, presumably by neighboring males who had not had the procedure. Perhaps vasectomies altered male territorial defense, increasing the frequency of extrapair copulations (EPCs). Thus, there was still the possibility that EPCs were rare for females mated to normal, untreated males.

The problem was not resolved until 1990, when a team of researchers published a DNA fingerprinting analysis of paternity in a population of redwings. Analysis of DNA from blood samples of nestlings and adults revealed that 45% of the nests had at least one offspring fathered by a male other than the one claiming that territory. The DNA fingerprinting even allowed the researchers to identify which males had ventured out of their territories and fathered offspring in the territories of neighbors. In FIGURE b, the green areas represent the territories of individual males. The fractions indicate the number of offspring sired by a male over the total number of young in his territory. The arrows show the direction of EPCs and the number of "extra" young sired by each male. Thus, male X fathered all three of the young in his territory, as well as two offspring in the neighboring territory of male Y and one offspring in the territory of male Z. A relatively large number of EPCs occur in the territories of males that have a larger-than-average number of females. Perhaps this is because a male with multiple mates cannot guard them all effectively.

In the past few years, DNA fingerprinting studies have shown that EPCs are common in many bird species that were previously assumed to be faithful in their pair-bonding. These techniques have reinvigorated the study of monogamous breeding systems by illuminating the variability in reproductive fitness that was previously undetectable.

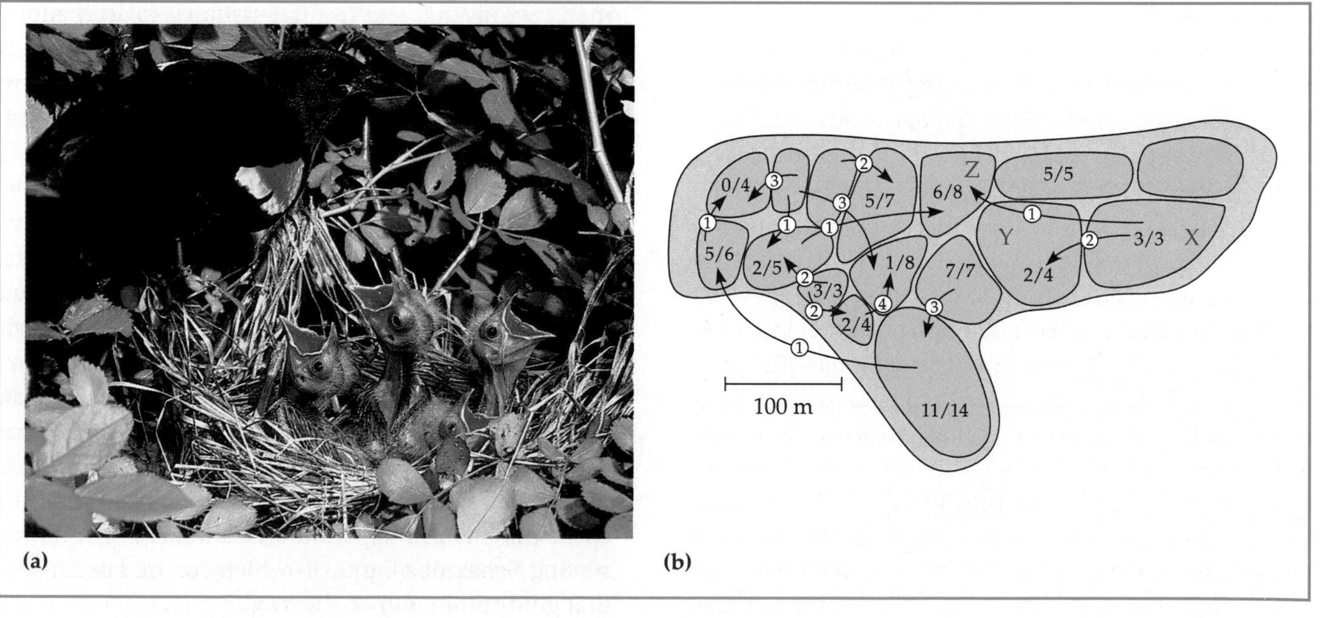

(a)

(b)

when they behave a certain way. Parental behavior correlated with certainty of paternity exists because it has been reinforced over generations by natural selection.

Social interactions depend on diverse modes of communication

Much of what we have discussed under competitive social interactions and mating behaviors involves instances in which animals intentionally, although not necessarily consciously, transmit information by special behaviors called *displays*. The intentional transmission of information between individuals is the usual definition of communication in behavioral ecology. Singing by male birds is, in effect, a display that transmits the information, "This is my territory. Keep out!" This is almost certainly an important message of singing; if we play the tape-recorded songs of another male in a male bird's territory, he becomes highly agitated, approaching and sometimes even attacking the speaker. Someone else has not only ignored his warning but has claimed the territory for himself. This simple experiment is so infallible that some bird watchers routinely use it to find and see secretive birds that would otherwise stay hidden. The playback procedure makes another important point. We cannot get into an animal's brain to determine whether it has received a message sent by another individual. How, then, do we know when communication has occurred? We usually say that communication has occurred when an act by a "sender" produces a detectable change in the behavior of another individual, the "receiver." Bird song is communication because it produces a response.

Most ethologists assumed that communications systems evolve in ways that maximize the quantity and accuracy of information. Behavioral ecologists take a different evolutionary perspective. They believe that these displays evolve because the fitness of the sender is increased by the effect the message has on the individuals receiving the message. This may explain adaptations in which the function of communication seems to be deception, especially when the message recipients belong to a different species. For example, male and female *Photinus* fireflies find each other and mate when females give a characteristic pattern of light flashes in response to flashes given by males. However, females of another firefly genus, *Photurus,* respond to these male *Photinus* flashes with the same flashing pattern. A male *Photinus* receives the incorrect message that a female *Photurus* is a suitable mate, and when he finds her she does not mate with him, but kills and eats him. Such cases of mimicry, in which communication is adaptive for the sender but maladaptive for the receiver, are widespread.

Deceit also occurs within species. In some mammals, after a dominant male assumes control of a social group, he kills young that are born too soon to be his offspring. Without dependent young, the females ovulate sooner, allowing the new dominant male to father their offspring. In an Asian monkey known as the hanuman langur, females in the early stages of pregnancy actively communicate receptiveness to copulations from new males. These females then give birth only shortly before young fathered by these males would appear. This may deceive a male into treating the female's young as if they were his own, when in fact they are not. Again, none of this implies that langurs have consciously shaped their behavior to maximize fitness. Instead, it simply implies that females with a tendency to solicit copulations from new males have had a selective advantage over other females without such a tendency.

Another important evolutionary consideration in communication is the sensory mode used to transmit information. Animals transmit information with visual, auditory, chemical (olfactory), tactile, and electrical signals. Which mode is used to transmit information is closely related to an animal's basic lifestyle. Most mammals are nocturnal, which makes visual displays relatively ineffective. But olfactory and auditory signals work as well in the dark as in the light, and most mammalian species stress these signals. Birds, by contrast, are mostly diurnal and use mostly visual and auditory signals. They almost never use olfactory signals, probably because they can fly faster than chemical signals can travel. (It is hard to imagine a system in which it would be adaptive for a messenger to arrive before its message.) Unlike most mammals, humans are diurnal and use the same visual and auditory communication modes as birds. Therefore, we can detect the songs and bright colors that birds use to communicate with each other. This may explain why bird watching is so popular. If humans had the well-developed olfactory abilities of most mammals and could detect their rich world of chemical cues, mammal sniffing might be as popular as bird watching.

Animals that communicate by odors emit chemical signals called **pheromones.** These are especially common among mammals and insects and often relate to reproductive behavior. Female silkworm moths, for example, emit a pheromone that can attract males from several kilometers away. Once the moths are together, pheromones are also important as releasers for specific courtship behaviors. Another example is the familiar trailing behavior of ants, in which scouts release scents that guide other ants to the food.

One of the most complex communication systems—certainly among invertebrates—is that of social, or hive, bees. For maximum foraging efficiency, workers must

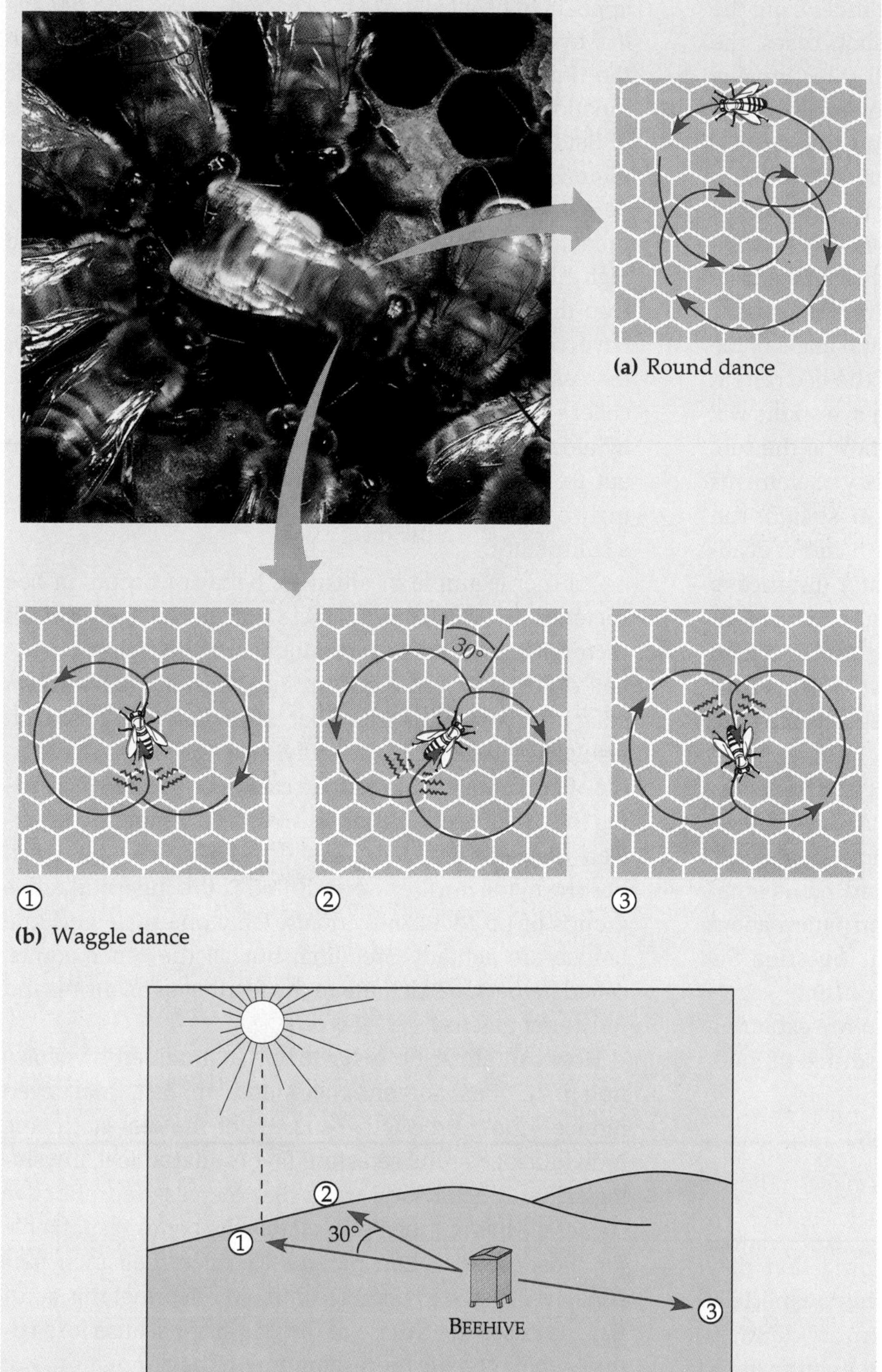

FIGURE 50.22

Communication in bees: one hypothesis. (**a**) The round dance indicates that food is near but provides no information on directionality or specific distance. (**b**) The waggle dance is performed when food is distant. This dance pattern resembles a figure eight, with a straight run between two semicircular movements. According to von Frisch's hypothesis, the waggle dance indicates both distance and direction. Distance is indicated by the duration of each waggle run or dance and the number of abdominal waggles performed per waggle run. Direction is indicated by the angle (in relation to the vertical surface of the hive) of the straight run that forms part of the dance itself. ① For instance, if the straight run is directly upward, this signals that food is in the same direction as the sun. ② If the angle is 30° to the right of vertical, the food is 30° to the right of the sun. ③ If the straight run is directly downward, the food is in the direction opposite the sun. Odor cues and sound may also convey information about the location and type of food.

convey to one another the location of good food sources, which may change frequently as various flowers bloom or new patches are discovered. How do bees communicate? The study of honeybee communication has a long and rich tradition of experimental research that continues to reveal new elements of the bees' language. The problem was first studied in the 1940s by Karl von Frisch, who carefully watched individual European honeybees (*Apis mellifera carnica*) as they returned to special observation hives. The returning bee would quickly become the center of attention by other bees, called followers, and begin to go through a repetitive behavior that von Frisch called a dance (FIGURE 50.22). If the food source was close to the hive (less than 50 m away), the returning bee moved in tight circles while waggling its abdomen from side to side. This dance was usually accompanied by the bee's regurgitating some of the acquired nectar. This behavior, which von Frisch called the round dance, had the effect of exciting the follower bees and motivating them to leave the hive and search for food that was nearby.

However, bees often forage at great distances from the hive, sometimes in excess of 5 km. In such cases, the round dance is insufficient, lacking both directionality and distance cues necessary for the followers to locate the food source efficiently. A worker returning from a longer distance does a "waggle dance": a half-circle swing in one direction, followed by a straight run, and then a half-circle swing in the other direction (FIGURE 50.22b). This dance indicates both direction and distance. The angle of the straight run in relation to the vertical surface of the open hive is the same as the horizontal angle of the food in relation to the sun. For example, if the bee runs at a 30° angle to the right of vertical, the other workers will fly 30° to the right of the horizontal direction of the sun. Distance to the food is indicated by a variety of elements of the waggle dance. For example, a longer straight run during the dance, and hence an increasing number of abdominal waggles per run, indicates a greater distance to the food source. Other features of the symbolic relationship between waggle run duration and distance to food source that have been discovered are that dialects exist among different subspecies and that the length of the run is innate.

During waggle dances, the bee also regurgitates nectar; thus, when bees leave to forage, they already "know" the type of food to seek, its distance, and its direction. There is also evidence that the sounds and odors emanating from the dancing bee provide information about the food source. In fact, some researchers question the strength of the evidence that the pattern of the waggle dance itself is the key communication and are exploring other hypotheses for how bees report a source of food.

The concept of inclusive fitness can account for most altruistic behavior

Many social behaviors are selfish, meaning that they benefit the individual at the expense of others, especially competitors. A bird that establishes a territory deprives other individuals of one, and if there is not enough habitat, these other individuals may be unable to breed. Even in species in which individuals do not engage in agonistic behavior, most adaptations that benefit one individual will indirectly harm others. For example, superior foraging ability by one individual may leave less food for others. It is easy to understand the pervasive nature of selfishness if natural selection shapes behavior. Behavior that maximizes an individual's reproductive success will be favored by selection, regardless of how much damage such behavior does to another individual, a local population, or even an entire species.

How, then, can we explain observed examples of what appears to be altruism? On occasion, animals do behave in ways that reduce their individual fitness and increase the fitness of the recipient of the behavior; this is our functional definition of **altruism.** Consider the example of Belding's ground squirrels, which live in some mountainous regions of the western United States and are vulnerable to predators such as coyotes and hawks. If a predator approaches, one of the squirrels often gives a high-pitched alarm call. This alerts unaware individuals, who then retreat to their burrows. Careful observations have confirmed that the conspicuous alarm behavior increases the risk of being killed, because it identifies the caller's location. How can a squirrel enhance its fitness by aiding other members of the population, which are apt to be its closest competitors? Fitness would seemingly be maximized by quietly allowing a predator to take a competitor.

Another example of altruistic behavior occurs in bee societies, in which the workers are sterile. The workers themselves will never reproduce, but they labor on behalf of a single fertile queen. Furthermore, the workers sting intruders, a behavior that helps defend the hive but results in the death of the individual. Still another example of apparent altruism is the case of cooperative breeding in some species of birds. This is illustrated by American crows, long-lived birds that have been observed nesting in the northeastern United States in reproductive groups of up to 14 individuals. Only one male and one female are actually breeding, but all the other adults, called helpers, participate in the feeding of offspring and territorial defense.

How can altruistic behavior be maintained by evolution if it does not enhance—and, in fact, may even reduce—the reproductive success of the self-sacrificing individuals? Natural selection favors anatomical, physiological, and behavioral traits that increase reproductive success, which in turn propagates the genes responsible for those traits. When parents sacrifice their own personal well-being to produce and aid offspring, this actually increases the fitness of the parents, because it maximizes their genetic representation in the population. But what about helping other close relatives? Like parents and offspring, siblings have half of their genes in common. Therefore, selection might also favor helping one's parents produce more siblings or even helping siblings directly. In the 1960s, W. D. Hamilton, then a graduate student in England, first realized that selection could result in an animal's increasing its genetic representation in the next generation by "altruistically" helping close relatives other than its own offspring. This realization led to the concept of **inclusive fitness,** which describes the total effect an individual has on proliferating its genes by producing its own offspring *and* by providing aid that enables other close relatives to increase the production of their offspring.

An important quantitative measure of inclusive fitness is the **coefficient of relatedness,** which is the proportion of genes that are identical in two individuals because of common ancestors. For example, the coefficient of relatedness for siblings is 0.5, meaning that they match in 50% of their genes. For cousins, the coefficient of relatedness is 0.125. We expect that the higher the coefficient of relatedness, the more likely an individual is to aid a relative: The individual's altruism can result in more genes identical to its own in the next generation if it aids a sibling rather than a cousin. This mechanism of increasing inclusive fitness is called **kin selection.** The contribution of kin selection to inclusive fitness varies among species. Kin selection may be rare to nonexistent in species that are never social or that disperse so widely that individuals never live near close relatives. In these species, an animal's inclusive fitness is essentially the same as its individual fitness—its fitness based on the production of only its own offspring.

British geneticist J. B. S. Haldane anticipated the concepts of inclusive fitness and kin selection by jokingly saying that he would lay down his life for two brothers or eight cousins. In today's terms, we would say that he would do this because either two brothers or eight cousins would result in as much representation of Haldane's genes as would two of his own offspring. But this assumes equal reproductive potential for all individuals. We assume that Haldane was as likely to have as many children as each of his individual brothers and cousins. Often, however, reproductive potentials differ. A sterile individual should theoretically sacrifice its life for one fertile brother or even one cousin, otherwise its inclusive fitness will be zero. Even differences in reproductive potentials that are not as great as between sterile and fertile individuals can still contribute to kin selection. For example, territories might be in limited supply, and one animal might have one while its brother does not. The homeless animal might have a greater increase in its inclusive fitness by helping its brother than by trying to win a territory of its own, if the second option has only a small chance of success. In trying to predict whether an individual animal will aid relatives, behavioral ecologists have derived a formula that combines coefficients of relatedness, costs to the altruist, and benefits to the recipient of aid. (Animals, of course, do not work out the kin selection formula for themselves or calculate behavior that increases their inclusive fitness.)

If kin selection explains altruism, then the examples of unselfish behavior we observe should involve close relatives. This expectation is met, but often in complex ways. Like most mammals, female Belding's squirrels settle close to their site of birth, while males settle at distant sites. Thus, only females are likely to live near close relatives, and nearly all alarm calls are given by females (FIGURE 50.23). However, if all of a female's close relatives are dead, she rarely gives alarm calls. In the case of worker bees, the individuals are sterile, and anything

(a)

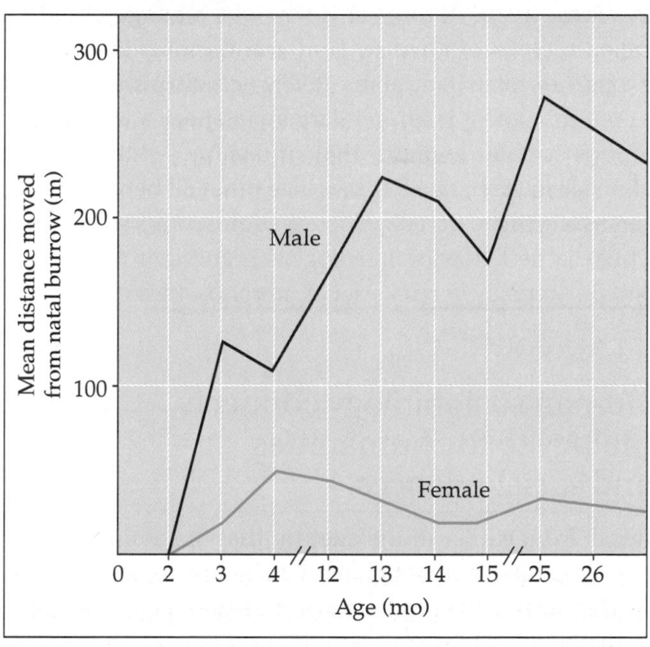

(b)

FIGURE 50.23
Altruistic behavior and a male-female difference in its expression. (**a**) By sounding an alarm call, this ground squirrel warns others of danger, such as an approaching predator. Nearly all alarm calls are given by females. (**b**) This graph helps explain the male-female difference in altruistic behavior of ground squirrels. After they are weaned, males disperse much farther from their birthplaces than do females. Therefore, females are more likely to live near close relatives, and warning these relatives increases the inclusive fitness of the altruist.

they do to help the entire hive benefits the only permanent member who is reproductively active—the queen, who is their mother.

In the case of the American crows mentioned earlier, the helpers are almost always older siblings that have not yet obtained a territory of their own. One long-term study of cooperative breeding in both rural and urban settings found that the size of crow family territories decreased and the number of individuals in each family group increased for crows breeding in urban settings. Ecological conditions in urban areas reduced the number of available territories and increased the competition for resources, thus selecting for increased helping behavior. For these nest assistants, helping a close relative may not be much of a sacrifice. And sometimes these helpers inherit the territory when their relatives die, which may be the surest way to secure a territory in a crowded landscape. This could mean that in some cases helping evolved primarily from a direct advantage to the individual, rather than because of kin selection.

Some animals occasionally behave altruistically toward others who are not relatives. A baboon may help an unrelated companion in a fight, or a wolf may offer food to another wolf even though they share no kinship. Such behavior can be adaptive if the aided individual returns the favor in the future. This sort of exchange of aid is called **reciprocal altruism** and is commonly invoked to explain altruism in humans. Reciprocal altruism is rare in other animals; it is limited largely to species with social groups stable enough that individuals have many chances to exchange aid. It is likely that all behavior that seems altruistic actually increases fitness in some way. Thus, some behavioral ecologists argue that true altruism never really occurs, except, perhaps, in humans.

Human sociobiology connects biology to the humanities and social sciences

Recall from earlier in the chapter that the main premise of sociobiology is that behavioral characteristics exist because they are expressions of genes that have been perpetuated by natural selection. In the last chapter of *Sociobiology,* Wilson speculates about the evolutionary basis of certain kinds of social behavior in humans. The book rekindled the nature-versus-nurture controversy, and the debate over sociobiology remains heated two decades after its publication.

Let's examine the debate in the context of a specific example: the avoidance of incest. Such avoidance is adaptive because inbreeding may increase the frequency of certain kinds of genetic disorders. Many birds and mammals clearly avoid incest. Similarly, nearly all human cultures have laws or taboos forbidding sexual relations or marriage between brother and sister. Is there an innate aversion to incest that we share with other species, or do we acquire this behavior as part of our socialization? Someone favoring the "nurture" side of the debate might argue that if this behavior were innate, cultural taboos would be superfluous. According to this argument, the avoidance of incest is a learned behavior, and the social stigma attached to incest may be based on experience; people who break the taboo are more likely to have children with congenital disorders. Some sociobiologists, on the other hand, would argue that the occurrence of a specific behavior in diverse cultures is evidence that the behavior has an innate component. According to this argument, incest taboos are simply proximate mechanisms that reinforce a behavior that ultimately evolved because of its effect on fitness.

Some sociobiologists have cited the experience of the Israeli communal societies known as kibbutzim as evidence that humans' innate repulsion against incest is strong enough to counter cultural factors. From the time of birth, kibbutz children used to spend most of their time in large child-care centers, which gave them the equivalent of a sibling relationship with all other children in the kibbutz. Early in the kibbutz movement, parents encouraged their children to marry within their own kibbutz. However, according to a study of more than 5000 people, such marriages were extremely rare, in spite of the wishes of parents and kibbutz leaders. Apparently, people who spend nearly all their time together as children have little sexual attraction for one another as adults. (In most situations, of course, individuals who cohabit throughout childhood are true siblings or other close relatives.) A sociobiologist might argue from this example that the human resistance to incest—in this case, with "artificial" siblings—overrides cultural pressure encouraging it.

Some sociobiologists, including Wilson, envision cultural and genetic components of social behavior as linked in a cycle of reinforcement. If most members of a society share an innate avoidance of incest, the aversion is likely to become formalized in laws and taboos. The cultural stigma, in turn, acts as an environmental factor in natural selection, amplifying the evolutionary component of the behavior. In the case of incest taboos, society may shun or imprison offenders, thus lowering their reproductive fitness. According to this view, genes and culture are integrated in human nature.

The spectrum of possible social behaviors may be circumscribed by our genetic potential, but this is very different from saying that genes are rigid determinants of behavior. This is at the core of the debate about sociobiology. Opponents fear that a sociobiological interpretation of human behavior can be used to justify the status quo in human society, thus rationalizing current social injustices. Sociobiologists argue that this is a gross oversimplification and misunderstanding of what the data tell us about human biology. Sociobiology does not reduce us to robots stamped out of rigid genetic molds. Individuals vary extensively in anatomical features, and we should expect inherent variations in behavior as well. Furthermore, though we are locked into our genotypes, our nervous systems are not "hard-wired." Environment intervenes in the pathway from genotype to phenotype for physical traits, and even more, in general, for behavioral traits. And because of our capacity for learning and versatility, human behavior is probably more plastic than that of any other animal. Over our recent evolutionary history, we have built up structured societies with governments, laws, cultural values, and religions that define what is acceptable behavior and what is not, even when unacceptable behavior might enhance an individual's Darwinian fitness. Perhaps it is our social and cultural institutions that make us truly unique and that provide the only feature in which there is no continuum between humans and other animals (FIGURE 50.24).

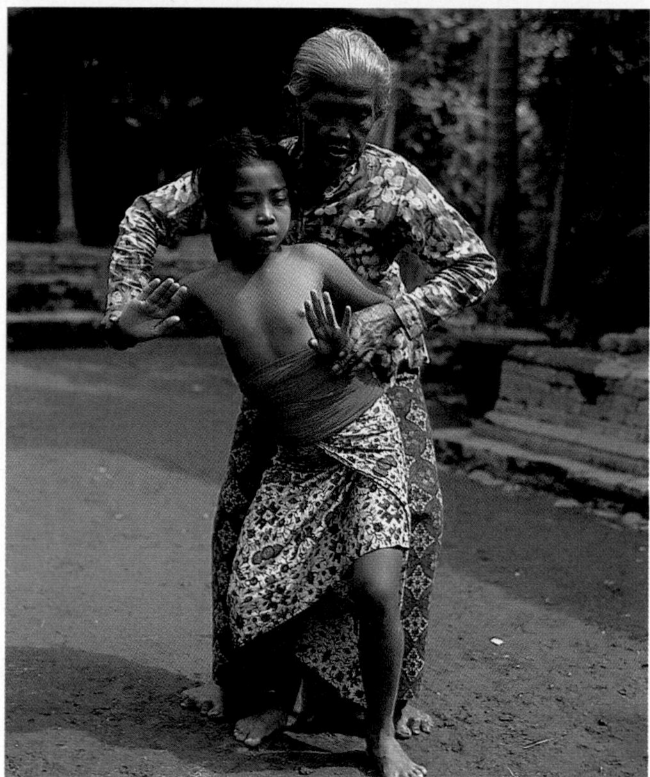

FIGURE 50.24
Both genes and culture build human nature. Teaching of the younger generation by the older is one of the basic ways in which all cultures are transmitted. Sociobiologists see tutoring as an innate tendency with adaptive value that has evolved in the human species.

*　*　*

The many facets of biology are reflected in the study of behavior; it is an intersection of biochemistry, genetics, physiology, evolutionary theory, and ecology. The study of behavior, especially social behavior, also connects biology to the social sciences and humanities. Through the study of life in general, it is inevitable that we will learn more about ourselves.

REVIEW OF KEY CONCEPTS (with page numbers and key figures)

- Behavior is what an animal does and how it does it (p. 1173, FIGURE 50.1)
  - Behavior includes observable motor activities and nonmotor processes, such as memory, that affect those activities.
  - Most animal behavior reflects some combination of innate and learned components.
- Behavioral ecology emphasizes evolutionary hypotheses: *science as a process* (pp. 1173–1174)
  - Behavioral ecology is based on the expectation that animals behave in ways that increase their Darwinian fitness (reproductive success).
- A behavior has both an ultimate and a proximate cause (pp. 1174–1176, FIGURE 50.4)
  - Questions about the stimuli that provoke behaviors and about the actual mechanisms of response address the proximate causes of animal behavior.
  - Evolutionary explanations in terms of why natural selection may have favored a particular behavior address the ultimate causes of animal behavior.
- Certain stimuli trigger innate behaviors called fixed action patterns (pp. 1176–1180, FIGURE 50.5)
  - A fixed action pattern (FAP) is a highly stereotypical, innate behavior that continues to completion after initiation by an external sign stimulus. Releasers are specific sign stimuli that function as communication signals between individuals of the same species.
  - Sign stimuli tend to be based on simple cues associated with the relevant object or activity.

- Learning is experience-based modification of behavior (pp. 1180–1186)
  - Some apparent learning is due mostly to inherent maturation.
  - Habituation is a simple kind of learning involving loss of sensitivity to unimportant stimuli.
  - Imprinting is a type of learned behavior acquired during a limited critical period.
  - Studies of song development in birds have revealed an interplay between the innate and learned components of imprinting.
  - Associative learning involves linking one stimulus with another, as demonstrated by classical conditioning.
  - Operant conditioning, or trial-and-error learning, is also a type of associative learning.
  - In observational learning, animals copy the behavior of other individuals in the population.
  - Insight learning involves the ability to reason by correctly performing a task on the first attempt in a situation in which the animal has had no earlier experience.
  - The animal cognition debate centers on whether non-human animals are conscious, thinking organisms.
- Rhythmic behaviors synchronize an animal's activities with daily and seasonal changes in the environment (pp. 1186–1187, FIGURE 50.13)
  - Various daily behaviors are governed by endogenous clocks, which in turn require exogenous cues to keep the behavior properly timed with the external environment.
  - Circannual behaviors, such as breeding and hibernation, seem to be controlled primarily by physiological and hormonal changes influenced by such exogenous factors as day length.
- Environmental cues guide animal movement (pp. 1187–1189, FIGURE 50.15)
  - Kinesis is a random movement displaying a stimulus-specific change in activity rate, whereas taxis is an automatic movement toward or away from a specific stimulus.
  - The most sophisticated mechanisms of orientation and navigation are used by migrating animals. Some species of migratory birds are guided by the sun or stars. Others may be able to use Earth's magnetic field. True navigation requires both a "compass sense" and a "map sense."
- Behavioral ecologists are using cost/benefit analysis to study foraging behavior (pp. 1189–1190, FIGURE 50.16)
  - Species may be specialists or generalists as foragers. Even generalists do not forage randomly, but form search images that can be changed if the food item becomes scarce.
  - According to the premise of optimal foraging, animals forage efficiently in nature by modifying their behavior in complex ways that favor a high ratio of energy intake to expenditure.
- Sociobiology places social behavior in an evolutionary context (pp. 1191–1192)
  - Social behavior encompasses the spectrum of interactions between two or more animals, usually of the same species. E. O. Wilson's *Sociobiology* provided a paradigm for the study of the adaptive nature of social behavior.
- Competitive social behaviors often represent contests for resources (pp. 1192–1194)
  - Agonistic behavior involves a contest in which one competitor gains an advantage in obtaining access to a limited resource, such as food or a mate. Natural selection has generally favored the evolution of symbolic, exaggerated ritual to resolve conflicts involving both aggressive and submissive behavior.
  - Some animals show dominance hierarchies, with differentially ranked individuals permitted options appropriate to their status in the "pecking order."
  - Territoriality is a behavior in which an animal defends a specific, fixed portion of its home range against intrusion by other animals of the same species through agonistic interactions.
- Mating behavior relates directly to an animal's fitness (pp. 1194–1198, FIGURE 50.21)
  - Courtship interactions are complex, species-specific behaviors that typically include a series of fixed action patterns and releasers, ensuring that the participating individuals are nonthreatening and of the proper species, sex, and physiological condition for mating.
  - Courtship may also involve the selection of a specific mate from an array of potential candidates.
  - Mating relationships, which vary widely among different species, include promiscuity, monogamy, and polygamy.
  - Evolution affects mating relationships, depending partly on parental investment in the next generation, which varies according to the degree of prenatal and postnatal input of the two parents.
- Social interactions depend on diverse modes of communication (pp. 1198–1200, FIGURE 50.22)
  - Animals communicate with one another through their various senses. Odors are particularly effective signals in many species, as demonstrated by the functions of pheromones.
  - Honeybees have a complex symbolic language of dances that scouts use to relay information to other foragers about the location and value of food resources.
- The concept of inclusive fitness can account for most altruistic behavior (pp. 1200–1202, FIGURE 50.23)
  - Altruistic behavior, which benefits a conspecific at the potential expense of the helpful individual, seems to be best explained by kin selection, a theory of inclusive fitness maintaining that genes enhance the survival of copies of themselves by directing organisms to care for others who share those genes.
  - Altruism commonly involves close relatives, although not always. Reciprocal altruism between unrelated individuals may be explained as ultimately advantageous to the altruistic individual in the future when another individual "returns the favor."
- Human sociobiology connects biology to the humanities and social sciences (pp. 1202–1203)
  - E. O. Wilson's *Sociobiology* suggests that human social behavior could be understood in evolutionary terms, igniting a heated and emotional debate in the scientific community.

- Sociobiologists acknowledge human behavioral plasticity, which is manifested in an impressive spectrum of social behaviors that is circumscribed, but not rigidly determined, by our genetic potential.

SELF-QUIZ

1. Bees can see colors we cannot see and detect minute amounts of chemicals we cannot smell. But unlike many insects, bees cannot hear very well. Which of the following statements best fits into the perspective of behavioral ecology?
   a. Bees are too small to have functional ears.
   b. Hearing must not contribute much to a bee's fitness.
   c. If a bee could hear, its tiny brain would be swamped with information.
   d. This is an example of a fixed action pattern.
   e. If bees could hear, the noise of the hive would distract the bees from their work.

2. The nature-versus-nurture controversy centers on
   a. the distinction between proximate and ultimate causes of behavior
   b. the role of genes in learning
   c. whether animals have conscious feelings or thoughts
   d. the extent to which an animal's behavior is innate or learned
   e. the importance of good parental care

3. Which of the following statements is *not* true of fixed action patterns?
   a. They are highly stereotypical, instinctive behaviors.
   b. They seem to lack adaptive significance.
   c. They are triggered by sign stimuli in the environment and, once begun, are continued to completion.
   d. They are often released by one or two simple cues associated with the relevant object or organism.
   e. A supernormal stimulus often produces a stronger response.

4. The return of salmon to their home streams to spawn is an example of
   a. olfactory imprinting
   b. insight
   c. associative learning
   d. operant conditioning
   e. habituation

5. Every morning Patricia turns on the light in the room and then feeds the fish in her aquarium. After a couple of weeks of this routine, she noticed that the fish came to the surface whenever she turned the light on, whether or not any food was present. This most likely illustrates
   a. habituation
   b. positive phototaxis
   c. imprinting
   d. associative learning
   e. observational learning

6. Which of the following is *not* true of agonistic behavior?
   a. It is most common among members of the same species.
   b. It may be used to establish and defend territories.
   c. It often involves symbolic conflict and often does not cause serious harm to either the winner or the loser in the encounter.
   d. It is a uniquely male behavior.
   e. It may be used to establish dominance hierarchies.

7. Which of the following suggests that some animals have internal clocks?
   a. Some birds can sense changes in Earth's magnetic field.
   b. A crab that has moved away from the shore can still sense the tides.
   c. A squirrel kept in darkness drifts away from a 24-hour cycle.
   d. Many animals become active at dawn and settle down at sunset.
   e. Rats kept in constant light show a daily rhythm of activity.

8. The core idea of sociobiology is that
   a. human behavior is rigidly predetermined by inheritance
   b. humans cannot learn to alter their social behavior
   c. many aspects of social behavior have an evolutionary basis
   d. the social behavior of humans is comparable to that of honeybees
   e. environment outweighs genes in human behavior

9. A biology teacher has set up a new aquarium in the lab, and the pump makes a noise that annoys the students. After a few days, the students no longer seem to hear or pay any attention to the noise. This is an example of
   a. conditioning
   b. imprinting
   c. habituation
   d. insight
   e. critical learning

10. A honeybee returning to the hive from a food source performs a waggle dance with the run oriented straight to the left on the vertical surface. This means that the food is located
    a. 90° left of the hive
    b. 90° left of the line from the hive to the sun
    c. in the opposite direction, straight to the right of the hive
    d. above the hive and slightly to the left
    e. very close to the hive

CHALLENGE QUESTIONS

1. European birds called wagtails eat insects. During winter when food is scarce, a wagtail may defend a territory where it captures an average of 20 insects per hour, or it may join a flock that ranges widely over the countryside. A bird in a flock averages 25 insects per hour. Discuss the two strategies in terms of optimal foraging. Why do you think the wagtails do not always travel in flocks?

2. Starting with the very first time a bee leaves the hive, it always flies in a circle around the hive before heading out on a foraging trip. If it is prevented from seeing the hive when it leaves or if the hive is moved while the bee is gone, the bee is not able to locate the hive when it returns. For this reason, beekeepers know that a hive should only be moved at night, when all the bees are inside. What component of this "orientation flight" behavior appears to be innate? What component shows learning?

3. Many species of gulls nest very close together on flat-topped islands, but herring gulls space their nests much farther apart. Kittiwakes nest in tiny pockets on the faces of cliffs. Suggest experiments to determine whether these birds are programmed to imprint on their own eggs or young. Which birds do you think will be found to imprint on their own eggs? Their own young? Why?

4. Explain how a bird's song is both programmed and learned.

5. Write a paragraph that frames the "nature-nurture debate" in your own words.

SCIENCE, TECHNOLOGY, AND SOCIETY

1. Researchers are very interested in studying identical twins who were separated at birth and raised apart. So far, data suggest that twins are much more alike than researchers would have predicted. They have similar personalities, mannerisms, habits, and interests. What kind of general question do you think researchers hope to answer by studying twins that have been raised apart? Why do twins make good subjects for this kind of research? What do the results suggest to you? What are the potential pitfalls of this research? What abuses might occur if the studies are not evaluated critically and the results are carelessly cited in support of a particular social agenda?

2. Chimpanzees, ants, lions, and wolves sometimes attack other groups of their own species and kill individuals, but human beings are the only animals that carry out warfare in an organized way on a mass scale. What does our knowledge of animal behavior tell us about human aggression and violence? Are humans more aggressive than other animals? Do other animals control their aggression in ways that we do not? What role does our technology play in the scale and ferocity of our wars? Is knowledge of animal behavior irrelevant when it comes to the subject of war? Why or why not?

3. Write an essay discussing why the subject of human sociobiology remains so controversial.

FURTHER READING

Alcock, J. *Animal Behavior: An Evolutionary Approach,* 5th ed. Sunderland, MA: Sinaur, 1993.

Buss, D. M. "The Strategies of Human Mating." *American Scientist,* May/June 1994. The perspective of behavioral ecology.

Davies, N., and M. Brooke. "Coevolution of the Cuckoo and Its Hosts." *Scientific American,* January 1991. The ultimate basis of a fascinating behavior.

Gordon, D. M. "The Development of Organization in an Ant Colony." *American Scientist,* January/February 1995. How group complexity can emerge from simple individual behavior.

Gould, J., and C. Gould. *The Animal Mind.* New York: Scientific American Library, 1994. Animal cognition and other current topics.

Grisham, J. "The Genetics of Timekeeping." *BioScience,* January 1995. Using mutations to study the biological clock and circadian rhythms.

Lohman, K. "How Sea Turtles Navigate." *Scientific American,* January 1992. One of the most remarkable homing phenomena in nature.

Nelson, B. "The Gannets of Cape Kidnappers." *Natural History,* January 1992. Different populations of the same species exhibit regional dialects in their ritualized displays.

Pfennig, D. W., and P. W. Sherman. "Kin Recognition." *Scientific American,* June 1995. How do animals identify relatives, and why?

Pool, R. "Putting Game Theory to the Test." *Science,* March 17, 1995. What do games have in common with animal behaviors such as aggression and cooperation?

Rosenthal, E. "The Forgotten Female." *Discover,* December 1991. When females choose mates, they have a powerful influence on evolution.

Wilson, E. O. *Naturalist.* Washington D.C.: Island Press, 1994. The autobiography of one of the twentieth century's most influential biologists.

Wilson, E. O. *Sociobiology: The New Synthesis.* Cambridge, MA: Harvard University Press, 1975. The book that launched a new field of research.

SELF-QUIZ ANSWERS

CHAPTER 2

1. b	6. c
2. a	7. c
3. b	8. a
4. c	9. b
5. c	10. b

CHAPTER 3

1. d	6. c
2. c	7. c
3. b	8. d
4. c	9. c
5. b	10. c

CHAPTER 4

1. b	6. b
2. c	7. b
3. d	8. d
4. d	9. d
5. a	10. b

CHAPTER 5

1. d	6. b
2. c	7. b
3. d	8. b
4. a	9. c
5. b	10. a

CHAPTER 6

1. b	6. a
2. c	7. c
3. b	8. c
4. b	9. d
5. d	10. c

CHAPTER 7

1. b	6. d
2. c	7. b
3. b	8. c
4. d	9. c
5. a	10. e

CHAPTER 8

1. b	6. fructose
2. c	7. glucose
3. a	8. cell contents
4. d	9. into cell
5. b	10. b

CHAPTER 9

1. d	6. a
2. b	7. b
3. c	8. b
4. d	9. b
5. a	10. b

CHAPTER 10

1. d	6. c
2. b	7. a
3. d	8. a
4. b	9. d
5. b	10. b

CHAPTER 11

1. c	6. c
2. b	7. See FIGURE 11.6
3. c	8. d
4. a	9. b
5. c	10. a

CHAPTER 12

1. d	6. d
2. b	7. a
3. d	8. b
4. d	9. c
5. c	10. d

CHAPTER 13

Genetics Problems

1. Incomplete dominance, with heterozygotes being gray in color. Mating a gray rooster with a black hen should yield approximately equal numbers of gray and black offspring.

2. F_1 cross is *AARR* × *aarr*. Genotype of progeny is *AaRr*, phenotype is all axial-pink. F_2 cross is *AaRr* × *AaRr*. Genotypes of progeny are 4 *AaRr* : 2 *AaRR* : 2 *AARr* : 2 *aaRr* : 2 *Aarr* : 1 *AARR* : 1 *aaRR* : 1 *AArr* : 1 *aarr*. Phenotypes of progeny are 6 axial-pink : 3 axial-red : 3 axial-white : 2 terminal-pink : 1 terminal-white : 1 terminal-red.

3. a. $^1/_{64}$ b. $^1/_{64}$ c. $^1/_8$ d. $^1/_{32}$

4. Albino is a recessive trait; black is dominant.

Parents	Gametes	Offspring
BB × *bb*	*B* and *b*	*All Bb*
bb × *Bb*	*b* and $^1/_2$*B*, $^1/_2$*b*	$^1/_2$*Bb*, $^1/_2$*bb*

5. a. *PPLl* × *PPLl*, *PpLl*, or *ppLl*
b. *ppLl* × *ppLl*
c. *PPLL* × any of the 9 possible genotypes
d. *PpLl* × *Ppll*
e. *PpLl* × *PpLl*

6. Man I^Ai; woman I^Bi; child *ii*. Other genotypes for children are $^1/_4$, I^AI^B, $^1/_4$ I^Ai, $^1/_4$ I^BI.

7. Four

8. a. $^3/_4 \times {}^3/_4 \times {}^3/_4 = {}^{27}/_{64}$
b. $1 - {}^{27}/_{64} = {}^{37}/_{64}$
c. $^1/_4 \times {}^1/_4 \times {}^1/_4 = {}^1/_{64}$
d. $1 - {}^1/_{64} = {}^{63}/_{64}$

9. a. $^1/_{256}$ b. $^1/_{16}$ c. $^1/_{256}$ d. $^1/_{64}$ e. $^1/_{128}$

10. If the "curl" allele is dominant, then the original mutant crossed with noncurl cats will produce both curl and noncurl offspring. If the mutation is recessive, then only curl offspring will result from curl × curl matings. You know that cats are true-breeding when curl × curl matings produce only curl offspring. A pure-bred curl cat is homozygous (as it turns out, for the dominant allele, which causes the curled ears).

11. a. 1 b. $^1/_{32}$ c. $^1/_8$ d. $^1/_2$

12. $^1/_9$ **13.** $^1/_{16}$

14. Twenty-five percent will be cross-eyed; all of the cross-eyed offspring will also be white.

15. The dominant allele *I* is epistatic to the *p* locus, and thus the F_1 generation will be:
9 *I_P_* : colorless
3 *I_pp* : colorless
3 *iiP_* : purple
1 *iipp* : red
Overall, 12 colorless : 3 purple : 1 red.

16. Recessive; George = *Aa*, Arlene = *aa*, Sandra = *AA* or *Aa*, Tom = *aa*, Sam = *Aa*, Wilma = *aa*, Ann = *Aa*, Michael = *Aa*, Daniel = *Aa*, Alan = *Aa*, Tina = *AA* or *Aa*, Carla = *aa*, Christopher = *AA* or *Aa*

17. $^1/_2$ **18.** $^1/_6$

19. 9 *B_A_* : agouti
3 *B_aa* : black
3 *bbA_* : white
1 *bbaa* : white
Overall, 9 agouti : 3 black : 4 white.

20. The gene is probably dominant if the condition is caused by a rare allele. Individual 1 is therefore heterozygous, and individuals 2 and 3 are homozygous recessive. (If the disorder were caused by a common allele, this pedigree is also consistent with recessive inheritance.)

CHAPTER 14

Genetics Problems

1. 0; $^1/_2$, $^1/_{16}$

2. Recessive. If the disorder were dominant, it would affect at least one parent of a child born with the disorder. For a girl to have the disorder, she would have to inherit recessive alleles from *both* parents. This would be very rare, especially since males with the allele die in their early teens.

3. $^1/_4$ for each daughter ($^1/_2$ chance that child will be female × $^1/_2$ chance of a homozygous recessive genotype); $^1/_2$ for first son.

4. 17%

5. 6%. Wild type (heterozygous for normal wings and red eyes) × recessive homozygote with vestigial wings and purple eyes.

6. Between *T* and *A*, 12%; between *A* and S, 5%.

7. Between *T* and *S*, 18%. Sequence of genes is *T-A-S*.

8. Both children would be blind; the son's children would have numbness, and the daughter's children would be blind.

9. The disorder would always be inherited from the mother.

10. *XXX*

11. *D–A–B–C*

12. In meiosis, the combined 14-21 chromosome will behave as one chromosome. If a gamete receives the combined 14-21 chromosome and a normal copy of chromosome 21, trisomy 21 will result when this gamete combines with a normal gamete.

13. Fifty percent of the offspring would show phenotypes that resulted from crossovers. These results would be the same as those from a cross where *A* and *B* were not linked. Further crosses involving other genes on the same chromosome would reveal the linkage and map distances.

14. One hypothesis is that a translocation has moved one of the genes to a different chromosome.

CHAPTER 15

1. c	6. a
2. d	7. d
3. b	8. d
4. c	9. d
5. b	10. b

CHAPTER 16

1. a	6. a
2. b	7. c
3. d	8. d
4. d	9. e
5. a	10. b

CHAPTER 17

1. e	6. d
2. d	7. a
3. b	8. c
4. d	9. a
5. e	10. e

CHAPTER 18

1. c	6. a
2. a	7. e
3. d	8. c
4. a	9. b
5. e	10. b

CHAPTER 19

1. b	6. c
2. b	7. e
3. c	8. c
4. b	9. d
5. e	10. b

CHAPTER 20

1. a	6. b
2. b	7. b
3. c	8. c
4. d	9. c
5. c	10. d

CHAPTER 21

1. b	6. a
2. d	7. c
3. c	8. b
4. a	9. e
5. c	10. b

CHAPTER 22

1. b	6. c
2. b	7. e
3. b	8. d
4. a	9. d
5. d	10. b

CHAPTER 23

1. d	6. b
2. b	7. a
3. a	8. d
4. e	9. c
5. c	10. b

CHAPTER 24

1. b	6. b
2. e	7. a
3. c	8. b
4. c	9. d
5. c	10. c

CHAPTER 25

1. b	6. a
2. a	7. e
3. d	8. c
4. c	9. a
5. b	10. c

CHAPTER 26

1. b	6. c
2. c	7. b
3. e	8. c
4. a	9. b
5. d	10. b

CHAPTER 27

1. c	6. d
2. a	7. a
3. a	8. b
4. b	9. a
5. a	10. a

CHAPTER 28

1. b	6. d
2. b	7. a
3. c	8. e
4. d	9. b
5. c	10. a

CHAPTER 29

1. c	6. a
2. e	7. d
3. d	8. a
4. c	9. e
5. c	10. b

CHAPTER 30

1. c	6. a
2. c	7. b
3. c	8. e
4. d	9. c
5. c	10. a

CHAPTER 31

1. d	6. a
2. a	7. c
3. c	8. d
4. b	9. a
5. b	10. d

CHAPTER 32

1. e	6. c
2. a	7. b
3. c	8. c
4. d	9. a
5. a	10. b

CHAPTER 33

1. b	6. c
2. b	7. b
3. e	8. b
4. c	9. b
5. d	10. b

CHAPTER 34

1. d	6. a
2. a	7. c
3. c	8. e
4. a	9. b
5. b	10. c

CHAPTER 35

1. b	6. a
2. a	7. b
3. b	8. b
4. d	9. c
5. b	10. a

CHAPTER 36

1. b	6. e
2. c	7. b
3. c	8. b
4. d	9. c
5. d	10. c

CHAPTER 37

1. b	6. d
2. b	7. c
3. c	8. d
4. c	9. b
5. d	10. c

CHAPTER 38

1. c	6. c
2. c	7. b
3. d	8. a
4. c	9. a
5. b	10. c

CHAPTER 39

1. d	6. c
2. b	7. b
3. d	8. b
4. c	9. b
5. a	10. c

CHAPTER 40

1. d	6. d
2. a	7. b
3. e	8. d
4. a	9. c
5. b	10. a

CHAPTER 41

1. c	6. b
2. a	7. e
3. e	8. c
4. d	9. c
5. c	10. c

CHAPTER 42

1. d	6. d
2. b	7. a
3. a	8. c
4. b	9. d
5. c	10. b

CHAPTER 43

1. a	6. c
2. b	7. c
3. d	8. a
4. c	9. d
5. a	10. d

CHAPTER 44

1. c	6. a
2. b	7. c
3. a	8. c
4. d	9. c
5. d	10. b

CHAPTER 45

1. e	6. e
2. d	7. b
3. a	8. d
4. b	9. c
5. c	10. b

CHAPTER 46

1. c	6. c
2. e	7. a
3. e	8. c
4. b	9. d
5. a	10. a

CHAPTER 47

1. c	6. b
2. d	7. c
3. c	8. d
4. d	9. a
5. c	10. c

CHAPTER 48

1. c	6. d
2. d	7. b
3. b	8. b
4. d	9. d
5. e	10. c

CHAPTER 49

1. c	6. a
2. e	7. b
3. b	8. c
4. e	9. c
5. d	10. c

CHAPTER 50

1. b	6. d
2. d	7. e
3. b	8. c
4. a	9. c
5. d	10. b

THE METRIC SYSTEM

MEASUREMENT	UNIT AND ABBREVIATION	METRIC EQUIVALENT	METRIC-TO-ENGLISH CONVERSION FACTOR	ENGLISH-TO-METRIC CONVERSION FACTOR
Length	1 kilometer (km)	= 1000 (10^3) meters	1 km = 0.62 mile	1 mile = 1.61 km
	1 meter (m)	= 100 (10^2) centimeters = 1000 millimeters	1 m = 1.09 yards 1 m = 3.28 feet 1 m = 39.37 inches	1 yard = 0.914 m 1 foot = 0.305 m
	1 centimeter (cm)	= 0.01 (10^{-2}) meter	1 cm = 0.394 inch	1 foot = 30.5 cm 1 inch = 2.54 cm
	1 millimeter (mm)	= 0.001 (10^{-3}) meter	1 mm = 0.039 inch	
	1 micrometer (μm) (formerly micron, μ)	= 10^{-6} meter (10^{-3}mm)		
	1 nanometer (nm) (formerly millimicron, mμ)	= 10^{-9} meter (10^{-3}μm)		
	1 angstrom (Å)	= 10^{-10} meter (10^{-4}μm)		
Area	1 hectare (ha)	= 10,000 square meters	1 ha = 2.47 acres	1 acre = 0.0405 ha
	1 square meter (m^2)	= 10,000 square centimeters	1 m^2 = 1.196 square yards 1 m^2 = 10.764 square feet	1 square yard = 0.8361 m^2 1 square foot = 0.0929 m^2
	1 square centimeter (cm^2)	= 100 square millimeters	1 cm^2 = 0.155 square inch	1 square inch = 6.4516 cm^2
Mass	1 metric ton (t)	= 1000 kilograms	1 t = 1.103 tons	1 ton = 0.907 t
	1 kilogram (kg)	= 1000 grams	1 kg = 2.205 pounds	1 pound = 0.4536 kg
	1 gram (g)	= 1000 milligrams	1 g = 0.0353 ounce 1 g = 15.432 grains	1 ounce = 28.35 g
	1 milligram (mg)	= 10^{-3} gram	1 mg = approx. 0.015 grain	
	1 microgram (μg)	= 10^{-6} gram		
Volume (solids)	1 cubic meter (m^3)	= 1,000,000 cubic centimeters	1m^3 = 1.308 cubic yards 1 m^3 = 35.315 cubic feet	1 cubic yard = 0.7646 m^3 1 cubic foot = 0.0283 m^3
	1 cubic centimeter (cm^3 or cc)	= 10^{-6} cubic meter	1 cm^3 = 0.061 cubic inch	1 cubic inch = 16.387 cm^3
	1 cubic millimeter (mm^3)	= 10^{-9} cubic meter (10^{-3} cubic centimeter)		
Volume (Liquids and Gases)	1 kiloliter (kl or kL)	= 1000 liters	1 kL = 264.17 gallons	1 gallon = 3.785 L
	1 liter (L)	= 1000 milliliters	1 L = 0.264 gallons 1 L = 1.057 quarts	1 quart = 0.946 L
	1 milliliter (mL)	= 10^{-3} liter = 1 cubic centimeter	1 mL = 0.034 fluid ounce 1 mL = approx. $\frac{1}{4}$ teaspoon 1 ml = approx. 15–16 drops (gtt.)	1 quart = 946 mL 1 pint = 473 mL 1 fluid ounce = 29.57 mL 1 teaspoon = approx. 5 mL
	1 microliter (μl or μL)	= 10^{-6} liter (10^{-3} milliliters)		
Time	1 second (s)	= $\frac{1}{60}$ minute		
	1 millisecond (ms)	= 10^{-3} second		
Temperature	Degrees Celsius (°C) (Absolute zero, when all molecular motion ceases, is –273 °C. The Kelvin (K) scale, which has the same size degrees as Celsius, has its zero point at absolute zero. Thus, 0° K = −273°C.)		°F = $\frac{9}{5}$°C + 32	°C = $\frac{5}{9}$(°F – 32)

CLASSIFICATION OF LIFE

This appendix presents the taxonomic classification used for the major groups of organisms discussed in this text; not all phyla are included. Plant and fungal divisions are the taxonomic equivalents of phyla. The classification reviewed here is based on the five-kingdom system; the rationale for alternative taxonomic systems is discussed in Unit Five of the text.

Kingdom Monera

(In some alternatives to the five-kingdom system, the archaebacteria and eubacteria are divided into two kingdoms and into two superkingdom taxa, the domain Archaea and the domain Bacteria—see Chapter 25.)

- Archaebacteria
  - Methanogens
  - Extreme halophiles
  - Thermoacidophiles
- Eubacteria
  - Proteobacteria
    - Purple bacteria
    - Chemoautotrophic proteobacteria
    - Chemoheterotrophic proteobacteria
  - Gram-positive eubacteria
  - Cyanobacteria
  - Spirochetes
  - Chlamydias

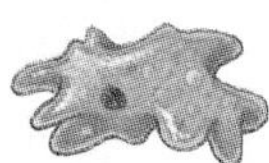

Kingdom Protista

(Some classification schemes split these organisms into two or more kingdoms—see Chapter 26.)

- Phylum Diplomonada (placed in separate kingdom, Archezoa, in some classification systems)
- Phylum Rhizopoda (amoebas)
- Phylum Actinopoda (heliozoans, radiolarians)
- Phylum Foraminifera (forams)
- Phylum Apicomplexa (apicomplexans)
- Phylum Zoomastigophora (zooflagellates)
- Phylum Ciliophora (ciliates)
- Phylum Myxomycota (plasmodial slime molds)
- Phylum Acrasiomycota (cellular slime molds)
- Phylum Oomycota (water molds)
- Phylum Euglenophyta (euglenoids)
- Phylum Dinoflagellata (dinoflagellates)
- Phylum Chrysophyta (golden algae)
- Phylum Bacillariophyta (diatoms)
- Phylum Phaeophyta (brown algae; grouped along with golden algae and diatoms in a separate kingdom, Chromista, in some classification systems)
- Phylum Rhodophyta (red algae)
- Phylum Chlorophyta (green algae)

Kingdom Plantae

- Division Bryophyta (mosses)
- Division Hepatophyta (liverworts)
- Division Anthocerophyta (hornworts)
- Division Psilophyta (whisk ferns)
- Division Lycophyta (club mosses)
- Division Sphenophyta (horsetails)
- Division Pterophyta (ferns)
- Division Coniferophyta (conifers)
- Division Cycadophyta (cycads)
- Division Ginkgophyta (ginkgos)
- Division Gnetophyta (gnetae)
- Division Anthophyta (flowering plants)
  - Class Monocotyledones (monocots)
  - Class Dicotyledones (dicots)

Kingdom Fungi

Division Chytridiomycota (chytrids)
Division Zygomycota (zygomycetes)
Division Ascomycota (sac fungi)
Division Basidiomycota (club fungi)
Division Deuteromycota (imperfect fungi)
Lichens (symbiotic associations of algae and fungi)

Kingdom Animalia

Phylum Porifera (sponges)
Phylum Cnidaria
- Class Hydrozoa (hydrozoans)
- Class Scyphozoa (jellyfishes)
- Class Anthozoa (sea anemones and coral animals)

Phylum Ctenophora (comb jellies)
Phylum Platyhelminthes (flatworms)
- Class Turbellaria (free-living flatworms)
- Class Trematoda (flukes)
- Class Monogenea (flukes)
- Class Cestoda (tapeworms)

Phylum Nemertea (proboscis worms)
Phylum Rotifera (rotifers)
Phylum Nematoda (roundworms)
Phylum Mollusca (mollusks)
- Class Polyplacophora (chitons)
- Class Gastropoda (gastropods: snails and their relatives)
- Class Bivalvia (bivalves)
- Class Cephalopoda (cephalopods; squids and octopuses)

Phylum Onychophora
Phylum Annelida (segmented worms)
- Class Oligochaeta (oligochaetes)
- Class Polychaeta (polychaetes)
- Class Hirudinea (leeches)

Phylum Arthropoda (arthropods)
- Subphylum Trilobitomorpha (trilobites)
- Subphylum Cheliceriformes (or Chelicerata, chelicerates)
  - Class Arachnida (arachnids: spiders, ticks, scorpions)
- Subphylum Uniramia (uniramians)
  - Class Diplopoda (millipedes)
  - Class Chilopoda (centipedes)
  - Class Insecta (insects)
- Subphylum Crustacea (crustaceans)

Phylum Phoronida (phoronids)
Phylum Bryozoa (bryozoans)
Phylum Brachiopoda (brachiopods: lamp shells)
Phylum Echinodermata (echinoderms)
- Class Asteroidea (sea stars)
- Class Ophiuroidea (brittle stars)
- Class Echinoidea (sea urchins and sand dollars)
- Class Crinoidea (sea lilies)
- Class Concentricycloidea (sea daisies)
- Class Holothuroidea (sea cucumbers)

Phylum Chordata (chordates)
- Subphylum Urochordata (urochordates: tunicates)
- Subphylum Cephalochordata (cephalochordates: lancelets)
- Subphylum Vertebrata (vertebrates)
  - Class Agnatha (jawless vertebrates)
  - Class Placodermi (extinct jawed fishes)
  - Class Chondrichthyes (cartilaginous fishes)
  - Class Osteichthyes (bony fishes)
  - Class Amphibia (amphibians)
  - Class Reptilia (reptiles)
  - Class Aves (birds)
  - Class Mammalia (mammals)

A COMPARISON OF THE LIGHT MICROSCOPE AND THE ELECTRON MICROSCOPE

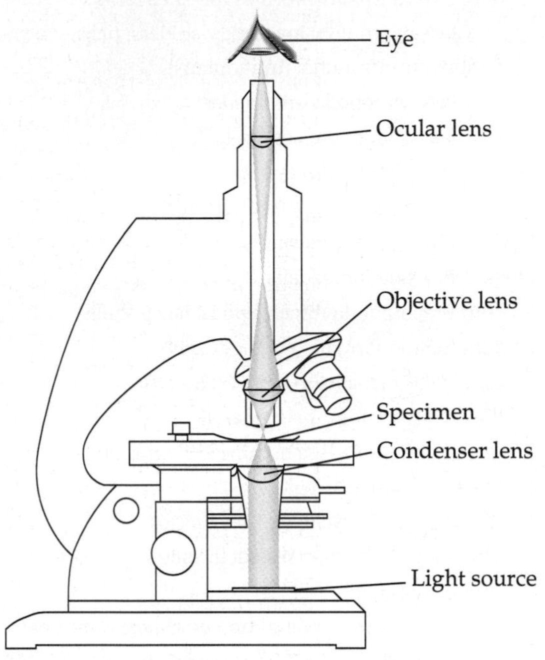

(a) LIGHT MICROSCOPE

In light microscopy, light is focused on a specimen by a glass condenser lens; the image is then magnified by an objective lens and an ocular lens, for projection on the eye or on photographic film.

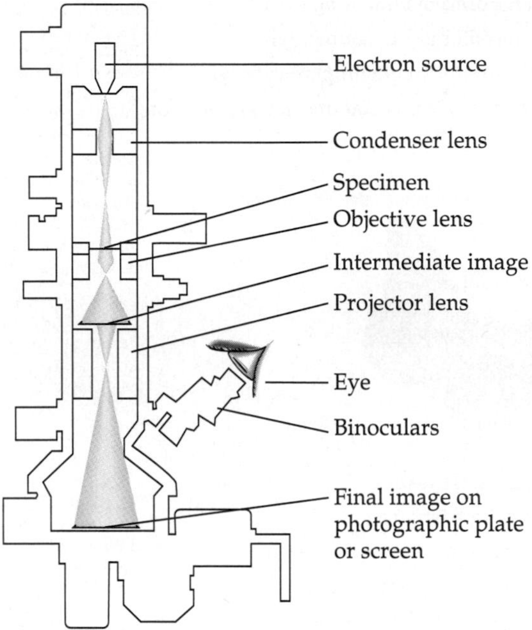

(b) ELECTRON MICROSCOPE

In electron microscopy, a beam of electrons (top of the microscope) is used instead of light, and electromagnets are used instead of glass lenses. The electron beam is focused on the specimen by a condenser lens; the image is magnified by an objective lens and a projector lens, for projection on a screen or on photographic film.

CREDITS

PHOTOGRAPHS

Title page: ©John Sexton 1977.

Front Matter *The Campbell Interviews:* Unit One: ©John Bohn. Unit Two: Mary DeChirico/Benjamin/Cummings. Unit Three: Kevin Keister/Benjamin/Cummings. Unit Four: Stuart Frawley/Benjamin/Cummings. Unit Five: ©John Bohn. Unit Six: Julian Kingma/Benjamin/-Cummings. Unit Seven: ©Rocky Thies. Unit Eight: Tom Foley/Benjamin/Cummings. *Table of Contents***:** Unit One ©Ken Eward/Biografx/Photo Researchers, Inc. Unit Two: ©A.B. Dowsett/SPL/ Photo Researchers, Inc. Unit Three: ©Omikron/ Science Source/Photo Researchers, Inc. Unit Four: ©Sinclair Stammers/SPL/Photo Researchers, Inc. Unit Five: ©Mark Moffett/Minden Pictures. Unit Six: ©Jim Brandenburg/Minden Pictures. Unit Seven: ©Michio Hoshino/Minden Pictures. Unit Eight: ©Dwight Kuhn.

Chapter 1 Opener: ©John Sexton 1977. 1.1a: Mary DeChirico/Benjamin/Cummings. 1.1b: ©Richard Howard. 1.1c: ©Mark Moffett/Minden Pictures. 1.1 d Courtesy of Margaret Davis, Univ. of Minnesota. 1.2a: ©Robert Fletterick. 1.2b: ©Dr. Jeremy Burgess/SPL/Photo Researchers, Inc. 1.2c: ©Manfred Kage/Peter Arnold, Inc. 1.2d: ©Dr. Jeremy Burgess/SPL/Photo Researchers, Inc. 1.2e: ©John Shaw/Tom Stack & Associates. 1.2f: ©Ric Ergenbright Photography. 1.3a: ©Bob Stovall/ Bruce Coleman, Inc. 1.3b: ©Michio Hoshino/ Minden Pictures. 1.3c: ©Michael Fogden/Bruce Coleman, Inc. 1.3d: ©Wolfgang Bayer/Bruce Coleman, Inc. 1.3e: ©Jeff Lepore/Photo Researchers, Inc. 1.3f: ©Gerald C. Kelley, 1986/Photo Researchers, Inc. 1.3g: ©Breck P. Kent/Animals Animals. 1.4 ©CNRI/SPL/Photo Researchers, Inc. 1.5 N.L. Max, University of California/Biological Photo Service. 1.6a: Courtesy of Ann Keller, "A Feeling for the Organism". 1.7a: ©Frans Lanting/Minden Pictures. 1.7b: ©Janice Sheldon. 1.7c: ©Don Fawcett/ Science Source/ Photo Researchers, Inc. 1.7d: Courtesy of N. Simionescu. 1.8 2 ©Ric Ergenbright Photography. 1.10a: 1 ©Ralph Robinson/Visuals Unlimited. 1.10a: 2 ©A.B. Dowsett/Science Source/ Photo Researchers 1.10b: ©D.P. Wilson/Photo Researchers, Inc. 1.10c: ©Cristina Taccone. 1.10d: ©Michael Fogden/Earth Scenes/Bruce Coleman, Inc. 1.10e: ©Gunter Ziesler/Peter Arnold, Inc. 1.11a: ©Manfred Kage/Peter Arnold, Inc. 1.11b: ©Manfred Kage/Peter Arnold, Inc. 1.11c: ©Omikron/Photo Researchers, Inc. 1.12: ©Chip Clark 1.14: Rudi Kuiter/©Discover Magazine. 1.15: David Reznick. 1.18a: Courtesy of Genentech. 1.18b: ©Peter Menzel.

Chapter 2 Opener: ©John Bennett. 2.2; left: ©Dr. Ed Degginger/Color Pic 2.2: middle, right; Stephen Frisch/Benjamin/Cummings. 2.3: ©Grant Heilman. 2.4: ©Ivan Polunin/Bruce Coleman, Inc. 2.6: Courtesy of M.E. Raichle. 2.6: Courtesy of Dr. Zhores Medvedev. 2.13: Stephen Frisch/Benjamin/ Cummings. 2.17b: Annalisa Kraft/Benjamin/ Cummings. 2.18: ©Runk/Schoenberger/Grant Heilman. Methods Box ©Terraphotographics/ Biological Photo Service. ; From M.C. Ratazzi et al., *Am J Human Genet* 28:143-154, 1976. Challenge Question: ©Phil Degginger.

Chapter 3 Opener: W. Reid Thompson, Cornell University. 3.2: ©Dr. E.R. Degginger/Color-Pic. 3.3: ©Stephen Dalton/NHPA 3.4: ©Mitchell Layton/Duomo. 3.6: ©Flip Nicklin 1990. 3.10: ©1991 Maresa Pryor/Earth Scenes. Challenge Question: ©G.I. Bernard/Animals Animals.

Chapter 4 Opener: Richard Wagner, UCSF/ Benjamin/Cummings. 4.1: ©Roger Ressmeyer/ Starlight 4.5: ©Manfred Kage/Peter Arnold, Inc. 4.8 left: ©W.J. Weber/Visuals Unlimited 4.8 right: ©Stephen J. Krasemann/Photo Researchers, Inc.

Chapter 5 Opener: ©Martin Shields 5.1a: Courtesy of Linus Pauling Institute. 5.1b: ©Louise Lockley/SCIRO/Science Photo Library/Photo Researchers, Inc. 5.6a: ©Dr. Jeremy Burgess/Photo Researchers,Inc. 5.6b: Courtesy of H. Shio and P.B. Lazerow. 5.8 ©J. Litray/Visuals Unlimited. 5.9a: ©F. Collet/Photo Researchers, Inc. 5.9b: ©George Disario/The Stock Market. 5.11a: ©Lara Hartley. 5.11b: Courtesy American Dairy Association. 5.17a: Courtesy of the Graphics Systems Research Group, IBM U.K. Scientific Centre. 5.19a: M. Murayama, Murayama Research Laboratory/Biological Photo Service. 5.21 left: ©Martin Shields 5.21 right: Vollrath & Edmunds, *Nature* 340:35-317. Methods Box Courtesy University of Califonia, Riverside.

Chapter 6 Opener: ©Ken Lucas/Biological Photo Service. 6.2 ©Eunice Harris 1987/Photo Researchers, Inc. 6.3a: ©Manfred Kage/Peter Arnold, Inc. 6.3b: ©John Cancalosi/Tom Stack & Associates. 6.11: Courtesy of Thomas Steitz. 6.17: ©R. Rodewald, University of Virginia/Biological Photo Service.

Chapter 7 Opener: ©M. Schliwa/Visuals Unlimited. 7.2: Courtesy of William L. Dentler, University of Kansas/Biological Photo Service. 7.4b: ©S.C. Holt, Univ. of Texas Health Center/ Biological Photo Service. 7.6a: Courtesy J. David Robertson. 7.9b: L. Orci and A. Perrelet, *Freeze-Etch Histology.* Heidelberg: Springer-Verlag, 1975. 7.9d: A.C. Faberge, *Cell Tiss. Res.* 151 1974:43. Springer-Verlag, 1974. 7.9e: U. Aebi. 7.10: ©D.W. Fawcett/Photo Researchers, Inc. 7.11: ©R. Bolender, D. Fawcett/Photo Researchers, Inc. 7.12: G.T. Cole, University of Texas, Austin/Biological Photo Service. 7.13a: ©R. Rodewald, University of Virginia/Biological Photo Service. 7.13b: Daniel S. Friend, Harvard Medical School. 7.15: Courtesy of E.H. Newcomb. 7.17: ©S.E. Frederick and E.H. Newcomb, *J. Cell Biol.* 43 1969 :343. Reproduced by copyright permission of The Rockefeller University Press. Provided by E.H. Newcomb. 7.18: Daniel S. Friend, Harvard Medical School. 7.19: Courtesy of WP Wergin and E.H. Newcomb, University of Wisconsin, Madison/Biological Photo Service. 7.20: Dr.John E.Heuser, Washington Univ. School of Medicine, St. Louis, Mo. 7.22: Courtesy of Kent McDonald. 7.23a: ©Richard Kessel/Visuals Unlimited. 7.23b: Courtesy D.D. Kunkel, University of Washington/Biological Photo Service. 7.24a: ©OMIKRON/Science Source/Photo Researchers, Inc. 7.24b,c: Courtesy W. L. Dentler, University of Kansas/Biological Photo Service. 7.26a: John Heuser 7.26b: *J. Cell Biol.* 94 1982:425, Hirokawa Nobutaka , Reproduced by copyright permission of Rockefeller Univ. Press. 7.28: ©G.F. Leedale/Photo Researchers, Inc. 7.30a: ©Douglas J. Kelly, *J. Cell Biol.* 28 1966 :51. Reproduced by copyright permission of the Rockefeller University Press. 7.30b: L. Orci and A. Perrelet, Freeze-Etch Histology, Heidelberg: Springer-Verlag, 1975. 7.30c: C. Peracchia and A.F. Dulhunty, *J. Cell Biol.* 70 1976:419. Reproduced by copyright permission of the Rockefeller University Press. 7.31: ©Boehringer Ingelheim International GmbH, photo Lennart Nilsson, *The Body Victorious,* Delacorte Press, Dell Publishing Co., Inc. Table 7.1ul: ©Biophoto Associates/Photo Researchers, Inc.; ur: ©Ed Reschke middle, ll: ©David M. Phillips/Visuals Unlimited.; lr: Courtesy of Noran Instruments. 7.2 left: Courtesy of Dr. Mary Osborn, Max Planck Institute.; middle: Courtesy of Drs. Frank Solomon and J. Dinsmore, Massachusetts Institute of Technology. ; right: Mark S. Ladinsky and J. Richard McIntosh, University of Colorado.

Chapter 8 8.11 ©Cabisco/Visuals Unlimited. 8.17a: R.N. Band and H.S. Pankratz, Michigan State University/Biological Photo Service. 8.17b: ©D.W. Fawcett/Photo Researchers, Inc. 8.17c: 1 M.M. Perry and A.B. Gilbert, *J. Cell Sci.* 39 1979 257. Copyright 1979 by The Company of Biologists Ltd. 8.17c: M.M. Perry and A.B. Gilbert, *J. Cell Sci.* 39 1979 257. Copyright 1979 by The Company of Biologists Ltd.

Chapter 9 Opener: ©Nuridsany & Perennou/ Photo Researchers, Inc. 9.17a: Egyptian Expedition of The Metropolitan Museum of Art, Rogers Fund, 1915 9.17b: Obester Winery, Half Moon Bay, CA. Photo by Kevin Schafer. Self-Testing: Courtesy of John A. Richardson, Southern Illinois University, Carbondale.

Chapter 10 Opener: ©David Muench. 10.1a: ©Renee Lynn/Photo Researchers, Inc. 10.1b: ©Bob Evans/Peter Arnold, Inc. 10.1c: ©Dwight Kuhn. 10.1 d J.R. Waaland,University of Washington/BPS 10.1 e ©Paul Johnson /Biological Photo Service. 10.2 2 ©M. Eichelberger/Visuals Unlimited 10.2 3 Courtesy of W.P. Wergin and E.H. Newcomb, University of Wisconsin/Biological Photo Service. 10.9b: ©Christine L. Case, Skyline College 10.18 left: ©C.F. Miescke/Biological Photo Service. 10.18 right: ©Phil Degginger.

Chapter 11 Opener: Courtesy of J.M. Murray, Department of Anatomy, University of Pennsylvania. 11.1a: ©Biophoto Associates/Photo Researchers, Inc. 11.1b: C.R. Wyttenbach, Univ. of Kansas/Biological Photo Service. 11.1c: ©Biophoto/Science Source/Photo Researchers, Inc. 11.3 Courtesy of J.M. Murray, Department of Anatomy, University of Pennsylvania. 11.4 2 ©Biophoto/Photo Researchers, Inc. 11.6 Ed Reschke 11.7b: 1 Courtesy of Richard Macintosh. 11.7b: 2 © Dr. Matthew Schibler, *Protoplasma* 137 1987 :29-44. 11.9a: ©David M. Phillips/Visuals Unlimited 11.9b: 1Micrograph by B.A. Palevitz. Courtesy of E.H. Newcomb, University of Wisconsin. 11.10 J.R. Waaland, University of Washington. Self Quiz Carolina Biological Supply Methods Box Courtesy of Dr. Gunter Albrecht-Buehler, Northwestern University.

Chapter 12 Opener: ©Jom Smeal/Galella Ltd. 12.1 ©Roland Birke/OKAPIA/Photo Researchers, Inc. 12.2 1 Courtesy of The Inouye and Yaneshiro Families 12.5a: ©Runk/Schoenberger/Grant Heilman, Inc. 12.5b: ©Stan Elems/Visuals Unlimited. Methods Box ©SIU/Visuals Unlimited, ©CNRI/SPL/Photo Researchers, Inc.

Chapter 13 Opener: The Bettmann Archive 13.8 Science Education Resources Pty. Ltd., Victoria, Australia. 13.14 1 ©Stan Goldblatt/Photo Researchers, Inc. 13.14 2 ©Gilbert Grant/Photo Researchers, Inc. 13.15 2 ©Jerry Koontz/The Picture Cube. 13.15 3 ©Elaine Sulle/The Image Bank 13.15 4,5 Anthony Loveday/Benjamin/ Cummings. 13.16 ©Bill Longcore/Photo Researchers, Inc. 13.17 Courtesy of Dr. Nancy Wexler, Columbia University. Self-quiz Breeder/owner: Patricia Speciale; photographer: Norma JubinVille.

Chapter 14 Opener: From Peter Lichter and David Ward*Science* 247 5 January 1990: cover. 14.2 ©Darwin Dale 14.10 ©Ron Kimball 14.13a: ©L. Willatt, East Anglian Regional Genetics

Svc./SPL/Photo Researchers, Inc. 14.13b: ©Science Photo Library/Photo Researchers, Inc. 14.15 ©John D. Cunningham/Visuals Unlimited.

Chapter 15 Opener: From J.D. Watson, The Double Helix, Atheneum, New York, 1968 p. 215 ©1968 by J.D. Watson. Courtesy of Cold Spring Harbor Laboratory Library Archives. 15.2a: ©Lee D. Simon/Photo Researchers, Inc. 15.4a: Cold Spring Harbor Archives 15.4b: From J.D. Watson, The Double Helix, Atheneum, New York, 1968 p. 215 ©1968 by J.D. Watson. 15.9b: From D. J. Burks and P. J. Stambrook, *J. Cell Biol.* 77 1978 : 762. Reproduced by permission of The Rockefeller University Press. Photos provided by P. J. Stambrook. 15.16 ©Conly L. Rieder/Biological Photo Service.

Chapter 16 6.6 Keith V. Wood 16.16b: B. Hamkalo and O.L. Miller, Jr. 16.18 From O. L. Miller , Jr., B. A. Hamkalo, and C. A Thomas, Jr., *Science* 169 1970 :392. Copyright ©1970 by the AAAS.

Chapter 17 17.2a,b: Robley C. Williams, Univ. of California, Berkeley/Biological Photo Service. 17.2c: John J. Cardamone, Jr., University of Pittsburgh/BPS 17.2 d R.C. Williams, University of California, Berkeley/Biological Photo Service. 17.7 ©C. Dauguet/Institute Pasteur/Photo Researchers, Inc. 17.8a: 1 ©Sherman Thomson/Visuals Unlimited. 17.8a: 2 ©Norm Thomas/Photo Researchers, Inc. 17.8b: 3 Photo provided by N. Obalka, M. Yeager, R. Beachy, and C. Fauquet, The Scripps Research Institute and ILTAB-ORSTOM. 17.11 ©Dennis Kunkel/Phototake NYC.

Chapter 18 Opener: Photo courtesy of M.B. Roth and J. Gall. 18.1(top to bottom, 1-6) : S.C. Holt, University of Texas Health Science Center, San Antonio/Biological Photo Service. A.L. Olins, University of Tennessee/Biological Photo Service. Courtesy of Barbara Hamkalo. Courtesy of J.R. Paulsen and U.K. Laemmli, *Cell.* G. F. Bahr, Armed Forces Institute of Pathology. 18.2: Courtesy of O.L. Miller Jr., Dept. of Biology, University of Virginia. 18.7 Courtesy of J.M. Amabis, Dept. of Biology, University of Sao Paulo. 18.9a: ©Evelyne Cudel-Epperson, University of Calif. Riverside 1995.

Chapter 19 Opener: ©Remi Benali /Gamma Liaison. Methods Box: ©Geoff Tompkinson/SPL/Photo Researchers, Inc.

Chapter 20 Opener: ©William E. Ferguson. 20.1:©David Schwimmer/Bruce Coleman, Inc. 20.2: ©Manfred Gottschalk/Tom Stack & Associates. 20.3b: ©Frans Lanting/Minden Pictures. 20.4a: ©Tui de Roy /Bruce Coleman, Inc. 20.4b: ©Tui de Roy /Bruce Coleman, Inc. 20.4c: ©Tui De Roy/Bruce Coleman, Inc. 20.5a: ©E.S. Ross, California Academy of Sciences. 20.5b: ©K.G. Preston Mafham/Animals Animals. 20.5c: ©P. and W. Ward/Animals Animals. 20.6a: ©Jack Wilburn/Earth Scenes/Animals Animals. 20.6b: ©Michael P. Gadomski/Photo Researchers, Inc. 20.6c-e: ©Dwight R. Kuhn. 20.6f: ©David Cavagnaro/Peter Arnold, Inc. 20.6g: ©Clyde H. Smith/Peter Arnold, Inc. 20.8: Michael Singer and Camille Parmesan. 20.9: Philip Gingerich © 1991 Discover Magazine. 20.11: ©Lennart Nilsson, *A Child Is Born,* Dell Publishing. 1990.

Chapter 21 Opener: ©E.S. Ross/Calif. Academy of Sciences. 21.1: ©J. Antonovics/Visuals Unlimited. 21.2a: ©David Cavagnaro. 21.2b: Produced from USAF DMSP Defense Meteorological Satellite Program film transparencies archived for NOAA/NESDIS at the University of Colorado, CIRES/National Sno and Ice Data Center. 21.6: ©1993 Time Magazine. 21.7: ©John Alcock. 21.8: Courtesy of E.D. Brodie III, University of California, Berkeley. 21.10: Courtesy of Thomas B. Smith, San Francisco State University. 21.11a: ©Dr. Michio Hori/Kyoto University. 21.13: ©Jane Burton/Bruce Coleman, Inc.

Chapter 22 Opener: Frans Lanting/Minden Pictures. 22.2a: 1 Don & Pat Valenti/Tom Stack & Associates 22.2a: 2 ©John Shaw/Tom Stack & Associates 22.2b: 1 ©John Eastcott/Yva Momatiuk/Woodfin Camp & Associates 22.2b: 2 ©Sven-Olof Lindblad/Photo Researchers, Inc. 22.2b: 3 ©Nita Winter 22.2b: 4 ©Geogrre Holton/Photo Researchers, Inc. 22.4 ©Joe McDonald/Visuals Unlimited. 22.5 ©John Eastcott/The Image Works. 22.6 ©Michael Fogden/Bruce Coleman, Inc. 22.7 1 ©John Shaw/Bruce Coleman, Inc. 22.7 2 ©Larry Ulrich/DRK Photo 22.7 3 ©M.P.L. Fogden/Bruce Coleman 22.8 ©Bob McKeever 1981/ Tom Stack & Assoc. 22.8 ©Steinhart Aquarium/Tom McHugh/ Photo Researchers, Inc. 22.9 ©David Cavagnaro. 22.12 Courtesy of J.D. Thompson. 22.12 Courtesy of Kenneth Kaneshiro Self-Quiz ©Russell C. Hansen

Chapter 23 Opener: ©Brian Parker/Tom Stack and Associates. 23.1a: Barbara J. Miller/Biological Photo Service. 23.1b: ©Margo Crabtree. 23.1c: ©Tom Till. 23.1d: ©Manfred Kage/Peter Arnold, Inc. 23.1e: ©W.H. Hodge/Peter Arnold, Inc. 23.1f: Dr. Martin Lockley. 23.1g: Courtesy Dr. David A. Grimald. Photo by Jacklyn Beckett/The American Museum of Natural History, N.Y. 23.3 top: ©John Shaw/NHPA. 23.3, bottom: ©John Shaw/NHPA. 23.4 ©Jane Burton/Photo Researchers, Inc. 23.13: ©Tom McHugh/Photo Researchers, Inc. 23.14 ©Gamma Liaison/Hanny Paul. 23.8b Self-Testing: ©Erwin & Peggy Bauer.

Chapter 24 Opener: ©Chip Clark. 24.1: S.M. Awramik, Univ. of California/Biological Photo Service, 24.3a: Courtesy John Stoltz. 24. b,c: S.M. Awramik, University of California/Biological Photo Service. 24.5a: Sidney Fox, University of Miami/ Biological Photo Service. 24.5b: Courtesy of F. M. Menger and Kurt Gabrielson, Emory University.

Chapter 25 Opener: ©Dr. Tony Brain and David Parker/Science Photo Library/Photo Researchers, Inc. 25.1a-c: ©David M. Phillips/ Visuals Unlimited. 25.2: ©Esther Angert, Indiana University. 25.3: S. Abraham and E.H. Beachey, VA Medical Center, Memphis/Biological Photo Service. 25.4: Courtesy J. Adler. 25.5a: Courtesy S.W. Watson ©Journal of Bacteriology, American Society of Microbiology/Biological Photo Service. 25.5b: N.J. Lang, University of California, Davis/Biological Photo Service. 25.6: ©John Durham/Science Photo Library/Photo Researchers, Inc. 25.8: Courtesy Wayne Carmichael/Wright State University. 25.8 inset: Courtesy Dr. Hans Paerl, Univ. of North Carolina. 25.9: Helen E. Carr/Biological Photo Service. Table 25.3 (top to bottom): ©Paul Johnson/ Biological Photo Service. ©L. Evans Roth/ Biological Photo Service; ©London School of Hygiene and Tropical Medicine/Science Photo Library/Photo Researchers, Inc; H.S. Pankratz, Michigan State University/Biological Photo Service; Michael Gabridge, University of Illinois/ Biological Photo Service; ©Biophoto Associates/ Science Source/Photo Researchers, Inc; P.W. Johnson and J. Mcn Sieburth, University of Rhode Island/Biological Photo Service; CNRI/SPL/Photo Researchers, Inc; ©Moredon Animal Health/ Science Photo Library/Photo Researchers, Inc. Methods Box ©Christine Case, Skyline College.

Chapter 26 Opener: ©M.I. Walker/Photo Researchers, Inc. 26.2: ©E. White/Visuals Unlimited. 26.5 ©Peter Parks/Animals. Animals. 26.6, 26.7a: ©Eric Grave/ Photo Researchers, Inc. 26.7b: ©Roland Birke/Peter Arnold, Inc. 26.8 inset: Masamichi Aikawa, *Science* 137 January 1990 :43. 26.9a: ©Eric Grave/ Photo Researchers, Inc. 26.9b: John Mansfield/University of Wisconsin. 26.10a: ©Eric Grave/ Photo Researchers, Inc. 26.10b: ©Manfred Kage/Peter Arnold Inc. 26.10c: 2 ©M. Abbey/Visuals Unlimited. 26.12 top: ©Ray Simons/Photo Researchers, Inc. 26.12, bottom: ©R. Calentine/Visuals Unlimited. 26.13: ©David Scharf/Peter Arnold, Inc. 26.14: ©Fred Rhoades/ Western Washington University. 26.15a: ©Biophoto Associates/Photo Researchers, Inc. 26.15b: ©Eric Grave/ Photo Researchers, Inc. 26.16a: ©Biophoto Associates/Photo Researchers, Inc. 26.16b: Courtesy of Fred Rhoades. 26.17 Courtesy of J. R. Waaland, University of Washington/ Biological Photo Service. 26.18: ©Anne Wertheim/ Animals Animals. 26.19: ©W. Lewis Trusty/ Animals. Animals. 26.20: Courtesy of J. R. Waaland, University of Washington/Biological Photo Service. 26.21a: Courtesy of J. R. Waaland, University of Washington/Biological Photo Service. 26.21b: ©D.P. Wilson and Eric and David Hosking/Photo Researchers, Inc. 26.21c: ©Gary R. Robinson/Visuals Unlimited. 26.22a: ©Manfred Kage/Peter Arnold, Inc. 26.22b: ©Estate of Dr. J. Metzner/Peter Arnold, Inc. 26.22c: ©Laurie Campbell/NHPA. 26.23: ©M.I. Walker/Science Source/Photo Researchers, Inc. 26.24: Courtesy of W.L. Dentler, University of Kansas. 26.25: Biological Photo Service.

Chapter 27 Opener: ©Carr Clifton. 27.1a: ©J.R. Waaland/Biological Photo Service. 27.1b: ©J.R. Waaland/Biological Photo Service. 27.4a: Martha Cook and Claudea Lipke. 27.4b: Linda Graham. 27.6 ©Stephen Kraseman/DRK Photo. 27.7 ©James W. Richardson/Visuals Unlimited. 27.8 ©John Shaw/Tom Stack & Associates. 27.10: ©Chip Clark 1988. 27.11: ©Dale and Marian Zimmerman/Bruce Coleman, Inc. 27.12: ©Robert and Linda Mitchell. 27.13: ©E.S. Ross 27.14: ©Milton Rand/Tom Stack and Associates; inset: ©Walter H. Hodge/Peter Arnold, Inc. 27.15: ©Milton Rand/Tom Stack & Associates. 27.16a: ©James Castner. 27.16b: ©Runk/ Schoenberger/Grant Heilman Photography. 27.16c: ©Micheal Fogden/Bruce Coleman, Inc. 27.16d: ©C.P. Hickman/Visuals Unlimited. 27.17: ©Dr. William M. Harlow/Photo Researchers, Inc. 27.18a: ©John Shaw/Tom Stack and Associates. 27.18b: ©W.H. Hodge/Peter Arnold, Inc. 27.21: ©1991 Gregory K. Scott/Photo Researchers, Inc. 27.23a: ©D. Wilder. 27.23b: ©Thomas C. Boyden. 27.23c: Merlin D. Tuttle, Bat Conservation International. 27.24: ©Lara Hartley.

Chapter 28 Opener: Ernst Roch, National Philatelic Centre, Canada Post Corporation, Antigonish, Nova Scotia. 28.1: ©Fred Rhoades/ Western Washington University. 28.3: ©Ed Reschke/Peter Arnold, Inc. 28.4a: ©Fred Rhoades/Western Washington University. 28.4b: ©Jacana/Photo Researchers, Inc. 28.4c: ©J.L. Lepore/Photo Researchers, Inc. 28.5: ©E.R. Degginger/Animals Animals. 28.6a: ©Kerry T. Givens/Tom Stack and Associates. 28.6b: ©Frans Lanting. 28.6c: ©A. Davies/Bruce Coleman, Inc. 28.7: ©Biophoto Associates/Photo Researchers, Inc. 28.8: ©Jack Bostrack/Visuals Unlimited; inset: ©M.F. Brown/Visuals Unlimited. 28.9: N. Allin and G.L. Barron, University of Guelph/Biological Photo Service. 28.10: Stephen J. Kron, University of Chicago. 28.11: ©Fred Rhoades/Western Washington University. 28.12: ©V. Ahmadjian/Visuals Unlimited. 28.13a: ©Bruce Iverson. 28.13b: ©Carolina Biological Supply/Phototake. 28.14 J.R. Waaland, Univ. of Washington/Biological Photo Service; inset: Courtesy of William Barstow, Dept. of Botany, Univ. Georgia, Athens.

Chapter 29 Opener: ©Kevin McCarthy/OffShoot Stock. 29.6, 29.10a: ©Andrew J. Martinez/Photo Researchers, Inc. 29.10b: Robert Brons/Biological Photo Service. 29.10c: ©Kevin McCarthy/OffShoot Stock. 29.10d: ©Chris Huss/Ellis Wildlife Collection. 29.11: ©Robert Brons/Biological Photo Service. 29.12: Fred Bavendam/Peter Arnold, Inc. 29.13: ©Bill Wood/Bruce Coleman, Inc. 29.15: Centers for Disease Control. 29.16: ©Drs. Kessel and Shih/Peter Arnold 29.17 ©Bill Wodd/NHPA. 29.18: ©M.I. Walker/Photo Researchers,Inc. 29.19: ©L.S. Stepanowicz/Photo Researchers, Inc. 29.21 ©Jeff Foott/Tom Stack & Associates. 29.23a: ©Kevin Schafer. 29.23b: ©Chris Huss. 29.24: ©H.W. Pratt/Biological Photo Service. 29.26a: ©Tom McHugh/Steinhart Aquarium/Photo Researchers, Inc. 29.26b: ©Fred Bavendam/Peter Arnold, Inc. 29.26c: ©Douglas Faulkner/Photo Researchers, Inc. 29.27b: ©Fred Bavendam/Peter Arnold, Inc. 29.29a: ©Sea Studios 29.29b: ©Kjell

Sandved/Sandved and Coleman. 29.29c: ©Robert and Linda Mitchell. 29.31: ©Cliff B. Frith/Bruce Coleman, Inc. 29.32: ©Chip Clark. 29.33: ©Milton H. Tierney, Jr./Visuals Unlimited. 29.34a: ©Robert and Linda Mitchell. 29.34b: ©David Scharf. 29.34c: Courtesy of Diana Sammatoro, Ohio State Univ., 29.35a: ©Paul Skelcher/Rainbow. 29.36a: ©Robert and Linda Mitchell. 29.36b: ©Dr. E.R. Degginger 29.37: ©Stephen Dalton/NHPA 29.39: ©John Shaw/Tom Stack & Associates. 29.40a: ©Marty Snyderman. 29.40b: ©Tom McHugh. 29.40c: ©C.R. Wyttenbach, University of Kansas/Biological Photo Service. 29.27a: ©Colin Milkins/Oxford Scientific Films/Animals Animals. 29.42a: ©Jeff Rotman/Peter Arnold, Inc. 29.42b: ©Gary Milburn/Tom Stack. 29.42c: ©Dave Woodward/Tom Stack & Associates. 29.42d: ©Marty Snyderman. 29.42e: ©Carl Roessler/Animals Animals. 29.42f: ©Jeff Rotman/Peter Arnold, Inc.

Chapter 30 Opener: ©Biophoto Associates/Photo Researchers, Inc. 30.2a: ©Scott Johnson/Animals Animals. 30.4a: Courtesy of M. Dale Stokes/Scripps. 30.4b: Courtesy Dr. Olson. 30.4c: ©Michael Neveux. 30.7a: ©Hervé Berthoule/Jacana/Photo Researchers, Inc. 30.7b,30.9a: ©Tom McHugh/Photo Researchers, Inc. 30.9b: ©Franklin Viola. 30.10a: ©E.R. Degginger/Bruce Coleman, Inc. 30.10b: ©J.M. Labat/Jacana/Photo Researchers, Inc. 30.12: ©Steve Martin/Tom Stack & Associates. 30.15a: ©Zig Leszczynski/Animals Animals. 30.15b: ©M.P.L. Fogden/Bruce Coleman, Inc. 30.15c: ©R. Andrew Odum/Peter Arnold, Inc. 30.16a: ©Hans Pfletschinger/Peter Arnold, Inc. 30.16b: ©Hans Pfletschinger/Peter Arnold, Inc. 30.16c: ©Hans Pfletschinger/Peter Arnold, Inc. 30.18: ©Jessie Cohen, Smithsonian Institution. 30.20: By Donna Braginetz, Courtesy of Natural History Magazine. 30.21a: ©John R. MacGregor/Peter Arnold, Inc. 30.21b: ©A.N.T./NHPA. 30.21c: ©R. Andrew Odum/Peter Arnold, Inc. 30.21d: ©Patricia Caulfield/Photo Researchers, Inc. 30.22: ©Claus Lotscher/Peter Arnold, Inc. 30.23 2 ©Janice Sheldon 30.25a: ©John Henry Dick, Academy of Natural Sciences, Philadelphia/Vireo. 30.25b: ©G. Neuchterlein, Academy of Natural Sciences, Philadelphia/Vireo. 30.25c: ©Bob and Clara Calhoun/Bruce Coleman, Inc. 30.25d: ©Art Wolfe. 30.27a: ©D. Parer & E. Parer-Cook/Auscape; inset: Courtesy of Mervyn Griffiths, *Australian Natural History* 17 September (1972):222-226; ©Ederic Slater. 30.27b: ©Tom McHugh/Photo Researchers, Inc. 30.27c: ©Mitch Reardon/Photo Researchers, Inc. 30.28: ©Erwin and Peggy Bauer/Bruce Coleman, Inc. 30.30a: ©Stephen Dalton/NHPA. 30.30b: ©Mickey Gibson/Animals Animals. 30.31a: ©E.R. Degginger/Animals Animals. 30.31b: ©Ulrich Nebelsiek/Peter Arnold, Inc. 30.31c: ©Nancy Adams E.P.I./Tom Stack & Associates 30.31d: ©Tom McHugh/Photo Researchers, Inc. 30.33a: ©Cleveland Museum of Natural History. 30.33b: ©John Reader 1982. 30.33c: ©Institute of Human Origins. Photo by Donald Johanson.

Chapter 31 Opener: ©Chip Clark. 31.1a: Elliot Meyerowitz, *Science* 254 11 October 1991:26 31.1 b,c Elliot Meyerowitz and John Bowman, Development 112 1991:1-20. 31.2: ©David Cavagnaro. 31.5: ©Dwight Kuhn. 31.8a: ©Dwight Kuhn. 31.8b: ©Larry Mellichamp/Visuals Unlimited. 31.8c: ©Kevin Schafer. 31.8 d ©E.S. Ross. 31.10a: ©Nels Lersten, University of Iowa. 31.10b: 1 ©Bruce Iverson. 31.10b: 2 ©Bruce Iverson. 31.10c: 1 ©Bruce Iverson 31.10c: 2 ©Ed Reschke/Peter Arnold, Inc. 31.10d: 1 ©George J. Wilder/Visuals Unlimited. 31.10d: 2 ©Randy Moore 31.14 1 ©Carolina Biological Supply/Phototake. 31.14 2 Courtesy of F.A.L. Clowes. 31.15a: 1 ©Ed Reschke. 31.15a: 2 ©Ed Reschke. 31.15b: ©Carolina Biological Supply. 31.16 ©Dwight Kuhn. 31.17 ©Ed Reschke 31.18 ©Ed Reschke. 31.21 1 ©Ed Reschke/Peter Arnold, Inc. 31.21 2 ©Ed Reschke/Peter Arnold, Inc.

Chapter 32 Opener: ©Branson Reynolds 32.11a-c: M.H. Zimmerman—supplied by Prof. P.B. Tomlinson, Biology Dept, Harvard Univ. Harvard Forest, Petersham,MA. Methods Box: top; Courtesy of Gary Tallman, Pepperdine University; middle; From T.D. Lamb, H.R. Matthews, and V. Torre, *J. Phys.* 372 1986:315-349.

Chapter 33 Opener: ©Kim Taylor/Bruce Coleman, Inc. 33.2a: Larry Lefever/Grant Heilman. 33.2b: ©John Colwell/Grant Heilman. 33.3: ©John H. Hoffman/Bruce Coleman, Inc. 33.4: ©Willam E. Ferguson. 33.6a: ©Science VU/Visuals Unlimited. 33.6b: ©Ken Wagner/Visuals Unlimited. 33.7: John Reganhold, U.S. Dept. of Agriculture Soil Conservation Service. 33.9a: ©William Ferguson. 33.9b: ©E.H. Newcomb, University of Wisconsin/Biological Photo Service. 33.11a: UN Food and Agriculture Organization. 33.11b: UN Food and Agriculture Organization, *Science* 250 9 November 1990 :748. 33.13a: ©Runk/Schoenberger/Grant Heilman. 33.13b: ©Kevin Schafer. 33.13c: ©Biophoto Associates/Photo Researchers, Inc. 33.14a: ©Jeff Lepore/Photo Researchers, Inc. 33.14b,c: ©John Shaw/Tom Stack and Associates. 33.15: ©R. Ronacordi/Visuals Unlimited.

Chapter 34 Opener: ©William Ferguson. 34.3a: ©Tom Branch 1978/Photo Researchers, Inc. 34.3b: ©Ed Reschke/Peter Arnold, Inc. 34.3c: ©Link, 1985/Visuals Unlimited. 34.3d: ©Stephen Dalton/Photo Researchers, Inc. 34.3e: ©William Ferguson. 34.3f left: ©Ray Coleman 1985/Photo Researchers, Inc; right: ©Ray Coleman 1985/Photo Researchers, Inc. 34.11a: ©James L. Castner. 34.11b: ©David Cavagnaro/DRK Photo. 34.13a: ©John C. Sanford/Cornell University. 34.14a: ©Jack Bostrack/Visuals Unlimited. 34.14b: ©John D. Cunnigham/Visuals Unlimited. 34.15: Courtesy of Susan Wick, University of Minnesota. 34.16: B. Wells and K. Roberts. Table top: ©Harold Taylor/OSF/Earth Scenes.; middle: ©William J. Weber/Visuals Unlimited; bottom: ©Bob Gossington/Bruce Coleman, Inc.

Chapter 35 Opener: Malcolm B. Wilkins, Regius Professor of Botany, Univ. of Glasgow. 35.6: ©Malcolm Wilkins. 35.7: ©Tugio Sasaki, Institute for Agricultural Research, Japan. 35.8: Courtesy of Stephen Gladfelter, Stanford University. 35.9: ©Ed Reschke. 35.10: Photo courtesy of Michael Evans, Ohio State University. 35.11a: ©John Kaprielian/Photo Researchers, Inc. 35.12: Courtesy Frank Salisbury. 35.17: Courtesy of J.L. Basq and M.C. Drew. 35.18: Courtesy of Janet Braam, from *Cell* 60 9 February 1990 :cover.

Chapter 36 Opener: ©Dwight R. Kuhn. 36.4: ©Ed Reschke. 36.8a: ©Jeff Foott/Tom Stack & Associaltes. 36.8b: ©Dr. E.R. Degginger. Methods Box: 1: Courtesy of Robert Full. Methods Box: 2 ©Yoav Levy/Phototake NYC. Challenge Question 3, a: Marion Rice. b: ©Bruce Iverson. c,e: ©Bruce Iverson; d: ©Eric Grave/Photo Researchers, Inc. f: ©Biophoto Associates/Photo Researchers, Inc.

Chapter 37 Opener: ©Marty Stouffer/Animals Animals. 37.1: ©C. Allan Morgan/Peter Arnold, Inc. 37.2: ©Tom Eisner, Cornell University. 37.3: ©Lennart Nilsson, Boehringer Ingelheim International GmbH. 37.4: ©Gunter Zeisler/Peter Arnold, Inc. 37.16 top: ©Brian Milne/Animals Animals; bottom: ©Hans and Judy Beste/Animals Animals. 37.18: ©Carol Hughes/Bruce Coleman, Inc. 37.20: ©Fred Bavendam/Peter Arnold, Inc.

Chapter 38 Opener: ©Stephen J. Kraseman/DRK Photo. 38.1: ©Norbert Wu. 38.14b: ©Lennart Nilsson/Boehringer Ingelheim International GmbH/*The Body Victorious,* Delacorte Press. 38.15 left ©Ed Reschke; right: ©W. Ober/Visuals Unlimited. 38.20b: Prepared by Dr. Hong Y. Yan, Univ. of Kentucky and Dr. Peng Chai, Univ. of Texas. 38.21 c2 ©David M. Phillips/Visuals Unlmited. 38.23: ©Professor Hans Rainer Dunker, Justus Liebig Univ., Giessen. 38.28: ©Kevin Schafer.

Chapter 39 Opener: ©Lennart Nilsson, Boehringer Ingelheim International GmbH. 39.2: ©Science Photo Library/Photo Researchers, Inc. 39.3: ©Lennart Nilsson, Boehringer Ingelheim International GmbH. 39.11c: ©A. J. Olson, Scripps Clinic and Research Foundation, 1986. 39.14b, 39.18: ©Lennart Nilsson, Boehringer Ingelheim International GmbH.

Chapter 40 Opener: ©John Shaw/Bruce Coleman, Inc. 40.3: ©John Crowe. 40.4a: ©Tom McHugh/Photo Researchers, Inc. 40.10b: Peter Andrews, Georgetown University, School of Medicine. 40.16a: ©Jeff Lepore/Photo Researchers, Inc. 40.16b: ©Dave B. Fleetham/Visuals Unlimited. 40.17a: ©E. Lyons/Bruce Coleman, Inc. 40.17b: ©Frans Lanting/Minden Pictures. 40.22 ©Warren Garst/Tom Stack & Associates.

Chapter 41 Opener: ©William E. Ferguson.

Chapter 42 Opener: ©Hans Pfletschinger/Peter Arnold, Inc. 42.1: ©David Wrobel 1992/Monterey Bay Aquarium. 42.2a: Courtesy of David Crews, Photo by P. DeVries. 42.3: ©Stephan Myers. 42.4: ©Dwight Kuhn. 42.5: ©William E. Ferguson. 42.18: ©Lennart Nilsson, Boehringer Ingelheim International GmbH. 42.21: ©Howard Sochurek/The Stock Market.

Chapter 43 Opener: Courtesy of W. Gehring, Univ. of Basel. *Science,* March 24, 1995. 43.1a: ©Dwight Kuhn. 43.1b: ©Hans Pfletschinger/Peter Arnold, Inc. 43.3 J.C. Gilkey, L.F. Jaffe, E.B. Ridgeway, and G.T. Reynolds, *J. Cell Biol.* 76:448-466, 1978. ©Rockefeller University Press. 43.6: George Watchmaker. . 43.7b: 43.8b: Courtesy of Robert F. Kalpin, William Sullivan and Douglas R. Daly, Univ. of Calif., Santa Cruz. 43.9: Courtesy of Charles A. Ettensohn, Carnegie Mellon University. 43.11: ©Cabisco/Visuals Unlimited. 43.11: Thomas Poole, SUNY Health Science Center. 43.13: Carolina Biological Supply. 43.16b: Hiroki Nishida, *Dev. Biol.* 121 1987 :526. 43.16c: J.E. Sulston and H.R. Horvitz, *Dev. Biol.* 56 1977 :110-156. 43.20a: Dr. Jean-Paul Thiery. From *J. Cell Bio.* 1983, Vol. 96, pp. 462-473 by Copyright permission of The Rockefeller University Press. 43.20: Richard Hynes, *Scientific American,* June 1986. 43.23 Kathryn Tosney, University of Michigan 43.24: Based on Honig & Summerbell 1985. Photograph courtesy of Dennis Summerbell. 43.25b: Courtesy of Dr. Ruth Lahmann, The Whitehead Institute. 43.25c-e: Drs. Jim Langeland, Steve Paddock, Sean Carroll, Univ. Wisconsin and The Howard Hughes Medical Institute. 43.26 ©F.R. Turner, Indiana University.

Chapter 44 Opener: ©John Stevens 1992/FPG. 44.2b: ©Manfred Kage/Peter Arnold, Inc. 44.11b: Dr. E.R. Lewis, ERL-EECS, University of California, Berkeley. 44.20a: Custom Medical Stock Photography. Methods Box:1b: ©William Thompson. 2a: ©Howard Sochurek/The Stock Market. 2b: Courtesy of Marcus Raichle, Washington University School of Medicine, St. Louis, MO.

Chapter 45 Opener: ©Stephen Dalton/NHPA 45.2: OSF/Animals Animals; inset: Dr. R.A. Steinbrecht, Max Planck Institut. 45.3a: ©Joe McDonald/Animals Animals. 45.3b: ©Russ Kinne/Photo Researchers, Inc. 45.5a: Janice Sheldon. 45.16: From Fundamentals of Entomology, Third Edition by Richard J. Elzinga 1987 . Reprinted by permission of Prentice-Hall Inc. 45.18: ©Joe Monroe/Photo Researchers, Inc. 45.25: Courtesy of Clara Franzini-Armstrong. 45.26: Courtesy of Dr. H.E. Huxley, Rosenstein Center, BrandeisUniversity, Waltham, MA. 45.31b: ©Eric Grave/Photo Researchers, Inc.

Chapter 46 Opener: Courtesy of NASA—LBJ Space Center. 46.1: Courtesy of Dan Simberloff. 46.2: ©Los Angeles Times. Photo by Associated Press. 46.13 : ©Norman Owen Tomalin/Bruce Coleman, Inc. 46.14: ©Jonathan Scott/Planet Earth Pictures. 46.15: ©Charlie Ott/Photo Researchers, Inc. 46.16: ©John D. Cunningham/Visuals Unlimited. 46.17: ©Frank Oberle. 46.18: ©Carr Clifton. 46.19: ©Lars Egede-Nissen/Biological Photo Service. 46.20: ©J. Warden/Superstock, Inc. 46.22: ©Bruce F. Molina/Terraphotographics.

46.23: Courtesy of D.N. Alstad, Univ. of Minnesota. 46.24: ©Steve Selum/Bruce Coleman, Inc. 46.25: ©M.E. Warre/Photo Researchers, Inc. 46.27: ©Chris Huss. 46.28: ©David Hall, 1984/Photo Researchers, Inc. 46.29: ©Dudley Foster/Woods Hole Oceanographic Institution.

Chapter 47 Opener: ©Will & Deni McIntyre/ Photo Researchers, Inc. 47.1: ©Victor Hutchinson/ Visuals Unlimited. 47.2, left: ©Sophie de Wilde/ Jacana/Photo Researchers, Inc; right: ©Frans Lanting/Minden Pictures; bottom: ©Will & Deni McIntrye/Photo Researchers, Inc. 47.7: ©Dr. E.R. Degginger. 47.9a: ©R. Calentine/Visuals Unlmited 47.9b: ©Max and Bea Hunn/Visuals Unlmited 47.19 ©Alan Carey/Photo Researchers, Inc. Methods Box: a: ©Frans Lanting; b: ©Frans Lanting

Chapter 48 Opener: ©Sven-Olof Linblad/Photo Researchers, Inc. 48.2: Courtesy of Paul Hertz. 48.3: ©L.E. Gilbert, University of Texas, Austin/ Biological Photo Service. 48.4: A. Morris/Academy of Natural Sciences©VIREO. 48.5: ©C. Allan Morgan/Peter Arnold, Inc. 48.6: ©John L. Tveten. 48.7: E.D. Brodie, Jr., Scientific American 263 July 1990: 28. 48.8a: ©Adrian Davies/Bruce Coleman, Inc. 48.8b: ©Peter J. Mayne. 48.10b: ©1985 Joseph T. Collins/Photo Researchers, Inc. 48.10c: ©Kevin deQueiroz, National Museum of Natural History. 48.13: ©Stephen Krasemen/Photo Researchers, Inc. 48.14: ©Robert and Linda Mitchell. 48.15: ©David M. Dennis/Tom Stack & Assoc. 48.16a: Courtesy of Karen Oberhauser. 48.17: Tom Bean. 48.18: ©Arthur C. Smith III/Grant Heilman Photography.

Chapter 49 Opener: 49.13a: ©John D. Cunningham/Visuals Unlimited 49.13b: Furnished by Dr. Robert Pierce/U.S. Department of Agriculture, Northeastern Experimental Forest Station, Durham, N.H. 49.15: D.W. Schindler, *Science* 184 24 May 1974:897-99. 49.19a,b: Matthew Meselson, Harvard University. 49.19c: ©Earth Observation Satellite Co, Lanham, Maryland/Science Photo Library/Photo Researchers, Inc. 49.20: ©William E. Ferguson. Methods Box: top: Courtesy of Sara Hotchkiss, Univ. of Minnesota; middle, bottom: Courtesy of Margaret Davis, University of Minnesota.

Chapter 50 Opener: ©William D. Griffin/Animals Animals. 50.7a: ©Ian Wyllie/Survival Anglia. 50.7b: ©Michael Leach/OSF/Animals Animals. 50.8: ©Lincoln P. Brower. 50.10: ©Ted Kerasote/Photo Researchers, Inc. 50.11: ©Bernd Heinrich/Univ. of Vermont. 50.12 ©Jeff Foott/Bruce Coleman, Inc. 50.13a: Courtesy of Pat De Coursey, University of South Carolina. 50.17a: ©G.R. Higbee/Photo Researchers, Inc. 50.17b: ©Bruce Davidson/ Animals Animals. 50.18: ©Visuals Unlimited. 50.19: ©Doug Wechsler/Animals Animals. 50.20a: ©Jonathan Scott/Seaphot Ltd./Planet Earth Pictures. 50.20b: ©Michael Dick/Animals Animals. 50.22: ©Kenneth Lorenzen, Univ of Calif., Davis. 50.23a: ©Stephen Kraseman/Peter Arnold, Inc. 50.24 ©Ivan Polunin/Bruce Coleman, Inc. Methods Box: a: ©E.R. Degginger/Animals Animals.

ILLUSTRATIONS

Contributing Artists to Previous Editions: Nea Bisek, Chris Carothers, Rachel Ciemma, Barbara Cousins, Pamela Drury-Wattenmaker, Cecile Duray-Bito, Janet Hayes, Darwen Hennings, Vally Hennings, Georg Klatt, Sandra McMahon, Linda McVay, Kenneth Miller, Fran Milner, Elizabeth Morales-Denney, Jackie Osborn, Carol Verbeeck, John Waller, and Judy Waller.

The following figures are adapted from Neil A. Campbell, Lawrence G. Mitchell, and Jane B. Reece, *Biology: Concepts and Connections* (Redwood City, CA: Benjamin/Cummings, 1994). ©1994 The Benjamin/Cummings Publishing Company:
Figures 3.9, 5.2, 9.15, 10.3, 11.4, 11.6, 11.7a, 11.9, 12.6, 15.5, 16.7, 20.12, 21.12, 22.11, 27.10, 29.9, 29.15, 29.16, 29.30, 29.43, 30.8, 31.23, 33.8, 33.12, 34.8, 34.13b, 35.5, 36.9, 36.10a, 37.7, 37.8, 37.9, 37.10, 37.11, 37.14, 37.17, 37.19, 38. Methods Box, 38.3, 38.4, 38.5, 38.6, 38.7, 38.9, 38.10, 38.11, 38.14a, 38.16, 38.18, 38.20a, 38.21a, 38.22, 38.24, 38.25, 38.27, 39.4, 40.2, 40.9, 40.14, 40.19b, 41.1, 41.9, 41.12, 41.13, 41.16, 42.8, 42.9, 42.11, 42.19a, 43.10, 44.1, 44.3, 44.18a, 44.20b, 44.21, 45.4, 45.8a, 45.22.

The following figures are adapted from Elaine N. Marieb, *Human Anatomy and Physiology,* 3rd ed. (Redwood City, CA: Benjamin/Cummings, 1995). ©1995 The Benjamin/Cummings Publishing Company:
Figures 7.11, 7.22, 44.2a, 45.23a, 45.26.

The following figures are adapted from C.K. Mathews and K.E. van Holde, *Biochemistry,* 2nd ed. (Menlo Park, CA: Benjamin/Cummings, 1996). ©1996 The Benjamin/Cummings Publishing Company:
Figures 4.5, 7.5, 9.9, 10.10.

The following figures are adapted from B. Alberts, D. Bray, J. Lewis, M. Raff, K. Roberts, J.D. Watson, *Molecular Biology of the Cell,* 2nd ed. (New York: Garland Publishing, 1989). ©1989 Garland Publishing:
Figures 6.1, 9.5, 11.11, 15.13, 17.13, 43.16b, c.

Figure 3.8 Adapted from an illustration by Michael Pique, The Scripps Research Institute.

Figure 4.6 Adapted from W.M. Becker, J.B. Reece, and M.F. Poenie, *The World of the Cell,* 3rd ed. (Menlo Park, CA: Benjamin/Cummings, 1996), Fig. 2.6. ©1996 The Benjamin/Cummings Publishing Co. **Figure 4.7** *Science News,* May 29, 1993. Clark Still, Columbia University.

Figure 5.12 From Robert A. Wallace, Gerald P. Sanders, Robert J. Ferl, *Biology: The Science of Life,* 3rd ed. (New York: HarperCollins, 1991). ©1991 HarperCollins Publishers, Inc. Reprinted by permission. **Figure 5.23** ©Irving Geis.

Figure 10.12 Adapted from David Walker, *Energy, Plants, and Man.* (Sheffield, England: University of Sheffield), Fig. 4.1, p. 69. ©David Walker and Richard Walker.

Chapter 17 opener James D. Watson et al., *Molecular Biology of the Gene,* 4th ed. (Menlo Park, CA: Benjamin/Cummings, 1987), Fig. 7.10, p. 187. © 1987 James D. Watson.

Figure 23.10 D. Raup, "Extinction: Bad Genes or Bad Luck?" *New Scientist,* September 14, 1991. **Figure 23.16** Cladogram by Peggy Conversano, illustrations by Ed Heck and Frank Ippolito. Courtesy of *Natural History,* June 1995, p. 34.

Figure 29.44 Adapted from "The Big Bang of Animal Evolution," *Scientific American,* November 1992, p. 85.

Figure 30.26 Gould et al., *The Book of Life* (New York: W.W. Norton), p. 96. **Figure 30.34** From an illustration by Laurie Grace, "The Recent African Genesis of Humans," A.C. Wilson, R.L. Cann, © April 1992 by Scientific American; illustration by Joe Lertola, "The Emergence of Modern Humans," C.B. Stringer, © December 1990 by Scientific American, Inc. All rights reserved.

Figure 39.11a,b Gerard J. Tortora, Berdell R. Funke, and Christine L. Case, *Microbiology: An Introduction,* 5th ed. (Redwood City, CA: Benjamin/Cummings, 1992). ©1992 The Benjamin/Cummings Publishing Company. **Figure 39.16** Adapted from Lennart Nilsson and Jan Lindberg, *The Body Victorious* (New York: Delacorte Press, 1987), p. 27. Urban Frank, Studio Frank & Co., illustrator.

Figure 40.18 Eckert, Randall and Augustine, *Animal Physiology,* 3rd ed. (New York: W.H. Freeman, 1988). ©1988 W.H. Freeman and Company. Used with permission.

Figure 43.27 Adapted from Peter Radetsky, "The Homeobox: Something Very Precious that We Share with Flies," *From Egg to Adult* (Bethesda, MD: Howard Hughes Medical Institute, 1992), p. 22

Figure 44.7 Adapted from G. Matthews, *Cellular Physiology of Nerve and Muscle* (Cambridge, MA: Blackwell Scientific Publications, 1986). Reprinted by permission of the publisher.

Figure 45.11 Adapted from an illustration by John Karapelou. ©1993 The Walt Disney Co. Reprinted with permission of Discover Magazine. **Figure 45.24** Lawrence G. Mitchell, John A. Mutchmor, Warren D. Dolphin, *Zoology* (Menlo Park, CA: Benjamin/Cummings, 1988). ©1988 The Benjamin/Cummings Publishing Company.

Figure 47.10 R.E. Ricklefs in D. S. Farner, *Breeding Biology of Birds* (Washington, DC: National Academy of Sciences, 1973), pp. 366–435.

Figure 48.10a A.S. Rand and E.E. Williams, "The Anoles of La Palma: Aspects of Their Ecological Relationships," *Breviora* no. 327, 1969. Museum of Comparative Zoology, Harvard.

Chapter 49 Methods Box (b) T. Webb, P.J. Bartlein, S.P.Harrison, and K.H. Anderson, "Vegetation, Lake Levels, and Climate in Eastern North America for the Past 18,000 years." In H.E. Wright et al., *Global Climates Since the Last Glacial Maximum* (Minneapolis: University of Minnesota Press, 1993). **Figure 49.9** From R.E. Ricklefs, *Ecology,* 2nd ed. ©1988 by Chiron Press. Reprinted by permission of W.H. Freeman. **Figure 49.10** P. Colinvaux, *Ecology* (New York: John Wiley & Sons, 1986). ©1986 John Wiley & Sons. Reprinted by permission of the publisher. **Figure 49.16** From G. Tyler Miller, *Living in the Environment,* 2nd ed. (Belmont, CA: Wadsworth Publishing Co., 1979), p. 87. ©1979 Wadsworth Publishing Company. **Figure 49.18** Tom Moore, ©1986 Discover Magazine.

Chapter 50 Methods Box (b) Lisle Gibbs, et al., "Realized Reproductive Successes of Polygonous Red-Winged Blackbirds Revealed by Markers," *Science* , v. 250, November/December 1990, pp. 1394–1397. **Figure 50.4** Lawrence G. Mitchell, John A. Mutchmor, Warren D. Dolphin, *Zoology* (Menlo Park, CA: Benjamin/Cummings, 1988). ©1988 The Benjamin/Cummings Publishing Company. **Figure 50.5** N. Tinbergen, *The Study of Instinct* (Oxford University Press, 1951). Reprinted by permission of Oxford University Press. **Figure 50.9** Courtesy of Masakazu Konishi.

GLOSSARY

A site Aminoacyl-tRNA site; the binding site on a ribosome that holds the tRNA carrying the next amino acid to be added to a growing polypeptide chain.

abscisic acid (ABA) *(ab-SIS-ik)* A plant hormone that generally acts to inhibit growth, promote dormancy, and help the plant tolerate stressful conditions.

absorption spectrum The range of a pigment's ability to absorb various wavelengths of light.

abyssal zone *(uh-BIS-ul)* The portion of the ocean floor where light does not penetrate and where temperatures are cold and pressures intense.

acclimatization *(uh-KLY-mih-ty-ZAY-shun)* Physiological adjustment to a change in an environmental factor.

accommodation The automatic adjustment of an eye to focus on near objects.

acetylcholine One of the most common neurotransmitters; functions by binding to receptors and altering the permeability of the postsynaptic membrane to specific ions, either depolarizing or hyperpolarizing the membrane.

acetyl CoA The entry compound for the Krebs cycle in cellular respiration; formed from a fragment of pyruvate attached to a coenzyme.

acid A substance that increases the hydrogen ion concentration in a solution.

acid precipitation Rain, snow, or fog that is more acidic than pH 5.6.

acoelomate *(a-SEEL-oh-mate)* A solid-bodied animal lacking a cavity between the gut and outer body wall.

acquired immunity The type of immunity achieved when antigens enter the body naturally or artificially. Acquired immunity is due to the stimulation of antibody production and the production of memory cells keyed to the antigen.

acrosome *(AK-ruh-some)* An organelle at the tip of a sperm cell that helps the sperm penetrate the egg.

actin *(AK-tin)* A globular protein that links into chains, two of which twist helically about each other, forming microfilaments in muscle and other contractile elements in cells.

action potential A rapid change in the membrane potential of an excitable cell, caused by stimulus-triggered, selective opening and closing of voltage-sensitive gates in sodium and potassium ion channels.

activation Induction by the sperm of increased respiration and protein synthesis in the egg; the earliest trigger of embryonic development.

active site The specific portion of an enzyme that attaches to the substrate by means of weak chemical bonds.

active transport The movement of a substance across a biological membrane against its concentration or electrochemical gradient, with the help of energy input and specific transport proteins.

adaptive peak An equilibrium state in a population when the gene pool has allele frequencies that maximize the average fitness of a population's members.

adaptive radiation The emergence of numerous species from a common ancestor introduced into an environment, presenting a diversity of new opportunities and problems.

adenylyl cyclase An enzyme that converts ATP to cyclic AMP in response to a chemical signal; a membrane protein that is the effector in a signal-transduction pathway.

adrenal gland *(uh-DREE-nul)* An endocrine gland located adjacent to the kidney in mammals; composed of two glandular portions: an outer cortex, which responds to endocrine signals in reacting to stress and effecting salt and water balance, and a central medulla, which responds to nervous inputs resulting from stress.

aerobic *(air-OH-bik)* Containing oxygen; referring to an organism, environment, or cellular process that requires oxygen.

age structure The relative number of individuals of each age in a population.

agnathan *(AG-naa-thun)* A member of a jawless class of vertebrates represented today by the lampreys and hagfishes.

agonistic behavior *(ag-on-IS-tik)* A type of behavior involving a contest of some kind that determines which competitor gains access to some resource, such as food or mates.

AIDS (acquired immunodeficiency syndrome) The name of the late stages of HIV infection; defined by a specified reduction of T cells and the appearance of characteristic secondary infections.

aldehyde *(AL-duh-hyde)* An organic molecule with a carbonyl group located at the end of the carbon skeleton.

aldosterone *(al-DAH-stair-own)* An adrenal hormone that acts on the distal tubules of the kidney to stimulate the reabsorption of sodium (Na^+) and the passive flow of water from the filtrate.

alga (plural, **algae**) A photosynthetic, plant-like protist.

all-or-none event An action that occurs either completely or not at all, such as the generation of an action potential by a neuron.

allantois *(AL-an-TOH-iss)* One of four extraembryonic membranes; serves as a repository for the embryo's nitrogenous waste.

allele *(uh-LEEL)* An alternative form of a gene.

allometric growth *(AL-oh-MET-rik)* The variation in the relative rates of growth of various parts of the body, which helps shape the organism.

allopatric speciation *(AL-oh-PAT-rik)* A mode of speciation induced when the ancestral population becomes segregated by a geographical barrier.

allopolyploid *(AL-oh-POL-ee-ploid)* A common type of polyploid species resulting from two different species interbreeding and combining their chromosomes.

allosteric site *(AL-oh-STEER-ik)* A specific receptor site on an enzyme molecule remote from the active site. Molecules bind to the allosteric site and change the shape of the active site, making it either more or less receptive to the substrate.

alpha (α) helix A spiral shape constituting one form of the secondary structure of proteins, arising from a specific hydrogen-bonding structure.

alternation of generations A life cycle in which there is both a multicellular diploid form, the sporophyte, and a multicellular haploid form, the gametophyte; characteristic of plants.

altruistic behavior *(AL-troo-IS-tik)* The aiding of another individual at one's own risk or expense.

alveolus *(al-VEE-oh-lus)* (plural, **alveoli**) (1) One of the deadend, multilobed air sacs that constitute the gas exchange surface of the lungs. (2) One of the milk-secreting sacs of epithelial tissue in the mammary glands.

amino acid *(uh-MEE-noh)* An organic molecule possessing both carboxyl and amino groups. Amino acids serve as the monomers of proteins.

amino group A functional group that consists of a nitrogen atom bonded to two hydrogen atoms; can act as a base in solution, accepting a hydrogen ion and acquiring a charge of $+1$.

aminoacyl-tRNA synthetases A family of enzymes, at least one for each amino acid, that catalyzes the attachment of an amino acid to its specific tRNA molecule.

amniocentesis *(AM-nee-oh-sen-TEE-sis)* A technique for determining genetic abnormalities in a fetus by the presence of certain chemicals or defective fetal cells in the amniotic fluid, obtained by aspiration from a needle inserted into the uterus.

amnion *(AM-nee-on)* The innermost of four extraembryonic membranes; encloses a fluid-filled sac in which the embryo is suspended.

amniote A vertebrate possessing an amnion surrounding the embryo; reptiles, birds, and mammals are amniotes.

amniotic egg *(AM-nee-AH-tik)* A shelled, water-retaining egg that enables reptiles, birds, and egg-laying mammals to complete their life cycles on dry land.

Amphibia The vertebrate class of amphibians, represented by frogs, salamanders, and caecilians.

amphipathic molecule A molecule that has both a hydrophilic region and a hydrophobic region.

anaerobic *(an-air-OH-bik)* Lacking oxygen; referring to an organism, environment, or cellular process that lacks oxygen and may be poisoned by it.

anagenesis *(AN-uh-JEN-eh-sis)* A pattern of evolutionary change involving the transformation of an entire population, sometimes to a state different enough from the ancestral population to justify renaming it as a separate species; also called phyletic evolution.

analogy The similarity of structure between two species that are not closely related; attributable to convergent evolution.

androgens *(AN-droh-jens)* The principal male steroid hormones, such as testosterone, which stimulate the development and maintenance of the male reproductive system and secondary sex characteristics.

aneuploidy *(AN-yoo-ploy-dee)* A chromosomal aberration in which certain chromosomes are present in extra copies or are deficient in number.

angiosperm *(AN-jee-oh-spurm)* A flowering plant, which forms seeds inside a protective chamber called an ovary.

anion *(AN-eye-on)* A negatively charged ion.

annual A plant that completes its entire life cycle in a single year or growing season.

anterior Referring to the head end of a bilaterally symmetrical animal.

anther *(AN-thur)* The terminal pollen sac of a stamen, inside which pollen grains with male gametes form in the flower of an angiosperm.

antheridium *(an-theh-RID-ee-um)* In plants, the male gametangium, a moist chamber in which gametes develop.

antibiotic A chemical that kills or inhibits the growth of bacteria, often via transcriptional or translational regulation.

antibody An antigen-binding immunoglobulin, produced by B cells, that functions as the effector in an immune response.

anticodon *(AN-tee-CO-don)* A specialized base triplet on one end of a tRNA molecule that recognizes a particular complementary codon on an mRNA molecule.

antidiuretic hormone (ADH) A hormone important in osmoregulation.

antigen *(AN-teh-jen)* A foreign macromolecule that does not belong to the host organism and that elicits an immune response.

aphotic zone *(ay-FOE-tik)* The part of the ocean beneath the photic zone, where light does not penetrate sufficiently for photosynthesis to occur.

apical dominance *(AY-pik-ul)* Concentration of growth at the tip of a plant shoot, where a terminal bud partially inhibits axillary bud growth.

apical meristem *(AY-pik-ul MARE-eh-stem)* Embryonic plant tissue in the tips of roots and in the buds of shoots that supplies cells for the plant to grow in length.

apomorphic character *(AP-oh-MORE-fik)* A derived phenotypic character, or homology, that evolved after a branch diverged from a phylogenetic tree.

apoplast *(AP-oh-plast)* In plants, the nonliving continuum formed by the extracellular pathway provided by the continuous matrix of cell walls.

aposematic coloration *(AP-oh-so-MAT-ik)* The bright coloration of animals with effective physical or chemical defenses that acts as a warning to predators.

aqueous solution *(AY-kwee-us)* A solution in which water is the solvent.

Archaea The domain name for the archaebacteria.

archaebacteria *(AR-kuh-bak-TEER-ee-uh)* An ancient lineage of prokaryotes, represented today by a few groups of bacteria inhabiting extreme environments. Some taxonomists place archaebacteria in their own kingdom, separate from the other bacteria.

archegonium *(ar-kih-GO-nee-um)* In plants, the female gametangium, a moist chamber in which gametes develop.

archenteron *(ark-EN-ter-on)* The endoderm-lined cavity, formed during the gastrulation process, that develops into the digestive tract of an animal.

Archezoa Primitive eukaryotic group that includes diplomonads, such as *Giardia;* some systematists assign kingdom status to archezoans.

artery A vessel that carries blood away from the heart to organs throughout the body.

arteriosclerosis A cardiovascular disease caused by the formation of hard plaques within the arteries.

artificial selection The selective breeding of domesticated plants and animals to encourage the occurrence of desirable traits.

ascus (plural, **asci**) A saclike spore capsule located at the tip of the ascocarp in dikaryotic hyphae; defining feature of the Ascomycota division of fungi.

asexual reproduction A type of reproduction involving only one parent that produces genetically identical offspring by budding or by the division of a single cell or the entire organism into two or more parts.

associative learning The acquired ability to associate one stimulus with another; also called classical conditioning.

assortative mating A type of nonrandom mating in which mating partners resemble each other in certain phenotypic characters.

asymmetric carbon A carbon atom covalently bonded to four different atoms or groups of atoms.

atomic number The number of protons in the nucleus of an atom, unique for each element and designated by a subscript to the left of the elemental symbol.

atomic weight The total atomic mass, which is the mass in grams of one mole of the atom.

ATP (adenosine triphosphate) *(uh-DEN-oh-sin try-FOS-fate)* An adenine-containing nucleoside triphosphate that releases free energy when its phosphate bonds are hydrolyzed. This energy is used to drive endergonic reactions in cells.

ATP synthase A protein complex that produces ATP.

atrioventricular valve A valve in the heart between each atrium and ventricle that prevents a backflow of blood when the ventricles contract.

atrium *(AY-tree-um)* (plural, **atria**) A chamber that receives blood returning to the vertebrate heart.

autogenesis model According to this model, eukaryotic cells evolved by the specialization of internal membranes originally derived from prokaryotic plasma membranes.

autoimmune disease An immunological disorder in which the immune system turns against itself.

autonomic nervous system *(AWT-uh-NAHM-ik)* A subdivision of the motor nervous system of vertebrates that regulates the internal environment; consists of the sympathetic and parasympathetic divisions.

autopolyploid *(AW-toe-POL-ee-ploid)* A type of polyploid species resulting from one species doubling its chromosome number to become tetraploid, which may self-fertilize or mate with other tetraploids.

autosome *(AW-tuh-some)* A chromosome that is not directly involved in determining sex, as opposed to the sex chromosomes.

autotroph *(AW-toh-TROHF)* An organism that obtains organic food molecules without eating other organisms. Autotrophs use energy from the sun or from the oxidation of inorganic substances to make organic molecules from inorganic ones.

auxins *(AWK-sins)* A class of plant hormones, including indoleacetic acid (IAA), having a variety of effects, such as phototropic response through the stimulation of cell elongation, stimulation of secondary growth, and the development of leaf traces and fruit.

auxotroph *(AWK-soh-trohf)* A nutritional mutant that is unable to synthesize and that cannot grow on media lacking certain essential molecules normally synthesized by wild-type strains of the same species.

Aves The vertebrate class of birds, characterized by feathers and other flight adaptations.

axillary bud *(AKS-ill-air-ee)* An embryonic shoot present in the angle formed by a leaf and stem.

axon *(AKS-on)* A typically long extension, or process, from a neuron that carries nerve impulses away from the cell body toward target cells.

B cell A type of lymphocyte that develops in the bone marrow and later produces antibodies, which mediate humoral immunity.

Bacteria The domain name for the eubacteria.

bacterium (plural, **bacteria**) A unicellular microorganism, also called a prokaryote, which has no true nucleus. Bacteria are classified into two groups based on a difference in cell walls, as determined by Gram staining.

balanced polymorphism A type of polymorphism in which the frequencies of the coexisting forms do not change noticeably over many generations.

bark All tissues external to the vascular cambium in a plant growing in thickness, consisting of phloem, phelloderm, cork cambium, and cork.

Barr body The dense object lying along the inside of the nuclear envelope in female mammalian cells, representing the one inactivated *X* chromosome.

basal metabolic rate (BMR) The minimal number of kilocalories a resting animal requires to fuel itself for a given time.

base A substance that reduces the hydrogen ion concentration in a solution.

basement membrane The floor of an epithelial membrane on which the basal cells rest.

base-pair substitution A point mutation; the replacement of one nucleotide and its partner from the complementary DNA strand by another pair of nucleotides.

basidium (plural, **basidia**) A reproductive appendage that produces sexual spores on the gills of mushrooms. The fungal division Basidiomycota is named for this structure.

Batesian mimicry *(BAYTZ-ee-un MIM-ih-kree)* A type of mimicry in which a harmless species looks like a different species that is poisonous or otherwise harmful to predators.

behavioral ecology A heuristic approach based on the expectation that Darwinian fitness (reproductive success) is improved by optimal behavior.

benthic zone The bottom surfaces of aquatic environments.

biennial *(by-EN-ee-ul)* A plant that requires two years to complete its life cycle.

bilateral symmetry Characterizing a body form with a central longitudinal plane that divides the body into two equal but opposite halves.

bilateria *(BY-leh-TEER-ee-uh)* Members of the branch of eumetazoans possessing bilateral symmetry.

binary fission The type of cell division by which prokaryotes reproduce; each dividing daughter cell receives a copy of the single parental chromosome.

binomial The two-part Latinized name of a species, consisting of genus and specific epithet.

bioenergetics The study of how organisms manage their energy resources.

biogeochemical cycles The various nutrient circuits, which involve both biotic and abiotic components of ecosystems.

biogeography The study of the past and present distribution of species.

biological magnification A trophic process in which retained substances become more concentrated with each link in the food chain.

biological species A population or group of populations whose members have the potential to interbreed.

biomass The dry weight of organic matter comprising a group of organisms in a particular habitat.

biome *(BY-ome)* One of the world's major communities, classified according to the predominant vegetation and characterized by adaptations of organisms to that particular environment.

biosphere *(BY-oh-sfeer)* The entire portion of Earth that is inhabited by life; the sum of all the planet's communities and ecosystems.

biotechnology The industrial use of living organisms or their components to improve human health and food production.

biotic *(by-OT-ik)* Pertaining to the living organisms in the environment.

blastocoel *(BLAS-toh-seel)* The fluid-filled cavity that forms in the center of the blastula embryo.

blastocyst An embryonic stage in mammals; a hollow ball of cells produced one week after fertilization in humans.

blastopore *(BLAS-toh-por)* The opening of the archenteron in the gastrula that develops into the mouth in protostomes and the anus in deuterostomes.

blastula *(BLAS-tyoo-la)* The hollow ball of cells marking the end stage of cleavage during early embryonic development.

blood-brain barrier A specialized capillary arrangement in the brain that restricts the passage of most substances into the brain, thereby preventing dramatic fluctuations in the brain's environment.

blood pressure The hydrostatic force that blood exerts against the wall of a vessel.

bond energy The quantity of energy that must be absorbed to break a particular kind of chemical bond; equal to the quantity of energy the bond releases when it forms.

book lungs Organs of gas exchange in spiders, consisting of stacked plates contained in an internal chamber.

bottleneck effect Genetic drift resulting from the reduction of a population, typically by a natural disaster, such that the surviving population is no longer genetically representative of the original population.

Bowman's capsule *(BOH-munz)* A cup-shaped receptacle in the vertebrate kidney that is the initial, expanded segment of the nephron where filtrate enters from the blood.

brain stem The hindbrain and midbrain of the vertebrate central nervous system. In humans, it forms a cap on the anterior end of the spinal cord, extending to about the middle of the brain.

bryophytes *(BRY-oh-fites)* The mosses, liverworts, and hornworts; a group of nonvascular plants that inhabit the land but lack many of the terrestrial adaptations of vascular plants.

budding An asexual means of propagation in which outgrowths from the parent form and pinch off to live independently or else remain attached to eventually form extensive colonies.

buffer A substance that consists of acid and base forms in solution and that minimizes changes in pH when extraneous acids or bases are added to the solution.

bulk flow The movement of water due to a difference in pressure between two locations.

C_3 plant A plant that uses the Calvin cycle for the initial steps that incorporate CO_2 into organic material, forming a three-carbon compound as the first stable intermediate.

C_4 plant A plant that prefaces the Calvin cycle with reactions that incorporate CO_2 into four-carbon compounds, the end-product of which supplies CO_2 for the Calvin cycle.

calcitonin *(kal-sih-TOH-nin)* A mammalian thyroid hormone that lowers blood calcium levels.

calmodulin *(kal-MOD-yoo-lin)* An intracellular protein to which calcium binds in its function as a second messenger in hormone action.

calorie (cal) The amount of heat energy required to raise the temperature of 1 g of water 1°C; the amount of heat energy that 1 g of water releases when it cools by 1°C. The Calorie (with a capital C), usually used to indicate the energy content of food, is a kilocalorie.

Calvin cycle The second of two major stages in photosynthesis (following the light reactions), involving atmospheric CO_2 fixation and reduction of the fixed carbon into carbohydrate.

CAM plant A plant that uses crassulacean acid metabolism, an adaptation for photosynthesis in arid conditions, first discovered in the family Crassulaceae. Carbon dioxide entering open stomata during the night is converted into organic acids, which release CO_2 for the Calvin cycle during the day, when stomata are closed.

5'cap During RNA processing, an "attach here" sign for small ribosomal subunits; at the 5' end of an mRNA molecule, the cap helps inhibit degradation and enhances translation.

capillary *(KAP-ill-air-ee)* A microscopic blood vessel that penetrates the tissues and consists of a single layer of endothelial cells that allows exchange between the blood and interstitial fluid.

capsid The protein shell that encloses the viral genome; rod-shaped, polyhedral, or more completely shaped.

carbohydrate *(KAR-bo-HY-drate)* A sugar (monosaccharide) or one of its dimers (disaccharides) or polymers (polysaccharides).

carbonyl group *(KAR-buh-nil)* A functional group present in aldehydes and ketones, consisting of a carbon atom double-bonded to an oxygen atom.

carboxyl group *(kar-BOX-ul)* A functional group present in organic acids, consisting of a single carbon atom double-bonded to an oxygen atom and also bonded to a hydroxyl group.

carcinogen *(kar-SIN-oh-jen)* A chemical agent that causes cancer.

cardiac muscle *(KAR-dee-ak)* A type of muscle that forms the contractile wall of the heart; its cells are joined by intercalated discs that relay each heartbeat.

cardiac output The volume of blood pumped per minute by the left ventricle of the heart.

carnivore An animal, such as a shark, hawk, or spider, that eats other animals.

carotenoids *(keh-ROT-en-oydz)* Accessory pigments, yellow and orange, in the chloroplasts of plants; by absorbing wavelengths of light that chlorophyll cannot, they broaden the spectrum of colors that can drive photosynthesis.

carpel *(KAR-pel)* The female reproductive organ of a flower, consisting of the stigma, style, and ovary.

carrying capacity The maximum population size that can be supported by the available resources, symbolized as *K*.

cartilage *(KAR-til-ij)* A type of flexible connective tissue with an abundance of collagenous fibers embedded in chondrin.

Casparian strip *(kas-PAR-ee-un)* A water-impermeable ring of wax around endodermal cells in plants that blocks the passive flow of water and solutes into the stele by way of cell walls.

catabolic pathway *(KAT-uh-BOL-ik)* A metabolic pathway that releases energy by breaking down complex molecules into simpler compounds.

catabolite activator protein (CAP) *(ka-TAB-ul-LITE)* In *E. coli*, a helper protein that stimulates gene expression by binding within the promoter region of an operon and enhancing the promoter's ability to associate with RNA polymerase.

cation *(KAT-eye-on)* An ion with a positive charge, produced by the loss of one or more electrons.

cation exchange A process in which positively charged minerals are made available to a plant when hydrogen ions in the soil displace mineral ions from the clay particles.

cell center A region in the cytoplasm near the nucleus from which microtubules originate and radiate.

cell cycle An ordered sequence of events in the life of a dividing cell, composed of the M, G_1, S, and G_2 phases.

cell fractionation The disruption of a cell and separation of its organelles by centrifugation.

cell-mediated immunity The type of immunity that functions in defense against fungi, protists, bacteria, and viruses inside host cells and against tissue transplants, with highly specialized cells that circulate in the blood and lymphoid tissue.

cell plate A double membrane across the midline of a dividing plant cell, between which the new cell wall forms during cytokinesis.

cell wall A protective layer external to the plasma membrane in plant cells, bacteria, fungi, and some protists. In the case of plant cells, the wall is formed of cellulose fibers embedded in a polysaccharide-protein matrix. The primary cell wall is thin and flexible, whereas the secondary cell wall is stronger and more rigid, and is the primary constituent of wood.

cellular differentiation The structural and functional divergence of cells as they become specialized during a multicellular organism's development; dependent on the control of gene expression.

cellular respiration The most prevalent and efficient catabolic pathway for the production of ATP, in which oxygen is consumed as a reactant along with the organic fuel.

cellulose *(SELL-yoo-lose)* A structural polysaccharide of cell walls, consisting of glucose monomers joined by β-1, 4-glycosidic linkages.

Celsius scale *(SELL-see-us)* A temperature scale (°C) equal to $\frac{5}{9}$ (°F – 32) that measures the freezing point of water at 0°C and the boiling point of water at 100°C.

central nervous system (CNS) In vertebrate animals, the brain and spinal cord.

centriole *(SEN-tree-ole)* One of two structures in the center of animal cells, composed of cylinders of nine triplet microtubules in a ring. Centrioles help organize microtubule assembly during cell division.

centromere *(SEN-troh-mere)* The centralized region joining two sister chromatids.

centrosome Material present in the cytoplasm of all eukaryotic cells and important during cell division; also called microtubule-organizing center.

cephalochordate A chordate without a backbone, represented by lancelets, tiny marine animals.

cerebellum *(SEH-reh-BELL-um)* Part of the vertebrate hindbrain (rhombencephalon) located dorsally; functions in unconscious coordination of movement and balance.

cerebral cortex *(seh-REE-brul)* The surface of the cerebrum; the largest and most complex part of the mammalian brain, containing sensory and motor nerve cell bodies of the cerebrum; the part of the vertebrate brain most changed through evolution.

cerebrum *(seh-REE-brum)* The dorsal portion, composed of right and left hemispheres, of the vertebrate forebrain; the integrating center for memory, learning, emotions, and other highly complex functions of the central nervous system.

chaparral *(SHAP-uh-RAL)* A scrubland biome of dense, spiny evergreen shrubs found at midlatitudes along coasts where cold ocean currents circulate offshore; characterized by mild, rainy winters and long, hot, dry summers.

chemical equilibrium In a reversible chemical reaction, the point at which the rate of the forward reaction equals the rate of the reverse reaction.

chemiosmosis *(KEE-mee-os-MOH-sis)* The ability of certain membranes to use chemical energy to pump hydrogen ions and then harness the energy stored in the H^+ gradient to drive cellular work, including ATP synthesis.

chemoautotroph *(KEE-moh-AW-toh-trohf)* An organism that needs only carbon dioxide as a carbon source but that obtains energy by oxidizing inorganic substances.

chemoheterotroph *(KEE-moh-HET-er-oh-trohf)* An organism that must consume organic molecules for both energy and carbon.

chemoreceptor A receptor that transmits information about the total solute concentration in a solution or about individual kinds of molecules.

chiasma *(KY-as-muh)* (plural, **chiasmata**) The X-shaped, microscopically visible region representing homologous chromatids that have exchanged genetic material through crossing over during meiosis.

chitin *(KY-tin)* A structural polysaccharide of an amino sugar found in many fungi and in the exoskeletons of all arthropods.

chlorophyll A green pigment located within the chloroplasts of plants; chlorophyll *a* can participate directly in the light reactions, which convert solar energy to chemical energy.

chloroplast *(KLOR-oh-plast)* An organelle found only in plants and photosynthetic protists that absorbs sunlight and uses it to drive the synthesis of organic compounds from carbon dioxide and water.

cholesterol *(kol-ESS-teh-rol)* A steroid that forms an essential component of animal cell membranes and acts as a precursor molecule for the synthesis of other biologically important steroids.

Chondrichthyes The vertebrate class of cartilaginous fishes, represented by sharks and their relatives.

chondrin A protein-carbohydrate complex secreted by chondrocytes; chondrin and collagen fibers form cartilage.

chordate *(KOR-date)* A member of a diverse phylum of animals that possess a notochord; a dorsal, hollow nerve cord; pharyngeal gill slits; and a postanal tail as embryos.

chorion *(KOR-ee-on)* The outermost of four extraembryonic membranes; contributes to the formation of the mammalian placenta.

chorionic villus sampling (CVS) *(KOR-ee-on-ik VILL-us)* A technique for diagnosing genetic and congenital defects while the fetus is in the uterus. A small sample of the fetal portion of the placenta is removed and analyzed.

chromatin *(KRO-muh-tin)* The aggregate mass of dispersed genetic material formed of DNA and protein and observed between periods of cell division in eukaryotic cells.

Chromista In some classification systems, a kingdom consisting of brown algae, golden algae, and diatoms.

chromosome *(KRO-muh-some)* A long, threadlike association of genes in the nucleus of all eukaryotic cells and most visible during mitosis and meiosis. Chromosomes consist of DNA and protein.

chytrid Fungus with flagellated stage; possible evolutionary link between fungi and protists.

cilium *(SILL-ee-um)* (plural, **cilia**) A short cellular appendage specialized for locomotion, formed from a core of nine outer doublet microtubules and two inner single microtubules ensheathed in an extension of plasma membrane.

circadian rhythm *(sur-KAY-dee-un)* A physiological cycle of about 24 hours, present in all eukaryotic organisms, that persists even in the absence of external cues.

cladistics *(kluh-DIS-tiks)* A taxonomic approach that classifies organisms according to the order in time at which branches arise along a phylogenetic tree, without considering the degree of morphological divergence.

cladogenesis *(KLAY-doh-GEN-eh-sis)* A pattern of evolutionary change that produces biological diversity by budding one or more new species from a parent species that continues to exist; also called branching evolution.

cladogram A dichotomous phylogenetic tree that branches repeatedly, suggesting a classification of organisms based on the time sequence in which evolutionary branches arise.

classical conditioning A type of associative learning; the association of a normally irrelevant stimulus with a fixed behavioral response.

cleavage The process of cytokinesis in animal cells, characterized by pinching of the plasma membrane; also, the succession of rapid cell divisions without growth during early embryonic development that converts the zygote into a ball of cells.

cleavage furrow The first sign of cleavage in an animal cell; a shallow groove in the cell surface near the old metaphase plate.

cline Variation in features of individuals in a population that parallels a gradient in the environment.

cloaca *(kloh-AY-kuh)* A common opening for the digestive, urinary, and reproductive tracts in all vertebrates except most mammals.

clonal selection *(KLOH-nul)* The mechanism that determines specificity and accounts for antigen memory in the immune system; occurs because an antigen introduced into the body selectively activates only a tiny fraction of inactive lymphocytes, which proliferate to form a clone of effector cells specific for the stimulating antigen.

clone A lineage of genetically identical individuals.

cloning vector An agent used to transfer DNA in genetic engineering, such as a plasmid that moves recombinant DNA from a test tube back into a cell, or a virus that transfers recombinant DNA by infection.

closed circulatory system A type of internal transport in which blood is confined to vessels.

cochlea *(KOH-klee-uh)* The complex, coiled organ of hearing that contains the organ of Corti.

codominance A phenotypic situation in which both alleles are expressed in the heterozygote.

codon *(KOH-don)* A three-nucleotide sequence of DNA or mRNA that specifies a particular amino acid or termination signal; the basic unit of the genetic code.

coelom *(SEE-lome)* A body cavity completely lined with mesoderm.

coelomate *(SEE-loh-mate)* An animal whose body cavity is completely lined by mesoderm, the layers of which connect dorsally and ventrally to form mesenteries.

coenocytic *(SEN-oh-SIT-ik)* Referring to a multinucleated condition resulting from the repeated division of nuclei without cytoplasmic division.

coenzyme *(ko-EN-zyme)* An organic molecule serving as a cofactor. Most vitamins function as coenzymes in important metabolic reactions.

coevolution The mutual influence on the evolution of two different species interacting with each other and reciprocally influencing each other's adaptations.

cofactor Any nonprotein molecule or ion that is required for the proper functioning of an enzyme. Cofactors can be permanently bound to the active site or may bind loosely with the substrate during catalysis.

cohesion The binding together of like molecules, often by hydrogen bonds.

collagen A glycoprotein in the extracellular matrix of animal cells that forms strong fibers, found extensively in connective tissue and bone; the most abundant protein in the animal kingdom.

collecting duct The location in the kidney where filtrate from renal tubules is collected; the filtrate is now called urine.

collenchyma cell *(koal-EN-keh-muh)* A flexible plant cell type that occurs in strands or cylinders that support young parts of the plant without restraining growth.

commensalism *(kuh-MEN-sul-iz-um)* A symbiotic relationship in which the symbiont benefits but the host is neither helped nor harmed.

community All the organisms that inhabit a particular area; an assemblage of populations of different species living close enough together for potential interaction.

companion cell A type of plant cell that is connected to a sieve-tube member by many plasmodesmata and whose nucleus and ribosomes may serve one or more adjacent sieve-tube members.

competitive exclusion principle The concept that when the populations of two species compete for the same limited resources, one population will use the resources more efficiently and have a reproductive advantage that will eventually lead to the elimination of the other population.

competitive inhibitor A substance that reduces the activity of an enzyme by entering the active site in place of the substrate whose structure it mimics.

complement system A group of at least 20 blood proteins that cooperate with other defense mechanisms; may amplify the inflammatory response, enhance phagocytosis, or directly lyse pathogens; activated by the onset of the immune response or by surface chemicals on microorganisms.

complementary DNA (cDNA) DNA that is identical to a native DNA containing a gene of interest, except that the cDNA lacks noncoding regions (introns) because it is synthesized in the laboratory using mRNA templates.

complete digestive tract A digestive tube that runs between a mouth and an anus; also called alimentary canal. An incomplete digestive tract has only one opening.

complete flower A flower that has sepals, petals, stamens, and carpels.

compound A chemical combination, in a fixed ratio, of two or more elements.

compound eye A type of multifaceted eye in insects and crustaceans consisting of up to several thousand light-detecting, focusing ommatidia; especially good at detecting movement.

condensation reaction A reaction in which two molecules become covalently bonded to each other through the loss of a small molecule, usually water; also called dehydration reaction.

cone cell One of two types of photoreceptors in the vertebrate eye; detects color during the day.

conidium (plural, **conidia**) A naked, asexual spore produced at the ends of hyphae in ascomycetes.

conifer A gymnosperm whose reproductive structure is the cone. Conifers include pines, firs, redwoods, and other large trees.

conjugation *(KON-joo-GAY-shun)* A recombination mechanism that results in the transfer of genetic material between two bacterial cells that are temporarily joined.

connective tissue Animal tissue that functions mainly to bind and support other tissues, having a sparse population of cells scattered through an extracellular matrix.

contraception The prevention of pregnancy.

convection The mass movement of warmed air or liquid to or from the surface of a body or object.

convergent evolution The independent development of similarity between species as a result of their having similar ecological roles and selection pressures.

cooperativity *(koh-OP-ur-uh-TIV-eh-tee)* An interaction of the constituent subunits of a protein causing a conformational change in one subunit to be transmitted to all the others.

cork cambium *(KAM-bee-um)* A cylinder of meristematic tissue in plants that produces cork cells to replace the epidermis during secondary growth.

corpus luteum *(KOR-pus LOO-tee-um)* A secreting tissue in the ovary that forms from the collapsed follicle after ovulation and produces progesterone.

cortex The region of the root between the stele and epidermis filled with ground tissue.

cotransport The coupling of the "downhill" diffusion of one substance to the "uphill" transport of another against its own concentration gradient.

cotyledons *(KOT-eh-LEE-dons)* The one (monocot) or two (dicot) seed leaves of an angiosperm embryo.

countercurrent exchange The opposite flow of adjacent fluids that maximizes transfer rates; for example, blood in the gills flows in the opposite direction in which water passes over the gills, maximizing oxygen uptake and carbon dioxide loss.

covalent bond *(koh-VAY-lent)* A type of strong chemical bond in which two atoms share one pair of electrons in a mutual valence shell.

crista *(KRIS-tuh)* (plural, **cristae**) An infolding of the inner membrane of a mitochondrion that houses the electron transport chain and the enzyme catalyzing the synthesis of ATP.

crossing over The reciprocal exchange of genetic material between nonsister chromatids during synapsis of meiosis I.

cryptic coloration *(KRIP-tik)* A type of camouflage that makes potential prey difficult to spot against its background.

cuticle *(KYOO-teh-kul)* (1) A waxy covering on the surface of stems and leaves that acts as an adaptation to prevent desiccation in terrestrial plants. (2) The exoskeleton of an arthropod, consisting of layers of protein and chitin that are variously modified for different functions.

cyanobacteria *(sy-AN-oh-bak-TEER-ee-uh)* Photosynthetic, oxygen-producing bacteria (formerly known as blue-green algae).

cyclic AMP (cAMP) (cyclic adenosine monophosphate) A small, ring-shaped molecule that acts as a chemical signal in slime molds, as an intracellular second messenger in vertebrate endocrine systems, and as a regulator of the *lac* operon.

cyclic electron flow A route of electron flow during the light reactions of photosynthesis that involves only photosystem I and produces ATP but not NADPH or oxygen.

cyclin *(SY-klin)* A regulatory protein whose concentration fluctuates cyclically.

cyclin-dependent kinase (Cdk) A protein kinase that is active only when attached to a particular cyclin.

cytochrome *(SY-toh-krome)* An iron-containing protein, a component of electron transport chains in mitochondria and chloroplasts.

cytokines In the vertebrate immune system, protein factors secreted by macrophages and helper T cells as regulators of neighboring cells.

cytokinesis *(SY-toh-kin-EE-sis)* The division of the cytoplasm to form two separate daughter cells immediately after mitosis.

cytokinins *(SY-toh-KY-nins)* A class of related plant hormones that retard aging and act in concert with auxins to stimulate cell division, influence the pathway of differentiation, and control apical dominance.

cytoplasm *(SY-toh-plaz-um)* The entire contents of the cell, exclusive of the nucleus, and bounded by the plasma membrane.

cytoplasmic streaming A circular flow of cytoplasm, involving myosin and actin filaments, that speeds the distribution of materials within cells.

cytoskeleton *(SY-toh-SKEL-eh-ton)* A network of microtubules, microfilaments, and intermediate filaments that branch throughout the cytoplasm and serve a variety of mechanical and transport functions.

cytosol *(SY-toh-sol)* The semifluid portion of the cytoplasm.

cytotoxic T cell (T_C) A type of lymphocyte that kills infected cells and cancer cells.

dalton *(DAWL-ton)* The atomic mass unit; a measure of mass for atoms and subatomic particles.

Darwinian fitness A measure of the relative contribution of an individual to the gene pool of the next generation.

day-neutral plant A plant whose flowering is not affected by photoperiod.

decomposers Saprotrophic fungi and bacteria that absorb nutrients from nonliving organic material such as corpses, fallen plant material, and the wastes of living organisms, and convert them into inorganic forms.

deletion (1) A deficiency in a chromosome resulting from the loss of a fragment through breakage. (2) A mutational loss of a nucleotide from a gene.

demography The study of statistics relating to births and deaths in populations.

denaturation A process in which a protein unravels and loses its native conformation, thereby becoming biologically inactive. Denaturation occurs under extreme conditions of pH, salt concentration, and temperature.

dendrite *(DEN-dryt)* One of usually numerous, short, highly branched processes of a neuron that conveys nerve impulses toward the cell body.

density The number of individuals per unit area or volume.

density-dependent factor Any factor influencing population regulation that has a greater impact as population density increases.

density-dependent inhibition The phenomenon observed in normal animal cells that causes them to stop dividing when they come into contact with one another.

density-independent factor Any factor influencing population regulation that acts to reduce population by the same percentage, regardless of size.

deoxyribonucleic acid (DNA) *(DEE-oks-ee-ry-boh-noo-KLAY-ik)* A double-stranded, helical nucleic acid molecule capable of replicating and determining the inherited structure of a cell's proteins.

deoxyribose The sugar component of DNA, having one less hydroxyl group than ribose, the sugar component of RNA.

depolarization An electrical state in an excitable cell whereby the inside of the cell is made less negative relative to the outside than at the resting membrane potential. A neuron membrane is depolarized if a stimulus decreases its voltage from the resting potential of –70 mV in the direction of zero voltage.

deposit-feeder A heterotroph, such as an earthworm, that eats its way through detritus, salvaging bits and pieces of decaying organic matter.

dermal tissue system The protective covering of plants; generally a single layer of tightly packed epidermal cells covering young plant organs formed by primary growth.

desmosome *(DEZ-muh-some)* A type of intercellular junction in animal cells that functions as an anchor.

determinate cleavage A type of embryonic development in protostomes that rigidly casts the developmental fate of each embryonic cell very early.

determinate growth A type of growth characteristic of animals, in which the organism stops growing after it reaches a certain size.

determination The progressive restriction of developmental potential, causing the possible fate of each cell to become more limited as the embryo develops.

detritus *(deh-TRY-tis)* Dead organic matter.

deuterostomes *(DOO-ter-oh-stomes)* One of two distinct evolutionary lines of coelomates, consisting of the echinoderms and chordates and characterized by radial, indeterminate cleavage, enterocoelous formation of the coelom, and development of the anus from the blastopore.

diaphragm A sheet of muscle that forms the bottom wall of the thoracic cavity in mammals; active in ventilating the lungs.

diastole *(dy-ASS-toh-lee)* The stage of the heart cycle in which the heart muscle is relaxed, allowing the chambers to fill with blood.

dicot *(DY-kot)* A subdivision of flowering plants whose members possess two embryonic seed leaves, or cotyledons.

differentiation *See* cellular differentiation.

diffusion The spontaneous tendency of a substance to move down its concentration gradient from a more concentrated to a less concentrated area.

digestion The process of breaking down food into molecules small enough for the body to absorb.

dihybrid cross *(DY-HY-brid)* A breeding experiment in which parental varieties differing in two traits are mated.

dikaryon *(dy-KAH-ree-on)* A mycelium of certain septate fungi that possesses two separate haploid nuclei per cell.

dioecious *(dy-EE-shus)* Referring to a plant species that has staminate and carpellate flowers on separate plants.

diploid cell *(DIP-loyd)* A cell containing two sets of chromosomes ($2n$), one set inherited from each parent.

directional selection Natural selection that favors individuals on one end of the phenotypic range.

disaccharide *(dy-SAK-ur-ide)* A double sugar, consisting of two monosaccharides joined by dehydration synthesis.

dispersion The distribution of individuals within geographical population boundaries.

diversifying selection Natural selection that favors extreme over intermediate phenotypes.

DNA ligase *(LY-gaze)* A linking enzyme essential for DNA replication; catalyzes the covalent bonding of the 3' end of a new DNA fragment to the 5' end of a growing chain.

DNA methylation The addition of methyl groups ($—CH_3$) to bases of DNA after DNA synthesis; may serve as a long-term control of gene expression.

DNA polymerase An enzyme that catalyzes the elongation of new DNA at a replication fork in the 5'⟶3' direction by the addition of nucleotides to the existing chain.

DNA probe A chemically synthesized, radioactively labeled segment of nucleic acid used to find a gene of interest by hydrogen-bonding to a complementary sequence.

domain A taxonomic category above the kingdom level; the three domains are archaebacteria, eubacteria, and eukaryotes.

dominance hierarchy A linear "pecking order" of animals, where position dictates characteristic social behaviors.

dominant allele In a heterozygote, the allele that is fully expressed in the phenotype.

double circulation A circulation scheme with separate pulmonary and systemic circuits, which ensures vigorous blood flow to all organs.

double fertilization A mechanism of fertilization in angiosperms, in which two sperm cells unite with two cells in the embryo sac to form the zygote and endosperm.

double helix The form of native DNA, referring to its two adjacent polynucleotide strands wound into a spiral shape.

Down syndrome A human genetic disease resulting from having an extra chromosome 21, characterized by mental retardation and heart and respiratory defects.

duodenum *(doo-oh-DEE-num)* The first section of the small intestine, where acid chyme from the stomach mixes with digestive juices from the pancreas, liver, gallbladder, and gland cells of the intestinal wall.

duplication An aberration in chromosome structure resulting from an error in meiosis or mutagens; duplication of a portion of a chromosome resulting from fusion with a fragment from a homologous chromosome.

dynein *(DY-nin)* A large contractile protein forming the sidearms of microtubule doublets in cilia and flagella.

ecdysone *(EK-deh-sone)* A steroid hormone that triggers molting in arthropods.

ecological efficiency The ratio of net productivity at one trophic level to net productivity at the next lower level.

ecological niche *(nich)* The sum total of an organism's utilization of the biotic and abiotic resources of its environment.

ecological succession Transition in the species composition of a biological community, often following ecological disturbance of the community; the establishment of a biological community in an area virtually barren of life.

ecology The study of how organisms interact with their environments.

ecosystem A level of ecological study that includes all the organisms in a given area as well as the abiotic factors with which they interact; a community and its physical environment.

ectoderm *(EK-tuh-durm)* The outermost of the three primary germ layers in animal embryos; gives rise to the outer covering and, in some phyla, the nervous system, inner ear, and lens of the eye.

ectotherm *(EK-toh-thurm)* An animal, such as a reptile, fish, or amphibian, that must use environmental energy and behavioral adaptations to regulate its body temperature.

effector cell A muscle cell or gland cell that performs the body's responses to stimuli; responds to signals from the brain or other processing center of the nervous system.

electrochemical gradient The diffusion gradient of an ion, representing a type of potential energy that accounts for both the concentration difference of the ion across a membrane and its tendency to move relative to the membrane potential.

electrogenic pump An ion transport protein generating voltage across the membrane.

electromagnetic spectrum The entire spectrum of radiation; ranges in wavelength from less than a nanometer to more than a kilometer.

electron microscope (EM) A microscope that focuses an electron beam through a specimen, resulting in resolving power a thousandfold greater than that of a light microscope. A transmission electron microscope (TEM) is used to study the internal structure of thin sections of cells. A scanning electron microscope (SEM) is used to study the fine details of cell surfaces.

electron transport chain A group of molecules in the inner membrane of a mitochondrion that synthesize ATP by means of an exergonic slide of electrons. Thylakoid membranes of chloroplasts are also equipped with electron transport chains.

electronegativity The tendency for an atom to pull electrons toward itself.

element Any substance that cannot be broken down to any other substance.

embryo sac The female gametophyte of angiosperms, formed from the growth and division of the megaspore into a multicellular structure with eight haploid nuclei.

enantiomer *(eh-NAN-she-uh-mer)* One of a pair of molecules that are mirror-image isomers of each other.

endergonic reaction *(EN-dur-GON-ik)* A nonspontaneous chemical reaction in which free energy is absorbed from the surroundings.

endocrine gland *(EN-doh-krin)* A ductless gland that secretes hormones directly into the bloodstream.

endocrine system The internal system of chemical communication involving hormones, the ductless glands that secrete hormones, and the molecular receptors on or in target cells that respond to hormones; functions in concert with the nervous system to effect internal regulation and maintain homeostasis.

endocytosis *(EN-doh-sy-TOH-sis)* The cellular uptake of macromolecules and particulate substances by localized regions of the plasma membrane that surround the substance and pinch off to form an intracellular vesicle.

endoderm *(EN-doh-durm)* The innermost of the three primary germ layers in animal embryos; lines the archenteron and gives rise to the liver, pancreas, lungs, and the lining of the digestive tract.

endodermis *(EN-doh-DUR-mis)* The innermost layer of the cortex in plant roots; a cylinder one cell thick that forms the boundary between the cortex and the stele.

endomembrane system The collection of membranes inside and around a eukaryotic cell, related either through direct physical contact or by the transfer of membranous vesicles.

endometrium *(EN-doh-MEE-tree-um)* The inner lining of the uterus, which is richly supplied with blood vessels.

endoplasmic reticulum (ER) *(EN-doh-plaz-mik reh-TIK-yoo-lum)* An extensive membranous network in eukaryotic cells, continuous with the outer nuclear membrane and composed of ribosome-studded (rough) and ribosome-free (smooth) regions.

endorphin *(en-DOR-fin)* A hormone produced in the brain and anterior pituitary that inhibits pain perception.

endoskeleton *(EN-doh-SKEL-eh-ton)* A hard skeleton buried within the soft tissues of an animal, such as the spicules of sponges, the plates of echinoderms, and the bony skeletons of vertebrates.

endosperm *(EN-doh-spurm)* A nutrient-rich tissue formed by the union of a sperm cell with two polar nuclei during double fertilization, which provides nourishment to the developing embryo in angiosperm seeds.

endospore A thick-coated, resistant cell produced within a bacterial cell exposed to harsh conditions.

endosymbiotic theory *(EN-doh-SIM-by-OT-ik)* A hypothesis about the origin of the eukaryotic cell, maintaining that the forerunners of eukaryotic cells were symbiotic associations of prokaryotic cells living inside larger prokaryotes.

endothelium *(EN-doh-THEEL-ee-um)* The innermost, simple squamous layer of cells lining the blood vessels; the only constituent structure of capillaries.

endotherm *(EN-doh-thurm)* An animal that uses metabolic energy to maintain a constant body temperature, such as a bird or mammal.

endotoxin *(EN-doh-TOKS-in)* A component of the outer membranes of certain gram-negative bacteria responsible for generalized symptoms of fever and ache.

energy The capacity to do work by moving matter against an opposing force.

enhancer A DNA sequence that recognizes certain transcription factors that can stimulate transcription of nearby genes.

entropy *(EN-truh-pee)* A quantitative measure of disorder or randomness, symbolized by *S*.

environmental grain An ecological term for the effect of spatial variation, or patchiness, relative to the size and behavior of an organism.

enzymes A class of proteins serving as catalysts, chemical agents that change the rate of a reaction without being consumed by the reaction.

epidermis *(EP-eh-DER-mis)* (1) The dermal tissue system in plants. (2) The outer covering of animals.

epigenesis *(EP-eh-JEN-eh-sis)* The progressive development of form in an embryo.

epiglottis A cartilaginous flap that blocks the top of the windpipe, the glottis, during swallowing, which prevents the entry of food or fluid into the respiratory system.

epinephrine A hormone produced as a response to stress; also called adrenaline.

epiphyte *(EP-eh-fite)* A plant that nourishes itself but grows on the surface of another plant for support, usually on the branches or trunks of tropical trees.

episome *(EP-eh-some)* A plasmid capable of integrating into the bacterial chromosome.

epistasis A phenomenon in which one gene alters the expression of another gene that is independently inherited.

epithelial tissue *(EP-eh-THEEL-ee-ul)* Sheets of tightly packed cells that line organs and body cavities.

epitope A localized region on the surface of an antigen that is chemically recognized by antibodies; also called antigenic determinant.

erythrocyte *(er-RITH-roh-site)* A red blood cell; contains hemoglobin, which functions in transporting oxygen in the circulatory system.

esophagus *(eh-SOF-eh-gus)* A channel that conducts food, by peristalsis, from the pharynx to the stomach.

essential amino acids The amino acids that an animal cannot synthesize itself and must obtain from food. Eight amino acids are essential in the human adult.

estivation *(ES-teh-VAY-shun)* A physiological state characterized by slow metabolism and inactivity, which permits survival during long periods of elevated temperature and diminished water supplies.

estrogens *(ES-troh-jens)* The primary female steroid sex hormones, which are produced in the ovary by the developing follicle during the first half of the cycle and in smaller quantities by the corpus luteum during the second half. Estrogens stimulate the development and maintenance of the female reproductive system and secondary sex characteristics.

estrous cycle *(ES-trus)* A type of reproductive cycle in all female mammals except higher primates, in which the nonpregnant endometrium is reabsorbed rather than shed, and sexual response occurs only during midcycle at estrus.

ethylene *(ETH-ul-een)* The only gaseous plant hormone, responsible for fruit ripening, growth inhibition, leaf abscission, and aging.

eubacteria *(YOO-bak-TEER-ee-uh)* The lineage of prokaryotes that includes the cyanobacteria and all other contemporary bacteria except archaebacteria.

euchromatin *(yoo-KROH-muh-tin)* The more open, unraveled form of eukaryotic chromatin, which is available for transcription.

eukaryotic cell *(YOO-kar-ee-OT-ik)* A type of cell with a membrane-enclosed nucleus and membrane-enclosed organelles, present in protists, plants, fungi, and animals; also called eukaryote.

eumetazoa *(YOO-met-uh-ZOH-uh)* Members of the subkingdom that includes all animals except sponges.

eutrophic lake A highly productive lake, having a high rate of biological productivity supported by a high rate of nutrient cycling.

evaporative cooling The property of a liquid whereby the surface becomes cooler during evaporation, owing to a loss of highly kinetic molecules to the gaseous state.

evolution All the changes that have transformed life on Earth from its earliest beginnings to the diversity that characterizes it today.

excitatory postsynaptic potential (EPSP) *(POST-sin-AP-tik)* An electrical change (depolarization) in the membrane of a postsynaptic neuron caused by the binding of an excitatory neurotransmitter from a presynaptic cell to a postsynaptic receptor; makes it more likely for a postsynaptic neuron to generate an action potential.

excretion The disposal of nitrogen-containing waste products of metabolism.

exergonic reaction *(EKS-ur-GON-ik)* A spontaneous chemical reaction in which there is a net release of free energy.

exocytosis *(EKS-oh-sy-TOH-sis)* The cellular secretion of macromolecules by the fusion of vesicles with the plasma membrane.

exon The coding region of a eukaryotic gene that is expressed. Exons are separated from each other by introns.

exoskeleton A hard encasement on the surface of an animal, such as the shells of mollusks or the cuticles of arthropods, that provides protection and points of attachment for muscles.

exotoxin *(EKS-oh-TOKS-in)* A toxic protein secreted by a bacterial cell that produces specific symptoms even in the absence of the bacterium.

exponential population growth The geometric increase of a population as it grows in an ideal, unlimited environment.

extraembryonic membranes *(EKS-truh-EM-bree-AHN-ik)* Four membranes (yolk sac, amnion, chorion, allantois) that support the developing embryo in reptiles, birds, and mammals.

F_1 generation The first filial or hybrid offspring in a genetic cross-fertilization.

F_2 generation Offspring resulting from interbreeding of the hybrid F_1 generation.

F plasmid The fertility factor in bacteria, a plasmid that confers the ability to form pili for conjugation and associated functions required for the transfer of DNA from donor to recipient.

facilitated diffusion The spontaneous passage of molecules and ions, bound to specific carrier proteins, across a biological membrane down their concentration gradients.

facultative anaerobe *(FAK-ul-tay-tiv AN-uh-robe)* An organism that makes ATP by aerobic respiration if oxygen is present but that switches to fermentation under anaerobic conditions.

fat (triacylglycerol) *(tri-AH-sil-GLIS-er-all)* A biological compound consisting of three fatty acids linked to one glycerol molecule.

fatty acid A long carbon chain carboxylic acid. Fatty acids vary in length and in the number and location of double bonds; three fatty acids linked to a glycerol molecule form fat.

feedback inhibition A method of metabolic control in which the end-product of a metabolic pathway acts as an inhibitor of an enzyme within that pathway.

fermentation A catabolic process that makes a limited amount of ATP from glucose without an electron transport chain and that produces a characteristic end-product, such as ethyl alcohol or lactic acid.

fertilization The union of haploid gametes to produce a diploid zygote.

fiber A lignified cell type that reinforces the xylem of angiosperms and functions in mechanical support; a slender, tapered sclerenchyma cell that usually occurs in bundles.

fibrin *(FY-brin)* The activated form of the blood-clotting protein fibrinogen, which aggregates into threads that form the fabric of the clot.

fibroblast *(FY-broh-blast)* A type of cell in loose connective tissue that secretes the protein ingredients of the extracellular fibers.

first law of thermodynamics *(THUR-moh-dy-NAM-iks)* The principle of conservation of energy. Energy can be transferred and transformed, but it cannot be created or destroyed.

fixed action pattern (FAP) A highly stereotypical behavior that is innate and must be carried to completion once initiated.

flaccid *(FLAS-id)* Limp; walled cells are flaccid in isotonic surroundings, where there is no tendency for water to enter.

flagellum *(fluh-JEL-um)* (plural, **flagella**) A long cellular appendage specialized for locomotion, formed from a core of nine outer doublet microtubules and two inner single microtubules, ensheathed in an extension of plasma membrane.

fluid-feeder An animal that lives by sucking nutrient-rich fluids from another living organism.

fluid mosaic model The currently accepted model of cell membrane structure, which envisions the membrane as a mosaic of individually inserted protein molecules drifting laterally in a fluid bilayer of phospholipids.

follicle *(FOL-eh-kul)* A microscopic structure in the ovary that contains the developing ovum and secretes estrogens.

food chain The pathway along which food is transferred from trophic level to trophic level, beginning with producers.

food web The elaborate, interconnected feeding relationships in an ecosystem.

founder effect A cause of genetic drift attributable to colonization by a limited number of individuals from a parent population.

fragile *X* syndrome A hereditary mental disorder, partially explained by genomic imprinting and the addition of nucleotides to a triplet repeat at the end of an *X* chromosome.

frameshift mutation A mutation occurring when the number of nucleotides inserted or deleted is not a multiple of 3, thus resulting in improper grouping into codons.

free energy A quantity of energy that interrelates entropy *(S)* and the system's total energy *(H)*; symbolized by *G*. The change in free energy of a system is calculated by the equation $G = \Delta H - T\Delta S$, where T is absolute temperature.

free energy of activation The initial investment of energy necessary to start a chemical reaction; also called activation energy.

frequency-dependent selection A decline in the reproductive success of a morph resulting from the morph's phenotype becoming too common in a population; a cause of balanced polymorphism in populations.

fruit A mature ovary of a flower that protects dormant seeds and aids in their dispersal.

functional group A specific configuration of atoms commonly attached to the carbon skeletons of organic molecules and usually involved in chemical reactions.

G protein A membrane protein that functions as a relay diary between hormone receptors in cell membranes and the enzyme adenylyl cyclase, which converts ATP to cAMP in the second messenger (cAMP) system in nonsteroid hormone action. Depending on the type of hormone, G proteins may increase or decrease the activity of adenylyl cyclase in producing cAMP.

G_1 phase The first growth phase of the cell cycle, consisting of the portion of interphase before DNA synthesis begins.

G_2 phase The second growth phase of the cell cycle, consisting of the portion of interphase after DNA synthesis occurs.

gametangium *(GAM-eh-TANJ-ee-um)* (plural, **gametangia**) The reproductive organ of bryophytes, consisting of the male antheridium and female archegonium; a multichambered jacket of sterile cells in which gametes are formed.

gamete *(GAM-eet)* A haploid egg or sperm cell; gametes unite during sexual reproduction to produce a diploid zygote.

gametophyte *(guh-MEE-toh-fite)* The multicellular haploid form in organisms undergoing alternation of generations, which mitotically produces haploid gametes that unite and grow into the sporophyte generation.

ganglion *(GANG-lee-un)* (plural, **ganglia**) A cluster (functional group) of nerve cell bodies in a centralized nervous system.

gap junction A type of intercellular junction in animal cells that allows the passage of material or current between cells.

gastrin A digestive hormone, secreted by the stomach, that stimulates the secretion of gastric juice.

gastrovascular cavity The central digestive compartment, usually with a single opening that functions as both mouth and anus.

gastrula *(GAS-troo-la)* The two-layered, cup-shaped embryonic stage.

gastrulation *(GAS-truh-LAY-shun)* The formation of a gastrula from a blastula.

gated ion channel A specific ion channel that opens and closes to allow the cell to alter its membrane potential.

gel electrophoresis *(JELL eh-LEK-troh-for-EE-sis)* The separation of nucleic acids or proteins, on the basis of their size and electrical charge, by measuring their rate of movement through an electrical field in a gel.

gene One of many discrete units of hereditary information located on the chromosomes and consisting of DNA.

gene amplification The selective synthesis of DNA, which results in multiple copies of a single gene, thereby enhancing expression.

gene cloning The formation by a bacterium, carrying foreign genes in a recombinant plasmid, of a clone of identical cells containing the replicated foreign genes.

gene flow The loss or gain of alleles from a population due to the emigration or immigration of fertile individuals, or the transfer of gametes, between populations.

gene pool The total aggregate of genes in a population at any one time.

genetic drift Changes in the gene pool of a small population due to chance.

genetic recombination The general term for the production of offspring that combine traits of the two parents.

genome *(JEE-nome)* The complete complement of an organism's genes; an organism's genetic material.

genomic imprinting The parental effect on gene expression. Identical alleles may have different effects on offspring, depending on whether they arrive in the zygote via the ovum or via the sperm.

genomic library A set of thousands of DNA segments from a genome, each carried by a plasmid or phage.

genotype *(JEE-noh-type)* The genetic makeup of an organism.

genus *(JEE-nus)* (plural, **genera**) A taxonomic category above the species level, designated by the first word of a species' binomial Latin name.

geographical range The geographic area in which a population lives.

geological time scale A time scale established by geologists that reflects a consistent sequence of historical periods, grouped into four eras: Precambrian, Paleozoic, Mesozoic, and Cenozoic.

gibberellins *(JIB-ur-EL-ins)* A class of related plant hormones that stimulate growth in the stem and leaves, trigger the germination of seeds and breaking of bud dormancy, and stimulate fruit development with auxin.

gill A localized extension of the body surface of many aquatic animals, specialized for gas exchange.

glial cell *(GLEE-ul)* A nonconducting cell of the nervous system that provides support, insulation, and protection for the neurons.

glomerulus *(glum-AIR-yoo-lus)* A ball of capillaries surrounded by Bowman's capsule in the nephron and serving as the site of filtration in the vertebrate kidney.

glucagon A peptide hormone secreted by pancreatic endocrine cells that raises blood glucose levels; an antagonistic hormone to insulin.

glucocorticoid A corticosteroid hormone secreted by the adrenal cortex that influences glucose metabolism and immune function.

glycocalyx *(GLY-koh-KAY-liks)* A fuzzy coat on the outside of animal cells, made of sticky oligosaccharides.

glycogen *(GLY-koh-jen)* An extensively branched glucose storage polysaccharide found in the liver and muscle of animals; the animal equivalent of starch.

glycolysis *(gly-KOL-eh-sis)* The splitting of glucose into pyruvate. Glycolysis is the one metabolic pathway that occurs in all living cells, serving as the starting point for fermentation or aerobic respiration.

Golgi apparatus *(GOAL-jee)* An organelle in eukaryotic cells consisting of stacks of membranes that modify, store, and route products of the endoplasmic reticulum.

gonadotropins *(goh-NAD-oh-TROH-pinz)* Hormones that stimulate the activities of the testes and ovaries; a collective term for follicle-stimulating and luteinizing hormones.

gonads *(GOH-nadz)* The male and female sex organs; the gamete-producing organs in most animals.

graded potential A local voltage change in a neuron membrane induced by stimulation of a neuron, with strength proportional to the strength of the stimulus and lasting about a millisecond.

gradualism A view of Earth's history that attributes profound change to the cumulative product of slow but continuous processes.

Gram stain A staining method that distinguishes between two different kinds of bacterial cell walls.

granum *(GRAN-um)* (plural, **grana**) A stacked portion of the thylakoid membrane in the chloroplast. Grana function in the light reactions of photosynthesis.

gravitropism *(GRAV-eh-TROH-piz-um)* A response of a plant or animal in relation to gravity.

greenhouse effect The warming of planet Earth due to the atmospheric accumulation of carbon dioxide, which absorbs infrared radiation and slows its escape from the irradiated Earth.

gross primary productivity (GPP) The total primary productivity of an ecosystem.

ground meristem A primary meristem that gives rise to ground tissue in plants.

ground tissue system A tissue of mostly parenchyma cells that makes up the bulk of a young plant and fills the space between the dermal and vascular tissue systems.

growth factor A protein that must be present in the extracellular environment (culture medium or animal body) for the growth and normal development of certain types of cells.

guard cell A specialized epidermal plant cell that forms the boundaries of the stomata.

guttation The exudation of water droplets caused by root pressure in certain plants.

gymnosperm *(JIM-noh-spurm)* A vascular plant that bears naked seeds not enclosed in any specialized chambers.

habituation A simple kind of learning involving a loss of sensitivity to unimportant stimuli, allowing an animal to conserve time and energy.

haploid cell *(HAP-loid)* A cell containing only one set of chromosomes *(n)*.

Hardy-Weinberg theorem An axiom maintaining that the sexual shuffling of genes alone cannot alter the overall genetic makeup of a population.

haustorium (plural, **haustoria**) In parasitic fungi, a nutrient-absorbing hyphal tip that penetrates the tissues of the host but remains outside the host cell membranes.

Haversian system *(ha-VER-shun)* One of many structural units of vertebrate bone, consisting of concentric layers of mineralized bone matrix surrounding lacunae, which contain osteocytes, and a central canal, which contains blood vessels and nerves.

heat The total amount of kinetic energy due to molecular motion in a body of matter. Heat is energy in its most random form.

heat-shock protein A protein that helps protect other proteins during heat stress, found in plants, animals, and microorganisms.

helper T cell (T_H) A type of T cell that is required by some B cells to help them make antibodies or that helps other T cells respond to antigens or secrete lymphokines or interleukins.

hemoglobin *(HEE-moh-gloh-bin)* An iron-containing protein in red blood cells that reversibly binds oxygen.

hemolymph In invertebrates with an open circulatory system, the body fluid that bathes tissues.

hepatic portal vessel A large circulatory channel that conveys nutrient-laden blood from the small intestine to the liver, which regulates the blood's nutrient content.

herbivore A heterotrophic animal that eats plants.

hermaphrodite *(her-MAF-roh-dite)* An individual that functions as both male and female in sexual reproduction by producing both sperm and eggs.

heterochromatin *(HET-ur-oh-KROH-muh-tin)* Nontranscribed eukaryotic chromatin that is so highly compacted that it is visible with a light microscope during interphase.

heterochrony Evolutionary changes in the timing or rate of development.

heterocyst *(HET-ur-oh-sist)* A specialized cell that engages in nitrogen fixation on some filamentous cyanobacteria.

heteromorphic *(HET-ur-oh-MOR-fik)* A condition in the life cycle of all modern plants in which the sporophyte and gametophyte generations differ in morphology.

heterosporous *(HET-ur-OS-pur-us)* Referring to plants in which the sporophyte produces two kinds of spores that develop into unisexual gametophytes, either female or male.

heterotroph *(HET-ur-oh-TROHF)* An organism that obtains organic food molecules by eating other organisms or their by-products.

heterozygote advantage *(HET-ur-oh-ZY-gote)* A mechanism that preserves variation in eukaryotic gene pools by conferring greater reproductive success on heterozygotes over individuals homozygous for any one of the associated alleles.

heterozygous *(HET-ur-oh-ZY-gus)* Having two different alleles for a given trait.

hibernation A physiological state that allows survival during long periods of cold temperatures and reduced food supplies, in which metabolism decreases, the heart and respiratory system slow down, and body temperature is maintained at a lower level than normal.

histamine *(HISS-tuh-meen)* A substance released by injured cells that causes blood vessels to dilate during an inflammatory response.

histone *(HISS-tone)* A small protein with a high proportion of positively charged amino acids that binds to the negatively charged DNA and plays a key role in its folding into chromatin.

HIV (human immunodeficiency virus) The infectious agent that causes AIDS; HIV is an RNA retrovirus.

holoblastic cleavage *(HOH-loh-BLAS-tik)* A type of cleavage in which there is complete division of the egg, as in eggs having little yolk (sea urchin) or a moderate amount of yolk (frog).

homeobox Specific sequences of DNA that regulate patterns of differentiation during the development of an organism.

homeosis Evolutionary alteration in the placement of different body parts.

homeostasis *(HOME-ee-oh-STAY-sis)* The steady-state physiological condition of the body.

homeotic genes *(HOME-ee-OT-ik)* Genes that control the overall body plan of animals by controlling the developmental fate of groups of cells.

homeotic mutation A mutation in genes regulated by positional information that results in the abnormal substitution of one type of body part for another.

homologous chromosomes *(home-OL-uh-gus)* Chromosome pairs of the same length, centromere position, and staining pattern that possess genes for the same traits at corresponding loci. One homologous chromosome is inherited from the organism's father, the other from the mother.

homologous structures Structures in different species that are similar because of common ancestry.

homology Similarity in characteristics resulting from a shared ancestry.

homosporous *(home-OS-pur-us)* Referring to plants in which a single type of spore develops into a bisexual gametophyte having both male and female sex organs.

homozygous *(HOME-oh-ZY-gus)* Having two identical alleles for a given trait.

hormone One of many types of circulating chemical signals in all multicellular organisms that are formed in specialized cells, travel in body fluids, and coordinate the various parts of the organism by interacting with target cells.

Human Genome Project An international collaborative effort to map and sequence the DNA of entire human genomes.

humoral immunity *(HYOO-mur-al)* The type of immunity that fights bacteria and viruses in body fluids with antibodies that circulate in blood plasma and lymph, fluids formerly called humors.

hybrid zone A region where two related populations that diverged after becoming geographically isolated make secondary contact and interbreed where their geographical ranges overlap.

hydrocarbon *(HY-droh-kar-bon)* An organic molecule consisting only of carbon and hydrogen.

hydrogen bond A type of weak chemical bond formed when the slightly positive hydrogen atom of a polar covalent bond in one molecule is attracted to the slightly negative atom of a polar covalent bond in another molecule.

hydrogen ion A single proton with a charge of +1. The dissociation of a water molecule (H_2O) leads to the generation of a hydroxide ion (OH^-) and a hydrogen ion (H^+).

hydrolysis *(hy-DROL-eh-sis)* A chemical process that lyses or splits molecules by the addition of water; an essential process in digestion.

hydrophilic *(HY-droh-FIL-ik)* Having an affinity for water.

hydrophobic *(HY-droh-FOH-bik)* Having an aversion to water; tending to coalesce and form droplets in water.

hydrophobic interaction A type of weak chemical bond formed when molecules that do not mix with water coalesce to exclude the water.

hydrostatic skeleton *(HY-droh-STAT-ik)* A skeletal system composed of fluid held under pressure in a closed body compartment; the main skeleton of most cnidarians, flatworms, nematodes, and annelids.

hydroxyl group *(hy-DROKS-ul)* A functional group consisting of a hydrogen atom joined to an oxygen atom by a polar covalent bond. Molecules possessing this group are soluble in water and are called alcohols.

hyperpolarization An electrical state whereby the inside of the cell is made more negative relative to the outside than at the resting membrane potential. A neuron membrane is hyperpolarized if a stimulus increases its voltage from the resting potential of −70 mV, reducing the chance that the neuron will transmit a nerve impulse.

hypertonic solution A solution with a greater solute concentration than another, a hypotonic solution.

hypha *(HY-fa)* (plural, **hyphae**) A filament that collectively makes up the body of a fungus.

hypothalmus *(HY-poh-THAL-uh-mus)* The ventral part of the vertebrate forebrain; functions in maintaining homeostasis, especially in coordinating the endocrine and nervous systems; secretes hormones of the posterior pituitary and releasing factors, which regulate the anterior pituitary.

hypotonic solution A solution with a lesser solute concentration than another, a hypertonic solution.

imaginal disk *(i-MAJ-in-ul)* An island of undifferentiated cells in an insect larva, which are committed (determined) to form a particular organ during metamorphosis to the adult.

immunoglobulin (Ig) *(IM-myoo-noh-GLOB-yoo-lin)* One of the class of proteins comprising the antibodies.

imprinting A type of learned behavior with a significant innate component, acquired during a limited critical period.

incomplete dominance A type of inheritance in which F_1 hybrids have an appearance that is intermediate between the phenotypes of the parental varieties.

incomplete flower A flower lacking sepals, petals, stamens, or carpels.

incomplete metamorphosis *(MET-uh-MOR-foh-sis)* A type of development in certain insects, such as grasshoppers, in which the larvae resemble adults but are smaller and have different body proportions. The animal goes through a series of molts, each time looking more like an adult, until it reaches full size.

indeterminate cleavage A type of embryonic development in deuterostomes, in which each cell produced by early cleavage divisions retains the capacity to develop into a complete embryo.

indeterminate growth A type of growth characteristic of plants, in which the organism continues to grow as long as it lives.

induced fit The change in shape of the active site of an enzyme so that it binds more snugly to the substrate, induced by entry of the substrate.

induction The ability of one group of embryonic cells to influence the development of another.

inflammatory response A line of defense triggered by penetration of the skin or mucous membranes, in which small blood vessels in the vicinity of an injury dilate and become leakier, enhancing the infiltration of leukocytes; may also be widespread in the body.

ingestion A heterotrophic mode of nutrition in which other organisms or detritus are eaten whole or in pieces.

inhibitory postsynaptic potential (IPSP) *(POST-sin-AP-tik)* An electrical charge (hyperpolarization) in the membrane of a postsynaptic neuron caused by the binding of an inhibitory neurotransmitter from a presynaptic cell to a postsynaptic receptor; makes it more difficult for a postsynaptic neuron to generate an action potential.

inner cell mass A cluster of cells in a mammalian blastocyst that protrudes into one end of the cavity and subsequently develops into the embryo proper and some of the extraembryonic membranes.

inositol trisphosphate (IP_3) *(in-NOS-i-tahl)* The second messenger, which functions as an intermediate between certain nonsteroid hormones and the third messenger, a rise in cytoplasmic Ca^{2+} concentration.

insertion A mutation involving the addition of one or more nucleotide pairs to a gene.

insertion sequence The simplest kind of a transposon, consisting of inserted repeats of DNA flanking a gene for transposase, the enzyme that catalyzes genetic transposition.

insight learning The ability of an animal to perform a correct or appropriate behavior on the first attempt in a situation with which it has had no prior experience.

insulin *(IN-sul-in)* A vertebrate hormone that lowers blood glucose levels by promoting the uptake of glucose by most body cells and the synthesis and storage of glycogen in the liver; also stimulates protein and fat synthesis; secreted by endocrine cells of the pancreas called islets of Langerhans.

interferon *(IN-tur-FEER-on)* A chemical messenger of the immune system, produced by virus-infected cells and capable of helping other cells resist the virus.

interleukin-1 *(IN-tur-loo-kin)* A chemical regulator (cytokin) secreted by macrophages that have ingested a pathogen or foreign molecule and have bound with a helper T cell; stimulates T cells to grow and divide and elevates body temperature. Interleukin-2, secreted by activated T cells, stimulates helper T cells to proliferate more rapidly.

intermediate filament A component of the cytoskeleton that includes all filaments intermediate in size between microtubules and microfilaments.

interneuron *(IN-tur-NOOR-ahn)* An association neuron; a nerve cell within the central nervous system that forms synapses with sensory and motor neurons and integrates sensory input and motor output.

internode The segment of a plant stem between the points where leaves are attached.

interphase The period in the cell cycle when the cell is not dividing. During interphase, cellular metabolic activity is high, chromosomes and organelles are duplicated, and cell size may increase. Interphase accounts for 90% of the time of each cell cycle.

interstitial cells *(IN-tur-STISH-ul)* Cells scattered among the seminiferous tubules of the vertebrate testis that secrete testosterone and other androgens, the male sex hormones.

interstitial fluid The internal environment of vertebrates, consisting of the fluid filling the spaces between cells.

intertidal zone The shallow zone of the ocean where land meets water.

intrinsic rate of increase The difference between the number of births and the number of deaths, symbolized as r_{max}; the maximum population growth rate.

introgression *(IN-troh-GRES-shun)* The transplantation of genes between species resulting from fertile hybrids mating successfully with one of the parent species.

intron *(IN-tron)* The noncoding, intervening sequence of a coding region (exon) in eukaryotic genes.

inversion An aberration in chromosome structure resulting from an error in meiosis or from mutagens; reattachment in a reverse orientation of a chromosomal fragment to the chromosome from which the fragment originated.

invertebrate An animal without a backbone; invertebrates make up 95% of animal species.

in vitro fertilization *(VEE-troh)* Fertilization of ova in laboratory containers followed by artificial implantation of the early embryo in the mother's uterus.

ion *(EYE-on)* An atom that has gained or lost electrons, thus acquiring a charge.

ionic bond *(eye-ON-ik)* A chemical bond resulting from the attraction between oppositely charged ions.

isogamy *(eye-SOG-uh-mee)* A condition in which male and female gametes are morphologically indistinguishable.

isomer *(EYE-sum-ur)* One of several organic compounds with the same molecular formula but different structures and therefore different properties. The three types are structural isomers, geometric isomers, and enantiomers.

isomorphic generations Alternating generations in which the sporophytes and gametophytes look alike, although they differ in chromosome number.

isotonic solutions Solutions of equal solute concentration.

isotope *(EYE-so-tope)* One of several atomic forms of an element, each containing a different number of neutrons and thus differing in atomic mass.

joule (J) A unit of energy: 1 J = 0.239 cal; 1 cal = 4.184 J.

juvenile hormone (JH) A hormone in arthropods, secreted by the corpora allata glands, that promotes the retention of larval characteristics.

juxtaglomerular apparatus (JGA) Specialized tissue located near the afferent arteriole that supplies blood to the kidney glomerulus; the JGA raises blood pressure by producing renin, which activates angiotensin.

***K*-selection** The concept that in certain (*K*-selected) populations, life history is centered around producing relatively few offspring that have a good chance of survival.

karyogamy The fusion of nuclei of two cells, as part of syngamy.

karyotype *(KAR-ee-oh-type)* A method of organizing the chromosomes of a cell in relation to number, size, and type.

keystone predator A predatory species that helps maintain species richness in a community by reducing the density of populations of the best competitors so that populations of less competitive species are maintained.

kilocalorie (kcal) A thousand calories; the amount of heat energy required to raise the temperature of 1 kg of water 1°C.

kin selection A phenomenon of inclusive fitness, used to explain altruistic behavior between related individuals.

kinesis *(kih-NEE-sis)* A change in activity rate in response to a stimulus.

kinetic energy *(kih-NET-ik)* The energy of motion, which is directly related to the speed of that motion. Moving matter does work by transferring some of its kinetic energy to other matter.

kinetochore *(kih-NET-oh-kor)* A specialized region on the centromere that links each sister chromatid to the mitotic spindle.

kingdom A taxonomic category, the second broadest after domain.

Koch's postulates A set of four criteria for determining whether a specific pathogen is the cause of a disease.

Krebs cycle A chemical cycle involving eight steps that completes the metabolic breakdown of glucose molecules to carbon dioxide; occurs within the mitochondrion; the second major stage in cellular respiration.

lacteal *(lak-TEEL)* A tiny lymph vessel extending into the core of an intestinal villus and serving as the destination for absorbed chylomicrons.

lagging strand A discontinuously synthesized DNA strand that elongates in a direction away from the replication fork.

larva *(LAR-vuh)* (plural, **larvae**) A free-living, sexually immature form in some animal life cycles that may differ from the adult in morphology, nutrition, and habitat.

lateral line system A mechanoreceptor system consisting of a series of pores and receptor units (neuromasts) along the sides of the body of fishes and aquatic amphibians; detects water movements made by an animal itself and by other moving objects.

lateral meristem *(MARE-eh-stem)* The vascular and cork cambium, a cylinder of dividing cells that runs most of the length of stems and roots and is responsible for secondary growth.

law of independent assortment Mendel's second law, stating that each allele pair segregates independently during gamete formation; applies when genes for two traits are located on different pairs of homologous chromosomes.

law of segregation Mendel's first law, stating that allele pairs separate during gamete formation, and then randomly re-form pairs during the fusion of gametes at fertilization.

leading strand The new continuous complementary DNA strand synthesized along the template strand in the mandatory 5' ⟶ 3' direction.

leukocyte *(LOO-koh-site)* A white blood cell; typically functions in immunity, such as phagocytosis or antibody production.

lichen *(LY-ken)* An organism formed by the symbiotic association between a fungus and a photosynthetic alga.

life table A table of data summarizing mortality in a population.

ligament A type of fibrous connective tissue that joins bones together at joints.

ligand *(LIG-und)* A molecule that binds specifically to a receptor site of another molecule.

light microscope (LM) An optical instrument with lenses that refract (bend) visible light to magnify images of specimens.

light reactions The steps in photosynthesis that occur on the thylakoid membranes of the chloroplast and convert solar energy to the chemical energy of ATP and NADPH, evolving oxygen in the process.

lignin *(LIG-nin)* A hard material embedded in the cellulose matrix of vascular plant cell walls that functions as an important adaptation for support in terrestrial species.

limbic system *(LIM-bik)* A group of nuclei (clusters of nerve cell bodies) in the lower part of the mammalian forebrain that interact with the cerebral cortex in determining emotions; includes the hippocampus and the amygdala.

linked genes Genes that are located on the same chromosome.

lipid *(LIH-pid)* One of a family of compounds, including fats, phospholipids, and steroids, that are insoluble in water.

lipoprotein A protein bonded to a lipid; includes the low-density lipoproteins (LDLs) and high-density lipoproteins (HDLs) that transport fats and cholesterol in blood.

locus *(LOH-kus)* (plural, **loci**) A particular place along the length of a certain chromosome where a given gene is located.

logistic population growth A model describing population growth that levels off as population size approaches carrying capacity.

long-day plant A plant that flowers, usually in late spring or early summer, only when the light period is longer than a critical length.

loop of Henle The long hairpin turn, with a descending and ascending limb, of the renal tubule in the vertebrate kidney; functions in water and salt reabsorption.

lungs The invaginated respiratory surfaces of terrestrial vertebrates, land snails, and spiders that connect to the atmosphere by narrow tubes.

lymph *(limf)* The colorless fluid, derived from interstitial fluid, in the lymphatic system of vertebrate animals.

lymphatic system *(lim-FAT-ik)* A system of vessels and lymph nodes, separate from the circulatory system, that returns fluid and protein to the blood.

lymphocyte A white blood cell. The lymphocytes that complete their development in the bone marrow are called B cells, and those that mature in the thymus are called T cells.

lysogenic cycle A type of viral replication cycle in which the viral genome becomes incorporated into the bacterial host chromosome as a prophage.

lysosome *(LY-so-some)* A membrane-enclosed bag of hydrolytic enzymes found in the cytoplasm of eukaryotic cells.

lysozyme *(LY-so-zime)* An enzyme in perspiration, tears, and saliva that attacks bacterial cell walls.

lytic cycle *(LIT-ik)* A type of viral replication cycle resulting in the release of new phages by death or lysis of the host cell.

M phase The mitotic phase of the cell cycle, which includes mitosis and cytokinesis.

macroevolution Evolutionary change on a grand scale, encompassing the origin of novel designs, evolutionary trends, adaptive radiation, and mass extinction.

macromolecule A giant molecule of living matter formed by the joining of smaller molecules, usually by condensation synthesis. Polysaccharides, proteins, and nucleic acids are macromolecules.

macrophage *(MAK-roh-fage)* An amoeboid cell that moves through tissue fibers, engulfing bacteria and dead cells by phagocytosis.

major histocompatibility complex (MHC) A large set of cell surface antigens encoded by a family of genes. Foreign MHC markers trigger T-cell responses that may lead to the rejection of transplanted tissues and organs.

Malpighian tubule *(mal-PIG-ee-un)* A unique excretory organ of insects that empties into the digestive tract, removes nitrogenous wastes from the blood, and functions in osmoregulation.

Mammalia The vertebrate class of mammals, characterized by body hair and mammary glands that produce milk to nourish the young.

mantle A heavy fold of tissue in mollusks that drapes over the visceral mass and may secrete a shell.

marsupial *(mar-SOOP-ee-ul)* A mammal, such as a koala, kangaroo, or opossum, whose young complete their embryonic development inside a maternal pouch called the marsupium.

matrix The nonliving component of connective tissue, consisting of a web of fibers embedded in homogeneous ground substance that may be liquid, jellylike, or solid.

matter Anything that takes up space and has mass.

mechanoreceptor A sensory receptor that detects physical deformations in the body's environment associated with pressure, touch, stretch, motion, and sound.

medulla oblongata *(meh-DOO-luh OBB-long-GAH-tuh)* The lowest part of the vertebrate brain; a swelling of the hindbrain dorsal to the anterior spinal cord that controls autonomic, homeostatic functions, including breathing, heart and blood vessel activity, swallowing, digestion, and vomiting.

medusa *(meh-DOO-suh)* The floating, flattened, mouth-down version of the cnidarian body plan. The alternate form is the polyp.

megapascal (MPa) *(MEG-uh-pass-KAL)* A unit of pressure equivalent to 10 atmospheres of pressure.

meiosis *(my-OH-sis)* A two-stage type of cell division in sexually reproducing organisms that results in gametes with half the chromosome number of the original cell.

membrane potential The charge difference between the cytoplasm and extracellular fluid in all cells, due to the differential distribution of ions. Membrane potential affects the activity of excitable cells and the transmembrane movement of all charged substances.

memory cell A clone of long-lived lymphocytes, formed during the primary immune response, that remains in a lymph node until activated by exposure to the same antigen that triggered its formation. Activated memory cells mount the secondary immune response.

menstrual cycle *(MEN-stroo-ul)* A type of reproductive cycle in higher female primates, in which the nonpregnant endometrium is shed as a bloody discharge through the cervix into the vagina.

meristem *(MARE-eh-stem)* Plant tissue that remains embryonic as long as the plant lives, allowing for indeterminate growth.

meroblastic cleavage *(MARE-oh-BLAS-tik)* A type of cleavage in which there is incomplete division of yolk-rich egg, characteristic of avian development.

mesentery *(MEZ-en-ter-ee)* A membrane that suspends many of the organs of vertebrates inside fluid-filled body cavities.

mesoderm *(MEZ-oh-durm)* The middle primary germ layer of an early embryo that develops into the notochord, the lining of the coelom, muscles, skeleton, gonads, kidneys, and most of the circulatory system.

mesophyll *(MEZ-oh-fil)* The ground tissue of a leaf, sandwiched between the upper and lower epidermis and specialized for photosynthesis.

messenger RNA (mRNA) A type of RNA synthesized from DNA in the genetic material that attaches to ribosomes in the cytoplasm and specifies the primary structure of a protein.

metabolism *(meh-TAB-oh-liz-um)* The totality of an organism's chemical processes, consisting of catabolic and anabolic pathways.

metamorphosis *(MET-uh-MOR-fuh-sis)* The resurgence of development in an animal larva that transforms it into a sexually mature adult.

metanephridium *(MET-uh-neh-FRID-ee-um)* (plural, **metanephridia**) In annelid worms, a type of excretory tubule with internal openings called nephrostomes that collect body fluids and external openings called nephridiopores.

metastasis *(meh-TAS-teh-sis)* The spread of cancer cells beyond their original site.

microevolution A change in the gene pool of a population over a succession of generations.

microfilament A solid rod of actin protein in the cytoplasm of almost all eukaryotic cells, making up part of the cytoskeleton and acting alone or with myosin to cause cell contraction.

microtubule A hollow rod of tubulin protein in the cytoplasm of all eukaryotic cells and in cilia, flagella, and the cytoskeleton.

microvillus (plural, **microvilli**) One of many fine, fingerlike projections of the epithelial cells in the lumen of the small intestine that increase its surface area.

middle lamella *(luh-MEL-uh)* A thin layer of adhesive extracellular material, primarily pectins, found between the primary walls of adjacent young plant cells.

mimicry A phenomenon in which one species benefits by a superficial resemblance to an unrelated species. A predator or species of prey may gain a significant advantage through mimicry.

mineralocorticoid A corticosteroid hormone secreted by the adrenal cortex that regulates salt and water homeostasis.

missense mutation The most common type of mutation involving a base-pair substitution within a gene that changes a codon, but the new codon makes sense in that it still codes for an amino acid.

mitochondrial matrix The compartment of the mitochondrion enclosed by the inner membrane and containing enzymes and substrates for the Krebs cycle.

mitochondrion *(MY-toh-KON-dree-un)* (plural, **mitochondria**) An organelle in eukaryotic cells that serves as the site of cellular respiration.

mitosis *(my-TOH-sis)* A process of cell division in eukaryotic cells conventionally divided into the growth period (interphase) and four stages: prophase, metaphase, anaphase, and telophase. The stages conserve chromosome number by equally allocating replicated chromosomes to each of the daughter cells.

modern synthesis A comprehensive theory of evolution emphasizing natural selection, gradualism, and populations as the fundamental units of evolutionary change; also called neo-Darwinism.

molarity A common measure of solute concentration, referring to the number of moles of solute in 1 L of solution.

mold A rapidly growing, asexually reproducing fungus.

mole The number of grams of a substance that equals its molecular weight in daltons and contains Avogadro's number of molecules.

molecular formula A type of molecular notation indicating only the quantity of the constituent atoms.

molecule Two or more atoms held together by chemical bonds.

molting A process in arthropods in which the exoskeleton is shed at intervals to allow growth by the secretion of a larger exoskeleton.

monoclonal antibody *(MON-oh-KLONE-ul)* A defensive protein produced by cells descended from a single cell; an antibody that is secreted by a clone of cells and, consequently, is specific for a single antigenic determinant.

monocot *(MON-oh-kot)* A subdivision of flowering plants whose members possess one embryonic seed leaf, or cotyledon.

monoculture Cultivation of large land areas with a single plant variety.

monoecious *(mon-EE-shus)* Referring to a plant species that has both staminate and carpellate flowers on the same individual.

monohybrid cross A breeding experiment that uses parental varieties differing in a single character.

monomer *(MON-uh-mer)* The subunit that serves as the building block of a polymer.

monophyletic *(MON-oh-fy-LEH-tik)* Pertaining to a taxon derived from a single ancestral species that gave rise to no species in any other taxa.

monosaccharide *(MON-oh-SAK-ur-ide)* The simplest carbohydrate, active alone or serving as a monomer for disaccharides and polysaccharides. Also known as simple sugars, the molecular formulas of monosaccharides are generally some multiple of CH_2O.

monotreme *(MON-uh-treem)* An egg-laying mammal, represented by the platypus and echidna.

morphogen A substance, such as bicoid protein, that provides positional information in the form of a concentration gradient along an embryonic axis.

morphogenesis *(MOR-foh-JEN-eh-sis)* The development of body shape and organization during ontogeny.

morphospecies A species defined by its anatomical features.

mosaic development A pattern of development, such as that of a mollusk, in which the early blastomeres each give rise to a specific part of the embryo. In some animals, the fate of the blastomeres is established in the zygote.

mosaic evolution The evolution of different features of an organism at different rates.

motor neuron A nerve cell that transmits signals from the brain or spinal cord to muscles or glands.

motor unit A single motor neuron and all the muscle fibers it controls.

MPF (M-phase promoting factor) A protein complex required for a cell to progress from late interphase to mitosis; the active form consists of cyclin and *cdc2*, a protein kinase.

Müllerian mimicry *(myoo-LER-ee-un)* A mutual mimicry by two unpalatable species.

multigene family A collection of genes with similar or identical sequences, presumably of common origin.

mutagen *(MYOOT-uh-jen)* A chemical or physical agent that interacts with DNA and causes a mutation.

mutagenesis *(MYOOT-uh-JEN-uh-sis)* The creation of mutations.

mutation *(myoo-TAY-shun)* A rare change in the DNA of genes that ultimately creates genetic diversity.

mutualism *(MYOO-choo-ul-iz-um)* A symbiotic relationship in which both the host and the symbiont benefit.

mycelium *(my-SEEL-ee-um)* The densely branched network of hyphae in a fungus.

mycorrhizae *(MY-koh-RY-zee)* Mutualistic associations of plant roots and fungi.

myelin sheath *(MY-eh-lin)* In a neuron, an insulating coat of cell membrane from Schwann cells that is interrupted by nodes of Ranvier where saltatory conduction occurs.

myofibril *(MY-oh-FY-brill)* A fibril collectively arranged in longitudinal bundles in muscle cells (fibers); composed of thin filaments of actin and a regulatory protein and thick filaments of myosin.

myoglobin *(MY-uh-glow-bin)* An oxygen-storing, pigmented protein in muscle cells.

myosin *(MY-uh-sin)* A type of protein filament that interacts with actin filaments to cause cell contraction.

NAD^+ (nicotinamide adenine dinucleotide) A coenzyme present in all cells that helps enzymes transfer electrons during the redox reactions of metabolism.

natural killer cell A nonspecific defensive cell that attacks tumor cells and destroys infected body cells, especially those harboring viruses.

natural selection Differential success in the reproduction of different phenotypes resulting from the interaction of organisms with their environment. Evolution occurs when natural selection causes changes in relative frequencies of alleles in the gene pool.

negative feedback A primary mechanism of homeostasis, whereby a change in a physiological variable that is being monitored triggers a response that counteracts the initial fluctuation.

nephron *(NEF-ron)* The tubular excretory unit of the vertebrate kidney.

neritic zone *(neh-RIT-ik)* The shallow regions of the ocean overlying the continental shelves.

net primary productivity (NPP) The gross primary productivity minus the energy used by the producers for cellular respiration; represents the storage of chemical energy in an ecosystem available to consumers.

neural crest A band of cells along the border where the neural tube pinches off from the ectoderm; the cells migrate to various parts of the embryo and form the pigment cells in the skin, bones of the skull, the teeth, the adrenal glands, and parts of the peripheral nervous system.

neuron *(NOOR-on)* A nerve cell; the fundamental unit of the nervous system, having structure and properties that allow it to conduct signals by taking advantage of the electrical charge across its cell membrane.

neurosecretory cells Hypothalamus cells that receive signals from other nerve cells, but instead of signaling to an adjacent nerve cell or muscle, they release hormones into the bloodstream.

neurotransmitter A chemical messenger released from the synaptic terminal of a neuron at a chemical synapse that diffuses across the synaptic cleft and binds to and stimulates the postsynaptic cell.

neutral variation Genetic diversity that confers no apparent selective advantage.

niche *See* ecological niche.

nitrogen fixation The assimilation of atmospheric nitrogen by certain prokaryotes into nitrogenous compounds that can be directly used by plants.

nitrogenase *(nih-TRAH-juh-nayz)* An enzyme, unique to certain prokaryotes, that reduces N_2 to NH_3.

node A point along the stem of a plant at which leaves are attached.

nodes of Ranvier *(ran-VEER)* The small gaps in the myelin sheath between successive glial cells along the axon of a neuron; also, the site of high concentration of voltage-gated ion channels.

noncompetitive inhibitor A substance that reduces the activity of an enzyme by binding to a location remote from the active site, changing its conformation so that it no longer binds to the substrate.

noncyclic electron flow A route of electron flow during the light reactions of photosynthesis that involves both photosystems and produces ATP, NADPH, and oxygen; the net electron flow is from water to $NADP^+$.

noncyclic photophosphorylation *(FO-toh-fos-FOR-eh-LAY-shun)* The production of ATP by noncyclic electron flow.

nondisjunction An accident of meiosis or mitosis, in which both members of a pair of homologous chromosomes or both sister chromatids fail to move apart properly.

nonpolar covalent bond A type of covalent bond in which electrons are shared equally between two atoms of similar electronegativity.

nonsense mutation A mutation that changes an amino acid codon to one of the three stop codons, resulting in a shorter and usually nonfunctional protein.

norm of reaction The range of phenotypic possibilities for a single genotype, as influenced by the environment.

notochord *(NO-toh-kord)* A longitudinal, flexible rod formed from dorsal mesoderm and located between the gut and the nerve cord in all chordate embryos.

nuclear envelope The membrane in eukaryotes that encloses the nucleus, separating it from the cytoplasm.

nucleic acid (polynucleotide) *(PAHL-ee-NOO-klee-o-tide)* A biological molecule (such as RNA or DNA) that allows organisms to reproduce; polymers composed of monomers called nucleotides joined by covalent bonds (phosphodiester linkages) between the phosphate of one nucleotide and the sugar of the next nucleotide.

nucleoid *(NOO-klee-oid)* A dense region of DNA in a prokaryotic cell.

nucleoid region The region in a prokaryotic cell consisting of a concentrated mass of DNA.

nucleolus *(noo-KLEE-oh-lus)* (plural, **nucleoli**) A specialized structure in the nucleus, formed from various chromosomes and active in the synthesis of ribosomes.

nucleoside *(NOO-klee-oh-side)* An organic molecule consisting of a nitrogenous base joined to a five-carbon sugar.

nucleosome *(NOO-klee-oh-some)* The basic, beadlike unit of DNA packaging in eukaryotes, consisting of a segment of DNA wound around a protein core composed of two copies of each of four types of histone.

nucleotide *(NOO-klee-oh-tide)* The building block of a nucleic acid, consisting of a five-carbon sugar covently bonded to a nitrogenous base and a phosphate group.

nucleus (1) An atom's central core, containing protons and neutrons. (2) The chromosome-containing organelle of a eukaryotic cell. (3) A cluster of neurons.

obligate aerobe *(OB-lig-it AIR-obe)* An organism that requires oxygen for cellular respiration and cannot live without it.

obligate anaerobe *(AN-ur-obe)* An organism that cannot use oxygen and is poisoned by it.

oceanic zone The region of water lying over deep areas beyond the continental shelf.

oligotrophic lake A nutrient-poor, clear, deep lake with minimum phytoplankton.

omnivore A heterotrophic animal that consumes both meat and plant material.

oncogene *(ON-koh-jeen)* A gene found in viruses or as part of the normal genome that is involved in triggering cancerous characteristics.

ontogeny *(on-TOJ-en-ee)* The embryonic development of an organism.

oogamy *(oh-OG-um-ee)* A condition in which male and female gametes differ, such that a small, flagellated sperm fertilizes a large, nonmotile egg.

oogenesis *(OO-oh-JEN-eh-sis)* The process in the ovary that results in the production of female gametes.

open circulatory system An arrangement of internal transport in which blood bathes the organs directly and there is no distinction between blood and interstitial fluid.

operant conditioning *(OP-ur-ent)* A type of associative learning that directly affects behavior in a natural context; also called trial-and-error learning.

operon *(OP-ur-on)* A unit of genetic function common in bacteria and phages, consisting of regulated clusters of genes with related functions.

organ A specialized center of body function composed of several different types of tissues.

organ-identity gene A plant gene in which a mutation causes a floral organ to develop in the wrong location.

organ of Corti The actual hearing organ of the vertebrate ear, located in the floor of the cochlear canal in the inner ear; contains the receptor cells (hair cells) of the ear.

organelle *(OR-guh-NEL)* One of several formed bodies with a specialized function, suspended in the cytoplasm and found in eukaryotic cells.

organic chemistry The study of carbon compounds (organic compounds).

organogenesis *(or-GAN-oh-JEN-eh-sis)* An early period of rapid embryonic development in which the organs take form from the primary germ layers.

orgasm Rhythmic, involuntary contractions of certain reproductive structures in both sexes during the human sexual response cycle.

osmoconformer An animal that does not actively adjust its internal osmolarity because it is isotonic with its environment.

osmolarity *(OZ-moh-LAR-eh-tee)* Solute concentration expressed as molarity.

osmoregulation Adaptations to control the water balance in organisms living in hypertonic, hypotonic, or terrestrial environments.

osmoregulator An animal whose body fluids have a different osmolarity than the environment, and that must either discharge excess water if it lives in a hypotonic environment or take in water if it inhabits a hypertonic environment.

osmosis *(oz-MOH-sis)* The diffusion of water across a selectively permeable membrane.

osmotic pressure *(oz-MOT-ik)* A measure of the tendency of a solution to take up water when separated from pure water by a selectively permeable membrane.

Osteichthyes The vertebrate class of bony fishes, characterized by a skeleton reinforced by calcium phosphate; the most abundant and diverse vertebrates.

ostracoderm *(os-TRAK-uh-durm)* An extinct agnathan; a fishlike creature encased in an armor of bony plates.

ovarian cycle *(oh-VAIR-ee-un)* The cyclic recurrence of the follicular phase, ovulation, and the luteal phase in the mammalian ovary, regulated by hormones.

ovary *(OH-vur-ee)* (1) In flowers, the portion of a carpel in which the egg-containing ovules develop. (2) In animals, the structure that produces female gametes and reproductive hormones.

oviduct *(OH-veh-dukt)* A tube passing from the ovary to the vagina in invertebrates or to the uterus in vertebrates.

oviparous *(oh-VIP-ur-us)* Referring to a type of development in which young hatch from eggs laid outside the mother's body.

ovoviviparous *(OH-voh-vy-VIP-ur-us)* Referring to a type of development in which young hatch from eggs that are retained in the mother's uterus.

ovulation The release of an egg from ovaries. In humans, an ovarian follicle releases an egg during each menstrual cycle.

ovule *(OV-yool)* A structure that develops in the plant ovary and contains the female gametophyte.

ovum *(OH-vum)* The female gamete; the haploid, unfertilized egg, which is usually a relatively large, nonmotile cell.

oxidation The loss of electrons from a substance involved in a redox reaction.

oxidative phosphorylation *(FOS-for-eh-LAY-shun)* The production of ATP using energy derived from the redox reactions of the electron transport chain.

oxidizing agent The electron acceptor in a redox reaction.

P site Peptidyl-tRNA site; the binding site on a ribosome that holds the tRNA carrying a growing polypeptide chain.

pacemaker A specialized region of the right atrium of the mammalian heart that sets the rate of contraction; also called the sinoatrial (SA) node.

paedogenesis *(pee-doh-JEN-eh-sis)* The precocious development of sexual maturity in a larva.

paedomorphosis *(PEE-doh-mor-FOH-sis)* The retention in an adult organism of the juvenile features of its evolutionary ancestors.

paleontology *(PAY-lee-un-TOL-uh-jee)* The scientific study of fossils.

Pangaea *(pan-JEE-uh)* The supercontinent formed near the end of the Paleozoic era when plate movements brought all the land masses of Earth together.

paraphyletic *(PAR-uh-FY-leh-tik)* Pertaining to a taxon that excludes some members that share a common ancestor with members included in the taxon.

parasite *(PAR-uh-site)* An organism that absorbs nutrients from the body fluids of living hosts.

parasitism A symbiotic relationship in which the symbiont (parasite) benefits at the expense of the host by living either within the host (endoparasite) or outside the host (ectoparasite).

parasympathetic division One of two divisions of the autonomic nervous system; generally enhances body activities that gain and conserve energy, such as digestion and reduced heart rate.

parathyroid glands Four endocrine glands, embedded in the surface of the thyroid gland, that secrete parathyroid hormone and raise blood calcium levels.

parazoa *(PAR-uh-ZOH-uh)* Members of the subkingdom of animals consisting of the sponges.

parenchyma *(pur-EN-kim-uh)* A relatively unspecialized plant cell type that carries most of the metabolism, synthesizes and stores organic products, and develops into more differentiated cell types.

parthenogenesis *(PAR-then-oh-JEN-eh-sis)* A type of reproduction in which females produce offspring from unfertilized eggs.

partial pressure The concentration of gases; a fraction of total pressure.

passive transport The diffusion of a substance across a biological membrane.

pattern formation The ordering of cells into specific three-dimensional structures, an essential part of shaping an organism and its individual parts during development.

pedigree A family tree describing the occurrence of heritable characters in parents and offspring across as many generations as possible.

pelagic zone *(pel-AY-jik)* The area of the ocean past the continental shelf, with areas of open water often reaching to very great depths.

peptide bond The covalent bond between two amino acid units, formed by condensation synthesis.

peptidoglycan *(PEP-tid-oh-GLY-kan)* A type of polymer in bacterial cell walls consisting of modified sugars cross-linked by short polypeptides.

perception The interpretation of sensations by the brain.

perennial *(pur-EN-ee-ul)* A plant that lives for many years.

pericycle *(PAIR-eh-sy-kul)* A layer of cells just inside the endodermis of a root that may become meristematic and begin dividing again.

periderm *(PAIR-eh-durm)* The protective coat that replaces the epidermis in plants during secondary growth, formed of the cork and cork cambium.

peripheral nervous system The sensory and motor neurons that connect to the central nervous system.

peristalsis *(PAIR-is-TAL-sis)* Rhythmic waves of contraction of smooth muscle that push food along the digestive tract.

peroxisome *(pur-OKS-eh-some)* A microbody containing enzymes that transfer hydrogen from various substrates to oxygen, producing and then degrading hydrogen peroxide.

petiole *(PET-ee-ole)* The stalk of a leaf, which joins the leaf to a node of the stem.

phage *(fage)* A virus that infects bacteria; also called a bacteriophage.

phagocytosis *(FAY-goh-sy-TOH-sis)* A type of endocytosis involving large, particulate substances.

pharynx *(FAH-rinks)* An area in the vertebrate throat where air and food passages cross; in flatworms, the muscular tube that protrudes from the ventral side of the worm and ends in the mouth.

phenetics *(feh-NEH-tiks)* An approach to taxonomy based entirely on measurable similarities and differences in phenotypic characters, without consideration of homology, analogy, or phylogeny.

phenotype *(FEE-nuh-type)* The physical and physiological traits of an organism.

pheromone *(FAIR-uh-mone)* A small, volatile chemical signal that functions in communication between animals and acts much like a hormone in influencing physiology and behavior.

phloem *(FLOH-um)* The portion of the vascular system in plants consisting of living cells arranged into elongated tubes that transport sugar and other organic nutrients throughout the plant.

phosphate group *(FOS-fate)* A functional group important in energy transfer.

phospholipids *(FOS-foh-LIP-ids)* Molecules that constitute the inner bilayer of biological membranes, having a polar, hydrophilic head and a nonpolar, hydrophobic tail.

photic zone *(FOH-tik)* The narrow top slice of the ocean, where light permeates sufficiently for photosynthesis to occur.

photoautotroph *(FOH-toh-AW-toh-trohf)* An organism that harnesses light energy to drive the synthesis of organic compounds from carbon dioxide.

photoheterotroph *(FOH-toh-HET-ur-oh-trohf)* An organism that uses light to generate ATP but that must obtain carbon in organic form.

photon *(FOH-tahn)* A quantum, or discrete amount, of light energy.

photoperiodism *(FOH-toh-PEER-ee-od-iz-um)* A physiological response to day length, such as flowering in plants.

photophosphorylation *(FOH-toh-fos-for-uh-LAY-shun)* The process of generating ATP from ADP and phosphate by means of a proton-motive force generated by the thylakoid membrane of the chloroplast during the light reactions of photosynthesis.

photorespiration A metabolic pathway that consumes oxygen, releases carbon dioxide, generates no ATP, and decreases photosynthetic output; generally occurs on hot, dry, bright days, when stomata close and the oxygen concentration in the leaf exceeds that of carbon dioxide.

photosynthesis The conversion of light energy to chemical energy that is stored in glucose or other organic compounds; occurs in plants, algae, and certain prokaryotes.

photosystem The light-harvesting unit in photosynthesis, located on the thylakoid membrane of the chloroplast and consisting of the antenna complex, the reaction-center chlorophyll *a*, and the primary electron acceptor. There are two types of photosystems, I and II; they absorb light best at different wavelengths.

phototropism *(FOH-toh-TROH-piz-um)* Growth of a plant shoot toward or away from light.

pH scale A measure of hydrogen ion concentration equal to $-\log [H^+]$ and ranging in value from 0 to 14.

phylogeny *(fih-LOJ-en-ee)* The evolutionary history of a species or group of related species.

phylum A taxonomic category; phyla are divided into classes.

phytoalexin *(fy-toh-ah-LEK-sin)* An antibiotic, produced by plants, that destroys microorganisms or inhibits their growth.

phytochrome *(FY-tuh-krome)* A pigment involved in many responses of plants to light.

pilus *(PILL-us)* (plural, **pili**) A surface appendage in certain bacteria that functions in adherence and the transfer of DNA during conjugation.

pineal gland *(PIN-ee-ul)* A small endocrine gland on the dorsal surface of the vertebrate forebrain; secretes the hormone melatonin, which regulates body functions related to seasonal day length.

pinocytosis *(PY-noh-sy-TOH-sis)* A type of endocytosis in which the cell ingests extracellular fluid and its dissolved solutes.

pith The core of the central vascular cylinder of monocot roots, consisting of parenchyma cells, which are ringed by vascular tissue; ground tissue interior to vascular bundles in dicot stems.

pituitary gland *(pih-TOO-ih-tair-ee)* An endocrine gland at the base of the hypothalamus; consists of a posterior lobe (neurohypophysis), which stores and releases two hormones produced by the hypothalamus, and an anterior lobe (adenohypophysis), which produces and secretes many hormones that regulate diverse body functions.

placenta *(pluh-SEN-tuh)* A structure in the pregnant uterus for nourishing a viviparous fetus with the mother's blood supply; formed from the uterine lining and embryonic membranes.

placental mammal A member of a group of mammals, including humans, whose young complete their embryonic development in the uterus, joined to the mother by a placenta.

placoderm *(PLAK-oh-durm)* A member of an extinct class of fishlike vertebrates that had jaws and were enclosed in a tough, outer armor.

plankton Mostly microscopic organisms that drift passively or swim weakly near the surface of oceans, ponds, and lakes.

plasma *(PLAZ-muh)* The liquid matrix of blood in which the cells are suspended.

plasma cell A derivative of B cells that secretes antibodies.

plasma membrane The membrane at the boundary of every cell that acts as a selective barrier, thereby regulating the cell's chemical composition.

plasmid *(PLAZ-mid)* A small ring of DNA that carries accessory genes separate from those of a bacterial chromosome.

plasmodesma *(PLAZ-moh-DEZ-muh)* (plural, **plasmodesmata**) An open channel in the cell wall of plants through which strands of cytoplasm connect from adjacent cells.

plasmogamy The fusion of the cytoplasm of cells from two individuals; occurs as one stage of syngamy.

plasmolysis *(plaz-MOL-eh-sis)* A phenomenon in walled cells in which the cytoplasm shrivels and the plasma membrane pulls away from the cell wall when the cell loses water to a hypertonic environment.

plastid One of a family of closely related plant organelles, including chloroplasts, chromoplasts, and amyloplasts (leucoplasts).

platelet A small enucleated blood cell important in blood clotting; derived from large cells in the bone marrow.

pleated sheet One form of the secondary structure of proteins in which the polypeptide chain folds back and forth, or where two regions of the chain lie parallel to each other and are held together by hydrogen bonds.

pleiotropy *(PLY-eh-troh-pee)* The ability of a single gene to have multiple effects.

plesiomorphic character *(PLEEZ-ee-oh-MOR-fik)* A primitive phenotypic character possessed by a remote ancestor.

pluripotent stem cell A cell within bone marrow that is a progenitor for any kind of blood cell.

point mutation A change in the chromosome at a single nucleotide within a gene.

polar covalent bond A type of covalent bond between atoms that differ in electronegativity. The shared electrons are pulled closer to the more electronegative atom, making it slightly negative and the other atom slightly positive.

polar molecule A molecule (such as water) with opposite charges on opposite sides.

pollen grain An immature male gametophyte that develops within the anthers of stamens in a flower.

pollination *(POL-eh-NAY-shun)* The placement of pollen onto the stigma of a carpel by wind or animal carriers, a prerequisite to fertilization.

polyandry *(POL-ee-AN-dree)* A polygamous mating system involving one female and many males.

poly-A tail During RNA processing, a nucleotide complex at the 3' end of an mRNA molecule that helps inhibit degradation and enhances translation.

polygenic inheritance *(POL-ee-JEN-ik)* An additive effect of two or more gene loci on a single phenotypic character.

polygyny *(pol-IJ-en-ee)* A polygamous mating system involving one male and many females.

polymer *(POL-eh-mur)* A large molecule consisting of many identical or similar monomers linked together.

polymerase chain reaction (PCR) A technique for amplifying DNA in vitro by incubating with special primers, DNA polymerase molecules and nucleotides.

polymorphic *(POL-ee-MOR-fik)* Referring to a population in which two or more physical forms are present in readily noticeable frequencies.

polymorphism *(POL-ee-MOR-fiz-um)* The coexistence of two or more distinct forms of individuals (polymorphic characters) in the same population.

polyp *(POL-ip)* The sessile variant of the cniderian body plan. The alternate form is the medusa.

polypeptide *(POL-ee-PEP-tide)* A polymer (chain) of many amino acids linked together by peptide bonds.

polyphyletic Pertaining to a taxon whose members were derived from two or more ancestral forms not common to all members.

polyploidy *(POL-ee-ploid-ee)* A chromosomal alteration in which the organism possesses more than two complete chromosome sets.

polyribosome *(POL-ee-RY-boh-some)* An aggregation of several ribosomes attached to one messenger RNA molecule.

polysaccharide *(POL-ee-SAK-ur-ide)* A polymer of up to over a thousand monosaccharides, formed by condensation synthesis.

population A group of individuals of one species that live in a particular geographic area.

positional information Signals, to which genes regulating development respond, indicating a cell's location relative to other cells in an embryonic structure.

positive feedback A physiological control mechanism in which a change in some variable triggers mechanisms that amplify the change.

postsynaptic membrane *(post-sin-AP-tik)* The surface of the cell on the opposite side of the synapse from the synaptic terminal of the stimulating neuron that contains receptor proteins and degradative enzymes for the neurotransmitter.

postzygotic barrier *(POST-zy-GOT-ik)* Any of several species-isolating mechanisms that prevent hybrids produced by two different species from developing into viable, fertile adults.

potential energy The energy stored by matter as a result of its location or spatial arrangement.

preadaptation A structure that evolves and functions in one environmental context but that can perform additional functions when placed in some new environment.

prezygotic barrier *(PREE-zy-GOT-ik)* A reproductive barrier that impedes mating between species or hinders fertilization of ova if interspecific mating is attempted.

primary consumer An herbivore; an organism in the trophic level of an ecosystem that eats plants or algae.

primary germ layers The three layers (ectoderm, mesoderm, endoderm) of the late gastrula, which develop into all parts of an animal.

primary growth Growth initiated by the apical meristems of a plant root or shoot.

primary immune response The initial immune response to an antigen, which appears after a lag of several days.

primary producer An autotroph, which collectively make up the trophic level of an ecosystem that ultimately supports all other levels; usually a photosynthetic organism.

primary productivity The rate at which light energy or inorganic chemical energy is converted to the chemical energy of organic compounds by autotrophs in an ecosystem.

primary structure The level of protein structure referring to the specific sequence of amino acids.

primary succession A type of ecological succession that occurs in an area where there were originally no organisms.

primer An already existing DNA chain bound to the template DNA to which nucleotides must be added during DNA synthesis.

principle of allocation The concept that each organism has an energy budget, or a limited amount of total energy available for all of its maintenance and reproductive needs.

prion An infectious form of protein that may increase in number by converting related proteins to more prions.

procambium *(pro-KAM-bee-um)* A primary meristem of roots and shoots that forms the vascular tissue.

prokaryotic cell *(pro-KAR-ee-OT-ik)* A type of cell lacking a membrane-enclosed nucleus and membrane-enclosed organelles; found only in the kingdom Monera; also called prokaryote.

promoter A specific nucleotide sequence in DNA, flanking the start of a gene; instructs RNA polymerase where to start transcribing RNA.

prophage *(PRO-fage)* A phage genome that has been inserted into a specific site on the bacterial chromosome.

prostaglandin (PG) *(PROS-tuh-GLAN-din)* One of a group of modified fatty acids secreted by virtually all tissues and performing a wide variety of functions as messengers.

protein *(PRO-teen)* A three-dimensional biological polymer constructed from a set of 20 different monomers called amino acids.

protein kinase An enzyme that regulates the activity of another protein by adding a phosphate group (phosphorylation).

proteoglycans *(pro-tee-oh-GLY-kanz)* A glycoprotein in the extracellular matrix of animal cells, rich in carbohydrate.

protoderm *(PRO-toh-durm)* The outermost primary meristem, which gives rise to the epidermis of roots and shoots.

proton-motive force The potential energy stored in the form of an electrochemical gradient, generated by the pumping of hydrogen ions across biological membranes during chemiosmosis.

proton pump *(PRO-tahn)* An active transport mechanism in cell membranes that consumes ATP to force hydrogen ions out of a cell and, in the process, generates a membrane potential.

protonephridium *(PRO-toh-nef-RID-ee-um)* An excretory system, such as the flame-cell system of flatworms, consisting of a network of closed tubules having external openings called nephridiopores and lacking internal openings.

proto-oncogene *(PRO-toh-ONK-oh-jeen)* A normal cellular gene corresponding to an oncogene; a gene with a potential to cause cancer, but that requires some alteration to become an oncogene.

protoplast The contents of a plant cell exclusive of the cell wall.

protostome *(PRO-toh-stome)* A member of one of two distinct evolutionary lines of coelomates, consisting of the annelids, mollusks, and arthropods, and characterized by spiral, determinate cleavage, schizocoelous formation of the coelom, and development of the mouth from the blastopore.

protozoan (plural, **protozoa**) A protist that lives primarily by ingesting food, an animal-like mode of nutrition.

provirus Viral DNA that inserts into a host genome.

proximate causation The hypothesis about why natural selection favored a particular animal behavior.

pseudocoelomate *(SOO-doh-SEEL-oh-mate)* An animal, such as a rotifer or roundworm, whose body cavity is not completely lined by mesoderm.

pseudopodium *(SOO-doh-POH-dee-um)* (plural, **pseudopodia**) A cellular extension of amoeboid cells used in moving and feeding.

punctuated equilibrium A theory of evolution advocating spurts of relatively rapid change followed by long periods of stasis.

quantitative character A heritable feature in a population that varies continuously as a result of environmental influences and the additive effect of two or more genes (polygenic inheritance).

quaternary structure *(KWAT-ur-nair-ee)* The particular shape of a complex, aggregate protein, defined by the characteristic three-dimensional arrangement of its constituent subunits, each a polypeptide.

quiescent center A region located within the zone of cell division in plant roots, containing meristematic cells that divide very slowly.

***r*-selection** The concept that in certain (*r*-selected) populations, a high reproductive rate is the chief determinant of life history.

radial cleavage A type of embryonic development in deuterostomes in which the planes of cell division that transform the zygote into a ball of cells are either parallel or perpendicular to the polar axis, thereby aligning tiers of cells one above the other.

radial symmetry Characterizing a body shaped like a pie or barrel, with many equal parts radiating outward like the spokes of a wheel; present in cnidarians and echinoderms.

radiata Members of the radially symmetrical animal phyla, including cnidarians.

radicle An embryonic root of a plant.

radioactive dating A method of determining the age of fossils and rocks using half-lives of radioactive isotopes.

radioactive isotope An isotope, an atomic form of a chemical element, that is unstable; the nucleus decays spontaneously, giving off detectable particles and energy.

radiometric dating A method paleontologists use for determining the ages of rocks and fossils on a scale of absolute time, based on the half-life of radioactive isotopes.

reaction center The location of one or a pair of specialized chlorophyll *a* molecules in the pigment assembly system of the light reactions of photosynthesis.

receptor-mediated endocytosis *(EN-doh-sy-TOH-sis)* The movement of specific molecules into a cell by the inward budding of membranous vesicles containing proteins with receptor sites specific to the molecules being taken in; enables a cell to acquire bulk quantities of specific substances.

receptor potential An initial response of a receptor cell to a stimulus, consisting of a change in voltage across the receptor membrane proportional to the stimulus strength. The intensity of the receptor potential determines the frequency of action potentials traveling to the nervous system.

recessive allele In a heterozygote, the allele that is completely masked in the phenotype.

reciprocal altruism *(AL-troo-IZ-um)* Altruistic behavior between unrelated individuals; believed to produce some benefit to the altruistic individual in the future when the current beneficiary reciprocates.

recognition concept of species A definition of species based on mate-recognition mechanisms; assumes that reproductive adaptations of a species consist of a set of features that maximize successful mating with members of the same population; an alternative to the biological species concept.

recombinant DNA A technique in which gene segments from different sources are recombined in vitro and transferred into cells, where the DNA may be expressed.

recombinant An offspring whose phenotype differs from that of the parents.

redox reaction *(REE-doks)* A chemical reaction involving the transfer of one or more electrons from one reactant to another; also called oxidation-reduction reaction.

reducing agent The electron donor in a redox reaction.

reduction The gaining of electrons by a substance involved in a redox reaction.

reflex An automatic reaction to a stimulus, mediated by the spinal cord or lower brain.

refractory period *(ree-FRAK-tor-ee)* The short time immediately after an action potential in which the neuron cannot respond to another stimulus, owing to an increase in potassium permeability.

regulative development A pattern of development, such as that of a mammal, in which the early blastomeres retain the potential to form the entire animal.

relative fitness The contribution of one genotype to the next generation compared to that of alternative genotypes for the same locus.

releaser A signal stimulus that functions as a communication signal between individuals of the same species.

releasing hormone A hormone produced by neurosecretory cells in the hypothalamus of the vertebrate brain that stimulates or inhibits the secretion of hormones by the anterior pituitary.

replication fork A Y-shaped point on a replicating DNA molecule where new strands are growing.

repressible enzyme An enzyme whose synthesis is inhibited by a specific metabolite.

repressor A protein that suppresses the expression of prophage or operon genes.

Reptilia The vertebrate class of reptiles, represented by lizards, snakes, turtles, and crocodilians.

resolving power A measure of the clarity of an image; the minimum distance that two points can be separated and still be distinguished as two separate points.

resource partitioning The division of environmental resources by coexisting species populations such that the niche of each species differs by one or more significant factors from the niches of all coexisting species populations.

resting potential The membrane potential characteristic of a nonconducting, excitable cell, with the inside of the cell more negative than the outside.

restriction enzyme A degradative enzyme that recognizes and cuts up DNA (including that of certain phages) that is foreign to a cell.

restriction fragment length polymorphisms (RFLPs) Differences in DNA sequence on homologous chromosomes that result in different patterns of restriction fragment lengths (DNA segments resulting from treatment with restriction enzymes); useful as genetic markers for making linkage maps.

restriction point During the G_1 phase of the cell cycle, the point at which the cell is committed to divide (called "start" in yeast).

restriction site A specific sequence on a DNA strand that is recognized as a "cut site" by a restriction enzyme.

retina *(REH-tin-uh)* The innermost layer of the vertebrate eye, containing photoreceptor cells (rods and cones) and neurons; transmits images formed by the lens to the brain via the optic nerve.

retinal The light-absorbing pigment in rods and cones of the vertebrate eye.

retrovirus *(REH-troh-VY-rus)* An RNA virus that reproduces by transcribing its RNA into DNA and then inserting the DNA into a cellular chromosome; an important class of cancer-causing viruses.

reverse transcriptase *(trans-KRIP-tase)* An enzyme encoded by some RNA viruses that uses RNA as a template for DNA synthesis.

rhodopsin A visual pigment consisting of retinal and opsin. When rhodopsin absorbs light, the retinal changes shape and dissociates from the opsin, after which it is converted back to its original form.

ribonucleic acid (RNA) *(RY-boh-noo-KLAY-ik)* A single-stranded nucleic acid molecule involved in protein synthesis, the structure of which is specified by DNA.

ribose The sugar component of RNA.

ribosomal RNA (rRNA) The most abundant type of RNA. Together with proteins, it forms the structure of ribosomes that coordinate the sequential coupling of tRNA molecules to the series of mRNA codons.

ribosome A cell organelle constructed in the nucleolus, consisting of two subunits and functioning as the site of protein synthesis in the cytoplasm.

ribozyme An enzymatic RNA molecule that catalyzes reactions during RNA splicing.

RNA polymerase *(pul-IM-ur-ase)* An enzyme that links together the growing chain of ribonucleotides during transcription.

RNA processing Modification of RNA before it leaves the nucleus, a process unique to eukaryotes.

RNA splicing The removal of noncoding portions of the RNA molecule after initial synthesis.

rod cell One of two kinds of photoreceptors in the vertebrate retina; sensitive to black and white and enables night vision.

root cap A cone of cells at the tip of a plant root that protects the apical meristem.

root hair A tiny projection growing just behind the root tips of plants, increasing surface area for the absorption of water and minerals.

root pressure The upward push of water within the stele of vascular plants, caused by active pumping of minerals into the xylem by root cells.

rough ER That portion of the endoplasmic reticulum studded with ribosomes.

R plasmid A bacterial plasmid whose resistance to certain antibiotics poses serious medical problems.

rubisco Ribulose carboxylase, the enzyme that catalyzes the first step (the addition of CO_2 to RuBP, or ribulose bisphosphate) of the Calvin cycle.

ruminant An animal, such as a cow or a sheep, with an elaborate, multicompartmentalized stomach specialized for an herbivorous diet.

S phase The synthesis phase of the cell cycle, constituting the portion of interphase during which DNA is replicated.

SA (sinoatrial) node The pacemaker of the heart, located in the wall of the right atrium. At the base of the wall separating the two atria is another patch of nodal tissue called the atrioventricular node (AV).

saltatory conduction *(SAHL-tuh-TOR-ee)* Rapid transmission of a nerve impulse along an axon resulting from the action potential jumping from one node of Ranvier to another, skipping the myelin-sheathed regions of membrane.

saprobe An organism that acts as a decomposer by absorbing nutrients from dead organic matter.

sarcomere *(SAR-koh-meer)* The fundamental, repeating unit of striated muscle, delimited by the Z lines.

sarcoplasmic reticulum *(SAR-koh-PLAZ-mik reh-TIK-yoo-lum)* A modified form of endoplasmic reticulum in striated muscle cells that stores calcium used to trigger contraction during stimulation.

saturated fatty acid A fatty acid in which all carbons in the hydrocarbon tail are connected by single bonds, thus maximizing the number of hydrogen atoms that can attach to the carbon skeleton.

savanna *(suh-VAN-uh)* A tropical grassland biome with scattered individual trees, large herbivores, and three distinct seasons based primarily on rainfall, maintained by occasional fires and drought.

Schwann cells A chain of supporting cells enclosing the axons of many neurons and forming an insulating layer called the myelin sheath.

sclereid *(SKLER-ee-id)* A short, irregular sclerenchyma cell in nutshells and seed coats and scattered through the parenchyma of some plants.

sclerenchyma cell *(skler-EN-kim-uh)* A rigid, supportive plant cell type usually lacking protoplasts and possessing thick secondary walls strengthened by lignin at maturity.

second law of thermodynamics The principle whereby every energy transfer or transformation increases the entropy of the universe. Ordered forms of energy are at least partly converted to heat, and in spontaneous reactions, the free energy of the system also decreases.

second messenger A chemical signal, such as calcium ions or cyclic AMP, that relays a hormonal message from a cell's surface to its interior.

secondary compound A chemical compound synthesized through the diversion of products of major metabolic pathways for use in defense by prey species.

secondary consumer A member of the trophic level of an ecosystem consisting of carnivores that eat herbivores.

secondary growth The increase in girth of the stems and roots of many plants, especially woody, perennial dicots.

secondary immune response The immune response elicited when an animal encounters the same antigen at some later time. The secondary immune response is more rapid, of greater magnitude, and of longer duration than the primary immune response.

secondary productivity The rate at which all the heterotrophs in an ecosystem incorporate organic material into new biomass, which can be equated to chemical energy.

secondary structure The localized, repetitive folding of the polypeptide backbone of a protein due to hydrogen bond formation between peptide linkages.

secondary succession A type of succession that occurs where an existing community has been severely cleared by some disturbance.

sedimentary rock *(SED-eh-MEN-tar-ee)* Rock formed from sand and mud that once settled in layers on the bottom of seas, lakes, and marshes. Sedimentary rocks are often rich in fossils.

seed An adaptation for terrestrial plants consisting of an embryo packaged along with a store of food within a resistant coat.

selection coefficient The difference between two fitness values, representing a relative measure of selection against an inferior genotype.

selective permeability A property of biological membranes that allows some substances to cross more easily than others.

self-incompatibility The capability of certain flowers to block fertilization by pollen from the same or a closely related plant.

semen *(SEE-men)* The fluid that is ejaculated by the male during orgasm; contains sperm and secretions from several glands of the male reproductive tract.

semicircular canals A three-part chamber of the inner ear that functions in maintaining equilibrium.

semilunar valve A valve located at the two exits of the heart, where the aorta leaves the left ventricle and the pulmonary artery leaves the right ventricle.

seminiferous tubules *(SEM-in-IF-er-us)* Highly coiled tubes in the testes in which sperm are produced.

sensation An impulse sent to the brain from activated receptors and sensory neurons.

sensory neuron A nerve cell that receives information from the internal and external environments and transmits the signals to the central nervous system.

sensory receptor A specialized structure that responds to specific stimuli from an animal's external or internal environment; transmits the information of an environmental stimulus to the animal's nervous system by converting stimulus energy to the electrochemical energy of action potentials.

sepal *(SEE-pul)* A whorl of modified leaves in angiosperms that encloses and protects the flower bud before it opens.

sex chromosomes The pair of chromosomes responsible for determining the sex of an individual.

sex-linked genes Genes located on one sex chromosome but not the other.

sexual dimorphism *(dy-MOR-fiz-um)* A special case of polymorphism based on the distinction between the secondary sex characteristics of males and females.

sexual reproduction A type of reproduction in which two parents give rise to offspring that have unique combinations of genes inherited from the gametes of the two parents.

sexual selection Selection based on variation in secondary sex characteristics, leading to the enhancement of sexual dimorphism.

shoot system The aerial portion of a plant body, consisting of stems, leaves, and flowers.

short-day plant A plant that flowers, usually in late summer, fall, or winter, only when the light period is shorter than a critical length.

sieve-tube member A chain of living cells that form sieve tubes in phloem.

sign stimulus An external sensory stimulus that triggers a fixed action pattern.

signal sequence A stretch of amino acids on polypeptides that targets proteins to specific destinations in eukaryotic cells.

signal-transduction pathway A mechanism linking a mechanical or chemical stimulus to a cellular response.

sister chromatids *(KROH-muh-tidz)* Replicated forms of a chromosome joined together by the centromere and eventually separated during mitosis or meiosis II.

skeletal muscle Striated muscle generally responsible for the voluntary movements of the body.

sliding-filament model The theory explaining how muscle contracts, based on change within a sarcomere, the basic unit of muscle organization, stating that thin (actin) filaments slide across thick (myosin) filaments, shortening the sarcomere; the shortening of all sarcomeres in a myofibril shortens the entire myofibril.

small nuclear ribonucleoprotein (snRNP) *(RY-boh-NOO-klee-oh-pro-teen)* One of a variety of small particles in the cell nucleus, composed of RNA and protein molecules; functions are not fully understood, but some form parts of spliceosomes, active in RNA splicing.

smooth ER That portion of the endoplasmic reticulum that is free of ribosomes.

smooth muscle A type of muscle lacking the striations of skeletal and cardiac muscle because of the uniform distribution of myosin filaments in the cell.

sociobiology The study of social behavior based on evolutionary theory.

sodium-potassium pump A special transport protein in the plasma membrane of animal cells that transports sodium out of and potassium into the cell against their concentration gradients.

solute *(SOL-yoot)* A substance that is dissolved in a solution.

solution A homogeneous, liquid mixture of two or more substances.

solvent The dissolving agent of a solution. Water is the most versatile solvent known.

somatic cell *(soh-MAT-ik)* Any cell in a multicellular organism except a sperm or egg cell.

somatic nervous system The branch of the motor division of the vertebrate peripheral nervous system composed of motor neurons that carry signals to skeletal muscles in response to external stimuli.

Southern blotting A hybridization technique that enables researchers to determine the presence of certain nucleotide sequences in a sample of DNA.

speciation *(SPEE-see-AY-shun)* The origin of new species in evolution.

species A particular kind of organism; members possess similar anatomical characteristics and have the ability to interbreed.

species diversity The number and relative abundance of species in a biological community.

species richness The number of species in a biological community.

species selection A theory maintaining that species living the longest and generating the greatest number of species determine the direction of major evolutionary trends.

specific heat The amount of heat that must be absorbed or lost for 1 g of a substance to change its temperature 1°C.

spectrophotometer An instrument that measures the proportions of light of different wavelengths absorbed and transmitted by a pigment solution.

spermatogenesis The continuous and prolific production of mature sperm cells in the testis.

sphincter *(SFINK-ter)* A ringlike valve, consisting of modified muscles in a muscular tube, such as a digestive tract; closes off the tube like a drawstring.

spindle An assemblage of microtubules that orchestrates chromosome movement during eukaryotic cell division.

spiral cleavage A type of embryonic development in protostomes, in which the planes of cell division that transform the zygote into a ball of cells occur obliquely to the polar axis, resulting in cells of each tier sitting in the grooves between cells of adjacent tiers.

spliceosome *(SPLY-see-oh-some)* A complex assembly that interacts with the ends of an RNA intron in splicing RNA; releases an intron and joins two adjacent exons.

sporangium (plural, **sporangia**) A capsule in fungi and plants in which meiosis occurs and haploid spores develop.

spore In the life cycle of a plant or alga undergoing alternation of generations, a meiotically produced haploid cell that divides mitotically, generating a multicellular individual, the gametophyte, without fusing with another cell.

sporophyte The multicellular diploid form in organisms undergoing alternation of generations that results from a union of gametes and that meiotically produces haploid spores that grow into the gametophyte generation.

stabilizing selection Natural selection that favors intermediate variants by acting against extreme phenotypes.

stamen The pollen-producing male reproductive organ of a flower, consisting of an anther and filament.

starch A storage polysaccharide in plants consisting entirely of glucose.

statocyst *(STAT-eh-SIST)* A type of mechanoreceptor that functions in equilibrium in invertebrates through the use of statoliths, which stimulate hair cells in relation to gravity.

stele The central vascular cylinder in roots where xylem and phloem are located.

stereoisomer A molecule that is a mirror image of another molecule with the same molecular formula.

steroids A class of lipids characterized by a carbon skeleton consisting of four rings with various functional groups attached.

stoma (plural, **stomata**) A microscopic pore surrounded by guard cells in the epidermis of leaves and stems that allows gas exchange between the environment and the interior of the plant.

strict aerobe *(AIR-obe)* An organism that can survive only in an atmosphere of oxygen, which is used in aerobic respiration.

strict anaerobe An organism that cannot survive in an atmosphere of oxygen. Other substances, such as sulfate or nitrate, are the terminal electron acceptors in the electron transport chains that generate their ATP.

stroma The fluid of the chloroplast surrounding the thylakoid membrane; involved in the synthesis of organic molecules from carbon dioxide and water.

stromatolite Rock made of banded domes of sediment in which are found the most ancient forms of life: prokaryotes dating back as far as 3.5 billion years.

structural formula A type of molecular notation in which the constituent atoms are joined by lines representing covalent bonds.

structural gene A gene that codes for a polypeptide.

substrate The substance on which an enzyme works.

substrate-level phosphorylation The formation of ATP by directly transferring a phosphate group to ADP from an intermediate substrate in catabolism.

summation A phenomenon of neural integration in which the membrane potential of the postsynaptic cell in a chemical synapse is determined by the total activity of all excitatory and inhibitory presynaptic impulses acting on it at any one time.

suppressor T cell (T_S) A type of T cell that causes B cells as well as other cells to ignore antigens.

surface tension A measure of how difficult it is to stretch or break the surface of a liquid. Water has a high surface tension because of the hydrogen bonding of surface molecules.

survivorship curve A plot of the number of members of a cohort that are still alive at each age; one way to represent age-specific mortality.

suspension-feeder An aquatic animal, such as a clam or a baleen whale, that sifts small food particles from the water.

sustainable agriculture Long-term productive farming methods that are environmentally safe.

swim bladder An adaptation, derived from a lung, that enables bony fishes to adjust their density and thereby control their buoyancy.

symbiont *(SIM-by-ont)* The smaller participant in a symbiotic relationship, living in or on the host.

symbiosis An ecological relationship between organisms of two different species that live together in direct contact.

sympathetic division One of two divisions of the autonomic nervous system of vertebrates; generally increases energy expenditure and prepares the body for action.

sympatric speciation A mode of speciation occurring as a result of a radical change in the genome that produces a reproductively isolated subpopulation in the midst of its parent population.

symplast In plants, the continuum of cytoplasm connected by plasmodesmata between cells.

synapomorphies Shared derived characters; homologies that evolved in an ancestor common to all species on one branch of a fork in a cladogram, but not common to species on the other branch.

synapse *(SIN-aps)* The locus where one neuron communicates with another neuron in a neural pathway; a narrow gap between a synaptic terminal of an axon and a signal-receiving portion (dendrite or cell body) of another neuron or effector cell. Neurotransmitter molecules released by synaptic terminals diffuse across the synapse, relaying messages to the dendrite or effector.

synapsis The pairing of replicated homologous chromosomes during prophase I of meiosis.

synaptic terminal A bulb at the end of an axon in which neurotransmitter molecules are stored and released.

syngamy *(SIN-gam-ee)* The process of cellular union during fertilization.

systematics The branch of biology that studies the diversity of life; encompasses taxonomy and is involved in reconstructing phylogenetic history.

systemic acquired resistance (SAR) A defensive response in infected plants that helps protect healthy tissue from pathogenic invasion.

systole *(SIS-toh-lee)* The stage of the heart cycle in which the heart muscle contracts and the chambers pump blood.

T cell A type of lymphocyte responsible for cell-mediated immunity that differentiates under the influence of the thymus.

taiga *(TY-guh)* The coniferous or boreal forest biome, characterized by considerable snow, harsh winters, short summers, and evergreen trees.

taxis *(TAKS-iss)* A movement toward or away from a stimulus.

taxon (plural, **taxa**) The named taxonomic unit at any given level.

taxonomy The branch of biology concerned with naming and classifying the diverse forms of life.

telomere The end of a chromosome. Repetitive DNA sequences at telomeres help conserve chromosome tips.

temperate deciduous forest A biome located throughout midlatitude regions where there is sufficient moisture to support the growth of large, broad-leaf deciduous trees.

temperate virus A virus that can reproduce without killing the host.

temperature A measure of the intensity of heat in degrees, reflecting the average kinetic energy of the molecules.

tendon A type of fibrous connective tissue that attaches muscle to bone.

tertiary structure *(TUR-shee-air-ee)* Irregular contortions of a protein molecule due to interactions of side chains involved in hydrophobic interactions, ionic bonds, hydrogen bonds, and disulfide bridges.

testcross Breeding of an organism of unknown genotype with a homozygous recessive individual to determine the unknown genotype. The ratio of phenotypes in the offspring determines the unknown genotype.

testis (plural, **testes**) The male reproductive organ, or gonad, in which sperm and reproductive hormones are produced.

testosterone The most abundant androgen hormone in the male body.

tetanus *(TET-un-us)* The maximal, sustained contraction of a skeletal muscle, caused by a very fast frequency of action potentials elicited by continual stimulation.

tetrapod A vertebrate possessing two pairs of limbs, such as amphibians, reptiles, birds, and mammals.

thalamus *(THAL-uh-mus)* One of two integrating centers of the vertebrate forebrain. Neurons with cell bodies in the thalamus relay neural input to specific areas in the cerebral cortex and regulate what information goes to the cerebral cortex.

thermoregulation The maintenance of internal temperature within a tolerable range.

thick filament A filament composed of staggered arrays of myosin molecules; a component of myofibrils in muscle fibers.

thigmomorphogenesis A response in plants to chronic mechanical stimulation, resulting from increased ethylene production; an example is thickening stems in response to strong winds.

thigmotropism *(THIG-moh-TROH-piz-um)* The directional growth of a plant in relation to touch.

threshold potential The potential an excitable cell membrane must reach for an action potential to be initiated.

thylakoid *(THY-luh-koid)* A flattened membrane sac inside the chloroplast, used to convert light energy to chemical energy.

thymus *(THY-mus)* An endocrine gland in the neck region of mammals that is active in establishing the immune system; secretes several messengers, including thymosin, that stimulate T cells.

thyroid gland An endocrine gland that secretes iodine-containing hormones (T_3 and T_4), which stimulate metabolism and influence development and maturation in vertebrates, and cacitonin, which lowers blood calcium levels in mammals.

thyroid-stimulating hormone (TSH) A hormone produced by the anterior pituitary that regulates the release of thyroid hormones.

Ti plasmid A plasmid of a tumor-inducing bacterium that integrates a segment of its DNA into the host chromosome of a plant; frequently used as a carrier for genetic engineering in plants.

tight junction A type of intercellular junction in animal cells that prevents the leakage of material between cells.

tissue An integrated group of cells with a common structure and function.

tonoplast A membrane that encloses the central vacuole in a plant cell, separating the cytosol from the cell sap.

totipotency The ability of embryonic cells to retain the potential to form all parts of the animal.

trace element An element indispensable for life but required in extremely minute amounts.

trachea *(TRAY-kee-uh)* The windpipe; that portion of the respiratory tube that has C-shaped cartilagenous rings and passes from the larynx to two bronchi.

tracheae *(TRAY-kee-ee)* Tiny air tubes that branch throughout the insect body for gas exchange.

tracheal system A gas exchange system of branched, chitin-lined tubes that infiltrate the body and carry oxygen directly to cells in insects.

tracheid *(TRAY-kee-id)* A water-conducting and supportive element of xylem composed of long, thin cells with tapered ends and walls hardened with lignin.

transcription The transfer of information from a DNA molecule into an RNA molecule.

transcription factor A regulatory protein that binds to DNA and stimulates transcription of specific genes.

transfer RNA (tRNA) An RNA molecule that functions as an interpreter between nucleic acid and protein language by picking up specific amino acids and recognizing the appropriate codons in the mRNA.

transformation (1) The conversion of a normal animal cell to a cancerous cell. (2) A phenomenon in which external genetic material is assimilated by a cell.

translation The transfer of information from an RNA molecule into a polypeptide, involving a change of language from nucleic acids to amino acids.

translocation (1) An aberration in chromosome structure resulting from an error in meiosis or from mutagens; attachment of a chromosomal fragment to a nonhomologous chromosome. (2) During protein synthesis, the third stage in the elongation cycle when the RNA carrying the growing polypeptide moves from the A site to the P site on the ribosome. (3) The transport via phloem of food in a plant.

transpiration The evaporative loss of water from a plant.

transposon *(trans-POH-son)* A transposable genetic element; a mobile segment of DNA that serves as an agent of genetic change.

triplet code A set of three-nucleotide-long words that specify the amino acids for polypeptide chains.

triploblastic Possessing three germ layers: the endoderm, mesoderm, and ectoderm. Most eumetazoa are triploblastic.

trophic level The division of species in an ecosystem on the basis of their main nutritional source. The trophic level that ultimately supports all others consists of autotrophs, or primary producers.

trophic structure The different feeding relationships in an ecosystem that determine the route of energy flow and the pattern of chemical cycling.

trophoblast The outer epithelium of the blastocyst, which forms the fetal part of the placenta.

tropic hormone A hormone that has another endocrine gland as a target.

tropical rain forest The most complex of all communities, located near the equator where rainfall is abundant; harbors more species of plants and animals than all other terrestrial biomes combined.

tropism A growth response that results in the curvature of whole plant organs toward or away from stimuli due to differential rates of cell elongation.

tumor A mass that forms within otherwise normal tissue, caused by the uncontrolled growth of a transformed cell.

tumor-suppressor gene A gene whose protein products inhibit cell division, thereby preventing uncontrolled cell growth (cancer).

tundra A biome at the extreme limits of plant growth; at the northernmost limits, it is called arctic tundra, and at high altitudes, where plant forms are limited to low shrubby or matlike vegetation, it is called alpine tundra.

turgid *(TUR-jid)* Firm; walled cells become turgid as a result of the entry of water from a hypotonic environment.

turgor pressure The force directed against a cell wall after the influx of water and the swelling of a walled cell due to osmosis.

ultimate causation The hypothetical evolutionary explanation for the existence of a certain pattern of animal behavior.

unsaturated fatty acid A fatty acid possessing one or more double bonds between the carbons in the hydrocarbon tail. Such bonding reduces the number of hydrogen atoms attached to the carbon skeleton.

urea A soluble form of nitrogenous waste excreted by mammals and most adult amphibians.

ureter A duct leading from the kidney to the urinary bladder.

urethra A tube that releases urine from the body near the vagina in females or through the penis in males; also serves in males as the exit tube for the reproductive system.

uric acid An insoluble precipitate of nitrogenous waste excreted by land snails, insects, birds, and some reptiles.

urochordate A chordate without a backbone, commonly called a tunicate, a sessile marine animal.

uterus A female reproductive organ where eggs are fertilized and/or development of the young occurs.

vaccine A harmless variant or derivative of a pathogen that stimulates a host's immune system to mount defenses against the pathogen.

vacuole A membrane-enclosed sac taking up most of the interior of a mature plant cell and containing a variety of substances important in plant reproduction, growth, and development.

valence shell The outermost energy shell of an atom, containing the valence electrons involved in the chemical reactions of that atom.

vascular cambium A continuous cylinder of meristematic cells surrounding the xylem and pith that produces secondary xylem and phloem.

vascular plants Plants with vascular tissue, consisting of all modern species except the mosses and their relatives.

vascular tissue Plant tissue consisting of cells joined into tubes that transport water and nutrients throughout the plant body.

vascular tissue system A system formed by xylem and phloem throughout the plant, serving as a transport system for water and nutrients, respectively.

vas deferens The tube in the male reproductive system in which sperm travel from the epididymis to the urethra.

vegetative reproduction Cloning of plants by asexual means.

vein A vessel that returns blood to the heart.

ventilation Any method of increasing contact between the respiratory medium and the respiratory surface.

vertebrate A chordate animal with a backbone: the mammals, birds, reptiles, amphibians, and various classes of fishes.

vessel element A specialized short, wide cell in angiosperms; arranged end to end, they form continuous tubes for water transport.

vestigial organ A type of homologous structure that is rudimentary and of marginal or no use to the organism.

viroid *(VY-roid)* A plant pathogen composed of molecules of naked RNA only several hundred nucleotides long.

visceral muscle Smooth muscle found in the walls of the digestive tract, bladder, arteries, and other internal organs.

visible light That portion of the electromagnetic spectrum detected as various colors by the human eye, ranging in wavelength from about 400 nm to about 700 nm.

vitalism The belief that natural phenomena are governed by a life force outside the realm of physical and chemical laws.

vitamin An organic molecule required in the diet in very small amounts; vitamins serve primarily as coenzymes or parts of coenzymes.

viviparous *(vy-VIP-er-us)* Referring to a type of development in which the young are born alive after having been nourished in the uterus by blood from the placenta.

voltage-gated channel Ion channel in a membrane that opens and closes in response to changes in membrane potential (voltage); the sodium and potassium channels of neurons are examples.

water potential The physical property predicting the direction in which water will flow, governed by solute concentration and applied pressure.

water vascular system A network of hydraulic canals unique to echinoderms that branches into extensions called tube feet, which function in locomotion, feeding, and gas exchange.

wavelength The distance between crests of waves, such as those of the electromagnetic spectrum.

wild type An individual with the normal phenotype.

wobble A violation of the base-pairing rules in that third nucleotide (5' end) of a tRNA anticodon can form hydrogen bonds with more than one kind of base in the third position (3' end) of a codon.

xylem *(ZY-lum)* The tube-shaped, nonliving portion of the vascular system in plants that carries water and minerals from the roots to the rest of the plant.

yeast A unicellular fungus that lives in liquid or moist habitats, primarily reproducing asexually by simple cell division or by budding of a parent cell.

yolk sac One of four extraembryonic membranes that supports embryonic development; the first site of blood cells and circulatory system function.

zygote The diploid product of the union of haploid gametes in conception; a fertilized egg.

INDEX

A *t* following a page number indicates a table; an *f* following a page number indicates a figure.